U0307381

Spark

大数据商业实战三部曲

内核解密 | 商业案例 | 性能调优

（第2版）

王家林　段智华◎编著

清華大學出版社
北　京

内 容 简 介

本书基于 Spark 2.4.X 版本，以 Spark 商业案例实战和 Spark 在生产环境下几乎所有类型的性能调优为核心，以 Spark 内核解密为基石，对企业生产环境下的 Spark 商业案例与性能调优抽丝剥茧地进行剖析。全书共分 4 篇，内核解密篇基于 Spark 源码，从一个实战案例入手，循序渐进地全面解析 Spark 2.4.X 版本的新特性及 Spark 内核源码；商业案例篇选取 Spark 开发中最具代表性的经典学习案例，在案例中综合介绍 Spark 的大数据技术；性能调优篇覆盖 Spark 在生产环境下的所有调优技术；Spark+AI 解密篇讲解深度学习实践、Spark+PyTorch 案例实战、Spark+TensorFlow 实战以及 Spark 上的深度学习内核解密。

本书适合所有 Spark 学习者和从业人员使用。对于有分布式计算框架应用经验的人员，本书也可作为 Spark 高手修炼的参考用书。本书还适合作为高等院校的大数据课程教材。

图书在版编目 (CIP) 数据

Spark 大数据商业实战三部曲：内核解密、商业案例、性能调优/王家林，段智华编著. —2 版. —北京：清华大学出版社，2020.4

ISBN 978-7-302-54453-1

Ⅰ. ①S… Ⅱ. ①王… ②段… Ⅲ. ①数据处理软件 Ⅳ. ①TP274

中国版本图书馆 CIP 数据核字 (2019) 第 265425 号

责任编辑： 袁金敏
封面设计： 刘新新
责任校对： 李建庄
责任印制： 丛怀宇

出版发行： 清华大学出版社
网　址：http://www.tup.com.cn, http://www.wqbook.com
地　址：北京清华大学学研大厦 A 座　　邮　编：100084
社 总 机：010-62770175　　邮　购：010-62786544
投稿与读者服务：010-62776969，c-service@tup.tsinghua.edu.cn
质 量 反 馈：010-62772015，zhiliang@tup.tsinghua.edu.cn
印 刷 者： 三河市铭诚印务有限公司
装 订 者： 三河市启晨纸制品加工有限公司
经　销： 全国新华书店
开　本： 185mm×260mm　　**印　张：** 83　　**字　数：** 2120 千字
版　次： 2018 年 2 月第 1 版　2020 年 6 月第 2 版　　**印　次：** 2020 年 6 月第 1 次印刷
定　价： 329.00 元

产品编号：086291-01

第 2 版前言

2019 年 4 月，在美国旧金山举办的 Spark+AI 峰会提出，大数据和人工智能需要统一，人工智能应用需要大量的、不断更新的训练数据来构建先进的模型。到目前为止，Apache Spark 是唯一一个将大规模数据处理与机器学习和人工智能算法完美结合的引擎。

Apache Spark 是大数据领域的通用计算平台。在 Full Stack 思想的指导下，Spark 中的 Spark SQL、Spark Streaming、MLLib、GraphX、R 五大子框架和库之间可以无缝地共享数据和操作，这不仅形成了 Spark 在当今大数据计算领域其他计算框架无可匹敌的优势，而且使其加速成为大数据处理中心首选的通用计算平台。

回顾 Spark 的历史可以发现，在任何规模的数据计算中，Spark 在性能和扩展性上都更具优势。

Hadoop 之父 Doug Cutting 指出：大数据项目的 MapReduce 引擎的使用将下降，由 Apache Spark 取代。

Hadoop 商业发行版本的市场领导者 Cloudera、HortonWorks、MapR 纷纷转投 Spark，并把 Spark 作为大数据解决方案的首选和核心计算引擎。

在 2014 年的 Sort Benchmark 测试中，Spark 秒杀 Hadoop，在使用其十分之一计算资源的情况下，对相同数据的排序，Spark 比 MapReduce 快 3 倍。在没有官方千万亿字节（PB）排序对比的情况下，首次利用 Spark 对 1PB 数据（10 万亿条记录）排序，在使用 190 个节点的情况下，工作负载在 4 小时内完成，同样远超雅虎之前使用 3800 台主机耗时 16 小时的记录。

2015 年 6 月，Spark 最大的集群——8000 个节点来自腾讯，单个最大 Job——1PB 来自阿里巴巴和 Databricks。

IBM 公司于 2015 年 6 月承诺大力推进 Apache Spark 项目，并称该项目为以数据为主导的、未来 10 年最重要的、新的开源项目。

2016 年，在有“计算界奥运会”之称的国际著名 Sort Benchmark 全球数据排序大赛中，由南京大学计算机科学与技术系 PASA 大数据实验室、阿里巴巴和 Databricks 公司组成的参赛团队 NADSort，使用 Apache Spark 大数据计算平台，以 144 美元的成本完成了 100TB 标准数据集的排序处理，创下了每万亿字节（TB）数据排序 1.44 美元成本的世界纪录。

2017 年，Spark Structured streaming 发布无缝整合流处理和其他计算范式。

2018 年，Spark 2.4.0 发布，成为全球最大的开源项目。

2019 年，任何个人和组织都可以基于 Spark 打造符合自己需求的基于大数据的 AI 全生态链计算引擎。

本书以 Spark 2.4.3 为基础，在第 1 版的基础上根据 Spark 的新版本全面更新源码，并以 TensorFlow 和 PyTorch 为核心，大幅度增加人工智能的内容及相应的实战案例。本书以 Spark 内核解密为基石，分为内核解密篇、商业案例篇、性能调优篇和 Spark+AI 内幕解密篇。虽然本书的内容增加了一篇，为了更好地与第 1 版延续，仍沿用三部曲的书名。

（1）内核解密篇。第 1 版基于 Spark 2.2.X 版本源码，从一个动手实战案例入手，循序渐进地全面解析了 Spark 新特性及 Spark 内核源码。第 2 版在第 1 版的基础上，将 Spark 2.2.X

源码更新为 Spark 2.4.X 源码，并对源码的版本更新做了详细解读，帮助读者学习 Spark 源码框架的演进及发展。

（2）商业案例篇。沿用第 1 版的案例内容，选取 Spark 开发中最具代表性的经典学习案例，深入浅出地介绍综合应用 Spark 的大数据技术。

（3）性能调优篇。第 1 版基于 Spark 2.2.X 源码，基本完全覆盖了 Spark 在生产环境下的所有调优技术；第 2 版在第 1 版的基础上，将 Spark 2.2.X 源码更新为 Spark 2.4.X 源码，基于 Spark 2.4.X 版本讲解 Spark 性能调优的内容。

（4）Spark+AI 内幕解密篇。本篇是第 2 版的全新内容，大幅度增加大数据在人工智能领域的应用内容，包括深度学习动手实践：人工智能下的深度学习、深度学习数据预处理、单节点深度学习训练、分布式深度学习训练；Spark+PyTorch 案例实战：PyTorch 在 Spark 上的安装、使用 PyTorch 实战图像识别、PyTorch 性能调优最佳实践；Spark+TensorFlow 实战：TensorFlow 在 Spark 上的安装、TensorBoard 解密、Spark TensorFlow 的数据转换；Spark 上的深度学习内核解密：使用 TensorFlow 进行图片的分布式处理、数据模型源码剖析、逻辑节点源码剖析、构建索引源码剖析、深度学习下 Spark 作业源码剖析、性能调优最佳实践。

在阅读本书的过程中，如发现任何问题或有任何疑问，可以加入本书的阅读群（QQ：418110145）讨论，会有专人答疑。同时，该群也会提供本书所用案例源码及本书的配套学习视频。作者的新浪微博是 http://weibo.com/ilovepains/，欢迎大家在微博上与作者进行互动。

由于时间仓促，书中难免存在不妥之处，请读者谅解，并提出宝贵意见。

王家林 2020 年 4 月于美国硅谷

第 1 版前言

大数据像当年的石油、人工智能（Artificial Intelligence），像当年的电力一样，正以前所未有的广度和深度影响所有的行业，现在及未来公司的核心壁垒是数据，核心竞争力来自基于大数据的人工智能的竞争。Spark 是当今大数据领域最活跃、最热门、最高效的大数据通用计算平台，2009 年诞生于美国加州大学伯克利分校 AMP 实验室，2010 年正式开源，2013 年成为 Apache 基金项目，2014 年成为 Apache 基金的顶级项目。基于 RDD，Spark 成功构建起了一体化、多元化的大数据处理体系。

在任何规模的数据计算中，Spark 在性能和扩展性上都更具优势。

（1）Hadoop 之父 Doug Cutting 指出：Use of MapReduce engine for Big Data projects will decline, replaced by Apache Spark（大数据项目的 MapReduce 引擎的使用将下降，由 Apache Spark 取代）。

（2）Hadoop 商业发行版本的市场领导者 Cloudera、HortonWorks、MapR 纷纷转投 Spark，并把 Spark 作为大数据解决方案的首选和核心计算引擎。

2014 年的 Sort Benchmark 测试中，Spark 秒杀 Hadoop，在使用十分之一计算资源的情况下，相同数据的排序上，Spark 比 MapReduce 快 3 倍！在没有官方千万亿字节（PB）排序对比的情况下，首次将 Spark 推到了 1PB 数据（10 万亿条记录）的排序，在使用 190 个节点的情况下，工作负载在 4 小时内完成，同样远超雅虎之前使用 3800 台主机耗时 16 个小时的记录。

2015 年 6 月，Spark 最大的集群——8000 个节点来自腾讯，单个 Job——1PB 最大分别是阿里巴巴和 Databricks，震撼人心！同时，Spark 的 Contributor 比 2014 年涨了 3 倍，达到 730 人；总代码行数也比 2014 年涨了 2 倍多，达到 40 万行。IBM 于 2015 年 6 月承诺大力推进 Apache Spark 项目，并称该项目为：以数据为主导的，未来十年最重要的新的开源项目。这一承诺的核心是将 Spark 嵌入 IBM 业内领先的分析和商务平台，并将 Spark 作为一项服务，在 IBMBluemix 平台上提供给客户。IBM 还将投入超过 3500 名研究和开发人员在全球 10 余个实验室开展与 Spark 相关的项目，并将为 Spark 开源生态系统无偿提供突破性的机器学习技术——IBM SystemML。同时，IBM 还将培养超过 100 万名 Spark 数据科学家和数据工程师。

2016 年，在有“计算界奥运会”之称的国际著名 Sort Benchmark 全球数据排序大赛中，由南京大学计算机科学与技术系 PASA 大数据实验室、阿里巴巴和 Databricks 公司组成的参赛团队 NADSort，以 144 美元的成本完成 100TB 标准数据集的排序处理，创下了每万亿字节（TB）数据排序 1.44 美元成本的最新世界纪录，比 2014 年夺得冠军的加州大学圣地亚哥分校 TritonSort 团队每万亿字节数据 4.51 美元的成本降低了近 70%，而这次比赛依旧使用 Apache Spark 大数据计算平台，在大规模并行排序算法以及 Spark 系统底层进行了大量的优化，以尽可能提高排序计算性能并降低存储资源开销，确保最终赢得比赛。

在 Full Stack 理想的指引下，Spark 中的 Spark SQL、SparkStreaming、MLLib、GraphX、R 五大子框架和库之间可以无缝地共享数据和操作，这不仅打造了 Spark 在当今大数据计算领域其他计算框架都无可匹敌的优势，而且使得 Spark 正在加速成为大数据处理中心首选通用计算平台，而 Spark 商业案例和性能优化必将成为接下来的重中之重！

本书根据王家林老师亲授课程及结合众多大数据项目经验编写而成，其中王家林、段智华编写了本书近 90%的内容，具体编写章节如下。

第 3 章　Spark 的灵魂：RDD 和 DataSet；
第 4 章　Spark Driver 启动内幕剖析；
第 5 章　Spark 集群启动原理和源码详解；
第 6 章　Spark Application 提交给集群的原理和源码详解；
第 7 章　Shuffle 原理和源码详解；
第 8 章　Job 工作原理和源码详解；
第 9 章　Spark 中 Cache 和 checkpoint 原理和源码详解；
第 10 章　Spark 中 Broadcast 和 Accumulator 原理和源码详解；
第 11 章　Spark 与大数据其他经典组件整合原理与实战；
第 12 章　Spark 商业案例之大数据电影点评系统应用案例；
第 13 章　Spark 2.2 实战之 Dataset 开发实战企业人员管理系统应用案例；
第 14 章　Spark 商业案例之电商交互式分析系统应用案例；
第 15 章　Spark 商业案例之 NBA 篮球运动员大数据分析系统应用案例；
第 16 章　电商广告点击大数据实时流处理系统案例；
第 17 章　Spark 在通信运营商生产环境中的应用案例；
第 18 章　使用 Spark GraphX 实现婚恋社交网络多维度分析案例；
第 23 章　Spark 集群中 Mapper 端、Reducer 端内存调优；
第 24 章　使用 Broadcast 实现 Mapper 端 Shuffle 聚合功能的原理和调优实战；
第 25 章　使用 Accumulator 高效地实现分布式集群全局计数器的原理和调优案例；
第 27 章　Spark 五大子框架调优最佳实践；
第 28 章　Spark 2.2.0 新一代钨丝计划优化引擎；
第 30 章　Spark 性能调优之数据倾斜调优一站式解决方案原理与实战；
第 31 章　Spark 大数据性能调优实战专业之路。

其中，段智华根据自身多年的大数据工作经验对本书的案例等部分进行了扩展。

除上述章节外，剩余内容由夏阳、郑采翎、闫恒伟三位作者根据王家林老师的大数据授课内容而完成。

在阅读本书的过程中，如发现任何问题或有任何疑问，可以加入本书的阅读群（QQ：418110145）讨论，会有专人答疑。同时，该群也会提供本书所用案例源码及本书的配套学习视频。

如果读者想要了解或者学习更多大数据相关技术，可以关注 DT 大数据梦工厂微信公众号 DT_Spark，也可以通过 YY 客户端登录 68917580 永久频道直接体验。

作者的新浪微博是 http://weibo.com/ilovepains/，欢迎大家在微博上与作者进行互动。

由于时间仓促，书中难免存在不妥之处，请读者谅解，并提出宝贵意见。

王家林 2017 年中秋之夜于美国硅谷

目　录

第1篇　内核解密篇

第 3 篇 性能调优篇

第 4 篇　Spark+AI 解密篇

第 1 篇　内核解密篇

第 1 章　电光石火间体验 Spark 2.4 开发实战

本章首先通过一个电影点评系统实战案例来体验一下 Spark 2.4 的程序代码特点。在 1.1 节中，我们将使用弹性分布式数据库（Resilient Distributed Datasets，RDD）的方式来编写 Spark 最基本的程序代码，而在第 1.2 节中，我们使用 DataFrame、DataSet 来感受另一种更易用的程序代码风格。

1.1　通过 RDD 实战电影点评系统入门及源码阅读

日常的数据来源有很多渠道，如网络爬虫、网页埋点、系统日志等。下面的案例中使用的是用户观看电影和点评电影的行为数据，数据来源于网络上的公开数据，共有 3 个数据文件：user.dat、ratings.dat 和 movies.dat。

其中，user.dat 的格式如下：

```
1.  UserID::Gender::Age::Occupation::Zip-code
```

这个文件里共有 6040 个用户的信息，每行中用“::”隔开的详细信息包括 ID、性别（F、M 分别表示女性、男性）、年龄（使用 7 个年龄段标记）、职业和邮编。

ratings.dat 的格式如下：

```
1.  UserID::MovieID::Rating::Timestamp
```

这个文件记录的是评分信息，即用户 ID、电影 ID、评分（满分是 5 分）和时间戳。

movies.dat 的格式如下：

```
1.  MovieID::Title::Genres
```

这个文件记录的是电影信息，即电影 ID、电影名称和电影类型。

1.1.1　Spark 核心概念图解

进入到案例实战前，首先来看几个至关重要的概念，这些概念承载着 Spark 集群运转和程序运行的重要使命。Spark 运行架构图如图 1-1 所示。

Master（图 1-1 中的 Cluster Manager）：就像 Hadoop 有 NameNode 和 DataNode 一样，Spark 有 Master 和 Worker。Master 是集群的领导者，负责管理集群资源，接收 Client 提交的作业，以及向 Worker 发送命令。

Worker（图 1-1 中的 Worker Node）：集群中的 Worker，执行 Master 发送的指令，来具体分配资源，并在这些资源中执行任务。

Driver：一个 Spark 作业运行时会启动一个 Driver 进程，也是作业的主进程，负责作业的解析、生成 Stage，并调度 Task 到 Executor 上。

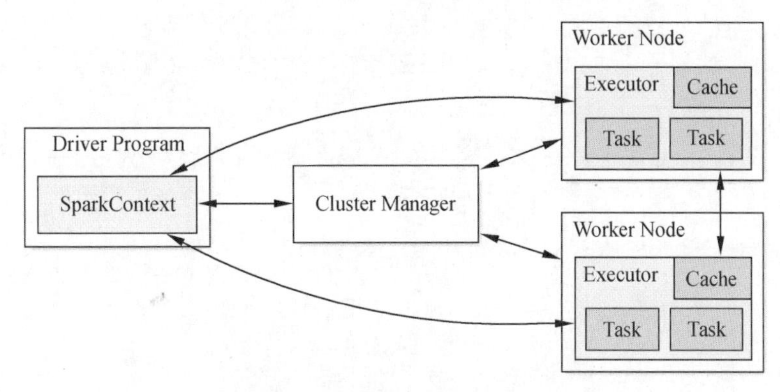

图 1-1　Spark 运行架构图

Executor：真正执行作业的地方。Executor 分布在集群中的 Worker 上，每个 Executor 接收 Driver 的命令来加载和运行 Task，一个 Executor 可以执行一到多个 Task。

SparkContext：是程序运行调度的核心，由高层调度器 DAGScheduler 划分程序的每个阶段，底层调度器 TaskScheduler 划分每个阶段的具体任务。SchedulerBackend 管理整个集群中为正在运行的程序分配的计算资源 Executor。

DAGScheduler：负责高层调度，划分 stage 并生成程序运行的有向无环图。

TaskScheduler：负责具体 stage 内部的底层调度，具体 task 的调度、容错等。

Job：（正在执行的叫 ActiveJob）是 Top-level 的工作单位，每个 Action 算子都会触发一次 Job，一个 Job 可能包含一个或多个 Stage。

Stage：是用来计算中间结果的 Tasksets。Tasksets 中的 Task 逻辑对于同一个 RDD 内的不同 partition 都一样。Stage 在 Shuffle 的地方产生，此时下一个 Stage 要用到上一个 Stage 的全部数据，所以要等到上一个 Stage 全部执行完才能开始。Stage 有两种：ShuffleMapStage 和 ResultStage，除了最后一个 Stage 是 ResultStage 外，其他 Stage 都是 ShuffleMapStage。ShuffleMapStage 会产生中间结果，以文件的方式保存在集群里，Stage 经常被不同的 Job 共享，前提是这些 Job 重用了同一个 RDD。

Task：任务执行的工作单位，每个 Task 会被发送到一个节点上，每个 Task 对应 RDD 的一个 partition。

RDD：是不可变的、Lazy 级别的、粗粒度的（数据集级别的而不是单个数据级别的）数据集合，包含了一个或多个数据分片，即 partition。

另外，Spark 程序中有两种级别的算子：Transformation 和 Action。Transformation 算子会由 DAGScheduler 划分到 pipeline 中，是 Lazy 级别的不会触发任务的执行；Action 算子会触发 Job 来执行 pipeline 中的运算。

介绍完上面的关键概念，下面开始进入到程序编写阶段。

首先写好 Spark 程序的固定框架，以便于在处理和分析数据的时候专注于业务逻辑本身。

创建一个 Scala 的 object 类，在 main 方法中配置 SparkConf 和 SparkContext，这里指定程序在本地运行，并且把程序名字设置为“RDD_Movie_Users_Analyzer”。

RDD_Movie_Users_Analyzer 代码如下：

```
1.   val conf = new SparkConf().setMaster("local[*]")
2.  .setAppName("RDD_Movie_Users_Analyzer")
3.   /**
```

```
   *Spark 2.0 引入 SparkSession 封装了 SparkContext 和 SQLContext，并且会在
   *builder 的 getOrCreate 方法中判断是否有符合要求的 SparkSession 存在，有则使用，
   *没有则进行创建
   */
4.  val spark = SparkSession.builder.config(conf).getOrCreate()
5.  //获取 SparkSession 的 SparkContext
6.  val sc = spark.sparkContext
7.  //把 Spark 程序运行时的日志设置为 warn 级别，以方便查看运行结果
8.  sc.setLogLevel("warn")
9.  //把用到的数据加载进来转换为 RDD，此时使用 sc.textFile 并不会读取文件，而是标记了有
    //这个操作，遇到 Action 级别算子时才会真正去读取文件
10. val usersRDD = sc.textFile(dataPath + "users.dat")
11. val moviesRDD = sc.textFile(dataPath + "movies.dat")
12. val ratingsRDD = sc.textFile(dataPath + "ratings.dat")
13. /**具体数据处理的业务逻辑*/
14. //最后关闭 SparkSession
15. spark.stop
```

1.1.2　通过 RDD 实战电影点评系统案例

首先我们来写一个案例计算，并打印出所有电影中评分最高的前 10 个电影名和平均评分。

第一步：从 ratingsRDD 中取出 MovieID 和 rating，从 moviesRDD 中取出 MovieID 和 Name，如果后面的代码重复使用这些数据，则可以把它们缓存起来。首先把使用 map 算子上面的 RDD 中的每一个元素（即文件中的每一行）以“::”为分隔符进行拆分，然后再使用 map 算子从拆分后得到的数组中取出需要用到的元素，并把得到的 RDD 缓存起来。

```
1.  println("所有电影中平均得分最高（口碑最好）的电影:")
2.  val movieInfo = moviesRDD.map(_.split("::")).map(x => (x(0), x(1))).cache()
3.  val ratings = ratingsRDD.map(_.split("::"))
4.  .map(x => (x(0), x(1), x(2))).cache()
```

第二步：从 ratings 的数据中使用 map 算子获取到形如（movieID,(rating,1)）格式的 RDD，然后使用 reduceByKey 把每个电影的总评分以及点评人数算出来。

```
1.  val moviesAndRatings = ratings.map(x => (x._2, (x._3.toDouble, 1)))
2.       .reduceByKey((x, y) => (x._1 + y._1, x._2 + y._2))
```

此时得到的 RDD 格式为（movieID,(Sum(ratings),Count(ratings))）。

第三步：把每个电影的 Sum(ratings)和 Count(ratings)相除，得到包含了电影 ID 和平均评分的 RDD：

```
1.  val avgRatings= moviesAndRatings.map(x => (x._1,x._2._1.toDouble / x._2._2))
```

第四步：把 avgRatings 与 movieInfo 通过关键字（key）连接到一起，得到形如（movieID,(MovieName,AvgRating)）的 RDD，然后格式化为（AvgRating,MovieName），并按照 key（也就是平均评分）降序排列，最终取出前 10 个并打印出来。

```
1.  avgRatings.join(movieInfo).map(item => (item._2._1,item._2._2))
2.       .sortByKey(false).take(10)
3.       .foreach(record => println(record._2+"评分为: "+record._1))
```

评分最高电影运行结果如图 1-2 所示。

```
Schlafes Bruder (Brother of Sleep) (1995)评分为：5.0
Gate of Heavenly Peace, The (1995)评分为：5.0
Lured (1947)评分为：5.0
```

图 1-2　评分最高电影运行结果

接下来我们来看另外一个功能的实现：分析最受男性喜爱的电影 Top10 和最受女性喜爱的电影 Top10。

首先来分析一下：单从 ratings 中无法计算出最受男性或者女性喜爱的电影 Top10，因为该 RDD 中没有 Gender 信息，如果需要使用 Gender 信息进行 Gender 的分类，此时一定需要聚合。当然，我们力求聚合使用的是 mapjoin（分布式计算的一大痛点是数据倾斜，map 端的 join 一定不会数据倾斜），这里是否可使用 mapjoin？不可以，因为 map 端的 join 是使用 broadcast 把相对小得多的变量广播出去，这样可以减少一次 shuffle，这里，用户的数据非常多，所以要使用正常的 join。

```
1.  Val usersGender = usersRDD.map(_.split("::")).map(x => (x(0), x(1)))
2.  val genderRatings = ratings.map(x => (x._1, (x._1, x._2, x._3)))
3.  .join(usersGender).cache()
4.  genderRatings.take(10).foreach(println)
```

使用 join 连接 ratings 和 users 之后，对分别过滤出男性和女性的记录进行处理：

```
1.  val maleFilteredRatings = genderRatings
2.  .filter(x => x._2._2.equals("M")).map(x => x._2._1)
3.  val femaleFilteredRatings = genderRatings
4.  .filter(x => x._2._2.equals("F")).map(x => x._2._1)
```

接下来对两个 RDD 进行处理，处理逻辑和上面的案例相同，最终打印出来的结果分别如图 1-3 和图 1-4 所示。

```
Schlafes Bruder (Brother of Sleep) (1995)评分为：5.0
Gate of Heavenly Peace, The (1995)评分为：5.0
Lured (1947)评分为：5.0
```

图 1-3　最受男性喜爱的电影运行结果

```
Big Combo, The (1955)评分为：5.0
Gate of Heavenly Peace, The (1995)评分为：5.0
Ayn Rand: A Sense of Life (1997)评分为：5.0
```

图 1-4　最受女性喜爱的电影运行结果

所有电影中最受男性喜爱的电影 Top10 业务代码如下：

```
1.  println("所有电影中最受男性喜爱的电影 Top10:")
2.  maleFilteredRatings.map(x => (x._2, (x._3.toDouble, 1)))
3.    .reduceByKey((x, y) => (x._1 + y._1, x._2 + y._2))
4.  .map(x => (x._1,x._2._1.toDouble / x._2._2))
5.  .join(movieInfo)
6.  .map(item => (item._2._1,item._2._2))
7.  .sortByKey(false) .take(10)
8.  .foreach(record => println(record._2+"评分为: "+record._1))
```

所有电影中最受女性喜爱的电影 Top10 业务代码如下：

```
1.   println("所有电影中最受女性喜爱的电影 Top10:")
2.  femaleFilteredRatings.map(x => (x._2, (x._3.toDouble, 1)))
3.    .reduceByKey((x, y) => (x._1 + y._1, x._2 + y._2))
4.  .map(x => (x._1,x._2._1.toDouble / x._2._2)) .join(movieInfo)
5.  .map(item => (item._2._1,item._2._2)).sortByKey(false) .take(10)
6.  .foreach(record => println(record._2+"评分为: "+record._1))
```

在现实业务场景中，二次排序非常重要，并且经常遇到。下面来模拟一下这些场景，实现对电影评分数据进行二次排序，以 Timestamp 和 Rating 两个维度降序排列，值得一提的是，Java 版本的二次排序代码非常烦琐，而使用 Scala 实现就会很简捷，首先我们需要一个继承自 Ordered 和 Serializable 的类。

```
1.   class SecondarySortKey(val first: Double, val second: Double)
2.  extends Ordered[SecondarySortKey] with Serializable {
3.  //在这个类中重写 compare 方法
4.  override def compare(other: SecondarySortKey): Int = {
5.  //既然是二次排序，那么首先要判断第一个排序字段是否相等，如果不相等，就直接排序
6.  if (this.first - other.first != 0) {
7.        (this.first - other.first).toInt
8.      } else {
9.       //如果第一个字段相等，则比较第二个字段，若想实现多次排序，也可以按照这个模式继
         //续比较下去
10.        if (this.second - other.second > 0) {
11.          Math.ceil(this.second - other.second).toInt
12.        } else if (this.second - other.second < 0) {
13.          Math.floor(this.second - other.second).toInt
14.        } else {
15.          (this.second - other.second).toInt
16.        }
17.      }
18.   }
19. }
```

然后再把 RDD 的每条记录里想要排序的字段封装到上面定义的类中作为 key，把该条记录整体作为 value。

```
1.    println("对电影评分数据以 Timestamp 和 Rating 两个维度进行二次降序排列:")
2.  val pairWithSortkey = ratingsRDD.map(line => {
3.       val splited = line.split("::")
4.       (new SecondarySortKey(splited(3).toDouble, splited(2).toDouble), line)
5.     })
6.  //直接调用 sortByKey，此时会按照之前实现的 compare 方法排序
7.  val sorted = pairWithSortkey.sortByKey(false)
8.
9.  val sortedResult = sorted.map(sortedline => sortedline._2)
10. sortedResult.take(10).foreach(println)
```

取出排序后的 RDD 的 value，此时这些记录已经是按照时间戳和评分排好序的，最终打印出的结果如图 1-5 所示，从图中可以看到已经按照 timestamp 和评分降序排列了。

```
4958::1924::4::1046454590
4958::3264::4::1046454548
4958::2634::3::1046454548
4958::1407::5::1046454443
```

图 1-5　电影系统二次排序运行结果

1.2　通过 DataFrame 和 DataSet 实战电影点评系统

DataFrameAPI 是从 Spark 1.3 开始就有的，它是一种以 RDD 为基础的分布式无类型数据集，它的出现大幅度降低了普通 Spark 用户的学习门槛。

DataFrame 类似于传统数据库中的二维表格。DataFrame 与 RDD 的主要区别在于，前者带有 schema 元信息，即 DataFrame 表示的二维表数据集的每一列都带有名称和类型。这使得 Spark SQL 得以解析到具体数据的结构信息，从而对 DataFrame 中的数据源以及对 DataFrame 的操作进行了非常有效的优化，从而大幅提升了运行效率。

DataSetAPI 是从 1.6 版本提出的，在 Spark 2.2 的时候，DataSet 和 DataFrame 趋于稳定，可以投入生产环境使用。与 DataFrame 不同的是，DataSet 是强类型的，而 DataFrame 实际上就是 DataSet[Row]（也就是 Java 中的 DataSet<Row>）。

DataSet 是 Lazy 级别的，Transformation 级别的算子作用于 DataSet 会得到一个新的 DataSet。当 Action 算子被调用时，Spark 的查询优化器会优化 Transformation 算子形成的逻辑计划，并生成一个物理计划，该物理计划可以通过并行和分布式的方式来执行。

反观 RDD，由于无从得知其中数据元素的具体内部结构，故很难被 Spark 本身自行优化，对于新手用户很不友好，但是，DataSet 底层是基于 RDD 的，所以最终的优化尽头还是对 RDD 的优化，这就意味着优化引擎不能自动优化的地方，用户在 RDD 上可能有机会进行手动优化。

1.2.1　通过 DataFrame 实战电影点评系统案例

现在我们通过实现几个功能来了解 DataFrame 的具体用法。先来看第一个功能：通过 DataFrame 实现某部电影观看者中男性和女性不同年龄分别有多少人。

```
1.  println("功能一：通过 DataFrame 实现某部电影观看者中男性和女性不同年龄人数")
2.  //首先把 Users 的数据格式化，即在 RDD 的基础上增加数据的元数据信息
3.  val schemaForUsers = StructType(
4.  "UserID::Gender::Age::OccupationID::Zip-code".split("::")
5.  .map(column => StructField(column, StringType, true)))
6.  //然后把我们的每一条数据变成以 Row 为单位的数据
7.  val usersRDDRows = usersRDD
8.  .map(_.split("::"))
9.  .map(line =>
10. Row(line(0).trim,line(1).trim,line(2).trim,line(3).trim,line(4).trim))
11. //使用 SparkSession 的 createDataFrame 方法，结合 Row 和 StructType 的元数据信息
    //基于 RDD 创建 DataFrame，这时 RDD 就有了元数据信息的描述
12. val usersDataFrame = spark.createDataFrame(usersRDDRows, schemaForUsers)
13. //也可以对 StructType 调用 add 方法来对不同的 StructField 赋予不同的类型
14. val schemaforratings = StructType("UserID::MovieID".split("::").
15. map(column => StructField(column, StringType, true)))
16. .add("Rating", DoubleType, true)
17. .add("Timestamp",StringType, true)
18.
```

```
19. val ratingsRDDRows = ratingsRDD
20. .map(_.split("::"))
21. .map(line =>
22.  Row(line(0).trim,line(1).trim,line(2).trim.toDouble,line(3).trim))
23. val ratingsDataFrame = spark.createDataFrame(ratingsRDDRows,
    schemaforratings)
24. //接着构建 movies 的 DataFrame
25. val schemaformovies = StructType("MovieID::Title::Genres".split("::")
26. .map(column => StructField(column, StringType, true)))
27. val moviesRDDRows = moviesRDD
28. .map(_.split("::"))
29. .map(line => Row(line(0).trim,line(1).trim,line(2).trim))
30. val moviesDataFrame = spark.createDataFrame(moviesRDDRows,
    schemaformovies)
31.
32. //这里能够直接通过列名 MovieID 为 1193 过滤出这部电影，这些列名都是在上面指定的
33. ratingsDataFrame.filter(s" MovieID = 1193")
34. //Join 的时候直接指定基于 UserID 进行 Join，这相对于原生的 RDD 操作而言更加方便快捷
35. .join(usersDataFrame, "UserID")
36. //直接通过元数据信息中的 Gender 和 Age 进行数据的筛选
37. .select("Gender", "Age")
38. //直接通过元数据信息中的 Gender 和 Age 进行数据的 groupBy 操作
39. .groupBy("Gender", "Age")
40. //基于 groupBy 分组信息进行 count 统计操作，并显示出分组统计后的前 10 条信息
41. .count().show(10)
```

最终打印结果如图 1-6 所示，类似一张普通的数据库表。

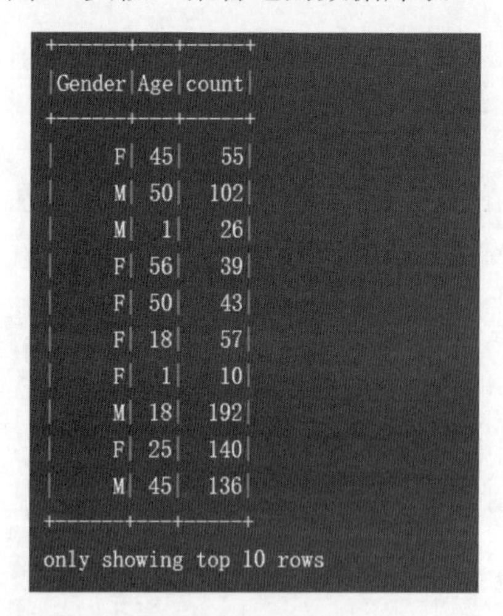

Gender	Age	count
F	45	55
M	50	102
M	1	26
F	56	39
F	50	43
F	18	57
F	1	10
M	18	192
F	25	140
M	45	136

only showing top 10 rows

图 1-6　电影观看者中男性和女性人数

上面案例中的代码无论是从思路上，还是从结构上都和 SQL 语句十分类似，下面通过写 SQL 语句的方式来实现上面的案例。

```
1.  println("功能二：用 LocalTempView 实现某部电影观看者中不同性别不同年龄分别有多少
    人？")
2.  //既然使用 SQL 语句，那么表肯定是要有的，所以需要先把 DataFrame 注册为临时表
3.  ratingsDataFrame.createTempView("ratings")
4.  usersDataFrame.createTempView("users")
```

```
5.  //然后写 SQL 语句，直接使用 SparkSession 的 sql 方法执行 SQL 语句即可
6.  val sql_local = "SELECT Gender, Age, count(*) from  users u join
7.  ratings as r on u.UserID = r.UserID where MovieID = 1193 group by Gender,
    Age"
8.  spark.sql(sql_local).show(10)
```

这样我们就可以得到与上面案例相同的结果，这对写 SQL 比较多的用户是十分友好的。但是有一个问题需要注意，这里调用 createTempView 创建的临时表是会话级别的，会话结束时这个表也会消失。那么，怎么创建一个 Application 级别的临时表呢？可以使用 createGlobalTempView 来创建临时表，但是这样就要在写 SQL 语句时在表名前面加上 global_temp，例如：

```
1.   ratingsDataFrame.createGlobalTempView("ratings")
2.  usersDataFrame.createGlobalTempView("users")
3.
4.  val sql = "SELECT Gender, Age, count(*) from  global_temp.users u join
    global_temp.ratings as r on u.UserID = r.UserID where MovieID = 1193 group
    by Gender, Age"
5.  spark.sql(sql).show(10)
```

第一个 DataFrame 案例实现了简单的类似 SQL 语句的功能，但这是远远不够的，我们要引入一个隐式转换来实现复杂的功能。

```
1.   import spark.sqlContext.implicits._
2.  ratingsDataFrame.select("MovieID", "Rating")
3.  .groupBy("MovieID").avg("Rating")
4.  //接着我们可以使用“$”符号把引号里的字符串转换成列来实现相对复杂的功能，例如，下面
    //我们把 avg(Rating)作为排序的字段降序排列
5.  .orderBy($"avg(Rating)".desc).show(10)
```

从图 1-7 的结果可以看到，求平均值的那一列列名和在 SQL 语句里使用函数时的列名一样变成了 avg(Rating)，程序中的 orderBy 里传入的列名要和这个列名一致，否则会报错，提示找不到列。

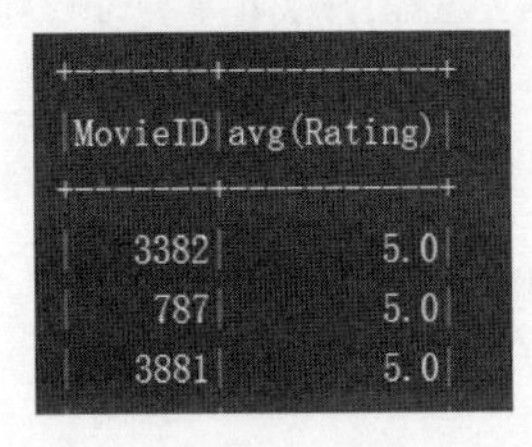

图 1-7　电影系统 SQL 运行结果

有时我们也可能会在使用 DataFrame 的时候在中间某一步转换到 RDD 里操作，以便实现更加复杂的逻辑。下面来看一下 DataFrame 和 RDD 的混合编程。

```
1.  ratingsDataFrame.select("MovieID", "Rating")
2.  .groupBy("MovieID").avg("Rating")
3.  //这里直接使用 DataFrame 的 rdd 方法转到 RDD 里操作
4.  .rdd.map(row =>(row(1),(row(0), row(1))))
5.  .sortBy(_._1.toString.toDouble, false)
6.  .map(tuple => tuple._2)
7.  .collect.take(10).foreach(println)
```

1.2.2　通过 DataSet 实战电影点评系统案例

前面提到的 DataFrame 其实就是 DataSet[Row]，所以只要学会了 DataFrame 的使用，就可以快速接入 DataSet，只不过在创建 DataSet 的时候要注意与创建 DataFrame 的方式略有不同。DataSet 可以由 DataFrame 转换而来，只需要用 yourDataFrame.as[yourClass]即可得到封装了 yourClass 类型的 DataSet，之后就可以像操作 DataFrame 一样操作 DataSet 了。接下来我们讲一下如何直接创建 DataSet，因为 DataSet 是强类型的，封装的是具体的类（DataFrame 其实封装了 Row 类型），而类本身可以视作带有 Schema 的，所以只需要把数据封装进具体的类，然后直接创建 DataSet 即可。

首先引入一个隐式转换，并创建几个 caseClass 用来封装数据。

```
1.  import spark.implicits._
2. case class User(UserID:String, Gender:String, Age:String, OccupationID:
   String, Zip_Code:String)
3. case  class  Rating(UserID:String,  MovieID:String,  Rating:Double,
   Timestamp:String)
4. 然后把数据封装进这些 Class:
5. val usersForDSRDD = usersRDD.map(_.split("::")).map(line =>
6.     User(line(0).trim,line(1).trim,line(2).trim,line(3).trim,line(4).trim))
7.     最后直接创建 DataSet:
8. val usersDataSet = spark.createDataset[User](usersForDSRDD)
9. usersDataSet.show(10)
10.
```

电影系统运行结果如图 1-8 所示，列名为 User 类的属性名。下面使用同样的方法创建 ratingsDataSet 并实现一个案例：找出观看某部电影的不同性别不同年龄的人数。

```
1.  val ratingsForDSRDD = ratingsRDD.map(_.split("::")).map(line =>
2. Rating(line(0).trim,line(1).trim,line(2).trim.toDouble,line(3).trim))
3. val ratingsDataSet = spark.createDataset(ratingsForDSRDD)
4. //下面的实现代码和使用 DataFrame 方法几乎完全一样（把 DataFrame 换成 DataSet 即可）
5. ratingsDataSet.filter(s" MovieID = 1193").join(usersDataSet, "UserID")
6.      .select("Gender", "Age").groupBy("Gender", "Age").count()
7.      .orderBy($"Gender".desc,$"Age").show()
```

观看电影性别、年龄统计结果如图 1-9 所示。

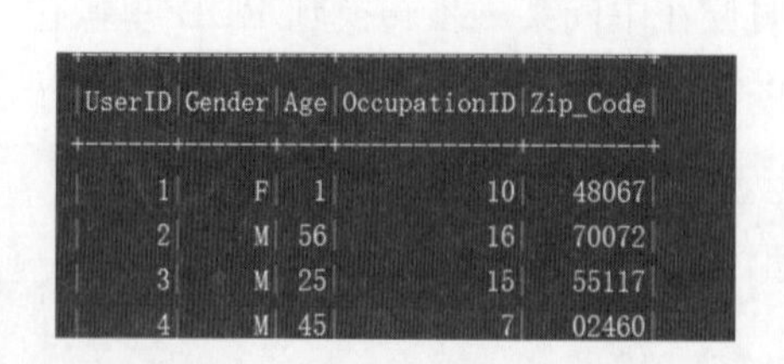

UserID	Gender	Age	OccupationID	Zip_Code
1	F	1	10	48067
2	M	56	16	70072
3	M	25	15	55117
4	M	45	7	02460

图 1-8　电影系统运行结果

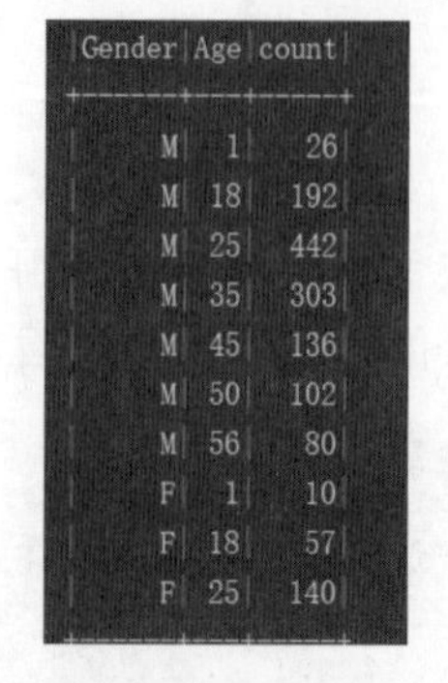

Gender	Age	count
M	1	26
M	18	192
M	25	442
M	35	303
M	45	136
M	50	102
M	56	80
F	1	10
F	18	57
F	25	140

图 1-9　观看电影性别、年龄统计结果

当然，也可以把 DataFrame 和 DataSet 混着用（这样做会导致代码混乱，故不建议这样做），得到的结果完全一样。

最后根据源码，有几点需要补充：

RDD 的 cache 方法等于 MEMORY_ONLY 级别的 persist，而 DataSet 的 cache 方法等于 MEMORY_AND_DISK 级别的 persist，因为重新计算的代价非常昂贵。如果想使用其他级别的缓存，可以使用 persist 并传入相应的级别。

RDD.scala 源码如下：

```
1.   /**
2.     *使用默认的存储级别持久化 RDD ('MEMORY_ONLY').
3.     */
4.    def cache(): this.type = persist()
```

Dataset.scala 源码如下：

```
1.   /**
2.     *使用默认的存储级别持久化 DataSet ('MEMORY_AND_DISK').
3.     *
4.     * @group basic
5.     * @since 1.6.0
6.     */
7.    def cache(): this.type = persist()
```

基于 DataSet 的计算会像 SQL 一样被 Catalyst 引擎解析生成执行查询计划，然后执行。我们可以使用 explain 方法来查看执行计划。

1.3　Spark 2.4 源码阅读环境搭建及源码阅读体验

对于 Spark 的应用，仅仅会使用其 API 来编程只能达到初级（助理）工程师或中级（熟练）工程师的水平，而学会调优则可以让你进阶为高级工程师。那么，怎么成为顶尖的工程师呢？源码！源码毫无保留地展示了 Spark 巧妙的实现原理和严谨的工作流程，同时也可能暴露出了 Spark 的缺陷；它不仅可以让我们深入地理解我们之前写过的代码的每一行的背后隐藏了什么，也可以让我们进一步改造 Spark 来更加完美地配合我们的业务。下面简单介绍一下源码阅读环境的搭建。

Spark 源码是使用 Scala 语言编写的，这是一个基于 JVM 的语言，与 Java 有很多类似之处，在决定读源码之前需先学会 Scala。

首先，准备好配置了 Scala 2.11 和 Maven 的集成开发环境（这里以 IntelliJ IDEA 为例），然后到 Spark 官网（http://spark.apache.org/downloads.html）下载并解压源码，如图 1-10 所示。

Spark 源码安装包解压之后，打开 IDEA，单击 Import Project，如图 1-11 所示选择 Maven 方式，然后单击 Next 按钮。

出现图 1-12，勾选自动导入 Maven 项目，自动下载 Source 和 Documentation，在右下角的 Environment settings 里面选择要使用的 Maven 和 Maven 配置文件。接着单击 Next 按钮。

图 1-10　Spark 官网下载示意图

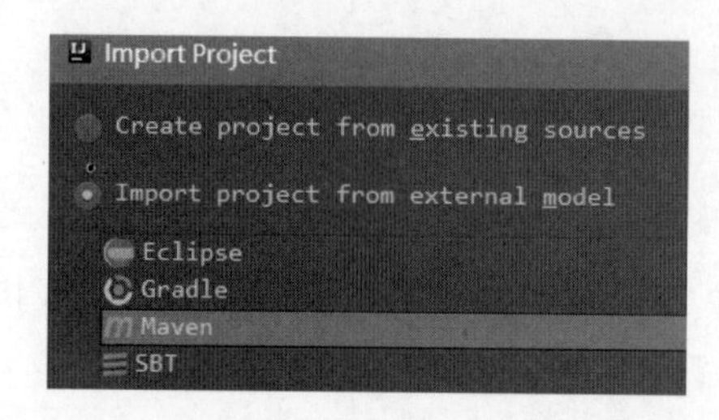

图 1-11　Maven 的方式导入源码

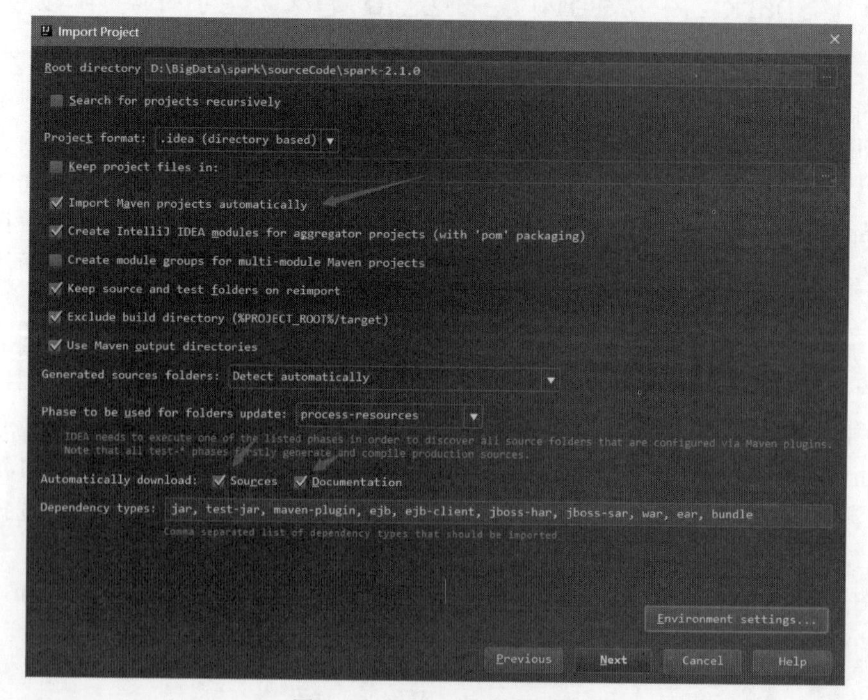

图 1-12　Maven 选项示意图

然后选择要加载的 profile，根据自己的需要勾选，然后一直单击 Next 按钮，直到出现 Finish。IEDA 打开后会引入源码并解析 pom 文件开始下载源码需要的 Jar 包（建议 Maven 配置文件的仓库选择比较快的镜像，否则会下载很长时间）。

最后把每个源码目录下面的/src/main/里的 scala 文件夹选中，在右键选项里选择 Mark Directory as : Sources Root，如图 1-13 所示。

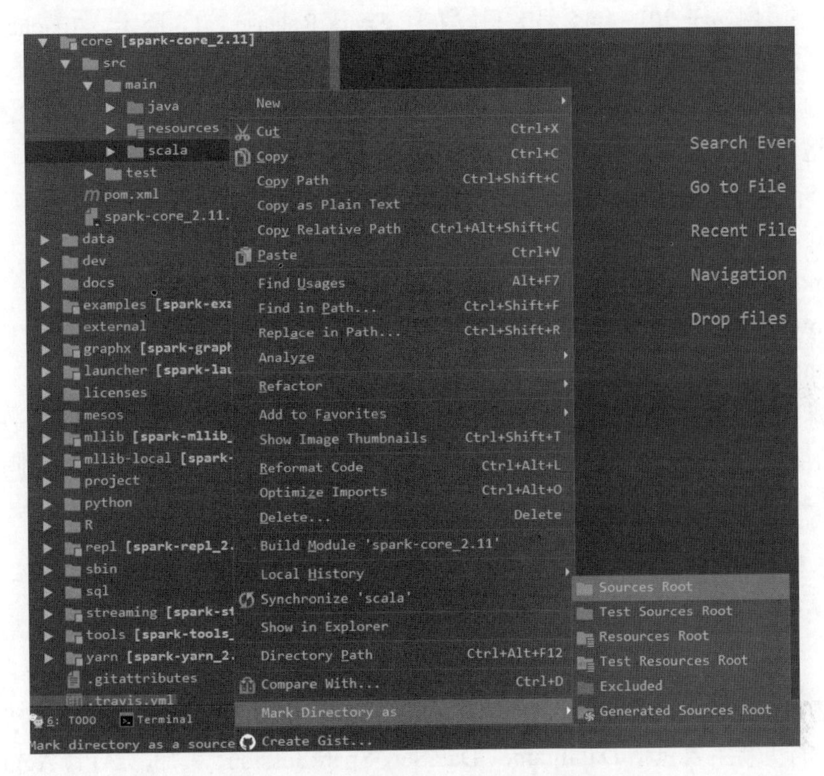

图 1-13　Mark Sources Root 示意图

如果在编写程序的时候随时查看源码，可以新建一个应用程序，创建完 SparkSession 之后，按住 Ctrl 键并单击 SparkSession，在右上角单击 attachsources，选中源码目录即可把源码关联进来。

搭建好源码环境后，可根据我们使用 Spark 的步骤阅读源码。

首先，使用 start-all.sh 启动集群。在 sbin 目录下可以找到这个脚本，打开后会看到 start-master.sh 和 start-salves.sh 两个脚本，前者会引导我们进入 Master 类（在 IDEA 中双击 Shift 键，输入 Master，打开 Master.scala），Master 继承自 RpcEndpoint，所以直接看其 onStart 方法即可，接着跟踪代码就可以明白 Master 的启动流程。在 start-salves.sh 脚本里最终会引导我们进入 Worker 类，同样可以直接看其 onStart 方法。

启动完集群之后，接着提交任务。在 bin 目录下有一个 spark-submit 文件用来提交程序，这个文件会引导我们进入 SparkSubmit 类，我们可以直接看它伴生类里的 main 方法，然后可跟着代码一步一步阅读。

这里有几个重点：资源分配（Master 里的 schedule 方法）、DAGScheduler、TaskScheduler、ScheduleBackend 以及 Shuffle 等。

第 2 章　Spark 2.4 技术及原理

Apache 官方网站于 2019 年 5 月 8 日发布了 Spark Release 2.4.3 版本。Apache Spark 2.4.3 版本是 Spark 2.4 系列上的第 3 个版本。 Spark 2.2.0 是 Spark 2.2 中第一个在生产环境可以使用的版本，对于 Spark 具有里程碑意义。Spark 2.4.X 版本在 Spark 2.2.X、Spark 2.3.X 的基础上进行了改进。

Apache Spark 2.4.X 版本的一些新变化如下。

- Core and Spark SQL 核心和 Spark SQL。
- Windows 性能和稳定性。
- Known Issues 已知的问题。
- Notable Changes 显著变化。

如无特殊说明，本书所有内容都基于最新最稳定的 Spark 2.4.3 版本的源码编写，为体现 Spark 源码的演进过程，部分核心源码在 Spark 1.5.X、Spark 1.6.X、Spark 2.2.X 源码的基础上，新增 Spark 2.4.3 版本的源码，便于读者系统比对、研习 Spark 源码。

2.1　Spark 2.4 综述

Spark 2.0 中更新发布了新的流处理框架（Structured Streaming）；对于 API 的更新，Spark 2.0 版本 API 的更新主要包括 DataFrame、DataSet、SparkSession、累加器 API、Aggregator API 等 API 的变动。

2.1.1　连续应用程序

自从 Spark 得到广泛使用以来，其流处理框架 Spark Streaming 也逐渐吸引到了很多用户，得益于其易用的高级 API 和一次性语义，使其成为使用最广泛的流处理框架之一。但是，我们不仅需要流处理来构建实时应用程序，很多时候我们的应用程序只有一部分需要用到流处理，对于这种应用程序，Databricks 公司把它称为 Continuous Application（实时响应数据的端到端的应用程序），也就是连续的应用程序。在 Continuous Application 中有许多难点，如数据交互的完整性、流数据与离线数据的结合使用、在线机器学习等。

Spark 2.0 最重磅的更新是新的流处理框架——Structured Streaming。它允许用户使用 DataFrame/DataSetAPI 编写与离线批处理几乎相同的代码，便可以作用到流数据和静态数据上，引擎会自动增量化流数据计算，同时保证了数据处理的一致性，并且提供了和存储系统的事务集成。

2.1.2　新的 API

在 Spark 2.0 版本的 API 中，共有如下几个 API 的变动。

（1）统一了 DataFrame 和 DataSet。现在 DataFrame 不再是一个独立的类，而是作为 DataSet[Row]的别名定义在 org.apache.spark.sql 这个包对象中。

sql\package.scala 源码如下：

```
1.    package object sql {
2.
3.   /**
      * 将一个逻辑计划转换为零个或多个 SparkPlans。这个 API 是查询计划实验使用，不是为
      * Spark 稳定发行版设计的。编写库的开发者应该考虑使用[[org.apache.spark.sql.
      * sources]]提供的稳定 APIs
4.    */
5.    @DeveloperApi
6.    @InterfaceStability.Unstable
7.    type Strategy = SparkStrategy
8.
9.    type DataFrame = Dataset[Row]
10. }
```

（2）加入了 SparkSession，用于替换 DataFrame 和 Dataset API 的 SQLContext 和 HiveContext（这两个 API 仍然可以使用）。

（3）为 SparkSession 和 SparkSQL 加入一个新的，精简的配置参数——RuntimeConfig，用来设置和获得与 SparkSQL 有关的 Spark 或者 Hadoop 设置。

SparkSession.scala 源码如下：

```
1.   /**
2.     * Spark 运行时的配置接口
       *
3.     *  这是用户可以获取并设置所有 Spark 和 Hadoop 的接口。将触发 Spark SQL 相关的配
       *  置。当获取配置值时，默认设置值在 SparkContext 里
4.     * @since 2.0.0
5.     */
6.    @transient lazy val conf: RuntimeConfig = new RuntimeConfig
      (sessionState.conf)
```

（4）更简单、更高性能的累加器 API。

（5）用于 DataSet 中类型化聚合的新的改进的 Aggregator API。

2.2　Spark 2.4 Core

本节讲解第二代 Tungsten 引擎、SparkSession 和累加器 API 的使用。

2.2.1　第二代 Tungsten 引擎

Spark 备受瞩目的原因之一在于它的高性能，Spark 开发者为了保持这个优势，一直在不

断地进行各种层次的优化，其中最令人兴奋的莫过于钨丝计划（Project Tungsten），因为钨丝计划的提出给 Spark 带来了极大的性能提升，并且在一定程度上引导了 Spark 的发展方向。

Spark 是使用 Scala 和 Java 语言开发的，不可避免地运行于 JVM 之上。当然，内存管理也是依赖于 JVM 的内存管理机制，而对于大数据量的基于内存的处理，JVM 对象模型对内存的额外开销，以及频繁的 GC 和 Full GC 都是非常致命的问题。另外，随着网络带宽和磁盘 I/O 的不断提升，内存和 CPU 又重新作为性能瓶颈受到关注，JVM 对象的序列化、反序列化带来的性能损耗亟待解决。Spark 1.5 版本加入的钨丝计划从 3 大方面着手解决这些问题：

（1）统一内存管理模型和二进制处理（Binary Processing）。统一内存管理模型代替之前基于 JVM 的静态内存管理，引入 Page 来管理堆内存和堆外内存（on-heap 和 off-heap），并且直接操作内存中的二进制数据，而不是 Java 对象，很大程度上摆脱了 JVM 内存管理的限制。

（2）基于缓存感知的计算（Cache-aware Computation）。Spark 内存读取操作也会带来一部分性能损耗，钨丝计划便设计了缓存友好的算法和数据结构来提高缓存命中率，充分利用 L1/L2/L3 三级缓存，大幅提高了内存读取速度，进而缩短了内存中整个计算过程的时间。

（3）代码生成（Code Generation）。在 JVM 中，所有代码的执行由解释器一步步地解释执行，CodeGeneration 这一功能则在 Spark 运行时动态生成用于部分算子求值的 bytecode，减少了对基础数据类型的封装，并且缓解了调用虚函数的额外开销。

Spark 2.0 升级了第二代 Tungsten 引擎。其中最重要的一点是把 CodeGeneration 作用于全阶段的 SparkSQL 和 DataFrame 之上（即全阶段代码生成 Whole Stage Code Generation），为常见的算子带来 10 倍左右的性能提升。

2.2.2　SparkSession

加入 SparkSession，取代原来的 SQLContext 和 HiveContext，为了兼容两者，仍然保留。SparkSession 使用方法如下：

```
1.    SparkSession.builder()
2.        .master("local")
3.        .appName("Word Count")
4.        .config("spark.some.config.option", "some-value")
5.        .getOrCreate()
```

首先获得 SparkSession 的 Builder，然后使用 Builder 为 SparkSession 设置参数，最后使用 getOrCreate 方法检测当前线程是否有一个已经存在的 Thread-local 级别的 SparkSession，如果有，则返回它；如果没有，则检测是否有全局级别的 SparkSession，有，则返回，没有，则创建新的 SparkSession。

在程序中如果要使用 SparkContext，调用 sparkSession.sparkContext 即可。在程序的最后我们需要调用 sparkContext.stop 方法，这个方法会调用 sparkContext.stop 来关闭 sparkContext。

从 Spark 2.0 开始，DataFrame 和 DataSet 既可以容纳静态、有限的数据，也可以容纳无限的流数据，所以用户也可以使用 SparkSession 像创建静态数据集一样来创建流式数据集，并且可以使用相同的操作算子。这样，整合了实时流处理和离线处理的框架，结合其他容错、扩展等特性就形成了完整的 Lambda 架构。

2.2.3 累加器 API

Spark 2.0 引入了一个更加简单和更高性能的累加器 API，如在 1.X 版本中可以这样使用累加器。

```
1.  //定义累加器，这里直接使用 SparkContext 内置的累加器，设置初始值为 0，名字为"My
    //Accumulator"
2.  val accum = sc.accumulator(0, "My Accumulator")
3.  //计算值
4.  sc.parallelize(Array(1, 2, 3, 4)).foreach(x => accum += x)
5.  //获取累加器的值，（Executor 上面只能对累加器进行累加操作，只有 Driver 才能读取累加
    //器的值，Driver 读取值的时候会把各个 Executor 上存储的本地累加器的值加起来），这里
    //的结果是 10
6.  accum.value
```

在 Spark 2.X 版本里使用 SparkContext 里内置的累加器。

```
1.   //与 Spark 1.X 不同的是，需要指定累加器的类型，目前 SparkContext 有 Long 类型和
     //Double 类型的累加器可以直接使用（不需要指定初始值）
2.  val accum = sc.longAccumulator("My Accumulator")
3.  sc.parallelize(Array(1, 2, 3, 4)).foreach(x => accum.add(x))
4.  print(accum.value)
```

只使用 SparkContext 里内置的累加器功能肯定不能满足略微复杂的业务类型，此时我们就可以自定义累加器。在 1.X 版本中的做法是（下面是官网的例子）：

```
1.   //继承 AccumulatorParam[Vector]，返回类型为 Vector
2.  object VectorAccumulatorParam extends AccumulatorParam[Vector] {
3.  //定义“零”值，这里把传入的初始值的 size 作为“零”值
4.  def zero(initialValue: Vector): Vector = {
5.      Vector.zeros(initialValue.size)
6.    }
7.  //定义累加操作的计算方式
8.   def addInPlace(v1: Vector, v2: Vector): Vector = {
9.      v1 += v2
10.   }
11. }
```

上面的累加器元素和返回类型是相同的，在 Scala 中还有另外一种方式来自定义累加器，用户只需要继承 Accumulable，就可以把元素和返回值定义为不同的类型，这样我们就可以完成添加操作（如往 Int 类型的 List 里添加整数，此时元素为 Int 类型，而返回类型为 List）。

在 Spark 2.X 中加入一个新的抽象类——AccumulatorV2，继承这个类要实现以下几种方法。

add 方法：指定元素相加操作。

copy 方法：指定对自定义累加器的复制操作。

isZero 方法：返回该累加器的值是否为“零”。

merge 方法：合并两个相同类型的累加器。

reset 方法：重置累加器。

value 方法：返回累加器当前的值。

重写这几种方法之后，只需实例化自定义累加器，并连同累加器名字一起传给 sparkContext.register 方法。

下面简单实现一个把字符串合并为数组的累加器。

```
1.   //首先要继承 AccumulatorV2，并指定输入为 String 类型，输出为 ArrayBuffer[String]
2.  class MyAccumulator extends AccumulatorV2[String, ArrayBuffer[String]] {
3.  //设置累加器的结果，类型为 ArrayBuffer[String]
4.    private var result = ArrayBuffer[String]()
5.
6.  //判断累加器当前值是否为“零值”，这里我们指定如果 result 的 size 为 0，则累加器的当
    //前值是“零值”
7.    override def isZero: Boolean = this.result.size == 0
8.
9.  //copy 方法设置为新建本累加器，并把 result 赋给新的累加器
10.   override def copy(): AccumulatorV2[String, ArrayBuffer[String]] = {
11.     val newAccum = new MyAccumulator
12.     newAccum.result = this.result
13.     newAccum
14.   }
15. //reset 方法设置为把 result 设置为新的 ArrayBuffer
16. override def reset(): Unit = this.result == new ArrayBuffer[String]()
17.
18. //add 方法把传进来的字符串添加到 result 内
19. override def add(v: String): Unit = this.result += v
20.
21. //merge 方法把两个累加器的 result 合并起来
22. override def merge(other: AccumulatorV2[String, ArrayBuffer[String]]):
    Unit = {
23.       result.++=:(other.value)
24.     }
25. //value 方法返回 result
26.   override def value: ArrayBuffer[String] = this.result
27. }
28. //接着在 main 方法里使用累加器
29. val Myaccum = new MyAccumulator()
30.
31. //向 SparkContext 注册累加器
32.     sc.register(Myaccum)
33.
34. //把“a”“b”“c”“d”添加进累加器的 result 数组并打印出来
35.     sc.parallelize(Array("a","b","c","d")).foreach(x => Myaccum.add(x))
36.     println(Myaccum.value)
```

运行结果显示的 ArrayBuffer 里的值顺序是不固定的，取决于各个 Executor 的值到达 Driver 的顺序。

2.3　Spark 2.4 SQL

Spark 2.0 通过对 SQL 2003 的支持增强了 SQL 功能，Catalyst 新引擎提升了 Spark 查询优化的速度；本节对 DataFrame 和 Dataset API、时间窗口进行了讲解。

在 Apache Spark 2.2.X 版本的基础上，Apache Spark 2.3.X 版本、Apache Spark 2.4.X 版本中核心和 Spark SQL 的主要更新如下。

1. API更新

- SPARK-18278：Spark 基于 Kubernetes。一个新的 Kubernetes 调度器后端，支持将 Spark 作业提交到 Kubernetes 进行集群管理。这种支持目前是实验性的，应该对配置、容器映像和入口点进行更改。
- SPARK-16060：矢量化 ORC 阅读器。增加了对新 ORC 阅读器的支持，通过矢量化（2～5 倍）大大提高了 ORC 扫描吞吐量。要启用阅读器，用户可以将 spark.sql.orc.impl 设置为 native。
- SPARK-18085：Spark History Server V2。一个新的 Spark History Server（SHS）后端，通过更高效的事件存储机制为大规模应用程序提供更好的可扩展性。
- SPARK-15689、SPARK-22386：Data Source API V2。在 Spark 中插入新数据源的实验 API。新的 API 试图解决 V1 版本 API 的几个局限性，旨在促进高性能、易于维护和可扩展的外部数据源的开发。此 API 仍在进行开发。
- SPARK-22216、SPARK-21187：通过快速数据序列化和矢量化执行显著提高了 Python 的性能和互操作性。
- SPARK-26266：对 Scala 2.12.8 更新（需要最近的 Java 8 版本）。
- SPARK-26188：Spark 2.4.0 分区行为打破向后兼容性。
- SPARK-27198：driver 和 executor 中的心跳间隔不匹配。
- SPARK-24374：屏障执行模式。在调度器中支持屏障执行模式，更好地与深度学习框架集成。
- SPARK-14220：可以使用 Scala 2.12 构建 Spark，并在 Scala 2.12 中编写 Spark 应用程序。
- SPARK-23899：高阶函数。添加许多新的内置函数，包括高阶函数，以便更容易地处理复杂的数据类型。
- SPARK-24768：内置 AVRO 数据源。内嵌 Spark-Avro 包，具有逻辑类型支持、更好的性能和可用性。
- SPARK-24035：透视转换的 SQL 语法。
- SPARK-24940：用于 SQL 查询的 Coalesce 和 Repartition 提示。
- SPARK-19602：支持完全限定列名的列解析。
- SPARK-21274：实现 EXCEPT ALL 及 INTERSECT ALL。

2. 性能优化和系统稳定性

- SPARK-21975：基于成本的优化器中的直方图支持。
- SPARK-20331：更好地支持 Hive 分区修剪的谓词下推。
- SPARK-19112：支持 ZStandard 标准压缩编解码器。
- SPARK-21113：支持预读输入流在溢出读取器中分摊磁盘 I/O 成本。
- SPARK-22510、SPARK-22692、SPARK-21871：进一步稳定 codegen 框架，以避免 Java 方法和 Java 编译器常数池受到 64KB JVM 字节码限制。
- SPARK-23207：修复了 Spark 中一个长期存在的 BUG，在这种情况下，数据帧上的连续洗牌+重新分区可能导致某些案例的不正确。

- SPARK-22062、SPARK-17788、SPARK-21907：解决 OOM 的各种原因。
- SPARK-22489、SPARK-22916、SPARK-22895、SPARK-20758、SPARK-22266、SPARK-19122、SPARK-22662、SPARK-21652：基于规则的优化器和规划器中的增强功能。
- SPARK-26080：无法在 Windows 上运行 worker.py。
- SPARK-27419：如果设置值 spark.executor.heartbeatinterval 少于 1 秒，将经常失败，因为值将被转换为 0，心跳将始终超时，最终杀死 executor。
- SPARK-25535：解决常见加密中的错误检查。
- SPARK-26891：RequestExecutors 反映节点黑名单并可序列化。
- SPARK-27496：RPC 返回致命错误。
- SPARK-27544：修复 Python 测试脚本以在 Scala 2.12 版本上工作。
- SPARK-27469：将 commons beanutils 更新为 1.9.3。
- SPARK-16406：大量列的引用解析应该更快。
- SPARK-23486：lookupFunctions 从外部目录缓存函数名以查找函数。
- SPARK-23803：支撑桶修剪。
- SPARK-24802：优化规则排除。
- SPARK-4502：Parquet 表的嵌套方案修剪。
- SPARK-24296：支持大于 2GB 的复制块。
- SPARK-24307：支持从内存发送超过 2GB 的消息。
- SPARK-23243：随机洗牌+对 RDD 重新分区可能导致错误答案。
- SPARK-25181：限制了 BlockManager 主线程池和从线程池的大小，在网络速度较慢时降低了内存开销。

3. 更新变化

- SPARK-22036：默认情况下，如果无法精确表示，小数之间的算术运算将返回舍入值（以前的版本中返回空值）。
- SPARK-22937：当所有输入都是二进制的时候，sql elt()返回一个二进制输出。否则将作为字符串返回。在以前的版本中，不管输入类型如何，总是以字符串形式返回。
- SPARK-22895：如果可能，在第一个非确定性谓词之后的连接/筛选器的确定性谓词也会向下通过子运算符推送。在以前的版本中，这些过滤器不符合谓词下推的条件。
- SPARK-22771：当所有输入都是二进制的时候，functions.concat()返回一个二进制输出。否则，它将作为字符串返回。在以前的版本中，不管输入类型如何，它总是以字符串形式返回。
- SPARK-22489：当任何一个 join 端可以广播时，广播提示中显式指定的表。
- SPARK-22165：分区列推断以前发现不同推断类型的公共类型不正确，例如，以前它以 double 类型作为 double 类型和 date 类型的公共类型结束，现在，它为此类冲突找到了正确的通用类型。
- SPARK-22100：percentile_approx 函数以前接受 numeric 数字类型输入，并输出 double 类型结果。现在支持日期类型、时间戳类型和数字类型作为输入类型，结果类型也更改为与输入类型相同，这对百分位数更为合理。

- SPARK-21610：当引用的列只包含内部损坏记录列（默认情况下名为“损坏记录”）时，不允许从原始 JSON/CSV 文件查询。相反，可以缓存或保存解析的结果，然后发送相同的查询。
- SPARK-23421：Spark 2.2.1 和 2.3.0，当数据源表具有同时存在于分区模式和数据模式中的列时，总是在运行时推断模式。推断模式没有分区列。读取表时，Spark 将查询这些重叠列的分区值，而不是数据源文件中存储的值。在 2.2.0 和 2.1.X 版本中，推断的模式是分区的，但用户看不到表的数据（结果集为空）。
- SPARK-19732：na.fill()或 fillna 也接受布尔值，并用布尔值替换空值。在以前的 Spark 版本中，pyspark 只是忽略它并返回原始的数据集/数据帧。
- SPARK-22395：Pandas 0.19.2 或更高版本需要使用 Pandas 相关功能，如 toPandas、从 Pandas DataFrame 创建数据帧等。
- SPARK-22395：与 panda 相关功能的时间戳值的行为已更改为会话时区，这在以前的版本中被忽略。
- SPARK-23328：在 to_replace 不是字典时，df.replace 不允许省略值。以前，值在其他情况下可以省略，默认情况下没有值，这是反直觉的，并且容易出错。
- SPARK-20236：支持 Hive 动态分区覆盖语义。
- SPARK-4131：支持 INSERT OVERWRITE DIRECTORY，直接从查询将数据写入文件系统。
- SPARK-19285、SPARK-22945、SPARK-21499、SPARK-20586、SPARK-20416、SPARK-20668： UDF 增强。
- SPARK-20463、SPARK-19951、SPARK-22934、SPARK-21055、SPARK-17729、SPARK-20962、SPARK-20963、SPARK-20841、SPARK-17642、SPARK-22475、SPARK-22934：改进了 ANSI SQL 兼容性和 Hive 兼容性。
- SPARK-20746：更全面的 SQL 内置函数。
- SPARK-21485：为内置函数生成 Spark SQL 文档。
- SPARK-19810：移除对 Scala 2.10 的支持。
- SPARK-22324：将 Arrow 升级到 0.8.0，neNetty 升级到 4.1.17。
- SPARK-25250：可能导致 job 永久挂起，将在 2.4.2 中恢复。
- SPARK-24935：从 Spark 2.2 开始执行 Hive UDF 时出现问题。
- SPARK-27539：修复包含空值的列的聚合输出行估计不准确的问题。
- SPARK-27563：自动获取 HiveExternalCatalogVersionsSuite 中的最新 Spark 版本。
- SPARK-24596：非级联缓存无效。
- SPARK-23880：不触发任何用于缓存数据的作业。
- SPARK-23510：支持 Hive 2.2 和 Hive 2.3 元存储。
- SPARK-23711：为 UnsafeProjection 添加 fallback 生成器。
- SPARK-24626：Analyze Table 命令中的并行位置大小计算。

4．Bug修复

- SPARK-26961：在 Spark Driver 中发现 Java 级死锁。
- SPARK-26998：在独立模式下执行 executor 进程时，spark.ssl.keyStorePassword 在

“ps -ef” 输出是明文形式。

- SPARK-27216：将 RoaringBitmap 升级到 0.7.45 以修复 kryo 不安全的 ser/dser 问题。
- SPARK-27244：使用选项 logconf=true 时修改密码。
- SPARK-27267：Snappy 1.1.7.1 在解压缩空的序列化数据时失败。
- SPARK-27275：EncryptedMessage.transferTo 的潜在损坏。
- SPARK-27301：DStreamCheckpointData 无法清除，因为它的文件系统已缓存。
- SPARK-27338：TaskMemoryManager 和 UnsafeExternalSorter$SpillableIterator 之间的死锁。
- SPARK-27351：AggregateEstification 之后的输出行估计错误，只有空值列。
- SPARK-27390：修复包名不匹配。
- SPARK-27394：当没有任务开始或完成时，UI 的过时可能会持续几分钟或几小时。
- SPARK-27403：修复 updateTableStats，更新表状态时始终使用新状态或不使用任何状态。
- SPARK-27406：当两台机器的 oops 大小不同时，UnsafeArrayData 序列化将中断。
- SPARK-27453：DSv1 将自动删除 DataFrameWriter.partitionBy。
- SPARK-25271：带 Hive parquet 的 CTA 应利用本地 parquet 源。
- SPARK-25879：选择嵌套字段和顶级字段时模式修剪失败。
- SPARK-25906：spark shell 无法正确处理-i 选项。
- SPARK-25921：Python worker 重用导致屏障任务在没有 BarrierTaskContext 的情况下运行。
- SPARK-25918：LOAD DATA LOCAL INPATH 应处理相对路径。

5. 改进

- SPARK-27346：更新断言 examples 表达式中的字段信息。
- SPARK-27358：将 jquery 更新为 1.12.X，以获取安全修补程序。
- SPARK-27479：隐藏 API 文档 org.apache.spark.util.kvstore。
- SPARK-23523：规则 OptimizeMetadataOnlyQuery 导致的结果不正确。
- SPARK-23406：stream-stream self-joins 的 Bugs。
- SPARK-25088：rest 服务器默认值和文档更新。
- SPARK-23549：比较时间戳和日期时强制转换为时间戳。
- SPARK-24324：Pandas Grouped Map UDF 应按名称分配结果列。
- SPARK-23425：使用通配符加载 HDFS 文件路径的数据工作不正常。
- SPARK-23173：from_json 可以为标记为不可为空的字段生成空值。
- SPARK-24966：为集合操作实现优先规则。
- SPARK-25708：Having 没有 group by 的应该是全局聚合。
- SPARK-24341：正确处理子查询中的多值。
- SPARK-19724：使用现有默认位置创建托管表应引发异常。

6. 任务

- SPARK-27382：更新 Spark 2.4.X 测试 HiveExternalCatalogVersionsSuite。

- ❑ SPARK-23972：将 Parquet 从 1.8.2 更新为 1.10.0。
- ❑ SPARK-25419：Parquet 谓词下推改进。
- ❑ SPARK-23456：本机 ORC 阅读器默认为打开状态。
- ❑ SPARK-22279：默认情况下，使用本机 ORC 阅读器读取 Hive serde 表。
- ❑ SPARK-21783：默认情况下打开 ORC 过滤器下推。
- ❑ SPARK-24959：JSON 和 CSV 的加速计数方法 count()。
- ❑ SPARK-24244：解析 csv 分析器所需的列。
- ❑ SPARK-23786：csv 模式验证不检查列名。
- ❑ SPARK-24423：用于指定从 JDBC 读取的查询的选项。
- ❑ SPARK-22814：支持 JDBC 分区列中的日期/时间戳。
- ❑ SPARK-24771：将 avro 从 1.7.7 更新为 1.8。
- ❑ SPARK-23984：用于 kubernetes 的 pyspark 绑定。
- ❑ SPARK-24433：kubernetes 的 R 绑定。
- ❑ SPARK-23146：支持 kubernetes 集群后端的客户端模式。
- ❑ SPARK-23529：支持安装 kubernetes 卷。
- ❑ SPARK-24215：实现对数据帧 API 的热切评估。
- ❑ SPARK-22274：Pandas UDF 的用户定义聚合函数。
- ❑ SPARK-22239：Pandas UDF 的用户定义窗口函数。
- ❑ SPARK-24396：为 Python 添加结构化流式 ForeachWriter。
- ❑ SPARK-23874：将 Apache Arrow 升级到 0.10.0。
- ❑ SPARK-25004：增加 spark.executor.pyspark.memory 限制。
- ❑ SPARK-23030：使用 Arrow stream 格式创建和收集 Pandas DataFrames。
- ❑ SPARK-24624：支持 Python UDF 和 scalar Pandas UDF 的混合。

2.3.1 Spark SQL

Spark 2.0 通过对 SQL 2003 的支持大幅增强了 SQL 功能，现在可以运行所有 99 个 TPC-DS 查询。这个版本中的 SparkSQL 主要有以下几点改进。

（1）引入了支持 ANSISQL 和 HiveSQL 的本地解析器。

（2）本地实现 DDL 命令。

（3）支持非相关标量子查询。

（4）在 Where 与 having 条件中，支持(not)in 和(not)exists。

（5）即使 Spark 没有和 Hive 集成搭建，SparkSQL 也支持它们一起搭建时的除了 Hive 连接、Hive UDF（User Defined Function 用户自定义函数）和脚本转换之外的大部分功能。

（6）Hive 式的分桶方式的支持。

另外，Catalyst 查询优化器对于常见的工作负载也有了很多提升，对如 nullability propagation 之类的查询做了更好的优化。Catalyst 查询优化器从最早的应用于 SparkSQL 到现在应用于 DataSetAPI，对 Spark 程序的高效率运行起到了非常重要的作用，并且随着 DataSetAPI 的流行，以及优化器自身的不断演进，未来肯定会对 Spark 的所有框架带来更高的执行效率。

2.3.2 DataFrame 和 Dataset API

在 Spark 1.X 版本中，DataFrame 的 API 存在很多问题，如 DataFrame 不是类型安全的（not type-safe）、不是面向对象的（not object-oriented），为了克服这些问题，Spark 在 1.6 版本引入了 Dataset，并在 2.0 版本的 Scala 和 Java 中将二者进行了统一（在 Python 和 R 中，由于缺少类型安全性，DataFrame 仍是主要的编程接口），DataFrame 成为 DataSet[Row]的别名，而且 Spark 2.0 版本为 DataSet 的类型化聚合加入了一个新的聚合器，让基于 DataSet 的聚合更加高效。

在 Spark 2.X 版本中，DataFrame 和 Dataset API 晋升为稳定的 API。也就是说，可以在生产实践中使用它们，且后续会基于向后兼容的前提不断强化。

DataSetAPI 是 High-LevelAPI，有更高的抽象级别，与 RDDAPI 这样的 Low-LevelAPI 相比更加易用，它对于提升用户的工作效率，以及提高程序的可读性而言意义非凡。由于 WholeStageCodeGeneration 的引入，SparkSQL 和 DataSetAPI 中的常见算子的性能提升了 2～10 倍。加上 Catalyst 查询优化器和 Tungsten 的帮助，用户不用过多地关注对程序的优化，也能获得很好的执行效率。

所以，毋庸置疑地，这样一种简单高效的 API 将成为 Spark 未来主流的编程接口。

2.3.3 Timed Window

对于经常用到复杂 SQL 的用户而言，窗口函数一直以来都是不可或缺的，在 Spark 2.X 版本中，通过对 Hive 中的窗口函数的本地化实现，使用 Spark 的内存管理机制，从而提升了窗口函数的性能。

2.4 Spark 2.4 Streaming

Spark 2.0 为我们带来了一个新的流处理框架 Structured Streaming，这是一个基于 Spark SQL 和 Catalyst 优化器构建的高级流 API。它允许用户使用与操作静态数据的 DataFrame/Dataset API 对流数据进行编程，利用 Catalyst 优化器自动地增量化查询计划，并且它不但支持流数据的不断写入，还支持其他的静态数据的插入。

在 Apache Spark 2.2.X 版本的基础上，Apache Spark 2.3.X 版本、Apache Spark 2.4.3 版本中 Structured Streaming 的更新如下。

- SPARK-27494：当处理键值为空时，修复了 Kafka source v2 中的一个错误。
- SPARK-27550：修复“test dependencies.sh”对 Scala 2.12 不使用“kafka-0-8”配置文件。
- Continuous Processing：一种新的执行引擎，只更改一行用户代码，可以以亚毫秒的端到端延迟执行流式查询。
- Stream-Stream Joins：能够连接两个数据 stream，缓冲行，直到匹配的元组到达另一个 stream。谓词可用于事件时间列，以绑定需要保留的状态量。

- Streaming API V2：一种用于新数据源和接收器的实验 API，适用于批量、微批量和连续执行。注意，这个 API 仍在进行积极的开发。
- SPARK-24565：使用 ForeachBatch (Python,Scala,Java),将每个微批次的输出为 DataFrame。
- SPARK-24396：为 foreach 和 ForeachWriter 添加了 Python API。
- SPARK-25005：支持“kafka.isolation.level”以只读模式使用事务生成器，从 Kafka 主题中提交记录。
- SPARK-24662：支持追加或完成模式的 LIMIT 运算符。
- SPARK-24763：从流聚合的值中删除冗余密钥数据。
- SPARK-24156：当输入流中没有数据时，使用有状态操作（mapGroupsWithState, stream-stream join, streaming aggregation, streaming dropDuplicates）更快生成输出结果和状态清理。
- SPARK-24730：支持在查询中有多个输入流时选择最小或最大水印。
- SPARK-25399：修复了一个错误，在该错误中，将连续处理的执行线程重新用于微批量流可能导致正确性问题。
- SPARK-18057：将 Kafka 客户端版本从 0.10.0.1 升级到 2.0.0。

Spark 2.3 引入了新的低延迟处理模式，称为连续处理，它可以在至少一次的情况下实现低至 1ms 的端到端延迟。在查询中不更改数据集/数据帧操作的情况下，可以根据应用程序要求选择模式。

2.4.1 Structured Streaming

Structured Streaming 会通过 Checkpoint（检查点）机制和预写日志的方式来确保对流数据的语义一致性 exactly-once 处理，让整个处理过程更加可靠。

在 Spark 2.2 版本中增加了对 kafka 0.10 的支持，为记录添加了可见的基于事件时间水印的延迟，并且现在 Structured Streaming 已经支持对所有格式的文件操作。

到 Spark 2.1 版本为止，Structured Streaming 还处于试验阶段，还需几个版本的迭代，才能稳定到可以在生产环境下使用的程度。

下面结合官方文档了解一下 Structured Streaming。

首先写一个程序来监听网络端口发来的内容，然后进行 WordCount。

第一步：创建程序入口 SparkSession，并引入 spark.implicits 来允许 Scalaobject 隐式转换为 DataFrame。

```
1.   val spark = SparkSession.builder
2.  .appName("StructuredNetworkCount").getOrCreate()
3.  import spark.implicits._
```

第二步：创建流。配置从 socket 读取流数据，地址和端口为 localhost:9999。

```
1.   val lines = spark.readStream.format("socket")
2.              .option("host","localhost").option("port","9999")
3.              .load()
```

第三步：进行单词统计。这里 lines 是 DataFrame，使用 as[String]给它定义类型转换为

DataSet。之后在 DataSet 里进行单词统计。

```
1.      val words = lines.as[String].flatMap(_.split(" "))
2.      val wordcount = words.groupBy("values").count()
```

第四步：创建查询句柄，定义打印结果方式并启动程序。这里使用 writeStream 方法，输出模式为全部输出到控制台。

```
1.    val query = wordcount.writeStream
2.      .outputMode("complete").format("console").start()
3.   //调用 awaitTermination 方法来防止程序在处理数据时停止
4.      query.awaitTermination()
```

接下来运行该程序，在 Linux 命令窗口运行 nc–lk 9999 开启 9999 端口。

然后在 Spark 的 Home 目录下提交程序，传入 IP 和端口号。

```
1.    .bin/run-example
2.   org.apache.spark.examples.sql.streaming.StructuredNetworkWordCount
3.   localhost 9999
```

程序启动之后，在前面的命令窗口中输入单词 apache spark，程序会运行并打印出结果，如图 2-1 所示。

最后再输入 hello spark，程序会再次运行并打印出如图 2-2 所示的结果。

```
+------+-----+
| value|count|
+------+-----+
|apache|    1|
| spark|    1|
+------+-----+
```

```
+------+-----+
| value|count|
+------+-----+
| hello|    1|
|apache|    1|
| spark|    2|
+------+-----+
```

图 2-1　Spark Streaming 运行示意图 1　　图 2-2　Spark Streaming 运行示意图 2

可见，融合了 DataSetAPI 的流处理框架的程序代码十分简捷，并且执行效率也会比原来的 SparkStreaming 更高。

Structured Streaming 的关键思想如图 2-3 所示：把数据流视为一张数据不断增加的表，这样用户就可以基于这张表进行数据处理，就好像使用批处理来处理静态数据一样，但实际上 Spark 底层是把新数据不断地增量添加到这张无界的表的下一行中。

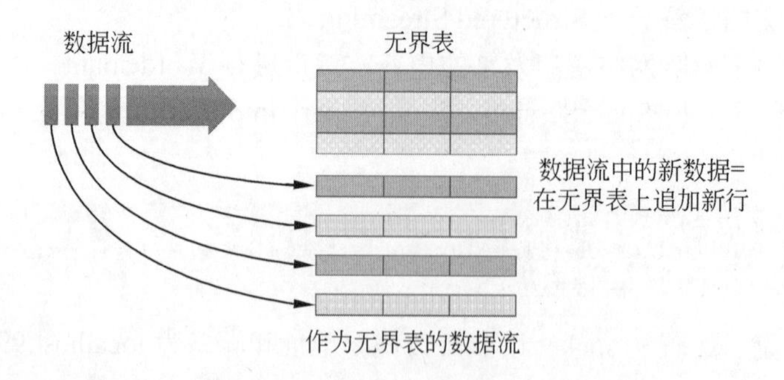

图 2-3　Structured Streaming 的关键思想

Structured Streaming 共有 3 种输出模式，这 3 种模式都只适用于某些类型的查询。

（1）CompleteMode：完整模式。整个更新的结果表将被写入外部存储器。由存储连接器决定如何处理整张表的写入。聚合操作以及聚合之后的排序操作支持这种模式。

（2）AppendMode：附加模式。只有自上次触发执行后在结果表中附加的新行会被写入外部存储器。这仅适用于结果表中的现有行不会更改的查询，如 select、where、map、flatMap、filter、join 等操作支持这种模式。

（3）UpdateMode：更新模式（这个模式将在以后的版本中实现）。只有自上次触发执行后在结果表中更新的行将被写入外部存储器（不输出未更改的行）。

2.4.2　增量输出模式

上面例子中使用的是 CompleteMode，程序中接收数据的输入表是 lines，它是一个 DataFrame，新来的数据会被添加进去。之后的 wordCounts 是结果表。当程序启动时，Spark 会不断检测是否有新数据加入到 lines 中，如果有新数据，则运行一个增量的查询，与上一次查询的结果合并，并且更新结果表，如图 2-4 所示。

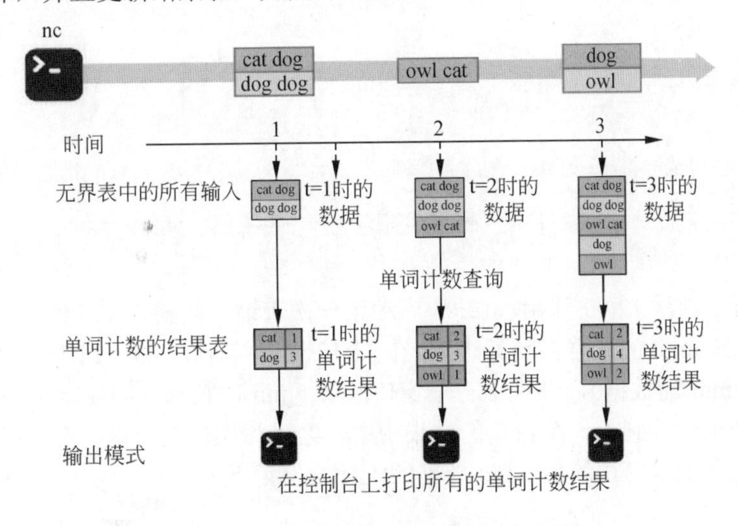

图 2-4　Spark Streaming 示意图

这个看似非常简单的设计，背后的实现逻辑却并不简单，因为其他的流处理框架用户是需要自己推理怎么解决容错和数据一致性的问题，如经典的 at-most-once、at-least-once、exactly-once 问题，而在上面的 CompleteMode 下，Spark 因为只在有新数据进来的时候才会更新结果，所以帮用户解决了这些问题。我们可以通过 StructuredStreaming 在基于 event-time 的操作和延迟数据的处理这两个问题上的解决方式来简单地了解背后的实现机制。

event-time 是嵌入事件本身的时间，记录了事件发生的时间。很多时候我们需要用这个时间来实现业务逻辑，例如，我们要获取 IOT 设备每分钟产生的事件数量，则可能需要使用生成数据的时间，而不是 Spark 接收它们的时间。在这个模式下，event-time 作为每行数据中的一列，可以用于基于时间窗口的聚合，也正因如此，基于 event-time 的窗口函数可以同样被定义在静态数据集上（如日志文件等）。

而关于容错方面，提供端到端的 exactly-once 语义是 Structured Streaming 的主要设计目标之一，要实现 exactly-once，就要考虑数据源（sources）、执行引擎（execution）和存储（sinks）3个方面。Structured Streaming 是这样实现的：假定每个数据源都有偏移量（与 kafka 的 offsets 类似）用来追溯数据在数据流中的位置；在执行引擎中会通过 checkpoint（检查点）和 WAL（writeaheadlogs 预写日志）记录包括被处理的数据的偏移量范围在内的程序运行进度信息；在存储层设计成多次处理结果幂等，即处理多次结果相同。这样确保了 Structured Streaming 端到端 exactly-once 的语义一致性。

上面提到的 event-time 列带来的另一个好处是，可以很自然地处理“迟到”的数据（如没有按照 event-time 的时间顺序被 Spark 接收），因为更新旧的结果表时，可以完全控制更新和清除旧的聚合结果来限制处于中间状态的数据（窗口函数中，这些数据可以是有状态的）的大小。Spark 2.1 引入的 watermarking 允许用户指定延迟数据的阈值，也允许引擎清除掉旧的状态。

先来看一下基础的窗口函数在 Structured Streaming 中的应用。例如，要对 10min 的单词进行计数，每 5min 统计一次，我们要把滑动窗口的大小设置为 10min，每 5min 滑动一次。具体代码如下：

```
1.  //假如输入的数据 words 格式是 timestamp: Timestamp, word: String
2.  val windowedCounts = words
3.  .groupBy(
4.  //设置窗口按照 timestamp 列为参照时间,10min 为窗口大小,5min 滑动一次,并且按照 word
    //进行分组计数
5.     window($"timestamp","10 minutes","5 minutes"),word
6.  ).count
```

如图 2-5 所示，从 12:00 开始，12:05 启动第一次查询，数据是 12:00—12:10 时间段的，当然，此时只有前 5min 的数据被统计进来。在 12:10 分时统计 12:05—12:15 时间段的数据（此时也是只有前 5min 的数据），并把上一个窗口的后 5min 的数据（即 12:05—12:10 时间段的数据）的统计结果合并到上一个窗口的结果中去，之后每次启动查询，都会把上一个窗口的查询结果补全，并把本窗口的前 5min 数据的统计结果记下来。

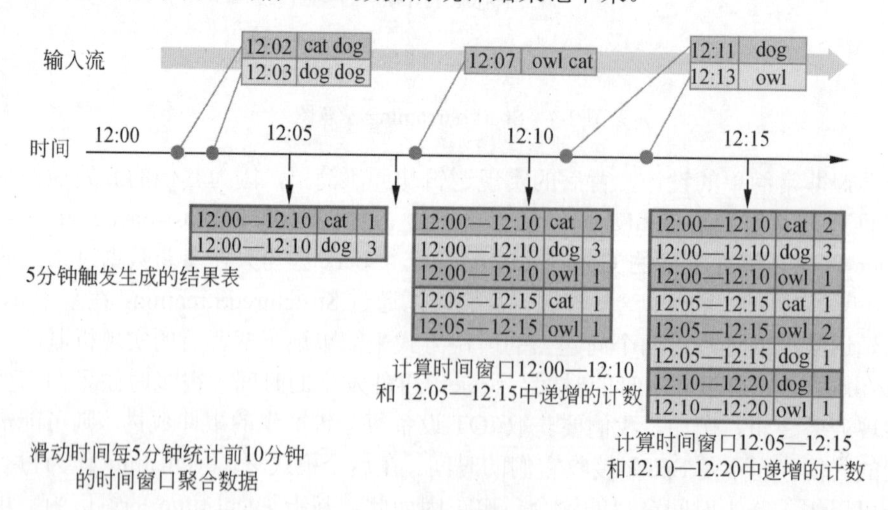

图 2-5 Spark 流处理示意图

但是，我们可能会遇到这种情况：如图 2-6 所示，12:04 产生的数据，一直到 12:10 之后才被程序接收到，此时数据依然会被正确地合并到对应的窗口中去，但是这样会导致查询结果长时间处于中间状态，而如果要长时间运行程序，就必须限制累积到内存中的中间状态结果的大小，这就要求系统知道什么时候清理这些中间状态的数据。

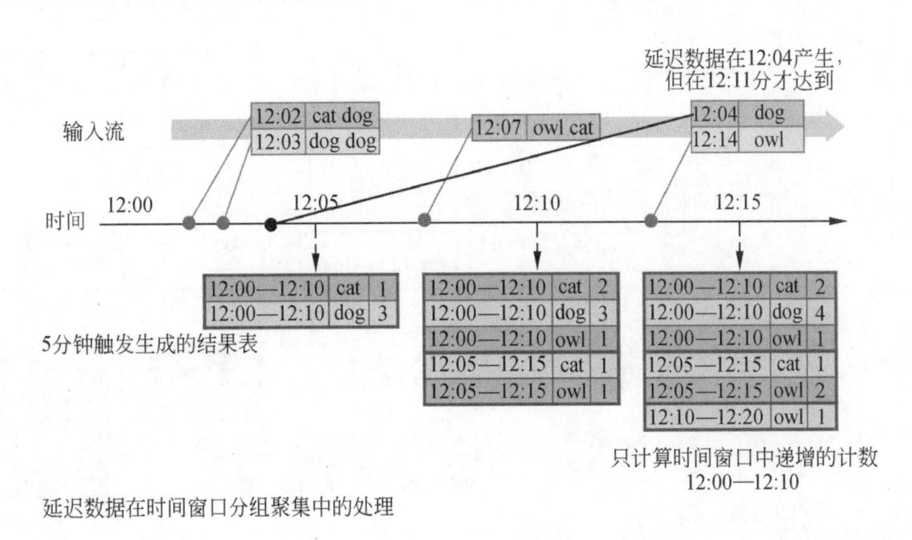

图 2-6　流处理 Watermark 数据示意图

在 Spark 2.1 中引入水印（watermarking），使得系统可以自动跟踪目前的 event-time，并按照用户指定的时间清理旧的状态。使用方法如下：

```
1.   val windowedCounts = words
2.  .withWatermark("timestamp", "10 minutes")
3.      .groupBy(window($"timestamp", "10 minutes", "5 minutes"),$"word")
4.      .count()
```

这里设置时间为 10min，在 Append 模式下，具体执行逻辑如图 2-7 所示。

图 2-7 中点线表示目前收到的数据的最大 event-time，实线是水印（计算方法是运算截止到触发点时收到的数据最大的 event-time 减去 latethreshold，也就是减去 10），当水印时间小于窗口的结束时间时，计算的数据都被保留为中间数据，当水印时间大于窗口结束时间时，就把这个窗口的运算结果加入到结果表中去，之后即使再收到属于这个窗口的数据，也不再进行计算，而直接忽略掉。图中在 12:15 时刻最大的 event-time 为 12:14，所以水印时间为 12:04，小于第一个窗口的 12:10，所以此时计算的数据处于中间状态，存放在内存中，在下一个触发点也就是 12:20 时，最大 event-time 为 12:21，水印时间是 12:11，大于 12:10，此时认为 12:00—12:10 窗口的数据已经完全到达，把中间结果中属于这个窗口的数据写入到结果表中，并且之后不再对这个结果进行更新。

需要注意的是，在 Append 模式下，系统在输出某个窗口的运行结果之前一定会根据设置等待延迟数据，这就意味着如果用户把延迟阈值设置为 1 天，那么在这一天内就无法看到这个窗口的（中间）结果。不过，在以后的版本中，Spark 会加入 Update 模式来解决这一问题。

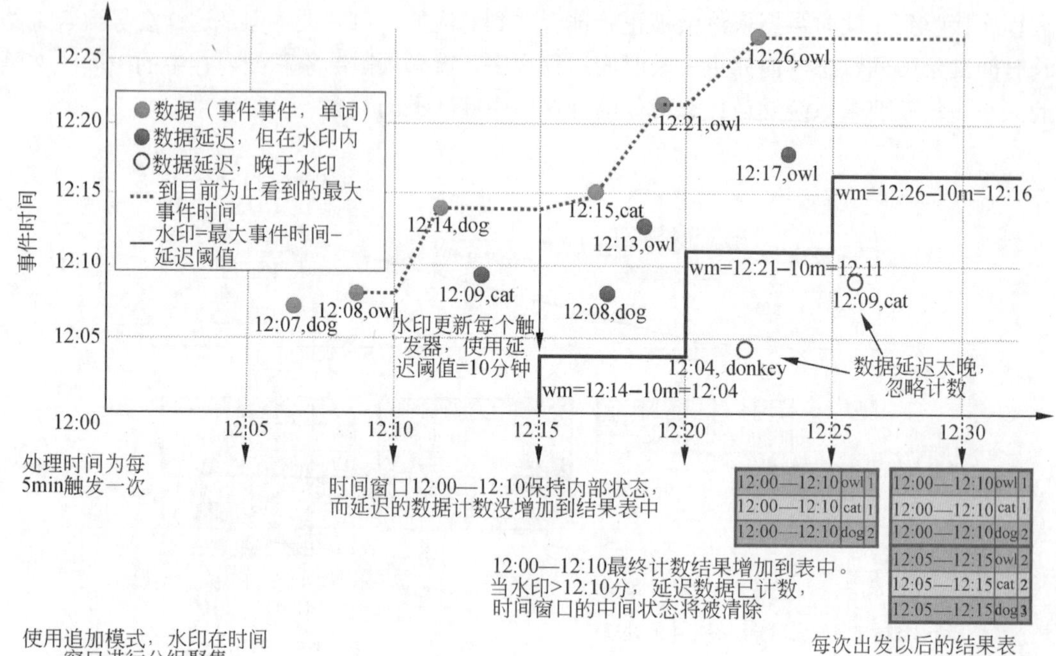

图 2-7 Append 逻辑示意图

Structured Streaming 目前有 4 种输出数据的处理方式。

（1）writeStream.format("parquet").start()，输出到文件中，Append 模式支持这种方式。这种方式自带容错机制。

（2）writeStream.format("console").start()，输出到控制台，用于调试。Append 模式和 Complete 模式支持这种方式。

（3）writeStream.format("memory").queryName("table").start()，以 table 的形式输出到内存，可以在之后的程序中使用 spark.sql(select * from table).show 来对结果进行处理。这种方式同样用于调试。Append 模式和 Complete 模式支持这种方式。

（4）writeStream.foreach(…).start()，对输出结果进行进一步处理。要使用它，用户必须传入一个实现 ForeachWriter 接口的 writer 类，这个类必须是可序列化的，因为稍后会把结果发送到不同的 executor 进行分布式计算。这个接口有 3 个方法要实现，即 open、process、close。

- open 方法：open 方法有两个入参：version、partitionId，其中 partitionId 是分区 ID，version 是重复数据删除的唯一 ID 标识。open 方法用于处理 Executor 分区的初始化（如打开连接启动事物等）。version 用于数据失败时重复数据的删除，当从失败中恢复时，一些数据可能生成多次，但它们具有相同的版本。如果此方法发现使用的 partitionId 和 version 这个分区已经处理后，可以返回 false，以跳过进一步的数据处理，但 close 仍然将被要求清理资源。
- process 方法：调用此方法处理 Executor 中的数据。此方法只在 open 时调用，返回 true。
- close 方法：停止执行 Executor 一个分区中的新数据时调用，保证被调用 open 时返回 true，或者返回 false。然而，在下列情况下，close 不被调用：JVM 崩溃，没有抛出异常 Throwable；open 方法抛出异常 Throwable。

最后要说明的是，为了程序在重启之后可以接着上次的执行结果继续执行，需要设置检查点，方式如下：

```
1.  aggDF.writeStream.outputMode("complete")
2.    .option("checkpointLocation", "path/to/HDFS/dir")
3.    .format("memory").start()
```

2.5　Spark 2.4 MLlib

Spark 2.2 版本中新增了基于 DataFrame 的机器学习；Spark 2.2 对 R 语言的机器学习新增了更多算法的支持，如 Random Forest（随机森林）、Gaussian Mixture Model（高斯混合模型）、Naive Bayes（朴素贝叶斯）、Survival Regression（生存回归分析）以及 K-Means（K-均值）等算法。

在 Apache Spark 2.2.X 版本的基础上，Apache Spark 2.3.X 版本、Apache Spark 2.4.X 版本中 MLlib 的主要更新如下。

1. 新API及更新

- SPARK-21866：内置的支持将图像读入 DataFrame（Scala/Java/Python）。
- SPARK-19634：用于向量列的描述汇总 DataFrame 的函数（Scala/Java）。
- SPARK-14516：聚类优化算法的聚类评估，用于支持聚类余弦轮廓和平方欧几里得轮廓度量（Scala/Java/Python）。
- SPARK-3181：具有 Huber 损失的健壮线性回归（Scala/Java/Python）。
- SPARK-13969：FeatureHasher 转换器（Scala/Java/Python）。
- SPARK-13030：独热编码评估器（Scala/Java/Python）。
- SPARK-22397：定量分解器（Scala/Java）。
- SPARK-20542：Bucketizer（Scala/Java/Python）。
- SPARK-21633、SPARK-21542：改进了对 Python 中自定义管道组件的支持。
- SPARK-26559：ML 图像不能与 1.9 之前的 Numpy 版本一起使用。
- SPARK-22666：图像格式的 Spark 数据源。
- SPARK-22119：KMeans/BisectingKMeans/Clustering evaluator 添加余弦距离度量。
- SPARK-10697：关联规则挖掘中的提升计算。
- SPARK-14682：提供 spark.ml GBTs 的 evaluateEachIteration 方法或等效方法。
- SPARK-7132：将 fit 和 validation set 添加到 spark.ml GBT。
- SPARK-15784：将 Power 迭代集群添加到 spark.ml。
- SPARK-15064：StopWordsRemover 中的本地支持。
- SPARK-21741：用于基于数据帧的多变量摘要的 Python API。
- SPARK-21898：MLLIB 中 KolmogorovSmirnovTest 的特征奇偶性。
- SPARK-10884：支持回归和分类相关模型的单实例预测。

- ❑ SPARK-23783：为 ML 管道添加新的通用出口特性。
- ❑ SPARK-11239：PMML 导出用于 ML 线性回归。

2．新特点

- ❑ SPARK-21087：交叉验证器和 TrainValidationSplit 可以在拟合时收集所有模型（Scala/Java），允许检查或保存所有已拟合的模型。
- ❑ SPARK-19357：Meta-algorithms CrossValidator, TrainValidationSplit, OneVsRest 支持并行参数，用于在并行 Spark 作业中拟合多个子模型。
- ❑ SPARK-17139：多项式逻辑回归模型（Scala/Java/Python）模型综述。
- ❑ SPARK-18710：在一般线性模型中添加偏移量。
- ❑ SPARK-20199：向 GBTClassifier 和 GBTRegressor 添加了 featureSubsetStrategy 参数。使用此功能进行子样本分析可以显著提高训练速度；此选项是 XGBoost 的一个关键优势。

3．其他显著变化

- ❑ SPARK-22156：修复了 Word2Vec 学习率缩放与 num 迭代。新的学习率被设置为与原来的 Word2Vec c 代码相匹配，并且从训练中获得更好的结果。
- ❑ SPARK-22289：添加对矩阵参数的 JSON 支持，修复了使用系数边界时使用 LogisticRegressionModel 进行 ML 持久性的错误。
- ❑ SPARK-22700：Bucketizer.transform 错误地删除了包含 NaN 的行。当参数 handleinvalid 设置为 skip 时，如果另一个（不相关的）列具有 NaN 值，则 Bucketizer 将在输入列中删除具有有效值的行。
- ❑ SPARK-22446：当 handleInvalid 设置为 error 时，Catalyst 优化器有时会导致 StringIndexerModel 引发错误的 Unsen Label 异常。由于谓词下推，筛选后的数据可能会发生这种情况，即使已从输入数据中筛选出无效行，也会导致错误。
- ❑ SPARK-21681：修正了多项式逻辑回归中的一个边缘情况错误，当某些特征方差为 0 时，会导致系数不正确。
- ❑ SPARK-22707：减少了交叉验证程序的内存消耗。
- ❑ SPARK-22949：减少了训练验证拆分的内存消耗。
- ❑ SPARK-21690：输入器应使用单次数据传递进行训练。
- ❑ SPARK-14371：OnlineLDAOptimizer 避免为每个小批量向 driver 收集统计信息。
- ❑ SPARK-17139：打破 API 的变化，逻辑回归模型摘要的类和特征层次被修改为更清晰，更好地适应多类摘要的添加。这是一个用户代码的突破性更改，它将 LogisticRegulationTrainingSummary 强制转换为 BinaryLogisticRegulationTrainingSummary。用户应改为使用 model.binarysummary 方法。
- ❑ SPARK-21806：BinaryClassificationMetrics.pr()：第一个点（0.0,1.0）具有误导性，已被（0.0,p）替换，其中精度 p 与最低召回点匹配。
- ❑ SPARK-16957：决策树在选择分割值时使用加权中点，可能会改变模型训练的结果。
- ❑ SPARK-14657：没有截获的 RFormula 在编码字符串项时输出引用类别，以便与本机 R 行为匹配，可能会改变模型训练的结果。

- SPARK-21027：OneVsRest 中使用的默认并行性设置为 1（即串行）。在 2.2 和早期版本中，并行度级别在 Scala 中设置为默认的 threadpool 大小，这可能会改变性能。
- SPARK-21523：Breeze 升级到 0.13.2，包括一个重要的错误修复——L-BFGS 算法中的线搜索 line-search 和强 Wolfe 条件。
- SPARK-15526：JPMML 依赖性被遮蔽了。

4．弃用

- SPARK-23122：为 SQLContext 中的 UDFs 和 PySpark 中的 Catalog 弃用 register。
- SPARK-13030：OneHotEncoder 已被弃用，将在 3.0 中删除。它已被新的 OneHotEncodeRestimator 取代。请注意，OneHotEncodeRestimator 将在 3.0 中重命名为 OneHotEncoder（但 OneHotEncodeRestimator 将保留为别名）。
- SPARK-23451：弃用 KMeans computeCost。
- SPARK-25345：从 ImageSchema 中弃用 readImages API。

2.5.1　基于 DataFrame 的 Machine Learning API

我们可以从下载的 Spark 源码中看到加入了 spark.ml 包（原来的 spark.mllib 仍然存在），这是在新版本中加入的基于 DataFrame 的机器学习代码包，存储在 DataFrames 中的向量和矩阵现在使用更高效的序列化，减少了调用 MLlib 算法的开销。现在 spark.ml 包代替基于 RDD 的 ML 成为主要的 SparkMLAPI。有一个很重要的功能是，现在可以保存和加载 Spark 支持的所有语言的 Machine Learning pipeline 和 model 了。

2.5.2　R 的分布式算法

在 Apache Spark 2.2.X 版本的基础上，Apache Spark 2.3.X 版本、Apache Spark 2.4.X 版本中 SparkR 的更新如下。

- SPARK-22933：用于 WithWatermark、Trigger、PartitionBy 和 stream-stream 连接的结构化流式 API。
- SPARK-21266：支持 DDL 格式架构的 SparkR UDF。
- SPARK-20726、SPARK-22924、SPARK-22843：几个新的 DataFrame API 包装器。
- SPARK-15767、SPARK-21622、SPARK-20917、SPARK-20307、SPARK-20906：几个新的 SparkML API 包装器。
- SPARK-26010：SparkR 在 CRAN 的 Java 11 上标记失败。
- SPARK-26422：不支持 Hadoop 版本时，无法禁用 SparkR 中的配置单元支持。
- SPARK-26910：SparkR 发布到 CRAN。
- SPARK-25393：从_csv()添加新函数。
- SPARK-21291：在数据框中添加 R partitionBy API。
- SPARK-25007：添加 array_intersect/array_except/array_union/shuffle 到 SparkR。
- SPARK-25234：避免并行中整数溢出。
- SPARK-25117：在 R 中添加 EXCEPT ALL 及 INTERSECT ALL 的支持。

- SPARK-24537：增加 array_remove / array_zip / map_from_arrays / array_distinct。
- SPARK-24187：在 SparkR 中添加 array_join 函数。
- SPARK-24331：在 SparkR 中添加 arrays_overlap, array_repeat, map_entries。
- SPARK-24198：在 SparkR 中添加 slice 函数。
- SPARK-24197：在 SparkR 中添加 array_sort 函数。
- SPARK-24185：在 SparkR 中添加 flatten 函数。
- SPARK-24069：增加 array_min / array_max 函数。
- SPARK-24054：增加 array_position / element_at 函数。
- SPARK-23770：在 SparkR 中增加 repartitionByRange API。

Spark 2.X 版本对 SparkR 的最大改进是用户自定义函数，包括 dapply、gapply 和 lapply，前两者可以用于执行基于分区的用户自定义函数（如分区域模型学习），而后者可用于超参数整定。

Spark 2.X 对 R 语言的机器学习增加了几种算法：Random Forest（随机森林）、Gaussian Mixture Model（高斯混合模型）、Naive Bayes（朴素贝叶斯）、Survival Regression（生存回归分析）以及 K-Means（K-均值）等。支持多项逻辑回归，来提供与 glmnet R 相似的功能。

同时对 Python 的机器学习也增加了一些算法，如 LDA（线性判别式分析 Linear Discriminant Analysis）、高斯混合模型、广义线性回归等。

2.6　Spark 2.4 GraphX

在 Apache Spark 2.2.X 版本的基础上，Apache Spark 2.3.X 版本、Apache Spark 2.4.X 版本中 GraphX 的更新如下。

- SPARK-5484：Pregel 现在定期检查以避免堆栈溢出错误。
- SPARK-21491：几个地方的性能改善很小。
- SPARK-26757：GraphX EdgeRDDImpl 及 VertexRDDImpl 的 count 方法不能处理空的 RDDs。
- SPARK-25268：运行 Parallel Personalized PageRank 抛出序列化异常。

第 3 章　Spark 的灵魂：RDD 和 DataSet

本章重点讲解 Spark 的 RDD 和 DataSet。3.1 节讲解 RDD 的定义、五大特性剖析及 DataSet 的定义和内部机制剖析；3.2 节对 RDD 弹性特性七个方面进行解析；3.3 节讲解 RDD 依赖关系，包括窄依赖、宽依赖；3.4 节解析 Spark 中 DAG 逻辑视图；3.5 节对 RDD 内部的计算机制及计算过程进行深度解析；3.6 节讲解 Spark RDD 容错原理及其四大核心要点解析；3.7 节对 Spark RDD 中 Runtime 流程进行解析；3.8 节通过一个 WordCount 实例，解析 Spark RDD 内部机制；3.9 节基于 DataSet 的代码，深入分析 DataSet 一步步转化成为 RDD 的过程。

3.1　为什么说 RDD 和 DataSet 是 Spark 的灵魂

Spark 建立在抽象的 RDD 上，使得它可以用一致的方式处理大数据不同的应用场景，把所有需要处理的数据转化成为 RDD，然后对 RDD 进行一系列的算子运算，从而得到结果。RDD 是一个容错的、并行的数据结构，可以将数据存储到内存和磁盘中，并能控制数据分区，且提供了丰富的 API 来操作数据。Spark 一体化、多元化的解决方案极大地减少了开发和维护的人力成本和部署平台的物力成本，并在性能方面有极大的优势，特别适合于迭代计算，如机器学习和图计算；同时，Spark 对 Scala 和 Python 交互式 shell 的支持也极大地方便了通过 shell 直接使用 Spark 集群来验证解决问题的方法，这对于原型开发至关重要，对数据分析人员有着无法拒绝的吸引力。

3.1.1　RDD 的定义及五大特性剖析

RDD 是分布式内存的一个抽象概念，是一种高度受限的共享内存模型，即 RDD 是只读的记录分区的集合，能横跨集群所有节点并行计算，是一种基于工作集的应用抽象。

RDD 底层存储原理：其数据分布存储于多台机器上，事实上，每个 RDD 的数据都以 Block 的形式存储于多台机器上，每个 Executor 会启动一个 BlockManagerSlave，并管理一部分 Block；而 Block 的元数据由 Driver 节点上的 BlockManagerMaster 保存，BlockManagerSlave 生成 Block 后向 BlockManagerMaster 注册该 Block，BlockManagerMaster 管理 RDD 与 Block 的关系，当 RDD 不再需要存储的时候，将向 BlockManagerSlave 发送指令删除相应的 Block。

BlockManager 管理 RDD 的物理分区，每个 Block 就是节点上对应的一个数据块，可以存储在内存或者磁盘上。而 RDD 中的 Partition 是一个逻辑数据块，对应相应的物理块 Block。本质上，一个 RDD 在代码中相当于数据的一个元数据结构，存储着数据分区及其逻辑结构映射关系，存储着 RDD 之前的依赖转换关系。

BlockManager 在每个节点上运行管理 Block（Driver 和 Executors），它提供一个接口检索本地和远程的存储变量，如 memory、disk、off-heap。使用 BlockManager 前必须先初始化。

BlockManager.scala 的部分源码如下：

```
1.   private[spark] class BlockManager(
2.      executorId: String,
3.      rpcEnv: RpcEnv,
4.      val master: BlockManagerMaster,
5.      val serializerManager: SerializerManager,
6.      val conf: SparkConf,
7.      memoryManager: MemoryManager,
8.      mapOutputTracker: MapOutputTracker,
9.      shuffleManager: ShuffleManager,
10.     val blockTransferService: BlockTransferService,
11.     securityManager: SecurityManager,
12.     numUsableCores: Int)
13.   extends BlockDataManager with BlockEvictionHandler with Logging {
```

BlockManagerMaster 会持有整个 Application 的 Block 的位置、Block 所占用的存储空间等元数据信息，在 Spark 的 Driver 的 DAGScheduler 中，就是通过这些信息来确认数据运行的本地性的。Spark 支持重分区，数据通过 Spark 默认的或者用户自定义的分区器决定数据块分布在哪些节点。RDD 的物理分区是由 Block-Manager 管理的，每个 Block 就是节点上对应的一个数据块，可以存储在内存或者磁盘。而 RDD 中的 partition 是一个逻辑数据块，对应相应的物理块 Block。本质上，一个 RDD 在代码中相当于数据的一个元数据结构（一个 RDD 就是一组分区），存储着数据分区及 Block、Node 等的映射关系，以及其他元数据信息，存储着 RDD 之前的依赖转换关系。分区是一个逻辑概念，Transformation 前后的新旧分区在物理上可能是同一块内存存储。

Spark 通过读取外部数据创建 RDD，或通过其他 RDD 执行确定的转换 Transformation 操作（如 map、union 和 groubByKey）而创建，从而构成了线性依赖关系，或者说血统关系（Lineage），在数据分片丢失时可以从依赖关系中恢复自己独立的数据分片，对其他数据分片或计算机没有影响，基本没有检查点开销，使得实现容错的开销很低，失效时只需要重新计算 RDD 分区，就可以在不同节点上并行执行，而不需要回滚（Roll Back）整个程序。落后任务（即运行很慢的节点）是通过任务备份，重新调用执行进行处理的。

因为 RDD 本身支持基于工作集的运用，所以可以使 Spark 的 RDD 持久化（persist）到内存中，在并行计算中高效重用。多个查询时，我们就可以显性地将工作集中的数据缓存到内存中，为后续查询提供复用，这极大地提升了查询的速度。在 Spark 中，一个 RDD 就是一个分布式对象集合，每个 RDD 可分为多个片（Partitions），而分片可以在集群环境的不同节点上计算。

RDD 作为泛型的抽象的数据结构，支持两种计算操作算子：Transformation（变换）与 Action（行动）。且 RDD 的写操作是粗粒度的，读操作既可以是粗粒度的，也可以是细粒度的。

RDD.scala 的源码如下：

```
1.  /** 每个 RDD 都有 5 个主要特性
2.     *-分区列表
3.     *-每个分区都有一个计算函数
4.     *-依赖于其他 RDD 的列表
5.     *- 数据类型(Key-Value)的 RDD 分区器
6.     *- 每个分区都有一个分区位置列表
7.     */
8.  abstract class RDD[T: ClassTag](
```

```
9.      @transient private var _sc: SparkContext,
10.     @transient private var deps: Seq[Dependency[_]]
11.   ) extends Serializable with Logging {
```

其中，SparkContext 是 Spark 功能的主要入口点，一个 SparkContext 代表一个集群连接，可以用其在集群中创建 RDD、累加变量、广播变量等，在每一个可用的 JVM 中只有一个 SparkContext，在创建一个新的 SparkContext 之前，必须先停止该 JVM 中可用的 SparkContext，这种限制可能最终会被修改。SparkContext 被实例化时需要一个 SparkConf 对象去描述应用的配置信息，在这个配置对象中设置的信息，会覆盖系统默认的配置。

RDD 有以下五大特性。

（1）分区列表（a list of partitions）。Spark RDD 是被分区的，每一个分区都会被一个计算任务（Task）处理，分区数决定并行计算数量，RDD 的并行度默认从父 RDD 传给子 RDD。默认情况下，一个 HDFS 上的数据分片就是一个 Partition，RDD 分片数决定了并行计算的力度,可以在创建 RDD 时指定 RDD 分片个数，如果不指定分区数量，当 RDD 从集合创建时，则默认分区数量为该程序所分配到的资源的 CPU 核数（每个 Core 可以承载 2～4 个 Partition），如果是从 HDFS 文件创建，默认为文件的 Block 数。

（2）每一个分区都有一个计算函数（a function for computing each split）。每个分区都会有计算函数，Spark 的 RDD 的计算函数是以分片为基本单位的，每个 RDD 都会实现 compute 函数，对具体的分片进行计算，RDD 中的分片是并行的，所以是分布式并行计算。有一点非常重要，就是由于 RDD 有前后依赖关系，遇到宽依赖关系，例如，遇到 reduceBykey 等宽依赖操作的算子，Spark 将根据宽依赖划分 Stage，Stage 内部通过 Pipeline 操作，通过 Block Manager 获取相关的数据，因为具体的 split 要从外界读数据，也要把具体的计算结果写入外界，所以用了一个管理器，具体的 split 都会映射成 BlockManager 的 Block，而具体 split 会被函数处理，函数处理的具体形式是以任务的形式进行的。

（3）依赖于其他 RDD 的列表（a list of dependencies on other RDDs）。RDD 的依赖关系，由于 RDD 每次转换都会生成新的 RDD，所以 RDD 会形成类似流水线的前后依赖关系，当然，宽依赖就不类似于流水线了，宽依赖后面的 RDD 具体的数据分片会依赖前面所有的 RDD 的所有的数据分片，这时数据分片就不进行内存中的 Pipeline，这时一般是跨机器的。因为有前后的依赖关系，所以当有分区数据丢失的时候，Spark 会通过依赖关系重新计算，算出丢失的数据，而不是对 RDD 所有的分区进行重新计算。RDD 之间的依赖有两种：窄依赖（Narrow Dependency）、宽依赖（Wide Dependency）。RDD 是 Spark 的核心数据结构，通过 RDD 的依赖关系形成调度关系。通过对 RDD 的操作形成整个 Spark 程序。

RDD 有 Narrow Dependency 和 Wide Dependency 两种不同类型的依赖，其中的 Narrow Dependency 指的是每一个 parent RDD 的 Partition 最多被 child RDD 的一个 Partition 所使用，而 Wide Dependency 指的是多个 child RDD 的 Partition 会依赖于同一个 parent RDD 的 Partition。可以从两个方面来理解 RDD 之间的依赖关系：一方面是该 RDD 的 parent RDD 是什么；另一方面是依赖于 parent RDD 的哪些 Partitions；根据依赖于 parent RDD 的 Partitions 的不同情况，Spark 将 Dependency 分为宽依赖和窄依赖两种。Spark 中宽依赖指的是生成的 RDD 的每一个 partition 都依赖于父 RDD 的所有 partition，宽依赖典型的操作有 groupByKey、sortByKey 等，宽依赖意味着 shuffle 操作，这是 Spark 划分 Stage 边界的依据，Spark 中宽依赖支持两种 Shuffle Manager，即 HashShuffleManager 和 SortShuffleManager，前者是基于 Hash

的 Shuffle 机制，后者是基于排序的 Shuffle 机制。Spark 2.2 现在的版本中已经没有 Hash Shuffle 的方式。

（4）key-value 数据类型的 RDD 分区器（-Optionally,a Partitioner for key-value RDDS），控制分区策略和分区数。每个 key-value 形式的 RDD 都有 Partitioner 属性，它决定了 RDD 如何分区。当然，Partition 的个数还决定每个 Stage 的 Task 个数。RDD 的分片函数，想控制 RDD 的分片函数的时候可以分区（Partitioner）传入相关的参数，如 HashPartitioner、RangePartitioner，它本身针对 key-value 的形式，如果不是 key-value 的形式，它就不会有具体的 Partitioner。Partitioner 本身决定了下一步会产生多少并行的分片，同时，它本身也决定了当前并行（parallelize）Shuffle 输出的并行数据，从而使 Spark 具有能够控制数据在不同节点上分区的特性，用户可以自定义分区策略，如 Hash 分区等。Spark 提供了“partitionBy”运算符，能通过集群对 RDD 进行数据再分配来创建一个新的 RDD。

（5）每个分区都有一个优先位置列表（-Optionally, a list of preferred locations to compute each split on）。它会存储每个 Partition 的优先位置，对于一个 HDFS 文件来说，就是每个 Partition 块的位置。观察运行 Spark 集群的控制台会发现 Spark 的具体计算，具体分片前，它已经清楚地知道任务发生在什么节点上，也就是说，任务本身是计算层面的、代码层面的，代码发生运算之前已经知道它要运算的数据在什么地方，有具体节点的信息。这就符合大数据中数据不动代码动的特点。数据不动代码动的最高境界是数据就在当前节点的内存中。这时有可能是 memory 级别或 Alluxio 级别的，Spark 本身在进行任务调度时候，会尽可能将任务分配到处理数据的数据块所在的具体位置。据 Spark 的 RDD.Scala 源码函数 getPreferredLocations 可知，每次计算都符合完美的数据本地性。

RDD 类源码文件中的 4 个方法和一个属性对应上述阐述的 RDD 的 5 大特性。

RDD.scala 的源码如下：

```
1.
2.   /**
3.     * :: DeveloperApi ::
4.     * 通过子类实现给定分区的计算
5.     */
6.    @DeveloperApi
7.    def compute(split: Partition, context: TaskContext): Iterator[T]
8.
9.   /**
     *  通过子类实现，返回一个 RDD 分区列表，这个方法仅只被调用一次，它是安全地执行一次
     *  耗时计算数组中的分区必须符合以下属性设置
10.    *  'rdd.partitions.zipWithIndex.forall { case (partition, index) =>
     *  partition.index == index }'
11.    */
12.   protected def getPartitions: Array[Partition]
13.
14.  /**
     *返回对父 RDD 的依赖列表，这个方法仅只被调用一次，它是安全地执行一次耗时计算
15.    */
16.   protected def getDependencies: Seq[Dependency[_]] = deps
17.
18.  /**
     * 可选的，指定优先位置，输入参数是 split 分片，输出结果是一组优先的节点位置
19.    */
```

```
20.   protected def getPreferredLocations(split: Partition): Seq[String] = Nil
21.
22.   /** 可选的，通过子类来实现。指定如何分区 */
23.   @transient val partitioner: Option[Partitioner] = None
```

其中，TaskContext 是读取或改变执行任务的环境，用 org.apache.spark.TaskContext.get()可返回当前可用的 TaskContext，可以调用内部的函数访问正在运行任务的环境信息。Partitioner 是一个对象，定义了如何在 key-Value 类型的 RDD 元素中用 Key 分区，从 0 到 numPartitions–1 区间内映射每一个 Key 到 Partition ID。Partition 是一个 RDD 的分区标识符。Partition.scala 的源码如下：

```
1.   trait Partition extends Serializable {
2.     /**
      * 获取父 RDD 的分区索引
3.     */
4.
5.     def index: Int
6.
7.     //最好默认实现 HashCode
8.     override def hashCode(): Int = index
9.     override def equals(other: Any): Boolean = super.equals(other)
10. }
```

3.1.2 DataSet 的定义及内部机制剖析

DataSet 是可以并行使用函数或关系操作转换特定域对象的强类型集合。每个 DataSet 有一个非类型化的 DataFrame。DataFrame 是 Dataset[Row]的别名。DataSet 中可用的算子分为转换算子和行动算子。转换算子可以产生新的 DataSet；行动算子将触发计算和返回结果。转换算子包括 map、filter、select 和聚集算子，如 groupBy。行动算子包括 count、show，或者将数据保存到文件系统中。

DataSet 是“懒加载”的，即只有在行动算子被触发时，才进行计算操作。本质上，DataSet 表示一个逻辑计划，它描述了生成数据所需的计算。当行动算子被触发时，Spark 查询优化器将优化逻辑计划，生成一个并行、分布式有效执行的物理计划。使用 explain 函数可以查看逻辑计划以及优化的物理计划。

为了有效地支持特定领域的对象，编码器[Encoder]是必需的。编码器将特定类型 T 转换为 Spark 的内部类型。例如，给定一个类 Person 有两个属性，包括‘名字’（string）和‘年龄’（int），编码器告诉 Spark 在运行生成代码时将 Person 对象序列化成二进制数据。二进制数据通常占用更少的内存以及更优化的数据处理效率（如列存储格式）。可以使用 schema 函数来查看了解数据的内部二进制结构。

通常有两种创建数据集 Dataset 的方法。

方法一：最常见的方式是 Spark 在 SparkSession 中使用 read 功能读入存储系统中的文件。例如，Scala 版本：可以使用 spark.read.parquet 方式读入 parquet 格式的文件，使用 as 方法转换为[Person]数据类型的 DataSet。Java 版本：使用 spark.read.parquet 方式读入 parquet 格式的文件，在 as 方法中使用编码器对 Person.class 数据类型进行编码，生成 DataSet。

```
1.   * {{{
2.   *   val people = spark.read.parquet("...").as[Person]  //Scala
```

```
3.   *   Dataset<Person> people = spark.read().parquet("...").as(Encoders
         .bean(Person.class)); //Java
4.   * }}}
```

方法二：DataSet 也可以通过现有 DataSet 进行转换创建。例如，在现有的 DataSet 中使用过滤算子，创建一个新的数据集。下面看一个生成新 DataSet 的例子。Scala 版本：在已有的 Dataset[Person]中使用 map 转换函数，获取 Person 中的姓名，将生成新的 Dataset[String]；Java 版本：在已有的数据集 Dataset<String>中使用 map 转换函数，通过(Person p) -> p.name 获取 Person 中的姓名，编码器指定姓名属性的类型为 String 类型，生成新的姓名的数据集 Dataset<String>。(Person p) -> p.name 这种写法为 Lambda 表达式，这是 Java 8 之后才有的新特性。

```
1.   ** {{{
2.   *   val names = people.map(_.name)
                                     //在 Scala 中，names 是一个 String 类型的 DataSet
3.   *   Dataset<String> names = people.map((Person p) -> p.name, Encoders
         .STRING));
4.       //in Java 8
5.   * }}}
```

通过各种特定领域的语言（DSL）定义的功能： Dataset（类），[Column]和[functions]等非类型化数据集的操作也可以。这些操作非常类似于 R 或 Python 语言在数据表中的抽象操作。在 scala 中，我们使用 apply 方法，从 people 的数据集中选择“年龄”这一列；在 Java 中使用"col"方法，通过 people.col("age")获取到年龄列。

```
1.   * 从 DataSet 中选择一列，在 Scala 中使用 apply 方法，在 Java 中使用 col 方法
2.   * {{{
3.   *   val ageCol = people("age")                 //在 Scala 中
4.   *   Column ageCol = people.col("age");         //在 Java 中
5.   * }}}
```

注意，[Column]类型也可以通过它的各种函数来操作。例如，以下代码在人员数据集中创建一个新的列，每个人的年龄增加 10。在 Scala 中使用的方法是 people("age") + 10；在 Java 中使用 plus 方法。

```
1.   * {{{
2.   *   //下面创建一个新的列，每个人的年龄增加 10
3.   *   people("age") + 10                //在 Scala 中
4.   *   people.col("age").plus(10);       //在 Java 中
5.   * }}}
```

下面是一个更具体的例子：使用 spark.read.parquet 分别读入 parquet 格式的人员数据及部门数据，过滤出年龄大于 30 岁的人员，根据部门 ID 和部门数据进行 join，然后按照姓名、性别分组，再使用 agg 方法，调用内置函数 avg 计算出部门中的平均工资、人员的最大年龄。Scala 版本代码如下：

```
1.   * {{{
2.   *   //使用 SparkSession 创建 Dataset[Row]
3.   *   val people = spark.read.parquet("...")
4.   *   val department = spark.read.parquet("...")
5.   *
```

```
6.  *   people.filter("age > 30")
7.  *     .join(department, people("deptId") === department("id"))
8.  *     .groupBy(department("name"), "gender")
9.  *     .agg(avg(people("salary")), max(people("age")))
10. * }}}
```

以上例子的 Java 版本代码如下：

```
1.  * {{{
2.  *   //To create Dataset<Row> using SparkSession
3.  *   Dataset<Row> people = spark.read().parquet("...");
4.  *   Dataset<Row> department = spark.read().parquet("...");
5.  *
6.  *   people.filter("age".gt(30))
7.  *     .join(department, people.col("deptId").equalTo(department("id")))
8.  *     .groupBy(department.col("name"), "gender")
9.  *     .agg(avg(people.col("salary")), max(people.col("age")));
10. * }}}
```

3.2 RDD 弹性特性 7 个方面解析

RDD 作为弹性分布式数据集，它的弹性具体体现在以下 7 个方面。

1. 自动进行内存和磁盘数据存储的切换

Spark 会优先把数据放到内存中，如果内存实在放不下，会放到磁盘里面，不但能计算内存放下的数据，也能计算内存放不下的数据。如果实际数据大于内存，则要考虑数据放置策略和优化算法。当应用程序内存不足时，Spark 应用程序将数据自动从内存存储切换到磁盘存储，以保障其高效运行。

2. 基于Lineage（血统）的高效容错机制

Lineage 是基于 Spark RDD 的依赖关系来完成的（依赖分为窄依赖和宽依赖两种形态），每个操作只关联其父操作，各个分片的数据之间互不影响，出现错误时只要恢复单个 Split 的特定部分即可。常规容错有两种方式：一个是数据检查点；另一个是记录数据的更新。数据检查点的基本工作方式，就是通过数据中心的网络链接不同的机器，然后每次操作的时候都要复制数据集，就相当于每次都有一个复制，复制是要通过网络传输的，网络带宽就是分布式的瓶颈，对存储资源也是很大的消耗。记录数据更新就是每次数据变化了就记录一下，这种方式不需要重新复制一份数据，但是比较复杂，消耗性能。Spark 的 RDD 通过记录数据更新的方式为何很高效？因为①RDD 是不可变的且 Lazy；②RDD 的写操作是粗粒度的。但是，RDD 读操作既可以是粗粒度的，也可以是细粒度的。

3. Task如果失败，会自动进行特定次数的重试

默认重试次数为 4 次，TaskSchedulerImpl 的源码如下所示。

Spark 2.2.1 版本的 TaskSchedulerImpl.scala 的源码如下：

```
1. private[spark] class TaskSchedulerImpl private[scheduler](
2.     val sc: SparkContext,
```

```
3.     val maxTaskFailures: Int,
4.     private[scheduler] val blacklistTrackerOpt: Option[BlacklistTracker],
5.     isLocal: Boolean = false)
6.   extends TaskScheduler with Logging {
7.
8.   import TaskSchedulerImpl._
9.
10.  def this(sc: SparkContext) = {
11.    this(
12.      sc,
13.      sc.conf.get(config.MAX_TASK_FAILURES),
14.      TaskSchedulerImpl.maybeCreateBlacklistTracker(sc))
15.  }
16.
17.  def this(sc: SparkContext, maxTaskFailures: Int, isLocal: Boolean) = {
18.    this(
19.      sc,
20.      maxTaskFailures,
21.      TaskSchedulerImpl.maybeCreateBlacklistTracker(sc),
22.      isLocal = isLocal)
23.  }
24. ……
25. //config\package.scala
26. ……
27.  private[spark] val MAX_TASK_FAILURES =
28.    ConfigBuilder("spark.task.maxFailures")
29.      .intConf
30.      .createWithDefault(4)
```

Spark 2.4.3 版本的 TaskSchedulerImpl.scala 源码与 Spark 2.2.1 版本相比具有如下特点。

- 上段代码中第 1 行取消了类 TaskSchedulerImpl 的访问权限限制。
- 上段代码中第 4 行删除黑名单列表跟踪变量 blacklistTrackerOpt。
- 上段代码中第 14 行删除 this 构造函数的 maybeCreateBlacklistTracker 参数。
- 上段代码中第 15 行之后新增变量 blacklistTrackerOpt，用于跟踪问题 executors 和 nodes 节点，延迟初始化 BlackListTrackerOpt 以避免获取空的 ExecutionAllocationClient，因为 ExecutorAllocationClient 是在此 TaskSchedulerImpl 之后创建的。
- 上段代码中第 17～23 行，删除带 sc、maxTaskFailures、isLocal 参数的 this 构造函数。

```
1. private[spark] class TaskSchedulerImpl(
2.    ……
3. private[scheduler] lazy val blacklistTrackerOpt =
   maybeCreateBlacklistTracker(sc)
4.  ……
```

TaskSchedulerImpl 是底层的任务调度接口 TaskScheduler 的实现，这些 Schedulers 从每一个 Stage 中的 DAGScheduler 中获取 TaskSet，运行它们，尝试是否有故障。DAGScheduler 是高层调度，它计算每个 Job 的 Stage 的 DAG，然后提交 Stage，用 TaskSets 的形式启动底层 TaskScheduler 调度在集群中运行。

4．Stage如果失败，会自动进行特定次数的重试

这样，Stage 对象可以跟踪多个 StageInfo（存储 SparkListeners 监听到的 Stage 的信息，将 Stage 信息传递给 Listeners 或 web UI）。默认重试次数为 4 次，且可以直接运行计算失败

的阶段，只计算失败的数据分片，Stage 的源码如下所示。

Stage.scala 的源码如下：

```
1.  private[scheduler] abstract class Stage(
2.      val id: Int,
3.      val rdd: RDD[_],
4.      val numTasks: Int,
5.      val parents: List[Stage],
6.      val firstJobId: Int,
7.      val callSite: CallSite)
8.    extends Logging {
9.
10.   val numPartitions = rdd.partitions.length
11.
12.   /** 属于这个工作集的 Stage */
13.   val jobIds = new HashSet[Int]
14.
15.   /**用于此 Stage 的下一个新 attempt 的标识 ID */
16.   private var nextAttemptId: Int = 0
17.
18.   val name: String = callSite.shortForm
19.   val details: String = callSite.longForm
20.
21.   /**
22.    * 最新的[StageInfo] object 指针，需要被初始化，
23.    *任何 attempts 都是被创造出来的，因为 DAGScheduler 使用 StageInfo
24.    *告诉 SparkListeners 工作何时开始（即发生前的任何阶段已经创建）
25.    */
26.   private var _latestInfo: StageInfo = StageInfo.fromStage(this, nextAttemptId)
27.
28.   /**
29.         *设置 stage attempt IDs 当失败时可以读取失败信息，
30.         *跟踪这些失败，为了避免无休止地重复失败
31.         *跟踪每一次 attempt，以便避免记录重复故障
32.         *如果从同一 stage 创建多任务失败（spark-5945）
33.    */
34.   val failedAttemptIds = new HashSet[Int]
35.
36.   private[scheduler] def clearFailures() : Unit = {
37.     failedAttemptIds.clear()
38.   }
39.
40.   /**  通过使用新 ID 创建新的 StageInfo，为此阶段创建新尝试。*/
41.   def makeNewStageAttempt(
42.       numPartitionsToCompute: Int,
43.       taskLocalityPreferences: Seq[Seq[TaskLocation]] = Seq.empty): Unit = {
44.     val metrics = new TaskMetrics
45.     metrics.register(rdd.sparkContext)
46.     _latestInfo = StageInfo.fromStage(
47.       this, nextAttemptId, Some(numPartitionsToCompute), metrics,
            taskLocalityPreferences)
48.     nextAttemptId += 1
49.   }
50.
51.   /** 返回此阶段最近尝试的 StageInfo。 */
52.   def latestInfo: StageInfo = _latestInfo
```

```
53.
54.   override final def hashCode(): Int = id
55.
56.   override final def equals(other: Any): Boolean = other match {
57.     case stage: Stage => stage != null && stage.id == id
58.     case _ => false
59.   }
60.
61.   /** 返回即将计算的分区 ID 序列（即需要计算）。 */
62.   def findMissingPartitions(): Seq[Int]
63. }
```

在 Stage 终止之前允许的 Stage 连续尝试的次数为 4 次，在 DAGScheduler.scala 的源码 object DAGScheduler 中进行定义。DAGScheduler.scala 的源码如下：

```
1.   /**
      *在终止之前允许的连续尝试的次数
2.    */
3.
4.    private[scheduler] val maxConsecutiveStageAttempts =
5.      sc.getConf.getInt("spark.stage.maxConsecutiveAttempts",
6.        DAGScheduler.DEFAULT_MAX_CONSECUTIVE_STAGE_ATTEMPTS)
7.  ……
8.
9.  private[spark] object DAGScheduler {
10.   //在毫秒级别，等待读取失败事件后就停止（在下一个检测到来之前）；这是一个避免重新提
      //交任务的简单方法，非读取数据的 map 中更多失败事件的到来
11.   val RESUBMIT_TIMEOUT = 200
12.
13.   //在终止之前允许连续尝试的次数
14.   val DEFAULT_MAX_CONSECUTIVE_STAGE_ATTEMPTS = 4
15. }
```

Stage 是 Spark Job 运行时具有相同逻辑功能和并行计算任务的一个基本单元。Stage 中所有的任务都依赖同样的 Shuffle，每个 DAG 任务通过 DAGScheduler 在 Stage 的边界处发生 Shuffle 形成 Stage，然后 DAGScheduler 运行这些阶段的拓扑顺序。每个 Stage 都可能是 ShuffleMapStage，如果是 ShuffleMapStage，则跟踪每个输出节点（nodes）上的输出文件分区，它的任务结果是输入其他的 Stage(s)，或者输入一个 ResultStage，若输入一个 ResultStage，这个 ResultStage 的任务直接在这个 RDD 上运行计算这个 Spark Action 的函数（如 count()、save() 等），并生成 shuffleDep 等字段描述 Stage 和生成变量，如 outputLocs 和 numAvailableOutputs，为跟踪 map 输出做准备。每个 Stage 会有 firstjobid，确定第一个提交 Stage 的 Job，使用 FIFO 调度时，会使得其前面的 Job 先行计算或快速恢复（失败时）。

ShuffleMapStage 是 DAG 产生数据进行 Shuffle 的中间阶段，它发生在每次 Shuffle 操作之前，可能包含多个 Pipelined 操作，ResultStage 阶段捕获函数在 RDD 的分区上运行 Action 算子计算结果，有些 Stage 不是运行在 RDD 的所有的分区上，例如，first()、lookup()等。SparkListener 是 Spark 调度器的事件监听接口。注意，这个接口随着 Spark 版本的不同会发生变化。

5．checkpoint和persist（检查点和持久化），可主动或被动触发

checkpoint 是对 RDD 进行的标记，会产生一系列的文件，且所有父依赖都会被删除，是整个依赖（Lineage）的终点。checkpoint 也是 Lazy 级别的。persist 后 RDD 工作时每个工作

节点都会把计算的分片结果保存在内存或磁盘中，下一次如果对相同的 RDD 进行其他的 Action 计算，就可以重用。

因为用户只与 Driver Program 交互，因此只能用 RDD 中的 cache()方法去 cache 用户能看到的 RDD。所谓能看到，是指经过 Transformation 算子处理后生成的 RDD，而某些在 Transformation 算子中 Spark 自己生成的 RDD 是不能被用户直接 cache 的。例如，reduceByKey()中会生成的 ShuffleRDD、MapPartitionsRDD 是不能被用户直接 cache 的。在 Driver Program 中设定 RDD.cache()后，系统怎样进行 cache?首先，在计算 RDD 的 Partition 之前就去判断 Partition 要不要被 cache，如果要被 cache，先将 Partition 计算出来，然后 cache 到内存。cache 可使用 memory，如果写到 HDFS 磁盘的话，就要检查 checkpoint。调用 RDD.cache()后，RDD 就变成 persistRDD 了，其 StorageLevel 为 MEMORY_ONLY，persistRDD 会告知 Driver 说自己是需要被 persist 的。此时会调用 RDD.iterator()。

RDD.scala 的 iterator()的源码如下：

```
1.   /**
      * RDD 的内部方法，将从合适的缓存中读取，否则计算它
      * 这不应该被用户直接使用，但可用于实现自定义的子 RDD
2.    */
3.
4.
5.    final def iterator(split: Partition, context: TaskContext): Iterator[T]
      = {
6.      if (storageLevel != StorageLevel.NONE) {
7.        getOrCompute(split, context)
8.      } else {
9.        computeOrReadCheckpoint(split, context)
10.     }
11.   }
```

当 RDD.iterator()被调用的时候，也就是要计算该 RDD 中某个 Partition 的时候，会先去 cacheManager 那里获取一个 blockId，然后去 BlockManager 里匹配该 Partition 是否被 checkpoint 了，如果是，那就不用计算该 Partition 了，直接从 checkpoint 中读取该 Partition 的所有 records 放入 ArrayBuffer 里面。如果没有被 checkpoint 过，先将 Partition 计算出来，然后将其所有 records 放到 cache 中。总体来说，当 RDD 会被重复使用（不能太大）时，RDD 需要 cache。Spark 自动监控每个节点缓存的使用情况，利用最近最少使用原则删除老旧的数据。如果想手动删除 RDD，可以使用 RDD.unpersist()方法。

此外，可以利用不同的存储级别存储每一个被持久化的 RDD。例如，它允许持久化集合到磁盘上，将集合作为序列化的 Java 对象持久化到内存中、在节点间复制集合或者存储集合到 Alluxio 中。可以通过传递一个 StorageLevel 对象给 persist()方法设置这些存储级别。cache()方法使用默认的存储级别-StorageLevel.MEMORY_ONLY。RDD 根据 useDisk、useMemory、useOffHeap、deserialized、replication 5 个参数的组合提供了常用的 12 种基本存储，完整的存储级别介绍如下。

StorageLevel.scala 的源码如下：

```
1.  object StorageLevel {
2.    val NONE = new StorageLevel(false, false, false, false)
3.    val DISK_ONLY = new StorageLevel(true, false, false, false)
4.    val DISK_ONLY_2 = new StorageLevel(true, false, false, false, 2)
```

```
5.    val MEMORY_ONLY = new StorageLevel(false, true, false, true)
6.    val MEMORY_ONLY_2 = new StorageLevel(false, true, false, true, 2)
7.    val MEMORY_ONLY_SER = new StorageLevel(false, true, false, false)
8.    val MEMORY_ONLY_SER_2 = new StorageLevel(false, true, false, false, 2)
9.    val MEMORY_AND_DISK = new StorageLevel(true, true, false, true)
10.   val MEMORY_AND_DISK_2 = new StorageLevel(true, true, false, true, 2)
11.   val MEMORY_AND_DISK_SER = new StorageLevel(true, true, false, false)
12.   val MEMORY_AND_DISK_SER_2 = new StorageLevel(true, true, false, false, 2)
13.   val OFF_HEAP = new StorageLevel(true, true, true, false, 1)
```

StorageLevel是控制存储RDD的标志，每个StorageLevel记录RDD是否使用memory，或使用ExternalBlockStore存储，如果RDD脱离了memory或ExternalBlockStore，是否扔掉RDD，是否保留数据在内存中的序列化格式，以及是否复制多个节点的RDD分区。另外，org.apache.spark.storage.StorageLevel是单实例（singleton）对象，包含了一些静态常量和常用的存储级别，且可用singleton对象工厂方法StorageLevel(…)创建定制化的存储级别。

Spark的多个存储级别意味着在内存利用率和CPU利用率间的不同权衡。推荐通过下面的过程选择一个合适的存储级别：①如果RDD适合默认的存储级别（MEMORY_ONLY），就选择默认的存储级别。因为这是CPU利用率最高的选项，会使RDD上的操作尽可能地快。②如果不适合用默认级别，就选择MEMORY_ONLY_SER。选择一个更快的序列化库提高对象的空间使用率，但是仍能够相当快地访问。③除非算子计算RDD花费较大或者需要过滤大量的数据，不要将RDD存储到磁盘上，否则重复计算一个分区，就会和从磁盘上读取数据一样慢。④如果希望更快地恢复错误，可以利用replicated存储机制，所有的存储级别都可以通过replicated计算丢失的数据来支持完整的容错。另外，replicated的数据能在RDD上继续运行任务，而不需要重复计算丢失的数据。在拥有大量内存的环境中或者多应用程序的环境中，Off_Heap（将对象从堆中脱离出来序列化，然后存储在一大块内存中，这就像它存储到磁盘上一样，但它仍然在RAM内存中。Off_Heap对象在这种状态下不能直接使用，须进行序列化及反序列化。序列化和反序列化可能会影响性能，Off_Heap 堆外内存不需要进行GC）。Off_Heap具有如下优势：Off_Heap运行多个执行者共享的Alluxio中相同的内存池，显著地减少GC。如果单个的Executor崩溃，缓存的数据也不会丢失。

6. 数据调度弹性，DAGScheduler、TASKScheduler和资源管理无关

Spark将执行模型抽象为通用的有向无环图计划（DAG），这可以将多Stage的任务串联或并行执行，从而不需要将Stage中间结果输出到HDFS中，当发生节点运行故障时，可有其他可用节点代替该故障节点运行。

7. 数据分片的高度弹性（coalesce）

Spark进行数据分片时，默认将数据放在内存中，如果内存放不下，一部分会放在磁盘上进行保存。

Spark 2.2.1版本的RDD.scala的coalesce算子代码如下：

```
1.   def coalesce(numPartitions: Int, shuffle: Boolean = false,
2.               partitionCoalescer: Option[PartitionCoalescer] = Option.empty)
3.              (implicit ord: Ordering[T] = null)
4.       : RDD[T] = withScope {
5.     require(numPartitions > 0, s"Number of partitions ($numPartitions)
       must be positive.")
```

```
6.     if (shuffle) {
7.       /**从随机分区开始，将元素均匀分布在输出分区上*/
8.       val distributePartition = (index: Int, items: Iterator[T]) => {
9.         var position = (new Random(index)).nextInt(numPartitions)
10.        items.map { t =>
11.          //注：Key 的哈希码是 Key 本身，HashPartitioner 分区器将它与总分区数进行
           //取模运算
12.
13.          position = position + 1
14.          (position, t)
15.        }
16.      } : Iterator[(Int, T)]
17.
18.      //包括一个 shuffle 步骤，使我们的上游任务仍然是分布式的
19.      new CoalescedRDD(
20.        new ShuffledRDD[Int, T, T](mapPartitionsWithIndex
           (distributePartition),
21.        new HashPartitioner(numPartitions)),
22.        numPartitions,
23.        partitionCoalescer).values
24.    } else {
25.      new CoalescedRDD(this, numPartitions, partitionCoalescer)
26.    }
27.  }
```

Spark 2.4.3 版本的 RDD.scala 源码与 Spark 2.2.1 版本相比具有如下特点。

- 上段代码中第 20 行 mapPartitionsWithIndex 方法调整为 mapPartitionsWithIndexInternal 方法，mapPartitionsWithIndexInternal 方法中传入的参数新增一个字段 isOrderSensitive。isOrderSensitive 用来标识函数是否区分顺序。如果它是对顺序敏感的，当输入顺序改变时，它可能返回完全不同的结果。大多数状态函数是对顺序敏感的。

```
1.    ......
2.    mapPartitionsWithIndexInternal(distributePartition,
      isOrderSensitive = true),
3.    ......
```

mapPartitionsWithIndexInternal 方法将构建一个 MapPartitionsRDD 类，构建类实例时传入 isOrderSensitive。

MapPartitionsRDD.scala 源代码如下：

```
1.    private[spark] def mapPartitionsWithIndexInternal[U: ClassTag](
2.        f: (Int, Iterator[T]) => Iterator[U],
3.        preservesPartitioning: Boolean = false,
4.        isOrderSensitive: Boolean = false): RDD[U] = withScope {
5.      new MapPartitionsRDD(
6.        this,
7.        (context: TaskContext, index: Int, iter: Iterator[T]) => f(index, iter),
8.        preservesPartitioning = preservesPartitioning,
9.        isOrderSensitive = isOrderSensitive)
10.   }
```

MapPartitionsRDD.scala 的 getOutputDeterministicLevel 方法将获取 DeterministicLevel 的 Value 值。

MapPartitionsRDD.scala 源代码如下：

```
1.  override protected def getOutputDeterministicLevel = {
2.      if (isOrderSensitive && prev.outputDeterministicLevel ==
          DeterministicLevel.UNORDERED) {
3.        DeterministicLevel.INDETERMINATE
4.      } else {
5.        super.getOutputDeterministicLevel
6.      }
7.    }
```

其中 DeterministicLevel 定义了 RDD 输出结果的确定级别（即“RDD compute”返回的值）。当 Spark RDD 重新运行任务时，输出将有所不同。包括以下级别：

（1）确定 DETERMINATE：在重新运行后，RDD 输出总是以相同顺序的相同数据集。

（2）无序：RDD 输出总是相同的数据集，但重新运行之后顺序可能不同。

（3）不确定的。重新运行后，RDD 输出可能不同。

注意，RDD 的输出通常依赖于父 RDD。当父 RDD 的输出是不确定的，很可能 RDD 的输出也是不确定的。

RDD.scala 的 DeterministicLevel 代码如下：

```
1. private[spark] object DeterministicLevel extends Enumeration {
2.     val DETERMINATE, UNORDERED, INDETERMINATE = Value
3. }
```

例如，在计算的过程中，会产生很多的数据碎片，这时产生一个 Partition 可能会非常小，如果一个 Partition 非常小，每次都会消耗一个线程去处理，这时可能会降低它的处理效率，需要考虑把许多小的 Partition 合并成一个较大的 Partition 去处理，这样会提高效率。另外，有可能内存不是那么多，而每个 Partition 的数据 Block 比较大，这时需要考虑把 Partition 变成更小的数据分片，这样让 Spark 处理更多的批次，但是不会出现 OOM。

3.3　RDD 依赖关系

RDD 依赖关系为成两种：窄依赖（Narrow Dependency）、宽依赖（Shuffle Dependency）。窄依赖表示每个父 RDD 中的 Partition 最多被子 RDD 的一个 Partition 所使用；宽依赖表示一个父 RDD 的 Partition 都会被多个子 RDD 的 Partition 所使用。

3.3.1　窄依赖解析

RDD 的窄依赖（Narrow Dependency）是 RDD 中最常见的依赖关系，用来表示每一个父 RDD 中的 Partition 最多被子 RDD 的一个 Partition 所使用，如图 3-1 窄依赖关系图所示，父 RDD 有 2～3 个 Partition，每一个分区都只对应子 RDD 的一个 Partition（join with inputs co-partitioned：对数据进行基于相同 Key 的数值相加）。

窄依赖分为两类：第一类是一对一的依赖关系，在 Spark 中用 OneToOneDependency 来表示父 RDD 与子 RDD 的依赖关系是一对一的依赖关系，如 map、filter、join with inputs co-partitioned；第二类是范围依赖关系，在 Spark 中用 RangeDependency 表示，表示父 RDD 与子 RDD 的一对一的范围内依赖关系，如 union。

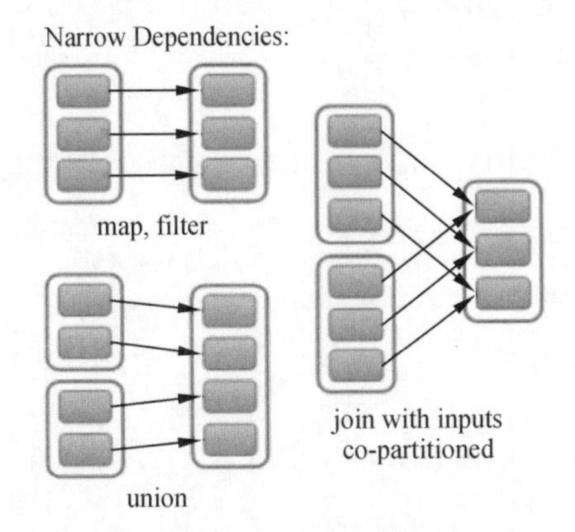

图 3-1　窄依赖关系图

OneToOneDependency 依赖关系的 Dependency.scala 的源码如下：

```
1.  class OneToOneDependency[T](rdd: RDD[T]) extends NarrowDependency[T]
    (rdd) {
2.   override def getParents(partitionId: Int): List[Int] = List(partitionId)
3.  }
```

OneToOneDependency 的 getParents 重写方法引入了参数 partitionId，而在具体的方法中也使用了这个参数，这表明子 RDD 在使用 getParents 方法的时候，查询的是相同 partitionId 的内容。也就是说，子 RDD 仅仅依赖父 RDD 中相同 partitionID 的 Partition。

Spark 窄依赖中第二种依赖关系是 RangeDependency。Dependency.scala 的 RangeDependency 的源码如下：

```
1.  class RangeDependency[T](rdd: RDD[T], inStart: Int, outStart: Int, length: Int)
2.   extends NarrowDependency[T](rdd) {
3.
4.   override def getParents(partitionId: Int): List[Int] = {
5.     if (partitionId >= outStart && partitionId < outStart + length) {
6.       List(partitionId - outStart + inStart)
7.     } else {
8.       Nil
9.     }
10.  }
11. }
```

RangeDependency 和 OneToOneDependency 最大的区别是实现方法中出现了 outStart、length、inStart，子 RDD 在通过 getParents 方法查询对应的 Partition 时，会根据这个 partitionId 减去插入时的开始 ID，再加上它在父 RDD 中的位置 ID，换而言之，就是将父 RDD 中的 Partition，根据 partitionId 的顺序依次插入到子 RDD 中。

分析完 Spark 中的源码，下边通过两个例子来讲解从实例角度去看 RDD 窄依赖输出的结果。

对于 OneToOneDependency，采用 map 操作进行实验，实验代码和结果如下：

```
1. def main (args: Array[String]) {
2. val num1 = Array(100,80,70)
3. val rddnum1 = sc.parallelize(num1)
4. val mapRdd = rddnum1.map(_*2)
```

```
5.  mapRdd.collect().foreach(println)
6.  }
```

结果为 200 160 140。

对于 RangeDependency，采用 union 操作进行实验，实验代码和结果如下：

```
1.  def main (args: Array[String]) {
2.   //创建数组 1
3.  val data1= Array("spark","scala","hadoop")
4.    //创建数组 2
5.  val data2=Array("SPARK","SCALA","HADOOP")
6.   //将数组 1 的数据形成 RDD1
7.  val rdd1 = sc.parallelize(data1)
8.    //将数组 2 的数据形成 RDD2
9.  val rdd2=sc.parallelize(data2)
10.   //把 RDD1 与 RDD2 联合
11. val unionRdd = rdd1.union(rdd2)
12.   //将结果收集并输出
13. unionRdd.collect().foreach(println)
14. }
```

结果为 spark scala hadoop SPARK SCALA HADOOP。

3.3.2　宽依赖解析

RDD 的宽依赖（Shuffle Dependency）是一种会导致计算时产生 Shuffle 操作的 RDD 操作，用来表示一个父 RDD 的 Partition 都会被多个子 RDD 的 Partition 使用，如图 3-2 宽依赖关系图中 groupByKey 算子操作所示，父 RDD 有 3 个 Partition，每个 Partition 中的数据会被子 RDD 中的两个 Partition 使用。

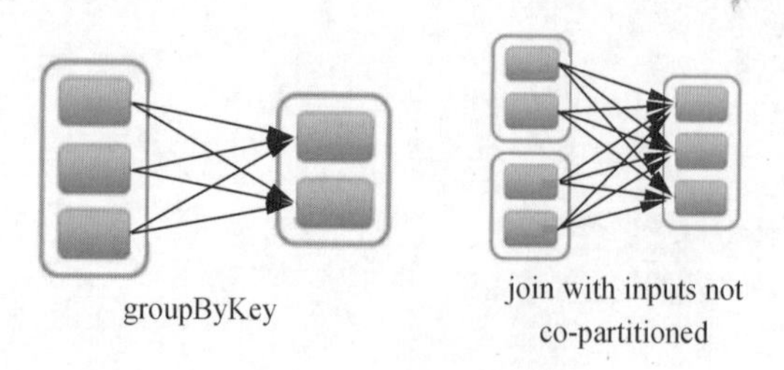

图 3-2　宽依赖关系图

宽依赖的源码位于 Dependency.scala 文件的 ShuffleDependency 方法中，newShuffleId() 产生了新的 shuffleId，表明宽依赖过程需要涉及 shuffle 操作，后续的代码表示宽依赖进行时的 shuffle 操作需要向 shuffleManager 注册信息。

Spark 2.2.1 版本的 Dependency.scala 的 ShuffleDependency 的源码如下：

```
1.   @DeveloperApi
2.  class ShuffleDependency[K: ClassTag, V: ClassTag, C: ClassTag](
3.      @transient private val _rdd: RDD[_ <: Product2[K, V]],
4.      val partitioner: Partitioner,
5.      val serializer: Serializer = SparkEnv.get.serializer,
6.      val keyOrdering: Option[Ordering[K]] = None,
7.      val aggregator: Option[Aggregator[K, V, C]] = None,
```

```
8.      val mapSideCombine: Boolean = false)
9.    extends Dependency[Product2[K, V]] {
10.
11.   override def rdd: RDD[Product2[K, V]] = _rdd.asInstanceOf[RDD[Product2
      [K, V]]]
12.
13.   private[spark] val keyClassName: String = reflect.classTag[K]
      .runtimeClass.getName
14.   private[spark] val valueClassName: String = reflect.classTag[V]
      .runtimeClass.getName
15.   //如果combineBykey使用PairRDDFunctions中的方法,而不是combineBykeyWithClassTag,
16.   //则 combiner 类标签可能是空的
17.   private[spark] val combinerClassName: Option[String] =
18.     Option(reflect.classTag[C]).map(_.runtimeClass.getName)
19.
20.   val shuffleId: Int = _rdd.context.newShuffleId()
21.
22.   val shuffleHandle: ShuffleHandle = _rdd.context.env.shuffleManager
      .registerShuffle(
23.     shuffleId, _rdd.partitions.length, this)
24.
25.   _rdd.sparkContext.cleaner.foreach(_.registerShuffleForCleanup(this))
26. }
```

Spark 2.4.3 版本的 Dependency.scala 源码与 Spark 2.2.1 版本相比具有如下特点。

- 上段代码中第 9 行后面新增加了对 mapSideCombine 的条件判断。

```
1.  ......
2.    if (mapSideCombine) {
3.      require(aggregator.isDefined, "Map-side combine without Aggregator
    specified!")
4.    }
5.  ......
```

Spark 中宽依赖关系非常常见，其中较经典的操作为 GroupByKey（将输入的 key-value 类型的数据进行分组，对相同 key 的 value 值进行合并，生成一个 tuple2，如图 3-3 所示），具体代码和操作结果如下所示。输入 5 个 tuple2 类型的数据，通过运行产生 3 个 tuple2 数据。

```
1.  def main (args: Array[String]) {
2.    //设置输入的 Tuple2 数组
3.  val data = Array(Tuple2("spark",100),Tuple2("spark",95),
4.  Tuple2("hadoop",99),Tuple2("hadoop",80),Tuple2("scala",75))
5.    //将数组内容转化为 RDD
6.  val rdd = sc.parallelize(data)
7.    //对 RDD 进行 groupByKey 操作
8.  val rddGrouped=rdd.groupByKey()
9.    //输出结果
10. rddGrouped.collect.foreach(println)
11. }
```

操作结果如图 3-3 所示。

```
(scala,CompactBuffer(75))
(spark,CompactBuffer(100, 95))
(hadoop,CompactBuffer(99, 80))
```

图 3-3　GroupByKey 结果

3.4 解析 Spark 中的 DAG 逻辑视图

本节讲解 DAG 生成的机制，通过 DAG，Spark 可以对计算的流程进行优化；通过 WordCounts 的示例对 DAG 逻辑视图进行解析。

3.4.1 DAG 生成的机制

在图论中，如果一个有向图无法从任意顶点出发经过若干条边回到该点，则这个图是一个有向无环图（DAG 图）。而在 Spark 中，由于计算过程很多时候会有先后顺序，受制于某些任务必须比另一些任务较早执行的限制，我们必须对任务进行排队，形成一个队列的任务集合，这个队列的任务集合就是 DAG 图，每一个定点就是一个任务，每一条边代表一种限制约束（Spark 中的依赖关系）。

通过 DAG，Spark 可以对计算的流程进行优化，对于数据处理，可以将在单一节点上进行的计算操作进行合并，并且计算中间数据通过内存进行高效读写，对于数据处理，需要涉及 Shuffle 操作的步骤划分 Stage，从而使计算资源的利用更加高效和合理，减少计算资源的等待过程，减少计算中间数据读写产生的时间浪费（基于内存的高效读写）。

Spark 中 DAG 生成过程的重点是对 Stage 的划分，其划分的依据是 RDD 的依赖关系，对于不同的依赖关系，高层调度器会进行不同的处理。对于窄依赖，RDD 之间的数据不需要进行 Shuffle，多个数据处理可以在同一台机器的内存中完成，所以窄依赖在 Spark 中被划分为同一个 Stage；对于宽依赖，由于 Shuffle 的存在，必须等到父 RDD 的 Shuffle 处理完成后，才能开始接下来的计算，所以会在此处进行 Stage 的切分。

在 Spark 中，DAG 生成的流程关键在于回溯，在程序提交后，高层调度器将所有的 RDD 看成是一个 Stage，然后对此 Stage 进行从后往前的回溯，遇到 Shuffle 就断开，遇到窄依赖，则归并到同一个 Stage。等到所有的步骤回溯完成，便生成一个 DAG 图。

DAG 生成的相关源码位于 Spark 的 DAGScheduler.scala。getOrCreateParentStages 获取或创建一个给定 RDD 的父 Stages 列表，getOrCreateParentStages 调用了 getShuffleDependencies（rdd），getShuffleDependencies 返回给定 RDD 的父节点中直接的 Shuffle 依赖。

Spark 2.2.1 版本的 DAGScheduler.scala 的 getOrCreateParentStages 的源码如下：

```
1.   private def getOrCreateParentStages(rdd: RDD[_], firstJobId: Int):
     List[Stage] = {
2.     getShuffleDependencies(rdd).map { shuffleDep =>
3.       getOrCreateShuffleMapStage(shuffleDep, firstJobId)
4.     }.toList
5.    }
6.
7.   ......
8.  private[scheduler] def getShuffleDependencies(
9.        rdd: RDD[_]): HashSet[ShuffleDependency[_, _, _]] = {
10.     val parents = new HashSet[ShuffleDependency[_, _, _]]
11.     val visited = new HashSet[RDD[_]]
12.     val waitingForVisit = new Stack[RDD[_]]
13.     waitingForVisit.push(rdd)
```

```
14.     while (waitingForVisit.nonEmpty) {
15.       val toVisit  = waitingForVisit.pop()
16.       if (!visited(toVisit)) {
17.         visited += toVisit
18.         toVisit.dependencies.foreach {
19.           case shuffleDep: ShuffleDependency[_, _, _] =>
20.             parents += shuffleDep
21.           case dependency =>
22.             waitingForVisit.push(dependency.rdd)
23.         }
24.       }
25.     }
26.     parents
27.   }
```

Spark 2.4.3 版本的 DAGScheduler.scala 源码与 Spark 2.2.1 版本相比具有如下特点。

- 上段代码中第 12 行，构建 Stack 实例修改为构建 ArrayStack 实例。

```
1.   ......
2.   val waitingForVisit = new ArrayStack[RDD[_]]
```

3.4.2 DAG 逻辑视图解析

本节通过一个简单计数案例讲解 DAG 具体的生成流程和关系。示例代码如下：

```
1.  val conf = new SparkConf()//创建 SparkConf
2.  conf.setAppName("Wow,My First Spark App")//设置应用名称
3.  conf.setMaster("local")//在本地运行
4.  val sc =new SparkContext(conf)
5.  val lines = sc.textFile ("C://Users//feng//IdeaProjects//WordCount//src
    //SparkText.txt",1)
6.  //操作 1，flatMap 由 lines 通过 flatMap 操作形成新的 MapPartitionRDD
7.  val words = lines.flatMap{ lines => lines.split(" ") }
8.  //操作 2，map 由 word 通过 Map 操作形成新的 MapPartitionRDD
9.  val pairs =words.map { word => (word,1) }
10. //操作 3，reduceByKey（包含 2 步 reduce）
11. //此步骤生成 MapPartitionRDD 和 ShuffleRDD
12. val WordCounts =pairs.reduceByKey(_+_)
13. WordCounts.collect.foreach(println)
14. sc.stop()
```

在程序正式运行前，Spark 的 DAG 调度器会将整个流程设定为一个 Stage，此 Stage 包含 3 个操作，5 个 RDD，分别为 MapPartitionRDD（读取文件数据时）、MapPartitionRDD（flatMap 操作）、MapPartitionRDD（map 操作）、MapPartitionRDD（reduceByKey 的 local 段的操作）、ShuffleRDD（reduceByKeyshuffle 操作）。

（1）回溯整个流程，在 shuffleRDD 与 MapPartitionRDD（reduceByKey 的 local 段的操作）中存在 shuffle 操作，整个 RDD 先在此切开，形成两个 Stage。

（2）继续向前回溯，MapPartitionRDD（reduceByKey 的 local 段的操作）与 MapPartitionRDD（map 操作）中间不存在 Shuffle（即两个 RDD 的依赖关系为窄依赖），归为同一个 Stage。

（3）继续回溯，发现往前的所有的 RDD 之间都不存在 Shuffle，应归为同一个 Stage。

（4）回溯完成，形成 DAG，由两个 Stage 构成。

- 第一个 Stage 由 MapPartitionRDD（读取文件数据时）、MapPartitionRDD（flatMap 操作）、MapPartitionRDD（map 操作）、MapPartitionRDD（reduceByKey 的 local 段的操作）构成，如图 3-4 所示。

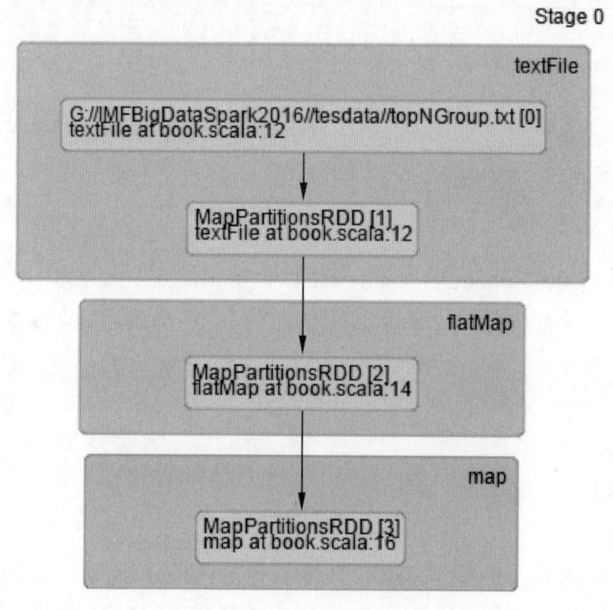

图 3-4　Stage 0 的构成

- 第二个 Stage 由 ShuffleRDD（reduceByKey Shuffle 操作）构成，如图 3-5 所示。

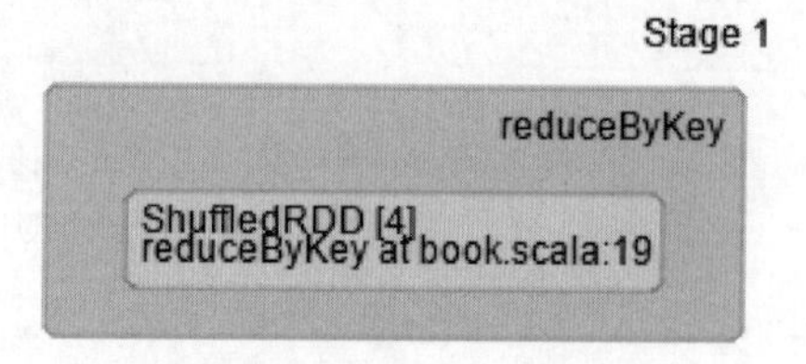

图 3-5　Stage 1 的构成

3.5　RDD 内部的计算机制

RDD 的多个 Partition 分别由不同的 Task 处理。Task 分为两类：shuffleMapTask、resultTask。本节基于源码对 RDD 的计算过程进行深度解析。

3.5.1　Task 解析

Task 是计算运行在集群上的基本计算单位。一个 Task 负责处理 RDD 的一个 Partition，一个 RDD 的多个 Partition 会分别由不同的 Task 去处理，通过之前对 RDD 的窄依赖关系的讲解，我们可以发现在 RDD 的窄依赖中，子 RDD 中 Partition 的个数基本都大于等于父 RDD 中 Partition 的个数，所以 Spark 计算中对于每一个 Stage 分配的 Task 的数目是基于该 Stage

中最后一个 RDD 的 Partition 的个数来决定的。最后一个 RDD 如果有 100 个 Partition，则 Spark 对这个 Stage 分配 100 个 Task。

Task 运行于 Executor 上，而 Executor 位于 CoarseGrainedExecutorBackend（JVM 进程）中。

Spark Job 中，根据 Task 所处 Stage 的位置，我们将 Task 分为两类：第一类为 shuffleMapTask，指 Task 所处的 Stage 不是最后一个 Stage，也就是 Stage 的计算结果还没有输出，而是通过 Shuffle 交给下一个 Stage 使用；第二类为 resultTask，指 Task 所处 Stage 是 DAG 中最后一个 Stage，也就是 Stage 计算结果需要进行输出等操作，计算到此已经结束；简单地说，Spark Job 中除了最后一个 Stage 的 Task 为 resultTask，其他所有 Task 都为 shuffleMapTask。

3.5.2　计算过程深度解析

Spark 中的 Job 本身内部是由具体的 Task 构成的，基于 Spark 程序内部的调度模式，即根据宽依赖的关系，划分不同的 Stage，最后一个 Stage 依赖倒数第二个 Stage 等，我们从最后一个 Stage 获取结果；在 Stage 内部，我们知道有一系列的任务，这些任务被提交到集群上的计算节点进行计算，计算节点执行计算逻辑时，复用位于 Executor 中线程池中的线程，线程中运行的任务调用具体 Task 的 run 方法进行计算，此时，如果调用具体 Task 的 run 方法，就需要考虑不同 Stage 内部具体 Task 的类型，Spark 规定最后一个 Stage 中的 Task 的类型为 resultTask，因为我们需要获取最后的结果，所以前面所有 Stage 的 Task 是 shuffleMapTask。

RDD 在进行计算前，Driver 给其他 Executor 发送消息，让 Executor 启动 Task，在 Executor 启动 Task 成功后，通过消息机制汇报启动成功信息给 Driver。Task 计算示意图如图 3-6 所示。

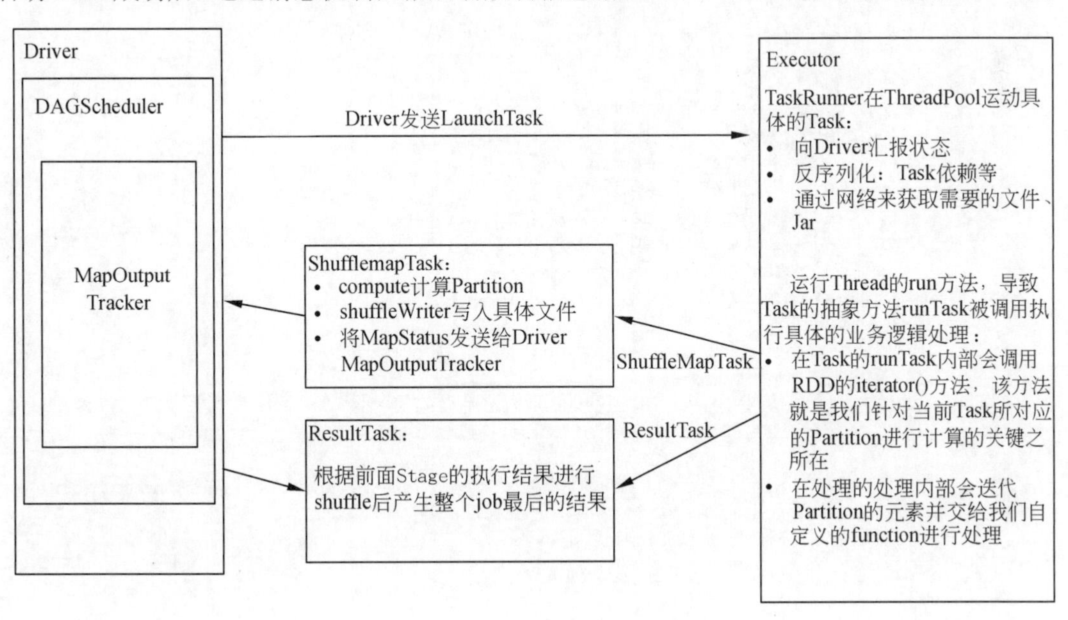

图 3-6　Task 计算示意图

详细情况如下：Driver 中的 CoarseGrainedSchedulerBackend 给 CoarseGrainedExecutorBackend 发送 LaunchTask 消息。

（1）首先反序列化 TaskDescription。

CoarseGrainedExecutorBackend.scala 的 receive 的源码如下：

```
1.  override def receive: PartialFunction[Any, Unit] = {
2.    ......
3.    case LaunchTask(data) =>
4.        if (executor == null) {
5.          exitExecutor(1, "Received LaunchTask command but executor was null")
6.        } else {
7.          val taskDesc = TaskDescription.decode(data.value)
8.          logInfo("Got assigned task " + taskDesc.taskId)
9.          executor.launchTask(this, taskDesc)
10.       }
```

launchTask 中调用了 decode 方法，解析读取 dataIn、taskId、attemptNumber、executorId、name、index 等信息，读取相应的 JAR、文件、属性，返回 TaskDescription 值。

Spark 2.2.1 版本的 TaskDescription.scala 的 decode 的源码如下：

```
1.      def decode(byteBuffer: ByteBuffer): TaskDescription = {
2.      val dataIn = new DataInputStream(new ByteBufferInputStream(byteBuffer))
3.      val taskId = dataIn.readLong()
4.      val attemptNumber = dataIn.readInt()
5.      val executorId = dataIn.readUTF()
6.      val name = dataIn.readUTF()
7.      val index = dataIn.readInt()
8.
9.      //读文件
10.     val taskFiles = deserializeStringLongMap(dataIn)
11.
12.     //读取 jars 包
13.     val taskJars = deserializeStringLongMap(dataIn)
14.
15.     //读取属性
16.     val properties = new Properties()
17.     val numProperties = dataIn.readInt()
18.     for (i <- 0 until numProperties) {
19.       val key = dataIn.readUTF()
20.       val valueLength = dataIn.readInt()
21.       val valueBytes = new Array[Byte](valueLength)
22.       dataIn.readFully(valueBytes)
23.       properties.setProperty(key, new String(valueBytes,
          StandardCharsets.UTF_8))
24.     }
25.
26.     //创建一个子缓冲用于序列化任务将其变成自己的缓冲区（被反序列化后）
27.     val serializedTask = byteBuffer.slice()
28.
29.     new TaskDescription(taskId, attemptNumber, executorId, name, index,
        taskFiles, taskJars,
30.       properties, serializedTask)
31.   }
32. }
```

Spark 2.4.3 版本的 TaskDescription.scala 源码与 Spark 2.2.1 版本相比具有如下特点。

- 上段代码中第 7 行之后新增加 partitionId 的变量。

```
1.  ......
2.  val partitionId = dataIn.readInt()
```

（2）Executor 会通过 launchTask 执行 Task。

（3）Executor 的 launchTask 方法创建一个 TaskRunner 实例在 threadPool 来运行具体的 Task。

Executor.scala 的 launchTask 的源码如下：

```
1.    def launchTask(context: ExecutorBackend, taskDescription: TaskDescription):
       Unit = {
2.      //调用 TaskRunner 句柄创建 TaskRunner 对象
3.       val tr = new TaskRunner(context, taskDescription)
4.      //将创建的 TaskRunner 对象放入即将进行的堆栈中
5.       runningTasks.put(taskDescription.taskId, tr)
6.      //从线程池中分配一条线程给 TaskRunner
7.       threadPool.execute(tr)
8.    }
```

在 TaskRunner 的 run 方法首先会通过 statusUpdate 给 Driver 发信息汇报自己的状态，说明自己处于 running 状态。同时，TaskRunner 内部会做一些准备工作，如反序列化 Task 的依赖，通过网络获取需要的文件、Jar 等；然后反序列化 Task 本身。

Spark 2.2.1 版本的 Executor.scala 的 run 方法的源码如下：

```
1.  override def run(): Unit = {
2.        threadId = Thread.currentThread.getId
3.        Thread.currentThread.setName(threadName)
4.        val threadMXBean = ManagementFactory.getThreadMXBean
5.        val taskMemoryManager = new TaskMemoryManager(env.memoryManager, taskId)
6.        val deserializeStartTime = System.currentTimeMillis()
7.        val deserializeStartCpuTime = if (threadMXBean.
          isCurrentThreadCpuTimeSupported) {
8.          threadMXBean.getCurrentThreadCpuTime
9.        } else 0L
10.       Thread.currentThread.setContextClassLoader(replClassLoader)
11.       val ser = env.closureSerializer.newInstance()
12.       logInfo(s"Running $taskName (TID $taskId)")
13.       //通过 statusUpdate 给 Driver 发信息汇报自己状态说明自己是 running 状态
14.       execBackend.statusUpdate(taskId, TaskState.RUNNING, EMPTY_BYTE_BUFFER)
15.       var taskStart: Long = 0
16.       var taskStartCpu: Long = 0
17.       startGCTime = computeTotalGcTime()
18.
19.       try {
20.         //必须在调用 UpdateDependencies()之前设置，获取依赖项
21.         //需要访问的属性（例如用于访问控制）。
22.
    Executor.taskDeserializationProps.set(taskDescription.properties)
23.
24.         updateDependencies(taskDescription.addedFiles, taskDescription
            .addedJars)
25.        //反序列化 Task 本身
26.         task = ser.deserialize[Task[Any]](
27.         taskDescription.serializedTask, Thread.currentThread
              .getContextClassLoader)
28.         task.localProperties = taskDescription.properties
29.         task.setTaskMemoryManager(taskMemoryManager)
```

Spark 2.4.3 版本的 Executor.scala 的 run 源码与 Spark 2.2.1 版本相比具有如下特点。

- 上段代码中第 15 行代码变量 taskStart 的名称调整为 taskStartTime。

```
1.    var taskStartTime: Long = 0
```

（4）调用反序列化后的 Task.run 方法来执行任务，并获得执行结果。

Spark 2.2.1 版本的 Executor.scala 的 run 方法的源码如下：

```
1.  //Task 计算开始时间.
2.          taskStart = System.currentTimeMillis()
3.          taskStartCpu = if (threadMXBean.isCurrentThreadCpuTimeSupported) {
4.            threadMXBean.getCurrentThreadCpuTime
5.          } else 0L
6.          var threwException = true
7.          val value = try {
8.        //运行 Task 的 run 方法
9.            val res = task.run(
10.             taskAttemptId = taskId,
11.             attemptNumber = taskDescription.attemptNumber,
12.             metricsSystem = env.metricsSystem)
13.           threwException = false
14.           res
15.         } finally {
16.           val releasedLocks = env.blockManager.releaseAllLocksForTask(taskId)
17.           val freedMemory = taskMemoryManager.cleanUpAllAllocatedMemory()
18.
19.           if (freedMemory > 0 && !threwException) {
20.             val errMsg = s"Managed memory leak detected; size = $freedMemory
              bytes, TID = $taskId"
21.             if (conf.getBoolean("spark.unsafe.exceptionOnMemoryLeak", false)) {
22.               throw new SparkException(errMsg)
23.             } else {
24.               logWarning(errMsg)
25.             }
26.           }
27.
28.           if (releasedLocks.nonEmpty && !threwException) {
29.             val errMsg =
30.               s"${releasedLocks.size} block locks were not released by TID =
                $taskId:\n" +
31.                 releasedLocks.mkString("[", ", ", "]")
32.             if (conf.getBoolean("spark.storage.exceptionOnPinLeak", false)) {
33.               throw new SparkException(errMsg)
34.             } else {
35.               logInfo(errMsg)
36.             }
37.           }
38.         }
39.         task.context.fetchFailed.foreach { fetchFailure =>
40.       //用户代码捕获了获取失败，但没有引发任何异常（尽管不可能），记录一个错误并继续.
41.           logError(s"TID ${taskId} completed successfully though internally it
              encountered " +
42.             s"unrecoverable fetch failures!  Most likely this means user
              code is incorrectly " +
43.             s"swallowing Spark's internal ${classOf[FetchFailedException]}",
              fetchFailure)
44.         }
45.        //计算完成时间
46.         val taskFinish = System.currentTimeMillis()
```

```
47.        val taskFinishCpu = if (threadMXBean.isCurrentThreadCpuTimeSupported) {
48.          threadMXBean.getCurrentThreadCpuTime
49.        }
```

Spark 2.4.3 版本的 Executor.scala 的 run 源码与 Spark 2.2.1 版本相比具有如下特点。

- 上段代码中第 2 行 taskStart 名称调整为 taskStartTime。
- 上段代码中第 7 行 try 方法调整为 Utils.tryWithSafeFinally 方法。

```
1.  ……
2.  taskStartTime = System.currentTimeMillis()
3.  ……
4.   val value = Utils.tryWithSafeFinally {……
```

task.run 方法调用了 runTask 的方法，而 runTask 方法是一个抽象方法，runTask 方法内部会调用 RDD 的 iterator()方法，该方法就是针对当前 Task 对应的 Partition 进行计算的关键所在，在处理的方法内部会迭代 Partition 的元素，并交给我们自定义的 function 进行处理。

Task.scala 的 run 方法的源码如下：

```
1.   final def run(
2.       taskAttemptId: Long,
3.       attemptNumber: Int,
4.       metricsSystem: MetricsSystem): T = {
5.     ……
6.     try {
7.       runTask(context)
8.     } catch
9.  ……
```

task 有两个子类，分别是 ShuffleMapTask 和 ResultTask，下面分别对两者进行讲解。

1. ShuffleMapTask

ShuffleMapTask.scala 的源码如下：

```
1.   override def runTask(context: TaskContext): MapStatus = {
2.     //使用广播变量反序列化 RDD
3.     val threadMXBean = ManagementFactory.getThreadMXBean
4.     val deserializeStartTime = System.currentTimeMillis()
5.     val deserializeStartCpuTime = if (threadMXBean.
       isCurrentThreadCpuTimeSupported) {
6.       threadMXBean.getCurrentThreadCpuTime
7.     } else 0L
8.  //创建序列化器
9.     val ser = SparkEnv.get.closureSerializer.newInstance()
10. //反序列化出 RDD 和依赖关系
11. val (rdd, dep) = ser.deserialize[(RDD[_], ShuffleDependency[_, _, _])](
12.       ByteBuffer.wrap(taskBinary.value), Thread.currentThread.
          getContextClassLoader)
13. //RDD 反序列化的时间
14.  _executorDeserializeTime = System.currentTimeMillis() - deserializeStartTime
15.     _executorDeserializeCpuTime = if (threadMXBean
       .isCurrentThreadCpuTimeSupported) {
16.       threadMXBean.getCurrentThreadCpuTime - deserializeStartCpuTime
17.     } else 0L
18.  //创建 Shuffle 的 writer 对象，用来将计算结果写入 Shuffle 管理器
```

```
19.     var writer: ShuffleWriter[Any, Any] = null
20.     try {
21. //实例化 shuffleManager
22.       val manager = SparkEnv.get.shuffleManager
23.   //对 writer 对象赋值
24.      writer = manager.getWriter[Any, Any](dep.shuffleHandle, partitionId,
       context)
25. //将计算结果通过 writer 对象的 write 方法写入 shuffle 过程
26.  writer.write(rdd.iterator(partition, context).asInstanceOf[Iterator[_
     <: Product2[Any, Any]]])
27.       writer.stop(success = true).get
28.     } catch {
29.       case e: Exception =>
30.         try {
31.           if (writer != null) {
32.             writer.stop(success = false)
33.           }
34.         } catch {
35.           case e: Exception =>
36.             log.debug("Could not stop writer", e)
37.         }
38.         throw e
39.     }
40.   }
```

首先，ShuffleMapTask 会反序列化 RDD 及其依赖关系，然后通过调用 RDD 的 iterator 方法进行计算，而 iterator 方法中进行的最终运算的方法是 compute()。

RDD.scala 的 iterator 方法的源码如下：

```
1.     final def iterator(split: Partition, context: TaskContext): Iterator[T]
       = {//判断此 RDD 的持久化等级是否为 NONE(不进行持久化)
2.      if (storageLevel != StorageLevel.NONE) {
3.        getOrCompute(split, context)
4.      } else {
5.        computeOrReadCheckpoint(split, context)
6.      }
7.    }
```

其中，RDD.scala 的 computeOrReadCheckpoint 的源码如下：

```
1.     private[spark] def computeOrReadCheckpoint(split: Partition, context:
       TaskContext): Iterator[T] =
2.    {
3.      if (isCheckpointedAndMaterialized) {
4.        firstParent[T].iterator(split, context)
5.      } else {
6.        compute(split, context)
7.      }
8.    }
```

RDD 的 compute 方法是一个抽象方法，每个 RDD 都需要重写的方法。

此时，选择查看 MapPartitionsRDD 已经实现的 compute 方法，可以发现 compute 方法的实现是通过 f 方法实现的，而 f 方法就是我们创建 MapPartitionsRDD 时输入的操作函数。

Spark 2.2.1 版本的 MapPartitionsRDD.scala 的源码如下：

```
1.     private[spark] class MapPartitionsRDD[U: ClassTag, T: ClassTag](
2.     var prev: RDD[T],
```

```
3.      f: (TaskContext, Int, Iterator[T]) => Iterator[U],  //(TaskContext,
        partition index, iterator)
4.      preservesPartitioning: Boolean = false)
5.    extends RDD[U](prev) {
6.
7.    override val partitioner = if (preservesPartitioning) firstParent[T]
      .partitioner else None
8.
9.    override def getPartitions: Array[Partition] = firstParent[T]
      .partitions
10.
11.   override def compute(split: Partition, context: TaskContext): Iterator
      [U] =
12.     f(context, split.index, firstParent[T].iterator(split, context))
13.
14.   override def clearDependencies() {
15.     super.clearDependencies()
16.     prev = null
17.   }
18. }
```

Spark 2.4.3 版本 MapPartitionsRDD.scala 源码与 Spark 2.2.1 版本相比具有如下特点。

- 上段代码中第 4 行之后新增 2 个参数 isFromBarrier、isOrderSensitive，isFromBarrier 参数指示此 RDD 是否从 RDDBarrier 转换，至少含有一个 RDDBarrier 的 Stage 阶段将转变为屏障阶段（BarrierS tage）。isOrderSensitive 参数指示函数是否区分顺序。
- 上段代码中第 18 行之后新增 isBarrier_、getOutputDeterministicLevel 方法。

```
1. ……
2.     isFromBarrier: Boolean = false,
3.     isOrderSensitive: Boolean = false)
4. ……
5.   @transient protected lazy override val isBarrier_ : Boolean =
6.     isFromBarrier || dependencies.exists(_.rdd.isBarrier())
7.
8.   override protected def getOutputDeterministicLevel = {
9.     if (isOrderSensitive && prev.outputDeterministicLevel ==
           DeterministicLevel.UNORDERED) {
10.      DeterministicLevel.INDETERMINATE
11.    } else {
12.      super.getOutputDeterministicLevel
13.    }
14.  }
15. }
```

注意：通过迭代器的不断叠加，将每个 RDD 的小函数合并成一个大的函数流。

然后在计算具体的 Partition 之后，通过 shuffleManager 获得的 shuffleWriter 把当前 Task 计算的结果根据具体的 shuffleManager 实现写入到具体的文件中，操作完成后会把 MapStatus 发送给 Driver 端的 DAGScheduler 的 MapOutputTracker。

2. ResultTask

Driver 端的 DAGScheduler 的 MapOutputTracker 把 shuffleMapTask 执行的结果交给 ResultTask，ResultTask 根据前面 Stage 的执行结果进行 shuffle 后产生整个 job 最后的结果。

ResultTask.scala 的 runTask 的源码如下：

```
1.    override def runTask(context: TaskContext): U = {
2.      //使用广播变量反序列化 RDD 及函数
3.      val threadMXBean = ManagementFactory.getThreadMXBean
4.      val deserializeStartTime = System.currentTimeMillis()
5.      val deserializeStartCpuTime = if (threadMXBean.
      isCurrentThreadCpuTimeSupported) {
6.        threadMXBean.getCurrentThreadCpuTime
7.      } else 0L
8.      //创建序列化器
9.      val ser = SparkEnv.get.closureSerializer.newInstance()
10.     //反序列 RDD 和 func 处理函数
11.     val (rdd, func) = ser.deserialize[(RDD[T], (TaskContext, Iterator[T])
      => U)](
12.       ByteBuffer.wrap(taskBinary.value), Thread.currentThread
        .getContextClassLoader)
13.     _executorDeserializeTime = System.currentTimeMillis() -
      deserializeStartTime
14.     _executorDeserializeCpuTime = if (threadMXBean
      .isCurrentThreadCpuTimeSupported) {
15.       threadMXBean.getCurrentThreadCpuTime - deserializeStartCpuTime
16.     } else 0L
17.
18.     func(context, rdd.iterator(partition, context))
19.   }
```

而 ResultTask 的 runTask 方法中反序列化生成 func 函数，最后通过 func 函数计算出最终的结果。

3.6　Spark RDD 容错原理及其四大核心要点解析

本节讲解 RDD 不同的依赖关系（宽依赖、窄依赖）的 Spark RDD 容错处理；对 Spark 框架层面容错机制的三大层面（调度层、RDD 血统层、Checkpoint 层）及 Spark RDD 容错四大核心要点进行深入解析。

3.6.1　Spark RDD 容错原理

RDD 不同的依赖关系导致 Spark 对不同的依赖关系有不同的处理方式。

对于宽依赖而言，由于宽依赖实质是指父 RDD 的一个分区会对应一个子 RDD 的多个分区，在此情况下出现部分计算结果丢失，单一计算丢失的数据无法达到效果，便采用重新计算该步骤中的所有数据，从而会导致计算数据重复；对于窄依赖而言，由于窄依赖实质是指父 RDD 的分区最多被一个子 RDD 使用，在此情况下出现部分计算的错误，由于计算结果的数据只与依赖的父 RDD 的相关数据有关，所以不需要重新计算所有数据，只重新计算出错部分的数据即可。

3.6.2　RDD 容错的四大核心要点

Spark 框架层面的容错机制，主要分为三大层面（调度层、RDD 血统层、Checkpoint 层），在这三大层面中包括 Spark RDD 容错四大核心要点。

- Stage 输出失败，上层调度器 DAGScheduler 重试。
- Spark 计算中，Task 内部任务失败，底层调度器重试。
- RDD Lineage 血统中窄依赖、宽依赖计算。
- Checkpoint 缓存。

1. 调度层（包含DAG生成和Task重算两大核心）

从调度层面讲，错误主要出现在两个方面，分别是在 Stage 输出时出错和在计算时出错。

1）DAG 生成层

Stage 输出失败，上层调度器 DAGScheduler 会进行重试，如下列源码所示。

DAGScheduler.scala 的 resubmitFailedStages 的源码如下：

```
1.      private[scheduler] def resubmitFailedStages() {
2.    //判断是否存在失败的 Stages
3.      if (failedStages.size > 0) {
4.        //失败的阶段可以通过作业取消删除，如果 ResubmitFailedStages 事件已调度，失
          //败将是空值
5.
6.        logInfo("Resubmitting failed stages")
7.        clearCacheLocs()
8.      //获取所有失败 Stage 的列表
9.        val failedStagesCopy = failedStages.toArray
10.     //清空 failedStages
11.       failedStages.clear()
12.     //对之前获取的所有失败的 Stage，根据 jobId 排序后逐一重试
13.       for (stage <- failedStagesCopy.sortBy(_.firstJobId)) {
14.         submitStage(stage)
15.       }
16.     }
17.   }
```

2）Task 计算层

Spark 计算过程中，计算内部某个 Task 任务出现失败，底层调度器会对此 Task 进行若干次重试（默认 4 次）。

Spark 2.2.1 版本的 TaskSetManager.scala 的 handleFailedTask 的源码如下：

```
1.  def  handleFailedTask(tid: Long, state: TaskState, reason:
    TaskFailedReason) {
2.      ......
3.    if (!isZombie && reason.countTowardsTaskFailures) {
4.        taskSetBlacklistHelperOpt.foreach(_.updateBlacklistForFailedTask(
5.          info.host, info.executorId, index))
6.        assert (null != failureReason)
7.  //对失败的 Task 的 numFailures 进行计数加一
8.        numFailures(index) += 1
9.  //判断失败的 Task 计数是否大于设定的最大失败次数，如果大于，则输出日志，并不再重试
10.       if (numFailures(index) >= maxTaskFailures) {
11.         logError("Task %d in stage %s failed %d times; aborting job".format(
12.           index, taskSet.id, maxTaskFailures))
13.         abort("Task %d in stage %s failed %d times, most recent failure:
            %s\nDriver stacktrace:"
14.           .format(index,  taskSet.id,  maxTaskFailures,  failureReason),
```

```
            failureException)
15.         return
16.       }
17.   ......}
18. //如果运行的 Task 为 0 时，则完成 Task 步骤
19.     maybeFinishTaskSet()
20.   }
21. ......
```

Spark 2.4.3 版本 TaskSetManager.scala 源码与 Spark 2.2.1 版本相比具有如下特点。

- 上段代码中第 3 行以后新增一个断言 failureReason 的判断，删除第 6 行的代码。
- 上段代码中第 4 行，执行 foreach 方法的时候新传入一个参数 failureReason。

```
1. ......
2.      assert (null != failureReason)
3.       taskSetBlacklistHelperOpt.foreach(_.updateBlacklistForFailedTask(
4.         info.host, info.executorId, index, failureReason))
```

2. RDD Lineage血统层容错

Spark 中 RDD 采用高度受限的分布式共享内存，且新的 RDD 的产生只能够通过其他 RDD 上的批量操作来创建，依赖于以 RDD 的 Lineage 为核心的容错处理，在迭代计算方面比 Hadoop 快 20 多倍，同时还可以在 5～7s 内交互式地查询 TB 级别的数据集。

Spark RDD 实现基于 Lineage 的容错机制，基于 RDD 的各项 transformation 构成了 compute chain，在部分计算结果丢失的时候可以根据 Lineage 重新恢复计算。

- 在窄依赖中，在子 RDD 的分区丢失，要重算父 RDD 分区时，父 RDD 相应分区的所有数据都是子 RDD 分区的数据，并不存在冗余计算。
- 在宽依赖情况下，丢失一个子 RDD 分区，重算的每个父 RDD 的每个分区的所有数据并不是都给丢失的子 RDD 分区用的，会有一部分数据相当于对应的是未丢失的子 RDD 分区中需要的数据，这样就会产生冗余计算开销和巨大的性能浪费。

3. checkpoint层容错

Spark checkpoint 通过将 RDD 写入 Disk 作检查点，是 Spark lineage 容错的辅助，lineage 过长会造成容错成本过高，这时在中间阶段做检查点容错，如果之后有节点出现问题而丢失分区，从做检查点的 RDD 开始重做 Lineage，就会减少开销。

checkpoint 主要适用于以下两种情况。

- DAG 中的 Lineage 过长，如果重算，开销太大，如 PageRank、ALS 等。
- 尤其适合于在宽依赖上作 checkpoint，这个时候就可以避免为 Lineage 重新计算而带来的冗余计算。

3.7　Spark RDD 中 Runtime 流程解析

本节讲解 Spark 的 Runtime 架构图，并从一个作业的视角通过 Driver、Master、Worker、Executor 等角色透视 Spark 的 Runtime 生命周期。

3.7.1 Runtime 架构图

（1）从 Spark Runtime 的角度讲，包括五大核心对象：Master、Worker、Executor、Driver、CoarseGrainedExecutorBackend。

（2）Spark 在做分布式集群系统设计的时候：最大化功能独立、模块化封装具体独立的对象、强内聚松耦合。Spark 运行架构图如图 3-7 所示。

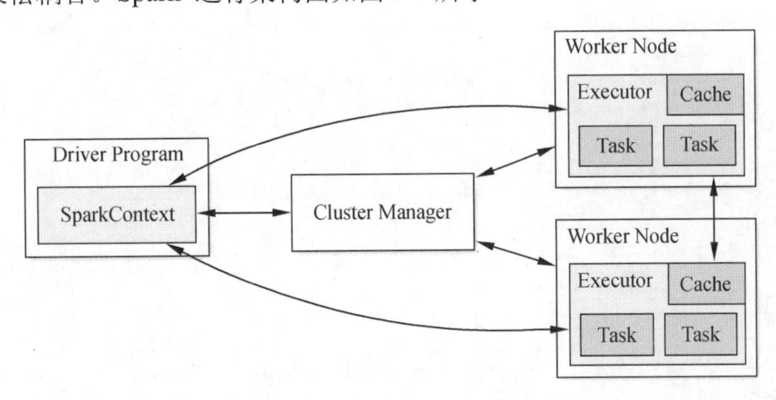

图 3-7　Spark 运行架构图

（3）当 Driver 中的 SparkContext 初始化时会提交程序给 Master，Master 如果接受该程序在 Spark 中运行，就会为当前的程序分配 AppID，同时会分配具体的计算资源。需要特别注意的是，Master 是根据当前提交程序的配置信息来给集群中的 Worker 发指令分配具体的计算资源，但是，Master 发出指令后并不关心具体的资源是否已经分配，换言之，Master 是发指令后就记录了分配的资源，以后客户端再次提交其他的程序，就不能使用该资源了。其弊端是可能会导致其他要提交的程序无法分配到本来应该可以分配到的计算资源；最终的优势是 Spark 分布式系统功能在耦合的基础上最快地运行系统（否则如果 Master 要等到资源最终分配成功后才通知 Driver，就会造成 Driver 阻塞，不能够最大化并行计算资源的使用率）。需要补充说明的是：Spark 在默认情况下由于集群中一般都只有一个 Application 在运行，所有 Master 分配资源策略的弊端就没有那么明显了。

3.7.2 生命周期

本节对 Spark Runtime（Driver、Master、Worker、Executor）内幕解密，从 Spark Runtime 全局的角度看 Spark 具体是怎么工作的，从一个作业的视角通过 Driver、Master、Worker、Executor 等角色来透视 Spark 的 Runtime 生命周期。

Job 提交过程源码解密中一个非常重要的技巧是通过在 spark-shell 中运行一个 Job 来了解 Job 提交的过程，然后再用源码验证这个过程。我们可以在 spark-shell 中运行一个程序，从控制台观察日志。

```
1.  sc.textFile("/library/dataforSortedShufffle").flatMap(_.split(" ")).map
    (word => (word, 1).reduceByKey(_+_).saveAsTextFile("/library/dataoutput2")
```

这里我们编写 WordCountJobRuntime.scala 代码，从 IDEA 中观察日志。读入的数据源文件内容如下：

```
1.  Hello Spark Hello Scala
2.  Hello Hadoop
3.  Hello Flink
4.  Spark is Awesome
```

WordCountJobRuntime.scala 的代码如下：

```
1.   package com.dt.spark.sparksql
2.
3.  import org.apache.log4j.{Level, Logger}
4.  import org.apache.spark.{SparkConf, SparkContext}
5.
6.  /**
7.    * 使用 Scala 开发本地测试的 Spark WordCount 程序
8.    * @author DT 大数据梦工厂
9.    * 新浪微博：http://weibo.com/ilovepains/
10.   */
11. object WordCountJobRuntime {
12.   def main(args: Array[String]){
13.     Logger.getLogger("org").setLevel(Level.ALL)
14.     /**
15.       * 第 1 步：创建 Spark 的配置对象 SparkConf，设置 Spark 程序运行时的配置信息，
          * 例如，通过 setMaster 来设置程序要链接的 Spark 集群的 Master 的 URL，如果设置
        * 为 local，则代表 Spark 程序在本地运行，特别适合于机器配置非常差（如只有 1GB 的内
        * 存）的初学者
16.     */
17.
18.
19.     val conf = new SparkConf() //创建 SparkConf 对象
20.     conf.setAppName("Wow,WordCountJobRuntime!")
                    //设置应用程序的名称，在程序运行的监控界面可以看到名称
21.     conf.setMaster("local") //此时，程序在本地运行，不需要安装 Spark 集群
22.
23.     /**
24.       * 第 2 步：创建 SparkContext 对象
25.       * SparkContext 是 Spark 程序所有功能的唯一入口，采用 Scala、Java、Python、
          * R 等，都必须有一个 SparkContext
26.       * SparkContext 核心作用：初始化 Spark 应用程序运行所需要的核心组件，包括
          * DAGScheduler、TaskScheduler、SchedulerBackend，同时还会负责 Spark 程序
          * 往 Master 注册程序等
27.       * SparkContext 是整个 Spark 应用程序中至关重要的一个对象
28.       */
29.     val sc = new SparkContext(conf)
        //创建 SparkContext 对象，通过传入 SparkConf
        //实例来定制 Spark 运行的具体参数和配置信息
30.
31.     /**
32.       * 第 3 步：根据具体的数据来源（如 HDFS、HBase、Local FS、DB、S3 等）通过
          * SparkContext 创建 RDD
33.       * RDD 的创建有 3 种方式：根据外部的数据来源（如 HDFS）、根据 Scala 集合、由其他
          * 的 RDD 操作
```

```
34.        * 数据会被 RDD 划分为一系列的 Partitions，分配到每个 Partition 的数据属于一
           * 个 Task 的处理范畴
35.        */
36.       val lines = sc.textFile("data/wordcount/helloSpark.txt")
37.      /**
38.        * 第 4 步：对初始的 RDD 进行 Transformation 级别的处理，如通过 map、filter 等
           * 高阶函数等的编程，进行具体的数据计算
39.        *  第 4.1 步：将每一行的字符串拆分成单个单词
40.        */
41.
42.      val words = lines.flatMap { line => line.split(" ")}
                                              //对每一行的字符串进行单词拆分，并把所有行的拆
                                              //分结果通过 flat 合并成为一个大的单词集合
43.
44.      /**
45.        * 第 4 步：对初始的 RDD 进行 Transformation 级别的处理，如通过 map、filter 等
           * 高阶函数等的编程，进行具体的数据计算
46.        *  第 4.2 步：在单词拆分的基础上对每个单词实例计数为 1，也就是 word => (word, 1)
47.        */
48.      val pairs = words.map { word => (word, 1) }
49.
50.      /**
51.        * 第 4 步：对初始的 RDD 进行 Transformation 级别的处理，如通过 map、filter 等
           * 高阶函数等的编程，进行具体的数据计算
52.        *  第 4.3 步：在每个单词实例计数为 1 的基础上统计每个单词在文件中出现的总次数
53.        */
54.      val wordCountsOdered = pairs.reduceByKey(_+_).saveAsTextFile("data/
         wordcount/wordCountResult.log")
55.
56.      while(true){
57.
58.      }
59.      sc.stop()
60.
61.    }
62. }
```

在 IDEA 中运行，WordCountJobRuntime 的运行结果保存在 data/wordcount/wordCountResult.log 目录的 part-00000 中。

```
1.   (Awesome,1)
2.  (Flink,1)
3.  (Spark,2)
4.  (is,1)
5.  (Hello,4)
6.  (Scala,1)
7.  (Hadoop,1)
```

在 IDEA 的控制台中观察 WordCountJobRuntime.scala 运行日志，这里 Spark 版本是 version 2.4.3。其中，MemoryStore 是从 Storage 内存角度来看的，Storage 是磁盘管理和内存管理。这里，Spark 读取了 Hadoop 的 HDFS，因此使用了 Hadoop 的内容，如 FileInputFormat，日志中显示 FileInputFormat: Total input paths to process : 1 说明有一个文件要处理。

```
1.  Using Spark's default log4j profile: org/apache/spark/log4j-defaults.properties
2.  19/05/18 13:32:05 INFO SparkContext: Running Spark version 2.4.3
```

```
3.  ......
4.  19/05/18 13:32:12 DEBUG DiskBlockManager: Adding shutdown hook
5.  19/05/18 13:32:12 DEBUG ShutdownHookManager: Adding shutdown hook
6.  19/05/18 13:32:12 INFO MemoryStore: MemoryStore started with capacity
    894.3 MB
7.  19/05/18 13:32:12 INFO SparkEnv: Registering OutputCommitCoordinator
8.  ......
9.  19/05/18 13:32:15 DEBUG FileInputFormat: Time taken to get FileStatuses:
    17
10. 19/05/18 13:32:15 INFO FileInputFormat: Total input paths to process : 1
11. 19/05/18 13:32:15 DEBUG FileInputFormat: Total # of splits generated by
    getSplits: 1, TimeTaken: 44
```

在 Spark 中，所有的 Action 都会触发至少一个 Job，在 WordCountJobRuntime.scala 代码中，是通过 saveAsTextFile 来触发 Job 的。在日志中查看 SparkContext: Starting job: saveAsTextFile 触发 saveAsTextFile。紧接着交给 DAGScheduler，日志中显示 DAGScheduler: Registering RDD，因为这里有两个 Stage，从具体计算的角度，前面 Stage 计算的时候保留输出。然后是 DAGScheduler 获得了 job 的 ID（job 0）。

```
1.  19/05/18 13:32:16 INFO SparkContext: Starting job: runJob at SparkHadoopWriter.
    scala:78
2.  19/05/18 13:32:17 DEBUG SortShuffleManager: Can't use serialized shuffle
    for shuffle 0 because we need to do map-side aggregation
3.  19/05/18 13:32:17 INFO DAGScheduler: Registering RDD 3 (map at
    WordCountJobRuntime.scala:52)
4.  19/05/18 13:32:17 INFO DAGScheduler: Got job 0 (runJob at SparkHadoopWriter
    .scala:78) with 1 output partitions
5.  ......
```

SparkContext 在实例化的时候会构造 StandaloneSchedulerBackend（Spark 2.0 版本将之前的 SparkDeploySchedulerBackend 名字更新为 StandaloneSchedulerBackend）、DAGScheduler、TaskSchedulerImpl、MapOutputTrackerMaster 等对象。

- 其中，StandaloneSchedulerBackend 负责集群计算资源的管理和调度，这是从作业的角度来考虑的，注册给 Master 的时候，Master 给我们分配资源，资源从 Executor 本身转过来向 StandaloneSchedulerBackend 注册，这是从作业调度的角度来考虑的，不是从整个集群来考虑，整个集群是 Master 来管理计算资源的。
- DAGScheduler 负责高层调度（如 Job 中 Stage 的划分、数据本地性等内容）。
- TaskSchedulerImple 负责具体 Stage 内部的底层调度（如具体每个 Task 的调度、Task 的容错等）。
- MapOutputTrackerMaster 负责 Shuffle 中数据输出和读取的管理。Shuffle 的时候将数据写到本地，下一个 Stage 要使用上一个 Stage 的数据，因此写数据的时候要告诉 Driver 中的 MapOutputTrackerMaster 具体写到哪里，下一个 Stage 读取数据的时候也要访问 Driver 的 MapOutputTrackerMaster 获取数据的具体位置。

MapOutputTrackerMaster 的源码如下：

```
1.  private[spark] class MapOutputTrackerMaster(conf: SparkConf,
2.     broadcastManager: BroadcastManager, isLocal: Boolean)
3.   extends MapOutputTracker(conf) {
```

DAGScheduler 是面向 Stage 调度的高层调度实现。它为每一个 Job 计算 DAG，跟踪 RDDS

及 Stage 输出结果进行物化，并找到一个最小的计划去运行 Job，然后提交 stages 中的 TaskSets 到底层调度器 TaskScheduler 提交集群运行，TaskSet 包含完全独立的任务，基于集群上已存在的数据运行（如从上一个 Stage 输出的文件），如果这个数据不可用，获取数据可能会失败。

Spark Stages 根据 RDD 图中 Shuffle 的边界来创建，如果 RDD 的操作是窄依赖，如 map() 和 filter()，在每个 Stages 中将一系列 tasks 组合成流水线执行。但是，如果是宽依赖，Shuffle 依赖需要多个 Stages（上一个 Stage 进行 map 输出写入文件，下一个 Stage 读取数据文件），每个 Stage 依赖于其他的 Stage，其中进行多个算子操作。算子操作在各种类型的 RDDS（如 MappedRDD、FilteredRDD）的 RDD.compute()中实际执行。

在 DAG 阶段，DAGScheduler 根据当前缓存状态决定每个任务运行的位置，并将任务传递给底层的任务调度器 TaskScheduler。此外，它处理 Shuffle 输出文件丢失的故障，在这种情况下，以前的 Stage 可能需要重新提交。Stage 中不引起 Shuffle 文件丢失的故障由任务调度器 TaskScheduler 处理，在取消整个 Stage 前，将重试几次任务。

当浏览这个代码时，有以下几个关键概念。

- Jobs 作业（表现为[ActiveJob]）作为顶级工作项提交给调度程序。当用户调用一个 action，如 count()算子，Job 将通过 submitJob 进行提交。每个作业可能需要执行多个 stages 来构建中间数据。
- Stages ([Stage])是一组任务的集合，在相同的 RDD 分区上，每个任务计算相同的功能，计算 Jobs 的中间结果。Stage 根据 Shuffle 划分边界，我们必须等待前一阶段 Stage 完成输出。有两种类型的 Stage：[ResultStage]是执行 action 的最后一个 Stage，[ShuffleMapStage]是 Shuffle Stages 通过 map 写入输出文件中的。如果 Jobs 重用相同的 RDDs，Stages 可以跨越多个 Jobs 共享。
- Tasks 任务是单独的工作单位，每个任务发送到一个分布式节点。
- 缓存跟踪：DAGScheduler 记录哪些 RDDS 被缓存，避免重复计算，以及记录 Shuffle map Stages 已经生成的输出文件，避免在 map 端重新计算。
- 数据本地化：DAGScheduler 基于 RDDS 的数据本地性、缓存位置，或 Shuffle 数据在 Stage 中运行每一个任务的 Task。
- 清理：当依赖于它们的运行作业完成时，所有数据结构将被清除，防止在长期运行的应用程序中内存泄漏。
- 为了从故障中恢复，同一个 Stage 可能需要运行多次，这被称为重试“attempts”。如在上一个 Stage 中的输出文件丢失，TaskScheduler 中将报告任务失败，DAGScheduler 通过检测 CompletionEvent 与 FetchFailed 或 ExecutorLost 事件重新提交丢失的 Stage。DAGScheduler 将等待看是否有其他节点或任务失败，然后在丢失计算任务的阶段 Stage 中重新提交 TaskSets。在这个过程中，可能须创建之前被清理的 Stage。旧 Stage 的任务仍然可以运行，但必须在正确的 Stage 中接收事件并进行操作。

做改变或者回顾时需要看的清单如下。

- Job 运行结束时，所有的数据结构将被清理，及清理程序运行中的状态。
- 添加一个新的数据结构时，在新结构中更新'DAGSchedulerSuite.assertDataStructuresEmpty'，包括新结构，将有助于捕获内存泄漏。

DAGScheduler.scala 的源码如下：

```
1.  private[spark]
2. class DAGScheduler(
3.     private[scheduler] val sc: SparkContext,
4.     private[scheduler] val taskScheduler: TaskScheduler,
5.     listenerBus: LiveListenerBus,
6.     mapOutputTracker: MapOutputTrackerMaster,
7.     blockManagerMaster: BlockManagerMaster,
8.     env: SparkEnv,
9.     clock: Clock = new SystemClock())
10.  extends Logging {
```

回到运行日志，SparkContext 在实例化的时候会构造 StandaloneSchedulerBackend、DAGScheduler、TaskSchedulerImpl、MapOutputTrackerMaster 四大核心对象，DAGScheduler 获得 Job ID，日志中显示 DAGScheduler：Final stage：ResultStage 1，Final stage 是 ResultStage；Parents of final stage 是 ShuffleMapStage，DAGScheduler 是面向 Stage 的。日志中显示两个 Stage：Stage 1 是 Final stage，Stage 0 是 ShuffleMapStage。

接下来序号改变，运行时最左侧从 0 开始，日志中显示 DAGScheduler：missing：List(ShuffleMapStage 0)，父 Stage 是 ShuffleMapStage，DAGScheduler 调度时必须先计算父 Stage，因此首先提交的是 ShuffleMapStage 0，这里 RDD 是 MapPartitionsRDD，只有 Stage 中的最后一个算子是真正有效的，Stage 0 中的最后一个操作是 map，因此生成了 MapPartitionsRDD。Stage 0 无父 Stage，因此提交，提交时进行广播等内容，然后提交作业。

```
1.  19/05/18 13:32:17 INFO DAGScheduler: Final stage: ResultStage 1 (runJob
    at SparkHadoopWriter.scala:78)
2.  19/05/18 13:32:17 INFO DAGScheduler: Parents of final stage:
    List(ShuffleMapStage 0)
3.  19/05/18 13:32:17 INFO DAGScheduler: Missing parents: List(ShuffleMapStage 0)
4.  19/05/18 13:32:17 DEBUG DAGScheduler: submitStage(ResultStage 1)
5.  19/05/18 13:32:17 DEBUG DAGScheduler: missing: List(ShuffleMapStage 0)
6.  19/05/18 13:32:17 DEBUG DAGScheduler: submitStage(ShuffleMapStage 0)
7.  19/05/18 13:32:17 DEBUG DAGScheduler: missing: List()
8.  19/05/18 13:32:17 INFO DAGScheduler: Submitting ShuffleMapStage 0
    (MapPartitionsRDD[3] at map at WordCountJobRuntime.scala:52), which has
    no missing parents
9.  19/05/18 13:32:17 DEBUG DAGScheduler: submitMissingTasks(ShuffleMapStage 0)
10. 19/05/18 13:32:18 TRACE BlockInfoManager: Task -1024 trying to put
    broadcast_1
11. ......
```

我们从 Web UI 的角度看一下，如图 3-8 所示，Web UI 中显示生成两个 Stage：Stage 0、Stage 1。

日志中显示 DAGScheduler: Submitting 1 missing tasks from ShuffleMapStage 0，DAGScheduler 提交作业，显示提交一个须计算的任务，ShuffleMapStage 在本地运行是一个并行度，交给 TaskSchedulerImpl 运行。这里是一个并行度，提交底层的调度器 TaskScheduler，TaskScheduler 收到任务后，就发布任务到集群中运行，由 TaskSetManager 进行管理：日志中显示 TaskSetManager：Starting task 0.0 in stage 0.0 (TID 0, localhost, executor driver, partition 0, PROCESS_LOCAL, 7415 bytes)，显示具体运行的位置，及 worker 运行了哪些任务。这里在本地只运行了一个任务。

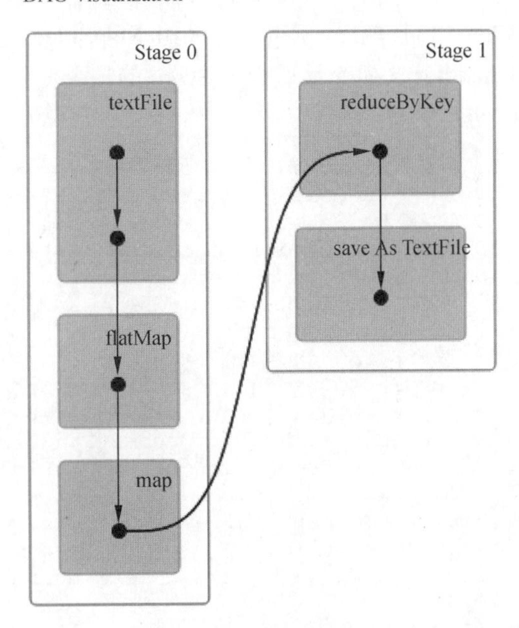

图 3-8　Stage 划分

```
1.  19/05/18 13:32:18 INFO DAGScheduler: Submitting 1 missing tasks from
    ShuffleMapStage 0 (MapPartitionsRDD[3] at map at WordCountJobRuntime.scala:52)
    (first 15 tasks are for partitions Vector(0))
2.  19/05/18 13:32:18 INFO TaskSchedulerImpl: Adding task set 0.0 with 1 tasks
3.  19/05/18 13:32:18 DEBUG TaskSetManager: Epoch for TaskSet 0.0: 0
4.  19/05/18 13:32:18 DEBUG TaskSetManager: Valid locality levels for TaskSet
    0.0: NO_PREF, ANY
5.  19/05/18 13:32:18 DEBUG TaskSchedulerImpl: parentName: , name: TaskSet_0.0,
    runningTasks: 0
6.  19/05/18 13:32:18 DEBUG TaskSetManager: Valid locality levels for TaskSet
    0.0: NO_PREF, ANY
7.  19/05/18 13:32:18 INFO TaskSetManager: Starting task 0.0 in stage 0.0 (TID
    0, localhost, executor driver, partition 0, PROCESS_LOCAL, 7415 bytes)
8.  19/05/18 13:32:18 INFO Executor: Running task 0.0 in stage 0.0 (TID 0)
......
```

然后是完成作业，日志中显示 TaskSetManager: Finished task 0.0 in stage 0.0 (TID 0) in 1529 ms on localhost (executor driver)，在本地机器上完成作业。当 Stage 的一个任务完成后，ShuffleMapStage 就已完成。Task 任务运行完后向 DAGScheduler 汇报，DAGScheduler 查看曾经提交了几个 Task，计算 Task 的数量如果等于 Task 的总数量，那 Stage 也就完成了。这个 Stage 完成以后，下一个 Stage 开始运行。

```
1.  19/05/18 13:32:19 INFO TaskSetManager: Finished task 0.0 in stage 0.0 (TID
    0) in 1529 ms on localhost (executor driver) (1/1)
2.  19/05/18 13:32:19 INFO TaskSchedulerImpl: Removed TaskSet 0.0, whose tasks
    have all completed, from pool
3.  19/05/18 13:32:19 DEBUG DAGScheduler: ShuffleMapTask finished on driver
4.  19/05/18 13:32:19 INFO DAGScheduler: ShuffleMapStage 0 (map at
    WordCountJobRuntime.scala:52) finished in 1.781 s
5.  ......
```

ShuffleMapStage 完成后，将运行下一个 Stage。日志中显示 DAGScheduler：looking for newly runnable stages，这里一共有两个 Stage，ShuffleMapStage 运行完成，那只有一个 ResultStage 将运行。DAGScheduler 又提交最后一个 Stage 的一个任务，默认并行度是继承的。同样，发布任务给 Executor 进行计算。

```
1.  19/05/18 13:32:19 INFO DAGScheduler: looking for newly runnable stages
2.  19/05/18 13:32:19 INFO DAGScheduler: running: Set()
3.  19/05/18 13:32:19 INFO DAGScheduler: waiting: Set(ResultStage 1)
4.  19/05/18 13:32:19 INFO DAGScheduler: failed: Set()
5.  19/05/18 13:32:19 DEBUG MapOutputTrackerMaster: Increasing epoch to 1
6.  19/05/18 13:32:19 TRACE DAGScheduler: Checking if any dependencies of
    ShuffleMapStage 0 are now runnable
7.  19/05/18 13:32:19 TRACE DAGScheduler: running: Set()
8.  19/05/18 13:32:19 TRACE DAGScheduler: waiting: Set(ResultStage 1)
9.  19/05/18 13:32:19 TRACE DAGScheduler: failed: Set()
10. 19/05/18 13:32:19 DEBUG DAGScheduler: submitStage(ResultStage 1)
11. 19/05/18 13:32:19 DEBUG DAGScheduler: missing: List()
12. 19/05/18 13:32:19 INFO DAGScheduler: Submitting ResultStage 1
    (MapPartitionsRDD[5] at saveAsTextFile at WordCountJobRuntime.scala:58),
    which has no missing parents
13. 19/05/18 13:32:19 DEBUG DAGScheduler: submitMissingTasks(ResultStage 1)
14. ......
15. 19/05/18 13:32:19 INFO DAGScheduler: Submitting 1 missing tasks from
    ResultStage 1 (MapPartitionsRDD[5] at saveAsTextFile at WordCountJobRuntime.
    scala:58) (first 15 tasks are for partitions Vector(0))
16. 19/05/18 13:32:19 INFO TaskSchedulerImpl: Adding task set 1.0 with 1 tasks
......
```

Task 任务运行完后向 DAGScheduler 汇报，DAGScheduler 计算曾经提交了几个 Task，如果 Task 的数量等于 Task 的总数量，ResultStage 也运行完成。然后进行相关的清理工作，两个 Stage（ShuffleMapStage、ResultStage）完成，Job 也就完成。

```
1.  19/05/18 13:32:20 DEBUG MapOutputTrackerMaster: Fetching outputs for shuffle 0,
    partitions 0-1
2.  19/05/18 13:32:20 DEBUG ShuffleBlockFetcherIterator: maxBytesInFlight:
    50331648, targetRequestSize: 10066329, maxBlocksInFlightPerAddress:
    2147483647
3.  19/05/18 13:32:20 INFO ShuffleBlockFetcherIterator: Getting 1 non-empty
    blocks including 1 local blocks and 0 remote blocks
4.  19/05/18 13:32:20 INFO ShuffleBlockFetcherIterator: Started 0 remote
    fetches in 16 ms
5.  19/05/18 13:32:20 DEBUG ShuffleBlockFetcherIterator: Start fetching local
    blocks: shuffle_0_0_0
6.  19/05/18 13:32:20 DEBUG ShuffleBlockFetcherIterator: Got local blocks in
    22 ms
7.  19/05/18 13:32:20 INFO HadoopMapRedCommitProtocol: Using output committer
    class org.apache.hadoop.mapred.FileOutputCommitter
8.  19/05/18 13:32:20 DEBUG TaskMemoryManager: Task 1 release 0.0 B from
    org.apache.spark.util.collection.ExternalAppendOnlyMap@55aac15d
9.  ......
10. 19/05/18 13:32:20 INFO Executor: Finished task 0.0 in stage 1.0 (TID 1).
    1465 bytes result sent to driver
11.19/05/18 13:32:20 DEBUG TaskSchedulerImpl: parentName: , name: TaskSet_1.0,
    runningTasks: 0
12. 19/05/18 13:32:20 INFO TaskSetManager: Finished task 0.0 in stage 1.0 (TID
    1) in 1082 ms on localhost (executor driver) (1/1)
13. 19/05/18 13:32:20 INFO TaskSchedulerImpl: Removed TaskSet 1.0, whose tasks
```

```
    have all completed, from pool
14. 19/05/18 13:32:20 INFO DAGScheduler: ResultStage 1 (runJob at SparkHadoopWriter
    .scala:78) finished in 1.139 s
15. 19/05/18 13:32:20 DEBUG DAGScheduler: After removal of stage 1, remaining
    stages = 1
16. 19/05/18 13:32:20 DEBUG DAGScheduler: After removal of stage 0, remaining
    stages = 0
17. 19/05/18 13:32:20 INFO DAGScheduler: Job 0 finished: runJob at SparkHadoopWriter
    .scala:78, took 4.583493 s
……
```

下面看一下 WebUI，ShuffleMapStage 中的任务交给 Executor，图 3-9 中显示了任务的相关信息，如 Shuffle 的输出等，第一个 Stage 肯定生成 Shuffle 的输出，可以看一下最右侧的 Shuffle Write Size/Records。图 3-9 中的 Input Size/Records 是从 Hdfs 中读入的文件数据。

Summary Metrics for 1 Completed Tasks

Metric	Min	25th percentile	Median	75th percentile	Max
Duration	0.4 s	0.4 s	0.4 s	0.4 s	0.4 s
GC Time	19 ms	19 ms	19 ms	19 ms	19 ms
Input Size / Records	68.0 B / 4	68.0 B / 4	68.0 B / 4	68.0 B / 4	68.0 B / 4
Shuffle Write Size / Records	104.0 B / 7	104.0 B / 7	104.0 B / 7	104.0 B / 7	104.0 B / 7

▾ Aggregated Metrics by Executor

Executor ID ▴	Address	Task Time	Total Tasks	Failed Tasks	Killed Tasks	Succeeded Tasks	Input Size / Records	Shuffle Write Size / Records	Blacklisted
driver	192.168.189.100:57007	2 s	1	0	0	1	68.0 B / 4	104.0 B / 7	false

▾ Tasks (1)

Index ▴	ID	Attempt	Status	Locality Level	Executor ID	Host	Launch Time	Duration	GC Time	Input Size / Records	Write Time	Shuffle Write Size / Records	Errors
0	0	0	SUCCESS	PROCESS_LOCAL	driver	localhost	2019/05/18 13:46:57	0.4 s	19 ms	68.0 B / 4	8 ms	104.0 B / 7	

图 3-9　ShuffleMapStage 运行

接下来看一下第二个 Stage。第二个 Stage 同样显示 Executor 的信息，图 3-10 最右侧显示 Shuffle Read Size/Records。如果在分布式集群运行，须远程读取数据，例如，原来是 4 个 Executor 计算，在第二个 Stage 中是两个 Executor 计算，因此一部分数据是本地的，一部分是远程的，或从远程节点拉取数据。ResultStage 最后要产生输出，输出到文件保存。

Task 的运行解密如下。

（1）Task 是运行在 Executor 中的，而 Executor 又是位于 CoarseGrainedExecutorBackend 中的，且 CoarseGrainedExecutorBackend 和 Executor 是一一对应的；计算运行于 Executor，而 Executor 位于 CoarseGrainedExecutorBackend 中，CoarseGrainedExecutorBackend 是进程。发任务消息也是在 CoarseGrainedExecutorBackend。

（2）当 CoarseGrainedExecutorBackend 接收到 TaskSetManager 发过来的 LaunchTask 消息后会反序列化 TaskDescription，然后使用 CoarseGrainedExecutorBackend 中唯一的 Executor 来执行任务。

CoarseGrainedExecutorBackend 收到 Driver 发送的 LaunchTask 任务消息，其中 LaunchTask 是 case class，而不是 case object，是因为每个消息是一个消息实例，每个消息状态不一样，

Summary Metrics for 1 Completed Tasks

Metric	Min	25th percentile	Median	75th percentile	Max
Duration	0.9 s	0.9 s	0.9 s	0.9 s	0.9 s
GC Time	0 ms	0 ms	0 ms	0 ms	0 ms
Output Size / Records	82.0 B / 7	82.0 B / 7	82.0 B / 7	82.0 B / 7	82.0 B / 7
Shuffle Read Size / Records	104.0 B / 7	104.0 B / 7	104.0 B / 7	104.0 B / 7	104.0 B / 7

▾ Aggregated Metrics by Executor

Executor ID ▴	Address	Task Time	Total Tasks	Failed Tasks	Killed Tasks	Succeeded Tasks	Output Size / Records	Shuffle Read Size / Records	Blacklisted
driver	192.168.189.100:57007	1 s	1	0	0	1	82.0 B / 7	104.0 B / 7	false

▾ Tasks (1)

Index ▴	ID	Attempt	Status	Locality Level	Executor ID	Host	Launch Time	Duration	GC Time	Output Size / Records	Shuffle Read Size / Records	Errors
0	1	0	SUCCESS	ANY	driver	localhost	2019/05/18 13:46:59	0.9 s		82.0 B / 7	104.0 B / 7	

图 3-10　ResultStage 运行

而 case object 是唯一的，因此使用 case class。

```
1.      //Driver 节点到 executors 节点
2.    case class LaunchTask(data: SerializableBuffer) extends
      CoarseGrainedClusterMessage
```

Executor.scala 的源码如下：

```
1.  //维护正在运行的任务列表
2.  private val runningTasks = new ConcurrentHashMap[Long, TaskRunner]
3.  ……
4.    def launchTask(context: ExecutorBackend, taskDescription: TaskDescription):
      Unit = {
5.      val tr = new TaskRunner(context, taskDescription)
6.      runningTasks.put(taskDescription.taskId, tr)
7.      threadPool.execute(tr)
8.    }
9.  ……
10.   class TaskRunner(
11.       execBackend: ExecutorBackend,
12.       private val taskDescription: TaskDescription)
13.     extends Runnable {
……
```

在 Executor.scala 中查看 launchTask 方法，运行任务 runningTasks 使用了 ConcurrentHashMap 数据结构，运行 launchTask 的时候构建了一个 TaskRunner，TaskRunner 是一个 Runnable，而 Runnable 是 Java 中的接口，Scala 可以直接调用 Java 的代码，run 方法中包括任务的反序列化等内容。通过 Runnable 封装任务，然后放入到 runningTasks 中，在 threadPool 中执行任务。threadPool 是一个 newDaemonCachedThreadPool。任务交给 Executor 的线程池中的线程去执行，执行的时候下载资源、数据等内容。

Executor.scala 的 threadPool 的源码如下：

```
1.     //启动 worker 线程池
2.    private val threadPool = {
```

```
3.      val threadFactory = new ThreadFactoryBuilder()
4.        .setDaemon(true)
5.        .setNameFormat("Executor task launch worker-%d")
6.        .setThreadFactory(new ThreadFactory {
7.          override def newThread(r: Runnable): Thread =
8.            //使用 UninterruptibleThread 运行任务，这样我们就可以允许运行代码
9.            //不被 Thread.interrupt()线程中断。例如，KAFKA-1894、HADOOP-10622,
              //如果某些方法被中断，程序将会一直挂起
10.           new UninterruptibleThread(r, "unused") //thread name will be set
                                                     //by ThreadFactoryBuilder
11.       })
12.       .build()
13.     Executors.newCachedThreadPool(threadFactory).asInstanceOf
        [ThreadPoolExecutor]
14.   }
```

3.8 通过 WordCount 实战解析 Spark RDD 内部机制

本节通过 Spark WordCount 动手实践，编写单词计数代码；在 wordcount.scala 的基础上，从数据流动的视角深入分析 Spark RDD 的数据处理过程。

3.8.1 Spark WordCount 动手实践

本节进行 Spark WordCount 动手实践。首先建立一个文本文件 helloSpark.txt，将文本文件放到文件目录 data/wordcount/中。helloSpark.txt 的文本内容如下：

```
1.  Hello Spark Hello Scala
2.  Hello Hadoop
3.  Hello Flink
4.  Spark is Awesome
```

在 IDEA 中编写 wordcount.scala 的代码如下：

```
1.   package com.dt.spark.sparksql
2.  import org.apache.spark.SparkConf
3.  import org.apache.spark.SparkContext
4.  import org.apache.spark.rdd.RDD
5.  /**
6.    * 使用 Scala 开发本地测试的 Spark WordCount 程序
7.    * @author DT 大数据梦工厂
8.    * 新浪微博：http://weibo.com/ilovepains/
9.    */
10. object WordCount {
11.   def main(args: Array[String]){
12.     /**
13.       * 第 1 步：创建 Spark 的配置对象 SparkConf，设置 Spark 程序运行时的配置信息，
          * 例如，通过 setMaster 设置程序要链接的 Spark 集群的 Master 的 URL，如果设置
          * 为 local，则代表 Spark 程序在本地运行，特别适合于机器配置非常差（如只有 1GB
          * 的内存）的初学者
14.       */
```

```
15.
16.     val conf = new SparkConf() //创建 SparkConf 对象
17.     conf.setAppName("Wow,My First Spark App!")
                              //设置应用程序的名称，在程序运行的监控界面可以看到名称
18.     conf.setMaster("local") //此时程序在本地运行，不需要安装 Spark 集群
19.
20.     /**
21.       * 第 2 步：创建 SparkContext 对象
22.       * SparkContext 是 Spark 程序所有功能的唯一入口，采用 Scala、Java、Python、
          * R 等都必须有一个 SparkContext
23.       * SparkContext 核心作用：初始化 Spark 应用程序，运行所需要的核心组件，包括
          * DAGScheduler、TaskScheduler、SchedulerBackend，同时还会负责 Spark 程
          * 序往 Master 注册程序等，SparkContext 是整个 Spark 应用程序中至关重要的一个对象
24.       */
25.     val sc = new SparkContext(conf)
                              //创建 SparkContext 对象，通过传入 SparkConf 实例来定
                              //制 Spark 运行的具体参数和配置信息
26.
27.     /**
28.       * 第 3 步：根据具体的数据来源（如 HDFS、HBase、Local FS、DB、S3 等）通过
          * SparkContext 来创建 RDD
29.       * RDD 的创建有 3 种方式：根据外部的数据来源（如 HDFS）、根据 Scala 集合、由其他
          * 的 RDD 操作
30.       * 数据会被 RDD 划分成为一系列的 Partitions，分配到每个 Partition 的数据属于
          * 一个 Task 的处理范畴
31.       */
32.
33.     val lines = sc.textFile("data/wordcount/helloSpark.txt", 1)
                                          //读取本地文件并设置为一个 Partition
34.
35.     /**
36.       * 第 4 步：对初始的 RDD 进行 Transformation 级别的处理，如通过 map、filter 等
          * 高阶函数等的编程，进行具体的数据计算
37.       *  第 4.1 步：将每一行的字符串拆分成单个单词
38.       */
39.     val words = lines.flatMap { line => line.split(" ")}
                                      //对每一行的字符串进行单词拆分，并把所有行的拆
                                      //分结果通过 flat 合并成为一个大的单词集合
40.     /**
41.       * 第 4 步：对初始的 RDD 进行 Transformation 级别的处理，如通过 map、filter 等
          * 高阶函数等的编程，进行具体的数据计算
42.       *  第 4.2 步：在单词拆分的基础上对每个单词实例计数为 1，也就是 word => (word, 1)
43.       */
44.     val pairs = words.map { word => (word, 1) }
45.
46.     /**
47.       * 第 4 步：对初始的 RDD 进行 Transformation 级别的处理，如通过 map、filter 等
          * 高阶函数等的编程，进行具体的数据计算
48.       *  第 4.3 步：在每个单词实例计数为 1 基础之上统计每个单词在文件中出现的总次数
49.       */
50.     val wordCountsOdered = pairs.reduceByKey(_+_).map(pair => (pair._2,
        pair._1)).sortByKey(false).map(pair => (pair._2, pair._1))
        //对相同的 Key，进行 Value 的累计（包括 Local 和 Reducer 级别，同时 Reduce）
```

```
51.     wordCountsOdered.collect.foreach(wordNumberPair => println
        (wordNumberPair._1 + " : " + wordNumberPair._2))
52.     sc.stop()
53.
54.   }
55. }
```

在 IDEA 中运行程序，wordcount.scala 的运行结果如下：

```
1.  19/05/18 15:16:57 INFO DAGScheduler: Job 1 finished: collect at WordCount
    .scala:63, took 13.969929 s
2.  Hello : 4
3.  Spark : 2
4.  Scala : 1
5.  Flink : 1
6.  is : 1
7.  Hadoop : 1
8.  Awesome : 1
……
```

3.8.2　解析 RDD 生成的内部机制

下面详细解析一下 wordcount.scala 的运行原理。

（1）从数据流动视角解密 WordCount，使用 Spark 作单词计数统计，搞清楚数据到底是怎么流动的。

（2）从 RDD 依赖关系的视角解密 WordCount。Spark 中的一切操作都是 RDD，后面的 RDD 对前面的 RDD 有依赖关系。

（3）DAG 与血统 Lineage 的思考。

在 wordcount.scala 的基础上，我们从数据流动的视角分析数据到底是怎么处理的。我们绘制一张 WordCount 数据处理过程图，由于图片较大，为了方便阅读，将原图分成两张图，如图 3-11 和图 3-12 所示。

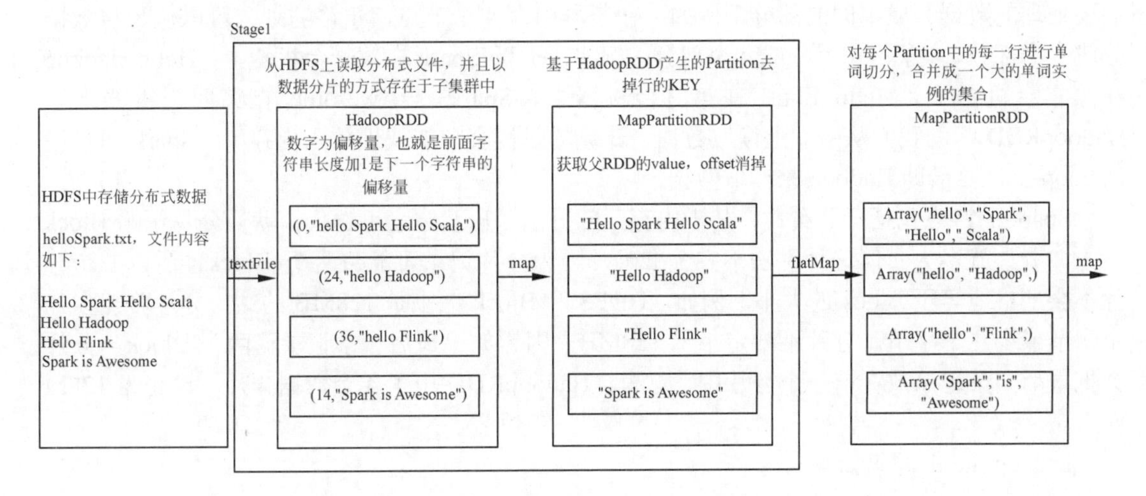

图 3-11　WordCount 数据处理过程图 1

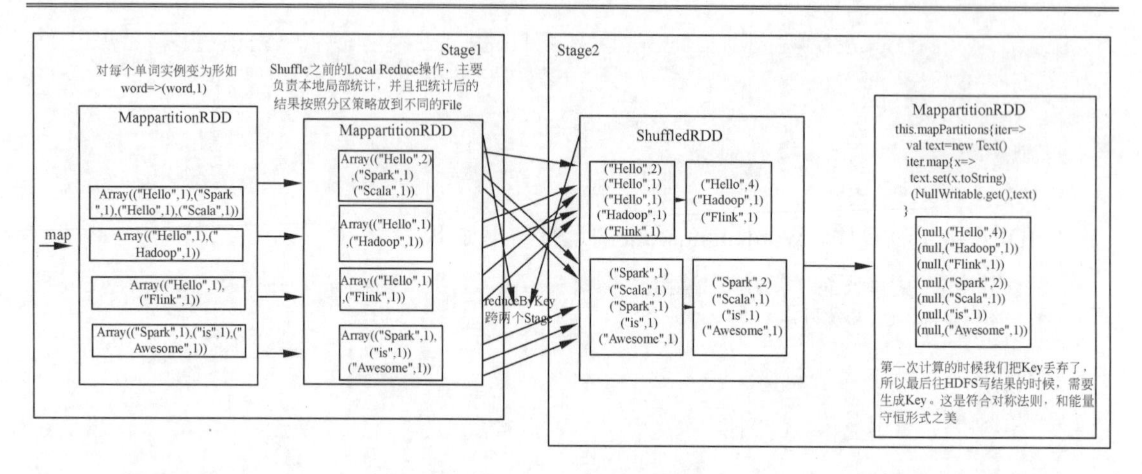

图 3-12　WordCount 数据处理过程图 2

数据在生产环境中默认在 HDFS 中进行分布式存储，如果在分布式集群中，我们的机器会分成不同的节点对数据进行处理，这里我们在本地测试，重点关注数据是怎么流动的。处理的第一步是获取数据，读取数据会生成 HadoopRDD。

在 WordCount.scala 中，单击 sc.textFile 进入 Spark 框架，SparkContext.scala 的 textFile 的源码如下：

```
1.    def textFile(
2.        path: String,
3.        minPartitions: Int = defaultMinPartitions): RDD[String] = withScope {
4.      assertNotStopped()
5.      hadoopFile(path,  classOf[TextInputFormat],  classOf[LongWritable],
      classOf[Text],
6.        minPartitions).map(pair => pair._2.toString).setName(path)
7.    }
```

下面看一下 hadoopFile 的源码，通过 new()函数创建一个 HadoopRDD，HadoopRDD 从 Hdfs 上读取分布式数据，并且以数据分片的方式存在于集群中。所谓的数据分片，就是把我们要处理的数据分成不同的部分，例如，在集群中有 4 个节点，粗略的划分可以认为将数据分成 4 个部分，4 条语句就分成 4 个部分。例如，Hello Spark 在第一台机器上，Hello Hadoop 在第二台机器上，Hello Flink 在第三台机器上，Spark is Awesome 在第四台机器上。HadoopRDD 帮助我们从磁盘上读取数据，计算的时候会分布式地放入内存中，Spark 运行在 Hadoop 上，要借助 Hadoop 来读取数据。

Spark 的特点包括：分布式、基于内存（部分基于磁盘）、可迭代；默认分片策略 Block 多大，分片就多大。但这种说法不完全准确，因为分片记录可能跨两个 Block，所以一个分片不会严格地等于 Block 的大小。例如，HDFS 的 Block 大小是 128MB 的话，分片可能多几个字节或少几个字节。分片不一定小于 128MB，因为如果最后一条记录跨两个 Block，分片会把最后一条记录放在前一个分片中。这里，HadoopRDD 用了 4 个数据分片，设想为 128M 左右。

hadoopFile 的源码如下：

```
1.   def hadoopFile[K, V](
2.       path: String,
3.       inputFormatClass: Class[_ <: InputFormat[K, V]],
```

```
4.      keyClass: Class[K],
5.      valueClass: Class[V],
6.      minPartitions: Int = defaultMinPartitions): RDD[(K, V)] = withScope {
7.    assertNotStopped()
8.
9.    //加载 hdfs-site.xml 配置文件
10.   //详情参阅 Spark-11227
11.   FileSystem.getLocal(hadoopConfiguration)
12.
13.   //Hadoop 配置文件大约有 10 KB，相当大，所以进行广播
14.   val confBroadcast = broadcast(new SerializableConfiguration
      (hadoopConfiguration))
15.   val setInputPathsFunc = (jobConf: JobConf) => FileInputFormat
      .setInputPaths(jobConf, path)
16.   new HadoopRDD(
17.     this,
18.     confBroadcast,
19.     Some(setInputPathsFunc),
20.     inputFormatClass,
21.     keyClass,
22.     valueClass,
23.     minPartitions).setName(path)
24.  }
```

SparkContext.scala 的 textFile 源码中，调用 hadoopFile 方法后进行了 map 转换操作，map 对读取的每一行数据进行转换，读入的数据是一个 Tuple，Key 值为索引，Value 值为每行数据的内容，生成 MapPartitionsRDD。这里，map(pair => pair._2.toString)是基于 HadoopRDD 产生的 Partition 去掉的行 Key 产生的 Value，第二个元素是读取的每行数据内容。MapPartitionsRDD 是 Spark 框架产生的，运行中可能产生一个 RDD，也可能产生两个 RDD。例如，textFile 中 Spark 框架就产生了两个 RDD，即 HadoopRDD 和 MapPartitionsRDD。下面是 map 的源码。

```
1.   def map[U: ClassTag](f: T => U): RDD[U] = withScope {
2.     val cleanF = sc.clean(f)
3.     new MapPartitionsRDD[U, T](this, (context, pid, iter) => iter.map
       (cleanF))
4.   }
```

我们看一下 WordCount 业务代码，对读取的每行数据进行 flatMap 转换。这里，flatMap 对 RDD 中的每一个 Partition 的每一行数据内容进行单词切分，如有 4 个 Partition 分别进行单词切分，将“Hello Spark”切分成单词“Hello”和“Spark”，对每一个 Partition 中的每一行进行单词切分并合并成一个大的单词实例的集合。flatMap 转换生成的仍然是 MapPartitionsRDD：

RDD.scala 的 flatMap 的源码如下：

```
1.   def flatMap[U: ClassTag](f: T => TraversableOnce[U]): RDD[U] = withScope {
2.     val cleanF = sc.clean(f)
3.     new MapPartitionsRDD[U, T](this, (context, pid, iter) => iter.flatMap
       (cleanF))
4.   }
```

继续 WordCount 业务代码，words.map { word => (word, 1) }通过 map 转换将单词切分以后单词计数为 1。例如，将单词“Hello”和“Spark”变成（Hello，1），(Spark，1)。这里生

成了 MapPartitionsRDD。

RDD.scala 的 map 的源码如下：

```
1.    def map[U: ClassTag](f: T => U): RDD[U] = withScope {
2.      val cleanF = sc.clean(f)
3.      new MapPartitionsRDD[U, T](this, (context, pid, iter) => iter.map
        (cleanF))
4.    }
```

继续 WordCount 业务代码，计数之后进行一个关键的 reduceByKey 操作，对全局的数据进行计数统计。reduceByKey 对相同的 Key 进行 Value 的累计（包括 Local 和 Reducer 级别，同时 Reduce）。reduceByKey 在 MapPartitionsRDD 之后，在 Local reduce 级别本地进行了统计，这里也是 MapPartitionsRDD。例如，在本地将（Hello，1），（Spark，1），（Hello，1），（Scala，1）汇聚成（Hello，2），（Spark，1），（Scala，1）。Shuffle 之前的 Local Reduce 操作主要负责本地局部统计，并且把统计以后的结果按照分区策略放到不同的 file。举一个简单的例子，如果下一个阶段 Stage 是 3 个并行度，每个 Partition 进行 local reduce 以后，将自己的数据分成 3 种类型，最简单的方式是根据 HashCode 按 3 取模。

PairRDDFunctions.scala 的 reduceByKey 的源码如下：

```
1.    def  reduceByKey(func: (V, V) => V): RDD[(K, V)] = self.withScope {
2.      reduceByKey(defaultPartitioner(self), func)
3.    }
```

至此，前面所有的操作都是一个 Stage，一个 Stage 意味着什么：完全基于内存操作。父 Stage：Stage 内部的操作是基于内存迭代的，也可以进行 Cache，这样速度快很多。不同于 Hadoop 的 Map Redcue，Hadoop Map Redcue 每次都要经过磁盘。

reduceByKey 在 Local reduce 本地汇聚以后生成的 MapPartitionsRDD 仍属于父 Stage；然后 reduceByKey 展开真正的 Shuffle 操作，Shuffle 是 Spark 甚至整个分布式系统的性能瓶颈，Shuffle 产生 ShuffleRDD，ShuffledRDD 就变成另一个 Stage，为什么是变成另外一个 Stage？因为要网络传输，网络传输不能在内存中进行迭代。

从 WordCount 业务代码 pairs.reduceByKey(_+_)中看一下 PairRDDFunctions.scala 的 reduceByKey 的源码。

```
1.     def reduceByKey(partitioner: Partitioner, func: (V, V) => V): RDD[(K,
       V)] = self.withScope {
2.      combineByKeyWithClassTag[V]((v: V) => v, func, func, partitioner)
3.    }
```

reduceByKey 内部调用了 combineByKeyWithClassTag 方法。下面看一下 PairRDDFunctions.scala 的 combineByKeyWithClassTag 的源码。

```
1.  def combineByKeyWithClassTag[C](
2.      createCombiner: V => C,
3.      mergeValue: (C, V) => C,
4.      mergeCombiners: (C, C) => C,
5.      partitioner: Partitioner,
6.      mapSideCombine: Boolean = true,
7.      serializer: Serializer = null)(implicit ct: ClassTag[C]): RDD[(K,
        C)] = self.withScope {
8.    require(mergeCombiners != null, "mergeCombiners must be defined")
      //required as of Spark 0.9.0
```

```
9.      if (keyClass.isArray) {
10.       if (mapSideCombine) {
11.         throw new SparkException("Cannot use map-side combining with array
            keys.")
12.       }
13.       if (partitioner.isInstanceOf[HashPartitioner]) {
14.         throw new SparkException("HashPartitioner cannot partition array
            keys.")
15.       }
16.     }
17.     val aggregator = new Aggregator[K, V, C](
18.       self.context.clean(createCombiner),
19.       self.context.clean(mergeValue),
20.       self.context.clean(mergeCombiners))
21.     if (self.partitioner == Some(partitioner)) {
22.       self.mapPartitions(iter => {
23.         val context = TaskContext.get()
24.         new InterruptibleIterator(context, aggregator.combineValuesByKey
            (iter, context))
25.       }, preservesPartitioning = true)
26.     } else {
27.       new ShuffledRDD[K, V, C](self, partitioner)
28.         .setSerializer(serializer)
29.         .setAggregator(aggregator)
30.         .setMapSideCombine(mapSideCombine)
31.     }
32.   }
```

在 combineByKeyWithClassTag 方法中就用 new()函数创建了 ShuffledRDD。

前面假设有 4 台机器并行计算，每台机器在自己的内存中进行迭代计算，现在产生 Shuffle，数据就要进行分类，MapPartitionsRDD 数据根据 Hash 已经分好类，我们就抓取 MapPartitionsRDD 中的数据。我们从第一台机器中获取的内容为（Hello，2），从第二台机器中获取的内容为（Hello，1），从第三台机器中获取的内容为（Hello，1），把所有的 Hello 都抓过来。同样，我们把其他的数据（Hadoop，1），（Flink，1）……都抓过来。

这就是 Shuffle 的过程，根据数据的分类拿到自己需要的数据。注意，MapPartitionsRDD 属于第一个 Stage，是父 Stage，内部基于内存进行迭代，不需要操作都要读写磁盘，所以速度非常快；从计算算子的角度讲，reduceByKey 发生在哪里？reduceByKey 发生的计算过程包括两个 RDD：一个是 MapPartitionsRDD；一个是 ShuffledRDD。ShuffledRDD 要产生网络通信。

reduceByKey 之后，我们将结果收集起来，进行全局级别的 reduce，产生 reduceByKey 的最后结果，如将（Hello，2），（Hello，1），（Hello，1）在内部变成（Hello，4），其他数据也类似统计。这里 reduceByKey 之后，如果通过 Collect 将数据收集起来，就会产生 MapPartitionsRDD。从 Collect 的角度讲，MapPartitionsRDD 的作用是将结果收集起来发送给 Driver；从 saveAsTextFile 输出到 Hdfs 的角度讲，例如输出（Hello，4），其中 Hello 是 key，4 是 Value 吗？不是！这里（Hello，4）就是 value，这就需要设计一个 key 出来。

下面是 RDD.scala 的 saveAsTextFile 方法。

```
1.     def saveAsTextFile(path: String): Unit = withScope {
2.      //https://issues.apache.org/jira/browse/SPARK-2075
3.      //NullWritable 在 Hadoop 1.+版本中是 Comparable，所以编译器无法发现隐式排
        //序，将使用默认的‘空’。然而，在 Hadoop 2.+中是 Comparable[NullWritable]，
        //编译器将调用隐式的“排序”方法来创建一个排序的 NullWritable。这就是为什么对
```

```
    //于 Hadoop 1.+版本和 Hadoop 2.+版本的 saveAsTextFile，编译器会生成不同的匿
    //名类。因此，这里提供了一个显式排序的“null”来确保编译器为 saveAsTextFile 生
    //成相同的字节码
4.
5.
6.      val nullWritableClassTag = implicitly[ClassTag[NullWritable]]
7.      val textClassTag = implicitly[ClassTag[Text]]
8.      val r = this.mapPartitions { iter =>
9.        val text = new Text()
10.       iter.map { x =>
11.         text.set(x.toString)
12.         (NullWritable.get(), text)
13.       }
14.     }
15.     RDD.rddToPairRDDFunctions(r)(nullWritableClassTag, textClassTag, null)
16.       .saveAsHadoopFile[TextOutputFormat[NullWritable, Text]](path)
17.   }
```

RDD.scala 的 saveAsTextFile 方法中的 iter.map {x=>text.set(x.toString) (NullWritable.get(), text)}，这里，key 转换成 Null，value 就是内容本身（Hello，4）。saveAsHadoopFile 中的 TextOutputFormat 要求输出的是 key-value 的格式，而我们处理的是内容。回顾一下，之前我们在 textFile 读入数据的时候，读入 split 分片将 key 去掉了，计算的是 value。因此，输出时，须将丢失的 key 重新弄进来，这里 key 对我们没有意义，但 key 对 Spark 框架有意义，只有 value 对我们有意义。第一次计算的时候我们把 key 丢弃了，所以最后往 HDFS 写结果的时候需要生成 key，这符合对称法则和能量守恒形式。

总结：

第一个 Stage 有哪些 RDD？HadoopRDD、MapPartitionsRDD、MapPartitionsRDD、MapPartitionsRDD、MapPartitionsRDD。

第二个 Stage 有哪些 RDD？ShuffledRDD、MapPartitionsRDD。

3.9　基于 DataSet 的代码如何转化为 RDD

基于 DataSet 的代码转换为 RDD 之前需要一个 Action 的操作，基于 Spark 中的新解析引擎 Catalyst 进行优化，Spark 中的 Catalyst 不仅限于 SQL 的优化，Spark 的五大子框架（Spark Cores、Spark SQL、Spark Streaming、Spark GraphX、Spark Mlib）将来都会基于 Catalyst 基础之上。

Dataset.scala 的 collect 方法的源码如下：

```
1.    def collect(): Array[T] = withAction("collect", queryExecution)
      (collectFromPlan)
```

withAction 方法是一个高阶函数，第一个参数包括两项，字符串“collect”、QueryExecution 类实例 queryExecution。第二个参数是一个函数（collectFromPlan 函数），collectFromPlan 函数输入一个 SparkPlan 计划，输出是一个数组。withAction 方法将 Dataset 的 action 包裹起来，这样可跟踪 QueryExecution 和时间成本，然后汇报给用户注册的回调函数。

Dataset.scala 的 withAction 源代码如下：

```
1.   private def withAction[U](name: String, qe: QueryExecution)(action:
     SparkPlan => U) = {
2.     try {
3.       qe.executedPlan.foreach { plan =>
4.         plan.resetMetrics()
5.       }
6.       val start = System.nanoTime()
7.       val result = SQLExecution.withNewExecutionId(sparkSession, qe) {
8.         action(qe.executedPlan)
9.       }
10.      val end = System.nanoTime()
11.      sparkSession.listenerManager.onSuccess(name, qe, end - start)
12.      result
13.    } catch {
14.      case e: Exception =>
15.        sparkSession.listenerManager.onFailure(name, qe, e)
16.        throw e
17.    }
18.  }
```

在第 9 行执行 action(qe.executedPlan)，将 QueryExecution 类 queryExecution.executedPlan 传给 collectFromPlan 函数，collectFromPlan 函数收集 executedPlan 执行计划的所有元素，返回一个数组。

QueryExecution 类是 Spark 执行关系查询的主要 workflow。查看 QueryExecution 类的执行计划 executedPlan，executedPlan 不用来初始化任何 SparkPlan，仅用于执行。其中 executedPlan.execute()是关键性的代码。

QueryExecution.scala 的源码如下：

```
1.  class QueryExecution(val sparkSession: SparkSession, val logical:
    LogicalPlan) {
2.  ……
3.  //executePlan 不应该被用来初始化任何 Spark Plan，executePlan 只用于执行
4.    lazy val executedPlan: SparkPlan = prepareForExecution(sparkPlan)
5.  ……
6.    lazy val toRdd: RDD[InternalRow] = executedPlan.execute()
……
```

接下来查看 collectFromPlan 方法，collectFromPlan 方法从 SparkPlan 中获取所有的数据。Dataset.scala 的源代码如下：

```
1.    private def collectFromPlan(plan: SparkPlan): Array[T] = {
2.  //此投影输出将写入 InternalRow，此投影应用不是线程安全的。在这个方法中创建投影，以
3.  //使 Dataset 线程安全
4.  val objProj = GenerateSafeProjection.generate(deserializer :: Nil)
5.  plan.executeCollect().map { row =>
6.  //SafeProjection 返回的行是“SpecificInternalRow”，忽略“get”方法参数的数
7.  //据类型，在这里使用 null 是安全的
8.  objProj(row).get(0, null).asInstanceOf[T]
9.  }
10.   }
```

在第 5 行 collectFromPlan 调用 plan.executeCollect()方法，plan 是 SparkPlan 类实例，executeCollect 方法执行查询，以数组形式返回结果。

SparkPlan.scala 的 executeCollect 源代码如下：

```
1.    def executeCollect(): Array[InternalRow] = {
2.      val byteArrayRdd = getByteArrayRdd()
3.
4.      val results = ArrayBuffer[InternalRow]()
5.      byteArrayRdd.collect().foreach { countAndBytes =>
6.        decodeUnsafeRows(countAndBytes._2).foreach(results.+=)
7.      }
8.      results.toArray
9.    }
```

其中，byteArrayRdd.collect()方法调用 RDD.scala 的 collect 方法。collect 方法最终通过 sc.runJob 提交 Spark 集群运行。

RDD.scala 的 collect 方法源码如下：

```
1.     def collect(): Array[T] = withScope {
2.      val results = sc.runJob(this, (iter: Iterator[T]) => iter.toArray)
3.      Array.concat(results: _*)
4.    }
```

回到 SparkPlan.scala 的 executeCollect 方法，getByteArrayRdd 方法将 UnsafeRows 打包到字节数组中，以便更快地序列化。

SparkPlan.scala 的 getByteArrayRdd 源代码如下：

```
1.  private def getByteArrayRdd(n: Int = -1): RDD[(Long, Array[Byte])] = {
2.      execute().mapPartitionsInternal { iter =>
3.        var count = 0
4.        val buffer = new Array[Byte](4 << 10)  // 4K
5.        val codec = CompressionCodec.createCodec(SparkEnv.get.conf)
6.        val bos = new ByteArrayOutputStream()
7.        val out = new DataOutputStream(codec.compressedOutputStream(bos))
8.        // iter.hasnext 可以生成一行并对其进行缓冲，只在没有命中时才调用它
9.        while ((n < 0 || count < n) && iter.hasNext) {
10.         val row = iter.next().asInstanceOf[UnsafeRow]
11.         out.writeInt(row.getSizeInBytes)
12.         row.writeToStream(out, buffer)
13.         count += 1
14.       }
15.       out.writeInt(-1)
16.       out.flush()
17.       out.close()
18.       Iterator((count, bos.toByteArray))
19.     }
20.   }
```

在第 2 行调用执行 execute().mapPartitionsInternal 方法，execute 方法内部调用 doExecute 方法，将此查询的结果作为 RDD[internalRow]返回。

SparkPlan.scala 的 execute 源代码如下：

```
1.    final def execute(): RDD[InternalRow] = executeQuery {
2.      if (isCanonicalizedPlan) {
3.        throw new IllegalStateException("A canonicalized plan is not supposed
          to be executed.")
4.      }
5.      doExecute()
6.    }
```

execute 返回的查询结果类型为 RDD[InternalRow]。SparkPlan.scala 的 doExecute()抽象方法没有具体实现，需通过 SparkPlan 子类重写此方法来进行具体实现。

```
1.      protected def doExecute(): RDD[InternalRow]
```

InternalRow 是通过语法树生成的一些数据结构。其子类包括 BaseGenericInternalRow、JoinedRow、Row、UnsafeRow。

InternalRow.scala 的源码如下：

```
1.  abstract class InternalRow extends SpecializedGetters with Serializable {
2.  ......
3.    def setBoolean(i: Int, value: Boolean): Unit = update(i, value)
4.    def setByte(i: Int, value: Byte): Unit = update(i, value)
5.    def setShort(i: Int, value: Short): Unit = update(i, value)
6.    def setInt(i: Int, value: Int): Unit = update(i, value)
7.    def setLong(i: Int, value: Long): Unit = update(i, value)
8.    def setFloat(i: Int, value: Float): Unit = update(i, value)
9.    def setDouble(i: Int, value: Double): Unit = update(i, value)
10. ......
```

DataSet 的代码转化成为 RDD 的内部流程如下：

Parse SQL(DataSet)→Analyze Logical Plan→Optimize Logical Plan→Generate Physical Plan→Prepareed Spark Plan→Execute SQL→Generate RDD。

基于 DataSet 的代码一步步转化成为 RDD：最终调用 execute()生成 RDD。

第 4 章　Spark Driver 启动内幕剖析

本章将对 Spark Driver 启动内幕进行剖析。4.1 节对 Spark Driver Program 进行解析，深度剖析 SparkContext；4.2 节解析 DAGScheduler，讲解 DAGScheduler 划分 Stage 及 Stage 内部 Task 获取最佳位置的算法；4.3 节对底层调度器 TaskScheduler 进行解析；4.4 节解析 SchedulerBackend；4.5 节整体讲解 Spark 系统运行内幕机制的循环流程。

4.1　Spark Driver Program 剖析

SparkContext 是通往 Spark 集群的唯一入口，是整个 Application 运行调度的核心。本节将深度剖析 SparkContext。

4.1.1　Spark Driver Program

Spark Driver Program（以下简称 Driver）是运行 Application 的 main 函数并且新建 SparkContext 实例的程序。其实，初始化 SparkContext 是为了准备 Spark 应用程序的运行环境，在 Spark 中，由 SparkContext 负责与集群进行通信、资源的申请、任务的分配和监控等。当 Worker 节点中的 Executor 运行完毕 Task 后，Driver 同时负责将 SparkContext 关闭。通常也可以使用 SparkContext 来代表驱动程序（Driver）。

Driver（SparkContext）整体架构图如图 4-1 所示。

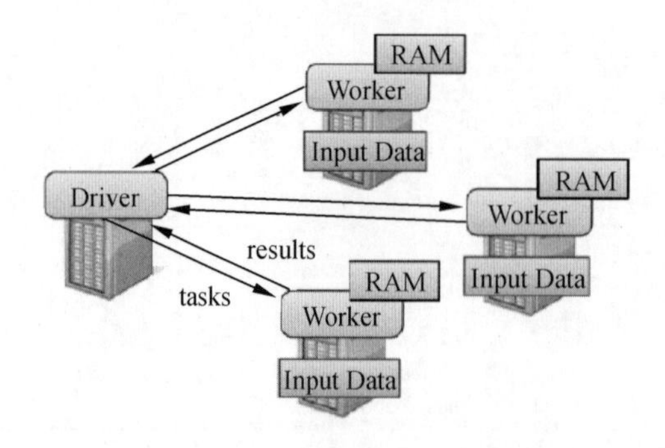

图 4-1　Driver（SparkContext）整体架构图

4.1.2　SparkContext 深度剖析

SparkContext 是通往 Spark 集群的唯一入口，可以用来在 Spark 集群中创建 RDDs、累加器（Accumulators）和广播变量（Broadcast Variables）。SparkContext 也是整个 Spark 应用程

序（Application）中至关重要的一个对象，可以说是整个 Application 运行调度的核心（不是指资源调度）。

SparkContext 的核心作用是初始化 Spark 应用程序运行所需要的核心组件，包括高层调度器（DAGScheduler）、底层调度器（TaskScheduler）和调度器的通信终端（SchedulerBackend），同时还会负责 Spark 程序向 Master 注册程序等。

一般而言，通常为了测试或者学习 Spark 开发一个 Application，在 Application 的 main 方法中，最开始几行编写的代码一般是这样的：首先，创建 SparkConf 实例，设置 SparkConf 实例的属性，以便覆盖 Spark 默认配置文件 spark-env.sh,spark-default.sh 和 log4j.properties 中的参数；然后，SparkConf 实例作为 SparkContext 类的唯一构造参数来实例化 SparkContext 实例对象。SparkContext 在实例化的过程中会初始化 DAGScheduler、TaskScheduler 和 SchedulerBackend，而当 RDD 的 action 触发了作业（Job）后，SparkContext 会调用 DAGScheduler 将整个 Job 划分成几个小的阶段（Stage），TaskScheduler 会调度每个 Stage 的任务（Task）进行处理。还有，SchedulerBackend 管理整个集群中为这个当前的 Application 分配的计算资源，即 Executor。

如果用一个车来比喻 Spark Application，那么 SparkContext 就是车的引擎，而 SparkConf 是关于引擎的配置参数。说明：只可以有一个 SparkContext 实例运行在一个 JVM 内存中，所以在创建新的 SparkContext 实例前，必须调用 stop 方法停止当前 JVM 唯一运行的 SparkContext 实例。

Spark 程序在运行时分为 Driver 和 Executor 两部分：Spark 程序编写是基于 SparkContext 的，具体包含两方面。

- Spark 编程的核心基础 RDD 是由 SparkContext 最初创建的（第一个 RDD 一定是由 SparkContext 创建的）。
- Spark 程序的调度优化也是基于 SparkContext，首先进行调度优化。
- Spark 程序的注册是通过 SparkContext 实例化时生产的对象来完成的（其实是 SchedulerBackend 来注册程序）。
- Spark 程序在运行时要通过 Cluster Manager 获取具体的计算资源，计算资源获取也是通过 SparkContext 产生的对象来申请的（其实是 SchedulerBackend 来获取计算资源的）。
- SparkContext 崩溃或者结束的时候，整个 Spark 程序也结束。

4.1.3 SparkContext 源码解析

SparkContext 是 Spark 应用程序的核心。我们运行 WordCount 程序，通过日志来深入了解 SparkContext。

WordCount.scala 的代码如下：

```
1.  package com.dt.spark.sparksql
2.
3. import org.apache.log4j.{Level, Logger}
4. import org.apache.spark.rdd.RDD
5. import org.apache.spark.{SparkConf, SparkContext}
6.
```

```
7.  /**
8.    * 使用 Scala 开发本地测试的 Spark WordCount 程序
9.    * @author DT 大数据梦工厂
10.   * 新浪微博：http://weibo.com/ilovepains/
11.   */
12. object WordCount {
13.   def main(args: Array[String]){
14.     Logger.getLogger("org").setLevel(Level.ALL)
15.     /**
16.       * 第 1 步：创建 Spark 的配置对象 SparkConf，设置 Spark 程序的运行时的配置信息，
          * 例如，通过 setMaster 设置程序要链接的 Spark 集群的 Master 的 URL，如果设置
          * 为 local，则代表 Spark 程序在本地运行，特别适合于机器配置非常差（如只有 1GB
          * 的内存）的初学者
17.       */
18.
19.     val conf = new SparkConf() //创建 SparkConf 对象
20.     conf.setAppName("Wow,WordCountJobRuntime!")
                                  //设置应用程序的名称，在程序运行的监控界面中可以看到名称
21.     conf.setMaster("local")    //此时，程序在本地运行，不需要安装 Spark 集群
22.
23.     /**
24.       * 第 2 步：创建 SparkContext 对象
25.       * SparkContext 是 Spark 程序所有功能的唯一入口，采用 Scala、Java、Python、
          * R 等都必须有一个 SparkContext
26.       * SparkContext 核心作用：初始化 Spark 应用程序运行所需要的核心组件，包括
          * DAGScheduler、TaskScheduler、SchedulerBackend
27.       * 同时还会负责 Spark 程序往 Master 注册程序等
28.       * SparkContext 是整个 Spark 应用程序中至关重要的一个对象
29.       */
30.     val sc = new SparkContext(conf)
                                      //创建 SparkContext 对象，通过传入 SparkConf
                                      //实例来定制 Spark 运行的具体参数和配置信息
31.     /**
32.       * 第 3 步：根据具体的数据来源（如 HDFS、HBase、Local FS、DB、S3 等）通过
          * SparkContext 来创建 RDD
33.       * RDD 的创建有 3 种方式：根据外部的数据来源（如 HDFS），根据 Scala 集合，由其他
          * 的 RDD 操作
34.       * 数据会被 RDD 划分成一系列的 Partitions，分配到每个 Partition 的数据属于一
          * 个 Task 的处理范畴
35.       */
36.     val lines = sc.textFile("data/wordcount/helloSpark.txt")
                                           //读取本地文件并设置为一个 Partition
37.
38.     /**
39.       * 第 4 步：对初始的 RDD 进行 Transformation 级别的处理，如通过 map、filter 等
          * 高阶函数等的编程，进行具体的数据计算
40.       *  第 4.1 步：将每一行的字符串拆分成单个单词
41.       */
42.
43.     val words = lines.flatMap { line => line.split(" ")}
                               //对每一行的字符串进行单词拆分，并把所有行的拆分结果通过
                               //flat 合并成为一个大的单词集合
```

```
44.
45.     /**
46.       * 第 4 步：对初始的 RDD 进行 Transformation 级别的处理，如通过 map、filter 等
          * 高阶函数等的编程，进行具体的数据计算
47.       *  第 4.2 步：在单词拆分的基础上对每个单词实例计数为 1，也就是 word => (word, 1)
48.       */
49.     val pairs: RDD[(String, Int)] = words.map { word => (word, 1) }
50.     pairs.cache()
51.     /**
52.       * 第 4 步：对初始的 RDD 进行 Transformation 级别的处理，如通过 map、filter 等
          * 高阶函数等的编程，进行具体的数据计算
53.       *  第 4.3 步：在每个单词实例计数为 1 的基础上统计每个单词在文件中出现的总次数
54.       */
55.     val wordCountsOdered = pairs.reduceByKey(_+_).saveAsTextFile("data/
        wordcount/wordCountResult.log")
56.
57.     while(true){
58.
59.     }
60.     sc.stop()
61.
62.   }
63. }
```

在 IDEA 中运行 WordCount.scala 代码，日志显示如下：

```
1.  Using Spark's default log4j profile: org/apache/spark/log4j-defaults.properties
2.  19/05/18 20:17:13 INFO SparkContext: Running Spark version 2.4.3
3.  ......
4.  19/05/18 20:17:17 INFO SparkEnv: Registering MapOutputTracker
5.  19/05/18 20:17:17 DEBUG MapOutputTrackerMasterEndpoint: init
6.  19/05/18 20:17:17 INFO SparkEnv: Registering BlockManagerMaster
7.  ......
8.  19/05/18  20:17:21  INFO  SparkContext:  Starting  job:  runJob  at
    SparkHadoopWriter.scala:78
9.  ......
```

程序一开始，日志里显示的是：INFO SparkContext: Running Spark version 2.4.3，日志中间部分是一些随着 SparkContext 创建而创建的对象，另一条比较重要的日志信息，作业启动了并正在运行：INFO SparkContext: Starting job: runJob at SparkHadoopWriter.scala:78。

在程序运行的过程中会创建 TaskScheduler、DAGScheduler 和 SchedulerBackend，它们有各自的功能。DAGScheduler 是面向 Job 的 Stage 的高层调度器；TaskScheduler 是底层调度器。SchedulerBackend 是一个接口，根据具体的 ClusterManager 的不同会有不同的实现。程序打印结果后便开始结束。日志显示：INFO SparkContext: Successfully stopped SparkContext。

```
1.  ......
2.  19/05/18 20:17:25 INFO BlockManagerMaster: BlockManagerMaster stopped
3.  19/05/18 20:17:25 INFO OutputCommitCoordinator$OutputCommitCoordinatorEndpoint:
    OutputCommitCoordinator stopped!
4.  19/05/18 20:17:25 INFO SparkContext: Successfully stopped SparkContext
5.  19/05/18 20:17:25 INFO ShutdownHookManager: Shutdown hook called
6.  ......
```

通过这个例子可以感受到 Spark 程序的运行到处都可以看到 SparkContext 的存在，我们

将 SparkContext 作为 Spark 源码阅读的入口，来理解 Spark 的所有内部机制。

图 4-2 是从一个整体去看 SparkContext 创建的实例对象。首先，SparkContext 构建的顶级三大核心为 DAGScheduler、TaskScheduler、SchedulerBackend，其中，DAGScheduler 是面向 Job 的 Stage 的高层调度器；TaskScheduler 是一个接口，是底层调度器，根据具体的 ClusterManager 的不同会有不同的实现，Standalone 模式下具体的实现是 TaskSchedulerImpl。SchedulerBackend 是一个接口，根据具体的 ClusterManager 的不同会有不同的实现。Standalone 模式下具体的实现是 StandaloneSchedulerBackend。

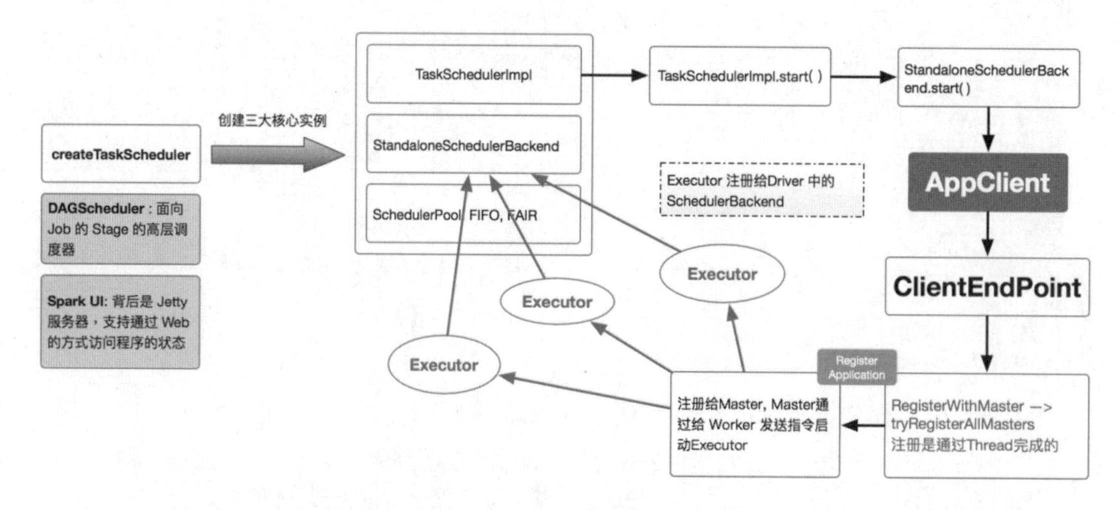

图 4-2　SparkContext 整体运行图

从整个程序运行的角度讲，SparkContext 包含四大核心对象：DAGScheduler、TaskScheduler、SchedulerBackend、MapOutputTrackerMaster。StandaloneSchedulerBackend 有三大核心功能：负责与 Master 连接，注册当前程序 RegisterWithMaster；接收集群中为当前应用程序分配的计算资源 Executor 的注册并管理 Executors；负责发送 Task 到具体的 Executor 执行。

第一步：程序一开始运行时会实例化 SparkContext 里的对象，所有不在方法里的成员都会被实例化！一开始实例化时第一个关键的代码是 createTaskScheduler，它位于 SparkContext 的 PrimaryConstructor 中，当它实例化时会直接被调用，这个方法返回的是 taskScheduler 和 dagScheduler 的实例，然后基于这个内容又构建了 DAGScheduler，最后调用 taskScheduler 的 start()方法。要先创建 taskScheduler，然后再创建 dagScheduler，因为 taskScheduler 是受 dagScheduler 管理的。

SparkContext.scala 的源码如下：

```
1.   //创建和启动调度器
2.      val (sched, ts) = SparkContext.createTaskScheduler(this, master,
        deployMode)
3.      _schedulerBackend = sched
4.      _taskScheduler = ts
5.      _dagScheduler = new DAGScheduler(this)
6.      _heartbeatReceiver.ask[Boolean](TaskSchedulerIsSet)
7.      //在 DAGScheduler 构造器中设置 taskScheduler 的引用以后，启动 TaskScheduler
8.      _taskScheduler.start()
9.   ……
```

第二步：调用 createTaskScheduler，这个方法创建了 TaskSchedulerImpl 和 StandaloneSchedulerBackend，createTaskScheduler 方法的第一个入参是 SparkContext，传入的 this 对象是在应用程序中创建的 sc，第二个入参是 master 的地址。

以下是 WordCount.scala 创建 SparkConf 和 SparkContext 的上下文信息。

```
1.   val conf = new SparkConf() //创建 SparkConf 对象
2.     conf.setAppName("Wow,WordCount")
           //设置应用程序的名称，在程序运行的监控界面中可以看到名称
3.     conf.setMaster("local")
4.   val sc = new SparkContext(conf)
```

当 SparkContext 调用 createTaskScheduler 方法时，根据集群的条件创建不同的调度器，例如，createTaskScheduler 第二个入参 master 如传入 local 参数，SparkContext 将创建 TaskSchedulerImpl 实例及 LocalSchedulerBackend 实例，在测试代码的时候，可以尝试传入 local[*]或者是 local[2]的参数，然后跟踪代码，看看创建了什么样的实例对象。

SparkContext 中的 SparkMasterRegex 对象定义不同的正则表达式，从 master 字符串中根据正则表达式适配 master 信息。

SparkContext.scala 的源码如下：

```
1.   private object SparkMasterRegex {
2.    //正则表达式 local[N]和 local[*] 用于 master 格式
3.    val LOCAL_N_REGEX = """local\[([0-9]+|\*)\]""".r
4.    //正则表达式 local[N, maxRetries]用于失败任务的测试
5.    val LOCAL_N_FAILURES_REGEX = """local\[([0-9]+|\*)\s*,\s*([0-9]+)\]""".r
6.    //正则表达式用于模拟 Spark 本地集群 [N, cores, memory]
7.    val LOCAL_CLUSTER_REGEX = """local-cluster\[\s*([0-9]+)\s*,\s*([0-9]+)
      \s*,\s*([0-9]+)\s*]""".r
8.    //用于连接到 Spark 部署集群的正则表达式
9.    val SPARK_REGEX = """spark://(.*)""".r
10. }
```

这是设计模式中的策略模式，它会根据实际需要创建出不同的 SchedulerBackend 的子类。

SparkContext.scala 的 createTaskScheduler 方法的源码如下：

```
1.   /**
      *基于给定的主 URL 创建任务调度器，返回一个二元调度程序的后台和任务调度
2.    */
3.
4.    private def createTaskScheduler(
5.        sc: SparkContext,
6.        master: String,
7.        deployMode: String): (SchedulerBackend, TaskScheduler) = {
8.      import SparkMasterRegex._
9.
10.     //当在本地运行时，不要试图在失败时重新执行任务
11.     val MAX_LOCAL_TASK_FAILURES = 1
12.
13.     master match {
14.       case "local" =>
15.         val scheduler = new TaskSchedulerImpl(sc, MAX_LOCAL_TASK_FAILURES,
          isLocal = true)
16.         val backend = new LocalSchedulerBackend(sc.getConf, scheduler, 1)
17.         scheduler.initialize(backend)
```

```
18.         (backend, scheduler)
19.
20.       case LOCAL_N_REGEX(threads) =>
21.         def localCpuCount: Int = Runtime.getRuntime.availableProcessors()
22.         //local[*]估计机器上的核数; local[N] 精确地使用 N 个线程
23.         val threadCount = if (threads == "*") localCpuCount else
            threads.toInt
24.         if (threadCount <= 0) {
25.           throw new SparkException(s"Asked to run locally with $threadCount
              threads")
26.         }
27.         val scheduler = new TaskSchedulerImpl(sc, MAX_LOCAL_TASK_FAILURES,
            isLocal = true)
28.         val backend = new LocalSchedulerBackend(sc.getConf, scheduler,
            threadCount)
29.         scheduler.initialize(backend)
30.         (backend, scheduler)
31.
32.       case LOCAL_N_FAILURES_REGEX(threads, maxFailures) =>
33.         def localCpuCount: Int = Runtime.getRuntime.availableProcessors()
34.         //local[*, M] 计算机核发生 M 个故障
35.         //local[N, M] 意味着 N 个线程 M 个故障
36.         val threadCount = if (threads == "*") localCpuCount else threads
            .toInt
37.         val scheduler = new TaskSchedulerImpl(sc, maxFailures.toInt,
            isLocal = true)
38.         val backend = new LocalSchedulerBackend(sc.getConf, scheduler,
            threadCount)
39.         scheduler.initialize(backend)
40.         (backend, scheduler)
41.
42.       case SPARK_REGEX(sparkUrl) =>
43.         val scheduler = new TaskSchedulerImpl(sc)
44.         val masterUrls = sparkUrl.split(",").map("spark://" + _)
45.         val backend = new StandaloneSchedulerBackend(scheduler, sc,
            masterUrls)
46.         scheduler.initialize(backend)
47.         (backend, scheduler)
48.
49.       case LOCAL_CLUSTER_REGEX(numSlaves, coresPerSlave, memoryPerSlave) =>
50.         //确认请求的内存<= memoryPerSlave，否则 Spark 将会挂起
51.         val memoryPerSlaveInt = memoryPerSlave.toInt
52.         if (sc.executorMemory > memoryPerSlaveInt) {
53.           throw new SparkException(
54.             "Asked to launch cluster with %d MB RAM / worker but requested
                %d MB/worker".format(
55.               memoryPerSlaveInt, sc.executorMemory))
56.         }
57.
58.         val scheduler = new TaskSchedulerImpl(sc)
59.         val localCluster = new LocalSparkCluster(
60.           numSlaves.toInt, coresPerSlave.toInt, memoryPerSlaveInt, sc.conf)
61.         val masterUrls = localCluster.start()
62.         val backend = new StandaloneSchedulerBackend(scheduler, sc,
            masterUrls)
63.         scheduler.initialize(backend)
64.         backend.shutdownCallback = (backend: StandaloneSchedulerBackend) => {
65.           localCluster.stop()
66.         }
```

```
67.         (backend, scheduler)
68.
69.       case masterUrl =>
70.         val cm = getClusterManager(masterUrl) match {
71.           case Some(clusterMgr) => clusterMgr
72.           case None => throw new SparkException("Could not parse Master URL:
            '" + master + "'")
73.         }
74.         try {
75.           val scheduler = cm.createTaskScheduler(sc, masterUrl)
76.           val backend = cm.createSchedulerBackend(sc, masterUrl, scheduler)
77.           cm.initialize(scheduler, backend)
78.           (backend, scheduler)
79.         } catch {
80.           case se: SparkException => throw se
81.           case NonFatal(e) =>
82.             throw new SparkException("External scheduler cannot be
              instantiated", e)
83.         }
84.     }
85.   }
```

在实际生产环境下，我们都是用集群模式，即以 spark://开头，此时在程序运行时，框架会创建一个 TaskSchedulerImpl 和 StandaloneSchedulerBackend 的实例，在这个过程中也会初始化 taskscheduler，把 StandaloneSchedulerBackend 的实例对象作为参数传入。StandaloneSchedulerBackend 被 TaskSchedulerImpl 管理，最后返回 TaskScheduler 和 StandaloneSchdeulerBackend。

SparkContext.scala 的源码如下：

```
1.      case SPARK_REGEX(sparkUrl) =>
2.          val scheduler = new TaskSchedulerImpl(sc)
3.          val masterUrls = sparkUrl.split(",").map("spark://" + _)
4.          val backend = new StandaloneSchedulerBackend(scheduler, sc,
            masterUrls)
5.          scheduler.initialize(backend)
6.          (backend, scheduler)
```

createTaskScheduler 方法执行完毕后，调用了 taskscheduler.start()方法来正式启动 taskscheduler，这里虽然调用了 taskscheduler.start 方法，但实际上是调用了 taskSchedulerImpl 的 start 方法，因为 taskSchedulerImpl 是 taskScheduler 的子类。

Task 默认失败重试次数是 4 次，如果任务不容许失败，就可以调大这个参数。调大 spark.task.maxFailures 参数有助于确保重要的任务失败后可以重试多次。

初始化 TaskSchedulerImpl：调用 createTaskScheduler 方法时会初始化 TaskSchedulerImpl，然后把 StandaloneSchedulerBackend 当作参数传进去，初始化 TaskSchedulerImpl 时首先是创建一个 Pool 来初定义资源分布的模式 Scheduling Mode，默认是先进先出（FIFO）的模式。

Spark 2.1.1 版本的 TaskSchedulerImpl.scala 的 initialize 的源码如下：

```
1.    def initialize(backend: SchedulerBackend) {
2.      this.backend = backend
3.      schedulableBuilder = {
4.        schedulingMode match {
5.          case SchedulingMode.FIFO =>
6.            new FIFOSchedulableBuilder(rootPool)
7.          case SchedulingMode.FAIR =>
```

```
8.          new FairSchedulableBuilder(rootPool, conf)
9.        case _ => throw new IllegalArgumentException(s"Unsupported $SCHEDULER_
10.         MODE_PROPERTY: " + s"$schedulingMode")
11.     }
12.   }
13.   schedulableBuilder.buildPools()
14. }
```

rootPool 作为 TaskSchedulerImpl 类的成员变量，在构建 TaskSchedulerImpl 时初始化。

```
1. val rootPool: Pool = new Pool("", schedulingMode, 0, 0)
2. ......
```

可以设置 spark.scheduler.mode 参数来定义资源调度池，例如 FAIR、FIFO，默认资源调度池是先进先出（FIFO）模式。

TaskSchedulerImpl.scala 的源码如下：

```
1. private[spark] object TaskSchedulerImpl {
2.   val SCHEDULER_MODE_PROPERTY = "spark.scheduler.mode"
3.   ......
4. private val schedulingModeConf = conf.get(SCHEDULER_MODE_PROPERTY,
5.   SchedulingMode.FIFO.toString)
6.   val schedulingMode: SchedulingMode =
7.     try {
8.       SchedulingMode.withName(schedulingModeConf.toUpperCase(Locale.ROOT))
9.     } catch {
10.      case e: java.util.NoSuchElementException =>
11.        throw new SparkException(s"Unrecognized $SCHEDULER_MODE_PROPERTY:
           $schedulingModeConf")
12.    }
13. ......
```

SchedulingMode.scala 的源代码如下：

```
1. object SchedulingMode extends Enumeration {
2.   type SchedulingMode = Value
3.   val FAIR, FIFO, NONE = Value
4. }
5. ......
```

回到 taskScheduler start 方法，taskScheduler.start 方法调用时会再调用 schedulerbackend 的 start 方法。

TaskSchedulerImpl.scala 的 start 方法的源码如下：

```
1. override def start() {
2.     backend.start()
3.
4.     if (!isLocal && conf.getBoolean("spark.speculation", false)) {
5.       logInfo("Starting speculative execution thread")
6.       speculationScheduler.scheduleWithFixedDelay(new Runnable {
7.         override def run(): Unit = Utils.tryOrStopSparkContext(sc) {
8.           checkSpeculatableTasks()
9.         }
10.      }, SPECULATION_INTERVAL_MS, SPECULATION_INTERVAL_MS, TimeUnit.MILLISECONDS)
11.    }
12.  }
13. ......
```

SchedulerBackend 包含多个子类，分别是 LocalSchedulerBackend、CoarseGrainedSchedulerBackend 和 StandaloneSchedulerBackend、MesosCoarseGrainedSchedulerBackend、YarnSchedulerBackend。

StandaloneSchedulerBackend 的 start 方法调用了 CoarseGraninedSchedulerBackend 的 start 方法，通过 StandaloneSchedulerBackend 注册程序把 command 提交给 Master：Command ("org.apache.spark.executor.CoarseGrainedExecutorBackend", args, sc.executorEnvs, classPathEntries ++ testingClassPath, libraryPathEntries, javaOpts)来创建一个 StandaloneAppClient 的实例。

StandaloneSchedulerBackend.scala 的 start 方法的源码如下：

```
1.  override def start() {
2.      super.start()
3.
4.     //SPARK-21159: 只有在 client 模式下 scheduler backend 才去连接 launcher
5.     //在 cluster 集群下，应用程序应提交给 Master
6.      if (sc.deployMode == "client") {
7.        launcherBackend.connect()
8.      }
9.
10.     // executors 节点与用户通信的端点
11.     val driverUrl = RpcEndpointAddress(
12.       sc.conf.get("spark.driver.host"),
13.       sc.conf.get("spark.driver.port").toInt,
14.       CoarseGrainedSchedulerBackend.ENDPOINT_NAME).toString
15.     val args = Seq(
16.       "--driver-url", driverUrl,
17.       "--executor-id", "{{EXECUTOR_ID}}",
18.       "--hostname", "{{HOSTNAME}}",
19.       "--cores", "{{CORES}}",
20.       "--app-id", "{{APP_ID}}",
21.       "--worker-url", "{{WORKER_URL}}")
22.     val extraJavaOpts = sc.conf.getOption("spark.executor.extraJavaOptions")
23.       .map(Utils.splitCommandString).getOrElse(Seq.empty)
24.     val classPathEntries = sc.conf.getOption("spark.executor.extraClassPath")
25.       .map(_.split(java.io.File.pathSeparator).toSeq).getOrElse(Nil)
26.     val libraryPathEntries = sc.conf.getOption("spark.executor.extraLibraryPath")
27.       .map(_.split(java.io.File.pathSeparator).toSeq).getOrElse(Nil)
28.
29.     //测试时，将父类路径公开给子对象，由 compute-classpath.{cmd,sh}计算路径
30.     //当“*-provided”配置启用，子进程可使用所有需要的 jar 包
31.     val testingClassPath =
32.       if (sys.props.contains("spark.testing")) {
33.         sys.props("java.class.path").split(java.io.File.pathSeparator).toSeq
34.       } else {
35.         Nil
36.       }
37.
38.     //使用注册调度必要的一些配置启动 executors
39.     val sparkJavaOpts = Utils.sparkJavaOpts(conf, SparkConf.isExecutorStartupConf)
40.     val javaOpts = sparkJavaOpts ++ extraJavaOpts
41.     val command = Command("org.apache.spark.executor.CoarseGrainedExecutorBackend",
42.       args, sc.executorEnvs, classPathEntries ++ testingClassPath,
          libraryPathEntries, javaOpts)
43.     val webUrl = sc.ui.map(_.webUrl).getOrElse("")
44.     val coresPerExecutor = conf.getOption("spark.executor.cores").map(_.toInt)
45.     //如果使用动态分配，现在将初始执行器限制设置为 0
```

```
46.     //ExecutorAllocationManager 将实际的初始限制发送给 Master 节点
47.     val initialExecutorLimit =
48.       if (Utils.isDynamicAllocationEnabled(conf)) {
49.         Some(0)
50.       } else {
51.         None
52.       }
53.     val appDesc = ApplicationDescription(sc.appName, maxCores,
          sc.executorMemory, command,
54.       webUrl, sc.eventLogDir, sc.eventLogCodec, coresPerExecutor,
            initialExecutorLimit)
55.     client = new StandaloneAppClient(sc.env.rpcEnv, masters, appDesc, this, conf)
56.     client.start()
57.     launcherBackend.setState(SparkAppHandle.State.SUBMITTED)
58.     waitForRegistration()
59.     launcherBackend.setState(SparkAppHandle.State.RUNNING)
60.   }
61. ……
```

Master 发指令给 Worker 去启动 Executor 所有的进程时加载的 Main 方法所在的入口类就是 command 中的 CoarseGrainedExecutorBackend，在 CoarseGrainedExecutorBackend 中启动 Executor（Executor 是先注册，再实例化），Executor 通过线程池并发执行 Task，然后再调用它的 run 方法。

CoarseGrainedExecutorBackend.scala 的源码如下：

```
1.    def main(args: Array[String]) {
2.      var driverUrl: String = null
3.      var executorId: String = null
4.      var hostname: String = null
5.      var cores: Int = 0
6.      var appId: String = null
7.      var workerUrl: Option[String] = None
8.      val userClassPath = new mutable.ListBuffer[URL]()
9.
10.     var argv = args.toList
11.     while (!argv.isEmpty) {
12.       argv match {
13.         case ("--driver-url") :: value :: tail =>
14.           driverUrl = value
15.           argv = tail
16.         case ("--executor-id") :: value :: tail =>
17.           executorId = value
18.           argv = tail
19.         case ("--hostname") :: value :: tail =>
20.           hostname = value
21.           argv = tail
22.         case ("--cores") :: value :: tail =>
23.           cores = value.toInt
24.           argv = tail
25.         case ("--app-id") :: value :: tail =>
26.           appId = value
27.           argv = tail
28.         case ("--worker-url") :: value :: tail =>
29.           //Worker url 用于 spark standalone 模式，以加强与 Worker 的分享
30.           workerUrl = Some(value)
31.           argv = tail
32.         case ("--user-class-path") :: value :: tail =>
```

```
33.          userClassPath += new URL(value)
34.          argv = tail
35.        case Nil =>
36.        case tail =>
37.          //scalastyle:off println
38.          System.err.println(s"Unrecognized options: ${tail.mkString(" ")}")
39.          //scalastyle:on println
40.          printUsageAndExit()
41.      }
42.    }
43.
44.    if (driverUrl == null || executorId == null || hostname == null || cores
       <= 0 ||
45.      appId == null) {
46.      printUsageAndExit()
47.    }
48.
49.    run(driverUrl, executorId, hostname, cores, appId, workerUrl,
       userClassPath)
50.    System.exit(0)
51.  }
```

CoarseGrainedExecutorBackend 的 main 入口方法中调用了 run 方法。

Spark 2.2.1 版本的 CoarseGrainedExecutorBackend 的 run 入口方法的源码如下：

```
1.  private def run(
2.        driverUrl: String,
3.        executorId: String,
4.        hostname: String,
5.        cores: Int,
6.        appId: String,
7.        workerUrl: Option[String],
8.        userClassPath: Seq[URL]) {
9.
10.     Utils.initDaemon(log)
11.
12.     SparkHadoopUtil.get.runAsSparkUser { () =>
13.      //Debug 代码
14.       Utils.checkHost(hostname)
15.
16.       // Bootstrap 去抓取 driver 节点的 Spark 属性
17.       val executorConf = new SparkConf
18.       val port = executorConf.getInt("spark.executor.port", 0)
19.       val fetcher = RpcEnv.create(
20.         "driverPropsFetcher",
21.         hostname,
22.         port,
23.         executorConf,
24.         new SecurityManager(executorConf),
25.         clientMode = true)
26.       val driver = fetcher.setupEndpointRefByURI(driverUrl)
27.       val cfg = driver.askSync[SparkAppConfig](RetrieveSparkAppConfig)
28.       val props = cfg.sparkProperties ++ Seq[(String,
          String)](("spark.app.id", appId))
29.       fetcher.shutdown()
30.
31.       // 从 driver 节点获取属性信息，创建 SparkEnv
32.       val driverConf = new SparkConf()
33.       for ((key, value) <- props) {
```

```
34.         //这是 SSL 在独立模式下需要的
35.         if (SparkConf.isExecutorStartupConf(key)) {
36.           driverConf.setIfMissing(key, value)
37.         } else {
38.           driverConf.set(key, value)
39.         }
40.       }
41.       if (driverConf.contains("spark.yarn.credentials.file")) {
42.         logInfo("Will periodically update credentials from: " +
43.           driverConf.get("spark.yarn.credentials.file"))
44.         SparkHadoopUtil.get.startCredentialUpdater(driverConf)
45.       }
46.
47.       val env = SparkEnv.createExecutorEnv(
48.       driverConf, executorId, hostname, port, cores, cfg.ioEncryptionKey,
          isLocal = false)
49.       env.rpcEnv.setupEndpoint("Executor", new CoarseGrainedExecutorBackend(
50.         env.rpcEnv, driverUrl, executorId, hostname, cores, userClassPath, env))
51.       workerUrl.foreach { url =>
52.         env.rpcEnv.setupEndpoint("WorkerWatcher",
            new WorkerWatcher(env.rpcEnv, url))
53.       }
54.       env.rpcEnv.awaitTermination()
55.       SparkHadoopUtil.get.stopCredentialUpdater()
56.     }
57.   }
58. ……
```

Spark 2.4.3 版本的 CoarseGrainedExecutorBackend.scala 源码与 Spark 2.2.1 版本相比具有如下特点。

- 删除上段代码中第 18 行 val port = executorConf.getInt("spark.executor.port", 0)。
- 上段代码中第 22 行将 port 修改为–1。
- 删除上段代码中第 41～45 行代码。
- 上段代码中第 40 行之后新增确认设置正确的 Hadoop 配置的代码。
- 上段代码中第 48 行去掉 port 端口的参数。
- 删除上段代码中第 55 行代码。

```
1.  -1,
2. ……
3.       cfg.hadoopDelegationCreds.foreach { tokens =>
4.         SparkHadoopUtil.get.addDelegationTokens(tokens, driverConf)
5.       }
6. ……
7.       val env = SparkEnv.createExecutorEnv(
8.         driverConf, executorId, hostname, cores, cfg.ioEncryptionKey,
           isLocal = false)
9.  ……
```

CoarseGrainedExecutorBackend 通过消息循环体向 driver 发送 RetrieveSparkAppConfig 消息，RetrieveSparkAppConfig 是一个 case object。Driver 端的 CoarseGrainedSchedulerBackend 消息循环体收到消息以后，将 Spark 的属性信息 sparkProperties 及加密 key 等内容封装成 SparkAppConfig 消息，将 SparkAppConfig 消息再回复给 CoarseGrainedExecutorBackend。

```
1.  ......
2.  val cfg = driver.askSync[SparkAppConfig](RetrieveSparkAppConfig)
3.  ......
```

回到 StandaloneSchedulerBackend.scala 的 start 方法：其中创建了一个很重要的对象，即 StandaloneAppClient 对象，然后调用它的 client.start()方法。

在 start 方法中创建一个 ClientEndpoint 对象。

StandaloneAppClient.scala 的 star 方法的源码如下：

```
1.    def start() {
2.      //启动 rpcEndpoint; it will call back into the listener.
3.      endpoint.set(rpcEnv.setupEndpoint("AppClient", new ClientEndpoint
      (rpcEnv)))
4.    }
```

ClientEndpoint 是一个 RpcEndPoint，首先调用自己的 onStart 方法，接下来向 Master 注册。

StandaloneAppClient.scala 的 ClientEndpoint 类的源码如下：

```
1.     private class ClientEndpoint(override val rpcEnv: RpcEnv) extends
       ThreadSafeRpcEndpoint
2.     ......
3.      override def onStart(): Unit = {
4.        try {
5.          registerWithMaster(1)
6.        } catch {
7.          case e: Exception =>
8.            logWarning("Failed to connect to master", e)
9.            markDisconnected()
10.           stop()
11.       }
12.     }
13. ......
```

调用 registerWithMaster 方法，从 registerWithMaster 调用 tryRegisterAllMasters，开一条新的线程来注册，然后发送一条信息（RegisterApplication 的 case class）给 Master。

StandaloneAppClient.scala 的 registerWithMaster 的源码如下：

```
1.     private def registerWithMaster(nthRetry: Int) {
2.       registerMasterFutures.set(tryRegisterAllMasters())
3.       registrationRetryTimer.set(registrationRetryThread.schedule(new
         Runnable {
4.         override def run(): Unit = {
5.           if (registered.get) {
6.             registerMasterFutures.get.foreach(_.cancel(true))
7.             registerMasterThreadPool.shutdownNow()
8.           } else if (nthRetry >= REGISTRATION_RETRIES) {
9.             markDead("All masters are unresponsive! Giving up.")
10.          } else {
11.            registerMasterFutures.get.foreach(_.cancel(true))
12.            registerWithMaster(nthRetry + 1)
13.          }
14.        }
15.      }, REGISTRATION_TIMEOUT_SECONDS, TimeUnit.SECONDS))
16.    }
17. ......
```

StandaloneAppClient.scala 的 tryRegisterAllMasters 的源码如下：

```
1.       private def tryRegisterAllMasters(): Array[JFuture[_]] = {
2.        for (masterAddress <- masterRpcAddresses) yield {
3.          registerMasterThreadPool.submit(new Runnable {
4.            override def run(): Unit = try {
5.              if (registered.get) {
6.                return
7.              }
8.              logInfo("Connecting to master " + masterAddress.toSparkURL +
              "...")
9.              val masterRef = rpcEnv.setupEndpointRef(masterAddress,
              Master.ENDPOINT_NAME)
10.             masterRef.send(RegisterApplication(appDescription, self))
11.           } catch {
12.             case ie: InterruptedException => //Cancelled
13.             case NonFatal(e) => logWarning(s"Failed to connect to master
              $masterAddress", e)
14.           }
15.         })
16.       }
17.     }
18. ......
```

Master 收到 RegisterApplication 信息后便开始注册，注册后再次调用 schedule()方法。Master.scala 的 receive 方法的源码如下：

```
1.    override def receive: PartialFunction[Any, Unit] = {
2.     ......
3.
4.     case RegisterApplication(description, driver) =>
5.       //待办事宜：防止某些 driver 重复注册
6.       if (state == RecoveryState.STANDBY) {
7.         //忽略，不要发送响应
8.       } else {
9.         logInfo("Registering app " + description.name)
10.        val app = createApplication(description, driver)
11.        registerApplication(app)
12.        logInfo("Registered app " + description.name + " with ID " + app.id)
13.        persistenceEngine.addApplication(app)
14.        driver.send(RegisteredApplication(app.id, self))
15.        schedule()
16.      }
17. ......
```

总结：从 SparkContext 创建 taskSchedulerImpl 初始化不同的实例对象来完成最终向 Master 注册的任务，中间包括调用 scheduler 的 start 方法和创建 StandaloneAppClient 来间接创建 ClientEndPoint 完成注册工作。

我们把 SparkContext 称为天堂之门，SparkContext 开启天堂之门：Spark 程序是通过 SparkContext 发布到 Spark 集群的；SparkContext 导演天堂世界：Spark 程序的运行都是在 SparkContext 为核心的调度器的指挥下进行的；SparkContext 关闭天堂之门：SparkContext 崩溃或者结束的时候整个 Spark 程序也结束。

4.2　DAGScheduler 解析

DAGScheduler 是面向 Stage 的高层调度器。本节讲解 DAG 的定义、DAG 的实例化、DAGScheduler 划分 Stage 的原理、DAGScheduler 划分 Stage 的具体算法、Stage 内部 Task 获取最佳位置的算法等内容。

4.2.1　DAG 的定义

DAGScheduler 是面向 Stage 的高层级的调度器，DAGScheduler 把 DAG 拆分成很多的 Tasks，每组的 Tasks 都是一个 Stage，解析时是以 Shuffle 为边界反向解析构建 Stage，每当遇到 Shuffle，就会产生新的 Stage，然后以一个个 TaskSet（每个 Stage 封装一个 TaskSet）的形式提交给底层调度器 TaskScheduler。DAGScheduler 需要记录哪些 RDD 被存入磁盘等物化动作，同时要寻求 Task 的最优化调度，如在 Stage 内部数据的本地性等。DAGScheduler 还需要监视因为 Shuffle 跨节点输出可能导致的失败，如果发现这个 Stage 失败，可能就要重新提交该 Stage。

为了更好地理解 Spark 高层调度器 DAGScheduler，须综合理解 RDD、Application、Driver Program、Job 内容，还需要了解以下概念。

（1）Stage：一个 Job 需要拆分成多组任务来完成，每组任务由 Stage 封装。与一个 Job 所有涉及的 PartitionRDD 类似，Stage 之间也有依赖关系。

（2）TaskSet：一组任务就是一个 TaskSet，对应一个 Stage。其中，一个 TaskSet 的所有 Task 之间没有 Shuffle 依赖，因此互相之间可以并行运行。

（3）Task：一个独立的工作单元，由 Driver Program 发送到 Executor 上去执行。通常情况下，一个 Task 处理 RDD 的一个 Partition 的数据。根据 Task 返回类型的不同，Task 又分为 ShuffleMapTask 和 ResultTask。

4.2.2　DAG 的实例化

在 Spark 源码中，DAGScheduler 是整个 Spark Application 的入口，即在 SparkContext 中声明并实例化。在实例化 DAGScheduler 之前，已经实例化了 SchedulerBackend 和底层调度器 TaskScheduler，而 SchedulerBackend 和 TaskScheduler 是通过 SparkContext 的方法 createTaskScheduler 实例化的。DAGScheduler 在提交 TaskSet 给底层调度器的时候是面向 TaskScheduler 接口的，这符合面向对象中依赖抽象，而不依赖具体实现的原则，带来底层资源调度器的可插拔性，以至于 Spark 可以运行在众多的部署模式上，如 Standalone、Yarn、Mesos、Local 及其他自定义的部署模式。

SparkContext.scala 的源码中相关的代码如下：

```
1.  class SparkContext(config: SparkConf) extends Logging {
2.  ......
3.  @volatile private var _dagScheduler: DAGScheduler = _
```

```
4.  ……
5.    private[spark] def dagScheduler: DAGScheduler = _dagScheduler
6.    private[spark] def dagScheduler_=(ds: DAGScheduler): Unit = {
7.      _dagScheduler = ds
8.    }
9.  ……
10.    val  (sched,  ts)  =  SparkContext.createTaskScheduler(this,  master,
       deployMode)
11.      _schedulerBackend = sched
12.      _taskScheduler = ts
13.      //实例化 DAGScheduler 时传入当前的 SparkContext 实例化对象
14.      _dagScheduler = new DAGScheduler(this)
15.      _heartbeatReceiver.ask[Boolean](TaskSchedulerIsSet)
16. ……
17.  _taskScheduler.start()
```

DAGScheduler.scala 的源码中相关的代码如下：

```
1.    private[spark]
2.   class DAGScheduler(
3.      private[scheduler] val sc: SparkContext,
4.      private[scheduler] val taskScheduler: TaskScheduler,
5.      listenerBus: LiveListenerBus,
6.      mapOutputTracker: MapOutputTrackerMaster,
7.      blockManagerMaster: BlockManagerMaster,
8.      env: SparkEnv,
9.      clock: Clock = new SystemClock())
10.   extends Logging {
11. ……
12.   def this(sc: SparkContext, taskScheduler: TaskScheduler) = {
13.     this(
14.       sc,
15.       taskScheduler,
16.       sc.listenerBus,
17.       sc.env.mapOutputTracker.asInstanceOf[MapOutputTrackerMaster],
18.       sc.env.blockManager.master,
19.       sc.env)
20.   }
21.
22.   def this(sc: SparkContext) = this(sc, sc.taskScheduler)
```

4.2.3 DAGScheduler 划分 Stage 的原理

Spark 将数据在分布式环境下分区，然后将作业转化为 DAG，并分阶段进行 DAG 的调度和任务的分布式并行处理。DAG 将调度提交给 DAGScheduler，DAGScheduler 调度时会根据是否需要经过 Shuffle 过程将 Job 划分为多个 Stage。

DAG 划分 Stage 及 Stage 并行计算示意图如图 4-3 所示。

其中，实线圆角方框标识的是 RDD，方框中的矩形块为 RDD 的分区。

在图 4-3 中，RDD A 到 RDD B 之间，以及 RDD F 到 RDD G 之间的数据需要经过 Shuffle 过程，因此 RDD A 和 RDD F 分别是 Stage 1 跟 Stage 3 和 Stage 2 跟 Stage 3 的划分点。而 RDD B 到 RDD G 之间，以及 RDD C 到 RDD D 到 RDD F 和 RDD E 到 RDD F 之间的数据不需要经过 Shuffle 过程，因此，RDD G 和 RDD B 的依赖是窄依赖，RDD B 和 RDD G 划分到同一个 Stage 3，RDD F 和 RDD D 和 RDD E 的依赖以及 RDD D 和 RDD C 的依赖是窄依赖，RDD

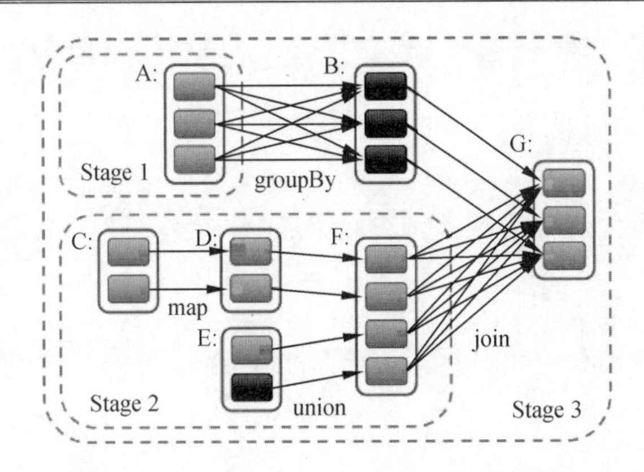

图 4-3　DAG 划分 Stage 及 Stage 并行计算示意图

C、RDD D、RDD E 和 RDD F 划分到同一个 Stage 2。Stage 1 和 Stage 2 是相互独立的，可以并发执行。而由于 Stage 3 依赖 Stage 1 和 Stage 2 的计算结果，所以 Stage 3 最后执行计算。

根据以上 RDD 依赖关系的描述，图 4-3 中的操作算子中，map 和 union 是窄依赖的操作，因为子 RDD（如 D）的分区只依赖父 RDD（如 C）的一个分区，其他常见的窄依赖的操作如 filter、flatMap 和 join（每个分区和已知的分区 join）等。groupByKey 和 join 是宽依赖的操作，其他常见的宽依赖的操作如 reduceByKey 等。

由此可见，在 DAGScheduler 的调度过程中，Stage 阶段的划分是根据是否有 Shuffle 过程，也就是当存在 ShuffleDependency 的宽依赖时，需要进行 Shuffle，这时才会将作业（Job）划分成多个 Stage。

4.2.4　DAGScheduler 划分 Stage 的具体算法

Spark 作业调度的时候，在 Job 提交过程中进行 Stage 划分以及确定 Task 的最佳位置。Stage 的划分是 DAGScheduler 工作的核心，涉及作业在集群中怎么运行，Task 最佳位置数据本地性的内容。Spark 算子的构建是链式的，涉及怎么进行计算，首先是划分 Stage，Stage 划分以后才是计算的本身；分布式大数据系统追求最大化的数据本地性。数据本地性是指数据进行计算的时候，数据就在内存中，甚至不用计算就直接获得结果。

Spark Application 中可以因为不同的 Action 触发众多的 Job。也就是说，一个 Application 中可以有很多的 Job，每个 Job 是由一个或者多个 Stage 构成的，后面的 Stage 依赖于前面的 Stage。也就是说，只有前面依赖的 Stage 计算完毕后，后面的 Stage 才会运行。

Stage 划分的根据是宽依赖。什么时候产生宽依赖呢？例如，reducByKey、groupByKey 等。

我们从 RDD 的 collect()方法开始，collect 算子是一个 Action，会触发 Job 的运行。

RDD.scala 的 collect 方法的源码调用了 runJob 方法。

```
1.    def collect(): Array[T] = withScope {
2.      val results = sc.runJob(this, (iter: Iterator[T]) => iter.toArray)
3.      Array.concat(results: _*)
4.    }
```

进入 SparkContext.scala 的 runJob 方法如下：

```
1.    def runJob[T, U: ClassTag](rdd: RDD[T], func: Iterator[T] => U): Array[U]
      = {
```

```
2.      runJob(rdd, func, 0 until rdd.partitions.length)
3.    }
```

继续重载 runJob 方法如下：

```
1.    def runJob[T, U: ClassTag](
2.        rdd: RDD[T],
3.        func: Iterator[T] => U,
4.        partitions: Seq[Int]): Array[U] = {
5.      val cleanedFunc = clean(func)
6.      runJob(rdd, (ctx: TaskContext, it: Iterator[T]) => cleanedFunc(it),
      partitions)
7.    }
```

继续重载 runJob 方法，SparkContext.scala 的源码如下：

```
1.     def runJob[T, U: ClassTag](
2.       rdd: RDD[T],
3.       processPartition: Iterator[T] => U,
4.       resultHandler: (Int, U) => Unit)
5.    {
6.      val processFunc = (context: TaskContext, iter: Iterator[T]) =>
      processPartition(iter)
7.      runJob[T, U](rdd, processFunc, 0 until rdd.partitions.length,
      resultHandler)
8.    }
```

继续重载 runJob 方法如下：

```
1.       def runJob[T, U: ClassTag](
2.       rdd: RDD[T],
3.       func: (TaskContext, Iterator[T]) => U,
4.       partitions: Seq[Int],
5.       resultHandler: (Int, U) => Unit): Unit = {
6.     if (stopped.get()) {
7.       throw new IllegalStateException("SparkContext has been shutdown")
8.     }
9.     val callSite = getCallSite
10.    val cleanedFunc = clean(func)
11.    logInfo("Starting job: " + callSite.shortForm)
12.    if (conf.getBoolean("spark.logLineage", false)) {
13.      logInfo("RDD's recursive dependencies:\n" + rdd.toDebugString)
14.    }
15.    dagScheduler.runJob(rdd, cleanedFunc, partitions, callSite,
       resultHandler, localProperties.get)
16.    progressBar.foreach(_.finishAll())
17.    rdd.doCheckpoint()
18.  }
```

进入 DAGScheduler.scala 的 runJob 方法，DAGScheduler.scala 的源码如下：

```
1.  def runJob[T, U](
2.       rdd: RDD[T],
3.       func: (TaskContext, Iterator[T]) => U,
4.       partitions: Seq[Int],
5.       callSite: CallSite,
6.       resultHandler: (Int, U) => Unit,
```

```
7.         properties: Properties): Unit = {
8.       val start = System.nanoTime
9.       val waiter = submitJob(rdd, func, partitions, callSite, resultHandler,
         properties)
10.      ThreadUtils.awaitReady(waiter.completionFuture, Duration.Inf)
11.      waiter.completionFuture.value.get match {
12.        case scala.util.Success(_) =>
13.          logInfo("Job %d finished: %s, took %f s".format
14.            (waiter.jobId, callSite.shortForm, (System.nanoTime - start) / 1e9))
15.        case scala.util.Failure(exception) =>
16.          logInfo("Job %d failed: %s, took %f s".format
17.            (waiter.jobId, callSite.shortForm, (System.nanoTime - start) / 1e9))
18.          // SPARK-8644: 源自 DagScheduler 的异常，包括用户堆栈跟踪
19.          val callerStackTrace = Thread.currentThread().getStackTrace.tail
20.          exception.setStackTrace(exception.getStackTrace ++ callerStackTrace)
21.          throw exception
22.      }
23.    }
24. ......
```

DAGScheduler runJob 的时候就交给了 submitJob，waiter 等待作业调度的结果，作业成功或者失败，打印相关的日志信息。进入 DAGScheduler 的 submitJob 方法如下：

```
1.       def submitJob[T, U](
2.         rdd: RDD[T],
3.         func: (TaskContext, Iterator[T]) => U,
4.         partitions: Seq[Int],
5.         callSite: CallSite,
6.         resultHandler: (Int, U) => Unit,
7.         properties: Properties): JobWaiter[U] = {
8.       //检查，以确保我们不在不存在的分区上启动任务
9.       val maxPartitions = rdd.partitions.length
10.      partitions.find(p => p >= maxPartitions || p < 0).foreach { p =>
11.        throw new IllegalArgumentException(
12.          "Attempting to access a non-existent partition: " + p + ". " +
13.            "Total number of partitions: " + maxPartitions)
14.      }
15.
16.      val jobId = nextJobId.getAndIncrement()
17.      if (partitions.size == 0) {
18.        //如果作业正在运行 0 个任务，则立即返回
19.        return new JobWaiter[U](this, jobId, 0, resultHandler)
20.      }
21.
22.      assert(partitions.size > 0)
23.      val func2 = func.asInstanceOf[(TaskContext, Iterator[_]) => _]
24.      val waiter = new JobWaiter(this, jobId, partitions.size,
         resultHandler)
25.      eventProcessLoop.post(JobSubmitted(
26.        jobId, rdd, func2, partitions.toArray, callSite, waiter,
27.        SerializationUtils.clone(properties)))
28.      waiter
29.    }
```

submitJob 方法中，submitJob 首先获取 rdd.partitions.length，校验运行的时候 partitions 是否存在。submitJob 方法关键的代码是 eventProcessLoop.post(JobSubmitted 的 JobSubmitted，

JobSubmitted 是一个 case class，而不是一个 case object，因为 application 中有很多的 Job，不同的 Job 的 JobSubmitted 实例不一样，如果使用 case object，case object 展示的内容是一样的，就像全局唯一变量，而现在我们需要不同的实例，因此使用 case class。JobSubmitted 的成员 finalRDD 是最后一个 RDD。

由 Action（如 collect）导致 SparkContext.runJob 的执行，最终导致 DAGScheduler 中的 submitJob 的执行，其核心是通过发送一个 case class JobSubmitted 对象给 eventProcessLoop。其中，JobSubmitted 的源码如下：

```
1.    private[scheduler] case class JobSubmitted(
2.     jobId: Int,
3.     finalRDD: RDD[_],
4.     func: (TaskContext, Iterator[_]) => _,
5.     partitions: Array[Int],
6.     callSite: CallSite,
7.     listener: JobListener,
8.     properties: Properties = null)
9.   extends DAGSchedulerEvent
```

JobSubmitted 是 private[scheduler]级别的，用户不可直接调用它。JobSubmitted 封装了 jobId，封装了最后一个 finalRDD，封装了具体对 RDD 操作的函数 func，封装了有哪些 partitions 要进行计算，也封装了作业监听器 listener、状态等内容。

DAGScheduler 的 submitJob 方法关键代码 eventProcessLoop.post(JobSubmitted 中，将 JobSubmitted 放入到 eventProcessLoop。post 就是 Java 中的 post，往一个线程中发一个消息。eventProcessLoop 的源码如下：

```
1.   private[scheduler] val eventProcessLoop = new DAGSchedulerEventProcessLoop
     (this)
```

DAGSchedulerEventProcessLoop 继承自 EventLoop。

```
1.    private[scheduler] class DAGSchedulerEventProcessLoop(dagScheduler:
      DAGScheduler)
2.   extends EventLoop[DAGSchedulerEvent]("dag-scheduler-event-loop") with
     Logging {
```

EventLoop 中开启了一个线程 eventThread，线程设置成 Daemon 后台运行的方式；run 方法里面调用了 onReceive(event)方法。post 方法就是往 eventQueue.put 事件队列中放入一个元素。EventLoop 的源码如下：

```
1.  private[spark] abstract class EventLoop[E](name: String) extends Logging {
2.
3.   private val eventQueue: BlockingQueue[E] = new LinkedBlockingDeque[E]()
4.
5.   private val stopped = new AtomicBoolean(false)
6.
7.   private val eventThread = new Thread(name) {
8.     setDaemon(true)
9.
10.    override def run(): Unit = {
11.      try {
12.        while (!stopped.get) {
13.          val event = eventQueue.take()
14.          try {
```

```
15.           onReceive(event)
16.         } catch {
17.           case NonFatal(e) =>
18.             try {
19.               onError(e)
20.             } catch {
21.               case NonFatal(e) => logError("Unexpected error in " + name, e)
22.             }
23.         }
24.       }
25.     } catch {
26.       case ie: InterruptedException => //即使 eventQueue 不为空，退出
27.       case NonFatal(e) => logError("Unexpected error in " + name, e)
28.     }
29.   }
30.
31. }
32.
33. def start(): Unit = {
34.   if (stopped.get) {
35.     throw new IllegalStateException(name + " has already been stopped")
36.   }
37.   //调用 OnStart 启动事件线程，确保其发生在 onReceive 方法前
38.   onStart()
39.   eventThread.start()
40. }
41. ......
42. def post(event: E): Unit = {
43.   eventQueue.put(event)
44. }
```

eventProcessLoop 是 DAGSchedulerEventProcessLoo 实例，DAGSchedulerEventProcessLoop 继承自 EventLoop，具体实现 onReceive 方法，onReceive 方法又调用 doOnReceive 方法。

doOnReceive 收到消息后开始处理。

Spark 2.2.1 版本的 DAGScheduler.scala 的源码如下：

```
1. private def doOnReceive(event: DAGSchedulerEvent): Unit = event match {
2.     case JobSubmitted(jobId, rdd, func, partitions, callSite, listener,
       properties) =>
3.       dagScheduler.handleJobSubmitted(jobId, rdd, func, partitions, callSite,
         listener, properties)
4.
5.     case MapStageSubmitted(jobId, dependency, callSite, listener, properties) =>
6.       dagScheduler.handleMapStageSubmitted(jobId, dependency, callSite,
         listener, properties)
7.
8.     case StageCancelled(stageId, reason) =>
9.       dagScheduler.handleStageCancellation(stageId, reason)
10.
11.    case JobCancelled(jobId, reason) =>
12.      dagScheduler.handleJobCancellation(jobId, reason)
13.
14.    case JobGroupCancelled(groupId) =>
15.      dagScheduler.handleJobGroupCancelled(groupId)
16.
17.    case AllJobsCancelled =>
```

```
18.       dagScheduler.doCancelAllJobs()
19.
20.     case ExecutorAdded(execId, host) =>
21.       dagScheduler.handleExecutorAdded(execId, host)
22.
23.     case ExecutorLost(execId, reason) =>
24.       val filesLost = reason match {
25.         case SlaveLost(_, true) => true
26.         case _ => false
27.       }
28.       dagScheduler.handleExecutorLost(execId, filesLost)
29.
30.     case BeginEvent(task, taskInfo) =>
31.       dagScheduler.handleBeginEvent(task, taskInfo)
32.
33.     case GettingResultEvent(taskInfo) =>
34.       dagScheduler.handleGetTaskResult(taskInfo)
35.
36.     case completion: CompletionEvent =>
37.       dagScheduler.handleTaskCompletion(completion)
38.
39.     case TaskSetFailed(taskSet, reason, exception) =>
40.       dagScheduler.handleTaskSetFailed(taskSet, reason, exception)
41.
42.     case ResubmitFailedStages =>
43.       dagScheduler.resubmitFailedStages()
44.   }
```

Spark 2.4.3 版本的 DAGScheduler.scala 的源码与 Spark 2.2.1 版本相比具有如下特点。

- 上段代码中第 24 行及第 28 行，将 filesLost 变量名称修改为 workerLost。
- 上段代码中第 29 行新增加收到 WorkerRemoved 消息的处理。
- 上段代码中第 32 行新增加收到 SpeculativeTaskSubmitted 消息的处理。

```
1.    val workerLost = reason match {
2.   ......
3.   dagScheduler.handleExecutorLost(execId, workerLost)
4.   ......
5.     case WorkerRemoved(workerId, host, message) =>
6.       dagScheduler.handleWorkerRemoved(workerId, host, message)
7.   ......
8.     case SpeculativeTaskSubmitted(task) =>
9.       dagScheduler.handleSpeculativeTaskSubmitted(task)
```

总结：EventLoop 里面开启一个线程，线程里面不断循环一个队列，post 的时候就是将消息放到队列中，由于消息放到队列中，在不断循环，所以可以拿到这个消息，转过来回调方法 onReceive(event)，在 onReceive 处理的时候就调用了 doOnReceive 方法。

关于线程的异步通信：为什么要新开辟一条线程？例如，在 DAGScheduler 发送消息为何不直接调用 doOnReceive，而需要一个消息循环器。DAGScheduler 这里自己给自己发消息，不管是自己发消息，还是别人发消息，都采用一条线程去处理，两者处理的逻辑是一致的，扩展性就非常好。使用消息循环器，就能统一处理所有的消息，保证处理的业务逻辑都是一致的。

eventProcessLoop 是 DAGSchedulerEventProcessLoop 的具体实例，而 DAGScheduler-EventProcessLoop 是 EventLoop 的子类，具体实现 EventLoop 的 onReceive 方法，onReceive 方法转过来回调 doOnReceive。

在 doOnReceive 中通过模式匹配的方式把执行路由到 case JobSubmitted，调用 dagScheduler.handleJobSubmitted 方法。

```
1.   private def doOnReceive(event: DAGSchedulerEvent): Unit = event match {
2.     case JobSubmitted(jobId, rdd, func, partitions, callSite, listener,
       properties) =>
3.       dagScheduler.handleJobSubmitted(jobId,  rdd,  func,  partitions,
         callSite, listener, properties)
```

Spark 2.2.1 版本的 DAGScheduler 的 handleJobSubmitted 的源码如下：

```
1.    private[scheduler] def handleJobSubmitted(jobId: Int,
2.        finalRDD: RDD[_],
3.        func: (TaskContext, Iterator[_]) => _,
4.        partitions: Array[Int],
5.        callSite: CallSite,
6.        listener: JobListener,
7.        properties: Properties) {
8.      var finalStage: ResultStage = null
9.      try {
10.       //如果作业运行在 HadoopRDD 上,而底层 HDFS 的文件已被删除,那么在创建新的 Stage
          //时将会跑出一个异常
11.       finalStage = createResultStage(finalRDD, func, partitions, jobId,
          callSite)
12.     } catch {
13.       case e: Exception =>
14.         logWarning("Creating new stage failed due to exception - job: " +
            jobId, e)
15.         listener.jobFailed(e)
16.         return
17.     }
18.
19.     val job = new ActiveJob(jobId, finalStage, callSite, listener,
        properties)
20.     clearCacheLocs()
21.     logInfo("Got job %s (%s) with %d output partitions".format(
22.       job.jobId, callSite.shortForm, partitions.length))
23.     logInfo("Final stage: " + finalStage + " (" + finalStage.name + ")")
24.     logInfo("Parents of final stage: " + finalStage.parents)
25.     logInfo("Missing parents: " + getMissingParentStages(finalStage))
26.
27.     val jobSubmissionTime = clock.getTimeMillis()
28.     jobIdToActiveJob(jobId) = job
29.     activeJobs += job
30.     finalStage.setActiveJob(job)
31.     val stageIds = jobIdToStageIds(jobId).toArray
32.     val stageInfos = stageIds.flatMap(id => stageIdToStage.get(id)
        .map(_.latestInfo))
33.     listenerBus.post(
34.       SparkListenerJobStart(job.jobId,  jobSubmissionTime,  stageInfos,
          properties))
35.     submitStage(finalStage)
36.   }
```

Spark 2.4.3 版本的 DAGScheduler.scala 的源码与 Spark 2.2.1 版本相比具有如下特点。

- 上段代码中第 13 行代码之后新增 BarrierJobSlotsNumberCheckFailed 异常情况的处理。
- 上段代码中第 19 行代码之前新增清除内部数据的代码。

```
1.        case e: BarrierJobSlotsNumberCheckFailed =>
2.          logWarning(s"The job $jobId requires to run a barrier stage that
            requires more slots " +
3.            "than the total number of slots in the cluster currently.")
4.          // 如果 jobId 在映射中不存在，scala 会自动将其值 null 转换为 0:Int
5.          val numCheckFailures =
    barrierJobIdToNumTasksCheckFailures.compute(jobId,
6.            new BiFunction[Int, Int, Int] {
7.              override def apply(key: Int, value: Int): Int = value + 1
8.            })
9.          if (numCheckFailures <= maxFailureNumTasksCheck) {
10.           messageScheduler.schedule(
11.             new Runnable {
12.               override def run(): Unit = eventProcessLoop.post(JobSubmitted(jobId,
                  finalRDD, func,
13.                 partitions, callSite, listener, properties))
14.             },
15.             timeIntervalNumTasksCheck,
16.             TimeUnit.SECONDS
17.           )
18.           return
19.         } else {
20.           // Job 失败，清理内部数据
21.           barrierJobIdToNumTasksCheckFailures.remove(jobId)
22.           listener.jobFailed(e)
23.           return
24.         }
25.       //提交作业，清除内部数据
26.       barrierJobIdToNumTasksCheckFailures.remove(jobId)
```

Stage 开始：每次调用一个 runJob 就产生一个 Job；finalStage 是一个 ResultStage，最后一个 Stage 是 ResultStage，前面的 Stage 是 ShuffleMapStage。

在 handleJobSubmitted 中首先创建 finalStage，创建 finalStage 时会建立父 Stage 的依赖链条。

通过 createResultStage 创建 finalStage，传入的参数包括最后一个 finalRDD，操作的函数 func，分区 partitions、jobId、callSite 等内容。创建过程中可能捕获异常。例如，在 Hadoop 上，底层的 hdfs 文件被删除了或者被修改了，就出现异常。

Spark 2.2.1 版本的 createResultStage 的源码如下：

```
1.   private def createResultStage(
2.       rdd: RDD[_],
3.       func: (TaskContext, Iterator[_]) => _,
4.       partitions: Array[Int],
5.       jobId: Int,
6.       callSite: CallSite): ResultStage = {
7.     val parents = getOrCreateParentStages(rdd, jobId)
8.     val id = nextStageId.getAndIncrement()
9.     val stage = new ResultStage(id, rdd, func, partitions, parents, jobId,
       callSite)
10.    stageIdToStage(id) = stage
11.    updateJobIdStageIdMaps(jobId, stage)
12.    stage
13.  }
```

Spark 2.4.3 版本的 DAGScheduler.scala 的源码与 Spark 2.2.1 版本相比具有如下特点。

- 上段代码中第 6 行代码之后新增检查屏障阶段的代码。

```
1.     checkBarrierStageWithDynamicAllocation(rdd)
```

```
2.        checkBarrierStageWithNumSlots(rdd)
3.        checkBarrierStageWithRDDChainPattern(rdd, partitions.toSet.size)
```

三个检查屏障阶段的方法如下。

（1）checkBarrierStageWithDynamicAllocation：如果在启用动态资源分配的情况下运行屏障阶段，将在作业提交时执行检查并快速失败。

（2）checkBarrierStageWithNumSlots：检查屏障阶段是否需要比当前活动插槽总数更多的插槽（以便能够启动屏障阶段的全部任务）。如果试图提交一个障碍阶段，需要比当前总数更多的插槽，检查会失败。如果检查连续失败，超过了作业的配置数，当前作业的提交将失败。

（3）checkBarrierStageWithRDDChainPattern：检查以确保我们不使用不支持的 RDD 链模式启动屏障阶段，以下模式不支持。

① 与 RDD 具有不同分区数的父 RDD（例如 union()/coalesce()/first()/take()/PartitionPruning-RDD）。

② 依赖多个屏障 RDD 的 RDD（例如 barrierRdd1.zip(barrierRdd2)）。

createResultStage 中，基于作业 ID，作业 ID（jobId）是作为第三个参数传进来的，创建了 ResultStage。

createResultStage 的 getOrCreateParentStages 获取或创建一个给定 RDD 的父 Stages 列表，新的 Stages 将提供 firstJobId 创建。

getOrCreateParentStages 的源码如下：

```
1.        private def getOrCreateParentStages(rdd: RDD[_], firstJobId: Int):
          List[Stage] = {
2.      getShuffleDependencies(rdd).map { shuffleDep =>
3.        getOrCreateShuffleMapStage(shuffleDep, firstJobId)
4.      }.toList
5.    }
```

getOrCreateParentStages 调用了 getShuffleDependencies(rdd)，getShuffleDependencies 返回给定 RDD 的父节点中直接的 shuffle 依赖。这个函数不会返回更远祖先节点的依赖。例如，如果 C shuffle 依赖于 B，B shuffle 依赖于 A：A <-- B <-- C。在 RDD C 中调用 getShuffleDependencies 函数，将只返回 B <-- C 的依赖。此功能可用作单元测试。

下面根据 DAG 划分 Stage 示意图，如图 4-4 所示。

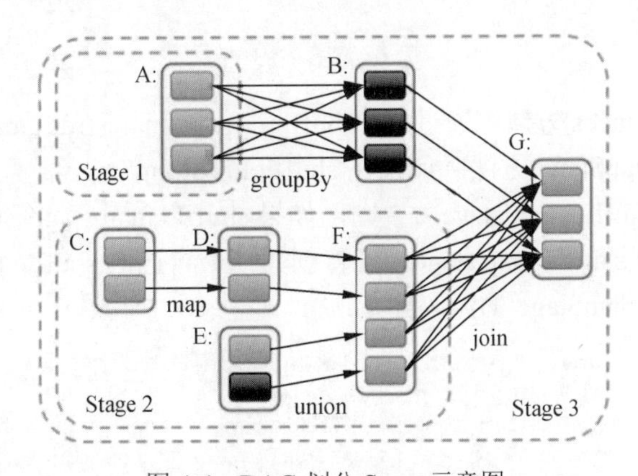

图 4-4　DAG 划分 Stage 示意图

RDD G 在 getOrCreateParentStages 的 getShuffleDependencies 的时候同时依赖于 RDD B,RDD F；看依赖关系，RDD G 和 RDD B 在同一个 Stage 里，RDD G 和 RDD F 不在同一个 Stage 里，根据 Shuffle 依赖产生了一个新的 Stage。如果不是 Shuffle 级别的依赖，就将其加入 waitingForVisit.push(dependency.rdd)，waitingForVisit 是一个栈 Stack，把当前依赖的 RDD push 进去。然后进行 while 循环，当 waitingForVisit 不是空的情况下，将 waitingForVisit.pop() 的内容弹出来放入到 toVisit，如果已经访问过的数据结构 visited 中没有访问记录，那么 toVisit.dependencies 再次循环遍历：如果是 Shuffle 依赖，就加入到 parents 数据结构；如果是窄依赖，就加入到 waitingForVisit。

例如，首先将 RDD G 放入到 waitingForVisit，然后看 RDD G 的依赖关系，依赖 RDD B、RDD F；RDD G 和 RDD F 构成的是宽依赖，所以就加入父 Stage 里，是一个新的 Stage。但如果是窄依赖，就把 RDD B 放入到栈 waitingForVisit 中，RDD G 和 RDD B 在同一个 Stage 中。栈 waitingForVisit 现在又有新的元素 RDD B，然后再次进行循环，获取到宽依赖 RDD A，将构成一个新的 Stage。RDD G 的 getShuffleDependencies 最终返回 HashSet (ShuffleDependency(RDD F),ShuffleDependency(RDD A))。然后 getShuffleDependencies(rdd) .map 遍历调用 getOrCreateShuffleMapStage 直接创建父 Stage。

getShuffleDependencies 的源码如下：

```
1.       private[scheduler] def getShuffleDependencies(
2.       rdd: RDD[_]): HashSet[ShuffleDependency[_, _, _]] = {
3.     val parents = new HashSet[ShuffleDependency[_, _, _]]
4.     val visited = new HashSet[RDD[_]]
5.     val waitingForVisit = new Stack[RDD[_]]
6.     waitingForVisit.push(rdd)
7.     while (waitingForVisit.nonEmpty) {
8.       val toVisit = waitingForVisit.pop()
9.       if (!visited(toVisit)) {
10.        visited += toVisit
11.        toVisit.dependencies.foreach {
12.          case shuffleDep: ShuffleDependency[_, _, _] =>
13.            parents += shuffleDep
14.          case dependency =>
15.            waitingForVisit.push(dependency.rdd)
16.        }
17.      }
18.    }
19.    parents
20.  }
```

getOrCreateParentStages 方法中通过 getShuffleDependencies(rdd).map 进行 map 转换时用了 getOrCreateShuffleMapStage 方法。如果在 shuffleIdToMapStage 数据结构中 shuffleId 已经存在，那就获取一个 shuffle map stage，否则，如果 shuffle map stage 不存在，除了即将进行计算的更远祖先节点的 shuffle map stage，还将创建一个自己的 shuffle map stage。

getOrCreateShuffleMapStage 的源码如下：

```
1.    private def getOrCreateShuffleMapStage(
2.        shuffleDep: ShuffleDependency[_, _, _],
3.        firstJobId: Int): ShuffleMapStage = {
4.      shuffleIdToMapStage.get(shuffleDep.shuffleId) match {
5.        case Some(stage) =>
6.          stage
```

```
7.
8.        case None =>
9.          //创建所有即将计算的祖先 shuffle 依赖的阶段
10.         getMissingAncestorShuffleDependencies(shuffleDep.rdd).foreach
            { dep =>
11.     //尽管 getMissingAncestorShuffleDependencies 只返回 shuffle 的依赖，其已
        //不在 shuffleIdToMapStage 中。我们在 foreach 循环中得到一个特定的依赖是可能
        //的，将被增加到 shuffleIdToMapStage 依赖中，其是通过早期的依赖关系创建的阶段，
        //参考 SPARK-13902
12.           if (!shuffleIdToMapStage.contains(dep.shuffleId)) {
13.             createShuffleMapStage(dep, firstJobId)
14.           }
15.         }
16.         //最后，为给定的 shuffle 依赖创建一个阶段
17.         createShuffleMapStage(shuffleDep, firstJobId)
18.     }
19.   }
```

getOrCreateShuffleMapStage 方法中：

- 如果根据 shuffleId 模式匹配获取到 Stage，就返回 Stage。首先从 shuffleIdToMapStage 中根据 shuffleId 获取 Stage。shuffleIdToMapStage 是一个 HashMap 数据结构，将 Shuffle dependency ID 对应到 ShuffleMapStage 的映射关系，shuffleIdToMapStage 只包含当前运行作业的映射数据，当 Shuffle Stage 作业完成时，Shuffle 映射数据将被删除，Shuffle 的数据将记录在 MapOutputTracker 中。
- 如果根据 shuffleId 模式匹配没有获取到 Stage，调用 getMissingAncestorShuffleDependencies 方法，createShuffleMapStage 创建所有即将进行计算的祖先 shuffle 依赖的 Stages。

getMissingAncestorShuffleDependencies 查找 shuffle 依赖中还没有进行 shuffleToMapStage 注册的祖先节点。

Spark 2.2.1 版本 DAGScheduler.scala 的 getMissingAncestorShuffleDependencies 源码如下：

```
1.      private def getMissingAncestorShuffleDependencies(
2.       rdd: RDD[_]): Stack[ShuffleDependency[_, _, _]] = {
3.     val ancestors = new Stack[ShuffleDependency[_, _, _]]
4.     val visited = new HashSet[RDD[_]]
5.     //手动维护堆栈来防止通过递归访问造成的堆栈溢出异常
6.     val waitingForVisit = new Stack[RDD[_]]
7.     waitingForVisit.push(rdd)
8.     while (waitingForVisit.nonEmpty) {
9.       val toVisit = waitingForVisit.pop()
10.      if (!visited(toVisit)) {
11.        visited += toVisit
12.        getShuffleDependencies(toVisit).foreach { shuffleDep =>
13.          if (!shuffleIdToMapStage.contains(shuffleDep.shuffleId)) {
14.            ancestors.push(shuffleDep)
15.            waitingForVisit.push(shuffleDep.rdd)
16.          } //依赖关系及其已经注册的祖先
17.        }
18.      }
19.    }
20.    ancestors
21.  }
```

Spark 2.4.3 版本的 DAGScheduler.scala 的源码与 Spark 2.2.1 版本相比具有如下特点。

- 上段代码中的第 2 行、第 3 行、第 6 行，将 Stack 实例修改为 ArrayStack 实例。

```
1.  ......
2.    rdd: RDD[_]): ArrayStack[ShuffleDependency[_, _, _]] = {
3.     val ancestors = new ArrayStack[ShuffleDependency[_, _, _]]
4.  ......
5.  val waitingForVisit = new ArrayStack[RDD[_]]
```

createShuffleMapStage 根据 Shuffle 依赖的分区创建一个 ShuffleMapStage，如果前一个 Stage 已生成相同的 Shuffle 数据，那 Shuffle 数据仍是可用的，createShuffleMapStage 方法将复制 Shuffle 数据的位置信息去获取数据，无须再重新生成一次数据。

Spark 2.2.1 版本的 DAGScheduler.scala 的 createShuffleMapStage 源码如下：

```
1.   def createShuffleMapStage(shuffleDep: ShuffleDependency[_, _, _], jobId:
     Int): ShuffleMapStage = {
2.     val rdd = shuffleDep.rdd
3.     val numTasks = rdd.partitions.length
4.     val parents = getOrCreateParentStages(rdd, jobId)
5.     val id = nextStageId.getAndIncrement()
6.     val stage = new ShuffleMapStage(id, rdd, numTasks, parents, jobId,
       rdd.creationSite, shuffleDep)
7.
8.     stageIdToStage(id) = stage
9.     shuffleIdToMapStage(shuffleDep.shuffleId) = stage
10.    updateJobIdStageIdMaps(jobId, stage)
11.
12.    if (mapOutputTracker.containsShuffle(shuffleDep.shuffleId)) {
13.      //以前运行的阶段为这个 shuffle 生成的分区，对于每个输出仍然可用，将输出位置的
         //信息复制到新阶段（所以没必要重新计算数据）
14.      val serLocs = mapOutputTracker.getSerializedMapOutputStatuses
         (shuffleDep.shuffleId)
15.      val locs = MapOutputTracker.deserializeMapStatuses(serLocs)
16.      (0 until locs.length).foreach { i =>
17.        if (locs(i) ne null) {
18.          //locs(i) will be null if missing
19.          stage.addOutputLoc(i, locs(i))
20.        }
21.      }
22.    } else {
23.      //这里需要注册 RDDS 与缓存和 map 输出跟踪器，不能在 RDD 构造函数实现，因为分区
         //是未知的
24.      logInfo("Registering RDD " + rdd.id + " (" + rdd.getCreationSite + ")")
25.      mapOutputTracker.registerShuffle(shuffleDep.shuffleId,
         rdd.partitions.length)
26.    }
27.    stage
28.  }
```

Spark 2.4.3 版本的 DAGScheduler.scala 的源码与 Spark 2.2.1 版本相比具有如下特点。

- 上段代码中第 2 行之后新增三个检查屏障阶段的方法：checkBarrierStageWithDynamicAllocation、checkBarrierStageWithNumSlots、checkBarrierStageWithRDDChainPattern。
- 上段代码中第 6 行构建 ShuffleMapStage 实例时，新增传入一个参数 mapOutputTracker。
- 上段代码中第 12 行将代码调整为!mapOutputTracker.containsShuffle(shuffleDep.shuffleId))。

❑ 删掉上段代码中第 13～22 行代码。

```
1.     checkBarrierStageWithDynamicAllocation(rdd)
2.     checkBarrierStageWithNumSlots(rdd)
3.     checkBarrierStageWithRDDChainPattern(rdd, rdd.getNumPartitions)
4.   ......
5.   val stage = new ShuffleMapStage(
6.        id, rdd, numTasks, parents, jobId, rdd.creationSite, shuffleDep,
        mapOutputTracker)
7.   ......
8.   if (!mapOutputTracker.containsShuffle(shuffleDep.shuffleId)) {
```

回到 handleJobSubmitted，创建 finalStage 以后将提交 finalStage。

```
1.    private[scheduler] def handleJobSubmitted(jobId: Int,
2.   ......
3.    finalStage  =  createResultStage(finalRDD,  func,  partitions,  jobId,
    callSite)
4.   ......
5.    submitStage(finalStage)
6.    }
```

submitStage 提交 Stage，首先递归提交即将计算的父 Stage。
submitStage 的源码如下：

```
1.    private def submitStage(stage: Stage) {
2.      val jobId = activeJobForStage(stage)
3.      if (jobId.isDefined) {
4.        logDebug("submitStage(" + stage + ")")
5.        if (!waitingStages(stage) && !runningStages(stage) && !failedStages
        (stage)) {
6.          val missing = getMissingParentStages(stage).sortBy(_.id)
7.          logDebug("missing: " + missing)
8.          if (missing.isEmpty) {
9.            logInfo("Submitting " + stage + " (" + stage.rdd + "), which has
            no missing parents")
10.           submitMissingTasks(stage, jobId.get)
11.         } else {
12.           for (parent <- missing) {
13.             submitStage(parent)
14.           }
15.           waitingStages += stage
16.         }
17.       }
18.     } else {
19.       abortStage(stage, "No active job for stage " + stage.id, None)
20.     }
21.   }
```

其中调用了 getMissingParentStages。
DAGScheduler.scala 的源码如下：

```
1.     private def getMissingParentStages(stage: Stage): List[Stage] = {
2.     val missing = new HashSet[Stage]
3.     val visited = new HashSet[RDD[_]]
4.     //人工维护堆栈来防止通过递归访问造成的堆栈溢出异常
5.     val waitingForVisit = new Stack[RDD[_]]
6.     def visit(rdd: RDD[_]) {
```

```
7.        if (!visited(rdd)) {
8.          visited += rdd
9.          val rddHasUncachedPartitions = getCacheLocs(rdd).contains(Nil)
10.         if (rddHasUncachedPartitions) {
11.           for (dep <- rdd.dependencies) {
12.             dep match {
13.               case shufDep: ShuffleDependency[_, _, _] =>
14.                 val mapStage = getOrCreateShuffleMapStage(shufDep, stage
                    .firstJobId)
15.                 if (!mapStage.isAvailable) {
16.                   missing += mapStage
17.                 }
18.               case narrowDep: NarrowDependency[_] =>
19.                 waitingForVisit.push(narrowDep.rdd)
20.             }
21.           }
22.         }
23.       }
24.     }
25.     waitingForVisit.push(stage.rdd)
26.     while (waitingForVisit.nonEmpty) {
27.       visit(waitingForVisit.pop())
28.     }
29.     missing.toList
30.   }
```

Spark 2.4.3 版本的 DAGScheduler.scala 的源码与 Spark 2.2.1 版本相比具有如下特点。

- 上段代码中第 5 行将构建 Stack 实例修改为构建 ArrayStack 实例。

```
1.  val waitingForVisit = new ArrayStack[RDD[_]]
```

接下来，我们结合 Spark DAG 划分 Stage 示意（见图 4-5）进行详细阐述。

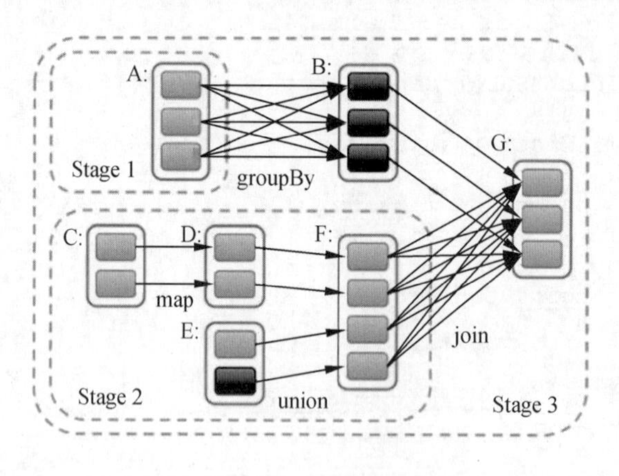

图 4-5　DAG 划分 Stage 示意图

RDD A 到 RDD B，以及 RDD F 到 RDD G 之间的数据需要经过 Shuffle 过程，因此，RDD A 和 RDD F 分别是 Stage 1 跟 Stage 3、Stage 2 跟 Stage 3 的划分点。而 RDD B 到 RDD G 没有 Shuffle，因此，RDD G 和 RDD B 的依赖是窄依赖，RDD B 和 RDD G 划分到同一个 Stage 3；RDD C 到 RDD D、RDD F，RDD E 到 RDD F 之间的数据不需要经过 Shuffle，RDD F 和 RDD D 加 RDD E 的依赖、RDD D 和 RDD C 的依赖是窄依赖，因此，RDD C、RDD D、RDD E 和 RDD F 划分到同一个 Stage 2。Stage 1 和 Stage 2 是相互独立的，可以并发执行。

而由于 Stage 3 依赖 Stage 1 和 Stage 2 的计算结果，所以 Stage 3 最后执行计算。

- createResultStage：基于作业 ID（jobId）创建 ResultStage。调用 getOrCreateParentStages 创建所有父 Stage，返回 parents: List[Stage]作为父 Stage，将 parents 传入 ResultStage，实例化生成 ResultStage。

在 DAG 划分 Stage 示意图中，对 RDD G 调用 createResultStage，通过 getOrCreateParentStages 获取所有父 List[Stage]：Stage 1、Stage 2，然后创建自己的 Stage 3。

- getOrCreateParentStages：获取或创建给定 RDD 的父 Stage 列表。将根据提供的 firstJobId 创建新的 Stages。

在 DAG 划分 Stage 示意图中，RDD G 的 getOrCreateParentStages 会调用 getShuffleDependencies 获得 RDD G 所有直接宽依赖集合 HashSet(ShuffleDependency(RDD F), ShuffleDependency(RDD A))，这里是 RDD F 和 RDD A 的宽依赖集合，然后遍历集合，对 (ShuffleDependency(RDD F), ShuffleDependency(RDD A))分别调用 getOrCreateShuffleMapStage。

- 对 ShuffleDependency(RDD A) 调用 getOrCreateShuffleMapStage，getOrCreateShuffleMapStage 中根据 shuffleDep.shuffleId 模式匹配调用 getMissingAncestorShuffleDependencies，返回为空；对 ShuffleDependency(RDD A)调用 createShuffleMapStage，RDD A 已无父 Stage，因此创建 Stage 1。
- 对 ShuffleDependency(RDD F)调用 getOrCreateShuffleMapStage，getOrCreateShuffleMapStage 中根据 shuffleDep.shuffleId 模式匹配调用 getMissingAncestorShuffleDependencies，返回为空；对 ShuffleDependency(RDD F)调用 createShuffleMapStage，RDD F 之前的 RDD C 到 RDD D、RDD F；RDD E 到 RDD F 之间都没有 Shuffle，没有宽依赖就不会产生 Stage。因此，RDD F 已无父 Stage，创建 Stage 2。
- 最后，把 List(Stage 1,Stage 2) 作为 Stage 3 的父 Stage，创建 Stage 3。Stage 3 是 ResultStage。

回到 DAGScheduler.scala 的 handleJobSubmitted 方法，首先通过 createResultStage 构建 finalStage。

handleJobSubmitted 的源码如下：

```
1.   private[scheduler] def handleJobSubmitted(jobId: Int,
2.     ......
3.       finalStage = createResultStage(finalRDD, func, partitions, jobId,
         callSite)
4.    ......
5.     val job = new ActiveJob(jobId, finalStage, callSite, listener,
       properties)
6.    ......
7.     logInfo("Missing parents: " + getMissingParentStages(finalStage))
8.     ......
9.     submitStage(finalStage)
10.  }
```

handleJobSubmitted 方法中的 ActiveJob 是一个普通的数据结构，保存了当前 Job 的一些信息。

```
1.   private[spark] class ActiveJob(
2.      val jobId: Int,
3.      val finalStage: Stage,
4.      val callSite: CallSite,
```

```
5.      val listener: JobListener,
6.      val properties: Properties) {
```

handleJobSubmitted 方法日志打印信息：getMissingParentStages(finalStage))，getMissing-ParentStages 根据 finalStage 找父 Stage，如果有父 Stage，就直接返回；如果没有父 Stage，就进行创建。

handleJobSubmitted 方法中的 submitStage 比较重要。submitStage 的源码如下：

```
1.   private def submitStage(stage: Stage) {
2.      val jobId = activeJobForStage(stage)
3.      if (jobId.isDefined) {
4.        logDebug("submitStage(" + stage + ")")
5.        if (!waitingStages(stage) && !runningStages(stage) && !failedStages
        (stage)) {
6.          val missing = getMissingParentStages(stage).sortBy(_.id)
7.          logDebug("missing: " + missing)
8.          if (missing.isEmpty) {
9.            logInfo("Submitting " + stage + " (" + stage.rdd + "), which has
            no missing parents")
10.           submitMissingTasks(stage, jobId.get)
11.         } else {
12.           for (parent <- missing) {
13.             submitStage(parent)
14.           }
15.           waitingStages += stage
16.         }
17.       }
18.     } else {
19.       abortStage(stage, "No active job for stage " + stage.id, None)
20.     }
21.   }
```

submitStage 首先从 activeJobForStage 中获得 JobID；如果 JobID 已经定义 isDefined，那就获得即将计算的 Stage(getMissingParentStages)，然后进行升序排列。如果父 Stage 为空，那么提交 submitMissingTasks，DAGScheduler 把处理的过程交给具体的 TaskScheduler 去处理。如果父 Stage 不为空，将循环递归调用 submitStage(parent)，从后往前回溯。后面的 Stage 依赖于前面的 Stage。也就是说，只有前面依赖的 Stage 计算完毕后，后面的 Stage 才会运行。submitStage 一直循环调用，导致的结果是父 Stage 的父 Stage……一直回溯到最左侧的父 Stage 开始计算。

4.2.5　Stage 内部 Task 获取最佳位置的算法

Task 任务本地性算法实现：DAGScheudler 的 submitMissingTasks 方法中体现了如何利用 RDD 的本地性得到 Task 的本地性，从而获取 Stage 内部 Task 的最佳位置。接下来看一下 submitMissingTasks 的源码，关注 Stage 本身的算法以及任务本地性。runningStages 是一个 HashSet[Stage]数据结构，表示正在运行的 Stages，将当前运行的 Stage 增加到 runningStages 中，根据 Stage 进行判断，如果是 ShuffleMapStage，则从 getPreferredLocs(stage.rdd, id)获取任务本地性信息；如果是 ResultStage，则从 getPreferredLocs(stage.rdd, p)获取任务本地性信息。

DAGScheduler.scala 的源码如下：

```
1.    private def submitMissingTasks(stage: Stage, jobId: Int) {
2.     ......
3.      runningStages += stage
4.     ......
5.      stage match {
6.        case s: ShuffleMapStage =>
7.          outputCommitCoordinator.stageStart(stage = s.id, maxPartitionId =
          s.numPartitions - 1)
8.        case s: ResultStage =>
9.          outputCommitCoordinator.stageStart(
10.           stage = s.id, maxPartitionId = s.rdd.partitions.length - 1)
11.     }
12.
```

在 submitMissingTasks 中会通过调用以下代码来获得任务的本地性。

```
1.      val taskIdToLocations: Map[Int, Seq[TaskLocation]] = try {
2.        stage match {
3.          case s: ShuffleMapStage =>
4.            partitionsToCompute.map { id => (id, getPreferredLocs(stage.rdd,
            id))}.toMap
5.          case s: ResultStage =>
6.            partitionsToCompute.map { id =>
7.              val p = s.partitions(id)
8.              (id, getPreferredLocs(stage.rdd, p))
9.            }.toMap
10.       }
```

partitionsToCompute 获得要计算的 Partitions 的 id。

```
1.      val partitionsToCompute: Seq[Int] = stage.findMissingPartitions()
```

如果 stage 是 ShuffleMapStage，在代码 partitionsToCompute.map { id => (id, getPreferredLocs (stage.rdd, id))}.toMap 中，id 是 partitions 的 id，使用匿名函数生成一个 Tuple，第一个元素值是数据分片的 id，第二个元素是把 rdd 和 id 传进去，获取位置 getPreferredLocs。然后通过 toMap 转换，返回 Map[Int, Seq[TaskLocation]]。第一个值是 partitions 的 id，第二个值是 TaskLocation。

具体一个 Partition 中的数据本地性的算法实现在下述 getPreferredLocs 代码中。

```
1.    private[spark]
2.    def getPreferredLocs(rdd: RDD[_], partition: Int): Seq[TaskLocation] = {
3.      getPreferredLocsInternal(rdd, partition, new HashSet)
4.    }
```

getPreferredLocsInternal 是 getPreferredLocs 的递归实现：这个方法是线程安全的，它只能被 DAGScheduler 通过线程安全方法 getCacheLocs()使用。

getPreferredLocsInternal 的源码如下：

```
1.    private def getPreferredLocsInternal(
2.        rdd: RDD[_],
3.        partition: Int,
4.        visited: HashSet[(RDD[_], Int)]): Seq[TaskLocation] = {
5.      //如果分区已被访问，则无须重新访问。这避免了路径探索。SPARK-695
6.      if (!visited.add((rdd, partition))) {
7.        //已访问的分区返回零
8.        return Nil
9.      }
```

```
10.     //如果分区已经缓存，返回缓存的位置
11.     val cached = getCacheLocs(rdd)(partition)
12.     if (cached.nonEmpty) {
13.       return cached
14.     }
15.     //如果 RDD 位置优先（输入 RDDs 的情况），就获取它
16.     val rddPrefs = rdd.preferredLocations(rdd.partitions(partition))
        .toList
17.     if (rddPrefs.nonEmpty) {
18.       return rddPrefs.map(TaskLocation(_))
19.     }
20.
21.     //如果 RDD 是窄依赖，将选择第一个窄依赖的第一分区作为位置首选项。理想情况下，我们
        //将基于传输大小选择
22.     rdd.dependencies.foreach {
23.       case n: NarrowDependency[_] =>
24.         for (inPart <- n.getParents(partition)) {
25.           val locs = getPreferredLocsInternal(n.rdd, inPart, visited)
26.           if (locs != Nil) {
27.             return locs
28.           }
29.         }
30.
31.       case _ =>
32.     }
33.
34.     Nil
35.   }
```

getPreferredLocsInternal 代码中：

在 visited 中把当前的 rdd 和 partition 加进去，visited 是一个 HashSet，如果已经存在了就会出错。

如果 partition 被缓存（partition 被缓存是指数据已经在 DAGScheduler 中），则在 getCacheLocs(rdd)(partition)传入 rdd 和 partition，获取缓存的位置信息。如果获取到缓存位置信息，就返回。

getCacheLocs 的源码如下：

```
1.    private[scheduler]
2.    def getCacheLocs(rdd: RDD[_]): IndexedSeq[Seq[TaskLocation]] =
      cacheLocs.synchronized {
3.      //注意：这个不用 getOrElse()，因为方法被调用 0(任务数)次
4.      if (!cacheLocs.contains(rdd.id)) {
5.        //注：如果存储级别为 NONE，我们不需要从块管理器获取位置信息
6.        val locs: IndexedSeq[Seq[TaskLocation]] = if (rdd.getStorageLevel
          == StorageLevel.NONE) {
7.          IndexedSeq.fill(rdd.partitions.length)(Nil)
8.        } else {
9.          val blockIds =
10.           rdd.partitions.indices.map(index => RDDBlockId(rdd.id,
              index)).toArray[BlockId]
11.         blockManagerMaster.getLocations(blockIds).map { bms =>
12.           bms.map(bm => TaskLocation(bm.host, bm.executorId))
13.         }
14.       }
```

```
15.      cacheLocs(rdd.id) = locs
16.    }
17.    cacheLocs(rdd.id)
18.  }
```

getCacheLocs 中的 cacheLocs 是一个 HashMap，包含每个 RDD 的分区上的缓存位置信息。map 的 key 值是 RDD 的 ID，Value 是由分区编号索引的数组。每个数组值是 RDD 分区缓存位置的集合。

```
1.    private val cacheLocs = new HashMap [Int, IndexedSeq[Seq[TaskLocation]]]
```

getPreferredLocsInternal 方法在具体算法实现的时候首先查询 DAGScheduler 的内存数据结构中是否存在当前 Partition 的数据本地性的信息，如果有，则直接返回；如果没有，首先会调用 rdd.getPreferedLocations。

如果自定义 RDD，那一定要写 getPreferedLocations，这是 RDD 的五大特征之一。例如，想让 Spark 运行在 HBase 上或者运行在一种现在还没有直接支持的数据库上面，此时开发者需要自定义 RDD。为了保证 Task 计算的数据本地性，最关键的方式是必须实现 RDD 的 getPreferedLocations。数据不动代码动，以 HBase 为例，Spark 要操作 HBase 的数据，要求 Spark 运行在 HBase 所在的集群中，HBase 是高速数据检索的引擎，数据在哪里，Spark 也需要运行在哪里。Spark 能支持各种来源的数据，核心就在于 getPreferedLocations。如果不实现 getPreferedLocations，就要从数据库或 HBase 中将数据抓过来，速度会很慢。

RDD.scala 的 getPreferedLocations 的源码如下：

```
1.    final def preferredLocations(split: Partition): Seq[String] = {
2.     checkpointRDD.map(_.getPreferredLocations(split)).getOrElse {
3.       getPreferredLocations(split)
4.     }
5.   }
```

这是 RDD 的 getPreferredLocations。

```
1.   protected def getPreferredLocations(split: Partition): Seq[String] = Nil
```

这样，数据本地性在运行前就已经完成，因为 RDD 构建的时候已经有元数据的信息。说明：本节代码基于 Spark 2.2 的源码版本。

DAGScheduler 计算数据本地性的时候巧妙地借助了 RDD 自身的 getPreferedLocations 中的数据，最大化地优化效率，因为 getPreferedLocations 中表明了每个 Partition 的数据本地性，虽然当前 Partition 可能被 persist 或者 checkpoint，但是 persist 或者 checkpoint 默认情况下肯定和 getPreferedLocations 中的 Partition 的数据本地性是一致的，所以这就极大地简化了 Task 数据本地性算法的实现和效率的优化。

4.3　TaskScheduler 解析

TaskScheduler 的核心任务是提交 TaskSet 到集群运算并汇报结果。

（1）为 TaskSet 创建和维护一个 TaskSetManager，并追踪任务的本地性以及错误信息。

（2）遇到 Straggle 任务时，会放到其他节点进行重试。

（3）向 DAGScheduler 汇报执行情况，包括在 Shuffle 输出丢失的时候报告 fetch failed 错误等信息。

4.3.1 TaskScheduler 原理剖析

DAGScheduler 将划分的一系列的 Stage（每个 Stage 封装一个 TaskSet），按照 Stage 的先后顺序依次提交给底层的 TaskScheduler 去执行。接下来我们分析 TaskScheduler 接收到 DAGScheduler 的 Stage 任务后，是如何来管理 Stage（TaskSet）的生命周期的。

首先，回顾一下 DAGScheduler 在 SparkContext 中实例化的时候，TaskScheduler 以及 SchedulerBackend 就已经先在 SparkContext 的 createTaskScheduler 创建出实例对象了。

虽然 Spark 支持多种部署模式（包括 Local、Standalone、YARN、Mesos 等），但是底层调度器 TaskScheduler 接口的实现类都是 TaskSchedulerImpl。并且，考虑到方便读者对 TaskScheduler 的理解，对于 SchedulerBackend 的实现，我们也只专注 Standalone 部署模式下的具体实现 StandaloneSchedulerBackend 来作分析。

TaskSchedulerImpl 在 createTaskScheduler 方法中实例化后，就立即调用自己的 initialize 方法把 StandaloneSchedulerBackend 的实例对象传进来，从而赋值给 TaskSchedulerImpl 的 backend。在 TaskSchedulerImpl 的 initialize 方法中，根据调度模式的配置创建实现了 SchedulerBuilder 接口的相应的实例对象，并且创建的对象会立即调用 buildPools 创建相应数量的 Pool 存放和管理 TaskSetManager 的实例对象。实现 SchedulerBuilder 接口的具体类都是 SchedulerBuilder 的内部类。

（1）FIFOSchedulableBuilder：调度模式是 SchedulingMode.FIFO，使用先进先出策略调度。先进先出（FIFO）为默认模式。在该模式下只有一个 TaskSetManager 池。

（2）FairSchedulableBuilder：调度模式是 SchedulingMode.FAIR，使用公平策略调度。

在 createTaskScheduler 方法返回后，TaskSchedulerImpl 通过 DAGScheduler 的实例化过程设置 DAGScheduler 的实例对象。然后调用自己的 start 方法。在 TaskSchedulerImpl 调用 start 方法的时候，会调用 StandaloneSchedulerBackend 的 start 方法，在 StandaloneSchedulerBackend 的 start 方法中，会最终注册应用程序 AppClient。TaskSchedulerImpl 的 start 方法中还会根据配置判断是否周期性地检查任务的推测执行。

TaskSchedulerImpl 启动后，就可以接收 DAGScheduler 的 submitMissingTasks 方法提交过来的 TaskSet 进行进一步处理。TaskSchedulerImpl 在 submitTasks 中初始化一个 TaskSetManager，对其生命周期进行管理，当 TaskSchedulerImpl 得到 Worker 节点上的 Executor 计算资源的时候，会通过 TaskSetManager 发送具体的 Task 到 Executor 上执行计算。

如果 Task 执行过程中有错误导致失败，会调用 TaskSetManager 来处理 Task 失败的情况，进而通知 DAGScheduler 结束当前的 Task。TaskSetManager 会将失败的 Task 再次添加到待执行 Task 队列中。Spark Task 允许失败的次数默认是 4 次，在 TaskSchedulerImpl 初始化的时候，通过 spark.task.maxFailures 设置该值。

如果 Task 执行完毕，执行的结果会反馈给 TaskSetManager，由 TaskSetManager 通知 DAGScheduler，DAGScheduler 根据是否还存在待执行的 Stage，继续迭代提交对应的 TaskSet 给 TaskScheduler 去执行，或者输出 Job 的结果。

通过下面的调度链，Executor 把 Task 执行的结果返回给调度器（Scheduler）。

（1）Executor.run。

（2）CoarseGrainedExecutorBackend.statusUpdate（发送 StatusUpdate 消息）。

（3）CoarseGrainedSchedulerBackend.receive（处理 StatusUpdate 消息）。

（4）TaskSchedulerImpl.statusUpdate。

（5）TaskResultGetter.enqueueSuccessfulTask 或者 enqueueFailedTask。

（6）TaskSchedulerImpl.handleSuccessfulTask 或者 handleFailedTask。

（7）TaskSetManager.handleSuccessfulTask 或者 handleFailedTask。

（8）DAGScheduler.taskEnded。

（9）DAGScheduler.handleTaskCompletion。

在上面的调度链中值得关注的是：第（7）步中，TaskSetManager 的 handleFailedTask 方法会将失败的 Task 再次添加到待执行 Task 队列中。在第（6）步中，TaskSchedulerImpl 的 handleFailedTask 方法在 TaskSetManager 的 handleFailedTask 方法返回后，会调用 CoarseGrainedSchedulerBackend 的 reviveOffers 方法给重新执行的 Task 获取资源。

4.3.2 TaskScheduler 源码解析

TaskScheduler 是 Spark 的底层调度器。底层调度器负责 Task 本身的调度运行。

下面编写一个简单的测试代码，setMaster("local-cluster[1, 1, 1024]")设置为 Spark 本地伪分布式开发模式，从代码的运行日志中观察 Spark 框架的运行情况。

```
1.   object SparkTest {
2.    def main(args: Array[String]): Unit = {
3.      Logger.getLogger("org").setLevel(Level.ALL)
4.      val conf = new SparkConf()                    //创建 SparkConf 对象
5.      conf.setAppName("Wow,My First Spark App!") //设置应用程序的名称，在程序
                                                  //运行的监控界面中可以看到名称
6.      conf.setMaster("local-cluster[1, 1, 1024]")
7.      conf.setSparkHome(System.getenv("SPARK_HOME"))
8.      val sc = new SparkContext(conf)
                               //创建 SparkContext 对象，通过传入 SparkConf
                               //实例来定制 Spark 运行的具体参数和配置信息
9.      sc.parallelize(Array("100","200"),1).count()
10.     sc.stop()
11.   }
12. }
```

在 IDEA 中运行代码，运行结果中打印的日志如下：

```
1.  Using Spark's default log4j profile: org/apache/spark/log4j-defaults.properties
2.  19/05/19 22:12:52 INFO SparkContext: Running Spark version 2.4.3
3.  ......
4.  19/05/19 22:13:00 INFO WorkerWebUI: Bound WorkerWebUI to 0.0.0.0, and
    started at http://windows10.microdone.cn:59745
5.  19/05/19 22:13:00 INFO Worker: Connecting to master 192.168.189.100:59722...
6.  ......
7.  19/05/19 22:13:01 INFO StandaloneAppClient$ClientEndpoint: Connecting to
    master spark://192.168.189.100:59722...
8.  ......
9.  19/05/19 22:13:01 INFO TransportClientFactory: Successfully created
```

```
    connection to /192.168.189.100:59722 after 127 ms (0 ms spent in bootstraps)
10. 19/05/19 22:13:01 DEBUG TransportClientFactory: Connection to /192.168.
    189.100:59722 successful, running bootstraps...
11. 19/05/19 22:13:01 INFO TransportClientFactory: Successfully created
    connection to /192.168.189.100:59722 after 33 ms (0 ms spent in bootstraps)
12. ......
13. 19/05/19 22:13:01 INFO Master: Registering worker 192.168.189.100:59744
    with 1 cores, 1024.0 MB RAM
14. ......
15. 19/05/19 22:13:01 INFO Worker: Successfully registered with master
    spark://192.168.189.100:59722
16. ......
17. 19/05/19 22:13:01 INFO Master: Registering app Wow,My First Spark App!
18. 19/05/19 22:13:01 INFO Master: Registered app Wow,My First Spark App! with
    ID app-20190519221301-0000
19. ......
20. 19/05/19 22:13:01 INFO StandaloneAppClient$ClientEndpoint: Executor added:
    app-20190519221301-0000/0 on worker-20190519221300-192.168.189.100-59744
    (192.168.189.100:59744) with 1 core(s)
21. ......
22. 19/05/19 22:13:01 INFO StandaloneSchedulerBackend: SchedulerBackend is
    ready for scheduling beginning after reached minRegisteredResourcesRatio:
    0.0
23. ......

24. 19/05/19 22:13:01 INFO StandaloneAppClient$ClientEndpoint: Executor updated:
    app-20190519221301-0000/0 is now RUNNING
```

日志中显示：StandaloneAppClient$ClientEndpoint: Connecting to master spark://192.168.189.100: 59722，表明 StandaloneAppClient 的 ClientEndpoint 注册给 master。日志中显示 StandaloneAppClient$ClientEndpoint: Executor added 获取了 Executor。具体是通过 Standalone-AppClient 的 ClientEndpoint 来管理 Executor。日志中显示 StandaloneSchedulerBackend: SchedulerBackend is ready for scheduling beginning after reached minRegisteredResourcesRatio: 0.0，说明 Standalone- SchedulerBackend 已经准备好。

这里是在 IDEA 本地伪分布式运行的（通过 count 的 action 算子启动了 Job）。如果是通过 Spark-shell 运行程序来观察日志，当启动 Spark-shell 本身的时候，命令终端反馈回来的主要是 ClientEndpoint 和 StandaloneSchedulerBackend，因为此时还没有任何 Job 的触发，这是启动 Application 本身而已，所以主要就是实例化 SparkContext，并注册当前的应用程序给 Master，且从集群中获得 ExecutorBackend 计算资源。

IDEA 本地伪分布式运行，Job 启动的日志如下：

```
1.  19/05/19 22:13:02 INFO DAGScheduler: Got job 0 (count at SparkTest.scala:23)
    with 1 output partitions
2.  19/05/19 22:13:02 INFO DAGScheduler: Final stage: ResultStage 0 (count
    at SparkTest.scala:23)
3.  19/05/19 22:13:02 INFO DAGScheduler: Parents of final stage: List()
4.  19/05/19 22:13:02 INFO DAGScheduler: Missing parents: List()
5.  19/05/19 22:13:02 DEBUG DAGScheduler: submitStage(ResultStage 0)
6.  19/05/19 22:13:02 DEBUG DAGScheduler: missing: List()
7.  19/05/19 22:13:02 INFO DAGScheduler: Submitting ResultStage 0
    (ParallelCollectionRDD[0] at parallelize at SparkTest.scala:23), which
    has no missing parents
8.  19/05/19 22:13:02 DEBUG DAGScheduler: submitMissingTasks(ResultStage 0)
```

```
9.  ......
10. 19/05/19 22:13:03 INFO DAGScheduler: Submitting 1 missing tasks from
    ResultStage 0 (ParallelCollectionRDD[0] at parallelize at SparkTest
    .scala:23) (first 15 tasks are for partitions Vector(0))
```

count 是 action 算子触发了 Job；然后 DAGScheduler 获取 Final stage: ResultStage，提交 Submitting ResultStage。最后提交任务给 TaskSetManager，启动任务。任务完成后，DAGScheduler 完成 Job。

DAGScheduler 划分好 Stage 后，会通过 TaskSchedulerImpl 中的 TaskSetManager 来管理当前要运行的 Stage 中的所有任务 TaskSet。TaskSetManager 会根据 locality aware 来为 Task 分配计算资源、监控 Task 的执行状态（如重试、慢任务进行推测式执行等）。

TaskSet 是一个数据结构，TaskSet 包含了一系列高层调度器交给底层调度器的任务的集合。第一个成员是 Tasks，第二个成员 task 属于哪个 Stage，stageAttemptId 是尝试的 ID，priority 优先级，调度的时候有一个调度池，调度归并调度的优先级。

```
1.   private[spark] class TaskSet(
2.      val tasks: Array[Task[_]],
3.      val stageId: Int,
4.      val stageAttemptId: Int,
5.      val priority: Int,
6.      val properties: Properties) {
7.    val id: String = stageId + "." + stageAttemptId
8.
9.    override def toString: String = "TaskSet " + id
10. }
```

TaskSetManager 实例化的时候完成 TaskSchedulerImpl 的工作，接收 TaskSet 任务的集合，maxTaskFailures 是任务失败重试的次数。

TaskSetManager.scala 的源码如下：

```
1.  private[spark] class TaskSetManager(
2.      sched: TaskSchedulerImpl,
3.      val taskSet: TaskSet,
4.      val maxTaskFailures: Int,
5.      blacklistTracker: Option[BlacklistTracker] = None,
6.      clock: Clock = new SystemClock()) extends Schedulable with Logging {
......
```

TaskSetManager 的 blacklistTracker 的成员变量，用于黑名单列表 executors 及 nodes 的跟踪。

TaskScheduler 与 SchedulerBackend 总体的底层任务调度的过程如下。

（1）TaskSchedulerImpl.submitTasks：主要作用是将 TaskSet 加入到 TaskSetManager 中进行管理。

DAGScheduler.scala 收到 JobSubmitted 消息，调用 handleJobSubmitted 方法。

```
1.      private def doOnReceive(event: DAGSchedulerEvent): Unit = event match {
2.      case JobSubmitted(jobId, rdd, func, partitions, callSite, listener,
        properties) =>
3.        dagScheduler.handleJobSubmitted(jobId, rdd, func, partitions,
          callSite, listener, properties)
```

在 handleJobSubmitted 方法中提交 submitStage。

```
1.    private[scheduler] def handleJobSubmitted(jobId: Int,
2.
3.    ......
4.      submitStage(finalStage)
5.    }
```

submitStage 方法调用 submitMissingTasks 提交 task。

```
1.    private def submitStage(stage: Stage) {
2.    ......
3.            submitMissingTasks(stage, jobId.get)
4.      ......
```

DAGScheduler.scala 的 submitMissingTasks 里面调用了 taskScheduler.submitTasks。

Spark 2.2.1 版本的 DAGScheduler.scala 的源码如下：

```
1.  private def submitMissingTasks(stage: Stage, jobId: Int) {
2.    ......
3.  val tasks: Seq[Task[_]] = try {
4.        stage match {
5.          case stage: ShuffleMapStage =>
6.            ......
7.            new ShuffleMapTask(stage.id, stage.latestInfo.attemptId,
8.            ......
9.          case stage: ResultStage =>
10.           ......
11.           new ResultTask(stage.id, stage.latestInfo.attemptId,
12.           ......
13.     if (tasks.size > 0) {
14.       logInfo(s"Submitting ${tasks.size} missing tasks from $stage
15.       (${stage.rdd}) (first 15 " +
16.       s"tasks are for partitions ${tasks.take(15).map(_.partitionId)})")
17.        taskScheduler.submitTasks(new TaskSet(
18.          tasks.toArray,  stage.id,  stage.latestInfo.attemptId,  jobId,
           properties))
19.        stage.latestInfo.submissionTime = Some(clock.getTimeMillis())
20.    ......
```

Spark 2.4.3 版本的 DAGScheduler.scala 的源码与 Spark 2.2.1 版本相比具有如下特点。

- 上段代码中第 7 行、第 11 行、第 18 行的 stage.latestInfo.attemptId 调整为 stage.latestInfo.attemptNumber，从 Spark 2.3.0 版本以后，将 attemptId 改为使用 attemptNumber。
- 上段代码中第 19 行删除。

```
1.  ......
2.  new ShuffleMapTask(stage.id, stage.latestInfo.attemptNumber,
3.  ......
4.  new ResultTask(stage.id, stage.latestInfo.attemptNumber,
5.  ......
6.    tasks.toArray, stage.id, stage.latestInfo.attemptNumber, jobId, properties))
7.  ......
```

taskScheduler 是一个接口 trait，这里没有具体的实现。

```
1.    //提交要运行的任务序列
2.    def submitTasks(taskSet: TaskSet): Unit
```

taskScheduler 的子类是 TaskSchedulerImpl，TaskSchedulerImpl 中 submitTasks 的具体实现如下。

Spark 2.2.1 版本的 TaskSchedulerImpl.scala 的 submitTasks 的源码。

```
1.    override def submitTasks(taskSet: TaskSet) {
2.     val tasks = taskSet.tasks
3.     logInfo("Adding task set " + taskSet.id + " with " + tasks.length +
       " tasks")
4.     this.synchronized {
5.       val manager = createTaskSetManager(taskSet, maxTaskFailures)
6.       val stage = taskSet.stageId
7.       val stageTaskSets =
8.         taskSetsByStageIdAndAttempt.getOrElseUpdate(stage, new HashMap
         [Int, TaskSetManager])
9.       stageTaskSets(taskSet.stageAttemptId) = manager
10.      val conflictingTaskSet = stageTaskSets.exists { case (_, ts) =>
11.        ts.taskSet != taskSet && !ts.isZombie
12.      }
13.      if (conflictingTaskSet) {
14.        throw new IllegalStateException(s"more than one active taskSet for
         stage $stage:" +
15.          s" ${stageTaskSets.toSeq.map{_._2.taskSet.id}.mkString(",")}")
16.      }
17.      schedulableBuilder.addTaskSetManager(manager, manager.taskSet
       .properties)
18.
19.      if (!isLocal && !hasReceivedTask) {
20.        starvationTimer.scheduleAtFixedRate(new TimerTask() {
21.          override def run() {
22.            if (!hasLaunchedTask) {
23.              logWarning("Initial job has not accepted any resources; " +
24.                "check your cluster UI to ensure that workers are registered " +
25.                "and have sufficient resources")
26.            } else {
27.              this.cancel()
28.            }
29.          }
30.        }, STARVATION_TIMEOUT_MS, STARVATION_TIMEOUT_MS)
31.      }
32.      hasReceivedTask = true
33.    }
34.    backend.reviveOffers()
35.  }
```

Spark 2.4.3 版本的 TaskSchedulerImpl.scala 的源码与 Spark 2.2.1 版本相比具有如下特点。

- 上段代码中第 8 行之后新增任务集管理器 TaskSetManagers，检查任务集是否为 Zombie 的代码。
- 删掉上段代码中第 10～16 行的代码。

```
8.  ……
9.   /**
10. *在添加新任务时，将此阶段的所有现有任务集管理器 TaskSetManagers 标记为 zombi，这是
    处理极端情况所必需的。假设一个 stage 有 10 个分区，2 个分区任务集管理器
    TaskSetManagers：TSM1(zombie)和 TSM2(active)。TSM1 的 10 个分区的任务都运行完
    成了。TSM2 完成了分区 1-9 的任务，认为它仍然活跃，因为分区 10 还没有完成。但是，
    DAGScheduler 获取到所有 10 个分区完成任务的事件，并认为该阶段已完成。如果是洗牌阶段
```

```
    shuffle 不知何故缺少映射输出，DAGScheduler 将重新提交它并创建一个 TSM3 。由于一个
    阶段不能有多个活动任务集管理器，因此必须标记 TSM2 是 zombie（实际上是）。
11. */
12.   stageTaskSets.foreach { case (_, ts) =>
13.         ts.isZombie = true
14.       }
```

高层调度器把任务集合传给了 TaskSet，任务可能是 ShuffleMapTask，也可能是 ResultTask。获得 taskSet.tasks 任务赋值给变量 tasks。然后使用了同步块 synchronized，在同步块中调用 createTaskSetManager，创建 createTaskSetManager。createTaskSetManager 代码如下。

TaskSchedulerImpl.scala 的源码如下：

```
1.    private[scheduler] def createTaskSetManager(
2.        taskSet: TaskSet,
3.        maxTaskFailures: Int): TaskSetManager = {
4.      new TaskSetManager(this, taskSet, maxTaskFailures, blacklistTrackerOpt)
5.    }
```

TaskSchedulerImpl.scala 的 createTaskSetManager 会调用 new()函数创建一个 TaskSetManager，传进来的 this 是其本身 TaskSchedulerImpl、任务集 taskSet、最大失败重试次数 maxTaskFailures。maxTaskFailures 是在构建 TaskSchedulerImpl 时传入的。

而 TaskSchedulerImpl 是在 SparkContext 中创建的。SparkContext 的源码如下：

```
1.          val (sched, ts) = SparkContext.createTaskScheduler(this, master,
            deployMode)
2.      _schedulerBackend = sched
3.      _taskScheduler = ts
4.      _dagScheduler = new DAGScheduler(this)
5.      _heartbeatReceiver.ask[Boolean](TaskSchedulerIsSet)
6.  ......
7.  private def createTaskScheduler(
8.  ......
9.   case SPARK_REGEX(sparkUrl) =>
10.         val scheduler = new TaskSchedulerImpl(sc)
11.         val masterUrls = sparkUrl.split(",").map("spark://" + _)
12.         val  backend  =  new  StandaloneSchedulerBackend(scheduler,  sc,
            masterUrls)
13.         scheduler.initialize(backend)
14.         (backend, scheduler)
```

在 SparkContext.scala 中，通过 createTaskScheduler 创建 taskScheduler，而在 createTaskScheduler 方法中，模式匹配到 Standalone 的模式，用 new 函数创建一个 TaskSchedulerImpl。

TaskSchedulerImpl 的构造方法将获取配置文件中的 config.MAX_TASK_FAILURES，MAX_TASK_FAILURES 默认的最大失败重试次数是 4 次。

回到 TaskSchedulerImpl，createTaskSetManager 创建了 TaskSetManager 后，非常关键的一行代码是 schedulableBuilder.addTaskSetManager(manager, manager.taskSet.properties)。

（2）SchedulableBuilder.addTaskSetManager：SchedulableBuilder 会确定 TaskSetManager 的调度顺序，然后按照 TaskSetManager 的 locality aware 来确定每个 Task 具体运行在哪个 ExecutorBackend 中。

schedulableBuilder 是应用程序级别的调度器。SchedulableBuilder 是一个接口 trait，建立

调度树。buildPools：建立树节点 pools。addTaskSetManager：建立叶子节点 TaskSetManagers。

```
1.   private[spark] trait SchedulableBuilder {
2.   def rootPool: Pool
3.   def buildPools(): Unit
4.   def addTaskSetManager(manager: Schedulable, properties: Properties):
     Unit
5. }
```

schedulableBuilder 支持两种调度模式：FIFOSchedulableBuilder、FairSchedulableBuilder。FIFOSchedulableBuilder 是先进先出调度模式。FairSchedulableBuilder 是公平调度模式。调度策略可以通过 spark-env.sh 中的 spark.scheduler.mode 进行具体设置，默认是 FIFO 的方式。

回到 TaskSchedulerImpl 的 submitTasks，看一下 schedulableBuilder.addTaskSetManager 中的调度模式 schedulableBuilder。

```
1.   var schedulableBuilder: SchedulableBuilder = null
```

schedulableBuilder 是 SparkContext 中 new TaskSchedulerImpl(sc)在创建 TaskSchedulerImpl 的时候通过 scheduler.initialize(backend)的 initialize 方法对 schedulableBuilder 进行了实例化。

具体调度模式有 FIFO 和 FAIR 两种，对应的 SchedulableBuilder 也有两种，即 FIFOSchedulableBuilder、FairSchedulableBuilder。initialize 方法中的 schedulingMode 模式默认是 FIFO。

回到 TaskSchedulerImpl 的 submitTasks，schedulableBuilder.addTaskSetManager 之后，关键的一行代码是 backend.reviveOffers()。

（3）CoarseGrainedSchedulerBackend.reviveOffers：给 DriverEndpoint 发送 ReviveOffers。SchedulerBackend.scala 的 reviveOffers 方法没有具体实现。

```
1.   private[spark] trait SchedulerBackend {
2.    private val appId = "spark-application-" + System.currentTimeMillis
3.    def start(): Unit
4.    def stop(): Unit
5.    def reviveOffers(): Unit
6.    def defaultParallelism(): Int
```

CoarseGrainedSchedulerBackend 是 SchedulerBackend 的子类。CoarseGrainedScheduler-Backend 的 reviveOffers 方法如下：

```
1.    override def reviveOffers() {
2.      driverEndpoint.send(ReviveOffers)
3.    }
```

CoarseGrainedSchedulerBackend 的 reviveOffers 方法中给 DriverEndpoint 发送 ReviveOffers 消息，而 ReviveOffers 本身是一个空的 case object 对象，ReviveOffers 本身是一个空的 case object 对象，只是起到触发底层资源调度的作用，在有 Task 提交或者计算资源变动的时候，会发送 ReviveOffers 这个消息作为触发器。

```
1.      case object ReviveOffers extends CoarseGrainedClusterMessage
```

TaskScheduler 中要负责为 Task 分配计算资源：此时程序已经具备集群中的计算资源了，根据计算本地性原则确定 Task 具体要运行在哪个 ExecutorBackend 中。

driverEndpoint.send(ReviveOffers)将 ReviveOffers 消息发送给 driverEndpoint，而不是发送

给 StandaloneAppClient，因为 driverEndpoint 是程序的调度器。driverEndpoint 的 receive 方法中模式匹配到 ReviveOffers 消息，就调用 makeOffers 方法。

```
1.       override def receive: PartialFunction[Any, Unit] = {
2.        case StatusUpdate(executorId, taskId, state, data) =>
3.     ......
4.        case ReviveOffers =>
5.          makeOffers()
```

（4）在 DriverEndpoint 接受 ReviveOffers 消息并路由到 makeOffers 具体的方法中：在 makeOffers 方法中首先准备好所有可以用于计算的 workOffers（代表了所有可用 ExecutorBackend 中可以使用的 Cores 等信息）。

Spark 2.2.1 版本的 CoarseGrainedSchedulerBackend.scala 的源码如下：

```
1.       private def makeOffers() {
2.        //确保在执行器上启动任务时没有 executor 被杀死
3.        val taskDescs = CoarseGrainedSchedulerBackend.this.synchronized {
4.          //过滤掉已被杀掉的 executors 节点
5.          val activeExecutors = executorDataMap.filterKeys(executorIsAlive)
6.          val workOffers = activeExecutors.map { case (id, executorData) =>
7.            new WorkerOffer(id, executorData.executorHost, executorData.freeCores)
8.          }.toIndexedSeq
9.          scheduler.resourceOffers(workOffers)
10.       }
11.       if (!taskDescs.isEmpty) {
12.         launchTasks(taskDescs)
13.       }
14.    }
```

Spark 2.4.3 版本的 CoarseGrainedSchedulerBackend.scala 的源码与 Spark 2.2.1 版本相比具有如下特点。

- 上段代码中第 7 行构建 WorkerOffer 实例时，新建一个地址 address 参数，address 是可选的 hostPort 字符串，在同一主机上启动多个执行器 executors 时，address 提供的信息比 host 更有用。

```
1.  new WorkerOffer(id, executorData.executorHost, executorData.freeCores,
             Some(executorData.executorAddress.hostPort))
```

其中的 executorData 类如下，包括 freeCores、totalCores 等信息。

```
1.    private[cluster] class ExecutorData(
2.     val executorEndpoint: RpcEndpointRef,
3.     val executorAddress: RpcAddress,
4.     override val executorHost: String,
5.     var freeCores: Int,
6.     override val totalCores: Int,
7.     override val logUrlMap: Map[String, String]
8.  ) extends ExecutorInfo(executorHost, totalCores, logUrlMap)
```

在 makeOffers 中首先找到可以利用的 activeExecutors，然后创建 workOffers。workOffers 是一个数据结构 case class，表示具体的 Executor 可能的资源。这里只考虑 CPU cores，不考虑内存，因为之前内存已经分配完成。

Spark 2.2.1 版本的 WorkerOffer.scala 的源码如下：

```
1.    private[spark]
2.  case class WorkerOffer(executorId: String, host: String, cores: Int)
```

Spark 2.4.3 版本的 WorkerOffer.scala 的源码与 Spark 2.2.1 版本相比具有如下特点。

- 上段代码中第 2 行 WorkerOffer 类增加一个 address 成员变量。

```
1.  private[spark]
2.  case class WorkerOffer(
3.      executorId: String,
4.      host: String,
5.      cores: Int,
6.      address: Option[String] = None)
```

makeOffers 方法中，TaskSchedulerImpl.resourceOffers 为每个 Task 具体分配计算资源，输入 offers: IndexedSeq[WorkerOffer]一维数组是可用的计算资源，ExecutorBackend 及其上可用的 Cores，输出 TaskDescription 的二维数组 Seq[Seq[TaskDescription]]定义每个任务的数据本地性及放在哪个 Executor 上执行。

TaskDescription 包括 executorId，TaskDescription 中已经确定好了 Task 具体要运行在哪个 ExecutorBackend 上。而确定 Task 具体运行在哪个 ExecutorBackend 上的算法由 TaskSetManager 的 resourceOffer 方法决定。

Spark 2.2.1 版本的 TaskDescription.scala 的源码如下：

```
1.  private[spark] class TaskDescription(
2.      val taskId: Long,
3.      val attemptNumber: Int,
4.      val executorId: String,
5.      val name: String,
6.      val index: Int,    //在该任务中的 TaskSet 的索引
7.      val addedFiles: Map[String, Long],
8.      val addedJars: Map[String, Long],
9.      val properties: Properties,
10.     val serializedTask: ByteBuffer) {
```

Spark 2.4.3 版本的 TaskDescription.scala 的源码与 Spark 2.2.1 版本相比具有如下特点。

- 上段代码中第 6 行之后新增一个成员变量 partitionId。

```
1.    ......
2.    val partitionId: Int,
```

resourceOffers 由群集管理器调用提供 slaves 的资源，根据优先级顺序排列任务，以循环的方式填充每个节点的任务，使得集群的任务运行均衡。

Spark 2.2.1 版本的 TaskSchedulerImpl.scala 的源码如下：

```
1.  def resourceOffers(offers: IndexedSeq[WorkerOffer]):
    Seq[Seq[TaskDescription]] = synchronized {
2.      //标记每一个 slave 节点活跃状态，记录主机名
3.     //如是新的 executor 节点增加，则进行跟踪
4.      var newExecAvail = false
5.      for (o <- offers) {
6.        if (!hostToExecutors.contains(o.host)) {
7.          hostToExecutors(o.host) = new HashSet[String]()
8.        }
9.        if (!executorIdToRunningTaskIds.contains(o.executorId)) {
10.         hostToExecutors(o.host) += o.executorId
```

```
11.         executorAdded(o.executorId, o.host)
12.         executorIdToHost(o.executorId) = o.host
13.         executorIdToRunningTaskIds(o.executorId) = HashSet[Long]()
14.         newExecAvail = true
15.       }
16.       for (rack <- getRackForHost(o.host)) {
17.         hostsByRack.getOrElseUpdate(rack, new HashSet[String]()) += o.host
18.       }
19.     }
20.
21.     //在进行任何 offers 之前，从黑名单中删除黑名单已过期的任何节点，这样做是为了
        //避免单独的线程和增加的同步开销，因为只有在提供任务时才更新黑名单
22.
23.
24.     blacklistTrackerOpt.foreach(_.applyBlacklistTimeout())
25.
26.     val filteredOffers = blacklistTrackerOpt.map { blacklistTracker =>
27.       offers.filter { offer =>
28.         !blacklistTracker.isNodeBlacklisted(offer.host) &&
29.           !blacklistTracker.isExecutorBlacklisted(offer.executorId)
30.       }
31.     }.getOrElse(offers)
32.
33.     val shuffledOffers = shuffleOffers(filteredOffers)
34.     //创建要分配给每个 worker 的任务列表.
35.     val tasks = shuffledOffers.map(o => new ArrayBuffer[TaskDescription](o.cores))
36.     val availableCpus = shuffledOffers.map(o => o.cores).toArray
37.     val sortedTaskSets = rootPool.getSortedTaskSetQueue
38.     for (taskSet <- sortedTaskSets) {
39.       logDebug("parentName: %s, name: %s, runningTasks: %s".format(
40.         taskSet.parent.name, taskSet.name, taskSet.runningTasks))
41.       if (newExecAvail) {
42.         taskSet.executorAdded()
43.       }
44.     }
45.
46.   //把每个 TaskSet 放在调度顺序中，然后提供它的每个节点本地性级别的递增顺序，以便它有
      //机会启动所有任务的本地任务
47.     //注意: 数据本地性优先级别顺序: PROCESS_LOCAL, NODE_LOCAL, NO_PREF, RACK_LOCAL,
        //ANY
48.
49.     for (taskSet <- sortedTaskSets) {
50.       var launchedAnyTask = false
51.       var launchedTaskAtCurrentMaxLocality = false
52.       for (currentMaxLocality <- taskSet.myLocalityLevels) {
53.         do {
54.           launchedTaskAtCurrentMaxLocality = resourceOfferSingleTaskSet(
55.             taskSet, currentMaxLocality, shuffledOffers, availableCpus, tasks)
56.           launchedAnyTask |= launchedTaskAtCurrentMaxLocality
57.         } while (launchedTaskAtCurrentMaxLocality)
58.       }
59.       if (!launchedAnyTask) {
60.         taskSet.abortIfCompletelyBlacklisted(hostToExecutors)
61.       }
62.     }
63.
64.     if (tasks.size > 0) {
65.       hasLaunchedTask = true
66.     }
```

```
67.     return tasks
68.   }
69. ……
```

Spark 2.4.3 版本的 TaskSchedulerImpl.scala 的源码与 Spark 2.2.1 版本相比具有如下特点。

- 上段代码中第 35 行将 o.cores 调整为 o.cores / CPUS_PER_TASK。
- 上段代码中第 35 行之后新增代码计算可用的槽位。
- 上段代码中第 49～68 行替换为新增的代码。

```
1.  val tasks = shuffledOffers.map(o => new ArrayBuffer[TaskDescription](o.cores /
    CPUS_PER_TASK))
2.  val availableSlots = shuffledOffers.map(o => o.cores / CPUS_PER_TASK).sum
3.  ……
4.
5.      for (taskSet <- sortedTaskSets) {
6.      //如果可用插槽少于挂起任务的数量，则跳过屏障任务集
7.      if (taskSet.isBarrier && availableSlots < taskSet.numTasks) {
8.        //跳过启动过程
9.        //spark-24819 如果作业需要的插槽多于可用插槽（繁忙和空闲槽），提交时作业失败
10.       logInfo(s"Skip current round of resource offers for barrier stage $
          {taskSet.stageId} " +
11.         s"because the barrier taskSet requires ${taskSet.numTasks} slots,
            while the total " +
12.         s"number of available slots is $availableSlots.")
13.     } else {
14.       var launchedAnyTask = false
15.       // 记录所有 executor ID 分配的障碍任务
16.       val addressesWithDescs = ArrayBuffer[(String, TaskDescription)]()
17.       for (currentMaxLocality <- taskSet.myLocalityLevels) {
18.         var launchedTaskAtCurrentMaxLocality = false
19.         do {
20.           launchedTaskAtCurrentMaxLocality = resourceOfferSingleTaskSet(taskSet,
21.             currentMaxLocality, shuffledOffers, availableCpus, tasks,
              addressesWithDescs)
22.           launchedAnyTask |= launchedTaskAtCurrentMaxLocality
23.         } while (launchedTaskAtCurrentMaxLocality)
24.       }
25.
26.       if (!launchedAnyTask) {
27.         taskSet.getCompletelyBlacklistedTaskIfAny(hostToExecutors)
            .foreach { taskIndex =>
28.   /**
29.     *如果任务集是不可计算的，将尝试查找现有的空闲黑名单 executor，如果找不到，就立即
      中止。否则就杀掉空闲的 executor，如果它没有在超时内调度任务，启动一个中止程序。
      如果无法从任务集中调度任何任务，它将中止任务集
30.    *注 1：我们跟踪每个任务集的可调度性，而不是每个任务的可调度性
31.    *注 2：当存在多个空闲黑名单 executors 并且启用动态分配时，任务集仍然可以中止。当
     一个被杀死的空闲 executor 没有及时被 ExecutorAllocationManager 替换时，就会
     发生这种情况，因为它依赖于挂起的任务，并且在空闲超时时不会杀死 executors，从而
     导致过期并中止任务集
32.    */
33.             executorIdToRunningTaskIds.find(x => !isExecutorBusy(x._1)) match {
34.               case Some ((executorId, _)) =>
35.           if (!unschedulableTaskSetToExpiryTime.contains(taskSet)) {
36.                 blacklistTrackerOpt.foreach(blt =>
```

```
    blt.killBlacklistedIdleExecutor(executorId))
37.
38.                    val timeout =
    conf.get(config.UNSCHEDULABLE_TASKSET_TIMEOUT) * 1000
39.                    unschedulableTaskSetToExpiryTime(taskSet) =
    clock.getTimeMillis() + timeout
40.                    logInfo(s"Waiting for $timeout ms for completely "
41.                      + s"blacklisted task to be schedulable again before aborting
                         $taskSet.")
42.                    abortTimer.schedule(
43.                      createUnschedulableTaskSetAbortTimer(taskSet,
                         taskIndex), timeout)
44.                  }
45.                case None => //立即中止
46.                  logInfo("Cannot schedule any task because of complete
    blacklisting. No idle" +
47.                    s" executors can be found to kill. Aborting $taskSet." )
48.                  taskSet.abortSinceCompletelyBlacklisted(taskIndex)
49.              }
50.          }
51.        } else {
52.   /**
53.   *只要有一个未列入黑名单的 executor，就要推迟杀掉任何任务集。可用于从任何活动任务集
     中调度任务。这确保了工作可以取得进展。注意：理论上，任务集可能永远不会在非黑名单
     executor 被调度，并且由于不断提交新的任务集，中止计时器不会启动
54.    */
55.          if (unschedulableTaskSetToExpiryTime.nonEmpty) {
56.            logInfo("Clearing the expiry times for all unschedulable
               taskSets as a task was " +
57.              "recently scheduled.")
58.            unschedulableTaskSetToExpiryTime.clear()
59.          }
60.        }
61.
62.        if (launchedAnyTask && taskSet.isBarrier) {
63.   /**
64.    检查屏障任务是否部分启动。 SPARK-24818 处理断言失败案例（当某些位置
65.    没有满足需求，应该恢复已启动的任务）
66.    */
67.          require(addressesWithDescs.size == taskSet.numTasks,
68.            s"Skip current round of resource offers for barrier stage
               ${taskSet.stageId} " +
69.              s"because only ${addressesWithDescs.size} out of a total
                 number of " +
70.              s"${taskSet.numTasks} tasks got resource offers. The resource
                 offers may have " +
71.              "been blacklisted or cannot fulfill task locality requirements.")
72.
73.          //实现屏障调度
74.          maybeInitBarrierCoordinator()
75.
76.          // 将 taskInfos 更新到所有屏障任务属性中
77.          val addressesStr = addressesWithDescs
78.            //按分区 ID 排序
79.            .sortBy(_._2.partitionId)
80.            .map(_._1)
81.            .mkString(",")
82.
```

```
    addressesWithDescs.foreach(_._2.properties.setProperty("addresses",
    addressesStr))
83.
84.          logInfo(s"Successfully scheduled all the ${addressesWithDescs.size}
             tasks for barrier " +
85.            s"stage ${taskSet.stageId}.")
86.        }
87.      }
88.    }
89.
90.    //spark-24823 如果屏障任务未在配置的时间内启动，则取消包含屏障阶段的作业
91.    if (tasks.size > 0) {
92.      hasLaunchedTask = true
93.    }
94.    return tasks
95.  }
```

resourceOffers 中：

- 标记每一个活着的 slave，记录它的主机名，并跟踪是否增加了新的 Executor。感知集群动态资源的状况。
- offers 是集群有哪些可用的资源，循环遍历 offers，hostToExecutors 是否包含当前的 host，如果不包含，就将 Executor 加进去。因为这里是最新请求，获取机器有哪些可用的计算资源。
- getRackForHost 是数据本地性，默认情况下，在一个机架 Rack 里面，生产环境中可能分若干个机架 Rack。
- 重要的一行代码 val shuffledOffers = shuffleOffers(filteredOffers)：将可用的计算资源打散。
- tasks 将获得洗牌后的 shuffledOffers 通过 map 转换，对每个 worker 用了 ArrayBuffer[TaskDescription]，每个 Executor 可以放几个[TaskDescription]，就可以运行多少个任务。即多少个 Cores，就可以分配多少任务。ArrayBuffer 是一个一维数组，数组的长度根据当前机器的 CPU 个数决定。

ArrayBuffer[TaskDescription](o.cores / CPUS_PER_TASK)说明当前 ExecutorBackend 上可以分配多少个 Task，并行运行多少 Task，和 RDD 的分区个数是两个概念：这里不是决定 Task 的个数，RDD 的分区数在创建 RDD 时就已经决定了。这里，具体任务调度是指 Task 分配在哪些机器上，每台机器上分配多少 Task，一次能分配多少 Task。

- 在 TaskSchedulerImpl 中的 initialize 中创建 rootPool，将 schedulingMode 调度模式传进去。rootPool 的叶子节点是 TaskSetManagers，按照一定的算法计算 Stage 的 TaskSet 调度的优先顺序。
- for 循环遍历 sortedTaskSets，如果有新的可用的 Executor，通过 taskSet.executorAdded()加入 taskSet。

TastSetManager 的 executorAdded 方法如下：

```
1.    def recomputeLocality() {
2.      val previousLocalityLevel = myLocalityLevels(currentLocalityIndex)
3.      myLocalityLevels = computeValidLocalityLevels()
4.      localityWaits = myLocalityLevels.map(getLocalityWait)
5.      currentLocalityIndex = getLocalityIndex(previousLocalityLevel)
6.    }
```

```
7.
8.    def executorAdded() {
9.      recomputeLocality()
10.   }
```

数据本地优先级从高到低依次为：PROCESS_LOCAL、NODE_LOCAL、NO_PREF、RACK_LOCAL、ANY。其中，NO_PREF 是指机器本地性，一台机器上有很多 Node，Node 的优先级高于机器本地性。

- resourceOffers 中追求最高级别的优先级本地性源码如下：

```
1.        for (taskSet <- sortedTaskSets) {
2.        var launchedAnyTask = false
3.        var launchedTaskAtCurrentMaxLocality = false
4.        for (currentMaxLocality <- taskSet.myLocalityLevels) {
5.        do {
6.           launchedTaskAtCurrentMaxLocality = resourceOfferSingleTaskSet(
7.             taskSet, currentMaxLocality, shuffledOffers, availableCpus,
             tasks)
8.           launchedAnyTask |= launchedTaskAtCurrentMaxLocality
9.         } while (launchedTaskAtCurrentMaxLocality)
10.       }
11.       if (!launchedAnyTask) {
12.         taskSet.abortIfCompletelyBlacklisted(hostToExecutors)
13.       }
14.     }
```

循环遍历 sortedTaskSets，对其中的每个 taskSet，首先考虑 myLocalityLevels 的优先性，myLocalityLevels 计算数据本地性的 Level，将 PROCESS_LOCAL、NODE_LOCAL、NO_PREF、RACK_LOCAL、ANY 循环一遍。myLocalityLevels 是通过 computeValidLocalityLevels 方法获取到的。

TaskSetManager.scala 的 computeValidLocalityLevels 的源码如下：

```
1.  private def computeValidLocalityLevels(): Array[TaskLocality.TaskLocality] = {
2.      import TaskLocality.{PROCESS_LOCAL, NODE_LOCAL, NO_PREF, RACK_LOCAL, ANY}
3.      val levels = new ArrayBuffer[TaskLocality.TaskLocality]
4.      if (!pendingTasksForExecutor.isEmpty &&
5.  pendingTasksForExecutor.keySet.exists(sched.isExecutorAlive(_))) {
6.        levels += PROCESS_LOCAL
7.      }
8.      if (!pendingTasksForHost.isEmpty &&
9.  pendingTasksForHost.keySet.exists(sched.hasExecutorsAliveOnHost(_))) {
10.       levels += NODE_LOCAL
11.     }
12.     if (!pendingTasksWithNoPrefs.isEmpty) {
13.       levels += NO_PREF
14.     }
15.     if (!pendingTasksForRack.isEmpty &&
16.         pendingTasksForRack.keySet.exists(sched.hasHostAliveOnRack(_))) {
17.       levels += RACK_LOCAL
18.     }
19.     levels += ANY
20.     logDebug("Valid locality levels for " + taskSet + ": " + levels.mkString(", "))
21.     levels.toArray
22.   }
```

Spark 2.2.1 版本的 TaskSchedulerImpl.scala 的 resourceOfferSingleTaskSet 的源码如下：

```
1.  private def resourceOfferSingleTaskSet(
2.        taskSet: TaskSetManager,
3.        maxLocality: TaskLocality,
4.        shuffledOffers: Seq[WorkerOffer],
5.        availableCpus: Array[Int],
6.        tasks: IndexedSeq[ArrayBuffer[TaskDescription]]) : Boolean = {
7.      var launchedTask = false
8.      //针对整个应用程序被列入黑名单的节点和 executors 已被此点过滤掉
9.      for (i <- 0 until shuffledOffers.size) {
10.       val execId = shuffledOffers(i).executorId
11.       val host = shuffledOffers(i).host
12.       if (availableCpus(i) >= CPUS_PER_TASK) {
13.         try {
14.           for (task <- taskSet.resourceOffer(execId, host, maxLocality)) {
15.             tasks(i) += task
16.             val tid = task.taskId
17.             taskIdToTaskSetManager(tid) = taskSet
18.             taskIdToExecutorId(tid) = execId
19.             executorIdToRunningTaskIds(execId).add(tid)
20.             availableCpus(i) -= CPUS_PER_TASK
21.             assert(availableCpus(i) >= 0)
22.             launchedTask = true
23.           }
24.         } catch {
25.           case e: TaskNotSerializableException =>
26.             logError(s"Resource offer failed, task set ${taskSet.name} was
                not serializable")
27.             //不为这个任务提供资源，但不抛出错误，以允许其他任务集提交任务
28.             return launchedTask
29.         }
30.       }
31.     }
32.     return launchedTask
33.   }
```

Spark 2.4.3 版本的 TaskSchedulerImpl.scala 的源码与 Spark 2.2.1 版本相比具有如下特点。

- 上段代码中第 6 行之后新增一个参数 addressesWithDescs: ArrayBuffer[(String, TaskDescription)])。
- 上段代码中第 21 行之后新增屏障任务主机更新的代码。

```
1.  addressesWithDescs: ArrayBuffer[(String, TaskDescription)]) : Boolean =
    {
2.  ......
3.      //只为屏障任务更新主机
4.      if (taskSet.isBarrier) {
5.      //executor 地址应为非空
6.      addressesWithDescs += (shuffledOffers(i).address.get -> task)
7.      }
```

resourceOfferSingleTaskSet 方法中的 CPUS_PER_TASK 是每个 Task 默认采用一个线程进行计算的。TaskSchedulerImpl.scala 中 CPUS_PER_TASK 的源码如下：

```
1.       //每个任务请求的 CPUs
2.    val CPUS_PER_TASK = conf.getInt("spark.task.cpus", 1)
```

resourceOfferSingleTaskSet 方法中的 taskSet.resourceOffer，通过调用 TaskSetManager 的

resourceOffer 最终确定每个 Task 具体运行在哪个 ExecutorBackend 的具体的 Locality Level。

Spark 2.2.1 版本的 TaskSetManager.scala 的源码如下：

```
1.  def resourceOffer(
2.        execId: String,
3.        host: String,
4.        maxLocality: TaskLocality.TaskLocality)
5.      : Option[TaskDescription] =
6.    {
7.  ……
8.       sched.dagScheduler.taskStarted(task, info)
9.         new TaskDescription(
10.          taskId,
11.          attemptNum,
12.          execId,
13.          taskName,
14.          index,
15.          addedFiles,
16.          addedJars,
17.          task.localProperties,
18.          serializedTask)
```

Spark 2.4.3 版本的 TaskSetManager.scala 的源码与 Spark 2.2.1 版本相比具有如下特点。

- 上段代码中第 14 行构建 TaskDescription 实例时，新增传入 task.partitionId 成员变量。

```
1.  ……
2.            task.partitionId,
3.  ……
```

以上内容都在做一件事情：获取 Locality Level 本地性的层次。DagScheduler 告诉我们任务运行在哪台机器上，DAGScheduler 是从数据层面考虑 preferedLocation 的，DAGScheduler 从 RDD 的层面确定就可以；而 TaskScheduler 是从具体计算 Task 的角度考虑计算的本地性，TaskScheduler 是更具体的底层调度。本地性的两个层面：①数据的本地性；②计算的本地性。

总结：scheduler.resourceOffers 确定了每个 Task 具体运行在哪个 ExecutorBackend 上；resourceOffers 到底是如何确定 Task 具体运行在哪个 ExecutorBackend 上的呢？

① 通过 Random.shuffle 方法重新洗牌所有的计算资源，以寻求计算的负载均衡。

② 根据每个 ExecutorBackend 的 cores 的个数声明类型为 TaskDescription 的 ArrayBuffer 数组。

③ 如果有新的 ExecutorBackend 分配给我们的 Job，此时会调用 executorAdded 来获得最新的、完整的可用计算资源。

④ 追求最高级别的优先级本地性。

⑤ 通过调用 TaskSetManager 的 resourceOffer 最终确定每个 Task 具体运行在哪个 ExecutorBackend 的具体的 Locality Level。

回到 CoarseGrainedSchedulerBackend.scala 的 launchTasks 方法。

Spark 2.2.1 版本的 CoarseGrainedSchedulerBackend.scala 的源码如下：

```
1.  private def launchTasks(tasks: Seq[Seq[TaskDescription]]) {
2.        for (task <- tasks.flatten) {
3.          val serializedTask = TaskDescription.encode(task)
4.          if (serializedTask.limit >= maxRpcMessageSize) {
5.            scheduler.taskIdToTaskSetManager.get(task.taskId).foreach
              { taskSetMgr =>
```

```
6.             try {
7.               var msg = "Serialized task %s:%d was %d bytes, which exceeds max
               allowed: " +
8.                 "spark.rpc.message.maxSize (%d bytes). Consider increasing " +
9.                 "spark.rpc.message.maxSize or using broadcast variables for
                 large values."
10.              msg = msg.format(task.taskId, task.index, serializedTask.limit,
               maxRpcMessageSize)
11.              taskSetMgr.abort(msg)
12.            } catch {
13.            case e: Exception => logError("Exception in error callback", e)
14.            }
15.          }
16.        }
17.        else {
18.          val executorData = executorDataMap(task.executorId)
19.          executorData.freeCores -= scheduler.CPUS_PER_TASK
20.
21.          logDebug(s"Launching task ${task.taskId} on executor id: ${task.
           executorId} hostname: " +
22.            s"${executorData.executorHost}.")
23.
24.          executorData.executorEndpoint.send(LaunchTask(new
           SerializableBuffer(serializedTask)))
25.        }
26.      }
27.    }
```

Spark 2.4.3 版本的 CoarseGrainedSchedulerBackend.scala 的源码与 Spark 2.2.1 版本相比具有如下特点。

- 上段代码中第 5 行调整为使用 Option 方法。

```
1.  ......
2.      Option(scheduler.taskIdToTaskSetManager.get(task.taskId)).foreach
        { taskSetMgr =>
3.  ......
```

（5）通过 launchTasks 把任务发送给 ExecutorBackend 去执行。

launchTasks 首先进行序列化，但序列化 Task 的大小不能太大，如果超过 maxRpcMessageSize，则提示出错信息。

RpcUtils.scala 中 maxRpcMessageSize 的定义，spark.rpc.message.maxSize 默认设置是 128MB。

```
1.      private val maxRpcMessageSize = RpcUtils.maxMessageSizeBytes(conf)
2.  ......
3.    def maxMessageSizeBytes(conf: SparkConf): Int = {
4.      val maxSizeInMB = conf.getInt("spark.rpc.message.maxSize", 128)
5.      if (maxSizeInMB > MAX_MESSAGE_SIZE_IN_MB) {
6.        throw new IllegalArgumentException(
7.          s"spark.rpc.message.maxSize should not be greater than $MAX_
          MESSAGE_SIZE_IN_MB MB")
8.      }
9.      maxSizeInMB * 1024 * 1024
10.   }
11. }
```

Task 进行广播时的 maxSizeInMB 大小是 128MB，如果任务大于等于 128MB，则 Task

直接被丢弃掉；如果小于 128MB，会通过 CoarseGrainedSchedulerBackend 去 launchTask 到具体的 ExecutorBackend 上。

CoarseGrainedSchedulerBackend.scala 的 launchTasks 方法：通过 executorData.executorEndpoint.send(LaunchTask(new SerializableBuffer(serializedTask)))交给 Task 要运行的 ExecutorBackend，给它发送一个消息 LaunchTask，发送序列化的 Task。

CoarseGrainedExecutorBackend 就收到了 launchTasks 消息，启动 executor.launchTask。

4.4 SchedulerBackend 解析

本节讲解 SchedulerBackend 原理剖析、SchedulerBackend 源码解析、Spark 程序的注册机制、Spark 程序对计算资源 Executor 的管理等内容。

4.4.1 SchedulerBackend 原理剖析

以 Spark Standalone 部署方式为例，StandaloneSchedulerBackend 在启动的时候构造了 StandaloneAppClient 实例，并在该实例 start 的时候启动了 ClientEndpoint 消息循环体，ClientEndpoint 在启动的时候会向 Master 注册当前程序。而 StandaloneSchedulerBackend 的父类 CoarseGrainedSchedulerBackend 在 start 的时候会实例化类型为 DriverEndPoint（这就是程序运行时的经典的对象 Driver）的消息循环体，StandaloneSchedulerBackend 专门负责收集 Worker 上的资源信息，当 ExecutorBackend 启动的时候，会发送 RegisteredExecutor 信息向 DriverEndpoint 注册，此时 StandaloneSchedulerBackend 就掌握了当前应用程序拥有的计算资源，TaskScheduler 就是通过 StandaloneSchedulerBackend 拥有的计算资源来具体运行 Task 的。

4.4.2 SchedulerBackend 源码解析

StandaloneSchedulerBackend 收集和分配资源给调度的 Task 使用。

StandaloneSchedulerBackend.scala 的源码如下：

```
1.  private[spark] class StandaloneSchedulerBackend(
2.  ......
3.   override def start() {
4.  ......
5.   val command = Command("org.apache.spark.executor.
     .CoarseGrainedExecutorBackend",
6.       args, sc.executorEnvs, classPathEntries ++ testingClassPath,
         libraryPathEntries, javaOpts)
7.  ......
8.    client = new StandaloneAppClient(sc.env.rpcEnv, masters, appDesc, this,
      conf)
9.      client.start()
10. ......
```

在 StandaloneAppClient 的 start 方法中调用 new()函数创建一个 ClientEndpoint，将在 ClientEndpoint 中向 Master 注册。

StandaloneAppClient.scala 的源码如下：

```
1.  ......
2.    def start() {
3.      //启动 rpcEndpoint；将直接回调到 listener
4.      endpoint.set(rpcEnv.setupEndpoint("AppClient",  new  ClientEndpoint
        (rpcEnv)))
5.    }
```

4.4.3 Spark 程序的注册机制

在上面的源码分析中，StandaloneAppClient 在启动的时候创建了 StandaloneAppClient 内部类 ClientEndpoint 的实例对象作为消息循环体，以便向 Master 注册当前的 Application。既然 ClientEndpoint 是 RpcEndpoint 的子类，那么就会有这样的生命周期：constructor -> onStart -> receive ->onStop。根据这个原理，我们来看 ClientEndpoint 的 onStart 方法代码。

StandaloneAppClient.scala 的源码如下：

```
1.      override def onStart(): Unit = {
2.        try {
3.          registerWithMaster(1)
4.        } catch {
5.       ......
```

ClientEndpoint 在启动时就立即调用 registerWithMaster 来注册 Application，继续查看 registerWithMaster 方法代码。

StandaloneAppClient.scala 的源码如下：

```
1.    private def registerWithMaster(nthRetry: Int) {
2.  //向所有 Master 异步地尝试注册 Application
3.      registerMasterFutures.set(tryRegisterAllMasters())
4.     ......
5.     }
```

ClientEndpoint 在 tryRegiesterAllMasters 方法中会向所有的 Master 尝试注册 Application。向 Master 发送 RegisterApplication 消息。

StandaloneAppClient.scala 的源码如下：

```
1.      private def tryRegisterAllMasters(): Array[JFuture[_]] = {
2.    ......
3.      val masterRef = rpcEnv.setupEndpointRef(masterAddress, Master
        .ENDPOINT_NAME)
4.            masterRef.send(RegisterApplication(appDescription, self))
5.    ......
```

Master 也是 RpcEndpoint 的子类，所以可以通过 receive 方法接收 DeployMessage 类型的消息 RegisterApplication。

Master.scala 的源码如下：

```
1.       override def receive: PartialFunction[Any, Unit] = {
2.     ......
3.   case RegisterApplication(description, driver) =>
4.     ......
5.       registerApplication(app)
```

```
6.        ......
7.          driver.send(RegisteredApplication(app.id, self))
8.     ......
```

ClientEndpoint 最后在 receive 方法中得到来自 Master 注册好 Application 的确认消息 RegisteredApplication。

StandaloneAppClient.scala 的源码如下：

```
1.      override def receive: PartialFunction[Any, Unit] = {
2.        case RegisteredApplication(appId_, masterRef) =>
3.     ......
4.          appId.set(appId_)
5.          registered.set(true)
6.      ......
```

至此，Application 向 Master 注册完毕。在上面的 RegisterApplication 中，调用了 schedule 方法，这个方法将完成 Application 的调度，并在 Worker 节点上启动分配好的 Executor 给 Application 使用。

4.4.4　Spark 程序对计算资源 Executor 的管理

从 TaskSchedulerImpl 的 submitTasks 的方法中我们知道，Spark Standalone 部署模式调用 StandaloneSchedulerBackend 的 reviveOffers 方法进行 TaskSet 所需要资源的分配，得到足够的资源后，将 TaskSet 中的 Task 逐个发送到 Executor 去执行。下面来看这里的资源，即 Executor 是如何得到和分配的。

StandaloneSchedulerBackend 的 reviveOffers 方法很简单，就是发送一个 ReviveOffers 消息给内部类 DriverEndpoint，代码如下所示。

CoarseGrainedSchedulerBackend.scala 的源码如下：

```
1.    override def reviveOffers() {
2.      driverEndpoint.send(ReviveOffers)
3.    }
```

DriverEndpoint 的 receive 方法处理 ReviveOffers 消息也很简单，就是调用 makeOffers 方法。receive 方法部分关键代码如下所示。

CoarseGrainedSchedulerBackend.scala 的源码如下：

```
1.    override def receive: PartialFunction[Any, Unit] = {
2.   ......
3.   case ReviveOffers =>
4.         makeOffers()
5.   ......
```

DriverEndpoint 的 makeOffers 方法首先过滤出 Alive 状态的 Executor 放到 activeExecutorsHahMap 变量中，然后使用 id、ExecutorData.ExecutorHost、ExecutorData.freeCores 构建代表 Executor 可用资源的 WorkerOffer。然后是最重要的两个方法调用。先是调用 TaskSchedulerImpl 的 resourceOffers 得到 TaskDescription 的二维数组，包含 Task ID、Executor ID、Task Index 等 Task 执行需要的信息。然后回调 DriverEndpoint 的 launchTask 给每个 Task 对应的 Executor 发执行 Task 的 LaunchTask 消息（其实是由 CourseGrainedExecutorBackend 转发 LauchTask 消息）。

TaskSchedulerImpl 的 resourceOffers 方法返回二维数组 TaskDescription 后作为 DriverEndpoint 的 launchTasks 方法的参数，DriverEndpoint 的 launchTasks 方法中首先对传入的 tasks 进行扁平化操作（例如，将多维数组降维成一维数组），得到所有的 Task，然后遍历所有的 Task。在遍历过程中，调用 serialize()方法对 task 进行序列化，得到 serializedTask。判断如果 serializedTask 大于等于 Akka 帧减去 Akka 预留空间大小，则调用 TaskSetManager 的 abort 方法终止该任务的执行，否则将 LaunchTask(new SerializableBuffer(serializedTask))消息发送到 CoarseGrainedExecutorBackend。

CoarseGrainedExecutorBackend 匹配到 LaunchTask(data)消息后，首先调用 deserialized 方法，反序列化出 task，然后调用 Executor 的 lauchTask 方法执行 Task 的处理。

4.5 打通 Spark 系统运行内幕机制循环流程

Spark 通过 DAGScheduler 面向整个 Job 划分出了不同的 Stage，划分 Stage 之后，Stage 从后往前划分，执行的时候从前往后执行，每个 Stage 内部有一系列的任务，Stage 里面的任务是并行计算，并行任务的逻辑是完全相同的，但处理的数据不同。DAGScheduler 以 TaskSet 的方式，把一个 DAG 构建的 Stage 中的所有任务提交给底层的调度器 TaskScheduler。TaskScheduler 是一个接口，与具体的任务解耦合，可以运行在不同的调度模式下，如可运行在 Standalone 模式，也可运行在 Yarn 上。

Spark 基础调度（见图 4-6）包括 RDD Objects、DAGScheduler、TaskScheduler、Worker 等内容。

DAGScheduler 在提交 TaskSet 给底层调度器的时候是面向接口 TaskScheduler 的，这符合面向对象中依赖抽象而不依赖具体的原则，带来底层资源调度器的可插拔性，导致 Spark 可以运行在众多的资源调度器模式上，如 Standalone、Yarn、Mesos、Local、EC2、其他自定义的资源调度器；在 Standalone 的模式下我们聚焦于 TaskSchedulerImpl。

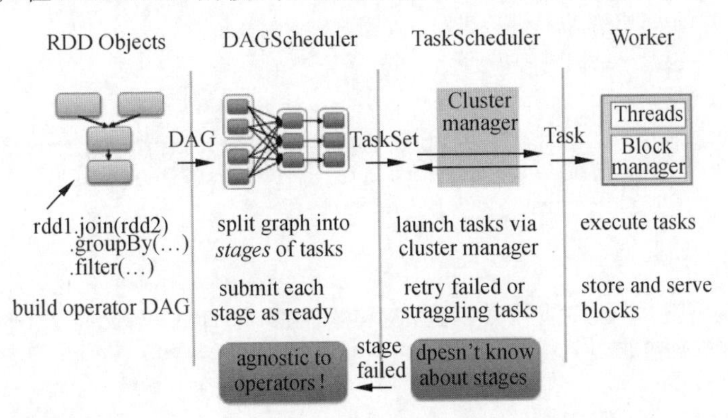

图 4-6　Spark 基础调度图

TaskScheduler 是一个接口 Trait，底层任务调度接口，由[org.apache.spark.scheduler.TaskSchedulerImpl]实现。这个接口允许插入不同的任务调度程序。每个任务调度器在单独的 SparkContext 中调度任务。任务调度程序从每个 Stage 的 DAGScheduler 获得提交的任务集，负责发送任务到集群运行，如果任务运行失败，将重试，返回 DAGScheduler 事件。

Spark 2.2.1 版本的 TaskScheduler.scala 的源码如下：

```
1.  private[spark] trait TaskScheduler {
2.
3.    private val appId = "spark-application-" + System.currentTimeMillis
4.
5.    def rootPool: Pool
6.
7.    def schedulingMode: SchedulingMode
8.
9.    def start(): Unit
10.
11.   //成功初始化后调用（通常在 Spark 上下文中）。Yarn 使用这个来引导基于优先位置的资源
12.   //分配，等待从节点登记等
13.   def postStartHook() { }
14.
15.   //从群集断开连接
16.   def stop(): Unit
17.
18.   //提交要运行的任务序列
19.   def submitTasks(taskSet: TaskSet): Unit
20.
21.   //取消 Stage
22.   def cancelTasks(stageId: Int, interruptThread: Boolean): Unit
23.
24.   /**
25.    * 杀掉尝试任务
26.    *
27.    * @return 任务是否成功终止
28.    */
29.   def killTaskAttempt(taskId: Long, interruptThread: Boolean, reason:
      String): Boolean
30.
31.   //系统为 upcalls 设置 DAG 调度，这是保证在 submitTasks 被调用前被设置
32.   def setDAGScheduler(dagScheduler: DAGScheduler): Unit
33.
34.   //获取集群中使用的默认并行级别，作为对作业的提示
35.   def defaultParallelism(): Int
36.
37.  /**
38.     更新正运行任务，让 master 知道 BlockManager 仍活着。如果 driver 知道给定的块
39.     管理器，则返回 true；否则返回 false，指示块管理器应重新注册
40.    */
41.
42.   def executorHeartbeatReceived(
43.       execId: String,
44.       accumUpdates: Array[(Long, Seq[AccumulatorV2[_, _]])],
45.       blockManagerId: BlockManagerId): Boolean
46.
47.   /**
48.    * 获取与作业相关联的应用程序 ID
49.    *
50.    * @return An application ID
51.    */
52.   def applicationId(): String = appId
53.
54.   /**
55.    * 处理丢失的 executor
56.    */
```

```
57.   def executorLost(executorId: String, reason: ExecutorLossReason): Unit
58.
59.   /**
60.    *获取与作业相关联的应用程序的尝试 ID
61.    *
62.    * @return 应用程序的尝试 ID
63.    */
64.   def applicationAttemptId(): Option[String]
65.
66. }
```

Spark 2.4.3 版本的 TaskScheduler.scala 的源码与 Spark 2.2.1 版本相比具有如下特点。

- 上段代码中第 29 行之后新增加了 killAllTaskAttempts 方法。
- 上段代码中第 57 行之后新增 workerRemoved 的方法。

```
1.  ......
2.  def killAllTaskAttempts(stageId: Int, interruptThread: Boolean, reason:
    String): Unit
3.  ......
4.  def workerRemoved(workerId: String, host: String, message: String): Unit
5.  ......
```

DAGScheduler 把 TaskSet 交给底层的接口 TaskScheduler，具体实现时有不同的方法。TaskScheduler 主要由 TaskSchedulerImpl 实现。

TaskSchedulerImpl 也有自己的子类 YarnScheduler。

```
1.    private[spark] class YarnScheduler(sc: SparkContext) extends
      TaskSchedulerImpl(sc) {
2.
3.   //RackResolver 记录 INFO 日志信息时，解析 rack 的信息
4.   if (Logger.getLogger(classOf[RackResolver]).getLevel == null) {
5.     Logger.getLogger(classOf[RackResolver]).setLevel(Level.WARN)
6.   }
7.
8.   //默认情况下，rack 是未知的
9.   override def getRackForHost(hostPort: String): Option[String] = {
10.    val host = Utils.parseHostPort(hostPort)._1
11.    Option(RackResolver.resolve(sc.hadoopConfiguration, host).
       getNetworkLocation)
12.  }
13. }
```

YarnScheduler 的子类 YarnClusterScheduler 实现如下：

```
1.   private[spark] class YarnClusterScheduler(sc: SparkContext) extends
     YarnScheduler(sc) {
2.   logInfo("Created YarnClusterScheduler")
3.
4.   override def postStartHook() {
5.     ApplicationMaster.sparkContextInitialized(sc)
6.     super.postStartHook()
7.     logInfo("YarnClusterScheduler.postStartHook done")
8.   }
9.
10. }
```

默认情况下，我们研究 Standalone 的模式，所以主要研究 TaskSchedulerImpl。

DAGScheduler 把 TaskSet 交给 TaskScheduler，TaskScheduler 中通过 TastSetManager 管理具体的任务。TaskScheduler 的核心任务是提交 TaskSet 到集群运算，并汇报结果。

- 为 TaskSet 创建和维护一个 TaskSetManager，并追踪任务的本地性以及错误信息。
- 遇到延后的 Straggle 任务，会放到其他节点重试。
- 向 DAGScheduler 汇报执行情况，包括在 Shuffle 输出 lost 的时候报告 fetch failed 错误等信息。

TaskSet 是一个普通的类，第一个成员是 tasks，tasks 是一个数组。TaskSet 的源码如下：

```
1.    private[spark] class TaskSet(
2.     val tasks: Array[Task[_]],
3.     val stageId: Int,
4.     val stageAttemptId: Int,
5.     val priority: Int,
6.     val properties: Properties) {
7.   val id: String = stageId + "." + stageAttemptId
8.
9.   override def toString: String = "TaskSet " + id
10. }
```

TaskScheduler 内部有 SchedulerBackend，SchedulerBackend 管理 Executor 资源。从 Standalone 的模式来讲，具体实现是 StandaloneSchedulerBackend（Spark 2.0 版本将之前的 AppClient 名字更新为 StandaloneAppClient）。

SchedulerBackend 本身是一个接口，是一个 trait。Spark 2.2.1 版本的 SchedulerBackend 的源码如下：

```
1.  private[spark] trait SchedulerBackend {
2.   private val appId = "spark-application-" + System.currentTimeMillis
3.
4.   def start(): Unit
5.   def stop(): Unit
6.   def reviveOffers(): Unit
7.   def defaultParallelism(): Int
8.
9.   def killTask(taskId: Long, executorId: String, interruptThread:
     Boolean, reason: String): Unit =
10.    throw new UnsupportedOperationException
11.  def isReady(): Boolean = true
12.
13.  /**
14.   *获取与作业关联的应用 ID
15.   *
16.   * @return 应用程序 ID
17.   */
18.  def applicationId(): String = appId
19.
20.  /**
21.   *如果集群管理器支持多个尝试，则获取此运行的尝试 ID，应用程序运行在客户端模式将没
      *有尝试 ID
22.   *
23.   * @return 如果可用，返回应用程序尝试 ID
24.   */
25.  def applicationAttemptId(): Option[String] = None
26.
```

```
27.    /**
28.     *得到 driver 日志的 URL。这些 URL 是用来在用户界面中显示链接 driver 的 Executors
        *选项卡
29.     * @return Map 包含日志名称和 URLs
30.     */
31.    def getDriverLogUrls: Option[Map[String, String]] = None
32.
33. }
```

Spark 2.4.3 版本的 SchedulerBackend.scala 的源码与 Spark 2.2.1 版本相比具有如下特点。

- 上段代码中第 31 行之后新增 maxNumConcurrentTasks 方法。获取当前可并发启动的最大任务数。注意，请不要缓存此方法返回的值，因为添加/删除 executors，数字可能会更改。

```
1. def maxNumConcurrentTasks(): Int
```

StandaloneSchedulerBackend：专门负责收集 Worker 的资源信息。接收 Worker 向 Driver 注册的信息，ExecutorBackend 启动的时候进行注册，为当前应用程序准备计算资源，以进程为单位。

StandaloneSchedulerBackend 的源码如下：

```
1.   private[spark] class StandaloneSchedulerBackend(
2.      scheduler: TaskSchedulerImpl,
3.      sc: SparkContext,
4.      masters: Array[String])
5.    extends CoarseGrainedSchedulerBackend(scheduler, sc.env.rpcEnv)
6.    with StandaloneAppClientListener
7.    with Logging {
8.    private var client: StandaloneAppClient = null
9.  ......
```

StandaloneSchedulerBackend 里有一个 Client: StandaloneAppClient。

```
1.    private[spark] class StandaloneAppClient(
2.      rpcEnv: RpcEnv,
3.      masterUrls: Array[String],
4.      appDescription: ApplicationDescription,
5.      listener: StandaloneAppClientListener,
6.      conf: SparkConf)
7.    extends Logging {
```

StandaloneAppClient 允许应用程序与 Spark standalone 集群管理器通信。获取 Master 的 URL、应用程序描述和集群事件监听器，当各种事件发生时可以回调监听器。masterUrls 的格式为 spark://host:port，StandaloneAppClient 需要向 Master 注册。

StandaloneAppClient 在 StandaloneSchedulerBackend.scala 的 start 方法启动时进行赋值，用 new()函数创建一个 StandaloneAppClient。

StandaloneSchedulerBackend.scala 的源码如下：

```
1.  private[spark] class StandaloneSchedulerBackend(
2.  ......
3.  override def start() {
4.  ......
5.  val appDesc = ApplicationDescription(sc.appName, maxCores, sc.executorMemory,
    command,
6.        webUrl, sc.eventLogDir, sc.eventLogCodec, coresPerExecutor,
```

```
        initialExecutorLimit)
7.      client = new StandaloneAppClient(sc.env.rpcEnv, masters, appDesc, this, conf)
8.      client.start()
9.      launcherBackend.setState(SparkAppHandle.State.SUBMITTED)
10.     waitForRegistration()
11.     launcherBackend.setState(SparkAppHandle.State.RUNNING)
```

StandaloneAppClient.scala 中，里面有一个类是 ClientEndpoint，核心工作是在启动时向 Master 注册。StandaloneAppClient 的 start 方法启动时，就调用 new 函数创建一个 ClientEndpoint。

StandaloneAppClient 的源码如下：

```
1.     private[spark] class StandaloneAppClient(
2.  ......
3.   private class ClientEndpoint(override val rpcEnv: RpcEnv) extends
     ThreadSafeRpcEndpoint
4.     with Logging {
5.  ......
6.  def start() {
7.     //启动 rpcEndpoint; 将回调 listener
8.     endpoint.set(rpcEnv.setupEndpoint("AppClient", new ClientEndpoint
       (rpcEnv)))
9.    }
```

StandaloneSchedulerBackend 在启动时构建 StandaloneAppClient 实例，并在 StandaloneAppClient 实例 start 时启动了 ClientEndpoint 消息循环体。ClientEndpoint 在启动时会向 Master 注册当前程序。

StandaloneAppClient 中 ClientEndpoint 类的 onStart()方法如下：

```
1.     override def onStart(): Unit = {
2.       try {
3.         registerWithMaster(1)
4.       } catch {
5.         case e: Exception =>
6.           logWarning("Failed to connect to master", e)
7.           markDisconnected()
8.           stop()
9.       }
10.    }
```

这是 StandaloneSchedulerBackend 的第一个注册的核心功能。StandaloneSchedulerBackend 继承自 CoarseGrainedSchedulerBackend。而 CoarseGrainedSchedulerBackend 在启动时就创建 DriverEndpoint，从实例的角度讲，DriverEndpoint 也属于 StandaloneSchedulerBackend 实例。

```
1.    private[spark]
2.  class CoarseGrainedSchedulerBackend(scheduler: TaskSchedulerImpl, val
    rpcEnv: RpcEnv)
3.    extends ExecutorAllocationClient with SchedulerBackend with Logging
4.  {
5.  ......
6.   class DriverEndpoint(override val rpcEnv: RpcEnv, sparkProperties:
     Seq[(String, String)])
7.     extends ThreadSafeRpcEndpoint with Logging {
8.  ......
```

StandaloneSchedulerBackend 的父类 CoarseGrainedSchedulerBackend 在 start 的时候会实例化类型为 DriverEndpoint（这就是我们程序运行时的经典对象 Driver）的消息循环体。

StandaloneSchedulerBackend 在运行时向 Master 注册申请资源，当 Worker 的 ExecutorBackend 启动时会发送 RegisteredExecutor 信息向 DriverEndpoint 注册，此时 StandaloneSchedulerBackend 就掌握了当前应用程序拥有的计算资源，TaskScheduler 就是通过 StandaloneSchedulerBackend 拥有的计算资源来具体运行 Task 的；StandaloneSchedulerBackend 不是应用程序的总管，应用程序的总管是 DAGScheduler、TaskScheduler，StandaloneSchedulerBackend 向应用程序的 Task 分配具体的计算资源，并把 Task 发送到集群中。

SparkContext、DAGScheduler、TaskSchedulerImpl、StandaloneSchedulerBackend 在应用程序启动时只实例化一次，应用程序存在期间始终存在这些对象。

这里基于 Spark 2.2 版本讲解如下。

Spark 调度器三大核心资源为 SparkContext、DAGScheduler 和 TaskSchedulerImpl。TaskSchedulerImpl 作为具体的底层调度器，运行时需要计算资源，因此需要 StandaloneSchedulerBackend。StandaloneSchedulerBackend 设计巧妙的地方是启动时启动 StandaloneAppClient，而 StandaloneAppClient 在 start 时有一个 ClientEndpoint 的消息循环体，ClientEndpoint 的消息循环体启动的时候向 Master 注册应用程序。

StandaloneSchedulerBackend 的父类 CoarseGrainedSchedulerBackend 在 start 启动的时候会实例化 DriverEndpoint，所有的 ExecutorBackend 启动的时候都要向 DriverEndpoint 注册，注册最后落到了 StandaloneSchedulerBackend 的内存数据结构中，表面上看是在 CoarseGrainedSchedulerBackend，但是实例化的时候是 StandaloneSchedulerBackend，注册给父类的成员其实就是子类的成员。

作为前提问题：TaskScheduler、StandaloneSchedulerBackend 是如何启动的？TaskSchedulerImpl 是什么时候实例化的？

TaskSchedulerImpl 是在 SparkContext 中实例化的。在 SparkContext 类实例化的时候，只要不是方法体里面的内容，都会被执行，(sched, ts)是 SparkContext 的成员，将调用 createTaskScheduler 方法。调用 createTaskScheduler 方法返回一个 Tuple，包括两个元素：sched 是我们的 schedulerBackend；ts 是 taskScheduler。

```
1. class SparkContext(config: SparkConf) extends Logging {
2.  ......
3. //创建启动调度器 scheduler
4.   val (sched, ts) = SparkContext.createTaskScheduler(this, master,
     deployMode)
5.   _schedulerBackend = sched
6.   _taskScheduler = ts
7.   _dagScheduler = new DAGScheduler(this)
8.   _heartbeatReceiver.ask[Boolean](TaskSchedulerIsSet)
```

createTaskScheduler 里有很多运行模式，这里关注 Standalone 模式，首先调用 new()函数创建一个 TaskSchedulerImpl，TaskSchedulerImpl 和 SparkContext 是一一对应的，整个程序运行的时候只有一个 TaskSchedulerImpl，也只有一个 SparkContext；接着实例化 StandaloneSchedulerBackend，整个程序运行的时候只有一个 StandaloneSchedulerBackend。createTaskScheduler 方法如下：

```
1.     private def createTaskScheduler(
2.       sc: SparkContext,
3.       master: String,
4.       deployMode: String): (SchedulerBackend, TaskScheduler) = {
```

```
5.     import SparkMasterRegex._
6.  ......
7.   master match {
8.  ......
9.       case SPARK_REGEX(sparkUrl) =>
10.        val scheduler = new TaskSchedulerImpl(sc)
11.        val masterUrls = sparkUrl.split(",").map("spark://" + _)
12.        val  backend  =  new  StandaloneSchedulerBackend(scheduler,  sc,
           masterUrls)
13.        scheduler.initialize(backend)
14.        (backend, scheduler)
15. ......
```

在 SparkContext 实例化的时候通过 createTaskScheduler 来创建 TaskSchedulerImpl 和 StandaloneSchedulerBackend。然后在 createTaskScheduler 中调用 scheduler.initialize(backend)。

initialize 的方法参数把 StandaloneSchedulerBackend 传进来，schedulingMode 模式匹配有两种方式：FIFO、FAIR。

initialize 的方法中调用 schedulableBuilder.buildPools()。buildPools 方法根据 FIFOSchedulableBuilder、FairSchedulableBuilder 不同的模式重载方法实现。

```
1.   private[spark] trait SchedulableBuilder {
2.    def rootPool: Pool
3.
4.    def buildPools(): Unit
5.
6.    def addTaskSetManager(manager: Schedulable, properties: Properties):
      Unit
7.  }
```

initialize 的方法把 StandaloneSchedulerBackend 传进来了，但还没有启动 StandaloneSchedulerBackend。在 TaskSchedulerImpl 的 initialize 方法中把 StandaloneSchedulerBackend 传进来，从而赋值为 TaskSchedulerImpl 的 backend；在 TaskSchedulerImpl 调用 start 方法时会调用 backend.start 方法，在 start 方法中会最终注册应用程序。

下面来看 SparkContext.scala 的 taskScheduler 的启动。

```
1.    val  (sched,  ts)  =  SparkContext.createTaskScheduler(this,  master,
      deployMode)
2.      _schedulerBackend = sched
3.      _taskScheduler = ts
4.      _dagScheduler = new DAGScheduler(this)
5.  ......
6.      _taskScheduler.start()
7.      _applicationId = _taskScheduler.applicationId()
8.      _applicationAttemptId = taskScheduler.applicationAttemptId()
9.      _conf.set("spark.app.id", _applicationId)
10. ......
```

其中调用了_taskScheduler 的 start 方法。

```
1.   private[spark] trait TaskScheduler {
2.  ......
3.
4.   def start(): Unit
5.  ......
```

TaskScheduler 的 start()方法没有具体实现。TaskScheduler 子类的 TaskSchedulerImpl 的 start()方法的源码如下：

```
1.    override def start() {
2.     backend.start()
3.
4.     if (!isLocal && conf.getBoolean("spark.speculation", false)) {
5.       logInfo("Starting speculative execution thread")
6.       speculationScheduler.scheduleAtFixedRate(new Runnable {
7.         override def run(): Unit = Utils.tryOrStopSparkContext(sc) {
8.           checkSpeculatableTasks()
9.         }
10.      }, SPECULATION_INTERVAL_MS, SPECULATION_INTERVAL_MS, TimeUnit.
       MILLISECONDS)
11.    }
12.  }
```

TaskSchedulerImpl 的 start 通过 backend.start 启动了 StandaloneSchedulerBackend 的 start 方法。

StandaloneSchedulerBackend 的 start 方法中，将 command 封装注册给 Master，Master 转过来要 Worker 启动具体的 Executor。command 已经封装好指令，Executor 具体要启动进程入口类 CoarseGrainedExecutorBackend。然后调用 new()函数创建一个 StandaloneAppClient，通过 client.start 启动 client。

StandaloneAppClient 的 start 方法中调用 new()函数创建一个 ClientEndpoint。

```
1.    def start() {
2.     //启动 rpcEndpoint; 将回调 listener
3.     endpoint.set(rpcEnv.setupEndpoint("AppClient", new ClientEndpoint
       (rpcEnv)))
4.   }
```

ClientEndpoint 的源码如下：

```
1.     private class ClientEndpoint(override val rpcEnv: RpcEnv) extends
       ThreadSafeRpcEndpoint
2.     with Logging {
3.  ......
4.     override def onStart(): Unit = {
5.       try {
6.         registerWithMaster(1)
7.       } catch {
8.         case e: Exception =>
9.           logWarning("Failed to connect to master", e)
10.          markDisconnected()
11.          stop()
12.      }
13.    }
```

ClientEndpoint 是一个 ThreadSafeRpcEndpoint。ClientEndpoint 的 onStart 方法中调用 registerWithMaster(1)进行注册，向 Master 注册程序。registerWithMaster 方法如下：

```
1.       private def registerWithMaster(nthRetry: Int) {
2.       registerMasterFutures.set(tryRegisterAllMasters())
3.       registrationRetryTimer.set(registrationRetryThread.schedule(new
         Runnable {
```

```
4.        override def run(): Unit = {
5.          if (registered.get) {
6.            registerMasterFutures.get.foreach(_.cancel(true))
7.            registerMasterThreadPool.shutdownNow()
8.          } else if (nthRetry >= REGISTRATION_RETRIES) {
9.            markDead("All masters are unresponsive! Giving up.")
10.         } else {
11.           registerMasterFutures.get.foreach(_.cancel(true))
12.           registerWithMaster(nthRetry + 1)
13.         }
14.       }
15.     }, REGISTRATION_TIMEOUT_SECONDS, TimeUnit.SECONDS))
16.   }
```

程序注册后，Master 通过 schedule 分配资源，通知 Worker 启动 Executor，Executor 启动的进程是 CoarseGrainedExecutorBackend，Executor 启动后又转过来向 Driver 注册，Driver 其实是 StandaloneSchedulerBackend 的父类 CoarseGrainedSchedulerBackend 的一个消息循环体 DriverEndpoint。

总结：在 SparkContext 实例化时调用 createTaskScheduler 来创建 TaskSchedulerImpl 和 StandaloneSchedulerBackend，同时在 SparkContext 实例化的时候会调用 TaskSchedulerImpl 的 start，在 start 方法中会调用 StandaloneSchedulerBackend 的 start，在该 start 方法中会创建 StandaloneAppClient 对象，并调用 StandaloneAppClient 对象的 start 方法，在该 start 方法中会创建 ClientEndpoint，创建 ClientEndpoint 时会传入 Command 来指定具体为当前应用程序启动的 Executor 的入口类的名称为 CoarseGrainedExecutorBackend，然后 ClientEndpoint 启动并通过 tryRegisterMaster 来注册当前的应用程序到 Master 中，Master 接收到注册信息后如果可以运行程序，为该程序生产 Job ID 并通过 schedule 来分配计算资源，具体计算资源的分配是通过应用程序的运行方式、Memory、cores 等配置信息决定的。最后，Master 会发送指令给 Worker，Worker 为当前应用程序分配计算资源时会首先分配 ExecutorRunner。ExecutorRunner 内部会通过 Thread 的方式构建 ProcessBuilder 来启动另外一个 JVM 进程，这个 JVM 进程启动时加载的 main 方法所在的类的名称就是在创建 ClientEndpoint 时传入的 Command 来指定具体名称为 CoarseGrainedExecutorBackend 的类，此时 JVM 在通过 ProcessBuilder 启动时获得了 CoarseGrainedExecutorBackend 后加载并调用其中的 main 方法，在 main 方法中会实例化 CoarseGrainedExecutorBackend 本身这个消息循环体，而 CoarseGrainedExecutorBackend 在实例化时会通过回调 onStart 向 DriverEndpoint 发送 RegisterExecutor 来注册当前的 CoarseGrainedExecutorBackend，此时 DriverEndpoint 收到该注册信息并保存在 StandaloneSchedulerBackend 实例的内存数据结构中，这样 Driver 就获得了计算资源。

CoarseGrainedExecutorBackend.scala 的 main 方法如下：

```
1.  def main(args: Array[String]) {
2.    var driverUrl: String = null
3.    var executorId: String = null
4.    var hostname: String = null
5.    var cores: Int = 0
6.    var appId: String = null
7.    var workerUrl: Option[String] = None
8.    val userClassPath = new mutable.ListBuffer[URL]()
9.
```

```
10.    var argv = args.toList
11.    ......
12.    run(driverUrl, executorId, hostname, cores, appId, workerUrl,
       userClassPath)
13.    System.exit(0)
14.  }
```

CoarseGrainedExecutorBackend 的 main 然后开始调用 run 方法。

```
1.     private def run(
2.        driverUrl: String,
3.        executorId: String,
4.        hostname: String,
5.        cores: Int,
6.        appId: String,
7.        workerUrl: Option[String],
8.        userClassPath: Seq[URL]) {
9.  ......
10.        env.rpcEnv.setupEndpoint("Executor", new CoarseGrainedExecutorBackend(
11.         env.rpcEnv, driverUrl, executorId, hostname, cores, userClassPath,
            env))
12. ......
```

在 CoarseGrainedExecutorBackend 的 main 方法中，通过 env.rpcEnv.setupEndpoint ("Executor", new CoarseGrainedExecutorBackend(env.rpcEnv, driverUrl, executorId, hostname, cores, userClassPath, env))构建了 CoarseGrainedExecutorBackend 实例本身。

4.6　本章总结

本章内容紧紧围绕 Spark 调度器（Scheduler）的运行机制，介绍了其中涉及的重要概念，如 Spark Driver Program、Spark Job、高层调度器（DAGScheduler）、底层调度器（TaskScheduler）和调度器的通信终端（SchedulerBackend）。同时，从外围的运行框架，到内部的调度器和通信终端，分别深度剖析了各自的运行原理。并且，每个原理都结合了 Spark 源码的解析，加深对整个 Spark 调度器运行机制的理解。

SparkContext、DAGScheduler、TaskScheduler、SchedulerBackend 在应用程序启动时只实例化一次，应用程序存在期间始终存在这些对象。

第 5 章　Spark 集群启动原理和源码详解

本章深入讲解 Spark 集群启动原理和源码。5.1 节讲解 Master 启动原理和源码；5.2 节讲解 Worker 启动原理和源码；5.3 节阐述了 ExecutorBackend 启动原理和源码、ExecutorBackend 接口与 Executor 的关系、ExecutorBackend 的不同实现、ExecutorBackend 中的通信及异常处理；5.4 节讲解 Executor 中任务的执行、加载、任务线程池、任务执行失败处理、TaskRunner 运行内幕；5.5 节讲解 Executor 执行结果的处理方式。

5.1　Master 启动原理和源码详解

本节讲解 Master 启动的原理和源码；Master HA 双机切换；Master 的注册机制和状态管理解密等内容。

5.1.1　Master 启动的原理详解

Spark 应用程序作为独立的集群进程运行，由主程序中的 SparkContext 对象（称为驱动程序）协调。Spark 集群部署组件如图 5-1 所示。

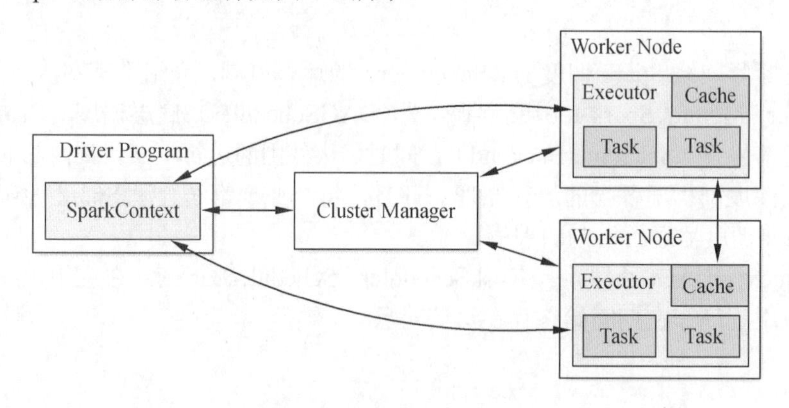

图 5-1　Spark 集群部署组件图

其中各个术语及相关术语的描述如下。

（1）Driver Program：运行 Application 的 main 函数并新建 SparkContext 实例的程序，称为驱动程序（Driver Program）。通常可以使用 SparkContext 代表驱动程序。

（2）Cluster Manager：集群管理器（Cluster Manager）是集群资源管理的外部服务。Spark 上现在主要有 Standalone、YARN、Mesos 3 种集群资源管理器。Spark 自带的 Standalone 模式能够满足绝大部分纯粹的 Spark 计算环境中对集群资源管理的需求，基本上只有在集群中运行多套计算框架的时候才建议考虑 YARN 和 Mesos。

（3）Worker Node：集群中可以运行 Application 代码的工作节点（Worker Node），相当于 Hadoop 的 Slave 节点。

（4）Executor：在 Worker Node 上为 Application 启动的一个工作进程，在进程中负责任务（Task）的运行，并且负责将数据存放在内存或磁盘上，在 Executor 内部通过多线程的方式（即线程池）并发处理应用程序的具体任务。

每个 Application 都有各自独立的 Executors，因此应用程序之间是相互隔离的。

（5）Task：任务（Task）是指被 Driver 送到 Executor 上的工作单元。通常，一个任务会处理一个 Partition 的数据，每个 Partition 一般是一个 HDFS 的 Block 块的大小。

（6）Application：是创建了 SparkContext 实例对象的 Spark 用户程序，包含了一个 Driver program 和集群中多个 Worker 上的 Executor。

（7）Job：和 Spark 的 action 对应，每个 action，如 count、savaAsTextFile 等都会对应一个 Job 实例，每个 Job 会拆分成多个 Stages，一个 Stage 中包含一个任务集（TaskSet），任务集中的各个任务通过一定的调度机制发送到工作单位（Executor）上并行执行。

Spark Standalone 集群的部署采用典型的 Master/Slave 架构。其中，Master 节点负责整个集群的资源管理与调度，Worker 节点（也可以称 Slave 节点）在 Master 节点的调度下启动 Executor，负责执行具体工作（包括应用程序以及应用程序提交的任务）。

5.1.2　Master 启动的源码详解

Spark 中各个组件是通过脚本来启动部署的。下面以脚本为入口点开始分析 Master 的部署。每个组件对应提供了启动的脚本，同时也会提供停止的脚本。停止脚本比较简单，在此仅分析启动脚本。

1. Master部署的启动脚本解析

首先看一下 Master 的启动脚本./sbin/start-master.sh，内容如下：

```
1.  # 在脚本的执行节点启动 Master 组件
2.  #如果没有设置环境变量 SPARK_HOME，会根据脚本所在位置自动设置
3.  if [ -z "${SPARK_HOME}" ]; then
4.    export SPARK_HOME="$(cd "'dirname "$0"'"/..; pwd)"
5.  fi
6.
7.  ##注：提取的类名必须和 SparkSubmit 的类相匹配。任何变化都需在类中进行反映
8.  # Master 组件对应的类
9.  CLASS="org.apache.spark.deploy.master.Master"
10.
11. #脚本的帮助信息
12. if [[ "$@" = *--help ]] || [[ "$@" = *-h ]]; then
13.   echo "Usage: ./sbin/start-master.sh [options]"
14.   pattern="Usage:"
15.   pattern+="\|Using Spark's default log4j profile:"
16.   pattern+="\|Registered signal handlers for"
17.
18. # 通过脚本 spark-class 执行指定的 Master 类，参数为--help
19.   "${SPARK_HOME}"/bin/spark-class $CLASS --help 2>&1 | grep -v "$pattern"
    1>&2
```

```
20.   exit 1
21. fi
22.
23. ORIGINAL_ARGS="$@"
24.
25. . "${SPARK_HOME}/sbin/spark-config.sh"
26.
27. . "${SPARK_HOME}/bin/load-spark-env.sh"
28.
29. #下面的一些参数对应的默认配置属性
30. if [ "$SPARK_MASTER_PORT" = "" ]; then
31.   SPARK_MASTER_PORT=7077
32. fi
33.
34. if [ "$SPARK_MASTER_HOST" = "" ]; then
35.   case 'uname' in
36.       (SunOS)
37.       SPARK_MASTER_HOST="'/usr/sbin/check-hostname | awk '{print $NF}''"
38.       ;;
39.       (*)
40.       SPARK_MASTER_HOST="'hostname -f'"
41.       ;;
42.   esac
43. fi
44.
45. if [ "$SPARK_MASTER_WEBUI_PORT" = "" ]; then
46.   SPARK_MASTER_WEBUI_PORT=8080
47. fi
48.
49. #通过启动后台进程的脚本spark-daemon.sh来启动Master组件
50. "${SPARK_HOME}/sbin"/spark-daemon.sh start $CLASS 1 \
51.   --host $SPARK_MASTER_HOST --port $SPARK_MASTER_PORT --webui-port
      $SPARK_MASTER_WEBUI_PORT \
52.   $ORIGINAL_ARGS
```

通过脚本的简单分析，可以看出Master组件是以后台守护进程的方式启动的，对应的后台守护进程的启动脚本spark-daemon.sh，在后台守护进程的启动脚本spark-daemon.sh内部，通过脚本spark-class启动一个指定主类的JVM进程，相关代码如下：

```
1.    case "$mode" in
2.  #这里对应的是启动一个Spark类
3.      (class)
4.        execute_command nice -n "$SPARK_NICENESS" "${SPARK_HOME}"/bin/
    spark-class "$command" "$@"
5.        ;;
6.  #这里对应提交一个Spark 应用程序
7.      (submit)
8.        execute_command nice -n "$SPARK_NICENESS" bash "${SPARK_HOME}"/bin/
          spark-submit --class "$command" "$@"
9.        ;;
10.
11.     (*)
12.       echo "unknown mode: $mode"
13.       exit 1
14.       ;;
```

通过脚本的分析，可以知道最终执行的是Master类（对应的代码为前面的CLASS="org.apache.spark.deploy.master.Master"），对应的入口点是Master伴生对象中的main

方法。下面以该方法作为入口点进一步解析 Master 部署框架。

部署 Master 组件时，最简单的方式是直接启动脚本，不带任何选项参数，命令如下：

```
1.  ./sbin/start-master.sh
```

如需设置选项参数，可以查看帮助信息，根据自己的需要进行设置。

2. Master的源码解析

首先查看 Master 伴生对象中的 main 方法。

Spark 2.2.1 版本的 Master.scala 的源码如下：

```
1.    def main(argStrings: Array[String]) {
2.      utils.initDaemon(log)
3.      val conf = new SparkConf
4.    //构建参数解析的实例
5.      val args = new MasterArguments(argStrings, conf)
6.     //启动 RPC 通信环境以及 Master 的 RPC 通信终端
7.      val (rpcEnv, _, _) = startRpcEnvAndEndpoint(args.host, args.port,
        args.webUiPort, conf)
8.      rpcEnv.awaitTermination()
9.    }
```

Spark 2.4.3 版本的源码与 Spark 2. 2.1 版本相比具有如下特点。

- 上段代码中第 2 行之前新增 SparkUncaughtExceptionHandler 处理的代码。

```
1.     Thread.setDefaultUncaughtExceptionHandler(new SparkUncaughtExceptionHandler(
2.       exitOnUncaughtException = false))
```

和其他类（如 SparkSubmit）一样，Master 类的入口点处也包含了对应的参数类 MasterArguments。MasterArguments 类包括 Spark 属性配置相关的一些解析。

MasterArguments.scala 的源码如下：

```
1.   private[master] class MasterArguments(args: Array[String], conf:
     SparkConf) extends Logging {
2.    var host = Utils.localHostName()
3.    var port = 7077
4.    var webUiPort = 8080
5.    var propertiesFile: String = null
6.
7.   //读取启动脚本中设置的环境变量
8.    if (System.getenv("SPARK_MASTER_IP") != null) {
9.      logWarning("SPARK_MASTER_IP is deprecated, please use SPARK_MASTER_
      HOST")
10.     host = System.getenv("SPARK_MASTER_IP")
11.   }
12.
13.   if (System.getenv("SPARK_MASTER_HOST") != null) {
14.     host = System.getenv("SPARK_MASTER_HOST")
15.   }
16.   if (System.getenv("SPARK_MASTER_PORT") != null) {
17.     port = System.getenv("SPARK_MASTER_PORT").toInt
18.   }
19.   if (System.getenv("SPARK_MASTER_WEBUI_PORT") != null) {
20.     webUiPort = System.getenv("SPARK_MASTER_WEBUI_PORT").toInt
21.   }
```

```
22.  //命令行选项参数的解析
23.   parse(args.toList)
24.
25.   //加载 SparkConf 文件，所有的访问必须经过此行
26.   propertiesFile = Utils.loadDefaultSparkProperties(conf, propertiesFile)
27.
28.   if (conf.contains("spark.master.ui.port")) {
29.     webUiPort = conf.get("spark.master.ui.port").toInt
30.   }
31. ......
```

MasterArguments 中的 printUsageAndExit 方法对应的就是命令行中的帮助信息。

MasterArguments.scala 的源码如下：

```
1.    private def printUsageAndExit(exitCode: Int) {
2.      //scalastyle:off println
3.      System.err.println(
4.        "Usage: Master [options]\n" +
5.        "\n" +
6.        "Options:\n" +
7.        "  -i HOST, --ip HOST     Hostname to listen on (deprecated, please
        use --host or -h) \n" +
8.        "  -h HOST, --host HOST   Hostname to listen on\n" +
9.        "  -p PORT, --port PORT   Port to listen on (default: 7077)\n" +
10.       "  --webui-port PORT      Port for web UI (default: 8080)\n" +
11.       "  --properties-file FILE Path to a custom Spark properties file.\n" +
12.       "                         Default is conf/spark-defaults.conf.")
13.     //scalastyle:on println
14.     System.exit(exitCode)
15.   }
```

解析完 Master 的参数后，调用 startRpcEnvAndEndpoin 方法启动 RPC 通信环境以及 Master 的 RPC 通信终端。

Master.scala 的 startRpcEnvAndEndpoint 的源码如下：

```
1.
2.   /**
3.     * 启动 Master 并返回一个三元组：
4.     *   (1) The Master RpcEnv
5.     *   (2) The web UI bound port
6.     *   (3) The REST server bound port, if any
7.     */
8.
9.  def startRpcEnvAndEndpoint(
10.       host: String,
11.       port: Int,
12.       webUiPort: Int,
13.       conf: SparkConf): (RpcEnv, Int, Option[Int]) = {
14.     val securityMgr = new SecurityManager(conf)
15.    //构建 RPC 通信环境
16.     val rpcEnv = RpcEnv.create(SYSTEM_NAME, host, port, conf, securityMgr)
17.    //构建 RPC 通信终端，实例化 Master
18.     val masterEndpoint = rpcEnv.setupEndpoint(ENDPOINT_NAME,
19.       new Master(rpcEnv, rpcEnv.address, webUiPort, securityMgr, conf))
20.
21.  //向 Master 的通信终端发送请求，获取绑定的端口号
22.  //包含 Master 的 Web UI 监听端口号和 REST 的监听端口号
```

```
23.
24.   val portsResponse = masterEndpoint.askSync[BoundPortsResponse]
      (BoundPortsRequest)
25.   (rpcEnv, portsResponse.webUIPort, portsResponse.restPort)
26. }
```

startRpcEnvAndEndpoint 方法中定义了 ENDPOINT_NAME。

Master.scala 的源码如下：

```
1.   private[deploy] object Master extends Logging {
2.   val SYSTEM_NAME = "sparkMaster"
3.   val ENDPOINT_NAME = "Master"
4. ......
```

startRpcEnvAndEndpoint 方法中通过 masterEndpoint.askWithRetry[BoundPortsResponse](BoundPortsRequest)给 Master 自己发送一个消息 BoundPortsRequest，是一个 case object。发送消息 BoundPortsRequest 给自己，确保 masterEndpoint 正常启动起来。返回消息的类型是 BoundPortsResponse，是一个 case class。

MasterMessages.scala 的源码如下：

```
1.  private[master] object MasterMessages {
2. ......
3.   case object BoundPortsRequest
4.   case class BoundPortsResponse(rpcEndpointPort: Int, webUIPort: Int,
     restPort: Option[Int])
5. }
```

Master 收到消息 BoundPortsRequest，发送返回消息 BoundPortsResponse。

Master.scala 的源码如下：

```
1.  override def receiveAndReply(context: RpcCallContext): PartialFunction
    [Any, Unit] = {
2. ......
3. case BoundPortsRequest =>
4. context.reply(BoundPortsResponse(address.port, webUi.boundPort,
   restServerBoundPort))
```

在 BoundPortsResponse 传入的参数 restServerBoundPort 是在 Master 的 onStart 方法中定义的。

Master.scala 的源码如下：

```
1.   ......
2.   private var restServerBoundPort: Option[Int] = None
3.
4.   override def onStart(): Unit = {
5. ......
6.    restServerBoundPort = restServer.map(_.start())
7. ......
```

而 restServerBoundPort 是通过 restServer 进行 map 操作启动赋值。下面看一下 restServer。

Master.scala 的源码如下：

```
1.   private var restServer: Option[StandaloneRestServer] = None
2. ......
3.    if (restServerEnabled) {
4.      val port = conf.getInt("spark.master.rest.port", 6066)
```

```
5.      restServer = Some(new StandaloneRestServer(address.host, port, conf,
        self, masterUrl))
6.    }
7.  ……
```

其中调用 new()函数创建一个 StandaloneRestServer。StandaloneRestServer 服务器响应请求提交的[RestSubmissionClient]，将被嵌入到 standalone Master 中，仅用于集群模式。服务器根据不同的情况使用不同的 HTTP 代码进行响应。

- 200 OK-请求已成功处理。
- 400 错误请求：请求格式错误，未成功验证，或意外类型。
- 468 未知协议版本：请求指定了此服务器不支持的协议。
- 500 内部服务器错误：服务器在处理请求时引发内部异常。

服务器在 HTTP 主体中总包含一个 JSON 表示的[SubmitRestProtocolResponse]。如果发生错误，服务器将包括一个[ErrorResponse]。如果构造了这个错误响应内部失败时，响应将由一个空体组成。响应体指示内部服务器错误。

StandaloneRestServer.scala 的源码如下：

```
1.  private[deploy] class StandaloneRestServer(
2.     host: String,
3.     requestedPort: Int,
4.     masterConf: SparkConf,
5.     masterEndpoint: RpcEndpointRef,
6.     masterUrl: String)
7.   extends RestSubmissionServer(host, requestedPort, masterConf) {
```

下面看一下 RestSubmissionClient 客户端。客户端提交申请[RestSubmissionServer]。在协议版本 V1 中，REST URL 以表单形式出现 http://[host:port]/v1/submissions/[action]，[action]可以是 create、kill 或状态中的其中一种。每种请求类型都表示为发送到以下前缀的 HTTP 消息。

（1）submit-POST to /submissions/create；

（2）kill-POST /submissions/kill/[submissionId]；

（3）status-GET /submissions/status/[submissionId]。

在（1）情况下，参数以 JSON 字段的形式发布到 HTTP 主体中。否则，URL 指定按客户端的预期操作。由于该协议预计将在 Spark 版本中保持稳定，因此现有字段不能添加或删除，但可以添加新的可选字段。如在少见的事件中向前或向后兼容性被破坏，Spark 须引入一个新的协议版本（如 V2）。客户机和服务器必须使用协议的同一版本进行通信。如果不匹配，服务器将用它支持的最高协议版本进行响应。此客户机的实现可以用指定的版本使用该信息重试。

RestSubmissionClient.scala 的源码如下：

```
1.  private[spark] class RestSubmissionClient(master: String) extends
    Logging {
2.   import RestSubmissionClient._
3.   private val supportedMasterPrefixes = Seq("spark://", "mesos://")
4.  ……
```

Restful 把一切都看成是资源。利用 Restful API 可以对 Spark 进行监控。程序运行的每一个步骤、Task 的计算步骤都可以可视化，对 Spark 的运行进行详细监控。

回到 startRpcEnvAndEndpoint 方法中，新创建了一个 Master 实例。Master 实例化时会对所有的成员进行初始化，如默认的 Cores 个数等。

Master.scala 的源码如下：

```
1.  private[deploy] class Master(
2.    override val rpcEnv: RpcEnv,
3.    address: RpcAddress,
4.    webUiPort: Int,
5.    val securityMgr: SecurityManager,
6.    val conf: SparkConf)
7.   extends ThreadSafeRpcEndpoint with Logging with LeaderElectable {
8.  ......
9.   //缺省 maxCores 时默认应用程序没有指定（通过 Int.MaxValue）
10.  private val defaultCores = conf.getInt("spark.deploy.defaultCores",
     Int.MaxValue)
11.  val reverseProxy = conf.getBoolean("spark.ui.reverseProxy", false)
12.  if (defaultCores < 1) {
13.    throw new SparkException("spark.deploy.defaultCores must be positive")
14.  }
15. ......
16.
```

Master 继承了 ThreadSafeRpcEndpoint 和 LeaderElectable，其中继承 LeaderElectable 涉及 Master 的高可用性（High Availability，HA）机制。这里先关注 ThreadSafeRpcEndpoint，继承该类后，Master 作为一个 RpcEndpoint，实例化后首先会调用 onStart 方法。

Spark 2.2.1 版本的 Master.scala 的源码如下：

```
1.   override def onStart(): Unit = {
2.     logInfo("Starting Spark master at " + masterUrl)
3.     logInfo(s"Running Spark version ${org.apache.spark.SPARK_VERSION}")
4.     //构建一个 Master 的 Web UI，查看向 Master 提交的应用程序等信息
5.     webUi = new MasterWebUI(this, webUiPort)
6.     webUi.bind()
7.     masterWebUiUrl = "http://" + masterPublicAddress + ":" + webUi.boundPort
8.     if (reverseProxy) {
9.       masterWebUiUrl = conf.get("spark.ui.reverseProxyUrl", masterWebUiUrl)
10.      logInfo(s"Spark Master is acting as a reverse proxy. Master, Workers
       and " +
11.       s"Applications UIs are available at $masterWebUiUrl")
12.    }
13.  //在一个守护线程中，启动调度机制，周期性检查 Worker 是否超时，当 Worker 节点超时后，
     //会修改其状态或从 Master 中移除其相关的操作
14.
15.    checkForWorkerTimeOutTask = forwardMessageThread.scheduleAtFixedRate
     (new Runnable {
16.      override def run(): Unit = Utils.tryLogNonFatalError {
17.        self.send(CheckForWorkerTimeOut)
18.      }
19.    }, 0, WORKER_TIMEOUT_MS, TimeUnit.MILLISECONDS)
20.
21.  //默认情况下会启动 Rest 服务，可以通过该服务向 Master 提交各种请求
22.    if (restServerEnabled) {
23.      val port = conf.getInt("spark.master.rest.port", 6066)
24.     restServer = Some(new StandaloneRestServer(address.host, port, conf,
      self, masterUrl))
25.    }
```

```
26.     restServerBoundPort = restServer.map(_.start())
27.   //度量（Metroics）相关的操作，用于监控
28.     masterMetricsSystem.registerSource(masterSource)
29.     masterMetricsSystem.start()
30.     applicationMetricsSystem.start()
31.    //度量系统启动后，将主程序和应用程序度量 handler 处理程序附加到 Web UI 中
32.     masterMetricsSystem.getServletHandlers.foreach(webUi.attachHandler)
33.     applicationMetricsSystem.getServletHandlers.foreach(webUi.
        attachHandler)
34.    //Master HA 相关的操作
35.     val serializer = new JavaSerializer(conf)
36.     val (persistenceEngine_, leaderElectionAgent_) = RECOVERY_MODE match {
37.       case "ZOOKEEPER" =>
38.         logInfo("Persisting recovery state to ZooKeeper")
39.         val zkFactory =
40.           new ZooKeeperRecoveryModeFactory(conf, serializer)
41.         (zkFactory.createPersistenceEngine(), zkFactory
            .createLeaderElectionAgent(this))
42.       case "FILESYSTEM" =>
43.         val fsFactory =
44.           new FileSystemRecoveryModeFactory(conf, serializer)
45.         (fsFactory.createPersistenceEngine(), fsFactory.createLeader-
            ElectionAgent(this))
46.       case "CUSTOM" =>
47.         val clazz = Utils.classForName(conf.get("spark.deploy
            .recoveryMode.factory"))
48.         val factory = clazz.getConstructor(classOf[SparkConf], classOf
            [Serializer])
49.           .newInstance(conf, serializer)
50.           .asInstanceOf[StandaloneRecoveryModeFactory]
51.         (factory.createPersistenceEngine(), factory.createLeaderElectionAgent
            (this))
52.       case _ =>
53.         (new BlackHolePersistenceEngine(), new MonarchyLeaderAgent(this))
54.     }
55.     persistenceEngine = persistenceEngine_
56.     leaderElectionAgent = leaderElectionAgent_
57.   }
```

Spark 2.4.3 版本的源码与 Spark 2.2.1 版本相比具有如下特点。

- 上段代码中第 9 行之后新增 webUi.addProxy()的代码。

```
1.         webUi.addProxy()
```

其中在 Master 的 onStart 方法中用 new()函数创建 MasterWebUI，启动一个 webServer。Master.scala 的源码如下：

```
1.    override def onStart(): Unit = {
2.   ......
3.   webUi = new MasterWebUI(this, webUiPort)
4.     webUi.bind()
5.     masterWebUiUrl = "http://" + masterPublicAddress + ":" + webUi
       .boundPort
```

如 MasterWebUI 的 spark.ui.killEnabled 设置为 True，可以通过 WebUI 页面把 Spark 的进程杀掉。

Spark 2.2.1 版本的 MasterWebUI.scala 的源码如下：

```
1.    private[master]
2.  class MasterWebUI(
3.      val master: Master,
4.      requestedPort: Int)
5.    extends  WebUI(master.securityMgr,  master.securityMgr.getSSLOptions
      ("standalone"),
6.      requestedPort, master.conf, name = "MasterUI") with Logging {
7.    val masterEndpointRef = master.self
8.    val killEnabled = master.conf.getBoolean("spark.ui.killEnabled", true)
9.  ......
10. initialize()
11.
12.   /**初始化所有的服务器组件 */
13.   def initialize() {
14.     val masterPage = new MasterPage(this)
15.     attachPage(new ApplicationPage(this))
16.     attachPage(masterPage)
17.     attachHandler(createStaticHandler(MasterWebUI.STATIC_RESOURCE_DIR,
        "/static"))
18.     attachHandler(createRedirectHandler(
19.       "/app/kill", "/", masterPage.handleAppKillRequest, httpMethods =
          Set("POST")))
20.     attachHandler(createRedirectHandler(
21.       "/driver/kill", "/", masterPage.handleDriverKillRequest,
          httpMethods = Set("POST")))
22.   }
```

Spark 2.4.3 版本的源码与 Spark 2. 2.1 版本相比具有如下特点。

- 上段代码中第 17 行将 attachHandler 方法调整为 addStaticHandler 方法。

```
1.    addStaticHandler(MasterWebUI.STATIC_RESOURCE_DIR)
```

MasterWebUI 中在初始化时用 new()函数创建 MasterPage，在 MasterPage 中通过代码去写 Web 页面。

MasterPage.scala 的源码如下：

```
1.   private[ui] class MasterPage(parent: MasterWebUI) extends WebUIPage("") {
2.   ......
3.    override def renderJson(request: HttpServletRequest): JValue = {
4.      JsonProtocol.writeMasterState(getMasterState)
5.    }
6.  ......
7.    val content =
8.          <div class="row-fluid">
9.            <div class="span12">
10.             <ul class="unstyled">
11.               <li><strong>URL:</strong> {state.uri}</li>
12.               {
13.                 state.restUri.map { uri =>
14.                   <li>
15.                     <strong>REST URL:</strong> {uri}
16.                     <span class="rest-uri"> (cluster mode)</span>
17.                   </li>
18.                 }.getOrElse { Seq.empty }
19.               }
20.               <li><strong>Alive Workers:</strong> {aliveWorkers.length}</li>
21.               <li><strong>Cores in use:</strong> {aliveWorkers.map
                (_.cores).sum} Total,
```

```
22.               {aliveWorkers.map(_.coresUsed).sum} Used</li>
23.             <li><strong>Memory in use:</strong>
24.               {Utils.megabytesToString(aliveWorkers.map(_.memory).sum)}
              Total,
25.               {Utils.megabytesToString(aliveWorkers.map(_.memoryUsed)
              .sum)} Used</li>
26.             <li><strong>Applications:</strong>
27.               {state.activeApps.length} <a href="#running-app">Running</a>,
28.               {state.completedApps.length}  <a  href="#completed-app">
              Completed</a> </li>
29.             <li><strong>Drivers:</strong>
30.               {state.activeDrivers.length} Running,
31.               {state.completedDrivers.length} Completed </li>
32.             <li><strong>Status:</strong> {state.status}</li>
33.           </ul>
34.         </div>
35.       </div>
36. ......
```

回到 MasterWebUI.scala 的 initialize()方法，其中调用了 attachPage 方法，在 WebUI 中增加 Web 页面。

Spark 2.2.1 版本的 WebUI.scala 的源码如下：

```
1.     def attachPage(page: WebUIPage) {
2.     val pagePath = "/" + page.prefix
3.     val renderHandler = createServletHandler(pagePath,
4.       (request: HttpServletRequest) => page.render(request),
       securityManager, conf, basePath)
5.     val  renderJsonHandler  =  createServletHandler(pagePath.stripSuffix
     ("/") + "/json",
6.       (request: HttpServletRequest) => page.renderJson(request),
       securityManager, conf, basePath)
7.     attachHandler(renderHandler)
8.     attachHandler(renderJsonHandler)
9.     val  handlers  =  pageToHandlers.getOrElseUpdate(page,  ArrayBuffer
     [ServletContextHandler]())
10.    handlers += renderHandler
11.  }
```

Spark 2.4.3 版本的 WebUI.scala 的源码与 Spark 2. 2.1 版本相比具有如下特点。

- 上段代码中第 1 行进行修改，增加 attachPage 方法的返回类型。

```
1.  def attachPage(page: WebUIPage): Unit = {
```

在 WebUI 的 bind 方法中启用了 JettyServer。

WebUI.scala 的 bind 的源码如下：

```
1.   def bind() {
2.     assert(!serverInfo.isDefined, s"Attempted to bind $className more than
     once!")
3.     try {
4.       val host = Option(conf.getenv("SPARK_LOCAL_IP")).getOrElse
       ("0.0.0.0")
5.      serverInfo = Some(startJettyServer(host, port, sslOptions, handlers,
      conf, name))
6.       logInfo(s"Bound $className to $host, and started at $webUrl")
7.     } catch {
8.       case e: Exception =>
```

```
9.         logError(s"Failed to bind $className", e)
10.        System.exit(1)
11.     }
12.   }
```

JettyUtils.scala 的 startJettyServer 尝试将 Jetty 服务器绑定到所提供的主机名、端口。startJettyServer 的源码如下：

```
1.  def startJettyServer(
2.        hostName: String,
3.        port: Int,
4.        sslOptions: SSLOptions,
5.        handlers: Seq[ServletContextHandler],
6.        conf: SparkConf,
7.        serverName: String = ""): ServerInfo = {
```

5.1.3　Master HA 双机切换

Spark 生产环境下一般采用 ZooKeeper 作 HA，且建议为 3 台 Master，ZooKeeper 会自动化管理 Masters 的切换。

采用 ZooKeeper 作 HA 时，ZooKeeper 会保存整个 Spark 集群运行时的元数据，包括 Workers、Drivers、Applications、Executors。

ZooKeeper 遇到当前 Active 级别的 Master 出现故障时会从 Standby Masters 中选取出一台作为 Active Master，但是要注意，被选举后到成为真正的 Active Master 之前需要从 ZooKeeper 中获取集群当前运行状态的元数据信息并进行恢复。

在 Master 切换的过程中，所有已经在运行的程序皆正常运行。因为 Spark Application 在运行前就已经通过 Cluster Manager 获得了计算资源，所以在运行时，Job 本身的调度和处理和 Master 是没有任何关系的。

在 Master 的切换过程中唯一的影响是不能提交新的 Job：一方面不能够提交新的应用程序给集群，因为只有 Active Master 才能接收新的程序提交请求；另一方面，已经运行的程序中也不能因为 Action 操作触发新的 Job 提交请求。

ZooKeeper 下 Master HA 的基本流程如图 5-2 所示。

ZooKeeper 下 Master HA 的基本流程如下。

（1）使用 ZooKeeperPersistenceEngine 读取集群的状态数据，包括 Drivers、Applications、Workers、Executors 等信息。

（2）判断元数据信息是否有空的内容。

（3）把通过 ZooKeeper 持久化引擎获得了 Drivers、Applications、Workers、Executors 等信息，重新注册到 Master 的内存中缓存起来。

（4）验证获得的信息和当前正在运行的集群状态的一致性。

（5）将 Application 和 Workers 的状态标识为 Unknown，然后向 Application 中的 Driver 以及 Workers 发送现在是 Leader 的 Standby 模式的 Master 的地址信息。

（6）当 Driver 和 Workers 收到新的 Master 的地址信息后会响应该信息。

（7）Master 接收到来自 Drivers 和 Workers 响应的信息后会使用一个关键的方法：completeRecovery()来对没有响应的 Applications (Drivers)、Workers (Executors) 进行处理。处理完毕后，Master 的 State 会变成 RecoveryState.ALIVE，从而开始对外提供服务。

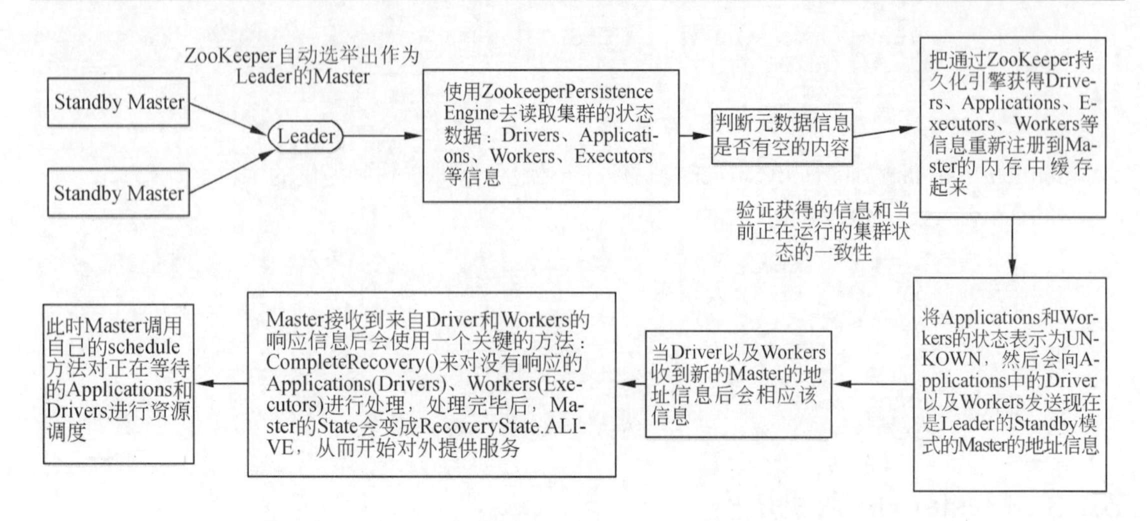

图 5-2　ZooKeeper 下 Master HA 的基本流程

（8）此时 Master 使用自己的 Schedule 方法对正在等待的 Application 和 Drivers 进行资源调度。

Master HA 的 4 大方式分别是 ZOOKEEPER、FILESYSTEM、CUSTOM、NONE。

需要说明的是：

（1）ZOOKEEPER 是自动管理 Master。

（2）FILESYSTEM 的方式在 Master 出现故障后需要手动重新启动机器，机器启动后会立即成为 Active 级别的 Master 来对外提供服务（接收应用程序提交的请求、接收新的 Job 运行的请求）。

（3）CUSTOM 的方式允许用户自定义 Master HA 的实现，这对高级用户特别有用。

（4）NONE，这是默认情况，Spark 集群中就采用这种方式，该方式不会持久化集群的数据，Master 启动后立即管理集群。

Master.scala 的 HA 的源码如下：

```
1.   override def onStart(): Unit = {
2.  ……
3.  val serializer = new JavaSerializer(conf)
4.     val (persistenceEngine_, leaderElectionAgent_) = RECOVERY_MODE match {
5.       case "ZOOKEEPER" =>
6.         logInfo("Persisting recovery state to ZooKeeper")
7.         val zkFactory =
8.           new ZooKeeperRecoveryModeFactory(conf, serializer)
9.         (zkFactory.createPersistenceEngine(), zkFactory.createLeader-
           ElectionAgent(this))
10.      case "FILESYSTEM" =>
11.        val fsFactory =
12.          new FileSystemRecoveryModeFactory(conf, serializer)
13.        (fsFactory.createPersistenceEngine(), fsFactory.
           createLeaderElectionAgent(this))
14.      case "CUSTOM" =>
15.        val clazz = Utils.classForName(conf.get("spark.deploy.
           recoveryMode.factory"))
16.        val factory = clazz.getConstructor(classOf[SparkConf], classOf
           [Serializer])
```

```
17.          .newInstance(conf, serializer)
18.          .asInstanceOf[StandaloneRecoveryModeFactory]
19.        (factory.createPersistenceEngine(), factory.createLeaderElectionAgent
           (this))
20.      case _ =>
21.        (new BlackHolePersistenceEngine(), new MonarchyLeaderAgent(this))
22.    }
23.    persistenceEngine = persistenceEngine_
24.    leaderElectionAgent = leaderElectionAgent_
25.  }
26. ……
```

Spark默认的HA方式是NONE。

```
1.  private val RECOVERY_MODE = conf.get("spark.deploy.recoveryMode", "NONE")
```

如使用ZOOKEEPER的HA方式，RecoveryModeFactory.scala的源码如下：

```
1.    private[master] class ZooKeeperRecoveryModeFactory(conf: SparkConf,
      serializer: Serializer)
2.    extends StandaloneRecoveryModeFactory(conf, serializer) {
3.
4.    def createPersistenceEngine(): PersistenceEngine = {
5.      new ZooKeeperPersistenceEngine(conf, serializer)
6.    }
7.
8.    def createLeaderElectionAgent(master: LeaderElectable):
      LeaderElectionAgent = {
9.      new ZooKeeperLeaderElectionAgent(master, conf)
10.   }
11. }
```

通过调用zkFactory.createPersistenceEngine()用new()函数创建一个ZooKeeper-PersistenceEngine。

ZooKeeperPersistenceEngine.scala的源码如下：

```
1.     private[master] class ZooKeeperPersistenceEngine(conf: SparkConf, val
       serializer: Serializer)
2.    extends PersistenceEngine
3.    with Logging {
4.
5.    private  val  WORKING_DIR  =  conf.get("spark.deploy.zookeeper.dir",
      "/spark") + "/master_status"
6.    private val zk: CuratorFramework = SparkCuratorUtil.newClient(conf)
7.
8.    SparkCuratorUtil.mkdir(zk, WORKING_DIR)
9.
10.
11.   override def persist(name: String, obj: Object): Unit = {
12.     serializeIntoFile(WORKING_DIR + "/" + name, obj)
13.   }
14.
15.   override def unpersist(name: String): Unit = {
16.     zk.delete().forPath(WORKING_DIR + "/" + name)
17.   }
18.
19.   override def read[T: ClassTag](prefix: String): Seq[T] = {
20.     zk.getChildren.forPath(WORKING_DIR).asScala
21.       .filter(_.startsWith(prefix)).flatMap(deserializeFromFile[T])
```

```
22.   }
23.
24.   override def close() {
25.     zk.close()
26.   }
27.
28.   private def serializeIntoFile(path: String, value: AnyRef) {
29.     val serialized = serializer.newInstance().serialize(value)
30.     val bytes = new Array[Byte](serialized.remaining())
31.     serialized.get(bytes)
32.     zk.create().withMode(CreateMode.PERSISTENT).forPath(path, bytes)
33.   }
34.
35.   private  def  deserializeFromFile[T](filename:  String)(implicit  m:
      ClassTag[T]): Option[T] = {
36.     val fileData = zk.getData().forPath(WORKING_DIR + "/" + filename)
37.     try {
38.       Some(serializer.newInstance().deserialize[T](ByteBuffer.wrap
          (fileData)))
39.     } catch {
40.       case e: Exception =>
41.         logWarning("Exception while reading persisted file, deleting", e)
42.         zk.delete().forPath(WORKING_DIR + "/" + filename)
43.         None
44.     }
45.   }
46. }
```

PersistenceEngine 中有至关重要的方法 persist 来实现数据持久化，readPersistedData 来恢复集群中的元数据。

PersistenceEngine.scala 的源码如下：

```
1.  def persist(name: String, obj: Object): Unit
2.  ......
3.   final def readPersistedData(
4.       rpcEnv: RpcEnv): (Seq[ApplicationInfo], Seq[DriverInfo], Seq
         [WorkerInfo]) = {
5.     rpcEnv.deserialize { () =>
6.       (read[ApplicationInfo]("app_"), read[DriverInfo]("driver_"), read
         [WorkerInfo]("worker_"))
7.     }
8.   }
```

下面来看 createdLeaderElectionAgent 方法。在 createdLeaderElectionAgent 方法中调用 new()函数创建 ZooKeeperLeaderElectionAgent 实例。

StandaloneRecoveryModeFactory.scala 的源码如下：

```
1.    def createLeaderElectionAgent(master: LeaderElectable):
      LeaderElectionAgent = {
2.      new ZooKeeperLeaderElectionAgent(master, conf)
3.    }
4.  }
```

ZooKeeperLeaderElectionAgent 的源码如下：

```
1.    private[master] class ZooKeeperLeaderElectionAgent(val masterInstance:
      LeaderElectable,
2.      conf: SparkConf) extends LeaderLatchListener with LeaderElectionAgent
        with Logging  {
```

```
3.
4.   val WORKING_DIR = conf.get("spark.deploy.zookeeper.dir", "/spark") +
     "/leader_election"
5.
6.   private var zk: CuratorFramework = _
7.   private var leaderLatch: LeaderLatch = _
8.   private var status = LeadershipStatus.NOT_LEADER
9.
10.  start()
11.
12.  private def start() {
13.    logInfo("Starting ZooKeeper LeaderElection agent")
14.    zk = SparkCuratorUtil.newClient(conf)
15.    leaderLatch = new LeaderLatch(zk, WORKING_DIR)
16.    leaderLatch.addListener(this)
17.    leaderLatch.start()
18.  }
19.
20.  override def stop() {
21.    leaderLatch.close()
22.    zk.close()
23.  }
24.
25.  override def isLeader() {
26.    synchronized {
27.      //可以取得领导权
28.      if (!leaderLatch.hasLeadership) {
29.        return
30.      }
31.
32.      logInfo("We have gained leadership")
33.      updateLeadershipStatus(true)
34.    }
35.  }
36.
37.  override def notLeader() {
38.    synchronized {
39.      //可以取得领导权
40.      if (leaderLatch.hasLeadership) {
41.        return
42.      }
43.
44.      logInfo("We have lost leadership")
45.      updateLeadershipStatus(false)
46.    }
47.  }
48.
49.  private def updateLeadershipStatus(isLeader: Boolean) {
50.    if (isLeader && status == LeadershipStatus.NOT_LEADER) {
51.      status = LeadershipStatus.LEADER
52.      masterInstance.electedLeader()
53.    } else if (!isLeader && status == LeadershipStatus.LEADER) {
54.      status = LeadershipStatus.NOT_LEADER
55.      masterInstance.revokedLeadership()
56.    }
57.  }
58.
59.  private object LeadershipStatus extends Enumeration {
60.    type LeadershipStatus = Value
```

```
61.     val LEADER, NOT_LEADER = Value
62.   }
63. }
```

FILESYSTEM 和 NONE 的方式采用 MonarchyLeaderAgent 的方式来完成 Leader 的选举，其实现是直接把传入的 Master 作为 Leader。

LeaderElectionAgent.scala 的源码如下：

```
1.   private[spark] class MonarchyLeaderAgent(val masterInstance:
     LeaderElectable)
2.    extends LeaderElectionAgent {
3.    masterInstance.electedLeader()
4.  }
```

FileSystemRecoveryModeFactory.scala 的源码如下：

```
1.  private[master] class FileSystemRecoveryModeFactory(conf: SparkConf,
    serializer: Serializer)
2.    extends StandaloneRecoveryModeFactory(conf, serializer) with Logging {
3.
4.    val RECOVERY_DIR = conf.get("spark.deploy.recoveryDirectory", "")
5.
6.    def createPersistenceEngine(): PersistenceEngine = {
7.      logInfo("Persisting recovery state to directory: " + RECOVERY_DIR)
8.      new FileSystemPersistenceEngine(RECOVERY_DIR, serializer)
9.    }
10.
11.   def createLeaderElectionAgent(master: LeaderElectable):
      LeaderElectionAgent = {
12.     new MonarchyLeaderAgent(master)
13.   }
14. }
```

如果 WorkerState 状态为 UNKNOWN（Worker 不响应），就把它删除，如果以集群方式运行，driver 失败后可以重新启动，最后把状态变回 ALIVE。注意，这里要加入--supervise 这个参数。

Master.scala 的源码如下：

```
1.    private def completeRecovery() {
2.      // 使用短同步周期确保“only-once”恢复一次语义
3.      if (state != RecoveryState.RECOVERING) { return }
4.      state = RecoveryState.COMPLETING_RECOVERY
5.
6.      //杀掉不响应消息的 workers 和 apps
7.      workers.filter(_.state == WorkerState.UNKNOWN).foreach(
8.        removeWorker(_, "Not responding for recovery"))
9.      apps.filter(_.state == ApplicationState.UNKNOWN).foreach(finishApplication)
10.
11.     //将恢复的应用程序的状态更新为正在运行
12.     apps.filter(_.state == ApplicationState.WAITING).foreach(_.state =
        ApplicationState.RUNNING)
13.
14.     //重新调度 drivers，其未被任何 workers 声明
15.     drivers.filter(_.worker.isEmpty).foreach { d =>
16.       logWarning(s"Driver ${d.id} was not found after master recovery")
17.       if (d.desc.supervise) {
18.         logWarning(s"Re-launching ${d.id}")
```

```
19.         relaunchDriver(d)
20.       } else {
21.         removeDriver(d.id, DriverState.ERROR, None)
22.         logWarning(s"Did not re-launch ${d.id} because it was not supervised")
23.       }
24.     }
25.
26.     state = RecoveryState.ALIVE
27.     schedule()
28.     logInfo("Recovery complete - resuming operations!")
29.   }
```

5.1.4　Master 的注册机制和状态管理解密

1．Master对其他组件注册的处理

Master 接收注册的对象主要是 Driver、Application、Worker；Executor 不会注册给 Master，Executor 是注册给 Driver 中的 SchedulerBackend 的。

Worker 是在启动后主动向 Master 注册的，所以如果在生产环境下加入新的 Worker 到正在运行的 Spark 集群上，此时不需要重新启动 Spark 集群就能够使用新加入的 Worker，以提升处理能力。假如在生产环境中的集群中有 500 台机器，可能又新加入 100 台机器，这时不需要重新启动整个集群，就可以将 100 台新机器加入到集群。

Spark 2.2.1 版本的 Worker 的源码如下：

```
1.  private[deploy] class Worker(
2.      override val rpcEnv: RpcEnv,
3.      webUiPort: Int,
4.      cores: Int,
5.      memory: Int,
6.      masterRpcAddresses: Array[RpcAddress],
7.      endpointName: String,
8.      workDirPath: String = null,
9.      val conf: SparkConf,
10.     val securityMgr: SecurityManager)
11.   extends ThreadSafeRpcEndpoint with Logging {
```

Spark 2.4.3 版本的 Worker.scala 源码与 Spark 2. 2.1 版本相比具有如下特点。

- 上段代码中第 10 行之后新增 externalShuffleServiceSupplier 的成员变量，externalShuffleServiceSupplier 是 ExternalShuffleService 类型，ExternalShuffleService 提供一个 server，Executors 可以从 server 中读取 shuffle 文件（而不是彼此之间直接读取），在 executors 被关闭或杀死的情况下提供对文件的不间断访问，需要 SASL 身份验证才能读取。

```
1.    externalShuffleServiceSupplier: Supplier[ExternalShuffleService] = null)
```

Worker 是一个消息循环体，继承自 ThreadSafeRpcEndpoint，可以收消息，也可以发消息。Worker 的 onStart 方法如下：

```
1.      override def onStart() {
2.    ......
3.      workerWebUiUrl = s"http://$publicAddress:${webUi.boundPort}"
4.      registerWithMaster()
```

```
5.      ......
6.    }
```

Worker 的 onStart 方法中调用了 registerWithMaster()。

```
1.      private def registerWithMaster() {
2.      ......
3.      registrationRetryTimer match {
4.        case None =>
5.          registered = false
6.          registerMasterFutures = tryRegisterAllMasters()
7.       ......
```

registerWithMaster 方法中调用了 tryRegisterAllMasters，向所有的 Master 进行注册。Worker.scala 的源码如下：

```
1.      private def tryRegisterAllMasters(): Array[JFuture[_]] = {
2.      masterRpcAddresses.map { masterAddress =>
3.        registerMasterThreadPool.submit(new Runnable {
4.          override def run(): Unit = {
5.            try {
6.              logInfo("Connecting to master " + masterAddress + "...")
7.              val masterEndpoint = rpcEnv.setupEndpointRef(masterAddress,
                Master.ENDPOINT_NAME)
8.              sendRegisterMessageToMaster(masterEndpoint)
9.            } catch {
10.             case ie: InterruptedException => //Cancelled
11.             case NonFatal(e) => logWarning(s"Failed to connect to master
                $masterAddress", e)
12.           }
13.         }
14.       })
15.     }
16.   }
```

tryRegisterAllMasters 方法中，由于实际运行时有很多 Master，因此使用线程池的线程进行提交，然后获取 masterEndpoint。masterEndpoint 是一个 RpcEndpointRef，通过 sendRegisterMessageToMaster (masterEndpoint)进行注册。

sendRegisterMessageToMaster 方法仅将 RegisterWorker 消息发送给 Master 消息循环体。sendRegisterMessageToMaster 方法内部不作其他处理。

```
1.    private def sendRegisterMessageToMaster(masterEndpoint: RpcEndpointRef):
      Unit = {
2.      masterEndpoint.send(RegisterWorker(
3.        workerId,
4.        host,
5.        port,
6.        self,
7.        cores,
8.        memory,
9.        workerWebUiUrl,
10.       masterEndpoint.address))
11.   }
```

sendRegisterMessageToMaster 方法中的 masterEndpoint.send 传进去的是 RegisterWorker。RegisterWorker 是一个 case class，包括 id、host、port、worker、cores、memory 等信息，这里 Worker 是自己的引用 RpcEndpointRef，Master 通过 Ref 通 worker 通信。

RegisterWorker.scala 的源码如下：

```
1.   case class RegisterWorker(
2.      id: String,
3.      host: String,
4.      port: Int,
5.      worker: RpcEndpointRef,
6.      cores: Int,
7.      memory: Int,
8.      workerWebUiUrl: String,masterAddress:RpcAddress)
9.    extends DeployMessage {
10.   Utils.checkHost(host, "Required hostname")
11.   assert (port > 0)
12.  }
```

Worker 通过 sendRegisterMessageToMaster 向 Master 发送了 RegisterWorker 消息，Master 收到 RegisterWorker 请求后，进行相应的处理。

Master.scala 的 receive 的源码如下：

```
1.  override def receive: PartialFunction[Any, Unit] = {
2.  ......
3.
4.   case RegisterWorker(
5.       id, workerHost, workerPort, workerRef, cores, memory, workerWebUiUrl,
    masterAddress) =>
6.       logInfo("Registering worker %s:%d with %d cores, %s RAM".format(
7.         workerHost, workerPort, cores, Utils.megabytesToString(memory)))
8.       if (state == RecoveryState.STANDBY) {
9.         workerRef.send(MasterInStandby)
10.      } else if (idToWorker.contains(id)) {
11.        workerRef.send(RegisterWorkerFailed("Duplicate worker ID"))
12.      } else {
13.        val worker = new WorkerInfo(id, workerHost, workerPort, cores, memory,
14.          workerRef, workerWebUiUrl)
15.        if (registerWorker(worker)) {
16.          persistenceEngine.addWorker(worker)
17.          workerRef.send(RegisteredWorker(self, masterWebUiUrl, masterAddress))
18.          schedule()
19.        } else {
20.          val workerAddress = worker.endpoint.address
21.          logWarning("Worker registration failed. Attempted to re-register
           worker at same " +
22.            "address: " + workerAddress)
23.          workerRef.send(RegisterWorkerFailed("Attempted to re-register
          worker at same address: "
24.            + workerAddress))
25.        }
26.          }
```

RegisterWorker 中，Master 接收到 Worker 的注册请求后，首先判断当前的 Master 是否是 Standby 的模式，如果是，就不处理；Master 的 idToWorker 包含了所有已经注册的 Worker 的信息，然后会判断当前 Master 的内存数据结构 idToWorker 中是否已经有该 Worker 的注册，如果有，此时不会重复注册；其中 idToWorker 是一个 HashMap，Key 是 String 代表 Worker 的字符描述，Value 是 WorkerInfo。

```
1.     private val idToWorker = new HashMap[String, WorkerInfo]
```

WorkerInfo 包括 id、host、port 、cores、memory、endpoint 等内容。

```
1.      private[spark] class WorkerInfo(
2.      val id: String,
3.      val host: String,
4.      val port: Int,
5.      val cores: Int,
6.      val memory: Int,
7.      val endpoint: RpcEndpointRef,
8.      val webUiAddress: String)
9.    extends Serializable {
```

Master 如果决定接收注册的 Worker，首先会创建 WorkerInfo 对象来保存注册的 Worker 的信息，然后调用 registerWorker 执行具体的注册的过程，如果 Worker 的状态是 DEAD 的状态，则直接过滤掉。对于 UNKNOWN 的内容，调用 removeWorker 进行清理（包括清理该 Worker 下的 Executors 和 Drivers）。其中，UNKNOWN 的情况：Master 进行切换时，先对 Worker 发 UNKNOWN 消息，只有当 Master 收到 Worker 正确的回复消息，才将状态标识为正常。

Spark 2.2.1 版本的 Master.scala 的 registerWorker 的源码如下：

```
1.       private def registerWorker(worker: WorkerInfo): Boolean = {
2.    //在同一节点上可能有一个或多个指向挂掉的 workers 节点的引用（不同 ID），须删除它们
3.      workers.filter { w =>
4.        (w.host == worker.host && w.port == worker.port) && (w.state ==
          WorkerState.DEAD)
5.      }.foreach { w =>
6.        workers -= w
7.      }
8.
9.      val workerAddress = worker.endpoint.address
10.     if (addressToWorker.contains(workerAddress)) {
11.       val oldWorker = addressToWorker(workerAddress)
12.       if (oldWorker.state == WorkerState.UNKNOWN) {
13.         //未知状态的 worker 意味着在恢复过程中 worker 被重新启动。旧的 worker 节点
            //挂掉，须删掉旧节点，接收新 worker 节点
14.         removeWorker(oldWorker)
15.       } else {
16.         logInfo("Attempted to re-register worker at same address: " +
            workerAddress)
17.         return false
18.       }
19.     }
20.
21.     workers += worker
22.     idToWorker(worker.id) = worker
23.     addressToWorker(workerAddress) = worker
24.     if (reverseProxy) {
25.        webUi.addProxyTargets(worker.id, worker.webUiAddress)
26.     }
27.     true
28.   }
```

Spark 2.4.3 版本的 Master.scala 源码与 Spark 2. 2.1 版本相比具有如下特点。

- 上段代码中第 14 行 removeWorker(oldWorker)新增一个消息说明信息。

```
1.  removeWorker(oldWorker, "Worker replaced by a new worker with same address")
```

在 registerWorker 方法中，Worker 注册完成后，把注册的 Worker 加入到 Master 的内存数据结构中。

```
1.  val workers = new HashSet[WorkerInfo]
2.  private val idToWorker = new HashMap[String, WorkerInfo]
3.    private val addressToWorker = new HashMap[RpcAddress, WorkerInfo]
4.  ......
5.
6.    workers += worker
7.      idToWorker(worker.id) = worker
8.      addressToWorker(workerAddress) = worker
```

回到Master.scala的receiveAndReply方法，Worker注册完成后，调用persistenceEngine.addWorker(worker)，PersistenceEngine 是持久化引擎，在 Zookeeper 下就是 Zookeeper 的持久化引擎，把注册的数据进行持久化。

PersistenceEngine.scala 的 addWorker 方法如下：

```
1.        final def addWorker(worker: WorkerInfo): Unit = {
2.      persist("worker_" + worker.id, worker)
3.    }
```

ZooKeeperPersistenceEngine 是 PersistenceEngine 的一个具体实现子类，其 persist 方法如下：

```
1.    private[master] class ZooKeeperPersistenceEngine(conf: SparkConf, val
      serializer: Serializer)
2.   extends PersistenceEngine
3.  ......
4.   override def persist(name: String, obj: Object): Unit = {
5.     serializeIntoFile(WORKING_DIR + "/" + name, obj)
6.   }
7.  ......
8.  private def serializeIntoFile(path: String, value: AnyRef) {
9.     val serialized = serializer.newInstance().serialize(value)
10.    val bytes = new Array[Byte](serialized.remaining())
11.    serialized.get(bytes)
12.    zk.create().withMode(CreateMode.PERSISTENT).forPath(path, bytes)
13.  }
```

回到Master.scala的receiveAndReply方法，注册的Worker数据持久化后，进行schedule()。至此，Worker 的注册完成。

同样，Driver 的注册过程：Driver 提交给 Master 进行注册，Master 会将 Driver 的信息放入内存缓存中，加入等待调度的队列，通过持久化引擎（如 ZooKeeper）把注册信息持久化，然后进行 Schedule。

Application 的注册过程：Application 提交给 Master 进行注册，Driver 启动后会执行 SparkContext 的初始化，进而导致 StandaloneSchedulerBackend 的产生，其内部有 StandaloneAppClient。StandaloneAppClient 内部有 ClientEndpoint。ClientEndpoint 来发送 RegisterApplication 信息给 Master。Master 会将 Application 的信息放入内存缓存中，把 Application 加入等待调度的 Application 队列，通过持久化引擎（如 ZooKeeper）把注册信息持久化，然后进行 Schedule。

2．Master对Driver和Executor状态变化的处理

1）对 Driver 状态变化的处理

如果 Driver 的各个状态是 DriverState.ERROR | DriverState.FINISHED | DriverState.KILLED | DriverState.FAILED，就将其清理掉。其他情况则报异常。

```
1.    override def receive: PartialFunction[Any, Unit] = {
2.  ......
3.   case DriverStateChanged(driverId, state, exception) =>
4.       state match {
5.         case DriverState.ERROR | DriverState.FINISHED | DriverState.KILLED
           | DriverState.FAILED =>
6.           removeDriver(driverId, state, exception)
7.         case _ =>
8.           throw new Exception(s"Received unexpected state update for driver
             $driverId: $state")
9.       }
```

removeDriver 清理掉 Driver 后，再次调用 schedule 方法，removeDriver 的源码如下：

```
1.      private def removeDriver(
2.       driverId: String,
3.       finalState: DriverState,
4.       exception: Option[Exception]) {
5.     drivers.find(d => d.id == driverId) match {
6.       case Some(driver) =>
7.         logInfo(s"Removing driver: $driverId")
8.         drivers -= driver
9.         if (completedDrivers.size >= RETAINED_DRIVERS) {
10.          val toRemove = math.max(RETAINED_DRIVERS / 10, 1)
11.          completedDrivers.trimStart(toRemove)
12.        }
13.        completedDrivers += driver
14.        persistenceEngine.removeDriver(driver)
15.        driver.state = finalState
16.        driver.exception = exception
17.        driver.worker.foreach(w => w.removeDriver(driver))
18.        schedule()
19.      case None =>
20.        logWarning(s"Asked to remove unknown driver: $driverId")
21.    }
22.  }
23. }
```

2）对 Executor 状态变化的处理

ExecutorStateChanged 的源码如下：

```
1.   override def receive: PartialFunction[Any, Unit] = {
2.   ......
3.  case ExecutorStateChanged(appId, execId, state, message, exitStatus) =>
4.       val execOption = idToApp.get(appId).flatMap(app => app.executors
         .get(execId))
5.       execOption match {
6.         case Some(exec) =>
7.           val appInfo = idToApp(appId)
8.           val oldState = exec.state
9.           exec.state = state
```

```
10.
11.          if (state == ExecutorState.RUNNING) {
12.            assert(oldState == ExecutorState.LAUNCHING,
13.              s"executor $execId state transfer from $oldState to RUNNING
               is illegal")
14.            appInfo.resetRetryCount()
15.          }
16.
17.          exec.application.driver.send(ExecutorUpdated(execId,    state,
           message, exitStatus, false))
18.
19.          if (ExecutorState.isFinished(state)) {
20.            //从 worker 和 app 中删掉 executor
21.            logInfo(s"Removing  executor  ${exec.fullId}  because  it  is
             $state")
22.            //如果应用程序已经完成，保存其状态及在 UI 上正确显示其信息
23.            if (!appInfo.isFinished) {
24.              appInfo.removeExecutor(exec)
25.            }
26.            exec.worker.removeExecutor(exec)
27.
28.            val normalExit = exitStatus == Some(0)
29.            //只重试一定次数，这样就不会进入无限循环。重要提示：这个代码路径不是通过
             //测试执行的，所以改变 if 条件时必须小心
30.            if (!normalExit
31.                && appInfo.incrementRetryCount() >= MAX_EXECUTOR_RETRIES
32.                && MAX_EXECUTOR_RETRIES >= 0) { //< 0 disables this
                 application-killing path
33.              val execs = appInfo.executors.values
34.              if (!execs.exists(_.state == ExecutorState.RUNNING)) {
35.                logError(s"Application ${appInfo.desc.name} with ID
                 ${appInfo.id} failed " +
36.                  s"${appInfo.retryCount} times; removing it")
37.                removeApplication(appInfo, ApplicationState.FAILED)
38.              }
39.            }
40.          }
41.          schedule()
42.        case None =>
43.          logWarning(s"Got  status  update  for  unknown  executor  $appId/
           $execId")
44.      }
```

Executor 挂掉时系统会尝试一定次数的重启（最多重启 10 次）。

```
1.      private val MAX_EXECUTOR_RETRIES = conf.getInt("spark.deploy.
        maxExecutorRetries", 10)
```

5.2　Worker 启动原理和源码详解

本节讲解 Worker 启动原理和源码。对于 Worker 的部署启动，我们以 Worker 的脚本为入口点进行分析。

5.2.1 Worker 启动的原理流程

Spark 中各个组件是通过脚本启动部署的。Worker 的部署以脚本为入口点开始分析。每个组件对应提供了启动的脚本，同时也会提供停止的脚本，停止脚本比较简单，在此仅分析启动的脚本。

部署 Worker 组件时，最简单的方式是通过配置 Spark 部署目录下的 conf/slaves 文件，然后以批量的方式启动集群中在该文件中列出的全部节点上的 Worker 实例。启动组件的命令如下所示：

```
1.  ./sbin/start-slaves.sh
```

或者动态地在某个新增节点上（注意是新增节点，如果之前已经部署过，可以参考后面对启动多个实例的进一步分析）启动一个 Worker 实例，此时可以在该新增的节点上执行如下启动命令。

```
1.  ./sbin/start-slave.sh  MasterURL
```

其中，参数 MasterURL 表示当前集群中 Master 的监听地址，启动后 Worker 会通过该地址动态注册到 Master 组件，实现为集群动态添加 Worker 节点的目的。

下面是 Worker 部署脚本的解析。

部署脚本根据单个节点以及多个节点的 Worker 部署，对应有两个脚本：start-slave.sh 和 start-slaves.sh。其中，start-slave.sh 负责在脚本执行节点启动一个 Worker 组件。start-slaves.sh 脚本则会读取配置的 conf/slaves 文件，逐个启动集群中各个 Slave 节点上的 Worker 组件。

1. 首先分析脚本start-slaves.sh

脚本 start-slaves.sh 提供了批量启动集群中各个 Slave 节点上的 Worker 组件的方法，即可以在配置好 Slave 节点（即配置好 conf/slaves 文件）后，通过该脚本一次性全部启动集群中的 Worker 组件。

脚本的代码如下：

```
1.  #在根据 conf/slaves 文件指定的每个节点上启动一个 Slave 实例，即 Worker 组件
2.  if [ -z "${SPARK_HOME}" ]; then
3.    export SPARK_HOME="$(cd "'dirname "$0"'"/..; pwd)"
4.  fi
5.
6.  . "${SPARK_HOME}/sbin/spark-config.sh"
7.  . "${SPARK_HOME}/bin/load-spark-env.sh"
8.
9.  # 找到主机的端口号
10. if [ "$SPARK_MASTER_PORT" = "" ]; then
11.   SPARK_MASTER_PORT=7077
12. fi
13.
14. if [ "$SPARK_MASTER_HOST" = "" ]; then
15.   case 'uname' in
16.       (SunOS)
17.       SPARK_MASTER_HOST="'/usr/sbin/check-hostname | awk '{print $NF}''"
18.       ;;
```

```
19.         (*)
20.         SPARK_MASTER_HOST="'hostname -f'"
21.         ;;
22.     esac
23. fi
24.
25. # 启动 slaves
26. "${SPARK_HOME}/sbin/slaves.sh" cd "${SPARK_HOME}" \; "${SPARK_HOME}/
    sbin/start-slave.sh" "spark://$SPARK_MASTER_HOST:$SPARK_MASTER_PORT"
```

其中，脚本 slaves.sh 通过 ssh 协议在指定的各个 Slave 节点上执行各种命令。

在 ssh 启动的 start-slave.sh 命令中，可以看到它的参数是"spark://$SPARK_MASTER_IP:$SPARK_MASTER_PORT"，即启动 slave 节点上的 Worker 进程时，使用的 Master URL 的值是通过两个环境变量（SPARK_MASTER_IP 和 SPARK_MASTER_PORT）拼接而成的。

2. 脚本start-slave.sh分析

从前面 start-slaves.sh 脚本的分析中可以看到，最终是在各个 Slave 节点上执行 start-slave.sh 脚本来部署 Worker 组件。对应地，就可以通过该脚本，动态地为集群添加新的 Worker 组件。

脚本的代码如下：

```
1.
2.  if [ -z "${SPARK_HOME}" ]; then
3.    export SPARK_HOME="$(cd "'dirname "$0"'"/..; pwd)"
4.  fi
5.
6.  #注：提取的类名必须和 SparkSubmit 的类相匹配，任何变化都需在类中反映
7.
8.  # Worker 组件对应的类
9.  CLASS="org.apache.spark.deploy.worker.Worker"
10.
11. #脚本的用法，其中 master 参数是必选的，Worker 需要与集群的 Master 通信
12. #这里的 master 对应 Master URL 信息
13. if [[ $# -lt 1 ]] || [[ "$@" = *--help ]] || [[ "$@" = *-h ]]; then
14.   echo "Usage: ./sbin/start-slave.sh [options] <master>"
15.   pattern="Usage:"
16.   pattern+="\|Using Spark's default log4j profile:"
17.   pattern+="\|Registered signal handlers for"
18.
19.   "${SPARK_HOME}"/bin/spark-class $CLASS --help 2>&1 | grep -v "$pattern"
      1>&2
20.   exit 1
21. fi
22.
23. . "${SPARK_HOME}/sbin/spark-config.sh"
24.
25. . "${SPARK_HOME}/bin/load-spark-env.sh"
26.
27. #第一个参数应该是 master，先保存它，因为我们可能需要在它和其他参数之间插入参数
28.
29. MASTER=$1
30. shift
31.
32. # Worker 的 Web UI 的端口号设置
```

```
33.
34. if [ "$SPARK_WORKER_WEBUI_PORT" = "" ]; then
35.   SPARK_WORKER_WEBUI_PORT=8081
36. fi
37.
38. #在节点上启动指定序号的Worker实例
39. #
40. #快速启动Worker的本地功能
41. function start_instance {
42. #指定的Worker实例的序号，一个节点上可以部署多个Worker组件，对应有多个实例
43.   WORKER_NUM=$1
44.   shift
45.
46.   if [ "$SPARK_WORKER_PORT" = "" ]; then
47.     PORT_FLAG=
48.     PORT_NUM=
49.   else
50.     PORT_FLAG="--port"
51.     PORT_NUM=$(( $SPARK_WORKER_PORT + $WORKER_NUM - 1 ))
52.   fi
53.   WEBUI_PORT=$(( $SPARK_WORKER_WEBUI_PORT + $WORKER_NUM - 1 ))
54.
55. #和Master组件一样，Worker组件也是使用启动守护进程的spark-daemon.sh脚本来启动
    #一个Worker实例的
56.
57.   "${SPARK_HOME}/sbin"/spark-daemon.sh start $CLASS $WORKER_NUM \
58.      --webui-port "$WEBUI_PORT" $PORT_FLAG $PORT_NUM $MASTER "$@"
59. }
60.
61. #一个节点上部署几个Worker组件是由SPARK_WORKER_INSTANCES环境变量控制#的，默
    #认情况下只部署一个实例，start_instance方法的第一个参数为实例的序号
62. if [ "$SPARK_WORKER_INSTANCES" = "" ]; then
63.   start_instance 1 "$@"
64. else
65.   for ((i=0; i<$SPARK_WORKER_INSTANCES; i++)); do
66.     start_instance $(( 1 + $i )) "$@"
67.   done
68. fi
```

手动启动 Worker 实例时，如果需要在一个节点上部署多个 Worker 组件，则需要配置 SPARK_WORKER_INSTANCES 环境变量，否则多次启动脚本部署 Worker 组件时会报错，其原因在于 spark-daemon.sh 脚本的执行控制，这里给出关键代码的简单分析。

首先，脚本中带了实例是否已经运行的判断，代码如下：

```
1.  run_command() {
2.    mode="$1"
3.    shift
4.
5.    mkdir -p "$SPARK_PID_DIR"
6.
7.  #检查记录对应实例的PID的文件，如果对应进程已经运行，则会报错
8.    if [ -f "$pid" ]; then
9.      TARGET_ID="$(cat "$pid")"
10.     if [[ $(ps -p "$TARGET_ID" -o comm=) =~ "java" ]]; then
11.       echo "$command running as process $TARGET_ID.  Stop it first."
12.       exit 1
13.     fi
14.   fi
```

其中，记录对应实例的 PID 的文件相关的代码如下：

```
1.  #这是 PID 文件所在的目录，如果没有设置，默认为/tmp
2.  #如果使用了默认目录，可能会出现停止组件失败的信息
3.  #原因在于该/tmp 下的文件可能会被系统自动删除
4.  if [ "$SPARK_PID_DIR" = "" ]; then
5.    SPARK_PID_DIR=/tmp
6.  fi
7.
8.  #这是指定实例编号对应的 pid 文件的路径
9.  #其中$instance 代表实例编号，因此如果编号相同，对应的就是同一个文件
10. pid="$SPARK_PID_DIR/spark-$SPARK_IDENT_STRING-$command-$instance.pid"
```

从上面的分析可以看出，如果不是通过设置 SPARK_WORKER_INSTANCES，然后一次性启动多个 Worker 实例，而是手动一个个地启动，对应的在脚本中每次启动时的实例编号都是 1，在后台守护进程的 spark-daemon.sh 脚本中生成的 pid 就是同一个文件。因此，第二次启动时，pid 文件已经存在，此时就会报错（对应停止时也是通过读取 pid 文件获取进程 ID 的，因此自动停止多个实例，也需要设置 SPARK_WORKER_INSTANCES）。

5.2.2　Worker 启动的源码详解

首先查看 Worker 伴生对象中的 main 方法，Spark 2.2.1 版本的 Worker.scala 源码如下：

```
1.  private[deploy] object Worker extends Logging {
2.   val SYSTEM_NAME = "sparkWorker"
3.   val ENDPOINT_NAME = "Worker"
4.
5.   def main(argStrings: Array[String]) {
6.     Utils.initDaemon(log)
7.     val conf = new SparkConf
8.    //构建解析参数的实例
9.     val args = new WorkerArguments(argStrings, conf)
10.   //启动 RPC 通信环境以及 Worker 的 RPC 通信终端
11.    val rpcEnv = startRpcEnvAndEndpoint(args.host, args.port, args.
    webUiPort, args.cores,
      args.memory, args.masters, args.workDir, conf = conf)
12.    rpcEnv.awaitTermination()
13.  }
```

Spark 2.4.3 版本的 Worker.scala 源码与 Spark 2.2.1 版本相比具有如下特点。

- 上段代码中第 3 行之后新增 SSL_NODE_LOCAL_CONFIG_PATTERN 的变量。
- 上段代码中第 5 行之后新增 SparkUncaughtExceptionHandler 的处理。
- 上段代码中第 11 行之后新增对外部 shuffle 服务的处理。

```
1.  private val SSL_NODE_LOCAL_CONFIG_PATTERN = """\-Dspark\.ssl\.useNodeLocalConf\
    =(.+)""".r
2.  ......
3.    Thread.setDefaultUncaughtExceptionHandler(new
    SparkUncaughtExceptionHandler(
4.      exitOnUncaughtException = false))
5.  ......
6.  /**启用外部 shuffle 服务后，如果请求在一台主机上启动多个 workers 进程，只能成功地启
    动第一个 worker 线程，其余的则失败，因为端口绑定后，只能在每个主机上启动不超过一个外
```

```
   部 shuffle 服务。当这种情况发生时，应该给出明确的失败原因，而不是默默地失败。更多详
   情见 SPARK-20989
7.  */
8.      val externalShuffleServiceEnabled = conf.get(config.SHUFFLE_SERVICE_ENABLED)
9.      val sparkWorkerInstances = scala.sys.env.getOrElse("SPARK_WORKER_
        INSTANCES", "1").toInt
10.     require(externalShuffleServiceEnabled == false || sparkWorkerInstances <= 1,
11.       "Starting multiple workers on one host is failed because we may launch
          no more than one " +
12.         "external shuffle service on each host, please set spark.shuffle
            .service.enabled to " +
13.         "false or set SPARK_WORKER_INSTANCES to 1 to resolve the conflict.")
```

可以看到，Worker 伴生对象中的 main 方法、格式和 Master 基本一致。通过参数的类型 WorkerArguments 来解析命令行参数。具体的代码解析可以参考 Master 节点部署时的 MasterArguments 的代码解析。

另外，MasterArguments 中的 printUsageAndExit 方法，对应的就是命令行中的帮助信息。

解析完 Worker 的参数后，调用 startRpcEnvAndEndpoint 方法启动 RPC 通信环境以及 Worker 的 RPC 通信终端。该方法的代码解析可以参考 Master 节点部署时使用的同名方法的代码解析。

最终会实例化一个 Worker 对象。Worker 也继承 ThreadSafeRpcEndpoint，对应的也是一个 RPC 的通信终端，实例化该对象后会调用 onStart 方法，该方法的代码如下所示。

Spark 2.2.1 版本的 Worker.scala 的源码如下：

```
1.   override def onStart() {
2.   //刚启动时 Worker 肯定是未注册的状态
3.      assert(!registered)
4.      logInfo("Starting Spark worker %s:%d with %d cores, %s RAM".format(
5.        host, port, cores, Utils.megabytesToString(memory)))
6.      logInfo(s"Running Spark version ${org.apache.spark.SPARK_VERSION}")
7.      logInfo("Spark home: " + sparkHome)
8.   //构建工作目录
9.     createWorkDir()
10.  //启动 Shuffle 服务
11.     shuffleService.startIfEnabled()
12.  //启动一个 Web UI
13.     webUi = new WorkerWebUI(this, workDir, webUiPort)
14.     webUi.bind()
15.
16.     workerWebUiUrl = s"http://$publicAddress:${webUi.boundPort}"
17.  //每个 Slave 节点上启动 Worker 组件时，都需要向集群中的 Master 注册
18.
19.    registerWithMaster()
20.
21.     metricsSystem.registerSource(workerSource)
22.     metricsSystem.start()
23.     //度量系统启动后，将 Worker 度量的 Servlet 处理程序附加到 Web 用户界面
24.     metricsSystem.getServletHandlers.foreach(webUi.attachHandler)
25.   }
```

Spark 2.4.3 版本的 Worker.scala 源码与 Spark 2.2.1 版本相比具有如下特点。

- 上段代码中第 11 行 shuffleService.startIfEnabled()代码替换为 startExternalShuffleService()。

```
1.   startExternalShuffleService()
```

其中，createWorkDir()方法对应构建了该 Worker 节点上的工作目录，后续在该节点上执行的 Application 相关信息都会存放在该目录下。

Worker.scala 的 createWorkDir 的源码如下：

```
1.    private def createWorkDir() {
2.    workDir = Option(workDirPath).map(new File(_)).getOrElse(new File
      (sparkHome, "work"))
3.    try {
4.      //这偶尔会失败，不知道原因 ... !workDir.exists() && !workDir.mkdirs()
5.      //因此，尝试创建并检查目录是否创建
6.      workDir.mkdirs()
7.      if ( !workDir.exists() || !workDir.isDirectory) {
8.        logError("Failed to create work directory " + workDir)
9.        System.exit(1)
10.     }
11.     assert (workDir.isDirectory)
12.   } catch {
13.     case e: Exception =>
14.       logError("Failed to create work directory " + workDir, e)
15.       System.exit(1)
16.   }
17. }
```

可以看到，如果 workDirPath 没有设置，默认使用的是 sparkHome 目录下的 work 子目录。对应的 workDirPath 在 Worker 实例化时传入，反推代码可以查到该变量在 WorkerArguments 中设置。相关代码有两处：一处在 WorkerArguments 的主构造体中，代码如下所示。

WorkerArguments.scala 的源码如下：

```
1.   if (System.getenv("SPARK_WORKER_DIR") != null) {
2.     workDir = System.getenv("SPARK_WORKER_DIR")
3.   }
```

即 workDirPath 由环境变量 SPARK_WORKER_DIR 设置。

另外一处在命令行选项解析时设置，代码如下所示。

WorkerArguments.scala 的源码如下：

```
1.  private def parse(args: List[String]): Unit = args match {
2.  .....
3.      case ("--work-dir" | "-d") :: value :: tail =>
4.        workDir = value
5.        parse(tail)
6.  .....
```

即 workDirPath 由启动 Worker 实例时传入的可选项--work-dir 设置。属性配置：通常由命令可选项来动态设置启动时的配置属性，此时配置的优先级高于默认的属性文件以及环境变量中设置的属性。

启动 Worker 后一个关键的步骤就是注册到 Master，对应的方法 registerWithMaster()的代码如下所示。

Worker.scala 的源码如下：

```
1.    private def registerWithMaster() {
2.      .....
3.        registerMasterFutures = tryRegisterAllMasters()
```

```
4.      .....
5.
```

继续查看 tryRegisterAllMasters 方法，代码如下：

```
1.      private def tryRegisterAllMasters(): Array[JFuture[_]] = {
2.  .....
3.    sendRegisterMessageToMaster(masterEndpoint)
4.  .....
```

其中，sendRegisterMessageToMaster(masterEndpoint)向特定 Master 的 RPC 通信终端发送消息 RegisterWorker。

Worker 接收到反馈消息后，进一步调用 handleRegisterResponse 方法进行处理。对应的处理代码如下所示。

Worker.scala 的源码如下：

```
1.  private def handleRegisterResponse(msg: RegisterWorkerResponse): Unit =
    synchronized {
2.    msg match {
3.      case RegisteredWorker(masterRef, masterWebUiUrl, masterAddress) =>
4.        if (preferConfiguredMasterAddress) {
5.          logInfo("Successfully registered with master " + masterAddress.toSparkURL)
6.        } else {
7.          logInfo("Successfully registered with master " + masterRef.address
          .toSparkURL)
8.        }
9.        registered = true
10.       changeMaster(masterRef, masterWebUiUrl, masterAddress)
11.
12. //启动周期性心跳发送调度器，在 Worker 生命周期中定期向 Worker 发送自己的心跳信息
13.       forwordMessageScheduler.scheduleAtFixedRate(new Runnable {
14.         override def run(): Unit = Utils.tryLogNonFatalError {
15.           self.send(SendHeartbeat)
16.         }
17.       }, 0, HEARTBEAT_MILLIS, TimeUnit.MILLISECONDS)
18.   //启动工作目录的定期清理调度器，默认情况下，该配置的属性为 False，需要手动设置，
      //对应属性名为 spark.worker.cleanup.enabled
19.
20.       if (CLEANUP_ENABLED) {
21.         logInfo(
22.           s"Worker cleanup enabled; old application directories will be
            deleted in: $workDir")
23.         forwordMessageScheduler.scheduleAtFixedRate(new Runnable {
24.           override def run(): Unit = Utils.tryLogNonFatalError {
25.             self.send(WorkDirCleanup)
26.           }
27.         }, CLEANUP_INTERVAL_MILLIS, CLEANUP_INTERVAL_MILLIS,
          TimeUnit.MILLISECONDS)
28.       }
29.
30.       val execs = executors.values.map { e =>
31.         new ExecutorDescription(e.appId, e.execId, e.cores, e.state)
32.       }
33.       masterRef.send(WorkerLatestState(workerId, execs.toList,
        drivers.keys.toSeq))
34.
```

```
35.     //注册失败，则退出
36.      case RegisterWorkerFailed(message) =>
37.        if (!registered) {
38.          logError("Worker registration failed: " + message)
39.          System.exit(1)
40.        }
41.     //注册的 Master 处于 Standby 状态
42.      case MasterInStandby =>
43.        //Ignore. Master not yet ready.
44.    }
45.  }
```

分析到这一步，已经明确了注册以及对注册的反馈信息的处理细节。下面进一步分析注册重试定时器的相关处理。注册重试定时器会定期向 Worker 本身发送 ReregisterWithMaster 消息，因此可以在 receive 方法中查看该消息的处理，具体代码如下：

```
1.   override def receive: PartialFunction[Any, Unit] = synchronized {
2.  .....
3.       case ReregisterWithMaster =>
4.       reregisterWithMaster()
5.  .....
```

5.3　ExecutorBackend 启动原理和源码详解

ExecutorBackend 是 Executor 向集群发送更新消息的一个可插拔的接口。ExecutorBackend 拥有不同的实现。Standalone 模式下 ExecutorBackend 的默认实现是 CoarseGrainedExecutorBackend；在 Local 模式下，ExecutorBackend 的默认实现是 LocalBackend。在 Mesos 调度模式下，ExecutorBackend 的默认实现是 MesosExecutorBackend。本节主要探索 Standalone 模式下的 ExecutorBackend，通过源码深入理解 ExecutorBackend 接口设计的精髓。

5.3.1　ExecutorBackend 接口与 Executor 的关系

本节将详细分析 Standalone 模式下 ExecutorBackend 和 Executor 的关系。在 StandaloneSchedulerBackend 中会实例化一个 StandaloneAppClient。StandaloneAppClient 中携带了 command 信息，command 信息中指定了要启动的 ExecutorBackend 的实现类，Standalone 模式下，该 ExecutorBackend 的实现类是 org.apache.spark.executor.CoarseGrainedExecutorBackend 类。

StandaloneSchedulerBackend.scala 的 start 方法中构建了一个 Command 对象，该对象的第一个参数是 mainClass，即进程的主类。该类在 Standalone 模式下为 org.apache.spark.executor.CoarseGrainedExecutorBackend。分别将 sparkJavaopts、javaOpts、command、appUiAddress、coresPerExecutor、appDes 传入 StandaloneAppClient 构造函数。StandaloneAppClient 将会向 Master 发送 RegisterApplication 注册请求，Master 受理后通过 launchExecutor 方法在 Worker 节点启动一个 ExecutorRunner 对象，该对象用于管理一个 Executor 进程。在 ExecutorRunner 中将通过 CommandUtil 构建一个 ProcessBuilder，调用 ProcessBuilder 的 start 方法将会以进程的方式启动 org.apache.spark.executor.CoarseGrainedExecutorBackend。在 CoarseGrainedExecotorBackend 的

onStart 方法中，将会向 Driver 端发送 RegisterExecutor(executorId, self, hostPort, cores, extractLogUrls)消息请求注册，完成注册后将立即返回一个 RegisteredExecutor(executorAddress.host)消息，CoarseGraiendExecutorBackend 收到该消息，马上实例化出一个 Executor。

CoarseGrainedExecutorBackend.scala 的源码如下：

```
1.     override def receive: PartialFunction[Any, Unit] = {
2.      case RegisteredExecutor =>
3.        logInfo("Successfully registered with driver")
4.        try {
5.          executor = new Executor(executorId, hostname, env, userClassPath,
          isLocal = false)
6.        } catch {
7.          case NonFatal(e) =>
8.            exitExecutor(1, "Unable to create executor due to " + e.getMessage, e)
9.        }
```

从这里可以看出，CoarseGrainedExecutorBackend 比 Executor 先实例化。CoarseGrained-ExecutorBackend 负责与集群通信，而 Executor 则专注于任务的处理，它们是一对一的关系，在集群中各司其职。

每个 Worker 节点上可以启动多个 CoarseGrainedExecutorBackend 进程，每个进程对应一个 Executor。

5.3.2　ExecutorBackend 的不同实现

ExecutorBackend 是与集群交互的接口，该接口在不同的调度模式下有不同的实现。图 5-3 是 ExecutorBackend 及其实现的关系类图。

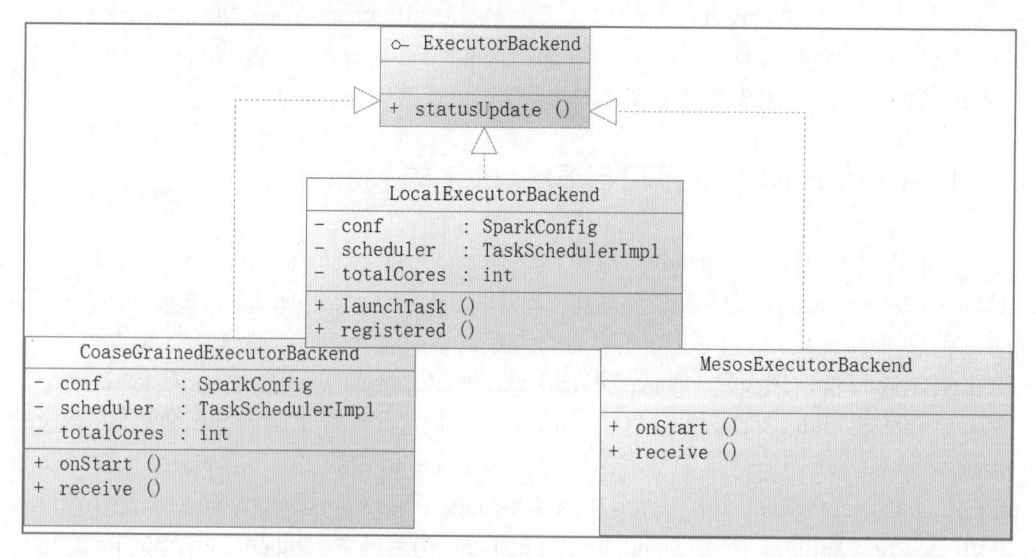

图 5-3　ExecutorBackend 及其实现的关系类图

不同模式下，ExecutorRunner 启动的进程不一样。在 Standalone 模式下启动的是 org.apache.spark.executor.CoarseGrainedExecutorBackend 进程；在 Local 模式下，启动的是 org.apache.spark.executor.LocalExecutorBackend 进程；在 Mesos 模式下，启动的是 org.apache

.spark.executor.MesosExecutorBackend 进程。

下面来看 Standalone 模式下 CoarseGrainedExecutorBackend 的启动。在 Standalone 模式下，会启动 org.apache.spark.deploy.Client 类，该类将向 Master 发送 RequestSubmitDriver (driverDescription)消息，Master 中匹配到 RequestSubmitDriver(driverDescription)后，将会调用 schedule 方法。

Master.scala 的 receiveAndReply 的源码如下：

```
1.  override def receiveAndReply(context: RpcCallContext): PartialFunction
    [Any, Unit] = {
2.  .....
3.   case RequestSubmitDriver(description) =>
4.   //若 state 不为 ALIVE，直接向 Client 返回 SubmitDriverResponse(self,false,
     //None,msg)消息
5.        if (state != RecoveryState.ALIVE) {
6.          val msg = s"${Utils.BACKUP_STANDALONE_MASTER_PREFIX}: $state. " +
7.            "Can only accept driver submissions in ALIVE state."
8.          context.reply(SubmitDriverResponse(self, false, None, msg))
9.        } else {
10.         logInfo("Driver submitted " + description.command.mainClass)
11. //使用 description 创建 driver，该方法返回 DriverDescription
12.         val driver = createDriver(description)
13.         persistenceEngine.addDriver(driver)
14.         waitingDrivers += driver
15.  //waitingDrivers 等待在调度数组中加入该 driver
16.         drivers.add(driver)
17.  //用 schedule 方法调度资源
18.         schedule()
19.  //向 ClientEndpoint 回复 SubmitDriverResponse 消息
20.
21.         context.reply(SubmitDriverResponse(self, true, Some(driver.id),
22.           s"Driver successfully submitted as ${driver.id}"))
23.       }
```

Master 的 receiveAndReply 收到 RequestSubmitDriver 消息后，调用 schedule 方法。

Master 的 schedule 的源码如下：

```
1.         private def schedule(): Unit = {
2.      if (state != RecoveryState.ALIVE) {
3.        return
4.      }
5.      //Drivers 优先于 executors
6.      val shuffledAliveWorkers = Random.shuffle(workers.toSeq.filter
        (_.state == WorkerState.ALIVE))
7.      val numWorkersAlive = shuffledAliveWorkers.size
8.      var curPos = 0
9.      for (driver <- waitingDrivers.toList) { //遍历 waitingDrivers
10.       //以循环的方式给每个等候的 driver 分配 Worker。对于每个 driver，我们从分配
          //给 driver 的最后一个 Worker 开始，继续前进，直到所有活跃的 Worker 节点
11.
12.       var launched = false
13.       var numWorkersVisited = 0
14.       while (numWorkersVisited < numWorkersAlive && !launched) {
```

```
15.         val worker = shuffledAliveWorkers(curPos)
16.         numWorkersVisited += 1
17.         if (worker.memoryFree >= driver.desc.mem && worker.coresFree >=
            driver.desc.cores) {
18.           launchDriver(worker, driver)
19.           waitingDrivers -= driver
20.           launched = true
21.         }
22.         curPos = (curPos + 1) % numWorkersAlive
23.       }
24.     }
25.     startExecutorsOnWorkers()
26.   }
```

上面代码中，RecoveryState 若不为 ALIVE，则直接返回，否则使用 Random.shuffle 将 Workers 集合打乱，过滤出 ALIVE 的 Worker，生成新的集合 shuffledAliveWorkers，尽量考虑到选择 Driver 的负载均衡。在 for 语句中遍历 waitingDrivers 队列，判断 Worker 剩余内存和剩余物理核是否满足 Driver 需求，如满足，则调用 launchDriver(worker,driver)方法在选中的 Worker 上启动 Driver 进程。

实例化 SparkContext 时，在 SparkContext 中将实例化出 DAGScheduler、StandaloneSchedulerBackend。Driver 在 Worker 节点上启动之后，在 StandaloneSchedulerBackend 中将会调用 new()函数创建一个 StandaloneAppClient。StandaloneAppClient 中有一个 ClientEndpoint，在其 onStart 方法中将向 Master 发送 RegisterApplication 请求注册 application，注册好 application 后，Master 又会调用 schedule 方法，在满足条件的 Worker 上为 application 启动 Executor，首先会启动 ExecutorRunner，在 ExecutorRunner 中启动 CoarseGrainedExecutorBackend，启动后将会实例化出 Executor。为什么在 Standalone 模式下会启动 CoarseGrainedExecutorBackend 呢？在什么地方设置要启动的 CoarseGrainedExecutorBackend 进程呢？其实，在实例化 StandaloneAppClient 的时候就已经传入了。

StandaloneSchedulerBackend.scala 的 start 方法代码中设置了 Command 对象。Command 对象的第一个参数是启动进程的 mainClass。因此，ExecutorRunner 中启动进程时，启动的是 org.apache.spark.executor.CoarseGrainedExecutorBackend。

5.3.3 ExecutorBackend 中的通信

ExecutorBackend 是一个被 Executor 使用的可插拔的与集群通信的接口。在 ExecutorBackend 中有 statusUpdate(taskId: Long, state: TaskState, data: ByteBuffer)方法，通过这个方法向集群发送 Task 执行的各种信息，如果任务执行失败，则返回失败的信息；如果执行成功，则返回任务执行的结果。本节重点讲解在 Standalone 模式下 CoarseGrainedExecutorBackend 中的通信。CoarseGrainedExecutorBackend 在整个集群中的通信如图 5-4 所示。

在图 5-4 中，Executor 与 CoarseGrainedExecutorBackend 协作，将任务计算的结果通过 CoarseGrainedExecutorBackend 的 statusUpdate 方法将 taskId、TaskState 以及结果数据发送给 Driver。Driver 收到 StatusUpdate(executorId,tasked,state,data)消息，通过判断 state 的不同状态，进行不同的处理。例如，当 state 的状态为 TaskState.LOST 时，Driver 端会移除 Executor；当 state 的状态为 TaskState.FINISHED 时，Driver 端会调用 enqueueSuccessfulTask 进行处理。

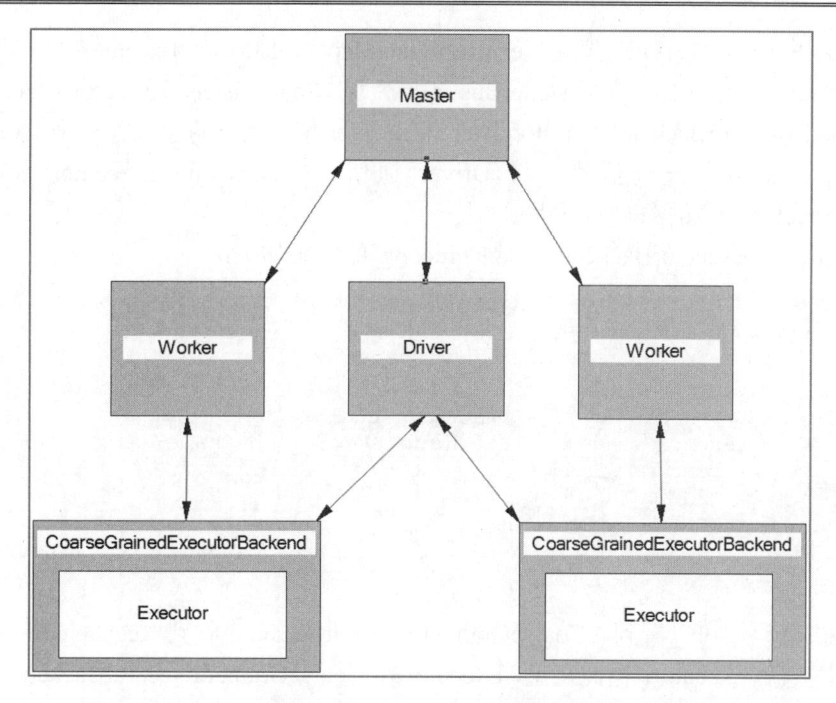

图 5-4　CoarseGrainedExecutorBackend 在整个集群中的通信

这里主要看 CoarseGrainedExecutorBackend 与 Driver 之间的通信。当 Worker 节点中启动 ExecutorRunner 时，ExecutorRunner 中会启动 CoarseGrainedExecutorBackend 进程，在 CoarseGrainedExecutorBackend 的 onStart 方法中，向 Driver 发出 RegisterExecutor 注册请求。

CoarseGrainedExecutorBackend 的 onStart 方法的源码如下：

```
1.     override def onStart() {
2.     logInfo("Connecting to driver: " + driverUrl)
3.     rpcEnv.asyncSetupEndpointRefByURI(driverUrl).flatMap { ref =>
4.       //这是一个非常快的行动，所以我们可以用"ThreadUtils.sameThread"
5.       driver = Some(ref)
6.   //向 Driver 发送 ask 请求，等待 Driver 回应
7.       ref.ask[Boolean](RegisterExecutor(executorId, self, hostname, cores,
       extractLogUrls))
8.     }(ThreadUtils.sameThread).onComplete {
9.       //这是一个非常快的行动，所以我们可以用"ThreadUtils.sameThread"
10.      case Success(msg) =>
11.        //经常收到 true，可以忽略
12.      case Failure(e) =>
13.        exitExecutor(1, s"Cannot register with driver: $driverUrl", e,
         notifyDriver = false)
14.    }(ThreadUtils.sameThread)
15.  }
```

上面的代码中，Some(ref)得到 Driver 的引用，通过 ask 方法返回 Future[Boolean]，然后在 Future 对象上调用 onComplete 方法进行额外的处理。Driver 端收到注册请求，将会注册 Executor 的请求，并向 ListenerBus 中发送 SparkListenerExecutorAdded 事件。

如果 executorDataMap 中已经存在该 Executor 的 id，就返回 RegisterExecutorFailed，如

果不存在该 Executor 的 id，则在 executorDataMap 中加入该 Executor 的 id，并返回 RegisteredExecutor 消息且向 listenerBus 中添加 SparkListenerExecutorAdded 事件。CoarseGrainedExecutorBackend 收到 RegisteredExecutor 消息后，将会新建一个 Executor 执行器，并为此 Executor 充当信使，与 Driver 通信。CoarseGrainedExecutorBackend 收到 RegisteredExecutor 消息的源码如下所示。

CoarseGrainedExecutorBackend.scala 的 receive 的源码如下：

```
1.   override def receive: PartialFunction[Any, Unit] = {
2.     case RegisteredExecutor =>
3.       logInfo("Successfully registered with driver")
4.       try {
5.   //收到 RegisteredExecutor 消息，立即创建 Executor
6.         executor = new Executor(executorId, hostname, env, userClassPath,
         isLocal = false)
7.       } catch {
8.         case NonFatal(e) =>
9.           exitExecutor(1, "Unable to create executor due to " + e.getMessage, e)
10.       }
```

从上面的代码中可以看到，CoarseGrainedExecutorBackend 收到 RegisteredExecutor 消息后，将会新建一个 Executor。由此可见，Executor 在 CoarseGrainedExecutorBackend 后实例化，这与 Executor 和 CoarseGrainedExecutorBackend 的不同职责有关，Executor 主要负责计算，而 CoarseGrainedExecutorBackend 主要负责通信，通信环境准备好了，架起同 CoarseGrainedSchedulerBackend 通信的桥梁，就可以接收 CoarseGrainedSchedulerBackend 中调用 launchTask 方法发送的 LaunchTask 消息了，因此通信在前，计算在后。

Executor 中的计算结果是通过 CoarseGrainedExecutorBackend 的 statusUpdate 方法返回给 CoarseGrainedExecutorBackend 的。statusUpdate 方法的代码如下所示。

CoarseGrainedExecutorBackend.scala 的源码如下：

```
1.     override def statusUpdate(taskId: Long, state: TaskState, data:
       ByteBuffer) {
2.     val msg = StatusUpdate(executorId, taskId, state, data)
3.     driver match {
4.  //向 Driver 发送 StatusUpdate 消息
5.       case Some(driverRef) => driverRef.send(msg)
6.       case None => logWarning(s"Drop $msg because has not yet connected
         to driver")
7.     }
8.   }
```

上面源码中，通过参数 taskId、state、data 构建一个 StatusUpdate 对象，该对象将被当作消息发送到 Driver 端，Driver 根据返回结果的需要，将会向 CoarseGrainedExecutorBackend 发送新的指令消息，如 LaunchTask、KillTask、StopExecutors、Shutdown 等。

5.3.4　ExecutorBackend 的异常处理

若 CoarseGrainedExecutorBackend 在运行中出现异常，将调用 exitExecutor 方法进行处理，处理以后，系统退出。exitExecutor 函数可以由其他子类重载来处理，Executor 执行的退出方式不同。例如，当 Executor 挂掉了，后台程序可能不会让父进程也挂掉。如果须通知 Driver，

Driver 将清理挂掉的 Executor 的数据。

Spark 2.2.1 版本的 CoarseGrainedExecutorBackend 的 exitExecutor 方法的源码如下：

```
1.     protected def exitExecutor(code: Int,
2.                                reason: String,
3.                                throwable: Throwable = null,
4.                                notifyDriver: Boolean = true) = {
5.     val message = "Executor self-exiting due to : " + reason
6.     if (throwable != null) {
7.       logError(message, throwable)
8.     } else {
9.       logError(message)
10.    }
11.
12.    if (notifyDriver && driver.nonEmpty) {
13.      driver.get.ask[Boolean](
14.        RemoveExecutor(executorId, new ExecutorLossReason(reason))
15.      ).onFailure { case e =>
16.        logWarning(s"Unable to notify the driver due to " + e.getMessage, e)
17.      }(ThreadUtils.sameThread)
18.    }
19.
20.    System.exit(code)
21.  }
22. }
```

Spark 2.4.3 版本的 CoarseGrainedExecutorBackend.scala 源码与 Spark 2. 2.1 版本相比具有如下特点。

- 上段代码中第 13～17 行替换为 driver.get.send 代码。

```
1.    driver.get.send(RemoveExecutor(executorId, new ExecutorLossReason(reason)))
```

CoarseGrainedExecutorBackend 在运行中一旦出现异常情况，将调用 exitExecutor 方法处理。

- Executor 向 Driver 注册 RegisterExecutor 失败。
- Executor 收到 Driver 的 RegisteredExecutor 注册成功消息以后，创建 Executor 实例失败。
- Driver 返回 Executor 注册失败消息 RegisterExecutorFailed。
- Executor 收到 Driver 的 LaunchTask 启动任务消息，但是 Executor 为 null。
- Executor 收到 Driver 的 KillTask 消息，但是 Executor 为 null。
- Executor 和 Driver 失去连接。

5.4　Executor 中任务的执行

本节讲解 Executor 中任务的加载，通过 launchTask()方法加载任务，将任务以 TaskRunner 的形式放入线程池中运行；Executor 中的任务线程池可以减少在创建和销毁线程上所花的时间和系统资源开销；TaskRunner 任务执行失败处理以及 TaskRunner 的运行内幕等内容。

5.4.1 Executor 中任务的加载

Executor 是基于线程池的任务执行器。通过 launchTask 方法加载任务，将任务以 TaskRunner 的形式放入线程池中运行。

DAGScheduler 划分好 Stage 通过 submitMissingTasks 方法分配好任务，并把任务交由 TaskSchedulerImpl 的 submitTasks 方法，将任务加入调度池，之后调用 CoarseGrainedScheduler-Backend 的 riviveOffers 方法为 Task 分配资源，指定 Executor。任务资源都分配好之后，CoarseGrainedSchedulerBackend 将向 CoarseGranedExecutorBackend 发送 LaunchTask 消息，将具体的任务发送到 Executor 上进行计算。

CoarseGranedExecutorBackend 匹配到 LaunchTask(data)消息之后，将会调用 Executor 的 launchTask 方法。launchTask 方法中将会构建 TaskRunner 对象，并放入线程池中执行。

Executor 中 Task 的加载时序图如图 5-5 所示。

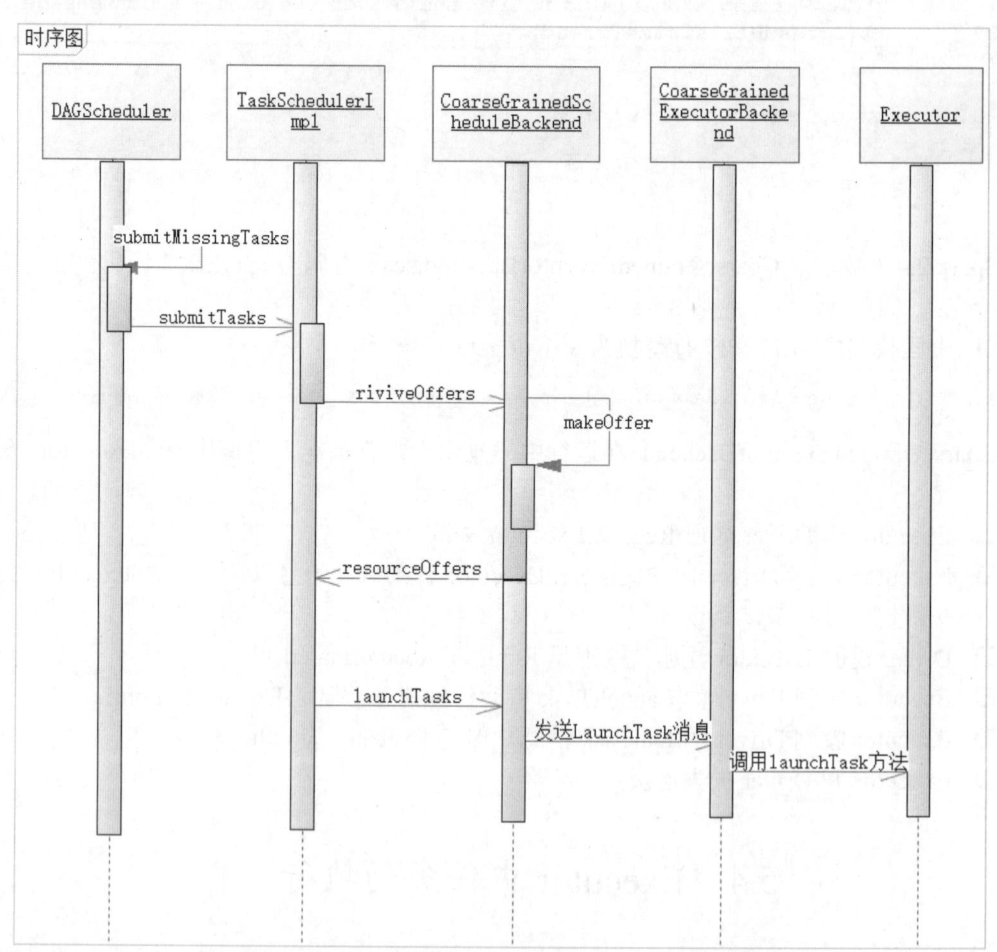

图 5-5　Executor 中 Task 的加载时序图

任务加载好后，在 Executor 中将会把构建好的 TaskRunner 放入线程池运行，至此任务完成加载，开始运行。

5.4.2 Executor 中的任务线程池

Executor 是构建在线程池之上的任务执行器。在 Executor 中使用线程池的好处是显而易见的，使用线程池可以减少在创建和销毁线程上所花的时间和系统资源开销。如果不使用线程池，可能造成系统创建大量的线程而导致消耗完系统内存以及出现“过度切换”。

为什么 Executor 中需要线程池？使用线程池基于以下原因：首先，在 Executor 端执行的任务处理时间都比较短，需要频繁地创建和销毁线程，这样就带来了巨大的创建和销毁线程的开销，造成额外的系统资源开销；其次，Executor 中处理的任务数量巨大，如果每个任务都创建一个线程，将导致消耗完系统内存，出现“过度切换”。

首先来看 Executor 中的线程池。Executor 中使用的是 CachedThreadPool，使用这种类型线程池的好处是：任务比较多时可以自动新增处理线程，而任务比较少时自动回收空闲线程。

CoarseGrainedExecutorBackend 调用 Executor 的 launchTask 方法，将会新建 TaskRunner，然后放入线程池进行处理。

从上面的源码中可以看到，新建的 TaskRunner 对象首先放入 runningTasks 这样一个 ConcurrentHashMap 里面，然后使用线程池的 Execute 方法运行 TaskRunner。Execute 方法将会调用 TaskRunner 的 run 方法。在 TaskRunner 的 run 方法中执行计算任务。

5.4.3 任务执行失败处理

TaskRunner 在计算的过程中可能发生各种异常，甚至错误，如抓取 shuffle 结果失败、任务被杀死、没权限向 HDFS 写入数据等。当 TaskRunner 的 run 方法运行的时候，代码中通过 try-catch 语句捕获这些异常，并通过调用 CoarseGrainedExecutorBackend 的 statusUpdate 方法向 CoarseGrainedSchedulerBackend 汇报。

下面是 CoarseGrainedExecutorBackend 的 statusUpdate 方法的源码如下：

```
1.      override def statusUpdate(taskId: Long, state: TaskState, data:
        ByteBuffer) {
2.    val msg = StatusUpdate(executorId, taskId, state, data)
3.    driver match {
4.      case Some(driverRef) => driverRef.send(msg)
5.      case None => logWarning(s"Drop $msg because has not yet connected
        to driver")
6.    }
7.  }
```

在 statusUpdate 方法中，通过方法的参数 taskId、state、data 构建一个 StatusUpdate 对象，并通过 driverRef 的 send 方法将该对象发送回 CoarseGrainedSheduleBackend。CoarseGrainedScheduleBackend 匹配到 StatusUpdate 时，将根据 StatusUpdate 对象中的 state 值对该 Task 的执行情况做出判断，并执行不同的处理逻辑。

从源码中可以发现有 TaskState 对象，其实这里的 TaskState 是一个枚举变量，该枚举变量中包括 LAUNCHING、RUNNING、FINISHED、FAILED、KILLED、LOST 这些枚举值，分别对应任务执行的不同状态。Executor 根据任务执行的不同状态，通过 statusUpdate 方法返回特定的 TaskState 值，该值通过 ExecutorBackend 返回给 SchedulerBackend，在

SchedulerBackend 中根据 TaskState 中的值进行处理。

TaskState.scala 的源码如下：

```
1.   private[spark] object TaskState extends Enumeration {
2.
3.    val LAUNCHING, RUNNING, FINISHED, FAILED, KILLED, LOST = Value
4.
5.    private val FINISHED_STATES = Set(FINISHED, FAILED, KILLED, LOST)
6.
7.    type TaskState = Value
8.
9.    def isFailed(state: TaskState): Boolean = (LOST == state) || (FAILED ==
      state)
10.
11.   def isFinished(state: TaskState): Boolean = FINISHED_STATES.contains
      (state)
12. }
13.
```

以 TaskState.FAILED 这种情况为例，在 Executor 的 run 方法中，如果发生 FetchFailed-Exeception、CommitDeniedExeception 或其他 Throwable 的子类的异常，就会返回 TaskState.FAILED 状态，该状态通过 CoaseGainedExecutorBackend 返回。在 CoaseGaiendScheduler-Backend 中，匹配到 StatusUpdate 消息，将进行相应的处理。

CoarseGrainedSchedulerBackend.scala 的 StatusUpdate 的源码如下：

```
1.     override def receive: PartialFunction[Any, Unit] = {
2.        case StatusUpdate(executorId, taskId, state, data) =>
3.  //调用 TaskSchedulerImpl 更新状态
4.          scheduler.statusUpdate(taskId, state, data.value)
5.  //若状态为 FINISHED,则从 executorDataMap 中取出 executorId 对应的 ExecutorInfo
6.          if (TaskState.isFinished(state)) {
7.            executorDataMap.get(executorId) match {
8.              case Some(executorInfo) =>
9.                executorInfo.freeCores += scheduler.CPUS_PER_TASK
10. //用 makeOffers 方法重新分配资源
11.               makeOffers(executorId)
12.             case None =>
13.               //因为不知道 executor，忽略更新
14.               logWarning(s"Ignored task status update ($taskId state $state) " +
15.                 s"from unknown executor with ID $executorId")
16.
17.           }
18.         }
```

上面的代码中，首先调用 TaskSchedulerImpl 的 statusUpdate 方法，该方法用于更新 taskId 对应任务的状态。完成更新之后，判断 state 状态是否 FINISHED，若状态为 FINISHED，则从 executorDataMap 这个哈希表中取出 executorId 对应的 ExecutorData 对象，修改该对象中的 freeCores。因为状态已经为 FINISHED，因此 ExecutorData 中的 freeCores 会增加 CPUS_PER_TASK 个，这里的 CPU_PER_TASK 为每个任务占用的 CPU 核的个数，该个数可以通过 spark.task.cpus 配置项进行配置。

更新完成 ExecutorData 上的可用 CPU 后，这些闲置的 CPU 通过 makeOffers 方法再次分配给其他任务使用。

Spark 2.2.1 版本的 CoarseGrainedSchedulerBackend.scala 的 makeOffers 的源码如下：

```
1.  private def makeOffers(executorId: String) {
2.        // 执行某项任务时，确保没有 Executor 被杀死
3.        val taskDescs = CoarseGrainedSchedulerBackend.this.synchronized {
4.          // 过滤掉被杀死的 executors
5.          if (executorIsAlive(executorId)) {
6.            val executorData = executorDataMap(executorId)
7.            val workOffers = IndexedSeq(
8.              new WorkerOffer(executorId, executorData.executorHost, executorData
              .freeCores))
9.            scheduler.resourceOffers(workOffers)
10.         } else {
11.           Seq.empty
12.         }
13.       }
14.       if (!taskDescs.isEmpty) {
15.         launchTasks(taskDescs)
16.       }
17.     }
```

Spark 2.4.3 版本的 CoarseGrainedSchedulerBackend.scala 的 makeOffers 源码与 Spark 2.2.1 版本相比具有如下特点。

- 上段代码中第 3 行修改为 withLock 方法，需要确保在 TaskSchedulerImpl 和 CoarseGrainedSchedulerBackend 对象之间有锁获取的顺序，以便修复 spark-27112 中暴露的死锁问题。
- 上段代码中第 8 行构建 WorkerOffer 实例时新增一个主机端口成员变量。

```
1. .....
2.          val taskDescs = withLock {
3. .....
4.    new WorkerOffer(executorId, executorData.executorHost, executorData.freeCores,
          Some(executorData.executorAddress.hostPort)))
```

每个 Executor 上的资源发生变动时，都将调用 makeOffers 方法，该方法的作用是为等待执行的任务分配资源，并通过 launchTasks 方法将这些任务发送到这些 Executor 上运行。这些任务将被包装成 TaskRenner 对象，运行于 Executor 上的线程池中。

5.4.4　揭秘 TaskRunner

TaskRunner 位于 Executor 中，继承自 Runnable 接口，代表一个可执行的任务。Driver 端下发的任务最终都要在 Executor 中封装成 TaskRunner。在 TaskRunner 的 run 方法中，将会进行任务的解析，并调用 Task 接口的 run 方法进行计算。

TaskRunner 定义的代码如下：

```
1.   class TaskRunner(
2.      execBackend: ExecutorBackend,
3.      private val taskDescription: TaskDescription)
4.    extends Runnable {
```

TaskRunner 的构造函数中传入 execBackend、taskDescription 成员变量，其中 taskDescription 是 TaskDescription 类型，包括 taskId、attemptNumber、taskName、serializedTask 等参数。其

中，execBackend 作为和 CoarseGrainedSchedulerBackend 通信的使者传入到 TaskRunner 中，在任务计算状态发生变化的时候，调用 execBackend 的 statusUpdate 方法向 CoarseGrainedScheduler-Backend 报告。传入 taskId 是为了使用 TaskMemoryManager 管理该 Task。attemptNumber 代表任务尝试执行的次数，serializedTask 是序列化的任务。序列化的任务通过序列化工具反序列化得到任务对象。

在 TaskRunner 中是如何运行任务的？我们知道，在线程池中启动 Runnable 任务会自动调用 Runnable 的 run 方法，TaskRunner 作为一个 Runnable 接口的实现类，启动时会自动调用其 run 方法。run 方法主要完成以下任务。

- 调用 ExecutorBackend 的 statusUpdate 方法向 SchedulerBackend 发送任务状态更新消息。
- 反序列化出 Task 和相关依赖 Jar 包。
- 调用 Task 上的 run 方法运行任务。
- 返回 Task 运行结果。

Task 是一个接口，ResultTask 和 ShffleMapTask 是其两种实现。Task 接口中提供了 run 方法，用于运行任务。TaskRunner 的 run 方法中，会通过反序列化器反序列化出 Task，并调用 Task 上的 run 方法运行任务，这里怎么知道是 ResultTask，还是 ShffleMapTask 呢？其实，这里不管是 ResultTask，还是 ShffleMapTask，都一视同仁，因为 ResultTask 和 ShffleMapTask 都实现了 Task 接口，都有 run 方法。这正是面向接口编程带来的最大的好处，灵活且最大限度地复用代码。

Task 运行结果的处理情况有 3 种：第一种情况是 resultSize 大于 maxResultSize，这种情况下构建 IndirectTaskResult 对象，并返回该 IndirectTaskResult 对象，IndirectTaskResult 对象中包含结果所在的 BlockId，在 SchedulerBackend 中可以通过 BlockManager 获得该 BlockId 对应的结果数据，这里的 maxResultSize 默认为 1GB；第二种情况是 resultSize 大于 Akka 帧的大小，这种情况下也是构建 IndirectTaskResult 对象，并返回该 IndirectTaskResult 对象，Akka 帧的大小为 128MB；第三种情况是直接返回 DirectTaskResult，这是在 resultSize 小于 Akka 帧大小的情况下采取的默认返回方式。

5.5 Executor 执行结果的处理方式

本节讲解 Executor 工作原理、ExecutorBackend 注册源码解密、Executor 实例化内幕、Executor 具体工作内幕 。

Master 让 Worker 启动，启动了一个 Executor 所在的进程。在 Standalone 模式中，Executor 所在的进程是 CoarseGrainedExecutorBackend。

- Master 侧：Master 发指令给 Worker，启动 Executor。
- Worker 侧：Worker 接收到 Master 发过来的指令，通过 ExecutorRunner 启动另外一个进程来运行 Executor。这里是指启动另外一个进程来启动 Executor，而不是直接启动 Executor。Master 向 Worker 发送指令，Worker 为什么启动另外一个进程？在另外一个进程中注册给 Driver，然后启动 Executor。因为 Worker 是管理机器上的资源的，所以机器上的资源变动时要汇报给 Master。Worker 不是用来计算的，不能在 Worker

中进行计算；Spark 集群中有很多应用程序，需要很多 Executor，如果不是给每个 Executor 启动一个对应的进程，而是所有的应用程序进程都在同一个 Executor 里面，那么一个程序崩溃将导致其他程序也崩溃。

- 启动 CoarseGrainedExecutorBackend。CoarseGrainedExecutorBackend 是 Executor 所在的进程。CoarseGrainedExecutorBackend 启动时，须向 Driver 注册。通过发送 RegisterExecutor 向 Driver 注册，注册的内容是 RegisterExecutor。

CoarseGrainedExecutorBackend.scala 的 onStart 方法的源码如下：

```
1.   override def onStart() {
2.      logInfo("Connecting to driver: " + driverUrl)
3.      rpcEnv.asyncSetupEndpointRefByURI(driverUrl).flatMap { ref =>
4.        //这是一个非常快的 Action，所以可以用 ThreadUtils.sameThread
5.        driver = Some(ref)
6.        ref.ask[Boolean](RegisterExecutor(executorId,   self,   hostname,
          cores, extractLogUrls))
7.      }(ThreadUtils.sameThread).onComplete {
8.        //这是一个非常快的 Action，所以可以用 ThreadUtils.sameThread
9.        case Success(msg) =>
10.         //经常收到 true，可忽略
11.       case Failure(e) =>
12.         exitExecutor(1, s"Cannot register with driver: $driverUrl", e,
            notifyDriver = false)
13.     }(ThreadUtils.sameThread)
14.   }
```

其中，RegisterExecutor 是一个 case class，源码如下：

```
1.      case class RegisterExecutor(
2.        executorId: String,
3.        executorRef: RpcEndpointRef,
4.        hostname: String,
5.        cores: Int,
6.        logUrls: Map[String, String])
7.      extends CoarseGrainedClusterMessage
```

CoarseGrainedExecutorBackend 启动时，向 Driver 发送 RegisterExecutor 消息进行注册；Driver 收到 RegisterExecutor 消息，在 Executor 注册成功后会返回消息 RegisteredExecutor 给 CoarseGrainedExecutorBackend。这里注册的 Executor 和真正工作的 Executor 没有任何关系，其实注册的是 RegisterExecutorBackend。可以将 RegisteredExecutor 理解为 RegisterExecutorBackend。

需要特别注意的是，在 CoarseGrainedExecutorBackend 启动时向 Driver 注册 Executor，其实质是注册 ExecutorBackend 实例，和 Executor 实例之间没有直接关系。

- CoarseGrainedExecutorBackend 是 Executor 运行所在的进程名称，CoarseGrained-ExecutorBackend 本身不会完成任务的计算。
- Executor 才是正在处理任务的对象。Executor 内部是通过线程池的方式来完成 Task 的计算的。Executor 对象运行于 CoarseGrainedExecutorBackend 进程。
- CoarseGrainedExecutorBackend 和 Executor 是一一对应的。
- CoarseGrainedExecutorBackend 是一个消息通信体(其具体实现了 ThreadSafeRPCEndpoint)，可以发送信息给 Driver，并可以接受 Driver 中发过来的指令，如启动 Task 等。

CoarseGrainedExecutorBackend 继承自 ThreadSafeRpcEndpoint，CoarseGrainedExecutor-Backend 是一个消息通信体，可以收消息，也可以发消息，源码如下：

```
1.      private[spark] class CoarseGrainedExecutorBackend(
2.      override val rpcEnv: RpcEnv,
3.      driverUrl: String,
4.      executorId: String,
5.      hostname: String,
6.      cores: Int,
7.      userClassPath: Seq[URL],
8.      env: SparkEnv)
9.    extends ThreadSafeRpcEndpoint with ExecutorBackend with Logging {
```

CoarseGrainedExecutorBackend 发消息给 Driver。Driver 在 StandaloneSchedulerBackend 里面（Spark 2.0 中已将 SparkDeploySchedulerBackend 更名为 StandaloneSchedulerBackend）。StandaloneSchedulerBackend 继承自 CoarseGrainedSchedulerBackend，start 启动时启动 StandaloneAppClient。StandaloneAppClient（Spark 2.0 中已将 AppClient 更名为 StandaloneAppClient）代表应用程序本身。

StandaloneAppClient.scala 的源码如下：

```
1.      private[spark] class StandaloneAppClient(
2.      rpcEnv: RpcEnv,
3.      masterUrls: Array[String],
4.      appDescription: ApplicationDescription,
5.      listener: StandaloneAppClientListener,
6.      conf: SparkConf)
7.    extends Logging {
8.   .....
9.   private  class  ClientEndpoint(override  val  rpcEnv:  RpcEnv)  extends
     ThreadSafeRpcEndpoint
10.     with Logging {
11.  .....
```

在 Driver 进程中有两个至关重要的 Endpoint。

- ClientEndpoint：主要负责向 Master 注册当前的程序，是 AppClient 的内部成员。
- DriverEndpoint：这是整个程序运行时的驱动器，是 CoarseGrainedExecutorBackend 的内部成员。

CoarseGrainedSchedulerBackend 的 DriverEndpoint 的源码如下：

```
1.      class DriverEndpoint(override val rpcEnv: RpcEnv, sparkProperties:
        Seq[(String, String)])
2.      extends ThreadSafeRpcEndpoint with Logging {
```

DriverEndpoint 会接收到 RegisterExecutor 消息，并完成在 Driver 上的注册。

RegisterExecutor 中有一个数据结构 executorDataMap，是 Key-Value 的方式。

```
1.          private val executorDataMap = new HashMap[String, ExecutorData]
```

ExecutorData 中的 executorEndpoint 是 RpcEndpointRef。ExecutorData 的源码如下：

```
1.   private[cluster] class ExecutorData(
2.     val executorEndpoint: RpcEndpointRef,
3.     val executorAddress: RpcAddress,
4.     override val executorHost: String,
5.     var freeCores: Int,
6.     override val totalCores: Int,
7.     override val logUrlMap: Map[String, String]
8.   ) extends ExecutorInfo(executorHost, totalCores, logUrlMap)
```

CoarseGrainedExecutorBackend.scala 的 RegisteredExecutor 的源码如下：

```
1.   override def receive: PartialFunction[Any, Unit] = {
2.      case RegisteredExecutor =>
3.        logInfo("Successfully registered with driver")
4.        try {
5.          executor = new Executor(executorId, hostname, env, userClassPath,
          isLocal = false)
6.        } catch {
7.          case NonFatal(e) =>
8.            exitExecutor(1, "Unable to create executor due to " + e.getMessage, e)
9.
```

CoarseGrainedExecutorBackend 收到 RegisteredExecutor 消息以后，用 new()函数创建一个 Executor，而 Executor 就是一个普通的类。

Executor.scala 的源码如下：

```
1.  private[spark] class Executor(
2.      executorId: String,
3.      executorHostname: String,
4.      env: SparkEnv,
5.      userClassPath: Seq[URL] = Nil,
6.      isLocal: Boolean = false,
7.      uncaughtExceptionHandler: UncaughtExceptionHandler = new
        SparkUncaughtExceptionHandler)
8.    extends Logging {
```

回到 ExecutorData.scala，其中的 RpcEndpointRef 是代理句柄，代理 CoarseGrainedExecutorBackend。在 Driver 中，通过 ExecutorData 封装并注册 ExecutorBackend 的信息到 Driver 的内存数据结构 executorMapData 中。

```
1.   private[cluster] class ExecutorData(
2.     val executorEndpoint: RpcEndpointRef,
3.     val executorAddress: RpcAddress,
4.     override val executorHost: String,
5.     var freeCores: Int,
6.     override val totalCores: Int,
7.     override val logUrlMap: Map[String, String]
8.  ) extends ExecutorInfo(executorHost, totalCores, logUrlMap)
```

Executor 注册消息提交给 DriverEndpoint，通过 DriverEndpoint 写数据给 CoarseGrainedSchedulerBackend 里面的数据结构 executorMapData。executorMapData 是 CoarseGrainedSchedulerBackend 的成员，因此最终注册给 CoarseGrainedSchedulerBackend。CoarseGrainedSchedulerBackend 获得 Executor（其实是 ExecutorBackend）的注册信息。

实际在执行的时候，DriverEndpoint 会把信息写入 CoarseGrainedSchedulerBackend 的内存数据结构 executorMapData 中，所以最终是注册给了 CoarseGrainedSchedulerBackend。也就是说，CoarseGrainedSchedulerBackend 掌握了为当前程序分配的所有的 ExecutorBackend 进程，而在每个 ExecutorBackend 进行实例中，会通过 Executor 对象负责具体任务的运行。在运行的时候使用 synchronized 关键字来保证 executorMapData 安全地并发写操作。

CoarseGrainedSchedulerBackend.scala 的 receiveAndReply 方法中 RegisterExecutor 注册的过程，源码如下所示。

Spark 2.2.1 版本的 CoarseGrainedSchedulerBackend.scala 的 receiveAndReply 方法的源码

如下：

```
1.  override def receiveAndReply(context: RpcCallContext):
    PartialFunction[Any, Unit] = {
2.
3.        case RegisterExecutor(executorId, executorRef, hostname, cores, logUrls) =>
4.       //检查 executorDataMap 中是否包含该 executorId，如果包含，就返回
                   //RegisterExecutorFailed 消息
5.          if (executorDataMap.contains(executorId)) {
6.            executorRef.send(RegisterExecutorFailed("Duplicate executor ID:
            " + executorId))
7.            context.reply(true)
8.          } else if (scheduler.nodeBlacklist != null &&
9.            scheduler.nodeBlacklist.contains(hostname)) {
10.      //如果集群管理器分配给用户一个 Executor，而这个 Executor 在黑名单节点列
11.      //表中（因为通知它是黑名单节点之前，集群已经开始分配这些资源了），如果集群
12.      //忽略了黑名单，那么立即拒绝 Executor
13.           logInfo(s"Rejecting $executorId as it has been blacklisted.")
14.           executorRef.send(RegisterExecutorFailed(s"Executor is blacklisted:
            $executorId"))
15.           context.reply(true)
16.         } else {
17.           //如果 executor 的 rpc env 未在监听连接，hostport 将为空，应使用客户端连
            //接与 executor 联系
18.           val executorAddress = if (executorRef.address != null) {
19.               executorRef.address
20.             } else {
21.               context.senderAddress
22.             }
23.           logInfo(s"Registered executor $executorRef ($executorAddress)
            with ID $executorId")
24. //在 addressToExecutorId 哈希表中加入 executorAddress 和 executorId 的
25. //对应关系
26.           addressToExecutorId(executorAddress) = executorId
27.          //totalCore 增加 cores 个
28.           totalCoreCount.addAndGet(cores)
29.           totalRegisteredExecutors.addAndGet(1)
30.         //创建 ExecutorData 对象
31.           val data = new ExecutorData(executorRef, executorRef.address, hostname,
            cores, cores, logUrls)
32.           //必须同步，因为在请求 executors 时读取此块中变化的变量
33.           CoarseGrainedSchedulerBackend.this.synchronized {
34.           //在 executorDataMap 中加入 executorId 和 ExecutorData 的对应关系
35.             executorDataMap.put(executorId, data)
36.             if (currentExecutorIdCounter < executorId.toInt) {
37.               currentExecutorIdCounter = executorId.toInt
38.             }
39.        //如果挂起的 Executors 的数量大于 0
40.             if (numPendingExecutors > 0) {
41.               numPendingExecutors -= 1
42.               logDebug(s"Decremented number of pending executors
              ($numPendingExecutors left)")
43.             }
44.           }
45.           executorRef.send(RegisteredExecutor)
46.           //注意：有些测试希望在 executor 放入映射之后得到响应
47.           //向 CoarseGrainedExecutorBackend 回复 true
```

```
48.          context.reply(true)
49.          listenerBus.post(
50.            SparkListenerExecutorAdded(System.currentTimeMillis(),
               executorId, data))
51.         //调用 makeOffers，给 Executor 发送执行任务
52.          makeOffers()
53.        }
54.
```

Spark 2.4.3 版本的 CoarseGrainedSchedulerBackend.scala 的 receiveAndReply 方法的源码与 Spark 2.2.1 版本相比具有如下特点。

- 上段代码中第 8 行 if 判断语句中去掉 scheduler.nodeBlacklist != null 的判断。
- 上段代码中第 31 行构建 ExecutorData 实例时，传入的 executorRef.address 调整为 executorAddress。

```
1.  ……
2.  } else if (scheduler.nodeBlacklist.contains(hostname)) {
3.  ……
4.  val data = new ExecutorData(executorRef, executorAddress, hostname,
5.            cores, cores, logUrls)
6.
```

CoarseGrainedSchedulerBackend.scala 中的 RegisterExecutor：

- 先判断 executorDataMap 是否已经包含 executorId，如果已经包含，就会发送注册失败的消息 RegisterExecutorFailed，因为已经有重复的 executor ID 的 Executor 在运行。
- 然后进行 Executor 的注册，获取到 executorAddress，在 executorRef.address 为空的情况下就获取到 senderAddress。
- 定义了 3 个数据结构：addressToExecutorId、totalCoreCount、totalRegisteredExecutors，其中，addressToExecutorId 是 DriverEndpoint 的数据结构，而 totalCoreCount、totalRegisteredExecutors 是 CoarseGrainedSchedulerBackend 的数据结构。addressToExecutorId、totalCoreCount、totalRegisteredExecutors 包含 Executors 注册的信息分别为：RPC 地址主机名和端口与 ExecutorId 的对应关系、集群中的总核数 Cores、当前注册的 Executors 总数等。

```
1.   protected val addressToExecutorId = new HashMap[RpcAddress, String]
2.   protected val totalCoreCount = new AtomicInteger(0)
3.   protected val totalRegisteredExecutors = new AtomicInteger(0)
```

- 然后调用 new()函数创建一个 ExecutorData，提取出 executorRef、executorRef.address、hostname、cores、cores、logUrls 等信息。
- 同步代码块 CoarseGrainedSchedulerBackend.this.synchronized：集群中很多 Executor 向 Driver 注册，为防止写冲突，因此设计一个同步代码块。在运行时使用 synchronized 关键字，来保证 executorMapData 安全地并发写操作。
- executorRef.send(RegisteredExecutor)发消息 RegisteredExecutor 给我们的 sender，sender 是 CoarseGrainedExecutorBackend。而 CoarseGrainedExecutorBackend 收到消息 RegisteredExecutor 以后，就调用 new()函数创建了 Executor。

CoarseGrainedExecutorBackend 收到 DriverEndpoint 发送过来的 RegisteredExecutor 消息后会启动 Executor 实例对象，而 Executor 实例对象事实上是负责真正 Task 计算的。

```
1.      override def receive: PartialFunction[Any, Unit] = {
2.      case RegisteredExecutor =>
3.        logInfo("Successfully registered with driver")
4.        try {
5.          executor = new Executor(executorId, hostname, env, userClassPath,
            isLocal = false)
6.        } catch {
7.          case NonFatal(e) =>
8.            exitExecutor(1, "Unable to create executor due to " + e.getMessage, e)
9.        }
```

下面来看一下 Executor.scala，其中的 threadPool 是一个线程池。

Executor 是真正负责 Task 计算的；其在实例化的时候会实例化一个线程池 threadPool 来准备 Task 的计算。threadPool 是一个 newDaemonCachedThreadPool。newDaemonCachedThreadPool 创建线程池，线程工厂按照需要的格式调用 new()函数创建线程。语法实现如下：

```
1.     def newDaemonCachedThreadPool(prefix: String): ThreadPoolExecutor = {
2.      val threadFactory = namedThreadFactory(prefix)
3.      Executors.newCachedThreadPool(threadFactory).asInstanceOf
        [ThreadPoolExecutor]
4.    }
```

namedThreadFactory 的源码如下：

```
1.     def namedThreadFactory(prefix: String): ThreadFactory = {
2.      new  ThreadFactoryBuilder().setDaemon(true).setNameFormat(prefix  +
        "-%d").build()
3.    }
```

newCachedThreadPool 创建一个线程池，根据需要创建新线程，线程池中的线程可以复用，使用提供的 ThreadFactory 创建新线程。newCachedThreadPool 的源码如下：

```
1.         public static ExecutorService newCachedThreadPool(ThreadFactory
           threadFactory) {
2.         return new ThreadPoolExecutor(0, Integer.MAX_VALUE,
                                         60L, TimeUnit.SECONDS,
                                         new SynchronousQueue<Runnable>(),
                                         threadFactory);
3.       }
```

创建的 threadPool 中以多线程并发执行和线程复用的方式来高效地执行 Spark 发过来的 Task。线程池创建好后，接下来是等待 Driver 发送任务给 CoarseGrainedExecutorBackend，不是直接发送给 Executor，因为 Executor 不是一个消息循环体。

Executor 具体是如何工作的?

当 Driver 发送过来 Task 的时候，其实是发送给了 CoarseGrainedExecutorBackend 这个 RpcEndpoint，而不是直接发送给了 Executor（Executor 由于不是消息循环体，所以永远也无法直接接收远程发过来的信息）。

Driver 向 CoarseGrainedExecutorBackend 发送 LaunchTask，转过来交给线程池中的线程去执行。先判断 Executor 是否为空，Executor 为空，则提示错误，进程就直接退出。如果 Executor 不为空，则反序列化任务调用 Executor 的 launchTask，其中，attemptNumber 是任务可以重试的次数。

ExecutorBackend 收到 Driver 发送的消息，调用 launchTask 方法，提交给 Executor 执行。

Executor.scala 的 launchTask 接收到 Task 执行的命令后，首先将 Task 封装在 TaskRunner 里面，然后放到 runningTasks。runningTasks 是一个简单的数据结构。

```
1.        private val runningTasks = new ConcurrentHashMap[Long, TaskRunner]
```

launchTask 最后交给 threadPool.execute(tr)，交给线程池中的线程执行任务。 TaskRunner 继承自 Runnable，是 Java 的一个对象。

TaskRunner 其实是 Java 中 Runnable 接口的具体实现，在真正工作时会交给线程池中的线程去运行，此时会调用 run 方法来执行 Task。

Executor.scala 中的 Run 方法最终调用 task.run 方法。

Spark 2.2.1 版本的 Executor.scala 的 run 方法的源码如下：

```
1.      override def run(): Unit = {
2.    .....
3.        var threwException = true
4.          val value = try {
5.            val res = task.run(
6.              taskAttemptId = taskId,
7.              attemptNumber = taskDescription.attemptNumber,
8.              metricsSystem = env.metricsSystem)
9.            threwException = false
10.           res
11.         } finally {
12.           val  releasedLocks  =  env.blockManager.releaseAllLocksForTask
              (taskId)
```

Spark 2.4.3 版本的 Executor.scala 的 run 方法源码与 Spark 2.2.1 版本相比具有如下特点。

- 上段代码中第 4 行的 try 语句调整为 Utils.tryWithSafeFinally 语句。

```
1.    ......
2.       val value = Utils.tryWithSafeFinally {
3.    ......
```

跟进 Task.scala 中的 run 方法，在里面调用 runTask。

```
1.      final def run(
2.        taskAttemptId: Long,
3.        attemptNumber: Int,
4.        metricsSystem: MetricsSystem): T = {
5.     .....
6.      try {
7.        runTask(context)
8.      } catch {
9.     .....
```

TaskRunner 在调用 run 方法时会调用 Task 的 run 方法，而 Task 的 run 方法会调用 runTask，实际上，Task 有 ShuffleMapTask 和 ResultTask。

5.6　本章总结

本章主要讲解了 Master、Worker 的启动原理和源码详解；讲解了 ExecutorBackend 通信接口，以及 ExecutorBackend 与 Executor 的关系。ExecutorBackend 负责与集群通信，而 Executor 则专注于任务的处理，它们是一对一的关系；讲解了 Executor 中任务执行的细节，包括任务的加载、任务线程池、任务执行失败处理、任务的载体 TaskRunner。然后，本章整体上贯通 Executor 工作原理、ExecutorBackend 注册源码、Executor 实例化内幕、Executor 具体工作内幕等内容。

第 6 章　Spark Application 提交给集群的原理和源码详解

本章讲解 Spark Application 提交给集群的原理和源码。6.1 节讲解 Spark Application 到底是如何提交给集群的。6.2 节讲解 Spark Application 是如何向集群申请资源的。6.3 节讲解从 Application 提交的角度重新审视 Driver；6.4 节讲解从 Application 提交的角度重新审视 Executor。6.5 节讲解 Spark 1.6 RPC 内幕解密：运行机制、源码详解、Netty 与 Akka 等内容。

6.1　Spark Application 是如何提交给集群的

本节讲解 Application 提交参数配置、Application 提交给集群原理、Application 提交给集群源码等内容，将彻底解密 Spark Application 到底是如何提交给集群的。

6.1.1　Application 提交参数配置详解

用户应用程序可以使用 bin/spark-submit 脚本来启动。spark-submit 脚本负责使用 Spark 及其依赖关系设置类路径，并可支持 Spark 支持的不同群集管理器和部署模式。

bin/spark-submit 脚本示例如下：

```
1.     ./bin/spark-submit \
2.    --class <main-class> \
3.    --master <master-url> \
4.    --deploy-mode <deploy-mode> \
5.    --conf <key>=<value> \
6.    ... # other options
7.    <application-jar> \
8.    [application-arguments]
```

spark-submit 脚本提交参数配置中一些常用的选项。

--class：应用程序的入口点（如 org.apache.spark.examples.SparkPi）。

--master：集群的主 URL（如 spark://23.195.26.187:7077）。

--deploy-mode：将 Driver 程序部署在集群 Worker 节点（cluster）；或作为外部客户端（client）部署在本地（默认值：client）。

--conf：任意 Spark 配置属性，使用 key = value 格式。对于包含空格的值，用引号括起来，如“key = value”。

application-jar：包含应用程序和所有依赖关系 Jar 包的路径。该 URL 必须在集群内全局可见。例如，所有节点上存在的 hdfs://路径或 file://路径。

application-arguments：传递给主类的 main 方法的参数。

6.1.2　Application 提交给集群原理详解

在 Spark 官网部署页面（http://spark.apache.org/docs/latest/cluster-overview.html），可以看到当前集群支持以下 4 种集群管理器（cluster manager）。

（1）Standalone：Spark 原生的简单集群管理器。使用 Standalone 可以很方便地搭建一个集群。

（2）Apache Mesos：一个通用的集群管理器，可以在上面运行 HadoopMapReduce 和一些服务型的应用。

（3）Hadoop YARN：在 Hadoop 2 中提供的资源管理器。

（4）Kubernetes：一个开源系统，用于自动化容器化应用程序的部署、扩张和管理。

另外，Spark 提供的 EC2 启动脚本，可以很方便地在 Amazon EC2 上启动一个 Standalone 集群。

实际上，除了上面这些通用的集群管理器外，Spark 内部也提供一些方便我们测试、学习的简单集群部署模式。为了更全面地理解，我们会从 Spark 应用程序部署点切入，也就是从提交一个 Spark 应用程序开始，引出并详细解析各种部署模式。

说明： 下面涉及类的描述时，如果可以通过类名唯一确定一个类，将直接给出类名，如果不能，会先给出全路径的类名，然后在不出现歧义的地方再简写为类名。

为了简化应用程序提交的复杂性，Spark 提供了各种应用程序提交的统一入口，即 spark-submit 脚本，应用程序的提交都间接或直接地调用了该脚本。下面简单分析几个脚本，包含./bin/spark-shell、./bin/pyspark、./bin/sparkR、./bin/spark-sql、./bin/run-example、./bin/speak-submit，以及所有脚本最终都调用到的一个执行 Java 类的脚本./bin/spark-class。

1. 脚本./bin/spark-shell

通过该脚本可以打开使用 Scala 语言进行开发、调试的交互式界面，脚本的代码如下：

```
1.  ......
2.  function main() {
3.  ......
4.  "${SPARK_HOME}"/bin/spark-submit  --class  org.apache.spark.repl.Main
    --name "Spark shell" "$@"
5.  stty-icanon echo > /dev/null 2>&1
6.  else
7.  export SPARK_SUBMIT_OPTS
8.  "${SPARK_HOME}"/bin/spark-submit  --class  org.apache.spark.repl.Main
    --name "Spark shell" "$@"
9.  fi
10. }
11. ......
```

对应在第 4 行和第 8 行处，调用了应用程序提交脚本./bin/spark-submit。脚本./bin/spark-shell 的基本用法如下：

```
1.  "Usage: ./bin/spark-shell [options]"
```

其他脚本类似。下面分别针对各个脚本的用法（具体用法可查看脚本的帮助信息，如通过--help 选项来获取）与关键执行语句等进行简单解析。了解工具（如脚本）如何使用，最根本的是先查看其帮助信息，然后在此基础上进行扩展。

2. 脚本./bin/pyspark

通过该脚本可以打开使用 Python 语言开发、调试的交互式界面。

（1）该脚本的用法如下：

```
1.  "Usage: ./bin/pyspark [options]"
```

（2）该脚本的执行语句如下：

```
1.  exec "${SPARK_HOME}"/bin/spark-submit pyspark-shell-main --name
    "PySparkShell" "$@"
```

3. 脚本./bin/sparkR

通过该脚本可以打开使用 sparkR 开发、调试的交互式界面。

（1）该脚本的用法如下：

```
1.  "Usage: ./bin/sparkR [options]"
```

（2）该脚本的执行语句如下：

```
1.  exec "${SPARK_HOME}"/bin/spark-submit sparkr-shell-main "$@"
```

4. 脚本./bin/spark-sql

通过该脚本可以打开使用 SparkSql 开发、调试的交互式界面。

（1）该脚本的用法如下：

```
1.  "Usage: ./bin/spark-sql [options] [cli option]"
```

（2）该脚本的执行语句如下：

```
1.  exec "${SPARK_HOME}"/bin/spark-submit --class org.apache.spark.sql.hive.
    thriftserver.SparkSQLCLIDriver "$@"
```

5. 脚本./bin/run-example

可以通过该脚本运行 Spark 2.4.3 自带的案例代码。该脚本中会自动补全案例类的路径。

（1）该脚本的用法如下：

```
1.  export _SPARK_CMD_USAGE="Usage: ./bin/run-example [options] example-class
    [example args]"
```

（2）该脚本的执行语句如下：

```
1. exec "${SPARK_HOME}"/bin/spark-submit run-example "$@"
```

6. 脚本./bin/spark-submit

./bin/spark-submit 是提交 Spark 应用程序最常用的一个脚本。从前面各个脚本的解析可以看出，各个脚本最终都调用了./bin/spark-submit 脚本。

（1）该脚本的用法。

该脚本的用法需要从源码中获取，具体源码位置参考 SparkSubmitArguments 类的方法 printUsageAndExit，代码如下：

```
1.  val command = sys.env.get("_SPARK_CMD_USAGE").getOrElse(
2.  """Usage: spark-submit [options] <app jar | python file> [app arguments]
3.      |Usage: spark-submit --kill [submission ID] --master [spark://...]
4.      |Usage: spark-submit --status [submission ID] --master [spark:
        //...]""".stripMargin)
```

（2）该脚本的执行语句如下：

```
1.  exec "${SPARK_HOME}"/bin/spark-class org.apache.spark.deploy.
    SparkSubmit "$@"
```

7. 脚本./bin/spark-class

该脚本是Spark 2.4.3所有其他脚本最终都调用到的一个执行Java类的脚本。其中关键的执行语句如下：

```
1.  CMD=()
2.  while IFS= read -d '' -r ARG; do
3.    CMD+=("$ARG")
4.  done < <(build_command "$@")
5.  ......
6.   build_command() {
7.    "$RUNNER" -Xmx128m -cp "$LAUNCH_CLASSPATH" org.apache.spark.launcher.Main "$@"
8.    printf "%d\0" $?
9.  }
```

其中，负责运行的RUNNER变量设置如下：

```
1.  # Find the java binary
2.  //将RUNNER设置为Java
3.  if [ -n "${JAVA_HOME}" ]; then
4.    RUNNER="${JAVA_HOME}/bin/java"
5.  else
6.    if [ 'command -v java' ]; then
7.      RUNNER="java"
8.    else
9.      echo "JAVA_HOME is not set" >&2
10.     exit 1
11.   fi
12. fi
```

在脚本中，LAUNCH_CLASSPATH变量对应Java命令运行时所需的classpath信息。最终Java命令启动的类是org.apache.spark.launcher.Main。Main类的入口函数main，会根据输入参数构建出最终执行的命令，即这里返回的${CMD[@]}信息，然后通过exec执行。

6.1.3 Application提交给集群源码详解

本节从应用部署的角度解析相关的源码，主要包括脚本提交时对应JVM进程启动的主类org.apache.spark.launcher.Main、定义应用程序提交的行为类型的类org.apache.spark.deploy.SparkSubmitAction、应用程序封装底层集群管理器和部署模式的类org.apache.spark.deploy.SparkSubmit，以及代表一个应用程序的驱动程序的类org.apache.spark.SparkContext。

1. Main解析

从前面的脚本分析，得出最终都是通过org.apache.spark.launcher.Main类（下面简称Main

类）启动应用程序的。因此，首先解析一下 Main 类。

在 Main 类的源码中，类的注释如下：

```
1.  /**
2.    * Spark 启动器的命令行接口。在 Spark 脚本内部使用
3.    *
4.    */
```

对应地，在 Main 对象的入口方法 main 的注释如下。

Main.java 源码如下：

```
1.
2.   /**
3.     * Usage: Main [class] [class args]
4.     * <p>
5.     * 命令行界面工作在两种模式下:
6.     * <ul>
7.     *   <li>"spark-submit": if <i>class</i> is "org.apache.spark.deploy.
       *   SparkSubmit", the {@link SparkLauncher} class is used to launch a
       *   Spark application. </li>
8.     *   <li>"spark-class": if another class is provided, an internal Spark
       *   class is run.</li>
9.     * </ul>
10.  .....
11. public static void main(String[] argsArray) throws Exception {
12. .....
```

Main 类主要有两种工作模式，分别描述如下。

（1）spark-submit

启动器要启动的类为 org.apache.spark.deploy.SparkSubmit 时，对应为 spark-submit 工作模式。此时，使用 SparkSubmitCommandBuilder 类来构建启动命令。

（2）spark-class

启动器要启动的类是除 SparkSubmit 之外的其他类时，对应为 spark-class 工作模式。此时使用 SparkClassCommandBuilder 类的 buildCommand 方法来构建启动命令。

Spark 2.2.1 版本的 Main.java 源码如下：

```
1.  public static void main(String[] argsArray) throws Exception {
2.  ......
3.   String className = args.remove(0);
4.  ......
5.   AbstractCommandBuilder builder;
6.      if (className.equals("org.apache.spark.deploy.SparkSubmit")) {
7.        try {
8.          builder = new SparkSubmitCommandBuilder(args);
9.       ......
10.     } else {
11.       builder = new SparkClassCommandBuilder(className, args);
12.     }
13.    ......
```

Spark 2.4.3 版本的 Main.java 源码与 Spark 2.2.1 版本相比具有如下特点。

- ❑ 上段代码中第 5 行删掉 AbstractCommandBuilder builder 的定义。
- ❑ 上段代码中第 8 行、第 11 行 builder 变量调整为 AbstractCommandBuilder builder。

```
1.   ......
2.  AbstractCommandBuilder builder = new SparkSubmitCommandBuilder(args);
3.  ......
4.  AbstractCommandBuilder builder = new SparkClassCommandBuilder(className,
    args);
```

以 spark-submit 工作模式为例，对应的在构建启动命令的 SparkSubmitCommandBuilder 类中，上述调用的 SparkClassCommandBuilder 构造函数定义如下。

Spark 2.2.1 版本的 SparkSubmitCommandBuilder.java 的源码如下：

```
1.     SparkSubmitCommandBuilder(List<String> args) {
2.      this.allowsMixedArguments = false;
3.      this.sparkArgs = new ArrayList<>();
4.      boolean isExample = false;
5.      List<String> submitArgs = args;
6.   //根据输入的第一个参数设置，包括主资源 appResource 等
7.      if (args.size() > 0) {
8.        switch (args.get(0)) {
9.          case PYSPARK_SHELL:
10.           this.allowsMixedArguments = true;
11.           appResource = PYSPARK_SHELL;
12.           submitArgs = args.subList(1, args.size());
13.           break;
14.
15.         case SPARKR_SHELL:
16.           this.allowsMixedArguments = true;
17.           appResource = SPARKR_SHELL;
18.           submitArgs = args.subList(1, args.size());
19.           break;
20.
21.         case RUN_EXAMPLE:
22.           isExample = true;
23.           submitArgs = args.subList(1, args.size());
24.       }
25.
26.       this.isExample = isExample;
27.       OptionParser parser = new OptionParser();
28.       parser.parse(submitArgs);
29.       this.isAppResourceReq = parser.isAppResourceReq;
30.     } else {
31.       this.isExample = isExample;
32.       this.isAppResourceReq = false;
33.     }
34.   }
```

Spark 2.4.3 版本的 SparkSubmitCommandBuilder.java 源码与 Spark 2.2.1 版本相比具有如下特点。

- ❑ 上段代码中第 5 行后新增加一行代码 this.userArgs = Collections.emptyList()。
- ❑ 上段代码中第 22 行后新增加一行代码 appResource = SparkLauncher.NO_RESOURCE。
- ❑ 上段代码中第 27 行构建 OptionParser 实例时，新增一个参数 True。
- ❑ 上段代码中第 29 行将 this.isAppResourceReq = parser.isAppResourceReq 调整为 this.isSpecialCommand = parser.isSpecialCommand。

- 上段代码中第 32 行将 this.isAppResourceReq = false 调整为 this.isSpecialCommand = true。

```
1.   ......
2.   this.userArgs = Collections.emptyList();
3.  ......
4.  appResource = SparkLauncher.NO_RESOURCE;
5.  ......
6.  OptionParser parser = new OptionParser(true);
7.  ......
8.  this.isSpecialCommand = parser.isSpecialCommand;
9.  ......
10. this.isSpecialCommand = true;
```

从这些初步的参数解析可以看出，前面脚本中的参数与最终对应的主资源间的对应关系见表 6-1。

表 6-1　脚本中的参数与主资源间的对应关系

脚 本 名	脚本中的参数	主 资 源
./bin/pyspark	PYSPARK_SHELL = "pyspark-shell-main"	PYSPARK_SHELL_RESOURCE = "pyspark-shell"
./bin/sparkR	SPARKR_SHELL = "sparkr-shell-main"	SPARKR_SHELL_RESOURCE = "sparkr-shell"

如果继续跟踪 appResource 赋值的源码，可以跟踪到一些特殊类的类名与最终对应的主资源间的对应关系，见表 6-2。

表 6-2　特殊类的类名与主资源间的对应关系

参考的脚本名	类　名	主 资 源
./bin/spark-shell	"org.apache.spark.repl.Main"	"spark-shell"
./bin/spark-sql	"org.apache.spark.sql.hive.thriftserver.SparkSQLCLIDriver"	"spark-internal"
./sbin/start-thriftserver	"org.apache.spark.sql.hive.thriftserver.HiveThriftServer2"	"spark-internal"

如果有兴趣，可以继续跟踪 SparkClassCommandBuilder 类的 buildCommand 方法的源码，查看构建的命令具体有哪些。

通过 Main 类的简单解析，可以将前面的脚本分析结果与后面即将进行分析的 SparkSubmit 类关联起来，以便进一步解析与应用程序提交相关的其他源码。

从前面的脚本分析可以看到，提交应用程序时，Main 启动的类，也就是用户最终提交执行的类是 org.apache.spark.deploy.SparkSubmit。因此，下面开始解析 SparkSubmit 相关的源码，包括提交行为的定义、提交时的参数解析以及最终提交运行的代码解析。

2. SparkSubmitAction解析

SparkSubmitAction 定义了提交应用程序的行为类型。SparkSubmit.scala 的源码如下：

```
1.   private[deploy] object SparkSubmitAction extends Enumeration {
2.   type SparkSubmitAction = Value
3.   val SUBMIT, KILL, REQUEST_STATUS = Value
4.  }
```

从源码中可以看到，分别定义了 SUBMIT、KILL、REQUEST_STATUS 这 3 种行为类型，对应提交应用、停止应用、查询应用的状态。

3. SparkSubmit解析

SparkSubmit 的全路径为 org.apache.spark.deploy.SparkSubmit。从 SparkSubmit 类的注释可以看出，SparkSubmit 是启动一个 Spark 应用程序的主入口点，这和前面从脚本分析得到的结论一致。首先看一下 SparkSubmit 类的注释，格式如下：

```
1.  /**
2.    *启动一个 Spark 应用程序的主入口点
3.    *
4.    *
5.    *这个程序处理与 Spark 依赖相关的类路径设置，提供 Spark 支持的在不同集群管理器的部
      *署模式
6.    *
7.    */
```

SparkSubmit 会帮助我们设置 Spark 相关依赖包的 classpath，同时，为了帮助用户简化提交应用程序的复杂性，SparkSubmit 提供了一个抽象层，封装了底层复杂的集群管理器与部署模式的各种差异点，即通过 SparkSubmit 的封装，集群管理器与部署模式对用户是透明的。

在 SparkSubmit 中体现透明性的集群管理器定义的源码如下所示。

Spark 2.2.1 版本的 SparkSubmit.scala 的源码如下：

```
1.  //集群管理器
2.  //Cluster managers
3.  private val YARN = 1
4.  private val STANDALONE = 2
5.  private val MESOS = 4
6.  private val LOCAL = 8
7.  private val ALL_CLUSTER_MGRS = YARN | STANDALONE | MESOS | LOCAL
```

Spark 2.4.3 版本的 SparkSubmit.scala 源码与 Spark 2.2.1 版本相比具有如下特点。

- 上段代码中第 6 行之后新增一行代码 private val KUBERNETES = 16。
- 上段代码中第 7 行新增加一个变量 KUBERNETES。

```
1.  ......
2.    private val KUBERNETES = 16
3.    private val ALL_CLUSTER_MGRS = YARN | STANDALONE | MESOS | LOCAL |
      KUBERNETES
```

在 SparkSubmit 中体现透明性的部署模式定义的源码如下：

```
1.  //部署模式
2.  //Deploy modes
3.  private val CLIENT = 1
4.  private val CLUSTER = 2
5.  private val ALL_DEPLOY_MODES = CLIENT | CLUSTER
```

作为提交应用程序的入口点，SparkSubmit 中根据具体的集群管理器进行参数转换、参数校验等操作，如对模式的检查，代码中给出了针对特定情况，不支持的集群管理器与部署模式，在这些模式下提交应用程序会直接报错退出。

Spark 2.2.1 版本的 SparkSubmit.scala 的源码如下：

```
1.    //不支持的集群管理器与部署模式
2.      (clusterManager, deployMode) match {
3.       case (STANDALONE, CLUSTER) if args.isPython =>
4.         printErrorAndExit("Cluster deploy mode is currently not supported
            for python " +"applications on standalone clusters.")
5.       case (STANDALONE, CLUSTER) if args.isR =>
6.         printErrorAndExit("Cluster deploy mode is currently not supported
            for R " +"applications on standalone clusters.")
7.       case (LOCAL, CLUSTER) =>
8.         printErrorAndExit("Cluster deploy mode is not compatible with
           master \"local\"")
9.       case (_, CLUSTER) if isShell(args.primaryResource) =>
10.        printErrorAndExit("Cluster deploy mode is not applicable to Spark
           shells.")
11.      case (_, CLUSTER) if isSqlShell(args.mainClass) =>
12.        printErrorAndExit("Cluster deploy mode is not applicable to Spark
           SQL shell.")
13.      case (_, CLUSTER) if isThriftServer(args.mainClass) =>
14.        printErrorAndExit("Cluster deploy mode is not applicable to Spark
           Thrift server.")
15.      case _ =>
16.    }
```

Spark 2.4.3 版本的 SparkSubmit.scala 源码与 Spark 2.2.1 版本相比具有如下特点。

- 上段代码中第 4、6、8、10、12、14 行的 printErrorAndExit 方法调整为 error 方法。

```
1.  ......
2.          error("Cluster deploy mode is currently not supported for python
            " + "applications on standalone clusters.")
3.  ......
4.          error("Cluster deploy mode is currently not supported for R " +
5.            "applications on standalone clusters.")
6.  ......
7.          error("Cluster deploy mode is not compatible with master \"local\"")
8.  ......
9.          error("Cluster deploy mode is not applicable to Spark shells.")
10. ......
11.         error("Cluster deploy mode is not applicable to Spark SQL shell.")
12. ......
13.         error("Cluster deploy mode is not applicable to Spark Thrift server.")
14. ......
```

首先，一个程序运行的入口点对应单例对象的 main 函数，因此在执行 SparkSubmit 时，对应的入口点是 objectSparkSubmit 的 main 函数。

Spark 2.2.1 版本的 SparkSubmit.scala 的源码如下：

```
1.  //入口点函数 main 的定义
2.  def main(args: Array[String]): Unit = {
3.    val appArgs = new SparkSubmitArguments(args)
4.    ......
5.    //根据 3 种行为分别进行处理
6.  appArgs.action match {
```

```
7.      case SparkSubmitAction.SUBMIT => submit(appArgs)
8.      case SparkSubmitAction.KILL => kill(appArgs)
9.      case SparkSubmitAction.REQUEST_STATUS =>requestStatus(appArgs)
10.   }
11. }
```

Spark 2.4.3 版本的 SparkSubmit.scala 源码与 Spark 2.2.1 版本相比具有如下特点。

- 上段代码中第 3 行代码调整为 parseArguments 方法。
- 上段代码中第 9 行之后新增一行代码 case SparkSubmitAction.PRINT_VERSION => printVersion()。

```
1.    val submit = new SparkSubmit() {
2.       self =>
3.
4.       override protected def parseArguments(args: Array[String]):
         SparkSubmitArguments = {
5.         new SparkSubmitArguments(args) {
6.           override protected def logInfo(msg: => String): Unit = self.logInfo(msg)
7.
8.           override protected def logWarning(msg: => String): Unit =
             self.logWarning(msg)
9.         }
10.      }
11. ……
12. case SparkSubmitAction.PRINT_VERSION => printVersion()
```

printVersion()方法用于打印 Spark 的版本信息。

```
1.    private def printVersion(): Unit = {
2.      logInfo("""Welcome to
3.        ____              __
4.       / __/__  ___ _____/ /__
5.      _\ \/ _ \/ _ `/ __/  '_/
6.     /___/ .__/\_,_/_/ /_/\_\   version %s
7.        /_/
8.                          """.format(SPARK_VERSION))
9.      logInfo("Using Scala %s, %s, %s".format(
10.       Properties.versionString, Properties.javaVmName, Properties.javaVersion))
11.     logInfo(s"Branch $SPARK_BRANCH")
12.     logInfo(s"Compiled by user $SPARK_BUILD_USER on $SPARK_BUILD_DATE")
13.     logInfo(s"Revision $SPARK_REVISION")
14.     logInfo(s"Url $SPARK_REPO_URL")
15.     logInfo("Type --help for more information.")
    }
```

其中，SparkSubmitArguments 类对应用户调用提交脚本 spark-submit 时传入的参数信息。对应的脚本的帮助信息（./bin/spark-submit --help），也是由该类的 printUsageAndExit 方法提供的。

找到上面的入口点代码之后，就可以开始分析其内部的源码。对应参数信息的 SparkSubmitArguments 可以参考脚本的帮助信息，来查看具体参数对应的含义。参数分析后，便是对各种提交行为的具体处理。SparkSubmit 支持 SparkSubmitAction 包含的 3 种行为，下面以行为 SparkSubmitAction.SUBMIT 为例进行分析，其他行为也可以通过各自的具体处理代码进行分析。

对应处理 SparkSubmitAction.SUBMIT 行为的代码入口点为 submit(appArgs)，进入该方法，即进入提交应用程序的处理方法。

Spark 2.2.1 版本的 SparkSubmit.scala 的源码如下：

```
1.    private def submit(args: SparkSubmitArguments): Unit = {
2.   //准备应用程序提交的环境，该步骤包含了内部封装的各个细节处理
3.     val (childArgs, childClasspath, sysProps, childMainClass) =
       prepareSubmitEnvironment(args)
4.
5.     def doRunMain(): Unit = {
6.       if (args.proxyUser != null) {
7.         val proxyUser = UserGroupInformation.createProxyUser
             (args.proxyUser,UserGroupInformation.getCurrentUser())
8.
9.         try {
10.          proxyUser.doAs(new PrivilegedExceptionAction[Unit]() {
11.            override def run(): Unit = {
12.              runMain(childArgs, childClasspath, sysProps, childMainClass,
                 args.verbose)
13.            }
14.          })
15.        } catch {
16.          case e: Exception =>
17.            //hadoop 的 AuthorizationException 抑制异常堆栈跟踪，通过 JVM 打印输
               //出的消息不是很有帮助。这里检测异常以及空栈，对其采用不同的处理
18.            if (e.getStackTrace().length == 0) {
19.              //scalastyle:off println
20.              printStream.println(s"ERROR:    ${e.getClass().getName()}:
                 ${e.getMessage()}")
21.              //scalastyle:on println
22.              exitFn(1)
23.            } else {
24.              throw e
25.            }
26.        }
27.      } else {
28.        runMain(childArgs,  childClasspath,  sysProps,  childMainClass,
           args.verbose)
29.      }
30.    }
31.
32.    //Standalone 集群模式下，有两种提交应用程序的方式
33.    //1.传统的 RPC 网关方式使用 o.a.s.deploy.Client 进行封装
34.    //2.Spark 1.3 使用新 REST-based 网关方式，作为 Spark 1.3 的默认方法，如果 Master
       //节点不是 REST 服务器节点，Spark 应用程序提交时会切换到传统的网关模式
35.     if (args.isStandaloneCluster && args.useRest) {
36.       try {
37.         //scalastyle:off println
38.         printStream.println("Running Spark using the REST application
            submission protocol.")
39.         //scalastyle:on println
40.         doRunMain()
41.       } catch {
42.           //如果失败，则使用传统的提交方式
43.         case e: SubmitRestConnectionException =>
44.           printWarning(s"Master endpoint ${args.master} was not a REST
              server. " + "Falling back to legacy submission gateway instead.")
45.
```

```
46.         //重新设置提交方式的控制开关
47.           args.useRest = false
48.           submit(args)
49.       }
50.     //在所有其他模式中，只要准备好主类就可以
51.     } else {
52.       doRunMain()
53.     }
54.   }
```

Spark 2.4.3 版本的 SparkSubmit.scala 源码与 Spark 2.2.1 版本相比具有如下特点。

- ❑ 上段代码中第 1 行 submit 方法新增加一个参数 uninitLog。
- ❑ 上段代码中第 3 行、12 行、28 行的变量名称 sysProps 调整为 sparkConf。
- ❑ 上段代码中第 20 行的日志打印语句 printStream.println 调整为 error 方法。
- ❑ 上段代码中第 30 行之后新增加代码，如果 uninitLog 为 True，则让主类在日志系统启动后重新初始化。
- ❑ 上段代码中第 38 行的日志打印语句 printStream.println 调整为 logInfo 方法。
- ❑ 上段代码中第 44 行的日志打印语句 printWarning 调整为 logWarning 方法。
- ❑ 上段代码中第 48 行的 submit(args)新增一个传入参数 false，调整为 submit(args, false)。

```
1.  private def submit(args: SparkSubmitArguments, uninitLog: Boolean): Unit
2.  = {
3.  ......
4.  val (childArgs, childClasspath, sparkConf, childMainClass) =
    prepareSubmitEnvironment(args)
5.  ......
6.  runMain(childArgs, childClasspath, sparkConf, childMainClass,
    args.verbose)
7.  ......
8.  runMain(childArgs, childClasspath, sparkConf, childMainClass,
    args.verbose)
9.  ......
10. error(s"ERROR: ${e.getClass().getName()}: ${e.getMessage()}")
11. ......
12.   if (uninitLog) {
13.     Logging.uninitialize()
14.   }
15. ......
16. logInfo("Running Spark using the REST application submission protocol.")
17. ......
18. logWarning(s"Master endpoint ${args.master} was not a REST server. " +
19.            "Falling back to legacy submission gateway instead.")
20. ......
21.  submit(args, false)
```

其中，最终运行所需的参数都由 prepareSubmitEnvironment 方法负责解析、转换，然后根据其结果执行。解析的结果包含以下 4 部分。

- ❑ 子进程运行所需的参数。
- ❑ 子进程运行时的 classpath 列表。
- ❑ 系统属性的映射。
- ❑ 子进程运行时的主类。

解析之后调用 runMain 方法，该方法中除了一些环境设置等操作外，最终会调用解析得到的 childMainClass 的 main 方法。下面简单分析一下 prepareSubmitEnvironment 方法，通过该方法来了解 SparkSubmit 是如何帮助底层的集群管理器和部署模式的封装的。里面涉及的各种细节比较多，这里以不同集群管理器和部署模式下最终运行的 childMainClass 类的解析为主线进行分析。

（1）当部署模式为 CLIENT 时，将 childMainClass 设置为传入的 mainClass，对应代码如下：

```
1.      //在 CLIENT 模式下，直接启动应用程序的主类
2.      //此外，在类路径中添加主应用程序 jar 和任何添加的 jar（如果有的话）。如果 yarn
        //客户机需要这些 jar，还应将主应用程序 jar 和任何添加的 jar 添加到类路径中
3.        if (deployMode == CLIENT) {
4.            childMainClass = args.mainClass
5.            if (localPrimaryResource != null && isUserJar(localPrimaryResource)) {
6.              childClasspath += localPrimaryResource
7.            }
8.            if (localJars != null) { childClasspath ++= localJars.split(",") }
9.          }
10.         //将主应用程序jar和任何添加的jar添加到类路径,以防yarn客户机需要这些jar。
11.         //这假设primaryResource 和user jar都是本地jar，否则它不会添加到yarn客户
            //机的类路径中
12.         if (isYarnCluster) {
13.           if (isUserJar(args.primaryResource)) {
14.             childClasspath += args.primaryResource
15.           }
16.           if (args.jars != null) { childClasspath ++= args.jars.split(",") }
17.         }
18.
19.         if (deployMode == CLIENT) {
20.           if (args.childArgs != null) { childArgs ++= args.childArgs }
21.     }
```

（2）当集群管理器为 STANDALONE、部署模式为 CLUSTER 时，根据提交的两种方式将 childMainClass 分别设置为不同的类，同时将传入的 args.mainClass（提交应用程序时设置的主类）及其参数根据不同集群管理器与部署模式进行转换，并封装到新的主类所需的参数中，对应的设置见表 6-3。

表 6-3　STANDALONE+CLUSTER时两种不同提交方式下的childMainClass封装

提 交 方 式	childMainClass
REST 方式（Spark 1.3+）	"org.apache.spark.deploy.rest.RestSubmissionClient"
传统方式	"org.apache.spark.deploy.Client"

其中，表述性状态传递（Representational State Transfer，REST）是 Roy Fielding 博士在 2000 年他的博士论文中提出来的一种软件架构风格。

这些设置的主类相当于封装了应用程序提交时的主类，运行后负责向 Master 节点申请启动提交的应用程序。

（3）当集群管理器为 YARN、部署模式为 CLUSTER 时，childMainClass 以及对应的 mainClass 的设置见表 6-4。

表 6-4　YARN+CLUSTER时childMainClass下的childMainClass封装

执 行 对 象	childMainClass	被封装的执行主类（mainClass）
isPython	"org.apache.spark.deploy.yarn.Client"	"org.apache.spark.deploy.PythonRunner"
isR		"org.apache.spark.deploy.RRunner"
其他		args.mainClass

（4）当集群管理器为 MESOS、部署模式为 CLUSTER 时，childMainClass 以及对应的 mainClass 的设置见表 6-5。

表 6-5　MESOS+CLUSTER时childMainClass下的childMainClass封装

执 行 对 象	childMainClass	被封装的执行主类（mainClass）
isPython	"org.apache.spark.deploy.rest.RestSubmissionClient"	""
其他		args.mainClass

从上面的分析中可以看到，使用 CLIENT 部署模式进行提交时，由于设置的 childMainClass 为应用程序提交时的主类，因此是直接在提交点执行设置的主类，即 mainClass，当使用 CLUSTER 部署模式进行提交时，则会根据具体集群管理器等信息，使用相应的封装类。这些封装类会向集群申请提交应用程序的请求，然后在由集群调度分配得到的节点上，启动所申请的应用程序。

以封装类设置为 org.apache.spark.deploy.Client 为例，从该类主入口 main 方法查看，可以看到构建了一个 ClientEndpoint 实例，该实例构建时，会将提交应用程序时设置的 mainClass 等信息封装到 DriverDescription 实例中，然后发送到 Master，申请执行用户提交的应用程序。

对应各种集群管理器与部署模式的组合，实际代码中的处理细节非常多。这里仅给出一种源码阅读的方式，和对应的大数据处理一样，通常采用化繁为简的方式去阅读复杂的源码。例如，这里在理解整个大框架的调用过程后，以 childMainClass 的设置作为主线去解读源码，对应地，在扩展阅读其他源码时，也可以采用这种方式，以某种集群管理器与部署模式为主线，详细阅读相关的代码。最后，在了解各种组合的处理细节之后，通过对比、抽象等方法，对整个 SparkSubmit 进行归纳总结。

提交的应用程序的驱动程序（Driver Program）部分对应包含了一个 SparkContext 实例。因此，接下来从该实例出发，解析驱动程序在不同的集群管理器的部署细节。

4. SparkContext解析

在详细解析 SparkContext 实例前，首先查看一下 SparkContext 类的注释部分，具体如下：

```
1.  /**
2.   * Spark 功能的主入口点。一个 SparkContext 代表连接到 Spark 集群，并可用于在集群中
     * 创建 RDDs、累加器和广播变量
3.   * .....
4.   * @param 描述应用程序配置的配置对象。在该配置的任何设置将覆盖默认的配置以及系统属性
5.   */
```

SparkContext 类是 Spark 功能的主入口点。一个 SparkContext 实例代表了与一个 Spark 集群的连接，并且通过该实例，可以在集群中构建 RDDs、累加器以及广播变量。SparkContext 实例的构建参数 config 描述了应用程序的 Spark 配置。在该参数中指定的配置属性会覆盖默

认的配置属性以及系统属性。

在 SparkContext 类文件中定义了一个描述集群管理器类型的单例对象 SparkMasterRegex，在该对象中详细给出了当前 Spark 支持的各种集群管理器类型。

SparkContext.scala 的源码如下：

```
1.  /**
2.   * 定义了从 Master 信息中抽取集群管理器类型的一个正则表达式集合
3.   *
4.   */
5.  private object SparkMasterRegex {
6.   //对应 Master 格式如 local[N] 和 local[*]的正则表达式
7.   //对应的 Master 格式如 local[N]和 local[*]的正则表达式
8.   val LOCAL_N_REGEX = """local\[([0-9]+|\*)\]""".r
9.
10.  //对应的 Master 格式如 local[N, maxRetries]的正则表达式
11.  //这种集群管理器类型用于具有任务失败尝试功能的测试
12.
13.   val LOCAL_N_FAILURES_REGEX = """local\[([0-9]+|\*)\s*,\s*([0-9]+)\]""".r
14.
15. //一种模拟 Spark 集群的本地模式的正则表达式,对应的 Master 格式如 local-cluster[N,
16. //cores, memory]
17.
18.                                         LOCAL_CLUSTER_REGEX = """local-cluster\
                                            [\s*([0-9]+)\s*,\s*([0-9]+)\s*,\s*
                                            ([0-9]+)\s*]""".r
19.
20. //连接 Spark 部署集群的正则表达式
21.
22.   val SPARK_REGEX = """spark://(.*)""".r
23. }
```

在 SparkContext 类中的主要流程可以归纳如下。

（1）createSparkEnv：创建 Spark 的执行环境对应的 SparkEnv 实例。

对应代码如下：

```
1.      //Create the Spark execution environment (cache, map output tracker,
        //etc)
2.      _env = createSparkEnv(_conf, isLocal, listenerBus)
3.  SparkEnv.set(_env)
```

（2）createTaskScheduler：创建作业调度器实例。

对应代码如下：

```
1.  //创建和启动调度器 scheduler
2.  val (sched, ts) = SparkContext.createTaskScheduler(this, master, deployMode)
3.  _schedulerBackend = sched
4.  _taskScheduler = ts
```

其中，TaskScheduler 是低层次的任务调度器，负责任务的调度。通过该接口提供可插拔的任务调度器。每个 TaskScheduler 负责调度一个 SparkContext 实例中的任务，负责调度上层 DAG 调度器中每个 Stage 提交的任务集（TaskSet），并将这些任务提交到集群中运行，在任务提交执行时，可以使用失败重试机制设置失败重试的次数。上述对应高层的 DAG 调度器的实例构建参见下一步。

（3）new DAGScheduler：创建高层 Stage 调度的 DAG 调度器实例。

对应代码如下：

```
1.   _dagScheduler = new DAGScheduler(this)
```

DAGScheduler 是高层调度模块，负责作业（Job）的 Stage 拆分，以及最终将 Stage 对应的任务集提交到低层次的任务调度器上。

下面基于这些主要流程，针对 SparkMasterRegex 单例对象中给出的各种集群部署模式进行解析。对应不同集群模式，这些流程中构建了包括 TaskScheduler 与 SchedulerBackend 的不同的具体子类，所构建的相关实例具体见表 6-6。

表 6-6　各种情况下TaskScheduler与SchedulerBackend的不同的具体子类

<table>
<tr><th>部署模式（Master）</th><th>实例对应的类</th><th>备　注</th></tr>
<tr><td>"local"</td><td rowspan="3">_taskScheduler：TaskSchedulerImpl
_schedulerBackend：LocalBackend</td><td>最简单的本地模式
这种本地模式下，任务的失败重试次数为 1，即失败不重试</td></tr>
<tr><td>local[*]、local[N]</td><td>指定线程个数的本地模式，指定方式及最终的线程数如下：
① local[*]：当前处理器个数
② local[N]：指定的 N
这种本地模式下，任务的失败重试次数为 1，即失败不重试</td></tr>
<tr><td>local[*, M]、
local[N, M]</td><td>指定线程个数以及失败重试次数的本地模式，仅比上一种本地模式多了一个失败重试次数的设置，对应为 M</td></tr>
<tr><td>local-cluster[numSlaves,
coresPerSlave,
memoryPerSlave]</td><td rowspan="2">_taskScheduler：TaskSchedulerImpl
_schedulerBackend：
StandaloneSchedulerBackend</td><td>本地伪分布式集群，由于本地模式下没有集群，因此需要构建一个用于模拟集群的实例：localCluster = new
LocalSparkCluster
对应的 3 个参数为
numSlaves：模拟集群的 Slave 节点个数。
coresPerSlave：模拟集群的各个 Slave 节点上的内核数。
memoryPerSlave：模拟集群的各个 Slave 节点上的内存大小</td></tr>
<tr><td>Spark Standalone</td><td>Spark Standalone 对应 Spark 原生的完全分布式集群因此，此种方式下不需要像上面的本地伪分布式集群那样，构建一个虚拟的本地集群</td></tr>
<tr><td>YARN Client</td><td>_taskScheduler：YarnScheduler
_schedulerBackend：
YarnClientSchedulerBackend</td><td>YARN 集群管理器 + Client 部署</td></tr>
<tr><td>YARN Cluster</td><td>_taskScheduler：YarnClusterScheduler
_schedulerBackend：
YarnClusterSchedulerBackend</td><td>YARN 集群管理器 + Cluster 部署</td></tr>
</table>

与 TaskScheduler 和 SchedulerBackend 不同的是，在不同集群模式中，应用程序的高层调度器 DAGScheduler 的实例是相同的，即对应在 Spark on YARN 与 Mesos 等集群管理器中，

应用程序内部的高层 Stage 调度是相同的。

6.2 Spark Application 是如何向集群申请资源的

本节讲解 Application 申请资源的两种类型：第一种是尽可能在集群的所有 Worker 上分配 Executor；第二种是运行在尽可能少的 Worker 上。本节讲解 Application 申请资源的源码内容，将彻底解密 Spark Application 是如何向集群申请资源的。

6.2.1 Application 申请资源的两种类型详解

Master 负责资源管理和调度。资源调度的方法 schedule 位于 Master.scala 类中，当注册程序或者资源发生改变时，都会导致 schedule 的调用。Schedule 调用的时机：每次有新的应用程序提交或者集群资源状况发生改变时（包括 Executor 增加或者减少、Worker 增加或者减少等）。

Spark 默认为应用程序启动 Executor 的方式是 FIFO 的方式，也就是所有提交的应用程序都放在调度的等待队列中，先进先出，只有在满足了前面应用程序的资源分配的基础上，才能够满足下一个应用程序资源的分配；在 FIFO 的情况下，默认是 spreadOutApps 来让应用程序尽可能多地运行在所有的 Node 上。为应用程序分配 Executors 有两种方式：第一种方式是尽可能在集群的所有 Worker 上分配 Executor，这种方式往往会带来潜在的、更好的数据本地性；第二种方式是尝试运行在尽可能少的 Worker 上。

为了更形象地描述 Master 的调度机制，下面通过图 6-1 介绍抽象的资源调度框架。

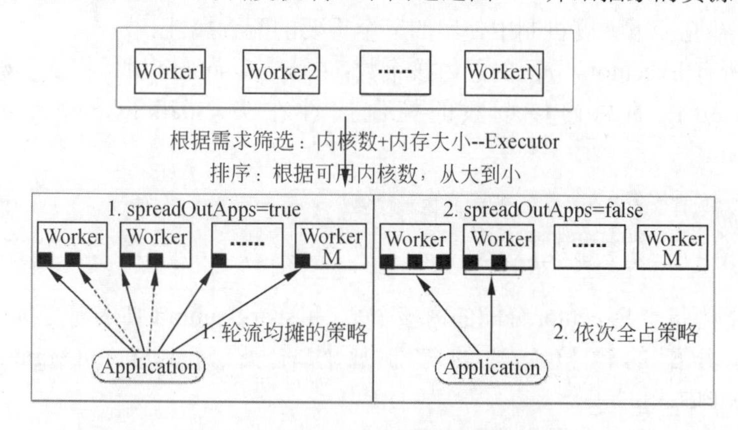

图 6-1　Master 中抽象的资源调度框架

其中，Worker1 到 WorkerN 是集群中全部的 Workers 节点，调度时，会根据应用程序请求的资源信息，从全部 Workers 节点中过滤出资源足够的节点，假设可以得到 Worker1 到 WorkerM 的节点。当前过滤的需求是内核数和内存大小足够启动一个 Executor，因为 Executor 是集群执行应用程序的单位组件（注意：和任务（Task）不是同一个概念，对应的任务是在 Executor 中执行的）。

选出可用 Workers 之后，会根据内核大小进行排序，这可以理解成是一种基于可用内核排序的、简单的负载均衡策略。然后根据设置的 spreadOutApps 参数，对应指定两种资源分配策略。

（1）当 spreadOutApps=true：使用轮流均摊的策略，也就是采用圆桌（round-robin）算法，图中的虚线表示第一次轮流摊派的资源不足以满足申请的需求，因此开始第二轮摊派，依次轮流均摊，直到符合资源需求。

（2）当 spreadOutApps=false：使用依次全占策略，依次从可用 Workers 上获取该 Worker 上可用的全部资源，直到符合资源需求。

对应图中 Worker 内部的小方块，在此表示分配的资源的抽象单位。对应资源的条件，理解的关键点在于资源是分配给 Executor 的，因此最终启动 Executor 时，占用的资源必须满足启动所需的条件。

前面描述了 Workers 上的资源是如何分配给应用程序的，之后正式开始为 Executor 分配资源，并向 Worker 发送启动 Executor 的命令了。根据申请时是否明确指定需要为每个 Executor 分配确定的内核个数，有：

（1）明确指定每个 Executor 需要分配的内核个数时：每次分配的是一个 Executor 所需的内核数和内存数，对应在某个 Worker 分配到的总的内核数可能是 Executor 的内核数的倍数，此时，该 Worker 节点上会启动多个 Executor，每个 Executor 需要指定的内核数和内存数（注意该 Worker 节点上分配到的总的内存大小）。

（2）未明确指定每个 Executor 需要分配的内核个数时：每次分配一个内核，最后所有在某 Worker 节点上分配到的内核都会放到一个 Executor 内（未明确指定内核个数，因此可以一起放入一个 Executor）。因此，最终该应用程序在一个 Worker 上只有一个 Executor（这里指的是针对一个应用程序，当该 Worker 节点上存在多个应用程序时，仍然会为每个应用程序分别启动相应的 Executor）。

在此强调、补充一下调度机制中使用的三个重要的配置属性。

① 指定为所有 Executors 分配的总内核个数：在 spark-submit 脚本提交参数时进行配置。所有 Executors 分配的总内核个数的控制属性在类 SparkSubmitArguments 的方法 printUsageAndExit 中。

```
1.  //指定为所有 Executors 分配的总内核个数
2.  | Spark standalone and Mesos only:
3.  |  --total-executor-cores NUM  Total cores for all executors.
```

② 指定需要为每个 Executor 分配的内核个数：在 spark-submit 脚本提交参数时进行配置。每个 Executor 分配的内核个数的控制属性在类 SparkSubmitArguments 的方法 printUsageAndExit 中。

SparkSubmitArguments.scala 的源码如下：

```
1.  // 指定需要为每个 Executor 分配的内核个数
2.  || Spark standalone and YARN only:
3.  |  --executor-cores NUM       Number of cores per executor. (Default: 1
    in YARN mode,
4.    or all available cores on the worker in standalone mode)
```

③ 资源分配策略：数据本地性（数据密集）与计算密集的控制属性，对应的配置属性在 Master 类中，代码如下：

```
1.    private val spreadOutApps = conf.getBoolean("spark.deploy.spreadOut",
      true)
```

6.2.2　Application 申请资源的源码详解

1. 任务调度与资源调度的区别

- 任务调度是通过 DAGScheduler、TaskScheduler、SchedulerBackend 等进行的作业调度。
- 资源调度是指应用程序如何获得资源。
- 任务调度是在资源调度的基础上进行的，如果没有资源调度，任务调度就成为无源之水，无本之木。

2. 资源调度内幕

（1）因为 Master 负责资源管理和调度，所以资源调度的方法 schedule 位于 Master.scala 类中，注册程序或者资源发生改变时都会导致 schedule 的调用，如注册程序时：

```
1.        case RegisterApplication(description, driver) =>
2.        //待办事项：防止重复注册 Driver
3.        if (state == RecoveryState.STANDBY) {
4.          //忽略，不要发送响应
5.        } else {
6.          logInfo("Registering app " + description.name)
7.          val app = createApplication(description, driver)
8.          registerApplication(app)
9.          logInfo("Registered app " + description.name + " with ID " + app.id)
10.         persistenceEngine.addApplication(app)
11.         driver.send(RegisteredApplication(app.id, self))
12.         schedule()
13.       }
```

（2）Schedule 调用的时机：每次有新的应用程序提交或者集群资源状况发生改变的时候（包括 Executor 增加或者减少、Worker 增加或者减少等）。

进入 schedule()，schedule 为当前等待的应用程序分配可用的资源。每当一个新的应用程序进来时，schedule 都会被调用。或者资源发生变化时（如 Executor 挂掉，Worker 挂掉，或者新增加机器），schedule 都会被调用。

（3）当前 Master 必须以 ALIVE 的方式进行资源调度，如果不是 ALIVE 的状态，就会直接返回，也就是 Standby Master 不会进行 Application 的资源调用。

```
1.         if (state != RecoveryState.ALIVE) {
2.        return
3.     }
```

（4）接下来通过 workers.toSeq.filter(_.state == WorkerState.ALIVE)过滤判断所有 Worker 中哪些是 ALIVE 级别的 Worker，ALIVE 才能够参与资源的分配工作。

```
1.     val shuffledAliveWorkers = Random.shuffle(workers.toSeq.filter(_.state
       == WorkerState.ALIVE))
```

（5）使用 Random.shuffle 把 Master 中保留的集群中所有 ALIVE 级别的 Worker 的信息随机打乱；Master 的 schedule()方法中：workers 是一个数据结构，打乱 workers 有利于负载均衡。例如，不是以固定的顺序启动 launchDriver。WorkerInfo 是 Worker 注册时将信息注册过来。

```
1.  val workers = new HashSet[WorkerInfo]
2.  ……
3.   val shuffledAliveWorkers = Random.shuffle(workers.toSeq.filter(_.state
    == WorkerState.ALIVE))
```

WorkerInfo.scala 的源码如下：

```
1.   private[spark] class WorkerInfo(
2.      val id: String,
3.      val host: String,
4.      val port: Int,
5.      val cores: Int,
6.      val memory: Int,
7.      val endpoint: RpcEndpointRef,
8.      val webUiAddress: String)
9.    extends Serializable {
```

随机打乱的算法：将 Worker 的信息传进来，先调用 new()函数创建一个 ArrayBuffer，将所有的信息放进去。然后将两个索引位置的内容进行交换。例如，如果有 4 个 Worker，依次分别为第一个 Worker 至第四个 Worker，第一个位置是第 1 个 Worker，第 2 个位置是第 2 个 Worker，第 3 个位置是第 3 个 Worker，第 4 个位置是第 4 个 Worker；通过 Shuffle 以后，现在第一个位置可能是第 3 个 Worker，第 2 个位置可能是第 1 个 Worker，第 3 个位置可能是第 4 个 Worker，第 4 个位置可能是第 2 个 Worker，位置信息打乱。

Random.scala 中的 shuffle 方法，其算法内部是循环随机交换所有 Worker 在 Master 缓存数据结构中的位置。

```
1.   def shuffle[T, CC[X] <: TraversableOnce[X]](xs: CC[T])(implicit bf:
     CanBuildFrom[CC[T], T, CC[T]]): CC[T] = {
2.      val buf = new ArrayBuffer[T] ++= xs
3.
4.      def swap(i1: Int, i2: Int) {
5.        val tmp = buf(i1)
6.        buf(i1) = buf(i2)
7.        buf(i2) = tmp
8.      }
9.
10.     for (n <- buf.length to 2 by -1) {
11.       val k = nextInt(n)
12.       swap(n - 1, k)
13.     }
14.
15.     (bf(xs) ++= buf).result
16.   }
```

（6）Master 的 schedule()方法中：循环遍历等待启动的 Driver，如果是 Client 模式，就不需要 waitingDrivers 等待；如果是 Cluster 模式，此时 Driver 会加入 waitingDrivers 等待列表。

当 SparkSubmit 指定 Driver 在 Cluster 模式的情况下，此时 Driver 会加入 waitingDrivers 等待列表中，在每个 DriverInfo 的 DriverDescription 中有要启动 Driver 时对 Worker 的内存及 Cores 的要求等内容。

```
1.  private val waitingDrivers = new ArrayBuffer[DriverInfo]
2. ......
```

DriverInfo 包括启动时间、ID、描述信息、提交时间等内容。

DriverInfo.scala 的源码如下：

```
1.  private[deploy] class DriverInfo(
2.    val startTime: Long,
3.    val id: String,
4.    val desc: DriverDescription,
5.    val submitDate: Date)
6.   extends Serializable {
```

其中，DriverInfo 的 DriverDescription 描述信息中包括 jarUrl、内存、Cores、supervise、command 等内容。如果在 Cluster 模式中指定 supervise 为 True，那么 Driver 挂掉时就会自动重启。

DriverDescription.scala 的源码如下：

```
1.    private[deploy] case class DriverDescription(
2.     jarUrl: String,
3.     mem: Int,
4.     cores: Int,
5.     supervise: Boolean,
6.     command: Command) {
```

在符合资源要求的情况下，采用随机打乱后的一个 Worker 来启动 Driver，worker 是 Master 中对 Worker 的一个描述。

Master.scala 的 launchDriver 方法如下：

```
1.      private def launchDriver(worker: WorkerInfo, driver: DriverInfo) {
2.     logInfo("Launching driver " + driver.id + " on worker " + worker.id)
3.     worker.addDriver(driver)
4.     driver.worker = Some(worker)
5.     worker.endpoint.send(LaunchDriver(driver.id, driver.desc))
6.     driver.state = DriverState.RUNNING
7.   }
```

Master 通过 worker.endpoint.send(LaunchDriver)发指令给 Worker，让远程的 Worker 启动 Driver，Driver 启动以后，Driver 的状态就变成 DriverState.RUNNING。

（7）先启动 Driver，才会发生后续的一切资源调度的模式。

（8）Spark 默认为应用程序启动 Executor 的方式是 FIFO 方式，也就是所有提交的应用程序都是放在调度的等待队列中的，先进先出，只有满足了前面应用程序的资源分配的基础，才能够满足下一个应用程序资源的分配。

Master 的 schedule()方法中，调用 startExecutorsOnWorkers()为当前的程序调度和启动 Worker 的 Executor，默认情况下排队的方式是 FIFO。

Spark 2.2.1 版本的 Master.scala 的 startExecutorsOnWorkers 的源码如下：

```
1.   private def startExecutorsOnWorkers(): Unit = {
2.     //这是一个非常简单的 FIFO 调度。我们尝试在队列中推入第一个应用程序，然后推入第二
       //个应用程序等
3.     for (app <- waitingApps if app.coresLeft > 0) {
4.       val coresPerExecutor: Option[Int] = app.desc.coresPerExecutor
5.       //筛选出 workers，其没有足够资源来启动 Executor
```

```
6.        val usableWorkers = workers.toArray.filter(_.state == WorkerState
          .ALIVE)
7.          .filter(worker => worker.memoryFree >= app.desc
            .memoryPerExecutorMB &&
8.            worker.coresFree >= coresPerExecutor.getOrElse(1))
9.          .sortBy(_.coresFree).reverse
10.       val assignedCores = scheduleExecutorsOnWorkers(app, usableWorkers,
          spreadOutApps)
11.
12.       //现在我们决定每个worker分配多少cores，进行分配
13.       for (pos <- 0 until usableWorkers.length if assignedCores(pos) > 0) {
14.         allocateWorkerResourceToExecutors(
15.           app, assignedCores(pos), coresPerExecutor, usableWorkers(pos))
16.       }
17.     }
18.   }
```

Spark 2.4.3 版本的 Master.scala 源码与 Spark 2.2.1 版本相比具有如下特点。

- 上段代码中第 3 行将 for 循环遍历语句调整为 for (app <- waitingApps)。
- 上段代码中第 4 行构建 coresPerExecutor 变量调整为 app.desc.coresPerExecutor.getOrElse(1)，如果剩余的核心小于 coresPerExecutor，则不会分配剩余的核心。
- 上段代码中第 8 行 coresPerExecutor.getOrElse(1)调整为 coresPerExecutor。
- 上段代码中第 15 行 coresPerExecutor 调整为 app.desc.coresPerExecutor。

```
1.  ......
2.  for (app <- waitingApps) {
3.  ......
4.        val coresPerExecutor = app.desc.coresPerExecutor.getOrElse(1)
5.        //如果剩余的核心小于coresPerExecutor，则不会分配剩余的核心
6.        if (app.coresLeft >= coresPerExecutor) {
7.  ......
8.  worker.coresFree >= coresPerExecutor)
9.  ......
10. app, assignedCores(pos), app.desc.coresPerExecutor, usableWorkers(pos))
```

（9）为应用程序具体分配 Executor 前要判断应用程序是否还需要分配 Core，如果不需要，则不会为应用程序分配 Executor。

startExecutorsOnWorkers 中的 coresLeft 是请求的 requestedCores 和可用的 coresGranted 的相减值。例如，如果整个程序要求 1000 个 Cores，但是目前集群可用的只有 100 个 Cores，如果 coresLeft 不为 0，就放入等待队列中；如果 coresLeft 是 0，那么就不需要调度。

```
1.    private[master] def coresLeft: Int = requestedCores - coresGranted
```

（10）Master.scala 的 startExecutorsOnWorkers 中，具体分配 Executor 之前，要求 Worker 必须是 ALIVE 的状态且必须满足 Application 对每个 Executor 的内存和 Cores 的要求，并且在此基础上进行排序，产生计算资源由大到小的 usableWorkers 数据结构。

```
1.        val usableWorkers = workers.toArray.filter(_.state == WorkerState
          .ALIVE)
2.          .filter(worker => worker.memoryFree >= app.desc
            .memoryPerExecutorMB &&
```

```
3.          worker.coresFree >= coresPerExecutor)
4.        .sortBy(_.coresFree).reverse
5. val assignedCores = scheduleExecutorsOnWorkers(app, usableWorkers,
   spreadOutApps)
```

然后调用 scheduleExecutorsOnWorkers，在 FIFO 的情况下，默认 spreadOutApps 让应用程序尽可能多地运行在所有的 Node 上。

```
1.    private val spreadOutApps = conf.getBoolean("spark.deploy.spreadOut",
      true)
```

scheduleExecutorsOnWorker 中，minCoresPerExecutor 表示每个 Executor 最小分配的 core 个数。scheduleExecutorsOnWorker 的源码如下：

```
1.    private def scheduleExecutorsOnWorkers(
2.      app: ApplicationInfo,
3.      usableWorkers: Array[WorkerInfo],
4.      spreadOutApps: Boolean): Array[Int] = {
5.     val coresPerExecutor = app.desc.coresPerExecutor
6.     val minCoresPerExecutor = coresPerExecutor.getOrElse(1)
7.     val oneExecutorPerWorker = coresPerExecutor.isEmpty
8.     val memoryPerExecutor = app.desc.memoryPerExecutorMB
9.     val numUsable = usableWorkers.length
10.    val assignedCores = new Array[Int](numUsable)
11.    val assignedExecutors = new Array[Int](numUsable)
12.    var coresToAssign = math.min(app.coresLeft, usableWorkers.map
       (_.coresFree).sum)
13. ……
```

（11）为应用程序分配 Executors 有两种方式：第一种方式是尽可能在集群的所有 Worker 上分配 Executor，这种方式往往会带来潜在的、更好的数据本地性；第二种方式是尝试运行在尽可能少的 Worker 上。

（12）具体在集群上分配 Cores 时会尽可能地满足我们的要求。math.min 用于计算最小值。coresToAssig 用于计算 app.coresLeft 与可用的 Worker 中可用的 Cores 的和的最小值。例如，应用程序要求 1000 个 Cores，但整个集群中只有 100 个 Cores，所以只能先分配 100 个 Cores。

scheduleExecutorsOnWorkers 方法如下：

```
1.      var coresToAssign = math.min(app.coresLeft, usableWorkers.map
        (_.coresFree).sum)
2. ……
```

（13）如果每个 Worker 下面只能为当前的应用程序分配一个 Executor，那么每次只分配一个 Core。scheduleExecutorsOnWorkers 方法如下：

```
1.        if (oneExecutorPerWorker) {
2.          assignedExecutors(pos) = 1
3.        } else {
4.          assignedExecutors(pos) += 1
5.        }
```

总结为两种情况：一种情况是尽可能在一台机器上运行程序的所有功能；另一种情况是尽可能在所有节点上运行程序的所有功能。无论是哪种情况，每次给 Executor 增加 Cores，是增加一个，如果是 spreadOutApps 的方式，循环一轮再下一轮。例如，有 4 个 Worker，第一次为每个 Executor 启动一个线程，第二次循环分配一个线程，第三次循环再分配一个线程……

scheduleExecutorsOnWorkers 方法如下：

```
1.            while (freeWorkers.nonEmpty) {
2.        freeWorkers.foreach { pos =>
3.          var keepScheduling = true
4.          while (keepScheduling && canLaunchExecutor(pos)) {
5.            coresToAssign -= minCoresPerExecutor
6.            assignedCores(pos) += minCoresPerExecutor
7.
8.            //如果每个 worker 上启动一个 Executor，那么每次迭代在 Executor 上分配一
              //个核，否则，每次迭代都将把内核分配给一个新的 Executor
9.            if (oneExecutorPerWorker) {
10.             assignedExecutors(pos) = 1
11.           } else {
12.             assignedExecutors(pos) += 1
13.           }
14.
15.           //展开应用程序意味着将 Executors 展开到尽可能多的 workers 节点。如果不展
              //开，将对这个 workers 的 Executors 进行调度，直到使用它的全部资源。否则，
              //只是移动到下一个 worker 节点
16.           if (spreadOutApps) {
17.             keepScheduling = false
18.           }
19.         }
20.       }
```

回到 Master.scala 的 startExecutorsOnWorkers，现在已经决定为每个 worker 分配多少个 cores，然后进行资源分配。

```
1.            for (pos <- 0 until usableWorkers.length if assignedCores(pos)
              > 0) {
2.          allocateWorkerResourceToExecutors(
3.            app, assignedCores(pos), app.desc.coresPerExecutor, usableWorkers(pos))
4.        }
```

allocateWorkerResourceToExecutors 的源码如下：

```
1.     private def allocateWorkerResourceToExecutors(
2.      app: ApplicationInfo,
3.      assignedCores: Int,
4.      coresPerExecutor: Option[Int],
5.      worker: WorkerInfo): Unit = {
6.   //如果指定了每个 Executor 的内核数，我们就将分配的内核无剩余地均分给 worker 节点的
     //Executors。否则，我们启动一个单一的 Executor，抓住这个 worker 节点所有的
     //assignedCores
7.    val numExecutors = coresPerExecutor.map { assignedCores / _ }.getOrElse(1)
8.    val coresToAssign = coresPerExecutor.getOrElse(assignedCores)
9.    for (i <- 1 to numExecutors) {
10.     val exec = app.addExecutor(worker, coresToAssign)
11.     launchExecutor(worker, exec)
12.     app.state = ApplicationState.RUNNING
13.   }
14. }
```

allocateWorkerResourceToExecutors 中的 app.addExecutor 增加一个 Executor，记录 Executor 的相关信息。

```
1.      private[master] def addExecutor(
2.        worker: WorkerInfo,
3.        cores: Int,
4.        useID: Option[Int] = None): ExecutorDesc = {
5.      val exec = new ExecutorDesc(newExecutorId(useID), this, worker, cores,
        desc.memoryPerExecutorMB)
6.      executors(exec.id) = exec
7.      coresGranted += cores
8.      exec
9.    }
```

回到 allocateWorkerResourceToExecutors 方法中，launchExecutor(worker, exec)启动 Executor。

```
1.    launchExecutor(worker, exec)
```

（14）准备具体要为当前应用程序分配的 Executor 信息后，Master 要通过远程通信发指令给 Worker 来具体启动 ExecutorBackend 进程。

launchExecutor 方法如下：

```
1.    private def launchExecutor(worker: WorkerInfo, exec: ExecutorDesc):
      Unit = {
2.      logInfo("Launching executor " + exec.fullId + " on worker " + worker.id)
3.      worker.addExecutor(exec)
4.      worker.endpoint.send(LaunchExecutor(masterUrl,
5.        exec.application.id, exec.id, exec.application.desc, exec.cores,
          exec.memory))
6.    ......
```

（15）紧接着给应用程序的 Driver 发送一个 ExecutorAdded 的信息。

launchExecutor 方法如下：

```
1.      exec.application.driver.send(
2.        ExecutorAdded(exec.id, worker.id, worker.hostPort, exec.cores,
          exec.memory))
3.    }
```

6.3　从 Application 提交的角度重新审视 Driver

本节从 Application 提交的角度重新审视 Driver，彻底解密 Driver 到底是什么时候产生的，以及 Driver 和 Master 交互原理、Driver 和 Master 交互源码。

6.3.1　Driver 到底是什么时候产生的

在 SparkContext 实例化时，通过 createTaskScheduler 来创建 TaskSchedulerImpl 和 StandaloneSchedulerBackend。

SparkContext.scala 的源码如下：

```
1.  class SparkContext(config: SparkConf) extends Logging {
2.  ......
```

```
3.  val (sched, ts) = SparkContext.createTaskScheduler(this, master,
    deployMode)
4.      _schedulerBackend = sched
5.      _taskScheduler = ts
6.
7.      _dagScheduler = new DAGScheduler(this)
8.      _heartbeatReceiver.ask[Boolean](TaskSchedulerIsSet)
9.  ......
10. private def createTaskScheduler(
11.    ......
12.       case SPARK_REGEX(sparkUrl) =>
13.         val scheduler = new TaskSchedulerImpl(sc)
14.         val masterUrls = sparkUrl.split(",").map("spark://" + _)
15.         val  backend  =  new  StandaloneSchedulerBackend(scheduler,  sc,
            masterUrls)
16.         scheduler.initialize(backend)
17.         (backend, scheduler)
18. ......
```

在 createTaskScheduler 中调用 scheduler.initialize(backend)，initialize 的方法参数把 StandaloneSchedulerBackend 传进来。

TaskSchedulerImpl 的 initialize 的源码如下：

```
1.       def initialize(backend: SchedulerBackend) {
2.     this.backend = backend
3.     ......
```

initialize 的方法把 StandaloneSchedulerBackend 传进来了，但还没有启动 StandaloneSchedulerBackend。在 TaskSchedulerImpl 的 initialize 方法中，把 StandaloneSchedulerBackend 传进来，赋值为 TaskSchedulerImpl 的 backend。

在 TaskSchedulerImpl 中调用 start 方法时，会调用 backend.start 方法，在 start 方法中会注册应用程序。

SparkContext.scala 的 taskScheduler 的源码如下：

```
1.    val  (sched,  ts)  =  SparkContext.createTaskScheduler(this,  master,
      deployMode)
2.      _schedulerBackend = sched
3.      _taskScheduler = ts
4.      _dagScheduler = new DAGScheduler(this)
5.  ......
6.      _taskScheduler.start()
7.      _applicationId = _taskScheduler.applicationId()
8.      _applicationAttemptId = taskScheduler.applicationAttemptId()
9.      _conf.set("spark.app.id", _applicationId)
10. ......
```

其中调用了_taskScheduler 的 start 方法。

```
1.   private[spark] trait TaskScheduler {
2.   ......
3.
4.    def start(): Unit
5.   ......
```

TaskScheduler 的 start()方法没具体实现，TaskScheduler 子类的 TaskSchedulerImpl 的 start()方法的源码如下：

```
1.    override def start() {
2.     backend.start()
3.  ……
```

TaskSchedulerImpl 的 start()通过 backend.start()启动了 StandaloneSchedulerBackend 的 start 方法。

StandaloneSchedulerBackend 的 start 方法中，将 command 封装注册给 Master，Master 转过来要 Worker 启动具体的 Executor。command 已经封装好指令，Executor 具体要启动进程入口类 CoarseGrainedExecutorBackend。然后调用 new()函数创建一个 StandaloneAppClient，通过 client.start()启动 client。

StandaloneAppClient 的 start 方法中调用 new()函数创建一个 ClientEndpoint。

```
1.    def start() {
2.     //启动一个 rpcEndpoint，它将回调到监听器
3.     endpoint.set(rpcEnv.setupEndpoint("AppClient",  new  ClientEndpoint
       (rpcEnv)))
4.    }
```

ClientEndpoint 的源码如下：

```
1.     private class ClientEndpoint(override val rpcEnv: RpcEnv) extends
       ThreadSafeRpcEndpoint
2.     with Logging {
3.  ……
4.     override def onStart(): Unit = {
5.       try {
6.         registerWithMaster(1)
7.       } catch {
8.         case e: Exception =>
9.           logWarning("Failed to connect to master", e)
10.          markDisconnected()
11.          stop()
12.      }
13.    }
```

ClientEndpoint 是一个 ThreadSafeRpcEndpoint。ClientEndpoint 的 onStart()方法中调用 registerWithMaster(1)进行注册，向 Master 注册程序。registerWithMaster 方法如下。

StandaloneAppClient.scala 的源码如下：

```
1.      private def registerWithMaster(nthRetry: Int) {
2.      registerMasterFutures.set(tryRegisterAllMasters())
3.     ……
```

registerWithMaster 中调用了 tryRegisterAllMasters 方法。在 tryRegisterAllMasters 方法中，ClientEndpoint 向 Master 发送 RegisterApplication 消息进行应用程序的注册。

StandaloneAppClient.scala 的源码如下：

```
1.     private def tryRegisterAllMasters(): Array[JFuture[_]] = {
2.    ……
3.           masterRef.send(RegisterApplication(appDescription, self))
4.     ……
```

程序注册以后，Master 通过 schedule()分配资源，通知 Worker 启动 Executor，Executor 启动的进程是 CoarseGrainedExecutorBackend，Executor 启动以后又转过来向 Driver 注册，

Driver 其实是 StandaloneSchedulerBackend 的父类 CoarseGrainedSchedulerBackend 的一个消息循环体 DriverEndpoint。

Master.scala 的 receive 方法的源码如下：

```
1.    override def receive: PartialFunction[Any, Unit] = {
2.    case RegisterApplication(description, driver) =>
3.      ......
4.          registerApplication(app)
5.          logInfo("Registered app " + description.name + " with ID " + app.id)
6.          persistenceEngine.addApplication(app)
7.          driver.send(RegisteredApplication(app.id, self))
8.          schedule()
9.        }
```

在 Master 的 receive 方法中调用了 schedule 方法。Schedule 方法在等待的应用程序中调度当前可用的资源。每次一个新的应用程序连接或资源发生可用性的变化时，此方法将被调用。

Master.scala 的 schedule 方法的源码如下：

```
1.    private def schedule(): Unit = {
2.    ......
3.          if (worker.memoryFree >= driver.desc.mem && worker.coresFree >=
          driver.desc.cores) {
4.            launchDriver(worker, driver)
5.            waitingDrivers -= driver
6.            launched = true
7.          }
8.          curPos = (curPos + 1) % numWorkersAlive
9.        }
10.     }
11.     startExecutorsOnWorkers()
12.   }
```

Master.scala 在 schedule 方法中调用 launchDriver 方法。launchDriver 方法给 Worker 发送 launchDriver 的消息。Master.scala 的 launchDriver 的源码如下：

```
1.     private def launchDriver(worker: WorkerInfo, driver: DriverInfo) {
2.       logInfo("Launching driver " + driver.id + " on worker " + worker.id)
3.       worker.addDriver(driver)
4.       driver.worker = Some(worker)
5.       worker.endpoint.send(LaunchDriver(driver.id, driver.desc))
6.       driver.state = DriverState.RUNNING
7.     }
```

launchDriver 本身是一个 case class，包括 driverId、driverDesc 等信息。

```
1.       case class LaunchDriver(driverId: String, driverDesc: DriverDescription)
         extends DeployMessage
```

DriverDescription 包含了 jarUrl、memory、cores、supervise、command 等内容。

```
1.   private[deploy] case class DriverDescription(
2.       jarUrl: String,
3.       mem: Int,
4.       cores: Int,
5.       supervise: Boolean,
6.       command: Command) {
```

```
7.
8.    override def toString: String = s"DriverDescription (${command.
      mainClass})"
9.  }
```

Master.scala 中 launchDriver 启动了 Driver，接下来，launchExecutor 启动 Executor。Master.scala 的 launchExecutor 的源码如下：

```
1.    private def launchExecutor(worker: WorkerInfo, exec: ExecutorDesc):
      Unit = {
2.      logInfo("Launching executor " + exec.fullId + " on worker " + worker.id)
3.      worker.addExecutor(exec)
4.      worker.endpoint.send(LaunchExecutor(masterUrl,
5.        exec.application.id, exec.id, exec.application.desc, exec.cores,
          exec.memory))
6.      exec.application.driver.send(
7.        ExecutorAdded(exec.id, worker.id, worker.hostPort, exec.cores,
          exec.memory))
8.    }
```

Master 给 Worker 发送一个消息 LaunchDriver 启动 Driver，然后是 launchExecutor 启动 Executor，launchExecutor 有自己的调度方式，资源调度后，也是给 Worker 发送了一个消息 LaunchExecutor。

Worker 就收到 Master 发送的 LaunchDriver、LaunchExecutor 消息。

图 6-2 是 Worker 原理内幕和流程机制。

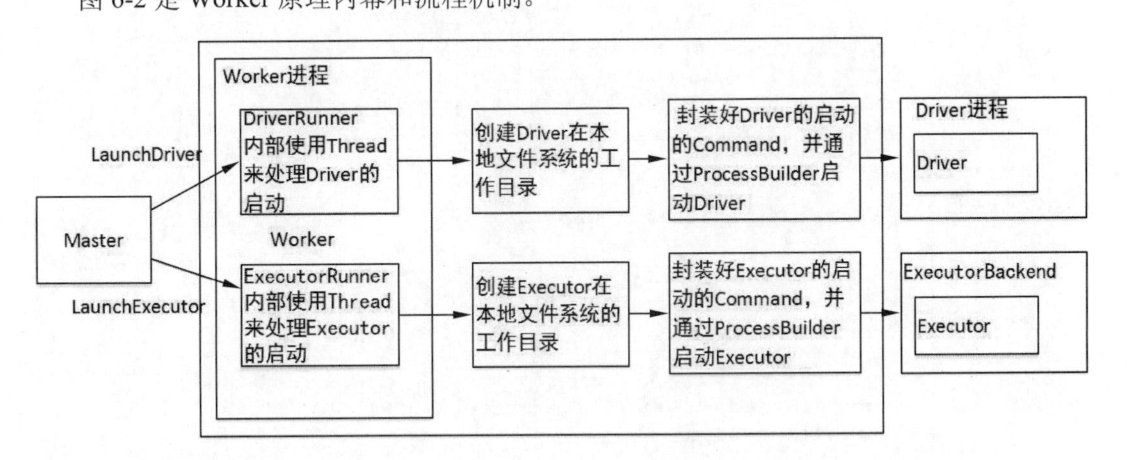

图 6-2　Worker 原理内幕和流程机制

Master、Worker 部署在不同的机器上，Master、Worker 为进程存在。Master 给 Worker 发两种不同的指令：一种指令是 LaunchDriver；另一种指令是 LaunchExecutor。

- Worker 收到 Master 的 LaunchDriver 消息以后，调用 new()函数创建一个 DriverRunner，然后启动 driver.start()方法。

Worker.scala 的源码如下：

```
1.  case LaunchDriver(driverId, driverDesc) =>
2.     ......
3.  val driver = new DriverRunner(
4.  ......
5.   driver.start()
```

- Worker 收到 Master 的 LaunchExecutor 消息以后，new()函数创建一个 ExecutorRunner，然后启动 manager.start()方法。

Worker.scala 的源码如下：

```
1.   case LaunchExecutor(masterUrl, appId, execId, appDesc, cores_, memory_) =>
2.  ......
3.  val manager = new ExecutorRunner(
4.  ......
5.  manager.start()
```

Worker 的 DriverRunner、ExecutorRunner 在调用 start 方法时，在 start 内部都启动了一条线程，使用 Thread 来处理 Driver、Executor 的启动。以 Worker 收到 LaunchDriver 消息，new 出 DriverRunnerDriverRunner 为例，DriverRunner.scala 的 start 的源码如下：

```
1.  /**启动一个线程来运行和管理 Driver*/
2.    private[worker] def start() = {
3.      new Thread("DriverRunner for " + driverId) {
4.        override def run() {
5.          var shutdownHook: AnyRef = null
6.          try {
7.            shutdownHook = ShutdownHookManager.addShutdownHook { () =>
8.              logInfo(s"Worker shutting down, killing driver $driverId")
9.              kill()
10.           }
11.
12.           //准备 Driver 的 jars 包，运行 Driver
13.           val exitCode = prepareAndRunDriver()
14.
15.           //设置的最终状态取决于是否强制删除，并处理退出代码
16.           finalState = if (exitCode == 0) {
17.             Some(DriverState.FINISHED)
18.           } else if (killed) {
19.             Some(DriverState.KILLED)
20.           } else {
21.             Some(DriverState.FAILED)
22.           }
23.         } catch {
24.           case e: Exception =>
25.             kill()
26.             finalState = Some(DriverState.ERROR)
27.             finalException = Some(e)
28.         } finally {
29.           if (shutdownHook != null) {
30.             ShutdownHookManager.removeShutdownHook(shutdownHook)
31.           }
32.         }
33.
34.         //通知 worker 节点 Driver 的最终状态及可能的异常
35.         worker.send(DriverStateChanged(driverId, finalState.get,
            finalException))
36.       }
37.     }.start()
38.   }
```

DriverRunner.scala 的 start 方法中调用了 prepareAndRunDriver 方法，准备 Driver 的 jar 包和启动 Driver。prepareAndRunDriver 的源码如下：

```
1.    private[worker] def prepareAndRunDriver(): Int = {
2.      val driverDir = createWorkingDirectory()
3.      val localJarFilename = downloadUserJar(driverDir)
4.
5.      def substituteVariables(argument: String): String = argument match {
6.        case "{{WORKER_URL}}" => workerUrl
7.        case "{{USER_JAR}}" => localJarFilename
8.        case other => other
9.      }
10.
11.      //待办事项：如果我们增加了提交多个 jars 包的能力，在这里也要增加
12.     val builder = CommandUtils.buildProcessBuilder(driverDesc.command,
        securityManager, driverDesc.mem, sparkHome.getAbsolutePath,
        substituteVariables)
13.
14.
15.     runDriver(builder, driverDir, driverDesc.supervise)
16.   }
```

LaunchDriver 的启动过程如下。

- Worker 进程：Worker 的 DriverRunner 调用 start 方法，内部使用 Thread 来处理 Driver 启动。DriverRunner 创建 Driver 在本地系统的工作目录（即 Linux 的文件目录），每次工作都有自己的目录，封装好 Driver 的启动 Command，通过 ProcessBuilder 启动 Driver。这些内容都属于 Worker 进程。
- Driver 进程：启动的 Driver 属于 Driver 进程。

LaunchExecutor 的启动过程如下。

- Worker 进程：Worker 的 ExecutorRunner 调用 start 方法，内部使用 Thread 来处理 Executor 启动。ExecutorRunner 创建 Executor 在本地系统的工作目录（即 Linux 的文件目录），每次工作都有自己的目录，封装好 Executor 的启动 Command，通过 ProcessBuilder 来启动 Executor。这些内容都属于 Worker 进程。
- Executor 进程：启动的 Executor 属于 Executor 进程。Executor 在 ExecutorBackend 里面，ExecutorBackend 在 Spark standalone 模式中是 CoarseGrainedExecutorBackend。CoarseGrainedExecutorBackend 继承自 ExecutorBackend。Executor 和 ExecutorBackend 是一对一的关系，一个 ExecutorBackend 有一个 Executor，在 Executor 内部是通过线程池并发处理的方式来处理 Spark 提交过来的 Task 的。
- Executor 启动后要向 Driver 注册，注册给 SchedulerBackend。

CoarseGrainedExecutorBackend 的源码如下：

```
1.     private[spark] class CoarseGrainedExecutorBackend(
2.      override val rpcEnv: RpcEnv,
3.      driverUrl: String,
4.      executorId: String,
5.      hostname: String,
6.      cores: Int,
7.      userClassPath: Seq[URL],
8.      env: SparkEnv)
9.    extends ThreadSafeRpcEndpoint with ExecutorBackend with Logging {
10.
11.   private[this] val stopping = new AtomicBoolean(false)
12.   var executor: Executor = null
13.   @volatile var driver: Option[RpcEndpointRef] = None
14. ......
```

再次看一下 Master 的 schedule 方法。

```
1.     private def schedule(): Unit = {
2.      ……
3.        if (worker.memoryFree >= driver.desc.mem && worker.coresFree >=
          driver.desc.cores) {
4.          launchDriver(worker, driver)
5.          waitingDrivers -= driver
6.          launched = true
7.        }
8.        curPos = (curPos + 1) % numWorkersAlive
9.      }
10.    }
11.    startExecutorsOnWorkers()
12.  }
```

Master 的 schedule 方法中，如果 Driver 运行在集群中，通过 launchDriver 来启动 Driver。launchDriver 发送一个消息交给 worker 的 endpoint，这是 RPC 的通信机制。

```
1.    private def launchDriver(worker: WorkerInfo, driver: DriverInfo) {
2.     logInfo("Launching driver " + driver.id + " on worker " + worker.id)
3.     worker.addDriver(driver)
4.     driver.worker = Some(worker)
5.     worker.endpoint.send(LaunchDriver(driver.id, driver.desc))
6.     driver.state = DriverState.RUNNING
7.   }
```

Master 的 schedule 方法中启动 Executor 的部分，通过 startExecutorsOnWorkers 启动，startExecutorsOnWorkers 也是通过 RPC 的通信方式。

Master.scala 的方法中调用 allocateWorkerResourceToExecutors 方法进行正式分配。

allocateWorkerResourceToExecutors 正式分配时就通过 launchExecutor 方法启动 Executor。

```
1.   private def launchExecutor(worker: WorkerInfo, exec: ExecutorDesc): Unit
     = {
2.     logInfo("Launching executor " + exec.fullId + " on worker " + worker.id)
3.     worker.addExecutor(exec)
4.     worker.endpoint.send(LaunchExecutor(masterUrl, exec.application.id,
       exec.id, exec.application.desc, exec.cores, exec.memory))
5.     exec.application.driver.send( ExecutorAdded(exec.id, worker.id,
       worker.hostPort, exec.cores, exec.memory))
6.   }
```

Master 发送消息给 Worker，发送两个消息：一个是 LaunchDriver；另一个是 LaunchExecutor。Worker 收到 Master 的 LaunchDriver、LaunchExecutor 消息。下面看一下 Worker。

Spark 2.2.1 版本的 Worker.scala 源代码如下：

```
1.    private[deploy] class Worker(
2.     override val rpcEnv: RpcEnv,
3.     webUiPort: Int,
4.     cores: Int,
5.     memory: Int,
6.     masterRpcAddresses: Array[RpcAddress],
7.     endpointName: String,
8.     workDirPath: String = null,
```

```
9.      val conf: SparkConf,
10.     val securityMgr: SecurityManager)
11.   extends ThreadSafeRpcEndpoint with Logging {
```

Spark 2.4.3 版本的 Worker.scala 源码与 Spark 2.2.1 版本相比具有如下特点。

- 上段代码中第 10 行之后新增加一行代码，新增加一个成员变量 externalShuffleServiceSupplier。

```
1.  externalShuffleServiceSupplier: Supplier[ExternalShuffleService] = null)
```

Worker 实现 RPC 通信，继承自 ThreadSafeRpcEndpoint。ThreadSafeRpcEndpoint 是一个 trait，其他的 RPC 对象可以给它发消息。

```
1.      private[spark] trait ThreadSafeRpcEndpoint extends RpcEndpoint
```

Worker 在 receive 方法中接收消息。就像一个邮箱，不断地循环邮箱接收邮件，我们可以把消息看成邮件。

```
1.     override def receive: PartialFunction[Any, Unit] = synchronized {
2.      case SendHeartbeat =>
3.      ......
4.      case WorkDirCleanup =>
5.      ......
6.      case MasterChanged(masterRef, masterWebUiUrl) =>
7.      ......
8.      case ReconnectWorker(masterUrl) =>
9.       ......
10.    case LaunchExecutor(masterUrl, appId, execId, appDesc, cores_, memory_)
       =>
11.     ......
12.    case executorStateChanged @ ExecutorStateChanged(appId, execId, state,
       message, exitStatus)
13.         ......
14.      case KillExecutor(masterUrl, appId, execId) =>
15.      ......
16.     case LaunchDriver(driverId, driverDesc) =>
17.      ......
```

Worker.scala 的 receive 方法 LaunchDriver 启动 Driver 的源码如下：

```
1.   case LaunchDriver(driverId, driverDesc) =>
2.       logInfo(s"Asked to launch driver $driverId")
3.       val driver = new DriverRunner(
4.         conf,
5.         driverId,
6.         workDir,
7.         sparkHome,
8.         driverDesc.copy(command = Worker.maybeUpdateSSLSettings
           (driverDesc.command, conf)),
9.         self,
10.        workerUri,
11.        securityMgr)
12.      drivers(driverId) = driver
13.      driver.start()
14.
15.      coresUsed += driverDesc.cores
16.      memoryUsed += driverDesc.mem
```

LaunchDriver 方法首先打印日志，传进来时肯定会告诉 driverId。启动 Driver 或者 Executor 时，Driver 或者 Executor 所在的进程一定满足内存级别的要求，但不一定满足 Cores 的要求，实际的 Cores 可能比期待的 Cores 多，也有可能少。

logInfo 方法打印日志使用了封装。

```
1.      protected def logInfo(msg: => String) {
2.      if (log.isInfoEnabled) log.info(msg)
3.    }
```

回到 LaunchDriver 方法，其中调用 new()函数创建一个 DriverRunner。DriverRunner 包括 driverId、工作目录（workDir）、spark 的路径（sparkHome）、driverDesc、workerUri、securityMgr 等内容。在代码 drivers(driverId) = driver 中，将 driver 交给一个数据结构 drivers，drivers 是一个 HashMap，是 Key-Value 的方式，其中 Key 是 Driver 的 ID，Value 是 DriverRunner。Worker 下可能启动很多 Executor，须根据具体的 ID 管理 DriverRunner。DriverRunner 内部通过线程的方式启动另外一个进程 Driver。DriverRunner 是 Driver 所在进程的代理。

```
1.       val drivers = new HashMap[String, DriverRunner]
```

回到 Worker.scala 的 LaunchDriver，Worker 在启动 driver 前，将相关的 DriverRunner 数据保存到 Worker 的内存数据结构中，然后进行 driver.start()。start 之后，将消耗的 cores、memory 增加到 coresUsed、memoryUsed。

接下来进入 DriverRunner.scala 的源码。DriverRunner 管理 Driver 的执行，包括在 Driver 失败的时候自动重启。如 Driver 运行在集群模式中，加入 supervise 关键字可以自动重启。

```
1.       private[deploy] class DriverRunner(
2.      conf: SparkConf,
3.      val driverId: String,
4.      val workDir: File,
5.      val sparkHome: File,
6.      val driverDesc: DriverDescription,
7.      val worker: RpcEndpointRef,
8.      val workerUrl: String,
9.      val securityManager: SecurityManager)
10.   extends Logging {
```

其中 DriverDescription 的源码如下。其中包括 DriverDescription 的成员 supervise，supervise 是一个布尔值，如果设置为 True，在集群模式中 Driver 运行失败的时候，Worker 会负责重新启动 Driver。

```
1.     private[deploy] case class DriverDescription(
2.      jarUrl: String,
3.      mem: Int,
4.      cores: Int,
5.      supervise: Boolean,
6.      command: Command) {
7.
8.    override def toString: String = s"DriverDescription (${command
    .mainClass})"
9.  }
```

回到 Worker.scala 的 LaunchDriver，DriverRunner 构造出后，调用其 start 方法，通过一个线程管理 Driver，包括启动 Driver 及关闭 Driver。其中，Thread("DriverRunner for " +

driverId)，DriverRunner for　driverId 是线程的名字，Thread 是 Java 的代码，scala 可以无缝连接 Java。

DriverRunner 的 start 方法调用 prepareAndRunDriver 来实现 driver jar 包的准备及启动 driver。

prepareAndRunDriver 方法中调用了 createWorkingDirectory 方法创建目录。通过 Java 的 new File 创建了 Driver 的工作目录，如果目录不存在而且创建不成功，就提示失败。在本地文件系统创建一个目录一般不会失败，除非磁盘满。createWorkingDirectory 的源码如下：

```
1.   private def createWorkingDirectory(): File = {
2.      val driverDir = new File(workDir, driverId)
3.      if (!driverDir.exists() && !driverDir.mkdirs()) {
4.        throw new IOException("Failed to create directory " + driverDir)
5.      }
6.      driverDir
7.    }
```

回到 DriverRunner.scala 的 prepareAndRunDriver 方法，其中采用 downloadUserJar 方法下载 jar 包。我们自己写的代码是一个 jar 包，这里下载用户的 jar 包到本地。jar 包在 Hdfs 中，开发人员需要从 Hdfs 中获取 Jar 包下载到本地。

downloadUserJar 方法的源码如下：

```
1.        private def downloadUserJar(driverDir: File): String = {
2.      val jarFileName = new URI(driverDesc.jarUrl).getPath.split("/").last
3.      val localJarFile = new File(driverDir, jarFileName)
4.      if (!localJarFile.exists()) { //如果在一个节点上运行多个 Worker，文件可能
                                        //已经存在
5.        logInfo(s"Copying user jar ${driverDesc.jarUrl} to $localJarFile")
6.        Utils.fetchFile(
7.          driverDesc.jarUrl,
8.          driverDir,
9.          conf,
10.         securityManager,
11.         SparkHadoopUtil.get.newConfiguration(conf),
12.         System.currentTimeMillis(),
13.         useCache = false)
14.       if (!localJarFile.exists()) { //验证复制成功
15.         throw new IOException(
16.           s"Can not find expected jar $jarFileName which should have been
              loaded in $driverDir")
17.       }
18.     }
19.     localJarFile.getAbsolutePath
20.   }
```

downloadUserJar 方法调用了 fetchFile，fetchFile 借助 Hadoop，从 Hdfs 中下载文件。我们提交文件时，将 jar 包上传到 Hdfs 上，提交一份，大家都可以从 Hdfs 中下载。

Spark 2.2.1 版本的 Utils.scala 源代码如下：

```
1.   def fetchFile(
2.       url: String,
3.       targetDir: File,
4.       conf: SparkConf,
5.       securityMgr: SecurityManager,
```

```
6.        hadoopConf: Configuration,
7.        timestamp: Long,
8.        useCache: Boolean) {
9.      val fileName = decodeFileNameInURI(new URI(url))
10.     val targetFile = new File(targetDir, fileName)
11.     val fetchCacheEnabled = conf.getBoolean("spark.files.useFetchCache",
        defaultValue = true)
12.     if (useCache && fetchCacheEnabled) {
13.       val cachedFileName = s"${url.hashCode}${timestamp}_cache"
14.       val lockFileName = s"${url.hashCode}${timestamp}_lock"
15.       val localDir = new File(getLocalDir(conf))
16.       val lockFile = new File(localDir, lockFileName)
17.       val lockFileChannel = new RandomAccessFile(lockFile, "rw").getChannel()
18.       //只有一个 executor 入口。FileLock 用来控制 executors 下载的文件同步，无论
          //锁类型是 mandatory 还是 advisory，它始终是安全的
19.       val lock = lockFileChannel.lock()
20.       val cachedFile = new File(localDir, cachedFileName)
21.       try {
22.         if (!cachedFile.exists()) {
23.           doFetchFile(url, localDir, cachedFileName, conf, securityMgr,
              hadoopConf)
24.         }
25.       } finally {
26.         lock.release()
27.         lockFileChannel.close()
28.       }
29.       copyFile(
30.         url,
31.         cachedFile,
32.         targetFile,
33.         conf.getBoolean("spark.files.overwrite", false)
34.       )
35.     } else {
36.       doFetchFile(url, targetDir, fileName, conf, securityMgr,
          hadoopConf)
37.     }
```

Spark 2.4.3 版本的 Utils.scala 源码与 Spark 2.2.1 版本相比具有如下特点。

- 上段代码中第 8 行新增加了函数的返回类型 File。
- 上段代码中第 15 行构建本地目录变量，调整为以下第 5～12 行代码。

```
1.  ……
2.    useCache: Boolean): File = {
3.  ……
4.        // 第一次设置 cachedlocaldir，稍后重新使用
5.        if (cachedLocalDir.isEmpty) {
6.          this.synchronized {
7.            if (cachedLocalDir.isEmpty) {
8.              cachedLocalDir = getLocalDir(conf)
9.            }
10.         }
11.       }
12.       val localDir = new File(cachedLocalDir)
13. ……
```

回到 DriverRunner.scala 的 prepareAndRunDriver 方法，driverDesc.command 表明运行什

么类，构建进程运行类的入口，然后是 runDriver 启动 Driver。

```
1.     private[worker] def prepareAndRunDriver(): Int = {
2.    ......
3.     val builder = CommandUtils.buildProcessBuilder(driverDesc.command,
       securityManager,
4.       driverDesc.mem, sparkHome.getAbsolutePath, substituteVariables)
5.
6.     runDriver(builder, driverDir, driverDesc.supervise)
7.   }
```

DriverRunner.scala 的 runDriver 方法如下。runDriver 中重定向输出文件和 err 文件，可以通过 log 文件查看执行的情况。最后是调用 runCommandWithRetry 方法。

```
1.   private def runDriver(builder: ProcessBuilder, baseDir: File, supervise:
     Boolean): Int = {
2.     builder.directory(baseDir)
3.     def initialize(process: Process): Unit = {
4.       //stdout 和 stderr 重定向到文件
5.       val stdout = new File(baseDir, "stdout")
6.       CommandUtils.redirectStream(process.getInputStream, stdout)
7.
8.       val stderr = new File(baseDir, "stderr")
9.       val formattedCommand = builder.command.asScala.mkString("\"", "\"
         \"", "\"")
10.      val header = "Launch Command: %s\n%s\n\n".format(formattedCommand,
         "=" * 40)
11.      Files.append(header, stderr, StandardCharsets.UTF_8)
12.      CommandUtils.redirectStream(process.getErrorStream, stderr)
13.    }
14.    runCommandWithRetry(ProcessBuilderLike(builder), initialize, supervise)
15.  }
```

runCommandWithRetry 中传入的参数是 ProcessBuilderLike(builder)，这里调用 new()函数创建一个 ProcessBuilderLike，在重载方法 start 中执行 processBuilder.start()。ProcessBuilderLike 的源码如下：

```
1.  private[deploy] object ProcessBuilderLike {
2.    def apply(processBuilder: ProcessBuilder): ProcessBuilderLike = new
      ProcessBuilderLike {
3.      override def start(): Process = processBuilder.start()
4.      override def command: Seq[String] = processBuilder.command().asScala
5.    }
6.  }
```

runCommandWithRetry 的源码如下：

```
1.   private[worker] def runCommandWithRetry(
2.       command: ProcessBuilderLike, initialize: Process => Unit, supervise:
         Boolean): Int = {
3.     var exitCode = -1
4.     //等待时间提交重试
5.     var waitSeconds = 1
6.     //运行一定秒的时间以后回退重置
7.     val successfulRunDuration = 5
8.     var keepTrying = !killed
9.
10.    while (keepTrying) {
```

```
11.       logInfo("Launch Command: " + command.command.mkString("\"", "\" \"",
          "\""))
12.
13.       synchronized {
14.         if (killed) { return exitCode }
15.         process = Some(command.start())
16.         initialize(process.get)
17.       }
18.
19.       val processStart = clock.getTimeMillis()
20.       exitCode = process.get.waitFor()
21.
22.       //如果尝试另一个运行检查
23.       keepTrying = supervise && exitCode != 0 && !killed
24.       if (keepTrying) {
25.         if (clock.getTimeMillis() - processStart > successfulRunDuration
            * 1000) {
26.           waitSeconds = 1
27.         }
28.         logInfo(s"Command exited with status $exitCode, re-launching after
            $waitSeconds s.")
29.         sleeper.sleep(waitSeconds)
30.         waitSeconds = waitSeconds * 2 //exponential back-off
31.       }
32.     }
33.
34.     exitCode
35.   }
36. }
```

runCommandWithRetry 第一次不一定能申请成功，因此循环遍历重试。DriverRunner 启动进程是通过 ProcessBuilder 中的 process.get.waitFor 来完成的。如果 supervise 设置为 True，exitCode 为非零退出码及 driver 进程没有终止，我们将 keepTrying 设置为 True，继续循环重试启动进程。

回到 DriverRunner.scala 的 LaunchDriver 方法如下：

```
1.      case LaunchDriver(driverId, driverDesc) =>
2.       ......
3.         drivers(driverId) = driver
4.         driver.start()
```

采用 driver.start 方法启动 Driver，进入 start 的源码如下：

```
1.    private[worker] def start() = {
2.      new Thread("DriverRunner for " + driverId) {
3.        override def run() {
4.        ......
5.          } catch {
6.            case e: Exception =>
7.              kill()
8.              finalState = Some(DriverState.ERROR)
9.              finalException = Some(e)
10.         } finally {
11.           if (shutdownHook != null) {
12.             ShutdownHookManager.removeShutdownHook(shutdownHook)
13.           }
14.         }
15.
```

```
16.         //通知 worker 节点 Driver 的最终状态及可能的异常
17.         worker.send(DriverStateChanged(driverId, finalState.get,
            finalException))
18.       }
19.     }.start()
20.   }
```

Start 启动时运行到了 finalState，可能是 Spark 运行出状况了，如 Driver 运行时 KILLED 或者 FAILED，出状况以后，通过 worker.send 给自己发一个消息，通知 DriverStateChanged 状态改变。下面是 Worker.scala 中的 driverStateChanged 的源码。

```
1.    case driverStateChanged @ DriverStateChanged(driverId, state, exception) =>
2.       handleDriverStateChanged(driverStateChanged)
```

在其中调用 handleDriverStateChanged 方法，handleDriverStateChanged 的源码如下：

```
1.    private[worker]   def   handleDriverStateChanged(driverStateChanged:
      DriverStateChanged): Unit = {
2.      val driverId = driverStateChanged.driverId
3.      val exception = driverStateChanged.exception
4.      val state = driverStateChanged.state
5.      state match {
6.        case DriverState.ERROR =>
7.          logWarning(s"Driver $driverId failed with unrecoverable exception:
            ${exception.get}")
8.        case DriverState.FAILED =>
9.          logWarning(s"Driver $driverId exited with failure")
10.       case DriverState.FINISHED =>
11.         logInfo(s"Driver $driverId exited successfully")
12.       case DriverState.KILLED =>
13.         logInfo(s"Driver $driverId was killed by user")
14.       case _ =>
15.         logDebug(s"Driver $driverId changed state to $state")
16.     }
17.     sendToMaster(driverStateChanged)
18.     val driver = drivers.remove(driverId).get
19.     finishedDrivers(driverId) = driver
20.     trimFinishedDriversIfNecessary()
21.     memoryUsed -= driver.driverDesc.mem
22.     coresUsed -= driver.driverDesc.cores
23.   }
```

Worker.scala 的 handleDriverStateChanged 方法中对于 state 的不同情况，打印相关日志。关键代码是 sendToMaster(driverStateChanged)，发一个消息给 Master，告知 Driver 进程挂掉。消息内容是 driverStateChanged。sendToMaster 的源码如下：

```
1.    private def sendToMaster(message: Any): Unit = {
2.      master match {
3.        case Some(masterRef) => masterRef.send(message)
4.        case None =>
5.          logWarning(
6.            s"Dropping $message because the connection to master has not yet
              been established")
7.      }
8.    }
```

下面来看一下 Master 的源码。Master 收到 DriverStateChanged 消息以后，无论 Driver 的状态是 DriverState.ERROR | DriverState.FINISHED | DriverState.KILLED | DriverState.FAILED

中的任何一个，都把Driver从内存数据结构中删掉，并把持久化引擎中的数据清理掉。

```
1.      case DriverStateChanged(driverId, state, exception) =>
2.      state match {
3.        case DriverState.ERROR | DriverState.FINISHED | DriverState.KILLED
          | DriverState.FAILED =>
4.          removeDriver(driverId, state, exception)
5.        case _ =>
6.          throw new Exception(s"Received unexpected state update for driver
            $driverId: $state")
7.      }
```

进入removeDriver的源码，清理掉相关数据以后，再次调用schedule方法。

```
1.   private def removeDriver(
2.       driverId: String,
3.       finalState: DriverState,
4.       exception: Option[Exception]) {
5.     drivers.find(d => d.id == driverId) match {
6.       case Some(driver) =>
7.         logInfo(s"Removing driver: $driverId")
8.         drivers -= driver
9.         if (completedDrivers.size >= RETAINED_DRIVERS) {
10.          val toRemove = math.max(RETAINED_DRIVERS / 10, 1)
11.          completedDrivers.trimStart(toRemove)
12.        }
13.        completedDrivers += driver
14.        persistenceEngine.removeDriver(driver)
15.        driver.state = finalState
16.        driver.exception = exception
17.        driver.worker.foreach(w => w.removeDriver(driver))
18.        schedule()
19.      case None =>
20.        logWarning(s"Asked to remove unknown driver: $driverId")
21.    }
22.  }
23. }
```

接下来看一下启动Executor。Worker.scala的LaunchExecutor方法的源码如下所示。

Worker.scala的源码如下：

```
1.     case LaunchExecutor(masterUrl, appId, execId, appDesc, cores_, memory_) =>
2.       if (masterUrl != activeMasterUrl) {
3.         logWarning("Invalid Master (" + masterUrl + ") attempted to launch
           executor.")
4.       } else {
5.         try {
6.           logInfo("Asked to launch executor %s/%d for %s".format(appId,
             execId, appDesc.name))
7.
8.           //创建executor节点的工作目录
9.           val executorDir = new File(workDir, appId + "/" + execId)
10.          if (!executorDir.mkdirs()) {
11.            throw new IOException("Failed to create directory " + executorDir)
12.          }
13.
14.         //创建executor的本地目录，通过SPARK_EXECUTOR_DIRS环境变量传递给
15.         //executor。应用程序完成后，这些目录将被Worker删除
16.          val appLocalDirs = appDirectories.getOrElse(appId, {
```

```
17.           val localRootDirs = Utils.getOrCreateLocalRootDirs(conf)
18.           val dirs = localRootDirs.flatMap { dir =>
19.             try {
20.               val appDir = Utils.createDirectory(dir, namePrefix = "executor")
21.               Utils.chmod700(appDir)
22.               Some(appDir.getAbsolutePath())
23.             } catch {
24.               case e: IOException =>
25.                 logWarning(s"${e.getMessage}. Ignoring this directory.")
26.                 None
27.             }
28.           }.toSeq
29.           if (dirs.isEmpty) {
30.             throw new IOException("No subfolder can be created in " +
31.               s"${localRootDirs.mkString(",")}.")
32.           }
33.           dirs
34.         })
35.         appDirectories(appId) = appLocalDirs
36.         val manager = new ExecutorRunner(
37.           appId,
38.           execId,
39.           appDesc.copy(command = Worker.maybeUpdateSSLSettings
              (appDesc.command, conf)),
40.           cores_,
41.           memory_,
42.           self,
43.           workerId,
44.           host,
45.           webUi.boundPort,
46.           publicAddress,
47.           sparkHome,
48.           executorDir,
49.           workerUri,
50.           conf,
51.           appLocalDirs, ExecutorState.RUNNING)
52.         executors(appId + "/" + execId) = manager
53.         manager.start()
54.         coresUsed += cores_
55.         memoryUsed += memory_
56.         sendToMaster(ExecutorStateChanged(appId, execId, manager.state,
            None, None))
57.       } catch {
58.         case e: Exception =>
59.           logError(s"Failed to launch executor $appId/$execId for
              ${appDesc.name}.", e)
60.           if (executors.contains(appId + "/" + execId)) {
61.             executors(appId + "/" + execId).kill()
62.             executors -= appId + "/" + execId
63.           }
64.           sendToMaster(ExecutorStateChanged(appId, execId, ExecutorState.FAILED,
65.             Some(e.toString), None))
66.       }
67.     }
```

直接看一下 manager.start 方法，启动一个线程 Thread，在 run 方法中调用 fetchAndRunExecutor。Spark 2.2.1 版本的 fetchAndRunExecutor 的源码如下：

```
1.      private def fetchAndRunExecutor() {
2.     try {
3.       //启动进程
4.       val builder = CommandUtils.buildProcessBuilder(appDesc.command, new
         SecurityManager(conf),
5.         memory, sparkHome.getAbsolutePath, substituteVariables)
6.       val command = builder.command()
7.       val formattedCommand = command.asScala.mkString("\"", "\" \"", "\"")
8.       logInfo(s"Launch command: $formattedCommand")
9.
10.      builder.directory(executorDir)
11.      builder.environment.put("SPARK_EXECUTOR_DIRS", appLocalDirs.mkString
         (File.pathSeparator))
12.      //如果在Spark Shell中运行，避免创建一个“Scala”的父进程执行executor命令
13.      builder.environment.put("SPARK_LAUNCH_WITH_SCALA", "0")
14.
15.      //增加WebUI日志网址
16.      val baseUrl =
17.        if (conf.getBoolean("spark.ui.reverseProxy", false)) {
18.          s"/proxy/$workerId/logPage/?appId=$appId&executorId=$execId&
           logType="
19.        } else {
20.          s"http://$publicAddress:$webUiPort/logPage/?appId=
           $appId&executorId=$execId&logType="
21.        }
22.      builder.environment.put("SPARK_LOG_URL_STDERR", s"${baseUrl}
         stderr")
23.      builder.environment.put("SPARK_LOG_URL_STDOUT", s"${baseUrl}
         stdout")
24.
25.      process = builder.start()
26.      val header = "Spark Executor Command: %s\n%s\n\n".format(
27.        formattedCommand, "=" * 40)
28.
29.      //重定向stdout和stderr文件
30.      val stdout = new File(executorDir, "stdout")
31.      stdoutAppender = FileAppender(process.getInputStream, stdout, conf)
32.
33.      val stderr = new File(executorDir, "stderr")
34.      Files.write(header, stderr, StandardCharsets.UTF_8)
35.      stderrAppender = FileAppender(process.getErrorStream, stderr, conf)
36.
37.      //等待它退出；执行器可以退出代码0（当driver 指示它关闭）或非零退出码
38.      val exitCode = process.waitFor()
39.      state = ExecutorState.EXITED
40.      val message = "Command exited with code " + exitCode
41.      worker.send(ExecutorStateChanged(appId, execId, state, Some
         (message), Some(exitCode)))
42.    } catch {
43.      case interrupted: InterruptedException =>
44.        logInfo("Runner thread for executor " + fullId + " interrupted")
45.        state = ExecutorState.KILLED
46.        killProcess(None)
47.      case e: Exception =>
48.        logError("Error running executor", e)
49.        state = ExecutorState.FAILED
50.        killProcess(Some(e.toString))
51.    }
52.  }
53. }
```

Spark 2.4.3 版本的 ExecutorRunner.scala 源码与 Spark 2.2.1 版本相比具有如下特点。

- 上段代码中第 3 行之后新增加代码，构建 subsCommand 变量。
- 上段代码中第 4 行 appDesc.command 调整为 subsCommand 变量。

```
1.  ……
2.      val subsOpts = appDesc.command.javaOpts.map {
3.        Utils.substituteAppNExecIds(_, appId, execId.toString)
4.      }
5.      val subsCommand = appDesc.command.copy(javaOpts = subsOpts)
6.  ……
7.      val builder = CommandUtils.buildProcessBuilder(subsCommand, new
        SecurityManager(conf),
8.        memory, sparkHome.getAbsolutePath, substituteVariables)
9.  ……
```

fetchAndRunExecutor 类似于启动 Driver 的过程，在启动 Executor 时首先构建 CommandUtils.buildProcessBuilder，然后是 builder.start()，退出时发送 ExecutorStateChanged 消息给 Worker。

Worker.scala 源码中的 executorStateChanged 如下：

```
1.    case executorStateChanged @ ExecutorStateChanged(appId, execId, state,
      message, exitStatus) =>
2.      handleExecutorStateChanged(executorStateChanged)
```

进入 handleExecutorStateChanged 源码，sendToMaster(executorStateChanged)发 executorStateChanged 消息给 Master。

Spark 2.2.1 版本的 Worker.scala 源码如下：

```
1.  private[worker] def handleExecutorStateChanged(executorStateChanged:
    ExecutorStateChanged):
2.      Unit = {
3.      sendToMaster(executorStateChanged)
4.      val state = executorStateChanged.state
5.      if (ExecutorState.isFinished(state)) {
6.        val appId = executorStateChanged.appId
7.        val fullId = appId + "/" + executorStateChanged.execId
8.        val message = executorStateChanged.message
9.        val exitStatus = executorStateChanged.exitStatus
10.       executors.get(fullId) match {
11.         case Some(executor) =>
12.           logInfo("Executor " + fullId + "finished with state" + state +
13.             message.map(" message " + _).getOrElse("") +
14.             exitStatus.map(" exitStatus " + _).getOrElse(""))
15.           executors -= fullId
16.           finishedExecutors(fullId) = executor
17.           trimFinishedExecutorsIfNecessary()
18.           coresUsed -= executor.cores
19.           memoryUsed -= executor.memory
20.         case None =>
21.           logInfo("Unknown Executor " + fullId + " finished with state "
              + state +
22.             message.map(" message " + _).getOrElse("") +
23.             exitStatus.map(" exitStatus " + _).getOrElse(""))
24.       }
25.       maybeCleanupApplication(appId)
26.     }
```

```
27.   }
28. }
```

Spark 2.4.3 版本的 Worker.scala 源码与 Spark 2.2.1 版本相比具有如下特点。

- 上段代码中第 19 行之后新增代码，如果 CLEANUP_NON_SHUFFLE_FILES_ENABLED 为 True，则删除 executor。

```
1.  ......
2.          if (CLEANUP_NON_SHUFFLE_FILES_ENABLED) {
3.
    shuffleService.executorRemoved(executorStateChanged.execId.toString,
    appId)
4.          }
5.  ......
6.  //是否清除 executor exits 的非洗牌文件
7.  private val CLEANUP_NON_SHUFFLE_FILES_ENABLED =
8.      conf.getBoolean("spark.storage.cleanupFilesAfterExecutorExit", true)
```

下面看一下 Master.scala。Master 收到 ExecutorStateChanged 消息。如状态发生改变，通过 exec.application.driver.send 给 Driver 也发送一个 ExecutorUpdated 消息，流程和启动 Driver 基本是一样的。ExecutorStateChanged 的源码如下：

```
1.   case ExecutorStateChanged(appId, execId, state, message, exitStatus) =>
2.       val execOption = idToApp.get(appId).flatMap(app => app.executors
         .get(execId))
3.       execOption match {
4.         case Some(exec) =>
5.           val appInfo = idToApp(appId)
6.           val oldState = exec.state
7.           exec.state = state
8.
9.           if (state == ExecutorState.RUNNING) {
10.            assert(oldState == ExecutorState.LAUNCHING,
11.              s"executor $execId state transfer from $oldState to RUNNING
                 is illegal")
12.            appInfo.resetRetryCount()
13.          }
14.
15.          exec.application.driver.send(ExecutorUpdated(execId, state,
             message, exitStatus, false))
16.
17.          if (ExecutorState.isFinished(state)) {
18.            //从 Worker 和应用程序中删除此 executor
19.            logInfo(s"Removing executor ${exec.fullId} because it is
               $state")
20.            //如果应用程序已经完成，保存应用程序状态，以在 UI 页面上正确显示信息
21.
22.            if (!appInfo.isFinished) {
23.              appInfo.removeExecutor(exec)
24.            }
25.            exec.worker.removeExecutor(exec)
26.
27.            val normalExit = exitStatus == Some(0)
28.            //只重试一定次数，这样就不会进入无限循环。重要提示：此代码路径不是通过测
               //试执行的，改变 if 条件时要小心
29.
30.
```

```
31.             if (!normalExit
32.                 && appInfo.incrementRetryCount() >= MAX_EXECUTOR_RETRIES
33.                 && MAX_EXECUTOR_RETRIES >= 0) { //< 0 disables this
                    application-killing path
34.               val execs = appInfo.executors.values
35.               if (!execs.exists(_.state == ExecutorState.RUNNING)) {
36.                 logError(s"Application ${appInfo.desc.name} with ID
                    ${appInfo.id} failed " +
37.                   s"${appInfo.retryCount} times; removing it")
38.                 removeApplication(appInfo, ApplicationState.FAILED)
39.               }
40.             }
41.           }
42.           schedule()
43.         case None =>
44.           logWarning(s"Got status update for unknown executor
              $appId/$execId")
45.       }
```

6.3.2 Driver 和 Master 交互原理解析

Driver 和 Master 交互，Master 是一个消息循环体。本节讲解 Driver 消息循环体的产生过程，Driver 消息循环体生成之后，就可以与 Master 互相通信了。

Spark 应用程序提交时，我们会提交一个 spark-submit 脚本。spark-submit 脚本中直接运行了 org.apache.spark.deploy.SparkSubmit 对象。Spark-submit 脚本内容如下所示。

```
#!/usr/bin/env bash
if [ -z "${SPARK_HOME}" ]; then
  source "$(dirname "$0")"/find-spark-home
fi

# disable randomized hash for string in Python 3.3+
export PYTHONHASHSEED=0

exec "${SPARK_HOME}"/bin/spark-class org.apache.spark.deploy.SparkSubmit "$@"
```

进入到 SparkSubmit 中，main 函数代码如下所示。

Spark 2.2.1 版本的 SparkSubmit.scala 的源码如下。

```
1. def main(args: Array[String]): Unit = {
2.     //由启动 main 函数传入的参数构建 SparkSubmitArguments 对象
3.     val appArgs = new SparkSubmitArguments(args)
4.     //打印参数信息
5.     if (appArgs.verbose) {
6.       printStream.println(appArgs)
7.     }
8.     appArgs.action match {
9.       //提交，调用 submit 方法
10.       case SparkSubmitAction.SUBMIT => submit(appArgs)
11.       //杀死，调用 kill 方法
12.       case SparkSubmitAction.KILL => kill(appArgs)
13.       //请求状态，调用 requestStatus 方法
14.       case SparkSubmitAction.REQUEST_STATUS => requestStatus(appArgs)
15.     }
16.   }
```

Spark 2.4.3 版本的 SparkSubmit.scala 源码与 Spark 2.2.1 版本相比具有如下特点。

❑ 上段代码中第 1～16 行整体替换为以下代码。

```
1.  override def main(args: Array[String]): Unit = {
2.      val submit = new SparkSubmit() {
3.        self =>
4.
5.        override protected def parseArguments(args: Array[String]):
          SparkSubmitArguments = {
6.          new SparkSubmitArguments(args) {
7.            override protected def logInfo(msg: => String): Unit = self.logInfo(msg)
8.
9.            override protected def logWarning(msg: => String): Unit =
              self.logWarning(msg)
10.         }
11.       }
12.
13.       override protected def logInfo(msg: => String): Unit = printMessage(msg)
14.
15.       override protected def logWarning(msg: => String): Unit =
          printMessage(s"Warning: $msg")
16.
17.       override def doSubmit(args: Array[String]): Unit = {
18.         try {
19.           super.doSubmit(args)
20.         } catch {
21.           case e: SparkUserAppException =>
22.             exitFn(e.exitCode)
23.         }
24.       }
25.
26.     }
27.
28.     submit.doSubmit(args)
29.   }
30. ......
31.   def doSubmit(args: Array[String]): Unit = {
32.    //如果尚未完成日志记录，则初始化日志记录。跟踪在应用程序启动之前是否需要重置日志
       //记录
33.     val uninitLog = initializeLogIfNecessary(true, silent = true)
34.
35.     val appArgs = parseArguments(args)
36.     if (appArgs.verbose) {
37.       logInfo(appArgs.toString)
38.     }
39.     appArgs.action match {
40.       case SparkSubmitAction.SUBMIT => submit(appArgs, uninitLog)
41.       case SparkSubmitAction.KILL => kill(appArgs)
42.       case SparkSubmitAction.REQUEST_STATUS => requestStatus(appArgs)
43.       case SparkSubmitAction.PRINT_VERSION => printVersion()
44.     }
45.   }
```

上面的代码中，spark-submit 脚本提交的命令行参数通过 main 函数的 args 获取，并将 args 参数传入 SparkSubmitArguments 中完成解析。最后通过匹配 appArgs 参数中的 action 类型，执行 submit、kill、requestStatus、PRINT_VERSION 操作。

进入到 SparkSubmitArguments 中，分析一下参数的解析过程。SparkSubmitArguments 中

的关键代码如下所示。

Spark 2.2.1 版本的 SparkSubmitArguments.scala 的源码如下：

```
1. //调用 parse 方法，从命令行解析出各个参数
2.   try {
3.     parse(args.asJava)
4.   } catch {
5.     //捕获到 IllegalArgumentException，打印错误并退出
6.     case e: IllegalArgumentException =>
7.       SparkSubmit.printErrorAndExit(e.getMessage())
8.   }
9.   //合并默认的 Spark 配置项，使用传入的配置覆盖默认的配置
10.   mergeDefaultSparkProperties()
11.   //从 sparkProperties 移除不是“spark.”为开始的配置
12.   ignoreNonSparkProperties()
13.   //加载系统环境变量中的配置信息
14.   loadEnvironmentArguments()
15.   //验证参数是否合法
16.   validateArguments()
```

Spark 2.4.3 版本的 SparkSubmitArguments.scala 源码与 Spark 2.2.1 版本相比具有如下特点。

- 上段代码中第 2～8 行删掉 try-catch 的处理，替换为 parse(args.asJava)。
- 上段代码中第 14 行之后新增一行代码，构建 useRest 变量。

```
1.  ……
2.  parse(args.asJava)
3.  ……
4.    useRest = sparkProperties.getOrElse("spark.master.rest.enabled",
      "false").toBoolean
```

在上面的代码中，parse(args.toList)将会解析命令行参数，通过 mergeDefaultSparkProperties 合并默认配置，调用 ignoreNonSparkProperties 方法忽略不是以“spark.”为开始的配置，方法 loadEnvironmentArguments 加载系统环境变量，最后调用 validateArguments 方法检验参数的合法性。这些配置如何提交呢？main 函数中由 case SparkSubmitAction.SUBMIT => submit(appArgs)这句代码判断是否提交参数并执行程序，如果匹配到 SparkSubmitAction.SUBMIT，则调用 submit(appArgs)方法，参数 appArgs 是 SparkSubmitArguments 类型，appArgs 中包含了提交的各种参数，包括命令行传入以及默认的配置项。

submit(appArgs)方法主要完成两件事情。

（1）准备提交环境。

（2）执行 main 方法，完成提交。

首先来看 Spark 中是如何准备环境的。在 submit(appArgs)方法中，有如下源码。

SparkSubmit.scala 的源码如下：

```
1.  private def submit(args: SparkSubmitArguments, uninitLog: Boolean): Unit = {
2.      val (childArgs, childClasspath, sparkConf, childMainClass) =
        prepareSubmitEnvironment(args)
3.  ……
4.      runMain(childArgs, childClasspath, sparkConf, childMainClass,
        args.verbose)
5.  ……
```

这段代码中，调用 prepareSubmitEnvironment(args)方法，完成提交环境的准备。该方法返回一个四元 Tuple，分别表示子进程参数、子进程 classpath 列表、系统属性 map、子进程 main 方法。完成了提交环境的准备工作后，接下来就启动子进程，在 Standalone 模式下，启动的子进程是 org.apache.spark.deploy.Client 对象。具体的执行过程在 runMain 函数中，关键代码如下所示。

Spark 2.2.1 版本的 SparkSubmit.scala 的源码如下：

```
1. private def runMain(
2.       childArgs: Seq[String],
3.       childClasspath: Seq[String],
4.       sysProps: Map[String, String],
5.       childMainClass: String,
6.       verbose: Boolean): Unit = {
7.    ......
8.     Thread.currentThread.setContextClassLoader(loader)//获得 classLoader
9.     for (jar <- childClasspath) {     //遍历 Classpath 列表
10.      addJarToClasspath(jar, loader)
                              //使用 loader 类加载器将 jar 包依赖加入 Classpath
11.     }
12.     for ((key, value) <- sysProps) {
                              //将 sysProps 中的配置全部设置到 System 全局变量中
13.      System.setProperty(key, value)
14.     }
15.    var mainClass: Class[_] = null
16.    try{
       mainClass = Utils.classForName(childMainClass)//获取启动的 MainClass
17.   ......//得到启动的对象的 main 方法
18.     val mainMethod = mainClass.getMethod("main", new Array[String]
       (0).getClass)
19.   ......//使用反射执行 main 方法，并将 childArgs 作为参数传入该 main 方法
20.     mainMethod.invoke(null, childArgs.toArray)
21.   }
```

Spark 2.4.3 版本的 SparkSubmit.scala 源码与 Spark 2.2.1 版本相比具有如下特点。

- 上段代码中第 4 行 sysProps 参数调整为 sparkConf。
- 上段代码中第 12～14 行删除。
- 上段代码中第 18 行调整为 new JavaMainApplication(mainClass)。
- 上段代码中第 20 行反射执行 main 方法调整为在 JavaMainApplication 实例的 start 方法里面执行。

```
1.   ......
2.     sparkConf: SparkConf,
3.   ......
4.   new JavaMainApplication(mainClass)
5.   ......
6.     app.start(childArgs.toArray, sparkConf)
7.   ......
8.   //SparkApplication.scala 代码
9.   private[deploy] class JavaMainApplication(klass: Class[_]) extends
     SparkApplication {
10.
11.    override def start(args: Array[String], conf: SparkConf): Unit = {
12.      val mainMethod = klass.getMethod("main", new Array[String](0).getClass)
13.      if (!Modifier.isStatic(mainMethod.getModifiers)) {
```

```
14.         throw new IllegalStateException("The main method in the given main
            class must be static")
15.       }
16.
17.       val sysProps = conf.getAll.toMap
18.       sysProps.foreach { case (k, v) =>
19.         sys.props(k) = v
20.       }
21.
22.       mainMethod.invoke(null, args)
23.     }
```

在上面的代码中，使用 Utils 工具提供的 classForName 方法，找到主类，然后在 mainClass 上调用 getMethod 方法得到 main 方法，最后在 mainMethod 上调用 invoke 执行 main 方法。需要注意的是，执行 invoke 方法同时传入了 childArgs 参数，这个参数中保留了配置信息。Utils.classForName(childMainClass)方法将会返回要执行的主类，这里的 childMainClass 是哪一个类呢?其实，这个参数在不同的部署模式下是不一样的，standalone 模式下，childMainClass 指的是 org.apache.spark.deploy.Client 类，从源码中可以找到依据，源码如下所示。

Spark 2.2.1 版本的 SparkSubmit.scala 的源码如下：

```
1.  //在 prepareSubmitEnvironment 方法中判断是否为 Standalone 集群模式
2.  if (args.isStandaloneCluster) {
3.        //判断使用 Rest，使用 Rest childMainClass 为 org.apache.spark.deploy
          //.rest.RestSubmissionClient
4.        if (args.useRest) {
5.          childMainClass = "org.apache.spark.deploy.rest.RestSubmissionClient"
6.          childArgs += (args.primaryResource, args.mainClass)
7.        } else {
8.          //非 Rest，childMainClass 为 org.apache.spark.deploy.Client
9.          childMainClass = "org.apache.spark.deploy.Client"
10.         if (args.supervise) { childArgs += "--supervise" }
11.         //设置 driver memory
12.         Option(args.driverMemory).foreach { m => childArgs += ("--memory", m) }
13.         //设置 driver cores
14.         Option(args.driverCores).foreach { c => childArgs += ("--cores", c) }
15.         childArgs += "launch"
16.         childArgs += (args.master, args.primaryResource, args.mainClass)
17.       }
18.       if (args.childArgs != null) {
19.         childArgs ++= args.childArgs
20.       }
21.     }
```

Spark 2.4.3 版本的 SparkSubmit.scala 源码与 Spark 2.2.1 版本相比具有如下特点。

- 上段代码中第 5 行将 childMainClass 调整为 REST_CLUSTER_SUBMIT_CLASS。
- 上段代码中第 9 行将 childMainClass 调整为 STANDALONE_CLUSTER_SUBMIT_CLASS。

```
1.  ......
2.   childMainClass = REST_CLUSTER_SUBMIT_CLASS
3.  ......
4.  childMainClass = STANDALONE_CLUSTER_SUBMIT_CLASS
5.  ......
6.    private[deploy] val YARN_CLUSTER_SUBMIT_CLASS =
7.      "org.apache.spark.deploy.yarn.YarnClusterApplication"
```

```
8.    private[deploy] val REST_CLUSTER_SUBMIT_CLASS =
      classOf[RestSubmissionClientApp].getName()
9.    private[deploy] val STANDALONE_CLUSTER_SUBMIT_CLASS =
      classOf[ClientApp].getName()
10.   private[deploy] val KUBERNETES_CLUSTER_SUBMIT_CLASS =
11.     "org.apache.spark.deploy.k8s.submit.KubernetesClientApplication"
```

在上面的代码中，程序首先根据 args.isStandaloneCluster 判断部署模式，如果是 standalone 模式并且不使用 REST 服务，childMainClass = STANDALONE_CLUSTER_SUBMIT_CLASS，获取的 childMainClass 其实是 org.apache.spark.deploy.Client。从上述代码中可以看出，childArgs 中存入了 Executor 的 memory 配置和 cores 配置。与 runMain 方法中描述一样，程序将启动 org.apache.spark.deploy.Client 类，并运行主方法。Client 类中做了哪些事情？先来看这个类是怎样完成调用的。下面是 Client 对象及主方法。

Spark 2.2.1 版本的 Client.scala 的源码如下：

```
1.        object Client {
2.    def main(args: Array[String]) {
3.     //若 sys 中不包含 SPARK_SUBMIT，则打印警告信息
4.      if (!sys.props.contains("SPARK_SUBMIT")) {
5.        println("WARNING: This client is deprecated and will be removed in
          a future version of Spark")
6.        println("Use  ./bin/spark-submit  with  \"--master  spark://host:
          port\"")
7.      }
8.      //scalastyle:on println
9.     //创建 SparkConf 对象
10.     val conf = new SparkConf()
11.    //创建 ClientArguments 对象，代表 Driver 端的参数
12.     val driverArgs = new ClientArguments(args)
13.
14.    //设置 RPC 请求超时时间为 10s
15.     if (!conf.contains("spark.rpc.askTimeout")) {
16.       conf.set("spark.rpc.askTimeout", "10s")
17.     }
18.     Logger.getRootLogger.setLevel(driverArgs.logLevel)
19.    //使用 RpcEnv 的 create 创建 RPC 环境
20.     val rpcEnv =
21.       RpcEnv.create("driverClient", Utils.localHostName(), 0, conf, new
          SecurityManager(conf))
22.
23.     //得到 master 的 URL 并得到 Master 的 Endpoints，用于同 Master 通信
24.     val masterEndpoints = driverArgs.masters.map(RpcAddress.fromSparkURL).
25.       map(rpcEnv.setupEndpointRef(_, Master.ENDPOINT_NAME))
26.    rpcEnv.setupEndpoint("client", new ClientEndpoint(rpcEnv, driverArgs,
       masterEndpoints, conf))
27. //等待 rpcEnv 的终止
28.     rpcEnv.awaitTermination()
29.   }
30. }
```

Spark 2.4.3 版本的 Client.scala 源码与 Spark 2.2.1 版本相比具有如下特点。

- 上段代码中第 8～30 行进行函数封装，整体替换为 ClientApp().start 方法。

```
1.  ......
2.   new ClientApp().start(args, new SparkConf())
```

```
3.  ......
4.  private[spark] class ClientApp extends SparkApplication {
5.
6.    override def start(args: Array[String], conf: SparkConf): Unit = {
7.      val driverArgs = new ClientArguments(args)
8.
9.      if (!conf.contains("spark.rpc.askTimeout")) {
10.       conf.set("spark.rpc.askTimeout", "10s")
11.     }
12.     Logger.getRootLogger.setLevel(driverArgs.logLevel)
13.
14.     val rpcEnv =
15.       RpcEnv.create("driverClient", Utils.localHostName(), 0, conf, new
          SecurityManager(conf))
16.
17.     val masterEndpoints = driverArgs.masters.map(RpcAddress.fromSparkURL)
18.       .map(rpcEnv.setupEndpointRef(_, Master.ENDPOINT_NAME))
19.     rpcEnv.setupEndpoint("client", new ClientEndpoint(rpcEnv, driverArgs,
        masterEndpoints, conf))
20.
21.     rpcEnv.awaitTermination()
22.   }
23.
24. }
```

上面的代码中，首先实例化出一个 SparkConfig 对象，通过这个配置对象，可以在代码中指定一些配置项，如 appName、Master 地址等。val driverArgs = new ClientArguments(args)使用传入的 args 参数构建一个 ClientArguments 对象，该对象同样保留传入的配置信息，如 Executor memory、Executor cores 等都包含在这个对象中。

使用 RpcEnv.create 工厂方法，创建一个 rpcEnv 成员，使用该成员设置好到 Master 的通信端点，通过该端点实现同 Master 的通信。Spark 2.0 中默认采用 Netty 框架来实现远程过程调用（Remote Precedure Call，RPC），通过使用 RPC 异步通信机制，完成各节点之间的通信。在 rpcEnv.setupEndpoint 方法中调用 new()函数创建一个 Driver ClientEndpoint。ClientEndpoint 是一个 ThreadSafeRpcEndpoint 消息循环体，至此就生成了 Driver ClientEndpoint。在 ClientEndpoint 的 onStart 方法中向 Master 提交注册。这里通过 masterEndpoint 向 Master 发送 RequestSubmitDriver(driverDescription)请求，完成 Driver 的注册。

Client.scala 的 onStart 的源码如下：

```
1.   private class ClientEndpoint(
2.      override val rpcEnv: RpcEnv,
3.      driverArgs: ClientArguments,
4.      masterEndpoints: Seq[RpcEndpointRef],
5.      conf: SparkConf)
6.    extends ThreadSafeRpcEndpoint with Logging {
7.
8.  override def onStart(): Unit = {
9.     driverArgs.cmd match {
10.    ......
11.        val driverDescription = new DriverDescription(
12.          driverArgs.jarUrl,
13.          driverArgs.memory,
14.          driverArgs.cores,
15.          driverArgs.supervise,
16.          command)
```

```
17.        ayncSendToMasterAndForwardReply[SubmitDriverResponse](
18.          RequestSubmitDriver(driverDescription))
19.
20. ......
```

Master 收到 Driver ClientEndpoint 的 RequestSubmitDriver 消息以后，就将 Driver 的信息加入到 waitingDrivers 和 drivers 的数据结构中。然后进行 schedule()资源分配，Master 向 Worker 发送 LaunchDriver 的消息指令。

Master.scala 的源码如下：

```
1.          case RequestSubmitDriver(description) =>
2.     ......
3.         val driver = createDriver(description)
4.         persistenceEngine.addDriver(driver)
5.         waitingDrivers += driver
6.         drivers.add(driver)
7.         schedule()
8.   ......
```

在 Client.scala 的 onStart 代码中，提交的配置参数始终在不同的对象、节点上传递。Master 把 Driver 加载到 Worker 节点并启动，Worker 节点上运行的 Driver 同样包含配置参数。当 Driver 端的 SparkContext 启动并实例化 DAGScheduler、TaskScheduler 时，StandaloneSchedulerBackend 在做另一件事情——实例化 StandaloneAppClient，StandaloneAppClient 中有 StandaloneAppClientPoint，也是一个 RPC 端口的引用，用于和 Master 进行通信。在 StandaloneAppClientPoint 的 onStart 方法中，向 Master 发送 RegisterApplication(appDescription,self)请求，Master 节点收到请求并调用 schedule 方法，向 Worker 发送 LaunchExecutor(masterUrl,exec.application.id, exec.id, exec.application.desc, exec.cores, exec.memory)请求，Worker 节点启动 ExecutorRunner。ExecutorRunner 中启动 CoarseGrainedExecutorBackend 并向 Driver 注册。

在 CoarseGrainedExecutorBackend 的 main 方法中，代码如下：

```
1. var argv = args.toList                               //将 args 转化成 List
2.    while (!argv.isEmpty) {                           //argv 不为空，则一直循环
3.      argv match {
4.        case ("--driver-url") :: value :: tail =>
5.          driverUrl = value                           //得到 driveRurl
6.          argv = tail
7.        case ("--executor-id") :: value :: tail =>
8.          executorId = value                          //得到 executorid
9.          argv = tail
10.        case ("--hostname") :: value :: tail =>
11.          hostname = value                           //得到 hostname
12.          argv = tail
13.        case ("--cores") :: value :: tail =>
14.          cores = value.toInt                        //得到配置的 Executor 核的个数
15.          argv = tail
16.        case ("--app-id") :: value :: tail =>
17.          appId = value                              //得到 application 的 id
18.          argv = tail
19.        case ("--worker-url") :: value :: tail =>
20.          workerUrl = Some(value)                    //得到 worker 的 url
21.          argv = tail
22.        case ("--user-class-path") :: value :: tail =>
```

```
23.         userClassPath += new URL(value)    //得到用户类路径
24.         argv = tail
25.       case Nil =>
26.       case tail =>
27.         System.err.println(s"Unrecognized options: ${tail.mkString(" ")}")
28.         printUsageAndExit()                  //打印并退出
29.     }
30.   }
```

从程序提交一直到 CoarseGrainedExecutorBackend 进程启动，配置参数一直被传递。在 CoarseGrainedExecutorBackend 中取出了 cores 配置信息，并通过 run(driverUrl, executorId, hostname, cores, appId, workerUrl, userClassPath)将 cores 传入 run 方法，CoarseGrainedExecutor-Backend 以进程的形式在 JVM 中启动，此时 JVM 的资源指占用资源的数量并启动起来。需要注意的是，在一个 Worker 节点上，只要物理内核的个数和内存大小能够满足 Executor 启动要求，一个 Worker 节点上就可以运行多个 Executor。

6.3.3　Driver 和 Master 交互源码详解

从 Spark-Submit 的脚本分析，提交应用程序时，Main 启动的类，也就是用户最终提交执行的类是 org.apache.spark.deploy.SparkSubmit。SparkSubmit 的全路径为 org.apache.spark.deploy .SparkSubmit。SparkSubmit 是启动一个 Spark 应用程序的主入口点。当集群管理器为 STANDALONE、部署模式为 CLUSTER 时，根据提交的两种方式将 childMainClass 分别设置为不同的类，同时将传入的 args.mainClass（提交应用程序时设置的主类）及其参数根据不同集群管理器与部署模式进行转换，并封装到新的主类所需的参数中。在 REST 方式（Spark 1.3+）方式中，childMainClass 是"org.apache.spark.deploy.rest.RestSubmissionClient"；在传统方式中，childMainClass 是"org.apache.spark.deploy.Client"。

接下来以 REST 方式讲解。当提交方式为 REST 方式（Spark 1.3+）时，会将应用程序的主类等信息封装到 RestSubmissionClient 类中，由该类负责向 RestSubmissionServer 发送提交应用程序的请求，而 RestSubmissionServer 接收到应用程序提交的请求后，会向 Master 发送 RequestSubmitDriver 消息，然后由 Master 根据资源调度策略，启动集群中相应的 Driver，执行提交的应用程序。Cluster 部署模式下的部署与执行框架如图 6-3 所示。

为了体现各个组件间的部署关系，这里以框架图的形式进行描述，对应地，可以从时序图的角度去理解各个类或组件之间的交互关系。其中，组件 Master 和 Worker 的标注在方框的左上角，其他方框表示一个具体的实例。

其中，RestSubmissionClient 是提交应用程序的客户端处，对提交的应用程序进行封装的类。之后各个组件间的交互流程分析如下。

（1）第 1 步 constructSubmitRequest，就是在 RestSubmissionClient 实例中，根据提交的应用程序信息，构建出提交请求。

（2）然后继续第 2 步 createSubmission，在该步骤中向 RestSubmissionServer 发送 post 请求，即图 6-3 中对应的第 3 步（注意，实际上是在第 2 步中调用）。

（3）RestSubmissionServer 接收到 post 请求后，由对应的 Servlet 进行处理，这里对应为 StandaloneSubmitRequestServlet，即开始第 4 步，调用 doPost，发送 Post 请求。

（4）doPost 中继续第 5 步 handleSubmit，开始处理提交请求。在处理过程中，向 Master

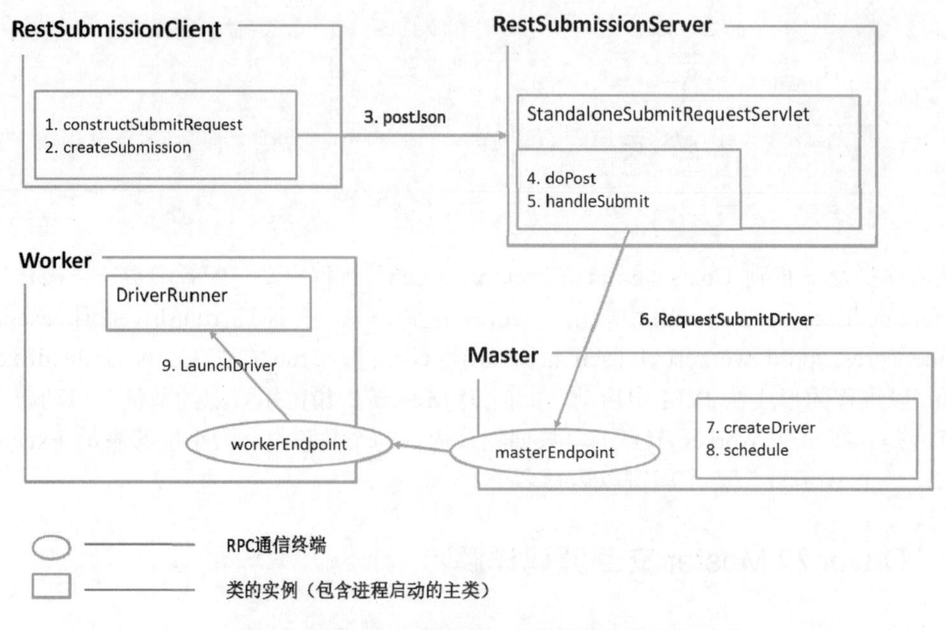

图 6-3　Cluster 部署模式下的部署与执行框架

的 RPC 终端发送消息 RequestSubmitDriver，对应图中的第 6 步。

（5）Master 接收到该消息后，执行第 7 步 createDriver，创建 Driver，需要由 Master 的调度机制创建，对应第 8 步 schedule，获取分配的资源后，向 Worker（这些 Worker 启动时会注册到 Master 上）的 RPC 终端发送 LaunchDriver 消息。

（6）Worker 在 RPC 终端接收到消息后开始处理，实例化一个 DriverRunner，并运行之前封装的应用程序。

注意：从上面部署框架及其术语解析部分可以知道，由于提交的应用程序在 main 部分包含了 SparkContext 实例，因此我们也称之为 Driver Program，即驱动程序。因此，在框架中，对应在 Master 和 Worker 处都使用 Driver，而不是 Application（应用程序）。

其中主要的源码及其分析如下。

（1）RestSubmissionClient 中的 run 方法。

RestSubmissionClient.scala 的源码如下：

```
1.   def run(
2.       appResource: String,
3.       mainClass: String,
4.       appArgs: Array[String],
5.       conf: SparkConf,
6.       env: Map[String, String] = Map()): SubmitRestProtocolResponse = {
7.     val master = conf.getOption("spark.master").getOrElse {
8.       throw new IllegalArgumentException("'spark.master' must be set.")
9.     }
10.    val sparkProperties = conf.getAll.toMap
11.     //创建一个 Rest 提交客户端
12.    val client = new RestSubmissionClient(master)
13.   //封装应用程序的相关信息，包括主资源、主类等
```

```
14.     val submitRequest = client.constructSubmitRequest(
15.       appResource, mainClass, appArgs, sparkProperties, env)
16.    //Rest 提交客户端开始创建 Submission，创建过程中向 RestSubmissionServer 发送
       //post 请求
17.
18.     client.createSubmission(submitRequest)
19.   }
```

（2）收到提交的 Post 请求之后，StandaloneSubmitRequestServlet 向 Master 的 RPC 终端发送 RequestSubmitDriver 请求。

StandaloneRestServer.scala 的源码如下：

```
1.   protected override def handleSubmit(
2.        requestMessageJson: String,
3.        requestMessage: SubmitRestProtocolMessage,
4.        responseServlet: HttpServletResponse): SubmitRestProtocolResponse
          = {
5.      requestMessage match {
6.        case submitRequest: CreateSubmissionRequest =>
7.
8.   //在这里开始构建驱动程序（也就是包含 SparkContext 的应用程序）的描述信息，
     //对应 DriverDescription 实例并向 Master 的 RPC 终端 masterEndpoint 发
     //送请求消息 RequestSubmitDriver
9.   val driverDescription = buildDriverDescription(submitRequest)
10.         val response = masterEndpoint.askSync[DeployMessages
            .SubmitDriverResponse](DeployMessages.RequestSubmitDriver
            (driverDescription))
11.
12.         val submitResponse = new CreateSubmissionResponse
13.         submitResponse.serverSparkVersion = sparkVersion
14.         submitResponse.message = response.message
15.         submitResponse.success = response.success
16.         submitResponse.submissionId = response.driverId.orNull
17. ……
```

（3）构建 DriverDescription 的 buildDriverDescription 方法。

StandaloneRestServer.scala 的源码如下：

```
1.  private def buildDriverDescription(request:
    CreateSubmissionRequest): DriverDescription = {
2.  ……
3.     //构建 Command 实例，将主类 mainClass 封装到 DriverWrapper（可以通过 jps 查看）
4.      val command = new Command(
5.        "org.apache.spark.deploy.worker.DriverWrapper",
6.  Seq("{{WORKER_URL}}", "{{USER_JAR}}", mainClass) ++ appArgs,
    //args to the DriverWrapper
7.        environmentVariables, extraClassPath, extraLibraryPath, javaOpts)
8.  ……
9.
10.  //构建驱动程序的描述信息 DriverDescription
11.  new DriverDescription(
12. appResource, actualDriverMemory, actualDriverCores, actualSuperviseDriver,
    command)
13.   }
```

（4）Master 接收 RequestSubmitDriver，处理消息并返回 SubmitDriverResponse 消息。

Master.scala 的源码如下：

```
1.   case RequestSubmitDriver(description) =>
2.        if (state != RecoveryState.ALIVE) {
3.          val msg = s"${Utils.BACKUP_STANDALONE_MASTER_PREFIX}: $state. " +
4.            "Can only accept driver submissions in ALIVE state."
5.          context.reply(SubmitDriverResponse(self, false, None, msg))
6.        } else {
7.          logInfo("Driver submitted " + description.command.mainClass)
8.          val driver = createDriver(description)
9.          persistenceEngine.addDriver(driver)
10.         waitingDrivers += driver
11.         drivers.add(driver)
12.         schedule()
13.
14.         //待办事项:让提交的客户端轮询master来确定driver的当前状态。目前使用fire
            //and forget方式发送消息
15.
16.
17.         context.reply(SubmitDriverResponse(self, true, Some(driver.id),
18.           s"Driver successfully submitted as ${driver.id}"))
19.       }
```

（5）Master 的 schedule()：调度机制的调度。

Master.scala 的源码如下：

```
1.      private def schedule(): Unit = {
2.
3.            launchDriver(worker, driver)
4.        ......
5.      startExecutorsOnWorkers()
6.    }
```

（6）Worker 上的 Driver 启动。

Worker.scala 的源码如下：

```
1.        case LaunchDriver(driverId, driverDesc) =>
2.        logInfo(s"Asked to launch driver $driverId")
3.        val driver = new DriverRunner(
4.          conf,
5.          driverId,
6.          workDir,
7.          sparkHome,
8.          driverDesc.copy(command = Worker.maybeUpdateSSLSettings
            (driverDesc.command, conf)),
9.          self,
10.         workerUri,
11.         securityMgr)
12.       drivers(driverId) = driver
13.       driver.start()
14.
15.       coresUsed += driverDesc.cores
16.       memoryUsed += driverDesc.mem
```

Driver Client 管理 Driver，包括向 Master 提交 Driver、请求 Kill Driver 等。Driver Client 与 Master 间的交互消息如下。

DeployMessages.scala 的源码如下：

```
1.      //DriverClient <-> Master
```

```
2.  //Driver Client 向 Master 请求提交 Driver
3.    case class RequestSubmitDriver(driverDescription: DriverDescription)
      extends DeployMessage
4.  //Master 向 Driver Client 返回注册是否成功的消息
5.    case class SubmitDriverResponse(
6.        master: RpcEndpointRef, success: Boolean, driverId: Option[String],
          message: String)
7.      extends DeployMessage
8.  //Driver Client 向 Master 请求 Kill Driver
9.    case class RequestKillDriver(driverId: String) extends DeployMessage
10. //Master 回复 Kill Driver 是否成功
11.   case class KillDriverResponse(
12.       master: RpcEndpointRef, driverId: String, success: Boolean, message:
          String)
13.     extends DeployMessage
14. //Driver Client 向 Master 请求 Driver 状态
15.   case class RequestDriverStatus(driverId: String) extends DeployMessage
16. //Master 向 Driver Client 返回状态请求信息
17.   case class DriverStatusResponse(found: Boolean, state: Option
      [DriverState],
18.     workerId: Option[String], workerHostPort: Option[String], exception:
        Option[Exception])
```

Driver 在 handleSubmit 方法中向 Master 请求提交 RequestSubmitDriver 消息。

Master 收到 Driver StandaloneSubmitRequestServlet 发送的消息 RequestSubmitDriver。Master 做相应的处理以后，返回 Driver StandaloneSubmitRequestServlet 消息 SubmitDriver-Response。

Master 的源码如下：

```
1.       case RequestSubmitDriver(description) =>
2.        if (state != RecoveryState.ALIVE) {
3.          val msg = s"${Utils.BACKUP_STANDALONE_MASTER_PREFIX}: $state. " +
4.            "Can only accept driver submissions in ALIVE state."
5.          context.reply(SubmitDriverResponse(self, false, None, msg))
6.        } else {
7.          logInfo("Driver submitted " + description.command.mainClass)
8.          val driver = createDriver(description)
9.          persistenceEngine.addDriver(driver)
10.         waitingDrivers += driver
11.         drivers.add(driver)
12.         schedule()
13.
14.         //待办事项:让提交的客户端轮询 master 来确定 driver 的当前状态,目前使用 fire
            //and forget 方式发送消息
15.
16.
17.         context.reply(SubmitDriverResponse(self, true, Some(driver.id),
18.           s"Driver successfully submitted as ${driver.id}"))
19.       }
```

类似地，Master 收到 Driver StandaloneKillRequestServlet 方法中发送的 RequestKillDriver 消息，Master 做相应的处理以后，返回 Driver StandaloneKillRequestServlet 消息 KillDriverResponse。

Master 收到 Driver StandaloneStatusRequestServlet 方法中发送的 RequestDriverStatus 更新消息，Master 做相应的处理以后，返回 Driver StandaloneStatusRequestServlet 消息 DriverStatusResponse。

6.4　从 Application 提交的角度重新审视 Executor

本节从 Application 提交的角度重新审视 Executor，彻底解密 Executor 到底是什么时候启动的，以及 Executor 如何把结果交给 Application。

6.4.1　Executor 到底是什么时候启动的

SparkContext 启动后，StandaloneSchedulerBackend 中会调用 new()函数创建一个 StandaloneAppClient，StandaloneAppClient 中有一个名叫 ClientEndPoint 的内部类，在创建 ClientEndpoint 时会传入 Command 来指定具体为当前应用程序启动的 Executor 进行的入口类的名称为 CoarseGrainedExecutorBackend。ClientEndPoint 继承自 ThreadSafeRpcEndpoint，其通过 RPC 机制完成和 Master 的通信。在 ClientEndPoint 的 start 方法中，会通过 registerWithMaster 方法向 Master 发送 RegisterApplication 请求，Master 收到该请求消息后，首先通过 registerApplication 方法完成信息登记，之后将会调用 schedule 方法，在 Worker 上启动 Executor。Master 对 RegisterApplication 请求处理源码如下：

```
1.        case RegisterApplication(description, driver) =>
2.        //待办事项：防止 driver 程序重复注册
3.      //Master 处于 STANDBY(备用)状态，不作处理
4.        if (state == RecoveryState.STANDBY) {
5.          //忽略，不发送响应
6.        } else {
7.          logInfo("Registering app " + description.name)
8.      //由 description 描述，构建 ApplicationInfo
9.          val app = createApplication(description, driver)
10.         registerApplication(app)
11.         logInfo("Registered app " + description.name + " with ID " + app.id)
12.      //在持久化引擎中加入 application
13.         persistenceEngine.addApplication(app)
14.         driver.send(RegisteredApplication(app.id, self))
15.   //调用 schedule 方法，在 worker 节点上启动 Executor
16.         schedule()
17.       }
```

在上面的代码中，Master 匹配到 RegisterApplication 请求，先判断 Master 的状态是否为 STANDBY（备用）状态，如果不是，说明 Master 为 ALIVE 状态，在这种状态下调用 createApplication(description,sender)方法创建 ApplicationInfo，完成之后调用 persistenceEngine.addApplication(app)方法，将新创建的 ApplicationInfo 持久化，以便错误恢复。完成这两步操作后，通过 driver.send(RegisteredApplication(app.id, self))向 StandaloneAppClient 返回注册成功后 ApplicationInfo 的 Id 和 master 的 url 地址。

ApplicationInfo 对象是对 application 的描述，下面先来看 createApplication 方法的源码。

Master.scala 的源码如下：

```
1.       private def createApplication(desc: ApplicationDescription, driver:
         RpcEndpointRef):
```

```
2.        ApplicationInfo = {
3.   //ApplicationInfo 创建时间
4.     val now = System.currentTimeMillis()
5.     val date = new Date(now)
6.    //由 date 生成 application id
7.     val appId = newApplicationId(date)
8.   //创建 ApplicationInfo
9.     new ApplicationInfo(now, appId, desc, date, driver, defaultCores)
10.   }
```

上面的代码中，createApplication 方法接收 ApplicationDescription 和 ActorRef 两种类型的参数，并调用 newApplicationId 方法生成 appId，关键代码如下：

```
1. val appId = "app-%s-%04d".format(createDateFormat.format(submitDate),
   nextAppNumber)
```

由代码所决定，appid 的格式形如：app-20160429101010-0001。desc 对象中包含一些基本的配置，包括从系统中传入的一些配置信息，如 appname、maxCores、memoryPerExecutorMB 等。最后使用 desc、date、driver、defaultCores 等作为参数构造一个 ApplicatiOnInfo 对象并返回。函数返回之后，调用 registerApplication 方法，完成 application 的注册，该方法是如何完成注册的？

Spark 2.2.1 版本的 Master.scala 的源码如下：

```
1.    private def registerApplication(app: ApplicationInfo): Unit = {
2.    //Driver 的地址，用于 Master 和 Driver 通信
3.      val appAddress = app.driver.address
4.    //如果 addressToApp 中已经有了该 Driver 地址，说明该 Driver 已经注册过了，直接
      //return
5.
6.      if (addressToApp.contains(appAddress)) {
7.        logInfo("Attempted to re-register application at same address: " +
          appAddress)
8.        return
9.      }
10.   //向度量系统注册
11.     applicationMetricsSystem.registerSource(app.appSource)
12.   //apps 是一个 HashSet，保存数据不能重复，向 HashSet 中加入 app
13.     apps += app
14.  //idToApp 是一个 HashMap，该 HashMap 用于保存 id 和 app 的对应关系
15.     idToApp(app.id) = app
16.  //endpointToApp 是一个 HashMap，该 HashMap 用于保存 driver 和 app 的对应关系
17.     endpointToApp(app.driver) = app
18.  //addressToApp 是一个 HashMap，记录 app Driver 的地址和 app 的对应关系
19.     addressToApp(appAddress) = app
20.  //waitingApps 是一个数组，记录等待调度的 app 记录
21.     waitingApps += app
22.     if (reverseProxy) {
23.       webUi.addProxyTargets(app.id, app.desc.appUiUrl)
24.     }
25.   }
```

Spark 2.4.3 版本的 Master.scala 源码与 Spark 2.2.1 版本相比具有如下特点。

- 上段代码中第 22～24 行删掉。

上面的代码中，首先通过 app.driver.path.address 得到 Driver 的地址，然后查看 appAddress 映射表中是否已经存在这个路径，如果存在，表示该 application 已经注册，直接返回；如果

不存在，则在 waitingApps 数组中加入该 application，同时在 idToApp、endpointToApp、addressToApp 映射表中加入映射关系。加入 waitingApps 数组中的 application 等待 schedule 方法的调度。

schedule 方法有两个作用：第一，完成 Driver 的调度，将 waitingDrivers 数组中的 Driver 发送到满足条件的 Worker 上运行；第二，在满足条件的 Worker 节点上为 application 启动 Executor。

Master.scala 的 schedule 方法的源码如下：

```
1.  private def schedule(): Unit = {
2.  ……
3.      launchDriver(worker, driver)
4.      ……
5.      startExecutorsOnWorkers()
6.    }
```

在 Master 中，schedule 方法是一个很重要的方法，每一次新的 Driver 的注册、application 的注册，或者可用资源发生变动，都将调用 schedule 方法。schedule 方法用于为当前等待调度的 application 调度可用的资源，在满足条件的 Worker 节点上启动 Executor。这个方法还有另外一个作用，就是当有 Driver 提交的时候，负责将 Driver 发送到一个可用资源满足 Driver 需求的 Worker 节点上运行。launchDriver(worker,driver)方法负责完成这一任务。

application 调度成功之后，Master 将会为 application 在 Worker 节点上启动 Executors，调用 startExecutorsOnWorkers 方法完成此操作。

在 scheduleExecutorsOnWorkers 方法中，有两种启动 Executor 的策略：第一种是轮流均摊策略（round-robin），采用圆桌算法依次轮流均摊，直到满足资源需求，轮流均摊策略通常会有更好的数据本地性，因此它是默认的选择策略；第二种是依次全占，在 usableWorkers 中，依次获取每个 Worker 上的全部资源，直到满足资源需求。

scheduleExecutorsOnWorkers 方法为 application 分配好逻辑意义上的资源后，还不能真正在 Worker 节点为 application 分配出资源，需要调用动作函数为 application 真正地分配资源。allocateWorkerResourceToExecutors 方法的调用，将会在 Worker 节点上实际分配资源。下面是 allocateWorkerResourceToExecutors 的源码。

Master.scala 的源码如下：

```
1. private def allocateWorkerResourceToExecutors(
2.    ……
3.      launchExecutor(worker, exec)
4.    ……
```

上面代码调用了 launchExecutor（worker,exec）方法，这个方法有两个参数：第一个参数是满足条件的 WorkerInfo 信息；第二个参数是描述 Executor 的 ExecutorDesc 对象。这个方法将会向 Worker 节点发送 LaunchExecutor 的请求，Worker 节点收到该请求之后，将会负责启动 Executor。launchExecutor 方法代码清单如下所示。

Master.scala 的源码如下：

```
1.      private def launchExecutor(worker: WorkerInfo, exec: ExecutorDesc):
        Unit = {
2.      logInfo("Launching executor " + exec.fullId + " on worker " + worker.id)
3.   //向 WorkerInfo 中加入 exec 这个描述 Executor 的 ExecutorDesc 对象
```

```
4.      worker.addExecutor(exec)
5.   //向 worker 发送 LaunchExecutor 消息，加载 Executor 消息中携带了 masterUrl 地址、
     //application id、Executor id、Executor 描述 desc、Executor 核的个数、Executor
     //分配的内存大小
6.
7.      worker.endpoint.send(LaunchExecutor(masterUrl,
8.        exec.application.id, exec.id, exec.application.desc, exec.cores,
          exec.memory))
9.     //向 Driver 发回 ExecutorAdded 消息，消息携带 worker 的 id 号、worker 的 host 和
       //port、分配的核的个数和内存大小
10.     exec.application.driver.send(
11.       ExecutorAdded(exec.id, worker.id, worker.hostPort, exec.cores,
          exec.memory))
12.   }
```

launchExecutor 有两个参数，第一个参数是 worker:WorkerInfo，代表 Worker 的基本信息；第二个参数是 exec：ExecutorDesc，这个参数保存了 Executor 的基本配置信息，如 memory、cores 等。此方法中有 worker.endpoint.send(LaunchExecutor(...))，向 Worker 发送 LaunchExecutor 请求，Worker 收到该请求之后，将会调用方法启动 Executor。

向 Worker 发送 LaunchExecutor 消息的同时，通过 exec.application.driver.send (ExecutorAdded(…))向 Driver 发送 ExecutorAdded 消息，该消息为 Driver 反馈 Master 都在哪些 Worker 上启动了 Executor，Executor 的编号是多少，为每个 Executor 分配了多少个核，多大的内存以及 Worker 的联系 hostport 等消息。

Worker 收到 LaunchExecutor 消息会做相应的处理。首先判断传过来的 masterUrl 是否和 activeMasterUrl 相同，如果不相同，说明收到的不是处于 ALIVE 状态的 Master 发送过来的请求，这种情况直接打印警告信息。如果相同，则说明该请求来自 ALIVE Master，于是为 Executor 创建工作目录，创建好工作目录之后，使用 appid、execid、appDes 等参数创建 ExecutorRunner。顾名思义，ExecutorRunner 是 Executor 运行的地方，在 ExecutorRunner 中有一个工作线程，这个线程负责下载依赖的文件，并启动 CoarseGaindExecutorBackend 进程，该进程单独在一个 JVM 上运行。下面是 ExecutorRunner 中的线程启动的源码。

ExecutorRunner.scala 的源码如下：

```
1.   private[worker] def start() {
2.       //创建线程
3.      workerThread = new Thread("ExecutorRunner for " + fullId) {
4.       //线程 run 方法中调用 fetchAndRunExecutor
5.        override def run() { fetchAndRunExecutor() }
6.      }
7.       //启动线程
8.      workerThread.start()
9.
10.      //终止回调函数，用于杀死进程
11.     shutdownHook = ShutdownHookManager.addShutdownHook { () =>
12.       //这是可能的，调用 fetchAndRunExecutor 之前，state 将是 ExecutorState
          //.RUNNING。在这种情况下，我们应该设置“状态”为“失败”
13.       if (state == ExecutorState.RUNNING) {
14.         state = ExecutorState.FAILED
15.       }
16.       killProcess(Some("Worker shutting down")) }
17.   }
```

上面代码中定义了一个 Thread，这个 Thread 的 run 方法中调用 fetchAndRunExecutor 方法，fetchAndRunExecutor 负责以进程的方式启动 ApplicationDescription 中携带的 org.apache.spark.executor.CoarseGrainedExecutorBackend 进程。

其中，fetchAndRunExecutor 方法中的 CommandUtils.buildProcessBuilder(appDesc.command，传入的入口类是"org.apache.spark.executor.CoarseGrainedExecutorBackend"，当 Worker 节点中启动 ExecutorRunner 时，ExecutorRunner 中会启动 CoarseGrainedExecutorBackend 进程，在 CoarseGrainedExecutorBackend 的 onStart 方法中，向 Driver 发出 RegisterExecutor 注册请求。

CoarseGrainedExecutorBackend 的 onStart 方法的源码如下：

```
1.     override def onStart() {
2.  ......
3.      driver = Some(ref)
4.  //向 driver 发送 ask 请求，等待 driver 回应
5.      ref.ask[Boolean](RegisterExecutor(executorId, self, hostname,
        cores, extractLogUrls))
6.    ......
```

Driver 端收到注册请求，将会注册 Executor 的请求。

CoarseGrainedSchedulerBackend.scala 的 receiveAndReply 方法的源码如下：

```
1.     override def receiveAndReply(context: RpcCallContext):
       PartialFunction[Any, Unit] = {
2.
3.      case RegisterExecutor(executorId, executorRef, hostname, cores,
        logUrls) =>
4.   ......
5.        executorRef.send(RegisteredExecutor)
6.   ......
```

如上面代码所示，Driver 向 CoarseGrainedExecutorBackend 发送 RegisteredExecutor 消息，CoarseGrainedExecutorBackend 收到 RegisteredExecutor 消息后将会新建一个 Executor 执行器，并为此 Executor 充当信使，与 Driver 通信。CoarseGrainedExecutorBackend 收到 RegisteredExecutor 消息的源码如下所示。

CoarseGrainedExecutorBackend.scala 的 receive 的源码如下：

```
1.  override def receive: PartialFunction[Any, Unit] = {
2.    case RegisteredExecutor =>
3.      logInfo("Successfully registered with driver")
4.      try {
5.   //收到 RegisteredExecutor 消息，立即创建 Executor
6.        executor = new Executor(executorId, hostname, env, userClassPath,
          isLocal = false)
7.      } catch {
8.        case NonFatal(e) =>
9.          exitExecutor(1, "Unable to create executor due to " + e.getMessage, e)
10.     }
```

从上面的代码中可以看到，CoarseGrainedExecutorBackend 收到 RegisteredExecutor 消息后，将会新创建一个 org.apache.spark.executor.Executor 对象，至此 Executor 创建完毕。

6.4.2 Executor 如何把结果交给 Application

CoarseGrainedExecutorBackend 给 DriverEndpoint 发送 StatusUpdate 传输执行结果，DriverEndpoint 会把执行结果传递给 TaskSchedulerImpl 处理，然后交给 TaskResultGetter 内部通过线程分别处理 Task 执行成功和失败的不同情况，然后告诉 DAGScheduler 任务处理结束的状况。

CoarseGrainedSchedulerBackend.scala 中 DriverEndpoint 的 receive 方法的源码如下。

```
1.    override def receive: PartialFunction[Any, Unit] = {
2.      case StatusUpdate(executorId, taskId, state, data) =>
3.        scheduler.statusUpdate(taskId, state, data.value)
4.        if (TaskState.isFinished(state)) {
5.          executorDataMap.get(executorId) match {
6.            case Some(executorInfo) =>
7.              executorInfo.freeCores += scheduler.CPUS_PER_TASK
8.              makeOffers(executorId)
9.            case None =>
10.             //忽略更新，因为我们不知道 Executor
11.             logWarning(s"Ignored task status update ($taskId state $state)"+
12.               s"from unknown executor with ID $executorId")
13.         }
14.       }
```

DriverEndpoint 的 receive 方法中的 StatusUpdate 调用 scheduler.statusUpdate，然后释放资源，再次进行资源调度 makeOffers(executorId)。

TaskSchedulerImpl 的 statusUpdate 中：

- 如果是 TaskState.LOST，则记录原因，将 Executor 清理掉。
- 如果是 TaskState.isFinished，则从 taskSet 中运行的任务中清除掉，调用 taskResultGetter .enqueueSuccessfulTask 处理。
- 如果是 TaskState.FAILED、TaskState.KILLED、TaskState.LOST，调用 taskResultGetter .enqueueFailedTask 处理。

6.5 Spark 1.6 RPC 内幕解密：运行机制、源码详解、Netty 与 Akka 等

Spark 1.6 推出了以 RpcEnv、RPCEndpoint、RPCEndpointRef 为核心的新型架构下的 RPC 通信方式，就目前的实现而言，其底层依旧是 Akka。Akka 是基于 Actor 的分布式消息通信系统，而在 Spark 1.6 中封装了 Akka，提供更高层的 Rpc 实现，目的是移除对 Akka 的依赖，为扩展和自定义 Rpc 打下基础。

Spark 2.0 版本中 Rpc 的变化情况如下。

- SPARK-6280：从 Spark 中删除 Akka systemName。
- SPARK-7995：删除 AkkaRpcEnv，并从 Core 的依赖中删除 Akka。

- SPARK-7997：删除开发人员 api SparkEnv.actorSystem 和 AkkaUtils。

RpcEnv 是一个抽象类 abstract class，传入 SparkConf。RPC 环境中[RpcEndpoint]需要注册自己的名字[RpcEnv]来接收消息。[RpcEnv]将处理消息发送到[RpcEndpointRef]或远程节点，并提供给相应的[RpcEndpoint]。[RpcEnv]未被捕获的异常，[RpcEnv]将使用[RpcCallContext.sendFailure]发送异常给发送者，如果没有这样的发送者，则记录日志 NotSerializableException。

RpcEnv.scala 的源码如下：

```
1.   private[spark] abstract class RpcEnv(conf: SparkConf) {
2.
3.   private[spark] val defaultLookupTimeout = RpcUtils.lookupRpcTimeout(conf)
4.  ……
```

RpcCallContext.scala 处理异常的方法包括 reply、sendFailure、senderAddress，其中 reply 是给发送者发送一个信息。如果发送者是[RpcEndpoint]，它的[RpcEndpoint.receive]将被调用。

其中，RpcCallContext 的地址 RpcAddress 是一个 case class，包括 hostPort、toSparkURL 等成员。

RpcAddress.scala 的源码如下：

```
1.   private[spark] case class RpcAddress(host: String, port: Int) {
2.   def hostPort: String = host + ":" + port
3.   /**返回一个字符串，该字符串的形式为：spark://host:port*/
4.   def toSparkURL: String = "spark://" + hostPort
5.   override def toString: String = hostPort
6.  }
```

RpcAddress 伴生对象 object RpcAddress 属于包 org.apache.spark.rpc，fromURIString 方法从 String 中提取出 RpcAddress；fromSparkURL 方法也是从 String 中提取出 RpcAddress。说明：case class RpcAddress 通过伴生对象 object RpcAddress 的方法调用，case class RpcAddress 也有自己的方法 fromURIString、fromSparkURL，而且方法 fromURIString、fromSparkURL 的返回值也是 RpcAddress。

伴生对象 RpcAddress 的源码如下：

```
1.  private[spark] object RpcAddress {
2.   /**返回[RpcAddress]为代表的 uri */
3.   def fromURIString(uri: String): RpcAddress = {
4.     val uriObj = new java.net.URI(uri)
5.     RpcAddress(uriObj.getHost, uriObj.getPort)
6.   }
7.   /**返回[RpcAddress]，编码的形式：spark://host:port */
8.   def fromSparkURL(sparkUrl: String): RpcAddress = {
9.     val (host, port) = Utils.extractHostPortFromSparkUrl(sparkUrl)
10.    RpcAddress(host, port)
11.  }
12. }
```

RpcEnv 解析如下。

（1）RpcEnv 是 RPC 的环境（相当于 Akka 中的 ActorSystem），所有的 RPCEndpoint 都需要注册到 RpcEnv 实例对象中（注册的时候会指定注册的名称，这样客户端就可以通过名称查询到 RpcEndpoint 的 RpcEndpointRef 引用，从而进行通信），在 RpcEndpoint 接收到消息后会调用 receive 方法进行处理。

（2）RpcEndpoint 如果接收到需要 reply 的消息，就会交给自己的 receiveAndReply 来处理（回复时是通过 RpcCallContext 中的 reply 方法来回复发送者的），如果不需要 reply，就交给 receive 方法来处理。

（3）RpcEnvFactory 是负责创建 RpcEnv 的，通过 create 方法创建 RpcEnv 实例对象，默认用 Netty。

RpcEnv 示意图如图 6-4 所示。

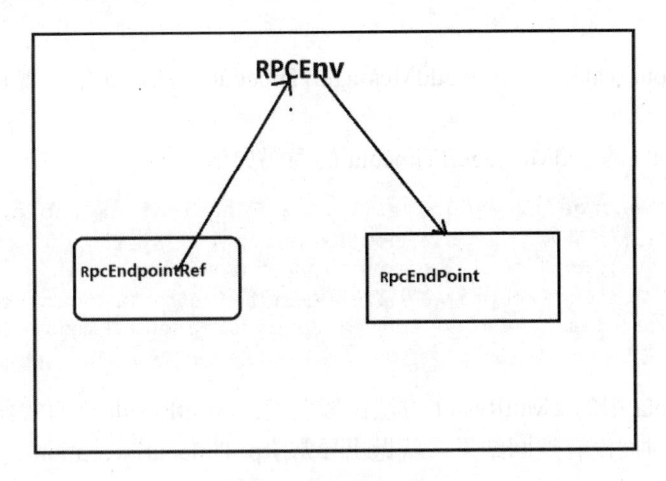

图 6-4　RPCEnv 示意图

回到 RpcEnv.scala 的源码，首先调用 RpcUtils.lookupRpcTimeout(conf)，返回 RPC 远程端点查找时默认 Spark 的超时时间。方法 lookupRpcTimeout 中构建了一个 RpcTimeout，定义 spark.rpc.lookupTimeout。spark.network.timeout 的超时时间是 120s。

RpcUtils.scala 的 lookupRpcTimeout 方法的源码如下：

```
1.    def lookupRpcTimeout(conf: SparkConf): RpcTimeout = {
2.      RpcTimeout(conf, Seq("spark.rpc.lookupTimeout", "spark.network
        .timeout"), "120s")
3.    }
```

进入 RpcTimeout，进行 RpcTimeout 关联超时的原因描述，当 TimeoutException 发生的时候，关于超时的额外的上下文将包含在异常消息中。

RpcTimeout.scala 的源码如下：

```
1.    private[spark] class RpcTimeout(val duration: FiniteDuration, val
      timeoutProp: String)
2.    extends Serializable {
3.
4.    /**修正 TimeoutException 标准的消息包括描述 */
5.    private def createRpcTimeoutException(te: TimeoutException):
      RpcTimeoutException = {
6.      new RpcTimeoutException(te.getMessage + ". This timeout is controlled
        by " + timeoutProp, te)
7.    }
```

其中的 RpcTimeoutException 继承自 TimeoutException。

```
1.    private[rpc] class RpcTimeoutException(message: String, cause:
      TimeoutException)
2.   extends TimeoutException(message) { initCause(cause) }
```

其中的 TimeoutException 继承自 Exception。

```
1.    public class TimeoutException extends Exception {
2.  ……
3.     public TimeoutException(String message) {
4.         super(message);
5.     }
6.  }
```

回到 RpcTimeout.scala，其中的 addMessageIfTimeout 方法，如果出现超时，将加入这些信息。

RpcTimeout.scala 的 addMessageIfTimeout 的源码如下：

```
1.   def addMessageIfTimeout[T]: PartialFunction[Throwable, T] = {
2.     //异常已被转换为一个 RpcTimeoutException，就抛出它
3.     case rte: RpcTimeoutException => throw rte
4.     //其他 TimeoutException 异常转换为修改的消息 RpcTimeoutException
5.     case te: TimeoutException => throw createRpcTimeoutException(te)
6.   }
```

RpcTimeout.scala 中的 awaitResult 方法比较关键：awaitResult 一直等结果完成并获得结果，如果在指定的时间没有返回结果，就抛出异常[RpcTimeoutException]。

RpcTimeout.scala 的源码如下：

```
1.    def awaitResult[T](future: Future[T]): T = {
2.     try {
3.       ThreadUtils.awaitResult(future, duration)
4.     } catch addMessageIfTimeout
5.   }
6.  }
```

其中的 future 是 Future[T]类型，继承自 Awaitable。

```
1.  trait Future[+T] extends Awaitable[T]
```

Awaitable 是一个 trait，其中的 ready 方法是指 Duration 时间片内，Awaitable 的状态变成 completed 状态，就是 ready。在 Await.result 中，result 本身是阻塞的。

Awaitable.scala 的源码如下：

```
1.   trait Awaitable[+T] {
2.  ……
3.  def ready(atMost: Duration)(implicit permit: CanAwait): this.type
4.  ……
5.   @throws(classOf[Exception])
6.   def result(atMost: Duration)(implicit permit: CanAwait): T
7.  }
8.
```

回到 RpcEnv.scala 中，其中 endpointRef 方法返回我们注册的 RpcEndpoint 的引用，是代理的模式。我们要使用 RpcEndpoint，是通过 RpcEndpointRef 来使用的。Address 方法是 RpcEnv 监听的地址；setupEndpoint 方法注册时根据 RpcEndpoint 名称返回 RpcEndpointRef。fileServer 返回用于服务文件的文件服务器实例。如果 RpcEnv 不以服务器模式运行，可能是 null 值。

RpcEnv.scala 的源码如下：

```
1.   private[spark] abstract class RpcEnv(conf: SparkConf) {
2.
3.    private[spark] val defaultLookupTimeout = RpcUtils.lookupRpcTimeout
     (conf)
4.  ......
5.    private[rpc] def endpointRef(endpoint: RpcEndpoint): RpcEndpointRef
6.  def address: RpcAddress
7.  def setupEndpoint(name: String, endpoint: RpcEndpoint): RpcEndpointRef
8.  ......
9.    def fileServer: RpcEnvFileServer
10. ......
```

RpcEnv.scala 中的 RpcEnvFileServer 方法中的 RpcEnvConfig 是一个 case class。Spark 2.2.1 版本的 RpcEnvFileServer 的源码如下：

```
1.    private[spark] trait RpcEnvFileServer {
2.    def addFile(file: File): String
3.  ......
4.  private[spark] case class RpcEnvConfig(
5.      conf: SparkConf,
6.      name: String,
7.      bindAddress: String,
8.      advertiseAddress: String,
9.      port: Int,
10.     securityManager: SecurityManager,
11.     clientMode: Boolean)
```

Spark 2.4.3 版本的 RpcEnv.scala 源码与 Spark 2.2.1 版本相比具有如下特点。

- 上段代码中第 10 行新增加一个参数 numUsableCores。

```
1.        ......
2.  numUsableCores: Int,
3.  ......
```

RpcEnv 是一个抽象类，其具体的子类是 NettyRpcEnv。Spark 1.6 版本中包括 AkkaRpcEnv 和 NettyRpcEnv 两种方式。Spark 2.0 版本中只有 NettyRpcEnv。

下面看一下 RpcEnvFactory。RpcEnvFactory 是一个工厂类，创建[RpcEnv]，必须有一个空构造函数，以便可以使用反射创建。create 根据具体的配置，反射出具体的实例对象。RpcEndpoint 方法中定义了 receiveAndReply 方法和 receive 方法。

RpcEndpoint.scala 的源码如下：

```
1.   private[spark] trait RpcEnvFactory {
2.
3.    def create(config: RpcEnvConfig): RpcEnv
4.  }
5.  private[spark] trait RpcEndpoint {
6.  ......
7.    val rpcEnv: RpcEnv
8.
9.    ......
10.   final def self: RpcEndpointRef = {
11.     require(rpcEnv != null, "rpcEnv has not been initialized")
12.     rpcEnv.endpointRef(this)
13.   }
```

```
14. ……
15.
16.   def receive: PartialFunction[Any, Unit] = {
17.     case _ => throw new SparkException(self + " does not implement
        'receive'")
18.   }
19. ……
20.   def  receiveAndReply(context:  RpcCallContext):  PartialFunction[Any,
      Unit] = {
21.     case _ => context.sendFailure(new SparkException(self + " won't reply
        anything"))
22.   }
23. ……
```

Master 继承自 ThreadSafeRpcEndpoint，接收消息使用 receive 方法和 receiveAndReply 方法。

其中，ThreadSafeRpcEndpoint 继承自 RpcEndpoint：ThreadSafeRpcEndpoint 是一个 trait，需要 RpcEnv 线程安全地发送消息给它。线程安全是指在处理下一个消息之前通过同样的[ThreadSafeRpcEndpoint]处理一条消息。换句话说，改变[ThreadSafeRpcEndpoint]的内部字段在处理下一个消息是可见的，[ThreadSafeRpcEndpoint]的字段不需要 volatile 或 equivalent，不能保证对于不同的消息在相同的[ThreadSafeRpcEndpoint]线程中来处理。

```
1.      private[spark] trait ThreadSafeRpcEndpoint extends RpcEndpoint
```

回到 RpcEndpoint.scala，重点看一下 receiveAndReply 方法和 receive 方法。receive 方法处理从[RpcEndpointRef.send]或者[RpcCallContext.reply]发过来的消息，如果收到一个不匹配的消息，[SparkException]会抛出一个异常 onError。receiveAndReply 方法处理从[RpcEndpointRef.ask]发过来的消息，如果收到一个不匹配的消息，[SparkException]会抛出一个异常 onError。receiveAndReply 方法返回 PartialFunction 对象。

RpcEndpoint.scala 的源码如下：

```
1.      def receive: PartialFunction[Any, Unit] = {
2.     case _ => throw new SparkException(self + " does not implement
       'receive'")
3.    }
4.
5.   ……
6.    def  receiveAndReply(context:  RpcCallContext):  PartialFunction[Any,
      Unit] = {
7.     case _ => context.sendFailure(new SparkException(self + " won't reply
       anything"))
8.    }
```

在 Master 中，Receive 方法中收到消息以后，不需要回复对方。

Master.scala 的 Receive 方法的源码如下：

```
1.     override def receive: PartialFunction[Any, Unit] = {
2.      case ElectedLeader =>
3.      ……
4.    recoveryCompletionTask = forwardMessageThread.schedule(new Runnable {
5.           override def run(): Unit = Utils.tryLogNonFatalError {
6.             self.send(CompleteRecovery)
7.           }
8.         }, WORKER_TIMEOUT_MS, TimeUnit.MILLISECONDS)
9.       }
10.
```

```
11.     case CompleteRecovery => completeRecovery()
12.
13.
14.     case RevokedLeadership =>
15.       logError("Leadership has been revoked -- master shutting down.")
16.       System.exit(0)
17.
18.     case RegisterApplication(description, driver) =>
19.      ......
20.         schedule()
21.
```

在 Master 中，receiveAndReply 方法中收到消息以后，都要通过 context.reply 回复对方。

在 Master 中，RpcEndpoint 如果接收到需要 reply 的消息，就会交给自己的 receiveAndReply 来处理（回复时是通过 RpcCallContext 中的 reply 方法来回复发送者的），如果不需要 reply，就交给 receive 方法来处理。

RpcCallContext 的源码如下：

```
1.  private[spark] trait RpcCallContext {
2.
3.   /**
      *回复消息的发送者。如果发送者是[RpcEndpoint]，其[RpcEndpoint.receive]
      *将被调用
5.    */
6.   def reply(response: Any): Unit
7.
8.   /**
      *向发送方报告故障
9.    */
10.  def sendFailure(e: Throwable): Unit
11.
12.  /**
      *此消息的发送者
13.   */
14.  def senderAddress: RpcAddress
15. }
```

回到 RpcEndpoint.scala，RpcEnvFactory 是一个 trait，负责创建 RpcEnv，通过 create 方法创建 RpcEnv 实例对象，默认用 Netty。

RpcEndpoint.scala 的源码如下：

```
1.   private[spark] trait RpcEnvFactory {
2.
3.    def create(config: RpcEnvConfig): RpcEnv
4.   }
```

RpcEnvFactory 的 create 方法没有具体的实现。下面看一下 RpcEnvFactory 子类 NettyRpcEnvFactory 中 create 的具体实现，使用的方式为 nettyEnv。

Spark 2.2.1 版本的 NettyRpcEnv.scala 的 create 方法的源码如下：

```
1.    def create(config: RpcEnvConfig): RpcEnv = {
2.      val sparkConf = config.conf
3.      //在多个线程中使用 JavaSerializerInstance 是安全的。然而，如果将来计划支持
        //KryoSerializer，必须使用 ThreadLocal 来存储 SerializerInstance
4.
```

```
5.      val javaSerializerInstance =
6.        new JavaSerializer(sparkConf).newInstance().asInstanceOf
          [JavaSerializerInstance]
7.      val nettyEnv =
8.        new NettyRpcEnv(sparkConf, javaSerializerInstance,
          config.advertiseAddress,
9.          config.securityManager)
10.     if (!config.clientMode) {
11.       val startNettyRpcEnv: Int => (NettyRpcEnv, Int) = { actualPort =>
12.         nettyEnv.startServer(config.bindAddress, actualPort)
13.         (nettyEnv, nettyEnv.address.port)
14.       }
15.       try {
16.         Utils.startServiceOnPort(config.port, startNettyRpcEnv, sparkConf,
            config.name)._1
17.       } catch {
18.         case NonFatal(e) =>
19.           nettyEnv.shutdown()
20.           throw e
21.       }
22.     }
23.     nettyEnv
24.   }
25. }
```

Spark 2.4.3 版本的 NettyRpcEnv.scala 源码与 Spark 2.2.1 版本相比具有如下特点。

- 上段代码中第 9 行新增加一个参数 config.numUsableCores。

```
1.   ......
2.      config.securityManager, config.numUsableCores)
3.   ......
```

在 Spark 2.0 版本中回溯一下 NettyRpcEnv 的实例化过程。在 SparkContext 实例化时调用 createSparkEnv 方法。

Spark 2.2.1 版本的 SparkContext.scala 的源码如下：

```
1.   ......
2.   _env = createSparkEnv(_conf, isLocal, listenerBus)
3.      SparkEnv.set(_env)
4.   ......
5.
6.    private[spark] def createSparkEnv(
7.        conf: SparkConf,
8.        isLocal: Boolean,
9.        listenerBus: LiveListenerBus): SparkEnv = {
10.     SparkEnv.createDriverEnv(conf, isLocal, listenerBus, SparkContext
        .numDriverCores(master))
11.   }
12.
13. ......
```

Spark 2.4.3 版本的 SparkContext.scala 源码与 Spark 2.2.1 版本相比具有如下特点。

- 上段代码中第 10 行 SparkContext.numDriverCores 新增加一个参数 conf。

```
1.   ......
2.      SparkEnv.createDriverEnv(conf, isLocal, listenerBus, SparkContext.
        numDriverCores(master, conf))
3.   ......
```

SparkContext 的 createSparkEnv 方法中调用了 SparkEnv.createDriverEnv 方法。下面看一下 createDriverEnv 方法的实现，其调用了 create 方法。

Spark 2.2.1 版本的 SparkEnv.scala 的 createDriverEnv 的源码如下：

```
1.   private[spark] def createDriverEnv(
2.     ……
3.     create(
4.       conf,
5.       SparkContext.DRIVER_IDENTIFIER,
6.       bindAddress,
7.       advertiseAddress,
8.       port,
9.       isLocal,
10.      numCores,
11.      ioEncryptionKey,
12.      listenerBus = listenerBus,
13.      mockOutputCommitCoordinator = mockOutputCommitCoordinator
14.    )
15.  }
16.
17. private def create(
18.    ……
19.    val rpcEnv = RpcEnv.create(systemName, bindAddress, advertiseAddress,
       port, conf,
20.      securityManager, clientMode = !isDriver)
21. ……
```

Spark 2.4.3 版本的 SparkEnv.scala 源码与 Spark 2.2.1 版本相比具有如下特点。

- 上段代码中第 8 行 port 参数调整为 Option(port)。

```
1.  ……
2.    Option(port),
3.  ……
```

在 RpcEnv.scala 中，creat 方法直接调用 new()函数创建一个 NettyRpcEnvFactory，调用 NettyRpcEnvFactory().create 方法，NettyRpcEnvFactory 继承自 RpcEnvFactory。在 Spark 2.0 中，RpcEnvFactory 直接使用 NettyRpcEnvFactory 的方式。

Spark 2.2.1 版本的 RpcEnv.scala 的源码如下：

```
1.   private[spark] object RpcEnv {
2.   ……
3.
4.    def create(
5.        name: String,
6.        bindAddress: String,
7.        advertiseAddress: String,
8.        port: Int,
9.        conf: SparkConf,
10.       securityManager: SecurityManager,
11.       clientMode: Boolean): RpcEnv = {
12.     val config = RpcEnvConfig(conf, name, bindAddress, advertiseAddress,
        port, securityManager,
13.       clientMode)
14.     new NettyRpcEnvFactory().create(config)
15.   }
```

Spark 2.4.3 版本的 RpcEnv.scala 源码与 Spark 2.2.1 版本相比具有如下特点。

- 上段代码中第 10 行之后新增一个参数 numUsableCores。
- 上段代码中第 13 行新增一个参数 numUsableCores。

```
1.  ……
2.    numUsableCores: Int,
3.  ……
4.     numUsableCores,
```

NettyRpcEnvFactory().create 的方法如下。

Spark 2.2.1 版本的 NettyRpcEnv.scala 的源码如下：

```
1.   private[rpc] class NettyRpcEnvFactory extends RpcEnvFactory with
     Logging {
2.
3.   def create(config: RpcEnvConfig): RpcEnv = {
4.   ……
5.     val nettyEnv =
6.       new NettyRpcEnv(sparkConf, javaSerializerInstance,
       config.advertiseAddress,
7.         config.securityManager)
8.     if (!config.clientMode) {
9.       val startNettyRpcEnv: Int => (NettyRpcEnv, Int) = { actualPort =>
10.        nettyEnv.startServer(config.bindAddress, actualPort)
11.        (nettyEnv, nettyEnv.address.port)
12.      }
13.      try {
14.        Utils.startServiceOnPort(config.port, startNettyRpcEnv,
         sparkConf, config.name)._1
15.      } catch {
16.        case NonFatal(e) =>
17.          nettyEnv.shutdown()
18.          throw e
19.      }
20.    }
21.    nettyEnv
22.  }
23. }
```

Spark 2.4.3 版本的 NettyRpcEnv.scala 源码与 Spark 2.2.1 版本相比具有如下特点。

- 上段代码中第 7 行新增加一个参数 config.numUsableCores。

```
1.  ……
2.    config.securityManager, config.numUsableCores)
3.  ……
```

在 NettyRpcEnvFactory().create 中调用 new()函数创建一个 NettyRpcEnv。NettyRpcEnv 传入 SparkConf 参数，包括 fileServer、startServer 等方法。

Spark 2.2.1 版本的 NettyRpcEnv.scala 的源码如下：

```
1.  private[netty] class NettyRpcEnv(
2.     val conf: SparkConf,
3.     javaSerializerInstance: JavaSerializerInstance,
4.     host: String,
5.     securityManager: SecurityManager) extends RpcEnv(conf) with Logging {
6.
7.  ……
8.   override def fileServer: RpcEnvFileServer = streamManager
```

```
9.   ......
10.   def startServer(bindAddress: String, port: Int): Unit = {
11.     val bootstraps: java.util.List[TransportServerBootstrap] =
12.       if (securityManager.isAuthenticationEnabled()) {
13.         java.util.Arrays.asList(new  SaslServerBootstrap(transportConf,
          securityManager))
14.       } else {
15.         java.util.Collections.emptyList()
16.       }
17.     server = transportContext.createServer(bindAddress, port, bootstraps)
18.     dispatcher.registerRpcEndpoint(
19.       RpcEndpointVerifier.NAME, new RpcEndpointVerifier(this, dispatcher))
20.   }
```

Spark 2.4.3 版本的 NettyRpcEnv.scala 源码与 Spark 2.2.1 版本相比具有如下特点。

- 上段代码中第 5 行新增加一个参数 numUsableCores。

```
1.   ......
2.        securityManager: SecurityManager,
3.     numUsableCores: Int) extends RpcEnv(conf) with Logging {
4.   ......
```

NettyRpcEnv.scala 的 startServer 中，通过 transportContext.createServer 创建 Server，使用 dispatcher.registerRpcEndpoint 方法 dispatcher 注册 RpcEndpoint。在 createServer 方法中调用 new()函数创建一个 TransportServer。

TransportContext 的 createServer 方法的源码如下：

```
1.    public TransportServer createServer(
2.        int port, List<TransportServerBootstrap> bootstraps) {
3.      return new TransportServer(this, null, port, rpcHandler, bootstraps);
4.    }
```

Spark 2.2.1 版本的 TransportServer.java 的源码如下：

```
1.  public TransportServer(
2.        TransportContext context,
3.        String hostToBind,
4.        int portToBind,
5.        RpcHandler appRpcHandler,
6.        List<TransportServerBootstrap> bootstraps) {
7.      this.context = context;
8.      this.conf = context.getConf();
9.      this.appRpcHandler = appRpcHandler;
10.     this.bootstraps = Lists.newArrayList(Preconditions.checkNotNull
        (bootstraps));
11.
12.     try {
13.       init(hostToBind, portToBind);
14.     } catch (RuntimeException e) {
15.       JavaUtils.closeQuietly(this);
16.       throw e;
17.     }
18.   }
```

Spark 2.4.3 版本的 TransportServer.scala 源码与 Spark 2.2.1 版本相比具有如下特点。

- 上段代码中第 10 行之后新增一行代码，增加一个布尔值变量 shouldClose。
- 上段代码中第 13 行之后新增一行代码，shouldClose = false。

- 上段代码中第14～16行调整为finally的异常处理代码。

```
1.  ......
2.  boolean shouldClose = true;
3.  ......
4.  shouldClose = false;
5.  ......
6.      } finally {
7.        if (shouldClose) {
8.          JavaUtils.closeQuietly(this);
9.        }
```

TransportServer.java中的关键方法是init，这是Netty本身的实现内容。

TransportServer.java中的init的源码如下：

```
1.         private void init(String hostToBind, int portToBind) {
2.
3.      IOMode ioMode = IOMode.valueOf(conf.ioMode());
4.      EventLoopGroup bossGroup =
5.        NettyUtils.createEventLoop(ioMode, conf.serverThreads(),
          conf.getModuleName() + "-server");
6.      EventLoopGroup workerGroup = bossGroup;
7.  ......
```

接下来，我们看一下RpcEndpointRef。RpcEndpointRef是一个抽象类，是代理模式。

RpcEndpointRef.scala的源码如下：

```
1.   private[spark] abstract class RpcEndpointRef(conf: SparkConf)
2.    extends Serializable with Logging {
3.
4.    private[this] val maxRetries = RpcUtils.numRetries(conf)
5.    private[this] val retryWaitMs = RpcUtils.retryWaitMs(conf)
6.    private[this] val defaultAskTimeout = RpcUtils.askRpcTimeout(conf)
7.  ......
8.  def send(message: Any): Unit
9.  def ask[T: ClassTag](message: Any, timeout: RpcTimeout): Future[T]
10. ......
```

NettyRpcEndpointRef是RpcEndpointRef的具体实现子类。ask 方法通过调用nettyEnv.ask传递消息。RequestMessage是一个case class。

NettyRpcEnv.scala的NettyRpcEndpointRef的源码如下：

```
1.   private[netty] class NettyRpcEndpointRef(
2.      @transient private val conf: SparkConf,
3.      private val endpointAddress: RpcEndpointAddress,
4.      @transient @volatile private var nettyEnv: NettyRpcEnv) extends
        RpcEndpointRef(conf) {
5.  ......
6.    override def ask[T: ClassTag](message: Any, timeout: RpcTimeout):
      Future[T] = {
7.      nettyEnv.ask(new RequestMessage(nettyEnv.address, this, message),
        timeout)
8.    }
```

下面从实例的角度来看RPC的应用。

RpcEndpoint的生命周期：构造（constructor）-> 启动（onStart）、消息接收（receive、receiveAndReply）、停止（onStop）。

Master 中接收消息的方式有两种：① receive 接收消息不回复；② receiveAndReply 通过 context.reply 的方式回复消息。例如，Worker 发送 Master 的 RegisterWorker 消息，当 Master 注册成功，Master 就返回 Worker RegisteredWorker 消息。

Worker 启动时，从生命周期的角度，Worker 实例化的时候提交 Master 进行注册。

Worker 的 onStart 的源码如下：

```
1.     override def onStart() {
2.    ......
3.     registerWithMaster()
4.
5.     metricsSystem.registerSource(workerSource)
6.     metricsSystem.start()
7.     //在度量系统启动后,将 Worker Metrics Servlet 处理程序附加到 Web UI
8.     metricsSystem.getServletHandlers.foreach(webUi.attachHandler)
9.    }
```

进入 registerWithMaster 方法。

Worker 的 registerWithMaster 的源码如下：

```
1.   private def registerWithMaster() {
2.      ......
3.         registerMasterFutures = tryRegisterAllMasters()
4.      ......
```

进入 tryRegisterAllMasters 方法：在 rpcEnv.setupEndpointRef 中根据 masterAddress、ENDPOINT_NAME 名称获取 RpcEndpointRef。

Worker.scala 的 tryRegisterAllMasters 的源码如下：

```
1.    private def tryRegisterAllMasters(): Array[JFuture[_]] = {
2.     ......
3.            val masterEndpoint = rpcEnv.setupEndpointRef(masterAddress,
              Master.ENDPOINT_NAME)
4.            sendRegisterMessageToMaster(masterEndpoint)
5.        ......
```

基于 masterEndpoint，使用 sendRegisterMessageToMaster 方法注册。

Worker.scala 的 sendRegisterMessageToMaster 的源码如下：

```
1.     private def sendRegisterMessageToMaster(masterEndpoint:
       RpcEndpointRef): Unit = {
2.      masterEndpoint.send(RegisterWorker(
3.        workerId,
4.        host,
5.        port,
6.        self,
7.        cores,
8.        memory,
9.        workerWebUiUrl,
10.       masterEndpoint.address))
11.   }
```

sendRegisterMessageToMaster 方法中的 Worker 发送 RegisterWorker 消息给 Master 以后，

就完成此次注册。Master 节点收到 RegisterWorker 消息另行处理，如果注册成功，Master 就发送 Worker 节点成功的 RegisteredWorker 消息；如果注册失败，Master 就发送 Worker 节点失败的 RegisterWorkerFailed 消息。

Worker.scala 的 handleRegisterResponse 源码如下：

```
1.   override def receive: PartialFunction[Any, Unit] = synchronized {
2.     case msg: RegisterWorkerResponse =>
3.       handleRegisterResponse(msg)
4.  ……
5.
6.   private def handleRegisterResponse(msg: RegisterWorkerResponse): Unit =
     synchronized {
7.     msg match {
8.       case RegisteredWorker(masterRef, masterWebUiUrl, masterAddress) =>
9.         if (preferConfiguredMasterAddress) {
10.          logInfo("Successfully registered with master " +
           masterAddress.toSparkURL)
11.        } else {
12.          logInfo("Successfully registered with master " + masterRef
           .address.toSparkURL)
13.        }
14.        registered = true
15.        changeMaster(masterRef, masterWebUiUrl, masterAddress)
16.        forwordMessageScheduler.scheduleAtFixedRate(new Runnable {
17.          override def run(): Unit = Utils.tryLogNonFatalError {
18.            self.send(SendHeartbeat)
19.          }
20.        }, 0, HEARTBEAT_MILLIS, TimeUnit.MILLISECONDS)
21.        if (CLEANUP_ENABLED) {
22.          logInfo(
23.            s"Worker cleanup enabled; old application directories will be
             deleted in: $workDir")
24.          forwordMessageScheduler.scheduleAtFixedRate(new Runnable {
25.            override def run(): Unit = Utils.tryLogNonFatalError {
26.              self.send(WorkDirCleanup)
27.            }
28.          }, CLEANUP_INTERVAL_MILLIS, CLEANUP_INTERVAL_MILLIS,
           TimeUnit.MILLISECONDS)
29.        }
30.
31.        val execs = executors.values.map { e =>
32.          new ExecutorDescription(e.appId, e.execId, e.cores, e.state)
33.        }
34.        masterRef.send(WorkerLatestState(workerId, execs.toList,
         drivers.keys.toSeq))
35.
36.      case RegisterWorkerFailed(message) =>
37.        if (!registered) {
38.          logError("Worker registration failed: " + message)
39.          System.exit(1)
40.        }
41.
42.      case MasterInStandby =>
43.        //忽略，Master 仍未准备好
44.    }
45.  }
```

6.6　本章总结

本章讲解了 Spark Application 提交给集群的原理和源码，主要内容包括 Spark Application 到底是如何提交给集群的？Spark Application 是如何向集群申请资源的？从 Application 提交的角度重新审视 Driver，Driver 到底是什么时候产生的以及 Driver 和 Master 交互通信。同时，本章也从 Application 提交的角度重新审视 Executor，Executor 到底是什么时候启动的？以及 Executor 如何把结果交给 Application。最后，本章还讲解了 Spark 1.6 RPC 内幕解密：运行机制、源码详解、Netty 与 Akka 等内容。

第7章　Shuffle原理和源码详解

本章对Shuffle原理和源码进行剖析。7.1节讲解MapReduce的Shuffle架构；7.2节讲解Shuffle的框架、框架内核、Shuffle数据读写的源码解析；7.3节讲解Hash Based Shuffle内核及Hash Based Shuffle数据读写的源码解析；7.4节讲解Sorted Based Shuffle内核及数据读写的源码解析；7.5节讲解Tungsten Sorted Based内核及Tungsten Sorted Based数据读写的源码解析；7.6节讲解Shuffle与Storage模块间的交互以及Shuffle注册的交互，Shuffle读写数据的交互，并讲解BlockManager架构原理及源码，解密BlockManager进阶的相关内容。

7.1　概　　述

在MapReduce框架中，Shuffle阶段是连接Map和Reduce之间的桥梁，Map阶段通过Shuffle过程将数据输出到Reduce阶段中。由于Shuffle涉及磁盘的读写和网络I/O，因此Shuffle性能的高低直接影响整个程序的性能。Spark本质上也是一种MapReduce框架，因此也会有自己的Shuffle过程实现。

在学习Shuffle的过程中，通常都会引用HadoopMapReduce框架中的Shuffle过程作为入门或比较，同时也会引用在Hadoop MapReduce框架中的Shuffle过程中常用的术语。下面是网络上描述该过程的经典的框架图，如图7-1所示。

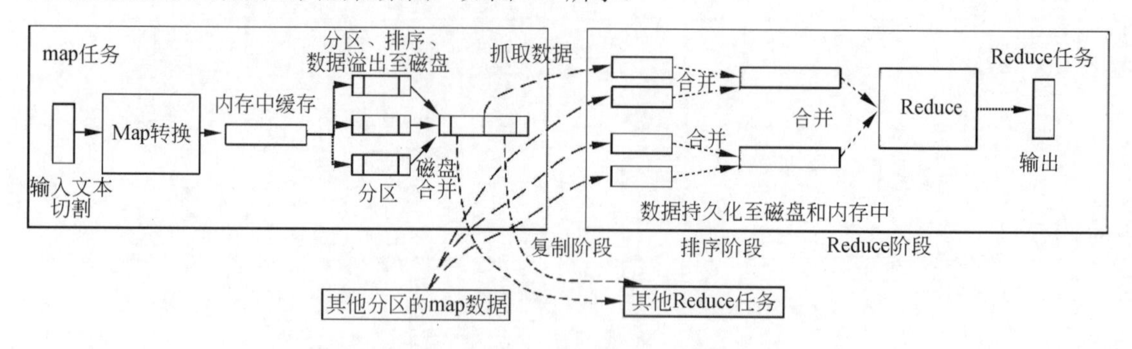

图7-1　Hadoop MapReduce框架中的Shuffle框架

其中，Shuffle是MapReduce框架中的一个特定的阶段，介于Map阶段和Reduce阶段之间。Map阶段负责准备数据，Reduce阶段则读取Map阶段准备的数据，然后进一步对数据进行处理，即Map阶段实现Shuffle过程中的数据持久化（即数据写），而Reduce阶段实现Shuffle过程中的数据读取。

在图7-1中，Mapper端与Reduce端之间的数据交互通常都伴随着一定的网络I/O，因此对应数据的序列化与压缩等技术也是Shuffle中必不可少的一部分。可以根据特定场景选取合适的序列化方式与压缩算法进行调优，在此仅给出相关的配置属性的简单描述，具体见表7-1所示。

表 7-1　Shuffle序列化方式与压缩算法的配置属性

配置属性	默认值	描述
spark.serializer	org.apache.spark.serializer.JavaSerializer(当使用 Spark SQL Thrift Server 时，默认为 org.apache.spark.serializer.KryoSerializer)	序列化器，当需要在网络中传输或以序列化方式缓存时，用于序列化对象所需的类
spark.shuffle.compress	true	是否对 Map 端输出文件进行压缩。为了减少网络 I/O 等，通常会使用压缩。对应的压缩算法由 spark.io.compression.codec 指定
spark.shuffle.spill.compress	true	在数据 Spill 过程中是否进行压缩的控制。对应的压缩算法由 spark.io.compression.codec 指定
spark.io.compression.codec	Snappy（Spark 2.0 版本默认是 lz4 压缩格式 DEFAULT_COMPRESSION_CODEC = "lz4"）	该 codec 用于压缩内部数据，如 RDD 分区数据、广播变量的数据以及 Shuffle 的输出数据。默认情况下，Spark 提供 3 种 codecs：lz4、lzf 和 snappy。指定时也可以指定完整类名。如：org.apache.spark.io.LZ4CompressionCodec、org.apache.spark.io.LZFCompressionCodec 和 org.apache.spark.io.SnappyCompressionCodec

可以基于图 7-1，抽象地去理解 Shuffle 阶段的实现过程，但在具体实现上，不同的 Shuffle 设计会有不同的实现细节。

7.2　Shuffle 的框架

本节讲解 Shuffle 的框架、Shuffle 的框架内核、Shuffle 数据读写的源码解析。Spark Shuffle 从基于 Hash 的 Shuffle，引入了 Shuffle Consolidate 机制（即文件合并机制），演进到基于 Sort 的 Shuffle 实现方式。随着 Tungsten 计划的引入与优化，引入了基于 Tungsten-Sort 的 Shuffle 实现方式。

7.2.1　Shuffle 的框架演进

Spark 的 Shuffle 框架演进历史可以从框架本身的演进、Shuffle 具体实现机制的演进两部分进行解析。

框架本身的演进可以从面向接口编程的原则出发，结合 Build 设计模式进行理解。整个 Spark 的 Shuffle 框架从 Spark 1.1 版本开始，提供便于测试、扩展的可插拔式框架。

而对应 Shuffle 的具体实现机制的演进部分，可以跟踪 Shuffle 实现细节在各个版本中的变更。具体体现在 Shuffle 数据的写入或读取，以及读写相关的数据块解析方式。下面简单描述一下整个演进过程。

在 Spark 1.1 之前，Spark 中只实现了一种 Shuffle 方式，即基于 Hash 的 Shuffle。在基于 Hash 的 Shuffle 的实现方式中，每个 Mapper 阶段的 Task 都会为每个 Reduce 阶段的 Task 生成一个文件，通常会产生大量的文件（即对应为 M×R 个中间文件，其中，M 表示 Mapper

阶段的 Task 个数，R 表示 Reduce 阶段的 Task 个数）。伴随大量的随机磁盘 I/O 操作与大量的内存开销。

为了缓解上述问题，在 Spark 0.8.1 版本中为基于 Hash 的 Shuffle 的实现引入了 Shuffle Consolidate 机制（即文件合并机制），将 Mapper 端生成的中间文件进行合并的处理机制。通过将配置属性 spark.shuffle.consolidateFiles 设置为 true，减少中间生成的文件数量。通过文件合并，可以将中间文件的生成方式修改为每个执行单位（类似于 Hadoop 的 Slot）为每个 Reduce 阶段的 Task 生成一个文件。其中，执行单位对应为：每个 Mapper 阶段的 Cores 数/每个 Task 分配的 Cores 数（默认为 1）。最终可以将文件个数从 M×R 修改为 E×C/T×R，其中，E 表示 Executors 个数，C 表示可用 Cores 个数，T 表示 Task 分配的 Cores 个数。

基于 Hash 的 Shuffle 的实现方式中，生成的中间结果文件的个数都会依赖于 Reduce 阶段的 Task 个数，即 Reduce 端的并行度，因此文件数仍然不可控，无法真正解决问题。为了更好地解决问题，在 Spark 1.1 版本引入了基于 Sort 的 Shuffle 实现方式，并且在 Spark 1.2 版本之后，默认的实现方式也从基于 Hash 的 Shuffle，修改为基于 Sort 的 Shuffle 实现方式，即使用的 ShuffleManager 从默认的 hash 修改为 sort。首先，每个 Mapper 阶段的 Task 不会为每个 Reduce 阶段的 Task 生成一个单独的文件；而是全部写到一个数据（Data）文件中，同时生成一个索引（Index）文件，Reduce 阶段的各个 Task 可以通过该索引文件获取相关的数据。避免产生大量文件的直接收益就是降低随机磁盘 I/O 与内存的开销。最终生成的文件个数减少到 2M，其中 M 表示 Mapper 阶段的 Task 个数，每个 Mapper 阶段的 Task 分别生成两个文件（1 个数据文件、1 个索引文件），最终的文件个数为 M 个数据文件与 M 个索引文件。因此，最终文件个数是 2×M 个。

随着 Tungsten 计划的引入与优化，从 Spark 1.4 版本开始（Tungsten 计划目前在 Spark 1.5 与 Spark 1.6 两个版本中分别实现了第一与第二两个阶段），在 Shuffle 过程中也引入了基于 Tungsten-Sort 的 Shuffle 实现方式，通过 Tungsten 项目所做的优化，可以极大提高 Spark 在数据处理上的性能。

为了更合理、更高效地使用内存，在 Spark 的 Shuffle 实现方式演进过程中，引进了外部排序等处理机制（针对基于 Sort 的 Shuffle 机制。基于 Hash 的 Shuffle 机制从最原始的全部放入内存改为记录级写入）。同时，为了保存 Shuffle 结果提高性能以及支持资源动态分配等特性，也引进了外部 Shuffle 服务等机制。

7.2.2 Shuffle 的框架内核

Shuffle 框架的设计可以从两方面理解：一方面，为了 Shuffle 模块更加内聚并与其他模块解耦；另一方面，为了更方便替换、测试、扩展 Shuffle 的不同实现方式。从 Spark 1.1 版本开始，引进了可插拔式的 Shuffle 框架（通过将 Shuffle 相关的实现封装到一个统一的对外接口，提供一种具体实现可插拔的框架）。Spark 框架中，通过 ShuffleManager 来管理各种不同实现机制的 Shuffle 过程，由 ShuffleManager 统一构建、管理具体实现子类来实现 Shuffle 框架的可插拔的 Shuffle 机制。

在详细描述 Shuffle 框架实现细节之前，先给出可插拔式 Shuffle 的整体架构的类图，如图 7-2 所示。

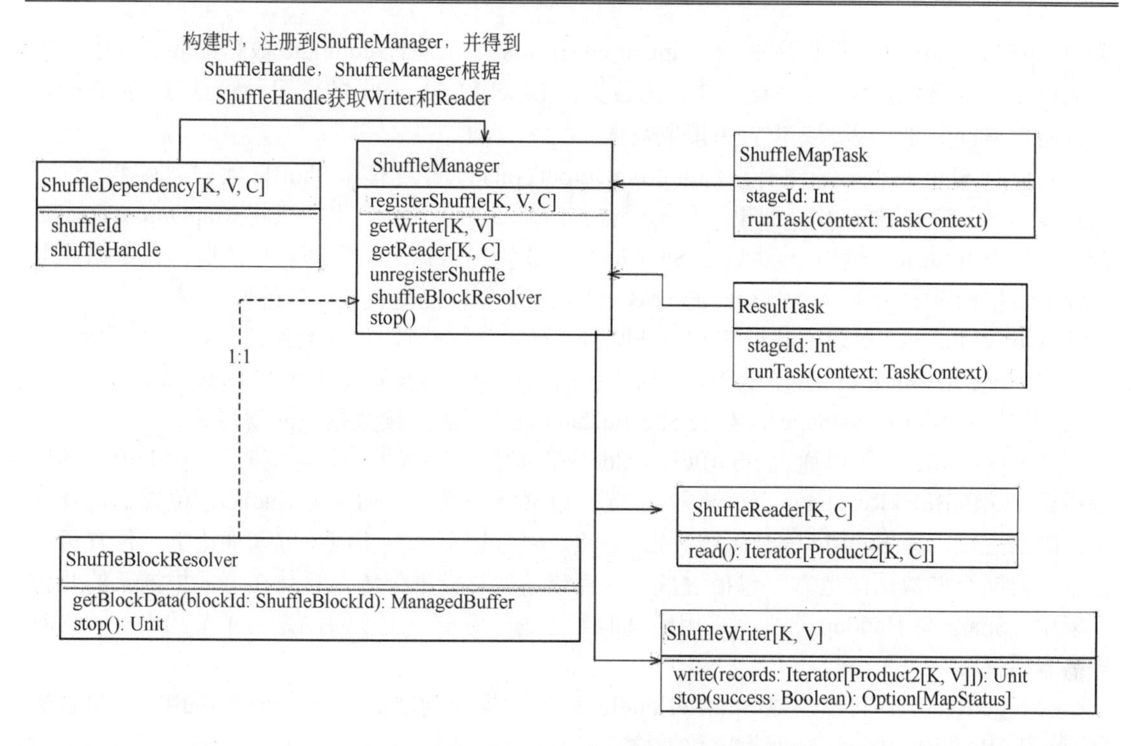

图 7-2　可插拔式 Shuffle 的整体架构的类图

在 DAG 的调度过程中，Stage 阶段的划分是根据是否有 Shuffle 过程，也就是当存在 ShuffleDependency 的宽依赖时，需要进行 Shuffle，这时会将作业（Job）划分成多个 Stage。对应地，在源码实现中，通过在划分 Stage 的关键点——构建 ShuffleDependency 时——进行 Shuffle 注册，获取后续数据读写所需的 ShuffleHandle。

Stage 阶段划分后，最终每个作业（Job）提交后都会对应生成一个 ResultStage 与若干个 ShuffleMapStage，其中 ResultStage 表示生成作业的最终结果所在的 Stage。ResultStage 与 ShuffleMapStage 中的 Task 分别对应了 ResultTask 与 ShuffleMapTask。一个作业，除了最终的 ResultStage，其他若干 ShuffleMapStage 中的各个 ShuffleMapTask 都需要将最终的数据根据相应的分区器（Partitioner）对数据进行分组（即将数据重组到新的各个分区中），然后持久化分组后的数据。对应地，每个 RDD 本身记录了它的数据来源，在计算（compute）时会读取所需数据，对于带有宽依赖的 RDD，读取时会获取在 ShuffleMapTask 中持久化的数据。

从图 7-2 中可以看到，外部宽依赖相关的 RDD 与 ShuffleManager 之间的注册交互，通过该注册，每个 RDD 自带的宽依赖（ShuffleDependency）内部会维护 Shuffle 的唯一标识信息 ShuffleId 以及与 Shuffle 过程具体读写相关的句柄 ShuffleHandle，后续在 ShuffleMapTask 中启动任务（Task）的运行时，可以通过该句柄获取相关的 Shuffle 写入器实例，实现具体的数据磁盘写操作。

而在带有宽依赖（ShuffleDependency）的 RDD 中，执行 compute 时会去读取上一 Stage 为其输出的 Shuffle 数据，此时同样会通过该句柄获取相关的 Shuffle 读取器实例，实现具体数据的读取操作。需要注意的是，当前 Shuffle 的读写过程中，与 BlockManager 的交互，是通过 MapOutputTracker 来跟踪 Shuffle 过程中各个任务的输出数据的。在任务完成等场景中，会将对应的 MapStatus 信息注册到 MapOutputTracker 中，而在 compute 数据读取过程中，也会通过该跟踪器来获取上一 Stage 的输出数据在 BlockManager 中的位置，然后通过 getReader 得到的数据读取器，从这些位置中读取数据。

目前对 Shuffle 的输出进行跟踪的 MapOutputTracker 并没有和 Shuffle 数据读写类一样，也封装到 Shuffle 的框架中。如果从代码聚合与解耦等角度出发，也可以将 MapOutputTracker 合并到整个 Shuffle 框架中，然后在 Shuffle 写入器输出数据之后立即进行注册，在数据读取器读取数据前获取位置等（但对应的 DAG 等调度部分，也需要进行修改）。

ShuffleManager 封装了各种 Shuffle 机制的具体实现细节，包含的接口与属性如下所示。

（1）registerShuffle：每个 RDD 在构建它的父依赖（这里特指 ShuffleDependency）时，都会先注册到 ShuffleManager，获取 ShuffleHandler，用于后续数据块的读写等。

（2）getWriter：可以通过 ShuffleHandler 获取数据块写入器，写数据时通过 Shuffle 的块解析器 shuffleBlockResolver，获取写入位置（通常将写入位置抽象为 Bucket，位置的选择则由洗牌的规则，即 Shuffle 的分区器决定），然后将数据写入到相应位置（理论上，位置可以位于任何能存储数据的地方，包括磁盘、内存或其他存储框架等，目前在可插拔框架的几种实现中，Spark 与 Hadoop 一样都采用磁盘的方式进行存储，主要目的是为了节约内存，同时提高容错性）。

（3）getReader：可以通过 ShuffleHandler 获取数据块读取器，然后通过 Shuffle 的块解析器 shuffleBlockResolver，获取指定数据块。

（4）unregisterShuffle：与注册对应，用于删除元数据等后续清理操作。

（5）shuffleBlockResolver：Shuffle 的块解析器，通过该解析器，为数据块的读写提供支撑层，便于抽象具体的实现细节。

7.2.3 Shuffle 框架的源码解析

用户可以通过自定义 ShuffleManager 接口，并通过指定的配置属性进行设置，也可以通过该配置属性指定 Spark 已经支持的 ShuffleManager 具体实现子类。

在 SparkEnv 源码中可以看到设置的配置属性，以及当前在 Spark 的 ShuffleManager 可插拔框架中已经提供的 ShuffleManager 具体实现。Spark 2.0 版本中支持 sort、tungsten-sort 两种方式。

SparkEnv.scala 的源码如下：

```
1. //用户可以通过短格式的命名来指定所使用的 ShuffleManager
2.     val shortShuffleMgrNames = Map(
3.       "sort" -> classOf[org.apache.spark.shuffle.sort.SortShuffleManager]
         .getName,
4.       "tungsten-sort" -> classOf[org.apache.spark.shuffle.sort
         .SortShuffleManager].getName)
5.     val shuffleMgrName = conf.get("spark.shuffle.manager", "sort")
6.     val shuffleMgrClass =
7.       shortShuffleMgrNames.getOrElse(shuffleMgrName
         .toLowerCase(Locale.ROOT), shuffleMgrName)
8.     val shuffleManager = instantiateClass[ShuffleManager](shuffleMgrClass)
9.
10.    val useLegacyMemoryManager = conf.getBoolean("spark.memory.useLegacyMode",
       false)
11.    val memoryManager: MemoryManager =
```

```
12.       if (useLegacyMemoryManager) {
13.         new StaticMemoryManager(conf, numUsableCores)
14.       } else {
15.         UnifiedMemoryManager(conf, numUsableCores)
16.       }
```

从代码中可以看出，ShuffleManager 是 Spark Shuffle 系统提供的一个可插拔式接口，可以通过 spark.shuffle.manager 配置属性来设置自定义的 ShuffleManager。

在 Driver 和每个 Executor 的 SparkEnv 实例化过程中，都会创建一个 ShuffleManager，用于管理块数据，提供集群块数据的读写，包括数据的本地读写和读取远程节点的块数据。

Shuffle 系统的框架可以以 ShuffleManager 为入口进行解析。在 ShuffleManager 中指定了整个 Shuffle 框架使用的各个组件，包括如何注册到 ShuffleManager，以获取一个用于数据读写的处理句柄 ShuffleHandle，通过 ShuffleHandle 获取特定的数据读写接口：ShuffleWriter 与 ShuffleReader，以及如何获取块数据信息的解析接口 ShuffleBlockResolver。下面通过源码分别对这几个比较重要的组件进行解析。

1. ShuffleManager的源码解析

由于 ShuffleManager 是 Spark Shuffle 系统提供的一个可插拔式接口，提供具体实现子类或自定义具体实现子类时，都需要重写 ShuffleManager 类的抽象接口。下面首先分析 ShuffleManager 的源码。

ShuffleManager.scala 的源码如下：

```
1.
2.  //Shuffle 系统的可插拔接口。在 Driver 和每个 Executor 的 SparkEnv 实例中创建
3.  private[spark] trait ShuffleManager {
4.
5.    /**
6.      *在 Driver 端向 ShuffleManager 注册一个 Shuffle，获取一个 Handle
7.      *在具体 Tasks 中会通过该 Handle 来读写数据
8.      */
9.    def registerShuffle[K, V, C](
10.       shuffleId: Int,
11.       numMaps: Int,
12.       dependency: ShuffleDependency[K, V, C]): ShuffleHandle
13.
14.   /**
      *获取对应给定的分区使用的 ShuffleWriter，该方法在 Executors 上执行各个 Map
      *任务时调用
15.     */
16.   def getWriter[K, V](handle: ShuffleHandle, mapId: Int, context:
    TaskContext): ShuffleWriter[K, V]
17.   /**
      * 获取在 Reduce 阶段读取分区的 ShuffleReader，对应读取的分区由[startPartition
      * to endPartition-1]区间指定。该方法在 Executors 上执行，在各个 Reduce 任务时调用
      *
18      */
19.   def getReader[K, C](
20
21.       handle: ShuffleHandle,
22.       startPartition: Int,
23.       endPartition: Int,
```

```
24.       context: TaskContext): ShuffleReader[K, C]
25.
26.   /**
27.     *该接口和 registerShuffle 分别负责元数据的取消注册与注册
28.     *调用 unregisterShuffle 接口时，会移除 ShuffleManager 中对应的元数据信息
29.     */
30.   def unregisterShuffle(shuffleId: Int): Boolean
31.
32.   /**
      *返回一个可以基于块坐标来获取 Shuffle 块数据的 ShuffleBlockResolver
33.     */
34.   def shuffleBlockResolver: ShuffleBlockResolver
35.
36.   /**终止 ShuffleManager */
37.   def stop(): Unit
38. }
```

2．ShuffleHandle的源码解析

```
1.  abstract class ShuffleHandle(val shuffleId: Int) extends Serializable {}
```

ShuffleHandle 比较简单，用于记录 Task 与 Shuffle 相关的一些元数据，同时也可以作为不同具体 Shuffle 实现机制的一种标志信息，控制不同具体实现子类的选择等。

3．ShuffleWriter的源码解析

ShuffleWriter.scala 的源码如下：

```
1.   private[spark] abstract class ShuffleWriter[K, V] {
2.    /** 将一系列记录写入此任务的输出 */
3.    @throws[IOException]
4.    def write(records: Iterator[Product2[K, V]]): Unit
5.
6.    /** 停止写入，传递 map 是否已完成 */
7.    def stop(success: Boolean): Option[MapStatus]
8.  }
```

继承 ShuffleWriter 的每个具体子类会实现 write 接口，给出任务在输出时的写记录的具体方法。

4．ShuffleReader的源码解析

ShuffleReader.scala 的源码如下：

```
1.   private[spark] trait ShuffleReader[K, C] {
2.    /** 读取 reduce 任务的组合键值 */
3.    def read(): Iterator[Product2[K, C]]
```

继承 ShuffleReader 的每个具体子类会实现 read 接口，计算时负责从上一阶段 Stage 的输出数据中读取记录。

5．ShuffleBlockResolver的源码解析

ShuffleBlockResolver 的源码如下：

```
1.  /**
      *该特质的具体实现子类知道如何通过一个逻辑 Shuffle 块标识信息来获取一个块数据。具体
      *实现可以使用文件或文件段来封装 Shuffle 的数据。这是获取 Shuffle 块数据时使用的抽
      *象接口，在 BlockStore 中使用
2.    */
3.
4.
5.  trait ShuffleBlockResolver {
6.    type ShuffleId = Int
7.
8.    /**
       *获取指定块的数据。如果指定块的数据无法获取，则抛出异常
9.     */
10.   def getBlockData(blockId: ShuffleBlockId): ManagedBuffer
11.
12.   def stop(): Unit
13. }
```

继承 ShuffleBlockResolver 的每个具体子类会实现 getBlockData 接口，给出具体的获取块数据的方法。

目前在 ShuffleBlockResolver 的各个具体子类中，除给出获取数据的接口外，通常会提供如何解析块数据信息的接口，即提供了写数据块时的物理块与逻辑块之间映射关系的解析方法。

7.2.4　Shuffle 数据读写的源码解析

1. Shuffle写数据的源码解析

从 Spark Shuffle 的整体框架中可以看到，ShuffleManager 提供了 Shuffle 相关数据块的写入与读取，即对应的接口 getWriter 与 getReader。

在解析 Shuffle 框架数据读取过程中，可以构建一个具有 ShuffleDependency 的 RDD，查看执行过程中，Shuffle 框架中的数据读写接口 getWriter 与 getReader 如何使用，通过这种具体案例的方式来加深对源码的理解。

Spark 中 Shuffle 具体的执行机制可以参考本书的其他章节，在此仅分析与 Shuffle 直接相关的内容。通过 DAG 调度机制的解析，可以知道 Spark 中一个作业可以根据宽依赖切分 Stages，而在 Stages 中，相应的 Tasks 也包含两种，即 ResultTask 与 ShuffleMapTask。其中，一个 ShuffleMapTask 会基于 ShuffleDependency 中指定的分区器，将一个 RDD 的元素拆分到多个 buckets 中，此时通过 ShuffleManager 的 getWriter 接口获取数据与 buckets 的映射关系。而 ResultTask 对应的是一个将输出返回给应用程序 Driver 端的 Task，在该 Task 执行过程中，最终都会调用 RDD 的 compute 对内部数据进行计算，而在带有 ShuffleDependency 的 RDD 中，在 compute 计算时，会通过 ShuffleManager 的 getReader 接口，获取上一个 Stage 的 Shuffle 输出结果作为本次 Task 的输入数据。

首先来看 ShuffleMapTask 中的写数据流程。ShuffleMapTask.scala 的源码如下：

```
1.  override def runTask(context: TaskContext): MapStatus = {
2.    ......
```

```
3.  //首先从 SparkEnv 获取 ShuffleManager
4.  //然后从 ShuffleDependency 中获取注册到 ShuffleManager 时得到的 shuffleHandle
5.  //根据 shuffleHandle 和当前 Task 对应的分区 ID，获取 ShuffleWriter
6.  //最后根据获取的 ShuffleWriter，调用其 write 接口，写入当前分区的数据
7.  var writer: ShuffleWriter[Any, Any] = null
8.      try {
9.        val manager = SparkEnv.get.shuffleManager
10.      writer = manager.getWriter[Any, Any](dep.shuffleHandle, partitionId,
         context)
11.       writer.write(rdd.iterator(partition, context).asInstanceOf
          [Iterator[_ <: Product2[Any, Any]]])
12.       writer.stop(success = true).get
13.     } catch {
14.       ......
15.     }
16.   }
```

2. Shuffle读数据的源码解析

对应的数据读取器，从 RDD 的 5 个抽象接口可知，RDD 的数据流最终会经过算子操作，即 RDD 中的 compute 方法。下面以包含宽依赖的 RDD、CoGroupedRDD 为例，查看如何获取 Shuffle 的数据。CoGroupedRDD.scala 的源码如下：

```
1.  //对指定分区进行计算的抽象接口，以下为 CoGroupedRDD 具体子类中该方法的实现
2.  override def compute(s: Partition, context: TaskContext): Iterator[(K,
    Array[Iterable[_]])] = {
3.      val split = s.asInstanceOf[CoGroupPartition]
4.      val numRdds = dependencies.length
5.
6.      //列表（RDD 迭代器、依赖项编号）
7.      val rddIterators = new ArrayBuffer[(Iterator[Product2[K, Any]], Int)]
8.      for ((dep, depNum) <- dependencies.zipWithIndex) dep match {
9.        case oneToOneDependency: OneToOneDependency[Product2[K, Any]]
          @unchecked =>
10.         val dependencyPartition = split.narrowDeps(depNum).get.split
11.         // 从父依赖读
12.         val it = oneToOneDependency.rdd.iterator(dependencyPartition, context)
13.         rddIterators += ((it, depNum))
14.
15.       case shuffleDependency: ShuffleDependency[_, _, _] =>
16. //首先从 SparkEnv 获取 ShuffleManager
17.        //然后从ShuffleDependency中获取注册到ShuffleManager时得到的shuffleHandle
18.        //根据 shuffleHandle 和当前 Task 对应的分区 ID，获取 ShuffleWriter
19.        //最后根据获取的 ShuffleReader，调用其 read 接口，读取 Shuffle 的 Map 输出
20.
21.         val it = SparkEnv.get.shuffleManager
22.           .getReader(shuffleDependency.shuffleHandle, split.index,
            split.index + 1, context)
23.           .read()
24.         rddIterators += ((it, depNum))
25.     }
26.
27.     val map = createExternalMap(numRdds)
28.     for ((it, depNum) <- rddIterators) {
29.       map.insertAll(it.map(pair => (pair._1, new CoGroupValue(pair._2,
```

```
        depNum)))
30.     }
31.     context.taskMetrics().incMemoryBytesSpilled(map.memoryBytesSpilled)
32.     context.taskMetrics().incDiskBytesSpilled(map.diskBytesSpilled)
33.     context.taskMetrics().incPeakExecutionMemory(map.peakMemoryUsedBytes)
34.     new InterruptibleIterator(context,
35.       map.iterator.asInstanceOf[Iterator[(K, Array[Iterable[_]])]])
36.   }
```

从代码中可以看到，带宽依赖的 RDD 的 compute 操作中，最终是通过 SparkEnv 中的 ShuffleManager 实例的 getReader 方法，获取数据读取器的，然后再次调用读取器的 read 读取指定分区范围的 Shuffle 数据。注意，是带宽依赖的 RDD，而非 ShuffleRDD，除了 ShuffleRDD 外，还有其他 RDD 也可以带上宽依赖的，如前面给出的 CoGroupedRDD。

目前支持的几种具体 Shuffle 实现机制在读取数据的处理上都是一样的。从源码角度可以看到，当前继承了 ShuffleReader 这一数据读取器的接口的具体子类只有 BlockStoreShuffleReader，因此，本章内容仅在此对各种 Shuffle 实现机制的数据读取进行解析，后续各实现机制中不再重复描述。

源码解析的第一步仍然是查看该类的描述信息，具体格式如下：

```
1.  /**
2.   *通过从其他节点上请求读取 Shuffle 数据来接收并读取指定范围[起始分区， 结束分区)
     *——对应为左闭右开区间
3.   *通过从其他节点上请求读取 Shuffle 数据来接收并读取指定范围[起始分区，结束分区]
4.   *——对应为左闭右开区间
5.   */
```

从注释上可以看出，读取器负责上一 Stage 为下一 Stage 输出数据块的读取。从前面对 ShuffleReader 接口的解析可知，继承的具体子类需要实现真正的数据读取操作，即实现 read 方法。因此，该方法便是需要重点关注的源码。

Spark 2.2.1 版本的 BlockStoreShuffleReader.scala 的源码如下：

```
1. //为该 Reduce 任务读取并合并 key-values 值
2. override def read(): Iterator[Product2[K, C]] = {
3. //真正的数据 Iterator 读取是通过 ShuffleBlockFetcherIterator 来完成的
4.     val wrappedStreams = new ShuffleBlockFetcherIterator(
5.       context,
6.       blockManager.shuffleClient,
7.       blockManager,
8.       //可以看到，当 ShuffleMapTask 完成后注册 mapOutputTracker 的元数据信息
9.       //同样会通过 mapOutputTracker 来获取，在此同时还指定了获取的分区范围
10.      //通过该方法的返回值类型
11.      mapOutputTracker.getMapSizesByExecutorId(handle.shuffleId,
   startPartition, endPartition),
12.      serializerManager.wrapStream,
13.      //注意：当没有为向后兼容性提供后缀时，使用 getSizeAsMb
14.      //默认读取时的数据大小限制为 48m，对应后续并行的读取，都是一种数据读取的控制策
15.      //略，一方面可以避免目标机器占用过多带宽，同时也可以启动并行机制，加快读取速度
16.
17.      SparkEnv.get.conf.getSizeAsMb("spark.reducer.maxSizeInFlight",
         "48m") * 1024 * 1024,
18.      SparkEnv.get.conf.getInt("spark.reducer.maxReqsInFlight", Int.MaxValue),
19.      SparkEnv.get.conf.get(config.REDUCER_MAX_BLOCKS_IN_FLIGHT_PER_ADDRESS),
```

```
20.    SparkEnv.get.conf.get(config.REDUCER_MAX_REQ_SIZE_SHUFFLE_TO_MEM),
21.    SparkEnv.get.conf.getBoolean("spark.shuffle.detectCorrupt", true))
22.
23.    val serializerInstance = dep.serializer.newInstance()
24.
25.    //为每个流 stream 创建一个键/值迭代器
26.    val recordIter = wrappedStreams.flatMap { case (blockId, wrappedStream) =>
27.      //注意：下面的 asKeyValueIterator 将键/值迭代器包装在 Nextiterator 内部
28.      //NextIterator 确保在
29.
30. //读取所有记录后的基础输入流
31. serializerInstance.deserializeStream(wrappedStream).asKeyValueIterator
32.    }
33.
34.    // 为每个记录更新上下文任务度量
35.    val readMetrics = context.taskMetrics.createTempShuffleReadMetrics()
36.    val metricIter = CompletionIterator[(Any, Any), Iterator[(Any, Any)]](
37.      recordIter.map { record =>
38.        readMetrics.incRecordsRead(1)
39.        record
40.      },
41.      context.taskMetrics().mergeShuffleReadMetrics())
42.
43.    //为了支持任务取消，这里必须使用可中断迭代器
44.    val interruptibleIter = new InterruptibleIterator[(Any, Any)](context,
       metricIter)
45.   //对读取到的数据进行聚合处理
46.    val aggregatedIter: Iterator[Product2[K, C]] = if (dep.aggregator.isDefined) {
47.      //如果在 Map 端已经做了聚合的优化操作，则对读取到的聚合结果进行聚合，
48.      //注意此时的聚合操作与数据类型和 Map 端未做优化时是不同的
49.
50.      if (dep.mapSideCombine) {
51.       //对读取到的数据进行聚合处理
52.        val combinedKeyValuesIterator = interruptibleIter
         .asInstanceOf[Iterator[(K, C)]]
53.      //Map 端各分区针对 Key 进行合并后的结果再次聚合，
54.      //Map 的合并可以大大减少网络传输的数据量
55.
56.        dep.aggregator.get.combineCombinersByKey(combinedKeyValuesIterator,
        context)
57.      } else {
58.       //我们无须关心值的类型，但应确保聚合是兼容的，其将把值的类型转化成聚合以后的
        //C 类型
59.
60.        val keyValuesIterator = interruptibleIter
         .asInstanceOf[Iterator[(K, Nothing)]]
61.        dep.aggregator.get.combineValuesByKey(keyValuesIterator, context)
62.      }
63.    } else {
64.      require(!dep.mapSideCombine, "Map-side combine without Aggregator
       specified!")
65.      interruptibleIter.asInstanceOf[Iterator[Product2[K, C]]]
66.    }
67.
68.      //在基于 Sort 的 Shuffle 实现过程中，默认基于 PartitionId 进行排序，
69.     //在分区的内部，数据是没有排序的，因此添加了 keyOrdering 变量，提供
70.     //是否需要针对分区内的数据进行排序的标识信息
```

```
71.
72.        //如果定义了排序，则对输出结果进行排序
73.
74.     dep.keyOrdering match {
75.       case Some(keyOrd: Ordering[K]) =>
76. //创建一个外部排序器来对数据进行排序
77. //为了减少内存的压力，避免 GC 开销，引入了外部排序器对数据进行排序，当内存不足
78. //以容纳排序的数据量时，会根据配置的 spark.shuffle.spill 属性来决定是否需要
79. //溢出到磁盘中，默认情况下会打开 spill 开关，若不打开 spill 开关，数据量比
80. //较大时会引发内存溢出问题（Out of Memory，OOM）
81.
82.         val sorter =
83.           new ExternalSorter[K, C, C](context, ordering = Some(keyOrd),
            serializer = dep.serializer)
84.         sorter.insertAll(aggregatedIter)
85.         context.taskMetrics().incMemoryBytesSpilled(sorter.memoryBytesSpilled)
86.         context.taskMetrics().incDiskBytesSpilled(sorter.diskBytesSpilled)
87.         context.taskMetrics().incPeakExecutionMemory(sorter
          .peakMemoryUsedBytes)
88.         CompletionIterator[Product2[K, C], Iterator[Product2[K, C]]]
          (sorter.iterator, sorter.stop())
89.       case None =>
90.       //不需要排序分区内部数据时直接返回
91.         aggregatedIter
92.     }
93.   }
```

Spark 2.4.3 版本的 BlockStoreShuffleReader.scala 源码与 Spark 2.2.1 版本相比具有如下特点。

- 上段代码中第 20 行将 config.REDUCER_MAX_REQ_SIZE_SHUFFLE_TO_MEM 调整为 config.MAX_REMOTE_BLOCK_SIZE_FETCH_TO_MEM。
- 上段代码中第 64 行删掉。
- 上段代码中第 74 行将 dep.keyOrdering match 调整为 val resultIter = dep.keyOrdering match。
- 上段代码中第 87 行之后的新增代码，如果任务已完成或取消，使用回调停止排序。
- 上段代码中第 92 行之后的新增代码，对 resultIter 的两种情况进行适配处理。

```
1.   ......
2.     SparkEnv.get.conf.get(config.MAX_REMOTE_BLOCK_SIZE_FETCH_TO_MEM),
3.   ......
4.   val resultIter = dep.keyOrdering match {
5.   ......
6.   //如果任务已完成或取消，使用回调停止排序
7.     context.addTaskCompletionListener[Unit](_ => {
8.             sorter.stop()
9.           })
10.  ......
11.      resultIter match {
12.        case _: InterruptibleIterator[Product2[K, C]] => resultIter
13.        case _ =>
14.          //这里使用另一个可中断迭代器来支持任务取消，因为聚合器排序器可能已经使用
           //了以前的可中断迭代器
15.          new InterruptibleIterator[Product2[K, C]](context, resultIter)
16.      }
17. ......
```

下面进一步解析数据读取的部分细节。首先是数据块获取、读取的 ShuffleBlockFetcherIterator 类，在类的构造体中调用了 initialize 方法（构造体中的表达式会在构造实例时执行），该方法中会根据数据块所在位置（本地节点或远程节点）分别进行读取。

ShuffleBlockFetcherIterator.scala 的源码如下：

```
1.     private[this] def initialize(): Unit = {
2.     //任务完成进行回调清理（在成功案例和失败案例中调用）
3.     context.addTaskCompletionListener(_ => cleanup())
4.     //本地与远程的数据读取方式不同，因此先进行拆分，注意拆分时会考虑一次获取的数据
       //大小（拆分时会同时考虑并行数）封装请求，最后会将剩余不足该大小的数据获取也封装
       //为一个请求
5.
6.
7.
8.     val remoteRequests = splitLocalRemoteBlocks()
9.     //存入需要远程读取的数据块请求信息
10.    fetchRequests ++= Utils.randomize(remoteRequests)
11.    assert ((0 == reqsInFlight) == (0 == bytesInFlight),
12.      "expected reqsInFlight = 0 but found reqsInFlight = " + reqsInFlight +
13.      ", expected bytesInFlight = 0 but found bytesInFlight = " +
         bytesInFlight)
14.
15.    //发送数据获取请求
16.    fetchUpToMaxBytes()
17.
18.    val numFetches = remoteRequests.size - fetchRequests.size
19.    logInfo("Started " + numFetches + " remote fetches in" +
       Utils.getUsedTimeMs(startTime))
20.
21.     //除了远程数据获取外，下面是获取本地数据块的方法调用
22.    fetchLocalBlocks()
23.    logDebug("Got local blocks in " + Utils.getUsedTimeMs(startTime))
24.  }
```

与 Hadoop 一样，Spark 计算框架也基于数据本地性，即移动数据而非移动计算的原则，因此在获取数据块时，也会考虑数据本地性，尽量从本地读取已有的数据块，然后再远程读取。

另外，数据块的本地性是通过 ShuffleBlockFetcherIterator 实例构建时所传入的位置信息来判断的，而该信息由 MapOutputTracker 实例的 getMapSizesByExecutorId 方法提供，可以参考该方法的返回值类型查看相关的位置信息，返回值类型为：Seq[(BlockManagerId, Seq[(BlockId, Long)])]。其中，BlockManagerId 是 BlockManager 的唯一标识信息，BlockId 是数据块的唯一信息，对应的 Seq[(BlockId, Long)]表示一组数据块标识 ID 及其数据块大小的元组信息。

最后简单分析一下如何设置分区内部的排序标识，当需要对分区内的数据进行排序时，会设置 RDD 中的宽依赖（ShuffleDependency）实例的 keyOrdering 变量。下面以基于排序的 OrderedRDDFunctions 提供的 sortByKey 方法给出解析。

OrderedRDDFunctions.scala 的源码如下：

```
1.    def sortByKey(ascending: Boolean = true, numPartitions: Int =
      self.partitions.length)
2.        : RDD[(K, V)] = self.withScope
```

```
3.     {
4.       //注意，这里设置了该方法构建的 RDD 使用的分区器
         //根据 Range 而非 Hash 进行分区，对应的 Range 信息需要计算并将结果
         //反馈到 Driver 端，因此对应调用 RDD 中的 Action，即会触发一个 Job 的执行
5.   val part = new RangePartitioner(numPartitions, self, ascending)
6.        //在构建 RDD 实例后，设置 Key 的排序算法，即 Ordering 实例
7.        new ShuffledRDD[K, V, V](self, part)
8.          .setKeyOrdering(if (ascending) ordering else ordering.reverse)
9.     }
```

当需要对分区内部的数据进行排序时，构建 RDD 的同时会设置 Key 值的排序算法，结合前面的 read 代码，当指定 Key 值的排序算法时，就会使用外部排序器对分区内的数据进行排序。

7.3　Hash Based Shuffle

本节讲解 Hash Based Shuffle，包括 Hash Based Shuffle 概述、Hash Based Shuffle 内核、Hash Based Shuffle 的数据读写的源码解析等内容。

7.3.1　概述

在 Spark 1.1 之前，Spark 中只实现了一种 Shuffle 方式，即基于 Hash 的 Shuffle。在 Spark 1.1 版本中引入了基于 Sort 的 Shuffle 实现方式，并且在 Spark 1.2 版本之后，默认的实现方式从基于 Hash 的 Shuffle，修改为基于 Sort 的 Shuffle 实现方式，即使用的 ShuffleManager 从默认的 hash 修改为 sort。说明在 Spark 2.0 版本中，Hash 的 Shuffle 方式已经不再使用。

Spark 之所以一开始就提供基于 Hash 的 Shuffle 实现机制，其主要目的之一就是为了避免不需要的排序（这也是 Hadoop Map Reduce 被人诟病的地方，将 Sort 作为固定步骤，导致许多不必要的开销）。但基于 Hash 的 Shuffle 实现机制在处理超大规模数据集的时候，由于过程中会产生大量的文件，导致过度的磁盘 I/O 开销和内存开销，会极大地影响性能。

但在一些特定的应用场景下，采用基于 Hash 的实现 Shuffle 机制的性能会超过基于 Sort 的 Shuffle 实现机制。关于基于 Hash 与基于 Sort 的 Shuffle 实现机制的性能测试方面，可以参考 Spark 创始人之一的 ReynoldXin 给的测试："sort-basedshuffle has lower memory usage and seems to outperformhash-based in almost all of our testing"。

相关数据可以参考 https://issues.apache.org/jira/browse/SPARK-3280。

因此，在 Spark 1.2 版本中修改为默认基于 Sort 的 Shuffle 实现机制时，同时也给出了特定应用场景下回退的机制。

7.3.2　Hash Based Shuffle 内核

1. 基于Hash的Shuffle实现机制的内核框架

基于 Hash 的 Shuffle 实现，ShuffleManager 的具体实现子类为 HashShuffleManager，对应的具体实现机制如图 7-3 所示。

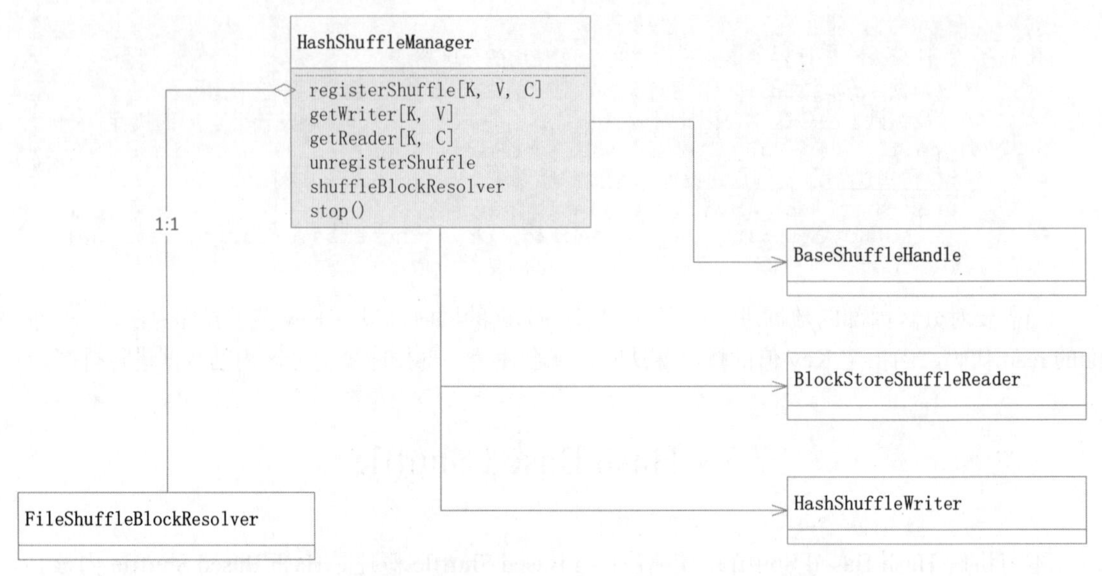

图 7-3　基于哈希算法的 Shuffle 实现机制的内核框架

其中，HashShuffleManager 是 ShuffleManager 的基于哈希算法实现方式的具体实现子类。数据块的读写分别由 BlockStoreShuffleReader 与 HashShuffleWriter 实现；数据块的文件解析器则由具体子类 FileShuffleBlockResolver 实现；BaseShuffleHandle 是 ShuffleHandle 接口的基本实现，保存 Shuffle 注册的信息。

HashShuffleManager 继承自 ShuffleManager，对应实现了各个抽象接口。基于 Hash 的 Shuffle，内部使用的各组件的具体子类如下所示。

（1）BaseShuffleHandle：携带了 Shuffle 最基本的元数据信息，包括 shuffleId、numMaps 和 dependency。

（2）BlockStoreShuffleReader：负责写入的 Shuffle 数据块的读操作。

（3）FileShuffleBlockResolver：负责管理，为 Shuffle 任务分配基于磁盘的块数据的 Writer。每个 ShuffleShuffle 任务为每个 Reduce 分配一个文件。

（4）HashShuffleWriter：负责 Shuffle 数据块的写操作。

在此与解析整个 Shuffle 过程一样，以 HashShuffleManager 类作为入口进行解析。

首先看一下 HashShuffleManager 具体子类的注释，如下所示。

Spark 1.6.0 版本的 HashShuffleManager.scala 的源码（Spark 2.4 版本已无 HashShuffleManager 方式）如下：

```
1.  /**
      *使用 Hash 的 ShuffleManager 具体实现子类，针对每个 Mapper 都会为各个 Reduce 分
      *区构建一个输出文件（也可能是多个任务复用文件）
2.    */
3.  private[spark]  class  HashShuffleManager(conf:  SparkConf)  extends
    ShuffleManager with Logging {
4.  ……
```

2. 基于Hash的Shuffle实现方式一

为了避免 Hadoop 中基于 Sort 方式的 Shuffle 所带来的不必要的排序开销，Spark 在开始

时采用了基于 Hash 的 Shuffle 方式。但这种方式存在不少缺陷，这些缺陷大部分是由于在基于 Hash 的 Shuffle 实现过程中创建了太多的文件所造成的。在这种方式下，每个 Mapper 端的 Task 运行时都会为每个 Reduce 端的 Task 生成一个文件，具体如图 7-4 所示。

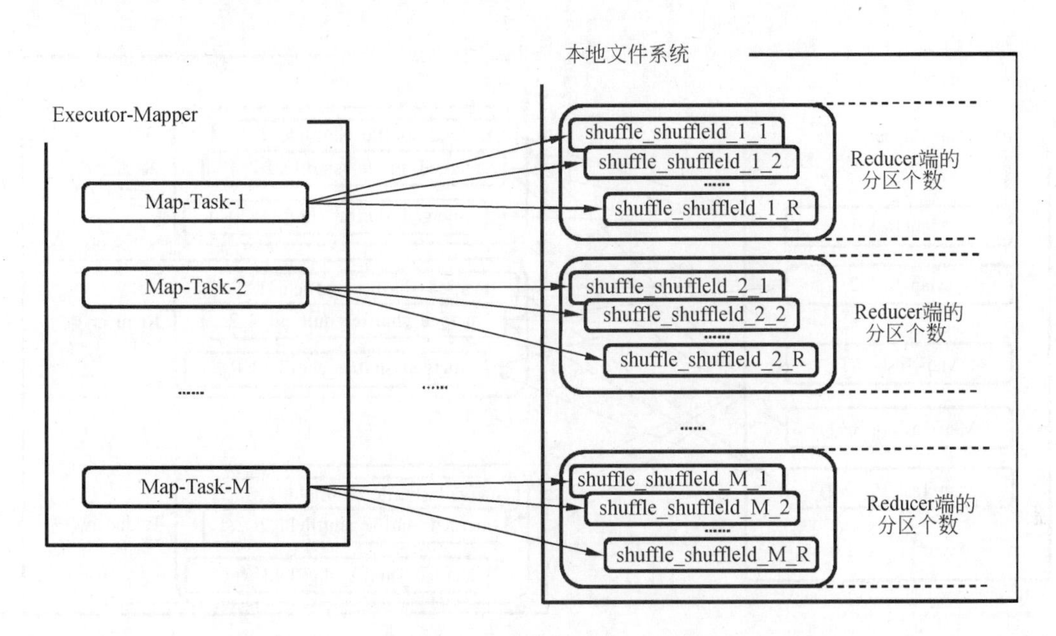

图 7-4　基于 Hash 的 Shuffle 实现方式——文件的输出细节图

Executor-Mapper 表示执行 Mapper 端的 Tasks 的工作点，可以分布到集群中的多台机器节点上，并且可以以不同的形式出现，如以 Spark Standalone 部署模式中的 Executor 出现，也可以以 Spark On Yarn 部署模式中的容器形式出现，关键是它代表了实际执行 Mapper 端的 Tasks 的工作点的抽象概念。其中，M 表示 Mapper 端的 Task 的个数，R 表示 Reduce 端的 Task 的个数。

对应在右侧的本地文件系统是在该工作点上所生成的文件，其中 R 表示 Reduce 端的分区个数。生成的文件名格式为：shuffle_shuffleId_mapId_reduceId，其中的 shuffle_shuffleId_1_1 表示 mapId 为 1，同时 reduceId 也为 1。

在 Mapper 端，每个分区对应启动一个 Task，而每个 Task 会为每个 Reducer 端的 Task 生成一个文件，因此最终生成的文件个数为 M×R。

由于这种实现方式下，对应生成文件个数仅与 Mapper 端和 Reducer 端各自的分区数有关，因此图中将 Mapper 端的全部 M 个 Task 抽象到一个 Executor-Mapper 中，实际场景中通常是分布到集群中的各个工作点中。

生成的各个文件位于本地文件系统的指定目录中，该目录地址由配置属性 spark.local.dir 设置。说明：分区数与 Task 数，一个是静态的数据分块个数，一个是数据分块对应执行的动态任务个数，因此，在特定的、描述个数的场景下，两者是一样的。

3. 基于Hash的Shuffle实现方式二

为了减少 Hash 所生成的文件个数，对基于 Hash 的 Shuffle 实现方式进行了优化，引入文件合并的机制，该机制设置的开关为配置属性 spark.shuffle.consolidateFiles。在引入文件合并的机制后，当设置配置属性为 true，即启动文件合并时，在 Mapper 端的输出文件会进行合并，在一定程度上可以大量减少文件的生成，降低不必要的开销。文件合并的实现方式可以参考图 7-5。

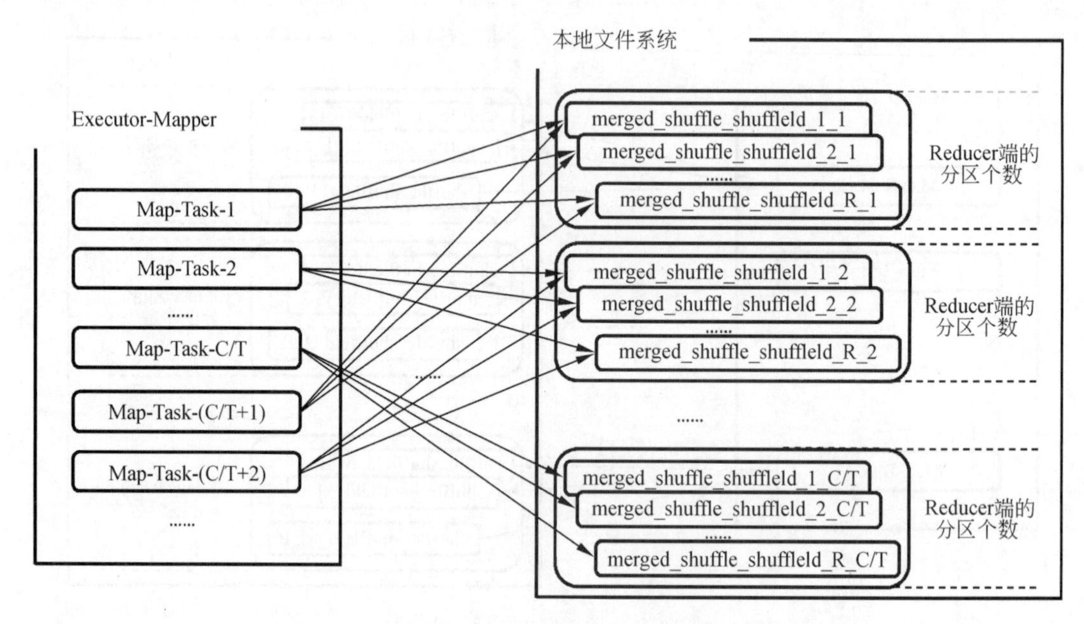

图 7-5　基于 Hash 的 Shuffle 的合并文件机制的输出细节图

Executor-Mapper 表示集群中分配的某个工作点，其中，C 表示在该工作点上所分配到的内核（Core）个数，T 表示在该工作点上为每个 Task 分配的内核个数。C/T 表示在该工作点上调度时最大的 Task 并行个数。

右侧的本地文件系统是在该工作点上所生成的文件，其中 R 表示 Reduce 端的分区个数。生成的文件名格式为：merged_shuffle_shuffleId_bucketId_fileId，其中的 merged_shuffle_shuffleId_1_1 表示 bucketId 为 1，同时 fileId 也为 1。

在 Mapper 端，Task 会复用文件组，由于最大并行个数为 C/T，因此文件组最多分配 C/T 个，当某个 Task 运行结束后，会释放该文件组，之后调度的 Task 则复用前一个 Task 所释放的文件组，因此会复用同一个文件。最终在该工作点上生成的文件总数为 C/T*R，如果设工作点个数为 E，则总的文件数为 E*C/T*R。

4. 基于Hash的Shuffle机制的优缺点

1）优点

- 可以省略不必要的排序开销。
- 避免了排序所需的内存开销。

2）缺点

- 生成的文件过多，会对文件系统造成压力。
- 大量小文件的随机读写会带来一定的磁盘开销。
- 数据块写入时所需的缓存空间也会随之增加，会对内存造成压力。

7.3.3　Hash Based Shuffle 数据读写的源码解析

1. 基于Hash的Shuffle实现方式一的源码解析

下面针对 Spark 1.6 版本中的基于 Hash 的 Shuffle 实现在数据写方面进行源码解析（Spark2.0 版本中已无 Hash 的 Shuffle 实现方式）。在基于 Hash 的 Shuffle 实现机制中，采用 HashShuffleWriter 作为数据写入器。在 HashShuffleWriter 中控制 Shuffle 写数据的关键代码如下所示。

Spark 1.6.0 版本的 HashShuffleWriter.scala 的源码（Spark 2.4 版本已无 HashShuffle-Manager 方式）如下：

```
1.  private[spark] class HashShuffleWriter[K, V](
2.      shuffleBlockResolver: FileShuffleBlockResolver,
3.      handle: BaseShuffleHandle[K, V, _],
4.  mapId: Int,
5.      context: TaskContext)
6.    extends ShuffleWriter[K, V] with Logging {
7.
8.    //控制每个 Writer 输出时的切片个数，对应分区个数
9.    private val dep = handle.dependency
10.   private val numOutputSplits = dep.partitioner.numPartitions
11.
12.   ……
13.   //获取数据读写的块管理器
14.   private val blockManager = SparkEnv.get.blockManager
15.   private val ser = Serializer.getSerializer(dep.serializer.getOrElse(null))
16.
17. //从 FileShuffleBlockResolver 的 forMapTask 方法中获取指定的 shuffleId 对应
    //的 mapId
18. //对应分区个数构建的数据块写的 ShuffleWriterGroup 实例
19.   private val shuffle = shuffleBlockResolver.forMapTask(dep.shuffleId,
      mapId, numOutputSplits, ser, writeMetrics)
20.
21.
22.   /** Task 输出时一组记录的写入 */
23.
24.   override def write(records: Iterator[Product2[K, V]]): Unit = {
25. //判断在写时是否需要先聚合，即定义了 Map 端 Combine 时，先对数据进行聚合再写入，否则
    //直接返回需要写入的一批记录
26.
27.   val iter = if (dep.aggregator.isDefined) {
28.       if (dep.mapSideCombine) {
29.         dep.aggregator.get.combineValuesByKey(records, context)
30.       } else {
31.         records
32.       }
33.     } else {
34.       require(!dep.mapSideCombine, "Map-side combine without Aggregator
          specified!")
35.       records
36.     }
37.
```

```
38.     //根据分区器，获取每条记录对应的 bucketId（即所在 Reduce 序号），根据 bucketId
        //从 FileShuffleBlockResolver 构建的 ShuffleWriterGroup 中，获取 DiskBlock-
        //ObjectWriter 实例，对应磁盘数据块的数据写入器
39.     for (elem<- iter) {
40.       val bucketId = dep.partitioner.getPartition(elem._1)
41.       shuffle.writers(bucketId).write(elem._1, elem._2)
42.     }
43.   }
44.     ......
45. }
```

当需要在 Map 端进行聚合时，使用的是聚合器（Aggregator）的 combineValuesByKey 方法，在该方法中使用 ExternalAppendOnlyMap 类对记录集进行处理，处理时如果内存不足，会引发 Spill 操作。早期的实现会直接缓存到内存，在数据量比较大时容易引发内存泄漏。

在 HashShuffleManager 中，ShuffleBlockResolver 特质使用的具体子类为 FileShuffleBlock-Resolver，即指定了具体如何从一个逻辑 Shuffle 块标识信息来获取一个块数据，对应为下面第 7 行调用的 forMapTask 方法。

Spark 1.6.0 版本的 FileShuffleBlockResolver.scala 的源码（Spark 2.4 版本已无 HashShuffleManager 方式）如下：

```
1.  /**
2.    *针对给定的 Map Task，指定一个 ShuffleWriterGroup 实例，在数据块写入器成功
3.    *关闭时，会注册为完成状态
4.    */
5.
6.
7.   def forMapTask(shuffleId: Int, mapId: Int, numReduces: Int, serializer:
     Serializer,
8.  writeMetrics: ShuffleWriteMetrics): ShuffleWriterGroup = {
9.      new ShuffleWriterGroup {
10.       //在 FileShuffleBlockResolver 中维护着当前 Map Task 对应 shuffleId 标识的
          //Shuffle 中，指定 numReduces 个数的 Reduce 的各个状态
11.
12. shuffleStates.putIfAbsent(shuffleId, new ShuffleState(numReduces))
13.       private val shuffleState = shuffleStates(shuffleId)
14.
15. ......
16.          //根据 Reduce 端的任务个数，构建元素类型为 DiskBlockObjectWriter 的数组，
             //DiskBlockObjectWriter 负责具体数据的磁盘写入
17.    //原则上，Shuffle 的输出可以存放在各种提供存储机制的系统上，但为了容错性等方面的
       //考虑，目前的 Shuffle 实行机制都会写入到磁盘中
18.
19.  val writers: Array[DiskBlockObjectWriter] = {
20.        //这里的逻辑 Bucket 的 Id 值即对应的 Reduce 的任务序号，或者说分区 ID
21.        Array.tabulate[DiskBlockObjectWriter](numReduces) { bucketId =>
22.        //针对每个 Map 端分区的 Id 与 Bucket 的 Id 构建数据块的逻辑标识
23.          val blockId = ShuffleBlockId(shuffleId, mapId, bucketId)
24.          val blockFile = blockManager.diskBlockManager.getFile(blockId)
25.          val tmp = Utils.tempFileWith(blockFile)
26.blockManager.getDiskWriter(blockId, tmp, serializerInstance, bufferSize,
   writeMetrics)
27.        }
28.      }
29. ......
```

```
30.       //任务完成时回调的释放写入器方法
31.       override def releaseWriters(success: Boolean) {
32. shuffleState.completedMapTasks.add(mapId)
33.       }
34.     }
35.   }
```

其中，ShuffleBlockId 实例构建的源码如下：

```
1.  case class ShuffleBlockId(shuffleId: Int, mapId: Int, reduceId: Int)
    extends BlockId {
2.    override def name: String = "shuffle_" + shuffleId + "_" + mapId + "_"
      + reduceId
3.  }
```

从 name 方法的重载上可以看出，后续构建的文件与代码中的 mapId、reduceId 的关系。当然，所有同一个 Shuffle 的输出数据块，都会带上 shuffleId 这个唯一标识的，因此全局角度上，逻辑数据块 name 不会重复（针对一些推测机制或失败重试机制之类的场景而已，逻辑 name 没有带上时间信息，因此缺少多次执行的输出区别，但在管理这些信息时会维护一个时间作为有效性判断）。

2. 基于Hash的Shuffle实现方式二的源码解析

下面通过详细解析 FileShuffleBlockResolver 源码来加深对文件合并机制的理解。

由于在 Spark 1.6 中，文件合并机制已经删除，因此下面基于 Spark 1.5 版本的代码对文件合并机制的具体实现细节进行解析。以下代码位于 FileShuffleBlockResolver 类中。

合并机制的关键控制代码如下所示。

Spark 1.5.0 版本的 FileShuffleBlockResolver.scala 的源码（Spark 2.4 版本已无 HashShuffleManager 方式）如下：

```
1.  /**
2.    *获取一个针对特定 Map Task 的 ShuffleWriterGroup
3.    */
4.
5.
6.    def forMapTask(shuffleId: Int, mapId: Int, numBuckets: Int, serializer:
      Serializer,
7.  writeMetrics: ShuffleWriteMetrics): ShuffleWriterGroup = {
8.      new ShuffleWriterGroup {
9.  ......
10.       val writers: Array[DiskBlockObjectWriter] = if
        (consolidateShuffleFiles) {
11. //获取未使用的文件组
12. fileGroup = getUnusedFileGroup()
13.         Array.tabulate[DiskBlockObjectWriter](numBuckets) { bucketId =>
14. val blockId = ShuffleBlockId(shuffleId, mapId, bucketId)
15. //注意获取磁盘写入器时，传入的第二个参数与未使用文件合并机制时的差异
16. //fileGroup(bucketId)：构造器方式调用，对应 apply 的方法调用
17.blockManager.getDiskWriter(blockId, fileGroup(bucketId), serializerInstance,
   bufferSize,
18. writeMetrics)
19.         }
20.       } else {
21.         Array.tabulate[DiskBlockObjectWriter](numBuckets) { bucketId =>
```

```
22. val blockId = ShuffleBlockId(shuffleId, mapId, bucketId)
23. //根据 ShuffleBlockId 信息获取文件名
24.         val blockFile = blockManager.diskBlockManager.getFile(blockId)
25.         val tmp = Utils.tempFileWith(blockFile)
26.blockManager.getDiskWriter(blockId, tmp, serializerInstance, bufferSize,
   writeMetrics)
27.       }
28.     }
29. ......
30. writeMetrics.incShuffleWriteTime(System.nanoTime - openStartTime)
31.      override def releaseWriters(success: Boolean) {
32. //带文件合并机制时，写入器在释放后的处理
33. //3 个关键信息 mapId、offsets、lengths
34.       if (consolidateShuffleFiles) {
35.         if (success) {
36.           val offsets = writers.map(_.fileSegment().offset)
37.           val lengths = writers.map(_.fileSegment().length)
38. fileGroup.recordMapOutput(mapId, offsets, lengths)
39.         }
40. //回收文件组，便于后续复用
41. recycleFileGroup(fileGroup)
42.       } else {
43. shuffleState.completedMapTasks.add(mapId)
44.       }
45.     }
```

其中，第 10 行中的 consolidateShuffleFiles 变量，是判断是否设置了文件合并机制，当设置 consolidateShuffleFiles 为 true 后，会继续调用 getUnusedFileGroup 方法，在该方法中会获取未使用的文件组，即重新分配或已经释放可以复用的文件组。

获取未使用的文件组（ShuffleFileGroup）的相关代码 getUnusedFileGroup 如下所示。

Spark 1.5.0 版本的 FileShuffleBlockResolver.scala 的源码（Spark 2.4 版本已无 HashShuffleManager 方式）如下：

```
1.  private def getUnusedFileGroup(): ShuffleFileGroup = {
2.  //获取已经构建但未使用的文件组，如果获取失败，则重新构建一个文件组
3.        val fileGroup = shuffleState.unusedFileGroups.poll()
4.        if (fileGroup != null) fileGroup else newFileGroup()
5.      }
6.  //重新构建一个文件组的源码
7.      private def newFileGroup(): ShuffleFileGroup = {
8.  //构建后会对文件编号进行递增，该文件编号最终用在生成的文件名中
9.        val fileId = shuffleState.nextFileId.getAndIncrement()
10.       val files = Array.tabulate[File](numBuckets) { bucketId =>
11. //最终的文件名，可以通过文件名的组成及取值细节，加深对实现细节在文件个数上的差异的理解
12.
13.         val filename = physicalFileName(shuffleId, bucketId, fileId)
14. blockManager.diskBlockManager.getFile(filename)
15.       }
16. //构建并添加到 shuffleState 中，便于后续复用
17.       val fileGroup = new ShuffleFileGroup(shuffleId, fileId, files)
18. shuffleState.allFileGroups.add(fileGroup)
19. fileGroup
20.     }
```

其中，第 13 行代码对应生成的文件名，即物理文件名，相关代码如下所示。

Spark 1.5.0 版本的 FileShuffleBlockResolver.scala 的源码（Spark 2.4 版本已无 HashShuffleManager 方式）如下：

```
1.  private def physicalFileName(shuffleId: Int, bucketId: Int, fileId: Int) = {
2.      "merged_shuffle_%d_%d_%d".format(shuffleId, bucketId, fileId)
3.    }
```

可以看到，与未使用文件合并时的基于 Hash 的 Shuffle 实现方式不同的是，在生成的文件名中没有对应的 mapId，取而代之的是与文件组相关的 fileId，而 fileId 则是多个 Mapper 端的 Task 所共用的，在此仅从生成的物理文件名中也可以看出文件合并的某些实现细节。

另外，对应生成的文件组既然是复用的，当一个 Mapper 端的 Task 执行结束后，便会释放该文件组（ShuffleFileGroup），之后继续调度时便会复用该文件组。对应地，调度到某个 Executor 工作点上同时运行的 Task 最大个数，就对应了最多分配的文件组个数。

而在 TaskSchedulerImpl 调度 Task 时，各个 Executor 工作点上 Task 调度控制的源码说明了在各个 Executor 工作点上调度并行的 Task 数，具体代码如下所示。

Spark 2.2.1 版本的 TaskSchedulerImpl.scala 的源码如下：

```
1.  private def resourceOfferSingleTaskSet(
2.  taskSet: TaskSetManager,
3.  maxLocality: TaskLocality,
4.  shuffledOffers: Seq[WorkerOffer],
5.  availableCpus: Array[Int],
6.      tasks: Seq[ArrayBuffer[TaskDescription]]) : Boolean = {
7.  var launchedTask = false
8.    for (i <- 0 until shuffledOffers.size) {
9.  val execId = shuffledOffers(i).executorId
10. val host = shuffledOffers(i).host
11. //判断当前 Executor 工作点上可用的内核个数是否满足 Task 所需的内核个数
12. //CPUS_PER_TASK：表示设置的每个 Task 所需的内核个数
13. if (availableCpus(i) >= CPUS_PER_TASK) {
14. try {
15. for (task <- taskSet.resourceOffer(execId, host, maxLocality)) {
16. ......
17. launchedTask = true
18. }
19.       } catch {
20. ......
21.         }
22.       }
23.   }
24. return launchedTask
25. }
```

Spark 2.4.3 版本的 TaskSchedulerImpl.scala 源码与 Spark 2.2.1 版本相比具有如下特点。

- 上段代码中第 6 行之后新增一个参数 addressesWithDescs。

```
1.   ......
2.     addressesWithDescs: ArrayBuffer[(String, TaskDescription)]) : Boolean = {
3.   ......
```

其中，设置每个 Task 所需的内核个数的配置属性如下：

```
1.  //每个任务请求的 CPU 个数
2.  val CPUS_PER_TASK = conf.getInt("spark.task.cpus", 1)
```

对于这些会影响 Executor 中并行执行的任务数的配置信息，设置时需要多方面考虑，包括内核个数与任务个数的合适比例，在内存模型中，为任务分配内存的具体策略等。任务分配内存的具体策略可以参考 Spark 官方给出的具体设计文档，以及文档中各种设计方式的权衡等内容。

7.4　Sorted Based Shuffle

在历史的发展中，为什么 Spark 最终还是放弃了 HashShuffle，使用了 Sorted-Based Shuffle，而且作为后起之秀的 Tungsten-based Shuffle 到底是在什么样的背景下产生的。Tungsten-Sort Shuffle 已经并入了 Sorted-Based Shuffle，Spark 的引擎会自动识别程序需要的是 Sorted-Based Shuffle，还是 Tungsten-Sort Shuffle，Spark 会检查相对的应用程序有没有 Aggregrate 的操作。Sorted-Based Shuffle 也有缺点，其缺点反而是它排序的特性，它强制要求数据在 Mapper 端必须先进行排序（注意，这里没有说对计算结果进行排序），所以导致它排序的速度有点慢。而 Tungsten-Sort Shuffle 对它的排序算法进行了改进，优化了排序的速度。

Spark 会根据宽依赖把它一系列的算子划分成不同的 Stage，Stage 的内部会进行 Pipeline、Stage 与 Stage 之间进行 Shuffle。Shuffle 的过程包含三部分，如图 7-6 所示。

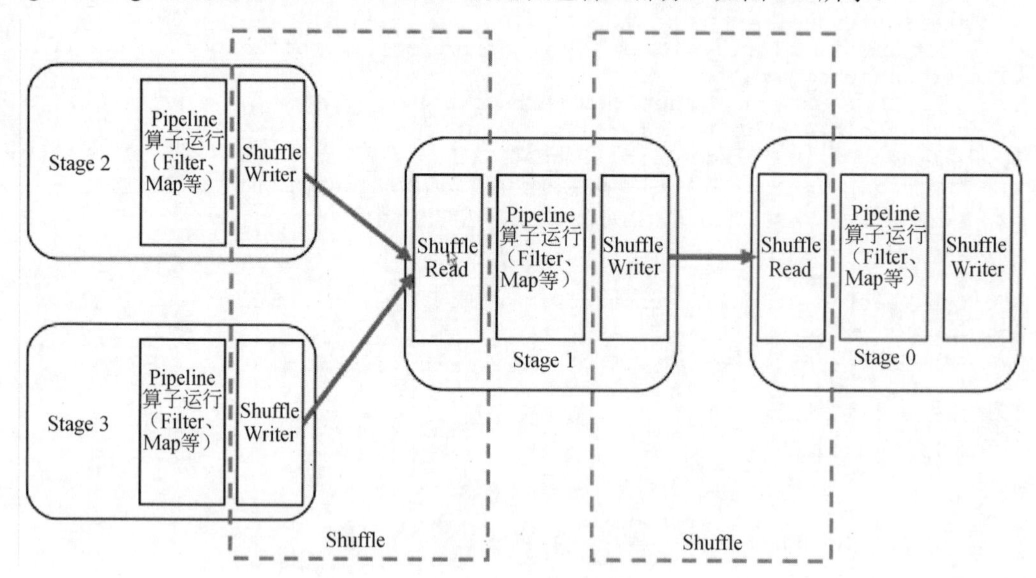

图 7-6　Shuffle 的过程示意图

第一部分是 Shuffle 的 Writer；第二部分是网络传输；第三部分是 Shuffle 的 Read，这三大部分设置了内存操作、磁盘 I/O、网络 I/O 以及 JVM 的管理。而这些东西是影响了 Spark 应用程序 95%以上效率的唯一原因。假设程序代码本身非常好，性能的 95%都消耗在 Shuffle 阶段的本地写磁盘文件、网络传输数据以及抓取数据这样的生命周期中，如图 7-7 所示。

在 Shuffle 写数据的时候，内存中有一个缓存区叫 Buffer，可以将其想象成一个 Map，同时在本地磁盘有对应的本地文件。如果本地磁盘有文件，在内存中肯定也需要有对应的管理句柄。也就是说，单从 ShuffleWriter 内存占用的角度讲，已经有一部分内存空间用在存储

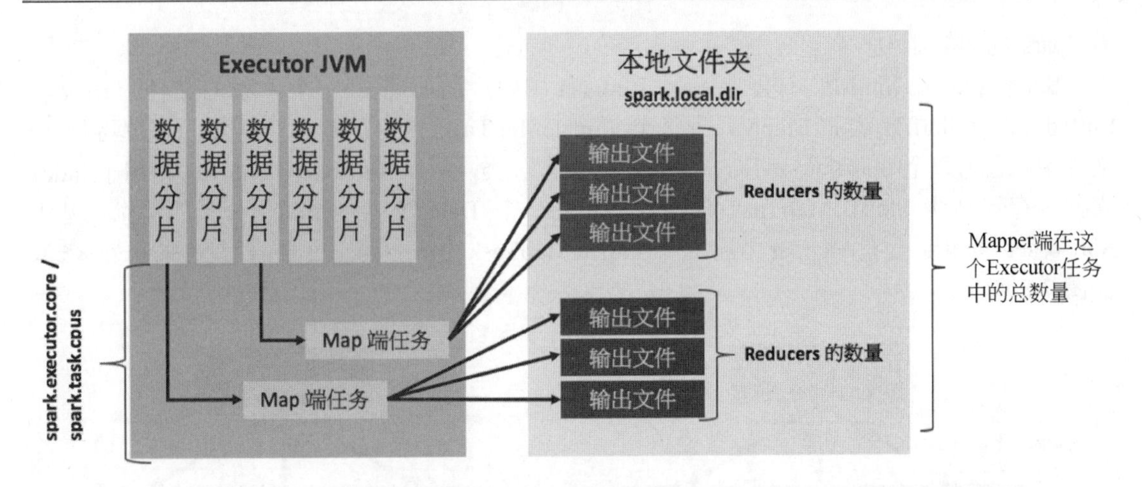

图 7-7　Shuffle 示意图

Buffer 数据，另一部分内存空间是用来管理文件句柄的，回顾 HashShuffle 所产生小文件的个数是 Mapper 分片数量×Reducer 分片数量（M×R）。例如，Mapper 端有 1000 个数据分片，Reducer 端也有 1000 个数据分片，在 HashShuffle 的机制下，它在本地内存空间中会产生 1000×1000=1000000 个小文件，结果可想而知，这么多的 I/O，这么多的内存消耗、这么容易产生 OOM，以及这么沉重的 CG 负担。再说，如果 Reducer 端去读取 Mapper 端的数据时，Mapper 端有这么多的小文件，要打开很多网络通道去读数据，打开 1000000 端口不是一件很轻松的事。这会导致一个非常经典的错误：Reducer 端下一个 Stage 通过 Driver 去抓取上一个 Stage 属于它自己的数据的时候，说文件找不到。其实，这个时候不是真的在磁盘上找不到文件，而是程序不响应，因为它在进行垃圾回收（GC）操作。

Spark 最根本要优化和迫切要解决的问题是：减少 Mapper 端 ShuffleWriter 产生的文件数量，这样便可以让 Spark 从几百台集群的规模瞬间变成可以支持几千台，甚至几万台集群的规模（一个 Task 背后可能是一个 Core 去运行，也可能是多个 Core 去运行，但默认情况下是用一个 Core 去运行一个 Task）。

减少 Mapper 端的小文件带来的好处如下。

（1）Mapper 端的内存占用变少了。

（2）Spark 不仅仅可以处理小规模的数据，即使处理大规模的数据，也不会很容易达到性能瓶颈。

（3）Reducer 端抓取数据的次数变少了。

（4）网络通道的句柄变少了。

（5）不仅仅减少了数据级别内存的消耗，更极大减少了 Spark 框架运行时必须消耗 Reducer 的内容。

7.4.1　概述

Sorted-Based Shuffle 的出现，最显著的优势是把 Spark 从只能处理中小规模数据的平台，变成可以处理无限大规模数据的平台。集群规模意味着 Spark 处理数据的规模，也意味着 Spark 的运算能力。

Sorted-Based Shuffle 不会为每个 Reducer 中的 Task 生产一个单独的文件，相反，Sorted-Based Shuffle 会把 Mapper 中每个 ShuffleMapTask 所有的输出数据 Data 只写到一个文件中，因为每个 ShuffleMapTask 中的数据会被分类，所以 Sort-based Shuffle 使用了 index 文件，存储具体 ShuffleMapTask 输出数据在同一个 Data 文件中是如何分类的信息。基于 Sort-based Shuffle 会在 Mapper 中的每个 ShuffleMapTask 中产生两个文件（并发度的个数×2），如图 7-8 所示。

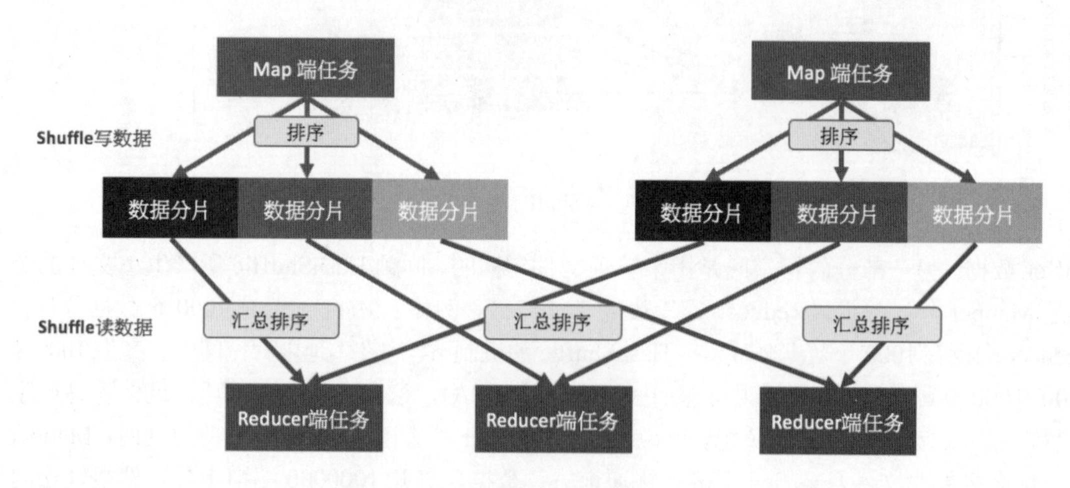

图 7-8　Sorted-Based Shuffle 示意图

图7-8会产生一个Data文件和一个Index文件。其中，Data 文件是存储当前Task的Shuffle输出的，而 Index 文件则存储了 Data 文件中的数据通过 Partitioner 的分类信息，此时下一个阶段的 Stage 中的 Task 就是根据这个 Index 文件获取自己所需要抓取的上一个 Stage 中 ShuffleMapTask 所产生的数据。

假设现在 Mapper 端有 1000 个数据分片，Reducer 端也有 1000 个数据分片，它的并发度是 100，使用 Sorted-Based Shuffle 会产生多少个 Mapper 端的小文件，答案是 100×2 = 200 个。它的 MapTask 会独自运行，每个 MapTask 在运行时写两个文件，运行成功后就不需要这个 MapTask 的文件句柄，无论是文件本身的句柄，还是索引的句柄，都不需要，所以如果它的并发度是 100 个 Core，每次运行 100 个任务，它最终只会占用 200 个文件句柄，这与 HashShuffle 的机制不一样，HashShuffle 最差的情况是 Hashed 句柄存储在内存中。

图 7-9 中，Sorted-Based Shuffle 主要在 Mapper 阶段，这个跟 Reducer 端没有任何关系，在 Mapper 阶段，Sorted-Based Shuffle 要进行排序，可以认为是二次排序，它的原理是有两个 Key 进行排序，第一个是 PartitionId 进行排序，第二个是本身数据的 Key 进行排序。它会把 PartitionId 分成 3 个，索引分别为 0、1、2，这个在 Mapper 端进行排序的过程其实是让 Reducer 去抓取数据的时候变得更高效。例如，第一个 Reducer，它会到 Mapper 端的索引为 0 的数据分片中抓取数据。具体而言，Reducer 首先找 Driver 去获取父 Stage 中每个 ShuffleMapTask 输出的位置信息，根据位置信息获取 Index 文件，解析 Index 文件，从解析的 Index 文件中获取 Data 文件中属于自己的那部分内容。

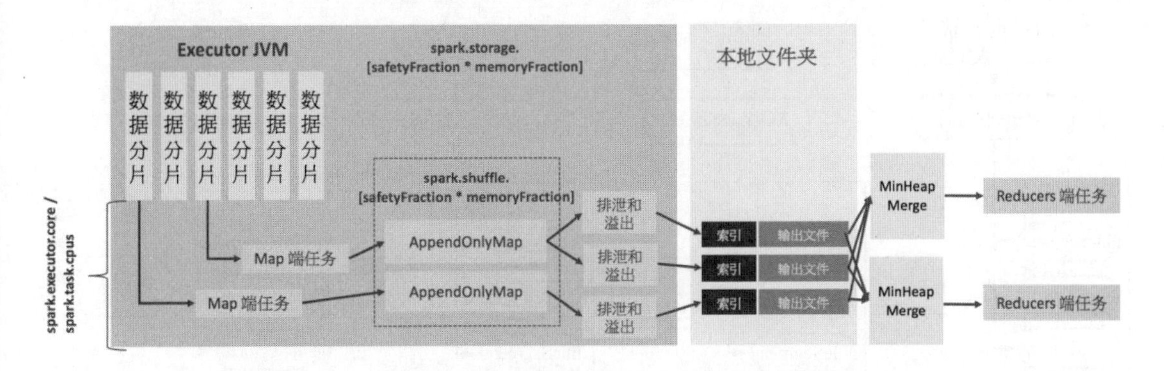

图 7-9　Sorted-Based Shuffle 流程图

一个 Mapper 任务除了有一个数据文件外，它也会有一个索引文件，Map Task 把数据写到文件磁盘的顺序是根据自身的 Key 写进去的，同时也是按照 Partition 写进去的，因为它是顺序写数据，记录每个 Partition 的大小。

Sort-Based Shuffle 的弱点如下。

（1）如果 Mapper 中 Task 的数量过大，依旧会产生很多小文件，此时在 Shuffle 传数据的过程中到 Reducer 端，Reducer 会需要同时大量地记录进行反序列化，导致大量内存消耗和 GC 负担巨大，造成系统缓慢，甚至崩溃！

（2）强制了在 Mapper 端必须要排序，这里的前提是数据本身不需要排序。

（3）如果在分片内也需要进行排序，此时需要进行 Mapper 端和 Reducer 端的两次排序。

（4）它要基于记录本身进行排序，这就是 Sort-Based Shuffle 最致命的性能消耗。

7.4.2　Sorted Based Shuffle 内核

Sorted-Based Shuffle 的核心是借助于 ExternalSorter 把每个 ShuffleMapTask 的输出排序到一个文件中（FileSegmentGroup），为了区分下一个阶段 Reducer Task 不同的内容，它还需要有一个索引文件（Index）来告诉下游 Stage 的并行任务，那一部分是属于下游 Stage 的，如图 7-10 所示。

图 7-10 中，在 Reducer 端有 4 个 Reducer Task，它会产生一组 File Group 和一个索引文件，File Group 里的 FileSegement 会进行排序，下游的 Task 很容易根据索引（index）定位到这个 File 中的那一部分。FileSegement 是属于下游的，相当于一个指针，下游的 Task 要向 Driver 去确定文件在哪里，然后到这个 File 文件所在的地方，实际上会与 BlockManager 进行沟通，BlockManager 首先会读一个 Index 文件，根据它的命名规则进行解析。例如，下一个阶段的第一个 Task，一般就是抓取第一个 Segment，这是一个指针定位的过程。

Sort-Based Shuffle 最大的意义是减少临时文件的输出数量，且只会产生两个文件：一个是包含不同内容，划分成不同 FileSegment 构成的单一文件 File；另外一个是索引文件 Index。图 7-10 中，Sort-Based Shuffle 展示了一个 Sort and Spill 的过程（它是 Spill 到磁盘的时候再进行排序的）。

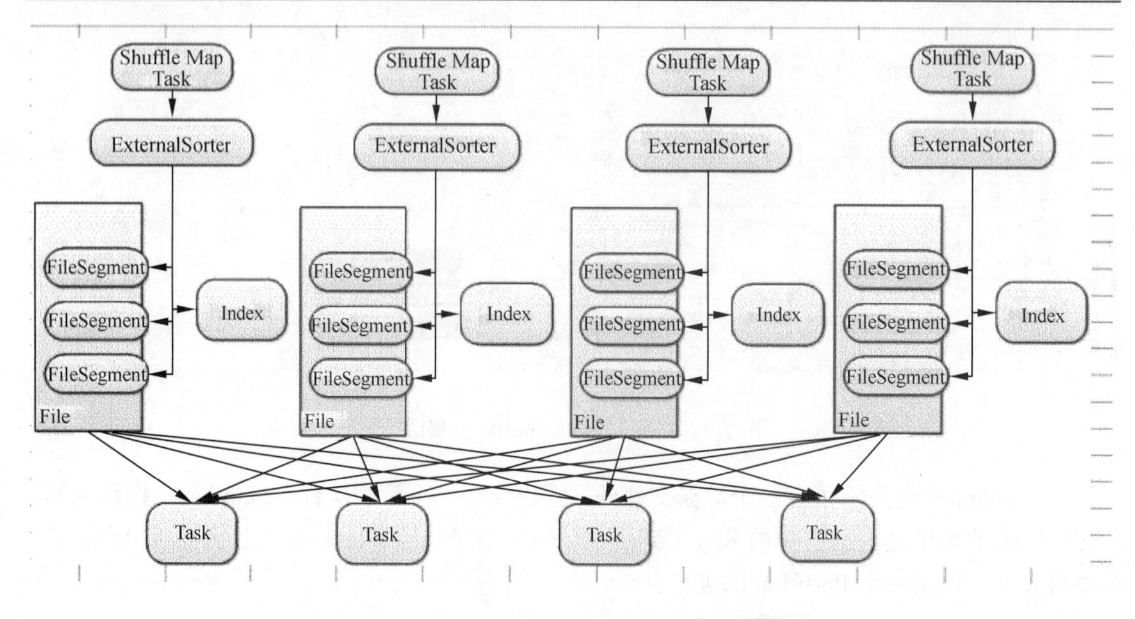

图 7-10　Sorted-Based Shuffle 的核心示意图

7.4.3　Sorted Based Shuffle 数据读写的源码解析

Sorted Based Shuffle，即基于 Sorted 的 Shuffle 实现机制，在该 Shuffle 过程中，Sorted 体现在输出的数据会根据目标的分区 Id（即带 Shuffle 过程的目标 RDD 中各个分区的 Id 值）进行排序，然后写入一个单独的 Map 端输出文件中。相应地，各个分区内部的数据并不会再根据 Key 值进行排序，除非调用带排序目的的方法，在方法中指定 Key 值的 Ordering 实例，才会在分区内部根据该 Ordering 实例对数据进行排序。当 Map 端的输出数据超过内存容纳大小时，会将各个排序结果 Spill 到磁盘上，最终再将这些 Spill 的文件合并到一个最终的文件中。在 Spark 的各种计算算子中到处体现了一种惰性的理念，在此也类似，在需要提升性能时，引入根据分区 Id 排序的设计，同时仅在指定分区内部排序的情况下，才会全局去排序。而 Hadoop 的 MapReduce 相比之下带有一定的学术气息，中规中矩，严格设计 Shuffle 阶段中的各个步骤。

基于 Hash 的 Shuffle 实现，ShuffleManager 的具体实现子类为 HashShuffleManager，对应的具体实现机制如 7-11 所示。

在图 7-11 中，各个不同的 ShuffleHandle 与不同的具体 Shuffle 写入器实现子类是一一对应的，可以认为是通过注册时生成的不同 ShuffleHandle 设置不同的 Shuffle 写入器实现子类。

从 ShuffleManager 注册的配置属性与具体实现子类的映射关系，即前面提及的在 SparkEnv 中实例化的代码，可以看出 sort 与 tungsten-sort 对应的具体实现子类都是 org.apache.spark.shuffle.sort.SortShuffleManager。也就是当前基于 Sort 的 Shuffle 实现机制与使用 Tungsten 项目的 Shuffle 实现机制都是通过 SortShuffleManager 类来提供接口，两种实现机制的区别在于，该类中使用了不同的 Shuffle 数据写入器。

SortShuffleManager 根据内部采用的不同实现细节，对应有两种不同的构建 Map 端文件输出的写方式，分别为序列化排序模式与反序列化排序模式。

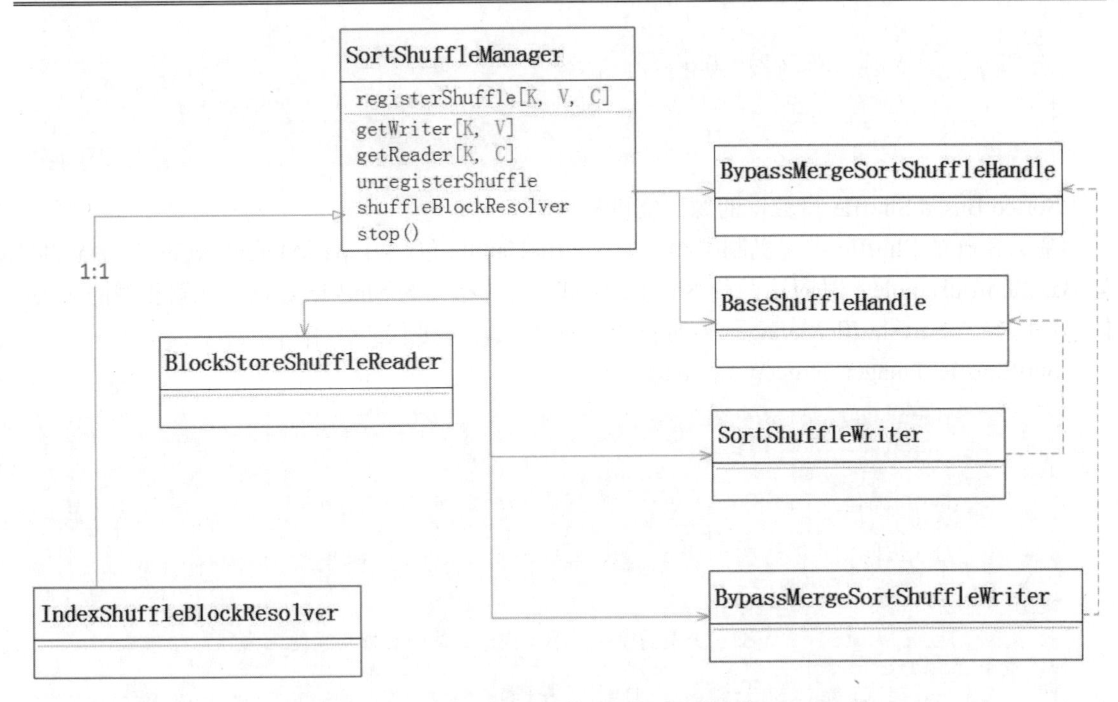

图 7-11　基于 Sorted 的 Shuffle 实现机制的框架类图

（1）序列化排序（Serialized sorting）模式：这种方式对应了新引入的基于 Tungsten 项目的方式。

（2）反序列化排序（Deserialized sorting）模式：这种方式对应除了前面这种方式之外的其他方式。

基于 Sort 的 Shuffle 实现机制采用的是反序列化排序模式。下面分析该实现机制下的数据写入器的实现细节。

基于 Sort 的 Shuffle 实现机制，具体的写入器的选择与注册得到的 ShuffleHandle 类型有关，参考 SortShuffleManager 类的 registerShuffle 方法，相关代码如下所示。

SortShuffleManager.scala 的源码如下：

```
1.   override def registerShuffle[K, V, C](
2.       shuffleId: Int,
3.       numMaps: Int,
4.       dependency: ShuffleDependency[K, V, C]): ShuffleHandle = {
5.     //通过 shouldBypassMergeSort 方法判断是否满足回退到 Hash 风格的 Shuffle 条件
6.     if (SortShuffleWriter.shouldBypassMergeSort(conf, dependency)) {
7.     //如果当前的分区个数小于设置的配置属性:
       //spark.shuffle.sort.bypassMergeThreshold,同时不需要在 Map 对数据进行聚合,
       //此时可以直接写文件，并在最后将文件合并
8.
9.   new BypassMergeSortShuffleHandle[K, V](
10.        shuffleId, numMaps, dependency.asInstanceOf[ShuffleDependency[K,
           V, V]])
11.     } else if (SortShuffleManager.canUseSerializedShuffle(dependency)) {
12.       //否则，试图 Map 输出缓冲区的序列化形式，因为这杨更高效
13.       new SerializedShuffleHandle[K, V](
14.         shuffleId, numMaps, dependency.asInstanceOf[ShuffleDependency[K,
            V, V]])
```

```
15.     } else {
16.       //否则，缓冲在反序列化形式 Map 输出
17.       new BaseShuffleHandle(shuffleId, numMaps, dependency)
18.     }
19.   }
```

Sorted Based Shuffle 写数据的源码解析如下。

基于 Sort 的 Shuffle 实现机制中相关的 ShuffleHandle 包含 BypassMergeSortShuffleHandle 与 BaseShuffleHandle。对应这两种 ShuffleHandle 及其相关的 Shuffle 数据写入器类型的相关代码可以参考 SortShuffleManager 类的 getWriter 方法，关键代码如下所示。

SortShuffleManager 的 getWriter 的源码如下：

```
1.      override def getWriter[K, V](
2.       handle: ShuffleHandle,
3.       mapId: Int,
4.       context: TaskContext): ShuffleWriter[K, V] = {
5.     numMapsForShuffle.putIfAbsent(
6.       handle.shuffleId, handle.asInstanceOf[BaseShuffleHandle[_, _,
       _]].numMaps)
7.     val env = SparkEnv.get
8.   //通过对 ShuffleHandle 类型的模式匹配，构建具体的数据写入器
9.     handle match {
10.      case unsafeShuffleHandle: SerializedShuffleHandle[K @unchecked, V
       @unchecked] =>
11.        new UnsafeShuffleWriter(
12.          env.blockManager,
13.          shuffleBlockResolver.asInstanceOf[IndexShuffleBlockResolver],
14.          context.taskMemoryManager(),
15.          unsafeShuffleHandle,
16.          mapId,
17.          context,
18.          env.conf)
19.      case bypassMergeSortHandle: BypassMergeSortShuffleHandle[K
       @unchecked, V @unchecked] =>
20.        new BypassMergeSortShuffleWriter(
21.          env.blockManager,
22.          shuffleBlockResolver.asInstanceOf[IndexShuffleBlockResolver],
23.          bypassMergeSortHandle,
24.          mapId,
25.          context,
26.          env.conf)
27.      case other: BaseShuffleHandle[K @unchecked, V @unchecked, _] =>
28.        new SortShuffleWriter(shuffleBlockResolver, other, mapId,
         context)
29.    }
30.  }
```

在对应构建的两种数据写入器类 BypassMergeSortShuffleWriter 与 SortShuffleWriter 中，都是通过变量 shuffleBlockResolver 对逻辑数据块与物理数据块的映射进行解析，而该变量使用的是与基于 Hash 的 Shuffle 实现机制不同的解析类，即当前使用的 IndexShuffleBlockResolver。

下面开始解析这两种写数据块方式的源码实现。

1. BypassMergeSortShuffleWriter写数据的源码解析

该类实现了带 Hash 风格的基于 Sort 的 Shuffle 机制，为每个 Reduce 端的任务构建一个

输出文件，将输入的每条记录分别写入各自对应的文件中，并在最后将这些基于各个分区的文件合并成一个输出文件。

在 Reducer 端任务数比较少的情况下，基于 Hash 的 Shuffle 实现机制明显比基于 Sort 的 Shuffle 实现机制要快，因此基于 Sort 的 Shuffle 实现机制提供了一个 fallback 方案，对于 Reducer 端任务数少于配置属性 spark.shuffle.sort.bypassMergeThreshold 设置的个数时，使用带 Hash 风格的 fallback 计划，由 BypassMergeSortShuffleWriter 具体实现。

使用该写入器的条件如下。

（1）不能指定 Ordering，从前面数据读取器的解析可以知道，当指定 Ordering 时，会对分区内部的数据进行排序。因此，对应的 BypassMergeSortShuffleWriter 写入器避免了排序开销。

（2）不能指定 Aggregator。

（3）分区个数小于 spark.shuffle.sort.bypassMergeThreshold 配置属性指定的个数。

和其他 ShuffleWriter 的具体子类一样，BypassMergeSortShuffleWriter 写数据的具体实现位于实现的 write 方法中，关键代码如下所示。

BypassMergeSortShuffleWriter.scala 的 write 的源码如下：

```
1.    public void write(Iterator<Product2<K, V>> records) throws IOException {
2.   //为每个 Reduce 端的分区打开的 DiskBlockObjectWriter 存放于 partitionWriters,
     //需要根据具体 Reduce 端的分区个数进行构建
3.
4.
5.     assert (partitionWriters == null);
6.     if (!records.hasNext()) {
7.       partitionLengths = new long[numPartitions];
8.    //初始化索引文件的内容，此时对应各个分区的数据量或偏移量需要在后续获取分区的真实
      //数据量时重写
9.
10.       shuffleBlockResolver.writeIndexFileAndCommit(shuffleId, mapId,
          partitionLengths, null);
11.      //下面代码的调用形式是对应在 Java 类中调用 Scala 提供的 object 中的 apply 方法
         //的形式，是由编译器编译 Scala 中的 object 得到的结果来决定的
12.
13.
14.       mapStatus = MapStatus$.MODULE$.apply(blockManager.shuffleServerId(),
          partitionLengths);
15.       return;
16.     }
17.     final SerializerInstance serInstance = serializer.newInstance();
18.     final long openStartTime = System.nanoTime();
19.   //对应每个分区各配置一个磁盘写入器 DiskBlockObjectWriter
20.     partitionWriters = new DiskBlockObjectWriter[numPartitions];
21.     partitionWriterSegments = new FileSegment[numPartitions];
22.     //注意，在该写入方式下，会同时打开 numPartitions 个 DiskBlockObjectWriter,
        //因此对应的分区数不应设置过大，避免带来过大的内存开销目前对应 DiskBlock-
        //ObjectWriter 的缓存大小默认配置为 32KB，比早先的 100KB 降低了很多，但也说明
        //不适合同时打开太多的 DiskBlockObjectWriter 实例
23.     for (int i = 0; i < numPartitions; i++) {
24.       final Tuple2<TempShuffleBlockId, File> tempShuffleBlockIdPlusFile =
25.         blockManager.diskBlockManager().createTempShuffleBlock();
26.       final File file = tempShuffleBlockIdPlusFile._2();
27.       final BlockId blockId = tempShuffleBlockIdPlusFile._1();
28.       partitionWriters[i] =
```

```
29.         blockManager.getDiskWriter(blockId, file, serInstance,
            fileBufferSize, writeMetrics);
30.     }
31.
32.     //创建文件写入和创建磁盘写入器都涉及与磁盘的交互，当打开许多文件时，磁盘写会花费
        //很长时间，所以磁盘写入时间应包含在 Shuffle 写入时间内
33.
34.     writeMetrics.incWriteTime(System.nanoTime() - openStartTime);
35.     //读取每条记录，并根据分区器将该记录交由分区对应的 DiskBlockObjectWriter，
        //写入各自对应的临时文件中
36.
37.     while (records.hasNext()) {
38.       final Product2<K, V> record = records.next();
39.       final K key = record._1();
40.       partitionWriters[partitioner.getPartition(key)].write(key,
          record._2());
41.     }
42.
43.     for (int i = 0; i < numPartitions; i++) {
44.       final DiskBlockObjectWriter writer = partitionWriters[i];
45.       partitionWriterSegments[i] = writer.commitAndGet();
46.       writer.close();
47.     }
48.      //获取最终合并后的文件名，对应格式为："shuffle_" + shuffleId + "_" + mapId
        // + "_" + reduceId + ".index"，并且其中的 reduceId 为 0，对应的含义就是
        //该文件包含所有为 Reduce 端输出的数据
49.     File output = shuffleBlockResolver.getDataFile(shuffleId, mapId);
50.     File tmp = Utils.tempFileWith(output);
51.     try {
52.     //在此合并前面生成的各个中间临时文件，并获取各个分区对应的数据量，由数据量可以得
        //到对应的偏移量
53.
54.       partitionLengths = writePartitionedFile(tmp);
55.  //主要是根据前面获取的数据量，重写 Index 文件中的偏移量信息
56.       shuffleBlockResolver.writeIndexFileAndCommit(shuffleId, mapId,
          partitionLengths, tmp);
57.     } finally {
58.       if (tmp.exists() && !tmp.delete()) {
59.         logger.error("Error while deleting temp file {}",
            tmp.getAbsolutePath());
60.       }
61.     }
62.  //封装并返回任务结果
63.    mapStatus = MapStatus$.MODULE$.apply(blockManager.shuffleServerId(),
       partitionLengths);
64.  }
```

其中调用的 createTempShuffleBlock 方法描述了各个分区生成的中间临时文件的格式与对应的 BlockId。

DiskBlockManager 的 createTempShuffleBlock 的源码如下：

```
1.   /**中间临时文件名的格式由前缀 temp_shuffle_与 randomUUID 组成，可以唯一标识
     BlockId*/
2.   def createTempShuffleBlock(): (TempShuffleBlockId, File) = {
3.      var blockId = new TempShuffleBlockId(UUID.randomUUID())
4.      while (getFile(blockId).exists()) {
```

```
5.        blockId = new TempShuffleBlockId(UUID.randomUUID())
6.      }
7.      (blockId, getFile(blockId))
8.    }
```

从上面的分析中可以知道，每个 Map 端的任务最终会生成两个文件，即数据（Data）文件和索引（Index）文件。

另外，使用 DiskBlockObjectWriter 写记录时，是以 32 条记录批次写入的，不会占用太大的内存。但由于对应不能指定聚合器（Aggregator），写数据时也是直接写入记录，因此对应后续的网络 I/O 的开销也会很大。

2. SortShuffleWriter写数据的源码解析

前面 BypassMergeSortShuffleWriter 的写数据是在 Reducer 端的分区个数较少的情况下提供的一种优化方式，但当数据集规模非常大时，使用该写数据方式不合适时，就需要使用 SortShuffleWriter 来写数据块。

和其他 ShuffleWriter 的具体子类一样，SortShuffleWriter 写数据的具体实现位于实现的 write 方法中。

Spark 2.2.1 版本的 SortShuffleWriter 的 write 的源码如下：

```
1.      override def write(records: Iterator[Product2[K, V]]): Unit = {
2.   //当需要在 Map 端进行聚合操作时，此时将会指定聚合器（Aggregator）
3.   //将 Key 值的 Ordering 传入到外部排序器 ExternalSorter 中
4.      sorter = if (dep.mapSideCombine) {
5.        require(dep.aggregator.isDefined, "Map-side combine without
          Aggregator specified!")
6.        new ExternalSorter[K, V, C](
7.          context, dep.aggregator, Some(dep.partitioner), dep.keyOrdering,
            dep.serializer)
8.      } else {
9.     //没有指定 Map 端使用聚合时，传入 ExternalSorter 的聚合器（Aggregator）
       //与 Key 值的 Ordering 都设为 None，即不需要传入，对应在 Reduce 端读取数据
       //时才根据聚合器分区数据进行聚合，并根据是否设置 Ordering 而选择是否对分区
       //数据进行排序
10.       new ExternalSorter[K, V, V](
11.         context, aggregator = None, Some(dep.partitioner), ordering = None,
            dep.serializer)
12.     }
13.      //将写入的记录集全部放入外部排序器
14.     sorter.insertAll(records)
15.
16.     //不要费心在 Shuffle 写时间中，包括打开合并输出文件的时间，因为它只打开一个文件，
        //所以通常太快，无法精确测量（见 Spark-3570）
17.     //和 BypassMergeSortShuffleWriter 一样，获取输出文件名和 BlockId
18.     val output = shuffleBlockResolver.getDataFile(dep.shuffleId, mapId)
19.     val tmp = Utils.tempFileWith(output)
20.     try {
21.       val blockId = ShuffleBlockId(dep.shuffleId, mapId,
          IndexShuffleBlockResolver.NOOP_REDUCE_ID)
22.  //将分区数据写入文件，返回各个分区对应的数据量
23.       val partitionLengths = sorter.writePartitionedFile(blockId, tmp)
24.    //和 BypassMergeSortShuffleWriter 一样，更新索引文件的偏移量信息
```

```
25.       shuffleBlockResolver.writeIndexFileAndCommit(dep.shuffleId, mapId,
          partitionLengths, tmp)
26.       mapStatus = MapStatus(blockManager.shuffleServerId, partitionLengths)
27.     } finally {
28.       if (tmp.exists() && !tmp.delete()) {
29.         logError(s"Error while deleting temp file ${tmp.getAbsolutePath}")
30.       }
31.     }
32.   }
```

Spark 2.4.3 版本的 SortShuffleWriter.scala 源码与 Spark 2.2.1 版本相比具有如下特点。

- 上段代码中第 5 行删掉。

在这种基于 Sort 的 Shuffle 实现机制中引入了外部排序器（ExternalSorter）。ExternalSorter 继承了 Spillable，因此内存使用达到一定阈值时，会 Spill 到磁盘，可以减少内存带来的开销。

外部排序器的 insertAll 方法内部在处理完（包含聚合和非聚合两种方式）每条记录时，都会检查是否需要 Spill。内部各种细节比较多，这里以 Spill 条件判断为主线，简单描述一下条件相关的代码。具体判断是否需要 Spill 的相关代码可以参考 Spillable 类中的 maybeSpill 方法（该方法的简单调用流程为：ExternalSorter #insterAll–>ExternalSorter #maybeSpillCollection –>Spillable#maybeSpill）。

Spillable.scala 的 maybeSpill 的源码如下：

```
1.      protected def maybeSpill(collection: C, currentMemory: Long): Boolean = {
2.    //判断是否需要 Spill
3.      var shouldSpill = false
4.    //1. 检查当前记录数是否是 32 的倍数——即对小批量的记录集进行 Spill
5.    //2. 同时，当前需要的内存大小是否达到或超过了当前分配的内存阈值
6.      if (elementsRead % 32 == 0 && currentMemory >= myMemoryThreshold) {
7.        //从 Shuffle 内存池中获取当前内存的两倍
8.        val amountToRequest = 2 * currentMemory - myMemoryThreshold
9.    //实际上会先申请内存，然后再次判断，最后决定是否 Spill
10.       val granted = acquireMemory(amountToRequest)
11.       myMemoryThreshold += granted
12.
13.       //内存很少时，如果准许内存进一步增长（tryToAcquire 返回 0，或者比
          //myMemoryThreshold 更多的内存），当前的 collection 将会溢出
14.       shouldSpill = currentMemory >= myMemoryThreshold
15.     }
16.     //当满足下列条件之一时，需要 Spill，条件如下所示：
17.     //1. 当前判断结果为 true
18.     //2. 从上次 Spill 之后所读取的记录数超过配置的阈值时
19.     //配置属性为：spark.shuffle.spill.numElementsForceSpillThreshold
20.     shouldSpill = shouldSpill || _elementsRead >
        numElementsForceSpillThreshold
21.     //Actually spill
22.     if (shouldSpill) {
23.       _spillCount += 1
24.       logSpillage(currentMemory)
25.       spill(collection)
26.       _elementsRead = 0
27.       _memoryBytesSpilled += currentMemory
28.       releaseMemory()
```

```
29.     }
30.     shouldSpill
31.   }
```

对于外部排序器，除了 insertAll 方法外，它的 writePartitionedFile 方法也非常重要。

ExternalSorter.scala 的 writePartitionedFile 的源码如下：

```
1.  def writePartitionedFile(
2.  blockId: BlockId,
3.  outputFile: File): Array[Long] = {
```

其中，BlockId 是数据块的逻辑位置，File 参数是对应逻辑位置的物理存储位置。这两个参数值的获取方法和使用 BypassMergeSortShuffleHandle 及其对应的 ShuffleWriter 是一样的。

在该方法中，有一个容易混淆的地方，与 Shuffle 的度量（Metric）信息有关，对应代码如下：

```
1.  context.taskMetrics().incMemoryBytesSpilled(memoryBytesSpilled)
2.  context.taskMetrics().incDiskBytesSpilled(diskBytesSpilled)
```

其中，第 1 行对应修改了 Spilled 的数据在内存中的字节大小，第 2 行则对应修改了 Spilled 的数据在磁盘中的字节大小。在内存中时，数据是以反序列化形式存放的，而存储到磁盘（默认会序列化）时，会对数据进行序列化。反序列化后的数据会远远大于序列化后的数据（也可以通过 UI 界面查看这两个度量信息的大小差异来确认，具体差异的大小和数据以及选择的序列化器有关，有兴趣的读者可以参考各序列器间的性能等比较文档）。

从这一点也可以看出，如果在内存中使用反序列化的数据，会大大增加内存的开销（也意味着增加 GC 负载），并且反序列化也会增加 CPU 的开销，因此引入了利用 Tungsten 项目的基于 Tungsten Sort 的 Shuffle 实现机制。Tungsten 项目的优化主要有三个方面，这里从避免反序列化的数据量会极大消耗内存这方面考虑，主要是借助 Tungsten 项目的内存管理模型，可以直接处理序列化的数据；同时，CPU 开销方面，直接处理序列化数据，可以避免数据反序列化的这部分处理开销。

7.5 Tungsten Sorted Based Shuffle

本节讲解 Tungsten Sorted Based Shuffle，包括 Tungsten Sorted Based Shuffle 概述、Tungsten Sorted Based 内核、Tungsten Sorted Based 数据读写的源码解析等内容。

7.5.1 概述

基于 Tungsten Sort 的 Shuffle 实现机制主要是借助 Tungsten 项目所做的优化来高效处理 Shuffle。

Spark 提供了配置属性，用于选择具体的 Shuffle 实现机制，但需要说明的是，虽然默认情况下 Spark 默认开启的是基于 Sort 的 Shuffle 实现机制（对应 spark.shuffle.manager 的默认值），但实际上，参考 Shuffle 的框架内核部分可知基于 Sort 的 Shuffle 实现机制与基于 Tungsten

Sort 的 Shuffle 实现机制都是使用 SortShuffleManager，而内部使用的具体的实现机制，是通过提供的两个方法进行判断的。对应非基于 Tungsten Sort 时，通过 SortShuffleWriter.shouldBypassMergeSort 方法判断是否需要回退到 Hash 风格的 Shuffle 实现机制，当该方法返回的条件不满足时，则通过 SortShuffleManager.canUseSerializedShuffle 方法判断是否需要采用基于 Tungsten Sort 的 Shuffle 实现机制，而当这两个方法返回都为 false，即都不满足对应的条件时，会自动采用常规意义上的基于 Sort 的 Shuffle 实现机制。

因此，当设置了 spark.shuffle.manager=tungsten-sort 时，也不能保证就一定采用基于 Tungsten Sort 的 Shuffle 实现机制。有兴趣的读者可以参考 Spark 1.5 及之前的注册方法的实现，该实现中 SortShuffleManager 的注册方法仅构建了 BaseShuffleHandle 实例，同时对应的 getWriter 中也只对应构建了 BaseShuffleHandle 实例。

7.5.2　Tungsten Sorted Based Shuffle 内核

基于 Tungsten Sort 的 Shuffle 实现机制的入口点仍然是 SortShuffleManager 类，与同样在 SortShuffleManager 类控制下的其他两种实现机制不同的是，基于 Tungsten Sort 的 Shuffle 实现机制使用的 ShuffleHandle 与 ShuffleWriter 分别为 SerializedShuffleHandle 与 UnsafeShuffleWriter。因此，对应的具体实现机制如图 7-12 所示。

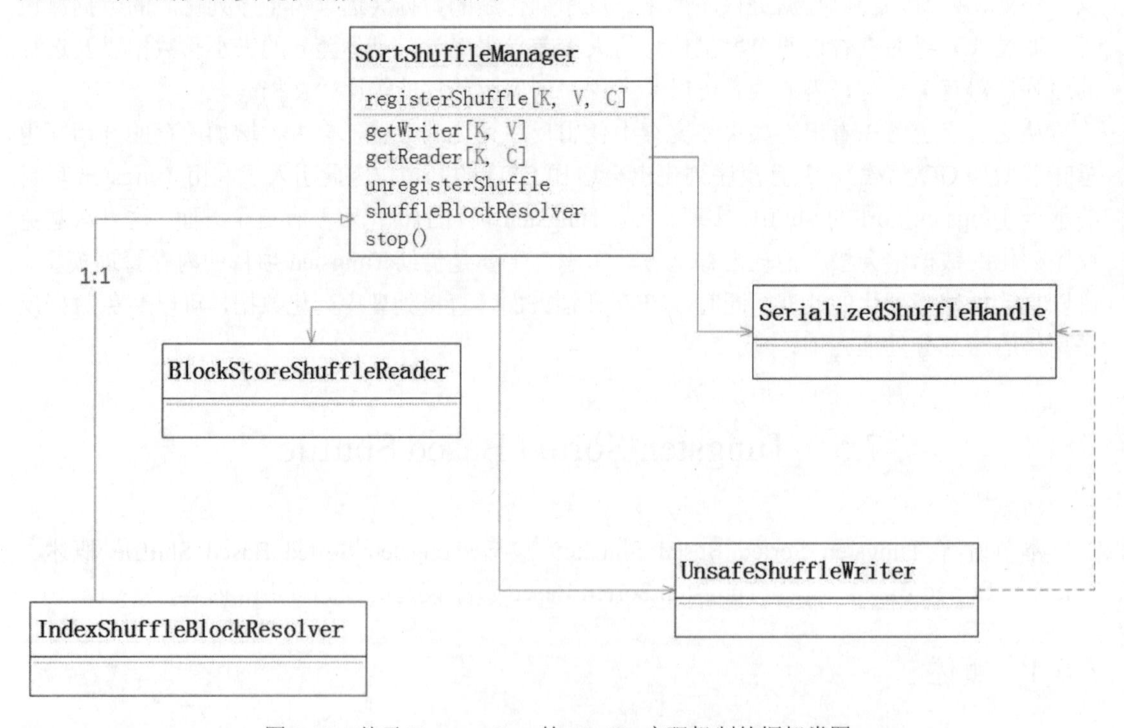

图 7-12　基于 TungstenSort 的 Shuffle 实现机制的框架类图

在 Sorted Based Shuffle 中，SortShuffleManager 根据内部采用的不同实现细节，分别给出两种排序模式，而基于 TungstenSort 的 Shuffle 实现机制对应的就是序列化排序模式。

从图 7-12 中可以看到基于 Sort 的 Shuffle 实现机制，具体的写入器的选择与注册得到的 ShuffleHandle 类型有关，参考 SortShuffleManager 类的 registerShuffle 方法。

registerShuffle 方法中会判断是否满足序列化模式的条件，如果满足，则使用基于 TungstenSort 的 Shuffle 实现机制，对应在代码中，表现为使用类型为 SerializedShuffleHandle 的 ShuffleHandle。上述代码进一步说明了在 spark.shuffle.manager 设置为 sort 时，内部会自动选择具体的实现机制。对应代码的先后顺序，就是选择的先后顺序。

对应的序列化排序（Serialized sorting）模式需要满足的条件如下所示。

（1）Shuffle 依赖中不带聚合操作或没有对输出进行排序的要求。

（2）Shuffle 的序列化器支持序列化值的重定位（当前仅支持 KryoSerializer 以及 Spark SQL 子框架自定义的序列化器）。

（3）Shuffle 过程中的输出分区个数少于 16 777 216 个。

实际上，使用过程中还有其他一些限制，如引入那个 Page 形式的内存管理模型后，内部单条记录的长度不能超过 128MB（具体内存模型可以参考 PackedRecordPointer 类）。另外，分区个数的限制也是该内存模型导致的（同样参考 PackedRecordPointer 类）。

所以，目前使用基于 TungstenSort 的 Shuffle 实现机制条件还是比较苛刻的。

7.5.3 Tungsten Sorted Based Shuffle 数据读写的源码解析

对应这种 SerializedShuffleHandle 及其相关的 Shuffle 数据写入器类型的相关代码，可以参考 SortShuffleManager 类的 getWriter 方法。

SortShuffleManager.scala 的源码如下：

```
1.      /** 为指定的分区提供一个数据写入器。该方法在 Map 端的 Tasks 中调用*/
2.  override def getWriter[K, V](
3.      handle: ShuffleHandle,
4.      mapId: Int,
5.      context: TaskContext): ShuffleWriter[K, V] = {
6.    numMapsForShuffle.putIfAbsent(
7.      handle.shuffleId,   handle.asInstanceOf[BaseShuffleHandle[_,   _,
        _]].numMaps)
8.    val env = SparkEnv.get
9.    handle match {
10.   //SerializedShuffleHandle 对应的写入器为 UnsafeShuffleWriter
      //使用的数据块逻辑与物理映射关系仍然为 IndexShuffleBlockResolver，对应
      //SortShuffleManager 中的变量，因此相同
11.     case unsafeShuffleHandle: SerializedShuffleHandle[K @unchecked, V
        @unchecked] =>
12.       new UnsafeShuffleWriter(
13.         env.blockManager,
14.         shuffleBlockResolver.asInstanceOf[IndexShuffleBlockResolver],
15.         context.taskMemoryManager(),
16.         unsafeShuffleHandle,
17.         mapId,
18.         context,
19.         env.conf)
20.     case bypassMergeSortHandle: BypassMergeSortShuffleHandle[K @
        unchecked, V @unchecked] =>
21.       new BypassMergeSortShuffleWriter(
22.         env.blockManager,
23.         shuffleBlockResolver.asInstanceOf[IndexShuffleBlockResolver],
24.         bypassMergeSortHandle,
25.         mapId,
```

```
26.           context,
27.           env.conf)
28.       case other: BaseShuffleHandle[K @unchecked, V @unchecked, _] =>
29.         new SortShuffleWriter(shuffleBlockResolver, other, mapId, context)
30
31.     }
32.   }
```

数据写入器类 UnsafeShuffleWriter 中使用 SortShuffleManager 实例中的变量 shuffleBlockResolver 来对逻辑数据块与物理数据块的映射进行解析，而该变量使用的是与基于 Hash 的 Shuffle 实现机制不同的解析类，即当前使用的 IndexShuffleBlockResolver。

UnsafeShuffleWriter 构建时传入了一个与其他两种基于 Sorted 的 Shuffle 实现机制不同的参数：context.taskMemoryManager()，在此构建了一个 TaskMemoryManager 实例并传入 UnsafeShuffleWriter。TaskMemoryManager 与 Task 是一对一的关系，负责管理分配给 Task 的内存。

下面开始解析写数据块的 UnsafeShuffleWriter 类的源码实现。首先来看其 write 的方法。

UnsafeShuffleWriter.scala 的源码如下：

```
1.       public void write(scala.collection.Iterator<Product2<K, V>> records)
         throws IOException {
2.     ......
3.       boolean success = false;
4.       try {
5.       //对输入的记录集 records，循环将每条记录插入到外部排序器
6.         while (records.hasNext()) {
7.           insertRecordIntoSorter(records.next());
8.         }
9.         closeAndWriteOutput();
10.        //生成最终的两个结果文件，和 Sorted Based Shuffle 的实现机制一样，每个 Map
           //端的任务对应生成一个数据（Data）文件和对应的索引（Index）文件
11.
12.
13.        success = true;
14.      } finally {
15.        if (sorter != null) {
16.          try {
17.            sorter.cleanupResources();
18.          } catch (Exception e) {
19.       ......
```

写过程的关键步骤有以下三步。

（1）通过 insertRecordIntoSorter(records.next())方法将每条记录插入外部排序器。

（2）closeAndWriteOutput 方法写数据文件与索引文件，在写的过程中，会先合并外部排序器在插入过程中生成的 Spill 中间文件。

（3）sorter.cleanupResources()最后释放外部排序器的资源。

首先查看将每条记录插入外部排序器（ShuffleExternalSorter）时所使用的 insertRecordIntoSorter 方法。

UnsafeShuffleWriter.scala 的源码如下：

```
1.         void insertRecordIntoSorter(Product2<K, V> record) throws IOException {
2.       assert(sorter != null);
3.     //对于多次访问的 Key 值，使用局部变量，可以避免多次函数调用
```

```
4.      final K key = record._1();
5.      final int partitionId = partitioner.getPartition(key);
6.   //先复位存放每条记录的缓冲区，内部使用 ByteArrayOutputStream 存放每条记录，容量
     //为 1MB
7.
8.
9.      serBuffer.reset();
10.  //进一步使用序列化器从 serBuffer 缓冲区构建序列化输出流，将记录写入到缓冲区
11.     serOutputStream.writeKey(key, OBJECT_CLASS_TAG);
12.     serOutputStream.writeValue(record._2(), OBJECT_CLASS_TAG);
13.     serOutputStream.flush();
14.
15.     final int serializedRecordSize = serBuffer.size();
16.     assert (serializedRecordSize > 0);
17.    //将记录插入到外部排序器中，serBuffer 是一个字节数组，内部数据存放的偏移量为
       //Platform.BYTE_ARRAY_OFFSET
18.
19.
20.     sorter.insertRecord(
21.       serBuffer.getBuf(), Platform.BYTE_ARRAY_OFFSET, serializedRecordSize,
          partitionId);
22.   }
```

下面继续查看第二步写数据文件与索引文件的 closeAndWriteOutput 方法。

closeAndWriteOutput 的源码如下：

```
1.     void closeAndWriteOutput() throws IOException {
2.      assert(sorter != null);
3.      updatePeakMemoryUsed();
4.      //设为 null，用于 GC 垃圾回收
5.      serBuffer = null;
6.      serOutputStream = null;
7.      //关闭外部排序器，并获取全部 Spill 信息
8.      final SpillInfo[] spills = sorter.closeAndGetSpills();
9.      sorter = null;
10.     final long[] partitionLengths;
11.     //通过块解析器获取输出文件名
12.     final File output = shuffleBlockResolver.getDataFile(shuffleId, mapId);
13.     //在后续合并 Spill 文件时先使用临时文件名，最终再重命名为真正的输出文件名，
        //即在 writeIndexFileAndCommit 方法中会重复通过块解析器获取输出文件名
14.     final File tmp = Utils.tempFileWith(output);
15.     try {
16.       try {
17.         partitionLengths = mergeSpills(spills, tmp);
18.       } finally {
19.         for (SpillInfo spill : spills) {
20.           if (spill.file.exists() && ! spill.file.delete()) {
21.             logger.error("Error while deleting spill file {}",
                spill.file.getPath());
22.           }
23.         }
24.       }
25.
26.  //将合并 Spill 后获取的分区及其数据量信息写入索引文件，并将临时数据文件重命名为
     //真正的数据文件名
27.
```

```
28.       shuffleBlockResolver.writeIndexFileAndCommit(shuffleId, mapId,
          partitionLengths, tmp);
29.     } finally {
30.       if (tmp.exists() && !tmp.delete()) {
31.         logger.error("Error while deleting temp file {}", tmp.getAbsolutePath());
32.       }
33.     }
34.    mapStatus = MapStatus$.MODULE$.apply(blockManager.shuffleServerId(),
       partitionLengths);
35.   }
```

closeAndWriteOutput 方法主要有以下三步。

（1）触发外部排序器，获取 Spill 信息。

（2）合并中间的 Spill 文件，生成数据文件，并返回各个分区对应的数据量信息。

（3）根据各个分区的数据量信息生成数据文件对应的索引文件。

writeIndexFileAndCommit 方法和 Sorted Based Shuffle 机制的实现一样，在此仅分析过程中不同的 Spill 文件合并步骤，即 mergeSpills 方法的具体实现。

UnsafeShuffleWriter.scala 的 mergeSpills 方法的源码如下：

```
1.   /**
      * 合并 0 个或多个 Spill 的中间文件，基于 Spills 的个数以及 I/O 压缩码选择最
      * 快速的合并策略。返回包含合并文件中各个分区的数据长度的数组。
2.    */
3.
4.
5.  private long[] mergeSpills(SpillInfo[] spills, File outputFile) throws
    IOException {
6.
7.
8.  //获取 Shuffle 的压缩配置信息
9.      final boolean compressionEnabled = sparkConf.getBoolean("spark
        .shuffle.compress", true);
10.     final CompressionCodec compressionCodec = CompressionCodec$.MODULE$
        .createCodec(sparkConf);
11.    //获取是否启动 unsafe 的快速合并
12.     final boolean fastMergeEnabled =
13.       sparkConf.getBoolean("spark.shuffle.unsafe.fastMergeEnabled", true);
14.
15.    //没有压缩或者当压缩码支持序列化流合并时，支持快速合并
16.     final boolean fastMergeIsSupported = !compressionEnabled ||
17.       CompressionCodec$.MODULE$.supportsConcatenationOfSerializedStreams
          (compressionCodec);
18.     final boolean encryptionEnabled = blockManager.serializerManager()
        .encryptionEnabled();
19.     try {
20.
21.         //没有中间的 Spills 文件时，创建一个空文件，并返回包含分区数据长度的
            //空数组。后续读取时会过滤掉空文件
22.
23.
24.       if (spills.length == 0) {
25.         new FileOutputStream(outputFile).close(); //Create an empty file
26.         return new long[partitioner.numPartitions()];
27.       } else if (spills.length == 1) {
28.
```

```
29.         //最后一个 Spills 文件已经更新 metrics 信息，因此不需要重复更新，直接
            //重命名 Spills 的中间临时文件为目标输出的数据文件，同时将该 Spills 中间
            //文件的各分区数据长度的数组返回即可
30.         Files.move(spills[0].file, outputFile);
31.         return spills[0].partitionLengths;
32.       } else {
33.         final long[] partitionLengths;
34.        //当存在多个 Spill 中间文件时，根据不同的条件，采用不同的文件合并策略
35.
36.         if (fastMergeEnabled && fastMergeIsSupported) {
37.          //由 spark.file.transferTo 配置属性控制，默认为 true
38.           if (transferToEnabled && !encryptionEnabled) {
39.             logger.debug("Using transferTo-based fast merge");
40.
41.             //通过 NIO 的方式合并各个 Spills 的分区字节数据
42.             //仅在 I/O 压缩码和序列化器支持序列化流的合并时安全
43.
44.             partitionLengths = mergeSpillsWithTransferTo(spills, outputFile);
45.           } else {
46.             logger.debug("Using fileStream-based fast merge");
47.             //使用 Java FileStreams 文件流的方式进行合并
48.             partitionLengths = mergeSpillsWithFileStream(spills, outputFile,
                null);
49.           }
50.         } else {
51.           logger.debug("Using slow merge");
52.          partitionLengths = mergeSpillsWithFileStream(spills, outputFile,
             compressionCodec);
53.         }
54.         //更新 Shuffle 写数据的度量信息
55.         writeMetrics.decBytesWritten(spills[spills.length - 1].file
            .length());
56.         writeMetrics.incBytesWritten(outputFile.length());
57.         return partitionLengths;
58.       }
59.     } catch (IOException e) {
60.       if (outputFile.exists() && !outputFile.delete()) {
61.
62.         logger.error("Unable to delete output file {}", outputFile
            .getPath());
63.       }
64.       throw e;
65.     }
66.   }
```

各种合并策略在性能上具有一定差异，会根据具体的条件采用，主要有基于 Java NIO(New I/O)和基于普通文件流合并文件的方式。下面简单描述一下基于文件合并流的处理过程。

Spark 2.2.1 版本的 UnsafeShuffleWriter.scala 的 mergeSpillsWithFileStream 方法的源码如下：

```
1.    /** 使用 Java FileStreams 文件流的方式合并*/
2.  private long[] mergeSpillsWithFileStream(
3.        SpillInfo[] spills,
4.        File outputFile,
5.        @Nullable CompressionCodec compressionCodec) throws IOException {
6.      assert (spills.length >= 2);
7.      final int numPartitions = partitioner.numPartitions();
```

```
8.      final long[] partitionLengths = new long[numPartitions];
9.
10. //对应打开的输入流的个数为 Spills 的临时文件个数
11.     final InputStream[] spillInputStreams = new FileInputStream
        [spills.length];
12.
13.     //使用计数输出流避免关闭基础文件并询问文件系统在每个分区写入文件大小
14.
15.     final CountingOutputStream mergedFileOutputStream = new
        CountingOutputStream(
16.       new FileOutputStream(outputFile));
17.
18.     boolean threwException = true;
19.     try {
20.       //为每个 Spills 中间文件打开文件输入流
21.       for (int i = 0; i < spills.length; i++) {
22.         spillInputStreams[i] = new FileInputStream(spills[i].file);
23.       }
24.       //遍历分区
25.       for (int partition = 0; partition < numPartitions; partition++) {
26.         final long initialFileLength = mergedFileOutputStream.getByteCount();
27.         //屏蔽底层输出流的 close()调用，以便能够关闭高层流，以确保所有数据都真正刷
            //新并清除内部状态
28.
29.         OutputStream partitionOutput = new CloseShieldOutputStream(
30.           new TimeTrackingOutputStream(writeMetrics, mergedFileOutputStream));
31.         partitionOutput = blockManager.serializerManager().wrapForEncryption
            (partitionOutput);
32.         if (compressionCodec != null) {
33.           partitionOutput = compressionCodec.compressedOutputStream
              (partitionOutput);
34.         }
35.
36.         //依次从各个 Spills 输入流中读取当前分区的数据长度指定个数的字节，到各个分
            //区对应的输出文件流中
37.         for (int i = 0; i < spills.length; i++) {
38.           final long partitionLengthInSpill = spills[i].partitionLengths
              [partition];
39.           if (partitionLengthInSpill > 0) {
40.             InputStream partitionInputStream = new LimitedInputStream
                (spillInputStreams[i],
41.               partitionLengthInSpill, false);
42.             try {
43.               partitionInputStream = blockManager.serializerManager()
                  .wrapForEncryption(
44.                 partitionInputStream);
45.               if (compressionCodec != null) {
46.                 partitionInputStream = compressionCodec.compressedInputStream
                    (partitionInputStream);
47.               }
48.               ByteStreams.copy(partitionInputStream, partitionOutput);
49.             } finally {
50.               partitionInputStream.close();
51.             }
52.           }
53.         }
54.         partitionOutput.flush();
55.         partitionOutput.close();
```

```
56.  //将当前写入的数据长度存入返回的数组中
57.        partitionLengths[partition] = (mergedFileOutputStream
           .getByteCount() - initialFileLength);
58.      }
59.      threwException = false;
60.    } finally {
61.      //为了避免屏蔽异常以后导致过早进入 finally 块的异常处理，只能在清理过程中抛出
         //异常
62.
63.      for (InputStream stream : spillInputStreams) {
64.        Closeables.close(stream, threwException);
65.      }
66.      Closeables.close(mergedFileOutputStream, threwException);
67.    }
68.    return partitionLengths;
69.  }
```

Spark 2.4.3 版本的 UnsafeShuffleWriter.scala 源码与 Spark 2.2.1 版本相比具有如下特点。

- 上段代码中第 15 行之前新增构建 BufferedOutputStream 实例 bos 的代码。
- 上段代码中第 16 行将 new FileOutputStream(outputFile)调整为 bos。
- 上段代码中第 22 行将构建 FileInputStream 实例，调整为构建 NioBufferedFileInputStream 实例。

```
1.  ......
2.      final OutputStream bos = new BufferedOutputStream(
3.              new FileOutputStream(outputFile),
4.              outputBufferSizeInBytes);
5.  ......
6.  final CountingOutputStream mergedFileOutputStream = new CountingOutputStream(bos);
7.  ......
8.    spillInputStreams[i] = new NioBufferedFileInputStream(
9.              spills[i].file,
10.             inputBufferSizeInBytes);
11.       }
```

基于 NIO 的文件合并流程基本类似，只是底层采用 NIO 的技术实现。

7.6　Shuffle 与 Storage 模块间的交互

在 Spark 中，存储模块被抽象成 Storage。顾名思义，Storage 是存储的意思，代表着 Spark 中的数据存储系统，负责管理和实现数据块（Block）的存放。其中存取数据的最小单元是 Block，数据由不同的 Block 组成，所有操作都是以 Block 为单位进行的。从本质上讲，RDD 中的 Partition 和 Storage 中的 Block 是等价的，只是所处的模块不同，看待的角度不一样而已。

Storage 抽象模块的实现分为两个层次，如图 7-13 所示。

（1）通信层：通信层是典型的 Master-Slave 结构，Master 和 Slave 之间传输控制和状态信息。通信层主要由 BlockManager、BlockManagerMaster、BlockManagerMasterEndpoint、BlockManagerSlaveEndpoint 等类实现。

（2）存储层：负责把数据存储到内存、磁盘或者堆外内存中，有时还需要为数据在远程节点上生成副本，这些都由存储层提供的接口实现。Spark 2.2.0 具体的存储层的实现类有

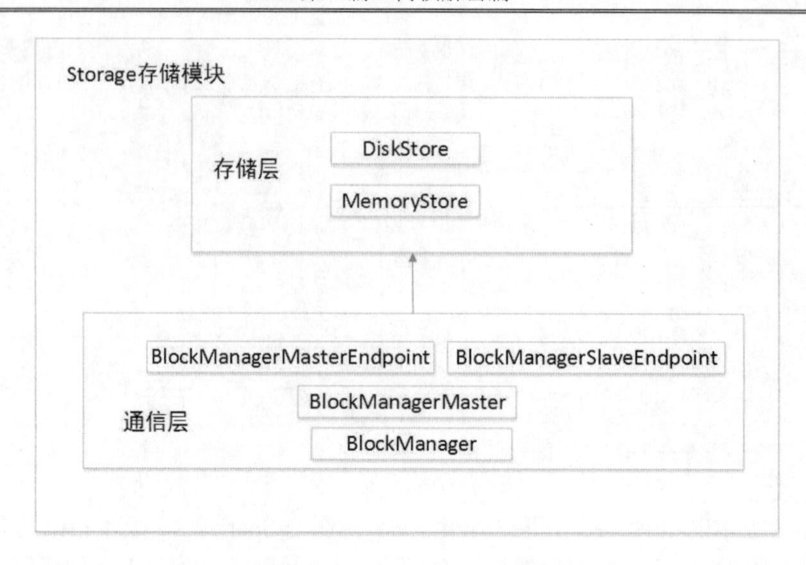

图 7-13　Storage 存储模块

DiskStore 和 MemoryStore。

Shuffle 模块若要和 Storage 模块进行交互，需要通过调用统一的操作类 BlockManager 来完成。如果把整个存储模块看成一个黑盒，BlockManager 就是黑盒上留出的一个供外部调用的接口。

7.6.1　Shuffle 注册的交互

Spark 中 BlockManager 在 Driver 端的创建，在 SparkContext 创建的时候会根据具体的配置创建 SparkEnv 对象。

Spark 2.2.1 版本的 SparkContext.scala 的源码如下：

```
1.   _env = createSparkEnv(_conf, isLocal, listenerBus)
2.     SparkEnv.set(_env)
3.  ......
4.  private[spark] def createSparkEnv(
5.        conf: SparkConf,
6.        isLocal: Boolean,
7.        listenerBus: LiveListenerBus): SparkEnv = {
8.  //创建 Driver 端的运行环境
9.     SparkEnv.createDriverEnv(conf, isLocal, listenerBus, SparkContext
       .numDriverCores(master))
10.  }
```

Spark 2.4.3 版本的 SparkContext.scala 源码与 Spark 2.2.1 版本相比具有如下特点。

- 上段代码中第 9 行在 SparkContext.numDriverCores 新增一个参数 conf。

```
1.  ......
2.  SparkEnv.createDriverEnv(conf, isLocal, listenerBus, SparkContext
    .numDriverCores(master, conf))
3.  ......
```

createSparkEnv 方法中，传入 SparkConf 配置对象、isLocal 标志，以及 LiveListenerBus，方法中使用 SparkEnv 对象的 createDriverEnv 方法创建 SparkEnv 并返回。在 SparkEnv 的

createDriverEvn 方法中，将会创建 BlockManager、BlockManagerMaster 等对象，完成 Storage 在 Driver 端的部署。

SparkEnv 中创建 BlockManager、BlockManagerMaster 关键源码如下：

```
1.     val blockTransferService =
2.        new NettyBlockTransferService(conf, securityManager, bindAddress,
          advertiseAddress, blockManagerPort, numUsableCores)
3.
4.  //创建 BlockManagerMasterEndpoint
5.      val blockManagerMaster = new BlockManagerMaster(registerOrLookupEndpoint(
6.        BlockManagerMaster.DRIVER_ENDPOINT_NAME,
7.    //创建 BlockManagerMasterEndpoint
8.        new BlockManagerMasterEndpoint(rpcEnv, isLocal, conf, listenerBus)),
9.        conf, isDriver)
10.  //创建 BlockManager
11.     //注：blockManager 无效，直到 initialize()被调用
12.     val blockManager = new BlockManager(executorId, rpcEnv,
        blockManagerMaster,
13.       serializerManager, conf, memoryManager, mapOutputTracker,
          shuffleManager,
14.       blockTransferService, securityManager, numUsableCores)
```

使用 new 关键字实例化出 BlockManagerMaster，传入 BlockManager 的构造函数，实例化出 BlockManager 对象。这里的 BlockManagerMaster 和 BlockManager 属于聚合关系。BlockManager 主要对外提供统一的访问接口，BlockManagerMaster 主要对内提供各节点之间的指令通信服务。

构建 BlockManager 时，传入 shuffleManager 参数，shuffleManager 是在 SparkEnv 中创建的，将 shuffleManager 传入到 BlockManager 中，BlockManager 就拥有 shuffleManager 的成员变量，从而可以调用 shuffleManager 的相关方法。

BlockManagerMaster 在 Driver 端和 Executors 中的创建稍有差别。首先来看在 Driver 端创建的情形。创建 BlockManagerMaster 传入的 isDriver 参数，isDriver 为 true，表示在 Driver 端创建，否则视为在 Slave 节点上创建。

当 SparkContext 中执行_env.blockManager.initialize(_applicationId)代码时，会调用 Driver 端 BlockManager 的 initialize 方法。Initialize 方法的源码如下所示。

SparkContext.scala 的源码如下：

```
1.      _env.blockManager.initialize(_applicationId)
```

BlockManager.scala 的源码如下：

```
1.   def initialize(appId: String): Unit = {
2.   //调用 blockTransferService 的 init 方法，blockTransferService 用于在不同节点
     //fetch 数据、传送数据
3.     blockTransferService.init(this)
4.   //shuffleClient 用于读取其他 Executor 上的 shuffle  files
5.      shuffleClient.init(appId)
6.
7.      blockReplicationPolicy = {
8.        val priorityClass = conf.get(
9.          "spark.storage.replication.policy", classOf
```

```
        [RandomBlockReplicationPolicy].getName)
10.       val clazz = Utils.classForName(priorityClass)
11.       val ret = clazz.newInstance.asInstanceOf[BlockReplicationPolicy]
12.       logInfo(s"Using $priorityClass for block replication policy")
13.       ret
14.     }
15.
16.     val id =
17.       BlockManagerId(executorId, blockTransferService.hostName,
          blockTransferService.port, None)
18.
19.    //向 blockManagerMaster 注册 BlockManager。在 registerBlockManager 方法中传
       //入了 slaveEndpoint，slaveEndpoint 为 BlockManager 中的 RPC 对象，用于和
       //blockManagerMaster 通信
20.     val idFromMaster = master.registerBlockManager(
21.       id,
22.       maxOnHeapMemory, maxOffHeapMemory,
23.       slaveEndpoint)
24.    //得到 blockManagerId
25.     blockManagerId = if (idFromMaster != null) idFromMaster else id
26.
27. //得到 shuffleServerId
28.     shuffleServerId = if (externalShuffleServiceEnabled) {
29.       logInfo(s"external shuffle service port = $externalShuffleServicePort")
30.       BlockManagerId(executorId, blockTransferService.hostName,
          externalShuffleServicePort)
31.     } else {
32.       blockManagerId
33.     }
34.   //注册 shuffleServer
35.     //如果存在，将注册 Executors 配置与本地 shuffle 服务
36.     if (externalShuffleServiceEnabled && !blockManagerId.isDriver) {
37.       registerWithExternalShuffleServer()
38.     }
39.
40.     logInfo(s"Initialized BlockManager: $blockManagerId")
41.   }
```

如上面的源码所示，initialize 方法使用 appId 初始化 BlockManager，主要完成以下工作。

（1）初始化 BlockTransferService。

（2）初始化 ShuffleClient。

（3）创建 BlockManagerId。

（4）将 BlockManager 注册到 BlockManagerMaster 上。

（5）若 ShuffleService 可用，则注册 ShuffleService。

在 BlockManager 的 initialize 方法上右击 Find Usages，可以看到 initialize 方法在两个地方得到调用：一个是 SparkContext；另一个是 Executor。启动 Executor 时，会调用 BlockManager 的 initialize 方法。Executor 中调用 initialize 方法的源码如下所示。

Spark 2.2.1 版本的 Executor.scala 的源码如下：

```
1.   //CoarseGrainedExecutorBackend 中实例化 Executor，isLocal 设置成 false，即
     //Executor 中 isLocal 始终为 false
2.
3.   if (!isLocal) {
4.   //向度量系统注册
```

```
5.      env.metricsSystem.registerSource(executorSource)
6.    //调用BlockManager的initialize方法,initialize方法将向BlockManagerMaster
      //注册，完成 Executor 中的 BlockManager 向 Driver 中的 BlockManager 注册
7.      env.blockManager.initialize(conf.getAppId)
8.    }
```

Spark 2.4.3 版本的 Executor.scala 源码与 Spark 2.2.1 版本相比具有如下特点。

- 上段代码中第 5 行和第 7 行互换一下顺序。
- 上段代码中第 7 行后新增代码，向度量系统进行 env.blockManager.shuffleMetricsSource 的注册。

```
1.   ......
2.      env.blockManager.initialize(conf.getAppId)
3.      env.metricsSystem.registerSource(executorSource)
4.   env.metricsSystem.registerSource(env.blockManager.shuffleMetricsSource)
5.   ......
```

上面代码中调用了 env.blockManager.initialize 方法。在 initialize 方法中，完成 BlockManger 向 Master 端 BlockManagerMaster 的注册。使用方法 master.registerBlockManager (id,maxMemory,slaveEndpoint)完成注册，registerBlockManager 方法中传入 Id、maxMemory、salveEndPoint 引用，分别表示 Executor 中的 BlockManager、最大内存、BlockManager 中的 BlockMangarSlaveEndpoint。BlockManagerSlaveEndpoint 是一个 RPC 端点，使用它完成同 BlockManagerMaster 的通信。BlockManager 收到注册请求后将 Executor 中注册的 BlockManagerInfo 存入哈希表中，以便通过 BlockManagerSlaveEndpoint 向 Executor 发送控制命令。

ShuffleManager 是一个用于 shuffle 系统的可插拔接口。在 Driver 端 SparkEnv 中创建 ShuffleManager，在每个 Executor 上也会创建。基于 spark.shuffle.manager 进行设置。Driver 使用 ShuffleManager 注册到 shuffles 系统，Executors（或 Driver 在本地运行的任务）可以请求读取和写入数据。这将被 SparkEnv 的 SparkConf 和 isDriver 布尔值作为参数。

ShuffleManager.scala 的源码如下：

```
1.   private[spark] trait ShuffleManager {
2.
3.    /**
4.      *注册一个 shuffle 管理器，获取一个句柄传递给任务
5.      */
6.    def registerShuffle[K, V, C](
7.        shuffleId: Int,
8.        numMaps: Int,
9.        dependency: ShuffleDependency[K, V, C]): ShuffleHandle
10.
11.   /**为给定分区获取一个写入器。Executors 节点通过 Map 任务调用*/
12.   def getWriter[K, V](handle: ShuffleHandle, mapId: Int, context:
      TaskContext): ShuffleWriter[K, V]
13.
14.   /**
        *获取读取器汇聚一定范围的分区（从 startPartition 到 endPartition-1）。在
        *Executors 节点，通过 reduce 任务调用
15.     */
16.
17.   def getReader[K, C](
18.       handle: ShuffleHandle,
```

```
19.         startPartition: Int,
20.         endPartition: Int,
21.         context: TaskContext): ShuffleReader[K, C]
22.
23.     /**
24.       *从 ShuffleManager 移除一个 shuffle 的元数据
25.       * @return 如果元数据成功删除，则返回 true，否则返回 false
26.       */
27.     def unregisterShuffle(shuffleId: Int): Boolean
28.
29.     /**
          * 返回一个能够根据块坐标来检索 shuffle 块数据的解析器
30.       */
31.     def shuffleBlockResolver: ShuffleBlockResolver
32.
33.     /** 关闭 ShuffleManager */
34.     def stop(): Unit
35. }
```

Spark Shuffle Pluggable 框架 ShuffleBlockManager 在 Spark 1.6.0 之后改成了 ShuffleBlockResolver。ShuffleBlockResolver 具体读取 shuffle 数据，是一个 trait。在 ShuffleBlockResolver 中已无 getBytes 方法。getBlockData（blockId: ShuffleBlockId）方法返回的是 ManagedBuffer，这是核心。

ShuffleBlockResolver 的源码如下：

```
1.      trait ShuffleBlockResolver {
2.      type ShuffleId = Int
3.
4.      /**
          *为指定的块检索数据。如果块数据不可用，则抛出一个未指明的异常
5.        */
6.      def getBlockData(blockId: ShuffleBlockId): ManagedBuffer
7.
8.      def stop(): Unit
9.    }
```

Spark 2.0 版本中通过 IndexShuffleBlockResolver 来具体实现 ShuffleBlockResolver（SortBasedShuffle 方式），已无 FileShuffleBlockManager（Hashshuffle 方式）。IndexShuffleBlockResolver 创建和维护逻辑块和物理文件位置之间的 shuffle blocks 映射关系。来自于相同 map task 任务的 shuffle blocks 数据存储在单个合并数据文件中；数据文件中的数据块的偏移量存储在单独的索引文件中。将 shuffleBlockId + reduce ID set to 0 + ".后缀"作为数据 shuffle data 的 shuffleBlockId 名字。其中，文件名后缀为".data"的是数据文件；文件名后缀为".index"的是索引文件。

7.6.2　Shuffle 写数据的交互

基于 Sort 的 Shuffle 实现的 ShuffleHandle 包含 BypassMergeSortShuffleHandle 与 BaseShuffleHandle。两种 ShuffleHandle 写数据的方法可以参考 SortShuffleManager 类的 getWriter 方法，关键代码如下所示。

SortShuffleManager 的 getWriter 的源码如下：

```
1.      override def getWriter[K, V](
2.    ......
3.      case bypassMergeSortHandle: BypassMergeSortShuffleHandle[K
        @unchecked, V @unchecked] =>
4.        new BypassMergeSortShuffleWriter(
5.          env.blockManager,
6.          shuffleBlockResolver.asInstanceOf[IndexShuffleBlockResolver],
7.        ......
8.      case other: BaseShuffleHandle[K @unchecked, V @unchecked, _] =>
9.        new SortShuffleWriter(shuffleBlockResolver, other, mapId, context)
10.     }
11.   }
```

在对应构建的两种数据写入器类 BypassMergeSortShuffleWriter 与 SortShuffleWriter 中，都是通过变量 shuffleBlockResolver 对逻辑数据块与物理数据块的映射进行解析。BypassMergeSortShuffleWriter 写数据的具体实现位于实现的 write 方法中，其中调用的 createTempShuffleBlock 方法描述了各个分区所生成的中间临时文件的格式与对应的 BlockId。SortShuffleWriter 写数据的具体实现位于实现的 write 方法中。

7.6.3　Shuffle 读数据的交互

SparkEnv.get.shuffleManager.getReader 是 SortShuffleManager 的 getReader，是获取数据的阅读器，getReader 方法中创建了一个 BlockStoreShuffleReader 实例。SortShuffleManager.scala 的 read()方法的源码如下：

```
1.    override def getReader[K, C](
2.        handle: ShuffleHandle,
3.        startPartition: Int,
4.        endPartition: Int,
5.        context: TaskContext): ShuffleReader[K, C] = {
6.      new BlockStoreShuffleReader(
7.        handle.asInstanceOf[BaseShuffleHandle[K, _, C]], startPartition,
          endPartition, context)
8.    }
```

BlockStoreShuffleReader 实例的 read 方法，首先实例化 new ShuffleBlockFetcherIterator。ShuffleBlockFetcherIterator 是一个阅读器，里面有一个成员 blockManager。blockManager 是内存和磁盘上数据读写的统一管理器；ShuffleBlockFetcherIterator.scala 的 initialize 方法中 splitLocalRemoteBlocks()划分本地和远程的 blocks，Utils.randomize(remoteRequests)把远程请求通过随机的方式添加到队列中，fetchUpToMaxBytes()发送远程请求获取我们的 blocks，fetchLocalBlocks()获取本地的 blocks。

7.6.4　BlockManager 架构原理、运行流程图和源码解密

BlockManager 是管理整个 Spark 运行时数据的读写，包含数据存储本身，在数据存储的基础上进行数据读写。由于 Spark 是分布式的，所以 BlockManager 也是分布式的，BlockManager 本身相对而言是一个比较大的模块，Spark 中有非常多的模块：调度模块、资

源管理模块等。BlockManager 是另外一个非常重要的模块。BlockManager 本身的源码量非常大。本节从 BlockManager 原理流程对 BlockManager 做深刻地讲解。在 Shuffle 读写数据的时候，我们需要读写 BlockManager。因此，BlockManager 是至关重要的内容。

编写一个业务代码 WordCount.scala，通过观察 WordCount 运行时 BlockManager 的日志来理解 BlockManager 的运行。

WordCount.scala 的代码如下：

```
1.  package com.dt.spark.sparksql
2.
3.  import org.apache.log4j.{Level, Logger}
4.  import org.apache.spark.SparkConf
5.  import org.apache.spark.SparkContext
6.  import org.apache.spark.internal.config
7.  import org.apache.spark.rdd.RDD
8.
9.  /**
10.   * 使用 Scala 开发本地测试的 Spark WordCount 程序
11.   *
12.   * @author DT 大数据梦工厂
13.   *        新浪微博：http://weibo.com/ilovepains/
14.   */
15. object WordCount {
16.   def main(args: Array[String]) {
17.     Logger.getLogger("org").setLevel(Level.ALL)
18.
19.     /**
20.       *第 1 步：创建 Spark 的配置对象 SparkConf，设置 Spark 程序的运行时的配置信息，
21.       *例如，通过 setMaster 设置程序要链接的 Spark 集群的 Master 的 URL，如果设置
22.       *为 local，则代表 Spark 程序在本地运行，特别适合于机器配置条件非常差（如只有
23.       *1GB 的内存）的初学者
24.       */
25.     val conf = new SparkConf() //创建 SparkConf 对象
26.     conf.setAppName("Wow,My First Spark App!") //设置应用程序的名称，在程序
                                                    //运行的监控界面可以看到名称
27.     conf.setMaster("local") //此时，程序在本地运行，不需要安装 Spark 集群
28.     /**
29.       * 第 2 步：创建 SparkContext 对象
30.       * SparkContext 是 Spark 程序所有功能的唯一入口，采用 Scala、Java、Python、
          * R 等都必须有一个 SparkContext
31.       * SparkContext 核心作用：初始化 Spark 应用程序运行所需要的核心组件，包括
          * DAGScheduler、TaskScheduler、SchedulerBackend
32.       * 同时还会负责 Spark 程序往 Master 注册程序等
33.       * SparkContext 是整个 Spark 应用程序中最重要的一个对象
34.       */
35.     val sc = new SparkContext(conf)
                            //创建 SparkContext 对象，通过传入 SparkConf
                            //实例来定制 Spark 运行的具体参数和配置信息
36.     /**
37.       * 第 3 步：根据具体的数据来源（如 HDFS、HBase、Local FS、DB、S3 等）通过
          * SparkContext 创建 RDD
38.       * RDD 的创建基本有三种方式：根据外部的数据来源（如 HDFS）、根据 Scala 集合、由
```

```
        * 其他的 RDD 操作
39.     * 数据会被 RDD 划分为一系列的 Partitions，分配到每个 Partition 的数据属于一
        * 个 Task 的处理范畴
40.     */
41.    //val lines: RDD[String] = sc.textFile("D://Big_Data_Software spark-
       1.6.0-bin-hadoop2.6README.md", 1) //读取本地文件并设置为一个 Partition
42.    // val lines = sc.textFile("D://Big_Data_Software spark-1.6.0-bin-
       hadoop2.6//README.md", 1) //读取本地文件并设置为一个 Partition
43.
44.    val lines = sc.textFile("data/wordcount/helloSpark.txt", 1)
                                        //读取本地文件并设置为一个 Partition
45.    /**
46.     * 第 4 步：对初始的 RDD 进行 Transformation 级别的处理，如通过 map、filter 等
        * 高阶函数等的编程，进行具体的数据计算
47.     * 第 4.1 步：将每一行的字符串拆分成单个单词
48.     */
49.
50.    val words = lines.flatMap { line => line.split(" ") }
  //对每一行的字符串进行单词拆分并把所有行的拆分结果通过 flat 合并成为一个大的单词集合
51.
52.    /**
53.     * 第 4 步：对初始的 RDD 进行 Transformation 级别的处理，如通过 map、filter 等
        * 高阶函数等的编程，进行具体的数据计算
54.     * 第 4.2 步：在单词拆分的基础上，对每个单词实例计数为 1，也就是 word => (word, 1)
55.     */
56.    val pairs = words.map { word => (word, 1) }
57.
58.    /**
59.     * 第 4 步：对初始的 RDD 进行 Transformation 级别的处理，如通过 map、filter 等
        * 高阶函数等的编程，进行具体的数据计算
60.     * 第 4.3 步：在每个单词实例计数为 1 基础上，统计每个单词在文件中出现的总次数
61.     */
62.    val wordCountsOdered = pairs.reduceByKey(_ + _).map(pair => (pair._2,
       pair._1)).sortByKey(false).map(pair => (pair._2, pair._1))
        //对相同的 Key，进行 Value 的累计（包括 Local 和 Reducer 级别同时 Reduce）
63.    wordCountsOdered.collect.foreach(wordNumberPair => println
       (wordNumberPair._1 + " : " + wordNumberPair._2))
64.    while (true) {
65.
66.    }
67.    sc.stop()
68.
69.  }
70. }
```

在 IDEA 中运行一个业务程序 WordCount.scala，日志中显示以下内容。

- SparkEnv：Registering MapOutputTracker，其中 MapOutputTracker 中数据的读写都和 BlockManager 关联。
- SparkEnv：Registering BlockManagerMaste，其中 Registering BlockManagerMaster 由 BlockManagerMaster 进行注册。
- DiskBlockManager：Created local directory C:\Users\dell\AppData\Local\Temp\blockmgr-... 其中 DiskBlockManager 是管理磁盘存储的，里面有我们的数据。可以访问 Temp 目

录下以 blockmgr-开头的文件的内容。

WordCount 运行结果如下：

```
1.  Using Spark's default log4j profile: org/apache/spark/log4j-defaults.properties
2.  19/07/13 17:22:31 INFO SparkContext: Running Spark version 2.4.3
3.  ......
4.  19/07/13 17:22:38 INFO SparkEnv: Registering MapOutputTracker
5.  19/07/13 17:22:38 INFO SparkEnv: Registering BlockManagerMaster
6.  19/07/13 17:22:38 INFO BlockManagerMasterEndpoint: Using org.apache
    .spark.storage.DefaultTopologyMapper for getting topology information
7.  19/07/13 17:22:38 INFO BlockManagerMasterEndpoint: BlockManagerMasterEndpoint
    up
8.  19/07/13 17:22:38 INFO DiskBlockManager: Created local directory at
    C:\Users\lenovo\AppData\Local\Temp\blockmgr-7b7206ac-5ab7-4057-b349-
    b5386796d8b4
9.  19/07/13 17:22:38 INFO MemoryStore: MemoryStore started with capacity
    894.3 MB
10. 19/07/13 17:22:38 INFO SparkEnv: Registering OutputCommitCoordinator
11. 19/07/13 17:22:39 INFO Utils: Successfully started service 'SparkUI' on
    port 4040.
12. ......
```

从 Application 启动的角度观察 BlockManager。

（1）Application 启动时会在 SparkEnv 中注册 BlockManagerMaster 以及 MapOutputTracker，其中：

① BlockManagerMaster：对整个集群的 Block 数据进行管理。

② MapOutputTrackerMaster：跟踪所有的 Mapper 的输出。

BlockManagerMaster 中有一个引用 driverEndpoint，isDriver 判断是否运行在 Driver 上。

BlockManagerMaster 的源码如下：

```
1.   private[spark]
2.  class BlockManagerMaster(
3.      var driverEndpoint: RpcEndpointRef,
4.      conf: SparkConf,
5.      isDriver: Boolean)
6.    extends Logging {
```

BlockManagerMaster 注册给 SparkEnv，SparkEnv 在 SparkContext 中。

SparkContext.scala 的源码如下：

```
1.     ......
2.    private var _env: SparkEnv = _
3.  ......
4.    _env = createSparkEnv(_conf, isLocal, listenerBus)
5.      SparkEnv.set(_env)
```

在 SparkContext.scala 的 createSparkEnv 方法中调用 SparkEnv.createDriverEnv 方法。

进入 SparkEnv.scala 的 createDriverEnv 方法。

Spark 2.2.1 版本的 SparkEnv.scala 的源码如下：

```
1.      private[spark] def createDriverEnv(
2.     ......
3.     create(
4.       conf,
5.       SparkContext.DRIVER_IDENTIFIER,
```

```
6.        bindAddress,
7.        advertiseAddress,
8.        port,
9.        isLocal,
10.       numCores,
11.       ioEncryptionKey,
12.       listenerBus = listenerBus,
13.       mockOutputCommitCoordinator = mockOutputCommitCoordinator
14.     )
15.   }
16. ......
```

Spark 2.4.3 版本的 SparkEnv.scala 源码与 Spark 2.2.1 版本相比具有如下特点。

❑ 上段代码中第 8 行将 port 调整为 Option(port)。

```
1.  ......
2.        Option(port),
3.  ......
```

SparkEnv.scala 的 createDriverEnv 中调用了 create 方法，判断是否是 Driver。create 方法的源码如下。

Spark 2.2.1 版本的 SparkEnv.scala 的源码如下：

```
1.    private def create(
2.        conf: SparkConf,
3.        executorId: String,
4.        bindAddress: String,
5.        advertiseAddress: String,
6.        port: Int,
7.        isLocal: Boolean,
8.        numUsableCores: Int,
9.        ioEncryptionKey: Option[Array[Byte]],
10.       listenerBus: LiveListenerBus = null,
11.       mockOutputCommitCoordinator:  Option[OutputCommitCoordinator]  =
          None): SparkEnv = {
12.
13.     val isDriver = executorId == SparkContext.DRIVER_IDENTIFIER
14.      ......
15.     if (isDriver) {
16.       conf.set("spark.driver.port", rpcEnv.address.port.toString)
17.     } else if (rpcEnv.address != null) {
18.       conf.set("spark.executor.port", rpcEnv.address.port.toString)
19.       logInfo(s"Setting spark.executor.port to: ${rpcEnv.address.port
          .toString}")
20.     }
21.   ......
22.    val mapOutputTracker = if (isDriver) {
23.       new MapOutputTrackerMaster(conf, broadcastManager, isLocal)
24.     } else {
25.       new MapOutputTrackerWorker(conf)
26.     }
27. ......
28. SparkContext.scala
29. private[spark] val DRIVER_IDENTIFIER = "driver"
30. ......
```

Spark 2.4.3 版本的 SparkEnv. scala 源码与 Spark 2.2.1 版本相比具有如下特点。

❑ 上段代码中第 6 行将 port: Int 调整为 port: Option[Int]。

```
1.  ……
2.        port: Option[Int],
3.  ……
```

在 SparkEnv.scala 的 createDriverEnv 中调用 new()函数创建一个 MapOutputTrackerMaster。MapOutputTrackerMaster 的源码如下。

```
1.   private[spark] class MapOutputTrackerMaster(conf: SparkConf,
2.      broadcastManager: BroadcastManager, isLocal: Boolean)
3.    extends MapOutputTracker(conf) {
4.  ……
```

然后看一下 blockManagerMaster。在 SparkEnv.scala 中调用 new()函数创建一个 blockManagerMaster。

```
1.       val blockManagerMaster = new BlockManagerMaster
         (registerOrLookupEndpoint(
2.        BlockManagerMaster.DRIVER_ENDPOINT_NAME,
3.        new BlockManagerMasterEndpoint(rpcEnv, isLocal, conf, listenerBus)),
4.        conf, isDriver)
```

BlockManagerMaster 对整个集群的 Block 数据进行管理，Block 是 Spark 数据管理的单位，与数据存储没有关系，数据可能存在磁盘上，也可能存储在内存中，还可能存储在 offline，如 Alluxio 上，源码如下：

```
1.  private[spark]
2.  class BlockManagerMaster(
3.      var driverEndpoint: RpcEndpointRef,
4.      conf: SparkConf,
5.      isDriver: Boolean)
6.    extends Logging {
7.  ……
```

构建 BlockManagerMaster 的时候调用 new()函数创建一个 BlockManagerMasterEndpoint，这是循环消息体。

```
1.      private[spark]
2.  class BlockManagerMasterEndpoint(
3.      override val rpcEnv: RpcEnv,
4.      val isLocal: Boolean,
5.      conf: SparkConf,
6.      listenerBus: LiveListenerBus)
7.    extends ThreadSafeRpcEndpoint with Logging {
```

（2）BlockManagerMasterEndpoint 本身是一个消息体，会负责通过远程消息通信的方式去管理所有节点的 BlockManager。

查看 WordCount 在 IDEA 中的运行日志，日志中显示 BlockManagerMasterEndpoint: Registering block manager，向 block manager 进行注册。

```
1.  19/07/13 17:22:40 INFO BlockManager: Using org.apache.spark.storage
    .RandomBlockReplicationPolicy for block replication policy
2.  19/07/13 17:22:40 INFO BlockManagerMaster: Registering BlockManager
    BlockManagerId (driver, windows10.microdone.cn, 50764, None)
3.  19/07/13 17:22:40 INFO BlockManagerMasterEndpoint: Registering block
    manager windows10.microdone.cn:50764 with 894.3 MB RAM, BlockManagerId
    (driver, windows10. microdone.cn, 50764, None)
```

```
4.  19/07/13 17:22:40 INFO BlockManagerMaster: Registered BlockManager
    BlockManagerId (driver, windows10.microdone.cn, 50764, None)
5.  19/07/13 17:22:40 INFO BlockManager: Initialized BlockManager: BlockManagerId
    (driver, windows10.microdone.cn, 50764, None)
6.  ......
```

（3）每启动一个 ExecutorBackend，都会实例化 BlockManager，并通过远程通信的方式注册给 BlockManagerMaster；实质上是 Executor 中的 BlockManager 在启动的时候注册给了 Driver 上的 BlockManagerMasterEndpoint。

（4）MemoryStore 是 BlockManager 中专门负责内存数据存储和读写的类。

查看 WordCount 在 IDEA 中的运行日志，日志中显示 MemoryStore: Block broadcast_0 stored as values in memory，数据存储在内存中。

```
1.  ......
2.  19/07/13 17:22:43 INFO MemoryStore: Block broadcast_0 stored as values
    in memory (estimated size 208.6 KB, free 894.1 MB)
3.  19/07/13 17:22:43 INFO MemoryStore: Block broadcast_0_piece0 stored as
    bytes in memory (estimated size 20.0 KB, free 894.1 MB)
4.  19/07/13 17:22:43 INFO BlockManagerInfo: Added broadcast_0_piece0 in
    memory on windows10.microdone.cn:50764 (size: 20.0 KB, free: 894.3 MB)
5.  ......
```

Spark 读写数据是以 block 为单位的，MemoryStore 将 block 数据存储在内存中。MemoryStore.scala 的源码如下：

```
1.   private[spark] class MemoryStore(
2.      conf: SparkConf,
3.      blockInfoManager: BlockInfoManager,
4.      serializerManager: SerializerManager,
5.      memoryManager: MemoryManager,
6.      blockEvictionHandler: BlockEvictionHandler)
7.    extends Logging {
8.  ......
```

（5）DiskStore 是 BlockManager 中专门负责基于磁盘的数据存储和读写的类。

DiskStore.scala 的源码如下：

```
1.  private[spark] class DiskStore(
2.      conf: SparkConf,
3.      diskManager: DiskBlockManager,
4.      securityManager: SecurityManager) extends Logging {
```

（6）DiskBlockManager：管理 Logical Block 与 Disk 上的 Physical Block 之间的映射关系并负责磁盘文件的创建、读写等。

查看 WordCount 在 IDEA 中的运行日志，日志中显示 INFO DiskBlockManager: Created local directory。DiskBlockManager 负责磁盘文件的管理。

```
1.  ......
2.  19/07/13 17:22:38 INFO BlockManagerMasterEndpoint: Using org.apache
    .spark.storage.DefaultTopologyMapper for getting topology information
3.  19/07/13 17:22:38 INFO BlockManagerMasterEndpoint:
    BlockManagerMasterEndpoint up
4.  19/07/13 17:22:38 INFO DiskBlockManager: Created local directory at
    C:\Users\lenovo\AppData\Local\Temp\blockmgr-7b7206ac-5ab7-4057-b349-
    b5386796d8b4
```

```
5.  19/07/13 17:22:38 INFO MemoryStore: MemoryStore started with capacity
    894.3 MB
6.  ......
```

DiskBlockManager 负责管理逻辑级别和物理级别的映射关系，根据 BlockID 映射一个文件。在目录 spark.local.dir 或者 SPARK_LOCAL_DIRS 中，Block 文件进行 hash 生成。通过 createLocalDirs 生成本地目录。DiskBlockManager 的源码如下：

```
1.   private[spark] class DiskBlockManager(conf: SparkConf, deleteFilesOnStop:
     Boolean) extends Logging {
2.  ......
3.  private def createLocalDirs(conf: SparkConf): Array[File] = {
4.      Utils.getConfiguredLocalDirs(conf).flatMap { rootDir =>
5.        try {
6.          val localDir = Utils.createDirectory(rootDir, "blockmgr")
7.          logInfo(s"Created local directory at $localDir")
8.          Some(localDir)
9.        } catch {
10.         case e: IOException =>
11.           logError(s"Failed to create local dir in $rootDir. Ignoring this
              directory.", e)
12.           None
13.       }
14.     }
15.   }
```

从 Job 运行的角度来观察 BlockManager：

查看 WordCount.scala 的运行日志：日志中显示 INFO BlockManagerInfo: Added broadcast_0_piece0 in memory，将 BlockManagerInfo 的广播变量加入到内存中。

```
1.  ......
2.  19/07/13 17:22:43 INFO MemoryStore: Block broadcast_0 stored as values
    in memory (estimated size 208.6 KB, free 894.1 MB)
3.  19/07/13 17:22:43 INFO MemoryStore: Block broadcast_0_piece0 stored as
    bytes in memory (estimated size 20.0 KB, free 894.1 MB)
4.  19/07/13 17:22:43 INFO BlockManagerInfo: Added broadcast_0_piece0 in
    memory on windows10.microdone.cn:50764 (size: 20.0 KB, free: 894.3 MB)
5.  ......
```

Driver 使用 BlockManagerInfo 管理 ExecutorBackend 中 BlockManager 的元数据，BlockManagerInfo 的成员变量包括 blockManagerId、系统当前时间 timeMs、最大堆内内存 maxOnHeapMem、最大堆外内存 maxOffHeapMem、slaveEndpoint。

BlockManagerMasterEndpoint.scala 的源码如下：

```
1.  private[spark] class BlockManagerInfo(
2.      val blockManagerId: BlockManagerId,
3.      timeMs: Long,
4.      val maxOnHeapMem: Long,
5.      val maxOffHeapMem: Long,
6.      val slaveEndpoint: RpcEndpointRef)
7.    extends Logging {
```

集群中每启动一个节点，就创建一个 BlockManager，BlockManager 是在每个节点（Driver 及 Executors）上运行的管理器，用于存放和检索本地和远程不同的存储块（内存、磁盘和堆外内存）。BlockManagerInfo 中的 BlockManagerId 标明是哪个 BlockManager，slaveEndpoint

是消息循环体，用于消息通信。

（1）首先通过 MemoryStore 存储广播变量。

（2）在 Driver 中是通过 BlockManagerInfo 来管理集群中每个 ExecutorBackend 中的 BlockManager 中的元数据信息的。

（3）当改变了具体的 ExecutorBackend 上的 Block 信息后，就必须发消息给 Driver 中的 BlockManagerMaster 来更新相应的 BlockManagerInfo。

（4）当执行第二个 Stage 的时候，第二个 Stage 会向 Driver 中的 MapOutputTracker-MasterEndpoint 发消息请求上一个 Stage 中相应的输出，此时 MapOutputTrackerMaster 会把上一个 Stage 的输出数据的元数据信息发送给当前请求的 Stage。图 7-14 是 BlockManager 工作原理和运行机制简图。

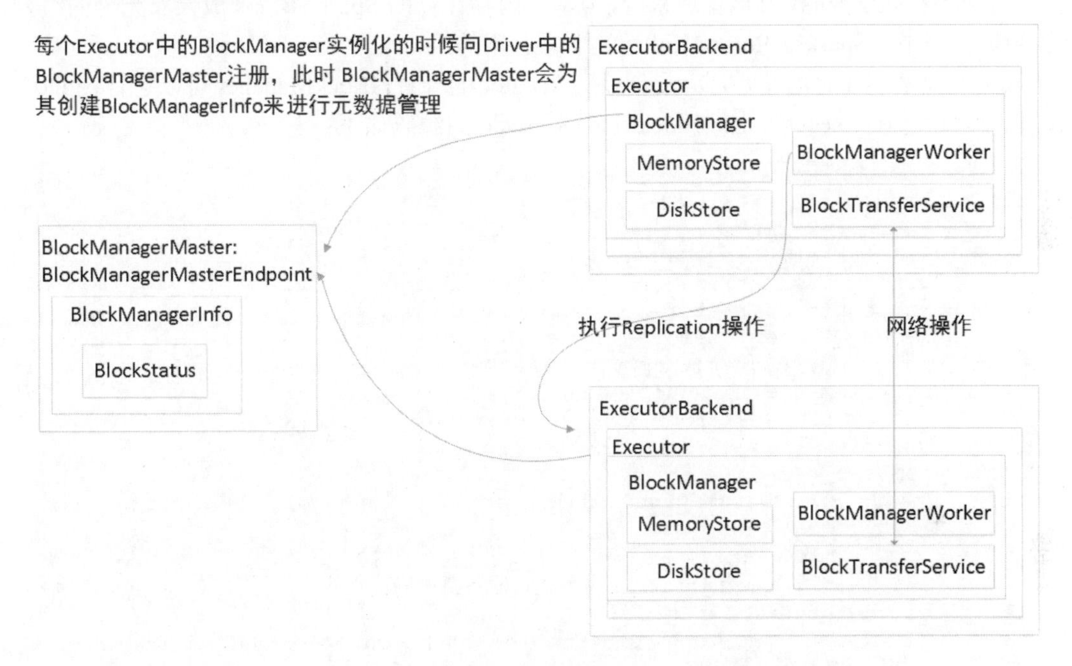

图 7-14　BlockManager 工作原理和运行机制简图

BlockManagerMasterEndpoint.scala 中 BlockManagerInfo 的 getStatus 方法如下：

```
1.    def getStatus(blockId: BlockId): Option[BlockStatus] = Option(_blocks.
      get(blockId))
```

其中的 BlockStatus 是一个 case class。

```
1.     case class BlockStatus(storageLevel: StorageLevel, memSize: Long,
       diskSize: Long) {
2.    def isCached: Boolean = memSize + diskSize > 0
3.  }
```

BlockTransferService.scala 进行网络连接操作，获取远程数据。

```
1.   private[spark]
2.  abstract class BlockTransferService extends ShuffleClient with Closeable
    with Logging {
```

7.6.5　BlockManager 解密进阶

本节讲解 BlockManager 初始化和注册解密、BlockManagerMaster 工作解密、BlockTransferService 解密、本地数据读写解密、远程数据读写解密。

BlockManager 既可以运行在 Driver 上，也可以运行在 Executor 上。在 Driver 上的 BlockManager 管理集群中 Executor 的所有的 BlockManager，BlockManager 分成 Master、Slave 结构，一切的调度、一切的工作由 Master 触发，Executor 在启动的时候一定会启动 BlockManager。BlockManager 主要提供了读和写数据的接口，可以从本地读写数据，也可以从远程读写数据。读写数据可以基于磁盘，也可以基于内存以及 OffHeap。OffHeap 就是堆外空间（如 Alluxio 是分布式内存管理系统，与基于内存计算的 Spark 系统形成天衣无缝的组合，在大数据领域中，Spark+Alluxio+Kafka 是非常有用的组合）。

从整个程序运行的角度看，Driver 也是 Executor 的一种，BlockManager 可以运行在 Driver 上，也可以运行在 Executor 上。BlockManager.scala 的源码如下：

```
1.   private[spark] class BlockManager(
2.      executorId: String,
3.      rpcEnv: RpcEnv,
4.      val master: BlockManagerMaster,
5.      val serializerManager: SerializerManager,
6.      val conf: SparkConf,
7.      memoryManager: MemoryManager,
8.      mapOutputTracker: MapOutputTracker,
9.      shuffleManager: ShuffleManager,
10.     val blockTransferService: BlockTransferService,
11.     securityManager: SecurityManager,
12.     numUsableCores: Int)
13.   extends BlockDataManager with BlockEvictionHandler with Logging {
14. ......
15. val diskBlockManager = {
16.     //如果外部服务不为 shuffle 文件提供服务执行清理文件
17.     val deleteFilesOnStop =
18.       !externalShuffleServiceEnabled || executorId == SparkContext.
        DRIVER_IDENTIFIER
19.     new DiskBlockManager(conf, deleteFilesOnStop)
20.   }
21. ......
22. private val futureExecutionContext = ExecutionContext.fromExecutorService(
23.     ThreadUtils.newDaemonCachedThreadPool("block-manager-future", 128))
24. ......
25.   private[spark] val memoryStore =
26.    new MemoryStore(conf, blockInfoManager, serializerManager, memoryManager,
       this)
27.   private[spark] val diskStore = new DiskStore(conf, diskBlockManager,
       securityManager)
28.   memoryManager.setMemoryStore(memoryStore)
29. ......
30.   def initialize(appId: String): Unit = {
31. ......
```

BlockManager 中的成员变量中：BlockManagerMaster 对整个集群的 BlockManagerMaster 进行管理；serializerManager 是默认的序列化器；MemoryManager 是内存管理；MapOutputTracker 是 Shuffle 输出的时候，要记录 ShuffleMapTask 输出的位置，以供下一个

Stage 使用，因此需要进行记录。BlockTransferService 是进行网络操作的，如果要连同另外一个 BlockManager 进行数据读写操作，就需要 BlockTransferService。Block 是 Spark 运行时数据的最小抽象单位，可能放入内存中，也可能放入磁盘中，还可能放在 Alluxio 上。

SecurityManager 是安全管理；numUsableCores 是可用的 Cores。

BlockManager 中 DiskBlockManager 管理磁盘的读写，创建并维护磁盘上逻辑块和物理块之间的逻辑映射位置。一个 block 被映射到根据 BlockId 生成的一个文件，块文件哈希列在目录 spark.local.dir 中（如果设置了 SPARK LOCAL DIRS），或在目录（SPARK LOCAL DIRS）中。

然后在 BlockManager 中创建一个缓存池：block-manager-future 以及 memoryStore、diskStore。

Shuffle 读写数据的时候是通过 BlockManager 进行管理的。

Spark 2.2.1 版本的 BlockManager.scala 的源码如下：

```
1.     var blockManagerId: BlockManagerId = _
2.
3.    //服务此 Executor 的 shuffle 文件的服务器的地址，这或者是外部的服务，或者只是我们
      //自己的 Executor 的 BlockManager
4.    private[spark] var shuffleServerId: BlockManagerId = _
5.
6.    //客户端读取其他 Executors 的 shuffle 文件。这或者是一个外部服务，或者只是
      //标准 BlockTransferService 直接连接到其他 Executors
7.
8.    private[spark] val shuffleClient = if (externalShuffleServiceEnabled) {
9.      val transConf = SparkTransportConf.fromSparkConf(conf, "shuffle",
        numUsableCores)
10.     new ExternalShuffleClient(transConf, securityManager, securityManager.
          isAuthenticationEnabled())
11.   } else {
12.     blockTransferService
13.   }
```

Spark 2.4.3 版本的 BlockManager.scala 的源码与 Spark 2.2.1 版本相比具有如下特点。

- 上段代码中第 12 行新增加一个参数。conf.get(config.SHUFFLE_REGISTRATION_TIMEOUT)。

```
1.  ......
2.   new ExternalShuffleClient(transConf, securityManager,
     securityManager.isAuthenticationEnabled(),
     conf.get(config.SHUFFLE_REGISTRATION_TIMEOUT))
3.  ......
```

BlockManager.scala 中，BlockManager 实例对象通过调用 initialize 方法才能正式工作，传入参数是 appId，基于应用程序的 ID 初始化 BlockManager。initialize 不是在构造器的时候被使用，因为 BlockManager 实例化的时候还不知道应用程序的 ID，应用程序 ID 是应用程序启动时，ExecutorBackend 向 Master 注册时候获得的。

BlockManager.scala 的 initialize 方法中的 BlockTransferService 进行网络通信。ShuffleClient 是 BlockManagerWorker 每次启动时向 BlockManagerMaster 注册。BlockManager.scala 的 initialize 方法中调用了 registerBlockManager，向 Master 进行注册，告诉 BlockManagerMaster 把自己注册进去。

BlockManagerMaster.scala 的 registerBlockManager 的源码如下：

```
1.    def registerBlockManager(
2.        blockManagerId: BlockManagerId,
3.        maxOnHeapMemSize: Long,
4.        maxOffHeapMemSize: Long,
5.        slaveEndpoint: RpcEndpointRef): BlockManagerId = {
6.      logInfo(s"Registering BlockManager $blockManagerId")
7.      val updatedId = driverEndpoint.askSync[BlockManagerId](
8.        RegisterBlockManager(blockManagerId, maxOnHeapMemSize, maxOffHeapMemSize,
          slaveEndpoint))
9.      logInfo(s"Registered BlockManager $updatedId")
10.     updatedId
11.   }
```

registerBlockManager 方法的 RegisterBlockManager 是一个 case class。

BlockManagerMessages.scala 的源码如下：

```
1.  case class RegisterBlockManager(
2.        blockManagerId: BlockManagerId,
3.        maxOnHeapMemSize: Long,
4.        maxOffHeapMemSize: Long,
5.        sender: RpcEndpointRef)
6.      extends ToBlockManagerMaster
```

在 Executor 实例化的时候，要初始化 blockManager。blockManager 在 initialize 中将应用程序 ID 传进去。

Executor.scala 中，Executor 每隔 10s 向 Master 发送心跳消息，如收不到心跳消息，blockManager 须重新注册。

Spark 2.1.2 版本的 Executor.scala 的源码如下：

```
1.  ……
2.  val message = Heartbeat(executorId, accumUpdates.toArray,
    env.blockManager.blockManagerId)
3.      try {
4.        val response = heartbeatReceiverRef.askRetry
          [HeartbeatResponse](
5.            message, RpcTimeout(conf, "spark.executor.heartbeatInterval",
            "10s"))
6.        if (response.reregisterBlockManager) {
7.          logInfo("Told to re-register on heartbeat")
8.          env.blockManager.reregister()
9.        }
10.       heartbeatFailures = 0
11.     } catch {
12.       case NonFatal(e) =>
13.         logWarning("Issue communicating with driver in heartbeater", e)
14.         heartbeatFailures += 1
15.         if (heartbeatFailures >= HEARTBEAT_MAX_FAILURES) {
16.           logError(s"Exit as unable to send heartbeats to driver " +
17.             s"more than $HEARTBEAT_MAX_FAILURES times")
18.           System.exit(ExecutorExitCode.HEARTBEAT_FAILURE)
19.         }
20.     }
```

Spark 2.4.3 版本的 Executor.scala 的源码与 Spark 2.1.1 版本相比具有如下特点。

- 上段代码中第 5 行 RpcTimeout 调整为以下代码：

```
1.  ……
2.          val response = heartbeatReceiverRef.askSync[HeartbeatResponse](
            message, new RpcTimeout(HEARTBEAT_INTERVAL_MS.millis, EXECUTOR_
            HEARTBEAT_INTERVAL.key))
```

回到 BlockManagerMaster.scala 的 registerBlockManager：

registerBlockManager 中 RegisterBlockManager 传入的 slaveEndpoint 是：具体的 Executor 启动时会启动一个 BlockManagerSlaveEndpoint，会接收 BlockManagerMaster 发过来的指令。在 initialize 方法中通过 master.registerBlockManager 传入 slaveEndpoint，而 slaveEndpoint 是在 rpcEnv.setupEndpoint 方法中调用 new()函数创建的 BlockManagerSlaveEndpoint。

总结如下。

（1）当 Executor 实例化的时候，会通过 BlockManager.initialize 来实例化 Executor 上的 BlockManager，并且创建 BlockManagerSlaveEndpoint 这个消息循环体来接受 Driver 中 BlockManagerMaster 发过来的指令，如删除 Block 等。

```
1.        env.blockManager.initialize(conf.getAppId)
```

BlockManagerSlaveEndpoint.scala 的源码如下：

```
1.    class BlockManagerSlaveEndpoint(
2.      override val rpcEnv: RpcEnv,
3.      blockManager: BlockManager,
4.      mapOutputTracker: MapOutputTracker)
5.    extends ThreadSafeRpcEndpoint with Logging {
```

（2）当 BlockManagerSlaveEndpoint 实例化后，Executor 上的 BlockManager 需要向 Driver 上的 BlockManagerMasterEndpoint 注册。

BlockManagerMaster 的 registerBlockManager 方法，其中的 driverEndpoint 是构建 BlockManagerMaster 时传进去的。

（3）BlockManagerMasterEndpoint 接收到 Executor 上的注册信息并进行处理。

BlockManagerMasterEndpoint.scala 的源码如下：

```
1. private[spark]
2. class BlockManagerMasterEndpoint(
3.     override val rpcEnv: RpcEnv,
4. ……
5. override def receiveAndReply(context: RpcCallContext): PartialFunction[Any,
   Unit] = {
6.     case RegisterBlockManager(blockManagerId, maxOnHeapMemSize,
       maxOffHeapMemSize, slaveEndpoint) =>
7.       context.reply(register(blockManagerId, maxOnHeapMemSize,
         maxOffHeapMemSize, slaveEndpoint))
```

BlockManagerMasterEndpoint 的 register 注册方法，为每个 Executor 的 BlockManager 生成对应的 BlockManagerInfo。BlockManagerInfo 是一个 HashMap[BlockManagerId, BlockManagerInfo]。

BlockManagerMasterEndpoint.scala 的 register 注册方法源码如下：

```
1.  private val blockManagerInfo = new mutable.HashMap[BlockManagerId,
    BlockManagerInfo]
2.   ……
3.  private def register(
4.        idWithoutTopologyInfo: BlockManagerId,
5.        maxOnHeapMemSize: Long,
```

```
6.        maxOffHeapMemSize: Long,
7.        slaveEndpoint: RpcEndpointRef): BlockManagerId = {
8.      //dummy id 不应包含拓扑信息
9.      //在这里得到信息和回应一个块标识符
10.
11.     val id = BlockManagerId(
12.       idWithoutTopologyInfo.executorId,
13.       idWithoutTopologyInfo.host,
14.       idWithoutTopologyInfo.port,
15.       topologyMapper.getTopologyForHost(idWithoutTopologyInfo.host))
16.
17.     val time = System.currentTimeMillis()
18.     if (!blockManagerInfo.contains(id)) {
19.       blockManagerIdByExecutor.get(id.executorId) match {
20.         case Some(oldId) =>
21.           //同一个 Executor 的块管理器已经存在，所以删除它（假设已挂掉）
22.           logError("Got two different block manager registrations on same
            executor - "
23.               + s" will replace old one $oldId with new one $id")
24.           removeExecutor(id.executorId)
25.         case None =>
26.       }
27.       logInfo("Registering block manager %s with %s RAM, %s".format(
28.         id.hostPort, Utils.bytesToString(maxOnHeapMemSize +
          maxOffHeapMemSize), id))
29.
30.       blockManagerIdByExecutor(id.executorId) = id
31.
32.       blockManagerInfo(id) = new BlockManagerInfo(
33.         id, System.currentTimeMillis(), maxOnHeapMemSize, maxOffHeapMemSize,
          slaveEndpoint)
34.     }
35.     listenerBus.post(SparkListenerBlockManagerAdded(time, id,
      maxOnHeapMemSize + maxOffHeapMemSize,
36.         Some(maxOnHeapMemSize), Some(maxOffHeapMemSize)))
37.     id
38.   }
```

BlockManagerMasterEndpoint 中，BlockManagerId 是一个 class，标明了 BlockManager 在哪个 Executor 中，以及 host 主机名、port 端口等信息。

BlockManagerId.scala 的源码如下：

```
1.   class BlockManagerId private (
2.       private var executorId_ : String,
3.       private var host_ : String,
4.       private var port_ : Int,
5.       private var topologyInfo_ : Option[String])
6.     extends Externalizable {
```

BlockManagerMasterEndpoint 中，BlockManagerInfo 包含内存、slaveEndpoint 等信息。

回到 BlockManagerMasterEndpoint 的 register 注册方法：如果 blockManagerInfo 没有包含 BlockManagerId，根据 BlockManagerId.executorId 查询 BlockManagerId，如果匹配到旧的 BlockManagerId，就进行清理。

BlockManagerMasterEndpoint 的 removeExecutor 方法如下：

```
1.      private def removeExecutor(execId: String) {
```

```
2.      logInfo("Trying to remove executor " + execId + " from
        BlockManagerMaster.")
3.      blockManagerIdByExecutor.get(execId).foreach(removeBlockManager)
4.    }
```

进入 removeBlockManager 方法，从 blockManagerIdByExecutor 数据结构中清理掉 block manager 信息，从 blockManagerInfo 数据结构中清理掉所有的 blocks 信息。removeBlockManager 源码如下。

BlockManagerMasterEndpoint.scala 的 removeBlockManager 的源码如下：

```
1.  private def removeBlockManager(blockManagerId: BlockManagerId) {
2.      val info = blockManagerInfo(blockManagerId)
3.
4.      //从blockManagerIdByExecutor删除块管理
5.      blockManagerIdByExecutor -= blockManagerId.executorId
6.
7.     //将它从blockManagerInfo 删除所有的块
8.      blockManagerInfo.remove(blockManagerId)
9.
10.     val iterator = info.blocks.keySet.iterator
11.     while (iterator.hasNext) {
12.       val blockId = iterator.next
13.       val locations = blockLocations.get(blockId)
14.       locations -= blockManagerId
15.    //如果没有块管理器，就注销这个块。否则，如果主动复制启用，块block是一个RDD
       //或测试块 block（后者用于单元测试），发送一条消息随机选择Executor的位置来
       //复制给定块block。注意，此处忽略了其他块block类型（如广播broadcast shuffle
16.    //blocks），因为复制在这种情况下没有多大意义
17.
18.       if (locations.size == 0) {
19.         blockLocations.remove(blockId)
20.         logWarning(s"No more replicas available for $blockId !")
21.       } else if (proactivelyReplicate && (blockId.isRDD || blockId
          .isInstanceOf[TestBlockId])) {
22.         //假设Executor未能找出故障前存在的副本数量
23.         val maxReplicas = locations.size + 1
24.         val i = (new Random(blockId.hashCode)).nextInt(locations.size)
25.         val blockLocations = locations.toSeq
26.         val candidateBMId = blockLocations(i)
27.         blockManagerInfo.get(candidateBMId).foreach { bm =>
28.           val remainingLocations = locations.toSeq.filter(bm => bm !=
              candidateBMId)
29.           val replicateMsg = ReplicateBlock(blockId, remainingLocations,
              maxReplicas)
30.           bm.slaveEndpoint.ask[Boolean](replicateMsg)
31.         }
32.       }
33.     }
34.
35.     listenerBus.post(SparkListenerBlockManagerRemoved(System
        .currentTimeMillis(), blockManagerId))
36.     logInfo(s"Removing block manager $blockManagerId")
37.
38.   }
```

removeBlockManager 中的一行代码 blockLocations.remove 的 remove 方法如下。

HashMap.java 的源码如下：

```
1.      public V remove(Object key) {
2.          Node<K,V> e;
3.          return (e = removeNode(hash(key), key, null, false, true)) == null ?
4.              null : e.value;
5.      }
```

回到 BlockManagerMasterEndpoint 的 register 注册方法：然后在 blockManagerIdByExecutor 中加入 BlockManagerId，将 BlockManagerId 加入 BlockManagerInfo 信息，在 listenerBus 中进行监听，函数返回 BlockManagerId，完成注册。

回到 BlockManager.scala，在 initialize 方法通过 master.registerBlockManager 注册成功以后，将返回值赋值给 idFromMaster。Initialize 初始化之后，看一下 BlockManager.scala 中其他的方法。

reportAllBlocks 方法：具体的 Executor 须向 Driver 不断地汇报自己的状态。

BlockManager.scala 的 reportAllBlocks 方法的源码如下：

```
1.        private def reportAllBlocks(): Unit = {
2.      logInfo(s"Reporting ${blockInfoManager.size} blocks to the master.")
3.      for ((blockId, info) <- blockInfoManager.entries) {
4.        val status = getCurrentBlockStatus(blockId, info)
5.        if (info.tellMaster && !tryToReportBlockStatus(blockId, status)) {
6.          logError(s"Failed to report $blockId to master; giving up.")
7.          return
8.        }
9.      }
10.   }
```

reportAllBlocks 方法中调用了 getCurrentBlockStatus，包括内存、磁盘等信息。

getCurrentBlockStatus 的源码如下：

```
1.   private def getCurrentBlockStatus(blockId: BlockId, info: BlockInfo):
     BlockStatus = {
2.     info.synchronized {
3.       info.level match {
4.         case null =>
5.           BlockStatus.empty
6.         case level =>
7.           val inMem = level.useMemory && memoryStore.contains(blockId)
8.           val onDisk = level.useDisk && diskStore.contains(blockId)
9.           val deserialized = if (inMem) level.deserialized else false
10.          val replication = if (inMem || onDisk) level.replication else 1
11.          val storageLevel = StorageLevel(
12.            useDisk = onDisk,
13.            useMemory = inMem,
14.            useOffHeap = level.useOffHeap,
15.            deserialized = deserialized,
16.            replication = replication)
17.          val memSize = if (inMem) memoryStore.getSize(blockId) else 0L
18.          val diskSize = if (onDisk) diskStore.getSize(blockId) else 0L
19.          BlockStatus(storageLevel, memSize, diskSize)
20.      }
21.    }
22.  }
```

getCurrentBlockStatus 方法中的 BlockStatus，包含存储级别 StorageLevel、内存大小、磁

盘大小等信息。

BlockManagerMasterEndpoint.scala 的 BlockStatus 的源码如下：

```
1.  case class BlockStatus(storageLevel: StorageLevel, memSize: Long,
    diskSize: Long) {
2.    def isCached: Boolean = memSize + diskSize > 0
3.  }
4.  ......
5.    object BlockStatus {
6.    def empty: BlockStatus = BlockStatus(StorageLevel.NONE, memSize = 0L,
      diskSize = 0L)
7.  }
```

回到 BlockManager.scala，其中的 getLocationBlockIds 方法比较重要，根据 BlockId 获取这个 BlockId 所在的 BlockManager。

BlockManager.scala 的 getLocationBlockIds 的源码如下：

```
1.    private def getLocationBlockIds(blockIds: Array[BlockId]): Array[Seq
      [BlockManagerId]] = {
2.     val startTimeMs = System.currentTimeMillis
3.     val locations = master.getLocations(blockIds).toArray
4.     logDebug("Got multiple block location in %s".format
       (Utils.getUsedTimeMs(startTimeMs)))
5.     locations
6.    }
```

getLocationBlockIds 方法中根据 BlockId 通过 master.getLocations 向 Master 获取位置信息，因为 master 管理所有的位置信息。getLocations 方法里的 driverEndpoint 是 BlockManagerMasterEndpoint，Executor 向 BlockManagerMasterEndpoint 发送 GetLocationsMultipleBlockIds 消息。

BlockManagerMaster.scala 的 getLocations 方法的源码如下：

```
1.   def getLocations(blockIds: Array[BlockId]): IndexedSeq[Seq[BlockManagerId]] = {
2.     driverEndpoint.askSync[IndexedSeq[Seq[BlockManagerId]]](
3.       GetLocationsMultipleBlockIds(blockIds))
4.   }
```

getLocations 中的 GetLocationsMultipleBlockIds 是一个 case class。

```
1.      case class GetLocationsMultipleBlockIds(blockIds: Array[BlockId])
        extends ToBlockManagerMaster
```

在 BlockManagerMasterEndpoint 侧接收 GetLocationsMultipleBlockIds 消息。

BlockManagerMasterEndpoint.scala 的 receiveAndReply 方法如下：

```
1.   override def receiveAndReply(context: RpcCallContext): PartialFunction
     [Any, Unit] = {
2.  ......
3.   case GetLocationsMultipleBlockIds(blockIds) =>
4.       context.reply(getLocationsMultipleBlockIds(blockIds))
```

进入 getLocationsMultipleBlockIds 方法，进行 map 操作，开始查询位置信息。

```
1.    private def getLocationsMultipleBlockIds(
2.        blockIds: Array[BlockId]): IndexedSeq[Seq[BlockManagerId]] = {
3.     blockIds.map(blockId => getLocations(blockId))
4.    }
```

进入 getLocations 方法，首先判断内存缓存结构 blockLocations 中是否包含 blockId，如果已包含，就获取位置信息，否则返回空的信息。

```
1.    private def getLocations(blockId: BlockId): Seq[BlockManagerId] = {
2.      if (blockLocations.containsKey(blockId)) blockLocations.get
        (blockId).toSeq else Seq.empty
3.    }
```

其中，blockLocations 是一个重要的数据结构，是一个 JHashMap。Key 是 BlockId。Value 是一个 HashSet[BlockManagerId]，使用 HashSet。因为每个 BlockId 在磁盘上有副本，不同机器的位置不一样，而且不同副本对应的 BlockManagerId 不一样，位于不同的机器上，所以使用 HashSet 数据结构。

BlockManagerMasterEndpoint.scala 的 blockLocations 的源码如下：

```
1.   private val blockLocations = new JHashMap[BlockId, mutable.HashSet
     [BlockManagerId]]
```

回到 BlockManager.scala，getLocalValues 是一个重要的方法，从 blockInfoManager 中获取本地数据。

- 首先根据 blockId 从 blockInfoManager 中获取 BlockInfo 信息。
- 从 BlockInfo 信息获取 level 级别，根据 level.useMemory && memoryStore.contains（blockId）判断是否在内存中，如果在内存中，就从 memoryStore 中获取数据。
- 根据 level.useDisk && diskStore.contains（blockId）判断是否在磁盘中，如果在磁盘中，就从 diskStore 中获取数据。

BlockManager.scala 的 getLocalValues 方法的源码如下：

```
1.  def getLocalValues(blockId: BlockId): Option[BlockResult] = {
2.      logDebug(s"Getting local block $blockId")
3.      blockInfoManager.lockForReading(blockId) match {
4.        case None =>
5.          logDebug(s"Block $blockId was not found")
6.          None
7.        case Some(info) =>
8.          val level = info.level
9.          logDebug(s"Level for block $blockId is $level")
10.         val taskAttemptId = Option(TaskContext.get()).map(_.taskAttemptId())
11.         if (level.useMemory && memoryStore.contains(blockId)) {
12.           val iter: Iterator[Any] = if (level.deserialized) {
13.             memoryStore.getValues(blockId).get
14.           } else {
15.             serializerManager.dataDeserializeStream(
16.               blockId, memoryStore.getBytes(blockId).get
                .toInputStream())(info.classTag)
17.           }
18.           //如果迭代器是从另一个没有 TaskContext 集的线程触发的，那么我们需要捕获当
              //前 taskID；请参阅 spark-18406
19.           val ci = CompletionIterator[Any, Iterator[Any]](iter, {
20.             releaseLock(blockId, taskAttemptId)
21.           })
22.           Some(new BlockResult(ci, DataReadMethod.Memory, info.size))
23.         } else if (level.useDisk && diskStore.contains(blockId)) {
24.           val diskData = diskStore.getBytes(blockId)
25.           val iterToReturn: Iterator[Any] = {
```

```
26.             if (level.deserialized) {
27.               val diskValues = serializerManager.dataDeserializeStream(
28.                 blockId,
29.                 diskData.toInputStream())(info.classTag)
30.               maybeCacheDiskValuesInMemory(info, blockId, level, diskValues)
31.             } else {
32.               val stream = maybeCacheDiskBytesInMemory(info, blockId, level,
                  diskData)
33.                 .map { _.toInputStream(dispose = false) }
34.                 .getOrElse { diskData.toInputStream() }
35.               serializerManager.dataDeserializeStream(blockId, stream)
                  (info.classTag)
36.             }
37.           }
38.           val ci = CompletionIterator[Any, Iterator[Any]](iterToReturn, {
39.             releaseLockAndDispose(blockId, diskData, taskAttemptId)
40.           })
41.           Some(new BlockResult(ci, DataReadMethod.Disk, info.size))
42.         } else {
43.           handleLocalReadFailure(blockId)
44.         }
45.     }
46.   }
```

回到 BlockManager.scala，getRemoteValues 方法从远程的 BlockManager 中获取 block 数据，在 JVM 中不需要去获取锁。

BlockManager.scala 的 getRemoteValues 方法的源码如下：

```
1.    private def getRemoteValues[T: ClassTag](blockId: BlockId): Option
      [BlockResult] = {
2.      val ct = implicitly[ClassTag[T]]
3.      getRemoteBytes(blockId).map { data =>
4.        val values =
5.          serializerManager.dataDeserializeStream(blockId, data.toInputStream
            (dispose = true))(ct)
6.        new BlockResult(values, DataReadMethod.Network, data.size)
7.      }
8.    }
```

getRemoteValues 方法中调用 getRemoteBytes，获取远程的数据，如果获取的失败次数超过最大的获取次数（locations.size），就提示失败，返回空值；如果获取到远程数据，就返回。

getRemoteBytes 方法调用 blockTransferService.fetchBlockSync 方法实现远程获取数据。

Spark 2.2.1 版本的 BlockTransferService.scala 的 fetchBlockSync 方法的源码如下：

```
1.   def fetchBlockSync(host: String, port: Int, execId: String, blockId:
     String): ManagedBuffer = {
2.      //线程等待的监视器
3.      val result = Promise[ManagedBuffer]()
4.      fetchBlocks(host, port, execId, Array(blockId),
5.        new BlockFetchingListener {
6.          override def onBlockFetchFailure(blockId: String, exception:
            Throwable): Unit = {
7.            result.failure(exception)
8.          }
9.          override def onBlockFetchSuccess(blockId: String, data:
            ManagedBuffer): Unit = {
```

```
10.          val ret = ByteBuffer.allocate(data.size.toInt)
11.          ret.put(data.nioByteBuffer())
12.          ret.flip()
13.          result.success(new NioManagedBuffer(ret))
14.        }
15.      }tempShuffleFileManager=null)
16.    ThreadUtils.awaitResult(result.future, Duration.Inf)
17.  }
```

Spark 2.4.3 版本的 BlockTransferService.scala 的 fetchBlockSync 方法的源码与 Spark 2.2.1 版本相比具有如下特点。

- 上段代码中第 1 行新增加一个 tempFileManager 参数。
- 上段代码中第 9～15 行 onBlockFetchSuccess 方法整体替换为以下的 onBlockFetchSuccess 代码。

```
1.  ……
2.  tempFileManager: DownloadFileManager): ManagedBuffer = {
3.  ……
4.    override def onBlockFetchSuccess(blockId: String, data: ManagedBuffer):
    Unit = {
5.          data match {
6.            case f: FileSegmentManagedBuffer =>
7.              result.success(f)
8.            case e: EncryptedManagedBuffer =>
9.              result.success(e)
10.           case _ =>
11.             try {
12.               val ret = ByteBuffer.allocate(data.size.toInt)
13.               ret.put(data.nioByteBuffer())
14.               ret.flip()
15.               result.success(new NioManagedBuffer(ret))
16.             } catch {
17.               case e: Throwable => result.failure(e)
18.             }
19.         }
20.       }
21.     }, tempFileManager)
```

fetchBlocks 方法用于从远程节点异步获取序列块，仅在调用[init]之后可用。注意，这个 API 需要一个序列，可以实现批处理请求，而不是返回一个 future，底层实现可以调用 onBlockFetchSuccess 来尽快获取块的数据，而不是等待所有块被取出来。

fetchBlockSync 中调用 fetchBlocks 方法，NettyBlockTransferService 继承自 BlockTransferService，是 BlockTransferService 实现子类。

Spark 2.2.1 版本的 NettyBlockTransferService 的 fetchBlocks 的源码如下：

```
1. override def fetchBlocks(
2.       host: String,
3.       port: Int,
4.       execId: String,
5.       blockIds: Array[String],
6.       listener: BlockFetchingListener,
7.       tempShuffleFileManager: TempShuffleFileManager): Unit = {
8.     logTrace(s"Fetch blocks from $host:$port (executor id $execId)")
9.     try {
10.      val blockFetchStarter = new RetryingBlockFetcher.BlockFetchStarter {
```

```
11.         override def createAndStart(blockIds: Array[String], listener:
            BlockFetchingListener) {
12.           val client = clientFactory.createClient(host, port)
13.           new OneForOneBlockFetcher(client, appId, execId, blockIds, listener,
14.             transportConf, tempShuffleFileManager).start()
15.         }
16.       }
17.
18.       val maxRetries = transportConf.maxIORetries()
19.       if (maxRetries > 0) {
20.         //注意，Fetcher 将正确处理 maxRetries 等于 0 的情况；避免它在代码中产生 Bug，
21.         //一旦确定了稳定性，就应该删除 if 语句
22.         new RetryingBlockFetcher(transportConf, blockFetchStarter, blockIds,
            listener).start()
23.       } else {
24.         blockFetchStarter.createAndStart(blockIds, listener)
25.       }
26.     } catch {
27.       case e: Exception =>
28.         logError("Exception while beginning fetchBlocks", e)
29.         blockIds.foreach(listener.onBlockFetchFailure(_, e))
30.     }
31.   }
```

Spark 2.4.3 版本的 NettyBlockTransferService.scala 源码与 Spark 2.2.1 版本相比具有如下特点。

- 上段代码中第 7、14 行 tempShuffleFileManager 调整为 tempFileManager。DownloadFileManager 用于创建临时块文件的管理器，用于获取远程数据以减少内存使用。当文件不再使用时，它将清除文件。

```
1.  ......
2.  tempFileManager: DownloadFileManager): Unit = {
3.  ......
4.  transportConf, tempFileManager).start()
```

回到 BlockManager.scala，无论是 doPutBytes()，还是 doPutIterator()方法中，都会使用 doPut 方法。

BlockManager.scala 的 doPut 方法的源码如下：

```
1.   private def doPut[T](
2.       blockId: BlockId,
3.       level: StorageLevel,
4.       classTag: ClassTag[_],
5.       tellMaster: Boolean,
6.       keepReadLock: Boolean)(putBody: BlockInfo => Option[T]): Option[T]
         = {
7.     require(blockId != null, "BlockId is null")
8.     require(level != null && level.isValid, "StorageLevel is null or
       invalid")
9.
10.    val putBlockInfo = {
11.      val newInfo = new BlockInfo(level, classTag, tellMaster)
12.      if (blockInfoManager.lockNewBlockForWriting(blockId, newInfo)) {
13.        newInfo
14.      } else {
15.        logWarning(s"Block $blockId already exists on this machine; not
           re-adding it")
```

```
16.         if (!keepReadLock) {
17.     //在现有的块上 lockNewBlockForWriting 返回一个读锁，所以我们必须释放它
18.           releaseLock(blockId)
19.         }
20.         return None
21.       }
22.     }
23.
24.     val startTimeMs = System.currentTimeMillis
25.     var exceptionWasThrown: Boolean = true
26.     val result: Option[T] = try {
27.       val res = putBody(putBlockInfo)
28.       exceptionWasThrown = false
29.      ......
30.      result
31.   }
```

doPut 方法中，lockNewBlockForWriting 写入一个新的块前先尝试获得适当的锁，如果我们是第一个写块，获得写入锁后继续后续操作。否则，如果另一个线程已经写入块，须等待写入完成，才能获取读取锁，调用 new()函数创建一个 BlockInfo 赋值给 putBlockInfo，然后通过 putBody(putBlockInfo)将数据存入。putBody 是一个匿名函数，输入 BlockInfo，输出的是一个泛型 Option[T]。putBody 函数体内容是 doPutIterator 方法（doPutBytes 方法也类似调用 doPut）调用 doPut 时传入的。

BlockManager.scala 的 doPutIterator 调用 doPut 方法，在其 putBody 匿名函数体中进行判断：

如果是 level.useMemory，则在 memoryStore 中放入数据。

如果是 level.useDisk，则在 diskStore 中放入数据。

如果 level.replication 大于 1，则在其他节点中存入副本数据。

其中，Spark 2.2.1 版本的 BlockManager.scala 的 replicate 方法的源码如下：

```
1.    private def replicate(
2.        blockId: BlockId,
3.        data: BlockData,
4.        level: StorageLevel,
5.        classTag: ClassTag[_],
6.        existingReplicas: Set[BlockManagerId] = Set.empty): Unit = {
7.   ......
8.
9.    while(numFailures <= maxReplicationFailures &&
10.       !peersForReplication.isEmpty &&
11.       peersReplicatedTo.size < numPeersToReplicateTo) {
12.       val peer = peersForReplication.head
13.       try {
14.         val onePeerStartTime = System.nanoTime
15.         logTrace(s"Trying to replicate $blockId of ${data.size} bytes to $peer")
16.         blockTransferService.uploadBlockSync(
17.           peer.host,
18.           peer.port,
19.           peer.executorId,
20.           blockId,
21.           new BlockManagerManagedBuffer(blockInfoManager, blockId, data, false),
22.           tLevel,
23.           classTag)
24. ......
```

Spark 2.4.3 版本的 BlockManager.scala 的 replicate 方法的源码与 Spark 2.2.1 版本相比具有如下特点。

- 上段代码中第 15 行之后新增代码，构建 BlockManagerManagedBuffer 的实例 buffer。
- 上段代码中第 21 行将 BlockManagerManagedBuffer 替换为 buffer。

```
1.  ……
2.  //这个线程在块上保持一个锁，所以我们不希望 netty 线程在块发送完消息后解锁它
3.          val buffer = new BlockManagerManagedBuffer(blockInfoManager, blockId,
            data, false,
4.            unlockOnDeallocate = false)
5.  ……
6.  buffer,
7.  ……
```

replicate 方法中调用了 blockTransferService.uploadBlockSync 方法。

BlockTransferService.scala 的 uploadBlockSync 的源码如下：

```
1.   def uploadBlockSync(
2.       hostname: String,
3.       port: Int,
4.       execId: String,
5.       blockId: BlockId,
6.       blockData: ManagedBuffer,
7.       level: StorageLevel,
8.       classTag: ClassTag[_]): Unit = {
9.     val future = uploadBlock(hostname, port, execId, blockId, blockData,
       level, classTag)
10.    ThreadUtils.awaitResult(future, Duration.Inf)
11.  }
12. }
```

uploadBlockSync 中又调用 uploadBlock 方法，BlockTransferService.scala 的 uploadBlock 方法无具体实现，NettyBlockTransferService 是 BlockTransferService 的子类，具体实现 uploadBlock 方法。

Spark 2.2.1 版本的 NettyBlockTransferService 的 uploadBlock 的源码如下：

```
1.   override def uploadBlock(
2.       hostname: String,
3.       port: Int,
4.       execId: String,
5.       blockId: BlockId,
6.       blockData: ManagedBuffer,
7.       level: StorageLevel,
8.       classTag: ClassTag[_]): Future[Unit] = {
9.     val result = Promise[Unit]()
10.    val client = clientFactory.createClient(hostname, port)
11.
12.    //使用 JavaSerializer 序列号器将 StorageLevel 和 ClassTag 序列化。其他一切都
       //用我们的二进制协议编码
13.    val metadata = JavaUtils.bufferToArray(serializer.newInstance()
       .serialize((level, classTag)))
14.
15.    //为了序列化，转换或复制 NIO 缓冲到数组
16.    val array = JavaUtils.bufferToArray(blockData.nioByteBuffer())
17.
18.    client.sendRpc(new  UploadBlock(appId,  execId,  blockId.toString,
```

```
        metadata, array).toByteBuffer,
19.       new RpcResponseCallback {
20.         override def onSuccess(response: ByteBuffer): Unit = {
21.           logTrace(s"Successfully uploaded block $blockId")
22.           result.success((): Unit)
23.         }
24.         override def onFailure(e: Throwable): Unit = {
25.           logError(s"Error while uploading block $blockId", e)
26.           result.failure(e)
27.         }
28.       })
29.
30.     result.future
31.   }
```

Spark 2.4.3 版本的 NettyBlockTransferService.scala 源码与 Spark 2.2.1 版本相比具有如下特点。

- 上段代码中删掉第 16～18 行。
- 上段代码中第 19 行之前新增代码，新建 asStream 变量，其将 blockData 与 MAX_REMOTE_ BLOCK_SIZE_FETCH_TO_MEM 进行比较。当块的大小高于此阈值时，远程块以字节为单位将被提取到磁盘。这是为了避免一个大的请求占用太多的内存。通过设置特定值（例如 200）可以启用这个配置。注意，此配置将影响 shuffle 获取和块管理器远程块获取。对于启用了外部 shuffle 服务，此功能只能在高于 Spark 2.2 的版本，在外部 shuffle 启用时使用服务。
- 上段代码中第 19 行新建一个变量 callback。
- 上段代码中第 21 行 logTrace 日志的内容进行调整。
- 上段代码中第 25 行 logError 日志的内容进行调整。
- 上段代码中第 29 行之后新增一段代码，如果 blockData 大于 MAX_REMOTE_BLOCK_SIZE_FETCH_TO_MEM，则调用 client.uploadStream 方法，将数据作为流发送到远程端，与 stream()方法的不同之处在于，这是一个请求向远程端发送数据，而不是从远程端接收数据。如果将 blockData 小于阈值，则将 NIO 缓冲区转换或复制到数组中，以便对其进行序列化。

```
1.   ......
2.     val asStream = blockData.size() > conf.get(config.MAX_REMOTE_BLOCK_
       SIZE_FETCH_TO_MEM)
3.   ......
4.   val callback = new RpcResponseCallback {
5.   ......
6.   logTrace(s"Successfully uploaded block $blockId${if (asStream) " as
     stream" else ""}")
7.   ......
8.   logError(s"Error while uploading $blockId${if (asStream) " as stream" else
     ""}", e)
9.   ......
10.    if (asStream) {
11.      val streamHeader = new UploadBlockStream(blockId.name, metadata)
         .toByteBuffer
12.      client.uploadStream(new NioManagedBuffer(streamHeader), blockData,
         callback)
13.    } else {
14.      // 将 NIO 缓冲区转换或复制到数组中，以便对其进行序列化
```

```
15.       val array = JavaUtils.bufferToArray(blockData.nioByteBuffer())
16.
17.       client.sendRpc(new UploadBlock(appId, execId, blockId.name,
          metadata, array).toByteBuffer,
18.         callback)
19.     }
20. ......
```

回到 BlockManager.scala，看一下 dropFromMemory 方法。如果存储级别定位为 MEMORY_AND_DISK，那么数据可能放在内存和磁盘中，内存够的情况下不会放到磁盘上；如果内存不够，就放到磁盘上，这时就会调用 dropFromMemory。如果存储级别不是定义为 MEMORY_AND_DISK，而只是存储在内存中，内存不够时，缓存的数据此时就会丢弃。如果仍需要数据，那就要重新计算。

BlockManager.scala 的 dropFromMemory 的源码如下：

```
1.    private[storage] override def dropFromMemory[T: ClassTag](
2.        blockId: BlockId,
3.        data: () => Either[Array[T], ChunkedByteBuffer]): StorageLevel = {
4.      logInfo(s"Dropping block $blockId from memory")
5.      val info = blockInfoManager.assertBlockIsLockedForWriting(blockId)
6.      var blockIsUpdated = false
7.      val level = info.level
8.
9.      // 如果存储级别要求，则保存到磁盘
10.     if (level.useDisk && !diskStore.contains(blockId)) {
11.       logInfo(s"Writing block $blockId to disk")
12.       data() match {
13.         case Left(elements) =>
14.           diskStore.put(blockId) { channel =>
15.             val out = Channels.newOutputStream(channel)
16.             serializerManager.dataSerializeStream(
17.               blockId,
18.               out,
19.               elements.toIterator)(info.classTag.asInstanceOf[ClassTag[T]])
20.           }
21.         case Right(bytes) =>
22.           diskStore.putBytes(blockId, bytes)
23.       }
24.       blockIsUpdated = true
25.     }
26.
27.     // 实际由内存存储
28.     val droppedMemorySize =
29.       if (memoryStore.contains(blockId)) memoryStore.getSize(blockId) else 0L
30.     val blockIsRemoved = memoryStore.remove(blockId)
31.     if (blockIsRemoved) {
32.       blockIsUpdated = true
33.     } else {
34.       logWarning(s"Block $blockId could not be dropped from memory as it
    does not exist")
35.     }
36.
37.     val status = getCurrentBlockStatus(blockId, info)
38.     if (info.tellMaster) {
39.       reportBlockStatus(blockId, status, droppedMemorySize)
40.     }
41.     if (blockIsUpdated) {
```

```
42.        addUpdatedBlockStatusToTaskMetrics(blockId, status)
43.      }
44.      status.storageLevel
45.    }
```

总结：dropFromMemory 是指在内存不够的时候，尝试释放一部分内存给要使用内存的应用，释放的这部分内存数据需考虑是丢弃，还是放到磁盘上。如果丢弃，如 5000 个步骤作为一个 Stage，前面 4000 个步骤进行了 Cache，Cache 时可能有 100 万个 partition 分区单位，其中丢弃了 100 个，丢弃的 100 个数据就要重新计算；但是，如果设置了同时放到内存和磁盘，此时会放入磁盘中，下次如果需要，就可以从磁盘中读取数据，而不是重新计算。

7.7　本章总结

本章阐述了 Shuffle 原理和源码，Shuffle 的框架、Shuffle 的框架演进、Shuffle 的框架内核及源码、Shuffle 数据读写的源码解析等内容；分别对 Hash Based Shuffle、Sorted Based Shuffle、Tungsten Sorted Based、Shuffle 与 Storage 模块间的交互进行了讲解。同时，本章着重阐述了 BlockManager 架构原理、运行流程和源码解密等内幕内容。

第 8 章　Job 工作原理和源码详解

本章对 Job 工作原理和源码进行详解。8.1 节讲解 Job 到底在什么时候产生；8.2 节讲解 Stage 划分内幕；8.3 节讲解 Task 全生命周期详解；8.4 节讲解 ShuffleMapTask 和 ResultTask 处理结果是如何被 Driver 管理的。

8.1　Job 到底在什么时候产生

典型的 Job 逻辑执行图如图 8-1 所示，经过下面四个步骤可以得到最终执行结果。

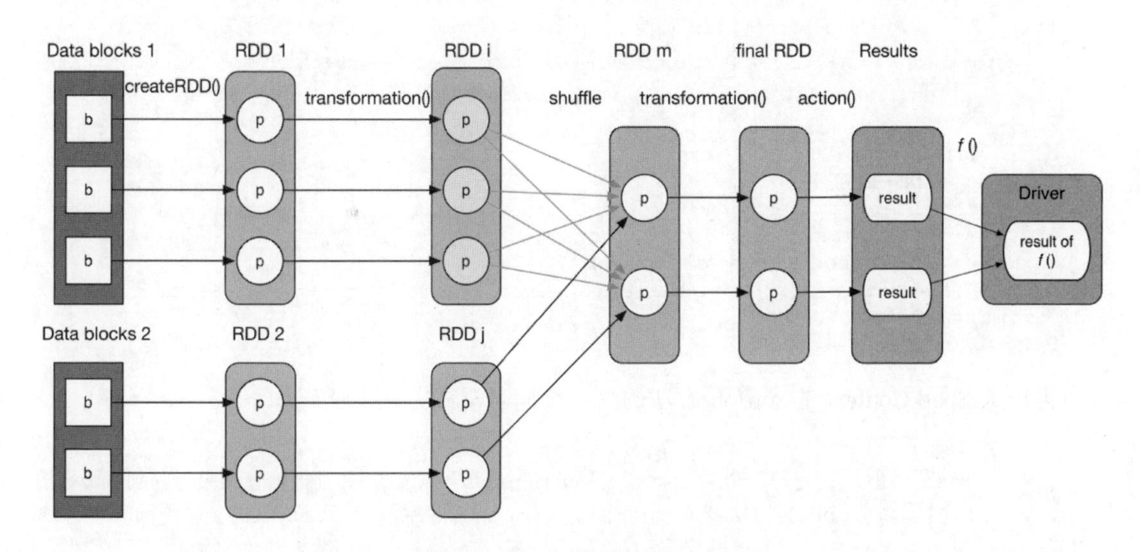

图 8-1　典型的 Job 逻辑执行图

（1）从数据源（可以是本地 file、内存数据结构、HDFS、HBase 等）读取数据创建最初的 RDD。

（2）对 RDD 进行一系列的 transformation() 操作，每个 transformation()会产生一个或多个包含不同类型 T 的 RDD[T]。T 可以是 Scala 里面的基本类型或数据结构，不限于（K, V）。

（3）对最后的 final RDD 进行 action()操作，每个 partition 计算后产生结果 result。

（4）将 result 回送到 driver 端，进行最后的 f(list[result])计算。RDD 可以被 Cache 到内存或者 checkpoint 到磁盘上。RDD 中的 partition 个数不固定，通常由用户设定。RDD 和 RDD 间的 partition 的依赖关系可以不是 1 对 1，如图 8-1 所示，既有 1 对 1 关系，也有多对多关系。

8.1.1　触发 Job 的原理和源码解析

对于 Spark Job 触发流程的源码，以 RDD 的 count 方法为例开始。RDD 的 count 方法代码如下：

```
1.  /**
     * 返回 RDD 中元素的数量
2.   */
3.  def count(): Long = sc.runJob(this, Utils.getIteratorSize _).sum
```

从上面的代码可以看出，count 方法触发 SparkContext 的 runJob 方法的调用。SparkContext 的 runJob 方法代码如下：

```
1.    /**
       * 触发一个 Job 处理一个 RDD 的所有 partitions，并且把处理结果返回到一个数组
2.     */
3.    def runJob[T, U: ClassTag](rdd: RDD[T], func: Iterator[T] => U): Array[U] ={
4.  runJob(rdd, func, 0 until rdd.partitions.length)
5.    }
```

进入 SparkContext 的 runJob 方法的同名重载方法，代码如下：

```
1.    /**
       *触发一个 Job 处理一个 RDD 的指定部分的 partitions，并且把处理结果返回到一个数组
       *比第一个 runJob 方法多了一个 partitions 数组参数
2.     */
3.    def runJob[T, U: ClassTag](
4.  rdd: RDD[T],
5.  func: Iterator[T] => U,
6.        partitions: Seq[Int]): Array[U] = {
7.      val cleanedFunc = clean(func)
8.  runJob(rdd,  (ctx:  TaskContext,  it:  Iterator[T])  =>cleanedFunc(it),
    partitions)
9.    }
```

再进入 SparkContext 的 runJob 方法的另一个同名重载方法，代码如下：

```
1.    /**
       *触发一个 Job 处理一个 RDD 的指定部分的 partitions，并且把处理结果返回到一个数组
       *比第一个 runJob 方法多了一个 partitions 数组参数，并且 func 的类型不同
2.     */
3.    def runJob[T, U: ClassTag](
4.  rdd: RDD[T],
5.  func: (TaskContext, Iterator[T]) => U,
6.        partitions: Seq[Int]): Array[U] = {
7.      val results = new Array[U](partitions.size)
8.  runJob[T, U](rdd, func, partitions, (index, res) => results(index) = res)
9.      results
10.   }
```

最后一次进入 SparkContext 的 runJob 方法的另一个同名重载方法，代码如下：

```
1.    /**
       *触发一个 Job 处理一个 RDD 的指定部分的 partitions，并把处理结果给指定的 handler
       *函数，这是 Spark 所有 Action 的主入口
2.     */
3.    def runJob[T, U: ClassTag](
4.  rdd: RDD[T],
5.  func: (TaskContext, Iterator[T]) => U,
6.        partitions: Seq[Int],
```

```
7.  resultHandler: (Int, U) => Unit): Unit = {
8.      if (stopped.get()) {
9.        throw new IllegalStateException("SparkContext has been shutdown")
10.     }
11. //记录了方法调用的方法栈
12.     val callSite = getCallSite
13. //清除闭包，为了函数能够序列化
14.     val cleanedFunc = clean(func)
15. logInfo("Starting Job: " + callSite.shortForm)
16.     if (conf.getBoolean("spark.logLineage", false)) {
17. logInfo("RDD's recursive dependencies:\n" + rdd.toDebugString)
18.     }
19.     //向高层调度器（DAGScheduler）提交 Job，从而获得 Job 执行结果
20. dagScheduler.runJob(rdd, cleanedFunc, partitions, callSite, resultHandler,
    localProperties.get)
21. progressBar.foreach(_.finishAll())
22. rdd.doCheckpoint()
23.   }
```

8.1.2 触发 Job 的算子案例

Spark Application 里可以产生一个或者多个 Job，例如，spark-shell 默认启动时内部就没有 Job，只是作为资源的分配程序，可以在 spark-shell 里面写代码产生若干个 Job，普通程序中一般可以有不同的 Action，每个 Action 一般也会触发一个 Job。

给定 Job 的逻辑执行图，如何生成物理执行图，也就是给定这样一个复杂数据依赖图，如何合理划分 Stage，并确定 Task 的类型和个数？

一个直观的想法是将前后关联的 RDDs 组成一个 Stage，每个 Stage 生成一个 Task。这样虽然可以解决问题，但效率不高。除了效率问题，这个想法还有一个更严重的问题：大量中间数据需要存储。对于 task 来说，其执行结果要么存到磁盘，要么存到内存，或者两者皆有。如果每个箭头都是 Task，每个 RDD 里面的数据都需要存起来，占用空间可想而知。

仔细观察一下逻辑执行图会发现：在每个 RDD 中，每个 Partition 是独立的。也就是说，在 RDD 内部，每个 Partition 的数据依赖各自不会相互干扰。因此，一个大胆的想法是将整个流程图看成一个 Stage，为最后一个 finalRDD 中的每个 Partition 分配一个 Task。

Spark 算法构造和物理执行时最基本的核心：最大化 Pipeline。基于 Pipeline 的思想，数据被使用的时候才开始计算，从数据流动的视角来说，是数据流动到计算的位置。实质上，从逻辑的角度看，是算子在数据上流动。从算法构建的角度而言：肯定是算子作用于数据，所以是算子在数据上流动；方便算法的构建。

从物理执行的角度而言：是数据流动到计算的位置；方便系统最为高效地运行。对于 Pipeline 而言，数据计算的位置就是每个 Stage 中最后的 RDD，一个震撼人心的内幕真相是：每个 Stage 中除了最后一个 RDD 算子是真实的外，前面的算子都是假的。计算的 Lazy 特性导致计算从后往前回溯，形成 Computing Chain，导致的结果是需要首先计算出具体一个 Stage 内部左侧的 RDD 中本次计算依赖的 Partition，如图 8-2 所示。

整个 Computing Chain 根据数据依赖关系自后向前建立，遇到 ShuffleDependency 后形成 Stage。在每个 Stage 中，每个 RDD 中的 compute()调用 parentRDD.iter()将 parent RDDs 中的 records 一个个 fetch 过来。

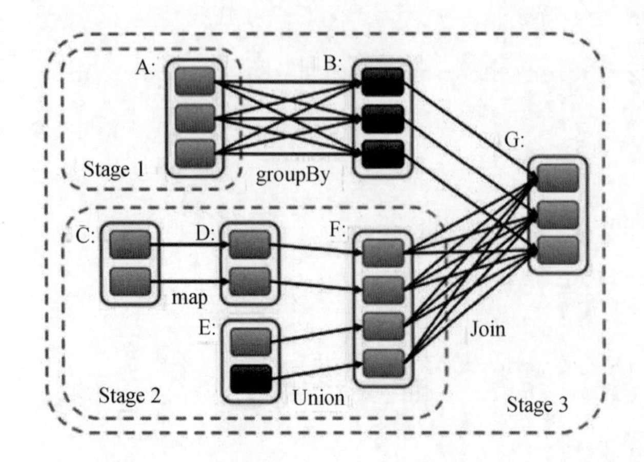

RDD算子计算：

- 图中三个Stage：Stage 1、Stage 2、Stage 3。
- Stage 3：内部左侧B的计算依赖于G的Partition，如G为3个Partition，那就有3个并行Task，从后往前回溯，B也是3个Partition，3个Task。
- Stage 1：内部A设置3个Partition，与Stage3的B进行groupBy操作。
- Stage 2：内部左侧C、D、F，左侧C的计算依赖于F的Partition，F设置两个Partition，两个Task并行计算，从后往前回溯，推出C、D、F也是两个Task，C、D、F形成了map的Pipeline；F设置两个Partition，与E是两个Task并行计算；将C、D与E的计算结果进行Union
- Stage 2：F计算的结果再与Stage 3的B进行Join。

图 8-2　Stage 示意图

例如，collect 前面的 RDD 是 transformation 级别的，不会立即执行。从后往前推，回溯时如果是窄依赖，则在内存中迭代，否则把中间结果写出到磁盘，暂存给后面的计算使用。

依赖分为窄依赖和宽依赖。例如，现实生活中，工作依赖一个对象，是窄依赖，依赖很多对象，是宽依赖。窄依赖除了一对一，还有 range 级别的依赖，依赖固定的个数，随着数据的规模扩大而改变。如果是宽依赖，DAGScheduler 会划分成不同的 Stage，Stage 内部是基于内存迭代的，也可以基于磁盘迭代，Stage 内部计算的逻辑是完全一样的，只是计算的数据不同而已。具体的任务就是计算一个数据分片，一个 Partition 的大小是 128MB。一个 partition 不是完全精准地等于一个 block 的大小，一般最后一条记录跨两个 block。

Spark 程序的运行有两种部署方式：Client 和 Cluster。

默认情况下建议使用 Client 模式，此模式下可以看到更多的交互性信息，及运行过程的信息。此时要专门使用一台机器来提交我们的 Spark 程序，配置和普通的 Worker 配置一样，而且要和 Cluster Manager 在同样的网络环境中，因为要指挥所有的 Worker 去工作，Worker 里的线程要和 Driver 不断地交互。由于 Driver 要驱动整个集群，频繁地和所有为当前程序分配的 Executor 去交互，频繁地进行网络通信，所以必须在同样的网络中。

也可以指定部署方式为 Cluster，这样真正的 Driver 会由 Master 决定在 Worker 中的某一台机器。Master 为你分配的第一个 Executor 就是 Driver 级别的 Executor。不推荐学习、开发的时候使用 Cluster，因为 Cluster 无法直接看到一些日志信息，所以建议使用 Client 方式。

8.2　Stage 划分内幕

本节讲解 Stage 划分原理及 Stage 划分源码。一个 Application 中，每个 Job 由一个或多个 Stage 构成，Stage 根据宽依赖（如 reducByKey、groupByKey 算子等）进行划分。

8.2.1　Stage 划分原理详解

Spark Application 中可以因为不同的 Action 触发众多的 Job。也就是说，一个 Application 中可以有很多的 Job，每个 Job 是由一个或者多个 Stage 构成的，后面的 Stage 依赖于前面的

Stage。也就是说，只有前面依赖的 Stage 计算完毕后，后面的 Stage 才会运行。

Stage 划分的依据就是宽依赖，什么时候产生宽依赖呢？例如，reducByKey、groupByKey 等；Action（如 collect）导致 SparkContext.runJob 的执行，最终导致 DAGScheduler 中的 submitJob 的执行，其核心是通过发送一个 case class JobSubmitted 对象给 eventProcessLoop。

eventProcessLoop 是 DAGSchedulerEventProcessLoop 的具体实例，而 DAGSchedulerEventProcessLoop 是 EventLoop 的子类，具体实现 EventLoop 的 onReceive 方法。onReceive 方法转过来回调 doOnReceive。在 doOnReceive 中通过模式匹配的方式把执行路由到 JobSubmitted，在 handleJobSubmitted 中首先创建 finalStage，创建 finalStage 时会建立父 Stage 的依赖链条。

8.2.2 Stage 划分源码详解

Spark 的 Action 算子执行 SparkContext.runJob，提交至 DAGScheduler 中的 submitJob，submitJob 发送 JobSubmitted 对象到 eventProcessLoop 循环消息队列，提交该任务。

DAGSchedulerEvent.scala 的源码如下：

```
1.  private[scheduler] case class JobSubmitted(
2.     jobId: Int,
3.     finalRDD: RDD[_],
4.     func: (TaskContext, Iterator[_]) => _,
5.     partitions: Array[Int],
6.     callSite: CallSite,
7.     listener: JobListener,
8.     properties: Properties = null)
9.   extends DAGSchedulerEvent
```

eventProcessLoop 是 DAGSchedulerEventProcessLoop 的具体实例，而 DAGSchedulerEventProcessLoop 是 EventLoop 的子类，具体实现 EventLoop 的 onReceive 方法，onReceive 方法转过来回调 doOnReceive。

DAGScheduler.scala 的源码如下：

```
1. private def doOnReceive(event: DAGSchedulerEvent): Unit = event match {
2.     case JobSubmitted(jobId, rdd, func, partitions, callSite, listener,
       properties) =>
3.       dagScheduler.handleJobSubmitted(jobId,  rdd,  func,  partitions,
         callSite, listener, properties)
4.
5.     case MapStageSubmitted(jobId, dependency, callSite, listener,
       properties) =>
6.       dagScheduler.handleMapStageSubmitted(jobId, dependency, callSite,
         listener, properties)
7.  ......
```

8.3 Task 全生命周期详解

本节讲解 Task 的生命过程，对 Task 在 Driver 和 Executor 中交互的全生命周期原理和源码进行详解。

8.3.1 Task的生命过程详解

Task的生命过程详解如下。

（1）当Driver中的CoarseGrainedSchedulerBackend给CoarseGrainedExecutorBackend发送LaunchTask之后，CoarseGrainedExecutorBackend收到LaunchTask消息后，首先会反序列化TaskDescription。

（2）Executor会通过launchTask执行Task，在launchTask方法中调用new()函数创建TaskRunner，TaskRunner继承自Runnable接口。

（3）TaskRunner在ThreadPool运行具体的Task，在TaskRunner的run方法中首先会通过调用statusUpdate给Driver发信息汇报自己的状态，说明自己是Running状态。其中execBackend是ExecutorBackend，ExecutorBackend是一个trait，其具体的实现子类是CoarseGrainedExecutorBackend，其中的statusUpdate方法中将向Driver提交StatusUpdate消息。

（4）TaskRunner内部会做一些准备工作：例如，反序列化Task的依赖，然后通过网络获取需要的文件、Jar等。

（5）然后是反序列Task本身。

（6）调用反序列化后的Task.run方法来执行任务，并获得执行结果。其中Task的run方法调用时会导致Task的抽象方法runTask的调用，在Task的runTask内部会调用RDD的iterator方法，该方法就是我们针对当前Task所对应的Partition进行计算的关键所在，在处理的内部会迭代Partition的元素，并交给我们自定义的function进行处理。

- 对于ShuffleMapTask，首先要对RDD以及其依赖关系进行反序列化，最终计算会调用RDD的compute方法。具体计算时有具体的RDD，例如，MapPartitionsRDD的compute。compute方法其中的f就是我们在当前的Stage中计算具体Partition的业务逻辑代码。
- 对于ResultTask：调用rdd.iterator方法，最终计算仍然会调用RDD的compute方法。

（7）把执行结果序列化，并根据大小判断不同的结果传回给Driver。

（8）CoarseGrainedExecutorBackend给DriverEndpoint发送StatusUpdate来传输执行结果。DriverEndpoint会把执行结果传递给TaskSchedulerImpl处理，然后交给TaskResultGetter内部通过线程去分别处理Task执行成功和失败时的不同情况，最后告诉DAGScheduler任务处理结束的状况。

说明：

① 在执行具体Task的业务逻辑前，会进行四次反序列。

- TaskDescription的反序列化。
- 反序列化Task的依赖。
- Task的反序列化。
- RDD反序列化。

② 在Spark 1.6中，AkkFrameSize是128MB，所以可以广播非常大的任务；而任务的执行结果最大可以达到1GB。Spark 2.2版本中，CoarseGrainedSchedulerBackend的launchTask方法中序列化任务大小的限制是maxRpcMessageSize为128MB。

8.3.2 Task 在 Driver 和 Executor 中交互的全生命周期原理和源码详解

在 Standalone 模式中，Driver 中的 CoarseGrainedSchedulerBackend 给 CoarseGrained-ExecutorBackend 发送 launchTasks 消息，CoarseGrainedExecutorBackend 收到 launchTasks 消息以后会调用 executor.launchTask。

CoarseGrainedExecutorBackend 的 receive 方法如下，模式匹配收到 LaunchTask 消息：

（1）LaunchTask 判断 Executor 是否存在，如果 Executor 不存在，则直接退出，然后会反序列化 TaskDescription。

CoarseGrainedExecutorBackend 的 receive 方法的源码如下：

```
1.      val taskDesc = ser.deserialize[TaskDescription](data.value)
```

（2）Executor 会通过 launchTask 来执行 Task，launchTask 方法中分别传入 taskId、尝试次数、任务名称、序列化后的任务本身。

CoarseGrainedExecutorBackend 的 receive 方法的源码如下：

```
1.      executor.launchTask(this, taskId = taskDesc.taskId, attemptNumber =
        taskDesc.attemptNumber,  taskDesc.name, taskDesc.serializedTask)
```

进入 Executor.scala 的 launchTask 方法，在 launchTask 方法中调用 new()函数创建一个 TaskRunner，传入的参数包括 taskId、尝试次数、任务名称、序列化后的任务本身。然后放入 runningTasks 数据结构，在 threadPool 中执行 TaskRunner。

TaskRunner 本身是一个 Runnable 接口。

下面看一下 TaskRunner 的 run 方法。TaskMemoryManager 是内存的管理，deserialize-StartTime 是反序列化开始的时间，setContextClassLoader 是 ClassLoader 加载具体的类。ser 是序列化器。

然后调用 execBackend.statusUpdate，statusUpdate 是 ExecutorBackend 的方法，Executor-Backend 通过 statusUpdate 给 Driver 发信息，汇报自己的状态。

```
1.     private[spark] trait ExecutorBackend {
2.    def statusUpdate(taskId: Long, state: TaskState, data: ByteBuffer): Unit
3.  }
```

其中，execBackend 是 ExecutorBackend，ExecutorBackend 是一个 trait，其具体的实现子类是 CoarseGrainedExecutorBackend。execBackend 实例是在 CoarseGrainedExecutorBackend 的 receive 方法收到 LaunchTask 消息，调用 executor.launchTask(this, taskId = taskDesc.taskId, attemptNumber = taskDesc.attemptNumber, taskDesc.name, taskDesc.serializedTask) 时将 CoarseGrainedExecutorBackend 自己本身的 this 实例传进来的。这里调用 CoarseGrained-ExecutorBackend 的 statusUpdate 方法。statusUpdate 方法将向 Driver 提交 StatusUpdate 消息。

CoarseGrainedExecutorBackend 的 statusUpdate 的源码如下：

```
1.       override def statusUpdate(taskId: Long, state: TaskState, data:
         ByteBuffer) {
2.     val msg = StatusUpdate(executorId, taskId, state, data)
3.     driver match {
```

```
4.         case Some(driverRef) => driverRef.send(msg)
5.         case None => logWarning(s"Drop $msg because has not yet connected
           to driver")
6.     }
7.   }
```

（3）TaskRunner 的 run 方法中，TaskRunner 在 ThreadPool 中运行具体的 Task，在 TaskRunner 的 run 方法中首先会通过调用 statusUpdate 给 Driver 发信息汇报自己的状态，说明自己是 Running 状态。

```
1.    execBackend.statusUpdate(taskId, TaskState.RUNNING, EMPTY_BYTE_BUFFER)
```

其中，EMPTY_BYTE_BUFFER 没有具体内容。

```
1.   private val EMPTY_BYTE_BUFFER = ByteBuffer.wrap(new Array[Byte](0))
```

接下来通过 Task.deserializeWithDependencies(serializedTask)反序列化 Task，得到一个 Tuple，获取到 taskFiles、taskJars、taskProps、taskBytes 等信息。

（4）Executor 会通过 TaskRunner 在 ThreadPool 中运行具体的 Task，TaskRunner 内部会做一些准备工作：反序列化 Task 的依赖。

Executor.scala 的源码如下：

```
1.   ……
2.   Executor.taskDeserializationProps.set(taskDescription.properties)
3.         updateDependencies(taskDescription.addedFiles,
           taskDescription.addedJars)
4.         task = ser.deserialize[Task[Any]](
5.           taskDescription.serializedTask, Thread.currentThread
             .getContextClassLoader)
6.         task.localProperties = taskDescription.properties
7.         task.setTaskMemoryManager(taskMemoryManager)
8.   ……
```

然后通过网络来获取需要的文件、Jar 等。

Executor.scala 的源码如下：

```
1.         updateDependencies(taskDescription.addedFiles, taskDescription
           .addedJars)
```

再来看一下 updateDependencies 方法。从 SparkContext 收到一组新的文件 JARs，下载 Task 运行需要的依赖 Jars，在类加载机中加载新的 JARs 包。updateDependencies 方法的源码如下所示。

Executor.scala 的源码如下：

```
1.   private def updateDependencies(newFiles: Map[String, Long], newJars:
     Map[String, Long]) {
2.     Lazy val hadoopConf = SparkHadoopUtil.get.newConfiguration(conf)
3.     synchronized {
4.       //获取将要计算的依赖关系
5.       for ((name, timestamp) <- newFiles if currentFiles.getOrElse(name,
         -1L) < timestamp) {
6.         logInfo("Fetching " + name + " with timestamp " + timestamp)
7.         //使用 useCache 获取文件，本地模式关闭缓存
8.         Utils.fetchFile(name, new File(SparkFiles.getRootDirectory()), conf,
9.           env.securityManager, hadoopConf, timestamp, useCache = !isLocal)
```

```
10.         currentFiles(name) = timestamp
11.       }
12.       for ((name, timestamp) <- newJars) {
13.         val localName = name.split("/").last
14.         val currentTimeStamp = currentJars.get(name)
15.           .orElse(currentJars.get(localName))
16.           .getOrElse(-1L)
17.         if (currentTimeStamp < timestamp) {
18.           logInfo("Fetching " + name + " with timestamp " + timestamp)
19.           //使用 useCache 获取文件，本地模式关闭缓存
20.           Utils.fetchFile(name, new File(SparkFiles.getRootDirectory()), conf,
21.             env.securityManager, hadoopConf, timestamp, useCache = !isLocal)
22.           currentJars(name) = timestamp
23.           //将它增加到类装入器中
24.           val url = new File(SparkFiles.getRootDirectory(), localName)
          .toURI.toURL
25.           if (!urlClassLoader.getURLs().contains(url)) {
26.             logInfo("Adding " + url + " to class loader")
27.             urlClassLoader.addURL(url)
28.           }
29.         }
30.       }
31.     }
32.   }
```

Executor 的 updateDependencies 方法中，Executor 运行具体任务时进行下载，下载文件使用 synchronized 关键字，因为 Executor 在线程中运行，同一个 Stage 内部不同的任务线程要共享这些内容，因此 ExecutorBackend 多条线程资源操作的时候，需要通过同步块加锁。

updateDependencies 方法的 Utils.fetchFile 将文件或目录下载到目标目录，支持各种方式获取文件，包括 HTTP，Hadoop 兼容的文件系统、标准文件系统的文件，基于 URL 参数。获取目录只支持从 Hadoop 兼容的文件系统。如果 usecache 设置为 true，第一次尝试取文件到本地缓存，执行同一应用程序进行共享。usecache 主要用于 executors，而不是本地模式。如果目标文件已经存在，并有不同于请求文件的内容，将抛出 SparkException 异常。

```
1.     def fetchFile(
2.         url: String,
3.         targetDir: File,
4.         conf: SparkConf,
5.         securityMgr: SecurityManager,
6.         hadoopConf: Configuration,
7.         timestamp: Long,
8.         useCache: Boolean) {
9.   ......
10.    doFetchFile(url, localDir, cachedFileName, conf, securityMgr, hadoopConf)
11. ......
```

Spark 2.2.1 版本的 doFetchFile 方法如下，包括 spark、http | https | ftp、file 各种协议方式的下载。

```
1.   private def doFetchFile(
2.       url: String,
3.       targetDir: File,
4.       filename: String,
5.       conf: SparkConf,
6.       securityMgr: SecurityManager,
7.       hadoopConf: Configuration) {
```

```
8.      val targetFile = new File(targetDir, filename)
9.      val uri = new URI(url)
10.     val fileOverwrite = conf.getBoolean("spark.files.overwrite",
        defaultValue = false)
11.     Option(uri.getScheme).getOrElse("file") match {
12.       case "spark" =>
13.       ……
14.         downloadFile(url, is, targetFile, fileOverwrite)
15.       case "http" | "https" | "ftp" =>
16.     ……
17.         downloadFile(url, in, targetFile, fileOverwrite)
18.       case "file" =>
19.       ……
20.       copyFile(url, sourceFile, targetFile, fileOverwrite)
21.       case _ =>
22.         val fs = getHadoopFileSystem(uri, hadoopConf)
23.         val path = new Path(uri)
24.         fetchHcfsFile(path, targetDir, fs, conf, hadoopConf, fileOverwrite,
25.                   filename = Some(filename))
26.     }
27.   }
```

Spark 2.4.3 版本的 Utils.scala 源码与 Spark 2.2.1 版本相比具有如下特点。

- 上段代码中第 7 行新增一个函数，返回类型 File。

```
1.  ……
2.  hadoopConf: Configuration): File = {
3.  ……
```

（5）回到 TaskRunner 的 run 方法，所有依赖的 Jar 都下载完成后，然后是反序列 Task 本身。Executor.scala 的源码如下：

```
1.      task = ser.deserialize[Task[Any]](
2.            taskDescription.serializedTask, Thread.currentThread
            .getContextClassLoader)
```

在执行具体 Task 的业务逻辑前会进行四次反序列。

① TaskDescription 的反序列化。

② 反序列化 Task 的依赖。

③ Task 的反序列化。

④ RDD 反序列化。

（6）回到 TaskRunner 的 run 方法，调用反序列化后的 Task.run 方法来执行任务并获得执行结果。

其中，Task 的 run 方法调用时会导致 Task 的抽象方法 runTask 的调用，在 Task 的 runTask 内部会调用 RDD 的 iterator 方法，该方法就是针对当前 Task 所对应的 Partition 进行计算的关键所在，在处理的内部会迭代 Partition 的元素并交给自定义的 function 进行处理。

进入 task.run 方法，在 run 方法里面再调用 runTask 方法。

Spark 2.2.1 版本的 Task.scala 的源码如下：

```
1.       final def run(
2.       taskAttemptId: Long,
3.       attemptNumber: Int,
4.       metricsSystem: MetricsSystem): T = {
```

```
5.      SparkEnv.get.blockManager.registerTask(taskAttemptId)
6.      context = new TaskContextImpl(
7.    ......
8.      TaskContext.setTaskContext(context)
9.     ......
10.     try {
11.       runTask(context)
12.  ......
```

Spark 2.4.3 版本的 Task.scala 源码与 Spark 2.2.1 版本相比具有如下特点。

- 上段代码中第 6 行 context 名称调整为 taskContext。

```
1.  ......
2.  //TODO Spark-24874 允许基于分区创建 BarrierTaskContext，取代判断阶段是否为屏障
    阶段
3.      val taskContext = new TaskContextImpl(
4.  ......
```

进入 Task.scala 的 runTask 方法，这里是一个抽象方法，没有具体的实现。

```
1.    def runTask(context: TaskContext): T
```

Task 包括两种 Task：ResultTask 和 ShuffleMapTask。抽象 runTask 方法由子类的 runTask 实现。先看一下 ShuffleMapTask 的 runTask 方法，runTask 实际运行的时候会调用 RDD 的 iterator，然后针对 partition 进行计算。

```
1.      override def runTask(context: TaskContext): MapStatus = {
2.   ......
3.      val ser = SparkEnv.get.closureSerializer.newInstance()
4.      val (rdd, dep) = ser.deserialize[(RDD[_], ShuffleDependency[_, _, _])](
5.      ......
6.       val manager = SparkEnv.get.shuffleManager
7.      writer = manager.getWriter[Any, Any](dep.shuffleHandle, partitionId,
        context)
8.       writer.write(rdd.iterator(partition, context).asInstanceOf[Iterator
         [_ <: Product2[Any, Any]]])
9.       writer.stop(success = true).get
10.    ......
```

ShuffleMapTask 在计算具体的 Partition 之后实际上会通过 shuffleManager 获得的 shuffleWriter 把当前 Task 计算内容根据具体的 shuffleManager 实现写入到具体的文件中。操作完成以后会把 MapStatus 发送给 DAGscheduler，Driver 的 DAGScheduler 的 MapOutputTracker 会收到注册的信息。

同样地，ResultTask 的 runTask 方法也是调用 RDD 的 iterator，然后针对 partition 进行计算。MapOutputTracker 会把 ShuffleMapTask 执行结果交给 ResultTask，ResultTask 根据前面 Stage 的执行结果进行 Shuffle，产生整个 Job 最后的结果。

```
1.     override def runTask(context: TaskContext): U = {
2.     ......
3.      val ser = SparkEnv.get.closureSerializer.newInstance()
4.      val (rdd, func) = ser.deserialize[(RDD[T], (TaskContext, Iterator[T])
        => U)](
5.   ......
6.      func(context, rdd.iterator(partition, context))
7.    }
```

ResultTask、ShuffleMapTask 的 runTask 方法真正执行的时候，调用 RDD 的 iterator，对 Partition 进行计算。ResultTask.scala 的 runTask 方法的源码如下：

```
1.    override def runTask(context: TaskContext): U = {
2.    ……
3.      val ser = SparkEnv.get.closureSerializer.newInstance()
4.      val (rdd, func) = ser.deserialize[(RDD[T], (TaskContext, Iterator[T])
        => U)](
5.   ……
6.      func(context, rdd.iterator(partition, context))
7.    }
```

RDD.scala 的 iterator 方法源码如下：

```
1.   final def iterator(split: Partition, context: TaskContext): Iterator[T] = {
2.     if (storageLevel != StorageLevel.NONE) {
3.       getOrCompute(split, context)
4.     } else {
5.       computeOrReadCheckpoint(split, context)
6.     }
7.   }
```

如果 storageLevel 不等于 NONE，就直接获取或者计算得到 RDD 的分区；如果 storageLevel 是空，就从 checkpoint 中读取或者计算 RDD 分区。

进入 computeOrReadCheckpoint：

```
1.      private[spark] def computeOrReadCheckpoint(split: Partition,
        context: TaskContext): Iterator[T] =
2.    {
3.     if (isCheckpointedAndMaterialized) {
4.       firstParent[T].iterator(split, context)
5.     } else {
6.       compute(split, context)
7.     }
8.   }
```

最终计算会调用 RDD 的 compute 方法。

```
1.    def compute(split: Partition, context: TaskContext): Iterator[T]
```

RDD 的 compute 方法中的 Partition 是一个 trait。

```
1.    trait Partition extends Serializable {
2.   def index: Int
3.   override def hashCode(): Int = index
4.   override def equals(other: Any): Boolean = super.equals(other)
5. }
```

RDD 的 compute 方法中的 TaskContext 里面有很多方法，包括任务是否完成、任务是否中断、任务是否在本地运行、任务运行完成时的监听器、任务运行失败的监听器、stageId、partitionId、重试的次数等。

```
1.  abstract class TaskContext extends Serializable {
2.  def isCompleted(): Boolean
3.  def isInterrupted(): Boolean
4.  def isRunningLocally(): Boolean
5.  def addTaskCompletionListener(listener: TaskCompletionListener): TaskContext
6.  def addTaskCompletionListener(f: (TaskContext) => Unit): TaskContext
```

```
7.  def addTaskFailureListener(listener: TaskFailureListener): TaskContext
8.  def addTaskFailureListener(f: (TaskContext, Throwable) => Unit): TaskContext
9.  def stageId()
10. def partitionId(): Int
11.  def attemptNumber(): Int
12. ......
```

下面看一下 TaskContext 具体的实现 TaskContextImpl。TaskContextImpl 维持了很多上下文信息，如 stageId、partitionId、taskAttemptId、重试次数、taskMemoryManager 等。

```
1.   private[spark] class TaskContextImpl(
2.      val stageId: Int,
3.      val partitionId: Int,
4.      override val taskAttemptId: Long,
5.      override val attemptNumber: Int,
6.      override val taskMemoryManager: TaskMemoryManager,
7.      localProperties: Properties,
8.      @transient private val metricsSystem: MetricsSystem,
9.      //默认值仅用于测试
10.     override val taskMetrics: TaskMetrics = TaskMetrics.empty)
11.   extends TaskContext
12.   with Logging {
13. ......
```

RDD 的 compute 方法具体计算的时候有具体的 RDD，如 MapPartitionsRDD 的 compute、传进去的 Partition 及 TaskContext 上下文。

MapPartitionsRDD.scala 的 compute 的源码如下：

```
1.   override def compute(split: Partition, context: TaskContext): Iterator[U] =
2.     f(context, split.index, firstParent[T].iterator(split, context))
```

MapPartitionsRDD.scala 的 compute 中的 f 就是我们在当前的 Stage 中计算具体 Partition 的业务逻辑代码。f 是函数，是我们自己写的业务逻辑。Stage 从后往前推，把所有的 RDD 合并变成一个，函数也会变成一个链条，展开成一个很大的函数。Compute 返回的是一个 Iterator。

Task 包括两种 Task：ResultTask 和 ShuffleMapTask。

先看一下 ShuffleMapTask 的 runTask 方法，从 ShuffleMapTask 的角度讲，rdd.iterator 获得数据记录以后，对 rdd.iterator 计算后的 Iterator 记录进行 write。

```
1.   val manager = SparkEnv.get.shuffleManager
2.      writer = manager.getWriter[Any, Any](dep.shuffleHandle, partitionId,
        context)
3.       writer.write(rdd.iterator(partition, context).asInstanceOf
         [Iterator[_ <: Product2[Any, Any]]])
4.       writer.stop(success = true).get
```

ResultTask.scala 的 runTask 方法较简单：在 ResultTask 中，rdd.iterator 获得数据记录以后，直接调用 func 函数。func 函数是 Task 任务反序列化后直接获得的 fun 函数。

```
1.  val (rdd, func) = ser.deserialize[(RDD[T], (TaskContext, Iterator[T]) => U)](
2.        ByteBuffer.wrap(taskBinary.value), Thread.currentThread.
          getContextClassLoader)
3.    ......
4.     func(context, rdd.iterator(partition, context))
```

（7）回到 TaskRunner 的 run 方法，把执行结果序列化，并根据大小判断不同的结果传回给 Driver。

- task.run 运行的结果赋值给 value。
- resultSer.serialize(value)把 task.run 的执行结果 value 序列化。
- maxResultSize > 0 && resultSize > maxResultSize 对任务执行结果的大小进行判断，并进行相应的处理。任务执行完以后，任务的执行结果最大可以达到 1GB。

如果任务执行结果特别大，超过 1GB，日志就会提示超出任务大小限制。返回元数据 ser.serialize(new IndirectTaskResult[Any](TaskResultBlockId(taskId), resultSize))。

如果任务执行结果小于 1GB，大于 maxDirectResultSize（128MB），就放入 blockManager，返回元数据 ser.serialize(new IndirectTaskResult[Any](blockId, resultSize))。

如果任务执行结果小于 128MB，就直接返回 serializedDirectResult。

TaskRunner 的 run 方法如下所示。

Executor.scala 的源码如下：

```
1.          override def run(): Unit = {
2.  ……
3.  val value = try {
4.            val res = task.run(
5.              taskAttemptId = taskId,
6.              attemptNumber = taskDescription.attemptNumber,
7.              metricsSystem = env.metricsSystem)
8.            threwException = false
9.            Res
10. ……
11.    val valueBytes = resultSer.serialize(value)
12. ……
13.  val directResult = new DirectTaskResult(valueBytes, accumUpdates)
14.         val serializedDirectResult = ser.serialize(directResult)
15.         val resultSize = serializedDirectResult.limit
16. ……
17.
18.  val serializedResult: ByteBuffer = {
19.           if (maxResultSize > 0 && resultSize > maxResultSize) {
20.          ……
21.        ser.serialize(new IndirectTaskResult[Any](TaskResultBlockId
         (taskId), resultSize))
22.           } else if (resultSize > maxDirectResultSize) {
23.             val blockId = TaskResultBlockId(taskId)
24.             env.blockManager.putBytes(
25.               blockId,
26.               new ChunkedByteBuffer(serializedDirectResult.duplicate()),
27.               StorageLevel.MEMORY_AND_DISK_SER)
28.           ……
29.             ser.serialize(new IndirectTaskResult[Any](blockId, resultSize))
30.           } else {
31.        ……
32.             serializedDirectResult
33.           }
34.         }
```

Spark 2.4.3 版本的 Executor.scala 的源码与 Spark 2.2.1 版本相比具有如下特点。

- 上段代码中第 3 行将 try 语句调整为 Utils.tryWithSafeFinally。tryWithSafeFinally 方法

执行代码块，然后执行 finally 块，但如果异常发生在 finally 块，不要抑制原来的异常。这主要是 finally out.close 块的问题，其中需要调用 Close 来清除，但如果在 Out.Write 发生异常，它很可能已损坏，Out.Close 也将失败，这将抑制原始的可能更有意义的 out.write 调用的异常。

```
1.  ……
2.     val value = Utils.tryWithSafeFinally {
```

其中的 maxResultSize 大小是 1GB，任务的执行结果最大可以达到 1GB。

Spark 2.2.1 版本的 Executor.scala 的源码如下：

```
1.    ……
2.  //对结果的总大小限制的字节数（默认为 1GB）
3.    private val maxResultSize = Utils.getMaxResultSize(conf)
4.  ……
5.  Utils.scala
6.    //对结果的总大小限制的字节数（默认为 1GB）
7.    def getMaxResultSize(conf: SparkConf): Long = {
8.      memoryStringToMb(conf.get("spark.driver.maxResultSize",
        "1g")).toLong << 20
9.    }
```

Spark 2.4.3 版本的 Executor.scala 源码与 Spark 2.2.1 版本相比具有如下特点。

- 上段代码中第 3 行 getMaxResultSize 方法调整为从配置文件读取 MAX_RESULT_SIZE，用于对结果的大小限制。

```
1.  ……
2.    private val maxResultSize = conf.get(MAX_RESULT_SIZE)
3.  ……
4.  //config/package.scala
5.    private[spark] val MAX_RESULT_SIZE =
      ConfigBuilder("spark.driver.maxResultSize")
6.      .doc("Size limit for results.")
7.      .bytesConf(ByteUnit.BYTE)
8.      .createWithDefaultString("1g")
9.  ……
```

其中的 Executor.scala 中的 maxDirectResultSize 大小，取 spark.task.maxDirectResultSize 和 RpcUtils.maxMessageSizeBytes 的最小值。其中 spark.rpc.message.maxSize 默认配置是 128MB。spark.task.maxDirectResultSize 在配置文件中进行配置。

```
1.    private val maxDirectResultSize = Math.min(
2.      conf.getSizeAsBytes("spark.task.maxDirectResultSize", 1L << 20),
3.      RpcUtils.maxMessageSizeBytes(conf))
4.  ……
5.  def maxMessageSizeBytes(conf: SparkConf): Int = {
6.      val maxSizeInMB = conf.getInt("spark.rpc.message.maxSize", 128)
7.      if (maxSizeInMB > MAX_MESSAGE_SIZE_IN_MB) {
8.        throw new IllegalArgumentException(
9.          s"spark.rpc.message.maxSize should not be greater than $MAX_
          MESSAGE_SIZE_IN_MB MB")
10.     }
11.     maxSizeInMB * 1024 * 1024
12.   }
```

Driver 发消息给 Executor，CoarseGrainedSchedulerBackend 的 launchTask 方法中序列化任务大小的限制是 maxRpcMessageSize，大小是 128MB。

CoarseGrainedSchedulerBackend.scala 的源码如下：

```
1.    private def launchTasks(tasks: Seq[Seq[TaskDescription]]) {
2.  ......
3.      if (serializedTask.limit() >= maxRpcMessageSize) {
4.  ......
5.    private val maxRpcMessageSize = RpcUtils.maxMessageSizeBytes(conf)
6.  ......
7.    def maxMessageSizeBytes(conf: SparkConf): Int = {
8.      val maxSizeInMB = conf.getInt("spark.rpc.message.maxSize", 128)
9.      if (maxSizeInMB > MAX_MESSAGE_SIZE_IN_MB) {
10.       throw new IllegalArgumentException(
11.         s"spark.rpc.message.maxSize should not be greater than $MAX_
          MESSAGE_SIZE_IN_MB MB")
12.     }
13.     maxSizeInMB * 1024 * 1024
14.   }
15. }
```

回到 TaskRunner 的 run 方法，execBackend.statusUpdate(taskId, TaskState.FINISHED, serializedResult)给 Driver 发送一个消息，消息中将 taskId、TaskState.FINISHED、serializedResult 放进去。

statusUpdate 方法的源码如下：

```
1.    override def statusUpdate(taskId: Long, state: TaskState, data:
      ByteBuffer) {
2.      val msg = StatusUpdate(executorId, taskId, state, data)
3.      driver match {
4.        case Some(driverRef) => driverRef.send(msg)
5.        case None => logWarning(s"Drop $msg because has not yet connected
          to driver")
6.      }
7.    }
```

（8）CoarseGrainedExecutorBackend 给 DriverEndpoint 发送 StatusUpdate 来传输执行结果，DriverEndpoint 会把执行结果传递给 TaskSchedulerImpl 处理，然后交给 TaskResultGetter 内部通过线程去分别处理 Task 执行成功和失败的不同情况，最后告诉 DAGScheduler 任务处理结束的状况。

CoarseGrainedSchedulerBackend.scala 中 DriverEndpoint 的 receive 方法如下：

```
1.    override def receive: PartialFunction[Any, Unit] = {
2.      case StatusUpdate(executorId, taskId, state, data) =>
3.        scheduler.statusUpdate(taskId, state, data.value)
4.        if (TaskState.isFinished(state)) {
5.          executorDataMap.get(executorId) match {
6.            case Some(executorInfo) =>
7.              executorInfo.freeCores += scheduler.CPUS_PER_TASK
8.              makeOffers(executorId)
9.            case None =>
10.             //忽略更新，因为我们不知道 Executor
11.             logWarning(s"Ignored task status update ($taskId state $state)" +
12.               s"from unknown executor with ID $executorId")
13.         }
14.       }
```

DriverEndpoint 的 receive 方法中，StatusUpdate 调用 scheduler.statusUpdate，然后释放资源，再次进行资源调度 makeOffers(executorId)。

TaskSchedulerImpl 的 statusUpdate 中：

- 如果是 TaskState.LOST，则记录下原因，将 Executor 清理掉。
- 如果是 TaskState.isFinished，则从 taskSet 中运行的任务中 remove 掉任务，调用 taskResultGetter.enqueueSuccessfulTask 处理。
- 如果是 TaskState.FAILED、TaskState.KILLED、TaskState.LOST，则调用 taskResultGetter.enqueueFailedTask 处理。

Spark 2.2.1 版本的 TaskSchedulerImpl 的 statusUpdate 的源码如下：

```
1.        def statusUpdate(tid: Long, state: TaskState, serializedData:
          ByteBuffer) {
2.     var failedExecutor: Option[String] = None
3.     var reason: Option[ExecutorLossReason] = None
4.     synchronized {
5.       try {
6.         taskIdToTaskSetManager.get(tid) match {
7.           case Some(taskSet) =>
8.             if (state == TaskState.LOST) {
9.      //TaskState.LOST 只被废弃的 Mesos 细粒度的调度模式使用，每个 Executor 对应单
       //个任务，因此将 Executor 标记为失败
10.              val execId = taskIdToExecutorId.getOrElse(tid, throw new
                 IllegalStateException(
11.                "taskIdToTaskSetManager.contains(tid) <=> taskIdToExecutorId
                   .contains(tid)"))
12.              if (executorIdToRunningTaskIds.contains(execId)) {
13.                reason = Some(
14.                  SlaveLost(s"Task $tid was lost, so marking the executor
                     as lost as well."))
15.                removeExecutor(execId, reason.get)
16.                failedExecutor = Some(execId)
17.              }
18.            }
19.            if (TaskState.isFinished(state)) {
20.              cleanupTaskState(tid)
21.              taskSet.removeRunningTask(tid)
22.              if (state == TaskState.FINISHED) {
23.                taskResultGetter.enqueueSuccessfulTask(taskSet, tid,
                   serializedData)
24.              } else if (Set(TaskState.FAILED, TaskState.KILLED, T
                 askState.LOST).contains(state)) {
25.                taskResultGetter.enqueueFailedTask(taskSet, tid, state,
                   serializedData)
26.              }
27.            }
28.          case None =>
29.            logError(
30.              ("Ignoring update with state %s for TID %s because its task
                 set is gone (this is " +
31.                "likely the result of receiving duplicate task finished
                   status updates) or its " +
32.                "executor has been marked as failed.")
33.                .format(state, tid))
34.        }
35.      } catch {
```

```
36.         case e: Exception => logError("Exception in statusUpdate", e)
37.       }
38.     }
39.     //更新 DAGScheduler 时没持有这个锁，所以可能导致死锁
40.     if (failedExecutor.isDefined) {
41.       assert(reason.isDefined)
42.       dagScheduler.executorLost(failedExecutor.get, reason.get)
43.       backend.reviveOffers()
44.     }
45.   }
```

Spark 2.4.3 版本的 TaskSchedulerImpl.scala 源码与 Spark 2.2.1 版本相比具有如下特点。

- 上段代码中第 6 行将根据任务 ID 获取的 TaskSetManager 调整为 Option(taskIdToTaskSetManager.get(tid))。

```
1.  ......
2.      Option(taskIdToTaskSetManager.get(tid)) match {
3.  ......
```

其中，taskResultGetter 是 TaskResultGetter 的实例化对象。

```
1.        private[spark] var taskResultGetter = new TaskResultGetter(sc.env,
          this)
```

TaskResultGetter.scala 的源码如下：

```
1.     private[spark] class TaskResultGetter(sparkEnv: SparkEnv, scheduler:
       TaskSchedulerImpl)
2.    extends Logging {
3.
4.    private val THREADS = sparkEnv.conf.getInt("spark.resultGetter
      .threads", 4)
5.
6.    //用于测试
7.    protected val getTaskResultExecutor: ExecutorService =
8.      ThreadUtils.newDaemonFixedThreadPool(THREADS, "task-result-getter")
9.  ......
10. def enqueueSuccessfulTask(
11.       taskSetManager: TaskSetManager,
12.       tid: Long,
13.       serializedData: ByteBuffer): Unit = {
14.     getTaskResultExecutor.execute(new Runnable {
15.       override def run(): Unit = Utils.logUncaughtExceptions {
16.         try {
17.          val (result, size) = serializer.get().deserialize[TaskResult[_]]
           (serializedData) match {
18.             case directResult: DirectTaskResult[_] =>
19.               if (!taskSetManager.canFetchMoreResults(serializedData
                .limit())) {
20.                 return
21.               }
22.         //反序列化“值”时不持有任何锁，所以不会阻止其他线程。我们在这里调用它，这样在
           //TaskSetManager.handleSuccessfulTask 中，当它再次被调用时，不需要反序列化值
23.               directResult.value(taskResultSerializer.get())
24.               (directResult, serializedData.limit())
25.             case IndirectTaskResult(blockId, size) =>
26.               if (!taskSetManager.canFetchMoreResults(size)) {
27.                 //如果大小超过 maxResultSize，将被 Executor 丢弃
```

```
28.                 sparkEnv.blockManager.master.removeBlock(blockId)
29.                 return
30.               }
31.               logDebug("Fetching indirect task result for TID %s".format(tid))
32.               scheduler.handleTaskGettingResult(taskSetManager, tid)
33.               val serializedTaskResult = sparkEnv.blockManager.getRemoteBytes
                  (blockId)
34.
35.               if (!serializedTaskResult.isDefined) {
36.                 /*如果运行任务的机器失败，我们将无法获得任务结果
                      当任务结束，我们试图取结果时，块管理器必须刷新结果*/
37.                 scheduler.handleFailedTask(
38.                   taskSetManager, tid, TaskState.FINISHED, TaskResultLost)
39.                 return
40.               }
41.               val deserializedResult = serializer.get().deserialize
                  [DirectTaskResult[_]](
42.                 serializedTaskResult.get.toByteBuffer)
43.               //反序列化获取值
44.               deserializedResult.value(taskResultSerializer.get())
45.               sparkEnv.blockManager.master.removeBlock(blockId)
46.               (deserializedResult, size)
47.           }
48.
49.       //从 Executors 接收的累加器更新中设置任务结果大小，我们需要在 Driver 上执行此操
          //作，因为如果我们在 Executors 上执行此操作，那么将结果更新大小后须进行序列化
50.           result.accumUpdates = result.accumUpdates.map { a =>
51.             if (a.name == Some(InternalAccumulator.RESULT_SIZE)) {
52.               val acc = a.asInstanceOf[LongAccumulator]
53.               assert(acc.sum == 0L, "task result size should not have been
                  set on the executors")
54.               acc.setValue(size.toLong)
55.               acc
56.             } else {
57.               a
58.             }
59.           }
60.
61.           scheduler.handleSuccessfulTask(taskSetManager, tid, result)
62.         } catch {
63.           case cnf: ClassNotFoundException =>
64.             val loader = Thread.currentThread.getContextClassLoader
65.             taskSetManager.abort("ClassNotFound with classloader: " +
                loader)
66.           //匹配 NonFatal，所以我们不从上面的 return 捕获 ControlThrowable 异常
67.           case NonFatal(ex) =>
68.             logError("Exception while getting task result", ex)
69.             taskSetManager.abort("Exception while getting task result:
                %s".format(ex))
70.         }
71.       }
72.     })
73.   }
```

TaskResultGetter.scala 的 enqueueSuccessfulTask 方法中，处理成功任务的时候开辟了一条新线程，先将结果反序列化，然后根据接收的结果类型 DirectTaskResult、IndirectTaskResult 分别处理。

如果是 DirectTaskResult，则直接获得结果并返回。

如果是 IndirectTaskResult，就通过 blockManager.getRemoteBytes 远程获取。获取以后再进行反序列化。

最后是 scheduler.handleSuccessfulTask。

TaskSchedulerImpl 的 handleSuccessfulTask 的源码如下：

```
1.    def handleSuccessfulTask(
2.      taskSetManager: TaskSetManager,
3.      tid: Long,
4.      taskResult: DirectTaskResult[_]): Unit = synchronized {
5.     taskSetManager.handleSuccessfulTask(tid, taskResult)
6.   }
```

TaskSchedulerImpl 中也有失败任务的相应处理。

TaskSchedulerImpl.scala 的源码如下：

```
1.     def handleFailedTask(
2.       taskSetManager: TaskSetManager,
3.       tid: Long,
4.       taskState: TaskState,
5.       reason: TaskFailedReason): Unit = synchronized {
6.     taskSetManager.handleFailedTask(tid, taskState, reason)
7.     if (!taskSetManager.isZombie && !taskSetManager.someAttemptSucceeded
       (tid)) {
8.       //任务集管理状态更新后，需要再次分配资源，失败的任务需要重新运行
9.       backend.reviveOffers()
10.    }
11.  }
```

TaskSchedulerImpl 的 handleSuccessfulTask 交给 TaskSetManager 调用 handleSuccessfulTask，告诉 DAGScheduler 任务处理结束的状况，并且 Kill 掉其他尝试的相同任务（因为一个任务已经尝试成功，其他的相同任务没必要再次去尝试）。

Spark 2.2.1 版本的 TaskSetManager 的 handleSuccessfulTask 的源码如下：

```
1.   def handleSuccessfulTask(tid: Long, result: DirectTaskResult[_]): Unit
2.  = {
3.     val info = taskInfos(tid)
4.     val index = info.index
5.     info.markFinished(TaskState.FINISHED, clock.getTimeMillis())
6.     if (speculationEnabled) {
7.       successfulTaskDurations.insert(info.duration)
8.     }
9.     removeRunningTask(tid)
10.
11.    //杀掉同一任务的任何其他尝试（因为现在不需要这些任务，所以一次尝试成功）
12.    for (attemptInfo <- taskAttempts(index) if attemptInfo.running) {
13.      logInfo(s"Killing attempt ${attemptInfo.attemptNumber} for task
         ${attemptInfo.id} " +
14.        s"in stage ${taskSet.id} (TID ${attemptInfo.taskId}) on
           ${attemptInfo.host} " +
15.        s"as the attempt ${info.attemptNumber} succeeded on ${info.host}")
16.      sched.backend.killTask(
17.        attemptInfo.taskId,
18.        attemptInfo.executorId,
```

```
19.          interruptThread = true,
20.          reason = "another attempt succeeded")
21.      }
22.      if (!successful(index)) {
23.        tasksSuccessful += 1
24.        logInfo(s"Finished task ${info.id} in stage ${taskSet.id} (TID
           ${info.taskId}) in" +
25.          s" ${info.duration} ms on ${info.host} (executor ${info.executorId})" +
26.          s" ($tasksSuccessful/$numTasks)")
27.        // Mark successful and stop if all the tasks have succeeded.
28.        successful(index) = true
29.        if (tasksSuccessful == numTasks) {
30.          isZombie = true
31.        }
32.      } else {
33.        logInfo("Ignoring task-finished event for " + info.id + " in stage " +
           taskSet.id +
34.          " because task " + index + " has already completed successfully")
35.      }
36.      //此方法由TaskSchedulerImpl.handleSuccessfulTask调用,将TaskSchedulerImpl
        //锁定，直到退出
37.      //为了避免SPARK-7655问题，不应该“反序列化”持有锁时的值，以避免阻塞其他线程
38.      //所以调用TaskResultGetter.enqueueSuccessfulTask中的result.value()
39.      //注意，“result.value()”只在第一次调用时反序列化该值，因此这里的“result.
         //value()”只返回值，不会阻塞其他线程
40.      sched.dagScheduler.taskEnded(tasks(index), Success, result.value(),
         result.accumUpdates, info)
41.      maybeFinishTaskSet()
42.    }
```

Spark 2.4.3 版本的 TaskSetManager.scala 的 handleSuccessfulTask 的源码与 Spark 2.2.1 版本相比具有如下特点。

- 上段代码中第 4 行之后新增代码，检查在此之前是否有其他尝试成功。如已完成，但由于另一次尝试成功而未提交，将此任务作为已终止的任务处理。
- 上段代码中第 16 行之前新增 killedByOtherAttempt 的代码。killedByOtherAttempt 是一个 HashSet，当任务被其他尝试任务杀死时，将任务的 tid 添加到此 HashSet 中。这是在将 spark.speculation 设置为 true 时发生的。被别人杀死的任务，executor 丢失时不应重新提交。

```
1.  ……
2.      //检查在此之前是否有其他尝试成功，并且此尝试尚未处理
3.      if (successful(index) && killedByOtherAttempt.contains(tid)) {
4.        //撤销之前calculatedTasks和totalResultSize的检查，是否可以获取更多结果
5.        calculatedTasks -= 1
6.        val resultSizeAcc = result.accumUpdates.find(a =>
7.          a.name == Some(InternalAccumulator.RESULT_SIZE))
8.        if (resultSizeAcc.isDefined) {
9.          totalResultSize -= resultSizeAcc.get.asInstanceOf[LongAccumulator].value
10.       }
11.
12.       //将此任务作为已终止的任务处理。已完成，但由于另一次尝试成功而未提交
13.       handleFailedTask(tid, TaskState.KILLED,
14.         TaskKilled("Finish but did not commit due to another attempt succeeded"))
```

```
15.       return
16.     }
17. ……
18. killedByOtherAttempt += attemptInfo.taskId
```

speculationEnabled 默认设置为 spark.speculation=false，用于推测执行慢的任务；如果设置为 true，successfulTaskDurations 使用 MedianHeap 记录成功任务的持续时间，这样就可以确定什么时候启动推测性任务，这种情况只在启用推测时使用，以避免不使用堆时增加堆中的开销。

TaskSetManager 的 handleSuccessfulTask 中调用了 maybeFinishTaskSet。maybeFinishTaskSet 的源码如下：

```
1.   private def maybeFinishTaskSet() {
2.     if (isZombie && runningTasks == 0) {
3.       sched.taskSetFinished(this)
4.       if (tasksSuccessful == numTasks) {
5.        blacklistTracker.foreach(_.updateBlacklistForSuccessfulTaskSet(
6.          taskSet.stageId,
7.          taskSet.stageAttemptId,
8.          taskSetBlacklistHelperOpt.get.execToFailures))
9.       }
10.    }
11.  }
```

TaskSetManager：单 TaskSet 的任务调度在 TaskSchedulerImpl 中进行。TaskSetManager 类跟踪每项任务，如果任务重试失败（超过有限的次数），对于 TaskSet 处理本地调度主要的接口是 resourceOffer，询问 TaskSet 是否要在一个节点上运行任务，进行状态更新 statusUpdate，告诉 TaskSet 的一个任务的状态发生了改变（如已完成）。线程：这个类被设计成只在具有锁的代码 TaskScheduler 上调用（如事件处理程序），不应该从其他线程调用。

总结：Task 执行及结果处理原理流程图如图 8-3 所示。任务从 Driver 上发送过来，CoarseGrainedSchedulerBackend 发送任务，CoarseGrainedExecutorBackend 收到任务后，交给

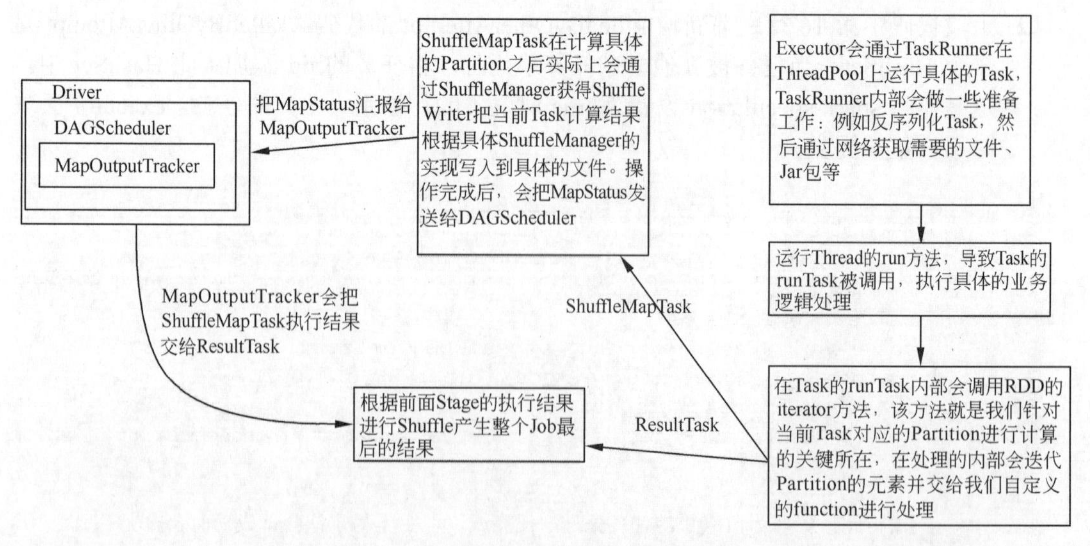

图 8-3 Task 执行及结果处理原理流程图

Executor 处理，Executor 会通过 launchTask 执行 Task。TaskRunner 内部会做很多准备工作：反序列化 Task 的依赖，通过网络获取需要的文件、Jar、反序列 Task 本身等待；然后调用 Task 的 runTask 执行，runTask 有 ShuffleMapTask、ResultTask 两种。通过 iterator()方法根据业务逻辑循环遍历，如果是 ShuffleMapTask，就把 MapStatus 汇报给 MapOutTracker；如果是 ResultTask，就从前面的 MapOutTracker 中获取信息。

8.4　Driver 如何管理 ShuffleMapTask 和 ResultTask 的处理结果

Spark Job 中，根据 Task 所处 Stage 的位置，我们将 Task 分为两类：第一类叫 shuffleMapTask，指 Task 所处的 Stage 不是最后一个 Stage，也就是 Stage 的计算结果还没有输出，而是通过 Shuffle 交给下一个 Stage 使用；第二类叫 resultTask，指 Task 所处 Stage 是 DAG 中最后的一个 Stage，也就是 Stage 计算结果需要进行输出等操作，计算到此为止已经结束。简单地说，Spark Job 中除了最后一个 Stage 的 Task 叫 resultTask，其他所有 Task 都叫 ShuffleMapTask。

8.4.1　ShuffleMapTask 执行结果和 Driver 的交互原理及源码详解

Driver 中的 CoarseGrainedSchedulerBackend 给 CoarseGrainedExecutorBackend 发送 launchTasks 消息，CoarseGrainedExecutorBackend 收到 launchTasks 消息以后会调用 executor.launchTask。通过 launchTask 执行 Task，launchTask 方法中根据传入的参数：taskId、尝试次数、任务名称、序列化后的任务创建一个 TaskRunner，在 threadPool 中执行 TaskRunner。TaskRunner 内部会先做一些准备工作，如反序列化 Task 的依赖，通过网络获取需要的文件、Jar 等；然后调用反序列化后的 Task.run 方法来执行任务并获得执行结果。

其中，Task 的 run 方法调用的时候会导致 Task 的抽象方法 runTask 的调用，Task.scala 的 runTask 方法是一个抽象方法。Task 包括 ResultTask、ShuffleMapTask 两种 Task，抽象 runTask 方法具体的实现由子类的 runTask 实现。ShuffleMapTask 的 runTask 实际运行的时候会调用 RDD 的 iterator，然后针对 Partition 进行计算。

ShuffleMapTask.scala 的源码如下：

```
1.     override def runTask(context: TaskContext): MapStatus = {
2.   ......
3.     val ser = SparkEnv.get.closureSerializer.newInstance()
4.     val  (rdd, dep)  = ser.deserialize[(RDD[_], ShuffleDependency[_, _,
       _])](
5.     ......
6.      val manager = SparkEnv.get.shuffleManager
7.     writer = manager.getWriter[Any, Any](dep.shuffleHandle, partitionId,
       context)
8.      writer.write(rdd.iterator(partition, context).asInstanceOf[Iterator
        [_ <: Product2[Any, Any]]])
9.      writer.stop(success = true).get
10.   ......
```

ShuffleMapTask 方法中调用 ShuffleManager 写入器 writer 方法，在 write 时最终计算会调用 RDD 的 compute 方法。通过 writer.stop(success = true).get，如果写入成功，就返回 MapStatus 结果值。

SortShuffleWriter.scala 的源码如下：

```
1.  override def write(records: Iterator[Product2[K, V]]): Unit = {
2.  ......
3.       val blockId = ShuffleBlockId(dep.shuffleId, mapId,
         IndexShuffleBlockResolver.NOOP_REDUCE_ID)
4.       val partitionLengths = sorter.writePartitionedFile(blockId, tmp)
5.      shuffleBlockResolver.writeIndexFileAndCommit(dep.shuffleId, mapId,
        partitionLengths, tmp)
6.       mapStatus = MapStatus(blockManager.shuffleServerId, partitionLengths)
7.  ......
8.   override def stop(success: Boolean): Option[MapStatus] = {
9.       ......
10.      if (success) {
11.        return Option(mapStatus)
12.      } else {
13.        return None
14.      }
15. ......
```

回到 TaskRunner 的 run 方法，把 task.run 执行结果通过 resultSer.serialize(value)序列化，生成一个 directResult。然后根据大小判断不同的结果赋值给 serializedResult，传回给 Driver。

（1）如果任务执行结果特别大，超过 1GB，日志就提示超出任务大小限制，返回元数据 ser.serialize(new IndirectTaskResult[Any](TaskResultBlockId(taskId), resultSize))。

Executor.scala 的源码如下：

```
1.    if (maxResultSize > 0 && resultSize > maxResultSize) {
2.            logWarning(s"Finished $taskName (TID $taskId). Result is larger
              than maxResultSize " + s"(${Utils.bytesToString(resultSize)} >
             ${Utils.bytesToString (maxResultSize)}), " + s"dropping it.")
3.
4.            ser.serialize(new IndirectTaskResult[Any](TaskResultBlockId
              (taskId), resultSize))
```

（2）如果任务执行结果小于 1GB，大于 maxDirectResultSize（128MB），就放入 blockManager，返回元数据 ser.serialize(new IndirectTaskResult[Any](blockId, resultSize))。

Executor.scala 的源码如下：

```
1.  ......
2.    } else if (resultSize > maxDirectResultSize) {
3.            val blockId = TaskResultBlockId(taskId)
4.            env.blockManager.putBytes(
5.              blockId,
6.              new ChunkedByteBuffer(serializedDirectResult.duplicate()),
7.              StorageLevel.MEMORY_AND_DISK_SER)
8.            logInfo(
9.              s"Finished $taskName (TID $taskId). $resultSize bytes result
                sent via BlockManager)")
10.           ser.serialize(new IndirectTaskResult[Any](blockId, resultSize))
```

（3）如果任务执行结果小于 128MB，就直接返回 serializedDirectResult。

Executor.scala 的源码如下：

```
1.  ......
2.  } else {
3.            logInfo(s"Finished $taskName (TID $taskId). $resultSize bytes
              result sent to driver")
4.            serializedDirectResult
5.  ......
```

接下来，TaskRunner 的 run 方法中调用 execBackend.statusUpdate(taskId, TaskState.FINISHED, serializedResult)给 Driver 发送一个消息，消息中将 taskId、TaskState.FINISHED、serializedResult 传进去。这里，execBackend 是 CoarseGrainedExecutorBackend。

Executor.scala 的源码如下：

```
1.     override def run(): Unit = {
2.   ......
3.        execBackend.statusUpdate(taskId, TaskState.FINISHED, serializedResult)
4.  ......
```

CoarseGrainedExecutorBackend 的 statusUpdate 方法的源码如下：

```
1.    override def statusUpdate(taskId: Long, state: TaskState, data:
      ByteBuffer) {
2.      val msg = StatusUpdate(executorId, taskId, state, data)
3.      driver match {
4.        case Some(driverRef) => driverRef.send(msg)
5.        case None => logWarning(s"Drop $msg because has not yet connected
          to driver")
6.      }
7.    }
```

CoarseGrainedExecutorBackend 给 DriverEndpoint 发送 StatusUpdate 来传输执行结果。DriverEndpoint 是一个 ThreadSafeRpcEndpoint 消息循环体，模式匹配收到 StatusUpdate 消息，调用 scheduler.statusUpdate(taskId, state, data.value)方法执行。这里的 scheduler 是 TaskSchedulerImpl。

CoarseGrainedSchedulerBackend.scala 的 DriverEndpoint 的源码如下：

```
1.        override def receive: PartialFunction[Any, Unit] = {
2.      case StatusUpdate(executorId, taskId, state, data) =>
3.        scheduler.statusUpdate(taskId, state, data.value)
```

DriverEndpoint 会把执行结果传递给 TaskSchedulerImpl 处理，交给 TaskResultGetter 内部，通过线程去分别处理 Task 执行成功和失败时的不同情况，然后告诉 DAGScheduler 任务处理结束的状况。

TaskSchedulerImpl.scala 的 statusUpdate 的源码如下：

```
1.  def statusUpdate(tid: Long, state: TaskState, serializedData: ByteBuffer) {
2.  ......
3.    if (TaskState.isFinished(state)) {
4.              cleanupTaskState(tid)
5.              taskSet.removeRunningTask(tid)
6.              if (state == TaskState.FINISHED) {
7.                taskResultGetter.enqueueSuccessfulTask(taskSet, tid,
                  serializedData)
8.              } else if (Set(TaskState.FAILED, TaskState.KILLED,
              TaskState.LOST).contains(state)) {
9.                taskResultGetter.enqueueFailedTask(taskSet, tid, state,
                  serializedData)
10.             }
11.         }
```

TaskResultGetter.scala 的 enqueueSuccessfulTask 方法中，开辟一条新线程处理成功任务，对结果进行相应的处理后调用 scheduler.handleSuccessfulTask。

TaskSchedulerImpl 的 handleSuccessfulTask 的源码如下：

```
1.     def handleSuccessfulTask(
2.       taskSetManager: TaskSetManager,
3.       tid: Long,
4.       taskResult: DirectTaskResult[_]): Unit = synchronized {
5.     taskSetManager.handleSuccessfulTask(tid, taskResult)
6.   }
```

TaskSchedulerImpl 的 handleSuccessfulTask 交给 TaskSetManager 调用 handleSuccessfulTask。

TaskSetManager 的 handleSuccessfulTask 的源码如下：

```
1.     def handleSuccessfulTask(tid: Long, result: DirectTaskResult[_]): Unit = {
2.       ……
3.     sched.dagScheduler.taskEnded(tasks(index), Success, result.value(),
       result.accumUpdates, info)
4.    ……
5.
```

handleSuccessfulTask 方法中调用 sched.dagScheduler.taskEnded，taskEnded 由 TaskSetManager 调用，汇报任务完成或者失败。将任务完成的事件 CompletionEvent 放入 eventProcessLoop 事件处理循环中。

DAGScheduler.scala 的源码如下：

```
1.   def taskEnded(
2.       task: Task[_],
3.       reason: TaskEndReason,
4.       result: Any,
5.       accumUpdates: Seq[AccumulatorV2[_, _]],
6.       taskInfo: TaskInfo): Unit = {
7.     eventProcessLoop.post(
8.       CompletionEvent(task, reason, result, accumUpdates, taskInfo))
9.   }
```

由事件循环线程读取消息，并调用 DAGSchedulerEventProcessLoop.onReceive 方法进行消息处理。

DAGScheduler.scala 的源码如下：

```
1.     override def onReceive(event: DAGSchedulerEvent): Unit = {
2.     val timerContext = timer.time()
3.     try {
4.       doOnReceive(event)
5.     } finally {
6.       timerContext.stop()
7.     }
8.   }
```

onReceive 中调用 doOnReceive(event)方法，模式匹配到 CompletionEvent，调用 dagScheduler.handleTaskCompletion 方法。

DAGScheduler.scala 的源码如下：

```
1.   private def doOnReceive(event: DAGSchedulerEvent): Unit = event match {
2.     case JobSubmitted(jobId, rdd, func, partitions, callSite, listener,
```

```
      properties) =>
3.        dagScheduler.handleJobSubmitted(jobId, rdd, func, partitions,
          callSite, listener, properties)
4.  ......
5.    case completion: CompletionEvent =>
6.        dagScheduler.handleTaskCompletion(completion)
7.  ......
```

DAGScheduler.handleTaskCompletion 中 task 执行成功的情况，根据 ShuffleMapTask 和 ResultTask 两种情况分别处理。其中，ShuffleMapTask 将 MapStatus 汇报给 MapOutTracker。

Spark 2.2.1 版本的 DAGScheduler 的 handleTaskCompletion 的源码如下：

```
1.    private[scheduler] def handleTaskCompletion(event: CompletionEvent) {
2.    ......
3.      val stage = stageIdToStage(task.stageId)
4.    ......
5.      event.reason match {
6.        case Success =>
7.         ......
8.          task match {
9.           ......
10.           case smt: ShuffleMapTask =>
11.             val shuffleStage = stage.asInstanceOf[ShuffleMapStage]
12.             val status = event.result.asInstanceOf[MapStatus]
13.             val execId = status.location.executorId
14.             logDebug("ShuffleMapTask finished on " + execId)
15.             if (stageIdToStage(task.stageId).latestInfo.attemptId == task.
                stageAttemptId) {
16.       //此任务用于当前正在运行的阶段尝试。从 TaskSetManager 的角度任务成功完成，
17.       //标记为不再挂起，TaskSetManager 甚至可能认为任务已完成，需要忽略输出，
18.       //因为任务的 epoch 太小。在这种情况下，当挂起的分区为空时，
19.       //仍然会丢失输出位置，这将导致 DAGScheduler 重新提交下面的阶段
20.               shuffleStage.pendingPartitions -= task.partitionId
21.             }
22.             if (failedEpoch.contains(execId) && smt.epoch <=
                failedEpoch(execId)) {
23.               logInfo(s"Ignoring possibly bogus $smt completion from
                  executor $execId")
24.             } else {
25.       //任务的 epoch 是可以接受的（任务是在执行者最近发现的失败之后启动的），因此将任
          //务的输出标记为可用
26.
27.               shuffleStage.addOutputLoc(smt.partitionId, status)
28.
29.       //从挂起的分区中删除任务的分区。这可能已完成，但在任务为阶段的早期尝试（不是当
          //前正在运行的尝试）
30.       //这允许 DagScheduler 在一个每个任务的副本已成功完成的情况下，标记阶段完成，即
          //使当前活动阶段任务仍在运行
31.
32.               shuffleStage.pendingPartitions -= task.partitionId
33.             }
34.
35.             if (runningStages.contains(shuffleStage) && shuffleStage
                .pendingPartitions.isEmpty) {
36.               markStageAsFinished(shuffleStage)
37.               logInfo("looking for newly runnable stages")
38.               logInfo("running: " + runningStages)
```

```
39.             logInfo("waiting: " + waitingStages)
40.             logInfo("failed: " + failedStages)
41.
42.      //我们设置为 true，来递增纪元编号，以防止 map 输出重新计算。在这种情况下，
43.      //一些节点可能已经缓存了损坏的位置（从我们检测到的错误），将需要纪元编号递增来
         //重取它们
44.      //待办事项：如果这不是第一次，那么只增加纪元编号，我们注册了 map 输出
45.
46.             mapOutputTracker.registerMapOutputs(
47.               shuffleStage.shuffleDep.shuffleId,
48.               shuffleStage.outputLocInMapOutputTrackerFormat(),
49.               changeEpoch = true)
50.
51.             clearCacheLocs()
52.
53.             if (!shuffleStage.isAvailable) {
54.      //有些任务已经失败了，重新提交 shuffleStage
55.      //待办事项：低级调度器也应该能处理这个问题
56.               logInfo("Resubmitting " + shuffleStage + " (" +
                  shuffleStage.name +
57.                 ") because some of its tasks had failed: " +
58.                 shuffleStage.findMissingPartitions().mkString(", "))
59.               submitStage(shuffleStage)
60.             } else {
61.      //标识任何 map 阶段的作业都在这个阶段等待完成
62.               if (shuffleStage.mapStageJobs.nonEmpty) {
63.                 val stats = mapOutputTracker.getStatistics
                    (shuffleStage.shuffleDep)
64.                 for (job <- shuffleStage.mapStageJobs) {
65.                   markMapStageJobAsFinished(job, stats)
66.                 }
67.               }
68.               submitWaitingChildStages(shuffleStage)
69.             }
70.           }
71.         }
72. ......
```

Spark 2.4.3 版本的 DAGScheduler 的 handleTaskCompletion 的源码与 Spark 2.2.1 版本相比具有如下特点。

- 上段代码中第 11 行之后新增代码，从 shuffleStage.pendingPartitions 中去掉分区 ID。
- 上段代码中删掉第 15～21 行。
- 上段代码中删掉第 25～32 行，新增 mapOutputTracker.registerMapOutput 进行注册的代码。
- 上段代码中删掉第 46～49 行，新增 mapOutputTracker.incrementEpoch 的代码。
- 上段代码中删掉第 61～67 行，将其封装为一个方法，新增 markMapStageJobsAsFinished(shuffleStage)的代码。

```
1. ......
2.   shuffleStage.pendingPartitions -= task.partitionId
3. ......
4.       mapOutputTracker.registerMapOutput(
5.               shuffleStage.shuffleDep.shuffleId, smt.partitionId, status)
6. ......
7. mapOutputTracker.incrementEpoch()
```

```
8. ……
9.   markMapStageJobsAsFinished(shuffleStage)
10. ……
11.   private[scheduler] def markMapStageJobsAsFinished(shuffleStage:
      ShuffleMapStage): Unit = {
12.     //将等待此阶段的任何 map 阶段作业标记为已完成
13.     if (shuffleStage.isAvailable && shuffleStage.mapStageJobs.nonEmpty) {
14.       val stats = mapOutputTracker.getStatistics(shuffleStage.shuffleDep)
15.       for (job <- shuffleStage.mapStageJobs) {
16.         markMapStageJobAsFinished(job, stats)
17.       }
18.     }
19.   }
```

8.4.2 ResultTask 执行结果与 Driver 的交互原理及源码详解

Task 的 run 方法调用的时候会导致 Task 的抽象方法 runTask 的调用，Task.scala 的 runTask 方法是一个抽象方法。Task 包括 ResultTask、ShuffleMapTask 两种 Task，抽象 runTask 方法具体的实现由子类的 runTask 实现。ResultTask 的 runTask 具体实现的源码如下。

ResultTask.scala 的 runTask 的源码如下：

```
1.   override def runTask(context: TaskContext): U = {
2.     ……
3. //反序列 RDD 和 func 处理函数
4.     val (rdd, func) = ser.deserialize[(RDD[T], (TaskContext, Iterator[T])
       => U)](
5.     ……
6.     func(context, rdd.iterator(partition, context))
7.   }
```

而 ResultTask 的 runTask 方法中反序列化生成 func 函数，最后通过 func 函数计算出最终的结果。

ResultTask 执行结果与 Driver 的交互过程同 ShuffleMapTask 类似，最终，DAGScheduler.handleTaskCompletion 中 Task 执行结果，根据 ShuffleMapTask 和 ResultTask 两种情况分别处理。其中，ResultTask 的处理结果如下所示。

DAGScheduler 的 handleTaskCompletion 的源码如下：

```
1.       case rt: ResultTask[_, _] =>
2.             //因为是 ResultTask 的一部分，所以对应为 ResultStage
3.             //待办事宜：这一功能进行重构，接受 ResultStage
4.
5.             val resultStage = stage.asInstanceOf[ResultStage]
6.             resultStage.activeJob match {
7.               case Some(job) =>
8.                 if (!job.finished(rt.outputId)) {
9.                   job.finished(rt.outputId) = true
10.                  job.numFinished += 1
11.                  // 如果整个作业完成，就删除
12.                  if (job.numFinished == job.numPartitions) {
13.                    markStageAsFinished(resultStage)
14.                    cleanupStateForJobAndIndependentStages(job)
15.                    listenerBus.post(
16.                      SparkListenerJobEnd(job.jobId, clock.getTimeMillis(),
```

```
                    JobSucceeded))
17.                 }
18.
19.                 //taskSucceeded 运行用户代码可能会抛出一个异常
20.                 try {
21.                   job.listener.taskSucceeded(rt.outputId, event.result)
22.                 } catch {
23.                   case e: Exception =>
24.                     //待办事项：标记 resultStage 失败
25.                     job.listener.jobFailed(new
                        SparkDriverExecutionException(e))
26.                 }
27.               }
28.             case None =>
29.               logInfo("Ignoring result from " + rt + " because its job has
                  finished")
30.         }
```

Driver 端的 DAGScheduler 的 MapOutputTracker 把 shuffleMapTask 执行的结果交给 ResultTask，ResultTask 根据前面 Stage 的执行结果进行 shuffle 后产生整个 Job 最后的结果。

第 9 章　Spark 中 Cache 和 checkpoint 原理和源码详解

本章讲解 Spark 中 Cache 和 checkpoint 原理和源码。9.1 节讲解 Spark 中 Cache 原理和源码，CacheManager 管理缓存，缓存可基于内存或者磁盘。CacheManager 通过 BlockManager 来操作数据；9.2 节对 Spark 中 checkpoint 原理和源码进行详解。Spark 在生产环境下，如果 Tranformations 的 RDD 非常多或者具体 Tranformation 产生的 RDD 本身计算特别复杂和耗时，我们就可以通过 checkpoint 对计算结果数据进行持久化。

9.1　Spark 中 Cache 原理和源码详解

本节对 Spark 中 Cache 原理及 Spark 中 Cache 源码进行详解。

9.1.1　Spark 中 Cache 原理详解

Spark 中 Cache 机制原理：首先，RDD 是通过 iterator 进行计算的。

（1）CacheManager 会通过 BlockManager 从 Local 或者 Remote 获取数据直接通过 RDD 的 compute 进行计算，有可能需要考虑 checkpoint。

（2）通过 BlockManager 首先从本地获取数据，如果得不到数据，就会从远程获取数据。

（3）首先查看当前的 RDD 是否进行了 checkpoint，如果进行了的话，就直接读取 checkpoint 的数据，否则必须进行计算；因为此时 RDD 需要缓存，所以计算如果需要，则通过 BlockManager 再次进行持久化。

（4）如果持久化的时候只是缓存到磁盘中，就直接使用 BlockManager 的 doPut 方法写入磁盘（需要考虑 Replication）。

（5）如果指定内存作缓存，优先保存到内存中，此时会使用 MemoryStore.unrollSafely 方法来尝试安全地将数据保存在内存中，如果内存不够，会使用一个方法来整理一部分内存空间，然后基于整理出来的内存空间放入我们想缓存的最新数据。

（6）直接通过 RDD 的 compute 进行计算，有可能需要考虑 checkpoint。

Spark 中，Cache 原理示意图如图 9-1 所示。

9.1.2　Spark 中 Cache 源码详解

CacheManager 管理是缓存，而缓存可以是基于内存的缓存，也可以是基于磁盘的缓存。CacheManager 需要通过 BlockManager 来操作数据。

Task 发生计算时要调用 RDD 的 compute 进行计算。下面看一下 MapPartitionsRDD 的

compute 方法。

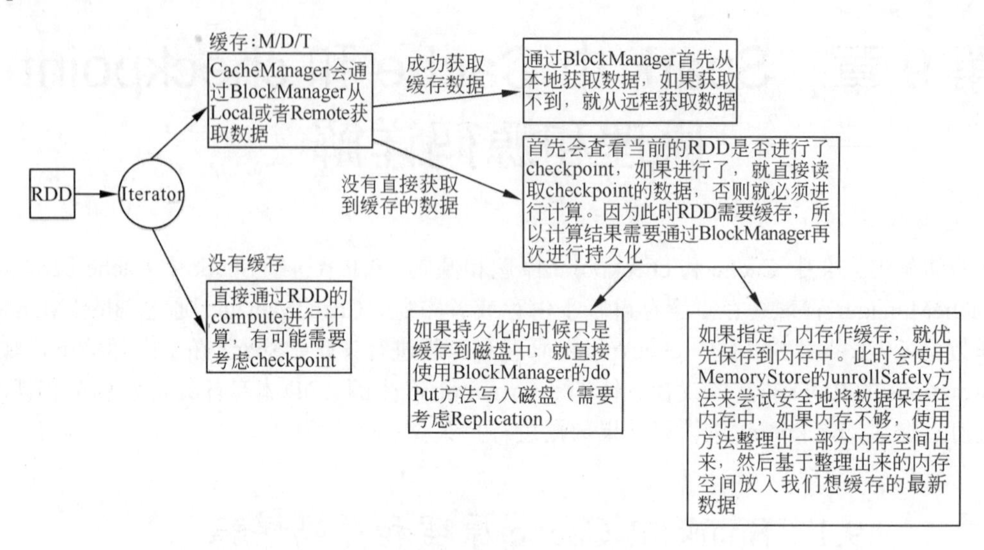

图 9-1　Cache 原理示意图

Spark 2.2.1 版本的 MapPartitionsRDD 的源码如下：

```
1.  private[spark] class MapPartitionsRDD[U: ClassTag, T: ClassTag](
2.    var prev: RDD[T],
3.    f: (TaskContext, Int, Iterator[T]) => Iterator[U],  //(TaskContext,
      partition index, iterator)
4.    preservesPartitioning: Boolean = false)
5.   extends RDD[U](prev) {
6.
7.   override val partitioner = if (preservesPartitioning) firstParent[T]
     .partitioner else None
8.
9.   override def getPartitions: Array[Partition] = firstParent[T]
     .partitions
10.
11.  override def compute(split: Partition, context: TaskContext):
     Iterator[U] =
12.    f(context, split.index, firstParent[T].iterator(split, context))
13.
14.  override def clearDependencies() {
15.    super.clearDependencies()
16.    prev = null
17.  }
18. }
```

Spark 2.4.3 版本的 MapPartitionsRDD.scala 源码与 Spark 2.2.1 版本相比具有如下特点。

- 上段代码中第 4 行之后新增 2 个参数 isFromBarrier、isOrderSensitive。isFromBarrier 指示此 RDD 是否从 RDDBarrier 转换，包含至少一个 RDDBarrier 的阶段应转换为屏障阶段。isOrderSensitive 指示函数是否区分顺序。如果它是顺序敏感的，当输入顺序改变时，它可能返回完全不同的结果。大多数状态函数是顺序敏感的。
- 上段代码中第 17 行之后新增代码，isBarrier_方法指示 RDD 是否处于屏障阶段。Spark 必须同时启动屏障阶段的所有任务。如果 RDD 处于屏障阶段，至少有一个父 RDD

或其自身映射自一个 RDDBarrier。对于 ShuffledRDD，此函数总是返回 false，因为 ShuffledRDD 表示新阶段的开始；对于 MapPartitionsRDD，可以从 RDDBarrier 转换，在这种情况下，MapPartitionsRDD 应标记为屏障。重写 GetOutputDeterministicLevel 方法实现自定义逻辑计算 RDD 输出的确定级别。

```
1.  ……
2.     isFromBarrier: Boolean = false,
3.     isOrderSensitive: Boolean = false)
4.  ……
5.    @transient protected lazy override val isBarrier_ : Boolean =
6.     isFromBarrier || dependencies.exists(_.rdd.isBarrier())
7.
8.    override protected def getOutputDeterministicLevel = {
9.     if (isOrderSensitive && prev.outputDeterministicLevel ==
       DeterministicLevel.UNORDERED) {
10.       DeterministicLevel.INDETERMINATE
11.     } else {
12.       super.getOutputDeterministicLevel
13.     }
```

compute 真正计算的时候通过 iterator 计算，MapPartitionsRDD 的 iterator 依赖父 RDD 计算。iterator 是 RDD 内部的方法，如有缓存，将从缓存中读取数据，否则进行计算。这不是被用户直接调用，但可用于实现自定义子 RDD。

RDD.scala 的 iterator 方法如下：

```
1.    final def iterator(split: Partition, context: TaskContext): Iterator[T] = {
2.     if (storageLevel != StorageLevel.NONE) {
3.       getOrCompute(split, context)
4.     } else {
5.       computeOrReadCheckpoint(split, context)
6.     }
7.    }
```

RDD.scala 的 iterator 方法中判断 storageLevel != StorageLevel.NONE，说明数据可能存放在内存、磁盘中，调用 getOrCompute(split, context)方法。如果之前计算过一次，再次计算可以找 CacheManager 要数据。

RDD.scala 的 getOrCompute 的源码如下：

```
1.   private[spark] def getOrCompute(partition: Partition, context:
     TaskContext): Iterator[T] = {
2.     val blockId = RDDBlockId(id, partition.index)
3.     var readCachedBlock = true
4.     //这种方法被 Executors 调用，所以我们需要调用 SparkEnv.get 代替 sc.env
5.     SparkEnv.get.blockManager.getOrElseUpdate(blockId,    storageLevel,
       elementClassTag, () => {
6.       readCachedBlock = false
7.       computeOrReadCheckpoint(partition, context)
8.     }) match {
9.       case Left(blockResult) =>
10.        if (readCachedBlock) {
11.          val existingMetrics = context.taskMetrics().inputMetrics
12.          existingMetrics.incBytesRead(blockResult.bytes)
13.          new InterruptibleIterator[T](context, blockResult.data.
             asInstanceOf[Iterator[T]]) {
14.            override def next(): T = {
```

```
15.              existingMetrics.incRecordsRead(1)
16.              delegate.next()
17.            }
18.          }
19.        } else {
20.          new InterruptibleIterator(context, blockResult.data.asInstanceOf
             [Iterator[T]])
21.        }
22.      case Right(iter) =>
23.        new InterruptibleIterator(context, iter.asInstanceOf[Iterator[T]])
24.    }
25.  }
```

在有缓存的情况下，缓存可能基于内存，也可能基于磁盘，getOrCompute获取缓存；如没有缓存，则需重新计算RDD。为何需要重新计算？如果数据放在内存中，假设缓存了100万个数据分片，下一个步骤计算的时候需要内存，因为需要进行计算的内存空间占用比之前缓存的数据占用内存空间重要，假设须腾出10000个数据分片所在的空间，因此从BlockManager中将内存中的缓存数据drop到磁盘上，如果不是内存和磁盘的存储级别，那10000个数据分片的缓存数据就可能丢失，99万个数据分片可以复用，而这10000个数据分片须重新进行计算。

Cache在工作的时候会最大化地保留数据，但是数据不一定绝对完整，因为当前的计算如果需要内存空间，那么Cache在内存中的数据必须让出空间，此时如何在RDD持久化的时候同时指定可以把数据放在Disk上，那么部分Cache的数据就可以从内存转入磁盘，否则数据就会丢失。

getOrCompute方法返回的是Iterator。进行Cache以后，BlockManager对其进行管理，通过blockId可以获得曾经缓存的数据。具体CacheManager在获得缓存数据的时候会通过BlockManager来抓到数据。

getOrElseUpdate方法中，如果block存在，检索给定的块block；如果不存在，则调用提供makeIterator方法计算块block，对块block进行持久化，并返回block的值。

BlockManager.scala的getOrElseUpdate的源码如下：

```
1.      def getOrElseUpdate[T](
2.        blockId: BlockId,
3.        level: StorageLevel,
4.        classTag: ClassTag[T],
5.        makeIterator: () => Iterator[T]): Either[BlockResult, Iterator[T]] = {
6.      //尝试从本地或远程存储读取块。如果它存在，那么我们就不需要通过本地get或put路
        //径获取
7.      get[T](blockId)(classTag) match {
8.        case Some(block) =>
9.          return Left(block)
10.       case _ =>
11.         //需要计算块
12.     }
13.     //需要计算blockInitially，在块上我们没有锁
14.     doPutIterator(blockId, makeIterator, level, classTag, keepReadLock =
        true) match {
15.       case None =>
16.         //doput()方法没有返回，所以块已存在或者已成功存储。因此，我们现在在块上持有
            //读取锁
17.         val blockResult = getLocalValues(blockId).getOrElse {
```

```
18.      //在 doPut()和 get()方法调用的时候，我们持有读取锁，块不应被驱逐，这样，get()
         //方法没返回块，表示发生一些内部错误
19.         releaseLock(blockId)
20.         throw new SparkException(s"get() failed for block $blockId even
            though we held a lock")
21.       }
22.     //我们已经持有调用 doPut()方法在块上的读取锁，getLocalValues()再一次获取锁，
        //所以我们需要调用 releaseLock()，这样获取锁的数量是 1(因为调用者只 release()一次)
23.        releaseLock(blockId)
24.        Left(blockResult)
25.       case Some(iter) =>
26.     //输入失败，可能是因为数据太大而不能存储在内存中，不能溢出到磁盘上。因此，我们需
        //要将输入迭代器传递给调用者，他们可以决定如何处理这些值（例如，不缓存它们）
27.        Right(iter)
28.     }
29.   }
```

BlockManager.scala 的 getOrElseUpdate 中根据 blockId 调用了 get[T](blockId)方法，get 方法从 block 块管理器（本地或远程）获取一个块 block。如果块在本地存储且没获取锁，则先获取块 block 的读取锁。如果该块是从远程块管理器获取的，当 data 迭代器被完全消费以后，那么读取锁将自动释放。get 的时候，如果本地有数据，从本地获取数据返回；如果没有数据，则从远程节点获取数据。

BlockManager.scala 的 get 方法的源码如下：

```
1.    def get[T: ClassTag](blockId: BlockId): Option[BlockResult] = {
2.      val local = getLocalValues(blockId)
3.      if (local.isDefined) {
4.        logInfo(s"Found block $blockId locally")
5.        return local
6.      }
7.      val remote = getRemoteValues[T](blockId)
8.      if (remote.isDefined) {
9.        logInfo(s"Found block $blockId remotely")
10.       return remote
11.     }
12.     None
13.   }
```

BlockManager 的 get 方法从 Local 的角度讲，如果数据在本地，get 方法调用 getLocalValues 获取数据。如果数据在内存中（level.useMemory 且 memoryStore 包含了 blockId），则从 memoryStore 中获取数据；如果数据在磁盘中（level.useDisk 且 diskStore 包含了 blockId），则从 diskStore 中获取数据。这说明数据在本地缓存，可以在内存中，也可以在磁盘上。

BlockManager 的 get 方法从 remote 的角度讲，get 方法中将调用 getRemoteValues 方法。

BlockManager.Scala 的 getRemoteValues 的源码如下：

```
1.      private def getRemoteValues[T: ClassTag](blockId: BlockId): Option
        [BlockResult] = {
2.      val ct = implicitly[ClassTag[T]]
3.      getRemoteBytes(blockId).map { data =>
4.        val values =
5.          serializerManager.dataDeserializeStream(blockId, data.toInputStream
            (dispose = true))(ct)
6.        new BlockResult(values, DataReadMethod.Network, data.size)
7.      }
8.    }
```

getRemoteValues 方法中调用 getRemoteBytes 方法，通过 blockTransferService.fetchBlockSync 从远程节点获取数据。

Spark 2.2.1 版本的 BlockManager.Scala 的 getRemoteBytes 的源码如下：

```
1.    def getRemoteBytes(blockId: BlockId): Option[ChunkedByteBuffer] = {
2.      logDebug(s"Getting remote block $blockId")
3.      require(blockId != null, "BlockId is null")
4.      var runningFailureCount = 0
5.      var totalFailureCount = 0
6.      val locations = getLocations(blockId)
7.      val maxFetchFailures = locations.size
8.      var locationIterator = locations.iterator
9.      while (locationIterator.hasNext) {
10.       val loc = locationIterator.next()
11.       logDebug(s"Getting remote block $blockId from $loc")
12.       val data = try {
13.         blockTransferService.fetchBlockSync(
14.           loc.host, loc.port, loc.executorId, blockId.toString).nioByteBuffer()
15.       } catch {
16.         case NonFatal(e) =>
17.           runningFailureCount += 1
18.           totalFailureCount += 1
19.
20.           if (totalFailureCount >= maxFetchFailures) {
21.             //放弃尝试的位置。要么我们已经尝试了所有的原始位置，或者我们已经从 master
              //节点刷新了位置列表，并且仍然在刷新列表中尝试位置后命中失败
              logWarning (s"Failed to fetch block after $totalFailureCount fetch
              failures." + s"Most recent failure cause:", e)
22.
23.             return None
24.           }
25.
26.           logWarning(s"Failed to fetch remote block $blockId " +
27.             s"from $loc (failed attempt $runningFailureCount)", e)
28.
29.     //如果有大量的 Executors，那么位置列表可以包含一个旧的条目造成大量重试，可能花
        //费大量的时间。在一定数量的获取失败之后，为去掉这些旧的条目，我们刷新块位置
30.           if (runningFailureCount >= maxFailuresBeforeLocationRefresh) {
31.             locationIterator = getLocations(blockId).iterator
32.             logDebug(s"Refreshed locations from the driver " +
33.               s"after ${runningFailureCount} fetch failures.")
34.             runningFailureCount = 0
35.           }
36.
37.           //此位置失败，所以我们尝试从不同的位置获取，这里返回一个 null
38.           null
39.       }
40.
41.       if (data != null) {
42.         return Some(new ChunkedByteBuffer(data))
43.       }
44.       logDebug(s"The value of block $blockId is null")
45.     }
46.     logDebug(s"Block $blockId not found")
47.     None
48.   }
```

Spark 2.4.3 版本的 BlockManager.scala 源码与 Spark 2.2.1 版本相比具有如下特点。

- 上段代码中第 5 行之后新增代码，因为所有的远程块都注册在 driver 中，所以不需要所有的从属执行器获取块状态。
- 上段代码中第 6 行将 getLocations 方法调整为 sortLocations 方法。sortLocations 方法返回给定块的位置列表，本地计算机的优先级从多个块管理器可以共享同一个主机，然后是同一机架上的主机。
- 上段代码中第 14 行新增一个 tempFileManager 参数。如果块大小超过阈值，将 FileManager 传递给 BlockTransferService，利用它来溢出块；如果没有，传递空值意味着块将持久存在内存中。
- 上段代码中第 31 行将 getLocations 方法调整为使用 sortLocations(master.getLocations (blockId)).iterator 方法，如果有大量执行者，则位置列表可以包含大量过时的条目导致大量重试，可能花大量的时间。除去这些陈旧的条目，在一定数量的提取失败后刷新块位置。
- 上段代码中第 42 行代码进行替换，对于未记录的 escape hatch，如果 ChunkedByteBuffer 时出现任何问题，返回到旧代码路径。如果新路径稳定，可在 Spark 2.4 以后的版本中清除。

```
1.  ……
2.  /* TODO SPARK-25905:如果将此方法更改为返回 ManagedBuffer，则 getRemoteValues
    只能在临时文件上使用 inputStream，而不是将该文件读取到内存中。在此之前，即使已经
3.  把数据读到磁盘上了，但副本可能会导致进程使用过多的内存并被终止*/
4.  ……
5.
6.      //因为所有的远程块都注册在 driver 中，所以不需要要求所有的从执行器获取块状态
7.      val locationsAndStatus = master.getLocationsAndStatus(blockId)
8.      val blockSize = locationsAndStatus.map { b =>
9.        b.status.diskSize.max(b.status.memSize)
10.     }.getOrElse(0L)
11.     val blockLocations = locationsAndStatus.map(_.locations)
        .getOrElse(Seq.empty)
12.
13.     //如果块大小超过阈值，则将 FileManager 传递给 BlockTransferService，利用
        //它来溢出块；如果没有，传递空值意味着块将持久存在内存中
14.     val tempFileManager = if (blockSize > maxRemoteBlockToMem) {
15.       remoteBlockTempFileManager
16.     } else {
17.       null
18.     }
19. ……
20.    val locations = sortLocations(blockLocations)
21. ……
22. loc.host, loc.port, loc.executorId, blockId.toString, tempFileManager)
23. ……
24. locationIterator = sortLocations(master.getLocations(blockId)).iterator
25. ……
26.         //SPARK-24307 未记录的 escape hatch，以防转换为 ChunkedByteBuffer 时出
            //现任何问题，返回到旧代码路径。如果新路径稳定，可在 Spark 2.4 以后的版本中清除
27.         if (remoteReadNioBufferConversion) {
28.           return Some(new ChunkedByteBuffer(data.nioByteBuffer()))
29.         } else {
30.           return Some(ChunkedByteBuffer.fromManagedBuffer(data))
31.         }
```

BlockManager 的 get 方法，如果本地有数据，则从本地获取数据返回；如果远程有数据，则从远程获取数据返回；如果都没有数据，就返回 None。get 方法的返回类型是 Option[BlockResult]，Option 的结果分为两种情况：①如果有内容，则返回 Some[BlockResult；②如果没有内容，则返回 None。这是 Option 的基础语法。

Option.scala 的源码如下：

```
1.    sealed abstract class Option[+A] extends Product with Serializable {
2.    self =>
3.  .....
4.  final case class Some[+A](x: A) extends Option[A] {
5.    def isEmpty = false
6.    def get = x
7.  }
8.
9.  ......
10. case object None extends Option[Nothing] {
11.   def isEmpty = true
12.   def get = throw new NoSuchElementException("None.get")
13. }
```

回到 BlockManager 的 getOrElseUpdate 方法，从 get 方法返回的结果进行模式匹配，如果有数据，则对 Some(block)返回 Left(block)，这是获取到 block 的情况；如果没数据，则是 None，须计算 block。

回到 RDD.scala 的 getOrCompute 方法，在 getOrCompute 方法中调用 SparkEnv.get.blockManager.getOrElseUpdate 方法时，传入 blockId、storageLevel、elementClassTag，其中第四个参数是一个匿名函数，在匿名函数中调用了 computeOrReadCheckpoint(partition, context)。然后在 getOrElseUpdate 方法中，根据 blockId 获取数据，如果获取到缓存数据，就返回；如果没有数据，就调用 doPutIterator(blockId, makeIterator, level, classTag, keepReadLock = true)进行计算，doPutIterator 其中第二个参数 makeIterator 就是 getOrElseUpdate 方法中传入的匿名函数，在匿名函数中获取到 Iterator 数据。

RDD.getOrCompute 中 computeOrReadCheckpoint 方法，如果 RDD 进行了 checkpoint，则从父 RDD 的 iterator 中直接获取数据；或者没有 Checkpoint 物化，则重新计算 RDD 的数据。

RDD.scala 的 computeOrReadCheckpoint 的源码如下：

```
1.     private[spark] def computeOrReadCheckpoint(split: Partition, context:
       TaskContext): Iterator[T] =
2.     {
3.      if (isCheckpointedAndMaterialized) {
4.        firstParent[T].iterator(split, context)
5.      } else {
6.        compute(split, context)
7.      }
8.    }
```

BlockManager.scala 的 getOrElseUpdate 方法中如果根据 blockID 没有获取到本地数据，则调用 doPutIterator 将通过 BlockManager 再次进行持久化。

BlockManager.scala 的 getOrElseUpdate 方法的源码如下：

```
1.   def getOrElseUpdate[T](
2.       blockId: BlockId,
3.       level: StorageLevel,
```

```
4.       classTag: ClassTag[T],
5.       makeIterator: () => Iterator[T]): Either[BlockResult, Iterator[T]] = {
6.     //尝试从本地或远程存储读取块。如果它存在，那么我们就不需要通过本地 GET 或 PUT 路
       //径获取
7.     get[T](blockId)(classTag) match {
8.       case Some(block) =>
9.         return Left(block)
10.      case _ =>
11.        //Need to compute the block.
12.    }
13.    //起初我们不锁这个块
14.    doPutIterator(blockId, makeIterator, level, classTag, keepReadLock =
       true) match {
15. .......
```

BlockManager.scala 的 getOrElseUpdate 方法中调用了 doPutIterator。doPutIterator 将 makeIterator 从父 RDD 的 checkpoint 读取的数据或者重新计算的数据存放到内存中，如果内存不够，就溢出到磁盘中持久化。

BlockManager.scala 的 doPutIterator 方法的源码如下：

```
1.     private def doPutIterator[T](
2.       blockId: BlockId,
3.       iterator: () => Iterator[T],
4.       level: StorageLevel,
5.       classTag: ClassTag[T],
6.       tellMaster: Boolean = true,
7.       keepReadLock: Boolean = false): Option[PartiallyUnrolledIterator[T]]={
8.     doPut(blockId, level, classTag, tellMaster = tellMaster, keepReadLock
       = keepReadLock) { info =>
9.       val startTimeMs = System.currentTimeMillis
10.      var iteratorFromFailedMemoryStorePut: Option[PartiallyUnrolledIterator
         [T]] = None
11.      //块的大小为字节
12.      var size = 0L
13.      if (level.useMemory) {
14.        //首先把它放在内存中，即使 useDisk 设置为 true；如果内存存储不能保存，我们
           //稍后会把它放在磁盘上
15.        if (level.deserialized) {
16.          memoryStore.putIteratorAsValues(blockId, iterator(), classTag)
             match {
17.            case Right(s) =>
18.              size = s
19.            case Left(iter) =>
20.              //没有足够的空间来展开块；如果适用，可以溢出到磁盘
21.              if (level.useDisk) {
22.                logWarning(s"Persisting block $blockId to disk instead.")
23.                diskStore.put(blockId) { fileOutputStream =>
24.                  serializerManager.dataSerializeStream(blockId,
                     fileOutputStream, iter)(classTag)
25.                }
26.                size = diskStore.getSize(blockId)
27.              } else {
28.                iteratorFromFailedMemoryStorePut = Some(iter)
29.              }
30.          }
31.        } else { //!level.deserialized
32.          memoryStore.putIteratorAsBytes(blockId, iterator(), classTag,
```

```
            level.memoryMode) match {
33.           case Right(s) =>
34.             size = s
35.           case Left(partiallySerializedValues) =>
36.             //没有足够的空间来展开块；如果适用，可以溢出到磁盘
37.             if (level.useDisk) {
38.              logWarning(s"Persisting block $blockId to disk instead.")
39.              diskStore.put(blockId) { channel =>
40.                val out = Channels.newOutputStream(channel)
                   partiallySerializedValues.finishWritingToStream(out)
41.              }
42.              size = diskStore.getSize(blockId)
43.             } else {
44.              iteratorFromFailedMemoryStorePut = Some
                 (partiallySerializedValues. valuesIterator)
45.             }
46.         }
47.       }
48.
49.     } else if (level.useDisk) {
50.      diskStore.put(blockId) { channel =>
51.        val out = Channels.newOutputStream(channel)
           serializerManager.dataSerializeStream(blockId, out,
           iterator())(classTag)
52.       }
53.       size = diskStore.getSize(blockId)
54.     }
55.
56.     val putBlockStatus = getCurrentBlockStatus(blockId, info)
57.     val blockWasSuccessfullyStored = putBlockStatus.storageLevel.isValid
58.     if (blockWasSuccessfullyStored) {
59.       //现在块位于内存或磁盘存储中，通知 master
60.       info.size = size
61.       if (tellMaster && info.tellMaster) {
62.         reportBlockStatus(blockId, putBlockStatus)
63.       }
64.       addUpdatedBlockStatusToTaskMetrics(blockId, putBlockStatus)
65.       logDebug("Put block %s locally took %s".format(blockId, Utils
          .getUsedTimeMs(startTimeMs)))
66.       if (level.replication > 1) {
67.         val remoteStartTime = System.currentTimeMillis
68.         val bytesToReplicate = doGetLocalBytes(blockId, info)
69.         //[SPARK-16550] 使用默认的序列化时擦除 classTag 类型，当反序列化类时
            //NettyBlockRpcServer 崩溃。待办事项（EKL）删除远程节点类装载器的问题
            //已经修复
              val remoteClassTag = if (!serializerManager.canUseKryo(classTag)) {
70.           scala.reflect.classTag[Any]
71.         } else {
72.           classTag
73.         }
74.         try {
75.           replicate(blockId, bytesToReplicate, level, remoteClassTag)
76.         } finally {
77.           bytesToReplicate.unmap()
78.         }
79.         logDebug("Put block %s remotely took %s"
80.           .format(blockId, Utils.getUsedTimeMs(remoteStartTime)))
81.       }
82.     }
```

```
83.      assert(blockWasSuccessfullyStored == iteratorFromFailedMemoryStorePut
         .isEmpty)
84.       iteratorFromFailedMemoryStorePut
85.     }
86.   }
```

9.2　Spark 中 checkpoint 原理和源码详解

本节对 Spark 中 checkpoint 原理及 Spark 中 checkpoint 源码进行详解。

9.2.1　Spark 中 checkpoint 原理详解

checkpoint 到底是什么？

（1）Spark 在生产环境下经常会面临 Tranformations 的 RDD 非常多（例如，一个 Job 中包含 10 000 个 RDD）或者具体 Tranformation 产生的 RDD 本身计算特别复杂和耗时（例如，计算时常超过 1h），此时我们必须考虑对计算结果数据的持久化。

（2）Spark 擅长多步骤迭代，同时擅长基于 Job 的复用，这时如果能够对曾经计算的过程产生的数据进行复用，就可以极大地提升效率。

（3）如果采用 persist 把数据放在内存中，虽然是最快速的，但是也是最不可靠的。如果放在磁盘上，也不是完全可靠的。例如，磁盘会损坏，管理员可能清空磁盘等。

（4）checkpoint 的产生就是为了相对更加可靠地持久化数据，checkpoint 可以指定把数据放在本地并且是多副本的方式，但是在正常的生产情况下是放在 HDFS，这就自然地借助 HDFS 高容错、高可靠的特征完成了最大化的、可靠的持久化数据的方式。

（5）为确保 RDD 复用计算的可靠性，checkpoint 把数据持久化到 HDFS 中，保证数据最大程度的安全性。

（6）checkpoint 就是针对整个 RDD 计算链条中特别需要数据持久化的环节（后面会反复使用当前环节的RDD）开始基于HDFS等的数据持久化复用策略，通过对RDD启动checkpoint机制来实现容错和高可用。

9.2.2　Spark 中 checkpoint 源码详解

1．checkpoint的运行原理和源码实现彻底详解

RDD 进行计算前须先看一下是否有 checkpoint，如果有 checkpoint，就不需要再进行计算了。RDD.scala 的 iterator 方法的源码如下：

```
1.       final def iterator(split: Partition, context: TaskContext):
         Iterator[T] = {
2.     if (storageLevel != StorageLevel.NONE) {
3.       getOrCompute(split, context)
4.     } else {
5.       computeOrReadCheckpoint(split, context)
6.     }
7.   }
```

进入 RDD.scala 的 getOrCompute 方法，源码如下：

```
1.  private[spark] def getOrCompute(partition: Partition, context: TaskContext):
    Iterator[T] = {
2.      val blockId = RDDBlockId(id, partition.index)
3.      var readCachedBlock = true
4.      //这种方法被 Executors 调用，所以我们需要调用 SparkEnv.get 代替 sc.env
5.      SparkEnv.get.blockManager.getOrElseUpdate(blockId,    storageLevel,
        elementClassTag, () => {
6.        readCachedBlock = false
7.        computeOrReadCheckpoint(partition, context)
8.      }) match {
```

getOrCompute 方法的 getOrElseUpdate 方法传入的第四个参数是匿名函数，调用 computeOrReadCheckpoint(partition, context)检查 checkpoint 中是否有数据。

RDD.scala 的 computeOrReadCheckpoint 的源码如下：

```
1.      private[spark]  def  computeOrReadCheckpoint(split:  Partition,
        context: TaskContext): Iterator[T] =
2.    {
3.      if (isCheckpointedAndMaterialized) {
4.        firstParent[T].iterator(split, context)
5.      } else {
6.        compute(split, context)
7.      }
8.    }
```

computeOrReadCheckpoint 方法中的 isCheckpointedAndMaterialized 是一个布尔值，判断这个 RDD 是否 checkpointed 和被物化，Spark 2.0 checkpoint 中有两种方式：reliably 或者 locally。computeOrReadCheckpoint 作为 isCheckpointed 语义的别名返回值。

isCheckpointedAndMaterialized 方法的源码如下：

```
1.     private[spark] def isCheckpointedAndMaterialized: Boolean =
2.    checkpointData.exists(_.isCheckpointed)
```

回到 RDD.scala 的 computeOrReadCheckpoint，如果已经持久化及物化 isCheckpointedAndMaterialized，就调用 firstParent[T]的 iterator。如果没有持久化，则进行 compute。

2．checkpoint原理机制

（1）通过调用 SparkContext.setCheckpointDir 方法指定进行 checkpoint 操作的 RDD 把数据放在哪里，在生产集群中是放在 HDFS 上的，同时为了提高效率，在进行 checkpoint 的使用时，可以指定很多目录。

SparkContext 为即将计算的 RDD 设置 checkpoint 保存的目录。如果在集群中运行，必须是 HDFS 的目录路径。

SparkContext.scala 的 setCheckpointDir 的源码如下：

```
1.    def setCheckpointDir(directory: String) {
2.
3.  /**
    *如果在集群上运行，如目录是本地的，则记录一个警告。否则，driver 可能会试图从它自己
    *的本地文件系统重建 RDD 的 checkpoint 检测点，因为 checkpoint 检查点文件不正确。
    *实际上是在 Executor 机器上
```

```
    */
4.    if (!isLocal && Utils.nonLocalPaths(directory).isEmpty) {
5.      logWarning("Spark is not running in local mode, therefore the
        checkpoint directory " +
6.        s"must not be on the local filesystem. Directory '$directory' " +
7.        "appears to be on the local filesystem.")
8.    }
9.
10.   checkpointDir = Option(directory).map { dir =>
11.     val path = new Path(dir, UUID.randomUUID().toString)
12.     val fs = path.getFileSystem(hadoopConfiguration)
13.     fs.mkdirs(path)
14.     fs.getFileStatus(path).getPath.toString
15.   }
16. }
```

RDD.scala 的 checkpoint 方法标记 RDD 的检查点 checkpoint。它将保存到 SparkContext#setCheckpointDir 的目录检查点内的文件中，所有引用它的父 RDDs 将被移除。须在任何作业之前调用此函数。建议 RDD 在内存中缓存，否则保存在文件中时需要重新计算。

RDD.scala 的 checkpoint 的源码如下：

```
1.  def checkpoint(): Unit = RDDCheckpointData.synchronized {
2.      //注意：我们在这里使用全局锁，原因是下游的复杂性：子 RDD 分区指向正确的父分区。未
        //来我们应该重新考虑这个问题
3.      if (context.checkpointDir.isEmpty) {
4.        throw new SparkException("Checkpoint directory has not been set in
          the SparkContext")
5.      } else if (checkpointData.isEmpty) {
6.        checkpointData = Some(new ReliableRDDCheckpointData(this))
7.      }
8.    }
```

其中的 checkpointData 是 RDDCheckpointData。

```
1.      private[spark] var checkpointData: Option[RDDCheckpointData[T]] = None
```

RDDCheckpointData 标识某个 RDD 要进行 checkpoint。如果某个 RDD 要进行 checkpoint，那在 Spark 框架内部就会生成 RDDCheckpointData。

```
1.   private[spark] abstract class RDDCheckpointData[T: ClassTag](@transient
     private val rdd: RDD[T])
2.    extends Serializable {
3.
4.    import CheckpointState._
5.
6.    //相关的 RDD 检查状态
7.    protected var cpState = Initialized
8.
9.    //RDD 包含检查点数据
10.   private var cpRDD: Option[CheckpointRDD[T]] = None
11.
12.   //待办事宜：确定需要在下面的方法中使用全局锁吗
13.
14.   /**
15.     *返回 RDD 的 checkpoint 数据是否已经持久化
16.     */
17.   def isCheckpointed: Boolean = RDDCheckpointData.synchronized {
```

```
18.     cpState == Checkpointed
19.   }
20.
21.   /**
22.     *物化 RDD 和持久化其内容
23.     *RDD 的第一个行动完成以后立即触发调用
24.     */
25.   final def checkpoint(): Unit = {
26.    //防止多个线程同时对相同 RDDCheckpointing，这 RDDCheckpointData 状态自动翻转
27.     RDDCheckpointData.synchronized {
28.       if (cpState == Initialized) {
29.         cpState = CheckpointingInProgress
30.       } else {
31.         return
32.       }
33.     }
34.
35.     val newRDD = doCheckpoint()
36.
37.     //更新我们的状态和截断 RDD 的血统
38.     RDDCheckpointData.synchronized {
39.       cpRDD = Some(newRDD)
40.       cpState = Checkpointed
41.       rdd.markCheckpointed()
42.     }
43.   }
44.
45.   /**
46.     *物化 RDD 和持久化其内容
47.     *
48.     *子类应重写此方法，以定义自定义检查点行为
49.     * @return the Checkpoint RDD 在进程中创建
50.     */
51.   protected def doCheckpoint(): CheckpointRDD[T]

52.   /**
      *返回包含我们的检查点数据。如果 checkpoint 的状态是 Checkpointed，才定义
      */
53.
54.   def checkpointRDD: Option[CheckpointRDD[T]] = RDDCheckpointData
      .synchronized { cpRDD }
55.   /**
      *返回 checkpoint RDD 的分区，仅用于测试
      */
56.
57.   def getPartitions: Array[Partition] = RDDCheckpointData.synchronized {
58.     cpRDD.map(_.partitions).getOrElse { Array.empty }
59.   }
60.
61. }

62. /**
     *同步检查点操作的全局锁
     */
63.
64. private[spark] object RDDCheckpointData
```

（2）在进行RDD的checkpoint的时候，其所依赖的所有的RDD都会从计算链条中清空掉。

（3）作为最佳实践，一般在进行checkpoint方法调用前都要进行persist把当前RDD的数据持久化到内存或者磁盘上，这是因为checkpoint是Lazy级别，必须有Job的执行，且在Job执行完成后，才会从后往前回溯哪个RDD进行了checkpoint标记，然后对标记过的RDD新启动一个Job执行具体的checkpoint过程。

（4）checkpoint改变了RDD的Lineage。

（5）当调用checkpoint方法要对RDD进行checkpoint操作，此时框架会自动生成RDDCheckpointData，当RDD上运行过一个Job后，就会立即触发RDDCheckpointData中的checkpoint方法，在其内部会调用doCheckpoint，实际上在生产时会调用ReliableRDDCheckpointData的doCheckpoint，在生产过程中会导致ReliableCheckpointRDD的writeRDDToCheckpointDirectory的调用，而在writeRDDToCheckpointDirectory方法内部，会触发runJob来执行把当前的RDD中的数据写到checkpoint的目录中，同时会产生ReliableCheckpointRDD实例。

RDDCheckpointData.scala的checkpoint方法进行真正的checkpoint：在RDDCheckpointData.synchronized同步块中先判断cpState的状态，然后调用doCheckpoint()。

RDDCheckpointData.scala的checkpoint方法的源码如下：

```
1.     final def checkpoint(): Unit = {
2.    //防止多个线程同时对相同RDDcheckpointing，这RDDCheckpointData状态自动翻转
3.  RDDCheckpointData.synchronized {
4.       if (cpState == Initialized) {
5.         cpState = CheckpointingInProgress
6.       } else {
7.         return
8.       }
9.     }
10.
11.     val newRDD = doCheckpoint()
12.
13.     //更新我们的状态和截断RDD的血统
14.     RDDCheckpointData.synchronized {
15.       cpRDD = Some(newRDD)
16.       cpState = Checkpointed
17.       rdd.markCheckpointed()
18.     }
19.   }
```

其中的doCheckpoint方法是RDDCheckpointData.scala中的方法，这里没有具体的实现。

```
1.      protected def doCheckpoint(): CheckpointRDD[T]
```

RDDCheckpointData的子类包括LocalRDDCheckpointData、ReliableRDDCheckpointData。ReliableRDDCheckpointData子类中doCheckpoint方法具体的实现，在方法中进行writeRDDToCheckpointDirectory的调用。

ReliableRDDCheckpointData.scala的doCheckpoint的源码如下：

```
1.     protected override def doCheckpoint(): CheckpointRDD[T] = {
2.     val newRDD = ReliableCheckpointRDD.writeRDDToCheckpointDirectory(rdd,
       cpDir)
```

```
3.
4.      //如果引用超出范围，则可选地清理检查点文件
5.      if (rdd.conf.getBoolean("spark.cleaner.referenceTracking
        .cleanCheckpoints", false)) {
6.        rdd.context.cleaner.foreach { cleaner =>
7.          cleaner.registerRDDCheckpointDataForCleanup(newRDD, rdd.id)
8.        }
9.      }
10.
11.     logInfo(s"Done checkpointing RDD ${rdd.id} to $cpDir, new parent is
        RDD ${newRDD.id}")
12.     newRDD
13.   }
14.
15. }
```

writeRDDToCheckpointDirectory 将 RDD 的数据写入到 checkpoint 的文件中，返回一个 ReliableCheckpointRDD。

- 首先找到 sparkContext，赋值给 sc 变量。
- 基于 checkpointDir 创建 checkpointDirPath。
- fs 获取文件系统的内容。
- 然后是广播 sc.broadcast，将路径信息广播给所有的 Executor。
- 接下来是 sc.runJob，触发 runJob 执行，把当前的 RDD 中的数据写到 checkpoint 的目录中。
- 最后返回 ReliableCheckpointRDD。无论是对哪个 RDD 进行 checkpoint，最终都会产生 ReliableCheckpointRDD，以 checkpointDirPath.toString 中的数据为数据来源；以 originalRDD.partitioner 的分区器 partitioner 作为 partitioner；这里的 originalRDD 就是要进行 checkpoint 的 RDD。

writeRDDToCheckpointDirectory 的源码如下：

```
1.    def writeRDDToCheckpointDirectory[T: ClassTag](
2.        originalRDD: RDD[T],
3.        checkpointDir: String,
4.        blockSize: Int = -1): ReliableCheckpointRDD[T] = {
5.      val checkpointStartTimeNs = System.nanoTime()
6.      val sc = originalRDD.sparkContext
7.
8.      //为检查点创建输出路径
9.      val checkpointDirPath = new Path(checkpointDir)
10.     val fs = checkpointDirPath.getFileSystem(sc.hadoopConfiguration)
11.     if (!fs.mkdirs(checkpointDirPath)) {
12.       throw new SparkException(s"Failed to create checkpoint path
          $checkpointDirPath")
13.     }
14.
15.     //保存文件，并重新加载它作为一个 RDD
16.     val broadcastedConf = sc.broadcast(
17.       new SerializableConfiguration(sc.hadoopConfiguration))
18.     //待办事项：这是代价昂贵的，因为它又一次计算 RDD 是不必要的（SPARK-8582）
        sc.runJob(originalRDD,
19.       writePartitionToCheckpointFile[T](checkpointDirPath.toString,
          broadcastedConf) _)
20.
```

```
21.     if (originalRDD.partitioner.nonEmpty) {
22.       writePartitionerToCheckpointDir(sc,  originalRDD.partitioner.get,
          checkpointDirPath)
23.     }
24.
25.     val checkpointDurationMs =
26.       TimeUnit.NANOSECONDS.toMillis(System.nanoTime() - checkpointStartTimeNs)
27.         logInfo(s"Checkpointing took $checkpointDurationMs ms.")
28.     val newRDD = new ReliableCheckpointRDD[T](
29.       sc, checkpointDirPath.toString, originalRDD.partitioner)
30.     if (newRDD.partitions.length != originalRDD.partitions.length) {
31.       throw new SparkException(
32.         s"Checkpoint RDD $newRDD(${newRDD.partitions.length}) has
            different " +
33.           s"number of partitions from original RDD $originalRDD
              (${originalRDD.partitions.length})")
34.     }
35.     newRDD
36.   }
```

ReliableCheckpointRDD 是读取以前写入可靠存储系统检查点文件数据的 RDD。其中的 partitioner 是构建 ReliableCheckpointRDD 的时候传进来的。其中的 getPartitions 是构建一个一个的分片。其中，getPreferredLocations 获取数据本地性，fs.getFileBlockLocations 获取文件的位置信息。compute 方法通过 ReliableCheckpointRDD.readCheckpointFile 读取数据。

ReliableCheckpointRDD.scala 的源码如下：

```
1.    private[spark] class ReliableCheckpointRDD[T: ClassTag](
2.      sc: SparkContext,
3.      val checkpointPath: String,
4.      _partitioner: Option[Partitioner] = None
5.    ) extends CheckpointRDD[T](sc) {
6.
7.    @transient private val hadoopConf = sc.hadoopConfiguration
8.    @transient private val cpath = new Path(checkpointPath)
9.    @transient private val fs = cpath.getFileSystem(hadoopConf)
10.   private val broadcastedConf = sc.broadcast(new SerializableConfiguration
      (hadoopConf))
11.   //如果检查点目录不存在，则快速失败
12.   require(fs.exists(cpath), s"Checkpoint  directory  does  not  exist:
      $checkpointPath")
13.   /**
        *返回 checkpoint 的路径，RDD 从中读取数据
        */
14.
15.   override val getCheckpointFile: Option[String] = Some(checkpointPath)
16.   override val partitioner: Option[Partitioner] = {
17.     _partitioner.orElse {
18.       ReliableCheckpointRDD.readCheckpointedPartitionerFile(context,
          checkpointPath)
19.     }
20.   }
21.  /**
       *返回检查点目录中的文件所描述的分区
       *由于原来的 RDD 可能属于一个之前的应用，没办法知道之前的分区数。此方法假定在应用
       *生命周期，原始集检查点文件完全保存在可靠的存储里面
       */
```

```
22.
23.   protected override def getPartitions: Array[Partition] = {
24.     //如果路径不存在，listStatus 就抛出异常
25.     val inputFiles = fs.listStatus(cpath)
26.       .map(_.getPath)
27.       .filter(_.getName.startsWith("part-"))
28.       .sortBy(_.getName.stripPrefix("part-").toInt)
29.     //如果输入文件无效，则快速失败
30.     inputFiles.zipWithIndex.foreach { case (path, i) =>
31.       if (path.getName != ReliableCheckpointRDD.checkpointFileName(i)) {
32.         throw new SparkException(s"Invalid checkpoint file: $path")
33.       }
34.     }
35.     Array.tabulate(inputFiles.length)(i => new CheckpointRDDPartition(i))
36.   }
37.  /**
      *返回与给定分区关联的检查点文件的位置
38.    */
39.   protected override def getPreferredLocations(split: Partition):
      Seq[String] = {
40.     val status = fs.getFileStatus(
41.       new Path(checkpointPath, ReliableCheckpointRDD.checkpointFileName
         (split.index)))
42.     val locations = fs.getFileBlockLocations(status, 0, status.getLen)
43.     locations.headOption.toList.flatMap(_.getHosts).filter(_ != "localhost")
44.   }
45.
46.   /**
      *读取与给定分区关联的检查点文件的内容
47.    */
48.   override def compute(split: Partition, context: TaskContext): I
      terator[T] = {
49.     val file = new Path(checkpointPath, ReliableCheckpointRDD
        .checkpointFileName(split.index))
50.     ReliableCheckpointRDD.readCheckpointFile(file, broadcastedConf, context)
51.   }
52.
53. }
54. ......
```

下面看一下 ReliableCheckpointRDD.scala 中 compute 方法中的 ReliableCheckpointRDD.readCheckpointFile。readCheckpointFile 读取指定检查点文件 checkpoint 的内容。readCheckpointFile 方法通过 deserializeStream 反序列化 fileInputStream 文件输入流，然后将 deserializeStream 变成一个 Iterator。

Spark 2.2.1 版本的 ReliableCheckpointRDD.scala 的 readCheckpointFile 的源码如下：

```
1.  def readCheckpointFile[T](
2.        path: Path,
3.        broadcastedConf: Broadcast[SerializableConfiguration],
4.        context: TaskContext): Iterator[T] = {
5.      val env = SparkEnv.get
6.      val fs = path.getFileSystem(broadcastedConf.value.value)
7.      val bufferSize = env.conf.getInt("spark.buffer.size", 65536)
8.      val fileInputStream = {
9.        val fileStream = fs.open(path, bufferSize)
10.       if (env.conf.get(CHECKPOINT_COMPRESS)) {
11.
```

```
   CompressionCodec.createCodec(env.conf).compressedInputStream(fileStream)
12.       } else {
13.         fileStream
14.       }
15.     }
16.     val serializer = env.serializer.newInstance()
17.     val deserializeStream = serializer.deserializeStream(fileInputStream)
18.
19.     //注册一个任务完成回调，关闭输入流
20.     context.addTaskCompletionListener(context => deserializeStream.close())
21.
22.     deserializeStream.asIterator.asInstanceOf[Iterator[T]]
23.   }
```

Spark 2.4.3 版本的 ReliableCheckpointRDD.scala 源码与 Spark 2.2.1 版本相比具有如下特点。

- 上段代码中第 20 行新增一个返回类型 Unit。

```
1.  ......
2.      context.addTaskCompletionListener[Unit](context =>
        deserializeStream.close())
```

ReliableRDDCheckpointData.scala 的 cleanCheckpoint 方法，清理 RDD 数据相关的 checkpoint 文件。

```
1.      def cleanCheckpoint(sc: SparkContext, rddId: Int): Unit = {
2.       checkpointPath(sc, rddId).foreach { path =>
3.         path.getFileSystem(sc.hadoopConfiguration).delete(path, true)
4.       }
5.     }
```

在生产环境中不使用 LocalCheckpointRDD。LocalCheckpointRDD 的 getPartitions 直接从 toArray 级别中调用 new()函数创建 CheckpointRDDPartition。LocalCheckpointRDD 的 compute 方法直接报异常。

LocalCheckpointRDD 的源码如下：

```
1.    private[spark] class LocalCheckpointRDD[T: ClassTag](
2.      sc: SparkContext,
3.      rddId: Int,
4.      numPartitions: Int)
5.    extends CheckpointRDD[T](sc) {
6.  ......
7.  protected override def getPartitions: Array[Partition] = {
8.      (0 until numPartitions).toArray.map { i => new CheckpointRDDPartition(i) }
9.    }
10. ......
11.   override def compute(partition: Partition, context: TaskContext):
      Iterator[T] = {
12.     throw new SparkException(
13.       s"Checkpoint block ${RDDBlockId(rddId, partition.index)} not found!
          Either the executor " +
14.       s"that originally checkpointed this partition is no longer alive,
          or the original RDD is " +
15.       s"unpersisted. If this problem persists, you may consider using
          'rdd.checkpoint()' " +
16.       s"instead, which is slower than local checkpointing but more fault-
          tolerant.")
17.   }
```

```
18.
19. }
```

checkpoint运行流程图如图9-2所示。

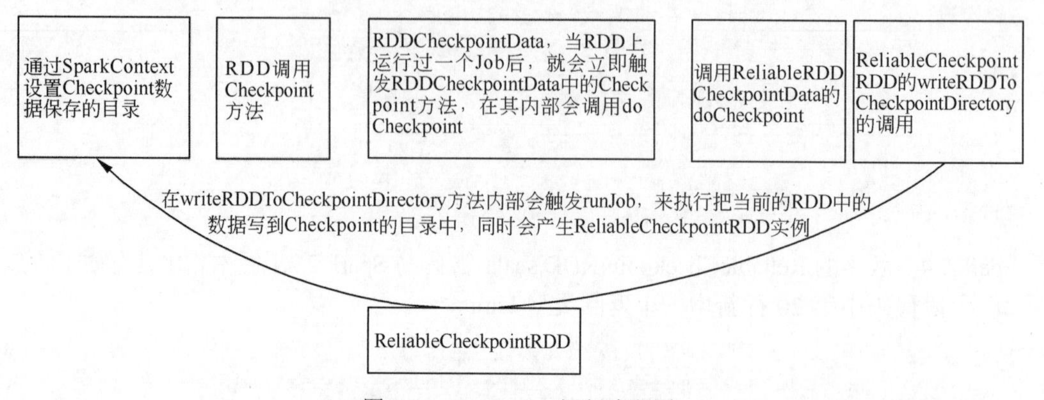

图9-2　Checkpoint运行流程图

通过SparkContext设置Checkpoint数据保存的目录，RDD调用checkpoint方法，生产RDDCheckpointData，当RDD上运行一个Job后，就会立即触发RDDCheckpointData中的checkpoint方法，在其内部会调用doCheckpoint；然后调用ReliableRDDCheckpointData的doCheckpoint；ReliableCheckpointRDD的writeRDDToCheckpointDirectory的调用；在writeRDDToCheckpointDirectory方法内部会触发runJob，来执行把当前的RDD中的数据写到Checkpoint的目录中，同时会产生ReliableCheckpointRDD实例。

checkpoint保存在HDFS中，具有多个副本；persist保存在内存中或者磁盘中。在Job作业调度的时候，checkpoint沿着finalRDD的“血统”关系lineage从后往前回溯向上查找，查找哪些RDD曾标记为要进行checkpoint，标记为checkpointInProgress；一旦进行checkpoint，RDD所有父RDD就被清空。

第 10 章　Spark 中 Broadcast 和 Accumulator 原理和源码详解

本章讲解 Spark 中 Broadcast 和 Accumulator 原理和源码。10.1 节中讲解 Spark 中 Broadcast 原理和源码，Broadcast 将数据从一个节点发送到其他节点上，一般用于处理共享配置文件、通用的 Dataset、常用的数据结构等；10.2 节对 Spark 中 Accumulator 原理和源码进行详解。Accumulator 是分布式全局只写的数据结构，用于数据的累加。

10.1　Spark 中 Broadcast 原理和源码详解

本节讲解 Spark 中 Broadcast 原理及 Spark 中 Broadcast 源码。

10.1.1　Spark 中 Broadcast 原理详解

Broadcast 在机器学习、图计算、构建日常的各种算法中到处可见。Broadcast 将数据从一个节点发送到其他节点上；例如，Driver 上有一张表，而 Executor 中的每个并行执行的 Task（100 万个 Task）都要查询这张表，那我们通过 Broadcast 的方式只需要往每个 Executor 发送一次这张表就行了，Executor 中的每个运行的 Task 查询这张唯一的表，而不是每次执行的时候都从 Driver 获得这张表。

Java 中的 Servlet 里有一个 ServletContext，是 JSP 或 Java 代码运行时的上下文，通过上下文可以获取各种资源。Broadcast 类似于 ServletContext 中的资源、变量或数据，Broadcast 广播出去是基于 Executor 的，里面的每个任务可以用上下文，Task 的上下文就是 Executor，可以抓取数据。这就好像 ServletContext 的具体作用，只是 Broadcast 是分布式的共享数据，默认情况下，只要程序在运行，Broadcast 变量就会存在，因为 Broadcast 在底层是通过 BlockManager 管理的。但是，你可以手动指定或者配置具体周期来销毁 Broadcast 变量。可以指定 Broadcast 的 unpersist 销毁 Broadcast 变量，因为 Spark 应用程序中可能运行很多 Job，可能一个 Job 需要很多 Broadcast 变量，但下一个 Job 不需要这些变量，但是应用程序还存在，因此需手工销毁 Broadcast 变量。

Broadcast 一般用于处理共享配置文件、通用的 Dataset、常用的数据结构等；但是在 Broadcast 中不适合存放太大的数据，Broadcast 不会内存溢出，因为其数据的保存的 StorageLevel 是 MEMORY_AND_DISK 的方式；虽然如此，我们也不可以放入太大的数据在 Broadcast 中，因为网络 I/O 和可能的单点压力会非常大（Spark 1.6 版本 Broadcast 有两种方式：HttpBroadcast、TorrentBroadcast。HttpBroadcast 可能有单点压力；TorrentBroadcast 下载没有单点压力，但可能有网络压力）。Spark 2.0 版本中已经去掉 HTTPBroadcast (SPARK-12588)了，Spark 2.0 版本的 TorrentBroadcast 是 Broadcast 唯一的广播实现方式。

广播 Broadcast 变量是只读变量，如果 Broadcast 不是只读变量而可以更新，那带来的问题是：①一个节点上 Broadcast 可以更新，其他的节点 Broadcast 也要更新；②如果多个节点 Broadcast 同时更新，如何确定更新的顺序，以及容错等内容。因此，广播 Broadcast 变量是只读变量，最为轻松保持了数据的一致性。

Broadcast 广播变量是只读变量，缓存在每个节点上，而不是每个 Task 去获取它的一份复制副本。例如，以高效的方式给每个节点发送一个 dataset 的副本。Spark 尝试在分布式发送广播变量时使用高效的广播算法减少通信的成本。

广播变量是由一个变量 V 通过调用[org.apache.spark.SparkContext#broadcast]创建的。广播变量是一个围绕 V 的包装器，它的值可以通过调用 value 方法来获取。例如：

```
1.    scala> val broadcastVar = sc.broadcast(Array(1, 2, 3))
2.    broadcastVar:org.apache.spark.broadcast.Broadcast[Array[Int]]=Broadcast(0)
3.
4.    scala> broadcastVar.value
5.    res0: Array[Int] = Array(1, 2, 3)
```

如果要更新广播变量，只有再广播一次，那就是一个新的广播变量，使用一个新的广播变量 ID。

广播变量创建后，在群集上运行时，V 变量不是在任何函数都使用，以便 V 传送到节点时不止一次。此外，对象 V 不应该被修改，是为了确保广播所有节点得到相同的广播变量值（例如，如果变量被发送到后来的一个新节点）。

Broadcast 的源码如下：

```
1.   @param id  广播变量的唯一标识符。
2.   @tparam T   广播变量的数据类型。
3.   abstract class Broadcast[T: ClassTag](val id: Long) extends Serializable
     with Logging {
4.
5.    @volatile private var _isValid = true
6.
7.    private var _destroySite = ""
8.
9.    /**获得广播值.*/
10.   def value: T = {
11.     assertValid()
12.     getValue()
13.   }
14. ……
```

Spark 1.6 版本的 HttpBroadcast 方式的 Broadcast，最开始的时候数据放在 Driver 的本地文件系统中，Driver 在本地会创建一个文件夹来存放 Broadcast 中的 data，然后启动 HttpServer 访问文件夹中的数据，同时写入到 BlockManager（StorageLevel 是 MEMORY_AND_DISK）中获得 BlockId(BroadcastBlockId)，当第一次 Executor 中的 Task 要访问 Broadcast 变量的时候，会向 Driver 通过 HttpServer 来访问数据，然后会在 Executor 中的 BlockManager 中注册该 Broadcast 中的数据 BlockManager，Task 访问 Broadcast 变量时，首先查询 BlockManager，如果 BlockManager 中已有此数据，Task 就可直接使用 BlockManager 中的数据（说明 SPARK-12588，HTTPBroadcast 方式在 Spark 2.0 版本中已经去掉）。

10.1.2　Spark 中 Broadcast 源码详解

BroadcastManager 是用来管理 Broadcast 的，该实例对象是在 SparkContext 创建 SparkEnv 的时候创建的。

SparkEnv.scala 的源码如下：

```
1.     val broadcastManager = new BroadcastManager(isDriver, conf, securityManager)
2.
3.     val mapOutputTracker = if (isDriver) {
4.       new MapOutputTrackerMaster(conf, broadcastManager, isLocal)
5.     } else {
6.       new MapOutputTrackerWorker(conf)
7.     }
```

BroadcastManager.scala 中 BroadcastManager 实例化的时候会调用 initialize 方法，initialize 方法就创建 TorrentBroadcastFactory 的方式。

BroadcastManager 的源码如下：

```
1.
2.  private[spark] class BroadcastManager(
3.      val isDriver: Boolean,
4.      conf: SparkConf,
5.      securityManager: SecurityManager)
6.    extends Logging {
7.
8.    private var initialized = false
9.    private var broadcastFactory: BroadcastFactory = null
10.
11.   initialize()
12.
13.   //使用广播前，被 SparkContext 或 Executor 调用
14.   private def initialize() {
15.     synchronized {
16.       if (!initialized) {
17.         broadcastFactory = new TorrentBroadcastFactory
18.         broadcastFactory.initialize(isDriver, conf, securityManager)
19.         initialized = true
20.       }
21.     }
22.   }
```

Spark 2.0 版本中的 TorrentBroadcast 方式：数据开始在 Driver 中，A 节点如果使用了数据，A 就成为供应源，这时 Driver 节点、A 节点两个节点成为供应源，如第三个节点 B 访问的时候，第三个节点 B 也成了供应源，同样地，第四个节点、第五个节点……都成了供应源，这些都被 BlockManager 管理，这样不会导致一个节点压力太大，从理论上讲，数据使用的节点越多，网络速度就越快。

TorrentBroadcast 按照 BLOCK_SIZE（默认是 4MB）将 Broadcast 中的数据划分成为不同的 Block，然后将分块信息（也就是 Meta 信息）存放到 Driver 的 BlockManager 中，同时会告诉 BlockManagerMaster，说明 Meta 信息存放完毕。

SparkContext.scala 的 broadcast 方法的源码如下：

```
1.        def broadcast[T: ClassTag](value: T): Broadcast[T] = {
2.      assertNotStopped()
3.      require(!classOf[RDD[_]].isAssignableFrom(classTag[T].runtimeClass),
4.        "Can not directly broadcast RDDs; instead, call collect() and
         broadcast the result.")
5.      val bc = env.broadcastManager.newBroadcast[T](value, isLocal)
6.      val callSite = getCallSite
7.      logInfo("Created broadcast " + bc.id + " from " + callSite.shortForm)
8.      cleaner.foreach(_.registerBroadcastForCleanup(bc))
9.      bc
10.   }
```

SparkContext.scala 的 broadcast 方法中调用 env.broadcastManager.newBroadcast。BroadcastManager.scala 的 newBroadcast 方法如下：

```
1.   def newBroadcast[T: ClassTag](value_ : T, isLocal: Boolean): Broadcast[T]
      = {
2.      broadcastFactory.newBroadcast[T](value_, isLocal, nextBroadcastId
        .getAndIncrement())
3.    }
```

newBroadcast 方法调用 new()函数创建一个 Broadcast，第一个参数是 Value，第三个参数是 BroadcastId。这里，BroadcastFactory 是一个 trait，没有具体的实现。

```
1.    private[spark] trait BroadcastFactory {
2.   ......
3.   def newBroadcast[T: ClassTag](value: T, isLocal: Boolean, id: Long):
     Broadcast[T]
4.   ......
```

TorrentBroadcastFactory 是 BroadcastFactory 的具体实现。

```
1.  private[spark] class TorrentBroadcastFactory extends BroadcastFactory {
2.  ......
3.  override def newBroadcast[T: ClassTag](value_ : T, isLocal: Boolean, id:
    Long): Broadcast[T] = {
4.      new TorrentBroadcast[T](value_, id)
5.    }
```

BroadcastFactory 的 newBroadcast 方法创建 TorrentBroadcast 实例。

TorrentBroadcast.scala 的源码如下：

```
1.   private[spark] class TorrentBroadcast[T: ClassTag](obj: T, id: Long)
2.     extends Broadcast[T](id) with Logging with Serializable {
3.   ......
4.   private def readBlocks(): Array[BlockData] = {
5.      //获取数据块。注意，所有这些块存储在 BlockManager 且向 driver 汇报，其他
6.      //Executors 也可以从这个 Executors 中提取这些块
7.      val blocks = new Array[BlockData](numBlocks)
8.      val bm = SparkEnv.get.blockManager
9.
10.     for (pid <- Random.shuffle(Seq.range(0, numBlocks))) {
11.       val pieceId = BroadcastBlockId(id, "piece" + pid)
12.       logDebug(s"Reading piece $pieceId of $broadcastId")
13.       //第一次尝试 getLocalBytes 从本地读取：因为以前试图获取广播块时已经获取了一些
14.       //块，在这种情况下，一些块将在本地（在 Executor 上）
15.       bm.getLocalBytes(pieceId) match {
16.         case Some(block) =>
```

```
17.            blocks(pid) = block
18.            releaseLock(pieceId)
19.          case None =>
20.            bm.getRemoteBytes(pieceId) match {
21.              case Some(b) =>
22.                if (checksumEnabled) {
23.                  val sum = calcChecksum(b.chunks(0))
24.                  if (sum != checksums(pid)) {
25.                    throw new SparkException(s"corrupt remote block $pieceId
                       of $broadcastId:" +
26.                      s" $sum != ${checksums(pid)}")
27.                  }
28.                }
29.                //从远程 Executors/driver 的 BlockManager 查找块，所以把块放在
30.              //Executor 的节点 BlockManager
31.                if (!bm.putBytes(pieceId, b, StorageLevel.MEMORY_AND_DISK_SER,
                   tellMaster = true)) {
32.                  throw new SparkException(
33.                    s"Failed to store $pieceId of $broadcastId in local BlockManager")
34.                }
35.                blocks(pid) = new ByteBufferBlockData(b, true)
36.              case None =>
37.                throw new SparkException(s"Failed to get $pieceId of $broadcastId")
38.            }
39.        }
40.      }
41.      blocks
42.    }
```

TorrentBroadcast.scala 的 readBlocks 方法中 Random.shuffle(Seq.range(0, numBlocks)进行随机洗牌，是因为数据有很多来源 DataServer，为了保持负载均衡，因此使用 shuffle。

TorrentBroadcast 将元数据信息存放到 BlockManager，然后汇报给 BlockManagerMaster。数据存放到 BlockManagerMaster 中就变成了全局数据，BlockManagerMaster 具有所有的信息，Driver、Executor 就可以访问这些内容。Executor 运行具体的 TASK 的时候，通过 TorrentBroadcast 的方式 readBlocks，如果本地有数据，就从本地读取，如果本地没有数据，就从远程读取数据。Executor 读取信息以后，通过 TorrentBroadcast 的机制通知 BlockManagerMaster 数据多了一份副本，下一个 Task 读取数据的时候，就有两个选择，分享的节点越多，下载的供应源就越多，最终变成点到点的方式。

Broadcast 可以广播 RDD，Join 操作性能优化之一也是采用 Broadcast。

10.2　Spark 中 Accumulator 原理和源码详解

本节讲解 Spark 中 Accumulator 原理及对 Spark 中 Accumulator 源码进行详解。

10.2.1　Spark 中 Accumulator 原理详解

Spark 的 Broadcast 和 Accumulator 很重要，在实际的企业级开发环境中一般会使用 Broadcast 和 Accumulator。Broadcast、Accumulator 和 RDD 是 Spark 中并列的三大基础数据结构。大家谈 Spark 的时候，首先谈 RDD。RDD 是一个并行的数据，关注在 JVM 中怎么处

理数据。很多时候可能忽略了 Broadcast 和 Accumulator，这两个变量都是全局级别的。例如，集群中有 1000 台机器，那 Broadcast 和 Accumulator 可以在 1000 台机器中共享。在分布式的基础上，如果有共享的数据结构，那是非常有用的。

分布式大数据系统中，进行编程的时候首先考虑数据结构。

- RDD：分布式私有数据结构。RDD 本身是一个并行化的、本地化的数据结构，运行时在一个个线程中运行，RDD 是私有的运行数据和私有的运行过程，但在一个 Stage 里面是一样的，一个线程一个时刻只处理一个数据分片，另一个线程一个时刻只处理另一个数据片。在设计业务逻辑的时候，我们通常考虑这个分片如何去处理。
- Broadcast：分布式全局只读数据结构。
- Accumulator：分布式全局只写的数据结构。我们不会在线程池中读取 Accumulator，但在 Driver 上可以读取 Accumulator。

在生产环境下，我们一定会自定义 Accumulator。

（1）自定义时可以让 Accumulator 非常复杂，基本上可以是任意类型的 Java 和 Scala 对象。

（2）自定义 Accumulator 时，可以实现一些“技术福利”。例如，在 Accumulator 变化的时候可以把数据同步到 MySQL 中。我们在进行流处理的时候，数据不断地流进来，如要查询用户点击量的趋势图，计算点击量以后须实时反馈到生产环境的 server 上。一个非常简单的实现方式是：每次发现累加的时候，就更新一下数据库，这是一个非常强大的同步机制和同步效果。

10.2.2　Spark 中 Accumulator 源码详解

Accumulator 是一个简单的 value 值[Accumulable]，相同类型的元素合并时结果可以累加，通过 added 到关联和交换操作，可以有效地支持并行，可以用来实现计数（如 MapReduce）或求和。Spark 原生支持数值类型的累加器，也可以自定义开发实现新类型的支持。

累加器由一个初始值 V 通过调用[SparkContext#accumulator SparkContext.accumulator]创建。在群集上运行的任务可以使用“+=”运算符写入，但是不能读取它的值。只有 Driver 程序使用[#value]方法可以读取累加器的值。例如：

```
1.    scala> val accum = sc.accumulator(0)
2.   accum: org.apache.spark.Accumulator[Int] = 0
3.   scala> sc.parallelize(Array(1, 2, 3, 4)).foreach(x => accum += x)
4.   ......
5.   10/09/29 18:41:08 INFO SparkContext: Tasks finished in 0.317106 s
6.
7.   scala> accum.value
8.   res2: Int = 10
```

Accumulator.scala 的源码如下：

```
1.    @deprecated("use AccumulatorV2", "2.0.0")
2.  class Accumulator[T] private[spark] (
3.      // SI-8813: 必须显式地定义 private val，否则 Scala 2.11 不编译
4.      @transient private val initialValue: T,
5.      param: AccumulatorParam[T],
6.      name: Option[String] = None,
7.      countFailedValues: Boolean = false)
```

```
8.    extends Accumulable[T, T](initialValue, param, name, countFailedValues)
9.  ……
```

Accumulator 是一个类，继承自 Accumulable。Accumulator 已经被标识为过时的（deprecated），在 Spark 2.0 版本中可以使用 AccumulatorV2。

```
1.   abstract class AccumulatorV2[IN, OUT] extends Serializable {
2.    private[spark] var metadata: AccumulatorMetadata = _
3.    private[this] var atDriverSide = true
4.  ……
```

可以通过继承创建自己的类型 AccumulatorV2。AccumulatorV2 抽象类有几种方法必须覆盖：reset 用于将累加器重置为零，add 用于将另一个值添加到累加器中，merge 用于将另一个相同类型的累加器合并到该累加器中。例如，假设有一个 MyVector 代表数学向量的类，代码如下：

```
1.   class VectorAccumulatorV2 extends AccumulatorV2[MyVector, MyVector] {
2.
3.    private val myVector: MyVector = MyVector.createZeroVector
4.
5.    def reset(): Unit = {
6.      myVector.reset()
7.    }
8.
9.    def add(v: MyVector): Unit = {
10.     myVector.add(v)
11.   }
12.   ……
13. }
14.
15. //创建一个这种类型的累加器
16. val myVectorAcc = new VectorAccumulatorV2
17. //然后，把它注册到 Spark 上下文中
18. sc.register(myVectorAcc, "MyVectorAcc1")
```

当自定义自己的 AccumulatorV2 类型时，生成的类型可能与添加的元素的类型不同。累加器更新仅在 Action 动作内执行，Spark 保证每个任务对累加器的更新只能应用一次，即重新启动的任务将不会更新该值。在 transformations 转换中，如果重新执行任务或作业阶段，则每个任务的更新可能会被多次执行。Accumulators 不会改变 Spark 的 Lazy 评估模型。如果它们在 RDD 的操作中更新，则只有在 RDD 作为操作的一部分进行计算时，才会更新其值。因此，累加器更新不能保证在 Lazy 变换中执行时执行 map()。

以下代码中，accum 仍然为 0，因为没有 action 算子触发 map 操作。

```
1.   val accum = sc.longAccumulator
2.  data.map { x => accum.add(x); x }
```

第 11 章　Spark 与大数据其他经典组件整合原理与实战

本章讲解 Spark 与大数据其他经典组件整合原理与实战。11.1 节中讲解 Spark 组件综合应用；11.2 节讲解 Spark 与 Alluxio 整合原理与实战；11.3 节讲解 Spark 与 Job Server 整合原理与实战；11.4 节通过生产环境实战案例讲解 Spark 与 Redis 整合原理与实战。

11.1　Spark 组件综合应用

Apache Spark 生态系统的外部软件项目，Spark 第三方项目组件的综合应用如下。

1. spark-packages.org

spark-packages.org 是一个外部社区管理的第三方库，是附加组件和与 Apache Spark 一起使用的应用程序的列表。只要有一个 GitHub 仓库，就可以添加一个包。

2. 基础项目

- Spark Job Server：用于在同一个群集上管理和提交 Spark 作业的 REST 接口。
- SparkR：Spark 的 R 前端。
- MLbase：Spark 机器学习研究项目。
- Apache Mesos：支持运行 Spark 的群集管理系统。
- Alluxio（Tachyon）：支持运行 Spark 的内存速度虚拟分布式存储系统。
- Spark Cassandra 连接器：轻松将 Cassandra 数据加载到 Spark 和 Spark SQL 中；来自 Datastax。
- FiloDB：一个 Spark 集成分析及列数据库，基于内存能够进行亚秒级并发查询。
- ElasticSearch：Spark SQL 集成。
- Spark-Scalding：轻松过渡 Cascading/Scalding 代码到 Spark。
- Zeppelin：类似于 IPython，还有 ISPark 和 Spark Notebook。
- IBM Spectrum Spark：集群管理软件与 Spark 集成。
- EclairJS：使 Node.js 开发人员可以对 Spark 进行编码，数据科学家可以在 Jupyter 中使用 Javascript。
- SnappyData：与同一个 JVM 集成的开源 OLTP + OLAP 数据库。
- GeoSpark：地理空间 RDD 和连接。
- Spark Cluster：部署 OpenStack 工具。

3．使用Spark的应用程序

- Apache Mahout：以前运行在 Hadoop MapReduce 上，Mahout 已经转向使用 Spark 作为后端。
- Apache MRQL：用于大规模，分布式数据分析的查询处理和优化系统，构建在 Apache Hadoop、Hama 和 Spark 上。
- BlinkDB：一个大规模并行的大致查询引擎，建立在 Shark 和 Spark 上。
- Spindle：基于 Spark / Parquet 的网络分析查询引擎。
- Spark Spatial：Spark 的空间连接和处理。
- Thunderain：是一个使用 Spark 和 Shark 的实时分析处理实例。
- DF from Ayasdi：类似 Pandas 的数据框架实现。
- Oryx：Apache Spark 上的 Oryx Lambda 架构，Apache Kafka 用于实时大规模机器学习。
- ADAM：使用 Apache Spark 加载，转换和分析基因组数据的框架和 CLI。

4．附加语言绑定

- C#/.NET：Spark 的 C# API 接口。
- Clojure：Spark 的 Clojure API 接口。
- Groovy：Groovy REPL 支持 Spark。

11.2　Spark 与 Alluxio 整合原理与实战

Alluxio 以前称为 Tachyon，是世界上第一个内存速度虚拟分布式存储系统。它统一数据访问、桥接计算框架和底层存储系统。应用程序只需要连接 Alluxio 来访问存储在任何底层存储系统中的数据。Alluxio 以内存为中心的架构使数据访问速度比现有解决方案更快。

本节讲解 Spark 与 Alluxio 整合原理及 Spark 与 Alluxio 整合实战。

11.2.1　Spark 与 Alluxio 整合原理

在大数据生态系统中，Alluxio 位于计算框架或作业 jobs 之间，如 Apache Spark、Apache MapReduce、Apache HBase、Apache Hive 或 Apache Flink，以及各种存储系统，如 Amazon S3、Google Cloud Storage、OpenStack Swift、GlusterFS、HDFS、MaprFS、Ceph、NFS 和 Alibaba OSS。Alluxio 为生态系统带来显著的性能改善。例如，百度使用 Alluxio 提升数据分析速度近 30 倍；Barclays（巴克莱）银行使用 Alluxio 把不可能变成了可能，从之前计算的小时级变成了秒级；Qunar（去哪儿网）在 Alluxio 之上进行实时数据分析。除了性能外，传统存储系统中的数据通过桥接存储在 Alluxio 中进行新的工作负载。用户可以使用其独立的集群模式运行 Alluxio。例如，在 Amazon EC2、Google Compute Engine 上，或者使用 Apache Mesos 或 Apache Yarn 启动 Alluxio。

Alluxio 兼容 Hadoop。现有的数据分析应用程序，如 Spark 和 MapReduce 程序，可以运行在 Alluxio 上，无须任何代码更改。Alluxio 项目是 Apache License 2.0 下的开源项目，部署在许多公司。它是增长速度最快的开源项目之一。Alluxio 拥有三年的开源历史，吸引了来自 150 多家机构的 600 多名参与者，包括阿里巴巴、Alluxio、百度、CMU、谷歌、IBM、英特尔、NJU、红帽、加州大学伯克利分校。Alluxio 项目是 Berkeley 数据分析堆栈（BDAS）的存储层，也是 Fedora 发行版的一部分。Alluxio 由 100 多个组织部署在生产中，并且运行在超过 1000 个节点的集群上。

Alluxio 功能如下。

- 灵活的文件 API： Alluxio 的原生 API 类似于 java.io.File 类，提供 InputStream 和 OutputStream 接口以及对内存映射 I/O 的高效支持。建议使用此 API 从 Alluxio 获得最佳性能。Alluxio 还提供了一个兼容 Hadoop 的 FileSystem 接口，允许 Hadoop MapReduce 和 Spark 使用 Alluxio 代替 HDFS。
- 提供容错能力的可插拔存储：Alluxio 将内存中的数据 checkpoints 到底层存储系统。Alluxio 具有通用接口，可以方便地插入不同的底层存储系统。Alluxio 目前支持 Amazon S3、Google Cloud Storage、OpenStack Swift、GlusterFS、HDFS、MaprFS、Ceph、NFS、Alibaba OSS 和单节点本地文件系统，并支持许多其他文件系统。
- 采用分层存储，除了内存外，Alluxio 还可以管理 SSD 和 HDD，允许将更大的数据集存储在 Alluxio 中。数据将自动在不同层之间进行管理，保持热数据。自定义策略插拔、引脚允许直接的用户控制。
- 统一命名空间： Alluxio 通过安装功能实现跨不同存储系统的有效数据管理。此外，透明命名可确保在将这些对象持久存储到底层存储系统时，保留在 Alluxio 中创建的对象的文件名和目录层次结构。
- Lineage 血统： Alluxio 可以实现高吞吐量写入，通过使用 Lineage 提供容错性，通过重新执行创建输出的作业恢复丢失的输出。使用 Lineage，应用程序将输出写入内存，Alluxio 会以异步方式定期检查输出到文件系统。如果出现故障，Alluxio 将启动重新计算，以恢复丢失的文件。
- Web UI 和命令行：用户可以通过 Web UI 轻松浏览文件系统。在调试模式下，管理员可以查看每个文件的详细信息，包括位置、检查点路径等。用户还可以使用./bin/alluxio fs 与 Alluxio 进行交互，例如，复制数据进出文件系统。

在 Alluxio 上运行 Apache Spark。HDFS 作为分布式存储系统，除了 HDFS 外，Alluxio 还支持许多其他存储系统，支持 Spark 等框架从任何数量的系统读取数据或写入数据。Alluxio 与 Spark 1.1 之后的新版本配合使用。

11.2.2 Spark 与 Alluxio 整合实战

本节根据 Alluxio 本地模式与 Spark 进行整合实战。

1. 在本地运行Alluxio部署的步骤

（1）在 Linux 系统上安装 JDK 7 或更高版本。

（2）安装部署 Alluxio 1.5.0。

下载 alluxio-1.5.0 的 Jar 安装包。

```
1.  wget http://alluxio.org/downloads/files/1.5.0/alluxio-1.5.0-bin.tar.gz
2.  tar xvfz alluxio-1.5.0-bin.tar.gz
3.  cd alluxio-1.5.0
```

在本地独立模式下运行，配置以下内容。

- 设置 alluxio.master.hostname：在 conf/alluxio-site.properties 配置为 localhost（即 alluxio.master.hostname=localhost）。
- 设置 alluxio.underfs.address：在 conf/alluxio-site.properties 配置本地文件系统中的 tmp 目录（例如，alluxio.underfs.address=/tmp）。
- 打开远程登录服务：登录 ssh localhost 成功。如无须重复输入密码，则可配置主机的公共 ssh 密钥 ~/.ssh/authorized_keys。

格式化 Alluxio 文件系统。注意：首次运行 Alluxio 时，才需要执行此步骤。如果为现有 Alluxio 群集运行此命令，则 Alluxio 文件系统中之前存储的所有数据和元数据将被删除。但是，存储中的数据将不会更改。

```
1.  ./bin/alluxio format
```

在本地启动 Alluxio 文件系统：运行以下命令启动 Alluxio 文件系统。在 Linux 上，为了设置 RAMFS，此命令可能需要输入密码，以获取 sudo 权限。

```
1.  ./bin/alluxio-start.sh local
```

（3）验证 Alluxio 正在运行。

要验证 Alluxio 是否正在运行，可以访问 http://localhost:19999，或查看 logs 文件夹中的日志。也可运行 runTests 命令进行检查。

```
1.  ./bin/alluxio runTests
```

（4）停止 Alluxio 运行。

```
1.  ./bin/alluxio-stop.sh local
```

2. Alluxio本地模式与Spark进行整合

（1）Alluxio 客户端使用 Spark 特定的配置文件进行编译。alluxio 使用以下命令从顶级目录构建整个项目。

```
1.  mvn clean package -Pspark -DskipTests
```

（2）添加以下行到 spark/conf/spark-defaults.conf。

```
1.  spark.driver.extraClassPath  /<PATH_TO_ALLUXIO>/core/client/runtime/target/
    alluxio-core-client-runtime-1.6.0-SNAPSHOT-jar-with-dependencies.jar
2.  spark.executor.extraClassPath  /<PATH_TO_ALLUXIO>/core/client/runtime/
    target/alluxio-core-client-runtime-1.6.0-SNAPSHOT-jar-with-dependenci
    es.jar
```

（3）HDFS 的附加设置：如果 Alluxio 运行在 Hadoop 1.x 群集上，则创建一个 spark/conf/core-site.xml 包含以下内容的新文件。

```
1.  <configuration>
2.   <property>
3.     <name>fs.alluxio.impl</name>
```

```
4.      <value>alluxio.hadoop.FileSystem</value>
5.    </property>
6.  </configuration>
```

（4）如果使用 zookeeper 在容错模式下运行 alluxio，并且 Hadoop 集群是 1.x，将以下内容添加到之前的 spark/conf/core-site.xml。

```
1.      <property>
2.    <name>fs.alluxio-ft.impl</name>
3.    <value>alluxio.hadoop.FaultTolerantFileSystem</value>
4.  </property>
```

增加以下内容到 spark/conf/spark-defaults.conf。

```
1.  spark.driver.extraJavaOptions
    -Dalluxio.zookeeper.address=zookeeperHost1:2181,zookeeperHost2:2181
    -Dalluxio.zookeeper.enabled=true
2.  spark.executor.extraJavaOptions
    -Dalluxio.zookeeper.address=zookeeperHost1:2181,zookeeperHost2:2181
    -Dalluxio.zookeeper.enabled=true
```

（5）使用 Alluxio 作为 Spark 应用程序的输入和输出源。

使用 Alluxio 中的数据。先把一些本地数据复制到 Alluxio 文件系统，将文件 LICENSE 放入 Alluxio 中，假设在 Alluxio 项目目录中。

```
1.   bin/alluxio fs copyFromLocal LICENSE /LICENSE
```

然后运行 spark-shell。Alluxio Master 在 localhost 模式下运行。

```
1.   > val s = sc.textFile("alluxio://localhost:19998/LICENSE")
2.  > val double = s.map(line => line + line)
3.  > double.saveAsTextFile("alluxio://localhost:19998/LICENSE2")
```

（6）我们已经在 Spark 应用程序中读入和保存了 Alluxio 系统中的文件，进行检查验证。

打开浏览器检查 http://localhost:19999/browse。应该有一个输出文件 LICENSE2，使 LICENSE 原文件中的每行内容都输出两次。

Alluxio 的更多内容，读者可以登录 Alluxio 的官网（http://www.alluxio.org/）进行学习。

11.3　Spark 与 Job Server 整合原理与实战

本节讲解 Spark 与 Job Server 整合原理及 Spark 与 Job Server 整合实战。

11.3.1　Spark 与 Job Server 整合原理

Spark-jobserver 提供了一个 RESTful 接口来提交和管理 spark 的 jobs、jars 和 job contexts。Spark-jobserver 项目包含了完整的 Spark job server 的项目，包括单元测试和项目部署脚本。

Spark-jobserver 的特性如下。

- Spark as Service：针对 job 和 contexts 的各个方面提供了 REST 风格的 api 接口进行管理。

- ❑ 支持 SparkSQL、Hive、Streaming Contexts/jobs 以及定制 job contexts。
- ❑ 通过集成 Apache Shiro 来支持 LDAP 权限验证。
- ❑ 通过长期运行的 job contexts 支持亚秒级别低延迟的任务。
- ❑ 可以通过结束 context 停止运行的作业（job）。
- ❑ 分割 jar 上传步骤，以提高 job 的启动。
- ❑ 异步和同步的 job API，其中同步 API 对低延时作业非常有效。
- ❑ 支持 Standalone Spark 和 Mesos、yarn。
- ❑ Job 和 jar 信息通过一个可插拔的 DAO 接口来持久化。
- ❑ 对 RDD 或 DataFrame 对象命名并缓存，通过该名称获取 RDD 或 DataFrame，这样可以提高对象在作业间的共享和重用。
- ❑ 支持 Scala 2.10 版本和 2.11 版本。

Spark-jobserver 的部署如下。

（1）复制 conf/local.sh.template 文件到 local.sh。备注：如果需要编译不同版本的 Spark，则须修改 SPARK_VERSION 属性。

（2）复制 config/shiro.ini.template 文件到 shiro.ini。备注：仅需 authentication = on 时执行这一步。

（3）复制 config/local.conf.template 到 <environment>.conf。

（4）bin/server_deploy.sh <environment>，这一步将 job-server 以及配置文件打包，并一同推送到配置的远程服务器上。

（5）在远程服务器上部署的文件目录下通过执行 server_start.sh 启动服务，如需关闭服务，可执行 server_stop.sh。

Spark-jobserver 的各种运行方式如下。

- ❑ Docker 模式：尝试使用作业服务器预先打包 Spark 分发的 Docker 容器，允许启动并部署。
- ❑ 本地模式：在 SBT 内以本地开发模式构建并运行 Job Server。注意：这不适用于 YARN，实际上仅推荐 spark.master 设置为 local[*]。
- ❑ 集群模式：将作业服务器部署到集群，有两种部署方式。
  - ➢ server_deploy.sh 将作业服务器部署到远程主机上的目录。
  - ➢ server_package.sh 将作业服务器部署到本地的目录，为 Mesos 或 YARN 部署创建.tar.gz。
- ❑ EC2 部署脚本：按照 EC2 中的说明，使用作业服务器和示例应用程序启动 Spark 群集。
- ❑ EMR 部署指令：按照 EMR 中的说明进行操作。

11.3.2　Spark 与 Job Server 整合实战

本节根据 Spark-jobserver 本地模式，在 SBT 内以本地开发模式构建并运行 Job Server。Spark 与 Job Server 本地模式的整合步骤如下。

1. Spark-jobserver本地模式服务的启动

（1）Linux 系统中要先安装 SBT。设置当前版本。

```
1.   export VER=`sbt version | tail -1 | cut -f2`
```

（2）在 Linux 系统中下载安装 Spark-jobserver。Spark-jobserver 的下载地址为 https://github.com/spark-jobserver/spark-jobserver#users。

（3）进入 spark-jobserver-master 的安装目录，在 Linux 系统提示符下输入 sbt，在 SBT shell 中键入 reStart，使用默认配置文件启动 Spark-jobserver 服务。可选参数是替代配置文件的路径，还可以在“---”之后指定 JVM 参数。包括所有选项如下：

```
1.   job-server-extras/reStart /path/to/my.conf --- -Xmx8g
```

（4）Spark-jobserver 服务启动测试验证：在浏览器中打开 Url 地址：http://localhost:8090，将显示 Spark Job Server UI 的 Web 页面。

2．Spark-jobserver的示例WordCountExample

（1）首先，将 WordCountExample 代码（WordCountExample 代码功能是进行单词计数：统计输入字符串中每个单词出现的次数）打成 Jar 包：sbt job-server-tests/package，然后上传 jar 包到作业服务器。

```
1.   curl--data-binary@job-server-tests/target/scala-2.10/job-server-   tests-
     $VER.jar localhost:8090/jars/test
```

（2）上述 job-server-tests-$VER.jar 的 jar 包作为应用程序 test 上传服务器。接下来，开始进行单词计数作业，作业服务器将创建自己的 SparkContext，并返回一个作业 ID，用于后续查询。

```
1.   curl -d "input.string = a b c a b see" "localhost:8090/jobs?appName=
     test&classPath=spark.jobserver.WordCountExample"
2.  {
3.    "duration": "Job not done yet",
4.    "classPath": "spark.jobserver.WordCountExample",
5.    "startTime": "2016-06-19T16:27:12.196+05:30",
6.    "context": "b7ea0eb5-spark.jobserver.WordCountExample",
7.    "status": "STARTED",
8.    "jobId": "5453779a-f004-45fc-a11d-a39dae0f9bf4"
9.  }
```

在 input.string 参数中传入字符串"a b c a b see"，将字符串"a b c a b see"提交给 job-server 服务器的 test 应用（即之前我们上传的 WordCountExample 服务），然后启动服务，应用程序的 jobId 号是 5453779a-f004-45fc-a11d-a39dae0f9bf4。

（3）根据 jobId 号查询 job-server 服务器作业的计算结果。在 curl 语句中输入的 jobs 查询参数就是上述应用程序的 jobId 号，是 5453779a-f004-45fc-a11d-a39dae0f9bf4。

```
1.    curl localhost:8090/jobs/5453779a-f004-45fc-a11d-a39dae0f9bf4
2.  {
3.    "duration": "6.341 secs",
4.    "classPath": "spark.jobserver.WordCountExample",
5.    "startTime": "2015-10-16T03:17:03.127Z",
6.    "context": "b7ea0eb5-spark.jobserver.WordCountExample",
7.    "result": {
8.      "a": 2,
9.      "b": 2,
10.     "c": 1,
```

```
11.     "see": 1
12.   },
13.   "status": "FINISHED",
14.   "jobId": "5453779a-f004-45fc-a11d-a39dae0f9bf4"
15. }
```

从 job-server 服务器返回计算结果为 "a": 2 次, "b": 2 次,"c": 1 次,"see": 1 次，计算结果准确。上述是异步模式获取计算结果，须根据 JobID 号查询结果。如果 Spark 分布式计算数据量不是很大，我们也可以在 curl 语句中配置参数&sync=true，在 POST 到 job-server 服务请求时同步返回结果。

```
1.   curl -d "input.string = a b c a b see" "localhost:8090/jobs?appName=
     test&classPath=spark.jobserver.WordCountExample&sync=true"
```

Job Server 的更多内容，读者可以登录 Job Server 的 github 网站（https://github.com/spark-jobserver/spark-jobserver#users）进行学习。

11.4　Spark 与 Redis 整合原理与实战

本节通过生产环境实战案例讲解 Spark 与 Redis 整合原理及 Spark 与 Redis 整合实战。

11.4.1　Spark 与 Redis 整合原理

Redis 是一个开源项目（BSD 许可）。Redis 以内存数据结构存储，用作数据库、缓存和消息代理。它支持数据结构，如字符串、散列、列表、集合、具有范围查询的排序集、位图、超文本和具有半径查询的地理空间索引等。Redis 内置复制、Lua 脚本、LRU eviction、事务和不同级别的磁盘持久性，通过 Redis Sentinel 提供高可用性，并通过 Redis Cluster 进行自动分区。

Redis 可以对这些类型运行原子操作，如附加到字符串、在哈希中增加值、将元素推送到列表中、计算集交集、联合与差异于一体；或者在排序集中获得最高排名的成员。

为了实现其卓越的性能，Redis 使用内存中的数据集。根据业务用例，可以通过将数据集一次性转储到磁盘中，或通过将每个命令附加到日志来持久化。如果只需功能丰富的网络内存缓存，则可以选择禁用持久性。

Redis 还支持简单的主从异步复制，第一次同步非阻塞的速度非常快，网络切分传输时可自动重连同步。

Redis 的其他功能包括：

- 事务 Transactions。
- 发布/订阅。
- Lua 脚本。
- Keys with a limited time-to-live。
- LRU eviction of keys。
- 自动故障切换。

大多数编程语言可以使用 Redis。

Redis 以 ANSI C 编写，适用于大多数 POSIX 系统，如 Linux、* BSD、OS X，无须外部依赖。Linux 和 OS X 是 Redis 开发和测试的两个操作系统，建议使用 Linux 进行部署。Redis 可能在诸如 SmartOS 的 Solaris 衍生系统中工作，但支持是尽力而为的。没有官方支持 Windows 版本，但是 Microsoft 开发并维护了 Redis 的 Win-64 端口。

11.4.2　Spark 与 Redis 整合实战

本节以生产环境中 Spark 与 Redis 整合实战案例来讲解。在通信运营商的 Spark 大数据项目中，Spark 每分钟实时读取 Hdfs 中的话单数据，经过业务逻辑代码分析转换后，将清洗以后的数据转换成一个 List 字符串列表，然后遍历 List 字符串列表，将每条记录拼接成一个 Key-Value 字符串放入 Redis 队列。

Spark 与 Redis 整合实战案例实现步骤如下。

（1）通过 Maven 方式下载 Redis 的 jedis 2.6.0 Jar 包。

pom.xml 文件增加的内容如下：

```
1.       <dependency>
2.            <groupId>redis.clients</groupId>
3.            <artifactId>jedis</artifactId>
4.            <version>2.6.0</version>
5.        </dependency>
```

在 Spark 中导入 Redis 的 Jar 包。

```
1.  import redis.clients.jedis.Jedis;
2.  import redis.clients.jedis.JedisPool;
3.  import redis.clients.jedis.JedisPoolConfig;
```

（2）在项目 config.properties 配置文件中增加 Redis 的主机地址、端口、密码，以及 Redis 连接池分配的连接数、等待时间等信息。

```
1.  ## REDIS
2.  redis.ip=100.*.*.100
3.  redis.port=6379
4.  redis.password=password
5.  ……
6.  ## redis
7.  #最大分配的对象数
8.  redis.pool.maxTotal=1024
9.  #最大能够保持 idle 状态的对象数
10. redis.pool.maxIdle=200
11. #当池内没有返回对象时，最大等待时间
12. redis.pool.maxWait=1000
13. #当调用 borrow Object 方法时，是否进行有效性检查
14. redis.pool.testOnBorrow=true
15. #当调用 return Object 方法时，是否进行有效性检查
16. redis.pool.testOnReturn=true
17.
```

（3）编写 RedisServiceImpl 实现类，从配置文件中获取 Redis 的主机地址、端口、密码等信息；编写 getFromPool 方法从 redis 访问池中获取 redis 实例；编写 getSingle 方法获取 redis

实例。

```
1.  public class RedisServiceImpl {
2.     private static final Map<String, String> REDIS_CONFIG = Config
       .getInstance().getRedisParams();
3.     private static final String REDIS_IP = REDIS_CONFIG.get("redis.ip");
4.     private static final String REDIS_PORT = REDIS_CONFIG.get("redis.port");
5.     private static final String REDIS_PASSWORD = REDIS_CONFIG.get("redis.
       password");
6.     private static final int REDIS_TIMEOUT = 2000;
7.     private static Logger log = LoggerFactory.getLogger(RedisServiceImpl.class);
8.     private static JedisPool pool;
9.     /**
        * 从 redis 访问池中获取 redis 实例
        *
        * @return redis 实例
10.     */
11.    public static Jedis getFromPool() {
12.       if (pool == null) {
13.          JedisPoolConfig config = new JedisPoolConfig();
14.          config.setMaxTotal(Integer.valueOf(REDIS_CONFIG.get("redis.
             pool.maxTotal")));
15.          config.setMaxIdle(Integer.valueOf(REDIS_CONFIG.get("redis.
             pool.maxIdle")));
16.          config.setMaxWaitMillis(Integer.valueOf(REDIS_CONFIG.get("redis.
             pool.maxWait")));
17.          config.setTestOnBorrow(Boolean.valueOf(REDIS_CONFIG.get("redis.
             pool.testOnBorrow")));
18.          config.setTestOnReturn(Boolean.valueOf(REDIS_CONFIG.get("redis.
             pool.testOnReturn")));
19.          pool = new JedisPool(config, REDIS_IP, Integer.valueOf(REDIS_
            PORT), REDIS_TIMEOUT, REDIS_PASSWORD);
20.       }
21.       return pool.getResource();
22.    }
23.
24.    public static void returnResource(Jedis redis) {
25.       pool.returnResource(redis);
26.    }
27.
28.    /**
        * 获取 redis 实例
        *
        * @return redis 实例
29.     */
30.
31.    public static Jedis getSingle() {
32.       Jedis redis = new Jedis(REDIS_IP, Integer.valueOf(REDIS_PORT),
          REDIS_TIMEOUT);
33.       redis.auth(REDIS_PASSWORD);
34.       return redis;
35.    }
36. }
```

（4）编写项目的业务代码，RedisBean 类数据结构用于要存放 Redis 的数据。

```
1.  public class RedisBean {
2.     public void setKey(String key) {
```

```
3.          this.key = key;
4.      }
5.
6.      public void setValue(String value) {
7.          this.value = value;
8.      }
9.
10.     public String getKey() {
11.         return key;
12.     }
13.
14.     public String getValue() {
15.         return value;
16.     }
17.
18.     protected String key;
19.     protected String value;
20.
21. }
```

根据项目的业务需求，将 Spark 中提取转换后的每条记录存入业务数据结构 RedisBean 的 key、value 中；然后加入到 RedisBeanList 列表中。RedisBeanList 是 List<RedisBean> RedisBeanList 类型。

```
1.          if ("RedisTestQUALINFO".equals(keyType)) {
2.                RedisBean.key = "RedisTestQUALINFO";
3.                RedisBean.value = RedisTestReslut.toString();
4.                RedisBeanList.add(RedisBean);
5.            }
```

（5）最终调用业务方法 addToRedis，通过 RedisServiceImpl.getSingle()获取 Redis 实例，然后循环遍历 List<RedisBean>的每个元素，调用 redis.lpush 方法分别将 Key 值、Value 值 lpush 到 Redis 中。lPush 完成以后，通过 redis.close 关闭连接。

```
1.          public static void addToRedis(List<RedisBean> redisBeanList) {
2.          Jedis redis = RedisServiceImpl.getSingle();
3.          for (RedisBean redisData : redisBeanList) {
4.             redis.lpush(redisData.getKey(), redisData.getValue());
5.          }
6.          redis.close();
7.      }
```

（6）Redis 业务验证：可以登录到 Redis 系统中，查询数据已持久化至 Redis。

```
1.   redis-cli -h 100.*.*.100  -p 6379 -a 'password'
2.  select 0  ---切换到 0 库
3.  keys *    ---列出所有的 key
4.  lrange CDNNODEQUALINFO 0 -1 查看所有的记录
```

第2篇 商业案例篇

第 12 章　Spark 商业案例之大数据电影点评系统应用案例

本章讲解 Spark 商业案例之大数据电影点评系统应用案例。电影点评系统可通过各种方式实现：如纯粹通过 RDD 的方式实现、通过 DataFrame 和 RDD 相结合的方式实现、纯粹使用 DataFrame 方式实现、纯粹通过 DataSet 方式实现等。综合应用实现大数据电影点评系统的功能：统计某特定电影观看者中男性和女性不同年龄的人数、计算所有电影中平均得分最高（口碑最好）的电影 TopN、计算最流行电影（即所有电影中粉丝或者观看人数最多）的电影 TopN、实现所有电影中最受男性、女性喜爱的电影 TopN、实现所有电影中 QQ 或者微信核心目标用户最喜爱电影 TopN 分析、实现所有电影中淘宝核心目标用户最喜爱电影 TopN 分析、电影点评系统实现 Java 和 Scala 版本的二次排序系统等。

12.1　通过 RDD 实现分析电影的用户行为信息

在本节中，我们首先搭建 IDEA 的开发环境。电影点评系统基于 IDEA 开发环境进行开发，本节对大数据电影点评系统中电影数据格式和来源进行了说明，然后通过 RDD 方式实现分析电影的用户行为信息的功能。

12.1.1　搭建 IDEA 开发环境

1．IntelliJ IDEA环境的安装

如图 12-1 所示，登录 IDEA 的官网，打开 http://www.jetbrains.com/idea/网站，单击 DOWNLOAD 进行 IDEA 的下载。IDEA 全称 IntelliJ IDEA，是 Java 语言开发的集成环境，具备智能代码助手、代码自动提示、重构、J2EE 支持、Ant、JUnit、CVS 整合、代码审查等方面的功能，支持 Maven、Gradle 和 STS，集成 Git、SVN、Mercurial 等，在 Spark 开发程序时通常使用 IDEA。单击 DOWNLOAD 下载，下载安装包以后根据 IDEA 安装提示一步步完成安装。

图 12-1　IDEA 的官网

IDEA 在本地计算机上安装完成以后，打开 IDEA 的默认显示主题风格是 Dracula 的主题格式，也是众多 IDEA 开发者喜欢的格式。但为了便于读者阅读，这里将 IDEA 的显示主题风格调整为 IntelliJ 格式，单击 File→Settings→Appearance→Theme→IntelliJ，这样书本纸质显示更清晰，如图 12-2 所示。

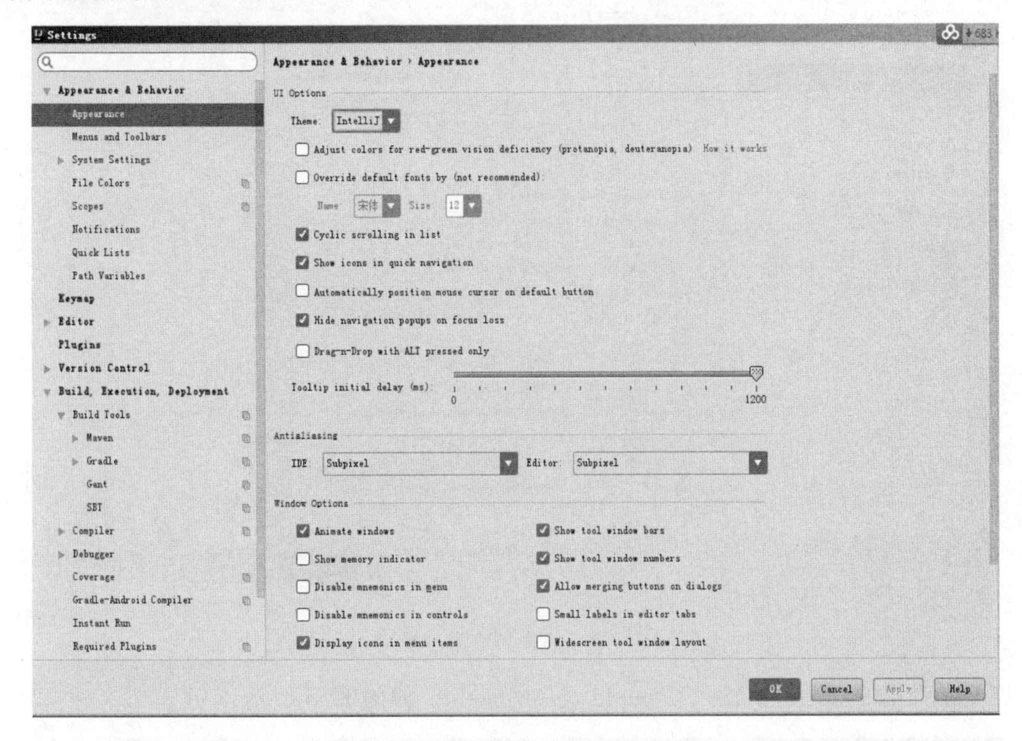

图 12-2　修改 IDEA 显示主题格式

IDEA 安装完成，在 Windows 系统中完成 Windows JDK 的安装与配置。安装和配置完成以后，测试验证 JDK 能否在设备上运行。选择“开始”→“运行”命令，在运行窗口中输入 CMD 命令，进入 DOS 环境，在命令行提示符中直接输入 java -version，按回车键，系统会显示 JDK 的版本，说明 JDK 已经安装成功，代码如下：

```
1.  C:\Windows\System32>java -version
2.  java version "1.8.0_121"
3.  Java(TM) SE Runtime Environment (build 1.8.0_121-b13)
4.  Java HotSpot(TM) 64-Bit Server VM (build 25.121-b13, mixed mode)
```

2. 新建Maven工程（SparkApps工程），导入Spark 2.0相关JAR包及源码

（1）在 IDEA 菜单栏中新建工程，单击 File→Project，如图 12-3 所示。

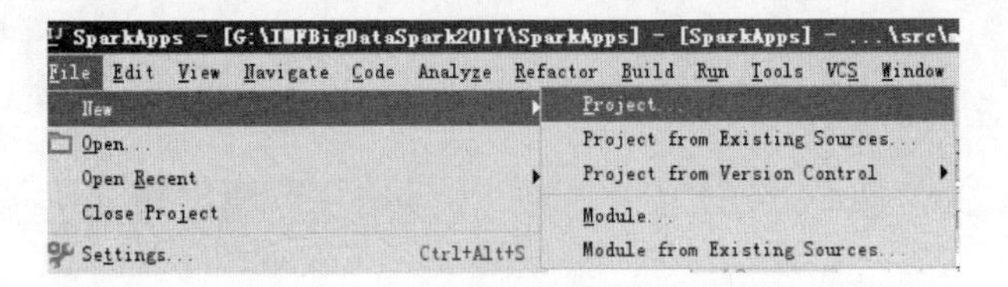

图 12-3　新建工程

（2）在弹出的 New Project 对话框中选择 Maven 方式，单击 Next 按钮，如图 12-4 所示。

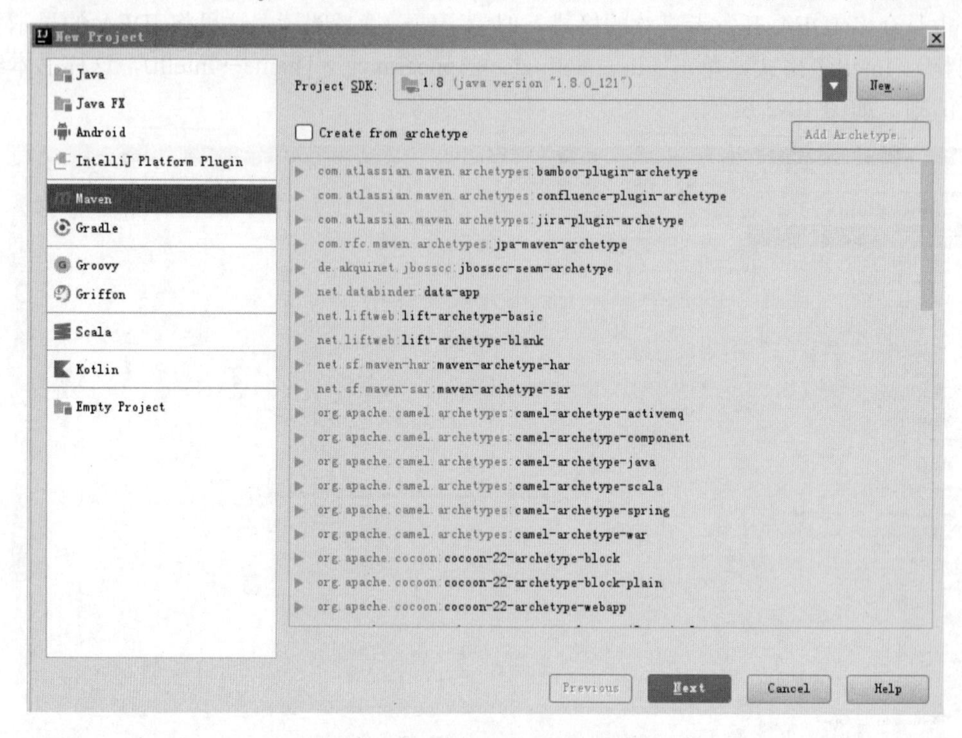

图 12-4　选择 Maven 方式

（3）在弹出的对话框中输入 GroupId 及 ArtifactId，如图 12-5 所示。

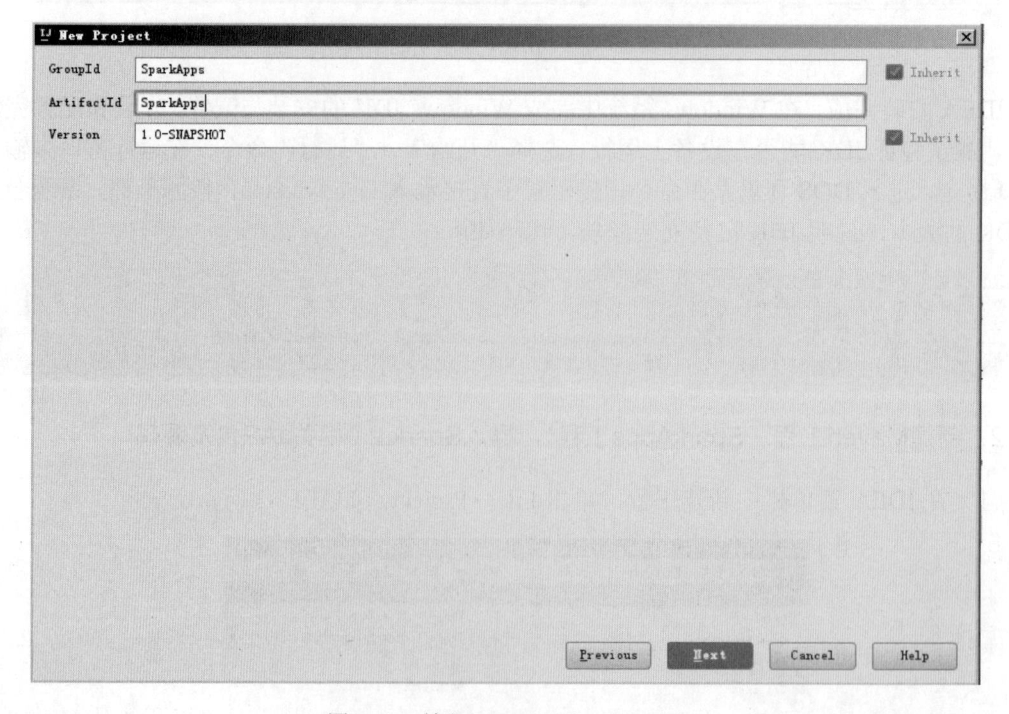

图 12-5　输入 GroupId 及 ArtifactId

（4）在弹出的对话框中输入工程名及工程保存位置，单击 Finish 按钮完成，如图 12-6 所示。

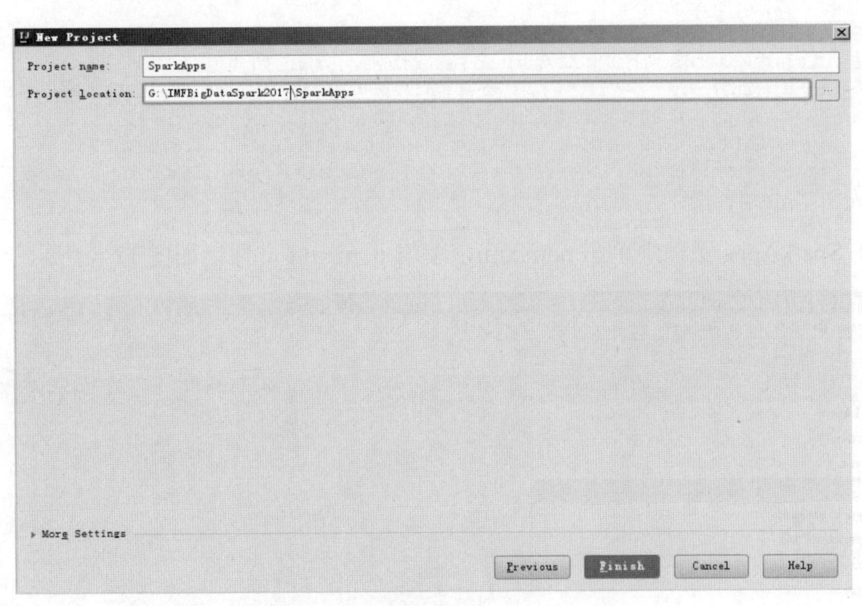

图 12-6　输入工程名及工程保存位置

（5）在 SparkApps 工程中设置 Maven 配置参数。单击 File→Settings→Build,Execution, Deployment→Maven→User settings file 及 Local repository，输入用户配置文件及本地库保存地址，如图 12-7 所示。

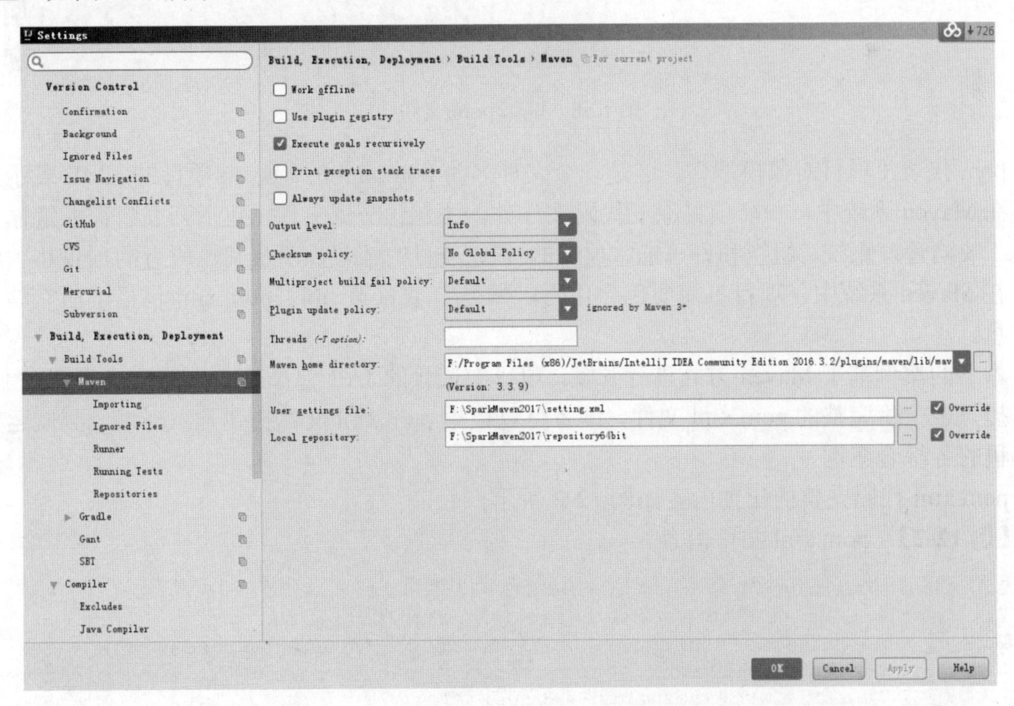

图 12-7　输入用户配置文件及本地库保存地址

其中，setting.xml 代码完整的配置内容如例 12-1 所示。

【例 12-1】　setting.xml 文件内容。

```
1.  <?xml version="1.0" encoding="UTF-8"?>
2.  <settings xmlns="http://maven.apache.org/SETTINGS/1.0.0"
3.          xmlns:xsi="http://www.w3.org/2001/XMLSchema-instance"
4.          xsi:schemaLocation="http://maven.apache.org/SETTINGS/1.0.0
            http://maven.apache.org/xsd/settings-1.0.0.xsd">
5.  <localRepository>F:\SparkMaven2017\repository64bit</localRepository>
6.  </settings>
```

（6）在 SparkApps 工程中单击 pom.xml，编辑 pom.xml 文件，如图 12-8 所示。

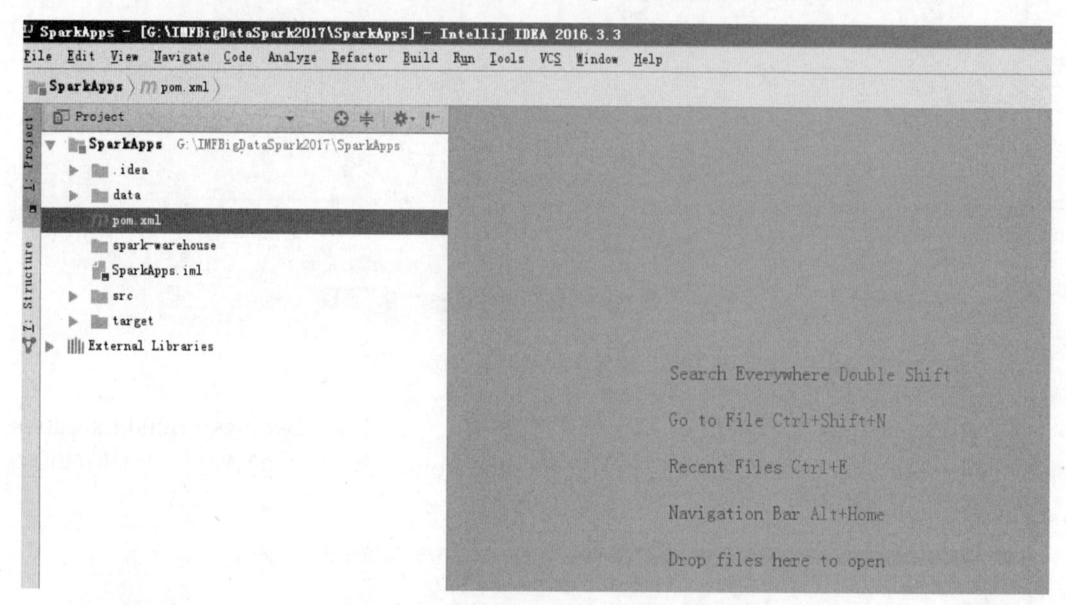

图 12-8　编辑 pom 文件

pom 代表“项目对象模型”。这是一个文件名为 pom.xml 的 Maven 项目的 XML 表示形式。在 Maven 系统中，一个项目除了代码文件外，还包含配置文件，包括开发者需要遵循的规则、缺陷管理系统、组织和许可证、项目的 url、项目的依赖，以及其他所有的项目相关因素。在 Maven 系统中，项目不需要包含代码，只是一个 pom.xml。pom.xml 包括了所有的项目信息。

本书的案例基于 Maven 方式进行开发，项目中依赖的 JAR 包都从 pom.xml 中下载获取，这里提供了一份完整的 pom.xml 文件，读者可以根据 pom.xml 文件搭建开发环境，测试运行本书稿的各综合案例。

pom.xml 代码完整的配置内容如例 12-2 所示。

【例 12-2】　pom.xml 文件内容。

```
1.  <?xml version="1.0" encoding="UTF-8"?>
2.  <project xmlns="http://maven.apache.org/POM/4.0.0"
3.          xmlns:xsi="http://www.w3.org/2001/XMLSchema-instance"
4.        xsi:schemaLocation="http://maven.apache.org/POM/4.0.0 http://maven.
          apache.org/xsd/maven-4.0.0.xsd">
5.      <modelVersion>4.0.0</modelVersion>
```

```
6.  <!-- 基础配置 -->
7.      <groupId>2017SparkCase100</groupId>
8.      <artifactId>2017SparkCase100</artifactId>
9.      <version>1.0-SNAPSHOT</version>
10.
11.     <properties>
12.         <scala.version>2.11.8</scala.version>
13.         <spark.version>2.1.0</spark.version>
14.         <jedis.version>2.8.2</jedis.version>
15.         <fastjson.version>1.2.14</fastjson.version>
16.         <jetty.version>9.2.5.v20141112</jetty.version>
17.         <container.version>2.17</container.version>
18.         <java.version>1.8</java.version>
19.     </properties>
20.
21.     <repositories>
22.         <repository>
23.             <id>scala-tools.org</id>
24.             <name>Scala-Tools Maven2 Repository</name>
25.             <url>http://scala-tools.org/repo-releases</url>
26.         </repository>
27.     </repositories>
28.
29.     <pluginRepositories>
30.         <pluginRepository>
31.             <id>scala-tools.org</id>
32.             <name>Scala-Tools Maven2 Repository</name>
33.             <url>http://scala-tools.org/repo-releases</url>
34.         </pluginRepository>
35.     </pluginRepositories>
36. <!-- 依赖关系-->
37.     <dependencies>
38.         <!-- put javax.ws.rs as the first dependency, it is important!!! -->
39.         <dependency>
40.             <groupId>javax.ws.rs</groupId>
41.             <artifactId>javax.ws.rs-api</artifactId>
42.             <version>2.0</version>
43.         </dependency>
44.
45.         <dependency>
46.             <groupId>org.scala-lang</groupId>
47.             <artifactId>scala-library</artifactId>
48.             <version>${scala.version}</version>
49.         </dependency>
50.         <dependency>
51.             <groupId>org.scala-lang</groupId>
52.             <artifactId>scala-compiler</artifactId>
53.             <version>${scala.version}</version>
54.         </dependency>
55.         <dependency>
56.             <groupId>org.scala-lang</groupId>
57.             <artifactId>scala-reflect</artifactId>
58.             <version>${scala.version}</version>
59.         </dependency>
60.
61.         <dependency>
62.             <groupId>org.scala-lang</groupId>
63.             <artifactId>scalap</artifactId>
64.             <version>${scala.version}</version>
```

```
65.         </dependency>
66.
67.         <dependency>
68.             <groupId>junit</groupId>
69.             <artifactId>junit</artifactId>
70.             <version>4.4</version>
71.             <scope>test</scope>
72.         </dependency>
73.         <dependency>
74.             <groupId>org.specs</groupId>
75.             <artifactId>specs</artifactId>
76.             <version>1.2.5</version>
77.             <scope>test</scope>
78.         </dependency>
79.         <dependency>
80.             <groupId>org.apache.spark</groupId>
81.             <artifactId>spark-core_2.11</artifactId>
82.             <version>${spark.version}</version>
83.         </dependency>
84.         <dependency>
85.             <groupId>org.apache.spark</groupId>
86.             <artifactId>spark-launcher_2.11</artifactId>
87.             <version>2.1.0</version>
88.         </dependency>
89.         <dependency>
90.             <groupId>org.apache.spark</groupId>
91.             <artifactId>spark-network-shuffle_2.11</artifactId>
92.             <version>2.1.0</version>
93.         </dependency>
94.         <dependency>
95.             <groupId>org.apache.spark</groupId>
96.             <artifactId>spark-sql_2.11</artifactId>
97.             <version>${spark.version}</version>
98.         </dependency>
99.         <dependency>
100.                <groupId>org.apache.spark</groupId>
101.                <artifactId>spark-hive_2.11</artifactId>
102.                <version>2.1.0</version>
103.            </dependency>
104.            <dependency>
105.                <groupId>org.apache.spark</groupId>
106.                <artifactId>spark-catalyst_2.11</artifactId>
107.                <version>2.1.0</version>
108.            </dependency>
109.            <dependency>
110.                <groupId>org.apache.spark</groupId>
111.                <artifactId>spark-streaming-flume-assembly_2.11</artifactId>
112.                <version>2.1.0</version>
113.            </dependency>
114.            <dependency>
115.                <groupId>org.apache.spark</groupId>
116.                <artifactId>spark-streaming-flume_2.11</artifactId>
117.                <version>2.1.0</version>
118.            </dependency>
119.            <dependency>
120.                <groupId>org.apache.spark</groupId>
121.                <artifactId>spark-streaming_2.11</artifactId>
122.                <version>${spark.version}</version>
123.            </dependency>
124.            <dependency>
```

```
125.            <groupId>org.apache.spark</groupId>
126.            <artifactId>spark-graphx_2.11</artifactId>
127.            <version>2.1.0</version>
128.        </dependency>
129.        <dependency>
130.            <groupId>org.scalanlp</groupId>
131.            <artifactId>breeze_2.11</artifactId>
132.            <version>0.11.2</version>
133.            <scope>compile</scope>
134.            <exclusions>
135.                <exclusion>
136.                    <artifactId>junit</artifactId>
137.                    <groupId>junit</groupId>
138.                </exclusion>
139.                <exclusion>
140.                    <artifactId>commons-math3</artifactId>
141.                    <groupId>org.apache.commons</groupId>
142.                </exclusion>
143.            </exclusions>
144.        </dependency>
145.        <dependency>
146.            <groupId>org.apache.commons</groupId>
147.            <artifactId>commons-math3</artifactId>
148.            <version>3.4.1</version>
149.            <scope>compile</scope>
150.        </dependency>
151.        <dependency>
152.            <groupId>org.apache.spark</groupId>
153.            <artifactId>spark-mllib_2.11</artifactId>
154.            <version>2.1.0</version>
155.        </dependency>
156.        <dependency>
157.            <groupId>org.apache.spark</groupId>
158.            <artifactId>spark-mllib-local_2.11</artifactId>
159.            <version>2.1.0</version>
160.            <scope>compile</scope>
161.        </dependency>
162.        <dependency>
163.            <groupId>org.apache.spark</groupId>
164.            <artifactId>spark-mllib-local_2.11</artifactId>
165.            <version>2.1.0</version>
166.            <type>test-jar</type>
167.            <scope>test</scope>
168.        </dependency>
169.        <dependency>
170.            <groupId>org.apache.spark</groupId>
171.            <artifactId>spark-repl_2.11</artifactId>
172.            <version>2.1.0</version>
173.        </dependency>
174.        <dependency>
175.            <groupId>org.apache.hadoop</groupId>
176.            <artifactId>hadoop-client</artifactId>
177.            <version>2.6.0</version>
178.        </dependency>
179.        <dependency>
180.            <groupId>org.apache.spark</groupId>
181.            <artifactId>spark-streaming-kafka-0-8_2.10</artifactId>
182.            <version>2.1.0</version>
183.        </dependency>
184.        <dependency>
```

```
185.                <groupId>org.apache.spark</groupId>
186.                <artifactId>spark-streaming-flume_2.11</artifactId>
187.                <version>${spark.version}</version>
188.            </dependency>
189.            <dependency>
190.                <groupId>mysql</groupId>
191.                <artifactId>mysql-connector-java</artifactId>
192.                <version>5.1.6</version>
193.            </dependency>
194.            <dependency>
195.                <groupId>org.apache.hive</groupId>
196.                <artifactId>hive-jdbc</artifactId>
197.                <version>1.2.1</version>
198.            </dependency>
199.            <dependency>
200.                <groupId>org.apache.httpcomponents</groupId>
201.                <artifactId>httpclient</artifactId>
202.                <version>4.4.1</version>
203.            </dependency>
204.            <dependency>
205.                <groupId>org.apache.httpcomponents</groupId>
206.                <artifactId>httpcore</artifactId>
207.                <version>4.4.1</version>
208.            </dependency>
209.
210.            <!-- https://mvnrepository.com/artifact/org.apache.hadoop/
            hadoop- common -->
211.            <dependency>
212.                <groupId>org.apache.hadoop</groupId>
213.                <artifactId>hadoop-common</artifactId>
214.                <version>2.6.0</version>
215.            </dependency>
216.
217.            <dependency>
218.                <groupId>org.apache.hadoop</groupId>
219.                <artifactId>hadoop-client</artifactId>
220.                <version>2.6.0</version>
221.            </dependency>
222.
223.            <!--https://mvnrepository.com/artifact/org.apache.hadoop/hadoop-hdfs-->
224.            <dependency>
225.                <groupId>org.apache.hadoop</groupId>
226.                <artifactId>hadoop-hdfs</artifactId>
227.                <version>2.6.0</version>
228.            </dependency>
229.
230.
231.            <dependency>
232.                <groupId>redis.clients</groupId>
233.                <artifactId>jedis</artifactId>
234.                <version>${jedis.version}</version>
235.            </dependency>
236.            <dependency>
237.                <groupId>org.json</groupId>
238.                <artifactId>json</artifactId>
239.                <version>20090211</version>
240.            </dependency>
241.            <dependency>
```

```
242.                <groupId>com.fasterxml.jackson.core</groupId>
243.                <artifactId>jackson-core</artifactId>
244.                <version>2.6.3</version>
245.            </dependency>
246.            <dependency>
247.                <groupId>com.fasterxml.jackson.core</groupId>
248.                <artifactId>jackson-databind</artifactId>
249.                <version>2.6.3</version>
250.            </dependency>
251.            <dependency>
252.                <groupId>com.fasterxml.jackson.core</groupId>
253.                <artifactId>jackson-annotations</artifactId>
254.                <version>2.6.3</version>
255.            </dependency>
256.            <dependency>
257.                <groupId>com.alibaba</groupId>
258.                <artifactId>fastjson</artifactId>
259.                <version>1.1.41</version>
260.            </dependency>
261.            <dependency>
262.                <groupId>fastutil</groupId>
263.                <artifactId>fastutil</artifactId>
264.                <version>5.0.9</version>
265.            </dependency>
266.            <dependency>
267.                <groupId>org.eclipse.jetty</groupId>
268.                <artifactId>jetty-server</artifactId>
269.                <version>${jetty.version}</version>
270.            </dependency>
271.
272.            <dependency>
273.                <groupId>org.eclipse.jetty</groupId>
274.                <artifactId>jetty-servlet</artifactId>
275.                <version>${jetty.version}</version>
276.            </dependency>
277.
278.            <dependency>
279.                <groupId>org.eclipse.jetty</groupId>
280.                <artifactId>jetty-util</artifactId>
281.                <version>${jetty.version}</version>
282.            </dependency>
283.
284.            <dependency>
285.                <groupId>org.glassfish.jersey.core</groupId>
286.                <artifactId>jersey-server</artifactId>
287.                <version>${container.version}</version>
288.            </dependency>
289.            <dependency>
290.                <groupId>org.glassfish.jersey.containers</groupId>
291.                <artifactId>jersey-container-servlet-core</artifactId>
292.                <version>${container.version}</version>
```

```
293.            </dependency>
294.            <dependency>
295.                <groupId>org.glassfish.jersey.containers</groupId>
296.                <artifactId>jersey-container-jetty-http</artifactId>
297.                <version>${container.version}</version>
298.            </dependency>
299.            <dependency>
300.                <groupId>org.apache.hadoop</groupId>
301.                <artifactId>hadoop-mapreduce-client-core</artifactId>
302.                <version>2.6.0</version>
303.            </dependency>
304.
305.            <dependency>
306.                <groupId>org.antlr</groupId>
307.                <artifactId>antlr4-runtime</artifactId>
308.                <version>4.5.3</version>
309.            </dependency>
310.
311.        </dependencies>
312.    <!-- 编译配置 -->
313.        <build>
314.            <plugins>
315.                <plugin>
316.                    <artifactId>maven-assembly-plugin</artifactId>
317.                    <configuration>
318.                        <classifier>dist</classifier>
319.                        <appendAssemblyId>true</appendAssemblyId>
320.                        <descriptorRefs>
321.                            <descriptor>jar-with-dependencies</descriptor>
322.                        </descriptorRefs>
323.                    </configuration>
324.                    <executions>
325.                        <execution>
326.                            <id>make-assembly</id>
327.                            <phase>package</phase>
328.                            <goals>
329.                                <goal>single</goal>
330.                            </goals>
331.                        </execution>
332.                    </executions>
333.                </plugin>
334.
335.                <plugin>
336.                    <artifactId>maven-compiler-plugin</artifactId>
337.                    <configuration>
338.                        <source>1.7</source>
339.                        <target>1.7</target>
340.                    </configuration>
341.                </plugin>
342.
343.                <plugin>
344.                    <groupId>net.alchim31.maven</groupId>
345.                    <artifactId>scala-maven-plugin</artifactId>
346.                    <version>3.2.2</version>
347.                    <executions>
348.                        <execution>
349.                            <id>scala-compile-first</id>
```

```
350.                 <phase>process-resources</phase>
351.                 <goals>
352.                     <goal>compile</goal>
353.                 </goals>
354.             </execution>
355.         </executions>
356.         <configuration>
357.             <scalaVersion>${scala.version}</scalaVersion>
358.             <recompileMode>incremental</recompileMode>
359.             <useZincServer>true</useZincServer>
360.             <args>
361.                 <arg>-unchecked</arg>
362.                 <arg>-deprecation</arg>
363.                 <arg>-feature</arg>
364.             </args>
365.             <jvmArgs>
366.                 <jvmArg>-Xms1024m</jvmArg>
367.                 <jvmArg>-Xmx1024m</jvmArg>
368.             </jvmArgs>
369.             <javacArgs>
370.                 <javacArg>-source</javacArg>
371.                 <javacArg>${java.version}</javacArg>
372.                 <javacArg>-target</javacArg>
373.                 <javacArg>${java.version}</javacArg>
374.                 <javacArg>-Xlint:all,-serial,-path</javacArg>
375.             </javacArgs>
376.         </configuration>
377.     </plugin>
378.
379.     <plugin>
380.         <groupId>org.antlr</groupId>
381.         <artifactId>antlr4-maven-plugin</artifactId>
382.         <version>4.3</version>
383.         <executions>
384.             <execution>
385.                 <id>antlr</id>
386.                 <goals>
387.                     <goal>antlr4</goal>
388.                 </goals>
389.                 <phase>none</phase>
390.             </execution>
391.         </executions>
392.         <configuration>
393.             <outputDirectory>src/test/java</outputDirectory>
394.             <listener>true</listener>
395.             <treatWarningsAsErrors>true</treatWarningsAsErrors>
396.         </configuration>
397.     </plugin>
398.   </plugins>
399.  </build>
400. </project>
```

（7）在 pom.xml 中，按 Ctrl+S 快捷键保存 pom.xml 文件，IDEA 会自动从网上下载各类 Jar 包，下载的时间根据网络带宽的情况可能需要几十分钟，也可能需要几个小时，全部下载好以后，可以看到工程中 External Libraries 已经加载了 Spark 相关的 Jar 包及源码，如图 12-9 所示。

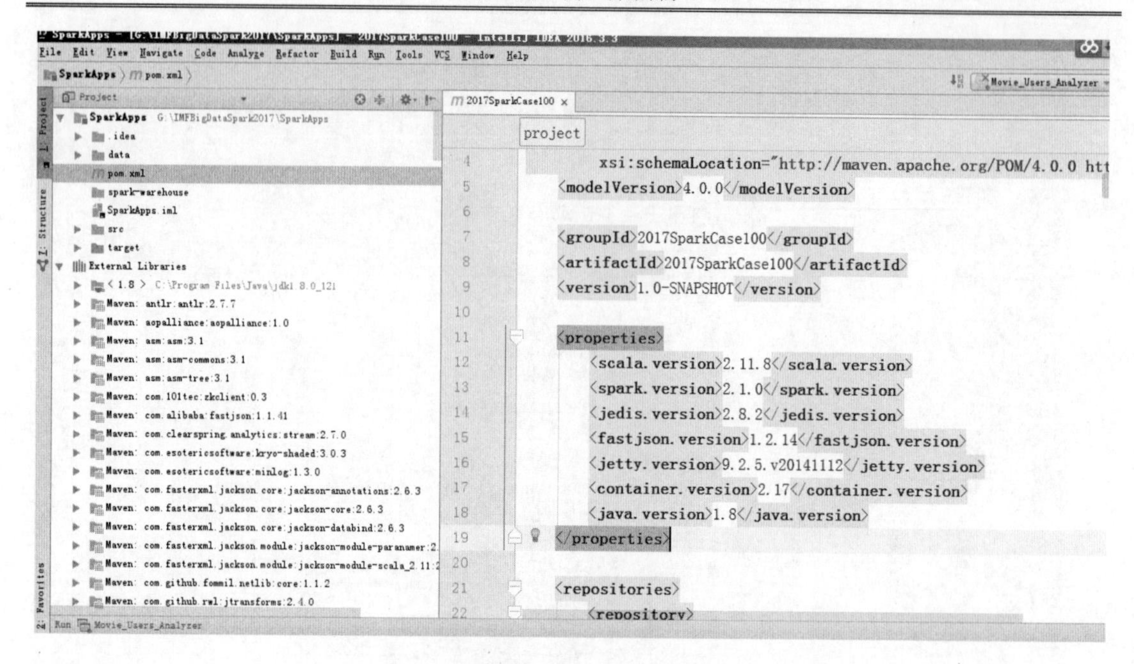

图 12-9　使用 Maven 方式加载 Jar 包

3. 在SparkApps工程中建立scala目录

单击 SparkApps 工程下的 src/main 目录，右击 main，从弹出的快捷菜单中选择 New→Directory 命令，新建一个目录 scala，如图 12-10 所示。

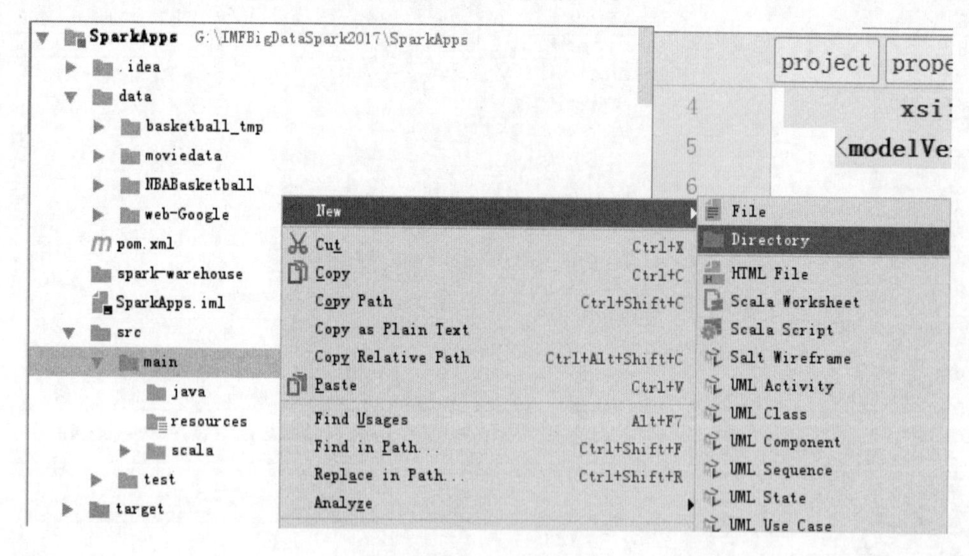

图 12-10　新建目录

单击 SparkApps 工程下的 src/main/scala 目录，右击 scala，从弹出的快捷菜单中选择 Mark Directory as→Resource Root 命令，标识目录 scala 为源码目录。至此，IDEA 本地开发环境搭建完成，如图 12-11 所示。

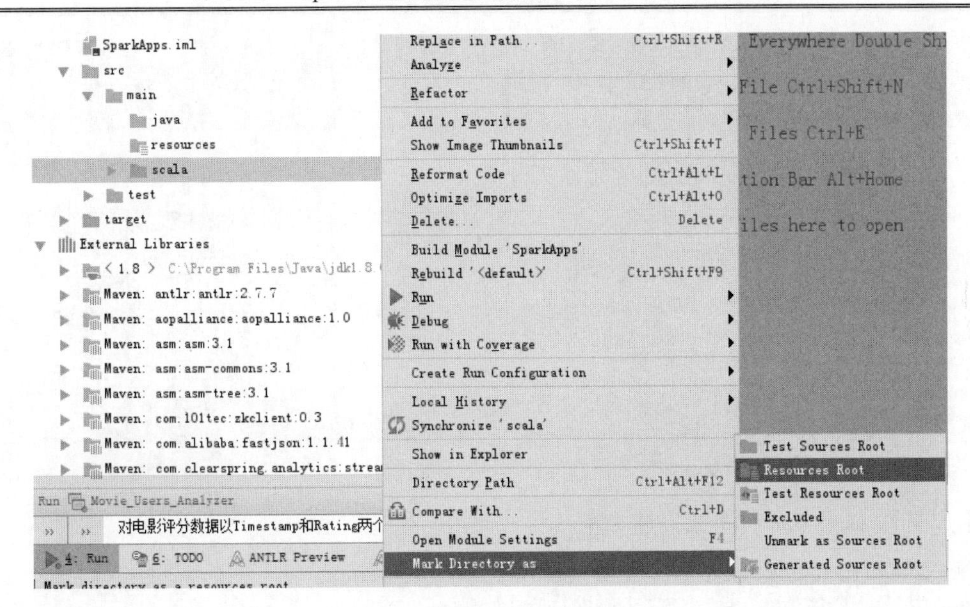

图 12-11　设置为代码目录

12.1.2　大数据电影点评系统中电影数据说明

1．大数据电影点评系统电影数据的来源

电影推荐系统（MovieLens）是美国明尼苏达大学（Minnesota）计算机科学与工程学院的 GroupLens 项目组创办的，是一个非商业性质的、以研究为目的的实验性站点。电影推荐系统主要使用协同过滤和关联规则相结合的技术，向用户推荐他们感兴趣的电影。这个项目是由 John Riedl 教授和 Joseph Konstan 教授领导的。该项目从 1992 年开始研究自动化协同过滤，在 1996 年使用自动化协同过滤系统应用于 USENET 新闻组中。自那以后，项目组扩大了研究范围，基于内容方法以及改进当前的协作过滤技术来研究所有的信息过滤解决方案。

电影推荐系统（MovieLens）的数据下载地址为: https://grouplens.org/datasets/movielens/。GroupLens 项目研究收集了从电影推荐系统 MovieLens 站点提供评级的数据集（http://MovieLens.org），收集了不同时间段的数据，我们可以根据电影分析业务需求下载不同规模大小的数据源文件。

2．大数据电影点评系统电影数据的格式说明

这里下载的是中等规模的电影推荐系统数据集。在本地目录 moviedata/medium 包含的电影点评系统数据源中提供了在 2000 年 6040 个用户观看约 3900 部电影发表的 1 000 209 条匿名评级数据信息。

评级文件 ratings.dat 的格式描述如下：

```
1.  UserID::MovieID::Rating::Timestamp
2.  用户 ID、电影 ID、评分数据、时间戳
3.  - 用户 ID 范围在 1～6040 之间
4.  - 电影 ID 范围在 1～3952 之间
```

```
5.   - 评级：使用五星评分方式
6.   - 时间戳表示系统记录的时间
7.   - 每个用户至少有 20 个评级
```

评级文件 ratings.dat 中摘取部分记录如下：

```
1.   1::1193::5::978300760
2.   1::661::3::978302109
3.   1::914::3::978301968
4.   1::3408::4::978300275
5.   1::2355::5::978824291
6.   1::1197::3::978302268
7.   1::1287::5::978302039
8.   1::2804::5::978300719
9.   1::594::4::978302268
10.  1::919::4::978301368
```

用户文件 users.dat 的格式描述如下：

```
1.   UserID::Gender::Age::Occupation::Zip-code
2.   用户 ID、性别、年龄、职业、邮编代码
3.   -所有的用户资料由用户自愿提供， GroupLens 项目组不会去检查用户数据的准确性。这个数
     据集中包含用户提供的用户数据
4.   -性别：“M”是男性、“F”是女性
5.   -年龄由以下范围选择：
6.   * 1:  "少于 18 岁"
7.   * 18:  "18 年龄段：从 18 岁到 24 岁"
8.   * 25:  "25 年龄段：从 25 岁到 34 岁"
9.   * 35:  "35 岁年龄段：从 35 岁到 44 岁"
10.  * 45:  "45 岁年龄段：从 45 岁到 49 岁"
11.  * 50:  "50 岁年龄段：从 50 岁到 55 岁"
12.  * 56:  "56 岁年龄段：大于 56 岁"
```

从用户文件 users.dat 中摘取部分记录如下：

```
1.   1::F::1::10::48067
2.   2::M::56::16::70072
3.   3::M::25::15::55117
4.   4::M::45::7::02460
5.   5::M::25::20::55455
6.   6::F::50::9::55117
7.   7::M::35::1::06810
8.   8::M::25::12::11413
9.   9::M::25::17::61614
10.  10::F::35::1::95370
```

电影文件 movies.dat 的格式描述如下：

```
1.   MovieID::Title::Genres
2.   电影 ID、电影名、电影类型
3.   -标题是由亚马逊公司的互联网电影资料库（IMDB）提供的，包括电影发布年份
4.   -电影类型包括以下类型：
5.   * Action：行动
6.   * Adventure：冒险
7.   * Animation：动画
8.   * Children's：儿童
9.   * Comedy：喜剧
10.  * Crime：犯罪
```

```
11. * Documentary：纪录片
12. * Drama：戏剧
13. * Fantasy：幻想
14. * Film-Noir：黑色电影
15. * Horror：恐怖
16. * Musical：音乐
17. * Mystery：神秘
18. * Romance：浪漫
19. * Sci-Fi：科幻
20. * Thriller：惊悚
21. * War：战争
22. * Western：西方
23. -由于偶然重复的电影记录或者电影记录测试，一些电影 ID 和电影名可能不一致
24. -电影记录大多是 GroupLens 手工输入的，因此不一定准确
```

电影文件 movies.dat 中摘取的部分记录如下：

```
1.  1::Toy Story (1995)::Animation|Children's|Comedy
2.  2::Jumanji (1995)::Adventure|Children's|Fantasy
3.  3::Grumpier Old Men (1995)::Comedy|Romance
4.  4::Waiting to Exhale (1995)::Comedy|Drama
5.  5::Father of the Bride Part II (1995)::Comedy
6.  6::Heat (1995)::Action|Crime|Thriller
7.  7::Sabrina (1995)::Comedy|Romance
8.  8::Tom and Huck (1995)::Adventure|Children's
9.  9::Sudden Death (1995)::Action
10. 10::GoldenEye (1995)::Action|Adventure|Thriller
```

职业文件 occupations.dat 的格式描述如下：

```
1.  OccupationID::Occupation
2.  职业 ID、职业名
3.  -职业包含如下选择：
4.  * 0：“其他”或未指定
5.  * 1：“学术/教育者”
6.  * 2：“艺术家”
7.  * 3：“文书/行政”
8.  * 4：“高校毕业生”
9.  * 5：“客户服务”
10. * 6：“医生/保健”
11. * 7：“行政/管理”
12. * 8：“农民”
13. * 9：“家庭主妇”
14. * 10：“中小学生”
15. * 11：“律师”
16. * 12：“程序员”
17. * 13：“退休”
18. * 14：“销售/市场营销”
19. * 15：“科学家”
20. * 16：“个体户”
21. * 17：“技术员/工程师”
22. * 18：“商人和工匠”
```

```
23. * 19: “失业”
24. * 20: “作家”
```

从职业文件 occupations.dat 中摘取部分记录如下：

```
1.  0::other or not specified
2.  1::academic/educator
3.  2::artist
4.  3::clerical/admin
5.  4::college/grad student
6.  5::customer service
7.  6::doctor/health care
8.  7::executive/managerial
9.  8::farmer
10. 9::homemaker
11. 10::K-12 student
12. 11::lawyer
13. 12::programmer
14. 13::retired
15. 14::sales/marketing
16. 15::scientist
17. 16::self-employed
18. 17::technician/engineer
19. 18::tradesman/craftsman
20. 19::unemployed
21. 20::writer
```

12.1.3　电影点评系统用户行为分析统计实战

在本节大数据电影点评系统用户行为分析统计实战中，我们需统计用户观看电影和点评电影行为数据的采集、过滤、处理和展示。对于用户行为的数据采集：在生产环境中，企业通常使用 Kafka 的方式实时收集前端服务器中发送的用户行为日志记录信息；对于用户行为的数据过滤：可以在前端服务器端进行用户行为数据的过滤和格式化，也可以采用 Spark SQL 的方式进行数据过滤。在大数据电影点评系统用户行为分析统计实战中，基于 GroupLens 项目组电影推荐系统（MovieLens）已经采集的用户电影观看和点评数据文件，我们直接基于 ratings.dat、users.dat、movies.dat、occupations.dat 文件进行用户行为实战分析。

用户行为分析统计的数据处理：① 一个基本的技巧是，先使用传统的 SQL 去实现一个数据处理的业务逻辑（自己可以手动模拟一些数据）；② 在 Spark2.x 的时候，再一次推荐使用 DataSet 去实现业务功能，尤其是统计分析功能；③ 如果想成为专家级别的顶级 Spark 人才，请使用 RDD 实现业务功能，为什么？原因很简单，因为使用 Spark DataSet 方式有一个底层的引擎 catalyst，基于 DataSet 的编程，catalyst 的引擎会对我们的代码进行优化，有很多优化的言外之意是你看不到问题到底是怎么来的，假设出错了，优化后的 RDD 跑在 Spark 上，打印的错误不是直接基于 DataSet 产生的错误，DataSet 是在内核上的封装，运行的时候是基于 RDD 的！因此，打印的错误是基于 RDD 的。DataSet 的优化引擎 catalyst 涉及 Spark 底层的代码封装。在 DataSet 的解析过程中，基于抽象语法树和语法规则的相互配合，引擎 catalyst 完成了词法分析、未解析的逻辑计划、解析以后的逻辑计划、优化后的逻辑计划、物理计划、可执行的物理计划、物理计划执行、生成 RDD 等一系列过程。如果使用 DataSet 出现问题，我们可能不知其所以然。而在业务代码编码中，如果我们直接使用 RDD，可以直接基于 RDD 来排查问题。在本节大数据电影点评系统用户行为分析统计实战中，我们通过

RDD 的方式直接统计分析用户的电影行为。

用户行为分析统计的数据格式：在生产环境中，强烈建议大家使用 Parquet 的文件格式。Parquet 是面向分析型业务的列式存储格式，由 Twitter 和 Cloudera 合作开发，2015 年 5 月成为 Apache 顶级项目。Parquet 是列式存储格式的一种文件类型，可以适配多种计算框架，是语言无关的，而且不与任何一种数据处理框架绑定在一起，适配多种语言和组件。在本节大数据电影点评系统用户行为分析统计实战中，我们研究试验中小规模的用户电影点评数据的分析，专注于大数据 Spark RDD 的算子实现，这里仍使用 GroupLens 项目组提供的文本文件格式，不进行 Parquet 格式的转换。

大数据电影点评系统用户行为分析统计的数据源格式如下：

```
1."ratings.dat": UserID::MovieID::Rating::Timestamp
2."users.dat": UserID::Gender::Age::OccupationID::Zip-code
3."movies.dat": MovieID::Title::Genres
4.  "occupations.dat": OccupationID::OccupationName
```

大数据电影点评系统用户行为分析统计实战，我们使用 Spark 本地模式进行开发，在 IDEA 开发环境的 SparkApps 工程中的 src/main/scala 目录中新建包 com.dt.spark.cores，然后在 com.dt.spark.cores 下新建 Movie_Users_Analyzer_RDD.scala 文件。

在 Movie_Users_Analyzer_RDD.scala 文件中导入电影点评数据。

```
1.      //设置打印日志的输出级别
2.       Logger.getLogger("org").setLevel(Level.ERROR)
3.
4.       var masterUrl = "local[4]" //默认程序运行在本地 Local 模式中，主要用于学习和测试
5.       var dataPath = "data/moviedata/medium/"  //数据存放的目录
6.
7.       /**
          *当我们把程序打包运行在集群上的时候，一般都会传入集群的 URL 信息，这里我们假设
          *如果传入参数，第一个参数只传入 Spark 集群的 URL，第二个参数传入的是数据的地址信息
8.        */
9.
10.      if(args.length > 0) {
11.        masterUrl = args(0)
12.      } else if (args.length > 1) {
13.        dataPath = args(1)
14.      }
15.
16.      /**
          *创建 Spark 集群上下文 sc，在 sc 中可以进行各种依赖和参数的设置等，大家可以通
          *过 SparkSubmit 脚本的 help 去看设置信息
17.       */
18.      val sc = new SparkContext(new SparkConf().setMaster(masterUrl).setAppName
         ("Movie_Users_Analyzer"))
19.
20.      /**
          *读取数据，用什么方式读取数据呢？这里使用的是 RDD
21.       */
22.      val usersRDD = sc.textFile(dataPath + "users.dat")
23.      val moviesRDD = sc.textFile(dataPath + "movies.dat")
24.      val occupationsRDD = sc.textFile(dataPath + "occupations.dat")
25.          val ratingsRDD = sc.textFile(dataPath + "ratings.dat")
26.      val ratingsRDD = sc.textFile("data/moviedata/large/" + "ratings.dat")
```

电影点评系统用户行为分析之一，统计具体某部电影观看的用户信息，如电影 ID 为 1193 的用户信息（用户的 ID、Age、Gender、Occupation）。为了便于阅读，我们在 Spark Driver 端 collect()获取到 RDD 的元素集合以后，使用 collect().take(2)算子打印输出 RDD 的两个元素，最后的用户信息的输出结果，我们使用 collect().take(20)显示 10 个元素。

```
1.      /**
          *电影点评系统用户行为分析之一：分析具体某部电影观看的用户信息，如电影 ID 为
          *1193 的用户信息（用户的 ID、Age、Gender、Occupation）
2.        */
3.
4.      val usersBasic: RDD[(String, (String, String, String))] = usersRDD
        .map(_.split("::")).map { user =>
5.        ( //UserID::Gender::Age::OccupationID
6.          user(3),
7.          (user(0), user(1), user(2))
8.        )
9.      }
10.     for (elem <- usersBasic.collect().take(2)) {
11.       println("usersBasicRDD (职业 ID,(用户 ID，性别，年龄)): " + elem)
12.     }
13.     val occupations: RDD[(String, String)] = occupationsRDD.map(_.split
        ("::")).map(job => (job(0), job(1)))
14.     for (elem <- occupations.collect().take(2)) {
15.       println("occupationsRDD(职业 ID，职业名): " + elem)
16.     }
17.     val userInformation: RDD[(String, ((String, String, String), String))]
        = usersBasic.join(occupations)
18.     userInformation.cache()
19.
20.     for (elem <- userInformation.collect().take(2)) {
21.       println("userInformationRDD (职业 ID,((用户 ID,性别,年龄),职业名)): " + elem)
22.     }
23.
24.     val  targetMovie:  RDD[(String,  String)]  =  ratingsRDD.map(_.split
        ("::")).map(x => (x(0), x(1))).filter(_._2.equals("1193"))
25.
26.     for (elem <- targetMovie.collect().take(2)) {
27.       println("targetMovie(用户 ID,电影 ID) : " + elem)
28.     }
29.
30.     val targetUsers: RDD[(String, ((String, String, String), String))] =
        userInformation.map(x => (x._2._1._1, x._2))
31.     for (elem <- targetUsers.collect().take(2)) {
32.       println("targetUsers (用户 ID, ((用户 ID,性别,年龄), 职业名)): " + elem)
33.     }
34.     println("电影点评系统用户行为分析,统计观看电影 ID 为 1193 的电影用户信息：用户
        的 ID、性别、年龄、职业名 ")
35.     val userInformationForSpecificMovie: RDD[(String, (String, ((String,
        String, String), String)))] = targetMovie.join(targetUsers)
36.     for (elem <- userInformationForSpecificMovie.collect().take(10)) {
37.       println("userInformationForSpecificMovie(用户 ID, (电影 ID, ((用户 ID,
          性别,年龄), 职业名))) : " + elem)
38.     }
```

在 IDEA 中运行代码，结果如下：

```
1.   Using Spark's default log4j profile: org/apache/spark/log4j-defaults
```

```
   .properties
2. [Stage 0:> (0 + 0) / 2]
3. usersBasicRDD (职业 ID,(用户 ID,性别,年龄)): (10,(1,F,1))
4. usersBasicRDD (职业 ID,(用户 ID,性别,年龄)): (16,(2,M,56))
5.
6. occupationsRDD(职业 ID，职业名)：(0,other or not specified)
7. occupationsRDD(职业 ID，职业名)：(1,academic/educator)
8.
9. userInformationRDD (职业 ID, ((用户 ID,性别,年龄), 职业名)): (4,((25,M,18),
   college/grad student))
10. userInformationRDD (职业 ID, ((用户 ID,性别,年龄), 职业名)): (4,((38,F,18),
   college/grad student))
11.
12. targetMovie(用户 ID,电影 ID) : (6,1193)
13. targetMovie(用户 ID,电影 ID) : (10,1193)
14.
15. targetUsers (用户 ID, ((用户 ID,性别,年龄), 职业名)): (25,((25,M,18),
   college/grad student))
16. targetUsers (用户 ID, ((用户 ID,性别,年龄), 职业名)): (38,((38,F,18),
   college/grad student))
17. 电影点评系统用户行为分析，统计观看电影 ID 为 1193 的电影用户信息：用户的 ID、性别、年
   龄、职业名
18. userInformationForSpecificMovie(用户 ID, (电影 ID, ((用户 ID,性别,年龄), 职
   业名))) : (3638,(1193,((3638,M,25),artist)))
19. userInformationForSpecificMovie(用户 ID, (电影 ID, ((用户 ID,性别,年龄), 职
   业名))) : (2060,(1193,((2060,M,1),academic/educator)))
20. userInformationForSpecificMovie(用户 ID, (电影 ID, ((用户 ID,性别,年龄), 职
   业名))) : (91,(1193,((91,M,35),executive/managerial)))
21. userInformationForSpecificMovie(用户 ID, (电影 ID, ((用户 ID,性别,年龄), 职
   业名))) : (4150,(1193,((4150,M,25),other or not specified)))
22. userInformationForSpecificMovie(用户 ID, (电影 ID, ((用户 ID,性别,年龄), 职
   业名))) : (3168,(1193,((3168,F,35),customer service)))
23. userInformationForSpecificMovie(用户 ID, (电影 ID, ((用户 ID,性别,年龄), 职
   业名))) : (2596,(1193,((2596,M,50),executive/managerial)))
24. userInformationForSpecificMovie(用户 ID, (电影 ID, ((用户 ID,性别,年龄), 职
   业名))) : (2813,(1193,((2813,M,25),writer)))
25. userInformationForSpecificMovie(用户 ID, (电影 ID, ((用户 ID,性别,年龄), 职
   业名))) : (5445,(1193,((5445,M,25),other or not specified)))
26. userInformationForSpecificMovie(用户 ID, (电影 ID, ((用户 ID,性别,年龄), 职
   业名))) : (3652,(1193,((3652,M,25),programmer)))
27. userInformationForSpecificMovie(用户 ID, (电影 ID, ((用户 ID,性别,年龄), 职
   业名))) : (3418,(1193,((3418,F,18),clerical/admin)))
```

12.2　通过 RDD 实现电影流行度分析

本节统计所有电影中平均得分最高（口碑最好）的电影及观看人数最多（流行度最高）的电影。

所有电影中平均得分最高的 Top10 电影实现思路：如果想算总的评分，一般肯定需要 reduceByKey 操作或者 aggregateByKey 操作。

评级文件 ratings.dat 的格式描述如下：

```
1.  UserID::MovieID::Rating::Timestamp
2.  用户 ID、电影 ID、评分数据、时间戳
```

第一步：把数据变成 Key-Value，大家想一下在这里什么是 Key，什么是 Value？把 MovieID 设置为 Key，把 Rating 设置为 Value。具体实现过程：将 ratingsRDD 中的每行数据按"::"分隔符进行分割，然后 map 格式化为（用户 ID，电影 ID，评分）元组；接下来对 ratings 进行 map 转换，取 ratings 元组的第 2 个元素"电影 ID"作为 Key，（第 3 个元素即评分，1）元组值作为 Value，格式化成为 Key-Value 的方式，即（电影 ID，（评分，1））。

第二步：通过 reduceByKey 操作或者 aggregateByKey 实现聚合，然后呢？具体实现过程：对两个具有相同 Key，而 Value 不同的元组，如（电影 ID，（x.评分，x.1）），（电影 ID，（y.评分，y.1）），我们使用 reduceByKey 算子对 Value 值进行汇聚转换，计算得出（x 的评分+y 的评分，x 的计数+y 的计数），转换以后 Key 是电影 ID，Value 是（总的评分，总的点评人数），格式化为 Key-Value，即（电影 ID，（总评分，总点评人数））。

第三步：排序，如何做？进行 Key 和 Value 的交换。上一步 reduceByKey 算子执行完毕，接下来进行 map 转换操作，交换 Key-Value 值，并且计算出电影平均分=总评分/总点评人数，即将（电影 ID，（总评分，总点评人数））转换成（（总评分/总点评人数），电影 ID），然后使用 sortByKey(false)算子按电影平均分降序排列，再通过 take(10)算子获取所有电影中平均得分最高的 Top10，打印输出。

所有电影中电影粉丝或者观看人数最多的电影实现思路：

第一步：把数据变成 Key-Value：取 ratings 元组的第 2 个元素电影 ID 作为 Key，计数 1 次作为 Value，格式化成为 Key-Value，即（电影 ID，1）。

第二步：通过 reduceByKey 操作实现聚合：对相同 Key 的 Value 值进行累加。生成 Key-Value，即（电影 ID，总次数）。

第三步：排序，进行 Key 和 Value 的交换。上一步 reduceByKey 算子执行完毕，然后进行 map 转换操作，交换 Key-Value 值，即将（电影 ID，总次数）转换成（总次数，电影 ID），然后使用 sortByKey(false)算子按总次数降序排列。

第四步：再次进行 Key 和 Value 的交换，打印输出。我们使用 map 转换函数将（总次数，电影 ID）进行交换，转换为（电影 ID，总次数），再通过 take(10)算子获取所有电影中粉丝或者观看人数最多的电影 Top10，打印输出。

大数据电影点评系统中，电影流行度分析须注意以下事项。

（1）转换数据格式的时候一般都会使用 map 操作，有时转换可能特别复杂，需要在 map 方法中调用第三方 jar 或者 so 库。

（2）RDD 从文件中提取的数据成员默认都是 String 类型，需要根据实际需要进行转换类型。

（3）RDD 如果要重复使用，一般都会进行 Cache 操作。

（4）重磅注意事项，RDD 的 Cache 操作之后不能直接再跟其他的算子操作，否则在一些版本中 Cache 不生效。

电影点评系统用户行为分析，统计所有电影中平均得分最高（口碑最好）的电影以及电影粉丝或者观看人数最多的电影的代码如下：

```
1.      println("所有电影中平均得分最高（口碑最好）的电影:")
2.      val  ratings= ratingsRDD.map(_.split("::")).map(x  =>  (x(0),  x(1),
        x(2))).cache()  //格式化出电影 ID 和评分
3.      ratings.map(x => (x._2, (x._3.toDouble, 1)))//格式化为 Key-Value 的方式
4.        .reduceByKey((x, y) => (x._1 + y._1,x._2 + y._2))
         //对 Value 进行 reduce 操作，分别得出每部电影的总的评分和总的点评人数
5.        .map(x => (x._2._1.toDouble / x._2._2, x._1))  //求出电影平均分
6.        .sortByKey(false)  //降序排列
7.        .take(10)          //取 Top10
8.        .foreach(println)  //打印到控制台
9.
10.     /**
11.       *上面的功能计算的是口碑最好的电影，接下来分析粉丝或者观看人数最多的电影
12.       */
13.     println("所有电影中粉丝或者观看人数最多的电影:")
14.     ratings.map(x => (x._2, 1)).reduceByKey(_+_).map(x => (x._2, x._1))
        .sortByKey(false)
15.       .map(x => (x._2, x._1)).take(10).foreach(println)
```

在 IDEA 中运行代码，结果如下：

```
1.  所有电影中平均得分最高（口碑最好）的电影:
2.  [Stage 17:==============================>           (4 + 4) / 8](5.0,33264)
3.  (5.0,64275)
4.  (5.0,42783)
5.  (5.0,53355)
6.  (5.0,51209)
7.  (4.75,26073)
8.  (4.75,26048)
9.  (4.75,65001)
10. (4.75,5194)
11. (4.75,4454)
12. 所有电影中粉丝或者观看人数最多的电影:
13. (296,34864)
14. (356,34457)
15. (593,33668)
16. (480,32631)
17. (318,31126)
18. (110,29154)
19. (457,28951)
20. (589,28948)
21. (260,28566)
22. (150,27035)
```

12.3　通过 RDD 分析各种类型的最喜爱电影 TopN 及性能优化技巧

通过 RDD 分析大数据电影点评系统各种类型的最喜爱电影 TopN。本节分析最受男性喜爱的电影 Top10 和最受女性喜爱的电影 Top10。

评级文件 ratings.dat 的格式描述如下：

```
1.  UserID::MovieID::Rating::Timestamp
```

```
2.  用户 ID、电影 ID、评分数据、时间戳
```

用户文件 users.dat 的格式描述如下：

```
1.  UserID::Gender::Age::Occupation::Zip-code
2.  用户 ID、性别、年龄、职业、邮编代码
```

单单从评分数据 ratings 中无法计算出最受男性或者女性喜爱的电影 Top10，因为该 RDD 中没有性别 Gender 信息，如果需要使用性别 Gender 信息进行性别 Gender 的分类，此时一定需要聚合。当然，我们力求聚合的使用是 mapjoin（分布式计算的杀手是数据倾斜，Mapper 端的 Join 是一定不会数据倾斜的），这里可否使用 mapjoin 呢？不可以，因为用户的数据非常多！所以，这里要使用正常的 Join，此处的场景不会数据倾斜，因为用户一般都均匀地分布。

最受男性喜爱的电影 Top10 和最受女性喜爱的电影 Top10 分析需注意以下事项。

（1）因为要再次使用电影数据的 RDD，所以复用了前面 Cache 的 ratings 数据。

（2）在根据性别过滤出数据后，关于 TopN 部分的代码直接复用前面的代码就行了。

（3）要进行 Join，需要 key-value。

（4）在进行 Join 的时候，通过 take 等方法注意 Join 后的数据格式(3319,((3319, 50, 4.5),F))。

（5）使用数据冗余来实现代码复用或者更高效地运行，这是企业级项目的一个非常重要的技巧。

大数据电影点评系统中，统计最受男性喜爱的电影 Top10 和最受女性喜爱的电影 Top10，我们先分别过滤出男性、女性相关的数据，具体实现思路如下。

（1）对 ratings 中的（用户 ID，电影 ID，评分）元组进行 map 转换，格式化成 Key-Value，即（用户 ID，（用户 ID，电影 ID，评分））。

（2）对 usersRDD 中的每行数据按“::”分隔符分割，然后进行 map 转换，格式化成 Key-Value，即（用户 ID，性别）。

（3）将（用户 ID，（用户 ID，电影 ID，评分））与（用户 ID，性别）进行 Join 生成新的 genderRatings RDD。格式为：（用户 ID，（（用户 ID，电影 ID，评分），性别）），性别，并且进行 Cache 缓存。

（4）对 genderRatings RDD 进行过滤转换，从元组（x._1 用户 ID，（x._2._1（用户 ID，电影 ID，评分），x._2._2 性别））过滤出 x._2._2 性别等于男性的数据。然后进行 map 转换为 x._2._1，即转换成（用户 ID，电影 ID，评分）格式，生成 maleFilteredRatings。

（5）对 genderRatings RDD 进行过滤转换，从元组（x._1 用户 ID，（x._2._1（用户 ID，电影 ID，评分），x._2._2 性别））过滤出 x._2._2 性别等于女性的数据。然后进行 map 转换为 x._2._1，即转换成（用户 ID，电影 ID，评分）格式，生成 femaleFilteredRatings。

从大数据电影点评系统中过滤男性、女性相关的数据的代码如下：

```
1.      val male = "M"
2.      val female = "F"
3.      val genderRatings = ratings.map(x => (x._1, (x._1, x._2, x._3))).join(
4.        usersRDD.map(_.split("::")).map(x => (x(0), x(1)))).cache()
5.      genderRatings.take(2).foreach(println)
6.      val maleFilteredRatings: RDD[(String, String, String)] = gender
        Ratings.filter(x => x._2._2.equals("M")).map(x => x._2._1)
7.      val femaleFilteredRatings = genderRatings.filter(x => x._2._2.equals
        ("F")).map(x => x._2._1)
```

在 IDEA 中运行代码，打印出 genderRatings 的数据，取 10 个数据，格式为：(用户 ID，((用户 ID，电影 ID，评分)，性别))，结果如下：

```
1.  (3319,((3319,32,5),F))
2.  (3319,((3319,50,4.5),F))
3.  (3319,((3319,163,4.5),F))
4.  (3319,((3319,180,5),F))
5.  (3319,((3319,296,5),F))
6.  (3319,((3319,318,5),F))
7.  (3319,((3319,405,4),F))
8.  (3319,((3319,914,4.5),F))
9.  (3319,((3319,1088,4),F))
10. (3319,((3319,1136,5),F))
```

电影点评系统用户行为分析，统计所有电影中最受男性喜爱的电影 Top10，具体实现思路如下：

（1）将性别为男的用户过滤以后的数据（用户 ID，电影 ID，评分）进行 map 转换，格式化成为 Key-Value 的方式，即（电影 ID，（评分，1））。

（2）使用 reduceByKey 算子对 Value 值进行汇聚转换，对两个具有相同 Key 值，而 Value 不同的元组，如（电影 ID，（x.评分，x.1）），（电影 ID，（y.评分，y.1）），计算得出（x 的评分+y 的评分，x 的计数+y 的计数），转换以后 Key 是电影 ID，Value 是（总的评分，总的点评人数），格式化成为 Key-Value，即（电影 ID，（总评分，总点评人数））。

（3）reduceByKey 算子执行完毕，接下来进行 map 转换操作，交换 Key-Value 值，并且计算出电影平均分=总评分/总点评人数，即将（电影 ID，（总评分，总点评人数））转换成（（总评分/总点评人数），电影 ID），然后使用 sortByKey(false)算子按电影平均分降序排列。

（4）再次进行 Key 和 Value 的交换，打印输出。使用 map 转换函数将（（总评分/总点评人数），电影 ID）进行交换，转换为（电影 ID，（总评分/总点评人数）），再通过 take(10)算子获取所有电影中最受男性喜爱的电影 Top10，进行打印输出。

在所有电影中分析最受男性喜爱的电影 Top10 的代码如下：

```
1.  println("所有电影中最受男性喜爱的电影 Top10:")
2.      maleFilteredRatings.map(x=>(x._2,(x._3.toDouble, 1)))//格式化成为 Key-Value
3.        .reduceByKey((x, y) => (x._1 + y._1,x._2 + y._2))
        //对 Value 进行 reduce 操作，分别得出每部电影的总的评分和总的点评人数
4.        .map(x => (x._2._1.toDouble / x._2._2, x._1))  //求出电影平均分
5.        .sortByKey(false)  //降序排列
6.        .map(x => (x._2, x._1))
7.        .take(10)           //取 Top10
8.        .foreach(println)  //打印到控制台
```

在 IDEA 中运行代码，结果如下：

```
1.  所有电影中最受男性喜爱的电影 Top10:
2.  (855,5.0)
3.  (6075,5.0)
4.  (1166,5.0)
5.  (3641,5.0)
6.  (1045,5.0)
7.  (4136,5.0)
8.  (2538,5.0)
9.  (7227,5.0)
```

```
10. (8484,5.0)
11. (5599,5.0)
```

同样地，在电影点评系统用户行为分析中，我们可以统计所有电影中最受女性喜爱的电影 Top10，具体实现思路和最受男性喜爱的电影 Top10 类似，这里不再赘述。

从所有电影中分析最受女性喜爱的电影 Top10 的代码如下：

```
1.      println("所有电影中最受女性喜爱的电影 Top10:")
2.      femaleFilteredRatings.map(x=>(x._2,(x._3.toDouble,1)))//格式化成为Key-Value
3.        .reduceByKey((x, y) => (x._1 + y._1,x._2 + y._2))
        //对 Value 进行 reduce 操作，分别得出每部电影的总的评分和总的点评人数
4.        .map(x => (x._2._1.toDouble / x._2._2, x._1))  //求出电影平均分
5.        .sortByKey(false)  //降序排列
6.        .map(x => (x._2, x._1))
7.        .take(10)          //取 Top10
8.        .foreach(println)  //打印到控制台
```

在 IDEA 中运行代码，结果如下：

```
1.  所有电影中最受女性喜爱的电影 Top10:
2.  [Stage 43:====================================>     7 + 1) / 8](789,5.0)
3.  (855,5.0)
4.  (32153,5.0)
5.  (4763,5.0)
6.  (26246,5.0)
7.  (2332,5.0)
8.  (503,5.0)
9.  (4925,5.0)
10. (8767,5.0)
11. (44657,5.0)
```

12.4　通过 RDD 分析电影点评系统仿 QQ 和微信等用户群分析及广播背后机制解密

通过 RDD 分析大数据电影点评系统仿 QQ 和微信等用户群分析，在本节统计最受不同年龄段人员欢迎的电影 TopN。

用户文件 users.dat 的格式描述如下：

```
1.  UserID::Gender::Age::Occupation::Zip-code
2.  用户 ID、性别、年龄、职业、邮编代码
```

电影文件 movies.dat 的格式描述如下：

```
1.  MovieID::Title::Genres
2.  电影 ID、电影名、电影类型
```

评级文件 ratings.dat 的格式描述如下：

```
1.  UserID::MovieID::Rating::Timestamp
2.  用户 ID、电影 ID、评分数据、时间戳
```

大数据电影点评系统仿 QQ 和微信等用户群分析实现思路：首先计算 TopN，但是这里的关注点有两个。

1．不同年龄阶段如何界定的问题

这个问题其实是业务问题。一般情况下，我们都是在原始数据中直接对要进行分组的年龄段提前进行数据清洗 ETL，例如，进行数据清洗 ETL 后产生以下数据。

```
1.  * 1:  "少于 18 岁"
2.      * 18:  "18 年龄段：从 18 岁到 24 岁"
3.      * 25:  "25 年龄段：从 25 岁到 34 岁"
4.      * 35:  "35 岁年龄段：从 35 岁到 44 岁"
5.      * 45:  "45 岁年龄段：从 45 岁到 49 岁"
6.      * 50:  "50 岁年龄段：从 50 岁到 55 岁"
7.      * 56:  "56 岁年龄段：大于 56 岁"
```

2．性能问题

第一点：在实现的时候可以使用 RDD 的 filter 算子，如 13 < age <18，但这样做会导致运行时进行大量的计算，因为要进行扫描，所以非常耗性能，通过提前进行数据清洗 ETL 把计算发生在 Spark 的业务逻辑运行前，用空间换时间，当然，这些实现也可以使用 Hive，因为 Hive 语法支持非常强悍且内置了最多的函数。

第二点：这里要使用 mapjoin，原因是 targetUsers 数据只有 UserID，数据量一般不会太大。

大数据电影点评系统仿 QQ 和微信等用户群分析，先过滤出仿 QQ 用户及微信用户 18 年龄段（从 18 岁到 24 岁）及仿淘宝用户 25 年龄段（从 25 岁到 34 岁）的用户数据，然后构建 Broadcast 数据结构类型的 targetQQUsersBroadcast、targetTaobaoUsersBroadcast 广播变量。具体实现方法如下。

（1）仿 QQ 用户及微信用户 18 年龄段用户 targetQQUsers 的创建：将 usersRDD 中的每行数据按“::”分隔符分割，然后 map 格式化为（用户 ID，年龄）元组，最后使用 filter 算子遍历过滤出元组的第二个元素等于 18 的数据。

（2）仿淘宝用户 25 年龄段用户 targetTaobaoUsers 的创建：将 usersRDD 中的每行数据按“::”分隔符分割，然后 map 格式化为（用户 ID，年龄）元组，然后使用 filter 算子遍历过滤出元组的第二个元素等于 25 的数据。

（3）构建仿 QQ 用户及微信用户数据集合 targetQQUsersSet。将 targetQQUsers（用户 ID，年龄）元组使用 map 转换函数获取第一个元素用户 ID，然后使用 collect()算子收集所有的仿 QQ 用户及微信用户的用户 ID。代码“HashSet() ++ targetQQUsers.map(_._1).collect()”，这里使用“++”运算符是因为：collect()算子返回的数据类型是数组类型 Array[T]，使用 HashSet() 进行“++”运算，结果将是 HashSet[String]类型，在之后的广播变量计算中将使用。而如果使用“+”运算符，结果将是 HashSet[Array[String]]类型，之后的广播变量计算编译器会提示类型不匹配。

（4）构建仿淘宝用户数据集合 targetTaobaoUsersSet。将 targetTaobaoUsers（用户 ID，年龄）元组使用 map 转换函数获取第一个元素用户 ID，然后使用 collect()算子收集所有的仿淘宝用户的用户 ID。

（5）构建广播变量数据结构 targetQQUsersBroadcast、targetTaobaoUsersBroadcast。在 Spark 中如何实现 mapjoin 呢？显然是要借助于 Broadcast，把数据广播到 Executor 级别，让该 Executor 上的所有任务共享该唯一的数据，而不是每次运行 Task 的时候都要发送一份数据的复制，这显著地降低了网络数据的传输和 JVM 内存的消耗。这里我们使用 sc.broadcast 分别创建了广播变量 targetQQUsersBroadcast、targetTaobaoUsersBroadcast。

```
1.        val  targetQQUsers  =  usersRDD.map(_.split("::")).map(x  =>  (x(0),
          x(2))).filter(_._2.equals("18"))
2.        val targetTaobaoUsers = usersRDD.map(_.split("::")).map(x => (x(0),
          x(2))).filter(_._2.equals("25"))
3.      /**
          * 在 Spark 中如何实现 mapjoin 呢？显然要借助于 Broadcast，把数据广播到 Executor
          * 级别，让该 Executor 上的所有任务共享该唯一的数据，而不是每次运行 Task 的时候
          * 都要发送一份数据的复制，这显著地降低了网络数据的传输和 JVM 内存的消耗
4.        */
5.      val targetQQUsersSet = HashSet() ++ targetQQUsers.map(_._1).collect()
6.      val targetTaobaoUsersSet = HashSet() ++ targetTaobaoUsers.map(_._1)
        .collect()
7.
8.      val targetQQUsersBroadcast = sc.broadcast(targetQQUsersSet)
9.      val targetTaobaoUsersBroadcast = sc.broadcast(targetTaobaoUsersSet)
```

所有电影中仿 QQ 或者微信核心目标用户最喜爱电影 TopN 分析，具体实现如下。

（1）将 moviesRDD 中的每行数据按“::”分隔符分割，然后 map 格式化为（电影 ID，电影名）元组，使用 collect 算子收集所有数据，通过 toMap 转换成 map 数据结构。

（2）将 ratingsRDD 中的每行数据按“::”分隔符分割，然后 map 格式化为（用户 ID，电影 ID）元组，最后使用 filter 算子遍历过滤。过滤条件为：元组的第一个元素用户 ID 包含在广播变量 targetQQUsersBroadcast 的集合 HashSet（用户 ID）中，即从 ratingsRDD 中过滤出 18 年龄段（从 18 岁到 24 岁）的用户的评分数据，数据格式为（用户 ID，电影 ID）。

（3）然后进行 map 转换，把数据变成 Key-Value，取 ratingsRDD 元组的第 2 个元素电影 ID 作为 Key，计数 1 次作为 Value，格式化为 Key-Value，即（电影 ID，1）。

（4）通过 reduceByKey 操作实现聚合：对相同 Key 的 Value 值进行累加。生成 Key-Value，即（电影 ID，总次数）。

（5）然后进行 Key 和 Value 的交换。上一步 reduceByKey 算子执行完毕，然后进行 map 转换操作，交换 Key-Value 值，即将（电影 ID，总次数）转换成（总次数，电影 ID）。

（6）接下来排序，我们使用 sortByKey(false)算子按总次数降序排列。

（7）排序完成，再次进行 Key 和 Value 的交换。我们使用 map 转换函数将（总次数，电影 ID）进行交换，转换为（电影 ID，总次数），再通过 take(10)算子获取 Top10 的记录。数据格式为（电影 ID，总次数）。

（8）为了使输出显示更加直观，我们使用 movieID2Name.getOrElse 获取电影 ID 对应的电影名，然后打印输出，格式为（电影名，总次数），即打印输出所有电影中仿 QQ 或者微信用户最喜爱电影的 Top10。

仿 QQ 或者微信核心目标用户最喜爱电影 TopN 分析代码如下：

```
1.      val movieID2Name = moviesRDD.map(_.split("::")).map(x => (x(0),
        x(1))).collect.toMap
2.     println("所有电影中 QQ 或者微信核心目标用户最喜爱电影 TopN 分析:")
3.     ratingsRDD.map(_.split("::")).map(x => (x(0), x(1))).filter(x =>
4.       targetQQUsersBroadcast.value.contains(x._1)
5.     ).map(x => (x._2, 1)).reduceByKey(_ + _).map(x => (x._2, x._1))
6.       .sortByKey(false).map(x => (x._2, x._1)).take(10).
7.       map(x => (movieID2Name.getOrElse(x._1, null), x._2)).foreach (println)
```

在 IDEA 中运行代码，结果如下：

```
1.  所有电影中 QQ 或者微信核心目标用户最喜爱电影 TopN 分析:
2.  (Silence of the Lambs, The (1991),524)
3.  (Pulp Fiction (1994),513)
4.  (Forrest Gump (1994),508)
5.  (Jurassic Park (1993),465)
6.  (Shawshank Redemption, The (1994),437)
7.  (Star Wars: Episode IV - A New Hope (1977),427)
8.  (Braveheart (1995),418)
9.  (Terminator 2: Judgment Day (1991),398)
10. (Toy Story (1995),396)
11. (Independence Day (ID4) (1996),393)
```

所有电影中淘宝核心目标用户最喜爱电影 TopN 分析，具体实现思路和仿 QQ 或者微信核心目标用户最喜爱电影 TopN 分析类似，这里不再赘述，实现代码如下：

```
1.     println("所有电影中淘宝核心目标用户最喜爱电影 TopN 分析:")
2.      ratingsRDD.map(_.split("::")).map(x => (x(0), x(1))).filter(x =>
3.        targetTaobaoUsersBroadcast.value.contains(x._1)
4.      ).map(x => (x._2, 1)).reduceByKey(_ + _).map(x => (x._2, x._1))
5.        .sortByKey(false).map(x => (x._2, x._1)).take(10).
6.        map(x => (movieID2Name.getOrElse(x._1, null), x._2)).foreach(println)
```

在 IDEA 中运行代码，结果如下：

```
1.  所有电影中淘宝核心目标用户最喜爱电影 TopN 分析:
2.  (Pulp Fiction (1994),959)
3.  (Silence of the Lambs, The (1991),949)
4.  (Forrest Gump (1994),935)
5.  (Jurassic Park (1993),894)
6.  (Shawshank Redemption, The (1994),859)
7.  (Star Wars: Episode IV - A New Hope (1977),853)
8.  (Toy Story (1995),801)
9.  (Independence Day (ID4) (1996),801)
10. (Braveheart (1995),798)
11. (Terminator 2: Judgment Day (1991),784)
```

12.5　通过 RDD 分析电影点评系统实现 Java 和 Scala 版本的二次排序系统

本节实现 RDD 分析大数据电影点评系统的二次排序功能，分别使用 Java 和 Scala 语言来实现。在实现大数据电影点评系统的二次排序前，我们先实现一个 Java 版本二次排序的例子，来阐述二次排序的实现。

数据准备：在 data/moviedata/medium/目录下新建文本文件 dataforsecondarysorting.txt，在文本文件中输入如下测试数据。

```
1.  2 4
2.  2 10
3.  3 6
4.  1 5
```

文本文件 dataforsecondarysorting.txt 中有两列数字，二次排序的含义是先按照第一列数字进行排序，如果第一列数字中有相同的数字，然后再按照第二列的数字进行第二次排序。如上述文本文件，进行二次排序的结果如下：

```
1.  3 6
2.  2 10
3.  2 4
4.  1 5
```

编写 MovieUsersAnalyzerTest.java 及 SecondarySortingKey.java 实现对 dataforsecondarysorting.txt 的文本文件内容进行二次排序。SecondarySortingKey.java 是自定义 Key 值类。MovieUsersAnalyzerTest.java 中实现了二次排序功能。

12.5.1　二次排序自定义 Key 值类实现（Java）

Spark 中可以使用 sortByKey 算子对数据的 Key 值进行排序，本节中，我们须进行二次排序，二次排序的关键是自定义 Key 值，在自定义 Key 中实现排序的功能。这里将每一行数据的第一个数据、第二个数据组合成一个 Key 值，然后在自定义 Key 中实现二次排序。同样的思路，如果业务需求需实现三次排序、四次排序，甚至更多维度的排序，重点也是自定义 Key 值，将多维数据组合成为自定义 Key，从而实现多维度的排序功能。

自定义 Key 值的 Java 类为 SecondarySortingKey.java，在类中定义需要二次排序的组合 key 值：first、second，然后重写$greater、$greater$eq、$less、$less$eq、compare、compareTo 方法，以及 hashCode、equals 方法。

compare 的方法是比较自身对象 this 与要比较对象 that 的比较结果。compare 方法用来确定如何将要排序的对象进行排序。假如返回的结果为 x，那么

- ❑ x < 0　表示　this < that
- ❑ x == 0　表示　this == that
- ❑ x > 0　表示　this > that

在 SecondarySortingKey.java 的重载方法 compareTo 中，分别将两个 SecondarySortingKey 对象进行比较，例如，dataforsecondarysorting.txt 的第一行（3　6）组合成 Key 的 this 对象，它将与第二行的（2　10）组合成 Key 的 that 对象进行比较，首先比较 this 和 that 的第一列数字，this 对象的第一列数字是 3，比 that 对象的第一列数字 2 大，因此相减值大于 0，表示 this（3　6）对象比 that 对象（2　10）大。如果 this 对象和 that 对象的第一列数字相等，那么再比较第二列数字的大小，进行二次排序。例如，Key（2　10）与 Key（2　4）第一列数字相等的情况下，再比较第二列数字，10 比 4 要大，因此 Key（2　10）、Key（2　4）二次排序以后，Key（2　10）比 Key（2　4）大。

方法$greater、$greater$eq、$less、$less$eq、compare 是同样的思路。以下是 Secondary

SortingKey.java 的重载方法 compare 的代码。

```
1.  @Override
2.     public int compareTo(SecondarySortingKey that) {
3.         if(this.first - that.getFirst() != 0){
4.            return this.first - that.getFirst();
5.         } else {
6.            return this.second - that.getSecond();
7.         }
8.     }
```

SecondarySortingKey.java 的完整代码如下：

```
1.   package com.dt.spark.cores;
2.  import scala.Serializable;
3.  import scala.math.Ordered;
4.  public class SecondarySortingKey implements Ordered<SecondarySortingKey>,
    Serializable{
5.     private int first;
6.     private int second;
7.     public int getFirst() {
8.        return first;
9.     }
10.    public void setFirst(int first) {
11.       this.first = first;
12.    }
13.    public int getSecond() {
14.       return second;
15.    }
16.    public void setSecond(int second) {
17.       this.second = second;
18.    }
19.    public SecondarySortingKey(int first, int second){
20.       this.first = first;
21.       this.second = second;
22.    }
23.    @Override
24.    public String toString() {
25.       return super.toString();
26.    }
27.    @Override
28.    public boolean equals(Object o) {
29.       if (this == o) return true;
30.       if (o == null || getClass() != o.getClass()) return false;
31.       SecondarySortingKey that = (SecondarySortingKey) o;
32.       if (first != that.first) return false;
33.       return second == that.second;
34.    }
35.    @Override
36.    public int hashCode() {
37.       int result = first;
38.       result = 31 * result + second;
39.       return result;
40.    }
41.    @Override
42.    public int compare(SecondarySortingKey that) {
43.       if(this.first - that.getFirst() != 0){
44.          return this.first - that.getFirst();
45.       } else {
```

```
46.            return this.second - that.getSecond();
47.        }
48.    }
49.    @Override
50.    public boolean $less(SecondarySortingKey that) {
51.        if(this.first < that.getFirst()) {
52.            return  true;
53.        } else if (this.first == that.getFirst() &&this.second < that
           .getSecond()){
54.            return true;
55.        }
56.        return false;
57.    }
58.    @Override
59.    public boolean $less$eq(SecondarySortingKey other) {
60.        if(SecondarySortingKey.this.$less(other)){
61.            return true;
62.        } else if (this.first == other.getFirst() && this.second == other
           .getSecond()){
63.            return true;
64.        }
65.        return false;
66.    }
67.    @Override
68.    public boolean $greater(SecondarySortingKey that) {
69.        if(this.first > that.getFirst()){
70.            return true;
71.        }else if(this.first == that.getFirst() && this.second > that
           .getSecond()) {
72.            return true;
73.        }
74.        return false;
75.    }
76.    @Override
77.    public boolean $greater$eq(SecondarySortingKey that) {
78.        if (SecondarySortingKey.this.$greater(that)){
79.            return true;
80.        } else if (this.first == that.getFirst() && this.second == that
           .getSecond()){
81.            return true;
82.        }
83.        return false;
84.    }
85.    @Override
86.    public int compareTo(SecondarySortingKey that) {
87.        if(this.first - that.getFirst() != 0){
88.            return this.first - that.getFirst();
89.        } else {
90.            return this.second - that.getSecond();
91.        }
92.    }
93. }
```

12.5.2　电影点评系统二次排序功能实现（Java）

我们已经实现了 SecondarySortingKey.java 自定义 Key 值类，接下来先编写一个 Java 测试类 MovieUsersAnalyzerTest.java，验证自定义 Key 值二次排序的功能。验证通过以后，只需将读入的 dataforsecondarysorting.txt 文本文件修改为读入电影点评系统文件，重新组合成电影点评系统的 Key，就可将二次排序代码移植到电影点评系统代码中。

下面先实现 Java 测试类 MovieUsersAnalyzerTest.java，具体实现思路如下。

（1）读入 dataforsecondarysorting.txt 每行的数据记录。

（2）对读入的数据进行 mapToPair 转换，格式为 Key-Value。先将每行数据按空格进行单词切分，返回一个 Key-Value 键值对。Key 为 SecondarySortKey 自定义类（每行数据切分后的第 0 个、第 1 个数据），value 为每行的原数据。格式化 Key-Value 的结果，如((2 4)，"2 4")。

（3）对 keyvalues 使用 sortByKey 算子按 SecondarySortKey 类型的 Key 值进行排序。通过 key 值 SecondarySortKey 的 compareTo 方法排序，实现了二次排序。

（4）上一步使用 sortByKey 算子进行 Key 值二次排序完毕，这里 Key 值不再需要了，我们使用 map 函数进行转换，直接返回 Key-Value 键值对的 Value。例如，((2 4)，"2 4") map 转换以后返回结果为"2 4"。

（5）打印输出二次排序后的结果。

MovieUsersAnalyzerTest.java 的完整代码如下：

```
1.  package com.dt.spark.cores;
2.  import org.apache.spark.SparkConf;
3.  import org.apache.spark.api.java.JavaPairRDD;
4.  import org.apache.spark.api.java.JavaRDD;
5.  import org.apache.spark.api.java.JavaSparkContext;
6.  import org.apache.spark.api.java.function.Function;
7.  import org.apache.spark.api.java.function.PairFunction;
8.  import scala.Tuple2;
9.  import java.util.List;
10. public class MovieUsersAnalyzerTest {
11.     public static void main(String[] args) {
12.         /**
            *创建 Spark 集群上下文 sc，在 sc 中可以进行各种依赖和参数的设置等，大家可以
            *通过 SparkSubmit 脚本的 help 去看设置信息
13.         */
14.         JavaSparkContext sc = new JavaSparkContext(new SparkConf().setMaster
            ("local[4]").setAppName("Movie_Users_Analyzer"));
15.         JavaRDD<String> lines = sc.textFile("data/moviedata/medium/" + "
           dataforsecondarysorting.txt");
16.         JavaPairRDD<SecondarySortingKey, String> keyvalues = lines.mapToPair
             (new PairFunction<String, SecondarySortingKey, String>() {
17.             private static final long serialVersionUID = 1L;
18.             @Override
19.             public Tuple2<SecondarySortingKey, String> call(String line)
                throws Exception {
20.                 String[] splited = line.split(" ");
21.                 SecondarySortingKey key = new SecondarySortingKey(Integer.
                    valueOf(splited[0]),
22.                         Integer.valueOf(splited[1])); //组合成 Key 值
23.                 return new Tuple2<SecondarySortingKey, String>(key, line);
24.             }
25.         });
26. //按 Key 值进行二次排序
```

```
27.         JavaPairRDD<SecondarySortingKey, String> sorted = keyvalues
            .sortByKey(false);
28.         JavaRDD<String> result = sorted.map(new Function<Tuple2<Secondary
            SortingKey, String>, String>() {
29.             private static final long serialVersionUID = 1L;
30.
31.             @Override
32.             public String call(Tuple2<SecondarySortingKey, String> tuple)
                throws Exception {
33.                 return tuple._2;//取第二个值 Value 值
34.             }
35.         });
36.         List<String> collected = result.take(10);
37.         for (String item : collected) {//打印输出二次排序后的结果
38.             System.out.println(item);
39.         }
40.     }
41. }
```

在 IDEA 中运行代码，结果如下：

```
1.  Using Spark's default log4j profile: org/apache/spark/log4j-defaults
    .properties
2.  17/03/01 11:07:16 INFO SparkContext: Running Spark version 2.1.0
3.  ......
4.  17/03/01 11:07:24 INFO DAGScheduler: Job 2 finished: take at MovieUsers
    AnalyzerTest.java:48, took 0.088430 s
5.  3 6
6.  2 10
7.  2 4
8.  1 5
```

Java 测试类 MovieUsersAnalyzerTest.java 已经测试验证通过，下面将读入的 dataforsecondarysorting.txt 文本文件修改为读入电影点评系统文件，重新组合成电影点评系统的 Key，从而实现电影点评系统的二次排序功能。在 MovieUsersAnalyzer.java 中读入电影评分数据 ratings.dat，对电影评分数据从时间、评分数二个维度进行排序，先按时间排序，然后按评分进行二次排序。

评级文件 ratings.dat 的格式描述如下：

```
1.  UserID::MovieID::Rating::Timestamp
2.  用户 ID、电影 ID、评分数据、时间戳
```

MovieUsersAnalyzer.java 在 MovieUsersAnalyzerTest.java 的基础上修改：①修改读入文件 ratings.dat；②读入的每行数据按“::”分隔符进行分割；③将时间戳、评分数据组合成 Key 值放入到 SecondarySortingKey 类。

MovieUsersAnalyzer.java 修改的地方如下：

```
1.      //"ratings.dat": UserID::MovieID::Rating::Timestamp
2.          JavaRDD<String> lines = sc.textFile("data/moviedata/medium/" +
            "ratings.dat");
3.          JavaPairRDD<SecondarySortingKey, String> keyvalues = lines. mapToPair
            (new PairFunction<String, SecondarySortingKey, String>() {
4.              private static final long serialVersionUID = 1L;
5.
6.              @Override
```

```
7.          public Tuple2<SecondarySortingKey, String> call(String line)
            throws Exception {
8.              String[] splited = line.split("::");
9.              SecondarySortingKey key = new SecondarySortingKey (Integer
                .valueOf(splited[3]),
10.                     Integer.valueOf(splited[2]));
11.             return new Tuple2<SecondarySortingKey, String>(key, line);
12.         }
13.     });
```

在 IDEA 中运行代码，显示结果"用户 ID、电影 ID、评分数据、时间戳"，先按时间戳排序，然后按评分排序，打印输出结果如下：

```
1.  Using Spark's default log4j profile: org/apache/spark/log4j- defaults.properties
2.  17/03/01 13:28:30 INFO SparkContext: Running Spark version 2.1.0
3.  ......
4.  17/03/01 13:28:38 INFO ShuffleBlockFetcherIterator: Started 0 remote
    fetches in 5 ms
5.  4958::1924::4::1046454590
6.  4958::3264::4::1046454548
7.  4958::2634::3::1046454548
8.  4958::1407::5::1046454443
9.  4958::2399::1::1046454338
10. 4958::3489::4::1046454320
11. 4958::2043::1::1046454282
12. 4958::2453::4::1046454260
13. 5312::3267::4::1046444711
14. 5948::3098::4::1046437932
```

12.5.3　二次排序自定义 Key 值类实现（Scala）

之前我们已使用 Java 实现了 RDD 分析大数据电影点评系统的二次排序功能，本节使用 Scala 语言来实现。在 Movie_Users_Analyzer_RDD.scala 代码文件中对电影评分数据进行二次排序，以时间戳 Timestamp 和评分 Rating 两个维度降序排列。

评级文件 ratings.dat 的格式描述如下。

```
1.  UserID::MovieID::Rating::Timestamp
2.  用户 ID、电影 ID、评分数据、时间戳
```

对电影评分数据进行二次排序完成的功能是最近时间中最新发生的点评信息。Scala 相对于 Java 代码实现更简洁优雅。具体实现方法如下。

（1）自定义 SecondarySortKey 类，在 SecondarySortKey 类构造函数中传入 first、second 两个参数。SecondarySortKey 类继承自 Ordered 排序接口及序列化器。

（2）在 SecondarySortKey 类中定义了 compare 方法，类似 Java 的二次排序方法，this、other 的比较和之前 Java 类中 this、that 的比较类似，这里不再赘述。和 Java 中代码实现不同的是，Scala 代码中引入了 Math.ceil 及 Math.floor 方法。

- Math.ceil 方法是向上取整计算，它返回的是大于或等于函数参数，并且与之最接近的整数。
- Math.floor 方法是求一个最接近它的整数，它的值小于或等于这个浮点数。

为什么这里需要调用 Math.ceil、Math.floor 方法呢？原因是 this 对象和 other 对象比较的

结果是一个 Int 整数。如果计算返回值是正小数，就需要实现大于 0 的小数，向上取整返回 1，表示第一个对象大于第二个对象；如果计算返回值是负小数，就需要实现小于 0 的小数向下取整返回-1，表示第一个对象小于第二个对象。实现的技巧就是通过 Math.ceil、Math.floor 方法来实现的。

compare 的方法是比较自身对象 this 与要比较对象 other 的比较结果。compare 方法用来确定如何将要排序的对象进行排序。假如计算 this.second - other.second 的值，那么：

- this.second - other.second = 0.5 返回值取 1，即取值 Math.ceil(0.5)。
- this.second - other.second = -0.5 返回值取-1，即取值 Math. floor(-0.5)。

SecondarySortKey 的代码如下：

```
1.  class  SecondarySortKey(val  first:Double,val  second:Double)  extends
    Ordered [SecondarySortKey] with Serializable {
2.       def compare(other:SecondarySortKey):Int = {
3.         if (this.first - other.first !=0) {
4.           (this.first - other.first).toInt
5.         } else {
6.           if (this.second - other.second > 0){
7.              Math.ceil(this.second - other.second).toInt
8.           } else if (this.second - other.second < 0){
9.             Math.floor(this.second - other.second).toInt
10.          } else {
11.            (this.second - other.second).toInt
12.          }
13.        }
```

12.5.4　电影点评系统二次排序功能实现（Scala）

接下来使用 Scala 实现电影点评系统的二次排序功能，实现对电影评分数据以时间戳 Timestamp 和评分 Rating 两个维度进行二次降序排列。

评级文件 ratings.dat 的格式描述如下：

```
1.  UserID::MovieID::Rating::Timestamp
2.  用户 ID、电影 ID、评分数据、时间戳
```

具体实现方法如下。

（1）读入电影点评系统文件，之前已读入 ratings.dat 文件生成 ratingsRDD。使用 map 函数对 ratingsRDD 进行转换，将每行数据按“::”分割，取到第 3 个元素时间戳，第 2 个元素评分数据，将时间戳、评分数据组合成为 Key 值，每行数据为 Value 值，形成格式化 Key-Value 键值对。格式为：((时间戳、评分数据)，“用户 ID、电影 ID、评分数据、时间戳”)。

（2）使用 sortByKey 算子按（时间戳、评分数据）Key 值排序。

（3）sortByKey 排序完毕，对于排序后的每行的元组数据（（时间戳、评分数据），“用户 ID、电影 ID、评分数据、时间戳”），取出元组的第 2 个元素，即“用户 ID、电影 ID、评分数据、时间戳”。

（4）使用 take 算子取出前 10 个数据，打印输出结果。

Movie_Users_Analyzer_RDD.scala 中电影点评系统二次排序功能实现代码如下：

```
1.        println("对电影评分数据以 Timestamp 和 Rating 两个维度进行二次降序排列:")
```

```
2.         val pairWithSortkey = ratingsRDD.map(line =>{
3.           val splited = line.split("::")
4.           (new SecondarySortKey( splited(3).toDouble,splited(2).toDouble),line )})
5.
6.         val sorted = pairWithSortkey.sortByKey(false)
7.
8.         val sortedResult = sorted.map(sortedline => sortedline._2)
9.         sortedResult.take(10).foreach(println)
```

在 IDEA 中运行代码，先按第 4 列时间戳排序，再按第 3 列评分数据排序，二次排序的结果如下：

```
1.  对电影评分数据以 Timestamp 和 Rating 两个维度进行二次降序排列：
2.  62510::34148::3::1231131736
3.  62510::4784::2.5::1231131303
4.  63100::2953::2::1231131142
5.  63100::7000::3.5::1231131137
6.  63100::2478::3.5::1231131132
7.  63100::231::3.5::1231131127
8.  63100::410::3.5::1231131118
9.  63100::4816::3.5::1231131112
10. 63100::2361::4.5::1231131107
11. 63100::36477::3::1231131103
```

12.6　通过 Spark SQL 中的 SQL 语句实现电影点评系统用户行为分析

之前我们已经详细阐述了通过 RDD 实现大数据电影点评系统中电影的用户行为信息分析，本节通过 Spark SQL 中的 SQL 语句实现大数据电影点评系统用户行为分析，通过 DataFrame 实现某特定电影观看者中男性和女性不同年龄分别有多少人？具体实现思路：① 从点评数据中获得观看者的信息 ID；② 把评分 ratings 表和用户 users 表进行 Join 操作，获得用户的性别信息；③ 使用内置函数（内部包含超过 200 个内置函数）进行信息统计和分析。

这里通过 DataFrame 来实现：首先通过 DataFrame 的方式表现评分 ratings 和用户 users 的数据，然后进行 Join 和统计操作。

用户文件 users.dat 的格式描述如下：

```
1.  UserID::Gender::Age::Occupation::Zip-code
2.  用户 ID、性别、年龄、职业、邮编代码
```

评级文件 ratings.dat 的格式描述如下：

```
1.  UserID::MovieID::Rating::Timestamp
2.  用户 ID、电影 ID、评分数据、时间戳
```

电影文件 movies.dat 的格式描述如下：

```
1.  MovieID::Title::Genres
2.  电影 ID、电影名、电影类型
```

通过 Spark SQL 中的 SQL 语句实现大数据电影点评系统用户行为分析，在 Movie_

Users_Analyzer_DateFrame.scala 代码文件中，首先读取电影点评系统的用户数据，用什么方式读取数据呢？这里使用 RDD 方式读入数据。

```
1.     val usersRDD = sc.textFile(dataPath + "users.dat")
2.     val moviesRDD = sc.textFile(dataPath + "movies.dat")
3.     val occupationsRDD = sc.textFile(dataPath + "occupations.dat")
4.     val ratingsRDD = sc.textFile(dataPath + "ratings.dat")
```

接下来使用 Spark SQL 中的 SQL 方式实现电影点评系统用户行为分析，具体实现步骤如下。

（1）使用 StructType 方式把用户的数据格式化，即在 RDD 的基础上增加数据的元数据信息，即 StructType(StructField(UserID, StringType, true), StructField(Gender, StringType, true), StructField(Age, StringType, true), StructField(Occupation, StringType, true), StructField(Zip-code, StringType, true))的格式。

（2）从用户 usersRDD 的每行数据按“::”分隔符分割，把每条数据变成以 Row 为单位的数据，Row 每行中的数据格式为（用户 ID、性别、年龄、职业、邮编代码）。

（3）创建 usersDataFrame：结合 usersRDDRows 和 schemaforusers StructType 的元数据信息基于 RDD 创建 DataFrame，这时 RDD 就有了元数据信息的描述。

```
1. val  schemaforusers  =  StructType("UserID::Gender::Age::OccupationID::
   Zip-code".split("::").
2.       map(column => StructField(column, StringType, true)))
       //使用 Struct 方式把 Users 的数据格式化，即在 RDD 的基础上增加数据的元数据信息
3.     val usersRDDRows: RDD[Row] = usersRDD.map(_.split("::")).map(line =>
       Row(line(0).trim,line(1).trim,line(2).trim,line(3).trim,line(4)
       .trim))//把我们的每条数据变成以 Row 为单位的数据
4.     val usersDataFrame = spark.createDataFrame(usersRDDRows, schemaforusers)
       //结合 Row 和 StructType 的元数据信息，基于 RDD 创建 DataFrame，这时 RDD 就有了
       //元数据信息的描述
```

（4）同样，使用 StructType 方式把评分的数据格式化，在 RDD 的基础上增加数据的元数据信息，即 StructType(StructField(UserID, StringType, true), StructField(MovieID, StringType, true)，StructField(Rating, DoubleType, true)，StructField(Timestamp, StringType, true))的格式。

（5）从评分 ratingsRDD 的每行数据按“::”分隔符分割，把评分的每条数据变成以 Row 为单位的数据，Row 每行中的数据格式为（用户 ID、电影 ID、评分数据、时间戳）。

（6）创建 ratingsDataFrame：结合 ratingsRDDRows 和 schemaforratings StructType 的元数据信息，基于 RDD 创建 ratingsDataFrame。

```
1.    val schemaforratings = StructType("UserID::MovieID".split("::")
2.       .map(column => StructField(column, StringType, true)))
3.       .add("Rating", DoubleType, true)
4.       .add("Timestamp",StringType, true)
5.
6.     val ratingsRDDRows = ratingsRDD.map(_.split("::")).map(line =>
       Row(line(0).trim,line(1)
7.       .trim,line(2).trim.toDouble,line(3).trim))
8.     val ratingsDataFrame = spark.createDataFrame(ratingsRDDRows, schemaforratings)
9.
```

（7）同样，使用 StructType 方式把电影的数据格式化，在 RDD 的基础上增加电影数据

的元数据信息，即 StructType(StructField(MovieID, StringType, true)，StructField(Title, StringType, true)，StructField(Genres, DoubleType, true))的格式。

（8）从电影 moviesRDD 的每行数据按“::”分隔符分割，把电影的每条数据变成以 Row 为单位的数据，Row 每行中的数据格式为（电影 ID、电影名、电影类型）。

（9）创建 moviesDataFrame：结合 moviesRDDRows 和 schemaformovies StructType 的元数据信息，基于 RDD 创建 moviesDataFrame。

```
1.      val schemaformovies = StructType("MovieID::Title::Genres".split("::")
2.        .map(column => StructField(column, StringType, true)))
         //使用 Struct 方式把 Users 的数据格式化，即在 RDD 的基础上增加数据的元数据信息
3.      val moviesRDDRows = moviesRDD.map(_.split("::")).map(line =>
        Row(line(0).trim,line(1)
4.        .trim,line(2).trim)) //把我们的每条数据变成以 Row 为单位的数据
5.      val moviesDataFrame = spark.createDataFrame(moviesRDDRows, schemaformovies)
        //结合 Row 和 StructType 的元数据信息，基于 RDD 创建 DataFrame，这时 RDD 就有了
        //元数据信息的描述
```

（10）在 ratingsDataFrame 中使用 filter 过滤算子，过滤出电影 ID 等于 1193 的数据，这里能够直接指定电影 ID 的原因是 DataFrame 中有该元数据信息。

（11）评分数据 ratingsDataFrame 和用户数据进行 Join 关联，Join 的时候直接指定基于用户 ID 进行 Join，这相对于原生的 RDD 操作而言更加方便、快捷。

（12）然后选择性别、年龄数据。可以直接通过元数据信息中的 Gender 和 Age 进行数据的筛选。

（13）使用 groupBy 算子对性别、年龄进行分组。可以直接通过元数据信息中的 Gender 和 Age 进行数据的 groupBy 操作。

（14）基于 groupBy 分组信息进行计数统计操作。

（15）显示出分组统计后的前 10 条信息。

代码如下：

```
1.      ratingsDataFrame.filter(s" MovieID = 1193")
        //这里能够直接指定 MovieID 的原因是 DataFrame 中有该元数据信息
2.            .join(usersDataFrame, "UserID")
            //Join 的时候直接指定基于 UserID 进行 Join，这相对于原生的 RDD 操作而言更
            //加方便、快捷
3.            .select("Gender", "Age")
             //直接通过元数据信息中的 Gender 和 Age 进行数据的筛选
4.            .groupBy("Gender", "Age")
             //直接通过元数据信息中的 Gender 和 Age 进行数据的 groupBy 操作
5.            .count()  //基于 groupBy 分组信息进行 count 统计操作
6.            .show(10) //显示出分组统计后的前 10 条信息
```

在 IDEA 中运行代码，结果如下：

```
1.   功能一：通过 DataFrame 实现某特定电影观看者中男性和女性不同年龄分别有多少人？
2.  +------+---+-----+
3.  |Gender|Age|count|
4.  +------+---+-----+
5.  |     F|45 |   55|
6.  |     M|50 |  102|
7.  |     M| 1 |   26|
8.  |     F| 56|   39|
```

```
9.  |     F | 50|   43|
10. |     F | 18|   57|
11. |     F |  1|   10|
12. |     M | 18|  192|
13. |     F | 25|  140|
14. |     M | 45|  136|
15. +------+---+-----+
16. only showing top 10 rows
```

接下来，通过 Spark SQL 中的 SQL 语句实现大数据电影点评系统用户行为分析，统计某特定电影观看者中男性和女性不同年龄分别有多少人。这里分别使用全局临时视图 GlobalTempView、临时视图 TempView 两种 SQL 语句方式来实现。具体实现步骤：从 DataFrame 创建全局临时视图或者临时视图，然后编写常规的 SQL 代码语句，在 Spark 执行 SQL 语句操作进行展示。全局临时视图 GlobalTempView、临时视图 LocalTempView 的区别参考本章的知识点内容。两种 SQL 方式实现的代码如下：

```
1.   println("功能二：用 GlobalTempView 的 SQL 语句实现某特定电影观看者中男性和女性不同年龄分别有多少人？")
2.      ratingsDataFrame.createGlobalTempView("ratings")
3.      usersDataFrame.createGlobalTempView("users")
4.
5.      spark.sql("SELECT Gender, Age, count(*) from  global_temp.users u join global_temp.ratings as r on u.UserID = r.UserID where MovieID = 1193" +
6.        " group by Gender, Age").show(10)
7.
8.      println("功能二：用 LocalTempView 的 SQL 语句实现某特定电影观看者中男性和女性不同年龄分别有多少人？")
9.      ratingsDataFrame.createTempView("ratings")
10.     usersDataFrame.createTempView("users")
11.
12.     spark.sql("SELECT Gender, Age, count(*) from  users u join  ratings as r on u.UserID = r.UserID where MovieID = 1193" +
13.       " group by Gender, Age").show(10)
```

在 IDEA 中运行代码，结果如下：

```
1.   功能二：用 GlobalTempView 的 SQL 语句实现某特定电影观看者中男性和女性不同年龄分别有多少人？
2.   +------+---+--------+
3.   |Gender|Age|count(1)|
4.   +------+---+--------+
5.   |     F | 45|      55 |
6.   |     M | 50|     102 |
7.   |     M |  1|      26 |
8.   |     F | 56|      39 |
9.   |     F | 50|      43 |
10.  |     F | 18|      57 |
11.  |     F |  1|      10 |
12.  |     M | 18|     192 |
13.  |     F | 25|     140 |
14.  |     M | 45|     136 |
15.  +------+---+--------+
16.  only showing top 10 rows
17.
18.  功能二：用 LocalTempView 的 SQL 语句实现某特定电影观看者中男性和女性不同年龄分别有多少人？
```

```
19. +------+---+--------+
20. |Gender|Age|count(1)|
21. +------+---+--------+
22. |     F | 45|      55 |
23. |     M | 50|     102 |
24. |     M |  1|      26 |
25. |     F | 56|      39 |
26. |     F | 50|      43 |
27. |     F | 18|      57 |
28. |     F |  1|      10 |
29. |     M | 18|     192 |
30. |     F | 25|     140 |
31. |     M | 45|     136 |
32. +------+---+--------+
33. only showing top 10 rows
```

12.7 通过 Spark SQL 下的两种不同方式实现口碑最佳电影分析

所有电影中平均得分最高(口碑最好)的电影 TopN 计算可以有多种实现方式,其中 Spark SQL 下有两种不同方式。

- ❑ Spark SQL 方式一：通过 DataFrame 和 RDD 结合的方式，本节将阐述。
- ❑ Spark SQL 方式二：通过纯粹使用 DataFrame 方式计算，本节将阐述。
- ❑ RDD 方式：纯粹通过 RDD 的方式实现，12.2 节已经详细阐述过。
- ❑ DataSet 方式：例如，纯粹通过 DataSet 对电影点评系统进行流行度和不同年龄阶段兴趣分析，将在 12.11 节阐述。

RDD 方式：在 12.2 节通过 RDD 实现分析电影流行度分析统计实战中，我们已经对纯粹通过 RDD 的方式实现所有电影中平均得分最高（口碑最好）的电影的代码实现进行过详细阐述，这里不再重复。通过 RDD 的方式实现的代码如下：

```
1.   println("纯粹通过 RDD 的方式实现所有电影中平均得分最高（口碑最好）的电影 TopN:")
2.      val  ratings=ratingsRDD.map(_.split("::")).map(x  =>  (x(0),  x(1),
     x(2))).cache()  //格式化出电影 ID 和评分
3.      ratings.map(x=>(x._2,(x._3.toDouble,1.toDouble)))//格式化成 Key-Value 的方式
4.        .reduceByKey((x, y) => (x._1 + y._1,x._2 + y._2))
       //对 Value 进行 reduce 操作，分别得出每部电影的总的评分和总的点评人数
5.        .map(x => (x._2._1.toDouble/x._2._2.toDouble,x._1))//求出电影平均分
6.        .sortByKey(false)  //降序排列
7.          .map(x => (x._2, x._1))
8.        .take(10)            //取 Top10
9.        .foreach(println)  //打印到控制台
```

Spark SQL 方式一：现在通过 DataFrame 和 RDD 相结合的方式计算所有电影中平均得分最高（口碑最好）的电影 TopN 以及所有电影中粉丝或者观看人数最多（最流行电影）的电影 TopN。

评级文件 ratings.dat 的格式描述如下：

```
1.  UserID::MovieID::Rating::Timestamp
```

```
2.  用户 ID、电影 ID、评分数据、时间戳
```

通过 DataFrame 和 RDD 相结合的方式计算所有电影中平均得分最高（口碑最好）的电影 TopN 的具体方法如下。

（1）之前我们已经结合 ratingsRDDRows 的行数据和 StructType 的 schemaforratings 元数据信息基于 RDD 创建了 ratingsDataFrame。打印 ratingsDataFrame 的结构体显示如下：

```
ratingsDataFrame.printSchema()
1.   |-- UserID: string (nullable = true)
2.   |-- MovieID: string (nullable = true)
3.   |-- Rating: double (nullable = true)
4.   |-- Timestamp: string (nullable = true)
```

（2）通过 DataFrame 方式从 ratingsDataFrame 中选择两列数据（电影 ID，评分数据），然后按照电影 ID 进行分组，使用 avg("Rating")算子计算出每部电影的平均分数。

（3）然后将 ratingsDataFrame 转换成 RDD，格式为（电影 ID，平均分）。

（4）转换为 RDD 以后，接下来结合 RDD 的方式进行统计分析。将 RDD 每行数据（电影 ID，平均分）格式化成 Key-Value，即（平均分，（电影 ID，平均分））。

（5）然后使用 RDD 的 sortBy 算子按 Key 平均分进行排序。排序以后的输出格式为（平均分，（电影 ID，平均分））。

（6）进行 map 转换，取元组的第 2 个数据（电影 ID，平均分）的前 10 名打印输出。

通过 DataFrame 和 RDD 相结合的方式计算所有电影中平均得分最高（口碑最好）的电影 TopN 代码如下：

```
1.   println("通过 DataFrame 和 RDD 相结合的方式计算所有电影中平均得分最高（口碑最好）
     的电影 TopN:")
2.      ratingsDataFrame.select("MovieID", "Rating").groupBy("MovieID")
3.        .avg("Rating").rdd.
4.        map(row=>(row(1),(row(0),row(1)))).sortBy(_._1.toString.toDouble,false)
5.        .map(tuple => tuple._2).collect.take(10).foreach(println)
```

在 IDEA 中运行代码，统计所有电影中平均得分最高（口碑最好）的电影 Top10，输出结果如下：

```
1.  通过 DataFrame 和 RDD 相结合的方式计算所有电影中平均得分最高（口碑最好）的电影 TopN:
2.  (3656,5.0)
3.  (3280,5.0)
4.  (989,5.0)
5.  (787,5.0)
6.  (3607,5.0)
7.  (3172,5.0)
8.  (3233,5.0)
9.  (3881,5.0)
10. (1830,5.0)
11. (3382,5.0)
```

Spark SQL 方式二：纯粹使用 DataFrame 方式计算所有电影中平均得分最高（口碑最好）的电影 TopN。

纯粹使用 DataFrame 方式计算所有电影中平均得分最高（口碑最好）的电影 TopN 代码如下：

```
import spark.sqlContext.implicits._
1.      println("通过纯粹使用 DataFrame 方式计算所有电影中平均得分最高（口碑最好）的电
        影 TopN:")
2.      ratingsDataFrame.select("MovieID", "Rating").groupBy("MovieID").
3.        avg("Rating").orderBy($"avg(Rating)".desc).show(10)
```

下面详细阐述一下纯粹使用 DataFrame 方式的具体实现思路。

（1）导入 SparkContext 隐式转换的方法。sqlContext.implicits 继承自 SQLImplicits，SQLImplicits 中包括一系列的隐式方法集，隐式方法可以将普通的 Scala 对象隐式转换成[Dataset]。

ratingsDataFrame 结构如下：

```
ratingsDataFrame.printSchema()
1.    |-- UserID: string (nullable = true)
2.    |-- MovieID: string (nullable = true)
3.    |-- Rating: double (nullable = true)
4.    |-- Timestamp: string (nullable = true)
```

（2）从 ratingsDataFrame 使用 select 方法选择电影 ID、评分数据两列。我们使用 ratingsDataFrame.select("MovieID", "Rating").explain(true)，查看 select explain 的解析过程。

① 经 sql 解析器词法分析生成未解析的逻辑计划，从[UserID#4, MovieID#5, Rating#6, Timestamp#7]中投影选择未解析的两列数据：电影 ID、评分数据。

② 通过语法分析器，形成解析以后的逻辑计划。取到电影 ID、评分数据，即[MovieID#5, Rating#6]。

③ 经优化器进行优化，生成优化以后的逻辑计划。这里仅做了 select 简单操作，不用优化。

④ 然后通过 Spark 计划，生成物理计划。

explain 的解析过程如下：

```
1.   == Parsed Loical Plan ==        //生成未解析的逻辑计划
2.  'Project [unresolvedalias('MovieID, None), unresolvedalias('Rating, None)]
3.  +- LogicalRDD [UserID#4, MovieID#5, Rating#6, Timestamp#7]
4.
5.  == Analyzed Logical Plan ==      //解析以后的逻辑计划
6.  MovieID: string, Rating: double
7.  Project [MovieID#5, Rating#6]
8.  +- LogicalRDD [UserID#4, MovieID#5, Rating#6, Timestamp#7]
9.
10. == Optimized Logical Plan ==     //生成优化以后的逻辑计划
11. Project [MovieID#5, Rating#6]
12. +- LogicalRDD [UserID#4, MovieID#5, Rating#6, Timestamp#7]
13.
14. == Physical Plan ==              //生成物理计划
15. *Project [MovieID#5, Rating#6]
16. +- Scan ExistingRDD[UserID#4,MovieID#5,Rating#6,Timestamp#7]
```

（3）使用 groupBy 方法按照电影 ID 进行分组。使用 groupBy 算子将返回的数据类型是 RelationalGroupedDataset。在 RelationalGroupedDataset 类中包括所有可用的聚集函数，如计数平均分的函数 avg 等。

Spark DataSet.scala 框架源码中的 groupBy 源码如下：

```
1.   /**
      * 使用指定列对数据集进行分组，可以在数据集上运行聚集函数。[RelationalGrouped
      * Dataset]中包括所有可用的聚集函数。只能利用现有列的列名（即不能构造表达式）进行
      * groupBy 操作
2.    * {{{
3.    *   // 例如：计算由各部门分组的所有数值列的平均值
4.    *   ds.groupBy("department").avg()
5.    *
6.    *   //计算最大的年龄和平均工资，按部门和性别分组
7.    *   ds.groupBy($"department", $"gender").agg(Map(
8.    *     "salary" -> "avg",
9.    *     "age" -> "max"
10.   *   ))
11.   * }}}
12.   * @group untypedrel
13.   * @since 2.0.0
14.   */
15.  @scala.annotation.varargs
16.  def groupBy(col1: String, cols: String*): RelationalGroupedDataset = {
17.    val colNames: Seq[String] = col1 +: cols
18.    RelationalGroupedDataset(
19.      toDF(), colNames.map(colName => resolve(colName)), Relational
       GroupedDataset.GroupByType)
20.  }
```

（4）使用 avg 方法计算评分数据的平均分。使用 ratingsDataFrame.select("MovieID", "Rating").groupBy("MovieID").avg("Rating").explain(true)查看 avg 的解析过程。

① 经 sql 解析器词法分析生成逻辑计划，在分组中使用聚集函数 Aggregate 形成：电影ID，评分平均分。

② 通过语法分析器，形成解析以后的逻辑计划。取到电影 ID，评分平均分。

③ 经优化器进行优化，生成优化以后的逻辑计划。这里不需要优化。

④ 然后通过 Spark 计划，生成物理计划。使用 hash 分区，并行度定义为 200。

avg 的解析过程如下：

```
== Parsed Logical Plan == //生成逻辑计划
1.  Aggregate [MovieID#5], [MovieID#5, avg(Rating#6) AS avg(Rating)#37]
2.  +- Project [MovieID#5, Rating#6]
3.     +- LogicalRDD [UserID#4, MovieID#5, Rating#6, Timestamp#7]
4.
5.  == Analyzed Logical Plan ==//解析以后的逻辑计划
6.  MovieID: string, avg(Rating): double
7.  Aggregate [MovieID#5], [MovieID#5, avg(Rating#6) AS avg(Rating)#37]
8.  +- Project [MovieID#5, Rating#6]
9.     +- LogicalRDD [UserID#4, MovieID#5, Rating#6, Timestamp#7]
10.
11. == Optimized Logical Plan ==  //生成优化以后的逻辑计划
12.
13. Aggregate [MovieID#5], [MovieID#5, avg(Rating#6) AS avg(Rating)#37]
14. +- Project [MovieID#5, Rating#6]
15.    +- LogicalRDD [UserID#4, MovieID#5, Rating#6, Timestamp#7]
16.
17. == Physical Plan == //生成物理计划
```

```
18. *HashAggregate(keys=[MovieID#5],  functions=[avg(Rating#6)],  output=[MovieID#5,
    avg(Rating)#37])
19. +- Exchange hashpartitioning(MovieID#5, 200)
20.    +- *HashAggregate(keys=[MovieID#5], functions=[partial_avg(Rating#6)],
       output=[MovieID#5, sum#44, count#45L])
21.       +- *Project [MovieID#5, Rating#6]
22.          +- Scan ExistingRDD[UserID#4,MovieID#5,Rating#6,Timestamp#7]
```

（5）使用 orderBy 方法按平均分进行降序排列，然后打印输出前 10 个数据。同样，经过逻辑计划、解析后的逻辑计划、优化后的逻辑计划、物理计划几个步骤，这里不再赘述。

orderBy 的解析过程如下：

```
== Parsed Logical Plan ==
1.  'Sort ['avg(Rating) DESC NULLS LAST], true
2.  +- Aggregate [MovieID#5], [MovieID#5, avg(Rating#6) AS avg(Rating)#54]
3.     +- Project [MovieID#5, Rating#6]
4.        +- LogicalRDD [UserID#4, MovieID#5, Rating#6, Timestamp#7]
5.
6.  == Analyzed Logical Plan ==
7.  MovieID: string, avg(Rating): double
8.  Sort [avg(Rating)#54 DESC NULLS LAST], true
9.  +- Aggregate [MovieID#5], [MovieID#5, avg(Rating#6) AS avg(Rating)#54]
10.    +- Project [MovieID#5, Rating#6]
11.       +- LogicalRDD [UserID#4, MovieID#5, Rating#6, Timestamp#7]
12.
13. == Optimized Logical Plan ==
14. Sort [avg(Rating)#54 DESC NULLS LAST], true
15. +- Aggregate [MovieID#5], [MovieID#5, avg(Rating#6) AS avg(Rating)#54]
16.    +- Project [MovieID#5, Rating#6]
17.       +- LogicalRDD [UserID#4, MovieID#5, Rating#6, Timestamp#7]
18.
19. == Physical Plan ==
20. *Sort [avg(Rating)#54 DESC NULLS LAST], true, 0
21. +- Exchange rangepartitioning(avg(Rating)#54 DESC NULLS LAST, 200)
22.    +- *HashAggregate(keys=[MovieID#5], functions=[avg(Rating#6)], output=
       [MovieID#5, avg(Rating)#54])
23.       +- Exchange hashpartitioning(MovieID#5, 200)
24.          +- *HashAggregate(keys=[MovieID#5], functions=[partial_avg(Rating#6)],
             output=[MovieID#5, sum#62, count#63L])
25.             +- *Project [MovieID#5, Rating#6]
26.                +- Scan ExistingRDD[UserID#4,MovieID#5,Rating#6,Timestamp#7]
```

在 IDEA 中运行代码，统计纯粹使用 DataFrame 方式计算所有电影中平均得分最高（口碑最好）的电影 TopN，输出结果如下：

```
1.  通过纯粹使用 DataFrame 方式计算所有电影中平均得分最高（口碑最好）的电影 TopN:
2.  +-------+-----------+
3.  |MovieID|avg(Rating)|
4.  +-------+-----------+
5.  |   3382 |       5.0 |
6.  |   1830 |       5.0 |
7.  |   3607 |       5.0 |
8.  |   3881 |       5.0 |
9.  |   3172 |       5.0 |
10. |   3280 |       5.0 |
11. |    989 |       5.0 |
12. |   3656 |       5.0 |
```

```
13. |    787 |        5.0 |
14. |   3233 |        5.0 |
15. +-------+-----------+
16. only showing top 10 rows
```

12.8　通过 Spark SQL 下的两种不同方式实现最流行电影分析

类似于所有电影中平均得分最高（口碑最好）的电影 TopN 计算思路，所有电影中粉丝或者观看人数最多（最流行电影）的电影 TopN 同样可有多种实现方式，其中 Spark SQL 下有两种不同方式。

- Spark SQL 方式一：通过 DataFrame 和 RDD 相结合的方式。
- Spark SQL 方式二：通过纯粹使用 DataFrame 方式计算。
- RDD 方式：纯粹通过 RDD 的方式实现，12.2 节已经详细阐述过。
- DataSet 方式：例如，纯粹通过 DataSet 对电影点评系统进行流行度和不同年龄阶段兴趣分析，将在 12.11 节阐述。

RDD 方式：在 12.2 节通过 RDD 实现分析电影流行度分析统计实战中，我们已经对纯粹通过 RDD 的方式实现所有电影中粉丝或者观看人数最多（最流行电影）的电影 TopN 的代码实现进行过详细阐述，这里不再重复。通过 RDD 的方式实现的代码如下：

```
1.   println("纯粹通过 RDD 的方式计算所有电影中粉丝或者观看人数最多(最流行电影)的电影
     TopN:")
2.      ratings.map(x => (x._2, 1)).reduceByKey(_+_).map(x => (x._2, x._1))
3.        .sortByKey(false).map(x => (x._2, x._1)).collect()
4.        .take(10).foreach(println)
```

Spark SQL 方式一：本节通过 DataFrame 和 RDD 相结合的方式计算所有电影中粉丝或者观看人数最多（最流行电影）的电影 TopN，具体实现思路如下。

（1）通过 DataFrame 方式从 ratingsDataFrame 中选择两列数据（电影 ID，时间戳），然后按照电影 ID 进行分组，使用 count()算子计算出每部电影观看的总次数。

（2）将 ratingsDataFrame 转换成 RDD，格式为（电影 ID，总次数）。

（3）转换为 RDD 后，接下来结合 RDD 的方式进行统计分析。将 RDD 每行数据（电影 ID，总次数）格式化成 Key-Value，即（总次数，（电影 ID，总次数））。

（4）然后使用 RDD 的 sortBy 算子按 Key 总次数进行排序。排序以后的输出格式为（总次数，（电影 ID，总次数））。

（5）进行 map 转换，取元组的第 2 个数据（电影 ID，总次数）的前 10 名打印输出。

通过 DataFrame 和 RDD 结合的方式计算最流行电影（即所有电影中粉丝或者观看人数最多）的电影 TopN 代码如下：

```
1.  println("通过 DataFrame 和 RDD 结合的方式计算最流行电影（即所有电影中粉丝或者观看
    人数最多）的电影 TopN:")
2.     ratingsDataFrame.select("MovieID","Timestamp").groupBy("MovieID")
       .count().rdd.
3.       map(row => (row(1).toString.toLong, (row(0), row(1)))).sortByKey(false)
4.       .map(tuple => tuple._2).collect().take(10).foreach(println)
```

在 IDEA 中运行代码，统计所有电影中粉丝或者观看人数最多（最流行电影）的电影 Top10，输出结果如下：

```
1.  通过 DataFrame 和 RDD 结合的方式计算最流行电影即（所有电影中粉丝或者观看人数最多）的
    电影 TopN:(2858,3428)
2.  (260,2991)
3.  (1196,2990)
4.  (1210,2883)
5.  (480,2672)
6.  (2028,2653)
7.  (589,2649)
8.  (2571,2590)
9.  (1270,2583)
10. (593,2578)
```

Spark SQL 方式二：纯粹通过 DataFrame 的方式计算最流行电影（即所有电影中粉丝或者观看人数最多）的电影 TopN，ratingsDataFrame 可以直接按照电影 ID 进行分组，再根据分组进行统计计数，然后打印输出统计列排序后的结果，代码如下：

```
1.      println("纯粹通过 DataFrame 的方式计算最流行电影(即所有电影中粉丝或者观看人数
        最多) 的电影 TopN:")
2.  //    ratingsDataFrame.select("MovieID","Timestamp").
3.  //    ratingsDataFrame.select("MovieID").
4.        ratingsDataFrame.groupBy("MovieID").count()
5.        .orderBy($"count".desc).show(10)
```

在 IDEA 中运行代码，统计纯粹通过 DataFrame 的方式计算最流行电影（即所有电影中粉丝或者观看人数最多）的电影 TopN，输出结果如下：

```
1.  纯粹通过 DataFrame 的方式计算最流行电影即所有电影中粉丝或者观看人数最多(最流行电影)
    的电影 TopN:
2.  [Stage 84:=============================> (1 + 1) / 2]+-------+-----+
3.  |MovieID|count|
4.  +-------+-----+
5.  |   2858| 3428|
6.  |    260| 2991|
7.  |   1196| 2990|
8.  |   1210| 2883|
9.  |    480| 2672|
10. |   2028| 2653|
11. |    589| 2649|
12. |   2571| 2590|
13. |   1270| 2583|
14. |    593| 2578|
15. +-------+-----+
16. only showing top 10 rows
```

12.9 通过 DataFrame 分析最受男性和女性喜爱电影 TopN

在 12.3 节通过 RDD 分析各种类型的最喜爱电影 TopN 及性能优化技巧中，我们已经对纯粹通过 RDD 的方式实现最受男性和女性喜爱电影 TopN 的代码实现进行过详细阐述，这里不再赘述。

纯粹使用 RDD 实现所有电影中最受男性喜爱的电影 Top10 的代码如下：

```
1.     println("纯粹使用 RDD 实现所有电影中最受男性喜爱的电影 Top10:")
2.      maleFilteredRatings.map(x=>(x._2,(x._3.toDouble,1)))//格式化为Key-Value的形式
3.        .reduceByKey((x, y) => (x._1 + y._1,x._2 + y._2))
        //对 Value 进行 reduce 操作，分别得出每部电影的总的评分和总的点评人数
4.        .map(x => (x._2._1.toDouble / x._2._2, x._1))  //求出电影平均分
5.        .sortByKey(false)  //降序排列
6.        .map(x => (x._2, x._1))
7.        .take(10)          //取 Top10
8.        .foreach(println) //打印到控制台
```

在本节，我们纯粹使用 DataFrame 实现所有电影中最受男性喜爱的电影 Top10，具体实现思路如下。

（1）12.6 节结合 Row 和 StructType 的元数据信息基于 RDD 创建 DataFrame，分别创建了 usersDataFrame 和 ratingsDataFrame。

（2）统计生成 genderRatingsDataFrame。根据评分 ratingsDataFrame 和用户数据 usersDataFrame 按用户 ID 号 Join，并进行缓存。

genderRatingsDataFrame 格式如下：

```
1.  +------+-------+------+---------+------+---+------------+--------+
2.  |UserID|MovieID|Rating|Timestamp|Gender|Age|OccupationID|Zip-code|
3.  +------+-------+------+---------+------+---+------------+--------+
4.  |  1090 |   1250|   3.0 |974931632|     M | 25|          7  |  94105 |
5.  |  1090 |   2997|   3.0 |974931925|     M | 25|          7  |  94105 |
6.  |  1090 |    589|   4.0 |974930654|     M | 25|          7  |  94105 |
7.  |  1090 |    593|   3.0 |974930569|     M | 25|          7  |  94105 |
8.  |  1090 |   1287|   4.0 |974931632|     M | 25|          7  |  94105 |
9.  |  1090 |   1293|   4.0 |974931575|     M | 25|          7  |  94105 |
10. |  1090 |    908|   3.0 |974930672|     M | 25|          7  |  94105 |
11. |  1090 |    909|   3.0 |974931763|     M | 25|          7  |  94105 |
12. |  1090 |    912|   4.0 |974931647|     M | 25|          7  |  94105 |
13. |  1090 |    913|   3.0 |974932091|     M | 25|          7  |  94105 |
14. +------+-------+------+---------+------+---+------------+--------+
15. only showing top 10 rows
```

（3）genderRatingsDataFrame 按性别为“男”进行过滤。

（4）过滤以后，按电影 ID 进行分组，按评分求平均分，然后按平均分进行排序，打印输出。

纯粹使用 DataFrame 实现所有电影中最受男性喜爱的电影 Top10 代码如下：

```
1.      val schemaforusers: StructType = StructType("UserID::Gender::Age::
        OccupationID::Zip-code".split("::")
2.        .map(column => StructField(column, StringType, true)))
        //使用 Struct 方式把 Users 的数据格式化，即在 RDD 的基础上增加数据的元数据信息
3.      val usersRDDRows: RDD[Row] = usersRDD.map(_.split("::")).map(line =>
        Row(line(0).trim,line(1)
4.        .trim,line(2).trim,line(3).trim,line(4).trim))
        //把我们的每一条数据变成以 Row 为单位的数据
5.      val usersDataFrame: DataFrame = spark.createDataFrame(usersRDDRows,
        schemaforusers)//结合 Row 和 StructType 的元数据信息，基于 RDD 创建 DataFrame,
                    //这时 RDD 就有了元数据信息的描述
6.  ......
```

```
7.   val schemaforratings = StructType("UserID::MovieID".split("::")
8.        .map(column => StructField(column, StringType, true)))
9.        .add("Rating", DoubleType, true)
10.       .add("Timestamp",StringType, true)
11.
12.     val ratingsRDDRows: RDD[Row] = ratingsRDD.map(_.split("::")).map(line=>
     Row(line(0).trim,line(1)
13.       .trim,line(2).trim.toDouble,line(3).trim))
14.     val ratingsDataFrame: DataFrame = spark.createDataFrame(ratingsRDDRows,
     schemaforratings)
15. ......
16. val genderRatingsDataFrame = ratingsDataFrame.join(usersDataFrame,
    "UserID").cache()
17. ......
18. val maleFilteredRatingsDataFrame = genderRatingsDataFrame.filter("Gender=
    'M'").select("MovieID", "Rating")
19. ......
20. maleFilteredRatingsDataFrame.groupBy("MovieID").avg("Rating").orderBy
    ($"avg(Rating)".desc).show(10):
```

在 IDEA 中运行代码，统计纯粹使用 DataFrame 实现所有电影中最受男性喜爱的电影 Top10 代码，输出结果如下：

```
1.  纯粹使用 DataFrame 实现所有电影中最受男性喜爱的电影 Top10:
2.  +-------+-----------+
3.  |MovieID|avg(Rating)|
4.  +-------+-----------+
5.  |   1830|        5.0|
6.  |    439|        5.0|
7.  |    130|        5.0|
8.  |   3233|        5.0|
9.  |   3172|        5.0|
10. |   3280|        5.0|
11. |   3656|        5.0|
12. |    989|        5.0|
13. |    787|        5.0|
14. |   3517|        5.0|
15. +-------+-----------+
16. only showing top 10 rows
```

接下来统计分析纯粹使用 DataFrame 实现所有电影中最受女性喜爱的电影 Top10。

在 12.3 节通过 RDD 分析各种类型的最受喜爱电影 TopN 及性能优化技巧中，我们已经对纯粹使用 RDD 实现所有电影中最受女性喜爱的电影 Top10 进行了阐述，代码如下：

```
1.    println("纯粹使用 RDD 实现所有电影中最受女性喜爱的电影 Top10:")
2.      femaleFilteredRatings.map(x=>(x._2,(x._3.toDouble,1)))
      //格式化成 Key-Value 的形式
3.        .reduceByKey((x, y) => (x._1 + y._1,x._2 + y._2))
        //对 Value 进行 reduce 操作，分别得出每部电影的总的评分和总的点评人数
4.        .map(x => (x._2._1.toDouble / x._2._2, x._1))  //求出电影平均分
5.        .sortByKey(false)  //降序排列
6.        .map(x => (x._2, x._1))
7.        .take(10)           //取 Top10
8.        .foreach(println)  //打印到控制台
```

现在通过纯粹使用 DataFrame 的方式，实现所有电影中最受女性喜爱的电影 Top10。

```
1.  femaleFilteredRatingsDataFrame.groupBy("MovieID").avg("Rating")
    .orderBy($"avg(Rating)".desc, $"MovieID".desc).show(10)
```

在 IDEA 中运行代码，统计纯粹使用 DataFrame 实现所有电影中最受女性喜爱的电影 Top10，输出结果如下：

```
1.  纯粹使用 DataFrame 实现所有电影中最受女性喜爱的电影 Top10:
2.  +-------+-----------+
3.  |MovieID|avg(Rating)|
4.  +-------+-----------+
5.  |    854 |      5.0  |
6.  |    787 |      5.0  |
7.  |    687 |      5.0  |
8.  |    681 |      5.0  |
9.  |     53 |      5.0  |
10. |    394 |      5.0  |
11. |   3888 |      5.0  |
12. |   3881 |      5.0  |
13. |   3817 |      5.0  |
14. |   3641 |      5.0  |
15. +-------+-----------+
16. only showing top 10 rows
```

12.10　纯粹通过 DataFrame 分析电影点评系统仿 QQ 和微信、淘宝等用户群

12.4 节已经详细阐述了通过 RDD 分析电影点评系统仿 QQ 和微信等用户群分析。本节将阐述通过 DataFrame 分析电影点评系统仿 QQ 和微信、淘宝等用户群。

RDD 方式：纯粹通过 RDD 的方式实现所有电影中 QQ 或者微信核心目标用户最喜爱电影 TopN 分析代码如下：

```
1.   println("纯粹通过 RDD 的方式实现所有电影中 QQ 或者微信核心目标用户最喜爱电影 TopN
     分析:")
2.      ratingsRDD.map(_.split("::")).map(x => (x(0), x(1))).filter(x =>
3.        targetQQUsersBroadcast.value.contains(x._1)
4.      ).map(x => (x._2, 1)).reduceByKey(_ + _).map(x => (x._2, x._1))
5.        .sortByKey(false).map(x => (x._2, x._1)).take(10)
6.        .map(x => (movieID2Name.getOrElse(x._1, null), x._2)).foreach
        (println)
```

在 IDEA 中运行代码，纯粹通过 RDD 的方式实现所有电影中 QQ 或者微信核心目标用户最喜爱电影 TopN 分析，输出结果如下：

```
1.  纯粹通过 RDD 的方式实现所有电影中 QQ 或者微信核心目标用户最喜爱电影 TopN 分析:
2.  (American Beauty (1999),715)
3.  (Star Wars: Episode VI - Return of the Jedi (1983),586)
4.  (Star Wars: Episode V - The Empire Strikes Back (1980),579)
5.  (Matrix, The (1999),567)
6.  (Star Wars: Episode IV - A New Hope (1977),562)
7.  (Braveheart (1995),544)
8.  (Saving Private Ryan (1998),543)
9.  (Jurassic Park (1993),541)
```

```
10. (Terminator 2: Judgment Day (1991),529)
11. (Men in Black (1997),514)
```

DataFrame 方式：纯粹通过 DataFrame 的方式实现所有电影中 QQ 或者微信核心目标用户最喜爱电影 TopN 分析，代码如下：

```
1.   ratingsDataFrame.join(usersDataFrame, "UserID").filter("Age = '18'").groupBy
     ("MovieID")
2.       .count().orderBy($"count".desc).printSchema()
3.   ratingsDataFrame.join(usersDataFrame, "UserID").filter("Age = '18'").groupBy
     ("MovieID").
4.       count().join(moviesDataFrame, "MovieID").select("Title", "count")
         .orderBy($"count".desc).show(10)
```

在 IDEA 中运行代码，纯粹通过 DataFrame 的方式实现所有电影中 QQ 或者微信核心目标用户最喜爱电影 TopN 分析，输出结果如下：

```
1.  纯粹通过 DataFrame 的方式实现所有电影中 QQ 或者微信核心目标用户最喜爱电影 TopN 分析:
2.  root
3.   |-- MovieID: string (nullable = true)
4.   |-- count: long (nullable = false)
5.
6.  +--------------------+-----+
7.  |               Title|count|
8.  +--------------------+-----+
9.  |American Beauty (...|  715|
10. |Star Wars: Episod...|  586|
11. |Star Wars: Episod...|  579|
12. |  Matrix, The (1999)|  567|
13. |Star Wars: Episod...|  562|
14. |    Braveheart (1995)|  544|
15. |Saving Private Ry...|  543|
16. |Jurassic Park (1993)|  541|
17. |Terminator 2: Jud...|  529|
18. | Men in Black (1997)|  514|
19. +--------------------+-----+
20. only showing top 10 rows
```

接下来，统计所有电影中淘宝核心目标用户最喜爱电影 TopN 分析。

RDD 方式：纯粹通过 RDD 的方式实现所有电影中淘宝核心目标用户最喜爱电影 TopN 分析，代码如下：

```
1.    println("纯粹通过RDD的方式实现所有电影中淘宝核心目标用户最喜爱电影TopN分析:")
2.      ratingsRDD.map(_.split("::")).map(x => (x(0), x(1))).filter(x =>
3.        targetTaobaoUsersBroadcast.value.contains(x._1)
4.      ).map(x => (x._2, 1)).reduceByKey(_ + _).map(x => (x._2, x._1))
5.        .sortByKey(false).map(x => (x._2, x._1)).take(10)
6.        .map(x => (movieID2Name.getOrElse(x._1, null), x._2)).foreach(println)
```

在 IDEA 中运行代码，纯粹通过 RDD 的方式实现所有电影中淘宝核心目标用户最喜爱电影 TopN 分析，输出结果如下：

```
1.  纯粹通过 RDD 的方式实现所有电影中淘宝核心目标用户最喜爱电影 TopN 分析:
2.  (American Beauty (1999),1334)
3.  (Star Wars: Episode V - The Empire Strikes Back (1980),1176)
4.  (Star Wars: Episode VI - Return of the Jedi (1983),1134)
5.  (Star Wars: Episode IV - A New Hope (1977),1128)
```

```
6.  (Terminator 2: Judgment Day (1991),1087)
7.  (Silence of the Lambs, The (1991),1067)
8.  (Matrix, The (1999),1049)
9.  (Saving Private Ryan (1998),1017)
10. (Back to the Future (1985),1001)
11. (Jurassic Park (1993),1000)
```

纯粹通过 DataFrame 的方式实现所有电影中淘宝核心目标用户最喜爱电影 TopN 分析，代码如下：

```
1.  println("纯粹通过 DataFrame 的方式实现所有电影中淘宝核心目标用户最喜爱电影 TopN 分析:")
2.     ratingsDataFrame.join(usersDataFrame, "UserID").filter("Age = '25'")
    .groupBy("MovieID")
3.       .count().join(moviesDataFrame, "MovieID").select("Title", "count").orderBy
       ($"count".desc).show(10)
```

在 IDEA 中运行代码，纯粹通过 DataFrame 的方式实现所有电影中淘宝核心目标用户最喜爱电影 TopN 分析，输出结果如下：

```
1.  纯粹通过 DataFrame 的方式实现所有电影中淘宝核心目标用户最喜爱电影 TopN 分析:
2.  +--------------------+-----+
3.  |               Title|count|
4.  +--------------------+-----+
5.  |American Beauty (...| 1334|
6.  |Star Wars: Episod...| 1176|
7.  |Star Wars: Episod...| 1134|
8.  |Star Wars: Episod...| 1128|
9.  |Terminator 2: Jud...| 1087|
10. |Silence of the La...| 1067|
11. |  Matrix, The (1999)| 1049|
12. |Saving Private Ry...| 1017|
13. |Back to the Futur...| 1001|
14. |Jurassic Park (1993)| 1000|
15. +--------------------+-----+
16. only showing top 10 rows
```

12.11　纯粹通过 DataSet 对电影点评系统进行流行度和不同年龄阶段兴趣分析等

前面的章节已经阐述了使用 Spark 的多种方式来实现大数据电影点评系统的功能，包括纯粹 RDD 方式、DataFrame 和 RDD 相结合的方式、纯粹使用 DataFrame 方式；本节将使用纯粹通过 DataSet 对电影点评系统进行流行度和不同年龄阶段兴趣分析。

纯粹使用 DataSet 方式对电影点评系统统计分析的具体思路如下。

（1）定义用户、评分、电影 case class 类。

（2）使用 as 方法将 DataFrame 转换为 DataSet 数据结构。用户、评分、电影类是一个 case class 类，case class 类的字段将被映射到 DataSet 中同名的列。

（3）DataSet 中具有用户、评分、电影的元数据信息，就可以直接通过元数据信息中的列名进行 join、select、groupBy 等算子操作。

12.11.1　通过 DataSet 实现某特定电影观看者中男性和女性不同年龄的人数

本节通过 DataSet 实现某特定电影（即观看的电影 ID 等于 1193）观看者中男性和女性不同年龄分别有多少人。

（1）首先在 Movie_Users_Analyzer_DateSet.scala 代码中定义用户、评分、电影 case class 类。

```
1.   case class User(UserID:String, Gender:String, Age:String, OccupationID:
     String, Zip_Code:String)
2.    case class Rating(UserID:String, MovieID:String, Rating:Double, Times
      tamp:String)
3.    case class Movie(MovieID:String, Title:String, Genres:String)
```

（2）将 usersDataFrame、ratingsDataFrame 分别转化为 usersDataSet、ratingsDataSet。

```
1.      val schemaforusers = StructType("UserID::Gender::Age::OccupationID::
        Zip_Code".split("::")
2.        .map(column => StructField(column, StringType, true)))
          //使用 Struct 方式把 Users 的数据格式化，即在 RDD 的基础上增加数据的元数据信息
3.      val usersRDDRows = usersRDD.map(_.split("::")).map(line => Row(line
        (0).trim,line(1)
4.        .trim,line(2).trim,line(3).trim,line(4).trim))
          //把我们的每条数据变成以 Row 为单位的数据
5.      val usersDataFrame: DataFrame = spark.createDataFrame(usersRDDRows,
        schemaforusers)//结合 Row 和 StructType 的元数据信息基于 RDD 创建 DataFrame，
                         //这时 RDD 就有了元数据信息的描述
6.      val usersDataSet = usersDataFrame.as[User]
7.  ......
8.      val ratingsRDDRows = ratingsRDD.map(_.split("::")).map(line =>
        Row(line(0).trim,line(1).
9.        trim,line(2).trim.toDouble,line(3).trim))
10.     val ratingsDataFrame = spark.createDataFrame(ratingsRDDRows, schemaforratings)
11.     val ratingsDataSet = ratingsDataFrame.as[Rating]
```

（3）将 ratingsDataFrame、usersDataFrame 根据电影 ID 关联，使用 select、groupBy、count 等算子操作，通过 DataSet 实现电影 ID 为 1193 的观看者中男性和女性不同年龄分别有多少人，代码如下：

```
1.  ratingsDataSet.filter(s" MovieID = 1193")
    //这里能够直接指定 MovieID 的原因是 DataFrame 中有该元数据信息
2.        .join(usersDataSet, "UserID")
          //Join 的时候直接指定基于 UserID 进行 Join，这相对于原生的 RDD 操作而言，更加
          //方便、快捷
3.        .select("Gender", "Age")
          //直接通过元数据信息中的 Gender 和 Age 进行数据的筛选
4.        .groupBy("Gender", "Age")
          //直接通过元数据信息中的 Gender 和 Age 进行数据的 groupBy 操作
5.        .count()  //基于 groupBy 分组信息进行 count 统计操作
6.        .show(10) //显示出分组统计后的前 10 条信息
```

在 IDEA 中运行代码，通过 DataSet 实现某特定电影观看者中男性和女性不同年龄的人数，输出结果如下：

```
1.   功能一：通过 DataSet 实现某特定电影观看者中男性和女性不同年龄分别有多少人？
```

```
2.  +------+---+-----+
3.  |Gender|Age|count|
4.  +------+---+-----+
5.  |     F |45 |   55|
6.  |     M |50 |  102|
7.  |     M | 1 |   26|
8.  |     F |56 |   39|
9.  |     F |50 |   43|
10. |     F |18 |   57|
11. |     F | 1 |   10|
12. |     M |18 |  192|
13. |     F |25 |  140|
14. |     M |45 |  136|
15. +------+---+-----+
16. only showing top 10 rows
```

12.11.2　通过 DataSet 方式计算所有电影中平均得分最高（口碑最好）的电影 TopN

本节通过 DataSet 实现通过 DataSet 方式计算所有电影中平均得分最高（口碑最好）的电影 TopN。

（1）在 12.11.1 节 Movie_Users_Analyzer_DateSet.scala 代码中已经定义了用户、评分、电影 case class 类。

```
1.   case class User(UserID:String, Gender:String, Age:String, OccupationID:
     String, Zip_Code:String)
2.   case class Rating(UserID:String, MovieID:String, Rating:Double,
     Timestamp:String)
3.   case class Movie(MovieID:String, Title:String, Genres:String)
```

（2）将评分数据 ratingsDataFrame 分别转化为 ratingsDataSet。

```
1.       ......
2.  val ratingsRDDRows = ratingsRDD.map(_.split("::")).map(line =>
    Row(line(0).trim,line(1).
3.      trim,line(2).trim.toDouble,line(3).trim))
4.     val ratingsDataFrame = spark.createDataFrame(ratingsRDDRows, schemaforratings)
5.     val ratingsDataSet = ratingsDataFrame.as[Rating]
```

（3）在评分数据集中查询电影 ID、评分两列，根据电影 ID 分组，计算评分的平均分，然后按照平均分降序排列，通过纯粹使用 DataSet 方式计算所有电影中平均得分最高（口碑最好）的电影 Top10，代码如下：

```
1.  println("通过 DataSet 方式计算所有电影中平均得分最高（口碑最好）的电影 TopN:")
2.     ratingsDataSet.select("MovieID", "Rating").groupBy("MovieID").
3.       avg("Rating").orderBy($"avg(Rating)".desc).show(10)
```

在 IDEA 中运行代码，通过纯粹使用 DataSet 方式计算所有电影中平均得分最高（口碑最好）的电影 TopN，输出结果如下：

```
1.  通过 DataSet 方式计算所有电影中平均得分最高（口碑最好）的电影 TopN:
2.  +-------+-----------+
3.  |MovieID|avg(Rating)|
4.  +-------+-----------+
5.  |   3382|        5.0 |
6.  |   3607|        5.0 |
```

```
7.  |   3881|       5.0  |
8.  |   3233|       5.0  |
9.  |   3172|       5.0  |
10. |   3280|       5.0  |
11. |   3656|       5.0  |
12. |    787|       5.0  |
13. |    989|       5.0  |
14. |   1830|       5.0  |
15. +-------+-----------+
16. only showing top 10 rows
```

12.11.3　通过 DataSet 方式计算所有电影中粉丝或者观看人数最多(最流行电影)的电影 TopN

本节通过 DataSet 实现通过 DataSet 的方式计算即所有电影中粉丝或者观看人数最多（最流行电影）的电影 TopN。

（1）在 12.11.1 节 Movie_Users_Analyzer_DateSet.scala 代码中已经定义了用户、评分、电影 case class 类。

```
1.    case class User(UserID:String, Gender:String, Age:String, OccupationID:
      String, Zip_Code:String)
2.    case class Rating(UserID:String, MovieID:String, Rating:Double,
      Timestamp:String)
3.    case class Movie(MovieID:String, Title:String, Genres:String)
```

（2）将评分数据 ratingsDataFrame 分别转化为 ratingsDataSet。

```
1.      ......
2.  val ratingsRDDRows = ratingsRDD.map(_.split("::")).map(line =>
    Row(line(0).trim,line(1)
3.      .trim,line(2).trim.toDouble,line(3).trim))
4.    val ratingsDataFrame = spark.createDataFrame(ratingsRDDRows, schemaforratings)
5.    val ratingsDataSet = ratingsDataFrame.as[Rating]
```

（3）在评分数据集中，根据电影 ID 使用 groupBy 先进行分组，然后使用 Count 算子计算每组的行数，使用 orderBy 算子根据 Count 计数进行降序排列。通过 DataSet 的方式计算所有电影中粉丝或者观看人数最多（最流行电影）的电影 Top10，代码如下：

```
1.   println("纯粹通过 DataSet 的方式计算所有电影中粉丝或者观看人数最多(最流行电影)的
     电影 TopN:")
2.     ratingsDataSet.groupBy("MovieID").count().
3.       orderBy($"count".desc).show(10)
```

在 IDEA 中运行代码，通过 DataSet 的方式计算所有电影中粉丝或者观看人数最多（最流行电影）的电影 TopN，输出结果如下：

```
1.    纯粹通过 DataSet 的方式计算所有电影中粉丝或者观看人数最多(最流行电影)的电影 TopN:
2.  +-------+-----+
3.  |MovieID|count|
4.  +-------+-----+
5.  |   2858| 3428|
6.  |    260| 2991|
```

```
7.  |    1196| 2990|
8.  |    1210| 2883|
9.  |     480| 2672|
10. |    2028| 2653|
11. |     589| 2649|
12. |    2571| 2590|
13. |    1270| 2583|
14. |     593| 2578|
15. +-------+-----+
16. only showing top 10 rows
```

12.11.4　纯粹通过 DataSet 的方式实现所有电影中最受男性、女性喜爱的电影 Top10

本节通过纯粹通过 DataSet 的方式实现所有电影中最受男性、女性喜爱的电影 Top10。

（1）在 12.11.1 节 Movie_Users_Analyzer_DateSet.scala 代码中已经定义了用户、评分、电影 case class 类。

```
1.    case class User(UserID:String, Gender:String, Age:String, OccupationID:
      String, Zip_Code:String)
2.    case class Rating(UserID:String, MovieID:String, Rating:Double, Timestamp:
      String)
3.    case class Movie(MovieID:String, Title:String, Genres:String)
```

（2）将评分数据集 ratingsDataSet 和用户数据集 usersDataSet 根据用户 ID 关联，生成 genderRatingsDataSet 以后，这里的 genderRatingsDataSet 其实是 DataFrame，可以使用 as 转换为 DataSet。然后再分别过滤出男性、女性的数据 maleFilteredRatingsDataSet、femaleFilteredRatingsDataSet，然后查询出电影 ID、评分两列数据，代码如下：

```
1.   ......
2.     val genderRatingsDataFrame = ratingsDataFrame.join(usersDataFrame,
       "UserID").cache()
3.     val genderRatingsDataSet = ratingsDataSet.join(usersDataSet, "UserID")
       .cache()
4.     val  maleFilteredRatingsDataFrame  =  genderRatingsDataFrame.filter
       ("Gender= 'M'").select("MovieID", "Rating")
5.     val  maleFilteredRatingsDataSet  =  genderRatingsDataSet.filter("Gender=
       'M'").select("MovieID", "Rating")
6.     val  femaleFilteredRatingsDataFrame  =  genderRatingsDataFrame.filter
       ("Gender= 'F'").select("MovieID", "Rating")
7.     val  femaleFilteredRatingsDataSet  =  genderRatingsDataSet.filter("Gender=
       'F'").select("MovieID", "Rating")
```

（3）将过滤后的男性、女性评分数据集 maleFilteredRatingsDataSet、femaleFiltered RatingsDataSet，根据电影 ID 进行分组，使用 avg 算子计算评分平均分，然后按平均分降序排列。纯粹通过 DataSet 的方式实现所有电影中最受男性、女性喜爱的电影 Top10，代码如下：

```
1.   println("纯粹通过 DataSet 的方式实现所有电影中最受男性喜爱的电影
    Top10:")maleFiltered  RatingsDataSet.groupBy("MovieID").avg("Rating").orderBy
    ($"avg(Rating)".desc).show(10)
2.    ......
3.   println("纯粹通过 DataSet 的方式实现所有电影中最受女性喜爱的电影 Top10:")female
     Filtered   RatingsDataSet.groupBy("MovieID").avg("Rating").   orderBy($
     "avg(Rating)".  desc, $"MovieID".desc).show(10)
```

在 IDEA 中运行代码，纯粹通过 DataSet 的方式实现所有电影中最受男性、女性喜爱的电影 Top10，输出结果如下：

```
1.  纯粹通过 DataSet 的方式实现所有电影中最受男性喜爱的电影 Top10:
2.  +-------+-----------+
3.  |MovieID|avg(Rating)|
4.  +-------+-----------+
5.  |    985|        5.0|
6.  |    439|        5.0|
7.  |   3233|        5.0|
8.  |    130|        5.0|
9.  |   3172|        5.0|
10. |   3280|        5.0|
11. |    989|        5.0|
12. |    787|        5.0|
13. |   3656|        5.0|
14. |   3517|        5.0|
15. +-------+-----------+
16. only showing top 10 rows
17. ......
18. 纯粹通过 DataSet 的方式实现所有电影中最受女性喜爱的电影 Top10:
19. +-------+-----------+
20. |MovieID|avg(Rating)|
21. +-------+-----------+
22. |    854|        5.0|
23. |    787|        5.0|
24. |    687|        5.0|
25. |    681|        5.0|
26. |     53|        5.0|
27. |    394|        5.0|
28. |   3888|        5.0|
29. |   3881|        5.0|
30. |   3817|        5.0|
31. |   3641|        5.0|
32. +-------+-----------+
33. only showing top 10 rows
```

12.11.5　纯粹通过 DataSet 的方式实现所有电影中 QQ 或者微信核心目标用户最喜爱电影 TopN 分析

本节我们纯粹通过 DataSet 的方式实现所有电影中 QQ 或者微信核心目标用户最喜爱电影 TopN 分析。

（1）在 12.11.1 节 Movie_Users_Analyzer_DateSet.scala 代码中已经定义了用户、评分、电影 case class 类。

```
1.     case class User(UserID:String, Gender:String, Age:String, OccupationID:
       String, Zip_Code:String)
2.     case class Rating(UserID:String, MovieID:String, Rating:Double,
       Timestamp:String)
3.     case class Movie(MovieID:String, Title:String, Genres:String)
```

（2）将评分数据 ratingsDataFrame 转化为 ratingsDataSet，电影数据 moviesDataFrame 转化为 moviesDataSet，用户数据 usersDataFrame 转换为 usersDataSet。

```
1.    ......
2.      val schemaforusers = StructType("UserID::Gender::Age::OccupationID::
        Zip_Code".split("::")
3.        .map(column => StructField(column, StringType, true)))
          //使用 Struct 方式把 Users 的数据格式化，即在 RDD 的基础上增加数据的元数据信息
4.      val usersRDDRows = usersRDD.map(_.split("::")).map(line => Row(line
        (0).trim,line(1)
5.        .trim,line(2).trim,line(3).trim,line(4).trim))
          //把我们的每条数据变成以 Row 为单位的数据
6.      val usersDataFrame: DataFrame = spark.createDataFrame(usersRDDRows,
        schemaforusers)  //结合 Row 和 StructType 的元数据信息，基于 RDD 创建
                        //DataFrame，这时 RDD 就有了元数据信息的描述
7.      val usersDataSet = usersDataFrame.as[User]
8.    ......
9.      val ratingsRDDRows=ratingsRDD.map(_.split("::")).map(line=>Row(line
        (0).trim,line(1).
10.       trim,line(2).trim.toDouble,line(3).trim))
11.     val ratingsDataFrame = spark.createDataFrame(ratingsRDDRows, s
        chemaforratings)
12.     val ratingsDataSet = ratingsDataFrame.as[Rating]
13.   ......
14.     val schemaformovies = StructType("MovieID::Title::Genres".split("::")
15.       .map(column => StructField(column, StringType, true)))
          //使用 Struct 方式把 Users 的数据格式化，即在 RDD 的基础上增加数据的元数据信息
16.     val moviesRDDRows = moviesRDD.map(_.split("::")).map(line =>
        Row(line(0).trim,line(1)
17.       .trim,line(2).trim)) //把我们的每条数据变成以 Row 为单位的数据
18.     val moviesDataFrame = spark.createDataFrame(moviesRDDRows, schemaformovies)
        //结合 Row 和 StructType 的元数据信息，基于 RDD 创建 DataFrame，这时 RDD 就有了
        //元数据信息的描述
19.     val moviesDataSet = moviesDataFrame.as[Movie]
```

（3）将评分 ratingsDataSet 和用户 usersDataSet 进行关联，过滤年龄为 18 岁年龄段的用户，根据电影 ID 进行分组，统计观看次数，然后再根据电影 ID 和电影 moviesDataSet 进行关联，查询电影名、观看次数两列数据，并且按观看次数降序排列，纯粹通过 DataSet 的方式实现所有电影中 QQ 或者微信核心目标用户最喜爱电影 TopN 分析，代码如下：

```
1.    println("纯粹通过 DataSet 的方式实现所有电影中 QQ 或者微信核心目标用户最喜爱电影
      TopN 分析:")
2.      ratingsDataSet.join(usersDataSet,  "UserID").filter("Age  =  '18'")
        .groupBy("MovieID").
3.        count().join(moviesDataSet,  "MovieID").select("Title",  "count")
          .sort($"count".desc).show(10)
```

在 IDEA 中运行代码，纯粹通过 DataSet 的方式实现所有电影中 QQ 或者微信核心目标用户最喜爱电影 TopN 分析，输出结果如下：

```
1.  纯粹通过 DataSet 的方式实现所有电影中 QQ 或者微信核心目标用户最喜爱电影 TopN 分析:
2.  +--------------------+-----+
3.  |       Title        |count|
4.  +--------------------+-----+
5.  |American Beauty (...|  715|
6.  |Star Wars: Episod...|  586|
7.  |Star Wars: Episod...|  579|
8.  | Matrix, The (1999) |  567|
9.  |Star Wars: Episod...|  562|
```

```
10. |Braveheart (1995)   |  544|
11. |Saving Private Ry...|  543|
12. |Jurassic Park (1993)|  541|
13. |Terminator 2: Jud...|  529|
14. | Men in Black (1997)|  514|
15. +--------------------+-----+
16. only showing top 10 rows
```

12.11.6　纯粹通过 DataSet 的方式实现所有电影中淘宝核心目标用户最喜爱电影 TopN 分析

本节我们纯粹通过 DataSet 的方式实现所有电影中淘宝核心目标用户最喜爱电影 TopN 分析。

（1）在 12.11.1 节 Movie_Users_Analyzer_DateSet.scala 代码中已经定义了用户、评分、电影 case class 类。

```
1.    case class User(UserID:String, Gender:String, Age:String, OccupationID:
      String, Zip_Code:String)
2.    case class Rating(UserID:String, MovieID:String, Rating:Double,
      Timestamp:String)
3.    case class Movie(MovieID:String, Title:String, Genres:String)
```

（2）将评分数据 ratingsDataFrame 转化为 ratingsDataSet，电影数据 moviesDataFrame 转化为 moviesDataSet，用户数据 usersDataFrame 转换为 usersDataSet。

```
1.   ......
2.      val schemaforusers = StructType("UserID::Gender::Age::OccupationID::
        Zip_Code".split("::").
3.        map(column => StructField(column, StringType, true)))
          //使用 Struct 方式把 Users 的数据格式化，即在 RDD 的基础上增加数据的元数据信息
4.      val usersRDDRows = usersRDD.map(_.split("::")).map(line =>
        Row(line(0).trim,line(1)
5.        .trim,line(2).trim,line(3).trim,line(4).trim))
          //把我们的每条数据变成以 Row 为单位的数据
6.      val usersDataFrame: DataFrame = spark.createDataFrame(usersRDDRows,
        schemaforusers)  //结合 Row 和 StructType 的元数据信息，基于 RDD 创建
                         //DataFrame，这时 RDD 就有了元数据信息的描述
7.      val usersDataSet = usersDataFrame.as[User]
8.   ......
9.      val ratingsRDDRows=ratingsRDD.map(_.split("::")).map(line=>
        Row(line(0).trim,line(1)
10.       .trim,line(2).trim.toDouble,line(3).trim))
11.     val ratingsDataFrame = spark.createDataFrame(ratingsRDDRows, schemaforratings)
12.     val ratingsDataSet = ratingsDataFrame.as[Rating]
13.  ......
14.     val schemaformovies = StructType("MovieID::Title::Genres".split("::")
15.       .map(column => StructField(column, StringType, true)))
          //使用 Struct 方式把 Users 的数据格式化，即在 RDD 的基础上增加数据的元数据信息
16.     val moviesRDDRows = moviesRDD.map(_.split("::")).map(line =>
        Row(line(0).trim,line(1)
17.       .trim,line(2).trim)) //把我们的每条数据变成以 Row 为单位的数据
18.     val moviesDataFrame = spark.createDataFrame(moviesRDDRows, schemaformovies)
        //结合 Row 和 StructType 的元数据信息，基于 RDD 创建 DataFrame，这时 RDD 就有了
        //元数据信息的描述
```

```
19.     val moviesDataSet = moviesDataFrame.as[Movie]
```

（3）将评分 ratingsDataSet 和用户 usersDataSet 进行关联，过滤年龄为 25 岁年龄段的用户，根据电影 ID 进行分组，统计观看次数，然后再根据电影 ID 和电影 moviesDataSet 进行关联，查询电影名、观看次数两列数据，并且按观看次数降序排列，纯粹通过 DataSet 的方式实现所有电影中淘宝核心目标用户最喜爱电影 TopN 分析，代码如下：

```
1.  println("纯粹通过 DataSet 的方式实现所有电影中淘宝核心目标用户最喜爱电影 TopN 分析:")
2.      ratingsDataSet.join(usersDataSet,  "UserID").filter("Age  =  '25'")
    .groupBy("MovieID")
3.        .count().join(moviesDataSet,  "MovieID").select("Title",  "count")
      .sort($"count".desc).limit(10).show()
```

在 IDEA 中运行代码，纯粹通过 DataSet 的方式实现所有电影中淘宝核心目标用户最喜爱电影 TopN 分析，输出结果如下：

```
1.  纯粹通过 DataSet 的方式实现所有电影中淘宝核心目标用户最喜爱电影 TopN 分析:
2.  +--------------------+-----+
3.  |Title               |count|
4.  +--------------------+-----+
5.  |American Beauty (...| 1334|
6.  |Star Wars: Episod...| 1176|
7.  |Star Wars: Episod...| 1134|
8.  |Star Wars: Episod...| 1128|
9.  |Terminator 2: Jud...| 1087|
10. |Silence of the La...| 1067|
11. |  Matrix, The (1999)| 1049|
12. |Saving Private Ry...| 1017|
13. |Back to the Futur...| 1001|
14. |Jurassic Park (1993)| 1000|
15. +--------------------+-----+
```

12.12　大数据电影点评系统应用案例涉及的核心知识点原理、源码及案例代码

本节大数据电影点评系统应用案例涉及的核心知识点主要包括以下几方面。

- 知识点：Broadcast。
- 知识点：SQL 全局临时视图及临时视图。

12.12.1　知识点：广播变量 Broadcast 内幕机制

在 12.4 节中，我们通过 RDD 分析电影点评系统仿 QQ 和微信等用户群分析中用到了广播变量 Broadcast。下面讲解一下广播变量 Broadcast 内幕机制的知识点。

广播变量就是一个用来广播的变量，允许 Spark 应用程序获取一个只读的变量，只读的广播变量缓存到每个分布式节点上，而不是给每个任务传送它的一份数据副本。Spark 以高效的方式给每个节点复制一个大的输入数据集，使用高效广播算法来分发广播变量，从而减少网络通信成本。

广播变量是由一个变量 v 通过调用[org.apache.spark.SparkContext#broadcast]创建的。广播变量将 v 封装起来。广播变量的值可以通过 value 方法来获取，例如：

```
1.    scala> val broadcastVar = sc.broadcast(Array(1, 2, 3))
2.   broadcastVar: org.apache.spark.broadcast.Broadcast[Array[Int]] = Broadcast(0)
3.
4.   scala> broadcastVar.value
5.   res0: Array[Int] = Array(1, 2, 3)
```

在实际的企业级开发项目中，基本上稍微复杂一点的项目，就一定会用到 Broadcast 广播变量和 Accumulator 计数器。Broadcast 广播变量和 Accumulator 计数器是和 RDD 并列的三大基础数据结构。Broadcast 广播变量和 Accumulator 计数器是全局级别的。从 Spark 数据结构的角度分析，RDD 是分布式私有数据结构，Broadcast 广播变量是分布式全局只读数据结构，Accumulator 计数器是分布式全局只写的数据结构。

下面通过图 12-12 说明 Broadcast 的工作机制。Spark Driver 中有一个变量 number，在 Executor 中有 4 个 Task，如没有使用广播变量，Driver 需给 Executor 的 4 个 Task 都发送一个变量 number 数据副本，需发送 4 次，这是对网络传输和内存消耗的极大浪费，如果变量比较大，则 Executor 内存占用大，极易出现 OOM。如果使用广播变量，Driver 发送广播变量到 Executor 的内存中，只有 1 份广播变量 number 数据副本，只要发送 1 次，4 个 Task 任务就可以共享唯一的一份广播变量，极大地减少了网络传输和内存消耗。

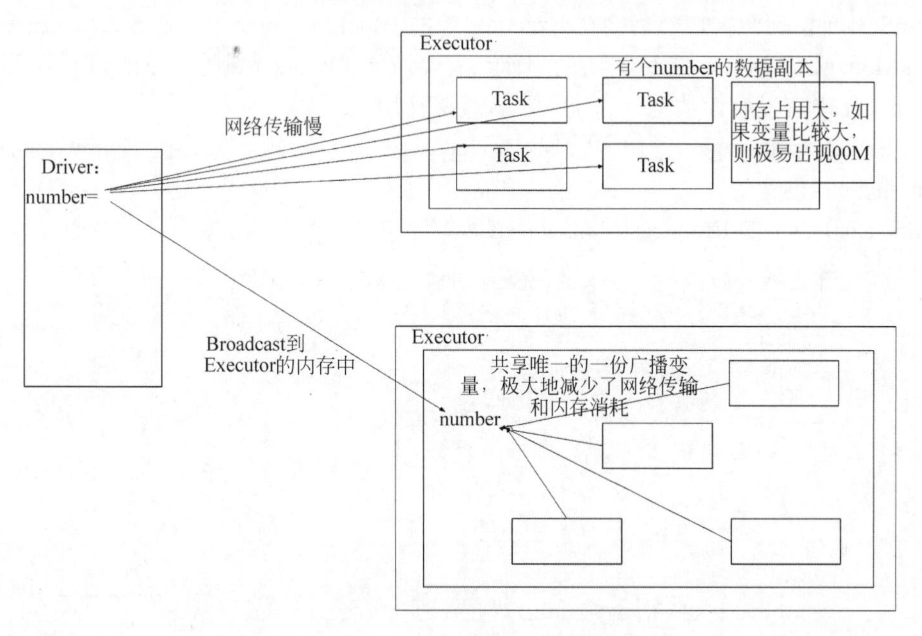

图 12-12　Broadcast 的工作机制图

现在我们深入思考 Spark Broadcast 的运行机制，Broadcast 广播变量内幕解密如下。

（1）Broadcast 就是将数据从一个节点发送到其他节点上。例如，Driver 上有一张表，而 Executor 中的每个并行执行的 Task（100 万个 Task）都要查询这张表，那我们通过 Broadcast 的方式往每个 Executor 把这张表发送一次就行了，Executor 中的每个运行的 Task 查询这张唯一的表，而不是每次执行的时候都从 Driver 获得这张表。

（2）这就好像 ServletContext 的具体作用，只是 Broadcast 是分布式的共享数据，默认情况下，只要程序在运行，Broadcast 变量就会存在，因为 Broadcast 在底层是通过 BlockManager 管理的。但是你可以手动指定或者配置具体周期来销毁 Broadcast 变量。

（3）Broadcast 一般用于处理共享配置文件、通用的 Dataset、常用的数据结构等；但是在 Broadcast 中，不适合存放太大的数据，Broadcast 不会内存溢出，因为其数据保存的 StorageLevel 是 MEMORY_AND_DISK 的方式；虽然如此，我们也不可以放入太大的数据在 Broadcast 中，因为网络 I/O 和可能的单点压力会非常大。

（4）广播 Broadcast 变量是只读变量，最为轻松保持了数据的一致性。

（5）Broadcast 的使用。

```
1.    scala> val broadcastVar = sc.broadcast(Array(1, 2, 3))
2.   broadcastVar: org.apache.spark.broadcast.Broadcast[Array[Int]] = Broadcast(0)
3.
4.   scala> broadcastVar.value
5.   res0: Array[Int] = Array(1, 2, 3)
```

（6）HttpBroadcast 方式的 Broadcast，最开始的时候，数据放在 Driver 的本地文件系统中，Driver 在本地会创建一个文件夹来存放 Broadcast 中的 data，然后启动 HttpServer 访问文件夹中的数据，同时写入到 BlockManager（StorageLevel 是 MEMORY_AND_DISK）中获得 BlockId（BroadcastBlockId），当第一次 Executor 中的 Task 要访问 Broadcast 变量的时候，会向 Driver 通过 HttpServer 访问数据，然后会在 Executor 中的 BlockManager 中注册该 Broadcast 中的数据 BlockManager，这样，之后需要的 Task 需要访问 Broadcast 的变量的时候会首先查询 BlockManager 中有没有该数据，如果有，就直接使用。

（7）BroadcastManager 是用来管理 Broadcast 的，该实例对象是在 SparkContext 创建 SparkEnv 的时候创建的。

Spark 1.6.0 版本的 BroadcastManager.scala 的源码如下：

```
1.   //使用广播变量前，被 SparkContext 或者 Executor 调用
2.    private def  initialize() {
3.      synchronized {
4.        if (!initialized) {
5.          val broadcastFactoryClass =
6.            conf.get("spark.broadcast.factory", "org.apache.spark.broadcast.
              TorrentBroadcastFactory")
7.
8.          broadcastFactory =
9.            Utils.classForName(broadcastFactoryClass).newInstance
               .asInstanceOf[BroadcastFactory]
10.
11.         //初始化相应的 BroadcastFactory 和 BroadcastObject
12.         broadcastFactory.initialize(isDriver, conf, securityManager)
13.
14.         initialized = true
15.       }
16.     }
17.   }
```

Spark 2.2.0 版本的 BroadcastManager.scala 的源码与 Spark 1.6.0 版本相比具有如下特点：Spark 2.2.0 版本的 BroadcastFactory 仅提供 TorrentBroadcastFactory 实现方式。

```
1.   private def initialize() {
2.     synchronized {
3.       if (!initialized) {
4.         broadcastFactory = new TorrentBroadcastFactory
5.         broadcastFactory.initialize(isDriver, conf, securityManager)
6.         initialized = true
7.       }
8.     }
9.   }
```

实例化 BroadcastManager 时会创建 BroadcastFactory 工厂来构建具体的 Broadcast 类型，默认情况下是 TorrentBroadcastFactory。

（8）HttpBroadcast 存在单点故障和网络 I/O 性能问题，所以默认使用 TorrentBroadcast 的方式，开始数据在 Driver 中，假设 A 节点用了数据，B 访问的时候 A 节点就变成数据源，依此类推，都是数据源，当然是被 BlockManger 进行管理的，数据源越多，节点压力会大大降低。

（9）TorrentBroadcast 按照 BLOCK_SIZE（默认是 4MB）将 Broadcast 中的数据划分成为不同的 Block，然后将分块信息（也就是 meta 信息）存放到 Driver 的 BlockManager 中，同时会告诉 BlockManagerMaster，说明 Meta 信息存放完毕。

12.12.2　知识点：SQL 全局临时视图及临时视图

在 12.6 节中，我们通过 Spark SQL 中的 SQL 语句实现电影点评系统用户行为分析，其中使用到 Spark SQL 的全局临时视图及临时视图。下面讲解一下 Spark SQL 的全局临时视图及临时视图。

Spark SQL 中的临时视图（Temporary views）是会话范围的，如果创建它的会话终止，临时视图将消失。如果需建立在所有会话之间共享的临时视图，并保持活动状态，直到 Spark 应用程序终止，那么可以创建一个全局临时视图（Global Temporary View）。全局临时视图绑定到 Spark 系统保留的数据库 global_temp，我们必须使用限定名称来引用它，例如 SELECT * FROM global_temp.view1。

全局临时视图的代码例子如下：

```
1.  //使用全局临时视图注册 DataFrame
2.  df.createGlobalTempView("people")
3.
4.  //全局临时视图绑定到 Spark 系统保留的数据库 global_temp
5.  spark.sql("SELECT * FROM global_temp.people").show()
6.  // +----+-------+
7.  // | age|   name|
8.  // +----+-------+
9.  // |null|Michael|
10. // |  30|   Andy |
11. // |  19| Justin|
12. // +----+-------+
13.
14. //全局临时视图是会话共享的
15. spark.newSession().sql("SELECT * FROM global_temp.people").show()
16. // +----+-------+
17. // | age|   name|
18. // +----+-------+
19. // |null|Michael|
```

```
20. // |  30|   Andy|
21. // |  19| Justin|
22. // +----+-------+
```

打开 Dataset.scala 框架源码，看一下创建全局临时视图 createGlobalTempView 方法，从源码中可以看到，这里 Spark createGlobalTempView 源码注释时出现了一个小错误，我们在创建 createGlobalTempView 的时候，DB 的名称实际为 global_temp，而在源码注释中写成了_global_temp，如果在自己编写的业务代码中使用_global_temp 限定词，编译时会报错，提示找不到表_global_temp。

```
1.    /**
       *  使用限定名称创建全局临时视图。全局临时视图的生命周期与 Spark 应用程序绑定。全
       * 局临时视图是跨会话的，全局临时视图的生命周期等同于 Spark 应用程序的生命周期，当
       * Spark 应用程序终止，全局临时视图会自动被终止。全局临时生命周期绑定于系统保留数
       * 据库_global_temp，我们必须使用限定名称指向全局临时视图
2.     *
3.     * view, e.g. `SELECT * FROM _global_temp.view1`.
4.     *
5.     * @throws AnalysisException if the view name already exists
6.     *
7.     * @group basic
8.     * @since 2.1.0
9.     */
10.   @throws[AnalysisException]
11.   def createGlobalTempView(viewName: String): Unit = withPlan {
12.     createTempViewCommand(viewName, replace = false, global = true)
13.   }
```

打开 Dataset.scala 框架源码，看一下创建临时视图 createTempView 方法，从源码中可以看到：

```
1.    /**
       *  使用给定名称创建本地临时视图，本地临时视图的生命周期是和[SparkSession]Spark
       *  会话绑定在一起的，sparksession 用于创建 Dataset 数据集。本地临时视图是会话作
       *  用域范围的，本地临时视图的生命周期等同于会话的生命周期，会话创建本地临时视图，
       *  当会话终止时，本地临时视图会自动被删除。本地临时视图不与任何数据库绑定，即我们
       *  不能用 db1.view1 引用本地临时视图
2.     *
3.     * @throws AnalysisException if the view name already exists
4.     *
5.     * @group basic
6.     * @since 2.0.0
7.     */
8.    @throws[AnalysisException]
9.    def createTempView(viewName: String): Unit = withPlan {
10.     createTempViewCommand(viewName, replace = false, global = false)
11.   }
```

12.12.3　大数据电影点评系统应用案例完整代码

1. RDD方式案例代码

Spark 商业案例之大数据电影点评系统应用案例代码 Movie_Users_Analyzer_RDD.scala 如例 12-3 所示。

【例 12-3】 Movie_Users_Analyzer_RDD.scala 代码。

```
1.   package com.dt.spark.cores
2.  import org.apache.spark.{SparkConf, SparkContext}
3.  import scala.collection.immutable.HashSet
4.  import org.apache.log4j.{Level, Logger}
5.  /**
6.    * 版权：DT 大数据梦工厂所有
7.    * 时间：2017 年 1 月 1 日；
8.    * 电影点评系统用户行为分析：用户观看电影和点评电影的所有行为数据的采集、过滤、处理
      * 和展示：
9.    *   数据采集：企业中一般越来越多地喜欢直接把 Server 中的数据发送给 Kafka，因为更加
      * 具备实时性；
10.   *   数据过滤：趋势是直接在 Server 端进行数据过滤和格式化，当然，采用 Spark SQL 进
      * 行数据的过滤也是一种主要形式；
11.   *   数据处理：
12.   *     1.一个基本的技巧是，先使用传统的 SQL 去实现一个数据处理的业务逻辑（自己可以
      * 手动模拟一些数据）；
13.   *     2.再一次推荐使用 DataSet 去实现业务功能，尤其是统计分析功能；
14.   *     3.如果你想成为专家级别的顶级 Spark 人才，请使用 RDD 实现业务功能，为什么？运
      *     行的时候是基于 RDD 的！
15.   *
16.   *  数据：强烈建议大家使用 Parquet
17.   *  1."ratings.dat": UserID::MovieID::Rating::Timestamp
18.   *  2."users.dat": UserID::Gender::Age::OccupationID::Zip-code
19.   *  3."movies.dat": MovieID::Title::Genres
20.   *  4."occupations.dat": OccupationID::OccupationName   一般情况下都会以
      * 程序中数据结构 Haskset 的方式存在，是为了做 mapjoin
21.   */
22. object Movie_Users_Analyzer {
23.   def main(args: Array[String]){
24.
25.     Logger.getLogger("org").setLevel(Level.ERROR)
26.
27.     var masterUrl = "local[4]" //默认程序运行在本地 Local 模式中，主要用于学习和测试
28.     var dataPath = "data/moviedata/medium/"  //数据存放的目录
29.
30.     /**
31.       * 当我们把程序打包运行在集群上时，一般都会传入集群的 URL 信息，这里我们假设如果传入
          * 参数，第一个参数只传入 Spark 集群的 URL，第二个参数传入的是数据的地址信息
31.       */
32.
33.
34.     if(args.length > 0) {
35.       masterUrl = args(0)
36.     } else if (args.length > 1) {
37.       dataPath = args(1)
38.     }
39.
40.     /**
          * 创建 Spark 集群上下文 sc，在 sc 中可以进行各种依赖和参数的设置等，大家可以通
          * 过 SparkSubmit 脚本的 help 去看设置信息
41.       */
42.
43.     val sc = new SparkContext(new SparkConf().setMaster(masterUrl)
        .setAppName("Movie_Users_Analyzer"))
```

```
44.
45.     /**
          * 读取数据，用什么方式读取数据呢？这里使用的是 RDD!
46.       */
47.
48.     val usersRDD = sc.textFile(dataPath + "users.dat")
49.     val moviesRDD = sc.textFile(dataPath + "movies.dat")
50.     val occupationsRDD = sc.textFile(dataPath + "occupations.dat")
51. //    val ratingsRDD = sc.textFile(dataPath + "ratings.dat")
52.     val ratingsRDD = sc.textFile("data/moviedata/large/" + "ratings.dat")
53.
54.     /**
          * 电影点评系统用户行为分析之一：分析具体某部电影观看的用户信息，如电影 ID 为
          * 1193 的用户信息（用户的 ID、Age、Gender、Occupation）
55.       */
56.
57.
58.     val usersBasic = usersRDD.map(_.split("::")).map{user => (//UserID::
        Gender::Age::OccupationID
59.       user(3),
60.       (user(0),user(1),user(2))
61.       )
62.     }
63.     val occupations = occupationsRDD.map(_.split("::")).map(job =>
        (job(0), job(1)))
64.
65.     val userInformation = usersBasic.join(occupations)
66.
67.     userInformation.cache()
68.
69. //    for (elem <- userInformation.collect()) {
70. //      println(elem)
71. //    }
72.
73.
74.     val targetMovie = ratingsRDD.map(_.split("::")).map(x => (x(0), x(1)))
       .filter(_._2.equals("1193"))
75.     //(11,((4882,M,45),lawyer))
76.     val targetUsers = userInformation.map(x => (x._2._1._1, x. 2))
77.
78.     val userInformationForSpecificMovie = targetMovie.join(targetUsers)
79.     for (elem <- userInformationForSpecificMovie.collect()) {
80.           println(elem)
81.         }
82.
83.
84.     /**
85.       * 电影流行度分析：所有电影中平均得分最高（口碑最好）的电影及观看人数最多的电影
          *（流行度最高）
86.       * "ratings.dat": UserID::MovieID::Rating::Timestamp
87.       * 得分最高的 Top10 电影实现思路：如果想算总的评分，一般需要 reduceByKey 操作
          * 或者 aggregateByKey 操作
88.       *   第一步：把数据变成 Key-Value，大家想一下在这里什么是 Key，什么是 Value。
          * 把 MovieID 设置为 Key，把 Rating 设置为 Value;
89.       *   第二步：通过 reduceByKey 操作或者 aggregateByKey 实现聚合，然后呢？
90.       *   第三步：排序，如何做？进行 Key 和 Value 的交换
91.       *
92.       *  注意：
93.       *   1. 转换数据格式时一般都会使用 map 操作，有时转换可能特别复杂，需要在 map 方
          * 法中调用第三方 jar 或者 so 库;
```

```
94.       *  2. RDD 从文件中提取的数据成员默认都是 String 方式，需要根据实际转换格式；
95.       *  3. RDD 如果要重复使用，一般都会进行 Cache；
96.       *  4. RDD 的 Cache 操作之后不能直接用与其他的算子操作，否则在一些版本中 Cache
          * 不生效
97.       */
98.       println("所有电影中平均得分最高（口碑最好）的电影:")
99.     val  ratings=  ratingsRDD.map(_.split("::")).map(x  =>  (x(0),  x(1),
        x(2))).cache()  //格式化出电影 ID 和评分
100.        ratings.map(x => (x._2, (x._3.toDouble, 1)))  //格式化成为 Key-Value
101.          .reduceByKey((x, y) => (x._1 + y._1,x._2 + y._2))
          //对 Value 进行 reduce 操作，分别得出每部电影的总的评分和总的点评人数
102.          .map(x => (x._2._1.toDouble / x._2._2, x._1))  //求出电影平均分
103.          .sortByKey(false) //降序排列
104.          .take(10) //取 Top10
105.          .foreach(println) //打印到控制台
106.
107.        /**
      * 上面的功能计算的是口碑最好的电影，接下来分析粉丝或者观看人数最多的电影
108.          */
109.
110.        println("所有电影中粉丝或者观看人数最多的电影:")
111.        ratings.map(x => (x._2, 1)).reduceByKey(_+_).map(x => (x._2,
          x._1)).sortByKey(false)
112.          .map(x => (x._2, x._1)).take(10).foreach(println)
113.
114.        /**
115.          * 今日作业：分析最受男性喜爱的电影 Top10 和最受女性喜爱的电影 Top10
116.          * 1."users.dat": UserID::Gender::Age::OccupationID::Zip-code
117.          * 2."ratings.dat": UserID::MovieID::Rating::Timestamp
118.          *   分析：单从 ratings 中无法计算出最受男性或者女性喜爱的电影 Top10，因为
          *     该 RDD 中没有 Gender 信息，如果需要使用 Gender 信息进行 Gender 的分
          *     类，此时一定需要聚合，当然，我们力求聚合使用的是 mapjoin（分布式计
          *     算的 Killer 是数据倾斜，Mapper 端的 Join 是一定不会数据倾斜的），这
          *     里可否使用 mapjoin 呢？不可以，因为用户的数据非常多！所以，这里要使
          *     用正常的 Join，此处的场景不会数据倾斜，因为用户一般都很均匀地分布（但
          *     是，系统信息搜集端要注意黑客攻击）
119.          * Tips:
120.          *   1. 因为要再次使用电影数据的 RDD，所以复用了前面 Cache 的 ratings 数据
121.          *   2. 在根据性别过滤出数据后，关于 TopN 部分的代码直接复用前面的代码就行了。
122.          *   3. 要进行 Join 的话，需要 key-value；
123.          *   4. 在进行 Join 的时候通过 take 等方法注意 Join 后的数据格式，例如
          *   (3319,((3319,50,4.5),F))
124.          *   5. 使用数据冗余来实现代码复用或者更高效地运行，这是企业级项目的一个非
          *   常重要的技巧！
125.          */
126.        val male = "M"
127.        val female = "F"
128.        val genderRatings = ratings.map(x => (x._1, (x._1, x._2,
          x._3))).join(
129.          usersRDD.map(_.split("::")).map(x => (x(0), x(1)))).cache()
130.        genderRatings.take(10).foreach(println)
131.        val maleFilteredRatings = genderRatings.filter(x =>
          x._2._2.equals("M")).map(x => x._2._1)
132.        val femaleFilteredRatings = genderRatings.filter(x => x._2._2
```

```
            .equals("F")).map(x => x._2._1)
133.
134.        /**
135.          * (855,5.0)
136.          * (6075,5.0)
137.          * (1166,5.0)
138.          * (3641,5.0)
139.          * (1045,5.0)
140.          * (4136,5.0)
141.          * (2538,5.0)
142.          * (7227,5.0)
143.          * (8484,5.0)
144.          * (5599,5.0)
145.          */
146.        println("所有电影中最受男性喜爱的电影 Top10:")
147.        maleFilteredRatings.map(x => (x._2, (x._3.toDouble, 1)))
            //格式化成为 Key-Value 的形式
148.          .reduceByKey((x, y) => (x._1 + y._1,x._2 + y._2))
              //对 Value 进行 reduce 操作，分别得出每部电影的总的评分和总的点评人数
149.          .map(x => (x._2._1.toDouble / x._2._2, x._1))  //求出电影平均分
150.          .sortByKey(false)  //降序排列
151.          .map(x => (x._2, x._1))
152.          .take(10)          //取 Top10
153.          .foreach(println)  //打印到控制台
154.
155.        /**
156.          * (789,5.0)
157.          * (855,5.0)
158.          * (32153,5.0)
159.          * (4763,5.0)
160.          * (26246,5.0)
161.          * (2332,5.0)
162.          * (503,5.0)
163.          * (4925,5.0)
164.          * (8767,5.0)
165.          * (44657,5.0)
166.          */
167.        println("所有电影中最受女性喜爱的电影 Top10:")
168.        femaleFilteredRatings.map(x => (x._2, (x._3.toDouble, 1)))
            //格式化成为 Key-Value 的形式
169.          .reduceByKey((x, y) => (x._1 + y._1,x._2 + y._2))
              //对 Value 进行 reduce 操作，分别得出每部电影的总的评分和总的点评人数
170.          .map(x => (x._2._1.toDouble / x._2._2, x._1))  //求出电影平均分
171.          .sortByKey(false)  //降序排列
172.          .map(x => (x._2, x._1))
173.          .take(10)          //取 Top10
174.          .foreach(println)  //打印到控制台
175.
176.
177.        /**
178.          * 最受不同年龄段人员欢迎的电影 TopN
179.          * "users.dat": UserID::Gender::Age::OccupationID::Zip-code
180.          * 思路：首先还是计算 TopN，但是这里的关注点有两个：
181.          *   1. 不同年龄阶段如何界定，这个问题其实是业务的问题，当然，实现时可以使
              *      用 RDD 的 filter，例如 13 < age <18，这样做会导致运行时进行大量的
              *      计算，因为要进行扫描，所以会非常耗性能。所以，一般情况下，我们都是在
```

```
          *      原始数据中直接对要进行分组的年龄段提前进行好 ETL，例如，进行 ETL 后
          *      产生以下数据：
182.      *      - Gender is denoted by a "M" for male and "F" for female
183.      *      - Age is chosen from the following ranges:
184.      *  1:  "Under 18"
185.      * 18:  "18-24"
186.      * 25:  "25-34"
187.      * 35:  "35-44"
188.      * 45:  "45-49"
189.      * 50:  "50-55"
190.      * 56:  "56+"
191.      *   2. 性能问题：
192.      *      第一点：实现时可以使用 RDD 的 filter，如 13 < age <18，这样做会导
          *      致运行时候进行大量的计算，因为要进行扫描，所以会非常耗性能，我们通过
          *      提前的 ETL 把计算发生在 Spark 业务逻辑运行以前，用空间换时间，当然这
          *      些实现也可以使用 Hive，因为 Hive 语法支持非常强悍且内置了最多的函数；
          *      第二点：这里要使用 mapjoin，原因是 targetUsers 数据只有 UserID，
          *      数据量一般不会太多
193.      */
194.
195.      val targetQQUsers = usersRDD.map(_.split("::")).map(x => (x(0),
          x(2))).filter(_._2.equals("18"))
196.      val targetTaobaoUsers = usersRDD.map(_.split("::")).map(x =>
          (x(0), x(2))).filter(_._2.equals("25"))
197.
198.    /**
          * 在 Spark 中如何实现 mapjoin 呢，显然是要借助于 Broadcast，把数据广播到
          * Executor 级别，让该 Executor 上的所有任务共享该唯一的数据，而不是每次
          * 运行 Task 的时候，都要发送一份数据的复制，这显著地降低了网络数据的传输
          * 和 JVM 内存的消耗
199.      */
200.    val targetQQUsersSet = HashSet() ++ targetQQUsers.map(_._1)
        .collect()
201.    val targetTaobaoUsersSet = HashSet() ++ targetTaobaoUsers
        .map(_._1).collect()
202.    val targetQQUsersBroadcast = sc.broadcast(targetQQUsersSet)
203.    val targetTaobaoUsersBroadcast = sc.broadcast(targetTaobaoUsersSet)
204.
205.    /**
206.      * QQ 或者微信核心目标用户最喜爱电影 TopN 分析
207.      * (Silence of the Lambs, The (1991),524)
208.      * (Pulp Fiction (1994),513)
209.      * (Forrest Gump (1994),508)
210.      * (Jurassic Park (1993),465)
211.      * (Shawshank Redemption, The (1994),437)
212.      */
213.    val movieID2Name = moviesRDD.map(_.split("::")).map(x => (x(0),
        x(1))).collect.toMap
214.    println("所有电影中 QQ 或者微信核心目标用户最喜爱电影 TopN 分析:")
215.    ratingsRDD.map(_.split("::")).map(x => (x(0), x(1))).filter(x =>
216.      targetQQUsersBroadcast.value.contains(x._1)
217.    ).map(x => (x._2, 1)).reduceByKey(_ + _).map(x => (x._2, x._1))
218.      .sortByKey(false).map(x => (x._2, x._1)).take(10)
219.      .map(x => (movieID2Name.getOrElse(x._1, null), x._2)).foreach
        (println)
```

```
220.
221.        /**
222.          * taobao 核心目标用户最喜爱电影 TopN 分析
223.          * (Pulp Fiction (1994),959)
224.          * (Silence of the Lambs, The (1991),949)
225.          * (Forrest Gump (1994),935)
226.          * (Jurassic Park (1993),894)
227.          * (Shawshank Redemption, The (1994),859)
228.          */
229.        println("所有电影中淘宝核心目标用户最喜爱电影 TopN 分析:")
230.        ratingsRDD.map(_.split("::")).map(x => (x(0), x(1))).filter(x =>
231.          targetTaobaoUsersBroadcast.value.contains(x._1)
232.        ).map(x => (x._2, 1)).reduceByKey(_ + _).map(x => (x._2, x._1))
233.          .sortByKey(false).map(x => (x._2, x._1)).take(10)
234.          .map(x => (movieID2Name.getOrElse(x._1, null), x._2))
         .foreach(println)
235.
236.
237.        /**
238.          * 对电影评分数据进行二次排序，以 Timestamp 和 Rating 两个维度降序排列
239.          * "ratings.dat": UserID::MovieID::Rating::Timestamp
240.          *
241.          * 完成的功能是最近时间中最新发生的点评信息
242.          */
243.        println("对电影评分数据以 Timestamp 和 Rating 两个维度进行二次降序排列:")
244.        val pairWithSortkey = ratingsRDD.map(line =>{
245.          val splited = line.split("::")
246.          (new SecondarySortKey( splited(3).toDouble,splited(2).toDouble),line )})
247.
248.        val sorted = pairWithSortkey.sortByKey(false)
249.
250.        val sortedResult = sorted.map(sortedline => sortedline._2)
251.        sortedResult.take(10).foreach(println)
252.
253.
254.        while(true){} //和通过 Spark Shell 运行代码可以一直看到 Web 终端的原理一
                      //样，因为 Spark Shell 内部有一个 LOOP 循环
255.
256.        sc.stop()
257.
258.
259.      }
260.    }
261.    class SecondarySortKey(val first:Double,val second:Double) extends
    Ordered [SecondarySortKey] with Serializable {
262.      def compare(other:SecondarySortKey):Int = {
263.        if (this.first - other.first !=0) {
264.          (this.first - other.first).toInt
265.        } else {
266.          if (this.second - other.second > 0){
267.              Math.ceil(this.second - other.second).toInt
268.          } else if (this.second - other.second < 0){
269.            Math.floor(this.second - other.second).toInt
270.          } else {
271.            (this.second - other.second).toInt
```

```
272.          }
273.        }
274.      }
275.    }
```

2. DateFrame方式案例代码

Spark 商业案例之大数据电影点评系统应用案例代码 Movie_Users_Analyzer_DateFrame.scala 如例 12-4 所示。

【例 12-4】 Movie_Users_Analyzer_DateFrame.scala 代码。

```
1.   package com.dt.spark.sparksql
2.
3.  import org.apache.spark.SparkConf
4.  import org.apache.log4j.{Level, Logger}
5.  import org.apache.spark.sql.{Row, SparkSession}
6.  import org.apache.spark.sql.types.{StringType, StructField, StructType,
    DoubleType}
7.  import scala.collection.immutable.HashSet
8.
9.
10.
11.
12. /**
13.   * 版权：DT 大数据梦工厂所有
14.   * 时间：2017 年 1 月 11 日；
15.   * 电影点评系统用户行为分析：用户观看电影和点评电影的所有行为数据的采集、过滤、处理
      * 和展示：
16.   *   数据采集：企业中一般越来越多地喜欢直接把 Server 中的数据发送给 Kafka，因为更加
      *   具备实时性；
17.   *   数据过滤：趋势是直接在 Server 端进行数据过滤和格式化，当然，采用 Spark SQL 进
      *   行数据过滤也是一种主要形式；
18.   *   数据处理：
19.   *     1. 一个基本的技巧是，先使用传统的 SQL 去实现一个数据处理的业务逻辑（自己可以
      *     手动模拟一些数据）；
20.   *     2. 再次推荐使用 DataSet 去实现业务功能，尤其是统计分析功能；
21.   *     3. 如果你想成为专家级别的顶级 Spark 人才，请使用 RDD 实现业务功能，为什么？
      *     运行的时候是基于 RDD 的！
22.   *
23.   *  数据：强烈建议大家使用 Parquet
24.   *  1."ratings.dat": UserID::MovieID::Rating::Timestamp
25.   *  2."users.dat": UserID::Gender::Age::OccupationID::Zip-code
26.   *  3."movies.dat": MovieID::Title::Genres
27.   *  4."occupations.dat": OccupationID::OccupationName   一般情况下都会以
      *  程序中数据结构 Haskset 的方式存在，是为了作 mapjoin
28.   */
29. object Movie_Users_Analyzer_DateFrame {
30.   def main(args: Array[String]){
31.
32.     Logger.getLogger("org").setLevel(Level.ERROR)
33.
34.     var masterUrl = "local[8]" //默认程序运行在本地 Local 模式中，主要用于学习和测试
35.     var dataPath = "data/moviedata/medium/"  //数据存放的目录
36.
```

```
37.     /**
          * 当我们把程序打包运行在集群上的时候，一般都会传入集群的 URL 信息，这里我们假设
          * 如果传入参数，第一个参数只传入 Spark 集群的 URL，第二个参数传入的是数据的地
          * 址信息
38.       */
39.
40.
41.     if(args.length > 0) {
42.       masterUrl = args(0)
43.     } else if (args.length > 1) {
44.       dataPath = args(1)
45.     }
46.
47.
48.     /**
          * 创建 Spark 会话上下文 SparkSession 和集群上下文 SparkContext，在 SparkConf 中可
          * 以进行各种依赖和参数的设置等，大家可以通过 SparkSubmit 脚本的 help 去看设置
          * 信息，其中 SparkSession 统一了 Spark SQL 运行的不同环境
49.       */
50.
51.
52.     val sparkConf = new SparkConf().setMaster(masterUrl).
        setAppName("Movie_Users_Analyzer_SparkSQL")
53.
54.     /**
          *SparkSession 统一了 Spark SQL 执行时的不同的上下文环境，也就是说，Spark SQL
          * 无论运行在哪种环境下，我们都可以只使用 SparkSession 这样一个统一的编程入口，
          * 来处理 DataFrame 和 DataSet 编程，不需要关注底层是否有 Hive 等
55.       */
56
57.
58.     val spark = SparkSession
59.       .builder()
60.       .config(sparkConf)
61.       .getOrCreate()
62.
63.     val sc = spark.sparkContext //从 SparkSession 获得的上下文，这是因为我们
                                    //读原生文件时或者实现一些 Spark SQL 目前还不
                                    //支持的功能时需要使用 SparkContext
64.
65.
66.     /**
          * 读取数据，用什么方式读取数据呢？这里使用的是 RDD!
67.       */
68.
69.     val usersRDD = sc.textFile(dataPath + "users.dat")
70.     val moviesRDD = sc.textFile(dataPath + "movies.dat")
71.     val occupationsRDD = sc.textFile(dataPath + "occupations.dat")
72.     val ratingsRDD = sc.textFile(dataPath + "ratings.dat")
73.
74.
75.     /**
76.       * 功能一：通过 DataFrame 实现某特定电影观看者中男性和女性不同年龄分别有多少人？
77.       *   1. 从点评数据中获得观看者的信息 ID;
78.       *   2. 把 ratings 和 users 表进行 Join 操作，获得用户的性别信息；
79.       *   3. 使用内置函数（内部包含超过 200 个内置函数）进行信息统计和分析
80.       *  这里我们通过 DataFrame 来实现：首先通过 DataFrame 的方式表现 ratings 和
          *  users 的数据，然后进行 Join 和统计操作
81.       */
82.
83.     println("功能一：通过 DataFrame 实现某特定电影观看者中男性和女性不同年龄分别有
```

```
      多少人? ")
84.   val schemaforusers = StructType("UserID::Gender::Age::OccupationID::
      Zip-code".split("::")
85.     .map(column => StructField(column, StringType, true)))
        //使用 Struct 方式把 Users 的数据格式化，即在 RDD 的基础上增加数据的元数据信息
86.   val usersRDDRows = usersRDD.map(_.split("::")).map(line =>
      Row(line(0).trim,line(1)
87.     .trim,line(2).trim,line(3).trim,line(4).trim))
          //把我们的每条数据变成以 Row 为单位的数据
88.   val usersDataFrame = spark.createDataFrame(usersRDDRows, schemaforusers)
      //结合 Row 和 StructType 的元数据信息，基于 RDD 创建 DataFrame，这时 RDD 就有了
      //元数据信息的描述
89.
90.   val schemaforratings = StructType("UserID::MovieID".split("::")
91.     .map(column => StructField(column, StringType, true)))
92.     .add("Rating", DoubleType, true)
93.     .add("Timestamp",StringType, true)
94.
95.   val ratingsRDDRows = ratingsRDD.map(_.split("::")).map(line =>
      Row(line(0).trim,line(1)
96.     .trim,line(2).trim.toDouble,line(3).trim))
97.   val ratingsDataFrame = spark.createDataFrame(ratingsRDDRows,
      schemaforratings)
98.
99.   val schemaformovies = StructType("MovieID::Title::Genres".split("::")
100.        .map(column => StructField(column, StringType, true)))
          //使用 Struct 方式把 Users 的数据格式化，即在 RDD 的基础上增加数据的元数据信息
101.      val moviesRDDRows = moviesRDD.map(_.split("::")).map(line =>
          Row(line(0).trim,line(1)
102.        .trim,line(2).trim)) //把我们的每条数据变成以 Row 为单位的数据
103.      val moviesDataFrame = spark.createDataFrame(moviesRDDRows,
          schemaformovies)  //结合 Row 和 StructType 的元数据信息基于 RDD 创建
                            //DataFrame，这时 RDD 就有了元数据信息的描述
104.
105.
106.      ratingsDataFrame.filter(s" MovieID = 1193")
          //这里能够直接指定 MovieID 的原因是 DataFrame 中有该元数据信息！
107.          .join(usersDataFrame, "UserID")
              //Join 的时候直接指定基于 UserID 进行 Join，这相对于原生的 RDD 操作而
              //言，更加方便、快捷
108.          .select("Gender", "Age")
              //直接通过元数据信息中的 Gender 和 Age 进行数据的筛选
109.          .groupBy("Gender", "Age")
              //直接通过元数据信息中的 Gender 和 Age 进行数据的 groupBy 操作
110.          .count()  //基于 groupBy 分组信息进行 count 统计操作
111.          .show(10) //显示出分组统计后的前 10 条信息
112.      /**
113.        * 功能二：用 SQL 语句实现某特定电影观看者中男性和女性不同年龄分别有多少人？
114.        * 1. 注册临时表，写 SQL 语句需要 Table;
115.        * 2. 基于上述注册的临时表写 SQL 语句;
116.        */
117.      println("功能二：用 GlobalTempView 的 SQL 语句实现某特定电影观看者中男性
          和女性不同年龄分别有多少人？")
118.      ratingsDataFrame.createGlobalTempView("ratings")
119.      usersDataFrame.createGlobalTempView("users")
120.
121.
122.      spark.sql("SELECT Gender, Age, count(*) from  global_temp.users u
          join  global_temp.ratings as r on u.UserID = r.UserID where MovieID
```

```
123.          = 1193" +" group by Gender, Age").show(10)
124.
125.        println("功能二：用 LocalTempView 的 SQL 语句实现某特定电影观看者中男性和
            女性不同年龄分别有多少人？")
126.        ratingsDataFrame.createTempView("ratings")
127.        usersDataFrame.createTempView("users")
128.
129.
130.        spark.sql("SELECT Gender, Age, count(*) from  users u join
            ratings as r on u.UserID = r.UserID where MovieID = 1193" +
131.          " group by Gender, Age").show(10)
132.
133.        /**
134.          * 功能三：使用 DataFrame 进行电影流行度分析：所有电影中平均得分最高（口碑
              * 最好）的电影及观看人数最多的电影（流行度最高）
135.          * "ratings.dat": UserID::MovieID::Rating::Timestamp
136.          * 得分最高的 Top10 电影实现思路：如果想算总的评分，一般需要 reduceByKey
              * 操作或者 aggregateByKey 操作
137.          *   第一步：把数据变成 Key-Value，大家想一下在这里什么是 Key，什么是
              *   Value。把 MovieID 设置为 Key，把 Rating 设置为 Value；
138.          *   第二步：通过 reduceByKey 操作或者 aggregateByKey 实现聚合，然后呢？
139.          *   第三步：排序，如何做？进行 Key 和 Value 的交换
140.          */
141.        println("纯粹通过 RDD 的方式实现所有电影中平均得分最高(口碑最好)的电影 TopN:")
142.        val ratings= ratingsRDD.map(_.split("::")).map(x => (x(0), x(1),
            x(2))).cache()  //格式化出电影 ID 和评分
143.        ratings.map(x => (x._2, (x._3.toDouble, 1.toDouble)))
            //格式化成为 Key-Value 的形式
144.          .reduceByKey((x, y) => (x._1 + y._1,x._2 + y._2))
              //对 Value 进行 reduce 操作，分别得出每部电影的总的评分和总的点评人数
145.          .map(x => (x._2._1.toDouble / x._2._2.toDouble, x._1))
              //求出电影平均分
146.          .sortByKey(false)  //降序排列
147.            .map(x => (x._2, x._1))
148.          .take(10)           //取 Top10
149.          .foreach(println)  //打印到控制台
150.
151.        println("通过 DataFrame 和 RDD 相结合的方式计算所有电影中平均得分最高（口
            碑最好）的电影 TopN:")
152.        ratingsDataFrame.select("MovieID", "Rating").groupBy("MovieID")
153.          .avg("Rating").rdd
154.          .map(row => (row(1),(row(0), row(1)))).sortBy(_._1.toString
              .toDouble, false)
155.          .map(tuple => tuple._2).collect.take(10).foreach(println)
156.
157.        import spark.sqlContext.implicits._
158.        println("通过纯粹使用 DataFrame 方式计算所有电影中平均得分最高（口碑最好）
            的电影 TopN:")
159.        ratingsDataFrame.select("MovieID", "Rating").groupBy("MovieID")
160.          .avg("Rating").orderBy($"avg(Rating)".desc).show(10)
161.        /**
              * 上面的功能计算的是口碑最好的电影，接下来分析粉丝或者观看人数最多的电影
162.          */
163.
```

```
164.        println("纯粹通过RDD的方式计算所有电影中粉丝或者观看人数最多(最流行电影)
            的电影 TopN:")
165.        ratings.map(x => (x._2, 1)).reduceByKey(_+_).map(x => (x._2, x._1))
166.          .sortByKey(false).map(x => (x._2, x._1)).collect()
167.          .take(10).foreach(println)
168.        println("通过 DataFrame 和 RDD 结合的方式计算最流行电影（即所有电影中粉丝
            或者观看人数最多）的电影 TopN:")
169.       ratingsDataFrame.select("MovieID","Timestamp").groupBy("MovieID")
           .count().rdd
170.          .map(row => (row(1).toString.toLong, (row(0), row(1))))
              .sortByKey(false)
171.          .map(tuple => tuple._2).collect().take(10).foreach(println)
172.
173.        println("纯粹通过 DataFrame 的方式计算最流行电影（即所有电影中粉丝或者观
            看人数最多）的电影 TopN:")
174.    //    ratingsDataFrame.select("MovieID","Timestamp")
175.    //    .ratingsDataFrame.select("MovieID")
176.          .ratingsDataFrame.groupBy("MovieID").count()
177.          .orderBy($"count".desc).show(10)
178.
179.
180.        /**
181.          * 功能四：分析最受男性喜爱的电影 Top10 和最受女性喜爱的电影 Top10
182.          * 1. "users.dat": UserID::Gender::Age::OccupationID::Zip-code
183.          * 2. "ratings.dat": UserID::MovieID::Rating::Timestamp 分析：单
              * 从 ratings 中无法计算出最受男性或者女性喜爱的电影 Top10，因为该 RDD 中
              * 没有 Gender 信息，如果我们需要使用 Gender 信息进行 Gender 的分类，此时
              * 一定需要聚合，当然，我们力求聚合使用的是 mapjoin（分布式计算的 Killer
              * 是数据倾斜,Mapper 端的 Join 是一定不会数据倾斜的),这里可否使用 mapjoin
              * 呢？不可以，因为用户的数据非常多！所以这里要使用正常的 Join，此处的场景
              * 不会数据倾斜，因为用户一般都很均匀地分布(但是系统信息搜集端要注意黑客攻击)
184.          * Tips:
185.          *   1. 因为要再次使用电影数据的 RDD,所以复用了前面 Cache 的 ratings 数据；
186.          *   2. 在根据性别过滤出数据后，关于 TopN 部分的代码，直接复用前面的代码就行了；
187.          *   3. 要进行 Join 的话，需要 key-value；
188.          *   4. 在进行 Join 的时候通过 take 等方法注意 Join 后的数据格式，例如
              *   (3319,((3319,50,4.5),F))
189.          *   5. 使用数据冗余实现代码复用或者更高效地运行，这是企业级项目的一个非常
              *   重要的技巧！
190.          */
191.        val male = "M"
192.        val female = "F"
193.        val genderRatings = ratings.map(x => (x._1, (x._1, x._2, x._
            3))).join(
194.          usersRDD.map(_.split("::")).map(x => (x(0), x(1)))).cache()
195.        val genderRatingsDataFrame = ratingsDataFrame.join(users
            DataFrame, "UserID").cache()
196.
197.        genderRatings.take(10).foreach(println)
198.        val maleFilteredRatings = genderRatings.filter(x => x._2._2
            .equals("M")).map(x => x._2._1)
199.        val maleFilteredRatingsDataFrame = genderRatingsDataFrame
            .filter("Gender= 'M'").select("MovieID", "Rating")
200.        val femaleFilteredRatings = genderRatings.filter(x => x._2._2
```

```
        .equals("F")).map(x => x._2._1)
201.    val femaleFilteredRatingsDataFrame = genderRatingsDataFrame
        .filter("Gender= 'F'").select("MovieID", "Rating")
202.
203.    /**
204.      * (855,5.0)
205.      * (6075,5.0)
206.      * (1166,5.0)
207.      * (3641,5.0)
208.      * (1045,5.0)
209.      * (4136,5.0)
210.      * (2538,5.0)
211.      * (7227,5.0)
212.      * (8484,5.0)
213.      * (5599,5.0)
214.      */
215.    println("纯粹使用 RDD 实现所有电影中最受男性喜爱的电影 Top10:")
216.    maleFilteredRatings.map(x => (x._2, (x._3.toDouble, 1)))
        //格式化成为 Key-Value 的形式
217.      .reduceByKey((x, y) => (x._1 + y._1,x._2 + y._2))
          //对 Value 进行 reduce 操作，分别得出每部电影的总的评分和总的点评人数
218.      .map(x => (x._2._1.toDouble / x._2._2, x._1))  //求出电影平均分
219.      .sortByKey(false)  //降序排列
220.      .map(x => (x._2, x._1))
221.      .take(10)          //取 Top10
222.      .foreach(println)  //打印到控制台
223.    println("纯粹使用 DataFrame 实现所有电影中最受男性喜爱的电影 Top10:")
224.    maleFilteredRatingsDataFrame.groupBy("MovieID").avg("Rating")
        .orderBy($"avg(Rating)".desc).show(10)
225.    /**
226.      * (789,5.0)
227.      * (855,5.0)
228.      * (32153,5.0)
229.      * (4763,5.0)
230.      * (26246,5.0)
231.      * (2332,5.0)
232.      * (503,5.0)
233.      * (4925,5.0)
234.      * (8767,5.0)
235.      * (44657,5.0)
236.      */
237.    println("纯粹使用 RDD 实现所有电影中最受女性喜爱的电影 Top10:")
238.    femaleFilteredRatings.map(x => (x._2, (x._3.toDouble, 1)))
        //格式化成为 Key-Value 的形式
239.      .reduceByKey((x, y) => (x._1 + y._1,x._2 + y._2))
      //对 Value 进行 reduce 操作，分别得出每部电影的总的评分和总的点评人数
240.      .map(x => (x._2._1.toDouble / x._2._2, x._1))  //求出电影平均分
241.      .sortByKey(false)  //降序排列
242.      .map(x => (x._2, x._1))
243.      .take(10)          //取 Top10
244.      .foreach(println)  //打印到控制台
245.    println("纯粹使用 DataFrame 实现所有电影中最受女性喜爱的电影 Top10:")
246.    femaleFilteredRatingsDataFrame.groupBy("MovieID").avg("Rating")
        .orderBy($"avg(Rating)".desc, $"MovieID".desc).show(10)
247.
248.    /**
```

```
         * 思考题：如果想让 RDD 和 DataFrame 计算的 TopN 的每次结果都一样，该如何
         * 保证？现在的情况是，例如计算 Top10，而其同样评分的不止 10 个，所以每次都会
         * 从中取出 10 个，这就导致 Top10 结果不一致，这时我们可以使用一个新的列参
         * 与排序，如果是 RDD，该怎么做呢？这时就要进行二次排序。
249.     *   如果是 DataFrame，该如何做呢？非常简单，只需要在 orderBy 函数中增加
         *   一个排序维度的字段即可
250.     */
251.
252.
253.
254.
255.    /**
256.     * 功能五：最受不同年龄段人员欢迎的电影 TopN
257.     * "users.dat": UserID::Gender::Age::OccupationID::Zip-code
258.     * 思路：首先计算 TopN，但是这里的关注点有两个：
259.     *   1.不同年龄阶段如何界定，这个问题其实是业务的问题，当然，实现时可以使
         *   用 RDD 的 filter，例如 13 < age <18，这样做会导致运行时进行大量的计
         *   算，因为要进行扫描，所以会非常耗性能。所以，一般情况下，我们都是在原始
         *   数据中直接对要进行分组的年龄段提前进行 ETL，例如，进行 ETL 后产生
         *   以下数据：
260.     *     - Gender is denoted by a "M" for male and "F" for female
261.     *     - Age is chosen from the following ranges:
262.     *  1:  "Under 18"
263.     * 18:  "18-24"
264.     * 25:  "25-34"
265.     * 35:  "35-44"
266.     * 45:  "45-49"
267.     * 50:  "50-55"
268.     * 56:  "56+"
269.     *   2. 性能问题：
270.     *     第一点：实现时可以使用 RDD 的 filter，例如 13 < age <18，这样做会
         *     导致运行时进行大量的计算，因为要进行扫描，所以会非常耗性能，我们通过
         *     提前的 ETL 把计算发生在 Spark 业务逻辑运行以前，用空间换时间，当然，
         *     这些实现也可以使用 Hive，因为 Hive 语法支持非常强悍且内置了最多的函数；
271.     *     第二点：这里要使用 mapjoin，原因是 targetUsers 数据只有 UserID，
         *     数据量一般不会太多
272.     */
273.
274.
275.    val targetQQUsers = usersRDD.map(_.split("::")).map(x => (x(0),
        x(2))).filter(_._2.equals("18"))
276.    val targetTaobaoUsers = usersRDD.map(_.split("::")).map(x =>
        (x(0), x(2))).filter(_._2.equals("25"))
277.
278.    /**
         * 在 Spark 中如何实现 mapjoin 呢，显然要借助于 Broadcast，把数据广播到
         * Executor 级别，让该 Executor 上的所有任务共享该唯一的数据，而不是每次
         * 运行 Task 时都要发送一份数据的复制，这显著地降低了网络数据的传输和 JVM
         * 内存的消耗
279.     */
280.
281.    val targetQQUsersSet = HashSet() ++ targetQQUsers.map(_._1)
        .collect()
282.    val targetTaobaoUsersSet = HashSet() ++ targetTaobaoUsers
```

```
        .map(_._1).collect()
283.
284.    val targetQQUsersBroadcast = sc.broadcast(targetQQUsersSet)
285.    val targetTaobaoUsersBroadcast = sc.broadcast(targetTaobaoUsersSet)
286.
287.    /**
288.      * QQ 或者微信核心目标用户最喜爱电影 TopN 分析
289.      * (Silence of the Lambs, The (1991),524)
290.      * (Pulp Fiction (1994),513)
291.      * (Forrest Gump (1994),508)
292.      * (Jurassic Park (1993),465)
293.      * (Shawshank Redemption, The (1994),437)
294.      */
295.    val movieID2Name = moviesRDD.map(_.split("::")).map(x => (x(0),
        x(1))).collect.toMap
296.    println("纯粹通过 RDD 的方式实现所有电影中 QQ 或者微信核心目标用户最喜爱电
        影 TopN 分析:")
297.    ratingsRDD.map(_.split("::")).map(x => (x(0), x(1))).filter(x =>
298.      targetQQUsersBroadcast.value.contains(x._1)
299.    ).map(x => (x._2, 1)).reduceByKey(_ + _).map(x => (x._2, x._1))
300.      .sortByKey(false).map(x => (x._2, x._1)).take(10)
301.      .map(x => (movieID2Name.getOrElse(x._1, null), x._2)).foreach
          (println)
302.
303.    println("纯粹通过 DataFrame 的方式实现所有电影中 QQ 或者微信核心目标用户最
        喜爱电影 TopN 分析:")
304.
305.    ratingsDataFrame.join(usersDataFrame, "UserID").filter("Age =
        '18'").groupBy("MovieID")
306.      .count().orderBy($"count".desc).printSchema()
307.
308.    /**
309.      * Tips:
310.      *   1. orderBy 操作需要在 Join 之后进行
311.      */
312.    ratingsDataFrame.join(usersDataFrame, "UserID").filter("Age =
        '18'").groupBy("MovieID")
313.      .count().join(moviesDataFrame, "MovieID").select("Title",
          "count").orderBy($"count".desc).show(10)
314.
315.    /**
316.      * 淘宝核心目标用户最喜爱电影 TopN 分析
317.      * (Pulp Fiction (1994),959)
318.      * (Silence of the Lambs, The (1991),949)
319.      * (Forrest Gump (1994),935)
320.      * (Jurassic Park (1993),894)
321.      * (Shawshank Redemption, The (1994),859)
322.      */
323.    println("纯粹通过 RDD 的方式实现所有电影中淘宝核心目标用户最喜爱电影 TopN
        分析:")
324.    ratingsRDD.map(_.split("::")).map(x => (x(0), x(1))).filter(x =>
325.      targetTaobaoUsersBroadcast.value.contains(x._1)
326.    ).map(x => (x._2, 1)).reduceByKey(_ + _).map(x => (x._2, x._1))
327.      .sortByKey(false).map(x => (x._2, x._1)).take(10)
328.      .map(x => (movieID2Name.getOrElse(x._1, null), x._2))
          .foreach(println)
329.
```

```
330.        println("纯粹通过 DataFrame 的方式实现所有电影中淘宝核心目标用户最喜爱电
            影 TopN 分析:")
331.        ratingsDataFrame.join(usersDataFrame, "UserID").filter("Age =
            '25'").groupBy("MovieID")
332.          .count().join(moviesDataFrame, "MovieID").select("Title", "count")
             .orderBy($"count".desc).show(10)
333.
334.
335.
336.
337.          while(true){} //和通过 Spark Shell 运行代码可以一直看到 Web 终端
                            //的原理一样，因为 Spark Shell 内部有一个 LOOP 循环
338.
339.        sc.stop()
340.
341.
342.      }
343.    }
```

3. DateSet方式案例代码

Spark 商业案例之大数据电影点评系统应用案例代码 Movie_Users_Analyzer_DateSet.scala 如例 12-5 所示。

【例 12-5】 Movie_Users_Analyzer_DateSet.scala 代码。

```
1.    package com.dt.spark.sparksql
2.
3.  import org.apache.log4j.{Level, Logger}
4.  import org.apache.spark.SparkConf
5.  import org.apache.spark.sql.types.{DoubleType, StringType, StructField,
    StructType}
6.  import org.apache.spark.sql.{Row, SparkSession}
7.
8.  import scala.collection.immutable.HashSet
9.
10.
11. /**
12.   * 版权：DT 大数据梦工厂所有
13.   * 时间：2017 年 1 月 19 日；
14.   * 电影点评系统用户行为分析：用户观看电影和点评电影的所有行为数据的采集、过滤、处理
      * 和展示：
15.   *    数据采集：企业中一般越来越多地喜欢直接把 Server 中的数据发送给 Kafka，因为更加
      *    具备实时性；
16.   *    数据过滤：趋势是直接在 Server 端进行数据过滤和格式化，当然，采用 Spark SQL 进
      *    行数据的过滤也是一种主要形式；
17.   *    数据处理：
18.   *       1. 一个基本的技巧是，先使用传统的 SQL 去实现一个数据处理的业务逻辑（自己可以
      *       手动模拟一些数据）；
19.   *       2. 再次推荐使用 DataSet 去实现业务功能尤其是统计分析功能；
20.   *       3. 如果你想成为专家级别的顶级 Spark 人才，请使用 RDD 实现业务功能，为什么？
      *       因为运行的时候是基于 RDD 的！
21.   *
22.   *   数据：强烈建议大家使用 Parquet
23.   *   1. "ratings.dat": UserID::MovieID::Rating::Timestamp
```

```
24.   *  2. "users.dat": UserID::Gender::Age::OccupationID::Zip-code
25.   *  3. "movies.dat": MovieID::Title::Genres
26.   *  4. "occupations.dat": OccupationID::OccupationName   一般情况下都会以
      *  程序中数据结构 Haskset 的方式存在，是为了做 mapjoin
27.   */
28. object Movie_Users_Analyzer_DateSet {
29.
30.   case class User(UserID:String, Gender:String, Age:String, OccupationID:
      String, Zip_Code:String)
31.   case class Rating(UserID:String, MovieID:String, Rating:Double,
      Timestamp:String)
32.   case class Movie(MovieID:String, Title:String, Genres:String)
33.
34.
35.   def main(args: Array[String]){
36.
37.
38.
39.
40.     Logger.getLogger("org").setLevel(Level.ERROR)
41.
42.     var masterUrl = "local[8]"
      //默认程序运行在本地 Local 模式中，主要用于学习和测试
43.     var dataPath = "data/moviedata/medium/"  //数据存放的目录
44.
45.     /**
        * 当我们把程序打包运行在集群上的时候，一般都会传入集群的 URL 信息，这里我们假设
        * 如果传入参数，第一个参数只传入 Spark 集群的 URL，第二个参数传入的是数据的地
        * 址信息
46.       */
47.
48.
49.     if(args.length > 0) {
50.       masterUrl = args(0)
51.     } else if (args.length > 1) {
52.       dataPath = args(1)
53.     }
54.
55.
56.     /**
        * 创建 Spark 会话上下文 SparkSession 和集群上下文 SparkContext，在 SparkConf 中
        * 可以进行各种依赖和参数的设置等，大家可以通过 SparkSubmit 脚本的 help 去看设
        * 置信息，其中 SparkSession 统一了 Spark SQL 运行的不同环境
57.       */
58.
59.
60.     val sparkConf = new SparkConf().setMaster(masterUrl)
      .setAppName("Movie_Users_Analyzer_DataSet")
61.
62.     /**
        * SparkSession 统一了 Spark SQL 执行时的不同的上下文环境，也就是说，Spark SQL
        * 无论运行在哪种环境下，我们都可以只使用 SparkSession 这样一个统一的编程入口
        * 来处理 DataFrame 和 DataSet 编程，不需要关注底层是否有 Hive 等
63.       */
64.
65.
66.     val spark = SparkSession
67.       .builder()
68.       .config(sparkConf)
69.       .getOrCreate()
70.
```

```
71.      val sc = spark.sparkContext
                //从 SparkSession 获得的上下文，这是因为我们读原生文件的时候或者实现一
                //些 Spark SQL 目前还不支持的功能的时候需要使用 SparkContext
72.
73.      import spark.implicits._
74.      /**
          * 读取数据，用什么方式读取数据呢？这里使用的是 RDD!
75.       */
76.
77.      val usersRDD = sc.textFile(dataPath + "users.dat")
78.      val moviesRDD = sc.textFile(dataPath + "movies.dat")
79.      val occupationsRDD = sc.textFile(dataPath + "occupations.dat")
80.      val ratingsRDD = sc.textFile(dataPath + "ratings.dat")
81.
82.
83.      /**
84.       * 功能一：通过 DataFrame 实现某特定电影观看者中男性和女性不同年龄分别有多少人？
85.       *   1. 从点评数据中获得观看者的信息 ID；
86.       *   2. 把 ratings 和 users 表进行 Join 操作获得用户的性别信息；
87.       *   3. 使用内置函数（内部包含超过 200 个内置函数）进行信息统计和分析
88.       *  这里我们通过 DataFrame 来实现：首先通过 DataFrame 的方式来表现 ratings 和
          *  users 的数据，然后进行 Join 和统计操作
89.       */
90.
91.      println("功能一：通过 DataFrame 实现某特定电影观看者中男性和女性不同年龄分别有
         多少人？")
92.      val schemaforusers = StructType("UserID::Gender::Age::OccupationID::
         Zip_Code".split("::")
93.        .map(column => StructField(column, StringType, true)))
        //使用 Struct 方式把 Users 的数据格式化，即在 RDD 的基础上增加数据的元数据信息
94.      val usersRDDRows = usersRDD.map(_.split("::")).map(line =>
         Row(line(0).trim,line(1)
95.        .trim,line(2).trim,line(3).trim,line(4).trim))
        //把我们的每条数据变成以 Row 为单位的数据
96.      val usersDataFrame = spark.createDataFrame(usersRDDRows, schemaforusers)
         //结合 Row 和 StructType 的元数据信息，基于 RDD 创建 DataFrame，这时 RDD 就有了
         //元数据信息的描述
97.      val usersDataSet = usersDataFrame.as[User]
98.
99.
100.         val schemaforratings = StructType("UserID::MovieID".split("::")
101.           .map(column => StructField(column, StringType, true)))
102.           .add("Rating", DoubleType, true)
103.           .add("Timestamp",StringType, true)
104.
105.         val ratingsRDDRows = ratingsRDD.map(_.split("::")).map(line =>
             Row(line(0).trim,line(1)
106.           .trim,line(2).trim.toDouble,line(3).trim))
107.         val ratingsDataFrame = spark.createDataFrame(ratingsRDDRows,
             schemaforratings)
108.         val ratingsDataSet = ratingsDataFrame.as[Rating]
109.
110.
111.         val schemaformovies = StructType("MovieID::Title::Genres".split("::")
112.           .map(column => StructField(column, StringType, true)))
            //使用 Struct 方式把 Users 的数据格式化，即在 RDD 的基础上增加数据的元数据信息
113.         val moviesRDDRows = moviesRDD.map(_.split("::")).map(line =>
             Row(line(0).trim,line(1)
114.          .trim,line(2).trim)) //把我们的每条数据变成以 Row 为单位的数据
115.         val moviesDataFrame = spark.createDataFrame(moviesRDDRows,
             schemaformovies)  //结合 Row 和 StructType 的元数据信息，基于 RDD 创建
```

```
                    //DataFrame，这时 RDD 就有了元数据信息的描述
116.        val moviesDataSet = moviesDataFrame.as[Movie]
117.
118.        println
119.        ratingsDataFrame.filter(s" MovieID = 1193")
        //这里能够直接指定 MovieID 的原因是 DataFrame 中有该元数据信息！
120.            .join(usersDataFrame, "UserID")
            //Join 的时候直接指定基于 UserID 进行 Join，这相对于原生的 RDD 操作而
            //言，更加方便、快捷
121.            .select("Gender", "Age")
            //直接通过元数据信息中的 Gender 和 Age 进行数据的筛选
122.            .groupBy("Gender", "Age")
            //直接通过元数据信息中的 Gender 和 Age 进行数据的 groupBy 操作
123.            .count()  //基于 groupBy 分组信息进行 count 统计操作
124.            .show(10) //显示出分组统计后的前 10 条信息
125.        println("功能一：通过 DataSet 实现某特定电影观看者中男性和女性不同年龄分别
        有多少人？")
126.        ratingsDataSet.filter(s" MovieID = 1193")
        //这里能够直接指定 MovieID 的原因是 DataFrame 中有该元数据信息
127.          .join(usersDataFrame, "UserID")
          //Join 的时候直接指定基于 UserID 进行 Join，这相对于原生的 RDD 操作而言，
          //更加方便、快捷
128.          .select("Gender", "Age")
          //直接通过元数据信息中的 Gender 和 Age 进行数据的筛选
129.          .groupBy("Gender", "Age")
          //直接通过元数据信息中的 Gender 和 Age 进行数据的 groupBy 操作
130.          .count()  //基于 groupBy 分组信息进行 count 统计操作
131.          .show(10) //显示出分组统计后的前 10 条信息
132.        /**
133.          * 功能二：用 SQL 语句实现某特定电影观看者中男性和女性不同年龄分别有多少人？
134.          * 1.注册临时表，写 SQL 语句需要 Table；
135.          * 2.基于上述注册的临时表写 SQL 语句
136.          */
137.        println("功能二：用 GlobalTempView 的 SQL 语句实现某特定电影观看者中男性
        和女性不同年龄分别有多少人？")
138.        ratingsDataFrame.createGlobalTempView("ratings")
139.        usersDataFrame.createGlobalTempView("users")
140.
141.
142.        spark.sql("SELECT Gender, Age, count(*) from  global_temp.users u
        join  global_temp.ratings as r on u.UserID = r.UserID where MovieID
        = 1193" +
143.          " group by Gender, Age").show(10)
144.
145.        println("功能二：用 LocalTempView 的 SQL 语句实现某特定电影观看者中男性和
        女性不同年龄分别有多少人？")
146.        ratingsDataFrame.createTempView("ratings")
147.        usersDataFrame.createTempView("users")
148.
149.
150.        spark.sql("SELECT Gender, Age, count(*) from  users u join
        ratings as r on u.UserID = r.UserID where MovieID = 1193" +
151.          " group by Gender, Age").show(10)
152.
153.        /**
154.          * 功能三：使用 DataFrame 进行电影流行度分析：所有电影中平均得分最高（口碑
          * 最好）的电影及观看人数最多的电影（流行度最高）
```

```
155.         * "ratings.dat": UserID::MovieID::Rating::Timestamp
156.         * 得分最高的 Top10 电影实现思路：如果想计算总的评分，一般需要 reduceByKey
             * 操作或者 aggregateByKey 操作
157.         *   第一步：把数据变成 Key-Value，大家想一下在这里什么是 Key，什么是
             *   Value。把 MovieID 设置为 Key，把 Rating 设置为 Value；
158.         *   第二步：通过 reduceByKey 操作或者 aggregateByKey 实现聚合，然后呢？
159.         *   第三步：排序，如何做？进行 Key 和 Value 的交换
160.         */
161.
162.       println("通过纯粹使用 DataFrame 方式计算所有电影中平均得分最高（口碑最好）
           的电影 TopN:")
163.       ratingsDataFrame.select("MovieID", "Rating").groupBy("MovieID")
164.         .avg("Rating").orderBy($"avg(Rating)".desc).show(10)
165.       println("通过纯粹使用 DataSet 方式计算所有电影中平均得分最高（口碑最好）的
           电影 TopN:")
166.       ratingsDataSet.select("MovieID", "Rating").groupBy("MovieID")
167.         .avg("Rating").orderBy($"avg(Rating)".desc).show(10)
168.       /**
             * 上面的功能计算的是口碑最好的电影，接下来分析粉丝或者观看人数最多的电影
169.         */
170.       println("纯粹通过 DataFrame 的方式计算最流行电影（即所有电影中粉丝或者观
           看人数最多）的电影 TopN:")
171.   //    ratingsDataFrame.select("MovieID","Timestamp")
172.   //    .ratingsDataFrame.select("MovieID")
173.         .ratingsDataFrame.groupBy("MovieID").count()
174.         .orderBy($"count".desc).show(10)
175.       println("纯粹通过 DataSet 的方式计算最流行电影(即所有电影中粉丝或者观看人
           数最多）的电影 TopN:")
176.       ratingsDataSet.groupBy("MovieID").count()
177.         .orderBy($"count".desc).show(10)
178.
179.
180.       /**
181.         * 功能四：分析最受男性喜爱的电影 Top10 和最受女性喜爱的电影 Top10
182.         * 1. "users.dat": UserID::Gender::Age::OccupationID::Zip-code
183.         * 2. "ratings.dat": UserID::MovieID::Rating::Timestamp
184.         *   分析：单从 ratings 中无法计算出最受男性或者女性喜爱的电影 Top10，因为
             *   该 RDD 中没有 Gender 信息，如果我们需要使用
185.         *     Gender 信息进行 Gender 的分类，此时一定需要聚合，当然，我们力求聚合
             *     使用的是 mapjoin（分布式计算的 Killer 是数据倾斜，Mapper 端的 Join
             *     是一定不会数据倾斜的），这里可否使用 mapjoin 呢？不可以，因为用户的
             *     数据非常多！所以，在这里要使用正常的 Join，此处的场景不会数据倾斜，
             *     因为用户一般都很均匀地分布（但是，系统信息搜集端要注意黑客攻击）
186.         * Tips:
187.         *   1. 因为要再次使用电影数据的 RDD,所以复用了前面 Cache 的 ratings 数据；
188.         *   2. 在根据性别过滤出数据后，关于 TopN 部分的代码，直接复用前面的代码就行；
189.         *   3. 要进行 Join，需要 key-value；
190.         *   4. 在进行 Join 的时候通过 take 等方法注意 Join 后的数据格式，例如
             *   (3319,((3319,50,4.5),F))
191.         *   5. 使用数据冗余来实现代码复用或者更高效地运行，这是企业级项目的一个非
```

```
         *  常重要的技巧
192.     */
193.
194.     val genderRatingsDataFrame = ratingsDataFrame.join(usersDataFrame,
         "UserID").cache()
195.     val genderRatingsDataSet = ratingsDataSet.join(usersDataSet,
         "UserID").cache()
196.     val maleFilteredRatingsDataFrame = genderRatingsDataFrame.filter
         ("Gender= 'M'").select("MovieID", "Rating")
197.     val maleFilteredRatingsDataSet = genderRatingsDataSet.filter
         ("Gender= 'M'").select("MovieID", "Rating")
198.     val femaleFilteredRatingsDataFrame = genderRatingsDataFrame
         .filter("Gender= 'F'").select("MovieID", "Rating")
199.     val femaleFilteredRatingsDataSet = genderRatingsDataSet.filter
         ("Gender= 'F'").select("MovieID", "Rating")
200.
201.     /**
202.      * (855,5.0)
203.      * (6075,5.0)
204.      * (1166,5.0)
205.      * (3641,5.0)
206.      * (1045,5.0)
207.      * (4136,5.0)
208.      * (2538,5.0)
209.      * (7227,5.0)
210.      * (8484,5.0)
211.      * (5599,5.0)
212.      */
213.
214.     println("纯粹使用 DataFrame 实现所有电影中最受男性喜爱的电影 Top10:")
215.     maleFilteredRatingsDataFrame.groupBy("MovieID").avg("Rating")
         .orderBy($"avg(Rating)".desc).show(10)
216.     println("纯粹使用 DataSet 实现所有电影中最受男性喜爱的电影 Top10:")
217.     maleFilteredRatingsDataSet.groupBy("MovieID").avg("Rating")
         .orderBy($"avg(Rating)".desc).show(10)
218.     /**
219.      * (789,5.0)
220.      * (855,5.0)
221.      * (32153,5.0)
222.      * (4763,5.0)
223.      * (26246,5.0)
224.      * (2332,5.0)
225.      * (503,5.0)
226.      * (4925,5.0)
227.      * (8767,5.0)
228.      * (44657,5.0)
229.      */
230.
231.     println("纯粹使用 DataFrame 实现所有电影中最受女性喜爱的电影 Top10:")
232.     femaleFilteredRatingsDataFrame.groupBy("MovieID").avg("Rating")
         .orderBy($"avg(Rating)".desc, $"MovieID".desc).show(10)
233.
234.     println("纯粹使用 DataSet 实现所有电影中最受女性喜爱的电影 Top10:")
235.     femaleFilteredRatingsDataSet.groupBy("MovieID").avg("Rating")
         .orderBy($"avg(Rating)".desc, $"MovieID".desc).show(10)
236.
237.     /**
```

```
238.        * 思考题：如果想让 RDD 和 DataFrame 计算的 TopN 的每次结果都一样，该如何
            * 保证？现在的情况是，例如计算 Top10，而其同样评分的不止 10 个，所以每次都会
            * 从中取出 10 个，这就导致大家的结果不一致，这个时候，我们可以使用一个新的
            * 列参与排序：
            *   如果是 RDD，该怎么做呢？这时就要进行二次排序，按照我们前面和大家讲解的
            *   二次排序的视频内容即可，如果是 DataFrame，该如何做呢？非常简单，我们
            *   只需要在 orderBy 函数中增加一个排序维度的字段即可
            */
239.      /**
240.        * 功能五：最受不同年龄段人员欢迎的电影 TopN
241.        * "users.dat": UserID::Gender::Age::OccupationID::Zip-code
242.        * 思路：首先还是计算 TopN，但是这里的关注点有两个：
243.        *   1. 不同年龄阶段如何界定，这个问题其实是业务的问题，当然，实现时可以使
            *   用 RDD 的 filter，例如 13 < age <18，这样做会导致运行时进行大量的计
            *   算，因为要进行扫描，所以会非常耗性能。所以，一般情况下，我们都是在原始
            *   数据中直接对要进行分组的年龄段提前进行 ETL，例如，进行 ETL 后产生以下
            *   数据：
244.        *     - Gender is denoted by a "M" for male and "F" for female
245.        *     - Age is chosen from the following ranges:
246.        *  1:  "Under 18"
247.        * 18:  "18-24"
248.        * 25:  "25-34"
249.        * 35:  "35-44"
250.        * 45:  "45-49"
251.        * 50:  "50-55"
252.        * 56:  "56+"
253.        *   2. 性能问题：
254.        *     第一点：实现时可以使用 RDD 的 filter，例如 13 < age <18，这样做会
            *     导致运行时进行大量的计算，因为要进行扫描，所以会非常耗性能，我们通过
            *     提前的 ETL 把计算发生在 Spark 业务逻辑运行以前，用空间换时间，当然，
            *     这些实现也可以使用 Hive，因为 Hive 语法支持非常强悍且内置了最多的函数；
255.        *     第二点：这里要使用 mapjoin，原因是 targetUsers 数据只有 UserID，
            *     数据量一般不会太多
256.        */
257.
258.     println("纯粹通过 DataFrame 的方式实现所有电影中 QQ 或者微信核心目标用户最
         喜爱电影 TopN 分析:")
259.
260.     ratingsDataFrame.join(usersDataFrame, "UserID").filter("Age =
         '18'").groupBy("MovieID")
261.       .count().orderBy($"count".desc).printSchema()
262.
263.
264.
265.     ratingsDataSet.join(usersDataSet, "UserID").filter("Age =
         '18'").groupBy("MovieID")
266.       .count().orderBy($"count".desc).printSchema()
267.
268.
269.     /**
270.       * Tips:
```

```
271.            *   1.orderBy 操作需要在 Join 之后进行
272.            */
273.          println("纯粹通过 DataFrame 的方式实现所有电影中 QQ 或者微信核心目标用户最
          喜爱电影 TopN 分析:")
274.          ratingsDataFrame.join(usersDataFrame, "UserID").filter("Age =
          '18'").groupBy("MovieID")
275.           .count().join(moviesDataFrame, "MovieID").select("Title", "count")
           .orderBy($"count".desc).show(10)
276.          println("纯粹通过 DataSet 的方式实现所有电影中 QQ 或者微信核心目标用户最喜
          爱电影 TopN 分析:")
277.          ratingsDataSet.join(usersDataSet, "UserID").filter("Age =
          '18'").groupBy("MovieID")
278.            .count().join(moviesDataSet, "MovieID").select("Title",
            "count").sort($"count".desc).show(10)
279.
280.
281.
282.
283.          /**
284.            * 淘宝核心目标用户最喜爱电影 TopN 分析
285.            * (Pulp Fiction (1994),959)
286.            * (Silence of the Lambs, The (1991),949)
287.            * (Forrest Gump (1994),935)
288.            * (Jurassic Park (1993),894)
289.            * (Shawshank Redemption, The (1994),859)
290.            */
291.
292.          println("纯粹通过 DataFrame 的方式实现所有电影中淘宝核心目标用户最喜爱电
          影 TopN 分析:")
293.          ratingsDataFrame.join(usersDataFrame, "UserID").filter("Age =
          '25'").groupBy("MovieID").
294.           count().join(moviesDataFrame, "MovieID").select("Title", "count")
           .orderBy($"count".desc).show(10)
295.
296.          println("纯粹通过 DataSet 的方式实现所有电影中淘宝核心目标用户最喜爱电影
          TopN 分析:")
297.          ratingsDataSet.join(usersDataSet, "UserID").filter("Age =
          '25'").groupBy("MovieID")
298.            .count().join(moviesDataSet, "MovieID").select("Title", "count")
            .sort($"count".desc).limit(10).show()
299.
300.
301.              while(true){}
              //和通过 Spark Shell 运行代码可以一直看到 Web 终端的原理一样,因为 Spark
              //Shell 内部有一个 LOOP 循环
302.
303.          sc.stop()
304.
305.
306.      }
307.    }
```

12.13　本 章 总 结

本章 Spark 商业案例之大数据电影点评系统应用案例从 Spark 的多种实现方式对电影点

评系统进行了统计分析，如纯粹通过 RDD 的方式实现、通过 DataFrame 和 RDD 相结合的方式实现、纯粹使用 DataFrame 方式计算、纯粹通过 DataSet 实现的方式，综合阐述了电影点评系统业务场景的实现。例如，统计某特定电影观看者中男性和女性不同年龄的人数、计算所有电影中平均得分最高（口碑最好）的电影 TopN、计算最流行电影（即所有电影中粉丝或者观看人数最多）的电影 TopN、实现所有电影中最受男性和女性喜爱的电影 TopN、实现所有电影中 QQ 或者微信核心目标用户最喜爱电影 TopN 分析、实现所有电影中淘宝核心目标用户最喜爱电影 TopN 分析、电影点评系统实现 Java 和 Scala 版本的二次排序系统等。读者深入掌握大数据电影点评系统应用案例全面综合的应用实现，就可以在生产环境类似系统中推广实现。

第 13 章　Spark 2.2 实战之 Dataset 开发实战企业人员管理系统应用案例

本章的企业人员管理系统应用案例将通过 Spark 2.2 进行 Dataset 开发实战。

Apache Spark 2.0.0 是 2.X 上的第一个版本。Spark 2.0.0 新版本的主要更新包括 API 可用性，SQL 2003 支持，性能改进，结构化流式处理，R 语言 UDF 自定义函数的支持、操作改进等。在 Apache Spark 2.0.0 版本中，300 多位源码贡献者修复了 2500 多个补丁。

Spark 2.0 较大的变化包括以下两个方面。

（1）统一了 DataFrame and Dataset：在 Scala 和 Java 中，DataFrame 和 Dataset 已经统一，DataFrame 只是 Dataset　Row 类型的别名。SparkSession 是统一的新入口点：用于替换 DataFrame 和 Dataset API 的旧 SQLContext 和 HiveContext，保留 SQLContext 和 HiveContext，以实现向后兼容。

（2）Spark 2.0 发布了 Structured Streaming 的实验版本。Structured Streaming 是基于 Spark SQL 和 Catalyst 优化器构建的高级流 API。结构化流允许用户使用与静态数据源相同的 DataFrame/Dataset API 对结构流数据进行编程，利用 Catalyst 优化器自动实现增量化查询计划。

13.1　企业人员管理系统应用案例业务需求分析

Spark 2.2 实战之 Dataset 开发实战企业人员管理系统应用案例的业务需求分析。

（1）企业人员管理系统应用案例中的企业人员信息包括姓名、年龄数据信息。在企业人员管理系统中，需对企业人员的年龄进行统计分析，如估算员工 10 年以后的工资数值等，涉及对员工的年龄进行计算。我们可以使用 map、flatMap、mapPartitions 来实现。

（2）在企业人员管理系统中，可通过 sort、join、joinWith 算子对人员信息中的年龄进行排序，将企业人员信息记录和企业人员评分数据进行关联，查询企业人员的年龄、姓名及评分数据信息。

（3）通过 randomSplit、sample、select 算子对人员信息中的人员进行随机采样分析；人员信息中如包括较多的属性信息，也可通过投影选择需要进行查询的属性进行展示。

（4）在企业人员管理系统中，对企业人员信息的姓名、年龄进行分组，然后使用 agg 中的 groupBy、agg、col 等内置函数进行计数、最大值、最小值、平均值等各种统计分析。

（5）通过 avg、sum、countDistinct 等一系列算子统计分析年龄总和、平均年龄、最大年龄、最小年龄、唯一年龄计数、平均年龄等人员信息数据。

13.2　企业人员管理系统应用案例数据建模

1. 企业人员管理系统应用案例数据的来源

企业人员管理系统应用案例中使用的是 Spark 官网提供的实例数据。从 Spark 官网（http://spark.apache.org/downloads.html）下载 Pre-built for Hadoop 2.6 预编译版本 spark-2.1.0-bin-hadoop2.6.tgz，将文件解压缩到本地文件目录。spark-2.1.0-bin-hadoop2.6\examples\src\main\resources 目录下包括 people.json 文件。

在工程 SparkApps 的 data 目录下新建 peoplemanagedata 目录，将上述解压缩生成的企业人员信息 people.json 文件保存到 data\peoplemanagedata 目录下。本节的重点是阐述 Spark 2.0 Dataset 在企业人员管理系统应用案例中的应用，企业人员分数评分信息 peopleScores.json 是人工创建的 json 文件，在 peopleScores.json 中模拟输入企业人员分数评分的数据。

2. 企业人员管理系统应用案例数据的格式说明

企业人员管理系统应用案例的数据包含两个数据文件：企业人员信息、企业人员分数评分信息。

首先查看企业人员信息的 JSON 格式描述信息。JSON 数据的书写格式是：名称/值对。名称/值对组合中的“名称”写在前面，“值对”写在后面，中间用冒号（：）隔开，企业人员信息文件 people.json 的格式描述如下：

```
1.  {" name ": name -Value, " age ": age -Value }
2.  姓名/值对、年龄/值对
3.  - name：用户姓名。
4.  - name-Value：用户姓名的值。
5.  - age：用户年龄。
6.  - age -Value：用户年龄的值。
```

从用户信息文件 people.json 中摘取部分记录如下：

```
1.  {"name":"Michael", "age":16}
2.  {"name":"Andy", "age":30}
3.  {"name":"Justin", "age":19}
4.  {"name":"Justin", "age":29}
5.  {"name":"Michael", "age":46}
```

接下来查看企业人员分数评分信息的 JSON 格式描述信息。JSON 数据的书写格式是：名称/值对。名称/值对组合中的“名称”写在前面，“值对”写在后面，中间用冒号（：）隔开。企业人员分数评分信息文件 peopleScores.json 的格式描述如下：

```
1.  {" n ": n -Value, " score ": score -Value}
2.  姓名/值对、分数评分/值对
3.  - n：用户姓名。
4.  - n-Value：用户姓名的值。
5.  - score：用户分数评分。
6.  - score -Value：用户分数评分的值。
```

从企业人员分数评分信息文件 peopleScores.json 中摘取部分记录如下：

```
1.  {"n":"Michael", "score":88}
2.  {"n":"Andy", "score":100}
3.  {"n":"Justin", "score":89}
```

13.3　通过 SparkSession 创建案例开发实战上下文环境

本节通过 SparkSession 创建企业人员管理系统应用案例开发实战上下文环境。

13.3.1　Spark 1.6.0 版本 SparkContext

在讲解 Spark 2.0.0 SparkSession 前，先回顾一下 Spark 之前版本 Spark 1.6.0 SparkContext 的使用。SparkContext 是 Spark 程序所有功能的唯一入口，采用 Scala、Java、Python、R 等都必须有一个 SparkContext。SparkContext 的核心作用：初始化 Spark 应用程序运行所需要的核心组件，包括 DAGScheduler、TaskScheduler、SchedulerBackend；同时还会负责 Spark 程序往 Master 注册程序等；SparkContext 是整个 Spark 应用程序中至关重要的一个对象。SparkContext 是迈入 Spark 的天堂之门！使用 val sc = new SparkContext(conf)时会新建一个 SparkContext 实例，在 SparkContext 构造器 class SparkContext(config: SparkConf){}类初始化的时候，SparkContext 除了类中的方法以外，其他所有语句都会执行！如加载的 import 包、定义的辅助构造器、定义的属性等。

Spark 1.6.0 中对数据的查询分成两个分支：hiveContext 和 sqlContext。两者之间的关系：hiveContext 继承自 sqlContext，除拥有 sqlContext 的特性外，还拥有自身的特性（支持 hive 的语法）。在 Spark Streaming 流式处理中，Spark 1.6.0 使用 StreamingContext 上下文。

（1）hiveContext 使用方式：在 Spark 1.6 版本中，HiveContext 的创建方式如下：

```
1.    val conf = new SparkConf() //创建 SparkConf 对象
2.    conf.setAppName("SparkSQLWindowsFuntionOps")
      //设置应用程序的名称，在程序运行的监控界面中可以看到名称
3.    conf.setMaster("local") //此时，程序在本地运行，不需要安装 Spark 集群
4.      val sc = new SparkContext(conf)
5.      val hiveContext =new HiveContext(sc)
6.       hiveContext.sql("use hive")
7.       hiveContext.sql("DROP TABLE IF EXISTS scores")
```

（2）sqlContext 使用方式：在 Spark 1.6 版本中，sqlContext 的创建方式如下：

```
1.      val conf = new SparkConf() //创建 SparkConf 对象
2.      conf.setAppName("SparkSQLInnerFunctions")
        //设置应用程序的名称，在程序运行的监控界面中可以看到名称
3.      conf.setMaster("local")
4.      val sc = new SparkContext(conf) //创建 SparkContext 对象,通过传入 SparkConf
                                     //实例来定制 Spark 运行的具体参数和配置信息
5.      val sqlContext = new SQLContext(sc)   //构建 SQL 上下文
6.   ......
7.   val userDataDF = sqlContext.createDataFrame(userDataRDDRow,structTypes)
8.   ......
```

（3）StreamingContext 使用方式：在 Spark 1.6 版本中，StreamingContext 的创建方式如下：

```
1.  val sparkConf = new SparkConf().setAppName("AdClickedStreamingStats")
2.      .setMaster("spark://192.168.189.1:7077").setJars(List(.......jar"))
3.     val ssc = new StreamingContext(sparkConf, Seconds(10))
4.     ssc.checkpoint("/usr/local/IMF_testdata/IMFcheckpoint114")
5. ......
```

13.3.2　Spark 2.0.0 版本 SparkSession

在 Spark 2.0 使用构造器模式（Builder）新建了一个新的入口 SparkSession，用于替换 DataFrame 和 Dataset API 的旧版本中的 SQLContext 和 HiveContext，但在 Spark 2.0 中保留了 SQLContext 和 HiveContext，实现向后兼容。Spark Session 整合了 SQLContext 和 HiveContext 的入口，内部创建的仍是 sparkContext。Spark Session 只是基于 sparkContext 基础之上的封装，Spark 程序所有功能的唯一入口仍然是 SparkContext。

SparkSession 使用方式：在 Spark 2.0 版本中，SparkSession 中的创建方式如下：

```
1.     val conf = new SparkConf().setMaster("local[3]").setAppName("HiSpark")
2.     val spark = SparkSession
3.         .builder()
4.         .config(conf)
5.         .config("spark.sql.warehouse.dir", "spark-warehouse")
6.         .getOrCreate()
```

SparkSession 使用构造器模式如下。

（1）SparkSession Builder 模式使用了链式调用，链式构建配置的键值对，配置的信息可读性更佳。

SparkSession.scala 的源码如下：

```
1. class Builder extends Logging {
2.
3.     private[this] val options = new scala.collection.mutable.HashMap
       [String, String]
4.     private[this] val extensions = new SparkSessionExtensions
5.     private[this] var userSuppliedContext: Option[SparkContext] = None
6.
7.     private[spark] def sparkContext(sparkContext: SparkContext): Builder
       = synchronized {
8.       userSuppliedContext = Option(sparkContext)
9.       this
10.    }
11.
12.    /**
13.     *为应用程序设置一个名称，应用名称将在 Spark Web UI 中显示。如果没有设置应
        *用程序名称，将使用随机生成的名称
14.     *
15.     * @since 2.0.0
16.     */
17.    def appName(name: String): Builder = config("spark.app.name", name)
18.
19.    /**
20.     * 设置配置选项，配置的选项集将自动应用到 sparkconf 和 sparksession 自身的配置中
21.     *
22      * @since 2.0.0
23.     */
```

```
24.
25.     def config(key: String, value: String): Builder = synchronized {
26.       options += key -> value
27.       this
28.     }
29.
30.     /**
31.      *设置配置选项，配置的选项集将自动应用到 sparkconf 和 sparksession 自身的配置中
32.      *
33.      * @since 2.0.0
34.      */
35.
36.     def config(key: String, value: Long): Builder = synchronized {
37.       options += key -> value.toString
38.       this
39.     }
40.
41.     /**
42.      *设置配置选项，配置的选项集将自动应用到 sparkconf 和 sparksession 自身的配置中
43.      *
44.      * @since 2.0.0
45.      */
46.     def config(key: String, value: Double): Builder = synchronized {
47.       options += key -> value.toString
48.       this
49.     }
50.
51.     /**
52.      *  设置配置选项，配置的选项集将自动应用到 sparkconf 和 sparksession 自身的配置中
53.      *
54.      * @since 2.0.0
55.      */
56.     def config(key: String, value: Boolean): Builder = synchronized {
57.       options += key -> value.toString
58.       this
59.     }
60.
61.     /**
62.      *根据给出的 sparkconf 设置列表选项的配置
63.      *
64.      * @since 2.0.0
65.      */
66.     def config(conf: SparkConf): Builder = synchronized {
67.       conf.getAll.foreach { case (k, v) => options += k -> v }
68.       this
69.     }
70.
71.     /**
72.      *设置 Spark master URL 链接，如 local 为本地运行，local[ 4 ]为在本地运行 4
         *个核，spark://master:7077 运行在 Spark standalone cluster 集群
73.      *
74.      * @since 2.0.0
75.      */
76.     def master(master: String): Builder = config("spark.master", master)
77.
78.     /**
79.      *  提供 Hive 的支持，包括 Hive 元数据 Hive metastore 的持久化连接，支持 Hive
         *  的用户自定义函数
80.      *
81.      * @since 2.0.0
82.      */
```

```
83.     def enableHiveSupport(): Builder = synchronized {
84.       if (hiveClassesArePresent) {
85.         config(CATALOG_IMPLEMENTATION.key, "hive")
86.       } else {
87.         throw new IllegalArgumentException(
88.           "Unable to instantiate SparkSession with Hive support because " +
89.             "Hive classes are not found.")
90.       }
91.     }
92.
93.     /**
94.
```

SparkSessionExtensions 是实验性质的，是[SparkSession]的注入点，这里我们对二进制兼容性和方法源兼容性的稳定不做任何保证。

SparkSessionExtensions 提供以下扩展点。

- Analyzer Rules：分析器规则。
- Check Analysis Rules：检查分析规则。
- Optimizer Rules：优化器规则。
- Planning Strategies：计划策略。
- Customized Parser：自定义解析器。
- (External) Catalog listeners：（外部）目录监听器。

SparkSessionExtensions 扩展可以在[SparkSession.Builder]中调用 withExtension，例如：

```
1.      SparkSession.builder()
2.        .master("...")
3.        .conf("...", true)
4.        .withExtensions { extensions =>
5.          extensions.injectResolutionRule { session =>
6.            ...
7.          }
8.          extensions.injectParser { (session, parser) =>
9.            ...
10.         }
11.       }
12.       .getOrCreate()
```

注意：没有注入的 builders 应该假设[SparkSession]是完全初始化的，不应接触会话的内部（如会话状态）。

（2）SparkSession Builder 中配置信息完成以后，返回的是 Builder 数据类型。在 Builder 中调用 getOrCreate 方法，如果已有 SparkSession，就返回已有的 SparkSession；如果没有 SparkSession，就新建立一个 sparksession，返回 SparkSession 数据类型的实例，在新建的 SparkSession 中创建 SparkContext。

Spark 2.2.1 版本的 SparkSession.scala 的 getOrCreate 方法的源码如下：

```
1.  /**获取已有的[SparkSession]，如果不存在，根据 Builder 的配置参数创建一个新的
    *SparkSession。该方法首先检查是否有一个有效的本地线程 SparkSession，如果已有
    *SparkSession，则返回 SparkSession，然后检查是否有有效的全局默认 SparkSession，
    *如果有，则返回 SparkSession。如果没有有效的全局默认 SparkSession，该方法创建一个
    *新的 SparkSession 和分配新创建的 SparkSession 为全局默认值。如果返回一个已有的
    *SparkSession，Builder 的配置参数将应用到已有的 SparkSession。@since 2.0.0
2.  */
```

```
3.  def getOrCreate(): SparkSession = synchronized {
4.        //从当前线程的活动会话中获取会话
5.        var session = activeThreadSession.get()
6.        if ((session ne null) && !session.sparkContext.isStopped) {
7.          options.foreach { case (k, v) => session.sessionState.conf
          .setConfString(k, v) }
8.          if (options.nonEmpty) {
9.            logWarning("Using an existing SparkSession; some configuration
            may not take effect.")
10.         }
11.         return session
12.       }
13.
14.       //全局同步变量，只设置一个默认会话
15.       SparkSession.synchronized {
16.         //如果本地现场没有会话，则从全局会话中获取会话
17.         session = defaultSession.get()
18.         if ((session ne null) && !session.sparkContext.isStopped) {
19.           options.foreach { case (k, v) => session.sessionState.conf
            .setConfString(k, v) }
20.           if (options.nonEmpty) {
21.             logWarning("Using an existing SparkSession; some configuration
              may not take effect.")
22.           }
23.           return session
24.         }
25.
26.         //如果没有全局默认会话，就创建一个新的会话
27.         val sparkContext = userSuppliedContext.getOrElse {
28.           //如果没有应用名，则设置一个应用名
29.           val randomAppName = java.util.UUID.randomUUID().toString
30.           val sparkConf = new SparkConf()
31.           options.foreach { case (k, v) => sparkConf.set(k, v) }
32.           if (!sparkConf.contains("spark.app.name")) {
33.             sparkConf.setAppName(randomAppName)
34.           }
35.           val sc = SparkContext.getOrCreate(sparkConf)
36.           //已有的 SparkContext，更新 SparkConf 配置参数
37.           options.foreach { case (k, v) => sc.conf.set(k, v) }
38.           if (!sc.conf.contains("spark.app.name")) {
39.             sc.conf.setAppName(randomAppName)
40.           }
41.           sc
42.         }
43.
44.         //如果用户定义了配置程序类，则初始化扩展
45.         val extensionConfOption = sparkContext.conf.get(StaticSQLConf
          .SPARK_SESSION_EXTENSIONS)
46.         if (extensionConfOption.isDefined) {
47.           val extensionConfClassName = extensionConfOption.get
48.           try {
49.             val extensionConfClass = Utils.classForName(extensionConfClassName)
50.             val extensionConf = extensionConfClass.newInstance()
51.               .asInstanceOf[SparkSessionExtensions => Unit]
52.             extensionConf(extensions)
53.           } catch {
54.             //如果找不到类或类的类型不正确，忽略该错误
55.             case e @ (_: ClassCastException |
56.                       _: ClassNotFoundException |
```

```
57.                 _: NoClassDefFoundError) =>
58.             logWarning(s"Cannot use $extensionConfClassName to configure
                session extensions.", e)
59.           }
60.         }
61.
62.         session = new SparkSession(sparkContext, None, None, extensions)
63.         options.foreach { case (k, v) => session.initialSessionOptions.put(k, v) }
64.         defaultSession.set(session)
65.
66.         //向 singleton 注册一个成功实例化的上下文，在类的末尾定义，以便只有在实例的
            //构造中没有异常时才更新 singleton
67.         sparkContext.addSparkListener(new SparkListener {
68.           override def onApplicationEnd(applicationEnd:
              SparkListenerApplicationEnd): Unit = {
69.             defaultSession.set(null)
70.             sqlListener.set(null)
71.           }
72.         })
73.       }
74.
75.       return session
76.     }
```

Spark 2.4.3 版本的 SparkSession.scala 的源码与 Spark 2.2.1 版本相比具有如下特点。

- 上段代码中第 3 行之后新增加 assertOnDriver()代码。
- 上段代码中第 29 行代码删掉。
- 上段代码中第 33 行调整应用的名称为 java.util.UUID.randomUUID().toString。
- 上段代码中第 35~42 行替换为 SparkContext.getOrCreate(sparkConf)。
- 上段代码中第 64 行调整为 setDefaultSession、setActiveSession 方法。

```
1.   ......
2.  assertOnDriver()
3.  ......
4.    sparkConf.setAppName(java.util.UUID.randomUUID().toString)
5.  .....
6.  SparkContext.getOrCreate(sparkConf)
7.  ......
8.          setDefaultSession(session)
9.          setActiveSession(session)
```

（3）在 Spark 2.0 版本中，SparkSession 的统一入口使用 hiveContext，调用 enableHive Support 方法提供对 Hive 的支持。使用 Hive 时，必须实例化 SparkSession 提供对 Hive 的支持，包括 Hive 元数据、Hive metastore 的元数据的持久连接、支持自定义函数功能和 Hive 序列化，代码如下：

```
1.       val conf = sys.env.get("SPARK_AUDIT_MASTER") match {
2.        case Some(master) => new SparkConf().setAppName("Simple Sql App")
          .setMaster(master)
3.        case None => new SparkConf().setAppName("Simple Sql App")
4.      }
5.      val sc = new SparkContext(conf)
6.      val sparkSession = SparkSession.builder
7.        .enableHiveSupport()
8.        .getOrCreate()
9.
```

下面看一下在 SparkSession 中执行 Spark 的 SQL 查询的示例。在 Spark 中执行 spark.sql() 语句返回的结果类型是 DataFrame，对 SQL 中的语句进行解析时，Spark SQL 解析器默认使用 spark.sql.dialect 方言来解析 SQL 语句。

执行 Spark 的 SQL 查询示例的具体实现方法如下。

（1）设置 spark.sql.warehouse.dir。作为 Windows 用户，由于不同的文件前缀跨操作系统，为避免潜在错误前缀的问题，当启动 sparksession 时，Spark 目前的解决方法是指定 spark.sql.warehouse.dir。这里我们设置 spark.sql.warehouse.dir 为本地文件目录的绝对路径。

（2）使用 spark.sql("CREATE TABLE IF NOT EXISTS src(key INT, value STRING)")SQL 语句创建一个表 src，表中包括 key 及 value 两个字段。

（3）使用 spark.sql("LOAD DATA LOCAL INPATH 'data/peoplemanagedata/kv1.txt' INTO TABLE src")加载外部相对目录中 kv1.txt 的文本文件中的数据。v1.txt 文本文件的数据内容（Key Value）键值对如下：

```
1.  238 val_238
2.  86 val_86
3.  311 val_311
4.  27 val_27
5.  165 val_165
6.  ......
```

（4）使用 spark.sql("SELECT * FROM src").show 查询出表 src 的内容进行展示。

在 SparkSession 中执行 Spark SQL 语句的代码如下：

```
1.  package com.dt.spark.cores
2.  import org.apache.spark.SparkConf
3.  import org.apache.spark.sql.{Row, SparkSession}
4.  object HiSpark {
5.    case class Person(name: String, age: Long)
6.    def main(args: Array[String]) {
7.      val conf = new SparkConf().setMaster("local[3]").setAppName("HiSpark")
8.      val spark = SparkSession
9.        .builder()
10.       .config(conf)
11.       .config("spark.sql.warehouse.dir",G:\\IMFBigDataSpark2017\\SparkApps
          \\spark-warehouse")
12.       .enableHiveSupport()
13.       .getOrCreate()
14.     spark.sql("CREATE TABLE IF NOT EXISTS src(key INT, value STRING)")
15.     spark.sql("LOAD DATA LOCAL INPATH 'data/peoplemanagedata/kv1.txt'
        INTO TABLE src")
16.     spark.sql("SELECT * FROM src").show()
17.     spark.sql("SELECT COUNT(*) FROM src").show()
18.   }
19. }
```

在 IDEA 中运行代码，在 SparkSession 中执行 Spark SQL 语句，输出结果如下：

```
1.  Using Spark's default log4j profile: org/apache/spark/log4j-defaults.
    properties
2.  17/03/13 20:50:46 INFO SparkContext: Running Spark version 2.1.0
3.  ......
4.  17/03/13 20:51:04 INFO SparkSqlParser: Parsing command: SELECT COUNT(*)
    FROM src
```

```
5.  +---+-------+
6.  |key| value|
7.  +---+-------+
8.  |238|val_238|
9.  | 86| val_86|
10. |311|val_311|
11. | 27| val_27|
12. |165|val_165|
13. |409|val_409|
14. |255|val_255|
15. |278|val_278|
16. | 98| val_98|
17. |484|val_484|
18. |265|val_265|
19. |193|val_193|
20. |401|val_401|
21. |150|val_150|
22. |273|val_273|
23. |224|val_224|
24. |369|val_369|
25. | 66| val_66|
26. |128|val_128|
27. |213|val_213|
28. +---+-------+
29. only showing top 20 rows
30. .......
31. 7/03/13 20:51:06 INFO CodeGenerator: Code generated in 15.186225 ms
32. +--------+
33. |count(1)|
34. +--------+
35. |    2500|
36. +--------+
```

13.3.3　DataFrame、DataSet 剖析与实战

Spark 2.0 基于 DataSet 实现开发，Spark 2.0 DataSet 背后会被 Tungsten 优化，而这里面会采用 Whole-Stage Code Generation 的技术，所以出错时定位错误和调优非常困难。例如，for 循环翻译成了自己的方式，出错的话，错误信息定位就非常困难，生产环境面临错误和调优，搞不定了还是要用 RDD，因此，RDD 是万能的，基于 RDD 的 Spark CORE 是王道。

Spark 2.0 DataSet 的类型如下。

- SQL 是无类型的：例如，写的 SQL 语法，数据类型是否对，列是否不存在，这些在编写 SQL 语句的时候判定不出来。只有运行时才能发现是什么问题。
- DataFrame 是弱类型的：相当于 DataSet[Row]，DataFrame 其实就是一张表，如在表中声明一些列，但实际运行中其中某些列不存在值，是弱耦合的。
- DataSet 是强类型的：必须严格声明，而且类型要匹配，在编译时期就决定了数据类型是否准确。

从 Spark 2.2 开始，Spark 不再支持 Python；Spark 2.2 DataSet 大部分使用 Scala 语言开发；Java 主要是 DataFrame 开发。因此，从 Spark 2.2 开始，编程语言逐渐转向 Scala 语言开发。这里，Spark 2.0　DataSet 开发实战中，我们使用 Scala 开发。

接下来，我们看一个 DataFrame、DataSet 使用的代码示例，具体实现如下。

（1）定义 Person 的 case class 类，包括姓名、年龄。

（2）通过 spark.read.json 读入 people.json 格式文件，生成 persons 的 DataFrame。

（3）在 DataFrame 中通过 as 方法转换为 Dataset[Person]。

（4）打印展示 Dataset 的结果和结构。

```
1.  package com.dt.spark.cores
2.  import org.apache.spark.SparkConf
3.  import org.apache.spark.sql.{DataFrame, Row, SparkSession}
4.  object HiSpark {
5.    case class Person(name: String, age: Long)
6.    def main(args: Array[String]) {
7.      val conf = new SparkConf().setMaster("local[3]").setAppName("HiSpark")
8.      val spark = SparkSession
9.        .builder()
10.       .config(conf)
11.       .config("spark.sql.warehouse.dir",  "G:\\IMFBigDataSpark2017\\SparkApps
        \\spark-warehouse")
12.       .enableHiveSupport()
13.       .getOrCreate()
14.     import spark.implicits._
15.     import org.apache.spark.sql.functions._
16.     val persons = spark.read.json("data/peoplemanagedata/people.json")
17.     val personsDS = persons.as[Person]
18.     personsDS.show()
19.     personsDS.printSchema()
20.     val personsDF = personsDS.toDF()
21.   }
22. }
```

在 IDEA 中运行代码，DataFrame、DataSet 代码示例的输出结果如下：

```
1.  Using Spark's default log4j profile: org/apache/spark/log4j-defaults
    .properties
2.  17/03/13 20:58:20 INFO SparkContext: Running Spark version 2.1.0
3.  ......
4.  17/03/13 20:58:38 INFO CodeGenerator: Code generated in 30.13774 ms
5.  +---+-------+
6.  |age|   name|
7.  +---+-------+
8.  | 16|Michael|
9.  | 30|   Andy|
10. | 19| Justin|
11. | 29| Justin|
12. | 46|Michael|
13. +---+-------+
14.
15. root
16.  |-- age: long (nullable = true)
17.  |-- name: string (nullable = true)
```

在上述 DataFrame、DataSet 示例代码中，通过 as 方法将 DataFrame 转换为 DataSet 结构。下面看一下 as 方法的源码。在 Spark 2.2 中，DataFrame 类型是 Dataset[Row]的别名。

```
1.      type DataFrame = Dataset[Row]
```

查看 Dataset.scala 框架源码 as 方法，将 DataFrame 转换为 DataSet 数据结构。

```
1.  :: 实验性的 ::
```

```
2.        当每个记录映射到标识的类型的时候，as 方法将依赖范型 U 的类型来转换列。
          - 当 U 是一个类，类的字段将被映射到同名的列。大小写敏感度由 spark.sql
          .caseSensitive 配置决定。
3.        - 当 U 是一个元组，列将按序号映射（例如，第一列将被分配到_1）。
4.        - 当 U 是一个原始类型（如 String 字符串、Int 整型等），那么第一列将被 DataFrame
          使用。
5.      如果数据集的结构与所需的 U 类型不匹配，可以使用 select 与 alias 或 as 重新排列或重
        命名
6.       @group basic
7.       @since 1.6.0
8.      /
9.    @Experimental
10.   @InterfaceStability.Evolving
11.   def as[U : Encoder]: Dataset[U] = Dataset[U](sparkSession, logicalPlan)
```

其中，U : Encoder 的源码解析如下：

```
1.  /**
2.   * :: 实验性的::
3.   * 用于将一个 JVM 对象类型 T 使用 Spark SQL 表示
4.   * == Scala ==
5.   * 通过 sparksession 的隐式转换，编码器可以自动创建，或者也可以调用静态方法
   * [Encoders]显式创建
6.   * {{{
7.   *   import spark.implicits._
8.   *   val ds = Seq(1, 2, 3).toDS()
   //隐式转换为 Dataset[Int]类型 (spark.implicits.newIntEncoder)
9.   * }}}
10.  *
11.  * == Java ==
12.  * 编码器调用静态方法[Encoders]
13.  * {{{
14.  *   List<String> data = Arrays.asList("abc", "abc", "xyz");
15.  *   Dataset<String> ds = context.createDataset(data, Encoders.STRING());
16.  * }}}
17.  *
18.  * 编码器可以使用元组
19.  *
20.  * {{{
21.  *   Encoder<Tuple2<Integer, String>> encoder2 = Encoders.tuple(Encoders
   *   .INT(), Encoders.STRING());
22.  *   List<Tuple2<Integer, String>> data2 = Arrays.asList(new scala.Tuple2(1,
   *"a");
23.  *   Dataset<Tuple2<Integer, String>> ds2 = context.createDataset(data2,
   *encoder2);
24.  * }}}
25.  *
26.  * 或者采用 JAVA Bean 来构建编码格式
27.  *
28.  * {{{
29.  *   Encoders.bean(MyClass.class);
30.  * }}}
31.  *
32.  * == 实现 ==
33.  * 编码器不需要是线程安全的，因此不需要使用锁来保护。针对并发访问，编码器重用内部缓
```

```
    *冲区来提高性能
34.  *
35.  * @since 1.6.0
36.  */
37. @Experimental 实验性的
38. @InterfaceStability.Evolving 演进
39. @implicitNotFound("Unable to find encoder for type stored in a Dataset
    .Primitive types " +
40.   "(Int, String, etc) and Product types (case classes) are supported by
    importing " +
41.   "spark.implicits._  Support for serializing other types will be added
    in future " +
42.   "releases.")
43. trait Encoder[T] extends Serializable {
44.
45.   /**返回行的编码格式*/
46.   def schema: StructType
47.
48.   /**
     *  ClassTag 构建包含 T 类型的数组集合
49.    */
50.   def clsTag: ClassTag[T]
51. }
```

13.4　通过 map、flatMap、mapPartitions 等分析企业人员管理系统

本节通过 map、flatMap、mapPartitions 等分析企业人员管理系统。企业人员管理系统应用案例中的企业人员信息包括姓名、年龄数据信息。在企业人员管理系统中，需对企业人员的年龄进行统计分析，如估算员工 10 年以后的工资数值等，涉及对员工的年龄进行计算。我们可以使用 map、flatMap、mapPartitions 来实现。

【例 13-1】 map 算子解析。

使用 map 算子将 personDS DataSet 转换为 Dataset[(String, Long)]类型，其中年龄值加 100，生成姓名、年龄元组类型的 DataSet。

```
1.     personsDS.map { person =>
2.       (person.name, person.age + 100L)
3.     }.show()
```

在 IDEA 中运行代码，map 算子示例的输出结果如下：

```
1.   Using Spark's default log4j profile: org/apache/spark/log4j-defaults
     .properties
2.   17/03/14 12:32:41 INFO SparkContext: Running Spark version 2.1.0
3.   ......
4.   17/03/14 12:33:03 INFO CodeGenerator: Code generated in 60.718305 ms
5.   +-------+---+
6.   |     _1 | _2|
7.   +-------+---+
8.   |Michael|116|
9.   |   Andy|130|
```

```
10. | Justin|119|
11. | Justin|129|
12. |Michael|146|
13. +-------+---+
14. ......
```

【例 13-2】 flatMap 算子解析。

使用 flatMap 算子对 personDS 进行转换，模式匹配如果姓名为 Andy，则年龄加上 70，其他员工年龄加上 30。

```
1.     personsDS.flatMap(persons => persons match {
2.       case Person(name, age) if (name == "Andy") => List((name, age + 70))
3.       case Person(name, age) => List((name, age+30))
4.     }).show()
```

在 IDEA 中运行代码，flatMap 算子示例的输出结果如下：

```
1.   Using Spark's default log4j profile: org/apache/spark/log4j-defaults.
     properties
2.  17/03/14 21:17:57 INFO SparkContext: Running Spark version 2.1.0
3.  ......
4.  17/03/14 21:18:07 INFO DAGScheduler: Job 2 finished: show at DatasetOps
    .scala:77, took 0.345143 s
5.  +-------+---+
6.  |     _1 | _2|
7.  +-------+---+
8.  |Michael| 46|
9.  |   Andy |100|
10. | Justin| 49|
11. | Justin| 59|
12. |Michael| 76|
13. +-------+---+
```

【例 13-3】 mapPartitions 算子解析。

使用 mapPartitions 算子对 personDS 执行 mapPartitions 算子，对于每个分区的记录集合，循环遍历 persons 记录集，如 persons 有元素，则将姓名、年龄值加上 1000 组成的元组值加入到 ArrayBuffer 列表，最终返回 result.iterator 的值。

```
1.     personsDS.mapPartitions { persons =>
2.       val result = ArrayBuffer[(String, Long)]()
3.       while (persons.hasNext) {
4.         val person = persons.next()
5.         result += ((person.name, person.age + 1000))
6.       }
7.       result.iterator
8.
9.     }.show
```

在 IDEA 中运行代码，mapPartitions 算子示例的输出结果如下：

```
1.   Using Spark's default log4j profile: org/apache/spark/log4j-defaults
     .properties
2.  17/03/14 14:46:04 INFO SparkContext: Running Spark version 2.1.0
3.  ......
4.  17/03/14 14:46:15 INFO DAGScheduler: Job 3 finished: show at DatasetOps.
    scala:67, took 0.109263 s
5.  +--------+----+
6.  |      _1 |  _2|
```

```
7.  +-------+----+
8.  |Michael|1016|
9.  |   Andy|1030|
10. | Justin|1019|
11. | Justin|1029|
12. |Michael|1046|
13. +-------+----+
```

13.5　通过 dropDuplicate、coalesce、repartition 等分析企业人员管理系统

本节通过 dropDuplicate、coalesce、repartition 等分析企业人员管理系统。企业人员管理系统应用案例中的企业人员信息包括姓名、年龄数据信息。在企业人员管理系统中，可能需对人员信息中重名的记录进行清除。Spark 2.0 计算的时候，我们也可以使用 coalesce、repartition 重新对人员信息数据记录进行分区。

【例 13-4】 dropDuplicate 算子解析。

使用 dropDuplicate 算子删除重复元素，例如，使用 personDS.dropDuplicates("name").show 删除姓名中重复员工的记录，之前 people.json 文本文件中包含重复的两条姓名为 Justin 的记录，去重以后结果只保留一条记录。

```
1.        println("使用 dropDuplicate 算子统计企业人员管理系统姓名无重复员工的记录：")
2.         personsDS.dropDuplicates("name").show()
```

在 IDEA 中运行代码，dropDuplicate 算子示例的输出结果如下：

```
1.     使用 dropDuplicate 算子统计企业人员管理系统姓名无重复员工的记录：
2.  +---+-------+
3.  |age|   name|
4.  +---+-------+
5.  | 16|Michael|
6.  | 30|   Andy|
7.  | 19| Justin|
8.  +---+-------+
```

dropDuplicate 算子与 distinct 算子比较：distinct 算子是从 Dataset 中返回一个新的数据集，新的数据集中包含唯一的记录行。distinct 是 dropDuplicates 的别名。PersonsDS 中包含两条同名员工 Justin 的记录，其年龄分别为 19 岁、29 岁，因为姓名、年龄组合的数据不重复，所示两条记录都打印输出。

```
1.       personsDS.distinct().show()
```

在 IDEA 中运行代码，distinct 算子示例的输出结果如下：

```
1.  +---+-------+
2.  |age|   name|
3.  +---+-------+
4.  | 46|Michael|
5.  | 30|   Andy|
6.  | 19| Justin|
7.  | 16|Michael|
```

```
8.  | 29| Justin|
9.  +---+-------+
```

【例 13-5】 repartition 算子解析。

使用 repartition 算子重新设置 personsDS 的分区数，将之前的 1 个分区调整为 4 个分区。

```
1.  println("使用 repartition 算子设置分区：")
2.      println("原分区数："+personsDS.rdd.partitions.size)
3.      val repartitionDs = personsDS.repartition(4)
4.      println("repartition 设置分区数："+repartitionDs.rdd.partitions.size)
```

在 IDEA 中运行代码，repartition 算子示例的输出结果如下：

```
1.  使用 repartition 算子设置分区：
2.  原分区数：1
3.  repartition 设置分区数：4
```

【例 13-6】 coalesce 算子解析。

把很大的分区变成很小的分区使用 coalesce 算子。例如，把 10 000 个分区变成 100 个分区。其他情况如把很小的分区变成很大的分区使用 repartition。不管是把分区数变多，还是变少，repartition 都会产生 shuffle。coalesce 算子在 Spark 集群中是符合它的分区数的，但在本地有一点问题，这里的分区数仍为 1。

```
1.      println("使用 coalesce 算子设置分区：")
2.      val coalesced = repartitionDs.coalesce(2)
3.      println("coalesce 设置分区数："+ coalesced.rdd.partitions.size)
4.      coalesced.show
```

在 IDEA 中运行代码，coalesce 算子本地模式运行的输出结果如下：

```
1.  使用 coalesce 算子设置分区：
2.  coalesce 设置分区数：1
3.  +---+-------+
4.  |age|   name |
5.  +---+-------+
6.  | 16|Michael|
7.  | 30|   Andy |
8.  | 19| Justin|
9.  | 29| Justin|
10. | 46|Michael|
11. +---+-------+
```

coalesce 算子源码解析如下：

```
1.    /**
2.      * 返回一个新的数据集，这个新的数据集有 numpartitions 个分区。在 RDD 中也有相似的
        * coalesce 定义。coalesce 算子操作的结果应用于窄依赖中，将 1000 个分区转换成 100 个
        * 分区，这个过程不会发生 shuffle，相反，如果 10 个分区转换成 100 个分区，将会发生
        * shuffle
3.      *
4.      * @group typedrel
5.      * @since 1.6.0
6.      */
7.    def coalesce(numPartitions: Int): Dataset[T] = withTypedPlan {
```

```
8.      Repartition(numPartitions, shuffle = false, logicalPlan)
9.    }
```

13.6　通过 sort、join、joinWith 等分析企业人员管理系统

本节通过 sort、join、joinWith 算子分析企业人员管理系统。企业人员管理系统应用案例中的企业人员信息包括姓名、年龄数据信息。在企业人员管理系统中，需对人员信息中的年龄进行排序，将企业人员信息记录和企业人员评分数据进行关联，查询企业人员的年龄、姓名及评分数据信息。

【例 13-7】 sort 算子解析。

使用 sort 算子对 personDS 按年龄进行降序排列。

```
1.      println("使用 sort 算子对年龄进行降序排列：")
2.        personsDS.sort($"age".desc).show
```

在 IDEA 中运行代码，sort 算子示例的输出结果如下：

```
1.   使用 sort 算子对年龄进行降序排列:
2.   +---+-------+
3.   |age|   name |
4.   +---+-------+
5.   | 46|Michael|
6.   | 30|   Andy |
7.   | 29| Justin|
8.   | 19| Justin|
9.   | 16|Michael|
10. +---+-------+
```

【例 13-8】 join 算子解析。

使用 join 算子将数据集根据提供的表达式进行关联。

```
1.     println("使用 join 算子关联企业人员信息、企业人员分数评分信息：")
2.       personsDS.join(personScoresDS, $"name" === $"n").show
```

在 IDEA 中运行代码，join 算子示例的输出结果如下：

```
1.    使用 join 算子关联企业人员信息、企业人员分数评分信息:
2.   +---+-------+-------+-----+
3.   |age|   name|      n |score|
4.   +---+-------+-------+-----+
5.   | 16|Michael|Michael|   88|
6.   | 30|   Andy|   Andy |  100|
7.   | 19| Justin| Justin|   89|
8.   | 29| Justin| Justin|   89|
9.   | 46|Michael|Michael|   88|
10. +---+-------+-------+-----+
```

【例 13-9】 joinWith 算子解析。

使用 joinWith 算子对数据集进行内关联，当关联的姓名相等时，返回一个 tuple2 键值对，格式分别为年龄，姓名及姓名，评分。

```
1.      println("使用 joinWith 算子关联企业人员信息、企业人员分数评分信息：")
2.         personsDS.joinWith(personScoresDS, $"name" === $"n").show
```

在 IDEA 中运行代码，joinWith 算子示例的输出结果如下：

```
1.   使用 joinWith 算子关联企业人员信息、企业人员分数评分信息：
2.  +------------+------------+
3.  |          _1 |          _2 |
4.  +------------+------------+
5.  |[16,Michael]|[Michael,88]|
6.  |   [30,Andy]|  [Andy,100]|
7.  | [19,Justin]| [Justin,89]|
8.  | [29,Justin]| [Justin,89]|
9.  |[46,Michael]|[Michael,88]|
10. +------------+------------+
```

13.7　通过 randomSplit、sample、select 等分析企业人员管理系统

本节通过 randomSplit、sample、select 算子分析企业人员管理系统。企业人员管理系统应用案例中的企业人员信息包括姓名、年龄数据信息。在企业人员管理系统中，可能需对人员信息中的人员进行随机采样分析。人员信息中如包括较多的属性信息，也可通过投影选择需要进行查询的属性进行展示。

【例 13-10】 randomSplit 算子解析。

使用 randomSplit 算子对 personDS 进行随机切分。randomSplit 的参数 weights 表示权重，传入的两个值将会切分成两个 Dataset[Person]，把原来的 Dataset[Person]按照权重 10，20 随机划分到两个 Dataset[Person]中，权重高的 Dataset[Person]，划分的概率就大。

```
1.   println("使用 randomSplit 算子进行随机切分：")
2.     personsDS.randomSplit(Array(10, 20)).foreach(dataset => dataset.show())
```

在 IDEA 中运行代码，randomSplit 算子示例的输出结果如下：

```
1.   使用 randomSplit 算子进行随机切分：
2.  +---+-------+
3.  |age|   name|
4.  +---+-------+
5.  | 16|Michael|
6.  | 29| Justin|
7.  +---+-------+
8.
9.  +---+-------+
10. |age|   name|
11. +---+-------+
12. | 19| Justin|
13. | 30|   Andy|
14. | 46|Michael|
15. +---+-------+
```

为体现随机性，再在 IDEA 中运行一次代码，随机切分结果和上次随机切分的结果不一样。randomSplit 算子示例的输出结果如下：

```
1.   使用 randomSplit 算子进行随机切分：
```

```
2.  +---+-------+
3.  |age|   name |
4.  +---+-------+
5.  | 16|Michael|
6.  | 19| Justin|
7.  | 29| Justin|
8.  | 30|   Andy |
9.  | 46|Michael|
10. +---+-------+
11.
12. +---+----+
13. |age|name|
14. +---+----+
15. +---+----+
```

【例 13-11】 sample 算子解析。

使用 sample 算子对 personDS 进行随机采样，第一个参数 true 表示有放回去的抽样，false 表示没有放回去的抽样； 第二个参数为采样率在 0～1 之间。例如，personDS.sample(false, 0.5).show()。

```
1.    println("使用 sample 算子进行随机采样：")
2.      personsDS.sample(false, 0.5).show()
```

在 IDEA 中运行代码，sample 算子示例的输出结果如下：

```
1.  使用 sample 算子进行随机采样：
2.  +---+------+
3.  |age|  name |
4.  +---+------+
5.  | 19|Justin|
6.  | 29|Justin|
7.  +---+------+
```

为体现随机性，再在 IDEA 中运行一次代码，结果和上次随机取样的结果不一样。sample 算子示例的输出结果如下：

```
1.  使用 sample 算子进行随机采样：
2.  +---+-------+
3.  |age|   name |
4.  +---+-------+
5.  | 16|Michael|
6.  | 30|   Andy |
7.  | 29| Justin|
8.  +---+-------+
```

【例 13-12】 select 算子解析。

使用 select 算子选择年龄进行显示。

```
1.     println("使用 select 算子选择列：")
2.      personsDS.select("name").show()
```

在 IDEA 中运行代码，select 算子示例的输出结果如下：

```
1.     使用 select 算子选择列：
2.  +-------+
3.  |   name |
4.  +-------+
```

```
5.  |Michael|
6.  |   Andy|
7.  | Justin|
8.  | Justin|
9.  |Michael|
10. +-------+
```

13.8　通过 groupBy、agg、col 等分析企业人员管理系统

本节通过 groupBy、agg、col 算子分析企业人员管理系统。企业人员管理系统应用案例中的企业人员信息包括姓名、年龄数据信息。在企业人员管理系统中，需要对企业人员信息的姓名、年龄进行分组，然后使用 agg 中的内置函数进行计数、最大值、最小值、平均值等各种统计分析。

【例 13-13】 groupBy 算子解析。

使用 groupBy 算子对姓名、年龄进行分组，统计分组后的计数。

```
1.      println("使用 groupBy 算子进行分组：")
2.      val personsDSGrouped = personsDS.groupBy($"name", $"age").count()
3.      personsDSGrouped.show()
```

在 IDEA 中运行代码，groupBy 算子示例的输出结果如下：

```
1.  使用 groupBy 算子进行分组：
2.  +-------+---+-----+
3.  |   name|age|count|
4.  +-------+---+-----+
5.  | Justin| 29|    1|
6.  |   Andy| 30|    1|
7.  |Michael| 16|    1|
8.  |Michael| 46|    1|
9.  | Justin| 19|    1|
10. +-------+---+-----+
```

【例 13-14】 agg 算子解析。

使用 agg 算子 concat 内置函数，将姓名、年龄连接在一起，成为单个字符串列。

```
1.     println("使用 agg 算子 concat 内置函数，将姓名、年龄连接在一起，成为单个字符串列：")
2.      personsDS.groupBy($"name", $"age").agg(concat($"name", $"age")).show
```

在 IDEA 中运行代码，agg 算子示例的输出结果如下：

```
1.  使用 agg 算子 concat 内置函数，将姓名、年龄连接在一起，成为单个字符串列：
2.  +-------+---+-----------------+
3.  |   name|age|concat(name, age)|
4.  +-------+---+-----------------+
5.  | Justin| 29|        Justin29 |
6.  |   Andy| 30|          Andy30 |
7.  |Michael| 16|       Michael16 |
8.  |Michael| 46|       Michael46 |
9.  | Justin| 19|        Justin19 |
10. +-------+---+-----------------+
```

【例 13-15】 col 算子解析。

使用 col 算子选择 personsDS 的姓名列以及 personScoresDS 的姓名列，如相等，则进行 joinWith 关联。

```
1.  println("使用 col 算子选择列 : ")
2.      personsDS.joinWith(personScoresDS,personsDS.col("name")===   person
        ScoresDS.col("n")).show
3.
```

在 IDEA 中运行代码，col 算子示例的输出结果如下：

```
1.  使用 col 算子选择列 :
2.  +------------+------------+
3.  |          _1|          _2|
4.  +------------+------------+
5.  |[16,Michael]|[Michael,88]|
6.  |   [30,Andy]|  [Andy,100]|
7.  | [19,Justin]| [Justin,89]|
8.  | [29,Justin]| [Justin,89]|
9.  |[46,Michael]|[Michael,88]|
10. +------------+------------+
```

13.9　通过 collect_list、collect_set 等分析企业人员管理系统

本节通过 collect_list、collect_set 算子分析企业人员管理系统。企业人员管理系统应用案例中的企业人员信息包括姓名、年龄数据信息。在企业人员管理系统中，对人员信息中的人员进行分组，根据人员信息管理的要求决定人员集合中重复的元素的去留。

【例 13-16】 collect_list、collect_set 函数解析。

将 personDS 按姓名分组，然后调用 agg 方法，collect_list 是分组以后的姓名集合，collect_set 是去重以后的姓名集合。collect_list 函数结果中包含重复元素；collect_set 函数结果中无重复元素。

```
1.    println("函数 collect_list、collect_set 比较，collect_list 函数结果中包含重
      复元素；collect_set 函数结果中无重复元素：")
2.      personsDS.groupBy($"name")
3.        .agg(collect_list($"name"), collect_set($"name"))
4.        .show()
```

在 IDEA 中运行代码，collect_list、collect_set 函数示例的输出结果如下：

```
1.   函数 collect_list、collect_set 比较，collect_list 函数结果中包含重复元素；
     collect_set 函数结果中无重复元素：
2.   +-------+------------------+----------------+
3.   |   name|collect_list(name)|collect_set(name)|
4.   +-------+------------------+----------------+
5.   |Michael|[Michael, Michael]|       [Michael]|
6.   |   Andy|            [Andy]|          [Andy]|
7.   | Justin|  [Justin, Justin]|        [Justin]|
8.   +-------+------------------+----------------+
```

13.10 通过 avg、sum、countDistinct 等分析企业人员管理系统

本节通过 avg、sum、countDistinct 算子分析企业人员管理系统。企业人员管理系统应用案例中的企业人员信息包括姓名、年龄数据信息。在企业人员管理系统中，对人员信息按照姓名进行分组，调用 agg 方法，统计分析年龄总和、平均年龄、最大年龄、最小年龄、唯一年龄计数、平均年龄等人员信息数据。

【例 13-17】 sum、avg、max、min、count、countDistinct、mean、current_date 函数解析。

将 personDS 按照姓名进行分组，调用 agg 方法，分别执行年龄求和、平均年龄、最大年龄、最小年龄、唯一年龄计数、平均年龄、当前时间等函数。

```
1.      println("使用 sum、avg 等函数计算年龄总和、平均年龄、最大年龄、最小年龄、唯一
        年龄计数、平均年龄、当前时间等数据 : ")
2.     personsDS.groupBy($"name").agg(sum($"age"), avg($"age"), max($"age"),
       min($"age"), count($"age"),countDistinct($"age"),mean($"age"),
       current_date()).show
```

在 IDEA 中运行代码，sum、avg、max、min、count、countDistinct、mean、current_date 函数示例的输出结果如下：

```
1.  使用 sum、avg 等函数计算年龄总和、平均年龄、最大年龄、最小年龄、唯一年龄计数、平均年
    龄、当前时间等数据 :
2.  +-----+------+------+------+----+-------+-----------+-----+---------+
3.  |   name|sum(age)|avg(age)|max(age)|min(age)|count(age)|count(DISTINCT
    age)|avg(age)|current_date()|
4.  +-----+------+------+------+----+-------+-----------+-----+---------+
5.  |Michael|      62|    31.0|      46|      16|         2|               2|
    31.0|    2017-03-16|
6.  |   Andy|      30|    30.0|      30|      30|         1|               1|
    30.0|    2017-03-16|
7.  | Justin|      48|    24.0|      29|      19|         2|               2|
    24.0|    2017-03-16|
8.  +-----+------+------+------+----+-------+-----------+-----+---------+
```

13.11 Dataset 开发实战企业人员管理系统应用案例代码

Spark 2.2 实战之 Dataset 开发实战企业人员管理系统应用案例如例 13-18 所示。

【例 13-18】 DatasetOps.scala 代码。

```
1.  package com.dt.spark.cores
2.
3.  import org.apache.log4j.{Level, Logger}
4.  import org.apache.spark.sql.{Dataset, Encoders, SparkSession}
5.
6.  import scala.collection.mutable.ArrayBuffer
7.
8.  object DatasetOps {
```

```
9.
10.  case class Person(name: String, age: Long)
11.
12.  case class Score(n: String, score: Long)
13.
14.
15.  def main(args: Array[String]) {
16.    Logger.getLogger("org").setLevel(Level.ERROR)
17.    val spark = SparkSession
18.      .builder
19.      .appName("DatasetOps").master("local[4]")
20.      .config("spark.sql.warehouse.dir", "G:\\IMFBigDataSpark2017\\SparkApps
       \\spark-warehouse")
21.      .getOrCreate()
22.
23.    import org.apache.spark.sql.functions._
24.    import spark.implicits._
25.
26.    /**
27.      * Dataset 中的 tranformation 和 Action 操作，Action 类型的操作有：
28.      * show、collect、first、reduce、take、count 等
29.      * 这些操作都会产生结果，也就是说会执行逻辑计算的过程
30.      */
31.
32.    val personsDF = spark.read.json("data/peoplemanagedata/people.json")
33.    val personScoresDF = spark.read.json("data/peoplemanagedata/people
     Scores.json")
34.
35.    val personsDS = personsDF.as[Person]
36.    val personScoresDS = personScoresDF.as[Score]
37.
38.    println("使用 groupBy 算子进行分组：")
39.    val personsDSGrouped = personsDS.groupBy($"name", $"age").count()
40.    personsDSGrouped.show()
41.
42.    println("使用 agg 算子 concat 内置函数，将姓名、年龄连接在一起，成为单个字符串
     列 ：")
43.    personsDS.groupBy($"name", $"age").agg(concat($"name", $"age")).show
44.
45.    println("使用 col 算子选择列 ：")
46.    personsDS.joinWith(personScoresDS, personsDS.col("name") === person
     ScoresDS.col("n")).show
47.
48.    println("使用 sum、avg 等函数计算年龄总和、平均年龄、最大年龄、最小年龄、唯一年
     龄计数、平均年龄、当前时间等数据 ：")
49.    personsDS.groupBy($"name").agg(sum($"age"), avg($"age"), max($"age"),
     min($"age"),count($"age"),countDistinct($"age"),mean($"age"),current_
     date()).show
50.
51.
52.    println("函数 collect_list、collect_set 比较，collect_list 函数结果中包含
```

```
    重复元素; collect_set 函数结果中无重复元素: ")
53.     personsDS.groupBy($"name")
54.       .agg(collect_list($"name"), collect_set($"name"))
55.       .show()
56.
57.     println("使用 sample 算子进行随机采样: ")
58.     personsDS.sample(false, 0.5).show()
59.
60.     println("使用 randomSplit 算子进行随机切分: ")
61.     personsDS.randomSplit(Array(10, 20)).foreach(dataset => dataset.show())
62.     println("使用 select 算子选择列: ")
63.     personsDS.select("name").show()
64.
65.     println("使用 joinWith 算子关联企业人员信息、企业人员分数评分信息: ")
66.     personsDS.joinWith(personScoresDS, $"name" === $"n").show
67.
68.     println("使用 join 算子关联企业人员信息、企业人员分数评分信息: ")
69.     personsDS.join(personScoresDS, $"name" === $"n").show
70.
71.     println("使用 sort 算子对年龄进行降序排列: ")
72.     personsDS.sort($"age".desc).show
73.
74.     import spark.implicits._
75.
76.     def myFlatMapFunction(myPerson: Person, myEncoder: Person): Dataset
        [Person] = {
77.       personsDS
78.     }
79.
80.     personsDS.flatMap(persons => persons match {
81.       case Person(name, age) if (name == "Andy") => List((name, age + 70))
82.       case Person(name, age) => List((name, age + 30))
83.     }).show()
84.
85.     personsDS.mapPartitions { persons =>
86.       val result = ArrayBuffer[(String, Long)]()
87.       while (persons.hasNext) {
88.         val person = persons.next()
89.         result += ((person.name, person.age + 1000))
90.       }
91.       result.iterator
92.
93.     }.show
94.
95.     println("使用 dropDuplicates 算子统计企业人员管理系统姓名无重复员工的记录:")
96.     personsDS.dropDuplicates("name").show()
97.     personsDS.distinct().show()
98.
99.     println("使用 repartition 算子设置分区: ")
100.        println("原分区数: " + personsDS.rdd.partitions.size)
101.        val repartitionDs = personsDS.repartition(4)
```

```
102.        println("repartition 设置分区数：" + repartitionDs.rdd.partitions.size)
103.
104.        println("使用 coalesce 算子设置分区：")
105.        val coalesced: Dataset[Person] = repartitionDs.coalesce(2)
106.        println("coalesce 设置分区数：" + coalesced.rdd.partitions.size)
107.        coalesced.show
108.
109.        spark.stop()
110.
111.     }
112.   }
```

13.12　本章总结

本章 Spark 2.2 实战之 Dataset 开发实战企业人员管理系统应用案例着重讲解了企业人员管理系统在 Spark 2.2 中的应用，通过对 Spark 2.2 中一系列算子的阐述，读者可以举一反三，触类旁通，从案例中熟练掌握 Spark 2.2 算子的应用。

第 14 章　Spark 商业案例之电商交互式分析系统应用案例

电商交互式分析系统应用案例：在实际生产环境下，一般都是以 J2EE+Hadoop+Spark+DB(Redis)的方式实现的综合技术栈，使用 Spark 进行电商用户行为分析时一般都是交互式的，什么是交互式的？也就是说，公司内部人员（如营销部门人员）向按照特定时间查询访问次数最多的用户或者购买金额最大的用户 TopN，这些分析结果对于公司的决策、产品研发和营销都是至关重要的，而且很多时候是立即想要结果的，如果此时使用 Hive 去实现，可能非常缓慢（如 1h），而在电商类企业中，经过深度调优后的 Spark 一般都会比 Hive 快 5 倍以上，此时的运行时间可能就是分钟级别，这个时候就可以达到即查即用的目的，也就是所谓的交互式，而交互式的大数据系统是未来的主流。

我们在这里分析电商用户的多维度的行为特征，如分析特定时间段访问人数的 TopN、特定时间段购买金额排名的 TopN、注册后一周内购买金额排名 TopN、注册后一周内访问次数排名 Top 等，但是这里的技术和业务场景同样适合于门户网站（如网易、新浪等），也同样适合于在线教育系统（如分析在线教育系统的学员的行为），当然也适用于 SNS 社交网络系统。

14.1　纯粹通过 DataSet 进行电商交互式分析系统中特定时段访问次数 TopN

14.1.1　电商交互式分析系统数据说明

1. 电商交互式分析系统数据的来源

本节的电商交互式分析系统的数据，由我们编写模拟代码来生成。在工程 SparkApps 的 data 目录下新建 sql 目录，将模拟生成的 log、user 的模拟数据文件保存到 data\sql 目录下。log、user 的模拟数据文件的文件格式包括 parquet、json 格式。

JSON（JavaScript Object Notation）是一种轻量级的数据交换格式，是基于 ECMAScript 的一个子集。JSON 是存储和交换文本信息的语法，采用完全独立于语言的文本格式，类似 XML，但 JSON 比 XML 更易解析。

Apache Parquet 是面向分析型业务的列式存储格式，由 Twitter 和 Cloudera 合作开发，并于 2015 年 5 月成为 Apache 顶级项目。Parquet 是一种语言无关列式存储格式的文件类型，可以适配多种计算框架，而且不与任何一种数据处理框架绑定在一起，适配多种语言和组件。

在本节电商交互式分析系统的编码实现中，我们使用在生产环境系统中广泛运用的 JSON、Parquet 文本格式，通过 DataSet 进行电商交互式分析系统中特定时段访问次数 TopN

的代码实战。

2. 电商交互式分析系统数据的格式说明

电商交互式分析系统的数据包含两个数据文件：用户信息、用户访问记录。

先查看用户信息的 JSON 格式描述信息。JSON 数据的书写格式是：名称/值对。名称/值对组合中的“名称”写在前面，“值对”写在后面，中间用冒号（:）隔开，用户信息文件 user.json 的格式描述如下：

```
1.  { "userID": userID-Value, "name": name-Value, "registeredTime": registeredTime--Value}
2.  用户 ID/值对、用户姓名/值对、注册时间/值对
3.  - userID：用户 ID。
4.  - userID-Value ：用户 ID 的值。
5.  - name：用户姓名。
6.  - name-Value：用户姓名的值。
7.  - registeredTime：用户注册时间。
8.  -registeredTime--Value：具体注册时间的值。
```

从用户信息文件 user.json 中摘取部分记录如下：

```
1.  {"userID": 0, "name": "spark0", "registeredTime": "2016-10-11 18:06:25"}
2.  {"userID": 1, "name": "spark1", "registeredTime": "2016-10-11 18:06:25"}
3.  {"userID": 2, "name": "spark2", "registeredTime": "2016-09-26 18:06:25"}
4.  {"userID": 3, "name": "spark3", "registeredTime": "2016-10-04 18:06:25"}
5.  {"userID": 4, "name": "spark4", "registeredTime": "2016-10-05 18:06:25"}
6.  {"userID": 5, "name": "spark5", "registeredTime": "2016-10-05 18:06:25"}
7.  {"userID": 6, "name": "spark6", "registeredTime": "2016-11-08 11:09:12"}
8.  {"userID": 7, "name": "spark7", "registeredTime": "2016-10-07 18:06:25"}
9.  {"userID": 8, "name": "spark8", "registeredTime": "2016-10-05 18:06:25"}
10. {"userID": 9, "name": "spark9", "registeredTime": "2016-10-07 18:06:25"}
```

接下来查看用户访问记录的 JSON 格式描述信息。用户访问记录文件 log.json 的格式描述如下：

```
1. {"logID":logID-Value,"userID":userID-Value, "time": time-Value, "typed": typed-Value, "consumed":consumed-Value}
2.   日志 ID/值对，用户 ID/值对，时间/值对，类型/值对，消费金额/值对
3.  - logID：日志 ID。
4.  - logID-Value ：日志 ID 的值。
5.  - userID：用户 ID。
6.  - userID-Value：用户 ID 的值。
7.  - time：注册时间。
8.  - time-Value：具体注册时间的值。
9.  - typed：类型。typed=0 为用户访问电商网站；typed=1 是用户在电商网站购买商品。
10. - typed-Value：类型的值。
11. - consumed：消费金额。
12. -consumed-Value：消费金额的值。
```

从用户访问记录文件 log.json 中摘取部分记录如下：

```
1.  {"logID": 00,"userID": 0, "time": "2016-10-04 15:42:45", "typed": 0, "consumed": 0.0}
2.  {"logID": 01,"userID": 0, "time":"2016-10-17 15:42:45","typed":1,"consumed": 33.36}
3.  {"logID":  02,"userID":  0,  "time":  "2016-10-18  15:42:45",  "typed":  0,
```

```
    "consumed": 0.0}
4.  {"logID": 03,"userID": 0, "time": "2016-10-14 15:42:45", "typed": 0,
    "consumed": 0.0}
5.  {"logID": 04,"userID": 0, "time": "2016-11-09 08:45:33", "typed": 1,
    "consumed": 664.35}
6.  {"logID": 05,"userID": 0, "time": "2016-10-13 15:42:45", "typed": 0,
    "consumed": 0.0}
7.  {"logID": 06,"userID": 0, "time": "2016-10-03 15:42:45", "typed": 1,
    "consumed": 606.34}
8.  {"logID": 07,"userID": 0, "time": "2016-10-04 15:42:45", "typed": 1,
    "consumed": 120.72}
9.  {"logID": 08,"userID": 0, "time": "2016-09-24 15:42:45", "typed": 1,
    "consumed": 264.96}
10. {"logID": 09,"userID": 0, "time": "2016-09-25 15:42:45", "typed": 0,
    "consumed": 0.0}
```

前面已经讲解了用户信息的JSON格式描述信息，接下来看一下用户信息的parquet格式。用户信息（parquet 格式）本地文件保存在 SparkApps\data\sql\userparquet.parquet 目录下，目录里共有 4 个文件。本地文件目录结构如下：

```
1.   ._SUCCESS.crc
2.  .part-r-00000-52b3efd4-5b4a-43a8-b2d7-0e5f94396d82.snappy.parquet.crc
3.  _SUCCESS
4.  part-r-00000-52b3efd4-5b4a-43a8-b2d7-0e5f94396d82.snappy
```

parquet 格式的文本文件无法用记事本直接打开，用记事本打开将看到二进制数据，我们在 Spark 应用程序中，使用 printSchema()方法打印显示用户的 userparquet.parquet 数据结构如下：

```
1.   User
2.       |-- name: string (nullable = true)
3.       |-- registeredTime: string (nullable = true)
4.       |-- userID: long (nullable = true)
```

用户访问记录（parquet 格式）本地文件保存在 SparkApps\data\sql\logparquet.parquet 目录下，目录里面共有 4 个文件。本地文件目录结构如下：

```
1.   ._SUCCESS.crc
2.  .part-r-00000-673b6323-2dab-4665-8c52-f91ef90ab9b5.snappy.parquet.crc
3.  _SUCCESS
4.  part-r-00000-673b6323-2dab-4665-8c52-f91ef90ab9b5.snappy.parquet
```

使用 printSchema 方法打印显示用户访问记录的 logparquet.parquet 数据结构如下：

```
1.       Log
2.      |-- consumed: double (nullable = true)
3.      |-- logID: long (nullable = true)
4.      |-- time: string (nullable = true)
5.      |-- typed: long (nullable = true)
6.      |-- userID: long (nullable = true)
```

14.1.2　特定时段内用户访问电商网站排名 TopN

本节纯粹通过 DataSet 实现电商交互式分析系统中特定时段访问次数 TopN 的功能。在实现特定时段访问次数 TopN 的功能前，请思考以下问题。

第一个问题：特定时段中的时间是从哪里来的？一般都来自于 J2EE 调度系统，如一个营销人员通过系统传入了 2017.01.01~2017.01.10。

第二个问题：计算的时候我们会使用哪些核心算子：join、groupBy、agg（在 agg 中可以使用大量的 functions.scala 中的函数，极大方便快速地实现业务逻辑系统）。

第三个问题：计算完成后，数据保存在哪里？在生产环境中，一般保存在 DB、HBase/Canssandra、Redis 中。

具体实现方法如下。

（1）在通过 DataSet 实现电商交互式分析系统中特定时段访问次数 TopN 的 EB_Users_Analyzer_DateSet.scala 代码中导入 sql 函数方法和隐式转换方法。org.apache.spark.sql.functions.scala 文件包含了大量的内置函数，尤其在 agg 中会广泛使用。

```
1.  import org.apache.spark.sql.functions._
2.  import spark.implicits._
```

（2）方式一：在 EB_Users_Analyzer_DateSet.scala 代码中使用 JSON 格式获取用户信息数据和用户访问记录数据。

```
1.      val userInfo = spark.read.format("json"). json("data/sql/user.json")
2.      val userLog = spark.read.format("json"). json("data/sql/log.json")
3.      println("用户信息及用户访问记录文件 JSON 格式 :")
4.      userInfo.printSchema()
5.      userLog.printSchema()
```

在 IDEA 中运行代码，打印用户信息数据和用户访问记录数据 JSON 格式，输出结果如下：

```
1.  用户信息及用户访问记录文件 JSON 格式:
2.  root
3.   |-- name: string (nullable = true)
4.   |-- registeredTime: string (nullable = true)
5.   |-- userID: long (nullable = true)
6.
7.  root
8.   |-- _corrupt_record: string (nullable = true)
     //源数据文件格式损坏，可检查双引号
9.   |-- consumed: double (nullable = true)
10.  |-- logID: long (nullable = true)
11.  |-- time: string (nullable = true)
12.  |-- typed: long (nullable = true)
13.  |-- userID: long (nullable = true)
```

（3）方式二：在 EB_Users_Analyzer_DateSet.scala 代码中也可使用 parquet 格式获取用户信息数据和用户访问记录数据。

```
1.      val userInfo = spark.read.format("parquet"). parquet("data/sql/
        userparquet.parquet")
2.      val userLog = spark.read.format("parquet"). parquet("data/sql/
        logparquet.parquet")
3.      println("用户信息及用户访问记录文件 parquet 格式 :")
4.      userInfo.printSchema()
5.      userLog.printSchema()
```

在 IDEA 中运行代码，打印用户信息数据和用户访问记录数据 parquet 格式，输出结果如下：

```
1.    用户信息及用户访问记录文件 parquet 格式：
2.  root
3.   |-- name: string (nullable = true)
4.   |-- registeredTime: string (nullable = true)
5.   |-- userID: long (nullable = true)
6.
7.  root
8.   |-- _corrupt_record: string (nullable = true)
9.   |-- consumed: double (nullable = true)
10.  |-- logID: long (nullable = true)
11.  |-- time: string (nullable = true)
12.  |-- typed: long (nullable = true)
13.  |-- userID: long (nullable = true)
```

（4）统计特定时段访问次数最多的 Top5，使用 filter 算子过滤出 10 月 1 日至 11 月 1 日时间范围内用户访问电商网站的数据，根据用户 ID 使用 join 算子和用户数据关联，按用户 ID 和用户姓名进行分组，然后使用 agg 算子，利用内置函数 count 统计日志 ID 的总访问次数，将用户的总访问次数取别名为 userLogCount，最后按别名进行降序排列，打印输出前 5 个元素。

统计特定时段访问次数最多的 Top5 的代码如下：

```
1.      val startTime =  "2016-10-01"
2.      val endTime = "2016-11-01"
3.    println("功能一：统计特定时段访问次数最多的Top5: 例如2016-10-01 ~ 2016-11-01 :")
4.      userLog.filter("time >= '" + startTime + "' and time <= '" + endTime
      + "' and typed = 0")
5.        .join(userInfo, userInfo("userID") === userLog("userID"))
6.        .groupBy(userInfo("userID"),userInfo("name"))
7.        .agg(count(userLog("logID")).alias("userLogCount"))
8.        .sort($"userLogCount".desc)
9.        .limit(5)
10.       .show()
```

在 IDEA 中运行代码，特定时段内用户访问电商网站排名 TopN，输出结果如下：

```
1.  功能一 ：统计特定时段访问次数最多的 Top5: 例如 2016-10-01 ~ 2016-11-01:
2.  +------+-------+------------+
3.  |userID|   name |userLogCount|
4.  +------+-------+------------+
5.  |    39 |spark39|         45  |
6.  |    11 |spark11|         39  |
7.  |     9 | spark9|         39  |
8.  |     4 | spark4|         38  |
9.  |    76 |spark76|         38  |
10. +------+-------+------------+
```

14.2　纯粹通过 DataSet 分析特定时段购买金额 Top10 和访问次数增长 Top10

本节纯粹通过 DataSet 分析特定时段购买金额 Top10 和访问次数增长 Top10。

（1）分析特定时段购买金额 Top10：使用 filter 算子过滤出 10 月 1 日至 11 月 1 日时间范围内用户访问电商网站的数据，根据用户 ID 使用 join 算子和用户数据关联，按用户 ID 和用户姓名进行分组，然后使用 agg 算子，利用内置函数 sum 求和，统计用户消费的总金额，保留 2 位小数取整，将用户消费的总金额取别名为 totalCount，最后按别名进行降序排列，打印输出前 10 个元素。

统计特定时段内用户购买总金额排名 TopN。

```
1.     println("功能二：统计特定时段内用户购买总金额排名 TopN:")
2.     userLog.filter("time >= '" + startTime + "' and time <= '" + endTime + "'")
3.       .join(userInfo, userInfo("userID") === userLog("userID"))
4.       .groupBy(userInfo("userID"),userInfo("name"))
5.       .agg(round(sum(userLog("consumed")),2).alias("totalCount"))
6.       .sort($"totalCount".desc)
7.       .limit(10)
8.       .show()
```

在 IDEA 中运行代码，统计特定时段内用户购买总金额排名 TopN，输出结果如下：

```
1.  功能二：统计特定时段内用户购买总金额排名 TopN:
2.  +------+-------+----------+
3.  |userID|   name |totalCount|
4.  +------+-------+----------+
5.  |    92 |spark92|   20109.1|
6.  |    14 |spark14|  19991.64|
7.  |    40 |spark40|  19098.22|
8.  |    64 |spark64|  19095.71|
9.  |    46 |spark46|  18895.19|
10. |    41 |spark41|  18878.94|
11. |    10 |spark10|  18489.67|
12. |    80 |spark80|  18003.31|
13. |    84 |spark84|  18002.74|
14. |    49 |spark49|  17902.01|
15. +------+-------+----------+
```

（2）纯粹通过 DataSet 分析特定时段访问次数增长 Top10： 在电商交互式分析系统应用案例中，统计本周比上周用户访问次数的增长排名。例如，上一周的时间范围为 2016 年 10 月 1 日至 7 日，本周的时间范围为 2016 年 10 月 8 日至 14 日，统计出访问电商网站次数本周比上周环比增长较快的用户。

电商交互式分析系统应用案例统计分析特定时段访问次数增长 Top10 的实现方法如下。

（1）定义 case class 类。定义 UserLog 用户类，使用 as[UserLog]方法将 DataFrame 转换为 DataSet。定义 LogOnce 类，用于存放用户访问网站的计数 1 次的信息，记录用户的日志 ID、用户 ID 及计数。定义 ConsumedOnce 类，用于存放用户购买商品金额的信息，记录用户的日志 ID、用户 ID 及消费金额。

case class 类定义如下：

```
1.  case class UserLog(logID: Long, userID: Long, time: String, typed: Long,
    consumed: Double)
```

```
2.  case class LogOnce(logID: Long, userID: Long, count: Long)
3.  case class ConsumedOnce(logID: Long, userID: Long, consumed: Double)
```

（2）分析特定时段访问次数增长 Top10。

统计访问电商网站次数本周比上周环比增长较快的用户的代码编写技巧如下。

- 在用户访问记录中过滤出本周（2016 年 10 月 8 日至 14 日）的访问数据，使用 map 转换函数，将数据集 Dataset[UserLog]的每行数据格式化为 LogOnce 类，生成新的数据集 Dataset[LogOnce]，数据格式为（日志 ID,用户 ID,1）。
- 在用户访问记录中过滤出上周（2016 年 10 月 1 日至 7 日）的访问数据，使用 map 转换函数，将数据集 Dataset[UserLog]的每行数据格式化为 LogOnce 类，生成新的数据集 Dataset[LogOnce]，数据格式为（日志 ID,用户 ID,–1）。

（3）将本周和上周的用户访问记录进行 union 合并，生成新的数据集 userAccessTemp，其中的计数正 1 表示本周的记录，负 1 表示上周的数据，数据格式为（日志 ID,用户 ID,计数）。

（4）根据用户 ID 将 userAccessTemp 和用户数据进行 join 关联，按用户 ID、姓名分组，使用 agg 算子，利用内置函数 sum 统计次数，这里有一个构思巧妙的小技巧：计数正 1 表示本周的记录，负 1 表示上周的数据，sum 求和就正负抵消巧妙地统计出同一用户本周比上周增长的访问次数，将用户本周比上周增长的访问次数取别名为 viewIncreasedTmp，最后按别名进行降序排列，打印输出前 10 个元素。

```
1.      println("功能三：统计特定时段内用户访问次数增长排名 TopN:")
2.      val userAccessTemp = userLog.as[UserLog].filter("time >= '2016-10-08'
        and time <= '2016-10-14' and typed = '0'")
3.        .map(log => LogOnce(log.logID, log.userID, 1))
4.        .union(userLog.as[UserLog].filter("time >= '2016-10-01' and time <=
        '2016-10-07' and typed = '0'")
5.          .map(log => LogOnce(log.logID, log.userID, -1)))
6.
7.      userAccessTemp.join(userInfo, userInfo("userID") === userAccessTemp
        ("userID"))
8.        .groupBy(userInfo("userID"), userInfo("name"))
9.        .agg(round(sum(userAccessTemp("count")), 2).alias("viewIncreasedTmp"))
10.       .sort($"viewIncreasedTmp".desc)
11.       .limit(10)
12.       .show()
```

在 IDEA 中运行代码，统计分析特定时段访问次数增长 Top10，输出结果如下：

```
1.  功能三：统计特定时段内用户访问次数增长排名 TopN:
2.  +------+-------+----------------+
3.  |userID|  name |viewIncreasedTmp|
4.  +------+-------+----------------+
5.  |    8 | spark8|          10    |
6.  |   52 |spark52|           9    |
7.  |   75 |spark75|           7    |
8.  |   78 |spark78|           7    |
9.  |   28 |spark28|           7    |
10. |   20 |spark20|           7    |
11. |   88 |spark88|           6    |
12. |   22 |spark22|           6    |
13. |   66 |spark66|           6    |
14. |   85 |spark85|           6    |
15. +------+-------+----------------+
```

14.3　纯粹通过 DataSet 进行电商交互式分析系统中各种类型 TopN 分析实战详解

纯粹通过 DataSet 进行电商交互式分析系统中各种类型 TopN 分析实战详解包括：统计特定时段购买金额最多的 Top5、访问次数增长最多的 Top5 用户、购买金额增长最多的 Top5 用户、注册之后前两周内访问最多的 Top10、注册之后前两周内购买总额最多的 Top10。

14.3.1　统计特定时段购买金额最多的 Top5 用户

统计特定时段购买次数最多的 Top5 在 14.2 节已经讲解过，使用 filter 算子过滤出 2016 年 10 月 1 日至 11 月 1 日时间范围内用户访问电商网站的数据，根据用户 ID 使用 join 算子和用户数据关联，按用户 ID 和用户姓名进行分组，然后使用 agg 算子，利用内置函数 sum 求和统计用户消费的总金额，round 函数保留 2 位小数取整，将用户消费的总金额取别名为 totalConsumed，最后按别名进行降序排列，打印输出前 5 个元素。

统计特定时段购买次数最多的 Top5 代码如下：

```
1.     val startTime = "2016-10-01"
2.     val endTime = "2016-11-01"
3.  ......
4.   println("统计特定时段购买金额最多的 Top5: 例如 2016-10-01 ~ 2016-11-01 :")
5.     userLog.filter("time >= '" + startTime + "' and time <= '" + endTime
       + "' and typed = 1")
6.       .join(userInfo, userInfo("userID") === userLog("userID"))
7.       .groupBy(userInfo("userID"), userInfo("name"))
8.       .agg(round(sum(userLog("consumed")), 2).alias("totalConsumed"))
9.       .sort($"totalConsumed".desc)
10.      .limit(5)
11.      .show
```

在 IDEA 中运行代码，统计特定时段购买金额最多的 Top5，输出结果如下：

```
1.  统计特定时段购买金额最多的 Top5: 例如 2016-10-01 ~ 2016-11-01:
2.  +------+-------+-------------+
3.  |userID|   name |totalConsumed|
4.  +------+-------+-------------+
5.  |    92 |spark92|     20109.1 |
6.  |    14 |spark14|    19991.64 |
7.  |    40 |spark40|    19098.22 |
8.  |    64 |spark64|    19095.71 |
9.  |    46 |spark46|    18895.19 |
10. +------+-------+-------------+
```

14.3.2　统计特定时段访问次数增长最多的 Top5 用户

统计特定时段访问次数增长最多的 Top5 用户，例如，这周比上周访问次数增长最快的 5 位用户。实现思路：一种非常直接的方式是计算这周每个用户的访问次数，同时计算出上周每个用户的访问次数，然后相减并进行排名，但是这种实现思路比较消耗性能，我们可以采用一种既能实现业务目标，又能够提升性能的方式，即把这周的每次用户访问计数为 1，把上周的每次用户访问计数为-1，在 agg 操作中采用 sum 即可巧妙地实现增长趋势的量化。

统计特定时段访问次数增长最多的 Top5 用户，代码如下：

```
1.      val userLogDS = userLog.as[UserLog].filter("time >= '2016-10-08' and
        time <= '2016-10-14' and typed = '0'")
2.        .map(log => LogOnce(log.logID, log.userID, 1))
3.        .union(userLog.as[UserLog].filter("time >= '2016-10-01' and time <=
          '2016-10-07' and typed = '0'")
4.          .map(log => LogOnce(log.logID, log.userID, -1)))
5.
6.      println("统计特定时段访问次数增长最多的 Top5 用户，例如这周比上周访问次数增
        长最快的 5 位用户： ")
7.      userLogDS.join(userInfo, userLogDS("userID") === userInfo("userID"))
8.        .groupBy(userInfo("userID"), userInfo("name"))
9.        .agg(sum(userLogDS("count")).alias("viewCountIncreased"))
10.       .sort($"viewCountIncreased".desc)
11.       .limit(5)
12.       .show()
```

在 IDEA 中运行代码，统计特定时段访问次数增长最多的 Top5 用户，输出结果如下：

```
1.   统计特定时段访问次数增长最多的 Top5 用户，例如这周比上周访问次数增长最快的 5 位用户：
2.   [Stage 19:============================================>(165 + 8)/200]
     +------+-------+------------------+
3.   |userID|  name |viewCountIncreased|
4.   +------+-------+------------------+
5.   |    8 | spark8|              10  |
6.   |   52 |spark52|               9  |
7.   |   75 |spark75|               7  |
8.   |   28 |spark28|               7  |
9.   |   78 |spark78|               7  |
10.  +------+-------+------------------+
```

14.3.3　统计特定时段购买金额增长最多的 Top5 用户

统计特定时段购买金额增长最多的 Top5 用户，例如这周比上周购买金额增长最多的 5 位用户。在电商交互式分析系统应用案例中，统计本周比上周用户购买金额增长最多的排名，例如上周的时间范围为 2016 年 10 月 1 日至 7 日，本周的时间范围为 2016 年 10 月 8 日至 14 日，统计出用户购买金额本周比上周环比增长最多的 5 位用户。

电商交互式分析系统应用案例统计特定时段购买金额增长最多的 Top5 用户实现方法如下。

（1）定义 case class 类。定义 UserLog 用户类，使用 as[UserLog]方法将 DataFrame 转换为 DataSet。定义 LogOnce 类，用于存放用户访问网站的计数 1 次的信息，记录用户的日志 ID、用户 ID 及计数。定义 ConsumedOnce 类，用于存放用户购买商品金额的信息，记录用户的日志 ID、用户 ID 及消费金额。

case class 类定义如下：

```
1.  case class UserLog(logID: Long, userID: Long, time: String, typed: Long,
    consumed: Double)
```

```
2.  case class LogOnce(logID: Long, userID: Long, count: Long)
3.  case class ConsumedOnce(logID: Long, userID: Long, consumed: Double)
```

（2）统计特定时段本周比上周购买金额增长最多的 Top5 用户的代码编写技巧。

- 在用户访问记录中过滤出本周（2016 年 10 月 8 日至 14 日）及购买商品行为（类型为 1）的访问数据，使用 map 转换函数，将数据集 Dataset[UserLog]的每行数据格式化为 ConsumedOnce 类，生成新的数据集 Dataset[ConsumedOnce]。数据格式为（日志 ID,用户 ID,消费金额）。
- 在用户访问记录中过滤出上周（2016 年 10 月 1 日至 7 日）及购买商品行为（类型为 1）的访问数据，使用 map 转换函数，将数据集 Dataset[UserLog]的每行数据格式化为 ConsumedOnce 类，生成新的数据集 Dataset[ConsumedOnce]。数据格式为（日志 ID,用户 ID,-消费金额）。

（3）将本周和上周的用户访问记录进行 union 合并，生成新的数据集 userLogConsumerDS，其中正的消费金额表示本周的记录，负的消费金额表示上周的数据。数据格式为（日志 ID,用户 ID,消费金额）。

（4）根据用户 ID 将 userLogConsumerDS 和用户数据进行 join 关联，按用户 ID、姓名分组，使用 agg 算子，利用内置函数 sum 统计求和，正的消费金额表示本周的记录，负的消费金额表示上周的数据，sum 求和就巧妙地统计出同一用户本周比上周增长的消费金额。round 函数保留 2 位小数取整，将用户本周比上周增长的消费金额取别名为 viewConsumedIncreased，最后按别名进行降序排列，打印输出前 5 个元素。

统计特定时段购买金额增长最多的 Top5 用户，代码如下：

```
1.   println("统计特定时段购买金额增长最多的 Top5 用户，例如这周比上周购买金额增长最
     多的 5 位用户：")
2.     val userLogConsumerDS = userLog.as[UserLog].filter("time >= '2016-
     10-08' and time <= '2016-10-14' and typed == 1")
3.       .map(log => ConsumedOnce(log.logID, log.userID, log.consumed))
4.       .union(userLog.as[UserLog].filter("time >= '2016-10-01' and time <=
     '2016-10-07' and typed == 1")
5.         .map(log => ConsumedOnce(log.logID, log.userID, -log.consumed)))
6.
7.     userLogConsumerDS.join(userInfo,  userLogConsumerDS("userID")  ===
     userInfo("userID"))
8.       .groupBy(userInfo("userID"), userInfo("name"))
9.       .agg(round(sum(userLogConsumerDS("consumed")),  2).alias("viewConsumed
       Increased"))
10.      .sort($"viewConsumedIncreased".desc)
11.      .limit(5)
12.      .show()
```

在 IDEA 中运行代码，统计特定时段购买金额增长最多的 Top5 用户，输出结果如下：

```
1.    统计特定时段购买金额增长最多的 Top5 用户，例如这周比上周购买金额增长最多的 5 位用户：
2.  +------+-------+---------------------+
3.  |userID|   name |viewConsumedIncreased|
4.  +------+-------+---------------------+
5.  |    47 |spark47|            6684.31   |
6.  |    62 |spark62|             5319.8   |
7.  |    83 |spark83|            5282.87   |
8.  |    45 |spark45|            4677.08   |
```

```
9.  |   11 |spark11|           4480.73   |
10. +------+-------+---------------------+
```

14.3.4　统计特定时段注册之后前两周内访问次数最多的 Top10 用户

在电商交互式分析系统应用案例中，统计注册之后前两周内访问次数最多的 Top10 的用户。例如，新用户的注册时间是 2016-10-01，在注册之后的两周（2016-10-01 至 2016-10-14）时间范围内，统计出新用户访问电商网站次数 Top10 的用户，电商网站就可对活跃的新用户进行营销推广。

电商交互式分析系统应用案例统计注册之后前两周内访问次数最多的 Top10 用户实现方法如下。

（1）根据用户 ID，将用户访问记录和用户信息进行 join 关联。

（2）过滤出 2016-10-01 至 2016-10-14 期间注册的新用户，用户访问电商网站的时间在用户新注册时间及注册时间两周内（注册时间加上 14 天）的时间范围内，而且是访问网站（类型等于 0）的用户记录。注意：这里使用了 functions.scala 中的内置函数 date_add,date_add 在开始时间 start 基础之加上 days 的时间，即加上 14 天。

（3）根据用户 ID、姓名分组。

（4）使用 agg 算子，按用户访问记录的日志 ID 统计计数，取别名为 logTimes。

（5）根据别名 logTimes 进行降序排列，取前 10 个数据，打印输出。

统计注册之后前两周内访问次数最多的 Top10，代码如下：

```
1.     println("统计注册之后前两周内访问次数最多的 Top10:")
2.      userLog.join(userInfo, userInfo("userID") === userLog("userID"))
3.        .filter(userInfo("registeredTime") >= "2016-10-01"
4.          && userInfo("registeredTime") <= "2016-10-14"
5.          && userLog("time") >= userInfo("registeredTime")
6.          && userLog("time") <= date_add(userInfo("registeredTime"), 14)
7.          && userLog("typed") === 0)
8.        .groupBy(userInfo("userID"), userInfo("name"))
9.        .agg(count(userLog("logID")).alias("logTimes"))
10.       .sort($"logTimes".desc)
11.       .limit(10)
12.       .show()
```

在 IDEA 中运行代码，统计注册之后前两周内访问次数最多的 Top10，输出结果如下：

```
1.  统计注册之后前两周内访问次数最多的 Top10:
2.  +------+-------+--------+
3.  |userID|   name |logTimes|
4.  +------+-------+--------+
5.  |    11 |spark11|      29 |
6.  |    66 |spark66|      26 |
7.  |    92 |spark92|      25 |
8.  |     8 | spark8|      25 |
9.  |    63 |spark63|      24 |
10. |     9 | spark9|      24 |
11. |     4 | spark4|      24 |
12. |    37 |spark37|      23 |
13. |    74 |spark74|      23 |
14. |    96 |spark96|      23 |
15. +------+-------+--------+
```

14.3.5　统计特定时段注册之后前两周内购买总额最多的 Top10 用户

在电商交互式分析系统应用案例中，统计注册之后前两周内购买总额最多的 Top10 的用户。例如，新用户的注册时间是 2016-10-01，在注册之后的两周（2016-10-01 至 2016-10-14）时间范围内，统计出两周内新用户购买总额最多 Top10 的用户。

电商交互式分析系统应用案例统计注册之后前两周内购买总额最多的 Top10 用户实现方法如下。

（1）根据用户 ID，将用户访问记录和用户信息进行 join 关联。

（2）过滤出在 2016-10-01 至 2016-10-14 期间注册的新用户，用户访问电商网站的时间在用户新注册时间及注册时间两周内（注册时间加上 14 天）的时间范围内，而且是购买商品（类型等于 1）的用户的记录。同样，这里使用了 functions.scala 中的内置函数 date_add，date_add 在开始时间 start 基础上加上 days 的时间，即加上 14 天。

（3）根据用户 ID、姓名分组。

（4）使用 agg 算子和 sum 函数对用户消费金额求和，round 函数保留 2 位小数取整，取别名为 totalConsumed。

（5）根据别名 totalConsumed 进行降序排列，取前 10 个数据，打印输出。

统计注册之后前两周内购买总额最多的 Top10，代码如下：

```
1.     println("统计注册之后前两周内购买总额最多的 Top10 :")
2.      userLog.join(userInfo, userInfo("userID") === userLog("userID"))
3.        .filter(userInfo("registeredTime") >= "2016-10-01"
4.          && userInfo("registeredTime") <= "2016-10-14"
5.          && userLog("time") >= userInfo("registeredTime")
6.          && userLog("time") <= date_add(userInfo("registeredTime"), 14)
7.          && userLog("typed") === 1)
8.        .groupBy(userInfo("userID"), userInfo("name"))
9.        .agg(round(sum(userLog("consumed")), 2).alias("totalConsumed"))
10.       .sort($"totalConsumed".desc)
11.       .limit(10)
12.       .show()
```

在 IDEA 中运行代码，统计注册之后前两周内购买总额最多的 Top10，输出结果如下：

```
1.    统计注册之后前两周内购买总额最多的 Top10:
2.  +------+-------+-------------+
3.  |userID|   name |totalConsumed|
4.  +------+-------+-------------+
5.  |    92 |spark92|     15555.91 |
6.  |    65 |spark65|     15191.59 |
7.  |    59 |spark59|     14238.02 |
8.  |    40 |spark40|     14176.72 |
9.  |    80 |spark80|     13328.77 |
10. |    91 |spark91|     13245.58 |
11. |    57 |spark57|     12997.61 |
12. |    64 |spark64|     12717.84 |
13. |    26 |spark26|     11667.84 |
14. |     5 | spark5|     11608.82 |
15. +------+-------+-------------+
```

14.4　电商交互式分析系统应用案例涉及的核心知识点原理、源码及案例代码

14.4.1　知识点：Functions.scala

Spark 框架源码 functions.scala 文件包含了大量的内置函数，尤其在 agg 中会广泛使用，请认真反复阅读该源码并实践。

首先看一下 functions.scala 包括的函数功能分类，如 UDF 自定义函数、聚合函数、日期时间函数、排序函数、非聚合函数、数学函数、窗口函数、字符串函数、集合函数、其他函数等，如下所示。

```
1.  Functions 函数功能可用于 DataFrame 的操作。
2.  @groupname  udf_funcs UDF 自定义函数
3.  @groupname  agg_funcs 聚合函数
4.  @groupname  datetime_funcs 日期时间函数
5.  @groupname  sort_funcs 排序功能
6.  @groupname  normal_funcs 非聚合函数
7.  @groupname  math_funcs 数学函数
8.  @groupname  misc_funcs 其他功能
9.  @groupname  window_funcs 窗口函数
10. @groupname  string_funcs 字符串函数
11. @groupname  collection_funcs 集合函数功能
12. @groupname  DataFrames 不分组支持功能
13. @since 自从 1.3.0
```

统计 Spark 框架源码中 functions.scala 中新增 API 函数的演进情况，从 Spark 1.3.0 演进到 Spark 2.2.0 API 函数共计 307 个。functions.scala 中 API 函数统计见表 14-1。

表 14-1　functions.scala中API函数统计

Spark 1.3.0	47	Spark 2.0.0	20
Spark 1.4.0	91	Spark 2.1.0	21
Spark 1.5.0	86	Spark 2.2.0	9
Spark 1.6.0	33		

下面讲解一下 Spark 2.0.0 中新增的时间窗口函数的应用。

Window 时间窗口函数是在 Spark2.0.0 中新增的 API 函数，根据 Window 函数中给定时间戳指定列，生成滚动时间窗口。时间窗口是左开右闭的，例如 12:05 将落在窗口[12:05,12:10)，不落在窗口[12:00,12:05)。Windows 可以支持微秒级精度。时间窗口不支持月份的顺序。Windows 开始时间为 1970-01-01 00:00:00 UTC。

Window 时间窗口函数一，包括时间列、窗口时间两个参数，源码如下：

```
1.     * 流式查询， 可以使用 current_timestamp 生成处理时间窗口
2.     *
3.     * @param timeColumn 列或表达式作为窗口时间的时间戳，时间列必须为 Timestamp 类型
4.     * @param windowDuration 指定窗口宽度的字符串，如“10 分钟”“1 秒钟”。使用
       * [org.apache.spark.unsafe.types.CalendarInterval]检查有效持续时间标识符
5.     *
6.     * @group 日期时间函数
7.     * @since 2.0.0 从 Spark 2.0.0 新增
8.     */
9.    @Experimental 试验性的
10.   @InterfaceStability.Evolving 进化演进
11.   def window(timeColumn: Column, windowDuration: String): Column = {
12.     window(timeColumn, windowDuration, windowDuration, "0 second")
13.   }
```

Window 时间窗口函数一的应用，如一分钟的滚动窗口的平均股价，代码如下：

```
1.     * {{{
2.     *   val df = ... // schema => timestamp: TimestampType, stockId:
       *   StringType, price: DoubleType
3.     *  df.groupBy(window($"time", "1 minute"), $"stockId")
4.     *    .agg(mean("price"))
5.     * }}}
```

Window 时间窗口函数一显示如下：

```
1.     * {{{
2.     *  09:00:00-09:01:00
3.     *  09:01:00-09:02:00
4.     *  09:02:00-09:03:00 ...
5.     * }}}
```

Window 时间窗口函数二，包括时间列、窗口时间、滑动时间窗口 3 个参数，源码如下：

```
1.     * 流式查询，可以使用 current_timestamp 生成处理时间窗口
2.     * @param timeColumn 列或表达式作为窗口时间的时间戳，时间列必须为 Timestamp 类型
3.     * @param windowDuration  指定窗口宽度的字符串，如“10 分钟”“1 秒钟”。使用
       * [org.apache.spark.unsafe.types.CalendarInterval]检查有效持续时间标识
       * 符。注意，持续时间是一个固定长度的时间，并不会随日历时间的变化而变化。例如，1 day
       * 总是意味着 86400000 ms，而不是日历日
4.     * @param slideDuration  指定滑动时间窗口的字符串，例如“1 分钟”，根据每个
       * slideDuration 生成新的时间窗口，slideDuration 必须小于或等于 windowduration。
       * 使用[org.apache.spark.unsafe.types.CalendarInterval]检查有效持续时间
       * 标识符。这个持续时间也是绝对的，不根据日历变化
5.     *
6.     * @group  日期时间函数
1.     * @since 2.0.0 从 Spark 2.0.0 新增
7.     */
8.    @Experimental 试验性的
9.    @InterfaceStability.Evolving 进化演进
10.   def window(timeColumn: Column, windowDuration: String, slideDuration:
       String): Column = {
11.     window(timeColumn, windowDuration, slideDuration, "0 second")
12.   }
```

Window 时间窗口函数二的应用，如滑动时间窗口每隔 10s，每一分钟时间窗口的平均股价，代码如下：

```
1.    * {{{
2.    *   val df = ... // schema => timestamp: TimestampType, stockId:
      *   StringType, price: DoubleType
3.    *  df.groupBy(window($"time", "1 minute", "10 seconds"), $"stockId")
4.    *    .agg(mean("price"))
5.    * }}}
```

Window 时间窗口函数二显示如下：

```
1.    * {{{
2.    *  09:00:00-09:01:00
3.    *  09:00:10-09:01:10
4.    *  09:00:20-09:01:20 ...
5.    * }}}
```

Window 时间窗口函数三，包括时间列、窗口时间、滑动时间窗口、开始时间 4 个参数，源码如下：

```
1.    * 流式查询，可以使用 current_timestamp 生成处理时间窗口
2.    * @param timeColumn 列或表达式作为窗口时间的时间戳，时间列必须为 Timestamp 类型。
3.    * @param windowDuration  指定窗口宽度的字符串，如“10 分钟”，“1 秒钟”。使
      * 用[org.apache.spark.unsafe.types.CalendarInterval]检查有效持续时间标识
      * 符。注意，持续时间是一个固定长度的时间，并不会随日历时间的变化而变化。例如，1 day
      * 总是意味着 86400000 ms，而不是日历日
4.    * @param slideDuration  指定滑动时间窗口的字符串，如“1 分钟”，根据每个
      * slideDuration 生成新的时间窗口，slideDuration 必须小于或等于 windowduration。
      * 使用[org.apache.spark.unsafe.types.CalendarInterval]检查有效持续时间标
      * 识符。这个持续时间也是绝对的，不根据日历变化
5.    * @param startTime 相对于 1970-01-01 00:00:00 UTC 开始偏移窗口的时间间隔。
      * 例如，每小时滚动时间窗口开始时间之后的 15 min，如 12:15-13:15，13:15-14:15…提供
6.    * 开始时间 startTime 为 15 min
7.    *
8.    * @group 日期时间函数
9.    * @since 2.0.0
10.   */
11.  @Experimental 实验性的
12.  @InterfaceStability.Evolving 演进进化
13.  def window(
14.      timeColumn: Column,
15.      windowDuration: String,
16.      slideDuration: String,
17.      startTime: String): Column = {
18.    withExpr {
19.      TimeWindow(timeColumn.expr, windowDuration, slideDuration, startTime)
20.    }.as("window")
21.  }
22.
```

Window 时间窗口函数三应用，如滑动时间窗口每 10s 开始时间的后 5s，每一分钟时间窗口的平均股价，代码如下：

```
1.     * {{{
2.     *   val df = ... // schema => timestamp: TimestampType, stockId:
       *   StringType, price: DoubleType
3.     *  df.groupBy(window($"time", "1 minute", "10 seconds", "5 seconds"),
       *  $"stockId")
4.     *    .agg(mean("price"))
5.     * }}}
```

Window 时间窗口函数三显示如下：

```
1.     *
2.     * {{{
3.     *  09:00:05-09:01:05
4.     *  09:00:15-09:01:15
5.     *  09:00:25-09:01:25 ...
6.     * }}}
7.     *
```

functions.scala 代码中的 API 函数清单见表 14-2。

表 14-2　functions.scala代码中的API函数清单

函 数 名	函 数 功 能
abs(Column e)	计算绝对值
acos(Column e)	计算给定值的反余弦；返回的角度为 0.0～π
acos(String columnName)	计算给定列的反余弦；返回的角度为 0.0～π
add_months(Column startDate, int numMonths)	返回在开始日期之后为 numMonths 的日期
approx_count_distinct(Column e)	聚合函数：返回组中不同记录的近似数量
approx_count_distinct(Column e, double rsd)	聚合函数：返回组中不同记录的近似数量
approx_count_distinct(String columnName)	聚合函数：返回组中不同记录的近似数量
approx_count_distinct(String columnName, double rsd)	聚合函数：返回组中不同记录的近似数量
approxCountDistinct(Column e)	已弃用。可使用 spark 2.1.0 中的 approx_count_distinct 函数
approxCountDistinct(Column e, double rsd)	已弃用。可使用 spark 2.1.0 中的 approx_count_distinct 函数
approxCountDistinct(String columnName)	已弃用。可使用 spark 2.1.0 中的 approx_count_distinct 函数
approxCountDistinct(String columnName, double rsd)	已弃用。可使用 spark 2.1.0 中的 approx_count_distinct 函数
array_contains(column: Column, value: Any): Column	如果数组包含 value，则返回 true
array(cols: Column*): Column	创建一个新的数组列。输入列必须具有相同的数据类型
array(colName: String, colNames: String*): Column	创建一个新的数组列。输入列必须具有相同的数据类型
asc_nulls_first(String columnName)	返回基于列的升序的排序表达式，空值在非空值之前返回
asc_nulls_last(String columnName)	根据列的升序返回排序表达式，空值显示在非空值后面
asc(String columnName)	根据列的升序返回排序表达式
ascii(Column e)	计算字符串列的第一个字符的数值，并将结果作为 int 列返回
asin(Column e)	计算给定值的正弦倒数；返回的角度为–π/2～π/2

续表

函 数 名	函 数 功 能
asin(String columnName)	计算给定值的正弦倒数；返回的角度为-π/2～π/2
atan(Column e)	计算给定值的正切倒数
atan(String columnName)	计算给定列的正切倒数
atan2(Column l, Column r)	返回从直角坐标（x，y）到极坐标（r，θ）的转换角度θ
atan2(Column l, double r)	返回从直角坐标（x，y）到极坐标（r，θ）的转换角度θ
atan2(Column l, String rightName)	返回从直角坐标（x，y）到极坐标（r，θ）的转换角度θ
atan2(double l, Column r)	返回从直角坐标（x，y）到极坐标（r，θ）的转换角度θ
atan2(double l, String rightName)	返回从直角坐标（x，y）到极坐标（r，θ）的转换角度θ
atan2(String leftName, Column r)	返回从直角坐标（x，y）到极坐标（r，θ）的转换角度θ
atan2(String leftName, double r)	返回从直角坐标（x，y）到极坐标（r，θ）的转换角度θ
atan2(String leftName, String rightName)	返回从直角坐标（x，y）到极坐标（r，θ）的转换角度θ
avg(Column e)	聚合函数：返回组中值的平均值
avg(String columnName)	聚合函数：返回组中值的平均值
base64(Column e)	计算二进制列的 BASE64 编码，并将其作为字符串列返回
bin(Column e)	返回给定长度的二进制值的字符串表达式。如 bin("12")返回"1100"
bin(String columnName)	返回给定长度的二进制值的字符串表达式。如 bin("12")返回"1100"
bitwiseNOT(Column e)	位不进行计算
broadcast[T](df: Dataset[T]): Dataset[T]	将 DataFrame 标记为足够小，以在广播中使用 joinKey 进行连接
bround(Column e)	使用 HALF_EVEN 向最接近数字方向舍入的模式返回小数点后 0 位的列值 e
bround(Column e, int scale)	如果 scale 大于或等于 0，用 HALF_EVEN 舍入模式将值舍入 e 为 scale 小数位，如果 scale 小于 0，则舍入为整数部分
callUDF(String udfName, Column... cols)	调用用户定义的函数
callUDF(String udfName, scala.collection.Seq<Column> cols)	调用用户定义的函数
cbrt(Column e)	计算给定值的立方根
cbrt(String columnName)	计算给定列的立方根
ceil(Column e)	计算给定值的上限
ceil(String columnName)	计算给定列的上限
coalesce(Column... e)	返回非空的第一列，如果所有输入为空，则返回 null
coalesce(scala.collection.Seq<Column> e)	返回非空的第一列，如果所有输入为空，则返回 null
col(String colName)	基于给定的列名返回列
collect_list(Column e)	聚合函数：返回具有重复的对象的列表
collect_list(String columnName)	聚合函数：返回具有重复的对象的列表
collect_set(Column e)	聚合函数：返回没有重复元素的对象
collect_set(String columnName)	聚合函数：返回没有重复元素的对象
column(String colName)	基于给定的列名返回列
concat_ws(String sep, Column... exprs)	使用给定的分隔符将多个输入字符串列连接在一起，成为单个字符串列

续表

函 数 名	函 数 功 能
concat_ws(String sep, scala.collection.Seq<Column> exprs)	使用给定的分隔符将多个输入字符串列连接在一起，成为单个字符串列
concat(Column... exprs)	将多个输入字符串列连接在一起，成为单个字符串列
concat(scala.collection.Seq<Column> exprs)	将多个输入字符串列连接在一起，成为单个字符串列
conv(Column num, int fromBase, int toBase)	将字符串列中的数字从一个基数转换为另一个基数
corr(Column column1, Column column2)	聚合函数：返回两列的皮尔森相关系数
corr(String columnName1, String columnName2)	聚合函数：返回两列的皮尔森相关系数
cos(Column e)	计算给定值的余弦值
cos(String columnName)	计算给定列的余弦
cosh(Column e)	计算给定值的双曲余弦值
cosh(String columnName)	计算给定列的双曲余弦值
count(Column e)	聚合函数：返回组中的记录数
count(columnName: String): TypedColumn[Any, Long]	聚合函数：返回组中的记录数
countDistinct(Column expr, Column... exprs)	聚合函数：返回组中不同记录的数量
countDistinct(Column expr, scala.collection.Seq<Column> exprs)	聚合函数：返回组中不同记录的数量
countDistinct(String columnName, scala.collection.Seq<String> columnNames)	聚合函数：返回组中不同记录的数量
countDistinct(String columnName, String... columnNames)	聚合函数：返回组中不同记录的数量
covar_pop(Column column1, Column column2)	聚合函数：返回两列的总体协方差
covar_pop(String columnName1, String columnName2)	聚合函数：返回两列的总体协方差
covar_samp(Column column1, Column column2)	聚合函数：返回两列的样本协方差
covar_samp(String columnName1, String columnName2)	聚合函数：返回两列的样本协方差
crc32(Column e)	计算二进制列的循环冗余校验值（CRC32），并将该值作为 bigint 返回
cume_dist()	窗口函数：返回窗口分区内的累积值分布，如在当前行下的行分片
current_date()	将当前日期作为日期列返回
current_timestamp()	将当前时间戳返回为时间戳列
date_add(Column start, int days)	返回是开始日期 days 之后的日期
date_format(Column dateExpr, String format)	将日期/时间戳/字符串转换为由第二个参数指定的日期格式的字符串值
date_sub(Column start, int days)	返回是开始日期 days 之前的日期
datediff(Column end, Column start)	返回从开始日期到终止日期的天数
dayofmonth(Column e)	从指定的日期/时间戳/字符串中提取一个月中的某一天为整数
dayofyear(Column e)	从给定的日期/时间戳/字符串中提取一年中的日期作为整数

续表

函 数 名	函 数 功 能
decode(Column value, String charset)	使用提供的字符集（'US-ASCII','ISO-8859-1','UTF-8', 'UTF-16BE','UTF-16LE','UTF-16'），将第一个参数从二进制计算为字符串
degrees(Column e)	将以弧度测量的角度转换为以度为单位测量的近似等效角度
degrees(String columnName)	将以弧度测量的角度转换为以度为单位测量的近似等效角度
dense_rank()	窗口函数：返回窗口分区中的行的排名，排名无间隔。密集排名和排名的区别：例如，第一名有 1 人，第二名并列有 3 人，那么在密集排名中，下一个人的排名是第三名；而在一般排名中，下一个人的排名是第五名
desc_nulls_first(String columnName)	返回基于列的降序的排序表达式，空值出现在非空值之前
desc_nulls_last(String columnName)	根据列的降序返回排序表达式，空值显示在非空值后面
desc(String columnName)	根据列的降序返回排序表达式
encode(Column value, String charset)	使用提供的字符集（'US-ASCII','ISO-8859-1','UTF-8', 'UTF-16BE', 'UTF-16LE','UTF-16'），将第一个参数从字符串转为二进制数据
exp(Column e)	计算给定值的指数
exp(String columnName)	计算给定列的指数
explode(Column e)	为给定数组或映射列中的每个元素创建一个新行
expm1(Column e)	计算给定值的指数减 1
expm1(String columnName)	计算给定列的指数
expr(String expr)	将表达式字符串解析到它所代表的列中，类似于 DataFrame.selectExpr
factorial(Column e)	计算给定值的阶乘
first(Column e)	聚合函数：返回组中的第一个值
first(Column e, boolean ignoreNulls)	聚合函数：返回组中的第一个值
first(String columnName)	聚合函数：返回组中列的第一个值
first(String columnName, boolean ignoreNulls)	聚合函数：返回组中列的第一个值
floor(Column e)	计算给定值的下限
floor(String columnName)	计算给定值的下限
format_number(Column x, int d)	将数字列 x 格式化为类似“#,###,###.##”的格式，四舍五入为 d 个小数位，并将结果作为字符串列返回
format_string(String format, Column... arguments)	格式化 printf 风格的参数，并将结果作为字符串列返回
format_string(String format, scala.collection.Seq<Column> arguments)	格式化 printf 风格的参数，并将结果作为字符串列返回
from_json(Column e, String schema, java.util.Map<String,String> options)	使用指定的模式将包含 JSON 字符串的列解析为 StructType 元素
from_json(Column e, StructType schema)	使用指定的模式将包含 JSON 字符串的列解析为 StructType 元素
from_json(Column e, StructType schema, scala.collection.immutable.Map<String,String> options)	（适用于 Scala）将包含 JSON 字符串的列解析为 StructType 具有指定模式的列

续表

函 数 名	函 数 功 能
from_json(Column e, StructType schema, java.util.Map<String,String> options)	（适用于 Java）将包含 JSON 字符串的列解析为 StructType 具有指定模式的字符串
from_unixtime(Column ut)	将从 UNIX 纪元（1970-01-01 00:00:00 UTC）的秒数转换为表示当前系统时区中给定格式的时间字符串
from_unixtime(Column ut, String f)	将从 UNIX 纪元（1970-01-01 00:00:00 UTC）的秒数转换为表示当前系统时区中给定格式的时间字符串
from_utc_timestamp(Column ts, String tz)	根据 UTC 中特定时间对应的时间戳，返回与给定时区中的相同时间对应的时间戳
get_json_object(Column e, String path)	从基于指定路径的 JSON 字符串中提取 JSON 对象，并返回提取的 JSON 对象的 JSON 字符串
greatest(Column... exprs)	返回值列表中最大的值，跳过空值
greatest(scala.collection.Seq<Column> exprs)	返回值列表中最大的值，跳过空值
greatest(String columnName, scala.collection.Seq<String> columnNames)	返回列名列表中最大的值，跳过空值
greatest(String columnName, String... columnNames)	返回列名列表中最大的值，跳过空值
grouping_id(scala.collection.Seq<Column> cols)	聚合函数：返回分组的级别，等同于(grouping(c1) <<; (n–1)) + (grouping(c2) <<; (n–2)) + … + grouping(cn)
grouping_id(String colName, scala.collection.Seq<String> colNames)	聚合函数：返回分组的级别，等同于(grouping(c1) <<; (n–1)) + (grouping(c2) <<; (n–2)) + … + grouping(cn)
grouping(Column e)	聚合函数：指示 GROUP BY 列表中的指定列是否已聚合，在结果集中返回 1 表示聚合，返回 0 表示未聚合
grouping(String columnName)	聚合函数：指示 GROUP BY 列表中的指定列是否已聚合，在结果集中返回 1 表示聚合，返回 0 表示未聚合
hash(Column... cols)	计算给定列的哈希码，并将结果作为 int 列返回
hash(scala.collection.Seq<Column> cols)	计算给定列的哈希码，并将结果作为 int 列返回
hex(Column column)	计算给定列的十六进制值
hour(Column e)	从给定的 date/timestamp/string 中提取小时作为整数
hypot(Column l, Column r)	计算 sqrt(a^2^ + b^2^)的值，无中间溢出或下溢
hypot(Column l, double r)	计算 sqrt(a^2^ + b^2^)的值，无中间溢出或下溢
hypot(Column l, String rightName)	计算 sqrt(a^2^ + b^2^)的值，无中间溢出或下溢
hypot(double l, Column r)	计算 sqrt(a^2^ + b^2^)的值，无中间溢出或下溢
hypot(double l, String rightName)	计算 sqrt(a^2^ + b^2^)的值，无中间溢出或下溢
hypot(String leftName, Column r)	计算 sqrt(a^2^ + b^2^)的值，无中间溢出或下溢
hypot(String leftName, double r)	计算 sqrt(a^2^ + b^2^)的值，无中间溢出或下溢
hypot(String leftName, String rightName)	计算'sqrt(a^2^ + b^2^)'的值，无中间溢出或下溢
initcap(Column e)	通过将每个单词的第一个字母转换为大写，返回一个新的字符串列。例如，将 hello world 转化为 Hello World
input_file_name()	为当前 Spark 任务的文件名创建一个字符串列
instr(Column str, String substring)	找到给定字符串中第一次出现截取部分字符串的位置
isnan(Column e)	如果列为非数字值的特殊值 NaN，则返回 true
isnull(Column e)	如果列为空，则返回 true

续表

函　数　名	函 数 功 能
json_tuple(Column json, scala.collection.Seq<String> fields)	根据给定的字段名称为 JSON 列创建一个新行
json_tuple(Column json, String... fields)	根据给定的字段名称为 JSON 列创建一个新行
kurtosis(Column e)	聚合函数：入参为列，返回组中值的峰度
kurtosis(String columnName)	聚合函数：入参为字符串，返回组中值的峰度
lag(Column e, int offset)	窗口函数：返回当前行 offset 之前的行的值，如果当前行之前的行少于 offset 行，则返回 null 值
lag(Column e, int offset, Object defaultValue)	窗口函数：返回当前行 offset 之前的行的值，如果当前行之前的行少于 offset 行，则返回 null 值
lag(String columnName, int offset)	窗口函数：返回当前行 offset 之前的行的值，如果当前行之前的行少于 offset 行，则返回 null 值
lag(String columnName, int offset, Object defaultValue)	窗口函数：返回当前行 offset 之前的行的值，如果当前行之前的行少于 offset 行，则返回 null 值
last_day(Column e)	给定日期列，返回给定日期所属的月份的最后一天。例如，输入“2015-07-27”，返回“2015-07-31”，7 月 31 日是 2015 年 7 月的最后一天
last(Column e)	聚合函数：返回组中的最后一个值
last(Column e, boolean ignoreNulls)	聚合函数：返回组中的最后一个值
last(String columnName)	聚合函数：返回组中列的最后一个值
last(String columnName, boolean ignoreNulls)	聚合函数：返回组中列的最后一个值
lead(Column e, int offset)	窗口函数：返回当前行 offset 之后的行的值，如果当前行之后的行少于 offset 行，则返回 null 值
lead(Column e, int offset, Object defaultValue)	窗口函数：返回当前行 offset 之后的行的值，如果当前行之后的行少于 offset 行，则返回 null 值
lead(String columnName, int offset)	窗口函数：返回当前行 offset 之后的行的值，如果当前行之后的行少于 offset 行，则返回 null 值
lead(String columnName, int offset, Object defaultValue)	窗口函数：返回当前行 offset 之后的行的值，如果当前行之后的行少于 offset 行，则返回 null 值
least(Column... exprs)	返回值列表的最小值，跳过空值
least(scala.collection.Seq<Column> exprs)	返回值列表的最小值，跳过空值
least(String columnName, scala.collection.Seq<String> columnNames)	返回同一行中多个列的最小值，跳过空值
least(String columnName, String... columnNames)	返回同一行中多个列的最小值，跳过空值
length(Column e)	计算给定字符串或二进制列的长度
levenshtein(Column l, Column r)	计算两个给定字符串列的编辑距离。含义为两个字符串之间，由一个字符串转换成另一个字符串所需的最少编辑操作次数
lit(Object literal)	创建[Column]的字面量值
locate(String substr, Column str)	找到第一次出现的截取字符串的位置
locate(String substr, Column str, int pos)	在位置 pos 后的字符串列中找到截取字符串的第一个位置
log(Column e)	计算给定值的自然对数
log(double base, Column a)	返回第二个参数的基于第一个参数的对数
log(double base, String columnName)	返回第二个参数的基于第一个参数的对数

续表

函　数　名	函 数 功 能
log(String columnName)	计算给定列的自然对数
log10(Column e)	计算以 10 为底的给定值的对数
log10(String columnName)	计算以 10 为底的给定值的对数
log1p(Column e)	计算给定值的自然对数加一
log1p(String columnName)	计算给定列的自然对数加一
log2(Column expr)	计算以 2 为底的给定列的对数
log2(String columnName)	计算以 2 为底的给定值的对数
lower(Column e)	将字符串列转换为小写
lpad(Column str, int len, String pad)	Left-pad 字符串列
ltrim(Column e)	修剪指定字符串值的左端空格
map(Column... cols)	将输入的 key-value 键值对转换为新的列
map(scala.collection.Seq<Column> cols)	将输入的 key-value 键值对转换为新的列
max(Column e)	聚合函数：返回组中表达式的最大值
max(String columnName)	聚合函数：返回组中列的最大值
md5(Column e)	计算二进制列的 MD5 摘要，并将该值作为 32 个字符的十六进制字符串返回
mean(Column e)	聚合函数：返回组中值的平均值
mean(String columnName)	聚合函数：返回组中值的平均值
min(Column e)	聚合函数：返回组中表达式的最小值
min(String columnName)	聚合函数：返回组中列的最小值
minute(Column e)	从给定的日期/时间戳/字符串中提取分钟作为整数
monotonically_increasing_id()	生成单调递增的 64 位整数的列表达式
monotonicallyIncreasingId()	已弃用
month(Column e)	从给定的日期/时间戳/字符串中提取月份作为整数
months_between(Column date1, Column date2)	返回从日期 date1 到日期 date2 之间的月数
nanvl(Column col1, Column col2)	如果 col1 不是 NaN，则返回 col1;如果 col1 为 NaN，则返回 col2
negate(Column e)	取相反数
next_day(Column date, String dayOfWeek)	给定日期列，返回比指定的星期列的值晚的第一个日期。例如，next_day('2015-07-27', "Sunday")返回 2015-08-02，因为 2015-08-02 是 2015-07-27 之后的第一个星期天
not(Column e)	布尔表达式的取反
ntile(int n)	窗口函数：在一个有序的窗口分区返回 ntile 组 ID（从 1 到 n）。例如，如果 n 是 4，第一部分的行将得到值 1，第二部分的行将获得值 2，第三部分将获得值 3，最后一部分将获得值 4
percent_rank()	窗口函数：返回相对排名等同于 SQL 的 percent_rank 功能
pmod(Column dividend, Column divisor)	返回被除数 mod 除数的正值
posexplode(Column e)	为给定数组或映射列中具有位置的每个元素创建一个新行
pow(Column l, Column r)	返回第一个参数的值增加到第二个参数的幂
pow(Column l, double r)	返回第一个参数的值增加到第二个参数的幂
pow(Column l, String rightName)	返回第一个参数的值增加到第二个参数的幂

续表

函 数 名	函 数 功 能
pow(double l, Column r)	返回第一个参数的值增加到第二个参数的幂
pow(double l, String rightName)	返回第一个参数的值增加到第二个参数的幂
pow(String leftName, Column r)	返回第一个参数的值增加到第二个参数的幂
pow(String leftName, double r)	返回第一个参数的值增加到第二个参数的幂
pow(String leftName, String rightName)	返回第一个参数的值增加到第二个参数的幂
quarter(Column e)	从给定的日期/时间戳/字符串中提取季度作为整数
radians(Column e)	将以度为单位测量的角度转换为以弧度为单位测量的近似等效角度
radians(String columnName)	将以度为单位测量的角度转换为以弧度为单位测量的近似等效角度
rand()	从 U[0.0, 1.0]生成独立相同分布样本的随机列
rand(long seed)	从 U[0.0, 1.0]生成独立相同分布样本的随机列
randn()	从标准正态分布生成独立相同分布样本列
randn(long seed)	从标准正态分布生成独立相同分布样本列
rank()	窗口函数：返回窗口分区中的行的排名
regexp_extract(Column e, String exp, int groupIdx)	从指定的字符串列中抽取由 Java 正则表达式匹配的特定组
regexp_replace(Column e, Column pattern, Column replacement)	替换正则表达式匹配的指定的字符串值的所有值
regexp_replace(Column e, String pattern, String replacement)	替换正则表达式匹配的指定的字符串值的所有值
repeat(Column str, int n)	重复一个字符串列 *n* 次，并返回一个新的字符串列
reverse(Column str)	反转字符串列，并将其作为新的字符串列返回
rint(Column e)	返回一个与参数最接近的 double 值，其等于一个整数
rint(String columnName)	返回一个与参数最接近的 double 值，其等于一个整数
round(Column e)	返回 e 四舍五入为小数点后 0 位的列的值
round(Column e, int scale)	如果大于或等于 0，则将值四舍五入 e 到 scale 小数位，如果 scale 小于 0，则舍入为整数部分
row_number()	窗口函数：返回在窗口分区中从 1 开始的序列号
rpad(Column str, int len, String pad)	将长度为 len 的列进行右填充
rtrim(Column e)	修剪指定字符串值的右端的空格
second(Column e)	从给定的日期/时间戳/字符串中提取秒数作为整数
sha1(Column e)	计算二进制列的 SHA-1 摘要，并将该值作为 40 个字符的十六进制字符串返回
sha2(Column e, int numBits)	计算二进制列的哈希函数的 SHA-2 系列，并将该值作为十六进制字符串返回
shiftLeft(Column e, int numBits)	将给定值左移 numBits
shiftRight(Column e, int numBits)	将给定值右移 numBits
shiftRightUnsigned(Column e, int numBits)	无符号移位将给定值右移 numBits
signum(Column e)	计算给定值的符号
signum(String columnName)	计算给定列的符号
sin(Column e)	计算给定值的正弦值

续表

函 数 名	函 数 功 能
sin(String columnName)	计算给定列的正弦
sinh(Column e)	计算给定值的双曲正弦值
sinh(String columnName)	计算给定列的双曲正弦值
size(Column e)	返回数组或map的长度
skewness(Column e)	聚合函数：返回组中值的偏度
skewness(String columnName)	聚合函数：返回组中值的偏度
sort_array(Column e)	根据数组元素的自然排序，按升序对给定列的输入数组进行排序
sort_array(Column e, boolean asc)	根据数组元素的自然排序，按升序或降序对给定列的输入数组进行排序
soundex(Column e)	返回指定表达式的soundex编码
spark_partition_id()	返回分区ID
split(Column str, String pattern)	根据正则表达式进行切分
sqrt(Column e)	计算指定浮点值的平方根
sqrt(String colName)	计算指定浮点值的平方根
stddev_pop(Column e)	聚合函数：返回组中表达式的总体标准偏差
stddev_pop(String columnName)	聚合函数：返回组中表达式的总体标准偏差
stddev_samp(Column e)	聚合函数：返回组中表达式的样本标准偏差
stddev_samp(String columnName)	聚合函数：返回组中表达式的样本标准偏差
stddev(Column e)	聚合函数：[stddev_samp]函数的别名
stddev(String columnName)	聚合函数：[stddev_samp]函数的别名
struct(Column... cols)	创建一个新的结构列
struct(scala.collection.Seq<Column> cols)	创建一个新的结构列
struct(String colName, scala.collection.Seq<String> colNames)	创建组成多个输入列的新结构列
struct(String colName, String... colNames)	创建组成多个输入列的新结构列
substring_index(Column str, String delim, int count)	返回按分隔符从字符串中截取计数长度的子字符串。如果计数长度是正数，则分隔符左侧截取返回；如果计数长度是负数，从分隔符右侧返回。匹配区分大小写
substring(Column str, int pos, int len)	当str是String类型，返回以字节开始的字节数组的片段；当str是二进制类型，返回从pos开始，长度是len的子串
sum(Column e)	聚合函数：返回表达式中所有值的总和
sum(String columnName)	聚合函数：返回给定列中所有值的总和
sumDistinct(Column e)	聚合函数：返回表达式中不同值的总和
sumDistinct(String columnName)	聚合函数：返回表达式中不同值的总和
tan(Column e)	计算给定值的正切值
tan(String columnName)	计算给定列的正切值
tanh(Column e)	计算给定值的双曲正切值
tanh(String columnName)	计算给定列的双曲正切值
to_date(Column e)	将列转换为日期类型
to_json(Column e)	将StructType格式的列转换为指定模式的JSON字符串

续表

函 数 名	函 数 功 能
to_json(Column e, scala.collection.immutable.Map<String,String> options)	（适用于 Scala）将 StructType 格式的列转换为指定模式的 JSON 字符串
to_json(Column e, java.util.Map<String, String> options)	（适用于 Java）将 StructType 格式的列转换为指定模式的 JSON 字符串
to_utc_timestamp(Column ts, String tz)	给定时间戳，其对应于给定时区中的一天中的特定时间，返回对应于 UTC 中的同一时刻的另一时间戳
toDegrees(Column e)	已弃用
toDegrees(String columnName)	已弃用
translate(Column src, String matchingString, String replaceString)	当字符串中的字符与 matchingString 匹配时，将源字符串使用 replaceString 替换掉匹配的字符串
trim(Column e)	修剪指定字符串列的两端的空格
trunc(Column date, String format)	返回截断到由格式指定的单位的日期
static <RT> UserDefinedFunction	将用户定义的 0 个参数的函数定义为用户定义函数（UDF）
static <RT,A1> UserDefinedFunction	定义 1 个参数的用户定义函数作为用户定义函数（UDF）
static <RT,A1,A2> UserDefinedFunction	定义 2 个参数的用户定义函数作为用户定义函数（UDF）
static <RT,A1,A2,A3> UserDefinedFunction	定义 3 个参数的用户定义函数作为用户定义函数（UDF）
static <RT,A1,A2,A3,A4> UserDefinedFunction	定义 4 个参数的用户定义函数作为用户定义函数（UDF）
static <RT,A1,A2,A3,A4,A5> UserDefinedFunction	定义 5 个参数的用户定义函数作为用户定义函数（UDF）
static <RT,A1,A2,A3,A4,A5,A6> UserDefinedFunction	定义 6 个参数的用户定义函数作为用户定义函数（UDF）
static <RT,A1,A2,A3,A4,A5,A6,A7> UserDefinedFunction	定义 7 个参数的用户定义函数作为用户定义函数（UDF）
static <RT,A1,A2,A3,A4,A5,A6,A7,A8> UserDefinedFunction	定义 8 个参数的用户定义函数作为用户定义函数（UDF）
static <RT,A1,A2,A3,A4,A5,A6,A7,A8,A9> UserDefinedFunction	定义 9 个参数的用户定义函数作为用户定义函数（UDF）
static <RT,A1,A2,A3,A4,A5,A6,A7,A8,A9,A10> UserDefinedFunction	定义 10 个参数的用户定义函数作为用户定义函数（UDF）
unbase64(Column e)	解码 BASE64 编码的字符串列，并将其作为二进制列返回
unhex(Column column)	十六进制取反
unix_timestamp()	获取当前 UNIX 时间戳（以秒为单位）
unix_timestamp(Column s)	将格式为 yyyy-MM-dd HH：mm：ss 的时间字符串转换为 UNIX 时间戳（以秒为单位），使用默认时区和默认语言环境，如果失败，则返回 null
unix_timestamp(Column s, String p)	将给定模式的时间字符串（请参见[http://docs.oracle.com/javase/tutorial/i18n/format/simpleDateFormat.html]）转换为 UNIX 时间戳（以秒为单位），如果失败，则返回 null
upper(Column e)	将字符串列转换为大写
var_pop(Column e)	聚合函数：返回组中值的总体方差
var_pop(String columnName)	聚合函数：返回组中值的总体方差
var_samp(Column e)	聚合函数：返回组中值的无偏方差
var_samp(String columnName)	聚合函数：返回组中值的无偏方差

续表

函 数 名	函 数 功 能
variance(Column e)	聚合函数：[var_samp]函数的别名
variance(String columnName)	聚合函数：[var_samp]函数的别名
weekofyear(Column e)	从给定的日期/时间戳/字符串中提取星期为整数
when(Column condition, Object value)	评估条件列表，并返回多个可能的结果表达式之一。例如 people.select(when(people("gender") === "male", 0) .when(people("gender") === "female", 1) .otherwise(2))
window(Column timeColumn, String windowDuration)	给定一个时间戳指定列，生成滚动时间窗口
window(Column timeColumn, String windowDuration, String slideDuration)	给定一个时间戳指定列，生成滚动时间窗口
window(Column timeColumn, String windowDuration, String slideDuration, String startTime)	给定一个时间戳指定列，生成滚动时间窗口
year(Column e)	从给定的日期/时间戳/字符串中提取年为整数

Spark 2.2.0 版本的 functions.scala 新增的函数见表 14-3。

表 14-3　Spark 2.2.0 版本中的functions.scala 源码新增的函数

函 数 名	函 数 功 能
typedLit(T literal, scala.reflect.api.TypeTags.TypeTag<T> evidence$1)	创建一个列的字面量值
to_timestamp(Column s, String fmt)	以指定格式将时间字符串转换为 UNIX 时间戳（以秒为单位）
to_timestamp(Column s)	将时间字符串转换为 UNIX 时间戳（以秒为单位）
to_date(Column e, String fmt)	将列转换成一个特定格式的日期类型
explode_outer(Column e)	为给定数组或映射列中的每个元素创建新行。与 explode 不同，如果数组或 map 为 null 或空，则生成 null
posexplode_outer(Column e)	使用给定数组或映射列中的位置为每个元素创建一个新行。与 posexplode 不同，如果数组或 map 为 null 或空，则生成 row (null, null)
from_json(Column e, DataType schema, scala.collection.immutable.Map<String,String> options)	（适用于 Scala）解析列将含有一个 JSON 字符串转换为 StructType 或数组类型指定模式的 StructTypes
from_json(Column e, DataType schema, java.util.Map<String,String> options)	（适用于 JAVA）解析列将含有一个 JSON 字符串转换为 StructType 或数组类型指定模式的 StructTypes
from_json(Column e, DataType schema)	解析列将含有一个JSON字符串转换为StructType或数组类型指定模式的 StructTypes

Spark 2.3.0 版本的 functions.scala 新增的函数见表 14-4。

表 14-4　Spark 2.3.0 版本中的functions.scala 源码新增的函数

函 数 名	函 数 功 能
ltrim(e: Column, trimString: String)	从指定字符串列的左端修剪指定的字符串
rtrim(e: Column, trimString: String)	从指定字符串列的右端修剪指定的字符串
trim(e: Column, trimString: String)	从指定字符串列的两端修剪指定字符
dayofweek(e: Column)	从给定日期/时间戳/字符串中提取一周中的一天作为整数

续表

函 数 名	函 数 功 能
date_trunc(format: String, timestamp: Column)	返回截断为格式中所指定单位的时间戳。例如，date_tunc("2018-11-19 12:01:19"，"year")返回 2018-01-01 00:00:00
from_json(e: Column, schema: String, options: Map[String, String])	（适用于 Scala）将包含 JSON 字符串的列解析为带有 StringType 的 MapType 作为键类型，"StructType"或"ArrayType"使用指定的方案。如果是不可解析的字符串，则返回"null"
map_keys(e: Column)	返回包含映射键的无序数组
map_values(e: Column)	返回包含映射值的无序数组
udf(f: UDF0[_], returnType: DataType) …… udf(f:UDF10[_,…..._],returnType: DataType)	分别将 Java UDF0～Java UDF10 的实例定义为用户定义函数（UDF）。调用方必须指定输出数据类型，并且不存在自动输入类型强制。默认情况下，返回的 UDF 是确定性的。要将其更改为不确定的，请调用 API 函数 UserDefinedFunction.asUndeterministic（）

Spark 2.4.0 版本的 functions.scala 新增的函数见表 14-5。

表 14-5 Spark 2.4.0 版本的functions.scala源码新增的函数

函 数 名	函 数 功 能
map_from_arrays(keys: Column, values: Column)	创建新的映射列。第一列中的数组用于键。第二列中的数组用于值。键的数组中的所有元素都不应为空
months_between(end: Column, start: Column, roundOff: Boolean)	返回日期 end 和 start 之间的月数。如果 roundOff 设置为 true，则结果四舍五入为 8 位；否则不进行四舍五入
from_utc_timestamp(ts: Column, tz: Column)	给定时间戳，如 2017-07-14 02:40:00.0，将其解释为 UTC 格式的时间，并呈现时间作为给定时区中的时间戳。例如，GMT+1 将产生 2017-07-14 03:40:00.0
to_utc_timestamp(ts: Column, tz: Column)	转换为时间戳
arrays_overlap(a1: Column, a2: Column)	如果 a1 和 a2 至少有一个共同的非空元素，则返回 true。如果数组不是空，任何数组都包含一个 null，返回 null。否则返回为 false
slice(x: Column, start: Int, length: Int)	返回一个数组，该数组包含从索引 start 开始的 x 中的所有元素（如果 start 为负数，则从结尾开始）以及指定的 length
array_join(column: Column, delimiter: String, nullReplacement: String)	使用 delimiter 连接 column 的元素。空值使用 nullReplacement 替换
array_join(column: Column, delimiter: String)	使用 delimiter 连接 column 的元素
array_position(column: Column, value: Any)	将值在给定数组中第一次出现的位置返回，如果其中一个参数为 null，则返回 null
element_at(column: Column, value: Any)	如果列是数组，则返回值中给定索引处的数组元素。如果列是映射，则返回值中给定键的值
array_sort(e: Column)	按升序排列输入数组。输入数组的元素必须是可排序的。空元素将放在返回数组的末尾
array_remove(column:Column, element: Any)	从给定数组中移除与元素相等的所有元素
array_distinct(e: Column)	从数组中删除重复值
array_intersect(col1:Column,col2: Column)	返回给定两个数组相交处的元素数组，不重复
array_union(col1: Column, col2: Column)	返回给定两个数组的并集中元素的数组，不重复
array_except(col1:Column,col2: Column)	返回第一个数组（不是第二个数组）中元素的数组，但不返回重复项，结果中元素的顺序未确定

续表

函　数　名	函 数 功 能
from_json(e: Column, schema: Column)	（适用于 Scala）将包含 JSON 字符串的列解析为带有 StringType 的 MapType 作为键类型，“StructType”或“ArrayType”使用指定的方案。如果是不可解析的字符串，则返回“null”
from_json(e: Column, schema: Column, options: java.util.Map[String, String])	（适用于 Scala）将包含 JSON 字符串的列解析为带有 StringType 的 MapType
schema_of_json(json: String)	解析 JSON 字符串并以 DDL 格式推断其格式
schema_of_json(json: Column)	解析 JSON 字符串并以 DDL 格式推断其格式
array_min(e: Column)	返回数组中的最小值
array_max(e: Column)	返回数组中的最大值
shuffle(e: Column)	返回给定数组的随机排列
flatten(e: Column)	从数组中创建单个数组。如果嵌套数组的结构深度超过两层，则只删除一层的嵌套
sequence(start: Column, stop: Column, step: Column)	从开始到结束生成一个整数序列，逐步递增
sequence(start: Column, stop: Column)	从开始到结束生成一个整数序列，如果 start 小于或等于 stop，则递增 1，否则为–1
array_repeat(left: Column, right: Column)	创建包含左参数的数组，重复右参数给定的次数
array_repeat(e: Column, count: Int)	创建包含左参数的数组，重复右参数给定的次数
map_from_entries(e: Column)	返回从给定项数组创建的映射
arrays_zip(e: Column*)	返回结构的合并数组，其中第 N 个结构包含输入数组的所有第 N 个值
map_concat(cols: Column*)	返回所有给定映射的并集

14.4.2　电商交互式分析系统应用案例完整代码

1. 电商交互式分析系统应用案例代码

Spark 商业案例之电商交互式分析系统应用案例完整代码 EB_Users_Analyzer_DateSet.scala，如例 14-1 所示。

【例 14-1】　EB_Users_Analyzer_DateSet.scala 代码。

```
1.   package com.dt.spark.sparksql
2.
3.  import org.apache.log4j.{Level, Logger}
4.  import org.apache.spark.SparkConf
5.  import org.apache.spark.sql.types.{DoubleType, StringType, StructField,
    StructType}
6.  import org.apache.spark.sql.{Row, SparkSession}
7.
8.
9.  /**
10.   * 版权：DT 大数据梦工厂所有
11.   * 时间：2017 年 1 月 21 日；
12.   * 电商用户行为分析系统：在实际生产环境下，一般都是以 J2EE+Hadoop+Spark+DB
    * (Redis)的方式实现的综合技术栈，使用 Spark 进行电商用户行为分析时一般都都会是交
    * 互式的，什么是交互式的？
```

```
13.   * 例如，营销部门人员按照特定时间查询访问次数最多的用户，或查询购买金额最大的前 TopN
      * 个用户
14.   * 这些分析结果对于公司的决策、产品研发和营销都至关重要，而且很多时候是立即想要结果
      * 的，如果此时使用 Hive 去实现，可能非常缓慢（如需 1h），而在电商类企业中，经过深度
      * 调优后的 Spark 一般都会比 Hive 快 5 倍以上，此时的运行时间可能就是分钟级别，这时就可以
      * 达到即查即用的目的，也就是所谓的交互式，而交互式的大数据系统是未来的主流！
15.   * 我们在这里是分析电商用户的多维度的行为特征，例如，分析特定时间段访问人数的 TopN、
      * 特定时间段购买金额排名的 TopN、注册后一周内购买金额排名 TopN、注册后一周内访问次
      * 数排名 TopN 等，但是这里的技术和业务场景同样适合于门户网站（如网易、新浪）等，也
      * 同样适合于在线教育系统（如分析在线教育系统的学员的行为），当然也适用于 SNS 社交
      * 网络系统，如对于婚恋网，我们可以通过这几节课讲的内容来分析最匹配的 Couple，再如，
      * 我们可以分析每周婚恋网站访问次数 TopN，这时就可以分析出迫切想找到对象的人，婚恋
      * 网站可以基于这些分析结果进行更精准和更有效（更挣钱）的服务
16.   *
17.   *
18.   * 具体数据结构如下所示：
19.   * User
20.      |-- name: string (nullable = true)
21.      |-- registeredTime: string (nullable = true)
22.      |-- userID: long (nullable = true)
23.
24.     Log
25.      |-- consumed: double (nullable = true)
26.      |-- logID: long (nullable = true)
27.      |-- time: string (nullable = true)
28.      |-- typed: long (nullable = true)
29.      |-- userID: long (nullable = true)
30.   *
31.   * 注意：
32.   *   1. 在实际生产环境下，要么是 Spark SQL+Parquet 的方式，要么是 Spark SQL+Hive
33.   *   2. functions.scala 文件包含了大量的内置函数，尤其在 agg 中会广泛使用，请反复
      *   阅读该源码并进行实践
34.   */
35. object EB_Users_Analyzer_DateSet {
36.   case class UserLog(logID: Long, userID: Long, time: String, typed: Long,
      consumed: Double)
37.   case class LogOnce(logID: Long, userID: Long, count: Long)
38.   case class ConsumedOnce(logID: Long, userID: Long, consumed: Double)
39.
40.   def main(args: Array[String]){
41.
42.
43.
44.
45.     Logger.getLogger("org").setLevel(Level.ERROR)
46.
47.     var masterUrl = "local[8]"//默认程序运行在本地 Local 模式中，主要用于学习和测试
48.
49.     /**
          * 当我们把程序打包运行在集群上的时候，一般都会传入集群的 URL 信息，这里我们假设传入
          * 参数，第一个参数只传入 Spark 集群的 URL，第二个参数传入的是数据的地址信息
50.       */
51.     if(args.length > 0) {
```

```
52.        masterUrl = args(0)
53.      }
54.
55.
56.      /**
         * 创建 Spark 会话上下文 SparkSession 和集群上下文 SparkContext，在 SparkConf
         * 中可以进行各种依赖和参数的设置等，大家可以通过 SparkSubmit 脚本的 help 去查
         * 看设置信息，其中 SparkSession 统一了 Spark SQL 运行的不同环境
57.        */
58.      val  sparkConf  =  new  SparkConf().setMaster(masterUrl).setAppName
         ("EB_Users_Analyzer_DateSet")
59.
60.      /**
         * SparkSession 统一了 Spark SQL 执行时的不同的上下文环境，也就是说，Spark SQL
         * 无论运行在哪种环境下，我们都可以只使用 SparkSession 这样一个统一的编程入口
         * 来处理 DataFrame 和 DataSet 编程，不需要关注底层是否有 Hive 等
61.        */
62.      val spark = SparkSession
63.        .builder()
64.        .config(sparkConf)
65.        .getOrCreate()
66.
67.      val sc = spark.sparkContext
                 //从 SparkSession 获得的上下文，这是因为我们读原生文件的时候或者实现一
                 //些 Spark SQL 目前还不支持的功能的时候需要使用 SparkContext
68.
69.      import org.apache.spark.sql.functions._
         //2017 年 1 月 21 的第一个作业：通读 functions.scala 的源码
70.      import spark.implicits._
71.
72.      //2017 年 1 月 21 的第二个作业：根据电商业务分析需要用户编写的数据，需要注意的是，
         //任何实际生产环境的系统都不止一个数据文件或者不止一张表
73.      //例如，这里的电商用户行为分析系统肯定至少有用户的信息 usersInfo，同时肯定至少
         //有用户访问行为信息 usersAccessLog
74.
75.      /**
76.        * 功能一：特定时段内用户访问电商网站排名 TopN：
77.        *    第一个问题：特定时段中的时间是从哪里来的？一般都来自于 J2EE 调度系统，例
           *    如一个营销人员通过系统传入了 2017.01.01~2017.01.10；
78.        *    第二个问题：计算的时候，我们会使用哪些核心算子？如 Join、groupBy、agg 等
           *    算子，在 agg 算子中可使用大量的 functions scala 函数，functions 的内置函
           *    数有助于快速实现业务计算；
79.        *    第三个问题：计算完成后，数据保存在哪里？现在生产环境下是保存在 DB、HBase/Canssandra、
           *    Redis 等
80.        *
81.        */
82.
83.      /**
         * 读取用户行文数据，建议使用 parquet 的方式
84.        */
85.      val userInfo = spark.read.format("parquet")
86.        .parquet("data/sql/userparquet.parquet")
87.      val userLog = spark.read.format("parquet")
88.        .parquet("data/sql/logparquet.parquet")
89.
```

```
90.
91.     /**
       * 统计特定时段访问次数最多的 Top5: 例如 2016-10-01 ~ 2016-11-01
92.    */
93.     val startTime =  "2016-10-01"
94.     val endTime = "2016-11-01"
95.
96.
97.     println("统计特定时段访问次数最多的Top5: 例如2016-10-01 ~ 2016-11-01 :")
98.     userLog.filter("time >= '" + startTime + "' and time <= '" + endTime
         + "' and typed = 0")
99.       .join(userInfo, userInfo("userID") === userLog("userID"))
100.        .groupBy(userInfo("userID"),userInfo("name"))
101.        .agg(count(userLog("logID")).alias("userLogCount"))
102.        .sort($"userLogCount".desc)
103.        .limit(5)
104.        .show()
105.
106.
107.      /**
108.        * 作业: 生成 parquet 方式的数据, 其自己实现时间函数, 然后测试整个代码
109.        * val peopleDF = spark.read.json("examples/src/main/resources/
          * people.json")
110.         // DataFrames 可以保存为 Parqnet 文件维护 schema 信息
111.         peopleDF.write.parquet("people.parquet")
112.
113.         // 读取上面创建的 Parquet 文件
114.         // Parquet 是自描述的, 模式 schema 将被保存
115.         // 加载 parget 文件的结果也是一个 DataFrame
116.         val parquetFileDF = spark.read.parquet("people.parquet")
117.       */
118.
119.      /**
        * 统计特定时段购买次数最多的 Top 5: 例如 2016-10-01 ~ 2016-11-01
120.       */
121.      println("统计特定时段购买次数最多的Top5: 例如2016-10-01 ~ 2016-11-01 :")
122.      userLog.filter("time >= '" + startTime + "' and time <= '" + endTime
      + "' and typed = 1")
123.        .join(userInfo, userInfo("userID") === userLog("userID"))
124.        .groupBy(userInfo("userID"),userInfo("name"))
125.        .agg(round(sum(userLog("consumed")), 2).alias("totalConsumed"))
126.        .sort($"totalConsumed".desc)
127.        .limit(5)
128.        .show
129.
130.      /**
131.        * 统计特定时段访问次数增长最多的 Top5 用户, 例如这周比上周访问次数增长最
          * 快的 5 位用户
132.        * 实现思路: 一种非常直接的方式是计算这周每个用户的访问次数, 同时计算出上周
          * 每个用户的访问次数, 然后相减并进行排名, 但是这种实现思路比较消耗性能, 我
          * 们可以采取一种既能实现业务目标, 又能提升性能的方式, 即把这周的每次用户访
          * 问计数为 1, 把上周的每次用户访问计数为-1, 再在 agg 操作中采用 sum 即可
133.        * 巧妙地实现增长趋势的量化
134.        */
135.      val userLogDS = userLog.as[UserLog].filter("time >= '2016-10-08'
      and time <= '2016-10-14' and typed = '0'")
136.        .map(log => LogOnce(log.logID, log.userID, 1) )
```

```
137.          .union(userLog.as[UserLog].filter("time >= '2016-10-01' and
               time <= '2016-10-07' and typed = '0'")
138.           .map(log => LogOnce(log.logID, log.userID, -1) ))
139.
140.      println("统计特定时段访问次数增长最多的 Top5 用户，例如这周比上周访问次
          数增长最快的 5 位用户：")
141.      userLogDS.join(userInfo, userLogDS("userID") === userInfo("userID"))
142.        .groupBy(userInfo("userID"),userInfo("name"))
143.        .agg(sum(userLogDS("count")).alias("viewCountIncreased"))
144.        .sort($"viewCountIncreased".desc)
145.        .limit(5)
146.        .show()
147.
148.
149.      /**
           * 统计特定时段购买金额增长最多的 Top5 用户，例如这周比上周访问次数增长最
           * 快的 5 位用户
150.       */
151.      println("统计特定时段购买金额增长最多的 Top5 用户，例如这周比上周访问次
          数增长最快的 5 位用户：")
152.      val userLogConsumerDS = userLog.as[UserLog].filter("time >=
          '2016-10-08' and time <= '2016-10-14' and typed == 1")
153.        .map(log => ConsumedOnce(log.logID, log.userID, log.consumed))
154.        .union(userLog.as[UserLog].filter("time >= '2016-10-01' and
             time <= '2016-10-07' and typed == 1")
155.          .map(log => ConsumedOnce(log.logID, log.userID, -log.consumed) ))
156.
157.      userLogConsumerDS.join(userInfo, userLogConsumerDS("userID")
           === userInfo("userID"))
158.        .groupBy(userInfo("userID"),userInfo("name"))
159.        .agg(round(sum(userLogConsumerDS("consumed")), 2).alias("view
            ConsumedIncreased"))
160.        .sort($"viewConsumedIncreased".desc)
161.        .limit(5)
162.        .show()
163.
164.      /**
           * 统计注册后前两周内访问最多的前 10 个人
165.       */
166.      println("统计注册后前两周内访问最多的前 Top10:")
167.      userLog.join(userInfo, userInfo("userID") === userLog("userID"))
168.        .filter(userInfo("registeredTime") >= "2016-10-01"
169.          && userInfo("registeredTime") <= "2016-10-14"
170.          && userLog("time") >= userInfo("registeredTime")
171.          && userLog("time") <= date_add(userInfo("registeredTime"), 14)
172.          && userLog("typed") === 0)
173.        .groupBy(userInfo("userID"),userInfo("name"))
174.        .agg(count(userLog("logID")).alias("logTimes"))
175.        .sort($"logTimes".desc)
176.        .limit(10)
177.        .show()
178.
179.
180.      /**
           * 统计注册后前两周内购买总额最多的前 10 个人
181.       */
```

```
182.        println("统计注册后前两周内购买总额最多的 Top 10 :")
183.        userLog.join(userInfo, userInfo("userID") === userLog("userID"))
184.          .filter(userInfo("registeredTime") >= "2016-10-01"
185.            && userInfo("registeredTime") <= "2016-10-14"
186.            && userLog("time") >= userInfo("registeredTime")
187.            &&  userLog("time")  <=  date_add(userInfo("registeredTime"),
14)
188.            && userLog("typed") === 1)
189.          .groupBy(userInfo("userID"),userInfo("name"))
190.          .agg(round(sum(userLog("consumed")),2).alias("totalConsumed"))
191.          .sort($"totalConsumed".desc)
192.          .limit(10)
193.          .show()
194.
195.        //while(true){} //和通过 Spark Shell 运行代码可以一直看到 Web 终端
                           //的原理一样，因为 Spark Shell 内部有一个 LOOP 循环
196.
197.        sc.stop()
198.
199.
200.      }
201.    }
```

2. 电商交互式分析系统应用模拟数据生成代码

电商交互式分析系统应用模拟数据生成代码，分别生成用户信息文件、用户访问记录文件，如例 14-2 所示。

【例 14-2】 Mock_EB_Users_Data.scala 代码。

```
1.   package com.dt.spark.SparkApps.sql;
2.
3.  import java.text.SimpleDateFormat;
4.  import java.util.Date;
5.  import java.io.FileOutputStream;
6.  import java.io.OutputStreamWriter;
7.  import java.io.PrintWriter;
8.  import java.text.ParseException;
9.  import java.util.Calendar;
10. import java.util.Random;
11.
12. /**
13.  * 电商数据自动生成代码，数据格式如下：
14.  * 用户信息文件：用户数据 {"userID": 0, "name": "spark0", "registeredTime":
     * "2016-10-11 18:06:25"}
15.  *用户访问记录文件：日志数据{"logID": 00,"userID": 0, "time": "2016-10-04
     *15:42:45", "typed": 0, "consumed": 0.0}
16.  */
17. public class Mock_EB_Users_Data {
18.
19.     public static void main(String[] args) throws ParseException {
20.         /**
             * 通过传递进来的参数生成指定大小规模的数据
21.          */
22.
23.         long numberItems = 1000;
24.         String dataPath = "data/Mock_EB_Users_Data/";
25.
26.         if (args.length > 1) {
27.             numberItems = Integer.valueOf(args[0]);
```

```
28.             dataPath = args[1];
29.         }
30.         System.out.println("User log number is : " + numberItems);
31.         mockUserData(numberItems, dataPath);
32.         mockLogData(numberItems, dataPath);
33.
34.     }
35.
36.     private static void mockLogData(long numberItems, String dataPath) {
37.        //{"logID": 00,"userID": 0, "time": "2016-10-04 15:42:45", "typed":
           0, "consumed": 0.0}
38.         StringBuffer mock_Log_Buffer = new StringBuffer("");
39.         Random random = new Random();
40.         for (int i = 0; i < numberItems; i++) { //userID
41.             for (int j = 0; j < numberItems; j++) {
42.                 String initData = "2016-10-";
43.      //拼接随机时间字符串 randomData
44.                 String  randomData  =  String.format("%s%02d%s%02d%s%02d%
                    s%02d", initData, random.nextInt(31)
45.                         , " ", random.nextInt(24)
46.                         , ":", random.nextInt(60)
47.                         , ":", random.nextInt(60));
48.                 String result = "{\"logID\": " + String.format("%02d", j)
                    + ", \"userID\":"+i+",\"time\":\"" + randomData + "\", \"" +
49.                         "typed\":"+String.format("%01d", random.nextInt(2)) +
50.                         ",\"consumed\":" + String.format("%.2f", random.
                            nextDouble() * 1000)+ "}";
51.
52.
53.                 mock_Log_Buffer.append(result)
54.                         .append("\n");
55.
56.             }
57.         }
58.         System.out.println(mock_Log_Buffer);
59.         PrintWriter printWriter = null;
60.         try {
61. //保存到 JSON 文件
62.             printWriter = new PrintWriter(new OutputStreamWriter(
63.                     new FileOutputStream(dataPath + "Mock_EB_Log_Data.json")));
64.             printWriter.write(mock_Log_Buffer.toString());
65.         } catch (Exception e) {
66.             //待办事项：自动生成 catch 块
67.             e.printStackTrace();
68.         } finally {
69.             printWriter.close();
70.         }
71.
72.
73.     }
74.
75.     private static void mockUserData(long numberItems, String dataPath) {
76.         StringBuffer mock_User_Buffer = new StringBuffer("");
77.         Random random = new Random();
78.         for (int i = 0; i < numberItems; i++) {
79.             String initData = "2016-10-";
80. //拼接随机时间字符串 randomData
81.             String randomData = String.format("%s%02d%s%02d%s%02d%s%02d",
                initData, random.nextInt(31)
82.                     , " ", random.nextInt(24)
```

```
83.                 , ":", random.nextInt(60)
84.                 , ":", random.nextInt(60));
85.             String result = "{\"userID\": " + i + ", \"name\": \"spark" +
            i + "\", \"registeredTime\": \"" + randomData + "\"}";
86.             mock_User_Buffer.append(result).append("\n");
87.
88.
89.         }
90.         System.out.println(mock_User_Buffer);
91.         PrintWriter printWriter = null;
92.         try {
93. //保存到 JSON 文件
94.             printWriter = new PrintWriter(new OutputStreamWriter(
95.                 new FileOutputStream(dataPath + "Mock_EB_Users_Data.json")));
96.             printWriter.write(mock_User_Buffer.toString());
97.         } catch (Exception e) {
98.             // 待办事项：自动生成 catch 块
99.             e.printStackTrace();
100.         } finally {
101.             printWriter.close();
102.         }
103.     }
104. }
```

14.5　本章总结

本章基于 Spark 2.2.0 框架，详细阐述了特定时间段内用户访问电商网站排名 TopN、特定时段购买金额 Top10 和访问次数增长 Top10 及纯粹通过 DataSet 进行电商交互式分析系统中各种类型的 TopN 实战，包括统计特定时间段购买金额最多的 Top5 用户、访问次数增长最多的 Top5 用户、购买金额增长最多的 Top5 用户，统计注册之后前两周内访问最多的 Top10 用户、前两周内购买总额最多的 Top10 用户。深入掌握电商交互式分析系统应用案例，将有助于读者跨入 Spark DataSet 编码殿堂。

第 15 章　Spark 商业案例之 NBA 篮球运动员大数据分析系统应用案例

本章讲解 Spark 商业案例之 NBA 篮球运动员大数据分析系统应用案例，基于 NBA 球员 1970 年至 2016 年的历史数据，统计分析每年 NBA 球员比赛的各项数据。

15.1　NBA 篮球运动员大数据分析系统架构和实现思路

NBA 篮球运动员大数据分析决策支持系统：基于 NBA 球员历史数据 1970 年至 2016 年各种表现，为全方位分析球员的技能，构建最强 NBA 篮球团队做数据分析支撑系统。曾经非常火爆的梦幻篮球是基于现实中的篮球比赛数据根据对手的情况制定游戏的先发阵容和比赛结果（也就是说，比赛结果是由实际结果决定的），游戏中可以管理球员，例如调整比赛的阵容，其中也包括裁员、签入和交易等。而这里的大数据分析系统可以被认为是游戏背后的数据分析系统。

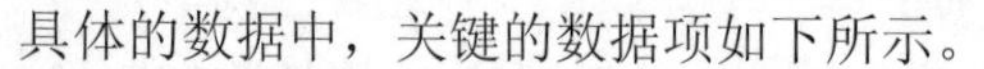

具体的数据中，关键的数据项如下所示。

- 3P：3 分命中。
- 3PA：3 分出手。
- 3P%：3 分命中率。
- 2P：2 分命中。
- 2PA：2 分出手。
- 2P%：2 分命中率。
- TRB：篮板球。
- STL：抢断。
- AST：助攻。
- BLK：盖帽。
- FT：罚球命中。
- TOV：失误。

基于球员的历史数据，如何对球员进行评价？也就是如何进行科学的指标计算，一个比较流行的算法是 Z-score：其基本的计算过程是基于球员的得分减去平均值后除以标准差。举一个简单的例子，某个球员在 2016 年的平均篮板数是 7.1，而所有球员在 2016 年的平均篮板数是 4.5，标准差是 1.3，那么该球员 Z-score 得分为 2。在计算球员的表现指标中可以计算 FT%、BLK、AST、FG%等。

具体如何通过 Spark 技术来实现呢？

第一步：数据预处理。例如，去掉不必要的标题等信息。

第二步：数据的缓存。为加速后面的数据处理打下基础。

第三步：基础数据项计算。方差、均值、最大值、最小值、出现次数等。

第四步：计算 Z-score，一般会进行广播，可以提升效率。

第五步：基于前面四步的基础，可以借助 Spark SQL 进行多维度 NBA 篮球运动员数据分析，可以使用 SQL 语句，也可以使用 DataSet（我们在这里可能会优先选择使用 SQL，为什么呢？其实原因非常简单，复杂的算法级别的计算已经在前面四步完成了且广播给了集群，我们在 SQL 中可以直接使用）。

第六步：把数据放在 Redis 或者 DB 中。

提示：（1）这里的一个非常重要的实现技巧是：通过 RDD 计算出来一些核心基础数据并广播出去，后面的业务基于 SQL 去实现，既简单，又可以灵活地应对业务变化需求，希望对大家能够有所启发；

（2）使用缓存和广播以及调整并行度等提升效率。

NBA 篮球运动员大数据分析系统应用数据说明如下。

1．NBA篮球运动员大数据分析系统应用数据的来源

美国职业篮球联赛（National Basketball Association，NBA）是由北美 30 支队伍组成的男子职业篮球联盟，汇集了世界上顶级的球员。NBA 一共有 30 支球队，东部分区和西部分区各有 15 支球队。其中，西部分区又被划分为西北赛区、太平洋赛区、西南赛区，每个赛区由 5 支球队组成。东部分区包括三大赛区：大西洋赛区、东南赛区、中部赛区，每个赛区也由 5 支球队组成。东部和西部联盟分别由前 8 名进入季后赛，对阵依据第一对第八，第二对第七，依此类推。每一轮系列赛均采取七局四胜的赛制，常规赛战绩占优的球队拥有主场优势。两大联盟的分区冠军进军决赛，同样为七局四胜。篮球参考网站（www.basketball-reference.com）提供了 NBA 篮球历届比赛球员、球队、季节赛、领先者、分数、季后赛、篮球指数的详细数据。Spark 商业案例之 NBA 篮球运动员大数据分析系统应用案例的数据来源就来自篮球参考网站。我们可以从网站上下载 NBA 篮球运动员历史数据。

（1）打开篮球参考网站的网页。

打开篮球参考网站（http://www.basketball-reference. com/leagues/NBA_2017_totals.html），查询 2016-2017 年度 NBA 赛季球员的比赛数据，如图 15-1 所示。

2016-17 NBA Season　Standings　Schedule and Results　Leaders　Player Stats ▾　Other ▾　2017 Playoffs Summary

Player Totals　Share & more ▾　Glossary　Hide Partial Rows

Rk	Player	Pos	Age	Tm	G	GS	MP	FG	FGA	FG%	3P	3PA	3P%	2P	2PA	2P%	eFG%	FT	FTA	FT%	ORB	DRB	TRB	AST	STL	BLK	TOV	PF	PTS
1	Alex Abrines	SG	23	OKC	68	6	1055	134	341	.393	94	247	.381	40	94	.426	.531	44	49	.898	18	68	86	40	37	8	33	114	406
2	Quincy Acy	PF	26	TOT	38	1	558	70	170	.412	37	90	.411	33	80	.413	.521	45	60	.750	20	95	115	18	14	15	21	67	222
2	Quincy Acy	PF	26	DAL	6	0	48	5	17	.294	1	7	.143	4	10	.400	.324	2	3	.667	2	6	8	0	0	0	2	9	13
2	Quincy Acy	PF	26	BRK	32	1	510	65	153	.425	36	83	.434	29	70	.414	.542	43	57	.754	18	89	107	18	14	15	19	58	209
3	Steven Adams	C	23	OKC	80	80	2389	374	655	.571	0	1	.000	374	654	.572	.571	157	257	.611	281	332	613	86	89	78	146	195	905
4	Arron Afflalo	SG	31	SAC	61	45	1580	185	420	.440	62	151	.411	123	269	.457	.514	83	93	.892	9	116	125	78	21	6	42	104	515

图 15-1　NBA 球员比赛数据

（2）Spark 商业案例之 NBA 篮球运动员大数据分析系统应用案例须抓取篮球参考网站的数据，而上述页面中网站没有导出数据的按钮。我们可以先在篮球参考网站上进行用户注册。

在网页右上角单击 Create Account 按钮，显示如图 15-2 所示的页面。

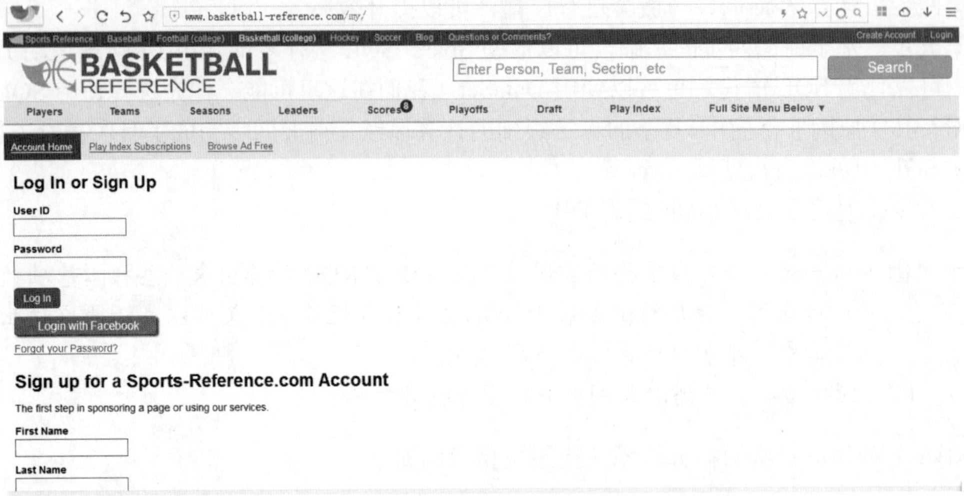

图 15-2　Create Account

（3）在图 15-3 所示的用户注册页面中填写用户的用户名、邮箱、密码相关信息。在用户邮箱中收到篮球参考网站发来的确认邮件，确认后就可以在篮球参考网站上进行登录。

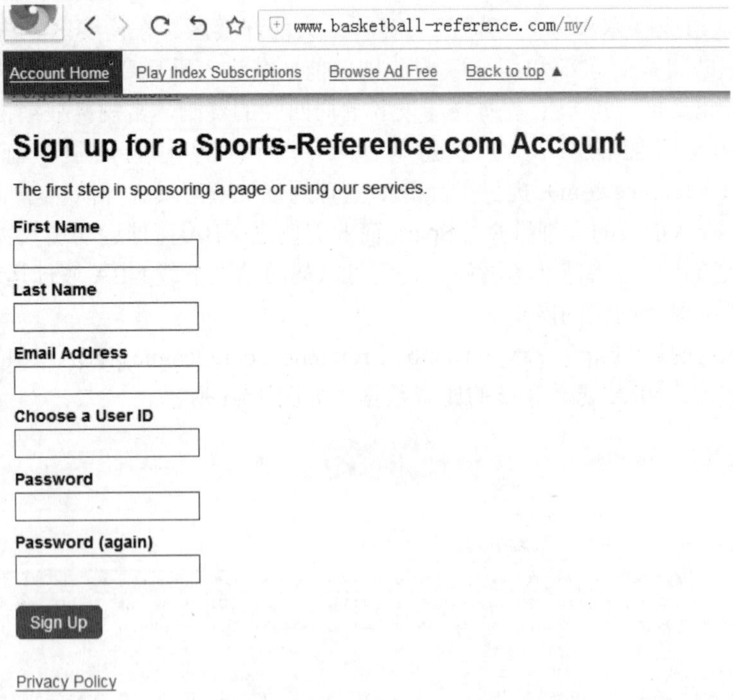

图 15-3　用户注册页面

（4）用户注册成功以后，在篮球参考网站进行登录，重新打开 2016—2017 年度 NBA 赛季球员的比赛数据的网页。如图 15-4 所示，在球员比赛数据上方出现工具栏，单击 share & more

按钮。

Player Totals　Share & more ▼　Glossary

Rk	Player	Pos	Age	Tm	G	GS	MP
1	Alex Abrines	SG	23	OKC	68	6	1055
2	Quincy Acy	PF	26	TOT	38	1	558
2	Quincy Acy	PF	26	DAL	6	0	48
2	Quincy Acy	PF	26	BRK	32	1	510
3	Steven Adams	C	23	OKC	80	80	2389
4	Arron Afflalo	SG	31	SAC	61	45	1580

图 15-4　单击 Share & more

（5）在 Share & more 列表中，通过 Excel、csv 等文本格式将 NBA 球员比赛的数据导出。这里单击 Get as Excel Workbook（experimental）导出为 Excel 格式，保存在本地文件 sportsref_download.xls 中，如图 15-5 所示。

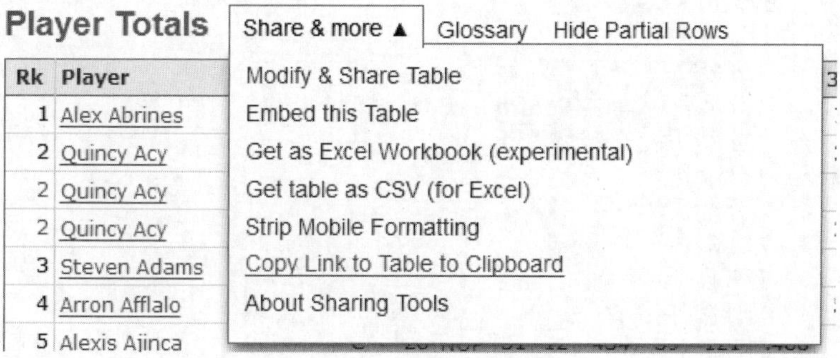

Player Totals　Share & more ▲　Glossary　Hide Partial Rows

Rk	Player
1	Alex Abrines
2	Quincy Acy
2	Quincy Acy
2	Quincy Acy
3	Steven Adams
4	Arron Afflalo
5	Alexis Ajinca

Modify & Share Table
Embed this Table
Get as Excel Workbook (experimental)
Get table as CSV (for Excel)
Strip Mobile Formatting
Copy Link to Table to Clipboard
About Sharing Tools

图 15-5　导出 NBA 球员数据

（6）打开 sportsref_download.xls 文件，从篮球参考网站上下载的球员比赛数据显示如图 15-6 所示。在 Spark 商业案例之 NBA 篮球运动员大数据分析系统应用案例中，我们从篮球参考网站中抓取 NBA 球员 1970—2016 年的历史数据进行统计分析。

Rk	Player	Pos	Age	Tm	G	GS	MP	FG	FGA	FG%	3P	3PA	3P%	2P	2PA	2P%	eFG%	FT	FTA	FT%	ORB	DRB	TRB	AST	STL	BLK	TOV	PF	PTS
1	Alex Abrines	SG	23	OKC	68	6	1055	134	341	0.393	94	247	0.381	40	94	0.426	0.531	44	49	0.898	18	68	86	40	37	8	33	114	406
2	Quincy Acy	PF	26	TOT	38	1	558	70	170	0.412	37	90	0.411	33	80	0.413	0.521	45	60	0.75	20	95	115	18	14	15	21	67	222
2	Quincy Acy	PF	26	DAL	6	0	48	5	17	0.294	1	7	0.143	4	10	0.4	0.324	2	3	0.667	2	6	8	0	0	0	2	9	13
2	Quincy Acy	PF	26	BRK	32	1	510	65	153	0.425	36	83	0.434	29	70	0.414	0.542	43	57	0.754	18	89	107	18	14	15	19	58	209
3	Steven Adams	C	23	OKC	80	80	2389	374	655	0.571	0	1	0	374	654	0.572	0.571	157	257	0.611	281	332	613	86	89	78	146	195	905
4	Arron Afflalo	SG	31	SAC	61	45	1580	185	420	0.44	62	151	0.411	123	269	0.457	0.514	83	93	0.892	9	116	125	78	21	6	42	104	515

图 15-6　NBA 球员比赛数据

2．NBA篮球运动员大数据分析系统应用数据的格式说明

Spark 商业案例之 NBA 篮球运动员大数据分析系统应用案例分析统计实战，我们使用 Spark 本地模式进行开发，在 IDEA 开发环境的 SparkApps 工程中的 data 目录中新建 NBABasketball 目录，将从篮球参考网站上抓取的球员历史数据文件转换为 csv 文件，将 leagues_NBA_1970_per_game_per_game.csv 至 leagues_NBA_2016_per_game_per_game.csv 文本文件保存到 data/NBABasketball 目录下。

NBA 篮球运动员文件 leagues_NBA_1970_per_game_per_game.csv 的格式描述如下：

```
1.  Rk      排名
2.  Player  球员名字
3.  Pos     打球位置
4.  Age     球员年龄
5.  Tm      效力球队
6.  G       上场次数
7.  GS      首发次数
8.  MP      比赛时间（分钟）
9.  FG      投篮命中次数
10. FGA     投篮出手次数
11. FG%     投篮命中率
12. 3P      3 分命中
13. 3PA     3 分出手
14. 3P%     3 分命中率
15. 2P      2 分命中
16. 2PA     2 分出手
17. 2P%     2 分命中率
18. eFG%    有效投篮命中率
19. FT      罚球命中
20. FTA     罚球出手次数
21. FT%     罚球命中率
22. ORB     进攻篮板球
23. DRB     防御篮板球
24. TRB     篮板球
25. AST     助攻
26. STL     抢断
27. BLK     盖帽
28. TOV     失误
29. PF      个人犯规
30. PTS     得分
```

从 NBA 篮球运动员文件 leagues_NBA_1970_per_game_per_game.csv 中摘取部分记录如下：

```
1.  Rk,Player,Pos,Age,Tm,G,GS,MP,FG,FGA,FG%,3P,3PA,3P%,2P,2PA,2P%,eFG%,
    FT,FTA,FT%,ORB,DRB,TRB,AST,STL,BLK,TOV,PF,PTS
2.  1,Kareem Abdul-Jabbar*,C,37,LAL,79,79,33.3,9.2,15.3,.599,0.0,0.0,.000,
    9.2,15.3,.600,.599,3.7,5.0,.732,2.1,5.8,7.9,3.2,0.8,2.1,2.5,3.0,22.0
3.  2,Alvan Adams,PF,30,PHO,82,69,26.0,5.8,11.2,.520,0.0,0.0,,5.8,11.2,
    .520,.520,3.0,3.5,.883,1.9,4.2,6.1,3.8,1.4,0.6,2.4,3.1,14.7
4.  3,Mark    Aguirre,SF,25,DAL,80,79,33.7,9.9,19.6,.506,0.3,1.1,.318,9.6,
    18.6,.517,.515,5.5,7.3,.759,2.4,3.6,6.0,3.1,0.8,0.3,3.2,3.1,25.7
```

```
5.  4,Danny Ainge,SG,25,BOS,75,73,34.2,5.6,10.6,.529,0.2,0.7,.268,5.4,
    9.8,.549,.539,1.6,1.8,.868,1.0,2.6,3.6,5.3,1.6,0.1,2.0,3.0,12.9
6.  ......
```

15.2　NBA 篮球运动员大数据分析系统代码实战：数据清洗和初步处理

本节 NBA 篮球运动员大数据分析系统代码实战：数据清洗和初步处理。首先构建 NBA 篮球运动员大数据分析系统 SparkSession 上下文运行环境，然后在 Spark 中读入 NBA 球员历史数据 1970—2017 年的数据，对原始的 NBA 球员数据进行清洗，并进行基础的计算处理。具体实现方法如下。

（1）构建 NBA 篮球运动员大数据分析系统实战 SparkSession 运行环境。SparkSession 统一了 Spark SQL 执行时的不同的上下文环境，Spark SQL 无论运行在哪种环境下，我们都可以只使用 SparkSession 统一的编程入口来处理 DataFrame 和 DataSet 编程，不需要关注底层是否有 Hive 等。有了 SparkSession 之后，就不再需要 SqlContext 或者 HiveContext 了。

构建 NBA 篮球运动员大数据分析系统实战 SparkSession 运行环境如下：

```
1.     def main(args: Array[String]) {
2.      Logger.getLogger("org").setLevel(Level.ERROR)
3.      //日志过滤级别，我们在控制台上只打印正确的结果或错误的信息
4.      var masterUrl = "local[8]" //默认程序运行在本地 Local 模式中，主要用于学习和测试
5.      /**
        * 当我们把程序打包运行在集群上的时候，一般都会传入集群的 URL 信息，这里我们假设传入
        * 参数，第一个参数只传入 Spark 集群的 URL，第二个参数传入的是数据的地址信息
6.      */
7.      if (args.length > 0) {
8.       masterUrl = args(0) //如果代码提交到集群上运行，传给 SparkSubmit 的第一个
                             //参数须是集群 master 的地址
9.      }
10.     /**
        * 创建 Spark 会话上下文 SparkSession 和集群上下文 SparkContext，在 SparkConf 中
        * 可以进行各种依赖和参数的设置等，大家可以通过 SparkSubmit 脚本的 help 去查看
        * 设置信息，其中 SparkSession 统一了 Spark SQL 运行的不同环境
11.     */
12.     val  sparkConf  =  new  SparkConf().setMaster(masterUrl).setAppName
        ("NBAPlayer_Analyzer_DateSet")
13.     /**
14.      * SparkSession 统一了 Spark SQL 执行时的不同的上下文环境，也就是说，Spark SQL
        * 无论运行在哪种环境下，我们都可以只使用 SparkSession 这样一个统一的编程入口
        * 来处理 DataFrame 和 DataSet 编程，不需要关注底层是否有 Hive 等
15.      * 有了 SparkSession 之后，就不再需要 SqlContext 或者 HiveContext 了
16.      */
17.     val spark = SparkSession
18.       .builder()
19.       .config(sparkConf)
20.       .getOrCreate()
```

```
21.     val sc = spark.sparkContext
                    //从 SparkSession 获得的上下文，这是因为我们读原生文件的时候或者实
                    //现一些 Spark SQL 目前还不支持的功能的时候，需要使用 SparkContext
22. ......
```

（2）数据清洗第一步：对原始的 NBA 球员数据进行初步处理，并进行缓存。

① 设置 NBA 球员历史数据存放的目录 data/NBABasketball；NBA 球员数据清洗以后保存的临时目录为 data/basketball_tmp。

② 循环遍历获取所有 NBA 球员文件的数据：SparkApps 工程中的 data 目录 data/NBABasketball 中存放了 2016—2017 年度 NBA 赛季球员的比赛数据，涉及的文件非常多，采用 for 语句循环遍历获取所有文件的数据。其中，2016 至 2017 年度 NBA 赛季球员数据文件名的格式均为 leagues_NBA_1970_per_game_per_game.csv，仅其中的年份（1970—2016）不同，我们通过输入年份变量读取每年 NBA 球员的数据信息。在 for 循环遍历中，读入的每年的 NBA 球员数据文本，每行中需包括“，”，筛选过滤清洗掉没有“，”的数据，并进行 map 转换，将读入的每行的球员数据格式化为 Key-Value 格式，Key 值为年份，Value 值为球员的比赛数据，即格式为（年份，球员数据）。将 Key-Value 格式化后的数据根据年份把每年数据保存到一个临时目录中，如 data/basketball_tmp/NBAStatsPerYear/1970。在生产环境中，转换以后的数据可以保存为 parquet 格式。

③ 读取已转换后的各年份目录下的文本文件（如 part-00000），part-00000 文本文件中第一行是（年份，标题栏）格式，其中标题栏中是 NBA 球员比赛数据对应的标题栏目，如 FG% 是投篮命中率，我们先把包含“FG%”标题的第一行过滤掉，然后再过滤出包含“，”的数据，即 NBA 球员每年的比赛数据。接下来进行数据清理，读入的每行数据中有的字段可能没有数据，原始数据记录中会出现多个“，”的情况，我们使用 map 转换函数将“，，”两个逗号的字符替换为“,0,”，这样进行格式转换，以便后续进行数据的清洗。

④ 接下来使用 persist 持久化算子进行数据缓存。为了加快后续的处理进度，我们一般对反复使用的数据进行缓存处理。这里我们使用 persist 算子，定义存储级别为 MEMORY_AND_DISK，首先将数据持久化到内存中，如果内存不足，就将 spill 溢出到磁盘，持久化到磁盘文件中。

NBA 篮球运动员大数据分析系统实战数据清洗代码如下：

```
1.      /**
          * 第一步：对原始的 NBA 球员数据进行初步的处理，并进行缓存
          *
2.        */
3.      val data_Path = "data/NBABasketball"  //数据存在的目录
4.      val data_Tmp = "data/basketball_tmp"
5.      /**
          * 因为文件非常多，此时我们需要采用循环读取所有文件的数据
6.        */
7.      FileSystem.get(new Configuration()).delete(new Path(data_Tmp),true)
        //如果临时文件夹已经存在，再删除其中的数据
8.      for(year <- 1970 to 2016){
9.        val statsPerYear = sc.textFile(s"${data_Path}/leagues_NBA_${year}
          *") //通过输入年份变量读取每年 NBA 球员的数据信息
10.       statsPerYear.filter(_.contains(",")).map(line => (year, line))
11.         .saveAsTextFile(s"${data_Tmp}/NBAStatsPerYear/${year}/")
12.     }
13.     val  NBAStats  =  sc.textFile(s"${data_Tmp}/NBAStatsPerYear/*/*")
```

```
    //读取所有 NBA 球员过去的历史数据
14.     /**
15.       * 进行数据初步的 ETL 清洗工作，实际产生的数据可能是不符合处理要求的数据，需要我
          * 们按照一定的规则进行清洗和格式化
16.       * 完成这个工作的关键是清晰地知道数据是如何产生的，以及我们需要什么样的数据
17.       */
18.     val filteredData = NBAStats.filter(line => !line.contains("FG%"))
        .filter(line => line.contains(","))
19.       .map(line => line.replace(",,", ",0,")) //数据清理的工作可能是持续的
20. filteredData.collect().take(10).foreach(println(_))
21.     /**
22.       * 数据缓存，为了加快后续的处理进度，我们一般对反复使用的数据都进行缓存处理
23.       * 推荐使用 StorageLevel.MEMORY_AND_DISK，因为这样可以更好地使用内存，且不
          * 让数据丢失
24.       */
25.     filteredData.persist(StorageLevel.MEMORY_AND_DISK)
26. ......
```

在 IDEA 中运行代码，NBA 篮球运动员大数据分析系统实战数据清洗代码，输出结果如下。

① 查看转换以后的 NBA 球员目录文件，如 1970 的目录结构如下：

```
1.  G:\IMFBigDataSpark2017\SparkApps\data\basketball_tmp\NBAStatsPerYear\1970
2.  ._SUCCESS.crc
3.  .part-00000.crc
4.  .part-00001.crc
5.  _SUCCESS
6.  part-00000
7.  part-00001
```

② 打开 part-00000 文本文件，显示 NBA 球员的比赛数据记录，第一行的数据包含的是标题栏，每行记录的 Key 值是年份值，Value 值是各年份 NBA 球员的比赛数据。

```
1.  (1970,Rk,Player,Pos,Age,Tm,G,GS,MP,FG,FGA,FG%,3P,3PA,3P%,2P,2PA,2P%,
    eFG%,FT,FTA,FT%,ORB,DRB,TRB,AST,STL,BLK,TOV,PF,PTS)
2.  (1970,1,Zaid Abdul-Aziz,PF,23,MIL,80,,20.5,3.0,6.8,.434,,,,3.0,6.8,
    .434,.434,1.5,2.3,.643,,,7.5,0.8,,,,2.1,7.4)
3.  (1970,2,Kareem Abdul-Jabbar*,C,22,MIL,82,,43.1,11.4,22.1,.518,,,,11.4,
    22.1,.518,.518,5.9,9.1,.653,,,14.5,4.1,,,,3.5,28.8)
4.  (1970,3,Mahdi Abdul-Rahman,PG,27,ATL,82,,33.6,6.0,12.9,.467,,,,6.0,
    12.9,.467,.467,3.3,4.0,.809,,,4.0,6.8,,,,3.2,15.3)
5.  (1970,4,Rick Adelman,PG,23,SDR,35,,20.5,2.7,7.1,.389,,,,2.7,7.1,.389,
    .389,1.9,2.6,.747,,,2.3,3.2,,,,2.6,7.4)
6.  (1970,5,Lucius Allen,PG,22,SEA,81,,22.4,3.8,8.5,.442,,,,3.8,8.5,.442,
    .442,2.2,3.1,.731,,,2.6,4.2,,,,2.5,9.8)
7.  (1970,6,Wally Anderzunas,SF,24,CIN,44,,8.4,1.5,3.8,.392,,,,1.5,3.8,
    .392,.392,0.7,1.0,.630,,,1.9,0.2,,,,1.1,3.6)
```

③ 对 NBA 球员比赛数据各行记录中包含连续两个逗号的记录清洗结果如下：

```
1.  (1970,1,Zaid Abdul-Aziz,PF,23,MIL,80,0,20.5,3.0,6.8,.434,0,,0,3.0,
    6.8,.434,.434,1.5,2.3,.643,0,,7.5,0.8,0,,0,2.1,7.4)
2.  (1970,2,Kareem Abdul-Jabbar*,C,22,MIL,82,0,43.1,11.4,22.1,.518,0,,0,
    11.4,22.1,.518,.518,5.9,9.1,.653,0,,14.5,4.1,0,,0,3.5,28.8)
3.  (1970,3,Mahdi Abdul-Rahman,PG,27,ATL,82,0,33.6,6.0,12.9,.467,0,,0,6.0,
```

```
    12.9,.467,.467,3.3,4.0,.809,0,,4.0,6.8,0,,0,3.2,15.3)
4.  (1970,4,Rick Adelman,PG,23,SDR,35,0,20.5,2.7,7.1,.389,0,,0,2.7,7.1,
    .389,.389,1.9,2.6,.747,0,,2.3,3.2,0,,0,2.6,7.4)
5.  (1970,5,Lucius Allen,PG,22,SEA,81,0,22.4,3.8,8.5,.442,0,,0,3.8,8.5,
    .442,.442,2.2,3.1,.731,0,,2.6,4.2,0,,0,2.5,9.8)
6.  (1970,6,Wally Anderzunas,SF,24,CIN,44,0,8.4,1.5,3.8,.392,0,,0,1.5,3.8,
    .392,.392,0.7,1.0,.630,0,,1.9,0.2,0,,0,1.1,3.6)
7.  (1970,7,John Arthurs,G,22,MIL,11,0,7.8,1.1,3.2,.343,0,,0,1.1,3.2,.343,
    .343,1.0,1.4,.733,0,,1.3,1.5,0,,0,1.4,3.2)
8.  (1970,8,Al Attles,PG,33,SFW,45,0,15.0,1.7,4.5,.386,0,,0,1.7,4.5,.386,
    .386,1.7,2.5,.664,0,,1.6,3.2,0,,0,2.3,5.1)
9.  (1970,9,Jim Barnes,C,28,BOS,77,0,13.6,2.3,5.6,.410,0,,0,2.3,5.6,.410,
    .410,1.2,1.7,.742,0,,4.5,0.7,0,,0,3.0,5.9)
10. (1970,10,Dick Barnett,SG,33,NYK,82,0,33.8,6.0,12.7,.475,0,,0,6.0,12.7,
    .475,.475,2.8,4.0,.714,0,,2.7,3.6,0,,0,2.7,14.9)
11. ……
```

（3）数据清洗第二步：对原始的 NBA 球员数据进行基础性处理，并进行基础数据项计算：方差、均值、最大值、最小值、出现次数等。

这里对 NBA 球员比赛数据进行了清洗，包括过滤清洗掉 NBA 球员比赛数据第一行的标题栏、星号清洗、两个逗号中插入数字 0 等清洗工作，部分清洗和第一步是重复的，读者可以直接从这里开始 NBA 球员比赛数据的数据清洗及处理。

NBA 篮球运动员数据清洗代码如下：

```
1.      //读入 NBA 球员的数据记录
2.      val stats = sc.textFile(s"${TMP_PATH}/BasketballStatsWithYear/*/*")
        .repartition(sc.defaultParallelism)
3.
4.      //过滤清洗掉 NBA 球员比赛数据第一行的标题栏、星号清洗、两个逗号中插入数字 0 等清洗工作
5.      val filteredStats: RDD[String] = stats.filter(x => !x.contains
        ("FG%")).filter(x => x.contains(",")) .map(x => x.replace("*", "")
        .replace(",,", ",0,"))
6.      filteredStats.cache() //清洗以后的数据进行缓存
7.      println("NBA 球员清洗以后的数据记录： ")
8.      filteredStats.take(10).foreach(println)
```

在 IDEA 中运行代码，NBA 篮球运动员数据清洗输出结果如下：

```
1.   NBA 球员清洗以后的数据记录：
2.  (1970,5,Lucius Allen,PG,22,SEA,81,0,22.4,3.8,8.5,.442,0,,0,3.8,8.5,
    .442,.442,2.2,3.1,.731,0,,2.6,4.2,0,,0,2.5,9.8)
3.  (1970,36,Dick Cunningham,C,23,MIL,60,0,6.9,0.9,2.4,.369,0,,0,0.9,2.4,
    .369,.369,0.4,0.6,.667,0,,2.7,0.5,0,,0,1.2,2.1)
4.  (1970,51,Dave Gambee,SF,32,SFW,73,0,13.0,2.5,6.4,.399,0,,0,2.5,6.4,
    .399,.399,2.1,2.5,.839,0,,3.3,0.8,0,,0,2.4,7.2)
5.  (1970,66,Clem Haskins,SG,26,CHI,82,0,39.2,8.1,18.1,.450,0,,0,8.1,18.1,
    .450,.450,4.0,5.2,.783,0,,4.6,7.6,0,,0,2.9,20.3)
6.  (1970,81,Wali Jones,PG,27,PHI,78,0,22.3,4.7,10.9,.430,0,,0,4.7,10.9,
    .430,.430,2.4,2.9,.841,0,,2.2,3.5,0,,0,2.7,11.8)
7.  (1970,96,Mike Lynn,SF,24,LAL,44,0,9.2,1.0,3.0,.331,0,,0,1.0,3.0,.331,
    .331,0.7,1.1,.646,0,,1.5,0.7,0,,0,2.0,2.7)
8.  (1970,111,Dorie Murrey,C,26,SEA,81,0,13.3,1.9,4.2,.446,0,,0,1.9,4.2,
```

```
   .446,.446,1.7,2.3,.731,0,,4.4,0.9,0,,0,2.4,5.5)
9.  (1970,126,Rick Roberson,C,22,LAL,74,0,27.1,3.5,7.9,.447,0,,0,3.5,7.9,
   .447,.447,1.6,2.9,.566,0,,9.1,1.2,0,,0,3.5,8.7)
10. (1970,141,Greg Smith,PF,23,MIL,82,0,28.9,4.1,8.1,.511,0,,0,4.1,8.1,
   .511,.511,1.5,2.1,.718,0,,8.7,1.9,0,,0,3.7,9.8)
11. (1970,157,Jimmy Walker,SG,25,DET,81,0,35.4,8.2,17.2,.478,0,,0,8.2,
   17.2,.478,.478,4.4,5.4,.807,0,,3.0,3.1,0,,0,2.5,20.8)
```

15.3　NBA 篮球运动员大数据分析代码实战之核心基础数据项编写

本节根据 NBA 球员 1970—2016 年的历史数据，按照表 15-1 中 NBA 球员比赛数据统计需求，统计出每年 NBA 球员比赛的各项统计数据。

表 15-1　NBA球员数据统计项

FT_min	最小罚球命中次数
DRB_max	最大防御篮板球数
FT%_count	罚球命中率
FG%_stdev	投篮命中率标准差
FTA_min	最小罚球出手次数
FT%_stdev	罚球命中率标准差
FG%_count	投篮命中率
FG_count	投篮命中次数
FG_stdev	投篮命中次数标准差
2P_stdev	2 分命中标准差
TOV_max	最大失误次数
AST_stdev	助攻标准差
BLK_stdev	盖帽标准差
FT_count	罚球命中次数
2P_count	2 分命中次数
DRB_avg	防御篮板球平均次数
3P_max	最大 3 分命中
2P%_count	2 分命中率
3P%_max	最大 3 分命中率
2P%_stdev	2 分命中率标准差
TOV_avg、3P_avg ……	平均失误次数、 平均 3 分命中 ……

15.3.1　NBA 球员数据每年基础数据项记录

根据 NBA 球员 1970—2016 年的历史数据，按照 NBA 球员比赛数据统计需求，对 NBA 篮球比赛感兴趣的读者，可以研究一下 NBA 比赛具体各个统计维度的业务含义。这里统计出每年 NBA 球员比赛的各项聚合统计数据。

具体实现方法如下。

（1）定义 NBA 球员数据统计维度的数组 txtStat，数组中包含 FG、FGA、FG%、3P、3PA、3P%、2P、2PA、2P%、eFG%、FT、FTA、FT%、ORB、DRB、TRB、AST、STL、BLK、TOV、PF、PTS 等字符串，其对应为投篮命中次数、投篮出手次数、投篮命中率、3 分命中、3 分出手、3 分命中率、2 分命中、2 分出手、2 分命中率、有效投篮命中率、罚球命中、罚球出手次数、罚球命中率、进攻篮板球、防御篮板球、篮板球、助攻、抢断、盖帽、失误、个人犯规、得分。

（2）通过函数 processStats 传入两个参数进行计算。第一个参数是清洗以后的 NBA 数据 filteredStats，第二个参数是 NBA 球员数据统计维度数组 txtStat。第三个参数 bStats 使用默认值空值，第四个参数 zStats 使用默认值空值。

processStats 函数的签名如下：

```
1.  def  processStats(stats0: RDD[String], txtStat: Array[String], bStats:
    Map[String, Double],
2.  zStats: Map[String, Double]) : RDD[(String, Double)]
```

（3）打印输出 NBA 球员基础数据项映射集。

NBA 每年球员比赛数据统计分析代码如下：

```
1.   //解析统计分析的结果保存为 map 结构
2.     val txtStat: Array[String] = Array("FG", "FGA", "FG%", "3P", "3PA",
       "3P%", "2P", "2PA", "2P%", "eFG%", "FT","FTA", "FT%", "ORB", "DRB", "TRB",
       "AST", "STL", "BLK", "TOV", "PF", "PTS")
3.
4.      println("NBA 球员数据统计维度: ")
5.      txtStat.foreach(println)
6.      val  aggStats:  Map[String,  Double]  =  processStats(filteredStats,
        txtStat).collectAsMap  //基础数据项，需要在集群中使用，因此会在后面广播出去
7.      println("NBA 球员基础数据项 aggStats MAP 映射集: ")
8.      aggStats.take(20).foreach{case (k,v) => println(" ("+k+","+v+" ) ")}
9.      //将 RDD 转换成 map 结构进行广播
10.     val broadcastStats = sc.broadcast(aggStats) //使用广播提升效率
```

在 IDEA 中运行代码，对原始的 NBA 球员数据进行基础性处理，输出结果如下：

```
1.   NBA 球员数据统计维度:
2.  FG
3.  FGA
4.  FG%
5.  3P
6.  3PA
7.  3P%
8.  2P
9.  2PA
10. 2P%
11. eFG%
12. FT
13. FTA
14. FT%
15. ORB
16. DRB
```

```
17. TRB
18. AST
19. STL
20. BLK
21. TOV
22. PF
23. PTS
24. ……
25. NBA 球员基础数据项 aggStats MAP 映射集:
26.   ( 1970_FT_min  , 0.0 )
27.   ( 1970_DRB_max  , 0.0 )
28.   ( 1970_FT%_count  , 171.0 )
29.   ( 1970_FG%_stdev  , 0.07887441636979554 )
30.   ( 1970_FTA_min  , 0.0 )
31.   ( 1970_FT%_stdev  , 0.14819090262313409 )
32.   ( 1970_FG%_count  , 171.0 )
33.   ( 1970_FG_count  , 171.0 )
34.   ( 1970_FG_stdev  , 2.745145972097947 )
35.   ( 1970_2P_stdev  , 2.745145972097947 )
36.   ( 1970_TOV_max  , 0.0 )
37.   ( 1970_AST_stdev  , 1.8797508579195414 )
38.   ( 1970_BLK_stdev  , 0.0 )
39.   ( 1970_FT_count  , 171.0 )
40.   ( 1970_2P_count  , 171.0 )
41.   ( 1970_DRB_avg  , 0.0 )
42.   ( 1970_3P_max  , 0.0 )
43.   ( 1970_2P%_count  , 171.0 )
44.   ( 1970_3P%_max  , 0.0 )
45.   ( 1970_2P%_stdev  , 0.07887441636979554 )
```

15.3.2　NBA 球员数据每年标准分 Z-Score 计算

根据 NBA 球员 1970—2016 年的历史数据，按照 NBA 球员比赛数据统计需求，统计出 NBA 球员数据每年标准分 Z-Score 计算。

具体实现方法如下。

（1）定义 NBA 球员数据标准分统计维度的数组 txtStatZ，数组中包含 FG、FT、3P、TRB、AST、STL、BLK、TOV、PTS 等字符串，其对应为投篮命中次数、罚球命中、3 分出手、篮板球、助攻、抢断、盖帽、失误、得分。

（2）通过函数 processStats 传入 3 个参数进行计算。第一个参数是清洗以后的 NBA 数据 filteredStats，第二个参数是 NBA 球员数据标准分统计维度的数组 txtStatZ，第三个参数 bStats 使用广播变量 broadcastStats.value，即在 15.3.1 节中计算出的 NBA 球员数据每年聚合统计数据，第四个参数 zStats 使用默认值空值。

processStats 函数的签名如下：

```
1.  def  processStats(stats0: RDD[String], txtStat: Array[String], bStats:
    Map[String, Double],
2.  zStats: Map[String, Double]) : RDD[(String, Double)]
```

（3）打印输出 NBA 球员 Z-Score 标准分映射集。

NBA 球员数据每年标准分 Z-Score 计算代码如下：

```
1.    //解析统计，跟踪权重
2.     val txtStatZ = Array("FG", "FT", "3P", "TRB", "AST", "STL", "BLK", "TOV",
       "PTS")
3.     val zStats: Map[String, Double] = processStats(filteredStats, txtStatZ,
       broadcastStats.value).collectAsMap
4.      println("NBA 球员 Z-Score 标准分 zStats  MAP 映射集: ")
5.      zStats.take(20).foreach{case (k,v) => println(" ("+k+", "+v  +")")}
6.      //将 RDD 转换为 map 结构，并使用广播变量广播到 Executor
7.      val zBroadcastStats = sc.broadcast(zStats)
```

在 IDEA 中运行代码，NBA 球员数据每年标准分 Z-Score 计算代码输出结果如下：

```
1.    NBA 球员 Z-Score 标准分 zStats  MAP 映射集:
2.    ( 1970_FT_min  , -3.6380461988304074 )
3.    ( 1970_FG_count  , 171.0 )
4.    ( 1970_FG_stdev  , 0.4780806983787295 )
5.    ( 1970_AST_stdev  , 1.0 )
6.    ( 1970_TOV_max  , -Infinity )
7.    ( 1970_3P_max  , -Infinity )
8.    ( 1970_FT_count  , 171.0 )
9.    ( 1970_BLK_stdev  , NaN )
10.   ( 1970_3P_avg  , 0.0 )
11.   ( 1970_TOV_avg  , 0.0 )
12.   ( 1970_FG_max  , 2.371563157894738 )
13.   ( 1970_FG_avg  , 0.1370776478232629 )
14.   ( 1970_FT_stdev  , 0.4094933599941332 )
15.   ( 1970_TRB_max  , 3.5104557688319065 )
16.   ( 1970_AST_min  , -1.247210200525836 )
17.   ( 1970_PTS_max  , 2.838208290871194 )
18.   ( 1970_TOV_count  , 0.0 )
19.   ( 1970_BLK_min  , Infinity )
20.   ( 1970_TOV_min  , Infinity )
```

15.3.3　NBA 球员数据每年归一化计算

根据 NBA 球员 1970—2016 年的历史数据，按照 NBA 球员比赛数据统计需求，统计出 NBA 球员数据每年归一化计算。

具体实现方法如下。

（1）对清洗后的 NBA 球员数据进行 map 转换。将读入的每行 NBA 球员数据调用 bbParse 函数，由 RDD[String]生成新的 RDD[BballData]，RDD 的元素类型是 BballData。

（2）将 nStats RDD[BballData]转换成 nPlayer RDD[Row]，后续我们可以将 nPlayer RDD[Row]再转换为 dataframe。在 map 转换过程中打印输出每行数据。

（3）NBA 球员比赛数据归一化处理。

```
1.     //解析统计，进行归一化处理
2.      val nStats: RDD[BballData] = filteredStats.map(x => bbParse(x,
        broadcastStats.value, zBroadcastStats.value))
3.      //转换 RDD 为 RDD[Row]，可以将其再转换为 dataframe
```

```
4.     val nPlayer: RDD[Row] = nStats.map(x => {
5.       val nPlayerRow: Row = Row.fromSeq(Array(x.name, x.year, x.age,
         x.position, x.team, x.gp, x.gs, x.mp)
6.         ++ x.stats ++ x.statsZ ++ Array(x.valueZ) ++ x.statsN ++ Array
          (x.valueN))
7.       println( nPlayerRow.mkString(" "))
8.       nPlayerRow
9.     })
```

在 IDEA 中运行代码，对 NBA 球员比赛数据进行归一化处理，在 map 转换过程中打印输出每行数据，分别截图显示输出结果，如图 15-7 和图 15-8 所示。

```
Vinny Del Negro 2000 33 PG MIL 67 0 18.1 2.3 4.9 0.471 0.1 0.4 0.333 2.2 4.5 0.482 0.483 0.5 0.6 0.897 0.1 1.5 1.6 2.4 0.5 0.0 0.7 1.2
 5.2 0.302247488803139 0.10301846983662138 -0.5551977334407006 -0.81418347444599 0.30488943735619084 -0.3638826077581982 -0
 .8070671514866472 0.6929313218203518 -0.4787406228362177 -1.615984872151451 0.04105621069430546 0.01432491946625797 -0
 .16020639315378787 -0.19196562192671113 0.06670532041193843 -0.0840722864519025 -0.13073442394709373 0.20463994678197225 -0
 .12764730551478282 -0.36789963363980394
LaPhonso Ellis 2000 29 PF ATL 58 8 22.6 3.6 8.0 0.45 0.1 0.4 0.143 3.6 7.6 0.465 0.454 1.1 1.6 0.695 1.7 3.3 5.0 1.0 0.6 0.4 0.9 2.3
 8.4 0.21251870052082839 -0.37035744683160193 -0.5551977334407006 0.53925260302153 -0.4707109277063714 -0.1506056186983l962 -0
 .05242928247144409 0.4457173029019164 0.07364937877390089 -0.3281630239302618 0.028867775145506914 -0.051498926435283296 -0
 .16020639315378787 0.12714328473083358 -0.10298462133134548 -0.0347962734313615 -0.008492864601461984 0.1316314651588224 0
 .019637240511637802 -0.05069931340643942
Lawrence Funderburke 2000 29 PF SAC 75 1 13.7 2.5 4.7 0.523 0.0 0.0 0.0 2.5 4.7 0.526 0.523 1.5 2.2 0.706 1.3 1.8 3.1 0.4 0.4 0.3 0.5
 1.2 6.4 0.8843130446410064 -0.29282674701582545 -0.7466602668110601 -0.21707932262208415 -0.8031110841617554 -0.5771595968180768
 -0.24108874972524494 0.940145340738787 -0.2715943722324232 -1.325061754006677 0.12012190018277118 -0.04071813117803504 -0
 .21545431663730155 -0.051182280754264946 -0.17570888207846716 -0.1333482994724435 -0.03905325443786993 0.27764842840512205 -0
 .07241560075487508 -0.330110436725364
```

图 15-7　NBA 球员比赛数据归一化处理结果 1

```
A.C. Green 2000 36 PF LAL 82 82 23.5 2.1 4.7 0.447 0.0 0.0 0.25 2.1 4.7 0.449 0.448 0.8 1.2 0.695 2.0 4.0 5.9 1.0 0.6 0.2 0.6 1.5 5.0
 0.002976921421987413 -0.3618598142864145 -0.7466602668110601 0.8975150941158737 -0.4707109277063714 -0.15060561869831962 -0
 .42974821697904564 0.8165383312795694 -0.5132649979368502 -0.9558194956006308 4.0437428812224865E-4 -0.05031731403066582 -0
 .21545431663730155 0.21161328943430133 -0.10298462133134548 -0.0347962734313615 -0.06961364427427785 0.24114418759354717 -0
 .13685258964143412 -0.1568569080304156
Hersey Hawkins 2000 33 SG CHI 61 49 26.6 2.6 6.1 0.424 0.9 2.3 0.39 1.7 3.8 0.444 0.497 1.8 2.0 0.899 0.5 2.4 2.9 2.2 1.2 0.2 1.6 2.4
 7.9 -0.27941733401388125 1.143172881056742 0.9765025335221765 -0.29669320953193834 0.1940893852043964 1.129056315660952 -0
 .42974821697904564 -0.4195317633126071 -0.012661558977680142 2.0047690326291145 -0.03795504466337155 0.1589604221759328 0
 .2817769947143218 -0.06995339291059112 0.0424639001628979 0.2608598046918845 -0.06961364427427785 -0.12389822052220205 -0
 .00337596980499042 0.43926484956960404
Larry Hughes 2000 21 SG TOT 82 37 28.3 5.6 14.0 0.4 0.4 1.5 0.232 5.2 12.5 0.421 0.413 3.4 4.6 0.74 1.4 2.9 4.3 2.5 1.4 0.3 2.4 2.3
 15.0 -1.1968924940358465 0.33811567135826986 0.01918986667037839 0.2606039988370405 0.36028946343208823 1.5556102937807095 -0
 .24108874972524494 -1.408387838986348 1.2129537570947704 0.9003939684258171 -0.16258156720559036 0.04701564457488364 0
 .005537377296753273 0.061444392183692 0.07882603053645873 0.3594118307329666 -0.03905325443786993 -0.4159321470148013 0
 .3234116166911303 0.2580799233576229
```

图 15-8　NBA 球员比赛数据归一化处理结果 2

（4）创建 DataFrame 的元数据结构 schemaN。

（5）根据元数据结构 schemaN StructType 和每行的 RDD 数据集 nPlayer: RDD[Row]创建 DataFrame。

（6）将 dfPlayersT 创建临时表视图 tPlayers。

（7）在 Spark SQL 中编写 SQL 语句代码，根据 NBA 球员比赛数据的业务查询需求编写 SQL 代码，然后展示打印输出。

（8）新建一个临时表 Players，将 SQL 查询转换后的 dfPlayers DataFrame 进行保存。

NBA 球员数据 DataFrame 的 spark sql 代码如下：

```
1.     //为 DataFrame 创建模式 schema
```

```
2.      val schemaN: StructType = StructType(
3.        StructField("name", StringType, true) ::
4.          StructField("year", IntegerType, true) ::
5.          StructField("age", IntegerType, true) ::
6.          StructField("position", StringType, true) ::
7.          StructField("team", StringType, true) ::
8.          StructField("gp", IntegerType, true) ::
9.          StructField("gs", IntegerType, true) ::
10.         StructField("mp", DoubleType, true) ::
11.         StructField("FG", DoubleType, true) ::
12.         StructField("FGA", DoubleType, true) ::
13.         StructField("FGP", DoubleType, true) ::
14.         StructField("3P", DoubleType, true) ::
15.         StructField("3PA", DoubleType, true) ::
16.         StructField("3PP", DoubleType, true) ::
17.         StructField("2P", DoubleType, true) ::
18.         StructField("2PA", DoubleType, true) ::
19.         StructField("2PP", DoubleType, true) ::
20.         StructField("eFG", DoubleType, true) ::
21.         StructField("FT", DoubleType, true) ::
22.         StructField("FTA", DoubleType, true) ::
23.         StructField("FTP", DoubleType, true) ::
24.         StructField("ORB", DoubleType, true) ::
25.         StructField("DRB", DoubleType, true) ::
26.         StructField("TRB", DoubleType, true) ::
27.         StructField("AST", DoubleType, true) ::
28.         StructField("STL", DoubleType, true) ::
29.         StructField("BLK", DoubleType, true) ::
30.         StructField("TOV", DoubleType, true) ::
31.         StructField("PF", DoubleType, true) ::
32.         StructField("PTS", DoubleType, true) ::
33.         StructField("zFG", DoubleType, true) ::
34.         StructField("zFT", DoubleType, true) ::
35.         StructField("z3P", DoubleType, true) ::
36.         StructField("zTRB", DoubleType, true) ::
37.         StructField("zAST", DoubleType, true) ::
38.         StructField("zSTL", DoubleType, true) ::
39.         StructField("zBLK", DoubleType, true) ::
40.         StructField("zTOV", DoubleType, true) ::
41.         StructField("zPTS", DoubleType, true) ::
42.         StructField("zTOT", DoubleType, true) ::
43.         StructField("nFG", DoubleType, true) ::
44.         StructField("nFT", DoubleType, true) ::
45.         StructField("n3P", DoubleType, true) ::
46.         StructField("nTRB", DoubleType, true) ::
47.         StructField("nAST", DoubleType, true) ::
48.         StructField("nSTL", DoubleType, true) ::
49.         StructField("nBLK", DoubleType, true) ::
50.         StructField("nTOV", DoubleType, true) ::
51.         StructField("nPTS", DoubleType, true) ::
52.         StructField("nTOT", DoubleType, true) :: Nil
53.     )
54.
55.     //创建 DaraFrame
56.     val dfPlayersT: DataFrame = spark.createDataFrame(nPlayer, schemaN)
57.
58.     //将所有数据保存为临时表
59.     dfPlayersT.createOrReplaceTempView("tPlayers")
60.
```

```
61.     //计算 exp 和 zdiff，ND2FF
62.     val dfPlayers: DataFrame = spark.sql("select age-min_age as exp,
        tPlayers.* from tPlayers join" +
63.       " (select name,min(age)as min_age from tPlayers group by name) as t1" +
64.       " on tPlayers.name=t1.name order by tPlayers.name, exp ")
65.     println("计算 exp and zdiff, ndiff")
66.     dfPlayers.show()
67.     //保存为表
68.     dfPlayers.createOrReplaceTempView("Players")
69.     //filteredStats.unpersist()
70.
```

在 IDEA 中运行代码，NBA 球员数据的 DataFrame 使用 dfPlayers.show 方法打印输出每行数据，分别截图显示部分输出结果，如图 15-9 和图 15-10 所示。

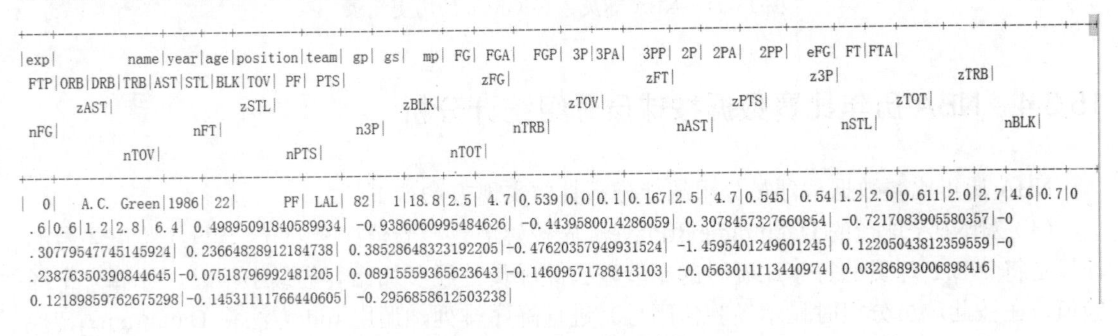

```
|exp|        name|year|age|position|team| gp| gs|  mp| FG| FGA|  FGP| 3P|3PA|  3PP| 2P| 2PA|  2PP|  eFG| FT|FTA|
FTP|ORB|DRB|TRB|AST|STL|BLK|TOV| PF| PTS|                 zFG|                 zFT|                 z3P|                zTRB|
      zAST|                 zSTL|                 zBLK|                 zTOV|                 zPTS|                 zTOT|
nFG|                 nFT|                 n3P|                 nTRB|                 nAST|                 nSTL|                 nBLK|
         nTOV|                 nPTS|                 nTOT|
| 0|   A.C. Green|1986| 22|      PF| LAL| 82|  1|18.8|2.5| 4.7|0.539|0.0|0.1|0.167|2.5| 4.7|0.545| 0.54|1.2|2.0|0.611|2.0|2.7|4.6|0.7|0
.6|0.6|1.2|2.8| 6.4| 0.49895091840589934| -0.9386060995484626| -0.4439580014286059| 0.3078457327660854| -0.7217083905580357|-0
.30779547745145924| 0.23664828912184738| 0.38528648323192205|-0.47620357949931524| -1.4595401249601245| 0.12205043812359559|-0
.23876350390844645|-0.07518796992481205| 0.08915559365623643|-0.14609571788413103| -0.0563011113440974| 0.03286893006898416|
0.12189859762675298|-0.14531111766440605| -0.2956858612503238|
```

图 15-9 dfPlayers 展示结果 1

```
| 1|   A.C. Green|1987| 23|      PF| LAL| 79| 72|28.4|4.0| 7.4|0.538|0.0|0.1|  0.0|4.0| 7.4|0.543|0.538|2.8|3.6| 0.78|2.7|5.1|7.8|1.1|0
.9|1.0|1.3|2.2|10.8|  1.0647443558788734|  0.4317423319030898| -0.4956343703118639| 1.4736712977283297| -0.5280946863340036|
0.29866739913350376|  0.9393031455733895|  0.2042513776981104| 0.23705460068644219|  3.6257054519558705| 0.20229884757986655|
0.08108703012520127|-0.10652353426919897|  0.3743133377279719|-0.11043091348881631|  0.0645866213427218|  0.1495373024322332|0
.053751503376815724|0.055923449487806434|  0.7645436443146016|
| 2|   A.C. Green|1988| 24|      PF| LAL| 82| 64|32.1|3.9| 7.8|0.503|0.0|0.0|  0.0|3.9| 7.8|0.505|0.503|3.6|4.6|0.773|3.0|5.7|8.7|1.1|1
.1|0.5|1.5|2.5|11.4|  0.5344617689896461|  0.4672660813119134|-0.48712024642979096|   1.884472699069965| -0.5043192405380542|
0.6765261358530328| 0.08403845256928431|-0.06968723861666007| 0.35920910809709483|   2.944847520306431| 0.11827604123363826|
0.11540827445867732|-0.07753335737468439|  0.5396323766526926| -0.0919921267999587| 0.14663402692778477|0.014288365188346563|-0
.02214891611687...| 0.08871830658292658|  0.8312829907525517|
| 3|   A.C. Green|1989| 25|      PF| LAL| 82| 82|30.6|4.9| 9.2|0.529|0.0|0.2|0.235|4.8| 9.0|0.536|0.532|3.4|4.4|0.786|3.1|5.9|9.0|1.3|1
.1|0.7|1.5|2.1|13.3|  1.3345675807499173|  0.6822090371092797| -0.5024920538018735| 1.9687421401290168|-0.38359131638997784|
0.5564611052881215|  0.4381116198347995|-0.05595318985683938|  0.6372690105802086|   4.675323933642653| 0.31023539020714025|
0.15279899364513802|-0.08206771631600944|   0.540098436595252|-0.07478092974899755| 0.13500583430571764| 0.06654914059056855|-0
.01913006305439...| 0.18019184014127934|   1.208900926365691|
```

图 15-10 dfPlayers 展示结果 2

从图中可以看出，NBA 球员 A.C. Green 参加了 NBA 1986—2001 年共计 16 年度的比赛。我们也可以将 dfPlayers 转换成 RDD 以后，通过 map 转换及 filter 过滤函数过滤出 NBA 球员 A.C. Green 的历史记录，然后打印输出结果。

```
1.    println("打印 NBA 球员的历年比赛记录：   ")
2.     dfPlayers.rdd.map(x =>
3.      (x.getString(1),x)).filter(_._1.contains("A.C. Green")).foreach(println)
```

在 IDEA 中运行代码，过滤出 NBA 球员 A.C. Green 的历史记录，输出结果如图 15-11 所示。

```
打印NBA球员的历年比赛记录：
[Stage 113:>                                                          (0 + 4) / 5](A.C.  Green, [0, A.C.  Green, 1986, 22, PF, LAL, 82, 1, 18.8, 2.5, 4
.7, 0.539, 0.0, 0.1, 0.167, 2.5, 4.7, 0.545, 0.54, 1.2, 2.0, 0.611, 2.0, 2.7, 4.6, 0.7, 0.6, 0.6, 1.2, 2.8, 6.4, 0.49895091840589934, -0.9386060995484626, -0
.4439580014286059, 0.3078457327660854, -0.7217083905580357, -0.30779547745145924, 0.23664828912184738, 0.38528648323192205, -0
.47620357949931524, -1.4595401249601245, 0.12205043812359559, -0.23876350390844645, -0.07518796992481205, 0.08915559365623643, -0
.14609571788413103, -0.0563011113440974, 0.03286893006898416, 0.12189859762675298, -0.14531111766440605, -0.295685861250323 8])
(A.C.  Green, [1, A.C.  Green, 1987, 23, PF, LAL, 79, 72, 28.4, 4.0, 7.4, 0.538, 0.0, 0.1, 0.0, 4.0, 7.4, 0.543, 0.538, 2.8, 3.6, 0.78, 2.7, 5.1, 7.8, 1.1, 0.9, 1.0, 1
.3, 2.2, 10.8, 1.0647443558788734, 0.4317423319030898, -0.4956343703118639, 1.4736712977283297, -0.5280946863340036, 0.29866739913350376, 0
.9393031455733895, 0.2042513776981104, 0.23705460068644219, 3.6257054519558705, 0.20229884757986655, 0.08108703012520127, -0
.10652353426919897, 0.3743133377279719, -0.11043091348881631, 0.0645866213427218, 0.1495373024322332, 0.053751503376815724, 0
.055923449487806434, 0.7645436443146016])
(A.C.  Green, [2, A.C.  Green, 1988, 24, PF, LAL, 82, 64, 32.1, 3.9, 7.8, 0.503, 0.0, 0.0, 0.0, 3.9, 7.8, 0.505, 0.503, 3.6, 4.6, 0.773, 3.0, 5.7, 8.7, 1.1, 1.1, 0.5,
1.5, 2.5, 11.4, 0.5344617689896461, 0.4672660813119134, -0.48712024642979096, 1.884472699069965, -0.5043192405380542, 0.6765261358530328, 0
.08403845256928431, -0.06968723861666007, 0.35920910809709483, 2.944847520306431, 0.11827604123363826, 0.11540827445867732, -0
.07753335737468439, 0.5396323766526926, -0.0919921267999587, 0.14663402692778477, 0.014288365188346563, -0.022148916116871076, 0
.08871830658292658, 0.8312829907525517])
```

图 15-11　NBA 球员 A.C. Green 的历史记录

15.3.4　NBA 历年比赛数据按球员分组统计分析

NBA 历年比赛数据按球员分组统计分析具体实现方法如下。

（1）对 NBA 球员的 DataFrame dfPlayers 先按姓名及经验值序号 exp（这里经验值序号的计算是将 NBA 球员的每年比赛时的年龄减去此球员曾经参加比赛的最小年龄，计算一个经验值，在按此球员分组时显示经验值序号）进行降序排列，调用 rdd 方法将 DataFrame 转换为 RDD，接下来对 RDD 根据 NBA 球员比赛数据的统计需求进行 map 转换，格式化为 Key-Value 的格式，其中 Key 为球员姓名，Value 为（Double, Double, Int, Int, Array[Double], Int）。map 转换新生成 RDD[(String, (Double, Double, Int, Int, Array[Double], Int))]。

（2）对 pStats 使用 groupByKey 算子根据 NBA 球员姓名进行分组。分组以后生成的 RDD pStats，类型为 RDD[(String, Iterable[(Double, Double, Int, Int, Array[Double], Int)])]，格式为 Key-Value，Key 是 NBA 球员姓名，Value 是 Iterable 迭代器。由于 dfPlayers 转换以后的 RDD 中包括 NBA 球员历年比赛的数据记录，因此按 NBA 球员姓名进行分组以后，将按照 NBA 球员姓名进行聚合，同一个 NBA 球员的比赛数据记录将聚合成一条记录。对于 NBA 球员历年的比赛数据，可以使用 for 循环语句打印 Iterable 中的迭代元素。

（3）对分组以后的 NBA 比赛数据进行缓存。

（4）打印输出 NBA 球员历年比赛数据。在对 pStats 的 Value 值（x._2）进行遍历时，x._2 是一个 Iterable 迭代器，我们将 Iterable 转换成数组 Array 以后，数组中的元素是一个元组，元组中包括 6 个元素，6 个元素的类型分别为：(Double, Double, Int, Int, Array[Double], Int)，分别获取每个元素使用“，”进行拼接打印输出；其中第 5 个元素仍是一个数组，我们调用数组的 mkString(" || ")方法，将第 5 个元素此数组中的元素使用“||”分隔以后打印输出。

```
1.    //*******************
2.      //统计分析
3.      //*******************
4.
5.      //按球员名字分组
6.      val dfPlayersRDD: RDD[(String, (Double, Double, Int, Int, Array
        [Double], Int))] =dfPlayers.sort(dfPlayers("name"), dfPlayers("exp")
        asc).rdd.map(x =>
```

```
7.        (x.getString(1), (x.getDouble(50), x.getDouble(40), x.getInt(2),
          x.getInt(3),
8.          Array(x.getDouble(31), x.getDouble(32), x.getDouble(33),
            x.getDouble(34), x.getDouble(35),
9.            x.getDouble(36), x.getDouble(37), x.getDouble(38), x.getDouble
              (39)), x.getInt(0))))
10.
11.     val pStats: RDD[(String, Iterable[(Double, Double, Int, Int, Array
        [Double], Int)])] = dfPlayers.sort(dfPlayers("name"), dfPlayers("exp")
        asc).rdd.map(x =>
12.       (x.getString(1), (x.getDouble(50), x.getDouble(40), x.getInt(2),
           x.getInt(3),
13.        Array(x.getDouble(31), x.getDouble(32), x.getDouble(33), x.getDouble(34),
           x.getDouble(35),
14.          x.getDouble(36), x.getDouble(37), x.getDouble(38), x.getDouble(39)),
             x.getInt(0))))
15.       .groupByKey
16.     pStats.cache
17.     println("**********根据 NBA 球员名字分组:     ")
18.     pStats.take(15).foreach(x => {
19.       val myx2: Iterable[(Double, Double, Int, Int, Array[Double], Int)] = x._2
20.       println("按 NBA 球员: " + x._1 + " 进行分组，组中元素个数为: " + myx2.size)
21.       for (i <- 1 to myx2.size) {
22.         val myx2size: Array[(Double, Double, Int, Int, Array[Double], Int)]
            = myx2.toArray
23.         val mynext: (Double, Double, Int, Int, Array[Double], Int) =
            myx2size(i - 1)
24.         println(i +":"+ x._1+",while " + mynext._1+","+mynext._2 + ","
25.           + mynext._3 + " , " + mynext._4 + " ,     " + mynext._5
              .mkString(" || ") + "     , "
26.           + mynext._6)
27.       }
28.
29.     })
```

在 IDEA 中运行代码，各 NBA 球员历年比赛数据显示输出结果如下：

```
1.   **********根据 NBA 球员名字分组:
2.  按 NBA 球员: Ticky Burden 进行分组，组中元素个数为: 2
3.  1 : Ticky Burden , while  NaN , NaN , 1977 , 23 ,    -0.6044576934502505 ||
    -0.8412145834216773 || NaN || -1.0255911476495823 || -0.6748427858591513 ||
    -0.02653087908155743 || -0.8073063498178507 || NaN     ||
    -0.6057684427416488     , 0
4.  2 : Ticky Burden , while   NaN , NaN , 1978 , 24 ,    -0.24519181462070969
    ||   -0.3625849339335772   ||   NaN   ||   -1.3750184794003877   ||
    -1.0771703003088466  ||  -0.661301276913946  ||  -0.8395535379341299  ||
    1.8623897720774216 || -1.3843610042164856      , 1
5.  按 NBA 球员: Mike Miller 进行分组，组中元素个数为: 16
6.  1 : Mike Miller , while   0.6850657516085638 , 2.7043002558967197 , 2001 ,
    20 ,    0.09942421490696489 || -0.22744626593747377 || 2.615968313590763
    || 0.1465279100082745 || -0.06819346117564598 || -0.1520055271653655 ||
    -0.4324268269064523 || 0.04210420956399251 || 0.6803476890116621     , 0
7.  2 : Mike Miller , while   0.8900247732778397 , 4.309214482353773 , 2002 ,
    21 ,    0.15211021867758653 || 0.3129797163416611 || 2.328193996192313
    || 0.26946346735406823 || 0.6857468344583921 || 0.06604153848660722 ||
    -0.09268751781179885 || -0.636820696475112 || 1.2241869251300554     , 1
8.  3 : Mike Miller , while   1.0832980841639197 , 4.722323449993903 , 2003 ,
    22 ,    0.2442504460819869 || 1.0968611638428813 || 2.2418027978864536
    || 0.6535403686491095 || 0.4830277706163585 || -0.1540658801895493 ||
```

```
    -0.2578077282339976 || -0.8650689937005274 || 1.2797835050411863    , 2
9.  4 : Mike Miller , while  0.6597686977121573 , 2.606620129894566 , 2004 ,
    23  ,          0.23960944027544986  ||  -0.14858114384165513  ||
    1.3342733513356602 || -0.11432468760563451 || 1.0774857600742487 ||
    0.5054852017284827 || -0.4115758747556613 || -0.4530062626577849 ||
    0.5772543453414607    , 3
10. 5 : Mike Miller , while  1.0161733339413734 , 4.590390252689687 , 2005 ,
    24 ,  1.7296411647848224 || -0.19205553968724373 || 2.1571870175598313
    || 0.13182088906215006 || 0.6421646603684523 || 0.11992240020228848 ||
    -0.22239768296112397 || -0.6534648462831214 || 0.8775721896436323  , 4
11. 6 : Mike Miller , while  0.9977947449783138 , 5.046645379184111 , 2006 ,
    25 ,    0.5531982944425201 || 0.6135777095588025 || 2.3190446880510933
    || 0.7926580931605534 || 0.5729116646487513 || 0.21667885050859406 ||
    0.006451111691680683 || -0.9363490309404241 || 0.9084739980625388   , 5
12. 7 : Mike Miller , while  1.650517474001914 , 7.134146483925148 , 2007 ,
    26 ,    0.5270017267319922 || 0.4798406988528217 || 3.7133431130431824
    || 0.7643576849738877 || 1.4001410217241135 || 0.42005377032412117 ||
    -0.18258263715507275 || -1.6260973802760275 || 1.6380884857061297   , 6
13. 8 : Mike Miller , while  1.3133699942353605 , 5.401135277847273 , 2008 ,
    27 ,    1.5529069893496565 || 0.4058208109132261 || 2.5559301197655686
    || 1.2326313536861488 || 0.8440957346766597 || -0.2684179047380797 ||
    -0.4306738978184909 || -1.8225967640068124 || 1.3314388360193972   , 7
14. 9 : Mike Miller , while  0.5734468850025483 , 2.5112393543942533 , 2009 ,
    28 ,    0.5189380631832368 || -0.2160265412705209 || 0.9802020907937623
    || 1.2026538942524694 || 1.4803860923736023 || -0.5294774294348025 ||
    -0.07176441430155282 || -1.0803050219732657 || 0.22663262077132446 , 8
15. 10 : Mike Miller , while  0.958449193125004 , 3.733433766444313 , 2010 ,
    29 ,    0.8612250838806804 || 0.3179326781864516 || 1.5173165363367993
    || 1.0137831299243059 || 1.1643930828314868 || 0.18020452329899148 ||
    -0.5011914666380154 || -1.2156977168795124 || 0.39546791550312577  , 9
16.11 : Mike Miller , while  -0.29347820905755756 , -1.763458223108538 , 2011 ,
    30  ,          -0.7013416635715775  ||  -0.42080520471622745  ||
    0.7571942743956656 || 0.3804268294575492 || -0.3087281640843859 ||
    -0.28427972338550167 || -0.9469791817476815 || 0.21225963181769425 ||
    -0.4512050212740733   , 10
17.12 : Mike Miller , while  0.07142224106269339 , -0.5221381704016791 , 2012 ,
    31  ,          -0.11954057342167189  ||  -0.40409034293903157  ||
    1.4046903328230618 || -0.11189396474350337 || -0.36610268870172846 ||
    -0.6127682429692385 || -0.49968733764574624 || 0.521516896855799 ||
    -0.3342622496596199   , 11
18. 13 : Mike Miller , while  -0.14384868876900747 , -1.0063327682068888 ,
    2013 , 32 ,      -0.17620651413761934 || -0.29112729830094763 ||
    0.9282731583962052 || -0.32548024406260884 || -0.059587675974944194 ||
    -0.5516399655790644 || -0.7034076570377812 || 0.7480110247641681 ||
    -0.5751675962742966   , 12
19. 14 : Mike Miller , while  -0.0903503537589224 , -0.49970776283493756 ,
    2014 , 33 ,      0.35109325571501176 || -0.05182514210961124 ||
    0.9826688443655516 || -0.41409681899060763 || -0.09913247529764337 ||
    -0.7622768313158717 || -0.6724319422017886 || 0.3355490304181267 ||
    -0.16925568341810523   , 13
20. 15 : Mike Miller , while  -0.8599519450026487 , -3.798801467447749 , 2015 ,
    34  ,          -0.6736851211371261  ||  -0.23317877480844645  ||
    -0.069583364878147 || -0.7202553668317317 || -0.5303790955293763 ||
    -0.81369967347111  || -0.6449637880550209 ||  0.98314941611857  ||
    -1.0962056988553603   , 14
21.16 : Mike Miller , while  -0.7933007669491381 , -3.9162696664446126 , 2016 ,
    35  ,          -0.34219721906458794  ||  -0.28418680944243174  ||
    -0.3490592074254263 || -0.8895932363798921 || -0.44828266728246297 ||
    -0.7306920273216803 || -0.6534242946356905 || 0.9323148901405814 || -
    1.151149095033022   , 15
```

```
22. ......
```

15.3.5　NBA 球员年龄值及经验值列表获取

基于 NBA 历届各年的历史比赛数据，计算每个 NBA 球员的 valueZ 及 valueN 球员价值的变化情况。将结果保存成两个列表：一个为年龄列表；一个为经验列表。这里由于 1980 年我们只有部分数据，1980 年原始记录统计不全，因此数据清洗时将 1980 年的数据过滤掉。

NBA 球员年龄值及经验值列表获取具体实现方法如下。

（1）从 dfPlayers 中过滤出比赛年份为 1980 年的数据，选择姓名列，通过 collect 算子收集以后，使用 mkstring 函数将参加 1980 年比赛的 NBA 球员的姓名用“，”拼接成字符串。

（2）将 NBA 球员姓名进行分组以后生成的 RDD pStats 通过 map 函数进行转换，pStats 中每行元素是一个元组(String, Iterable[(Double, Double, Int, Int, Array[Double], Int)])，将其进行模式匹配，元组中的第一个元素是姓名 name，元组中的第二个元素是 stats，其数据类型为 Iterable[(Double, Double, Int, Int, Array[Double], Int)]。使用 foreach 方法遍历 stats 的 Iterable 的每个元素，根据 NBA 业务统计需求进行计算。

（3）pStats 进行 map 转换时，创建生成年龄列表 aList、经验值列表 eList，其类型为 ListBuffer[(Int, Array[Double])]，最后将（年龄列表 aList, 经验值列表 eList）组合成一个元组返回，生成 pStats1 的 RDD，其元素类型为 RDD[(ListBuffer[(Int, Array[Double])], ListBuffer[(Int, Array[Double])])]。

（4）pStats1 使用 cache 算子进行缓存。

（5）按 NBA 球员的年龄及经验值打印输出。其中，x._1 是年龄列表 aList 的值，x._1(i-1) 遍历 aList 列表中的每个元素，其中 aList 列表每一个元素的第一个值是年龄 x._1(i-1)._1，第二个元素是一个数组 x._1(i-1)._2，我们使用分隔符("||")进行拼接。其中，x._2 是经验值列表 eList 的值，x._2（i-1）遍历 eList 列表中的每个元素，其中 eList 列表每个元素的第一个值是经验值 exp x._2(i-1)._1，第二个元素是一个数组 x._2(i-1)._2，我们使用“||”进行拼接。

NBA 球员年龄值及经验值代码如下：

```
1.  import spark.implicits._
2.      // 计算每个 NBA 球员的 valueZ 及 valueN 球员价值的变化情况。将结果保存成两个列表：
        // 一个为年龄列表；一个为经验列表
3.      val excludeNames: String = dfPlayers.filter(dfPlayers("year") ===
        1980).select(dfPlayers("name"))
4.        .map(x => x.mkString).collect().mkString(",")
5.
6.      val pStats1: RDD[(ListBuffer[(Int, Array[Double])], ListBuffer[(Int,
        Array[Double])])] = pStats.map { case (name, stats) =>
7.        var last = 0
8.        var deltaZ = 0.0
9.        var deltaN = 0.0
10.       var valueZ = 0.0
11.       var valueN = 0.0
12.       var exp = 0
13.       val aList = ListBuffer[(Int, Array[Double])]()
14.       val eList = ListBuffer[(Int, Array[Double])]()
15.       stats.foreach(z => {
16.         if (last > 0) {
17.           deltaN = z._1 - valueN
18.           deltaZ = z._2 - valueZ
```

```
19.        } else {
20.          deltaN = Double.NaN
21.          deltaZ = Double.NaN
22.        }
23.        valueN = z._1
24.        valueZ = z._2
25.        last = z._4
26.        aList += ((last, Array(valueZ, valueN, deltaZ, deltaN)))
27.        if (!excludeNames.contains(z._1)) {
28.          exp = z._6
29.          eList += ((exp, Array(valueZ, valueN, deltaZ, deltaN)))
30.        }
31.      })
32.      (aList, eList)
33.    }
34.
35.    pStats1.cache
36.
37.
38.    println("按 NBA 球员的年龄及经验值进行统计：    ")
39.    pStats1.take(10).foreach(x => {
40.      //pStats1: RDD[(ListBuffer[(Int, Array[Double])], ListBuffer[(Int,
       Array[Double])])]
41.      for (i <- 1 to x._1.size) {
42.        println("年龄：" + x._1(i - 1)._1+"," + x._1(i - 1)._2.mkString("||") +
43.          "  经验：" + x._2(i-1)._1 + " , " + x._2(i - 1)._2.mkString("||"))
44.      }
45.    })
```

在 IDEA 中运行代码，NBA 球员年龄值及经验值列表获取显示输出结果如图 15-12 所示。

```
按NBA球员的年龄及经验值进行统计：
年龄：23 , NaN||NaN||NaN||NaN  经验：0 , NaN||NaN||NaN||NaN
年龄：24 , NaN||NaN||NaN||NaN  经验：1 , NaN||NaN||NaN||NaN
年龄：20 , 2.7043002558967197||0.6850657516085638||NaN||NaN  经验：0 , 2.7043002558967197||0.6850657516085638||NaN||NaN
年龄：21 , 4.309214482353773||0.8900247732778397||1.604914226457053||0.20495902166927593  经验：1 , 4.309214482353773||0
.8900247732778397||1.604914226457053||0.20495902166927593
年龄：22 , 4.722323449993903||1.0832980841639197||0.41310896764013005||0.19327331088608002  经验：2 , 4.722323449993903||1
.0832980841639197||0.41310896764013005||0.19327331088608002
年龄：23 , 2.606620129894566||0.6597686977121573||-2.1157033200993367||-0.4235293864517624  经验：3 , 2.606620129894566||0
.6597686977121573||-2.1157033200993367||-0.4235293864517624
年龄：24 , 4.590390252689687||1.0161733339413734||1.983770122795121||0.35640463622921614  经验：4 , 4.590390252689687||1
.0161733339413734||1.983770122795121||0.35640463622921614
年龄：25 , 5.046645379184111||0.9977947449783138||0.45625512649442346||-0.01837858896305966  经验：5 , 5.046645379184111||0
.9977947449783138||0.45625512649442346||-0.01837858896305966
```

图 15-12　NBA 球员年龄值及经验值列表获取

15.3.6　NBA 球员年龄值及经验值统计分析

NBA 球员年龄值及经验值统计分析的具体实现方法如下。

（1）对已获取的年龄列表 aList 及经验值列表 eList 的 RDD pStats1 通过 flatMap 算子进行转换，提取年龄列表数据生成新的 RDD pStats2。

数据类型 RDD[(ListBuffer[(Int, Array[Double])], ListBuffer[(Int, Array[Double])])]转换为 RDD[(Int, Array[Double])]。

（2）将 RDD pStats2 通过 processStatsAgeOrExperience 函数处理，传入参数 age，生成年

龄统计的 DataFrame dfAge。

（3）打印展示 dfAge。

（4）对已获取的年龄列表 aList 及经验值列表 eList 的 RDD pStats1 通过 flatMap 算子进行转换，提取经验列表数据生成新的 RDD pStats3。

数据类型 RDD[(ListBuffer[(Int, Array[Double])], ListBuffer[(Int, Array[Double])])]转换为 RDD[(Int, Array[Double])]。

（5）将 RDD pStats3 通过 processStatsAgeOrExperience 函数处理，传入参数 Experience，生成经验统计的 DataFrame ddfExperience。

（6）打印展示 dfExperience。

NBA 球员年龄值及经验值代码如下：

```
1.   //********************
2.     //计算年龄统计分析
3.     //*********************
4.
5.     //抽取年龄列表 age list
6.     val pStats2: RDD[(Int, Array[Double])]=pStats1.flatMap { case (x, y) => x }
7.
8.     //创建年龄 dataframe
9.     val dfAge: DataFrame = processStatsAgeOrExperience(pStats2, "age")
10.    dfAge.show() //展示
11.    //保存为表
12.    dfAge.createOrReplaceTempView("Age")
13.
14.    //抽取经验列表 experience list
15.    val pStats3: RDD[(Int, Array[Double])] = pStats1.flatMap { case (x, y)=>y}
16.
17.    //创建经验 dataframe
18.    val dfExperience: DataFrame = processStatsAgeOrExperience(pStats3,
       "Experience")
19.    dfExperience.show()
20.    //保存为表
21.    dfExperience.createOrReplaceTempView("Experience")
22.    //去掉持久化
23.    pStats1.unpersist()
```

在 IDEA 中运行代码，NBA 球员年龄列表 DataFrame dfAge.show()的显示结果如图 15-13 所示。

```
+---+------------+------------------+------------------+------------------+------------------+------------+------------------+----
|age|valueZ_count|       valueZ_mean|      valueZ_stdev|        valueZ_max|        valueZ_min|valueN_count|       valueN_mean|
   valueN_stdev|        valueN_max|        valueN_min|deltaZ_count|       deltaZ_mean|      deltaZ_stdev|        deltaZ_max|
     deltaZ_min|deltaN_count|       deltaN_mean|      deltaN_stdev|        deltaN_max|        deltaN_min|
+---+------------+------------------+------------------+------------------+------------------+------------+------------------+----
| 25|      1455.0| 0.25379208325747654| 4.469784981759089|  18.37868106915914| -9.768147167470113|      1455.0| 0.05356755942526729|
 0.9363648516284568|   4.084154283162466|-2.0434554117927335|         1297.0| 0.28261995970634324|2.614627372691107 3|  13.000726189142348|
  -9.877047197045487|        1297.0|0.060896748790134854| 0.5579199143095767|  2.9988850416463215|-1.6748235925501425|
| 30|       883.0|  0.6386628373762322| 4.150782078317183| 19.250691022504732| -8.908506726387039|       883.0| 0.13588879424834538|
 0.8678142345638947|   3.953292476287818| -1.786976053934938|          856.0| -0.6245408433328367| 2.563521963997515|    13
 .46284963135733|-11.626289129299085|          856.0|-0.13836178606647828| 0.5473719505010776|   2.155319710971554|-2.9701318159216368|
| 35|       251.0|-0.13235133435419635| 3.961448143850996| 11.225378864018998| -8.576621367534942|       251.0|-0.03176256936783...|
 0.8259434133041167|    2.37338783765769|-1.7177478005272369|          248.0| -1.0531438304210574|2.0348301829775175|     3
 .839124523312493|-11.394073219134894|          248.0| -0.2275861148364875| 0.4374962873798658|  0.8761850410735177|-1.8979816577642625|
```

图 15-13　NBA 球员年龄列表 DataFrame dfAge.show()的显示结果

NBA 球员经验列表 DataFrame dfExperience.show()的显示结果如图 15-14 所示。

```
+----------+------------+-------------------+------------------+------------------+-------------------+------------+
|Experience|valueZ_count|        valueZ_mean|       valueZ_stdev|        valueZ_max|         valueZ_min|valueN_count|
 valueN_mean|        valueN_stdev|          valueN_max|         valueN_min|deltaZ_count|        deltaZ_mean|      deltaZ_stdev|
  deltaZ_max|          deltaZ_min|deltaN_count|         deltaN_mean|      deltaN_stdev|        deltaN_max|        deltaN_min|
+----------+------------+-------------------+------------------+------------------+-------------------+------------+
|        15|        90.0|  -0.348780200560152|   4.146224123151|  11.21414315439408|  -6.36267200315876|        90.0|-0
.09838034502652812| 0.8597100317394312|   2.37338783765769|-1.3618458180628608|        90.0|   -1.32156390899947| 2.494686233791554|
 4.949390589183601| -9.713975160916561|        90.0|-0.29985188278545294| 0.5012939060825078|  0.8761850410735177| -1.8523304625735544|
|         0|      2502.0|  -2.699725685968074|2.9097378003192222| 13.307255142897896|-12.426726255774238|      2502.0| -0
.5709560010594176|  0.606056640307347|  2.9231669384755223|-2.5456706028768226|         6.0|  0.9449536820938823| 4.340782917423647|
 8.547256226143105|  -4.46240737700405|         6.0| 0.26506254886106556| 1.0257103369016858|  2.2219561872156377| -0.9483858042334511|
|        20|         3.0| -2.7881137061708317| 1.180383156651202|-1.141397509766164|-3.848539138969675|         3.0| -0
.5800732817179493| 0.3031146424370221|-0.15140860346456714|-0.7960411339228379|         3.0| -1.4685559595246707|1.0389147731704078|-0
.0823613251839248| -2.583392752747825|         3.0| -0.2867980479922298|0.23379500428993494|0.043164217649615644| -0.470050291501099 4|
|        10|       562.0|  0.5781598777498813| 4.224673728993368| 15.779549978808273|  -7.49564814660065|       562.0| 0
.13113052602508968|  0.889720273904276|  3.1317470994374585|-1.6978042762106786|       551.0|  -1.042677772297701|2.3445804672158435|
 11.130624597433679| -9.996872630906662|       551.0|-0.21897173557180127| 0.5084908560909539|  2.4110473116912563| -2.1962330901489002|
```

图 15-14　NBA 球员经验列表 DataFrame dfExperience.show()的显示结果

15.3.7　NBA 球员系统内部定义的函数、辅助工具类

NBA 球员系统内部定义的函数包括 statNormalize、bbParse、processStatsAgeOrExperience、processStats；辅助工具类包括 BballData、BballStatCounter。

NBA 球员系统内部定义的函数、辅助工具类跟 NBA 篮球运动员大数据分析系统的业务需求相关，如读者对 NBA 篮球比赛系统感兴趣，可以深入研究 NBA 篮球系统具体的指标算法的计算。这里不对 NBA 业务需求实现展开叙述。

NBA 球员系统内部定义的函数、辅助工具类如下：

```
1.   //********************
2.     //类，辅助函数、变量
3.     //********************
4.     import org.apache.spark.sql.Row
5.     import org.apache.spark.sql.types._
6.     import org.apache.spark.util.StatCounter
7.
8.     import scala.collection.mutable.ListBuffer
9.
10.    //计算归一化的辅助函数
11.    def statNormalize(stat: Double, max: Double, min: Double) = {
12.      val newmax = math.max(math.abs(max), math.abs(min))
13.      stat / newmax
14.    }
15.
16.    //初始化 + 权重统计+归一统计
17.    case class BballData(val year: Int, name: String, position: String,
       age: Int, team: String, gp: Int, gs: Int, mp: Double, stats: Array
       [Double], statsZ: Array[Double] = Array[Double](), valueZ: Double =
       0, statsN: Array[Double] = Array[Double](), valueN: Double = 0,
       experience: Double = 0)
18.
19.    //解析转换为 BBallDataZ 对象
20.    def bbParse(input: String, bStats: scala.collection.Map[String,
```

```
    Double] = Map.empty, zStats: scala.collection.Map[String, Double] =
    Map.empty) = {
21.     val line = input.replace(",,", ",0,")
22.     val pieces = line.substring(1, line.length - 1).split(",")
23.     val year = pieces(0).toInt
24.     val name = pieces(2)
25.     val position = pieces(3)
26.     val age = pieces(4).toInt
27.     val team = pieces(5)
28.     val gp = pieces(6).toInt
29.     val gs = pieces(7).toInt
30.     val mp = pieces(8).toDouble
31.     val stats = pieces.slice(9, 31).map(x => x.toDouble)
32.     var statsZ: Array[Double] = Array.empty
33.     var valueZ: Double = Double.NaN
34.     var statsN: Array[Double] = Array.empty
35.     var valueN: Double = Double.NaN
36.
37.     if (!bStats.isEmpty) {
38.       val fg = (stats(2) - bStats.apply(year.toString + "_FG%_avg")) *
          stats(1)
39.       val tp = (stats(3) - bStats.apply(year.toString + "_3P_avg")) /
          bStats.apply(year.toString + "_3P_stdev")
40.       val ft = (stats(12) - bStats.apply(year.toString + "_FT%_avg")) *
          stats(11)
41.       val trb = (stats(15) - bStats.apply(year.toString + "_TRB_avg"))
          / bStats.apply(year.toString + "_TRB_stdev")
42.       val ast = (stats(16) - bStats.apply(year.toString + "_AST_avg"))
          / bStats.apply(year.toString + "_AST_stdev")
43.       val stl = (stats(17) - bStats.apply(year.toString + "_STL_avg"))
          / bStats.apply(year.toString + "_STL_stdev")
44.       val blk = (stats(18) - bStats.apply(year.toString + "_BLK_avg"))
          / bStats.apply(year.toString + "_BLK_stdev")
45.       val tov = (stats(19) - bStats.apply(year.toString + "_TOV_avg"))
          / bStats.apply(year.toString + "_TOV_stdev") * (-1)
46.       val pts = (stats(21) - bStats.apply(year.toString + "_PTS_avg"))
          / bStats.apply(year.toString + "_PTS_stdev")
47.       statsZ = Array(fg, ft, tp, trb, ast, stl, blk, tov, pts)
48.       valueZ = statsZ.reduce(_ + _)
49.
50.       if (!zStats.isEmpty) {
51.         val zfg = (fg - zStats.apply(year.toString + "_FG_avg")) /
            zStats.apply(year.toString + "_FG_stdev")
52.         val zft = (ft - zStats.apply(year.toString + "_FT_avg")) /
            zStats.apply(year.toString + "_FT_stdev")
53.         val fgN = statNormalize(zfg, (zStats.apply(year.toString +
            "_FG_max") - zStats.apply(year.toString + "_FG_avg"))
54.          / zStats.apply(year.toString + "_FG_stdev"), (zStats.apply
             (year.toString + "_FG_min")
55.          - zStats.apply(year.toString + "_FG_avg")) / zStats.apply
             (year.toString + "_FG_stdev"))
56.         val ftN = statNormalize(zft, (zStats.apply(year.toString +
            "_FT_max") - zStats.apply(year.toString + "_FT_avg"))
57.          / zStats.apply(year.toString + "_FT_stdev"), (zStats.apply
             (year.toString + "_FT_min")
58.          - zStats.apply(year.toString + "_FT_avg")) / zStats.apply
             (year.toString + "_FT_stdev"))
59.         val tpN = statNormalize(tp, zStats.apply(year.toString +
            "_3P_max"), zStats.apply(year.toString + "_3P_min"))
60.         val trbN = statNormalize(trb, zStats.apply(year.toString +
```

```
          "_TRB_max"), zStats.apply(year.toString + "_TRB_min"))
61.          val  astN  =  statNormalize(ast,  zStats.apply(year.toString  +
          "_AST_max"), zStats.apply(year.toString + "_AST_min"))
62.          val  stlN  =  statNormalize(stl,  zStats.apply(year.toString  +
          "_STL_max"), zStats.apply(year.toString + "_STL_min"))
63.          val  blkN  =  statNormalize(blk,  zStats.apply(year.toString  +
          "_BLK_max"), zStats.apply(year.toString + "_BLK_min"))
64.          val  tovN  =  statNormalize(tov,  zStats.apply(year.toString  +
          "_TOV_max"), zStats.apply(year.toString + "_TOV_min"))
65.          val  ptsN  =  statNormalize(pts,  zStats.apply(year.toString  +
          "_PTS_max"), zStats.apply(year.toString + "_PTS_min"))
66.          statsZ = Array(zfg, zft, tp, trb, ast, stl, blk, tov, pts)
67.          valueZ = statsZ.reduce(_ + _)
68.          statsN = Array(fgN, ftN, tpN, trbN, astN, stlN, blkN, tovN, ptsN)
69.          valueN = statsN.reduce(_ + _)
70.        }
71.      }
72.      BballData(year, name, position, age, team, gp, gs, mp, stats, statsZ,
      valueZ, statsN, valueN)
73.    }
74.
75.
76.    //该类是一个辅助工具类，在后面编写业务代码的时候会反复使用其中的方法
77.    class BballStatCounter extends Serializable {
78.      val stats: StatCounter = new StatCounter()
79.      var missing: Long = 0
80.
81.      def add(x: Double): BballStatCounter = {
82.        if (x.isNaN) {
83.          missing += 1
84.        } else {
85.          stats.merge(x)
86.        }
87.        this
88.      }
89.
90.      def merge(other: BballStatCounter): BballStatCounter = {
91.        stats.merge(other.stats)
92.        missing += other.missing
93.        this
94.      }
95.
96.      def printStats(delim: String): String = {
97.        stats.count + delim + stats.mean + delim + stats.stdev + delim +
        stats.max + delim + stats.min
98.      }
99.
100.        override def toString: String = {
101.          "stats: " + stats.toString + " NaN: " + missing
102.        }
103.      }
104.
105.      object BballStatCounter extends Serializable {
106.        def apply(x: Double) = new BballStatCounter().add(x)
       //这里使用了 Scala 语言的一个编程技巧，借助于 apply 工厂方法，在构造该对象
       //的时候就可以执行结果
107.      }
108.
109.      //处理原始数据为 zScores and nScores
```

```
110.        def processStats(stats0: org.apache.spark.rdd.RDD[String],
        txtStat: Array[String],
111.                        bStats:  scala.collection.Map[String,  Double]  =
                            Map.empty,
112.                        zStats:  scala.collection.Map[String,  Double]  =
                            Map.empty) = {
113.          //解析 stats
114.          val stats1 = stats0.map(x => bbParse(x, bStats, zStats))
115.
116.          //按年份进行分组
117.          val stats2 = {
118.            if (bStats.isEmpty) {
119.              stats1.keyBy(x => x.year).map(x => (x._1, x._2.stats))
                 .groupByKey()
120.            } else {
121.              stats1.keyBy(x => x.year).map(x => (x._1, x._2.statsZ))
                 .groupByKey()
122.            }
123.          }
124.
125.          //转换成 StatCounter
126.          val stats3 = stats2.map { case (x, y) => (x, y.map(a => a.map(b
          => BballStatCounter(b)))) }
127.
128.          //合并
129.          val stats4 = stats3.map { case (x, y) => (x, y.reduce((a, b) =>
          a.zip(b).map { case (c, d) => c.merge(d) })) }
130.
131.          //combine 合并聚合
132.          val stats5 = stats4.map { case (x, y) => (x, txtStat.zip(y)) }.map
{
133.            x => (x._2.map {
134.            case (y, z) => (x._1, y, z) }) }
135.
136.          //使用逗号分隔符打印输出
137.          val stats6 = stats5.flatMap(x => x.map(y => (y._1, y._2,
          y._3.printStats(","))))
138.
139.          //转换为 key-value 键值对
140.          val stats7 = stats6.flatMap { case (a, b, c) => {
141.            val pieces = c.split(",")
142.            val count = pieces(0)
143.            val mean = pieces(1)
144.            val stdev = pieces(2)
145.            val max = pieces(3)
146.            val min = pieces(4)
147.            Array((a + "_" + b + "_" + "count", count.toDouble),
148.              (a + "_" + b + "_" + "avg", mean.toDouble),
149.              (a + "_" + b + "_" + "stdev", stdev.toDouble),
150.              (a + "_" + b + "_" + "max", max.toDouble),
151.              (a + "_" + b + "_" + "min", min.toDouble))
152.          }
153.          }
154.          stats7
155.        }
156.
157.        //处理经验值函数
158.        def processStatsAgeOrExperience(stats0: org.apache.spark.rdd
        .RDD[(Int, Array[Double])], label: String) = {
```

```
159.
160.
161.        //按年龄分组
162.        val stats1 = stats0.groupByKey()
163.
164.        //转换为 StatCounter 对象
165.        val stats2 = stats1.map { case (x, y) => (x, y.map(z => z.map(a
            => BballStatCounter(a)))) }
166.
167.        //通过合并 StatCounter 对象汇聚行数据
168.        val stats3 = stats2.map { case (x, y) => (x, y.reduce((a, b) =>
            a.zip(b).map { case (c, d) => c.merge(d) })) }
169.
170.        //转换为 RDD[Row] 对象
171.        val stats4 = stats3.map(x => Array(Array(x._1.toDouble),
172.          x._2.flatMap(y => y.printStats(",").split(",")).map(y =>
              y.toDouble)).flatMap(y => y))
173.          .map(x =>
174.            Row(x(0).toInt, x(1), x(2), x(3), x(4), x(5), x(6), x(7), x(8),
175.             x(9), x(10), x(11), x(12), x(13), x(14), x(15), x(16), x(17),
               x(18), x(19), x(20)))
176.
177.        //创建年龄列表的元数据
178.        val schema = StructType(
179.          StructField(label, IntegerType, true) ::
180.            StructField("valueZ_count", DoubleType, true) ::
181.            StructField("valueZ_mean", DoubleType, true) ::
182.            StructField("valueZ_stdev", DoubleType, true) ::
183.            StructField("valueZ_max", DoubleType, true) ::
184.            StructField("valueZ_min", DoubleType, true) ::
185.            StructField("valueN_count", DoubleType, true) ::
186.            StructField("valueN_mean", DoubleType, true) ::
187.            StructField("valueN_stdev", DoubleType, true) ::
188.            StructField("valueN_max", DoubleType, true) ::
189.            StructField("valueN_min", DoubleType, true) ::
190.            StructField("deltaZ_count", DoubleType, true) ::
191.            StructField("deltaZ_mean", DoubleType, true) ::
192.            StructField("deltaZ_stdev", DoubleType, true) ::
193.            StructField("deltaZ_max", DoubleType, true) ::
194.            StructField("deltaZ_min", DoubleType, true) ::
195.            StructField("deltaN_count", DoubleType, true) ::
196.            StructField("deltaN_mean", DoubleType, true) ::
197.            StructField("deltaN_stdev", DoubleType, true) ::
198.            StructField("deltaN_max", DoubleType, true) ::
199.            StructField("deltaN_min", DoubleType, true) :: Nil
200.        )
201.
202.        //创建 data frame
203.        spark.createDataFrame(stats4, schema)
204.      }
```

15.4　NBA 篮球运动员大数据分析完整代码测试和实战

Spark 商业案例之 NBA 篮球运动员大数据分析系统应用案例 NBABasketball_

Analysis.scala，如例 15-1 所示。

【例 15-1】 NBABasketball_Analysis.scala 代码。

```
1.  package com.dt.spark.sparksql
2.
3.  import scala.language.postfixOps
4.  import org.apache.hadoop.conf.Configuration
5.  import org.apache.hadoop.fs.{FileSystem, Path}
6.  import org.apache.log4j.{Level, Logger}
7.  import org.apache.spark.SparkConf
8.  import org.apache.spark.broadcast.Broadcast
9.  import org.apache.spark.rdd.RDD
10. import org.apache.spark.sql.{DataFrame, SparkSession}
11.
12. import scala.collection.{Map, mutable}
13.
14. /**
15.   * 版权：DT 大数据梦工厂所有
16.   * 时间：2017 年 1 月 26 日；
17.   * NBA 篮球运动员大数据分析决策支持系统：
18.   * 基于 NBA 球员历史数据 1970—2017 年各种表现，全方位分析球员的技能，构建最强
      * NBA 篮球团队做数据分析支撑系统
19.   * 曾经非常火爆的梦幻篮球是基于现实中的篮球比赛数据根据对手的情况制定游戏的先发阵
      * 容和比赛结果（也就是说，比赛结果是由实际结果决定的），游戏中可以管理球员，例如，
      * 调整比赛的阵容，其中也包括裁员、签入和交易等
20.   *
21.   *
22.   * 这里的大数据分析系统可以被认为是游戏背后的数据分析系统
23.   * 具体关键的数据项如下所示
24.   * 3P：3 分命中
25.   * 3PA：3 分出手
26.   * 3P%：3 分命中率
27.   * 2P：2 分命中
28.   * 2PA：2 分出手
29.   * 2P%：2 分命中率
30.   * TRB：篮板球
31.   * STL：抢断
32.   * AST：助攻
33.   * BLT：盖帽
34.   * FT：罚球命中
35.   * TOV：失误
36.   *
37.   *
38.   * 基于球员的历史数据，如何对球员进行评价？也就是如何进行科学的指标计算，一个比较流
      * 行的算法是 Z-score：其基本的计算过程是：
39.   * 基于球员的得分减去平均值后除以标准差，举一个简单的例子，某个球员在 2016 年的平均
      * 篮板数是 7.1，而所有球员在 2016 年的平均篮板数是 4.5
40.   * 标准差是 1.3，那么该球员 Z-score 得分为 2
41.   *
42.   * 在计算球员的表现指标中，可以计算 FT%、BLK、AST、FG%等
43.   *
44.   *
45.   * 具体如何通过 Spark 技术来实现呢？
46.   * 第一步：数据预处理。例如，去掉不必要的标题等信息
47.   * 第二步：数据的缓存。为加速后面的数据处理打下基础
```

```
48.   * 第三步：基础数据项计算：方差、均值、最大值、最小值、出现次数等
49.   * 第四步：计算 Z-score，一般会进行广播，可以提升效率
50.   * 第五步：基于前面四步的基础，可以借助 Spark SQL 进行多维度 NBA 篮球运动员数据分析，
      * 可以使用 SQL 语句，也可以使用 DataSet（我们在这里会优先选择使用 SQL，为什么呢？
      * 原因非常简单，复杂的算法级别的计算已经在前面四步完成了且广播给了集群，我们在 SQL
      * 中可以直接使用）
51.   * 第六步：把数据放在 Redis 或者 DB 中
52.   *
53.   *
54.   *
55.   * Tips:
56.   * 1. 这里的一个非常重要的实现技巧是:通过 RDD 计算出一些核心基础数据，并广播出去，后面
57.   * 的业务基于 SQL 去实现，既简单，又可以灵活地应对业务变化需求，希望对大家有所启发；
58.   * 2. 使用缓存和广播以及调整并行度等提升效率
59.   *
60.   */
61. object NBABasketball_Analysis {
62.
63.   def main(args: Array[String]) {
64.     Logger.getLogger("org").setLevel(Level.ERROR)
65.     var masterUrl = "local[4]"
66.     if (args.length > 0) {
67.       masterUrl = args(0)
68.     }
69.
70.     // 根据给定的 master URL 地址创建 SparContext
71.     /**
72.       * Spark SQL 默认情况下 Shuffle 的时候并行度是 200，如果数据量不是非常大的情
        * 况下，设置 200 的 Shuffle 并行度会拖慢速度，所以这里我们根据实际情况进行了调
        * 整，因为 NBA 的篮球运动员的数据并不是那么多，这样做同时也可以使机器得到更有效
        * 的使用（如内存等）
73.       */
74.
75.     val  conf  =  new  SparkConf().setMaster(masterUrl).set("spark.sql.
      shuffle.partitions", "5").setAppName("FantasyBasketball")
76.     val spark = SparkSession
77.       .builder()
78.       .appName("NBABasketball_Analysis")
79.       .config(conf)
80.       .getOrCreate()
81.
82.     val sc = spark.sparkContext
83.
84.     //********************
85.     //SET-UP
86.     //********************
87.
88.
89.     val DATA_PATH = "data/NBABasketball"
90.     //数据存在的目录
91.     val TMP_PATH = "data/basketball_tmp"
92.
93.     val fs = FileSystem.get(new Configuration())
94.     fs.delete(new Path(TMP_PATH), true)
95.
96.     //处理文件使每行包含一年的数据
97.     for (i <- 1970 to 2016) {
98.       println(i)
```

```
99.      val yearStats = sc.textFile(s"${DATA_PATH}/leagues_NBA_$i*")
         .repartition(sc.defaultParallelism)
100.        yearStats.filter(x => x.contains(",")).map(x => (i, x)).saveAs
            TextFile(s"${TMP_PATH}/BasketballStatsWithYear/$i/")
101.      }
102.
103.
104.      //********************
105.      //CODE
106.      //********************
107.      //剪切和粘贴到 Spark Sheu。输入：paste 进入“剪切和粘贴”模式，然后输入 Ctrl+D
          //组合键进行处理
108.      //输入 spark-shell 命令：spark-shell --master yarn-client
109.      //********************
110.
111.
112.      //********************
113.      //类、辅助函数+变量
114.      //********************
115.      import org.apache.spark.sql.Row
116.      import org.apache.spark.sql.types._
117.      import org.apache.spark.util.StatCounter
118.
119.      import scala.collection.mutable.ListBuffer
120.
121.      //用辅助函数计算归一化值
122.      def statNormalize(stat: Double, max: Double, min: Double) = {
123.        val newmax = math.max(math.abs(max), math.abs(min))
124.        stat / newmax
125.      }
126.
127.      //保持初始化的篮球统计加权及归一化数据统计
128.      case class BballData(val year: Int, name: String, position: String,
129.                           age: Int, team: String, gp: Int, gs: Int, mp: Double,
130.                           stats: Array[Double], statsZ: Array[Double] =
                               Array[Double](),
131.                           valueZ: Double = 0, statsN: Array[Double] =
                               Array[Double](),
132.                           valueN: Double = 0, experience: Double = 0)
133.
134.      //解析数据为 BBallDataZ 对象
135.      def bbParse(input: String, bStats: scala.collection.Map[String,
           Double] = Map.empty,
136.                  zStats: scala.collection.Map[String, Double] = Map
                      .empty): BballData = {
137.        val line = input.replace(",,", ",0,")
138.        val pieces = line.substring(1, line.length - 1).split(",")
139.        val year = pieces(0).toInt
140.        val name = pieces(2)
141.        val position = pieces(3)
142.        val age = pieces(4).toInt
143.        val team = pieces(5)
144.        val gp = pieces(6).toInt
145.        val gs = pieces(7).toInt
146.        val mp = pieces(8).toDouble
147.
148.        val stats: Array[Double] = pieces.slice(9, 31).map(x => x.toDouble)
149.        var statsZ: Array[Double] = Array.empty
```

```
150.        var valueZ: Double = Double.NaN
151.        var statsN: Array[Double] = Array.empty
152.        var valueN: Double = Double.NaN
153.
154.        if (!bStats.isEmpty) {
155.          val fg: Double = (stats(2) - bStats.apply(year.toString +
              "_FG%_avg")) * stats(1)
156.          val tp = (stats(3) - bStats.apply(year.toString + "_3P_avg"))
              / bStats.apply(year.toString + "_3P_stdev")
157.          val ft = (stats(12) - bStats.apply(year.toString + "_FT%_avg"))
              * stats(11)
158.         val trb = (stats(15) - bStats.apply(year.toString + "_TRB_avg"))
             / bStats.apply(year.toString + "_TRB_stdev")
159.         val ast = (stats(16) - bStats.apply(year.toString + "_AST_avg"))
             / bStats.apply(year.toString + "_AST_stdev")
160.         val stl = (stats(17) - bStats.apply(year.toString + "_STL_avg"))
             / bStats.apply(year.toString + "_STL_stdev")
161.         val blk = (stats(18) - bStats.apply(year.toString + "_BLK_avg"))
             / bStats.apply(year.toString + "_BLK_stdev")
162.         val tov = (stats(19) - bStats.apply(year.toString + "_TOV_avg"))
             / bStats.apply(year.toString + "_TOV_stdev") * (-1)
163.         val pts = (stats(21) - bStats.apply(year.toString + "_PTS_avg"))
             / bStats.apply(year.toString + "_PTS_stdev")
164.          statsZ = Array(fg, ft, tp, trb, ast, stl, blk, tov, pts)
165.          valueZ = statsZ.reduce(_ + _)
166.
167.          if (!zStats.isEmpty) {
168.            val zfg = (fg - zStats.apply(year.toString + "_FG_avg")) /
                zStats.apply(year.toString + "_FG_stdev")
169.            val zft = (ft - zStats.apply(year.toString + "_FT_avg")) /
                zStats.apply(year.toString + "_FT_stdev")
170.            val fgN = statNormalize(zfg, (zStats.apply(year.toString +
                "_FG_max") - zStats.apply(year.toString + "_FG_avg"))
171.              / zStats.apply(year.toString + "_FG_stdev"), (zStats.apply
                  (year.toString + "_FG_min")
172.              - zStats.apply(year.toString + "_FG_avg")) / zStats.apply
                  (year.toString + "_FG_stdev"))
173.            val ftN = statNormalize(zft, (zStats.apply(year.toString +
                "_FT_max") - zStats.apply(year.toString + "_FT_avg"))
174.              / zStats.apply(year.toString + "_FT_stdev"), (zStats.apply
                  (year.toString + "_FT_min")
175.              - zStats.apply(year.toString + "_FT_avg")) / zStats.apply
                  (year.toString + "_FT_stdev"))
176.            val tpN = statNormalize(tp, zStats.apply(year.toString +
                "_3P_max"), zStats.apply(year.toString + "_3P_min"))
177.            val trbN = statNormalize(trb, zStats.apply(year.toString +
                "_TRB_max"), zStats.apply(year.toString + "_TRB_min"))
178.            val astN = statNormalize(ast, zStats.apply(year.toString +
                "_AST_max"), zStats.apply(year.toString + "_AST_min"))
179.            val stlN = statNormalize(stl, zStats.apply(year.toString +
                "_STL_max"), zStats.apply(year.toString + "_STL_min"))
180.            val blkN = statNormalize(blk, zStats.apply(year.toString +
                "_BLK_max"), zStats.apply(year.toString + "_BLK_min"))
181.            val tovN = statNormalize(tov, zStats.apply(year.toString +
                 "_TOV_max"), zStats.apply(year.toString + "_TOV_min"))
182.            val ptsN = statNormalize(pts, zStats.apply(year.toString +
                "_PTS_max"), zStats.apply(year.toString + "_PTS_min"))
183.            statsZ = Array(zfg, zft, tp, trb, ast, stl, blk, tov, pts)
184.            //  println("bbParse 函数中打印 statsZ:  " + statsZ.foreach
```

```
          (println(_)) )
185.          valueZ = statsZ.reduce(_ + _)
186.          statsN = Array(fgN, ftN, tpN, trbN, astN, stlN, blkN, tovN, ptsN)
187.          //   println("bbParse函数中打印statsN:  " + statsN.foreach
          (println(_)) )
188.          valueN = statsN.reduce(_ + _)
189.        }
190.      }
191.      BballData(year, name, position, age, team, gp, gs, mp, stats,
      statsZ, valueZ, statsN, valueN)
192.    }
193.
194.    //统计类counter，需要printStats方法打印出统计
195.    //该类是一个辅助工具类，在后面编写业务代码的时候会反复使用其中的方法
196.    class BballStatCounter extends Serializable {
197.      val stats: StatCounter = new StatCounter()
198.      var missing: Long = 0
199.
200.      def add(x: Double): BballStatCounter = {
201.        if (x.isNaN) {
202.          missing += 1
203.        } else {
204.          stats.merge(x)
205.        }
206.        this
207.      }
208.
209.      def merge(other: BballStatCounter): BballStatCounter = {
210.        stats.merge(other.stats)
211.        missing += other.missing
212.        this
213.      }
214.
215.      def printStats(delim: String): String = {
216.        stats.count + delim + stats.mean + delim + stats.stdev + delim
        + stats.max + delim + stats.min
217.      }
218.
219.      override def toString: String = {
220.        "stats: " + stats.toString + " NaN: " + missing
221.      }
222.    }
223.
224.    object BballStatCounter extends Serializable {
225.      def apply(x: Double) = new BballStatCounter().add(x)
      //这里使用了Scala语言的一个编程技巧，借助于apply工厂方法，在构造该对象
      //的时候就可以执行结果
226.    }
227.
228.    //处理原始数据为zScores和nScores
229.    def processStats(stats0: org.apache.spark.rdd.RDD[String],
    txtStat: Array[String],
230.                     bStats: scala.collection.Map[String, Double] =
                     Map.empty,
231.                     zStats: scala.collection.Map[String, Double] =
                     Map.empty): RDD[(String, Double)] = {
232.      //解析数据
233.      val stats1: RDD[BballData] = stats0.map(x => bbParse(x, bStats,
```

```
        zStats))
234.
235.    //按年度分组
236.    val stats2 = {
237.      if (bStats.isEmpty) {
238.        stats1.keyBy(x => x.year).map(x => (x._1, x._2.stats))
          .groupByKey()
239.      } else {
240.        stats1.keyBy(x => x.year).map(x => (x._1, x._2.statsZ))
          .groupByKey()
241.      }
242.    }
243.
244.    //转换每个 stat 到 StatCounter
245.    val stats3 = stats2.map { case (x, y) => (x, y.map(a => a.map(b
        => BballStatCounter(b)))) }
246.
247.    //合并所有统计数据
248.    val stats4 = stats3.map { case (x, y) => (x, y.reduce((a, b) =>
        a.zip(b).map { case (c, d) => c.merge(d) })) }
249.
250.    //将统计数据打上标签，并解析标签
251.    val stats5 = stats4.map{case (x, y) => (x, txtStat.zip(y)) }.map {
252.      x =>
253.        (x._2.map {
254.          case (y, z) => (x._1, y, z)
255.        })
256.    }
257.
258.    //将每个统计数据分离到自己的行上，并将统计数据打印到一个字符串中
259.    val stats6 = stats5.flatMap(x => x.map(y => (y._1, y._2,
        y._3.printStats(","))))
260.
261.    //打开属性元组与相应的属性的键/值对
262.    val stats7: RDD[(String, Double)] = stats6.flatMap { case (a, b,
        c) => {
263.      val pieces = c.split(",")
264.      val count = pieces(0)
265.      val mean = pieces(1)
266.      val stdev = pieces(2)
267.      val max = pieces(3)
268.      val min = pieces(4)
269.      /*   println("processStats 函数的返回结果 array" +
270.           (a + "_" + b + "_" + "count", count.toDouble),
271.           (a + "_" + b + "_" + "avg", mean.toDouble),
272.           (a + "_" + b + "_" + "stdev", stdev.toDouble),
273.           (a + "_" + b + "_" + "max", max.toDouble),
274.           (a + "_" + b + "_" + "min", min.toDouble))*/
275.
276.
277.      Array((a + "_" + b + "_" + "count", count.toDouble),
278.        (a + "_" + b + "_" + "avg", mean.toDouble),
279.        (a + "_" + b + "_" + "stdev", stdev.toDouble),
280.        (a + "_" + b + "_" + "max", max.toDouble),
281.        (a + "_" + b + "_" + "min", min.toDouble))
282.    }
283.    }
284.    stats7
```

```
285.        }
286.
287.        //年龄或经验值的设计
288.        def processStatsAgeOrExperience(stats0: org.apache.spark.rdd
            .RDD[(Int, Array[Double])], label: String): DataFrame = {
289.
290.
291.          //按年龄分组
292.          val stats1: RDD[(Int, Iterable[Array[Double]])] = stats0
              .groupByKey()
293.
294.          val stats2: RDD[(Int, Iterable[Array[BballStatCounter]])] =
              stats1.map {
295.            case (x: Int, y: Iterable[Array[Double]]) =>
296.              (x, y.map((z: Array[Double]) => z.map((a: Double) =>
                  BballStatCounter(a))))
297.          }
298.          //合并 StatCounter 对象进行汇聚
299.          val stats3: RDD[(Int, Array[BballStatCounter])] = stats2.map
              { case (x, y) => (x, y.reduce((a, b) => a.zip(b).map { case (c,
              d) => c.merge(d) })) }
300.          //将 DataFrame 数据转化为 RDD[Row]
301.          val stats4 = stats3.map(x => Array(Array(x._1.toDouble),
302.            x._2.flatMap(y => y.printStats(",").split(",")).map(y =>
              y.toDouble)).flatMap(y => y))
303.            .map(x =>
304.              Row(x(0).toInt, x(1), x(2), x(3), x(4), x(5), x(6), x(7), x(8),
305.               x(9), x(10), x(11), x(12), x(13), x(14), x(15), x(16), x(17),
                  x(18), x(19), x(20)))
306.
307.          //为年龄表创建 schema 模式
308.          val schema = StructType(
309.            StructField(label, IntegerType, true) ::
310.              StructField("valueZ_count", DoubleType, true) ::
311.              StructField("valueZ_mean", DoubleType, true) ::
312.              StructField("valueZ_stdev", DoubleType, true) ::
313.              StructField("valueZ_max", DoubleType, true) ::
314.              StructField("valueZ_min", DoubleType, true) ::
315.              StructField("valueN_count", DoubleType, true) ::
316.              StructField("valueN_mean", DoubleType, true) ::
317.              StructField("valueN_stdev", DoubleType, true) ::
318.              StructField("valueN_max", DoubleType, true) ::
319.              StructField("valueN_min", DoubleType, true) ::
320.              StructField("deltaZ_count", DoubleType, true) ::
321.              StructField("deltaZ_mean", DoubleType, true) ::
322.              StructField("deltaZ_stdev", DoubleType, true) ::
323.              StructField("deltaZ_max", DoubleType, true) ::
324.              StructField("deltaZ_min", DoubleType, true) ::
325.              StructField("deltaN_count", DoubleType, true) ::
326.              StructField("deltaN_mean", DoubleType, true) ::
327.              StructField("deltaN_stdev", DoubleType, true) ::
328.              StructField("deltaN_max", DoubleType, true) ::
329.              StructField("deltaN_min", DoubleType, true) :: Nil
330.          )
331.
332.          //创建 DataFrame
333.          spark.createDataFrame(stats4, schema)
334.        }
335.
```

```
336.         //********************
337.         //处理+转换
338.         //********************
339.
340.
341.         //********************
342.         //计算每年总统计数量
343.         //********************
344.
345.         //读入所有统计信息
346.         val stats = sc.textFile(s"${TMP_PATH}/BasketballStatsWithYear/*
             /*").repartition(sc.defaultParallelism)
347.
348.         //过滤掉垃圾行，清理录入错误的数据
349.         val filteredStats: RDD[String] = stats.filter(x => !x.contains
             ("FG%")).filter(x => x.contains(","))
350.           .map(x => x.replace("*", "").replace(",,", ",0,"))
351.         filteredStats.cache()
352.         println("NBA 球员清洗以后的数据记录： ")
353.         filteredStats.take(10).foreach(println)
354.
355.         //统计数据，并保存为 map
356.         val txtStat: Array[String] = Array("FG", "FGA", "FG%", "3P", "3PA",
             "3P%", "2P", "2PA", "2P%", "eFG%", "FT",
357.           "FTA", "FT%", "ORB", "DRB", "TRB", "AST", "STL", "BLK", "TOV",
               "PF", "PTS")
358.         println("NBA 球员数据统计维度： ")
359.         txtStat.foreach(println)
360.         val aggStats: Map[String, Double] = processStats(filteredStats,
             txtStat).collectAsMap//基础数据项，需要在集群中使用，因此会在后面广播出去
361.         println("NBA 球员基础数据项 aggStats MAP 映射集： ")
362.         aggStats.take(60).foreach { case (k, v) => println(" （ " + k + " ，
             " + v + " ） ") }
363.
364.         //收集 RDD 为 map，并创建广播变量
365.         val broadcastStats: Broadcast[Map[String, Double]] =
             sc.broadcast(aggStats) //使用广播提升效率
366.
367.
368.         //********************
369.         //统计计算每年的 Z-Score 值
370.         //********************
371.
372.         //解析统计信息，并跟踪权重
373.         val txtStatZ = Array("FG", "FT", "3P", "TRB", "AST", "STL", "BLK",
             "TOV", "PTS")
374.         val zStats: Map[String, Double] = processStats(filteredStats,
             txtStatZ, broadcastStats.value).collectAsMap
375.         println("NBA 球员 Z-Score 标准分 zStats  MAP 映射集： ")
376.         zStats.take(10).foreach { case (k, v) => println(" （ " + k + " ，
             " + v + " ） ") }
377.         //收集 RDD 为 map，并创建广播变量
378.         val zBroadcastStats = sc.broadcast(zStats)
379.
380.
381.         //********************
```

```
382.        //计算每年规范化的统计数据
383.        //******************
384.
385.        //解析统计信息并规范化
386.        val nStats: RDD[BballData] = filteredStats.map(x => bbParse(x,
            broadcastStats.value, zBroadcastStats.value))
387.
388.        //转换 RDD 为 RDD[Row]，我们可以把它转换成一个 DataFrame
389.
390.        val nPlayer: RDD[Row] = nStats.map(x => {
391.          val nPlayerRow: Row = Row.fromSeq(Array(x.name, x.year, x.age,
              x.position, x.team, x.gp, x.gs, x.mp)
392.            ++ x.stats ++ x.statsZ ++ Array(x.valueZ) ++ x.statsN ++
                Array(x.valueN))
393.          //println( nPlayerRow.mkString(" "))
394.          nPlayerRow
395.        })
396.
397.        //为 DataFrame 创建 Schema 模式
398.        val schemaN: StructType = StructType(
399.          StructField("name", StringType, true) ::
400.            StructField("year", IntegerType, true) ::
401.            StructField("age", IntegerType, true) ::
402.            StructField("position", StringType, true) ::
403.            StructField("team", StringType, true) ::
404.            StructField("gp", IntegerType, true) ::
405.            StructField("gs", IntegerType, true) ::
406.            StructField("mp", DoubleType, true) ::
407.            StructField("FG", DoubleType, true) ::
408.            StructField("FGA", DoubleType, true) ::
409.            StructField("FGP", DoubleType, true) ::
410.            StructField("3P", DoubleType, true) ::
411.            StructField("3PA", DoubleType, true) ::
412.            StructField("3PP", DoubleType, true) ::
413.            StructField("2P", DoubleType, true) ::
414.            StructField("2PA", DoubleType, true) ::
415.            StructField("2PP", DoubleType, true) ::
416.            StructField("eFG", DoubleType, true) ::
417.            StructField("FT", DoubleType, true) ::
418.            StructField("FTA", DoubleType, true) ::
419.            StructField("FTP", DoubleType, true) ::
420.            StructField("ORB", DoubleType, true) ::
421.            StructField("DRB", DoubleType, true) ::
422.            StructField("TRB", DoubleType, true) ::
423.            StructField("AST", DoubleType, true) ::
424.            StructField("STL", DoubleType, true) ::
425.            StructField("BLK", DoubleType, true) ::
426.            StructField("TOV", DoubleType, true) ::
427.            StructField("PF", DoubleType, true) ::
428.            StructField("PTS", DoubleType, true) ::
429.            StructField("zFG", DoubleType, true) ::
430.            StructField("zFT", DoubleType, true) ::
431.            StructField("z3P", DoubleType, true) ::
432.            StructField("zTRB", DoubleType, true) ::
433.            StructField("zAST", DoubleType, true) ::
434.            StructField("zSTL", DoubleType, true) ::
435.            StructField("zBLK", DoubleType, true) ::
436.            StructField("zTOV", DoubleType, true) ::
437.            StructField("zPTS", DoubleType, true) ::
438.            StructField("zTOT", DoubleType, true) ::
```

```
439.            StructField("nFG", DoubleType, true) ::
440.            StructField("nFT", DoubleType, true) ::
441.            StructField("n3P", DoubleType, true) ::
442.            StructField("nTRB", DoubleType, true) ::
443.            StructField("nAST", DoubleType, true) ::
444.            StructField("nSTL", DoubleType, true) ::
445.            StructField("nBLK", DoubleType, true) ::
446.            StructField("nTOV", DoubleType, true) ::
447.            StructField("nPTS", DoubleType, true) ::
448.            StructField("nTOT", DoubleType, true) :: Nil
449.        )
450.
451.        //创建 DataFrame
452.        val dfPlayersT: DataFrame = spark.createDataFrame(nPlayer, schemaN)
453.
454.        //将所有统计数据保存为临时表
455.        dfPlayersT.createOrReplaceTempView("tPlayers")
456.
457.        //计算 exp 和 zdiff, NDIFF
458.        val dfPlayers: DataFrame = spark.sql("select age-min_age as
           exp,tPlayers.* from tPlayers join"+" (select name,min(age)as
           min_age from tPlayers group by name)as t1" + " on tPlayers.name=
           t1.name order by tPlayers.name, exp  ")
459.
460.
461.        println("计算 exp and zdiff, ndiff")
462.        dfPlayers.show()
463.        //保存为表格
464.        dfPlayers.createOrReplaceTempView("Players")
465.        //filteredStats.unpersist()
466.
467.        //********************
468.        //ANALYSIS
469.        //********************
470.        println("打印 NBA 球员的历年比赛记录：    ")
471.        dfPlayers.rdd.map(x =>
472.          (x.getString(1), x)).filter(_._1.contains("A.C. Green"))
           .foreach (println)
473.
474.        val pStats: RDD[(String, Iterable[(Double, Double, Int, Int,
           Array[Double], Int)])] = dfPlayers.sort(dfPlayers("name"),
           dfPlayers("exp") asc).rdd.map(x =>
475.          (x.getString(1), (x.getDouble(50), x.getDouble(40),
           x.getInt(2), x.getInt(3),
476.            Array(x.getDouble(31), x.getDouble(32), x.getDouble(33),
           x.getDouble(34), x.getDouble(35),
477.              x.getDouble(36), x.getDouble(37), x.getDouble(38),
             x.getDouble(39)), x.getInt(0))))
478.          .groupByKey
479.        pStats.cache
480.
481.        println("**********根据 NBA 球员名字分组：    ")
482.        pStats.take(15).foreach(x => {
483.         val myx2: Iterable[(Double, Double, Int, Int, Array[Double], Int)]
           = x._2
484.          println("按 NBA 球员： " + x._1 + " 进行分组，组中元素个数为：" +
           myx2.size)
485.          for (i <- 1 to myx2.size) {
486.            val myx2size: Array[(Double, Double, Int, Int, Array[Double],
```

```
           Int)] = myx2.toArray
487.           val mynext: (Double, Double, Int, Int, Array[Double], Int) =
           myx2size(i - 1)
488.           println(i + " : " + x._1 + " , while   " + mynext._1 + " , "
           + mynext._2 + " , "
489.             + mynext._3 + " , " + mynext._4 + " ,     " + mynext._5
             .mkString(" || ") + "     , "
490.             + mynext._6)
491.         }
492.
493.       })
494.
495.
496.       import spark.implicits._
497.       //每个球员计算这些年份 valneZ 值和 valneN 值的变化，保存成两个列表，一个是
       //年龄，一个是经验值。不包括 1980 年份的球员数据，因为我们只有他们的部分数据
498.
499.
500.       val excludeNames: String = dfPlayers.filter(dfPlayers("year") ===
       1980).select(dfPlayers("name"))
501.         .map(x => x.mkString).collect().mkString(",")
502.
503.       val pStats1: RDD[(ListBuffer[(Int, Array[Double])], ListBuffer
       [(Int, Array[Double])])] = pStats.map { case (name, stats) =>
504.         var last = 0
505.         var deltaZ = 0.0
506.         var deltaN = 0.0
507.         var valueZ = 0.0
508.         var valueN = 0.0
509.         var exp = 0
510.         val aList = ListBuffer[(Int, Array[Double])]()
511.         val eList = ListBuffer[(Int, Array[Double])]()
512.         stats.foreach(z => {
513.           if (last > 0) {
514.             deltaN = z._1 - valueN
515.             deltaZ = z._2 - valueZ
516.           } else {
517.             deltaN = Double.NaN
518.             deltaZ = Double.NaN
519.           }
520.           valueN = z._1
521.           valueZ = z._2
522.           last = z._4
523.           aList += ((last, Array(valueZ, valueN, deltaZ, deltaN)))
524.           if (!excludeNames.contains(z._1)) {
525.             exp = z._6
526.             eList += ((exp, Array(valueZ, valueN, deltaZ, deltaN)))
527.           }
528.         })
529.         (aList, eList)
530.       }
531.
532.       pStats1.cache
533.
534.
535.       println("按 NBA 球员的年龄及经验值进行统计:     ")
536.       pStats1.take(10).foreach(x => {
537.         //pStats1: RDD[(ListBuffer[(Int, Array[Double])], ListBuffer
         [(Int, Array[Double])])]
538.         for (i <- 1 to x._1.size) {
```

```
539.            println("年龄: " + x._1(i - 1)._1 + " , " + x._1(i - 1)._2
               .mkString("||") +
540.              "  经验: " + x._2(i - 1)._1 + " , " + x._2(i - 1)._2
                .mkString("||"))
541.          }
542.        })
543.
544.
545.        //*******************
546.        //计算年龄统计
547.        //*******************
548.
549.        //提取出年龄列表
550.        val pStats2: RDD[(Int, Array[Double])] = pStats1.flatMap {
            case(x, y) => x }
551.
552.        //创建年龄及 DataFrame
553.        val dfAge: DataFrame = processStatsAgeOrExperience(pStats2, "age")
554.        dfAge.show()
555.        //作为表保存
556.        dfAge.createOrReplaceTempView("Age")
557.
558.        //提取出经验值列表
559.        val pStats3: RDD[(Int, Array[Double])] = pStats1.flatMap {
            case(x, y) => y }
560.
561.        //创造经验值 DataFrame
562.       val dfExperience: DataFrame = processStatsAgeOrExperience(pStats3,
          "Experience")
563.        dfExperience.show()
564.        //作为表保存
565.        dfExperience.createOrReplaceTempView("Experience")
566.
567.        pStats1.unpersist()
568.
569.        //while(true){}
570.      }
571.
572.    }
```

15.5　NBA 篮球运动员大数据分析系统应用案例涉及的核心知识点、原理、源码

15.5.1　知识点：StatCounter 源码分析

NBA 篮球运动员大数据分析系统应用案例中定义了一个 BballStatCounter 类。BballStatCounter 类是一个辅助工具类，在 NBA 篮球运动员大数据分析系统编写业务代码的时候会反复使用 BballStatCounter 类中的方法，在 BballStatCounter 类中涉及 StatCounter 类的创建。StatCounter 类具备统计级数、平均数和方差等功能，下面看一下其 Stats 方法的应用。

在 SparkContext 上下文中使用 parallelize 算子，将 100.1，200，300 的数据集创建生成

RDD[Double]，然后调用 stats()方法，分别计算 100.1，200，300 三个元素的计数、平均数、标准差、最大值、最小值，打印输出结果。接着，将 statCounter 合并加入一个新元素 5000，然后重新计算 100.1，200，300，5000 四个元素的计数、平均数、标准差、最大值、最小值，打印输出结果。

```
1.      println("stats 方法调用： ")
2.      val doubleRDD: RDD[Double] =sc.parallelize(Seq(100.1,200.0,300.0))
3.      val statCounter: StatCounter = doubleRDD.stats()
4.      println(statCounter.merge(5000))
5.      println(statCounter)
```

在 IDEA 中运行代码，Stats 方法的应用的输出结果如下：

```
1.  stats 方法调用：
2.  (count: 3, mean: 200.033333, stdev: 81.608837, max: 300.000000, min:
    100.100000)
3.  (count: 4, mean: 1400.025000, stdev: 2079.647807, max: 5000.000000, min:
    100.100000)
```

上述代码中，doubleRDD 变量的类型是 RDD[Double]，但我们查看 RDD 抽象类的源码，在 RDD 抽象类中是没有 stats 这个方法的，doubleRDD 本身不能调用 stats 方法，但在 RDD 抽象类中定义了隐式转换，RDD[Double]通过隐式转化 DoubleRDDFunctions 就可以获得超人的力量，新增一些函数功能，如 stats 方法。DoubleRDDFunctions 新功能使用完毕，又返回到普通的 RDD[Double]，超人变回了普通人。

隐式转化 DoubleRDDFunctions 的代码如下：

```
1.  在包 org.apache.spark.rdd 的 RDD 抽象类中定义
2.  implicit def doubleRDDToDoubleRDDFunctions(rdd: RDD[Double]):
    DoubleRDDFunctions = {
3.      new DoubleRDDFunctions(rdd)
4.    }
5.  ......
6.  在包 org.apache.spark.rdd 的 DoubleRDDFunctions 类中定义
7.    def stats(): StatCounter = self.withScope {
8.     self.mapPartitions(nums => Iterator(StatCounter(nums))).reduce((a, b)
     => a.merge(b))
9.    }
```

上述的 stats 方法执行完毕以后，返回[[org.apache.spark.util.StatCounter]] 对象，StatCounter 位于 org.apache.spark.util 包中，用于统计分析。

StatCounter 的源码如下：

```
1.  package org.apache.spark.util
2.
3.  import org.apache.spark.annotation.Since
4.
5.  /**
6.   * StatCounter 用于跟踪一组数字的统计(计数、均值和方差),可以支持两个 statcounters
     * 类的 merge 合并
7.   * @constructor 根据传入的参数初始化 StatCounter
8.   */
9.  class StatCounter(values: TraversableOnce[Double]) extends Serializable {
10.   private var n: Long = 0     // 对业务数据进行计数
```

```
11.   private var mu: Double = 0  //计算业务数据的平均值
12.   private var m2: Double = 0  //计算 (sum of (x - mean)^2)
13.   private var maxValue: Double = Double.NegativeInfinity
                                   //计算业务数据的最大值
14.   private var minValue: Double = Double.PositiveInfinity
                                   //计算业务数据的最小值
15.
16.   merge(values)
17.
18.   /**StatCounter 构造器，无传入参数*/
19.   def this() = this(Nil)
20.
21.   /**增加一个变量值到 StatCounter，更新内部的各统计值 */
22.   def merge(value: Double): StatCounter = {
23.     val delta = value - mu
24.     n += 1
25.     mu += delta / n
26.     m2 += delta * (value - mu)
27.     maxValue = math.max(maxValue, value)
28.     minValue = math.min(minValue, value)
29.     this
30.   }
31.
32.   /**增加多个数据集到 StatCounter，更新内部各统计值*/
33.   def merge(values: TraversableOnce[Double]): StatCounter = {
34.     values.foreach(v => merge(v))
35.     this
36.   }
37.
38.   /**两个 StatCounter 进行合并，增加更新各内部统计值*/
39.   def merge(other: StatCounter): StatCounter = {
40.     if (other == this) {
41.       merge(other.copy())  //避免以乱序覆盖值
42.     } else {
43.       if (n == 0) {
44.         mu = other.mu
45.         m2 = other.m2
46.         n = other.n
47.         maxValue = other.maxValue
48.         minValue = other.minValue
49.       } else if (other.n != 0) {
50.         val delta = other.mu - mu
51.         if (other.n * 10 < n) {
52.           mu = mu + (delta * other.n) / (n + other.n)
53.         } else if (n * 10 < other.n) {
54.           mu = other.mu - (delta * n) / (n + other.n)
55.         } else {
56.           mu = (mu * n + other.mu * other.n) / (n + other.n)
57.         }
58.         m2 += other.m2 + (delta * delta * n * other.n) / (n + other.n)
59.         n += other.n
60.         maxValue = math.max(maxValue, other.maxValue)
61.         minValue = math.min(minValue, other.minValue)
62.       }
63.       this
64.     }
65.   }
```

```
66.
67.   /**复制一个 StatCounter*/
68.   def copy(): StatCounter = {
69.     val other = new StatCounter
70.     other.n = n
71.     other.mu = mu
72.     other.m2 = m2
73.     other.maxValue = maxValue
74.     other.minValue = minValue
75.     other
76.   }
77.
78.   def count: Long = n
79.
80.   def mean: Double = mu
81.
82.   def sum: Double = n * mu
83.
84.   def max: Double = maxValue
85.
86.   def min: Double = minValue
87.
88.   /**返回值的总体方差*/
89.   def variance: Double = popVariance
90.
91.   /**
92.    *返回值的总体方差
93.    */
94.   @Since("2.1.0")
95.   def popVariance: Double = {
96.     if (n == 0) {
97.       Double.NaN
98.     } else {
99.       m2 / n
100.       }
101.      }
102.
103.       /**
104.        *返回样本方差，通过除以 N-1 而不是 N，使用方差估测来校正偏差
105.        */
106.       def sampleVariance: Double = {
107.         if (n <= 1) {
108.           Double.NaN
109.         } else {
110.           m2 / (n - 1)
111.         }
112.       }
113.
114.       /**返回值的总体标准差*/
115.       def stdev: Double = popStdev
116.
117.       /**
118.        *返回值的总体标准差
119.        */
120.       @Since("2.1.0")
121.       def popStdev: Double = math.sqrt(popVariance)
122.
123.       /**
124.        *返回值的样本标准偏差，通过除以 N-1，而不是 N，修正了预估的偏差
```

```
125.         */
126.       def sampleStdev: Double = math.sqrt(sampleVariance)
127.
128.       override def toString: String = {"(count: %d, mean: %f, stdev: %f,
           max: %f, min: %f)".format(count, mean, stdev, max, min)
129.       }
130.     }
131.
132.     object StatCounter {
133.       /**从变量集合中构建 StatCounter*/
134.       def apply(values: TraversableOnce[Double]): StatCounter = new
           StatCounter(values)
135.
136.       /**列表中的值为可变长度参数，传入参数构建 StatCounter*/
137.       def apply(values: Double*): StatCounter = new StatCounter(values)
138.     }
139.
```

15.5.2　知识点：StatCounter 应用案例

本节结合 NBA 篮球运动员大数据分析系统应用案例实现一个 StatCounter 应用案例。

（1）模拟定义一条 NBA 的比赛数据，类型为(Int, Iterable[Array[Double]])，其中第一个元素是年份，第二个元素是 Iterable 迭代器，里面包括 5 个 Array 数组，每一个数组中包括 9 个数字。在 SparkContext 上下文中使用 parallelize 算子，将定义的一条 NBA 的比赛数据集创建生成 NBAdata RDD，其类型为 RDD[(Int, Iterable[Array[Double]])]。

（2）然后通过 map 转换，case (x, y) => (x, y.map(a => a.map((b: Double) => BballStatCounter(b))))，将 NBAdata RDD 转换为 stats3:RDD[(Int, Iterable[Array[BballStatCounter]])]。

（3）循环遍历获取到 BballStatCounter，调用 BballStatCounter 的 stats 方法打印输出统计分析结果。

```
1.   object NBAStatsMytest {
2.    def main(args: Array[String]) {
3.      Logger.getLogger("org").setLevel(Level.ERROR)
4.      var masterUrl = "local[8]"
5.      var dataPath = "data/NBABasketball"
6.      if (args.length > 0) {
7.        masterUrl = args(0)
8.      } else if (args.length > 1) {
9.        dataPath = args(1)
10.     }
11.     val sparkConf = new SparkConf().setMaster(masterUrl)
        .setAppName("NBAStatsMytest")
12.     val spark = SparkSession
13.       .builder()
14.       .config(sparkConf)
15.       .getOrCreate()
16.     org.apache.spark.deploy.SparkHadoopUtil.get.conf.set("parquet.
        block.size", "new value")
17.     val sc = spark.sparkContext
18.  ……
```

```
19. //构建 NBA 测试数据
20.     val NBAdata: RDD[(Int, Iterable[Array[Double]])] =
        sc.parallelize(Seq((1984, Iterable(
21.       Array(-0.0682906451612 9113,  0.08352774193548394,  0.4160606568195 0233,
          -0.7778113892745171,  0.027867830846394284,  -0.26666016861312614,
          -0.6326720941933309, 0.6165415556370011, -0.595782979157253),
22.       Array(-0.0399512903225 8082, -0.0023312903225805592, -0.4107265458346371,
          -1.2289977869201123,   -0.8065836950655141,   -1.4532341246428297,
          -0.6326720941933309, 0.6165415556370011, -1.211374091693702),
23.       Array(-0.08202483870967833,  0.05875064516129 0485,  0.4160606568195 0233,
          -0.7778113892745171,   -0.4629860079253165,   -0.8599471466279779,
          -0.6326720941933309, 0.8489158805579382, -0.4726647566499632),
24.       Array(-0.3832293548387099,   -0.19493129032258052,   -0.4107265458346371,
          -0.8906079886859158,   -1.0520106144513695,   -1.2554717986378792,
          -0.6326720941933309, 1.0812902054788756, -1.3191025363875806),
25.       Array(0.07404774193548126, 0.393021935483872, 0.41606065681950233,
          0.19975913895760572,    -0.2666444724166322,    0.7221514614116271,
          -0.8151170701932682, -1.0100787188095592, 1.2971596918923274)
26.     ))))
27. println("NBA 测试数据进行统计分析测试: ")
28.     val stats3: RDD[(Int, Iterable[Array[BballStatCounter]])] = NBAdata
        .map { case (x, y) => (x, y.map(a => a.map((b: Double) =>
        BballStatCounter(b)))) }
29.     stats3.take(1).foreach(x => {
30.       val myX2: Iterable[Array[BballStatCounter]] = x._2
31.       for (i <- 1 to myX2.size) {
32.         val myX2size: Array[Array[BballStatCounter]] = myX2.toArray
33.         val myNext: Array[BballStatCounter] = myX2size(i - 1)
34.         for (j <- 1 to myNext.size) {
35.           println("第" + i + "个元素第 " + j + "个数字统计 : " + myNext(j - 1).stats)
36.         }
37.       }
38.     })
39.   class BballStatCounter extends Serializable {
40.     val stats: StatCounter = new StatCounter()
41.     var missing: Long = 0
42.
43.     def add(x: Double): BballStatCounter = {
44.       if (x.isNaN) {
45.         missing += 1
46.       } else {
47.         stats.merge(x)
48.       }
49.       this
50.     }
51.
52.     def merge(other: BballStatCounter): BballStatCounter = {
53.       stats.merge(other.stats)
54.       missing += other.missing
55.       this
56.     }
57.
58.     def printStats(delim: String): String = {
```

```
59.       stats.count + delim + stats.mean + delim + stats.stdev + delim +
          stats.max + delim + stats.min
60.     }
61.     override def toString: String = {
62.       "NBA  stats: " + stats.toString + " NBA NaN: " + missing
63.     }
64.   }
65.   object BballStatCounter extends Serializable {
66.     def apply(x: Double) = new BballStatCounter().add(x)
67.   }
```

在 IDEA 中运行代码，StatCounter 应用案例循环遍历调用 BballStatCounter 的 stats 方法的输出结果如下：

```
1.  NBA 测试数据进行统计分析测试:
2.  第 1 个元素第 1 个数字统计 : (count: 1, mean: -0.068291, stdev: 0.000000, max:
    -0.068291, min: -0.068291)
3.  第 1 个元素第 2 个数字统计 : (count: 1, mean: 0.083528, stdev: 0.000000, max:
    0.083528, min: 0.083528)
4.  第 1 个元素第 3 个数字统计 : (count: 1, mean: 0.416061, stdev: 0.000000, max:
    0.416061, min: 0.416061)
5.  第 1 个元素第 4 个数字统计 : (count: 1, mean: -0.777811, stdev: 0.000000, max:
    -0.777811, min: -0.777811)
6.  第 1 个元素第 5 个数字统计 : (count: 1, mean: 0.027868, stdev: 0.000000, max:
    0.027868, min: 0.027868)
7.  第 1 个元素第 6 个数字统计 : (count: 1, mean: -0.266660, stdev: 0.000000, max:
    -0.266660, min: -0.266660)
8.  第 1 个元素第 7 个数字统计 : (count: 1, mean: -0.632672, stdev: 0.000000, max:
    -0.632672, min: -0.632672)
9.  第 1 个元素第 8 个数字统计 : (count: 1, mean: 0.616542, stdev: 0.000000, max:
    0.616542, min: 0.616542)
10. 第 1 个元素第 9 个数字统计 : (count: 1, mean: -0.595783, stdev: 0.000000, max:
    -0.595783, min: -0.595783)
11. 第 2 个元素第 1 个数字统计 : (count: 1, mean: -0.039951, stdev: 0.000000, max:
    -0.039951, min: -0.039951)
12. 第 2 个元素第 2 个数字统计 : (count: 1, mean: -0.002331, stdev: 0.000000, max:
    -0.002331, min: -0.002331)
13. 第 2 个元素第 3 个数字统计 : (count: 1, mean: -0.410727, stdev: 0.000000, max:
    -0.410727, min: -0.410727)
14. 第 2 个元素第 4 个数字统计 : (count: 1, mean: -1.228998, stdev: 0.000000, max:
    -1.228998, min: -1.228998)
15. 第 2 个元素第 5 个数字统计 : (count: 1, mean: -0.806584, stdev: 0.000000, max:
    -0.806584, min: -0.806584)
16. 第 2 个元素第 6 个数字统计 : (count: 1, mean: -1.453234, stdev: 0.000000, max:
    -1.453234, min: -1.453234)
17. 第 2 个元素第 7 个数字统计 : (count: 1, mean: -0.632672, stdev: 0.000000, max:
    -0.632672, min: -0.632672)
18. 第 2 个元素第 8 个数字统计 : (count: 1, mean: 0.616542, stdev: 0.000000, max:
    0.616542, min: 0.616542)
19. 第 2 个元素第 9 个数字统计 : (count: 1, mean: -1.211374, stdev: 0.000000, max:
    -1.211374, min: -1.211374)
20. ......
```

15.6 本章总结

本章详细阐述了 Spark 商业案例之 NBA 篮球运动员大数据分析系统应用案例的实战实现。NBA 篮球运动员大数据分析系统综合 NBA 球员的多个分析维度，基于 NBA 球员历史数据 1970 年至 2017 年的各种表现，全方位分析球员的技能，统计分析 NBA 球员的方差、均值、最大值、最小值、出现次数等统计值。案例中综合应用 Spark 缓存、广播变量、并行度等优化技巧来提升并行计算效率。通过对 NBA 篮球运动员大数据分析系统的深入掌握，读者在生产环境中对于业务系统的复杂需求，借鉴 NBA 球员系统的实现思路触类旁通，可简单、灵活地应对业务的变化。

第 16 章 电商广告点击大数据实时流处理系统案例

本章详细阐述电商广告点击综合案例需求分析和技术架构、在线点击统计实战、黑名单过滤实现、底层数据层的建模和编码实现（基于 MySQL）、从数据库中获取黑名单封装成 RDD 及过滤、动态黑名单基于数据库 MySQL 的操作、通过 updateStateByKey 等实现广告点击流量的在线更新统计、实现每个省份点击排名 Top5、实现广告点击 Trend 趋势计算、模拟点击数据的生成和数据表 SQL 的建立，最终运行整个电商广告点击综合案例，计算出运行结果。同时，本章还提供了电商广告点击综合案例的业务源码及代码关注点。

16.1 电商广告点击综合案例需求分析和技术架构

在大数据流处理时代，Spark Streaming 有强大的吸引力，而且 Spark Streaming 发展前景广阔，在 Hadoop 大数据生态系统中，Hadoop 是大数据分布式存储系统，Spark 将取代 Hadoop Map Reduce 成为事实上的大数据分布式计算框架。Spark 分布式系统的生态系统也非常丰富，Spark Streaming 可以方便地调用其他基于 Spark Core 上的 Spark SQL、Spark Mllib、Spark Graph 等强大的子框架。因此，大数据流处理时代 Spark Streaming 必将一统天下。

Kafka 是分布式发布-订阅消息系统，与传统消息系统相比，Kafka 是分布式系统，易于扩展节点。Kafka 在内核空间处理消息数据，为分布式发布和订阅消息提供高吞吐量。Kafka 将消息存储到磁盘，消息在磁盘上可以持久化，可以重复消费；Kafka 是实时的动态数据来源，Spark Streaming 可以维护消息的状态；Kafka 能动态增减节点，节点信息被 Zookeeper 管理。

在电商广告点击大数据实时流处理系统中，我们综合应用 Spark Streaming+Kafka+Spark SQL+TopN+MySQL 大数据处理技术，从系统架构、整体部署、系统软件设计、核心代码、系统运行等方面全面、深入阐述电商广告点击系统的设计及实现。

16.1.1 电商广告点击综合案例需求分析

电商广告点击综合案例可以淘宝、京东网站为案例。以京东网站为例，用户登录京东网站（见图 16-1），点击广告立即购买，广告点击系统就会记录用户的广告点击信息。广告点击系统是整个电商系统的一部分，电商系统包括用户行为分析，页面浏览率、跳转率，用户登录信息，什么商户比较受欢迎，用户广告点击信息，个性化推荐系统等内容（推荐系统与机器学习、图计算相关）。

图 16-1　京东网站广告点击示意图

广告点击系统实时分析的意义：因为可以在线实时地看见广告的投放效果，所以为广告的更大规模地投入和调整打下了坚实的基础，从而为公司带来最大化的经济回报。

电商广告点击综合案例的核心需求如下。

（1）实时黑名单动态过滤出有效的用户广告点击行为；因为黑名单用户可能随时出现，所以需要动态更新。

（2）在线计算广告点击流量。

（3）Top3 热门广告。

（4）每个广告的流量趋势。

（5）广告点击用户的区域分布分析。

（6）最近一分钟的广告点击量。

（7）整个广告需 7×24 小时运行。

16.1.2　电商广告点击综合案例技术架构

电商广告点击综合案例在线实时处理技术架构如下。

（1）电商广告点击综合案例数据来源。电商广告点击综合案例的数据来源有多个渠道：如网站、App、设备等，互联网等以京东网站为例，京东网站进行广告的推送，当用户点击广告的时候，肯定有日志 Log 发送到服务器 Server，或者当用户使用 Android、iOS 等中的 App 的时候，京东网站系统都会设置用户数据记录的关键点（埋点）。如果是网站，经典的方式是通过 JS（JavaScript）及 Ajax（Asynchronous JavaScript And XML，异步 JavaScript 和 XML）把日志传回到服务器上，如果是移动 App，一般是通过 Socket，其他的传感器或者工业设备可以通过自己的通信协议把数据传回到服务器端。

（2）为了应对高并发访问，电商广告点击系统一般采用 Nginx 等作为服务器 Server 前端，通过服务器 Server 的分布式集群做负载均衡。Web 服务器在生产环境中可以使用 Tomcat、Apache 等第三方服务器。

（3）企业中可使用 Crontab 等定时工具通过日志整理工具把当天的日志采集、合并和初步的处理性能生成一份日志文件，然后将日志文件发送到 Flume 监控目录中。当 Flume 发现有新的日志文件进来时，会按照配置把数据通过通道 Channel 传递 Sink 到目的地，这里是传送 Sink 到 Kafka 集群中。

（4）Kafka 集群中的数据被消费。Spark Streaming 主动从 Kafka 集群抓到数据在线进行处理（处理过程中可能使用机器学习和图计算等非常复杂的算法和功能）。例如，在线广告推荐，这个时候就需要结合机器学习 ML 来实现。Spark Streaming 处理后的数据可以存储到 Kafka，供 Kafka 上的应用系统再次进行数据处理。

（5）对于 Spark Streaming 集群，可以使用 Ganglia 来监控 Spark 分布式集群节点的系统性能，如 CPU、内存、硬盘利用率，I/O 负载、网络流量情况等，统计展示每个节点的工作状态，对合理调整、分配系统资源，提高系统整体性能起到重要作用。

（6）Spark Streaming 在线处理的数据持久化到 DB/Redis 数据库中。实际生产环境下，在大项目中使用 Redis 比较多，因为 QPS 可以非常高，尤其是在并发和实时要求比较高的场景特别有用。

（7）数据持久化到 DB/Redis 数据库中后，数据的展示可通过 JDBC 接口用 Java 企业级组件获取数据库 DB 数据并通过报表展示，提供营销、运营、产品研发改进等。

电商广告点击综合案例离线批处理技术架构：

（1）企业中可使用 Crontab 等定时工具通过日志整理工具把当天的日志采集、合并和初步的处理性能生成一份日志文件，然后将日志文件发送到 Flume 监控目录中。当 Flume 发现有新的日志文件进来时，将日志文件传递到 Hadoop HDFS 系统。

（2）在 HDFS 生产系统中：①使用 MapReduce 作业对数据进行初步清洗，并写入新的 HDFS 文件中；②清洗后的数据一般导入到 Hive 数据仓库中，在 Hive 中可以采用分区表；③通过 Hive 中的 SQL 语句在数据仓库的基础上进行数据清洗 ETL，此时数据清洗 ETL 会把原始的数据生成很多张目标表 Table。

（3）Spark 对 HDFS 中的批量数据进行离线处理。市场营销人员、管理运营人员、决策人员、产品研发人员会将分析后的结果用于提升营业额、利润和市场占有率。

电商广告点击综合案例整体技术架构图如图 16-2 所示。

电商广告点击综合案例实现的技术细节如下。

- 数据格式：时间、用户、广告、地点等。
- 在线计算用户点击的次数分析、屏蔽 IP 等。
- 使用 updateStateByKey 或者 mapWithState 进行不同地区广告点击排名的计算。
- Spark Streaming+Spark SQL+Spark Core 等综合分析数据。
- 使用 Window 类型的操作。
- 高可用和性能调优。
- 流量趋势一般会结合 DB 等。

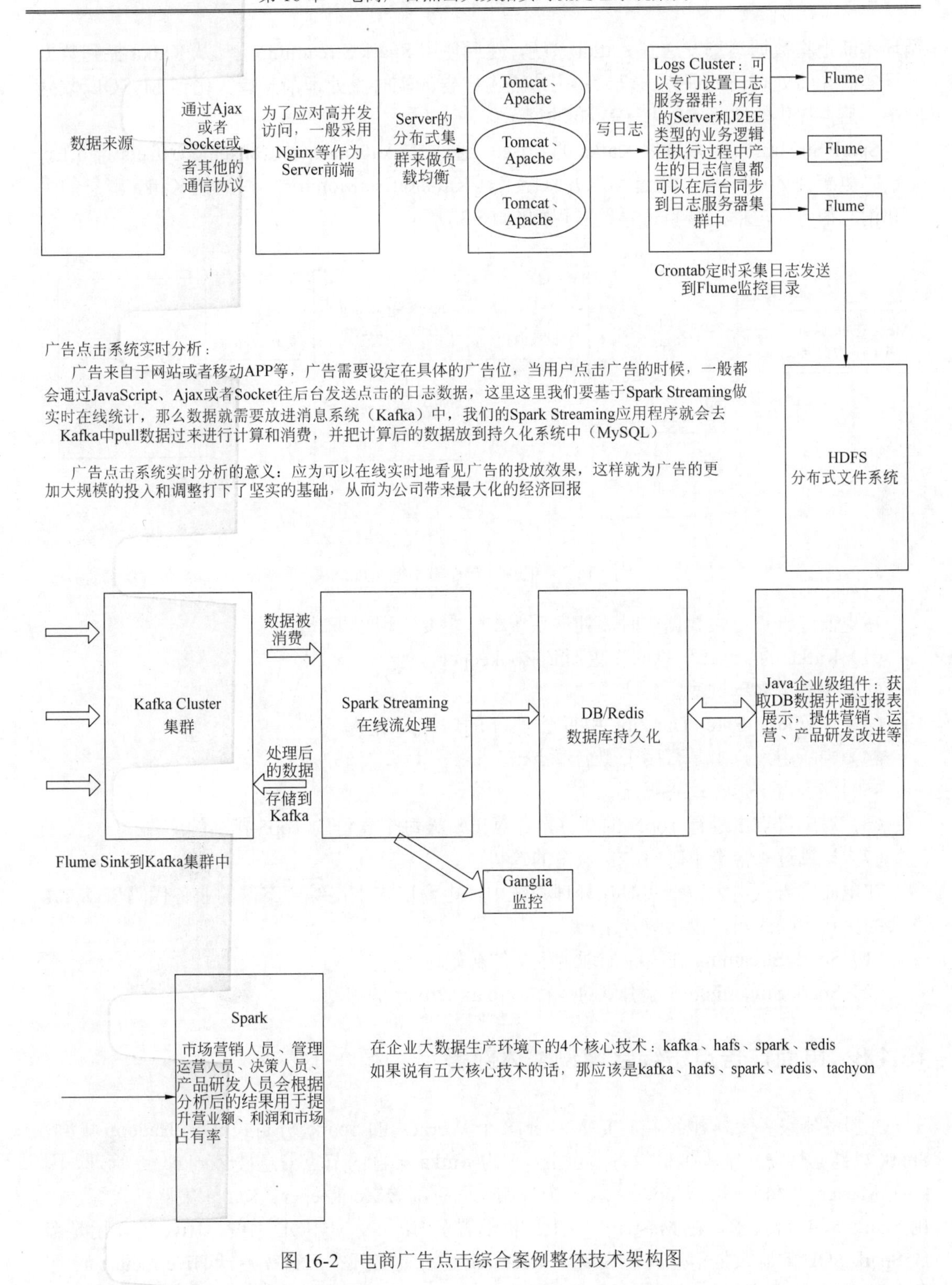

图 16-2 电商广告点击综合案例整体技术架构图

在电商广告点击大数据实时流处理系统中，我们将 Kafka 作为电商广告点击大数据实时流处理系统的动态数据源，采用 MockAdClickedStats 脚本生产模拟的电商广告点击数据消息，循环不断地将数据消息发送到Kafka系统，然后使用Spark Streaming主动从Kafka抓到数据进行实时在线处理，处理完成以后将用户点击广告的实时流处理消息持久化到MySQL数据库中，后期提供给Java EE进行Web的展示。

Spark Streaming直接读取Kafka的offset，通过KafkaCluster.setConsumerOffsets将offset等信息更新到Zookeeper，通过第三方监控工具KafkaOffsetMonitor实时监控Kafka消息的生产和消息信息。电商广告点击系统实现如图16-3所示。

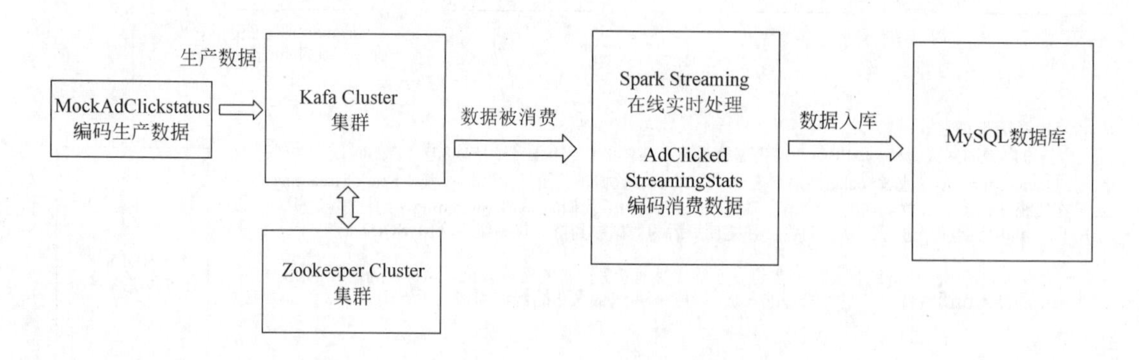

图16-3　电商广告点击系统实现

在电商广告点击大数据实时流处理系统中，主要实现以下功能。

（1）Kafka的offset信息同步更新到Zookeeper。

（2）在线黑名单过滤。

（3）计算每个Batch Duration中每个User的广告点击量。

（4）判断用户点击是否属于黑名单点击。

（5）广告点击累计动态更新。

（6）对广告点击进行TopN的计算，计算出每天每个省份的Top5排名的广告。

（7）计算过去半个小时内广告点击的趋势。

在电商广告点击大数据实时流处理系统中，电商广告点击综合案例的业务代码与Spark框架的交互代码的实现包括两方面。

（1）Spark Streaming在线流处理上下文的初始化。

（2）Spark Streaming执行引擎的运行，Spark Driver的启动。

16.1.3　电商广告点击综合案例整体部署

在服务器设备上搭建部署1个Master、8个Worker的Spark分布式集群，Hadoop集群、Spark集群运行在9个虚拟机设备节点上；因为Kafka集群的节点信息被Zookeeper管理，因此在Master、Worker1、Worker2这3个节点上同时部署Zookeeper、Kafka集群；为综合应用Spark SQL等技术，在Master节点上单机部署了Hive及MySQL系统，Hive的作用是存放Spark SQL的源数据信息，在Spark Streaming运行前必须在后台先启动Hive metastore服务，MySQL数据库的作用是持久化保存Spark Streaming的运行结果；在Master节点上也安装部署了Eclipse集成开发环境。Eclipse的作用是在Master虚拟机本地调试Spark Streaming

代码，快速测试，定位代码 Bug。

脚本代码部署：AdClickedStreamingStats.sh 脚本在 Master 节点上运行，通过运行 spark-submit 向 Spark 集群提交应用；为减轻 Master 系统运行负荷，将 MockAdClickedStats.sh 脚本部署在 Worker1 节点上运行，源源不断地向 Spark 集群发送数据；同时，为了监控 Kafka 集群生产者、消费者的实时数据消费情况，还部署了 KafkaOffsetMonitor 进行监控。

电商广告点击系统整体部署如图 16-4 所示。

<table>
<tr><td rowspan="10">整体部署</td><td>节点</td><td>IP地址</td><td>部署集群</td><td>部署服务</td><td>Mock数据来源</td><td>Spark集群提交应用</td></tr>
<tr><td>Master</td><td>192.168.189.1</td><td>Hadoop、Spark</td><td>Zookeeper、Kakfa、Hive、MySQL、Eclipse</td><td></td><td rowspan="9">AdClicked StreamingStats</td></tr>
<tr><td>Worker1</td><td>192.168.189.2</td><td>Hadoop、Spark</td><td>Zookeeper、Kakfa</td><td>MockAdClickedStats</td></tr>
<tr><td>Worker2</td><td>192.168.189.3</td><td>Hadoop、Spark</td><td>Zookeeper、Kakfa</td><td>KafkaOffsetMonitor</td></tr>
<tr><td>Worker3</td><td>192.168.189.4</td><td>Hadoop、Spark</td><td></td><td></td></tr>
<tr><td>Worker4</td><td>192.168.189.5</td><td>Hadoop、Spark</td><td></td><td></td></tr>
<tr><td>Worker5</td><td>192.168.189.6</td><td>Hadoop、Spark</td><td></td><td></td></tr>
<tr><td>Worker6</td><td>192.168.189.7</td><td>Hadoop、Spark</td><td></td><td></td></tr>
<tr><td>Worker7</td><td>192.168.189.8</td><td>Hadoop、Spark</td><td></td><td></td></tr>
<tr><td>Worker8</td><td>192.168.189.9</td><td>Hadoop、Spark</td><td></td><td></td></tr>
<tr><td rowspan="5">MySQL数据库表</td><td>adclicked</td><td colspan="5">表的字段：timestamp、ip、userID、adID、province、city、clickedCount</td></tr>
<tr><td>adclickedtrend</td><td colspan="5">表的字段：date、hour、minute、adID、clickedCount</td></tr>
<tr><td>adprovincetopn</td><td colspan="5">表的字段：timestamp、adID、province、clickedCount</td></tr>
<tr><td>blacklisttable</td><td colspan="5">表的字段：name</td></tr>
<tr><td>adclickedcount</td><td colspan="5">表的字段：timestamp、adID、province、city、clickedCount</td></tr>
<tr><td rowspan="8">业务流程</td><td colspan="6">Spark初始化</td></tr>
<tr><td colspan="6">在线黑名单过滤</td></tr>
<tr><td colspan="6">计算每个Batch Duration中每个User的广告点击量</td></tr>
<tr><td colspan="6">判断用户点击是否属于黑名单点击</td></tr>
<tr><td colspan="6">广告点击累计动态更新</td></tr>
<tr><td colspan="6">对广告点击进行TopN的计算，计算出每天每个省份的Top5排名的广告</td></tr>
<tr><td colspan="6">计算过去半个小时内广告点击的趋势</td></tr>
<tr><td colspan="6">Spark Streaming执行引擎 Driver开始运行</td></tr>
</table>

图 16-4　电商广告点击系统整体部署

16.1.4　生产数据业务流程及消费数据业务流程

电商广告点击大数据实时流处理系统案例系统整体上分成生产数据、消费数据两部分。

（1）生产数据业务流程：编写 MockAdClickStatus 的代码向 Kafka 集群发送数据。

（2）消费数据业务流程：编写 Spark Streaming　AdClickedStreamingStats 的代码从 Kafka 集群中抓取数据，将 Spark 实时处理的数据持久化到数据库中。

16.1.5　Spark JavaStreamingContext 初始化及启动

电商广告点击综合案例的业务代码在 Spark 集群分布式集群中运行，我们首先要编写 Spark Driver 应用程序的代码。

- Spark Streaming 在线流处理上下文的初始化。
- Spark Streaming 执行引擎的运行，Spark Driver 的启动。

1．Spark Streaming在线流处理上下文的初始化具体实现

第一步：配置 SparkConf。

（1）Spark Streaming 运行至少需要两条线程：因为 Spark Streaming 应用程序在运行的时

候，至少有一条线程用于不断地循环接收数据，并且至少有一条线程用于处理接收的数据（否则无法有线程用于处理数据，随着时间的推移，内存和磁盘都会不堪重负）。

（2）对于集群而言，每个Executor一般肯定不止一个Thread，那对于处理Spark Streaming的应用程序而言，每个Executor一般分配多少个Core比较合适？根据经验，5个左右的Core是最佳的（一般分配为奇数个Core表现最佳，如3个、5个、7个Core等）。

```
1.    SparkConf conf = new SparkConf().setMaster("spark://192.
      168.189.1:7077")
2.              .setAppName("114-AdClickedStreamingStats");
```

第二步：创建SparkStreamingContext。

（1）SparkStreamingContext是Spark Streaming应用程序所有功能的起始点和程序调度的核心。SparkStreamingContext的构建可以基于SparkConf参数，也可基于持久化的SparkStreamingContext的内容来恢复（典型的场景是Driver崩溃后重新启动，由于Spark Streaming具有连续7×24小时不间断运行的特征，所以需要在Driver重新启动后继续上一次的状态，此时的状态恢复需要基于曾经的checkpoint）。

（2）在一个Spark Streaming应用程序中可以创建若干个SparkStreamingContext对象，使用下一个SparkStreamingContext之前需要把前面正在运行的SparkStreamingContext对象关闭掉，由此，我们获得一个重大的启发：Spark Streaming框架也只是Spark Core上的一个应用程序而已，只不过Spark Streaming框架箱运行的话需要Spark工程师写业务逻辑处理代码。

```
1.    JavaStreamingContext jsc = new JavaStreamingContext(conf, Durations.
      seconds(10));
2.    jsc.checkpoint("/usr/local/IMF_testdata/IMFcheckpoint114");
```

第三步：创建Spark Streaming输入数据来源input Stream。

（1）数据输入来源可以基于File、HDFS、Flume、Kafka、Socket等。

（2）Spark Streaming可以指定数据来源于网络Socket端口，Spark Streaming连接上该端口并在运行的时候，一直监听该端口的数据（当然，该端口服务首先必须存在），并且在后续会根据业务需要不断地有数据产生（当然，对于Spark Streaming应用程序的运行而言，有无数据其处理流程都是一样的）。电商广告点击综合案例中，Spark Streaming指定数据来源是Kafka集群中的数据。

（3）如果经常在每间隔5s没有数据的话就不断地启动空的Job，其实会造成调度资源的浪费，因为并没有数据需要发生计算，所以实际的企业级生成环境的代码在具体提交Job前会判断是否有数据，如果没有，就不再提交Job。

（4）在电商广告点击综合案例中，Spark Streaming连接Kafka集群的KafkaUtils工具类调用createDirectStream方法的具体参数的含义如下。

第一个参数是StreamingContext实例。

第二个参数是keyClass，Kafka记录的Key值的类型。

第三个参数是valueClass，Kafka记录的Value值的类型。

第四个参数是keyDecoderClass，Key值的解码器类型。

第五个参数是valueDecoderClass，Value值的解码器类型。

第六个参数是kafkaParams，Kafka集群的参数配置信息，要求使用metadata.broker.list或者bootstrap.servers来配置Kafka集群的分布式节点信息，格式规范为host1:port1, host2:port2。如果不是从检查点开始，可以设置 auto.offset.reset 为最大值 largest 或者最小值 smallest（默认是最大值），确定流处理开始处理数据的位置。

```
1.      /**
          * 创建Kafka元数据，让Spark Streaming这个Kafka Consumer利用
2.        */
3.
4.      Map<String, String> kafkaParameters = new HashMap<String, String>();
5.      kafkaParameters.put("metadata.broker.list", "Master:9092,Worker1:
        9092,Worker2:9092");
6.
7.      Set<String> topics = new HashSet<String>();
8.      topics.add("AdClicked");
9.
10.     JavaPairInputDStream<String, String> adClickedStreaming = KafkaUtils
        .createDirectStream(jsc,String.class,String.class,StringDecoder
        .class, StringDecoder.class,kafkaParameters, topics);
```

第四步：接下来就像对于 RDD 编程一样基于 DStream 进行编程！原因是 DStream 是 RDD 产生的模板（或者说类），在 Spark Streaming 具体发生计算前，其实质是把每个 Batch 的 DStream 的操作翻译成为对 RDD 的操作！对初始的 DStream 进行 Transformation 级别的处理，如 map、filter 等高阶函数等的编程，来进行具体的数据计算。这里进行电商广告点击综合案例的一系列业务代码的实现：在线黑名单过滤、计算每个 Batch Duration 中每个 User 的广告点击量、判断用户点击是否属于黑名单点击、广告点击累计动态更新、对广告点击进行 TopN 的计算，计算出每天每个省份的 Top5 排名的广告、计算过去半个小时内广告点击的趋势。

2. Spark Streaming执行引擎的运行，Spark Driver的启动

Spark Streaming 执行引擎也就是 Driver 开始运行，Driver 启动的时候是位于一条新的线程中的，当然，其内部有消息循环体，用于接收应用程序本身或者 Executor 中的消息。

```
1.      jsc.start();
2.          jsc.awaitTermination();
3.          jsc.close();
```

16.1.6　Spark Streaming 使用 No Receivers 方式读取 Kafka 数据及监控

Spark Streaming 读取 Kafka 数据支持有两种方式：Receiver 方式和 No Receivers 方式。

（1）Receiver 方式：Spark Streaming kafkautil 使用 createStream 方法。

（2）No Receivers 方式：Spark Streaming kafkautil 使用 createDirectStream 方法。

目前，No Receivers 方式在企业中使用得越来越多，具有更强的自由度控制、语义一致性。No Receivers 方式更符合数据读取和数据操作，是我们操作数据来源的自然方式。

下面 No Receivers 方式直接抓取 Kafka 数据带来的好处。

好处一：

No Receivers 方式直接抓取 Kafka 的数据，没有缓存，不会出现内存溢出的问题。如果使用 kafka Receiver 方式读取数据，会存在缓存的问题，需要设置 kafka Receiver 读取的频率和 block interval 等信息。

好处二：

如果采用 Receivers 方式，Receivers 默认情况需要和 Worker 的 Executor 绑定，不方便做分布式。如果采用 No Receivers direct 方式，默认情况下数据会存在多个 Worker 上的 Executor，数据天然就是分布式的，默认分布在多个 Executor 上。而 Receivers 方式就不方便计算。

好处三：

数据消费的问题，在实际操作的时候采用 Receivers 的方式有一个弊端，消费数据来不及处理，如果延迟多次，Spark Streaming 程序就有可能崩溃。但如果是采用 No Receivers direct 方式访问 Kafka 数据，就不会存在此问题。因为 No Receivers direct 方式直接读取 Kafka 数据，如果数据有延迟 delay，那就不进行下一个处理，因此，No Receivers direct 方式就不会存在来不及消费、程序崩溃的问题。

好处四：

No Receivers direct 方式实现完全的语义一致性，不会重复消费数据，而且保证数据一定被消费。No Receivers direct 方式与 kafka 进行交互，只有数据真正执行成功后才会记录下来。

因此，在生产环境中强烈建议大家采用 No Receivers direct 的方式。本章电商广告点击综合案例采用 No Receivers direct 方式开发 Spark Streaming 应用程序。

```
1.      JavaPairInputDStream<String, String> adClickedStreaming = KafkaUtils
        .createDirectStream(jsc, String.class,String.class, StringDecoder
        .class, StringDecoder.class,kafkaParameters, topics);
```

下面是 createDirectStream 的源码。

```
1.   /**
2.     * 创建输入流 input stream 直接从 Kafka 集群节点拉取消息
3.     ......
4.     */
5.     * @param jssc JavaStreamingContext object
6.     * @param keyClass Class of the keys in the Kafka records
7.     * @param valueClass Class of the values in the Kafka records
8.     * @param keyDecoderClass Class of the key decoder
9.     * @param valueDecoderClass Class type of the value decoder
10.    * @param kafkaParams Kafka <a href="http://kafka.apache.org/documentation
       * .html#configuration">
11.    *  configuration parameters</a>. Requires "metadata.broker.list" or
       *  "bootstrap.servers"
12.    *  to be set with Kafka broker(s) (NOT zookeeper servers), specified in
13.    *  host1:port1,host2:port2 form.
14.    *  If not starting from a checkpoint, "auto.offset.reset" may be set
       *  to "largest" or "smallest"
15.    *  to determine where the stream starts (defaults to "largest")
16.    * @param topics Names of the topics to consume
17.    * @tparam K type of Kafka message key
18.    * @tparam V type of Kafka message value
19.    * @tparam KD type of Kafka message key decoder
20.    * @tparam VD type of Kafka message value decoder
21.    * @return DStream of (Kafka message key, Kafka message value)
22.    */
23.   def createDirectStream[K, V, KD <: Decoder[K], VD <: Decoder[V]](
24.      jssc: JavaStreamingContext,
25.      keyClass: Class[K],
```

```
26.       valueClass: Class[V],
27.       keyDecoderClass: Class[KD],
28.       valueDecoderClass: Class[VD],
29.       kafkaParams: JMap[String, String],
30.       topics: JSet[String]
31.     ): JavaPairInputDStream[K, V] = {
32.     implicit val keyCmt: ClassTag[K] = ClassTag(keyClass)
33.     implicit val valueCmt: ClassTag[V] = ClassTag(valueClass)
34.     implicit val keyDecoderCmt: ClassTag[KD] = ClassTag(keyDecoderClass)
35.     implicit val valueDecoderCmt: ClassTag[VD] = ClassTag(valueDecoderClass)
36.     createDirectStream[K, V, KD, VD](
37.       jssc.ssc,
38.       Map(kafkaParams.asScala.toSeq: _*),
39.       Set(topics.asScala.toSeq: _*)
40.     )
41.   }
42. }
```

Spark Streaming No Receivers 方式的 createDirectStream 方法不使用接收器，而是创建输入流直接从 Kafka 集群节点拉取消息。输入流保证每个消息从 Kafka 集群拉取以后只完全转换一次，保证语义一致性。

（1）无接收器：createDirectStream 方式不使用任何接收器，createDirectStream 直接从 Kafka 集群进行查询。

（2）偏移量：createDirectStream 方式不使用 Zookeeper 存储偏移量。消费的偏移量由 Stream 本身进行跟踪记录。Kafka 的监控依赖于 Zookeeper，为监控 createDirectStream 的消费信息，我们可以从 Spark Streaming 应用程序中编写代码来更新 Kafka/Zookeeper 的偏移量，偏移量可以从每一批流处理中生成的 RDDS 偏移量来获取[org.apache.spark.streaming.kafka.HasOffsetRanges]。

（3）容错恢复：如果是 Spark Driver 容错恢复，可以在 StreamingContext 中启用检查点。消费的偏移量消息可以从检查点恢复。

（4）端到端语义：createDirectStream 方式确保每个记录有效接收和转换一次，但不保证转换以后的数据只输出一次。对于端到端的语义一致性，必须确保输出操作是幂等的（注意多次执行所产生的影响均为与一次执行的影响相同），或使用交易记录自动输出。

本章电商广告点击综合案例采用 No Receivers direct 方式，通过 Spark Streaming direct 方式直接读取 Kafka 的数据，此时数据消费的偏移量没有自动更新到 Zookeeper 中。而我们使用的第三方工具 KafkaOffsetMonitor 监控 Kafka 消费信息的时候是从 Zookeeper 中获取消费记录信息的，因此在 KafkaOffsetMonitor 中监控不到 Kafka 消费的数据。

为了在第三方工具 KafkaOffsetMonitor 监控到 Kafka 的消费信息，我们需在 Spark Streaming 应用程序代码编码操作偏移量，实现 Kafka 集群的 kafkacluster 自动将偏移量更新到 Zookeeper 中，这样 KafkaOffsetMonitor 就能实时监控到 Kafka 的消费信息，并在 Web 页面展示。

Kafka 的偏移量信息同步更新到 Zookeeper。

- 从 adClickedStreaming.transformToPair 中获取 offsetRanges 信息。
- 在 foreachRDD 中循环遍历 offsetRanges 的内容，读取 topicAndPartition 及 untilOffset 数据。

- 通过 kafkaCluster.setConsumerOffsets 更新 Zookeeper 中的偏移量，使用第三方软件 KafkaOffsetMonitor 监控 Kafka 的生产数据、消费数据的情况。

Kafka 的 offset 信息同步更新到 Zookeeper 代码如下：

```
1.  scala.collection.immutable.Map scalaImmutablekafkaParameters= scala
    .collection.JavaConverters.mapAsScalaMapConverter(kafkaParameters)
    .asScala().toMap(scala.Predef.conforms());
2.          final AtomicReference<OffsetRange[]> offsetRanges = new
            AtomicReference<OffsetRange[]>();
3.          adClickedStreaming.transformToPair( new Function<JavaPairRDD<
            String, String>, JavaPairRDD<String, String>>() {
4.              private static final long serialVersionUID = 1L;
5.
6.              @Override
7.              public JavaPairRDD<String, String> call(JavaPairRDD<String,
S               tring> rdd) throws Exception {
8.                    OffsetRange[] offsets = ((HasOffsetRanges) rdd
                      .rdd()).offsetRanges();
9.                    offsetRanges.set(offsets);
10.                   return rdd;
11.             }
12.
13.         }).foreachRDD( new Function<JavaPairRDD<String, String>, Void>(){
14.
15.             private static final long serialVersionUID = 1L;
16.
17.             @Override
18.             public Void call(JavaPairRDD<String,String>rdd) throws Exception {
19.                 KafkaCluster kafkaCluster= new KafkaCluster(scala
                    ImmutablekafkaParameters);
20.                 for (OffsetRange o : offsetRanges.get()) {
21.                     TopicAndPartition topicAndPartition=new  TopicAndPartition
                        ("AdClicked",o.partition());
22.                    Map<TopicAndPartition,Long> offsetsmap = new HashMap
                       <TopicAndPartition, Long>();
23.                    offsetsmap.put(topicAndPartition, o.untilOffset());
24.                    scala.collection.immutable.Map scalaImmutableoffsetsmap=
                       scala.collection.JavaConverters.mapAsScalaMap
                       Converter(offsetsmap).asScala().toMap(scala
                       .Predef.conforms());
25.                     kafkaCluster.setConsumerOffsets("Kafka-groupId",
                        scalaImmutableoffsetsmap);
26.                           );
27.                  }
28.                 return null;
29.             }
30.         });
```

16.2　电商广告点击综合案例在线点击统计实战

电商广告点击综合案例在线点击统计实战，对黑名单用户的过滤要进行两次判断。

- 基于电商广告点击综合案例数据库中获取的黑名单进行过滤（黑名单的过滤在 16.3 节阐述），第 1 次过滤掉黑名单用户以后，初步统计点击了多少条广告。

- ❑ 基于初步统计的广告点击次数，动态判断用户点击的次数是否超过阈值，如果判断为黑名单用户，则更新一下黑名单的数据表；第 2 次过滤掉黑名单用户点击的无效广告数，计算出有效的广告点击。
- ❑ 电商广告点击综合案例在线点击统计的数据持久化到 MySQL 数据库。
- ❑ 广告点击的基本数据格式：timestamp、ip、userID、adID、province、city。

电商广告点击综合案例在线点击统计实战中，KafkaUtils 调用 createDirectStream 生成 JavaPairInputDStream 类型的 adClickedStreaming，adClickedStreaming 的格式为 Key-Value 键值对<String, String>，Key 为 Kafka 生成的 key 值，Value 值是 Spark Streaming 从 Kafka 中读取的一行行的数据。adClickedStreaming 经过黑名单的动态过滤后（黑名单的过滤在 16.3 节阐述），转换生成了 JavaPairDStream<String, String>类型的 filteredadClickedStreaming。

首先实现第 1 次过滤掉黑名单用户以后，初步统计点击了多少条广告。

（1）对 filteredadClickedStreaming 使用 mapToPair 算子进行转换。mapToPair 的函数签名如下，传入 PairFunction 函数的第一个参数是元组 Tuple2<String, String>，元组的第一个元素是 Kafka 提供的 key 值，元组的第二个元素是黑名单过滤以后读入的一行行广告点击的数据；PairFunction 函数的第二个参数是 mapToPair 转换以后的返回结果（Key-Value）的 key 值，类型是 String 类型，PairFunction 函数的第三个参数是 mapToPair 转换以后的返回结果（Key-Value）的 Value 值，类型是 Long 类型。

```
1.  public  <K2,  V2>  JavaPairDStream<java.lang.String,  java.lang.Long>
    mapToPair(PairFunction<scala.Tuple2<java.lang.String,  java.lang.String>,
    java.lang.String,java.lang.Long> f)
```

（2）对 filteredadClickedStreaming 做 mapToPair 转换，读入 filteredadClickedStreaming 的每一行数据是一个元组，元组第二个元素是黑名单过滤以后读入的一行行广告点击的数据，我们使用“\t”分隔符进行切分，依次获取时间戳、IP 地址、用户 ID、广告 ID、省份、城市等数据，然后使用“_”下画线分隔符重新组拼成一条新的记录 clickedRecord。重新组拼的目的是定义 Key 值，方便后续按 Key 进行聚合汇总统计总计数，最终生成新的 JavaPairDStream<String, Long>类型的 pairs，其格式是 Key-Value，Key 值为组拼的广告点击记录 clickedRecord（timestamp_ip_userID_adID_province_city），Value 值为计数 1 次，即（组拼的广告点击记录，计算 1 次）。

（3）对 pairs 使用 reduceByKey 算子，计算某个用户在某时间、IP、省份、城市点击某一条广告的总次数，按 Key（timestamp_ip_userID_adID_province_city）进行统计汇总。

电商广告点击综合案例在线点击统计代码如下：

```
1.   JavaPairDStream<String, Long> pairs = filteredadClickedStreaming
2.                  .mapToPair(new PairFunction<Tuple2<String, String>,
                   String, Long>() {
3.                      @Override
4.                      public Tuple2<String, Long> call(Tuple2<String,
                        String> t) throws Exception {
5.                          String[] splited = t._2.split("\t");
6.                          String timestamp = splited[0]; // yyyy-MM-dd
7.                          String ip = splited[1];
8.                          String userID = splited[2];
9.                          String adID = splited[3];
```

```
10.                             String province = splited[4];
11.                             String city = splited[5];
12.
13.                             String clickedRecord = timestamp + "_" + ip + "_"
                                + userID + "_" + adID + "_" + province + "_"
14.                                     + city;
15.
16.                             return new Tuple2<String, Long>(clickedRecord, 1L);
17.                         }
18.                     });
19.
20.         /**
21.          *对初始的 DStream 进行 Transformation 级别的处理，如通过 map、filter 等
             *高阶函数等的编程，进行具体的数据计算
22.          * 计算每个 Batch Duration 中每个 User 的广告点击量
23.          */
24.         JavaPairDStream<String, Long> adClickedUsers = pairs
            .reduceByKey(new Function2<Long, Long, Long>() {
25.
26.             @Override
27.             public Long call(Long v1, Long v2) throws Exception {
28.                 // TODO Auto-generated method stub
29.                 return v1 + v2;
30.             }
31.
32.         });
```

接下来进行第2次的黑名单点击过滤，基于之前初步统计的广告点击次数，动态判断用户点击的次数是否超过阈值，如果判断为黑名单用户，则更新一下黑名单的数据表；过滤掉黑名单用户点击的无效广告数，计算出有效的广告点击。

什么叫有效的广告点击？

（1）复杂化的有效广告点击的处理：一般都是采用机器学习训练好模型，直接使用机器学习模型在线进行过滤。

（2）简单的有效广告点击的处理：可以通过一个批处理时间中 Batch Duration 中的点击次数来判断是不是非法广告点击。例如，在某个时间用户连续点击广告几十次或上百次，但是实际上，非法广告点击程序会尽可能模拟真实的广告点击行为，不会在某个时间点连续地点击多次，所以通过一个批处理时间 Batch 来判断是否完整，我们需要对例如一天（也可以是每一个小时）的数据进行判断。

（3）有效广告点击的处理如下。例如，一段时间内，同一个 IP（MAC 地址）有多个用户的账号访问；可以统计一天内一个用户点击广告的次数，如果一天点击同样的广告操作 50 次，就列入黑名单；黑名单有一个重点的特征：动态生成！所以，每个批处理时间 Batch Duration 都要考虑是否有新的黑名单加入，此时黑名单需要存储起来，具体存储在什么地方呢，存储在 DB/Redis 中即可；例如，邮件系统中的“黑名单”，可以采用 Spark Streaming 不断地监控每个用户的操作，如果用户发送邮件的频率超过设定的值，就可以暂时把用户列入“黑名单”，从而阻止用户过度频繁的发送邮件。

本章节电商广告点击综合案例着重于 Spark Streaming 的在线流处理的开发，对有效广告点击的处理做了简化，简单地判断用户点击广告的次数大于 1，就假设为黑名单用户，更新黑名单的数据库表，做黑名单测试使用，这里省略了数据库表黑名单的更新代码；同时，为

测试方便，我们直接返回 false，对用户的广告点击记录都不进行过滤，以便后续应用的测试。生成 JavaPairDStream<String, Long>类型的 filteredClickInBatch。

有效广告点击处理的代码如下：

```
1.      JavaPairDStream<String, Long> filteredClickInBatch = adClickedUsers
2.                  .filter(new Function<Tuple2<String, Long>, Boolean>() {
3.
4.                      @Override
5.                      public Boolean call(Tuple2<String,Long>v1)throws Exception{
6.                          if (1 < v1._2) {
7.                              //更新黑名单的数据表
8.                              return false;
9.                          } else {
10.                             return true;
11.                         }
12.
13.                     }
14.                 });
```

接下来，我们对 filteredClickInBatch 使用 foreachRDD 算子，将 foreachRDD 中用户广告点击记录持久化到 MySQL 数据库中。对 filteredClickInBatch 进行 foreachRDD 函数触发，基于设置的 Duration 时间间隔启动 Job。将 record._1 按“_”进行分割，提取出 timestamp、ip、userID、adID、province、city、ClickedCount 各个字段，并在数据表中查询表 adclicked，如没记录，则插入；有记录，则更新。

```
1.  filteredClickInBatch.foreachRDD(new Function<JavaPairRDD<String, Long>,
    Void>() {
2.
3.              @Override
4.              public Void call(JavaPairRDD<String, Long> rdd) throws E
                xception {
5.
6.                  if (rdd.isEmpty()) {
7.
8.                  }
9.                  rdd.foreachPartition(new VoidFunction<Iterator<Tuple2<
                    String, Long>>>() {//Todo 插入数据库
10. }
```

16.3　电商广告点击综合案例黑名单过滤实现

本节阐述电商广告点击综合案例中黑名单的过滤实现。

因为 Spark Streaming 应用程序要对黑名单进行在线过滤，Spark Streaming 从 Kafka 集群中抓取的数据是在 Spark RDD 中的，所以在 Spark 中必然使用 transform 函数；但是，在这里我们必须使用 transformToPair，原因是读取进来的 Kafka 的数据是 Pair<String,String>类型的，另外一个原因是，过滤后的数据要进行进一步处理，所以必须是读进来的 Kafka 数据的原始类型 DStream<String, String>。每个 Batch Duration 中实际上将输入的数据被一个且仅仅被一个 RDD 封装，可以有多个 inputDstream，但是在产生 Job 的时候，这些不同的 InputDstream 在 Batch Duration 中就相当于 Spark 基于 HDFS 数据操作的不同文件来源。

根据电商广告点击综合案例的业务需求，黑名单过滤具体思路步骤如下。

（1）从 MySQL 数据库中查询到黑名单数据，并且将黑名单数据转换成 RDD（详见 16.5 节“对电商广告点击综合案例动态黑名单过滤真正的实现代码”）。

（2）Spark Streaming 从 Kafka 集群中抓取用户广告点击数据，将黑名单的 RDD 和流处理的 RDD 进行左关联，将每一批流处理的用户广告点击数据过滤掉黑名单用户的数据（16.5 节“对电商广告点击综合案例动态黑名单过滤真正的实现代码”）。

（3）基于用户广告点击数据表，动态过滤黑名单用户。黑名单生成来自两个方面：一方面，从数据库中查询到已有的黑名单用户进行过滤；另一方面，黑名单是动态生成的，Spark Streaming 持续不断地一直运行，在用户点击广告的时候，会动态产生黑名单用户。每个流处理时，都要考虑是否有新的黑名单加入。例如，某一时间用户广告点击的次数超过阈值，就判定该用户为黑名单用户。

（4）动态生成的黑名单可能包括重复的黑名单用户，我们需要首先将黑名单用户去重转换，然后将黑名单用户持久化到 MySQL 数据库中。

（5）使用数据库连接池的高效读写数据库的方式把唯一的不重复的黑名单用户数据写入数据库 MySQL，入库以后的黑名单数据又可以提供给第二步 Spark Streaming 每一批流处理 RDD 和黑名单 RDD 关联使用。

16.3.1　基于用户广告点击数据表及动态过滤黑名单用户

在电商广告点击综合案例之前，我们对有效广告点击的处理做了简化，简单地由用户点击广告的次数来判断用户是否为黑名单用户，生成黑名单过滤以后的 JavaPairDStream<String, Long>类型的 filteredClickInBatch，其格式是 Key-Value，Key 值为组拼的广告点击记录 clickedRecord(timestamp_ip_userID_adID_province_city)，Value 值为汇聚的点击总次数。

（1）对 filteredClickInBatch 使用算子 filter 进行过滤，filteredClickInBatch 每一行数据的第一个元素 v1._1 是组拼的广告点击记录 clickedRecord(timestamp_ip_userID_adID_province_city)，我们使用“_”分隔符进行切分，然后分别取出时间、用户 ID、广告 ID。

（2）接下来根据时间 date、用户 ID userID、广告 ID adID 等条件去查询用户点击广告的数据表，获得总的点击次数，基于点击次数判断是否属于黑名单点击。这里我们简化了电商广告点击综合案例的代码编写，省略了从数据库查询用户广告点击次数的步骤，直接赋予变量 clickedCountTotalToday 的值等于 81，来模拟某用户在某个时间点击了 81 次广告。clickedCountTotalToday 模拟了用户这个时间总的点击次数，我们设定一个黑名单用户点击的阈值，如可以设置 50 次，超过 50 次就可以判定用户为黑名单用户，if 表达式就返回 true，获取到黑名单用户的点击数据。

（3）生成 JavaPairDStream<String, Long> 类型的 blackListBasedOnHistory，blackListBasedOnHistory 是黑名单用户广告点击数据，其格式为 Key-Value。Key 值为黑名单用户的广告点击记录 clickedRecord(timestamp_ip_userID_adID_province_city)，Value 值为黑名单用户汇聚的点击总次数。

基于用户广告点击数据表，动态过滤黑名单用户代码如下：

```
1.  JavaPairDStream<String, Long> blackListBasedOnHistory = filteredClickInBatch
2.                  .filter(new Function<Tuple2<String, Long>, Boolean>() {
```

```
3.                      @Override
4.                      public Boolean call(Tuple2<String, Long> v1) throws
                        Exception {
5.          //广告点击的基本数据格式：timestamp、ip、userID、adID、province、city
6.                          String[] splited = v1._1.split("_");
7.                          String date = splited[0];
8.                          String userID = splited[2];
9.                          String adID = splited[3];
10.                         /**
                             *接下来根据 date、userID、adID 等条件查询用户点击广告
                             *的数据表，获得总的点击次数，这时基于点击次数判断是否
                             *属于黑名单点击
11.                          */
12.
13.                         int clickedCountTotalToday = 81;
14.                         if (clickedCountTotalToday > 50) {
15.                             return true;
16.                         } else {
17.                             return false;
18.                         }
19.                     }
20.                 });
```

16.3.2　黑名单的整个 RDD 进行去重操作

黑名单用户 blackListBasedOnHistory 的数据包括重复的黑名单用户，例如，某个用户在不同时间的点击次数都超过了设定阈值，那么用户就属于黑名单，而且出现了多次。我们需要将黑名单用户去重转换，然后将黑名单用户持久化到 MySQL 数据库中。

（1）blackListBasedOnHistory 格式为 Key-Value，blackListBasedOnHistory 每行数据的第一个元素 v1._1 的 Key 值为黑名单用户的广告点击记录 clickedRecord(timestamp_ip_ userID_ adID_province_city)，根据“_”分隔符进行切分，返回用户 ID，生成 JavaDStream<String> 格式的 blackListuserIDtBasedOnHistory，其值为黑名单用户 ID。

（2）blackListuserIDtBasedOnHistory 使用 transform 算子进行转换，将读入的每行黑名单用户广告点击数据 JavaRDD<String>，通过 distinct 方法进行去重，返回不重复的唯一的黑名单用户列表。生成 JavaDStream<String>类型的 blackListUniqueuserIDtBasedOnHistory，其值为去重的黑名单用户 ID。

黑名单用户去重操作的代码如下：

```
1.        /**
           *必须对黑名单的整个 RDD 进行去重操作！
2.         */
3.
4.        JavaDStream<String> blackListuserIDtBasedOnHistory = blacklist
5.            BasedOnHistory.map(new Function<Tuple2<String,Long>,String>() {
6.
7.                        @Override
8.                        public String call(Tuple2<String, Long> v1) throws
                          Exception {
9.                            //待办事项：自动生成方法存根
10.                           return v1._1.split("_")[2];
11.                       }
12.                   });
13.           JavaDStream<String> blackListUniqueuserIDtBasedOnHistory =
```

```
        blackListuserIDtBasedOnHistory
14.             .transform(new Function<JavaRDD<String>, JavaRDD<String>>() {
15.
16.                 @Override
17.                 public JavaRDD<String> call(JavaRDD<String> rdd)
                    throws Exception {
18.                     //待办事项：自动生成方法存根
19.                     return rdd.distinct();
20.                 }
21.             });
```

16.3.3　将黑名单写入到黑名单数据表

接下来我们将 blackListUniqueuserIDtBasedOnHistory 使用 foreachRDD 遍历每个 RDD，对于 RDD，使用 foreachPartition 遍历每个分区。

这里，我们使用数据库连接池的高效读写数据库的方式把数据写入数据库 MySQL；由于传入的参数是一个 Iterator 类型的集合，所以为了更加高效地操作，我们需要批量处理。例如，一次性插入 1000 条 Record，使用 insertBatch 或者 updateBatch 类型的操作；插入的用户信息可以只包含用户 ID useID，此时直接插入黑名单数据表即可。

将黑名单写入到黑名单数据表的代码如下：

```
1.  blackListUniqueuserIDtBasedOnHistory.foreachRDD(new Function<JavaRDD<
    String>, Void>() {
2.
3.              @Override
4.              public Void call(JavaRDD<String> rdd) throws Exception {
5.                  rdd.foreachPartition(new VoidFunction<Iterator<String>>(){
6.
7.                      @Override
8.                      public void call(Iterator<String> t) throws Exception {
9.      List<Object[]> blackList = new ArrayList<Object[]>();
10.
11.                         while (t.hasNext()) {
12.                             blackList.add(new Object[]{(Object)t.next()});
13.                         }
14.                         JDBCWrapper jdbcWrapper = JDBCWrapper.getJDBC
                            Instance();
15.                         jdbcWrapper.doBatch("INSERT INTO blacklisttable
                            VALUES (?) ", blackList);
16.                     }
17.                 });
18.                 return null;
19.             }
20.         });
```

16.4　电商广告点击综合案例底层数据层的建模和编码实现（基于 MySQL）

本节阐述电商广告点击综合案例实现底层数据层的建模和编码实现（基于 MySQL）。

在生产环境中，底层数据层的建模至关重要。传统的关系型数据库都是伪分布式，在 100 个使用了底层数据持久化层的案例中，至少有 90 个底层都是基于数据库的。当然，可以选择基于 Hbase 或者 Redis，但是实质上大多数人底层都会基于数据库。有两个非常重要的原因：第一，数据库可以非常高效地读和写，尤其是写的层次，频繁地、批量地插入，数据库特别适合；第二，传统的 IT 系统大多数都是基于数据库的，Spark 处理的结果如果扔给了数据库，传统的 Java EE 系统（如 spring），可以直接基于数据库的驱动，或者基于第三方框架去操作数据库中的数据。而数据库中的数据是 Spark 计算后的结果，可以操作这些数据绘制趋势图等。如股票一直变化的趋势图，使用 Spark Streaming+数据库 +Java EE 技术就特别适合。如每隔十秒刷新一次，Spark Streaming 是流式的功能，不断地更新数据，就会发现图一直在动。例如，发现纽约证券交易所、中国香港证券交易所的大屏幕上不同公司的股票一直在变化，变化的背后是经过了复杂的处理，此时我们基于 Spark Streaming 就特别适合完成诸如此类的事情。

在生产环境中，Spark 将计算处理的数据写到数据层中，Spark 在计算的过程中，有时也需要从数据库中读取数据，如黑名单数据的过滤，需要从数据库中读取黑名单的数据生成 RDD，然后执行 Join 操作。Java EE 系统从数据库中抓取数据，然后通过 Web 系统渲染数据，进行可视化展示。图 16-5 展示了数据层的重要性。数据层通常使用数据库，如 MySQL 数据库。底层数据层的建模示意图如图 16-5 所示。

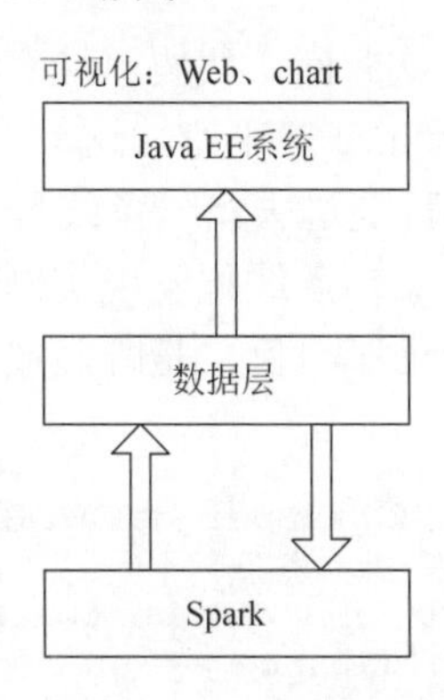

图 16-5　底层数据层的建模示意图

16.4.1　电商广告点击综合案例数据库链接单例模式实现

电商广告点击综合案例不是严格地采用 BAT（百度、阿里、腾讯公司）数据层真正的架构设计（BAT 公司规范了数据层设计、数据接口、数据库工厂，数据库的配置不能使用硬编码，一定会给一个配置）。我们主要是实现数据层的功能，将相关功能放到一个文件中，定义 JDBCWrapper。

（1）第一件事情，使用 static 代码块将数据库加载进来。Class.forName("com.mysql.jdbc.Driver")如果加载 MySQL 数据库有异常，可使用 try… catch 捕获异常。

（2）使用单例模式获取 jdbcInstance 实例。在 getJDBCInstance 方法中，只要获得一个实例就可以了。

首先按以下的代码编写，如果 jdbcInstance 等于空，就创建返回一个 JDBCWrapper。这样写单例不行，举一个很简单的例子，如果有 3 条线程同时来请求，3 条线程同时都发现没有实例，那就会同时创建出 3 个 jdbcInstance 实例。

```
1.   public static JDBCWrapper  getJDBCInstance() {
2.           if (jdbcInstance == null) {
3.               jdbcInstance = new  JDBCWrapper();
4.             }
5.           return jdbcInstance;
6.           }
```

对单例代码继续完善，有两个地方可以加 synchronized：第一个地方是在 getJDBCInstance 方法名前加 synchronized；第二个地方是在 synchronized (JDBCWrapper.class)方法名前加 synchronized。但这样写还不行，里面还要再进行一次判断。

```
1.       public  static  JDBCWrapper  getJDBCInstance() {
2.           if (jdbcInstance == null) {
3.               synchronized (JDBCWrapper.class) {
4.                   jdbcInstance = new  JDBCWrapper();
5.               }
6.           }
7.           return  jdbcInstance;
8.       }
```

接下来在单例代码里面再进行一次判断，代码已经有点技术含量了，虽然只是一个小单例。如果有 3 条线程来请求，jdbcInstance 等于空，这个地方有一个同步关键字 synchronized (JDBCWrapper.class)，那只有一条线程能进去，一条线程进去之后，运行完成，确实会生成一个实例；第二条线程获得这把锁，第二条线程发现实例不为空，就不运行了，直接返回；第三条线程也是直接返回。这样就保证了单例，是既高效、又轻松的方式。单例我们先做到这个程度。

```
1.           public static JDBCWrapper getJDBCInstance() {
2.           if (jdbcInstance == null) {
3.               synchronized (JDBCWrapper.class) {
4.                   if (jdbcInstance == null) {
5.                       jdbcInstance = new JDBCWrapper();
6.                   }
7.               }
8.           }
9.           return jdbcInstance;
10.      }
```

（3）在单例模式中构造私有的构造器 private JDBCWrapper()，只有 JDBCWrapper 类内才可以调用构造器。通过 for 循环遍历 10 次，当创建 new JDBCWrapper()时，一次就可以创建 10 个数据库链接，每次链接都使用 DriverManager.getConnection 获得一个数据库链接的实例，我们将数据库链接实例放入到 LinkedBlockingQueue<Connection>连接池队列中。

（4）写一个方法 getConnection()获得数据库句柄，从数据库池中返回 Connection。这里使用 poll 的方式，将数据库链接从队列中拎出来。考虑线程的问题，方法 getConnection()需加上关键字 synchronized，这样一个线程就获得一个实例。

```
1.       public synchronized Connection getConnection() {
2.
```

```
3.          return dbConnectionPool.poll();
4.      }
```

这里有一个问题，如果数据库连接池中没有东西，数据库连接池大小等于 0，那就获取不到连接，解决的方法是让线程等一会儿，睡眠 20ms，一直循环，确保可以获取到数据库链接。

```
1.      public synchronized Connection getConnection() {
2.          while (0 == dbConnectionPool.size()) {
3.              try {
4.                  Thread.sleep(20);
5.              } catch (InterruptedException e) {
6.                  //待办事项：自动生成 catch 块
7.                  e.printStackTrace();
8.              }
9.          }
10.
11.         return dbConnectionPool.poll();
12.     }
```

电商广告点击综合案例数据库链接单例模式实现代码如下：

```
1.      class  JDBCWrapper {
2.
3.      private static LinkedBlockingQueue<Connection> dbConnectionPool = new
        LinkedBlockingQueue<Connection>();
4.
5.      private static JDBCWrapper jdbcInstance = null;
6.
7.      static {
8.          try {
9.              Class.forName("com.mysql.jdbc.Driver");
10.         } catch (ClassNotFoundException e) {
11.             //待办事项：自动生成 catch 块
12.             e.printStackTrace();
13.         }
14.     }
15.
16.     public static JDBCWrapper getJDBCInstance() {
17.         if (jdbcInstance == null) {
18.
19.             synchronized (JDBCWrapper.class) {
20.                 if (jdbcInstance == null) {
21.                     jdbcInstance = new JDBCWrapper();
22.                 }
23.             }
24.
25.         }
26.
27.         return jdbcInstance;
28.     }
29.
30.     private JDBCWrapper() {
31.
32.         for (int i = 0; i < 10; i++) {
33.
34.             try {
35.                 Connection conn = DriverManager.getConnection("jdbc:
                    mysql://Master:3306/sparkstreaming", "root",
```

```
36.                     "root");
37.                 dbConnectionPool.put(conn);
38.             } catch (Exception e) {
39.                 //待办事项：自动生成 catch 块
40.                 e.printStackTrace();
41.             }
42.
43.         }
44.
45.     }
46.
47.     public synchronized Connection getConnection() {
48.         while (0 == dbConnectionPool.size()) {
49.             try {
50.                 Thread.sleep(20);
51.             } catch (InterruptedException e) {
52.                 //待办事项：自动生成 catch 块
53.                 e.printStackTrace();
54.             }
55.         }
56.
57.         return dbConnectionPool.poll();
58.     }
59.
```

16.4.2　电商广告点击综合案例数据库操作实现

电商广告点击综合案例获得数据库连接后，我们要对数据库进行操作：插入、查询、修改、删除。我们编写一个通用的方式来实现，首先考虑的问题是，在 Spark 中进行操作，操作的是一批又一批的数据，因此要考虑批量数据的处理，而不考虑一条数据的插入、查询、修改等。其次要考虑的是，在数据库批处理中有可能来一批参数，因此传入 paramsList。BAT（百度、阿里、腾讯）公司一般都用数据库框架，代码都差不多，我们这里主要实现数据库的相关操作。

电商广告点击综合案例数据库批量处理函数，批处理的 SQL 语句可以是插入、更新等。

```
1.  public int[] doBatch(String sqlText, List<Object[]> paramsList) {
2.          Connection conn = getConnection();//获取数据库链接
3.          PreparedStatement preparedStatement = null;
4.          int[] result = null;
5.          try {
6.              conn.setAutoCommit(false);
7.              preparedStatement = conn.prepareStatement(sqlText);
8.
9.              for (Object[] parameters : paramsList) {
                //循环遍历 paramsList，传入参数
10.                 for (int i = 0; i < parameters.length; i++) {
11.                     preparedStatement.setObject(i + 1, parameters[i]);
12.                 }
13.
14.                 preparedStatement.addBatch();
15.             }
16.
17.             result = preparedStatement.executeBatch();
18.
```

```
19.             conn.commit();
20.
21.         } catch (Exception e) {
22.             //待办事项：自动生成 catch 块
23.             e.printStackTrace();
24.         } finally {
25.             if (preparedStatement != null) {
26.                 try {
27.                     preparedStatement.close();//释放掉 preparedStatement
28.                 } catch (SQLException e) {
29.                     //待办事项：自动生成 catch 块
30.                     e.printStackTrace();
31.                 }
32.             }
33.
34.             if (conn != null) {
35.                 try {
36.                     dbConnectionPool.put(conn);
37.                 } catch (InterruptedException e) {
38.                     //待办事项：自动生成 catch 块
39.                     e.printStackTrace();
40.                 }
41.             }
42.         }
43.
44.         return result;
45.     }
```

电商广告点击综合案例数据库查询函数，数据库处理的结果有一个回调函数，回调函数一定是一个接口，将数据库查询的结果传进去，查询函数如下：

```
1.  public void doQuery(String sqlText, Object[] paramsList, ExecuteCallBack
    callBack) {
2.
3.          Connection conn = getConnection();
4.          PreparedStatement preparedStatement = null;
5.          ResultSet result = null;
6.
7.          try {
8.
9.              preparedStatement = conn.prepareStatement(sqlText);
10.
11.             if (paramsList != null) {
12.                 for (int i = 0; i < paramsList.length; i++) {
13.                     preparedStatement.setObject(i + 1, paramsList[i]);
14.
15.                 }
16.             }
17.             result = preparedStatement.executeQuery();
18.
19.             callBack.resultCallBack(result);
20.
21.         } catch (Exception e) {
22.             //待办事项：自动生成 catch 块
23.             e.printStackTrace();
24.         } finally {
25.             if (preparedStatement != null) {
26.                 try {
27.                     preparedStatement.close();
```

```
28.                 } catch (SQLException e) {
29.                     //待办事项：自动生成 catch 块
30.                     e.printStackTrace();
31.                 }
32.             }
33.
34.             if (conn != null) {
35.                 try {
36.                     dbConnectionPool.put(conn);
37.                 } catch (InterruptedException e) {
38.                     //待办事项：自动生成 catch 块
39.                     e.printStackTrace();
40.                 }
41.             }
42.         }
43.
44.     }
45. }
```

电商广告点击综合案例数据库查询回调函数，传入的参数类型是 ResultSet，数据库查询的业务逻辑在 resultCallBack 重载方法中实现。

```
1.   interface ExecuteCallBack {
2.       void  resultCallBack(ResultSet result) throws Exception;
3.   }
```

16.5　电商广告点击综合案例动态黑名单过滤真正的实现代码

在 16.3 节电商广告点击综合案例黑名单过滤实现中，我们阐述了黑名单过滤的具体思路步骤。其中，从 MySQL 数据库中查询到黑名单数据，并且将黑名单数据转换成 RDD 以及 Spark Streaming 从 Kafka 集群中抓取用户广告点击数据，将黑名单的 RDD 和流处理的 RDD 进行左关联，将每一批流处理的用户广告点击数据过滤掉黑名单用户的数据，这些内容将在本节详细阐述。

16.5.1　从数据库中获取黑名单封装成 RDD

从数据库中获取黑名单转换成 RDD，即新的 RDD 实例封装黑名单数据。黑名单的表中只有用户 ID，但是如果要进行 Join 操作，就必须是 Key-Value，所以这里我们需要基于数据表中的数据产生 Key-Value 类型的数据集合。

从数据库中查询黑名单表 blacklisttable，将查询到的黑名单放入 blackListNames 列表 List<String>，此时的 blackListNames 列表中只有一个元素用户 ID，而我们之后需将黑名单记录进行 Join 操作，因此需将列表中的元组转化为 Key-Value 的格式。循环遍历 blackListNames 列表，取出黑名单，默认定义黑名单为 true 值，生成元组（用户 ID，布尔值），将元组 Tuple2<String, Boolean>放入到 blackListFromDB 列表中，然后使用 jsc.parallelizePairs 方法创建 JavaPairRDD<String, Boolean> blackListRDD。blackListRDD 用于封装黑名单数据。

从数据库中获取黑名单转换成 RDD 的代码如下：

```
1.   JavaPairDStream<String, String> filteredadClickedStreaming = adClicked
                    Streaming.transformToPair(new Function<JavaPairRDD<String,
                    String>,JavaPairRDD<String, String>>() {
2.
3.
4.                       @Override
5.                       public JavaPairRDD<String, String> call(JavaPairRDD<
                         String, String> rdd) throws Exception {
6.
7.                           final List<String> blackListNames = new ArrayList<
                             String>();
8.                           JDBCWrapper jdbcWrapper = JDBCWrapper.getJDBC
                             Instance();
9.                           jdbcWrapper.doQuery("SELECT * FROM blacklisttable",
                             null, new ExecuteCallBack() {
10.
11.                              @Override
12.                              public void resultCallBack(ResultSet result)
                                 throws Exception {
13.      while (result.next()) {
         blackListNames.add(result.getString(1));
14.                                  }
15.                              }
16.
17.                          });
18.
19.                          List<Tuple2<String, Boolean>> blackListTuple =
                             new ArrayList<Tuple2<String, Boolean>>();
20.
21.                          for (String name : blackListNames) {
22.                              blackListTuple.add(new Tuple2<String, Boolean>
                                 (name, true));
23.                          }
24.
25.                          List<Tuple2<String, Boolean>> blackListFromDB =
                             blackListTuple; //数据来自于查询的黑名单表，并且映射
                                              //成为<String,Boolean>
26.
27.                          JavaSparkContext jsc = new JavaSparkContext
                             (rdd.context());
28.
29.                          /**
                              * 黑名单的表中只有 userID，但是如果要进行 Join 操作，就
                              * 必须是 Key-Value，所以这里我们需要基于数据表中的数据产
                              * 生 Key-Value 类型的数据集合
30.                           */
31.                          JavaPairRDD<String, Boolean> blackListRDD = jsc.
                             parallelizePairs(blackListFromDB);
32. ......
```

16.5.2　黑名单 RDD 和批处理 RDD 进行左关联及过滤掉黑名单

从数据库中获取黑名单封装成 RDD，代表黑名单的 RDD 的实例 blackListRDD 和 Batch Duration 产生的 RDD 进行 Join 操作时，Batch Duration 产生的 RDD 操作的时候肯定是基于用户 ID 进行 Join 的，所以必须把 adClickedStreaming transformToPair 传入的 RDD 进行 mapToPair 操作，将其转化成为符合格式的 RDD。之前，KafkaUtils 调用 createDirectStream 生

成 JavaPairInputDStream 类型的 adClickedStreaming，adClickedStreaming 的格式为 Key-Value 键值对<String, String>，第一个元素 Key 键值为 Kafka 生成的键值，第二个元素 Value 值是 Spark Streaming 从 Kafka 中读取的一行行的数据，我们对 RDD 使用 mapToPair 方法，读取第二个元素，即广告点击数据，其广告点击的数据格式为 timestamp、ip、userID、adID、province、city，我们使用“\t”分隔符进行切分，取到第 2 个值，即用户 ID，并且将用户 ID 作为 Key 值，Value 值仍是 Spark Streaming 从 Kafka 中读取的一行行的数据，创建生成 JavaPairRDD<String, Tuple2<String, String>> rdd2Pair，其格式为（用户 ID，每行广告点击数据）。

传入的 RDD 进行 mapToPair 操作转化成为符合格式的 RDD 的代码。

```
1.   JavaPairRDD<String, Tuple2<String, String>> rdd2Pair = rdd.mapToPair(new
     PairFunction<Tuple2<String, String>,String, Tuple2<String, String>>() {
2.   @Override
3.       public Tuple2<String, Tuple2<String, String>> call(Tuple2<String, String>t)
4.           throws Exception {
5.                                 String userID = t._2.split("\t")[2];
6.                                 return new Tuple2<String, Tuple2<String,
                                   String>>(userID, t);
7.                             }
8.                       });
9.   ......
```

我们把代表黑名单的 RDD 的实例 blackListRDD 和 Batch Duration 产生的 RDD 进行 Join 操作，准确地说是进行 leftOuterJoin 左关联操作。也就是说，使用 Batch Duration 产生的 rdd 和代表黑名单的 RDD 的实例进行 leftOuterJoin 左关联操作，如果两者都有内容，就会是 true，否则就是 false；我们要留下的是 leftOuterJoin 操作结果为 false 的消息记录。

- 将 rdd2Pair 与 blackListRDD 进行左关联，创建生成 JavaPairRDD<String, Tuple2<Tuple2<String, String>, Optional<Boolean>>> joined。其中，joined RDD 的每行记录的第一个元素 v1._1（即 Key 值）为用户 ID，第二个元素 v1._2（即 Value 值）为一个元组 Tuple2。
- 第二个元素 Tuple2 v1._2 中第一个元素 v1._2._1 对应 rdd2Pair 中的 Tuple2<String, String>，其 v1._2._1._1 中第一个元素 Key 为 Kafka 生成的键值，其 v1._2._1._2 中第二个元素 Value 值是 Spark Streaming 从 Kafka 中读取的一行行的数据。
- 第二个元素 Tuple2 v1._2 中第二个元素 v1._2._2 对应 blackListRDD 中的 Optional<Boolean>，即 blackListRDD 中定义的布尔值是否为黑名单。Optional 是可选的，Batch　Duration 产生的 RDD 和代表黑名单的 RDD 的实例进行 leftOuterJoin 左关联操作，如果 Batch　Duration 产生的 RDD 和代表黑名单的 RDD 两者的用户 ID 相等，则 Optional 值存在，在 if 表达式中返回 true；如果 Batch　Duration 产生的 RDD 中用户 ID 左关联时不能匹配上代表黑名单的 RDD 的用户 ID，此时 Optional 值是空值，if 表达式返回 false。

leftOuterJoin 左关联操作及黑名单过滤代码如下：

```
1.   JavaPairRDD<String, Tuple2<Tuple2<String, String>, Optional<Boolean>>>
     joined = rdd2Pair.leftOuterJoin(blackListRDD);
2.
3.
4.                             JavaPairRDD<String, String> result = joined.filter(
5.                                   new Function<Tuple2<String, Tuple2<Tuple2<
                                     String, String>, Optional<Boolean>>>,
                                     Boolean>() {
```

```
6.
7.                          @Override
8.                          public Boolean call(Tuple2<String, Tuple2<
                            Tuple2<String,String>,Optional<Boolean>>>v1)
9.                                  throws Exception {
10.                             Optional<Boolean> optional = v1._2._2;
11.
12.                             if(optional.isPresent()&&optional.get()){
13.                                 return false;
14.                             } else {
15.                                 return true;
16.                             }
17.
18.                         }
19.                     }).mapToPair(
20.                             new PairFunction<Tuple2<String, Tuple2<
                                Tuple2<String, String>,Optional<Boolean>>>,
                                String, String>() {
21.
22.                         @Override
23.                         public Tuple2<String, String> call(
24.                                 Tuple2<String, Tuple2<Tuple2<String,
                                     String>, Optional<Boolean>>> t)
25.                                         throws Exception {
26.                             //待办事项：自动生成方法存根
27.                             return t._2._1;
28.                         }
29.                     });
30.
31.                     return result;
32.                 }
33.             });
34.
35.         filteredadClickedStreaming.print();
36. ......
```

上述代码中的 filteredadClickedStreaming.print()说明：此处的 print 并不会直接触发 Job 的执行，因为现在的一切都在 Spark Streaming 框架的控制下，对于 Spark Streaming 而言，具体是否触发真正的 Job 运行是基于设置的 Duration 时间间隔的。注意，Spark Streaming 应用程序要想执行具体的 Job，对 Stream 就必须有 output Stream 操作， output Stream 有很多类型的函数触发，类 print、saveAsTextFile、saveAsHadoopFiles 等，最重要的一个方法是 foreachRDD，因为 Spark Streaming 处理的结果一般都会放在 Redis、DB、DashBoard 等上面，foreachRDD 主要用用来完成这些功能，而且可以随意地自定义具体数据到底放在哪里。

16.6　动态黑名单基于数据库 MySQL 的真正操作代码实战

电商广告点击综合案例中动态黑名单的入库，我们基于 MySQL 数据库来实现。

16.6.1　MySQL 数据库操作的架构分析

默认情况下，将 RDD 中的数据插入 MySQL 中是一条条地插入的，也就是说，去遍历每个 Partition 的 Iterator 中的每条记录，每次都要建立一次数据库的链接，当我们使用 foreachRDD 的时候，操作的对象是 RDD，然后我们使用 rdd 的 foreachPartition，此时操作的对象是分区 Partition，而不是一条条的记录。也就是说，每次读取的是整个分区 Partition，读取数据时效率非常高，然后我们采用 executeBatch 的方式插入或者更新数据，此时也是数据库更加高效的链接和更新方式。不过，一次读取一个分区 Partition 的弊端是有可能内存 OOM，所以此时需要关注内存的使用。

MySQL 数据库入库如图 16-6 所示。

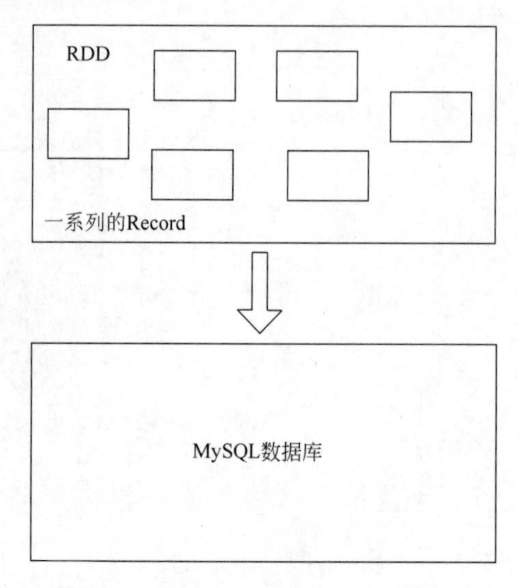

图 16-6　MySQL 数据库入库

16.6.2　MySQL 数据库操作的代码实战

本节基于动态黑名单内容进行数据库 MySQL 的代码实战，包括三部分：黑名单数据库查询，插入黑名单，插入、更新用户广告点击数据等内容。

1. MySQL数据库操作（黑名单数据库查询）

16.5.1 节实现了从数据库中获取黑名单封装成 RDD 的功能。这里进行 MySQL 数据库中黑名单数据库查询的操作。

MySQL 数据库黑名单数据库查询的步骤如下。

（1）获取数据库连接器 JDBCWrapper getJDBCInstance()。

（2）执行 doQuery 查询函数，传入查询的 SQL 语句，这里查询所有的黑名单，因此不需要传入 paramsList 参数。数据库查询的结果传到回调函数的返回值 result，然后循环遍历 result 的 next 值，通过 getString(1)获取到黑名单数据。

黑名单数据库查询代码如下：

```
1.          final List<String> blackListNames = new ArrayList<String>();
2.              JDBCWrapper jdbcWrapper = JDBCWrapper.getJDBCInstance();
3.                 jdbcWrapper.doQuery("SELECT * FROM blacklisttable", null,
```

```
                    new ExecuteCallBack() {
4.
5.                              @Override
6.                              public void resultCallBack(ResultSet result)
                                throws Exception {
7.                                  while (result.next()) {
8.                                      blackListNames.add(result.getString(1));
9.                                  }
10.                             }
11.
12.                         });
13.
14.                         List<Tuple2<String, Boolean>> blackListTuple =
                            new ArrayList<Tuple2<String, Boolean>>();
15.
16.                         for (String name : blackListNames) {
17.                             blackListTuple.add(new Tuple2<String, Boolean>
                                (name, true));
18.                         }
19.
20.                         List<Tuple2<String, Boolean>> blackListFromDB =
                            blackListTuple;
21. ......
```

2. MySQL数据库操作（插入黑名单）

16.3.3节实现了将黑名单写入到黑名单数据表的功能。MySQL数据库操作（插入黑名单）比较简单，这里就是插入一个用户ID。我们定义一个列表blackList，这只是功能性的实现，企业级对任何实体的封装肯定会创建一个JavaBean，相当于scala中的case class。然后while循环遍历Iterator的数据，拿到一批黑名单数据，加入到blackList。

MySQL数据库操作插入黑名单的步骤如下。

（1）获取数据库连接器JDBCWrapper getJDBCInstance()。

（2）批量插入Insert Batch，doBatch函数传入插入的SQL语句。

```
1.      blackListUniqueuserIDtBasedOnHistory.foreachRDD(new  Function<JavaRDD<
        String>, Void>() {
2.              @Override
3.              public Void call(JavaRDD<String> rdd) throws Exception {
4.                  rdd.foreachPartition(new VoidFunction<Iterator<String>>(){
5.                      @Override
6.                      public void call(Iterator<String> t) throws Exception
{
7.                  //定义一个列表
8.                          List<Object[]> blackList = new ArrayList<
                            Object[]>();
9.                          while (t.hasNext()) { //while 循环遍历
10.                             blackList.add(new Object[]{(Object)t.next()});
11.                         }
12.             //获得数据库连接器
13.                 JDBCWrapper jdbcWrapper = JDBCWrapper.getJDBCInstance();
14.         //批量插入数据
15.                 jdbcWrapper.doBatch("INSERT INTO blacklisttable VALUES
                    (?) ", blackList);
16.                     }
17.                 });
18.                 return null;
```

```
19.             }
20.         });
```

3. MySQL数据库操作（插入、更新用户广告点击数据）

在 16.2 节电商广告点击综合案例在线点击统计实战中，须实现将用户在线广告点击的数据持久化保存到 MySQL 数据库中。这里，我们使用数据库连接池的高效读写数据库的方式把数据写入数据库 MySQL。由于传入的参数是一个 Iterator 类型的集合，所以为了更加高效地操作，我们需要批量处理。例如，一次性插入 1000 条 Record，使用 insertBatch 或者 updateBatch 类型的操作；插入的用户信息可以只包含 timestamp、ip、userID、adID、province、city。里面有一个问题：可能出现两条记录的 Key 是一样的，此时就需要更新累加操作。

MySQL 数据库插入在线广告点击数据的具体实现如下。

（1）定义用户广告点击数据 UserAdClicked 的 JavaBean。插入的用户信息包含 timestamp、ip、userID、adID、province、city，我们定义一个 UserAdClicked 的 JavaBean，用于存放用户广告点击的数据。

UserAdClicked 的 JavaBean 代码如下：

```
1.  class  UserAdClicked  implements  Serializable {
2.      private String timestamp;
3.      private String ip;
4.      private String userID;
5.      private String adID;
6.      private String province;
7.      private String city;
8.      private Long clickedCount;
9.
10.     @Override
11.     public String toString() {
12.         return "UserAdClicked [timestamp=" + timestamp + ", ip=" + ip +
            ", userID=" + userID + ", adID=" + adID+ ", province=" + province
            + ", city=" + city + ",clickedCount=" + clickedCount + "]";
13.
14.     }
15.
16.     public Long getClickedCount() {
17.         return clickedCount;
18.     }
19.
20.     public void setClickedCount(Long clickedCount) {
21.         this.clickedCount = clickedCount;
22.     }
23.
24.     public String getTimestamp() {
25.         return timestamp;
26.     }
27.
28.     public void setTimestamp(String timestamp) {
29.         this.timestamp = timestamp;
30.     }
31.
32.     public String getIp() {
33.         return ip;
34.     }
35.
```

```
36.     public void setIp(String ip) {
37.         this.ip = ip;
38.     }
39.
40.     public String getUserID() {
41.         return userID;
42.     }
43.
44.     public void setUserID(String userID) {
45.         this.userID = userID;
46.     }
47.
48.     public String getAdID() {
49.         return adID;
50.     }
51.
52.     public void setAdID(String adID) {
53.         this.adID = adID;
54.     }
55.
56.     public String getProvince() {
57.         return province;
58.     }
59.
60.     public void setProvince(String province) {
61.         this.province = province;
62.     }
63.
64.     public String getCity() {
65.         return city;
66.     }
67.
68.     public void setCity(String city) {
69.         this.city = city;
70.     }
71. }
72. ......
```

（2）对 filteredClickInBatch 使用 foreachRDD 算子，将 foreachRDD 中用户广告点击记录持久化到 MySQL 数据库中。filteredClickInBatch 每行数据记录的格式为 Key-Value：Key 值为组拼的广告点击记录 clickedRecord(timestamp_ip_userID_adID_province_city)，Value 值为汇聚的广告点击次数。对 filteredClickInBatch 进行 foreachRDD 函数触发，将 filteredClickInBatch 每行记录的第一个元素 record._1 按“_”进行分割，提取出 timestamp、ip、userID、adID、province、city，ClickedCount 各个字段，并在数据表中查询表广告点击 adclicked，如果广告点击表中没有用户的点击记录，则插入，如果广告点击表中已经有用户的广告点击记录，则更新数据。

具体实现如下。

（1）将 filteredClickInBatch 每行记录的第一个元素 record._1 按“_”分割，提取出 timestamp、ip、userID、adID、province、city，ClickedCount 各个字段，存放到 UserAdClicked 的 Javabean 数据结构中，然后循环遍历用户广告点击数据，放入到 List<UserAdClicked>类型列表 userAdClicked。

（2）根据时间、用户 ID、广告 ID 在 MySQL 数据库进行点击次数查询。

- 获取数据库连接器 JDBCWrapper getJDBCInstance()。
- 执行 doQuery 查询函数，传入查询的 SQL 语句，传入的 paramsList 参数为时间、用户 ID、广告 ID。将数据库查询的结果传到回调函数的返回值 result，然后循环遍历 result 的 next 值，通过 getString(1)获取到黑名单数据。

（3）根据数据库广告点击查询的结果 result，将用户广告点击数据分成两类。

第一类：需更新的数据，List<UserAdClicked> updating，把从数据库中能查询到的数据 UserAdClicked clicked 放入到 updating 列表。

第二类：需插入的数据，List<UserAdClicked> inserting，若从数据库中查询不到记录，就将 UserAdClicked clicked 放入到 inserting 列表。

根据数据库查询的结果进行数据分类，代码如下：

```
1.     filteredClickInBatch.foreachRDD(new Function<JavaPairRDD<String,
       Long>, Void>() {
2.
3.             @Override
4.             public Void call(JavaPairRDD<String, Long> rdd) throws
               Exception {
5.
6.                 if (rdd.isEmpty()) {
7.
8.                 }
9.                 rdd.foreachPartition(new VoidFunction<Iterator<Tuple2<
                   String, Long>>>() {
10.                    @Override
11.                    public void call(Iterator<Tuple2<String, Long>> partition)
                       throws Exception {
12.                    /** 这里我们使用数据库连接池的高效读写数据库的方式把数据写
                         * 入数据库 MySQL; 由于传入的参数是一个 Iterator 类型的
                         * 集合，所以为了更加高效地操作，我们需要批量处理。例如，
                         * 一次性插入 1000 条 Record，使用 insertBatch 或者
                         * updateBatch 类型的操作；插入的用户信息可以只包含
                         * timestamp、ip、userID、adID、province、city，这
                         * 里面有一个问题：可能出现两条记录的 Key 是一样的，此时就
                         * 需要更新累加操作
13.                      */
14.                        List<UserAdClicked> userAdClickedList = new
                           ArrayList<UserAdClicked>();
15.
16.                        while (partition.hasNext()) {
17.                            Tuple2<String, Long> record = partition.next();
18.
19.                            String[] splited = record._1.split("_");
20.                            UserAdClicked userClicked = new UserAdClicked();
21.
22.                            userClicked.setTimestamp(splited[0]);
23.                            userClicked.setIp(splited[1]);
24.                            userClicked.setUserID(splited[2]);
25.                            userClicked.setAdID(splited[3]);
26.                            userClicked.setProvince(splited[4]);
27.                            userClicked.setCity(splited[5]);
28.                            userAdClickedList.add(userClicked);
29.
30.                        }
31.
32.                        final List<UserAdClicked> inserting = new ArrayList<
```

```
                         UserAdClicked>();
33.                      final List<UserAdClicked> updating = new ArrayList<
                         UserAdClicked>();
34.
35.                      JDBCWrapper jdbcWrapper = JDBCWrapper. getJDBC
                         Instance();
36.
37.                      //点击
38.              //表的字段：timestamp、ip、userID、adID、province、city、
                 //clickedCount
39.                      for (final UserAdClicked clicked : userAdClickedList){
40.                          jdbcWrapper.doQuery(
41.                                  "SELECT count(1) FROM adclicked WHERE "
42.                                          + " timestamp = ? AND userID
                                             = ? AND adID = ?",
43.                                  new Object[] { clicked.getTimestamp(),
                                     clicked.getUserID(),clicked.getAdID()},
44.                                  new ExecuteCallBack() {
45.
46.                              @Override
47.                              public void resultCallBack(ResultSet
                                 result) throws Exception {
48.
49.                                  if (result.getRow() != 0) {
50.                                      long count = result.getLong(1);
51.                                      clicked.setClickedCount(count);
52.                                      updating.add(clicked);
53.
54.                                  } else {
55.                                      clicked.setClickedCount(0L);
56.                                      inserting.add(clicked);
57.
58.                                  }
59.
60.                              }
61.                          });
62.                      }
```

（4）按照用户广告点击的两类数据分别执行数据库操作。

第一类：在 MySQL 数据库中更新数据。

- 循环遍历 updating 列表，取出 UserAdClicked inserRecord 中的时间、用户 IP、用户 ID、广告 ID、省份、城市、点击次数放入到 SQL 参数列表 insertParametersList。
- 执行 doBatch 批处理函数，传入更新的 SQL 语句，根据传入的 paramsList 参数更新时间、用户 IP、用户 ID、广告 ID、省份、城市、点击次数等用户广告点击数据。

第二类：在 MySQL 数据库中插入数据。

- 循环遍历 inserting 列表，取出 UserAdClicked inserRecord 中的时间、用户 IP、用户 ID、广告 ID、省份、城市、点击次数放入到 SQL 参数列表 insertParametersList。
- 执行 doBatch 批处理函数，传入插入的 SQL 语句，传入的 paramsList 参数为时间、用户 IP、用户 ID、广告 ID、省份、城市、点击次数，插入用户广告点击数据。

按照用户广告点击的两类数据分别执行数据库操作代码。

```
1.   //adclicked
2.       //表的字段：timestamp、ip、userID、adID、province、city、clickedCount
3.   //在 MySQL 数据库中插入数据
```

```
4.                      ArrayList<Object[]> insertParametersList = new
                        ArrayList<Object[]>();
5.                      for (UserAdClicked inserRecord : inserting) {
6.                          insertParametersList.add(new Object[]{ inserRecord
                            .getTimestamp(), inserRecord.getIp(),
7.                                  inserRecord.getUserID(), inserRecord
                                    .getAdID(), inserRecord.getProvince(),
8.                                  inserRecord.getCity(), inserRecord
                                    .getClickedCount() });
9.                      }
10.                     jdbcWrapper.doBatch("INSERT INTO adclicked
                        VALUES(?,?,?,?,?,?,?)", insertParametersList);
11.
12.                     //点击
13.                     //表的字段：timestamp、ip、userID、adID、province、
                        //city、clickedCount
14.                     //在 MySQL 数据库中更新数据
15.                     ArrayList<Object[]> updateParametersList = new
                        ArrayList<Object[]>();
16.                     for (UserAdClicked updateRecord : updating) {
17.                         updateParametersList.add(new Object[]{updateRecord
                            .getTimestamp(), updateRecord.getIp(),
18.                                 updateRecord.getUserID(), updateRecord
                                    .getAdID(),updateRecord.getProvince(),
19.                                 updateRecord.getCity(), updateRecord
                                    .getClickedCount() });
20.                     }
21.                     jdbcWrapper.doBatch("UPDATE adclicked set
                        clickedCount = ? WHERE "
22.                             + " timestamp = ? AND ip = ? AND userID
                                = ? AND adID = ? AND province = ? "
23.                             + "AND city = ? ", updateParametersList);
24.
25.                 }
26.             });
27.             return null;
28.         }
29.
30.     });
```

16.7　通过 updateStateByKey 等实现广告点击流量的在线更新统计

在 16.2 节电商广告点击综合案例在线点击统计实战中，我们使用 reduceByKey 算子计算的是每一个批处理时间 Batch Duration 中每个 User 的广告点击量。我们要不断在 Spark Streaming 历史的基础上进行更新，这时就需要 updateStateByKey。updateStateByKey 遵循 RDD 的不变性，采样的是 cogroup 的方式，cogroup 方式是根据 Key 对数据进行聚合操作，每次操作时都要进行全量的扫描，随着时间的推移，其性能会越来越差。可能在开始的时候，Spark Streaming 每个 Batch Duration 处理进行 updateStateByKey，响应时间是 5s，过一天就会发现变成 1 min 了。所以，Spark 1.6.X 推出了 MapWithState，MapWithState 是实验性的 AP，MapWithState 遵循 RDD 的不变性，其是基于一个数据结构，数据结构不变，但里面的内容

改变，而且 MapWithState 的内存类似于 HashMap，在历史的基础上进行更新。

在本节中，我们基于 updateStateByKey 等实现广告点击流量的在线更新统计，这是在生产环境中的应用方式。以后在 Spark 2.0 中，MapWithState 方式更加成熟。

电商广告点击综合案例中广告点击累计动态更新，每个 updateStateByKey 都会在 Batch Duration 时间间隔的基础上进行点击次数的更新，更新之后我们一般都会持久化到外部存储设备上，在这里我们存储到 MySQL 数据库中。

（1）定义用户广告点击数据 AdClicked 的 JavaBean。AdClicked 的用户信息包含时间、广告 ID、省份、城市、点击次数（timestamp、adID、province、city、clickedCount），用于存放广告点击累计动态更新的用户广告点击数据。

AdClicked 的 JavaBean 代码如下：

```
1.  class AdClicked implements Serializable {
2.      private String timestamp;
3.      private String adID;
4.      private String province;
5.      private String city;
6.      private Long clickedCount;
7.
8.      @Override
9.      public String toString() {
10.         return "AdClicked [timestamp="+ timestamp + ", adID=" + adID + ",
            province=" + province + ", city=" + city   +", clickedCount=" +
            clickedCount + "]";
11.     }
12.
13.     public String getTimestamp() {
14.         return timestamp;
15.     }
16.
17.     public void setTimestamp(String timestamp) {
18.         this.timestamp = timestamp;
19.     }
20.
21.     public String getAdID() {
22.         return adID;
23.     }
24.
25.     public void setAdID(String adID) {
26.         this.adID = adID;
27.     }
28.
29.     public String getProvince() {
30.         return province;
31.     }
32.
33.     public void setProvince(String province) {
34.         this.province = province;
35.     }
36.
37.     public String getCity() {
38.         return city;
39.     }
40.
41.     public void setCity(String city) {
42.         this.city = city;
```

```
43.     }
44.
45.     public Long getClickedCount() {
46.         return clickedCount;
47.     }
48.
49.     public void setClickedCount(Long clickedCount) {
50.         this.clickedCount = clickedCount;
51.     }
52.
53. }
```

（2）使用 updateStateByKey 算子进行广告点击累计动态更新。

电商广告点击综合案例在线点击统计实战中，KafkaUtils 调用 createDirectStream 生成 JavaPairInputDStream 类型的 adClickedStreaming，adClickedStreaming 经过黑名单的动态过滤以后，转换生成 JavaPairDStream<String, String> 类型的 filteredadClickedStreaming。filteredadClickedStreaming 的格式为 Key-Value 键值对<String, String>，Key 为 Kafka 生成的 key 值，Value 值是 Spark Streaming 从 Kafka 中读取的一行行的数据。

① 对 filteredadClickedStreaming 每进行 mapToPair 转换，其每行数据的第二个元素 t._2 是 park Streaming 从 Kafka 中读取的一行行的数据，按照分隔符“\t”对每行的数据进行切分，分别取出时间、IP、用户 ID、广告 ID、省份、城市等信息，然后将其重新组拼成 clickedRecord(timestamp_adID_province_city)，形成 Key-Value 键值对。Key 值为 clickedRecord (timestamp_adID_province_city)，Value 为计算 1 次。

② 使用 updateStateByKey 算子进行广告点击累计动态更新。updateStateByKey 的传入参数是一个函数 Function2，Function2 的第一个参数 List<Long> v1 代表当前 Key(timestamp_adID_province_city)在当前的 Batch Duration 中出现次数的集合，例如{1,1,1,1,1,1}；Function2 的第二个参数 Optional<Long> v2 代表当前 Key(timestamp_ adID_province_city)在以前的 Batch Duration 中积累下来的结果。Function2 的返回结果是 Optional<Long>>，即动态更新累计的状态值。

其中，v2 类型是可选的 Optional<Long>。

- 如果 Key 值在以前的 Batch　Duration 中没出现过，v2 值不存在，默认定义累计次数 clickedTotalHistory 等于 0 次。
- 如果 Key 值在以前的 Batch　Duration 中存在，那就由 v2.get()获取到以前的 Batch Duration 中积累下来的结果，再循环遍历 v1 中的元素，将当前的 Batch　Duration 批处理中 Key 值出现的次数，在以前的 Batch　Duration 中积累下来的结果的基础上进行累加，最终返回动态累计更新的广告点击次数，创建生成 JavaPairDStream<String, Long> updateStateByKeyDStream。

使用 updateStateByKey 算子进行广告点击累计动态更新。

```
1.          /**
             * 广告点击累计动态更新，每个 updateStateByKey 都会在 Batch Duration 的
             * 时间间隔的基础上进行更高点击次数的更新，更新之后，我们一般都会持久化到外
             * 部存储设备上，这里我们存储到 MySQL 数据库中
2.           */
3.
4.
5.          JavaPairDStream<String, Long> updateStateByKeyDStream =
```

```
        filteredadClickedStreaming
6.                .mapToPair(new PairFunction<Tuple2<String, String>,
                  String, Long>() {
7.
8.                    @Override
9.                    public Tuple2<String, Long> call(Tuple2<String,
                      String> t) throws Exception {
10.                       String[] splited = t._2.split("\t");
11.                       String timestamp = splited[0]; // yyyy-MM-dd
12.                       String ip = splited[1];
13.                       String userID = splited[2];
14.                       String adID = splited[3];
15.                       String province = splited[4];
16.                       String city = splited[5];
17.                       String clickedRecord = timestamp + "_" + adID +
                          "_" + province + "_" + city;
18.                       return new Tuple2<String, Long>(clickedRecord, 1L);
19.                   }
20.               }).updateStateByKey(new Function2<List<Long>, Optional<Long>,
                  Optional<Long>>() {
21.
22.                   @Override
23.                   public Optional<Long> call(List<Long> v1, Optional<
                      Long> v2) throws Exception {
24.                       /**
25.                        *v1:代表当前 Key 在当前的 Batch  Duration 中出现次
                           *数的集合，如{1,1,1,1,1,1}
26.                        *v2:代表当前 Key 在以前的 Batch  Duration 中积累下来的结果
27.                        */
28.                       Long clickedTotalHistory = 0L;
29.                       if (v2.isPresent()) {
30.                           clickedTotalHistory = v2.get();
31.                       }
32.                       for (Long one : v1) {
33.                           clickedTotalHistory += one;
34.                       }
35.                       return Optional.of(clickedTotalHistory);
36.                   }
37.               });
```

（3）广告点击累计动态更新的 MySQL 数据库入库操作。

updateStateByKeyDStream 使用 foreachRDD 算子遍历每一个 RDD，然后 rdd 使用 foreachPartition 方法对每个分区的元素进行数据库操作。将 updateStateByKeyDStream 每行记录的第一个元素 record._1 按“_”进行分割，提取出 timestamp、adID、province、city 各个字段；updateStateByKeyDStream 每行记录的第二个元素 record._2(即累计点击次数 ClickedCount 字段)，存放到 AdClicked 的 Javabean 数据结构中，然后循环遍历用户广告点击数据放入到 List< AdClicked>类型的列表 adClicked 中。

然后根据 timestamp、adID、province、city 在 MySQL 数据库进行点击次数查询，分别进行数据更新、数据插入操作。

广告点击累计动态更新的 MySQL 数据库入库操作代码如下：

```
1.  updateStateByKeyDStream.foreachRDD(new    Function<JavaPairRDD<String,
    Long>, Void>() {
2.          @Override
3.          public Void call(JavaPairRDD<String, Long> rdd) throws
```

```
            Exception {
4.              rdd.foreachPartition(new VoidFunction<Iterator<Tuple2<
                String, Long>>>() {
5.
6.                  @Override
7.                  public void call(Iterator<Tuple2<String, Long>>
                    partition) throws Exception {
8.  /**
9.    * 这里我们使用数据库连接池的高效读写数据库的方式把数据写入数据库 MySQL;
10.   * 由于传入的参数是一个 Iterator 类型的集合，所以为了更加高效地操作，我们需要批量处理，
      * 例如，一次性插入 1000 条 Record，使用 insertBatch 或者 updateBatch 类型的操作;
11.   * 插入的用户信息可以只包含 timestamp、adID、province、city
12.   * 这里面有一个问题：可能出现两条记录的 Key 一样，此时就需要更新累加操作
13.   */
14.       List<AdClicked> adClickedList = new ArrayList<AdClicked>();
15.
16.                     while (partition.hasNext()) {
17.                         Tuple2<String, Long> record = partition.next();
18.
19.                         String[] splited = record._1.split("_");
20.
21.                         AdClicked adClicked = new AdClicked();
22.                         adClicked.setTimestamp(splited[0]);
23.                         adClicked.setAdID(splited[1]);
24.                         adClicked.setProvince(splited[2]);
25.                         adClicked.setCity(splited[3]);
26.                         adClicked.setClickedCount(record._2);
27.
28.                         adClickedList.add(adClicked);
29.
30.                     }
31.
32.         JDBCWrapper jdbcWrapper = JDBCWrapper.getJDBCInstance();
33.
34.     final List<AdClicked> inserting = new ArrayList<AdClicked>();
35.     final List<AdClicked> updating = new ArrayList<AdClicked>();
36.
37.       //点击
38.       //表的字段：timestamp、ip、userID、adID、province、city、clickedCount
39.                     for (final AdClicked clicked : adClickedList) {
40.                         jdbcWrapper.doQuery(
41.                             "SELECT count(1) FROM adclickedcount WHERE "
42.                                     + " timestamp = ? AND adID = ? AND
                                        province = ? AND city = ? ",
43.                                 new Object[]{clicked.getTimestamp(), clicked
                                .getAdID(), clicked.getProvince(),
44.                                     clicked.getCity()},
45.                                 new ExecuteCallBack() {
46.
47.                                     @Override
48.                 public void resultCallBack(ResultSet result) throws Exception {
49.                                     if (result.getRow() != 0) {
50.                                         long count = result.getLong(1);
51.                                         clicked.setClickedCount(count);
52.                                         updating.add(clicked);
53.                                     } else {
54.                                         inserting.add(clicked);
```

```
55.                                             }
56.                                         }
57.                                     });
58.                        }
59.         //点击
60.         //表的字段：timestamp、ip、userID、adID、province、city、clickedCount
61.          ArrayList<Object[]> insertParametersList = new ArrayList<
            Object[]>();
62.                        for (AdClicked inserRecord : inserting) {
63.                            insertParametersList.add(new
              Object[]{inserRecord .getTimestamp(), inserRecord.getAdID(),
64.           inserRecord.getProvince(), inserRecord.getCity(), inserRecord
              .getClickedCount()});
65.                        }
66.                        jdbcWrapper.doBatch("INSERT INTO adclickedcount
                           VALUES(?,?,?,?,?)", insertParametersList);
67.
68.         //点击
69.         //表的字段：timestamp、ip、userID、adID、province、city、clickedCount
70.          ArrayList<Object[]> updateParametersList = new ArrayList<
            Object[]>();
71.                        for (AdClicked updateRecord : updating) {
72.                            updateParametersList.add(new Object[]{updateRecord
                               .getClickedCount(),
73.               updateRecord.getTimestamp(), updateRecord.getAdID(), updateRecord
                  .getProvince(),updateRecord.getCity()});
74.
75.                        }
76.                        jdbcWrapper.doBatch(
77.                                "UPDATE adclickedcount set clickedCount = ?
                                   WHERE "+ " timestamp = ? AND adID = ? AND
                              province = ? AND city = ? ",updateParametersList);
78.
79.                    }
80.                });
81.                return null;
82.            }
83.        });
```

16.8　实现每个省份点击排名 Top5 广告

电商广告点击综合案例实现每个省份广告点击排名 Top5。在广告点击排名中，可以按省份排名，也可以按城市排名，在上海市可以按照不同的区进行划分，如浦东区、静安区等，按不同的区进行细分排名，TopN 排名代码的思路是一样的。也可以进行多维度的划分，即所谓的二次排序，这个意义非常重要，我们看广告点击的 TopN，看哪些广告受人欢迎，对于广告策略的调整或者公司的营销非常重要，广告点击的 TopN 排名对于互联网和电商非常重要，因为大部分的收入来自于广告。在实际的生产环境下，我们一般会进行二次排序，二次排序不是排序 2 次就行了，可能进行 3 个维度，4 个维度，甚至 10 个维度的排序，这些都是 2 次排序。本节实现每天每个省份 Top5 的广告点击排名。

（1）定义用户不同省份 Top5 的广告点击排名数据 AdProvinceTopN 的 JavaBean。

AdProvinceTopN 的用户信息包含时间、广告 ID、省份、点击次数（timestamp、adID、province、clickedCount），用于存放用户不同省份 Top5 的广告点击排名数据。

AdProvinceTopN 的 JavaBean 代码如下。

```
1.  class AdProvinceTopN  implements Serializable {
2.     private String timestamp;
3.     private String adID;
4.
5.     public String getTimestamp() {
6.        return timestamp;
7.     }
8.
9.     public void setTimestamp(String timestamp) {
10.       this.timestamp = timestamp;
11.    }
12.
13.    public String getAdID() {
14.       return adID;
15.    }
16.
17.    public void setAdID(String adID) {
18.       this.adID = adID;
19.    }
20.
21.    public String getProvince() {
22.       return province;
23.    }
24.
25.    public void setProvince(String province) {
26.       this.province = province;
27.    }
28.
29.    public Long getClickedCount() {
30.       return clickedCount;
31.    }
32.
33.    public void setClickedCount(Long clickedCount) {
34.       this.clickedCount = clickedCount;
35.    }
36.
37.    private String province;
38.
39.    private Long clickedCount;
40. }
```

（2）在广告点击累计动态更新 updateStateByKeyDStream 的基础上进行 transform 转换。updateStateByKeyDStream 的每行数据的 Key 值 t._1 按"_"进行切分，提取出时间、广告 ID、省份按"_"重新组拼成 timestamp_adID_province；每行数据的 Value 值 t._2 是累计动态更新的计数值。重新生成 Key-Value 键值对（clickedRecord, t._2），然后使用 reduceByKey 算子汇总统计某时间、某省份的广告点击次数。

接下来使用 map 转换，目的是生成一行行的 JavaRDD<Row> 数据。实现也很简单，将上述汇总的某时间、某省份的广告点击次数重新拆分，v1._1 按"_"进行切分，提取时间、广告 ID、省份字段；v1._2 为汇总的点击次数，然后使用 RowFactory.create(timestamp, adID, province, v1._2)就创建生成了行数据：JavaRDD<Row> rowRDD。

广告点击累计动态更新数据转换生成 Row 行数据代码。

```
1.  updateStateByKeyDStream.transform(new Function<JavaPairRDD<String, Long>,
    JavaRDD<Row>>() {
2.
3.              @Override
4.              public JavaRDD<Row> call(JavaPairRDD<String, Long> rdd) throws
                Exception {
5.
6.                  JavaRDD<Row> rowRDD = rdd.mapToPair(new PairFunction<
                    Tuple2<String, Long>, String, Long>() {
7.
8.                      @Override
9.                      public Tuple2<String, Long> call(Tuple2<String, Long> t)
                        throws Exception {
10.                         String[] splited = t._1.split("_");
11.                         String timestamp = "2016-07-10"; // yyyy-MM-dd
12.                         String adID = splited[1];
13.                         String province = splited[2];
14.                         String clickedRecord = timestamp + "_" + adID + "_"
                            + province;
15.                         return new Tuple2<String, Long>(clickedRecord, t._2);
16.                     }
17.                 }).reduceByKey(new Function2<Long, Long, Long>() {
18.
19.                     @Override
20.                     public Long call(Long v1, Long v2) throws Exception {
21.                         //待办事项：自动生成方法存根
22.                         return v1 + v2;
23.                     }
24.                 }).map(new Function<Tuple2<String, Long>, Row>() {
25.
26.                     @Override
27.                     public Row call(Tuple2<String, Long> v1) throws Exception {
28.                         String[] splited = v1._1.split("_");
29.                         String timestamp = "2016-07-10"; // yyyy-MM-dd
30.                         String adID = splited[1];
31.                         String province = splited[2];
32.                         return RowFactory.create(timestamp, adID, province,
                            v1._2);
33.                     }
34.                 });
```

（3）创建 df：结合 rowRDD 和 structType 的元数据信息，基于 RDD 创建 df（hiveContext.createDataFrame(rowRDD, structType)）。然后，hiveContext 使用开窗函数进行 TopN 的查询。

DataFrame 的创建及开窗函数进行 TopN 的查询代码如下：

```
1.  StructType  structType = DataTypes.createStructType(
2.                          Arrays.asList(DataTypes.createStructField("timstamp",
                            DataTypes.StringType, true),
3.          DataTypes.createStructField("adID", DataTypes.StringType, true),
4.          DataTypes.createStructField("province", DataTypes.StringType, true),
5.          DataTypes.createStructField("clickedCount",DataTypes.LongType, true)));
6.
7.                  HiveContext hiveContext = new HiveContext(rdd.context());
8.
9.                  DataFrame df = hiveContext.createDataFrame(rowRDD, structType);
10.
11.                 df.registerTempTable("topNTableSource");
12. //开窗函数进行 TopN 的查询
```

```
13.                String IMFsqlText = "SELECT timstamp,adID,province,
                   clickedCount FROM "   + " ( SELECT timstamp,adID,province,
                   clickedCount, row_number() "
14.        + " OVER ( PARTITION BY province ORDER BY clickedCount DESC)rank "
15.         + " FROM topNTableSource ) subquery " + " WHERE rank <= 5 ";
16.
17.                DataFrame result = hiveContext.sql(IMFsqlText);
18.
19.                return result.toJavaRDD();
20.
21.            }
```

（4）将每个省份 Province 广告点击排名 Top5 的数据保存到 MySQL 数据库，提供后续进行 Java EE 的查询。这里将每个省份 Province 广告点击排名 Top5 的数据循环遍历到 List<AdProvinceTopN> adProvinceTopN，使用 set 集合清理掉 Timestamp_Province 重复的元素，然后从数据库中删除某个时间点的省份旧的排名数据，再 for 循环遍历插入每个省份 Province 广告点击新的排名 Top5 的数据。数据库操作和之前章节的操作基本一样，这里不再赘述。

数据库 MySQL 的操作代码如下：

```
1.    }).foreachRDD(new Function<JavaRDD<Row>, Void>() {
2.
3.            @Override
4.            public Void call(JavaRDD<Row> rdd) throws Exception {
5.
6.                rdd.foreachPartition(new VoidFunction<Iterator<Row>>() {
7.
8.                    @Override
9.                    public void call(Iterator<Row> t) throws Exception {
10.
11.                       List<AdProvinceTopN> adProvinceTopN = new ArrayList<
                          AdProvinceTopN>();
12.
13.                       while (t.hasNext()) {
14.                           Row row = t.next();
15.
16.                           AdProvinceTopN item = new AdProvinceTopN();
17.                           item.setTimestamp(row.getString(0));
18.                           item.setAdID(row.getString(1));
19.                           item.setProvince(row.getString(2));
20.
21.                           item.setClickedCount(row.getLong(3));
22.
23.                           adProvinceTopN.add(item);
24.                       }
25.
26.           JDBCWrapper jdbcWrapper = JDBCWrapper.getJDBCInstance();
27.
28.                       Set<String> set = new HashSet<String>();
29.                       for (AdProvinceTopN item : adProvinceTopN) {
30.                           set.add(item.getTimestamp()+"_"+item.getProvince());
31.                       }
32.
33.                       //点击
34.       //表的字段：timestamp、ip、userID、adID、province、city、clickedCount
35.           ArrayList<Object[]> deleteParametersList = new ArrayList<
```

```
        Object[]>();
36.                 for (String deleteRecord : set) {
37.                    String[] splited = deleteRecord.split("_");
38.       deleteParametersList.add(new Object[]{splited[0], splited[1]});
39.                 }
40.                 jdbcWrapper.doBatch("DELETE FROM adprovincetopn
                    WHERE timestamp = ? AND province = ?",
41.                         deleteParametersList);
42.
43.                 //不同省份 Top5 的广告点击排名
44.                 //表的字段：timestamp、adID、province、clickedCount
45.                 ArrayList<Object[]> insertParametersList = new
                    ArrayList<Object[]>();
46.                 for(AdProvinceTopN updateRecord : adProvinceTopN) {
47.                    insertParametersList.add(new Object[]{updateRecord
                       .getTimestamp(), updateRecord.getAdID(),
48.                            updateRecord.getProvince(), updateRecord
                               .getClickedCount()});
49.                 }
50.     jdbcWrapper.doBatch("INSERT INTO adprovincetopn VALUES (?,?,?,?) ",
        insertParametersList);
51.              }
52.           });
53.           return null;
54.        }
55.     });
```

16.9　实现广告点击 Trend 趋势计算实战

广告点击 Trend 趋势对于投放广告的效果至关重要。广告点击 Trend 趋势也可以用来分析用户的心理和行为特征。本节实现广告点击 Trend 趋势计算实战。

（1）定义广告点击 Trend 趋势数据 AdTrendStat 的 JavaBean。AdTrendStat 的用户信息包含日期、小时、分钟、广告 ID、点击次数（date、hour、minute、adID、clickedCount），用于存放广告点击 Trend 趋势的数据。

AdTrendStat 的 JavaBean 代码如下：

```
1.  class  AdTrendStat  implements Serializable {
2.     private String _date;
3.     private String _hour;
4.     private String _minute;
5.
6.     public String get_date() {
7.        return _date;
8.     }
9.
10.    @Override
11.    public String toString() {
12.       return "AdTrendStat [_date=" + _date + ", _hour=" + _hour + ", _minute
          =" + _minute + ", adID=" + adID
13.             + ", clickedCount=" + clickedCount + "]";
14.    }
15.
16.    public void set_date(String _date) {
17.       this._date = _date;
```

```
18.     }
19.
20.     public String get_hour() {
21.        return _hour;
22.     }
23.
24.     public void set_hour(String _hour) {
25.        this._hour = _hour;
26.     }
27.
28.     public String get_minute() {
29.        return _minute;
30.     }
31.
32.     public void set_minute(String _minute) {
33.        this._minute = _minute;
34.     }
35.
36.     public String getAdID() {
37.        return adID;
38.     }
39.
40.     public void setAdID(String adID) {
41.        this.adID = adID;
42.     }
43.
44.     public Long getClickedCount() {
45.        return clickedCount;
46.     }
47.
48.     public void setClickedCount(Long clickedCount) {
49.        this.clickedCount = clickedCount;
50.     }
51.
52.     private String adID;
53.     private Long clickedCount;
54. }
```

定义广告点击 Trend 趋势数据的历史数据 AdTrendCountHistory 的 JavaBean。AdTrendCountHistory 的信息包含历史点击次数（clickedCountHistory），用于存放广告点击 Trend 趋势的历史数据。

AdTrendCountHistory 的 JavaBean 代码如下：

```
1.   class AdTrendCountHistory implements Serializable {
2.      private Long clickedCountHistory;
3.      public Long getClickedCountHistory() {
4.         return clickedCountHistory;
5.      }
6.      public void setClickedCountHistory(Long clickedCountHistory) {
7.         this.clickedCountHistory = clickedCountHistory;
8.      }
9.   }
```

（2）算子 reduceByKeyAndWindow 讲解。算子 reduceByKeyAndWindow 根据 Key 值及时间窗口进行汇总统计，按时间窗口长度、滑动时间窗口进行统计。

reduceByKeyAndWindow 窗口转换操作算子。

- 第一个参数是聚合函数。

- 第二个参数是可逆聚合函数。
- 第三个参数是时间窗口长度 windowDuration。
- 第四个参数是滑动时间窗口 slideDuration，每隔多长时间滑动一次窗口。

reduceByKeyAndWindow 的源码如下：

```
1.   def  reduceByKeyAndWindow(
2.       reduceFunc: JFunction2[V, V, V],
3.       invReduceFunc: JFunction2[V, V, V],
4.       windowDuration: Duration,
5.       slideDuration: Duration
6.     ): JavaPairDStream[K, V] = {
7.     dstream.reduceByKeyAndWindow(reduceFunc, invReduceFunc, windowDuration,
       slideDuration)
8.   }
```

下面看一个采用 scala 语言写的 reduceByKeyAndWindow 代码，每隔 2s 时间统计过去 5s 时间内所有输入数据的统计信息，滑动窗口是 2s，窗口长度是 5s。

```
1.  filteredadClickedStreaming.mapToPair (...).reduceByKeyAndWindow(_ + _,
    _ - _, Seconds (5), Seconds (2))
```

假设在第一个时间窗口读入的数据为（time0 time1 time2 time3 time4）

时间窗口 1：

time0	time1	**time2**	**time3**	**time4**

假设在第二个时间窗口读入的数据为（time2 time3 time4 time5 time6）

时间窗口 2：

time2	time3	time4	time5	time6

接下来看一下 reduceByKeyAndWindow 的计算过程。

- 执行_+_操作：时间窗口 1（time0 time1 time2 time3 time4）加下一个时间窗口新读入的数据（time5 time6）计算得出的结果（time0 time1 time2 time3 time4 time5 time6）。
- 执行_-_ 操作：在上述结果的基础上，再减去上一个时间窗口旧的数据（time0 time1）计算得出的结果正好是第二个时间窗口的数据（time2 time3 time4 time5 time6）。基于重叠的时间窗口数据，只对增加、减少的数据进行计算，提升了计算效率。

电商广告点击综合案例实现广告点击 Trend 趋势计算实战。其中，filteredad ClickedStreaming 的格式为 Key-Value 键值对<String, String>，Key 为 Kafka 生成的 Key 值，Value 值是 Spark Streaming 从 Kafka 中读取的一行行的数据。从 filteredadClickedStreaming 每行数据的第二个元素 t._2 广告点击数据中提取出广告 ID、时间，按“_”组合成新的 Key 值 time_adID，Value 值为计数 1 次。然后将每行数据(time_adID,1)执行 reduceByKey AndWindow 算子，每隔 1 min 统计过去 30 min 的用户广告点击次数。

广告点击 Trend 趋势计算实战 reduceByKeyAndWindow 代码如下：

```
1.      filteredadClickedStreaming.mapToPair(new  PairFunction<Tuple2<String,
        String>, String, Long>() {
```

```
2.
3.             @Override
4.             public Tuple2<String, Long> call(Tuple2<String, String> t)
               throws Exception {
5.                String[] splited = t._2.split("\t");
6.
7.                String adID = splited[3];
8.
9.                String time = splited[0];
                  //Todo：后续需要重构代码实现时间戳和分钟的转换提取，此处需要提取出
                  //该广告的点击分钟单位
10.
11.               return new Tuple2<String, Long>(time + "_" + adID, 1L);
12.            }
13.         }).reduceByKeyAndWindow(new Function2<Long, Long, Long>() {
14.
15.            @Override
16.            public Long call(Long v1, Long v2) throws Exception {
17.               //待办事项：自动生成方法存根
18.               return v1 + v2;
19.            }
20.         }, new Function2<Long, Long, Long>() {
21.
22.            @Override
23.            public Long call(Long v1, Long v2) throws Exception {
24.               //待办事项：自动生成方法存根
25.               return v1 - v2;
26.            }
27.         }, Durations.minutes(30), Durations.minutes(1)).foreachRDD(new
            Function<JavaPairRDD<String, Long>, Void>() {
28. ......
29.
```

（3）广告点击 Trend 趋势的数据插入到数据库时具体包括字段：time、adID、clickedCount，我们通过 J2EE 技术进行趋势绘图时肯定是需要年、月、日、时、分这个维度的，所以我们在这里需要年、月、日、小时、分钟这些时间维度；然后根据日期、小时、分钟查询广告点击 Trend 趋势表，如查询出相关的记录，则进行更新；如查询没有记录，则插入新的记录。数据库操作和之前章节的操作基本一样，这里不再赘述。

广告点击 Trend 趋势进行数据库 MySQL 入库的操作代码如下：

```
1.       @Override
2.             public Void call(JavaPairRDD<String, Long> rdd) throws Exception{
3.                rdd.foreachPartition(new VoidFunction<Iterator<Tuple2<String,
                  Long>>>() {
4.
5.                   @Override
6.                   public void call(Iterator<Tuple2<String, Long>> partition)
                     throws Exception {
7.
8.                      List<AdTrendStat> adTrend = new ArrayList<
                        AdTrendStat>();
9.
10.                     while (partition.hasNext()) {
11.                        Tuple2<String, Long> record = partition.next();
12.                        String[] splited = record._1.split("_");
13.                        String time = splited[0];
14.                        String adID = splited[1];
15.                        Long clickedCount = record._2;
```

```
16.
17.                    /**
                        * 插入数据到数据库时，具体需要哪些字段？time、adID、
                        * clickedCount；而我们通过 J2EE 技术进行趋势绘图的时
                        * 候肯定是需要年、月、日、时、分这些维度的，所以这里需
                        * 要年、月、日、小时、分钟这些时间维度
18.                     */
19.
20.
21.                    AdTrendStat adTrendStat = new AdTrendStat();
22.                    adTrendStat.setAdID(adID);
23.                    adTrendStat.setClickedCount(clickedCount);
24.                    adTrendStat.set_date(time);     //Todo:获取年月日
25.                    adTrendStat.set_hour(time);     //Todo:获取小时
26.                    adTrendStat.set_minute(time);   //Todo:获取分钟
27.
28.                    adTrend.add(adTrendStat);
29.
30.
31.                }
32.
33.                final List<AdTrendStat> inserting = new ArrayList<
                   AdTrendStat>();
34.                final List<AdTrendStat> updating = new ArrayList<
                   AdTrendStat>();
35.
36.                JDBCWrapper jdbcWrapper = JDBCWrapper.getJDBCInstance();
37.
38.                //广告点击 Trend 趋势
39.                //表的字段：date、hour、minute、adID、clickedCount
40.                for (final AdTrendStat clicked : adTrend) {
41.                    final AdTrendCountHistory adTrendCountHistory =
                       new AdTrendCountHistory();
42.
43.                    jdbcWrapper.doQuery(
44.                            "SELECT count(1)FROM adclickedtrend WHERE"
45.                                    + " date = ? AND hour = ? AND minute
                                   = ? AND adID = ?",
46.                            new Object[]{clicked.get_date(), clicked
                           .get_hour(), clicked.get_minute(),
47.                                    clicked.getAdID()},
48.                            new ExecuteCallBack() {
49.
50.                                @Override
51.                                public void resultCallBack(ResultSet
                                result) throws Exception {
52.
53.                                    if (result.getRow() != 0) {
54.                                        long count = result.getLong(1);
55.                                        adTrendCountHistory.setClicked
                                       CountHistory(count);
56.                                        updating.add(clicked);
57.                                    } else {
58.
59.                                        inserting.add(clicked);
60.                                    }
61.
62.                                }
```

```
63.                     });
64.                 }
65.                 //广告点击 Trend 趋势
66.                 //表的字段：date、hour、minute、adID、clickedCount
67.                 ArrayList<Object[]> insertParametersList = new ArrayList<
                    Object[]>();
68.                 for (AdTrendStat inserRecord : inserting) {
69.
70.
71.                     insertParametersList
72.                             .add(new Object[]{inserRecord.get_date(),
73.
74.                                 inserRecord.get_hour(), inserRecord. get_minute(),
                                    inserRecord.getAdID(),
75.                                    inserRecord.getClickedCount()});
76.                 }
77.
78.                 jdbcWrapper.doBatch("INSERT INTO adclickedtrend VALUES
                    (?,?,?,?,?)", insertParametersList);// IMF
79.                 //BUG
80.                 //FIXED
81.
82.                 //广告点击趋势
83.                 //表的字段：date、hour、minute、adID、clickedCount
84.                 ArrayList<Object[]> updateParametersList = new  ArrayList<
                    Object[]>();
85.                 for (AdTrendStat updateRecord : updating) {
86.                     updateParametersList.add(new Object[]{updateRecord. get
                         ClickedCount(),
87.                             updateRecord.get_date(),updateRecord.get_ hour(),
                                updateRecord.get_minute(),
88.                              updateRecord.getAdID()});
89.                 }
90.                 jdbcWrapper.doBatch("UPDATE adclickedtrend set  clicked
                    Count=? WHERE "+ " date = ? AND hour = ? AND minute = ? AND
91.                         adID = ?", updateParametersList);
92.
93.             }
94.         });
95.         return null;
96.     }
97. });
```

16.10　实战模拟点击数据的生成和数据表 SQL 的建立

本节模拟用户在电商网站点击广告，模拟生成电商广告点击数据；在 MySQL 数据库中建立数据库表。

16.10.1　电商广告点击综合案例模拟数据的生成

电商广告点击大数据实时流处理系统案例整体上分为生产数据、消费数据两部分。

（1）编写 MockAdClickStatus 的代码向 Kafka 集群发送数据。

（2）编写 Spark Streaming　AdClickedStreamingStats 的代码从 Kafka 集群中提取消费数据，将 Spark 实时处理的数据持久化到数据库中。

本节，我们编写 MockAdClickStatus 的代码向 Kafka 集群发送数据，具体实现方法如下。

（1）在 MockAdClickStatus 类中，模拟发送电商广告点击数据信息，其发送的数据内容为 timestamp、ip、userID、adID、province、city，即时间戳、IP 地址、用户 ID、广告 ID、省份、城市。

```
1.      final Random random = new Random();
2.          final String[] provinces = new String[]{"Guangdong","Zhejiang",
            "Jiangsu","Fujian"};
3.      //省份、城市数据
4.          final Map<String,String[]> cities = new HashMap<String,
            String[]>();
5.          cities.put("Guangdong",new String[]{"Guangzhou","Shenzhen","DongGuan"});
6.          cities.put("Zhejiang", new String[]{"Hangzhou","Wenzhou","Ningbo"});
7.          cities.put("Jiangsu", new String[]{"Nanjing","Suzhou","WuXi"});
8.          cities.put("Fujian", new String[]{"Fuzhou","Ximen","Sanming"});

9.          //IP 地址
10.         final String[] ips = new String[]{
11.                 "192.168.112.240",
12.                 "192.168.112.239",
13.                 "192.168.112.245",
14.                 "192.168.112.246",
15.                 "192.168.112.247",
16.                 "192.168.112.248",
17.                 "192.168.112.249",
18.                 "192.168.112.250",
19.                 "192.168.112.251",
20.                 "192.168.112.252",
21.                 "192.168.112.253",
22.                 "192.168.112.254"
23. };
```

（2）接下来配置 Kafka 相关的信息。配置序列化类、Kafka 集群各主机名、端口等信息。

```
1.          /**
             * Kafka 相关的基本配置信息
2.           */
3.
4.          Properties kafkaConf = new Properties();
5.          kafkaConf.put("serializer.class", "kafka.serializer.StringEncoder");
6.          kafkaConf.put("metadata.broker.list", "master:9092,worker1:9092,
            worker2:9092");
7.          ProducerConfig producerConfig = new ProducerConfig(kafkaConf);
```

（3）创建 kafka 的 Producer 生产者实例，通过启动线程，在线程中调用 Kafka 生产者实例的 send 发送方法，循环不断地向 Kafka 集群发送数据。发送的数据内容为电商广告点击的模拟数据：拼接成 clickedAd 字符串，格式为 timestamp + "\t" + ip + "\t" + userID + "\t" + adID +"\t" + province + "\t" + city，即时间戳 IP 地址 用户 ID 广告 ID 省份 城市。

```
1. final Producer<Integer, String> producer = new Producer<Integer, String>
   (producerConfig);
2.
3.          new Thread(new Runnable() {
4.
```

```
5.                @Override
6.                public void run() {
7.                    while(true){
8.                        //在线处理广告点击流 广告点击的基本数据格式：timestamp、ip、
                          //userID、adID、province、city
9.                        Long timestamp = new Date().getTime();
10.                       String ip = ips[random.nextInt(12)];
                          //可以采用网络上免费提供的 IP 库
11.                       int userID = random.nextInt(10000);
12.                       int adID = random.nextInt(100);
13.                       String province = provinces[random.nextInt(4)];
14.                       String city = cities.get(province)[random.nextInt(3)];
15.
16.                       String clickedAd = timestamp + "\t" + ip + "\t" + userID
                          + "\t" + adID + "\t" + province + "\t" + city ;
17.
18.
19.                       System.out.println(clickedAd);
20.
21.                       producer.send( new KeyedMessage("AdClicked",clickedAd));
22.
23.                       try {
24.                           Thread.sleep(2000);
25.                       } catch (InterruptedException e) {
26.                           // TODO Auto-generated catch block
27.                           e.printStackTrace();
28.                       }
29.                   }
30.               }
31.           }).start();
```

在 IDEA 中运行代码，在本地测试生产者发送的数据内容如下：

```
1.  1490331571280   192.168.112.239   1746   12   Jiangsu     Nanjing
2.  1490331573280   192.168.112.249   6204   77   Jiangsu     WuXi
3.  1490331575280   192.168.112.252   8372   80   Guangdong   Guangzhou
4.  1490331577280   192.168.112.246   1587   70   Fujian      Ximen
5.  1490331579280   192.168.112.239   9409   36   Zhejiang    Ningbo
6.  1490331581280   192.168.112.250   345    37   Guangdong   Shenzhen
7.  1490331583281   192.168.112.250   1946   29   Zhejiang    Wenzhou
8.  1490331585281   192.168.112.248   537    8    Guangdong   DongGuan
9.  1490331587281   192.168.112.239   2146   81   Fujian      Fuzhou
10. 1490331589281   192.168.112.253   5547   96   Fujian      Sanming
11. 1490331591286   192.168.112.254   897    23   Zhejiang    Ningbo
12. 1490331593287   192.168.112.245   3430   78   Fujian      Ximen
13. 1490331595287   192.168.112.245   5651   77   Guangdong   Shenzhen
14. 1490331597287   192.168.112.252   8551   6    Fujian      Fuzhou
```

（4）在本地测试验证以后，我们将本地的代码打包上传到 Kafka 分布式集群中运行。在 Eclipse 集成开发环境下将 MockAdClickStatus 导出 JAR 包，上传到 Worker1 节点上。为方便测试，使生产者循环不断地向 Kafka 集群发送数据，我们编写一个脚本来执行。在脚本中配置需导入的 JAR 包。

MockAdClickedStats.sh 脚本如下：

```
1. java -Xbootclasspath/a:/usr/local/kafka_2.10-0.8.2.1/libs/kafka_2.10-0.
   8.2.1.jar:/usr/local/scala-2.10.4/lib/scala-library.jar:/usr/local/ka
   fka_2.10-0.8.2.1/libs/log4j-1.2.16.jar:/usr/local/kafka_2.10-0.8.2.1/
```

```
libs/metrics-core-2.2.0.jar:/usr/local/spark-1.6.1-bin-hadoop2.6/lib/
spark-streaming_2.10-1.6.1.jar:/usr/local/kafka_2.10-0.8.2.1/libs/kaf
ka-clients-0.8.2.1.jar:/usr/local/kafka_2.10-0.8.2.1/libs/slf4j-log4j
12-1.6.1.jar:/usr/local/kafka_2.10-0.8.2.1/libs/slf4j-api-1.7.6.jar
-jar /usr/local/IMF_testdata/MockAdClickedStats.jar
```

（5）在 Worker1 的 Linux 命令提示符中输入 chmod 修改 MockAdClickedStats.sh 的权限，执行 MockAdClickedStats.sh 脚本即可向 Kafka 集群循环不断地发送数据。

```
1.  root@worker1:/usr/local/setup_scripts# chmod u+x MockAdClickedStats114.sh
2.  root@worker1:/usr/local/setup_scripts#MockAdClickedStats114.sh
```

16.10.2　电商广告点击综合案例数据表 SQL 的建立

电商广告点击综合案例使用 MockAdClickStatus 代码循环不断向 Kafka 集群发送电商模拟数据，然后使用 Spark Streaming AdClickedStreamingStats 代码从 Kafka 集群中提取消费数据，Spark Streaming 实时处理的数据最终持久化到 MySQL 数据库中，后续可以通过 J2EE Web 网站读取 MySQL 中的数据进行可视化展示。本节阐述电商广告点击系统在 MySQL 的建表操作。

电商广告点击综合案例 MySQL 数据库表设计：在 MySQL 数据库表中创建表 16-1，分别存放广告点击、广告点击趋势、广告点击每省前 5 名排名数据、黑名单数据、广告点击计数等数据信息。

表 16-1　电商广告点击系统数据库表设计

广告点击	timestamp、ip、userID、adID、province、city、clickedCount
广告点击趋势	date、hour、minute、adID、clickedCount
广告点击每省前 5 名排名数据	timestamp、adID、province、clickedCount
黑名单数据	name
广告点击计数	timestamp、adID、province、city、clickedCount

（1）在 Master 节点连接 MySQL 数据库，登录 sparkstreaming 数据库，查询数据库表。

```
1.  root@master:~# mysql -uroot -proot
2.  ......
3.  mysql> show databases;
4.  +--------------------+
5.  | Database           |
6.  +--------------------+
7.  | information_schema |
8.  | hive               |
9.  | mysql              |
10. | performance_schema |
11. | spark              |
12. | sparkstreaming     |
13. +--------------------+
14. 6 rows in set (1.13 sec)
15.
16. mysql> use sparkstreaming;
17. Reading table information for completion of table and column names
18. You can turn off this feature to get a quicker startup with -A
19.
20. Database changed
```

```
21. mysql> show tables;
22. +--------------------------+
23. | Tables_in_sparkstreaming |
24. +--------------------------+
25. | categorytop3             |
26. +--------------------------+
27. 1 row in set (0.00 sec)
```

（2）在 MySQL sparkstreaming 数据库，依次建立电商广告点击综合案例需要的表。

建立 adclicked 广告点击表：表的字段包括时间戳、IP 地址、用户 ID、广告 ID、省份、城市、点击次数（timestamp、ip、userID、adID、province、city、clickedCount）。

```
1.  CREATE TABLE IF NOT EXISTS adclicked (
2.  timestamp VARCHAR(255),
3.  ip VARCHAR(255),
4.  userID VARCHAR(255),
5.  adID VARCHAR(255),
6.  province VARCHAR(255),
7.  city VARCHAR(255),
8.  clickedCount int(10) DEFAULT '0'
9.  );
```

建立 adclickedtrend 广告点击趋势表：表的字段包括日期、小时、分钟、广告 ID、点击次数（date、hour、minute、adID、clickedCount）。

```
1.   CREATE TABLE IF NOT EXISTS adclickedtrend (
2.  date VARCHAR(255),
3.  hour VARCHAR(255),
4.  minute VARCHAR(255),
5.  adID VARCHAR(255),
6.  clickedCount int(10) DEFAULT '0'
7.  );
```

建立 adprovincetopn 各省广告点击 TopN 表：表的字段包括时间戳、广告 ID、省份、点击次数（timestamp、adID、province、clickedCount）。

```
1.  CREATE TABLE IF NOT EXISTS adprovincetopn (
2.  timestamp VARCHAR(255),
3.  adID VARCHAR(255),
4.  province VARCHAR(255),
5.  clickedCount int(10) DEFAULT '0'
6.  );
```

建立 blacklisttable 黑名单表：表的字段包括姓名（name）。

```
1.   CREATE TABLE IF NOT EXISTS blacklisttable(
2.       name VARCHAR(255)
3.       );
```

建立 adclickedcount 广告点击统计表：表的字段包括时间戳、广告 ID、省份、城市、点击次数（timestamp、adID、province、city、clickedCount）。

```
1.   CREATE TABLE IF NOT EXISTS adclickedcount (
2.       timestamp VARCHAR(255),
3.       adID VARCHAR(255),
4.       province VARCHAR(255),
5.       city VARCHAR(255),
6.       clickedCount int(10) DEFAULT '0'
```

```
7.       );
```

（3）在 MySQL sparkstreaming 数据库中通过 desc 查询创建表的各个字段的属性，验证电商广告点击综合案例的表已经建立成功。

```
1.  mysql> desc adclickedtrend;
2.  +--------------+--------------+------+-----+---------+-------+
3.  | Field        | Type         | Null | Key | Default | Extra |
4.  +--------------+--------------+------+-----+---------+-------+
5.  | date         | varchar(255) | YES  |     | NULL    |       |
6.  | hour         | varchar(255) | YES  |     | NULL    |       |
7.  | minute       | varchar(255) | YES  |     | NULL    |       |
8.  | adID         | varchar(255) | YES  |     | NULL    |       |
9.  | clickedCount | int(10)      | YES  |     | 0       |       |
10. +--------------+--------------+------+-----+---------+-------+
11. 5 rows in set (0.00 sec)
12.
13. mysql> desc  adclicked;
14. +--------------+--------------+------+-----+---------+-------+
15. | Field        | Type         | Null | Key | Default | Extra |
16. +--------------+--------------+------+-----+---------+-------+
17. | timestamp    | varchar(255) | YES  |     | NULL    |       |
18. | ip           | varchar(255) | YES  |     | NULL    |       |
19. | userID       | varchar(255) | YES  |     | NULL    |       |
20. | adID         | varchar(255) | YES  |     | NULL    |       |
21. | province     | varchar(255) | YES  |     | NULL    |       |
22. | city         | varchar(255) | YES  |     | NULL    |       |
23. | clickedCount | int(10)      | YES  |     | 0       |       |
24. +--------------+--------------+------+-----+---------+-------+
25. 7 rows in set (0.00 sec)
26.
27. mysql> desc adprovincetopn;
28. +--------------+--------------+------+-----+---------+-------+
29. | Field        | Type         | Null | Key | Default | Extra |
30. +--------------+--------------+------+-----+---------+-------+
31. | timestamp    | varchar(255) | YES  |     | NULL    |       |
32. | adID         | varchar(255) | YES  |     | NULL    |       |
33. | province     | varchar(255) | YES  |     | NULL    |       |
34. | clickedCount | int(10)      | YES  |     | 0       |       |
35. +--------------+--------------+------+-----+---------+-------+
36. 4 rows in set (0.00 sec)
37.
38. mysql> show tables;
39. +--------------------------+
40. | Tables_in_sparkstreaming |
41. +--------------------------+
42. | adclicked                |
43. | adclickedtrend           |
```

```
44. | adprovincetopn           |
45. | categorytop3             |
46. +--------------------------+
47. 4 rows in set (0.00 sec)
```

16.11 电商广告点击综合案例运行结果

本节比较简单，根据16.1.3节电商广告点击综合案例整体部署，依次启动Hadoop、Spark、Zookeeper、Kafka集群及Hive metastore服务、KafkaOffsetMonitor监控，然后运行电商广告点击综合案例的代码，将Spark Streaming的运行结果持久化到MySQL数据库中，验证测试电商广告点击综合案例的运行结果。

16.11.1 电商广告点击综合案例Hadoop集群启动

电商广告点击综合案例Hadoop集群部署在1个Master、8个Worker 分布式节点上。登录Hadoop集群的Master节点，进入Hadoop的bin目录，启动Hadoop集群。

输入# cd /usr/local/hadoop-2.6.0/sbin。

输入# start-dfs.sh。电商广告点击综合案例只需使用Hadoop的分布式文件系统，因此没有启动start-all.sh，仅使用start-dfs.sh启动Hdfs文件系统。

电商广告点击综合案例Hadoop集群启动。

```
1.  root@master:/usr/local/hadoop-2.6.0/sbin# start-dfs.sh
2.  Starting namenodes on [master]
3.  master: starting namenode, logging to /usr/local/hadoop-2.6.0/logs/
    hadoop-root-namenode-master.out
4.  worker4: starting datanode, logging to /usr/local/hadoop-2.6.0/logs/
    hadoop-root-datanode-worker4.out
5.  worker3: starting datanode, logging to /usr/local/hadoop-2.6.0/logs/
    hadoop-root-datanode-worker3.out
6.  worker2: starting datanode, logging to /usr/local/hadoop-2.6.0/logs/
    hadoop-root-datanode-worker2.out
7.  worker6: starting datanode, logging to /usr/local/hadoop-2.6.0/logs/
    hadoop-root-datanode-worker6.out
8.  worker1: starting datanode, logging to /usr/local/hadoop-2.6.0/logs/
    hadoop-root-datanode-worker1.out
9.  worker8: starting datanode, logging to /usr/local/hadoop-2.6.0/logs/
    hadoop-root-datanode-worker8.out
10. worker5: starting datanode, logging to /usr/local/hadoop-2.6.0/logs/
    hadoop-root-datanode-worker5.out
11. worker7: starting datanode, logging to /usr/local/hadoop-2.6.0/logs/
    hadoop-root-datanode-worker7.out
12. Starting secondary namenodes [0.0.0.0]
13. 0.0.0.0: starting secondarynamenode, logging to /usr/local/hadoop-2.6.0/
    logs/hadoop-root-secondarynamenode-master.out
```

在浏览器中输入 http://192.168.189.1:50070，查看 Hadoop 集群的相关信息，如图 16-7所示。

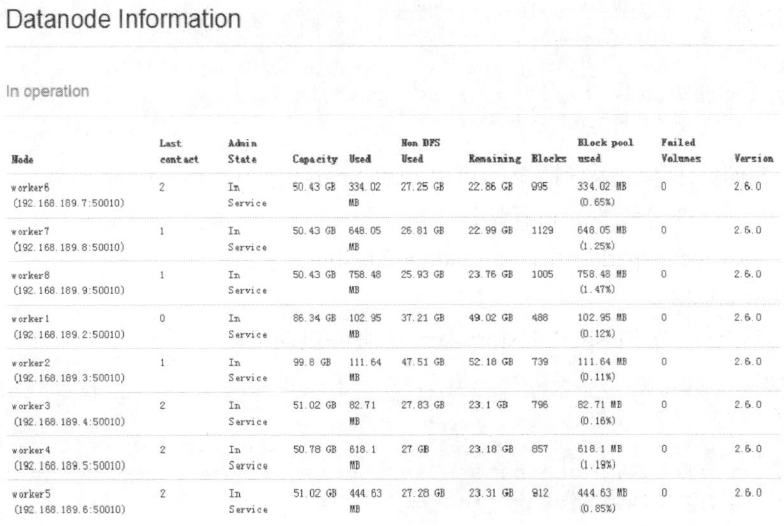

Datanode Information

In operation

Node	Last contact	Admin State	Capacity	Used	Non DFS Used	Remaining	Blocks	Block pool used	Failed Volumes	Version
worker6 (192.168.189.7:50010)	2	In Service	50.43 GB	334.02 MB	27.25 GB	22.86 GB	995	334.02 MB (0.65%)	0	2.6.0
worker7 (192.168.189.8:50010)	1	In Service	50.43 GB	648.05 MB	26.81 GB	22.99 GB	1129	648.05 MB (1.25%)	0	2.6.0
worker8 (192.168.189.9:50010)	1	In Service	50.43 GB	758.48 MB	25.93 GB	23.76 GB	1005	758.48 MB (1.47%)	0	2.6.0
worker1 (192.168.189.2:50010)	0	In Service	86.34 GB	102.95 MB	37.21 GB	49.02 GB	488	102.95 MB (0.12%)	0	2.6.0
worker2 (192.168.189.3:50010)	1	In Service	99.8 GB	111.64 MB	47.51 GB	52.18 GB	739	111.64 MB (0.11%)	0	2.6.0
worker3 (192.168.189.4:50010)	2	In Service	51.02 GB	82.71 MB	27.83 GB	23.1 GB	796	82.71 MB (0.16%)	0	2.6.0
worker4 (192.168.189.5:50010)	2	In Service	50.78 GB	618.1 MB	27 GB	23.18 GB	857	618.1 MB (1.19%)	0	2.6.0
worker5 (192.168.189.6:50010)	2	In Service	51.02 GB	444.63 MB	27.28 GB	23.31 GB	912	444.63 MB (0.85%)	0	2.6.0

图 16-7　Hadoop 集群的 Web 页面

16.11.2　电商广告点击综合案例 Spark 集群启动

电商广告点击综合案例 Spark 集群部署在 1 个 Master、8 个 Worker 分布式节点上。登录 Spark 集群的 Master 节点，进入 Spark 的 sbin 目录，启动 Spark 集群。

输入# cd　/usr/local/spark-1.6.1-bin-hadoop2.6/sbin。

输入#　start-all.sh。

电商广告点击综合案例 Spark 集群启动。

```
1.  启动 spark
2.  root@master:/usr/local/spark-1.6.1-bin-hadoop2.6/sbin# start-all.sh
3.  starting org.apache.spark.deploy.master.Master, logging to /usr/local/
    spark-1.6.1-bin-hadoop2.6/logs/spark-root-org.apache.spark.deploy.mas
    ter.Master-1-master.out
4.  worker4:  starting  org.apache.spark.deploy.worker.Worker,  logging  to
    /usr/local/spark-1.6.1-bin-hadoop2.6/logs/spark-root-org.apache.spark
    .deploy.worker.Worker-1-worker4.out
5.  worker8:  starting  org.apache.spark.deploy.worker.Worker,  logging  to
    /usr/local/spark-1.6.1-bin-hadoop2.6/logs/spark-root-org.apache.spark
    .deploy.worker.Worker-1-worker8.out
6.  worker2:  starting  org.apache.spark.deploy.worker.Worker,  logging  to
    /usr/local/spark-1.6.1-bin-hadoop2.6/logs/spark-root-org.apache.spark
    .deploy.worker.Worker-1-worker2.out
7.  worker7:  starting  org.apache.spark.deploy.worker.Worker,  logging  to
    /usr/local/spark-1.6.1-bin-hadoop2.6/logs/spark-root-org.apache.spark
    .deploy.worker.Worker-1-worker7.out
8.  worker6:  starting  org.apache.spark.deploy.worker.Worker,  logging  to
    /usr/local/spark-1.6.1-bin-hadoop2.6/logs/spark-root-org.apache.spark
    .deploy.worker.Worker-1-worker6.out
9.  worker3:  starting  org.apache.spark.deploy.worker.Worker,  logging  to
    /usr/local/spark-1.6.1-bin-hadoop2.6/logs/spark-root-org.apache.spark
    .deploy.worker.Worker-1-worker3.out
10. worker5:  starting  org.apache.spark.deploy.worker.Worker,  logging  to
```

```
   /usr/local/spark-1.6.1-bin-hadoop2.6/logs/spark-root-org.apache.spark
   .deploy.worker.Worker-1-worker5.out
11. worker1: starting org.apache.spark.deploy.worker.Worker, logging to
   /usr/local/spark-1.6.1-bin-hadoop2.6/logs/spark-root-org.apache.spark
   .deploy.worker.Worker-1-worker1.out
```

为了监控 Spark 任务的运行情况，可以开启 Spark 的 history-server 服务，在 Spark 任务执行完毕后，仍可以查看任务的执行情况。

输入# cd　/usr/local/spark-1.6.1-bin-hadoop2.6/sbin。

输入#　start-all.sh。

电商广告点击综合案例 Spark history-server 服务启动。

```
1.  root@master:/usr/local/spark-1.6.1-bin-hadoop2.6/sbin#  start-history-
    server.sh
2.  starting org.apache.spark.deploy.history.HistoryServer, logging to /usr/
    local/spark-1.6.1-bin-hadoop2.6/logs/spark-root-org.apache.spark.depl
    oy.history.HistoryServer-1-master.out
```

电商广告点击综合案例 Spark 集群的 Web 页面如图 16-8 所示。

192.168.189.1:8080

Spark 1.6.1 Spark Master at spark://192.168.189.1:7077

URL: spark://192.168.189.1:7077
REST URL: spark://192.168.189.1:6066 *(cluster mode)*
Alive Workers: 8
Cores in use: 8 Total, 0 Used
Memory in use: 16.0 GB Total, 0.0 B Used
Applications: 0 Running, 6 Completed
Drivers: 0 Running, 0 Completed
Status: ALIVE

Workers

Worker Id	Address	State	Cores	Memory
worker-20160606175720-192.168.189.3-55653	192.168.189.3:55653	ALIVE	1 (0 Used)	2.0 GB (0.0 B Used)
worker-20160606175720-192.168.189.9-59273	192.168.189.9:59273	ALIVE	1 (0 Used)	2.0 GB (0.0 B Used)
worker-20160606175721-192.168.189.5-45068	192.168.189.5:45068	ALIVE	1 (0 Used)	2.0 GB (0.0 B Used)
worker-20160606175723-192.168.189.2-36175	192.168.189.2:36175	ALIVE	1 (0 Used)	2.0 GB (0.0 B Used)
worker-20160606175740-192.168.189.8-40248	192.168.189.8:40248	ALIVE	1 (0 Used)	2.0 GB (0.0 B Used)
worker-20160606175741-192.168.189.6-60643	192.168.189.6:60643	ALIVE	1 (0 Used)	2.0 GB (0.0 B Used)
worker-20160606175741-192.168.189.7-48754	192.168.189.7:48754	ALIVE	1 (0 Used)	2.0 GB (0.0 B Used)
worker-20160606175742-192.168.189.4-33967	192.168.189.4:33967	ALIVE	1 (0 Used)	2.0 GB (0.0 B Used)

Running Applications

Application ID	Name	Cores	Memory per Node	Submitted Time	User	State	Duration

Completed Applications

Application ID	Name	Cores	Memory per Node	Submitted Time	User	State	Duration
app-20160606202042-0005	IMFAdClickedStreamingStats	8	1024.0 MB	2016/06/06 20:20:42	root	FINISHED	4.5 min
app-20160606201922-0004	IMFAdClickedStreamingStats	8	1024.0 MB	2016/06/06 20:19:22	root	FINISHED	39 s
app-20160606201436-0003	IMFAdClickedStreamingStats	8	1024.0 MB	2016/06/06 20:14:36	root	FINISHED	1.9 min
app-20160606200306-0002	IMFAdClickedStreamingStats	8	1024.0 MB	2016/06/06 20:03:06	root	FINISHED	9 s
app-20160606194522-0001	IMFAdClickedStreamingStats	8	1024.0 MB	2016/06/06 19:45:22	root	FINISHED	24 s
app-20160606192329-0000	IMFAdClickedStreamingStats	8	1024.0 MB	2016/06/06 19:23:29	root	FINISHED	1.3 min

图 16-8　Spark 集群的 Web 页面

16.11.3　电商广告点击综合案例 Zookeeper 集群启动

电商广告点击综合案例 Zookeeper 集群部署在 Master、Worker01、Worker02 3 个分布式

节点上。分别登录 Master、Worker01、Worker02 三个节点，在系统提示符下：

输入# zkServer.sh start 启动 Zookeeper。

输入# zkServer.sh status 查看 zkServer 的状态。其中，Worker01 是领导者 leader，Master、Worker02 是跟随者 follower。

电商广告点击综合案例 Zookeeper 集群启动。

```
1.  root@master:/usr/local/spark-1.6.1-bin-hadoop2.6/sbin# zkServer.sh start
2.  JMX enabled by default
3.  Using config: /usr/local/zookeeper-3.4.6/bin/../conf/zoo.cfg
4.  Starting zookeeper ... STARTED
5.  root@master:/usr/local/spark-1.6.1-bin-hadoop2.6/sbin# zkServer.sh status
6.  JMX enabled by default
7.  Using config: /usr/local/zookeeper-3.4.6/bin/../conf/zoo.cfg
8.  Mode: follower
9.  root@master:/usr/local/spark-1.6.1-bin-hadoop2.6/sbin# jps
10. 3633 NameNode
11. 3890 SecondaryNameNode
12. 5075 Jps
13. 4997 QuorumPeerMain
14. 4262 Master
15. 4383 HistoryServer
16. root@master:/usr/local/spark-1.6.1-bin-hadoop2.6/sbin#
17.
18. root@worker1:~# zkServer.sh start
19. JMX enabled by default
20. Using config: /usr/local/zookeeper-3.4.6/bin/../conf/zoo.cfg
21. Starting zookeeper ... STARTED
22. root@worker1:~# zkServer.sh status
23. JMX enabled by default
24. Using config: /usr/local/zookeeper-3.4.6/bin/../conf/zoo.cfg
25. Mode: leader
26. root@worker1:~# jps
27. 2448 DataNode
28. 2804 Worker
29. 3211 Jps
30. 3151 QuorumPeerMain
31. root@worker1:~#
32.
33. root@worker2:~# zkServer.sh start
34. JMX enabled by default
35. Using config: /usr/local/zookeeper-3.4.6/bin/../conf/zoo.cfg
36. Starting zookeeper ... STARTED
37. root@worker2:~# zkServer.sh status
38. JMX enabled by default
39. Using config: /usr/local/zookeeper-3.4.6/bin/../conf/zoo.cfg
40. Mode: follower
41. root@worker2:~# jps
42. 3203 Jps
43. 2453 DataNode
44. 2809 Worker
45. 3149 QuorumPeerMain
46. root@worker2:~#
```

16.11.4　电商广告点击综合案例 Kafka 集群启动

电商广告点击综合案例 Kafka 集群部署在 Master、Worker01、Worker02 三个分布式节点上。分别登录 Master、Worker01、Worker02 三个节点，在系统提示符下：

输入# nohup /usr/local/kafka_2.10-0.9.0.1/bin/kafka-server-start.sh/usr/local/kafka_2.10-0.9.0.1/config/server.properties & 启动 Kafka，&表示在系统后台运行。

输入# jps 查看 Kafka 的状态。在 3 个分布式节点上都启动了 Kafka 进程。

电商广告点击综合案例 Kafka 集群启动。

```
1.  root@master:/usr/local/spark-1.6.1-bin-hadoop2.6/sbin#  nohup  /usr/
    local/kafka_2.10-0.9.0.1/bin/kafka-server-start.sh  /usr/local/kafka_
    2.10-0.9.0.1/config/server.properties &
2.  [1] 5132
3.  root@master:/usr/local/spark-1.6.1-bin-hadoop2.6/sbin#  root@master:/
    usr/local/spark-1.6.1-bin-hadoop2.6/sbin#  nohup:  ignoring  input  and
    appending output to 'nohup.out'
4.
5.  root@master:/usr/local/spark-1.6.1-bin-hadoop2.6/sbin# jps
6.  5200 Jps
7.  3633 NameNode
8.  3890 SecondaryNameNode
9.  4997 QuorumPeerMain
10. 4262 Master
11. 5132 Kafka
12. 4383 HistoryServer
13. root@master:/usr/local/spark-1.6.1-bin-hadoop2.6/sbin#
14.
15. root@worker1:~#  nohup  /usr/local/kafka_2.10-0.9.0.1/bin/kafka-server-
    start.sh  /usr/local/kafka_2.10-0.9.0.1/config/server.properties &
16. [1] 3225
17.
18.
19. root@worker1:~# jps
20. 2448 DataNode
21. 3249 Jps
22. 2804 Worker
23. 3225 Kafka
24. 3151 QuorumPeerMain
25. root@worker1:~#
26.
27. root@worker2:~#  nohup  /usr/local/kafka_2.10-0.9.0.1/bin/kafka-server-
    start.sh  /usr/local/kafka_2.10-0.9.0.1/config/server.properties &
28. [1] 3216
29.
30.
31. root@worker2:~# jps
32. 3216 Kafka
33. 3235 Jps
34. 2453 DataNode
35. 2809 Worker
36. 3149 QuorumPeerMain
37. root@worker2:~#
```

电商广告点击综合案例 Kafka 集群测试验证如下。

（1）在电商广告点击综合案例 Kafka 集群使用 kafka-topics.sh --create 创建一个主题 topic，主题名称为 AdClicked。

```
1.  root@master:/usr/local/spark-1.6.1-bin-hadoop2.6/sbin# kafka-topics.sh
    --create --zookeeper master:2181,worker1:2181,worker2:2181 --replication-
    factor 1 --partitions 1 --topic  AdClicked
```

（2）在电商广告点击综合案例 Kafka 集群使用 kafka-topics.sh --describe 查看已有主题的详细情况。

```
1.  root@master:/usr/local/spark-1.6.1-bin-hadoop2.6/sbin# kafka-topics.sh
    --describe --zookeeper master:2181,worker1:2181,worker2:2181
2.  Topic:AdClicked PartitionCount:1        ReplicationFactor:1    Configs:
3.          Topic: AdClicked    Partition: 0    Leader: 0    Replicas: 0    Isr: 0
4.  Topic:IMFHelloKafka   PartitionCount:1     ReplicationFactor:1  Configs:
5.          Topic: IMFHelloKafka    Partition: 0    Leader: 1     Replicas: 1   Isr: 1
6.  Topic:SparkStreamingDirected    PartitionCount:1          ReplicationFactor:1
    Configs:  Topic:  SparkStreamingDirected     Partition:  0      Leader:  1
    Replicas: 1   Isr: 1
```

（3）在电商广告点击综合案例 Kafka 集群 Master 节点编辑 MockAdClickedStats114.sh 脚本，配置相应的 JAR 包，运行 16.10.1 节电商广告点击综合案例模拟数据生成的 MockAdClickedStats jar 代码，模拟生产者循环不断地广告点击数据发送给 Kafka 集群。

```
1.  root@master:/usr/local/setup_scripts# cat MockAdClickedStats114.sh
2.  java  -Xbootclasspath/a:/usr/local/kafka_2.10-0.8.2.1/libs/kafka_2.10-
    0.8.2.1.jar:/usr/local/scala-2.10.4/lib/scala-library.jar:/usr/local/
    kafka_2.10-0.8.2.1/libs/log4j-1.2.16.jar:/usr/local/kafka_2.10-0.8.2.
    1/libs/metrics-core-2.2.0.jar:/usr/local/spark-1.6.1-bin-hadoop2.6/li
    b/spark-streaming_2.10-1.6.1.jar:/usr/local/kafka_2.10-0.8.2.1/libs/k
    afka-clients-0.8.2.1.jar:/usr/local/kafka_2.10-0.8.2.1/libs/slf4j-log
    4j12-1.6.1.jar:/usr/local/kafka_2.10-0.8.2.1/libs/slf4j-api-1.7.6.jar
    -jar /usr/local/IMF_testdata/MockAdClickedStats.jar
3.
4.  root@master:/usr/local/setup_scripts#  MockAdClickedStats114.sh
5.  log4j:WARN  No  appenders  could  be  found  for  logger  (kafka.utils.
    VerifiableProperties).
6.  log4j:WARN Please initialize the log4j system properly.
7.  log4j:WARN  See  http://logging.apache.org/log4j/1.2/faq.html#noconfig
    for more info.
8.   ......
```

（4）输入 Kafka 消费者客户端命令 kafka-console-consumer.sh --zookeeper，查看广告点击数据信息。

```
1.   root@worker2:~#  kafka-console-consumer.sh  --zookeeper  master:2181,
     worker1:2181,worker2:2181 --from-beginning --topic  AdClicked
2.  1465210351273   192.168.112.250 4856    98     Jiangsu Nanjing
3.  1465210351830   192.168.112.248 4008    64     Zhejiang        Wenzhou
4.  1465210351884   192.168.112.254 4356    85     Zhejiang        Ningbo
5.  1465210351938   192.168.112.245 5451    19     Fujian  Sanming
6.  1465210351993   192.168.112.253 571     59     Jiangsu Suzhou
7.  1465210352051   192.168.112.254 2778    33     Guangdong       Guangzhou
8.  1465210352106   192.168.112.252 8072    2      Jiangsu Suzhou
9.  1465210352161   192.168.112.247 5699    70     Fujian  Sanming
10. ......
```

16.11.5　电商广告点击综合案例 Hive metastore 集群启动

电商广告点击综合案例实现每个 Province 点击排名 Top5 广告时，使用 hiveContext 的上下文，因此我们须启动 Hive 的元数据服务，Spark 框架会读取 Hive 的元数据服务信息。

在系统提示符下：

输入~# hive --service metastore &启动 Hive metastore 服务。

电商广告点击综合案例 Hive metastore 服务集群启动。

```
1.  root@master:~# hive --service metastore &
2.  [1] 5757
3.  ......
4.  SLF4J: See http://www.slf4j.org/codes.html#multiple_bindings for an
    explanation.
5.  SLF4J: Actual binding is of type [org.slf4j.impl.Log4jLoggerFactory]
6.  Starting Hive Metastore Server
7.  ......
8.  SLF4J: See http://www.slf4j.org/codes.html#multiple_bindings for an
    explanation.
9.  SLF4J: Actual binding is of type [org.slf4j.impl.Log4jLoggerFactory]
```

16.11.6　电商广告点击综合案例程序运行

电商广告点击大数据实时流处理系统案例系统整体上分成生产数据、消费数据两部分。

（1）MockAdClickedStats114.sh 脚本执行，循环不断地向 Kafka 集群发送数据：时间戳、IP 地址、用户 ID、广告 ID、省份、城市。

```
1.  root@master:/usr/local/setup_scripts# MockAdClickedStats114.sh
```

（2）AdClickedStreamingStats114.sh 脚本执行，将 AdClickedStreamingStats 类代码打成 JAR 包，上传到 Spark 分布式集群 Master 节点上，编写脚本运行 AdClickedStreamingStats.jar 代码，从 Kafka 集群中消费数据，将 Spark 实时处理的数据持久化到数据库中。

```
1.  root@master:/usr/local/setup_scripts# vi AdClickedStreamingStats114.sh
2.  /usr/local/spark-1.6.1-bin-hadoop2.6/bin/spark-submit --files /usr/
    local/apache-hive-1.2.1/conf/hive-site.xml --class com.dt.spark.
    sparkstreaming.AdClickedStreamingStats --master spark://192.168.189.1:7077
    --jars /usr/local/spark-1.6.1-bin-hadoop2.6/lib/mysql-connector-java-
    5.1.13-bin.jar,/usr/local/spark-1.6.1-bin-hadoop2.6/lib/spark-streami
    ng_2.10-1.6.1.jar,/usr/local/spark-1.6.1-bin-hadoop2.6/lib/spark-asse
    mbly-1.6.1-hadoop2.6.0.jar/usr/local/IMF_testdata/AdClickedStreamingStats.jar
3.
4.  root@master:/usr/local/setup_scripts# chmod u+x AdClickedStreaming
    Stats114.sh
5.
6.  root@master:/usr/local/setup_scripts# cat AdClickedStreamingStats114.sh
7.  /usr/local/spark-1.6.1-bin-hadoop2.6/bin/spark-submit --files /usr/
    local/apache-hive-1.2.1/conf/hive-site.xml --class com.dt.spark.sparkstreaming.
    AdClickedStreamingStats --master spark://192.168.189.1:7077 --jars
    /usr/local/spark-1.6.1-bin-hadoop2.6/lib/mysql-connector-java-5.1.13-
    bin.jar,/usr/local/spark-1.6.1-bin-hadoop2.6/lib/spark-streaming_2.10
    -1.6.1.jar,/usr/local/spark-1.6.1-bin-hadoop2.6/lib/spark-assembly-1.
```

```
    6.1-hadoop2.6.0.jar /usr/local/IMF_testdata/AdClickedStreamingStats.jar
8.  root@master:/usr/local/setup_scripts#
```

16.11.7　电商广告点击综合案例运行结果

电商广告点击大数据实时流处理系统案例运行，登录 Master 节点的 MySQL 数据库，分别查询广告点击、广告点击趋势、广告点击每省前 5 名排名数据、黑名单数据、广告点击计数等数据库表，MySQL 表中实时展示 Spark Streaming 处理保存的用户广告点击数据，测试验证电商广告点击综合案例运行成功。

```
1.  mysql> select * from adclicked;
2.  | 1467551315985 | 192.168.112.253|496 |66|Guangdong| Guangzhou| 0 |
3.  | 1467551328011 | 192.168.112.252|4490|0 |Guangdong| Shenzhen | 0 |
4.  | 1467551322000 | 192.168.112.251|1763|89|Guangdong| Guangzhou| 0 |
5.  | 1467551326007 | 192.168.112.247| 472|70|Fujian   | Fuzhou   | 0 |
6.  | 1467551330016 | 192.168.112.254|8216| 4|Zhejiang | Ningbo   | 0 |
7.  | 1467551338074 | 192.168.112.246|5954|87|Zhejiang | Wenzhou  | 0 |
8.  ......
9.  +---------------+---------------+----+--+---------+----------+---+
10. 1362 rows in set (0.00 sec)
11.
12. mysql> select * from adclickedcount;
13.
14. | 1467551273830 | 1    | Zhejiang  | Ningbo    |          1   |
15. | 1467551311971 | 99   | Guangdong | Guangzhou |          1   |
16. | 1467551305954 | 19   | Zhejiang  | Wenzhou   |          1   |
17. | 1467551269807 | 0    | Fujian    | Fuzhou    |          1   |
18. | 1467551315985 | 66   | Guangdong | Guangzhou |          1   |
19. | 1467551338074 | 87   | Zhejiang  | Wenzhou   |          1   |
20. | 1467551271816 | 72   | Jiangsu   | WuXi      |          1   |
21. | 1467551303943 | 85   | Fujian    | Sanming   |          1   |
22. ......
23. +---------------+------+-----------+-----------+--------------+
24. 14471 rows in set (0.02 sec)
25.
26. mysql> select * from adclickedtrend;
27. +---------------+---------------+---------------+-----+------------+
28. | date          | hour          |        minute |adID |clickedCount|
29. +---------------+---------------+---------------+-----+------------+
30. | 1467551271816 | 1467551271816 | 1467551271816 | 72  |          1 |
31. | 1467551283868 | 1467551283868 | 1467551283868 | 74  |          1 |
32. | 1467551285873 | 1467551285873 | 1467551285873 | 27  |          1 |
33. | 1467551289887 | 1467551289887 | 1467551289887 | 64  |          1 |
34. | 1467551301937 | 1467551301937 | 1467551301937 | 17  |          1 |
35. | 1467551319992 | 1467551319992 | 1467551319992 | 83  |          1 |
36. | 1467551315985 | 1467551315985 | 1467551315985 | 66  |          1 |
37. | 1467551281860 | 1467551281860 | 1467551281860 | 0   |          1 |
38. | 1467551303943 | 1467551303943 | 1467551303943 | 85  |          1 |
39. ......
40. | 1467551309964 | 1467551309964 | 1467551309964 | 37  |          1 |
41. +---------------+---------------+---------------+------+------------+
42. 23 rows in set (0.00 sec)
```

```
43.
44. mysql>
45. mysql> select * from adprovincetopn;
46. +------------+------+----------+--------------+
47. | timestamp  | adID | province | clickedCount |
48. +------------+------+----------+--------------+
49. | 2016-06-26 | 3    | Zhejiang |        3     |
50. ……
51. | 2016-06-26 | 16   | Fujian   |       16     |
52. | 2016-06-26 | 21   | Fujian   |       21     |
53. +------------+------+----------+--------------+
54. 21 rows in set (0.00 sec)
55.
56. mysql>
57.
58.
59. mysql>  select * from blacklisttable;
60. ……
61. | 2755        |
62. | 6514        |
63. | 4828        |
64. | 1664        |
65. | 1359        |
66. | 2329        |
67. | 5312        |
68. +-------------+
69. 1223 rows in set (0.00 sec)
70.
```

电商广告点击综合案例 Spark 监控图如图 16-9 所示。

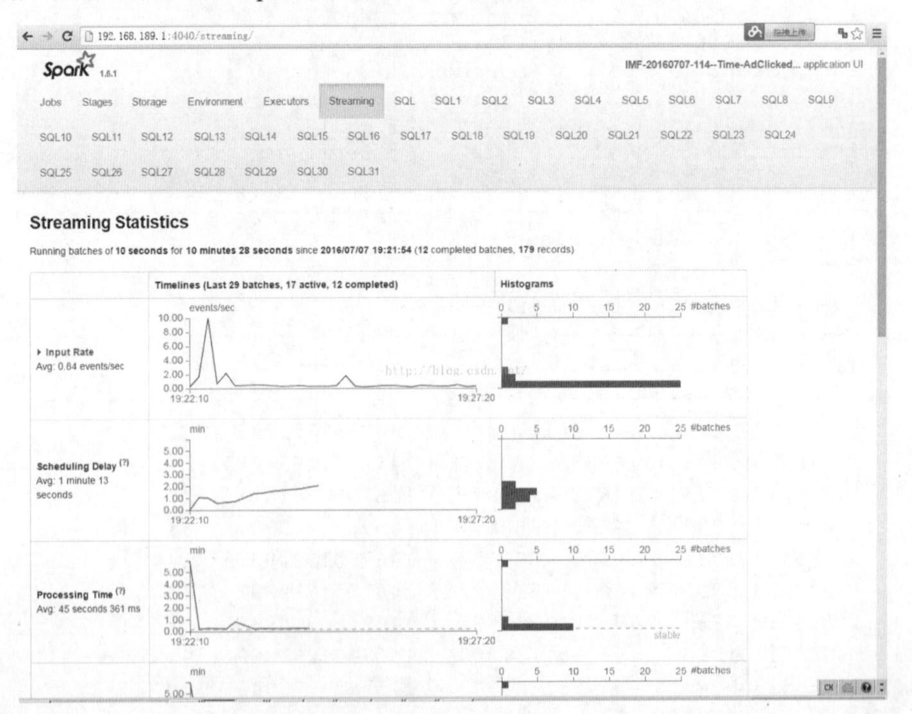

图 16-9　电商广告点击综合案例 Spark 监控图

Kafka 集群的 Web UI 页面监控 KafkaOffsetMonitor。

下载安装 Kafka 集群运行的第三方监控工具 KafkaOffsetMonitor，编写脚本 IMFKafkaOffsetMon.sh 运行 KafkaOffsetMonitor 工具。

```
1.  root@master:/usr/local/kafka_monitor# vi IMFKafkaOffsetMon.sh
2.  #! /bin/bash
3.  java -cp  /usr/local/kafka_monitor/KafkaOffsetMonitor-assembly-0.2.0.
    jar com.quantifind.kafka.offsetapp.OffsetGetterWeb  --zk master:2181,
    worker1:2181,worker2:2181  --port 8089  --refresh 10.seconds  --retain
    1.days
4.
5.  root@master:/usr/local/kafka_monitor# chmod u+x IMFKafkaOffsetMon.sh
6.  root@master:/usr/local/kafka_monitor# ls
7.  IMFKafkaOffsetMon.sh  KafkaOffsetMonitor-assembly-0.2.0.jar
```

Kafka 集群的 Web UI 页面监控图如图 16-10 所示。

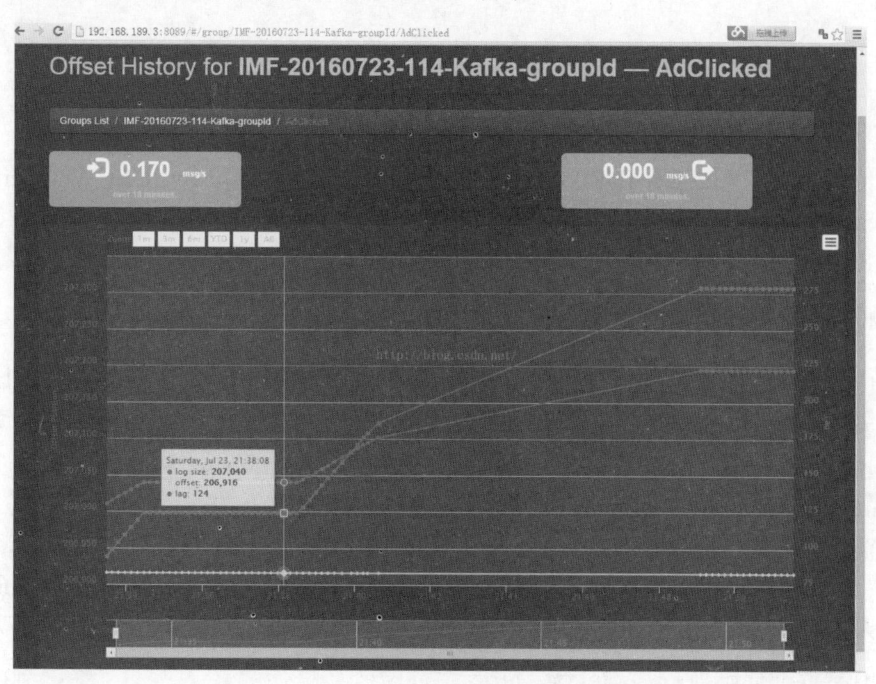

图 16-10　Kafka 集群的 Web UI 页面监控图

16.12　电商广告点击综合案例 Scala 版本关注点

电商广告点击综合案例之前是用 Java 语言开发的。本节使用 Scala 语言重新改写 Java 代码。Scala 开发中，我们关注一下 MySQL 数据库的相关开发内容。

1. 数据库的链接

电商广告点击综合案例 Java 版本中定义了一个数据库连接类 JDBCWrapper，在数据库连接类 JDBCWrapper 中定义了 MySQL 连接池、连接驱动、连接地址端口等内容；Java 中使用

static 关键字定义静态变量、静态代码块及静态方法。

电商广告点击综合案例 Scala 代码实现中，Scala 使用 Object 伴生对象实现 Java 的 static 静态对象内容。在 Scala 代码中使用单例模式创建 JDBCWrapper 类的伴生对象 object JDBCWrapper，新建一个 JDBCWrapper 类型的变量 jdbcInstance，在 getInstance 方法中，使用 synchronized 同步块锁定多线程中 new JDBCWrapper()的创建，jdbcInstance 如果为空，就创建一个 JDBCWrapper 实例，jdbcInstance 如果不为空，就返回之前已经创建的 jdbcInstance 实例。

```
1.  object  JDBCWrapper {
2.      private var  jdbcInstance: JDBCWrapper = _
3.      def getInstance(): JDBCWrapper = {
4.        synchronized {
5.        if (jdbcInstance == null) {
6.          jdbcInstance = new JDBCWrapper()
7.          }
8.        }
9.        jdbcInstance
10.     }
11.   }
```

如不使用伴生对象 object JDBCWrapper 单例模式，直接创建 new JDBCWrapper()实例，在 rdd.foreachPartition 中连接 MySQL 数据库就会提示“Too many connections”异常，改用上述单例模式可解决这个问题。

2. MySQL数据库批量数据增、删、改、查

电商广告点击综合案例 Java 版本中，doBatch 传入的第一个参数是 sqlText 语句，第二个参数是对象数组列表，paramsList 列表中的元素是一个对象数组，然后在 doBatch 方法中用 for 循环遍历 paramsList 列表，读入对象数组的数值，写入到 preparedStatement 批量参数列表中。preparedStatement.executeBatch()批量执行，获取数据库操作结果。

在电商广告点击综合案例 Scala 版本中，定义一个类 class paramsList 传入参数，params_Type 定义为参数类型，由于无法直接定义一个 Object 类型的对象，因此只好硬编码来实现不同的传入类型：params1～params7 传入 String 类型， params10_Long 传入 Long 类型。

```
1.  class paramsList extends Serializable {
2.      var params1: String = _
3.      var params2: String = _
4.      var params3: String = _
5.      var params4: String = _
6.      var params5: String = _
7.      var params6: String = _
8.      var params7: String = _
9.      var params10_Long: Long = _
10.      var params_Type : String = _
11.     var length: Int = _
12.
13.   }
```

在 JDBCWrapper 的 doBatch 方法中，根据传入 paramsList 参数列表的 params_Type 字段模式匹配，批量插入不同的 SQL 参数，如 params_Type 为 adclickedInsert，插入广告点击次数表的相关参数。

```
1.  def doBatch(sqlText: String, paramsList: ListBuffer[paramsList]): Array
    [Int] = {
2.        val conn: Connection = getConnection();
3.        var preparedStatement: PreparedStatement = null;
4.        val result: Array[Int] = null;
5.        try {
6.          conn.setAutoCommit(false);
7.          preparedStatement = conn.prepareStatement(sqlText)
8.          for (parameters <- paramsList) {
9.            println("====doBatch parameters.params_Type:   " + parameters.
              params_Type)
10.           parameters.params_Type match {
11.             case "adclickedInsert" => {
12.               println("adclickedInsert")
13.               preparedStatement.setObject(1, parameters.params1)
14.               preparedStatement.setObject(2, parameters.params2)
15.               preparedStatement.setObject(3, parameters.params3)
16.               preparedStatement.setObject(4, parameters.params4)
17.               preparedStatement.setObject(5, parameters.params5)
18.               preparedStatement.setObject(6, parameters.params6)
19.               preparedStatement.setObject(7, parameters.params10_Long)
20.             }
21. ……
22.           preparedStatement.addBatch();
23.         }
24.
25.         val result = preparedStatement.executeBatch();
26.
27.         conn.commit();
28.       } catch {
29.         //待办事项：自动生成 catch 块
30.         case e: Exception => e.printStackTrace()
31.       } finally {
32.         if (preparedStatement != null) {
33.           try {
34.             preparedStatement.close();
35.           } catch {
36.             //待办事项：自动生成 catch 块
37.             case e: SQLException => e.printStackTrace()
38.           }
39.         }
40.
41.         if (conn != null) {
42.           try {
43.             dbConnectionPool.put(conn);
44.           } catch {
45.             //待办事项：自动生成 catch 块
46.             case e: InterruptedException => e.printStackTrace()
47.           }
48.         }
49.       }
50.
51.       result;
52.     }
```

16.13　电商广告点击综合案例课程的 Java 源码

Spark Streaming 电商广告点击综合案例课程消费广告点击数据的 Java 源码，如例 16-1 所示。

【例 16-1】 AdClickedStreamingStats.java 代码。

```
1.     package com.dt.spark.SparkApps.SparkStreaming;
2.
3.  import java.io.Serializable;
4.  import java.sql.Connection;
5.  import java.sql.DriverManager;
6.  import java.sql.PreparedStatement;
7.  import java.sql.ResultSet;
8.  import java.sql.SQLException;
9.  import java.util.ArrayList;
10. import java.util.Arrays;
11. import java.util.HashMap;
12. import java.util.HashSet;
13. import java.util.Iterator;
14. import java.util.List;
15. import java.util.Map;
16. import java.util.Set;
17. import java.util.concurrent.LinkedBlockingQueue;
18.
19. import org.apache.spark.SparkConf;
20. import org.apache.spark.SparkContext;
21. import org.apache.spark.api.java.JavaPairRDD;
22. import org.apache.spark.api.java.JavaRDD;
23. import org.apache.spark.api.java.JavaSparkContext;
24. import org.apache.spark.api.java.function.Function;
25. import org.apache.spark.api.java.function.Function2;
26. import org.apache.spark.api.java.function.PairFunction;
27. import org.apache.spark.api.java.function.VoidFunction;
28. import org.apache.spark.sql.DataFrame;
29. import org.apache.spark.sql.Row;
30. import org.apache.spark.sql.RowFactory;
31. import org.apache.spark.sql.hive.HiveContext;
32. import org.apache.spark.sql.types.DataTypes;
33. import org.apache.spark.sql.types.StructField;
34. import org.apache.spark.sql.types.StructType;
35. import org.apache.spark.streaming.Durations;
36. import org.apache.spark.streaming.api.java.JavaDStream;
37. import org.apache.spark.streaming.api.java.JavaPairDStream;
38. import org.apache.spark.streaming.api.java.JavaPairInputDStream;
39. import org.apache.spark.streaming.api.java.JavaStreamingContext;
40. import org.apache.spark.streaming.kafka.KafkaUtils;
41.
42. import   org.apache.hadoop.hive.ql.parse.HiveParser_IdentifiersParser.
    function_return;
43.
44. import com.google.common.base.Optional;
45.
46. import kafka.serializer.StringDecoder;
47. import scala.Tuple2;
```

```
48. import scala.collection.Seq;
49.
50. /**
51.  *
52.  * 在线处理广告点击，广告点击的基本数据格式：timestamp、ip、userID、adID、province、city
53.  *
54.  * @author wangjialin
55.  *
56.  */
57. public class AdClickedStreamingStats {
58.
59.     public static void main(String[] args) {
60.
61.         /**
62.          * 第一步：配置 SparkConf：
63.          *① 至少 2 条线程：因为 Spark Streaming 应用程序在运行的时候，至少有一条
            *线程用于不断地循环接收数据，并且至少有一条线程用于处理接收的数据（否则无法有
            *线程用于处理数据，随着时间的推移，内存和磁盘都会不堪重负）；
64.          * ② 对于集群而言，每个 Executor 一般不止一个 Thread，那对于处理 Spark
            * Streaming 的应用程序而言，每个 Executor 一般分配多少 Core 比较合适？根据经
            * 验，5 个左右的 Core 最佳（一个段分配为奇数个 Core 表现最佳，如 3 个、5 个、
            * 7 个 Core 等）
            *
65.          */
66.
         SparkConf conf = new SparkConf().setMaster("spark://192.168.
         189.1:7077")
67.                 .setAppName("IMF-20160710-114-AdClickedStreamingStats");
68.
69.         // SparkConf conf = new
70.         //  SparkConf().setMaster("local[5]").setAppName("IMF-20160626-
         114-AdClickedStreamingStats")
71.
72.         /**
73.          * SparkConf conf = new SparkConf().setMaster("local[5]").setAppName(
74.          * "IMF-20160706-114-local-AdClickedStreamingStats") .setJars(new
75.          * String[] {
76.          * "/usr/local/spark-1.6.1-bin-hadoop2.6/lib/spark-streaming-
            * kafka_2.10-1.6.1.jar",
77.          * "/usr/local/kafka_2.10-0.8.2.1/libs/kafka-clients-0.8.2.1.jar",
78.          * "/usr/local/kafka_2.10-0.8.2.1/libs/kafka_2.10-0.8.2.1.jar",
89.          * "/usr/local/spark-1.6.1-bin-hadoop2.6/lib/spark-streaming_2.10-
            * 1.6.1.jar",
80.          * "/usr/local/kafka_2.10-0.8.2.1/libs/metrics-core-2.2.0.jar",
81.          * "/usr/local/kafka_2.10-0.8.2.1/libs/zkclient-0.3.jar",
82.          * "/usr/local/spark-1.6.1-bin-hadoop2.6/lib/spark-assembly-
            * 1.6.1-hadoop2.6.0.jar",
83.          * "/usr/local/spark-1.6.1-bin-hadoop2.6/lib/mysql-connector-java-
            * 5.1.13-bin.jar"
84.          *
85.          *
86.          * });
87.          */
88.
89.         /**
90.          * 第二步：创建 SparkStreamingContext：
```

```
91.          * ① 这是 SparkStreaming 应用程序所有功能的起始点和程序调度的核心
92.          * SparkStreamingContext 的构建可以基于 SparkConf 参数，也可基于持久化
             * 的 SparkStreamingContext 的内容来恢复(典型的场景是 Driver 崩溃后重新
             * 启动，由于 Spark Streaming 具有连续 7×24 小时不间断运行的特征，所以需要
             * 在 Driver 重新启动后继续之前系统的状态，此时的状态恢复需要基于曾经的
             * Checkpoint)；
93.          * ② 在一个 Spark Streaming 应用程序中，可以创建若干个 SparkStreaming
             * Context 对象，使用下一个 SparkStreamingContext 之前需要把前面正在运
             * 行的 SparkStreamingContext 对象关闭掉，由此，我们获得一个重大的启发，
             * Spark Streaming 框架也只是 Spark Core 上的一个应用程序而已，只不过
             * Spark Streaming 框架箱运行的话，需要 Spark 工程师写业务逻辑处理代码
94.          */
95.         JavaStreamingContext jsc = new JavaStreamingContext(conf,
                Durations.seconds(10));
96.         jsc.checkpoint("/usr/local/IMF_testdata/IMFcheckpoint114");
97.         /**
98.          * 第三步：创建 Spark Streaming 输入数据来源 input Stream：
99.          * ① 数据输入来源可以基于 File、HDFS、Flume、Kafka、Socket 等；
100.         * ② 这里我们指定数据来源于网络 Socket 端口，Spark Streaming 连接上
             * 该端口并在运行的时候一直监听该端口的数据（当然，该端口服务首先必须
             * 存在)，并且在后续会根据业务需要不断有数据产生(当然，对于 Spark
             * Streaming 应用程序的运行而言，有无数据其处理流程都一样)；
101.         * ③ 如果经常每间隔 5s 没有数据，不断地启动空的 Job 其实会造成调度资源
             * 的浪费，因为并没有数据需要发生计算，所以实例的企业级生成环境的代码
             * 在具体提交 Job 前会判断是否有数据，如果没有，就不再提交 Job；
102.         * ④ 在本案例中，具体参数的含义：
103.         * 第一个参数是 StreamingContext 实例；
104.         * 第二个参数是 ZooKeeper 集群信息(接收 Kafka 数据时会从 ZooKeeper
             * 中获得 Offset 等元数据信息)；
105.         * 第三个参数是 Consumer Group；
106.         * 第四个参数是消费的 Topic 以及并发读取 Topic 中 Partition 的线程数
107.         */
108.        /**
             * 创建 Kafka 元数据，让 Spark Streaming 这个 Kafka Consumer 利用
109.         */
110.            Map<String, String> kafkaParameters = new HashMap<String, String>();
111.            kafkaParameters.put("metadata.broker.list", "Master:9092,Worker1:
                9092,Worker2:9092");
112.
113.            Set<String> topics = new HashSet<String>();
114.            topics.add("AdClicked");
115.
116.            JavaPairInputDStream<String, String> adClickedStreaming =
                KafkaUtils.createDirectStream(jsc, String.class,
117.                    String.class, StringDecoder.class, StringDecoder.
                        class, kafkaParameters, topics);
118.            /**
                 * 因为要对黑名单进行在线过滤，而数据是在 RDD 中的，所以必然使用 transform
                 * 函数；但是，这里我们必须使用 transformToPair，原因是读取进来的
                 * Kafka 的数据是 Pair<String,String>类型的，另外一个原因是过滤后
                 * 的数据要进行进一步处理，所以必须是读进来的 Kafka 数据的原始类型 DStream<
```

```
            * String, String>
119.     *
120.     * 在此再次说明：每个 Batch Duration 中输入的数据就是被一个且仅仅被一个 RDD 封
         * 装的，你可以有多个 InputDstream，但是其实在产生 Job 的时候，这些不同的 InputDstream
         * 在 BatchDuration 中相当于 Spark 基于 HDFS 数据操作的不同文件来源
121.     */
122.    JavaPairDStream<String, String> filteredadClickedStreaming = adClickedStreaming
123.            .transformToPair(new Function<JavaPairRDD<String, String>,
                JavaPairRDD<String, String>>() {
124.
125.                @Override
126.                public JavaPairRDD<String, String> call(JavaPair RDD<
                    String, String> rdd) throws Exception {
127.                    /**
128.                     * 在线黑名单过滤思路步骤：
129.                     * ① 从数据库中获取黑名单转换成 RDD，即新的 RDD 实例封装黑名
                         * 单数据；
130.                     * ② 把代表黑名单的 RDD 的实例和 BatchDuration 产生 * 的 RDD
                         * 进行 Join 操作，准确地说，是进行 leftOuterJoin 操作，也就是
                         * 使用 BatchDuration 产生的 RDD 和代表黑名单的 RDD 的实例进行
                         * leftOuterJoin 操作，如果两者都有内容，就会是 true，否则
                         * 就是 false；
131.                     * 我们要留下的是 leftOuterJoin 操作结果，为 false；
132.                     *
133.                     */
134.
135.                    final List<String> blackListNames = new ArrayList<
                        String>();
136.                    JDBCWrapper jdbcWrapper = JDBCWrapper.getJDBCInstance();
137.                    jdbcWrapper.doQuery("SELECT * FROM blacklisttable", null,
                        new ExecuteCallBack() {
138.
139.                        @Override
140.                        public void resultCallBack(ResultSet result) throws
                                Exception{
141.
142.                            while (result.next()) {
143.
144.                            blackListNames.add(result.getString(1));
145.                            }
146.                        }
147.
148.                    });
149.
150.                    List<Tuple2<String,Boolean>> blackListTuple=new ArrayList<
                        Tuple2<String, Boolean>>();
151.
152.                    for (String name : blackListNames) {
153.                        blackListTuple.add(new Tuple2<String,
                        Boolean>(name, true));
154.                    }
155.
```

```
156.            List<Tuple2<String,Boolean>>blackListFromDB=blackListTuple;
                //数据来自于查询的黑名单表并且映射成为<String,Boolean>
157.
158.
159.                    JavaSparkContext jsc = new JavaSpark Context(rdd.
                        context());
160.
161.                    /**
                         * 黑名单的表中只有userID，但是如果要进行Join操作，就必须
                         * 是Key-Value，所以这里我们需要基于数据表中的数据产生
                         * Key-Value类型的数据集合
162.                     */
163.                    JavaPairRDD<String, Boolean> blackListRDD =
                        jsc.parallelizePairs(blackListFromDB);
164.
165.                    /**
                         * 进行操作的时候肯定是基于userID进行Join的，所以必须把传
                         * 入的RDD进行mapToPair操作转化成为符合格式的RDD
166.                     *
167.                     * 广告点击的基本数据格式：timestamp、ip、userID、adID、
                         * province、city
168.                     */
169.
170.                    JavaPairRDD<String, Tuple2<String, String>>
                        rdd2Pair = rdd
171.                            .mapToPair(new PairFunction<Tuple2< String,
                                String>,String, Tuple2< String, String>>() {
172.
173.                        @Override
174.                        public Tuple2<String, Tuple2<String,String>> call
                            (Tuple2<String, String> t)
175.                                throws Exception {
176.                            String userID = t._2.split("\t")[2];
177.                            return new Tuple2<String, Tuple2<
                                String, String>>(userID, t);
178.                        }
179.                    });
180.
181.                    JavaPairRDD<String, Tuple2<Tuple2<String, String>,
                        Optional<Boolean>>>joined=rdd2Pair
182.                            .leftOuterJoin(blackListRDD);
183.
184.                    JavaPairRDD<String,String> result= joined.filter(
185.                            new Function<Tuple2<String, Tuple2<
                                Tuple2<String, String>, Optional<
                                Boolean>>>, Boolean>() {
186.
187.                        @Override
188.                        public Boolean call(Tuple2<String, Tuple2<
                            Tuple2<String, String>, Optional< Boolean>>> v1)
189.                                throws Exception {
190.                            Optional<Boolean> optional = v1._2._2;
```

```
191.
192.                    if(optional.isPresent()&&optional.get()){
193.                        return false;
194.                    } else {
195.                            return true;
196.                        }
197.
198.                    }
199.                }).mapToPair(
200.                        new PairFunction<Tuple2<String, Tuple2< Tuple2<
                        String,String>,Optional<Boolean>>>, String,
                        String>() {
201.
202.                    @Override
203.                    public Tuple2<String, String> call(
204.                            Tuple2<String, Tuple2<Tuple2< String,
                            String>, Optional<Boolean>>> t)
205.                                    throws Exception {
206.                        // TODO Auto-generated method stub
207.                        return t._2._1;
208.                    }
209.                });
210.
211.                return result;
212.            }
213.        });
214.
215.    filteredadClickedStreaming.print();
216.
217.    /*
218.     * 第四步：接下来就像对于 RDD 编程一样，基于 DStream 进行编程！！！原因是 DStream
     * 是 RDD 产生的模板（或者说类），在 Spark Streaming 具体发生计算前，其实质是把每
     * 个 Batch 的 DStream 的操作翻译成为对 RDD 的操作
219.     * 对初始的 DStream 进行 Transformation 级别的处理，如通过 map、filter 等高阶
     * 函数等的编程，进行具体的数据计算
220.     * 广告点击的基本数据格式：timestamp、ip、userID、adID、province、city
221.     */
222.
223.    JavaPairDStream<String, Long> pairs = filteredadClickedStreaming
224.            .mapToPair(new PairFunction<Tuple2<String, String>,
            String, Long>() {
225.
226.                @Override
227.                public Tuple2<String, Long> call(Tuple2<String,
                String> t) throws Exception {
228.                    String[] splited = t._2.split("\t");
229.
230.                    String timestamp = splited[0]; // yyyy-MM-dd
231.                    String ip = splited[1];
232.                    String userID = splited[2];
233.                    String adID = splited[3];
234.                    String province = splited[4];
235.                    String city = splited[5];
236.
237.                    String clickedRecord = timestamp + "_" + ip
                    +"_"+userID+"_" + adID + "_" + province + "_"
238.                        + city;
```

```
239.
240.                return new Tuple2<String,Long>(clickedRecord,1L);
241.            }
242.        });
243.
244.    /**
245.     * 第四步：对初始的 DStream 进行 Transformation 级别的处理，如通过 map、filter
       * 等高阶函数等的编程，进行具体的数据计算
246.     * 计算每个 Batch Duration 中每个 User 的广告点击量
247.     */
248.    JavaPairDStream<String, Long> adClickedUsers = pairs.reduce
        ByKey(new Function2<Long, Long, Long>() {
249.
250.        @Override
251.        public Long call(Long v1, Long v2) throws Exception {
252.            //待办事项：自动生成方法存根
253.            return v1 + v2;
254.        }
255.
256.    });
257.
258.    /**
259.     *
260.     * 计算出什么叫有效的点击？
261      * ① 对于复杂化的，一般都采用机器学习训练好模 261.型直接在线进行过滤；
262.     * ② 对于简单的，可以通过一个 BatchDuration 中的点击次数判断是不是非法广告点
       * 击，但是实际上，非法广告点击程序会尽可能模拟真实的广告点击行为，所以通过一个
       * Batch 来判断是不完整的，我们需要对例如一天（也可以是每个小时）的数据进行判断！
263.     * ③ 比在线机器学习退而求其次的做法如下：例如，一段时间内，同一个 IP（MAC 地址）
       * 有多个用户的账号访问；
264.     * 例如：可以统一一天内一个用户点击广告的次数，如果一天内点击同样的广告操作 50
       * 次，就列入黑名单；
       *
265.     * 黑名单有一个重要的特征：动态生成！！！所以，每个 Batch Duration 都要考虑是
       * 否有新的黑名单加入，此时黑名单需要存储起来
266.     * 具体存储在什么地方呢？存储在 DB/Redis 中即可；
267.     * 例如，邮件系统中的“黑名单”，可以采用 Spark Streaming 不断地监控每个用户
       * 的操作，如果用户发送邮件的频率超过了设定的值，则可以暂时把用户列入“黑名单”，从
       * 而阻止用户过度频繁的发送邮件
268.     *
269.     */
270.
271.    JavaPairDStream<String, Long> filteredClickInBatch = adClickedUsers
272.            .filter(new Function<Tuple2<String,Long>,Boolean>(){
273.
274.                @Override
275.                public Boolean call(Tuple2<String, Long> v1) throws
                    Exception {
276.                    if (1 < v1._2) {
277.                        //更新黑名单的数据表
278.                        return false;
279.                    } else {
280.                        return true;
281.                    }
282.
283.                }
```

```
284.           });
285.
286.     /*
       * 此处的print并不会直接发出Job的执行，因为现在的一切都在Spark Streaming
       * 框架的控制下，对于Spark Streaming而言，具体是否触发真正的Job运行是基于设置
       * 的Duration时间间隔的
       *
       *
287.   * 要注意，Spark Streaming应用程序要想执行具体的Job，对DStream就必须有
       * output Stream操作，output Stream有很多类型的函数触发，如print、
       * saveAsTextFile、saveAsHadoopFiles等，最重要的一个方法是foraeachRDD，
       * 因为Spark Streaming处理的结果一般都会放在Redis、DB、DashBoard等上面，
       * foreachRDD
288.   *
289.   *
290.   *
291.   *
292.   * 主要就是用来完成这些功能的，而且可以随意地自定义具体数据到底放在哪里
293.   *
294.   */
295.
296.   filteredClickInBatch.foreachRDD(new Function<JavaPairRDD<String,
       Long>, Void>() {
297.
298.       @Override
299.       public Void call(JavaPairRDD<String, Long> rdd) throws Exception{
300.
301.           if (rdd.isEmpty()) {
302.
303.           }
304.           rdd.foreachPartition(new VoidFunction<Iterator<Tuple2< String,
               Long>>>() {
305.
306.               @Override
307.               public void call(Iterator<Tuple2<String, Long>> partition)
                   throws Exception {
308.                   /**
309.                    * 这里我们使用数据库连接池的高效读写数据库的方式把数据写入
                        * 数据库MySQL;
310.                    * 由于传入的参数是一个Iterator类型的集合,所以为了更加高效
                        * 地操作，我们需要批量处理。
311.                    * 例如，一次性插入1000条Record，使用insertBatch或者
                        * updateBatch类型的操作;
312.                    * 插入的用户信息可以只包含timestamp、ip、userID、adID、
                        * province、city
313.                    * 这里有一个问题：可能出现两条记录的Key是一样的情况，此时就
                        * 需要更新累加操作
314.                    */
315.
316.                   List<UserAdClicked> userAdClickedList = new
                       ArrayList<UserAdClicked>();
317.
318.                   while (partition.hasNext()) {
319.                       Tuple2<String, Long> record = partition.next();
320.
```

```
321.            String[] splited = record._1.split("_");
322.            UserAdClicked userClicked = new UserAdClicked();
323.
324.                userClicked.setTimestamp(splited[0]);
325.                userClicked.setIp(splited[1]);
326.                userClicked.setUserID(splited[2]);
327.                userClicked.setAdID(splited[3]);
328.                userClicked.setProvince(splited[4]);
329.                userClicked.setCity(splited[5]);
330.                userAdClickedList.add(userClicked);
331.
332.          }
333.
334.          final List<UserAdClicked> inserting = new ArrayList<
          UserAdClicked>();
335.          final List<UserAdClicked> updating = new ArrayList<
          UserAdClicked>();
336.
337.          JDBCWrapper jdbcWrapper = JDBCWrapper.getJDBCInstance();
338.
339.          //点击
340.          //表的字段：timestamp、ip、userID、adID、province、city、
          //clickedCount
341.          for (final UserAdClicked clicked:userAdClickedList) {
342.              jdbcWrapper.doQuery(
343.                      "SELECT count(1) FROM adclickedWHERE"
344.                              + " timestamp = ? AND userID = ?
                              AND adID = ?",
345.                      new Object[] { clicked.get Timestamp(),
                      clicked.getUserID(), clicked.getAdID() },
346.                      new ExecuteCallBack() {
347.
348.                  @Override
349.                  public void resultCallBack(ResultSet result)
                  throws Exception {
350.
351.                      if (result.getRow() != 0) {
352.                          long count = result.getLong(1);
353.                          clicked.setClickedCount(count);
354.                          updating.add(clicked);
355.
356.                      } else {
357.                          clicked.setClickedCount(0L);
358.                          inserting.add(clicked);
359.
360.                      }
361.
362.                  }
363.              });
364.          }
365.          //点击
366.          //表的字段：timestamp、ip、userID、adID、province、city、
          //clickedCount
367.          ArrayList<Object[]>insertParametersList=newArrayList<
          Object[]>();
368.          for (UserAdClicked inserRecord : inserting) {
```

```
369.                    insertParametersList.add(new Object[] {inserRecord.
                        getTimestamp(), inserRecord.getIp(),
370.                                inserRecord.getUserID(),inserRecord. getAdID(),
                                    inserRecord.getProvince(),
371.                                inserRecord.getCity(), inserRecord.
                                    getClickedCount()});
372.                    }
373.                    jdbcWrapper.doBatch("INSERT INTO adclicked VALUES
                        (?,?,?,?,?,?,?)", insertParametersList);
374.
375.                    //点击
376.                    //表的字段：timestamp、ip、userID、adID、province、city、
                        //clickedCount
377.                    ArrayList<Object[]>updateParametersList=newArrayList<
                        Object[]>();
378.                    for (UserAdClicked updateRecord : updating) {
379.                        updateParametersList.add(new Object[]
                            {updateRecord.getTimestamp(),updateRecord.
                            getIp(),
380.                                updateRecord.getUserID(),updateRecord. getAdID(),
                                    updateRecord.getProvince(),
381.                                updateRecord.getCity(),
                                    updateRecord.getClickedCount()});
382.                    }
383.                    jdbcWrapper.doBatch("UPDATE adclicked set
                        clickedCount = ? WHERE "
384.                            + " timestamp = ? AND ip = ? AND userID
                                = ? AND adID = ? AND province = ? "
385.                            + "AND city = ? ", updateParametersList);
386.
387.                }
388.            });
389.            return null;
390.        }
391.
392.    });
393.
394.    JavaPairDStream<String, Long> blackListBasedOnHistory = filtered
        ClickInBatch
395.            .filter(new Function<Tuple2<String, Long>, Boolean>() {
396.
397.                @Override
398.                public Boolean call(Tuple2<String, Long> v1) throws Exception{
399.                    //广告点击的基本数据格式：timestamp、ip、userID、adID、
                        //province、city
400.
401.                    String[] splited = v1._1.split("_");
402.                    String date = splited[0];
403.
404.                    String userID = splited[2];
405.                    String adID = splited[3];
406.
407.                    /**
408.                     * 接下来根据 date、userID、adID 等条件查询用户点击广告的数
                         * 据表，获得总的点击次数
```

```
409.                    * 这时基于点击次数判断是否属于黑名单点击 *
410.                    */
411.
412.                   int clickedCountTotalToday = 81;
413.
414.                   if (clickedCountTotalToday > 50) {
415.                       return true;
416.                   } else {
417.                       return false;
418.                   }
419.
420.               }
421.           });
422.
423.   /**
       * 必须对黑名单的整个 RDD 进行去重操作
424.    */
425.
426.   JavaDStream<String> blackListuserIDtBasedOnHistory = blackListBased
       OnHistory
427.           .map(new Function<Tuple2<String, Long>, String>() {
428.
429.               @Override
430.               public String call(Tuple2<String, Long> v1)throws Exception{
431.                   //待办事项：自动生成方法存根
432.
433.                   return v1._1.split("_")[2];
434.               }
435.           });
436.
437.   JavaDStream<String> blackListUniqueuserIDtBasedOnHistory =
       blackListuserIDtBasedOnHistory
438.           .transform(new Function<JavaRDD<String>, JavaRDD<String>>(){
439.
440.               @Override
441.               public JavaRDD<String> call(JavaRDD<String> rdd)
               throws Exception {
442.                   //待办事项：自动生成方法存根
443.                   return rdd.distinct();
444.               }
445.           });
446.
447.   //下一步写入黑名单数据表中
448.
449.   blackListUniqueuserIDtBasedOnHistory.foreachRDD(new Function<JavaRDD<
       String>, Void>() {
450.
451.       @Override
452.       public Void call(JavaRDD<String> rdd) throws Exception {
453.           rdd.foreachPartition(new VoidFunction<Iterator<String>>() {
454.
455.               @Override
456.               public void call(Iterator<String> t) throws Exception {
457.                   /**
458.                    * 这里我们使用数据库连接池的高效读写数据库的方式把数据写入
                    * 数据库 MySQL
459.                    * 由于传入的参数是一个 Iterator 类型的集合，所以为了更加高效
```

```
                        * 地操作，我们需要批量处理
460.                    * 例如，一次性插入1000条Record，使用insertBatch或者
                        * updateBatch类型的操作
461.                    * 插入的用户信息可以只包含useID,此时直接插入黑名数据表即可
462.                            */
463.
464.                    List<Object[]> blackList = new ArrayList<
                        Object[]>();
465.
466.                    while (t.hasNext()) {
467.                        blackList.add(new Object[]{(Object)t.next()});
468.                    }
469.                    JDBCWrapper jdbcWrapper = JDBCWrapper.getJDBCInstance();
470.                    jdbcWrapper.doBatch("INSERT INTO blacklisttable
                        VALUES (?) ", blackList);
471.                }
472.            });
473.            return null;
474.        }
475.    });
476.
477.    /**
     * 广告点击累计动态更新，每个updateStateByKey都会在BatchDuration的时间间
     * 隔的基础上进行更高点击次数的更新，更新之后，我们一般都会持久化到外部存储设备
     * 上，这里我们存储到MySQL数据库中
478.     */
479.    JavaPairDStream<String, Long> updateStateByKeyDStream =
        filteredadClickedStreaming
480.            .mapToPair(new PairFunction<Tuple2<String, String>,
            String, Long>() {
481.
482.                @Override
483.                public Tuple2<String, Long> call(Tuple2<String,
                String> t) throws Exception {
484.                    String[] splited = t._2.split("\t");
485.
486.                    String timestamp = splited[0]; // yyyy-MM-dd
487.                    String ip = splited[1];
488.                    String userID = splited[2];
489.                    String adID = splited[3];
490.                    String province = splited[4];
491.                    String city = splited[5];
492.
493.                    String clickedRecord = timestamp + "_" + adID
                    + "_" + province + "_" + city;
494.
495.                    return new Tuple2<String, Long>(clicked Record, 1L);
496.                }
497.            }).updateStateByKey(new Function2<List<Long>,Optional<Long>,
            Optional<Long>>() {
498.
499.                @Override
500.                public Optional<Long> call(List<Long> v1,Optional<Long>
                v2) throws Exception {
```

```
501.                    /**
                         * v1:代表当前的 Key 在当前的 Batch Duration 中出现次数的集
                         * 合，如{1,1,1,1,1,1}
                         * v2:代表当前 Key 在以前的 BatchDuration 中积累下来的结果
502.                     */
503.                    Long clickedTotalHistory = 0L;
504.                    if (v2.isPresent()) {
505.                    clickedTotalHistory = v2.get();
506.                    }
507.
508.                    for (Long one : v1) {
509.                        clickedTotalHistory += one;
510.                    }
511.
512.                    return Optional.of(clickedTotalHistory);
513.                }
514.            });
515.
516.    updateStateByKeyDStream.foreachRDD(new Function<JavaPairRDD< String,
        Long>, Void>() {
517.
518.        @Override
519.        public Void call(JavaPairRDD<String, Long> rdd) throws
            Exception {
520.            rdd.foreachPartition(new VoidFunction<Iterator< Tuple2<
                String, Long>>>() {
521.
522.                @Override
523.                public void call(Iterator<Tuple2<String, Long>>
                    partition) throws Exception {
524.                    /**
525.                     * 这里我们使用数据库连接池的高效读写数据库的方式把数据写入
                         * 数据库 MySQL
526.                     * 由于传入的参数是一个 Iterator 类型的集合,所以为了更加高效
                         * 地操作，我们需要批量处理
527.                     * 例如，一次性插入 1000 条 Record，使用 insertBatch 或者
                         * updateBatch 类型的操作
528.                     * 插入的用户信息可以只包含 timestamp、adID、province、city
529.                     * 这里有一个问题：可能出现两条记录的 Key 是一样的情况，此时就
                         * 需要更新累加操作
530.                     */
531.
532.                    List<AdClicked> adClickedList = new ArrayList<
                        AdClicked>();
533.
534.                    while (partition.hasNext()) {
535.                        Tuple2<String, Long> record = partition.next();
536.
537.                        String[] splited = record._1.split("_");
538.
539.                        AdClicked adClicked = new AdClicked();
540.                        adClicked.setTimestamp(splited[0]);
541.                        adClicked.setAdID(splited[1]);
542.                        adClicked.setProvince(splited[2]);
543.                        adClicked.setCity(splited[3]);
```

```
544.                            adClicked.setClickedCount(record._2);
545.
546.                            adClickedList.add(adClicked);
547.
548.                        }
549.
550.                        JDBCWrapper jdbcWrapper = JDBCWrapper.getJDBCInstance();
551.
552.                        final List<AdClicked> inserting = new ArrayList<
                            AdClicked>();
553.                        final List<AdClicked> updating = new ArrayList<
                            AdClicked>();
554.
555.                        //点击
556.                        //表的字段：timestamp、ip、userID、adID、province、city、
                            //clickedCount
557.                        for (final AdClicked clicked : adClickedList){
558.                            jdbcWrapper.doQuery(
559.                                    "SELECT count(1) FROM adclickedcount
                                    WHERE "
560.                                            + " timestamp = ? AND adID = ? AND
                                            province = ? AND city = ? ",
561.                                   new Object[] { clicked.getTimestamp(), clicked.
                                   getAdID(), clicked. getProvince(),
562.                                            clicked.getCity() },
563.                                    new ExecuteCallBack() {
564.
565.                                @Override
566.                                public void resultCallBack(ResultSet
                                result) throws Exception {
567.
568.                                    if (result.getRow() != 0) {
569.
570.                                        long count = result.getLong(1);
571.                                        clicked.setClickedCount(count);
572.                                        updating.add(clicked);
573.
574.                                    } else {
575.                                        inserting.add(clicked);
576.
577.                                    }
578.
579.                                }
580.                            });
581.                        }
582.                        //点击
583.                        //表的字段：timestamp、ip、userID、adID、province、city、
                            //clickedCount
584.                        ArrayList<Object[]> insertParametersList = new
                            ArrayList<Object[]>();
585.                        for (AdClicked inserRecord : inserting) {
586.                            insertParametersList.add(new Object[]
                                { inserRecord.getTimestamp(), inserRecord.
                                getAdID(),
587.                                        inserRecord.getProvince(), inserRecord.
                                        getCity(),inserRecord.getClickedCount()});
```

```
588.                    }
589.                    jdbcWrapper.doBatch("INSERT INTO adclickedcount
                        VALUES(?,?,?,?,?)", insertParametersList);
590.
591.                    //点击
592.                    //表的字段：timestamp、ip、userID、adID、province、city、
                        //clickedCount
593.                    ArrayList<Object[]> updateParametersList = new ArrayList<
                        Object[]>();
594.                    for (AdClicked updateRecord : updating) {
595.                        updateParametersList.add(new Object[]
                            { updateRecord.getClickedCount(),
596.                                updateRecord.getTimestamp(),updateRecord.
                                getAdID(),updateRecord.getProvince(),
597.                                updateRecord.getCity() });
598.                    }
599.                    jdbcWrapper.doBatch(
600.                            "UPDATE adclickedcount set clickedCount = ?
                            WHERE "
601.                                    + " timestamp = ? AND adID = ?
                                    AND province=? AND city = ? ",
602.                            updateParametersList);
603.
604.                }
605.            });
606.            return null;
607.        }
608.    });
609.
610.    /**
         * 对广告点击进行 TopN 的计算，计算出每天每个省份的 Top5 排名的广告；因为我们直
         * 接对 RDD 进行操作，所以使用了 transform 算子
611.     */
612.
613.   updateStateByKeyDStream.transform(new Function<JavaPairRDD< String, Long>,
        JavaRDD<Row>>() {
614.
615.        @Override
616.        public JavaRDD<Row> call(JavaPairRDD<String, Long> rdd)
            throws Exception {
617.
618.            JavaRDD<Row> rowRDD = rdd.mapToPair(new PairFunction< Tuple2<
                String, Long>, String, Long>() {
619.
620.                @Override
621.                public Tuple2<String, Long> call(Tuple2<String, Long> t)
                    throws Exception {
622.                    String[] splited = t._1.split("_");
623.
624.                    String timestamp = "2016-07-10"; // yyyy-MM-dd
625.
626.                    String adID = splited[1];
627.                    String province = splited[2];
628.
629.                    String clickedRecord = timestamp + "_" + adID + "_"
                        + province;
630.
```

```
631.                    return new Tuple2<String,Long>(clickedRecord,t._2);
632.                }
633.            }).reduceByKey(new Function2<Long, Long, Long>() {
634.
635.                @Override
636.                public Long call(Long v1, Long v2) throws Exception {
637.                    //待办事项：自动生成方法存根
638.                    return v1 + v2;
639.                }
640.            }).map(new Function<Tuple2<String, Long>, Row>() {
641.
642.                @Override
643.                public Row call(Tuple2<String,Long>v1)throws Exception {
644.                    String[] splited = v1._1.split("_");
645.
646.                    String timestamp = "2016-07-10"; // yyyy-MM-dd
647.                    String adID = splited[1];
648.                    String province = splited[2];
649.
650.                    return RowFactory.create(timestamp, adID,province,
                    v1._2);
651.                }
652.            });
653.
654.            StructType structType = DataTypes.createStructType(
655.                    Arrays.asList(DataTypes.createStructField
                    ("timstamp", DataTypes.StringType, true),
656.                            DataTypes.createStructField("adID",
                            DataTypes.StringType, true),
657.                            DataTypes.createStructField("province",
                            DataTypes.StringType, true),
658.
659.            DataTypes.createStructField("clickedCount", Data Types
            .LongType, true)));
660.
661.            HiveContext hiveContext = new HiveContext(rdd.context());
662.
663.            DataFrame df = hiveContext.createDataFrame(rowRDD,structType);
664.
665.            df.registerTempTable("topNTableSource");
666.
667.            String IMFsqlText = "SELECT timstamp,adID,province,
            clickedCount FROM "
668.                    + " ( SELECT timstamp,adID,province,clickedCount,
                    row_number() "
669.                    + " OVER ( PARTITION BY province ORDER BY clickedCount
                    DESC ) rank "
670.                    + " FROM topNTableSource ) subquery " + " WHERE rank
                    <= 5 ";
671.
672.            DataFrame result = hiveContext.sql(IMFsqlText);
673.
674.            return result.toJavaRDD();
675.
676.        }
677.
678.    }).foreachRDD(new Function<JavaRDD<Row>, Void>() {
```

```
679.
680.        @Override
681.        public Void call(JavaRDD<Row> rdd) throws Exception {
682.
683.            rdd.foreachPartition(new VoidFunction<Iterator<Row>>(){
684.
685.                @Override
686.                public void call(Iterator<Row> t)throws Exception{
687.
688.                    List<AdProvinceTopN> adProvinceTopN = new
                    ArrayList<AdProvinceTopN>();
689.
690.                    while (t.hasNext()) {
691.                        Row row = t.next();
692.
693.                        AdProvinceTopN item = new AdProvinceTopN();
694.                        item.setTimestamp(row.getString(0));
695.                        item.setAdID(row.getString(1));
696.                        item.setProvince(row.getString(2));
697.
698.                        item.setClickedCount(row.getLong(3));
699.
700.                        adProvinceTopN.add(item);
701.                    }
702.
703.                    JDBCWrapper jdbcWrapper = JDBCWrapper.getJDBCInstance();
704.
705.                    Set<String> set = new HashSet<String>();
706.                    for (AdProvinceTopN item : adProvinceTopN) {
707.                        set.add(item.getTimestamp()+"_"+item.getProvince());
708.                    }
709.
710.                    //点击
711.                    //表的字段：timestamp、ip、userID、adID、province、city、
                    //clickedCount
712.                    ArrayList<Object[]> deleteParametersList =
                    new ArrayList<Object[]>();
713.                    for (String deleteRecord : set) {
714.                        String[] splited = deleteRecord.split("_");
715.                        deleteParametersList.add(new Object[] { splited[0],
                        splited[1] });
716.                        }
717.                    jdbcWrapper.doBatch("DELETE FROM adprovincetopn WHERE
                    timestamp = ? AND province = ?",
718.                            deleteParametersList);
719.
720.                    //广告点击每个省份的 TopN 排名
721.                    //表的字段：timestamp、adID、province、clickedCount
722.                    ArrayList<Object[]> insertParametersList =
                    new ArrayList<Object[]>();
723.                    for (AdProvinceTopN updateRecord:adProvinceTopN) {
724.                        insertParametersList.add(new Object[]{updateRecord
                        .getTimestamp(),updateRecord.getAdID(),
725.                                updateRecord.getProvince(),updateRecord
                                .getClickedCount()});
```

```
726.                  }
727.                  jdbcWrapper.doBatch("INSERT INTO adprovincetopn VALUES
                      (?,?,?,?) ", insertParametersList);
728.
729.              }
730.          });
731.
732.          return null;
733.
734.      }
735.  });
736.
737.  /**
       * 计算过去半个小时内广告点击的趋势 用户广告点击信息可以只包含timestamp、ip、
       * userID、adID、province、city
738.   */
739.
740.  filteredadClickedStreaming.mapToPair(new PairFunction< Tuple2<
      String, String>, String, Long>() {
741.
742.      @Override
743.      public Tuple2<String, Long> call(Tuple2<String, String>t) throws
          Exception {
744.          String[] splited = t._2.split("\t");
745.
746.          String adID = splited[3];
747.
748.          String time = splited[0];
              //Todo：后续需要重构代码实现时间戳和分钟的转换提取，此处需要提取出该
              //广告的点击分钟单位
749.
750.          return new Tuple2<String, Long>(time + "_" + adID, 1L);
751.      }
752.  }).reduceByKeyAndWindow(new Function2<Long, Long, Long>() {
753.
754.      @Override
755.      public Long call(Long v1, Long v2) throws Exception {
756.          //待办事项：自动生成方法存根
757.          return v1 + v2;
758.      }
759.  }, new Function2<Long, Long, Long>() {
760.
761.      @Override
762.      public Long call(Long v1, Long v2) throws Exception {
763.          //待办事项：自动生成方法存根
764.          return v1 - v2;
765.      }
766.  }, Durations.minutes(30), Durations.minutes(1)).foreachRDD
      (new Function<JavaPairRDD<String, Long>, Void>() {
677.
768.      @Override
769.      public Void call(JavaPairRDD<String, Long> rdd) throws Exception{
770.          rdd.foreachPartition(new  VoidFunction<Iterator<Tuple2<String,
              Long>>>() {
771.
772.              @Override
773.              public void call(Iterator<Tuple2<String, Long>>
                  partition) throws Exception {
774.
```

```
775.                 List<AdTrendStat> adTrend = new ArrayList<AdTrendStat>();
776.
777.                 while (partition.hasNext()) {
778.                     Tuple2<String, Long> record = partition.next();
779.                     String[] splited = record._1.split("_");
780.                     String time = splited[0];
781.                     String adID = splited[1];
782.                     Long clickedCount = record._2;
783.
784.                     /**
785.                       * 在插入数据到数据库的时候具体需要哪些字段？time、
                          * adID、clickedCount；
786.                       * 而我们通过 J2EE 技术进行趋势绘图的时候肯定是需要年、
                          * 月、日、时、分这些维度的，所有我们在这里需要年、月、日、
                          * 小时、分钟这些时间维度
787.                       */
788.
789.                     AdTrendStat adTrendStat = new AdTrendStat();
790.                     adTrendStat.setAdID(adID);
791.                     adTrendStat.setClickedCount(clickedCount);
792.                     adTrendStat.set_date(time);//Todo:获取年月日
793.                     adTrendStat.set_hour(time);//Todo:获取小时
794.                     adTrendStat.set_minute(time);//Todo:获取分钟
795.
796.                     adTrend.add(adTrendStat);
797.
798.
799.                 }
800.
801.                 final List<AdTrendStat> inserting = new ArrayList<
                 AdTrendStat>();
802.                 final List<AdTrendStat> updating = new ArrayList<
                 AdTrendStat>();
803.
804.                 JDBCWrapper jdbcWrapper = JDBCWrapper.getJDBCInstance();
805.
806.                 //广告点击趋势
807.                 //表的字段：date、hour、minute、adID、clickedCount
808.                 for (final AdTrendStat clicked : adTrend) {
809.                     final AdTrendCountHistory adTrendCount
                     History = new AdTrendCountHistory();
810.
811.                     jdbcWrapper.doQuery(
812.                             "SELECT count(1) FROM adclickedtrend
                             WHERE "
813.                                     + " date = ? AND hour = ?
                                     AND minute=?AND adID =?",
814.                             new Object[] { clicked.get_ date(),
                             clicked.get_hour(),clicked.get_minute(),
815.                                     clicked.getAdID() },
```

```
816.                                   new ExecuteCallBack() {
817.
818.                               @Override
819.                               public void resultCallBack(ResultSet
                                   result) throws Exception {
820.
821.                                   if (result.getRow() != 0) {
822.                                       long count = result.getLong(1);
823.                                       adTrendCountHistory.set ClickedCount
                                           History(count);
824.                                       updating.add(clicked);
825.                                   } else {
826.                                       //
827.                                       inserting.add(clicked);
828.                                   }
829.
830.                               }
831.                           });
832.                       }
833.                       //广告点击趋势
834.                       //表的字段：date、hour、minute、adID、clickedCount
835.                       ArrayList<Object[]> insertParametersList = new
                           ArrayList<Object[]>();
836.                       for (AdTrendStat inserRecord : inserting) {
837.
838.
839.                           insertParametersList
840.                                   .add(new Object[] { inserRecord.get_date(),
841.
842.                                   inserRecord.get_hour(),inserRecord.
                                   get_minute(),inserRecord.getAdID(),
843.                                           inserRecord.getClicked Count()});
844.                       }
845.
846.                       jdbcWrapper.doBatch("INSERT INTO adclickedtrend
                           VALUES(?,?,?,?,?)",insertParametersList);  //IMF
847.                                               //BUG
848.                                               //FIXED
849.
850.                       //广告点击趋势
851.                       //表的字段：date、hour、minute、adID、clickedCount
852.                       ArrayList<Object[]> updateParametersList =
                           new ArrayList<Object[]>();
853.                       for (AdTrendStat updateRecord : updating) {
854.                           updateParametersList.add(new Object[]
                           { updateRecord.getClickedCount(),
855.                                   updateRecord.get_date(),updateRecord.
                                   get_hour(),updateRecord.get_minute(),
856.                                   updateRecord.getAdID() });
857.                       }
858.                               jdbcWrapper.doBatch("UPDATE  adclickedtrend
```

```
                    set clickedCount = ? WHERE "
859.                                + " date = ? AND hour = ? AND minute
                                    =?AND adID=?",updateParametersList);
860.
861.                        }
862.                    });
863.                    return null;
864.                }
865.            });
866.
867.            /**
             *Spark Streaming 执行引擎也就是 Driver 开始运行，Driver 启动时是位
             *于一条新的线程中的，当然其内部有消息循环体，用于接收应用程序本身或者
             *Executor 中的消息
868.             */
869.            jsc.start();
870.
871.            jsc.awaitTermination();
872.            jsc.close();
873.
874.        }
875.
876.    }
877.
878.    class JDBCWrapper {
879.
880.        private static LinkedBlockingQueue<Connection> dbConnectionPool
            = new LinkedBlockingQueue<Connection>();
881.
882.        private static JDBCWrapper jdbcInstance = null;
883.
884.        static {
885.            try {
886.                Class.forName("com.mysql.jdbc.Driver");
887.            } catch (ClassNotFoundException e) {
888.                //待办事项：自动生成 catch 块
889.                e.printStackTrace();
890.            }
891.        }
892.
893.        public static JDBCWrapper getJDBCInstance() {
894.            if (jdbcInstance == null) {
895.
896.                synchronized (JDBCWrapper.class) {
897.                    if (jdbcInstance == null) {
898.                        jdbcInstance = new JDBCWrapper();
899.                    }
900.                }
901.
902.            }
903.
904.            return jdbcInstance;
905.        }
906.
907.        private JDBCWrapper() {
908.
909.            for (int i = 0; i < 10; i++) {
910.
```

```
911.                try {
912.                    Connection conn = DriverManager. getConnection
                        ("jdbc:mysql://Master:3306/sparkstreaming", "root",
913.                            "root");
914.                    dbConnectionPool.put(conn);
915.                } catch (Exception e) {
916.                    //待办事项：自动生成 catch 块
917.                    e.printStackTrace();
918.                }
919.
920.            }
921.
922.        }
923.
924.        public synchronized Connection getConnection() {
925.            while (0 == dbConnectionPool.size()) {
926.                try {
927.                    Thread.sleep(20);
928.                } catch (InterruptedException e) {
929.                    //待办事项：自动生成 catch 块
930.                    e.printStackTrace();
931.                }
932.            }
933.
934.            return dbConnectionPool.poll();
935.        }
936.
937.        public int[] doBatch(String sqlText, List<Object[]> paramsList)
{
938.
939.            Connection conn = getConnection();
940.            PreparedStatement preparedStatement = null;
941.            int[] result = null;
942.            try {
943.                conn.setAutoCommit(false);
944.                preparedStatement = conn.prepareStatement(sqlText);
945.
946.                for (Object[] parameters : paramsList) {
947.                    for (int i = 0; i < parameters.length; i++) {
948.                        preparedStatement.setObject(i + 1, parameters[i]);
949.                    }
950.
951.                    preparedStatement.addBatch();
952.                }
953.
954.                result = preparedStatement.executeBatch();
955.
956.                conn.commit();
957.
958.            } catch (Exception e) {
959.                //待办事项：自动生成 catch 块
960.                e.printStackTrace();
961.            } finally {
962.                if (preparedStatement != null) {
963.                    try {
964.                        preparedStatement.close();
965.                    } catch (SQLException e) {
966.                        //待办事项：自动生成 catch 块
967.                        e.printStackTrace();
968.                    }
```

```
969.                }
970.
971.                if (conn != null) {
972.                    try {
973.                        dbConnectionPool.put(conn);
974.                    } catch (InterruptedException e) {
975.                        //待办事项：自动生成 catch 块
976.                        e.printStackTrace();
977.                    }
978.                }
979.            }
980.
981.            return result;
982.        }
983.
984.        public void doQuery(String sqlText, Object[] paramsList,
        ExecuteCallBack callBack) {
985.
986.            Connection conn = getConnection();
987.            PreparedStatement preparedStatement = null;
988.            ResultSet result = null;
989.
990.            try {
991.
992.                preparedStatement = conn.prepareStatement(sqlText);
993.
994.                if (paramsList != null) {
995.                    for (int i = 0; i < paramsList.length; i++) {
996.                        preparedStatement.setObject(i + 1, paramsList[i]);
997.
998.                    }
999.                }
1000.               result = preparedStatement.executeQuery();
1001.
1002.               callBack.resultCallBack(result);
1003.
1004.           } catch (Exception e) {
1005.               //待办事项：自动生成 catch 块
1006.               e.printStackTrace();
1007.           } finally {
1008.               if (preparedStatement != null) {
1009.                   try {
1010.                       preparedStatement.close();
1011.                   } catch (SQLException e) {
1012.                       //待办事项：自动生成 catch 块
1013.                       e.printStackTrace();
1014.                   }
1015.               }
1016.
1017.               if (conn != null) {
1018.                   try {
1019.                       dbConnectionPool.put(conn);
1020.                   } catch (InterruptedException e) {
1021.                       //待办事项：自动生成 catch 块
1022.                       e.printStackTrace();
1023.                   }
1024.               }
1025.           }
1026.
1027.       }
```

```
1028.   }
1029.
1030.   interface ExecuteCallBack {
1031.       void resultCallBack(ResultSet result) throws Exception;
1032.   }
1033.
1034.   class UserAdClicked implements Serializable {
1035.       private String timestamp;
1036.       private String ip;
1037.       private String userID;
1038.       private String adID;
1039.       private String province;
1040.       private String city;
1041.       private Long clickedCount;
1042.
1043.       @Override
1044.       public String toString() {
1045.           return"UserAdClicked [timestamp=" + timestamp + ", ip=" + ip + ",
                userID="+userID+",adID=" + adID+", province=" + province + ",
                city=" + city + ",clickedCount=" + clickedCount + "]";
1046.       }
1047.
1048.       public Long getClickedCount() {
1049.           return clickedCount;
1050.       }
1051.
1052.       public void setClickedCount(Long clickedCount) {
1053.           this.clickedCount = clickedCount;
1054.       }
1055.
1056.       public String getTimestamp() {
1057.           return timestamp;
1058.       }
1059.
1060.       public void setTimestamp(String timestamp) {
1061.           this.timestamp = timestamp;
1062.       }
1063.
1064.       public String getIp() {
1065.           return ip;
1066.       }
1067.
1068.       public void setIp(String ip) {
1069.           this.ip = ip;
1070.       }
1071.
1072.       public String getUserID() {
1073.           return userID;
1074.       }
1075.
1076.       public void setUserID(String userID) {
1077.           this.userID = userID;
1078.       }
1079.
1080.       public String getAdID() {
1081.           return adID;
1082.       }
1083.
1084.       public void setAdID(String adID) {
```

```
1085.            this.adID = adID;
1086.        }
1087.
1088.        public String getProvince() {
1089.            return province;
1090.        }
1091.
1092.        public void setProvince(String province) {
1093.            this.province = province;
1094.        }
1095.
1096.        public String getCity() {
1097.            return city;
1098.        }
1099.
1100.        public void setCity(String city) {
1101.            this.city = city;
1102.        }
1103.    }
1104.
1105.    class AdClicked implements Serializable {
1106.        private String timestamp;
1107.        private String adID;
1108.        private String province;
1109.        private String city;
1110.        private Long clickedCount;
1111.
1112.        @Override
1113.        public String toString() {
1114.            return "AdClicked [timestamp=" + timestamp + ", adID=" + adID
             + ", province=" + province + ", city=" + city+", clickedCount="
             + clickedCount + "]";
1115.        }
1116.
1117.        public String getTimestamp() {
1118.            return timestamp;
1119.        }
1120.
1121.        public void setTimestamp(String timestamp) {
1122.            this.timestamp = timestamp;
1123.        }
1124.
1125.        public String getAdID() {
1126.            return adID;
1127.        }
1128.
1129.        public void setAdID(String adID) {
1130.            this.adID = adID;
1131.        }
1132.
1133.        public String getProvince() {
1134.            return province;
1135.        }
1136.
1137.        public void setProvince(String province) {
1138.            this.province = province;
1139.        }
1140.
1141.        public String getCity() {
1142.            return city;
```

```
1143.         }
1144.
1145.         public void setCity(String city) {
1146.             this.city = city;
1147.         }
1148.
1149.         public Long getClickedCount() {
1150.             return clickedCount;
1151.         }
1152.
1153.         public void setClickedCount(Long clickedCount) {
1154.             this.clickedCount = clickedCount;
1155.         }
1156.
1157.     }
1158.
1159.     class AdProvinceTopN implements Serializable {
1160.         private String timestamp;
1161.         private String adID;
1162.
1163.         public String getTimestamp() {
1164.             return timestamp;
1165.         }
1166.
1167.         public void setTimestamp(String timestamp) {
1168.             this.timestamp = timestamp;
1169.         }
1170.
1171.         public String getAdID() {
1172.             return adID;
1173.         }
1174.
1175.         public void setAdID(String adID) {
1176.             this.adID = adID;
1177.         }
1178.
1179.         public String getProvince() {
1180.             return province;
1181.         }
1182.
1183.         public void setProvince(String province) {
1184.             this.province = province;
1185.         }
1186.
1187.         public Long getClickedCount() {
1188.             return clickedCount;
1189.         }
1190.
1191.         public void setClickedCount(Long clickedCount) {
1192.             this.clickedCount = clickedCount;
1193.         }
1194.
1195.         private String province;
1196.
1197.         private Long clickedCount;
1198.     }
1199.
1200.     class AdTrendStat implements Serializable {
1201.         private String _date;
1202.         private String _hour;
```

```
1203.        private String _minute;
1204.
1205.        public String get_date() {
1206.            return _date;
1207.        }
1208.
1209.        @Override
1210.        public String toString() {
1211.            return "AdTrendStat [_date=" + _date + ", _hour=" + _hour + ",
                 _minute=" + _minute + ", adID=" + adID+",clickedCount=" +
                 clickedCount + "]";
1212.        }
1213.
1214.        public void set_date(String _date) {
1215.            this._date = _date;
1216.        }
1217.
1218.        public String get_hour() {
1219.            return _hour;
1220.        }
1221.
1222.        public void set_hour(String _hour) {
1223.            this._hour = _hour;
1224.        }
1225.
1226.        public String get_minute() {
1227.            return _minute;
1228.        }
1229.
1230.        public void set_minute(String _minute) {
1231.            this._minute = _minute;
1232.        }
1233.
1234.        public String getAdID() {
1235.            return adID;
1236.        }
1237.
1238.        public void setAdID(String adID) {
1239.            this.adID = adID;
1240.        }
1241.
1242.        public Long getClickedCount() {
1243.            return clickedCount;
1244.        }
1245.
1246.        public void setClickedCount(Long clickedCount) {
1247.            this.clickedCount = clickedCount;
1248.        }
1249.
1250.        private String adID;
1251.        private Long clickedCount;
1252.    }
1253.
1254.    class AdTrendCountHistory implements Serializable {
1255.        private Long clickedCountHistory;
1256.
1257.        public Long getClickedCountHistory() {
1258.            return clickedCountHistory;
1259.        }
1260.
```

```
1261.        public void setClickedCountHistory(Long clickedCountHistory) {
1262.            this.clickedCountHistory = clickedCountHistory;
1263.        }
1264.    }
```

Spark Streaming 电商广告点击综合案例课程的模拟数据生成的 Java 源码，如例 16-2 所示。

【例 16-2】 MockAdClickedStats.java 代码。

```
1.  package com.dt.spark.SparkApps.SparkStreaming;
2.
3.  import java.util.Date;
4.  import java.util.HashMap;
5.  import java.util.Map;
6.  import java.util.Properties;
7.  import java.util.Random;
8.
9.  import kafka.javaapi.producer.Producer;
10. import kafka.producer.KeyedMessage;
11. import kafka.producer.ProducerConfig;
12.
13. public class MockAdClickedStats {
14.
15.     public static void main(String[] args) {
16.         final Random random = new Random();
17.         final String[] provinces = new String[]{"Guangdong","Zhejiang",
            "Jiangsu","Fujian"};
18.         final Map<String,String[]> cities = new HashMap<String,
            String[]>();
19.         cities.put("Guangdong", new String[]{"Guangzhou","Shenzhen",
            "DongGuan"});
20.         cities.put("Zhejiang", new String[]{"Hangzhou","Wenzhou","Ningbo"});
21.         cities.put("Jiangsu", new String[]{"Nanjing","Suzhou","WuXi"});
22.         cities.put("Fujian", new String[]{"Fuzhou","Ximen","Sanming"});
23.
24.         final String[] ips = new String[]{
25.                 "192.168.112.240",
26.                 "192.168.112.239",
27.                 "192.168.112.245",
28.                 "192.168.112.246",
29.                 "192.168.112.247",
30.                 "192.168.112.248",
31.                 "192.168.112.249",
32.                 "192.168.112.250",
33.                 "192.168.112.251",
34.                 "192.168.112.252",
35.                 "192.168.112.253",
36.                 "192.168.112.254"
37.         };
38.         /**
             * Kafka 相关的基本配置信息
39.          */
40.
41.         Properties kafkaConf = new Properties();
42.         kafkaConf.put("serializer.class", "kafka.serializer.StringEncoder");
43.         kafkaConf.put("metadata.broker.list", "master:9092,worker1:9092,
            worker2:9092");
44.         ProducerConfig producerConfig = new ProducerConfig(kafkaConf);
45.
46.         final Producer<Integer, String> producer = new Producer<Integer,
```

```
            String>(producerConfig);
47.
48.         new Thread(new Runnable() {
49.
50.             @Override
51.             public void run() {
52.                 while(true){
53.                     //在线处理广告点击流，广告点击的基本数据格式：timestamp、ip、
                    userID、adID、province、city
54.                     Long timestamp = new Date().getTime();
55.                     String ip = ips[random.nextInt(12)];
                    //可以采用网络上免费提供的 IP 库
56.                     int userID = random.nextInt(10000);
57.                     int adID = random.nextInt(100);
58.                     String province = provinces[random.nextInt(4)];
59.                     String city = cities.get(province)[random.nextInt(3)];
60.
61.                     String clickedAd = timestamp + "\t" + ip + "\t" + userID
                    + "\t" + adID + "\t" + province +
62.                             "\t" + city ;
63.
64.                     System.out.println(clickedAd);
65.
66.                     producer.send( new KeyedMessage("AdClicked",clickedAd));
67.
68.                     try {
69.                         Thread.sleep(2000);
70.                     } catch (InterruptedException e) {
71.                         // TODO Auto-generated catch block
72.                         e.printStackTrace();
73.                     }
74.                 }
75.             }
76.         }).start();
77.
78.
79.     }
80.
81. }
82.
```

16.14　电商广告点击综合案例课程的 Scala 源码

Spark Streaming 电商广告点击综合案例课程的 Scala 源码，如例 16-3 所示。

【例 16-3】 AdClickedStreamingStats.scala 代码。

```
1.   package com.dt.spark.streaming114
2.
3.  import java.sql.Connection
4.  import java.sql.DriverManager
5.  import java.sql.PreparedStatement
6.  import java.sql.ResultSet
7.  import java.sql.SQLException
8.  import java.util.concurrent.LinkedBlockingQueue
9.
```

```
10. import scala.collection.mutable
11. import scala.collection.mutable.ListBuffer
12.
13. import org.apache.spark.SparkConf
14. import org.apache.spark.sql.Row
15. import org.apache.spark.sql.hive.HiveContext
16. import org.apache.spark.sql.hive.HiveQLDialect
17. import org.apache.spark.sql.types.LongType
18. import org.apache.spark.sql.types.StringType
19. import org.apache.spark.sql.types.StructType
20. import org.apache.spark.sql.types.StructType
21. import org.apache.spark.streaming.Seconds
22. import org.apache.spark.streaming.StreamingContext
23. import org.apache.spark.streaming.kafka._
24.
25. import kafka.serializer.StringDecoder
26. import org.apache.spark.SparkContext
27.
28. object AdClickedStreamingStats {
29.   def main(args: Array[String]): Unit = {
30.
31.     val sparkConf = new SparkConf().setAppName("scala-20160903-
        blacklist-ok-114-AdClickedStreamingStats")
32.       .setMaster("spark://192.168.189.1:7077").setJars(List(
33.         // .setMaster("local[5]").setJars(List(
34.         "/usr/local/spark-1.6.1-bin-hadoop2.6/lib/spark-streaming-
           kafka_2.10-1.6.1.jar",
35.         "/usr/local/kafka_2.10-0.8.2.1/libs/kafka-clients-0.8.2.1.jar",
36.         "/usr/local/kafka_2.10-0.8.2.1/libs/kafka_2.10-0.8.2.1.jar",
37.         "/usr/local/spark-1.6.1-bin-hadoop2.6/lib/spark-streaming_2.10-
           1.6.1.jar",
38.         "/usr/local/kafka_2.10-0.8.2.1/libs/metrics-core-2.2.0.jar",
39.         "/usr/local/kafka_2.10-0.8.2.1/libs/zkclient-0.3.jar",
40.         //  "/usr/local/spark-1.6.1-bin-hadoop2.6/lib/spark-assembly-1.
           6.1-hadoop2.6.0.jar",
41.         "/usr/local/spark-1.6.1-bin-hadoop2.6/lib/mysql-connector-java-
            5.1.13-bin.jar",
42.         "/usr/local/IMF_testdata/AdClickedStreamingStats.jar"))
43.
44.     val ssc = new StreamingContext(sparkConf, Seconds(10))
45.     ssc.checkpoint("/usr/local/IMF_testdata/IMFcheckpoint114")
46.     val kafkaParameters = Map[String, String]("metadata.broker.list" ->
       "Master:9092,Worker1:9092,Worker2:9092")
47.     val topics = Set[String]("IMFScalaAdClicked")
48.     val adClickedStreaming = KafkaUtils.createDirectStream[String,
        String, StringDecoder, StringDecoder](ssc, kafkaParameters, topics)
49.
50.     val filteredadClickedStreaming=adClickedStreaming.transform(rdd =>{
51.       val blackListNames = ListBuffer[String]()
52.       val jdbcWrapper = JDBCWrapper.getInstance()
53.
54.       def querycallBack(result: ResultSet): Unit = {
55.
56.         while (result.next()) {
57.           result.getString(1)
58.           blackListNames += result.getString(1)
59.         }
60.       }
61.
62.       jdbcWrapper.doQuery("SELECT * FROM blacklisttable", null,
```

```
      querycallBack)
63.
64.       val blackListTuple = ListBuffer[(String, Boolean)]()
65.       for (name <- blackListNames) {
66.         val nameBoolean = (name, true)
67.         blackListTuple += nameBoolean
68.       }
69.
70.       val blackListFromDB = blackListTuple
71.
72.       val jsc = rdd.sparkContext
73.
74.       val blackListRDD = jsc.parallelize(blackListFromDB)
75.
76.       val rdd2Pair = rdd.map(t => {
77.         val userID = t._2.split("\t")(2)
78.         (userID, t)
79.
80.       })
81.
82.       val joined = rdd2Pair.leftOuterJoin(blackListRDD)
83.       val result = joined.filter(v1 => {
84.         val optional = v1._2._2;
85.         if (optional.isDefined && optional.get) {
86.           false
87.         } else {
88.           true
89.         }
90.
91.       }).map(_._2._1) //test
92.
93.       result
94.
95.     })
96.
97.     filteredadClickedStreaming.print()
98.
99.     /**
      * 第四步：接下来就像对于 RDD 编程一样，基于 DStream 进行编程！！！原因是 DStream
      * 是 RDD 产生的模板（或者说类），在 Spark Streaming 具体发生计算前，其实质是把
      * 每个 Batch 的 DStream 的操作翻译成为对 RDD 的操作
      * 对初始的 DStream 进行 Transformation 级别的处理，如通过 map、filter 等高
      * 阶函数等的编程，进行具体的数据计算
100.      * 广告点击的基本数据格式：timestamp、ip、userID、adID、province、city
101.      */
102.
103.       val pairs = filteredadClickedStreaming.map(t => {
104.         val splited = t._2.split("\t")
105.         val timestamp = splited(0) // yyyy-MM-dd
106.         val ip = splited(1)
107.         val userID = splited(2)
108.         val adID = splited(3)
109.         val province = splited(4)
110.         val city = splited(5)
111.         val clickedRecord = timestamp + "_" + ip + "_" + userID + "_" +
          adID + "_" + province + "_" + city
112.         (clickedRecord, 1L)
113.       })
```

```
114.
115.         /**
116.          * 第四步：对初始的 DStream 进行 Transformation 级别的处理，如通过 map、
             * filter 等高阶函数等的编程，进行具体的数据计算
117.          * 计算每个 Batch Duration 中每个 User 的广告点击量
118.          */
119.
120.         val adClickedUsers = pairs.reduceByKey(_ + _)
121.
122.         /**
123.          * 计算出什么叫有效的点击
124.          * ① 复杂化的一般都是采用机器学习训练好模型直接在线进行过滤
125.          * ② 简单的？可以通过一个 Batch Duration 中的点击次数来判断是不是非法广告
             * 点击，但是实际上，非法广告点击程序会尽可能模拟真实的广告点击行为，所以通
             * 过一个 Batch 来判断是不完整的，我们需要对例如一天（也可以是每个小时）的数
             * 据进行判断
126.          * ③ 比在线机器学习退而求其次的做法如下：例如，一段时间内，同一个 IP（MAC 地
             * 址）有多个用户账号访问
127.          * 例如，可以统一一天内一个用户点击广告的次数，如果一天点击同样的广告操作 50
             * 次，就列入黑名单
             *
128.          * 黑名单有一个重要的特征：动态生成。所以，每个 Batch Duration 都要考虑是
             * 否有新的黑名单加入，此时黑名单需要存储起来，具体存储在什么地方呢，存储在
             * DB/Redis 中即可
129.          * 例如，邮件系统中的“黑名单”，可以采用 Spark Streaming 不断地监控每个用
             * 户的操作，如果用户发送邮件的频率超过设定的值，可以暂时把用户列入“黑名单”，
             * 从而阻止用户过度频繁地发送邮件
130.          */
131.
132.         val filteredClickInBatch = adClickedUsers.filter(v1 => {
133.           if (1 < v1._2) { //更新黑名单的数据表
134.             false
135.           } else {
136.             true
137.           }
138.         })
139.
140.         /**
              * 此处的 print 并不会直接发出 Job 的执行，因为现在的一切都在 Spark Streaming 框
              * 架的控制下，对于 SparkStreaming 而言，具体是否触发真正的 Job 运行是基
              * 间隔的于设置的 Duration 时间
141.           * 注意，Spark Streaming 应用程序要想执行具体的 Job，对 DStream 就必须有
              * output Stream 操作，output Stream 由很多类型的函数触发，如
              * printsaveAsTextFile、saveAsHadoopFiles 等，最重要的一个方法是
              * foraeachRDD，因为 Spark Streaming 处理的结果一般都会放在 Redis、
              * DB、DashBoard 等上面，foreachRDD 主要用来完成这些功能，而且可以
              * 随意地自定义具体数据到底放在哪里
142.           */
143.
144.         filteredClickInBatch.foreachRDD(rdd => {
```

```
145.          if (rdd.isEmpty()) {}
146.          rdd.foreachPartition(partition => {
147.            /**
148.             * 这里我们使用数据库连接池高效读写数据库的方式把数据写入数据库 MySQL;
149.             * 由于传入的参数是一个 Iterator 类型的集合，所以为了更加高效地操作，我
               * 们需要批量处理
150.             * 例如，一次性插入 1000 条 Record，使用 insertBatch 或者 updateBatch
               * 类型的操作;
151.             * 插入的用户信息可以只包含 timestamp、ip、userID、adID、province、city
152.             * 这里有一个问题：可能出现两条记录的 Key 是一样的情况，此时就需要更新累
               * 加操作
153.             */
154.            val userAdClickedList = ListBuffer[UserAdClicked]()
155.            while (partition.hasNext) {
156.              val record = partition.next()
157.              val splited = record._1.split("_")
158.              val userClicked = new UserAdClicked()
159.              userClicked.timestamp = splited(0)
160.              userClicked.ip = splited(1)
161.              userClicked.userID = splited(2)
162.              userClicked.adID = splited(3)
163.              userClicked.province = splited(4)
164.              userClicked.city = splited(5)
165.              userAdClickedList += userClicked
166.            }
167.
168.            val inserting = ListBuffer[UserAdClicked]()
169.            val updating = ListBuffer[UserAdClicked]()
170.            val jdbcWrapper = JDBCWrapper.getInstance()
171.
172.            //点击
173.            //表的字段：timestamp、ip、userID、adID、province、city、
                //clickedCount
174.
175.            for (clicked <- userAdClickedList) {
176.              def clickedquerycallBack(result: ResultSet): Unit = {
177.                while (result.next()) {
178.                  if ((result.getRow - 1) != 0) {
179.                    val count = result.getLong(1)
180.                    clicked.clickedCount = count
181.                    updating += clicked
182.
183.                  } else {
184.                    clicked.clickedCount = 0L
185.                    inserting += clicked
186.
187.                  }
188.                }
189.              }
190.
191.              jdbcWrapper.doQuery("SELECT count(1) FROM adclicked WHERE "
192.                + " timestamp = ? AND userID = ? AND adID = ?",
                  Array(clicked.timestamp,
193.                clicked.userID, clicked.adID), clickedquerycallBack)
194.            }
195.
196.            //广告点击
197.            //表的字段：timestamp、ip、userID、adID、province、city、clickedCount
```

```
198.
199.            val insertParametersList = ListBuffer[paramsList]()
200.            for (inserRecord <- inserting) {
201.              val paramsListTmp = new paramsList()
202.              paramsListTmp.params1 = inserRecord.timestamp
203.              paramsListTmp.params2 = inserRecord.ip
204.              paramsListTmp.params3 = inserRecord.userID
205.              paramsListTmp.params4 = inserRecord.adID
206.              paramsListTmp.params5 = inserRecord.province
207.              paramsListTmp.params6 = inserRecord.city
208.              paramsListTmp.params10_Long = inserRecord.clickedCount
209.              paramsListTmp.params_Type = "adclickedInsert"
210.              insertParametersList += paramsListTmp
211.            }
212.
213.            jdbcWrapper.doBatch("INSERT INTO adclicked VALUES
            (?,?,?,?,?,?,?)", insertParametersList)
214.
215.            //广告点击
216.            //表的字段：timestamp、ip、userID、adID、province、city、
            //clickedCount
217.            val updateParametersList = ListBuffer[paramsList]()
218.            for (updateRecord <- updating) {
219.              val paramsListTmp = new paramsList()
220.              paramsListTmp.params1 = updateRecord.timestamp
221.              paramsListTmp.params2 = updateRecord.ip
222.              paramsListTmp.params3 = updateRecord.userID
223.              paramsListTmp.params4 = updateRecord.adID
224.              paramsListTmp.params5 = updateRecord.province
225.              paramsListTmp.params6 = updateRecord.city
226.              paramsListTmp.params10_Long = updateRecord.clickedCount
227.              paramsListTmp.params_Type = "adclickedUpdate"
228.              updateParametersList += paramsListTmp
229.            }
230.
231.            jdbcWrapper.doBatch("UPDATE adclicked set clickedCount = ? WHERE"
232.              + " timestamp = ? AND ip = ? AND userID = ? AND adID = ? AND
              province = ? "+ "AND city = ? ", updateParametersList)
233.
234.          })
235.        })
236.
237.        val blackListBasedOnHistory = filteredClickInBatch.filter(v1=>{
238.          val splited = v1._1.split("_")
239.          val date = splited(0)
240.          val userID = splited(2)
241.          val adID = splited(3)
242.          /**
243.            * 接下来根据 date、userID、adID 等条件去查询用户点击广告的数据表，获得
            * 总的点击次数
244.            * 这时基于点击次数判断是否属于黑名单点击
245.            */
246.          val clickedCountTotalToday = 81
247.          if (clickedCountTotalToday > 1) {
248.            true
249.          } else {
250.            false
251.          }
```

```
252.         })
253.         /**
              * 必须对黑名单的整个RDD进行去重操作！！！return v1._1.split("_")[2]
254.          */
255.         val blackListuserIDtBasedOnHistory = blackListBasedOnHistory.
             map(_._1.split("_")(2))
256.
257.         val blackListUniqueuserIDtBasedOnHistory = blackListuser
             IDtBasedOnHistory.transform(_.distinct())
258.         //下一步写入黑名单数据表中
259.         blackListUniqueuserIDtBasedOnHistory.foreachRDD(rdd => {
260.           /**
261.            * 这里使用数据库连接池高效读写数据库的方式把数据写入数据库MySQL；
262.            * 由于传入的参数是一个Iterator类型的集合，所以为了更加高效地操作，
               * 需要批量处理
263.            * 例如，一次性插入1000条Record，使用insertBatch或者updateBatch
               * 类型的操作；
264.            * 插入的用户信息可以只包含useID，此时直接插入黑名单数据表即可
265.            */
266.           rdd.foreachPartition(t => {
267.             val blackList = ListBuffer[paramsList]()
268.             while (t.hasNext) {
269.               //blackList += Array(t.next())
270.               val paramsListTmp = new paramsList()
271.               paramsListTmp.params1 = t.next()
272.
273.               paramsListTmp.params_Type = "blacklisttableInsert"
274.               blackList += paramsListTmp
275.
276.             }
277.             val jdbcWrapper = JDBCWrapper.getInstance()
278.             jdbcWrapper.doBatch("INSERT INTO blacklisttable VALUES (?) ",
                 blackList)
279.           })
280.         })
281.         /**
             * 广告点击累计动态更新，每个updateStateByKey都会在Batch Duration的
             * 时间间隔的基础上进行更高点击次数的更新，更新之后，一般都会持久化到外部存储
             * 设备上，这里存储到MySQL数据库中
282.          */
283.         val filteredadClickedStreamingmappair = filteredadClicked
             Streaming.map(t => {
284.           val splited = t._2.split("\t")
285.           val timestamp = splited(0) // yyyy-MM-dd
286.           val ip = splited(1)
287.           val userID = splited(2)
288.           val adID = splited(3)
289.           val province = splited(4)
290.           val city = splited(5)
291.
292.           val clickedRecord = timestamp + "_" + adID + "_" + province + "_" + city
293.
294.           (clickedRecord, 1L)
295.
296.         })
297.         val updateFunc = (values: Seq[Long], state: Option[Long]) => {
298.
```

```
299.          Some[Long](values.sum + state.getOrElse(0L))
300.
301.        }
302.
303.        val updateStateByKeyDStream = filteredadClickedStreamingmappair.
            updateStateByKey(updateFunc)
304.
305.        updateStateByKeyDStream.foreachRDD(rdd => {
306.          rdd.foreachPartition(partition => {
307.            /**
308.             * 这里使用数据库连接池高效读写数据库的方式把数据写入数据库 MySQL
309.             * 由于传入的参数是一个 Iterator 类型的集合，所以为了更加高效地操作，
               * 需要批量处理
310.             * 例如，一次性插入 1000 条 Record，使用 insertBatch 或者 updateBatch
               * 类型的操作
311.             * 插入的用户信息可以只包含 timestamp、adID、province、city
312.             * 这里面有一个问题：可能出现两条记录的 Key 是一样的情况，此时就需要更新
               * 累加操作
313.             */
314.
315.            val adClickedList = ListBuffer[AdClicked]()
316.            while (partition.hasNext) {
317.              val record = partition.next()
318.              val splited = record._1.split("_")
319.              val adClicked = new AdClicked()
320.              adClicked.timestamp = splited(0)
321.              adClicked.adID = splited(1)
322.              adClicked.province = splited(2)
323.              adClicked.city = splited(3)
324.              adClicked.clickedCount = record._2
325.              adClickedList += adClicked
326.            }
327.
328.            val inserting = ListBuffer[AdClicked]()
329.            val updating = ListBuffer[AdClicked]()
330.            val jdbcWrapper = JDBCWrapper.getInstance()
331.            //点击
332.            //表的字段：timestamp、ip、userID、adID、province、city、
                //clickedCount
333.            for (clicked <- adClickedList) {
334.
335.              def adClickedquerycallBack(result: ResultSet): Unit = {
336.                while (result.next()) {
337.                  if ((result.getRow - 1) != 0) {
338.                    val count = result.getLong(1)
339.                    clicked.clickedCount = count
340.                    updating += clicked
341.                  } else {
342.                    // clicked.clickedCount = 0L
343.                    inserting += clicked
344.                  }
345.                }
346.              }
347.              jdbcWrapper.doQuery(
348.                "SELECT count(1) FROM adclickedcount WHERE "
349.                  +"timestamp = ? AND adID = ? AND province = ? AND city =?",
350.                Array(clicked.timestamp, clicked.adID, clicked.province,
351.                  clicked.city), adClickedquerycallBack)
```

```
352.          }
353.          //点击
354.          //表的字段：timestamp、ip、userID、adID、province、city、
              //clickedCount
355.          val insertParametersList = ListBuffer[paramsList]()
356.          for (inserRecord <- inserting) {
357.            val paramsListTmp = new paramsList()
358.            paramsListTmp.params1 = inserRecord.timestamp
359.            paramsListTmp.params2 = inserRecord.adID
360.            paramsListTmp.params3 = inserRecord.province
361.            paramsListTmp.params4 = inserRecord.city
362.            paramsListTmp.params10_Long = inserRecord.clickedCount
363.            paramsListTmp.params_Type = "adclickedcountInsert"
364.            insertParametersList += paramsListTmp
365.
366.          }
367.          jdbcWrapper.doBatch("INSERT INTO adclickedcount VALUES
              (?,?,?,?,?)", insertParametersList)
368.          //点击
369.          //表的字段：timestamp、ip、userID、adID、province、city、
              //clickedCount
370.          val updateParametersList = ListBuffer[paramsList]()
371.          for (updateRecord <- updating) {
372.            val paramsListTmp = new paramsList()
373.            paramsListTmp.params1 = updateRecord.timestamp
374.            paramsListTmp.params2 = updateRecord.adID
375.            paramsListTmp.params3 = updateRecord.province
376.            paramsListTmp.params4 = updateRecord.city
377.            paramsListTmp.params10_Long = updateRecord.clickedCount
378.            paramsListTmp.params_Type = "adclickedUpdate"
379.            updateParametersList += paramsListTmp
380.
381.          }
382.          jdbcWrapper.doBatch(
383.            "UPDATE adclickedcount set clickedCount = ? WHERE "
384.              + " timestamp = ?AND adID= ? AND province=? AND city = ? ",
385.            updateParametersList)
386.
387.        })
388.      })
389.      /**
           * 对广告点击进行 TopN 的计算，计算出每天每个省份的 Top5 排名的广告；因为
           * 直接对 RDD 进行操作，所以使用了 transform 算子
390.       */
391.      val updateStateByKeyDStreamrdd = updateStateByKeyDStream.
          transform(rdd => {
392.
393.        val rowRDD = rdd.map(t => {
394.
395.          val splited = t._1.split("_")
396.          val timestamp = "2016-09-03" //yyyy-MM-dd
397.          val adID = splited(1)
398.          val province = splited(2)
399.          val clickedRecord = timestamp + "_" + adID + "_" + province
400.          (clickedRecord, t._2)
401.
402.        }).reduceByKey(_ + _).map(v1 => {
403.          val splited = v1._1.split("_")
```

```
404.            val timestamp = "2016-09-03" // yyyy-MM-dd
405.            val adID = splited(1)
406.            val province = splited(2)
407.            Row(timestamp, adID, province, v1._2)
408.
409.          })
410.          val structType = new StructType()
411.            .add("timstamp", StringType)
412.            .add("adID", StringType)
413.            .add("province", StringType)
414.            .add("clickedCount", LongType)
415.
416.          val hiveContext = new HiveContext(rdd.sparkContext)
417.          val df = hiveContext.createDataFrame(rowRDD, structType)
418.          df.registerTempTable("topNTableSource")
419.          val sqlText= "SELECT timstamp,adID,province,clickedCount FROM "+
420.            "(SELECT timstamp,adID,province,clickedCount, row_number() "+
421.            " OVER ( PARTITION BY province ORDER BY clickedCount DESC) rank"+
422.            " FROM topNTableSource ) subquery " + " WHERE rank <= 5 "
423.          val result = hiveContext.sql(sqlText)
424.          result.rdd
425.
426.        })
427.
428.        updateStateByKeyDStreamrdd.foreachRDD(rdd => {
429.          rdd.foreachPartition(t => {
430.            val adProvinceTopN = ListBuffer[AdProvinceTopN]()
431.            while (t.hasNext) {
432.              val row = t.next()
433.              val item = new AdProvinceTopN();
434.              item.timestamp = row.getString(0)
435.              item.adID = row.getString(1)
436.              item.province = row.getString(2)
437.              item.clickedCount = row.getLong(3)
438.              adProvinceTopN += item
439.            }
440.            val jdbcWrapper = JDBCWrapper.getInstance()
441.            val set = new mutable.HashSet[String]()
442.            for (itemTopn <- adProvinceTopN) {
443.              set += itemTopn.timestamp + "_" + itemTopn.province
444.            }
445.            //点击
446.            //表的字段：timestamp、ip、userID、adID、province、city、clickedCount
447.            val deleteParametersList = ListBuffer[paramsList]()
448.            for (deleteRecord <- set) {
449.              val splited = deleteRecord.split("_")
450.              val paramsListTmp = new paramsList()
451.              paramsListTmp.params1 = splited(0)
452.              paramsListTmp.params2 = splited(1)
453.              paramsListTmp.params_Type = "adprovincetopnDelete"
454.              deleteParametersList += paramsListTmp
455.
456.            }
457.            jdbcWrapper.doBatch("DELETE FROM adprovincetopn WHERE
            timestamp = ? AND province = ?",
458.              deleteParametersList);
459.            val insertParametersList = ListBuffer[paramsList]()
460.            for (updateRecord <- adProvinceTopN) {
461.              val paramsListTmp = new paramsList()
```

```
462.             paramsListTmp.params1 = updateRecord.timestamp
463.             paramsListTmp.params2 = updateRecord.adID
464.             paramsListTmp.params3 = updateRecord.province
465.             paramsListTmp.params10_Long = updateRecord.clickedCount
466.             paramsListTmp.params_Type = "adprovincetopnInsert"
467.             insertParametersList += paramsListTmp
468.
469.           }
470.           jdbcWrapper.doBatch("INSERT INTO adprovincetopn VALUES
               (?,?,?,?) ", insertParametersList)
471.
472.         })
473.       })
474.       /**
        * 计算过去半个小时内广告点击的趋势，用户广告点击信息可以只包含timestamp、
        * ip、userID、adID、province、city
475.        */
476.       val filteredadClickedStreamingpair = filteredadClickedStreaming.
       map(t => {
477.         val splited = t._2.split("\t")
478.         val adID = splited(3)
479.         val time = splited(0) //Todo：后续需要重构代码，实现时间戳和分钟的
                                   //转换提取，此处需要提取出该广告的点击分钟单位
480.         (time + "_" + adID, 1L)
481.
482.       })
483.       filteredadClickedStreamingpair.reduceByKeyAndWindow(_ + _, _ - _,
       Seconds(1800), Seconds(60))
484.         .foreachRDD(rdd => {
485.           rdd.foreachPartition(partition => {
486.             val adTrend = ListBuffer[AdTrendStat]()
487.             while (partition.hasNext) {
488.               val record = partition.next()
489.               val splited = record._1.split("_")
490.               val time = splited(0)
491.               val adID = splited(1)
492.               val clickedCount = record._2
493.               /**
494.                * 在插入数据到数据库的时候具体需要哪些字段？time、adID、
                * clickedCount；
495.                * 而通过J2EE技术进行趋势绘图的时候肯定是需要年、月、日、时、分
                * 这些维度的，所以我们在这里需要年、月、日、小时、分钟这些时间维度
496.                */
497.               val adTrendStat = new AdTrendStat()
498.               adTrendStat.adID = adID
499.               adTrendStat.clickedCount = clickedCount
500.               adTrendStat._date = time      //Todo:获取年、月、日
501.               adTrendStat._hour = time      //Todo:获取小时
502.               adTrendStat._minute = time    //Todo:获取分钟
503.               adTrend += adTrendStat
504.             }
505.             val inserting = ListBuffer[AdTrendStat]()
506.             val updating = ListBuffer[AdTrendStat]()
507.             val jdbcWrapper = JDBCWrapper.getInstance()
508.             //广告点击
509.             //表的字段：date、hour、minute、adID、clickedCount
510.             for (clicked <- adTrend) {
```

```
511.              val adTrendCountHistory = new AdTrendCountHistory()
512.
513.              def adTrendquerycallBack(result: ResultSet): Unit = {
514.                while (result.next()) {
515.                  if ((result.getRow - 1) != 0) {
516.                    val count = result.getLong(1)
517.                    adTrendCountHistory.clickedCountHistory = count
518.                    updating += clicked
519.                  } else {
520.                    inserting += clicked
521.                  }
522.                }
523.              }
524.
525.              jdbcWrapper.doQuery("SELECT count(1) FROM adclickedtrend
              WHERE "+ " date = ? AND hour = ? AND minute = ? AND adID = ?",
              Array(clicked._date,clicked._hour,clicked._minute,clicked.adID),
              adTrendquerycallBack)
526.
527.            }
528.            val insertParametersList = ListBuffer[paramsList]()
529.
530.            for (inserRecord <- inserting) {
531.
532.              val paramsListTmp = new paramsList()
533.              paramsListTmp.params1 = inserRecord._date
534.              paramsListTmp.params2 = inserRecord._hour
535.              paramsListTmp.params3 = inserRecord._minute
536.              paramsListTmp.params4 = inserRecord.adID
537.              paramsListTmp.params10_Long = inserRecord.clickedCount
538.              paramsListTmp.params_Type = "adclickedtrendInsert"
539.              insertParametersList += paramsListTmp
540.
541.            }
542.            jdbcWrapper.doBatch("INSERT INTO adclickedtrend
            VALUES(?,?,?,?,?)", insertParametersList)
543.
544.            val updateParametersList = ListBuffer[paramsList]()
545.            for (updateRecord <- updating) {
546.
547.              val paramsListTmp = new paramsList()
548.              paramsListTmp.params1 = updateRecord._date
549.              paramsListTmp.params2 = updateRecord._hour
550.              paramsListTmp.params3 = updateRecord._minute
551.              paramsListTmp.params4 = updateRecord.adID
552.              paramsListTmp.params10_Long = updateRecord.clickedCount
553.              paramsListTmp.params_Type = "adclickedtrendUpdate"
554.              updateParametersList += paramsListTmp
555.            }
556.            jdbcWrapper.doBatch("UPDATE adclickedtrend set clickedCount
            = ? WHERE " + " date = ? AND hour = ? AND minute = ? AND adID
            = ?",updateParametersList)
557.
558.          })
559.
560.        })
561.
```

```
562.        ssc.start()
563.        ssc.awaitTermination()
564.
565.      }
566.
567.      object JDBCWrapper {
568.
569.        private var jdbcInstance: JDBCWrapper = _
570.
571.        def getInstance(): JDBCWrapper = {
572.          synchronized {
573.
574.            if (jdbcInstance == null) {
575.
576.              jdbcInstance = new JDBCWrapper()
577.            }
578.          }
579.          jdbcInstance
580.        }
581.
582.      }
583.
584.      class JDBCWrapper {
585.
586.        val dbConnectionPool = new LinkedBlockingQueue[Connection]()
587.        try {
588.          Class.forName("com.mysql.jdbc.Driver")
589.        } catch {
590.          case e: ClassNotFoundException => e.printStackTrace()
591.        }
592.
593.        for (i <- 1 to 10) {
594.          try {
595.            val conn = DriverManager.getConnection("jdbc:mysql://Master:
            3306/sparkstreaming", "root","root");
596.            dbConnectionPool.put(conn);
597.          } catch {
598.            case e: Exception => e.printStackTrace()
599.          }
600.        }
601.
602.        def getConnection(): Connection = synchronized {
603.          while (0 == dbConnectionPool.size()) {
604.            try {
605.              Thread.sleep(20);
606.            } catch {
607.              case e: InterruptedException => e.printStackTrace()
608.            }
609.          }
610.          dbConnectionPool.poll();
611.        }
612.
613.        def doBatch(sqlText: String, paramsList: ListBuffer[paramsList]):
        Array[Int] = {
614.
615.          val conn: Connection = getConnection();
616.          var preparedStatement: PreparedStatement = null;
617.          val result: Array[Int] = null;
618.          try {
```

```
619.          conn.setAutoCommit(false);
620.          preparedStatement = conn.prepareStatement(sqlText)
621.          for (parameters <- paramsList) {
622.
623.            parameters.params_Type match {
624.
625.              case "adclickedInsert" => {
626.                println("adclickedInsert")
627.
628.                preparedStatement.setObject(1, parameters.params1)
629.                preparedStatement.setObject(2, parameters.params2)
630.                preparedStatement.setObject(3, parameters.params3)
631.                preparedStatement.setObject(4, parameters.params4)
632.                preparedStatement.setObject(5, parameters.params5)
633.                preparedStatement.setObject(6, parameters.params6)
634.                preparedStatement.setObject(7, parameters.params10_Long)
635.              }
636.
637.              case "blacklisttableInsert" => {
638.                println("blacklisttableInsert")
639.                preparedStatement.setObject(1, parameters.params1)
640.              }
641.              case "adclickedcountInsert" => {
642.                println("adclickedcountInsert")
643.                preparedStatement.setObject(1, parameters.params1)
644.                preparedStatement.setObject(2, parameters.params2)
645.                preparedStatement.setObject(3, parameters.params3)
646.                preparedStatement.setObject(4, parameters.params4)
647.                preparedStatement.setObject(5, parameters.params10_Long)
648.              }
649.              case "adprovincetopnInsert" => {
650.                println("adprovincetopnInsert")
651.                preparedStatement.setObject(1, parameters.params1)
652.                preparedStatement.setObject(2, parameters.params2)
653.                preparedStatement.setObject(3, parameters.params3)
654.                preparedStatement.setObject(4, parameters.params10_Long)
655.              }
656.              case "adclickedtrendInsert" => {
657.                println("adclickedtrendInsert")
658.                preparedStatement.setObject(1, parameters.params1)
659.                preparedStatement.setObject(2, parameters.params2)
660.                preparedStatement.setObject(3, parameters.params3)
661.                preparedStatement.setObject(4, parameters.params4)
662.                preparedStatement.setObject(5, parameters.params10_Long)
663.              }
664.              case "adclickedUpdate" => {
665.                println("adclickedUpdate")
666.                preparedStatement.setObject(1, parameters.params10_Long)
667.                preparedStatement.setObject(2, parameters.params1)
668.                preparedStatement.setObject(3, parameters.params2)
669.                preparedStatement.setObject(4, parameters.params3)
670.                preparedStatement.setObject(5, parameters.params4)
671.                preparedStatement.setObject(6, parameters.params5)
672.                preparedStatement.setObject(7, parameters.params6)
673.
674.              }
675.
676.              case "blacklisttableUpdate" => {
677.                println("blacklisttableUpdate")
678.                preparedStatement.setObject(1, parameters.params1)
```

```
679.            }
680.            case "adclickedcountUpdate" => {
681.              println("adclickedcountUpdate")
682.              preparedStatement.setObject(1, parameters.params10_Long)
683.              preparedStatement.setObject(2, parameters.params1)
684.              preparedStatement.setObject(3, parameters.params2)
685.              preparedStatement.setObject(4, parameters.params3)
686.              preparedStatement.setObject(5, parameters.params4)
687.
688.            }
689.            case "adprovincetopnUpdate" => {
690.              println("adprovincetopnUpdate")
691.              preparedStatement.setObject(1, parameters.params10_Long)
692.              preparedStatement.setObject(2, parameters.params1)
693.              preparedStatement.setObject(3, parameters.params2)
694.              preparedStatement.setObject(4, parameters.params3)
695.
696.            }
697.
698.            case "adprovincetopnDelete" => {
699.              println("adprovincetopnDelete")
700.
701.              preparedStatement.setObject(1, parameters.params1)
702.              preparedStatement.setObject(2, parameters.params2)
703.
704.            }
705.
706.            case "adclickedtrendUpdate" => {
707.              println("adclickedtrendUpdate")
708.              preparedStatement.setObject(1, parameters.params10_Long)
709.              preparedStatement.setObject(2, parameters.params1)
710.              preparedStatement.setObject(3, parameters.params2)
711.              preparedStatement.setObject(4, parameters.params3)
712.              preparedStatement.setObject(5, parameters.params4)
713.            }
714.
715.          }
716.
717.          preparedStatement.addBatch();
718.        }
719.
720.        val result = preparedStatement.executeBatch();
721.
722.        conn.commit();
723.      } catch {
724.        //待办事项：自动生成 catch 块
725.        case e: Exception => e.printStackTrace()
726.      } finally {
727.        if (preparedStatement != null) {
728.          try {
729.            preparedStatement.close();
730.          } catch {
731.            //待办事项：自动生成 catch 块
732.            case e: SQLException => e.printStackTrace()
733.          }
734.        }
735.
736.        if (conn != null) {
737.          try {
```

```
738.            dbConnectionPool.put(conn);
739.          } catch {
740.            //待办事项：自动生成 catch 块
741.            case e: InterruptedException => e.printStackTrace()
742.          }
743.        }
744.      }
745.
746.      result;
747.    }
748.    def doQuery(sqlText: String, paramsList: Array[_], callBack:
    ResultSet => Unit) {
749.
750.      val conn: Connection = getConnection();
751.      var preparedStatement: PreparedStatement = null;
752.      var result: ResultSet = null;
753.      try {
754.        preparedStatement = conn.prepareStatement(sqlText)
755.        if (paramsList != null) {
756.
757.          for (i <- 0 to paramsList.length - 1) {
758.            preparedStatement.setObject(i + 1, paramsList(i))
759.
760.          }
761.        }
762.        result = preparedStatement.executeQuery()
763.        callBack(result)
764.
765.      } catch {
766.        //待办事项：自动生成 catch 块
767.        case e: Exception => e.printStackTrace()
768.
769.      } finally {
770.        if (preparedStatement != null) {
771.          try {
772.            preparedStatement.close();
773.          } catch {
774.            //待办事项：自动生成 catch 块
775.            case e: SQLException => e.printStackTrace()
776.
777.          }
778.        }
779.
780.        if (conn != null) {
781.          try {
782.            dbConnectionPool.put(conn);
783.          } catch {
784.            //待办事项：自动生成 catch 块
785.            case e: InterruptedException => e.printStackTrace()
786.
787.          }
788.        }
789.      }
790.
791.    }
792.  }
793.
794.  def resultCallBack(result: ResultSet, blackListNames:
  List[String]): Unit = {
```

```
795.
796.      }
797.
798.      class paramsList extends Serializable {
799.        var params1: String = _
800.        var params2: String = _
801.        var params3: String = _
802.        var params4: String = _
803.        var params5: String = _
804.        var params6: String = _
805.        var params7: String = _
806.        var params10_Long: Long = _
807.        var params_Type: String = _
808.        var length: Int = _
809.
810.      }
811.
812.      class UserAdClicked extends Serializable {
813.        var timestamp: String = _
814.        var ip: String = _
815.        var userID: String = _
816.        var adID: String = _
817.        var province: String = _
818.        var city: String = _
819.        var clickedCount: Long = _
820.
821.        override def toString: String = "UserAdClicked [timestamp=" +
822.          timestamp + ", ip=" + ip + ", userID=" + userID + ", adID=" +
823.          adID + ", province=" + province + ", city=" + city + ",
          clickedCount=" +clickedCount + "]";
824.
825.      }
826.
827.      class AdClicked extends Serializable {
828.        var timestamp: String = _
829.        var adID: String = _
830.        var province: String = _
831.        var city: String = _
832.        var clickedCount: Long = _
833.        override def toString: String = "AdClicked [timestamp=" +
        timestamp + ", adID=" + adID + ", province=" + province + ", city="
        + city + ", clickedCount=" +clickedCount + "]"
834.      }
835.
836.      class AdProvinceTopN extends Serializable {
837.        var timestamp: String = _
838.        var adID: String = _
839.        var province: String = _
840.        var clickedCount: Long = _
841.      }
842.
843.      class AdTrendStat extends Serializable {
844.        var _date: String = _
845.        var _hour: String = _
846.        var _minute: String = _
847.        var adID: String = _
848.        var clickedCount: Long = _
849.        override def toString: String = "AdTrendStat [_date=" +
```

```
        _date+",_hour=" + _hour + ", _minute=" + _minute + ", adID=" +
        adID + ", clickedCount=" + clickedCount + "]"
850.     }
851.
852.     class AdTrendCountHistory extends Serializable {
853.       var clickedCountHistory: Long = _.
854.     }
855.   }
856.
```

16.15　本章总结

本章电商广告点击综合案例应用 Spark Streaming+Kafka+Spark SQL+TopN+MySQL 大数据处理技术，是一个综合、全面、实战型的 Spark 实时在线计算的案例。读者如能搭建一套 Spark 分布式集群环境，自己实践一遍案例代码并解决实际运行中遇到的各种问题，可极大地提升读者通过 Spark 解决问题的综合应用能力。

第 17 章　Spark 在通信运营商生产环境中的应用案例

17.1　Spark 在通信运营商融合支付系统日志统计分析中的综合应用案例

本章阐述 Spark 在通信运营商融合支付系统日志统计分析的综合应用案例，通过 Spark 核心计算框架及 Splunk 大数据 Web 展示系统在融合支付平台的综合应用，统计分析通信运营商生产环境中融合支付系统日志中调用外部系统接口的返回码、系统超时等日志数据。

17.1.1　融合支付系统日志统计分析综合案例需求分析

本章案例中的通信运营商集团公司是中国特大型国有通信企业，连续多年入选“世界 500 强企业”，主要经营固定电话、移动通信、卫星通信、互联网接入及应用等综合信息服务。中国特大型城市本地公司拥有集团公司内最大的城市电信网络，为超过 2200 万用户提供包括移动通信、宽带互联网接入、信息化应用及固定电话等产品在内的综合信息解决方案，始终保持中国特大型城市通信市场的领先地位。

首先看一下本案例涉及的业务场景：用户登录支付宝客户端查询、缴付通信运营商电信账单，流程示意图如图 17-1 所示。

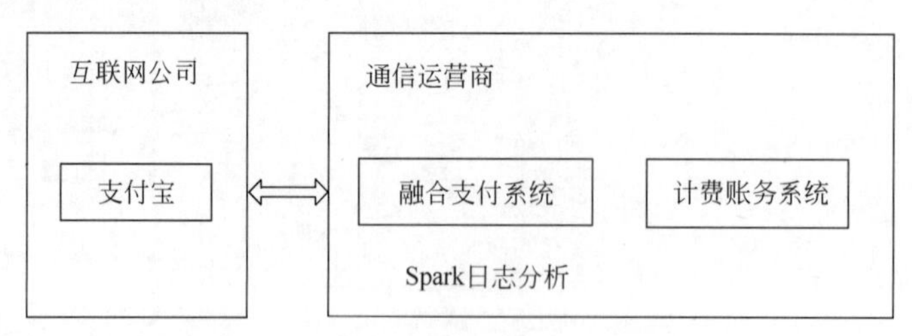

图 17-1　支付宝通信运营商账单支付

用户登录支付宝客户端缴费账单业务流程如下。

（1）用户登录支付宝客户端进行通信运营商账单查询缴费。支付宝系统侧将用户分账序号发送到通信运营商侧融合支付系统进行账单查询。

（2）融合支付平台收到查询请求，对账单相关的参数进行校验。

（3）如校验通过，融合支付平台调用通信运营商内部计费账务系统接口查询账单，若查询失败，则返回失败结果。

（4）通信运营商内部计费账务系统查询账单返回成功，融合支付侧把相应的账单查询结果返回给支付宝系统侧。

（5）支付宝系统侧收到查询结果，根据账单信息发起销账请求。

（6）融合支付平台收到销账请求，对账单相关的参数进行校验。

（7）如校验通过，融合支付平台调用通信运营商内部计费账务系统账单校验接口对分账序号进行确认，若校验失败，则返回失败结果。

（8）内部计费账务系统计费校验返回成功，融合支付平台将订单（会先作订单的交易状态校验，因为同一订单在支付失败后，允许再次支付）信息进行入库。

（9）融合支付平台调用通信运营商内部计费账务系统销账单。销账完毕，融合支付侧将销账结果返回给支付宝系统侧。

需求来源：

Spark 在通信运营商融合支付系统日志统计分析的业务需求来自通信运营商业务部门账务中心：用户登录支付宝客户端查询、缴付通信运营商电信账单，用户有时会遇到账单缴费失败、系统超时的情况，用户须知晓账单缴费失败的原因。整个业务流程经过支付宝公司（支付宝系统）、通信运营商（融合支付系统、计费账务系统），为了提升用户账单缴付体验，通信运营商业务部门召集支付宝和融合支付的技术部门一起共同查证。

对于支付宝公司：支付宝客户端直接面对客户入口，须了解账单支付请求到融合支付系统的失败原因。

- 系统超时：支付宝通过公网连接融合支付平台，从支付宝系统公网连接融合支付平台超时的现象较少，融合支付平台也提供相应的查询接口给支付宝系统，支付宝系统可以对没收到返回消息的账单缴费请求调用融合支付的查询接口，如查询不到记录，则再次重发一次账单销账请求。
- 账单销账失败的原因。从支付宝公司的角度，支付宝侧收到融合支付侧返回的系统返回码，对于某一时刻出现的返回码峰值，需要清楚返回码的原因。如图 17-2 所示，全球领先的独立第三方支付平台支付宝公司，从支付宝的监控系统中发现某一个时刻，支付宝侧到融合支付平台的返回失败码突然上升，支付宝公司不知晓具体的情况，及时报告通信运营商融合支付平台配合协查。

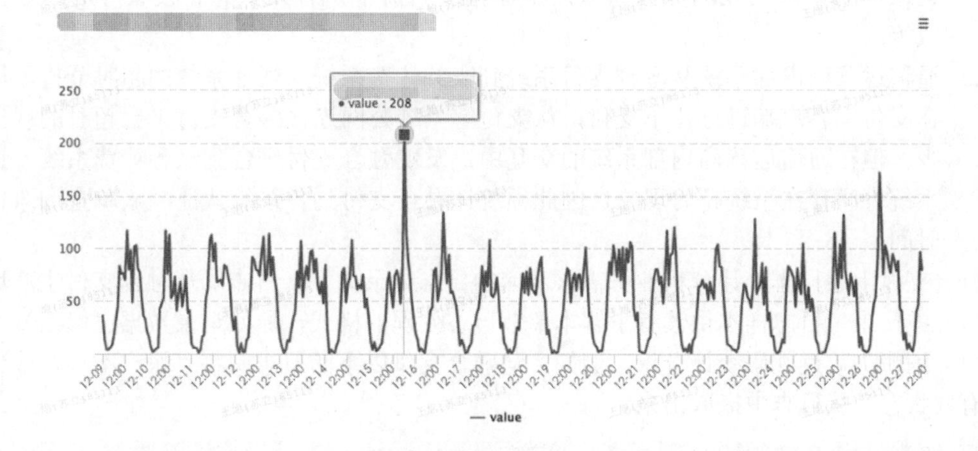

图 17-2　支付宝提供的分析图

对于通信运营商，为配合支付宝公司的账单销账失败码查证，首先在通信运营商内部进行排查：电信账单缴费经过了融合支付系统、计费账务系统，我们需分析通信运营商内部系统销账失败的情况，融合支付系统发送账单销账请求到计费账务系统，如计费系统销账失败，计费账务系统将销账失败的返回码返回给融合支付平台。融合支付系统收到计费系统的返回消息记录，将其失败返回码记录在日志中，在融合支付 Oracle 数据库系统中是没有入库的，因此，在数据库中查询不到相关的记录信息；失败返回码散布在各个日志文件中，每个日志文件大小在 2GB 左右，如人工查证日志，需人工使用第三方的文本切割工具切分日志，一条条地去看，人工难以统计出整个日志的失败返回码情况。因此，本案例就尝试使用 Spark 技术来分析融合支付系统发起请求到内部计费账务系统的日志记录，内部排查计费账务系统返回的错误码情况。

关于日志分析的技术选型，可能也有其他的一些技术能实现相同的功能，我们考虑使用 Spark 技术的原因在于：

- Spark 技术很适合于日志文件的处理。融合支付日志文件打印的日志记录非格式化，日志中包括各种支付交易的记录信息，Spark 读入一行行的数据 Iterator 进行处理，有利于根据正则规则进行日志数据清洗提取。
- Spark 技术具备日志离线、在线实时处理扩展性。目前，Spark 技术应用于融合支付日志系统的离线数据分析。在以后的应用中，相同的业务逻辑代码稍作修改就可以推广到融合支付日志实时在线处理系统中。
- 在电影点评、人员管理、电商分析、NBA 球员分析、电商广告案例的综合应用中积累的 Spark 经验知识，触类旁通就能将 Spark 技术应用到通信运营商的生产环境中，充分发挥 Spark 的作用，根据融合支付日志分析的业务需求实现相应的功能。

17.1.2　融合支付系统日志统计分析数据说明

Spark 在通信运营商生产环境中的应用案例统计分析以下日志数据：

- 配合支付宝公司侧电信销账失败返回码查证的原因分析：在日志中提取销账失败的记录，如销账失败的原因为账户不存在或当前不是有效状态、设备号对应账号为空等。
- 通信运营商内部系统从融合支付系统到账务计费系统、认证系统的超时分析。用户在支付宝客户端进行账单支付，从支付宝系统公网连接融合支付平台超时的现象较少。但在通信运营商内部系统的交互中，发现融合支付平台到账务计费系统、认证系统有时存在系统超时现象，因此须统计融合支付到各系统认证、账单查询缴付的时间。

融合支付上述日志统计分析业务需求对应的日志记录如下。日志均从融合支付生产环境中的日志中摘录（日志样本供读者了解本章节日志处理的格式，原始记录数据已修改）。

（1）销账失败返回码统计分析。例如，从融合支付以下的日志中将“账户不存在或当前不是有效状态”从日志中提取出来。

```
1. 2016-12-20 23:07:27,625 [DEBUG] [SocketUtils.java] : 94 -- Service
   socketReceiveMsgServe() received msg is 000416<?xml version="1.0"
   encoding="UTF-8"?>
```

```
2.  <MsgRequest><Head><RequestSeq>28*db3ba80cb45697</RequestSeq>
    <Command>Out*Invoice</Command><RequestModule>ALIPAY</RequestModule></
    Head><Body><MQAccountNo>*</MQAccountNo><MQStatus>0</MQStatus><MQFromD
    ate>2016-05</MQFromDate><MQToDate>2016-12</MQToDate></Body></MsgReque
    st>
3.  2016-12-20 23:07:27,656 [DEBUG] [MQQueryInvoiceBusiness.java] : 298 --
    流 水 号 :28*3ba80cb45697,queryInvoice  response:<soap:Envelope xmlns:
    soap="http://*/soap/envelope/"><soap:Body><ns2:queryInvoiceResponse
    xmlns:ns2="http://services.*.cn/"><response><head><errCode>00000001</
    errCode><errDesc>com.ctsh.payment.common.exception.ApplicationExcepti
    on: 账户不存在或当前不是有效状态
4.  at
5.  cn.*</errDesc></head></response></ns2:queryInvoiceResponse>
    </soap:Body></soap:Envelope>
6.
7.  2016-12-20 23:07:27,672 [ERROR] [MQQueryInvoiceBusiness.java] : 406 --
    流水号:281*697 MQQueryInvoiceBusiness queryInvoiceNew() webservice 账单余
    额列表查询 失败! errCode:00000001
8.  2016-12-20 23:07:27,672 [DEBUG] [NewZWService.java] : 605 -- ************
    最终响应消息: 000528<?xml version="1.0" encoding="UTF-8"?>
9.  <MsgResponse><Head><RespCode>0001</RespCode><ErrCode>0040</ErrCode>
    <ReplyCommand>Out*Invoice</ReplyCommand><ReplySeq>2811*a80cb45697</Re
    plySeq><ReplyModule></ReplyModule></Head><Body><MQCount/><MQPaymentMo
    de/><MQReseller/><MQResult/><MQInvoiceList/></Body></MsgResponse>
10. 2016-12-20 23:07:27,672 [DEBUG] [SocketUtils.java] : 29 -- SocketUtils
    socketSentMsg()  return  message  >>>>  000138<?xml  version="1.0"
    encoding="UTF-8"?>......
11.
```

融合支付销账失败返回码统计分析（账户不存在或当前不是有效状态）见表 17-1。

表 17-1　融合支付销账失败返回码统计分析（账户不存在或当前不是有效状态）

序号	记　录	字 段 说 明
1	时间戳	融合支付日志文件中记录的时间
2	日志级别	日志文件记录的级别
3	代码类	融合支付调用的 Java 类代码的名称
4	行号	Java 类代码中在行号位置打的日志
5	流水号描述	流水号描述
6	流水号	流水号
7	方法描述	queryInvoice 描述
8	MQ 返回报文	融合支付收到 MQ 计费系统的返回报文消息记录
9	MQ 计费系统返回代码描述	errCode 代码描述
10	MQ 计费系统返回代码值	00000001
11	errDesc 描述	errDesc 描述
12	errDesc 中文返回含义	errDesc 中文返回含义

（2）销账失败返回码统计分析。例如，从融合支付以下的日志中将“设备号对应账号为空”从日志中提取出来。

```
1.  2016-12-10 23:38:36,359 [DEBUG] [SocketUtils.java] : 94 -- Service
    socketReceiveMsgServe() received msg is 000355<?xml version="1.0"
    encoding="UTF-8"?>
```

```
2.   <MsgRequest><Head><RequestSeq>1008*4846</RequestSeq><Command>
     Out*AccountNo</Command><RequestModule>shup</RequestModule></Head><Bod
     y><MQServiceId>5*9</MQServiceId></Body></MsgRequest>
3.   2016-12-10 23:38:36,484 [WARN ] [MQQueryAccountNoBusiness.java] : 278 --
     流 水 号 :1*46,queryAccountNo   response:<soap:Envelope   xmlns:soap
     ="http://*/soap/envelope/"><soap:Body><ns2:queryAccountNoResponse
     xmlns:ns2="http://services.*.cn/"><response><head><errCode>00000001</
     errCode><errDesc>java.lang.Exception: 设备号对应账号为空
4.      at cn.*.cmd.CTSH*Cmd.execute(CTSHQue</errDesc></head></response>
        </ns2:queryAccountNoResponse></soap:Body></soap:Envelope>
5.
6.   2016-12-10 23:38:36,484 [ERROR] [MQQueryAccountNoBusiness.java] : 654 --
     流水号:10*4846 MQQueryAccountNoBusiness queryAccountNo() 账号查询失败!
     errCode:00000001
7.   2016-12-10 23:38:36,484 [DEBUG] [NewZWService.java] : 405 -- *************
     最终响应消息：000491<?xml version="1.0" encoding="UTF-8"?>
8.   ......
```

融合支付销账失败返回码统计分析（设备号对应账号为空）如表 17-2 所示。

表 17-2　融合支付销账失败返回码统计分析（设备号对应账号为空）

序号	记　录	字 段 说 明
1	时间戳	融合支付日志文件中记录的时间
2	日志级别	日志文件记录的级别
3	代码类	融合支付调用的 Java 类代码的名称
4	行号	Java 类代码中在行号位置打的日志
5	流水号描述	流水号描述
6	流水号	流水号
7	方法描述	queryAccountNo 描述
8	MQ 返回报文	融合支付收到 MQ 计费系统的返回报文消息记录
9	MQ 计费系统返回代码描述	errCode 代码描述
10	MQ 计费系统返回代码值	00000001
11	errDesc 描述	errDesc 描述
12	errDesc 中文返回含义	errDesc 中文返回含义

（3）通信运营商内部融合支付平台到 UAM 认证系统超时分析。例如，从融合支付以下的日志中将“去 UAM 认证的业务的时间 225ms”从日志中提取出来。

```
1.    2016-12-25 23:41:40,062 [ERROR] [MQAuthOperateBusiness.java] : 85 -- 去
      UAM 认证的业务的时间: >>>> Auth >>>> [225]ms
2.   2016-12-25 23:41:40,062 [DEBUG] [NewZWService.java] : 205 -- *************
     最终响应消息：000447<?xml version="1.0" encoding="UTF-8"?>
3.   <MsgResponse><Head><RespCode>0001</RespCode><ErrCode>9996</ErrCode>
     <ReplyCommand>Out*Operate</ReplyCommand><ReplySeq>001*524</ReplySeq><
     ReplyModule></ReplyModule></Head></MsgResponse>
4.   2016-12-25 23:41:40,062 [DEBUG] [SocketUtils.java] : 59 --  SocketUtils
     socketSentMsg()   return   message   >>>>   000947<?xml   version="1.0"
     encoding="UTF-8"?>......
```

通信运营商内部融合支付平台到 UAM 认证系统超时分析如表 17-3 所示。

表 17-3　到UAM认证系统超时分析

序　号	记　录	字段说明
1	时间戳	融合支付日志文件中记录的时间
2	日志级别	日志文件记录的级别
3	代码类	融合支付调用的 Java 类代码的名称
4	行号	Java 类代码中在行号位置打的日志
5	日志说明	日志记录的中文描述说明
6	接口方法	融合支付 Java 代码中使用的接口方法
7	时间	融合支付到 UAM 认证的时间

（4）通信运营商内部融合支付平台到账务计费系统 MQ 校验订单业务的超时分析。例如，从融合支付以下的日志中将“MQ 校验订单业务的时间 57 ms”从日志中提取出来。

```
1.  2016-12-05 23:41:45,484 [ERROR] [MQBillCheckBusiness.java] : 233 --
    流水号:2016*418 MQ 校验订单业务的时间: >>>> CheckBill >>>> [57]ms
2.  2016-12-05 23:41:45,484 [DEBUG] [MQXMLParser.java] : 318 -- xmlStrrrr ==
    <?xml version="1.0" encoding="UTF-8" ?><MsgResponse><Head><ErrCode>
    00000000</ErrCode><ErrDesc>Success</ErrDesc><ReplyCommand>CheckBill</
    ReplyCommand><ReplyID>2016*18</ReplyID><ReplyModule>gateway</ReplyMod
    ule></Head><Body><Status>2</Status><BillDate>2016.12.01</BillDate><Bi
    llType>01</BillType><BillAmount>11910</BillAmount><InvoiceNo>1*0</Inv
    oiceNo><Result>校验成功</Result></Body></MsgResponse>
3.  2016-12-05 23:41:45,547 [DEBUG] [MQXMLParser.java] : 139 -- docccc ===
    false
```

通信运营商内部融合支付平台到账务计费系统 MQ 校验订单超时分析如表 17-4 所示。

表 17-4　到账务计费系统MQ校验订单超时分析

序号	记　录	字段说明
1	时间戳	融合支付日志文件中记录的时间
2	日志级别	日志文件记录的级别
3	代码类	融合支付调用的 Java 类代码的名称
4	行号	Java 类代码中在行号位置的日志
5	流水号描述	流水号描述
6	流水号	流水号
7	MQ 校验描述	MQ 校验订单业务描述
8	接口方法	融合支付 Java 代码中使用的接口方法
9	时间	融合支付到 MQ 校验的时间

（5）通信运营商内部融合支付平台到账务计费系统 MQ 账单支付业务的超时分析。例如，从融合支付以下的日志中将“MQ 账单支付业务的时间 507 ms”从日志中提取出来。

```
1.  2016-12-28 23:41:46,047 [ERROR] [MQBillPaymentBusiness.java] : 182 --
    流水号:2016*9 MQ 账单支付业务的时间: >>>> Payment >>>> [507]ms
2.  2016-12-28 23:41:46,047 [DEBUG] [MQXMLParser.java] : 138 -- xmlStrrrr ==
    <?xml version="1.0" encoding="UTF-8" ?><MsgResponse><Head><ErrCode>
    00000000</ErrCode><ErrDesc>Success</ErrDesc><ReplyCommand>Payment</Re
    plyCommand><ReplyID>2016*19</ReplyID><ReplyModule>gateway</ReplyModul
```

```
    e></Head><Body><Result>支付成功</Result></Body></MsgResponse>
3.  2016-12-28 23:41:46,047 [DEBUG] [MQXMLParser.java] : 239 -- docccc ===
    false
```

通信运营商内部融合支付平台到账务计费系统 MQ 账单支付超时分析见表 17-5。

表 17-5　到账务计费系统MQ账单支付超时分析

序号	记　　录	字 段 说 明
1	时间戳	融合支付日志文件中记录的时间
2	日志级别	日志文件记录的级别
3	代码类	融合支付调用的 Java 类代码的名称
4	行号	Java 类代码中在行号位置打的日志
5	流水号描述	流水号描述
6	流水号	流水号
7	MQ 账单支付	MQ 账单支付业务描述
8	接口方法	融合支付 Java 代码中使用的接口方法
9	时间	融合支付到 MQ 校验的时间

17.1.3　融合支付系统日志清洗中 Scala 正则表达式与模式匹配结合的代码实战

本节 Spark 在通信运营商融合支付系统日志统计分析的综合应用案例中首先定义融合支付系统日志清洗的正则表达式规则集，然后根据正则表达式规则集进行模式匹配，将符合融合支付日志模式匹配的结果保存到本地文本文件中。

1．融合支付日志分析正则表达式规则集

融合支付日志分析正则表达式规则集如下。

- ❑ 提取融合支付到 UAM 认证的时间。
- ❑ 提取融合支付到计费账务系统销账单失败返回码（账户不存在）。
- ❑ 提取融合支付到计费账务系统 MQ 校验订单业务的时间。
- ❑ 提取融合支付到计费账务系统 MQ 账单支付业务的时间。
- ❑ 提取融合支付到计费账务系统销账失败返回码（设备号对应账号为空）。

融合支付日志分析正则表达式规则集代码如下：

```
1.        package com.noc.ronghezhifu.cores
2.
3.  /**
      * 融合支付日志正则表达式解析规则集
4.    */
5.
6.
7.  object rhzfLogRegex {
8.    /**
9.      * 正则表达式解析场景：
10.     * 日志分析：提取融合支付到 UAM 的日志
11.     * 日志样本：2017-01-15 17:47:56,242 [ERROR] [MQAuthOperateBusiness.
      * java] : 45 -- 去 UAM 认证的业务的时间：>>>> Auth >>>> [31]ms
12.     */
```

```
13.  val RHZF_UAM_TIME_REGEX =
14.    """^(\S+) (\S+) (\S+) (\S+) : (\S+) -- (\S+) >>>> (\S+) >>>>
     (\S+)](\S+)$""".r
15.  /**
16.    *正则表达式解析场景:
17.    *日志分析: 提取融合支付到计费账务系统销账单失败返回码
18.    *日志样本: 2017-01-15 17:47:58,430 [DEBUG] [MQQueryInvoiceBusiness
    *.java] : 498 -- 流水号:372@@8698,queryInvoice response:<soap:Envelope
    *xmlns:soap="http://@@/soap/envelope/"><soap:Body><ns2:query InvoiceResponse
    *xmlns:ns2="http://@@.sh.cn/"><response><head><errCode>00000001</
    *errCode><errDesc>com.@@.exception.ApplicationException: 账户不存在
    *或当前不是有效状态
19.    */
20.  val RHZF_MQ_QueryInvoice_ErrCode_REGEX =
21.    """^(\S+)(\S+)(\S+)(\S+) : (\S+) -- (\S+):(\S+),(\S+) (\S+) (\S+) (.*)
    <response><head><errCode>([0-9]*)</errCode><errDesc>(.*): (\S+)$""".r
22.
23.  /**
24.    *正则表达式解析场景:
25.    * 1: 日志分析: 提取融合支付到计费账务系统 MQ 校验订单业务的时间
26.    * 日志样本: 2017-01-15 17:41:21,654 [ERROR] [MQBillCheckBusiness
    * .java] : 233 -- 流水号:20170@@636 MQ 校验订单业务的时间: >>>> CheckBill
    * >>>>[554]ms
27.    * 2: 日志分析: 提取融合支付到计费账务系统 MQ 账单支付业务的时间
28.    *日志样本: 2016-12-15 15:14:30,263 [ERROR] [MQBillPaymentBusiness
    *.java] : 182 -- 流水号:2016@@41202 MQ 账单支付业务的时间: >>>> Payment >>>>
    *[423]ms
29.    *
30.    */
31.  val RHZF_MQ_CHECK_ORDER_TIME_REGEX=
32.    """^(\S+) (\S+) (\S+) (\S+) : (\S+) -- (\S+):(\S+) (\S+) >>>> (\S+)
    >>>> (\S+)](\S+)$""".r
33.
34.  /**
35.    *正则表达式解析场景:
36.    *日志分析: 提取融合支付到计费账务系统销账失败返回码
37.    *日志样本: 2016-12-15 21:31:52,015 [WARN ] [MQQueryAccountNoBusiness
    *.java] : 278 -- 流水号:1008@@27,queryAccountNo response:<soap:Envelope
    *xmlns:soap="http://@@/soap/envelope/"><soap:Body><ns2:queryAccount
    *NoResponse  xmlns:ns2="http://@@.sh.cn/"><response><head><errCode>
    *00000001</errCode><errDesc>java.lang.Exception: 设备号对应账号为空
38.    */
39.  val RHZF_MQ_queryAccountNo_ErrCode_REGEX =
40.    """^(\S+) (\S+) (\S+) (\S+) (\S+) : (\S+) -- (\S+):(\S+),(\S+) (\S+)
    (\S+) (.*)<response><head><errCode>([0-9]*)</errCode><errDesc>(.*):
    (\S+)$""".r
41. }
```

2. 融合支付日志分析Spark实战

融合支付日志分析 Spark 实战实现如下。

（1）Spark 上下文初始化。类似于之前各商业案例的 Spark 初始化，配置相应的参数初始化 Spark。

```
1.   var masterUrl = "local[4]"
2.     var dataPath = "G://IMFBigDataSpark2017//IMFSparkAppsnew2017//data//
```

```
       ronghezhifualipaylog//logs20161215//zw_out_core.log.2016-12-15"
3.     if (args.length > 0) {
4.       masterUrl = args(0)
5.     } else if (args.length > 1) {
6.       dataPath = args(1)
7.     }
8.     val sparkConf = new SparkConf().setMaster(masterUrl).setAppName
       ("rhzfAlipayLogAnalysis")
9.     val spark = SparkSession
10.      .builder()
11.      .config(sparkConf)
12.      .getOrCreate()
13.    val sc = spark.sparkContext
```

（2）读入融合支付日志文件中的中文字符的处理。使用 SparkContext 上下文的 hadoopFile 方法，创建生成一个 InputFormat 输入格式的 HadoopRDD，对于文本中的每行记录，Hadoop 的 RecordReader 类重用了相同的可写对象，将为同一对象创建多个引用，直接缓存返回 RDD 或直接传递到聚集 aggregation，或 Shuffle 操作中。如果计划直接使用缓存、排序或聚集操作 Hadoop 可写对象，可以首先使用 map 转换功能。这里我们进行 map 转换，读入的每行数据是 Key-Value 格式，Key 值（line._1）是 Hadoop 文件每行的序号，Value 值（line._2）是读入的每行融合支付的日志数据，将读入融合支付日志的字节码按照 GBK 的方式读取变成字符串，就能顺利地读入融合支付日志中的中文字符。

```
1.  val linesConvertToGbkRDD: RDD[String] = sc.hadoopFile(dataPath, classOf
    [TextInputFormat], classOf[LongWritable], classOf[Text], 4)
2.        .map(line => {
3.          new String(line._2.getBytes, 0, line._2.getLength, "GBK")
4.
5.        })
```

（3）根据融合支付日志分析正则表达式规则集中的各种情况进行模式匹配，如果能匹配，则直接返回融合支付每行的数据（日志记录包括融合支付到账务计费系统的失败返回码记录、各种超时的日志记录）；如果匹配不上，则是正则表达式规则之外的融合支付日志，这些日志不是融合支付需要的，因此就直接返回字符 MatcheIsNull。创建生成 linesRegexd 的 RDD，其数据类型为 RDD[String]，数据格式为经过清洗的融合支付的每行日志记录。

```
1.   val linesRegexd: RDD[String] = linesConvertToGbkRDD.map(rhzfline => {
2.        rhzfline match {
3.          case RHZF_UAM_TIME_REGEX(uamLog1, uamLog2, uamLog3, uamLog4,
            uamLog5, uamLog6, uamLog7, uamLog8, uamLog9)
4.          => rhzfline
5.          case RHZF_MQ_QueryInvoice_ErrCode_REGEX(mqLog1, mqLog2, mqLog3,
            mqLog4, mqLog5, mqLog6, mqLog7, mqLog8, mqLog9, mqLog10, mqLog11,
            mqLog12, mqLog13, mqLog14)
6.          => rhzfline
7.          case RHZF_MQ_CHECK_ORDER_TIME_REGEX(mqTimeLog1, mqTimeLog2,
            mqTimeLog3, mqTimeLog4, mqTimeLog5, mqTimeLog6, mqTimeLog7,
            mqTimeLog8, mqTimeLog9, mqTimeLog10, mqTimeLog11)
8.          => rhzfline
9.         case RHZF_MQ_queryAccountNo_ErrCode_REGEX(mqWarnLog1, mqWarnLog2,
           mqWarnLog3, mqWarnLog4, mqWarnLog5, mqWarnLog6, mqWarnLog7,
           mqWarnLog8, mqWarnLog9, mqWarnLog10, mqWarnLog11, mqWarnLog12,
```

```
        mqWarnLog13, mqWarnLog14 ,mqWarnLog15)
10.         => rhzfline
11.         case _ => "MatcheIsNull"
12.       }
13.     })
```

（4）接下来对清洗后的 linesRegexd 的 RDD，将其中的 MatcheIsNul 字符串进行过滤，过滤完毕后保存为本地文件，为了减少本地文件的数量，可以使用 coalesce 算子将计算的结果合并成一个分区，生成一个文件，后续就可以加载文件到 Splunk 中进行展示。

```
1.    val linedFilterd: RDD[String] = linesRegexd.filter(!_.contains("Matche
      IsNull"))
2.      linedFilterd.coalesce(1).saveAsTextFile("G://IMF*2017//ronghezhifu
        LogAnalysis2017//data//ronghezhifualipaylogresult//zw_*_20170207")
```

在 IDEA 中运行代码，Spark 再对融合支付系统日志进行清洗，保存到本地文件的结果 part-00000 如下：

```
1.  2016-12-15 00:02:55,619 [ERROR] [MQBillPaymentBusiness.java] : 282 --
    流水号:20*60 MQ 账单支付业务的时间: >>>> Payment >>>> [419]ms
2.  2016-12-15 00:02:59,228 [ERROR] [MQBillCheckBusiness.java] : 233 --
    流水号:20*66 MQ 校验订单业务的时间: >>>> CheckBill >>>> [147]ms
3.  2016-12-15 00:02:59,525 [ERROR] [MQBillPaymentBusiness.java] : 282 --
    流水号:20*67 MQ 账单支付业务的时间: >>>> Payment >>>> [287]ms
4.  2016-12-15 00:03:21,417 [DEBUG] [MQQueryInvoiceBusiness.java] : 98 --
    流水号:15c*59de61576,queryInvoice response:<soap:Envelope xmlns:soap=
    "http://*/soap/envelope/"><soap:Body><ns2:queryInvoiceResponse
    xmlns:ns2="http://*.sh.cn/"><response><head><errCode>00000001</errCod
    e><errDesc>*t.common.exception.ApplicationException: 账户不存在或当前不是
    有效状态
5.  2016-12-15 00:03:32,855 [DEBUG] [MQQueryInvoiceBusiness.java] : 98 --
    流水号:1156*4f70,queryInvoice response:<soap:Envelope xmlns:soap=
    "http://*/envelope/"><soap:Body><ns2:queryInvoiceResponse
    xmlns:ns2="http://*.sh.cn/"><response><head><errCode>00000001</errCod
    e><errDesc>*.exception.ApplicationException: 账户不存在或当前不是有效状态
6.  2016-12-15 00:03:33,027 [ERROR] [MQBillCheckBusiness.java] : 233 --
    流水号:1008*43 MQ 校验订单业务的时间: >>>> CheckBill >>>> [110]ms
7.  2016-12-15 00:03:33,183 [ERROR] [MQQueryAccIdBusiness.java] : 53 --
    流水号:1008*5643 MQ 条码号查询分账序号业务的时间: >>>> QueryAccExternalId >>>>
    [247]ms
8.  2016-12-15 00:03:33,417 [ERROR] [MQBillPaymentBusiness.java] : 282 --
    流水号:100*5643 MQ 账单支付业务的时间: >>>> Payment >>>> [109]ms
9.  2016-12-15 00:03:49,215 [ERROR] [MQBillCheckBusiness.java] : 233 --
    流水号:201*861 MQ 校验订单业务的时间: >>>> CheckBill >>>> [178]ms
10. 2016-12-15 00:03:49,512 [ERROR] [MQBillPaymentBusiness.java] : 282 --
    流水号:20*828863 MQ 账单支付业务的时间: >>>> Payment >>>> [287]ms
```

本地文件的目录结构如下：

```
1.   ._SUCCESS.crc
2.  .part-00000.crc
3.  _SUCCESS
4.  part-00000
```

17.1.4　融合支付系统日志在大数据 Splunk 中的可视化展示

Splunk 是第三方的运维智能平台，通过监控和分析客户的点击流、交易数据、信息安全事件和网络活动，获取有价值的数据。作为第三方的日志文件管理工具，其主要功能包括日志聚合、搜索、字段提取、数据格式化、可视化、电子邮件提醒。

Spark 在通信运营商融合支付系统日志统计分析中的综合应用案例，将应用第三方 Splunk 技术进行可视化展示，使用 Web 图表方式直观地展示结果。

1．Splunk工具安装及搜索应用

在 Splunk 官网（https://www.splunk.com）注册新用户，然后登录 Splunk 官网的下载页面（https://www.splunk.com/en_us/download/splunk-enterprise.html），下载 Splunk 的免费版 splunk-6.5.3-36937ad027d4-x64-release。Splunk 免费版可免费使用 60 天，免费期间每天最多可对 500 MB 数据建立索引。本节使用免费版本完成融合支付案例的展示，如图 17-3 所示。

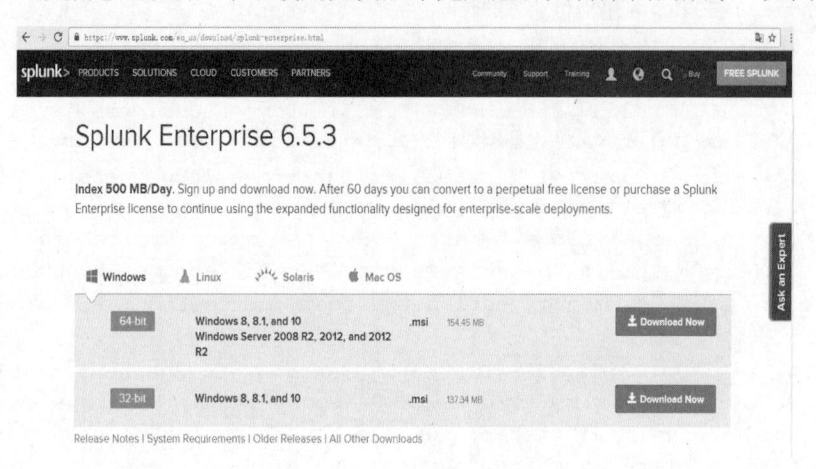

图 17-3　Splunk 下载页面

下载 splunk-6.5.3-36937ad027d4-x64-release 到本地，双击程序，一步步按照提示进行安装，如图 17-4 所示。

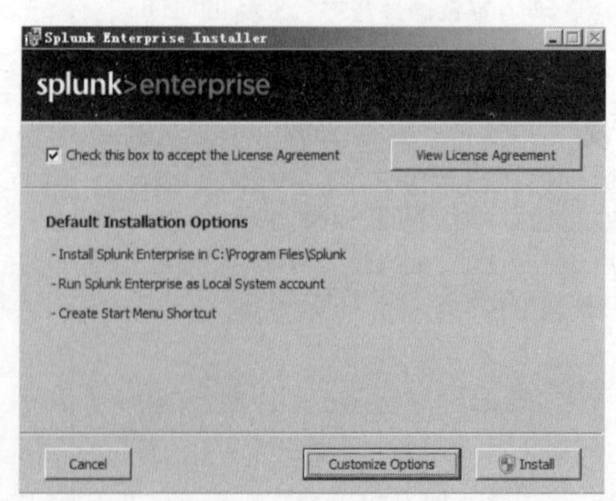

图 17-4　Splunk 安装

Splunk 安装完成以后，在浏览器中输入 http://localhost:8000/，打开 Splunk 登录页面，首次登录须修改密码，输入用户名、密码，进入 Splunk 系统，如图 17-5 所示。

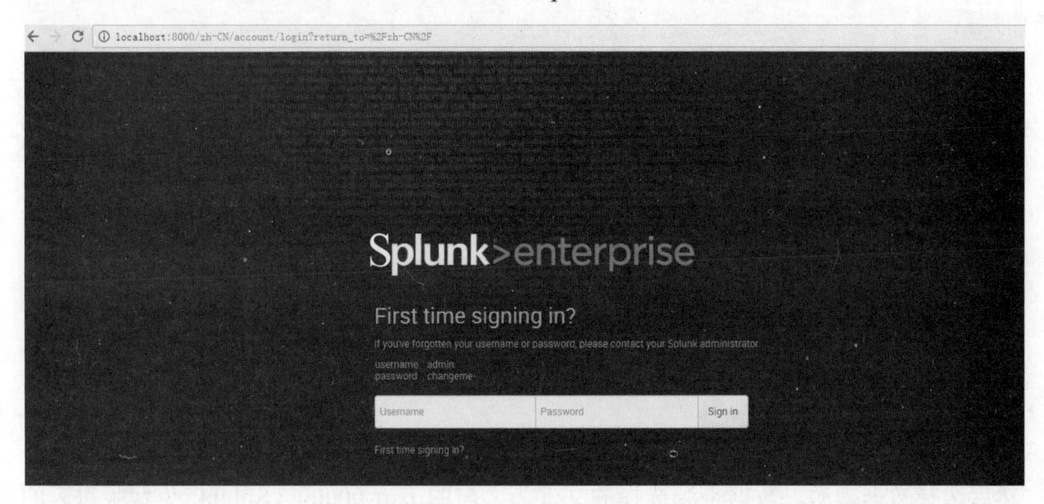

图 17-5　Splunk 登录首页

接下来为 Splunk 添加数据来源，在 Spark 通信运营商融合支付系统日志统计分析的综合应用案例中，Splunk 的数据来源是之前经过 Spark 数据清洗以后保存的本地文件 part-00000，如图 17-6 所示。

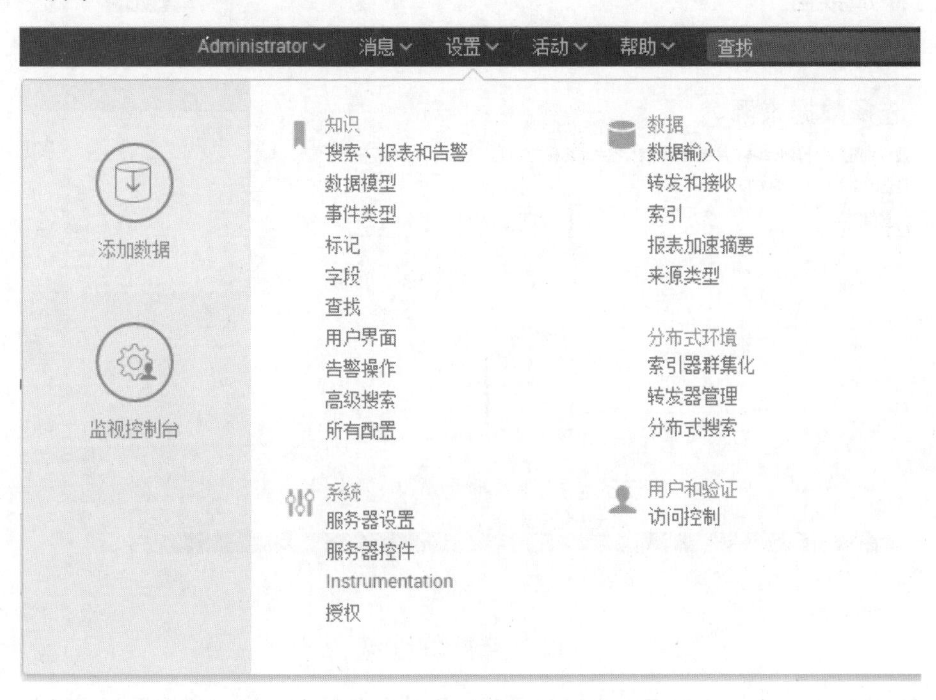

图 17-6　Splunk 添加数据

单击“添加数据”按钮，弹出新的页面，在新的页面中单击“上载”按钮，如图 17-7 所示。

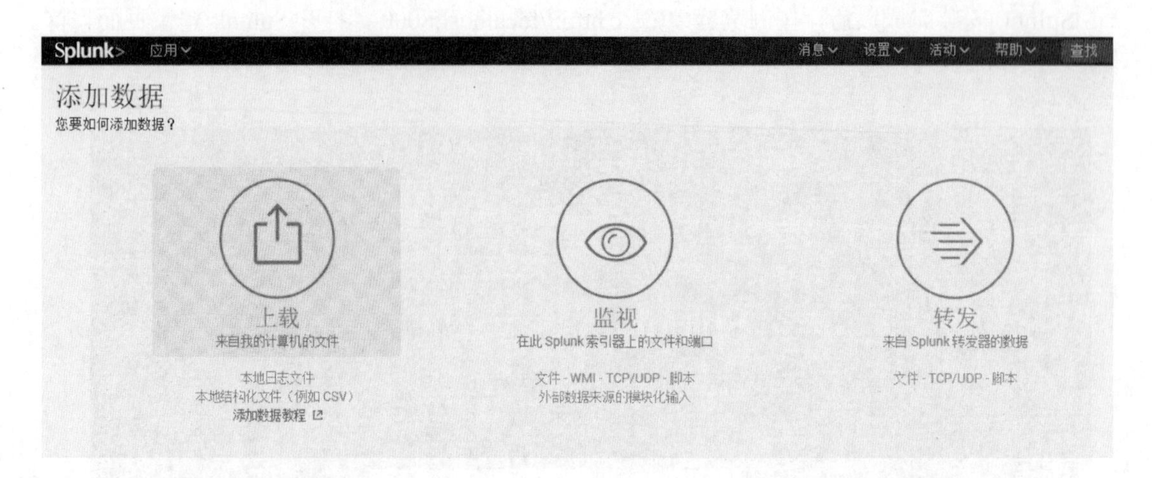

图 17-7　上传数据

然后选择数据来源。数据来源选择的文件为之前的 part-00000（Splunk 免费版每天的索引数据源不超过 500MB），如图 17-8 所示。

图 17-8　选择数据来源

选择好数据来源，接下来配置来源的类型来区分各种数据，这里设置数据来源的名称为 rhzf2017，单击“保存”按钮进行保存，如图 17-9 所示。

对于 Splunk 导入的数据来源，可以设置主机字段值，说明数据来自哪台主机设备，这里设置为 rhzf61，如图 17-10 所示。

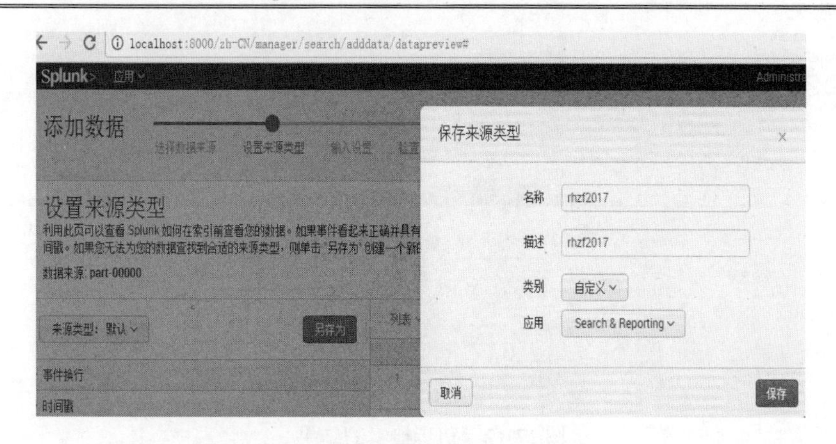

图 17-9　设置来源类型

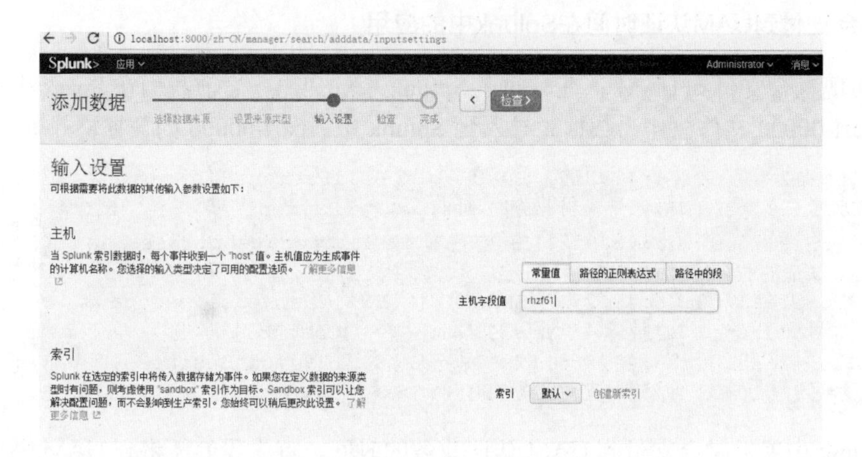

图 17-10　设置主机名

单击“下一步”按钮，Splunk 系统检查之前的配置情况，如图 17-11 所示。

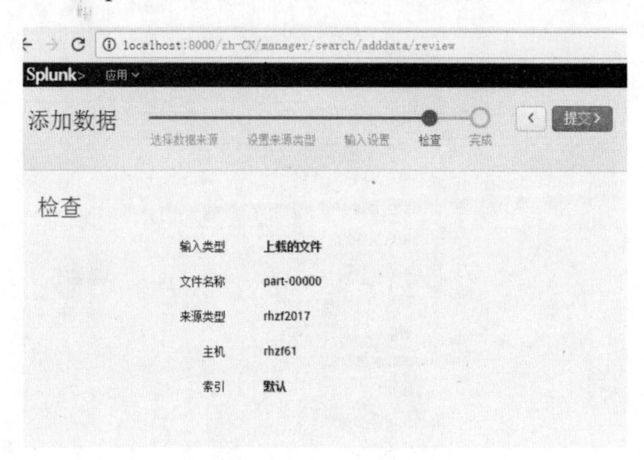

图 17-11　检查页面

Splunk 检查完成后，Splunk 打开搜索页面，在页面中搜索数据来源是 part-00000，主机名称为 rhzf61，数据来源类型是 rhzf2017 的数据。至此，Splunk 初步的安装应用完成，简单直观地显示出融合支付日志的时间索引图，如图 17-12 所示。

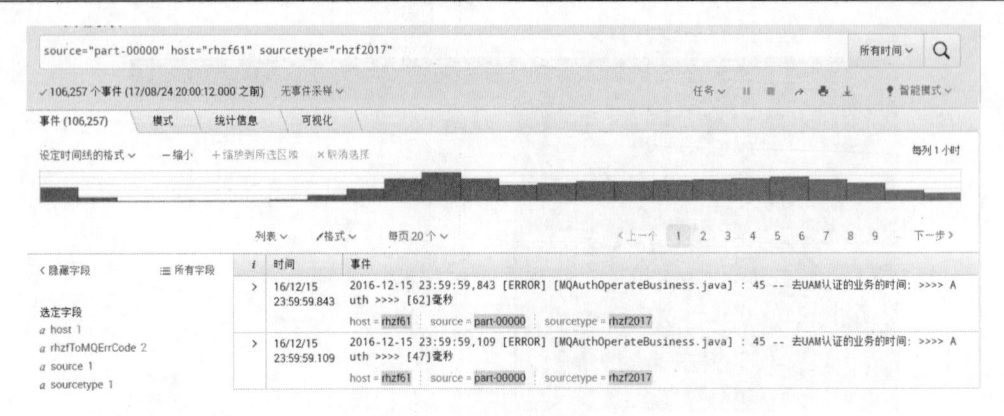

图 17-12　初步检索的结果

2．融合支付到UAM认证时间在Splunk中的应用

本节实现融合支付到 UAM 认证时间在 Splunk 中的应用。之前我们在 Spark 中已进行数据清洗，part-00000 文件已作为数据源导入到 Splunk 中。part-00000 的数据格式如下：

```
1.  2016-12-15 00:06:41,552 [ERROR] [MQBillPaymentBusiness.java] : 82 --
    流水号:201*209 MQ 账单支付业务的时间: >>>> Payment >>>> [156]ms
2.  2016-12-15 00:06:42,912 [ERROR] [MQAuthOperateBusiness.java] : 45 -- 去
    UAM 认证的业务的时间: >>>> Auth >>>> [219]ms
3.  2016-12-15 00:06:49,271 [ERROR] [MQBillCheckBusiness.java] : 133 --
    流水号:2016*20 MQ 校验订单业务的时间: >>>> CheckBill >>>> [62]ms
4.  2016-12-15 00:06:49,568 [ERROR] [MQBillPaymentBusiness.java] : 82 --
    流水号:2016*22 MQ 账单支付业务的时间: >>>> Payment >>>> [203]ms
```

在 Splunk 中展示融合支付到 UAM 认证业务的时间，首先我们需将到 UAM 系统认证的时间提取出来作为一个单独的字段，然后在 Splunk 中根据 Splunk 的查询语句进行业务需求的查询。在图 17-13 所示页面中点击“提取新字段”链接，进入提取新字段的页面。

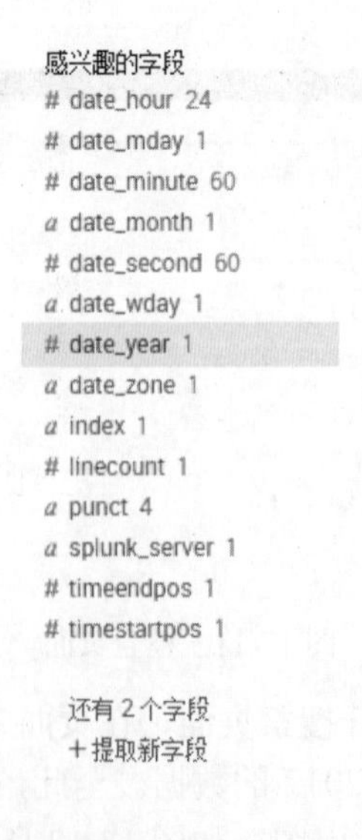

图 17-13　提取字段

在提取字段页面的事件列表中选择示例事件，这里点击融合支付到 UAM 认证的时间记录，然后单击“下一步”按钮进入下一个步骤，如图 17-14 所示。

图 17-14　提取字段选择记录

选择正则表达式作为提取融合支付到 UAM 认证时间字段的方法。单击“下一步”按钮继续，如图 17-15 所示。

图 17-15　选择正则表达式

Splunk 工具提供强大的自动生成正则表达式功能。在选择字段的时候，突出显示我们要提取的数据，这里选择“62”ms（融合支付到 UAM 认证的时间），则 Splunk 会自动弹出输入框，将提取的“62”ms 定义为一个字段名称 RhzfToUAM。RhzfToUAM 字段的示例值为 62，RhzfToUAM 字段代表在导入 part-00000 文件中匹配的每行记录融合支付到 UAM 的认证时间，其实现是根据 Splunk 生成的正则表达式自动识别的，如图 17-16 所示。

图 17-16　提取到 UAM 的时间

Splunk 提供了校验验证的功能。在上述自动提取字段的过程中，Splunk 可能将类似的时间记录都提取出来了，需要根据融合支付的业务场景进一步清理过滤。如在提取融合支付到 UAM 认证的时间过程中，也可能提取到融合支付到 MQ 账单支付业务的时间: [484]ms。Splunk 提供了便捷的功能，在页面的事件列表中，每个时间后面有一个“×”符号，直接点击一个 MQ 账单支付业务的时间: [484]ms 后面的“×”符号，就能将 part-00000 文件类似的记录从 UAM 认证时间匹配的正则表达式中清洗出去。这样，通过正则表达式过滤出融合支付到 UAM 认证的时间，如图 17-17 所示。

图 17-17　过滤掉不符合规则的数据

然后保存提取字段的结果，就完成了融合支付到 UAM 认证时间字段的提取，如图 17-18 所示。

图 17-18　字段提取完成

在 Splunk 搜索页面中，根据融合支付的业务场景，输入 Splunk 的搜索语句 index=_* OR index=* sourcetype=rhzf2017 | timechart max(RhzfToUAM) span=1m，检索出：时间间隔每隔 1min，数据来源是 rhzf2017，按照时间轴生成图表，统计每一分钟融合支付到 UAM 认证最大时间的时间数值。通过 Splunk 强大的可视化展示功能，就能轻松地展示出 Spark 数据处理的成果，如图 17-19 所示。

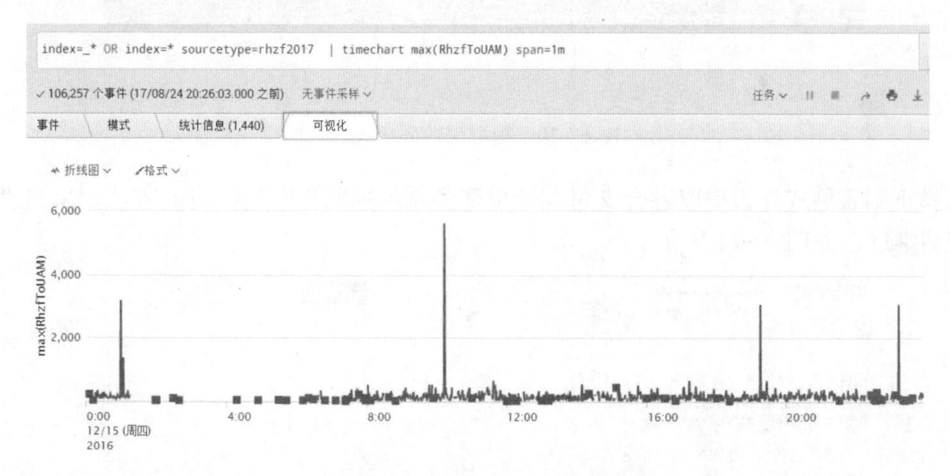

图 17-19　融合支付至 UAM 系统时间

3. 融合支付到计费账务系统销账失败返回码统计在Splunk中的应用

本节实现融合支付到计费账务系统销账失败返回码统计。同样，使用已导入 Splunk 的 part-00000 文件，part-00000 的数据格式如下：

```
1.  2016-12-15 00:09:00,856 [DEBUG] [MQQueryInvoiceBusiness.java] : 198 --
    流 水 号 :*ecc,queryInvoice  response:<soap:Envelope xmlns:soap="http:
    //*/soap/envelope/"><soap:Body><ns2:queryInvoiceResponse   xmlns:ns2=
    "http://*.sh.cn/"><response><head><errCode>00000001</errCode>
```

```
   <errDesc>*.exception.ApplicationException: 账户不存在或当前不是有效状态
2. 2016-12-15 00:09:01,278 [ERROR] [MQBillCheckBusiness.java] : 133 --
   流水号:2016*76 MQ 校验订单业务的时间: >>>> CheckBill >>>> [78]ms
3. 2016-12-15 00:09:01,575 [ERROR] [MQBillPaymentBusiness.java] : 82 --
   流水号:2016*78 MQ 账单支付业务的时间: >>>> Payment >>>> [188]ms
4. 2016-12-15 00:09:13,404 [DEBUG] [MQQueryInvoiceBusiness.java] : 198 --
   流 水 号 :*5f17,queryInvoice response:<soap:Envelope xmlns:soap=
   "http://*.org/soap/envelope/"><soap:Body><ns2:queryInvoiceResponse
   xmlns:ns2="http://*.cn/"><response><head><errCode>00000001</errCode><
   errDesc>*.exception.ApplicationException: 账户不存在或当前不是有效状态
```

在 Splunk 中展示融合支付到计费账务系统销账失败返回码统计，我们需将融合支付到计费账务系统销账失败返回码提取出来作为一个单独的字段，然后在 Splunk 中根据 Splunk 的查询语句进行业务需求的查询。单击“提取新字段”，进入提取字段的页面。

在提取字段页面的事件列表中选择示例事件，这里点击融合支付到计费账务系统销账失败返回码的记录，然后单击“下一步”按钮进入下一个步骤，如图 17-20 所示。

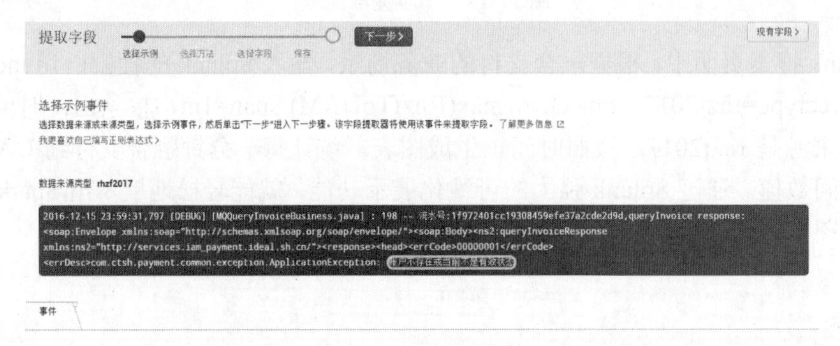

图 17-20　提取失败码字段

选择正则表达式作为提取融合支付到计费账务系统销账失败返回码的方法。单击“下一步”按钮继续，如图 17-21 所示。

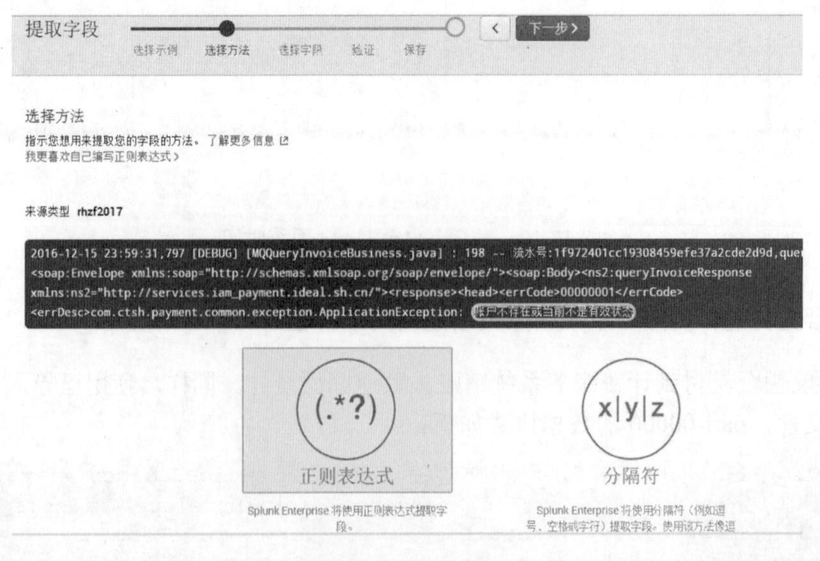

图 17-21　提取字段

Splunk 工具提供强大的自动生成正则表达式的功能。在选择字段的时候，突出显示我们要提取的数据，如图 17-22 所示。这里选择“账户不存在或当前不是有效状态”（融合支付到计费账务系统销账失败返回码含义），则 Splunk 会自动弹出输入框，将提取的“账户不存在或当前不是有效状态”定义为一个字段名称 rhzfToMQErrCode。rhzfToMQErrCode 字段的示例值为“账户不存在或当前不是有效状态”，RhzfToUAM 字段代表融合支付到计费账务系统销账失败返回码含义，失败返回码可能包含多种含义。

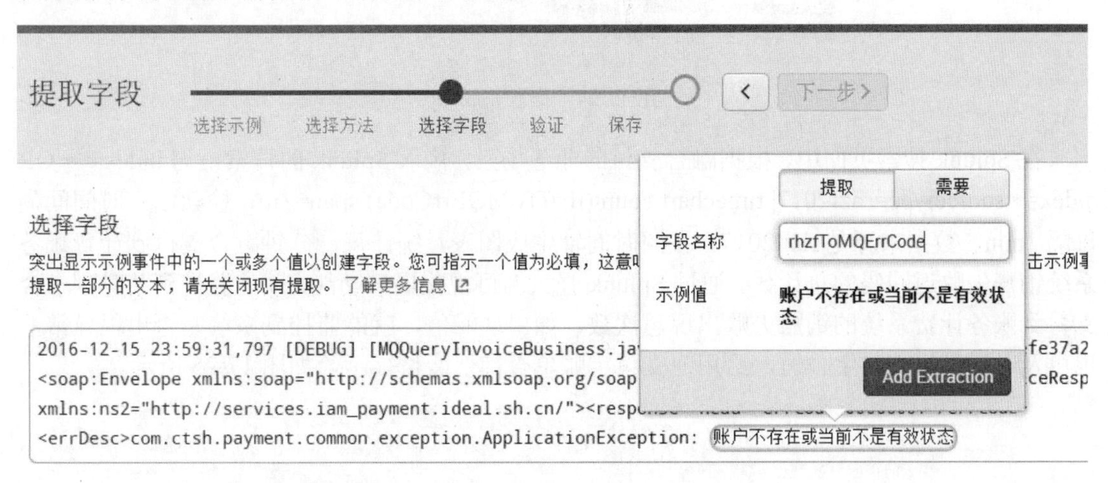

图 17-22　配置字段名

- ❑ 账户不存在或当前不是有效状态。
- ❑ 不可以查询军政用户账单。

Splunk 提供了校验验证的功能，这里检查过滤的记录都为账户不存在或当前不是有效状态”（融合支付到计费账务系统销账失败返回码含义），如图 17-23 所示。

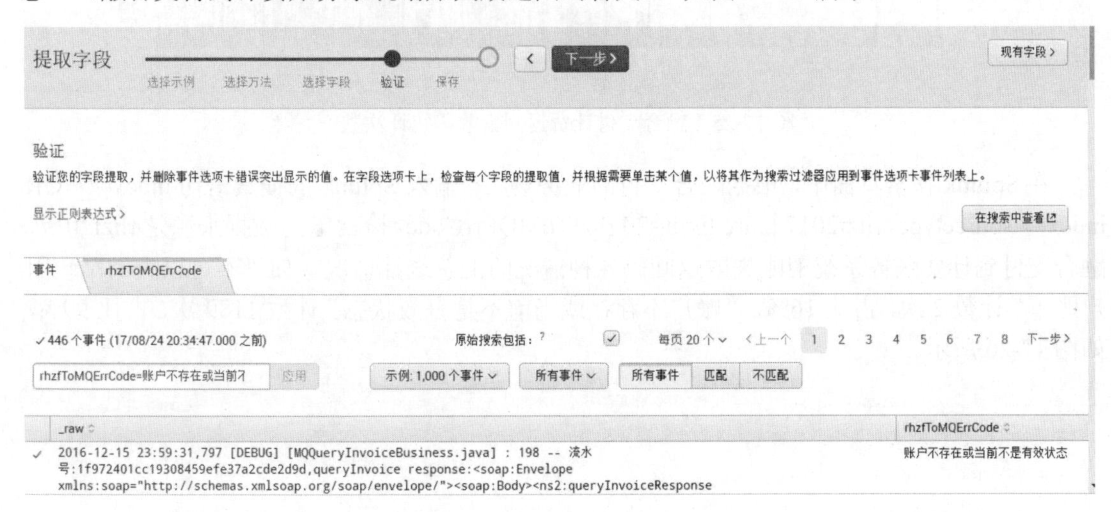

图 17-23　校验验证

然后保存提取字段的结果，就完成了融合支付到计费账务系统销账失败返回码字段的提取，如图 17-24 所示。

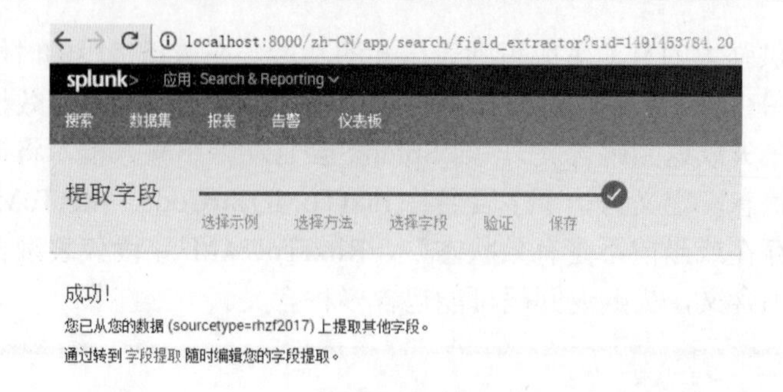

图 17-24 提取完成

在 Splunk 搜索页面中，根据融合支付的业务场景，输入 Splunk 的搜索语句 index=_* OR index=* sourcetype=rhzf2017| timechart count(rhzfToMQErrCode) span=1m，检索出：时间间隔每隔 1min，数据来源是 rhzf2017，按照时间轴生成图表，统计每一分钟融合支付到计费账务系统销账失败返回码的总计数。通过 Splunk 强大的可视化展示功能，就能监控到某时刻融合支付到账务计费系统的销账失败出现总次数，如出现峰值，就能监控到系统是否出现异常，也可及时提供给支付宝公司，给用户支付宝账单查询及缴付提示，如图 17-25 所示。

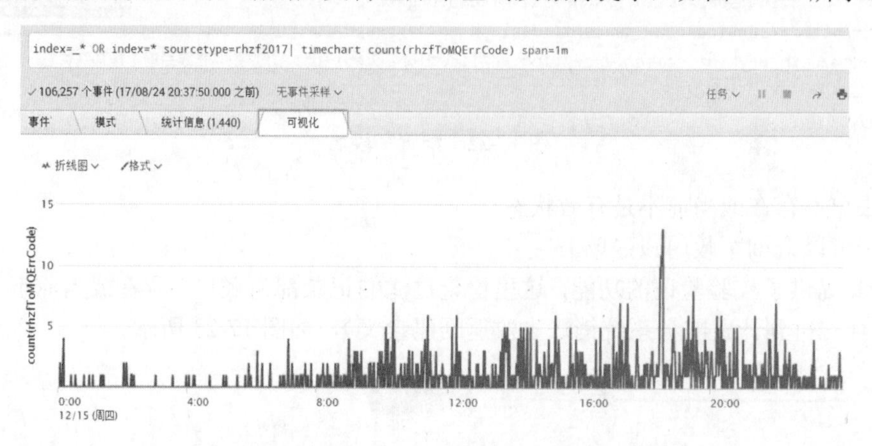

图 17-25 融合支付计费返回失败码统计次数

在 Splunk 搜索页面中，根据融合支付的业务场景，输入 Splunk 的搜索语句 index=_* OR index=* sourcetype=rhzf2017 | rare limit=20 rhzfToMQErrCode，检索出：数据来源是 rhzf2017，融合支付到计费账务系统销账失败返回码各种情况的汇总统计情况。如“不可以查询军政用户账单”计数 2 次，占比 16%，“账户不存在或当前不是有效状态”计数 1189 次，占比 99%，如图 17-26 所示。

index=_* OR index=* sourcetype=rhzf2017| rare limit=20 rhzfToMQErrCode

所有时间

✓106,257 个事件 (17/08/24 20:40:15.000 之前) 无事件采样

任务 智能模式

事件 模式 统计信息 (2) 可视化

每页 20 个 格式 预览

rhzfToMQErrCode	count	percent
不可以查询军政用户帐单	2	0.167926
账户不存在或当前不是有效状态	1189	99.832074

图 17-26 融合支付到计费账务系统销账失败返回码

同样，融合支付日志统计的其他情况也可在 Splunk 中展示，这里不再赘述。

17.1.5　融合支付系统日志统计分析案例涉及的正则表达式知识点及案例代码

Spark 在通信运营商融合支付系统日志统计分析中的综合应用案例，无论是 Spark 对日志文件的解析，还是 Splunk 自动提取字段，都应用到正则表达式的知识。本节主要对正则表达式的概念及在大数据系统中的应用进行讲解。

1．正则表达式概念及日志分析案例

在文本检索、日志分析等大数据处理过程中，正则表达式起着非常重要的作用，是必须掌握的基础知识。在正则表达式的知识基础上，可拓展学习词法分析器（Lexer）、语法分析器（Parser）、语法树分析器（Ast parser）的相关内容。

正则表达式（Regular Expression）在代码中常简写为 regex、regexp 或 RE。正则表达式通常用来检索、替换符合某个规则的文本。正则表达式最初从 UNIX 中的工具软件（sed 和 grep）普及。

要用好正则表达式，必须正确地理解元字符。表 17-6 列出了本节案例涉及的一些元字符及描述。

表 17-6　正则表达式

元字符	描　述
^	匹配输入字符串的开始位置。如果设置了 RegExp 对象的 Multiline 属性，^也匹配“\n”或“\r”之后的位置
$	匹配输入字符串的结束位置。如果设置了 RegExp 对象的 Multiline 属性，$也匹配“\n”或“\r”之前的位置
*	匹配前面的子表达式任意次。例如，zo*能匹配“z”，也能匹配“zo”以及“zoo”
+	匹配前面的子表达式一次或多次(大于等于 1 次）。例如，“zo+”能匹配“zo”以及“zoo”，但不能匹配“z”。+等价于{1,}
?	匹配前面的子表达式零次或一次。例如，“do(es)?”可以匹配“do”或“does”中的“do”，?等价于{0,1}
{n}	n 是一个非负整数。匹配确定的 n 次。例如，“o{2}”不能匹配“Bob”中的“o”，但是能匹配“food”中的两个 o
{n,}	n 是一个非负整数。至少匹配 n 次。例如，“o{2,}”不能匹配“Bob”中的“o”，但能匹配“fooood”中的所有 o。“o{1,}”等价于“o+”。“o{0,}”则等价于“o*”
{n,m}	m 和 n 均为非负整数，其中 n<=m。最少匹配 n 次且最多匹配 m 次。例如，“o{1,3}”将匹配“fooooood”中的前三个 o。“o{0,1}”等价于“o?”。注意，逗号和两个数之间不能有空格
?	当该字符紧跟在任何一个其他限制符（*,+,?，{n}，{n,}，{n,m}）后面时，匹配模式是非贪婪的。非贪婪模式尽可能少地匹配所搜索的字符串，而默认的贪婪模式则尽可能多地匹配所搜索的字符串。例如，对于字符串“oooo”，“o+”将匹配每个“o”，即 4 次匹配，而“o+?”将只匹配 1 次，即匹配“oooo”
(pattern)	匹配 pattern 并获取这一匹配。所获取的匹配可以从产生的 Matches 集合得到，在 VBScript 中使用 SubMatches 集合，在 JScript 中则使用$0…$9 属性。要匹配圆括号字符，请使用“”或“”

续表

<table>
<tr><th>元字符</th><th>描　述</th></tr>
<tr><td>(?:pattern)</td><td>非获取匹配，匹配 pattern 但不获取匹配结果，不进行存储供以后使用。这在使用或字符“(|)”来组合一个模式的各个部分时很有用。例如，“industr(?:y|ies)”就是一个比“industry|industries”更简略的表达式</td></tr>
<tr><td rowspan="2">(?=pattern)</td><td>非获取匹配，正向肯定预查，在任何匹配 pattern 的字符串开始处匹配查找字符串，在一个匹配发生后，在最后一次匹配之后立即开始下一次匹配的搜索</td></tr>
<tr><td>例如，“Windows(?=95|98|NT|2000)”能匹配“Windows2000”中的“Windows”，但不能匹配“Windows3.1”中的“Windows”。预查不消耗字符，也就是说，在一个匹配发生后，在最后一次匹配之后立即开始下一次匹配的搜索，而不是从包含预查的字符之后开始</td></tr>
<tr><td>(?!pattern)</td><td>非获取匹配，正向否定预查，在任何不匹配 pattern 的字符串开始处匹配查找字符串，该匹配不需要获取供以后使用。例如，“Windows(?!95|98|NT|2000)”能匹配“Windows3.1”中的“Windows”，但不能匹配“Windows2000”中的“Windows”</td></tr>
<tr><td>(?<=pattern)</td><td>非获取匹配，反向肯定预查，与正向肯定预查类似，只是方向相反。例如，“(?<=95|98|NT|2000)Windows”能匹配“2000Windows”中的“Windows”，但不能匹配“3.1Windows”中的“Windows”</td></tr>
<tr><td>(?<!pattern)</td><td>非获取匹配，反向否定预查，与正向否定预查类似，只是方向相反。例如，“(?<!95|98|NT|2000)Windows”能匹配“3.1Windows”中的“Windows”，但不能匹配“2000Windows”中的“Windows”。如果有多项，则任意一项都不能超过 2 位，如“(?<!95|98|NT|20)Windows 正确，“(?<!95|980|NT|20)Windows 报错，若是单独使用，则无限制，如(?<!2000)Windows 正确匹配</td></tr>
</table>

接下来看一下使用正则表达式进行日志分析的 Java 代码案例。

（1）首先定义一个字符串列表 exampleApacheLogs，存放模拟的 Apache 的日志记录。Apache 的日志格式如下：

```
1.  public static final List<String> exampleApacheLogs = Lists.newArrayList(
2.  "10.10.10.10 - \"FRED\" [18/Jan/2013:17:56:07 +1100] \"GET http://images.
    com/2013/Generic.jpg " +
3.  "HTTP/1.1\" 304 315 \"http://referall.com/\" \"Mozilla/4.0 (compatible;
    MSIE 7.0; " +
4.  "Windows NT 5.1; GTB7.4; .NET CLR 2.0.50727; .NET CLR 3.0.04506.30; .NET
    CLR 3.0.04506.648; " +
5.  ".NET CLR 3.5.21022; .NET CLR 3.0.4506.2152; .NET CLR 1.0.3705; .NET CLR
    1.1.4322; .NET CLR " +
6.  "3.5.30729; Release=ARP)\" \"UD-1\" - \"image/jpeg\" \"whatever\" 0.350
    \"-\" - \"\" 265 923 934 \"\" " +
7.  "62.24.11.25 images.com 1358492167 - Whatup",
8.  ……
```

（2）定义日志解析的正则表达式规则 apacheLogRegex，解析 Apache 的日志记录，按照正则表达式的规则进行匹配。

```
1.   public static final Pattern apacheLogRegex = Pattern.compile(
2.            "^([\\d.]+) (\\S+) (\\S+) \\[([\\w\\d:/]+\\s[+\\-]\\d{4})\\]
              \"(.+?)\" (\\d{3}) ([\\d\\-]+) \"([^\"]+)\" \"([^\"]+)\".*");
```

（3）定义一个 Stats 类，包括计数和字节数两个属性，Stats 类中提供了两个方法；其中 merge 方法用于将两个 Stats 类的计数、字节数分别进行汇总累加；toString()方法返回该 Stats 对象的字符串表示。

```
1.  public static class Stats implements Serializable {
2.
3.         private final int count;
4.         private final int numBytes;
5.
6.         public Stats(int count, int numBytes) {
7.             this.count = count;
8.             this.numBytes = numBytes;
9.         }
10.
11.        public Stats merge(Stats other) {
12.            return new Stats(count + other.count, numBytes + other.numBytes);
13.        }
14.
15.        public String toString() {
16.            return String.format("bytes=%s\tn=%s", numBytes, count);
17.        }
18.    }
```

（4）定义两个函数：extractKey 函数用于将传入的每行 Apache 日志记录使用正则表达式进行匹配，按照间隔的空格进行分组，提取的第一个分组是 IP 地址，提取的第二个分组是用户信息，提取的第三个分组是查询 URL，组合返回三元组（IP 地址、用户、查询 URL）；extractStats 函数用于将传入的每行 Apache 日志记录使用正则表达式进行匹配，按照间隔的空格进行分组，提取的第 7 个分组是 Apache 日志中记录的字节数，如果和 Apache 日志记录匹配，则计数为 1，字节数为日志中的字节数，如果匹配不上，则计数为 1，字节数为 0，组合返回 Stats 类（计数、字节数）。

```
1.    //提取 Apache 日志中的三元组：IP 地址、用户、查询 URL
2.   public static Tuple3<String, String, String> extractKey(String line) {
3.         Matcher m = apacheLogRegex.matcher(line);
4.         if (m.find()) {
5.            String ip = m.group(1);
6.            String user = m.group(3);
7.            String query = m.group(5);
8.            if (!user.equalsIgnoreCase("-")) {
9.               return new Tuple3<String, String, String>(ip, user, query);
10.           }
11.        }
12.        return new Tuple3<String, String, String>(null, null, null);
13.    }
14.  //提取 Apache 日志中的字节数
15.   , public static Stats extractStats(String line) {
16.        Matcher m = apacheLogRegex.matcher(line);
17.        if (m.find()) {
18.           int bytes = Integer.parseInt(m.group(7));
19.           return new Stats(1, bytes);
20.        } else {
21.           return new Stats(1, 0);
22.        }
23.    }
```

（5）使用 Spark 上下文的 parallelize 算子将 exampleApacheLogs 创建生成 JavaRDD <String> dataSet，其每行内容为 Apache 的日志记录。

```
1.        SparkConf sparkConf = new SparkConf().setAppName("JavaLogQuery")
          .setMaster("local[4]");
```

```
2.          JavaSparkContext jsc = new JavaSparkContext(sparkConf);
3.
4.          JavaRDD<String> dataSet = (args.length == 1) ? jsc.textFile
            (args[0]) : jsc.parallelize(exampleApacheLogs);
5.
```

（6）将 dataSet 进行 mapToPair 转换，将读入的每行 Apache 日志记录提取出三元组（IP 地址、用户、查询 URL）及 Stats（计数、字节数），格式化为 Key-Value 值 extracted，Key 值为三元组（IP 地址、用户、查询 URL），Value 值为 Stats（计数、字节数）。

```
1.     JavaPairRDD<Tuple3<String, String, String>, Stats> extracted =
       dataSet.mapToPair(new PairFunction<String, Tuple3<String, String,
       String>, Stats>() {
2.             @Override
3.             public Tuple2<Tuple3<String, String, String>, Stats> call
              (String s) {
4.                 return new Tuple2<Tuple3<String, String, String>, Stats>
                   (extractKey(s), extractStats(s));
5.             }
6.         });
```

（7）将 extracted 使用 reduceByKey 算子，对于具有相同 Key 值（IP 地址、用户、查询 URL）的 Stats，则将其 Stats 进行汇总，计数值进行累加，字节数也进行累加。

```
1.          JavaPairRDD<Tuple3<String, String, String>, Stats> counts =
            extracted.reduceByKey(new Function2<Stats, Stats, Stats>() {
2.             @Override
3.             public Stats call(Stats stats, Stats stats2) {
4.                 return stats.merge(stats2);
5.             }
6.          });
```

（8）然后使用 collect 算子收集全部的输出结果，循环遍历打印输出。

```
1.          List<Tuple2<Tuple3<String, String, String>, Stats>> output =
            counts.collect();
2.          for (Tuple2<?, ?> t : output) {
3.             System.out.println(t._1() + "\t" + t._2());
4.          }
```

正则表达式进行日志分析的案例的完整代码如下：

```
1.      package com.noc.ronghezhifu.cores;
2.
3.  import com.google.common.collect.Lists;
4.  import scala.Tuple2;
5.  import scala.Tuple3;
6.  import org.apache.spark.SparkConf;
7.  import org.apache.spark.api.java.JavaPairRDD;
8.  import org.apache.spark.api.java.JavaRDD;
9.  import org.apache.spark.api.java.JavaSparkContext;
10. import org.apache.spark.api.java.function.Function2;
11. import org.apache.spark.api.java.function.PairFunction;
12.
13. import java.io.Serializable;
14. import java.util.Collections;
15. import java.util.List;
16. import java.util.regex.Matcher;
17. import java.util.regex.Pattern;
```

```
18.
19.
20. /**
21.  * Executes a roll up-style query against Apache logs.
22.  * <p>
23.  * Usage: JavaLogQuery [logFile]
24.  */
25.
26. public final class JavaLogQuery {
27.
28.     public static final List<String> exampleApacheLogs = Lists
        .newArrayList(
29.             "10.10.10.10 - \"FRED\" [18/Jan/2013:17:56:07 +1100] \"GET http:
            //images. com/2013/Generic.jpg " +
30.                     "HTTP/1.1\" 304 315 \"http://referall.com/\" \"Mozilla/
                     4.0 (compatible; MSIE 7.0; " +
31.                     "Windows NT 5.1; GTB7.4; .NET CLR 2.0.50727; .NET CLR
                    3.0.04506.30; .NET CLR 3.0.04506.648; " +
32.                     ".NET CLR 3.5.21022; .NET CLR 3.0.4506.2152; .NET CLR
                    1.0.3705; .NET CLR 1.1.4322; .NET CLR " +
33.                     "3.5.30729; Release=ARP)\" \"UD-1\" - \"image/jpeg\"
                    \"whatever\" 0.350 \"-\" - \"\" 265 923 934 \"\" " +
34.                     "62.24.11.25 images.com 1358492167 - Whatup",
35.             "10.10.10.110 - \"FRED\" [18/Jan/2013:17:56:07 +1100] \"GET
            http://images.com/2013/Generic.jpg " +
36.                     "HTTP/1.1\" 304 315 \"http://referall.com/\" \"Mozilla/
                    4.0 (compatible; MSIE 7.0; " +
37.                     "Windows NT 5.1; GTB7.4; .NET CLR 2.0.50727; .NET CLR
                     3.0.04506.30; .NET CLR 3.0.04506.648; " +
38.                     ".NET CLR 3.5.21022; .NET CLR 3.0.4506.2152; .NET CLR
                    1.0.3705; .NET CLR 1.1.4322; .NET CLR " +
39.                     "3.5.30729; Release=ARP)\" \"UD-1\" - \"image/jpeg\"
                    \"whatever\" 0.350 \"-\" - \"\" 265 923 934 \"\" " +
40.                     "62.24.11.25 images.com 1358492167 - Whatup",
41.             "10.10.10.10 - \"FRED\" [18/Jan/2013:18:02:37 +1100] \"GET
            http://images.com/2013/Generic.jpg " +
42.                     "HTTP/1.1\" 304 306 \"http:/referall.com\" \"Mozilla/
                     4.0 (compatible; MSIE 7.0; Windows NT 5.1; " +
43.                     "GTB7.4; .NET CLR 2.0.50727; .NET CLR 3.0.04506.30; .NET
                    CLR 3.0.04506.648; .NET CLR " +
44.                     "3.5.21022; .NET CLR 3.0.4506.2152; .NET CLR 1.0.
                    3705; .NET CLR 1.1.4322; .NET CLR " +
45.                     "3.5.30729; Release=ARP)\" \"UD-1\" - \"image/jpeg\"
                    \"whatever\" 0.352 \"-\" - \"\" 256 977 988 \"\" " +
46.                     "0 73.23.2.15 images.com 1358492557 - Whatup");
47.
48.     public static final Pattern apacheLogRegex = Pattern.compile(
49.             "^([\\d.]+) (\\S+) (\\S+) \\[([\\w\\d:/]+\\s[+\\-]\\d{4})\\]
            \"(.+?)\" (\\d{3}) ([\\d\\-]+) \"([^\"]+)\" \"([^\"]+)\".*");
50.
51.
52.
53.     public static class Stats implements Serializable {
54.
55.         private final int count;
56.         private final int numBytes;
57.
58.         public Stats(int count, int numBytes) {
59.             this.count = count;
60.             this.numBytes = numBytes;
```

```
61.         }
62.
63.         public Stats merge(Stats other) {
64.            return new Stats(count + other.count, numBytes + other.numBytes);
65.         }
66.
67.         public String toString() {
68.            return String.format("bytes=%s\tn=%s", numBytes, count);
69.         }
70.     }
71.
72.     public static Tuple3<String, String, String> extractKey(String line) {
73.         Matcher m = apacheLogRegex.matcher(line);
74.         if (m.find()) {
75.            String ip = m.group(1);
76.            String user = m.group(3);
77.            String query = m.group(5);
78.            if (!user.equalsIgnoreCase("-")) {
79.               return new Tuple3<String, String, String>(ip, user, query);
80.            }
81.         }
82.         return new Tuple3<String, String, String>(null, null, null);
83.     }
84.
85.     public static Stats extractStats(String line) {
86.         Matcher m = apacheLogRegex.matcher(line);
87.         if (m.find()) {
88.            int bytes = Integer.parseInt(m.group(7));
89.            return new Stats(1, bytes);
90.         } else {
91.            return new Stats(1, 0);
92.         }
93.     }
94.
95.     public static void main(String[] args) {
96.
97.         SparkConf sparkConf = new SparkConf().setAppName("JavaLogQuery")
            .setMaster("local[4]");
98.         JavaSparkContext jsc = new JavaSparkContext(sparkConf);
99.
100.           JavaRDD<String> dataSet = (args.length == 1) ? jsc.textFile
               (args[0]) : jsc.parallelize(exampleApacheLogs);
101.
102.           JavaPairRDD<Tuple3<String, String, String>, Stats> extracted
               = dataSet.mapToPair(new PairFunction<String, Tuple3<String,
               String, String>, Stats>() {
103.              @Override
104.              public Tuple2<Tuple3<String, String, String>, Stats>
                  call(String s) {
105.                 return new Tuple2<Tuple3<String, String, String>,
                     Stats>(extractKey(s), extractStats(s));
106.              }
107.           });
108.
109.           JavaPairRDD<Tuple3<String, String, String>, Stats> counts =
               extracted.reduceByKey(new Function2<Stats, Stats, Stats>() {
110.              @Override
111.              public Stats call(Stats stats, Stats stats2) {
112.                 return stats.merge(stats2);
113.              }
```

```
114.            });
115.
116.            List<Tuple2<Tuple3<String, String, String>, Stats>> output =
                counts.collect();
117.            for (Tuple2<?, ?> t : output) {
118.               System.out.println(t._1() + "\t" + t._2());
119.            }
120.            jsc.stop();
121.        }
122.    }
123.
```

在 IDEA 中运行代码，正则表达式日志分析的输出结果如下。

```
1.      Using Spark's default log4j profile: org/apache/spark/log4j-
        defaults.properties
2.  17/04/06 21:20:23 INFO SparkContext: Running Spark version 2.1.0
3.  .......
4.  17/04/06 21:20:31 INFO DAGScheduler: Job 0 finished: collect at
    JavaLogQuery.java:116, took 1.432762 s
5.  (10.10.10.10,"FRED",GET http://images.com/2013/Generic.jpg HTTP/1.1)
    bytes=621   n=2
6.  (10.10.10.110,"FRED",GET http://images.com/2013/Generic.jpg HTTP/1.1)
    bytes=315   n=1
7.  ……
```

2. Spark在通信运营商生产环境中的应用案例代码

Spark 在通信运营商生产环境中的应用案例正则规则集代码如例 17-1 所示。

【例 17-1】 rhzfLogRegex.scala 代码。

```
1.      package com.noc.ronghezhifu.cores
2.
3.  /**
     * A collection of regexes for Ronghezhifu LOG extracting information from
     * the rhzfline string.
4.   */
5.
6.
7.  object rhzfLogRegex {
8.    /**
9.      * Regular expression used for Scene :
10.     * Log analysis:extraction the total time of the RHZF system to UAM
11.     *2017-01-15 17:47:56,242 [ERROR] [MQAuthOperateBusiness.java] : 45 --
        *去 UAM 认证业务的时间: >>>> Auth >>>> [31]ms
12.     */
13.   val RHZF_UAM_TIME_REGEX =
14.     """^(\S+) (\S+) (\S+) (\S+) : (\S+) -- (\S+) >>>> (\S+) >>>> (\S+)]
        (\S+)$""".r
15.   /**
16.     * Regular expression used for Scene :
17.     * Log analysis:extraction the error code and the error description of
        * the RHZF system  received from  IT's charging system
18.     * 2017-01-15 17:47:58,430 [DEBUG] [MQQueryInvoiceBusiness.java] : 498
        * -- 流水号:372@@8698,queryInvoice response:<soap:Envelope xmlns:
        * soap="http://@@/soap/envelope/"><soap:Body><ns2:queryInvoiceResponse
        * xmlns:ns2="http://@@.sh.cn/"><response><head><errCode>00000001< /errCode>
        * <errDesc>com.@@.exception.ApplicationException: 账户不存在或当前不是
        * 有效状态
```

```
19.     */
20.   val RHZF_MQ_QueryInvoice_ErrCode_REGEX =
21.     """^(\S+) (\S+) (\S+) (\S+) : (\S+) -- (\S+):(\S+),(\S+) (\S+) (\S+)
      (.*)<response><head><errCode>([0-9]*)</errCode><errDesc>(.*):
       (\S+)$""".r
22.
23.   /**
24.     * Regular expression used for Scene :
25.     * 1: Log analysis:extraction the MQ check order time of the RHZF system
      * to IT's MQ system
26.     *  2017-01-15 17:41:21,654 [ERROR] [MQBillCheckBusiness.java] : 233
      * -- 流水号:20170@@636 MQ校验订单业务的时间: >>>> CheckBill >>>> [554]ms
27.     * 2:Log analysis:extraction the MQ billing payment time of the RHZF
      * system to IT's MQ system
28.     * 2016-12-15 15:14:30,263 [ERROR] [MQBillPaymentBusiness.java] : 182
      * -- 流水号:2016@@41202 MQ账单支付业务的时间: >>>> Payment >>>> [423]ms
29.     *
30.     */
31.   val RHZF_MQ_CHECK_ORDER_TIME_REGEX=
32.     """^(\S+) (\S+) (\S+) (\S+) : (\S+) -- (\S+):(\S+) (\S+) >>>> (\S+)
      >>>> (\S+)](\S+)$""".r
33.
34.   /**
35.     * Regular expression used for Scene 4:
36.     *Log analysis:extraction the error code and the error description of
      * the RHZF system  received from  IT's charging system
37.     * 2016-12-15 21:31:52,015 [WARN ] [MQQueryAccountNoBusiness.java] :
      * 278 -- 流水号:1008@@27,queryAccountNo response:<soap:Envelope xmlns:
      *soap="http://@@/soap/envelope/"><soap:Body><ns2:queryAccountNo
      *Response    xmlns:ns2="http://@@.sh.cn/"><response><head><errCode>
      * 00000001</errCode><errDesc>java.lang.Exception: 设备号对应帐号为空
38.     */
39.   val RHZF_MQ_queryAccountNo_ErrCode_REGEX =
40.     """^(\S+) (\S+) (\S+) (\S+) (\S+) : (\S+) -- (\S+):(\S+),(\S+) (\S+)
      (\S+)  (.*)<response><head><errCode>([0-9]*)</errCode><errDesc>(.*):
      (\S+)$""".r
41.
42. }
43.
```

Spark 在通信运营商生产环境中的应用案例 Spark 实战代码如例 17-2 所示。

【例 17-2】 rhzfAlipayLogAnalysis.scala 代码。

```
1.   package com.noc.ronghezhifu.cores
2.
3.  import org.apache.spark.SparkConf
4.  import org.apache.log4j.{Level, Logger}
5.  import org.apache.spark.sql.SparkSession
6.  import org.apache.hadoop.mapred.TextInputFormat
7.  import org.apache.hadoop.io.LongWritable
8.  import org.apache.hadoop.io.Text
9.  import org.apache.spark.rdd.RDD
10. import rhzfLogRegex._
11. import scala.collection.immutable.HashSet
12. import scala.util.matching.Regex
13.
14. object rhzfAlipayLogAnalysis {
15.   def main(args: Array[String]): Unit = {
16.     //Logger.getLogger("org").setLevel(Level.ERROR)
```

```
17.     var masterUrl = "local[4]"
18.     var dataPath = "G://IMFBigDataSpark2017//IMFSparkAppsnew2017//data//
        ronghezhifualipaylog//logs20161215//zw_out_core.log.2016-12-15"
19.     if (args.length > 0) {
20.       masterUrl = args(0)
21.     } else if (args.length > 1) {
22.       dataPath = args(1)
23.     }
24.     val  sparkConf  =  new  SparkConf().setMaster(masterUrl).setAppName
        ("rhzfAlipayLogAnalysis")
25.     val spark = SparkSession
26.       .builder()
27.       .config(sparkConf)
28.       .getOrCreate()
29.     val sc = spark.sparkContext
30.
31.     val  linesConvertToGbkRDD:  RDD[String]  =  sc.hadoopFile(dataPath,
        classOf [TextInputFormat], classOf[LongWritable], classOf[Text], 4)
32.       .map(line => {
33.         new String(line._2.getBytes, 0, line._2.getLength, "GBK")
34.
35.       })
36.
37.     val linesRegexd: RDD[String] = linesConvertToGbkRDD.map(rhzfline => {
38.       rhzfline match {
39.         case  RHZF_UAM_TIME_REGEX(uamLog1,  uamLog2,  uamLog3,  uamLog4,
          uamLog5, uamLog6, uamLog7, uamLog8, uamLog9)
40.           => rhzfline
41.         case RHZF_MQ_QueryInvoice_ErrCode_REGEX(mqLog1, mqLog2, mqLog3,
          mqLog4, mqLog5, mqLog6, mqLog7, mqLog8, mqLog9, mqLog10, mqLog11,
          mqLog12, mqLog13, mqLog14)
42.           => rhzfline
43.         case   RHZF_MQ_CHECK_ORDER_TIME_REGEX(mqTimeLog1,   mqTimeLog2,
          mqTimeLog3,  mqTimeLog4,  mqTimeLog5,  mqTimeLog6,  mqTimeLog7,
          mqTimeLog8, mqTimeLog9, mqTimeLog10, mqTimeLog11)
44.           => rhzfline
45.         case RHZF_MQ_queryAccountNo_ErrCode_REGEX(mqWarnLog1, mqWarnLog2,
          mqWarnLog3, mqWarnLog4, mqWarnLog5, mqWarnLog6, mqWarnLog7,
          mqWarnLog8,  mqWarnLog9,  mqWarnLog10,  mqWarnLog11,  mqWarnLog12,
          mqWarnLog13, mqWarnLog14 ,mqWarnLog15)
46.           => rhzfline
47.         case _ => "MatcheIsNull"
48.       }
49.     })
50.
51.     val linedFilterd: RDD[String] = linesRegexd.filter(!_.contains
        ("MatcheIsNull"))
52.
53.
54.     linedFilterd.coalesce(1).saveAsTextFile("G://IMFBigDataSpark2017
        //ronghezhifuLogAnalysis2017//data//ronghezhifualipaylogresult
        //zw_out_core_result_splunkzw_out_core_20170207-ok")
55.   }
56. }
57.
```

17.2　Spark 在光宽用户流量热力分布 GIS 系统中的综合应用案例

本节阐述 Spark 在中国特大型城市通信运营商光宽用户流量热力分布 GIS 系统生产环境中的综合应用案例。

17.2.1　光宽用户流量热力分布 GIS 系统案例需求分析

目前，在中国特大型城市已发展几百万光宽用户，根据专业网管采集全网几十万无源光纤网络设备端口 PON 口的流量数据，其中几十万 PON 口发生用户流量，出向峰值流量最高的 PON 口达几百兆位每秒；按普遍情况下 1:64 分光，每个 PON 口下 64 家光宽用户；基于 PON 口可以预测 64 家用户区域的上网流量分布情况；若以 PON 口出向峰值 100Mb/s 流量为颗粒度标签颜色，使用不同颜色区分流量高低情况。

中国特大型城市通信运营商具备特大城市通信网络大屏幕监控系统，7×24 监控整个城市的通信网络运行情况，需要监控光宽用户流量热力分布的情况。根据光宽用户编号与 PON 口资源数据的映射关系，即可对 PON 口下具体 64 家用户编号按上述流量高低标签颜色；再根据用户编号与 GPS 坐标的映射关系在城市地图上展现光宽用户流量热力分布，由此可洞察哪些区域存在高流量使用情况，即挖潜存在发展高带宽产品商机的区域；同时，也可进一步根据 PON 口流量设定预警门限，如出向峰值超过 1000Mb/s 的 PON 口，挖掘其下用户分布情况，能直接在地图上展现 AD 编号位置信息，以便与用户申告投诉上网拥塞等潜在风险关联分析。

Spark 分布式集群计算框架在通信运营商光宽用户流量热力分布 GIS 系统案例中处于核心计算的位置，将数据源从 Hadoop 分布式文件系统中进行关联计算处理，经过 Spark 计算转换以后在 Oracle 数据库中持久化入库，提供给第三方地图 GIS 系统进行大屏幕展示，取得了很好的成果。

17.2.2　光宽用户流量热力分布 GIS 应用的数据说明

光宽用户流量热力分布 GIS 应用数据来源包括：

- ❑ PON 口的流量数据。
- ❑ 光网络设备的经纬度位置信息。

特大型城市通信运营商从现网运行网络设备上采集无源光纤网络设备端口 PON 口的流量数据，数据格式包括端口类型、区局名称、设备端口编号、采集时间、入向峰值、入向平均值、出向峰值、出向平均值。PON 口的流量数据经通信运营商网管系统每天采集以后，上传至大数据平台 Hadoop 系统。

PON 口的流量数据的格式描述如下。

```
1.  端口类型 PonPort
2.  区局 区局名称
3.  OLT 名称 分局 T1/ZTEC300-OLT04
4.  PON 口编号    epon_1/2/4
5.  采集时间 201701071458-   2.01604E+11
6.  入向峰值 6.17308
7.  入向平均值    4.3376384
8.  出向峰值 15.217123
9.  出向平均值    5.3401234
```

登录 HDFS 文件系统查看 PON 口的流量数据文件。

```
1.  [hdfs@master1 root]$ hdfs dfs -ls hdfs:///bigdata/Olt*/ |more
2.  Java HotSpot(TM) 64-Bit Server VM warning: Insufficient space for shared
    memory file:
3.    30246
4.  Try using the -Djava.io.tmpdir= option to select an alternate temp
    location.
5.
6.  Found 73 items
7.  -rw-r--r--      3  hdfs  hdfs                  608  2016-12-16  10:36
    hdfs:///big*/Olt*/Olt*_20161216_221005.txt
8.  -rw-r--r--      3  hdfs  hdfs                  608  2016-12-20  10:23
    hdfs:///big*/Olt*/Olt*_20161220_221005.txt
9.  -rw-r--r--      3  hdfs  hdfs                  648  2016-12-23  20:36
    hdfs:///big*/Olt*/Olt*_20161223_221005.txt
10. -rw-r--r--      3  hdfs  hdfs         32604216  2016-12-26  13:01
    hdfs:///big*/Olt*/Olt*_20161225_221005.txt
11. -rw-r--r--      3  hdfs  hdfs         32607740  2016-12-28  16:11
    hdfs:///big*/Olt*/Olt*_20161227_221004.txt
```

在 HDFS 文件系统中查看 PON 口的流量数据内容。

```
1.  [hdfs@master1 root]$ hdfs dfs -cat  hdfs:///big*/Olt*/Olt*_20170130.txt
    | more
2.  Java HotSpot(TM) 64-Bit Server VM warning: Insufficient space for shared
    memory file:
3.    29901
4.  Try using the -Djava.io.tmpdir= option to select an alternate temp
    location.
5.
6.  PonPort, 宝山, 宝山宾馆北楼 R1/HW5680T-OLT01, EPON 0/1/5,
    201701302058-201703302210, 0, 0, 0.22320748, 0.063800128
7.  ……
```

从 PON 口的流量数据摘取部分记录如下（样本供读者了解以下数据字段的格式，原始记录数据已修改）：

```
1.  PonPort, 宝山, 宝山宾馆北楼 T1/HW5680T-OLT01, EPON 0/1/0,
    201701302058-201701302210, 0, 0, 0.42321234, 0.063600122
2.  PonPort, 宝山, 宝山宾馆北楼 T1/HW5680T-OLT01, EPON 0/1/1,
    201701302058-201701302210, 0, 0, 0.42321234, 0.063600122
3.  PonPort, 宝山, 宝山宾馆北楼 T1/HW5680T-OLT01, EPON 0/1/2,
    201701302058-201701302210, 0, 0, 0.25320722, 0.063600128
4.  PonPort, 宝山, 宝山宾馆北楼 T1/HW5680T-OLT01, EPON 0/1/4,
    201701302058-201701302210, 0.22133228, 0.099186062, 0.46746829,
    0.265229958
```

```
5.  PonPort, 宝 山 , 宝 山 宾 馆 北 楼 T1/HW5680T-OLT01, EPON 0/1/3,
    201701302058-201701302210, 0.21802262, 0.167416384, 8.96763648,
    8.544179202
6.  PonPort, 宝 山 , 宝 山 宾 馆 北 楼 T1/HW5680T-OLT01, EPON 0/1/3,
    201701302058-201701302210, 0.8519842, 0.4448745, 3.82706225, 2.286044314
```

光网络设备的位置信息：

特大型城市通信运营商从现网运行网络设备上入库记录光网络设备的位置信息，数据格式包括光网络设备端口信息、设备位置经纬度。光网络设备的位置信息存放在特大型城市通信运营商大数据平台的 Hive 中。

查看 Hive 中表 OLTPORT_LOCATION 的信息。

```
1.   hive> desc OLTPORT_LOCATION;
2.  OK
3.  olt_port              string
4.  ad_location           string
5.  Time taken: 0.989 seconds, Fetched: 2 row(s)
6.  hive>
```

光网络设备的位置信息的格式描述如下：

```
1.  olt 端口信息
2.  olt 位置经纬度     按：分隔
```

从光网络设备摘取部分记录如下（样本供读者了解以下数据字段的格式，原始记录数据已修改）：

```
1.  崇明港沿 T1/ZTEC500-OLT04|03|07  124.670679:31.591234
2.  南汇惠南 D1/ZTEC220-OLT01|05|04  124.750808:31.041234
3.  松江方塔 T1/ZTEC600-OLT11|02|07  124.243126:31.002079
4.  奉贤肖塘 D1/HW5680T-OLT07|01|03  122.458160:30.971234
5.  黄兴 Z1/ZTEC220-OLT05|01|04      121.430594:41.801234
6.  长宁 T1/HW5680T-OLT04|08|02      121.426483:31.225152
7.  海宁 T2/HW5680T-OLT3803|00       124.455410:31.258940
8.  江苏 T2/HW5680T-OLT11|03|02      122.442038:31.222126
9.  ……
```

17.2.3 光宽用户流量热力分布 GIS 应用 Spark 实战

本节进行光宽用户流量热力分布 GIS 应用 Spark 实战。使用 Spark 对 PON 口的流量数据、光网络设备经纬度位置信息进行关联，并将结果持久化到 Oracle 数据库。

具体实现步骤如下。

（1）使用 StructType 方式把 PON 口的流量数据格式化，即在 RDD 的基础上增加 PON 口的流量数据的元数据信息。格式化的时候区分一下数据类型，入向流量、入向流量均值、出向流量、出向流量均值 4 个字段的类型是 DoubleType；OLT 端口、开始时间、结束时间 3 个字段的类型是 StringType，创建生成 StructType 类型的 structType_ponData。

（2）从特大型城市通信运营商大数据平台 Hadoop 分布式文件系统中读入 PON 口的流量数据进行 Row 行数据格式化。PON 口流量数据的每行数据包括端口类型、区局名称、设备端口编号、采集时间、入向峰值、入向平均值、出向峰值、出向平均值。

- 例如，某行数据“PonPort, 宝山, 宝山宾馆北楼 R1/HW5680T-OLT01, EPON 0/1/3, 201702302058-201702302210, 0.42133228, 0.099181234, 0.16746829, 0.265229958”。对读入的每行数据按","分隔符进行分割，生成变量 array。
- 获取端口值：array 第三个元素是端口信息，如 EPON 0/1/3，再按"/"进行分割，取出具体的端口值，端口第 0 个值是 0，端口第 1 个值是 1，端口第 2 个值是 3。然后将第二个元素，如“宝山宾馆北楼 R1/HW5680T-OLT01”+"|0"和第 1 个端口值 1+"|0"+第二个端口值 3 组拼成 oltPortTemp，除掉空格，即“宝山宾馆北楼 R1/HW5680T-OLT01|01|03”。
- 获取开始时间：array 第四个元素，如 201702302058-201702302210，按“-”分割以后取到第 0 个值，即开始时间 201702302058。
- 获取结束时间：array 第四个元素，如 201702302058-201702302210，按“-”分割以后取到第 1 个值，即结束时间 201702302210。

经过上述格式处理，最终把 PON 口的流量数据的每条数据变成以 Row 为单位的数据，Row 每行中的数据格式为 Row（olt 端口、开始时间、结束时间、入向峰值、入向平均值、出向峰值、出向平均值），创建生成 RDD[Row]格式的 ponData_rdd。其中，每行数据，如（宝山宾馆北楼 R1/HW5680T-OLT01|01|03，201702302058，201702302210，0.42133228, 0.099181234, 0.16746829, 0.265229958）。

（3）创建 df_ponData DataFrame：结合 ponData_rdd 和 structType_ponData StructType 的元数据信息，基于 RDD 创建 DataFrame，这时 RDD 就有了元数据信息的描述。

通过 DataFrame 的方式实现 PON 口的流量数据分析，代码如下：

```
1.     val  schemaString2  =  "OLTPORT2,FirstTime,LastTime,I_Mbps,IA_Mbps,
       O_Mbps,OA_Mbps"
2.     var fields2: Seq[StructField] = List[StructField]()
3.     for (columnName <- schemaString2.split(",")) {
4.       //分别规定 String 和 Double 类型
5.       if (columnName == "I_Mbps" || columnName == "IA_Mbps" || columnName
       == "O_Mbps" || columnName == "OA_Mbps")
6.         fields2 = fields2 :+ StructField(columnName, DoubleType, true)
7.       else
8.         fields2 = fields2 :+ StructField(columnName, StringType, true)
9.     }
10.    val structType_ponData = StructType.apply(fields2)
11.
12.    //结合数据给出要处理的时间格式为 yyyyMMdd
13.    val  fileName  =  "hdfs:///bigdata/OltPonPort/OltPonPort_FlowData_"  +
       args(0) + "*"
14.    var ponData = spark.read.textFile(fileName).rdd
15.
16.    //找出相关格式，进行关联
17.    val ponData_rdd = ponData.map(x => {
18.      val array = x.split(",")
19.      val port = array(3).split("/")
20.
21.      //使用 replace 方法去掉空格
22.      val oltPortTemp = array(2) + "|0" + port(1) + "|0" + port(2)
23.      val oltPort = oltPortTemp.replace(" ", "")
24.      val firstTime = array(4).split("-")(0).replace(" ", "")
25.      val lastTime = array(4).split("-")(1)
```

```
26.
27.       Row(oltPort, firstTime, lastTime, array(5).toDouble, array(6).
          toDouble, array(7).toDouble, array(8).toDouble)
28.
29.     })
30.
31.     var df_ponData: DataFrame = spark.createDataFrame(ponData_rdd,
        structType_ponData)
```

（4）Spark 从大数据平台 Hive 中读取光网络设备位置表 OLTPORT_LOCATION，创建生成 DataFrame 类型的 df_gisData。df_gisData 的每行数据包括光网络设备端口、设备经纬度位置信息（olt_port、ad_location）。

将 PON 口的流量数据的 DataFrame df_ponData 和光网络设备经纬度位置 DataFrame df_gisData 根据设备端口号进行关联 Join，去除重复的端口列，计算出的格式为（olt 端口数据、开始时间、结束时间、入向峰值、入向平均值、出向峰值、出向平均值，经纬度位置），最终创建生成关联以后的结果 result: DataFrame。

PON 口的流量数据及光网络设备经纬度位置关联代码如下：

```
1.      //在此读 Hive，读取相关的表，进行定期操作
2.      val df_gisData = spark.sql("SELECT * FROM OLTPORT_LOCATION")
3.      //进行关联
4.      val result = df_ponData.join(df_gisData, df_ponData.col("OLTPORT2")
        === df_gisData.col("olt_port")).drop("OLTPORT2")
5.      result.show(20)
6.      val printRdd = result.rdd.map(x => {
7.        var result = x.apply(0)
8.        for (i <- 1 to x.length - 1) {
9.          result = result + "," + x.apply(i)
10.       }
11.       result
12.     })
```

（5）将 PON 口的流量数据及光网络设备经纬度位置关联的结果（olt 端口数据、开始时间、结束时间、入向峰值、入向平均值、出向峰值、出向平均值，经纬度位置）持久化保存至 Oracle 数据库。关联的结果 result: DataFrame 使用 foreachPartition 算子按分区进行转换。

首先加载 Oracle 驱动引擎建立与 Oracle 数据库的连接。

接下来对读取的每行关联后的行数据 Row。

- 计算经纬度数据：获取 ad_location 字段，按“:”进行切分，其第 0 个数据是经度数据 jd，第 1 个数据是纬度数据 wd。然后经过转换，计算出转换后的经度 jd1、纬度 wd1。
- 获取当前时间：DRRQ。
- 获取开始时间：YWRQ。PON 口的流量数据中记录的开始时间。
- 获取光网络设备 OLTPORT 的分隔符“/”切分的第一个名称，如“宝山宾馆北楼 R1/HW5680T-OLT01|01|03”按分隔符“/”切分以后，其第 0 个元素是宝山宾馆北楼 R1。
- 依次获取开始时间、结束时间、入向峰值、入向平均值、出向峰值、出向平均值等字段数据。

最后批量将上述数据插入到 Oracle 数据库表 DX_LLXXB 中，插入的字段包括序号、开

始时间、结束时间、入向峰值、入向平均值、出向峰值、出向平均值、olt 光网设备端口号、经纬度位置、经度、纬度、当前时间、开始时间、转换后的经度、转换后的纬度等数据。

PON 口的流量数据及光网络设备经纬度位置关联结果入库 Oracle 的代码如下：

```
1.  //再次清洗，给出 JD WD
2.      result.foreachPartition(partition => {
3.
4.        //分区域进行数据库链接，减少连接池的压力
5.        val url = "jdbc:oracle:thin:@10.*.*.*:1521:ORCL"
6.        val user = "**"
7.        val password = "**"
8.        var conn: Connection = null
9.        var ps: PreparedStatement = null
10.
11.       //插入数据库语句
12.       val sql = "insert into ogg.DX_LLXXB(xh,firsttime,lasttime, i_mbps,
          ia_mbps,o_mbps,oa_mbps,oltport,olmc,l_local,jd,wd,drrq,ywrq,jd1,
          wd1) values(SEQ_DXLL.nextval,?,?,?,?,?,?,?,?,?,?,?,?,?,?,?)"
13.       Class.forName("oracle.jdbc.driver.OracleDriver").newInstance()
14.
15.       try {
16.
17.         //给出链接
18.         conn = DriverManager.getConnection(url, user, password)
19.         conn.setAutoCommit(false)
20.         //拼接 SQL
21.         ps = conn.prepareStatement(sql)
22.
23.         //每个分区内部批处理
24.         partition.toList.foreach(x => {
25.
26.           //给出经纬度计算
27.           val jd = x.getAs[String]("ad_location").split(":")(0)
28.           val wd = x.getAs[String]("ad_location").split(":")(1)
29.
30.           //转换后的经纬度
31.           val jd1 = jd.toDouble * 20037508.34 / 180
32.           var wd1 = math.log(Math.tan((90 + wd.toDouble) * math.Pi / 360))
              /(math.Pi/180)
33.           wd1 = wd1 * 20037508.34 / 180
34.
35.           //当前的时间
36.           //给出当前时间
37.           var now = new Date()
38.           val sdf = new SimpleDateFormat("yyyy-MM-dd HH:mm:ss")
39.           val DRRQ = sdf.format(now)
40.           val YWRQ = x.getAs[String]("FirstTime")
41.
42.           //对 SQL 进行赋值
43.           ps.setString(1, x.getAs[String]("FirstTime"))
44.           ps.setString(2, x.getAs[String]("LastTime"))
45.           ps.setDouble(3, x.getAs[Double]("I_Mbps"))
46.           ps.setDouble(4, x.getAs[Double]("IA_Mbps"))
47.           ps.setDouble(5, x.getAs[Double]("O_Mbps"))
48.           ps.setDouble(6, x.getAs[Double]("OA_Mbps"))
49.           ps.setString(7, x.getAs[String]("olt_port"))
50.
```

```
51.         //获取 OLT 的名称
52.         ps.setString(8, x.getAs[String]("olt_port").split("/")(0))
53.
54.         ps.setString(9, x.getAs[String]("ad_location"))
55.         ps.setString(10, jd)
56.         ps.setString(11, wd)
57.         ps.setString(12, DRRQ)
58.         ps.setString(13, YWRQ)
59.
60.         //进行格式的规范化，保留 7 位小数
61.         val df = new DecimalFormat("#.0000000")
62.
63.         ps.setString(14, df.format(jd1))
64.         ps.setString(15, df.format(wd1))
65.
66.         //进行批处理
67.         ps.addBatch()
68.       })
69.
70.       //进行批处理
71.       ps.executeBatch()
72.       conn.commit()
73.     } //结束尝试
74.     catch {
75.       case e: Exception => e.printStackTrace
76.     }
77.
78.     finally {
79.       //关闭两个管道
80.       if (ps != null) {
81.         ps.close()
82.       }
83.
84.       if (conn != null) {
85.         conn.close()
86.       }
87.     }
88.
89.
90.   })
```

17.2.4 光宽用户流量热力分布 GIS 应用 Spark 实战成果

在通信运营商大数据平台生产环境中运行 Spark 代码，Spark 运行完成以后，登录到 Oracle 数据库，检查数据库表 DX_LLXXB，数据字段格式为序号、开始时间、结束时间、入向峰值、入向平均值、出向峰值、出向平均值、olt 光网设备端口号、经纬度位置、经度、纬度、当前时间、开始时间、转换后的经度、转换后的纬度，从数据库表中可以查询到数据，验证测试成功（原始记录数据已修改）。

```
1. 37  201611302058    201611302210    2.98793000  1.875177838 65.75281337
   49.3881015540   中原 T1/HW5680T-OLT30|01|08  中 原  D1122.530336:33.334756
   122.530336  33.334756   2016-12-07 22:21:10 201611302058
   13528695.1175722    3676300.31930068
2. 38  201611302058    201611302210    10.50303314 6.5187775380
   47.80156946 35.46381150 中原 T1/HW5680T-OLT33|02|02  中 原  D1122.536467:
```

```
31.329279   122.536467  31.329279   2016-12-07 22:21:10 201611302058
13529377.6173702     3675586.52921234
```

基于 Spark 将 PON 口的流量数据及光网络设备经纬度位置关联的结果，中国特大型城市通信运营商可在 GIS 地图监控系统中成功地监控到用户流量热力分布信息。

17.2.5　光宽用户流量热力分布 GIS 应用 Spark 案例代码

光宽用户流量热力分布 GIS 应用案例如例 17-3 所示。

【例 17-3】 GisDataMap.scala 代码。

```
1.   package NOCGis.GisData.src.main.scala.Noc
2.
3.  import java.sql.{Connection, DriverManager, PreparedStatement}
4.  import java.text.{DecimalFormat, SimpleDateFormat}
5.  import java.util.{Calendar, Date}
6.  import org.apache.spark.sql.{Row, SQLContext, SparkSession}
7.  import org.apache.spark.sql.types._
8.  import org.apache.spark.sql.types.{StringType, StructField, StructType}
9.
10. object GisDataMap {
11.
12.   def main(args: Array[String]) {
13.     val warehouslocation = "/spark-warehouse"
14.     val spark = SparkSession
15.       .builder.appName("GisDataMap")
16.       .enableHiveSupport()
17.       .config("spark.sql.warehouse.dir", warehouslocation)
18.       .getOrCreate()
19.
20.     val  schemaString2  =  "OLTPORT2,FirstTime,LastTime,I_Mbps,IA_Mbps,
        O_Mbps,OA_Mbps"
21.     var fields2: Seq[StructField] = List[StructField]()
22.     for (columnName <- schemaString2.split(",")) {
23.       //分别规定 String 和 Double 类型
24.       if (columnName == "I_Mbps" || columnName == "IA_Mbps" || columnName
          == "O_Mbps" || columnName == "OA_Mbps")
25.         fields2 = fields2 :+ StructField(columnName, DoubleType, true)
26.       else
27.         fields2 = fields2 :+ StructField(columnName, StringType, true)
28.     }
29.     val structType_ponData = StructType.apply(fields2)
30.
31.     //结合数据给出要处理的时间，格式为 yyyyMMdd
32.     val fileName = "hdfs:///bigdata/OltPonPort/OltPonPort_FlowData_" +
        args(0) + "*"
33.     var ponData = spark.read.textFile(fileName).rdd
34.
35.     //找出相关格式，进行关联
36.     val ponData_rdd = ponData.map(x => {
37.       val array = x.split(",")
38.       val port = array(3).split("/")
39.
40.       //使用 replace 方法去掉空格
41.       val oltPortTemp = array(2) + "|0" + port(1) + "|0" + port(2)
42.       val oltPort = oltPortTemp.replace(" ", "")
```

```
43.        val firstTime = array(4).split("-")(0).replace(" ", "")
44.        val lastTime = array(4).split("-")(1)
45.
46.        Row(oltPort, firstTime, lastTime, array(5).toDouble, array(6)
           .toDouble, array(7).toDouble, array(8).toDouble)
47.
48.      })
49.
50.      var df_ponData = spark.createDataFrame(ponData_rdd, structType_ponData)
51.
52.      //在此读 Hive，读取相关的表，进行定期操作
53.      val df_gisData = spark.sql("SELECT * FROM OLTPORT_LOCATION")
54.
55.
56.      //进行关联
57.      val result = df_ponData.join(df_gisData, df_ponData.col("OLTPORT2")
          === df_gisData.col("olt_port")).drop("OLTPORT2")
58.
59.      result.show(20)
60.
61.      val printRdd = result.rdd.map(x => {
62.
63.        var result = x.apply(0)
64.        for (i <- 1 to x.length - 1) {
65.          result = result + "," + x.apply(i)
66.        }
67.        result
68.      })
69.
70.      //针对以上情况进行处理
71.      //再次清洗，给出 JD WD
72.      result.foreachPartition(partition => {
73.
74.        //分区域进行数据库链接，减少连接池的压力
75.        val url = "jdbc:oracle:thin:@10.100.100.109:1521:ORCL"
76.        val user = "ogg"
77.        val password = "ogg"
78.        var conn: Connection = null
79.        var ps: PreparedStatement = null
80.
81.        //插入数据库语句
82.        val sql = "insert into ogg.DX_LLXXB(xh,firsttime,lasttime,i_mbps,
           ia_mbps,o_mbps,oa_mbps,oltport,olmc,l_local,jd,wd,drrq,ywrq,jd1,
           wd1) values(SEQ_DXLL.nextval,?,?,?,?,?,?,?,?,?,?,?,?,?,?,?)"
83.        Class.forName("oracle.jdbc.driver.OracleDriver").newInstance()
84.
85.        try {
86.
87.          //给出链接
88.          conn = DriverManager.getConnection(url, user, password)
89.          conn.setAutoCommit(false)
90.          //拼接 SQL
91.          ps = conn.prepareStatement(sql)
92.
93.          //每个分区内部批处理
94.          partition.toList.foreach(x => {
95.            //给出经纬度计算
96.            val jd = x.getAs[String]("ad_location").split(":")(0)
```

```
97.          val wd = x.getAs[String]("ad_location").split(":")(1)
98.
99.          //转换后的经纬度
100.         val jd1 = jd.toDouble * 20037508.34 / 180
101.         var wd1 = math.log(Math.tan((90 + wd.toDouble) * math.Pi / 360))
             /(math.Pi/180)
102.         wd1 = wd1 * 20037508.34 / 180
103.
104.         //当前的时间
105.         //给出当前时间
106.         var now = new Date()
107.         val sdf = new SimpleDateFormat("yyyy-MM-dd HH:mm:ss")
108.         val DRRQ = sdf.format(now)
109.         val YWRQ = x.getAs[String]("FirstTime")
110.
111.         //对 SQL 进行赋值
112.         ps.setString(1, x.getAs[String]("FirstTime"))
113.         ps.setString(2, x.getAs[String]("LastTime"))
114.         ps.setDouble(3, x.getAs[Double]("I_Mbps"))
115.         ps.setDouble(4, x.getAs[Double]("IA_Mbps"))
116.         ps.setDouble(5, x.getAs[Double]("O_Mbps"))
117.         ps.setDouble(6, x.getAs[Double]("OA_Mbps"))
118.         ps.setString(7, x.getAs[String]("olt_port"))
119.
120.         //获取 OLT 的名称
121.         ps.setString(8, x.getAs[String]("olt_port").split("/")(0))
122.
123.         ps.setString(9, x.getAs[String]("ad_location"))
124.         ps.setString(10, jd)
125.         ps.setString(11, wd)
126.         ps.setString(12, DRRQ)
127.         ps.setString(13, YWRQ)
128.
129.         //进行格式的规范化，保留 7 位小数
130.         val df = new DecimalFormat("#.0000000")
131.
132.         ps.setString(14, df.format(jd1))
133.         ps.setString(15, df.format(wd1))
134.
135.         //进行批处理
136.         ps.addBatch()
137.       })
138.
139.       //进行批处理
140.       ps.executeBatch()
141.       conn.commit()
142.     } //结束尝试
143.     catch {
144.       case e: Exception => e.printStackTrace
145.     }
146.
147.     finally {
148.       //关闭两个管道
149.       if (ps != null) {
150.         ps.close()
151.       }
152.
153.       if (conn != null) {
```

```
154.         conn.close()
155.       }
156.     }
157.
158.
159.   })
160.
161.
162.  }
163.}
```

17.3　本章总结

本章详细阐述了 Spark 在通信运营商生产环境中的两个应用案例，第一个案例是 Spark 在通信运营商融合支付系统日志统计分析中的综合应用；第二个案例是 Spark 在光宽用户流量热力分布 GIS 系统中的综合应用。在通信运营商生产环境中，Spark 的作用在于对业务系统提出的需求，运用 Spark 大数据技术结合生产实际需求开发，实现了业务需求的各个功能。

第 18 章　使用 Spark GraphX 实现婚恋社交网络多维度分析案例

图计算广泛应用于社交网络、电子商务，地图等领域。Spark GraphX 是图计算领域的屠龙宝刀，对 Pregel API 的支持更是让 Spark GraphX 如虎添翼。Spark GraphX 可以轻而易举地完成基于度分布的中枢节点发现、基于最大连通图的社区发现、基于三角形计数的关系衡量、基于随机游走的用户属性传播等。得益于 Spark 的 RDD 抽象，Spark GraphX 可以无缝地与 Spark SQL、MLLib 等结合使用。例如，可以使用 Spark SQL 进行数据的 ETL 之后交给 Spark GraphX 进行处理，而 Spark GraphX 在计算的时候又可以和 MLLib 结合使用，来共同完成深度数据挖掘等人工智能化的操作，这些特性都是其他图计算平台无法比拟的。

在淘宝，Spark GraphX 不仅广泛应用于用户网络的社区发现、用户影响力、能量传播、标签传播等，而且也越来越多地应用到推荐领域的标签推理、人群划分、年龄段预测、商品交易时序跳转等。从技术层面讲，Spark GraphX 非常适合于微信、微博、社交网络、电子商务、地图导航等类型的产品，所以可以期待 Spark GraphX 在 Facebook、Twitter、Linkedin、腾讯、百度等的大规模应用。

18.1　Spark GraphX 发展演变历史和在业界的使用案例

Spark GraphX 是一个分布式图处理框架，基于 Spark 平台提供对图计算和图挖掘简洁易用而丰富多彩的接口，极大地方便了分布式图处理的需求。

社交网络中人与人之间有很多关系链，如 Twitter、Facebook、微博、微信，这些都是大数据产生的地方，都需要图计算。现在的图处理基本都是分布式的图处理，而并非单机处理。Spark GraphX 由于底层是基于 Spark 处理的，所以天然就是一个分布式的图处理系统。

图的分布式或者并行处理其实是把这张图拆分成很多的子图，然后分别对这些子图进行计算，计算时可以分别迭代进行分阶段的计算，即对图进行并行计算。

下面看一下图计算的简单示例，如图 18-1 所示。

从图 18-1 中可以看出：拿到 Wikipedia（维基百科）的文档后，可以变成 Table（表）形式的视图，然后基于 Table（表）形式的视图，我们可以分析 Hyperlinks（超链接），也可以分析 Term-Doc Graph（文本分词图），然后经过主题模型算法（Latent Dirichlet Allocation，LDA）之后进入 WordTopics（单词主题），对于上面的 Hyperlinks，我们可以使用 PageRank（网页排名算法）分析，在下面的 Editor Graph（图编辑）到 Community（社区），这个过程可以称为 Triangle Computation（三角计算），这是计算三角形的一个算法。基于此，会发现一个社区，从上面的分析中可以发现图计算有很多的做法和算法，同时也发现图和表格可以互相转换。不过，并非所有的图计算框架都支持图与表格的互相转换。

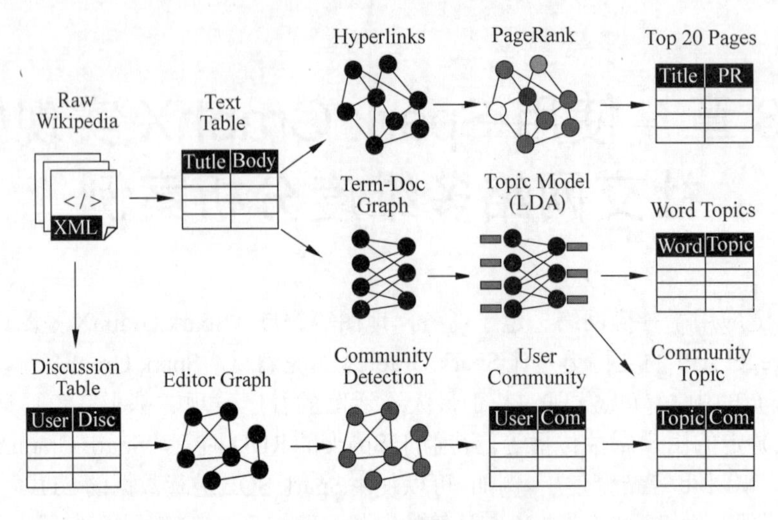

图 18-1　图计算示例

Spark GraphX 的优势在于能够把表格和图进行互相转换，这一点可以带来非常多的优势，现在很多框架也在渐渐地往这方面发展。例如，GraphLib 已经实现了可以读取 Graph 中的 Data，也可以读取 Table 中的 Data，还可以读取 Text 中的 data，即文本中的内容等。与此同时，Spark GraphX 基于 Spark 也为 GraphX 增添了额外的很多优势，如和 mllib、Spark SQL 协作等。

当今图计算领域对图的计算大多数只考虑邻居节点的计算，也就是说，一个节点计算的时候只会考虑其邻居节点，对于非邻居节点并不关心，如图 18-2 所示。

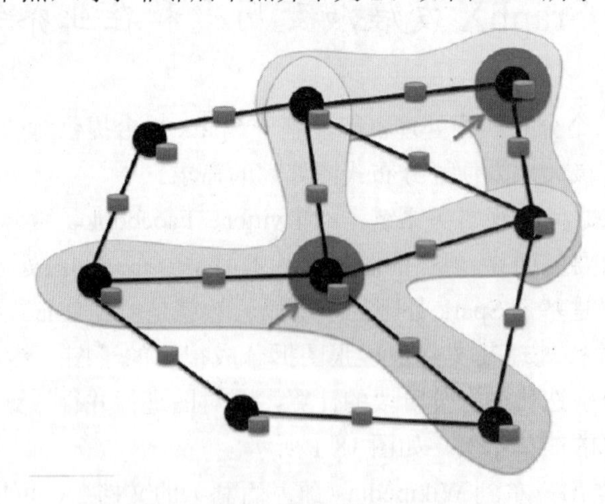

图 18-2　图计算节点

目前基于图的并行计算框架已经有很多，比如来自 Google 的 Pregel、来自 Apache 开源的图计算框架 Giraph，以及最著名的 GraphLab，当然也包含 HAMA，其中，Pregel、HAMA、Giraph 都是非常类似的，都是基于 BSP（Bulk Synchronous Parallel）整体同步并行模型，BSP 模型实现了 SuperStep（超步），BSP 首先进行本地计算，然后进行全局的通信，最后进行全局的 Barrier（栅栏）。BSP 最大的好处是编程简单，而其问题在于一些情况下 BSP 运算的性能非常差，因为我们有一个全局 Barrier 的存在，所以系统速度取决于最慢的计算，也是木桶原理的体现；另一方面，很多现实生活中的网络是符合幂律分布的，也就是定点、边等分布式很不均匀，所以这种情况下，BSP 的木桶原理导致性能问题会得到放大，对这个问题的解决，以 GraphLab（卡内基梅隆大学实验室提出的基于图像处理模型的开源图计算框架）为例，使用了一种异步的概念，而没有全部的 Barrier；最后，不得不提的一点是在 Spark GraphX 中可以用极简洁的代码非常方便地使用 Pregel 的 API。

基于图的计算框架的共同特点，抽象出了一批 API 来简化基于图的编程，这往往比一般的 data-parellel（数据并行）系统的性能高出很多倍。

传统的图计算往往需要不同的系统支持不同的 View（视图）。例如，在 Table View（表视图）下可能需要 Spark 的支持或者 Hadoop 的支持，而在 Graph View（图视图）下可能需要 Pregel 或者 GraphLab 的支持，也就是把图和表分别在不同的系统中进行拉练处理，如图 18-3 所示。

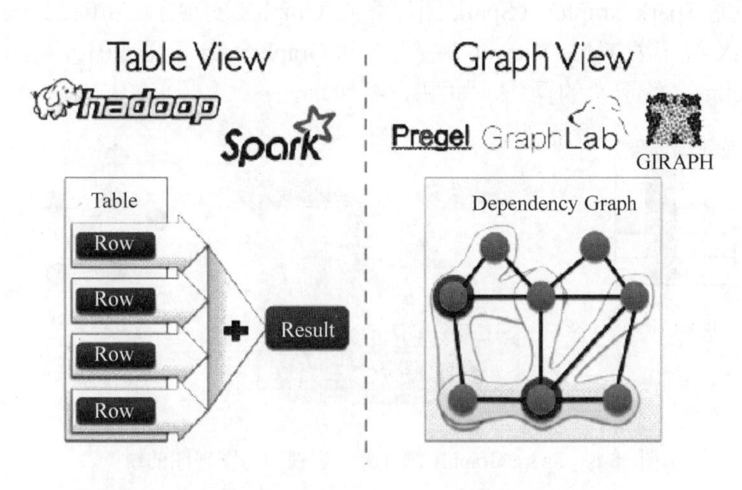

图 18-3　传统图的计算：图、表使用不同的视图处理

上面描述的图计算处理方式是传统的计算方式，当然，除了 Spark GraphX 外，图计算框架也在考虑这个问题；不同系统带来的问题之一是需要学习、部署和管理不同的系统，例如，要同时学习、部署和管理 Hadoop、Hive、Spark、Giraph、GraphLab 等。我们需要用更少的框架解决更多的问题，如图 18-4 所示。

图 18-4　大数据部署的各个系统

其次最关键的问题是效率问题。在不同的转换中间，如果每步转换都要落地，数据转换和复制带来的开销（如序列化带来的开销）也非常大，同时中间结果和相应的结构无法重用，特别是一些结构性的东西。例如，顶点或者边的结构一直没有变，这种情况下结构内部的 Structure（结构）是不需要改变的，而如果每次都重新构建，就算不变，也无法重用，这会

导致性能非常差，如图 18-5 所示。

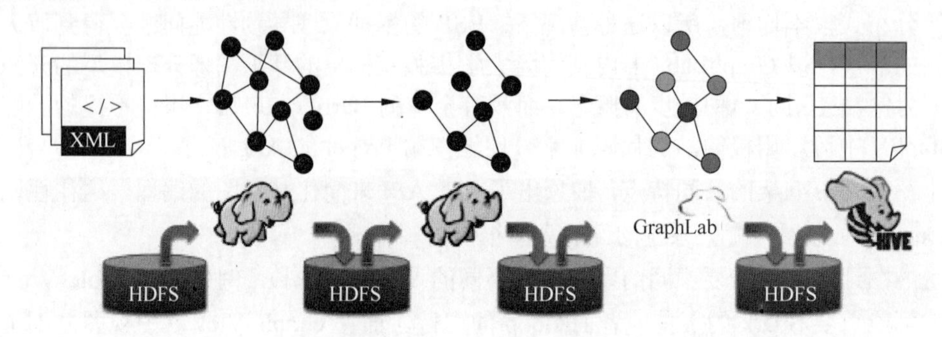

图 18-5　传统图的计算：每步转换落地带来的开销较大

解决方案就是 Spark GraphX（Spark 图计算），GraphX 实现了 Unified Representation（统一表示），GraphX 统一了 Table View（表视图）和 Graph View（图视图），基于 Spark 可以非常轻松地做 pipeline（管道）的操作，如图 18-6 所示。

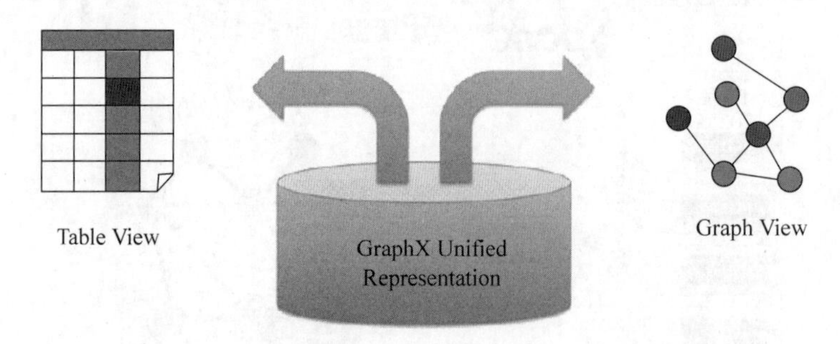

图 18-6　Spark GraphX 图计算：表视图与图视图的统一

如果和 Spark SQL 结合，可以用 SQL 语句进行 ETL（数据清洗），然后放入 GraphX（图计算）处理，这是非常方便的。

在 Spark GraphX 中的 Graph 其实是 Property Graph（属性图），也就是说，图的每个顶点和边都是有属性的，如图 18-7 所示。

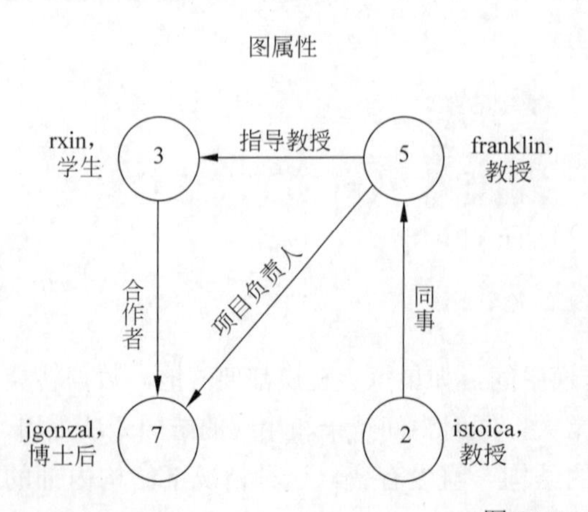

顶点表

ID	属性（顶点）
3	（rxin，学生 student）
7	（jgonzal，博士后 postdoc）
5	（franklin，教师 prof）
2	（istoica，教师 prof）

边表

源 ID	目的 ID	属性（顶点）
3	7	合作者
5	3	指导教授
2	5	同事
5	7	项目负责人

图 18-7　图计算示例

例如，顶点为 3 的名称为 rxin，是学生 stu.，顶点为 5 的名称是 franlin，是教授 prof.，边 5 到 3 表明 5 是 3 的 Advisor（指导教授），图 18-7 中（rxin，学生）表示的是相应顶点的 Property（属性），而“指导教授、合作者、项目负责人、同事”表示的是边的 Property，边和顶点都是有 ID 的，对于顶点而言，有自身的 ID，而对于边来说，有 SourceID（源 ID）和 DestinationID（目的 ID），即对于边而言，会有两个 ID 表达从哪个顶点出发到哪个顶点结束，来表明边的方向，这就是 Property Graph（属性图）的表示方法；如果把 Property 反映到表上，例如，在 Vertex Table（顶点表）中，Id 为 3 的 Property 就是（rxin, student），而在 Edge Table（边表）中，3 到 7 表明的边的 Property 是 Collaborator（合作者）的关系，2 到 5 是 Colleague（同事）的关系；更重要的是，Property Graph（属性图）和 Table（表）之间是可以相互转换的。在 GraphX 中，所有操作的基础是 table operator（表操作）和 graph operator（图操作），其继承自 Spark 中的 RDD，都是针对集合进行操作。

18.2　Spark GraphX 设计实现的核心原理

Spark GraphX 是基于 Spark 的，如何使得 GraphX 进行分布式计算呢？答案是要进行图的切分。

切分有两种：一种是对边进行切分；一种是对顶点进行切分，如图 18-8 所示。

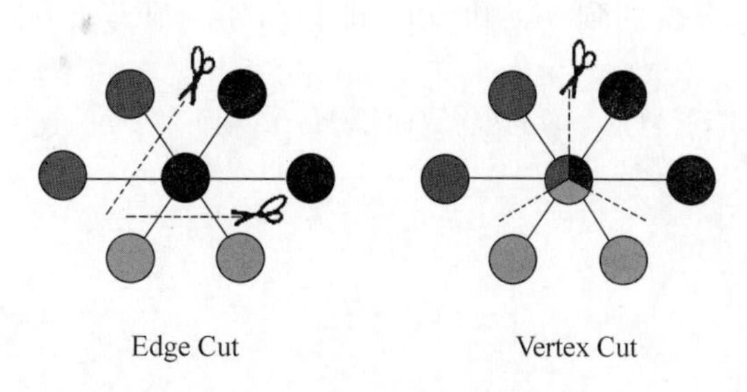

图 18-8　图计算图的切分

GraphX（图计算）使用的 Vertex Cut（顶点切分），即对顶点进行切分。GraphX 在进行切分的时候有几种不同的 Partition（分区）策略，PartitionStrategy（分区策略）专门定义了这些不同的策略，在 PartitionStrategy 的 object（对象）中定义了 4 种不同的 Partition 策略。

第一种分区策略是 RandomVertextCut。

```
1.    /**
        *通过将源顶点 ID 和目标顶点 ID 进行哈希计算分区，在随机顶点切分中，将两个顶点之间
        *所有相同方向的边进行聚合协作
2.      */
3.
4.    case object RandomVertexCut extends PartitionStrategy {
5.      override def getPartition(src: VertexId, dst: VertexId, numParts:
        PartitionID): PartitionID = {
6.        math.abs((src, dst).hashCode()) % numParts
7.      }
8.    }
```

可以看到，RandomVertextCut 是通过对源顶点 ID 和目标顶点 ID 运行 hash 计算来实现的。

第二种分区策略是 CanonicalRandomVertexCut。

```
1.   /**
     *对位于典型方向中的源顶点 ID 和目标顶点 ID 进行哈希计算分区。在随机顶点切分中，不
     *管是什么方向，将两个顶点之间所有的边进行聚合协作
2.   */
3.
4.   case object CanonicalRandomVertexCut extends PartitionStrategy {
5.     override def getPartition(src: VertexId, dst: VertexId, numParts:
       PartitionID): PartitionID = {
6.       if (src < dst) {
7.         math.abs((src, dst).hashCode()) % numParts
8.       } else {
9.         math.abs((dst, src).hashCode()) % numParts
10.      }
11.    }
12.  }
```

从实现上看，CanonicalRandomVertexCut 和 RandomVertextCut 没有本质上的差别。

第三种和第四种分区策略分别是 EdgePartition1D 和 EdgePartition2D。

在四种分区策略中，关键的是 EdgePartition1D 和 EdgePartition2D，其中 EdgePartition1D 只考虑源顶点的 ID，而在 EdgePartition2D 中，源顶点的 ID 和目标顶点的 ID 都会用到，在 EdgePartition2D 的时候，会考虑到 row 和 col，即行和列，源码如下：

```
1.   /**
2.     *使用稀疏边邻接矩阵的 2D 分区将边计算分区，保证 2 sqrt(numParts)绑定在复制的
       *顶点。假设有一个图，图有 12 个顶点，在超过 9 台机器的节点上进行分区，我们可以使用
       *以下稀疏矩阵表示
3.     *
4.     *       __________________________________
5.     *  V0   | P0 *     | P1       | P2    *  |
6.     *  V1   |  ****    |  *       |          |
7.     *  V2   |  ******* |      **  |  ****    |
8.     *  V3   |  *****   |  *  *    |       *  |
9.     *       ----------------------------------
10.    *  V4   | P3 *     | P4 ***   | P5 **  * |
11.    *  V5   |  *  *    |  *       |          |
12.    *  V6   |       *  |      **  |  ****    |
13.    *  V7   |  * * *   |  *  *    |       *  |
14.    *       ----------------------------------
15.    *  V8   | P6   *   | P7    *  | P8  *   * |
16.    *  V9   |     *    |  *    *  |          |
17.    *  V10  |       *  |      **  |  *  *    |
18.    *  V11  | * <-E    |  ***     |      **  |
19.    *       ----------------------------------
20.    *
21.    *边用 E 表示，连接顶点 V11 与顶点 V1，将其分配给处理器 P6 。根据获取的处理器数量，
       *通过 sqrt（numparts）块将矩阵分为 sqrt（numparts）。注意：顶点 V11 的相邻边
       *只能在块（P0，P3，P6）的第一列或者块（P6、P7、P8）的最后一行，因此可以保证
       *顶点 V11 被复制到最多 2 sqrt(numParts)个机器节点上
22.    *
23.
```

```
24.     * 注意， P0 有许多边，因此这个分区计算不均衡。为了提高平衡性，我们首先对每个顶点
        * ID 乘以一个大的素数，将顶点的位置重新打乱洗牌 Shuffle
25.     *
26.     *当请求的分区数不是一个完全平方数，我们使用不同的方法计算：其中最后一列可以有不
        *同数量的行，而其他列仍然保持相同大小的块
27.     */
28.   case object EdgePartition2D extends PartitionStrategy {
29.     override def getPartition(src: VertexId, dst: VertexId, numParts:
        PartitionID): PartitionID = {
30.       val ceilSqrtNumParts: PartitionID = math.ceil(math.sqrt(numParts))
          .toInt
31.       val mixingPrime: VertexId = 1125899906842597L
32.       if (numParts == ceilSqrtNumParts * ceilSqrtNumParts) {
33.         //如果 numParts 等于 ceilSqrtNumParts * ceilSqrtNumParts，我们仍使用
            //老的方法，以确保得到相同的结果
34.         val col: PartitionID = (math.abs(src * mixingPrime) % ceilSqrtNumParts)
            .toInt
35.         val row: PartitionID = (math.abs(dst * mixingPrime) % ceilSqrtNumParts)
            .toInt
36.         (col * ceilSqrtNumParts + row) % numParts
37.
38.       } else {
39.         //否则使用新方法
40.         val cols = ceilSqrtNumParts
41.         val rows = (numParts + cols - 1) / cols
42.         val lastColRows = numParts - rows * (cols - 1)
43.         val col = (math.abs(src * mixingPrime) % numParts / rows).toInt
44.         val row = (math.abs(dst * mixingPrime) % (if (col < cols - 1) rows
            else lastColRows)).toInt
45.         col * rows + row
46.
47.       }
48.     }
49.   }
50.
51.   /**
        *指定边仅使用源顶点 ID 计算分区，使用相同的源顶点协调计算
52.     */
53.
54.   case object EdgePartition1D extends PartitionStrategy {
55.     override def getPartition(src: VertexId, dst: VertexId, numParts:
        PartitionID): PartitionID = {
56.       val mixingPrime: VertexId = 1125899906842597L
57.       (math.abs(src * mixingPrime) % numParts).toInt
58.     }
59.   }
60.
61.
```

在 EdgePartition2D 中可以看到有一个 mixingPrime（素数），这个非常大的素数主要是为了取得计算平衡，但本质上讲，它是无法解决这个问题的，只能缓减这个问题。至此，我们阐述了 PartitionStrategy（分区策略）中定义的 4 种不同的 Partition（分区）策略。

接下来以图 18-9 为例，对 Spark GraphX 图计算的分区进行说明。

图 18-9 中左侧的 Property Graph（属性图）被分成两个 Partition，Vertex Table（顶点表）的信息表明其中 A、B、C 在一个 Partition 中，D、E、F 在另外一个 Partition 中；而右侧的

Edge Table（边表）表明每个 Partition 中的不同的 edge（边）。Spark GraphX 中一个非常棒的地方是 Routing Table（路由表）部分。Routing Table 记录了节点的路由，如图 18-9 中 A 在 Part 1 和 Part 2 中出现，也就是说，在做 mapVertices 和 mapEdges 等操作的时候，其内部结构是不会变的，所以我们可以重用，这个不变的内部结构同时也给 pipeline（管道）带来天然的优势。

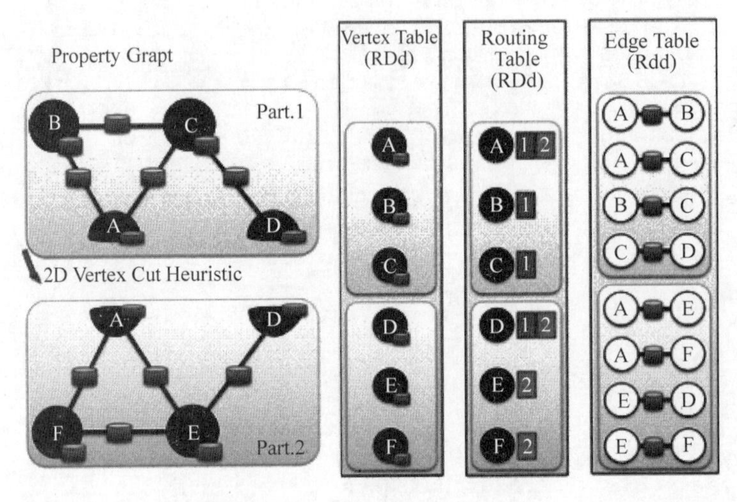

图 18-9　图计算分区示意图

18.3　Table Operator 和 Graph Operator

Spark GraphX 图计算源码中有两个非常重要的类：Graph 类、GraphOps 类。

先看一下 Graph 类。在 IDEA 中，在 Graph.scala 代码中使用 Ctrl+F12 查看 Graph 包含的属性及方法，如图 18-10 所示。

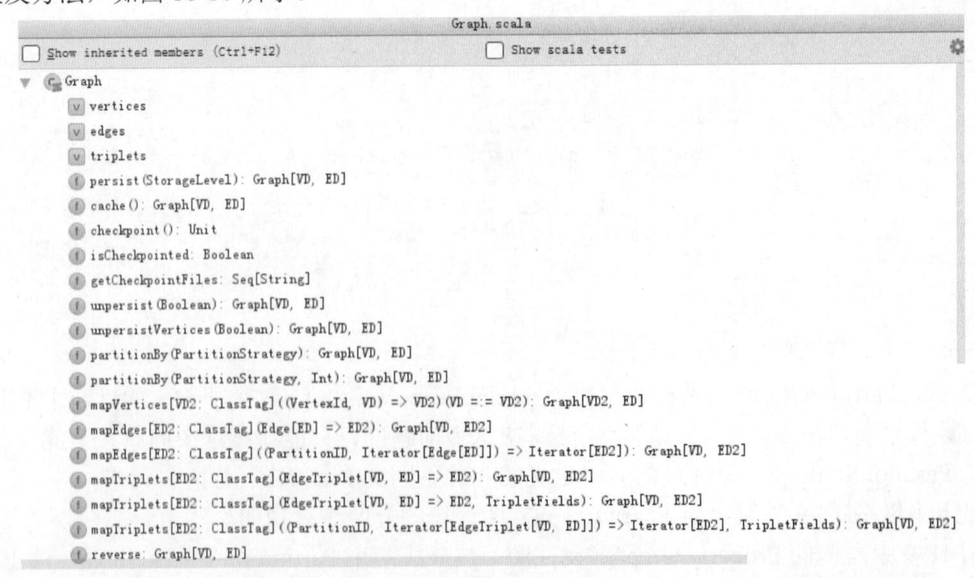

图 18-10　Graph 的属性及方法

在 Graph 中，我们看到 groupEdges、mapTriplets 等最重要的方法。Graph 类的说明正如其注释所示：图 Graph 抽象地表示任意图对象中顶点和边的关系。图提供基本操作去管理和操作顶点、边以及底层结构的相关数据。类似于 Spark RDD，图是一种函数式数据结构，可以通过转换操作返回一个新的图。

```
1.  /**
2.   *图 Graph 抽象地表示任意图对象中顶点和边的关系。图提供基本操作去管理和操作顶点、边
     *以及底层结构的相关数据。类似于 Spark RDD，图是一种函数式数据结构，可以通过转换操
     *作返回一个新的图
3.   *
4.   * @note [[GraphOps]] 包含附加的操作和图算法
5.   *
6.   * @tparam VD 顶点属性类型
7.   *
8.   * @tparam ED  边属性类型
9.   */
10. abstract class Graph[VD: ClassTag, ED: ClassTag] protected () extends
    Serializable {
11.
12.   /**
13.    *  顶点 RDD 包含顶点及相关的属性
14.    *
15.    * @note 顶点 ID 是唯一的
16.    * @return an RDD 返回 RDD 包含图中的顶点
17.    */
18.   val vertices: VertexRDD[VD]
19. ......
```

Graph 类是一个抽象类，很多内容都没有具体实现，具体实现是由 GraphImpl 完成的，源码如下：

```
1.  /**
     *图[org.apache.spark.graphx.Graph]的具体实现。图用两组 RDD 表示：vertices 顶
     *点包含顶点属性和将顶点属性传送到边分区的路由信息；replicatedVertexView 包含边、
     *及边涉及的顶点属性
2.   */
3.
4.  class GraphImpl[VD: ClassTag, ED: ClassTag] protected (
5.      @transient val vertices: VertexRDD[VD],
6.      @transient val replicatedVertexView: ReplicatedVertexView[VD, ED])
7.    extends Graph[VD, ED] with Serializable {
8.
9.    /**构造函数，支持序列化*/
10.   protected def this() = this(null, null)
11.
12.   @transient override val edges: EdgeRDDImpl[ED, VD] = replicatedVertexView.
      edges
13.
14.   /**返回 RDD，边中包含源顶点及目标顶点*/
15.   @transient override lazy val triplets: RDD[EdgeTriplet[VD, ED]] = {
16.     replicatedVertexView.upgrade(vertices, true, true)
17. ......
```

Spark GraphX 图计算源码中另外一个重要的类是 GraphOps 类。GraphOps 类是 Graph 非

常重要的协同工作的类。

```
1.  /**
2.    *包含图[Graph]的附加功能。为每个图对象隐式构造该类，所有操作都可使用 API
3.    *
4.    * @tparam VD 顶点属性类型
5.    * @tparam ED  边属性类型
6.    */
7.  class GraphOps[VD: ClassTag, ED: ClassTag](graph: Graph[VD, ED]) extends
    Serializable {
8.
9.    /** 图中边的数量. */
10.   @transient lazy val numEdges: Long = graph.edges.count()
11.
12.   /** 图中顶点的数量 */
13.   @transient lazy val numVertices: Long = graph.vertices.count()
14.
15.   /**
16.    * 图中每个顶点的度
17.    * @note Vertices with no in-edges are not returned in the resulting RDD.
18.    */
19.   @transient lazy val inDegrees: VertexRDD[Int] =
20.     degreesRDD(EdgeDirection.In).setName("GraphOps.inDegrees")
```

18.4　Vertices、edges 和 triplets

Vertices、edges 和 triplets 是 Spark GraphX 中 3 个最重要的概念。

Vertices 对应的 RDD 名称为 VertexRDD，属性有顶点 ID 和顶点属性。VertexRDD 的源码如下：

```
1.  /**
2.    *扩展 RDD[(VertexId, VD)]确保对于每个顶点，只有一个实体，对于每个实体，建立预
    *索引，这样可以建立快速高效的连接。两个顶点 RDDVertexRDDs 如有相同的索引，可以有
    *效地连接。所有的操作除了[reindex]外，都会保存索引。可以使用[org.apache.spark
    *.graphx.VertexRDD$ VertexRDD object]构建 VertexRDD
3.    *
4.    *
5.    *此外，存储路由信息可以使顶点属性与边 RDD[EdgeRDD]进行关联
6.    **
7.    *@example  从一个普通的 RDD 构建 VertexRDD
8.    *{{{
9.    *// 构建初始化顶点集
10.   *val  someData: RDD[(VertexId, SomeType)] = loadData(someFile)
11.   *val  vset = VertexRDD(someData)
12.   *//如果在 someData 中有一些冗余值，我们使用 reduceFunc 函数进行聚合
13.   *val  vset2 = VertexRDD(someData, reduceFunc)
14.   *//最后，可以转换 VertexRDD 去索引另一个数据集
15.   *val  otherData: RDD[(VertexId, OtherType)] = loadData(otherFile)
16.   *val  vset3 = vset2.innerJoin(otherData) { (vid, a, b) => b }
17.   *//现在我们可以构造两个集合之间的快速连接
```

```
18.   *val  vset4: VertexRDD[(SomeType, OtherType)] = vset.leftJoin(vset3)
19.   *}}}
20.   *
21.   *@tparam VD  每个顶点关联的顶点属性
22.   */
23. abstract  class  VertexRDD[VD](
24.     sc: SparkContext,
25.     deps: Seq[Dependency[_]]) extends RDD[(VertexId, VD)](sc, deps) {
26.   implicit protected def vdTag: ClassTag[VD]
27. ......
```

从源码中可以看出 VertexRDD 继承自 RDD[(VertexId, VD)]，RDD 的类型是 VertexId 和 VD，其中 VD 是属性的类型，也就是说，VertexRDD 有 ID 和顶点属性。

边 Edges 对应的是 EdgeRDD，属性有 3 个：源顶点的 ID、目标顶点的 ID、边属性。EdgeRDD 的源码如下：

```
1.  /**
    *EdgeRDD[ED, VD]继承自 RDD[Edge[ED]]，通过在每个分区以列格式存储边的数据，
    *还可以存储与边关联的顶点属性，提供边的三元组视图。顶点属性的传送由 impl.Replicated
    *VertexView 进行管理
2.    */
3.
4.  abstract class EdgeRDD[ED](
5.      sc: SparkContext,
6.      deps: Seq[Dependency[_]]) extends RDD[Edge[ED]](sc, deps) {
7.
8.    // scalastyle:关闭 structural.type
9.   private[graphx] def  partitionsRDD: RDD[(PartitionID, EdgePartition[ED,
     VD])] forSome { type VD }
10.   // scalastyle:打开 structural.type
11.
12.  override protected def  getPartitions: Array[Partition] = partitionsRDD.
     partitions
```

从源码中可以看出 EdgeRDD 继承的 RDD 的类型是 Edge[ED]。

Triplets 的属性有源顶点 ID、源顶点属性、边属性、目标顶点 ID、目标顶点属性，Triplets 其实是对 Vertices 和 Edges 做了 Join 操作，如图 18-11 所示。

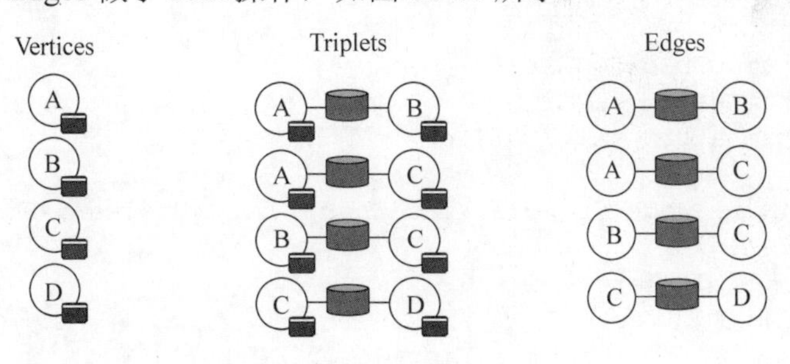

图 18-11　Triplets 属性

顶点 Vertices 具有顶点 ID 和属性，边 Edges 具有源顶点 ID、目标顶点 ID 和自己的属性，顶点 Triplets 对顶点 Vertices 和边 Edges 的 Join 操作使得 Triplets 具备源顶点的 ID 和属性、目标顶点的 ID 和属性以及自己的属性。也就是说，Triplets 把包含源顶点和目标顶点的属性以及自身的属性全部连在一起了，如果我们需要使用顶点及自己的属性以及和顶点关联的边

的属性，那就必须使用 Triplets。Triplets 的 RDD 的类型是 EdgeTriplet，其源码实现如下：

```
1.  /**
2.    *边三元组表示一条边以及边相邻顶点的顶点属性
3.    *
4.    *@tparam VD 顶点属性类型
5.    *@tparam ED  边属性类型
6.    */
7.  class EdgeTriplet[VD, ED] extends Edge[ED] {
8.    /**
      *源顶点属性
9.      */
10.
11.   var srcAttr: VD = _ // nullValue[VD]
12.
13.   /**
      * 目标顶点属性
14.    */
15.
16.   var dstAttr: VD = _ // nullValue[VD]
17.
18.   /**
      * 设置三元组中边的属性
19.    */
20.
21.   protected[spark] def  set(other: Edge[ED]): EdgeTriplet[VD, ED] = {
22.     srcId = other.srcId
23.     dstId = other.dstId
24.     attr = other.attr
25.     this
26.   }
```

从源码中可以看到 EdgeTriplet 继承自 Edge[ED]，同时，EdgeTriplet 用 srcAttr 表示源顶点的属性，用 dstAttr 表示目标顶点的属性。其父类 Edge 的源码如下：

```
1.  /**
2.    *由源 ID、目标 ID 组成的单个有向边，与边相关的数据。
3.    *
4.    * @tparam ED  边属性类型
5.    *
6.    * @param srcId  源顶点 ID
7.    * @param dstId  目标顶点 ID
8.    * @param attr   关联的边的属性
9.    */
10.  case class Edge[@specialized(Char, Int, Boolean, Byte, Long, Float,
     Double) ED] (
11.     var srcId: VertexId = 0,
12.     var dstId: VertexId = 0,
13.     var attr: ED = null.asInstanceOf[ED])
14.   extends Serializable {
15.
16.  /**
17.    *给定边中的一个顶点，返回另一个顶点
18.    *
19.    * @param vid 边上两个顶点，其中一个顶点的 ID
20.    * @return 边上另一个顶点的 ID
```

```
21.     */
22.   def otherVertexId(vid: VertexId): VertexId =
23.     if (srcId == vid) dstId else { assert(dstId == vid); srcId }
```

从边 Edge 的源码中可以看到，构造方法中用 srcId 标志源顶点的 ID，用 dstId 标志目标顶点的 ID，用 attr 表示 edge 的属性，这样，作为 Edge 的继承者 EdgeTriplet 就拥有 5 个属性，而前面分析的 VertexRDD 只有两个属性集，即 Vertex 的 ID 和属性。

18.5　以最原始的方式构建 Graph

以最原始的方式构建 Graph 其实使用的是 Graph 的伴生对象的 apply 的方式，其源码如下所示：

```
1.   /**
2.     *从顶点集合和边属性构建图。任意选择复制的顶点，在边集合中找到顶点，而不是在输入顶
3.     *点中查找，将顶点指定为默认属性
4.     *
5.     * @tparam VD  顶点属性类型
6.     * @tparam ED  边属性类型
7.     * @param vertices  顶点及其属性的“集合”
8.     * @param edges    图中边的集合
9.     * @param defaultVertexAttr 用于在边中但不在顶点中提到的顶点，设置默认顶点属性
10.    * @param edgeStorageLevel  必要的存储级别，在必要时缓存边
11.    * @param vertexStorageLevel  如果需要，缓存顶点的存储级别
12.    */
13.  def apply[VD: ClassTag, ED: ClassTag](
14.      vertices: RDD[(VertexId, VD)],
15.      edges: RDD[Edge[ED]],
16.      defaultVertexAttr: VD = null.asInstanceOf[VD],
17.      edgeStorageLevel: StorageLevel = StorageLevel.MEMORY_ONLY,
18.      vertexStorageLevel: StorageLevel = StorageLevel.MEMORY_ONLY): Graph
         [VD, ED] = {
19.    GraphImpl(vertices, edges, defaultVertexAttr, edgeStorageLevel, vertex
       StorageLevel)
20.  }
```

从源码中可以看到，构建图 Graph 首先需要顶点 Vertices 的 RDD，此时的顶点 Vertices 的类型是 RDD[(VertexId, VD)]，当然，也可以直接使用 VertexRDD，因为 VertexRDD 继承了 RDD[(VertexId, VD)]，第二点是要放入边 edges，第三个元素是放入默认顶点的属性，当然可以 null。需要注意的是，默认的顶点属性只是给我们使用，并不在顶点 Vertices 里面，接下来是 edgeStorageLevel 和 vertexStorageLevel，用来指定存储策略。图的边和顶点并不是在一起 Cache 的，这里默认的存储策略都是 StorageLevel.MEMORY_ONLY，最后我们发现创建一个 Graph 实例是使用 GraphImpl 来完成的。

18.6　第一个 Graph 代码实例并进行 Vertices、edges 和 triplets 操作实战

在这一节中，构建 Graph 的数据来源有两种：一种是本地的数据集；一种是来自 Google

的数据集。下面先看一下本地数据集构建 Graph 的操作，该内容也来自 Spark 1.0.2 GraphX 的官方文档：http://spark.apache.org/docs/latest/graphx-programming-guide.html。

我们在 IDEA 中已导入 Spark 7.1.0 的 Jar 包，将在 IDEA 中编码实现 Spark Graph 的案例。首先把相关应该导入的类导入进来。

```
1.  import org.apache.spark._
2.  import org.apache.spark.graphx._
3.  //在图计算的例子运行中我们需要导入 RDD 的包
4.  import org.apache.spark.rdd.RDD
```

可以看出我们导入了 org.apache.spark 下面的内容，同时也导入了 org.apache.spark.graphx 下面的内容，最后为了方便后续的 RDD 操作，还导入了 org.apache.spark.rdd.RDD。作为第一个例子，我们用的是图 18-7 中表示的数据。

Spark GraphX 官方文档提供的代码如下：

```
1.  //假设已经构建 SparkContext
2.  val sc: SparkContext
3.  //创建顶点的 RDD
4.  val users: RDD[(VertexId, (String, String))] =
5.    sc.parallelize(Array((3L,("rxin", "student")), (7L, ("jgonzal", "postdoc")),
6.                        (5L, ("franklin", "prof")), (2L, ("istoica", "prof"))))
7.  //创建边的 RDD
8.  val relationships: RDD[Edge[String]] =
9.    sc.parallelize(Array(Edge(3L, 7L, "collab"),   Edge(5L, 3L, "advisor"),
10.                        Edge(2L, 5L, "colleague"), Edge(5L, 7L, "pi")))
11. //定义案例中顶点的默认使用者，在缺失使用者的情况下，图可以建立关联
12. val defaultUser = ("John Doe", "Missing")
13. //初始化图
14. val graph = Graph(users, relationships, defaultUser)
```

对于图中的顶点 users 这个 RDD 而言，其每个元素包含一个 ID 和属性，属性是由姓名（name）和职业（occupation）构成的元组，因为要生产 RDD，所以使用 sc.parallelize 函数把 Array 转换一下。parallelize 方法的源码如下：

```
1.    /** 分配本地 Scala 集合形成一个 RDD
2.      *注：并行化操作是懒加载。如果 SEQ 是一个可变的集合，在 RDD 的第一个执行动作之前
        *调用并行化方法数据会发生改变，计算的结果 RDD 将反映变化以后的集合。传递一个复制
        *的参数来避免这个情况。
3.      *注：避免使用并行(seq())创建一个空的 RDD。考虑到 emptyrdd RDD 没有分区，或
        *并行(SEQ [t]())是空分区的情况。
4.      */
5.    def parallelize[T: ClassTag](
6.        seq: Seq[T],
7.        numSlices: Int = defaultParallelism): RDD[T] = withScope {
8.      assertNotStopped()
9.      new ParallelCollectionRDD[T](this, seq, numSlices, Map[Int, Seq
        [String]]())
10.   }
```

从官方给出的第一个图 GraphX 的源码同样可以看出，关联（relationships）每个元素由源顶点 ID、目标顶点 ID 和边的属性三部分构成。

接下来一个非常重要的对象为默认使用者（defaultUser），其主要作用就在于，当想描述

一种关联中不存在的目标顶点的时候，就会使用这个默认使用者。例如，5 到 0 这个关联是不存在的，那就会默认指向默认使用者，这就是默认使用者的用途。如果在编码中不想要默认使用者，这也是可以的。

在 IDEA 中编码首先生成使用者（users）这个 RDD。

```
1.   val users: RDD[(VertexId, (String, String))] =
2.       sc.parallelize(Array((3L, ("rxin", "student")),
3.      (7L, ("jgonzal", "postdoc")),
4.   (5L, ("franklin", "prof")),
5.  (2L, ("istoica", "prof"))))
```

上述代码生成的是顶点表（Vertex Table），如表 18-1 所示。

表 18-1　顶点表

ID	属性（顶点）	ID	属性（顶点）
3	（rxin，student）	5	（franklin，prof）
7	（jgonzal，postdoc）	2	（istoica，prof）

接下来生成关联这个 RDD。

```
1.   val relationships: RDD[Edge[String]] =
2.       sc.parallelize(Array(Edge(3L, 7L, "collab"),
3.                        Edge(5L, 3L, "advisor"),
4.                    Edge(2L, 5L, "colleague"),
5.                        Edge(5L, 7L, "pi")))
```

上述代码生成的是边表（Edge Table），如表 18-2 所示

表 18-2　边表

源 ID	目的 ID	属性（边）
3	7	collab
5	3	advisor
2	5	colleague
5	7	PI

代码中使用了 Edge 这个 case class，其源码如下：

```
1.  /**
2.    *  由源 ID、目标 ID 及边相关的数据组成的单条有向边。
3.    *
4.    * @tparam ED 边属性的类型
5.    *
6.    * @param srcId 源顶点的顶点 ID
7.    * @param dstId 目标顶点的顶点 ID
8.    * @param attr 边的关联属性
9.    */
10.case class Edge[@specialized(Char, Int, Boolean, Byte, Long, Float, Double)
   ED] (
11.     var srcId: VertexId = 0,
12.     var dstId: VertexId = 0,
13.     var attr: ED = null.asInstanceOf[ED])
14.   extends Serializable {
```

接下来放入默认使用者：

```
1.      val defaultUser = ("John Doe", "Missing")
```

此时我们使用 Graph 的 Graph 方法（调用 object Graph 伴生对象的 apply 方法）即可构造出图，代码如下所示。

```
1.  val graph = Graph(users, relationships, defaultUser)
```

再次看一下 Graph 实例构造的源码。

```
1.   /**
2.     *从顶点集合以及边属性集构建一个图，将边集合中可以找到的顶点进行复制，如发现顶点不
     *在输入的顶点中，将被分配默认属性
3.     *
4.     * @tparam VD 顶点的属性类型
5.     * @tparam ED 边的属性类型
6.     * @param vertices 顶点以及属性集
7.     * @param edges 图中边的集合
8.     * @param defaultVertexAttr  用于顶点的默认顶点属性，这些顶点在边中涉及，但不
     *在顶点中
9.     * @param edgeStorageLevel   定义的存储级别，必要时缓存边的数据
10.    * @param vertexStorageLevel 定义的存储级别，必要时缓存顶点的数据
11.    */
12.   def apply[VD: ClassTag, ED: ClassTag](
13.      vertices: RDD[(VertexId, VD)],
14.      edges: RDD[Edge[ED]],
15.      defaultVertexAttr: VD = null.asInstanceOf[VD],
16.      edgeStorageLevel: StorageLevel = StorageLevel.MEMORY_ONLY,
17.      vertexStorageLevel: StorageLevel = StorageLevel.MEMORY_ONLY): Graph[VD,
     ED] = {
18.    GraphImpl(vertices,  edges,  defaultVertexAttr,  edgeStorageLevel,
     vertexStorageLevel)
19.   }
```

从源码中可以看出使用 Graph 类的伴生对象的 apply 方法产生了 Graph 实例对象，具体的实现为 GraphImpl。在代码中依旧使用了 GraphImpl 伴生对象的 apply 方法，如下所示：

```
1.  /**
2.    * 从顶点和边数据构建一个图,设置默认的顶点为 defaultVertexAttr.
3.    */
4.    def apply[VD: ClassTag, ED: ClassTag](
5.       vertices: RDD[(VertexId, VD)],
6.       edges: RDD[Edge[ED]],
7.       defaultVertexAttr: VD,
8.       edgeStorageLevel: StorageLevel,
9.       vertexStorageLevel: StorageLevel): GraphImpl[VD, ED] = {
10.     val edgeRDD = EdgeRDD.fromEdges(edges)(classTag[ED], classTag[VD])
11.       .withTargetStorageLevel(edgeStorageLevel)
12.     val vertexRDD = VertexRDD(vertices, edgeRDD, defaultVertexAttr)
13.       .withTargetStorageLevel(vertexStorageLevel)
14.     GraphImpl(vertexRDD, edgeRDD)
15.   }
```

继续跟踪源码，在以下源码中可以看到第 8 行进行了顶点的缓存，第 13 行进行了边的缓存。

```
1.   /**
     *从顶点 RDD（VertexRDD）以及边 RDD（EdgeRDD）中任意复制顶点构建一个图。通过调
     *用 VertexRDD.withEdges 或者适当的 VertexRDD 构造器，顶点 RDD（VertexRDD）
     *将和边 RDD（EdgeRDD）建立有效的关联
2.    */
3.   def apply[VD: ClassTag, ED: ClassTag](
4.       vertices: VertexRDD[VD],
5.       edges: EdgeRDD[ED]): GraphImpl[VD, ED] = {
6.
7.     vertices.cache()
8.
9.     //将边中顶点分区转换为正确类型
10.    val newEdges = edges.asInstanceOf[EdgeRDDImpl[ED, _]]
11.      .mapEdgePartitions((pid, part) => part.withoutVertexAttributes[VD])
12.      .cache()
13.
14.    GraphImpl.fromExistingRDDs(vertices, newEdges)
15.  }
```

进入 fromExistingRDDs 中：

```
1.   /**
     *使用同样的复制顶点的类型，从顶点 RDD（VertexRDD）以及边 RDD（EdgeRDD）构建
     *一个图，通过调用 VertexRDD.withEdges 或者适当的 VertexRDD 构造器，顶点 RDD
2.    *（VertexRDD）将和边 RDD（EdgeRDD）建立有效关联
3.    */
4.   def fromExistingRDDs[VD: ClassTag, ED: ClassTag](
5.       vertices: VertexRDD[VD],
6.       edges: EdgeRDD[ED]): GraphImpl[VD, ED] = {
7.     new GraphImpl(vertices, new ReplicatedVertexView(edges.asInstanceOf
     [EdgeRDDImpl[ED, VD]]))
8.   }
```

此时使用 GraphImpl 的类构造出了 Graph 对象实例。

```
1.  /**
     *通过[org.apache.spark.graphx.Graph]的实现支持图计算
2.   *
3.
4.   *图表示为两个 RDD：vertices 包含了顶点的属性以及用于将顶点属性传送到边分区的路由
     *信息；replicatedVertexView,其中包含每个边所涉及的边和顶点属性
5.   */qqqqqqqqqqq
6.  class GraphImpl[VD: ClassTag, ED: ClassTag] protected (
7.      @transient val vertices: VertexRDD[VD],
8.      @transient val replicatedVertexView: ReplicatedVertexView[VD, ED])
9.    extends Graph[VD, ED] with Serializable {
10.
11.   /** Default constructor is provided to support serialization */
12.   protected def this() = this(null, null)
13.
14.   @transient override val edges: EdgeRDDImpl[ED, VD] = replicatedVertex
      View.edges
15.
16.   /** Return an RDD that brings edges together with their source and
      destination vertices. */
17.   @transient override lazy val triplets: RDD[EdgeTriplet[VD, ED]] = {
```

```
18.    replicatedVertexView.upgrade(vertices, true, true)
19.    replicatedVertexView.edges.partitionsRDD.mapPartitions(_.flatMap {
20.      case (pid, part) => part.tripletIterator()
21.    })
22.  }
```

其中 ReplicatedVertexView 的代码如下：

```
1.  /**
     *通过[org.apache.spark.graphx.EdgeRDD]管理顶点属性传送到边分区。顶点属性可以
     *部分传送，来构造只有一个顶点属性的三元组，它们可以被更新。一个活跃的顶点还可以将设置
     *传送到边分区。注意，不要将引用存储到 edges，因为它可能在属性传送级别更新的时候被修改
2.   */
3.  private[impl]
4.  class ReplicatedVertexView[VD: ClassTag, ED: ClassTag](
5.      var edges: EdgeRDDImpl[ED, VD],
6.      var hasSrcId: Boolean = false,
7.      var hasDstId: Boolean = false) {
8.
9.    /**
      *返回一个 ReplicatedVertexView 与指定的 EdgeRDD,必须使用相同的传送级别
10.   */
11.   def withEdges[VD2: ClassTag, ED2: ClassTag](
12.       _edges: EdgeRDDImpl[ED2, VD2]): ReplicatedVertexView[VD2, ED2] = {
13.     new ReplicatedVertexView(_edges, hasSrcId, hasDstId)
14.   }
```

可以看到，实际使用 GraphImpl 构造出了图的实例，此时就构造出了一张构建图，如图 18-12 所示。

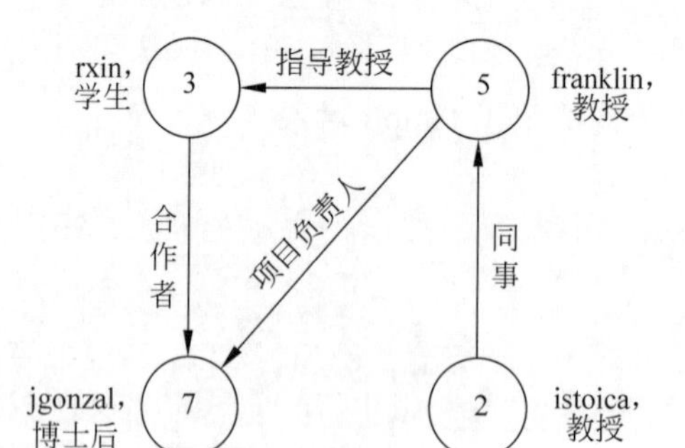

图 18-12　构建图

下面按照 Spark GraphX 官方文档提供的方式对构造出来的 Graph 对象进行操作。例如，要看职业（occupation）为博士后（postdoc）的顶点数目，使用如下代码即可。

```
1.     //val graph: Graph[(String, String), String] 在之前构建的图实例
2.     //计算图中职业是博士后的节点数量
3.     val filteredVertices: VertexId = graph.vertices.filter { case (id,
       (name, pos)) => pos == "postdoc" }.count
4.     System.out.println("图中职业是博士后的节点数量计数为:  " + filteredVertices)
```

在 IDEA 中运行代码结果如下：

```
1.  Using Spark's default log4j profile: org/apache/spark/log4j-defaults.
    properties
2.  17/02/22 12:20:40 INFO SparkContext: Running Spark version 7.1.0
3.  17/02/22 12:20:41 WARN NativeCodeLoader: Unable to load native-hadoop
    library for your platform... using builtin-java classes where applicable
4.  ......
5.  17/02/22 12:20:48 INFO DAGScheduler: ResultStage 2 (reduce at
    VertexRDDImpl.scala:90) finished in 0.207 s
6.  17/02/22 12:20:48 INFO DAGScheduler: Job 0 finished: reduce at
    VertexRDDImpl.scala:90, took 7.069002 s
7.  图中职业是博士后的节点数量计数为:    1
```

通过计算发现顶点使用者的职业是博士后的节点，此时从图的属性可以发现顶点为 7 的这个使用者，如图 18-12 所示。

例如，要计算生成的 graph 中源顶点 ID 大于目标顶点 ID 的数量，直接使用如下代码即可。

```
1.  //统计边中源 ID 大于目的 ID 的数量
2.      val filteredSrcDstId: VertexId = graph.edges.filter(e => e.srcId >
        e.dstId).count
3.      System.out.println("图中边的源 ID 大于目的 ID 的数量是:    " + filteredSrcDstId)
```

在 IDEA 中运行代码结果如下：

```
1.  Using Spark's default log4j profile: org/apache/spark/log4j-defaults.
    properties
2.  17/02/22 12:33:21 INFO SparkContext: Running Spark version 7.1.0
3.  ......
4.  17/02/22 12:33:28 INFO TaskSetManager: Finished task 0.0 in stage 3.0 (TID
    24) in 113 ms on localhost (executor driver) (8/8)
5.  17/02/22 12:33:28 INFO TaskSchedulerImpl: Removed TaskSet 3.0, whose tasks
    have all completed, from pool
6.  17/02/22 12:33:28 INFO DAGScheduler: ResultStage 3 (count at Graphx_
    VerticesEdgesTriplets.scala:62) finished in 0.114 s
7.  17/02/22 12:33:28 INFO DAGScheduler: Job 1 finished: count at Graphx_
    VerticesEdgesTriplets.scala:62, took 0.132638 s
8.  图中边的源 ID 大于目的 ID 的数量是:    1
```

计算结果表明只有一个源 ID 大于目的 ID 的情况，此时我们看一下前面生成的图 18-12。

从图 18-12 中可以发现只有 5 大于 3，其他都是源 ID 小于目的 ID。其实，对构建出来的图（Graph）除了可以对其顶点（Vertices）和边（edges）进行操作外，还可以对它的三元组（triplets）进行操作。当图被构建出来的时候，其本身就有 3 个非常重要的属性，源码如下：

```
1.  abstract class Graph[VD: ClassTag, ED: ClassTag] protected () extends
    Serializable {
2.
3.    /**
4.     * RDD 包含顶点以及顶点相关的属性
5.     *
6.     * @note 顶点 ID 是唯一的。
7.     * @return 返回图中包含顶点的 RDD
8.     */
9.    val vertices: VertexRDD[VD]
10.
```

```
11.   /**
12.    * RDD 包含边以及边相关的属性。RDD 中包含源 ID 和目的 ID 以及边的数据
13.    *
14.    * @return RDD 包含图中的边
15.    *
16.    * @see 'Edge' 边的类型
17.    * @see 'Graph#triplets'获取得到一个 RDD，RDD 包含所有的边以及连同边的顶点数据
18.    */
19.   val edges: EdgeRDD[ED]
20.
21.
22.
23.  /**
24.    *  RDD 包含边的三元组，边连同相邻的顶点的数据
25.    *  如果不需要顶点数据，调用方应该使用[edges]
26.    *  适用于需要边和相邻顶点 ID 的数据
27.    *
28.    * @return 返回 RDD 包含边的三元组
29.    *
30.    * @example  例子：此操作可用于评估图形着色，用于检查两个顶点是否为不同的颜色
31.    * {{{
32.    * type Color = Int
33.    * val graph: Graph[Color, Int] = GraphLoader.edgeListFile("hdfs://file
       *.tsv")
34.    * val numInvalid = graph.triplets.map(e => if (e.src.data == e.dst.data)
       *1 else 0).sum
35.    * }}}
36.    */
37.   val triplets: RDD[EdgeTriplet[VD, ED]]
```

接下来我们查看一下 triplets。

```
1.    val resTriplets: RDD[EdgeTriplet[(String, String), String]] = graph
      .triplets
```

为什么 EdgeTriplet 的类型是 EdgeTriplet[(String, String),String]呢？源码如下：

```
1.  /**
2.    * 边三元组表示一个边，以及它相邻顶点的顶点属性
3.    *
4.    * @tparam VD 顶点属性的类型
5.    * @tparam ED 边属性的类型
6.    */
7.  class EdgeTriplet[VD, ED] extends Edge[ED] {
```

从源码中可以看出第一个元素是顶点属性类型，在我们的例子中就是（name, occupation）的元组，第二个元素是边属性类型。因为其数量比较少，接下来进行 collect 操作。

```
1.  val resTriplets: RDD[EdgeTriplet[(String, String), String]] = graph
    .triplets
2.      for (elem <- resTriplets.collect()) {
3.        println("三元组:     " + elem)
4.      }
```

在 IDEA 中运行代码结果如下：

```
1.  Using Spark's default log4j profile: org/apache/spark/log4j-defaults
    .properties
2.  17/02/22 13:23:23 INFO SparkContext: Running Spark version 7.1.0
3.  ......
4.  17/02/22 13:23:29 INFO ShuffleBlockFetcherIterator: Started 0 remote
    fetches in 0 ms
5.  三元组:     ((3,(rxin,student)),(7,(jgonzal,postdoc)),collab)
6.  三元组:     ((5,(franklin,prof)),(3,(rxin,student)),advisor)
7.  三元组:     ((2,(istoica,prof)),(5,(franklin,prof)),colleague)
8.  三元组:     ((5,(franklin,prof)),(7,(jgonzal,postdoc)),pi)
```

下面看一下顶点：

```
1.      for (elem <- graph.vertices.collect()) {
2.        println("graph.vertices:    " + elem)
3.      }
```

在 IDEA 中运行代码结果如下：

```
1.  Using Spark's default log4j profile: org/apache/spark/log4j-defaults
    .properties
2.  17/02/22 13:26:31 INFO SparkContext: Running Spark version 7.1.0
3.  ......
4.  17/02/22 13:26:37 INFO DAGScheduler: Job 3 finished: collect at Graphx_
    VerticesEdgesTriplets.scala:69, took 0.079005 s
5.  graph.vertices:     (2,(istoica,prof))
6.  graph.vertices:     (3,(rxin,student))
7.  graph.vertices:     (5,(franklin,prof))
8.  graph.vertices:     (7,(jgonzal,postdoc))
```

也可以直接查看 graph 边的情况。

```
1.     for (elem <- graph.edges.collect()) {
2.         println("graph.edges:    " + elem)
3.       }
```

在 IDEA 中运行代码结果如下：

```
1.  Using Spark's default log4j profile: org/apache/spark/log4j-defaults
    .properties
2.  17/02/22 13:26:31 INFO SparkContext: Running Spark version 7.1.0
3.  ......
4.  17/02/22 13:26:37 INFO DAGScheduler: Job 4 finished: collect at EdgeRDDImpl
    .scala:50, took 0.082806 s
5.  graph.edges:     Edge(3,7,collab)
6.  graph.edges:     Edge(5,3,advisor)
7.  graph.edges:     Edge(2,5,colleague)
8.  graph.edges:     Edge(5,7,pi)
```

本节例子较简单，完整的代码如例 18-1 所示。

【例 18-1】 Vertices、edges、triplets 操作。

```
1.  package com.dt.spark.graphx
2.  import org.apache.spark.SparkConf
3.  import org.apache.spark.sql.SparkSession
4.  import org.apache.spark._
```

```
5.  import org.apache.spark.graphx._
6.  //在图计算的例子运行中需要导入 RDD 的包
7.  import org.apache.spark.rdd.RDD
8.
9.  object Graphx_VerticesEdgesTriplets {
10.   def main(args: Array[String]) {
11.     var masterUrl = "local[8]"
        //默认程序运行在本地 Local 模式中，主要用于学习和测试
12.     /**
          * 当我们把程序打包运行在集群上的时候，一般都会传入集群的 URL 信息，这里假设传入
          * 参数，第一个参数只传入 Spark 集群的 URL，第二个参数传入的是数据的地址信息
13.       */
14.
15.
16.     if (args.length > 0) {
17.       masterUrl = args(0)
18.     }
19.     /**
          *创建 Spark 会话上下文 SparkSession 和集群上下文 SparkContext，在 SparkConf
          *中可以进行各种依赖和参数的设置等，可以通过 SparkSubmit 脚本的 help 看设置信息，
          *其中 SparkSession 统一了 Spark SQL 运行的不同环境
20.       */
21.
22.     val  sparkConf  =  new  SparkConf().setMaster(masterUrl).setAppName
       ("Graphx_VerticesEdgesTriplets")
23.
24.     /**
          * SparkSession 统一了 Spark SQL 执行时的不同的上下文环境。也就是说，Spark SQL
          *无论运行在哪种环境下，我们都可以只使用 SparkSession 这样一个统一的编程入口来
          *处理 DataFrame 和 DataSet 编程，不需要关注底层是否有 Hive 等
25.       */
26.
27.     val spark = SparkSession
28.       .builder()
29.       .config(sparkConf)
30.       .getOrCreate()
31.
32.     val sc = spark.sparkContext
         //从 SparkSession 获得的上下文，这是因为我们读原生文件的时候或者实现一些
         //Spark SQL 目前还不支持的功能的时候，需要使用 SparkContext
33.
34.     //创建顶点的 RDD
35.     val users: RDD[(VertexId, (String, String))] =
36.       sc.parallelize(Array((3L, ("rxin", "student")), (7L, ("jgonzal",
        "postdoc")),     (5L, ("franklin", "prof")),(2L, ("istoica", "prof"))))
37.
38.  //创建边的 RDD
39.     val relationships: RDD[Edge[String]] =
40.       sc.parallelize(Array(Edge(3L, 7L, "collab"), Edge(5L, 3L, "advisor"),
41.         Edge(2L, 5L, "colleague"), Edge(5L, 7L, "pi")))
42.
43.     //定义案例中顶点的默认使用者，在缺失使用者的情况下，图可以建立关联
44.     val defaultUser = ("John Doe", "Missing")
45.     //初始化图
46.     val graph: Graph[(String, String), String] = Graph(users, relationships,
        defaultUser)
```

```
47.
48.     //统计图中职业是博士后的节点
49.     val filteredVertices: VertexId = graph.vertices.filter { case (id, (name,
        pos)) => pos == "postdoc" }.count
50.     System.out.println("图中职业是博士后的节点数量计数为:     " + filteredVertices)
51.
52.     //统计图中边的源 ID 大于目的 ID 的数量
53.     val filteredSrcDstId: VertexId = graph.edges.filter(e => e.srcId >
        e.dstId).count
54.     System.out.println("图中边的源 ID 大于目的 ID 的数量是:     " + filteredSrcDstId)
55.
56.     val resTriplets: RDD[EdgeTriplet[(String, String), String]] = graph
        .triplets
57.     for (elem <- resTriplets.collect()) {
58.       println("三元组:     " + elem)
59.     }
60.
61.     for (elem <- graph.vertices.collect()) {
62.       println("graph.vertices:     " + elem)
63.     }
64.
65.     for (elem <- graph.edges.collect()) {
66.       println("graph.edges:     " + elem)
67.     }
68.
69.   }
70.
71. }
```

18.7　数据加载成为 Graph 并进行操作实战

一般情况下，数据都在文件中（如在日志文件中），Spark GraphX 提供了非常方便的从文件中读取数据来构建 Graph 的接口，这个接口方法是 edgeListFile，位于 GraphLoader 这个 object 中。

```
1.  /**
     *提供工具从文件加载至[Graph]s
2.   */
3.
4.  object GraphLoader extends Logging {
5.
6.   /**
7.     *从格式化文件中的边表加载一个图，其中每行包含两个整数：源 ID 和目标 ID，由#开头的
       *就跳行
8.     *
9.     *通过设置 canonicalorientation 为 true，边可以自动面向正方向（源 ID 小于目
10.    *标 ID）
11.    *
12.    * @example 下列的格式加载文件：
13.    * {{{
14.    * # Comment Line
15.    * # Source Id <\t> Target Id
```

```
16.     * 1   -5
17.     * 1    2
18.     * 2    7
19.     * 1    8
20.     * }}}
21.     *
22.     * @param sc SparkContext
23.     * @param 文件路径(如/home/data/file 或者 hdfs://file)
24.     * @param canonicalOrientation  是否在正方向
25.     * @param numEdgePartitions 边 RDD 的分区数，设置为-1，将使用默认并行度
26.     * @param edgeStorageLevel 边分区的存储级别
27.     * @param vertexStorageLevel 顶分区的存储级别
28.     */
29.    def edgeListFile(
30.        sc: SparkContext,
31.        path: String,
32.        canonicalOrientation: Boolean = false,
33.        numEdgePartitions: Int = -1,
34.        edgeStorageLevel: StorageLevel = StorageLevel.MEMORY_ONLY,
35.        vertexStorageLevel: StorageLevel = StorageLevel.MEMORY_ONLY)
36.      : Graph[Int, Int] =
37.    {
38.      val startTime = System.currentTimeMillis
39.
40.      //将边数据表直接解析到边分区
41.      val lines =
42.        if (numEdgePartitions > 0) {
43.          sc.textFile(path, numEdgePartitions).coalesce(numEdgePartitions)
44.        } else {
45.          sc.textFile(path)
46.        }
47.      val edges = lines.mapPartitionsWithIndex { (pid, iter) =>
48.        val builder = new EdgePartitionBuilder[Int, Int]
49.        iter.foreach { line =>
50.          if (!line.isEmpty && line(0) != '#') {
51.            val lineArray = line.split("\\s+")
52.            if (lineArray.length < 2) {
53.              throw new IllegalArgumentException("Invalid line: " + line)
54.            }
55.            val srcId = lineArray(0).toLong
56.            val dstId = lineArray(1).toLong
57.            if (canonicalOrientation && srcId > dstId) {
58.              builder.add(dstId, srcId, 1)
59.            } else {
60.              builder.add(srcId, dstId, 1)
61.            }
62.          }
63.        }
64.        Iterator((pid, builder.toEdgePartition))
65.      }.persist(edgeStorageLevel).setName("GraphLoader.edgeListFile - edges
       (%s)".format(path))
66.      edges.count()
67.
68.      logInfo("It took %d ms to load the edges".format(System.currentTimeMillis
       - startTime))
69.
70.      GraphImpl.fromEdgePartitions(edges,defaultVertexAttr =1,edgeStorageLevel=
       edgeStorageLevel,
```

```
71.        vertexStorageLevel = vertexStorageLevel)
72.   } // edgeListFile 结束
```

在源码中我们需要注意的是，edgeListFile 的第三个参数 canonicalOrientation，如果是有方向的，即 canonicalOrientation 的值设置为 true，那么只有在源顶点的 ID 大于目标顶点的 ID 的时候才算一个。

我们发现，在从文件中构建 Graph 的时候，GraphX 默认把顶点和边的属性都设置为 1，同时，顶点与顶点之间的分隔可以是空格，也可以是 tab 键。当然，在源码中发现注释的时候会直接略过，关于 minEdgePartitions 的值，默认设置为 1，edgeStorageLevel 和 vertexStorageLevel 默认都设置为 MEMORY_ONLY 的方式，这些参数都可以根据需要进行调整。

本节测试使用 Google Contest 提供的数据，下载地址为 http://snap.stanford.edu/data/web-Google.html，下载 web-Google.txt.gz 文件到本地，在 IDEA 中新建一个文件夹 data，data 和 src 是同一级的目录，然后在 data 目录下建立 web-Google 子目录，将 web-Google.txt.gz 文件解压缩以后保存至相对路径的 data\web-Google 目录。如果具备 hdfs 环境，也可以将 web-Google.txt 文件上传至 hdfs 分布式文件系统中，在 Spark Graph 中导入加载 hdfs 系统中的 web-Google.txt 文件。这里我们加载本地的 web-Google.txt 文件。

打开 web-Google.txt 文本文件，其内容显示如下，第 1 行至第 4 行是注释部分，每行文件以“#”开头，Spark Graph 图计算导入文件时会自动跳过注释行。web-Google.txt 文件中，数据部分每行内容的第一列是源网页的 ID，第二列是目标网页的 ID。

```
1.  # 有向图（节点每个无序对保存一次）：web-Google.txt
2.  # 从 Google programming contest 下载的图书籍，2002
3.  # 节点数：875713 边数：5105039
4.  # 源节点 Id   目标节点 Id
5.  0              11342
6.  0              824020
7.  0              867923
8.  0              891835
9.  11342          0
10. 11342          27469
11. 11342          38716
12. 11342          309564
13. 11342          322178
14. 11342          387543
15. 11342          427436
16. 11342          538214
17. ……
```

在 IDEA 中新建一个 object 类 Graphx_webGoogle，导入 Spark　Graph 的相关 JAR 包。

```
1.  import org.apache.spark._
2.  import org.apache.spark.graphx._
3.  import org.apache.spark.rdd.RDD
```

然后加载本地 web-Google.txt 文件。

```
1.  //数据存放的目录
```

```
2.     var dataPath = "data/web-Google/"
3.     val graphFromFile: Graph[PartitionID, PartitionID] = GraphLoader
       .edgeListFile(sc, dataPath + "web-Google.txt")
4.  while (true) {}
```

为了查看 Spark Graph 运行时 Spark 系统资源的使用情况，在 IDEA 中的代码中加一个 While 循环语句，程序就会一直运行，我们可以在浏览器中打开 Spark 的 WebUI 控制台页面（http://127.0.0.1:4040/jobs/）查看 Spark 的资源使用情况。

点击 Stages 标题栏，从图 18-13 中可以看到 Spark 有 3 个任务在运行，即对应 3 个分区。

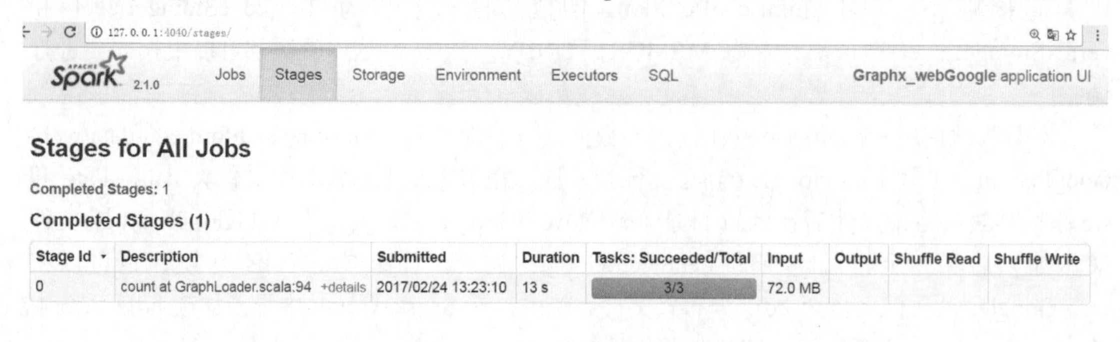

图 18-13　任务运行图

继续点击 count at GraphLoader.scala:94，进入以下页面，可以看到 3 个任务在执行，3 个输入的数据分别 32.1MB、32.1MB、7.9MB，如图 18-14 所示。

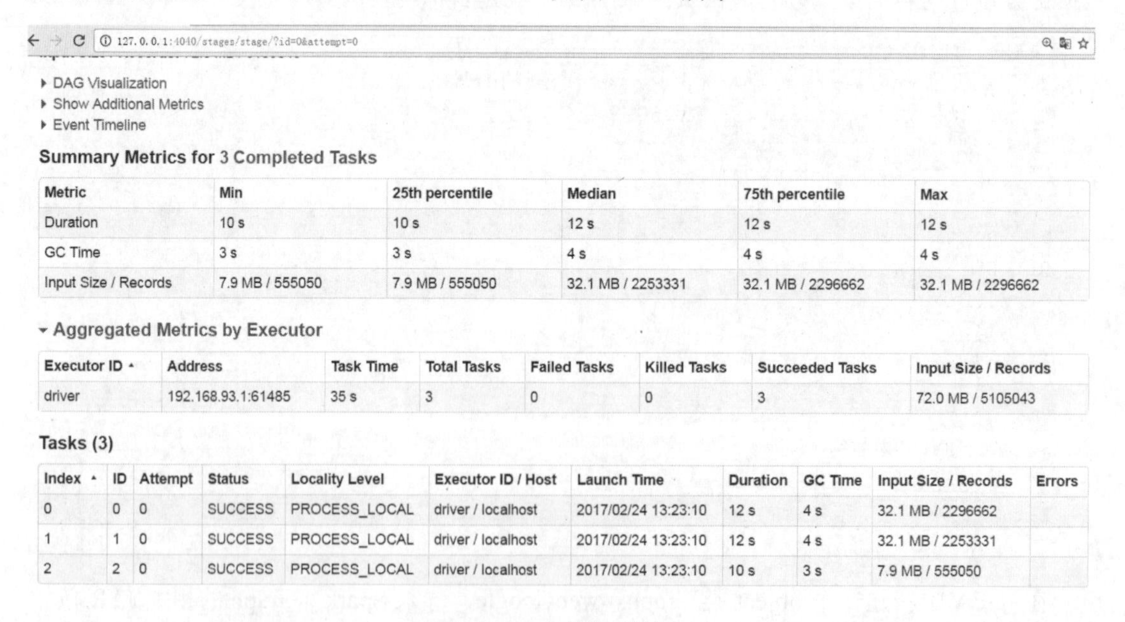

图 18-14　任务执行情况

点击 Storage 标题栏，从图中可以验证 Spark 图计算加载文件时就有 3 个分区。当 web-Google.txt 数据加载完成后，整个 web-Google.txt 文本数据就缓存保存在内存中。在本地模式下运行 Spark，默认的分区块是 32MB，而 web-Google.txt 文本的大小是 75MB，因此按照分区，块被切成 3 个分区，如图 18-15 所示。

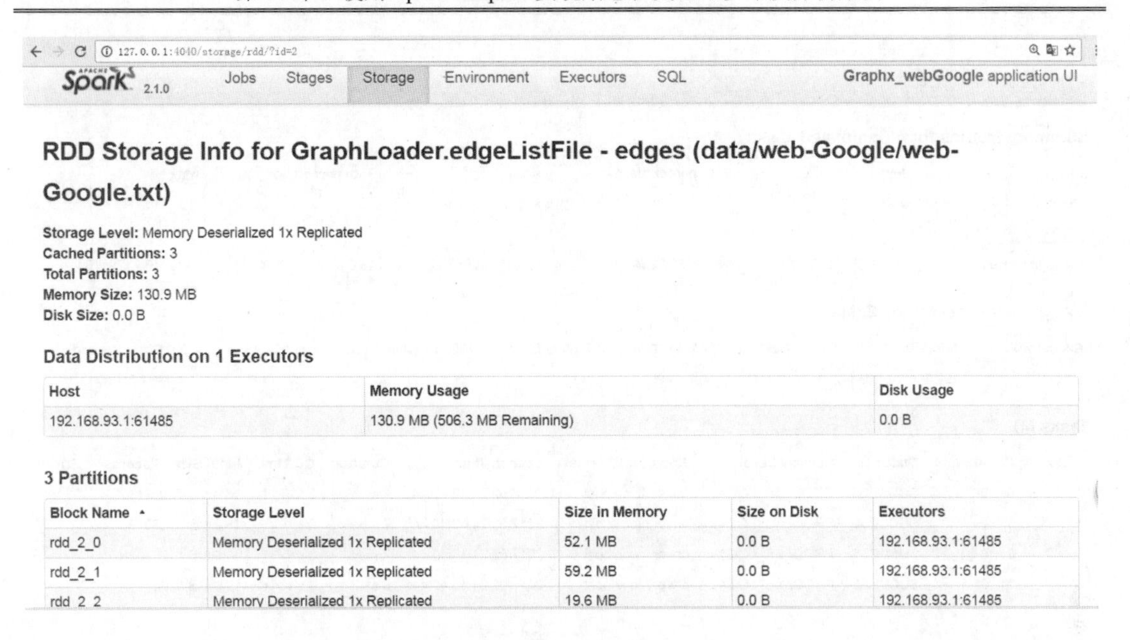

图 18-15　图的存储分区

在 IDEA 代码中，将 edgeListFile 的参数 minEdgePartitions 设置为 4，即定义分区数为 4。

```
1.  //数据存放的目录
2.      var dataPath = "data/web-Google/"
3.      val graphFromFile: Graph[PartitionID, PartitionID] = GraphLoader
        .edgeListFile(sc, dataPath + "web-Google.txt",numEdgePartitions =4)
```

在浏览器中打开 Spark 的 WebUI 控制台页面（http://127.0.0.1:4040/jobs/），查看 Spark 将 Graph 设置为 4 个分区的资源使用情况。

点击 Stages 标题栏，从图 18-16 中可以看到 Spark 有 4 个任务在运行，即对应 4 个分区。

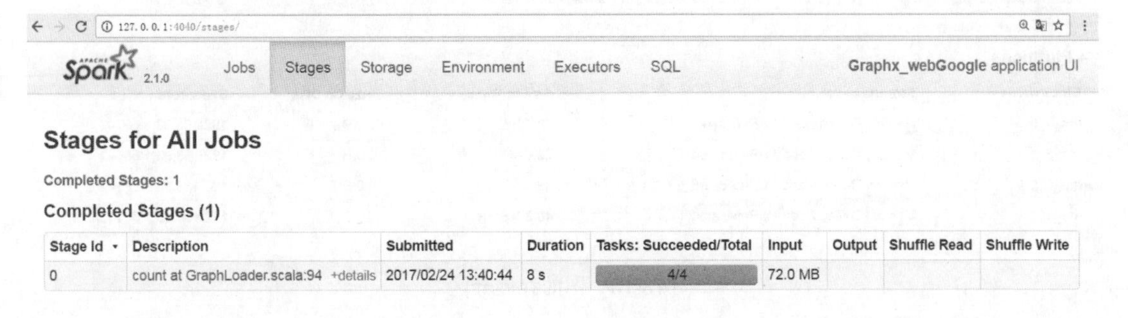

图 18-16　任务执行

继续点击 count at GraphLoader.scala:94，进入以下页面，可以看到 4 个任务在执行，输入的 4 个数据均为 18.0MB，如图 18-17 所示。

点击 Storage 标题栏，从图中可以验证 Spark 图计算加载文件时根据我们的设置，切分成了 4 个分区。将 GraphLoader.edgeListFile - edges 数据缓存保存在内存中，如图 18-18 所示。

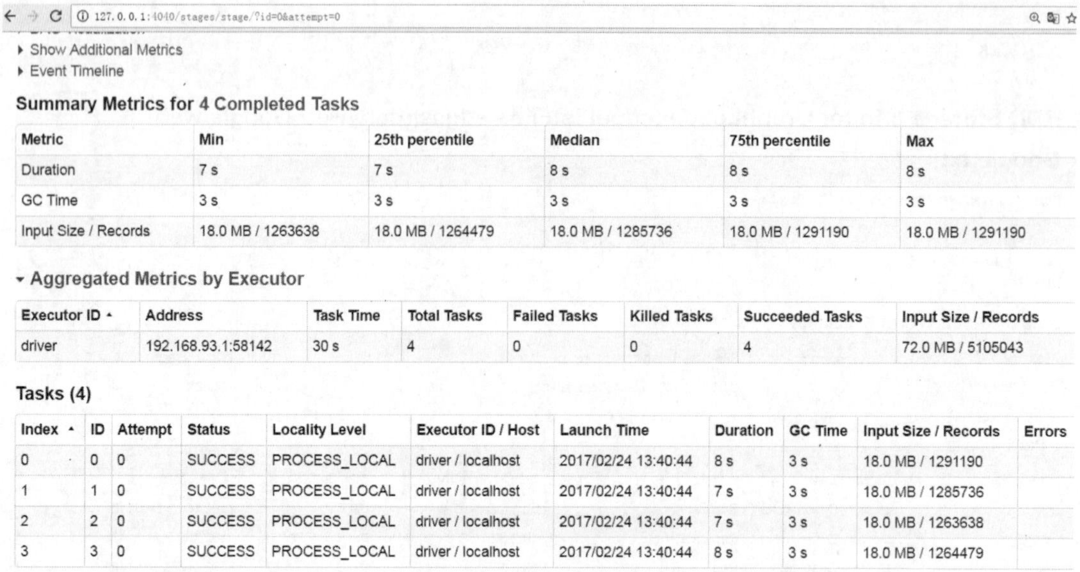

127.0.0.1:4040/stages/stage/?id=0&attempt=0

▸ Show Additional Metrics
▸ Event Timeline

Summary Metrics for 4 Completed Tasks

Metric	Min	25th percentile	Median	75th percentile	Max
Duration	7 s	7 s	8 s	8 s	8 s
GC Time	3 s	3 s	3 s	3 s	3 s
Input Size / Records	18.0 MB / 1263638	18.0 MB / 1264479	18.0 MB / 1285736	18.0 MB / 1291190	18.0 MB / 1291190

▾ **Aggregated Metrics by Executor**

Executor ID ▴	Address	Task Time	Total Tasks	Failed Tasks	Killed Tasks	Succeeded Tasks	Input Size / Records
driver	192.168.93.1:58142	30 s	4	0	0	4	72.0 MB / 5105043

Tasks (4)

Index ▴	ID	Attempt	Status	Locality Level	Executor ID / Host	Launch Time	Duration	GC Time	Input Size / Records	Errors
0	0	0	SUCCESS	PROCESS_LOCAL	driver / localhost	2017/02/24 13:40:44	8 s	3 s	18.0 MB / 1291190	
1	1	0	SUCCESS	PROCESS_LOCAL	driver / localhost	2017/02/24 13:40:44	7 s	3 s	18.0 MB / 1285736	
2	2	0	SUCCESS	PROCESS_LOCAL	driver / localhost	2017/02/24 13:40:44	7 s	3 s	18.0 MB / 1263638	
3	3	0	SUCCESS	PROCESS_LOCAL	driver / localhost	2017/02/24 13:40:44	8 s	3 s	18.0 MB / 1264479	

图 18-17　任务执行情况

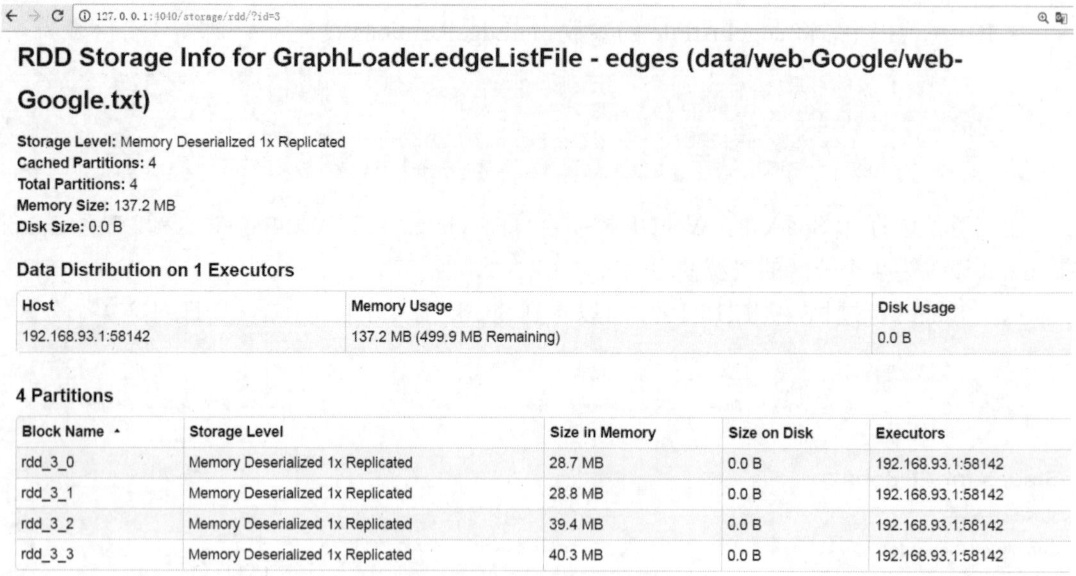

127.0.0.1:4040/storage/rdd/?id=3

RDD Storage Info for GraphLoader.edgeListFile - edges (data/web-Google/web-Google.txt)

Storage Level: Memory Deserialized 1x Replicated
Cached Partitions: 4
Total Partitions: 4
Memory Size: 137.2 MB
Disk Size: 0.0 B

Data Distribution on 1 Executors

Host	Memory Usage	Disk Usage
192.168.93.1:58142	137.2 MB (499.9 MB Remaining)	0.0 B

4 Partitions

Block Name ▴	Storage Level	Size in Memory	Size on Disk	Executors
rdd_3_0	Memory Deserialized 1x Replicated	28.7 MB	0.0 B	192.168.93.1:58142
rdd_3_1	Memory Deserialized 1x Replicated	28.8 MB	0.0 B	192.168.93.1:58142
rdd_3_2	Memory Deserialized 1x Replicated	39.4 MB	0.0 B	192.168.93.1:58142
rdd_3_3	Memory Deserialized 1x Replicated	40.3 MB	0.0 B	192.168.93.1:58142

图 18-18　图的存储分区

然后统计一下文本文件 web-Google.txt 中包含多少个顶点。

```
1.  println("graphFromFile.vertices.count:  " + graphFromFile.vertices.count())
```

在 IDEA 中运行代码，结果如下：

```
1.  Using Spark's default log4j profile: org/apache/spark/log4j-defaults
    .properties
2.  17/02/24 14:09:42 INFO SparkContext: Running Spark version 7.1.0
3.  ......
```

```
4.  17/02/24 14:09:59 INFO DAGScheduler: Job 1 finished: reduce at VertexRDDImpl
    .scala:90, took 3.212063 s
5.  graphFromFile.vertices.count:    875713
6.  ......
```

从以上的运行结果可以看出，加载的 web-Google.txt 文件，其构建的图数据中共包含 875 713 个顶点。然后再统计一下图中共有多少条边。

```
1.  println("graphFromFile.edges.count:    " + graphFromFile.edges.count())
```

在 IDEA 中运行代码，结果如下：

```
1.  Using Spark's default log4j profile: org/apache/spark/log4j-defaults
    .properties
2.  17/02/24 14:14:45 INFO SparkContext: Running Spark version 7.1.0
3.  ......
4.  17/02/24 14:15:20 INFO DAGScheduler: ResultStage 3 (reduce at EdgeRDDImpl
    .scala:90) finished in 0.033 s
5.  graphFromFile.edges.count:    5105039
6.  17/02/24 14:15:20 INFO DAGScheduler: Job 2 finished: reduce at EdgeRDDImpl
    .scala:90, took 0.049847 s
7.  ......
```

从以上的运行结果可以看出，加载的 web-Google.txt 文件，其构建的图数据中共包含 5 105 039 条边。可以看出，web-Google.txt 文件大小非常理想，很适合读者在本地环境中测试学习。

本节完整的代码如例 18-2 所示。

【**例 18-2**】 web-Google 文件操作。

```
1.  package com.dt.spark.graphx
2.  import org.apache.spark.SparkConf
3.  import org.apache.spark.graphx.{Graph, GraphLoader, PartitionID}
4.  import org.apache.spark.sql.SparkSession
5.
6.  object Graphx_webGoogle {
7.    def main(args: Array[String]) {
8.      var masterUrl = "local[8]"
9.      if (args.length > 0) {
10.       masterUrl = args(0)
11.     }
12.     val sparkConf = new SparkConf().setMaster(masterUrl).setAppName("Graphx_
        webGoogle")
13.     val spark = SparkSession
14.       .builder()
15.       .config(sparkConf)
16.       .getOrCreate()
17.     val sc = spark.sparkContext
18.     //数据存放的目录
19.     var dataPath = "data/web-Google/"
20.     val graphFromFile: Graph[PartitionID, PartitionID] = GraphLoader
        .edgeListFile(sc, dataPath + "web-Google.txt", numEdgePartitions = 4)
21.     //统计顶点的数量
22.    println("graphFromFile.vertices.count:    " + graphFromFile.vertices
       .count())
23.     //统计边的数量
24.     println("graphFromFile.edges.count:    " + graphFromFile.edges
```

```
        .count())
25.     while (true) {}
26.   }
27. }
```

18.8　图操作之 Property Operators 实战

本节专注于集群上 Property Operator 的内容，其中比较重要的是 mapVertices、mapEdges 和 mapTriplets，即对顶点进行 map、对边进行 map、对 Triplets 进行 map。在 Graph 中，其方法分别如下所示：

```
1.    /**
2.     * 使用 map 函数转换图中的每个顶点的属性
3.     *
4.     * @note 新的图具有相同的结构，底层结构可重用
5.     * @param 函数从顶点对象转换为新顶点值
6.     *
7.     * @tparam VD2 新顶点的类型
8.     *
9.     * @example 可以使用这个操作来改变顶点值，初始化算法从一种类型到另一种类型
10.    * {{{
11.    * val rawGraph: Graph[(), ()] = Graph.textFile("hdfs://file")
12.    * val root = 42
13.    * var bfsGraph = rawGraph.mapVertices[Int]((vid, data) => if (vid ==
       * root) 0 else Math.MaxValue)
14.    * }}}
15.    *
16.    */
17.   def mapVertices[VD2: ClassTag](map: (VertexId, VD) => VD2)
18.     (implicit eq: VD =:= VD2 = null): Graph[VD2, ED]
```

mapEdges 的方法如下：

```
1.    /**
2.     *使用 map 函数转换图中的每条边的属性。map 函数不是转换相邻边的顶点值。如果需要顶
       *点值，须使用 maptriplets
3.     *
4.     * @note 图不会更改，新的图具有相同结构，且底层结构可以复用
5.     *
6.     * @param map 函数将边对象映射到新的边值
7.     *
8.     * @tparam ED2 新的边的数据类型
9.     *
10.    * @example  函数可用于初始化边的属性
11.    *
12.    */
13.   def mapEdges[ED2: ClassTag](map: Edge[ED] => ED2): Graph[VD, ED2] = {
14.     mapEdges((pid, iter) => iter.map(map))
15.   }
```

mapTriplets 的方法如下：

```
1.   /**
2.     *使用 map 函数转换每条边的属性，使用函数转换相邻顶点属性。如果不需要相邻顶点值，
       *考虑使用 mapedges
3.     * @note 并没有改变图或修改此图的值，因此可以重用底层索引结构
4.     *
5.     * @param map 函数从边的对象转换到新的边值
6.     *
7.     * @tparam ED2 新的边的类型
8.     *
9.     * @example  此函数可用于初始化边的属性，属性基于与每个顶点相关联的属性
10.    * {{{
11.    * val rawGraph: Graph[Int, Int] = someLoadFunction()
12.    * val graph = rawGraph.mapTriplets[Int]( edge =>
13.    *   edge.src.data - edge.dst.data)
14.    * }}}
15.    *
16.    */
17.   def  mapTriplets[ED2:  ClassTag](map:  EdgeTriplet[VD,  ED]  =>  ED2):
      Graph[VD, ED2] = {
18.     mapTriplets((pid, iter) => iter.map(map), TripletFields.All)
19.   }
```

下面看一下 Graph 实例中的 10 个元素的具体值。

```
1.      for (elem <- graphFromFile.vertices.take(10)) {
2.        println("graphFromFile.vertices:     " + elem)
3.      }
```

在 IDEA 中运行代码，结果如下：

```
1.  Using Spark's default log4j profile: org/apache/spark/log4j-defaults
    .properties
2.  17/02/24 20:41:17 INFO SparkContext: Running Spark version 7.1.0
3.  ......
4.  17/02/24 20:41:35 INFO DAGScheduler: Job 3 finished: take at Graphx_
    webGoogle.scala:27, took 0.456157 s
5.  graphFromFile.vertices:    (185012,1)
6.  graphFromFile.vertices:    (612052,1)
7.  graphFromFile.vertices:    (354796,1)
8.  graphFromFile.vertices:    (182316,1)
9.  graphFromFile.vertices:    (199516,1)
10. graphFromFile.vertices:    (627804,1)
11. graphFromFile.vertices:    (170792,1)
12. graphFromFile.vertices:    (307248,1)
13. graphFromFile.vertices:    (512760,1)
14. graphFromFile.vertices:    (386896,1)
```

可以看到这 10 个顶点元素中每个顶点元素的属性值都是 1，这是源码设定的。下面我们把每个顶点的元素的值都变成 2，当然这样做没有实际意义，只是试验用途。

```
1.       val  tmpGraph:  Graph[PartitionID,  PartitionID]  =graphFromFile.map
         Vertices((vid, attr) => attr.toInt*2)
2.      for (elem <- tmpGraph.vertices.take(10)) {
3.        println("tmpGraph.vertices:     " + elem)
4.      }
```

在 IDEA 中运行代码，结果如下：

```
1.  Using Spark's default log4j profile: org/apache/spark/log4j-defaults
    .properties
2.  17/02/24 20:58:18 INFO SparkContext: Running Spark version 7.1.0
3.  ......
4.  17/02/24 20:58:35 INFO DAGScheduler: Job 4 finished: take at Graphx_
    webGoogle.scala:32, took 0.403449 s
5.  tmpGraph.vertices:     (185012,2)
6.  tmpGraph.vertices:     (612052,2)
7.  tmpGraph.vertices:     (354796,2)
8.  tmpGraph.vertices:     (182316,2)
9.  tmpGraph.vertices:     (199516,2)
10. tmpGraph.vertices:     (627804,2)
11. tmpGraph.vertices:     (170792,2)
12. tmpGraph.vertices:     (307248,2)
13. tmpGraph.vertices:     (512760,2)
14. tmpGraph.vertices:     (386896,2)
```

18.9　图操作之 Structural Operators 实战

Spark GraphX 中属于 Structural Operators 的操作主要有 reverse、subgraph、mask、groupEdges 等几种函数，它们在 Graph 中的源码分别如下所示。

reverse 函数的源码如下：

```
1.  /**
     *反转图中的所有边。如果此图包含从 a 到 b 的边，则返回图包含一个从 b 到 a 的边
2.   */
3.   def reverse: Graph[VD, ED]
```

subgraph 函数的源码如下：

```
1.    /**
2.     *  将图限制为满足谓词的顶点和边,获取达到条件的子图
3.     *
4.     * {{{
5.     * V' = {v : for all v in V where vpred(v)}
6.     * E' = {(u,v): for all (u,v) in E where epred((u,v)) && vpred(u) &&
       *vpred(v)}
7.     * }}}
8.     *
9.     * @param epred  边谓词需要一个三元组，如果边保持在子图中，则计算为真。注意：
10.    *只有两个顶点满足谓词的边才被考虑
11.    * @param vpred  顶点谓词需要一个顶点对象，如果顶点包含在子图中，则计算为 true
12.    *
13.    * @return  返回的子图只包含满足谓词的顶点和边
14.    */
15.   def subgraph(
16.      epred: EdgeTriplet[VD, ED] => Boolean = (x => true),
17.      vpred: (VertexId, VD) => Boolean = ((v, d) => true))
18.     : Graph[VD, ED]
```

mask 函数的源码如下：

```
1.  /**
2.     *限制图的顶点和边在 other，保持图的属性
3.     * @param 将图投影到其他图
4.     * @return  返回一个图对象，在当前图和其他图 other 中存在的顶点和边，从当前图获取
       * 这些顶点和边的数据
5.     */
6.   def mask[VD2: ClassTag, ED2: ClassTag](other: Graph[VD2, ED2]): Graph[VD,ED]
7.  ......
```

groupEdges 函数的源码如下：

```
1.   /**
2.     * 将两个顶点之间的多条边合并为一条边。对于正确的结果，图必须使用[partitionBy]
       * 进行分区
3.     *
4.     * @param merge  通过用户提供的关联函数复制边，合并边属性
5.     *
6.     * @return  返回的图中每条单边具备 (源顶点,目标顶点)顶点对
7.     */
8.   def groupEdges(merge: (ED, ED) => ED): Graph[VD, ED]
```

上述函数中用得比较多的是 subGraph。下面看一下如何使用 subGraph。首先来看基于 web-Google.txt 构建的 Graph 有多少个顶点 vertices。

```
1.    //数据存放的目录
2.     var dataPath = "data/web-Google/"
3.     val graphFromFile: Graph[PartitionID, PartitionID] = GraphLoader
       .edgeListFile(sc, dataPath + "web-Google.txt", numEdgePartitions = 4)
4.     //统计顶点的数量
5.    println("graphFromFile.vertices.count:    " + graphFromFile.vertices
      .count())
```

在 IDEA 中运行代码，图中统计顶点的数量结果如下：

```
1.  Using Spark's default log4j profile: org/apache/spark/log4j-defaults
    .properties
2.  17/04/09 20:19:08 INFO SparkContext: Running Spark version 2.1.0
3.  17/04/09 20:19:09 WARN NativeCodeLoader: Unable to load native-hadoop
    library for your platform... using builtin-java classes where applicable
4.  ......
5.  17/04/09  20:19:29  INFO  DAGScheduler:  Job  1  finished:  reduce  at
    VertexRDDImpl.scala:90, took 2.518919 s
6.  graphFromFile.vertices.count:    875713
7.  17/04/09 20:19:29 INFO SparkContext: Starting job: reduce at EdgeRDDImpl
    .scala:90
8.  ......
```

从执行结果上可以看出有 875 713 个顶点。

再看一下基于 web-Google.txt 构建的 Graph 有多少条边。

```
1.      //数据存放的目录
2.     var dataPath = "data/web-Google/"
3.     val graphFromFile: Graph[PartitionID, PartitionID] = GraphLoader
       .edgeListFile(sc, dataPath + "web-Google.txt", numEdgePartitions = 4)
4.  ......
```

```
5.     //统计边的数量
6.     println("graphFromFile.edges.count:    " + graphFromFile.edges.count())
```

在 IDEA 中运行代码，图中统计边的数量结果如下：

```
1.  Using Spark's default log4j profile: org/apache/spark/log4j-defaults
    .properties
2.  17/04/09 20:19:08 INFO SparkContext: Running Spark version 2.1.0
3.  17/04/09 20:19:09 WARN NativeCodeLoader: Unable to load native-hadoop
    library for your platform... using builtin-java classes where applicable
4.  ......
5.  17/04/09  20:19:29  INFO  DAGScheduler:  Job  2  finished:  reduce  at
    EdgeRDDImpl.scala:90, took 0.043325 s
6.  graphFromFile.edges.count:    5105039
7.  17/04/09  20:19:29  INFO  SparkContext:  Starting  job:  take  at  Graphx_
    webGoogle.scala:27
8.
```

可以发现有 5 105 039 条边。

接下来使用 subGraph，我们首先只考虑第一个参数，也就是说，第二个参数使用默认值，不参与判断。

```
1.      val subGraph: Graph[PartitionID, PartitionID] =graphFromFile.subgraph
        (epred = e  => e.srcId > e.dstId)//e 是 epred 的别名
2.     for(elem <- subGraph.edges.take(10)) {
3.      println("subGraph.edges:    "+  elem)
4.     }
```

在 IDEA 中运行代码，验证 subGraph 的结果。

```
1.   Using Spark's default log4j profile: org/apache/spark/log4j-defaults
     .properties
2.  17/04/09 20:41:42 INFO SparkContext: Running Spark version 2.1.0
3.  17/04/09 20:41:42 WARN NativeCodeLoader: Unable to load native-hadoop
    library for your platform... using builtin-java classes where applicable
4.  ......
5.  17/04/09  20:42:00  INFO  DAGScheduler:  ResultStage  6  (take  at  Graphx_
    webGoogle.scala:28) finished in 0.622 s
6.  subGraph.edges:    Edge(1122,429,1)
7.  subGraph.edges:    Edge(1300,606,1)
8.  subGraph.edges:    Edge(1436,409,1)
9.  subGraph.edges:    Edge(1509,1401,1)
10. subGraph.edges:    Edge(1513,1406,1)
11. subGraph.edges:    Edge(1624,827,1)
12. subGraph.edges:    Edge(1705,693,1)
13. subGraph.edges:    Edge(1825,1717,1)
14. subGraph.edges:    Edge(1985,827,1)
15. subGraph.edges:    Edge(2135,600,1)
16. 17/04/09 20:42:00 INFO DAGScheduler: Job 3 finished: take at Graphx_webGoogle.
    scala:28, took 1.703074 s
17. ......
```

执行结果表明源顶点 ID 都是大于目标顶点 ID 的。

下面看一下我们构建的 subGraph 的顶点的个数。

```
1.   println("subGraph.vertices.count():    "+  subGraph.vertices.count())
```

在 IDEA 中运行代码，验证 subGraph 的顶点的个数。

```
1.  Using Spark's default log4j profile: org/apache/spark/log4j-defaults
    .properties
2.  17/04/09 20:48:33 INFO SparkContext: Running Spark version 2.1.0
3.  17/04/09 20:48:35 WARN NativeCodeLoader: Unable to load native-hadoop
    library for your platform... using builtin-java classes where applicable
4.  ......
5.  17/04/09 20:49:19 INFO DAGScheduler: Job 4 finished: reduce at VertexRDDImpl
    .scala:90, took 0.156681 s
6.  subGraph.vertices.count():    875713
7.  17/04/09 20:49:19 INFO SparkContext: Starting job: take at Graphx_
    webGoogle.scala:34
```

此时发现顶点个数依旧是 875 713 个，和原图一样。

使用 subGraph.edges.count 看一下 subGraph 的边的个数。

```
1.   println("subGraph.edges.count():    "+ subGraph.edges.count())
```

在 IDEA 中运行代码，验证一下 subGraph 的边的个数。

```
1.  Using Spark's default log4j profile: org/apache/spark/log4j-defaults
    .properties
2.  17/04/09 20:52:23 INFO SparkContext: Running Spark version 2.1.0
3.  17/04/09 20:52:24 WARN NativeCodeLoader: Unable to load native-hadoop
    library for your platform... using builtin-java classes where applicable
4.  ......
5.  17/04/09 20:52:49 INFO TaskSetManager: Finished task 0.0 in stage 11.0
    (TID 25) in 1587 ms on localhost (executor driver) (3/4)
6.  subGraph.edges.count():    2420548
7.  17/04/09 20:52:49 WARN Executor: 1 block locks were not released by TID = 27:
8.  ......
```

从执行结果上可以看出，此时边的个数是 2 420 548，而原来的 Graph 的边的个数是 5 105 039，所以，通过过滤 subGraph 的边减少了。

接下来我们也加入对顶点过滤的子图构建。

```
1.   val subGraph2: Graph[PartitionID, PartitionID] = graphFromFile.subgraph
     (epred = e  => e.srcId > e.dstId,vpred= (id, _) => id>500000)
     //顶点 ID 大于 500000
2.    println("subGraph2.vertices.count():    "+ subGraph2.vertices.count())
3.
```

在 IDEA 中运行代码，验证 subGraph2 的顶点过滤后的个数。

```
1.  Using Spark's default log4j profile: org/apache/spark/log4j-defaults
    .properties
2.  17/04/09 21:00:43 INFO SparkContext: Running Spark version 2.1.0
3.  17/04/09 21:00:43 WARN NativeCodeLoader: Unable to load native-hadoop
    library for your platform... using builtin-java classes where applicable
4.  ......
5.  17/04/09 21:01:04 INFO DAGScheduler: ResultStage 13 (reduce at
    VertexRDDImpl.scala:90) finished in 0.182 s
6.  subGraph2.vertices.count():    400340
7.  17/04/09 21:01:04 INFO DAGScheduler: Job 6 finished: reduce at
    VertexRDDImpl.scala:90, took 0.209837 s
8.  ......
```

执行结果表明顶点的个数是 400 340，原来的顶点个数是 875 713，顶点在过滤条件的作用下减少了。

下面看一下顶点大于 1 000 000 的顶点个数。

```
1.      val subGraph2: Graph[PartitionID, PartitionID] = graphFromFile
        .subgraph(epred = e  => e.srcId > e.dstId,vpred= (id, _) => id>1000000)
2.     println("subGraph2.vertices.count():   "+ subGraph2.vertices.count())
3.     ......
```

在 IDEA 中运行代码，此时使用 subGraph2.vertices.count 查看顶点的个数。

```
1.   ......
2.   17/04/09 21:05:25 INFO TaskSetManager: Finished task 1.0 in stage 13.0
     (TID 30) in 155 ms on localhost (executor driver) (3/4)
3.   subGraph2.vertices.count():    0
4.   17/04/09 21:05:25 INFO Executor: Finished task 2.0 in stage 13.0 (TID 31).
     1112 bytes result sent to driver
```

从结果中可以发现不存在顶点的 ID 大于 1 000 000 的情况。

18.10　图操作之 Computing Degree 实战

度(Degree)是离散数学的概念。在 Spark GraphX 中把 Degree 分为 inDegrees、outDegrees、degrees 这 3 种不同的 degree，以图 18-12 为例。

在图 18-12 中，顶点 5 的 inDegrees 是 1、outDegrees 是 2、degrees 是 3。Degree 是 GraphOps 中的成员，源码如下：

```
1.  /**
2.     * 图中每个顶点的入度
3.     * @note 没有入边的顶点不在 RDD 中返回
4.     */
5.    @transient lazy val  inDegrees: VertexRDD[Int] =
6.      degreesRDD(EdgeDirection.In).setName("GraphOps.inDegrees")
7.
8.    /**
9.     * 图中每个顶点的出度
10.    * @note 没有出边的顶点不在 RDD 中返回
11.    */
12.   @transient lazy val  outDegrees: VertexRDD[Int] =
13.     degreesRDD(EdgeDirection.Out).setName("GraphOps.outDegrees")
14.
15.   /**
16.    *图中每个顶点的度
17.    * @note 无边的顶点不在 RDD 中返回
18.    */
19.   @transient lazy val degrees: VertexRDD[Int] =
20.     degreesRDD(EdgeDirection.Either).setName("GraphOps.degrees")
```

方向控制的时候是由 EdgeDirection 决定的，其源码如下：

```
1.  /**
```

```
2.   *  [EdgeDirection]边方向集合
3.   */
4.  object EdgeDirection {
5.    /**到达顶点的边 */
6.    final val In: EdgeDirection = new EdgeDirection("In")
7.
8.    /** 起源于顶点的边*/
9.    final val Out: EdgeDirection = new EdgeDirection("Out")
10.
11.   /**到达顶点的边或者起源于顶点的边 */
12.   final val Either: EdgeDirection = new EdgeDirection("Either")
13.
14.   /**到达顶点的边并且起源于顶点的边 */
15.   final val Both: EdgeDirection = new EdgeDirection("Both")
```

下面看一下基于 web-Google.txt 构建的 Graph 的 inDegrees。

```
1.   val    tmp: VertexRDD[PartitionID] =graphFromFile.inDegrees
2.  for (elem <- tmp.take(10)) {
3.    println("graphFromFile.inDegrees:     " + elem)
4.  }
```

在 IDEA 中运行代码，查看 Graph 的 inDegrees。

```
1.   Using Spark's default log4j profile: org/apache/spark/log4j-defaults
     .properties
2.  17/04/10 12:58:43 INFO SparkContext: Running Spark version 2.1.0
3.  17/04/10 12:58:44 WARN NativeCodeLoader: Unable to load native-hadoop
    library for your platform... using builtin-java classes where applicable
4.  ......
5.  17/04/10 12:59:11 INFO DAGScheduler: Job 8 finished: take at Graphx_
    webGoogle.scala:40, took 1.449161 s
6.  graphFromFile.inDegrees:     (185012,1)
7.  graphFromFile.inDegrees:     (354796,2)
8.  graphFromFile.inDegrees:     (199516,28)
9.  graphFromFile.inDegrees:     (627804,2)
10. graphFromFile.inDegrees:     (170792,1)
11. graphFromFile.inDegrees:     (307248,2)
12. graphFromFile.inDegrees:     (512760,2)
13. graphFromFile.inDegrees:     (386896,3)
14. graphFromFile.inDegrees:     (32676,3)
15. graphFromFile.inDegrees:     (513652,1)
16. 17/04/10 12:59:11 INFO SparkContext: Starting job: take at Graphx_
    webGoogle.scala:45
17. ......
```

也可以对 outDegrees 和 degrees 进行计算。

```
1.  val tmp1: VertexRDD[PartitionID]=graphFromFile.outDegrees
2.  for (elem <- tmp1.take(10)) {
3.    println("graphFromFile.outDegrees:     " + elem)
4.  }
5.
6.  val tmp2: VertexRDD[PartitionID]=graphFromFile.degrees
7.  for (elem <- tmp2.take(10)) {
8.    println("graphFromFile.degrees:     " + elem)
9.  }
```

在 IDEA 中运行代码，查看 Graph 的 outDegrees 和 degrees。

```
1.  ......
2.  17/04/10 19:48:41 INFO DAGScheduler: ResultStage 24 (take at Graphx_
    webGoogle.scala:44) finished in 0.141 s
3.  graphFromFile.outDegrees:     (612052,2)
4.  graphFromFile.outDegrees:     (354796,1)
5.  17/04/10 19:48:41 INFO DAGScheduler: Job 9 finished: take at Graphx_
    webGoogle.scala:44, took 0.861334 s
6.  graphFromFile.outDegrees:     (182316,2)
7.  graphFromFile.outDegrees:     (199516,20)
8.  graphFromFile.outDegrees:     (627804,8)
9.  graphFromFile.outDegrees:     (170792,9)
10. graphFromFile.outDegrees:     (307248,2)
11. graphFromFile.outDegrees:     (512760,7)
12. graphFromFile.outDegrees:     (386896,18)
13. graphFromFile.outDegrees:     (32676,15)
14. 17/04/10 19:48:41 INFO SparkContext: Starting job: take at Graphx_
    webGoogle.scala:49
15. ......
16. 17/04/10 19:48:42 INFO DAGScheduler: Job 10 finished: take at Graphx_
    webGoogle.scala:49, took 1.616604 s
17. graphFromFile.degrees:     (185012,1)
18. graphFromFile.degrees:     (612052,2)
19. graphFromFile.degrees:     (354796,3)
20. graphFromFile.degrees:     (182316,2)
21. graphFromFile.degrees:     (199516,48)
22. graphFromFile.degrees:     (627804,10)
23. graphFromFile.degrees:     (170792,10)
24. graphFromFile.degrees:     (307248,4)
25. graphFromFile.degrees:     (512760,9)
26. graphFromFile.degrees:     (386896,21)
27. ......
```

如果想算哪个顶点的 inDegrees 或者 outDegrees 或者 Degrees 最大，是非常简单的，首先定义一个比较两个顶点 Degree 中较大值的函数如下：

```
1.  def max(a:(VertexId,Int),b:(VertexId,Int)):(VertexId,Int)=if(a._2 > b._2)
    a else b
2.     println("graphFromFile.degrees.reduce(max):      " + graphFromFile.
       degrees.reduce(max))
3.  ......
```

在 IDEA 中运行代码，查看哪个顶点 Degrees 最大的结果如下：

```
1.  ......
2.  17/04/10 19:55:16 INFO DAGScheduler: ResultStage 32 (reduce at Graphx_
    webGoogle.scala:54) finished in 0.867 s
3.  graphFromFile.degrees.reduce(max):     (537039,6353)
4.  17/04/10 19:55:16 INFO DAGScheduler: Job 11 finished: reduce at Graphx_
    webGoogle.scala:54, took 0.880735 s
5.  17/04/10 19:55:16 INFO SparkContext: Starting job: take at Graphx_
    webGoogle.scala:56
6.  ......
```

计算结果表明 ID 为 537 039 的顶点的 Degrees 最大，有 6353 条边与之相连。

下面使用 graphFromFile.inDegrees.reduce(max) 看一下哪个节点具有最大的 inDegrees。

```
1.  def  max(a:(VertexId,Int),b:(VertexId,Int)):(VertexId,Int)=if(a._2  >
    b._2) a else b
2.      ......
3.  println("graphFromFile.inDegrees.reduce(max):    " + graphFromFile.inDegrees
    .reduce(max))
```

在 IDEA 中运行代码，查看哪个节点具有最大的 inDegrees 的结果如下：

```
1.   ......
2.  17/04/10 20:03:52 INFO TaskSetManager: Finished task 1.0 in stage 36.0
    (TID 57) in 354 ms on localhost (executor driver) (3/4)
3.  graphFromFile.inDegrees.reduce(max):    (537039,6326)
4.  17/04/10 20:03:52 WARN Executor: 1 block locks were not released by TID
    = 58:
5.  ......
```

此时发现同样是 ID 为 537 039 的顶点，如果这是一张网页，就表明是一个质量非常不错的网页。

下面使用 graphFromFile.outDegrees.reduce(max)看一下哪个节点具有最大的 outDegrees。

```
1.  def max(a:(VertexId,Int),b:(VertexId,Int)):(VertexId,Int)=if(a._2 > b._2) a
    else b
2.      ......
3.      println("graphFromFile.outDegrees.reduce(max):    " + graphFromFile
      .outDegrees.reduce(max))
```

在 IDEA 中运行代码，查看哪个节点具有最大的 outDegrees 的结果如下：

```
1.   ......
2.   17/04/10 20:09:19 INFO DAGScheduler: Job 13 finished: reduce at Graphx_
     webGoogle.scala:56, took 0.240818 s
3.  graphFromFile.outDegrees.reduce(max):    (506742,456)
4.  17/04/10  20:09:19  INFO  SparkContext:  Starting  job:  take  at  Graphx_
    webGoogle.scala:59
5.  ......
```

结果显示 ID 为 506 742 的顶点具有最大的 outDegrees，如果是网页，则其具有指向 456 个外部网页的链接，这一般都是导航网站。

18.11　图操作之 Collecting Neighbors 实战

计算方法主要有 collectNeighborIds 和 collectNeighbors 两个，源码都位于 GraphOps 中。collectNeighborIds 方法的源码如下：

```
1.   /**
2.     * 对于每个顶点，收集邻居顶点 ID
3.     * @param edgeDirection 边的方向，收集相邻顶点的方向
4.     * @return 每个顶点的相邻 ID 集
5.     */
6.   def collectNeighborIds(edgeDirection: EdgeDirection): VertexRDD[Array
     [VertexId]] = {
```

```
7.      val nbrs =
8.        if (edgeDirection == EdgeDirection.Either) {
9.          graph.aggregateMessages[Array[VertexId]](
10.           ctx => { ctx.sendToSrc(Array(ctx.dstId)); ctx.sendToDst(Array
              (ctx.srcId)) },
11.           _ ++ _, TripletFields.None)
12.       } else if (edgeDirection == EdgeDirection.Out) {
13.         graph.aggregateMessages[Array[VertexId]](
14.           ctx => ctx.sendToSrc(Array(ctx.dstId)),
15.           _ ++ _, TripletFields.None)
16.       } else if (edgeDirection == EdgeDirection.In) {
17.         graph.aggregateMessages[Array[VertexId]](
18.           ctx => ctx.sendToDst(Array(ctx.srcId)),
19.           _ ++ _, TripletFields.None)
20.       } else {
21.         throw new SparkException("It doesn't make sense to collect neighbor
            ids without a " +
22.           "direction. (EdgeDirection.Both is not supported; use EdgeDirection.
              Either instead.)")
23.       }
24.     graph.vertices.leftZipJoin(nbrs) { (vid, vdata, nbrsOpt) =>
25.       nbrsOpt.getOrElse(Array.empty[VertexId])
26.     }
27.   } // collectNeighborIds 结束
```

collectNeighbors 方法的源码如下：

```
1.    /**
2.     *为每个顶点收集邻居顶点属性。
3.     *
4.     * @note  这个函数在幂律图（幂律图是小部分顶点的度很大，大部分顶点的度很低）中可
       * 能效率很低，高度顶点可能会使大量的信息收集到一个位置
5.     *
6.     * @param edgeDirection  边的方向，收集相邻顶点的方向
7.     *
8.     * @return 每个顶点的邻近顶点属性的顶点集
9.     */
10.   def collectNeighbors(edgeDirection: EdgeDirection): VertexRDD[Array
      [(VertexId, VD)]] = {
11.     val nbrs = edgeDirection match {
12.       case EdgeDirection.Either =>
13.         graph.aggregateMessages[Array[(VertexId, VD)]](
14.           ctx => {
15.             ctx.sendToSrc(Array((ctx.dstId, ctx.dstAttr)))
16.             ctx.sendToDst(Array((ctx.srcId, ctx.srcAttr)))
17.           },
18.           (a, b) => a ++ b, TripletFields.All)
19.       case EdgeDirection.In =>
20.         graph.aggregateMessages[Array[(VertexId, VD)]](
21.           ctx => ctx.sendToDst(Array((ctx.srcId, ctx.srcAttr))),
22.           (a, b) => a ++ b, TripletFields.Src)
23.       case EdgeDirection.Out =>
24.         graph.aggregateMessages[Array[(VertexId, VD)]](
25.           ctx => ctx.sendToSrc(Array((ctx.dstId, ctx.dstAttr))),
26.           (a, b) => a ++ b, TripletFields.Dst)
27.       case EdgeDirection.Both =>
28.         throw new SparkException("collectEdges does not support
            EdgeDirection.Both. Use" +
```

```
29.         "EdgeDirection.Either instead.")
30.     }
31.     graph.vertices.leftJoin(nbrs) { (vid, vdata, nbrsOpt) =>
32.       nbrsOpt.getOrElse(Array.empty[(VertexId, VD)])
33.     }
34.   } // end of collectNeighbor
```

从上述两个方法的源码中可以看出 EdgeDirection.Both 是不被支持的，这一点在使用的时候需要注意。

18.12　图操作之 Join Operators 实战

Join Operators 是非常重要的图操作，其有两个核心方法：joinVertices 和 outerJoinVertices。其中，joinVertices 只作用于有 ID 链接的地方，而 outerJoinVertices 会作用于所有的 ID。

joinVertices 的源码位于 GraphOps 中，代码如下：

```
1.   /**
2.     * 关联顶点 RDD，从顶点和 RDD 转换应用一个新的顶点值。输入表最多包含每个顶点的一个
       * 实体。如果没有提供转换函数，则使用原来的值
3.     *
4.     * @tparam U 更新表中的实体类型
5.     * @param table  与图中顶点关联的表。表最多应包含每个顶点的一个实体
6.     * @param mapFunc  用于计算新顶点的方法。map 函数用于表中对应的实体，否则将使用
       * 原来的顶点值
7.     *
8.     * @example  此方法基于外部数据更新顶点，例如，对每个顶点，可以添加出度的数据
9.     *
10.    * {{{
11.    * val rawGraph: Graph[Int, Int] = GraphLoader.edgeListFile(sc, "webgraph")
12.    *   .mapVertices((_, _) => 0)
13.    * val outDeg = rawGraph.outDegrees
14.    * val graph = rawGraph.joinVertices[Int](outDeg)
15.    *   ((_, _, outDeg) => outDeg)
16.    * }}}
17.    *
18.    */
19.  def joinVertices[U: ClassTag](table: RDD[(VertexId, U)])(mapFunc: (VertexId,
     VD, U) => VD)
20.     : Graph[VD, ED] = {
21.     val uf = (id: VertexId, data: VD, o: Option[U]) => {
22.       o match {
23.         case Some(u) => mapFunc(id, data, u)
24.         case None => data
25.       }
26.     }
27.     graph.outerJoinVertices(table)(uf)
28.   }
```

outerJoinVertices 的源码位于 Graph 中，代码如下：

```
1.   /**
2.     *在 table RDD 关联顶点实体，使用 mapfunc 合并函数。输入表最多包含每个顶点的一个
       *实体。如果没有图中的某个特定顶点提供的 other 实体，map 函数就接收 None
```

```
3.     *
4.     * @tparam U 表更新实体的类型
5.     * @tparam VD2 新顶点的类型
6.     *
7.     * @param other  与图中顶点关联的表，最多包含每个顶点的一个实体
8.     * @param mapFunc  用于计算新顶点值的函数。所有顶点调用 map 函数，即使表中没有
9.     * 相应的项
10.    *
11.    * @example  此函数用于更新基于外部数据的新的顶点值。例如，可以对每个顶点增加出度
       * 记录
12.    *
13.    * {{{
14.    * val rawGraph: Graph[_, _] = Graph.textFile("webgraph")
15.    * val outDeg: RDD[(VertexId, Int)] = rawGraph.outDegrees
16.    * val graph = rawGraph.outerJoinVertices(outDeg) {
17.    *   (vid, data, optDeg) => optDeg.getOrElse(0)
18.    * }
19.    * }}}
20.    */
21.   def outerJoinVertices[U: ClassTag, VD2: ClassTag](other: RDD[(VertexId, U)])
22.       (mapFunc: (VertexId, VD, Option[U]) => VD2)(implicit eq: VD =:= VD2
        = null)
23.     : Graph[VD2, ED]
```

接下来演示 joinVertices 和 outerJoinVertices 的使用。

首先把所有顶点的属性都变成 0，代码如下：

```
1.   val rawGraph: Graph[PartitionID, PartitionID]=graphFromFile
     .mapVertices((id, attr) =>0 )
2.   for (elem <- rawGraph.vertices.take(10)) {
3.     println("rawGraph.vertices:     " + elem)
4.   }
```

在 IDEA 中运行代码，通过 rawGraph.vertices.take(10)看一下执行结果。

```
1.   ......
2.   17/04/11 12:38:06 INFO DAGScheduler: ResultStage 42 (take at Graphx_
     webGoogle.scala:60) finished in 0.040 s
3.   17/04/11 12:38:06 INFO DAGScheduler: Job 14 finished: take at Graphx_
     webGoogle.scala:60, took 0.046316 s
4.   rawGraph.vertices:     (185012,0)
5.   rawGraph.vertices:     (612052,0)
6.   rawGraph.vertices:     (354796,0)
7.   rawGraph.vertices:     (182316,0)
8.   rawGraph.vertices:     (199516,0)
9.   rawGraph.vertices:     (627804,0)
10.  rawGraph.vertices:     (170792,0)
11.  rawGraph.vertices:     (307248,0)
12.  rawGraph.vertices:     (512760,0)
13.  rawGraph.vertices:     (386896,0)
14.  17/04/11 12:38:07 INFO SparkContext: Starting job: take at Graphx_
     webGoogle.scala:64
15.  ......
```

发现此时成功地让顶点的属性值变成了 0。

接下来找到所有 outDegrees 的顶点的集合。

```
1.   val outDeg: VertexRDD[PartitionID] =rawGraph.outDegrees
```

下面执行 joinVertices 的操作。

```
1.   val tmpJoinVertices: Graph[PartitionID, PartitionID]=rawGraph
     .joinVertices[Int](outDeg)((_, _, optDeg) => optDeg)
2.  for (elem <- tmpJoinVertices.vertices.take(10)) {
3.    println("tmpJoinVertices.vertices:     " + elem)
4.  }
```

在 IDEA 中运行代码， joinVertices 执行结果如下：

```
1.   ......
2.  17/04/11 13:33:05 INFO DAGScheduler: Job 15 finished: take at Graphx_
    webGoogle.scala:67, took 1.634308 s
3.  tmpJoinVertices.vertices:     (185012,0)
4.  tmpJoinVertices.vertices:     (612052,2)
5.  tmpJoinVertices.vertices:     (354796,1)
6.  tmpJoinVertices.vertices:     (182316,2)
7.  tmpJoinVertices.vertices:     (199516,20)
8.  tmpJoinVertices.vertices:     (627804,8)
9.  tmpJoinVertices.vertices:     (170792,9)
10. tmpJoinVertices.vertices:     (307248,2)
11. tmpJoinVertices.vertices:     (512760,7)
12. tmpJoinVertices.vertices:     (386896,18)
13. 17/04/11  13:33:05  INFO  SparkContext:  Starting  job:  take  at  Graphx_
    webGoogle.scala:73
14.
```

可以看到 ID 为 185012 的顶点的属性值为 0，这是因为这个顶点没有参加 Join 操作。接下来看 outerJoinVertices 的操作。

```
1.   val tmpouterJoinVertices: Graph[PartitionID, PartitionID]=rawGraph
     .outerJoinVertices[Int,Int](outDeg)((_, _, optDeg) =>optDeg.getOrElse(0))
2.  for (elem <- tmpouterJoinVertices.vertices.take(10)) {
3.    println("tmpouterJoinVertices.vertices:     " + elem)
4.  }
```

在 IDEA 中运行代码，outerJoinVertices 执行结果如下：

```
1.   ......
2.   17/04/11 13:47:13 INFO DAGScheduler: Job 16 finished: take at Graphx_
     webGoogle.scala:72, took 0.242373 s
3.  tmpouterJoinVertices.vertices:     (185012,0)
4.  tmpouterJoinVertices.vertices:     (612052,2)
5.  tmpouterJoinVertices.vertices:     (354796,1)
6.  tmpouterJoinVertices.vertices:     (182316,2)
7.  tmpouterJoinVertices.vertices:     (199516,20)
8.  tmpouterJoinVertices.vertices:     (627804,8)
9.  tmpouterJoinVertices.vertices:     (170792,9)
10. tmpouterJoinVertices.vertices:     (307248,2)
11. tmpouterJoinVertices.vertices:     (512760,7)
12. tmpouterJoinVertices.vertices:     (386896,18)
13. 17/04/11  13:47:13  INFO  SparkContext:  Starting  job:  take  at  Graphx_
    webGoogle.scala:77
14.
```

18.13　图操作之 aggregateMessages 实战

许多图分析任务的关键步骤是聚合每个顶点邻域的信息。例如，我们可能想知道每个用户拥有的关注者数量或每个用户的追随者的平均年龄。许多迭代图算法（如 PageRank，最短路径和连接组件）重复聚合相邻顶点的属性（如当前 PageRank Value，源的最短路径和最小可达顶点 id）。为了提高性能，主要聚集操作从老的 API 函数 graph.mapReduceTriplets 改变为使用新的 API 函数 graph.AggregateMessages。

在 Spark 早期版本的 GraphX 中，邻域聚合是使用旧的 API 函数 mapReduceTriplets 运算符完成的。

```
1.  class Graph[VD, ED] {
2.    def mapReduceTriplets[Msg](
3.        map: EdgeTriplet[VD, ED] => Iterator[(VertexId, Msg)],
4.        reduce: (Msg, Msg) => Msg)
5.      : VertexRDD[Msg]
6.  }
```

在 mapReduceTriplets 操作中需要定义 map 转换函数应用到每个三元组，产生用户定义的映射函数的消息，使用用户定义的聚合 reduce 函数进行聚合，但是，返回的迭代器 Iterator 不能进行额外的优化（如局部顶点重新编号等）。

Spark 2.X 中已经不再使用 mapReduceTriplets 函数，取而代之的是新的函数 aggregateMessages。在 aggregateMessages 函数中引入 EdgeContext，定义三元组字段，提供了向源顶点和目标顶点发送消息的功能，同时删除了字节码检查。定义三元组实际需要哪些字段？

以下代码块使用旧的函数 mapReduceTriplets。

```
1.   val graph: Graph[Int, Float] = ...
2.  def msgFun(triplet: Triplet[Int, Float]): Iterator[(Int, String)] = {
3.    Iterator((triplet.dstId, "Hi"))
4.  }
5.  def reduceFun(a: String, b: String): String = a + " " + b
6.  val result = graph.mapReduceTriplets[String](msgFun, reduceFun)
```

在 Spark 2.X 中，使用新的函数 aggregateMessages。

```
1.   val graph: Graph[Int, Float] = ...
2.  def msgFun(triplet: EdgeContext[Int, Float, String]) {
3.    triplet.sendToDst("Hi")
4.  }
5.  def reduceFun(a: String, b: String): String = a + " " + b
6.  val result = graph.aggregateMessages[String](msgFun, reduceFun)
```

Spark 2.2 GraphX 中的核心聚合操作是 aggregateMessages。该操作将用户定义的 sendMsg 函数应用于图中的每个边三元组，然后使用 mergeMsg 函数在其目标顶点聚合这些消息。aggregateMessages 源码在 Graph 中，源码如下：

```
1.   /**
2.     *  从每个顶点的相邻边和顶点进行聚合。用户提供的 sendMsg 函数在图中的每条边上调用，
```

```
     * 生成 0 个或更多的消息发送到边的任一顶点。mergeMsg 用来将发送到同一个顶点的消息
     * 进行合并
3.   *
4.   * @tparam A 发送到每个顶点的消息类型
5.   *
6.   * @param sendMsg 运行在每条边上，发送消息到相邻顶点使用[EdgeContext]。
7.   * @param mergeMsg  用于合并 sendMsg 发送的目标顶点是同一个顶点的信息，合并可以
     * 交换和关联
8.   * @param tripletFields 定义哪些字段在[EdgeContext]传递给 sendMsg 函数。如果
     * 不是所有字段都需要，指定字段可以提高计算性能
9.   *
10.  * @example  可以使用这个函数来计算每个顶点
11.  * {{{
12.  * val rawGraph: Graph[_, _] = Graph.textFile("twittergraph")
13.  * val inDeg: RDD[(VertexId, Int)] =
14.  *   rawGraph.aggregateMessages[Int](ctx => ctx.sendToDst(1), _ + _)
15.  * }}}
16.  *
17.  * @note  通过计算边层次，实现最大并行度。这是 Graph API 的核心功能之一，实现邻
     * 域计算，函数可以用来对符合条件的邻居进行计数或实现 PageRank 网页排名
18.  *
19.  */
20. def aggregateMessages[A: ClassTag](
21.     sendMsg: EdgeContext[VD, ED, A] => Unit,
22.     mergeMsg: (A, A) => A,
23.     tripletFields: TripletFields = TripletFields.All)
24.   : VertexRDD[A] = {
25.   aggregateMessagesWithActiveSet(sendMsg, mergeMsg, tripletFields, None)
26. }
```

用户定义的 sendMsg 函数采用 EdgeContext，将源和目标属性以及边属性、函数（sendToSrc，sendToDst）一起发送到源和目标属性，可以将 sendMsg 函数理解成 map-reduce 中的 map 函数。用户定义的 mergeMsg 函数需要两个发往同一顶点的消息，并产生单个消息，可以将 mergeMsg 函数理解成 map-reduce 中的 reduce 函数。aggregateMessages 操作返回一个 VertexRDD[Msg]，将该聚合消息（Msg 的类型）发往每个顶点。没有收到消息的顶点不包括在返回的 VertexRDD 中。

aggregateMessages 使用一个可选的三元组字段 tripletsFields，表明哪个数据在 EdgeContext 中被访问（例如，是源顶点属性，而不是目标顶点属性）。tripletsFields 的可能选项在 tripletsFields 中定义，TripletFields 的默认值是 TripletFields.All，表明用户定义的 sendMsg 函数可以访问 EdgeContext 的任何字段。该 tripletFields 参数可用于通知 GraphX，只有部分 EdgeContext 允许 GraphX 选择优化的关联策略。例如，如果计算每个用户的追随者的平均年龄，只需要源字段，因此我们将用于 TripletFields.Src 表示我们只需要源字段。

在早期版本的 GraphX 中，我们使用字节码检测来推断 TripletFields，但是发现字节码检测有时不可靠，而是选择了更明确的用户控制。

在下面的例子中，我们使用 aggregateMessages 操作来计算所有比这个用户年龄大的用户的个数以及比这个用户年龄大的用户的平均年龄。首先随机生成一张图：

```
1.  import org.apache.spark.graphx.{Graph, VertexRDD}
2.  import org.apache.spark.graphx.util.GraphGenerators
3.  ……
```

```
4.  //创建一个图，以年龄作为顶点属性
5.  //这里，我们使用一个简单的随机图
6.      val graph: Graph[Double, Int] =
7.      GraphGenerators.logNormalGraph(sc, numVertices = 100).mapVertices
        ( (id, _) => id.toDouble )
8.  ……
9.  //打印显示随机生成的内容
10.     for(elem <- graph.vertices.take(10)) {
11.       println("graph.vertices:    "+ elem)
12.     }
13.     for(elem <- graph.edges.take(10)) {
14.       println("graph.edges:    "+ elem)
15.     }
16. ……
```

在 IDEA 中运行代码，打印显示随机生成的内容。

```
1.  ……
2.  17/04/12 08:38:56 INFO DAGScheduler: Job 0 finished: take at Graphx_
    aggregateMessages.scala:29, took 2.208011 s
3.  graph.vertices:    (96,96.0)
4.  graph.vertices:    (56,56.0)
5.  graph.vertices:    (16,16.0)
6.  graph.vertices:    (80,80.0)
7.  graph.vertices:    (48,48.0)
8.  graph.vertices:    (32,32.0)
9.  graph.vertices:    (0,0.0)
10. graph.vertices:    (24,24.0)
11. graph.vertices:    (64,64.0)
12. graph.vertices:    (40,40.0)
13. 17/04/12 08:38:56 INFO SparkContext: Starting job: take at Graphx_
    aggregateMessages.scala:32
14. ……
15. 17/04/12 08:38:56 INFO DAGScheduler: Job 1 finished: take at Graphx_
    aggregateMessages.scala:32, took 0.063005 s
16. graph.edges:    Edge(0,1,1)
17. graph.edges:    Edge(0,4,1)
18. graph.edges:    Edge(0,4,1)
19. graph.edges:    Edge(0,12,1)
20. graph.edges:    Edge(0,13,1)
21. graph.edges:    Edge(0,14,1)
22. graph.edges:    Edge(0,15,1)
23. graph.edges:    Edge(0,23,1)
24. graph.edges:    Edge(0,27,1)
25. graph.edges:    Edge(0,36,1)
26. 17/04/12 08:38:56 INFO SparkContext: Starting job: collect at Graphx_
    aggregateMessages.scala:52
27. ……
```

此时假设图顶点的 ID 为用户年龄，下面的代码可以计算出所有比这个用户年龄大的用户的个数以及比这个用户年龄大的用户的平均年龄。

```
1.    // 计算出比这个用户年龄大的用户的个数以及比这个用户年龄大的用户的总年龄
2.      val olderFollowers: VertexRDD[(Int, Double)] = graph.aggregateMessages
        [(Int, Double)](
```

```
3.       triplet => { // Map 函数
4.         if (triplet.srcAttr > triplet.dstAttr) {
5.           //发送消息到目标顶点包含计数和年龄
6.           triplet.sendToDst(1, triplet.srcAttr)
7.         }
8.       },
9.       //计数和年龄分别相加
10.      (a, b) => (a._1 + b._1, a._2 + b._2) // Reduce 函数
11.    )
12.    //比这个用户年龄大的用户的个数除总年龄，获得比这个用户年龄大的追随者的平均年龄
13.    val avgAgeOfOlderFollowers: VertexRDD[Double] =
14.      olderFollowers.mapValues( (id, value) =>
15.        value match { case (count, totalAge) => totalAge / count } )
16.    //结果展示
17.    avgAgeOfOlderFollowers.collect.foreach(println(_))
```

在 IDEA 中运行代码， 所有比这个用户年龄大的用户的平均年龄的内容结果如下：

```
1.  ……
2.  17/04/12 08:38:57 INFO DAGScheduler: Job 2 finished: collect at Graphx_
    aggregateMessages.scala:52, took 1.455939 s
3.  (56,77.22222222222223)
4.  (16,57.0)
5.  (80,85.66666666666667)
6.  (48,69.5)
7.  (32,65.28571428571429)
8.  (0,53.041666666666664)
9.  (24,58.54545454545455)
10. (64,81.77777777777777)
11. (40,67.29411764705883)
12. (72,81.66666666666667)
13. (8,47.84615384615385)
14. (88,94.5)
15. ……
```

18.14　图算法之 Pregel API 原理解析与实战

为什么 Spark GraphX 会提供 Pregel API 呢？主要是便于迭代操作。因为在 GraphX 里面，Graph 这张图并没有自动 Cache，而是手动 Cache，但是，在每次迭代中，为了加快速度，需要手动 Cache，每次迭代完就需要把没用的删除掉，而把有用的保留，这是非常难以控制的，因为 Graph 中的点和边是分开进行 Cache 的，而 Pregel 能够帮助我们做这件事情，能够把一些细节屏蔽掉，所以如果你有很多迭代操作，如 PageRank，就非常适合使用 Pregel 来做。

Pregel 的 API 的代码位于 GraphOps 类中，其处理过程是异步的，性能非常出色，源码如下：

```
1.  /**
2.      * 执行一个 Pregel-like 迭代并行顶点的抽象。用户定义的顶点程序 vprog 是并行执行的，
        * 对每个顶点接收的消息计算新的顶点值。sendMsg 函数在所有出向边上被调用，并用于计
        * 算一个可选的消息到目标顶点；mergeMsg 函数用于在相同的顶点上将消息进行聚合
3.      * 第一次迭代时，所有顶点接收 initialMsg；在随后的迭代中，如果顶点不接收消息，顶
```

```
     * 点程序将不会被调用
4.   * 函数一直迭代到没有剩余的消息，或达到最大 maxiterations 迭代次数
5.   * @tparam A  Pregel 消息类型
6.   * @param initialMsg  第一次迭代时，每个顶点将收到消息
7.   * @param maxIterations 运行的最大迭代次数
8.   *
9.   * @param activeDirection  边的方向事件，对于顶点，收到上一轮 sendMsg 函数发送
     * 的消息。例如，如果是 EdgeDirection.Out，只有出向边的顶点在上一轮中收到消息的
     * 才运行
10.  *
11.  * @param vprog  用户定义的顶点程序，运行在每个顶点上，根据接收的消息计算新的顶
     * 点值。第一次迭代时，顶点程序通过默认消息应用到所有顶点；在随后的迭代中，顶点程序
     * 只在这些已经收到消息的顶点上调用
12.  *
13.  * @param sendMsg 用户定义的函数应用到在当前迭代中收到消息的出向边的顶点中
14.  *
15.  * @param mergeMsg  用户定义的函数，需要两个传入类型 A 的消息，并将其合并到单个
     * A 类型的消息中。函数可以结合和交换，理想情况下，A 类型的大小不应该增加
16.  *
17.  * @return  计算结束时得到的图
18.  *
19.  */
20. def pregel[A: ClassTag](
21.     initialMsg: A,
22.     maxIterations: Int = Int.MaxValue,
23.     activeDirection: EdgeDirection = EdgeDirection.Either)(
24.     vprog: (VertexId, VD, A) => VD,
25.     sendMsg: EdgeTriplet[VD, ED] => Iterator[(VertexId, A)],
26.     mergeMsg: (A, A) => A)
27.   : Graph[VD, ED] = {
28.   Pregel(graph, initialMsg, maxIterations, activeDirection)(vprog,
    sendMsg, mergeMsg)
29. }
```

构造 Pregel 对象使用的是其 object 对象的 apply 方法。

Pregel.scala 源码如下：

```
1.  /**
2.  * 实现了一个 Pregel-like 批量同步消息传递接口。不同于以前的 Pregel-like API，
    * GraphX Pregel API 在边上应用 sendMessage 计算，使发送的消息计算读取顶点属性和
    * 约束图结构。这些变化在基于图的计算时大大提高了分布式执行的效率，也更有灵活性
3.  *
4.  * @example  例如，可以使用 Pregel 抽象实现 PageRank
5.  * {{{
6.  * val  pagerankGraph: Graph[Double, Double] = graph
7.  *   //与每个顶点的关联度
8.  *   .outerJoinVertices(graph.outDegrees) {
9.  *     (vid, vdata, deg) => deg.getOrElse(0)
10. *   }
11. *   //根据度设置边的权重
12. *   .mapTriplets(e => 1.0 / e.srcAttr)
13. *   //将顶点属性设置为初始 PageRank 值
14. *   .mapVertices((id, attr) => 1.0)
```

```
15.  *
16.  * def  vertexProgram(id: VertexId, attr: Double, msgSum: Double):
     * Double =
17.  *   resetProb + (1.0 - resetProb) * msgSum
18.  * def sendMessage(id: VertexId, edge: EdgeTriplet[Double, Double]):
     Iterator[(VertexId, Double)] =
19.  *   Iterator((edge.dstId, edge.srcAttr * edge.attr))
20.  * def  messageCombiner(a: Double, b: Double): Double = a + b
21.  * val  initialMessage = 0.0
22.  * //执行 Pregel 固定次数的迭代
23.  * Pregel(pagerankGraph, initialMessage, numIter)(
24.  *   vertexProgram, sendMessage, messageCombiner)
25.  * }}}
26.  *
27.  */
28. object Pregel extends Logging {
29.
30.   /**
31.     *  执行 Pregel-like 迭代顶点的并行抽象。用户定义的顶点程序 vprog 并行执行，
32.     *  每个顶点接收消息并计算新的顶点值。sendMsg 函数在所有的出向边被调用，用于在目
        *  标顶点计算可选的消息。mergeMsg 函数是可交换的，用于在相同的顶点将消息进行合
        *  并
33.     *
34.     * 第一次迭代时，所有顶点收到 initialMsg 消息；在随后的迭代中，如果顶点没有接收到
        * 消息，则顶点程序不会调用运行
35.     *
36.     *  函数一直迭代到没有剩余的消息，或迭代到最大的次数 maxiterations
37.     *
38.     * @tparam VD 顶点数据类型
39.     * @tparam ED  边数据类型
40.     * @tparam A   Pregel 消息类型
41.     *
42.     * @param 输入图
43.     *
44.     * @param initialMsg  第一次迭代时，每个顶点收到的初始消息
45.     *
46.     * @param maxIterations 运行的最大迭代次数
47.     *
48.     * @param activeDirection 边的方向事件，顶点收到上一轮 sendMsg 函数发送的消
        * 息。如果是 EdgeDirection.Out，只有在上一轮中收到消息的出向边的顶点才运行。
        * 默认值是 EdgeDirection.Either，在上一轮中收到消息的边会运行 sendMsg；如果
        * 是 EdgeDirection.Both，sendMsg 在边和顶点都收到消息才会运行
49.     *
50.     * @param vprog  用户定义的顶点程序，运行在每个顶点上，根据接收的消息计算新顶
        * 点的值。第一次迭代时，顶点程序通过默认消息应用到所有顶点；在随后的迭代中，顶点程序
        * 只在这些已经收到消息的顶点上调用
51.     *
52.     * @param sendMsg  用户定义的函数应用到在当前迭代中收到消息的出向边的顶点中
53.     * @param mergeMsg  用户定义的函数，需要两个传入类型 A 的消息，并将其合并到单
        * 个 A 类型的消息中。函数可以结合和交换，理想情况下，A 类型的大小不应该增加
54.     *
55.     * @return 在计算结束时得到的图
```

```
56.     */
57.   def apply[VD: ClassTag, ED: ClassTag, A: ClassTag]
58.      (graph: Graph[VD, ED],
59.       initialMsg: A,
60.       maxIterations: Int = Int.MaxValue,
61.       activeDirection: EdgeDirection = EdgeDirection.Either)
62.      (vprog: (VertexId, VD, A) => VD,
63.       sendMsg: EdgeTriplet[VD, ED] => Iterator[(VertexId, A)],
64.       mergeMsg: (A, A) => A)
65.     : Graph[VD, ED] =
66.   {
67.     require(maxIterations > 0, s"Maximum number of iterations must be
      greater than 0," +
68.       s" but got ${maxIterations}")
69.     val checkpointInterval = graph.vertices.sparkContext.getConf
70.       .getInt("spark.graphx.pregel.checkpointInterval", -1)
71.     var g = graph.mapVertices((vid, vdata) => vprog(vid, vdata, initialMsg))
72.     val graphCheckpointer = new PeriodicGraphCheckpointer[VD, ED](
73.       checkpointInterval, graph.vertices.sparkContext)
74.     graphCheckpointer.update(g)
75.
76.     // 计算消息
77.     var messages = GraphXUtils.mapReduceTriplets(g, sendMsg, mergeMsg)
78.     val messageCheckpointer = new PeriodicRDDCheckpointer[(VertexId, A)](
79.       checkpointInterval, graph.vertices.sparkContext)
80.     messageCheckpointer.update(messages.asInstanceOf[RDD[(VertexId, A)]])
81.     var activeMessages = messages.count()
82.
83.     // 循环
84.     var prevG: Graph[VD, ED] = null
85.     var i = 0
86.     while (activeMessages > 0 && i < maxIterations) {
87.       // 接收消息并更新顶点
88.       prevG = g
89.       g = g.joinVertices(messages)(vprog)
90.       graphCheckpointer.update(g)
91.
92.       val oldMessages = messages
93.       //发送新消息，跳过没有收到消息的边。必须缓存，因此消息可以在下一行被物化
94.       // 允许去掉上一次迭代的缓存
95.       messages = GraphXUtils.mapReduceTriplets(
96.         g, sendMsg, mergeMsg, Some((oldMessages, activeDirection)))
97.       //  调用 count()计算 messages 和 g 的顶点，隐藏了旧的消息 oldmessages
98.       //  （取决于 g 的顶点）和 prevG 顶点（依靠旧的消息 oldmessages 和 g 的顶点）
99.       messageCheckpointer.update(messages.asInstanceOf[RDD[(VertexId, A)]])
100.          activeMessages = messages.count()
101.
102.          logInfo("Pregel finished iteration " + i)
103.
104.          // 将隐藏在新物化的 RDD 去掉缓存
105.          oldMessages.unpersist(blocking = false)
106.          prevG.unpersistVertices(blocking = false)
107.          prevG.edges.unpersist(blocking = false)
108.          // 迭代计算
109.          i += 1
110.        }
111.        messageCheckpointer.unpersistDataSet()
```

```
112.        graphCheckpointer.deleteAllCheckpoints()
113.        messageCheckpointer.deleteAllCheckpoints()
114.        g
115.      } //  apply 结束
116.
117.    } // class Pregel 结束
```

接下来使用 Pregel API 来做一个例子，车载地图导航软件一般都有最短路径和最佳路径，这里以计算两个点之间的最短路径为例来说明 Pregel API 的使用，依旧使用 web-Google 提供的数据，此时就是计算任何两个网页之间的最短路径，当然，任何两个网页之间不一定都存在最短路径。

首先定义 sourceId，如下所示。

```
1.    val  sourceId: VertexId = 0
```

接下来通过 mapVertices 操作来获得新的 Graph：

```
1.    val g:=graphFromFile.mapVertices((id , _) =>if(id == sourceId) 0.0 else
      Double.PositiveInfinity)
```

下面开始用 pregel 进行处理。

```
1.  val  sssp = g.pregel(Double.PositiveInfinity)(
2.    (id, dist, newDist) => math.min(dist, newDist), //顶点程序
3.    triplet => {  //发送消息
4.      if (triplet.srcAttr + triplet.attr < triplet.dstAttr) {
5.        Iterator((triplet.dstId, triplet.srcAttr + triplet.attr))
6.      } else {
7.        Iterator.empty
8.      }
9.    },
10.   (a, b) => math.min(a, b) //合并消息
11. )
12. println(sssp.vertices.collect.mkString("\n"))
```

在 IDEA 中运行代码，结果如下：

```
1.  ......
2.  17/04/13  19:43:54  INFO  DAGScheduler:  Job  52  finished:  collect  at
    Graphx_webGoogle.scala:100, took 2.309694 s
3.  sssp.vertices.collect:  (185012,12.0)
4.  (612052,Infinity)
5.  (354796,11.0)
6.  (182316,Infinity)
7.  (199516,8.0)
8.  (627804,15.0)
9.  (170792,11.0)
10. (307248,Infinity)
11. (512760,12.0)
12. (386896,11.0)
```

前面我们设置 sourceId 为 0，那么从 0 开始到 185 012 这个网页时需要 12 步，而从 0 这个网页永远无法到达 612 052。

上述就是求最短路径的方法，此时你也可以把它看作是地图导航，不同的是，我们这里把每条边的值都看成 1，而地图导航两个点之间会有里程的不同，其计算过程是一致的。

18.15　图算法之 ShortestPaths 原理解析与实战

最短路径在地图、电子商务、社交网络中等都有广泛的应用。

关于最短路径的计算，前面已经有分析和代码案例，在此不再赘述。其源码如下：

```
1.  /**
    *计算给定的地标顶点的最短路径，返回一个图顶点属性，每个顶点属性的映射包含到每个可
    *达地标的最短路径距离
2.   */
3.
4.  object ShortestPaths {
5.    /** 存储一个映射：从地标的顶点 ID 到该地标的距离*/
6.    type SPMap = Map[VertexId, Int]
7.
8.    private def makeMap(x: (VertexId, Int)*) = Map(x: _*)
9.
10.   private def incrementMap(spmap: SPMap): SPMap = spmap.map { case (v, d)
    => v -> (d + 1) }
11.
12.   private def addMaps(spmap1: SPMap, spmap2: SPMap): SPMap =
13.     (spmap1.keySet ++ spmap2.keySet).map {
14.       k => k -> math.min(spmap1.getOrElse(k, Int.MaxValue), spmap2.getOr
        Else(k, Int.MaxValue))
15.     }.toMap
16.
17.  /**
18.    * 计算给定地标顶点集的最短路径
19.    *
20.    * @tparam ED 边的属性类型，不用于计算
21.    *
22.    * @param graph 图用于计算最短路径
23.    * @param landmarks  地标顶点 ID 列表。最短路径将被计算到每个地标
24.    *
25.    * @return   返回图的每个顶点属性是一个映射，包含距离每个可达地标顶点的最短路径
26.    */
27.   def run[VD, ED: ClassTag](graph: Graph[VD, ED], landmarks: Seq
    [VertexId]): Graph[SPMap, ED] = {
28.     val spGraph = graph.mapVertices { (vid, attr) =>
29.       if (landmarks.contains(vid)) makeMap(vid -> 0) else makeMap()
30.     }
31.
32.     val initialMessage = makeMap()
33.
34.     def vertexProgram(id: VertexId, attr: SPMap, msg: SPMap): SPMap = {
35.       addMaps(attr, msg)
36.     }
37.
38.     def sendMessage(edge: EdgeTriplet[SPMap, _]): Iterator[(VertexId,
    SPMap)] = {
39.       val newAttr = incrementMap(edge.dstAttr)
40.       if (edge.srcAttr != addMaps(newAttr, edge.srcAttr)) Iterator((edge
        .srcId, newAttr))
41.       else Iterator.empty
```

```
42.     }
43.
44.     Pregel(spGraph, initialMessage)(vertexProgram, sendMessage, addMaps)
45.   }
46. }
```

18.16　图算法之 PageRank 原理解析与实战

非常著名的 PageRank 的用途非常广泛，如社交网络的推荐等。PageRank 图如图 18-19 所示。

图 18-19　PageRank 图

Spark GraphX 的 GraphOps 中提供了 PageRank 方法，使用 PageRank 方法可以让你即使在不了解 PageRank 实现算法的情况下，通过一行代码也可使用 PageRank，其源码如下：

```
1.  /**
2.    * PageRank 算法的实现有两种方法
3.    *
4.    *  第一个实现方法：使用独立的 Graph 接口运行 PageRank 算法，迭代固定次数
5.    * {{{
6.    * var PR = Array.fill(n)( 1.0 )
7.    * val oldPR = Array.fill(n)( 1.0 )
8.    * for( iter <- 0 until numIter ) {
9.    *   swap(oldPR, PR)
10.   *   for( i <- 0 until n ) {
11.   *     PR[i] = alpha + (1 - alpha) * inNbrs[i].map(j => oldPR[j] / outDeg[j]).sum
12.   *   }
13.   * }
14.   * }}}
15.   *
16.   *  第二个实现方法：使用 Pregel 接口运行 PageRank 算法，直到聚合
17.   *
18.   * {{{
19.   * var PR = Array.fill(n)( 1.0 )
20.   * val oldPR = Array.fill(n)( 0.0 )
21.   * while( max(abs(PR - oldPr)) > tol ) {
22.   *   swap(oldPR, PR)
23.   *   for( i <- 0 until n if abs(PR[i] - oldPR[i]) > tol ) {
24.   *     PR[i] = alpha + (1 - \alpha) * inNbrs[i].map(j => oldPR[j] /
      *     outDeg[j]).sum
25.   *   }
26.   * }
27.   * }}}
```

```
28.   *
29.   *  alpha 是随机重置的概率(通常为 0.15),inNbrs[i]是邻居集,其关联到 i 和 outDeg[j],
      *  outDeg[j]是顶点 J 的出度
30.   *
31.   * @note  这不是“标准化”的 PageRank 算法，作为结果页，没有反向链接具备 PageRank
      * 的 alpha
32.   */
33. object PageRank extends Logging {
34.
35.
36.  /**
37.    *  PageRank 进行固定次数的迭代,返回一个图的顶点属性包含 PageRank 和标准化边权重
       *  的边属性
38.    *
39.    * @tparam VD 原始的顶点属性(不使用)
40.    * @tparam ED  原始的边属性(不使用)
41.
42.    *
43.    * @param graph  计算 PageRank 的图
44.    * @param numIter 运行的 PageRank 迭代次数
45.    * @param resetProb   随机重置概率（alpha）
46.    *
47.    * @return 图每个顶点包含 PageRank 和每条边包含标准化权重
48.    */
49.   def run[VD: ClassTag, ED: ClassTag](graph: Graph[VD, ED], numIter: Int,
50.     resetProb: Double = 0.15): Graph[Double, Double] =
51.   {
52.     runWithOptions(graph, numIter, resetProb)
53.   }
```

另一种 PageRank 的内部具体方法的实现如下：

```
1.   /**
2.     *运行一个动态版本的 PageRank，返回图的顶点属性包含标准化边权重的 PageRank 和边属性
3.     *
4.     * @tparam VD  原始的顶点属性 (不使用)
5.     * @tparam ED  原始的边属性 (不使用)
6.     *
7.     * @param graph 计算 PageRank 的图
8.     * @param tol  聚集时允许的容错（更小 => 更准确）
9.     * @param resetProb 随机重置概率（alpha）
10.    *
11.    * @return 图包含每个顶点的 PageRank 和每条边包含标准化权重
12.    */
13.   def  runUntilConvergence[VD: ClassTag, ED: ClassTag](
14.     graph: Graph[VD, ED], tol: Double, resetProb: Double = 0.15): Graph
      [Double, Double] =
15.   {
16.       runUntilConvergenceWithOptions(graph, tol, resetProb)
17.   }
```

下面看一下 PageRank 的使用，我们依旧基于 web-Google.txt 的数据。

```
1.  val  rank: VertexRDD[Double] = graphFromFile.pageRank(0.01).vertices
```

在 IDEA 中运行代码，PageRank 执行的结果如下所示。

```
1.  ......
2.  17/04/13 20:59:41 INFO DAGScheduler: Job 107 finished: take at Graphx_
    webGoogle.scala:102, took 0.129050 s
3.  rank:    (185012,0.18083095391383722)
4.  rank:    (612052,0.15)
5.  rank:    (354796,0.18810908106478783)
6.  ......
```

我们可以注意到在使用 PageRank 方法的时候需要传入一个参数，传入的这个参数的值越小，PageRank 计算的值就越精确，如果数据量特别大而传入的参数值又特别小，就会导致巨大的计算任务和太长的计算时间。

18.17　图算法之 TriangleCount 原理解析与实战

TriangleCount 的主要用途之一是用于社区发现，如图 18-20 所示。

图 18-20　社区发现

例如，在微博上你关注的人也互相关注，大家的关注关系中就会有很多三角形，这说明社区很强，很稳定，大家的联系都比较紧密；如果只是你一个人关注很多人，这说明你的社交群体非常小。

triangleCount 的源码位于 GraphOps 中，如下所示。

```
1.  /**
2.    *计算通过每个顶点的三角形数。
3.    *
4.    * @see [[org.apache.spark.graphx.lib.TriangleCount$#run]]
5.    */
6.  def  triangleCount(): Graph[Int, ED] = {
7.    TriangleCount.run(graph)
8.  }
```

进入 triangleCount 的 run 的源码实现如下：

```
1.  /**
2.    * 计算通过每个顶点的三角形数
3.    *
4.    * 该算法相对比较简单，计算分为三个步骤：
5.    * 1. 计算每个顶点的邻居集
6.    * 2. 对于每条边计算集的交集，发送计数到两个顶点
7.    * 3. 计算每个顶点的和除以 2，因为每个三元组被计算两次
8.  
```

```
9.    *  这里有两种实现方法,默认的实现方法TriangleCount.run首先移除自循环和规范化图,
      *  要确保以下条件:
10.   * 1. 没有自己的边
11.   * 2. 所有边的方向(src 顶点大于 dst 目标顶点)
12.   * 3. 没有重复的边
13.   *
14.   *  规范化过程代价较高，因为需要对图重新分区。如果输入的数据已符合“规范范式”，而且
      *  去掉了自循环，那取而代之可使用 TriangleCount.runPreCanonicalized
15.   *
16.   * {{{
17.   * val canonicalGraph = graph.mapEdges(e => 1).removeSelfEdges()
      * .canonicalizeEdges()
18.   * val counts = TriangleCount.runPreCanonicalized(canonicalGraph)
      * .vertices
19.   * }}}
20.   *
21.   */
22. object TriangleCount {
23.
24.   def run[VD: ClassTag, ED: ClassTag](graph: Graph[VD, ED]): Graph[Int,
      ED] = {
25.     //转换边的数据，进行 shuffle，然后规范化
26.     val canonicalGraph = graph.mapEdges(e => true).removeSelfEdges()
        .convertToCanonicalEdges()
27.     //获得三角形计数
28.     val counters = runPreCanonicalized(canonicalGraph).vertices
29.     //与原始的图关联
30.     graph.outerJoinVertices(counters) { (vid, _, optCounter: Option[Int]) =>
31.       optCounter.getOrElse(0)
32.     }
33.   }
```

从源码中可以清晰地看出，如果进行 Triangle 计算，需要保持 sourceId 小于 destId，所以此时创建 Graph 时必须指定 GraphLoader.edgeListFile 的 canonicalOrientation 参数为 true，如下所示。

```
1.     val graphFortriangleCount: Graph[PartitionID, PartitionID] = GraphLoader
       .edgeListFile(sc, dataPath + "web-Google.txt", true)
2.      val c: VertexRDD[PartitionID] = graphFromFile.triangleCount().vertices
3.      for (elem <- c.take(10)) {
4.        println("triangleCount:     " + elem)
5.      }
```

在 IDEA 中运行代码，一直要运行几个小时，读者可以自行测试一下。

18.18　使用 Spark GraphX 实现婚恋社交网络多维度分析实战

我们登录一个婚恋社交网站，如珍爱网、世纪佳缘网等，登录网站须进行注册，当用户在婚恋网站上注册时有一个用户 ID，注册用户的职业、年龄、婚姻状况、收入、学习背景、

性格等。例如，性格分成金、木、水、火、土，假设郭靖是水型性格，黄蓉是火型性格等，这个时候婚恋网有一套推断逻辑，哪类性格类型匹配什么类型。如果是土型性格加木型性格，那推荐的最佳的性格是水型性格加金型性格。

用户在婚恋网注册时的用户 ID 对应图计算中的顶点，顶点及相关的属性就形成了顶点的表。顶点的属性定义了（用户名称、用户年龄）。

在婚恋网站上，当男生点击某个女生，在点击的时候，点击的行为就会被婚恋网站记录下来，这就形成了图计算中一个又一个边的关系，用户点击关注的行为就形成了边的表。边的属性为源顶点用户点击了目标顶点的用户几次。

这就构建了一张顶点的表，一张边的表。顶点的表记录了用户的基本信息。边的表描述发生了什么社交网络关系，如图 18-21 所示。

婚恋社交图

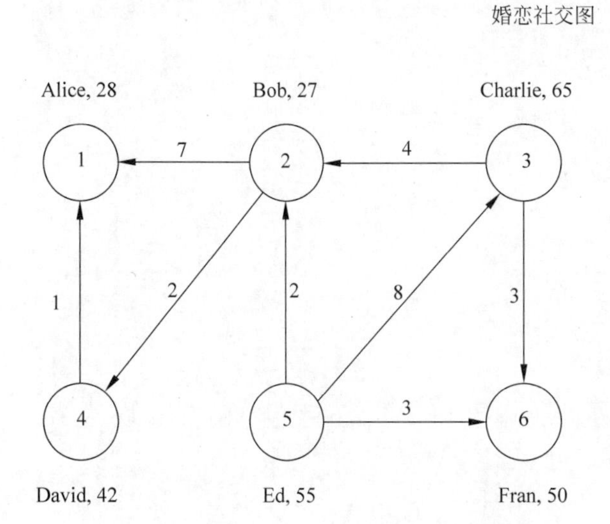

顶点表	
ID	属性（顶点）
1	（Alice, 28）
2	（Bob, 28）
3	（Charlie. 65）
4	（David, 42）
5	（Ed, 55）
6	（Fran, 50）

边表		
源 ID	目的 ID	属性（顶点）
2	1	7
2	4	2
3	2	4
3	6	3
4	1	1
5	2	2
5	3	8
5	6	3

图 18-21　婚恋社交图

假如男生关注一个女生，婚恋网站就会发送给用户这个女生相关的信息，提示用户这个女生喜欢什么类型的男生。根据图计算就能很好地计算这种情况，计算出用户的出度（关注的女生）。当男生、女士开始交往，婚恋网站就会分析记录双方的行为，使用机器学习进行行为分析。

作为女生，也很关注男生的职业分布，哪些男生关注了女生，追求男生的最大年龄是多少，最小年龄是多少，离家最近的男生，等等。所有这些要完成大数据系统分析，就需要结合 Spark SQL、图计算和机器学习的内容来实现。

本节简化需求，使用 Spark GraphX 实现婚恋社交网络多维度分析实战。

首先在 Spark 开发代码中屏蔽日志，设置运行环境。

```
1.    //屏蔽日志
2.      Logger.getLogger("org.apache.spark").setLevel(Level.WARN)
3.      Logger.getLogger("org.eclipse.jetty.server").setLevel(Level.OFF)
4.
5.      //设置运行环境
6.      val conf = new SparkConf().setAppName("SNSAnalysisGraphX").setMaster
        ("local[4]")
7.      val sc = new SparkContext(conf)
```

婚恋社交网络多维度分析实战先设置顶点和边，顶点和边是用元组定义的 Array，然后使用 SparkContext 的 parallelize 方法分别构造 vertexRDD 和 edgeRDD，最终构造图 Graph[VD,ED]。

```
1.  //设置顶点和边，注意顶点和边都是用元组定义的 Array
2.    //顶点的数据类型是 VD:(String,Int)
3.    val   vertexArray = Array(
4.      (1L, ("Alice", 28)),
5.      (2L, ("Bob", 27)),
6.      (3L, ("Charlie", 65)),
7.      (4L, ("David", 42)),
8.      (5L, ("Ed", 55)),
9.      (6L, ("Fran", 50))
10.   )
11.   //边的数据类型 ED:Int
12.   val   edgeArray = Array(
13.     Edge(2L, 1L, 7),
14.     Edge(2L, 4L, 2),
15.     Edge(3L, 2L, 4),
16.     Edge(3L, 6L, 3),
17.     Edge(4L, 1L, 1),
18.     Edge(5L, 2L, 2),
19.     Edge(5L, 3L, 8),
20.     Edge(5L, 6L, 3)
21.   )
22.   //构造 vertexRDD 和 edgeRDD
23.   val  vertexRDD: RDD[(Long, (String, Int))] = sc.parallelize(vertexArray)
24.   val  edgeRDD: RDD[Edge[Int]] = sc.parallelize(edgeArray)
25.
26.   //构造图 Graph[VD,ED]
27.   val  graph: Graph[(String, Int), Int] = Graph(vertexRDD, edgeRDD)
```

18.18.1　婚恋社交网络多维度分析实战图的属性演示

本节婚恋社交网络多维度分析实战图的属性演示中实现以下需求。

- 顶点操作：列出婚恋社交网络图中年龄大于 30 岁的用户。
- 边操作：列出婚恋社交网络图中用户点击他人超过 5 次的用户。
- 三元组操作：列出婚恋社交网络图中三元组中所有的源顶点用户喜欢目标顶点的用户。
- 三元组操作：列出婚恋社交网络图中三元组中用户点击他人超过 5 次的用户。
- 度操作：列出婚恋社交网络图中谁关注最多的人（用户关注了多少个其他用户），关注度最高的人（多少个人关注这个用户）、总关注度最大的人。

婚恋社交网络多维度分析实战图的属性演示具体方法实现如下。

（1）顶点操作：列出婚恋社交网络图中年龄大于 30 岁的用户。有两种实现方法：

方法 1，在图顶点的过滤方法中使用模式匹配，过滤年龄大于 30 岁的用户。图顶点的类型是(VertexId, VD)，顶点属性 VD 在这里是（姓名,年龄），图顶点 VertexRDD 类型格式为（图顶点 ID、（姓名,年龄）)，将大于 30 岁的年龄过滤掉就可以了。

方法 2，在图顶点的过滤方法中使用 v._2._2 > 30 进行过滤，VertexRDD 类型格式为（图顶点 ID、（姓名,年龄）)，每个图顶点的 v._1 第一个元素是图顶点 ID，每个图顶点的 v._2 第二个元素是图属性（姓名,年龄），v._2._1 是姓名，v._2._2 就获取到图顶点属性中的第二个元素年龄数据，过滤掉大于 30 岁的用户，然后打印输出。

（2）边操作：列出婚恋社交网络图中用户点击他人超过 5 次的用户。找出图中属性大于 5 的边，在图的边方法中过滤掉 graph.edges.filter(e => e.attr > 5)就可以。

（3）三元组操作：列出婚恋社交网络图中三元组中所有的源顶点用户喜欢目标顶点的用户。三元组包含五个元素，格式为((srcId, srcAttr), (dstId, dstAttr), attr)，在婚恋社交网络图中格式为((源顶点 ID，源顶点属性（姓名,年龄)),（目标顶点 ID，目标顶点属性（姓名,年龄)),边属性点击次数)。要列出用户关注的情况，使用图中的 triplet.srcAttr._1 获取源顶点属性 triplet.srcAttr（姓名，年龄）的第一个元素姓名，使用图中的 triplet.dstAttr._1 获取目标顶点属性 triplet.dstAttr（姓名，年龄）的第一个元素姓名，然后打印输出。

（4）三元组操作：列出婚恋社交网络图中三元组中用户点击他人超过 5 次的用户。首先在图的三元组中进行过滤，在婚恋社交网络图中三元组的格式为（(源顶点 ID，源顶点属性（姓名,年龄)),（目标顶点 ID,目标顶点属性（姓名，年龄)),边属性点击次数），t.attr 就是边属性点击次数，使用 t.attr > 5 就可以过滤掉用户点击超过 5 次的用户。然后打印输出源顶点、目标顶点姓名就可以。

（5）度操作：列出婚恋社交网络图中谁关注最多的人（用户关注了多少个其他用户），关注度最高的人（多少个人关注这个用户）、总关注度最大的人。计算婚恋社交网络图哪个顶点的 inDegrees 或者 outDegrees 或者 Degrees 最大，scala 代码是很强悍的，我们定义一个比较两个顶点 Degree 中较大值的函数 max，然后调用 graph.outDegrees.reduce(max)、graph.inDegrees.reduce(max)、graph.degrees.reduce(max)即可，就这么简单。

婚恋社交网络多维度分析实战图的属性操作代码如下：

```
1.    //*************************************************************
2.    //*************************** 图的属性******************************
      //*************************************************************
3.      println("*******************************************************")
4.      println("属性演示")
5.      println("*******************************************************")
6.      //方法一
7.      println("找出图中年龄大于 30 的顶点方法一：")
8.
9.      /**
         * 其实这里还可以加入性别等信息，例如我们可以看年龄大于 30 岁且是 female 的人
10.      */
11.     graph.vertices.filter { case (id, (name, age)) => age > 30}.collect
        .foreach {
12.       case (id, (name, age)) => println(s"$name is $age")
13.     }
14.     //方法二
15.     println("找出图中年龄大于 30 的顶点方法二：")
16.     graph.vertices.filter(v => v._2._2 > 30).collect.foreach(v =>
```

```
      println(s"${v._2._1} is ${v._2._2}"))
17.   println
18.
19.   //边操作：找出图中属性大于 5 的边
20.   println("找出图中属性大于 5 的边：")
21.   graph.edges.filter(e => e.attr > 5).collect.foreach(e => println(s"$
      {e.srcId} to ${e.dstId} att ${e.attr}"))
22.   println
23.
24.   //triplets 操作，((srcId, srcAttr), (dstId, dstAttr), attr)
25.   println("列出所有的 triplets：")
26.   for (triplet <- graph.triplets.collect) {
27.     println(s"${triplet.srcAttr._1} likes ${triplet.dstAttr._1}")
28.   }
29.   println
30.
31.   println("列出边属性>5 的 triplets：")
32.   for (triplet <- graph.triplets.filter(t => t.attr > 5).collect) {
33.     println(s"${triplet.srcAttr._1} likes ${triplet.dstAttr._1}")
34.   }
35.   println
36.
37.   //Degrees 操作
38.   println("找出图中最大的出度、入度、度数：")
39.   def max(a: (VertexId, Int), b: (VertexId, Int)): (VertexId, Int) = {
40.     if (a._2 > b._2) a else b
41.   }
42.   println("max of outDegrees:" + graph.outDegrees.reduce(max) + " max
      of inDegrees:" + graph.inDegrees.reduce(max) + " max of Degrees:" +
      graph.degrees.reduce(max))
43.   println
```

在 IDEA 中运行代码，婚恋社交网络多维度分析实战图的属性演示结果如下：

```
1.  Using Spark's default log4j profile: org/apache/spark/log4j-defaults
    .properties
2.  17/04/14 13:32:58 WARN NativeCodeLoader: Unable to load native-hadoop
    library for your platform... using builtin-java classes where applicable
3.  ********************************************************
4.  属性演示
5.  ********************************************************
6.  找出图中年龄大于 30 的顶点方法一：
7.  David is 42
8.  Ed is 55
9.  Fran is 50
10. Charlie is 65
11. 找出图中年龄大于 30 的顶点方法二：
12. David is 42
13. Ed is 55
14. Fran is 50
15. Charlie is 65
16.
17. 找出图中属性大于 5 的边：
18. 2 to 1 att 7
```

```
19. 5 to 3 att 8
20.
21. 列出所有的 triplets:
22. Bob likes Alice
23. Bob likes David
24. Charlie likes Bob
25. Charlie likes Fran
26. David likes Alice
27. Ed likes Bob
28. Ed likes Charlie
29. Ed likes Fran
30.
31. 列出边属性>5 的 triplets:
32. Bob likes Alice
33. Ed likes Charlie
34.
35. 找出图中最大的出度、入度、度数:
36. max of outDegrees:(5,3) max of inDegrees:(2,2) max of Degrees:(2,4)
```

18.18.2　婚恋社交网络多维度分析实战图的转换操作

婚恋社交网络多维度分析实战图的转换操作对婚恋社交图的顶点、边进行转换操作。

- 对婚恋社交图的顶点进行转换操作，将年龄加上 10。
- 对婚恋社交图的边进行转换操作，将用户点击的次数乘以 2。

婚恋社交网络多维度分析实战图的转换操作具体方法实现如下。

（1）对婚恋社交图的顶点进行转换操作，将年龄加上 10。使用图的 mapVertices 方法对图顶点进行转换操作，婚恋社交网的顶点格式为（图顶点 ID、（姓名,年龄）），转换时将年龄加上 10，graph.mapVertices{case (id, (name, age)) => (id, (name, age+10))}.vertices.collect 处理后返回的结果类型是 Array[(VertexId, (VertexId, (String, PartitionID)))]。使用 foreach 遍历 Array 数组中每个元素时，其类型是 (VertexId, (VertexId, (String, PartitionID)), 数组中的每个元素的 v._1 是图顶点 ID，数组中的每个元素的 v._2 是(VertexId, (String, PartitionID)，v._2._1 就获取到图顶点 ID，v._2._2 就获取到图顶点的属性（姓名，年龄），然后打印输出。

（2）对婚恋社交图的边进行转换操作，将用户点击的次数乘以 2。使用图的 mapEdges 进行转换，将 e.attr*2 边点击的次数乘以 2 就可以了。

```
1.  //*********************************************************************
    ********** 转换操作
    //*********************************************************************
2.      println("**********************************************************")
3.      println("转换操作")
4.      println("**********************************************************")
5.      println("顶点的转换操作，顶点 age + 10: ")
6.      graph.mapVertices{ case (id, (name, age)) => (id, (name, age+10))}
        .vertices.collect.foreach(v => println(s"${v._2._1} is ${v._2._2}"))
7.      println
8.      println("边的转换操作，边的属性*2: ")
9.      graph.mapEdges(e=>e.attr*2).edges.collect.foreach(e => println(s"$
```

```
     {e.srcId} to ${e.dstId} att ${e.attr}"))
10.    println
11.
```

在 IDEA 中运行代码，婚恋社交网络多维度分析实战图的转换操作结果如下：

```
1.  *******************************************************
2.  转换操作
3.  *******************************************************
4.  顶点的转换操作，顶点 age + 10:
5.  4 is (David,52)
6.  1 is (Alice,38)
7.  5 is (Ed,65)
8.  6 is (Fran,60)
9.  2 is (Bob,37)
10. 3 is (Charlie,75)
11.
12. 边的转换操作，边的属性*2:
13. 2 to 1 att 14
14. 2 to 4 att 4
15. 3 to 2 att 8
16. 3 to 6 att 6
17. 4 to 1 att 2
18. 5 to 2 att 4
19. 5 to 3 att 16
20. 5 to 6 att 6
```

18.18.3　婚恋社交网络多维度分析实战图的结构操作

婚恋社交网络多维度分析实战图的结构操作实现以下需求。

- ❑ 婚恋社交图中用户的年龄大于 30 岁的子图。
- ❑ 列出婚恋社交图中用户的年龄大于 30 岁的子图的所有顶点。
- ❑ 列出婚恋社交图中用户的年龄大于 30 岁的子图的所有边。

婚恋社交网络多维度分析实战图的结构操作具体方法实现如下。

（1）婚恋社交图中使用 graph.subgraph 方法对年龄大于 30 岁的用户进行结构转换。subgraph 方法中有两个传入参数，第一个参数是边的谓词函数 epred，这里没有传入参数；第二个参数是顶点的谓词函数 vpred，它是一个匿名函数，vpred 函数的传入参数类型是(VertexId, VD)，即（图顶点 ID,图顶点属性）。在婚恋社交图中，图顶点及属性为（VertexId 图顶点 ID、VD（姓名,年龄）），因此 vd._2 是图顶点属性的第二个元素（年龄），vd._2 >=30 就获取到图顶点属性中年龄大于 30 岁的用户。vpred 函数的返回结果为布尔值，最终生成年龄大于 30 岁的用户的子图。

（2）列出婚恋社交图中用户的年龄大于 30 岁的子图的所有顶点。在大于 30 岁的用户的子图中遍历打印顶点信息。这里 subGraph.vertices.collect 的类型是 Array[(VertexId, (String, PartitionID))]，即（顶点 ID，（姓名，年龄））。使用 foreach 遍历数组的元素，v._1 是图顶点 ID，v._2 是图顶点属性，v._2._1 是图顶点属性的姓名，v._2._2 是图顶点属性的年龄，最后打印输出。

（3）列出婚恋社交图中用户的年龄大于 30 岁的子图的所有边。在大于 30 岁的子图遍历打印边的信息。使用 subGraph.edges.collect 结果的类型是 Array[Edge[PartitionID]]，e.srcId 获取边的源顶点，e.dstId 获取边的目标顶点，e.attr 获取边的属性用户点击的次数，使用 foreach 循环遍历打印输出。

婚恋社交网络多维度分析实战图的结构操作代码如下：

```
1.  //*************************************************************************
2.  //**************************结构操作
3.  //*************************************************************************
4.      println("************************************************************")
5.      println("结构操作")
6.      println("************************************************************")
7.      println("顶点年纪>30 的子图：")
8.      val subGraph = graph.subgraph(vpred = (id, vd) => vd._2 >= 30)
9.      println("子图所有顶点：")
10.     subGraph.vertices.collect.foreach(v => println(s"${v._2._1} is
        ${v._2._2}"))
11.     println
12.     println("子图所有边：")
13.     subGraph.edges.collect.foreach(e => println(s"${e.srcId} to ${e.dstId}
        att ${e.attr}"))
14.     println
```

在 IDEA 中运行代码，婚恋社交网络多维度分析实战图的结构操作结果如下：

```
1.  ************************************************************
2.  结构操作
3.  ************************************************************
4.  顶点年纪>30 的子图：
5.  子图所有顶点：
6.  David is 42
7.  Ed is 55
8.  Fran is 50
9.  Charlie is 65
10.
11. 子图所有边：
12. 3 to 6 att 3
13. 5 to 3 att 8
14. 5 to 6 att 3
```

18.18.4　婚恋社交网络多维度分析实战图的连接操作

婚恋社交网络多维度分析实战图的连接操作实现以下需求。

- 婚恋社交网络中连接图的属性计算：用户被多少人关注，及用户关注了多少人。
- 婚恋社交网络中连接图的属性计算：列出用户关注人次和被人关注的人次相同的用户。

婚恋社交网络多维度分析实战图的连接操作具体方法实现如下。

（1）定义一个 User 的 case　class，用于之后创建新用户图的顶点的顶点属性信息，用户图顶点的属性包括（姓名,年龄,入度,出度），其中的入度表示多少人关注了用户，出度表示用户关注了多少人。

（2）创建一个新图 initialUserGraph，顶点 VD 的数据类型为 User，并从原始 graph 做顶点类型转换，将原来的顶点属性信息（姓名, 年龄）转换应用到 User 的 case　class 上，转换为新的顶点属性信息（姓名,年龄,入度,出度），入度、出度的初始值设置为 0。

（3）新图 initialUserGraph 与 inDegrees、outDegrees（RDD）进行连接，并修改 initialUserGraph 中的 inDeg 值、outDeg 值，创建新的用户图 userGraph，用户图 userGraph 的顶点属性格式为（姓名,年龄,入度,出度），其中，入度、出度不再为初始值 0，已经计算为实际的值。

- 计算入度（多少人关注了用户）：对于图 initialUserGraph，使用 outerJoinVertices 算子对 initialUserGraph.inDegrees 进行关联，其中 initialUserGraph.inDegrees 的数据类型是 VertexRDD[Int]，同时 VertexRDD[Int]继承自 RDD[(VertexId, VD)]；使用模式匹配，转换计算图的新的顶点属性（姓名,年龄,入度,出度），其中，姓名、年龄、出度从图 initialUserGraph 中直接沿用原来的值；入度（inDegOpt）是图 initialUserGraph 与 inDegrees 计算后的值，inDegOpt 是一个可选值 Option，如果 inDegOpt 能取到值，就填入到图顶点属性的入度里面，如果 inDegOpt 没有值，就仍然赋值 0。
- 计算出度（用户关注了多少人）：上述已在图顶点属性中填好了入度，接下来在图顶点属性中填写出度信息。对于图 initialUserGraph，使用 outerJoinVertices 算子对 initialUserGraph.outDegrees 进行关联，其中，initialUserGraph.outDegrees 的数据类型是 VertexRDD[Int]，同时 VertexRDD[Int]继承自 RDD[(VertexId, VD)]；使用模式匹配，转换计算图的新的顶点属性(姓名,年龄,入度,出度），其中姓名、年龄、入度从图 initialUserGraph 中直接沿用原来的值；出度（outDegOpt）是图 initialUserGraph 与 outDegrees 计算后的值，outDegOpt 是一个可选值 Option，如果 outDegOpt 能取到值，就填入到图顶点属性的出度里面，如果 outDegOpt 没有值，就仍然赋值 0。

（4）打印输出婚恋社交网络中连接图的属性：用户被多少人关注，及用户关注了多少人。userGraph.vertices.collect 的结果类型是 Array[(VertexId, User)]，使用 foreach 遍历数组中的每一个元素，其中 v._1 是图顶点 ID，v._2 是图顶点属性 case class 用户（User），依次使用 v._2.name、v._2.inDeg、v._2.outDeg 获取姓名、入度（用户被多少人关注）、出度（用户关注了多少人），打印输出。

（5）打印输出婚恋社交网络中连接图的属性：列出用户关注人次和被人关注的人次相同的用户。首先使用 u.inDeg == u.outDeg 过滤出图顶点属性中入度、出度相同的顶点，过滤以后，collect 算子计算的结果类型是 Array[(VertexId, User)]，使用 foreach 循环遍历数组中的元素，通过模式匹配到顶点属性信息用户 property (姓名,年龄,入度,出度），property.name 打印出用户的姓名。

婚恋社交网络多维度分析实战图的连接操作代码如下：

```
1.   //****************************************************************************
2.  //*******************************          连接操作
3.  //****************************************************************************
4.    println("*************************************************************")
```

```
5.      println("连接操作")
6.      println("***********************************************************")
7.      val  inDegrees: VertexRDD[Int] = graph.inDegrees
8.      case  class User(name: String, age: Int, inDeg: Int, outDeg: Int)
9.
10.     //创建一个新图，顶点 VD 的数据类型为 User，并从 Graph 做类型转换
11.     val initialUserGraph: Graph[User, Int] = graph.mapVertices { case (id,
        (name, age)) => User(name, age, 0, 0)}
12.
13.     //initialUserGraph 与 inDegrees、outDegrees（RDD）进行连接，并修改
        //initialUserGraph 中的 inDeg 值和 outDeg 值
14.    val  userGraph = initialUserGraph.outerJoinVertices(initialUserGraph
       .inDegrees) {
15.       case (id, u, inDegOpt) => User(u.name, u.age, inDegOpt.getOrElse(0),
         u.outDeg)
16.     }.outerJoinVertices(initialUserGraph.outDegrees) {
17.       case  (id, u, outDegOpt) => User(u.name, u.age, u.inDeg, outDegOpt
         .getOrElse(0))
18.     }
19.
20.     println("连接图的属性：")
21.    userGraph.vertices.collect.foreach(v => println(s"${v._2.name} inDeg:
       ${v._2.inDeg}  outDeg: ${v._2.outDeg}"))
22.     println
23.
24.     println("出度和入度相同的人员：")
25.     userGraph.vertices.filter {
26.       case  (id, u) => u.inDeg == u.outDeg
27.     }.collect.foreach {
28.       case  (id, property) => println(property.name)
29.     }
30.     println
```

在 IDEA 中运行代码，婚恋社交网络多维度分析实战图的连接操作结果如下：

```
1.  ***************************************************************
2.  连接操作
3.  ***************************************************************
4.  连接图的属性：
5.  David inDeg: 1  outDeg: 1
6.  Alice inDeg: 2  outDeg: 0
7.  Ed inDeg: 0  outDeg: 3
8.  Fran inDeg: 2  outDeg: 0
9.  Bob inDeg: 2  outDeg: 2
10. Charlie inDeg: 1  outDeg: 2
11.
12. 出度和入度相同的人员：
13. David
14. Bob
```

18.18.5　婚恋社交网络多维度分析实战图的聚合操作

婚恋社交网络多维度分析实战图的聚合操作实现以下需求。

- 在婚恋社交图中找出年龄最大的追求者。
- 在婚恋社交图中找出年龄最小的追求者。

- ❑ 在婚恋社交图中找出追求者的平均年龄。

婚恋社交网络多维度分析实战图的聚合操作具体方法实现如下。

（1）在婚恋社交图中找出年龄最大的追求者：使用图的 aggregateMessages 方法，在婚恋社交图中，aggregateMessages 算子发送到其他每个图顶点的消息类型是(String, Int)，其含义是（姓名，年龄）。

- ❑ 把 sendMsg 函数理解成 map-reduce 中的 map：将源顶点的属性发送给目标顶点，map 过程。别名为 triplet 三元组的数据类型是 EdgeContext[(String, PartitionID), PartitionID, (String, PartitionID)]，EdgeContext 中每个元素的含义分别是顶点属性（姓名，年龄），边属性（用户点击的次数），aggregateMessages 是发送到其他每个图顶点的消息类型（姓名，年龄）。EdgeContext 包含 5 个属性方法：源顶点 IDsrcId、目标顶点 IDdstId、源顶点属性 srcAttr、目标顶点属性 dstAttr、边属性 attr。triplet.srcAttr 获取源顶点的属性，其格式为（姓名，年龄），因此，triplet.srcAttr._1 是源顶点属性的姓名，triplet.srcAttr._2 是源顶点属性的年龄。通过 triplet.sendToDst 将（姓名，年龄）发送消息到图的每个顶点进行计算。
- ❑ 把 mergeMsg 函数理解成 map-reduce 中的 reduce：汇聚计算得到最大追求者，reduce 过程。mergeMsg 函数将收到的消息（姓名，年龄）进行合并汇聚，比较两个消息的第二个元素年龄，将年龄大的用户返回。最终计算出每个顶点年龄最大的追求者的 VertexRDD，创建生成 oldestFollower。

接下来将用户图 userGraph 的顶点 vertices 和 oldestFollower VertexRDD[(String, Int)] 进行左关联 leftJoin。将（id, user, optOldestFollower）进行模式匹配，其中，id 是 userGraph 图的图顶点 ID，user 是 userGraph 图的图顶点属性（姓名,年龄,入度,出度），optOldestFollower 的类型是可选类型 Option[(String, PartitionID)]，格式为（姓名，年龄）；如果模式匹配结果为空，则说明此用户顶点没有一个追求者，就将字符串“${user.name} does not have any followers.”作为新的图顶点的属性返回；如果(name, age)能匹配上，就将年龄最大的追求者 name 和 userGraph 图的图顶点属性 user.name 的用户姓名拼接成字符串“${name} is the oldest follower of ${user.name}”，作为新的图顶点的属性返回。因此，userGraph.vertices.leftJoin(oldestFollower)创建生成的图顶点格式为 VertexRDD[String]，这里 String 的内容就是上述的字符串。使用 collect 算子收集数据，然后使用 foreach 循环遍历打印输出婚恋社交图中每个用户年龄最大的追求者。

（2）在婚恋社交图中找出年龄最小的追求者，实现思路与在婚恋社交图中找出年龄最大的追求者相同，在 reduce 过程中，将最小追求者返回就可以，这里不再赘述。

（3）在婚恋社交图中找出追求者的平均年龄：使用图（graph）的 aggregateMessages 方法，在婚恋社交图中，这里的消息发送数据类型不同于找出年龄最大、最小的追求者的消息类型，因为我们需要对用户追求者进行计数，还需要获取用户追求者的年龄信息，因此，aggregateMessages 算子发送到其他每个图顶点的消息类型是(Int, Double)，其含义是（计数 1 次，年龄）。

- ❑ 把 sendMsg 函数理解成 map-reduce 中的 map：将源顶点的属性发送给目标顶点，map 过程。别名为 triplet 三元组的数据类型是 EdgeContext[(String, PartitionID), PartitionID, (PartitionID, Double)]，EdgeContext 中每个元素的含义分别是顶点属性（姓名，年龄），边属性（用户点击的次数），aggregateMessages 是发送到其他每个图顶

点的消息类型（计数 1 次，年龄）。EdgeContext 包含 5 个属性方法：源顶点 IDsrcId、目标顶点 IDdstId、源顶点属性 srcAttr、目标顶点属性 dstAttr、边属性 attr。triplet.srcAttr 获取源顶点的属性，其格式为（姓名，年龄），因此，triplet.srcAttr._1 是源顶点属性的姓名，triplet.srcAttr._2 是源顶点属性的年龄。通过 triplet.sendToDst 将（计数 1 次，年龄）发送消息到图的每个顶点进行计算。

- 把 mergeMsg 函数理解成 map-reduce 中的 reduce：汇聚计算，得到追求者的数量和总年龄。mergeMsg 函数将收到的消息（计数 1 次，年龄）进行合并汇聚，比较两个消息，分别将两个消息的第一个元素计数次数进行累加 a._1 + b._1；将两个消息的第二个元素年龄进行累加 a._2 + b._2，返回用户追求者的个数和追求者的总年龄，内容为（计数总次数，累加总年龄）。

graph.aggregateMessages 计算以后生成的数据类型是 VertexRDD[(PartitionID, Double)]，然后使用 mapValues 算子对(id, p)进行转换，从(图顶点 VertexId, 图顶点属性 p（计数总次数，累加总年龄））中获取 p._1 计数总次数及 p._2 累加总年龄，使用 p._2 / p._1 计算出追求者的平均年龄。最终创建生成 averageAge: VertexRDD[Double]。

接下来将用户图 userGraph 的顶点 vertices 和 averageAge: VertexRDD[Double] 进行左关联 leftJoin。将(id, user, optAverageAge)进行模式匹配，其中 id 是 userGraph 图的图顶点 ID，user 是 userGraph 图的图顶点属性（姓名,年龄,入度,出度），optAverageAge 的类型是可选类型 Option[Double]，格式为（追求者的平均年龄）；如果模式匹配结果为空，则说明此用户顶点没有一个追求者，就将字符串“${user.name} does not have any followers.”作为新的图顶点的属性返回；如果 Some(avgAge)能匹配上，就将追求者的平均年龄 avgAge 和 userGraph 图的图顶点属性 user.name 的用户姓名拼接成字符串 "The average age of ${user.name}\'s followers is $avgAge."，作为新的图顶点的属性返回。因此，userGraph.vertices.leftJoin(averageAge)创建生成的图顶点格式为：VertexRDD[String]，这里，String 的内容就是上述的字符串。使用 collect 算子收集数据，然后使用 foreach 循环遍历打印输出婚恋社交图中每个用户追求者的平均年龄。

婚恋社交网络多维度分析实战图的聚合操作代码如下：

```
1.    println("*********************************************************")
2.        println("聚合操作")
3.        println("*********************************************************")
4.        println("找出年龄最大的追求者：")
5.        val oldestFollower: VertexRDD[(String, Int)] = graph.aggregateMessages
          [(String, Int)](
6.          //将源顶点的属性发送给目标顶点，map 过程
7.          triplet => { // Map 方法
8.            //将消息发送到包含姓名和年龄的目标顶点
9.            triplet.sendToDst(triplet.srcAttr._1, triplet.srcAttr._2)
10.         },
11.         //得到年龄最大的追求者，reduce 过程
12.         (a, b) => if (a._2 > b._2) a else b
13.       )
14.
15.
16.       userGraph.vertices.leftJoin(oldestFollower) { (id, user, optOldest
          Follower) =>
17.         optOldestFollower match {
```

```
18.             case None => s"${user.name} does not have any followers."
19.             case Some((name, age)) => s"${name} is the oldest follower of
                ${user.name}."
20.           }
21.         }.collect.foreach { case (id, str) => println(str)}
22.         println
23.
24.     println("****************************************************")
25.     println("找出年龄最小的追求者：")
26.     val youngestFollower: VertexRDD[(String, Int)] = graph.aggregateMessages
        [(String, Int)](
27.       //将源顶点的属性发送给目标顶点，map 过程
28.       triplet => { //Map Function
29.         //Send message to destination vertex containing name and age
30.         triplet.sendToDst(triplet.srcAttr._1, triplet.srcAttr._2)
31.       },
32.       //得到年龄最小的追求者，reduce 过程
33.       (a, b) => if (a._2 > b._2) b else a
34.     )
35.
36.
37.     userGraph.vertices.leftJoin(youngestFollower) { (id, user,
        optYoungestFollower) =>
38.       optYoungestFollower  match {
39.         case None => s"${user.name} does not have any followers."
40.         case Some((name, age)) => s"${name} is the youngest follower of
             ${user.name}."
41.       }
42.     }.collect.foreach { case (id, str) => println(str)}
43.     println
44.
45. //找出追求者的平均年龄
46.         println("找出追求者的平均年龄：")
47.         val averageAge: VertexRDD[Double] = graph.aggregateMessages[(Int,
             Double)](
48.           //将源顶点的属性 (1, Age)发送给目标顶点，map 过程
49.           triplet => { //Map Function
50.             //将消息发送到包含姓名和年龄的目标顶点
51.             triplet.sendToDst((1, triplet.srcAttr._2.toDouble))
52.           },
53.           //得到追求者的数量和总年龄
54.           (a, b) => ((a._1 + b._1), (a._2 + b._2))
55.         ).mapValues((id, p) => p._2 / p._1)
56.
57.         userGraph.vertices.leftJoin(averageAge) { (id, user, optAverageAge)
            =>
58.           optAverageAge match {
59.             case None => s"${user.name} does not have any followers."
60.             case Some(avgAge) => s"The average age of ${user.name}\'s
                followers is $avgAge."
61.           }
62.         }.collect.foreach { case (id, str) => println(str)}
63.         println
```

在 IDEA 中运行代码，婚恋社交网络多维度分析实战图的聚合操作结果如下：

```
1.   ****************************************************
2.   聚合操作
```

```
3.  *******************************************************
4.  找出年龄最大的追求者：
5.  Bob is the oldest follower of David.
6.  David is the oldest follower of Alice.
7.  Ed does not have any followers.
8.  Charlie is the oldest follower of Fran.
9.  Charlie is the oldest follower of Bob.
10. Ed is the oldest follower of Charlie.
11.
12. *******************************************************
13. 找出年龄最小的追求者：
14. Bob is the youngest follower of David.
15. Bob is the youngest follower of Alice.
16. Ed does not have any followers.
17. Ed is the youngest follower of Fran.
18. Ed is the youngest follower of Bob.
19. Ed is the youngest follower of Charlie.
20.
21. 找出追求者的平均年龄：
22. The average age of David's followers is 27.0.
23. The average age of Alice's followers is 34.5.
24. Ed does not have any followers.
25. The average age of Fran's followers is 60.0.
26. The average age of Bob's followers is 60.0.
27. The average age of Charlie's followers is 55.0.
```

18.18.6　婚恋社交网络多维度分析实战图的实用操作

婚恋社交网络多维度分析实战图的实用操作：找出图中某个顶点 5 到图中各顶点的最短距离。

婚恋社交网络多维度分析实战图的实用操作具体方法实现如下。

（1）首先定义一个源顶点（第 5 个顶点），初始化顶点 5 到各顶点的距离。图使用 mapVertices 对顶点进行转换，如果顶点 ID 等于源顶点，则将图顶点第 5 个顶点的属性设置为 0，否则图顶点属性为无穷大，创建生成图 initialGraph: Graph[Double, PartitionID]。

（2）对图 initialGraph 使用 pregel 方法计算顶点 5 到各顶点的最短距离。

- 消息类型为 Double 类型。
- 顶点谓词函数 vprog：顶点程序，运行在每个顶点上，根据接收的消息计算新的顶点值。(id, dist, newDist)的含义为（顶点 ID，图顶点属性距离，收到的消息新的距离），按照 math.min(dist, newDist) 获取源顶点到图顶点的距离及收到消息中距离的最小值。
- sendMsg 函数，相当于 map-reduce 中的 map，这里计算权重值：triplet 的数据类型为 EdgeTriplet[Double, PartitionID]，EdgeTriplet 类中包含源顶点属性 srcAttr、目标顶点属性 dstAttr，EdgeTriplet[VD, ED] 继承了 Edge[ED]类，因此 EdgeTriplet 类还包含了边属性 attr。如果 triplet.srcAttr 加上边属性 triplet.attr 小于 triplet.dstAttr，则返回一个 Iterator（目标顶点 ID，消息（最短距离））。否则返回 Iterator.empty 空值。
- mergeMsg 函数，相当于 map-reduce 中的 reduce：这里使用(a, b) => math.min(a, b)计算出两个消息中的最短距离。

（3）打印输出图中顶点 5 到图中各顶点的最短距离。

婚恋社交网络多维度分析实战图的实用操作代码如下：

```
1.  //************************************************************************
2.  //******************************实用操作
3.  //************************************************************************
4.      println("**************************************************************")
5.      println("聚合操作")
6.      println("**************************************************************")
7.      println("找出顶点 5 到各顶点的最短距离：")
8.      val sourceId: VertexId = 5L //定义源点
9.      val initialGraph = graph.mapVertices((id, _) => if (id == sourceId)
        0.0 else Double.PositiveInfinity)
10.     val sssp = initialGraph.pregel(Double.PositiveInfinity)(
11.       (id, dist, newDist) => math.min(dist, newDist),
12.       triplet => {  //计算权重
13.         if (triplet.srcAttr + triplet.attr < triplet.dstAttr) {
14.           Iterator((triplet.dstId, triplet.srcAttr + triplet.attr))
15.         } else {
16.           Iterator.empty
17.         }
18.       },
19.       (a,b) => math.min(a,b) // 最短距离
20.     )
21.     println(sssp.vertices.collect.mkString("\n"))
```

在 IDEA 中运行代码，婚恋社交网络多维度分析实战图的实用操作结果如下：

```
1.  **************************************************************
2.  聚合操作
3.  **************************************************************
4.  找出顶点 5 到各顶点的最短距离：
5.  (4,4.0)
6.  (1,5.0)
7.  (5,0.0)
8.  (6,3.0)
9.  (2,2.0)
10. (3,8.0)
```

18.19　婚恋社交网络多维度分析案例代码

Spark GraphX 图操作的案例代码如例 18-3 所示。

【例 18-3】 Graphx_webGoogle.scala 代码。

```
1.   package com.dt.spark.graphx
2.
3.  import org.apache.log4j.{Level, Logger}
4.  import org.apache.spark.SparkConf
5.  import org.apache.spark.graphx._
6.  import org.apache.spark.sql.SparkSession
7.
8.  object Graphx_webGoogle {
```

```
9.    def main(args: Array[String]) {
10.     Logger.getLogger("org").setLevel(Level.ERROR)
11.     var masterUrl = "local[8]"
12.     if (args.length > 0) {
13.       masterUrl = args(0)
14.     }
15.     val  sparkConf  =  new  SparkConf().setMaster(masterUrl).setAppName
        ("Graphx_webGoogle")
16.     val spark = SparkSession
17.       .builder()
18.       .config(sparkConf)
19.       .getOrCreate()
20.     val sc = spark.sparkContext
21.     //数据存放的目录
22.     var dataPath = "data/web-Google/"
23.     val graphFromFile: Graph[PartitionID, PartitionID] = GraphLoader
        .edgeListFile(sc, dataPath + "web-Google.txt", numEdgePartitions = 4)
24.     //统计顶点的数量
25.    println("graphFromFile.vertices.count:     " + graphFromFile.vertices
       .count())
26.     //统计边的数量
27.     println("graphFromFile.edges.count:     " + graphFromFile.edges
        .count())
28.
29.     val subGraph: Graph[PartitionID, PartitionID] =graphFromFile.subgraph
        (epred = e  => e.srcId > e.dstId)
30.     for(elem <- subGraph.edges.take(10)) {
31.       println("subGraph.edges:     "+  elem)
32.     }
33.
34.     println("subGraph.vertices.count():     "+  subGraph.vertices
        .count())
35.     println("subGraph.edges.count():     "+  subGraph.edges.count())
36.
37.     val subGraph2: Graph[PartitionID, PartitionID] = graphFromFile.subgraph
        (epred = e  => e.srcId > e.dstId,vpred= (id, _) => id>1000000)
38.     println("subGraph2.vertices.count():     "+  subGraph2.vertices
        .count())
39.     println("subGraph2.edges.count():     "+  subGraph2.edges.count())
40.
41.     val tmp: VertexRDD[PartitionID] =graphFromFile.inDegrees
42.     for (elem <- tmp.take(10)) {
43.       println("graphFromFile.inDegrees:     " + elem)
44.     }
45.     val tmp1: VertexRDD[PartitionID] =graphFromFile.outDegrees
46.     for (elem <- tmp1.take(10)) {
47.       println("graphFromFile.outDegrees:     " + elem)
48.     }
49.
50.     val tmp2: VertexRDD[PartitionID] =graphFromFile.degrees
51.     for (elem <- tmp2.take(10)) {
52.       println("graphFromFile.degrees:     " + elem)
53.     }
54.
55.     def max(a:(VertexId,Int),b:(VertexId,Int)):(VertexId,Int)=if(a._2 >
        b._2) a else b
56.     println("graphFromFile.degrees.reduce(max):     " + graphFromFile
```

```
     .degrees.reduce(max) )
57.    println("graphFromFile.inDegrees.reduce(max):     " + graphFromFile
     .inDegrees.reduce(max) )
58.    println("graphFromFile.outDegrees.reduce(max):     " + graphFromFile
     .outDegrees.reduce(max) )
59.
60.
61.    val rawGraph: Graph[PartitionID, PartitionID] =graphFromFile
     .mapVertices((id, attr) =>0 )
62.    for (elem <- rawGraph.vertices.take(10)) {
63.      println("rawGraph.vertices:     " + elem)
64.    }
65.
66.    val outDeg: VertexRDD[PartitionID] =rawGraph.outDegrees
67.
68.    val tmpJoinVertices: Graph[PartitionID, PartitionID] =rawGraph.join
     Vertices[Int](outDeg)((_, _, optDeg) => optDeg)
69.    for (elem <- tmpJoinVertices.vertices.take(10)) {
70.      println("tmpJoinVertices.vertices:     " + elem)
71.    }
72.
73.    val tmpouterJoinVertices: Graph[PartitionID, PartitionID] =rawGraph
     .outerJoinVertices[Int,Int](outDeg)((_, _, optDeg) =>optDeg.getOrElse(0))
74.    for (elem <- tmpouterJoinVertices.vertices.take(10)) {
75.      println("tmpouterJoinVertices.vertices:     " + elem)
76.    }
77.
78.
79.    for (elem <- graphFromFile.vertices.take(10)) {
80.      println("graphFromFile.vertices:     " + elem)
81.    }
82.
83.    val tmpGraph: Graph[PartitionID, PartitionID] =graphFromFile
     .mapVertices((vid, attr) => attr.toInt*2)
84.    for (elem <- tmpGraph.vertices.take(10)) {
85.      println("tmpGraph.vertices:     " + elem)
86.    }
87.
88.    //Pregel API 例子
89.    val sourceId: VertexId = 0
90.    val g: Graph[Double, PartitionID] =graphFromFile.mapVertices((id, _)
     =>if(id == sourceId) 0.0 else Double.PositiveInfinity)
91.    val sssp: Graph[Double, PartitionID] = g.pregel(Double.Positive
     Infinity)(
92.      (id, dist, newDist) => math.min(dist, newDist), //顶点程序
93.      triplet => {  //发送消息
94.        if (triplet.srcAttr + triplet.attr < triplet.dstAttr) {
95.          Iterator((triplet.dstId, triplet.srcAttr + triplet.attr))
96.        } else {
97.          Iterator.empty
98.        }
99.      },
100.        (a, b) => math.min(a, b) //合并消息
101.      )
102.      println("sssp.vertices.collect: "+ sssp.vertices.collect.take(10)
       .mkString("\n"))
103.      val rank: VertexRDD[Double] = graphFromFile.pageRank(0.01)
```

```
          .vertices
104.      for (elem <- rank.take(10)) {
105.        println("rank:     " + elem)
106.      }
107.
108.      val graphFortriangleCount: Graph[PartitionID, PartitionID] =
          GraphLoader.edgeListFile(sc, dataPath + "web-Google.txt", true)
109.      val c: VertexRDD[PartitionID] = graphFromFile.triangleCount()
          .vertices
110.      for (elem <- c.take(10)) {
111.        println("triangleCount:     " + elem)
112.      }
113.
114.      while (true) {}
115.    }
116.  }
117.
```

婚恋社交网络多维度分析的案例代码如例 18-4 所示。

【例 18-4】 SNSAnalysisGraphX.scala 代码。

```
1.   package com.dt.spark.graphx
2.
3.  import org.apache.log4j.{Level, Logger}
4.  import org.apache.spark.{SparkContext, SparkConf}
5.  import org.apache.spark.graphx._
6.  import org.apache.spark.rdd.RDD
7.
8.  object SNSAnalysisGraphX {
9.    def main(args: Array[String]) {
10.     //屏蔽日志
11.     Logger.getLogger("org.apache.spark").setLevel(Level.ERROR)
12.     Logger.getLogger("org.eclipse.jetty.server").setLevel(Level.OFF)
13.
14.     //设置运行环境
15.     val conf = new SparkConf().setAppName("SNSAnalysisGraphX")
          .setMaster("local[4]")
16.     val sc = new SparkContext(conf)
17.
18.     //设置顶点和边，注意顶点和边都是用元组定义的 Array
19.     //顶点的数据类型是 VD:(String,Int)
20.     val vertexArray = Array(
21.       (1L, ("Alice", 28)),
22.       (2L, ("Bob", 27)),
23.       (3L, ("Charlie", 65)),
24.       (4L, ("David", 42)),
25.       (5L, ("Ed", 55)),
26.       (6L, ("Fran", 50))
27.     )
28.     //边的数据类型是 ED:Int
29.     val edgeArray = Array(
30.       Edge(2L, 1L, 7),
31.       Edge(2L, 4L, 2),
32.       Edge(3L, 2L, 4),
33.       Edge(3L, 6L, 3),
34.       Edge(4L, 1L, 1),
35.       Edge(5L, 2L, 2),
```

```
36.       Edge(5L, 3L, 8),
37.       Edge(5L, 6L, 3)
38.     )
39.
40.     //构造 vertexRDD 和 edgeRDD
41.     val vertexRDD: RDD[(Long, (String, Int))] = sc.parallelize
    (vertexArray)
42.     val edgeRDD: RDD[Edge[Int]] = sc.parallelize(edgeArray)
43.
44.     //构造图 Graph[VD,ED]
45.     val graph: Graph[(String, Int), Int] = Graph(vertexRDD, edgeRDD)
46.
47.     //***********************************************************************
48.     //*********************          图的属性          *********************
49.     //***********************************************************************
50.     println("**********************************************************")
51.     println("属性演示")
52.     println("**********************************************************")
53.     //方法一
54.     println("找出图中年龄大于 30 的顶点方法一：")
55.
56.     /**
      *其实，这里还可以加入性别等信息，例如，可以看年龄大于 30 岁且是 female 的人
57.       */
58.     graph.vertices.filter { case (id, (name, age)) => age > 30 }.collect
        .foreach {
59.       case (id, (name, age)) => println(s"$name is $age")
60.     }
61.     //方法二
62.     println("找出图中年龄大于 30 的顶点方法二：")
63.     graph.vertices.filter(v => v._2._2 > 30).collect.foreach(v =>
          println(s"${v._2._1} is ${v._2._2}"))
64.     println
65.
66.     //边操作：找出图中属性大于 5 的边
67.     println("找出图中属性大于 5 的边：")
68.     graph.edges.filter(e => e.attr > 5).collect.foreach(e =>
    println(s"${e.srcId} to ${e.dstId} att ${e.attr}"))
69.     println
70.
71.
72.     //triplets 操作，((srcId, srcAttr), (dstId, dstAttr), attr)
73.     println("列出所有的 tripltes：")
74.     for (triplet <- graph.triplets.collect) {
75.       println(s"${triplet.srcAttr._1} likes ${triplet.dstAttr._1}")
76.     }
77.     println
78.
79.     println("列出边属性>5 的 tripltes：")
80.     for (triplet <- graph.triplets.filter(t => t.attr > 5).collect) {
81.       println(s"${triplet.srcAttr._1} likes ${triplet.dstAttr._1}")
82.     }
83.     println
84.
85.     //Degrees 操作
```

```
86.    println("找出图中最大的出度、入度、度数：")
87.
88.    def max(a: (VertexId, Int), b: (VertexId, Int)):(VertexId, Int)= {
89.      if (a._2 > b._2) a else b
90.    }
91.
92.    println("max of outDegrees:" + graph.outDegrees.reduce(max) +
       "max of inDegrees:" + graph.inDegrees.reduce(max) + " max of
           Degrees:" + graph.degrees.reduce(max))
93.    println
94.
95.    //*************************************************************************
96.    //**********************         转换操作     **************************
97.    //*************************************************************************
98.    println("**********************************************************")
99.    println("转换操作")
100.   println("**********************************************************")
101.   println("顶点的转换操作，顶点 age + 10：")
102.   graph.mapVertices { case (id, (name, age)) => (id, (name, age +10))}
       .vertices.collect.foreach(v=>println(s"${v._2._1}is ${v._2._2}"))
103.   println
104.   println("边的转换操作，边的属性*2：")
105.   graph.mapEdges(e => e.attr * 2).edges.collect.foreach(e =>
       println(s"${e.srcId} to ${e.dstId} att ${e.attr}"))
106.   println
107.
108.
109.   //*************************************************************************
110.   //************************         结构操作     ***************************
111.   //*************************************************************************
112.   println("**********************************************************")
113.   println("结构操作")
114.   println("**********************************************************")
115.   println("顶点年龄>30 的子图：")
116.   val subGraph = graph.subgraph(vpred = (id, vd) => vd._2 >= 30)
117.   println("子图所有顶点：")
118.   subGraph.vertices.collect.foreach(v => println(s"${v._2._1} is
       ${v._2._2}"))
119.   println
120.   println("子图所有边：")
121.   subGraph.edges.collect.foreach(e => println(s"${e.srcId} to
       ${e.dstId} att ${e.attr}"))
122.   println
123.
124.   //*************************************************************************
125.   //**********************           连接操作     ***********************
126.   //*************************************************************************
127.   println("**********************************************************")
128.   println("连接操作")
129.   println("**********************************************************")
130.   val inDegrees: VertexRDD[Int] = graph.inDegrees
131.   case class User(name: String, age: Int, inDeg: Int, outDeg: Int)
132.
133.   //创建一个新图，顶点 VD 的数据类型为 User，并从 Graph 做类型转换
134.   val initialUserGraph: Graph[User, Int] = graph.mapVertices { case
```

```
    (id, (name, age)) => User(name, age, 0, 0) }
135.
136.  //initialUserGraph 与 inDegrees、outDegrees（RDD）进行连接，并修改
      //initialUserGraph 中的 inDeg 值和 outDeg 值
137.  val userGraph = initialUserGraph.outerJoinVertices(initialUserGraph
      .inDegrees) {
138.    case (id, u, inDegOpt) => User(u.name, u.age, inDegOpt.getOr
        Else(0), u.outDeg)
139.  }.outerJoinVertices(initialUserGraph.outDegrees) {
140.    case (id, u, outDegOpt) => User(u.name, u.age, u.inDeg, outDegOpt
        .getOrElse(0))
141.  }
142.
143.  println("连接图的属性：")
144.  userGraph.vertices.collect.foreach(v => println(s"${v._2.name}
      inDeg: ${v._2.inDeg}  outDeg: ${v._2.outDeg}"))
145.  println
146.
147.      println("出度和入度相同的人员：")
148.  userGraph.vertices.filter {
149.    case (id, u) => u.inDeg == u.outDeg
150.  }.collect.foreach {
151.    case (id, property) => println(property.name)
152.  }
153.  println
154.
155.  //***********************************************************************
       ***************************
156.  //**************  聚合操作  使用旧版本的 mapReduceTriplets 操作 *************
157.  //***********************************************************************
       ***************************
158.  //   println("******************************************************")
159.  //   println("聚合操作")
160.  //   println("******************************************************")
161.  //   println("找出年龄最大的追求者：")
162.  //   val oldestFollower: VertexRDD[(String, Int)] = userGraph
          .mapReduceTriplets[(String, Int)](
163.  //     // 将源顶点的属性发送给目标顶点，map 过程
164.  //     edge => Iterator((edge.dstId, (edge.srcAttr.name, edge
          .srcAttr.age))),
165.  //     // 得到年龄最大的追求者，reduce 过程
166.  //     (a, b) => if (a._2 > b._2) a else b
167.  //   )
168.  //
169.  //
170.  //   userGraph.vertices.leftJoin(oldestFollower) { (id, user,
         optOldestFollower) =>
171.  //     optOldestFollower match {
172.  //       case None => s"${user.name} does not have any followers."
173.  //       case Some((name, age)) => s"${name} is the oldest follower
           of ${user.name}."
174.  //     }
175.  //   }.collect.foreach { case (id, str) => println(str)}
176.  //   println
177.
178.  //找出追求者的平均年龄
```

```
179.  //   println("找出追求者的平均年龄：")
180.  //   val averageAge: VertexRDD[Double] = userGraph.mapReduce
          Triplets[(Int, Double)](
181.  //     //将源顶点的属性 (1, Age)发送给目标顶点，map 过程
182.  //     edge => Iterator((edge.dstId, (1, edge.srcAttr.age.toDouble))),
183.  //     //得到追求者的数量和总年龄
184.  //     (a, b) => ((a._1 + b._1), (a._2 + b._2))
185.  //   ).mapValues((id, p) => p._2 / p._1)
186.
187.  //   userGraph.vertices.leftJoin(averageAge) {(id,user,opt AverageAge) =>
188.  //     optAverageAge match {
189.  //       case None => s"${user.name} does not have any followers."
190.  //       case Some(avgAge) => s"The average age of ${user.name}\'s
               followers is $avgAge."
191.  //     }
192.  //   }.collect.foreach {case (id, str) => println(str)}
193.  //   println
194.
195.   //    ***************************************************************
          ****************************
196.  //   ********  聚合操作   使用 Spark 2.0.2 版本的 aggregateMessages *******
        *******************************************************************
197.   //    ***************************************************************
        ****************************
198.  println("*******************************************************")
199.  println("聚合操作")
200.  println("*******************************************************")
201.  println("找出年龄最大的追求者：")
202.  val oldestFollower: VertexRDD[(String, Int)] = graph.aggregate
       Messages[(String, Int)](
203.    //将源顶点的属性发送给目标顶点，map 过程
204.    triplet => {
205.      //Map 方法
206.      //将消息发送到包含姓名和年龄的目标顶点
207.      triplet.sendToDst(triplet.srcAttr._1, triplet.srcAttr._2)
208.    },
209.    //得到年龄最大的追求者，reduce 过程
210.    (a, b) => if (a._2 > b._2) a else b
211.  )
212.
213.
214.  userGraph.vertices.leftJoin(oldestFollower) { (id, user, optOldest
       Follower) =>
215.    optOldestFollower match {
216.      case None => s"${user.name} does not have any followers."
217.      case Some((name, age)) => s"${name} is the oldest follower of
          ${user.name}."
218.    }
219.  }.collect.foreach { case (id, str) => println(str) }
220.  println
221.
222.  println("*******************************************************")
223.  println("找出年龄最小的追求者：")
224.  val youngestFollower: VertexRDD[(String, Int)] = graph.aggregate
       Messages[(String, Int)](
225.    //将源顶点的属性发送给目标顶点，map 过程
```

```
226.    triplet => {
227.      //Map Function
228.      //Send message to destination vertex containing name and age
229.      triplet.sendToDst(triplet.srcAttr._1, triplet.srcAttr._2)
230.    },
231.    //得到年龄最小的追求者，reduce 过程
232.    (a, b) => if (a._2 > b._2) b else a
233.  )
234.
235.
236.  userGraph.vertices.leftJoin(youngestFollower) { (id, user,
      optYoungestFollower) =>
237.    optYoungestFollower match {
238.      case None => s"${user.name} does not have any followers."
239.      case Some((name, age)) => s"${name} is the youngest follower
         of ${user.name}."
240.    }
241.  }.collect.foreach { case (id, str) => println(str) }
242.  println
243.
244.  //找出追求者的平均年龄
245.  println("找出追求者的平均年龄：")
246.   val averageAge: VertexRDD[Double] = graph.aggregateMessages[(Int,
      Double)](
247.    //将源顶点的属性 (1, Age)发送给目标顶点，map 过程
248.    triplet => {
249.      //Map 方法
250.      //将消息发送到包含姓名和年龄的目标顶点
251.      triplet.sendToDst((1, triplet.srcAttr._2.toDouble))
252.    },
253.    //得到追求者的数量和总年龄
254.    (a, b) => ((a._1 + b._1), (a._2 + b._2))
255.  ).mapValues((id, p) => p._2 / p._1)
256.
257.  userGraph.vertices.leftJoin(averageAge) { (id, user,optAverageAge)=>
258.    optAverageAge match {
259.      case None => s"${user.name} does not have any followers."
260.      case Some(avgAge) => s"The average age of ${user.name}\'s
            followers is $avgAge."
261.    }
262.  }.collect.foreach { case (id, str) => println(str) }
263.  println
264.
265.  //*****************************************************************
      **************************
266.  //*************************      实用操作     ****************************
267.  //*****************************************************************
      **************************
268.  println("********************************************************")
269.  println("实用操作")
270.  println("********************************************************")
271.  println("找出顶点 5 到各顶点的最短距离：")
272.  val sourceId: VertexId = 5L
273.  //定义源点
274.  val initialGraph = graph.mapVertices((id, _) => if (id == sourceId)
        0.0 else Double.PositiveInfinity)
```

```
275.   val sssp = initialGraph.pregel(Double.PositiveInfinity)(
276.     (id, dist, newDist) => math.min(dist, newDist),
277.     triplet => {
278.       //计算权重
279.       if (triplet.srcAttr + triplet.attr < triplet.dstAttr) {
280.         Iterator((triplet.dstId, triplet.srcAttr + triplet.attr))
281.       } else {
282.         Iterator.empty
283.       }
284.     },
285.     (a, b) => math.min(a, b) //最短距离
286.   )
287.   println(sssp.vertices.collect.mkString("\n"))
288.
289.   while (true) {}
290.
291.   sc.stop()
292. }
293.     }
```

18.20　本章总结

本章详细阐述了使用 Spark GraphX 实现婚恋社交网络多维度分析案例，阐述了图的一些基本操作：属性演示、转换、结构、连接、聚合及相关的实用操作。

基于 Spark 的 RDD 抽象，Spark GraphX 可以无缝地与 Spark SQL、MLLib 等结合使用。我们可以使用 Spark SQL 进行 ETL 数据清洗，清洗之后交给 Spark GraphX 进行处理，Spark GraphX 在计算时又可以和 MLLib 结合使用，共同完成深度数据挖掘等人工智能化的操作，充分发挥 Spark 集成计算框架的优势。

第3篇　性能调优篇

第 19 章　对运行在 YARN 上的 Spark 进行性能调优

本章主要讲解运行在 YARN 上的 Spark 如何进行性能调优。19.1 节讲解运行环境 Jar 包管理及数据本地性原理调优实践；19.2 节讲解 Spark on YARN 两种不同的调度模型及其调优；19.3 节讲解 YARN 队列资源不足引起的 Spark 应用程序失败的原因及调优方案；19.4 节讲解 Spark on YARN 模式下 Executor 经常被杀死的原因及最佳调优方案；19.5 节讲解 YARN-Client 模式下网卡流量激增的原因及调优方案；19.6 节讲解 YARN-Cluster 模式下 JVM 栈内存溢出的原因及调优方案。

19.1　运行环境 Jar 包管理及数据本地性原理调优实践

本节首先讲解运行环境 Jar 包管理及数据本地性原理，然后讲解运行环境 Jar 包管理及数据本地性调优实践。

19.1.1　运行环境 Jar 包管理及数据本地性原理

在 YARN 上运行 Spark 需要在 Spark-env.sh 或环境变量中配置 HADOOP_CONF_DIR 或 YARN_CONF_DIR 目录指向 Hadoop 的配置文件。

在 Spark-default.conf 中配置 Spark.YARN.jars 指向 hdfs 上的 Spark 需要的 jar 包。如果不配置该参数，每次启动 Spark 程序会将 Driver 端的 SPARK_HOME 打包上传分发到各个节点。配置如下：

```
1.   spark.yarn.jars hdfs://clustername/spark/spark210/jars/*
```

对于分布式系统来说，由于数据可能也是分布式的，所以数据处理往往也是分布式的。要想保证性能，尽量保证数据本地性很重要。

分布式计算系统的精粹在于移动计算，而非移动数据，但是，在实际的计算过程中，总存在移动数据的情况，除非是在集群的所有节点上都保存数据的副本。移动数据将数据从一个节点移动到另一个节点进行计算，不但消耗了网络 I/O，也消耗了磁盘 I/O，降低了整个计算的效率。为了提高数据的本地性，除了优化算法（也就是修改 Spark 内存，难度有点大），就是合理设置数据的副本。设置数据的副本，这是需要通过配置参数并长期观察运行状态，才能获取的一个经验值。

Spark 中的数据本地性有以下 5 种。

- PROCESS_LOCAL：进程本地化。代码和数据在同一个进程中，也就是在同一个

Executor 中；计算数据的 Task 由 Executor 执行，数据在 Executor 的 BlockManager 中；性能最好。

- NODE_LOCAL：节点本地化。代码和数据在同一个节点中，数据作为一个 HDFS block 块，就在节点上，而 Task 在节点上某个 Executor 中运行；或者是数据和 Task 在一个节点上的不同 Executor 中；数据需要在进程间进行传输。也就是说，数据虽然在同一 Worker 中，但不是同一 JVM 中。这隐含着进程间移动数据的开销。
- NO_PREF：数据没有局部性首选位置。它能从任何位置同等访问。对于 Task 来说，数据从哪里获取都一样，无好坏之分。
- RACK_LOCAL：机架本地化。数据在不同的服务器上，但在相同的机架。数据需要通过网络在节点之间进行传输。
- ANY：数据在不同的服务器及机架上面。这种方式性能最差。

Spark 应用程序本身包含代码和数据两部分，单机版本一般情况下很少考虑数据本地性的问题，因为数据在本地。单机版本的程序，数据本性有 PROCESS_LOCAL 和 NODE_LOCAL 之分，但也应尽量让数据处于 PROCESS_LOCAL 级别。

通常，读取数据要尽量使数据以 PROCESS_LOCAL 或 NODE_LOCAL 方式读取。其中，PROCESS_LOCAL 还和 Cache 有关，如果 RDD 经常用，应将该 RDD Cache 到内存中。注意，由于 Cache 是 Lazy 级别的，所以必须通过一个 Action 的触发，才能真正地将该 RDD Cache 到内存中。

19.1.2　运行环境 Jar 包管理及数据本地性调优实践

启动 Spark 程序时，其他节点会自动下载 Jar 包并进行缓存，下次启动时如果包没有变化，则会直接读取本地缓存的包。缓存清理间隔在 YARN-site.xml 通过以下参数配置。

```
1.    yarn.nodemanager.localizer.Cache.cleanip.interval-ms
```

启动 Spark-Shell，指定 master 为 YARN，默认为 client 模式。在 cluster 模式下可以使用 --supervise，如果 Driver 异常退出，将会自行重启。

```
1.  /usr/install/Spark210/bin/Spark-Shell
2.  --master YARN
3.  --deploy-mode client
4.  --Executor-memory 1g
5.  --Executor-cores 1
6.  --Driver-memory 1g
7.  --queue test (默认队列是 default)
```

初次讲到参数配置，需要提醒的是，在代码中的 SparkConf 中的配置参数具有最高优先级，其次是 Spark-Submit 或 Spark-Shell 的参数，最后是配置文件（conf/Spark-defaults.conf）中的参数。

本地性级别以最近的级别开始，以最远的级别结束。Spark 调度任务执行是要让任务获得最近的本地性级别的数据。然而，有时它不得不放弃最近本地性级别，因为一些资源上的原因，有时不得不选择更远的。在数据与 Executor 在同一机器但数据处理比较忙，且存在一些空闲的 Executor 远离数据的情况下，Spark 会让 Executor 等特定的时间，看其能否完成正在处理的工作。如果该 Executor 仍不可用，一个新的任务将启用在更远的可用节点上，即数据被传送给那个节点。

可以给 Spark 中的 Executor 配置单种数据本地性级别可以等待的空闲时长，如下例所示。

```
1.  Val SparkConf = new SparkConf()
2.  SparkConf.set("Spark.locality.wait", "3s")
3.  SparkConf.set("Spark.locality.wait.node", "3s")
4.  SparkConf.set("Spark.locality.wait.process", "3s")
5.  SparkConf.set("Spark.locality.wait.rack", "3s")
```

默认情况下，这些等待时长都是 3s。下面讲解调节这个参数的方式。推荐大家在测试的时候，先用 Client 模式，在本地就直接可以看到比较全的日志。观察日志 Spark 作业的运行日志，里面会显示如下日志。

```
1.   17/02/11 21:59:45 INFO scheduler.TaskSetManager: Starting Task 0.0 in
     Stage 0.0 (TID 0, sandbox, PROCESS_LOCAL, 1260 bytes)
```

观察大部分 Task 的数据本地化级别：如果大多都是 PROCESS_LOCAL，那就不用调节了；如果发现很多的级别都是 NODE_LOCAL、ANY，那么最好调节一下数据本地化的等待时长。调节往往不可能一次到位，应该反复调节，每次调节完以后，再运行，观察日志。看看大部分的 Task 的本地化级别有没有提升；也看看整个 Spark 作业的运行时间有没有缩短，但别本末倒置，如果本地化级别提升了，但是因为大量的等待时长而导致 Spark 作业的运行时间反而增加，那也是不好的调节。

19.2　Spark on YARN 两种不同的调度模型及其调优

本节讲解 Spark on YARN 的两种不同类型模型（YARN-Client 模式、YARN-Cluster 模式）优劣分析，以及 Spark on YARN 的两种不同类型调优实践。

19.2.1　Spark on YARN 的两种不同类型模型优劣分析

按照 Spark 应用程序中的 Driver 分布方式的不同，Spark on YARN 有两种模式：YARN-Client 模式、YARN-Cluster 模式。

不论是在 Spark-Shell 或者 Spark-Submit 中，Driver 都运行在启动 Spark 应用的机器上。在这种情形下，YARN Application Master 仅负责从 YARN 中请求资源，这就是 YARN-Client 模式。

另一种方式，Driver 自动运行在 YARN Container（容器）里，客户端可以从集群中断开，或者用于其他作业。这叫作 YARN-Cluster 模式。

理解 YARN-Client 和 YARN-Cluster 深层次的区别之前先清楚一个概念：Application Master。在 YARN 中，每个 Application 实例都有一个 Application Master 进程，它是 Application 启动的第一个容器。它负责和 ResourceManager 打交道并请求资源，获取资源后告诉 NodeManager 为其启动 Container。从深层次的含义讲，YARN-Cluster 和 YARN-Client 模式

的区别其实就是 Application Master 进程的区别。

YARN-Client 模式下，Application Master 仅向 YARN 请求 Executor，Client 会和请求的 Container 通信来调度它们工作。YARN-Client 模式适合调试 Spark 程序，能在控制台输出一些调试信息。

YARN-Cluster 模式下，Driver 运行在 AM（Application Master）中，负责向 YARN 申请资源，并监督作业的运行状况。当用户提交了作业后，就可以关掉 Client，作业会继续在 YARN 上运行，因而 YARN-Cluster 模式不适合运行交互类型的作业。应用的运行结果不能在客户端显示（可以在 History Server 中查看），所以最好将结果保存在 HDFS，而非 Stdout 输出，客户端的终端显示的是作为 YARN 的 Job 的简单运行状况。但企业生产环境下会用 YARN-Cluster 模式来运行 Spark 应用程序。

19.2.2　Spark on YARN 的两种不同类型调优实践

Spark 在 YARN 上运行包括两种类型。

- YARN-Client 模式：Spark 向 YARN 集群提交应用程序（如 Spark-Shell 或 Spark-Submit），Driver 就运行在提交应用的节点上。
- YARN-Cluster 模式：Spark 向 YARN 集群提交应用程序，YARN 在集群内分配一个 YARN Container（容器）运行 Driver。

在 YARN-Cluster 上运行作业的简单例子如下：

```
1.   $spark_home/bin/spark-submit -master yarn-cluster -num-Executors 4
2.  --Executor-cores 3 -class main.class myjar.jar
```

另外，我们能在 YARN 上运行 Spark-Shell：

```
1.   $spark_home/bin/spark-shell -master yarn-client -num-Executors 4
     --Executor-cores 3
```

通过 YARN 资源管理器 WebUI 可看到活动的 YARN 作业。有时在 YARN 上运行会产生一些复杂情况。

一个常见的问题是 YARN Resource Manager 对 Spark 的申请资源的限制。例如，当启动 Spark 应用时，可能试图去分配比 YARN 集群中可用资源更多的资源，在这种情形下，Spark 将提出申请，而不是直接否定请求，YARN 将等资源可用，即使它们可能永远不会有足够的资源。

另一个通用的问题是单个 YARN Container 的可用资源是固定的。单个 Container 里面资源有限，即使分配多个 Container，当 Spark 应用在一个 YARN Container 里面超过了可用内存，就会出现 OOM 问题。在这种情形下，YARN 将终结容器并抛出错误，而且问题的底层原因很难追踪。

Driver 可在 YARN-Client 模式分配 1GB 内存及一个核。如果应用程序需要，有时也可增加可用的内存，特别是对于长时间运行的 Spark 工作很有意义，因为在执行过程中可能积累数据。下面是一个例子。

```
1.  spark-shell --num-Executors 8 -Executor-cores 5 -Driver-memory 2g
```

19.3　YARN 队列资源不足引起的 Spark 应用程序失败的原因及调优方案

本节讲解 YARN 队列资源不足引起的 Spark 应用程序失败的原因，以及 YARN 队列资源不足引起的 Spark 应用程序失败的 3 个解决方案。

19.3.1　失败的原因剖析

ResourceManager 会接收你提交的请求吗？YARN 一般把自己的资源分成不同的类型，我们提交的时候会专门提交到分配给 Spark 的那一组资源。例如，我们提交信息资源：Memory 1000G，Cores 800 个。此时你要提交的 Spark 应用程序可能需要 900GB 的内存和 700 个 Core，一定会没有问题吗？不一定！

另一种情况是当前的作业可以提交运行，已经消耗了 900GB 的内存和 700 个 Core，然后又提交了一个消耗 500GB 的内存和 300 个 Core 的 Spark 应用程序，这时资源不够，无法提交。

19.3.2　调优方案

YARN 队列资源不足引起的 Spark 应用程序失败有以下解决方案。

第一个方法：在 J2EE 中间层，通过线程池技术实现顺利地提交，可让线程池的大小设定为 1。

第二个办法：如果提交的 Spark 应用程序比较耗时，如均超过 10min，而其他的 Spark 程序都在 2min 内执行完成，这时可以把 Spark 拥有的资源进行分类（耗时任务和快速任务）。此时可以使用两个线程池，每个线程池都有一个线程。

第三种办法：只有一个程序运行的时候，可以把 Memory 和 Cores 都调整到最大，这样最大化地使用资源来最快速地完成程序的计算，同时也简化了集群的运维和故障解决。

19.4　Spark on YARN 模式下 Executor 经常被杀死的原因及调优方案

本节讲解 Spark on YARN 模式下 Executor 经常被杀死的原因，以及调优方案。

19.4.1　原因剖析

如果出现以下异常信息：

```
1.    ExecutorLostFailure (Executor 3 exited caused by one of the running Tasks)
      Reason: Container killed by YARN for exceeding memory limits. 52.6 GB of
```

```
50 GB physical memory used. Consider boosting Spark.YARN.Executor
.memoryOverhead
```

很明显，内存被用完。提示建议考虑增加 Spark.YARN.Executor.memoryOverhead 的配置。

19.4.2　调优方案

Spark on YARN 模式下 Executor 经常被杀死的调优方案可考虑：

- ❑ 移除 RDD 缓存操作。
- ❑ 增加该 Job 的 Spark.storage.memoryFraction 系数值。

Spark 配置参数如下：

```
1.   Spark.storage.memoryFraction
```

- ❑ 增加该 Job 的 spark.yarn.Executor.memoryoverhead 值。

Spark 配置参数如下：

```
1.   spark.yarn.Executor.memoryoverhead
```

19.5　YARN-Client 模式下网卡流量激增的原因及调优方案

本节讲解 YARN-Client 下网卡流量激增的原因，及 YARN-Client 下网卡流量激增的解决方案：在生产环境中使用 YARN-Cluster 模式去提交 Spark 作业。

19.5.1　原因剖析

YARN 集群分成两种节点。

- ❑ ResourceManager 负责资源的调度。
- ❑ NodeManager 负责资源的分配、应用程序执行。

通过 Spark-Submit 脚本使用 YARN-Client 方式提交，这种模式其实会在本地启动 Driver 程序。

我们将写的 Spark 程序打成 Jar 包，使用 Spark-Submit 提交，将 Jar 包中的 main 类通过 JVM 命令启动。JVM 中的进程其实就是 Driver 进程，Driver 进程启动后，执行我们写的 main 函数。

应该正确认识 ApplicationMaster 这个 YARN 中的核心概念，任何要在 YARN 上启动的作业类型（MR、Spark），都必须有一个 ApplicationMaster。每种计算框架（MapReduce、Spark），如果想在 YARN 上执行自己的计算应用，就必须自己实现和提供一个 ApplicationMaster，这就相当于是实现了 YARN 提供的接口，这是 Spark 自己开发的一个类。

Spark 在 YARN-Client 模式下，Application 的注册（Executor 的申请）和计算 Task 的调度，是分离开的。Standalone 模式下，这两个操作都是 Driver 负责的。

ApplicationMaster(ExecutorLauncher)负责 Executor 的申请：Driver 负责 Job 和 Stage 的划分，以及 Task 的创建、分配和调度。

YARN-Client 模式下，会产生什么样的问题呢？

由于我们的 Driver 是启动在本地机器的，而且 Driver 是全权负责所有任务的调度的，也就是说，要与 YARN 集群上运行的多个 Executor 进行频繁的通信（中间有 Task 的启动消息、Task 的执行统计消息、Task 的运行状态、Shuffle 的输出结果）。

例如，Executor 有 100 个，Stage 有 10 个，Task 有 1000 个。每个 Stage 运行的时候，都有 1000 个 Task 提交到 Executor 上面去运行，平均每个 Executor 有 10 个 Task。接下来问题来了，Driver 要频繁地与 Executor 上运行的 1000 个 Task 进行通信。通信消息特别多，通信的频率特别高。运行完一个 Stage，接着运行下一个 Stage，又是频繁的通信。

在整个 Spark 运行的生命周期内，都会频繁地进行通信和调度。所有的通信和调度都是在本地机器上发出去和接收到的。本地机器很可能在 30min 内（Spark 作业运行的周期内）进行频繁大量的网络通信。此时本地机器的网络通信负载非常高，会导致本地机器的网卡流量激增。

在一些拥有庞大计算机集群的公司，对每台机器的使用情况，都是有监控的。本地机器的网卡流量激增，可能对公司网络运行造成影响。因此，运维人员不会允许单个机器出现耗费大量网络带宽资源的情况。

19.5.2　调优方案

YARN-Client 下网卡流量激增问题的解决方法很简单，首先须清楚 YARN-Client 模式是在什么情况下使用的。

YARN-Client 模式通常只会使用在测试环境中。我们写好某个 Spark 作业打成 Jar 包，在某台测试机器上使用 YARN-Client 模式去提交测试。我们不会长时间连续提交大量的 Spark 作业去测试，YARN-Client 模式提交后可以在本地机器观察到详细全面的 Log。通过查看 Log，可以解决线上报错的故障（troubleshooting）、对性能进行观察并进行性能调优。

实际上线后，在生产环境中都得用 YARN-Cluster 模式去提交 Spark 作业。因此，YARN-Cluster 模式与本地机器引起的网卡流量激增的问题没有关系。即使在 YARN-Client 测试模式下网卡流量激增，也应该是 YARN 运维团队和基础运维团队去考虑 YARN 集群里每台机器是虚拟机，还是物理机？网卡流量激增后会不会对其他东西产生影响？如果网络流量激增，要不要给 YARN 集群增加一些网络带宽等。

使用 YARN-Cluster 模式后，就不是本地机器运行 Driver 负责 Task 调度了。YARN 集群中，有某个节点会运行 Driver 进程，负责 Task 调度。

19.6　YARN-Cluster 模式下 JVM 栈内存溢出的原因及调优方案

本节讲解 YARN-Cluster 模式下 JVM 栈内存溢出的原因及调优方案：在 Spark-Submit 脚本中设置 PermGen。

19.6.1　原因剖析

有些 Spark 作业在 YARN-Client 模式下是可以运行的，但在 YARN-Cluster 模式下，会报出 JVM 的 PermGen（永久代）的内存溢出（OOM）。

出现以上问题的原因是：YARN-Client 模式下，Driver 运行在本地机器上，Spark 使用 JVM 的 PermGen 的配置，是本地的默认配置 128MB；但在 YARN-Cluster 模式下，Driver 运行在集群的某个节点上，Spark 使用的 JVM 的 PermGen 是没有经过配置的，默认是 82MB，故有时会出现 PermGen Out Of Memory error Log。

19.6.2　调优方案

YARN-Cluster 模式下 JVM 栈内存溢出问题的调优方案如下。

（1）在 Spark-Submit 脚本中设置 PermGen。

```
1.   - conf Spark.Driver.extraJavaOptions="-XX:PermSize=128M -XX:MaxPermSize=
     256M" (最小 128M，最大 256M)
```

（2）如果使用 Spark SQL，SQL 中使用大量的 or 语句，也可能会报出 JVM stack overflow，JVM 栈内存溢出，此时可以把复杂的 SQL 简化为多个简单的 SQL 进行处理。

第 20 章　Spark 算子调优最佳实践

本章对 Spark 算子调优最佳实践进行讲解。20.1 节讲解使用 mapPartitions 或者 mapPartitionWithIndex 取代 map 操作；20.2 节讲解使用 foreachPartition 把 Spark 数据持久化到外部存储介质；20.3 节讲解使用 coalesce 取代 rePartition 操作；20.4 节讲解使用 repartitionAndSortWithinPartitions 取代 repartition 和 sort 的联合操作；20.5 节讲解使用 treeReduce 取代 reduce 的原理和源码；20.6 节讲解使用 treeAggregate 取代 Aggregate 的原理和源码；20.7 节讲解 reduceByKey 高效运行的原理和源码解密；20.8 节讲解使用 AggregateByKey 取代 groupByKey 的原理和源码；20.9 节讲解 Join 不产生 Shuffle 的情况及案例实战；20.10 节讲解 RDD 复用性能调优最佳实践。

20.1　使用 mapPartitions 或者 mapPartitionWithIndex 取代 map 操作

本节讲解 mapPartitions 内部工作机制和源码解析、mapPartitionWithIndex 内部工作机制和源码解析；然后讲解使用 mapPartitions 取代 map 案例和性能测试。

20.1.1　mapPartitions 内部工作机制和源码解析

mapPartitions 和 map 函数类似，只不过映射函数的参数由 RDD 中的每个元素变成了 RDD 中每个分区的迭代器。如果在映射的过程中需要频繁创建额外的对象，使用 mapPartitions 要比使用 map 高效。

例如，将 RDD 中的所有数据通过 JDBC 连接写入数据库，如果使用 map 函数，可能要为每个元素都创建一个 Connection，这样开销很大；如果使用 mapPartitions，那么只需要针对每个分区建立一个 Connection。

```
1.  def mapPartitions[U: ClassTag](
2.      f: Iterator[T] => Iterator[U],
3.      preservesPartitioning: Boolean = false): RDD[U]
```

20.1.2　mapPartitionWithIndex 内部工作机制和源码解析

mapPartitionsWithIndex 与 mapPartitions 基本相同，只是处理函数的参数是一个二元元组，元组的第一个元素是当前处理的分区的 index，元组的第二个元素是当前处理的分区元素组成的 Iterator。

RDD.scala 中的 mapPartitionsWithIndex 的源码如下：

```
1.    def mapPartitionsWithIndex[U: ClassTag](
2.        f: (Int, Iterator[T]) => Iterator[U],
3.        preservesPartitioning: Boolean = false): RDD[U] = withScope {
4.      val cleanedF = sc.clean(f)
5.      new MapPartitionsRDD(
6.        this,
7.        (context: TaskContext, index: Int, iter: Iterator[T]) => cleanedF(index,
          iter),
8.        preservesPartitioning)
9.    }
```

RDD.scala 中的 mapPartitions 的源码如下：

```
1.  def mapPartitions[U: ClassTag](
2.      f: Iterator[T] => Iterator[U],
3.      preservesPartitioning: Boolean = false): RDD[U] = withScope {
4.    val cleanedF = sc.clean(f)
5.    new MapPartitionsRDD(
6.      this,
7.      (context: TaskContext, index: Int, iter: Iterator[T]) => cleanedF(iter),
8.      preservesPartitioning)
9.  }
```

从源码中可以看到：其实 mapPartitions 已经获得了当前处理的分区的 index，只是没有传入分区处理函数，而 mapPartitionsWithIndex 将其传入分区处理函数。

20.1.3　使用 mapPartitions 取代 map 案例和性能测试

RDD 的 mapPartitions 是 map 的一个变种，它们都可进行分区的并行处理。

两者的主要区别是调用的粒度不一样：map 的输入变换函数应用于 RDD 中每个元素，而 mapPartitions 的输入函数应用于每个分区。

假设一个 RDD 有 10 个元素，分成 3 个分区。如果使用 map 方法，map 中的输入函数会被调用 10 次；而使用 mapPartitions 方法，其输入函数只会被调用 3 次，每个分区调用 1 次。

编写业务测试代码 mapPartitionsSuit.scala，调用 map 函数中的自定义函数 myfuncPerElement，对其中的每个元素进行处理，10 个元素调用了 10 次；调用 mapPartitions 函数中的自定义函数 myfuncPerPartition，对分区中的元素进行处理，3 个分区调用 3 次，mapPartitionsSuit.scala 代码如下：

```
1.   //生成 10 个元素 3 个分区的 RDD a，元素值为 1~10 的整数（1 2 3 4 5 6 7 8 9 10），
     //sc 为 SparkContext 对象
2.  val a = sc.parallelize(1 to 10, 3)
3.  //定义两个输入变换函数，它们的作用均是将 RDD a 中的元素值翻倍
4.  //map 的输入函数，其参数 e 为 RDD 元素值
5.  def myfuncPerElement(e:Int):Int = {
6.  println("e="+e)
7.  e*2
8.    }
9.   //mapPartitions 的输入函数。iter 是分区中元素的迭代子，返回类型也应是迭代子
10. def myfuncPerPartition ( iter : Iterator [Int] ) : Iterator [Int] = {
11. println("run in Partition")
```

```
12. var res = for (e <- iter ) yield e*2
13. res
14. }
15.
16. val b = a.map(myfuncPerElement).collect
17. val c =  a.mapPartitions(myfuncPerPartition).collect
```

在 IDEA 中运行代码，mapPartitionsSuit 的运行结果如下：

```
1.   Using Spark's default Log4j profile: org/apache/spark/Log4j-defaults
     .properties
2.  17/06/06 17:01:17 INFO SparkContext: Running Spark version 2.1.0
3.  ......
4.  e=1
5.  e=2
6.  e=3
7.  e=4
8.  e=5
9.  e=6
10. e=7
11. e=8
12. e=9
13. e=10
14. ……
15. run in Partition
16. run in Partition
17. run in Partition
18. ……
```

从输入函数（myfuncPerElement、myfuncPerPartition）层面看，map 是推模式，数据被推到 myfuncPerElement 中；mapPartitons 是拉模式，myfuncPerPartition 通过迭代子从分区中拉数据。

这两个方法的另一个区别是，在大数据集情况下的资源初始化开销和批处理：如果在 myfuncPerPartition 和 myfuncPerElement 中都要初始化一个耗时的资源，然后使用，如数据库链接。在上面的例子中，myfuncPerPartition 只需初始化 3 个资源（3 个分区每个 1 次），而 myfuncPerElement 要初始化 10 次（10 个元素每个 1 次）。显然，在大数据集情况下（数据集中元素个数远大于分区数），mapPartitons 的开销要小很多，也便于进行批处理操作。

20.2　使用 foreachPartition 把 Spark 数据持久化到外部存储介质

本节讲解 foreachPartition 内部工作机制和源码解析；使用 foreachPartition 写数据到 MySQL 中案例和性能测试。

20.2.1　foreachPartition 内部工作机制和源码解析

下面先看一下 RDD.scala 的 foreach 方法。在 foreach 方法中传入一个函数，在函数中提交 runJob，对于 iter 中的每个元素，使用 iter.foreach 遍历进行计算。

RDD.scala 的 foreach 的源码如下：

```
1.  def foreach(f: T => Unit): Unit = withScope {
2.    val cleanF = sc.clean(f)
3.    sc.runJob(this, (iter: Iterator[T]) => iter.foreach(cleanF))
4.  }
```

foreachPartition 函数根据传入的 function 进行处理，和 foreach 函数的不同之处在于：foreachPartition 中 function 的传入参数是一个 Partition 对应数据的 Iterator，而不是直接使用 iterator 的 foreach，然后对分区的数据进行计算。

RDD.scala 的 foreachPartition 的源码如下：

```
1.  def foreachPartition(f: Iterator[T] => Unit): Unit = withScope {
2.     val cleanF = sc.clean(f)
3.     sc.runJob(this, (iter: Iterator[T]) => cleanF(iter))
4.   }
```

20.2.2　使用 foreachPartition 写数据到 MySQL 中案例和性能测试

如果使用 foreach 算子，将在每个 RDD 的每条记录中都进行 Connection 的建立和关闭，这会导致不必要的高负荷，并且会降低整个系统的吞吐量，所以一个更好的方式是使用 RDD.foreachPartition，即对于每个 RDD 的 Partition，建立唯一的连接（每个 Partition 的 RDD 是运行在同一个 worker 上的），降低了频繁建立连接的负载，通常我们在链接数据库时会使用连接池。

foreachPartition 的数据库链接代码如下：

```
1.    RDD.foreachPartition { PartitionOfRecords =>
2.    //ConnectionPool 连接池是一个静态的，懒加载初始化连接池
3.  val Connection = ConnectionPool.getConnection()
4.  PartitionOfRecords.foreach(record => Connection.send(record))
5.  ConnectionPool.returnConnection(Connection)  //返回连接池，用于以后链接的复用
6.  }
```

通过持有一个静态连接池对象，可以重复利用 Connection，进一步优化了链接建立的开销，从而降低了负载。另外值得注意的是，同数据库的连接池类似，这里说的连接池同样应该是 Lazy 的按需建立链接，并且及时地收回超时的链接。

20.3　使用 coalesce 取代 rePartition 操作

本节讲解 coalesce 和 rePartition 工作机制和源码剖析；以及通过测试对比 coalesce 和 rePartition 的性能。

20.3.1　coalesce 和 repartition 工作机制和源码剖析

在 Spark 的 RDD 中，RDD 是分区的。

有时需要重新设置 RDD 的分区数量，如 RDD 中 RDD 分区比较多，但是每个 RDD 的数据量比较小，此时就需要设置一个比较合理的分区。或者需要把 RDD 的分区数量调大。还有就是通过设置一个 RDD 的分区来达到设置生成的文件的数量。

有两种方法可以重设 RDD 的分区。

- coalesce 方法。
- repartition 方法。

（1）Spark 2.2.1 版本的 RDD.scala 和 coalesce 方法的源码如下：

```
1.  def coalesce(numPartitions: Int, shuffle: Boolean = false,
2.               partitionCoalescer: Option[PartitionCoalescer] = Option.empty)
3.              (implicit ord: Ordering[T] = null)
4.       : RDD[T] = withScope {
5.     require(numPartitions > 0, s"Number of partitions ($numPartitions) must be
       positive.")
6.     if (shuffle) {
7.       /**从随机分区开始，将元素均匀分布在输出分区上*/
8.       val distributePartition = (index: Int, items: Iterator[T]) => {
9.         var position = (new Random(index)).nextInt(numPartitions)
10.        items.map { t =>
11.          //注意，密钥的哈希代码本身就是密钥。HashPartitioner 将根据总的分区数进行
           //取模
12.          position = position + 1
13.          (position, t)
14.        }
15.      } : Iterator[(Int, T)]
16.
17.      //包括一个 shuffle 步骤，以便我们的上游任务仍然是分布式的
18.      new CoalescedRDD(
19.        new ShuffledRDD[Int, T, T](mapPartitionsWithIndex(distributePartition),
20.        new HashPartitioner(numPartitions)),
21.        numPartitions,
22.        partitionCoalescer).values
23.    } else {
24.      new CoalescedRDD(this, numPartitions, partitionCoalescer)
25.    }
26.  }
```

Spark 2.4.3 版本的 RDD.scala 源码与 Spark 2.2.1 版本相比具有如下特点。

- 上段代码中第 9 行 index 调整为 hashing.byteswap32(index)。
- 上段代码中第 19 行 mapPartitionsWithIndexInternal 方法新增一个参数 isOrderSensitive。isOrderSensitive 指示函数是否区分顺序。如果它对顺序是敏感的，当输入顺序改变时，可能返回完全不同的结果。大多数状态函数是顺序敏感的。

```
1.  ......
2.      var position = new Random(hashing.byteswap32(index)).nextInt(numPartitions)
3.  ......
4.   mapPartitionsWithIndexInternal(distributePartition, isOrderSensitive = true),
5.  ......
```

coalesce 方法的作用是返回指定一个新的指定分区的 RDD。默认情况下，shuffle 设置为 False。如果是生成一个窄依赖的结果，那么不会发生 Shuffle。例如，1000 个分区被重新设置成 10 个分区，这样不会发生 Shuffle。如果分区的数量发生激烈的变化，如设置 numPartitions = 1，这可能会造成运行计算的节点比你想象的要少，为了避免这个情况，可以设置 Shuffle=True，

那么，这会增加 Shuffle 操作。

（2）repartition 方法的源码如下：

```
1.  def repartition(numPartitions: Int)(implicit ord: Ordering[T] = null):
    RDD[T] = withScope {
2.    coalesce(numPartitions, shuffle = true)
3.  }
```

repartition 方法就是 coalesce 方法 Shuffle 为 True 的情况。如果只是要减少父 RDD 的分区数量，并且要设置的分区数量变化并不是很激烈，则可以考虑直接使用 coalesce 方法来避免执行 Shuffle 操作，以提高效率。

20.3.2 通过测试对比 coalesce 和 repartition 的性能

通常对一个 RDD 执行 filter 算子过滤掉 RDD 中较多数据后（如 30%以上的数据），建议使用 coalesce 算子，手动减少 RDD 的 Partition 数量，将 RDD 中的数据压缩到更少的 Partition 中去。因为 filter 之后，RDD 的每个 Partition 中都会有很多数据被过滤掉，此时如果照常进行后续的计算，其实每个 Task 处理的 Partition 中的数据量并不是很多，有点资源浪费，而且此时处理的 Task 越多，速度反而越慢。因此，用 coalesce 减少 Partition 数量，将 RDD 中的数据压缩到更少的 Partition 之后，只要使用更少的 Task 即可处理完所有的 Partition。在某些场景下，对性能的提升会有一定帮助。

20.4 使用 repartitionAndSortWithinPartitions 取代 repartition 和 sort 的联合操作

本节讲解 repartitionAndSortWithinPartitions 工作机制和源码解析；repartitionAndSortWithinPartitions 性能测试。

20.4.1 repartitionAndSortWithinPartitions 的工作原理和源码

JavaPairRDD.scala 的 repartitionAndSortWithinPartitions 的源码如下：

```
1.    def repartitionAndSortWithinPartitions(partitioner: Partitioner): JavaPairRDD
      [K, V] = {
2.    val comp = com.google.common.collect.Ordering.natural().asInstanceOf
      [Comparator[K]]
3.    repartitionAndSortWithinPartitions(partitioner, comp)
4.  }
5.  ......
6.  def repartitionAndSortWithinPartitions(partitioner: Partitioner, comp:
    Comparator[K])
7.    : JavaPairRDD[K, V] = {
8.    implicit val ordering = comp //允许比较器做隐式转换进行排序
9.    fromRDD(
```

```
10.     new OrderedRDDFunctions[K, V, (K, V)](rdd).repartitionAndSortWithin
        Partitions(partitioner))
11. }
```

OrderedRDDFunctions.scala 的 repartitionAndSortWithinPartitions 的源码如下：

```
1.   def repartitionAndSortWithinPartitions(partitioner: Partitioner): RDD[(K,
     V)] = self.withScope {
2.     new ShuffledRDD[K, V, V](self, partitioner).setKeyOrdering(ordering)
3.   }
```

从源码中可以看出，该方法依据 Partitioner 对 RDD 进行分区，并且在每个结果分区中按 key 进行排序；通过对比 sortByKey 发现，这种方式比先分区，然后在每个分区中进行排序效率高，这是因为它可以将排序融入 Shuffle 阶段。

20.4.2　repartitionAndSortWithinPartitions 性能测试

repartitionAndSortWithinPartitions 是 Spark 官网推荐的一个算子。官方建议，如果 repartition 重分区之后，还要进行排序，建议直接使用 repartitionAndSortWithinPartitions 算子。因为该算子可以一边进行重分区的 Shuffle 操作，一边进行排序。Shuffle 与 sort 两个操作同时进行，比先 Shuffle 再 sort 性能要高。

20.5　使用 treeReduce 取代 reduce 的原理和源码

本节讲解使用 treeReduce 取代 reduce 的原理和源码；以及使用 treeReduce 进行性能测试。

20.5.1　treeReduce 进行 reduce 的工作原理和源码

treeReduce 类似于 treeAggregate，利用在 Executor 端进行多次 Aggregate 来缩小 Driver 的计算开销。

RDD.scala 的 treeReduce 的源码如下：

```
1.   def treeReduce(f: (T, T) => T, depth: Int = 2): T = withScope {
2.     require(depth >= 1, s"Depth must be greater than or equal to 1 but got
       $depth.")
3.     val cleanF = context.clean(f)
4.     val reducePartition: Iterator[T] => Option[T] = iter => {
5.       if (iter.hasNext) {
6.         Some(iter.reduceLeft(cleanF))
7.       } else {
8.         None
9.       }
10.    }
11.    val partiallyReduced = mapPartitions(it => Iterator(reducePartition
       (it)))
12.    val op: (Option[T], Option[T]) => Option[T] = (c, x) => {
13.      if (c.isDefined && x.isDefined) {
14.        Some(cleanF(c.get, x.get))
```

```
15.       } else if (c.isDefined) {
16.         c
17.       } else if (x.isDefined) {
18.         x
19.       } else {
20.         None
21.       }
22.     }
23.     partiallyReduced.treeAggregate(Option.empty[T])(op, op, depth)
24.       .getOrElse(throw new UnsupportedOperationException("empty collection"))
25.   }
```

treeReduce 函数先针对每个分区利用 scala 的 reduceLeft 函数进行计算；最后，再将局部合并的 RDD 进行 treeAggregate 计算，这里的 seqOp 和 combOp 一样，初值均为空。在实际应用中，可以用 treeReduce 代替 reduce，主要用于单个 reduce 操作开销比较大的情况，而 treeReduce 可以通过调整深度来控制每次 reduce 的规模。

20.5.2　使用 treeReduce 进行性能测试

reduceByKey 仅在 key-value 键值对的 RDD 上可用，而 treeReduce 是对任何 RDD 的泛化操作。reduceByKey 用于实现 treeReduce，reduceByKey 对每个键执行汇聚，reduceByKey 不是一个动作 Action，结果返回 Shuffled RDD。treeReduce 使用 reduceByKey 来执行并行汇聚，由于 treeReduce 不是 key-value 类型，因此我们将树的深度作为 Key，将 treeReduce 转换成一个 key-value 类型的 RDD。

许多机器学习 MLib 的算法使用 treeAggregate，在高斯混合模型 GaussianMixture 中，使用 treeAggregate 算子替代 aggregate 算子性能提升了 20%，而在 Online Variational Bayes for LDA 使用 treeAggregate 算子替代 reduce 算子来聚合预期的单词计数矩阵（可能是一个非常大的矩阵），不会发生扩展性问题。 MLlib 的“梯度下降”的实现使用的是 treeAggregate。

测试中使用 treeAggregate 替代 reduce 在实现反向传播算法中计算梯度下降。在测试中，具有 100 个特征的数据集、10M 实例（instance）及 96 个分区，在一个集群上执行，由 3 个 Worker 工作节点和 1 个应用 Master 主节点（每个具有 16 个 CPU 和 52 GB 内存）组成，神经网络只执行了 100 个训练次数，用了 36min，而替代前运行需花费几个小时的时间。

以下 Scala 代码示例生成两个随机 RDD，其中包含 100 万个值，并使用 map-reduce 模式，treeReduce 和 treeAggregate 计算欧氏距离如下：

```
1.  import org.apache.commons.lang.SystemUtils
2.  import org.apache.spark.mllib.random.RandomRDDs._
3.  import org.apache.spark.sql.SQLContext
4.  import org.apache.spark.{SparkConf, SparkContext}
5.
6.  import scala.math.sqrt
7.
8.  object Test{
9.
10.   def main(args: Array[String]) {
11.
12.     var mapReduceTimeArr : Array[Double]= Array.ofDim(20)
13.     var treeReduceTimeArr : Array[Double]= Array.ofDim(20)
14.     var treeAggregateTimeArr : Array[Double]= Array.ofDim(20)
```

```
15.
16.     //Spark 初始化
17.     val config = new SparkConf().setAppName("TestStack")
18.     val sc: SparkContext = new SparkContext(config)
19.     val sql: SQLContext = new SQLContext(sc)
20.
21.     //生成一个随机的 RDD，包含 100 万个来自标准正态分布 N(0, 1)的独立同分布的值，
        //均匀分布在 5 个分区中
22.     val input1 = normalRDD(sc, 1000000L, 5)
23.
24.    //生成一个随机的 RDD，包含 100 万个来自标准正态分布 N(0, 1)的独立同分布的值，
       //均匀分布在 5 个分区中
25.     val input2 = normalRDD(sc, 1000000L, 5)
26.
27.     val xy = input1.zip(input2).cache()
28.     //RDD 计算
29.     xy.count()
30.
31.     for(i:Int <-0 until 20){
32.       val t1 = System.nanoTime()
33.       val euclideanDistanceMapRed = sqrt(xy.map { case (v1, v2) => (v1 -
          v2) * (v1 - v2) }.reduce(_ + _))
34.       val t11 = System.nanoTime()
35.       println("Map-Reduce - Euclidean Distance "+euclideanDistanceMapRed)
36.       mapReduceTimeArr(i)=(t11 - t1)/1000000.0
37.       println("Map-Reduce - Elapsed time: " + (t11 - t1)/1000000.0  + "ms")
38.     }
39.
40.     for(i:Int <-0 until 20) {
41.       val t2 = System.nanoTime()
42.       val euclideanDistanceTreeRed = sqrt(xy.map { case (v1, v2) => (v1
          - v2) * (v1 - v2) }.treeReduce(_ + _))
43.       val t22 = System.nanoTime()
44.       println("TreeReduce - Euclidean Distance "+euclideanDistanceTreeRed)
45.       treeReduceTimeArr(i)=(t22 - t2) / 1000000.0
46.       println("TreeReduce - Elapsed time: " + (t22 - t2) / 1000000.0 + "ms")
47.     }
48.
49.     for(i:Int <-0 until 20) {
50.       val t3 = System.nanoTime()
51.       val euclideanDistanceTreeAggr = sqrt(xy.treeAggregate(0.0)(
52.         seqOp = (c, v) => {
53.           (c + ((v._1 - v._2) * (v._1 - v._2)))
54.         },
55.         combOp = (c1, c2) => {
56.           (c1 + c2)
57.         }))
58.       val t33 = System.nanoTime()
59.       println("TreeAggregate - Euclidean Distance " + euclideanDistance
          TreeAggr)
60.       treeAggregateTimeArr(i) = (t33 - t3) / 1000000.0
61.       println("TreeAggregate - Elapsed time: " + (t33 - t3) / 1000000.0
          + "ms")
62.     }
63.
64.     val mapReduceAvgTime = mapReduceTimeArr.sum / mapReduceTimeArr.length
65.     val treeReduceAvgTime = treeReduceTimeArr.sum / treeReduceTimeArr
```

```
        .length
66.     val treeAggregateAvgTime = treeAggregateTimeArr.sum / treeAggregate
        TimeArr.length
67.
68.     val mapReduceMinTime = mapReduceTimeArr.min
69.     val treeReduceMinTime = treeReduceTimeArr.min
70.     val treeAggregateMinTime = treeAggregateTimeArr.min
71.
72.     val mapReduceMaxTime = mapReduceTimeArr.max
73.     val treeReduceMaxTime = treeReduceTimeArr.max
74.     val treeAggregateMaxTime = treeAggregateTimeArr.max
75.
76.     println("Map-Reduce - Avg:" + mapReduceAvgTime+ "ms "+ "Max:"
        +mapReduceMaxTime+ "ms "+ "Min:" +mapReduceMinTime+ "ms ")
77.     println("TreeReduce - Avg:" + treeReduceAvgTime + "ms "+ "Max:"
        +treeReduceMaxTime+ "ms "+ "Min:" +treeReduceMinTime+ "ms ")
78.     println("TreeAggregate - Avg:" + treeAggregateAvgTime + "ms "+ "Max:"
        +treeAggregateMaxTime+ "ms "+ "Min:" +treeAggregateMinTime+ "ms ")
79.   }
80. }
```

20.6　使用 treeAggregate 取代 Aggregate 的原理和源码

本节讲解 treeAggregate 进行 Aggregate 的工作原理和源码；以及使用 treeAggregate 进行性能测试。

20.6.1　treeAggregate 进行 Aggregate 的工作原理和源码

treeAggregate 分层进行 Aggregate，由于 Aggregate 的时候，其分区的计算结果是传输到 Driver 端再进行合并的，如果分区比较多，计算结果返回的数据量比较大，那么 Driver 端需要缓存大量的中间结果，这样就会加大 Driver 端的计算压力，因此 treeAggregate 把分区计算结果的合并仍旧放在 Executor 端进行，将结果在 Executor 端不断合并缩小返回 Driver 的数据量，最后在 Driver 端进行合并。

Spark 2.2.1 版本的 RDD.scala 的 treeAggregate 的源码如下：

```
1.    /**
2.     *聚集 Aggregates 多级树状模式的 RDD 的元素
3.     *
4.     * @param depth 建议树的深度（默认值：2）
5.     * @see [[org.apache.spark.rdd.RDD#aggregate]]
6.     */
7.    def treeAggregate[U: ClassTag](zeroValue: U)(
8.        seqOp: (U, T) => U,
9.        combOp: (U, U) => U,
10.       depth: Int = 2): U = withScope {
11.     require(depth >= 1, s"Depth must be greater than or equal to 1 but got
        $depth.")
12.     if (partitions.length == 0) {
13.       Utils.clone(zeroValue, context.env.closureSerializer.newInstance())
14.     } else {
```

```
15.         val cleanSeqOp = context.clean(seqOp)
16.         val cleanCombOp = context.clean(combOp)
17.         //针对初始分区的聚合函数
18.         val aggregatePartition =
19.           (it: Iterator[T]) => it.aggregate(zeroValue)(cleanSeqOp, cleanCombOp)
20.         //针对初始的各分区先进行部分聚合
21.         var partiallyAggregated = mapPartitions(it => Iterator(aggregate
            Partition(it)))
22.         var numPartitions = partiallyAggregated.partitions.length
23.         //根据传入的 depth 计算出需要迭代计算的程度
24.         val scale = math.max(math.ceil(math.pow(numPartitions, 1.0 /
            depth)).toInt, 2)
25.       //如果创建一个额外的级别并不能帮助减少 wall-clock 时间，我们将停止树的聚集
26.
27.         //当不保存 wall-clock 时间时，将不触发树聚集 TreeAggregation
28.         while (numPartitions > scale + math.ceil(numPartitions.toDouble /
            scale)) {
29.        //计算迭代的程度
30.           numPartitions /= scale
31.           val curNumPartitions = numPartitions
32.        //减少分区个数，合并部分分区的结果
33.           partiallyAggregated = partiallyAggregated.mapPartitionsWithIndex {
34.             (i, iter) => iter.map((i % curNumPartitions, _))
35.           }.reduceByKey(new HashPartitioner(curNumPartitions), cleanCombOp)
              .values
36.         }
37.        //执行最后一次 reduce，返回最终结果
38.         partiallyAggregated.reduce(cleanCombOp)
39.       }
40.     }
```

Spark 2.4.3 版本的 RDD.scala 源码与 Spark 2.2.1 版本相比具有如下特点。

- 上段代码中第 21 行新增变量 partiallyAggregated 的类型说明。
- 上段代码中第 35 行将 reduceByKey 算子调整为 foldByKey 算子。
- 上段代码中第 36 行之后新增一行代码，使用 Spark 序列化程序克隆对象。
- 上段代码中第 38 行将 reduce 算子调整为 fold 算子。

```
1.  ……
2.     var partiallyAggregated: RDD[U] = mapPartitions(it =>
       Iterator(aggregatePartition(it)))
3.  ……
4.       }.foldByKey(zeroValue, new HashPartitioner(curNumPartitions))
         (cleanCombOp).values
5.  ……
6.  val copiedZeroValue = Utils.clone(zeroValue, sc.env.closureSerializer
    .newInstance())
7.  partiallyAggregated.fold(copiedZeroValue)(cleanCombOp)
8.  ……
```

20.6.2　使用 treeAggregate 进行性能测试

在 Spark（和原始 MapReduce）中的 reduce 算子或 aggregate 算子中，所有分区都必须将其汇聚的值发送到 Driver 节点，并且 Driver 节点对分区数量花费线性时间（由于分区计算结

果和网络带宽限制)。当有很多分区，每个分区的数据很大时，它成为一个瓶颈。

Spark 1.1 引入了基于多级聚合树的聚合模式。在此设置中，将数据组合在 Executors 上，然后将其发送到 Driver 节点，大大降低了 Driver 节点必须处理的负载。测试表明，这些功能将聚合时间减少一个数量级，特别是在具有大量分区的数据集上。

在 treeReduce 和 treeAggregate 中，分区以对数的轮数相互通信。在 treeAggregate 的情况下，假设在其叶子具有所有分区的 n-ary tree 树，树根节点将包含最终的汇聚值。这样就没有一台瓶颈节点，如图 20-1 所示。

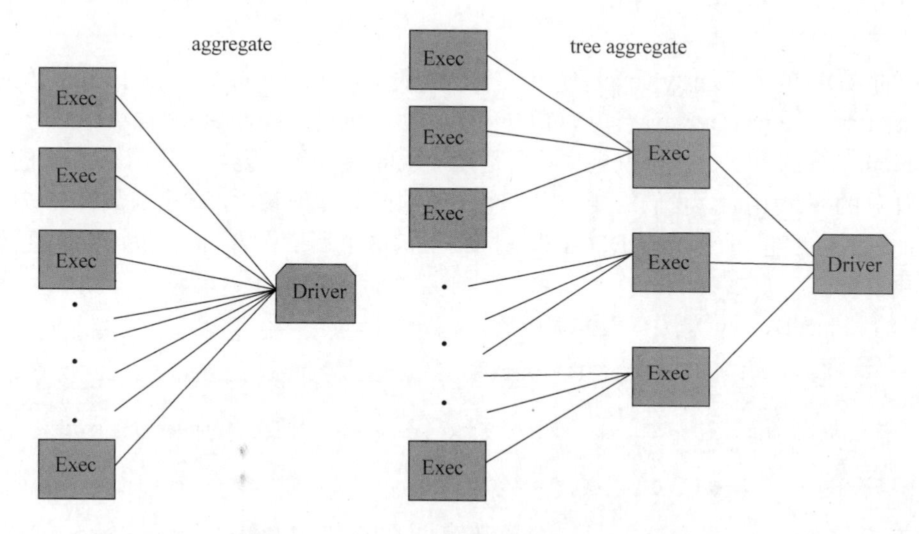

图 20-1　treeReduce 和 treeAggregate 对比图

下面是 treeReduce 和 treeAggregate 代码示例，需在具有大量分区的群集上运行。

```
1.   public void treeReduce() {
2.     JavaRDD<Integer> rdd = sc.parallelize(Arrays.asList(-5, -4, -3, -2, -1,
       1, 2, 3, 4), 10);
3.       Function2<Integer, Integer, Integer> add = new Function2<Integer,
         Integer, Integer>() {
4.         @Override
5.         public Integer call(Integer a, Integer b) {
6.           return a + b;
7.         }
8.       };
9.       for (int depth = 1; depth <= 10; depth++) {
10.        int sum = rdd.treeReduce(add, depth);
11.        assertEquals(-5, sum);
12.      }
13.  }
14.
15.
16. public void treeAggregate() {
17.      JavaRDD<Integer> rdd = sc.parallelize(Arrays.asList(-5, -4, -3, -2,
         -1, 1, 2, 3, 4), 10);
18.      Function2<Integer, Integer, Integer> add = new Function2<Integer,
         Integer, Integer>() {
19.        @Override
20.        public Integer call(Integer a, Integer b) {
21.          return a + b;
```

```
22.         }
23.      };
24.      for (int depth = 1; depth <= 10; depth++) {
25.        int sum = rdd.treeAggregate(0, add, add, depth);
26.        assertEquals(-5, sum);
27.      }
28.    }
```

20.7　reduceByKey 高效运行的原理和源码解密

Spark 的 RDD 的 reduceByKey 是使用一个相关的函数来合并每个 key 的 value 值的一个算子。本质上讲，reduceByKey 函数（算子）只作用于包含 key-value 的 RDD 上，它是 transformation 类型的算子，这也就意味着它是懒加载的（也就是说，不调用 Action 的方法，是不会去计算的）。使用时，我们需要传递一个相关的函数作为参数，这个函数将会被应用到源 RDD 上，并且创建一个新的 RDD 作为返回结果，这个算子作为 data Shuffling 在分区时被广泛使用。

Spark 的 reduceByKey 方法有 3 种重载形式。

```
1.  def reduceByKey(Partitioner: Partitioner, func: JFunction2[V, V, V]):
   JavaPairRDD[K, V]
2. def  reduceByKey(func:  JFunction2[V,  V,  V],  numPartitions:  Int):
   JavaPairRDD[K, V]
3. def reduceByKey(func: JFunction2[V, V, V]): JavaPairRDD[K, V]
```

前两种形式除了允许用户传入聚合函数外，还允许用户指定 Partitioner 或者指定 reduceByKey 后生成的 RDD 的 Partition 个数。

RDD.scala 的源码如下：

```
1.  def reduceByKey(func: JFunction2[V, V, V]): JavaPairRDD[K, V] = {
2.      fromRDD(reduceByKey(defaultPartitioner(RDD), func))
3.  }
```

其中，当用户没有指定 Partitioner 以及 Partition 的个数时，Spark 会调用 defaultPartitioner (RDD)函数去获取一个默认的 Partitioner。

Spark 2.2.1 版本的 Partitioner.scala 的源码如下：

```
1.       def defaultPartitioner(rdd: RDD[_], others: RDD[_]*): Partitioner = {
2.     val rdds = (Seq(rdd) ++ others)
3.     val hasPartitioner = rdds.filter(_.partitioner.exists(_.numPartitions >
       0))
4.     if (hasPartitioner.nonEmpty) {
5.       hasPartitioner.maxBy(_.partitions.length).partitioner.get
6.     } else {
7.       if (rdd.context.conf.contains("spark.default.parallelism")) {
8.         new HashPartitioner(rdd.context.defaultParallelism)
9.       } else {
10.        new HashPartitioner(rdds.map(_.partitions.length).max)
11.      }
12.    }
13.  }
14. }
```

Spark 2.4.3 版本的 Partitioner.scala 源码与 Spark 2.2.1 版本相比具有如下特点。

- 上段代码中第 4～14 行代码整体替换为以下代码，选择要用于多个 RDD 之间的类似 cogroup 操作的分区器。如果设置了 spark.default.parallelism，我们将使用 SparkContext defaultParallelism 的值作为默认分区数，否则我们将使用上游分区的最大数量。如果可用，我们将从具有最大分区数的 RDD 中选择分区器。如果这个分区器符合条件（分区数在最大分区数内），或者分区数高于默认分区数，我们使用这个分区器。否则，我们将使用具有默认分区数的新 HashPartitioner。除非设置了 spark.default.parallelism，否则分区数将取上游 RDD 中的最大分区数，因为这最不可能导致内存不足错误，将使用两个方法参数（rdd, others）来强制调用方至少传递一个 RDD。

```
1.  ……
2.  val hasMaxPartitioner: Option[RDD[_]] = if (hasPartitioner.nonEmpty) {
3.        Some(hasPartitioner.maxBy(_.partiti  def defaultPartitioner(rdd:
          RDD[_], others: RDD[_]*): Partitioner = {
4.      val rdds = (Seq(rdd) ++ others)
5.      val hasPartitioner = rdds.filter(_.partitioner
        .exists(_.numPartitions > 0))
6.      if (hasPartitioner.nonEmpty) {
7.        hasPartitioner.maxBy(_.partitions.length).partitioner.get
8.      } else {
9.        if (rdd.context.conf.contains("spark.default.parallelism")) {
10.         new HashPartitioner(rdd.context.defaultParallelism)
11.       } else {
12.         new HashPartitioner(rdds.map(_.partitions.length).max)
13.       }
14.     }
15.   }ons.length))
16.     } else {
17.       None
18.     }
19.
20.     val defaultNumPartitions = if (rdd.context.conf
        .contains("spark.default.parallelism")) {
21.       rdd.context.defaultParallelism
22.     } else {
23.       rdds.map(_.partitions.length).max
24.     }
25.
26.     //如果现有的max分区是可用分区，或者其分区数比默认分区数更大，使用现有的分区数
27.     if (hasMaxPartitioner.nonEmpty &&
        (isEligiblePartitioner(hasMaxPartitioner.get, rdds) ||
28.         defaultNumPartitions < hasMaxPartitioner.get.getNumPartitions)) {
29.       hasMaxPartitioner.get.partitioner.get
30.     } else {
31.       new HashPartitioner(defaultNumPartitions)
32.     }
33.   }
```

defaultPartitioner 方法允许传入两个 RDD，调用该方法时最少需要传入一个 RDD。方法首先会将传入的两个 RDD 合并成一个数组，遍历这个数组，从有 Partitioner 的 RDD 中挑选出 Partition 个数最多的 RDD，将其 Partitioner 返回。如果传入的 RDD 都没有 Partitioner，那么就会返回一个 HashPartitioner，其中，如果 Spark 配置了 Spark.default.parallelism 参数，则 Partition 的个数为该参数的值。否则，新生成的 RDD 中 Partition 的个数取与其依赖的父 RDD

中 Partition 个数的最大值。

再进一步，我们来看 HashPartitioner（HashPartitioner 是 Partitioner 的一个内部类）的划分规则是怎么样的。

Partitioner.scala 的源码如下：

```
1.   class HashPartitioner(partitions: Int) extends Partitioner {
2.    require(partitions >= 0, s"Number of partitions ($partitions) cannot be
      negative.")
3.
4.    def numPartitions: Int = partitions
5.
6.    def getPartition(key: Any): Int = key match {
7.      case null => 0
8.      case _ => Utils.nonNegativeMod(key.hashCode, numPartitions)
9.    }
10.
11.   override def equals(other: Any): Boolean = other match {
12.     case h: HashPartitioner =>
13.       h.numPartitions == numPartitions
14.     case _ =>
15.       false
16.   }
17.
18.   override def hashCode: Int = numPartitions
19. }
```

HashPartitioner 是一个基于 Java 的 Object.HashCode 实现，基于 Hash 的 Partitioner。由于 Java arrays 的 Hash code 是基于 arrays 的标识，而不是它的内容，所以如果使用 HashPartitioner 对 RDD[Array[_]]或者 RDD[(Array[_],_)]进行 Partition，可能会得到不正确的结果。也就是说，如果 RDD 中保存的数据类型是 arrays，这时默认的 HashPartitioner 是不可用的，用户在调用 reduceByKey 时需要自行实现一个 Partitioner，否则方法会抛出异常。

从源码中可以看出，HashPartitioner 的划分规则根据的是 Utils.nonNegativeMod (key.HashCode, numPartitions)方法，而这个方法也很简单。

Utils.scala 的源码如下：

```
1.   def nonNegativeMod(x: Int, mod: Int): Int = {
2.     val rawMod = x % mod
3.     rawMod + (if (rawMod < 0) mod else 0)
4.   }
```

nonNegativeMod 就是直接用对象的 HashCode 对 numPartition 取模。

回到 reduceByKey 方法，所有的 reduceByKey 重载最终都会调用以下代码。

```
1.  /**
2.     *使用关联和交换汇聚函数合并每个 Key 键的 Value 值。在每个 Mapper 端将结果发送到
       *Reducer 端之前，进行本地合并，类似于 MapReduce 的本地聚合 combiner
3.     */
4.  def reduceByKey(partitioner: Partitioner, func: (V, V) => V): RDD[(K, V)]
    = self.withScope {
5.    combineByKeyWithClassTag[V]((v: V) => v, func, func, partitioner)
6.  }
```

接下来看一下 combineByKeyWithClassTag 函数的实现。

- 如果执行这个操作，RDD 的 key 是一个数组类型时，同时设置 Mapper 端执行 combine 操作，提示错误。
- 如果 RDD 的 key 是一个数组类型，同时分区算子是默认的哈希算子时，提示错误。

PairRDDFunctions.scala 的源码如下：

```
1.   def combineByKeyWithClassTag[C](
2.   ......
3.      if (keyClass.isArray) {
4.        if (mapSideCombine) {
5.          throw new SparkException("Cannot use map-side combining with array
            keys.")
6.        }
7.        if (partitioner.isInstanceOf[HashPartitioner]) {
8.          throw new SparkException("HashPartitioner cannot partition array
             keys.")
9.        }
10.     }
11. ......
```

根据传入的前 3 个参数，生成 Aggregator，设置 mapSideCombine 时，Aggregator 必须存在。

PairRDDFunctions.scala 的源码如下：

```
1.  val aggregator = new Aggregator[K, V, C](
2.        self.context.clean(createCombiner),
3.        self.context.clean(mergeValue),
4.        self.context.clean(mergeCombiners))
5.     ......
```

如果执行当前操作传入的 Partitioner 与执行这个操作对应的 RDD 是相同的算子时，这时不对当前的操作生成新的 RDD，也就是这个操作不再执行 Shuffle 操作，直接使用当前操作的 RDD 的 Iterator。

PairRDDFunctions.scala 的源码如下：

```
1.   if (self.partitioner == Some(partitioner)) {
2.        self.mapPartitions(iter => {
3.          val context = TaskContext.get()
4.          new InterruptibleIterator(context, aggregator.combineValuesByKey
            (iter, context))
5.        }, preservesPartitioning = true)
6.  ......
```

如果当前操作传入的 Partitioner 与执行这个操作对应的 RDD 不相同时，执行这个操作的 Partitioner 是一个新生成的，或者说与当前要执行这个操作的 RDD 的 Partitioner 不是相同的实例，表示这个操作需要执行 Shuffle 操作，生成一个 ShuffledRDD 实例。

PairRDDFunctions.scala 的源码如下：

```
1.  ......
2.   } else {
3.        new ShuffledRDD[K, V, C](self, partitioner)
4.          .setSerializer(serializer)
5.          .setAggregator(aggregator)
6.          .setMapSideCombine(mapSideCombine)
7.      }
8.    }
```

ShuffledRDD 的实例生成：先看 ShuffledRDD 实例的生成部分。这里传入的 prev 是生成 ShuffleRDD 的上层的 RDD，在实例生成时设置对上层的 RDD 的依赖为 Nil，表示对上层 RDD 的依赖是 Nil。

```
1.    class ShuffledRDD[K: ClassTag, V: ClassTag, C: ClassTag](
2.    @transient var prev: RDD[_ <: Product2[K, V]],
3.    part: Partitioner)
4.  extends RDD[(K, C)](prev.context, Nil) {
```

接下来 ShuffledRDD 处理上层 RDD 的依赖部分，在 ShuffledRDD 中会重写 getDependencies 函数。

ShuffledRDD.scala 的源码如下：

```
1.    override def getDependencies: Seq[Dependency[_]] = {
2.    ......
3.  //这里，生成对 ShuffledRDD 的依赖为 ShuffleDependency 实例。这个依赖的 RDD
4.  //就是生成这个 ShuffledRDD 的上层的 RDD 的实例
5.     List(new ShuffleDependency(prev, part, serializer, keyOrdering, aggregator,
       mapSideCombine))
6.    }
```

在 getDependencies 方法中构建生成 ShuffleDependency 实例，每生成一个 ShuffleDependency 的实例时，会对每个 Shuffle 的依赖生成一个唯一的 ShuffleId，用于对此 Stage 中每个 Task 的结果集的跟踪。

Dependency.scala 的源码如下：

```
1.    class ShuffleDependency[K: ClassTag, V: ClassTag, C: ClassTag](
2.  ......
3.    val shuffleId: Int = _rdd.context.newShuffleId()
4.    val shuffleHandle: ShuffleHandle = _rdd.context.env.shuffleManager
      .registerShuffle(
5.      shuffleId, _rdd.partitions.length, this)
6.  ......
```

向 ShuffleManager 注册这个 Shuffle 的依赖。Task 的结果集向 Driver 通知时，首先需要这个 Shuffle 是一个注册的 Shuffle。reduceByKey 适合使用在大数据集上。因为 Spark 知道它可以在每个分区移动数据前将输出数据与一个共用的 key 结合。

20.8　使用 AggregateByKey 取代 groupByKey 的原理和源码

本节讲解使用 AggregateByKey 取代 groupByKey 的原理和源码；以及使用 AggregateByKey 取代 groupByKey 性能测试。

20.8.1　使用 AggregateByKey 取代 groupByKey 的工作原理

AggregateByKey 函数对 PairRDD 中相同 Key 的值进行聚合操作，在聚合过程中同样使用了一个中立的初始值。和 Aggregate 函数类似，AggregateByKey 返回值的类型不需要和 RDD 中 value 的类型一致。因为 AggregateByKey 是对相同 Key 中的值进行聚合操作，所以 AggregateByKey 函数最终返回的类型还是 Pair RDD，对应的结果是 Key 和聚合好的值；而 Aggregate 函数是直接返回非 RDD 的结果，这点需要注意。在实现过程中定义了 3 个 AggregateByKey 函数原型，但最终调用的 AggregateByKey 函数都一致。

函数原型如下：

```
1.   def AggregateByKey[U: ClassTag](zeroValue: U, Partitioner: Partitioner)
2.          (seqOp: (U, V) => U, combOp: (U, U) => U): RDD[(K, U)]
3.  def AggregateByKey[U: ClassTag](zeroValue: U, numPartitions: Int)
4.          (seqOp: (U, V) => U, combOp: (U, U) => U): RDD[(K, U)]
5.  def AggregateByKey[U: ClassTag](zeroValue: U)
6.          (seqOp: (U, V) => U, combOp: (U, U) => U): RDD[(K, U)]
```

第一个 AggregateByKey 函数我们可以自定义 Partitioner。除了这个参数外，其函数声明和 Aggregate 很类似；其他的 AggregateByKey 函数实现最终都是调用 AggregateByKey 函数。

第二个 AggregateByKey 函数可以设置分区的个数(numPartitions)，最终用的是 HashPartitioner。

最后一个 AggregateByKey 实现先判断当前 RDD 是否定义了分区函数，如果定义了，则用当前 RDD 的分区；如果当前 RDD 并未定义分区，则使用 HashPartitioner。

20.8.2　源码剖析

如使用 AggregateByKey 取代 groupByKey，先看一下 AggregateByKey 源码的主要代码。

PairRDDFunctions.scala 的源码如下：

```
1.  def aggregateByKey[U: ClassTag](zeroValue: U, partitioner: Partitioner)
    (seqOp: (U, V) => U,
2.        combOp: (U, U) => U): RDD[(K, U)] = self.withScope {
3.     ……
4.      val cleanedSeqOp = self.context.clean(seqOp)
5.      //seqOp 传入 createCombiner, mergeValue。combOp 传入 mergeCombiners
6.      combineByKeyWithClassTag[U]((v: V) => cleanedSeqOp(createZero(), v),
7.        cleanedSeqOp, combOp, partitioner)
8.    }
```

其中调用的 combineByKeyWithClassTag 方法源码中的 mapSideCombine: Boolean 默认为 true 值，在 mapSide 端进行聚合。

PairRDDFunctions.scala 的源码如下：

```
1.  def combineByKeyWithClassTag[C](
2.        createCombiner: V => C,
3.        mergeValue: (C, V) => C,
4.        mergeCombiners: (C, C) => C,
5.        partitioner: Partitioner,
6.        mapSideCombine: Boolean = true,
7.        serializer: Serializer = null)(implicit ct: ClassTag[C]): RDD[(K,
          C)] = self.withScope {
8.       ……
```

在 aggregateByKey 方法中调用 combineByKeyWithClassTag 时：

- ❑ createCombiner: cleanedSeqOp(createZero(), v)是 createCombiner，也就是传入的 seqOp 函数，只不过其中一个值是传入的 zeroValue。
- ❑ mergeValue：seqOp 函数同样是 mergeValue, createCombiner 和 mergeValue 函数都是 AggregateByKey 函数的关键。
- ❑ mergeCombiners：combOp 函数。

因此，当 createCombiner 和 mergeValue 函数的操作相同时，AggregateByKey 更合适。

当调用 groupByKey 时，所有的键值对（key-value pair）都会被移动，在网络上传输这些数据非常没必要，因此避免使用 GroupByKey。

为了确定将数据对移到哪个主机，Spark 会对数据对的 Key 调用一个分区算法。当移动的数据量大于单台执行机器内存总量时，Spark 会把数据保存到磁盘上。不过，在保存时每次会处理一个 Key 的数据，所以当单个 Key 的键值对超过内存容量时，会存在内存溢出的异常。应避免将数据保存到磁盘上，这会严重影响性能。

如果需要按 Key 分组聚合（如 sum 或 average），推荐使用 reduceByKey 或者 AggregateByKey，以获得更好的性能。

20.8.3　使用 AggregateByKey 取代 groupByKey 性能测试

计算计数时有两种不同的方式，如分别通过 reduceByKey 算子和 groupByKey 算子进行计算。

```
1.   val words = Array("one", "two", "two", "three", "three", "three")
2.  val wordPairsRDD = sc.parallelize(words).map(word => (word, 1))
3.
4.  val wordCountsWithReduce = wordPairsRDD .reduceByKey(_ + _) .collect()
5.  val wordCountsWithGroup = wordPairsRDD .groupByKey().map(t =>(t._1,
    t._2.sum)).collect()
```

reduceByKey 将在 Shuffle 前在本地将相同 Key 值的记录先进行汇聚（例如：（a，1）出现 2 次，在本地先汇聚（a，2）），如图 20-2 所示，groupByKey 本地不进行汇聚，通过 Shuffle 进行汇聚，如图 20-3 所示。

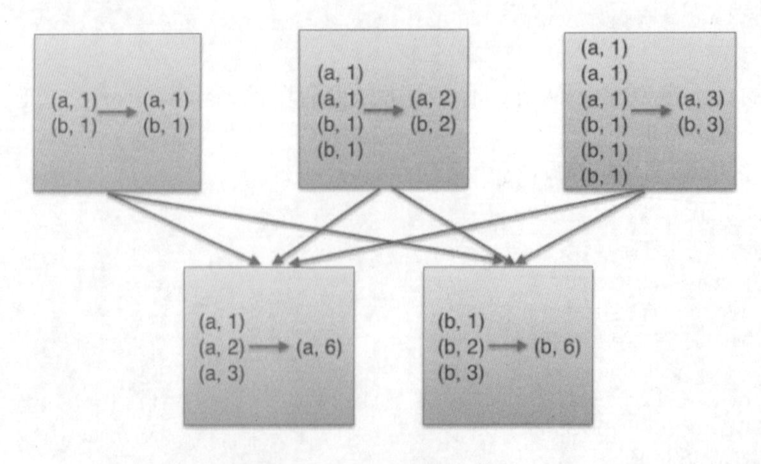

图 20-2　reduceByKey 算子

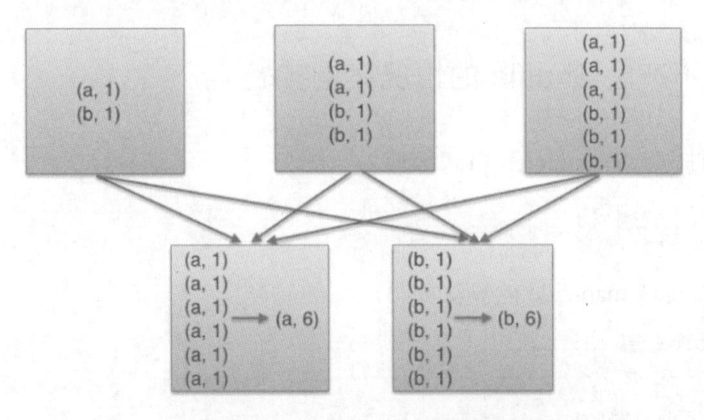

图 20-3　groupByKey 算子

AggregateByKey 算子类似于 reduceByKey 算子，在算子里面调用 combineByKey 方法。使用 AggregateByKey 取代 groupByKey 对大量数据的计算速度的提升是显而易见的。

20.9　Join 不产生 Shuffle 的情况及案例实战

本节讲解 Join 在什么情况下不产生 Shuffle 及其运行原理；Join 不产生 Shuffle 的情况案例实战。

20.9.1　Join 在什么情况下不产生 Shuffle 及其运行原理

在大数据处理场景中，多表 Join 是常见的一类运算。为了便于求解，通常会将多表 Join 问题转为多个两表连接问题。两表 Join 的实现算法非常多，一般我们会根据两表的数据特点选取不同的 Join 算法，其中，最常用的两个算法是 map-side join 和 reduce-side join。map-side join 也就是 Join 不产生 Shuffle。

Map-side Join 使用场景是一个大表和一个小表的连接操作，其中，“小表”是指文件足够小，可以加载到内存中。该算法可以将 Join 算子执行在 Mapper 端，无须经历 Shuffle 和 reduce 等阶段，因此效率非常高。

在 Hadoop MapReduce 中，map-side Join 是借助 DistributedCache 实现的。DistributedCache 可以帮我们将小文件分发到各个节点的 Task 工作目录下，这样，我们只需在程序中将文件加载到内存中（例如，保存到 Map 数据结构中），然后借助 Mapper 的迭代机制，遍历另一个大表中的每一条记录，并查找是否在小表中，如果在，则输出，否则跳过。

在 Apache Spark 中，同样存在类似于 DistributedCache 的功能，称为“广播变量”(Broadcast variable)。其实现原理与 DistributedCache 非常类似，但提供了更多的数据/文件广播算法，包括高效的 P2P 算法，该算法在节点数目非常多的场景下，效率远远好于 DistributedCache 这种基于 HDFS 共享存储的方式，具体比较可参考“Performance and Scalability of Broadcast in Spark”。使用 MapReduce DistributedCache 时，用户需要显式地使用 File API 编写程序从本地读取小表数据，而 Spark 则不用，它借助 Scala 语言强大的函数闭包特性，可以隐藏数据/文件广播过程，使用户编写程序更加简单。

20.9.2　Join 不产生 Shuffle 的情况案例实战

假设两个文件，一小一大，且格式类似为

```
1.  Key,value,value
2.  Key,value,value
```

则利用 Spark 实现 map-side 的算法如下。

```
1.  var table1 = sc.textFile(args(1))
2.  var table2 = sc.textFile(args(2))
3.
4.  //table1 是一个小表，因此广播这个小表作为 map<String, String>
5.  var pairs = table1.map { x =>
6.  var pos = x.indexOf(',')
7.    (x.substring(0, pos), x.substring(pos + 1))
8.  }.collectAsMap
9.  var broadCastMap = sc.broadcast(pairs) //保存表 table1 作为 map，然后进行广播
10.
11. //table2 和 table1 在 map 端进行 Join
12. var result = table2.map { x =>
13. var pos = x.indexOf(',')
14.   (x.substring(0, pos), x.substring(pos + 1))
15. }.mapPartitions({ iter =>
16. var m = broadCastMap.value
17. for{
18.     (key, value) <- iter
19. if(m.contains(key))
20.   } yield (key, (value, m.get(key).getOrElse("")))
21. })
22.
23. result.saveAsTextFile(args(3)) //保存结果到本地文件或者 HDFS 系统
24.
```

20.10　RDD 复用性能调优最佳实践

本节讲解什么时候需要复用 RDD，以及如何复用 RDD 算子。

20.10.1　什么时候需要复用 RDD

通常，开发一个 Spark 作业时，首先是基于某个数据源（如 Hive 表或 HDFS 文件）创建一个初始的 RDD；接着对这个 RDD 执行某个算子操作，然后得到下一个 RDD；以此类推，循环往复，直到计算出最终我们需要的结果。在这个过程中，多个 RDD 会通过不同的算子操作（如 map、reduce 等）串起来，这个“RDD 串”就是 RDD lineage，也就是“RDD 的血缘关系链”。

在开发过程中要注意：对于同一份数据，只应该创建一个 RDD，不能创建多个 RDD 来代表同一份数据。

一些 Spark 初学者在刚开始开发 Spark 作业时，或者是有经验的工程师在开发 RDD

lineage 极其冗长的 Spark 作业时，可能会忘了自己之前对于某一份数据已经创建过一个 RDD 了，从而导致对于同一份数据，创建了多个 RDD。这就意味着，我们的 Spark 作业会进行多次重复计算，来创建多个代表相同数据的 RDD，进而增加了作业的性能开销。

除了要避免在开发过程中对一份完全相同的数据创建多个 RDD 外，在对不同的数据执行算子操作时还要尽可能地复用一个 RDD。例如，有一个 RDD 的数据格式是 key-value 类型的，另一个是单 value 类型的，这两个 RDD 的 value 数据是完全一样的。那么，此时我们可以只使用 key-value 类型的那个 RDD，因为其中已经包含了另一个数据。对于类似这种多个 RDD 的数据有重叠或者包含的情况，我们应该尽量复用一个 RDD，这样可以尽可能地减少 RDD 的数量，从而尽可能减少算子执行的次数。

Spark 中对于一个 RDD 执行多次算子的默认原理：每次对一个 RDD 执行一个算子操作时，都会重新从源头处计算一遍，计算出那个 RDD，然后再对这个 RDD 执行算子操作。这种方式的性能很差。

因此对于这种情况，建议：对多次使用的 RDD 进行持久化。此时 Spark 就会根据持久化策略，将 RDD 中的数据保存到内存或者磁盘中。以后每次对这个 RDD 进行算子操作时，都会直接从内存或磁盘中提取持久化的 RDD 数据，然后执行算子操作，而不会从源头处重新计算一遍 RDD，再执行算子操作。

20.10.2　如何复用 RDD 算子

复用 RDD 一般基于以下几个原则。

原则一：避免创建重复的 RDD。

一个简单的例子：需要对名为“hello.txt”的 HDFS 文件进行一次 map 操作，再进行一次 reduce 操作。也就是说，需要对一份数据执行两次算子操作。错误的做法：对于同一份数据执行多次算子操作时，创建多个 RDD。这里执行了两次 textFile 方法，针对同一个 HDFS 文件，创建了两个 RDD，然后分别对每个 RDD 都执行一个算子操作。这种情况下，Spark 需要从 HDFS 上两次加载 hello.txt 文件的内容，并创建两个单独的 RDD；第二次加载 HDFS 文件以及创建 RDD 的性能开销，很明显是白白浪费掉的。

```
1.  val RDD1 = sc.textFile("hdfs://192.168.0.1:9000/hello.txt")
2.  RDD1.map(...)
3.  val RDD2 = sc.textFile("hdfs://192.168.0.1:9000/hello.txt")
4.  RDD2.reduce(...)
```

正确的用法：对于一份数据执行多次算子操作时，只使用一个 RDD。这种写法明显比上一种写法好多了，因为我们对于同一份数据只创建了一个 RDD，然后对这个 RDD 执行了多次算子操作。但是要注意，到这里为止，优化还没有结束，由于 RDD1 被执行了两次算子操作，第二次执行 reduce 操作的时候，还会再次从源头处重新计算一次 RDD1 的数据，因此还是会有重复计算的性能开销。

```
1.   val RDD1 = sc.textFile("hdfs://192.168.0.1:9000/hello.txt")
2.  RDD1.map(...)
3.  RDD1.reduce(...)
```

要彻底解决这个问题，必须结合“原则三：对多次使用的 RDD 进行持久化”，才能保证一个 RDD 被多次使用时只被计算一次。

原则二：尽可能复用同一个 RDD。

一个简单的例子：有一个<Long, String>格式的 RDD，即 RDD1。由于业务需要，对 RDD1 执行了一个 map 操作，创建了一个 RDD2，而 RDD2 中的数据仅仅是 RDD1 中的 value 值而已。也就是说，RDD2 是 RDD1 的子集。错误的做法：

```
1.  JavaPairRDD<Long, String> RDD1 = ...
2.  JavaRDD<String> RDD2 = RDD1.map(...)
3.  //分别对 RDD1 和 RDD2 执行了不同的算子操作
4.  RDD1.reduceByKey(...)
5.  RDD2.map(...)
```

上面这个 case 中，其实 RDD1 和 RDD2 的区别无非就是数据格式不同而已，RDD2 的数据完全就是 RDD1 的子集而已，却创建了两个 RDD，并对两个 RDD 都执行了一次算子操作。此时会因为对 RDD1 执行 map 算子来创建 RDD2，而多执行一次算子操作，进而增加性能开销。其实，在这种情况下完全可以复用同一个 RDD。可以使用 RDD1，既做 reduceByKey 操作，也做 map 操作。进行第二个 map 操作时，只使用每个数据的 tuple._2，也就是 RDD1 中的 value 值。正确的做法：

```
1.   JavaPairRDD<Long, String> RDD1 = ...
2.  RDD1.reduceByKey(...)
3.  RDD1.map(tuple._2...)
```

第二种方式相较第一种方式而言，很明显减少了一次 RDD2 的计算开销。但是，到这里为止，优化还没有结束，对 RDD1 我们还是执行了两次算子操作，RDD1 实际上还是会被计算两次。因此还需要配合“原则三：对多次使用的 RDD 进行持久化”进行使用，才能保证一个 RDD 被多次使用时只被计算一次。

原则三：对多次使用的 RDD 进行持久化。

对多次使用的 RDD 进行持久化的代码示例：如果要对一个 RDD 进行持久化，只要对这个 RDD 调用 Cache()和 Persist()即可。Cache 方法表示：使用非序列化的方式将 RDD 中的数据全部尝试持久化到内存中。此时再对 RDD1 执行两次算子操作时，只有在第一次执行 map 算子时，才会将这个 RDD1 从源头处计算一次。第二次执行 reduce 算子时，就会直接从内存中提取数据进行计算，不会重复计算一个 RDD。正确的做法：

```
1.   val RDD1 = sc.textFile("hdfs://192.168.0.1:9000/hello.txt").Cache()
2.  RDD1.map(...)
3.  RDD1.reduce(...)
```

Persist 方法表示：手动选择持久化级别，并使用指定的方式进行持久化。例如，StorageLevel.MEMORY_AND_DISK_SER 表示，内存充足时优先持久化到内存中，内存不充足时持久化到磁盘文件中。而且其中的_SER 后缀表示，使用序列化的方式来保存 RDD 数据，此时 RDD 中的每个 Partition 都会序列化成一个大的字节数组，然后再持久化到内存或磁盘中。序列化的方式可以减少持久化的数据对内存/磁盘的占用量，进而避免内存被持久化数据占用过多，从而发生频繁的垃圾回收（Garbage Collection，GC）。

```
1.  val  RDD1  =  sc.textFile("hdfs://192.168.0.1:9000/hello.txt").Persist
    (StorageLevel.MEMORY_AND_DISK_SER)
2.  RDD1.map(...)
3.  RDD1.reduce(...)
```

第 21 章 Spark 频繁遇到的性能问题及调优技巧

本章主要讲解 Spark 频繁遇到的性能问题及调优技巧。21.1 节讲解使用 BroadCast 广播大变量和业务配置信息原理和案例实战；21.2 节讲解使用 Kryo 取代 Scala 默认的序列器原理和案例；21.3 节讲解使用 FastUtil 优化 JVM 数据格式解析和案例；21.4 节讲解 Persist 及 checkpoint 使用时的正误方式；21.5 节讲解序列化导致的报错原因解析和调优；21.6 节讲解算子返回 NULL 产生的问题及解决办法。

21.1 使用 BroadCast 广播大变量和业务配置信息原理和案例实战

本节讲解使用 BroadCast 广播大变量和业务配置信息原理；使用 BroadCast 广播大变量和业务配置信息案例实战。

21.1.1 使用 BroadCast 广播大变量和业务配置信息原理

一个 Spark Application 的 Driver 进程其实就是我们写的 Spark 作业，打包成 Jar 运行起来的主进程。如果有一个 100MB 的 map（随机抽取的 map），创建 10 个副本，网络传输，分到 10 个机器上，则占用了 1GB 内存。这是不必要的网络消耗和内存消耗。如果你是从哪个表里面读取了一些维度数据，比方说，所有商品的品类的信息，在某个算子函数中使用到 100MB。如果有 1000 个 Task，就会有 100GB 的数据，要进行网络传输，集群瞬间性能下降。

广播变量允许程序员将一个只读的变量缓存在每台机器上，而不用在任务之间传递变量。广播变量可被用于有效地给每个节点一个大输入数据集的副本。Spark 还尝试使用高效的广播算法来分发变量，进而减少通信开销。Spark 的动作通过一系列的步骤执行，这些步骤由分布式的洗牌操作分开。Spark 自动地广播每个步骤、每个任务需要的通用数据。这些广播数据被序列化地缓存，在运行任务前被反序列化出来。这意味着当我们需要在多个阶段的任务之间使用相同的数据，或者以反序列化形式缓存数据是十分重要的时候，显式地创建广播变量才有用。

广播变量的好处：不是每个 Task 一份变量副本，而是变成每个节点的 Executor 才一份副本。这样，就可以让变量产生的副本大大减少。

广播变量，初始的时候，就在 Driver 上有一份副本。Task 在运行的时候，想要使用广播变量中的数据，此时首先会在自己本地的 Executor 对应的 BlockManager 中，尝试获取变量副本；如果本地没有，BlockManager 也许会从远程的 Driver 上去获取变量副本；也有可能从距离比较近的其他节点的 Executor 的 BlockManager 上去获取，并保存在本地的 BlockManager 中；BlockManager 负责管理某个 Executor 对应的内存和磁盘上的数据，此后这个 Executor 上的 Task，都会直接使用本地的 BlockManager 中的副本。

例如，50 个 Executor，1000 个 Task。一个 map 有 10MB。默认情况下，1000 个 Task，1000 份副本。10GB 的数据，网络传输，在集群中耗费 10GB 的内存资源。

如果使用了广播变量。50 个 Executor，50 个副本。500MB 的数据，网络传输，而且不一定都是从 Driver 传输到每个节点，还可能是就近从最近的节点的 Executor 的 BlockManager 上拉取变量副本，网络传输速度大大增加；内存消耗为 500MB。

21.1.2　使用 BroadCast 广播大变量和业务配置信息案例实战

广播变量使用 SparkContext 的 broadcast 方法，传入要广播的变量即可。

```
1.  Final Broadcast<Map<String, Map<String, IntList>>> broadcast =
    sc.broadcast(fastutilDateHourExtractMap);
```

使用广播变量的时候，直接调用广播变量（Broadcast 类型）的 value() / getValue()，可以获取到之前封装的广播变量。

```
1.  Map<String, Map<String, IntList>> dateHourExtractMap = broadcast
    .value();
```

21.2　使用 Kryo 取代 Scala 默认的序列器原理和案例实战

本节讲解使用 Kryo 取代 Scala 默认的序列器原理；使用 Kryo 取代 Scala 默认的序列器案例实战。

21.2.1　使用 Kryo 取代 Scala 默认的序列器原理

序列化对于提高分布式程序的性能起到非常重要的作用。一个不好的序列化方式（如序列化模式的速度非常慢或者序列化结果非常大）会极大地降低计算速度。很多情况下，这是优化 Spark 应用的第一选择。Spark 试图在方便和性能之间获取一个平衡。Spark 提供了两个序列化类库。

Java 序列化：默认情况下，Spark 采用 Java 的 ObjectOutputStream 序列化一个对象。该方式适用于所有实现了 java.io.Serializable 的类。通过继承 java.io.Externalizable，能进一步控制序列化的性能。Java 序列化非常灵活，但是速度较慢，在某些情况下序列化的结果也比较大。

Kryo 序列化：Spark 也能使用 Kryo（版本 2）序列化对象。Kryo 不但速度极快，而且产生的结果更紧凑（通常能提高 10 倍）。Kryo 的缺点是不支持所有类型，为了更好的性能，需要提前注册程序中使用的类（class）。

可以在创建 SparkContext 前，通过调用 System.setProperty("Spark.serializer", "Spark.Kryo Serializer")，将序列化方式切换成 Kryo。Kryo 不能成为默认方式的唯一原因是需要用户进行注册；但是，对于任何“网络密集型”（network-intensive）的应用，建议都采用该方式。

21.2.2　使用 Kryo 取代 Scala 默认的序列器案例实战

KryoSerialization 速度快，可以配置为任何 org.apache.Spark.serializer 的子类，但 Kryo 也不支持所有实现了 java.io.Serializable 接口的类型，它需要在程序中 register 需要序列化的类型，以得到最佳性能。

SparkConf 初始化时使用以下语句来使用 Kryo。

```
1.  conf.set("Spark.serializer", "org.apache.Spark.serializer.KryoSerializer")
```

这个设置不仅控制各个 Worker 节点之间的混洗数据序列化格式，同时还控制 RDD 存到磁盘上的序列化格式。需要在使用时注册需要序列化的类型，建议在对网络敏感的应用场景下使用 Kryo。

如果自定义类型需要使用 Kryo 序列化，可以用 registerKryoClasses 方法先注册。

```
1.   val conf = new SparkConf.setMaster(...).setAppName(...)
2.  conf.registerKryoClasses(Array(classOf[MyClass1], classOf[MyClass2]))
3.  val sc = new SparkContext(conf)
```

最后，如果不注册需要序列化的自定义类型，Kryo 也能工作，不过，每个对象实例的序列化结果都会包含一份完整的类名，这有点浪费空间。

21.3　使用 FastUtil 优化 JVM 数据格式解析和案例实战

本节讲解使用 FastUtil 优化 JVM 数据格式解析以及使用 FastUtil 优化 JVM 数据格式案例实战。

21.3.1　使用 FastUtil 优化 JVM 数据格式解析

FastUtil 是扩展了 Java 标准集合框架（Map、List、Set；HashMap、ArrayList、HashSet）的类库，提供了特殊类型的 map、set、list 和 queue。

FastUtil 能够提供更小的内存占用，更快的存取速度；我们使用 FastUtil 提供的集合类，来替代自己平时使用的 JDK 的原生的 Map、List、Set，好处在于，FastUtil 集合类可以减小内存的占用，并且在进行集合的遍历、根据索引（或者 key）获取元素的值和设置元素的值时，提供更快的存取速度。

FastUtil 也提供了 64 位的 array、set 和 list，以及高性能快速的、实用的 I/O 类，来处理二进制和文本类型的文件。

FastUtil 最新版本要求 Java 7 以及以上版本。

FastUtil 的每种集合类型，都实现了对应的 Java 中的标准接口（如 FastUtil 的 map，实现了 Java 的 Map 接口），因此可以直接放入已有系统的任何代码中。

FastUtil 还提供了一些 JDK 标准类库中没有的额外功能（如双向迭代器）。

FastUtil 除了对象和原始类型为元素的集合，FastUtil 也提供引用类型的支持，但是对引用类型是使用等于号（=）进行比较的，而不是 equals 方法。

FastUtil 尽量提供了在任何场景下都是速度最快的集合类库。

在 Spark 中应用 FastUtil 的场景如下。

（1）如果算子函数使用了外部变量：第一，可以使用 Broadcast 广播变量优化；第二，可以使用 Kryo 序列化类库，提升序列化性能和效率；第三，如果外部变量是某种比较大的集合，那么可以考虑使用 FastUtil 改写外部变量，首先从源头上就减少内存的占用，通过广播变量进一步减少内存占用，再通过 Kryo 序列化类库进一步减少内存占用。

（2）在算子函数里，也就是 Task 要执行的计算逻辑里面，如果逻辑中要创建比较大的 Map、List 等集合，可能会占用较大的内存空间，而且可能涉及消耗性能的遍历、存取等集合操作；那么，此时可以考虑将这些集合类型使用 FastUtil 类库重写，使用了 FastUtil 集合类以后，就可以在一定程度上减少 Task 创建出来的集合类型的内存占用，避免 Executor 内存频繁占满，频繁唤起 GC，导致性能下降。

FastUtil 调优虽然能起到一些作用，但作用并不是那么强大，性能不会得到惊人的提升。

21.3.2　使用 FastUtil 优化 JVM 数据格式案例实战

使用 FastUtil 优化 JVM 数据格式案例实战首先在 pom.xml 中引用 FastUtil 的包，拉取 Jar 包的时间可能比较长。

```
1.   <dependency>
2.  <groupId>fastutil</groupId>
3.  <artifactId>fastutil</artifactId>
4.  <version>5.0.9</version>
5.  </dependency>
```

FastUtil 优化基本都类似于 IntList 的格式。

```
1.   List<Integer> => IntList
```

前缀就是集合的元素类型；特殊的就是 Map 和 Int2IntMap，代表了 key-value 映射的元素类型。除此之外，还支持 object、reference。

FastUtil 的示例代码如下：

```
1.  import it.unimi.dsi.fastutil.ints.IntArrayList;
2.  import it.unimi.dsi.fastutil.ints.IntList;
3.
4.  SparkConf conf = new SparkConf().setAppName(Constants.SPARK_APP_NAME_
    SESSION)
5.  .registerKryoClasses(new Class[]{
6.  CategorySortKey.class,
7.  IntList.class});
8.  //注意，要在 SparkConf 里注册自定义的类
9.
```

```
10. /**
     * FastUtil 的使用很简单，如 List<Integer>的 list，对应到 FastUtil，就是 IntList
11.   */
12.
13. Map<String, Map<String, IntList>> fastutilDateHourExtractMap = new
    HashMap<String, Map<String, IntList>>();
14. for(Map.Entry<String, Map<String, List<Integer>>> dateHourExtractEntry :
    dateHourExtractMap.entrySet()) {
15.   String date = dateHourExtractEntry.getKey();
16.   Map<String, List<Integer>> hourExtractMap = dateHourExtractEntry
      .getValue();
17.   Map<String, IntList> fastutilHourExtractMap = new HashMap<String,
      IntList>();
18.  for(Map.Entry<String, List<Integer>> hourExtractEntry : hourExtractMap
     .entrySet()) {
19.     String hour = hourExtractEntry.getKey();
20.      List<Integer> extractList = hourExtractEntry.getValue();
21.      IntList fastutilExtractList = new IntArrayList();
22.      for(int i = 0; i < extractList.size(); i++) {
23.         fastutilExtractList.add(extractList.get(i));
24.         }
25.      fastutilHourExtractMap.put(hour, fastutilExtractList);
26.       }
27.    fastutilDateHourExtractMap.put(date, fastutilHourExtractMap);
28. }
```

21.4　Persist 及 checkpoint 使用时的正误方式

1. Persist的正误使用方式

Spark 支持缓存中间结果到内存。当高速缓存 RDD，Spark 分区能存于计算节点的内存或磁盘（依赖于请求方式）。Spark 将从持久化分区中返回数据快速做进一步 action（行动）处理。

调用 Cache 或 Persist 方法将 RDD 持久化。当执行第一个 action（行动）时 RDD 被缓存。因此在下面例子中，仅仅 collect action（行动）将从预先计算的值获利。

```
1.  myRDD.Cache()
2.  myRDD.count()
3.  myRDD.collect()
```

前面提到 Persist 或 Cache 两个方法，让 Spark 知道在计算后去临时存储 RDD。它们默认存储 RDD 到内存中。它们之间的不同是，Persist 也提供了 API 指定存储级别，方便在某种情形下改变默认行为。

Spark 有许多种持久化 RDD 的方式。

- MEMORY_ONLY：当用这种类型的存储级别时，Spark 将 RDD 作为未序列化的 Java 对象存于内存。如果 Spark 估算不是所有的分区能在内存中容纳，Spark 将不保存数据。如果在后续的处理管道中需要数据，将基于 RDD 血缘关系重新计算。此方式有一个缺点：比较其他储存级，将使用大量内存，如果缓存过多小对象，将对垃圾回收产生压力。用这个存储级别缓存 RDD，能用下面的方法。

```
1.   myRDD.Cache()
```

```
2.  myRDD.Persist()
3.  myRDD.Persist(StorageLevel.MEMORY_ONLY)
```

- MEMORY_ONLY_SER：RDD 也仅存于内存，但是以 Java 序列对象的形式存储。它在空间上更加高效，因为数据将更加紧凑，因此将能缓存更多数据。MEMORY_ONLY_SER 存储级别的缺点是 CPU 密集的（要进行密集的读操作，更耗费 CPU 资源），因为对象在读写时会序列化与解序列化。序列化的 Java 对象按分区存为字节数组。缓冲 RDD 时使用的序列化器的选择很重要。能用下面的方法以序列化形式仅存于内存。

```
1.  myRDD.Persist(StorageLevel.MEMORY_ONLY_SER)
```

- MEMORY_AND_DISK：这种存储级别与仅存于内存相似，Spark 将尝试在内存中缓存整个 RDD 作为未序列化的 Java 对象，但是这次如果有分区不能完全容纳于内存，将写到磁盘。如果这些数据稍后在其他操作中使用，它们将不用重新计算，而是直接从磁盘读取。这个存储级别依然使用大量内存，除此之外，它也隐含了 CPU 与磁盘 I/O 上的载荷。我们需要思考哪种方式更加昂贵：是写内存容纳不下的分区到硬盘需要时读取，还是每次使用它们再重新计算。

用下面的代码将 RDD 缓存于内存与磁盘。

```
1.  myRDD.Persist(StorageLevel.MEMORY_AND_DISK)
```

- MEMORY_AND_DISK_SER：此存储级别与 MEMORY_AND_DISK 相似；它们的不同仅仅在于数据以序列化形式存于内存。这次 RDD 更多的分区将能被内存容纳，因为它们是更加紧凑的，写于磁盘也会占更少空间。此选项对比 MEMORY_AND_DISK 是更加 CPU 密集的。MEMORY_AND_DISK_SER 优先尝试将数据缓存在内存中，内存缓存不下才写入磁盘，以序列化对象格式保存于内存及磁盘的方法如下。

```
1.   myRDD.Persist(StorageLevel.MEMORY_AND_DISK_SER)
```

- DISK_ONLY：用此选项将避免内存消耗，空间消耗也很低，因为数据是序列化格式的。因为整个数据集不得不解序列化、序列化、从磁盘读写，使得 CPU 载荷比较高，也产生磁盘 io 压力。例如：

```
1.   myRDD.Persist(StorageLevel.DISK_ONLY)
```

- 两节点缓存。

所有上面的存储级别能应用到集群的两结点。RDD 的每个分区将被复制到两节点的内存或硬盘。API 用法：

```
1.  myRDD.Persist(StorageLevel.MEMORY_ONLY_2)
2.  myRDD.Persist(StorageLevel.MEMORY_ONY_SER_2)
3.  myRDD.Persist(StorageLevel.MEMORY_AND_DISK_2)
4.  myRDD.Persist(StorageLevel.MEMROY_AND_DISK_SER_2)
5.  myRDD.Persist(StorageLevel.DISK_ONLY_2)
```

- 堆外存储。

在这种情形下，序列化 RDD 将存于 Alluxio(Tachyon)的堆外存储。这种选项有很多好处。最重要的是，能在 Executor 及其他应用中共享大量内存，垃圾回收带来的消耗被减少。用堆外持久化技术避免在 Executor 崩溃的情形下丢失内存缓存。

```
1.   myRDD.Persist(StorageLevel.OFF_HEAP)
```

如果 Spark 应用持久化需要的内存多过实际提供的，使用的分区会尽可能最少地从内存中驱逐。然而，Spark Cache（缓存）是故障容忍的，它将重新计算丢失的分区，不需要担心应用会崩溃。但是需要特别在意缓存什么数据及缓存多少。当使用缓存的数据集去提升应用性能，最终可能会增加执行时间。换一种方式讲，如果缓存了过多不需要的数据，有用的分区将被驱逐，再进一步执行 action（行动）时，它将被重新计算。

2．使用checkpoint错误及正确的方式

建议在执行 checkpoint()方法前先对 RDD 进行 Persist 操作。

为什么要这样呢？因为 checkpoint 会触发一个 Job，如果执行 checkpoint 的 RDD 是由其他 RDD 经过许多计算转换过来的，如果没有 Persist 这个 RDD，那么又要从头开始计算该 RDD，也就是做了重复的计算工作，所以建议先 Persist RDD，然后再 checkpoint，checkpoint 会丢弃该 RDD 的以前的依赖关系，使该 RDD 成为顶层父 RDD，这样在失败的时候只需要恢复该 RDD，而不需要重新计算该 RDD，这在迭代计算中是很有用的。假设你在迭代 1000 次的计算中在第 999 次失败了，而若没有 checkpoint，则只能重新开始恢复，如果恰好在第 998 次迭代的时候做了一个 checkpoint，那么只需要恢复第 998 次产生的 RDD，然后再执行 2 次迭代完成总共 1000 次的迭代，这样效率就很高，比较适用于迭代计算非常复杂的情况，也就是恢复计算代价非常高的情况，适当进行 checkpoint 会有很大的好处。

21.5　序列化导致的报错原因解析和调优实战

本节讲解序列化导致的报错原因解析以及调优实战。

21.5.1　报错原因解析

如果出现“org.apache.Spark.SparkException: Task not serializable”错误，一般是因为在 map、filter 等的参数使用了外部的变量，但是这个变量不能序列化（不是说不可以引用外部变量，只是要做好序列化工作）。其中最普遍的情形是：当引用了某个类（经常是当前类）的成员函数或变量时，会导致这个类的所有成员（整个类）都需要支持序列化。虽然许多情形下，当前类使用了“extends Serializable”声明支持序列化，但是由于某些字段不支持序列化，仍然会导致整个类序列化时出现问题，最终导致出现 Task 未序列化问题。

21.5.2　调优实战

由于 Spark 程序中的 map、filter 等算子内部引用了类成员函数或变量导致需要该类所有成员都支持序列化，又由于该类某些成员变量不支持序列化，最终引发 Task 无法序列化问题。为了验证上述原因，我们编写了一个实例程序，如下所示。该类的功能是从域名列表中（RDD）

过滤得到特定顶级域名（rootDomain，如.com,.cn,.org）的域名列表，而该特定顶级域名需在要函数调用时指定。

```
1.  class MyTest1(conf:String) extends Serializable{
2.   val list = List("a.com", "www.b.com", "a.cn", "a.com.cn", "a.org");
3. private val SparkConf = new SparkConf().setAppName("AppName");
4. private val sc = new SparkContext(SparkConf);
5. val RDD = sc.parallelize(list);
6. private val rootDomain = conf
7. def getResult(): Array[(String)] = {
8. val result = RDD.filter(item => item.contains(rootDomain))
9. result.take(result.count().toInt)
10.  }
11. }
```

依据上述分析的原因，由于依赖了当前类的成员变量，所以导致当前类全部需要序列化。当前类的某些字段未做好序列化，导致出错。实际情况与分析的原因一致，运行过程中出现的错误如下所示。分析下面的错误报告，可知错误是由于 sc（SparkContext）引起的。

```
1.  Exception in thread "main" org.apache.Spark.SparkException: Task not
   serializable
2. at org.apache.Spark.util.ClosureCleaner$.ensureSerializable(ClosureCleaner
   .scala:166)
3. at org.apache.Spark.util.ClosureCleaner$.clean(ClosureCleaner.scala:158)
4. at org.apache.Spark.SparkContext.clean(**SparkContext**.scala:1435)
5. ......
6. Caused by: java.io.NotSerializableException: org.apache.Spark.SparkContext
7.     - field (class "com.ntci.test.MyTest1", name: "sc", type: "class
        org.apache.Spark.SparkContext")
8.     - object (class "com.ntci.test.MyTest1", com.ntci.test.MyTest1@63700353)
9.     - field (class "com.ntci.test.MyTest1$$anonfun$1", name: "$outer",
         type: "class com.ntci.test.MyTest1")
```

为了验证上述结论，将不需要序列化的成员变量使用关键字“@transient”标注，表示不序列化当前类中的这两个成员变量，再次执行函数，同样报错。

```
1.   Exception in thread "main" org.apache.Spark.SparkException: Task not
     serializable
2. at org.apache.Spark.util.ClosureCleaner$.ensureSerializable(ClosureCleaner
   .scala:166)
3. ......
4.  Caused by: java.io.NotSerializableException: org.apache.Spark.SparkConf
5.     - field (class "com.ntci.test.MyTest1", name: "SparkConf", type:
       "class org.apache.Spark.**SparkConf**")
6.     - object (class "com.ntci.test.MyTest1", com.ntci.test.MyTest1@
       6107799e)
```

虽然错误原因相同，但是这次导致错误的字段是 SparkConf（SparkConf)。使用同样的“@transient”标注方式，将 sc（SparkContext）和 SparkConf（SparkConf）都标注为不需序列化，再次执行时，程序正常。

```
1.  class MyTest1(conf:String) extends Serializable{
2.   val list = List("a.com", "www.b.com", "a.cn", "a.com.cn", "a.org");
3.   @transient
4. private val SparkConf = new SparkConf().setAppName("AppName");
5.   @transient
6. private val sc = new SparkContext(SparkConf);
```

```
7.  val RDD = sc.parallelize(list);
8.
9.  private val rootDomain = conf
10.
11. def getResult(): Array[(String)] = {
12.
13. val result = RDD.filter(item => item.contains(rootDomain))
14. result.take(result.count().toInt)
15.   }
16. }
```

所以，通过上面的例子可以得到结论：由于 Spark 程序中的 map、filter 等算子内部引用了类成员函数或变量，导致该类所有成员都需要支持序列化，又由于该类某些成员变量不支持序列化，最终引发 Task 无法序列化问题。相反，对类中那些不支持序列化问题的成员变量标注后，使得整个类能够正常序列化，最终消除 Task 未序列化问题。

引用成员函数的实例分析：

成员变量与成员函数对序列化的影响相同，即引用了某类的成员函数，会导致该类所有成员都支持序列化。为了验证这个假设，我们在 map 中使用了当前类的一个成员函数，作用是如果当前域名没有以“www.”开头，那么就在域名头部添加“www.”前缀（注：由于 rootDomain 是在 getResult 函数内部定义的，所以就不存在引用类成员变量的问题，也就不存在和排除了上一个例子讨论和引发的问题。因此，这个例子主要讨论成员函数引用的影响；此外，不直接引用类成员变量也是解决这类问题的一个手段，如本例中为了消除成员变量的影响而在函数内部定义变量的这种做法）。下面的代码同样会报错，同上面的例子一样，由于当前类中的 sc(SparkContext) 和 SparkConf(SparkConf) 两个成员变量没有做好序列化处理，导致当前类的序列化出现问题。

```
1.   class MyTest1(conf:String)  extends Serializable{
2.    val list = List("a.com", "www.b.com", "a.cn", "a.com.cn", "a.org");
3.  private val SparkConf = new SparkConf().setAppName("AppName");
4.  private val sc = new SparkContext(SparkConf);
5.  val RDD = sc.parallelize(list);
6.
7.  def getResult(): Array[(String)] = {
8.  val rootDomain = conf
9.  val result = RDD.filter(item => item.contains(rootDomain))
10.     .map(item => addWWW(item))
11. result.take(result.count().toInt)
12.   }
13. def addWWW(str:String):String = {
14. if(str.startsWith("www."))
15. str
16. else
17.       "www."+str
18.   }
19. }
```

如同前面的做法，将 sc(SparkContext) 和 SparkConf(SparkConf) 两个成员变量使用“@transient”标注后，使当前类不序列化这两个变量，则程序可以正常运行。此外，与成员变量稍有不同的是，由于该成员函数不依赖特定的成员变量，因此可以定义在 scala 的 object 中（类似于 Java 中的 static 函数），这样也取消了对特定类的依赖。如下面的例子所示，将 addWWW 放到一个 object 对象（UtilTool）中，在 filter 操作中直接调用，这样处理以后，程序能够正常运行。

```
1.  def getResult(): Array[(String)] = {
2.  val rootDomain = conf
3.  val result = RDD.filter(item => item.contains(rootDomain))
4.     .map(item => UtilTool.addWWW(item))
5.  result.take(result.count().toInt)
6.   }
7.  object UtilTool {
8.  def addWWW(str:String):String = {
9.  if(str.startsWith("www."))
10. str
11. else
12.      "www."+str
13.   }
14. }
```

如上所述，引用了某类成员函数，会导致该类及所有成员都需要支持序列化。因此，对于使用了某类成员变量或函数的情形，首先该类需要序列化（extends Serializable），同时需要对某些不需要序列化的成员变量进行标记，以避免对序列化造成影响。对于上面两个例子，由于引用了该类的成员变量或函数，所以导致该类以及所有成员支持序列化，为了消除某些成员变量对序列化的影响，使用“@transient”进行标注。

为了进一步验证关于整个类需要序列化的假设，这里在上面例子使用“@transient”标注后并且能正常运行的代码基础上，将类序列化的相关代码删除（去掉 extends Serializable），这样程序执行会报该类为序列化的错误，如下所示。所以，这个实例说明了上面的假设。

```
1.  Caused by: java.io.NotSerializableException: com.ntci.test.MyTest1
2.     - field (class "com.ntci.test.MyTest1$$anonfun$1", name: "$outer",
       type: "class com.ntci.test.MyTest1")
```

通过上面的例子可以说明：map 等算子内部可以引用外部变量和某类的成员变量，但是要做好该类的序列化处理。首先是该类需要继承 Serializable 类，此外，对类中某些序列化会出错的成员变量做好处理，这也是 Task 未序列化问题的主要原因。出现这类问题，首先查看未能序列化的成员变量是哪个，对于可以不需要序列化的成员变量，可使用“@transient”标注。

此外，也不是 map 操作所在的类必须序列化不可（继承 Serializable 类），对于不需要引用某类成员变量或函数的情形，就不会要求相应的类必须实现序列化，如下面的例子所示。filter 操作内部没有引用任何类的成员变量或函数，因此当前类不用序列化，程序可正常执行。

```
1.  class MyTest1(conf:String) {
2.   val list = List("a.com", "www.b.com", "a.cn", "a.com.cn", "a.org");
3.  private val SparkConf = new SparkConf().setAppName("AppName");
4.  private val sc = new SparkContext(SparkConf);
5.  val RDD = sc.parallelize(list);
6.
7.  def getResult(): Array[(String)] = {
8.  val rootDomain = conf
9.  val result = RDD.filter(item => item.contains(rootDomain))
10. result.take(result.count().toInt)
11.   }
12. }
```

解决办法与编程建议：承上所述，这个问题主要是引用了某类的成员变量或函数，并且

相应的类没有做好序列化处理导致的。解决这个问题有以下两种方法：不在（或不直接在）map 等闭包内部直接引用某类（通常是当前类）的成员函数或成员变量；如果引用了某类的成员函数或变量，则需对相应的类做好序列化处理。

（1）不在（或不直接在）map 等闭包内部直接引用某类成员函数或成员变量。

① 对于依赖某类成员变量的情形。

- 如果程序依赖的值相对固定，可取固定的值，或定义在 map、filter 等操作内部，或定义在 scala object 对象中（类似于 Java 中的 static 变量）。
- 如果依赖值需要程序调用时动态指定（以函数参数形式），则在 map、filter 等操作时，可不直接引用该成员变量，而是在类似上面例子的 getResult 函数中根据成员变量的值重新定义一个局部变量，这样，map 等算子就无须引用类的成员变量。

② 对于依赖某类成员函数的情形。

如果函数功能独立，可定义在 scala object 对象中（类似于 Java 中的 static 方法），这样就无须特定的类。

（2）如果引用了某类的成员函数或变量，则需对相应的类做好序列化处理。

对于这种情况，需对该类做好序列化处理，首先该类继承序列化类，然后对不能序列化的成员变量使用“@transient”标注，告诉编译器不需要序列化。

此外，如果可以，可将依赖的变量独立放到一个小的 Class 中，让这个 Class 支持序列化，这样做可以减少网络传输量，提高效率。

21.6　算子返回 NULL 产生的问题及解决办法

有些场景下并不需要返回具体的值，这时往往会返回 NULL 值，但有时在下一步的 RDD 操作中要求 RDD 的元素不能为 NULL。如果是 NULL，就会抛出异常。这时可以在返回 NULL 的基础上，在下一步的时候通过 Option 进行模式匹配。

还有一种方法，可以返回一个特定的值，然后在下一步的业务逻辑操作前进行 filter 操作，把该特定的值过滤掉，这样就在无形中化解了 NULL 值的问题。

第 22 章　Spark 集群资源分配及并行度调优最佳实践

本章讲解 Spark 集群资源分配及并行度调优最佳实践。22.1 节讲解实际生产环境下每个 Executor 内存及 CPU 的具体配置及原因；22.2 节讲解 Spark 并行度设置最佳实践。

22.1　实际生产环境下每个 Executor 内存及 CPU 的具体配置及原因

本节讲解内存的具体配置及原因；以及实际生产环境下一般每个 Executor 的 CPU 的具体配置及原因。

22.1.1　内存的具体配置及原因

YARN Container 里实际的内存结构，即 YARN-Cluster 模式下 Executor 内存使用的实现方式，如图 22-1 所示。

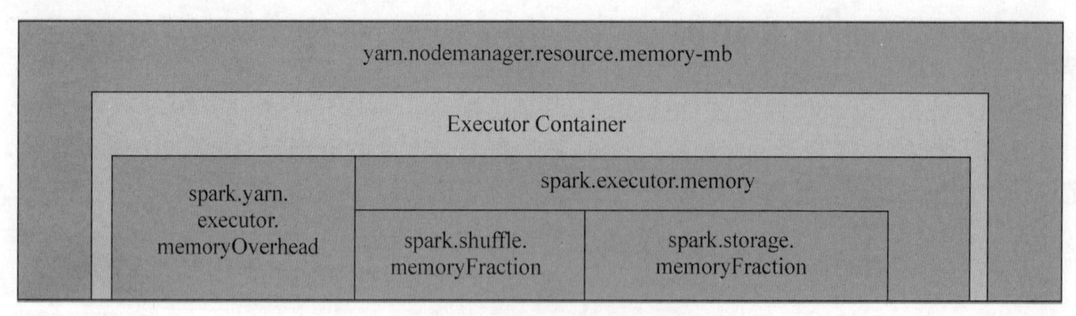

图 22-1　YARN 内存示意图

YARN.nodemanager.resource.memory-mb 控制在每个节点上 Container 能够使用的最大内存。

可以用 Spark.Executor.memory 来配置每个 Executor 使用的内存总量。例如：

```
--Executor-memory 6GB
```

Executor 可使用的内存中，又主要分为以下三块。

第一块是让 Task 执行用户编写的代码使用，默认占 Executor 总内存的 20%。

第二块是让 Task 通过 Shuffle 过程拉取了上一个 Stage 的 Task 的输出后，进行聚合等操作时使用，默认也占 Executor 总内存的 20%；用 Spark.Shuffle.memoryFraction 可配置比例。

第三块是让 RDD Cache 使用，默认占 Executor 总内存的 60%；用 Spark.storage .memoryFraction 可配置比例。官方文档建议这个比值不要超过 JVM Old Gen 区域的比值。因为 RDD Cache 数据通常都长期驻留于内存，理论上最终会被转移到 Old Gen 区域（如果该 RDD 还没有被删除），如果这部分数据允许的尺寸太大，势必把 Old Gen 区域占满，造成频繁地 Full GC。如何调整这个比值，取决于用户的应用对数据的使用模式和数据的规模，粗略来说，如果频繁发生 Full GC，可以考虑降低这个比值，这样 RDD Cache 可用的内存空间减少（剩下的部分 Cache 数据就需要通过 Disk Store 写到磁盘上），会带来一定的性能损失，但是腾出了更多的内存空间用于执行任务，减少了 Full GC 发生的次数，反而可能改善程序运行的整体性能。如果有过多的 Minor GC，而不是 Full GC，那么可以为 Eden 分配更大的内存。如果 Eden 的大小为 E，那么可以通过设置-Xmn = 4/3×E（将内存扩大到 4/3 是考虑到 Survivor 所需要的空间）。举一个例子，如任务从 HDFS 读取数据，那么任务需要的内存空间可以从读取的 block 数量估算出来。解压后的 block 通常为解压前的 2～3 倍，所以，如果同时执行 3～4 个任务，block 的大小为 64MB，我们可以估算出 Eden 的大小为 4×3×64MB。

还有一个 YARN 特定的参数非常重要。即 Spark.YARN.Executor.memoryOverhead 参数。系统分配可选的堆外存储。因此，当用 Spark.Executor.memory 分配 8GB 的内存时，YARN 因为 Spark.YARN.Executor.memoryOverhead 实际还分配给 Container 更多的内存，默认为 Spark.Executor.memory 的 7%及 384MB 的最大值。

虽然我们定义了分配给 Executor 的总内存，但是内存使用方式并不明显。Spark 对于不同的内部功能，需要不同的内存池。有两个主要分支，即堆内存和堆外内存。在 JVM 管理的堆范围内，Spark 对 3 个独立的内存池进行权衡。图 22-2 所示为 Spark 内存示意图。

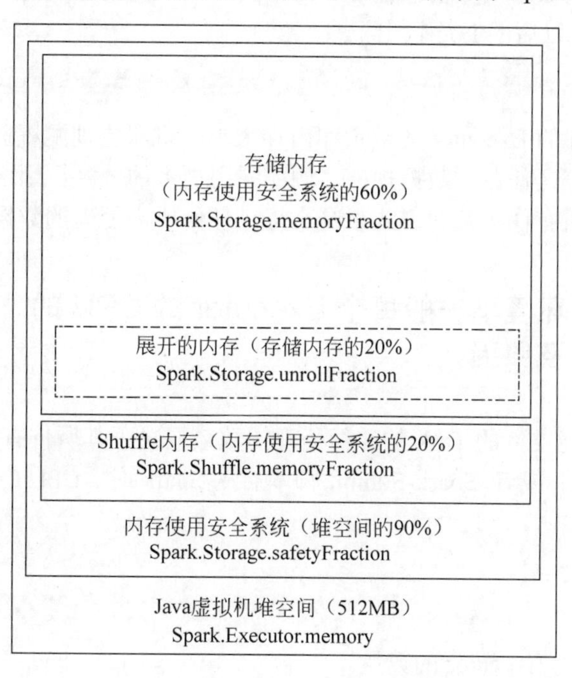

图 22-2　Spark 内存示意图

Spark 中可用内存第一部分用于持久化 RDDS 的内存存储。持久化 RDD 是 RDD 通过调用 Cache 或 Persist 被安放在内存中的 RDD，依赖于函数的参数，可能仅存储 RDD 部分至内存。将 RDD 持久化于内存，而不是每次从磁盘读写，对性能有显著的提升，但由于可用内存量一般小于处理的数据量，Spark 支持持久化数据到内存的不同配置项，例如允许溢出磁盘。在第一部分的内存池内有一块叫 Unroll 的内存，这块内存在数据从序列化到非序列化转换操作中使用。

Executor 内存的大小，和性能本身并没有直接关系，但是几乎所有运行时性能相关的内容都或多或少间接和内存大小相关。这个参数最终会被设置到 Executor 的 JVM 的 Heap 尺寸上，对应的是 Xmx 和 Xms 的值。

理论上，Executor 内存当然是多多益善，但是实际受机器配置、运行环境、资源共享、JVM GC 效率等因素的影响，还是有可能需要为它设置一个合理的大小。多大算合理，要看实际情况。

Executor 的内存基本上是由 Executor 内部的所有任务共享的，而每个 Executor 上可以支持的任务的数量取决于 Executor 管理的 CPU Core 资源的多少，因此需要了解每个任务的数据规模的大小，从而推算出每个 Executor 大致需要多少内存即可满足基本需求。

如何知道每个任务所需内存的大小呢？这个很难进行统一的衡量，因为除了数据集本身的开销，还包括算法所需各种临时内存空间的使用，而根据具体的代码算法等不同，临时内存空间的开销也不同。但是，数据集本身的大小对最终所需内存的大小还是有一定参考意义的。

通常，每个分区的数据集在内存中的大小，可能是其在磁盘上源数据大小的若干倍（不考虑源数据压缩，Java 对象相对于原始裸数据，也要算上用于管理数据的数据结构的额外开销），需要准确地知道大小，可以将 RDD Cache 在内存中，从 BlockManager 的 Log 输出可以看到每个 Cache 分区的大小（其实也是估算出来的，并不完全准确）。

例如，Cache 分区以后的 Log 输出记录：

```
1.   BlockManagerInfo: Added RDD_0_1 on disk on sr438:41134 (size: 495.3 MB)
```

反过来说，如果你的 Executor 的数量和内存大小受机器物理配置影响相对固定，那么你就需要合理规划每个分区任务的数据规模，例如采用更多的分区，用增加任务数量（进而需要更多的批次来运算所有任务）的方式来减小每个任务所需处理的数据大小。

22.1.2　实际生产环境下一般每个 Executor 的 CPU 的具体配置及原因

每个 Executor 所能支持的 Task 的并行处理的数量取决于其所持有的 CPU Core 的数量。

当通过 Spark-Shell 或者 Spark-Submit 脚本启动 Spark 时，CPU Core 数量的参数设置可以在命令行进行配置。

例如：

```
--num-Executors 10
```

10 是集群上启动的 Executor 的数量。

```
--Executor-cores 2
```

2 是每个 Executor 运行的核数，即 Executor 能最大运行的并行 Task（任务）数。Executor 的每个核共享 Executor 分配到的总内存。

22.2 Spark 并行度设置最佳实践

本节讲解并行度设置的原理和影响因素以及并行度设置最佳实践。

22.2.1 并行度设置的原理和影响因素

并行度就是 Spark 作业中，各个 Stage 的 Task 数量，也就代表了 Spark 作业在各个阶段（Stage）的并行度。

如果不调节并行度，导致并行度过低，会怎么样？

假设现在已经在 Spark-Submit 脚本里面，给我们的 Spark 作业分配了足够多的资源，例如 50 个 Executor，每个 Executor 有 10GB 内存，每个 Executor 有 3 个 CPU core。基本已经达到了集群或者 YARN 队列的资源上限。

Task 没有设置，或者设置的很少，例如就设置了 100 个 Task，50 个 Executor，每个 Executor 有 3 个 CPU core。也就是说，Application 的任何一个 Stage 运行的时候，都有总数在 150 个 CPU core 可以并行运行。但是现在只有 100 个 Task，平均分配一下，每个 Executor 分配到两个 Task，那么同时运行的 Task 只有 100 个，每个 Executor 只会并行运行两个 Task。每个 Executor 剩下的一个 CPU core 就浪费掉了。

合理的并行度的设置，应该是设置得足够大，大到可以完全合理地利用集群资源；例如上面的例子中，集群共有 150 个 CPU core，可以并行运行 150 个 Task。那么就应该将 Application 的并行度至少设置成 150，才能完全有效地利用集群资源，让 150 个 Task 并行执行；而且 Task 增加到 150 个以后，既可同时并行运行，还可让每个 Task 要处理的数据量变少；例如，总共 150GB 的数据要处理，如果是 100 个 Task，每个 Task 计算 1.5GB 的数据；现在增加到 150 个 Task，可以并行运行，而且每个 Task 主要处理 1GB 的数据就可以。

通过调整参数，可以提高集群并行度，让系统同时执行的任务更多，那么，对于相同的任务，并行度高了，可以减少轮询次数。举例说明：如果一个 Stage 有 100 个 Task，并行度为 50，那么执行完这次任务，需要轮询两次才能完成，如果并行度为 100，那么轮询一次就可以了。

22.2.2 并行度设置最佳实践

Task 数量至少设置成与 Spark Application 的总 CPU core 数量相同（最理想情况，例如总共 150 个 CPU core，分配了 150 个 Task，一起运行，差不多同一时间运行完毕）。

如果集群不能有效地被利用，则要为每个操作都设置足够高的并行度。Spark 会根据每个文件的大小自动设置运行在该文件 Map 任务的个数（也可以通过 SparkContext 的配置参数来控制）；对于分布式 Reduce 任务（如 groupByKey 或者 reduceByKey），则利用最大 RDD 的分区数。

可以通过第二个参数传入并行度（阅读文档 Spark.PairRDDFunctions）或者通过设置系统参数 Spark.default.parallelism 来改变默认值。通常，在集群中，建议为每个 CPU 核（core）分配 2～3 个任务。

事实上，官方推荐也是，把 Task 数量设置成 Spark Application 总 CPU core 数量的 2～3 倍，例如，150 个 CPU core，基本要设置 Task 数量为 300～500。

例如：

```
1.   SparkConf conf = new SparkConf()
2.  conf.set("Spark.default.parallelism", "500")
```

第 23 章 Spark 集群中 Mapper 端、Reducer 端内存调优

本章主要讲解两个方面。23.1 节讲解 Spark 集群中 Mapper 端内存调优实战；23.2 节讲解 Spark 集群中 Reducer 端内存调优实战。

23.1 Spark 集群中 Mapper 端内存调优实战

本节讲解 Spark 集群中 Mapper 端内存使用详解以及内存性能调优实战。

23.1.1 内存使用详解

Spark 集群 Shuffle 分为两部分：Mapper 端和 Reducer 端。Spark 集群中的 Shuffle 非常重要，Shuffle 的特殊之处在于它依赖于所有的数据，RDD 依赖是后面的 RDD 依赖前面的 RDD，当发生 Shuffle RDD 的时候，Reducer 端的 RDD 的每个 Partition 依赖于父 RDD 的所有的 Partition，不是固定依赖于某一个 RDD 的数据，或者某几个 Partition 的数据，它的依赖是不确定的，因此是依赖于所有的数据。假如有 100 万个 Partition，用户不会知道是依赖于其中的 50 万个 Partition，还是其中的一个 Partition，这个时候从所有 Partition 的角度考虑，就产生了 Shuffle 网络通信。

Spark 集群中 Mapper 端内存性能调优示意图如图 23-1 所示。

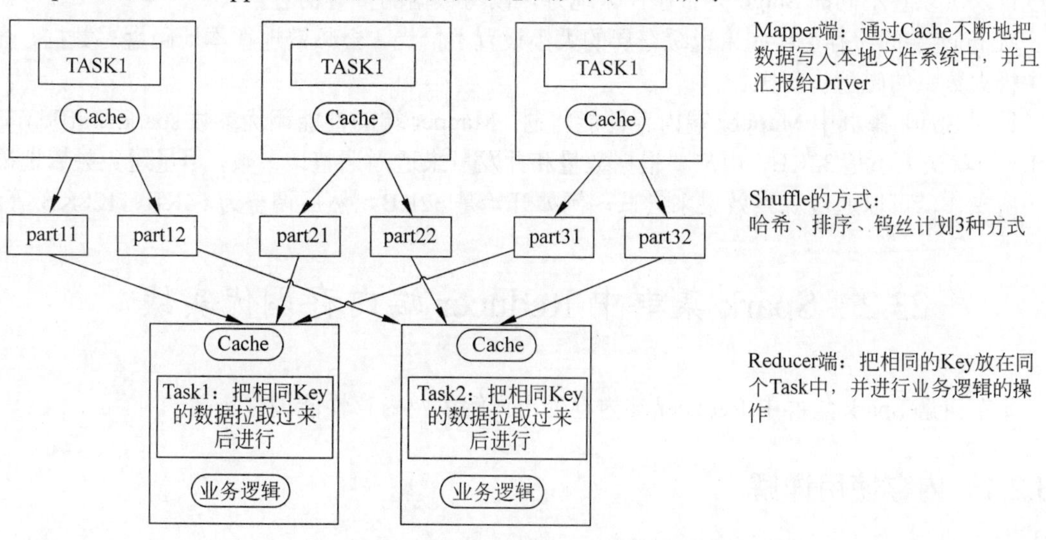

图 23-1 Spark 集群中 Mapper 端内存性能调优示意图

假设 Mapper 端有 3 个 Task：Task1、Task2、Task3，Reducer 端有两个 Task：Task1、Task2，数据传输到 Reducer 端的时候首先进行 Mapper 端的处理，Mapper 端的处理很简单，Mapper 端有一个缓存，Mapper 端会产生文件，这里将文件分成两部分，分别为 part11、part12；part21、part22；part31、part32；Shuffle 可以是哈希、排序、钨丝计划 3 种方式之一（其中，哈希方式在 Spark 2.2.0 中已不使用）。数据从 Mapper 缓存层写入文件中。Mapp 缓存层根据 Reducer 端的需要，将数据分成不同的部分，part11、part12 可能在一个文件中，也可能在两个文件中。然后在 Reducer 端抓取属于自己的数据进行 reduce 操作，Reducer 端操作的时候也有一个缓存层，是定义业务逻辑运行的地方。

Mapper 端：通过缓存层不断地把数据写入到文件系统中，并汇报给 Driver，Driver 须知道把数据写在什么地方。

Reducer 端：把相同的 Key 放在同一个 Task 中，并进行业务逻辑的操作。Reducer 端抓取数据的时候也有一个小的缓存层。

针对 Shuffle Mapper 端的过程，Shuffle Mapper 端怎么进行性能调优？性能调优点在什么地方？Mapper 端内存性能调优点在于缓存层：假设 Mapper 端的数据非常大，有 100 万个 Key，Mapper 端的缓存层大小是 16KB，每个 Task 的数据是 16GB，这时进行磁盘读写的次数会是一个非常大的数字。

Reducer 端将数据抓取过来，如果缓存空间不够，数据将溢出到磁盘上。Reducer 端也有一个缓存层，也需要在 Reducer 端进行缓存层的调优。

23.1.2　内存性能调优实战

Spark 集群中 Mapper 端内存性能调优示意图如图 23-1 所示，那么，用户怎么判断 Mapper 端要不要调优？什么时候进行 Mapper 端调优呢？

这个要看 log 和 Web UI 上面的信息来判断。log 信息可以给出重要参考；从 Web UI 的角度讲，可以看不同的 Stage 分布在什么地方，读写数据的量等内容。

Mapper 端的缓存层：如果说缓存层的大小设置不恰当，会频繁地往本地磁盘写数据，会产生极大数量的磁盘访问操作。

针对 Spark 集群中 Mapper 端内存性能问题：Mapper 端的性能调优参数 spark.shuffle.file.buffer 的默认大小是 32KB，用户要根据数量和并发量来适当调整该参数，尽量避免频繁地磁盘访问操作，可以通过观察效果来尝试，例如开始是 32KB，然后调整为 64KB、128KB 等。

23.2　Spark 集群中 Reducer 端内存调优实战

本节讲解 Spark 集群中 Reducer 端内存使用详解以及内存性能调优实战。

23.2.1　内存使用详解

进行 Shuffle 时，Mapper 端有一些文件按照某种规则给 Reducer 端，在整个 Shuffle 的过程中，Mapper 端有很多任务，Reducer 端也有很多任务，Shuffle 有很多不同的类型，不同类型的核心区别在于 Mapper 端的数据怎么交给 Reducer 端。

Spark 集群中 Reducer 端内存性能调优示意图如图 23-2 所示。

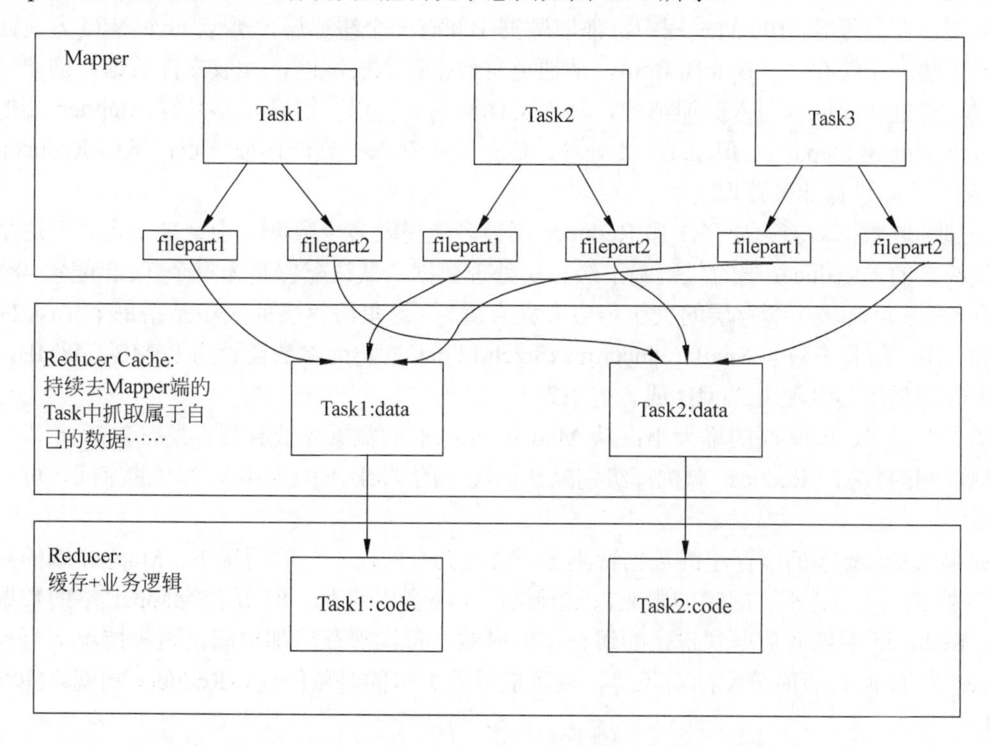

图 23-2　Spark 集群中 Reducer 端内存性能调优示意图

假设在 Mapper 端有 3 个 Task：Task1、Task2、Task3。在 Reducer 端有两个任务：Task1、Task2。

从 Reducer 端的角度考虑，每个 Task 生成几个部分的文件，因为在 Shuffle 的时候有不同的 Shuffle 策略：哈希方式、排序方式等。

在 Mapper 端和 Reducer 端中间加一个缓存层，Reducer 端的 Task 有两个，所以文件会分为两个小的文件：filepart1、filepart2。这里不是指第一个文件、第二个文件，而是文件的第一部分、第二部分。Mapper 端 Task 的数据也有两部分，不同的 Shuffle 策略会说明怎么划分这两部分。

缓存层分别从不同 Task 的 filepart1、filepart2 抓取属于自己的数据，然后把数据放到缓存层中，再从缓存层中把数据抓取到 Reducer 端，在 Reducer 端里面对 RDD 进行一系列业务逻辑的处理。

具体流程如下：整个 Spark 的每个 Job 分成 Mapper 端和 Reducer 端，Mapper 端产生的数据会分成若干个部分，具体分成几部分由 Reducer 端的并行度决定。例如，排序的方式就在一个文件中，而哈希的方式则进行了文件压缩。Reducer 端获取具体数据的时候，Reducer 端的前端有一个缓存，持续从 Mapper 端的 Task 输出中去抓取属于自己的数据，Reducer 端通过 transformation 业务逻辑代码对抓到的数据进行处理。

从上面整个过程来看，Spark 集群中 Reducer 端在如下环节可能出问题。

第一个是 Reducer 端的缓存层：Mapper 端不断地输出数据，根据不同的作业以及作业不同的阶段，数据可能很多，也可能很少；Reducer 端要运行 Task，是否要等到 Mapper 端将所有的数据都写到磁盘中之后，Reducer 端才从 Mapper 端抓取数据？不是。这里是一边 Shuffle，一边处理，在进行 Shuffle 的过程中，抓取数据中间有一个缓存层，类似 Java NIO 方式读写文件：创建一个缓存区（ByteBuffer），从通道将数据读入缓冲区，读取文件数据；创建一个缓存区，通过从缓冲区写入到通道中，写入文件数据。所以，缓存层不是等 Mapper 端把所有的数据都放到 filepart1、filepart2 才处理，而是数据存入一点就读取一点，然后 Reducer 端的 Task 的代码进行业务处理。

一边读取数据，一边处理，那 Reducer 端最多能提取多少数据、由谁来决定呢？这是由缓存层决定的。Reducer 端的代码基于缓存层处理数据，默认配置是为每个 Task 配置 48MB 的缓存，那实际环境中缓存层的大小一般多大合适？（这里缓存层的大小是指每个 Task 的缓存层的大小，设置参数为 Spark.reducer.maxSizeInFlight），将此参数设置为 96MB、48MB，或者更大？还是将其设置为 24MB 或者更小？

第二个是 Reducer 端的堆大小：从 Mapper 端抓取的数据先放到缓存层，然后才用 Task 执行抓取到的数据，Reducer 端执行级别默认情况下的 Task 堆的大小是 20%的空间，可以进行调整。

如果 Reducer 端的缓存层的数据特别多，会不会有问题？一般情况下，Mapper 端的数据不是特别多，远远达不到配置的限额，这种情况下不会出问题。但如果 Mapper 端的数据特别多，Reducer 端抓取数据到自己的缓存层的时候，每次缓存层都填满，这种情况下再加上 Reducer 端 Task 运行时分配的对象等，就可能导致大量的对象创建，Reducer 端就可能发生 OOM。

在企业生产环境中通常会遇到这个问题，在这种情况下该怎么办呢？可能有人提出增加 Executor 或者增加内存，但在实际生产环境中，资源被严格限制，所以先从技能的层面，在不改变资源的情况下，考虑如何去处理。这时比较简单的方式就是将 Reducer 端的缓存层减少。例如，原先缓存层在 48MB 时发生了 OOM，将参数调整成 24MB 就行了。这是最简单、最直接的解决方案。先让程序运行起来，然后才考虑让程序跑得更快。这个想法与平时单机版本的想法完全一样，如果单机版本发生了 OOM，是调大缓存大小，还是调小缓存大小？确实要调小缓存大小。

23.2.2　内存性能调优实战

问题 1：Reducer 端的业务逻辑（Business Logic）运行的空间分配不够，业务逻辑运行的时候被迫把数据溢出到磁盘上面，一方面造成了业务逻辑处理的时候需要读写磁盘，另一方面也会导致不安全（数据读写故障）。

针对问题 1 的调优办法：Reducer 端的性能调优参数 spark.shuffle.memoryFraction 默认大小是 0.2，Reducer 端的业务逻辑运行占用 Executor 的内存大小的 20%，很多公司的 Executor 中线程的并行度在 5 个左右，调整的时候可以从 0.2 调到 0.3、0.4 等。调整得越大，Spill 到磁盘的次数就越少，次数越少，从磁盘中读取文件的时候数量也会越小。

问题 2：发生 Reducer 端的 OOM，Reducer 端如果出现 OOM，一般由于内存中数据太多，无法容纳活跃的对象。

针对问题 2 的调优办法：调小 Reducer 端的缓存层。因为分配的内存有限，如果占用了太多的缓存，将导致太多的对象数据产生，会产生 OOM，将缓存层减少，OOM 的症状就很可能消失。这个办法也有代价，那就是缓存层变小了，向 Mapper 层提取数据的次数就变多了，Shuffle 的次数变得更多了，性能也就降低了。这样修改的目的是让程序先运行起来，然后再慢慢调，如增加 Executor，分配更多的内存，然后可以再调大缓存。

如果内存足够大，可以增加缓存的大小，例如，从 48MB 提升到 96MB，这样可以减少网络传输的次数，从而提高性能，配置参数是 spark.reducer.maxSizeInFlight。

问题 3：shuffle file not found 的问题。原因有可能是 GC，无论是 Minor GC，还是 final GC，只要有 GC，就有可能在 Mapper 端 GC 的时候无法把数据抓取过来。

针对问题 3 的调优方法：一般情况下，当 Executor 进行 GC 的时候，所有的线程都停止工作，包括进行数据传输的 Netty 中的线程，所以就暂时无法获取数据。

当 Reducer 端根据 Driver 端提供的信息到 Mapper 中指定的位置去获取数据时，首先会去定位数据所在的文件，此时可能发生 shuffle file not found 的错误。这个错误的出现一般是由于 Mapper 端正在进行 GC，对数据请求没有响应，默认情况为：

```
spark.shuffle.io.maxRetries =3
spark.shuffle.io.retryWait =5s
```

所以 15s 还没有抓取到数据就会出现 shuffle file not found 的错误。解决办法是调大上述参数，例如：

```
spark.shuffle.io.maxRetries =30
spark.shuffle.io.retryWait =30s
```

第 24 章 使用 Broadcast 实现 Mapper 端 Shuffle 聚合功能的原理和调优实战

本章讲解使用 Broadcast 实现 Mapper 端 Shuffle 聚合功能的原理和调优实战。

24.1 使用 Broadcast 实现 Mapper 端 Shuffle 聚合功能的原理

Shuffle 分为两部分：一个是 Mapper 端的 Shuffle；另外一个是 Reducer 端的 Shuffle。性能调优有一个很重要的总结就是，尽量不使用 Shuffle 类的算子，因为一般进行 Shuffle 的时候，它会把集群中多个节点上的同一个 Key 汇聚在同一个节点上，如 reduceByKey。然后会优先把结果数据放在内存中，但如果内存不够，就会放到磁盘上。Shuffle 在进行数据抓取前，为了整个集群的稳定性，它的 Mapper 端会把数据写到本地文件系统。这可能导致大量磁盘文件的操作。

在大数据处理场景中，多表 Join 是非常常见的一类运算。为了便于求解，通常将多表 Join 问题转为多个两表连接问题。

普通的 Join 是会走 Shuffle 过程的，而一旦 Shuffle，就相当于会将相同 Key 的数据拉取到一个 Shuffle read Task 中再进行 Join，此时就是 reduce Join。但是，如果一个 RDD 比较小，则可以采用广播变量方式，将较小 RDD 的数据进行全量广播，再利用 map 算子操作来实现与 Join 同样的效果，也就是 map Join，此时就不会发生 Shuffle 操作，也就不会发生数据倾斜。

24.2 使用 Broadcast 实现 Mapper 端 Shuffle 聚合功能调优实战

使用 Broadcast 实现 Mapper 端 Shuffle 聚合调优实战。本节使用电影点评系统中最受不同年龄段人员欢迎的电影 TopN 的功能来讲解。

电影点评系统的 Movie_Users_Analyzer_DateFrame.scala 业务代码如下：

```
1.  功能五：最受不同年龄段人员欢迎的电影 TopN
2.  ......
3.    val  targetQQUsers  =  usersRDD.map(_.split("::")).map(x  =>  (x(0),
      x(2))).filter(_._2.equals("18"))
4.     val targetTaobaoUsers = usersRDD.map(_.split("::")).map(x => (x(0),
       x(2))).filter(_._2.equals("25"))
5.
6.     /**
```

```
7.        * 在 Spark 中如何实现 map Join 呢，显然是要借助于 Broadcast，把数据广播到
          * Executor 级别，让该 Executor 上的所有任务共享
8.        * 该唯一的数据，而不是每次运行 Task 的时候都要发送一份数据的复制，这显著地降低
          * 了网络数据的传输和 JVM 内存的消耗
9.        */
10.     val targetQQUsersSet = HashSet() ++ targetQQUsers.map(_._1).collect()
11.     val targetTaobaoUsersSet = HashSet() ++ targetTaobaoUsers.map(_._1)
        .collect()
12.
13.     val targetQQUsersBroadcast = sc.broadcast(targetQQUsersSet)
14.     val targetTaobaoUsersBroadcast = sc.broadcast(targetTaobaoUsersSet)
15.
16.     /**
17.       * QQ 或者微信核心目标用户最喜爱电影 TopN 分析
18.       * (Silence of the Lambs, The (1991),524)
19.       * (Pulp Fiction (1994),513)
20.       * (Forrest Gump (1994),508)
21.       * (Jurassic Park (1993),465)
22.       * (Shawshank Redemption, The (1994),437)
23.       */
24.     val movieID2Name = moviesRDD.map(_.split("::")).map(x => (x(0),
        x(1))).collect.toMap
25.     println("纯粹通过 RDD 的方式实现所有电影中 QQ 或者微信核心目标用户最喜爱电影
        TopN 分析:")
26.     ratingsRDD.map(_.split("::")).map(x => (x(0), x(1))).filter(x =>
27.       targetQQUsersBroadcast.value.contains(x._1)
28.     ).map(x => (x._2, 1)).reduceByKey(_ + _).map(x => (x._2, x._1)).
29.       sortByKey(false).map(x => (x._2, x._1)).take(10).
30.       map(x => (movieID2Name.getOrElse(x._1, null), x._2)).foreach(println)
```

电影点评系统借助于 Broadcast，把不同年龄的用户数据广播到 Executor，让 Executor 上的所有任务共享该唯一的数据，在业务代码中没有 Join 操作，也就避免了 Shuffle 操作。

第 25 章　使用 Accumulator 高效地实现分布式集群全局计数器的原理和调优案例

本章讲解如何使用 Accumulator 高效地实现分布式集群全局计数器的原理和调优案例。

25.1　Accumulator 内部工作原理

累加器是仅仅被相关操作累加的全局共享变量，因此可以在并行计算中有效使用，从 Worker 节点聚合数值回 Driver。它可以用来实现计数器的总和。Spark 原生地只支持数字类型的累加器，用户可以添加新类型的支持。如果创建累加器时指定了名字，可以在 Spark 的 UI 看到。这有利于理解每个执行阶段的进程。

累加器通过对一个初始化的变量 v 调用 SparkContext.accumulator(v)来创建。在集群上运行的任务可以通过 add 或者“+=”方法在累加器上进行累加操作。但是，它们不能读取它的值。只有驱动程序通过累加器的 value 方法能够读取它的值。

下面是 Spark-Shell 下累加器的使用例子。

```
1.   scala> val accum = sc.accumulator(0)
2.  accum: Spark.Accumulator[Int] = 0
3.
4.  scala> sc.parallelize(Array(1, 2, 3, 4)).foreach(x => accum += x)
5.
6.  scala> accum.value
7.  res2: Int = 10
```

25.2　Accumulator 自定义实现原理和源码解析

Accumulator 是分布式全局只写的数据结构。在生产环境下，一定要自定义 Accumulator。自定义时可以让 Accumulator 非常复杂，基本上可以是任意类型的 Java 和 Scala 对象；自定义 Accumulator 时，可以实现一些“技术福利”，如在 Accumulator 变化的时候可以把数据同步到 MySQL 中。

下面来看一下源码：

```
1.   @deprecated("use AccumulatorV2", "2.0.0")
2.  class Accumulator[T] private[spark] (
3.      //SI-8813: 必须显式定义一个私有变量，否则 Scala 2.11 不进行编译
4.      @transient private val initialValue: T,
5.      param: AccumulatorParam[T],
6.      name: Option[String] = None,
7.      countFailedValues: Boolean = false)
```

```
8.    extends Accumulable[T, T](initialValue, param, name, countFailedValues)
9.
```

被累加的值的类型是泛型，可以对值的类型进行自定义。

25.3　Accumulator 作全局计数器案例实战

开发人员也可以通过继承 AccumulatorParam 类创建它们自己的累加器类型。AccumulatorParam 接口有两个方法。

（1）zero 方法为类型提供一个 0 值。

（2）addInPlace 方法将两个值相加。

假设有一个代表数学 vector 的 Vector 类，累加器实现如下：

```
1.   object VectorAccumulatorParam extends AccumulatorParam[Vector] {
2.  def zero(initialValue: Vector): Vector = {
3.  Vector.zeros(initialValue.size)
4.    }
5.  def addInPlace(v1: Vector, v2: Vector): Vector = {
6.      v1 += v2
7.    }
8.  }
9.
10. //然后创建一个此类型的累加器 Accumulator
11. val vecAccum = sc.accumulator(new Vector(…))(VectorAccumulatorParam)
```

在 Scala 里，Spark 提供更通用的累加器接口来累加数据，尽管结果的类型和累加的数据类型可能不一致（例如通过收集在一起的元素来创建一个列表）。同时，使用 SparkContext..accumulableCollection 方法来累加通用的 Scala 的集合类型。

累加器仅仅在动作操作内部被更新，Spark 保证每个任务在累加器上的更新操作只被执行一次，也就是说，重启任务也不会更新。在转换操作中，用户必须意识到每个任务对累加器的更新操作可能不止一次执行，例如重新执行了任务和作业的阶段。

累加器并没有改变 Spark 的惰性求值模型。如果它们被 RDD 上的操作更新，它们的值只有当 RDD 因为动作操作被计算时，才被更新。因此，当执行一个惰性的转换操作（如 map）时，不能保证对累加器值的更新被实际执行了。下面的代码片段演示了此特性。

```
1.   val accum = sc.accumulator(0)
2.  data.map { x => accum += x; f(x) }
```

这里，accum 的值仍然是 0，因为没有 Action 动作操作引起 map 实际的计算。

第 26 章　Spark 下 JVM 性能调优最佳实践

本章讲解 Spark 下 JVM 性能调优最佳实践。26.1 节讲解 JVM 内存架构详解及调优；26.2 节讲解 Spark 中对 JVM 使用的内存原理图详解及调优；26.3 节讲解 Spark 下 JVM 的 On-Heap 和 Off-Heap 解密；26.4 节讲解 Spark 下的 JVM GC 导致的 Shuffle 拉取文件失败及调优方案；26.5 节中讲解 Spark 下的 Executor 对 JVM 堆外内存连接等待时长调优；26.6 节讲解 Spark 下的 JVM 内存降低 Cache 内存占比的调优。

26.1　JVM 内存架构详解及调优

本节讲解 JVM 内存架构详解及调优，内容包括 JVM 的堆区、栈区、方法区等详解；JVM 线程引擎及内存共享区域详解；JVM 中年轻代和老年代及元空间原理详解；JVM 进行 GC 的具体工作流程详解；JVM 常见调优参数详解。

26.1.1　JVM 的堆区、栈区、方法区等详解

JVM 主要由类加载器子系统、运行时数据区（内存空间）、执行引擎以及与本地接口等组成。其中，运行时数据区又由方法区、堆、Java 栈、程序计数器、本地方法栈组成，如图 26-1 所示。

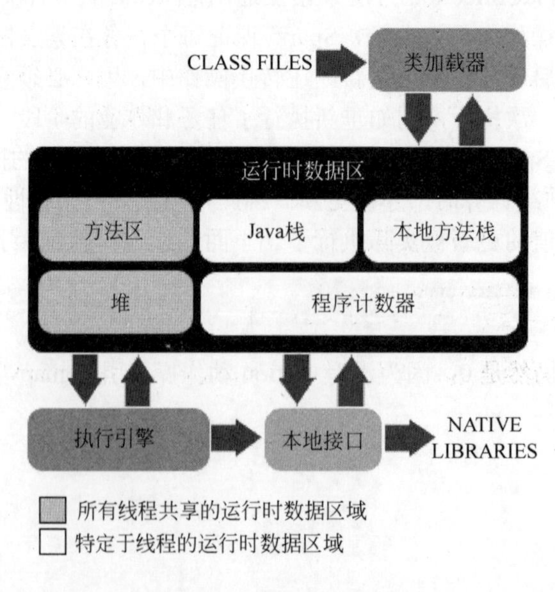

图 26-1　JVM 内存示意图

每个 JVM 中的方法区和堆区被 JVM 中的线程共享。当 JVM 加载了一个 class 文件后，

则 class 中的参数、类型等信息会存储在方法区中。程序运行时所有创建的对象都存储在堆中。

当每个新线程启动时，会有自己的程序计数器和栈，如果线程调用方法，则程序计数器表明下一条执行的指令。线程栈存储线程的方法调用状态（包括局部变量、被调用的参数、返回值、中间结果）。本地方法调用存储在独立的本地方法栈中，或其他独立的内存区域中。

栈区是由栈桢组成，每个栈桢就是每个调用的方法的栈。当方法调用结束时，JVM 会弹栈，即抛弃此方法的栈桢。

1. 堆区

（1）堆区存储的全部是对象，每个对象都包含一个与之对应的 class 的信息（class 的目的是得到操作指令）。

（2）JVM 只有堆区和方法区被所有线程共享，堆中不存放基本类型和对象引用，只存放对象本身。

（3）一般由程序员分配释放，若程序员不释放，程序结束时可能由 OS 回收。

2. 栈区

（1）每个线程包含一个栈区，栈中只保存基础数据类型的对象和自定义对象的引用（不是对象），对象都存放在堆区中。

（2）每个栈中的数据（原始类型和对象引用）都是私有的，其他栈不能访问。

（3）栈分为 3 部分：基本类型变量区、执行环境上下文、操作指令区（存放操作指令）。

（4）由编译器自动分配释放，存放函数的参数值、局部变量的值等。

3. 静态区/方法区

（1）方法区又称静态区，与堆一样，被所有的线程共享。方法区包含所有的 class 和 static 变量。

（2）方法区中包含的都是在整个程序中永远唯一的元素，如 class、static 变量。

（3）全局变量和静态变量的存储是放在一块的，初始化的全局变量和静态变量在一块区域，未初始化的全局变量和未初始化的静态变量在相邻的另一块区域。

26.1.2　JVM 线程引擎及内存共享区域详解

多线程的 Java 应用程序：为了让每个线程正常工作，提出了程序计数器（Program Counter Register），每个线程都有自己的程序计数器，这样，当线程执行切换的时候，就可以在上次执行的基础上继续执行。仅仅从一条线程线性执行的角度而言，代码是一条一条地往下执行的，这时就是程序计数器，JVM 就是通过读取程序计数器的值来决定该线程下一条需要执行的字节码指令，进而进行选择语句、循环、异常处理等。

JVM 线程引擎及内存共享区域示意图如图 26-2 所示。对于每个线程，从 OOP 角度而言，相当于一个对象。该对象中具有执行代码，同时也有要处理的数据。数据包含 Thread 工作时要访问的数据，同时也包含现在的 Stack，Stack 中包含了 Thread 本地的数据，也包含了复制的全局数据。从面向过程的角度而言：线程=代码+数据。

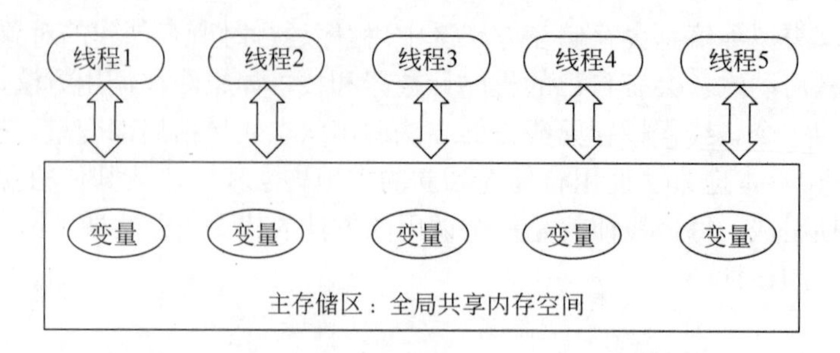

图 26-2 JVM 线程引擎及内存共享区域示意图

线程的执行，与全局内存共享区域交互的时候，涉及从全局内存共享区域复制数据到线程的工作内存中。复制过去后，交给线程的操作代码去处理数据。不是直接操作全局的数据，这是 JVM 实现的基本机制，防止发生状态不一致，所以实现复制机制。

程序计数器如图 26-2 所示，有 5 条线程，在工作的时候，线程工作的背后是 core，涉及 core 在线程之间的切换，线程的正常工作是指令一条条执行和状态的保持。为了让线程正常工作，提出了程序计数器。

字节码解释器是在程序计数器的基础上工作的，通过读取程序计数器的值决定该线程下一条需要执行的字节码指令，进而进行选择语句、循环、异常处理等。

ThreadLocal，线程本地存储，数据属于线程私有，不同于私有成员，其他线程是不能访问的。在各大框架中，都会有 ThreadLocal 的身影。

26.1.3 JVM 中年轻代和老年代及元空间原理详解

JVM 7 中的内存共划分为 3 个代：年轻代（Young Generation）、老年代（Old Generation）和持久代（Permanent Generation）。其中，持久代主要存放的是 Java 类的类信息，与垃圾收集要收集的 Java 对象关系不大，而年轻代和老年代对垃圾收集影响比较大，如图 26-3 所示。

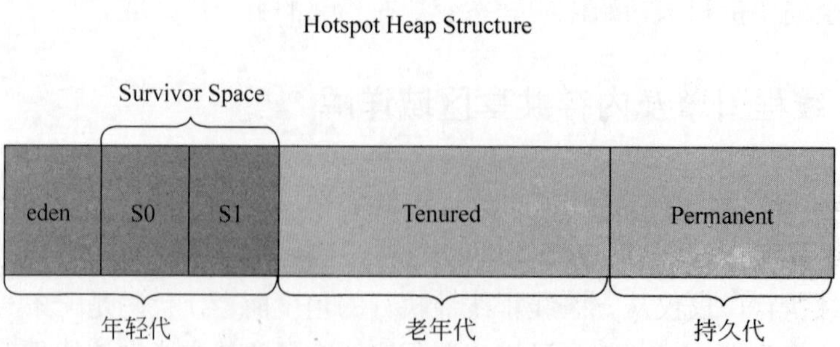

Hotspot Heap Structure

Survivor Space

eden S0 S1 Tenured Permanent

年轻代 老年代 持久代

图 26-3 JVM 中内存年轻代、老年代、持久代的划分

1. 年轻代

所有新生成的对象首先都放在年轻代。年轻代的目标就是尽可能快速地收集那些生命周期短的对象。年轻代分 3 个区：一个 Eden 区，两个 Survivor 区（一般而言）。大部分对象在

Eden 区中生成。当 Eden 区满时，还存活的对象将被复制到 Survivor 区（两个中的一个），当这个 Survivor 区满时，此区的存活对象将被复制到另外一个 Survivor 区，当这个 Survivor 区也满了的时候，从第一个 Survivor 区复制过来的并且此时还存活的对象，将被复制到“年老区（Tenured）”。注意，Survivor 的两个区是对称的，没有先后关系，所以同一个区中可能同时存在从 Eden 区复制过来的对象和从前一个 Survivor 区复制过来的对象，而复制到年老区的只有从第一个 Survivor 区复制过来的对象，而且 Survivor 区总有一个是空的。同时，根据程序需要，Survivor 区是可以配置为多个的（多于两个），这样可以增加对象在年轻代中的存在时间，减少被放到老年代的可能。

2．老年代

在年轻代中经历了 *N* 次垃圾回收后仍然存活的对象，就会被放到老年代中。因此，可以认为老年代中存放的都是一些生命周期较长的对象。

3．持久代

持久代用于存放静态文件、Java 类、方法等。持久代对垃圾回收没有显著影响，但是有些应用可能动态生成或者调用一些 class，如 hibernate 等，这时需要设置一个比较大的持久代空间来存放这些运行过程中新增的类。持久代的大小通过-XX：MaxPermSize=<N>进行设置。

4．Scavenge GC

一般情况下，当新对象生成，并且在 Eden 区申请空间失败时，就会触发 Scavenge GC，对 Eden 区进行 GC，清除非存活对象，并且把尚且存活的对象移动到 Survivor 区。然后整理 Survivor 的两个区。这种方式的 GC 只是对年轻代的 Eden 区进行，不会影响到老年代。因为大部分对象都是从 Eden 区开始的，同时 Eden 区不会分配得很大，所以 Eden 区的 GC 会频繁进行。因而，一般在这里需要使用速度快、效率高的算法，使 Eden 区能尽快空闲出来。

5．Full GC

对整个堆进行整理，包括 Young、Tenured 和 Perm。Full GC 因为需要对整个堆进行回收，所以比 Scavenge GC 慢，因此应该尽可能减少 Full GC 的次数。在对 JVM 调优的过程中，很大一部分工作是对 Full GC 的调节。

有如下原因可能导致 Full GC。

- 老年代（Tenured）被写满。
- 持久代（Perm）被写满。
- System.GC()被显式调用。
- 上一次 GC 之后，Heap 的各域分配策略动态变化。

```
at java.util.concurrent.ThreadPoolExecutor$Worker.runTask
(ThreadPoolExecutor.java: 886)
at java.util.concurrent.ThreadPoolExecutor$Worker.run
(ThreadPoolExecutor.java: 908)
```

```
at java.lang.Thread.run(Thread.java: 662)
Exception in thread "http-bio-17788-exec-74"
java.lang.OutOfMemoryError:  PermGen space
Exception in thread "http-bio-17788-exec-75"
java.lang.OutOfMemoryError:  PermGen space
Exception in thread "http-bio-17788-exec-76"
java.lang.OutOfMemoryError:  PermGen space
```

从以下日志明显可以看出是老年代的内存溢出，说明在容器下的静态文件过多，例如，编译的字节码，JSP 编译成 Servlet，或者 Jar 包。

```
1.   at java.util.concurrent.ThreadPoolExecutor$Worker.runTask
2.  (ThreadPoolExecutor.java: 886)
3.  at java.util.concurrent.ThreadPoolExecutor$Worker.run
4.  (ThreadPoolExecutor.java: 908)
5.  at java.lang.Thread.run(Thread.java: 662)
6.  Exception in thread "http-bio-17788-exec-74"
7.  java.lang.OutOfMemoryError:  PermGen space
8.  Exception in thread "http-bio-17788-exec-75"
9.  java.lang.OutOfMemoryError:  PermGen space
10. Exception in thread "http-bio-17788-exec-76"
11. java.lang.OutOfMemoryError:  PermGen space
```

要解决此问题，修改 JVM 的参数 PermSize 即可，PermSize 初始默认为 64MB。

6．JVM内存参数

```
1.   -vmargs -Xms128MB -Xmx512MB -XX: PermSize=64MB -XX: MaxPermSize=128MB
2.   -vmargs 说明后面是 VM 的参数
3.   -Xms128MB JVM 初始分配的堆内存
4.   -Xmx512MB JVM 最大允许分配的堆内存，按需分配
5.   -XX: PermSize=64MB JVM 初始分配的非堆内存
6.   -XX: MaxPermSize=128MB JVM 最大允许分配的非堆内存，按需分配
```

默认，新生代（Young）与老年代（Old）的比值为 1:2（该值可以通过参数 – XX: NewRatio 指定），即新生代（Young）= 1/3 的堆空间大小。

老年代（Old）=2/3 的堆空间大小。其中，新生代（Young）被细分为 Eden 区和两个 Survivor 区，这两个 Survivor 区分别被命名为 from 和 to 以示区分。

默认的配置：Edem:from:to = 8:1:1（可以通过参数 – XX：SurvivorRatio 来设定），即 Eden = 8/10 的新生代空间大小，from = to = 1/10 的新生代空间大小。

JVM 每次只会使用 Eden 区和其中的一块 Survivor 区来为对象服务，所以无论什么时候，总有一块 Survivor 区是空闲着的，如图 26-4 所示。

因此，新生代实际可用的内存空间为 9/10（即 90%）的新生代空间。

7．堆（Heap）和非堆（Non-heap）内存

按照官方的说法：“Java 虚拟机具有一个堆，堆是运行时的数据区域，所有类实例和数组的内存均从此处分配。堆是在 Java 虚拟机启动时创建的”；“在 JVM 中，堆之外的内存称为非堆内存（Non-heap Memory）”。

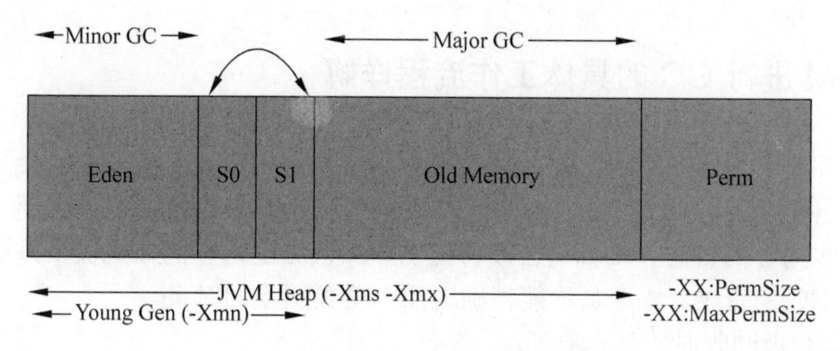

图 26-4　新生代内存空间示意图

可以看出，JVM 主要管理两种类型的内存：堆和非堆。简单来说，堆就是 Java 代码可及的内存，是留给开发人员使用的；非堆就是 JVM 留给自己用的。

Java 8 的一个特性就是完全地移除持久代（Permanent Generation，PermGen），这从 JDK 7 开始 Oracle 就行动了。例如，本地化的 String 从 JDK 7 开始就被移除了持久代。JDK 8 让它最终退役了。– XX：PermSize 和 – XX：MaxPermSize 选项会被忽略，如图 26-5 所示。

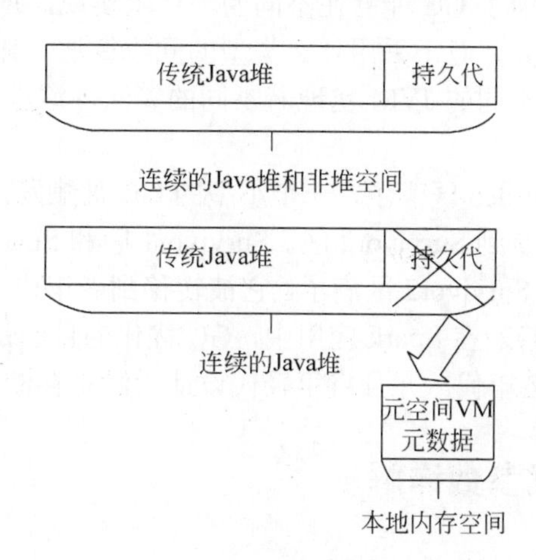

图 26-5　持久代变迁示意图

JDK 8.HotSpot JVM 开始使用本地化的内存存放类的元数据，这个空间叫作元空间（Metaspace）。类的元数据信息（Metadata）还在，只不过不再是存储在连续的堆空间上，而是移动到了元空间中。

类的元数据信息转移到元空间的原因是元空间很难调整。元空间中类的元数据信息在每次 Full GC 的时候可能会被收集，但成绩很难令人满意。而且应该为元空间分配多大的空间很难确定，因为 PermSize 的大小依赖于很多因素，如 JVM 加载的类的总数、常量池的大小、方法的大小等。

此外，在 HotSpot 中的每个垃圾收集器需要专门的代码来处理存储在元空间中的类的元数据信息。从元空间分离类的元数据信息到元空间，由于元空间的分配具有和 Java Heap 相同的地址空间，因此元空间和 Java Heap 可以无缝地管理，而且简化了 Full GC 的过程，以至将来可以并行地对元数据信息进行垃圾收集，而没有 GC 暂停。

26.1.4　JVM 进行 GC 的具体工作流程详解

内存处理器是编程人员容易出现问题的地方，忘记或者错误的内存回收会导致程序或系统的不稳定，甚至崩溃。Java 语言提供的 GC 功能可以自动地检测对象是否超过作用域，从而达到自动回收内存的目的。Java 语言没有提供释放已分配内存的显式操作方法，资源回收工作全部交由 GC 来完成，程序员不能精确地控制垃圾回收的时机。

GC 在实现垃圾回收时的基本原理：

Java 的内存管理实际就是对象的管理，其中包括对象的分配和释放。对于程序员来说，分配对象使用 new 关键字，释放对象时只是将对象赋值为 null，让程序员不能够再访问到这个对象，该对象被称为“不可达”。GC 将负责回收所有“不可达”对象的内存空间。

对于 GC 来说，当程序员创建对象时，GC 就开始监控这个对象地址、大小以及使用情况。通常，GC 采用有向图的方式记录并管理堆中的所有对象，通过这种方式确定哪些对象是“可达”的，哪些对象是“不可达”的。当 GC 确定一些对象为“不可达”时，GC 就有责任回收这些内存空间，但为了 GC 能够在不同的平台上实现，Java 规范对 GC 的很多行为都没有进行严格的规定。例如，对于采用什么类型的回收算法、什么时候进行回收等重要问题，都没有明确规定，因此不同的 JVM 实现着不同的算法，这也给 Java 程序员的开发带来了很多不确定性。

内存碎片整理步骤：当 Eden 区满了，一个小型的 GC 被触发，Eden 区和 Survivor1 区中幸存的仍被使用的对象被复制到 Survivor2 区。Survivor1 区和 Survivor2 区进行交换。当一个对象生存的时间足够长或者 Survivor2 区满了，它被转移到老年代。最终，当老年代的空间快满时，一个全面的 GC 被召唤。在 Spark 应用中，GC 优化的目的是使 Spark 保证只有长生存周期的 RDD 才会被存储在老年代，并且将年轻代设计为满足存储短生命周期的对象。

26.1.5　JVM 常见调优参数详解

针对 MetaSpace，JVM 8 增加了新的 Flag。

```
1.  -XX：MetaspaceSize 初始化元空间的大小（默认 12MB 在 32bit client VM and 16MB 在
    32bit server VM，在 64bit VM 上会更大些）
2.  -XX：MaxMetaspaceSize 最大元空间的大小（默认本地内存）
3.  -XX：MinMetaspaceFreeRatio 扩大空间的最小比率，当 GC 后，内存占用超过这一比率，
    就会扩大空间
4.  -XX：MaxMetaspaceFreeRatio 缩小空间的最小比率，当 GC 后，内存占用低于这一比率，
    就会缩小空间
```

默认的元空间只会受限于本地内存大小。当元空间达到 MetaspaceSize 的当前大小时，就会触发 GC。当然，可以设置 MetaspaceSize 为一个更大的值以延迟触发 GC。

Spark GC 调优的目标是确保老年代只保存长生命周期 RDD，同时，年轻代的空间又能足够保存短生命周期的对象。这样就能在任务执行期间，避免启动 Full GC。以下是 GC 调优的主要步骤。

- 从 GC 的统计日志中观察 GC 是否启动太多。如果某个任务结束前，多次启动了 Full GC，则意味着用以执行该任务的内存不够。

- 如果 GC 统计信息中显示，老年代的内存空间已经接近存满，可以通过降低 Spark.memory.storageFraction 来减少 RDD 缓存占用的内存；减少缓存对象总比任务执行缓慢强。
- 如果 Major GC 比较少，但 Minor GC 很多，就可以多分配一些 Eden 区内存。可以把 Eden 区的大小设为高于各个任务执行所需的工作内存。如果把 Eden 区大小设为 E，则可以这样设置年轻代区的大小：-Xmn=4/3×E（放大 4/3 倍，主要是为了给 Survivor 区保留空间）。例如，如果你的任务会从 HDFS 上读取数据，那么单个任务的内存需求可以用其读取的 HDFS 数据块的大小来评估。需要特别注意的是，解压后的 HDFS 块是解压前的 2～3 倍。所以，如果希望保留 3～4 个任务并行的工作内存，并且 HDFS 块的大小为 64MB，那么评估 Eden 区的大小可以设置为 4×3×64MB。
- 最后再观察一下垃圾回收的启动频率和总耗时有没有变化。

很多经验表明，GC 调优的效果和程序代码以及可用的总内存相关。网上还有不少调优的选项说明（many more tuning options），但总体来说，只要控制好 Full GC 的启动频率，就能有效地减少垃圾回收开销。

26.2　Spark 中对 JVM 使用的内存原理图详解及调优

Spark 作为分布式计算框架，由于其优先使用内存，故内存的管理对性能有重大影响。因此，了解其内存管理机制和对内存的使用情况，也将大大有助于程序的优化。

26.2.1　Spark 中对 JVM 使用的内存原理图说明

Spark 1.6.0 前使用的内存管理模式由 StaticMemoryManager 实现，如图 26-6 所示；而 Spark 1.6.0 后使用的内存管理模式由 UnifiedMemoryManager 实现，如图 26-7 所示。当然，也可以通过参数 spark.memory.useLegacyMode 来配置使用哪种内存管理模式。

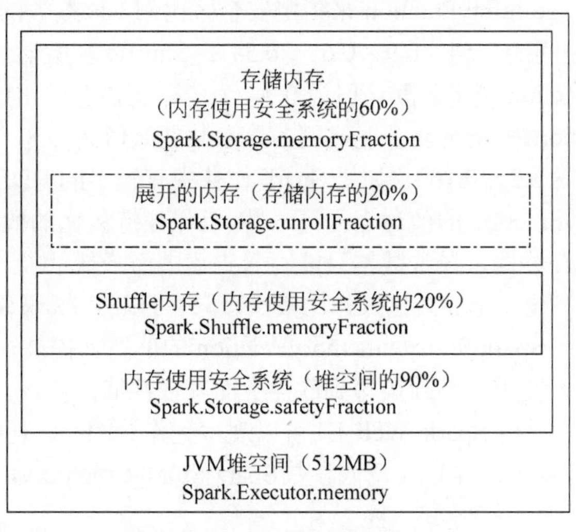

图 26-6　StaticMemoryManager

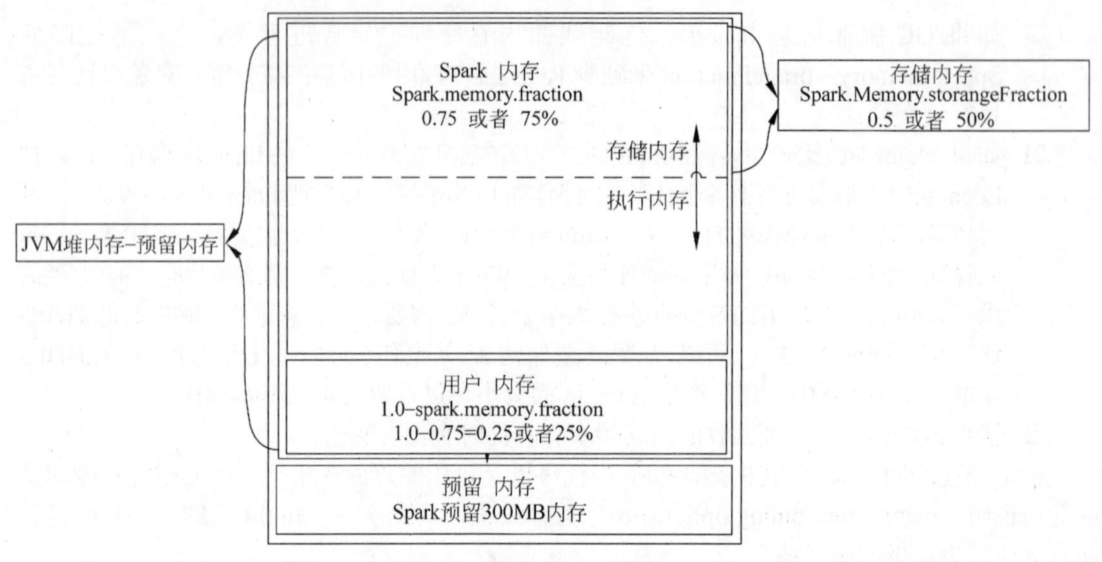

图 26-7 UnifiedMemoryManager

26.2.2 Spark 中对 JVM 使用的内存原理图内幕详解

Executor 中对内存的使用涉及以下几点。

（1）RDD 存储。当对 RDD 调用 Persist 或 Cache 方法时，RDD 的 Partitions 会被存储到内存里。

（2）Shuffle 操作。Shuffle 时，需要缓冲区来存储 Shuffle 的输出和聚合的中间结果。

（3）用户代码。用户编写的代码能使用的内存空间是整个堆空间除了上述两点之后剩下的空间。

StaticMemoryManage 模式下，堆空间分为 Storage 区和 Shuffle 区。Storage 区能使用的堆空间的比例由 spark.storage.memoryFraction 指定，默认值为 0.6，为了避免内存溢出的风险，还有一个参数 spark.storage.safetyFraction 来指定安全区比例，该参数的默认值是 0.9，故实际可用的 Storage 区为堆空间的 0.54（0.9×0.6 = 0.54）；Shuffle 区所能使用的堆空间的比例由 Spark.Shuffle.memoryFraction 指定，默认值为 0.2，为了避免内存溢出的风险，还有一个参数 Spark.Shuffle.safetyFraction 来指定安全区比例，该参数默认值是 0.8，故实际可用的 Shuffle 区为堆空间的 0.16（0.8×0.2 = 0.16）。如果 Spark 作业中有较多的 RDD 持久化操作，则可以将 Spark.Storage.memoryFraction 的值适当提高一些，以保证持久化的数据能够容纳在内存中，避免因内存不够缓存所有数据，导致数据只能写入磁盘中而降低性能。但是，如果 Spark 作业中的 Shuffle 类操作比较多，而持久化操作比较少，那么可以将 Spark.Storage.memoryFraction 的值适当降低一些，而将 Spark.Shuffle.memoryFraction 的值适当提高一些，以避免 Shuffle 过程中因数据过多时内存不够用，必须溢写到磁盘上而降低性能；此外，如果发现作业由于频繁地 GC 导致运行缓慢（通过 Spark Web UI 可以观察到作业的 GC 耗时），意味着 task 执行用户代码的内存不够用，那么同样建议调低参数 Spark.Storage.memoryFraction 和 Spark.Shuffle.memoryFraction 的值。

UnifiedMemoryManager 模式下，整个堆空间分为 Spark Memory 和 User Memory，在 Spark

Memory 内部又分为 Storage Memory 和 Execution Memory。Storage Memory 和 Execution Memory 并没有硬界限，可以相互借用空间。可以通过参数 Spark.Memory.fraction（默认为 0.75）来设置 Spark Memory 所占的整个堆空间的比例，剩下的空间就是 User Memory（默认为 1−0.75=0.25）；通过 Spark.Memory.storageFraction（默认为 0.5）设置 Storage Memory 所占的 Spark Memory 的比例。根据实际的程序中 Cache 的多少，Shuffle 的多少，对象的多少等，可以调整上述各个参数来调整各个内存区的大小，进而优化程序。

下面重点看一下 UnifiedMemoryManager 模式下的相关参数。

Spark.Storage.memoryFraction：该参数用于设置 RDD 持久化数据在 Executor 内存中能占的比例，默认是 0.6。也就是说，默认占 Executor 60%的内存，可以用来保存持久化的 RDD 数据。根据选择的不同的持久化策略，如果内存不够时，可能数据就不会持久化，或者数据会写入磁盘。参数调优建议：如果 Spark 作业中有较多的 RDD 持久化操作，该参数的值可以适当提高一些，保证持久化的数据能够容纳在内存中，避免内存不够缓存所有的数据，导致数据只能写入磁盘中，降低了性能。但是，如果 Spark 作业中的 Shuffle 类操作比较多，而持久化操作比较少，那么适当降低一些这个参数的值比较合适。此外，如果发现作业由于频繁的 GC 导致运行缓慢（通过 Spark Web UI 可以观察到作业的 GC 耗时），则意味着 Task 执行用户代码的内存不够用，那么同样建议调低这个参数的值。

Spark.Shuffle.memoryFraction：该参数用于设置 Shuffle 过程中一个 Task 拉取到上一个 stage 的 task 的输出后，进行聚合操作时能够使用的 Executor 内存的比例，默认是 0.2。也就是说，Executor 默认只有 20%的内存用来进行该操作。Shuffle 操作在进行聚合时，如果发现使用的内存超出了 20%的限制，那么多余的数据就会溢写到磁盘文件中，此时就会极大地降低性能。参数调优建议：如果 Spark 作业中的 RDD 持久化操作较少，Shuffle 操作较多时，建议降低持久化操作的内存占比，提高 Shuffle 操作的内存占比比例，避免 Shuffle 过程中数据过多时内存不够用，必须溢写到磁盘上而降低性能。此外，如果发现作业由于频繁的 GC 导致运行缓慢，则意味着 Task 执行用户代码的内存不够用，那么同样建议调低这个参数的值。

1．确定内存消耗

可以在程序中通过 Cache 方法将 RDD Cache 到内存中，然后通过 Web UI 的 Storage 页面查看该 RDD 占用了多少内存，或查看 Driver 的日志。

有很多工具可以帮助我们了解内存的消耗，如 JVM 自带的众多的内存消耗诊断工具 JMap、JSonsole 等，第三方工具如 IBM JVM Profile Tools 等。

通过这些方法，可以了解各种数据结构、广播变量等对内存的占用，从而选择使用对内存更友好的数据结构。

2．数据结构调优

Spark 程序的调优同普通的 Java 程序一样，都要关注数据结构的合理使用，又由于 Spark 优先基于内存的特点，程序员就更要注意选取合适的内存和友好的数据结构，以减少内存消耗。

3．采用合适的序列化器序列化要持久化的RDD

由于 Spark 默认的序列化器 org.apache.spark.serializer.JavaSerializer，相比 Kryo 序列化器

org.apache.spark.serializer.KryoSerializer，性能和空间表现都比较差，所以我们在持久化 RDD 时，应该优先使用压缩率更高、更快的 Kryo 序列化器。

26.2.3　Spark 下常见的 JVM 内存调优参数最佳实践

JVM 的垃圾回收在某些情况下可能造成瓶颈。例如，RDD 存储经常需要“换入换出”（新 RDD 抢占了老 RDD 内存，不过，如果程序没有这种情况，那么 JVM 垃圾回收一般不是问题。例如，RDD 只是载入一次，后续只是在这一个 RDD 上操作）。当 Java 需要把老对象逐出内存的时候，JVM 需要跟踪所有的 Java 对象，并找出哪些对象已经没有用了。概括起来就是：垃圾回收的开销和对象个数成正比，所以减少对象的个数（如用 Int 数组取代 LinkedList），就能大大减少垃圾回收的开销。当然，一个更好的方法是以序列化形式存储数据，这时每个 RDD 分区都只包含有一个对象（一个巨大的字节数组）。在尝试其他技术方案前，首先可以试试用序列化 RDD 的方式评估 GC 是不是一个瓶颈。

如果作业中各个任务需要的工作内存和节点上存储的 RDD 缓存占用的内存产生冲突，那么 GC 很可能会出现问题。下面讨论如何控制好 RDD 缓存使用的内存空间，以减少这种冲突。

GC 调优的第一步是统计一下 GC 启动的频率以及 GC 使用的总时间。给 JVM 设置参数：-verbose：GC -XX：+PrintGCDetails，就可以在后续 Spark 作业的 Worker 日志中看到每次 GC 花费的时间。注意，这些日志是在集群 Worker 节点上（在各节点的工作目录下的 Stdout 文件中），而不是在驱动器所在节点上。

为避免全面 GC 去收集 Spark 运行期间产生的临时对象，归纳一些实用技巧如下。

首先检查 GC 日志中是否有过于频繁的 GC。如果在一个任务完成前，全量 GC 被唤醒了多次，意味着对于执行任务来说，没有分配足够的内存。如果有太多的小型垃圾收集，但全量 GC 出现并不多，给 Eden 区分配更多的内存会很有帮助。可以为每个任务设置一个高于其所需内存的值。假设 Eden 区的内存需求量为 E，可以设置年轻代的内存为-Xmn=4/3×E。（这一设置同样也会导致 Survivor 区同时扩张）在 GC 打印的日志中，如果 OldGen 接近满时，可以通过降低 Spark.memory.fraction 减少用于缓存的空间。更好的方式是缓存更少的对象，而不是降低作业执行时间。一个可选的方案是减少年轻代的规模。如果设置了-Xmn，可以降低-Xmn。如果没有设置，可以尝试改变 JVM 的 NewRatio 参数。很多 JVM 的 NewRation 参数默认值是 2，这意味着老年代申请 2/3 的堆空间。它的值应足够大，以至可以超过 Spark.memory.fraction。尝试使用 G1GC 垃圾收集选项：-XX：+UseG1GC。当 GC 存在瓶颈时，采用这一选项在某些情况下可以提升性能。当执行器的堆空间比较大时，提升 G1 Region size（-XX：G1HeapRegionSize）是一种重要的选择。如果你的任务需要从 HDFS 读取数据，可以通过估计 HDFS 文件的大小来预估任务所需的内存量。需要注意的是，解压后的块大小是原大小的 2～3 倍。因此我们需要设置 3～4 倍的工作空间用于作业执行，例如 HDFS 的块大小为 128MB，我们需要预估 Eden 区的大小为 4×3×128MB。监控在新变化和设置生效后，GC 的频率和耗费的时间。

我们的经验建议是：GC 优化的成效依赖于你的应用和可用内存的多少。网上也有许多优化策略，但是需要更深的知识基础，例如通过控制全量 GC 发生的频率来降低总开销。

通过设置 Spark.Executor.extraJavaOptions 可以实现对 Executor 中 GC 的优化调整。

（1）Spark 的钨丝计划是专门用来解决 JVM 性能问题的，在 Spark 2.0 以前，钨丝计划功能不稳定且不完善，且只能在特定的情况下发生作用，包括 Spark 1.6.0 在内的 Spark 及其以前的版本，大多数情况下没有使用钨丝计划的功能，所以此时就必须关注 JVM 性能调优。

（2）JVM 性能调优的关键是调优 GC。为什么 GC 如此重要？主要是因为 Spark 热衷于 RDD 的持久化。GC 本身的性能开销是和数据量成正比的。

（3）初步可以考虑的是尽量多地使用 array 和 String，并且在序列化机制方面尽可能地采用 Kryo，让每个 Partition 都成为字节数组。

（4）监控 GC 的方式有两种。

① 配置 JVM 参数。

```
1.   Spark.Executor.extraJavaOptions = -verbose: GC -XX: +PrintGCDetails -XX:
     + PrintGCDateTimeStamps
```

例如：

```
1.    ./bin/Spark-Submit --name "My app" --master local[4] --conf Spark.Shuffle
      .spill=false  --conf "Spark.Executor.extraJavaOptions=-XX: +PrintGCDetails
      -XX: +PrintGCTimeStamps" myApp.jar
```

② SparkUI 的 4040 端口是通过 Web UI 页面进行监控的。

（5）Spark 默认情况下使用 60%的空间进行缓存 RDD 的内容。也就是说，Task 在执行的时候，只能使用剩下的 40% 的空间，如果空间不够用，就会触发（频繁的）GC。可以设置 Spark.memory.fraction 参数来调整空间的使用，例如，降低 Cache 的空间，让 Task 使用更多的空间创建对象和完成计算；再次强烈建议从 RDD 进行 Cache 的时候使用 Kryo 序列化，从而给 Task 分配更大的空间，来顺利完成计算（避免频繁的 GC）。

（6）因为在老年代空间满的时候会发生 FULL GC 操作，而老年代空间中基本都是存活得比较久的对象（经历数次 GC 依旧存在），此时会停下所有的程序线程，进入 FULL GC，对 OLD 区中的对象进行整理，严重影响性能，此时可以考虑：

① 设置 Spark.memory.fraction 参数（当前的内存管理器的最大内存使用比例，默认值 0.75）进行调整空间的使用，来给年轻代更多的空间，用于存放短时间存活的对象。

② -Xmn 调整 Eden 区域：对 RDD 中操作的对象和数据进行大小评估，如果在 HDFS 上解压后，一般体积可能会变成原有体积的 3 倍左右，根据数据的大小设置 Eden，如果有 10 个 Task，每个 Task 处理的 HDFS 上的数据为 128MB，则需要设置-Xmn 为 10×128×3×4/3 的大小。

③ -XX：SupervisorRatio。

④ -XX：NewRatio。

SupervisorRatio、NewRatio 正常情况下不用随便调，前提是在对 JVM 非常了解的情况下。但是，数据级别到 PB 级别之后，就完全不是这么一回事了，就要去研究 JVM 了。

26.3　Spark 下 JVM 的 On-Heap 和 Off-Heap 解密

本节讲解 Spark 下 JVM 的 On-Heap 和 Off-Heap 解密，内容包括：JVM 的 On-Heap 和

Off-Heap 详解；Spark 是如何管理 JVM 的 On-Heap 和 Off-Heap；Spark 下 JVM 的 On-Heap 和 Off-Heap 调优最佳实践。

26.3.1　JVM 的 On-Heap 和 Off-Heap 详解

JVM 可以使用的内存分为两种：堆内存和堆外内存。

JVM 堆内存分为两块：Permanent Space 和 Heap Space。

Permanent 即持久代（Permanent Generation），主要存放的是 Java 类定义信息，与垃圾收集器要收集的 Java 对象关系不大。

Heap = { Old + NEW = {Eden, from, to} }，Old 即老年代，New 即年轻代。老年代和年轻代的划分对垃圾收集影响比较大。

堆内存完全由 JVM 负责分配和释放，如果程序没有缺陷代码，导致内存泄漏，那么就不会遇到 java.lang.OutOfMemoryError 错误。

使用堆外内存，就是为了能直接分配和释放内存，提高效率。JDK5.0 之后，代码中能直接操作本地内存的方式有两种：使用未公开的 Unsafe 和 NIO 包下 ByteBuffer。

广义的堆外内存：

说到堆外内存，大家肯定想到堆内内存，这也是我们大家接触最多的，我们在 JVM 参数里通常设置-Xmx 来指定堆的最大值，不过，这还不是我们理解的 Java 堆，-Xmx 的值是新生代和老生代的和的最大值，我们在 JVM 参数里通常还会加一个参数-XX：MaxPermSize 来指定持久代的最大值，那么我们认识的 Java 堆的最大值其实是-Xmx 和-XX：MaxPermSize 的总和。在分代算法下，新生代、老生代和持久代是连续的虚拟地址，因为它们是一起分配的，那么剩下的都可以认为是堆外内存（广义的）了，这些包括了 JVM 本身在运行过程中分配的内存、代码缓冲（codeCache）、jni 里分配的内存、DirectByteBuffer 分配的内存等。

狭义的堆外内存：

而作为 Java 开发者，我们常说的堆外内存溢出了，其实是狭义的堆外内存，这主要指 java.nio.DirectByteBuffer 在创建的时候分配的内存，我们主要是讲狭义的堆外内存，因为它和我们平时碰到的问题关系比较密切。

DirectByteBuffer 在创建的时候会通过 Unsafe 的 native 方法直接使用 malloc 分配一块内存，这块内存是 heap 之外的，那么自然也不会对 GC 造成什么影响（System.GC 除外），因为 GC 耗时的操作主要是针对 heap 之内的对象，对这块内存的操作也是直接通过 Unsafe 的 native 方法来操作的，相当于 DirectByteBuffer 仅仅是一个壳，还有在通信过程中，如果数据是在 Heap 里的，最终也还是会复制一份到堆外，然后再进行发送，所以为什么不直接使用堆外内存呢？对于需要频繁操作的内存，并且仅仅是临时存在一会的，都建议使用堆外内存，并且做成缓冲池，不断循环利用这块内存。

如果大面积使用堆外内存并且没有限制，那迟早会导致内存溢出，毕竟程序是跑在一台资源受限的机器上，因为这块内存的回收不是直接能控制的，当然可以通过别的途径，如反射，直接使用 Unsafe 接口等，但是这些务必会带来一些烦恼，Java 与生俱来的优势被完全抛弃了（开发不需要关注内存的回收，由 GC 算法自动去实现）。另外，上面的 GC 机制与堆外内存的关系也说了，如果一直触发不了 cms GC 或者 Full GC，那么后果可能很严重。

26.3.2　Spark 是如何管理 JVM 的 On-Heap 和 Off-Heap 的

Spark 使用 sun.misc.Unsafe 进行 Off-heap 级别的内存分配、指针使用及内存释放。Spark 为了统一管理 Off-Heap 和 On-Heap，提出了 Page。

- On-heap 方式：由一个 64bit 的 Object 的引用和一个 64bit 的在 Object 中的 OffSet 指定具体数据。堆内内存空间 GC 的时候，对堆内结构重新组织。如果在运行的时候分配 Java Object 对象，地址不可以改变，JVM 对内存管理的负担远远大于 Off-Heap 方式，因为 GC 的时候会对 Heap 进行相应的重组。
- Off-Heap 方式：采用 C 语言的方式，一个指针直接指向数据实体。

Spark 对内存进行了封装抽象，访问数据的时候，数据可能在堆内，也可能在堆外，Spark 提供了内存管理器。内存管理器可以根据数据在堆内，还是在堆外进行具体寻址，但是，从程序运行的角度或者框架的角度看，堆内或堆外寻址对程序是封装不可见的，管理器会自动完成具体地址和寻址到具体数据的映射过程。

Page 会针对堆外内存和堆内内存两种情况进行具体的适配。Page 寻址包括两部分：数据在哪个 Page，Page 的具体偏移量 OffSet 值。

（1）Off-Heap 方式，内存就直接是一个 64bit 的 long 的指针指向具体数据。

寻址：一个指针直接指向数据结构。

（2）On-Heap 方式，堆内的情况有两部分，一部分是 64bit 的 Object 的引用，另外一部分是用 64bit 的 Object 内的 Offset 来表示具体的数据。

寻址：GC 会导致 Heap 重组。重组之后要确保 Object 的地址不变。

Page 是一个 Table。Spark 通过 Page Table 的方式进行内存管理。把内存分为很多页。页只是一个单位，和分配数组差不多，具体通过 TaskMemoryManager 对内存进行管理，根据 allocatePage 来分配页。地址和数据之间有一个映射，即将逻辑地址映射到实际的物理地址，这是钨丝计划内部的管理机制。由于逻辑地址是一个 64bit 的长整数，其前 13 个 bit 表示第几页，后 51bit 表示在页内的偏移量。所以，寻址的方式就是先找到 page，然后根据后面的 51bit，加上偏移量就找到了具体的数据在内存中的物理地址。

MemoryLocation：封装了两种逻辑地址寻址的方式。

MemoryLocation.java 的源码如下：

```
1.  /**
      *内存位置。通过内存地址跟踪（堆外内存分配），或通过 JVM 对象的偏移量（堆内存分配）
      *进行跟踪
2.    */
3.
4.  public class MemoryLocation {
5.
6.    @Nullable
7.    Object obj;
8.
9.    long offset;
10.
11.   public MemoryLocation(@Nullable Object obj, long offset) {
```

```
12.     this.obj = obj;
13.     this.offset = offset;
14.   }
15.
16.   public MemoryLocation() {
17.     this(null, 0);
18.   }
19.
20.   public void setObjAndOffset(Object newObj, long newOffset) {
21.     this.obj = newObj;
22.     this.offset = newOffset;
23.   }
24.
25.   public final Object getBaseObject() {
26.     return obj;
27.   }
28.
29.   public final long getBaseOffset() {
30.     return offset;
31.   }
32. }
```

TaskMemoryManager 管理 Page，管理 Off-heap 和 On-heap 的方式。其中的 allocatePage 方法分配内存页，分配以后加入 pageTable 中。

Spark 2.2.1 版本的 TaskMemoryManager.java 的源码如下：

```
1.   /**
       * 分配内存块，将在 MemoryManager 的页表记录；用于分配 Tungsten 内存模式下的内存
       * 块，这些内存将在操作算子之间共享。如果没有足够的内存来分配页面，则返回 null。可能
       * 返回一页，其包含的字节比请求的字节少，因此调用者应该验证返回页面的大小
2.     */
3.
4.    public MemoryBlock allocatePage(long size, MemoryConsumer consumer) {
5.      assert(consumer != null);
6.      assert(consumer.getMode() == tungstenMemoryMode);
7.      if (size > MAXIMUM_PAGE_SIZE_BYTES) {
8.        throw new IllegalArgumentException(
9.          "Cannot allocate a page with more than " + MAXIMUM_PAGE_SIZE_BYTES
              + " bytes");
10.     }
11. ......
```

Spark 2.4.3 版本的 TaskMemoryManager.java 源码与 Spark 2.2.1 版本相比具有如下特点。

- 上段代码中第 8 行抛出 IllegalArgumentException，异常调整为抛出 TooLargePageException 异常。

```
1.  ......
2.  throw new TooLargePageException(size);
3.  ......
4. public class TooLargePageException extends RuntimeException {
5.   TooLargePageException(long size) {
6.     super("Cannot allocate a page of " + size + " bytes.");
7.   }
8. }
```

26.3.3　Spark 下 JVM 的 On-Heap 和 Off-Heap 调优最佳实践

Java 内存分为堆和栈等，其中堆由两部分组成：分别是 Old Generation 和 Young Generation。Young Generation 又由三部分组成：一个 Eden 区域和两个 Survivor 区域，每次放对象的时候，都是放入 Eden 区域，和其中一个 Survivor 区域；另外一个 Survivor 区域是空闲的。

当 Eden 区域和一个 Survivor 区域放满了以后（Spark 运行过程中产生的对象实在太多了），就会触发 Minor GC，把不再使用的对象从内存中清空，给后面新创建的对象腾出内存空间。清理掉不再使用的对象后，将存活下来的对象（还要继续使用的），放入之前空闲的那个 Survivor 区域中。这里可能会出现一个问题。默认 Eden、Survivor1 和 Survivor2 的内存占比是 8:1:1。如果存活下来的对象是 1.5，一个 Survivor 区域放不下，此时就可能通过 JVM 的担保机制（不同 JVM 版本可能对应不同的行为），将多余的对象直接放入老年代。

如果 JVM 内存不够大，可能导致频繁的年轻代内存满溢，频繁地进行 Minor GC。频繁地 Minor GC 会导致短时间内，有些存活的对象多次垃圾回收都没有回收掉，导致这种短生命周期（其实不一定是要长期使用的）对象，年龄过大及垃圾回收次数太多还没有回收，而将这些对象移到老年代。老年代中可能会因为内存不足，囤积一大堆短生命周期的，本来应该在年轻代中的、可能马上就要被回收掉的对象。此时可能导致老年代频繁满溢，频繁进行 Full GC（全局/全面垃圾回收），Full GC 会去回收老年代中的对象。

Full GC 算法的设计是针对老年代中的对象数量很少，满溢进行 Full GC 频率很小的情况，采取了不太复杂，但是耗费性能和时间的垃圾回收算法，因此 Full GC 运行很慢。Full GC/Minor GC，无论是快，还是慢，都会导致 JVM 的工作线程停止工作。简言之，GC 的时候，Spark 停止工作了，等着垃圾回收结束。

内存不足时存在的问题如下。

- 频繁 Minor GC，也会导致 Spark 频繁停止工作。
- 老年代囤积大量活跃对象（短生命周期的对象），导致频繁 Full GC，Full GC 时间很长，短则数十秒，长则数分钟，甚至数小时。可能导致 Spark 长时间停止工作。
- 严重影响 Spark 的性能和运行的速度。

Spark 下 JVM 的 On-Heap 和 Off-Heap 调优方式：

Spark 中的堆内存被划分成两块：一块是专门用来给 RDD 的 Cache、Persist 操作进行 RDD 数据缓存用的；另一块是用来给 Spark 算子函数的运行使用的，存放函数中自己创建的对象。默认情况下，给 RDD Cache 操作的内存占比是 0.6，60%的内存都给 Cache 操作了。

如果某些情况下，Cache 不是那么的紧张， Task 算子函数中创建的对象过多，然而内存又不太大，导致频繁地 Minor GC，甚至频繁 Full GC，导致 Spark 频繁地停止工作，性能影响会很大。针对这种情况，可以在 Spark We bUI 页面（如在 YARN 中运行，可以通过 YARN 的 Web 页面）查看 Spark 作业的运行统计，从 Web UI 页面中一层一层点击进去可以看到每个 Stage 的运行情况，包括每个 Task 的运行时间、GC 时间等。如果发现 GC 太频繁，时间太长。此时就可以适当调节这个比例。

降低 Cache 操作的内存占比，可使用 Persist 操作，选择将一部分缓存的 RDD 数据写入磁盘；或者使用序列化方式，配合 Kryo 序列化类，减少 RDD 缓存的内存占用。降低 Cache 操作内存占比，对应的算子函数的内存占比就提升了。这时，可能降低 Minor GC 的频率，同时减少 Full GC 的频率，对 Spark 性能的提升是有一定帮助的。

调优方式：让 Task 执行算子函数时，有更多的内存可以使用。

StaticMemoryManager.scala 的源码的 spark.storage.memoryFraction 默认配置是 0.6：

```
1.   private def getMaxStorageMemory(conf: SparkConf): Long = {
2.     val systemMaxMemory = conf.getLong("spark.testing.memory", Runtime.
       getRuntime.maxMemory)
3.     val memoryFraction = conf.getDouble("spark.storage.memoryFraction", 0.6)
4.     val safetyFraction = conf.getDouble("spark.storage.safetyFraction", 0.9)
5.     (systemMaxMemory * memoryFraction * safetyFraction).toLong
6.   }
```

调整以后的参数设置方式如下，将 Spark 的存储空间从 0.6 调整为 0.5：

```
1.   SparkContext().set("Spark.storage.memoryFraction", "0.5")
```

数据量很大时，导致一个 Stage 内存溢出而挂掉，Block Manager 也没有了，使得后面的 Stage 取不到数据而出错，如 Shuffle file cannot find，Executor Task lost，Out Of Memory（内存溢出）。

解决方式：修改堆外内存最大值。

```
1.   --conf spark.yarn.executor.memoryoverhead=2048
```

有时 Task 需要的数据在其他节点，此时需要去拉取数据，而刚好此时那个节点正在执行垃圾回收而无法回应数据，导致连接超时（默认是 60s）而出现以下错误：uuid（dsfsfd-2342vs--sdf--sdfsd）not found、file lost。

此时应该增加连接等待时间。

```
1.   --conf Spark.core.Connection.ack.wait.timeout=300
```

例如，提交 Spark 应用程序的参数配置。

```
1.   /usr/local/spark/bin/spark-submit \
2.  --class com.ibeifeng.sparkstudy.wordcount \
3.  --num-executors 80 \
4.  --driver-memory 6g \
5.  --executor-memory 6g \
6.  --executor-cores 3 \
7.  --master yarn-cluster \
8.  --queue root.default \
9.  --conf spark.yarn.executor.memoryoverhead=2048 \
10. --conf spark.core.Connection.ack.wait.timeout=300 \
11. /usr/local/spark/spark.jar \
```

26.4 Spark 下 JVM GC 导致的 Shuffle 拉取文件失败及调优方案

本节讲解 Spark 下的 JVM GC 导致的 Shuffle 拉取文件失败原因解密；Spark 下的 JVM GC 导致的 Shuffle 拉取文件失败时候调优。

26.4.1　Spark 下 JVM GC 导致的 Shuffle 拉取文件失败原因解密

Spark 运行时有时出现 Shuffle 拉取文件失败的情况，如 Shuffle output file lost。真正的原因是 GC 导致的！如果 GC 尤其是 Full GC 产生，通常会导致线程停止工作，这个时候下一个 Stage 的 Task 在默认情况下就会重试来获取数据。一般重试 3 次，每次重试的时间为 5s。也就是说，默认情况下，15s 内如果还是无法抓到数据，就会出现 Shuffle output file lost 等情况，进而导致 Task 重试，甚至会导致 Stage 重试，最严重的是会导致 App 失败；在这个时候首先就要采用高效的内存数据结构和序列化机制、JVM 调优来减少 Full GC 的产生。

26.4.2　Spark 下 JVM GC 导致的 Shuffle 拉取文件失败时调优

Spark 下的 JVM GC 导致的 Shuffle 拉取文件失败时调优措施如下。

（1）在 Shuffle 的时候，Reducer 端获取数据会有一个指定大小的缓存空间，在内存足够大的情况下，可以适当地增大该缓存空间，否则会 Spill 到磁盘上，影响效率。此时可以调整（增大）spark.reducer.maxSizeInFlight 参数。

（2）在 ShuffleMapTask 端通常也会增大 Map 任务的写磁盘的缓存，默认情况下是 32KB，即 spark.shuffle.file.buffer 32k。

（3）调整获取 Shuffle 数据的重试次数，默认是 3 次，通常建议增大重试次数。

（4）调整获取 Shuffle 数据重试的时间间隔，默认为 5s，即 spark.shuffle.io.retryWait 5s，强烈建议增大该时间间隔。

（5）在 Reducer 端 Aggregation 的时候，默认是 20%的内存用来 Aggregation，如果超出这个大小，就会溢出到磁盘上，建议调大百分比来提高性能。

26.5　Spark 下 Executor 对 JVM 堆外内存连接等待时长调优

本节讲解 Spark 下的 Executor 对 JVM 堆外内存连接等待时长调优，内容包括：Executor 对堆外内存等待工作过程；Executor 对堆外内存等待时长调优。

26.5.1　Executor 对堆外内存等待工作过程

有时如果 Spark 作业处理的数据量特别大，如几亿数据量，Spark 作业运行时不时地报错，如 Shuffle file cannot find、Executor Task lost、Out Of Memory（内存溢出）等。可能是 Executor 的堆外内存不太够用，导致 Executor 在运行的过程中内存溢出，而后续的 Stage 的 Task 运行时需从一些 Executor 中去拉取 Shuffle map output 文件，如果此时 Stage0 的 Executor 挂了，Block Manager 也没有了，Stage1 的 Executor 的 Task 虽然通过 Driver 的 MapOutputTrakcer 获取到了自己数据的地址，但是去 Executor 的 Block Manager 是获取不到数据的，可能会报 Shuffle output file not found、DAGScheduler resubmitting Task、Executor lost 等各种错误，直

到挂掉，反复挂掉几次，反复报错几次，Spark 作业彻底崩溃。

上述情况下，就可以去考虑调节一下 Executor 的堆外内存。也许就可以避免报错；此外，有时堆外内存调节得比较大，对于性能来说，也会带来一定的提升。

默认情况下，堆外内存的上限为 300MB。真正处理大数据的时候，这里都会出现问题，导致 Spark 作业反复崩溃，无法运行；此时就会去调节这个参数到至少 1GB（1024MB），甚至 2GB、4GB。通常这个参数调上去以后，就会避免掉某些 JVM OOM 的异常问题，同时会让整体 Spark 作业的性能得到较大提升。

解决方式：修改堆外内存最大值。

```
1.   --conf spark.yarn.executor.memoryoverhead=2048
```

26.5.2　Executor 对堆外内存等待时长调优

Shuffle 的时候，Task 需从其他节点拉取数据，如果刚好此时那个节点正在执行垃圾回收而无法回应数据，导致连接超时（默认为 60s）出现以下错误：uuid（dsfsfd-2342vs--sdf--sdfsd）not found、file lost 等。这种情况下，可能是有数据的 Executor 在运行 JVM GC 垃圾回收，所以在拉取数据的时候建立不了连接，然后超过默认 60s 以后直接宣告失败。如果报错几次，几次都拉取不到数据，可能会导致 DAGScheduler 反复提交几次 Stage；或 TaskScheduler 反复提交几次 Task，大大延长 Spark 作业的运行时间，最终可能导致 Spark 作业的崩溃。

Executor 对堆外内存等待时长的问题，可以考虑调节连接的超时时长。

```
1.   --conf spark.core.Connection.ack.wait.timeout=300
```

注意：这里的配置需在 Spark-Submit 脚本里面使用--conf 的方式去添加，而不是在 Spark 作业代码中使用 new SparkConf().set()的方式去设置。SparkConf().set()的方式没有起作用。

```
1.  /usr/local/spark/bin/spark-submit \
2.  --class com.ibeifeng.sparkstudy.wordcount \
3.  --num-executors 80 \
4.  --driver-memory 6g \
5.  --executor-memory 6g \
6.  --executor-cores 3 \
7.  --master yarn-cluster \
8.  --queue root.default \
9.  --conf spark.yarn.executor.memoryoverhead=2048 \
10. --conf spark.core.Connection.ack.wait.timeout=300 \
11. /usr/local/spark/spark.jar \
```

26.6　Spark 下 JVM 内存降低 Cache 内存占比的调优

本节讲解 Spark 下 JVM 内存降低 Cache 内存占比的调优，内容包括：

- 什么时候需要降低 Cache 的内存占用。
- 降低 Cache 的内存占比调优最佳实践。

26.6.1 什么时候需要降低 Cache 的内存占用

Spark 中的堆内存划分如下。

- storageMemory： RDD 的 Cache、Persist 操作进行 RDD 数据缓存使用；默认情况下，RDD Cache 操作的内存占比是 0.6，60%的内存都给了 Cache 操作。
- shuffleMemory： Spark 算子函数的运行使用，存放函数中自己创建的对象。

如果某些情况下，Cache 不是那么紧张， Task 算子函数中创建的对象过多，然而内存又不太大，导致频繁的 Minor GC，甚至频繁 Full GC，导致 Spark 频繁地停止工作，性能影响会很大。针对这种情况，可以在 Spark Web UI 页面（如在 YARN 中运行，可以通过 YARN 的 Web 页面）查看 Spark 作业的运行统计，从 Web UI 页面中一层一层点击进去可以看到每个 Stage 的运行情况，包括每个 Task 的运行时间、GC 时间等。如果发现 GC 太频繁，时间太长，就可以适当调节这个比例。

26.6.2 降低 Cache 的内存占比调优最佳实践

降低 Cache 操作的内存占比调优最佳实践。

- 可使用 Persist 操作，选择将一部分缓存的 RDD 数据写入磁盘。
- 或者使用序列化方式，配合 Kryo 序列化类，减少 RDD 的缓存的内存占用。
- 降低 Cache 操作内存占比，提升对应的算子函数的内存占比。

调优方式：调整 spark.storage.memoryFraction 参数。spark.storage.memoryFraction 默认配置参数是 0.6。

调整以后的参数设置方式如下，如将 Spark 的存储空间从 0.6 调整为 0.5。

```
1.   SparkContext().set("Spark.storage.memoryFraction", "0.5")
```

第 27 章　Spark 五大子框架调优最佳实践

本章讲解 Spark 五大子框架调优最佳实践。27.1 节讲解 Spark SQL 调优原理调优参数及调优最佳实践；27.2 节讲解 Spark Streaming 调优原理及调优最佳实践；27.3 节讲解 Spark GraphX 调优原理及调优最佳实践；27.4 节讲解 Spark ML 调优原理及调优最佳实践；27.5 节讲解 SparkR 调优原理及调优最佳实践。

27.1　Spark SQL 调优原理及调优最佳实践

本节讲解 Spark SQL 调优原理；以及 Spark SQL 调优参数及调优最佳实践。

27.1.1　Spark SQL 调优原理

Spark SQL 是 Spark 用来处理结构化数据的一个模块。与基础的 Spark RDD API 不同，Spark SQL 提供了更多数据与执行计算的信息，在其实现中会使用这些额外信息进行优化。可以使用 SQL 语句和 Dataset API 与 Spark SQL 模块交互。无论使用哪种语言或 API 来执行计算，都会使用相同的引擎。可以选择熟悉的语言（支持 Scala、Java、R、Python）以及在不同场景下选择不同的方式进行计算。

（1）SQL：一种使用 Spark SQL 的方式是使用 SQL。Spark SQL 也支持从 Hive 中读取数据。使用编码方式执行 SQL 将会返回一个 Dataset/DataFrame。也可以使用命令行、JDBC/ODBC 与 Spark SQL 进行交互。

（2）Datasets 和 DataFrames：Dataset 是一个分布式数据集合。Dataset 是自 Spark 1.6 开始提供的新接口，能同时享受到 RDD 的优势（Dataset 是强类型，能使用强大的 lambda 函数）以及 Spark SQL 优化过的执行引擎。Dataset 可以从 JVM 对象创建而来，并且可以使用各种 transform 操作（如 map、flatMap、filter 等）。目前，Dataset API 支持 Scala 和 Java。Python 暂不支持 Dataset API。不过，得益于 Python 的动态属性，可以享受到许多 DataSet API 的益处。R 也是类似情况。

DataFrame 是具有名字的列，概念上相当于关系数据库中的表或 R/Python 下的 data frame，但有更多的优化。DataFrames（Dataset 亦如此）可以从很多数据中构造，例如，结构化文件、Hive 中的表、数据库、已存在的 RDD。DataFrame API 可在 Scala、Java、Python 和 R 中使用。在 Scala 和 Java 中，DataFrame 由一个元素为 Row 的 Dataset 表示。在 Scala API 中，DataFrame 只是 Dataset[Row] 的别名。在 Java API 中，类型为 Dataset<Row>。

1. Spark SQL运行架构

类似于关系型数据库，SparkSQL 也是语句，由 Projection（a1，a2，a3）、Data Source（tableA）、Filter（condition）组成，分别对应 SQL 查询过程中的 Result、Data Source、Operation。

也就是说，SQL 语句是按 Result→Data Source→Operation 的次序描述的，如图 27-1 所示。

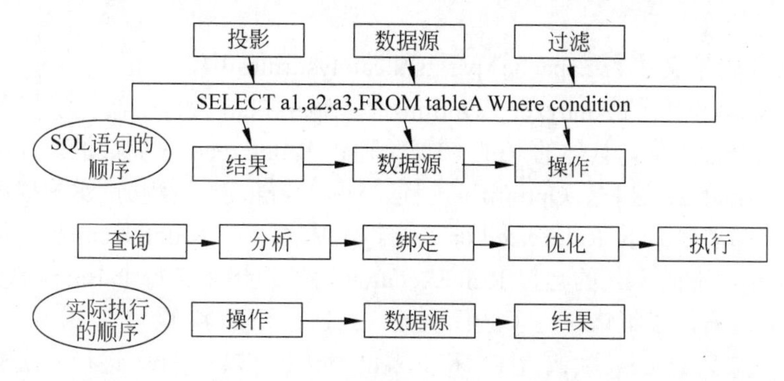

图 27-1　SQL 语句的表达顺序与 SQL 实际执行顺序的对比图

执行 SparkSQL 语句的顺序为：

（1）对读入的 SQL 语句进行解析（Parse），分辨出 SQL 语句中哪些词是关键词（如 SELECT、FROM、WHERE），哪些是表达式，哪些是 Projection，哪些是 Data Source 等，从而判断 SQL 语句是否规范。

（2）将 SQL 语句和数据库的数据字典（如列、表、视图等）进行绑定（Bind），如果相关的 Projection、Data Source 等都存在，就表示这个 SQL 语句是可以执行的。

（3）一般的数据库会提供几个执行计划，这些计划一般都有运行统计数据，数据库会在这些计划中选择一个最优计划（Optimize）。

（4）计划执行（Execute），按 Operation→Data Source→Result 的次序进行，在执行过程中有时甚至不需要读取物理表，就可以返回结果，如重新运行刚运行过的 SQL 语句，可能直接从数据库的缓冲池中获取返回结果。

1）Tree 和 Rule

SparkSQL 对 SQL 语句的处理和关系型数据库对 SQL 语句的处理采用了类似的方法，首先会将 SQL 语句进行解析（Parse），然后形成一个 Tree，在后续的（如绑定、优化等）处理过程中都是对 Tree 的操作，而操作的方法是采用 Rule，通过模式匹配，对不同类型的节点采用不同的操作。在整个 SQL 语句的处理过程中，Tree 和 Rule 相互配合，完成了解析、绑定（在 SparkSQL 中称为 Analysis）、优化、物理计划等过程，最终生成可以执行的物理计划。

（1）Tree。

- Tree 的相关代码定义在 org.apache.spark.sql.catalyst 中。
- Logical Plans、Expressions、Physical Operators 都可以使用 Tree 表示。
- Tree 的具体操作是通过 TreeNode 来实现的。
- TreeNode 可以使用 Scala 的集合操作方法（如 foreach、map、flatMap、collect 等）进行操作。有了 TreeNode，通过 Tree 中各个 TreeNode 之间的关系，可以对 Tree 进行遍历操作，如使用 transformDown、transformUp 将 Rule 应用到给定的树段，然后用结果替代旧的树段；也可以使用 transformChildrenDown、transformChildrenUp 对一个给定的节点进行操作，通过迭代将 Rule 应用到该节点以及子节点。

TreeNode 可以细分成 3 种类型的 Node。

- UnaryNode 一元节点，即只有一个子节点，如 Limit、Filter 操作。
- BinaryNode 二元节点，即有左右子节点的二叉节点，如 Join、Union 操作。

- LeafNode 叶子节点，没有子节点的节点，主要用于命令类操作，如 SetCommand。

（2）Rule。

Rule 的相关代码定义在 org.apache.spark.sql.catalyst.rules 中。

- Rule 在 SparkSQL 的 Analyzer、Optimizer、SparkPlan 等各个组件中都会应用到。
- Rule 是一个抽象类，具体的 Rule 实现是通过 RuleExecutor 完成的。凡需要处理执行计划树（Analyze 过程、Optimize 过程、SparkStrategy 过程），实施规则匹配和节点处理的，都需要继承 RuleExecutor[TreeType]抽象类。RuleExecutor 内部提供了一个 Seq[Batch]，里面定义的是该 RuleExecutor 的处理步骤。每个 Batch 代表一套规则，配备一个策略，该策略说明了迭代次数（一次，还是多次）。
- Rule 通过定义 Once（默认为 1）和 FixedPoint，可以对 Tree 进行一次操作或多次操作（如对某些 Tree 进行多次迭代操作的时候，达到 FixedPoint 次数迭代或达到前后两次的树结构没变化才停止操作，具体参看 RuleExecutor.apply。

2）Spark SQL 的运行过程

接下来看一下 Spark SQL 的运行过程。Spark SQL 的运行架构如图 27-2 所示。

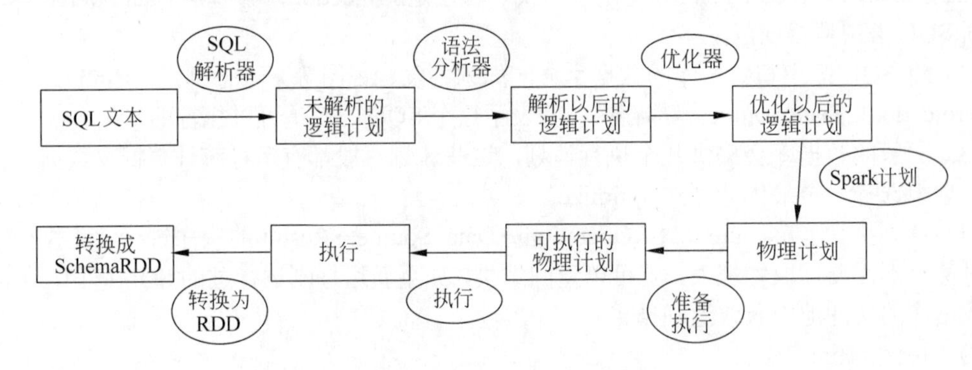

图 27-2　Spark SQL 的运行架构

通过初步解析不同来源的数据变为 UnresolvedLogical Plan（此过程会提取关键字，检查基本的语法，如果有问题，则下一步直接不能运行），进一步解析语法树生成 Logical Plan，进行 CombineFilters、CombineLimits 等优化策略，产生 Physical Plan，把需要执行的操作转换为 Spark 可以真正执行的 RDD。

先概括一下，其执行流程是：

Parse SQL→Analyze Logical Plan→Optimize Logical Plan→Generate Physical Plan→Prepareed Spark Plan→Execute SQL→Generate RDD。

SQLContext 里对 SQL 的解析和执行流程如下。

（1）第一步 Parse SQL (SQL: String)，simple SQL parser 做词法语法解析，生成 LogicalPlan。

（2）第二步 Analyzer(logicalPlan)，把做完词法语法解析的执行计划进行初步分析和映射。

在构建 SessionState 类时，作为参数传入 SessionState。

Spark 2.2.1 版本的 SessionState.scala 的源码如下：

```
1.    private[sql] class SessionState(
2.      sharedState: SharedState,
```

```
3.     val conf: SQLConf,
4.     val experimentalMethods: ExperimentalMethods,
5.     val functionRegistry: FunctionRegistry,
6.     val udfRegistration: UDFRegistration,
7.     val catalog: SessionCatalog,
8.     val sqlParser: ParserInterface,
9.     val analyzer: Analyzer,
10.    val optimizer: Optimizer,
11.    val planner: SparkPlanner,
12.    val streamingQueryManager: StreamingQueryManager,
13.    val listenerManager: ExecutionListenerManager,
14.    val resourceLoader: SessionResourceLoader,
15.    createQueryExecution: LogicalPlan => QueryExecution,
16.    createClone: (SparkSession, SessionState) => SessionState) {
```

Spark 2.4.3 版本的 SessionState.scala 源码与 Spark 2.2.1 版本相比具有如下特点。

- 上段代码中第 7 行将 catalog 成员变量调整为 catalogBuilder 方法，创建用于管理表和数据库状态的内部目录的函数。
- 上段代码中第 9 行将 analyzer 成员变量调整为 analyzerBuilder 方法，用于创建逻辑查询计划分析器的函数，用来解决未解析的属性和关系。
- 上段代码中第 10 行将 optimizer 成员变量调整为 optimizerBuilder 方法，用于创建逻辑查询计划优化器的函数。
- 上段代码中第 14 行将 resourceLoader 成员变量调整为 resourceLoaderBuilder 方法，创建会话共享资源加载程序以加载 jar、文件等的函数。

```
1.  ……
2.   catalogBuilder: () => SessionCatalog,
3.  ……
4.   analyzerBuilder: () => Analyzer,
5.  ……
6.   optimizerBuilder: () => Optimizer,
7.  ……
8.  resourceLoaderBuilder: () => SessionResourceLoader,
```

其中，SessionState 在构建 SparkSession 时进行初始化，SessionState 是跨会话隔离状态，包括 SQL 配置、临时表、已注册功能，以及一切接受[org.apache.spark.sql.internal.SQLConf]的配置内容。如果 parentSessionState 不为空，SessionState 将从父节点复制。这是 Spark 内部使用。

SparkSession.scala 的源码如下：

```
1.     ……
2.      @InterfaceStability.Unstable
3.      @transient
4.      lazy val sessionState: SessionState = {
5.        parentSessionState
6.          .map(_.clone(this))
7.          .getOrElse {
8.            val state = SparkSession.instantiateSessionState(
9.              SparkSession.sessionStateClassName(sparkContext.conf),
10.             self)
11.           initialSessionOptions.foreach { case (k, v) =>
              state.conf.setConfString(k, v) }
12.           state
13.         }
14.     }
15. ……
```

其中，instantiateSessionState 是辅助方法来创建一个基于配置 className 的 SessionState 实例，结果是 SessionState，或者是基于 Hive 的 SessionState。

SparkSession.scala 的 instantiateSessionState 方法的源码如下：

```
1.  ......
2.    private def instantiateSessionState(
3.        className: String,
4.        sparkSession: SparkSession): SessionState = {
5.      try {
6.        //触发 new [Hive]SessionStateBuilder(SparkSession, Option[SessionState])
7.        val clazz = Utils.classForName(className)
8.        val ctor = clazz.getConstructors.head
9.        ctor.newInstance(sparkSession, None).asInstanceOf[BaseSession
          StateBuilder]. build()
10.     } catch {
11.       case NonFatal(e) =>
12.         throw new IllegalArgumentException(s"Error while instantiating
            '$className':", e)
13.     }
14.   }
15. ......
```

其中，调用 ctor.newInstance(sparkSession, None).asInstanceOf[BaseSessionStateBuilder].build 方法，构建 Analyzer 实例对象。

Spark 2.2.1 版本的 BaseSessionStateBuilder.scala 的源码如下：

```
1.  ......
2.  //解析仍未解析属性和关系的逻辑查询计划分析器。说明：这取决于 conf 和 catalog 字段
3.  protected def analyzer: Analyzer = new Analyzer(catalog, conf) {
4.      override val extendedResolutionRules: Seq[Rule[LogicalPlan]] =
5.        new FindDataSourceTable(session) +:
6.          new ResolveSQLOnFile(session) +:
7.          customResolutionRules
8.
9.      override val postHocResolutionRules: Seq[Rule[LogicalPlan]] =
10.       PreprocessTableCreation(session) +:
11.         PreprocessTableInsertion(conf) +:
12.         DataSourceAnalysis(conf) +:
13.         customPostHocResolutionRules
14.
15.     override val extendedCheckRules: Seq[LogicalPlan => Unit] =
16.       PreWriteCheck +:
17.         HiveOnlyCheck +:
18.         customCheckRules
19.   }
20. ......
21. def build(): SessionState = {
22.     new SessionState(
23.       session.sharedState,
24.       conf,
25.       experimentalMethods,
26.       functionRegistry,
27.       udfRegistration,
28.       catalog,
29.       sqlParser,
30.       analyzer,
31.       optimizer,
32.       planner,
```

```
33.       streamingQueryManager,
34.       listenerManager,
35.       resourceLoader,
36.       createQueryExecution,
37.       createClone)
38.   }
39. ......
```

Spark 2.4.3 版本的 BaseSessionStateBuilder.scala 源码与 Spark 2.2.1 版本相比具有如下特点。

- 上段代码中第 28 行 catalog 调整为匿名函数，返回 catalog。
- 上段代码中第 30 行 analyzer 调整为匿名函数，返回 analyzer。
- 上段代码中第 31 行 optimizer 调整为匿名函数，返回 optimizer。
- 上段代码中第 35 行 resourceLoader 调整为匿名函数，返回 resourceLoader。

```
1.  ......
2.        () => catalog,
3.  ......
4.        () => analyzer,
5.        () => optimizer,
6.  ......
7.        () => resourceLoader,
```

Analyzer 传入的 catalog 为 SessionCatalog，catalog 用来注册 table 和查询 relation。catalog 在构建 SessionState 类时已经实例化，作为参数传入 SessionState。构建 SparkSession 时调用 instantiateSessionState 进行初始化，instantiateSessionState 方法调用 ctor.newInstance(sparkSession, None).asInstance Of[BaseSessionStateBuilder].build 方法，在 BaseSessionStateBuilder 类初始化时构建 SessionCatalog 实例，然后赋值给 catalog。SessionCatalog 管理表和数据库状态的目录 catalog。如果预先存在目录 catalog，则该目录 catalog 的状态（临时表和当前数据库）将被复制到新目录 catalog 中。这取决于 conf、functionRegistry 及 sqlParser 字段。

Spark 2.2.1 版本的 BaseSessionStateBuilder.scala 的源码如下：

```
1.   ......
2.   protected lazy val catalog: SessionCatalog = {
3.     val catalog = new SessionCatalog(
4.       session.sharedState.externalCatalog,
5.       session.sharedState.globalTempViewManager,
6.       functionRegistry,
7.       conf,
         SessionState.newHadoopConf(session.sparkContext.hadoopConfiguration,
         conf),
8.       sqlParser,
9.       resourceLoader)
10.    parentState.foreach(_.catalog.copyStateTo(catalog))
11.    catalog
12.  }
```

Spark 2.4.3 版本的 BaseSessionStateBuilder.scala 源码与 Spark 2.2.1 版本相比具有如下特点。

- 上段代码中第 4 行 externalCatalog 调整为匿名函数，返回 externalCatalog。
- 上段代码中第 5 行 globalTempViewManager 调整为匿名函数，返回 globalTempView-Manager。

```
1.  ......
2.        () => session.sharedState.externalCatalog,
```

```
3.        () => session.sharedState.globalTempViewManager,
4.   ......
```

其中，FunctionRegistry 是注册函数，由于存在 lookupFunction 方法，所以该 analyzer 支持 Function 注册，即 UDF 自定义函数。

Spark 2.2.1 版本的 FunctionRegistry.scala 的源码如下：

```
1.   trait FunctionRegistry {
2.
3.    final def registerFunction(name: String, builder: FunctionBuilder):
      Unit = {
4.      registerFunction(name, new ExpressionInfo(builder.getClass.getCanonicalName,
        name), builder)
5.    }
6.
7.    def registerFunction(name: String, info: ExpressionInfo, builder:
      FunctionBuilder): Unit
8.
9.    @throws[AnalysisException]("If function does not exist")
10.   def lookupFunction(name: String, children: Seq[Expression]): Expression
```

Spark 2.4.3 版本的 FunctionRegistry.scala 源码与 Spark 2.2.1 版本相比具有如下特点。

- 上段代码中第 3、7、10 行 name 的类型调整为 FunctionIdentifier。FunctionIdentifier 是一个 case class，用于标识数据库中的函数。如果未定义“database”，则使用当前数据库。
- 上段代码中第 4 行 registerFunction 函数的第二个参数调整为 info 变量。

```
1.  ......
2.     final def registerFunction(name: FunctionIdentifier, builder: FunctionBuilder):
       Unit = {
3.      val info = new ExpressionInfo(
4.        builder.getClass.getCanonicalName, name.database.orNull, name.funcName)
5.      registerFunction(name, info, builder)
6.  ......
7.    def registerFunction(
8.      name: FunctionIdentifier,
9.      info: ExpressionInfo,
10.     builder: FunctionBuilder): Unit
11. ......
12. def lookupFunction(name: FunctionIdentifier, children: Seq[Expression]):
    Expression
13. ......
14. case class FunctionIdentifier(funcName: String, database: Option[String])
15.   extends IdentifierWithDatabase {
16.
17.   override val identifier: String = funcName
18.
19.   def this(funcName: String) = this(funcName, None)
20.
21.   override def toString: String = unquotedString
22. }
23.
24. object FunctionIdentifier {
25.   def apply(funcName: String): FunctionIdentifier = new
      FunctionIdentifier(funcName)
26. }
```

Analyzer 内定义了的逻辑计划解析规则。Spark 2.2.1 版本的 Analyzer.scala 的源码如下：

```
1.    lazy val batches: Seq[Batch] = Seq(
2.      Batch("Hints", fixedPoint,
3.        new ResolveHints.ResolveBroadcastHints(conf),
4.        ResolveHints.RemoveAllHints),
5.      Batch("Simple Sanity Check", Once,
6.        LookupFunctions),
7.      Batch("Substitution", fixedPoint,
8.        CTESubstitution,
9.        WindowsSubstitution,
10.       EliminateUnions,
11.       new SubstituteUnresolvedOrdinals(conf)),
12.     Batch("Resolution", fixedPoint,
13.       ResolveTableValuedFunctions ::
14.       ResolveRelations ::
15.       ResolveReferences ::
16.       ResolveCreateNamedStruct ::
17.       ResolveDeserializer ::
18.       ResolveNewInstance ::
19.       ResolveUpCast ::
20.       ResolveGroupingAnalytics ::
21.       ResolvePivot ::
22.       ResolveOrdinalInOrderByAndGroupBy ::
23.       ResolveAggAliasInGroupBy ::
24.       ResolveMissingReferences ::
25.       ExtractGenerator ::
26.       ResolveGenerate ::
27.       ResolveFunctions ::
28.       ResolveAliases ::
29.       ResolveSubquery ::
30.       ResolveWindowOrder ::
31.       ResolveWindowFrame ::
32.       ResolveNaturalAndUsingJoin ::
33.       ExtractWindowExpressions ::
34.       GlobalAggregates ::
35.       ResolveAggregateFunctions ::
36.       TimeWindowing ::
37.       ResolveInlineTables(conf) ::
38.       ResolveTimeZone(conf) ::
39.       TypeCoercion.typeCoercionRules ++
40.       extendedResolutionRules : _*),
41.     Batch("Post-Hoc Resolution", Once, postHocResolutionRules: _*),
42.     Batch("View", Once,
43.       AliasViewChild(conf)),
44.     Batch("Nondeterministic", Once,
45.       PullOutNondeterministic),
46.     Batch("UDF", Once,
47.       HandleNullInputsForUDF),
48.     Batch("FixNullability", Once,
49.       FixNullability),
50.     Batch("Subquery", Once,
51.       UpdateOuterReferences),
52.     Batch("Cleanup", fixedPoint,
53.       CleanupAliases)
54.   )
```

Spark 2.4.3 版本中的 Analyzer.scala 源码，与 Spark 2.2.1 版本相比具有如下特点。

- 上段代码中第 3 行之后新增一行代码，新增加 ResolveHints.ResolveCoalesceHints，coalesce 提示接受名称“coalesce”和“repartition”，它的参数包括一个分区号。

- 上段代码中第29行之后新增一行代码ResolveSubqueryColumnAliases，使用投影替换子查询的未解析列别名。
- 上段代码中第32行之后新增一行代码ResolveOutputRelation，从逻辑计划中的数据输出表解析出列。此规则将按名称写入时重新排序列；数据类型不匹配时插入安全强制转换；当列名不匹配时插入别名；检测与输出表不兼容的计划并引发AnalysisException。
- 上段代码中第37行之后新增两行代码ResolveHigherOrderFunctions和ResolveLambda-Variables。

  ResolveHigherOrderFunctions规则用于从目录中解析高阶函数。这与常规函数解析不同，因为lambda函数只能在函数解析之后才能解析。因此，当解析所有子函数或lambda函数时，我们需要解析高阶函数。

  ResolveLambdaVariables规则解析由高阶函数提供的lambda变量。此规则分两步工作：第1步，将高阶函数提供的匿名变量绑定到lambda函数的参数；这将创建命名和类型化的lambda变量。将检查参数名称是否重复，并在此步骤中检查参数数量。第2步，解析lambda函数的函数表达式树中使用的lambda变量。注意，我们允许使用来自当前lambda外部的变量，这可以是在外部范围中定义的lambda函数，也可以是由计划的子级生成的属性。如果名称重复，则使用在最内部作用域中定义的名称。
- 上段代码中第38行之后新增一行代码ResolveRandomSeed，为随机数生成设置种子。

```
1.  ……
2.  ResolveHints.ResolveCoalesceHints,
3.  ……
4.  ResolveSubqueryColumnAliases ::
5.  ……
6.  ResolveOutputRelation ::
7.  ……
8.  ResolveHigherOrderFunctions(catalog) ::
9.  ResolveLambdaVariables(conf) ::
10. ……
11. ResolveRandomSeed ::
```

（3）从第二步得到的是初步的logicalPlan，接下来是optimizer(plan)。Optimizer里面也定义了几批规则，会按序对执行计划进行优化操作。

Spark 2.2.1版本的Optimizer.scala的源码如下：

```
1.  /**
2.    *抽象类继承的所有优化器，包含标准的批处理（扩展优化器可以重写）
3.    */
4.  abstract class Optimizer(sessionCatalog: SessionCatalog, conf: SQLConf)
5.    extends RuleExecutor[LogicalPlan] {
6.
7.    protected val fixedPoint = FixedPoint(conf.optimizerMaxIterations)
8.
9.    def batches: Seq[Batch] = {
10.     /*严格说来， Finish Analysis 中的一些规则不是优化器规则，更多地属于分析，这是
    因为它们需要的正确性（如 ComputeCurrentTime）。然而，因为我们也使用 analyzer 的规
    范化查询（如视图定义），故无须消除子查询或计算分析当前时间*/
11.     Batch("Finish Analysis", Once,
```

```
12.       EliminateSubqueryAliases,
13.       EliminateView,
14.       ReplaceExpressions,
15.       ComputeCurrentTime,
16.       GetCurrentDatabase(sessionCatalog),
17.       RewriteDistinctAggregates,
18.       ReplaceDeduplicateWithAggregate) ::
19.     ////////////////////////////////////////////////////////////////////////////////////////////////
20.     //优化器规则从这里开始:
21.     ////////////////////////////////////////////////////////////////////////////////////////////////
22.      //在开始主要的优化规则前，首先调用 CombineUnions，因为它可以减少迭代次数，其
         //他规则可以增加/移动两个相邻的联合运算符之间的额外运算符
23.      //在批处理规则 Batch("Operator Optimizations")后再次调用 CombineUnions,
         //因为其他规则可能使两个独立的联合操作符相邻
24.
25.     Batch("Union", Once,
26.       CombineUnions) ::
27.     Batch("Pullup Correlated Expressions", Once,
28.       PullupCorrelatedPredicates) ::
29.     Batch("Subquery", Once,
30.       OptimizeSubqueries) ::
31.     Batch("Replace Operators", fixedPoint,
32.       ReplaceIntersectWithSemiJoin,
33.       ReplaceExceptWithAntiJoin,
34.       ReplaceDistinctWithAggregate) ::
35.     Batch("Aggregate", fixedPoint,
36.       RemoveLiteralFromGroupExpressions,
37.       RemoveRepetitionFromGroupExpressions) ::
38.     Batch("Operator Optimizations", fixedPoint, Seq(
39.       //运算符下推
40.       PushProjectionThroughUnion,
41.       ReorderJoin(conf),
42.       EliminateOuterJoin(conf),
43.       PushPredicateThroughJoin,
44.       PushDownPredicate,
45.       LimitPushDown(conf),
46.       ColumnPruning,
47.       InferFiltersFromConstraints(conf),
48.       //运算符组合
49.       CollapseRepartition,
50.       CollapseProject,
51.       CollapseWindow,
52.       CombineFilters,
53.       CombineLimits,
54.       CombineUnions,
55.       //常量折叠和强度降低
56.       NullPropagation(conf),
57.       FoldablePropagation,
58.       OptimizeIn(conf),
59.       ConstantFolding,
60.       ReorderAssociativeOperator,
61.       LikeSimplification,
62.       BooleanSimplification,
63.       SimplifyConditionals,
64.       RemoveDispensableExpressions,
65.       SimplifyBinaryComparison,
66.       PruneFilters(conf),
67.       EliminateSorts,
```

```
68.        SimplifyCasts,
69.        SimplifyCaseConversionExpressions,
70.        RewriteCorrelatedScalarSubquery,
71.        EliminateSerialization,
72.        RemoveRedundantAliases,
73.        RemoveRedundantProject,
74.        SimplifyCreateStructOps,
75.        SimplifyCreateArrayOps,
76.        SimplifyCreateMapOps) ++
77.        extendedOperatorOptimizationRules: _*) ::
78.      Batch("Join Reorder", Once,
79.        CostBasedJoinReorder(conf)) ::
80.      Batch("Decimal Optimizations", fixedPoint,
81.        DecimalAggregates(conf)) ::
82.      Batch("Object Expressions Optimization", fixedPoint,
83.        EliminateMapObjects,
84.        CombineTypedFilters) ::
85.      Batch("LocalRelation", fixedPoint,
86.        ConvertToLocalRelation,
87.        PropagateEmptyRelation) ::
88.      Batch("Check Cartesian Products", Once,
89.        CheckCartesianProducts(conf)) ::
90.      Batch("OptimizeCodegen", Once,
91.        OptimizeCodegen(conf)) ::
92.      Batch("RewriteSubquery", Once,
93.        RewritePredicateSubquery,
94.        CollapseProject) :: Nil
95.    }
```

Spark 2.4.3 版本的 Optimizer.scala 的源码与 Spark 2.2.1 版本相比具有如下特点。

- 将上段代码整体替换为以下代码：

```
1.  ......
2.   abstract class Optimizer(sessionCatalog: SessionCatalog)
3.    extends RuleExecutor[LogicalPlan] {
4.
5.     //在测试模式下检查计划的结构完整性，目前只检查在执行每个规则之后计划是否被解决
6.     override protected def isPlanIntegral(plan: LogicalPlan): Boolean = {
7.       !Utils.isTesting || plan.resolved
8.     }
9.
10.    protected def fixedPoint = FixedPoint(SQLConf.get.optimizerMaxIterations)
11.
12.    /**
13.      *在优化器中定义默认规则批处理。此类的实现应重写此方法，并在必要时使用
         nonExcludableRules，而不是使用 batches
14.      *最终在优化器中运行的规则批处理，由 batches 返回的规则批将是(defaultBatches -
         (excludedRules - nonExcludableRules))
15.      */
16.    def defaultBatches: Seq[Batch] = {
17.      val operatorOptimizationRuleSet =
18.        Seq(
19.          //运算符下推
20.          PushProjectionThroughUnion,
21.          ReorderJoin,
22.          EliminateOuterJoin,
23.          PushPredicateThroughJoin,
24.          PushDownPredicate,
```

```
25.         LimitPushDown,
26.         ColumnPruning,
27.         InferFiltersFromConstraints,
28.         // 运算符组合
29.         CollapseRepartition,
30.         CollapseProject,
31.         CollapseWindow,
32.         CombineFilters,
33.         CombineLimits,
34.         CombineUnions,
35.         //常量折叠和强度降低
36.         NullPropagation,
37.         ConstantPropagation,
38.         FoldablePropagation,
39.         OptimizeIn,
40.         ConstantFolding,
41.         ReorderAssociativeOperator,
42.         LikeSimplification,
43.         BooleanSimplification,
44.         SimplifyConditionals,
45.         RemoveDispensableExpressions,
46.         SimplifyBinaryComparison,
47.         PruneFilters,
48.         EliminateSorts,
49.         SimplifyCasts,
50.         SimplifyCaseConversionExpressions,
51.         RewriteCorrelatedScalarSubquery,
52.         EliminateSerialization,
53.         RemoveRedundantAliases,
54.         RemoveRedundantProject,
55.         SimplifyExtractValueOps,
56.         CombineConcats) ++
57.         extendedOperatorOptimizationRules
58.
59.     val operatorOptimizationBatch: Seq[Batch] = {
60.       val rulesWithoutInferFiltersFromConstraints =
61.         operatorOptimizationRuleSet.filterNot(_ == InferFiltersFromConstraints)
62.       Batch("Operator Optimization before Inferring Filters", fixedPoint,
63.         rulesWithoutInferFiltersFromConstraints: _*) ::
64.       Batch("Infer Filters", Once,
65.         InferFiltersFromConstraints) ::
66.       Batch("Operator Optimization after Inferring Filters", fixedPoint,
67.         rulesWithoutInferFiltersFromConstraints: _*) :: Nil
68.     }
69.
70.     (Batch("Eliminate Distinct", Once, EliminateDistinct) ::
71.
72.     //从技术上讲，完成分析中的一些规则不是优化器规则，更多地属于分析器，因为它们是
        //正确性所必需的（例如 computecurrentTime）
73.     //但是，因为我们也使用分析器来规范化查询（用于视图定义），所以我们不会消除子查
        //询或计算分析器中的当前时间
74.
75.     Batch("Finish Analysis", Once,
76.       EliminateSubqueryAliases,
77.       EliminateView,
78.       ReplaceExpressions,
79.       ComputeCurrentTime,
80.       GetCurrentDatabase(sessionCatalog),
81.       RewriteDistinctAggregates,
```

```
82.         ReplaceDeduplicateWithAggregate) ::
83.       ////////////////////////////////////////////////////////////////////////////////////////////////////
84.       //优化器规则从这里开始
85.       ////////////////////////////////////////////////////////////////////////////////////////////////////
86.       //-在开始主要的优化规则前，首先调用 CombineUnions，因为它可以减少迭代次数，
87.       //其他规则可以增加/移动两个相邻的联合运算符之间的额外运算符
88.       //在批处理规则 Batch("Operator Optimizations")再次调用 CombineUnions，
89.       //因为其他规则可能使两个独立的联合操作符相邻
90.
91.
92.       Batch("Union", Once,
93.         CombineUnions) ::
94.       //提前运行一次，这可能会简化计划并降低优化器的成本
95.       //例如，如果存在一个筛选器（如 InferFiltersFromConstraints），
          //filter（LocalRelation）查询将遍历触发所有重要优化器规则
96.       //如果早些运行这个批处理，查询将变为
97.
98.       //LocalRelation，并且不会触发许多规则
99.       Batch("LocalRelation early", fixedPoint,
100.        ConvertToLocalRelation,
101.        PropagateEmptyRelation) ::
102.      Batch("Pullup Correlated Expressions", Once,
103.        PullupCorrelatedPredicates) ::
104.      Batch("Subquery", Once,
105.        OptimizeSubqueries) ::
106.      Batch("Replace Operators", fixedPoint,
107.        RewriteExceptAll,
108.        RewriteIntersectAll,
109.        ReplaceIntersectWithSemiJoin,
110.        ReplaceExceptWithFilter,
111.        ReplaceExceptWithAntiJoin,
112.        ReplaceDistinctWithAggregate) ::
113.      Batch("Aggregate", fixedPoint,
114.        RemoveLiteralFromGroupExpressions,
115.        RemoveRepetitionFromGroupExpressions) :: Nil ++
116.      operatorOptimizationBatch) :+
117.      Batch("Join Reorder", Once,
118.        CostBasedJoinReorder) :+
119.      Batch("Remove Redundant Sorts", Once,
120.        RemoveRedundantSorts) :+
121.      Batch("Decimal Optimizations", fixedPoint,
122.        DecimalAggregates) :+
123.      Batch("Object Expressions Optimization", fixedPoint,
124.        EliminateMapObjects,
125.        CombineTypedFilters) :+
126.      Batch("LocalRelation", fixedPoint,
127.        ConvertToLocalRelation,
128.        PropagateEmptyRelation) :+
129.      Batch("Extract PythonUDF From JoinCondition", Once,
130.        PullOutPythonUDFInJoinCondition) :+
131.      //以下批处理应在“Join Reorder”“LocalRelation”和“Extract PythonUDF From
          //JoinCondition”批处理后执行
132.      Batch("Check Cartesian Products", Once,
133.        CheckCartesianProducts) :+
134.      Batch("RewriteSubquery", Once,
135.        RewritePredicateSubquery,
136.        ColumnPruning,
137.        CollapseProject,
```

```
138.     RemoveRedundantProject) :+
139.    Batch("UpdateAttributeReferences", Once,
140.     UpdateNullabilityInAttributeReferences)
141.  }
142.
143.  /**
144.    *定义不能从优化器中排除的规则，即使在 SQL 配置“excludedrules”。如果需要，此
      类的实现可以重写此方法
145.    *规则批处理最终在优化器中运行，即由 batches 返回
      (defaultBatches - (excludedRules - nonExcludableRules))
146.    */
147.  def nonExcludableRules: Seq[String] =
148.    EliminateDistinct.ruleName ::
149.     EliminateSubqueryAliases.ruleName ::
150.     EliminateView.ruleName ::
151.     ReplaceExpressions.ruleName ::
152.     ComputeCurrentTime.ruleName ::
153.     GetCurrentDatabase(sessionCatalog).ruleName ::
154.     RewriteDistinctAggregates.ruleName ::
155.     ReplaceDeduplicateWithAggregate.ruleName ::
156.     ReplaceIntersectWithSemiJoin.ruleName ::
157.     ReplaceExceptWithFilter.ruleName ::
158.     ReplaceExceptWithAntiJoin.ruleName ::
159.     RewriteExceptAll.ruleName ::
160.     RewriteIntersectAll.ruleName ::
161.     ReplaceDistinctWithAggregate.ruleName ::
162.     PullupCorrelatedPredicates.ruleName ::
163.     RewriteCorrelatedScalarSubquery.ruleName ::
164.     RewritePredicateSubquery.ruleName ::
165.     PullOutPythonUDFInJoinCondition.ruleName :: Nil
166.
167.  /**
168.    *优化表达式中的所有子查询
169.    */
170.  object OptimizeSubqueries extends Rule[LogicalPlan] {
171.    private def removeTopLevelSort(plan: LogicalPlan): LogicalPlan = {
172.     plan match {
173.      case Sort(_, _, child) => child
174.      case Project(fields, child) => Project(fields, removeTopLevelSort(child))
175.      case other => other
176.     }
177.    }
178.    def apply(plan: LogicalPlan): LogicalPlan = plan transformAllExpressions {
179.     case s: SubqueryExpression =>
180.      val Subquery(newPlan) = Optimizer.this.execute(Subquery(s.plan))
181.      //现在有了一个优化的子查询计划，将附加到这个子查询表达式上。在这里可以安全
182.      //地删除计划中的任何顶级排序，因为子查询生成的元组是无序的
183.      s.withNewPlan(removeTopLevelSort(newPlan))
184.    }
185.  }
186.
187.  /**
188.    *重写以提供运算符优化批处理的其他规则
189.    */
190.  def extendedOperatorOptimizationRules: Seq[Rule[LogicalPlan]] = Nil
191.
192.  /**
```

```
193.    *返回 (defaultBatches - (excludedRules - nonExcludableRules))，规则
        批处理最终在优化器中运行，此类实现应重写
194.    *defaultBatches 和 nonExcludableRules，如有必要，请不要使用此方法。
195.    */
196.  final override def batches: Seq[Batch] = {
197.    val excludedRulesConf =
198.
   SQLConf.get.optimizerExcludedRules.toSeq.flatMap(Utils.stringToSeq)
199.    val excludedRules = excludedRulesConf.filter { ruleName =>
200.      val nonExcludable = nonExcludableRules.contains(ruleName)
201.      if (nonExcludable) {
202.        logWarning(s"Optimization rule '${ruleName}' was not excluded
          from the optimizer " +
203.          s"because this rule is a non-excludable rule.")
204.      }
205.      !nonExcludable
206.    }
207.    if (excludedRules.isEmpty) {
208.      defaultBatches
209.    } else {
210.      defaultBatches.flatMap { batch =>
211.        val filteredRules = batch.rules.filter { rule =>
212.          val exclude = excludedRules.contains(rule.ruleName)
213.          if (exclude) {
214.            logInfo(s"Optimization rule '${rule.ruleName}' is excluded
              from the optimizer.")
215.          }
216.          !exclude
217.        }
218.        if (batch.rules == filteredRules) {
219.          Some(batch)
220.        } else if (filteredRules.nonEmpty) {
221.          Some(Batch(batch.name, batch.strategy, filteredRules: _*))
222.        } else {
223.          logInfo(s"Optimization batch '${batch.name}' is excluded from
          the optimizer " +
224.            s"as all enclosed rules have been excluded.")
225.          None
226.        }
227.      }
228.    }
229.  }
230.    }
```

（4）优化后的执行计划提交给 SparkPlanner 处理，SparkPlanner 里面定义了一些策略，目的是根据逻辑执行计划树生成最后可以执行的物理执行计划树，即得到 SparkPlan。

（5）在最终真正执行物理执行计划前，还要进行 prepareForExecution 规则处理。QueryExecution 里定义这个过程叫 prepareForExecution。

QueryExecution.scala 的源码如下：

```
1.  /**
     *准备一个计划[SparkPlan]，用于执行 Shuffle 算子和内部行格式转换的需要
2.   */
3.
4.  protected def prepareForExecution(plan: SparkPlan): SparkPlan = {
5.    preparations.foldLeft(plan) { case (sp, rule) => rule.apply(sp) }
6.  }
```

（6）最后调用 SparkPlan 的 execute()执行计算。先调用 executeCollect 方法，在 executeCollect 方法中调用 getByteArrayRdd 方法，最终调用 execute()进行计算。这个 execute()在每种 SparkPlan 的实现里定义，会触发整棵 Tree 的计算。

Spark 2.2.1 版本的 SparkPlan.scala 的源码如下：

```
1.  /**
     *运行此查询，将结果作为数组返回
2.   */
3.
4.  def executeCollect(): Array[InternalRow] = {
5.    val byteArrayRdd = getByteArrayRdd()
6.
7.    val results = ArrayBuffer[InternalRow]()
8.    byteArrayRdd.collect().foreach { bytes =>
9.      decodeUnsafeRows(bytes).foreach(results.+=)
10.   }
11.   results.toArray
12. }
```

Spark 2.4.3 版本的 SparkPlan.scala 源码与 Spark 2.2.1 版本相比具有如下特点。

- 上段代码中第 8 行 bytes 名称调整为 countAndBytes。
- 上段代码中第 9 行 bytes 调整为 countAndBytes._2。上段代码第 5 行代码 byteArrayRdd 的类型是 RDD[Array[Byte]]，以下代码第 2 行代码 byteArrayRdd 的类型是 RDD[(Long, Array[Byte])]，因此获取 countAndBytes 的第二个元素传入 decodeUnsafeRows 方法，将字节数组解码为 UnsafeRows，并将其放入缓冲区。

```
1.  ......
2.    byteArrayRdd.collect().foreach { countAndBytes =>
3.      decodeUnsafeRows(countAndBytes._2).foreach(results.+=)
4.  ......
```

其中，executeCollect 方法中调用了 getByteArrayRdd 方法，getByteArrayRdd 方法封装 unsaferows 到更快的序列化的字节数组。字节数组的格式如下：[size] [bytes of UnsafeRow] [size] [bytes of UnsafeRow] … [−1]。UnsafeRow 是高度可压缩的（任何列至少 8B），字节数组也是可压缩的。

Spark 2.2.1 版本的 SparkPlan.scala 的源码如下：

```
1.  /**
2.    *包装 UnsafeRows 到更快的序列化的字节数组。字节数组的格式如下：
3.    *[size] [bytes of UnsafeRow] [size] [bytes of UnsafeRow] ... [-1]，UnsafeRow
      *是高度可压缩的（任何列至少 8B）字节数组也是可压缩的
4.    */
5.    private def getByteArrayRdd(n: Int = -1): RDD[Array[Byte]] = {
6.      execute().mapPartitionsInternal { iter =>
7.        var count = 0
8.        val buffer = new Array[Byte](4 << 10)  //4K
9.        val codec = CompressionCodec.createCodec(SparkEnv.get.conf)
10.       val bos = new ByteArrayOutputStream()
11.       val out = new DataOutputStream(codec.compressedOutputStream(bos))
12.       while (iter.hasNext && (n < 0 || count < n)) {
13.         val row = iter.next().asInstanceOf[UnsafeRow]
14.         out.writeInt(row.getSizeInBytes)
```

```
15.         row.writeToStream(out, buffer)
16.         count += 1
17.       }
18.       out.writeInt(-1)
19.       out.flush()
20.       out.close()
21.       Iterator(bos.toByteArray)
22.     }
23.   }
```

Spark 2.4.3 版本的 SparkPlan.scala 源码与 Spark 2.2.1 版本相比具有如下特点。

- 上段代码中第 5 行将 getByteArrayRdd 方法返回类型 RDD[Array[Byte]]调整为 RDD[(Long, Array[Byte])]。
- 上段代码中第 12 行条件表达式调整了前后顺序。
- 上段代码中第 21 行将返回 Array[Byte]，调整为返回(Long, Array[Byte])元组。

```
1.  ……
2.   private def getByteArrayRdd(n: Int = -1): RDD[(Long, Array[Byte])] = {
3.  ……
4.    while ((n < 0 || count < n) && iter.hasNext) {
5.  ……
6.  Iterator((count, bos.toByteArray))
7.  ……
```

SparkPlan 的 execute 方法的源码如下：

```
1.    /**
       *返回查询的结果为 RDD[InternalRow]，准备工作完成后委托给 doExecute,
       *SparkPlan 的具体实现须重写 doExecute
2.     */
3.    final def execute(): RDD[InternalRow] = executeQuery {
4.      doExecute()
5.    }
```

Spark 2.4.3 版本的 SparkPlan.scala 源码与 Spark 2.2.1 版本相比具有如下特点。

- 上段代码中第 4 行之前新增对 isCanonicalizedPlan 的判断。isCanonicalizedPlan 用于指示此计划是否是规范化的结果，一个规范化的计划不应该被执行。

```
1.  ……
2.      if (isCanonicalizedPlan) {
3.       throw new IllegalStateException("A canonicalized plan is not supposed
         to be executed.")
4.      }
```

（7）生成 RDD。

在整个运行过程中涉及多个 SparkSQL 的组件，如 SqlParse、analyzer、optimizer、SparkPlan 等。

HiveContext：是将存储在 Hive 中的数据集成 Spark SQL 执行引擎的实例。

配置文件 Hive 从 classpath 中的 hive-site.xml 读取。HiveContext 继承自 SQLContext 类。hiveContext 执行架构如图 27-3 所示。

① SQL 语句经过 HiveQl.parseSql 解析成 Unresolved LogicalPlan，在这个解析过程中对 HiveQL 语句使用 getAst()获取 AST 树，然后再进行解析。

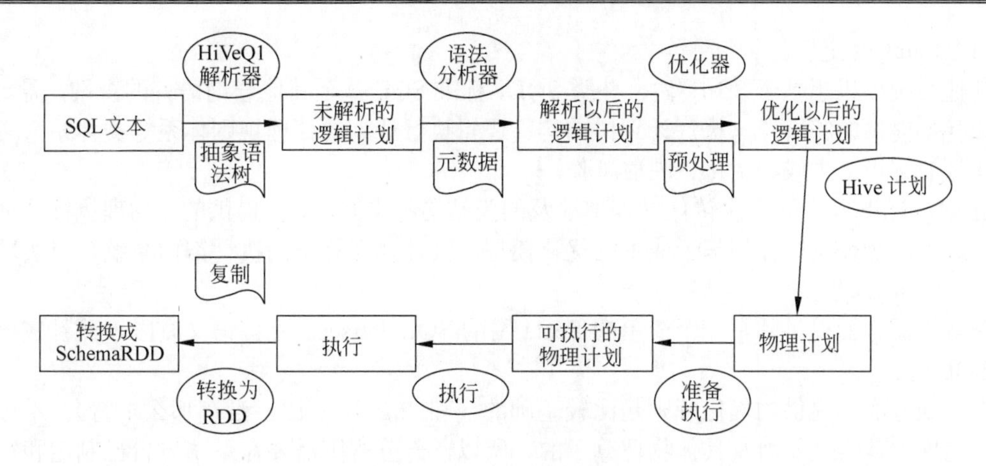

图 27-3　hiveContext 执行架构

② 使用 analyzer 结合数据 Hive 源数据 Metastore（新的 catalog）进行绑定，生成 resolved LogicalPlan。

③ 使用 optimizer 对 resolved LogicalPlan 进行优化，生成 optimized LogicalPlan，优化前使用了 ExtractPythonUdfs(catalog.PreInsertionCasts(catalog.CreateTables(analyzed)))进行预处理。

④ 使用 hivePlanner 将 LogicalPlan 转换成 PhysicalPlan。

⑤ 使用 prepareForExecution()将 PhysicalPlan 转换成可执行物理计划。

⑥ 使用 execute()执行可执行物理计划。

⑦ 执行后，使用 map(_.copy)将结果导入 SchemaRDD。

3）Catalyst 优化器

Spark 2.2 版本中，Spark SQL、Spark DataSet、Hive SQL、SQL 流式处理等组件的底层运行引擎都是 Catalyst。

Catalyst 是与 Spark 解耦的一个独立库，是一个 impl-free 的执行计划的生成和优化框架。以下是 Catalyst 较早期的架构图，展示的是代码结构和处理流程，如今的版本在核心功能上没有改变，我们仍然用这张图进行说明，如图 27-4 所示。

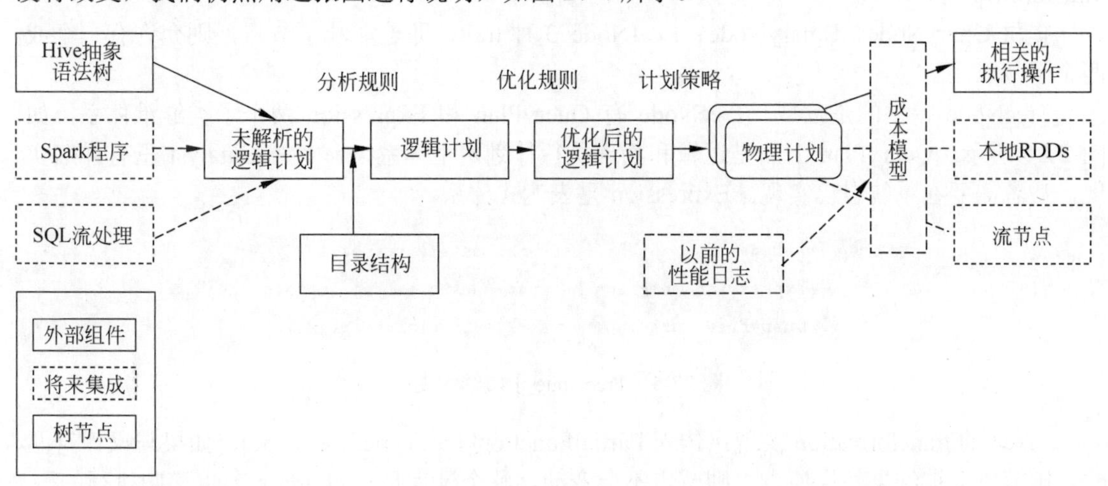

图 27-4　Catalyst 引擎运行架构图

（1）Catalyst 定位。

其他系统如果想基于 Spark 做一些类 SQL、标准 SQL 甚至其他查询语言的查询，需要基于 Catalyst 提供的解析器、执行计划树结构、逻辑执行计划的处理规则体系等类体系来实现执行计划的解析、生成、优化、映射工作。

TreeNodelib 及中间三次转化过程中涉及的类结构都是 Catalyst 提供的。物理执行计划映射生成过程，物理执行计划基于成本的优化模型，具体物理算子的执行都由系统自己实现。

（2）Catalyst 现状。

在解析器方面提供的是一个简单的 Scala 写的 SQL Parser，支持语义有限，而且应该是标准 SQL 的。

在规则方面，提供的优化规则是比较基础的（和 Pig/Hive 比，没有那么丰富）。不过，一些优化规则其实是要涉及具体物理算子的，所以，部分规则需要在系统方自己制定和实现（如 spark-sql 里的 SparkStrategy）。

（3）Catalyst 的类结构。

TreeNode 体系：TreeNode 是 Catalyst 执行计划表示的数据结构，是一个树结构，具备一些 scala collection 的操作能力和树遍历能力。这棵树一直在内存里维护，不会 dump 到磁盘以某种格式的文件存在，且无论在映射逻辑执行计划阶段，还是优化逻辑执行计划阶段，树的修改都是以替换已有节点的方式进行的。

TreeNode 内部带一个 children: Seq[BaseType]表示孩子节点。

TreeNode.scala 的源码如下：

```
1.  /**
      *返回该节点的 children 子序列，children 子节点不应改变，对于包含 child ren 子节
      *点的优化，须保持不变性
2.    */
3.  def children: Seq[BaseType]
```

TreeNode 类提供的方法包括：针对节点操作的方法，如 foreach、map、collect 等；遍历树上的节点，对匹配节点实施变化的方法，如 transformDown（默认，前序遍历）、transformUp 等。

提供 UnaryNode、BinaryNode、LeafNode 3 种 trait，即若非叶子节点，则允许有一个或两个子节点。

TreeNode 提供的是范型。TreeNode 有 QueryPlan 和 Expression 两个子类继承体系，如图 27-5 所示。QueryPlan 下面是逻辑和物理执行计划两个体系，前者在 Catalyst 里有详细实现，后者需要在系统自己实现。Expression 是表达式体系。

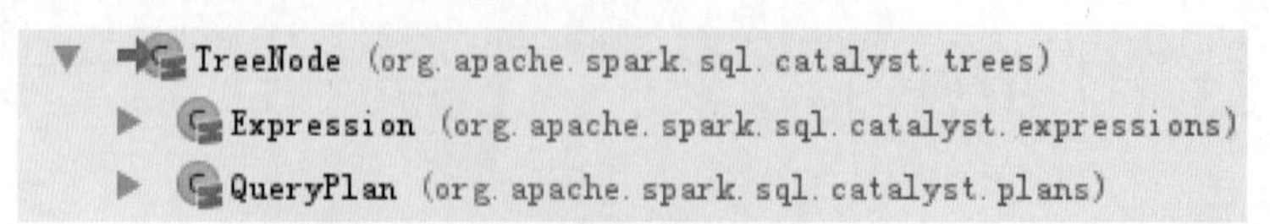

图 27-5　TreeNode 子类继承类

Tree 的 transformation 实现：传入 PartialFunction[TreeType,TreeType]，如果与操作符匹配，则节点会被结果替换掉，否则节点不会变动。整个过程是对 children 节点递归执行的。

执行计划表示模型：逻辑执行计划 QueryPlan 继承自 TreeNode，内部带一个 output: Seq[Attribute]，具备 transformExpressionDown、transformExpressionUp 方法。

在 Catalyst 中，QueryPlan 的主要子类体系是 LogicalPlan，即逻辑执行计划表示，如图 27-6 所示。其物理执行计划表示由使用方实现（Spark-SQL 项目中）。

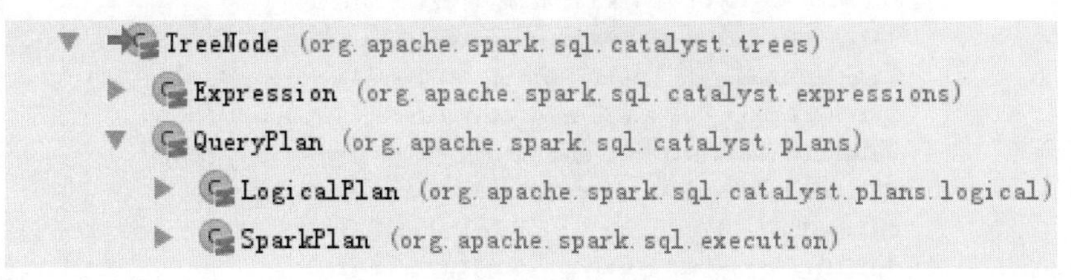

图 27-6　QueryPlan 子类继承类

LogicalPlan 继承自 QueryPlan，内部带一个 reference:Set[Attribute]，主要方法为 resolve(name:String): Option[NamedExpression]，用于分析生成对应的 NamedExpression。LogicalPlan 有许多具体子类，如 UnaryNode、BinaryNode、LeafNode 等类，如图 27-7 所示，具体在 org.apache.spark.sql.catalyst.plans.logical 路径下。

```
TreeNode (org.apache.spark.sql.catalyst.trees)
  QueryPlan (org.apache.spark.sql.catalyst.plans)
    LogicalPlan (org.apache.spark.sql.catalyst.plans.logical)
      BinaryNode (org.apache.spark.sql.catalyst.plans.logical)
      EventTimeWatermark (org.apache.spark.sql.catalyst.plans.logical)
      InsertIntoTable (org.apache.spark.sql.catalyst.plans.logical)
      LeafNode (org.apache.spark.sql.catalyst.plans.logical)
      ObjectProducer (org.apache.spark.sql.catalyst.plans.logical)
      UnaryNode (org.apache.spark.sql.catalyst.plans.logical)
      Union (org.apache.spark.sql.catalyst.plans.logical)
```

图 27-7　LogicalPlan 子类继承类

逻辑执行计划实现：LeafNode 子类（部分）如图 27-8 所示。

```
LeafNode (org.apache.spark.sql.catalyst.plans.logical)
  Command (org.apache.spark.sql.catalyst.plans.logical)
  ComplexPlan (org.apache.spark.sql.catalyst.trees)
  ExternalRDD (org.apache.spark.sql.execution)
  InMemoryRelation (org.apache.spark.sql.execution.columnar)
  JsonTestTreeNode (org.apache.spark.sql.catalyst.trees)
  LocalRelation (org.apache.spark.sql.catalyst.plans.logical)
  LogicalRDD (org.apache.spark.sql.execution)
  LogicalRelation (org.apache.spark.sql.execution.datasources)
  MemoryPlan (org.apache.spark.sql.execution.streaming)
  MetastoreRelation (org.apache.spark.sql.hive)
  NeverPlanned$ in SparkPlannerSuite (org.apache.spark.sql.execution)
  OneRowRelation$ (org.apache.spark.sql.catalyst.plans.logical)
  Range (org.apache.spark.sql.catalyst.plans.logical)
  SimpleCatalogRelation (org.apache.spark.sql.catalyst.catalog)
  SQLTable in SQLBuilder (org.apache.spark.sql.catalyst)
  StreamingExecutionRelation (org.apache.spark.sql.execution.streaming)
  StreamingRelation (org.apache.spark.sql.execution.streaming)
```

图 27-8　LeafNode 子类（部分）

UnaryNode 子类（部分）如图 27-9 所示。

```
▼ UnaryNode (org.apache.spark.sql.catalyst.plans.logical)
    Aggregate (org.apache.spark.sql.catalyst.plans.logical)
    AppendColumns (org.apache.spark.sql.catalyst.plans.logical)
    BroadcastHint (org.apache.spark.sql.catalyst.plans.logical)
    DeserializeToObject (org.apache.spark.sql.catalyst.plans.logical)
    Distinct (org.apache.spark.sql.catalyst.plans.logical)
    Expand (org.apache.spark.sql.catalyst.plans.logical)
    Filter (org.apache.spark.sql.catalyst.plans.logical)
    FlatMapGroupsInR (org.apache.spark.sql.catalyst.plans.logical)
    Generate (org.apache.spark.sql.catalyst.plans.logical)
    GlobalLimit (org.apache.spark.sql.catalyst.plans.logical)
    GroupingSets (org.apache.spark.sql.catalyst.plans.logical)
    LocalLimit (org.apache.spark.sql.catalyst.plans.logical)
    MapGroups (org.apache.spark.sql.catalyst.plans.logical)
  ▶ ObjectConsumer (org.apache.spark.sql.catalyst.plans.logical)
    Pivot (org.apache.spark.sql.catalyst.plans.logical)
    Project (org.apache.spark.sql.catalyst.plans.logical)
    Repartition (org.apache.spark.sql.catalyst.plans.logical)
    RepartitionByExpression (org.apache.spark.sql.catalyst.plans.logical)
    ReturnAnswer (org.apache.spark.sql.catalyst.plans.logical)
    Sample (org.apache.spark.sql.catalyst.plans.logical)
    ScriptTransformation (org.apache.spark.sql.catalyst.plans.logical)
    Sort (org.apache.spark.sql.catalyst.plans.logical)
```

图 27-9　UnaryNode 子类（部分）

BinaryNode 子类如图 27-10 所示。

```
▼ BinaryNode (org.apache.spark.sql.catalyst.plans.logical)
    CoGroup (org.apache.spark.sql.catalyst.plans.logical)
    Join (org.apache.spark.sql.catalyst.plans.logical)
  ▶ SetOperation (org.apache.spark.sql.catalyst.plans.logical)
    TestBinaryRelation in LogicalPlanSuite (org.apache.spark.sql.catalyst.plans)
```

图 27-10　BinaryNode 子类

物理执行计划：物理执行计划节点在具体系统里实现，如 spark-sql 工程里的 SparkPlan 继承体系。每个子类都要实现 execute 方法，作为 trait 接口的有以下 3 类：LeafNode 的子类、UnaryNode 的子类、BinaryNode 的子类。

执行计划映射：Catalyst 提供了一个 QueryPlanner[Physical <: TreeNode[PhysicalPlan]]抽象类，需要子类制定一批 strategies：Seq[Strategy]，其 apply 方法也是类似根据制定的具体策略来把逻辑执行计划算子映射成物理执行计划算子。由于物理执行计划的节点是在具体系统里实现的，所以 QueryPlanner 及里面的 strategies 也需要在具体系统里实现。

在 Spark-SQL 项目中，SparkStrategies 继承了 QueryPlanner[SparkPlan]，内部制定了 LeftSemiJoin、HashJoin、PartialAggregation、BroadcastNestedLoopJoin、CartesianProduct 等几种策略，每种策略接受的都是一个 LogicalPlan，生成的是 Seq[SparkPlan]，每个 SparkPlan 理解为具体 RDD 的算子操作。

2. Spark SQL核心组件

Spark SQL 核心组件包括 LogicalPlan 组件中的 LeafNode、UnaryNode、BinaryNode；SqlParser 组件中的 sqlPraser；Analyzer 组件中的 Analyzer 类。在 sparkSQL 的运行架构中，LogicalPlan 贯穿了大部分过程，其中 catalyst 中的 SqlParser、Analyzer、Optimizer 都要对 LogicalPlan 进行操作。

1）LogicalPlan

LogicalPlan 的定义如下。

在 LogicalPlan 里维护着一套统计数据和属性数据，也提供了解析方法，同时延伸了 3 种类型的 LogicalPlan。

- LeafNode：对应于 trees.LeafNode 的 LogicalPlan。
- UnaryNode：对应于 trees.UnaryNode 的 LogicalPlan。
- BinaryNode：对应于 trees.BinaryNode 的 LogicalPlan。

而对于 SQL 语句解析时，会调用和 SQL 匹配的操作方法进行解析；这些操作分 4 大类，最终生成 LeafNode、UnaryNode、BinaryNode 中的一种。

- basicOperators：一些数据基本操作，如 Join、Union、Filter、Project、Sort。
- commands：一些命令操作，如 SetCommand、CacheCommand。
- partitioning：一些分区操作，如 RedistributeData。
- ScriptTransformation：对脚本的处理，如 ScriptTransformation。

LogicalPlan 类的总体架构如图 27-11 所示。

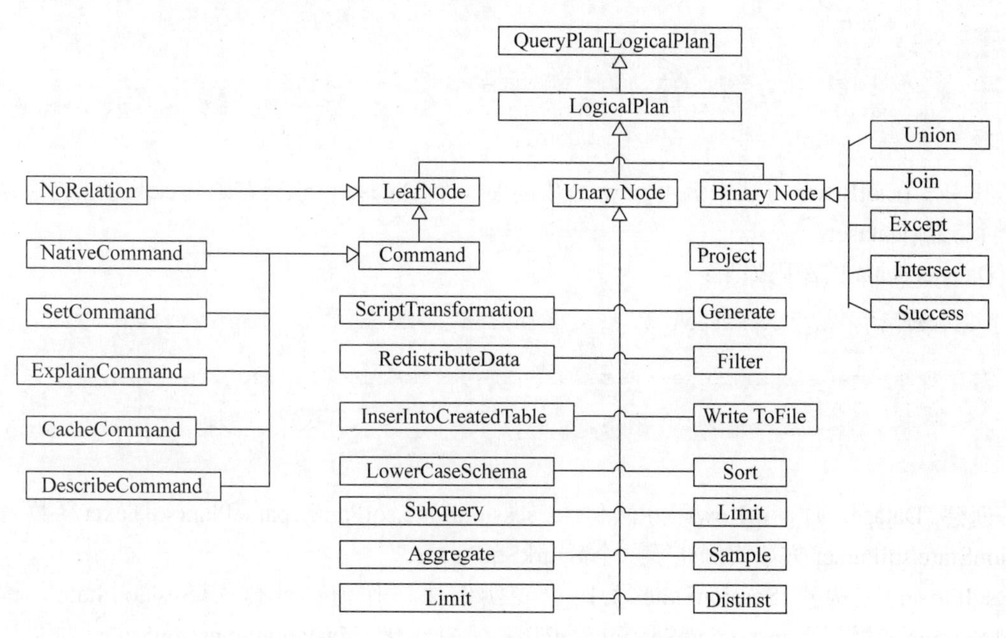

图 27-11　LogicalPlan 类的总体架构

2）SqlParser

SqlParser 的功能是将 SQL 语句解析成逻辑上的 UnesolvedLogicalPlan。

SqlParser 解析流程图如图 27-12 所示。

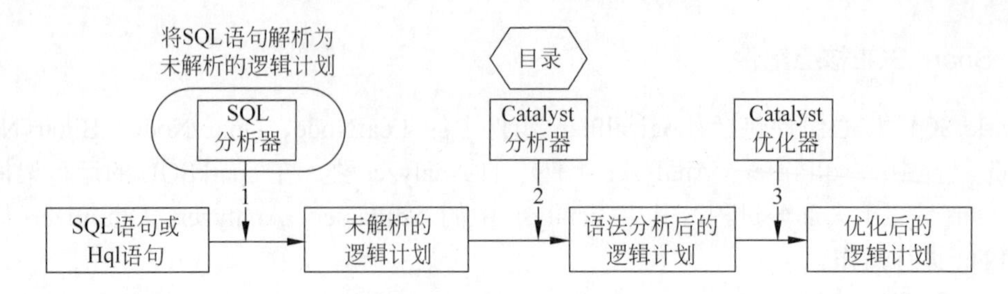

图 27-12　SqlParser 解析流程图

首先看一下源码中的继承结构，如图 27-13 所示。

AbstractSqlParser (org.apache.spark.sql.catalyst.parser)
CatalystSqlParser$ (org.apache.spark.sql.catalyst.parser)
SparkSqlParser (org.apache.spark.sql.execution)

图 27-13　SparkSQLParser 继承结构

其中，每个类中都创建了自己需要的关键字和方法。

当调用 sql("select name,value from temp_shengli")时，源码中执行如下语句。

SparkSession.scala 的源码如下：

```
1.  /**
2.    *使用 Spark 执行 SQL 查询，返回结果为 DataFrame。用于 SQL 解析的方言可以用 spark.
    *sql.dialect 来配置
3.    *
4.    * @since 2.0.0
5.    */
6.    def sql(sqlText: String): DataFrame = {
7.      Dataset.ofRows(self, sessionState.sqlParser.parsePlan(sqlText))
8.    }
```

其中，在 ofRows 方法中调用 new 函数创建一个 Dataset，也就是通过 sql 方法构造器产生一个新的 Dataset。

Dataset.scala 的源码如下：

```
1.  def ofRows(sparkSession: SparkSession, logicalPlan: LogicalPlan): DataFrame
    = {
2.    val qe = sparkSession.sessionState.executePlan(logicalPlan)
3.    qe.assertAnalyzed()
4.    new Dataset[Row](sparkSession, qe, RowEncoder(qe.analyzed.schema))
5.  }
```

创建 Dataset 时，ofRows 方法传入 sessionState.sqlParser.parsePlan(sqlText)参数值，sessionState.sqlParser 方法实例化了一个 SparkSqlParser。

sqlParser 在构建 SessionState 类时已经实例化，作为参数传入 SessionState。构建 SparkSession 时调用 instantiateSessionState 进行初始化，instantiateSessionState 方法调用 ctor.newInstance(sparkSession,None).asInstanceOf[BaseSessionStateBuilder].build 方法，在 BaseSessionStateBuilder 类初始化时构建 SparkSqlParser 实例，然后赋值给 sqlParser。

其中，new SparkSqlParser 新构建一个 SparkSqlParser。

SparkSqlParser.scala 的源码如下：

```
1.  /**
     *  Spark SQL 语句的具体分析器
2.   */
3.
4.  class SparkSqlParser(conf: SQLConf) extends AbstractSqlParser {
5.    val astBuilder = new SparkSqlAstBuilder(conf)
6.
7.    private val substitutor = new VariableSubstitution(conf)
8.
9.    protected override def parse[T](command: String)(toResult: SqlBaseParser =>
            T): T = {
10.     super.parse(substitutor.substitute(command))(toResult)
11.   }
12. }
```

回到 SparkSession.scala，sessionState.sqlParser.parsePlan(sqlText)语句调用 parsePlan 方法，ParserInterface 是一个 trait，其中的 parsePlan 是一个抽象方法，没有具体的实现。

ParserInterface.scala 的源码如下：

```
1.  /**
2.   *分析器 parser 的接口
3.   */
4.  @DeveloperApi
5.  trait ParserInterface {
6.    /**
7.     *解析字符串为逻辑计划[LogicalPlan]
8.     */
9.    @throws[ParseException]("Text cannot be parsed to a LogicalPlan")
10.   def parsePlan(sqlText: String): LogicalPlan
11.
12.   /**
13.     *解析字符串为表达式[Expression]
14.     */
15.   @throws[ParseException]("Text cannot be parsed to an Expression")
16.   def parseExpression(sqlText: String): Expression
17.
18.   /**
19.     *解析字符串为表标识[TableIdentifier]
20.     */
21.   @throws[ParseException]("Text cannot be parsed to a TableIdentifier")
22.   def parseTableIdentifier(sqlText: String): TableIdentifier
23.
24.   /**
25.     *解析字符串为函数标识[FunctionIdentifier]
26.     */
27.   @throws[ParseException]("Text cannot be parsed to a FunctionIdentifier")
28.   def parseFunctionIdentifier(sqlText: String): FunctionIdentifier
29.
30.   /**
31.     *解析字符串为结构类型[StructType]，SQL 字符串应该使用逗号分隔的列表字段定义，
       该字段定义将保存正确的 Hive 元数据
32.     */
33.   @throws[ParseException]("Text cannot be parsed to a schema")
34.   def parseTableSchema(sqlText: String): StructType
35.
```

```
36.    /**
37.      *Parse a string to a[DataType]
38.      */
39.    @throws[ParseException]("Text cannot be parsed to a DataType")
40.    def parseDataType(sqlText: String): DataType
41. }
```

其具体子类 AbstractSqlParser 的 parsePlan 的实现方法如下，其中调用了 parse 方法。ParseDriver.scala 的源码如下：

```
1.     /**根据给定的 SQL 字符串创建一个逻辑计划 LogicalPlan */
2.     override def parsePlan(sqlText: String): LogicalPlan = parse(sqlText)
       { parser =>
3.       astBuilder.visitSingleStatement(parser.singleStatement()) match {
4.         case plan: LogicalPlan => plan
5.         case _ =>
6.           val position = Origin(None, None)
7.          throw new ParseException(Option(sqlText), "Unsupported SQL statement",
           position, position)
8.       }
9.     }
```

Spark 2.2.1 版本的 ParseDriver.scala 源码中 parse 方法的源码如下：

```
1.  protected def parse[T](command: String)(toResult: SqlBaseParser => T):
    T = {
2.      logInfo(s"Parsing command: $command")
3.
4.      val lexer = new SqlBaseLexer(new ANTLRNoCaseStringStream(command))
5.      lexer.removeErrorListeners()
6.      lexer.addErrorListener(ParseErrorListener)
7.
8.      val tokenStream = new CommonTokenStream(lexer)
9.      val parser = new SqlBaseParser(tokenStream)
10.     parser.addParseListener(PostProcessor)
11.     parser.removeErrorListeners()
12.     parser.addErrorListener(ParseErrorListener)
13.
14.     try {
15.       try {
16.         //首先，使用可能更快的 SLL 模式尝试分析
17.         parser.getInterpreter.setPredictionMode(PredictionMode.SLL)
18.         toResult(parser)
19.       }
20.       catch {
21.         case e: ParseCancellationException =>
22.           //如果解析失败，使用 LL 模式进行解析
23.           tokenStream.reset() //rewind input stream
24.           parser.reset()
25.
26.           //重新试一次
27.           parser.getInterpreter.setPredictionMode(PredictionMode.LL)
28.           toResult(parser)
29.       }
30.     }
31.     catch {
32.       case e: ParseException if e.command.isDefined =>
33.         throw e
```

```
34.        case e: ParseException =>
35.          throw e.withCommand(command)
36.        case e: AnalysisException =>
37.          val position = Origin(e.line, e.startPosition)
38.          throw new ParseException(Option(command), e.message, position,
             position)
39.      }
40.    }
```

Spark 2.4.3 版本的 ParseDriver.scala 源码与 Spark 2.2.1 版本相比具有如下特点。

- 上段代码中第 4 行 ANTLRNoCaseStringStream 调整为 UpperCaseCharStream。UpperCaseCharStream 方法中字符串流仅为 lexer 提供大写字符，简化了词法分析，同时可以维护原始命令。
- 上段代码中第 6 行之后新增代码，当 LEGACY_SETOPS_PRECEDENCE_ENABLED 设置为 true 且计算顺序未由括号指定，将从左到右执行集合操作，如同在查询中的显示。当设置为 false 且括号中未指定计算顺序时，在任何 UNION、EXCEPT、MINUS 之前都会执行 INTERSECT 操作。
- 上段代码中第 12 行之后新增代码，设置 parser 的 legacy_setops_precedence_enbled。
- 上段代码中第 23 行调整为 tokenStream.seek(0)。

```
1.  ......
2.      val lexer = new SqlBaseLexer(new UpperCaseCharStream
        (CharStreams.fromString(command)))
3.  ......
4.    lexer.legacy_setops_precedence_enbled = SQLConf.get.setOpsPrecedenceEnforced
5.  ......
6.  parser.legacy_setops_precedence_enbled = SQLConf.get.setOpsPrecedenceEnforced
7.  ......
8.     tokenStream.seek(0)
```

其中，parse 方法的 SqlBaseLexer、SqlBaseParser 类是 ANTLR4 自动生成的。后面将研究 Spark SQL 中 ANTLR4 的应用。

3）Analyzer

Spark SQL 的执行流程中另一个核心的组件是 Analyzer。本节介绍 Analyzer 在 Spark SQL 里起到的作用。Analyzer 位于 Catalyst 的 analysis package 下，主要职责是将 Sql Parser 未能 Resolved 的 Logical Plan 进行 Resolved 解析。

Analyzer 会使用 Catalog 和 FunctionRegistry 将 UnresolvedAttribute 和 UnresolvedRelation 转换为 catalyst 里全类型的对象。

Analyzer 里面有 fixedPoint 对象，一个 Seq[Batch]。

Analyzer.scala 的源码如下：

```
1.  /**
2.    *提供了一个逻辑查询计划分析器，利用[SessionCatalog]和[FunctionRegistry]的信
      息将[UnresolvedAttribute]和[UnresolvedRelation]解析成完全类型化的对象
3.    */
4.
5.  class Analyzer(
6.      catalog: SessionCatalog,
7.      conf: SQLConf,
8.      maxIterations: Int)
9.    extends RuleExecutor[LogicalPlan] with CheckAnalysis {
```

```
10.
11.   def this(catalog: SessionCatalog, conf: SQLConf) = {
12.     this(catalog, conf, conf.optimizerMaxIterations)
13.   }
14.
15.   def resolver: Resolver = conf.resolver
16.
17.   protected val fixedPoint = FixedPoint(maxIterations)
18.
19.   /**
20.     *覆盖重写，为“Resolution”批处理提供额外的规则
21.     */
22.   val extendedResolutionRules: Seq[Rule[LogicalPlan]] = Nil
23.
24.    /**
25.      *覆盖重写，以提供 post-hoc 完成后处理的规则。注意，这些规则将被单独执行。
         *此批操作将在正常解析批处理之后运行，将执行一次规则
26.      */
27.
28.   val postHocResolutionRules: Seq[Rule[LogicalPlan]] = Nil
29. ……
```

Analyzer 里的一些对象解释如下。

- FixedPoint：相当于迭代次数的上限。

RuleExecutor.scala 的源码如下：

```
1.   Analyzer.scala
2.   protected val fixedPoint = FixedPoint(maxIterations)
3.  ……
4.  RuleExecutor.scala
5.    /**策略将一直运行到固定次数或最大迭代次数，以先达到的次数为准*/
6.    case class FixedPoint(maxIterations: Int) extends Strategy
```

- Batch：批次，Batch 对象由一系列 Rule 组成，采用一个策略。

RuleExecutor.scala 的源码如下：

```
1.   Analyzer.scala
2.  lazy val batches: Seq[Batch] = Seq(
3.     Batch("Substitution", fixedPoint,
4.  ……
5.  RuleExecutor.scala
6.    /**一批规则*/
7.    protected case class Batch(name: String, strategy: Strategy, rules:
      Rule[TreeType]*)
8.  ……
```

- Rule：理解为一种规则，这种规则会应用到 Logical Plan，从而将 UnResolved 转变为 Resolved。

Rule.scala 的源码如下：

```
1.   Analyzer.scala
2.   /**
       *覆盖重写，为 Resolution 批处理提供额外的规则
3.     */
4.
5.    val extendedResolutionRules: Seq[Rule[LogicalPlan]] = Nil
```

```
6.  ......
7.  Rule.scala
8.  abstract class Rule[TreeType <: TreeNode[_]] extends Logging {
9.
10.   /**根据类名称自动推断此规则的名称*/
11.   val ruleName: String = {
12.     val className = getClass.getName
13.     if (className endsWith "$") className.dropRight(1) else className
14.   }
15.
16.   def apply(plan: TreeType): TreeType
17. }
```

- Strategy：最大的执行次数，如果执行次数在最大迭代次数之前，就达到了 FixedPoint，策略就会停止，不再应用了。

RuleExecutor.scala 的源码如下：

```
1.   /**
      *表示最大执行次数的规则的执行策略。执行达到固定次数（即收敛最大迭代次数）之前，它
      *将停止迭代
2.    */
3.   abstract class Strategy { def maxIterations: Int }
4.
5.   /**策略只运行一次*/
6.   case object Once extends Strategy { val maxIterations = 1 }
7.
8.   /**策略将一直运行到固定次数或最大迭代次数，以先达到的次数为准*/
9.   case class FixedPoint(maxIterations: Int) extends Strategy
```

Analyzer 解析主要是根据这些 Batch 里面定义的策略和 Rule 来对 Unresolved 的逻辑计划进行解析。这里 Analyzer 类本身并没有定义执行的方法，而是从它的父类 RuleExecutor [LogicalPlan]中寻找。

4）Optimizer

Optimizer 的主要职责是将 Analyzer 给 Resolved 的 Logical Plan 根据不同的优化策略 Batch，来对语法树进行优化。优化逻辑计划节点（Logical Plan）以及表达式（Expression），也是转换成物理执行计划的前置。

（1）Optimizer。

Optimizer 类是在 Catalyst 里的 Optimizer 包下的一个类，继承结构如图 27-14 所示。Optimizer 的工作方式类似于 Analyzer，因为它们都继承自 RuleExecutor[LogicalPlan]，都是执行一系列的 Batch 操作，生成 Optimized LogicalPlan。

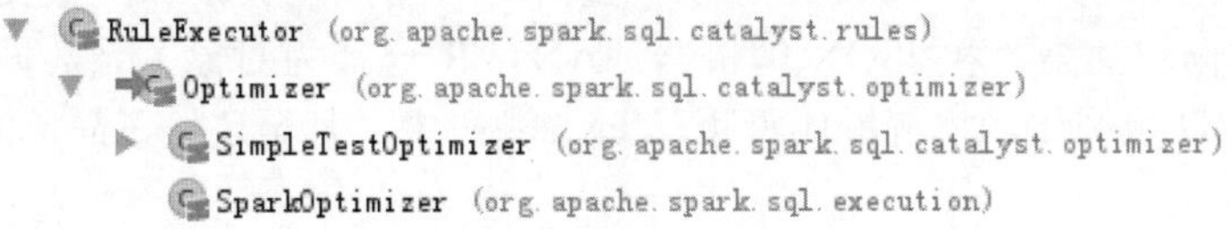

RuleExecutor (org.apache.spark.sql.catalyst.rules)
Optimizer (org.apache.spark.sql.catalyst.optimizer)
SimpleTestOptimizer (org.apache.spark.sql.catalyst.optimizer)
SparkOptimizer (org.apache.spark.sql.execution)

图 27-14　Optimizer 继承结构

（2）优化策略详解。

Optimizer 优化就是在 Catalyst 里将语法分析后的逻辑计划（Analyzed Logical Plan）通过对逻辑计划（Logical Plan）和表达式（Expression）进行规则（Rule）的应用及转换（Transform），

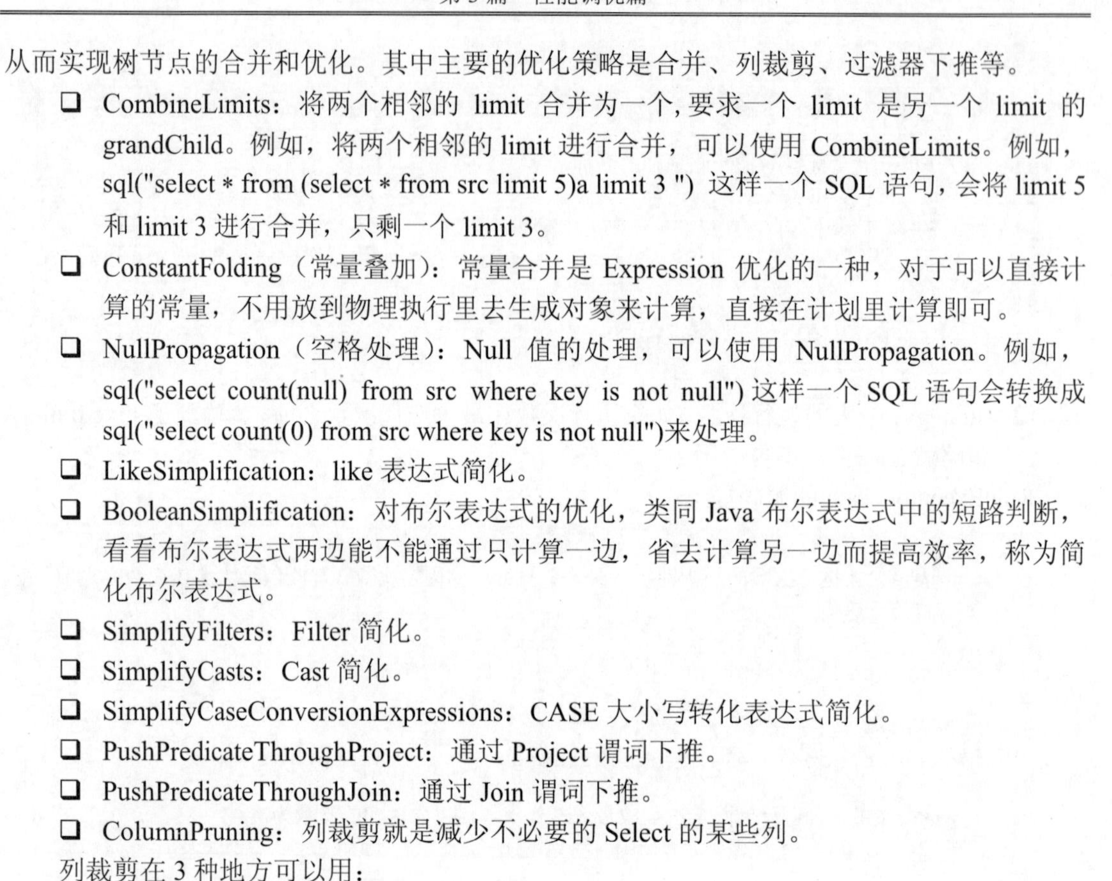

从而实现树节点的合并和优化。其中主要的优化策略是合并、列裁剪、过滤器下推等。

- ❑ CombineLimits：将两个相邻的 limit 合并为一个, 要求一个 limit 是另一个 limit 的 grandChild。例如，将两个相邻的 limit 进行合并，可以使用 CombineLimits。例如，sql("select * from (select * from src limit 5)a limit 3 ") 这样一个 SQL 语句，会将 limit 5 和 limit 3 进行合并，只剩一个 limit 3。
- ❑ ConstantFolding（常量叠加）：常量合并是 Expression 优化的一种，对于可以直接计算的常量，不用放到物理执行里去生成对象来计算，直接在计划里计算即可。
- ❑ NullPropagation（空格处理）：Null 值的处理，可以使用 NullPropagation。例如，sql("select count(null) from src where key is not null") 这样一个 SQL 语句会转换成 sql("select count(0) from src where key is not null")来处理。
- ❑ LikeSimplification：like 表达式简化。
- ❑ BooleanSimplification：对布尔表达式的优化，类同 Java 布尔表达式中的短路判断，看看布尔表达式两边能不能通过只计算一边，省去计算另一边而提高效率，称为简化布尔表达式。
- ❑ SimplifyFilters：Filter 简化。
- ❑ SimplifyCasts：Cast 简化。
- ❑ SimplifyCaseConversionExpressions：CASE 大小写转化表达式简化。
- ❑ PushPredicateThroughProject：通过 Project 谓词下推。
- ❑ PushPredicateThroughJoin：通过 Join 谓词下推。
- ❑ ColumnPruning：列裁剪就是减少不必要的 Select 的某些列。

列裁剪在 3 种地方可以用：

（1）在聚合操作中，可以做列裁剪。

（2）在 Join 操作中，左右孩子可以做列裁剪。

（3）合并相邻的 Project 的列时可以做列裁剪。

3. Spark SQL中ANTLR4的应用

ANTLR 是一个强大的解析器生成器，可用于读取、处理、执行或翻译结构化文本或二进制文件。它广泛应用于学术界和工业界，建立各种语言，工具和框架。例如，Twitter 搜索使用 ANTLR 进行查询解析，每天有超过 2 亿次查询。Hive 和 Pig 语言，Hadoop 的数据仓库和分析系统都使用 ANTLR。Lex Machina 使用 ANTLR 从法律文本中提取信息。Oracle 在 SQL Developer IDE 及其迁移工具中使用 ANTLR。NetBeans IDE 使用 ANTLR 解析 C++。Hibernate 对象关系映射框架中的 HQL 语言使用 ANTLR 构建。

除了这些大型的项目，ANTLR 还可以构建各种有用的工具，如配置文件读取器、旧代码转换器、Wiki 标记渲染器和 JSON 解析器。ANTLR 已经为对象关系数据库映射建立了一些工具，描述了 3D 可视化，将解析代码注入 Java 源码中，甚至还做了简单的 DNA 模式匹配示例。

从称为语法的形式语言描述中，ANTLR 生成可以自动构建解析树的语言的解析器，解释语法如何匹配输入的数据结构。ANTLR 还会自动生成 tree walkers，以使用它们访问这些树的节点来执行特定于应用程序的代码。

ANTLR 被广泛使用，ANTLR 易于理解、强大、灵活，能够生成可读的输出，具有 BSD

许可证下的完整源码，因此得到积极的支持。

ANTLR 对解析的理论和实践做出了贡献，包括：

- linear approximate lookahead。
- semantic and syntactic predicates。
- ANTLRWorks。
- tree parsing。
- LL(*)。
- Adaptive LL(*) in ANTLR v4。

Terence Parr 是 ANTLR 的发明者，自 1989 年以来一直在 ANTLR 工作，他是旧金山大学的计算机科学教授。

ANTLR 做两件事：① 将语法转换为 Java（或其他目标语言）的语法分析器/词法分析器的工具。② 运行期间生成解析器/词法分析器。如使用 ANTLR IntelliJ 插件或 ANTLRWorks 来运行 ANTLR 工具，生成的代码仍将需要运行时库。

首先，下载并安装 ANTLR 开发工具插件。然后，获取系统运行环境，以运行生成的解析器/词法分析器。下面将在 UNIX 系统和 Windows 系统中分别安装部署 antlr-4.5.3-complete.jar。

在 UNIX 系统中安装部署 antlr-4.5.3-complete.jar。

（1）安装 Java（1.6 或更高版本）。

（2）下载安装 antlr 的 Jar 包。

```
1.  $ cd /usr/local/lib
2.  $ curl -O http://www.antlr.org/download/antlr-4.5.3-complete.jar
```

或者从浏览器中下载：http://www.antlr.org/download.html，并将其放在/usr/local/lib 位置。将 antlr-4.5.3-complete.jar 附加到系统的 CLASSPATH。

```
1.   $ export CLASSPATH=".:/usr/local/lib/antlr-4.5.3-complete.jar:$CLASSPATH"
```

创建 ANTLR 工具别名和 TestRig。

```
1.  $ alias antlr4='java -Xmx500M -cp "/usr/local/lib/antlr-4.5.3-complete.jar:
    $CLASSPATH" org.antlr.v4.Tool'
2.  $ alias grun='java org.antlr.v4.gui.TestRig'
```

在 Windows 系统中安装部署 antlr-4.5.3-complete.jar。

（1）安装 Java（1.6 或更高版本）。

（2）从 http://www.antlr.org/download/下载 antlr-4.5.3-complete.jar，保存到 windows 本地目录中，如 C:\Javalib。

（3）将 antlr-4.5-complete.jar 到 CLASSPATH，或者使用系统属性对话框>环境变量>创建或附加到 CLASSPATH 变量。

```
1.  SET CLASSPATH=.;C:\Javalib\antlr-4.5.3-complete.jar;%CLASSPATH%
```

使用批处理文件或 doskey 命令为 ANTLR 工具和 TestRig 创建快捷命令：

批处理文件（系统 PATH 中的目录）antlr4.bat 和 grun.bat。

```
1.   java org.antlr.v4.Tool %*
2.  java org.antlr.v4.gui.TestRig %*
```

或者使用 doskey 命令：

```
1.  doskey antlr4=java org.antlr.v4.Tool $*
2. doskey grun =java org.antlr.v4.gui.TestRig $*
```

ANTLR 安装测试。

可以直接启动 org.antlr.v4.Tool。

```
1. $ java org.antlr.v4.Tool
2. ANTLR Parser Generator Version 4.5.3
3. -o ___ specify output directory where all output is generated
4. -lib ___ specify location of .tokens files
5. ……
```

或者在 Java 上使用-jar 选项。

```
1. $ java -jar /usr/local/lib/antlr-4.5.3-complete.jar
2. ANTLR Parser Generator Version 4.5.3
3. -o ___ specify output directory where all output is generated
4. -lib ___ specify location of .tokens files
5. ……
```

ANTLR 的第一个例子。

用 ANTLR 开发一个语法分析器，可以分成 3 个步骤。

- 写出要分析内容的文法。
- 用 ANTLR 生成相对该文法的语法分析器的代码。
- 编译运行语法分析器。

首先在临时目录中将以下语法放在 Hello.g4 文件中。

```
1. //定义一个名称为 Hello 的文法解析器
2. grammar Hello;
3. r  : 'hello' ID ;              //匹配关键词 hello，之后跟随标识符
4. ID : [a-z]+ ;                  //匹配小写的标识符
5. WS : [ \t\r\n]+ -> skip ;   //忽略空格、tab 符、换行符
```

Hello.g4 文件的第一行 grammar Hello 的 Hello 为文法的名称，Hello 与文件名一致，文件名后缀以.g4 结尾。Hello.g4 第二行开始是文法定义部分，文法是用扩展的巴科斯范式（EBNF1）推导式来描述的，每行都是一个规则（rule）或叫作推导式、产生式，每个规则的左边是文法中的一个名字，代表文法中的一个抽象概念。中间用一个“:”表示推导关系，右边是该名字推导出的文法形式。

Hello.g4 文法的规则定义如下。

- “r　: 'hello' ID ;”代表表达式语句，表示以'hello'开头的字符串，'hello'之后跟随的字符须符合表达式 ID 的规则，即任意小写字母的字符串。表达式本身是合法的语句，表达式也可以出现在赋值表达式中组成赋值语句，语句以“;”字符结束。
- “ID : [a-z]+ ;”以小写形式表达的词法描述部分，ID 表示变量由一个或多个小写字母组成。
- “WS : [\t\r\n]+ -> skip ;”WS 表示空白，它的作用是过滤掉空格、TAB 符号、回车换行的无意义字符。skip 的作用是跳过空白字符。

接下来运行 ANTLR 工具，使用 antlr4 批处理文件来编译文法。

```
1. $ cd /tmp
```

```
2.  $ antlr4 Hello.g4
3.  $ javac Hello*.java
```

然后进行测试，antlr4 提供了一个可以测试文法的工具类 java org.antlr.v4.gui.TestRig，使用 TestRig 来测试新开发的文法，使用 grun 批处理文件测试运行。

```
1.   $ grun Hello r -tree
2.  hello parrt
3.  ^D
4.  (r hello parrt)
5.  (That ^D means EOF on unix; it's ^Z in Windows.) The -tree option prints
    the parse tree in LISP notation.
6.  It's nicer to look at parse trees visually.
7.  $ grun Hello r -gui
8.  hello parrt
9.  ^D
```

使用-gui 参数可以让工具类显示一个窗口来显示解析后的树形结构。antlr4 运行结果示意图如图 27-15 所示。弹出一个对话框，显示 r 匹配关键字 hello 后跟标识符 parrt。

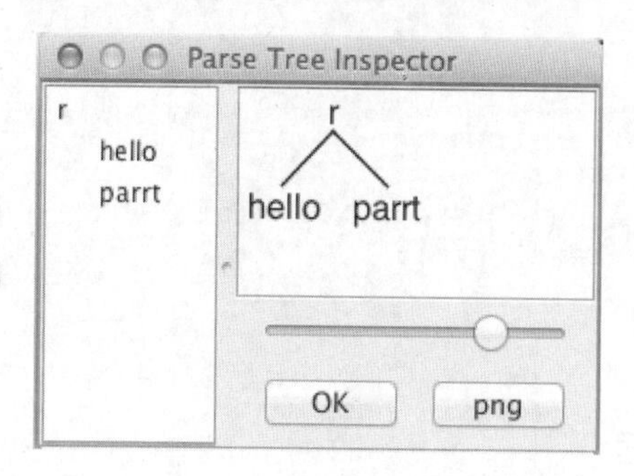

图 27-15　antlr4 运行结果示意图

Spark SQL 中 ANTLR4 的应用：

Spark 1.X 版本使用的是 Scala 原生的 parser 语法解析器，Spark 2.2 版本不同于 Spark 1.X 版本的 parser 语法解析器，Spark 2.2 版本使用的是第三方语法解析器工具 ANTLR4。

Spark 2.2 SQL 语句的解析采用的是 ANTLR4，ANTLR4 根据 spark-2.4.3\sql\catalyst\src\main\antlr4\org\apache\spark\sql\catalyst\parser\SqlBase.g4 文件自动解析生成的 Java 类：词法解析器 SqlBaseLexer 和语法解析器 SqlBaseParser。

SqlBaseLexer 和 SqlBaseParser 均使用 ANTLR4 自动生成的 Java 类。使用这两个解析器将 SQL 语句解析成 ANTLR 4 的语法树结构 ParseTree。然后在 parsePlan 中使用 AstBuilder（AstBuilder.scala）将 ANTLR4 语法树结构转换成 Catalyst 表达式逻辑计划。

ANTLR4 自动生成 Spark SQL 语法解析器 SqlBaseParser 类等 Java 代码。摘录部分代码如下。

SqlBaseParser.java 的源码如下：

```
1.   @SuppressWarnings({"all", "warnings", "unchecked", "unused", "cast"})
2.  public class SqlBaseParser extends Parser {
3.      static { RuntimeMetaData.checkVersion("4.7", RuntimeMetaData.VERSION); }
4.
5.      protected static final DFA[] _decisionToDFA;
```

```
6.      protected static final PredictionContextCache _sharedContextCache =
7.          new PredictionContextCache();
8.      public static final int
9.        T__0=1, T__1=2, T__2=3, T__3=4, T__4=5, T__5=6, T__6=7, SELECT=8,
          FROM=9,
10. ......
11.     public static final int
12.         RULE_singleStatement = 0, RULE_singleExpression = 1, RULE_single
            TableIdentifier = 2,
13. ......
14. public static final String[] ruleNames = {
15.         "singleStatement", "singleExpression", "singleTableIdentifier",
            "singleDataType",
16. ......
17.     private static final String[] _LITERAL_NAMES = {
18.         null, "'('", "')'", "','", "'.'", "'['", "']'", "':'", "'SELECT'",
            "'FROM'",
19.         "'ADD'", "'AS'", "'ALL'", "'DISTINCT'", "'WHERE'", "'GROUP'",
            "'BY'",
20.         "'GROUPING'", "'SETS'", "'CUBE'", "'ROLLUP'", "'ORDER'", "'HAVING'",
            "'LIMIT'",
21.         "'AT'", "'OR'", "'AND'", "'IN'", null, "'NO'", "'EXISTS'",
      "'BETWEEN'",
22. ......
23. public SqlBaseParser(TokenStream input) {
24.         super(input);
25.         _interp = new ParserATNSimulator(this,_ATN,_decisionToDFA,_
            sharedContextCache);
26.     }
27. ......
28.
```

4. Spark SQL工作流程

本节对一个简单的 Spark SQL 程序进行剖析，讲解在业务代码中编写的 SQL 语句，在 Spark SQL Catalyst 中最终是在哪里提交 Spark 框架进行执行的。

1）Spark SQL 示例剖析

先来看一个简单的 Spark SQL 程序。

```
1.  import sqlContext._
2.  val sqlContext = new org.apache.spark.sql.SQLContext(sc)
3.  case class Student(name: String, age: Int)
4.   val student = sc.textFile("examples/src/main/resources/Student.txt").
    map(_.split(",")).map(p => Student(p(0), p(1).trim.toInt))
5.  student.registerAsTable("student")
6.  val agers = sql("SELECT name FROM student WHERE age >= 13 AND age <= 19")
7.  agers.map(t => "Name: " + t(0)).collect().foreach(println)
```

程序逻辑解释如下。

第 1 行代码，导入 sqlContext 下面的 all。

第 2 行代码，生成 SQLContext，也就是运行 SparkSQL 的上下文环境。

第 3、4、5 行代码，是加载数据源注册 table。

第 6 行代码是真正的入口，是 SQL 函数，传入一个 SQL 语句，返回一个 SchemaRDD。这一步是 lazy 级别的代码，直到遇到第 7 行代码的 collect 这个 action 时，sql 才会被真正执行。

调用 sql 函数，返回的结果是创建一个 DataFrame，DataFrame 是 Dataset[Row]的别名，

DataFrame 在生成的时候就调用 sessionState.sqlParser.parsePlan 方法，parsePlan 的结果会生成一个逻辑计划。

当调用 Dataset 里面的 collect 方法时，会初始化 QueryExecution，开始启动执行。

Dataset.scala 的源码如下：

```
1.     def collect(): Array[T] = withAction("collect", queryExecution)
       (collectFromPlan)
2.   ……
3.   private def collectFromPlan(plan: SparkPlan): Array[T] = {
4.     plan.executeCollect().map(boundEnc.fromRow)
5.   }
6.   ……
```

通过 withAction 方法包裹 Dataset 的执行，跟踪 QueryExecution 查询执行及时间成本，并汇报给用户注册的回调函数。collect 方法返回一个数组，此数组包含数据集所有行的数据，运行 collect 算子收集所有数据到应用程序的 Driver 节点中，并且是一个非常大的数据集，collect 方法可能导致 OutOfMemoryError。

QueryExecution 类对象中定义了 SQL 执行过程中的关键步骤，是 SQL 执行的关键类。QueryExecution 类中的成员都是 lazy 级别的，被调用时才会执行。只有等到程序中出现 action 算子时，才会调用 QueryExecution 类中的 executedPlan 成员，原先生成的逻辑执行计划才会被优化器优化，并转换成物理执行计划真正地被系统调用执行。

QueryExecution 类的主要成员定义了 analyzer、optimizer 以及生成物理执行计划的 SparkPlan。最终调用 executedPlan.execute 生成 RDD。

Spark 2.2.1 版本的 QueryExecution.scala 的源码如下：

```
1.  /**
2.    *执行关联查询主要的工作流程：利用 Spark 的设计使开发人员很容易
3.    *进入查询执行的中间阶段。虽然 QueryExecution 不是一个公共类，但要避免
4.    *更改它们，因为很多开发人员都使用这个特性进行调试
5.    */
6.  class QueryExecution(val sparkSession: SparkSession, val logical:
    LogicalPlan) {
7.
8.    //将来要实现的：在这里从 SessionState 会话状态移动优化器
9.    protected def planner = sparkSession.sessionState.planner
10.
11.   def assertAnalyzed(): Unit = {
12.     //Analyzer 尝试 try 块外部调用，以避免在下面的 catch 块中再次调用它
13.     analyzed
14.     try {
15.       sparkSession.sessionState.analyzer.checkAnalysis(analyzed)
16.     } catch {
17.       case e: AnalysisException =>
18.         val ae = new AnalysisException(e.message, e.line, e.startPosition,
            Option(analyzed))
19.         ae.setStackTrace(e.getStackTrace)
20.         throw ae
21.     }
22.   }
23.
24.   def assertSupported(): Unit = {
25.     if (sparkSession.sessionState.conf.isUnsupportedOperationCheckEnabled) {
26.       UnsupportedOperationChecker.checkForBatch(analyzed)
```

```
27.     }
28.   }
29.  //调用 analyzer 解析器
30.   lazy val analyzed: LogicalPlan = {
31.     SparkSession.setActiveSession(sparkSession)
32.     sparkSession.sessionState.analyzer.execute(logical)
33.   }
34.
35.   lazy val withCachedData: LogicalPlan = {
36.     assertAnalyzed()
37.     assertSupported()
38.     sparkSession.sharedState.cacheManager.useCachedData(analyzed)
39.   }
40.   //调用 optimizer 优化器
41.   lazy val optimizedPlan: LogicalPlan = sparkSession.sessionState
      .optimizer.execute(withCachedData)
42.   //将优化后的逻辑执行计划转换成物理执行计划
43.   lazy val sparkPlan: SparkPlan = {
44.     SparkSession.setActiveSession(sparkSession)
45.     //将来要实现的：使用 next 方法，例如，从计划器 planner 返回第一个计划 plan，现
46.     //在在这里选择最佳的计划
47.     planner.plan(ReturnAnswer(optimizedPlan)).next()
48.   }
49.
50.   //executedPlan 不应该被用来初始化任何 SparkPlan，它应该只用于执行
51.   lazy val executedPlan: SparkPlan = prepareForExecution(sparkPlan)
52.
53.   /** 转换为内部的 RDD，避免复制及没有编目概要 */
54.   lazy val toRdd: RDD[InternalRow] = executedPlan.execute()
55.
56.   /**
57.    *  准备一个计划[SparkPlan]，用于执行 Shuffle 算子和内部行格式转换的需要
58.    */
59.   protected def prepareForExecution(plan: SparkPlan): SparkPlan = {
60.     preparations.foldLeft(plan) { case (sp, rule) => rule.apply(sp) }
61.   }
62.
63.   /**为了执行物理计划而应用的一系列规则*/
64.   protected def preparations: Seq[Rule[SparkPlan]] = Seq(
65.     python.ExtractPythonUDFs,
66.     PlanSubqueries(sparkSession),
67.     EnsureRequirements(sparkSession.sessionState.conf),
68.     CollapseCodegenStages(sparkSession.sessionState.conf),
69.     ReuseExchange(sparkSession.sessionState.conf),
70.     ReuseSubquery(sparkSession.sessionState.conf))
71.
72.   protected def stringOrError[A](f: => A): String =
73.     try f.toString catch { case e: AnalysisException => e.toString }
74.
75.
76.   /**
77.    *作为一个 Hive 兼容序列的字符串返回结果。对于本地命令，执行通过 Hive 返回
78.    */
79.   def hiveResultString(): Seq[String] = executedPlan match {
80.     case ExecutedCommandExec(desc: DescribeTableCommand) =>
81.       //如果是一个描述 Hive 表的描述命令，输出的格式与 Hive 相似
82.       desc.run(sparkSession).map {
83.         case Row(name: String, dataType: String, comment) =>
84.           Seq(name, dataType,
85.             Option(comment.asInstanceOf[String]).getOrElse(""))
```

```
86.            .map(s => String.format(s"%-20s", s))
87.            .mkString("\t")
88.        }
89.     //SHOW TABLES 在 Hive 中仅输出表的名称，此处输出数据库、表名等，是临时的内容
90.     case command @ ExecutedCommandExec(s: ShowTablesCommand) if !s.isExtended =>
91.       command.executeCollect().map(_.getString(1))
92.     case other =>
93.       val result: Seq[Seq[Any]] = other.executeCollectPublic().map(_.toSeq).toSeq
94.       //需要的类型，可以输出结构的字段名称
95.       val types = analyzed.output.map(_.dataType)
96.       //重新格式化，匹配 Hive 的制表符分隔的输出
97.       result.map(_.zip(types).map(toHiveString)).map(_.mkString("\t"))
98.   }
99.
100.      /**格式化数据（基于给定的数据类型），并返回字符串表示 */
101.      private def toHiveString(a: (Any, DataType)): String = {
102.        val primitiveTypes = Seq(StringType, IntegerType, LongType,
        DoubleType, FloatType,
103.          BooleanType, ByteType, ShortType, DateType, TimestampType, BinaryType)
104.
105.        def formatDecimal(d: java.math.BigDecimal): String = {
106.          if (d.compareTo(java.math.BigDecimal.ZERO) == 0) {
107.            java.math.BigDecimal.ZERO.toPlainString
108.          } else {
109.            d.stripTrailingZeros().toPlainString
110.          }
111.        }
112.
113.        /**Hive 输出的结构字段与顶级属性稍有不同 */
114.        def toHiveStructString(a: (Any, DataType)): String = a match {
115.          case (struct: Row, StructType(fields)) =>
116.            struct.toSeq.zip(fields).map {
117.              case (v, t) => s""""${t.name}":${toHiveStructString(v,
              t.dataType)}"""
118.            }.mkString("{", ",", "}")
119.          case (seq: Seq[_], ArrayType(typ, _)) =>
120.            seq.map(v => (v, typ)).map(toHiveStructString).mkString("[", ",", "]")
121.          case (map: Map[_, _], MapType(kType, vType, _)) =>
122.            map.map {
123.              case (key, value) =>
124.                toHiveStructString((key, kType)) + ":" +
                toHiveStructString((value, vType))
125.            }.toSeq.sorted.mkString("{", ",", "}")
126.          case (null, _) => "null"
127.          case (s: String, StringType) => "\"" + s + "\""
128.          case (decimal, DecimalType()) => decimal.toString
129.          case (other, tpe) if primitiveTypes contains tpe =>
          other.toString
130.        }
131.
132.        a match {
133.          case (struct: Row, StructType(fields)) =>
134.            struct.toSeq.zip(fields).map {
135.              case (v, t) => s""""${t.name}":${toHiveStructString(v,
              t.dataType)}"""
136.            }.mkString("{", ",", "}")
137.          case (seq: Seq[_], ArrayType(typ, _)) =>
138.            seq.map(v => (v, typ)).map(toHiveStructString).mkString("[",
            ",", "]")
139.          case (map: Map[_, _], MapType(kType, vType, _)) =>
```

```
140.          map.map {
141.            case (key, value) =>
142.              toHiveStructString((key, kType)) + ":" +
                  toHiveStructString((value, vType))
143.          }.toSeq.sorted.mkString("{", ",", "}")
144.        case (null, _) => "NULL"
145.        case (d: Date, DateType) =>
146.          DateTimeUtils.dateToString(DateTimeUtils.fromJavaDate(d))
147.        case (t: Timestamp, TimestampType) =>
148.          DateTimeUtils.timestampToString(DateTimeUtils.fromJavaTimestamp(t),
149.            DateTimeUtils.getTimeZone(sparkSession.sessionState.conf
                .sessionLocalTimeZone))
150.        case (bin: Array[Byte], BinaryType) => new String(bin,
            StandardCharsets.UTF_8)
151.        case (decimal: java.math.BigDecimal, DecimalType()) =>
            formatDecimal(decimal)
152.        case (other, tpe) if primitiveTypes.contains(tpe) => other.toString
153.      }
154.    }
155.
156.    def simpleString: String = {
157.      s"""== Physical Plan ==
158.         |${stringOrError(executedPlan.treeString(verbose = false))}
159.        """.stripMargin.trim
160.    }
161.
162.    override def toString: String = completeString(appendStats = false)
163.
164.    def toStringWithStats: String = completeString(appendStats = true)
165.
166.    private def completeString(appendStats: Boolean): String = {
167.      def output = Utils.truncatedString(
168.        analyzed.output.map(o => s"${o.name}: ${o.dataType.simpleString}"), ", ")
169.      val analyzedPlan = Seq(
170.        stringOrError(output),
171.        stringOrError(analyzed.treeString(verbose = true))
172.      ).filter(_.nonEmpty).mkString("\n")
173.
174.      val optimizedPlanString = if (appendStats) {
175.        //触发器计算逻辑计划的统计信息
176.        optimizedPlan.stats(sparkSession.sessionState.conf)
177.        optimizedPlan.treeString(verbose = true, addSuffix = true)
178.      } else {
179.        optimizedPlan.treeString(verbose = true)
180.      }
181.
182.      s"""== Parsed Logical Plan ==
183.         |${stringOrError(logical.treeString(verbose = true))}
184.         |== Analyzed Logical Plan ==
185.         |$analyzedPlan
186.         |== Optimized Logical Plan ==
187.         |${stringOrError(optimizedPlanString)}
188.         |== Physical Plan ==
189.         |${stringOrError(executedPlan.treeString(verbose = true))}
190.      """.stripMargin.trim
191.    }
192.
193.    /**用于调试查询执行命令的特殊命名空间 */
194.    //scalastyle:关
195.    object debug {
```

```
196.      //scalastyle:打开
197.
198.        /**
199.         *打印在这个计划中所有生成的代码到标准输出（即每个 WholeStageCodegen 子树
           的输出）
200.         */
201.        def codegen(): Unit = {
202.          //scalastyle:关闭打印
203.
   println(org.apache.spark.sql.execution.debug.codegenString(executedPl
   an))
204.          //scalastyle:打开打印
205.        }
206.      }
```

Spark 2.4.3 版本的 QueryExecution.scala 源码与 Spark 2.2.1 版本相比具有如下特点。

- 上段代码中第 11～22 行删除。
- 上段代码中第 32 行 execute 方法调整为 executeAndCheck 方法，对逻辑计划执行子类定义的批处理规则，对规则进行转换和排序，以便捕获第一个可能的失败，级联解析失败的结果。
- 上段代码中第 65 行删除。
- 上段代码中第 151 行之后新增代码，新增对 interval 日历间隔的处理。
- 上段代码中第 156～191 行替换为以下第 8～50 行代码。

```
1.   ......
2.    def assertAnalyzed(): Unit = analyzed
3.    ......
4.  sparkSession.sessionState.analyzer.executeAndCheck(logical)
5.  ......
6.    case (interval, CalendarIntervalType) => interval.toString
7.  ......
8.  def simpleString: String = withRedaction {
9.      s"""== Physical Plan ==
10.        |${stringOrError(executedPlan.treeString(verbose = false))}
11.       """.stripMargin.trim
12.   }
13.
14.   override def toString: String = withRedaction {
15.     def output = Utils.truncatedString(
16.       analyzed.output.map(o => s"${o.name}: ${o.dataType.simpleString}"), ", ")
17.     val analyzedPlan = Seq(
18.       stringOrError(output),
19.       stringOrError(analyzed.treeString(verbose = true))
20.     ).filter(_.nonEmpty).mkString("\n")
21.
22.     s"""== Parsed Logical Plan ==
23.        |${stringOrError(logical.treeString(verbose = true))}
24.        |== Analyzed Logical Plan ==
25.        |$analyzedPlan
26.        |== Optimized Logical Plan ==
27.        |${stringOrError(optimizedPlan.treeString(verbose = true))}
28.        |== Physical Plan ==
29.        |${stringOrError(executedPlan.treeString(verbose = true))}
30.     """.stripMargin.trim
31.   }
32.
```

```
33.   def stringWithStats: String = withRedaction {
34.     //触发器计算逻辑计划的统计信息
35.     optimizedPlan.stats
36.
37.     //仅显示优化的逻辑计划和物理计划
38.     s"""== Optimized Logical Plan ==
39.         |${stringOrError(optimizedPlan.treeString(verbose = true,
        addSuffix = true))}
40.         |== Physical Plan ==
41.         |${stringOrError(executedPlan.treeString(verbose = true))}
42.     """.stripMargin.trim
43.   }
44.
45.   /**
46.    *修改给定字符串中的敏感信息
47.    */
48.   private def withRedaction(message: String): String = {
49.     Utils.redact(sparkSession.sessionState.conf.stringRedactionPattern, message)
50.   }
```

2）Spark SQL 执行流程

Spark SQL 执行流程图如图 27-16 所示。

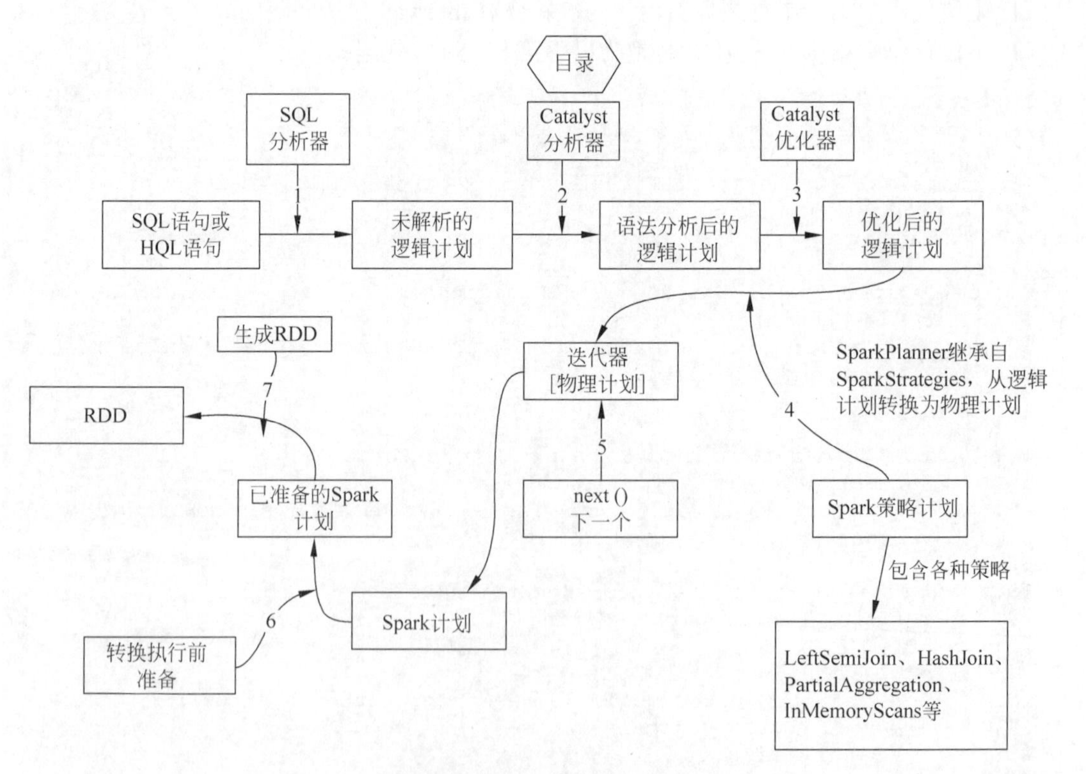

图 27-16　Spark SQL 执行流程图

以上整个流程可以总结为：

（1）sql parser(parse)生成 unresolved logical plan。

（2）analyzer(analysis)生成 analyzed logical plan。

（3）optimizer(optimize)optimized logical plan。

（4）spark planner(use strategies to plan)生成 physical plan。

（5）迭代器进行 next 迭代，采用不同 Strategies 生成 spark plan。

（6）spark plan(prepare) prepared spark plan。

（7）call toRDD（execute()函数调用）执行 SQL 生成 RDD。

27.1.2　Spark SQL 调优参数及调优最佳实践

对于某些工作负载，可以通过缓存内存中的数据或打开一些实验选项来提高性能。

缓存内存中的数据（Caching Data In Memory）：Spark SQL 可以通过调用 spark.catalog.cacheTable("tableName")或使用 dataFrame.cache()以内存中的列格式来缓存表。然后，Spark SQL 将仅扫描所需的列，并将自动调整压缩以最小化内存使用来减少 GC 压力。可以调用 spark.catalog.uncacheTable("tableName")从内存中删除表。

内存缓存的配置可以使用 SparkSession 的 setConf 方法，或在 SQL 语句中运行 SET key=value 来配置内存中的缓存。表 27-1 为内存缓冲配置表。

表 27-1　内存缓冲配置表

属　性　名	默认值	含　　义
spark.sql.inMemoryColumnarStorage.compressed	true	如果设置为 true，Spark SQL 将会根据数据统计信息，自动为每列选择单独的压缩编码方式
spark.sql.inMemoryColumnarStorage.batchSize	10000	控制列式缓存批量的大小。增大批量大小可以提高内存利用率和压缩率，但同时也会带来 OOM（Out Of Memory）的风险

以下选项也可用于调整查询执行的性能。这些选项有可能在将来的版本中被废弃，因为更多的优化是自动执行的。表 27-2 为 SQL 查询配置表。

表 27-2　SQL查询配置表

属　性　名	默　认　值	含　　义
Spark.sql.files.maxPartitionBytes	134217728 (128 MB)	读取文件并放入一个分区中的最大字节数
Spark.sql.files.openCostInBytes	4 194 304 (4MB)	小文件合并阈值。打开文件的估算成本，按照同一时间能够扫描的字节数来测量。将多个文件放入分区时使用。最好过度估计，那么具有小文件的分区将比具有较大文件的分区（先被调度）更快
Spark.sql.broadcastTimeout	300	广播变量做 Join 时的等待时间（s）
Spark.sql.autoBroadcastJoinThreshold	10 485 760 (10 MB)	配置 Join 操作时，能够作为广播变量的最大 table 的大小。设置为-1，表示禁用广播。注意，目前的元数据统计仅支持 Hive metastore 中的表，并且需要运行命令：ANALYSE TABLE <tableName> COMPUTE STATISTICS noscan
Spark.sql.Shuffle.Partitions	200	配置数据混洗（Shuffle）时（Join 或者聚合操作）使用的分区数

常用的一些 Spark SQL 设置：

（1）Spark.sql.Shuffle.Partitions=200

并行度的优化，这要根据自己集群的配置来调节。

（2）Spark.sql.files.maxPartitionBytes=128MB

调节每个 Partition 大小，默认 Partition 是 128MB，可以调大一点。

（3）Spark.sql.files.openCostInBytes=4MB

很多小文件浪费 Task，合并成一个 Task，提高处理性能，可以将值调大。

（4）Spark.sql.autoBroadcastJoinThreshold=10MB

两个表 Shuffle，如 Join。这个最有用，是经常使用的。

默认是 10MB，调成 100MB，甚至是 1GB。

综合实践例程：

```
1.   val conf = new SparkConf()
2.          .setAppName("moviesRatings")
3.          .setMaster("local")
4.          //可调优参数
5.          .set("Spark.sql.Shuffle.Partitions","4")    //调节并行度
6.          .set("Spark.sql.files.maxPartitionBytes","256")//调节每个Partition大小
7.          .set("Spark.sql.files.openCostInBytes","4")//小文件合并
8.          .set("Spark.sql.autoBroadcastJoinThreshold","100")
                                                     //Join时小表的大小
```

27.2　Spark Streaming 调优原理及调优最佳实践

本节讲解 Spark Streaming 调优原理及 Spark Streaming 调优参数及调优最佳实践。

27.2.1　Spark Streaming 调优原理

Spark Streaming 提供了高效便捷的流式处理模式，但是在有些场景下，使用默认的配置达不到最优，甚至无法实时处理来自外部的数据，这时我们就需要对默认的配置进行相关修改。由于现实中场景和数据量的多样性，所以我们无法设置一些通用的配置，我们需要根据数据量、场景的不同设置不一样的配置，这里只是给出建议，这些调优不一定适用于你的程序，一个好的配置是需要慢慢地尝试的。

1．设置合理的批处理时间（batchDuration）

构建 StreaminGContext 时，需要传进一个参数，用于设置 Spark Streaming 批处理的时间间隔。Spark 会每隔 batchDuration 时间去提交一次 Job，如果你的 Job 处理的时间超过了 batchDuration 的设置，就会导致后面的作业无法按时提交，随着时间的推移，越来越多的作业被拖延，最后导致整个 Streaming 作业被阻塞，这就间接地导致无法实时处理数据，这肯定不是我们想要的。

另外，虽然 batchDuration 的单位可以达到毫秒级别，但是经验告诉我们，如果这个值过小，将会导致因频繁提交作业，从而给整个 Streaming 带来负担，所以尽量不要将这个值设置为小于 500ms。在很多情况下，设置为 500ms 性能就很不错了。

那么，如何设置一个好的值呢？可以先将这个值置为比较大的值（如 10s），如果我们发现作业很快被提交完成，可以进一步减小这个值，直到 Streaming 作业刚好能够及时处理完

上一个批处理的数据，那么这个值就是我们要的最优值。

2．增加Job并行度

我们需要充分地利用集群的资源，尽可能地将 Task 分配到不同的节点，一方面可以充分利用集群资源；另一方面还可以及时地处理数据。例如，我们使用 Streaming 接收来自 Kafka 的数据，可以对每个 Kafka 分区设置一个接收器，这样可以达到负载均衡，及时处理数据。再如，类似 reduceByKey()和 Join 函数，都可以设置并行度参数。

3．使用Kryo系列化

Spark 默认的是使用 Java 内置的系列化类，虽然可以处理所有继承自 java.io.Serializable 的系列化的类，但是其性能不佳，如果这个成为性能瓶颈，可以使用 Kryo 系列化类。使用系列化数据可以很好地改善 GC 行为。

4．缓存需要经常使用的数据

对一些经常使用到的数据，我们可以显式地调用 RDD.Cache()来缓存数据，这样也可以加快数据的处理，但是我们需要更多的内存资源。

5．清除不需要的数据

随着时间的推移，有些数据是不需要的，但是这些数据缓存在内存中，会消耗宝贵的内存资源，我们可以通过配置 Spark.cleaner.ttl 为一个合理的值；但是这个值不能过小，因为如果后面计算需要用的数据被清除，会带来不必要的麻烦。而且，我们还可以配置选项 Spark.streaming.unPersist 为 true（默认是 true）来更智能地去持久化（unPersist）RDD。这个配置使系统找出那些不需要经常保留的 RDD，然后去持久化它们。这可以减少 Spark RDD 的内存使用，也可能改善垃圾回收的行为。

6．设置合理的GC

GC 是程序中最难调的一块。不合理的 GC 行为会给程序带来很大影响。在集群环境下，我们可以使用并行 Mark-Sweep 垃圾回收机制，虽然这消耗更多的资源，但是我们还是建议开启，配置如下：

```
1.   Spark.Executor.extraJavaOptions=-XX：+UseConcMarkSweepGC
```

7．设置合理的CPU资源数

很多情况下，Streaming 程序需要的内存不是很多，但是需要的 CPU 很多。在 Streaming 程序中，CPU 资源的使用可以分为两大类：① 用于接收数据；② 用于处理数据。我们需要设置足够的 CPU 资源，使得有足够的 CPU 资源用于接收和处理数据，这样才能及时高效地处理数据。

27.2.2　Spark Streaming 调优参数及调优最佳实践

Spark Streaming 调优参数之并行度解析，在 Spark 集群资源允许的前提下，可以提高数据接收、数据处理的并行度。

数据接收的并行度调优，有多个方面。

1. InputDStream的并行度

Spark Streaming 应用程序中涉及数据接收的第一个 DStream 是 InputDStream。Spark Streaming 数据接收方式可分别使用 Receiver 方式和 No Receiver 方式（生产环境建议使用）。

对 Receiver 方式：每个 InputDStream 都会在某个 Worker 节点上创建一个 Receiver。写应用程序时，可以创建多个 InputDStream，来接收同一数据源数据。还可以通过配置，让这些 DStream 分别接收数据源的不同分区的数据，最大 DStream 个数可以达到数据源提供的分区数。例如，一个接收两个 Kafka Topic 数据的输入 DStream，可以被拆分成两个接收不同 Topic 数据的 DStream。最后，可以在程序中把多个 InputDStream 再合并为一个 DStream，进行后续处理。下面给出基于 Kafka 的 Java 代码。

多个 InputDStream 合并为一个 DStream 的 Java 代码如下：

```
1.  int numStreams = 5;
2.  List<JavaPairDStream<String, String>> kafkaStreams = new ArrayList<
    JavaPairDStream<String, String>>(numStreams);
3.  for (int i = 0; i < numStreams; i++) {
4.  kafkaStreams.add(KafkaUtils.createStream(...));
5.  }
6.  JavaPairDStream<String, String> unifiedStream = streaminGContext.
    union(kafkaStreams.get(0), kafkaStreams.subList(1, kafkaStreams.size()));
7.
```

2. Task的并行度

数据接收使用的 BlockGenerator 里有一个 RecurringTimer 类型的对象 blockIntervalTimer，会周期性地发送 BlockGenerator 消息，进而周期性地生成和存储一个 Block。这个周期对应有一个配置参数 Spark.streaming.blockInterval。这个时间周期的缺省值是 200ms。

读写 Block，会用到 BlockManager。BlockManager 定义于 Spark Core 中，是 Storage 模块与其他模块交互最主要的类，提供了读和写 Block 的接口。这里的 Block，实际上就对应了 RDD 中提到的 Partition，每个 Partition 都会对应一个 Block。而 Spark Streaming 按 Batch Interval 来进行一次数据接收和处理，所以 Batch Interval 内的 Block 个数，就是 RDD 的 Partition 数，也就是 RDD 的并行 Task 数。因此，Task 的并行度等于 Batch Interval / Block Interval。例如，Batch Interval 是 2s，Block Interval 是 200ms，则 Task 并行度为 10。

通过调小 Block Interval，可以提高 Task 并行度。但建议一般不让 Block Interval 低于 50ms。

3. 数据处理前的重分区

多输入流或者多 Receiver 的一个可选方法是：明确地重新分配输入数据流，即在进一步处理数据前，利用 DStream.rePartition(<number of Partitions>)重新分区，把接收的数据分发到集群上。

4. 数据处理的并行度

如果运行在计算 Stage 上的并发任务数不足够大，就不会充分利用集群的资源。例如，对于分布式 reduce 操作，如 reduceByKey 和 reduceByKeyAndWindow，默认的并发任务数通过配置属性 Spark.default.parallelism 来确定。可以通过参数 PairDStreamFunctions 传递并行度，或者设置参数 Spark.default.parallelism 修改默认值。

5. 内存

Spark Streaming 应用需要的集群内存资源，是由使用的转换操作类型决定的。例如，如果想使用一个窗口长度为 10min 的窗口操作，那么集群就必须有足够的内存来保存 10min 内的数据。如果想使用 updateStateByKey 来维护许多 Key 的 state，那么内存就必须足够大。反过来说，如果想做一个简单的 map-filter-store 操作，需要使用的内存就很少。

通常，通过 Receiver 接收到的数据会使用 StorageLevel.MEMORY_AND_DISK_SER_2 持久化级别来进行存储，因此无法保存在内存中的数据会溢写到磁盘上，而溢写到磁盘上，是会降低应用的性能的。因此，通常建议为应用提供它需要的足够的内存资源。建议在一个小规模的场景下测试内存的使用量，并进行评估。

6. 序列化

Spark Streaming 默认将接收到的数据序列化存储，以减少内存使用。序列化和反序列化需要更多的 CPU 时间、更加高效的序列化方式（Kryo）。自定义的序列化接口可以更高效地使用 CPU。使用 Kryo 时，一定要考虑注册自定义的类，并且禁用对应引用的 tracking（Spark.kryo.referenceTracking）。

在流式计算的场景下，有两种类型的数据需要序列化。

（1）输入数据：默认情况下，接收到的输入数据是存储在 Executor 的内存中的，使用的持久化级别是 StorageLevel.MEMORY_AND_DISK_SER_2。这意味着，数据被序列化为字节，从而减小 GC 开销，并且会复制，以进行 Executor 失败的容错。因此，数据首先会存储在内存中，然后在内存不足时会溢写到磁盘上，从而为流式计算保存所有需要的数据。这里的序列化有明显的性能开销——Receiver 必须反序列化从网络接收到的数据，然后再使用 Spark 的序列化格式序列化数据。

（2）流式计算操作生成的持久化 RDD：流式计算操作生成的持久化 RDD，可能会持久化到内存中。例如，窗口操作默认就会将数据持久化在内存中，因为这些数据后面可能会在多个窗口中被使用，并被处理多次。然而，不像 Spark Core 的默认持久化级别，StorageLevel.MEMORY_ONLY，DStream 默认持久化级别是 MEMORY_ONLY_SER，默认就会减小 GC 开销。

在一些特殊的场景中，例如需要为流式应用保持的数据总量并不是很多，也许可以将数据以非序列化的方式进行持久化，从而减少序列化和反序列化的 CPU 开销，而且又不会有太昂贵的 GC 开销。举例来说，如果是数秒的 batch interval，并且没有使用 Window 操作，那么可以考虑通过显式地设置持久化级别，来禁止持久化时对数据进行序列化。这样就可以减少用于序列化和反序列化的 CPU 性能开销，并且不用承担太多的 GC 开销。

7. Batch Interval

要确保 Spark Streaming 应用程序在集群环境下稳定运行，系统必须尽快处理接收的数据。处理数据的速度要跟上数据流入的速度，即批次处理时间必须小于批次间隔时间。通过查看日志可以了解 Total delay，如果 delay 小于 Batch Interval，则系统稳定运行。

如果 Delay 一直增加，则说明系统的处理速度跟不上数据输入速度，要做调整。要想有一个稳定的配置，可以尝试提升数据处理的速度，或者增加 Batch Interval。临时性的数据增长导致暂时的延迟增长，可以被认为是合理的，只要延迟情况可以在短时间内恢复即可。

8. Task

任务的提交和分发都会有延迟，需要注意 Batch Interval 不能太小。可采取以下措施：

- 任务 Kryo 序列化可以减少任务大小，减少发送到 Worker 节点的时间。
- Standalone 模式、粗粒度 Mesos 模式下的 Spark，比细粒度模式有更低短延迟。

9. JVM GC

GC 会影响任务的运行。采取不同策略，以减小 GC 对 Job 运行的影响。降低系统吞吐量，就能减小 GC 的停顿。

对于流式应用来说，如果要获得低延迟，肯定不想有因为 JVM 垃圾回收导致的长时间延迟。有很多参数可以帮助降低内存使用和 GC 开销。

（1）DStream 的持久化：输入数据和某些操作生产的中间 RDD，默认持久化时都会序列化为字节。与非序列化的方式相比，这会降低内存和 GC 开销。使用 Kryo 序列化机制可以进一步减少内存使用和 GC 开销。进一步降低内存使用率，可以对数据进行压缩，由 Spark.RDD .compress 参数控制（默认为 false）。

（2）清理旧数据：默认情况下，所有输入数据和通过 DStream 的转换操作生成的持久化 RDD，会自动被清理。Spark Streaming 何时清理这些数据，取决于转换操作类型。例如，在使用窗口长度为 10min 内的 Window 操作，Spark 会保持 10min 以内的数据，时间过了以后，就会清理旧数据。但是，在某些特殊场景下，如 Spark SQL 和 Spark Streaming 整合使用时，在异步开启的线程中，使用 Spark SQL 针对 Batch RDD 进行查询。那么就需要让 Spark 保存更长时间的数据，直到 Spark SQL 查询结束。可以使用 StreaminGContext.remember 方法来实现。

（3）CMS 垃圾回收器：CMS 垃圾回收器使用并行的 mark-sweep 垃圾回收机制，用来保持 GC 低开销，所以推荐使用。虽然并行的 GC 会降低吞吐量，但是还是建议使用它来减少 Batch 的处理时间（降低处理过程中的 GC 开销）。如果要使用，那么要在 Driver 端和 Executor 端都开启。具体做法是用 Spark-Submit 提交应用程序执行时，增加如下两个设置。

```
1.  --Driver-java-options "-XX: +UseConcMarkSweepGC"
2.  --conf "Spark.Executor.extraJavaOptions=-XX: +UseConcMarkSweepGC"
```

27.3　Spark GraphX 调优原理及调优最佳实践

本节讲解 Spark GraphX 调优原理；以及 Spark GraphX 调优参数及调优最佳实践。

27.3.1　Spark GraphX 调优原理

在分布式系统中，往往倾向于用大量的小的低配机器，来完成巨大的计算任务。其思想是：即便再复杂的计算，只要将大数据分解为足够小的数据片，总能在足够多的机器上通过性能的降低和时间的拖延，来完成计算任务。但是，很遗憾，在图计算这样的场景下，尤其是 GraphX 的设计框架下，这是行不通的。

要发挥 GraphX 的最佳性能，最少要有 128GB 以上的内存。

主要原因有以下两个。

1. 节点复制——越小越浪费

GraphX 使用了点切割的方式，这是一种用空间换时间的方法，通过浪费一定的内存，将点和它的邻居放到一起，减少 Executor 之间的通信。如果用小内存的 Executor 来运行图算法，假设一个节点，需要 10 个 Executor 才能放下它的邻居，那么它就需要被复制 10 份，才能进行计算。如果用大内存的 Executor，一个 Executor 就能放下它的所有邻居，理论上它就只需要被复制一次，大大减少了空间占用。

2. 节点膨胀——越小越慢

图计算中，常常会进行消息的扩散和收集，并把最终的结果汇总到单个节点上。以共同好友数模型为例，第一步需要将节点的好友都收集到该节点上。即便根据邓巴的“150 定律”，将好友的个数限制在 150 之内。那么，图的占用空间还是很可能会膨胀 150 倍。这个时候，如果内存空间不够，GraphX 为了容下所有的数据，会需要在节点之间进行大量的 Shuffle 和 Spill 操作，使得后续的计算变得非常慢。

其实，这两个问题在 Spark 的其他机器学习算法中或多或少都会有，也是分布式计算系统中经常面临的问题。但是，在图计算中，它们是无法被忽略的问题，而且非常严重。所以，这决定了 GraphX 需要大的内存，才能有良好的性能。

在正常情况下，128GB 内存，减掉 8GB 的系统占用，剩下 120GB。这时配置给每个 Executor 60GB 内存，两个 Core，每个 Core 分到 30GB 的内存。这时不需要申请太多的 Executor，经过合理的性能优化，全量关系链计算，可以运行成功。

27.3.2　Spark GraphX 调优参数及调优最佳实践

Spark GraphX 调优方法如下。

1. 图缓存

Spark 和 GraphX 原本设计的精妙之处，亮点之一便在于 Cache，也就是 Persist (MEMORY_ONLY)，或者 Persist(MEMORY_AND_DISK)。可以把 RDD 和 Graph 的数据 Cache 到内存中，方便多次调用，而无须重新计算。那么，是否对所有的 RDD 或者图都 Cache 一下，Cache 是最佳的选择呢？答案是未必。

判断是否要 Cache 一个 Graph 或 RDD，最简单和重要的标准，就是该 Graph 是否会在后续的过程中被直接使用多次，包括迭代。

如果会，那么这个 Graph 就要被 Persist，然后通过 action 触发。如果不会，那么反过来，最好把这个 Graph 直接 unPersist 掉。一个 Graph 被 Cache，一般最终体现为两个 RDD 的 Cache，一个是 Edge，一个是 Vertex，其占用量非常巨大。在整体空间有限的情况下，Cache 会导致内存的使用量大大加剧，引发多次 GC 和重算，反而会拖慢速度。在 QQ 全量的关系链计算，一个全量图是非常大的，因此如果一个图没被多次使用，那么先 unPersist，再返回给下一个计算步骤，反而成了最佳实践。当然，既然 unPersist 了，那么它只能被再用一次。

示例代码如下。

```
1.  val oneNbrGraph = computeOneNbr(originalGraph)
2.  oneNbrGraph.unPersist()
```

```
3.  val resultRDD = oneNbrGraph.triplets.map {
4.      ......
5.  }
```

2. 分区策略：EdgePartition2D

GraphX 有 4 种分区的策略，其中性能最好的莫过于 EdgePartition2D 这种边分区策略。但是，由于 QQ 全量的关系链非常大，所以，如果先用默认策略，构造了图，再调用 PartitionBy 的方法来改变分区策略，那么就会多一步代价非常高的计算。

因此，为了减少不必要的计算步骤，建议在构造图前，先对 Edge 使用该策略进行划分，再用划分好的 Edge RDD 进行图构建。

示例代码如下。

```
1.   val edgesRePartitionRDD = edgeRDD.map { case (src, dst) => val pid =
     PartitionStrategy.EdgePartition2D.getPartition(src, dst, PartitionNum)
     (pid, (src, dst)) }.PartitionBy(new HashPartitioner(PartitionNum)).map
     { case (pid, (src, dst)) => Edge(src, dst, 1) }
2.  val g = Graph.fromEdges(
3.  edgesRePartitionRDD,
4.    ......
5.  )
```

注意：PartitionNum 分区数必须是平方数，才能达到最佳的性能。

GraphX 的代码，从 1.3 版本开始，便已经没有变动，基本是靠 Core 的优化来提高性能，没有任何实质性的改进，如果要继续使用，在核心上必须有所提升才行。

27.4　Spark ML 调优原理及调优最佳实践

本节讲解 Spark ML 调优原理；以及 Spark ML 调优参数及调优最佳实践。

27.4.1　Spark ML 调优原理

模型选择（又名超参数调优）：在 ML 中一个重要的任务就是模型选择，或者使用给定的数据为给定的任务寻找最适合的模型或参数。这也叫作调优。调优可以是对单个的 Estimator，如 LogisticRegression，或者是包含多个算法、向量化和其他步骤的整个 Pipline。用户可以一次性对整个 Pipline 进行调优，而不必对 Pipline 中的每个元素进行单独调优。

MLlib 支持使用像 CrossValidator 和 TrainValidationSplit 这样的工具进行模型选择。这些工具需要以下组件。

- Estimator：用户调优的算法或 Pipline。
- ParamMap 集合：提供参数选择，有时也叫作用户查找的“参数网格”。
- Evaluator：衡量模型在测试数据上的拟合程度。

在上层，这些模型选择工具的工作方式如下。

- 将输入数据切分成训练数据集和测试数据集。
- 对于每个（训练数据，测试数据）对，通过 ParamMap 集合进行迭代。
- 对于每个 ParamMap，使用它提供的参数对 Estimator 进行拟合，给出拟合模型，然后使用 Evaluator 评估模型的性能。

- 选择表现最好的参数集合生成的模型。

针对回归问题，Evaluator 可以是一个 RegressionEvaluator；针对二进制数据，可以是 BinaryClassificationEvaluator，或者是对于分类问题的 MulticlassClassificationEvaluator。用于选择最佳 ParamMap 的默认度量方式可以通过评估器的 setMetricName 方法进行覆盖。为了方便构造参数网格，用户可以使用通用的 ParamGridBuilder。

27.4.2　Spark ML 调优参数及调优最佳实践

1. 交叉验证

交叉验证（CrossValidator）将数据集切分成 *K* 折数据集合，并被分别用于训练和测试。例如，*K*=3 折时，CrossValidator 会生成 3 个（训练数据，测试数据）对，每个数据对的训练数据占 2/3，测试数据占 1/3。为了评估一个 ParamMap，CrossValidator 会计算这 3 个不同的（训练，测试）数据集对在 Estimator 拟合出的模型上的平均评估指标。

找出最好的 ParamMap 后，CrossValidator 会使用这个 ParamMap 和整个的数据集重新拟合 Estimator。

使用交叉验证进行模型选择。下面示例示范了使用 CrossValidator 从整个网格的参数中选择合适的参数。注意，在整个参数网格中进行交叉验证比较耗时。例如，在下面的例子中，参数网格有 3 个 HashingTF.numFeatures 值和两个 lr.regParam 值，CrossValidator 使用 2 折切分数据。最终将有(3×2)×2 = 12 个不同的模型被训练。在真实场景中，很可能使用更多的参数和进行更多折切分（*k*=3 和 *k*=10 都很常见）。换句话说，使用 CrossValidator 的代价可能会非常大。然而，对比启发式的手动调优，这是选择参数的行之有效的方法。

```
1.
2.  import org.apache.Spark.ml.Pipeline
3.  import org.apache.Spark.ml.classification.LogisticRegression
4.  import org.apache.Spark.ml.evaluation.BinaryClassificationEvaluator
5.  import org.apache.Spark.ml.feature.{HashingTF, Tokenizer}
6.  import org.apache.Spark.ml.linalg.Vector
7.  import org.apache.Spark.ml.tuning.{CrossValidator, ParamGridBuilder}
8.  import org.apache.Spark.sql.Row
9.
10. //从（ID、文本、标签）的元组列表数据中准备训练数据
11. val training = Spark.createDataFrame(Seq(
12.   (0L, "a b c d e Spark", 1.0),
13.   (1L, "b d", 0.0),
14.   (2L, "Spark f g h", 1.0),
15.   (3L, "hadoop mapreduce", 0.0),
16.   (4L, "b Spark who", 1.0),
17.   (5L, "g d a y", 0.0),
18.   (6L, "Spark fly", 1.0),
19.   (7L, "was mapreduce", 0.0),
20.   (8L, "e Spark program", 1.0),
21.   (9L, "a e c l", 0.0),
22.   (10L, "Spark compile", 1.0),
23.   (11L, "hadoop software", 0.0)
24. )).toDF("id", "text", "label")
25.
26. //配置一个机器学习 ML 的工作流，包括三个阶段：标记、HashingTF 和 LR
27. val tokenizer = new Tokenizer()
```

```
28.   .setInputCol("text")
29.   .setOutputCol("words")
30. val HashingTF = new HashingTF()
31.   .setInputCol(tokenizer.getOutputCol)
32.   .setOutputCol("features")
33. val lr = new LogisticRegression()
34.   .setMaxIter(10)
35. val pipeline = new Pipeline()
36.   .setStages(Array(tokenizer, HashingTF, lr))
37.
38. //用一个 ParamGridBuilder 构建网格参数进行搜索
39. //使用 HashingTF.numFeatures 的 3 个值，以及 lr.regParam 的两个值
40. //该网格将有 3×2 = 6 参数设置为 CrossValidator
41. val paramGrid = new ParamGridBuilder()
42.   .addGrid(HashingTF.numFeatures, Array(10, 100, 1000))
43.   .addGrid(lr.regParam, Array(0.1, 0.01))
44.   .build()
45. /**
    * 现在把工作流作为一种估计，它包裹在一个 CrossValidator 实例里
    *这将使用户能够共同选择所有的工作流阶段的参数。CrossValidator 需要进行估计
    *Estimator ParamMaps、Evaluator。注意,这里的评估是一个 inaryClassification
    *Evaluator,其默认度量是 areaUnderROC
46.   */
47. val cv = new CrossValidator()
48.   .setEstimator(pipeline)
49.   .setEvaluator(new BinaryClassificationEvaluator)
50.   .setEstimatorParamMaps(paramGrid)
51.   .setNumFolds(2)  //Use 3+ in practice
52.
53. //运行交叉验证，并选择最佳的参数集
54. val cvModel = cv.fit(training)
55.
56. //准备测试文件，这是未标记的（ID、文本）的元组
57. val test = Spark.createDataFrame(Seq(
58.   (4L, "Spark i j k"),
59.   (5L, "l m n"),
60.   (6L, "mapreduce Spark"),
61.   (7L, "apache hadoop")
62. )).toDF("id", "text")
63.
64. //对测试文档进行预测。cvModel 模型采用最好的模式（lrmodel）
65. cvModel.transform(test)
66.   .select("id", "text", "probability", "prediction")
67.   .collect()
68.   .foreach { case Row(id: Long, text: String, prob: Vector, prediction:
    Double) =>
69. println(s"($id, $text) --> prob=$prob, prediction=$prediction")
70.   }
```

2. 训练-验证切分

作为 CrossValidator 的附加，Spark 同样为超参数调优提供了 TrainValidationSplit。相对于 CrossValidator 的 K 次评估，TrainValidationSplit 只对每个参数组合评估一次。因此，它的评估代价没有这么高，但是，当训练数据集不够大的时候，其结果相对不够可信。

不同于 CrossValidator，TrainValidationSplit 创建单一的（训练，测试）数据集对。它使

用 trainRatio 参数将数据集切分成两部分。例如，当设置 trainRatio=0.75 时，TrainValidationSplit 会将数据切分 75%作为数据集，25%作为验证集，来生成训练、测试集对。

与 CrossValidator 相似，TrainValidationSplit 最终使用最好的 ParamMap 和完整的数据集来拟合 Estimator。

示例：通过训练/验证切分选择模型。

```
1.    import org.apache.Spark.ml.evaluation.RegressionEvaluator
2.  import org.apache.Spark.ml.regression.LinearRegression
3.  import org.apache.Spark.ml.tuning.{ParamGridBuilder, TrainValidationSplit}
4.
5.  //准备训练和测试数据
6.  val  data  =  Spark.read.format("libsvm").load("data/mllib/sample_linear_
    regression_data.txt")
7.  val Array(training, test) = data.randomSplit(Array(0.9, 0.1), seed =
    12345)
8.
9.  val lr = new LinearRegression()
10.     .setMaxIter(10)
11.
12. //我们用一个 ParamGridBuilder 构建网格参数进行搜索，TrainValidationSplit 将尝
    //试所有的组合和最佳值确定计算模型
13. val paramGrid = new ParamGridBuilder()
14.   .addGrid(lr.regParam, Array(0.1, 0.01))
15.   .addGrid(lr.fitIntercept)
16.   .addGrid(lr.elasticNetParam, Array(0.0, 0.5, 1.0))
17.   .build()
18.
19. //在这种情况下，评估是简单线性回归，TrainValidationSplit 需要评估 ParamMaps、
    //Evaluator
20. val trainValidationSplit = new TrainValidationSplit()
21.   .setEstimator(lr)
22.   .setEvaluator(new RegressionEvaluator)
23.   .setEstimatorParamMaps(paramGrid)
24.   //80%的数据将用于训练，剩下的 20%用于验证
25.   .setTrainRatio(0.8)
26.
27. //运行训练验证分离，并选择最佳的参数集
28. val model = trainValidationSplit.fit(training)
29.
30. //对试验数据进行预测，模型与参数的组合的模型是表现最好的
31. model.transform(test)
32.   .select("features", "label", "prediction")
33.   .show()
```

27.5　SparkR 调优原理及调优最佳实践

27.5.1　SparkR 调优原理

R 是数据科学家中最流行的编程语言和环境之一，在 Spark 中加入对 R 的支持是社区中较受关注的话题。作为增强 Spark 对数据科学家群体吸引力的最新举措，SparkR 使得熟悉 R 的用户可以在 Spark 的分布式计算平台基础上结合 R 本身强大的统计分析功能和丰富的第三

方扩展包，对大规模数据集进行分析和处理。

SparkR 架构主要由两部分组成：SparkR 包和 JVM 后端。SparkR 包是一个 R 扩展包，安装到 R 中之后，在 R 的运行时环境里提供了 RDD 和 DataFrame API，如图 27-17 所示。

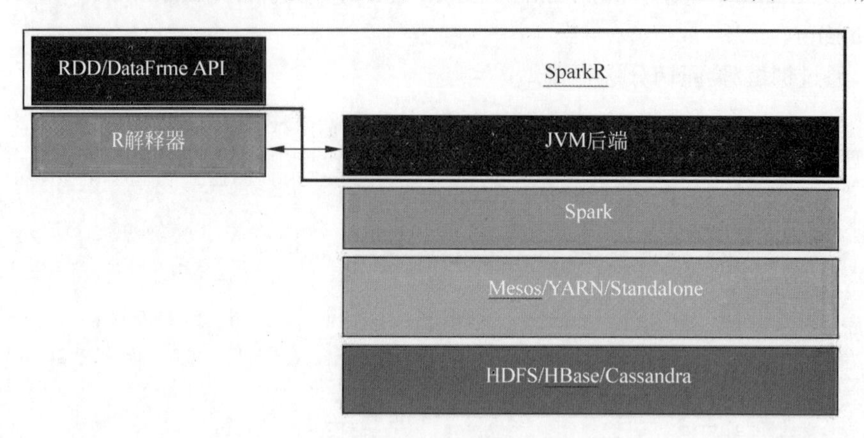

图 27-17 SparkR 部署图

SparkR 的整体架构如图 27-18 所示。

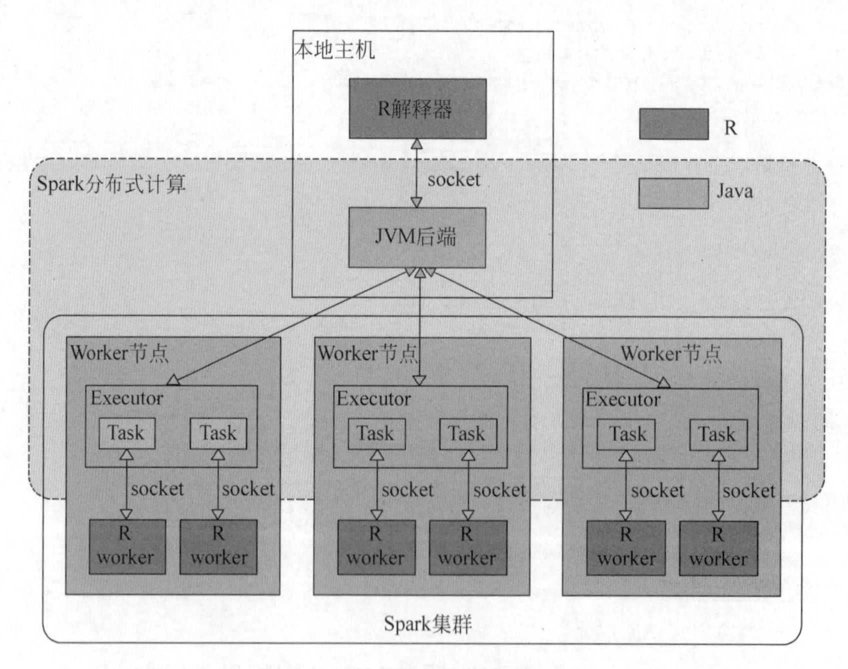

图 27-18 SparkR 的整体架构

R JVM 后端 SparkR API 运行在 R 解释器中，而 Spark Core 运行在 JVM 中，因此必须有一种机制能让 SparkR API 调用 Spark Core 的服务。R JVM 后端是 Spark Core 中的一个组件，提供了 R 解释器和 JVM 虚拟机之间的桥接功能，能够让 R 代码创建 Java 类的实例，调用 Java 对象的实例方法或者 Java 类的静态方法。JVM 后端基于 Netty 实现，和 R 解释器之间用 TCP socket 连接，用自定义的简单高效的二进制协议通信。

SparkR RDD API 和 Scala RDD API 相比：SparkR RDD 是 R 对象的分布式数据集，SparkR

RDD transformation 操作应用的是 R 函数。SparkR RDD API 的执行依赖于 Spark Core，但运行在 JVM 上的 Spark Core 既无法识别 R 对象的类型和格式，又不能执行 R 的函数，因此，如何在 Spark 的分布式计算核心的基础上实现 SparkR RDD API 是 SparkR 架构设计的关键。

SparkR 设计了 Scala RRDD 类，除了从数据源创建的 SparkR RDD 外，每个 SparkR RDD 对象概念上在 JVM 端有一个对应的 RRDD 对象。RRDD 派生自 RDD 类，改写了 RDD 的 compute 方法，在执行时会启动一个 R worker 进程，通过 socket 连接将父 RDD 的分区数据、序列化后的 R 函数以及其他信息传给 R worker 进程。R worker 进程反序列化接收到的分区数据和 R 函数，将 R 函数应到用分区数据上，再把结果数据序列化成字节数组传回 JVM 端。

从这里可以看出，与 Scala RDD API 相比，SparkR RDD API 的实现多了几项开销：启动 R worker 进程，将分区数据传给 R worker 和 R worker 将结果返回，分区数据的序列化和反序列化。这也是 SparkR RDD API 相比 Scala RDD API 有较大性能差距的原因。

27.5.2　SparkR 调优参数及调优最佳实践

SparkR 是一个 R 包，提供了一个轻量级的前端，使用 R 中的 Apache Spark。在 Spark 2.4.3 中，SparkR 提供了支持选择、过滤、聚合等操作的分布式数据 data frame 的实现（类似于 R 数据 data frame,dplyr）。在大数据集上，SparkR 还支持使用 MLlib 进行分布式机器学习。

SparkR 目前支持以下机器学习算法。

1. 分类

- spark.logit: Logistic Regression（逻辑回归）。
- spark.mlp: Multilayer Perceptron（MLP）（多层感知器）。
- spark.naiveBayes: Naive Bayes（朴素贝叶斯）。
- spark.svmLinear: Linear Support Vector Machine（线性支持向量机）。

2. 回归

- spark.survreg: Accelerated Failure Time (AFT) Survival Model（加速失效时间（AFT）生存模型）。
- spark.glm or glm: Generalized Linear Model (GLM)（广义线性模型）。
- spark.isoreg: Isotonic Regression（保序回归）。

3. 树

- spark.gbt: Gradient Boosted Trees for Regression and Classification（回归和分类的梯度增强树）。
- spark.randomForest: Random Forest for Regression and Classification（回归和分类的随机森林）。

4. 聚类

- spark.bisectingKmeans:　Bisecting k-means（二分 *k* 均值）。
- spark.gaussianMixture: Gaussian Mixture Model（GMM）（高斯混合模型）。

- spark.kmeans: K-Means（聚类算法）。
- spark.lda: Latent Dirichlet Allocation (LDA)（潜在狄利克雷分布）。

5．协同过滤

spark.als: Alternating Least Squares (ALS)（交替最小二乘法）。

6．频繁模式挖掘

spark.fpGrowth : FP-growth（关联分析算法）。

7．统计

spark.kstest: Kolmogorov-Smirnov Test（柯尔莫诺夫-斯米尔诺夫检验）。

SparkR 的性能调优可借助于 Spark 核心框架引擎的升级，Spark 旧版本升级到 Spark 2.X 新版本提升 SparkR 的运行性能。

（1）从 SparkR 1.5.X 升级到 1.6.X。在 Spark 1.6.0 之前写入的默认模式是 append。在 Spark 1.6.0 中更改 error 为与 Scala API 相匹配；SparkSQL 将 R 中的类型 NA 转换为 null，反之亦然。

（2）从 SparkR 1.6.X 升级到 2.0。

- table 方法已被删除，并替换为 tableToDF。
- 为避免名称冲突，类 DataFrame 已重命名 SparkDataFrame。
- Spark SQLContext 和 HiveContext 已过时，将被代替为 SparkSession。而不是 sparkR.init()调用 sparkR.session()来实例化 SparkSession 的位置。一旦完成，当前活动的 SparkSession 将用于 SparkDataFrame 操作。
- sparkExecutorEnv 参数不支持 sparkR.session。要为执行程序设置环境，使用前缀“spark.executorEnv.VAR_NAME”设置 Spark 配置属性，如“spark.executorEnv.PATH”。
- sqlContext 参数不再需要这些功能：createDataFrame，as.DataFrame，read.json，jsonFile，read.parquet，parquetFile，read.text，sql，tables，tableNames，cacheTable，uncacheTable，clearCache，dropTempTable，read.df，loadDF，createExternalTable。
- 方法 registerTempTable 已被弃用，替代为 createOrReplaceTempView。
- 方法 dropTempTable 已被弃用，替代为 dropTempView。
- sc 的 SparkContext 参数不再需要这些功能：setJobGroup，clearJobGroup，cancelJobGroup。

（3）升级到 SparkR 2.1.0。Join 默认情况下不再执行笛卡尔乘积，改用 crossJoin。

第 28 章　Spark 2.2.0 新一代钨丝计划优化引擎

本章讲解 Spark 2.2.0 新一代钨丝计划（Tungsten）优化引擎。28.1 节讲解钨丝计划概述；28.2 节讲解内存管理与二进制处理；28.3 节讲解缓存感知计算；28.4 节讲解代码生成。

28.1　概　　述

Spark 作为一个一体化、多元化的大数据处理通用平台，性能一直是其根本性的追求之一。Spark 基于内存迭代（部分基于磁盘迭代）的模型极大地满足了人们对分布式系统处理性能的渴望。Spark 是采用 Scala+ Java 语言编写的，所以运行在 JVM 平台。当然，JVM 是一个绝对伟大的平台，因为 JVM 让整个离散的主机融为一体（网络即 OS），但是 JVM 的死穴 GC 反过来限制了 Spark（也就是说，平台限制了 Spark），所以 Tungsten 聚焦于 CPU 和 Memory 使用，以达到对分布式硬件潜能的终极压榨！

对 Memory 的使用，Tungsten 使用了 Off-Heap，也就是在 JVM 之外的内存空间（这就好像 C 语言对内存的分配、使用和销毁），此时 Spark 实现了自己的独立的内存管理，就避免了 JVM 的 GC 引发的性能问题，其实也避免了序列化和反序列化。

对 Memory 的管理，Tungsten 提出了 Cache-aware computation。也就是说，使用对缓存友好的算法和数据结构来完成数据的存储和复用。

对 CPU 的使用，Tungsten 提出了 Code Generation，其首先在 Spark SQL 使用，通过 Tungsten 把该功能普及到 Spark 的所有功能中（这里的 CG 类似于 Android 的 art 模式）。

Java 操作的数据一般是自己的对象，但像 C、C++语言可以操作内存中的二进制数据一样，Java 同样也可以。Tungsten 的内存管理机制独立于 JVM，所以 Spark 操作数据时具体操作的是 Binary Data，而不是 JVM Object，而且还免去了序列化和反序列化的过程！

1．内存管理和二进制处理

（1）避免使用非 transient 的 Java 对象（它们以二进制格式存储），这样可以减少 GC 的开销。

（2）通过使用基于内存的密集数据格式，可以减少内存的使用。

（3）更好的内存计算（字节的大小），而不是依赖启发式。

（4）对于知道数据类型的操作（如 DataFrame 和 SQL），可以直接对二进制格式进行操作，这样我们就不需要进行系列化和反系列化的操作。

2．缓存感知计算

对 aggregations、Joins 和 Shuffle 操作进行快速排序和哈希操作。

3．代码生成

（1）更快的表达式求值和 DataFrame/SQL 操作。

（2）快速系列化。

Tungsten 的实施已有两个阶段：1.6.X 基于内存的优化；2.X 基于 CPU 的优化。还优化 Disk I/O、Network I/O，主要针对 Shuffle。

Apache Spark 已经非常快了，但是我们能不能让它再快 10 倍？这个问题使我们从根本上重新思考 Spark 物理执行层的设计。当调查一个现代数据引擎，会发现大部分的 CPU 周期都花费在无用的工作上，如虚函数的调用；或者读取/写入中间数据到 CPU 高速缓存或内存中。通过减少花在这些无用功的 CPU 周期一直是现代编译器长期性能优化的重点。

Apache Spark 2.0 中附带了第二代 Tungsten engine。这一代引擎是建立在现代编译器的想法上，并且把它们应用于数据的处理过程中。主要想法是，通过在运行期间优化那些拖慢整个查询的代码到一个单独的函数中，消除虚拟函数的调用以及利用 CPU 寄存器来存放那些中间数据。作为这种流线型策略的结果，我们显著提高 CPU 效率并且获得了性能提升，我们把这些技术统称为“整段代码生成”。

28.2　内存管理与二进制处理

本节讲解内存管理与二进制处理，主要包括：JVM 对象模型的内存开销和 GC 的开销；内存管理的模型及其实现类的解析；二进制处理及其实现类的解析。

28.2.1　概述

Spark 计算框架是基于 Scala 与 Java 语言开发的，其底层都使用了 Java 虚拟机（Java Virtual Machine，JVM）。而在 JVM 上运行的应用程序是依赖 JVM 的垃圾回收机制来管理内存的，随着 Spark 应用程序性能的不断提升，JVM 对象和 GC 开销产生的影响（包括内存不足、频繁 GC 或 Full GC 等）将非常致命，即引进新的 Tungsten 内存管理机制的主要原因在于，JVM 在内存方面和 GC 方面的开销。主要包含不必要的内存开销和 GC 的开销。

1．JVM对象模型的内存开销

可以通过 Java Object Layout 工具来查看在 JVM 上 Java 对象所占用的内存空间（可以参考 http://openjdk.java.net/projects/code-tools/jol/）。需要注意的是，在 32bit 与 64bit 的操作系统，占用空间会有所差异。下面是在 64bit 操作系统上对 String 的分析结果。

```
1.   java.lang.String object internals:
2.    OFFSET  SIZE        TYPE DESCRIPTION                     VALUE
```

```
3.        0    12          (object header)                    N/A
4.       12     4     char[] String.value                     N/A
5.       16     4           int String.hash                   N/A
6.       20     4         int String.hash32                   N/A
7.  Instance size: 24 bytes (estimated, the sample instance is not available)
8.  Space losses: 0 bytes internal + 0 bytes external = 0 bytes total
```

一个简单的 String 对象会额外占用一个 12B 的 Header 和 8B 的哈希信息。这是开启(-XX:+UseCompressedOops，默认）指针压缩方式（-XX:+UseCompressedOops）的结果，如果不开启（-XX:+UseCompressedOops）指针压缩（-XX:+UseCompressedOops），则内存更大，如下所示。

```
1.  java.lang.String object internals:
2.   OFFSET  SIZE       TYPE DESCRIPTION                      VALUE
3.        0    16          (object header)                    N/A
4.       16     8    char[] String.value                      N/A
5.       24     4          int String.hash                    N/A
6.       28     4        int String.hash32                    N/A
7.  Instance size: 32 bytes (estimated, the sample instance is not available)
8.  Space losses: 0 bytes internal + 0 bytes external = 0 bytes total
```

其中，Header 会占用 16B 的内存。另外，JVM 内存模型会采用 8B 对齐，因此也可能会增加一部分的内存开销。

以上仅仅是 String 对象在 JVM 内存模型中占用的内存大小，实际计算时需要考虑其内部的引用对象所占的内存。通过分析，char[]中默认的指针压缩情况下会占用 16B 内存，因此仅仅是一个空字符串，也会占用 24B + 16B =40B 的内存。

另外，在 JVM 内存模型中，为了更加通用，它重新定制了自己的储存机制，使用 UTF-16 方式编码每个字符（2B）。参考 http://www.javaworld.com/article/2077408/core-java/sizeof-for-java.html，java 对象的内存占用大小如下：

```
1.  //java.lang.Object shell size in bytes:
2.      public static final int OBJECT_SHELL_SIZE= 8;
3.      public static final int OBJREF_SIZE = 4;
4.      public static final int LONG_FIELD_SIZE = 8;
5.      public static final int INT_FIELD_SIZE = 4;
6.      public static final int SHORT_FIELD_SIZE = 2;
7.      public static final int CHAR_FIELD_SIZE = 2;
8.      public static final int BYTE_FIELD_SIZE = 1;
9.      public static final int BOOLEAN_FIELD_SIZE = 1;
10.     public static final int DOUBLE_FIELD_SIZE = 8;
11.     public static final int FLOAT_FIELD_SIZE = 4;
```

其中，CHAR_FIELD_SIZE 为 2，即每个字符串在 JVM 中采用了 UTF-16 编码，一个字符会占用 2B。

2. GC（垃圾回收机制）的开销

JVM 对象带来的另一个问题是 GC。通常情况下，JVM 内存模型中堆会分成两大块：Young Generation（年轻代）和 Old Generation（老年代），其中年轻代会有很高的分配/释放，通过利用年轻代对象的瞬时特性，垃圾收集器可以更有效率地对其进行管理。老年代的状态则非常稳定。GC 的开销在所有基于 JVM 的应用程序中都是不可忽视的，而且对应的调优也非常烦琐，在类似 Spark 这样的基于内存迭代处理的框架中，直接在底层对内存进行管理可以极大地提高效率。因此，对应引入 Project Tungsten 也就很合情合理了。

下面会详细解析 Project Tungsten 的内存模型及其源码实现，并且对基于该模型的 Shuffle 写数据过程中的二进制数据处理也给出了详细解析。

28.2.2　内存管理的模型及其实现类的解析

Spark 1.6 版本中提出了一个新的内存管理模型，即统一内存管理模型，对应在 Spark 1.5 及之前的版本则使用静态的内存管理模型。关于新的统一内存管理模型，可以参考 https://issues.apache.org/jira/secure/attachment/12765646/unified-memory-management-spark-10000.pdf。该文档详细描述了各种可能的设计以及各设计的优缺点。

为了解决现有基于 JVM 托管方式的内存模型存在的缺陷，Project Tungsten 设计了一套新的内存管理机制。在新的内存管理机制中，Spark 的 operation 可以直接使用分配的二进制数据，而不是 JVM objects，避免了数据处理过程中不必要的序列化与反序列化的开销，同时基于 Off-Heap 方式管理内存，降低了 GC 带来的开销。

Project Tungsten 通过 sun.misc.Unsafe 管理内存，关于 sun.misc.Unsafe（从命名上可知该工具不能滥用）及其使用等内容，可以参考官网文档（http://www.docjar.com/docs/api/sun/misc/Unsafe.html）。这里主要分析 Project Tungsten 中的内存管理模型的具体实现。

1．Project Tungsten的内存模型整体描述

Project Tungsten 内存管理模型主要的类图结构如图 28-1 所示。

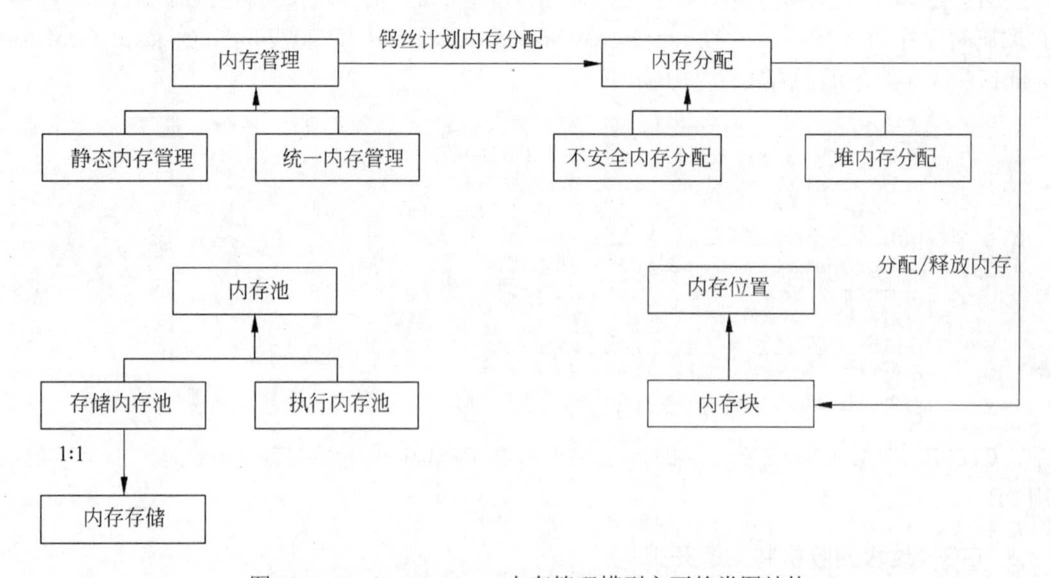

图 28-1　Project Tungsten 内存管理模型主要的类图结构

在图 28-1 中，基类 MemoryManager 封装了静态内存管理模型与统一内存管理模型，即分别对应两个具体实现子类 StaticMemoryManger 与 UnitedMemoryManager。对应的内存分配由 MemoryManager 的成员 tungstenMemoryMode 决定，即由基类 MemoryAllocator 负责具体内存分配，对应 Off-Heap 与 On-Heap 两种内存模式，分别实现了两个具体子类

UnsafeMemoryAllocator 与 HeapMemoryAllocator。MemoryAllocator 进行了 allocate 与 free 两个成员函数来提供内存的分配与释放，分配的内存以 MemoryBlock 表示。

另外，根据内存使用目的的不同，将内存分为 Storage 和 Execution 两大部分，对应的 MemoryPool 的两个具体实现子类 StorageMemoryPool 与 ExecutionMemoryPool 对其进行管理。实际上，除了这两部分，总的内存还包括为系统预留的 OtherMemory。

内存分类及其对应管理的主要类之间的关系可以通过图 28-2 来描述。

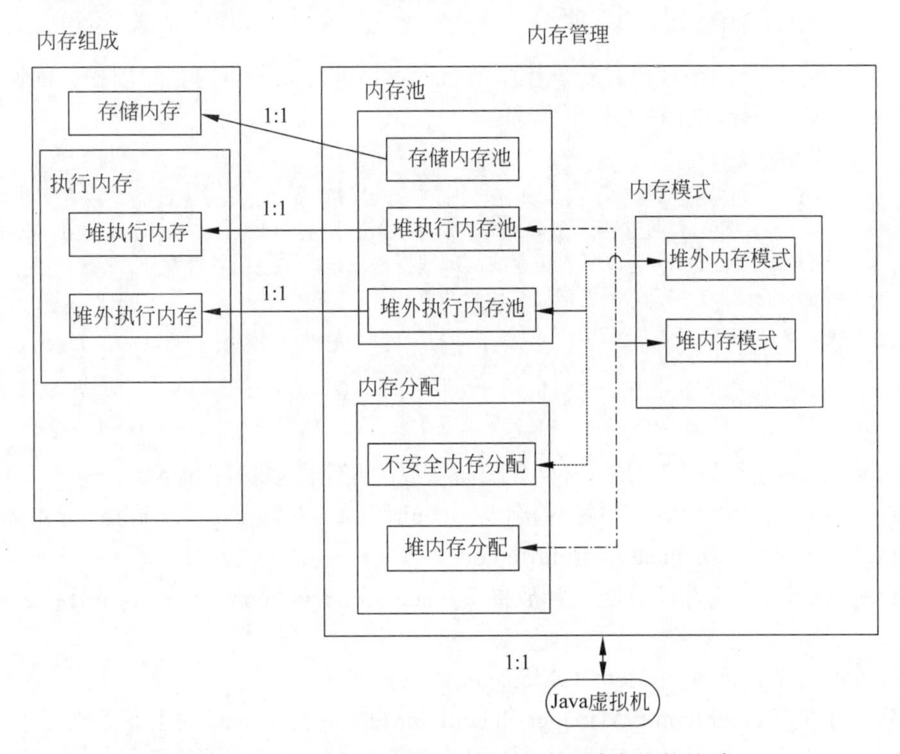

图 28-2　内存分类及其对应管理的主要类之间的关系

在 Worker 上运行的每个 Execution 进程（抽象描述，实际对应各部署场景下的具体 ExecutorBackend 实现子类），对应由一个内存管理负责管理其内存，即图 28-2 中内存管理与 Java 虚拟机的对应关系为 1∶1。

Storage 部分的内存由 StorageMemoryPool 负责管理，Execution 部分的内存根据不同的内存模式分为 On-Heap 与 Off-Heap 两种，分别由 onHeapExecutionMemoryPool 与 offHeapExecutionMemoryPool 进行管理。管理内存主要是通过内存使用量进行控制的，不涉及内存的分配与释放。

2. MemoryManager的实现及其源码解析

MemoryManager 目前实现了两种具体的内存管理模型，从 Spark 1.6 版本开始，默认使用统一内存管理模型，对应的配置属性为 spark.memory.useLegacyMode，控制代码位于 SparkEnv 类中，代码如下：

```
1.  //Spark 1.5 及之前的版本使用的内存管理模型对应配置属性 spark.memory.useLegacy
    //Mode 为 true。Spark 1.6 及之后的版本默认设置为 false
```

```
2.   val useLegacyMemoryManager = conf.getBoolean("spark.memory.useLegacyMode",
     false)
3.      val memoryManager: MemoryManager =
4.        if (useLegacyMemoryManager) {
5.          //使用静态内存管理模型
6.          new StaticMemoryManager(conf, numUsableCores)
7.        } else {
8.  //Spark 1.6 及之后的版本默认使用统一内存管理模型
9.          UnifiedMemoryManager(conf, numUsableCores)
10.       }
```

以上是选择具体采用哪种内存管理模型的代码，下面开始分析内存管理相关的源码，首先查看 MemoryManager 的注释，如下所示。

```
1.  /**
2.    *内存管理的抽象接口，用于指定如何在 Execution 与 Storage 间共享内存
3.    *Execution Memory 是指用于计算的内容，包括 Shuffles、Joins、sorts 以及 aggregations
4.    * Storage Memory 是指用于缓存或内部数据传输过程中使用的内存
5.    *MemoryManager 与 JVM 进程的对应关系为 1：1，即一个 JVM 进程中的内存由一个
6.    *MemoryManager 进行管理
7.    */
8.  private[spark] abstract class MemoryManager(
9.  ……
```

在 MemoryManager 类中提供的内存分配与释放的几个主要接口如下。

（1）Storage 部分内存的分配与释放接口：acquireStorageMemory、acquireUnrollMemory、releaseStorageMemory 以及 releaseUnrollMemory。

（2）Execution 部分内存的分配与释放接口：acquireExecutionMemory 与 releaseExecution Memory。

具体分配与释放的实现由 MemoryManager 的具体子类提供。

两大实现子类（StaticMemoryManager 和 UnifiedMemoryManager）的主要差别在于 Storage 与 Execution 内存之间的边界是静态的，还是动态可变的。下面分别简单描述了两大子类的实现细节。

StaticMemoryManager 类的注释如下：

```
1.  /**
2.    *静态划分 Storage 与 Execution 内存之间的边界的一种内存管理实现
3.    * Storage 与 Execution 内存大小分别由配置属性 spark.shuffle.memoryFraction 与
4.    *spark.storage.memoryFraction 各自指定，由于是静态划分边界，因此这两者之间不
      *能互相借用多余的内存
5.    */
6.  private[spark] class StaticMemoryManager(
```

静态内存管理模型中各部分内存的分配可以通过以下几个接口或成员变量查看。

（1）maxUnrollMemory：Unroll 过程中可用的内存，占最大可用 Storage 内存的 0.2（占比）。

（2）getMaxStorageMemory：获取分配给 Storage 使用的最大内存大小。

（3）getMaxExecutionMemory：获取分配给 Execution 使用的最大内存大小。

其中，getMaxStorageMemory 对应用于 Storage 的最大内存，具体配置如下所示。

```
1.    private def getMaxStorageMemory(conf: SparkConf): Long = {
2.      val systemMaxMemory = conf.getLong("spark.testing.memory", Runtime.
        getRuntime.maxMemory)
3.      val memoryFraction = conf.getDouble("spark.storage.memoryFraction",
        0.6)
4.      val safetyFraction = conf.getDouble("spark.storage.safetyFraction",
        0.9)
5.      (systemMaxMemory * memoryFraction * safetyFraction).toLong
6.    }
```

其中，配置属性 spark.storage.memoryFraction 表示 Storage 内存占用全部内存（除预留给系统的内存外）的占比，spark.storage.safetyFraction 对应为 Storage 内存的安全系数。

对应的 getMaxExecutionMemory 方法指明了用于 Execution 内存的相关配置属性，与 Storage 内存一样，包含占总内存的占比（0.2）及对应的安全系数。

另外，除了 Storage 内存（占用 60%）与 Execution 内存（占用 20%）之外的剩余内存，作为系统预留内存。

通过 StaticMemoryManager 类简单分析静态内存管理模型后，继续查看统一内存管理模型，首先查看其类注释。

```
1.  /**
2.    *[MemoryManager]管理执行内存和存储内存的软边界执行，用于执行内存和存储之间互相
      *借用内存。执行和存储之间的区域是总的堆空间减去 300MB 预留内存的一部分内存，可以通
      *过配置 spark.memory.fraction 参数（默认值为 0.6）实现，边界位置通过spark.memory.
      *storageFraction 参数（默认为 0.5）进一步确定。这意味着默认情况下存储内存的大小
      *是堆空间的 0.6×0.5 = 0.3
3.    *存储内存能借用执行内存（如果执行内存有空闲的内存），直到执行内存重新占用它的空间
      *为止。当发生这种情况时，缓存块将被从内存中清除，直到足够的借入内存被释放，满足执行
      *内存、请求内存的需要
4.    *类似地，执行内存可以借用尽可能多空闲的存储内存。然而，执行内存由于执行此操作所涉及
      *的复杂性，执行内存永远不会被存储区逐出。其含义是，如果执行内存已占用存储内存的大部
      *分空间，则缓存块的尝试可能失败。在这种情况下，根据存储级别，新的块将立即被逐出内存
5.    *@paramstorageRegionSize 存储区的大小，以字节为单位。这个区域是不是静态保留的，
      *如果必要，执行内存可以借用存储内存的空间；如果实际存储内存使用超出这个区域，缓存块
      *将被驱逐
6.    */
7.  private[spark] class UnifiedMemoryManager private[memory]
8.  ......
```

UnifiedMemoryManager 与 StaticMemoryManager 一样实现了 MemoryManager 的几个内存分配、释放的接口，对应分配与释放接口的实现，在 StaticMemoryManager 中相对比较简单，而在 UnifiedMemoryManager 中，由于考虑到动态借用的情况，实现相对比较复杂，具体细节可以参考官方提供的统一内存管理设计文档以及相关源码。例如，针对各个 Task 如何保证其最小分配的内存（最少为 $1/2N$，其中 N 表示当前活动状态的 Task 个数，最大的 Task 个数可以从 Executor 分配的内核个数/每个 Task 占用的内核个数得到）等等。

下面简单分析一下统一内存管理模型中，Storage 内存与 Execution 内存等相关的配置。

UnifiedMemoryManager 的 getMaxMemory 方法，在 Spark 1.6 版本中，spark.memory.fraction 的默认值是 0.75；在 Spark 2.2.0 版本中，spark.memory.fraction 的默认值是 0.6。

UnifiedMemoryManager.scala 的 getMaxMemory 的源码如下：

```
1.  /**
2.    *返回 Execution 与 Storage 共享的最大内存
3.    */
4.  private def getMaxMemory(conf: SparkConf): Long = {
5.      val systemMemory = conf.getLong("spark.testing.memory", Runtime.get
      Runtime.maxMemory)
6.    //系统预留的内存大小，默认为 300MB
7.      val reservedMemory = conf.getLong("spark.testing.reservedMemory",
8.        if (conf.contains("spark.testing")) 0 else RESERVED_SYSTEM_MEMORY_
        BYTES)
9.     //当前最小的内存需要 300×1.5，即 450MB，不满足该条件时会报错退出
10.     val minSystemMemory = (reservedMemory * 1.5).ceil.toLong
11.     if (systemMemory < minSystemMemory) {
12.       throw new IllegalArgumentException(s"System memory $systemMemory
       must " +
13.         s"be at least $minSystemMemory. Please increase heap size using the
         --driver-memory " +
14.         s"option or spark.driver.memory in Spark configuration.")
15.     }
16.     //SPARK-12759 如果内存不足，就检查 Executor 内存是否失败
17.     if (conf.contains("spark.executor.memory")) {
18.       val executorMemory = conf.getSizeAsBytes("spark.executor.memory")
19.       if (executorMemory < minSystemMemory) {
20.         throw new IllegalArgumentException(s"Executor memory $executor
         Memory must be at least " +
21.           s"$minSystemMemory. Please increase executor memory using the " +
22.           s"--executor-memory option or spark.executor.memory in Spark
           configuration.")
23.       }
24.     }
25.     val usableMemory = systemMemory - reservedMemory
26.
27.     //当前 Execution 与 Storage 共享的最大内存占比默认为 0.6，即
28.     //Execution 与 Storage 内存为可用内存的 0.6
29.     //用户内存为可用内存的（1-0.6）= 0.4
30.
31.     val memoryFraction = conf.getDouble("spark.memory.fraction", 0.6)
32.     (usableMemory * memoryFraction).toLong
33.   }
34. }
```

另外，虽然 Execution 与 Storage 之间共享内存，但仍然存在一个初始边界值，可以参考伴生对象 UnifiedMemoryManager 的 apply 工厂方法，具体代码如下所示。

UnifiedMemoryManager.scala 的源码如下：

```
1.     def apply(conf: SparkConf, numCores: Int): UnifiedMemoryManager = {
2.      val maxMemory = getMaxMemory(conf)
3.      new UnifiedMemoryManager(
4.        conf,
5.        maxHeapMemory = maxMemory,
6.       //通过配置属性 spark.memory.storageFraction，可以设置 Execution 与 Storage
7.       //共享内存的初始边界值，即默认初始时，各占总内存的一半
8.        onHeapStorageRegionSize =
9.          (maxMemory * conf.getDouble("spark.memory.storageFraction", 0.5)).toLong,
10.       numCores = numCores)
11.   }
```

另外，需要注意的是，前面 Execution 内存指的是 On-Heap 部分的内存，在 Project Tungsten 中引入了 Off-Heap（堆外）内存，这部分内存大小的设置在基类 MemoryManager 中，对应代码如下所示。

Spark 2.2.1 版本的 MemoryManager.scala 的源码如下：

```
1.    //1. Storage 部分的内存池初始大小设置
2.    onHeapStorageMemoryPool.incrementPoolSize(onHeapStorageMemory)
3.     //2. On-Heap 部分的 Execution 内存池初始大小设置
4.    onHeapExecutionMemoryPool.incrementPoolSize(onHeapExecutionMemory)
5.
6.    protected[this] val maxOffHeapMemory = conf.getSizeAsBytes("spark.
      memory.offHeap.size", 0)
7.    protected[this] val offHeapStorageMemory =
8.      (maxOffHeapMemory * conf.getDouble("spark.memory.storageFraction",
        0.5)).toLong
9.    //3. 计算获取 Off-Heap 内存池的初始内存大小
10.   offHeapExecutionMemoryPool.incrementPoolSize(maxOffHeapMemory - offHeap
      StorageMemory)
11.   offHeapStorageMemoryPool.incrementPoolSize(offHeapStorageMemory)
```

Spark 2.4.3 版本的 MemoryManager.scala 源码与 Spark 2.2.1 版本相比具有如下特点。

- 上段代码中第 6 行调整为使用 MEMORY_OFFHEAP_SIZE。MEMORY_OFFHEAP_SIZE 可用于堆外内存分配的绝对内存量（字节）。此设置对堆内存使用没有影响，如果 executors 的总内存消耗必须符合某个限制，那么一定要相应地缩小 JVM 堆的大小。当 spark.memory.offsheap.enabled=true 时，必须将其设置为正值。

```
1.   ……
2.          protected[this] val maxOffHeapMemory = conf.get(MEMORY_OFFHEAP_SIZE)
3.   ……
4.    private[spark] val MEMORY_OFFHEAP_SIZE =
      ConfigBuilder("spark.memory.offHeap.size")
5.      .doc("The absolute amount of memory in bytes which can be used for
        off-heap allocation. " +
6.        "This setting has no impact on heap memory usage, so if your executors'
          total memory " +
7.        "consumption must fit within some hard limit then be sure to shrink
          your JVM heap size " +
8.        "accordingly. This must be set to a positive value when spark.memory
          .offHeap.enabled=true.")
9.      .bytesConf(ByteUnit.BYTE)
10.     .checkValue(_ >= 0, "The off-heap memory size must not be negative")
11.     .createWithDefault(0)
```

当需要使用 Off-Heap 内存时，需要注意的是，除了需要修改 Off-Heap 内存池（offHeapExecutionMemoryPool）的内存初始值（默认为 0），还需要打开对应的控制开关，具体代码参考内存分配 MemoryManager 中内存模式的设置（该内存模式可以控制用于内存分配 MemoryAllocator 的具体子类）。

Spark 2.2.1 版本的 MemoryManager.scala 的源码如下：

```
1.     final val tungstenMemoryMode: MemoryMode = {
2.      if (conf.getBoolean("spark.memory.offHeap.enabled", false)) {
3.        require(conf.getSizeAsBytes("spark.memory.offHeap.size", 0) > 0,
4.          "spark.memory.offHeap.size must be > 0 when spark.memory.offHeap
           .enabled == true")
```

```
5.        require(Platform.unaligned(),
6.          "No support for unaligned Unsafe. Set spark.memory.offHeap.enabled
            to false.")
7.        MemoryMode.OFF_HEAP
8.      } else {
9.        MemoryMode.ON_HEAP
10.     }
11.   }
```

Spark 2.4.3 版本的 MemoryManager.scala 的源码与 Spark 2.2.1 版本相比具有如下特点。

- 上段代码中第 2 行调整使用 MEMORY_OFFHEAP_ENABLED，如果 spark.memory.offHeap.enabled 为 true，Spark 将试图在某些操作中使用堆外内存。如果启用了堆外内存使用，则 spark.memory.offsheap.size 必须为正。
- 上段代码中第 3 行调整使用 MEMORY_OFFHEAP_SIZE。

```
1.  ……
2.      if (conf.get(MEMORY_OFFHEAP_ENABLED)) {
3.            require(conf.get(MEMORY_OFFHEAP_SIZE) > 0,
4.  ……
```

从图 28-2 可以看出，Execution 内存根据不同的内存模式（On-Heap 或 Off-Heap）可以有两种内存池管理方式，对应可以查看一下 Execution 内存分配的方法（方法注释中给出了为 Task 分配内存的实现细节，有兴趣的读者可以查看源码注释），关键的代码如下所示。

UnifiedMemoryManager.scala 的源码如下：

```
1.     override private[memory] def acquireExecutionMemory(
2.       numBytes: Long,
3.       taskAttemptId: Long,
4.       memoryMode: MemoryMode): Long = synchronized {
5.     assertInvariants()
6.     assert(numBytes >= 0)
7.     val (executionPool, storagePool, storageRegionSize, maxMemory) =
       memoryMode match {
8.       case MemoryMode.ON_HEAP => (
9.   //当内存模式为 On-Heap 时，使用 onHeapExecutionMemoryPool 内存池管理
10.        onHeapExecutionMemoryPool,
11.        onHeapStorageMemoryPool,
12.        onHeapStorageRegionSize,
13.        maxHeapMemory)
14.      case MemoryMode.OFF_HEAP => (
15. //当内存模式为 Off-Heap 时，使用 offHeapExecutionMemoryPool 内存池管理
16.        offHeapExecutionMemoryPool,
17.        offHeapStorageMemoryPool,
18.        offHeapStorageMemory,
19.        maxOffHeapMemory)
20.    }
```

MemoryMode 是二选一，因此在启动 Off-Heap 内存模式时，可以将 Storage 的内存占比（对应配置属性 spark.memory.storageFraction）设置高一点，虽然在具体分配过程中，Storage 也可以向 On-Heap 这部分 Execution 借用内存。

关于内存池部分，可以阅读 Spark 内存管理的相关源码加深理解。内存池相关类图如图 28-3 所示。

主要是通过内部池大小和使用的内存大小等进行控制，对应统一内存管理模型，需要考虑内存借用等具体实现（关键代码可以查看 UnitedMemoryManager 对 StorageMemoryPool 类

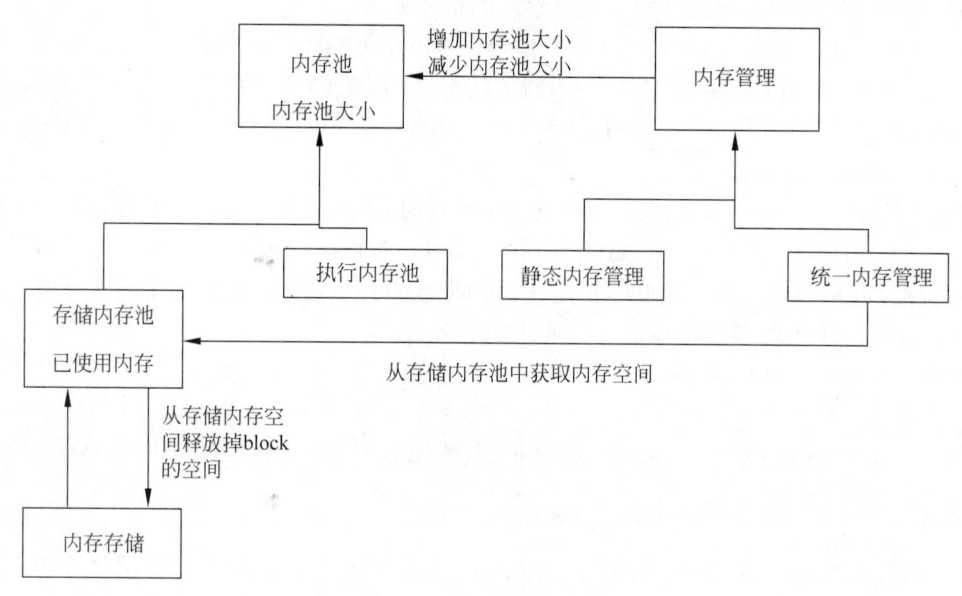

图 28-3　内存池相关类图

的 shrinkPoolToFreeSpace 方法的调用）。

以上是对 Tungsten 的两种内存管理模型的简单解析，下面开始对内存管理模型的内部组织结构进行解析。

3. 内存管理模型中对内存描述的封装

关于 Project Tungsten 的相关内容，可以参考 https://github.com/hustnn/TungstenSecret。其中对 Page Table 给出了描述非常详细的说明图。

下面从最基本的源码开始逐步分析内存管理模型中内存描述的封装，主要包含内存地址的封装和内存块的封装，分别对应 MemoryLocation 与 MemoryBlock。

在 ProjectTungsten 中，为了统一管理 On-Heap 与 Off-Heap 两种内存模式，引入了统一的地址表示形式，即通过 MemoryLocation 类来表示 On-Heap 或 Off-Heap 两种内存模式下的地址。

首先查看该类的注释信息。MemoryLocation.scala 的源码如下：

```
1.  /**
      *一个内存地址，用于跟踪 Off-Heap 模式下的内存地址或 On-Heap 模式下的内存地址
2.    */
3.  public class MemoryLocation {
```

当使用 Off-Heap 内存模式时，内存地址可以通过 64b 的绝对地址来描述，对应地，当使用 On-Heap 内存模式时，由于 GC 过程中会对堆（heap）内存进行重组，因此地址的定位需要通过对象在堆内存的引用以及在该对象内的偏移量来表示，此时便需要对象引用和一个偏移量来表示内存地址。因此，在 MemoryLocation 中定义了两个成员变量，具体代码如下所示。

MemoryLocation.scala 的源码如下：

```
1.  @Nullable
2.  Object obj;
3.  long offset;
```

对应两种不同的内存模式，两个成员变量的描述如下。

（1）Off-Heap 内存模式：obj 为 null，地址由 64b 的 offset 唯一标识。

（2）On-Heap 内存模式：obj 为堆中该对象的引用，offset 对应数据在该对象中的偏移量。

由以上分析可知，通过 MemoryLocation 类可以统一定位一个 Off-Heap 与 On-Heap 两种内存模式下的内存地址。

对应 MemoryLocation 类的继承子类为 MemoryBlock。顾名思义，该子类表示一个内存块，不管是 Off-Heap，还是 On-Heap 内存模式，在 ProjectTungsten 内存管理时，都使用一块连续的内存空间来存储数据，因此即使是在 On-Heap 模式下，也可以降低 GC 的开销。下面查看一下 MemoryBlock 类的注释信息，具体如下所示。

MemoryBlock.scala 的源码如下：

```
1.  /**
      * 一个连续的内存块，继承自描述内存地址的 MemoryLocation 类，同时提供内存块的大小
2.    */
3.  public class MemoryBlock extends MemoryLocation {
```

在代码复用方式上存在继承与组合两种形式。目前在 MemoryBlock 中使用继承的方式包含内存块的地址信息。在实现上，也可以采用组合这种复用方式，指定内存块的地址，以及内存块本身的内存大小。

下面简单介绍一下 MemoryBlock 类中除了继承自 MemoryLocation 类之外的部分成员。

（1）private final long length：表示内存块的长度。

（2）public int pageNumber：表示内存块对应的 page 号。

（3）public static MemoryBlock　fromLongArray(final long[] array)：这是提供的一个将 long 型数组转换为 MemoryBlock 内存块的接口。

提供了内存块之后，进一步就是如何去组织这些内存块，在 Project Tungsten 中采用了类似操作系统的内存管理模式，即使用 Page Table 方式来管理内存。因此，下面开始对 Page Table 管理方式进行解析。

4．内存管理模型中的内存组织、管理模式

Spark 是一个技术框架，数据以分区粒度进行处理，即每个分区对应一个处理的任务（Task），因此内存的组织与管理等可以通过与 Task 一一对应的 TaskMemoryManager 来理解。

下面首先给出任务内存管理与内存管理间的关系图，如图 28-4 所示。

在图 28-4 中，各个内存使用是具体处理时需要使用（消耗）内存块的实体，内存使用通过任务内存管理提供的接口向内存管理申请或释放内存资源，即申请或释放内存块，任务内存管理类中会管理全部内存使用，并对这些内存消耗实体申请的内存块进行组织与管理，具体是通过 PageTable 的方式实现的。

首先查看一下类的注释信息，原注释信息比较多，在此仅给出简单的中文描述，具体代码如下：

```
1.  /**
2.    * ……
3.    * 管理为单个 Task 分配的内存
4.    * 内存地址在不同的内存模式下的表示：
```

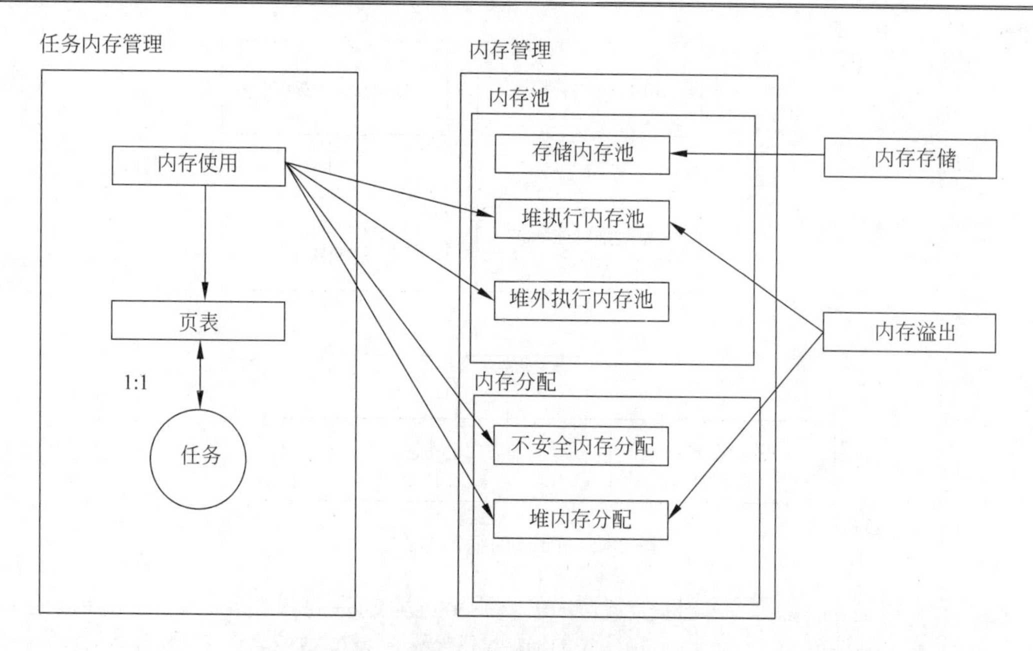

图 28-4　任务内存管理与内存管理间的关系图

```
5.      * 1. Off-Heap : 直接使用 64b 表示内存地址
6.      * 2. On-Heap : 通过 base object 和该对象中 64b 的偏移量表示
7.      * 通过封装类 MemoryBlock 统一表示内存块信息:
8.      * 1. Off-Heap : MemoryBlock 的 base object 为 null，偏移量对应 64b 的
        * 绝对地址
9.      * 2. On-Heap : MemoryBlock 的 base object 保存对象的引用（该引用可以由 page
        * 的索引从 pageTable 获取）
10.     *     偏移量对应数据在该对象中的偏移量
11.     *
12.     * 通过这两种内存模式对应的编码方式，最终对外提供的编码格式为 13b-pageNumber +
        * 51b-offset
13.     */
14. public class TaskMemoryManager {
```

下面从 3 个方面对任务内存管理进行解析，包含内存地址的编码与解码、PageTable 的组织与管理，以及内存的分配与释放。

首先解析内存地址的编码与解码部分。从任务内存管理类的注释部分可知，Off-Heap 与 On-Heap 两种内存模式最终对外都是采用一致的编码格式，即对应 13b 的 Page Number（页码）和 51b 的 offset（偏移量），可以通过图 28-5 来描述对应的编码方式。

下面分别对 TaskMemoryManager 类中与编码和解码相关的几个接口进行解析。编码接口主要有两个，encodePageNumberAndOffset 和 decodePageNumber、decodeOffset，其源码与解析如下所示。

Spark 2.2.1 版本的 TaskMemoryManager.scala 的 encodePageNumberAndOffset 的源码如下：

```
1.   /**
2.     * 给定该页中的内存页和偏移量，将此地址编码为 64b 长。只要对应的页面没有被释放，
       *这个地址仍然有效
3.     *
```

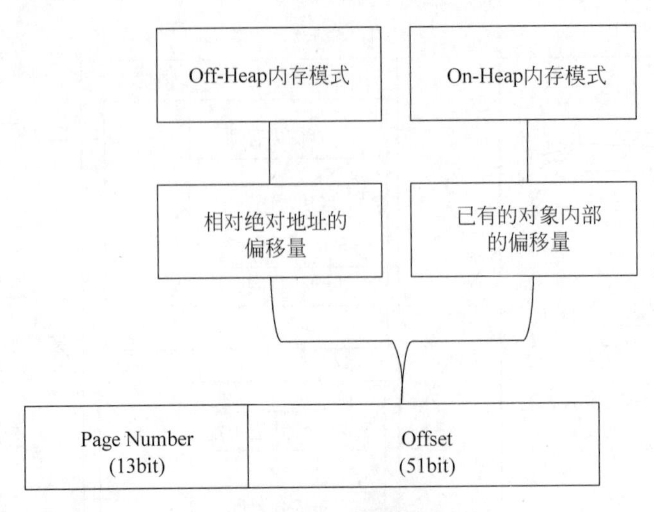

图 28-5　Page 的编码方式

```
4.     * @param page 一个数据页被其分配 {@link TaskMemoryManager#allocatePage}/
5.     * @param offsetInPage 页中的偏移量包含基偏移量。换言之，这是作为基偏移量传递
       * 给不安全的调用的值（例如，page.baseOffset()加上某地址）
6.     * @return 编码页地址
7.     *
8.     * 将针对某个 Page 的地址进行编码：
9.     * On-Heap : offsetInPage 是针对 base object 的偏移量。
10.    * Off-Heap : 此时，offsetInPage 是绝对地址，因此编码到 Page 方式的地址时，
11.    *            需要将绝对地址转换为相对于已有的 Page（MemoryBlock）中的绝对地址
       *            offset 的相对地址，最后将得到的两个偏移量和 Page Number 一起组装
       *            到（13 + 51)b 的 64b 中
12.    */
13.
14.
15. public long encodePageNumberAndOffset(MemoryBlock page, long offsetInPage) {
16.     if (tungstenMemoryMode == MemoryMode.OFF_HEAP) {
17.      //如果是 Off-Heap，则对应的 offsetInPage 为 64b 的绝对地址，需要转换为 Page
18.      //编码能容纳的 51b 编码中，因此此时需要将其转换为 Page 内的相对地址，即页内的
19.      //偏移地址
20.
21.       offsetInPage -= page.getBaseOffset();
22.     }
23.     return encodePageNumberAndOffset(page.pageNumber, offsetInPage);
24.   }
25. ......
26.   @VisibleForTesting
27.   public static long encodePageNumberAndOffset(int pageNumber, long
      offsetInPage) {
28.     assert (pageNumber != -1) : "encodePageNumberAndOffset called with
        invalid page";
29.      //将 13bit 的页码与 51bit 的页内偏移量组装成 64bit 的编码地址
30.     return (((long) pageNumber) << OFFSET_BITS) | (offsetInPage & MASK_
        LONG_LOWER_51_BITS);
31.   }
```

Spark 2.4.3 版本的 TaskMemoryManager.scala 源码与 Spark 2.2.1 版本相比具有如下特点。

❑ 上段代码中第 28 行调整判断条件为 pageNumber 大于等于 0。

```
1.  ……
2.  assert (pageNumber >= 0) : "encodePageNumberAndOffset called with invalid
    page";
3.  ……
```

通过 pageNumber 可以找到最终的 Page，Page 内部会根据 Off-Heap 或 On-Heap 两种模式分别存储 Page 对应内存块的起始地址（或对象内偏移地址），因此编码后的地址可以通过查找到 Page，最终解码出原始地址。

TaskMemoryManager.scala 的 decodePageNumber、decodeOffset 的源码如下：

```
1.       @VisibleForTesting
2.    public static int decodePageNumber(long pagePlusOffsetAddress) {
3.   //解析出编码地址中的页码信息
4.      return (int) (pagePlusOffsetAddress >>> OFFSET_BITS);
5.    }
6.
7.    private static long decodeOffset(long pagePlusOffsetAddress) {
8.   //通过 51bit 掩码解析出编码地址中的页码信息，即对应的低 51bit 内容
9.      return (pagePlusOffsetAddress & MASK_LONG_LOWER_51_BITS);
10.   }
```

在 TaskMemoryManager 类中另外还提供了针对 On-Heap 内存模式下，获取 base object 的接口，对应的源码及其解析如下所示。

TaskMemoryManager.scala 的 getPage 的源码如下：

```
1.   /**
2.     *  获取编码地址相关的 base object
3.     * {@link TaskMemoryManager#encodePageNumberAndOffset(MemoryBlock, long)}
4.     */
5.    public Object getPage(long pagePlusOffsetAddress) {
6.      if (tungstenMemoryMode == MemoryMode.ON_HEAP) {
7.        //首先从地址中解析出页码
8.        final int pageNumber = decodePageNumber(pagePlusOffsetAddress);
9.        assert (pageNumber >= 0 && pageNumber < PAGE_TABLE_SIZE);
10.      //根据页码从 pageTable 变量中获取对应的内存块
11.       final MemoryBlock page = pageTable[pageNumber];
12.       assert (page != null);
13.       assert (page.getBaseObject() != null);
14.      //获取内存块对应的 BaseObject
15.       return page.getBaseObject();
16.     } else {
17.      //Off-Heap 内存模式下 MemoryBlock 只需要保存一个绝对地址，因此对应 base
       object 为 null
18.       return null;
19.     }
20.   }
```

下面开始解析 Page Table 的组织与管理方面的内容，在解析前先给出内存以 Page Table 方式进行组织与管理的大致描述图，如图 28-6 所示。

在图 28-6 中，右侧是分配的内存块，即当前需要管理的 Page，在 TaskMemoryManager 中，通过 Page Table 存放内存块，同时，通过变量 allocatedPages 中指定值为 Page Number（页码）的下标（索引）对应的值是否为 1 来表示当前 Page Number 对应的 Page Table 中的 Page

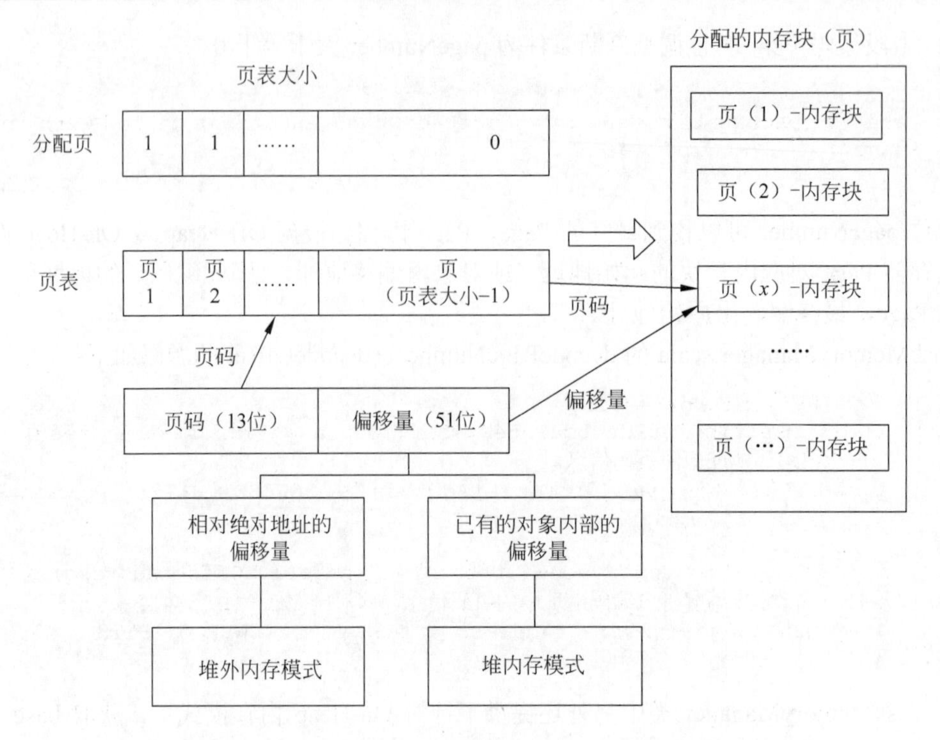

图 28-6　内存以 Page Table 方式组织与管理描述图

是否已经存放了对应的内存块，即每当分配到一个内存块时，从 allocatedPages 获取一个值为 0 的位置（页码），并将该位置作为内存块放入到 Page Table 中的位置。

简单描述的话，就是用 allocatedPages 中各个位置上的值为 1 或 0 来表示在 Page Table 中的相同位置是否已经放置了内存块（Page）。

而对应在 Page Table 中已经存放的内存块，实际上就是对应了右侧已经分配的内存块。

当针对一个 Page Encode（页编码）时，首先从中获取 Page Number（页码），根据该值从 Page Table（页表）中获取确定的内存块（MemoryBlock 或 Page），找到确定内存块之后，再通过页编码中的偏移量（具体两种内存模式下的概念如图 28-6 所示）确定内存块中的相关偏移量，如果是 Off-Heap，则该偏移量是相对于内存块（从前面分析可知，内存块本身的信息也与内存模式相关）中的绝对地址的相对地址，如果是 On-Heap，则该偏移量是相对于内存块的 base object 中的偏移量。

相关的源码主要涉及 TaskMemoryManager 类的两个成员变量，TaskMemoryManager.scala 的源码如下：

```
1.  //对应图中的页表
2.  private final MemoryBlock[] pageTable = new MemoryBlock[PAGE_TABLE_SIZE];
3.
4.  //对应图中的分配页
5.  private final BitSet  allocatedPages = new BitSet(PAGE_TABLE_SIZE);
```

PageTable 的组织与管理中关于页码的偏移量已经在上一部分给出详细描述，而对应的具体的管理操作则与实际的内存分配与解析部分相关。下面通过内存分配与解析部分来详细解析具体的管理细节。

下面开始分析 TaskMemoryManager 类提供的内存分配与解析部分。关于这部分内容，主要参考 allocatePage 与 freePage 两个方法，对于 allocatePage 内部如何申请内存，以及申请内存时采用的 Spill 策略等细节，大家可以继续深入，例如查看 acquireExecutionMemory 的具体源码来加深理解。

Spark 2.2.1 版本的 TaskMemoryManager.scala 的 allocatePage 的源码如下：

```
1.   /**
2.     * 分配内存块，将在 MemoryManager 的页表中记录；这是用于分配大型共享 Tungsten 内
       * 存块，这些内存将在操作算子之间共享。如果没有足够的内存来分配页面，则返回 null。返
       * 回页包含的字节可以比请求的字节少，因此调用者应该验证返回页面的大小
3.     *
4.     * 分配一块内存，并通过 MemoryManager（实际上是在 TaskMemoryManager 中）的 Page
       * Table 进行跟踪；分配的是 Execution 部分的内存
5.     * Project Tungsten 的内存包含 Off-Heap 和 On-Heap 两种模式，由底层 tungsten
       * MemoryMode（在 MemoryManager 中设置）控制具体分配的 MemoryAllocator 子类
6.     */
7.
8.    public MemoryBlock allocatePage(long size, MemoryConsumer consumer) {
9.      assert(consumer != null);
10.     assert(consumer.getMode() == tungstenMemoryMode);
11.     //页大小的限制
12.     if (size > MAXIMUM_PAGE_SIZE_BYTES) {
13.       throw new IllegalArgumentException(
14.         "Cannot allocate a page with more than " + MAXIMUM_PAGE_SIZE_BYTES
          + " bytes");
15.     }
16.      //申请一定的内存量
17.     long acquired = acquireExecutionMemory(size, consumer);
18.     if (acquired <= 0) {
19.       return null;
20.     }
21.
22.     final int pageNumber;
23.     synchronized (this) {
24.      //获取当前未被占用的页码
25.       pageNumber = allocatedPages.nextClearBit(0);
26.       if (pageNumber >= PAGE_TABLE_SIZE) {
27.         releaseExecutionMemory(acquired, consumer);
28.         throw new IllegalStateException(
29.           "Have already allocated a maximum of " + PAGE_TABLE_SIZE +
            "pages");
30.       }
31.      //设置该页码已经被占用（即设置对应页码位置的值）
32.       allocatedPages.set(pageNumber);
33.     }
34.     MemoryBlock page = null;
35.     try {
36.
37.     //开始通过 MemoryAllocator 真正分配内存
38.     //注意 : acquireExecutionMemory 中通过 ExecutionMemoryPool 进行分配时，
39.     //仅仅是内存使用大小上的控制，并没有真正分配内存
40.     //有兴趣的话，可以查看对 acquireExecutionMemory 的调用点（其中
41.     //可以指定与 tungstenMemoryMode 不同的其他内存模式，
42.     //此时是不存在真正的内存分配的）
43.
44.
```

```
45.       page = memoryManager.tungstenMemoryAllocator().allocate(acquired);
46.     } catch (OutOfMemoryError e) {
47.       logger.warn("Failed to allocate a page ({} bytes), try again.",
        acquired);
48.       //实际上没有足够的内存，这意味着实际的空闲内存小于 MemoryManager 管理的内存，
          //我们应该保持获得的内存
49.       synchronized (this) {
50.         acquiredButNotUsed += acquired;
51.         allocatedPages.clear(pageNumber);
52.       }
53.       //这可能会触发溢出，释放一些页面
54.       return allocatePage(size, consumer);
55.     }
56.
57.     //分配得到内存块之后，会设置该内存块对应的 pageNumber，即此时设置 MemoryBlock
        //在其管理的 Page Table 中的位置
58.
59.
60.     page.pageNumber = pageNumber;
61.     pageTable[pageNumber] = page;
62.     if (logger.isTraceEnabled()) {
63.       logger.trace("Allocate page number {} ({} bytes)", pageNumber,
        acquired);
64.     }
65.     return page;
66.   }
```

Spark 2.4.3 版本的 TaskMemoryManager.scala 源码与 Spark 2.2.1 版本相比具有如下特点。

- 上段代码中第 13 行抛出异常调整为抛出 TooLargePageException。

```
1.  ……
2.    throw new TooLargePageException(size);
3.  ……
4.  public class TooLargePageException extends RuntimeException {
5.    TooLargePageException(long size) {
6.      super("Cannot allocate a page of " + size + " bytes.");
7.    }
8.  }
```

其中，MAXIMUM_PAGE_SIZE_BYTES 是页内数据量的最大限制，从之前 MemoryBlock 提供的 long 型数组转换得到 MemoryBlock 接口，可以知道当前连续的内存块是通过 long 型数组来获取的，因此对应的内存块的大小也会受到数组的最大长度的限制。

至于对应在具体的处理过程中，对页内的数据量大小是否还有其他限制，可以参考具体的处理细节。下一节会给出一个具体处理过程的源码解析，其中会包含这部分内容。

由于分配的细节比较多，这里只给出主要的过程描述。

（1）首先通过 acquireExecutionMemory 方法，向 ExecutionMemoryPool 申请内存（根据统一或静态两种具体实现给出）：这一部分主要是判断当前可用内存是否满足申请需求，并根据申请结果修改当前内存池可用内存信息（实际是当前使用内存量信息）。

（2）从当前 Page Table 中找出一个可用位置，用于存放所申请的内存块（MemoryBlock 或 Page）。

（3）准备好前两步后，开始通过 MemoryAllocator 真正分配内存块。

（4）将分配的内存块放入 Page Table。

在整个过程中，allocatedPages 与 pageTable 这两个成员变量的使用是体现 Page Table 组织与管理的关键所在。

下面解析 freePage 的源码，如下所示。

Spark 2.2.1 版本的 TaskMemoryManager.scala 的 freePage 的源码如下：

```
1.   /**
2.     * Free a block of memory allocated via {@link TaskMemoryManager#allocate
       * Page}.
3.     * 更新 Page Table 相关信息，通过 MemoryAllocator 释放 Page 的内存，最后通过
4.     * MemoryManager 修改 ExecutorManagerPool 中的内存使用量（即释放）
5.     */
6.
7.    public void freePage(MemoryBlock page, MemoryConsumer consumer) {
8.   //首先确认当前释放的内存块在 Page Table 的管理中，即页码必须有效
9.      assert (page.pageNumber != -1) :
10.       "Called freePage() on memory that wasn't allocated with allocatePage()";
11.     assert(allocatedPages.get(page.pageNumber));
12.     pageTable[page.pageNumber] = null;
13.     //allocatedPages 是控制 Page Table 中对应位置是否可用的，需要考虑释放与分配时
        //的并发性，因此需同步处理
14.
15.
16.     synchronized(this) {
17.       allocatedPages.clear(page.pageNumber);
18.     }
19.     if (logger.isTraceEnabled()) {
20.       logger.trace("Freed page number {} ({} bytes)", page.pageNumber,
          page.size());
21.     }
22.    //通过当前内存模式对应的 MemoryAllocator 真正释放该内存块
23.     long pageSize = page.size();
24.     memoryManager.tungstenMemoryAllocator().free(page);
25.     //对应 ExecutionMemoryPool 部分的内存释放，参考前面 acquireExecutionMemory
        //解析一起了解
26.
27.
28.     releaseExecutionMemory(pageSize, consumer);
29.   }
```

Spark 2.4.3 版本的 TaskMemoryManager.scala 源码与 Spark 2.2.1 版本相比具有如下特点。

- 上段代码中第 9 行断言中的判断条件将–1 调整为 MemoryBlock.NO_PAGE_NUMBER 。
- 上段代码中第 10 行之后新增加代码，新增对 MemoryBlock.FREED_IN_ALLOCATOR_PAGE_NUMBER、MemoryBlock.FREED_IN_TMM_PAGE_NUMBER 的判断。
- 上段代码中第 23 行之后新增加代码，将块传递给 MemoryAllocator 的 free 方法之前清除页码。

```
1.  //将块传递给 MemoryAllocator 的 free 方法之前清除页码，这样做可以让 MemoryLocator
      检测任务内存管理器 TaskMemoryManager 未调用 TMM.freePage 方法就直接释放了页。
2.  page.pageNumber = MemoryBlock.FREED_IN_TMM_PAGE_NUMBER;
3.  ......
4.  assert (page.pageNumber != MemoryBlock.NO_PAGE_NUMBER) :
5.  ......
6.      assert (page.pageNumber != MemoryBlock.FREED_IN_ALLOCATOR_PAGE_NUMBER) :
7.        "Called freePage() on a memory block that has already been freed";
```

```
8.      assert (page.pageNumber != MemoryBlock.FREED_IN_TMM_PAGE_NUMBER) :
9.              "Called freePage() on a memory block that has already been freed";
10. ……
11. // MemoryBlock.java
12. public class MemoryBlock extends MemoryLocation {
13.     public static final int NO_PAGE_NUMBER = -1;
14.     public static final int FREED_IN_TMM_PAGE_NUMBER = -2;
15.     public static final int FREED_IN_ALLOCATOR_PAGE_NUMBER = -3;
```

释放 Page 的逻辑实际上可以参考申请 Page，大部分都是步骤相反而已。

28.2.3　二进制处理及其实现类的解析

就目前来说，Project Tungsten 的二进制数据处理主要用在 Shuffle、SQL 的 aggregation（聚合）（以及其他操作）的数据上，像为其他非 JVM 的本地类库（如 C++类库）等提供内存访问已经解决（有兴趣的读者可以参考 https://issues.apache.org/jira/browse/SPARK-10399 部分）。

本书基于 Shuffle 过程，解析在源码中具体如何使用 Project Tungsten 来处理数据，对应其他操作的处理细节，可以参考对应的 issues 及其设计文档。例如，在聚合方面使用 Project Tungsten 内存模型的详细设计可以参考 https://issues.apache.org/jira/browse/SPARK-7080，对应的设计文档为 https://github.com/apache/spark/pull/5725。读者可以基于这些设计文档，然后参考本节的源码解析过程加深理解。

在前面 Tungsten Sorted Based Shuffle 写数据的源码解析中已经提到，写数据时，会使用一个外部排序器 ShuffleExternalSorter 对 Shuffle 数据进行排序，该外部排序器中的数据处理就是建立在 Project Tungsten 内存模型基础之上的。因此，本节继续深入解析 Shuffle 写数据的过程，从 Project Tungsten 内存模型的使用角度结合源码进行解析。

首先，从前面对 TaskMemoryManager 的源码解析可以知道，所有内存申请与释放的请求都是通过 MemoryConsumer 提交的，因此首先需要了解在 Shuffle 的写过程中，外部排序器 ShuffleExternalSorter 与内存消费者 MemoryConsumer 之间的关系。为了了解这一点，可以先查看 ShuffleExternalSorter 类的注释与类定义，具体源码如下：

```
1.  /**
2.    * 为 sort-base Shuffle 定制的外部排序器
3.    * 输入的记录会附加到数据 pages 中。当所有数据插入后（或当前线程分配的 Shuffle 内
      * 存达到极限时），内存中（in-memory）的记录会根据分区 ID 进行排序（使用 ShuffleIn
      * MemorySorter）。排序后的记录会写入一个输出文件（或多个文件，如果发生 spilled）。
      * 输出文件的格式与 SortShuffleWriter 的输出文件格式相同：
4.    *每个输出的分区记录通过一个序列化的、压缩的流写入，之后再使用解压的、反序列化的流读取
5.    * 与 ExternalSorter 不同，ExternalSorter 排序器不会进行 Spill 文件的合并
6.    * 而 ShuffleExternalSorter 排序器在最后会进行合并，使用一个可以避免序列化/反序
      * 列化的特定的合并过程
7.    * ShuffleExternalSorter 继承了 MemoryConsumer，会向 TaskMemoryManager 申
      * 请释放 Execution 内存
8.    */
9.  final class ShuffleExternalSorter extends MemoryConsumer {
```

可以看到，ShuffleExternalSorter 继承了 MemoryConsumer，因此在数据处理时可以向 TaskMemoryManager 申请/释放 Execution 内存。

对应地，其他建立在 Project Tungsten 内存模型基础上的数据处理，也可以通过查看

MemoryConsumer 的子类来获取，例如，前面 SPARK-7080 提到的设计文档中的 BytesToBytesMap 子类。本节旨在通过某个具体子类的源码解析，提供一种源码阅读方式。

在详细解析源码之前，同样先给出 ShuffleExternalSorter 在内存处理上的流程，如图 28-7 所示。

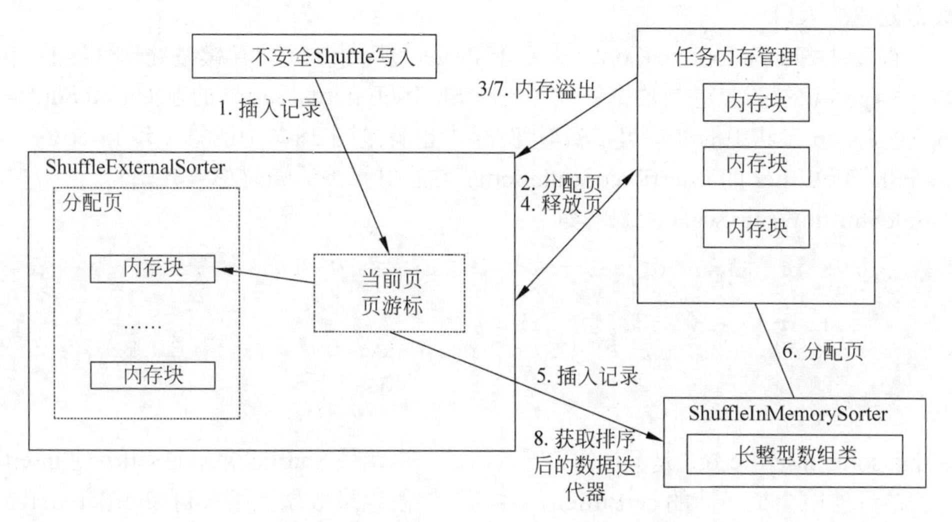

图 28-7　ShuffleExternalSorter 处理流程图

流程的步骤如下。

（1）首先插入记录，由 UnsafeShuffleWriter 调用 ShuffleExternalSorter 的 insertRecord 方法，向 currentPage 插入一条记录，即图 28-7 中的第 1 步 insertRecord。

（2）此时，如果当前页内存未分配，或剩余空间不足以容纳记录数据，则向 TaskMemory Manager 申请内存，即图 28-7 中的第 2 步 allocatePage。

（3）在申请内存时，有可能由于内存压力而产生 MemoryConsumer（即这里的 Shuffle ExternalSorter）的 Spill 操作，即图 28-7 中的第 3 步 Spill。

（4）触发 Spill 操作时，会获取 ShuffleInMemorySorter 的排序数据的迭代器，将排序后的数据 Spill 到文件中，即图 28-7 中的第 8 步 getSortedIterator；同时会再释放 ShuffleExternalSorter 占用的内存（通过由记录的内存页 allocatedPages 实现），即图 28-7 中的第 4 步 freePage。

（5）当 currentPage 能够容纳记录数据时，将数据插入到内存页中，同时会将记录的编码地址插入到 ShuffleInMemorySorter 的 LongArray 中，即图 28-7 中的第 5 步 insertRecord。

（6）在插入编码地址的过程中，也可能会由于 LongArray 内存不足而向 TaskMemory Manager 申请内存，即图 28-7 中的第 6 步 allocatedPage。在申请过程中也可能会触发 spill，这和前面内存申请时描述的过程一样。

（7）完成记录插入后也会调用第 8 步的 getSortedIterator，获取在 ShuffleInMemorySorter 的 LongArray 中未 Spill 到文件的内存数据，然后写入最后一个文件（所以，这个文件写入的数据量不作为 Spill 的 Metric 度量信息）。

在整个处理流程中，比较难以理解的是数据或地址在内存中的存储与处理（处理实际是用与存储相反的过程来读取数据进行处理的，所以本质上还是理解数据是以何种方式存入内存页的）。

插入记录的二进制数据处理：

ShuffleExternalSorter 类的内部处理过程，可以先从该类在 UnsafeShuffleWriter 中的使用开始去理解。如果已经熟悉了 UnsafeShuffleWriter 的写数据过程，可以直接忽略；如果不熟悉，也可以简单地在 UnsafeShuffleWriter 类中搜索 ShuffleExternalSorter 类的调用点来获取二进制数据处理的入口。

下面直接从 ShuffleExternalSorter 类基于 Project Tungsten 内存模型处理数据的关键入口点开始解析插入记录的二进制数据处理（对应 ShuffleExternalSorter 的在 UnsafeShuffleWriter 类的 open 与 stop 方法中的处理可以暂时忽略），也就是图 28-7 中的第 1 步 insertRecord，即从 UnsafeShuffleWriter 的 insertRecordIntoSorter 方法中相关的源码部分开始。

UnsafeShuffleWriter.scala 的源码如下：

```
1.   void insertRecordIntoSorter(Product2<K, V> record) throws IOException {
2.  ……
3.     sorter.insertRecord(
4.  serBuffer.getBuf(), Platform.BYTE_ARRAY_OFFSET, serializedRecordSize,
    partitionId);
5.   }
```

其中，sorter.insertRecord 是解析的关键入口点，也就是 ShuffleExternalSorter 的 insertRecord 方法。在该方法中会把当前的 serBuffer 内容（即一条记录数据）插入到 ShuffleExternalSorter 中。因此，接下来首先分析 insertRecord 方法，对应的源码及其解析如下所示。

Spark 2.2.1 版本的 ShuffleExternalSorter.scala 的 insertRecord 的源码如下：

```
1.    /**
       * 向 Shuffle 排序器写入一条记录
2.     */
3.
4.  public void insertRecord(Object recordBase, long recordOffset, int length,
    int partitionId)
5.     throws IOException {
6.
7.     //测试使用
8.     assert(inMemSorter != null);
9.     if (inMemSorter.numRecords() >= numElementsForSpillThreshold) {
10.      logger.info("Spilling data because number of spilledRecords crossed
         the threshold " + numElementsForSpillThreshold);
11.      spill();
12.    }
13.  //如果需要，增加内存中排序所需的 LongArray 内存大小
14.    growPointerArrayIfNecessary();
15.
16.    //需要 4 个字节来存储记录长度
17.    //记录插入时，以记录长度作为起始信息，然后是对应该长度的记录数据
18.    //因此申请的内存大小需要考虑长度本身占的 4 个字节
19.    final int required = length + 4;
20.    acquireNewPageIfNecessary(required);
21.
22.    assert(currentPage != null);
23.    //获取当前页的 base object
24.    final Object base = currentPage.getBaseObject();
25.    //针对 currentPage 内的页游标（当前起始地址）进行编码
26.    //该地址对应该记录的起始地址（格式：4 字节的长度 + 记录数据）
```

```
27.    final long recordAddress = taskMemoryManager.encodePageNumberAndOffset
       (currentPage, pageCursor);
28.
29.    //首先将记录的长度（Int，4 字节）放入 base + pageCursor 对应的内存地址，
       //更新当前页内地址 pageCursor
30.    Platform.putInt(base, pageCursor, length);
31.    pageCursor += 4;
32.   //将记录数据 recordBase + recordOffset 复制到 base + pageCursor（已更新 4 字节）
      //中，长度为记录长度，更新当前页内地址 pageCursor
33.    Platform.copyMemory(recordBase, recordOffset, base, pageCursor, length);
34.    pageCursor += length;
35.    //将记录的编码地址（PageNumber + 页内 Offset）及其分区号插入到内存排序器中
36.    inMemSorter.insertRecord(recordAddress, partitionId);
37.  }
```

Spark 2.4.3 版本的 ShuffleExternalSorter.scala 源码与 Spark 2.2.1 版本相比具有如下特点。

- 上段代码中第 19、31 行将 4 调整为 uaoSize，需要 4 字节或 8 字节来存储记录长度。

```
1.  ……
2.     final int uaoSize = UnsafeAlignedOffset.getUaoSize();
3.     //需要 4 或 8 个字节来存储记录长度
4.     final int required = length + uaoSize;
5.  ……
6.  pageCursor += uaoSize;
7.  ……
8.  //UnsafeAlignedOffset.java
9.  public class UnsafeAlignedOffset {
10.    private static final int UAO_SIZE = Platform.unaligned() ? 4 : 8;
11. ……
```

这里主要解析记录如何存储在内存页，即内存页中记录的组织形式，可以通过图 28-8 来描述。

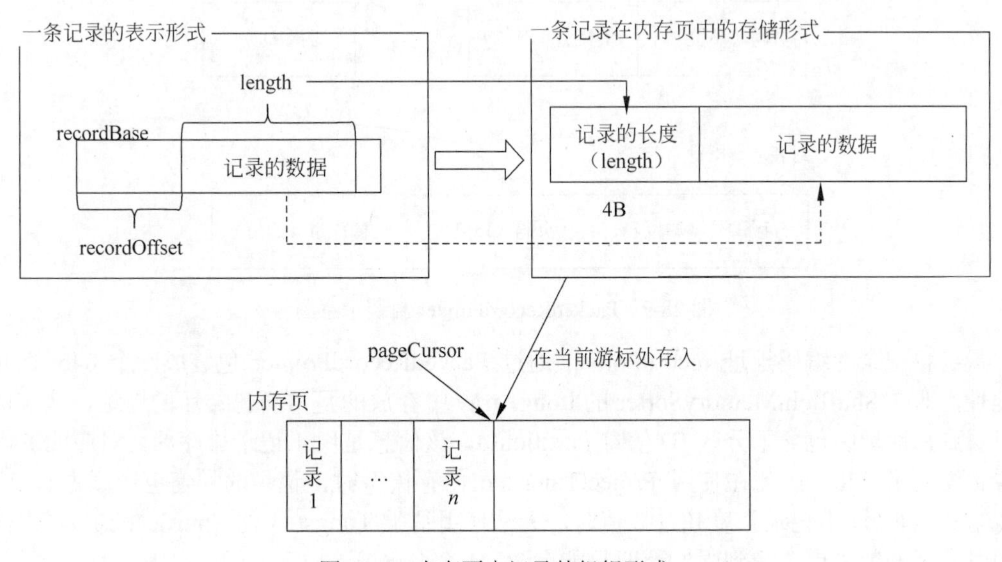

图 28-8　内存页中记录的组织形式

将记录存入内存页时，首先是在内存页当前游标（pageCursor）所在位置存放该记录的数据长度（长度 long 类型对应 4B，因此占用的空间是记录数据的长度+4B），然后通过描述

记录数据信息的三元素，将记录数据复制到数据长度后面的内存页空间中。

对记录数据信息的三元素描述如下：

（1）recordBase：记录数据所在的对象。

（2）recordOffset：记录数据在对象中的偏移量。

（3）length：记录数据的长度。

在处理过程中，通过 TaskMemoryManager 的 encodePageNumberAndOffset 方法，将记录在内存页中的存储地址进行编码（存储时已经包含记录的长度和数据，因此只需要起始地址即可），并将该编码地址和所在分区 ID 一起插入到 inMemSorter(ShuffleInMemorySorter)变量中。继续查看插入的信息如何在 ShuffleInMemorySorter 中组织，具体源码如下所示。

ShuffleInMemorySorter.scala 的 insertRecord 的源码如下：

```
1.     public void insertRecord(long recordPointer, int partitionId) {
2.    if (!hasSpaceForAnotherRecord()) {
3.      throw new IllegalStateException("There is no space for new record");
4.     }
5.    //以 PackedRecordPointer 封装记录数据的编码地址
6.    //编码格式为： 24b-PartitionId + 13b-PageNumber + 27b-offset
7.    //说明：array 对应元素为 Long 类型，因此长度为 64b
8.
9.    array.set(pos, PackedRecordPointer.packPointer(recordPointer, partitionId));
10.   pos++;
11. }
```

插入到 ShuffleInMemorySorter 时，会将信息重新封装为 PackedRecordPointer，然后存放到 LongArray 中。PackedRecordPointer 封装示意图如图 28-9 所示。

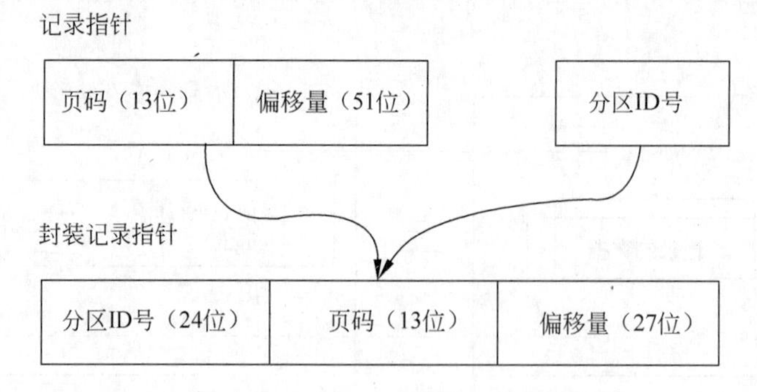

图 28-9　PackedRecordPointer 封装示意图

最终将记录的编码地址 recordPointer 通过 PackedRecordPointer 包装成一个 64b 的 long 型地址，即在 ShuffleInMemorySorter 的 LongArray 中存放的是重新包装后的地址，从该地址可以看到，地址中包含了分区 ID 信息 PartitionId，该信息是用于记录排序的。对应的页内偏移量也缩成了 27b，因此在使用 ProjectTungsten 内存模型时，记录的长度也从原先的 $2^{31}-1$ 变成了 $2^{27}-1$（即当记录长度超过该值时，无法使用基于 Tungsten 的 Shuffle 机制）。同样，使用时分区的数量也会受到限制，即只能有 $2^{24}-1$ 个分区。

对应地，在插入数据的过程中使用的两个方法在某些细节上可能不容易理解，因此这里给出简单的源码及其解析。

首先是 growPointerArrayIfNecessary 方法。

Spark 2.2.1 版本的 UnsafeExternalSorter.scala 的 growPointerArrayIfNecessary 的源码如下：

```
1.    private void growPointerArrayIfNecessary() throws IOException {
2.     assert(inMemSorter != null);
3.     if (!inMemSorter.hasSpaceForAnotherRecord()) {
4.       long used = inMemSorter.getMemoryUsage();
5.       LongArray array;
6.       try {
7.         //将触发溢出
8.         //内部会通过 TaskMemoryManager 的 allocatePage 方法申请内存
9.         //在申请时遇到内存不足时，会采用一定的策略进行 Spill
10.        array = allocateArray(used / 8 * 2);
11.      } catch (OutOfMemoryError e) {
12.        //应触发溢出
13.        if (!inMemSorter.hasSpaceForAnotherRecord()) {
14.          logger.error("Unable to grow the pointer array");
15.          throw e;
16.        }
17.        return;
18.      }
19.       //如果在申请内存过程中触发了溢出，使得 inMemSorter 有空间容纳另一条记录，
          //则释放刚申请的 array （LongArray 内部组合了申请的内存块 MemoryBlock）
20.      if (inMemSorter.hasSpaceForAnotherRecord()) {
21.        freeArray(array);
22.      } else {
23.  //如果没有触发 Spill，则 inMemSorter 使用新的 array（内部会有旧数据的迁移）
24.        inMemSorter.expandPointerArray(array);
25.      }
26.    }
27.  }
```

Spark 2.4.3 版本的 UnsafeExternalSorter.scala 源码与 Spark 2.2.1 版本相比具有如下特点。

- 上段代码中第 10 行新增捕获一个 TooLargePageException 异常，指针数组太大，无法在单个页中修复，发生溢出。

```
1.  ......
2.       } catch (TooLargePageException e) {
3.         //指针数组太大，无法在单个页中修复，溢出
4.         spill();
5.         return;
6.  ......
```

其次是 acquireNewPageIfNecessary 方法。

UnsafeExternalSorter.scala 的 acquireNewPageIfNecessary 的源码如下：

```
1.   /**
       *为了插入一些新的记录而申请更多的内存会从内存管理处申请所需的内存，并在申请失败时
       *触发 Spill
2.     */
3.   private void acquireNewPageIfNecessary(int required) {
4.     //当初始情况或当前页（currentPage）剩余空间不足以容纳所要求的大小时，申请新的
       //内存页，并更新当前页的游标（指向当前页中可用内存的起始位置），同时将当前内存也
       //放入 allocatedPages，以便后续在 Spill 或 Stop 等情况下释放全部页内存
5.     if (currentPage == null ||
6.       pageCursor + required > currentPage.getBaseOffset() + currentPage
         .size()) {
```

```
7.        //待办事项 TODO：尝试在前一页找到空间
8.        currentPage = allocatePage(required);
9.        pageCursor = currentPage.getBaseOffset();
10.       allocatedPages.add(currentPage);
11.     }
12.   }
```

Spill 时二进制数据的处理：

可以将继承 MemoryConsumer 类的子类作为二进制解析数据的关键入口点，从前面对 TaskMemoryManager 类的解析源码中已经知道，当内存不足时，会采用某种策略调用 MemoryConsumer 的 Spill 方法（对应策略比较简单，可以直接参考源码），因此，Spill 方法也是理解 ShuffleExternalSorter 类内部处理过程的一个关键入口点。

下面是 Spill 的关键代码及其解析。

ShuffleExternalSorter.scala 的 spill 的源码如下：

```
1.    /**
       * 在内存压力下，排序并 spill 当前记录集
2.     */
3.
4.  public long spill(long size, MemoryConsumer trigger) throws IOException {
5.      ......
6.      writeSortedFile(false);
7.   //writeSortedFile 不会释放内存，因此需要手动释放
8.      final long spillSize = freeMemory();
9.      inMemSorter.reset();
10.     //重置内存中的排序器指针数组，释放内存页记录。否则，如果任务是过度分配内存，则不
        //释放内存页，我们可能无法获得指针数组的内存
11.     taskContext.taskMetrics().incMemoryBytesSpilled(spillSize);
12.     return spillSize;
13.   }
```

在 Spill 的时候就会调用 writeSortedFile 方法（记录集处理完成后也会调用，只是参数不同），之后便释放占用的全部内存。对应 writeSortedFile 方法的解析，在理解了记录在内存页中存储的组织形式后，相对比较好理解，因此在此仅给出该方法的注释信息，具体如下：

```
1.   /**
2.     * 对内存记录进行排序，并将已排序的记录写入磁盘文件。此方法不释放排序数据结构
3.     *
4.     * @paramisLastFile
5.     *
6.     * 在内存中对记录进行排序，并写入磁盘的一个文件中
7.     * 该方法并不会释放排序的数据结构（即需要手动释放内存）
8.     * 当 isLastFile 为：
9.     *  1. true 时，表示最后一个输出文件，此时写的数据量统计在 Shuffle write 的度
       *  量信息中
10.    *  2. false 时，则同时统计在 Shuffle write 和 Shuffle spill 的度量信息中
11.    */
12.  private void writeSortedFile(boolean isLastFile) throws IOException {
```

至此基本上解析了基于 Project Tungsten 内存模型的整个 Shuffle 数据处理过程。下面简单描述一下内存中的排序（对应 ShuffleInMemorySorter 类），排序时主要使用了 timSort 排序

算法（封装 Java 上的实现），所采用的比较器为 SortComparator，对应的定义如下所示。

ShuffleInMemorySorter 的 SortComparator 的源码如下：

```
1.    private static final class SortComparator implements Comparator
      <PackedRecordPointer> {
2.     @Override
3.     public int compare(PackedRecordPointer left, PackedRecordPointer right) {
4.       int leftId = left.getPartitionId();
5.       int rightId = right.getPartitionId();
6.       return leftId < rightId ? -1 : (leftId > rightId ? 1 : 0);
7.     }
8.   }
```

从源码中可以看到，比较时，使用的是 PackedRecordPointer 对象中的分区 ID，所以在基于 Tungsten 的 Shuffle 机制中，记录是按分区 ID 进行排序，并没有对分区内部的记录进行排序。

基于 Project Tungsten 内存模型的数据处理，可以参考本节的源码阅读方式，从 MemoryConsumer 角度出发（包含那些内部使用了 MemoryConsumer 的类，如 UnsafeExternalSorter），逐个理解内部的数据结构组织与二进制数据处理等方式。

说明： 通常，代码中有 open 方法，也会同时对应有 stop 方法，就像资源有申请，同时也会有释放，或者元数据信息有创建，也会有销毁一样。这些配对的处理方式可以在阅读源码时引起注意。

28.3 缓存感知计算

本节讲解缓存感知计算，主要包括：缓存感知计算的解析；缓存感知计算类的解析。

28.3.1 概述

在解释缓存感知计算（Cache-aware Computation）前，我们先回顾一下“内存计算”，也是 Spark 广为业内知晓的优势。对于 Spark 来说，它可以更好地利用集群中的内存资源，提供比基于磁盘解决方案更快的速度。然而，Spark 同样可以处理超过内存大小的数据，自动地外溢到磁盘，并执行额外的操作，如排序和哈希。

类似的情况，缓存感知计算通过使用 L1/ L2/L3 CPU 缓存来提升速度，同样也可以处理超过寄存器大小的数据。在给用户 Spark 应用程序作性能分析时，我们发现大量的 CPU 时间因为等待从内存中读取数据而浪费。在 Tungsten 项目中，我们设计了更加友好的缓存算法和数据结构，从而让 Spark 应用程序可以花费更少的时间等待 CPU 从内存中读取数据，也给有关工作提供了更多的计算时间。

28.3.2 缓存感知计算的解析

我们看一个对记录排序的例子。一个标准的排序步骤需要为记录储存一组指针，并使用

quicksort 互换指针，直到所有记录被排序。基于顺序扫描的特性，排序通常能获得一个不错的缓存命中率。然而，排序一组指针的缓存命中率却很低，因为每个比较运算都需要对两个指针解引用，而这两个指针对应的却是内存中两个随机位置的数据。

那么该如何提高排序中的缓存本地性？其中一种方法就是通过指针顺序地储存每个记录的 sort key。举个例子，如果 sort key 是一个 64 位的整型，那么需要在指针阵列中使用 128 位（64 位指针，64 位 sort key）来存储每条记录。这种途径下，每个 quicksort 对比操作只需要线性地查找每对 pointer-key，从而不会产生任何的随机扫描。图 28-10 可以使我们对提高缓存本地性的方法有一定的了解。

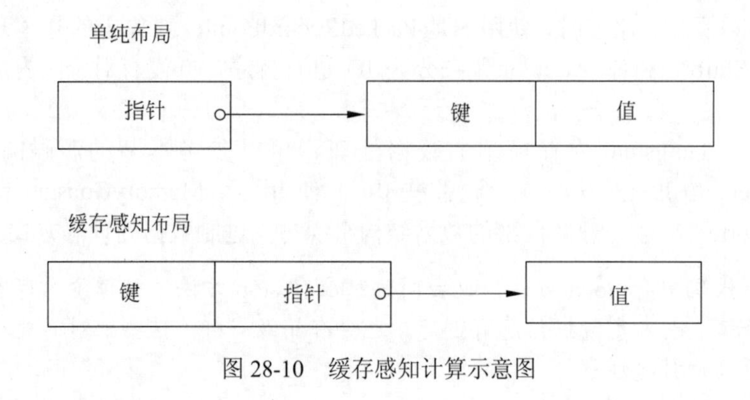

图 28-10　缓存感知计算示意图

28.3.3　缓存感知计算类的解析

缓存感知计算如何适用于 Spark？大多数分布式数据处理都可以归结为多个操作组成的一个小列表，如聚合、排序和 Join。因此，通过提升这些操作的效率，可以从整体上提升 Spark。我们已经为排序操作建立了一个具有缓存感知功能的排序版本，它比老版本的速度快 3 倍。这个新的 sort 将会被应用到 sort-based Shuffle、high cardinality aggregations 和 sort-merge join operator。将来所有 Spark 上的低等级算法都将升级为 Cache-aware，从而让所有应用程序的效率都得到提高——从机器学习到 SQL。

28.4　代码生成

本节讲解代码生成，主要包括：新型解析器的解析；代码生成的解析；表达式代码生成的应用解析。

28.4.1　概述

Spark 为 SQL 和 DataFrames 中的表达式评估引入了代码生成。表达式评估是在如 age > 35 && age < 40 特定记录上计算表达式的值的过程。运行时，Spark 会动态生成用于评估这些表

达式的字节码，而不是为每行代码进行交互解释运行。与解释器相比，代码生成减少了原始数据类型的打包，更重要的是，避免了昂贵的多态函数调度。

代码生成可以将许多 TPC-DS 查询加速几乎一个数量级。Spark 现在正在将代码生成覆盖扩展到大多数内置表达式。此外，计划将代码生成级别从一次性表达式评估增加到向量化表达式评估，利用 JIT 的功能，在现代 CPU 中利用更好的指令流水线，可以一次处理多条记录。

28.4.2　新型解析器的解析

我们还将代码生成应用于表达式评估之外的领域，以优化内部组件的 CPU 效率。Spark 对应用代码生成非常兴奋的一个领域是加快数据从内存二进制格式转换为 wire-protocol for shuffle。Shuffle 通常是数据序列化，而不是底层网络的瓶颈。通过代码生成，可以提高序列化的吞吐量，从而提高网络吞吐量。

图 28-11 比较了使用 Kryo 序列化器和代码生成的自定义序列化器在一个线程中运行 Shuffle 操作 800 万复杂行的计算性能。代码生成的序列化程序利用单个 Shuffle 中的所有行都具有相同的模式的特点，并为此生成专门的代码。这使得生成的版本比 Kryo 版本快 2 倍以上。

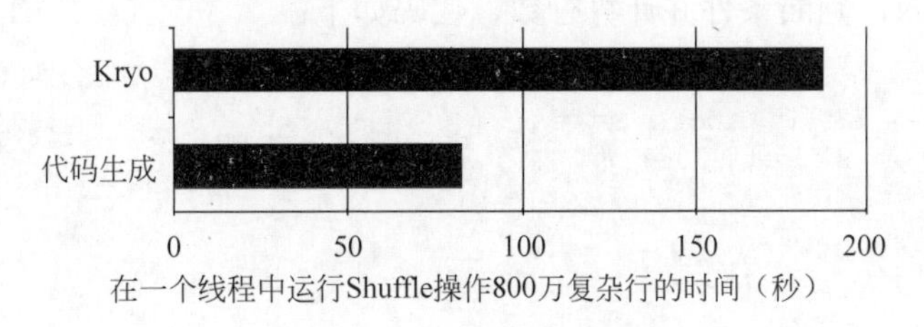

图 28-11　代码生成性能比较图

28.4.3　代码生成的解析

在深入介绍 whole-Stage code generation 前，先回顾一下现在的 Spark（以及大多数数据库系统）是如何运行的。首先看一下图 28-12。扫描一个表，然后计算出满足给定条件属性值的总行数。

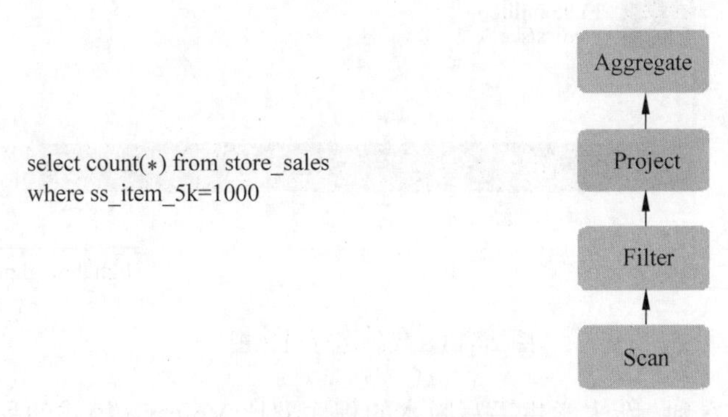

图 28-12　查询表达式示意图

为了计算这个查询，旧版本的 Spark(1.x)会利用基于迭代模型的经典查询评估策略（通常被称为 Volcano model）。在这个模型中，一个查询由多个算子（operators）组成，每个算子都提供了 next()接口，该接口每次只返回一个元组（tuple）给嵌套树中的下一个算子。例如，上面查询中的 Filter 算子大致可以翻译成下面的代码：

```
1.  class Filter(child: Operator, predicate: (Row => Boolean))
2.  extends Operator {
3.  def next(): Row = {
4.  var current = child.next()
5.  while (current == null || predicate(current)) {
6.  current = child.next()
7.     }
8.  return current
9.    }
10. }
```

让每个算子实现迭代器接口允许查询执行引擎来优雅地组合任意的算子，而不必担心每个算子提供的数据类型。结果 Volcano 模型在过去的 20 年间变成数据库系统的标准，这个也是 Spark 使用的架构。

如果给一个新大学生 10min 的时间使用 Java 来实现上面的查询，他很可能会想出一段迭代代码来循环遍历输入，判断条件并计算行数，代码如下：

```
1.   var count = 0
2.  for (ss_item_sk in store_sales) {
3.    if (ss_item_sk == 1000) {
4.      count += 1
5.    }
6.  }
```

上面的代码仅仅是专门解决一个给定的查询，而且很明显不能和其他算子组合。但是，这两种实现（Volcano 与手写代码）方式在性能上有什么重要区别呢？一方面，Spark 和大多数关系型数据库选择这种可以对不同算子进行组合的结构；另一方面，我们有一个由新手在 10min 内编写的程序。我们运行了一个简单的基准测试，对比了大学新生手写版的程序和 Spark 版的程序在使用单个线程的情况下运行上面同一份查询，并且这些数据存储在磁盘上，格式为 Parquet。它们之间的对比如图 28-13 所示。

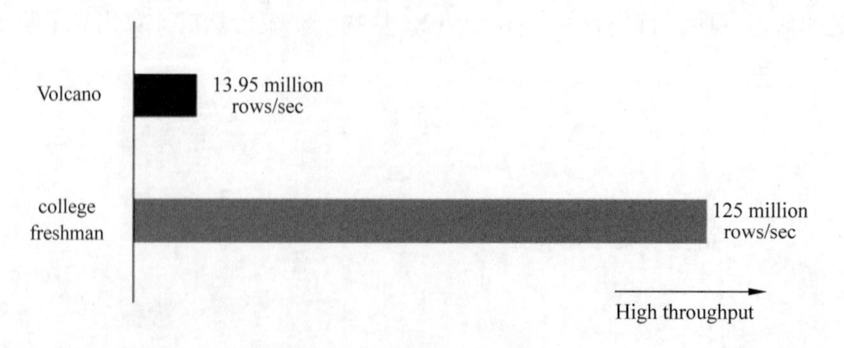

图 28-13 代码运行对比图

从图 28-13 可以看到，新大学生手写版本的程序要比 Volcano 模式的程序快一个数量级！因为 6 行的 Java 代码是被优化过的，其原因如下。

（1）没有虚函数调用：在 Volcano 模型中，处理一个元组（tuple）最少需要调用一次 next()函数。这些函数的调用是由编译器通过虚函数调度（通过 vtable）实现的；而手写版本的代码没有一个函数调用。虽然虚函数调度是现代计算机体系结构中的重点优化部分，它仍然需要消耗很多 CPU 指令而且相当慢，特别是调度数达十亿次。

（2）内存和 CPU 寄存器中的临时数据：在 Volcano 模型中，每次一个算子给另外一个算子传递元组的时候，都需要将这个元组存放在内存中；而在手写版本的代码中，编译器（这个例子中是 JVM JIT）实际上是将临时数据存放在 CPU 寄存器中。访问内存中的数据需要的 CPU 时间比直接访问寄存器中的数据要大一个数量级！

（3）循环展开（Loop unrolling）和 SIMD：运行简单的循环时，现代编译器和 CPU 的效率高得令人难以置信。编译器会自动展开简单的循环，甚至在每个 CPU 指令中产生 SIMD 指令来处理多个元组。CPU 的特性，如管道（pipelining）、预取（prefetching）以及指令重排序（instruction reordering）使得运行简单的循环非常高效。然而，这些编译器和 CPU 对复杂函数调用的优化极少，而这些函数正是 Volcano 模型依赖的。

这里的关键点是手写版本代码的编写正对上面的查询，所以它充分利用到已知的所有信息，导致消除了虚函数的调用，将临时数据存放在 CPU 寄存器中，并且可以通过底层硬件进行优化。

28.4.4　表达式代码生成的应用解析

未来整个阶段的代码生成：从上述观察中，我们的下一步是探索在运行时自动生成这个手写代码的可能性，我们称之为“全阶段代码生成”。这个想法的灵感来自于托马斯·诺依曼的 VLDB 2011 的论文 *Efficiently Compiling Efficient Query Plans for Modern Hardware*。

Spark 的目标是利用整个阶段的代码生成，引擎可以实现手写代码的性能，同时提供通用引擎的功能。运行时，这些运算符不是依赖运算符来处理数据，而是一起在运行时生成代码，并将所有的查询片段组成到单个函数中，仅需要运行生成的代码。

例如，在上一节的查询中，整个查询是一个单一的阶段，而 Spark 将生成图 28-14 所示的 JVM 字节码（以 Java 代码的形式显示）。更复杂的查询将导致多个阶段，从而导致由 Spark 生成多个不同的功能。

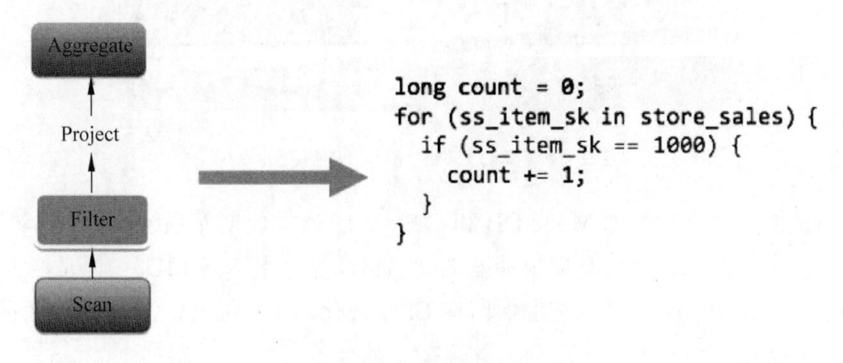

图 28-14　Java 代码图

全阶段代码生成模型：explain()表达式中的函数扩展到全阶段代码生成。在输出的结果里，当算子前面有一个*时，全阶段代码生成被启用。在以下情况下，Range、Filter 和两个 Aggregates 算子都运行全阶段代码生成。但是，Exchange 算子并没有实现整段代码生成，因为它需要通过网络发送数据。

```
1.    spark.range(1000).filter("id > 100").selectExpr("sum(id)").explain()
2.
3.  == Physical Plan ==
4.  *Aggregate(functions=[sum(id#201L)])
5.  +- Exchange SinglePartition, None
6.     +- *Aggregate(functions=[sum(id#201L)])
7.        +- *Filter (id#201L > 100)
8.           +- *Range 0, 1, 3, 1000, [id#201L]
```

已经关注 Spark 发展的人可能会问下面的问题：Apache Spark 1.1 的代码生成和新的代码生成有何区别？过去与其他 MPP 查询引擎类似，Spark 只将代码生成应用于表达式求值，并限于少量运算符（如 Project、Filter）。也就是说，过去的代码生成只是加快了诸如“1 + a”的表达式的评估，而今天的全阶段代码生成实际上是为整个查询计划生成代码。

矢量（Vectorization）：整个阶段的代码生成技术特别适用于对大型数据集执行简单、可预测的查询的大量操作。然而，存在生成代码，以将整个查询融合为单个功能是不可行的情况。操作可能很复杂（如 CSV 解析或拼接解码），或者可能是与第三方组件集成，无法将其代码集成到生成的代码中（例如，从调用到 Python / R 将计算卸载到 GPU）。

为了提高这些情况下的性能，我们采用了另一种称为“向量化”的技术。引擎一次只能处理一行数据，引擎会以多个列的格式将多个行进行批处理，每个操作符都使用简单的循环来迭代批量内的数据。每次调用 next()函数，next()将返回一批元组，以分摊虚拟功能调度的成本。这些简单的循环还将使编译器和 CPU 更有效地执行。

例如，对于具有 3 列（id，name，score）的表，以行、列为格式的内存布局如图 28-15 所示。

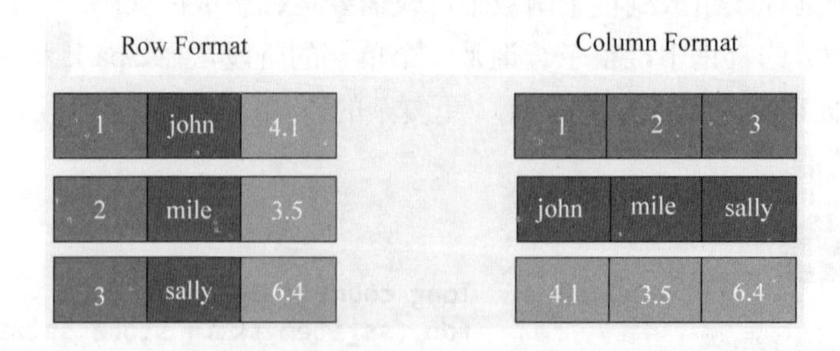

图 28-15　以行、列为格式的内存布局

行和列格式的内存布局：如 MonetDB 和 C-Store 的列状数据库系统的处理方式将会实现前面提到的三点中的两点：（没有虚拟功能调度和自动循环展开/SIMD）。但是，它仍然需要将中间数据放入内存中，而不是将它们保留在 CPU 寄存器中。因此，仅当不可能进行全阶段代码生成时，才使用向量化。

例如，我们已经实现了一个新的矢量化的 Parquet 读取器，它在列批量中进行解压缩和解码。当解码整数列（在磁盘上）时，这个新的阅读器比非向量化的读取器大约快 9 倍，如

图 28-16 所示。

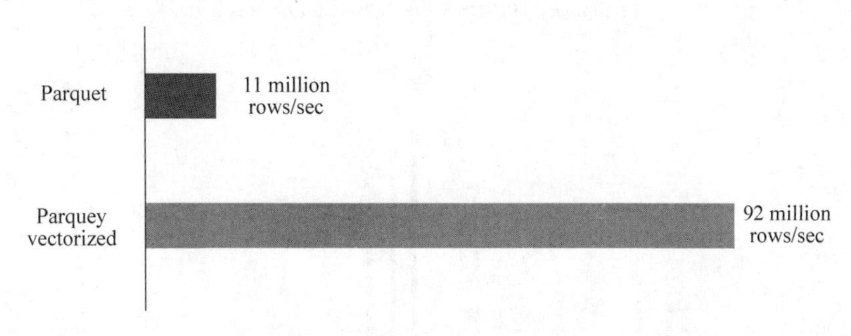

图 28-16 矢量化的 Parquet 示意图

将来，我们计划在更多的代码路径中使用向量化，如 Python / R 中的 UDF 支持。

性能基准：我们测量了在 Apache Spark 1.6 与 Apache Spark 2.0 中的一个核上处理一行的操作时间（单位是 ns），表 28-1 描述了新的 Tungsten engine 的性能提升。Spark 1.6 提供代码生成技术，也在当今一些商业数据库中使用。

每行成本（单位为 ms）如表 28-1 所示。

表 28-1 Tungsten engine的性能

成　　本	Spark 1.6/ns	Spark 2.0/ns
filter	15	1.1
sum w/o group	14	0.9
sum w/ group	79	10.7
hash join	115	4.0
sort (8-bit entropy)	620	5.3
sort (64-bit entropy)	620	40
sort-merge join	750	700
Parquet decoding (single int column)	120	13

最常用的运算符（如 filter、aggregate 以及 Hash Join）实现了整个阶段的代码生成。许多核心算子在整个阶段的代码生成中都快了一个数量级。然而，如 sort-merge join 的一些运算符本质上较慢，并且难以优化。我们在单个机器上执行 10 亿条记录的聚合和连接。在 Databricks 平台（使用英特尔 Haswell 处理器，包含 3 个内核）以及 Macbook Pro 上对 10 亿个元组执行 Hash Join 连接操作需要不到 1s 的时间。

Spark 的新引擎如何在端到端进行查询工作？除了整个阶段的代码生成和向量化外，还进行了大量的工作，改进了 Catalyst 优化器，用于一般查询优化，如 nullability propagation。图 28-17 为使用 TPC-DS 查询进行了一些初步分析来比较 Spark 1.6 和 Spark 2.0。

这是否意味着一旦升级到 Spark 2.0，计算就会比之前快 10 倍？不是。虽然我们相信新的 Tungsten 引擎在数据处理中实现了性能工程的最佳架构，但并不是所有的工作负载都能受益于同样的程度。例如，字符串的可变长度数据类型操作就很昂贵，并且一些工作负载受 I/O 吞吐量、元数据操作的其他因素限制。之前受 CPU 效率限制的工作负载将观察到最大的收益，并转向更多的 I/O 绑定，而之前 I/O 绑定的工作负载不太可能获得收益。

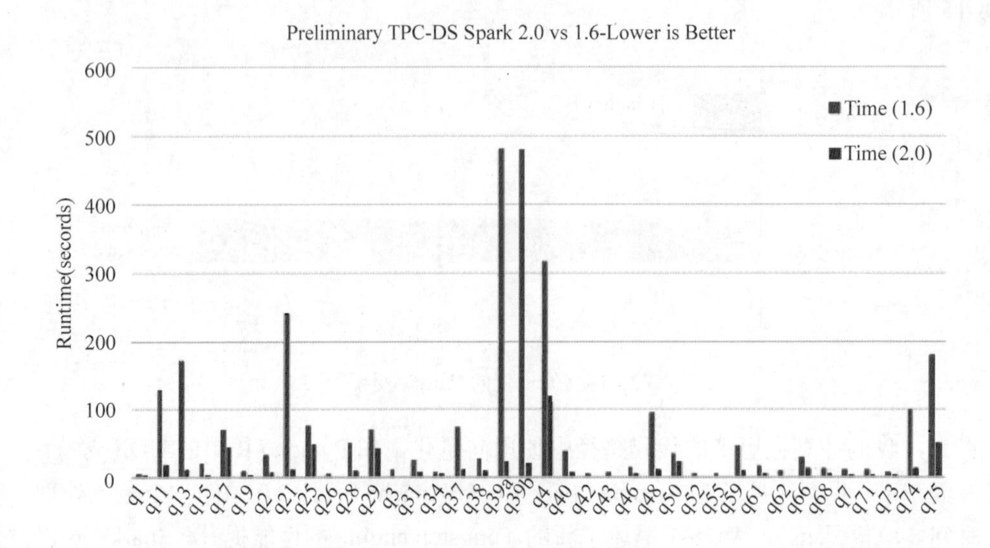

图 28-17　TPC-DS 查询比较

第二代钨执行引擎称为全阶段代码生成技术，引擎将实现以下功能。

（1）消除虚拟功能调度。

（2）将中间数据从存储器移动到 CPU 寄存器。

（3）利用现代 CPU 功能循环展开和使用 SIMD。通过 vectorization 技术，引擎将加快对复杂操作代码生成运行的速度。对于许多数据处理的核心算子，新引擎的运行速度要提升一个数量级。未来，考虑到执行引擎的效率，我们的大部分性能工作将转向优化 I/O 效率和更好地查询规划。

28.5　本章总结

本章内容主旨在于抛砖引玉，在结合 Databricks 公司发布的博文和 Spark issues 及其相关的设计文档的基础上，从源码角度对目前已经实现的部分 Project Tungsten 进行解析，主要内容包含 Project Tungsten 的内存模型，以及在该内存模型基础上，以 Shuffle 写数据的过程为例，详细解析在 Project Tungsten 的内存模型上对数据结构进行组织以及基于二进制处理数据等方面的内容。同时，本章也讲解了缓存感知计算及代码生成的相关内容。

第 29 章　Spark Shuffle 调优原理及实践

本章主要讲解如下几项内容。

- Shuffle 对性能消耗的原理详解。
- Spark.Shuffle.manager 参数调优原理及实践。
- Spark.Shuffle.blockTransferService 参数调优原理及实践。
- Spark.Shuffle.compress 参数调优原理及最佳实践。
- Spark.io.compression.codec 参数调优原理及实践。
- Spark.Shuffle.consolidateFiles 参数调优原理及实践。
- Spark.Shuffle.file.buffer 参数调优原理及实践。
- Spark.Shuffle.io.maxRetries 参数调优原理及实践。
- Spark.Shuffle.io.retryWait 参数调优原理及实践。
- Spark.Shuffle.io.numConnectionsPerPeer 参数调优原理及实践。
- Spark.reducer.maxSizeInFlight 参数调优原理及实践。
- Spark.Shuffle.io.preferDirectBufs 参数调优原理及实践。
- Spark.Shuffle.memoryFraction 参数调优原理及实践。
- Spark.Shuffle.service.enabled 参数调优原理及实践。
- Spark.Shuffle.service.port 参数调优原理及实践。
- Spark.Shuffle.sort.bypassMergeThreshold 参数调优原理及实践。
- Spark.Shuffle.spill 参数调优原理及实践。
- Spark.Shuffle.spill.compress 参数调优原理及实践。

29.1　Shuffle 对性能消耗的原理详解

在分布式系统中，数据分布在不同的节点上，每个节点计算一部分数据，如果不对各个节点上独立的部分进行汇聚，计算就得不到最终的结果。我们需要利用分布式来发挥 Spark 本身并行计算的能力，而后续又需要计算各节点上最终的结果，所以需要把数据汇聚集中，这就会导致 Shuffle，这也是为什么 Shuffle 是分布式不可避免的。因为 Shuffle 的过程中会产生大量的磁盘 I/O、网络 I/O，以及压缩、解压缩、序列化和反序列化的操作，这一系列操作对性能都是一个很大的负担。

调优是一个动态的过程，需要根据业务数据的特性和硬件设备的条件，经过不断地测试，才能达到一个最优化的水平。下面是一些 Spark 参数的介绍，以及一些调优的最佳实战。参数调优是其中一种减少 Shuffle 带来的性能负担的方法。

Spark 官网 https://spark.apache.org/docs/latest/configuration.html 提供的 Spark 集群 Shuffle 行为属性配置参数见表 29-1。

表 29-1　Spark集群Shuffle行为属性配置参数表

属 性 参 数	缺 省 值	含　义
spark.reducer.maxSizeInFlight	48MB	从每个 reduce 任务同时抓取 map 输出数据的最大大小。由于每个输出数据需要创建一个缓冲区来接收，是每个 reduce 任务固定的内存开销，因此须设置一个较小的值，除非有大量的内存
spark.reducer.maxReqsInFlight	Int.MaxValue	此配置限制了在任何给定点获取块的远程请求数。当集群中的主机数量增加时，可能会导致一个或多个节点的入站连接数量非常多，导致工作负载失败。通过限制获取请求的数量，可以减轻这种情况
spark.shuffle.compress	true	是否压缩 map 输出文件通常是一个好主意，压缩时使用 spark.io.compression.codec
spark.shuffle.file.buffer	32KB	每个 Shuffle 文件输出流的内存缓冲区大小。这些缓冲区减少了创建中间 Shuffle 文件时进行的磁盘查找和系统调用次数
spark.shuffle.io.maxRetries	3	（仅限 Netty）如果将其设置为非零值，在 I/O 相关异常导致获取数据失败，将自动重试。重试将有助于保障长时间 GC 停顿或瞬时网络连接问题情况下 Shuffle 的稳定性
spark.shuffle.io.numConnectionsPerPeer	1	（仅 Netty）重新使用主机之间的连接数，以减少大型集群的连接建立。对于具有多个硬盘和少量主机的集群，这可能导致并发性不足，以使所有磁盘饱和，因此用户可考虑增加此值
spark.shuffle.io.preferDirectBufs	true	（仅 Netty）非堆缓冲区用于在 Shuffle 和缓存块传输过程中减少垃圾回收。对于非堆内存严格限制的环境，用户可能希望将其关闭，以强制 Netty 的所有分配都在堆上
spark.shuffle.io.retryWait	5s	（仅 Netty）获取重试之间等待多长时间。默认情况下，重试引起的最大延迟时间为 15s，计算方式为 maxRetries ×retryWait
spark.shuffle.service.enabled	false	启用外部 Shuffle 服务。此服务保留由 Executors 写入的 Shuffle 文件，以便 Executors 可以安全地删除。如果 spark.dynamicAllocation.enabled 设置为 true，则必须启用，必须设置外部 Shuffle 服务才能启用它
spark.shuffle.service.port	7337	外部 Shuffle 服务运行的端口
spark.shuffle.service.index.cache.entries	1024	在 Shuffle 服务的索引缓存中的最大条目数
spark.shuffle.sort.bypassMergeThreshold	200	（高级）在基于排序的 Shuffle 管理器中，如果没有 map 侧聚合，则避免合并排序数据，可以减少分区
spark.shuffle.spill.compress	true	是否压缩在 Shuffle 期间溢出的数据。压缩将使用 spark.io.compression.codec

续表

属 性 参 数	缺 省 值	含 义
spark.shuffle.accurateBlockThreshold	100×1024×1024	当压缩 HighlyCompressedMapStatus 中的 Shuffle 块的大小时，如果超过此配置，我们将准确记录大小。这有助于通过避免在 Shuffle 获取块时低估 Shuffle 块的大小来防止 OOM
spark.io.encryption.enabled	false	启用 I/O 加密。目前支持除 Mesos 外的所有模式。建议在使用此功能时启用 RPC 加密
spark.io.encryption.keySizeBits	128	I/O 加密密钥大小以位为单位。支持的值为 128 192 和 256
spark.io.encryption.keygen.algorithm	HmacSHA1	生成 I/O 加密密钥时使用的算法。支持的算法在 Java 加密体系结构标准算法名称文档的 KeyGenerator 部分进行了描述

29.2　Spark.Shuffle.manager 参数调优原理及实践

Spark.Shuffle.manager 默认值：Sort。

参数说明：该参数用于设置 ShuffleManager 的类型。Spark 1.5 以后，有 3 个可选项：Hash、Sort 和 Tungsten-Sort。HashShuffleManager 是 Spark 1.2 以前的默认选项，但是 Spark 1.2 以及之后的版本默认都是 SortShuffleManager。Tungsten-Sort 与 Sort 类似，但是使用了 Tungsten 计划中的堆外内存管理机制，内存使用效率更高。

调优建议：由于 SortShuffleManager 默认会对数据进行排序，因此如果你的业务逻辑中需要该排序机制，则使用默认的 SortShuffleManager 就可以；而如果你的业务逻辑不需要对数据进行排序，那么建议参考后面的几个参数调优，通过 bypass 机制或优化的 HashShuffleManager 来避免排序操作，同时提供较好的磁盘读写性能。这里要注意的是，Tungsten-Sort 要慎用，因为之前发现了一些相应的 Bug。

Spark.Shuffle.manager 参数用于设置 ShuffleManager 的类型，Spark 2.0 以后，只有 Sort 和 Tungsten-Sort 两个可选项，从源码中查看，以前的 Hash-Based Shuffle 算法在新版本中已经被废弃。

SparkEnv.scala 的源码如下：

```
1.      val shortShuffleMgrNames = Map(
2.        "sort" -> classOf[org.apache.spark.shuffle.sort.SortShuffleManager]
          .getName,
3.        "tungsten-sort" -> classOf[org.apache.spark.shuffle.sort
          .SortShuffleManager].getName)
4.      val shuffleMgrName = conf.get("spark.shuffle.manager", "sort")
5.      val shuffleMgrClass =
6.        shortShuffleMgrNames.getOrElse(shuffleMgrName.toLowerCase
          (Locale.ROOT), shuffleMgrName)
7.      val shuffleManager = instantiateClass[ShuffleManager](shuffleMgrClass)
```

Spark 2.0 版本默认是 SortShuffleManager。Tungsten-Sort 与 Sort 类似，SortShuffleManager 默认对数据进行排序，因此如果用户的业务逻辑中需要该排序机制，则使用默认的 SortShuffle-Manager；如果需要使用 Tungsten-Sort，则把 Spark.Shuffle.manager 设置成 Tungsten-Sort。

29.3　Spark.Shuffle.blockTransferService 参数调优原理及实践

Spark.Shuffle.blockTransferService 参数用来实现在 Executor 之间传递 Shuffle 缓存块。有 Netty 和 Nio 两种可用的实现。基于 Netty 的块传递在具有相同效率的情况下更简单，默认值是 Netty。

在 Spark 1.2.0 中，这个配置的默认值是 Netty，而之前是 Nio。Netty 用于在各个 Executor 之间传输 Shuffle 数据。Netty 的实现更加简洁，但实际上用户不用太关心这个选项。除非有特殊的需求，否则采用默认配置就可以。

29.4　Spark.Shuffle.compress 参数调优原理及实践

Spark.Shuffle.compress 参数是判断是否对 mapper 端的聚合输出进行压缩，默认是 true，表示在每个 Shuffle 的过程中都会对 mapper 端的输出进行压缩。例如，说几千台或者上万台的机器进行汇聚计算，数据量和网络传输会非常大，这会造成大量内存消耗、磁盘 I/O 消耗和网络 I/O 消耗。此时如果在 Mapper 端进行了压缩，就会减少 Shuffle 过程中下一个 Stage 向上一个 Stage 抓数据的网络开销，大大地减轻 Shuffle 的压力。

UnsafeShuffleWriter.scala 的 mergeSpills 方法的源码如下：

```
1.      private long[] mergeSpills(SpillInfo[] spills, File outputFile) throws
        IOException {
2.      final boolean compressionEnabled = sparkConf.getBoolean("spark.Shuffle
        .compress", true);
3.      final CompressionCodec compressionCodec = CompressionCodec$.MODULE$
        .createCodec(sparkConf);
4.      final boolean fastMergeEnabled =
5.        sparkConf.getBoolean("spark.Shuffle.unsafe.fastMergeEnabled", true);
6.      final boolean fastMergeIsSupported = !compressionEnabled ||
7.        CompressionCodec$.MODULE$.supportsConcatenationOfSerializedStreams
          (compressionCodec);
8.      final boolean encryptionEnabled = blockManager.serializerManager()
        .encryptionEnabled();
9.      try {
10.       if (spills.length == 0) {
11.         new FileOutputStream(outputFile).close(); //Create an empty file
12.         return new long[Partitioner.numPartitions()];
13.       } else if (spills.length == 1) {
14.         //这里，我们不需要执行任何度量更新，因为写入这个字节，输出文件将被计算为已
            //写入的 Shuffle 字节数
15.         Files.move(spills[0].file, outputFile);
16.         return spills[0].PartitionLengths;
17.       } else {
18.         final long[] PartitionLengths;
19.
20. //这里有多个溢出合并，因此，对于 Shuffle 写入计数或 Shuffle 写入时间，这些溢出文
    //件的长度没有被计数。如果使用慢合并路径，最终输出文件的大小不一定等于溢出文件总大小
```

```
   //为了防止这种情况发生，我们查看输出文件的实际大小，在计算 Shuffle 字节数时，允许单
   //个合并方法报告来自不同的合并后的 I/O 时间
21. //既然不同的合并策略使用不同的 I/O 技术，对于 Shuffle 写入时间，我们在合并计算 I/O
   //时经常发现在 ExternalSorter 外部排序时，not bypassing merge-sort 是一致的
22.          if (fastMergeEnabled && fastMergeIsSupported) {
23.            //压缩被禁用，或者我们使用支持的 I/O 压缩编解码器。解压缩级联的压缩流，这
               //样我们就可以执行快速溢出合并，不需要体现溢出的字节
24.            if (transferToEnabled && !encryptionEnabled) {
25.              Logger.debug("Using transferTo-based fast merge");
26.              PartitionLengths = mergeSpillsWithTransferTo(spills, outputFile);
27.            } else {
28.              Logger.debug("Using fileStream-based fast merge");
29.              PartitionLengths = mergeSpillsWithFileStream(spills, outputFile,
                 null);
30.            }
31.          } else {
32.            Logger.debug("Using slow merge");
33.           PartitionLengths = mergeSpillsWithFileStream(spills, outputFile,
              compressionCodec);
34.          }
35.          //当关闭一个 UnsafeShuffleExternalSorter 不安全的外部排序器时，其已溢出
             //一次，但数据也在内存记录中，我们将内存中的记录写入一个文件，但不计算它，最后
             //写入字节溢出（被视为 Shuffle write 写入）。合并的需求被认为是 Shuffle 写入，
             //但这将导致最终 SpillInfo 的字节进行双倍计数
36.          writeMetrics.decBytesWritten(spills[spills.length - 1].file.length());
37.          writeMetrics.incBytesWritten(outputFile.length());
38.          return PartitionLengths;
39.        }
40.      } catch (IOException e) {
41.        if (outputFile.exists() && !outputFile.delete()) {
42.          Logger.error("Unable to delete output file {}", outputFile.getPath());
43.        }
44.        throw e;
45.      }
46.    }
```

29.5　Spark.io.compression.codec 参数调优原理及实践

Spark.io.compression.codec 参数用来压缩内部数据，如 RDD 分区、广播变量和 Shuffle 输出的数据等，所采用的压缩器有 lz4、Lzf 和 Snappy 这 3 种选择，默认是 Snappy，但和 Snappy 比较，Lzf 的压缩率较高，故在有大量 Shuffle 的情况下，使用 Lzf 可以提高 Shuffle 性能，进而提高程序的整体效率。

Spark 2.2 版本中，默认的压缩方式是 lz4。

Spark 2.2.1 版本的 CompressionCodec.scala 的源码如下：

```
1.    private val configKey = "spark.io.compression.codec"
2.  ......
3.   private val shortCompressionCodecNames = Map(
4.     "lz4" -> classOf[LZ4CompressionCodec].getName,
5.     "lzf" -> classOf[LZFCompressionCodec].getName,
6.     "snappy" -> classOf[SnappyCompressionCodec].getName)
7.  ......
```

```
8.    def getCodecName(conf: SparkConf): String = {
9.      conf.get(configKey, DEFAULT_COMPRESSION_CODEC)
10.   }
11. ……
12.   val FALLBACK_COMPRESSION_CODEC = "snappy"
13.   val DEFAULT_COMPRESSION_CODEC = "lz4"
14.   val ALL_COMPRESSION_CODECS = shortCompressionCodecNames.values.toSeq
```

Spark 2.4.3 版本的 CompressionCodec.scala 源码与 Spark 2.2.1 版本相比具有如下特点。

- 上段代码中第 6 行之后新增一行代码，ZStandard 实现 org.apache.spark.io.CompressionCodec，更多详情请参见 http://facebook.github.io/zstd/。注意，此编解码器的协议不能保证跨 Spark 版本兼容，这是为了在单个 Spark 应用程序中用作内部压缩工具。

```
1.  ……
2.  "zstd" -> classOf[ZStdCompressionCodec].getName)
```

29.6　Spark.Shuffle.consolidateFiles 参数调优原理及实践

Spark.Shuffle.consolidateFiles 默认值：false。

参数说明：如果使用 HashShuffleManager，该参数有效。如果设置为 true，那么就会开启 consolidate 机制，会大幅度合并 Shuffle Write 的输出文件。在 Shuffle Read Task 数量特别多的情况下，这种方法可以极大地减少磁盘 I/O 开销，提升性能。

调优建议：如果的确不需要 SortShuffleManager 的排序机制，那么除了使用 bypass 机制，还可以尝试将 Spark.shuffle.manager 参数手动指定为哈希，使用 HashShuffleManager 同时开启 consolidate 机制。在实践中尝试过，发现其性能比开启了 bypass 机制的 SortShuffleManager 要高出 10%~30%。

这个配置参数仅适用于 HashShuffleMananger 的实现，同样是为了解决生成过多文件的问题，采用的方式是在不同批次运行的 Map 任务之间重用 Shuffle 输出文件。也就是说，合并的是不同批次的 Map 任务的输出数据，但是每个 Map 任务需要的文件还是取决于 Reduce 分区的数量。因此，它并不减少同时打开的输出文件的数量，对内存使用量的减少并没有帮助。只是 HashShuffleManager 里的一个折中的解决方案。

说明：在 Spark 2.0 版本中已没有 HashShuffleManager 方式。

29.7　Spark.Shuffle.file.buffer 参数调优原理及实践

在 ShuffleMapTask 端通常也会增大 Map 任务的写磁盘的缓存，默认情况下是 32KB。

Spark.Shuffle.file.buffer 默认配置是 32KB，为什么默认情况下这么小呢？考虑在最小的硬件情况下都把它部署成功。

Spark.Shuffle.file.buffer 参数用于设置 Shuffle Write Task 的 BufferedOutputStream 的 buffer 缓冲大小。将数据写到磁盘文件之前，先写入 buffer 缓冲中，待缓冲写满之后，才会溢写到

磁盘。如果作业可用的内存资源较为充足，则可以适当增加这个参数的大小（如 64KB），从而减少 Shuffle Write 过程中溢写磁盘文件的次数，也就可以减少磁盘 I/O 次数，进而提升性能。在实践中发现，合理调节该参数，性能会有 1%～5%的提升。

Spark 2.2 版本中，spark.shuffle.file.buffer 的默认配置是 32KB。

Spark 2.2.1 版本的 ShuffleExternalSorter.java 的源码如下：

```
1.    final class ShuffleExternalSorter extends MemoryConsumer {
2.
3.    private static final Logger Logger = LoggerFactory.getLogger(Shuffle
      ExternalSorter.class);
4.
5.    @VisibleForTesting
6.    static final int DISK_WRITE_BUFFER_SIZE = 1024 * 1024;
7.
8.    private final int numPartitions;
9.    private final TaskMemoryManager taskMemoryManager;
10.   private final BlockManager blockManager;
11.   private final TaskContext taskContext;
12.   private final ShuffleWriteMetrics writeMetrics;
13.   /*当内存中有许多元素，强迫数据排序时溢出到磁盘，默认值大小是
       *1GB(1024×1024×1024)，指针数组最大值是 8GB
14.    */
15.   private final long numElementsForSpillThreshold;
16.
17.   /**缓冲区大小，溢出时使用 DiskBlockObjectWriter 写入*/
18.   private final int fileBufferSizeBytes;
19.
20.   /**
        *保存正在排序的记录的内存页。页列表中的页溢出时将释放内存，虽然原则上我们可以回
        *收这些页面溢出（另一方面，保持 TaskMemoryManager 本身可重用的页面池，可能不是
        *必要的）
21.     */
22.   private final LinkedList<MemoryBlock> allocatedPages = new LinkedList<>();
23.
24.   private final LinkedList<SpillInfo> spills = new LinkedList<>();
25.
26.   /** 排序器使用的峰值内存单位为字节*/
27.   private long peakMemoryUsedBytes;
28.
29.   //这些变量在溢出后重置
30.   @Nullable private ShuffleInMemorySorter inMemSorter;
31.   @Nullable private MemoryBlock currentPage = null;
32.   private long pageCursor = -1;
33.
34.   ShuffleExternalSorter(
35.       TaskMemoryManager memoryManager,
36.       BlockManager blockManager,
37.       TaskContext taskContext,
38.       int initialSize,
39.       int numPartitions,
40.       SparkConf conf,
41.       ShuffleWriteMetrics writeMetrics) {
42.     super(memoryManager,
43.      (int) Math.min(PackedRecordPointer.MAXIMUM_PAGE_SIZE_BYTES, memoryManager.
         pageSizeBytes()),
44.       memoryManager.getTungstenMemoryMode());
```

```
45.     this.taskMemoryManager = memoryManager;
46.     this.blockManager = blockManager;
47.     this.taskContext = taskContext;
48.     this.numPartitions = numPartitions;
49.     //使用 getSizeAsKb（不是字节），如果没有提供单位，则保持向后兼容
50.     this.fileBufferSizeBytes = (int) conf.getSizeAsKb("spark.shuffle
        .file.buffer", "32k") * 1024;
51.     this.numElementsForSpillThreshold =
52.       conf.getLong("spark.shuffle.spill.numElementsForceSpillThreshold",
          1024 * 1024 * 1024);
53.     this.writeMetrics = writeMetrics;
54.     this.inMemSorter = new ShuffleInMemorySorter(
55.       this, initialSize, conf.getBoolean("spark.shuffle.sort.useRadixSort",
          true));
56.     this.peakMemoryUsedBytes = getMemoryUsage();
57.   }
```

Spark 2.4.3 版本的 ShuffleExternalSorter.java 源码与 Spark 2.2.1 版本相比具有如下特点。

- 将上段代码中第 15 行 numElementsForSpillThreshold 类型调整为整型。
- 上段代码中第 18 行之后新增定义 diskWriteBufferSize 变量。
- 上段代码中第 50 行调整为获取 SHUFFLE_FILE_BUFFER_SIZE 的大小，spark.shuffle.file.buffer 是每个 shuffle 文件输出流的内存缓冲区的大小，以 KiB 为单位，这些缓冲区减少了创建中间 shuffle 文件时进行的磁盘查找和系统调用的数量。
- 上段代码中第 51 行调整为获取 SHUFFLE_SPILL_NUM_ELEMENTS_FORCE_SPILL_THRESHOLD，spark.shuffle.spill.numElementsForceSpillThreshold 是强制 shuffle 排序溢出前内存中元素的最大数目。默认情况下，它是 Integer.MAX_VALUE，这意味着在达到某些限制（如排序器中指针数组的最大页面大小限制）之前，将不会强制排序器溢出。
- 上段代码中第 56 行之后新增加 diskWriteBufferSize 代码，diskWriteBufferSize 是将已排序的记录写入磁盘文件时要使用的缓冲区大小。

```
1.  ......
2.  private final int numElementsForSpillThreshold;
3.  ......
4.  private final int diskWriteBufferSize;
5.  ......
6.  this.fileBufferSizeBytes =
7.          (int) (long) conf.get(package$.MODULE$.SHUFFLE_FILE_BUFFER_SIZE()) * 1024;
8.      this.numElementsForSpillThreshold =
9.          (int) conf.get(package$.MODULE$.SHUFFLE_SPILL_NUM_ELEMENTS_
            FORCE_SPILL_THRESHOLD());
10. ......
11.    this.diskWriteBufferSize =
12.         (int) (long) conf.get(package$.MODULE$.SHUFFLE_DISK_WRITE_BUFFER_SIZE());
13. ......
14.
15. //package.scala
16.   private[spark] val SHUFFLE_FILE_BUFFER_SIZE =
17.     ConfigBuilder("spark.shuffle.file.buffer")
18.       .doc("Size of the in-memory buffer for each shuffle file output stream,
          in KiB unless " +
19.         "otherwise specified. These buffers reduce the number of disk seeks
            and system calls " +
20.         "made in creating intermediate shuffle files.")
```

```
21.       .bytesConf(ByteUnit.KiB)
22.       .checkValue(v => v > 0 && v <= ByteArrayMethods.MAX_ROUNDED_ARRAY_LENGTH / 1024,
23.         s"The file buffer size must be positive and less than or equal to" +
24.           s" ${ByteArrayMethods.MAX_ROUNDED_ARRAY_LENGTH / 1024}.")
25.       .createWithDefaultString("32k")
26. ......
27.   private[spark] val SHUFFLE_SPILL_NUM_ELEMENTS_FORCE_SPILL_THRESHOLD =
28.
ConfigBuilder("spark.shuffle.spill.numElementsForceSpillThreshold")
29.       .internal()
30.       .doc("The maximum number of elements in memory before forcing the
        shuffle sorter to spill. " +
31.         "By default it's Integer.MAX_VALUE, which means we never force the
          sorter to spill, " +
32.         "until we reach some limitations, like the max page size limitation
          for the pointer " +
33.         "array in the sorter.")
34.       .intConf
35.       .createWithDefault(Integer.MAX_VALUE)
36. ......
37.   private[spark] val SHUFFLE_DISK_WRITE_BUFFER_SIZE =
38.     ConfigBuilder("spark.shuffle.spill.diskWriteBufferSize")
39.       .doc("The buffer size, in bytes, to use when writing the sorted records
        to an on-disk file.")
40.       .bytesConf(ByteUnit.BYTE)
41.       .checkValue(v => v > 0 && v <= Int.MaxValue,
42.         s"The buffer size must be greater than 0 and less than ${Int.MaxValue}.")
43.       .createWithDefault(1024 * 1024)
```

29.8　Spark.Shuffle.io.maxRetries 参数调优原理及实践

Spark.Shuffle.io.maxRetries 默认值：3。

参数说明：Shuffle Read Task 从 Shuffle Write Task 所在节点拉取属于自己的数据时，如果因为网络异常导致拉取失败，是会自动进行重试的。该参数就代表了可以重试的最大次数。如果在指定次数内拉取还是没有成功，就可能导致作业执行失败。

调优建议：对于那些包含了特别耗时的 Shuffle 操作的作业，建议增加重试最大次数（如 60 次），以避免由于 JVM 的 Full GC 或者网络不稳定等因素导致的数据拉取失败。在实践中发现，对于针对超大数据量（数十亿至上百亿）的 Shuffle 过程，调节该参数可以大幅度提升稳定性。

在 Spark 2.2 版本中，io.maxRetries 默认重试次数是 3 次。调整获取 Shuffle 数据的重试次数，通常建议增大重试次数到 8～10 次。

TransportConf.java 的源码如下：

```
1.  public TransportConf(String module, ConfigProvider conf) {
2.  ......
3.     SPARK_NETWORK_IO_MAXRETRIES_KEY = getConfKey("io.maxRetries");
4.     SPARK_NETWORK_IO_RETRYWAIT_KEY = getConfKey("io.retryWait");
5.  ......
6.  public int maxIORetries() { return conf.getInt(SPARK_NETWORK_IO_MAXRETRIES_
    KEY, 3); }
7.  ......
```

29.9　Spark.Shuffle.io.retryWait 参数调优原理及实践

Spark.Shuffle.io.retryWait 默认值：5s。

参数说明：代表了每次重试拉取数据的等待间隔，默认是 5s，在 Spark-conf 配置文件中配置。

调优建议：默认情况下，重试 3 次，每次重试的间隔时间为 5s，重试引起的最大延迟为 15s，以 maxRetries×retryWait 计算。建议加大间隔时长（如 60s），以增加 Shuffle 操作的稳定性。

TransportConf.java 的源码如下：

```
1.  public TransportConf(String module, ConfigProvider conf) {
2.  ......
3.     SPARK_NETWORK_IO_MAXRETRIES_KEY = getConfKey("io.maxRetries");
4.     SPARK_NETWORK_IO_RETRYWAIT_KEY = getConfKey("io.retryWait");
5.  ......
6.  public int ioRetryWaitTimeMs() {
7.      return  (int)  JavaUtils.timeStringAsSec(conf.get(SPARK_NETWORK_IO_
      RETRYWAIT_KEY, "5s")) * 1000;
8.    }
9.
```

29.10　Spark.Shuffle.io.numConnectionsPerPeer 参数调优原理及实践

Spark.Shuffle.io.numConnectionsPerPeer（仅 Netty 使用）重新使用主机之间的连接，以减少大型集群的连接建立。对于具有多个硬盘和少量主机的集群，这可能导致并发性不足，以使所有磁盘饱和，因此用户可考虑增加此值，默认是 1 次。

TransportConf.java 中的 numConnectionsPerPeer 用于获取数据的两个节点之间的并发连接数。TransportConf.java 的源码如下：

```
1.  public TransportConf(String module, ConfigProvider conf) {
2.  ......
3.  private final String SPARK_NETWORK_IO_NUMCONNECTIONSPERPEER_KEY;
4.  ......
5.   SPARK_NETWORK_IO_NUMCONNECTIONSPERPEER_KEY =  getConfKey
    ("io.numConnectionsPerPeer");
6.  ......
7.    public int numConnectionsPerPeer() {
8.      return conf.getInt(SPARK_NETWORK_IO_NUMCONNECTIONSPERPEER_KEY, 1);
9.    }
10. ......
```

29.11　Spark.reducer.maxSizeInFlight 参数调优原理及实践

Spark.reducer.maxSizeInFlight 默认值：48m。

参数说明：该参数用于设置 Shuffle Read Task 的 Buffer 大小，而这个 Buffer 决定了每次能够拉取多少数据。

调优建议：如果作业可用的内存资源较为充足，可以适当增加这个参数的大小（如 96m），从而减少拉取数据的次数，也就可以减少网络传输的次数，进而提升性能。在实践中发现，合理调节该参数，性能会有 1%～5%的提升。

BlockStoreShuffleReader.scala 的源码如下：

```
1.  private[spark] class BlockStoreShuffleReader[K, C](
2.  ……
3.   SparkEnv.get.conf.getSizeAsMb("spark.reducer.maxSizeInFlight", "48m")
     * 1024 * 1024,
4.    SparkEnv.get.conf.getInt("spark.reducer.maxReqsInFlight",
      Int.MaxValue))
5.  ……
```

29.12　Spark.Shuffle.io.preferDirectBufs 参数调优原理及实践

Spark.Shuffle.io.preferDirectBufs 参数仅 Netty 使用：堆外缓存可以有效减少垃圾回收和缓存复制。对于堆外内存紧张的用户来说，可以考虑禁用这个选项，以迫使 Netty 所有内存都分配在堆上，默认是 true。

```
1.    public TransportConf(String module, ConfigProvider conf) {
2.  ……
3.    private final String SPARK_NETWORK_IO_PREFERDIRECTBUFS_KEY;
4.  ……
5.    SPARK_NETWORK_IO_PREFERDIRECTBUFS_KEY = getConfKey("io.preferDirectBufs");
6.  ……
7.  public boolean preferDirectBufs() {
8.      return conf.getBoolean(SPARK_NETWORK_IO_PREFERDIRECTBUFS_KEY, true);
9.    }
```

29.13　Spark.Shuffle.memoryFraction 参数调优原理及实践

Spark.Shuffle.memoryFraction 参数的默认值为 20%。

参数说明：该参数代表了 Executor 内存中，分配给 Shuffle Read Task 进行聚合操作的内存比例，默认是 20%。

调优建议：如果内存充足，而且很少使用持久化操作，建议调高这个比例，给 Shuffle Read 的聚合操作更多内存，以避免由于内存不足导致聚合过程中频繁读写磁盘。在实践中发现，合理调节该参数可以将性能提升 10%左右。将存储 Mapper 端的输出结果存储在 JVM 的堆空间中，这个空间的大小取决于 Spark.Shuffle.memoryFraction 和 Spark.Shuffle.safetyFraction 这两个参数。

StaticMemoryManager.scala 的源码如下：

```
1.   private def getMaxExecutionMemory(conf: SparkConf): Long = {
```

```
2.      val systemMaxMemory = conf.getLong("spark.testing.memory", Runtime
        .getRuntime.maxMemory)
3.
4.      if (systemMaxMemory < MIN_MEMORY_BYTES) {
5.        throw new IllegalArgumentException(s"System memory $systemMaxMemory
          must " +
6.          s"be at least $MIN_MEMORY_BYTES. Please increase heap size using
            the --driver-memory " +
7.          s"option or spark.driver.memory in Spark configuration.")
8.      }
9.      if (conf.contains("spark.executor.memory")) {
10.       val executorMemory = conf.getSizeAsBytes("spark.executor.memory")
11.       if (executorMemory < MIN_MEMORY_BYTES) {
12.         throw new IllegalArgumentException(s"Executor memory $executor
            Memory must be at least " +
13.           s"$MIN_MEMORY_BYTES. Please increase executor memory using the " +
14.           s"--executor-memory option or spark.executor.memory in Spark
              configuration.")
15.       }
16.     }
17.     val memoryFraction = conf.getDouble("spark.shuffle.memoryFraction", 0.2)
18.     val safetyFraction = conf.getDouble("spark.shuffle.safetyFraction", 0.8)
19.     (systemMaxMemory * memoryFraction * safetyFraction).toLong
20.   }
```

默认的计算公式是 Spark.Shuffle.memoryFraction (0.2)×Spark.Shuffle.safetyFraction (0.8) = 0.16。也就是说，是 JVM HeapSize 的 16%。通过 Spark.Shuffle.memoryFraction 可以调整 Spill 的触发条件，即 Shuffle 占用内存的大小，进而调整 Spill 的频率和 GC 的行为。总的来说，如果 Spill 太过频繁，可以适当增加 Spark.Shuffle.memoryFraction 的大小，增加用于 Shuffle 的内存，减少 Spill 的次数。

29.14　Spark.Shuffle.service.enabled 参数调优原理及实践

Spark.Shuffle.service.enabled 必须配置为 true，默认为 false。如果这个配置设置为 true，BlockManager 实例生成时，需要读取 Spark.Shuffle.service.port 配置的 Shuffle 端口，同时对应 BlockManager 的 ShuffleClient 不再是默认的 BlockTransferService 实例，而是 ExternalShuffleClient 实例。

BlockManager.scala 中客户端读取其他 Executor 上的 Shuffle 文件有两个方式：一种方式是在 spark.shuffle.service.enabled 设置为 true 时，创建 shuffleClient 为 ExternalShuffleClient；另一种方式是在 spark.shuffle.service.enabled 设置为 false 时，创建 shuffleClient 为 BlockTransferService，直接读取其他 Executors 的数据。

Spark 2.2.1 版本的 BlockManager.scala 的源码如下：

```
1.     private[spark] class BlockManager(
2.      executorId: String,
3.   ......
4.    private[spark] val externalShuffleServiceEnabled =
5.      conf.getBoolean("spark.shuffle.service.enabled", false)
6.   ......
```

```
7.   private[spark] val shuffleClient = if (externalShuffleServiceEnabled) {
8.      val  transConf  =  SparkTransportConf.fromSparkConf(conf,  "shuffle",
        numUsableCores)
9.      new ExternalShuffleClient(transConf, securityManager, securityManager
        .isAuthenticationEnabled(),
10.       securityManager.isSaslEncryptionEnabled())
11.   } else {
12.     blockTransferService
13.   }
```

Spark 2.4.3 版本的 BlockManager.scala 源码与 Spark 2.2.1 版本相比具有如下特点。

- 上段代码中第 10 行 ExternalShuffleClient 新增一个参数 SHUFFLE_REGISTRATION_TIMEOUT，表示注册到外部 shuffle 服务的超时时间（毫秒）。

```
1.  ......
2.      new ExternalShuffleClient(transConf, securityManager,
3.        securityManager.isAuthenticationEnabled(),
          conf.get(config.SHUFFLE_REGISTRATION_TIMEOUT))
4.  ......
5.  //package.scala
6.    private[spark] val SHUFFLE_REGISTRATION_TIMEOUT =
7.      ConfigBuilder("spark.shuffle.registration.timeout")
8.        .doc("Timeout in milliseconds for registration to the external
          shuffle service.")
9.        .timeConf(TimeUnit.MILLISECONDS)
10.       .createWithDefault(5000)
```

启用外部 Shuffle Service，Shuffle Service 保留由 Executor 写入的 Shuffle 文件，以便 Executors 可以安全地删除。必须首先把 Spark.dynamicAllocation.enabled 设置为 true，才可以启动这个外部 Shuffle Service。NodeManager 中一个长期运行的辅助服务，用于提升 Shuffle 计算性能。Shuffle Service 默认为 false，表示不启用该功能。

Spark 2.2.1 版本的 ExternalShuffleService.scala 的源码如下：

```
1.    private[deploy]
2.  class  ExternalShuffleService(sparkConf:  SparkConf,  securityManager:
    SecurityManager)
3.    extends Logging {
4.    ......
5.    private val enabled = sparkConf.getBoolean("spark.shuffle.service
      .enabled", false)
6.  ......
7.    def startIfEnabled() {
8.      if (enabled) {
9.        start()
10.     }
11.   }
12. ......
```

Spark 2.4.3 版本的 ExternalShuffleService.scala 源码与 Spark 2.2.1 版本相比具有如下特点。

- 上段代码中第 5 行调整为获取 SHUFFLE_SERVICE_ENABLED 配置。

```
1.  ......
2.  private val enabled = sparkConf.get(config.SHUFFLE_SERVICE_ENABLED)
3.  ......
4.  //package.scala
5.    private[spark] val SHUFFLE_SERVICE_ENABLED =
```

```
    ConfigBuilder("spark.shuffle.service.enabled").booleanConf
    .createWithDefault(false)
```

Spark 系统在运行包含 Shuffle 过程的应用时，Executor 进程除了运行 Task，还要负责写 Shuffle 数据，给其他 Executor 提供 Shuffle 数据。当 Executor 进程任务过重，导致 GC 不能为其他 Executor 提供 Shuffle 数据时，会影响任务运行。

External Shuffle Service 是长期存在于 NodeManager 进程中的一个辅助服务。通过该服务抓取 Shuffle 数据，减少了 Executor 的压力，在 Executor GC 的时候也不会影响其他 Executor 的任务运行。

在 YARN-site.xml 中添加如下配置项。

```
1.   <property>
2.  <name>YARN.nodemanager.aux-services</name>
3.  <value>Spark_Shuffle</value>
4.  </property>
5.  <property>
6.  <name>YARN.nodemanager.aux-services.Spark_Shuffle.class</name>
7.  <value>org.apache.Spark.network.YARN.YARNShuffleService</value>
8.  </property>
9.  <property>
10. <name>Spark.Shuffle.service.port</name>
11. <value>7337</value>
12. </property>
```

29.15　Spark.Shuffle.service.port 参数调优原理及实践

Spark.Shuffle.service.port 是 Shuffle 服务监听数据获取请求的端口，可选配置，默认值为 7337。

Spark 2.2.1 版本的 BlockManager.scala 的源码如下：

```
1.     private val externalShuffleServicePort = {
2.      val tmpPort = Utils.getSparkOrYarnConfig(conf, "spark.shuffle.service
       .port", "7337").toInt
3.      if (tmpPort == 0) {
4.        //为了测试，我们在 Yarn 配置中设置 spark.shuffle.service.port 为 0，这样，
          //Yarn 就可找到开放的端口。但我们仍然需要告诉 Spark 应用程序使用的正确端口，
          //所以只有当 Yarn 配置将端口设置为 0 时，我们优先使用 config 配置中的值
5.        conf.get("spark.shuffle.service.port").toInt
6.      } else {
7.        tmpPort
8.      }
9.    }
```

Spark 2.4.3 版本的 BlockManager.scala 源码与 Spark 2.2.1 版本相比具有如下特点。

- 上段代码中第 2 行调整为获取 SHUFFLE_SERVICE_PORT.key、SHUFFLE_SERVICE_PORT.defaultValueString 配置。
- 上段代码中第 5 行调整为获取 SHUFFLE_SERVICE_PORT.key 配置。

```
1.  ......
2.     val tmpPort = Utils.getSparkOrYarnConfig(conf, config
       .SHUFFLE_SERVICE_PORT.key,
```

```
3.        config.SHUFFLE_SERVICE_PORT.defaultValueString).toInt
4.   ......
5.   conf.get(config.SHUFFLE_SERVICE_PORT.key).toInt
6.   ......
7.   // package.scala
8.   private[spark] val SHUFFLE_SERVICE_PORT = ConfigBuilder("spark.
     shuffle.service.port").intConf.createWithDefault(7337)
9.   ......
```

在 Spark-defaults.conf 中必须添加如下配置项。

```
1.   Spark.Shuffle.service.enabled   true
2.   Spark.Shuffle.service.port      7337
```

29.16　Spark.Shuffle.Sort.bypassMergeThreshold 参数调优原理及实践

Spark.Shuffle.Sort.bypassMergeThreshold 参数的默认值：200。

参数说明：当 ShuffleManager 为 SortShuffleManager 时，如果 Shuffle Read Task 的数量小于这个阈值（默认是 200），则 Shuffle Write 过程中不会进行排序操作，而是直接按照未经优化的 HashShuffleManager 方式去写数据，但是最后会将每个 Task 产生的所有临时磁盘文件都合并成一个文件，并会创建单独的索引文件。

调优建议：当使用 SortShuffleManager 时，如果的确不需要排序操作，那么建议将这个参数调大一些，大于 Shuffle Read Task 的数量。那么，此时就会自动启用 bypass 机制，map-side 就不会进行排序了，减少了排序的性能开销。但是，这种方式下，依然会产生大量的磁盘文件，因此 Shuffle Write 性能有待提高。

这个参数仅适用于 SortShuffleManager。SortShuffleManager 在处理不需要排序的 Shuffle 操作时，由于排序使性能下降。这个参数决定了在这种情况下，当 Reduce 分区的数量小于多少的时候，在 SortShuffleManager 内部不使用 Merge Sort 的方式处理数据，而是与 Hash Shuffle 类似，直接将分区文件写入单独的文件。不同的是，在最后一步还是会将这些文件合并成一个单独的文件。这样，通过去除 Sort 步骤来加快处理速度，代价是需要并发打开多个文件，所以内存消耗量增加，本质上是相对 HashShuffleMananger 的一个折中方案。这个配置的默认值是 200，用于设置在 Reducer 的 Partition 数目少于多少的时候，Sort Based Shuffle 内部不使用 Merge Sort 方式处理数据，而是直接将每个 Partition 写入单独的文件。这个方式和 Hash Based 方式类似，区别就是最后这些文件还是会合并成一个单独的文件，并通过一个 index 索引文件来标记不同 Partition 的位置信息。从 Reducer 看来，数据文件和索引文件的格式和内部是否做过 Merge Sort 是完全相同的。

这个可以看作 SortBased Shuffle 在 Shuffle 量比较小的时候对于 Hash Based Shuffle 的一种折中。当然，它和 Hash Based Shuffle 一样，也存在同时打开文件过多导致内存占用增加的问题。因此，如果 GC 比较严重或者内存比较紧张，可以适当地降低这个值。

SortShuffleWriter.scala 中的 shouldBypassMergeSort 方法中，如果分区个数小于 spark.shuffle.sort.bypassMergeThreshold（200），就返回 true，不需要进行排序。

Spark 2.2.1 版本的 SortShuffleWriter.scala 的 shouldBypassMergeSort 的源码如下。

```
1.   private[spark] object SortShuffleWriter {
2.    def shouldBypassMergeSort(conf: SparkConf, dep: ShuffleDependency[_,
      _, _]): Boolean = {
3.      //如果需要执行 map 端的聚合，就不能绕过排序
4.      if (dep.mapSideCombine) {
5.        require(dep.aggregator.isDefined, "Map-side combine without Aggregator
          specified!")
6.        false
7.      } else {
8.        val bypassMergeThreshold: Int = conf.getInt("spark.shuffle.sort
          .bypassMergeThreshold", 200)
9.        dep.Partitioner.numPartitions <= bypassMergeThreshold
10.     }
11.   }
12. }
```

Spark 2.4.3 版本的 SortShuffleWriter.scala 源码与 Spark 2.2.1 版本相比具有如下特点。

- 上段代码中删掉第 5 行代码。

29.17　Spark.Shuffle.spill 参数调优原理及实践

Shuffle 的过程中，如果涉及排序、聚合等操作，势必会需要在内存中维护一些数据结构，进而占用额外的内存。如果内存不够用，那只有两条路可以走：一就是 Out Of Memory 出错了；二就是将部分数据临时写到外部存储设备中，最后再合并到最终的 Shuffle 输出文件中。

这里，Spark.Shuffle.Spill 决定是否 Spill 到外部存储设备（默认打开），如果你的内存足够，或者数据集足够小，当然也就不需要 Spill，毕竟 Spill 带来了额外的磁盘操作。默认情况下，这个参数是 true，在 Shuffle 期间通过溢出数据到磁盘降低了内存使用总量，溢出阈值是由 Spark.Shuffle.memoryFraction 指定的。

Hash BasedShuffle 的 Shuffle Write 过程中使用的 org.apache.Spark.util.collection.AppendOnlyMap 就是全内存的方式，而 org.apache.Spark.util.collection.ExternalAppendOnlyMap 对 org.apache.Spark.util.collection.AppendOnlyMap 有了进一步的封装，在内存使用超过阈值时，会将它 Spill 到外部存储，最后会对这些临时文件进行 Merge。而 Sort BasedShuffle Write 使用到的 org.apache.Spark.util.collection.ExternalSorter 也会有类似的 Spill。

而对于 ShuffleRead，如果需要做 Aggregate，也可能在 Aggregate 的过程中将数据 Spill 到外部存储。

SortShuffleManager.scala 的源码如下：

```
1.   private[spark] class SortShuffleManager(conf: SparkConf) extends
     ShuffleManager with Logging {
2.
3.    if (!conf.getBoolean("spark.shuffle.spill", true)) {
4.      LogWarning(
5.        "spark.shuffle.spill was set to false, but this configuration is
          ignored as of Spark 1.6+." +
6.          " Shuffle will continue to spill to disk when necessary.")
7.    }
```

29.18　Spark.Shuffle.spill.compress 参数调优原理及实践

理论上，Spark.Shuffle.compress 设置为 true 通常都是合理的，因为如果使用千兆以下的网卡，网络带宽往往最容易成为瓶颈。此外，目前的 Spark 任务调度实现中，以 Shuffle 划分 Stage，下一个 Stage 的任务要等待上一个 Stage 的任务全部完成后，才能开始执行，所以 Shuffle 数据的传输和 CPU 计算任务之间通常不会重叠，这样，Shuffle 数据传输量的大小和所需的时间就直接影响到整个任务的完成速度。但是，压缩也是要消耗大量 CPU 资源的，所以打开压缩选项会增加 Map 任务的执行时间。因此，如果在 CPU 负载的影响远大于磁盘和网络带宽的影响的场合下，也可能将 Spark.Shuffle.compress 设置为 false 才是最佳的方案。

对于 Spark.Shuffle.spill.compress 而言，情况类似，但是 Spill 数据不会被发送到网络中，仅仅是临时写入本地磁盘，而且在一个任务中同时需要执行压缩和解压缩两个步骤，所以对 CPU 负载的影响会更大一些，而磁盘带宽（如果标配 12HDD 的话）可能往往不会成为 Spark 应用的主要问题，所以这个参数相对而言，或许更需要设置为 false。

总之，在 Shuffle 过程中数据是否应该压缩，取决于 CPU、磁盘、网络的实际能力和负载，应该综合考虑。

SerializerManager.scala 的源码如下：

```
1.  private[spark] class SerializerManager(
2.     defaultSerializer: Serializer,
3.     conf: SparkConf,
4.     encryptionKey: Option[Array[Byte]]) {
5. ……
6.
7. //是否压缩存储的广播变量
8.  private[this] val compressBroadcast = conf.getBoolean("spark.broadcast
    .compress", true)
9.   //是否压缩存储的 Shuffle 输出
10.  private[this] val compressShuffle = conf.getBoolean("spark.shuffle
     .compress", true)
11.  //是否压缩 RDD 分区存储序列化
12.  private[this] val compressRdds = conf.getBoolean("spark.rdd.compress",
     false)
13.  //是否压缩 Shuffle 临时输出溢出到磁盘
14.  private[this] val compressShuffleSpill = conf.getBoolean("spark.shuffle
     .spill.compress", true)
```

第 30 章　Spark 性能调优之数据倾斜调优一站式解决方案原理与实战

本章深入讲解 Spark 性能调优之数据倾斜调优一站式解决方案原理与实战。30.1 节讲解为什么数据倾斜是分布式大数据系统的性能噩梦；30.2 节讲解数据倾斜解决方案之一：对源数据进行聚合并过滤掉导致倾斜的 Keys；30.3 节讲解数据倾斜解决方案之二：适当提高 Reducer 端的并行度；30.4 节讲解数据倾斜解决方案之三：使用随机 Key 实现双重聚合；30.5 节讲解数据倾斜解决方案之四：使用 Mapper 端进行 Join 操作；30.6 节讲解数据倾斜解决方案之五：对倾斜的 Keys 采样后进行单独的 Join 操作；30.7 节讲解数据倾斜解决方案之六：使用随机数进行 Join；30.8 节讲解数据倾斜解决方案之七：通过扩容进行 Join；30.9 节结合电影点评系统进行数据倾斜解决方案的小结。

30.1　为什么数据倾斜是分布式大数据系统的性能噩梦

本节讲解数据倾斜为何是分布式大数据系统的性能噩梦：讲解什么是数据倾斜，数据倾斜对性能的巨大影响，如何判断 Spark 程序运行中出现了数据倾斜以及如何定位数据倾斜等内容。

30.1.1　什么是数据倾斜

何谓数据倾斜？数据倾斜是指并行处理数据集的某一部分（如 Spark 或 Kafka 的一个 Partition）的数据显著多于其他部分，从而使得该部分的处理速度成为整个数据集处理的瓶颈。数据倾斜的基本特征：个别任务处理大量数据，符合二八定律，约 20%的任务处理 80%的数据，数据中基本上都存在业务热点问题，这是现实问题，如图 30-1 所示。

举一个例子：在 Spark 中，同一个 Stage 不同的 Partition 可以并行处理，而具体依赖关系在不同 Stage 之间是串行处理的。假设某个 Spark Job 分为 Stage 0 和 Stage 1 两个 Stage，且 Stage 1 依赖于 Stage 0，那 Stage 0 完全处理结束之前不会处理 Stage 1。而 Stage 0 可能包含 *N* 个 Task，这 *N* 个 Task 可以并行进行。如果其中 *N*–1 个 Task 都在 10s 内完成，而另外一个 Task 却耗时 1min，那该 Stage 的总时间至少为 1min。换句话说，一个 Stage 耗费的时间主要由最慢的那个 Task 决定，由于同一个 Stage 内的所有 Task 执行相同的计算，在排除不同计算节点计算能力差异的前提下，不同 Task 之间耗时的差异主要由该 Task 处理的数据量决定。

导致数据倾斜的原因很简单，数据分配给不同的 Task，一般就是 Shuffle 的过程。在 Shuffle 的过程中，同样一个 Key 一般都会交给一个 Task 去处理，可能有时候运气很不好，同样一个

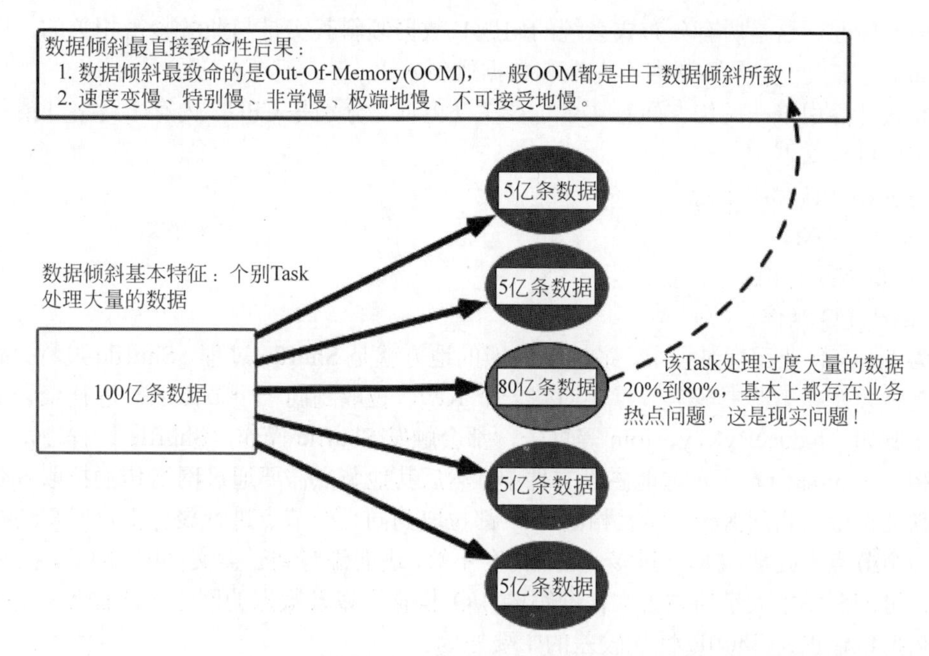

图 30-1　数据倾斜示意图

Key 的 Value 太多了。假设图 30-1 所示的例子有 5 个 Key，分别是 K1、K2、K3、K4、K5；同样一个 Key 会分成一个 Task，现在 K3 中的 Value 特别多，这会导致很多数据集中在 K3 这个 Task 中。

30.1.2　数据倾斜对性能的巨大影响

大数据基本有 3 个特性：第一是数据多样化，有着不同类型的数据，其中包括结构化和非结构化数据；第二是庞大的数据量；第三就是数据的流动性，从批处理到流处理。一般在处理大数据的时候，都会面对这 3 个特性的问题，而 Spark 就是基于内存的分布式计算引擎，以处理高效和稳定著称，是目前处理大数据的一个非常好的选择。然而在实际的应用开发过程中，开发者还是会遇到种种问题，其中一大类问题就是和性能相关的问题。

在分布式系统中，数据分布在不同的节点上，每个节点计算一部分数据，如果不对各个节点上独立的部分进行汇聚，我们就计算不到最终的结果。这就是因为我们需要利用分布式来发挥它本身并行计算的能力，而后续又需要计算各节点上最终的结果，所以需要把数据汇聚集中，这就会导致 Shuffle，而 Shuffle 又会导致数据倾斜。

数据倾斜最致命的就是 Out-Of-Memory（OOM）。一般 OOM 都是由数据倾斜所致！如果应用程序在运行时速度变得非常慢，就有可能出现数据倾斜。它带来的结果是原本程序可以在 10min 内运行完毕，因为数据倾斜的原因，其中有一个任务要处理的数据特别多，这个时候，当其他程序都运行完成时，就因为这个数据量特大的任务还在运行，导致这个程序原本可以用 10min 完成，最后用了 1h。这极大地降低了工作效率！

所有编程高手无论做什么类型的编程，最终思考的都是硬件方面的问题！最终思考的都是在 1s、1ms，甚至 1ns 到底是如何运行的，并且基于此进行算法实现和性能调优，最后都回到了硬件！大数据最怕的就是数据本地性（内存中）和数据倾斜或者叫数据分布不均衡、数据转输的问题，这是所有分布式系统的问题！数据倾斜其实是与业务紧密相关的。所以，调优 Spark 的重点一定是从数据本地性和数据倾斜入手。

分布式计算引擎在调优方面有 4 个主要关注方向，分别是 CPU、内存、网络开销和 I/O，其具体调优目标如下。

- 提高 CPU 利用率。
- 避免 OOM。
- 降低网络开销。
- 减少 I/O 操作。

因为 Spark 作业运行过程中，最消耗性能的地方就是 Shuffle 过程。Shuffle 过程，简单来说，就是将分布在集群中多个节点上的同一个 Key，拉取到同一个节点上，进行聚合或 Join 等操作。例如，reduceByKey、join 等算子，都会触发 Shuffle 操作。Shuffle 过程中，各个节点上的相同 Key 都会先写入本地磁盘文件中，然后其他节点需要通过网络传输拉取各个节点上的磁盘文件中的相同 Key。而且相同 Key 都拉取到同一个节点进行聚合操作时，还有可能会因为一个节点上处理的 Key 过多，导致内存不够，进而溢写到磁盘文件中。因此，在 Shuffle 过程中，可能会发生大量的磁盘文件读写的 I/O 操作，以及数据的网络传输操作。磁盘 I/O 和网络数据传输也是 Shuffle 性能较差的主要原因。

如果非要做 Shuffle，就要注意是否有数据倾斜的情况存在。出现数据倾斜的时候，Spark 作业看起来会运行得非常缓慢，甚至可能因为某个 Task 处理的数据量过大导致内存溢出。

30.1.3　如何判断 Spark 程序运行中出现了数据倾斜

分布式系统常见的一个性能问题是倾斜，即若干个 Task 相对于其他大多数 Task 来说消耗了相当长的时间。我们可以通过查看 Task 的 metrics 来判断是否有倾斜。Task 运行了多久，是否有些 Task 相比其他 Task 需要多得多的运行时间，如果是，就需要进一步分析这些 Task 执行慢的原因；是否有些 Task 相比其他 Task，读或写了多得多的数据；是否某些节点上的 Task 都运行得特别慢，等等。这些都是我们性能诊断时首先要关注的。我们还要关注 Task 在读、计算和写的各个阶段消耗了多少时间，如果 Task 读写数据消耗的时间不多，而整体消耗时间较多，则可能是因为应用程序代码的问题，就需要考虑代码的优化；也可能有些 Task 几乎所有的时间都消耗在从外部存储系统读取数据上了，这时瓶颈在输入的读取上，单纯优化 Spark 可能就没多大帮助了。

Stages 主页面如图 30-2 所示。

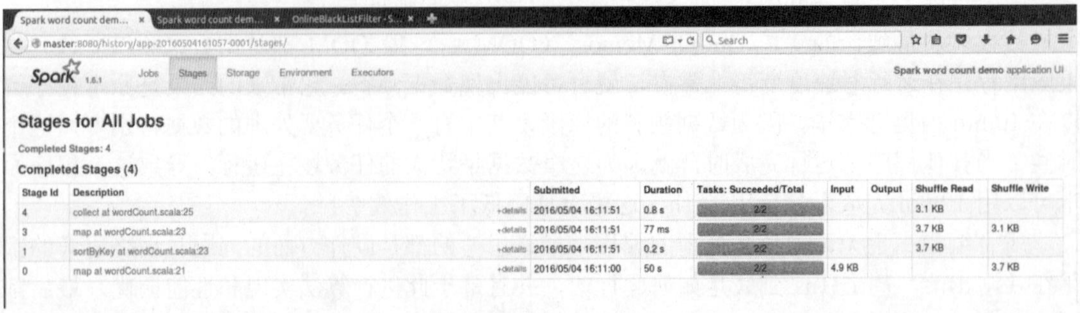

Stage Id	Description		Submitted	Duration	Tasks: Succeeded/Total	Input	Output	Shuffle Read	Shuffle Write
4	collect at wordCount.scala:25	+details	2016/05/04 16:11:51	0.8 s	2/2			3.1 KB	
3	map at wordCount.scala:23	+details	2016/05/04 16:11:51	77 ms	2/2			3.7 KB	3.1 KB
1	sortByKey at wordCount.scala:23	+details	2016/05/04 16:11:51	0.2 s	2/2			3.7 KB	
0	map at wordCount.scala:21	+details	2016/05/04 16:11:00	50 s	2/2	4.9 KB			3.7 KB

图 30-2　Stages 主页面

单击某个 Stage 即可进入到 Stages 的细节页面，如图 30-3 所示。

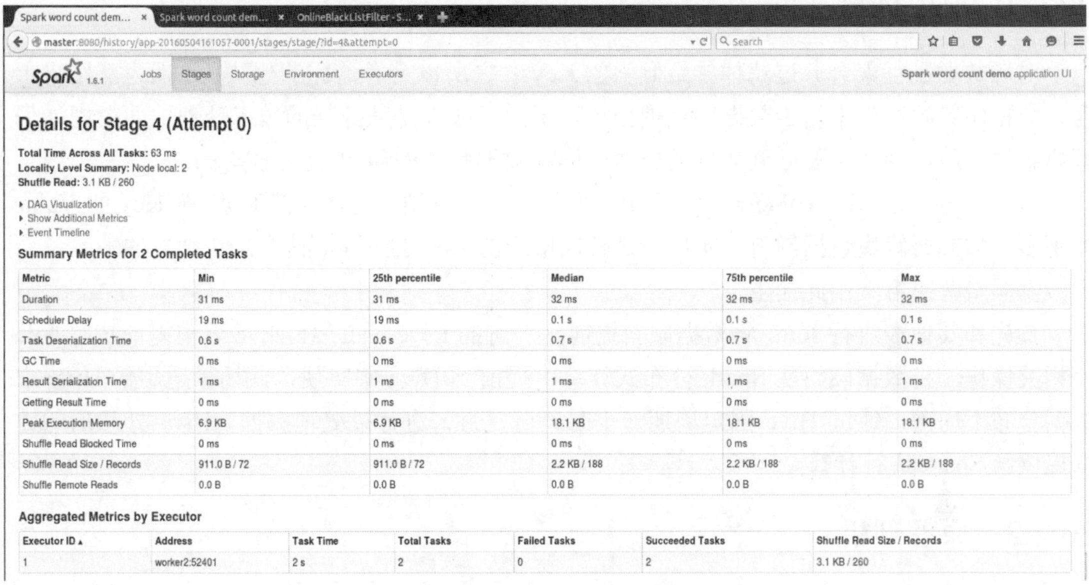

Details for Stage 4 (Attempt 0)

Total Time Across All Tasks: 63 ms
Locality Level Summary: Node local: 2
Shuffle Read: 3.1 KB / 260

- DAG Visualization
- Show Additional Metrics
- Event Timeline

Summary Metrics for 2 Completed Tasks

Metric	Min	25th percentile	Median	75th percentile	Max
Duration	31 ms	31 ms	32 ms	32 ms	32 ms
Scheduler Delay	19 ms	19 ms	0.1 s	0.1 s	0.1 s
Task Deserialization Time	0.6 s	0.6 s	0.7 s	0.7 s	0.7 s
GC Time	0 ms	0 ms	0 ms	0 ms	0 ms
Result Serialization Time	1 ms	1 ms	1 ms	1 ms	1 ms
Getting Result Time	0 ms	0 ms	0 ms	0 ms	0 ms
Peak Execution Memory	6.9 KB	6.9 KB	18.1 KB	18.1 KB	18.1 KB
Shuffle Read Blocked Time	0 ms	0 ms	0 ms	0 ms	0 ms
Shuffle Read Size / Records	911.0 B / 72	911.0 B / 72	2.2 KB / 188	2.2 KB / 188	2.2 KB / 188
Shuffle Remote Reads	0.0 B	0.0 B	0.0 B	0.0 B	0.0 B

Aggregated Metrics by Executor

Executor ID ▴	Address	Task Time	Total Tasks	Failed Tasks	Succeeded Tasks	Shuffle Read Size / Records
1	worker2:52401	2 s	2	0	2	3.1 KB / 260

图 30-3　Stages 的细节页面

30.1.4　如何定位数据倾斜

定位数据倾斜的方法如下。

- ❑ Spark Web UI 页面，可以清晰地看见 Task 运行的数据量大小。
- ❑ Log 日志：Log 的一个好处是可以清晰地显示哪一行出现 OOM 问题，同时可以清晰地看到具体在哪个 Stage 出现数据倾斜（数据倾斜一般是在 Shuffle 过程中产生的），从而定位具体 Shuffle 的代码，也可能发现绝大多数 Task 非常快，但是个别 Task 非常慢。
- ❑ 代码走读，重点看 Join、groupByKey、reduceByKey 等的关键代码。
- ❑ 对数据特征分布进行分析。

30.2　数据倾斜解决方案之一：对源数据进行聚合并过滤掉导致倾斜的 Keys

本节对源数据进行聚合并过滤掉导致倾斜的 Keys 进行解析。首先对源数据进行聚合并过滤掉导致倾斜的 Keys 的适用场景进行分析，然后对源数据清洗倾斜的 Keys 原理进行剖析；通过使用 Hive 等 ETL 工具对源数据进行聚合，使用 Spark SQL 对源数据进行清洗过滤等方式解决数据倾斜的问题。

30.2.1　适用场景分析

对源数据进行聚合并过滤掉导致倾斜的 Keys：之所以会有这样的想法，是因为从结果上看，数据倾斜的产生来自于数据的处理技术，用户一般都是从数据的技术处理层面考虑如何解决数据倾斜，现在需要回到数据的层面去解决数据倾斜的问题。

数据本身就是 Key-Value 的存在方式。所谓的数据倾斜，就是说某（几）个 Key 的 Value 特别多，如果要解决数据倾斜，实质上是解决单一的 Key 的 Value 的个数特别多的情况，新的数据倾斜解决方案由此诞生。

预先和其他表进行 Join，将数据倾斜提前到上游的 Hive ETL；在 Hive 表中提前执行 Join 等相关操作，将数据倾斜从 Spark 分布式计算中提前到 Hive 中实现，此场景适用于 Hive 中数据分布不均衡，通过 Hive 固定周期（小时、日、月）进行批处理数据，Hive 处理后的数据提供给 Spark 进行计算。

30.2.2　原理剖析

对源数据进行聚合并过滤掉导致倾斜的 Keys：在 Spark 分布式计算业务代码中读入数据源数据，如数据源（Hdfs、本地文件、Kafka 等）记录中导致数据倾斜的 Key 只有很少几个，这些 Key 不影响对业务的计算。例如，某些无效的 Key 数据（–1 值、null 值等），汇聚以后对应的 Value 值却很多，可能会导致数据倾斜。我们可以使用 filter 算子将这些倾斜的 Key 值过滤掉，这样，其他 Key 的值都对应均衡的 Value 值，从而不会产生数据倾斜。

RDD.scala 的 filter 的源码如下：

```
1.  /**
     * 返回一个新的 RDD，包含满足谓词的元素
2.   */
3.  def filter(f: T => Boolean): RDD[T] = withScope {
4.    val cleanF = sc.clean(f)
5.    new MapPartitionsRDD[T, T](
6.      this,
7.      (context, pid, iter) => iter.filter(cleanF),
8.      preservesPartitioning = true)
9.  }
```

30.2.3　使用 Hive 等 ETL 工具对源数据进行聚合并过滤掉导致倾斜的 Keys

Hive 是底层封装了 Hadoop 的数据仓库处理工具，使用 HiveQL 语言实现数据查询，所有 Hive 的数据都存储在 Hadoop 兼容的文件系统（Amazon S3、HDFS）中。Hive 基于 Hadoop 系统进行 SQL 查询，由于 Hadoop 延迟较高、作业提交及调度性能开销较大，因此，Hive 并不能够在大规模数据集上实现低延迟快速的查询，Hive 不适合那些需要低延迟的应用，但适用于大数据集的批处理作业，如网络日志分析。

如果 Hive 表中的数据存在数据倾斜，Hive 表中的数据分布很不均衡，如果采用 Spark 读

取 Hive 表中数据倾斜的数据，进行 Shuffle 算子计算时，由于 Hive 系统中数据倾斜，所以会导致 Spark 某些任务运行特别缓慢或者运行时出现 OOM 现象。此时我们可以将数据倾斜的操作前移到 Hive 中进行，在 Hive 中对数据倾斜的记录进行预处理，这样就从数据根源上解决了数据倾斜的问题，Spark 分布式计算时就不用执行 Shuffle 算子，从而也就避免了数据倾斜。

30.2.4　使用 Spark SQL 对源数据进行聚合并过滤掉导致倾斜的 Keys

对源数据进行聚合并过滤掉导致倾斜的 Keys，我们可以使用多种过滤清洗方式。

- 如果在 Spark SQL 中查询发生数据倾斜，可以在 Spark SQL 中使用 where 子句过滤掉数据倾斜的 Key，然后再进行 groupBy 等表关联操作。
- 加一个中间适配层，当数据进来的时候进行 Key 的统计和动态排名，基于该排名动态地调整 Key 分布；这种情况表示数据倾斜特别严重，此时可以使用内存级别的数据库，不断统计 Key 值并进行排名，如果 Key 值特别多，则可以对 Key 值进行调整，如触发一个过程，将 Key 值加上一个时间戳，以时间为考虑因素改变 Key 的分布。
- 采用 Spark 的 sample 算子，对 RDD 的 Key 值进行采样，如果某些倾斜的 Key 值不是业务需要的，则通过 filter 算子对倾斜的 Key 值进行过滤。

30.3　数据倾斜解决方案之二：适当提高 Reducer 端的并行度

本节谈一个大家既熟悉又陌生的并行度的使用，来解决数据倾斜的问题。本节分成两部分：①基本关于并行度的使用；②相对高级的内容：并行度的深度使用。数据倾斜的数据表现，某一个任务的数据量特别多，导致内存无法装载这个数据或者导致某个任务的执行特别缓慢。如果能够改变并行度，如将并行度变大，往往数据倾斜的问题也会有所改善。

30.3.1　适用场景分析

改善并行度之所以能改变数据倾斜的原因在于，如果某个 Task 有 100 个 Key 且数据量特别大，就极有可能导致 OOM 或者任务运行特别慢，此时如果把并行度变大，则可以分解该 Task 的数据量，例如，把原本 Task 的 100 个 Key 分解给 10 个 Task，这就可以减少每个 Task 的数据量，从而有可能解决 OOM 和任务运行慢的问题。

对于 reduceByKey，可以传入并行度的参数 numPartitions:int，该参数就设置了这个 Shuffle 算子执行时 Shuffle read Task 的数量，也可以自定义 Partitioner，增加 Executor 改变计算资源，从数据倾斜的角度来看，并不能直接去解决数据倾斜的问题，但是也有好处，好处是同时可以并发运行更多的 Task，结果是可能加快了运行速度。增加 Shuffle read Task 的数量，可以让原本分配给一个 Task 的多个 Key 分配给多个 Task，从而让每个 Task 处理比原来更少的数据。

30.3.2　原理剖析

Spark 中有很多 Shuffle 操作，如 groupByKey、reduceByKey 等。在 RDD.scala 源码中找不到 reduceByKey 算子源码，因为 RDD.scala 不是 Key-Value 类型的。我们在 PairRDDFunctions.scala 中去找 reduceByKey 算子。reduceByKey 有 3 个重载的方法，其中一个重载方法 reduceByKey(partitioner: Partitioner, func: (V, V) => V): RDD[(K, V)] 中有 Partitioner。下面是最简单的 reduceByKey 重载方法。

PairRDDFunctions.scala 的 reduceByKey 的源码如下：

```
1.   /**
       * 使用关联和交换汇聚函数合并每个键的值。在将结果发送到 Reducer 端前，在每个 map
       * 端执行本地合并，类似 MapReduce 的 combiner，输出将根据分区器/并行度进行重新分区
2.     */
3.    def reduceByKey(func: (V, V) => V): RDD[(K, V)] = self.withScope {
4.      reduceByKey(defaultPartitioner(self), func)
5.    }
```

其中，defaultPartitioner 是传入进来的分区器。在方法里面又调用了 reduceByKey 方法。

PairRDDFunctions.scala 的 reduceByKey 的源码如下：

```
1.   /**
       * 使用关联和交换汇聚函数合并每个键的值。在将结果发送到 Reducer 端前，在每个 map
       * 端执行本地合并，类似 MapReduce 的 combiner
2.     */
3.    def reduceByKey(partitioner: Partitioner, func: (V, V) => V): RDD[(K,
      V)] = self.withScope {
4.      combineByKeyWithClassTag[V]((v: V) => v, func, func, partitioner)
5.    }
```

这里如果要改变并行度，可以自定义一个 Partitioner，如果并行度是 1000，进行 Shuffle 的时候并行度也是 1000，除非没有 1000 个 Key。reduceByKey 传入 numPartitions 参数，numPartitions 是直接传入并行度的数字，这里采用 HashPartitioner 的分区方式。

PairRDDFunctions.scala 的 reduceByKey 的源码如下：

```
1.   /**
       *使用关联和交换汇聚函数合并每个键的值。在将结果发送到 Reducer 端前，在每个 map 端
       *上执行本地合并，类似 MapReduce 的 combiner，输出将根据分区器数量重新分区
2.     */
3.    def reduceByKey(func: (V, V) => V, numPartitions: Int): RDD[(K, V)] =
      self.withScope {
4.      reduceByKey(new HashPartitioner(numPartitions), func)
5.    }
```

30.3.3　案例实战

用并行度解决数据倾斜的基本应用：如 reduceByKey。

改变并行度之所以能够改善数据倾斜的原因在于，如果某个 Task 有 100 个 Key 且数据量特别大，就极有可能导致 OOM 或者任务运行特别缓慢，此时如果把并行度变大，则可以

分解该 Task 的数据量。例如，把原本该 Task 的 100 个 Key 分解给 10 个 Task，这就可以减少每个 Task 的数据量，从而有可能解决 OOM 和任务慢的问题。

对于 reduceByKey 而言，可以传入并行度的参数，也可以自定义 Partitioner。

增加 Executor：改变计算资源，仅从数据倾斜的角度看，并不能直接解决数据倾斜的问题，但是也有好处，好处是可以并发运行更多的 Task，结果是可能加快运行速度。

用并行度解决数据倾斜的深度应用：可参阅 30.4 节“使用随机 Key 实现双重聚合”，通过双重聚合（局部聚合+全局聚合）的方式将原本倾斜的 Key 值通过分而治之方案分散开，最后再进行全局聚合，其本质还是通过改变并行度去解决数据倾斜的问题。

30.3.4　注意事项

并行度指的就是 RDD 的分区数。由于一个分区对应一个 Task，并行度也是一个 Stage 中的 Task 数，这些 Task 被并行地处理。RDD 是以 Partition（即分区）的形式散落在集群上的，每个分区都包含了一部分待处理的数据，Spark 程序运行时，会为每个待处理的分区创建一个 Task，且默认情况下每个 Task 占用一个 CPU Core 来处理。

Spark 有一套自己自动推导出默认的分区数的机制。当我们在程序中通过操作算子（如 textFile 等）读取外部数据源，以获得 Input RDD 时，Spark 会自动根据外部数据源的大小推导出一个合适的、默认的分区数，如 HDFS 文件的每个 block 就对应一个分区；在对 RDD 进行 map 类不涉及 Shuffle 的操作时，由于分区数具有遗传性，新产生的 RDD 的分区数由 parent RDD 中最大的分区数决定；在对 RDD 进行 reduce 类涉及 Shuffle 操作的算子时（如 groupByKey、reduceByKey 等各种 reduce 操作算子），由于分区数具有遗传性，新产生的 RDD 的分区数也由 parent RDD 中最大的分区数决定。如果是在 spark-shell 交互式命令终端下，则可以通过方法 rdd.partitions.size 获得某个 RDD 的分区数，而在 Spark 1.6.0 以后的版本中，也可以通过 rdd.getNumPartitions 获得某个 RDD 的分区数。

并行度对性能的影响有两方面，当并行度不够大时，会存在资源的闲置与浪费，例如，一个应用程序分配到了 1000 个 Core，但是一个 Stage 里只有 30 个 Task，这时就可以提高并行度，以提升硬件利用率；而当并行度太大时，Task 常常几微秒就执行完毕，或 Task 读写的数据量很小，这种情况下，Task 频繁地开辟与销毁的不必要的开销就太大，我们就需要调小并行度。

由于 Spark 自动推导出的默认的分区数很多时候是不理想的，所以我们必须人为地加以控制，来改变并行度。Spark 提供了 4 种改变并行度的方式。

第一种：使用读取外部数据源的 textFile 类算子时，可以通过可选的参数 minPartitions 显式指定最小的分区数。

第二种：针对已经存在的 RDD，可以通过方法 repartition()或 coalesce()来改变并行度。repartition()和 coalesce()的区别在于，前者会产生 Shuffle，而后者默认不会产生 Shuffle。事实上，当有大量小任务（任务处理的数据量小且耗时短）时，如某个 RDD 在 filter 操作后，由于过滤掉了大量数据，每个分区都只剩下了很少量的数据，这时我们常用 coalesce()来合并分区，调小并行度，减少不必要的任务的开辟与销毁的消耗；而当任务耗时长且处理的数据量大时，如果计算只发生在部分 Executor 上，我们常用 repartition()重新分区，提高并行度，开辟更多的并行计算的任务来完成计算。

第三种：在对 RDD 进行 reduce 类涉及 shuffle 操作的算子时，这些算子大都可以接受一个显式指定的参数，来确定新产生的 RDD 的分区数，我们可以显示地指定这类参数来改变 Shuffle 后新产生的 RDD 的分区数，而不是采用系统推导出的默认的分区数。

第四种：也可以配置参数 spark.default.parallelism 来设置默认的并行度。该参数其实指定的是在对 RDD 进行 reduce 类涉及 Shuffle 操作的算子时，如果没有对这些算子显式指定参数，来确定新产生的 RDD 的分区数时，这类 reduce 类涉及 Shuffle 操作的算子产生新的 RDD 的 partition 数量。该参数也指定了 parallelize 等没有 parent RDD 类操作的算子所产生的新的 RDD 的分区数。

一个最佳实践是，并行度设置为集群的总的 CPU Cores 的个数的 2～3 倍，如 Executor 的总 CPU Core 数量为 400 个，那么设置 1000 个 Task 是可以的，此时可以充分利用 Spark 集群的资源；每个分区的大小在 128MB 左右。

需要说明的是，通过以上方式确定了任务的并行度，就确定了理论上能够并行执行的任务的数量，而实际执行时真正并发执行的任务的数量还要受应用分配到的实际资源数量的限制，要想改变应用程序获得的资源数目，就会涉及资源参数的调优。

30.4　数据倾斜解决方案之三：使用随机 Key 实现双重聚合

本节讲解使用随机 Key 实现双重聚合。首先讲解什么是随机 Key 双重聚合，接下来讲解使用随机 Key 实现双重聚合解决数据倾斜的适用场景分析、原理分析、案例实战，以及使用随机 Key 实现双重聚合解决数据倾斜注意事项等内容。

30.4.1　什么是随机 Key 双重聚合

随机 Key 双重聚合是指 Spark 分布式计算对 RDD 调用 reduceByKey 各算子进行计算，使用对 Key 值随机数前缀的处理技巧，对 Key 值进行二次聚合。

（1）第一次聚合（局部聚合）：对每个 Key 值加上一个随机数，执行第一次 reduceByKey 聚合操作。

（2）第二次聚合（双重聚合）：去掉 Key 值的前缀随机数，执行第二次 reduceByKey 聚合，最终得到全局聚合的结果。

30.4.2　适用场景分析

随机 Key 适用于 groupByKey、reduceByKey 等算子操作数据时某些 Key 值发生数据倾斜的情况。例如，电商广告点击系统中，如果根据用户点击的省份进行汇聚，原来的 Key 值是省份，如果某些省份的 Value 值特别多，发生了数据倾斜，可以将每个 Key 拆分成多个

Key，加上随机数前缀将 Key 值打散，组拼成 random_省份的新的 Key 值，调用 reduceByKey 做局部聚合，然后再将 random_前缀去掉，形成的 Key 值仍为省份，再调用 reduceByKey，进行全局聚合。

30.4.3　原理剖析

使用随机 Key 实现双重聚合，解决数据倾斜原理剖析：如 reduceByKey。

假设有倾斜的 Key，我们给所有的 Key 加上一个随机数，然后进行 reduceByKey 操作；此时同一个 Key 会有不同的随机数前缀，进行 reduceByKey 操作时使用的原来的一个非常大的倾斜的 Key 就分而治之变成若干个更小的 Key，不过，此时的结果和原来不一样，怎么办？进行 map 操作。map 操作的目的是把随机数前缀去掉，然后再次进行 reduceByKey 操作，（当然，可以再次做随机数前缀），这样就可以把原本倾斜的 Key 通过分而治之方案分散开，最后再进行了全局聚合。在这里的本质还是通过改变并行度去解决数据倾斜的问题。

30.4.4　案例实战

使用随机 Key 实现双重聚合解决数据倾斜案例实战：我们看一个 reduceByKey(_+_)的示例。假设 RDD 不同，Partition 的数据内容分别为（1，1）（1，2）（1，3）以及（2，1）（1，2）（1，3），我们通过 map 操作加上随机数，将数据转换为（1_1，1）（2_1，2）（3_1，3）以及（1_2，1）（2_1，2）（3_1，3；然后进行 reduceByKey 的累加操作，汇聚的结果为（1_1，1）（2_1，4）以及（3_1，6）（1_2，1），之后通过 map 转换去掉随机数前缀，转换为（1，1）（1，4）以及（1，6）（2，1），再进行一次 reduceByKey 操作，转换成（1，11）以及（2，1）。最终将结果输出到 Output file1、Output file2，如图 30-4 所示。

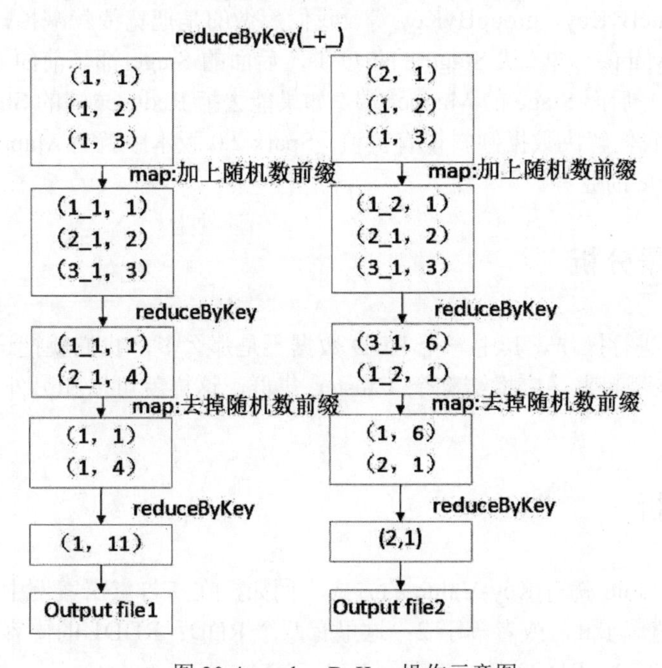

图 30-4　reduceByKey 操作示意图

30.4.5 注意事项

使用随机 Key 实现双重聚合解决数据倾斜方案局限于单个 RDD 的 reduceByKey、groupByKey 等算子。如果两个 RDD 的数据量都特别大，而且倾斜的 Key 特别多，就无法采用分而治之进行双重聚合的方法，我们须综合应用各种数据倾斜的解决方案，如通过扩容进行 Join 操作，解决数据倾斜问题。

30.5 数据倾斜解决方案之四：使用 Mapper 端进行 Join 操作

本节讲解使用 Mapper 端进行 Join 操作，首先讲解为什么要在 Mapper 端进行 Join 操作，然后讲解使用 Mapper 端进行 Join 操作解决数据倾斜问题的使用场景、原理流程、案例实战、注意事项等内容。

30.5.1 为什么要在 Mapper 端进行 Join 操作

解决数据倾斜有一个技巧：把 Reducer 端的操作变成 Mapper 端的 Reduce，通过这种方式不需要发生 Shuffle。如果把 Reducer 端的操作放在 Mapper 端，就避免了 Shuffle。避免了 Shuffle，在很大程度上就化解掉了数据倾斜的问题。Spark 是 RDD 的链式操作，DAGScheduler 根据 RDD 的不同类型的依赖关系划分成不同的 Stage，所谓不同类型的依赖关系，就是宽依赖、窄依赖。当发生宽依赖的时候，把 Stage 划分成更小的 Stage。划分的依据就是宽依赖。宽依赖的算子如 reducByKey、groupByKey 等。我们想做的是把宽依赖减掉，避免掉 Shuffle，把操作直接发生在 Mapper 端。从 Stage 的角度讲，后面的 Stage 都是前面 Stage 的 Reducer 端，前面的 Stage 都是后面 Stage 的 Mapper 端。如果能去掉 Reducer 端的 Shuffle 操作，将其放在 Mapper 端，对我们解决数据倾斜很有价值。Spark 2.0 版本中就有 Mapper 端聚合，只有 Mapper 端完成 Shuffle 的业务。

30.5.2 适用场景分析

如果两个 RDD 进行操作，其中一个 RDD 数据不是那么多，我们就把这个 RDD 的数据以广播变量的形式包裹起来，广播给整个 Cluster 集群，这样就可以和另外一个 RDD 进行 map 操作了。

30.5.3 原理剖析

进行 Join 操作，Join 就有 Key-Value 的方式。例如，在广告点击案例中，假设有两个不同的 Key 值，Key 值等于 1，或者等于 2；这里有两个 RDD：RDD1 的内容是（1,1）（1,2）（1,3）（2,1）（2,2）（2,3）；RDD2 的内容是（1,1）（2,1）。

RDD1 和 RDD2 进行 Join 操作，产生的 RDD3，假设这里有两个 Task，因为这里有两个 Key，那产生的结果是 Task1：(1,（1，1))（1,（2，1))（1,（3，1)）Task2：(2,（1，1))（2,（2，1))（2,（3，1)），如图 30-5 所示。

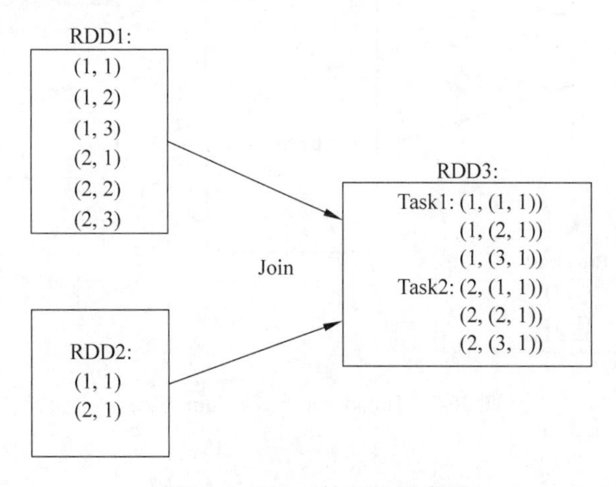

图 30-5　RDD 的 Join 示意图

在 Join 的过程中可能产生数据倾斜。图 30-5 是一个比较友好的例子，Key 比较均衡，没有数据倾斜。接下来演示一下数据倾斜的例子，假设 Key 等于 2 的记录现在只有一条，(2，1)。那么在两个 RDD 中，RDD1 的内容是（1,1）（1,2）（1,3）（2,1）；RDD2 的内容是（1,1）（2,1），那就产生数据倾斜，如图 30-6 所示。

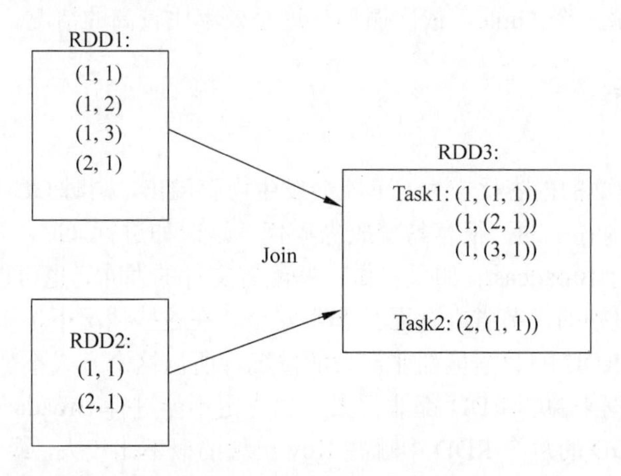

图 30-6　RDD 的数据倾斜示意图

在这种情况下会产生数据倾斜，那我们对 RDD1、RDD2 如何进行处理呢？RDD1 仍然是 RDD1，但 RDD2 此时有一个变化，RDD2 进行 Broadcast 变成广播变量，Broadcast 是进程级别的，Broadcast 将 RDD2 的数据（1,1）（2,1）广播出去。将数据（1,1）（2,1）广播后，如何完成 Join 的业务逻辑操作？Join 操作是对 Key 相同的 Value 进行 Join，这里将数据广播出去以后，没有进行 Shuffle，数据被 BlockManager 管理，在 Executor 中都有 Broadcast 的数据。Join 的业务逻辑通过对 RDD1 进行 map 操作实现，遍历 Broadcast 中的值，然后变成一个 Tuple。Map 操作是一条记录一条记录地读取数据，如进行性能优化，可以使用 mapPartitions 算子，读取分区中的一批数据进行转换，如图 30-7 所示。

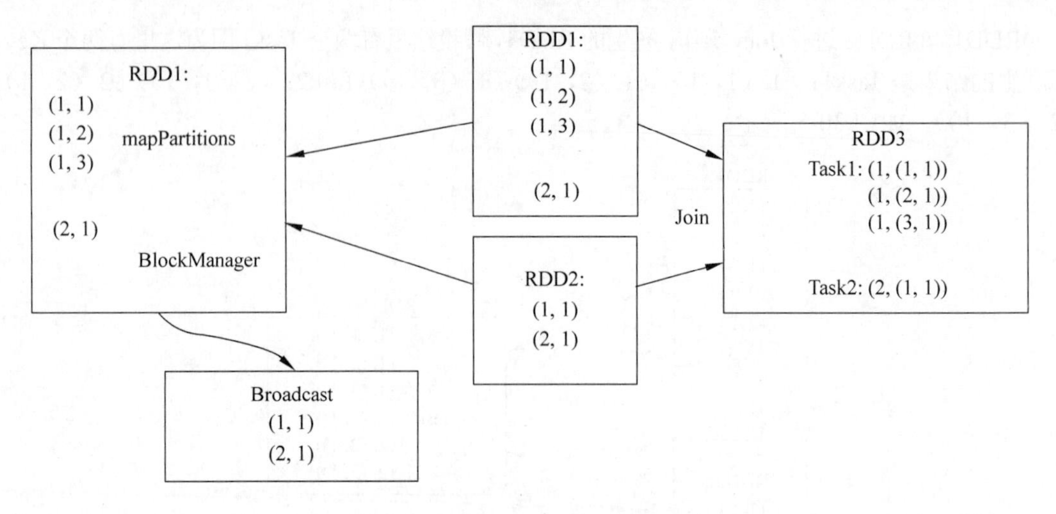

图 30-7　Broadcast 方式、Join 方式

30.5.4　案例实战

Join 的时候，RDD1 Join RDD2，如将 Join 去掉，可以将数据量小的部分做成 Broadcast，在 RDD1 不同的 Task 中，通过 mapPartitions 的方式每次读取一个 Partitions 的数据，在内部遍历 mapPartitions 里面的元素，和 Broadcast 里面的元素进行 Key 的比较，如果 Key 相同，就将它们的 Value 变成一个 Tuple，依次循环，这是效率比较高的情况。

30.5.5　注意事项

如果 Broadcast 里面的内容比较大，可能会发生内存溢出，同时 GC 也会非常麻烦，因为广播的数据是常驻在内存中的，很容易变成老年代。这时如进行 GC，是非常致命的。如果有一张小表，可以进行 Broadcast；如果要做一些配置文件的发布，也可以采用 Broadcast，因为 Broadcast 是进程级别的，节省内存而且非常安全。在这些情况下，Broadcast 方式非常有效，但不适用于两个 RDD 的数据量都非常大的情况，所以这个方式不是万能的。

如果代表两个数据来源的 RDD 都非常大，那肯定不能进行 Broadcast；数据倾斜采样的思路是采样出两个 RDD 的每个 RDD 中哪些 Key 出现的概率比较大，或者次数比较多，然后基于 Key 值，将数据独立抽取出来，进行 Join 操作，进行数据规模上的改变。利用 Map 操作增加或减少数据的规模。经过过滤以后，原来的两个 RDD 变成 4 个 RDD，剩下的两个 RDD 进行 Join 操作，然后进行 Union 操作。

30.6　数据倾斜解决方案之五：对倾斜的 Keys 采样后进行单独的 Join 操作

本节首先讲解为什么对倾斜的 Keys 采样进行单独的 Join 操作，然后讲解如何对倾斜的

Keys 进行采样，最后讲解对倾斜的 Keys 采样后进行单独的 Join 操作的使用场景、案例实战、注意事项等内容。

30.6.1　为什么对倾斜的 Keys 采样后进行单独的 Join 操作

本节采用采样算法的思想来解决数据倾斜。数据倾斜的时候如果能把 Join 的方式去除，在 Mapper 端就能完成 Join 的操作，这是最好的，但有一个前提条件：要进行 Join 的 RDD，其中有一个 RDD 的数据比较少。而在实际的生产环境下，有时不具备这样的前提条件，如果两个 RDD 的数据都比较多，我们将尝试采取进一步的做法来解决这个问题。

首先我们谈采样。采样是有一个数据的全量，假如有 100 亿条数据，采取一个规则来选取 100 亿条数据中的一部分数据，如 5%、10%、15%，采样通常不可能超过 30%的数据。采样算法的优劣决定了采样的效果。所谓采样的效果，即我们采样的结果能否代表全局的数据（100 亿条数据）。在 Spark 中，我们可以直接采用采样算法 Sample。采样算法对解决数据倾斜的作用：数据产生数据倾斜是由于某个 Key 或者某几个 Key，数据的 Value 特别多，进行 Shuffle 的时候，Key 是进行数据分类的依据。如果能够精准地找出是哪个 Key 或者哪几个 Key 导致了数据倾斜，这是解决问题的第一步：找出谁导致数据倾斜，就可以进行分而治之。

30.6.2　如何对倾斜的 Keys 进行采样

例如，RDD1 和 RDD2 进行 Join 操作，我们采用采样的方式发现 RDD1 中有严重的数据倾斜的 Key。

第一步：采用 Spark RDD 中提供的采样接口，可以很方便地对全体（如 100 亿条）数据进行采样，然后基于采样的数据可以计算出哪个（哪些）Key 的 Values 个数最多。

第二步：把全体数据分成两部分，即把原来的一个 RDD1 变成 RDD11 和 RDD12，其中 RDD11 代表导致数据倾斜的 Key，RDD12 中包含的是不会产生数据倾斜的 Key。

第三步：把 RDD11 和 RDD2 进行 Join 操作，且把 RDD12 和 RDD2 进行 Join 操作，然后把 Join 操作后的结果进行 Union 操作，从而得出和 RDD1 与 RDD2 直接进行 Join 相同的结果。

这样就解决了数据倾斜的问题：RDD12 和 RDD2 进行 Join 操作不会产生数据倾斜，因为里面没有特别的 Key，即没有哪个 Key 的 Value 特别多；RDD11 和 RDD2 进行 Join 操作，假设 RDD11 中只有一个 Key，其 Key 值的 Value 特别多，利用 Spark Core 天然的并行机制对 RDD11 的 Key 的数据进行拆分。

30.6.3　适用场景分析

两个 RDD 进行 Join 操作，如果一个 RDD 有严重的数据倾斜，那我们可以通过采样的方式发现 RDD1 中有严重的数据倾斜的 Key，然后将原来一个 RDD1 拆分成 RDD11（产生倾斜 Key 的数据）和 RDD12（不产生倾斜 Key 的数据），把 RDD11、RDD12 分别和 RDD2

进行 Join 操作，然后把 Join 操作后的结果进行 Union 操作。此外，倾斜的 Key 也可加上随机数处理。

30.6.4　案例实战

例如，RDD1 中的元素（1，1）（1，2）（1，3）（2，1），其中 Key 值等于 1 的 Value 值特别多；RDD2 中的元素（1，1）（2，1）。思路是：将 RDD1 拆分成 2 份，RDD1 变成了两个 RDD，即 RDD11、RDD12；RDD11 的元素为（1，1）（1，2）（1，3），RDD12 的元素为（2，1）。然后，RDD11、RDD12 分别和 RDD2 进行 Join，Join 以后产生 Result1、Result2 两个结果。Result1 的元素为（1，（1，1），（1，（2，1）），（1，（3，1））；Result2 的元素为（2，（1，1））；然后 Result1、Result2 进行 Union 操作，最终结果 FinalResult：（1，（1，1）），（1，（2，1）），（1，（3，1））（2，（1，1）），如图 30-8 所示。

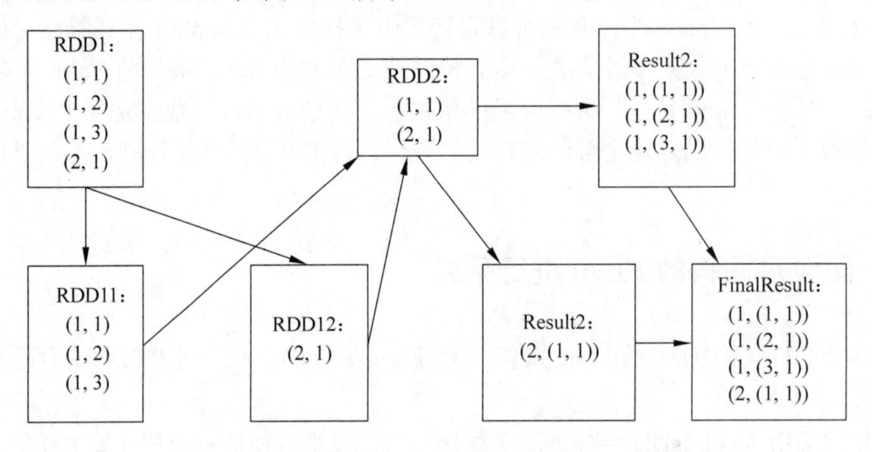

图 30-8　分而治之解决数据倾斜

为什么这样做缓解了数据倾斜的问题：将 RDD1 一分为二，将数据单独和 RDD2 进行 Join，基本上就不会产生 Shuffle，这是 Spark 本身的机制保证的。RDD11 中只有一个 Key，和 RDD2 进行 Join 操作，RDD2 中有若干个 Key。

这时有两种情况。

第一种情况：如果 RDD11 中的数据量不是很多，可以采用广播的方式，把 Reducer 端操作放在 Mapper 端，采用 Mapper 端的 Join 操作，避免了 Shuffle 和数据倾斜。

第二种情况：如果 RDD11 中的数据量特别多，此时之所以能够缓解数据倾斜，是因为采用了 Spark Core 天然的并行机制对 RDD11 中的同样一个 Key 的数据进行了拆分，从而达到让原本倾斜的 Key 分散到不同的 Task 的目的，就缓解了数据倾斜。

30.6.5　注意事项

对倾斜的 Keys 采样进行单独的 Join 操作步骤有点复杂：首先对 RDD1 进行采样，例如 RDD1 进行 Sample 抽样（15%）可以计算出一个结果，其是一个 RDD，采样之后进行 Map 操作，通过 reduceByKey 操作计数，然后对 Key 和 Value 进行置换，通过 SortByKey 进行排序，再进行 Map 置换操作，从而找出哪一个 Key 值倾斜比较严重，对其进行过滤，提取到 RDD11 中，剩下的提取到 RDD12 中。

如果倾斜的 Key 特别多，如 50 多个倾斜的 Key，我们可以一个一个地对 Key 进行过滤处理。

如果导致倾斜的 Key 特别多，如成千上万个 Key 都导致数据倾斜，那么这种方式也不适合。

30.7　数据倾斜解决方案之六：使用随机数进行 Join

本节讲解如何使用随机数，使用随机数进行 Join 来解决数据倾斜问题使用场景、案例实战、注意事项等内容。

30.7.1　如何使用随机数

倾斜的 Key 加上随机数：对 Key 进行 Map 操作加上随机数，计算出结果以后再次进行 Map，将随机数去掉，这样计算得到的结果和原先不加随机数的结果是一样的，好处是可以控制并行度。

随机数的使用如下。

（1）对 RDD1 使用 mapToPair 算子进行转换，使用 Random random = new Random();构建一个随机数，例如，任意一个 100 以内的随机数 random.nextInt(100)，将其赋值给 prefix；然后将 prefix 加上原来的 Key 值组拼成新的 Key 值：prefix_Key。

（2）对 RDD2 中使用 mapToPair 算子进行转换，对于 RDD1 相同的 Key 值，同样构建随机数及加上前缀，将 prefix 加上原来的 Key 值组拼成新的 Key 值：prefix_Key。

（3）RDD1 和 RDD2 根据 prefix_Key 进行 Join。

（4）RDD1 和 RDD2 进行 Join 的结果再次通过 Map 算子进行转换，去掉随机数，得到最终计算结果。

30.7.2　适用场景分析

两个 RDD 中的某一个 Key 或者某几个 Key 值的数据量特别大，在进行 Join 的时候发生了数据倾斜。我们可以将 RDD1 中一个或者几个 Key 值加上随机数前缀，然后 RDD2 中的相同的 Key 值也做同样的处理，加上随机数的前缀。Key 值加上随机数进行 Join 来解决数据倾斜问题。

30.7.3　案例实战

使用随机数进行 Join 来解决数据倾斜问题案例实战：加上随机数，并行 Task 数量可能增加。

例如，RDD1 中的元素（1，1）（1，2）（1，3）（2，1），其中 Key 值等于 1 的 Value

值特别多；RDD2 中的元素（1，1）（2，1）。我们先将 RDD1 拆分成 2 份，RDD1 变成了两个 RDD，即 RDD11、RDD12；RDD11 中的元素为（1，1）（1，2）（1，3），RDD12 中的元素为（2，1）。

如果 RDD11 和 RDD2 的数据规模特别大，可以将 RDD11 中的倾斜的 Key 加上 1000 以内的随机数，这时能直接和 RDD2 进行 Join 操作？不行！此时我们一定需要把 RDD11 中的 Key 在 RDD2 中的相同的 Key 加上 1000 以内的随机数，然后再进行 Join 操作，这样做的好处：让倾斜的 Key 更加不倾斜，在实际生产环境下会极大地解决在两个进行 Join 的 RDD 数量都很大且其中一个 RDD 有一个或者两三个明显倾斜的 Key 的情况下的数据倾斜问题。

30.7.4　注意事项

使用随机数进行 Join 来解决数据倾斜问题案例实战：其中，随机数须选取合适的数值，我们对 RDD1 和 RDD2 中相同的 Key 值加上随机数，加上随机数须确保 RDD1 和 RDD2 的新的 Key 值 prefix_Key 在 RDD1 和 RDD2 中都能匹配上，随机数 *n* 的具体数值根据数据规模来设置。

如果两个 RDD 数据都特别多且倾斜的 Key 有成千上万个，仅仅使用随机数进行 Join 来解决数据倾斜问题并不适用。我们可以考虑进行数据扩容，将其中一个 RDD 的数据进行扩容 *N* 倍后再加上前缀 *n*，另外一个 RDD 的每条数据都打上一个 *n* 以内的随机前缀，最后将两个处理后的 RDD 进行 Join。

30.8　数据倾斜解决方案之七：通过扩容进行 Join

本节讲解如何进行扩容，通过扩容进行 Join 操作解决数据倾斜问题使用场景、案例实战、注意事项等内容。

30.8.1　如何进行扩容

两个 RDD 数据都特别多且倾斜的 Key 有成千上万个，该如何解决数据倾斜的问题？初步的想法：在倾斜的 Key 上加上随机数。该想法的原因：Shuffle 的时候把 Key 的数据可以分到不同的 Task 里。加随机数有一个前提：必须知道哪些是倾斜的 Key。但是，现在的倾斜的 Key 非常多，成千上万，所以如果说采样找出倾斜的 Key 的话，并不是一个非常好的想法。

下一个想法是考虑进行扩容。

首先，什么是扩容？扩容就是把该 RDD 中的每条数据变成 5 条、10 条、20 条等。例如，RDD 中原来是 10 亿条数据，扩容后可能变成 1000 亿条数据。

其次，如何扩容？flatMap 中对要进行扩容的每条数据都通过 0～*N*−1 个不同的前缀变成 *N* 条数据。

问题：*N* 的值可以随便取吗？需要考虑当前程序能够使用的 Core 的数目。

答案：*N* 的数字一般不能取得特别大，通常都会小于 50，否则对磁盘、内存和网络都会形成极大的负担，如会造成 OOM。

最后，

（1）将另外一个 RDD 的每条数据都打上一个 *n* 以内的随机前缀。

（2）最后将两个处理后的 RDD 进行 Join 即可。

N 这个数字取成 10 和取成 1000 除了 OOM 等不同外，是否还有其他影响呢？

其实，*N* 的数字的大小还会对数据倾斜的解决程度构成直接影响！*N* 越大，越不容易倾斜，但是也会占用更多的内存、磁盘、网络以及消耗更多的 CPU 时间。

30.8.2　适用场景分析

如果两个 RDD 数据特别多，而且倾斜的 Key 成千上万个，则我们可以考虑进行数据扩容，将其中一个 RDD 的数据进行扩容 *N* 倍，另外一个 RDD 的每条数据都打上一个 *n* 以内的随机前缀，最后将两个处理后的 RDD 进行 Join。

30.8.3　案例实战

通过扩容进行 Join 操作解决数据倾斜问题：RDD1 和 RDD2 进行 Join 操作，我们对 RDD2 进行 flatMap 操作，由于 *n* 越大，就越容易造成 OOM，假设 *n* 等于 1000，分配到不同的 Task 中，数据规模变大，就容易 OOM，因此 *n* 的数字不能取得特别大。这里假设 *n* 等于 10，循环遍历 *n*，对 item 加上前缀 i；然后对 RDD1 进行 map 操作，加上 10 以内的随机数前缀。最后将 RDD11 和 RDD22 进行 Join，对 result 进行 Map 操作去掉前缀，得到结果。

示例代码如下：

```
1.  RDD1 join RDD2
2.
3.  rdd2 2= RDD2.flatMap {
4.      for(1  to 10) {
5.          1_item
6.      }
7.  }
8.
9.  rdd11 = RDD1.map{
10.     Random(10)
11.
12.     random_item
13.
14. }
15.
16. result = rdd11.join(rdd22)
17.
18. result.map{
19.     item_1.split   去掉前缀
20.
21. }
```

30.8.4　注意事项

n 的值可以随便定吗？需要考虑当前程序能够使用的 Core 的数目，扩容的问题是来解

决从程序运行不了的问题，从无法运行到能运行的结果。因为扩容想把小规模数据变成大规模数据，结果是对机器的内存和 CPU 造成很大的消耗。该方案更多的是缓解数据倾斜，而不是彻底避免数据倾斜，而且需要对整个 RDD 进行扩容，对内存资源要求很高。

30.9　结合电影点评系统进行数据倾斜解决方案的小结

在 Spark 商业案例之大数据电影点评系统应用案例中，我们通过 RDD、DataFrame、DataSet 等各种方式综合实现大数据电影点评系统的功能。结合本章节 Spark 性能调优数据倾斜解决方案，我们可以对电影点评系统应用进行性能调优。

（1）电影点评源数据进行 ETL 预处理。例如，TL 最受不同年龄段人员欢迎的电影 TopN 中，不同年龄阶段如何界定，这个问题其实是业务问题，实际在实现的时候可以使用 RDD 的 filter 算子，对于 13 < age <18 的计算，因为要进行全量扫描，所以会非常耗性能。一般情况下，我们都是在原始数据中直接对要进行分组的年龄段提前进行好 ETL，根据不同年龄预先计算出年龄范围的特征值。通过提前的 ETL 把计算发生在 Spark 业务逻辑运行前，用空间换时间。当然，这些实现也可以用 Hive，因为 Hive 语法支持非常强悍且内置了最多的函数。

（2）电影点评系统使用 Mapper 端进行 Join 操作。在电影点评系统中使用 Mapjoin，原因是目标用户 targetUsers 数据只有用户 UserID，数据量一般不会太多。Mapjoin 实行借助于广播变量 Broadcast，把数据广播到 Executor 级别，让该 Executor 上的所有任务共享该唯一的数据，而不是每次运行 Task 的时候都要发送一份数据的复制，这显著降低了网络数据的传输和 JVM 内存的消耗。

下面对 Spark 数据倾斜解决方案进行整体回顾和总结。

（1）数据倾斜运行的症状和危害。如果发现数据倾斜，往往发现作业任务运行特别缓慢，出现 OOM 内存溢出等现象。

（2）如果两个 RDD 进行操作，其中一个 RDD 数据不是那么多，我们把这个 RDD 的数据以广播变量的形式包裹起来，广播给整个 Cluster 集群，这样就可以和另外一个 RDD 进行 map 操作。

（3）两个 RDD 进行 Join 操作，如果一个 RDD 有严重的数据倾斜，那么我们可以通过采样的方式发现 RDD1 中有严重的数据倾斜的 Key，然后将原来一个 RDD1 拆分成 RDD11（产生倾斜 Key 的数据）和 RDD12（不产生倾斜 Key 的数据），把 RDD11、RDD12 分别和 RDD2 进行 Join 操作，然后把 Join 操作后的结果进行 Union 操作。此外，倾斜的 Key 也可加上随机数处理。

（4）如果两个 RDD 数据特别多，而且倾斜的 Key 成千上万个，我们可以考虑进行数据扩容，将其中一个 RDD 的数据扩容 *N* 倍，另外一个 RDD 的每条数据都打上一个 *n* 以内的随机前缀，最后将两个处理后的 RDD 进行 Join。

（5）从并行度的角度考虑。

- 初级级别的并行度解决方案：例如，reduceByKey 在第二个参数中指定并行度来改善数据倾斜。
- 进阶级别的并行度解决方案：以 reduceByKey 为例，在原有数据分片的基础上加上随机数前缀，进行 reduceByKey，然后进行 map 操作，去掉随机数前缀，再次进行

reduceByKey，这种方式很有效。

讲解了数据倾斜这么多的解决方式，几乎涵盖了数据倾斜方方面面的内容。改变并行度、Key 值加上随机数、进行广播变量、数据扩容等，都是类似的解决思路，这些方案是否足以解决数据倾斜的问题？作为数据倾斜解决方案的“银弹”，我们须穷尽一下解决方案，能不能将数据倾斜消灭在问题产生以前？Spark 是处理数据的，在处理数据的过程中产生了数据倾斜，由于数据的特征导致数据倾斜。如果在数据来源的根源上解决了数据倾斜，Spark 本身就不会面临数据倾斜的问题。在实际生产环境中，不能单一地考虑 Spark 本身的问题。

逃离 Spark 技术本身之外如何解决数据倾斜的问题？

之所以会有这样的想法，是因为从结果上看，数据倾斜的产生来自于数据和数据的处理技术，之前我们都是从数据的技术处理层面考虑如何解决数据倾斜，现在我们需要回到数据的层面去解决数据倾斜的问题。

数据本身就是 Key-Value 的存在方式。所谓的数据倾斜，就是说某（几）个 Key 的 Values 特别多，如果要解决数据倾斜，实质上是解决单一的 Key 的 Values 的个数特别多情况，新的的数据倾斜解决方案由此诞生了。

（1）把一个大的 Key-Values 的数据分解成为 Key-subKey-Values 的方式。例如，数据原来的 Key 值是 ID，ID 下面包括省份、城市、社区等很多数据，现在把 Key 变成 ID+省份+城市+社区的方式，这样就把原来庞大的 Value 的集合变成了更细化的数据。

（2）预先和其他表进行 Join，将数据倾斜提前到上游的 Hive ETL。

（3）可以把大的 Key-Values 中的 Values 组拼成为一个字符串，从而形成只有一个元素的 Key-Value。

（4）加一个中间适配层，当数据进来时进行 Key 的统计和动态排名，基于该排名动态地调整 Key 分布；这种情况表示数据倾斜特别严重，此时可以使用内存级别的数据库，不断统计 Key 值并进行排名，如果 Key 值特别多，可以对 Key 值进行调整，如触发一个过程，将 Key 值加上一个时间戳，以时间为考虑因素改变 Key 的分布。

（5）假如 10 万个 Key 都发生了数据倾斜，如何解决呢？此时一般是加内存和 Cores。如出现 OOM，就加内存，如运行特别慢，就加 Cores。

第31章 Spark大数据性能调优实战专业之路

本章主要讲解如下内容。

- 大数据性能调优的本质和Spark性能调优要点分析。
- Spark性能调优之系统资源使用原理和调优最佳实践。
- Spark性能调优之使用更高性能算子及其源码剖析。
- Spark旧版本中性能调优之HashShuffle剖析及调优。
- Shuffle是如何成为Spark性能杀手的及调优点思考。
- Spark Hash Shuffle源码解读与剖析。
- Sort-Based Shuffle产生的内幕及其tungsten-sort背景解密。
- Spark Shuffle令人费解的6大经典问题。
- Spark Sort-Based Shuffle排序具体实现内幕和源码详解。
- Spark 1.6.X以前Shuffle中JVM内存使用及配置内幕详情。
- Spark 2.4.X中Shuffle中内存管理源码解密：Static Memory和Unified Memory。
- Spark 2.4.X中Shuffle中JVM Unified Memory内幕详情。
- Spark 2.4.X中Shuffle下Task视角内存分配管理。
- Spark 2.4.X中Shuffle中Mapper端的源码实现。
- Spark 2.4.X中Shuffle中SortShuffleWriter排序源码内幕解密。
- Spark 2.4.X中Sort Shuffle中TimSort排序源码具体实现。
- Spark 2.4.X中Sort Shuffle中Reducer端源码内幕。

31.1 大数据性能调优的本质和Spark性能调优要点分析

我们谈大数据性能调优，到底在谈什么，它的本质是什么，以及Spark在性能调优部分的要点，这几点在进入性能调优之前都是至关重要的问题，它的本质限制了我们调优到底要达到一个什么样的目标，或者说我们是从什么本源上进行调优的。

Spark官网的性能优化指南（http://spark.apache.org/docs/latest/tuning.html）包括以下内容。

- 数据序列化。
- 内存调优。
  - 内存管理。
  - 内存消耗。
  - 调整数据结构。
  - 序列化RDD存储。

- ➢ 垃圾回收。
- ❑ 其他考量点。
  - ➢ 并行度。
  - ➢ 减少任务的内存使用。
  - ➢ 广播大变量。
  - ➢ 数据本地性。

Spark 官网性能优化指南的内容是冰山一角，接下来我们分析大数据性能调优的本质和 Spark 性能调优要点。

1．大数据性能调优的本质

编程的时候发现一个惊人的规律：软件是不存在的。所有编程高手级别的人无论做什么类型的编程，最终思考的都是硬件方面的问题。最终思考的都是硬件在一秒、一毫秒，甚至一纳秒到底是如何运行的，并且基于此进行算法实现和性能调优！最后都回到了硬件。

那么，我们回归到问题：大数据性能调优的本质是什么？答案是基于硬件的调优，即基于 CPU、Memory、I/O(Disk/Network)基础上构建算法和性能调优！无论是 Hadoop，还是 Spark，还是其他技术，都无法逃脱。

- ❑ CPU：计算。
- ❑ Memory：存储。
- ❑ I/O：数据交互。

2．Spark性能调优要点

读者可以讨论一下 Spark 在哪些方面进行调优：并行度、压缩和序列化、数据倾斜、JVM 调优（如 JVM 数据结构优化）、内存调优（如内存消耗诊断等）、Task 性能调优（例如包含 Mapper 和 Reducer 两种类型的 Task）、Shuffle 网络调优（如小文件合并等）、RDD 算子调优（如 RDD 复用、RDD 自定义）、数据本地性、容错调优等。

但忽略了一个非常重要的东西：大数据（Spark）系统最怕什么？

- ❑ 数据不在本地（数据不在内存中，数据本地性是和框架技术及开发者紧密相关的）。
- ❑ 数据倾斜（这是分布式系统的问题，这和数据的业务紧密相关）。

所以，调优 Spark 的重点肯定是从数据本地性和数据倾斜两方面入手！

参照“基于 CPU、Memory、I/O(Disk/Network)基础上构建算法和性能调优”的思路，Spark 性能调优如下所示。

（1）资源分配和使用：你能够申请到多少计算资源及如何最优化地使用计算资源。

（2）开发调优：如何基于 Spark 框架内核原理和运行机制最优化地实现代码功能。

（3）Shuffle 调优：分布式系统必然面临的“杀手级别”的问题。

（4）数据倾斜调优：这个问题的本质是分布式系统业务本身有数据倾斜，那么在我们现有的框架上该如何调优呢？有一套完整的分布式系统数据倾斜调优的解决方案。

基本上来讲，性能调优都无法逃脱 CPU、Memory、Disk I/O、Net I/O、JVM、Data Locality、Data Skew 等。

31.2　Spark 性能调优之系统资源使用原理和调优最佳实践

我们从 Spark 资源的角度讲解性能调优的原因，算子调优、Shuffle、数据倾斜等实质上都涉及资源的使用。从 Spark 官网（http：//spark.apache.org/docs/latest/cluster-overview.html）看一下 Spark 运行架构图（见图 31-1）。

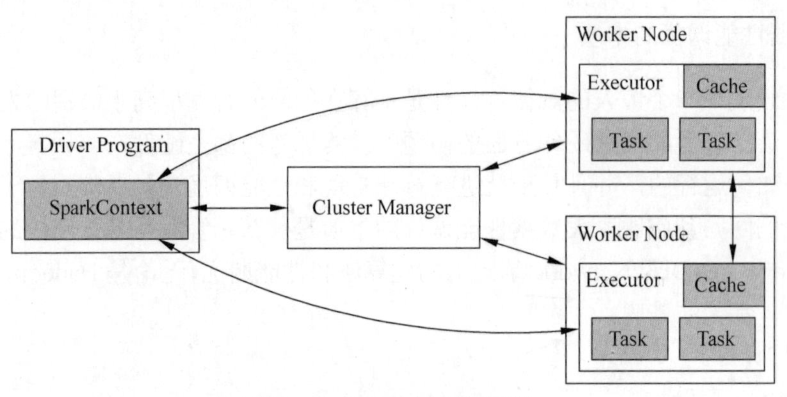

图 31-1　Spark 运行架构图

从程序运行的角度看，Spark 的组件分成三部分。

- ❑ Driver：类似 Linux 的驱动程序，驱动程序的运行。
- ❑ Executor：具体的处理数据的部分。
- ❑ Cluster Manager：资源调度器。

我们的应用程序运行本身在 Driver、Executor 上，运行的时候需要从 Cluster Manager 中获取资源，Cluster Manager 在 Spark 中可以是插拔的，在 Spark 中可以使用不同的集群资源管理器。集群资源管理器管理内存和 CPU。现在业界比较常用的、经典的资源管理器是 Yarn 和 Mesos，国内大部分使用的是 Yarn，因为 Spark 运作在 Hadoop 上，而 Hadoop 自带的资源管理器是 Yarn。

在实际生产环境中，Cluster Manager 一般都是 Yarn 的 ResourceManager，Driver 会向 ResourceManager 申请计算资源（一般情况下都是在发生计算前一次性申请请求），分配的计算资源就是 CPU、Cores 和 Memory，我们的具体的 Job 的 Task 就是基于这些分配的内存和 Cores 构建的线程池来运行 Tasks 的。

当然，在 Task 运行的时候需要消耗内存，而 Task 又分为 Mapper 和 Reducer 两种不同类型的 Task，也就是 ShuffleMapTask 和 ResultTask 两种类型。

在一个 Task 运行的时候，在 Spark 1.6.X 以前版本中默认是占用 Executor 内存的 20%，Shuffle 拉取数据和进行聚合操作等占用了另外 20%的内存，剩下 60%用于 RDD 的持久化（例如，Cache 数据到内存），Task 在运行的时候是跑在 Core 上的，比较理想的情况是有足够的 Cores，同时数据分布比较均匀，这时往往能够充分利用集群的资源。

核心调优参数如下。

- ❑ num-executors：该参数一定会被设置，Yarn 会按照 Driver 的申请最终为当前的

Application 生产指定个数的 Executors。实际生产环境下分配 80 个左右的 Executors 比较合适。

- executor-memory：与 JVM OOM 紧密相关，很多时候甚至决定了 Spark 运行的性能。实际生产环境下建议 8GB 左右，很多时候 Spark 是运行在 Yarn 的，内存占用量不超过 Yarn 的内存资源的 50%。
- executor-cores：决定了在 Executor 中能够并行执行的 Task 的个数。实际生产环境建议 4 个左右，一般情况下不要超过 Yarn 队列中 Cores 总数的 50%。
- driver-memory：作为驱动，默认是 1GB，老师在生产环境下都设置为 4GB。
- spark.default.parallelism：建议至少设置为 100 个，最好是 700 个左右。
- spark.storage.memoryFraction：默认占用 60%，如果计算比较依赖于历史数据，则可以适当调高该参数，但是如果计算严重依赖于 Shuffle，则需要降低该比例。
- spark.shuffle.memoryFraction：默认占用 20%，如果计算严重依赖于 Shuffle，则需要提高该比例。
- supervise：配置这个参数，当 Driver 运行在 Cluster 集群，如果出问题了，可自动重新启动。

31.3　Spark 性能调优之使用更高性能算子及其源码剖析

Spark 性能调优之使用更高性能算子的重要性在于：同样情况下，如果使用更高性能的算子，从算子级别给我们带来更高的效率。Spark 现在主推的是 DataSet 这个 API 接口，越来越多的算子可以基于 DataSet 去做，DataSet 基于天然自带的优化引擎，理论上讲比 RDD 的性能更高，DataSet 的弱点是无法自定义很多功能。从平时使用来讲，使用的最基本的是 Shuffle 的算子。Shuffle 分为两部分：Mapper 端和 Reducer 端。性能调优的准则是尽量不使用 Shuffle 类的算子，尽量避免 Shuffle。进行 Shuffle 的时候，将多个节点的同一个 Key 的数据汇聚到同样一个节点进行 Shuffle 操作，基本思路是将数据放入内存中，若内存中放不下，就放入磁盘中。

如果要避免 Shuffle，所有的分布式计算框架是产生 Mapper 端的 Join，两个 RDD 进行操作，先把 RDD 的数据收集过来，然后通过 Spark Context 进行 BroadCast 广播出去， 假设原先是 RDD1、RDD2，我们把 RDD2 广播出去，原来是进行 Join，Join 的过程肯定是进行某种计算，此时 RDD2 其实已经不是 RDD 了，就是一个数据结构体包裹着数据本身，对 RDD1 进行 Join 操作，就是一条条遍历数据跟另一个集合里的数据产生某种算法操作。

如果不能避免 Shuffle，我们退而求其次，需要更多的机器承担 Shuffle 的工作，充分利用 Mapper 端和 Reducer 端机器的计算资源，尽量让 Mapper 端承担聚合的任务。如果在 Mapper 端进行 Aggregate 的操作，在 Mapper 端的进程中就会合并相同的 Key。

（1）reduceByKey 和 aggregateByKey 取代 groupByKey。

PairRDDFunctions.scala 的 aggregateByKey，aggregateByKey 算子使用给定的组合函数和一个中立的“零值”聚合每个 Key 的值。这个函数可以返回不同的结果类型 U，不同于 RDD 的值的类型 V。在 scala.TraversableOnce 中，我们需要一个操作将 V 合并成 U 和一个操作合并两个 U，前者的操作用于合并分区内的值，后者用于在分区之间合并值。为了避免内存分配，这两个函数都允许修改和返回它们的第一个参数，而不是创造一个新的 U。aggregateByKey 的代码如下：

```
1.   def aggregateByKey[U: ClassTag](zeroValue: U, partitioner: Partitioner)
    (seqOp:  (U, V) => U,
2.       combOp:  (U, U) => U): RDD[(K, U)] = self.withScope {
3.     //序列化零值字节数组，我们可以在每个 Key 值进行新的克隆
4.     val  zeroBuffer  =  SparkEnv.get.serializer.newInstance().serialize
       (zeroValue)
5.     val zeroArray = new Array[Byte](zeroBuffer.limit)
6.     zeroBuffer.get(zeroArray)
7.
8.     Lazy val cachedSerializer = SparkEnv.get.serializer.newInstance()
9.     val createZero = () => cachedSerializer.deserialize[U](ByteBuffer
       .wrap(zeroArray))
10.
11.    //combinebykey 之后，进行清理关闭
12.    val cleanedSeqOp = self.context.clean(seqOp)
13.    combineByKeyWithClassTag[U]((v:  V) => cleanedSeqOp(createZero(), v),
14.      cleanedSeqOp, combOp, partitioner)
15.  }
```

例如，groupByKey 会不会进行 Mapper 的聚合操作呢？不会。groupByKey 重载函数都没有指定函数操作的功能。相对于 groupByKey 而言，我们倾向于采用 reduceByKey 和 aggregateByKey 来取代 groupByKey，因为 groupByKey 不会进行 Mapper 端的 aggregate 的操作，所有数据会通过网络传输传到 Reducer 端，性能会比较差。而我们进行 aggregateByKey 的时候，可以自定义 Mapper 端的操作和 Reducer 端的操作，当然，reduceByKey 和 aggregateByKey 算子是一样的。

PairRDDFunctions.scala 的 groupByKey 的代码如下：

```
1.   def groupByKey(partitioner:  Partitioner):  RDD[(K, Iterable[V])] =
     self.withScope {
2.     //groupByKey 不使用 Mapper 端聚合，因为 Mapper 端聚合不汇聚 Shuffled 的数据；
       //要求所有 Mapper 端的数据插入到哈希表中，导致生成更多对象
3.     val createCombiner = (v:  V) => CompactBuffer(v)
4.     val mergeValue = (buf:  CompactBuffer[V], v:  V) => buf += v
5.     val mergeCombiners = (c1:  CompactBuffer[V], c2:  CompactBuffer[V]) =>
       c1 ++= c2
6.     val bufs = combineByKeyWithClassTag[CompactBuffer[V]](
7.       createCombiner, mergeValue, mergeCombiners, partitioner, mapSideCombine
         = false)
8.     bufs.asInstanceOf[RDD[(K, Iterable[V])]]
9.   }
```

reduceByKey 和 aggregateByKey 在正常情况下取代 groupByKey 的两个问题如下。

- groupByKey 可进行分组，reduceByKey 和 aggregateByKey 怎么进行分组？可采用算法控制。
- reduceByKey 和 aggregateByKey 都可以取代 groupByKey。reduceByKey 和 aggregateByKey 有什么区别？区别很简单，aggregateByKey 给予我们更多的控制，可以定义 Mapper 端的 aggregate 函数和 Reducer 端的 aggregate 函数。

（2）批量处理数据 mapPartitions 算子取代 map 算子。

我们看一下 RDD.scala 的源码，RDD 在处理一块又一块的写数据的时候，不使用 map 算子，可以使用 mapPartitions 算子，但 mapPartitions 有一个弊端，会出现 OOM 的问题，因为每次处理掉一个 Partitions 的数据，对 JVM 也是一个负担。

RDD.scala 的 mapPartitions 的代码如下：

```
1.  def  mapPartitions[U:  ClassTag](
2.        f:  Iterator[T] => Iterator[U],
3.        preservesPartitioning:  Boolean = false):  RDD[U] = withScope {
4.      val cleanedF = sc.clean(f)
5.      new MapPartitionsRDD(
6.        this,
7.        (context:  TaskContext, index:  Int, iter:  Iterator[T]) => cleanedF
        (iter),
8.        preservesPartitioning)
9.    }
```

（3）批量数据处理 foreachPartition 取代 foreach。

foreach 处理一条条的数据，foreachPartition 将一批数据写入数据库或 Hbase，至少提升 50%的性能。RDD.scala 的 foreachPartition foreach 的源码如下：

```
1.    def foreach(f:  T => Unit):  Unit = withScope {
2.      val cleanF = sc.clean(f)
3.      sc.runJob(this, (iter:  Iterator[T]) => iter.foreach(cleanF))
4.    }
5.
6.    /**
7.     * RDD 的每个分区应用这个函数
8.     */
9.    def foreachPartition(f:  Iterator[T] => Unit):  Unit = withScope {
10.     val cleanF = sc.clean(f)
11.     sc.runJob(this, (iter:  Iterator[T]) => cleanF(iter))
12.   }
```

（4）使用 coalesce 算子整理碎片文件。coalesce 默认不产生 Shuffle，基本工作机制把更多并行度的数据变成更少的并行度。例如，10 000 个并行度的数据变成 100 个并行度。coalesce 算子返回一个新的 RDD，汇聚为 numPartitions 个分区。这将导致一个窄依赖。例如，如果从 1000 个分区变成 100 个分区，将不会产生 Shuffle，而不是从当前分区的 10 个分区变成 100 个新分区。然而，如果做一个激烈的合并，如 numPartitions = 1，这可能导致计算发生在更少的节点。（例如，设置 numPartitions = 1，将在一个节点上进行计算）。为了避免这个情况，可以设置 Shuffle = true。这将增加一个 Shuffle 的步骤，但意味着当前的上游分区将并行执行。Shuffle = true，可以汇聚到一个更大的分区，对于少量的分区，这是有用的，例如，100 个分区，可能有几个分区数据非常大。那使用 coalesce 算子合并（1000，shuffle = true），将导致使用哈希分区器将数据分布在 1000 个分区。注意，可选的分区 coalescer 必须是可序列化的。

Spark 2.2.1 版本的 RDD.scala 的 coalesce 算子代码如下：

```
1.    def coalesce(numPartitions:  Int, shuffle:  Boolean = false,
2.               partitionCoalescer:  Option[PartitionCoalescer] = Option.empty)
```

```
3.            (implicit ord:  Ordering[T] = null)
4.         : RDD[T] = withScope {
5.       require(numPartitions > 0, s"Number of partitions ($numPartitions)
         must be positive.")
6.       if (shuffle) {
7.         /**从随机分区开始，将元素均匀分布在输出分区上*/
8.         val distributePartition = (index: Int, items: Iterator[T]) => {
9.           var position = (new Random(index)).nextInt(numPartitions)
10.          items.map { t =>
11.     //注意，Key 的哈希代码仅对 Key 进行，HashPartitioner 将它与总分区数进行取模
12.            position = position + 1
13.            (position, t)
14.          }
15.        } : Iterator[(Int, T)]
16.
17.        //包括一个 Shuffle 步骤，以便我们的上游任务仍然是分布式的
18.        new CoalescedRDD(
19.          new ShuffledRDD[Int, T, T](mapPartitionsWithIndex(distributePartition),
20.          new HashPartitioner(numPartitions)),
21.          numPartitions,
22.          partitionCoalescer).values
23.      } else {
24.        new CoalescedRDD(this, numPartitions, partitionCoalescer)
25.      }
26.    }
```

Spark 2.4.3 版本的 RDD.scala 源码与 Spark 2.2.1 版本相比具有如下特点。

- 上段代码中第 9 行将 index 调整为 hashing.byteswap32(index)。
- 上段代码中第 19 行 mapPartitionsWithIndexInternal 新增加一个参数 isOrderSensitive，指明函数是否区分顺序。如果它是顺序敏感的，当输入顺序改变时，它可能返回完全不同的结果。大多数状态函数是顺序敏感的。

```
1.  ......
2.      var position = new Random(hashing.byteswap32(index)).nextInt(numPartitions)
3.  ......
4.          mapPartitionsWithIndexInternal(distributePartition,
            isOrderSensitive = true),
5.  ......
```

从最优化的角度讲，使用 coalesce 一般在使用 filter 算子之后。因为 filter 算子会产生数据碎片，Spark 的并行度会从上游传到下游，我们在 filter 算子之后一般会使用 coalesce 算子。

（5）使用 repartition 算子，其背后使用的仍是 coalesce。但是，Shuffle 值默认设置为 true，repartition 算子会产生 Shuffle。repartition 的代码如下：

```
1.      def repartition(numPartitions: Int)(implicit ord: Ordering[T] =
        null): RDD[T] = withScope {
2.      coalesce(numPartitions, shuffle = true)
3.    }
```

（6）repartition 算子碎片整理以后会进行排序，Spark 官方提供了一个 repartition AndSortWithinPartitions 算子。JavaPairRDD 的 repartitionAndSortWithinPartitions 方法的代码如下：

```
1.    def repartitionAndSortWithinPartitions(partitioner: Partitioner):
      JavaPairRDD[K, V] = {
```

```
2.     val comp = com.google.common.collect.Ordering.natural().asInstanceOf
       [Comparator[K]]
3.     repartitionAndSortWithinPartitions(partitioner, comp)
4.   }
```

（7）persist：数据复用时使用持久化算子。

```
1.  //设置这个 RDD 的存储级别，第一次计算以后持久化值，如果 RDD 没有一个存储级别，将分配
    //一个新的存储级别。局部检查点是一个例外
2.   def persist(newLevel: StorageLevel): this.type = {
3.     if (isLocallyCheckpointed) {
4.   //之前称为 localCheckpoint，这标记 RDD 已经持久化。这里我们要覆盖重写旧的存储级别，
     //这是由用户明确要求（使用磁盘以后）的
5.       persist(LocalRDDCheckpointData.transformStorageLevel(newLevel),
         allowOverride = true)
6.     } else {
7.       persist(newLevel, allowOverride = false)
8.     }
9.   }
```

（8）mapPartitionsWithIndex 算子：推荐使用，每个分区有一个 index，实际运行时看 RDD 上面有数字，如果对数字感兴趣，可以使用 mapPartitionsWithIndex 算子。

mapPartitionsWithIndex 算子的代码如下：

```
1.   def mapPartitionsWithIndex[U: ClassTag](
2.       f: (Int, Iterator[T]) => Iterator[U],
3.       preservesPartitioning: Boolean = false): RDD[U] = withScope {
4.     val cleanedF = sc.clean(f)
5.     new MapPartitionsRDD(
6.       this,
7.       (context: TaskContext, index: Int, iter: Iterator[T]) => cleanedF
         (index, iter),
8.       preservesPartitioning)
9.   }
```

（9）推荐使用 tree 开头的算子，如 treeReduce()和 treeAggregate()。

treeReduce 的源码如下：

```
1.   def treeReduce(f: (T, T) => T, depth: Int = 2): T = withScope {
2.     require(depth >= 1, s"Depth must be greater than or equal to 1 but got
       $depth.")
3.     val cleanF = context.clean(f)
4.     val reducePartition: Iterator[T] => Option[T] = iter => {
5.       if (iter.hasNext) {
6.         Some(iter.reduceLeft(cleanF))
7.       } else {
8.         None
9.       }
10.    }
11.    val partiallyReduced = mapPartitions(it => Iterator(reducePartition
       (it)))
12.    val op: (Option[T], Option[T]) => Option[T] = (c, x) => {
13.      if (c.isDefined && x.isDefined) {
14.        Some(cleanF(c.get, x.get))
15.      } else if (c.isDefined) {
16.        c
17.      } else if (x.isDefined) {
```

```
18.         x
19.       } else {
20.         None
21.       }
22.     }
23.     partiallyReduced.treeAggregate(Option.empty[T])(op, op, depth)
24.       .getOrElse(throw new UnsupportedOperationException("empty collection"))
25.   }
```

Spark 2.2.1 版本的 treeAggregate 算子的源码如下：

```
1.    def treeAggregate[U: ClassTag](zeroValue: U)(
2.        seqOp: (U, T) => U,
3.        combOp: (U, U) => U,
4.        depth: Int = 2): U = withScope {
5.      require(depth >= 1, s"Depth must be greater than or equal to 1 but got
      $depth.")
6.      if (partitions.length == 0) {
7.        Utils.clone(zeroValue, context.env.closureSerializer.newInstance())
8.      } else {
9.        val cleanSeqOp = context.clean(seqOp)
10.       val cleanCombOp = context.clean(combOp)
11.       val aggregatePartition =
12.         (it: Iterator[T]) => it.aggregate(zeroValue)(cleanSeqOp, cleanCombOp)
13.       var partiallyAggregated = mapPartitions(it => Iterator(aggregate
        Partition(it)))
14.       var numPartitions = partiallyAggregated.partitions.length
15.       val scale = math.max(math.ceil(math.pow(numPartitions, 1.0 /
        depth)).toInt, 2)
16.       //如果创建一个额外的级别并不能帮助减少 wall-clock 时间，停止聚合树，当不保存
        //wall-clock 时间时，不触发 TreeAggregation 聚合
17.       while (numPartitions > scale + math.ceil(numPartitions.toDouble /
        scale)) {
18.         numPartitions /= scale
19.         val curNumPartitions = numPartitions
20.         partiallyAggregated = partiallyAggregated.mapPartitionsWithIndex {
21.           (i, iter) => iter.map((i % curNumPartitions, _))
22.         }.reduceByKey(new HashPartitioner(curNumPartitions), cleanCombOp).
          values
23.       }
24.       partiallyAggregated.reduce(cleanCombOp)
25.     }
26.   }
```

Spark 2.4.3 版本的 RDD.scala 源码与 Spark 2.2.1 版本相比具有如下特点。

- 上段代码中第 22～24 行替换为以下代码，使用 foldByKey 方法关联函数和零值，合并每个键的值，零值可以添加到结果中任意次数，并且不能更改结果（例如，列表串联为 Nil，加法加 0，重复为 1 等）。

```
1.  ......
2.        }.foldByKey(zeroValue, new HashPartitioner(curNumPartitions))
        (cleanCombOp).values
3.        }
4.        val copiedZeroValue = Utils.clone(zeroValue, sc.env.closureSerializer
        .newInstance())
5.        partiallyAggregated.fold(copiedZeroValue)(cleanCombOp)
6.  ......
```

31.4　Spark 旧版本中性能调优之 HashShuffle 剖析及调优

大数据是分布式的，分布式大多情况下涉及 Shuffle。Spark 内核引擎是树根，Spark Shuffle 就相当于整个运行的树干，树枝相当于在 Mapper 端怎么表现，在 Reducer 端怎么表现，内部的 JVM 又是怎么做。HashShuffle 虽然在 Spark 新版本中已经不用了，但温习一下会了解 Spark Shuffle 到底是怎么回事。

我们看一下 Spark 运行架构图，回顾一下 Spark 运行的方式：Spark 本身运行分成两部分：第一部分是 Driver Program，里面的核心是 Spark Context，Driver 是驱动，指挥工作怎么做；另一部分是 Worker 节点上的 Task，具体工作运作在 Task 上。Task 运行在 JVM 的线程中，当程序运行时，不间断地由 Driver 与 Executor 所在的进程进行交互，交互的内容：第一、哪些 Task 发送消息给 Driver；第二、告诉 Task 数据源在哪里，例如说第三个 Stage 会找第二个 Stage 拿数据，怎么知道数据在哪里？向 Driver 拿数据位置信息这个过程，一方面是 Driver 跟 Executor 进行网络传输，另一方面是 Task 要从 Driver 抓取其他上游的 Task 执行的数据结果，所以在这个过程中就不断地产生网络传输。其中，下一个 Stage 向上一个 Stage 要数据这个过程，我们称之为 Shuffle。

Shuffle 从很多 Stage 的角度讲，第一个 Stage 是一个 Mapper，第二个 Stage 较第一个 Stage 而言，相当于第一个 Stage 的 Reducer，第二个 Stage 相当于第三个 Stage 的 Mapper，第三个 Stage 相当于第二个 Stage 的 Reducer，依此迭代。

由于上游阶段的数据是并行运行的，下游要进行汇总，我们把需要的那类数据抓到 Reducer 中，如图 31-2 中数据分为 3 类，上游的每个任务都要把自己的数据分成 3 类，但可能有一类一个数据也没有，里面的数据为空，但我们按照这种规则获取，这是最原始的 Shuffle。在这最原始的、也是最重要的 Shuffle 的基础上，我们才考虑内存调优及网络传输等内容。这里是 4 个 Task，每个 Task 分 3 种类型有 3 个 Buffer，每个 Buffer 对应一个文件，共有 12 个 Buffer，12 个文件。第一个 Task 开辟了 Buffer0、Buffer1、Buffer2，第二个 Task 仍需要再开辟 Buffer0、Buffer1、Buffer2，内存消耗很大。如果每个 Task 3 个 Buffer，那 400 个 Task 就有 1200 个 Buffer，1200 个本地文件。

优化后的 HashShuffle 如图 31-3 所示。这里是 4 个 Task，数据根据 Key 的分类分成 3 种类型，取模以后我们编号为 0、1、2。进程中无论有多少 Task，都把 Task 同样的 Key 放到同一个 Buffer 中，并把同一个 Task 的数据写到同一个本地文件中。优化以后，4 个 Task 变成 6 个 Buffer，这里因为每两个 Task 属于同一个进程，进程中所有的 Task 共用不同的 Buffer。例如，第一个 Task 开辟了 Buffer0、Buffer1、Buffer2，第二个 Task 不需要再开辟 Buffer0、Buffer1、Buffer2。同时，在运行的时候，现在是 2 个 Task，也可能是 20 个 Task，都往 3 个 Buffer 中根据取模算法写入标记为 0、1、2 的数据。例如，0 号写入标记为 0 的文件中，其优势在于，如果进程中有 200 个 Task，还是 3 个 Buffer，本地还是 3 个文件；另外的一个进程也是 200 个 Task，还是 3 个 Buffer，本地还是 3 个文件；那一共 400 个 Task，6 个 Buffer，6 个本地文件。同样的 Key 的数据合并到同一个 Buffer 及同一个数据文件，而且从抓取数据的角度讲简化了很多，以前要抓几百次数据及建很多的 Socket 端口等，现在只是抓取一次数据，和之前完全是不同的概念，减少了本地磁盘读取 I/O 的时间，以及网络传输的负担。

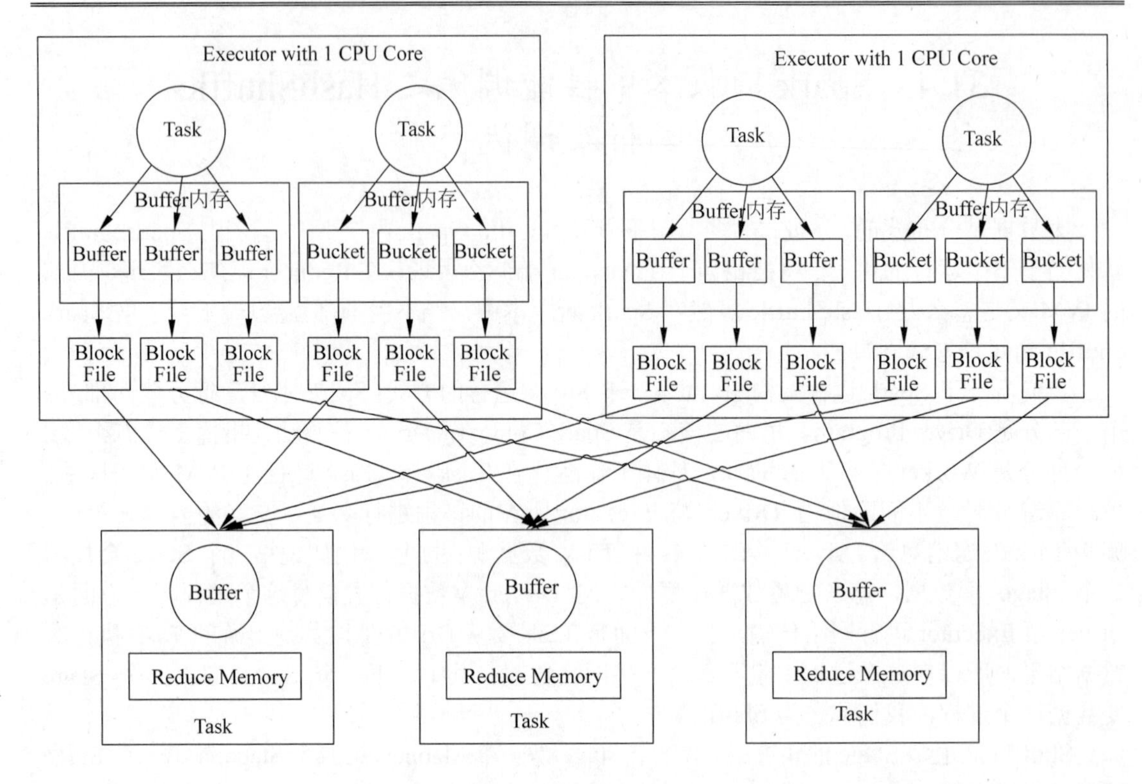

图 31-2　HashShuffle

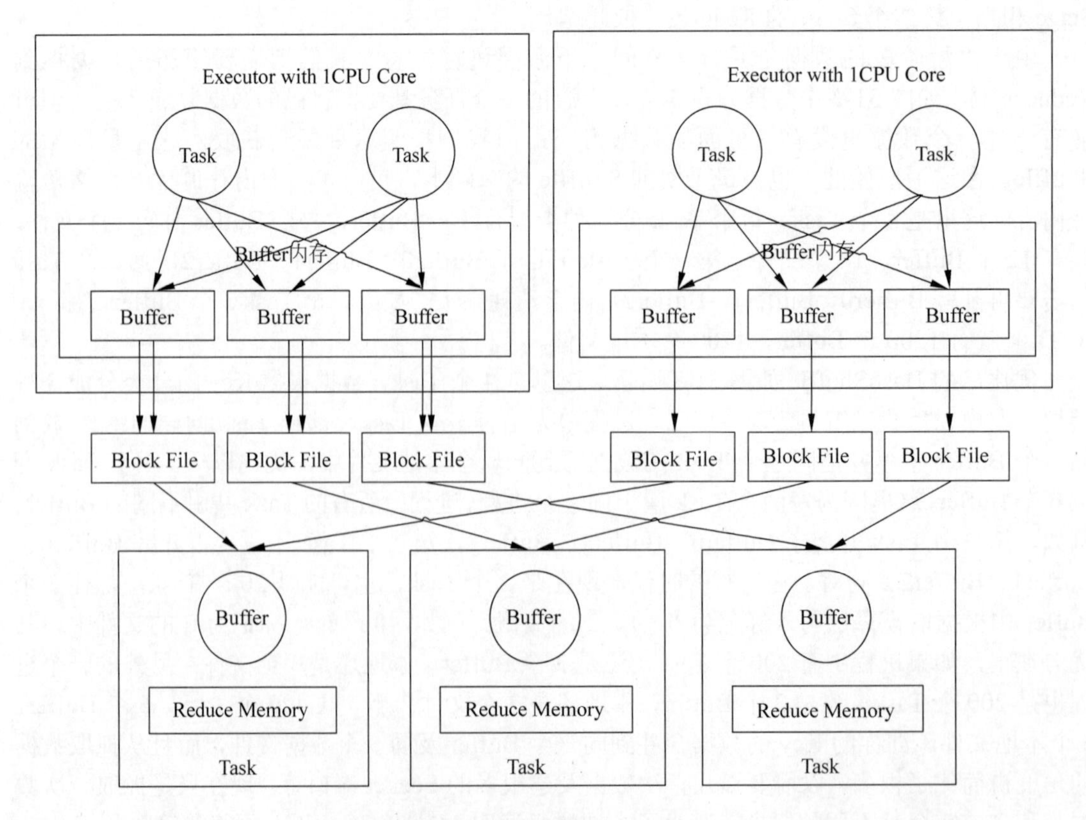

图 31-3　优化后的 HashShuffle

优化后的 HashShuffle 是 ConsolidateShuffle。Spark 1.6.X 以前版本可以设置以下参数：spark.shuffle.consolidateFiles=true。这里是两个 CPU 并行度。假设并行度是 2，Task 任务数是 200，那 Task 进行排队，不断地往 Buffer 里面写入数据。文件数量的计算方式基于 CPU Cores 的个数和下一个 Stage 的 Task 个数的乘积：CPU Cores×Reducer Task；这里 Key 分成 3 份，两个 CPU Cores 共计 2×3=6 个文件。现在的 Spark 2.4.X 中已经把这种方式废弃了。

31.5　Shuffle 如何成为 Spark 性能杀手

人们对 Spark 的第一印象往往是 Spark 基于内存进行计算。但实质上讲，Spark 基于内存进行计算，也可以基于磁盘进行计算，或者基于第三方的存储空间进行计算。背后的两层含义：①Spark 架构框架的实现模式是倾向于在内存中计算数据的，可以从 Storage、算法、库的不同方式看出来；②我们计算数据的时候，数据就在内存中。Shuffle 和 Spark 完全基于内存的计算愿景相违背，Shuffle 就变成了 Spark 的性能杀手，从现在的计算机硬件和软件的发展水平看，这是不可抗拒的。

Shuffle 不可以避免，是因为在分布式系统中把一个很大的任务/作业分成一百份或者一千份文件，这一百份或一千份文件在不同的机器上独自完成各自不同的部分，完成后针对整个作业要结果，在后面进行汇聚，从前一阶段到后一阶段及网络传输的过程就叫 Shuffle。

在 Spark 中，为了完成 Shuffle 的过程，会把一个作业划分为不同的 Stage，这个 Stage 的划分是根据依赖关系决定的。Shuffle 是整个 Spark 中最消耗性能的地方。试想，如果没有 Shuffle，Spark 几乎可以完成一个纯内存的操作，因为不需要网络传输，又有迭代的概念，那数据每次在内存中计算，如内存不够，也可以每次算一点，完成纯内存的操作。

那 Shuffle 如何破坏了这一点？因为在不同机器、不同节点上，我们要进行数据传输，如 reduceByKey，它会把每个 Key 对应的 Value 聚合成一个 Value，然后生成新的 RDD。下游 Stage 的 Task 拿到上一个 Stage 相同的 key，然后进行 Value 的 reduce 操作，这里分为 Mapper 端的 reduce，和 Reducer 端的 reduce。数据在通过网络发送之前，要先存储在内存中，内存达到一定的程度，它会写到本地磁盘（以前 Spark 的版本中没有 Buffer 的限制，会不断地写入 Buffer，内存满了就写入本地文件，Buffer 没有限制有很大的弊端，容易出现 OOM，但有一个好处，如以前的版本可以使用 48GB 的内存，可以减少 I/O 操作；现在的 Spark 版本对 Buffer 大小设定了限制，如限制 Buffer 为 48KB，48KB 刷新一次 Buffer，以防止出现 OOM。每个版本的变化都有原因，不能简单地说好，还是不好）。Mapper 端写入内存 Buffer，这个关乎到 GC 的问题，然后 Mapper 端的 Block 要写入本地磁盘文件，Spark 以前版本实现的时候曾经只要 RDD 存在，本地的临时文件并不会被删除掉，这个好处是计算到一半或者 95% 的时候，可以复用 Mapper 端本地磁盘的数据；但这个不好的地方是一直在运行，如流处理程序运行了 5 天的时间，缓存越来越大，最终造成系统崩溃。

Spark 一再鼓励数据就在内存中，我们就在内存中计算数据，但 Spark 是分布式的，不可避免地要进行网络传输，不可避免地进行内存与磁盘 I/O 的操作，以及磁盘 I/O 与网络 I/O 的操作，大量的磁盘与 I/O 的操作和磁盘与网络 I/O 的操作，这就构成了分布式的性能杀手。

如果要对最终计算结果进行排序，一般都会进行 sortByKey，如果以最终结果思考，可以认为是产生了一个很大很大的 partition，例如，使用 reduceByKey 的时候指定它的并行度，把 reduceByKey 的并行度变成 1，数据切片就变成 1，理论上讲，排序一般都会牵涉很多节点，如果把很多节点变成一个节点，然后进行排序，有时会取得更好的效果，因为数据就在一个节点上，就依靠一个进程进行排序。以前的一个项目就用到这一点：reduceByKey 之后变成一个数据分片，然后进行 mapPartitions，reduceByKey 是 Key-Value 的方式，对于 Key 值业务清楚其内容，mapPartitions 这里就是一个 Partition，Partition 内部可以按照 Key 进行比较，按照业务需要进行排序。还有另外一种算子 repartitionAndSortWithinPartitions，它在 Partitions 的过程中进行 Sort。SortByKey 默认是全局级别的，我们还是使用 reduceByKey 的时候指定它的并行度，把 reduceByKey 的并行度变为 1，这也是全局级别的。

Shuffle 还有一个很危险的地方，就是数据倾斜。例如，一个 ReduceTask 抓一部分数据，另一个 ReduceTask 抓一部分数据，Key 的分发时可能第一个节点处理了 99%的数据，第二个节点、第三个节点处理了 1%的数据。什么时候会导致数据倾斜？Shuffle 的时候会导致数据倾斜。读者可能会问，在计算的时候可能有些节点数据特别多，这不必担心，有两个层面：①Spark 的数据一般来源于 Hdfs，Hdfs 自动分配的数据尽可能地均匀分配。即使原始数据有数据倾斜，就算百亿级别的数据量计算，也是很快的，Spark 就很擅长在本地节点计算，很凸显 Spark 计算性能的地方；②但是，Shuffle 产生的数据倾斜，就须向上帝祈祷了。能否彻底掌握 Shuffle 及 Shuffle 中数据倾斜的所有内容是判断一个 Spark 高手的最直接的方式。

Shuffle 数据倾斜的问题会牵扯很多其他问题，例如，网络带宽、各种硬件故障、内存消耗 OOM、文件掉失。因为 Shuffle 的过程中会产生大量的磁盘 I/O、网络 I/O 以及压缩、解压缩、序列化和反序列化等。

Shuffle 的性能消耗包括磁盘消耗、内存消耗、网络消耗，以及分发数据过程中产生的数据倾斜。围绕 Shuffle 和数据倾斜有很多调优点。

- Mapper 端的 Buffer 设置为多大？Buffer 设置得大，可提升性能，减少磁盘 I/O，但是对内存有要求，对 GC 有压力；Buffer 设置得小，可能不占用那么多内存，但是可能频繁的磁盘 I/O、频繁的网络 I/O。spark.shuffle.file.buffer 默认是 32KB。
- Reducer 端的 Buffer 设置为多大？Buffer 设置得小，限制了每次拉取多少数据过来，spark.reducer.maxSizeInFlight 默认情况是 48MB，一次最大可以拉取 48MB 数据，如把数据变成 1GB，那拉取数据的次数变少了，拉取的效率变高了。假如变成 48KB，原来 10 次就够了，现在需要 1 万次。
- 在数据传输的过程中是否有压缩，以及使用什么方式压缩，默认使用 snappy 的压缩方式。使用 snappy 方式的好处是压缩速度快，但占用空间，可以根据业务需求使用配置其他的压缩方式。
- 网络传输失败重试的次数：网络传输有很大的风险，可能出故障。一般需要重试，那重试多少次，每次重试之间间隔多少时间也是非常重要的调优参数。腾讯曾经在做一个很大规模的图计算的时候，就是调优的时候调整了这两个参数，才让整个图计算分析工作得以顺利进行。例如，默认是重试 3 次，如果抓取 3 次还没有弄好，就宣告失败。在数据本来有，但重试 3 次抓不到数据的情况下，是上游可能在进行 GC，不响应请求。我们一般考虑将 3 次变成 30 次、50 次，或者变成 100 次。我们至少应将参数 spark.shuffle.io.maxRetries =30 次。
- 每次重试之间的时间间隔默认是 5s，为了减少重试次数，可以调大等待的时间。建议将时间间隔设置为 60s。

- 现在 Spark 是排序的 Shuffle，那作为业务，是否需要排序呢？不一定。SortShuffle 的弊端是浪费时间，SortShuffle 排序用的是全量的数据。在一定情况下，HashShuffle 比 SortShuffle 的性能更好，因为不需要排序，部分数据就可以进行 Shuffle。因此有一个参数开关 spark.shuffle.sort.bypassMergeThreshold，并行度多大的时候采用 SortShuffle 方式，什么时候采用 HashShuffle 方式，对性能的影响非常大。

31.6　Spark Hash Shuffle 源码解读与剖析

Spark 2.4.X 现在的版本已经没有 Hash Shuffle 的方式，那为什么我们还要讲解 Hash Shuffle 源码的内容呢？原因有 3 点：①在现在的实际生产环境下，很多人在用 Spark 1.5.X，实际上是在使用 Hash Shuffle 的方式；②Hash Shuffle 的方式是后续 Sort Shuffle 的基础；③在实际生产环境下，如果不需要排序，数据规模不是那么大，HashShuffle 的方式是性能比较好的一种方式，Spark Shuffle 是可以插拔的，我们可以进行配置。

本节基于 Spark 1.5.2 版本讲解 Hash Shuffle；Spark 1.6.3 是 Spark 1.6.0 中的一个版本，如果在生产环境中使用 Spark 1.x，最终都会转向 Spark 1.6.3。Spark 1.6.3 是 1.X 版本中的最后一个版本，也是最稳定、最强大的一个版本；Spark 2.4.3 是 Spark 最新版本，可以在生产环境中实验。

Shuffle 的过程是 Mapper 和 Reducer 以及网络传输构成的，Mapper 端会把自己的数据写入本地磁盘，Reducer 端会通过网络把数据抓取过来。Mapper 会先把数据缓存在内存中。默认情况下，缓存空间是 32KB，数据从内存到本地磁盘的一个过程就是写数据的一个过程。

这里有两个 Stage，上一个 Stage 叫 ShuffleMapTask，下一个 Stage 可能是 ShuffleMapTask，也有可能是 ResultsTask，取决于它这个任务是不是最后一个 Stage 产生的。ShuffleMapTask 会把我们处理的 RDD 的数据分成若干个 Bucket，即一个又一个的 Buffer。一个 Task 怎么去切分，具体要看你的 partitioner，ShuffleMapTask 肯定是属于具体的 Stage。

下面看一下 Spark 1.5.2 版本的 ShuffleMapTask，里面创建了一个 ShuffleWriter，它是负责把缓存中的数据写入本地磁盘，ShuffleWriter 写入入本地磁盘时，还有一个非常重要的工作，就是要跟 Spark 的 Driver 通信，告诉 Driver 把数据写到了什么地方，这样，下一个 Stage 找上一个 Stage 的数据的时候，通过 Driver(blockManagerMaster)去获取数据的位置信息，Driver(blockManagerMaster)会告诉下一个 Stage 中的 Task 需要的数据在哪里。ShuffleMapTask 的核心代码是 runTask。

Spark 1.5.2 版本的 ShuffleMapTask.scala 的源码如下：

```
1.   override def runTask(context: TaskContext): MapStatus = {
2.     //使用广播变量的 RDD 进行反序列化
3.     val deserializeStartTime = System.currentTimeMillis()
4.     val ser = SparkEnv.get.closureSerializer.newInstance()
5.     val (rdd, dep) = ser.deserialize[(RDD[_], ShuffleDependency[_, _,
       _])](
6.       ByteBuffer.wrap(taskBinary.value), Thread.currentThread
         .getContextClassLoader)
7.     _executorDeserializeTime = System.currentTimeMillis() -
       deserializeStartTime
8.
```

```
9.      metrics = Some(context.taskMetrics)
10.     var writer: ShuffleWriter[Any, Any] = null
11.     try {
12.       val manager = SparkEnv.get.shuffleManager
13.      writer = manager.getWriter[Any, Any](dep.shuffleHandle, partitionId,
         context)
14.       writer.write(rdd.iterator(partition,   context).asInstanceOf[Iterator
          [_ <: Product2[Any, Any]]])
15.       writer.stop(success = true).get
16.     } catch {
17.       case e: Exception =>
18.         try {
19.           if (writer != null) {
20.             writer.stop(success = false)
21.           }
22.         } catch {
23.           case e: Exception =>
24.             log.debug("Could not stop writer", e)
25.         }
26.         throw e
27.     }
28.   }
```

Spark 2.4.3 版本的 ShuffleMapTask.scala 的源码与 Spark 1.5.2 版本相比具有如下特点。

- 上面代码的第 2 行之后新增 threadMXBean 变量，构建系统托管 bean，用于 Java 虚拟机的线程系统。新增统计反序列化的开始时间 deserializeStartCpuTime，以及执行反序列化的时间_executorDeserializeCpuTime。
- 上段代码的第 9 行删除。

```
1.  ......
2.  val threadMXBean = ManagementFactory.getThreadMXBean
3.  ......
4.    val deserializeStartCpuTime = if (threadMXBean.isCurrentThreadCpuTime
      Supported) {
5.        threadMXBean.getCurrentThreadCpuTime
6.      } else 0L
7.  ......
8.  _executorDeserializeCpuTime = if (threadMXBean.isCurrentThreadCpuTime
    Supported) {
9.        threadMXBean.getCurrentThreadCpuTime - deserializeStartCpuTime
10.     } else 0L
11. ......
```

从 SparkEnv.get.shuffleManager 获取哈希的方式，查看 SparkEnv.scala 的 shuffleManager，在 Spark 1.5.2 版本中，有 3 种方式：HashShuffleManager、SortShuffleManager、UnsafeShuffleManager；在 Spark 1.5.2 版本中，默认也变成 SortShuffleManager 的方式。在配置 Spark Conf 的时候，可以进行配置。

Spark 1.5.2 版本的 SparkEnv.scala 的源码如下：

```
1.    val shortShuffleMgrNames = Map(
2.        "hash" -> "org.apache.spark.shuffle.hash.HashShuffleManager",
3.        "sort" -> "org.apache.spark.shuffle.sort.SortShuffleManager",
4.        "tungsten-sort" -> "org.apache.spark.shuffle.unsafe.UnsafeShuffle
          Manager")
```

```
5.      val shuffleMgrName = conf.get("spark.shuffle.manager", "sort")
6.      val shuffleMgrClass = shortShuffleMgrNames.getOrElse(shuffleMgrName
        .toLowerCase, shuffleMgrName)
7.      val shuffleManager = instantiateClass[ShuffleManager](shuffleMgrClass)
```

Spark 2.4.3 版本的 SparkEnv.scala 的源码与 Spark 1.5.2 版本相比具有如下特点。

- 上段代码第 2 行删除，Spark 2.4.3 版本已无哈希方式。
- 上段代码第 3 行、第 4 行使用 classOf[org.apache.spark.shuffle.sort.SortShuffleManager].getName 获取类名。
- 上段代码第 6 行 shuffleMgrName.toLowerCase 方法新增传入参数 Locale.ROOT。

```
1.     ......
2.     "sort" -> classOf[org.apache.spark.shuffle.sort.SortShuffleManager]
       .getName,
3.     "tungsten-sort" -> classOf[org.apache.spark.shuffle.sort.SortShuffleManager]
       .getName)
4.   ......
5.   val shuffleMgrClass =
6.         shortShuffleMgrNames.getOrElse(shuffleMgrName.toLowerCase(Locale
         .ROOT), shuffleMgrName)
7.
```

下面查看一下 HashShuffleManager.scala 的 getWriter 方法。

Spark 1.5.2 版本的 HashShuffleManager.scala 的源码如下：

```
1.       override def getWriter[K, V](handle: ShuffleHandle, mapId: Int,
         context: TaskContext)
2.       : ShuffleWriter[K, V] = {
3.     new HashShuffleWriter(
4.       shuffleBlockResolver, handle.asInstanceOf[BaseShuffleHandle[K, V,
         _]], mapId, context)
5.   }
```

Spark 2.4.3 版本已无 HashShuffleManager 代码。

从 getWriter 方式创建了 HashShuffleWriter 的实例对象，如果需要看它具体怎么写数据，必须看 HashShuffleWriter 类，然后它也必须有一个 write 方法。

HashShuffleWriter 类的 write 方法首先判断一下是否在 Mapper 端进行 aggregate 操作，也就是说，是否进行 Map Reduce 计算模型的 Local Reduce 本地聚合，如果有本地聚合操作，就循环遍历 Buffer 里面的数据，基于 records 进行聚合。例如，reduceByKey 操作。怎么进行聚合？取决于 reduceByKey 中传入的算子，如是加操作，还是乘操作。reduceByKey 将数据放到 Buffer 中聚合后，再写入本地 Block，在本地的聚合显现带来的好处是减少磁盘 I/O 的数据、操作磁盘 I/O 的次数、网络传输的数据量，以及这个 Reduce Task 抓取 Mapper Task 数据的次数，这个意义肯定是非常重大的。

Spark 1.5.2 版本的 HashShuffleWriter.scala 的 write 方法的源码如下：

```
1.   override def write(records: Iterator[Product2[K, V]]): Unit = {
2.     val iter = if (dep.aggregator.isDefined) {
3.       if (dep.mapSideCombine) {
4.         dep.aggregator.get.combineValuesByKey(records, context)
5.       } else {
6.         records
```

```
7.         }
8.       } else {
9.         require(!dep.mapSideCombine, "Map-side combine without Aggregator
         specified!")
10.        records
11.      }
12.
13.      for (elem <- iter) {
14.        val bucketId = dep.partitioner.getPartition(elem._1)
15.        shuffle.writers(bucketId).write(elem._1, elem._2)
16.      }
17.    }
```

Spark 2.4.3 版本已无 HashShuffleWriter 代码。

通过 HashShuffleWriter 的 write 代码可以看见，如果有本地聚合，就会在内存中完成聚合。例如，说 reduceByKey 是累加的话，就往累加上写数据，因为它是线性执行的；后面才是本地文件写数据，先获取 partitioner.getPartition，一个分片一个分片地写，如图 31-3 所示 bucketId 可以认为是内存操作的句柄，我们需要将 bucketId 传进去，然后使用 shuffle.writers (bucketId).write(elem._1, elem._2)写数据。

HashShuffleWriter 类中 write 方法的 shuffleBlockResolver.forMapTask 代码中，FileShuffle BlockResolver 类的 forMapTask 方法中的 ShuffleWriterGroup：如果启动了文件合并机制，写数据时，将很多的不同 Task 的相同 Key 的数据合并在同一个文件中，这就是 ShuffleWriter Group。里面会有一个判断 consolidateShuffleFiles，判断是否需要合并的过程。判断是否启动压缩机制，如果启动了压缩机制，就会有一个 fileGroup，否则就 getFile。

Spark 1.5.2 版本的 FileShuffleBlockResolver.scala 的 forMapTask 方法的源码如下：

```
1.     def forMapTask(shuffleId: Int, mapId: Int, numBuckets: Int, serializer:
       Serializer,
2.       writeMetrics: ShuffleWriteMetrics): ShuffleWriterGroup = {
3.     new ShuffleWriterGroup {
4.       shuffleStates.putIfAbsent(shuffleId, new ShuffleState(numBuckets))
5.       private val shuffleState = shuffleStates(shuffleId)
6.       private var fileGroup: ShuffleFileGroup = null
7.
8.       val openStartTime = System.nanoTime
9.       val serializerInstance = serializer.newInstance()
10.      val writers: Array[DiskBlockObjectWriter] = if (consolidateShuffleFiles) {
11.        fileGroup = getUnusedFileGroup()
12.        Array.tabulate[DiskBlockObjectWriter](numBuckets) { bucketId =>
13.          val blockId = ShuffleBlockId(shuffleId, mapId, bucketId)
14.          blockManager.getDiskWriter(blockId, fileGroup(bucketId), serializer
           Instance, bufferSize,
15.            writeMetrics)
16.        }
17.      } else {
18.        Array.tabulate[DiskBlockObjectWriter](numBuckets) { bucketId =>
19.          val blockId = ShuffleBlockId(shuffleId, mapId, bucketId)
20.          val blockFile = blockManager.diskBlockManager.getFile(blockId)
21.          //由于先前的失败，这个机器节点上可能已经存在 Shuffle 文件了，如果是，则删
           //除此文件
22.          if (blockFile.exists) {
23.            if (blockFile.delete()) {
24.              logInfo(s"Removed existing shuffle file $blockFile")
```

```
25.            } else {
26.              logWarning(s"Failed to remove existing shuffle file $blockFile")
27.            }
28.          }
29.         blockManager.getDiskWriter(blockId, blockFile, serializerInstance,
            bufferSize,
30.           writeMetrics)
31.         }
32.       }
33.       //创建文件和磁盘写入都涉及与磁盘的交互，所以应包括 Shuffle 写入时间
34.       writeMetrics.incShuffleWriteTime(System.nanoTime - openStartTime)
35.
36.       override def releaseWriters(success: Boolean) {
37.         if (consolidateShuffleFiles) {
38.           if (success) {
39.             val offsets = writers.map(_.fileSegment().offset)
40.             val lengths = writers.map(_.fileSegment().length)
41.             fileGroup.recordMapOutput(mapId, offsets, lengths)
42.           }
43.           recycleFileGroup(fileGroup)
44.         } else {
45.           shuffleState.completedMapTasks.add(mapId)
46.         }
47.       }
48.
49.       private def getUnusedFileGroup(): ShuffleFileGroup = {
50.         val fileGroup = shuffleState.unusedFileGroups.poll()
51.         if (fileGroup != null) fileGroup else newFileGroup()
52.       }
53.
54.       private def newFileGroup(): ShuffleFileGroup = {
55.         val fileId = shuffleState.nextFileId.getAndIncrement()
56.         val files = Array.tabulate[File](numBuckets) { bucketId =>
57.           val filename = physicalFileName(shuffleId, bucketId, fileId)
58.           blockManager.diskBlockManager.getFile(filename)
59.         }
60.         val fileGroup = new ShuffleFileGroup(shuffleId, fileId, files)
61.         shuffleState.allFileGroups.add(fileGroup)
62.         fileGroup
63.       }
64.
65.       private def recycleFileGroup(group: ShuffleFileGroup) {
66.         shuffleState.unusedFileGroups.add(group)
67.       }
68.     }
69.   }
```

Spark 2.4.3 版本已无 FileShuffleBlockResolver 代码。

是否进行 consolidateShuffleFiles，无论是哪种情况，最终都要写数据。写数据通过 blockManager 来实现，blockManager.getDiskWriter 把数据写到本地磁盘，就是很基本的 I/O 操作。

Spark 1.5.2 版本的 BlockManager.scala 的 getDiskWriter 的源码如下：

```
1.      def getDiskWriter(
2.        blockId: BlockId,
3.        file: File,
4.        serializerInstance: SerializerInstance,
```

```
5.        bufferSize:  Int,
6.        writeMetrics:  ShuffleWriteMetrics): DiskBlockObjectWriter = {
7.      val compressStream:  OutputStream => OutputStream = wrapForCompression
        (blockId,_)
8.      val syncWrites = conf.getBoolean("spark.shuffle.sync", false)
9.      new DiskBlockObjectWriter(blockId, file, serializerInstance, bufferSize,
        compressStream,
10.       syncWrites, writeMetrics)
11.   }
```

Spark 2.4.3 版本的 BlockManager.scala 的 getDiskWriter 的源码与 Spark 1.5.2 版本相比具有如下特点。

- 上段代码第 7 行代码已删除。
- 构建 DiskBlockObjectWriter 实例的参数顺序和名称进行了微调。

```
1.   ......
2.    new DiskBlockObjectWriter(file, serializerManager, serializerInstance,
      bufferSize,syncWrites, writeMetrics, blockId)
3.   ......
```

回到 HashShuffleWriter.scala，shuffleBlockResolver.forMapTask 传入的参数要注意：第一个参数是 shuffleId；第二个参数是 mapId；第三个参数是输出的 Split 个数；第四个参数是序列化器；第五个参数是 metric 来统计它的一些基本信息。

```
1.    private val shuffle = shuffleBlockResolver.forMapTask(dep.shuffleId,
      mapId, numOutputSplits, ser, writeMetrics)
```

HashShuffleWriter.scala 中先进行 forMapTask，然后进行 writer 操作。例如，reduceByKey 在本地进行了聚合，假设相同 Key 的 Value 有 10 000 个，原本需要写 10 000 次，objOut.writeKey(key)是写 key，objOut.writeValue(value)是写 value；但是，如果进行了本地聚合，将 10 000 个 Value 进行聚合，那只需要写 1 次。writer 写入部分的源码如下：

Spark 1.5.2 版本的 HashShuffleWriter.scala 的 writer 的源码如下：

```
1.    for (elem <- iter) {
2.    val bucketId = dep.partitioner.getPartition(elem._1)
3.    shuffle.writers(bucketId).write(elem._1, elem._2)
4.  }
```

Spark 2.4.3 版本已无 HashShuffleWriter 代码。

下面看一下 DiskBlockObjectWriter.scala 的 write 方法，这个 write 就是 Disk 级别的 write。

```
1.    def write(key:  Any, value:  Any) {
2.    if (!initialized) {
3.      open()
4.    }
5.
6.    objOut.writeKey(key)
7.    objOut.writeValue(value)
8.    recordWritten()
9.  }
```

HashShuffleWriter.scala 的 write 方法中 shuffle.writers(bucketId).write(elem._1, elem._2)中的 writers，把 bucketId 传进去，指明数据具体写在什么地方。

Spark 1.5.2 版本的 FileShuffleBlockResolver.scala 的源码如下：

```
1.   private[spark] trait ShuffleWriterGroup {
2.    val writers:  Array[DiskBlockObjectWriter]
3.
4.    /** @param success  表示所有写入成功。如果为 false，则没有块将被记录*/
5.    def releaseWriters(success:  Boolean)
6.  }
```

Spark 2.4.3 版本已无 FileShuffleBlockResolver 代码。

HashShuffle 在内存中有 bucket 缓存，在本地有磁盘文件，在调优的时候须注意内存和磁盘 I/O 的操作。再回看一下 HashShuffleWriter.scalad 的 write 写入代码 shuffle.writers (bucketId).write(elem._1, elem._2)，根据关联的 bucketId 将数据写入到本地文件中。循环遍历 iter 迭代器获取元素，write 方法写数据的时候传入 elem._1 和 elem._2 两个参数，其中 elem 的第一个元素 elem._1 是 Key 值，elem 的第二个元素 elem._2 是 value 值，即具体内容本身。

```
1.    for (elem <- iter) {
2.    val bucketId = dep.partitioner.getPartition(elem._1)
3.    shuffle.writers(bucketId).write(elem._1, elem._2)
4.  }
```

其中的 getPartition 代码，基于它的 key 分发到不同的 bucketId 上，Partitioner.scala 的 getPartition 是抽象类 Partitioner 的方法，没有具体实现。

```
1.   abstract class Partitioner extends Serializable {
2.    def numPartitions:  Int
3.    def getPartition(key:  Any):  Int
4.  }
```

下面看一下 Partitioner.scala 中 HashPartitioner 类的 getPartition 方法，把 key 值传进来。

Spark 默认的并行度会遗传，从上一个 Stage 传递到下一个 Stage。例如，如果上游有 4 个并行任务，下游也会有 4 个。

HashPartitioner 类的 getPartition 方法的源码如下：

```
1.   class HashPartitioner(partitions:  Int) extends Partitioner {
2.    require(partitions >= 0, s"Number of partitions ($partitions) cannot be
      negative.")
3.
4.    def numPartitions:  Int = partitions
5.
6.    def getPartition(key:  Any):  Int = key match {
7.      case null => 0
8.      case _ => Utils.nonNegativeMod(key.hashCode, numPartitions)
9.    }
10.
11.   override def equals(other:  Any):  Boolean = other match {
12.     case h:  HashPartitioner =>
13.       h.numPartitions == numPartitions
14.     case _ =>
15.       false
16.   }
17.
18.   override def hashCode:  Int = numPartitions
19. }
```

HashPartitioner 类的 getPartition 方法中调用了 nonNegativeMod 方法，定义了一个计算方式，传入两个参数：key 的 hashCode，以及要多少分片 numPartitions，就是普通的求模运算。

```
1.   def nonNegativeMod(x:  Int, mod:  Int):  Int = {
2.    val rawMod = x % mod
3.    rawMod + (if (rawMod < 0) mod else 0)
4.  }
```

基于 writer 的基础，看一下 reader。HashShuffleReader.scala 重点是看它的 Read 方法，首先会创建一个 ShuffleBlockFetcherIterator，这里有一个很重要的调优的参数 spark.reducer.maxSizeInFlight，也就是说，一次能最大抓取多少数据过来，在 Spark 1.5.2 默认情况下是 48MB，如果在内存足够大以及把 Shuffle 内存空间分配足够的情况下（Shuffle 默认占用 20%的内存空间），可以尝试调大这个参数，如可将 spark.reducer.maxSizeInFlight 调成 96MB，甚至更高。调大这个参数的好处是减少抓取次数，因为网络 I/O 的开销来建立新的连接其实很耗时。

HashShuffleReader.scala 的 Read 方法进行一个判断 mapSideCombine，是否需要聚合 aggregate，分别实现需要聚合及不需要聚合的操作；从 reducer 端借助 HashShuffleReader，从远程抓取数据，抓取数据过来后进行 aggregate 操作，至于汇聚之后进行分组或者还是 reduce 及其他一些操作，这是开发者决定的。

对于 Spark 1.5.2 版本 HashShuffleReader.scala，Spark 2.4.3 版本已将 HashShuffleReader 更名为 BlockStoreShuffleReader，相关功能在 BlockStoreShuffleReader 中实现。

这里谈到聚合，我们深入看一下 reduceByKey。reduceByKey 和 Hadoop 的 Map reduce 相比，有一个缺点：Hadoop 的 Map reduce 中无论业务是什么类型，Map reduce 都可以自定义，Map reduce 的业务逻辑都可以不一样。但 reduceByKey 有一个好处，可以很好地操作上一个 Stage 的算子，前面 Mapper 的算子也可以很好地操作下一个 Stage，具体 reduce 的算子。

PairRDDFunctions.scala 的 reduceByKey 方法的源码如下：

```
1.    def reduceByKey(partitioner:  Partitioner, func:  (V, V) => V):  RDD[(K,
      V)] = self.withScope {
2.      combineByKey[V]((v:  V) => v, func, func, partitioner)
3.    }
```

reduceByKey 方法里会调用 combineByKey。

combineByKey 方法中：

第一个参数 createCombiner，是所谓的 Combiner；例如，建立一个元素列表，将 V 类型转换为 C 类型。

第二个参数 mergeValue，在元素列表末尾追加元素，将 V 类型合并进 C 类型。

第三个参数 mergeCombiners，将两个 C 类型合并成 1 个。

第四个参数 partitioner，指定分区器。

其中，mapSideCombine 默认为 true，默认在 Mapper 端进行聚合。这里要注意 key 的类型不能是数组。第二个参数、第三个参数从 reduceByKey 的角度看是一样的。

```
1.   def combineByKey[C](createCombiner:  V => C,
2.       mergeValue:  (C, V) => C,
3.       mergeCombiners:  (C, C) => C,
4.       partitioner:  Partitioner,
```

```
5.       mapSideCombine:  Boolean = true,
6.       serializer:  Serializer = null):  RDD[(K, C)] = self.withScope {
7.
```

回到 HashShuffleReader.scala 的 read 方法，在 reducer 端抓取数据，需要进行网络通信的过程，那网络通信发生在什么时候呢？网络通信肯定由 read 方法中的 ShuffleBlockFetcherIterator 完成。

Spark 1.5.2 版本的 ShuffleBlockFetcherIterator.scala 的源码如下：

```
1.    final class ShuffleBlockFetcherIterator(
2.    context:  TaskContext,
3.    shuffleClient:  ShuffleClient,
4.    blockManager:  BlockManager,
5.    blocksByAddress:  Seq[(BlockManagerId, Seq[(BlockId, Long)])],
6.    maxBytesInFlight:  Long)
7.  extends Iterator[(BlockId, InputStream)] with Logging {
```

Spark 2.4.3 版本的 ShuffleBlockFetcherIterator.scala 的源码与 Spark 1.5.2 版本相比具有如下特点。

- 上段代码中第 5 行将 Seq 调整为 Iterator。
- 上段代码中第 5 行之后新增一行 streamWrapper 的代码，streamWrapper 用于封装返回输入流的函数。
- 上段代码中第 6 行之后新增代码 maxReqsInFlight、maxBlocksInFlightPerAddress、maxReqSizeShuffleToMem、detectCorrupt。maxReqsInFlight 是在任何给定点获取块的最大远程请求数；maxBlocksInFlightPerAddress 是在给定远程主机端口的任何给定点上获取的 shuffle 块的最大数目；maxReqSizeShuffleToMem 是 shuffle 到内存的请求的最大大小（字节）；detectCorrupt 是否检测获取块中的任何损坏。
- 上段代码中第 7 行调整继承自 DownloadFileManager。

```
1.    ......
2.    blocksByAddress: Iterator[(BlockManagerId, Seq[(BlockId, Long)])],
3.    ......
4.    streamWrapper: (BlockId, InputStream) => InputStream,
5.    ......
6.    maxReqsInFlight: Int,
   maxBlocksInFlightPerAddress: Int,
   maxReqSizeShuffleToMem: Long,
   detectCorrupt: Boolean)
7.    ......
8.    extends Iterator[(BlockId, InputStream)] with DownloadFileManager with
      Logging {
9.   ......
```

下面看一下 ShuffleBlockFetcherIterator.scala 的 initialize 方法。

Spark 1.5.2 版本的 ShuffleBlockFetcherIterator.scala 的源码如下：

```
1.    private[this] def initialize(): Unit = {
2.      //将任务完成回调（在成功案例和失败案例中调用）增加清理处理
3.      context.addTaskCompletionListener(_ => cleanup())
4.
5.      //切分本地和远程块
6.      val remoteRequests = splitLocalRemoteBlocks()
```

```
7.      //以随机顺序将远程请求增加到队列中
8.      fetchRequests ++= Utils.randomize(remoteRequests)
9.
10.     //发出块初始请求，达到 maxBytesInFlight
11.     while (fetchRequests.nonEmpty &&
12.       (bytesInFlight == 0 || bytesInFlight + fetchRequests.front.size <=
          maxBytesInFlight)) {
13.       sendRequest(fetchRequests.dequeue())
14.     }
15.
16.     val numFetches = remoteRequests.size - fetchRequests.size
17.     logInfo("Started " + numFetches + " remote fetches in" + Utils
        .getUsedTimeMs(startTime))
18.
19.     //获取本地块
20.     fetchLocalBlocks()
21.     logDebug("Got local blocks in " + Utils.getUsedTimeMs(startTime))
22.   }
```

Spark 2.4.3 版本的 ShuffleBlockFetcherIterator.scala 的源码与 Spark 1.5.2 版本相比具有如下特点。

- 将上段代码的第 10～14 行代码删除。
- 新增 0 == reqsInFlight 及 0 == bytesInFlight 的断言逻辑判断。
- 新增函数 fetchUpToMaxBytes()，发出块初始请求，达到我们的 maxBytesInFlight。

```
1.    ......
2.  assert ((0 == reqsInFlight) == (0 == bytesInFlight),
3.       "expected reqsInFlight = 0 but found reqsInFlight = " + reqsInFlight
         "+, expected bytesInFlight = 0 but found bytesInFlight = " +
         bytesInFlight)
4.
5.     //发出块初始请求，达到 maxBytesInFlight
6.     fetchUpToMaxBytes()
7.  ......
```

在 ShuffleBlockFetcherIterator.scala 的 initialize 方法中循环遍历，发生请求拉取数据，每次最大可以拉取 48MB 数据。下面看一下 ShuffleBlockFetcherIterator.scala 的 sendRequest 方法。

Spark 1.5.2 版本的 ShuffleBlockFetcherIterator.scala 的 sendRequest 方法的源码如下：

```
1.   private[this] def sendRequest(req: FetchRequest) {
2.      logDebug("Sending request for %d blocks (%s) from %s".format(
3.        req.blocks.size, Utils.bytesToString(req.size), req.address.hostPort))
4.      bytesInFlight += req.size
5.
6.      //这样，我们可以看到每个 blockId 大小
7.      val sizeMap = req.blocks.map { case (blockId, size) => (blockId
        .toString, size) }.toMap
8.      val blockIds = req.blocks.map(_._1.toString)
9.
10.     val address = req.address
11.    shuffleClient.fetchBlocks(address.host, address.port, address.executorId,
       blockIds.toArray,
12.      new BlockFetchingListener {
13.        override def onBlockFetchSuccess(blockId: String, buf: Managed
           Buffer): Unit = {
```

```
14.          //如果迭代器仍然处于活动状态，只在结果队列中添加缓冲区，cleanup()还没被
             //调用
15.          if (!isZombie) {
16.            //增加引用计数，因为我们需要把这个传递给不同的线程，需要在使用后释放
17.            buf.retain()
18.            results.put(new SuccessFetchResult(BlockId(blockId), address,
               sizeMap(blockId), buf))
19.            shuffleMetrics.incRemoteBytesRead(buf.size)
20.            shuffleMetrics.incRemoteBlocksFetched(1)
21.          }
22.          logTrace("Got remote block " + blockId + " after " + Utils.getUsed
             TimeMs(startTime))
23.        }
24.
25.        override def onBlockFetchFailure(blockId: String, e: Throwable):
           Unit = {
26.          logError(s"Failed to get block(s) from ${req.address.host}:${req
             .address.port}", e)
27.          results.put(new FailureFetchResult(BlockId(blockId),address, e))
28.        }
29.      }
30.    )
31.  }
```

Spark 2.4.3 版本的 ShuffleBlockFetcherIterator.scala 的 sendRequest 方法的源码与 Spark 1.5.2 版本相比具有如下特点。

- 上段代码中第 4 行之后新增一行代码 reqsInFlight += 1。
- 上段代码中第 8 行之前新增 remainingBlocks 的代码。
- 删除上段代码中的第 11 行和第 12 行，替换为新建 blockFetchingListener 的代码。
- 上段代码中第 15 行之前新增 ShuffleBlockFetcherIterator 同步块的代码。
- 上段代码中第 18 行之前新增 remainingBlocks -= blockId 的代码。
- 上段代码中第 18 行 SuccessFetchResult 实例中新增 remainingBlocks.isEmpty 的参数，并新增 remainingBlocks 的日志。
- 删除上段代码中的第 19 行和第 20 行。
- 上段代码中第 29 行之后新增代码，当请求大于 maxReqSizeShuffleToMem，获取远程 Shuffle 块保存到磁盘中。

```
1.  ......
2.  reqsInFlight += 1
3.  ......
4.  val remainingBlocks = new HashSet[String]() ++= sizeMap.keys
5.  ......
6.  val blockFetchingListener = new BlockFetchingListener {
7.  ......
8.  ShuffleBlockFetcherIterator.this.synchronized {
9.  ......
10. remainingBlocks -= blockId
11. results.put(new SuccessFetchResult(BlockId(blockId), address,
12.             sizeMap(blockId), buf,remainingBlocks.isEmpty))
13. logDebug("remainingBlocks: " + remainingBlocks)
14. ......
15.   //当请求太大时，将远程 Shuffle 块提取到磁盘，Shuffle 数据已经通过加密和压缩（根据
      //相关配置），我们可以直接获取数据并将其写入文件
```

```
16. if (req.size > maxReqSizeShuffleToMem) {
17.     shuffleClient.fetchBlocks(address.host, address.port,
    address.executorId, blockIds.toArray,
18.         blockFetchingListener, this)
19.     } else {
20.       shuffleClient.fetchBlocks(address.host, address.port,
    address.executorId, blockIds.toArray,
21.         blockFetchingListener, null)
22.     }
23.   }
```

sendRequest 方法中的 shuffleClient.fetchBlocks(address.host, address.port, address.executorId, blockIds.toArray,blockFetchingListener,null)代码中就有 host、port、机器的域名和端口、executorId、blockIds 等相关信息，抓到信息后会有 BlockFetchingListener 进行结果的处理。

下面看一下 ShuffleClient .scala 的 fetchBlocks 方法，从远程的节点中同步读取数据。

Spark 1.5.2 版本的 ShuffleClient .scala 的 fetchBlocks 方法的源码如下：

```
1.      public abstract void fetchBlocks(
2.       String host,
3.       int port,
4.       String execId,
5.       String[] blockIds,
6.       BlockFetchingListener listener);
7.  }
```

Spark 2.4.3 版本的 ShuffleClient.scala 的源码与 Spark 1.5.2 版本相比具有如下特点：上段代码中的第 6 行新增传入 downloadFileManager 的参数，downloadFileManager 参数用于创建和清理临时文件，如果不是 null，则远程块将被移动到临时 shuffle 文件中以减少内存使用，否则，它们将保留在内存中。

```
1.      ……
2.  DownloadFileManager downloadFileManager);
3.  ……
```

fetchBlocks 具体实现的代码是 BlockTransferService.scala，里面仍没有具体实现方法。

Spark 1.5.2 版本的 BlockTransferService.scala 的 fetchBlocks 方法的源码如下：

```
1.   override def fetchBlocks(
2.       host: String,
3.       port: Int,
4.       execId: String,
5.       blockIds: Array[String],
6.       listener: BlockFetchingListener): Unit
```

Spark 2.4.3 版本的 BlockTransferService.scala 的 fetchBlocks 的源码与 Spark 1.5.2 版本相比具有如下特点：新增传入 shuffleFiles 的参数。

```
1.      ……
2.   tempFileManager: DownloadFileManager): Unit
3.  ……
```

继续查看 BlockTransferService.scala 子类的实现方式，是 NettyBlockTransferService.scala，Netty 是基于 NIO 的理念进行网络通信，互联网公司进行不同进程的通信一般都使用 Netty。说明它底层有一套通信框架。我们基于这套通信框架进行数据的请求和传输，查看

NettyBlockTransferService.scala 的 fetchBlocks 的源码。

Spark 1.5.2 版本的 NettyBlockTransferService.scala 的 fetchBlocks 方法的源码如下：

```
1.   override def fetchBlocks(
2.       host: String,
3.       port: Int,
4.       execId: String,
5.       blockIds: Array[String],
6.       listener: BlockFetchingListener): Unit = {
7.     logTrace(s"Fetch blocks from $host: $port (executor id $execId)")
8.     try {
9.       val blockFetchStarter = new RetryingBlockFetcher.BlockFetchStarter {
10.        override def createAndStart(blockIds: Array[String], listener:
         BlockFetchingListener) {
11.          val client = clientFactory.createClient(host, port)
12.          new OneForOneBlockFetcher(client, appId, execId, blockIds.toArray,
             listener).start()
13.        }
14.      }
15.
16.      val maxRetries = transportConf.maxIORetries()
17.      if (maxRetries > 0) {
18.        //注意，Fetcher 将正确处理 maxRetries 等于 0 的情况；避免它在代码中发生 Bug,
           //一旦确定了稳定性，就应该删除 if 语句
19.        new RetryingBlockFetcher(transportConf, blockFetchStarter, blockIds,
           listener).start()
20.      } else {
21.        blockFetchStarter.createAndStart(blockIds, listener)
22.      }
23.    } catch {
24.      case e: Exception =>
25.        logError("Exception while beginning fetchBlocks", e)
26.        blockIds.foreach(listener.onBlockFetchFailure(_, e))
27.    }
28.  }
```

Spark 2.4.3 版本的 NettyBlockTransferService.scala 的 fetchBlocks 的源码与 Spark 1.5.2 版本相比具有如下特点。

- 上段代码中的第 6 行之后，fetchBlocks 方法中新增传入 tempFileManager 参数。
- 上段代码中的第 12 行构建 OneForOneBlockFetcher 实例时，新增 transportConf、tempFileManager 参数。

```
1. ......
2. tempFileManager: DownloadFileManager): Unit = {
3. ......
4.        new OneForOneBlockFetcher(client, appId, execId, blockIds, listener,
          transportConf, tempFileManager).start()
5. ......
```

fetchBlocks 方法中第 16 行的 maxRetries 是最大的重试次数， createAndStart 是底层的 Netty 实现方法。

其中，OneForOneBlockFetcher(client, appId, execId, blockIds.toArray, listener).start()就开始了通信过程。下面看一下 OneForOneBlockFetcher.java 的 start 方法的源码。

Spark 1.5.2 版本的 OneForOneBlockFetcher.java 的 start 方法的源码如下：

```
1.    public void start() {
2.      if (blockIds.length == 0) {
3.        throw new IllegalArgumentException("Zero-sized blockIds array");
4.      }
5.
6.      client.sendRpc(openMessage.toByteArray(), new RpcResponseCallback() {
7.        @Override
8.        public void onSuccess(byte[] response) {
9.          try {
10.           streamHandle = (StreamHandle) BlockTransferMessage.Decoder
              .fromByteArray(response);
11.           logger.trace("Successfully opened blocks {}, preparing to fetch
              chunks.", streamHandle);
12.
13.           //立即请求所有的块——我们期望请求的总大小是合理的，因为是在
              //[ShuffleBlockFetcherIterator]中更高层次的分块
14.           for (int i = 0; i < streamHandle.numChunks; i++) {
15.             client.fetchChunk(streamHandle.streamId, i, chunkCallback);
16.           }
17.         } catch (Exception e) {
18.           logger.error("Failed while starting block fetches after success", e);
19.           failRemainingBlocks(blockIds, e);
20.         }
21.       }
22.
23.       @Override
24.       public void onFailure(Throwable e) {
25.         logger.error("Failed while starting block fetches", e);
26.         failRemainingBlocks(blockIds, e);
27.       }
28.     });
29.   }
```

Spark 2.4.3 版本的 OneForOneBlockFetcher.java 的 start 方法的源码与 Spark 1.5.2 版本相比具有如下特点。

- 上段代码的第8行 onSuccess 方法的 response 参数类型由 byte[]类型调整为 ByteBuffer 类型。
- 新增 shuffleFiles 文件是否为空的逻辑判断处理。

```
1.  ......
2.  public void onSuccess(ByteBuffer response) {
3.  ......
4.  if (shuffleFiles != null) {
5.            client.stream(OneForOneStreamManager.genStreamChunkId(streamHandle
              .streamId, i),
6.                new DownloadCallback(shuffleFiles[i], i));
7.            } else {
8.  ......
```

如果进一步查看 fetch 内容，一般情况下会提供 HashMap 的数据结构，为了将数据聚合起来，总结如下。

（1）Shuffle 存数据和抓取数据，就是普通的 Scala 和 Java 编程，思想是一样的，有个缓存，然后往磁盘写，不过，Spark 是分布式系统，要跟 Driver 的管理器进行合作，或者说受 Driver 的控制，如写数据的时候告诉 Driver 数据写在哪里，然后下一个阶段要去读数据，到

Driver 中去要数据。Driver 会清晰地告诉你要读取的数据在哪里。具体读数据的过程是 Netty 的 rpc 框架，是基本的 I/O 操作。

（2）Reducer 端如果内存不够写磁盘，代价是双倍的。Mapper 如果内存不够，要写磁盘，不管够不够，都只写 1 次；而 Reducer 端如果内存不够，将数据存到磁盘，在计算数据的时候，又再次将数据从磁盘上抓回来。这时有一个很重要的调优参数，就是将 Shuffle 的内存适当调大。Shuffle 的内存默认占 20%，可以调大到 30%，但也不能太大，因为要进行 Persist，Persist 在磁盘中占用的空间会越来越小。

31.7　Sort-Based Shuffle 产生的内幕及其 tungsten-sort 背景解密

在历史的发展中，为什么 Spark 最终放弃了 HashShuffle，而使用 Sorted-Based Shuffle，而且作为后起之秀的 Tungsten-based Shuffle 到底是在什么样的背景下产生的。Tungsten-Sort Shuffle 已经并入了 Sorted-Based Shuffle，Spark 的引擎会自动识别程序需要原生的 Sorted-Based Shuffle，还是用 Tungsten-Sort Shuffle，那识别的依据是什么。其实，Spark 会检查相对的应用程序有没有 Aggregate 的操作。Sorted-Based Shuffle 也有缺点，其缺点反而是它排序的特性，它强制要求数据在 Mapper 端必须先进行排序（注意，这里没有说对计算结果进行排序），所以导致它排序的速度有点慢。而 Tungsten-Sort Shuffle 对它的排序算法进行了改进，优化了排序的速度。

Spark Sorted-Based Shuffle 的诞生：为什么 Spark 用 Sorted-Based Shuffle，而放弃了 Hash-Based Shuffle？在 Spark 里，为什么最终是 Sorted-Based Shuffle 成为核心，基本了解过 Spark 的学习者都会知道，如图 31-4 所示，Spark 会根据宽依赖把它一系列的算子划分成不同的 Stage，Stage 的内部会进行 Pipeline，Stage 与 Stage 之间进行 Shuffle。Shuffle 的过程包含 3 部分。

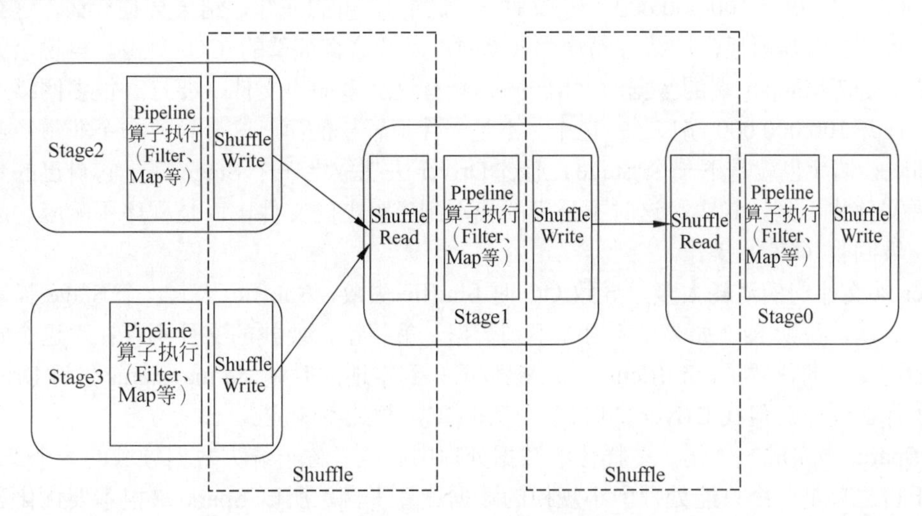

图 31-4　Shuffle 的过程

第一部分是 Shuffle 的 Write；第二部分是网络的传输；第三部分是 Shuffle 的 Read，这三大部分设置了内存操作、磁盘 I/O、网络 I/O 以及 JVM 的管理。而这些影响了 Spark 应用程序在 95%以上的效率，假设程序代码是非常好的情况下，Spark 性能的 95%都消耗在 Shuffle 阶段的本地写磁盘文件、网络传输数据以及抓取数据这样的生命周期中，如图 31-5 所示。

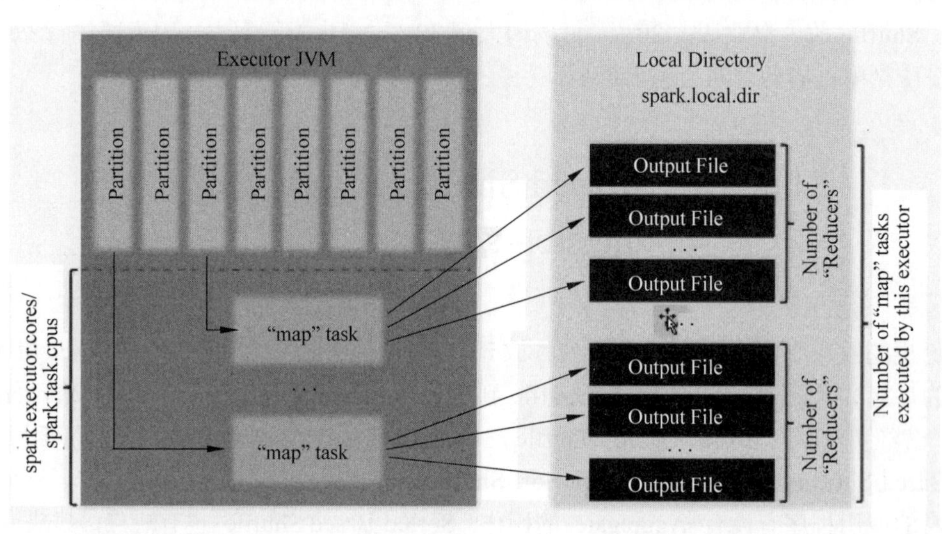

图 31-5　Shuffle 示意图

在 Shuffle 写数据的时候，内存中有一个缓存区叫 Buffer，可以想象成一个 Map，同时在本地磁盘有对应的本地文件。如果本地磁盘有文件，你在内存中肯定也需要有相应的管理句柄。也就是说，单从 ShuffleWrite 内存占用的角度讲，已经有一部分内存空间是用在存储 Buffer 数据的，另一部分内存空间是用来管理文件句柄的，回顾 HashShuffle 产生小文件的个数是 Mapper 分片数量 × Reducer 分片数量（M×R）。例如，Mapper 端有 10 000 个数据分片，Reducer 端也有 10 000 个数据分片，在 HashShuffle 的机制下，它在本地内存空间中会产生 10 000×10 000 = 100 000 000 个小文件（1 亿个），可想而知，结果会是什么，这么多的 I/O，这么多的内存消耗、这么容易产生 OOM，以及这么沉重的 CG 负担。再说，如果 Reducer 端去读取 Mapper 端的数据时，Mapper 端有这么多的小文件，要打开很多网络通道去读数据，打开 100 000 000 端口（1 亿个）不是一件很轻松的事。这会导致一个非常经典的错误：Reducer 端（也就是下一个 Stage）通过 Driver 去抓取上一个 Stage 属于它自己的数据的时候，说文件找不到。其实，这时不是真的找不到磁盘上的文件，而是程序不响应，因为它在进行垃圾回收（GC）操作。

Mapper 端文件句柄消耗太多，导致 GC 时 Shuffle 失败，Shuffle 从上一个 Stage 抓取数据，默认是 3 次，每次 5s，如果 15s 内抓不到数据，就报错，要找的文件不存在。这个时候其实已经知道文件的位置了，shufflemaptask 把数据写到本地的时候，将 mapstatus 告诉 Driver，数据放到了什么地方。但找 Driver 的时候，正在 GC，导致不响应。

因为 Spark 想完成一体化、多样化的数据处理中心或者称一统大数据领域的一个美梦，肯定不甘于自己只是一个只能处理中小规模的数据计算平台，所以 Spark 最根本要优化和逼切解决的问题是：减少 Mapper 端 ShuffleWriter 产生的文件数量，这样便可以能让 Spark 从几百台集群的规模中瞬间变成可以支持几千台，甚至几万台集群的规模（一个 Task 背后可能是一个 Core 去运行，也可能是多个 Core 去运行，但默认情况下是用一个 Core 去运行一个 Task）。

Spark shuffle 改进的根本在于：减少 Mapper 端产生的 shuffle Writer 文件的数量，这是精髓之所在，所有的学习 Spark Shuffle 从这个角度出发，才能够最直接地理解 Spark shuffle 不同版本的精髓，并进行最大程度的性能调优！

减少 Mapper 端的小文件带来的好处如下。

- Mapper 端的内存占用变少了。
- Spark 不仅仅可以处理小规模的数据，处理大规模的数据也不会很容易达到性能瓶颈。
- Reducer 端抓取数据的次数变少了。
- 网络通道的句柄也变少。
- 极大地减少 Reducer 的内存不仅是因为数据级别的消耗，而且是框架时要运行的必须消耗。

Sorted-Based Shuffle 的出现，最显著的优势是把 Spark 从只能处理中小规模的数据平台，变成可以处理无限大规模的数据平台。可能你会问规模真这么重要吗？当然，集群规模意味着它处理数据的规模，也意味着它的运算能力。

Sorted-Based Shuffle 不会为每个 Reducer 中的 Task 生产一个单独的文件，相反，Sorted-Based Shuffle 会把 Mapper 中每个 ShuffleMapTask 所有的输出数据 Data 只写到一个文件中，因为每个 ShuffleMapTask 中的数据会被分类，所以 Sort-based Shuffle 使用 index 文件存储具体 ShuffleMapTask 输出数据在同一个 Data 文件中是如何分类的信息。所以，基于 Sort-based Shuffle 会在 Mapper 中的每个 ShuffleMapTask 中产生两个文件（并发度的个数×2），如图 31-6 所示。

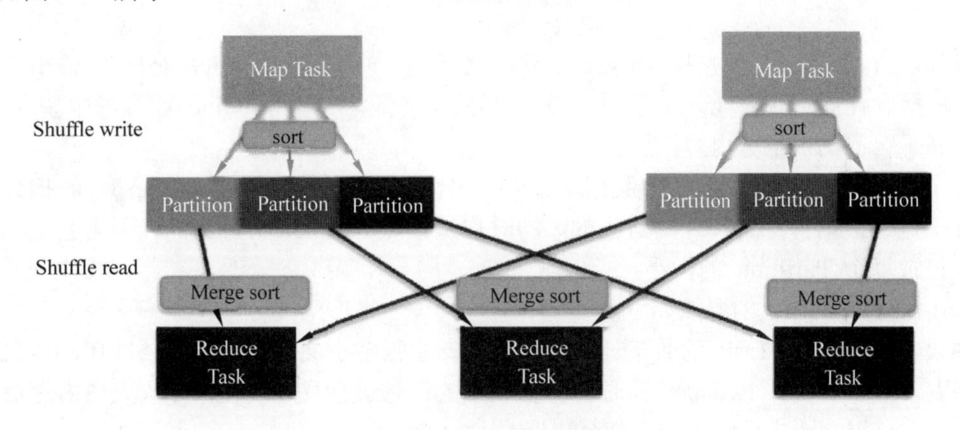

图 31-6 Sorted-Based Shuffle

它会产生一个 Data 文件和一个 Index 文件，其中 Data 文件是存储当前 Task 的 Shuffle 输出的，而 Index 文件则存储了 Data 文件中的数据通过 Partitioner 的分类信息，此时下一个阶段的 Stage 中的 Task 就是根据这个 Index 文件获取自己需要抓取的上一个 Stage 中 ShuffleMapTask 产生的数据。

假设现在 Mapper 端有 10 000 个数据分片，Reducer 端也有 10 000 个数据分片，它的并发度是 100，使用 Sorted-Based Shuffle 会产生多少个 Mapper 端的小文件，答案是 100×2 = 200 个。它的 MapTask 会独自运行，每个 MapTask 在运行的时候写两个文件，运行成功后就不需要这个 MapTask 的文件句柄，无论是文件本身的句柄，还是索引的句柄，都不需要，所以如果它的并发度是 100 个 Core，每次运行 100 个任务，它最终只会占用 200 个文件句柄，这与 HashShuffle 的机制不一样，HashShuffle 最差的情况是 Hashed 句柄存储在内存中的。

Sorted-Based Shuffle 主要在 Mapper 阶段，这个与 Reducer 端没有任何关系，在 Mapper 阶段，它要进行排序，你可以认为是二次排序，它的原理是有两个 Key 进行排序，第一个是 PartitionId 进行排序，第二个是本身数据的 Key 进行排序。如图 31-7 所示，它会把 PartitionId 分成 3 个，索引分别为 0、1、2，这个在 Mapper 端进行排序的过程其实是让 Reducer 去抓取数据的时候变得更高效。例如，说第一个 Reducer，它会到 Mapper 端的索引为 0 的数据分片中抓取数据。

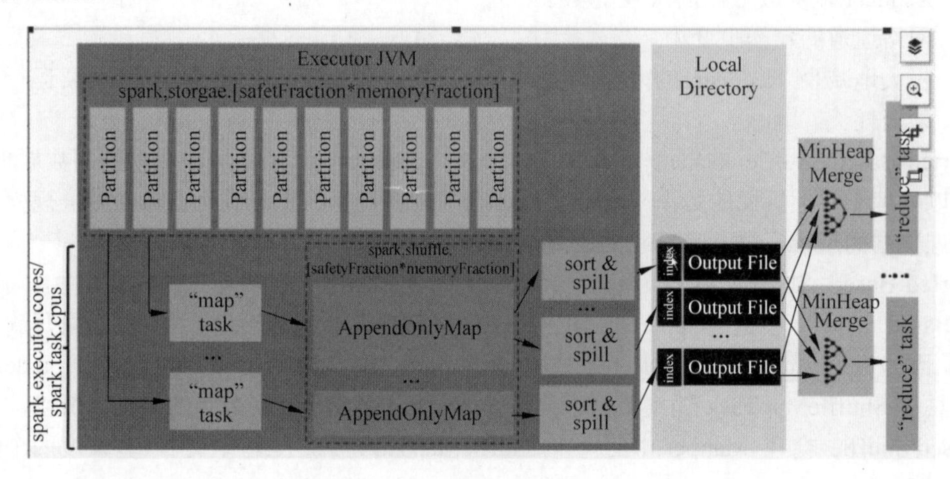

图 31-7　Sorted-Based Shuffle 示意图

具体而言，Reducer 首先找 Driver 去获取父 Stage 中每个 ShuffleMapTask 输出的位置信息，根据位置信息获取 Index 文件，解析 Index 文件，从解析的 Index 文件中获取 Data 文件中属于自己的那部分内容。

一个 Mapper 任务除了有一个数据文件外，它也会有一个索引文件，Map Task 把数据写到文件磁盘是根据自身的 Key 写进去的，同时也是按照 Partition 写进去的，因为它是顺序写数据，记录每个 Partition 的大小。

Sort-Based Shuffle 的弱点如下。

如果 Mapper 中 Task 的数量过大，依旧会产生很多小文件，此时在 Shuffle 传数据的过程中到 Reducer 端，Reducer 同时会需要大量的记录进行反序列化，导致大量内存消耗和 GC 的巨大负担，造成系统缓慢，甚至崩溃！

强制在 Mapper 端必须排序，这里的前提是本身数据根本不需要排序。

如果在分片内也进行排序，此时需要进行 Mapper 端和 Reducer 端的两次排序！

它要基于记录本身进行排序，这就是 Sort-Based Shuffle 最致命的性能消耗。

Spark 2.4.X 之后，现在只有一种 SortShuffleManager，废弃了 HashShuffleManager；后起之秀 tungsten-sort 并入了 SortShuffleManager（根据是否是 aggregate）。aggregate 的情况不适用于 tungsten-sort。

31.8　Spark Shuffle 令人费解的 6 大经典问题

（1）Shuffle 的第一大问题：什么时候进行 Shuffle 的抓取操作？Shuffle 具体在什么时候开始运行（是在一边 Mapper 的 Map 操作，同时进行 Reducer 端的 Shuffle 的 Reduce 操作吗）？

错误的观点：Spark 是一遍 Mapper 一遍 Shuffle，而 Hadoop 的 MapReduce 是先完成 Mapper，然后才进行 Reducer 的 Shuffle。

事实：Spark 一定先完成 Mapper 端所有的 Tasks，才会进行 Reducer 端的 Shuffle 过程。

原因：Spark 的 Job 是按照 stage 线性执行的，前面的 Stage 必须执行完，才能够执行后面的 Reducer 的 Shuffle 过程。

补充说明：Spark 的 Shuffle 是边拉取数据，边进行 Aggregate 操作。其实与 Hadoop MapReduce 相比，优势确实在速度上。但是也会导致一些算法不容易实现，如求平均值等（但是，Spark 提供了一些内置函数）。

（2）Shuffle 的第二大问题：Shuffle 抓过来的数据到底放到了哪里？

抓过来的数据首先肯定是放在 Reducer 端的内存缓冲区中的，Spark 曾经有版本要求只能放在内存缓存中，数据结构类似于 HashMap（AppendOnlyMap），显然特别消耗内存和极易出现 OOM，同时也从 Reducer 端极大地限制了 Spark 集群的规模，现在的实现都是内存+磁盘的方式（数据结构使用 ExternalAppendOnlyMap）。当然，大家也可以通过 spark.shuffle.spill=false 设置只能使用内存，使用 ExternalAppendOnlyMap 的方式时如果内存使用达到一定临界值后，会首先尝试在内存中扩大 ExternalAppendOnlyMap（内部有实现算法），如果不能扩容，才会 Spill 到磁盘。

（3）Shuffle 的第三大问题：Shuffle 的数据在 Mapper 端如何存储，在 Reducer 端又是如何知道数据具体在哪里的？

在 Spark 的实现中，每个 Stage（里面是 ShuffleMapTask）中的 Task 在 Stage 的最后一个 RDD 上一定会注册给 Driver 上的 MapOutputTrackerMaster，Mapper 和 MapOutputTrackerMaster 汇报 ShuffleMapTask 具体数据的位置（具体的输出文件及内容和 Reduce 有关）。Reducer 是向 Driver 中的 MapOutputTrackerMaster 请求数据的元数据，然后和 Mapper 所在的 Executor 进行通信。

（4）Shuffle 的第四大问题：仅从 HashShuffle 的角度讲，我们在 Shuffle 的时候到底可以产生多少 Mapper 端的中间文件？

例如，有 M 个 Mapper、R 个 Reducer 和 C 个 Core，那么 HashShuffle 可以产生多少个 Mapper 的中间文件？如果回答是 $M\times R$ 个临时中间文件，就是有问题的，我们从另外一个角度来说明一下。例如，在实际生产环境中有 Executors（如 100 个），每个 Executor 上有 C 个 Cores（如 10 个），同时有 R 个 Reducer。在 Hash Shuffle 情况下会产生多少 Mapper 端的中间文件呢？是否可以回答 $E\times C\times R$ 个临时文件呢？

答案：在没有 Consolidation 机制的情况下，第一个问题会产生 $M\times R$ 个中间文件，第二个问题的答案是实际的 Task 的个数×R；在有 Consolidation 机制的情况下，第一个问题会产生 $C\times R$ 个文件吗？不一定。这取决于一个越来越重要的配置参数 spark.task.cpus（该参数决定了运行 Spark 的每个 task 需要多少个 Cores，默认情况是 1 个），假如 Spark.task.cpus 为 T，那么第一个问题的答案是 $C/T \times R$，第二个问题的答案是 $E \times C/T \times R$；如何理解 Consolidation 机制，可以认为是文件池的复用。

（5）Shuffle 的第五大问题：Spark 中 Sorted-Based Shuffle 数据结果默认是排序的吗？Sorted-Based Shuffle 采用什么排序算法？这个排序算法的好处是什么？

Spark Sorted-Based Shuffle 在 Mapper 端是排序的，包括 Partition 的排序和每个 Partition 内部元素的排序！但在 Reducer 端是没有进行排序的，所以 Job 的结果默认不是排序的。Sorted-Based Shuffle 采用 Tim-Sort 排序算法的，好处是可以极为高效地使用 Mapper 端的排序成果全局排序。

（6）Shuffle 的第六大问题：Spark Tungsten-Sorted Shuffle 在 Mapper 中会对内部元素进行排序吗？Tungsten-Sorted Shuffle 不适用于什么情况？

Tungsten-Sorted Shuffle 在 Mapper 中不会对内部元素进行排序（它只会对 Partition 进行排序），原因是它自己管理的二进制序列化后的数据，问题来啦：数据是进入 Buffer 时或者是进入磁盘时才进行排序？答案是数据的排序是发生在 Buffer 要满了 Spill 到磁盘时才进行排序的。所以，Tungsten-Sorted Shuffle 对内部不会进行排序。

Tungsten-Sorted Shuffle 什么时候会退化成为 Sorted-Based Shuffle？它是在程序有 Aggregate 操作的时候；或者是 Mapper 端输出的 Partition 大于 16 777 216；或者是一条 Record 大于 128MB 的时候，原因也是因为它自己管理的二进制序列化后的数据以及数组指针管理范围。

31.9　Spark Sort-Based Shuffle 排序具体实现内幕和源码详解

为什么讲解 Sorted-Based shuffle？有两方面的原因。

（1）可能有些朋友看到 Sorted-Based Shuffle 的时候，会有一个误解，认为 Spark 基于 Sorted-Based Shuffle 产出的结果是有序的。

（2）Sorted-Based Shuffle 要排序，涉及一个排序算法。这部分内容可选学。

Sorted-Based Shuffle 的核心是借助于 ExternalSorter 把每个 ShuffleMapTask 的输出排序到一个文件中（FileSegmentGroup），为了区分下一阶段 Reducer Task 不同的内容，它还需要有一个索引文件（Index）来告诉下游 Stage 的并行任务，那一部分是属于你的，如图 31-8 所示。

Shuffle Map Task 在 ExternalSorter 溢出到磁盘的时候，产生一组 File（File Group 是 hashShuffle 中的概念，理解为一个 file 文件池，这里为区分，使用 File 的概念，FileSegment 根据 PartionID 排序）和一个索引文件，File 里的 FileSegement 会进行排序，在 Reducer 端有 4 个 Reducer Task，下游的 Task 可以很容易根据索引（index）定位到这个 File 中的哪部分 FileSegement 是属于下游的，它相当于一个指针，下游的 Task 要向 Driver 确定文件在哪里，然后到了这个 File 文件所在的地方，实际上会与 BlockManager 进行沟通，BlockManager 首先会读一个 Index 文件，根据它的命名规则进行解析，例如说下一个阶段的第一个 Task，一般就是抓取第一个 Segment，这是一个指针定位的过程。

再次强调，Sort-Based Shuffle 最大的意义是减少临时文件的输出数量，且只会产生两个文件：一个是包含不同内容划分成不同 FileSegment 构成的单一文件 File；另一个是索引文

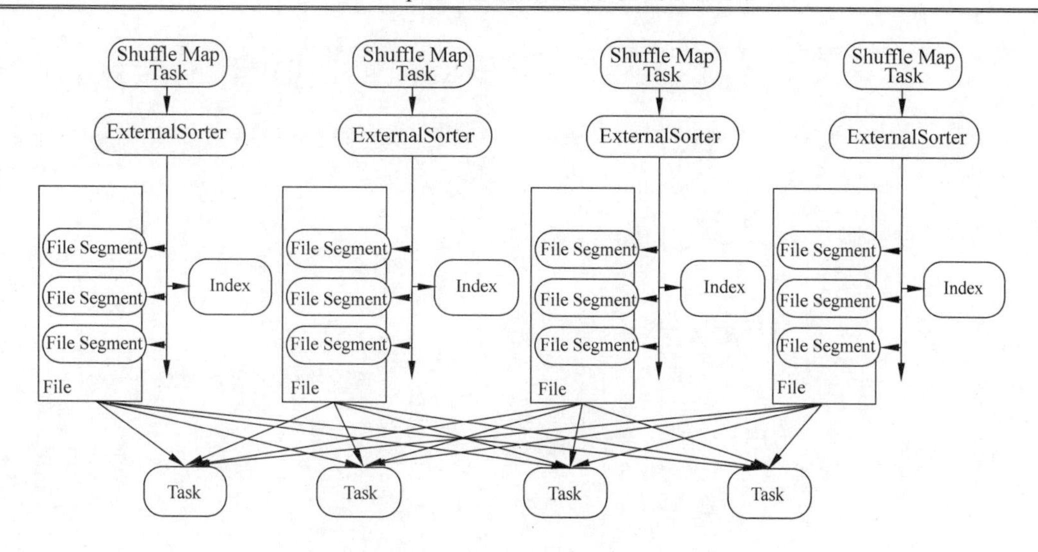

图 31-8　Sorted-Based Shuffle 图

件 Index。

一件很重要的事情：在 Sorted-Shuffle 中会排序吗？从测试结果看，一般不排序（例如，可以在 Spark 2.4.3 中做一个 wordcount 测试，结果是不排序的）。

Sort-Based Shuffle　Mapper 端的 Sort and Spill 的过程中，AppendOnlyMap 时不进行排序，Spill 到磁盘时再进行排序。

现在从源码的角度看一下 Sorted-Based Shuffle 的排序，默认情况是 sort 类型，全称 org.apache.spark.shuffle.sort.SortShuffleManager。

进入 org.apache.spark.shuffle.sort.SortShuffleManager，在 SortShuffleManager 中没找到 ExternalSorter，那我们从 ShuffleMapTask 中去看是怎么写数据的。

ShuffleMapTask.scala 的源码如下：

```
1.   override def runTask(context: TaskContext): MapStatus = {
2.     //使用广播变量的 RDD 反序列化
3.     val threadMXBean = ManagementFactory.getThreadMXBean
4.     val deserializeStartTime = System.currentTimeMillis()
5.     val deserializeStartCpuTime = if (threadMXBean.isCurrentThread
       CpuTimeSupported) {
6.       threadMXBean.getCurrentThreadCpuTime
7.     } else 0L
8.     val ser = SparkEnv.get.closureSerializer.newInstance()
9.     val (rdd, dep) = ser.deserialize[(RDD[_], ShuffleDependency[_, _,
        _])](
10.      ByteBuffer.wrap(taskBinary.value), Thread.currentThread
         .getContextClassLoader)
11.    _executorDeserializeTime = System.currentTimeMillis() - deserialize
       StartTime
12.    _executorDeserializeCpuTime = if (threadMXBean.isCurrentThreadCpuTime
       Supported) {
13.      threadMXBean.getCurrentThreadCpuTime - deserializeStartCpuTime
14.    } else 0L
15.
16.    var writer: ShuffleWriter[Any, Any] = null
17.    try {
```

```
18.         val manager = SparkEnv.get.shuffleManager
19.        writer = manager.getWriter[Any, Any](dep.shuffleHandle, partitionId,
           context)
20.         writer.write(rdd.iterator(partition,context).asInstanceOf[Iterator
            [_<: Product2[Any, Any]]])
21.         writer.stop(success = true).get
22.       } catch {
23.         case e: Exception =>
24.           try {
25.             if (writer != null) {
26.               writer.stop(success = false)
27.             }
28.           } catch {
29.             case e: Exception =>
30.               log.debug("Could not stop writer", e)
31.           }
32.           throw e
33.       }
34.   }
```

其中，manager = SparkEnv.get.shuffleManager 是从 SparkEnv 中通过反射获取的 shuffleManager，就是 SortShuffleManager。那 manager.getWriter 是 SortShuffleManager 的 getWriter。

SortShuffleManager.scala 的 getWriter 的源码如下：

```
1.    override def getWriter[K, V](
2.        handle: ShuffleHandle,
3.        mapId: Int,
4.        context: TaskContext): ShuffleWriter[K, V] = {
5.      numMapsForShuffle.putIfAbsent(
6.        handle.shuffleId, handle.asInstanceOf[BaseShuffleHandle[_, _,_]].
          numMaps)
7.      val env = SparkEnv.get
8.      handle match {
9.        case unsafeShuffleHandle: SerializedShuffleHandle[K @unchecked, V
          @unchecked] =>
10.         new UnsafeShuffleWriter(
11.           env.blockManager,
12.           shuffleBlockResolver.asInstanceOf[IndexShuffleBlockResolver],
13.           context.taskMemoryManager(),
14.           unsafeShuffleHandle,
15.           mapId,
16.           context,
17.           env.conf)
18.       case bypassMergeSortHandle: BypassMergeSortShuffleHandle[K @unchecked,
          V @unchecked] =>
19.         new BypassMergeSortShuffleWriter(
20.           env.blockManager,
21.           shuffleBlockResolver.asInstanceOf[IndexShuffleBlockResolver],
22.           bypassMergeSortHandle,
23.           mapId,
24.           context,
25.           env.conf)
26.       case other: BaseShuffleHandle[K @unchecked, V @unchecked, _] =>
27.         new SortShuffleWriter(shuffleBlockResolver, other, mapId, context)
28.     }
29.   }
```

SortShuffleManager getWriter Handle 提供了 3 种方式。

- unsafeShuffleHandle：tungsten 深度优化的方式。
- bypassMergeSortHandle：Sorted-Shuffle 在一定程度上可以退化为 hashShuffle 的方式。
- BaseShuffleHandle：是 SortShuffleWriter。

再回到之前 ShuffleMapTask 中，获取 shuflemanager getWriter 之后，要 write 写数据。

ShuffleMapTask.scala 的源码如下：

```
1.     var writer: ShuffleWriter[Any, Any] = null
2.     try {
3.       val manager = SparkEnv.get.shuffleManager
4.      writer = manager.getWriter[Any, Any](dep.shuffleHandle, partitionId,
        context)
5.       writer.write(rdd.iterator(partition, context).asInstanceOf[Iterator
         [_ <: Product2[Any, Any]]])
6.       writer.stop(success = true).get
```

SortShuffleWriter 的 write 方法的代码非常清晰、简洁。我们终于看到了 ExternalSorter。

SortShuffleWriter.scala 的源码如下：

```
1.    override def write(records: Iterator[Product2[K, V]]): Unit = {
2.     ……
3.       new ExternalSorter[K, V, C](
4.         context, dep.aggregator, Some(dep.partitioner), dep.keyOrdering,
           dep.serializer)
5.       ……
```

ExternalSorter.scala 中有两个很重要的数据结构。

（1）在 Mapper 端进行 combine：PartitionedAppendOnlyMap 是 map 类型的数据结构，map 是 key-value，在本地进行聚合，在本地 Key 值不变，Value 不断更新；Partitioned AppendOnlyMap 底层还是一个数组，基于数组实现 map 的原因是更节省空间，效率更高。那么，直接基于数组怎么实现 map：把数组的偶数标记设置为 map 的 Key 值，把奇数标记设置为 map 的 value 值。

（2）在 Mapper 端没有 combine：使用 PartitionedPairBuffer。

ExternalSorter .scala 的源码如下：

```
1.    @volatile private var map = new PartitionedAppendOnlyMap[K, C]
2.  @volatile private var buffer = new PartitionedPairBuffer[K, C]
```

下面看一下 insertAll 方法。

ExternalSorter.scala 的源码如下：

```
1.     def insertAll(records: Iterator[Product2[K, V]]): Unit = {
2.       //待办事项 TODO：如果发现汇聚系数不高，就停止合并
3.      val shouldCombine = aggregator.isDefined
4.
5.      if (shouldCombine) {
6.        //使用 AppendOnlyMap，首先在内存中组合值
7.        val mergeValue = aggregator.get.mergeValue
8.        val createCombiner = aggregator.get.createCombiner
9.        var kv: Product2[K, V] = null
10.       val update = (hadValue: Boolean, oldValue: C) => {
11.         if (hadValue) mergeValue(oldValue, kv._2) else createCombiner(kv._2)
```

```
12.       }
13.       while (records.hasNext) {
14.         addElementsRead()
15.         kv = records.next()
16.         map.changeValue((getPartition(kv._1), kv._1), update)
17.         maybeSpillCollection(usingMap = true)
18.       }
19.     } else {
20.       //将值插入缓冲区
21.       while (records.hasNext) {
22.         addElementsRead()
23.         val kv = records.next()
24.         buffer.insert(getPartition(kv._1), kv._1, kv._2.asInstanceOf[C])
25.         maybeSpillCollection(usingMap = false)
26.       }
27.     }
28.   }
```

首先判断是否聚合 shouldCombine。

（1）如果聚合，map.changeValue 此时 Key 不变，在历史 Value 基础上进行 combine。

（2）如果没有聚合，就直接在 Buffer 数据结构中插入一条记录。

注意：这时没有排序。

继续回到 SortShuffleWriter 的 write 方法：根据 dep.shuffleId, mapId 获取输出文件 output 写数据，根据 dep.shuffleId, mapId, partitionLengths, tmp，tmp 是中间临时文件写入文件和更新索引。task 运行结束后返回的 mapStatus 数据结构，告诉数据放在哪里。

SortShuffleWriter.scala 的 write 的源码如下：

```
1.   ......
2.   val output = shuffleBlockResolver.getDataFile(dep.shuffleId, mapId)
3.     val tmp = Utils.tempFileWith(output)
4.     try {
5.       val blockId = ShuffleBlockId(dep.shuffleId, mapId, IndexShuffleBlock
         Resolver.NOOP_REDUCE_ID)
6.       val partitionLengths = sorter.writePartitionedFile(blockId, tmp)
7.      shuffleBlockResolver.writeIndexFileAndCommit(dep.shuffleId, mapId,
        partitionLengths, tmp)
8.       mapStatus = MapStatus(blockManager.shuffleServerId, partitionLengths)
```

writePartitionedFile 方法，实现了 spill 和不 spill 怎么做。

ExternalSorter.sala 的源码如下：

```
1.  def writePartitionedFile(
2.        blockId: BlockId,
3.        outputFile: File): Array[Long] = {
4.
5.      //跟踪输出文件中每个范围的位置
6.      val lengths = new Array[Long](numPartitions)
7.      val writer = blockManager.getDiskWriter(blockId, outputFile, serInstance,
        fileBufferSize,context.taskMetrics().shuffleWriteMetrics)
8.
9.      if (spills.isEmpty) {
10.       //在只有内存中数据的情况下
11.       val collection = if (aggregator.isDefined) map else buffer
12.       val it = collection.destructiveSortedWritablePartitionedIterator
          (comparator)
```

```
13.       while (it.hasNext) {
14.         val partitionId = it.nextPartition()
15.         while (it.hasNext && it.nextPartition() == partitionId) {
16.           it.writeNext(writer)
17.         }
18.         val segment = writer.commitAndGet()
19.         lengths(partitionId) = segment.length
20.       }
21.     } else {
22.       //我们必须执行合并排序；通过分区获得一个迭代器，并直接写入所有内容
23.       for ((id, elements) <- this.partitionedIterator) {
24.         if (elements.hasNext) {
25.           for (elem <- elements) {
26.             writer.write(elem._1, elem._2)
27.           }
28.           val segment = writer.commitAndGet()
29.           lengths(id) = segment.length
30.         }
31.       }
32.     }
33.
34.     writer.close()
35.     context.taskMetrics().incMemoryBytesSpilled(memoryBytesSpilled)
36.     context.taskMetrics().incDiskBytesSpilled(diskBytesSpilled)
37.     context.taskMetrics().incPeakExecutionMemory(peakMemoryUsedBytes)
38.
39.     lengths
40.   }
```

里面有一句很关键的代码：val it = collection.destructiveSortedWritablePartitionedIterator(comparator)，生成一个it WritablePartitionedIterator写数据。

WritablePartitionedPairCollection.scala的源码如下：

```
1.  private[spark] trait WritablePartitionedPairCollection[K, V] {
2.    /**
3.     *将分区中的key-value键值对插入到集合中
4.     */
5.    def insert(partition: Int, key: K, value: V): Unit
6.
7.    /**
8.     * 按分区ID顺序遍历数据，然后按给定比较器进行迭代，这可能破坏底层集合
9.     */
10.   def partitionedDestructiveSortedIterator(keyComparator: Option[Comparator
    [K]])
11.     : Iterator[((Int, K), V)]
```

从这个地方看到了排序：以partition ID进行排序，实现快速的写、方便的读操作；关键是对Key进行操作。

下面看一下继承结构PartitionedAppendOnlyMap。

PartitionedAppendOnlyMap.scala的源码如下：

```
1.  private[spark] class PartitionedAppendOnlyMap[K, V]
2.    extends SizeTrackingAppendOnlyMap[(Int, K), V] with WritablePartitioned
    PairCollection[K, V] {
3.
4.    def partitionedDestructiveSortedIterator(keyComparator: Option[Comparator
    [K]])
```

```
5.      : Iterator[((Int, K), V)] = {
6.      val comparator = keyComparator.map(partitionKeyComparator).getOrElse
        (partitionComparator)
7.      destructiveSortedIterator(comparator)
8.    }
9.
10.   def insert(partition: Int, key: K, value: V): Unit = {
11.     update((partition, key), value)
12.   }
13. }
```

点击 destructiveSortedIterator。

AppendOnlyMap.scala 的源码如下：

```
1.    def destructiveSortedIterator(keyComparator: Comparator[K]): Iterator
      [(K, V)] = {
2.      destroyed = true
3.      //将 key-values 键值对插入到底层数组的前面
4.      var keyIndex, newIndex = 0
5.      while (keyIndex < capacity) {
6.        if (data(2 * keyIndex) != null) {
7.          data(2 * newIndex) = data(2 * keyIndex)
8.          data(2 * newIndex + 1) = data(2 * keyIndex + 1)
9.          newIndex += 1
10.       }
11.       keyIndex += 1
12.     }
13.     assert(curSize == newIndex + (if (haveNullValue) 1 else 0))
14.
15.     new Sorter(new KVArraySortDataFormat[K, AnyRef]).sort(data, 0, newIndex,
        keyComparator)
16.
17.     new Iterator[(K, V)] {
18.       var i = 0
19.       var nullValueReady = haveNullValue
20.       def hasNext: Boolean = (i < newIndex || nullValueReady)
21.       def next(): (K, V) = {
22.         if (nullValueReady) {
23.           nullValueReady = false
24.           (null.asInstanceOf[K], nullValue)
25.         } else {
26.           val item = (data(2 * i).asInstanceOf[K], data(2 * i + 1).as
              InstanceOf[V])
27.           i += 1
28.           item
29.         }
30.       }
31.     }
32.   }
```

其中关键的地方有一个 new Sorter。

AppendOnlyMap.scala 的源码如下：

```
1.    new Sorter(new KVArraySortDataFormat[K, AnyRef]).sort(data, 0, newIndex,
      keyComparator)
```

Sorter 里面使用的是 timSort 算法。

Sorter.scala 的源码如下：

```
1.  private[spark]
2. class Sorter[K, Buffer](private val s: SortDataFormat[K, Buffer]) {
3.
4.   private val timSort = new TimSort(s)
5.
6.   /**
     *对范围内[lo, hi)的输入缓冲区进行排序
7.    */
8.   def sort(a: Buffer, lo: Int, hi: Int, c: Comparator[_ >: K]): Unit = {
9.     timSort.sort(a, lo, hi, c)
10.  }
11. }
```

31.10　Spark 1.6.X 以前 Shuffle 中 JVM 内存使用及配置内幕详情

Spark 1.6.X 以前 Shuffle 中 JVM 内存使用及配置内幕详情：Spark 到底能够缓存多少数据，Shuffle 到底占用了多少数据，磁盘的数据远远比内存小却还是报告内存不足？本节将讲解以下内容。

- JVM 内存使用架构剖析。
- Spark 集群在 1.6.X 以前中 JVM 到底可以缓存多少数据。
- Spark 集群在 1.6.X 以前中 Shuffle JVM 到底缓存多少数据。
- Spark on Yarn 实际计算对内存的使用案例。

1. JVM内存使用架构剖析

JVM 有很多不同的区，最开始的时候，它会通过类装载器把类加载进来，在运行期数据区中有“本地方法栈”“程序计数器”“Java 栈”“Java 堆”“方法区”以及本地方法接口和它的本地方法库。从 Spark 的角度来谈代码的运行和数据的处理，主要是谈 Java 堆（Heap）空间的运用。

JVM 的体现架构如下。

- 本地方法栈：在梯归的时候至关重要。
- 程序计数器：这是一个全区计数器，对于线程切换至关重要。
- Java 栈（Stack）：Stack 区属于线程私有，高效的程序一般都是并发的，每个线程都会包含一个 Stack 区域，Stack 区域中含有基本的数据类型以及对象的引用，其他线程均不能直接访问该区域。Java 栈分为三大部分：基本数据类型区域、操作指令区域、上下文。
- Java 堆（Heap）：存储的全部都是 Object 对象实例，对象实例中一般都包含了其数据成员以及与该对象对应类的信息，它会指向类的引用一个，不同线程肯定要操作这个对象；一个 JVM 实例在运行的时候只有一个 Heap 区域，而且该区域被所有线程共享。

- 方法区：又名静态成员区域，包含整个程序的 class、static 成员等。类本身的字节码是静态的，它会被所有的线程共享，是全区级别的。

JVM 内存使用示意图如图 31-9 所示。

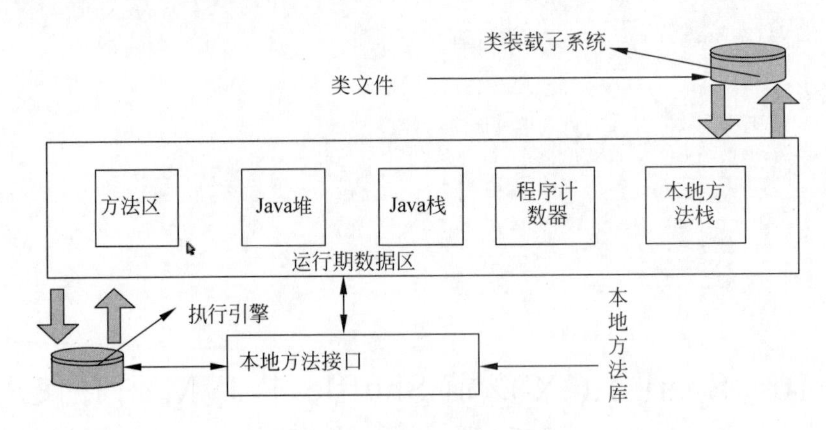

图 31-9　JVM 内存使用示意图

对于 Spark，我们这里研究的是 JVM 的 Heap 堆。对象放在堆中，堆上才有 Spark 的对象。关于 Heap Area：① 存储的全部都是 OBJECT 对象实例，对象实例中一般包含了其数据成员及对象对应的 Class 信息；② 一个 JVM 实例在运行的时候只有一个 Heap 区域，该区域被所有的线程共享。注意，GC 回收的是堆上的内容。

Spark 内存示意图如图 31-10 所示。左侧图是 Spark 1.6.X 之后的内存管理分配图；右侧图是 Spark 1.6.X 之前的内存示意图。

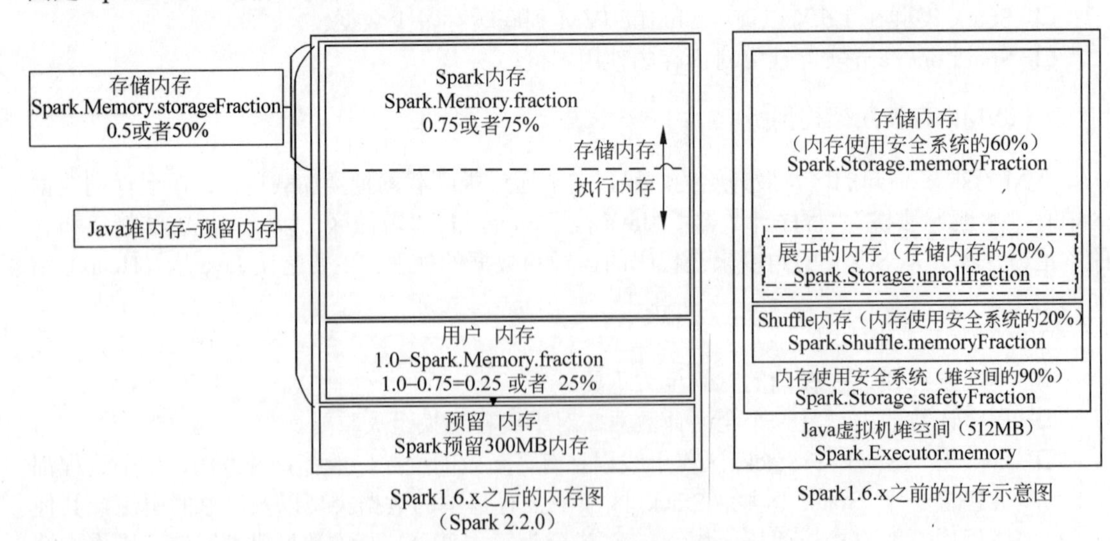

图 31-10　Spark 内存示意图

在回答 Spark JVM 到底可以缓存多少数据这个问题前，首先了解一下 JVM Heap 在 Spark 中是如何分配内存比例的。无论你定义 Spark.Executor.memory 的内存空间有多大，Spark 必然会定义一个安全空间，默认情况下只会使用 Java 堆上的 90% 作为安全空间，从单个 Executor 的角度讲，就是 Heap Size x 90%。

JVM Heap 默认是 512MB，机器的内存如果是 128GB，那么机器的内存就没有用起来，因此，实际运行时，JVM Heap 需要配置！配置参数为 Spark.Executor.memory。例如，Executor 的可用 Heap 大小是 10GB，实际上 Spark 只能使用 90%，也就是 9GB 的大小。（这里是指一个 Executor，还有 1GB 被 JVM 使用了。90%这个参数是由 spark.storage.safetyFraction 控制的（假如 Executor 配置使用了 100GB，剩余了 10GB 的内存没使用，那可以调高到 95%，不过一般不会去调整））。

2．Spark集群在1.6.X以前中JVM到底可以缓存多少数据

单个 Executor 的 Cache 数据量的计算公式：

Heap Size×park.Storage.safetyFraction×Spark.Storage.memoryFraction

从单个 Executor 的角度讲：Heap Size × 90%(Spark.Storage.safetyFraction) × 60%（Spark.Storage.memoryFraction)=Heap Size×54%。如果我们的 Executor Heap 的大小是 10GB，从理论上讲，单个 Executor 可以缓存的数据大小是 5.4GB，差不多占一半。缓存大小的控制参数由 spark.storage.safetyFraction 和 spark.storage.memoryFraction 共同决定。如果是 100 个 Executor，每个的 Heap 是 10GB，那么从理论上讲是可以缓存 540GB 的数据，普通规模的计算基本都可以满足了。

StaticMemoryManager.scala 的 getMaxStorageMemory 方法的源码如下：

```
1.   /**
      *返回存储区域可用的内存总量，以字节为单位
2.    */
3.   private def getMaxStorageMemory(conf: SparkConf): Long = {
4.     val systemMaxMemory = conf.getLong("spark.testing.memory", Runtime.
       getRuntime.maxMemory)
5.     val memoryFraction = conf.getDouble("spark.storage.memoryFraction", 0.6)
6.     val safetyFraction = conf.getDouble("spark.storage.safetyFraction", 0.9)
7.     (systemMaxMemory * memoryFraction * safetyFraction).toLong
8.   }
```

3．Spark集群在1.6.X以前中Shuffle JVM到底缓存多少数据

Shuffle 在一个 Executor 的 Heap 占用大小计算公式：Heap Size×Spark.Storage .safetyFraction×Spark.Shuffle.memoryFraction，默认情况下是 Heap Size×90%×20%=Heap Size×18%。

StaticMemoryManager.scala 的 getMaxExecutionMemory 方法的源码如下：

```
1.   private def getMaxExecutionMemory(conf: SparkConf): Long = {
2.     val systemMaxMemory = conf.getLong("spark.testing.memory", Runtime
       .getRuntime.maxMemory)
3.
4.     if (systemMaxMemory < MIN_MEMORY_BYTES) {
5.       throw new IllegalArgumentException(s"System memory $systemMaxMemory
         must " +
6.         s"be at least $MIN_MEMORY_BYTES. Please increase heap size using
           the --driver-memory " +
7.         s"option or spark.driver.memory in Spark configuration.")
8.     }
9.     if (conf.contains("spark.executor.memory")) {
```

```
10.        val executorMemory = conf.getSizeAsBytes("spark.executor.memory")
11.        if (executorMemory < MIN_MEMORY_BYTES) {
12.          throw new IllegalArgumentException(s"Executor memory $executorMemory
             must be at least " +
13.            s"$MIN_MEMORY_BYTES. Please increase executor memory using the " +
14.            s"--executor-memory option or spark.executor.memory in Spark
               configuration.")
15.        }
16.      }
17.      val memoryFraction = conf.getDouble("spark.shuffle.memoryFraction", 0.2)
18.      val safetyFraction = conf.getDouble("spark.shuffle.safetyFraction", 0.8)
19.      (systemMaxMemory * memoryFraction * safetyFraction).toLong
20.    }
21.
22. }
```

Unroll 占用空间计算及对缓存数据的影响是什么？Unroll 是反序列化的过程。需要反序列化的时候，占 20%，占用 Cache 的空间。

计算公式： Heap Size×Spark.Storage.safetyFraction×Spark.Storage.memoryFraction×Spark.Storage.unrollFraction，默认情况是 Heap Size×10.08%。

对 Cache 缓存数据的影响，由于 Unroll 是一个优先级高的操作，进行 Unroll 操作的时候会占用 Cache 空间，又可以挤掉缓存在内存中的数据，如果该数据的存储级别是 MemoryOnly，则该数据丢失。（有时运行好好的，却发现有数据丢失，就是这个原因。）这里有一个细节：Spark 集群在 1.6.X 以前中 Shuffle JVM 使用的内存其实还不到 Heap Size 的 18%。需考虑 Spark.Shuffle.safetyFraction 参数，那 Shuffle 在一个 Executor 的 Heap 占用大小计算公式调整为：

Heap Size×Spark.Storage.safetyFraction×Spark.Shuffle.memoryFraction×Spark.Shuffle.safetyFraction=Heap Size×90%×20×80%=14.4%。

StaticMemoryManager.scala 的 maxUnrollMemory 的源码如下：

```
1.      //数据展开时的最大数量的字节块
2.     private val maxUnrollMemory: Long = {
3.      (maxOnHeapStorageMemory * conf.getDouble("spark.storage.unrollFraction",
        0.2)).toLong
4.     }
```

4. Spark on Yarn实际计算对内存的使用案例

Spark on Yarn 内存使用示意图如图 31-11 所示。

Spark 运行在 Yarn 上，它有 Driver 和 Executor 两部分，在 Driver 部分有一个内存控制参数，Spark 1.6.X 以前是 spark.driver.memory，在实际生产环境下建议配置成 2GB。如果 Driver 比较繁忙或者是经常把某些数据收集到 Driver 上，建议把这个参数调大。

图 31-11 的左边是 Executor 部分，它是被 Yarn 管理的，每台机器上都有一个 Node Manager；Node Manager 是被 Resources Manager 管理的，Resources Manager 的工作主要是管理全区级别的计算资源，计算资源核心就是内存和 CPU，每台机器上都有一个 Node Manager 来管理当前内存和 CPU 等资源。Yarn 一般与 Hadoop 耦合，它底层会有 HDFS Node Manager，主要负责管理当前机器进程上的数据，并且与 HDFS Name Node 进行通信。

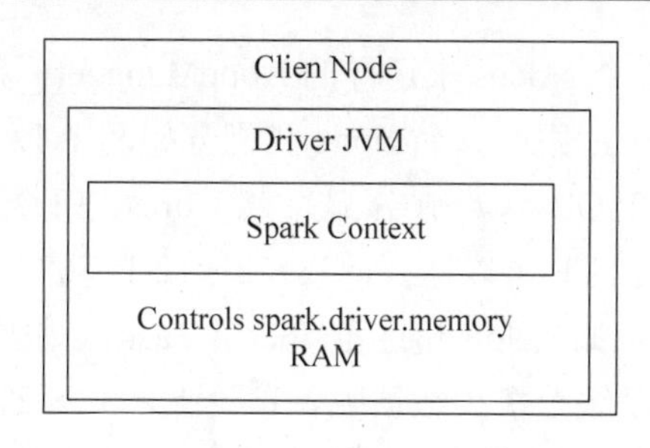

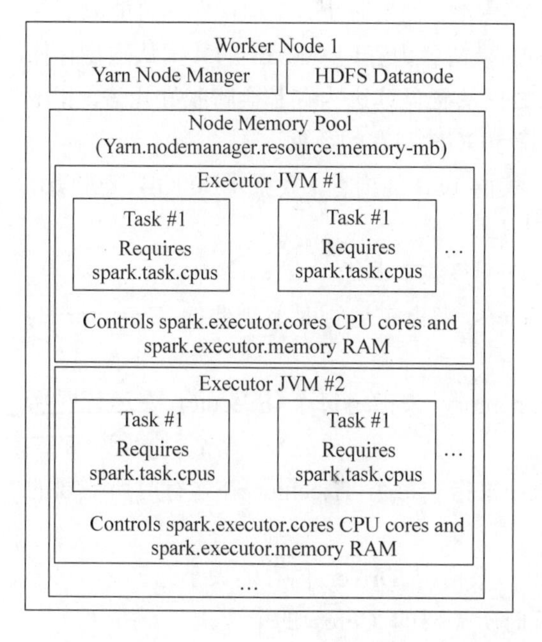

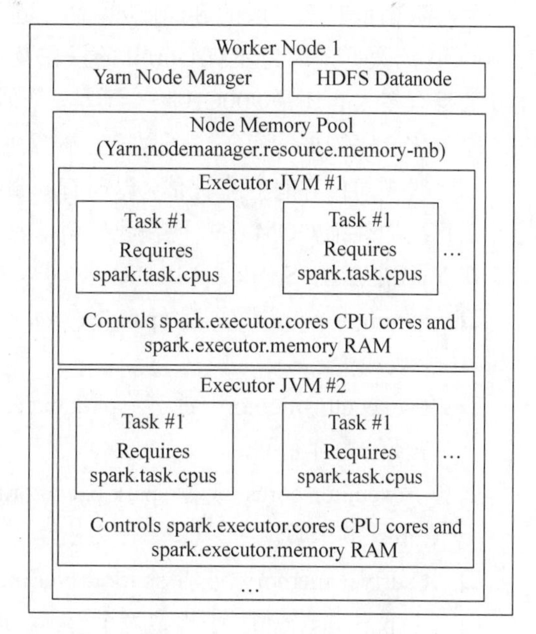

图 31-11 Spark on Yarn 内存使用示意图

每个节点上至少有两个进程：一个是 HDFS Data Node，负责管理磁盘上的数据，另一个是 Yarn Node Manager，负责管理执行进程。在这两个 Node 的下面有两个 Executors，每个 Executor 里运行的都是 Tasks。从 Yarn 的角度讲，会配置每个 Executor 所占用的空间，以防止资源竞争。Yarn 里有一个叫 Node Memory Pool 的概念，可以配置 64GB 或者是 128GB，Node Memory Pool 是当前节点上总共能够使用的内存大小。

图 31-11 中，这两个 Executors 在两个不同的进程（JVM#1 和 JVM#2）中，里面的 Task 是并行运行的，Task 运行在线程中，但你可以配置 Task 使用线程的数量，例如，2 条线程或者是 4 条线程，默认都是一条线程去处理一个 Task，你也可以用 spark.executor.cores 去配置可用的 Core 以及 spark.executor.memory 去配置可用的 RAM 的大小。

在 Yarn 上启动 Spark Application 时，我们会通过 num-executor 或者 spark.executor.instance 在 Yarn 上指定会有多少 Executor 实例运行当前的程序，同时也会指定每个 Executor 会使用多少内存：-executor-memory 或者 spark.executor.memory；同时也会通过-executor-cores 或者 spark.executor.cores 来指定每个 Executor 可以使用多少个 Cores，同时会通过 spark.task.cpus 来指定每个 Task 的运行需要多少个 Cores。对于 Driver，一般会通过 driver-memory 或者 spark.driver.memory 来指定 Driver 内存的大小。

例如，Yarn 集群上有 32 个 Nodes 来运行的 NodeManager，每个 Node 的内存是 64GB，每个 Node 的 Cores 是 32Cores，假如每个 Node 我们分配两个 Executors，那么我们就可以把每个 Executor 的内存分配为 28GB，Cores 分配为 12 Cores，每个 Spark Task 在运行的时候只需要一个 Core，那么 32nodes 同时可以运行：32×2×12/1=768 task slots。也就是说，这个集群可以并行运行 768 个 Task，如果 Job 超过了 768 个 Task，则需要排队。

那么，这个集群规模可以缓存多少数据呢？理论上，32×2×28×0.9×0.6=967.68GB，这个缓存数据量对于普通的 Spark Job 而言是完全够用的。而实际上，在运行中你可能只能缓存 900GB 的数据，因为还有 Unroll 的操作及其他的事情。内存中 900GB 的数据从磁盘存储角度数据有多大？还是 900GB 吗？不是，一般的数据从磁盘读进内存都会膨胀好几倍，和压缩、序列化反序列框架有关，所以，在磁盘上也就 300GB 左右的数据。

这个时候对内存有一个评估，以前有同学问：集群 1TB 的内存，数据才 500GB、600GB，为什么每次加载都 OOM 呢？现在我们就非常清楚了！

在 Yarn 上启动 Spark Application 时可以通过以下参数来调优。

- 用-num-executor 或者 spark.executor.instances 来指定运行时所需要的 Executor 的个数。
- 用-executor-memory 或者 spark.executor.memory 来指定每个 Executor 在运行时所需要的内存空间。
- 用-executor-cores 或者 spark.executor.cores 来指定每个 Executor 在运行时所需要的 Cores 的个数。
- 用-driver-memory 或者 spark.driver.memory 来指定 Driver 内存的大小。
- 用 spark.task.cpus 来指定每个 Task 运行时所需要的 Cores 的个数。

31.11　Spark 2.4.X 下 Shuffle 中内存管理源码解密：StaticMemory 和 UnifiedMemory

本节从源码的角度了解 Spark 内存管理是怎么设计的，从而知道应该配置哪个参数，让程序运行更适合你的实际需要。

- 了解 MemoryManger：Unified Memory Manager、Static Memory Manager 以及它们的核心功能与方法。
- 了解 MemoryPool：StorageMemoryPool、ExecutionMemoryPool 以及它们的核心功能与方法。

Spark Shuffle 的内存管理有两种：一种是联合内存管理器（Spark Unified Memory）；一种是静态内存管理器（Spark Static Memory）。首先，这两个类都是继承 MemoryManager，MemoryManager 是一个抽象类，我们在应用程序中使用接口就是为了包容未来的变化，因为现在只有两个内存管理器，将来可能会有好几种内存控制器，如图 31-12 所示。

MemoryManager 主要有以下几个功能。

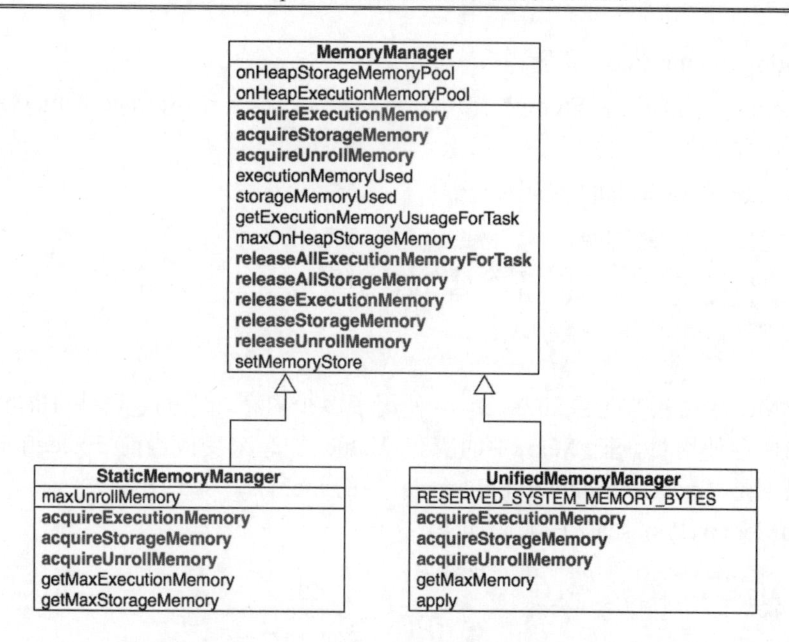

图 31-12　MemoryManager 示意图

- 记录用了多少 StorageMemory 和 ExecutionMemory。
- 申请 Storage、Execution 和 Unroll Memory：acquireStorageMemory、acquireExection Memory、acquireUnrollMemory。
- 释放 Storage 和 Execution Memory。

抽象类 MemoryManager 介绍如下。

- MemoryManger 强制管理储存（Storage）和执行（Execution）之间的内存使用，从 MemoryManager 申请可以把剩余空间借给对方。所有 Task 的运行就是 ShuffleTask 的运行，ExecutionMemory 是指 Shuffles、joins、sorts 和 aggregation 的操作；而 StorageMemory 是缓存和广播数据相关的，每个 JVM 会产生一个 MemoryManager 来负责管理内存。MemoryManager 构造时，需要指定 onHeapStorageMemeory 和 onHeapExecutionMemory 的参数。

MemoryManager.scala 的源码如下：

```
1.  /**
     *一种抽象内存管理器，用于在执行内存和存储内存之间共享内存。在这种情况下，执行内存是
     *指用于计算在 Shuffles、Joins、排序和聚合，而存储内存是指用于缓存集群内部数据的
     *存内。每个 JVM 都有一个 MemoryManager
2.   */
3.  private[spark] abstract class MemoryManager(
4.      conf: SparkConf,
5.      numCores: Int,
6.      onHeapStorageMemory: Long,
7.      onHeapExecutionMemory: Long) extends Logging {
```

构造 MemoryManager 对象时创建 StorageMemoryPool 和 ExecutionMemoryPool 对象，用来管理 Storage 和 Execution 的内存分配。MemoryManager.scala 中定义了 OnHeapStorageMemoryPool、OffHeapStorageMemoryPool、OnHeapExecutionMemoryPool、

OffHeapExecutionMemoryPool 变量。

StorageMemory 用来记录 Storage 使用了多少内存。以下是 StorageMemoryPool.scala 中的 memoryUsed 方法。

StorageMemoryPool.scala 的源码如下：

```
1.    @GuardedBy("lock")
2.   private[this] var _memoryUsed: Long = 0L
3.   override def memoryUsed: Long = lock.synchronized {
4.     _memoryUsed
5.   }
```

ExecutionMemory 用来记录 Execution 使用了多少内存，它创建一些 HashMap 来存储每个 Task 的内存使用量，把 Map 中的所有 Value 加起来变成当前 ExecutionMemory 的总使用量。以下是 ExecutionMemoryPool.scala 中的 memoryUsed 方法。

ExecutionMemoryPool.scala 的源码如下：

```
1.   /**
       * Map 数据结构：taskAttemptId -> 内存消耗的字节数
2.     */
3.   @GuardedBy("lock")
4.     private val memoryForTask = new mutable.HashMap[Long, Long]()
5.
6.     override def memoryUsed: Long = lock.synchronized {
7.       memoryForTask.values.sum
8.     }
```

MemoryStore 是被 BlockManager 管理的，以下是其中一个 MemoryStore.scala 中的 putBytes 方法。

MemoryStore.scala 的 putBytes 方法的源码如下：

```
1.     /**
2.       *使用 size 测试在 MemoryStore 内存中是否有足够的空间。如果有，创建 ByteBuffer
       *放入 MemoryStore 中。否则，ByteBuffer 缓冲区不会被创建。调用者应保证 size 是正
       *确的
3.       *
4.       * @return true  如果 put 方法是成功的，则返回 true，否则返回 false
5.       */
6.     def putBytes[T: ClassTag](
7.         blockId: BlockId,
8.         size: Long,
9.         memoryMode: MemoryMode,
10.        _bytes: () => ChunkedByteBuffer): Boolean = {
11.      require(!contains(blockId), s"Block $blockId is already present in the
         MemoryStore")
12.      if (memoryManager.acquireStorageMemory(blockId, size, memoryMode)) {
13.        //为这个块获得了足够的内存，所以把它放进去
14.        val bytes = _bytes()
15.        assert(bytes.size == size)
16.        val entry = new SerializedMemoryEntry[T](bytes, memoryMode,
           implicitly[ClassTag[T]])
17.        entries.synchronized {
18.          entries.put(blockId, entry)
19.        }
```

```
20.        logInfo("Block %s stored as bytes in memory (estimated size %s, free
           %s)".format(
21.          blockId, Utils.bytesToString(size), Utils.bytesToString(maxMemory
             - blocksMemoryUsed)))
22.        true
23.      } else {
24.        false
25.      }
26.    }
```

Spark 2.4.3 默认的 MemoryManager 是 UnifiedMemoryManager，以下源码里有一段条件判断的逻辑，如果 spark.memory.userLegacyMode 是 true，MemeoryManager 便是 StaticMemoryManager，否则就是 Spark Unified Memory。

SparkEnv.scala 的 memoryManager 的源码如下：

```
1.     val useLegacyMemoryManager = conf.getBoolean("spark.memory.useLegacy
       Mode", false)
2.     val memoryManager: MemoryManager =
3.       if (useLegacyMemoryManager) {
4.         new StaticMemoryManager(conf, numUsableCores)
5.       } else {
6.         UnifiedMemoryManager(conf, numUsableCores)
7.       }
```

在 MemoryManager 中有一个很关键的代码，如果想使用 OffHeap 作为储存的话，必须设置 spark.memory.offHeap.enabled 为 true，还要确定 offHeap 系统的空间必须大于 0。以下是 MemoryManager.scala 中的 tungstenMemoryMode 变量源码。

Spark 2.2.1 版本的 MemoryManager.scala 的 tungstenMemoryMode 的源码如下：

```
1.   //与钨丝计划管理内存相关的内容
2.    /**
       * 跟踪是否将钨丝计划内存分配到 JVM 堆或 Off-Heap 堆外使用（sun.misc.Unsafe）
3.     */
4.    final val tungstenMemoryMode: MemoryMode = {
5.      if (conf.getBoolean("spark.memory.offHeap.enabled", false)) {
6.        require(conf.getSizeAsBytes("spark.memory.offHeap.size", 0) > 0,
7.          "spark.memory.offHeap.size must be > 0 when spark.memory.offHeap
            .enabled == true")
8.        require(Platform.unaligned(),
9.          "No support for unaligned Unsafe. Set spark.memory.offHeap.enabled
            to false.")
10.       MemoryMode.OFF_HEAP
11.     } else {
12.       MemoryMode.ON_HEAP
13.     }
14.   }
```

Spark 2.4.3 版本的 MemoryManager.scala 源码与 Spark 2.2.1 版本相比具有如下特点。

- 上段代码中第 5 行调整为 MEMORY_OFFHEAP_ENABLED 参数，如果 spark.memory.offHeap.enabled 为 true，Spark 将试图在某些操作中使用堆外内存。如果启用了堆外内存使用，则 spark.memory.offSheap.size 必须为正。
- 上段代码中第 6 行调整为 MEMORY_OFFHEAP_SIZE 参数。spark.memory.offHeap.size 可用于堆外分配的绝对内存量（字节），此设置对堆内存使用没有影响，因此，如果 executors 的总内存消耗必须符合某个限制，那么一定要相应地缩小 JVM 堆的大小。

当 spark.memory.offSheap.enabled=true 时，必须将其设置为正值。

```
1.  ......
2.    if (conf.get(MEMORY_OFFHEAP_ENABLED)) {
3.      require(conf.get(MEMORY_OFFHEAP_SIZE) > 0,
4.  ......
5.  //package.scala
6.   private[spark] val MEMORY_OFFHEAP_ENABLED = ConfigBuilder("spark
     .memory.offHeap.enabled")
7.     .doc("If true, Spark will attempt to use off-heap memory for certain
       operations. " +
8.       "If off-heap memory use is enabled, then spark.memory.offHeap.size
         must be positive.")
9.     .withAlternative("spark.unsafe.offHeap")
10.    .booleanConf
11.    .createWithDefault(false)
12. ......
13.  private[spark] val MEMORY_OFFHEAP_SIZE = ConfigBuilder("spark.memory
     .offHeap.size")
14.    .doc("The absolute amount of memory in bytes which can be used for
       off-heap allocation. " +
15.      "This setting has no impact on heap memory usage, so if your executors'
         total memory " +
16.      "consumption must fit within some hard limit then be sure to shrink
         your JVM heap size " +
17.      "accordingly. This must be set to a positive value when
         spark.memory.offHeap.enabled=true.")
18.    .bytesConf(ByteUnit.BYTE)
19.    .checkValue(_ >= 0, "The off-heap memory size must not be negative")
20.    .createWithDefault(0)
```

内存管理（MemoryManager）属于 Spark 框架内部，包含两种类型。

- 统一内存管理（UnifiedMemoryManager），属于框架内部 private[memory]。
- 静态内存管理（StaticMemoryManager）。

其中 UnifiedMemoryManager 的源码如下：

```
1.  private[spark] class UnifiedMemoryManager private[memory] (
2.     conf: SparkConf,
3.     val maxHeapMemory: Long,
4.     onHeapStorageRegionSize: Long,
5.     numCores: Int)
6.   extends MemoryManager(
7.     conf,
8.     numCores,
9.     onHeapStorageRegionSize,
10.    maxHeapMemory - onHeapStorageRegionSize) {
```

StaticMemoryManager 的源码如下：

```
1.  private[spark] class StaticMemoryManager(
2.     conf: SparkConf,
3.     maxOnHeapExecutionMemory: Long,
4.     override val maxOnHeapStorageMemory: Long,
5.     numCores: Int)
6.   extends MemoryManager(
7.     conf,
8.     numCores,
9.     maxOnHeapStorageMemory,
```

```
10.     maxOnHeapExecutionMemory) {
```

UnifiedMemoryManager 和 StaticMemoryManager 继承自 MemoryManager。MemoryManager 是一个抽象类，预留未来的变化。MemoryManager 强制管理 Execution 和 Storage 的内存的使用：Storage 是存储层面，负责 Persist、Unroll 及 Broadcast 的数据；Execution 是 Shuffle 的数据。所有 Task 的运行就是 Shuffle Task 的运行！

在上下文中，Execution 内存使用于 Shuffles、joins、sorts、aggregations；而 Storage 内存适用于 caching、propagating internal data。每个 JVM 都有一个 MemoryManager。

MemoryManager 传入的参数包括：

- onHeapStorageMemory。
- onHeapExecutionMemory。

MemoryManager.scala 的源码如下：

```
1.   private[spark] abstract class MemoryManager(
2.     conf: SparkConf,
3.     numCores: Int,
4.     onHeapStorageMemory: Long,
5.     onHeapExecutionMemory: Long) extends Logging {
```

MemoryManager 定义了一些数据结构。

```
1.     @GuardedBy("this")
2.    protected val onHeapStorageMemoryPool = new StorageMemoryPool(this,
      MemoryMode.ON_HEAP)
3.    @GuardedBy("this")
4.    protected val offHeapStorageMemoryPool = new StorageMemoryPool(this,
      MemoryMode.OFF_HEAP)
5.    @GuardedBy("this")
6.    protected val onHeapExecutionMemoryPool = new ExecutionMemoryPool(this,
      MemoryMode.ON_HEAP)
7.    @GuardedBy("this")
8.   protected val offHeapExecutionMemoryPool = new ExecutionMemoryPool(this,
     MemoryMode.OFF_HEAP)
```

我们看一个结构 StorageMemoryPool，其使用模式匹配，定义了两种内存的方式。

- case MemoryMode.ON_HEAP => "on-heap storage"堆内内存。
- case MemoryMode.OFF_HEAP => "off-heap storage"堆外内存。

StorageMemoryPool.scala 的源码如下：

```
1.   private[memory] class StorageMemoryPool(
2.     lock: Object,
3.     memoryMode: MemoryMode
4.    ) extends MemoryPool(lock) with Logging {
5.
6.    private[this] val poolName: String = memoryMode match {
7.      case MemoryMode.ON_HEAP => "on-heap storage"
8.      case MemoryMode.OFF_HEAP => "off-heap storage"
9.    }
```

StorageMemoryPool 管理 Storage 的空间，memoryUsed 是已经使用的内存空间，其中 MemoryStore 非常重要。

StorageMemoryPool.scala 的源码如下：

```
1.   override def memoryUsed: Long = lock.synchronized {
2.      _memoryUsed
3.    }
4.
5.    private var _memoryStore: MemoryStore = _
6.    def memoryStore: MemoryStore = {
7.      if (_memoryStore == null) {
8.        throw new IllegalStateException("memory store not initialized yet")
9.      }
10.     _memoryStore
11.   }
```

MemoryStore：内存数据被 MemoryManger 管理，其实最终还是被 BlockManger 管理，MemoryStore 在构造的时候，根据上下文创建了 block、hashmap 等数据结构。

MemoryStore.scala 的源码如下：

```
1.  /**
     *内存中的存储块，无论是作为反序列化的 Java 对象的数组，还是作为序列化的字节缓冲区
2.   */
3.  private[spark] class MemoryStore(
4.      conf: SparkConf,
5.      blockInfoManager: BlockInfoManager,
6.      serializerManager: SerializerManager,
7.      memoryManager: MemoryManager,
8.      blockEvictionHandler: BlockEvictionHandler)
9.    extends Logging {
```

回到 StorageMemoryPool，acquireMemory 是申请内存。

StorageMemoryPool.scala 的源码如下：

```
1.    /**
2.     *获得 N 个字节的内存去缓存给定块，如果有必要，将驱逐现在的内存
3.     *
4.     * @return 是否所有 N 个字节都成功地被分配
5.     */
6.    def acquireMemory(blockId: BlockId, numBytes: Long): Boolean = lock
      .synchronized {
7.      val numBytesToFree = math.max(0, numBytes - memoryFree)
8.      acquireMemory(blockId, numBytes, numBytesToFree)
9.    }
```

判断一下内存是否足够，内存足够的话就分配内存：_memoryUsed += numBytesToAcquire。

StorageMemoryPool.scala 的源码如下：

```
1.    def acquireMemory(
2.        blockId: BlockId,
3.        numBytesToAcquire: Long,
4.        numBytesToFree: Long): Boolean = lock.synchronized {
5.      assert(numBytesToAcquire >= 0)
6.      assert(numBytesToFree >= 0)
7.      assert(memoryUsed <= poolSize)
8.      if (numBytesToFree > 0) {
9.        memoryStore.evictBlocksToFreeSpace(Some(blockId), numBytesToFree,
        memoryMode)
10.     }
```

```
11.     //注意：如果内存存储时驱逐块，驱逐将同步回调到 StorageMemoryPool，为了其释放
        //内存，这些变量已经更新
12.     val enoughMemory = numBytesToAcquire <= memoryFree
13.     if (enoughMemory) {
14.       _memoryUsed += numBytesToAcquire
15.     }
16.     enoughMemory
```

回到 MemoryManager：tungsten 内存分配的方式有两种。

- case MemoryMode.ON_HEAP => MemoryAllocator.HEAP。
- case MemoryMode.OFF_HEAP => MemoryAllocator.UNSAFE。

MemoryManager.scala 的源码如下：

```
1.   /**
      *为不安全/钨丝计划代码分配内存
2.    */
3.   private[memory] final val tungstenMemoryAllocator: MemoryAllocator = {
4.     tungstenMemoryMode match {
5.       case MemoryMode.ON_HEAP => MemoryAllocator.HEAP
6.       case MemoryMode.OFF_HEAP => MemoryAllocator.UNSAFE
7.     }
8.   }
9.  }
```

进入 MemoryAllocator：构建一个 HeapMemoryAllocator。MemoryAllocator HEAP = new HeapMemoryAllocator()。

MemoryAllocator.scala 的源码如下：

```
1.   /**
      * 分配一个连续的内存块。注意，分配的内存没有保证被清零（如果有必要，调用 fill(0)
      * 方法填充）
2.    */
3.   MemoryBlock allocate(long size) throws OutOfMemoryError;
4.
5.   void free(MemoryBlock memory);
6.
7.   MemoryAllocator UNSAFE = new UnsafeMemoryAllocator();
8.
9.   MemoryAllocator HEAP = new HeapMemoryAllocator();
10. }
```

进入 HeapMemoryAllocator：查看里面的 allocate 方法，先查询缓冲池是否为空，如果不为空，则获取 MemoryBlock，然后使用 memory.fill 进行分配。

Spark 2.2.1 版本的 HeapMemoryAllocator.scala 的源码如下：

```
1.    public MemoryBlock allocate(long size) throws OutOfMemoryError {
2.     if (shouldPool(size)) {
3.       synchronized (this) {
4.         final LinkedList<WeakReference<MemoryBlock>> pool = bufferPoolsBySize
           .get(size);
5.         if (pool != null) {
6.           while (!pool.isEmpty()) {
7.             final WeakReference<MemoryBlock> blockReference = pool.pop();
8.             final MemoryBlock memory = blockReference.get();
9.             if (memory != null) {
```

```
10.              assert (memory.size() == size);
11.              return memory;
12.            }
13.          }
14.          bufferPoolsBySize.remove(size);
15.        }
16.      }
17.    }
18.    long[] array = new long[(int) ((size + 7) / 8)];
19.    MemoryBlock memory = new MemoryBlock(array, Platform.LONG_ARRAY_
       OFFSET, size);
20.    if (MemoryAllocator.MEMORY_DEBUG_FILL_ENABLED) {
21.      memory.fill(MemoryAllocator.MEMORY_DEBUG_FILL_CLEAN_VALUE);
22.    }
23.    return memory;
24.  }
```

Spark 2.4.3版本的HeapMemoryAllocator.scala源码与Spark 2.2.1版本相比具有如下特点。

- 上段代码中第2行之前新增numWords、alignedSize变量。
- 上段代码中第2、4行size修改为alignedSize。
- 上段代码中第7～18行替换为以下第9～24行代码。

```
1.  ......
2.     int numWords = (int) ((size + 7) / 8);
3.     long alignedSize = numWords * 8L;
4.     assert (alignedSize >= size);
5.     if (shouldPool(alignedSize)) {
6.  ......
7.  final LinkedList<WeakReference<long[]>> pool = bufferPoolsBySize
    .get(alignedSize);
8.  ......
9.     final WeakReference<long[]> arrayReference = pool.pop();
10.            final long[] array = arrayReference.get();
11.            if (array != null) {
12.              assert (array.length * 8L >= size);
13.              MemoryBlock memory = new MemoryBlock(array, Platform
                 .LONG_ARRAY_OFFSET, size);
14.              if (MemoryAllocator.MEMORY_DEBUG_FILL_ENABLED) {
15.                memory.fill(MemoryAllocator.MEMORY_DEBUG_FILL_CLEAN_VALUE);
16.              }
17.              return memory;
18.            }
19.          }
20.          bufferPoolsBySize.remove(alignedSize);
21.        }
22.      }
23.    }
24.    long[] array = new long[numWords];
25. ......
```

其中的memory.fill方法内存分配借助JVM的Platform来分配内存。

MemoryBlock.scala的源码如下：

```
1.    /**
       * 用指定字节值填充内存块
2.     */
3.    public void fill(byte value) {
4.      Platform.setMemory(obj, offset, length, value);
```

```
5.    }
6.  }
```

Platform 类使用 JVM 提供的接口方法。底层 JVM 的操作是很原始的操作。

Platform.java 的源码如下：

```
1.   package org.apache.spark.unsafe;
2.
3.  import java.lang.reflect.Constructor;
4.  import java.lang.reflect.Field;
5.  import java.lang.reflect.Method;
6.  import java.nio.ByteBuffer;
7.
8.  import sun.misc.Cleaner;
9.  import sun.misc.Unsafe;
10.
11. public final class Platform {
12. ......
13.   public static void setMemory(Object object, long offset, long size, byte
      value) {
14.     _UNSAFE.setMemory(object, offset, size, value);
15.   }
```

现在看几个关键点： Unified Memory 是 Spark 1.6.X 之后引入的，那么怎么看 Spark 是哪种内存管理器呢？每台机器上都有 MemoryManger，那么 Driver 上也有 MemoryManger，是一个 master-slave 的结构。下面看一下 SparkEnv：val memoryManager: MemoryManager。

SparkEnv.scala 的源码如下：

```
1.  class SparkEnv (
2.      val executorId: String,
3.      private[spark] val rpcEnv: RpcEnv,
4.      val serializer: Serializer,
5.      val closureSerializer: Serializer,
6.      val serializerManager: SerializerManager,
7.      val mapOutputTracker: MapOutputTracker,
8.      val shuffleManager: ShuffleManager,
9.      val broadcastManager: BroadcastManager,
10.     val blockManager: BlockManager,
11.     val securityManager: SecurityManager,
12.     val metricsSystem: MetricsSystem,
13.     val memoryManager: MemoryManager,
14.     val outputCommitCoordinator: OutputCommitCoordinator,
15.     val conf: SparkConf) extends Logging {
```

在 SparkEnv.Scala 中，MemoryManager: conf.getBoolean("spark.memory.useLegacy Mode", false)设置为 false 表示遗弃了，是 static 级别的。那么，在 Spark 2.4.X 中，如果要使用旧版本的内存管理，在配置文件中设置为 true 就可以了。

- conf.getBoolean("spark.memory.useLegacyMode")，设置为 false，使用 StaticMemory Manager。
- conf.getBoolean("spark.memory.useLegacyMode")，设置为 true，使用 UnifiedMemory Manager。

SparkEnv.scala 的源码如下：

```
1.   val useLegacyMemoryManager = conf.getBoolean("spark.memory.useLegacy Mode",
     false)
```

```
2.   val memoryManager: MemoryManager =
3.     if (useLegacyMemoryManager) {
4.       new StaticMemoryManager(conf, numUsableCores)
5.     } else {
6.       UnifiedMemoryManager(conf, numUsableCores)
7.     }
```

Spark 2.4.3 中，我们使用的是 UnifiedMemoryManager，进入类 UnifiedMemoryManager，里面的预留空间是 300MB RESERVED_SYSTEM_MEMORY_BYTES = 300×1024×1024。

reservedMemory 要做事情，也可以将它调大。

UnifiedMemoryManager.scala 的源码如下：

```
1.   object UnifiedMemoryManager {
2.
3.     //为非存储内存、非执行内存的用途预留内存量，提供类似于 spark.memory.fraction 的
       //功能，但保证我们预留充足的系统内存，即使是很小的堆内存。例如，如果有一个 1GB 的 JVM，
       //用于执行内存和存储内存，默认为 (1024̄300)×0.6 = 434MB
4.     private val RESERVED_SYSTEM_MEMORY_BYTES = 300 * 1024 * 1024
5.
6.     def apply(conf: SparkConf, numCores: Int): UnifiedMemoryManager = {
7.       val maxMemory = getMaxMemory(conf)
8.       new UnifiedMemoryManager(
9.         conf,
10.        maxHeapMemory = maxMemory,
11.        onHeapStorageRegionSize =
12.          (maxMemory * conf.getDouble("spark.memory.storageFraction", 0.5))
             .toLong,
13.        numCores = numCores)
14.    }
```

Spark 新型的 JVM Heap 分为三大部分：Reserved Memory、User Memory、Spark Memory。图 31-10 中的 Reserved Memory（预留内存）是 300MB，Spark Memory 包括 Storage Memory 和 Execution Memory，里面所有的参数都可以调整。

下面看一下 maxMemory。maxMemory 是 Storage Memory 和 Execution Memory 需要的部分。val maxMemory = getMaxMemory(conf)；val memoryFraction = conf.getDouble ("spark.memory.fraction", 0.6)，Spark 2.1.X 的源码默认配置为 60%。

UnifiedMemoryManager 的 getMaxMemory 的源码如下：

```
1.   /**
      *返回执行内存和存储内存之间共享的内存总量，以字节为单位
2.    */
3.
4.   private def getMaxMemory(conf: SparkConf): Long = {
5.     val systemMemory = conf.getLong("spark.testing.memory", Runtime
       .getRuntime.maxMemory)
6.     val reservedMemory = conf.getLong("spark.testing.reservedMemory",
7.       if (conf.contains("spark.testing")) 0 else RESERVED_SYSTEM_MEMORY_
         BYTES)
8.     val minSystemMemory = (reservedMemory * 1.5).ceil.toLong
9.     if (systemMemory < minSystemMemory) {
10.      throw new IllegalArgumentException(s"System memory $systemMemory
         must " +
11.        s"be at least $minSystemMemory. Please increase heap size using the
           --driver-memory " +
```

```
12.         s"option or spark.driver.memory in Spark configuration.")
13.     }
14.     //SPARK-12759 如果内存不足，就检查 Executor 内存是否失败
15.     if (conf.contains("spark.executor.memory")) {
16.       val executorMemory = conf.getSizeAsBytes("spark.executor.memory")
17.       if (executorMemory < minSystemMemory) {
18.         throw new IllegalArgumentException(s"Executor memory $executor
            Memory must be at least " +
19.           s"$minSystemMemory. Please increase executor memory using the " +
20.           s"--executor-memory option or spark.executor.memory in Spark
              configuration.")
21.       }
22.     }
23.     val usableMemory = systemMemory - reservedMemory
24.     val memoryFraction = conf.getDouble("spark.memory.fraction", 0.6)
25.     (usableMemory * memoryFraction).toLong
26.   }
```

回到 Memory Manager，这里有很多成员。看一个比较关键的 tungstenMemoryMode，定义 Tungsten Memory 使用在 JVM 的 Heap 上面，还是使用 sun.misc.Unsafe OFF-HEAP。

- "spark.memory.offHeap.enabled"，默认是 false，使用 MemoryMode.ON_HEAP。
- "spark.memory.offHeap.enabled"，设置为 true，使用 MemoryMode.OFF_HEAP。

Memory Manager 定义了 pageSizeBytes。

MemoryManager.scala 的源码如下：

```
1.   /**
      * 默认页面大小，以字节为单位。如果用户没有显式地设置 spark.buffer.pageSize，我
      * 们计算出默认值，通过进程查看可用的内核数量和内存总量，然后除以一个安全系数
2.    */
3.
4.    val pageSizeBytes: Long = {
5.      val minPageSize = 1L * 1024 * 1024   //1MB
6.      val maxPageSize = 64L * minPageSize  //64MB
7.      val cores = if (numCores > 0) numCores else Runtime.getRuntime
        .availableProcessors()
8.      //取得下一个 2 的幂，可能在最坏情况下的安全系数为 8
9.      val safetyFactor = 16
10.     val maxTungstenMemory: Long = tungstenMemoryMode match {
11.       case MemoryMode.ON_HEAP => onHeapExecutionMemoryPool.poolSize
12.       case MemoryMode.OFF_HEAP => offHeapExecutionMemoryPool.poolSize
13.     }
14.     val size = ByteArrayMethods.nextPowerOf2(maxTungstenMemory / cores /
        safetyFactor)
15.     val default = math.min(maxPageSize, math.max(minPageSize, size))
16.     conf.getSizeAsBytes("spark.buffer.pageSize", default)
17.   }
```

Spark 1.6.X 版本之前的旧版本使用的内存管理方式是静态内存 StaticMemoryManager 方式。StaticMemoryManager.scala 的源码如下：

```
1.  private val maxUnrollMemory: Long = {
2.      (maxOnHeapStorageMemory * conf.getDouble("spark.storage.unrollFraction",
        0.2)).toLong
3.    }
```

Unroll 是反序列化的过程。数据在内存中是序列化的，要使用数据，必须反序列化。需要反序列化的时候，Unroll 占 20%的空间，占用 Cache 的空间。计算公式：Heap Size × Spark.Storage.safetyFraction × Spark.Storage.memoryFraction × Spark.Storage.unrollFraction，即 Heap Size × 90% × 60% × 20%，默认情况是 HeapSize × 10.08%。

这里定义了默认的 Storage Memory 的一些配置信息。Spark.Testing.memory 是系统运行时的最大内存大小。其中 spark.storage.safetyFraction 默认设置为 0.9；spark.storage.memory Fraction 默认设置为 0.6。

StaticMemoryManager.scala 的源码如下：

```
1.   /**
      *返回存储区域可用的内存总量，以字节为单位
2.    */
3.   private def getMaxStorageMemory(conf: SparkConf): Long = {
4.     val systemMaxMemory = conf.getLong("spark.testing.memory", Runtime
       .getRuntime.maxMemory)
5.     val memoryFraction = conf.getDouble("spark.storage.memoryFraction", 0.6)
6.     val safetyFraction = conf.getDouble("spark.storage.safetyFraction", 0.9)
7.     (systemMaxMemory * memoryFraction * safetyFraction).toLong
8.   }
```

下面看一下 Execution 级别的内存管理，即 Shuffle 的内存。Shuffle 在一个 Executor 的 Heap 占用大小计算公式为：

Heap Size × Spark.Storage.safetyFraction × Spark.Shuffle.memoryFraction × Spark.Shuffle.safety Fraction=Heap size × 90% × 20 × 80%=14.4%，这里涉及的参数为 spark.shuffle.memoryFraction，默认设置为 0.2；spark.shuffle.safetyFraction 默认设置为 0.8。

StaticMemoryManager.scala 的源码如下：

```
1.   /**
      *返回执行内存可用的内存总量，以字节为单位
2.    */
3.
4.   private def getMaxExecutionMemory(conf: SparkConf): Long = {
5.     val systemMaxMemory = conf.getLong("spark.testing.memory", Runtime
       .getRuntime.maxMemory)
6.
7.     if (systemMaxMemory < MIN_MEMORY_BYTES) {
8.       throw new IllegalArgumentException(s"System memory $systemMaxMemory
         must " +
9.         s"be at least $MIN_MEMORY_BYTES. Please increase heap size using
           the --driver-memory " +
10.        s"option or spark.driver.memory in Spark configuration.")
11.    }
12.    if (conf.contains("spark.executor.memory")) {
13.      val executorMemory = conf.getSizeAsBytes("spark.executor.memory")
14.      if (executorMemory < MIN_MEMORY_BYTES) {
15.        throw new IllegalArgumentException(s"Executor memory $executorMemory
           must be at least " +
16.          s"$MIN_MEMORY_BYTES. Please increase executor memory using the " +
17.          s"--executor-memory option or spark.executor.memory in Spark
             configuration.")
18.      }
19.    }
20.    val memoryFraction = conf.getDouble("spark.shuffle.memoryFraction", 0.2)
```

```
21.     val safetyFraction = conf.getDouble("spark.shuffle.safetyFraction", 0.8)
22.     (systemMaxMemory * memoryFraction * safetyFraction).toLong
23.   }
```

最后看一下 ExecutionMemoryPool、StorageMemoryPool。

首先看一下 ExecutionMemoryPool，里面有一个变量 memoryForTask。memoryForTask 是 HashMap 类型。memoryForTask 记录 Task 对内存的使用情况，Task 运行的时候怎么使用。

ExecutionMemoryPool.scala 的源码如下：

```
1.  /**
     * Map 数据结构：taskAttemptId -> 内存消耗的字节数
2.    */
3.  @GuardedBy("lock")
4.  private val memoryForTask = new mutable.HashMap[Long, Long]()
```

ExecutionMemoryPool 有一个重要方法 acquireMemory，申请内存可能申请不到，也可能机器比较忙，申请时会重试若干次。如果无法满足，就会循环一些次数。

ExecutionMemoryPool.scala 的源码如下：

```
1.    private[memory] def acquireMemory(
2.        numBytes: Long,
3.        taskAttemptId: Long,
4.        maybeGrowPool: Long => Unit = (additionalSpaceNeeded: Long) => Unit,
5.        computeMaxPoolSize: () => Long = () => poolSize): Long =
        lock.synchronized {
6.      assert(numBytes > 0, s"invalid number of bytes requested: $numBytes")
7.
8.      //待办事项 TODO：清理这个笨拙的方法签名
9.
10.     //将任务添加到 taskMemory，这样我们就可以保持活动任务的准确计数，调用 acquireMemory
        //让其他任务降低内存
11.     if (!memoryForTask.contains(taskAttemptId)) {
12.       memoryForTask(taskAttemptId) = 0L
13.       //这将导致等待中的任务被唤醒，再次检查任务数量
14.       lock.notifyAll()
15.     }
16.
17.     //继续循环，直到确认不批准这个请求（因为这个任务会超过 1/numActiveTasks 的内
        //存）或者我们有足够的空闲内存给它（让每个任务得到至少 1 / (2×numActiveTasks))
18.     //待办事项 TODO：简化此操作，以将每个任务限制到自己的槽中
19.     while (true) {
20.       val numActiveTasks = memoryForTask.keys.size
21.       val curMem = memoryForTask(taskAttemptId)
22.
23.        //在循环的每次迭代中，应该首先尝试回收任何从存储空间借来的执行空间。这是必要的，
           //因为在可能的情况条件下，新的存储块可能会获取该任务正在等待的空闲执行内存
24.       maybeGrowPool(numBytes - memoryFree)
25.
26.        //在池增长以后可能达到的池的最大大小。这是用来计算每个任务可以占用多少内存的上
           //限。这必须考虑到潜在的空闲内存以及当前池的占用数量。否则，我们可能会遇到 SPARK-
           //12155 的情况，在统一存储管理中，我们没有考虑到的空间可能已被逐出了缓存块
27.       val maxPoolSize = computeMaxPoolSize()
28.       val maxMemoryPerTask = maxPoolSize / numActiveTasks
```

```
29.       val minMemoryPerTask = poolSize / (2 * numActiveTasks)
30.
31.       //如何分配这个任务：保持其份额在 0 <= X <= 1 / numActiveTasks
32.       val maxToGrant = math.min(numBytes, math.max(0, maxMemoryPerTask -
          curMem))
33.       //给它尽可能多的空闲内存，如果没有达到 1 / numTasks
34.       val toGrant = math.min(maxToGrant, memoryFree)
35.
36.       //要让每个任务在阻塞前至少得到 1 / (2×numActiveTasks)；如果我们现在不能
          //给它这么多，就等待其他任务释放内存（如果旧任务在N增长之前分配了大量内存）
37.       if (toGrant < numBytes && curMem + toGrant < minMemoryPerTask) {
38.         logInfo(s"TID $taskAttemptId waiting for at least 1/2N of $poolName
          pool to be free")
39.         lock.wait()
40.       } else {
41.         memoryForTask(taskAttemptId) += toGrant
42.         return toGrant
43.       }
44.     }
45.     0L //Never reached
46.   }
```

下面看一下 StorageMemoryPool 申请内存。acquireMemory 申请时判断有没有内存，首先计算 Storage 空闲的内存量和申请的内存量，以及需要释放多少内存。如果申请的内存大于内存池的剩余内存，Storage 内存池就尝试释放一部分内存，如果释放后能满足，就将申请的内存分配给运行的 Task，让 Task 去使用。

问题：MemoryManger 在哪里被调用？每个 Executor 启动的时候肯定有一个 MemoryManger。很清楚，肯定是在 BlockManger 中。

BlockManager.scala 的源码如下：

```
1.   private[spark] class BlockManager(
2.     executorId: String,
3.     rpcEnv: RpcEnv,
4.     val master: BlockManagerMaster,
5.     val serializerManager: SerializerManager,
6.     val conf: SparkConf,
7.     memoryManager: MemoryManager,
8.     mapOutputTracker: MapOutputTracker,
9.     shuffleManager: ShuffleManager,
10.    val blockTransferService: BlockTransferService,
11.    securityManager: SecurityManager,
12.    numUsableCores: Int)
13.  extends BlockDataManager with BlockEvictionHandler with Logging {
```

BlockManager 里面有 MemoryManager，MemoryManager 成员是被传入进来的。那我们就看 BlockManager 是什么时候被实例化的。在 BlockManager 上按 Ctrl 键，选择 ShuffleExternalSorter，查看 ShuffleExternalSorter 的源码。

ShuffleExternalSorter.java 的源码如下：

```
1.   ShuffleExternalSorter(
2.       TaskMemoryManager memoryManager,
3.       BlockManager blockManager,
```

```
4.        TaskContext taskContext,
5.        int initialSize,
6.        int numPartitions,
7.        SparkConf conf,
8.        ShuffleWriteMetrics writeMetrics) {
```

继续跟踪至 UnsafeShuffleWriter.java 的 ShuffleExternalSorter，这个时候传的是 MemoryManager。

Spark 2.2.1 版本的 UnsafeShuffleWriter.scala 的源码如下：

```
1.   private void open() throws IOException {
2.     assert (sorter == null);
3.     sorter = new ShuffleExternalSorter(
4.       memoryManager,
5.       blockManager,
6.       taskContext,
7.       initialSortBufferSize,
8.       partitioner.numPartitions(),
9.       sparkConf,
10.      writeMetrics);
11.    serBuffer = new MyByteArrayOutputStream(1024 * 1024);
12.    serOutputStream = serializer.serializeStream(serBuffer);
13.  }
```

Spark 2.4.3 版本的 UnsafeShuffleWriter.java 源码与 Spark 2.2.1 版本相比具有如下特点。

- 上段代码中第 1 行去掉 IOException 的异常。
- 上段代码中第 11 行调整为 DEFAULT_INITIAL_SER_BUFFER_SIZE 的参数。

```
1.   private void open() {
2.  ......
3.  serBuffer = new MyByteArrayOutputStream(DEFAULT_INITIAL_SER_BUFFER_SIZE);
4.  ......
5.  static final int DEFAULT_INITIAL_SER_BUFFER_SIZE = 1024 * 1024;
```

我们看一下 MemoryManager。

UnsafeShuffleWriter.scala 的源码如下：

```
1.  public class UnsafeShuffleWriter<K, V> extends ShuffleWriter<K, V> {
2.  ......
3.   private final TaskMemoryManager memoryManager;
```

MemoryManager 是什么时候实例化的？MemoryManager 是在 UnsafeShuffleWriter 构造化的时候实例化的，那继续跟踪下去。

UnsafeShuffleWriter.scala 的源码如下：

```
1.   public UnsafeShuffleWriter(
2.       BlockManager blockManager,
3.       IndexShuffleBlockResolver shuffleBlockResolver,
4.       TaskMemoryManager memoryManager,
5.       SerializedShuffleHandle<K, V> handle,
6.       int mapId,
7.       TaskContext taskContext,
8.       SparkConf sparkConf) throws IOException {
```

这时就到 SortShuffleManager.scala，SortShuffleManager 的 getWriter 方法的第三个参数 context: TaskContext，这里是指 Task 的 Context，在 Task 反序列化时获得的，从 context .taskMemoryManager()就可以从上下文 Context 中获得 taskmemoryManager。

SortShuffleManager.scala 的源码如下：

```
1.   override def getWriter[K, V](
2.     ……
3.      handle match {
4.        case unsafeShuffleHandle: SerializedShuffleHandle[K @unchecked, V
          @unchecked] =>
5.          new UnsafeShuffleWriter(
6.            env.blockManager,
7.            shuffleBlockResolver.asInstanceOf[IndexShuffleBlockResolver],
8.            context.taskMemoryManager(),
9.            unsafeShuffleHandle,
10.           mapId,
11.           context,
12.           env.conf)
13. ……
```

taskcontext 上下文是 Task 启动过程中给内存分配内存管理器！Task 被序列化到 Executor 中，之后反序列化运行，构建 MemoryManager 的一个实例。

Unified 机制下 Execution 向 Storage 借空间源码解析。

Unified Memory Manager 有两个核心方法：acquiredExecutionMemeory 和 acquireStorage Memory。当 ExecutionMemory 有剩余空间时，可以借给 StorageMemory，然后通过调用 StorageMemoryPool 的 acquireMemory 方法向 storageMemoryPool 申请空间。

UnifiedMemoryManager.scala 的 acquireStorageMemory 的源码如下：

```
1.      override def acquireStorageMemory(
2.        blockId: BlockId,
3.        numBytes: Long,
4.        memoryMode: MemoryMode): Boolean = synchronized {
5.      assertInvariants()
6.      assert(numBytes >= 0)
7.      val (executionPool, storagePool, maxMemory) = memoryMode match {
8.        case MemoryMode.ON_HEAP => (
9.          onHeapExecutionMemoryPool,
10.         onHeapStorageMemoryPool,
11.         maxOnHeapStorageMemory)
12.       case MemoryMode.OFF_HEAP => (
13.         offHeapExecutionMemoryPool,
14.         offHeapStorageMemoryPool,
15.         maxOffHeapMemory)
16.     }
17.     if (numBytes > maxMemory) {
18.       //如果块不合适，很快就失败
19.       logInfo(s"Will not store $blockId as the required space ($numBytes
          bytes) exceeds our " +
20.         s"memory limit ($maxMemory bytes)")
21.       return false
22.     }
23.     if (numBytes > storagePool.memoryFree) {
24.       //存储池中没有足够的空闲内存，因此尝试从执行内存中借用空闲内存
25.       val memoryBorrowedFromExecution = Math.min(executionPool.memoryFree,
          numBytes-storagePool. memoryFree)
```

```
26.       executionPool.decrementPoolSize(memoryBorrowedFromExecution)
27.       storagePool.incrementPoolSize(memoryBorrowedFromExecution)
28.     }
29.     storagePool.acquireMemory(blockId, numBytes)
30.   }
```

acquiredExecutionMemory 主要是为当前的执行任务获得的执行空间，它首先会根据 onHeap 和 offHeap 方式进行分配。

在 MemoryManager 构造的时候，也分配一定的内存空间 poolSize。

MemoryManager.scala 的源码如下：

```
1.   offHeapExecutionMemoryPool.incrementPoolSize(maxOffHeapMemory -
     offHeapStorageMemory)
2.    offHeapStorageMemoryPool.incrementPoolSize(offHeapStorageMemory)
```

MemoryPool.scala 的 incremenPoolSize 方法的源码如下：

```
1.  /**
      * 通过 delta 字节扩大池
2.    */
3.  final def incrementPoolSize(delta: Long): Unit = lock.synchronized {
4.    require(delta >= 0)
5.    _poolSize += delta
6.  }
```

调用 computeMaxExecutionPoolSize 方法向 ExecutionPool 申请资源。过程中会调用 maybeGrowExecutionPool 来判断需要多少内存，包括计算内存空间的空闲资源与 Storage 曾经占用的空间。

UnifiedMemoryManager.scala 的 computeMaxExecutionPoolSize 方法的源码如下：

```
1.  def computeMaxExecutionPoolSize(): Long = {
2.     maxMemory - math.min(storagePool.memoryUsed, storageRegionSize)
3.   }
```

maybeGrowExecutionPool 方法首先判断申请的内存申请资源大于 0，然后判断剩余空间和 Storage 曾经占用的空间多，把需要的内存资源量提交给 StorageMemoryPool 的 freeSpaceToShrinkPool 方法。

最终的结果是调用方法 executionPool.acquireMemory。UnifiedMemoryManager.scala 的源码如下：

```
1.      executionPool.acquireMemory(
2.        numBytes, taskAttemptId, maybeGrowExecutionPool, computeMaxExecution
          PoolSize)
3.    }
```

然后判断当前 FreeSpace 能否满足 Execution 的需要，如果无法满足，则调用 MemoryStore 的 evictVlocksToFreeSpace 方法在 StorageMemoryPool 中挤掉一部分数据。

StorageMemoryPool.scala 的 freeSpaceToShrinkPool 方法的源码如下：

```
1.   def freeSpaceToShrinkPool(spaceToFree: Long): Long = lock.synchronized {
2.      val spaceFreedByReleasingUnusedMemory = math.min(spaceToFree,
        memoryFree)
3.      val remainingSpaceToFree = spaceToFree - spaceFreedByReleasingUnused
        Memory
```

```
4.      if (remainingSpaceToFree > 0) {
5.        //如果回收内存没有充分收缩池，则开始驱逐块
6.        val spaceFreedByEviction =
7.          memoryStore.evictBlocksToFreeSpace(None, remainingSpaceToFree,
          memoryMode)
8.        //当一个块被释放，BlockManager.dropFromMemory()调用 releaseMemory 方法，
          //这里我们不需要 decrement _memoryUsed。但是，我们确实需要减少池的大小
9.        spaceFreedByReleasingUnusedMemory + spaceFreedByEviction
10.     } else {
11.       spaceFreedByReleasingUnusedMemory
12.     }
13.   }
```

调用 ExecutionPool 的 acquireMemory 方法向 ExecutionPool 申请内存资源，每个 Task 理论上讲一般能使用的大小是从 poolSize/(2 × numActiveTasks) 到 maxPoolSize/numActiveTasks。

了解 Spark Shuffle 中的 JVM 内存使用空间对一个 Spark 应用程序的内存调优是至关重要的。根据不同的内存控制原理分别对存储和执行空间进行参数调优：spark.executor.memory、spark.storage.safetyFraction、spark.storage.memoryFraction、spark.storage.unrollFraction、spark.shuffle.memoryFraction、spark.shuffle.safteyFraction。

Spark 1.6 以前的版本使用的是固定的内存分配策略，把 JVM Heap 中的 90% 分配为安全空间，然后将这 90%的安全空间中的 60% 作为存储空间，例如进行 Persist、Unroll 以及 Broadcast 的数据。然后再把这 60%的 20%作为支持一些序列化和反序列化的数据工作。其次，当程序运行时，JVM Heap 会把其中的 80% 作为运行过程中的安全空间，这 80%的其中 20%是用来负责 Shuffle 数据传输的空间。

Spark 2.4.X 中推出了联合内存的概念，最主要的改变是存储和运行的空间可以动态移动。需要注意的是，执行比存储有更大的优先值，当空间不够时，可以向对方借空间，但前提是对方有足够的空间或者是 Execution 可以强制把 Storage 一部分空间挤掉。Excution 向 Storage 借空间有两种方式：第一种方式是 Storage 曾经向 Execution 借了空间，它缓存的数据可能非常多，当 Execution 需要空间时，可以强制拿回来；第二种方式是 Storage Memory 不足 50% 的情况下，Storage Memory 会很乐意地把剩余空间借给 Execution。

如果是你的计算比较复杂的情况，使用新型的内存管理（Unified Memory Management）会取得更高的效率，但是如果计算的业务逻辑需要更大的缓存空间，此时使用老版本的固定内存管理（StaticMemoryManagement）效果会更好。

31.12　Spark 2.4.X 下 Shuffle 中 JVM Unified Memory 内幕详情

Spark 2.4.X 中 Shuffle 中 JVM Unified Memory 内幕详情：Spark Unified Memory 的运行原理和机制是什么？Spark JVM 最小配置是什么？用户空间什么时候会出现 OOM？Spark 中的 Broadcast 到底存储在什么空间？ShuffleMapTask 使用的数据到底在什么地方？

- ❑ Spark Unified Memory 的运行原理和机制是什么？Spark Unified Memory，这是统一或者联合的意思，但是 Spark 没有用 Shared，例如，A 和 B 进行 Unified，A，B 进行 Shared 其实是两个不同的概念。

- Spark JVM 最小配置是什么？
- 用户空间什么时候会出现 OOM？Spark 2.4.X 中用户空间 OOM，首先要确定 user space memory 是什么，举一个很简单的例子，假如 Executor 是 100GB 的内存，那 user space memory 是什么，这个问题不是所有人能回答出来的，你的 user space memory 是 50GB、80GB、20GB，还是 25GB？为什么这件事情很重要？例如，在 Spark 中使用算子 mapPartition，一般要使用中间数据和临时对象，你这个时候使用的中间数据和临时对象，就是 user space 里面用户操作的数据空间，那这个空间的数据大小什么时候导致 OOM？
- Spark 中的 Broadcast 到底存储在什么空间？
- ShuffleMapTask 使用的数据到底在什么地方？是存在 Cache 空间中吗？

本节彻底解密 Spark 2.4.X 中 Shuffle 中 JVM Unified Memory 内幕详情。

（1）Spark 2.4.X 新型的 JVM Heap 分为三大部分。

- Reserved Memory。
- User Memory。
- Spark Memory。

（2）预留内存 Reserved Memory：系统运行时至少 Heap 的大小为 300MB×1.5=450MB。一般本地开发，例如在 Windows 系统上，建议 Windows 系统至少为 2GB。

（3）User Memory：从 Spark 的程序讲，什么是 user memory？什么时候接触到 memory 的空间？基于 RDD 编程，user memory 就是在 RDD 具体实现的方法中（如 map、mapPartitions 等方法），开发者在方法中写自己的代码，实现怎么处理 RDD 的数据举一个很简单的例子，你可能使用一个数据结构，mapPartitions 处理一个 Partition 的时候有 5000 个 Record 记录要处理，在这个过程中建立了一些临时数据，这些数据实际意义上讲和 Spark 本身不太相关，作为开发者 User 引入的中间数据，这些数据不缓存在 Spark 的 Storage Memory 中，存放在 User Memory 中。

5000 个 record 记录产生的中间数据放在哪里是一件很重要的事情，例如，产生的中间数据是 10G 的大小，或者 1M 的大小，对系统的运行有什么影响？单机版本的 Java、Python 就是这种级别的内存，分配一个 list、map，或者 ARRAY 写一个循环进行处理；例如，读数据库对 Key 进行匹配，或者累加一个数据，因此我们须考虑自己维护的数据是否会导致出现 OOM。在 Spark 中，人们对算子里面 RDD 数据以外的数据及数据结构却不那么重视，而在单机版本编程的时候，我们重视的主要就是这些数据结构及数据。

在 mapPartitions 算子中间使用的数据，也就是说不是 RDD 数据的数据，是存储在 Spark Memory 中吗？Spark Memory 是 Spark 框架使用的内存空间，相当于编程的时候涉及两个方面：一方面是框架方面，例如 tensorFlow 进行特征提取，解决了图像、声音特征的识别；另一方面是用户的部分，例如用户只要开发声音、图像、深度学习内容。

从 Spark 的角度讲，也分 Spark 框架的部分和用户空间的部分。

- Spark Memory 就是 Spark 运行时可以主导哪些空间。
- User Memory 就是你可以主导哪些空间：你在什么时候主导空间呢？唯一主导的时候就是 map、mapPartioion、groupByKey、aggregate 操作的时候中间会产生一些数据结构，这些数据结构就在 userMemory 中。而系统的空间不可侵犯的！Spark 在新版本空间划分非常明智，将用户空间、Spark 空间完全分离开！也是从安全方面的考量！

用户空间（User Memory）的计算公式：（Heap size-Reserverd Memory）×25%（默认情况下是 25%），这是用户可以支配的空间，从编程的角度讲，必须思考一件事情，用户可以支配多少空间，否则会出现 OOM。举一个简单的例子，例如有 4GB 大小，那么默认情况下 User Memory 大小是（4G–300M）×25%=949MB，所以在算子中，从 Stage 的 Task 的角度讲，Task 运行的时候展开每个 RDD 算子的内部，从 4GB 大小的角度讲，Stage 中可能有很多算子，如 mapPartitions、map 等，最大的使用空间不能超过 949MB，这不是指 Task 的运行占用的空间不能超过 949MB，而是指 Stage 内部的 Task 的所有算子在运行的时候中间的数据不能超出 949MB，这完全是两码事。Task 的运行是被 Spark 框架使用的，用户无能为力！用户只不过把自己的逻辑和数据嵌入到后来框架称之为 Task 一个又一个 RDD 的算子而已，我们谈的是数据在 RDD 算子中一个 Stage 内部一个 Task 的运行这些算子占用多大空间，就像 mapPartitions 中循环遍历数据库的 Record，要累加数据，中间使用 array，一下使用 2GB 就会出现 OOM。如果工程师使用如 mapPartitions 等一个 Task 内的所有算子使用的数据空间的大小大约 949MB，那么就会出现 OOM。

问题：100 个 Executor 每个为 4GB，处理 10GB 的大小，是否出现 OOM？这不好说。每个 Executor 的内存是 4GB，然后有 100 个 Executor，处理的磁盘数据一共才 100GB，理论上分配在 100 个 Executor，每个 Executor 分配 1GB 的数据，远远小于 4GB 的大小，为什么会出现 OOM？别说 100GB 的磁盘数据，10GB 的磁盘数据也会出现 OOM，因为在 mapPartitions 算子，你使用的是算子数据，而不是 RDD 的本身数据超过了 User Memory 的大小。

（4）Spark Memory 是框架空间，由 Storage Memory 和 Execution Memory 两部分构成。

Storage Memory：相当于 Spark 1.6.X 之前旧版本的 Storage 空间，旧版本的 Storage 占了 54%的 Heap 空间。

Execution Memory：相当于 Spark 1.6.X 之前旧版本的 Shuffle 空间，命名非常科学，执行分成许多 Stage，分析的就是 Shuffle。代码运行的时候也是谈 Shuffle。从整个运行的流程讲，涉及从上一个 Stage 抓数据，涉及聚合的操作。旧版本的 Storage 占了 54%的 Heap 空间，Shuffle 在旧版本中占了 14.4%的 Heap 空间。你认为这个数据合理吗？是数据缓存重要，还是执行重要？即是程序先跑起来重要呢，还是性能调优重要？当然是程序跑起来重要，Shuffle 肯定是最重要的！而遗憾的是，Spark 1.6.X 之前的版本中没有体现 Execution 最重要的位置。现在 Storage 和 Shuffle 其实是采用 Unified 的方式共同使用（Heap size–300M）×75%。默认情况下，Storage 和 Execution 各占该空间的 50%，默认情况下平分，但是这里并没有体现 unified，从箭头的角度看，一个往上，一个往下。补充：Storage Memory 新版本和旧版本有一个存储是一样的，Storage 中会负责 Persist、Unroll 及 Broadcast 的数据，广播是广播到这里面，大变量地广播出去是有道理的，因为这个空间还很大，从默认的情况讲，假设 4GB 的空间减去 300MB，乘以 75%，再乘以 50%，算下来大约 1.5GB 的空间，确实可以广播一些大变量。如果内存足够大，可以广播足够大的变量，对性能有很大的提升，对于线程共享的，Heap 的对象是线程共享的。而线程私有的是 Stack。这和 JVM 联系在一块了。

也可以从另外一个角度讲：这个是 Execution Memory，也就是 Task 运行级别的东西，Execution Memory 是从 shuffleMapTask 的角度来看的，作为 Task 运行的，运行在线程之上，这个 User Memory 是用户的数据，相当于是私有的，私有是从算子的角度讲。但是，整体上讲，大家都是这种类型的，也可以认为是公有的。这和 Heap、Stack 是两个完全不同层面的东西。这个数据为什么非常重要？是因为从算子运行的角度来讲，尽可能倾向于从 Storage

Memory 中拿到数据，这是所谓的内存计算，我们的 Persist、Unroll 及 Broadcast 的数据都在 Storage Memory 空间。如果说是 4GB 的堆大小，Storage Memory 占到 1.5GB 左右，Execution Memory 也是占到 1.5GB 左右。

Unified 统一内存内幕。

- Storage Memory：相当于旧版本的 Storage 空间。在旧版本中，Storage 占了 54%的 Heap 空间，这个空间会负责存储 Persist、Unroll 以及 Broadcast 的数据。假设 Executor 有 4GB，那么 Storage 空间是：(4GB−300MB)×75%×50% = 1423.5MB，也就是说，如果你的内存足够大，你可以扩播足够大的变量，扩播对性能提升是一件很重要的事情，因为它所有的线程都是共享的。从算子运行的角度讲，Spark 会倾向于数据直接从 Storage Memory 中抓取过来，这也就是所谓的内存计算。
- Execution Memory：相当于旧版本的 Shuffle 空间，这个空间会负责存储 ShuffleMapTask 的数据。例如，从上一个 Stage 抓取数据和一些聚合的操作，等等。在旧版本中，Shuffle 占了 16%的 Heap 空间。Execution 如果在空间不足的情况下，除了选择向 Storage Memory 借空间外，也可以把一部分数据 Spill 到磁盘上，但很多时候基于性能调优方面的考虑，都不想把数据 Spill 到磁盘上。

Storage Memory 以及 Execution Memory 互借空间。

第一点，Storage 和 Execution 在适当的时候可以借用彼此的 Memory。Execution Memory 可以使用 Storage Memory。Storage Memory 也可以使用 Execution Memory。

第二点，当 Execution 空间不足而且 Storage 空间也不足的时候，Storage 空间会被强制 Drop 掉一部分数据，来解决 Execution 的空间不足问题。举一个例子，如 4GB 的空间，两个加一起空间不到 3GB，非要用 4GB 的空间，怎么解决也解决不了。Storage 空间不足的情况下，Execution 空间也不足，这时放弃掉 Storage 的一部分空间来满足 Execution 的空间需求，为什么这么做？原因很简单，执行是更重要的事情！运行都运行不起来了，还管什么缓存？这里使用的是 drop，没有说丢失，drop 可能 drop 到 Disk 中，看 Persist 的 level 级别。

UnifiedMemoryManager 伴生对象里的 apply 方法中设置 spark.memory.storageFraction 为 0.5。spark.memory.storageFraction 的源码如下：

```
1.     def apply(conf: SparkConf, numCores: Int): UnifiedMemoryManager = {
2.     val maxMemory = getMaxMemory(conf)
3.     new UnifiedMemoryManager(
4.       conf,
5.       maxHeapMemory = maxMemory,
6.       onHeapStorageRegionSize =
7.        (maxMemory * conf.getDouble("spark.memory.storageFraction", 0.5)).
         toLong,
8.       numCores = numCores)
9.   }
```

Execution 向 Storage 借空间分成两种情况。

（1）Execution 向 Storage 能借到空间的情况：Storage 曾经向 Execution 借了空间，由于缓存的时候，数据非常多，Execution 又不需要那么多空间，原先各自平分 50%；Storage 默认占了 75%的空间或者 80%，如 Execution 使用了 20%的空间，那么空间不足了，Execution 需要将你曾经占用的空间数据强制 drop 掉，Execution 会向内存管理器发信号，若原先 Storage Memory 占据了共用空间的 80%的话，那现在 Execution Memory 需要占用更多的空间，那是否是说 Execution Memory 可以将 Storage Memory80%的数据空间都拿过来呢？假设 Execution

数据特别庞大。肯定不是的，这种情况下（Storage Memory 曾经超过了 50%的空间），Execution 这个时候需要更多的空间，Execution 把曾经 Storage Memory 占用的超过 50%的内容挤掉，但是剩下的，Storage Memory 还是占据 50%的空间，Execution Memory 不能继续把别人（Storage Memory）的数据挤掉了，这是有限度的。

Execution 向 Storage 借空间的第一种情况如图 31-13 所示。

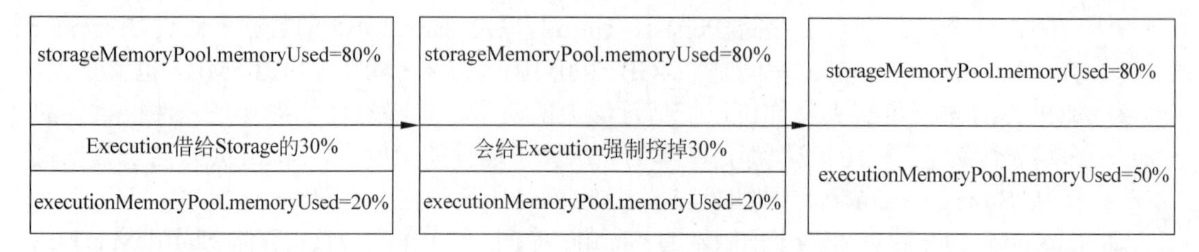

图 31-13　Execution 向 Storage 借空间的第一种情况

Storage 曾经向 Execution 借了空间，它缓存的数据可能非常多，然后 Execution 又不需要那么大的空间（默认情况下各占 50%），假设现在 Storage 占了 80%，Execution 占了 20%，然后 Execution 说自己空间不足，Execution 会向内存管理器发信号把 Storage 曾经占用的超过 50%数据的那部分强制挤掉，在这个例子中挤掉了 30%。

（2）Execution 向 Storage 拿 Memory，Memory 不足 50%的空间，这个时候就需要拿 Memory 的空间了。这是 Execution 可以强制要空间的两种情况。在 Execution 有剩余空间的时候，Storage Memory 可以向 Execution 借空间，而如果 Execution 没有空间，那么 Storage 是不能借的。

Execution 向 Storage 借空间的第二种情况如图 31-14 所示。

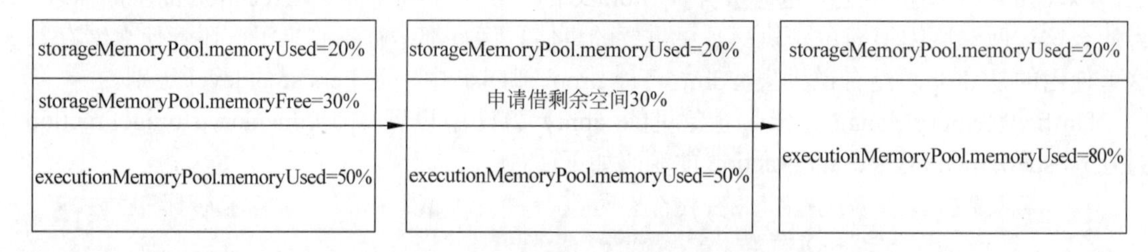

图 31-14　Execution 向 Storage 借空间的第二种情况

Execution 可以向 Storage Memory 借空间，在 Storage Memory 不足 50% 的情况下，Storage Memory 会把剩余空间借给 Execution。相反，当 Execution 有剩余空间的时候，Storage 也可以找 Execution 借空间。

总结：

（1）Execution 如果已经占据了 Execution 和 Storage 的 80%的共用空间，这个时候 Storage 需要更多的空间，能不能向 Execution 借空间？ 把默认超过 50%的剩余 30%的空间拿回去呢？那 Storage 是拿不回去的。Execution 是最重要的。

（2）无论是 Execution，还是 Storage 的空间，在 Execution 空间不足的情况下，都可以 Spill 到磁盘上，但 Shuffle 最影响性能。解决了 Shuffle 的问题，就解决了 95%的 Spark 的性能问题。所以，Execution 很强势，Execution 可以强制使用空间，这是有道理的。因为 Execution 与贡献成正比。Execution 将 Storage 的空间占用了，Storage 曾经没有占用的空间被 Execution

空间占用了，Storage 如果需要空间，Execution 也不给 Storage！Execution 是最重要的！而如果 Storage 占用了超过 50%的空间，Execution 需要更多的空间，那 Execution 将 Storage 空间挤掉。

（3）ShuffleMapTask 使用的数据到底在什么地方？

放在 Execution Memory 中，例如，聚合数据过来，从 500 个 Executor 中抓数据过来，就放在 Execution Memory 中。

31.13　Spark 2.4.X 下 Shuffle 中 Task 视角内存分配管理

Spark 2.4.X 内存管理包含两种类型：统一内存管理（UnifiedMemoryManager）、静态内存（StaticMemoryManager）。这两种内存的管理方式最终要落实到 Task 的运行。我们先从源码角度对 Spark 内存管理进行回顾，从 Spark Task 的视角解析 Task 运行内存管理源码。在 Spark 2.4.X 中默认使用 UnifiedMemoryManager 方式，从 Task 运行的角度来讲是 Execution 级别的，也就是 UnifiedMemoryManager 的核心是 Execution Memory。Execution Memory 主要做的事情是 Shuffle、Sort、Aggregate 等。为什么将 Execution Memory 的内存调大，理论上讲，Execution Memory 的内存越大，I/O 就越来越少。如图 31-15 所示，Execution Memory 相对于内存的占用比较强势。当 Execution Memory 空间不足，而且 Storage Memory 空间也不足的时候，Storage Memory 空间会被强制 drop 掉一部分数据，来解决 Execution 空间不足的问题。

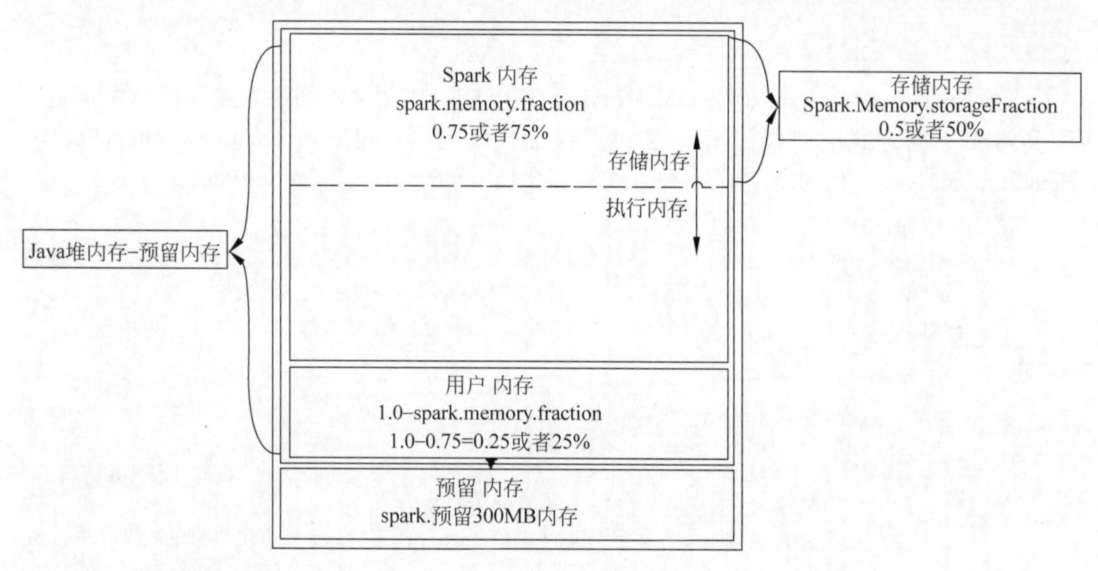

图 31-15　Spark 统一内存管理

Spark 2.4.X 内存管理 UnifiedMemoryManager 的管理方式从源码的角度看有以下两个核心方法。

- acquireExecutionMemory。
- acquireStorageMemory。

下面看一下 acquireExecutionMemory 的源码。acquireExecutionMemory 为当前的运行任务 Task 获得的内存空间。

```
1.    /**
      * 尝试为当前任务 Task 获取 numBytes 的执行内存（Execution Memory），并返回获
      * 得的字节数；如果没有可以分配的内存，就返回 0。此调用可能会阻塞，直到在某些情况下
      * 有足够的空闲内存，以确保每个 Task 在被强制 Spill 溢出前有机会上升到至少 1/2N 的
      * 总内存池（其中 N 是#的活动任务）。如果任务数量增加，这可能会发生，但旧的 Task
      * 已经分配了内存
2.    */
3.
4.   override private[memory] def acquireExecutionMemory(
5.       numBytes: Long,
6.       taskAttemptId: Long,
7.       memoryMode: MemoryMode): Long = synchronized {
8.     assertInvariants()
9.     assert(numBytes >= 0)
10.    val (executionPool, storagePool, storageRegionSize, maxMemory) =
     memoryMode match {
11.      case MemoryMode.ON_HEAP => (
12.        onHeapExecutionMemoryPool,
13.        onHeapStorageMemoryPool,
14.        onHeapStorageRegionSize,
15.        maxHeapMemory)
16.      case MemoryMode.OFF_HEAP => (
17.        offHeapExecutionMemoryPool,
18.        offHeapStorageMemoryPool,
19.        offHeapStorageMemory,
20.        maxOffHeapMemory)
21.    }
```

acquireExecutionMemory 根据 ON_HEAP、OFF_HEAP 两种不同的方式进行模式匹配，匹配到两种不同方式的时候，有自己的执行方式。内部 onHeapExecutionMemoryPool，onHeapStorageMemoryPool 无论是哪种方式，都会调 maybeGrowExecutionPool。

```
1.    /**
      *增加执行池（execution pool）的大小，缓存中的数据可能被清理掉，从而减少了存储
      *池 storage pool。当获取 Task 的内存时，执行池可能需要多个尝试，每一次尝试，必
      *须能够减少另一个 Task 的存储，并在尝试之间缓存一个大的块
2.    */
3.
4.     def maybeGrowExecutionPool(extraMemoryNeeded: Long): Unit = {
5.       if (extraMemoryNeeded > 0) {
6.         //执行池中没有足够的空闲内存，所以尝试从 storage pool 内存中回收内存。可
           //以从存储池中回收任意内存。如果存储池已经超过 storageRegionSize，可以回收
           //storage memory 存储已从执行内存 execution memory 中借来的内存
7.         val memoryReclaimableFromStorage = math.max(
8.           storagePool.memoryFree,
9.           storagePool.poolSize - storageRegionSize)
10.        if (memoryReclaimableFromStorage > 0) {
11.          //回收尽可能多的空间是必要的
12.          val spaceToReclaim = storagePool.freeSpaceToShrinkPool(
13.            math.min(extraMemoryNeeded, memoryReclaimableFromStorage))
14.          storagePool.decrementPoolSize(spaceToReclaim)
15.          executionPool.incrementPoolSize(spaceToReclaim)
```

```
16.         }
17.       }
18.     }
```

每个 Task 能够使用的内存理论上是多大呢？这是很重要的一件事情，因为能推测每个 Task 能使用的大小，计算公式为 poolSize/(2×numActiveTasks)到 maxPollSize/numActiveTasks。

在 maybeGrowExecutionPool 方法中，通过比较 storagePool.memoryFree 和 storagePool.poolSize - storageRegionSize 的大小，计算一个最大值：Storage Memory 剩余的内存和 Storage Memory 从 Execution Memory 借来的内存哪个大。Storage 和 Execution 在适当的时候可以借用彼此的 Memory，Storage Memory 在 Execution 执行的时候，Execution Memory 有空间，Storage Memory 缓存的时候向 Execution Memory 借用一些空间。如果 Execution Memory 内存不够，这时进行一个比较，整个 Execution Memory 能借到的最大内存是 Storage Memory 曾经找 Execution memory 借的内存+Storage Memory 空闲内存；根据 math.min(extraMemoryNeeded, memoryReclaimableFromStorage)，如果 Execution Memory 需要内存的大小小于能借到的最大内存，就以实际需要的内存为准。

在 maybeGrowExecutionPool 中调用 storagePool.decrementPoolSize(spaceToReclaim)方法减少内存，调用 executionPool.incrementPoolSize(spaceToReclaim)方法增加内容。

- storagePool.decrementPoolSize(spaceToReclaim)。
- executionPool.incrementPoolSize(spaceToReclaim)。

```
1.   /**
      *通过 delta 字节扩大池
2.    */
3.   final def incrementPoolSize(delta: Long): Unit = lock.synchronized {
4.     require(delta >= 0)
5.     _poolSize += delta
6.   }
7.
8.   /**
      *通过 delta 字节收缩池
9.    */
10.  final def decrementPoolSize(delta: Long): Unit = lock.synchronized {
11.    require(delta >= 0)
12.    require(delta <= _poolSize)
13.    require(_poolSize - delta >= memoryUsed)
14.    _poolSize -= delta
15.  }
```

接下来阐述 Storage Memory。Storage Memory 只有一种情况：在 Execution Memory 空闲的时候，Storage Memory 能借走 Execution Memory 的空闲内存。acquireStorageMemory 中有 On-Heap、Off-Heap 两种方式。如果使用的内存大于最大内存，就返回 false。其中 val memoryBorrowedFromExecution = Math.min(executionPool.memoryFree, numBytes-storagePool.memoryFree)，因此 memoryBorrowedFromExecution 取 executionPool.memoryFree，(numBytes - storagePool.memoryFree)的最小值，为多大内存取决于 executionPool 有多大空间。

无论是 storagePool，还是 executionPool，都要和 memoryPoll 打交道。下面看一下 memoryPool，其子类是 ExecutionMemoryPool 和 StorageMemoryPool。

```
1.  /**
2.    * 管理可调节 memory 内存大小的记录。这个类是内部的[MemoryManager]，具体的实现
      * 参见子类
3.    * @param lock a [MemoryManager] 用于同步。我们擦去类型 Object，以避免编程
      * 错误，因为这个对象只能用于同步目的
4.  private[memory] abstract class MemoryPool(lock: Object) {
5.
6.    @GuardedBy("lock")
7.    private[this] var _poolSize: Long = 0
8.
9.   /**
      *返回池的当前大小，以字节为单位
10.   */
11.   final def poolSize: Long = lock.synchronized {
12.     _poolSize
13.   }
14.
15.  /**
      *返回池中空闲内存的数量，以字节为单位
16.   */
17.   final def memoryFree: Long = lock.synchronized {
18.     _poolSize - memoryUsed
19.   }
20.
21.  /**
      *通过 delta 字节扩大池
22.   */
23.   final def incrementPoolSize(delta: Long): Unit = lock.synchronized {
24.     require(delta >= 0)
25.     _poolSize += delta
26.   }
27.
28.  /**
      *通过 delta 字节收缩池
29.   */
30.   final def decrementPoolSize(delta: Long): Unit = lock.synchronized {
31.     require(delta >= 0)
32.     require(delta <= _poolSize)
33.     require(_poolSize - delta >= memoryUsed)
34.     _poolSize -= delta
35.   }
36.
37.  /**
      *返回此池中使用内存的数量（以字节为单位）
38.   */
39.   def memoryUsed: Long
40. }
```

我们看一下 StorageMemoryPool，其对于内存的记录和管理，一方面是内存使用的记录，一方面是可调整大小内存（要么借进新的内存，要么借出内存）的管理。StorageMemoryPool 有 acquireMemory、releaseMemory 等方法。我们在 Shuffle 的时候，如果使用了 UnifiedMemoryManager 的新型内存管理方式，同时开启了 Off-Heap，对我们内存的使用有很大影响，如果有 offHeapExecutionMemoryPool，这时还存不存在一个概念，从 StorageMemory

中获取内存？实际上不需要找 StorageMemory 要内存。

每个 Task 能够分配的内存大小：poolSize/(2×numActiveTasks)到 maxPollSize/numActiveTasks。

- val maxMemoryPerTask = maxPoolSize / numActiveTasks。
- val minMemoryPerTask = poolSize / (2×numActiveTasks)。

如果 Shuffle Task 计算比较复杂，业务逻辑比较复杂，使用新型的 UnifiedMemoryManager 内存管理方式能取得较好的效果；但是，如果计算的业务逻辑不复杂，Shuffle 计算比较简单，这时可以回退到旧的内存管理方式，使用 Static Memory Management 效果会更好，因为缓存空间更大。

在这个基础上，从 Task 的角度看，Task 有一个 TaskMemoryManager。TaskMemoryManager 管理单个任务分配的内存。TaskMemoryManager 类中的大多数复杂情况涉及 64 位长的 Off-Heap 非堆地址编码。在 Off-Heap 非堆模式下，内存可以直接寻址 64 位。在堆模式下，内存是对象内的基本对象引用和 64 位偏移的组合寻址。这是一个问题，当我们想存储指针的数据结构内的其他结构，如在 hashmaps 或排序缓冲区的记录内的指针。即使我们决定使用 128 位来解决内存问题，我们不能只存储基本对象的地址，由于堆内存存在 GC，因此它不能保证保持稳定。相反，使用 64 位地址方式来编码记录指针：Off-Heap 非堆模式，只需存储原始地址，并在堆模式上使用地址的上 13 位存储"页码"和较低的 51 位来存储此页内的偏移量。这使我们能够解决 8192 页。在堆模式下，最大页大小受最大长度的数组限制，我们能够解决 8192×2^{32} 位的地址，大约 35 百万兆字节的内存。

TaskMemoryManager 肯定要申请内存，管理内存以及释放内存，需要一个 MemoryBlock[] 类型的内存的 pageTable，对堆的内存进行分配、释放，这个都是 JVM 进行管理的，我们调用 new 函数创建一个 MemoryBlock，只是对象的引用，不是具体内存空间的地址。堆内存有 GC 的问题，是 JVM 的死穴，那我们做堆外内存，JAVA 提供了 ByteBufferBlock 的工具类。

接下来看一下 MemoryLocation。MemoryLocation 是内存位置：跟踪内存地址（off-heap 非堆地址分配），或者 JVM 对象的偏移量（in-heap 堆分配）。

```
1.   public class MemoryLocation {
2.
3.    @Nullable
4.    Object obj;
5.
6.    long offset;
7.
8.    public MemoryLocation(@Nullable Object obj, long offset) {
9.      this.obj = obj;
10.     this.offset = offset;
11.   }
12.
13.   public MemoryLocation() {
14.     this(null, 0);
15.   }
16.
17.   public void setObjAndOffset(Object newObj, long newOffset) {
18.     this.obj = newObj;
19.     this.offset = newOffset;
20.   }
21.
22.   public final Object getBaseObject() {
```

```
23.     return obj;
24.   }
25.
26.   public final long getBaseOffset() {
27.     return offset;
28.   }
29. }
```

MemoryBlock 继承自 MemoryLocation。MemoryBlock 是一个连续的内存块，从{@link MemoryLocation} 开始具有固定大小的存储单元。

Spark 2.2.1 版本的 MemoryBlock.java 的源码如下：

```
1.  public class MemoryBlock extends MemoryLocation {
2.
3.   private final long length;
4.
5.   /**
     * pageNumber 变量是可选的页码；当 TaskMemoryManager 分配内存时，使用 MemoryBlock
     * 表示页，pageNumber 是公共变量，在不同的包里可以通过 TaskMemoryManager 修改
6.    */
7.   public int pageNumber = -1;
8.
9.   public MemoryBlock(@Nullable Object obj, long offset, long length) {
10.    super(obj, offset);
11.    this.length = length;
12.  }
13.
14.  /**
     *返回内存块的大小
15.   */
16.  public long size() {
17.    return length;
18.  }
19.
20.  /**
     *创建指向长数组使用内存的内存块
21.   */
22.  public static MemoryBlock fromLongArray(final long[] array) {
23.    return new MemoryBlock(array, Platform.LONG_ARRAY_OFFSET, array.length
     * 8L);
24.  }
25.
26.  /**
     *用指定字节值填充内存块
27.   */
28.  public void fill(byte value) {
29.    Platform.setMemory(obj, offset, length, value);
30.  }
31. }
```

Spark 2.4.3 版本的 MemoryBlock.java 源码与 Spark 2.2.1 版本相比具有如下特点。

- 上段代码中第 1 行之后新增 NO_PAGE_NUMBER、FREED_IN_TMM_PAGE_NUMBER、FREED_IN_ALLOCATOR_PAGE_NUMBER 变量。NO_PAGE_NUMBER 为由 Task-MemoryManagers 分配的页面的 pageNumber 值，初始化设置为−1；FREED_IN_TMM_PAGE_NUMBER 用于标记在 TaskMemoryManager 中释放的页的特殊

pageNumber 值，初始化设置为–2。在 TaskMemoryManager.freepage()中将 pageNumber 设置为该值，以便 MemoryAllocator 可以检测由 TaskMemoryManager 分配的页面是否已在 TMM 中释放，在传递给 MemoryAllocator.free()之前（在 TaskMemoryManager 中分配一个页面，然后在 MemoryAllocator 中直接释放该页面，而不执行 TMM freePage()，这是一个错误）；

FREED_IN_ALLOCATOR_PAGE_NUMBER 是 MemoryAllocator 释放的页面的 pageNumber 值，初始化设置为–3。这使用户能够检测到双重释放。

- 上段代码中第 7 行将–1 调整为 NO_PAGE_NUMBER 变量。

```
1.  ......
2.   public static final int NO_PAGE_NUMBER = -1;
3.   ......
4.   public static final int FREED_IN_TMM_PAGE_NUMBER = -2;
5.  ......
6.   public static final int FREED_IN_ALLOCATOR_PAGE_NUMBER = -3;
7.  ......
8.   public int pageNumber = NO_PAGE_NUMBER;
```

接下来看一下 UnsafeMemoryAllocator，一个简单的 {@link MemoryAllocator} 使用{@code Unsafe}分配 off-heap 非堆内存。通过 JVM 的 Platform 管理分配内存。

Spark 2.2.1 版本的 UnsafeMemoryAllocator.java 的源码如下：

```
1.  public class UnsafeMemoryAllocator implements MemoryAllocator {
2.
3.   @Override
4.   public MemoryBlock allocate(long size) throws OutOfMemoryError {
5.     long address = Platform.allocateMemory(size);
6.     MemoryBlock memory = new MemoryBlock(null, address, size);
7.     if (MemoryAllocator.MEMORY_DEBUG_FILL_ENABLED) {
8.       memory.fill(MemoryAllocator.MEMORY_DEBUG_FILL_CLEAN_VALUE);
9.     }
10.    return memory;
11.  }
12.
13.  @Override
14.  public void free(MemoryBlock memory) {
15.    assert (memory.obj == null) :
16.      "baseObject not null; are you trying to use the off-heap allocator
       to free on-heap memory?";
17.    if (MemoryAllocator.MEMORY_DEBUG_FILL_ENABLED) {
18.      memory.fill(MemoryAllocator.MEMORY_DEBUG_FILL_FREED_VALUE);
19.    }
20.    Platform.freeMemory(memory.offset);
21.  }
22. }
```

Spark2.4.3 版本的 UnsafeMemoryAllocator.java 源码与 Spark 2.2.1 版本相比具有如下特点。

- 上段代码中第 17 行前新增代码，增加 memory.pageNumber 判断的断言。
- 上段代码中第 20 行之后新增代码，作为一个额外的防御层，防止在释放之后出现 bugs，改变 MemoryBlock 来重置其指针；将页面标记为已释放（这样就可以检测到双重释放）。

```
1.  ......
2.     assert (memory.pageNumber != MemoryBlock.FREED_IN_ALLOCATOR_PAGE_NUMBER) :
```

```
3.         "page has already been freed";
4.       assert ((memory.pageNumber == MemoryBlock.NO_PAGE_NUMBER)
5.              || (memory.pageNumber ==
    MemoryBlock.FREED_IN_TMM_PAGE_NUMBER)) :
6.         "TMM-allocated pages must be freed via TMM.freePage(), not directly
           in allocator free()";
7.   ......
8.       memory.offset = 0;
9.   ......
10.      memory.pageNumber = MemoryBlock.FREED_IN_ALLOCATOR_PAGE_NUMBER;
11.  ......
```

再回到TaskMemoryManager，关注以下关键点。

（1）pageTable：与操作系统的页表类似，此数组将页编号映射为基准对象指针，支持哈希表的内部64位地址，以及baseObject+offset偏移量，这样可以支持堆地址和非堆地址。当使用“非堆地址分配”时，此映射中的每个条目将为null。使用堆分配器时，此映射中的项将指向页的基准对象。当新数据页被分配时，条目将被添加到该map。

```
1.     private final MemoryBlock[] pageTable = new MemoryBlock[PAGE_TABLE_
       SIZE];
```

（2）PAGE_TABLE_SIZE是页表中的条目数。

```
1.   private static final int PAGE_TABLE_SIZE = 1 << PAGE_NUMBER_BITS;
```

（3）PAGE_NUMBER_BITS：用于处理页表的位数。

OFFSET_BITS：用于在数据页中编码偏移量的位数，实际为51。

```
1.    /**用于页表地址的位数*/
2.    private static final int PAGE_NUMBER_BITS = 13;
3.
4.    /**用于数据页中编码偏移量的位数*/
5.    @VisibleForTesting
6.    static final int OFFSET_BITS = 64 - PAGE_NUMBER_BITS;  //51
7.
8.    /**页表条目数量*/
9.    private static final int PAGE_TABLE_SIZE = 1 << PAGE_NUMBER_BITS;
10.
11.  /**
      *最大支持的数据页大小（以字节为单位）。原则上，最大可寻址页大小是
      *（1L << OFFSET_BITS）字节，这是 2+petabytes。但是，在堆分配器上，最大页大
      *小受存储long[]数组的最大数据量的限制。这是 (2^32 - 1)×8 字节（或16 gigabytes）。
      *因此，最大是16 gigabytes
12.   */
13.   public static final long MAXIMUM_PAGE_SIZE_BYTES = ((1L << 31) - 1)×8L;
14.
15.   /**低51位的位掩码*/
16.   private static final long MASK_LONG_LOWER_51_BITS = 0x7FFFFFFFFFFFFL;
```

31.14　Spark 2.4.X下Shuffle中Mapper端的源码实现

Spark是MapReduce思想的实现之一，在一个作业中，会把不同的计算按照不同的依赖关系分成不同的Stage，前面的Stage是后面Stage的Mapper构建的一个有向无环图。我们

研究 Shuffle，实际上要研究 Mapper 端怎么实现，Reducer 端怎么实现，以及连接 Mapper 端、Reducer 端的过程，思路是非常清晰的。

我们回顾一下 MapReduce 思想在 Spark 的具体实现，到底是如何进行 Shuffle 的，主要根据依赖关系，如果有宽依赖，把我们的 Stage 进行划分，划分的时候就构成了 MapReduce，当然，可以有很多的 Stage，构建出很多 MapReduce 的关系。从源码的角度，要思考一件事情：我们写 Spark 业务代码的时候是基于 RDD 进行编程，当然也可能基于 DataSet、DataFrame，但它们背后也是 RDD，编程从 Shuffle 的层面讲，最终肯定会落到 RDD 的计算部分！

只有在 RDD 的计算部分，我们才能看出 Shuffle 的 Map 和 Reduce，为什么呢？因为我们写的 RDD，如果是窄依赖，在同一个 Stage 中进行计算，如果是最后一个 Stage，那将前面的计算结果最终汇聚起来，得出我们业务想要的结果；如果不是最后一个 Stage，是前面的 Stage，假设是第一个 Stage，那从外部或本地磁盘读取具体的数据信息，变成整个依赖关系的 Mapper。如果下面还有其他的 Stage，如果整个作业有 100 个 Stage，作为 Mapper，就有一个输出，这个输出就是 Shuffle 的输出过程，下一个 Stage 假设是第二个 Stage，就有一个输入过程，其实就是 Shuffle 的输入过程。所以，要研究源码具体在什么地方进行 Mapper 的过程，如果了解 Spark 的内核实现，就明确一件事情，肯定是在 RDD 的 Compute 方法中做的。我们看一个 RDD，如 HadoopRDD。

使用 HadoopRDD 可以读取 Hadoop 支持的文件系统或数据来源（例如，HDFS 文件、Hbase 数据源，或者 S3 数据源），使用 MapReduce API 接口（org.apache.hadoop.mapred）。主要看一下 HadoopRDD 的计算方法 compute，以及与 Shuffle 相关的部分，读取数据是看 Reader 部分，看 Reader 怎么读取数据。

```
1.  override def compute(theSplit:  Partition, context:  TaskContext):
    InterruptibleIterator[(K, V)] = {
2.    val iter = new NextIterator[(K, V)] {
3.  ......
4.  private var reader:  RecordReader[K, V] = null
5.      private val inputFormat = getInputFormat(jobConf)
6.      HadoopRDD.addLocalConfiguration(
7.        new SimpleDateFormat("yyyyMMddHHmmss", Locale.US)
        .format(createTime),
8.        context.stageId, theSplit.index, context.attemptNumber, jobConf)
9.
10.     reader =
11.       try {
12.         inputFormat.getRecordReader(split.inputSplit.value, jobConf,
            Reporter.NULL)
13.       } catch {
14.         case e:  IOException if ignoreCorruptFiles =>
15.           logWarning(s"Skipped the rest content in the corrupted file:
              ${split.inputSplit}", e)
16.           finished = true
17.           null
18.       }
19. ......
20. new InterruptibleIterator[(K, V)](context, iter)
21.  }
```

HadoopRDD 是从磁盘文件读取数据，作为整个 Stage 依赖关系的开端，读取数据进行计算的时候肯定要采用函数，采用函数计算肯定要用 Write 方法，Write 方法基于 ShuffleWriter，因为数据读取进来要进行处理，无论是简单的，还是复杂的，但 ShuffleWriter 进行 Writer 的时候，从这个角度来讲，我们认为 Stage 是 Mapper 部分，因为 Mapper 部分会将自己的数据保存在本地或者其他地方，默认保存在本地，供下一个 Stage 或 Reduce 读取数据，Stage 读取数据是从 Driver 读取数据，那我们看一下 ShuffleWriter。

```
1.  /**
     *在 map 任务中获得数据，将记录写入到 Shuffle 系统中
2.   */
3.  private[spark] abstract class ShuffleWriter[K, V] {
4.    /**将一个记录序列写入该任务的输出 */
5.    @throws[IOException]
6.    def write(records: Iterator[Product2[K, V]]): Unit
7.
8.    /**map 任务完成，关闭写入 */
9.    def stop(success: Boolean): Option[MapStatus]
10. }
```

ShuffleWriter 肯定是抽象的，为什么是抽象的？根据面向对象的设计法则，我们设计一个抽象的 ShuffleWriter 就可以有不同 Shuffle 的实现。例如，Spark 2.2.1 是默认的 SortShuffleWriter，我们看一下 ShuffleWriter 的继承结构。ShuffleWriter 的子类包括 SortShuffleWriter、UnsafeShuffleWriter（基于钨丝计划）、BypassMergeSortShuffleWriter（在数据规模不大的时候，退化为 HashShuffle 的方式）。现在核心做的是 SortShuffleWriter，这取决于 ShuffleManger。

SortShuffleWriter 写数据有一个过程。SortShuffleWriter 写数据是 Writer 方法，把数据一条一条的输出。

Spark 2.2.1 版本的 SortShuffleWriter.scala 的源码如下：

```
1.  /**写入记录到任务的输出 */
2.    override def write(records: Iterator[Product2[K, V]]): Unit = {
3.      sorter = if (dep.mapSideCombine) {
4.        require(dep.aggregator.isDefined, "Map-side combine without Aggregator
        specified!")
5.        new ExternalSorter[K, V, C](
6.          context, dep.aggregator, Some(dep.partitioner), dep.keyOrdering,
          dep.serializer)
7.      } else {
8.        //在这种情况下，我们既不聚合，也不进行排序，因为我们不关注每个分区中的 Key 键是
        //否被排序，如果正在运行的操作是 sortByKey，这将在 Reducer 端完成
9.        new ExternalSorter[K, V, V](
10.         context, aggregator = None, Some(dep.partitioner), ordering = None,
          dep.serializer)
11.     }
12.     sorter.insertAll(records)
13.
14.     //不要在 Shuffle 写入时间中计算打开合并输出文件的时间，因为它只打开一个文件，所
      //以通常太快，无法精确测量（见 SPARK-3570）
15.     val output = shuffleBlockResolver.getDataFile(dep.shuffleId, mapId)
16.     val tmp = Utils.tempFileWith(output)
17.     try {
18.       val blockId = ShuffleBlockId(dep.shuffleId, mapId, IndexShuffle
        BlockResolver.NOOP_REDUCE_ID)
19.       val partitionLengths = sorter.writePartitionedFile(blockId, tmp)
```

```
20.        shuffleBlockResolver.writeIndexFileAndCommit(dep.shuffleId, mapId,
           partitionLengths, tmp)
21.        mapStatus = MapStatus(blockManager.shuffleServerId, partitionLengths)
22.      } finally {
23.        if (tmp.exists() && !tmp.delete()) {
24.          logError(s"Error while deleting temp file ${tmp.getAbsolutePath}")
25.        }
26.      }
27.    }
```

Spark 2.4.3 版本的 SortShuffleWriter.scala 源码与 Spark 2.2.1 版本相比具有如下特点。

- 删掉上段代码中第 4 行代码。

SortShuffleWriter 的 Writer 方法中：

（1）sorter 的类型是 ExternalSorter[K, V, _]，我们要创建 ExternalSorter 实例，sorter 是要进行排序的。这里直接 new ExternalSorter，我们创建的 sorter 用于 Mapper 的 Task 的输出结果进行排序，如果需要对输出结果进行 Mapper 端的 mapSideCombine，就需要对 External 级别外部排序进行聚合。外部排序是指中间的数据非常多，基于中间数据生成一个个临时的小文件，基于小文件进行排序。如果不进行 mapSideCombine，那 ExternalSorter 进行外部排序，但不进行聚合。

（2）sorter.insertAll(records)是关键的代码，基于 sorter 具体排序的实现方式，将数据写入缓冲区中。如果 records 数据特别多，可能会导致内存溢出，Spark 现在的实现方式是 Spill 溢出写到磁盘中。

（3）val output = shuffleBlockResolver.getDataFile(dep.shuffleId, mapId)是一行非常关键的代码，从 shuffleBlockResolver 中获取输出结果，传进来的是 shuffleId、mapId，根据 Shuffle 的编号和 map 的编号获取数据文件，因为要写数据。然后通过工具类 Utils.tempFileWith(output) 创建了临时文件。

（4）val blockId = ShuffleBlockId(dep.shuffleId, mapId, IndexShuffleBlockResolver.NOOP_REDUCE_ID)，其中 ShuffleBlockId 是 case class，实质上根据 Shuffle 的 ID 和 map 的 ID 来获取 ShuffleBlock 的具体的 ID 编号。

（5）val partitionLengths = sorter.writePartitionedFile(blockId, tmp)，writePartitionedFile 通过 ExternalSorter 获取的所有数据写入到磁盘文件上。因为要进行外部排序，有些内容在内存中，还有些内容在磁盘文件上，有很多临时的磁盘文件，这时我们获取 PartitionedFile，需要将内存中的数据和磁盘上的文件数据组合成一个更大的文件。

（6）writeIndexFileAndCommit 按照 sort Shuffle 的方式分成两部分：一部分为创建索引的部分；另外一部分是我们创建了索引文件以后，在写数据的时候，把每个 Partition 数据文件中的起始和结束位置写入到我们创建的索引文件中，这样我们的 Reducer 获取位置的时候，首先根据索引文件确定属于这个 Partition 的文件起始和结束位置，抓到属于我们的数据。

Spark 2.2.1 版本的 IndexShuffleBlockResolver.scala 的 writeIndexFileAndCommit 方法如下：

```
1.    def writeIndexFileAndCommit(
2.        shuffleId: Int,
3.        mapId: Int,
4.        lengths: Array[Long],
5.        dataTmp: File): Unit = {
6.      val indexFile = getIndexFile(shuffleId, mapId)
7.      val indexTmp = Utils.tempFileWith(indexFile)
```

```
8.      try {
9.        val out = new DataOutputStream(new BufferedOutputStream(new
          FileOutputStream(indexTmp)))
10.       Utils.tryWithSafeFinally {
11.         //获取每个块的长度，需要转换成偏移量
12.         var offset = 0L
13.         out.writeLong(offset)
14.         for (length <- lengths) {
15.           offset += length
16.           out.writeLong(offset)
17.         }
18.       } {
19.         out.close()
20.       }
21.
22.       val dataFile = getDataFile(shuffleId, mapId)
23.     //每次执行 IndexShuffleBlockResolver，同步确认下面的检查和重命名是原子的
24.       synchronized {
25.         val existingLengths = checkIndexAndDataFile(indexFile, dataFile,
            lengths.length)
26.         if (existingLengths != null) {
27.       //同一任务的另一次尝试已经成功地写入了 map 输出，只需使用现有的分区长度，并删除
          //临时 map 输出
28.           System.arraycopy(existingLengths, 0, lengths, 0, lengths.length)
29.           if (dataTmp != null && dataTmp.exists()) {
30.             dataTmp.delete()
31.           }
32.           indexTmp.delete()
33.         } else {
34.           //这是为这个任务写入 map 输出的首次成功尝试，将覆盖我们写的所有索引和数据文件
35.           if (indexFile.exists()) {
36.             indexFile.delete()
37.           }
38.           if (dataFile.exists()) {
39.             dataFile.delete()
40.           }
41.           if (!indexTmp.renameTo(indexFile)) {
42.             throw new IOException("fail to rename file " + indexTmp + " to "
                + indexFile)
43.           }
44.           if (dataTmp != null && dataTmp.exists() && !dataTmp.renameTo
              (dataFile)) {
45.             throw new IOException("fail to rename file " + dataTmp + " to "
                + dataFile)
46.           }
47.         }
48.       }
49.     } finally {
50.       if (indexTmp.exists() && !indexTmp.delete()) {
51.         logError(s"Failed to delete temporary index file at ${indexTmp
            .getAbsolutePath}")
52.       }
53.     }
54.   }
```

Spark 2.4.3 版本的 IndexShuffleBlockResolver.scala 源码与 Spark 2.2.1 版本相比具有如下特点。

- 上段代码中第 8 行以后新增以下第 2～12 行代码。

- 删除上段代码中第 22～33 行代码。

```
1. ......
2.   val dataFile = getDataFile(shuffleId, mapId)
3.       //每个 executor 只有一个 IndexShuffleBlockResolver，此同步确保以下检查和
         //重命名是原子的
4.       synchronized {
5.         val existingLengths = checkIndexAndDataFile(indexFile, dataFile,
           lengths.length)
6.         if (existingLengths != null) {
7.           //对同一任务的另一个尝试已经成功地编写了 map 输出，因此只需使用现有的分区
             //长度并删除临时的 map 输出
8.           System.arraycopy(existingLengths, 0, lengths, 0, lengths.length)
9.           if (dataTmp != null && dataTmp.exists()) {
10.            dataTmp.delete()
11.          }
12.        } else {
13. ......
```

回到 SortShuffleWriter 中：

（7）mapStatus = MapStatus(blockManager.shuffleServerId, partitionLengths)：mapStatus = MapStatus(blockManager.shuffleServerId, partitionLengths)：Map 在最后的时候将我们的元数据写入到 MapStatus，MapStatus 返回给 Driver，这里是元数据信息，Driver 根据这个信息告诉下一个 Stage，你的上一个 Mapper 的数据写在什么地方。下一个 Stage 就根据 MapStatus 得到上一个 Stage 的处理结果。

回顾一下，使用 sort Shuffle 的方式写磁盘数据的时候，Mapper 本身有一个数据文件，也有 Index 文件。我们把相同的 Partition 放到一个文件中，Reducer 端拉取数据的时候，基于 Shuffle 读取数据。

SortShuffleWriter 中的核心代码肯定是 sorter.insertAll(records)，涉及数据写入到内存缓冲区及进行排序、内存和磁盘文件的管理关系，对内存的使用进行自己的管理。

- 在 insertAll 之前有判断，在 Mapper 端是否要进行聚合，如果没有进行聚合，将按照 Partition 写入到不同的文件中，最后按照 Partition 顺序合并到同样一个文件中。在这种情况下，适合 Partition 的数据比较少的情况；那我们将很多的 bucket 合并到一个文件，减少了 Mapper 端输出文件的数量，减少了磁盘 I/O，提升了性能。
- 除了既不想排序，又不想聚合的情况，也可能在 Mapper 端不进行聚合，但可能进行排序，这在缓存区中根据 PartitionID 进行排序，也可能根据 Key 进行排序。最后需要根据 PartitionID 进行排序，比较适合 Partition 比较多的情况。如果内存不够用，就会溢写到磁盘中。
- 第三种情况，既需要聚合，也需要排序，这时肯定先进行聚合，后进行排序。实现时，根据 Key 值进行聚合，在缓存中根据 PartitionID 进行排序，也可能根据 Key 进行排序，默认情况不需要根据 Key 进行排序。最后需要根据 PartitionID 进行合并，如果内存不够用，就会溢写到磁盘中。

insertAll 方法如下。

（1）shouldCombine 判断是否需要聚合。一个基本的问题：怎么知道是否需要聚合？算子和算子的配置参数决定了是否需要聚合。如果为 true，使用 AppendOnlyMap 首先在内存对值进行组合。

（2）如果需要聚合：其中的关键函数为 val update = (hadValue：Boolean, oldValue：C) => {if (hadValue) mergeValue(oldValue, kv._2) else createCombiner(kv._2) }。从 scala 语法角度讲，update 是偏函数。如果有值，就将新的 value 和旧的 value 进行合并；如果 hadValue 是 false，则新建 combiner，相当于没有旧的值。从 Hadoop 的角度讲，Merge 相当于 hadoop 的 combiner，相同 Key 的 Value 进行聚合。

然后进行循环遍历，map.changeValue((getPartition(kv._1), kv._1), update) 调用了偏函数 update，又更新了 Value 值。其中，map 是 PartitionedAppendOnlyMap[K, C]类型，是一个存储数据的内存结构。maybeSpillCollection 中如果超过内存设定的临界值，就溢写到磁盘中。

（3）如果不需要进行聚合，循环遍历时，buffer.insert(getPartition(kv._1), kv._1, kv._2 .asInstanceOf[C])直接把数据写入到缓存区中。

changeValue 调用的是 SizeTrackingAppendOnlyMap 的 changeValue 方法。

```
1.    override def changeValue(key: K, updateFunc: (Boolean, V) => V): V = {
2.     val newValue = super.changeValue(key, updateFunc)
3.     super.afterUpdate()
4.     newValue
5.   }
```

跟到 AppendOnlyMap 类的 changeValue 方法，使用聚合算法获得新的 Value。var pos = rehash(k.hashCode) & mask 获取位置，根据 Key 的 hashCode 以及掩码获得位置。curKey 获取 Key 的位置，偶数位(2×pos)表示 Key 的内容，奇数位(2×pos + 1)表示 Value 的内容。

```
1.  def  changeValue(key: K, updateFunc: (Boolean, V) => V): V = {
2.     assert(!destroyed, destructionMessage)
3.     val k = key.asInstanceOf[AnyRef]
4.     if (k.eq(null)) {
5.       if (!haveNullValue) {
6.         incrementSize()
7.       }
8.       nullValue = updateFunc(haveNullValue, nullValue)
9.       haveNullValue = true
10.      return nullValue
11.    }
12.    var pos = rehash(k.hashCode) & mask
13.    var i = 1
14.    while (true) {
15.      val curKey = data(2 * pos)
16.      if (curKey.eq(null)) {
17.        val newValue = updateFunc(false, null.asInstanceOf[V])
18.        data(2 * pos) = k
19.        data(2 * pos + 1) = newValue.asInstanceOf[AnyRef]
20.        incrementSize()
21.        return newValue
22.      } else if (k.eq(curKey) || k.equals(curKey)) {
23.        val newValue = updateFunc(true, data(2 * pos + 1).asInstanceOf[V])
24.        data(2 * pos + 1) = newValue.asInstanceOf[AnyRef]
25.        return newValue
26.      } else {
27.        val delta = i
28.        pos = (pos + delta) & mask
29.        i += 1
30.      }
31.    }
```

```
32.     null.asInstanceOf[V] //从不达到此语句，但需要进行编译
33.   }
```

下面看一下 incrementSize 方法，如果超过临界值，则增加空间。

```
1.   private def incrementSize() {
2.     curSize += 1
3.     if (curSize > growThreshold) {
4.       growTable()
5.     }
6.   }
```

继续跟踪 growThreshold 函数，其扩容双倍的表的大小（capacity×2）和重新散列一切。

```
1.       protected def growTable() {
2.     //容量< MAXIMUM_CAPACITY (2 ^ 29)，所以容量×2 不会溢出
3.     val newCapacity = capacity * 2
4.     require(newCapacity <= MAXIMUM_CAPACITY, s"Can't contain more than
       ${growThreshold} elements")
5.     val newData = new Array[AnyRef](2 * newCapacity)
6.     val newMask = newCapacity - 1
7.     //将所有旧值插入新数组中。注意，因为旧的 Key 值是唯一的，所以在插入时不需要检查相
       //等性
8.     var oldPos = 0
9.     while (oldPos < capacity) {
10.      if (!data(2 * oldPos).eq(null)) {
11.        val key = data(2 * oldPos)
12.        val value = data(2 * oldPos + 1)
13.        var newPos = rehash(key.hashCode) & newMask
14.        var i = 1
15.        var keepGoing = true
16.        while (keepGoing) {
17.          val curKey = newData(2 * newPos)
18.          if (curKey.eq(null)) {
19.            newData(2 * newPos) = key
20.            newData(2 * newPos + 1) = value
21.            keepGoing = false
22.          } else {
23.            val delta = i
24.            newPos = (newPos + delta) & newMask
25.            i += 1
26.          }
27.        }
28.      }
29.      oldPos += 1
30.    }
31.    data = newData
32.    capacity = newCapacity
33.    mask = newMask
34.    growThreshold = (LOAD_FACTOR * newCapacity).toInt
35.  }
```

回到 SizeTrackingAppendOnlyMap 的 changeValue 方法，super.changeValue(key, updateFunc)以后执行 super.afterUpdate()操作，每次更新后调用回调，在其中会调用 takeSample 方法，取一个新样本的当前集合的大小，其中采用 estimate 方法进行估值。

```
1.   private def takeSample(): Unit = {
2.     samples.enqueue(Sample(SizeEstimator.estimate(this), numUpdates))
```

```
3.      //只使用最后两个样本进行推算。
4.      if (samples.size > 2) {
5.        samples.dequeue()
6.      }
7.      val bytesDelta = samples.toList.reverse match {
8.        case latest :: previous :: tail =>
9.          (latest.size - previous.size).toDouble / (latest.numUpdates -
          previous.numUpdates)
10.       //如果少于 2 个样品，假设没有变化
11.       case _ => 0
12.     }
13.     bytesPerUpdate = math.max(0, bytesDelta)
14.     nextSampleNum = math.ceil(numUpdates * SAMPLE_GROWTH_RATE).toLong
15.   }
```

再次回到 SortShuffleWriter 的 Write 方法中的关键代码 sorter.insertAll(records)，如果不进行聚合，则直接将数据写入到 buffer 中。

```
1.   while (records.hasNext) {
2.         addElementsRead()
3.         val kv = records.next()
4.         buffer.insert(getPartition(kv._1), kv._1, kv._2.asInstanceOf[C])
5.         maybeSpillCollection(usingMap = false)
```

其中的 inert 方法：

```
1.   def insert(partition: Int, key: K, value: V): Unit = {
2.      if (curSize == capacity) {
3.        growArray()
4.      }
5.      data(2 * curSize) = (partition, key.asInstanceOf[AnyRef])
6.      data(2 * curSize + 1) = value.asInstanceOf[AnyRef]
7.      curSize += 1
8.      afterUpdate()
9.    }
```

如果已经达到容量，再扩容 2 倍的数组大小。

Spark 2.2.1 版本的 PartitionedPairBuffer.scala 的源码如下：

```
1.  private def growArray(): Unit = {
2.      if (capacity >= MAXIMUM_CAPACITY) {
3.        throw new IllegalStateException(s"Can't insert more than ${MAXIMUM_
        CAPACITY} elements")
4.      }
5.      val newCapacity =
6.        if (capacity * 2 < 0 || capacity * 2 > MAXIMUM_CAPACITY) { //溢出
7.          MAXIMUM_CAPACITY
8.        } else {
9.          capacity * 2
10.       }
11.     val newArray = new Array[AnyRef](2 * newCapacity)
12.     System.arraycopy(data, 0, newArray, 0, 2 * capacity)
13.     data = newArray
14.     capacity = newCapacity
15.     resetSamples()
16.   }
```

Spark 2.4.3 版本的 PartitionedPairBuffer.scala 源码与 Spark 2.2.1 版本相比具有如下特点。

❑ 上段代码中第 6 行调整为容量*2 大于最大容量的溢出判断。

```
1.  ......
2.  if (capacity * 2 > MAXIMUM_CAPACITY) { // Overflow
3.  ......
```

再次回到 SortShuffleWriter 的 Write 方法中的关键代码 sorter.insertAll(records)，在 insertAll 方法中无论是否聚合，都可能溢写到磁盘。下面看一下 maybeSpillCollection。

```
1.      private def maybeSpillCollection(usingMap:  Boolean):  Unit = {
2.      var estimatedSize = 0L
3.      if (usingMap) {
4.        estimatedSize = map.estimateSize()
5.        if (maybeSpill(map, estimatedSize)) {
6.          map = new PartitionedAppendOnlyMap[K, C]
7.        }
8.      } else {
9.        estimatedSize = buffer.estimateSize()
10.       if (maybeSpill(buffer, estimatedSize)) {
11.         buffer = new PartitionedPairBuffer[K, C]
12.       }
13.     }
14.
15.     if (estimatedSize > _peakMemoryUsedBytes) {
16.       _peakMemoryUsedBytes = estimatedSize
17.     }
18.   }
```

其中的 maybeSpill 是如何实现的？如果大于阈值 myMemoryThreshold，就要申请内存空间 acquireMemory。一种情况，如果分配的内存太小，就返回 0；另外一种情况，如果超过了阈值，就导致当前需要的内存大小需要 Spill。但在 Spill 之前会先扩容一次，源码如下：

```
1.      protected def maybeSpill(collection: C, currentMemory: Long): Boolean
        = {
2.      var shouldSpill = false
3.      if (elementsRead % 32 == 0 && currentMemory >= myMemoryThreshold) {
4.        //从 Shuffle 内存池中获取当前内存的两倍
5.        val amountToRequest = 2 * currentMemory - myMemoryThreshold
6.        val granted = acquireMemory(amountToRequest)
7.        myMemoryThreshold += granted
8.        //如果分配太少的内存（无论是 tryToAcquire 返回 0，还是已超过 myMemory
          Threshold 阈值内存），都将溢出当前的集合
9.        shouldSpill = currentMemory >= myMemoryThreshold
10.     }
11.     shouldSpill = shouldSpill || _elementsRead > numElementsForce
        SpillThreshold
12.     //实际上溢出
13.     if (shouldSpill) {
14.       _spillCount += 1
15.       logSpillage(currentMemory)
16.       spill(collection)
17.       _elementsRead = 0
18.       _memoryBytesSpilled += currentMemory
19.       releaseMemory()
20.     }
21.     shouldSpill
22.   }
```

点击进入 Spillable.scala 的 spill 方法，这里没有具体的实现。

```
1.    protected def spill(collection: C): Unit
```

查看 Spillable.scala 的子类 ExternalSorter 中的 spill 方法，生成 spillFile 文件。

```
1.    def spill(): Boolean = SPILL_LOCK.synchronized {
2.      ......
3.        val spillFile = spillMemoryIteratorToDisk(inMemoryIterator)
4.        forceSpillFiles += spillFile
5.        val spillReader = new SpillReader(spillFile)
6.        nextUpstream = (0 until numPartitions).iterator.flatMap { p =>
7.          val iterator = spillReader.readNextPartition()
8.          iterator.map(cur => ((p, cur._1), cur._2))
9.        }
10.       hasSpilled = true
11.       true
12.     }
13.   }
```

继续跟踪 spillMemoryIteratorToDisk 方法：val (blockId, file) = diskBlockManager .createTempShuffleBlock()：生成临时文件的 block 和临时文件本身；batchSizes：按写入磁盘的顺序，记录本身的大小。elementsPerPartition：记录分区有多少元素。flush：将数据写入磁盘。一批一批地写入数据，达到序列化大小的时候进行 flush 操作。

```
1.      private[this] def spillMemoryIteratorToDisk(inMemoryIterator :
        WritablePartitionedIterator)
2.        : SpilledFile = {
3.    //因为这些文件可能在 Shuffle 过程中被读取，所以它们的压缩必须由 spark.shuffle
      //.compress 参数（取代 spark.shuffle.spill.compress）控制，所以这里我们需要使
      //用 createTempShuffleBlock；更多的上下文参考 SPARK-3426
4.    val (blockId, file) = diskBlockManager.createTempShuffleBlock()
5.
6.    //这些变量在每次刷新后重置
7.    var objectsWritten: Long = 0
8.    val spillMetrics: ShuffleWriteMetrics = new ShuffleWriteMetrics
9.    val writer: DiskBlockObjectWriter =
10.     blockManager.getDiskWriter(blockId, file, serInstance, fileBufferSize,
        spillMetrics)
11.
12.   //按写入磁盘顺序的列表空间大小（字节）
13.   val batchSizes = new ArrayBuffer[Long]
14.
15.   //每个分区中有多少元素
16.   val elementsPerPartition = new Array[Long](numPartitions)
17.
18.   //将磁盘写入程序的内容刷新到磁盘，并更新相关变量
19.   //在处理结束时提交写入
20.   def flush(): Unit = {
21.     val segment = writer.commitAndGet()
22.     batchSizes += segment.length
23.     _diskBytesSpilled += segment.length
24.     objectsWritten = 0
25.   }
26.
```

```
27.     var success = false
28.     try {
29.       while (inMemoryIterator.hasNext) {
30.         val partitionId = inMemoryIterator.nextPartition()
31.         require(partitionId >= 0 && partitionId < numPartitions,
32.           s"partition Id: ${partitionId} should be in the range [0,
            ${numPartitions})")
33.         inMemoryIterator.writeNext(writer)
34.         elementsPerPartition(partitionId) += 1
35.         objectsWritten += 1
36.
37.         if (objectsWritten == serializerBatchSize) {
38.           flush()
39.         }
40.       }
41.       if (objectsWritten > 0) {
42.         flush()
43.       } else {
44.         writer.revertPartialWritesAndClose()
45.       }
46.       success = true
47.     } finally {
48.       if (success) {
49.         writer.close()
50.       } else {
51.         //在设置成功前，如文件路径发生异常，关闭内容，让异常进一步抛出
52.         writer.revertPartialWritesAndClose()
53.         if (file.exists()) {
54.           if (!file.delete()) {
55.             logWarning(s"Error deleting ${file}")
56.           }
57.         }
58.       }
59.     }
60.
61.     SpilledFile(file, blockId, batchSizes.toArray, elementsPerPartition)
62.   }
```

再次回到 SortShuffleWriter，我们将在后面继续研究 sorter.writePartitionedFile 方法。

31.15　Spark 2.4.X 下 Shuffle 中 SortShuffleWriter 排序源码内幕解密

Spark Shuffle 一个至关重要的内容，是我们的 SortShuffle 内部到底怎么排序？这里的排序是从整个框架的角度讲，SortShuffle 在不考虑业务排序的情况下是怎么进行排序的？SortShuffle 最原始的排序是按照 Partition 进行的。

关于 SortShuffle 的排序，我们主要看它的 Write 方法，因为只有进行输出的时候，才涉及排序，涉及排序中非常关键的一行代码。

```
1.   val partitionLengths = sorter.writePartitionedFile(blockId, tmp)
```

writePartitionedFile 的基本工作机制是 ExternalSorter 在进行排序的时候可能一部分数据在内存中，一部分数据在磁盘上。在磁盘上的数据可能是一个，也可能是若干个，假设磁盘上有很多小文件，那就会将小文件 Merge 成一个大文件。

（1）数组 lengths：跟踪输出文件中每个范围的位置。

（2）通过 blockManager 获得一个 writer，blockManager 管理了内存和磁盘的读写。

（3）判断 spills.isEmpty，如仅在内存中有数据的处理方法。如果数据在磁盘中，则必须执行合并排序；得到一个迭代器的分区和直接写入数据。

（4）数据仅在内存中的情况：其中重要的一行代码是 val it：WritablePartitionedIterator = collection.destructiveSortedWritablePartitionedIterator(comparator)，生成了一个迭代器 Iterator，这个迭代器非常重要，因为排序的时候需要迭代器。那么我们来看一下 WritablePartitionedPairCollection 类中的 destructiveSortedWritablePartitionedIterator 方法，该方法遍历数据并写出元素，而不是返回元素，记录根据 PartitionID 及给定的比较器进行排序。这可能会破坏基础集合。

```
1.  def  destructiveSortedWritablePartitionedIterator(keyComparator: Option
    [Comparator[K]])
2.      :  WritablePartitionedIterator = {
3.      val it = partitionedDestructiveSortedIterator(keyComparator)
4.      new WritablePartitionedIterator {
5.        private[this] var cur = if (it.hasNext) it.next() else null
6.
7.        def writeNext(writer:  DiskBlockObjectWriter):  Unit = {
8.          writer.write(cur._1._2, cur._2)
9.          cur = if (it.hasNext) it.next() else null
10.       }
11.
12.       def hasNext():  Boolean = cur != null
13.
14.       def nextPartition():  Int = cur._1._1
15.     }
16.   }
17. }
18.
```

在 destructiveSortedWritablePartitionedIterator 方法中：

- partitionedDestructiveSortedIterator 生成迭代器本身。按照分区 ID 和给定比较器迭代数据，这可能破坏基础集合。
- 构建一个对象 WritablePartitionedIterator，迭代器写元素到一个 DiskBlockObjectWriter，而不是返回元素。每个元素有关联分区。WritablePartitionedIterator 从设计模式讲，相当于一个代理类，代理的是迭代器的功能。其中，cur 基于迭代器 partitionedDestructiveSortedIterator 的 it 生成一个私有变量。

partitionedDestructiveSortedIterator 方法没有具体实现。

```
1.      def partitionedDestructiveSortedIterator(keyComparator: Option
        [Comparator[K]])
2.      :  Iterator[((Int, K), V)]
```

下面看 WritablePartitionedPairCollection 子类 PartitionedAppendOnlyMap，Partitioned-AppendOnlyMap 封装了一个 map，Key 值是(partition ID, K)，其中的 partitionedDestructive-SortedIterator 实现如下：

```
1.   def  partitionedDestructiveSortedIterator(keyComparator:  Option
     [Comparator[K]])
2.     :  Iterator[((Int, K), V)] = {
3.     val comparator = keyComparator.map(partitionKeyComparator)
       .getOrElse(partitionComparator)
4.     destructiveSortedIterator(comparator)
5.   }
```

partitionedDestructiveSortedIterator 的 keyComparator.map 操作传入的是 partitionKeyComparator，然后调用 destructiveSortedIterator 方法。

destructiveSortedIterator 方法按排序顺序返回映射的迭代器。这提供了一种方法来排序 map 没有使用额外的内存，以破坏 map 的有效性为代价。

destructiveSortedIterator 实现了一个算法，里面涉及较多的内容，重新回到 WritablePartitionedPairCollection.scala。

```
1.    def   destructiveSortedWritablePartitionedIterator(keyComparator  :
Option[Comparator[K]])
2.     :  WritablePartitionedIterator = {
3.     val it = partitionedDestructiveSortedIterator(keyComparator)
4.     new WritablePartitionedIterator {
5.    ......
```

destructiveSortedWritablePartitionedIterator 传入了一个参数 keyComparator，这个参数是从哪里来的？回到之前的 SortShuffleWriter 的 write 方法中。

```
1.   ......
2.  val blockId = ShuffleBlockId(dep.shuffleId, mapId, IndexShuffleBlock-
    Resolver.NOOP_REDUCE_ID)
3.      val partitionLengths = sorter.writePartitionedFile(blockId, tmp)
4.      shuffleBlockResolver.writeIndexFileAndCommit(dep.shuffleId, mapId,
        partitionLengths, tmp)
5.      mapStatus = MapStatus(blockManager.shuffleServerId, partitionLengths)
6.    ......
```

从 writePartitionedFile 跟进去，ExternalSorter.scala 的 writePartitionedFile 方法中。

```
1.   if (spills.isEmpty) {
2.       //只有内存数据的情况下
3.       val collection = if (aggregator.isDefined) map else buffer
4.       val it = collection.destructiveSortedWritablePartitionedIterator
         (comparator)
5.       while (it.hasNext) {
6.
```

这里就传入了一个 comparator，comparator 是从哪里来的？跟踪源码 comparator 本身是一个比较器，是一个函数。

```
1.         private def comparator:  Option[Comparator[K]] = {
2.       if (ordering.isDefined || aggregator.isDefined) {
3.         Some(keyComparator)
4.       } else {
5.         None
6.       }
7.     }
```

回到 WritablePartitionedPairCollection 类的 destructiveSortedWritablePartitionedIterator 方法，生成了迭代器 it 之后使用了一个代理类 WritablePartitionedIterator，接下来怎么跟源码呢？因为在实际运行的时候要进行排序，那我们回到 ExternalSorter 类的 writePartitionedFile 方法，框架本身会基于 Partition 进行排序。

writePartitionedFile 里面有一个 partitionedIterator 方法，这个方法非常重要。

```
1.    def  writePartitionedFile(
2.        ……
3.       if (spills.isEmpty) {
4.         //只有内存数据的情况下
5.       ……
6.       } else {
7.         //我们必须执行合并排序；通过分区获得一个迭代器，并直接写入所有内容
8.         for ((id, elements) <- this.partitionedIterator) {
9.           if (elements.hasNext) {
10.            for (elem <- elements) {
11.              writer.write(elem._1, elem._2)
12.            }
13.            val segment = writer.commitAndGet()
14.            lengths(id) = segment.length
15.          }
16.      ……
```

下面看一下 ExternalSorter 类的 partitionedIterator 方法：返回一个迭代器，在其中写入所有数据，迭代器按分区和聚合函数进行分组。对于每个分区，基于其中的内容有一个迭代器，这些将按顺序访问（没有读到上一个数据前，不能跳过一个分区）。对于每个分区，根据分区 partition ID 的顺序返回一个 key-value。我们一次合并所有溢出的文件，也可以修改为支持层次合并。

```
1.      def  partitionedIterator:  Iterator[(Int, Iterator[Product2[K,
        C]])] = {
2.      val usingMap = aggregator.isDefined
3.      val collection:  WritablePartitionedPairCollection[K, C] = if
        (usingMap) map else buffer
4.      if (spills.isEmpty) {
5.        //特殊情况：如果只在内存数据中，就不需要合并流，甚至不需要使用分区 ID 以外的任
          //何键来排序
6.        if (!ordering.isDefined) {
7.          //用户没有请求排序键，所以只能按分区 ID 排序，而不是按键 Key 排序
8.          groupByPartition(destructiveIterator(collection.partitioned-
            DestructiveSortedIterator(None)))
9.        } else {
10.         //我们需要通过分区 ID 和键 Key 进行排序
11.         groupByPartition(destructiveIterator(
```

```
12.         collection.partitionedDestructiveSortedIterator(Some(key-
            Comparator)))
13.       }
14.     } else {
15.       //合并溢出和内存数据
16.       merge(spills, destructiveIterator(
17.         collection.partitionedDestructiveSortedIterator(comparator)))
18.     }
19.   }
```

partitionedIterator 本身返回的是一个 Iterator，在 Mapper 端进行 aggregate 操作。特殊情况：如果只有内存数据，就不需要合并，甚至不需要按分区 ID 排序。

（1）如果没有 spills，则 spills 为空。

- 如果 ordering.isDefined 为 false，在不进行排序 ordering 的情况下：用户没有请求排序 Keys，所以 groupByPartition 只能按分区 ID 排序，而不是按 Key 排序；groupByPartition 传入的参数值是 None，即不对 Key 进行排序。
- 如果 ordering.isDefined 为 true，则我们需要根据 partition ID 和 Key 进行排序。

（2）在有 spills 的情况下，一部分数据在磁盘上，合并溢出和内存数据。

下面看一下 groupByPartition 方法，给定一个 stream((partition, key), combiner)，假定要按分区 ID 排序，将每个分区的键值对组合为子迭代器。

```
1.      private def groupByPartition(data:  Iterator[((Int, K), C)])
2.        :  Iterator[(Int, Iterator[Product2[K, C]])] =
3.    {
4.      val buffered = data.buffered
5.      (0 until numPartitions).iterator.map(p => (p, new IteratorForPartition
        (p, buffered)))
6.    }
```

groupByPartition 根据 partition 进行排序后，还根据 partition 进行聚合。IteratorForPartition 是单个 partition 的迭代器。IteratorForPartition 代码如下：

```
1.      private[this] class IteratorForPartition(partitionId:  Int, data:
        BufferedIterator[((Int, K), C)])
2.      extends Iterator[Product2[K, C]]
3.    {
4.      override def hasNext:  Boolean = data.hasNext && data.head._1._1 ==
        partitionId
5.
6.      override def next():  Product2[K, C] = {
7.        if (!hasNext) {
8.          throw new NoSuchElementException
9.        }
10.       val elem = data.next()
11.       (elem._1._2, elem._2)
12.     }
13.   }
```

IteratorForPartition 仅从底层缓冲区读取给定分区 ID 的元素的迭代器流，假设这个分区是下一个被读取，更容易返回来自内存集合的分区迭代器。

回到 ExternalSorter 的 partitionedIterator 方法，groupByPartition(destructiveIterator(collection.partitionedDestructiveSortedIterator(Some(keyComparator)))), 这里就是根据 partition ID 和 Key 进行排序的情况。partitionedDestructiveSortedIterator 是天然的排序，字母按照字母排序，数字按照数字排序，当然也可以自定义业务类的排序。

```
1.          //我们需要通过分区 ID 和键 Key 进行排序
2.         groupByPartition(destructiveIterator(
3.           collection.partitionedDestructiveSortedIterator(Some(key-
             Comparator))))
```

下面看一下 keyComparator 的代码，根据 Key 的 hashCode 进行排序。

```
1.          private val keyComparator: Comparator[K] = ordering.getOrElse
            (new Comparator[K] {
2.      override def compare(a: K, b: K): Int = {
3.        val h1 = if (a == null) 0 else a.hashCode()
4.        val h2 = if (b == null) 0 else b.hashCode()
5.        if (h1 < h2) -1 else if (h1 == h2) 0 else 1
6.      }
7.    })
```

回到 ExternalSorter 的 partitionedIterator 方法，查看 partitionedDestructiveSortedIterator 方法，按照分区 ID 和给定比较器迭代数据。这可能破坏基础集合。其继承者为 PartitionedAppendOnlyMap 和 PartitionedPairBuffer。

```
1.    def partitionedDestructiveSortedIterator(keyComparator: Option
      [Comparator[K]])
2.      : Iterator[((Int, K), V)]
```

回到 ExternalSorter 的 partitionedIterator 方法，根据是否进行 aggregator 操作区分两种情况。

（1）如果需要聚合，则使用 PartitionedAppendOnlyMap。

（2）如果不需要进行聚合，则使用 PartitionedPairBuffer。

```
1.  val usingMap = aggregator.isDefined
2.   val collection: WritablePartitionedPairCollection[K, C] = if (usingMap)
     map else buffer
```

看一下类的申明，数据结构在溢出前存储在内存对象中。取决于是否有聚合器，我们可以把数据放入 AppendOnlyMap，或者将它们存储在数组缓冲区 buffer 中。

```
1.  @volatile private var map: PartitionedAppendOnlyMap[K, C] = new
    PartitionedAppendOnlyMap[K, C]
2.    @volatile private var buffer: PartitionedPairBuffer[K, C] = new
      PartitionedPairBuffer[K, C]
```

PartitionedPairBuffer 和 PartitionedAppendOnlyMap 类中，我们主要看 partitioned-DestructiveSortedIterator 方法。

下面看一下 PartitionedPairBuffer 的 partitionedDestructiveSortedIterator 方法。

```
1.    override def partitionedDestructiveSortedIterator(keyComparator:
      Option[Comparator[K]])
2.      : Iterator[((Int, K), V)] = {
```

```
3.      val comparator = keyComparator.map(partitionKeyComparator).getOrElse
        (partitionComparator)
4.      new Sorter(new KVArraySortDataFormat[(Int, K), AnyRef]).sort(data, 0,
        curSize, comparator)
5.      iterator
6.    }
```

使用 partitionKeyComparator 将原有的 Comparator 进行了替换，partitionKeyComparator 是 partitionKey 的二次排序。两种情况：如果 keyComparator 传入了值，则根据 partitionID 和 Key 进行排序；如果 keyComparator 没有传入值，则只根据 partitionID 进行排序。

我们看一下 partitionKeyComparator，通过 partition ID 和 Key 对它们进行排序。

```
1.    def partitionKeyComparator[K](keyComparator: Comparator[K]):
      Comparator[(Int, K)] = {
2.      new Comparator[(Int, K)] {
3.        override def compare(a: (Int, K), b: (Int, K)): Int = {
4.          val partitionDiff = a._1 - b._1
5.          if (partitionDiff != 0) {
6.            partitionDiff
7.          } else {
8.            keyComparator.compare(a._2, b._2)
9.          }
10.       }
```

回到 PartitionedPairBuffer 的 partitionedDestructiveSortedIterator 方法， 这里调用 new 函数创建一个 Sorter。Sorter 内部使用 timSort 排序。

下面看一下 PartitionedAppendOnlyMap。

```
1.   private[spark] class PartitionedAppendOnlyMap[K, V]
2.     extends SizeTrackingAppendOnlyMap[(Int, K), V] with WritableParti-
       tionedPairCollection [K, V] {
3.
4.     def partitionedDestructiveSortedIterator(keyComparator: Option
       [Comparator[K]])
5.       : Iterator[((Int, K), V)] = {
6.       val comparator = keyComparator.map(partitionKeyComparator)
         .getOrElse(partitionComparator)
7.       destructiveSortedIterator(comparator)
8.     }
9.
10.    def insert(partition: Int, key: K, value: V): Unit = {
11.      update((partition, key), value)
12.    }
13. }
```

PartitionedAppendOnlyMap 中最关键的是 destructiveSortedIterator，destructiveSortedIterator 方法以排序的顺序返回 map 的迭代器，这以牺牲 map 的有效性为代价，提供了一种不需要额外的内存对 map 进行排序的方法。

PartitionedAppendOnlyMap 的 destructiveSortedIterator 方法中调用 new 函数创建一个 Sorter，Sorter 内部也是使用 timSort 排序。

31.16　Spark 2.4.X 下 Sort Shuffle 中 timSort 排序源码具体实现

timSort 排序方式是一种相对权衡了各方面的排序方式，假如排序的数据分成很多不同的块，timSort 有很好的排序性能上的表现。因此，有必要彻底研究一下 timSort 是怎么实现的。

回顾一下，我们跟踪代码是从 Sorter.scala 中跟到 timSort 的，也就是进行 ExternalSorter 的时候要进行排序，默认情况下基于 PartitionID 进行排序，对 PartitionID 进行排序并不意味着对数据本身进行排序，我们在 Sorter.scala 中用到了 timSort。研究一下 timSort 的源码，会发现 timSort 和 MergeSort 有一点类似，实质上有很大的区别，我们可以初步感知 timSort 的排序方式。MergeSort 排序的方式把数据分成很多片，开始分成很多小文件，最终把小文件合并成大文件。timSort 可以认为是 MergeSort 排序的改良。

TimSort 优化 MergeSort 排序，把它变成稳定的、适应的、迭代的排序，timSort 基于分布式的排序，效率有很大的提升。

MergeSort 排序默认长度是 1，归并的时候自动生成归并元素；timSort 是连续递增的，将其中的一块数据 run 进行反转，run 有自己具体的实现算法，run 可以认为是一块固定大小的数据，如果插入一段数据，数据的长度小于 run 的长度，timSort 就会采用二分的 insertSort，进行一些局部的优化。MergeSort 排序归并是固定的，而 timSort 是随机的，会有判断条件。timSort 在很多地方都有使用，如安卓等。

TimSort.java 位于 org.apache.spark.util.collection 包里面，其中还有一个测试类 TestTimSort，如创建测试数组等。TimSort.java 阅读源码的技巧先看 sort 排序，然后将其他的成员、方法关联起来。TimSort.java 代码如下：

```
1.    public void sort(Buffer a, int lo, int hi, Comparator<? super K> c) {
2.      assert c != null;
3.
4.      int nRemaining  = hi - lo;
5.      if (nRemaining < 2)
6.        return;  //大小为 0 和 1 的数组总是排序的
7.
8.      //如果数组是小的数组，则执行一个"mini-TimSort”，其没有做合并
9.      if (nRemaining < MIN_MERGE) {
10.       int initRunLen = countRunAndMakeAscending(a, lo, hi, c);
11.       binarySort(a, lo, hi, lo + initRunLen, c);
12.       return;
13.     }
14.
15.     /**
        *在数组中从左向右遍历，寻找 natural runs，扩展短的 natural runs 到 minRun
        *元素中，合并 runs 的时候保持堆栈不变
16.     */
17.     SortState sortState = new SortState(a, c, hi - lo);
18.     int minRun = minRunLength(nRemaining);
19.     do {
20.       //标识下一个 run
21.       int runLen = countRunAndMakeAscending(a, lo, hi, c);
```

```
22.
23.       //如果 run 是短的，就扩展到 min(minRun, nRemaining)
24.       if (runLen < minRun) {
25.         int force = nRemaining <= minRun ? nRemaining :  minRun;
26.         binarySort(a, lo, lo + force, lo + runLen, c);
27.         runLen = force;
28.       }
29.
30.       //压送 run 到等待运行堆栈，并可能合并
31.       sortState.pushRun(lo, runLen);
32.       sortState.mergeCollapse();
33.
34.       //查找下一个 run
35.       lo += runLen;
36.       nRemaining -= runLen;
37.     } while (nRemaining != 0);
38.
39.     //合并所有剩余的 run，以完成排序
40.     assert lo == hi;
41.     sortState.mergeForceCollapse();
42.     assert sortState.stackSize == 1;
43.   }
```

在 TimSort.java 代码中：

- nRemaining 是未排序的数组的长度，是从数组的角度考虑的，不过 timSort 是分布式的。
- 如果 nRemaining 小于 2，数组大小为 0、1 时，此时数据已经排好序，通过 return 语句直接返回。数组大小为 0、1 是已经排序的，那就不用排序。
- 如果 nRemaining 小于 MIN_MERGE，就变成 mini-TimSort，就是不使用归并排序。countRunAndMakeAscending 计算得到递增数据的长度，然后使用 binarySort 二分排序法，这个是基本的排序法。
- 之后是 SortState。SortState 是构建一个栈，创建一个 timSort 实例，维护我们排序的状态信息。
- minRunLength：获得最小的 run 长度。
- do while 循环首先得到递增数列的长度，如果 runLen 小于 minRun，则使用 binarySort 二分插入。sortState.pushRun(lo, runLen)是入栈，把即将运行的 run 放入栈中。sortState.mergeCollapse()：可能进行归并排序，内部视不同的情况进行判断。lo += runLen：下一个要进行的 run。
- 循环结束后，所有剩余的 run 完成排序。

TimSort.java 从源码实现的角度讲，第一个比较关键的一行代码是 int initRunLen = countRunAndMakeAscending(a, lo, hi, c)；我们看一下 countRunAndMakeAscending，首先找到 run 的尾部，在 while 中进行判断，反转我们的 run，最后返回 run 的长度。

countRunAndMakeAscending：返回 run 的长度。run 在指定的开始位置，如果它是递减的，则反转运行。如一个 run 是最长的升序序列：a[lo] <= a[lo + 1] <= a[lo + 2] <= …或者是最长的递减序列：a[lo] > a[lo + 1] > a[lo + 2] > …一个稳定的归并排序中严格的降序定义是必要的，能安全调用进行反转降序序列，而不破坏稳定性。

- @param a：数组中的 run 将被计数，并可能反转。

- @param lo：run 第一个元素的索引。
- @param hi：run 可能包含的最后一个元素的索引，需要 {@code lo < hi}。
- @param c：用于排序的比较器。
- @return：返回 run 的长度。

countRunAndMakeAscending 代码如下：

```
1.   private int countRunAndMakeAscending(Buffer a, int lo, int hi,
     Comparator<? super K> c) {
2.     assert lo < hi;
3.     int runHi = lo + 1;
4.     if (runHi == hi)
5.       return 1;
6.
7.     K key0 = s.newKey();
8.     K key1 = s.newKey();
9.
10.    //查找 run 的末尾，如果递减，则反转范围
11.    if (c.compare(s.getKey(a, runHi++, key0), s.getKey(a, lo, key1)) < 0)
       { //降序
12.      while (runHi < hi && c.compare(s.getKey(a, runHi, key0), s.getKey(a,
         runHi - 1, key1)) < 0)
13.        runHi++;
14.      reverseRange(a, lo, runHi);
15.    } else {                              //升序
16.      while (runHi < hi && c.compare(s.getKey(a, runHi, key0), s.getKey(a,
         runHi - 1, key1)) >= 0)
17.        runHi++;
18.    }
19.
20.    return runHi - lo;
21.  }
```

回到 TimSort.java 的 sort 方法，我们看一下 binarySort 的代码。二分法排序将指定数组的指定部分进行插入排序，小数量数据排序的最好情况需要进行 $O(n\log n)$次比较，但最坏情况下需移动 $O(n^2)$次数据。如果指定范围的初始部分已排序，此方法可以利用它：该方法假定包含索引{@code lo}的元素，包括到{@code start}，排除已排序的数据。

- @param a：需进行排序的数组范围。
- @param lo：索引中的第一个元素进行排序的范围。
- @param hi：索引中的最后一个元素之后的范围进行排序。
- @param start：索引中的第一个元素的范围是未知排序的（{@code lo <= start <= hi}）。
- @param C：比较器用于排序。

TimSort.java 的 binarySort 代码如下：

```
1.   private void binarySort(Buffer a, int lo, int hi, int start, Comparator<?
     super K> c) {
2.     assert lo <= start && start <= hi;
3.     if (start == lo)
4.       start++;
5.
6.     K key0 = s.newKey();
7.     K key1 = s.newKey();
8.
9.     Buffer pivotStore = s.allocate(1);
```

```
10.     for ( ; start < hi; start++) {
11.       s.copyElement(a, start, pivotStore, 0);
12.       K pivot = s.getKey(pivotStore, 0, key0);
13.
14.       //将 left (right)设置为一个 a[start] (pivot) 的索引
15.       int left = lo;
16.       int right = start;
17.       assert left <= right;
18.       /**
19.        * 不变量
20.        *   pivot >= all in [lo, left).
21.        *   pivot <  all in [right, start).
22.        */
23.       while (left < right) {
24.         int mid = (left + right) >>> 1;
25.         if (c.compare(pivot, s.getKey(a, mid, key1)) < 0)
26.           right = mid;
27.         else
28.           left = mid + 1;
29.       }
30.       assert left == right;
31.
32.       /**
           * 不变量仍然保持不变：pivot >= all in [lo, left)以及 pivot < all in [left,
           * start)，所以 pivot 属于 left。注意：如果元素等于 pivot，则 left 指向在它们
           * 之后的第一个 slot--这就是为什么这种类型是稳定的：滑动元素给 pivot 腾出空间
33.        */
34.       int n = start - left;  //要移动的元素的数目
35.       //默认情况下，Switch 切换是对数组复制优化
36.       switch (n) {
37.         case 2:  s.copyElement(a, left + 1, a, left + 2);
38.         case 1:  s.copyElement(a, left, a, left + 1);
39.           break;
40.         default: s.copyRange(a, left, a, left + 1, n);
41.       }
42.       s.copyElement(pivotStore, 0, a, left);
43.     }
44.   }
```

回到 TimSort.java 的 sort 方法，这里有一个 minRunLength，这个方法得到最小 run 的长度，minRunLength 里面是一个 while 循环，循环条件是 *n* 大于等于 MIN_MERGE，这里 MIN_MERGE 是 32，即 2 的 5 次方，然后进行基本的移位运算。minRunLength 的代码如下：

```
1.      private int minRunLength(int n) {
2.      assert n >= 0;
3.      int r = 0;      //如果有 1 位被移位，则为 1
4.      while (n >= MIN_MERGE) {
5.        r |= (n & 1);
6.        n >>= 1;
7.      }
8.      return n + r;
9.    }
```

TimSort.java 的 sort 方法中，sortState.pushRun(lo, runLen)中的 pushRun 就是一个栈。

```
1.        private void pushRun(int runBase, int runLen) {
```

```
2.         this.runBase[stackSize] = runBase;
3.         this.runLen[stackSize] = runLen;
4.         stackSize++;
5.       }
```

TimSort.java 的 sort 方法中有一句很关键的代码 sortState.mergeCollapse()。它的源码如下：

```
1.         private void mergeCollapse() {
2.         while (stackSize > 1) {
3.           int n = stackSize - 2;
4.           if ( (n >= 1 && runLen[n-1] <= runLen[n] + runLen[n+1])
5.             || (n >= 2 && runLen[n-2] <= runLen[n] + runLen[n-1])) {
6.             if (runLen[n - 1] < runLen[n + 1])
7.               n--;
8.           } else if (runLen[n] > runLen[n + 1]) {
9.             break; //建立不变量
10.          }
11.          mergeAt(n);
12.        }
13.      }
```

mergeCollapse 据说 openJDK 在实现 mergeCollapse 时有 Bug，在插入数据的时候插入的顺序可能有问题。但 Spark 进行过充分测试，mergeCollapse 没有 Bug。其中的关键代码是 mergeAt。我们看一下 mergeAt 的实现，runLen[i] 如果是栈顶的第 3 个位置，则将被交换为栈顶的第二个位置。gallopRight 从我们的 run1 找到 run2 中第一个元素的位置。在此基础上，run1 中的元素可以被忽略，将从 run2 找到 run1 中最后一个元素的位置，然后 run2 的元素被忽略。

```
1.       private void mergeAt(int i) {
2.         assert stackSize >= 2;
3.         assert i >= 0;
4.         assert i == stackSize - 2 || i == stackSize - 3;
5.
6.         int base1 = runBase[i];
7.         int len1 = runLen[i];
8.         int base2 = runBase[i + 1];
9.         int len2 = runLen[i + 1];
10.        assert len1 > 0 && len2 > 0;
11.        assert base1 + len1 == base2;
12.
13.        /**
            *记录组合 runs 的长度；如果 i 是倒数第三个 run，在最后一个 run 滑动时（在这个
            *合并中没有涉及），当前的 run (i+1)在任何情况下都会消失
14.         */
15.        runLen[i] = len1 + len2;
16.        if (i == stackSize - 3) {
17.          runBase[i + 1] = runBase[i + 2];
18.          runLen[i + 1] = runLen[i + 2];
19.        }
20.        stackSize--;
21.
22.        K key0 = s.newKey();
23.
24.        /**
            *找到 run2 的第一个元素在 run1 中的位置， run1 的前一个元素可以忽略（因为它
```

```
         *们已经到位）
25.      */
26     int k = gallopRight(s.getKey(a, base2, key0), a, base1, len1, 0, c);
27.    assert k >= 0;
28.    base1 += k;
29.    len1 -= k;
30.    if (len1 == 0)
31.      return;
32.
33.    /**
         *找到 run1 的最后一个元素在 run2 中的位置，run2 的后一个元素可以忽略（因为它
         *们已经到位）
34.      */
35.    len2 = gallopLeft(s.getKey(a, base1 + len1 - 1, key0), a, base2, len2,
       len2 - 1, c);
36.    assert len2 >= 0;
37.    if (len2 == 0)
38.      return;
39.
40.    //合并剩余的 runs，使用 min(len1, len2)临时数组的元素
41.    if (len1 <= len2)
42.      mergeLo(base1, len1, base2, len2);
43.    else
44.      mergeHi(base1, len1, base2, len2);
45.  }
```

其中有一个方法 gallopRight，类似于 gallopleft，除非包含相等的元素 key，gallopRight 返回最右边的相等元素的索引。

- @param key：关键的搜索插入点。
- @param a：需要搜索的数组。
- @param base：第一个元素的索引范围。
- @param len：范围的长度须大于 0。
- @param hint：开始搜索的索引，0 <= hint < n 结果越接近 hint，方法运行得越快。
- @param c：用于排序和搜索范围的比较器。
- @return k：返回 k，0 <= k <= n 这样 a[b + k−1] <= key < a[b + k]。

gallopRight 的代码如下：

```
1.     private int gallopRight(K key, Buffer a, int base, int len, int hint,
       Comparator<? super K> c) {
2.     assert len > 0 && hint >= 0 && hint < len;
3.
4.     int ofs = 1;
5.     int lastOfs = 0;
6.     K key1 = s.newKey();
7.
8.     if (c.compare(key, s.getKey(a, base + hint, key1)) < 0) {
9.       //飞奔模式 向左归并直到 a[b+hint - ofs] <= key < a[b+hint - lastOfs]
10.      int maxOfs = hint + 1;
11.      while (ofs < maxOfs && c.compare(key, s.getKey(a, base + hint - ofs,
         key1)) < 0) {
12.        lastOfs = ofs;
13.        ofs = (ofs << 1) + 1;
14.        if (ofs <= 0)   //整数溢出
15.          ofs = maxOfs;
```

```
16.        }
17.        if (ofs > maxOfs)
18.          ofs = maxOfs;
19.
20.        //计算相对于 B 的偏移量
21.        int tmp = lastOfs;
22.        lastOfs = hint - ofs;
23.        ofs = hint - tmp;
24.      } else { //a[b + hint] <= key
25.        //Gallop right 直到 a[b+hint + lastOfs] <= key < a[b+hint + ofs]
26.        int maxOfs = len - hint;
27.        while (ofs < maxOfs && c.compare(key, s.getKey(a, base + hint + ofs,
           key1)) >= 0) {
28.          lastOfs = ofs;
29.          ofs = (ofs << 1) + 1;
30.          if (ofs <= 0)   //整数溢出
31.            ofs = maxOfs;
32.        }
33.        if (ofs > maxOfs)
34.          ofs = maxOfs;
35.
36.        //计算相对于 B 的偏移量
37.        lastOfs += hint;
38.        ofs += hint;
39.      }
40.      assert -1 <= lastOfs && lastOfs < ofs && ofs <= len;
41.
42.      /**
          * 现在，a[b + lastOfs] <= key < a[b + ofs]，所以关键是 lastOfs 的 right，
          * 而不是 ofs 的 right，使用不变量 a[b + lastOfs - 1] <= key < a[b + ofs]
          * 做二元搜索
43.       */
44.      lastOfs++;
45.      while (lastOfs < ofs) {
46.        int m = lastOfs + ((ofs - lastOfs) >>> 1);
47.
48.        if (c.compare(key, s.getKey(a, base + m, key1)) < 0)
49.          ofs = m;              //key < a[b + m]
50.        else
51.          lastOfs = m + 1;     //a[b + m] <= key
52.      }
53.      assert lastOfs == ofs; //这样，a[b + ofs - 1] <= key < a[b + ofs]
54.      return ofs;
55.    }
```

前面的代码中还有一个 gallopLeft，定位将指定 Key 插入到指定的排序范围；如果该范围包含等于 Key 的元素，则返回最左边的相等的元素的索引。

- @param key：关键的搜索插入点。
- @param a：搜索的数组。
- @param base：第一个元素的索引范围。
- @param len：范围的长度需大于 0。
- @param hint：开始搜索的索引，0 <= hint < n 结果越接近 hint，方法运行得越快。
- @param c：用于排序和搜索范围的比较器。
- @return k：返回 k，0 <= k <= n 这样 a[b + k – 1] < key <= a[b + k]，假设 a[b−1]是负

无穷大，a[b + n]是无穷大。关键属于索引 b + k，换句话说，a 的第一个 k 元素应先于 Key，最后 n-k 元素应该排在其之后。

gallopLeft 代码如下：

```
1.    private int gallopLeft(K key, Buffer a, int base, int len, int hint,
      Comparator<? super K> c) {
2.      assert len > 0 && hint >= 0 && hint < len;
3.      int lastOfs = 0;
4.      int ofs = 1;
5.      K key0 = s.newKey();
6.
7.      if (c.compare(key, s.getKey(a, base + hint, key0)) > 0) {
8.        //Gallop right 直到 a[base+hint+lastOfs] < key <= a[base+hint+ofs]
9.        int maxOfs = len - hint;
10.       while (ofs < maxOfs && c.compare(key, s.getKey(a, base + hint + ofs,
          key0)) > 0) {
11.         lastOfs = ofs;
12.         ofs = (ofs << 1) + 1;
13.         if (ofs <= 0)   //整数溢出
14.           ofs = maxOfs;
15.       }
16.       if (ofs > maxOfs)
17.         ofs = maxOfs;
18.
19.       //计算相对于基底偏移量
20.       lastOfs += hint;
21.       ofs += hint;
22.     } else { //key <= a[base + hint]
23.       //Gallop left 直到 a[base+hint-ofs] < key <= a[base+hint-lastOfs]
24.       final int maxOfs = hint + 1;
25.       while (ofs < maxOfs && c.compare(key, s.getKey(a, base + hint - ofs,
          key0)) <= 0) {
26.         lastOfs = ofs;
27.         ofs = (ofs << 1) + 1;
28.         if (ofs <= 0)   //整数溢出
29.           ofs = maxOfs;
30.       }
31.       if (ofs > maxOfs)
32.         ofs = maxOfs;
33.
34.       //计算相对于基底偏移量
35.       int tmp = lastOfs;
36.       lastOfs = hint - ofs;
37.       ofs = hint - tmp;
38.     }
39.     assert -1 <= lastOfs && lastOfs < ofs && ofs <= len;
40.
41.     /**
.         * 现在， a[base+lastOfs] < key <= a[base+ofs]，所以关键是 lastOfs 的
          * right，而不是 ofs 的 right，使用不变量 a[base + lastOfs - 1] < key <= a[base
          * + ofs]做二元搜索
42.       */
43.     lastOfs++;
44.     while (lastOfs < ofs) {
45.       int m = lastOfs + ((ofs - lastOfs) >>> 1);
46.
47.       if (c.compare(key, s.getKey(a, base + m, key0)) > 0)
```

```
48.          lastOfs = m + 1;        //a[base + m] < key
49.        else
50.          ofs = m;                //key <= a[base + m]
51.      }
52.      assert lastOfs == ofs;      //这样，a[base + ofs - 1] < key <= a[base
                                     //+ ofs]
53.      return ofs;
54.    }
```

回到TimSort.java的mergeAt方法，下面看一个关键代码mergeLo(base1, len1, base2, len2); mergeLo 方法以稳定的方式合并两个相邻 run。第一个 run 的第一个元素必须大于第二个 run 的第一个元素 (a[base1] > a[base2])，第一个 run 的最后一个元素(a[base1 + len1−1])必须大于第二个 run 的所有元素。这种方法只有当 len1 <= len2 的时候被调用；另一个类似的方法 mergeHi 在 len1 > = len2 的情况下被调用。（如果 len1 == len2，任何一种方法都可被调用）。

- @param base1：第一个 run 的第一个元素被合并的索引。
- @param len1：第一个 run 被合并的长度（必须大于 0）。
- @param base2：第二个 run 被合并的第一个元素的索引（必须是 aBase + aLen）。
- @param len2：第二个 run 被合并的长度（必须大于 0）。

mergeLo 代码如下：

```
1.    private void mergeLo(int base1, int len1, int base2, int len2) {
2.        assert len1 > 0 && len2 > 0 && base1 + len1 == base2;
3.
4.        //复制第一个 run 到临时数组
5.        Buffer a = this.a;     //为了性能
6.        Buffer tmp = ensureCapacity(len1);
7.        s.copyRange(a, base1, tmp, 0, len1);
8.
9.        int cursor1 = 0;        //临时数组的索引
10.       int cursor2 = base2;   //整数 a 的索引
11.       int dest = base1;      //整数 a 的索引
12.
13.       //移动第二个 run 的第一个元素，处理退化情况
14.       s.copyElement(a, cursor2++, a, dest++);
15.       if (--len2 == 0) {
16.         s.copyRange(tmp, cursor1, a, dest, len1);
17.         return;
18.       }
19.       if (len1 == 1) {
20.         s.copyRange(a, cursor2, a, dest, len2);
21.         s.copyElement(tmp, cursor1, a, dest + len2);
                                           //最后的 run 1 到合并结束
22.         return;
23.       }
24.
25.       K key0 = s.newKey();
26.       K key1 = s.newKey();
27.
28.       Comparator<? super K> c = this.c;       //使用本地变量提升性能
29.       int minGallop = this.minGallop;         //"     "       "     "     "
30.       outer:
31.       while (true) {
32.         int count1 = 0;                       //第一个 run 胜出的次数
33.         int count2 = 0;                       //第二个 run 胜出的次数
34.
35.         /**
```

```
          *循环遍历执行，直到一个 run 胜出
36.       */
37.
38.     do {
39.       assert len1 > 1 && len2 > 0;
40.       if (c.compare(s.getKey(a, cursor2, key0), s.getKey(tmp, cursor1,
          key1)) < 0) {
41.         s.copyElement(a, cursor2++, a, dest++);
42.         count2++;
43.         count1 = 0;
44.         if (--len2 == 0)
45.           break outer;
46.       } else {
47.         s.copyElement(tmp, cursor1++, a, dest++);
48.         count1++;
49.         count2 = 0;
50.         if (--len1 == 1)
51.           break outer;
52.       }
53.     } while ((count1 | count2) < minGallop);
54.
55.     /**
          *一个 run 持续胜出，galloping 将胜出，继续尝试运行，直到没有 run 再持续胜出
56.       */
57.
58.     do {
59.       assert len1 > 1 && len2 > 0;
60.       count1 = gallopRight(s.getKey(a, cursor2, key0), tmp, cursor1,
          len1, 0, c);
61.       if (count1 != 0) {
62.         s.copyRange(tmp, cursor1, a, dest, count1);
63.         dest += count1;
64.         cursor1 += count1;
65.         len1 -= count1;
66.         if (len1 <= 1) //len1 == 1 || len1 == 0
67.           break outer;
68.       }
69.       s.copyElement(a, cursor2++, a, dest++);
70.       if (--len2 == 0)
71.         break outer;
72.
73.       count2 = gallopLeft(s.getKey(tmp, cursor1, key0), a, cursor2,
          len2, 0, c);
74.       if (count2 != 0) {
75.         s.copyRange(a, cursor2, a, dest, count2);
76.         dest += count2;
77.         cursor2 += count2;
78.         len2 -= count2;
79.         if (len2 == 0)
80.           break outer;
81.       }
82.       s.copyElement(tmp, cursor1++, a, dest++);
83.       if (--len1 == 1)
84.         break outer;
85.       minGallop--;
86.     } while (count1 >= MIN_GALLOP | count2 >= MIN_GALLOP);
87.     if (minGallop < 0)
88.       minGallop = 0;
89.     minGallop += 2;  //Penalize for leaving gallop mode
90.   } //“outer”循环结束
```

```
91.         this.minGallop = minGallop < 1 ? 1 :  minGallop;   //回写字段
92.
93.         if (len1 == 1) {
94.           assert len2 > 0;
95.           s.copyRange(a, cursor2, a, dest, len2);
96.           s.copyElement(tmp, cursor1, a, dest + len2);
                                                //最后的 run 1 到合并结束
97.         } else if (len1 == 0) {
98.           throw new IllegalArgumentException(
99.               "Comparison method violates its general contract!");
100.          } else {
101.            assert len2 == 0;
102.            assert len1 > 1;
103.            s.copyRange(tmp, cursor1, a, dest, len1);
104.          }
105.        }
```

TimSort.java 的 sort 方法中还有一行关键代码 sortState.mergeForceCollapse(), 合并堆栈上所有的 run，直到只有一个 run。这种方法是调用一次，完成排序。

mergeForceCollapse 方法如下：

```
1.         private void mergeForceCollapse() {
2.         while (stackSize > 1) {
3.           int n = stackSize - 2;
4.           if (n > 0 && runLen[n - 1] < runLen[n + 1])
5.             n--;
6.           mergeAt(n);
7.          }
8.       }
```

总结：timSort 会预先按连续递增的 run 的片段归并元素，进入插入的时候，如果长度小于 run，就会使用 insert 进行排序的实现。与 mergesort 相比，mergesort 归并是预先定义好的，而 timSort 比较灵活。如果第三个 run 小于栈顶的 run，那先归并第 2 个、第 3 个 run，从而得出需要归并的片段；如果 run1 的头部和 run2 的尾部有不用进行归并的部分，timSort 在截取的基础上进行二分排序，得到需要归并的起始位置。如果 run 的长度为 1，就会进行一些优化。

关于 timSort 排序：

timSort 是结合了合并排序（merge sort）和插入排序（insertion sort）而得出的排序算法，它在现实中有很好的效率。Tim Peters 在 2002 年设计了该算法，并在 Python 中使用（timSort 是 Python 中 list.sort 的默认实现）。该算法找到数据中已经排好序的块-分区，每个分区叫一个 run，然后按规则合并这些 run。Python 自从 2.3 版以来一直采用 timSort 算法排序，现在 Java SE7 和 Android 也采用 timSort 算法对数组排序。

timSort 的核心过程：

timSort 算法为了减少对升序部分的回溯和对降序部分的性能倒退，将输入按其升序和降序特点进行了分区。排序的输入不是一个个单独的数字，而是一个个的块分区。其中每个分区叫一个 run。针对这些 run 序列，每次拿一个 run 出来按规则进行合并。每次合并会将两个 run 合并成一个 run。合并的结果保存到栈中。合并直到消耗掉所有的 run，这时将栈上剩余的 run 合并到只剩一个 run 为止。这时这个仅剩的 run 便是排好序的结果。

综合上述过程，timSort 算法的过程包括：

- 如果数组长度小于某个值，则直接用二分插入排序算法。
- 找到各个 run，并入栈。

- 按规则合并 run。

现实中的大部分数据通常已经是排好序的，timSort 利用了这一特点。timSort 排序的输入的单位不是一个个单独的数字，而是一个个分区。其中每个分区叫一个 run。针对这个 run 序列，每次拿一个 run 出来进行归并。每次归并会将两个 run 合并成一个 run。每个 run 最少要有两个元素。timSort 按照升序和降序划分出各个 run：run 如果是升序的，那么 run 中的后一元素要大于或等于前一元素（a[lo] <= a[lo + 1] <= a[lo + 2] <= …）；如果 run 是严格降序的，即 run 中的前一元素大于后一元素（a[lo] > a[lo + 1] > a[lo + 2] > …），就需要将 run 中的元素翻转(这里注意降序的部分必须是“严格”降序，才能进行翻转。因为 timSort 的一个重要目标是保持稳定性 stability。如果在>= 的情况下进行翻转，这个算法就不再是 stable）。

run 的最小长度：run 是已经排好序的一块分区。run 可能有不同的长度，timSort 根据 run 的长度选择排序的策略。例如，如果 run 的长度小于某一个值，则会选择插入排序算法来排序。run 的最小长度（minrun）取决于数组的大小。当数组元素少于 64 个时，那么 run 的最小长度便是数组的长度，这时 timSort 用插入排序算法来排序。

优化 run 的长度：优化 run 的长度是指当 run 的长度小于 minrun 时，为了使这样的 run 的长度达到 minrun 的长度，会从数组中选择合适的元素插入 run 中。这样做使大部分的 run 的长度达到均衡，有助于后面 run 的合并操作。

合并 run：划分 run 和优化 run 长度后，然后就是对各个 run 进行合并。合并 run 的原则是：run 合并的技术要保证有最高的效率。当 timSort 算法找到一个 run 时，会将该 run 在数组中的起始位置和 run 的长度放入栈中，然后根据先前放入栈中的 run 决定是否应该合并 run。timSort 不会合并在栈中不连续的 run。

timSort 会合并栈中两个连续的 run。X、Y、Z 代表栈最上方的 3 个 run 的长度(见图 31-16)，当不能同时满足① X>Y+Z；② Y>Z 两个条件时，X、Y 这两个 run 会被合并，直到同时满足这两个条件，则合并结束。例如，如果 X<Y+Z，那么 X+Y 合并为一个新的 run，然后入栈。重复上述步骤，直到同时满足上述两个条件。当合并结束后，timSort 会继续找下一个 run，找到以后入栈，重复上述步骤，每次 run 入栈都会检查是否需要合并两个 run。

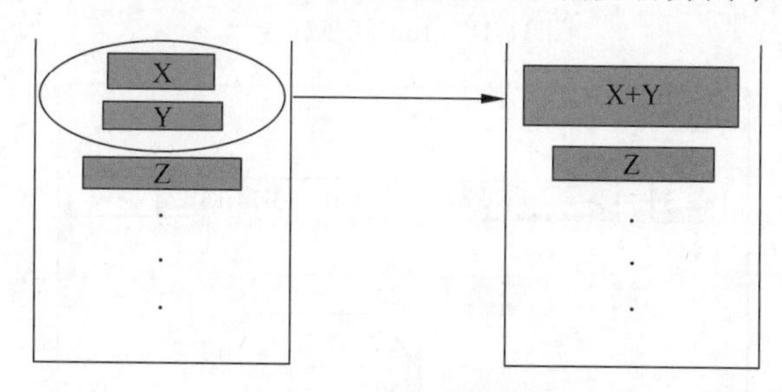

图 31-16　run 计算图

合并 run 的步骤：合并两个相邻的 run 需要临时存储空间。临时存储空间的大小是两个 run 中较小的 run 的大小。timSort 算法先将较小的 run 复制到这个临时存储空间，然后用原先存储这两个 run 的空间来存储合并后的 run，如图 31-17 所示。

简单的合并算法是用简单插入算法，依次从左到右或从右到左比较，然后合并两个 run。为了提高效率，timSort 采用二分插入算法。先用二分查找算法/折半查找算法（binary search）找到插入的位置，然后再插入。

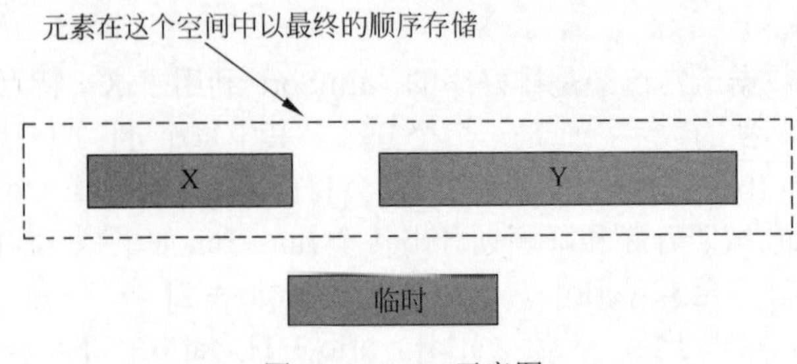

图 31-17　run 示意图

例如，要将A和B这两个run合并，且A是较小的run。因为A和B已经分别排好序了，二分查找会找到 B 的第一个元素在 A 中何处插入。同样，A 的最后一个元素找到在 B 的何处插入，找到以后，B 在这个元素之后的元素就不需要比较了。这种查找可能在随机数中效率不会很高，但是在其他情况下有很高的效率。

run 合并过程如图 31-18 和图 31-19 所示。

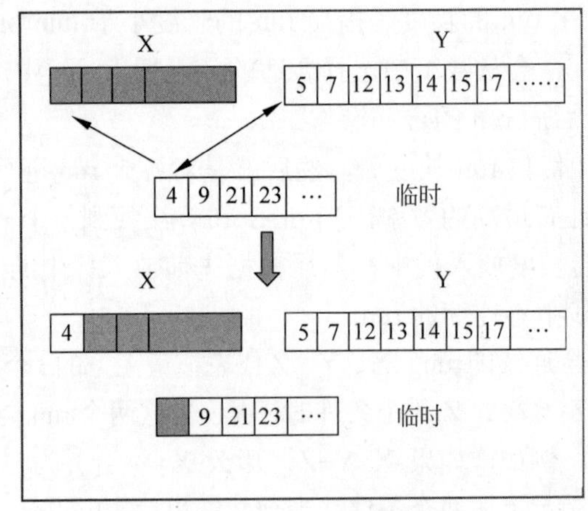

（5，7，12，13，14，15，17）
与4进行比较，更小的元素4移动到其最终位置

图 31-18　run 合并过程一

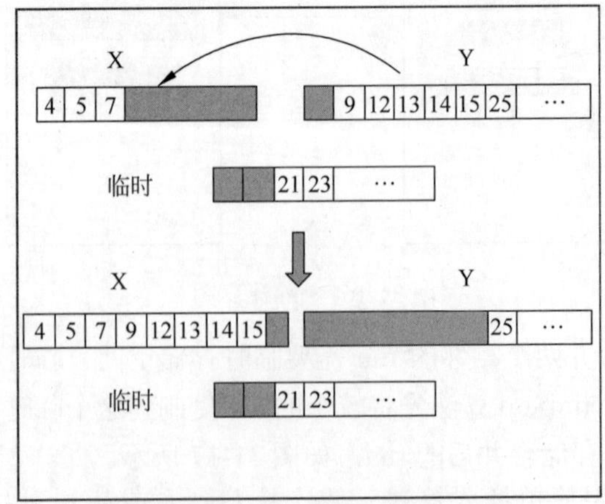

元素（9，12，13，14，15）都小于元素21，
因此可被移动到最终数组中

图 31-19　run 合并过程二

timSort 是稳定的算法，当待排序的数组中已经有排序好的数，它的时间复杂度就会小于 $n \log n$。与其他合并排序一样，timSrot 是稳定的排序算法，最坏时间复杂度是 $O(n \log n)$。在最坏情况下，timSort 算法需要的临时空间是 $n/2$，在最好情况下，它只需要一个很小的临时存储空间。

31.17　Spark 2.4.X 下 Sort Shuffle 中 Reducer 端的源码内幕

本节讲解 Spark 2.4.X 中 Sort Shuffle 中 Reducer 端的源码内幕，Spark 是 MapReduce 思想的一种实现，相对于 Hadoop 的 MapReduce，Spark 作业 Job 根据算子的依赖关系，当是宽依赖的时候，会产生 Shuffle，这时划分成不同的 Stage，前面的 Stage 是后面 Stage 的 Mapper，后面 Stage 是前面 Stage 的 Reducer。研究的核心是 MapReduce 以及中间网络传输的过程。从 Reduce 的角度讲，肯定有拉取数据的过程，这与原始的大数据分布式思想完全一致。在 Hadoop 的 MapReduce 中是链式的，Map、Reduce，接着 Map、Reduce，从 Hadoop 的角度讲，前面是 Map，Map，Map……后面是 Reduce，Hadoop 借助 Oozie 工具来实现将多个 Map/Reduce 作业连接到一起，而 Spark 基于 DAG 的模型天然可迭代。

Spark 研究 Reducer 端的 Stage 时，从 ShuffledRDD 去谈。从 RDD 的运行角度讲，Shuffled 的 RDD 是 RDD 的具体实现，因此关键是分析 compute 方法。

ShuffledRDD 的 compute 方法的代码如下：

```
1.   override def compute(split: Partition, context: TaskContext): Iterator
     [(K, C)] = {
2.     val dep = dependencies.head.asInstanceOf[ShuffleDependency[K, V, C]]
3.     SparkEnv.get.shuffleManager.getReader(dep.shuffleHandle, split.index,
       split.index + 1, context)
4.       .read()
5.       .asInstanceOf[Iterator[(K, C)]]
6.   }
7.
```

按照面向对象的机制，首先调用 RDD 的 compute 方法，而 RDD 的 compute 是空实现，是一个抽象方法，子类具体实现去计算一个分区 partition。RDD.scala 的 compute 如下：

```
1.   def compute(split: Partition, context: TaskContext): Iterator[T]
```

从 RDD 的角度讲，运行的是并行任务的集合，在 ShuffledRDD 的 compute 方法中，dependencies.head.asInstanceOf[ShuffleDependency[K, V, C]]是获得 Shuffle 级别的依赖关系，然后在 SparkEnv.get.shuffleManager 获得 SortShuffleManage，因为在构建 SparkEnv 的时候，Spark 2.4.X 中默认是 org.apache.spark.shuffle.sort.SortShuffleManager。

在 ShuffledRDD 的 compute 方法中，SparkEnv.get.shuffleManager.getReader 是获取 ShuffledRDD 的阅读器，阅读器的核心作用是读取分布在不同节点上 Map 中的 task 计算的结果的数据，传入的参数是 shuffleHandle，根据 Dependency 获得的。compute 获得的是一个迭代器 Iterator[(K, C)]，类型是泛型，C 是计算以后的类型，要进行聚合。

SparkEnv.get.shuffleManager.getReader 是 SortShuffleManager 的 getReader，是获取数据的阅读器。Reducer 端 Task 在运行的时候，会抓取自己想要的数据。

```
1.    override def getReader[K, C](
2.        handle: ShuffleHandle,
3.        startPartition: Int,
4.        endPartition: Int,
5.        context: TaskContext): ShuffleReader[K, C] = {
6.      new BlockStoreShuffleReader(
7.        handle.asInstanceOf[BaseShuffleHandle[K, _, C]], startPartition,
          endPartition, context)
8.    }
```

getReader 方法很简单，就调用 new 函数创建一个 BlockStoreShuffleReader 实例，进入 BlockStoreShuffleReader 实例中，看一下 BlockStoreShuffleReader.scala 的 read 方法。

Spark 2.2.1 版本的 BlockStoreShuffleReader.scala 的 read 的源码如下：

```
1.     override def read(): Iterator[Product2[K, C]] = {
2.      val wrappedStreams = new ShuffleBlockFetcherIterator(
3.        context,
4.        blockManager.shuffleClient,
5.        blockManager,
6.        mapOutputTracker.getMapSizesByExecutorId(handle.shuffleId,
          startPartition, endPartition),
7.        serializerManager.wrapStream,
8.        //注意：为保持后向兼容性，当没有后缀时，使用 getSizeAsMb
9.        SparkEnv.get.conf.getSizeAsMb("spark.reducer.maxSizeInFlight",
          "48m") * 1024 * 1024,
10.       SparkEnv.get.conf.getInt("spark.reducer.maxReqsInFlight",
          Int.MaxValue))
11.
12.       SparkEnv.get.conf.get(config.REDUCER_MAX_BLOCKS_IN_FLIGHT_PER_ADDRESS),
13.       SparkEnv.get.conf.get(config.REDUCER_MAX_REQ_SIZE_SHUFFLE_TO_MEM),
14.       SparkEnv.get.conf.getBoolean("spark.shuffle.detectCorrupt", true))
15.
16.
17.     val serializerInstance = dep.serializer.newInstance()
18.
19.     //为每个流创建一个键 Key/值 Value 迭代器
20.     val recordIter = wrappedStreams.flatMap {case (blockId, wrappedStream)=>
21.       //注意： asKeyValueIterator 在迭代器 NextIterator 内部包裹键 Key/值
          //Value。当下层 InputStream 所有记录已读，NextIterator 确保 close 方法被调用
22.       serializerInstance.deserializeStream(wrappedStream).asKeyValueIterator
23.     }
24.
25.     //为每个记录更新上下文任务度量
26.     val readMetrics = context.taskMetrics.createTempShuffleReadMetrics()
27.     val metricIter = CompletionIterator[(Any, Any), Iterator[(Any, Any)]](
28.       recordIter.map { record =>
29.         readMetrics.incRecordsRead(1)
30.         record
31.       },
32.       context.taskMetrics().mergeShuffleReadMetrics())
33.
34.     //为了支持任务取消，这里必须使用可中断迭代器
35.     val interruptibleIter = new InterruptibleIterator[(Any, Any)](context,
        metricIter)
36.
37.     val aggregatedIter: Iterator[Product2[K, C]] = if (dep.aggregator.
        isDefined) {
```

```
38.         if (dep.mapSideCombine) {
39.           //读取已经组合的值
40.           val combinedKeyValuesIterator = interruptibleIter.asInstanceOf
              [Iterator[(K, C)]]
41.           dep.aggregator.get.combineCombinersByKey(combinedKeyValues-
              Iterator, context)
42.         } else {
43.           //不知道值类型，但也不关注值类型——依赖已经确定了聚合，将 Value 值类型转换
              //为组合 C 类型
44.           val keyValuesIterator = interruptibleIter.asInstanceOf[Iterator
              [(K, Nothing)]]
45.           dep.aggregator.get.combineValuesByKey(keyValuesIterator,
              context)
46.         }
47.       } else {
48.         require(!dep.mapSideCombine, "Map-side combine without Aggregator
            specified!")
49.         interruptibleIter.asInstanceOf[Iterator[Product2[K, C]]]
50.       }
51.
52.       //如果定义排序，则对输出进行排序
53.       dep.keyOrdering match {
54.         case Some(keyOrd: Ordering[K]) =>
55.           //创建一个 ExternalSorter 对数据排序。注意，如果 spark.shuffle.spill 禁
              //用，则 ExternalSorter 不会溢出到磁盘
56.           val sorter =
57.             new ExternalSorter[K, C, C](context, ordering = Some(keyOrd),
                serializer = dep.serializer)
58.           sorter.insertAll(aggregatedIter)
59.           context.taskMetrics().incMemoryBytesSpilled(sorter.memoryBytes-
              Spilled)
60.           context.taskMetrics().incDiskBytesSpilled(sorter.diskBytes-
              Spilled)
61.           context.taskMetrics().incPeakExecutionMemory(sorter.peakMemory-
              UsedBytes)
62.           CompletionIterator[Product2[K, C], Iterator[Product2[K, C]]]
              (sorter.iterator, sorter.stop())
63.         case None =>
64.           aggregatedIter
65.       }
66.     }
```

Spark 2.4.3 版本的 BlockStoreShuffleReader.scala 的源码与 Spark 2.2.1 版本相比具有如下特点。

- 上段代码中的第 12 行调整为获取 MAX_REMOTE_BLOCK_SIZE_FETCH_TO_MEM 参数。spark.maxRemoteBlockSizeFetchToMem 参数用于当块的大小高于此阈值时，远程块将以字节为单位被提取到磁盘。这是为了避免一个巨大的请求占用太多的内存。我们可以启用这个配置，如设置特定值 200m 进行配置。注意，此配置将影响 shuffle 获取和块管理器远程块获取。对于启用了外部 shuffle 服务，此功能只能用于 Spark 2.2 以后的版本，在外部 shuffle 时使用。
- 删除上段代码中的第 47 行代码。
- 上段代码中的第 61 行之前新增代码，增加对 addTaskCompletionListener 任务完成的监听。addTaskCompletionListener 方法以 scala 闭包的形式添加一个侦听器，在任务完成时执行。这将在所有情况下调用：成功、失败或取消。将侦听器添加到已完成

的任务将导致立即调用该侦听器。HadoopRDD 的一个示例用法是注册回调以关闭输入流。侦听器引发的异常将导致任务失败。

- 上段代码中的第 64 行之后新增 resultIter 的处理代码，这里使用另一个可中断迭代器来支持任务取消，因为聚合器或排序器可能已经使用了以前的可中断迭代器。

```
1.    ......
2.    SparkEnv.get.conf.get(config.MAX_REMOTE_BLOCK_SIZE_FETCH_TO_MEM),
3.  ......
4.        context.addTaskCompletionListener[Unit](_ => {
5.          sorter.stop()
6.        })
7.  ......
8.      resultIter match {
9.        case _: InterruptibleIterator[Product2[K, C]] => resultIter
10.       case _ =>
11.         ......
12.         new InterruptibleIterator[Product2[K, C]](context, resultIter)
13.     }
14. ......
15. // package.scala
16.   private[spark] val MAX_REMOTE_BLOCK_SIZE_FETCH_TO_MEM =
17.     ConfigBuilder("spark.maxRemoteBlockSizeFetchToMem")
18.       .doc("Remote block will be fetched to disk when size of the block
        is above this threshold " +
19.         "in bytes. This is to avoid a giant request takes too much memory.
          We can enable this " +
20.         "config by setting a specific value(e.g. 200m). Note this
          configuration will affect " +
21.         "both shuffle fetch and block manager remote block fetch. For users
          who enabled " +
22.         "external shuffle service, this feature can only be worked when
          external shuffle" +
23.         "service is newer than Spark 2.2.")
24.       .bytesConf(ByteUnit.BYTE)
```

BlockStoreShuffleReader.scala 的 read 方法中，从 Reduce Task 的角度讲，我们读取 Key-value 的集合，因为同样的 Key 分布在很多节点上，这是 Map Reduce 实现思想。首先实例化 new ShuffleBlockFetcherIterator，这个非常重要。

- ShuffleBlockFetcherIterator 是一个阅读器，里面有一个成员 blockManager，blockManager 是内存和磁盘上数据读写的统一管理器；mapOutputTracker 是上一阶段 Mapper 输出的数据位置被 mapOutputTracker 跟踪。从 Driver 角度讲，Driver 中有一个 mapOutputTrackerMaster，其他节点上有一个 mapOutputTracker，是 master-slave 的结构。根据 mapOutputTracker.getMapSizesByExecutorId 方法通过发送消息获取 ShuffleMapTask 存储数据的具体位置，消息是发送给 Driver，在上一个 Mapper 阶段数据写到本地磁盘的时候，会告诉 Driver 数据写在什么地方，方法是和 Driver 通信，获得上一个 Stage 或 Mapper 的具体存储数据位置的元数据信息。
- spark.reducer.maxSizeInFlight：每次最大可以读取多少数据，默认是 48MB，网络好的情况下，参数可以适当调大一点，如增大为 96MB，甚至更高，这样可以减少抓取数据的次数。
- spark.reducer.maxReqsInFlight：是从远程节点请求块的数量，这里是无限次。最好减少其次数。

- blockFetcherItr 实例类通过 map 转换，基于压缩和加密，构建一个 wrappedStreams 包装流。
- serializerInstance 构建序列化器。
- 对于每个 wrappedStreams 生成一个 key-value 类型的迭代器 recordIter。其中 deserializeStream 是反序列化。
- readMetrics：每条记录被读取后，要更新 Metrics，这样在 Web UI 控制台或者交互式控制台上就会看到相关的信息。Shuffle 的运行肯定需要相关的 Metrics。metricIter 是一个迭代器。
- interruptibleIter 传入的是 metricIter，运行的时候可能要中断，需取消我们的迭代。
- aggregatedIter 判断在 Mapper 端进行聚合怎么做；不在 Mapper 端聚合怎么做。首先判断 aggregator 是否被定义，如果已经定义 aggregator，再判断 map 端是否需聚合，我们谈的是Reducer端，为什么这里需在Mapper端进行聚合呢？原因很简单：Reducer 可能还有下一个 Stage，如果还有下一个 Stage，那这个 Reducer 对于下一个 Stage 而言，其实就是 Mapper，是 Mapper 就须考虑在本地是否进行聚合。迭代是一个 DAG 图，假设如果有 100 个 Stage，这里是第 10 个 Stage，作为第 9 个 Stage 的 Reducer 端，但是作为第 11 个 Stage 是 Mapper 端，作为 Shuffle 而言，现在的 Reducer 端相对于 Mapper 端。Mapper 端需要聚合，则进行 combineCombinersByKey。Mapper 端也可能不需要聚合，只需要进行 Reducer 端的操作。如果 aggregator.isDefined 没定义，则出错提示。
- dep.keyOrdering 是否对输出结果进行排序：根据传进来的 keyOrdering 判断，如果进行排序，就调用 new 函数创建一个 ExternalSorter。我们之前在 Mapper 端已讲解过 ExternalSorter 的源码，ExternalSorter 里面是 timSort。接下来是 insertAll，insertAll 与 Mapper 端的代码一样，也须判断磁盘是否进行 Spill。

再次回到 ShuffledRDD 的 compute 方法，获得 getReader 以后，须进行 read，跟踪至 BlockStoreShuffleReader 的 read()，BlockStoreShuffleReader 的 read 里面最重要的内容是 ShuffleBlockFetcherIterator，须深入到 ShuffleBlockFetcherIterator 类。

ShuffleBlockFetcherIterator 中传入了 mapOutputTracker.getMapSizesByExecutorId(handle.shuffleId, startPartition, endPartition)，下面看一下 getMapSizesByExecutorId。

Spark 2.2.1 版本的 MapOutputTracker.scala 的源码如下：

```
1.   def getMapSizesByExecutorId(shuffleId: Int, startPartition: Int,
     endPartition: Int)
2.       : Seq[(BlockManagerId, Seq[(BlockId, Long)])] = {
3.     logDebug(s"Fetching outputs for shuffle $shuffleId, partitions
       $startPartition-$endPartition")
4.     val statuses = getStatuses(shuffleId)
5.     //在返回的数组上进行同步块锁，因为在 Driver 上它会发生变化
6.     statuses.synchronized {
7.       return MapOutputTracker.convertMapStatuses(shuffleId, startParti-
         tion, endPartition, statuses)
8.     }
9.   }
```

Spark 2.4.3 版本的 MapOutputTracker.scala 源码与 Spark 2.2.1 版本相比具有如下特点。

- 上段代码中第 2 行 Seq 调整为 Iterator。
- 上段代码中第 6～9 行替换为以下代码的第 4～12 行，新增加对 MetadataFetchFailedException 异常的处理。

```
1.  ……
2.  : Iterator[(BlockManagerId, Seq[(BlockId, Long)])] = {
3.  ……
4.      try {
5.      MapOutputTracker.convertMapStatuses(shuffleId, startPartition,
        endPartition, statuses)
6.     } catch {
7.      case e: MetadataFetchFailedException =>
8.        //遇到了一个获取失败，因此 mapStatuses 缓存已过时，清除它
9.        mapStatuses.clear()
10.       throw e
11.    }
12.  }
```

getMapSizesByExecutorId 是我们的 Executor 根据 URI 获取元数据，然后返回 Seq[(BlockManagerId, Seq[(BlockId, Long)])]。getStatuses 的代码如下：

```
1.   private def getStatuses(shuffleId: Int): Array[MapStatus] = {
2.     val statuses = mapStatuses.get(shuffleId).orNull
3.     if (statuses == null) {
4.       logInfo("Don't have map outputs for shuffle " + shuffleId + ", fetching
         them")
5.       val startTime = System.currentTimeMillis
6.       var fetchedStatuses: Array[MapStatus] = null
7.       fetching.synchronized {
8.         //其他任务在获取它，等任务执行完毕
9.         while (fetching.contains(shuffleId)) {
10.          try {
11.            fetching.wait()
12.          } catch {
13.            case e: InterruptedException =>
14.          }
15.        }
16.
17.        //或者在等待的时候，获取数据成功了；或者在同步块获取数据的时候，其他任务获
           //取到它
18.        fetchedStatuses = mapStatuses.get(shuffleId).orNull
19.        if (fetchedStatuses == null) {
20.          //去获取数据，其他任务等待
21.          fetching += shuffleId
22.        }
23.      }
24.
25.      if (fetchedStatuses == null) {
26.        //获取到状态
27.        logInfo("Doing the fetch; tracker endpoint = " + trackerEndpoint)
28.        //try-finally 防止由于超时挂掉
29.        try {
30.          val fetchedBytes = askTracker[Array[Byte]](GetMapOutputStatuses
             (shuffleId))
31.          fetchedStatuses = MapOutputTracker.deserializeMapStatuses
```

```
            (fetchedBytes)
32.           logInfo("Got the output locations")
33.           mapStatuses.put(shuffleId, fetchedStatuses)
34.         } finally {
35.           fetching.synchronized {
36.             fetching -= shuffleId
37.             fetching.notifyAll()
38.           }
39.         }
40.       }
41.       logDebug(s"Fetching map output statuses for shuffle $shuffleId took "+
42.         s"${System.currentTimeMillis - startTime} ms")
43.
44.       if (fetchedStatuses != null) {
45.         return fetchedStatuses
46.       } else {
47.         logError("Missing all output locations for shuffle " + shuffleId)
48.         throw new MetadataFetchFailedException(
49.           shuffleId, -1, "Missing all output locations for shuffle " +
            shuffleId)
50.       }
51.     } else {
52.       return statuses
53.     }
54.   }
```

MapOutputTracker.scala 中的 getStatuses 方法要读取上一个 Stage 的数据的元数据。

- mapStatuses.get(shuffleId)首先查看本地是否有数据，如果本地有数据，就返回；如果本地没有数据，则从远程节点获得数据。
- 这里有一个同步代码块：使用 while 进行循环遍历，因为从远程节点获取数据，由于网络因素，或者远程机器在进行 GC 时可能会获取数据不成功，或者别人占用了资源须等待。
- 等待之后再次尝试获取数据，这里使用了和之前同样的一行代码 fetchedStatuses = mapStatuses.get(shuffleId).orNull。因为在等待的过程中，同一个 Executor 的其他任务把数据拉到了本地，那就不用从远程拉数据，通过 mapStatuses.get 即可获取本地数据。如果 fetchedStatuses 为空，还是没有获取到数据，就需从远程获取，将要获取的数据 shuffleId 加入到 fetching。
- askTracker 及 GetMapOutputStatuses(shuffleId)真正从远程获取数据，fetchedBytes 获得数据后进行反序列化得到 fetchedStatuses，这里的数据是元数据。然后将数据加入到 mapStatuses 里面。如果 fetchedStatuses 不为空，则直接获取结果。如果 fetchedStatuses 还为空，则提示报错。

MapOutputTracker.scala 的 getStatuses 方法中的 val fetchedBytes = askTracker[Array[Byte]] (GetMapOutputStatuses(shuffleId))代码是关键，GetMapOutputStatuses 是一个 case class，本身是一个消息，继承自 MapOutputTrackerMessage，而 MapOutputTrackerMessage 是一个 trait。askTracker 通过 RPC 发送消息。根据传入的 shuffleId 不断地去获取信息。下面看一下 askTracker 的代码，这里使用了 rpc 的 trackerEndpoint。

MapOutputTracker.scala 的源码如下：

```
1.   protected def askTracker[T: ClassTag](message: Any): T = {
2.     try {
```

```
3.        trackerEndpoint.askWithRetry[T](message)
4.      } catch {
5.        case e:  Exception =>
6.          logError("Error communicating with MapOutputTracker", e)
7.          throw new SparkException("Error communicating with MapOutput-
            Tracker", e)
8.      }
9.    }
```

trackerEndpoint 是 RpcEndpointRef，Executor 在执行任务的时候实例化 Endpoint，从 Worker 的角度，发消息给 Driver。

```
1.   /** MapOutputTrackerMaster 的 RpcEndpoint 类*/
2.  private[spark] class MapOutputTrackerMasterEndpoint(
3.      override val rpcEnv:  RpcEnv, tracker:  MapOutputTrackerMaster, conf:
        SparkConf)
4.    extends RpcEndpoint with Logging {
5.
6.    logDebug("init") //logger 记录日志
7.
8.    override def receiveAndReply(context: RpcCallContext): PartialFunction
      [Any, Unit] = {
9.      case GetMapOutputStatuses(shuffleId:  Int) =>
10.       val hostPort = context.senderAddress.hostPort
11.       logInfo("Asked to send map output locations for shuffle " + shuffleId
          + " to " + hostPort)
12.       val mapOutputStatuses = tracker.post(new GetMapOutputMessage
          (shuffleId, context))
13.
14.     case StopMapOutputTracker =>
15.       logInfo("MapOutputTrackerMasterEndpoint stopped!")
16.       context.reply(true)
17.       stop()
18.   }
19. }
```

MapOutputTrackerMasterEndpoint 从运行角度讲，仍在 Executor 中。在 receiveAndReply 方法中，MapOutputTrackerMasterEndpoint 收到信息，模式匹配 GetMapOutputStatuses。tracker.post 方法是 Master 级别的，是 Driver 中的，tracker 本身是 MapOutputTrackerMaster，发消息 GetMapOutputMessage。GetMapOutputMessage 其实是一个简单的 case class。

那么，从哪里获取消息处理的代码呢？肯定是在 MapOutputTrackerMaster 中。MapOutputTrackerMaster 代码运行在 Driver 中。

Spark 2.2.1 版本的 MapOutputTracker.scala 的源码如下：

```
1.   private[spark] class MapOutputTrackerMaster(conf:  SparkConf,
2.      broadcastManager:  BroadcastManager, isLocal:  Boolean)
3.    extends MapOutputTracker(conf) {
4.
5.
6.  ......
7.   private class MessageLoop extends Runnable {
8.      override def run():  Unit = {
9.        try {
10.         while (true) {
11.           try {
```

```
12.            val data = mapOutputRequests.take()
13.             if (data == PoisonPill) {
14.              //放回 PoisonPill，其他 MessageLoops 可以看到它
15.              mapOutputRequests.offer(PoisonPill)
16.              return
17.            }
18.            val context = data.context
19.            val shuffleId = data.shuffleId
20.            val hostPort = context.senderAddress.hostPort
21.            logDebug("Handling request to send map output locations for
               shuffle " + shuffleId +
22.              " to " + hostPort)
23.            val mapOutputStatuses = getSerializedMapOutputStatuses
               (shuffleId)
24.            context.reply(mapOutputStatuses)
25.          } catch {
26.            case NonFatal(e) => logError(e.getMessage, e)
27.          }
28.        }
29.      } catch {
30.        case ie: InterruptedException => //退出
31.      }
32.    }
33.  }
```

Spark 2.4.3 版本的 MapOutputTracker.scala 源码与 Spark 2.2.1 版本相比具有如下特点。

- 上段代码中第 23 行调整为 shuffleStatus 变量。
- 上段代码中第 24 行调整回复发送者的信息。serializedMapStatus 方法将 mapStatuses 数组序列化为有效的压缩格式。查看 MapOutputTracker.serializeMapStatuses()，了解有关序列化格式的详细信息。此方法被设计为多次调用并实现缓存以加快后续请求。如果缓存为空并且多个线程同时尝试序列化 map 状态，序列化将只在单个线程中执行，所有其他线程将阻塞，直到填充缓存为止。

```
1.  ......
2.             val shuffleStatus = shuffleStatuses.get(shuffleId).head
3.             context.reply(
4.               shuffleStatus.serializedMapStatus(broadcastManager,
                 isLocal, minSizeForBroadcast))
5.  ......
```

MapOutputTrackerMaster 里面有很多 HashMap，采样线程池不断地接收消息，其中的 MessageLoop 中包含了业务，使用 while 不断循环，和 Spark 1.6.X 有差别，这里，while 循环的是 mapOutputRequests。mapOutputRequests 是一个链式的阻塞队列 LinkedBlockingQueue，我们发消息就发给 LinkedBlockingQueue。LinkedBlockingQueue 中的消息类型就是 GetMapOutputMessage，与刚才发的消息类型一样。run 方法中循环不断地从数据结构中获取数据，如果有数据，则解析数据。解析数据获取 data.shuffleId，读的数据通过 getSerializedMapOutputStatuses 进行序列化,然后返回消息进行 reply。

在 MapOutputTracker.scala 中，通过 val fetchedBytes = askTracker[Array[Byte]](GetMapOutputStatuses(shuffleId))调用 askTracker 方法。

回到代码的主线，获得数据之后，从实际理论上讲，必须对数据进行处理。从 MapOutputTracker.scala 中找一下哪里调用了 getStatuses 方法。

```
1.    override def getMapSizesByExecutorId(shuffleId: Int, startPartition:
      Int, endPartition: Int)
2.     ......
3.     val statuses = getStatuses(shuffleId)
4.     try {
5.       MapOutputTracker.convertMapStatuses(shuffleId, startPartition,
         endPartition, statuses)
6.  ......
```

getStatuses 获取数据后，使用同步代码块，MapOutputTracker.convertMapStatuses 要做的一件事情是什么呢？我们看一下 MapOutputTracker.scala 的 convertMapStatuses 代码，把返回的数据按照一定的格式，获得 status.location 具体的位置信息。

Spark 2.2.1 版本的 MapOutputTracker.scala 的源码如下：

```
1.      private def convertMapStatuses(
2.        shuffleId: Int,
3.        startPartition: Int,
4.        endPartition: Int,
5.        statuses: Array[MapStatus]): Seq[(BlockManagerId, Seq[(BlockId,
          Long)])] = {
6.      assert (statuses != null)
7.      val splitsByAddress = new HashMap[BlockManagerId,
        ArrayBuffer[(BlockId, Long)]]
8.      for ((status, mapId) <- statuses.zipWithIndex) {
9.        if (status == null) {
10.         val errorMessage = s"Missing an output location for shuffle
            $shuffleId"
11.         logError(errorMessage)
12.         throw new MetadataFetchFailedException(shuffleId, startPartition,
            errorMessage)
13.       } else {
14.         for (part <- startPartition until endPartition) {
15.           splitsByAddress.getOrElseUpdate(status.location, ArrayBuffer()) +=
16.             ((ShuffleBlockId(shuffleId, mapId, part), status.getSizeFor-
                Block(part)))
17.         }
18.       }
19.     }
20.
21.     splitsByAddress.toSeq
22.   }
23. }
```

Spark 2.4.3 版本的 MapOutputTracker.scala 源码与 Spark 2.2.1 版本相比具有如下特点。

- 上段代码中第 5 行将 Seq 调整为 Iterator。
- 上段代码中第 8 行调整为 statuses.iterator.zipWithIndex。
- 上段代码中第 15～21 行调整为以下代码第 7～15 行。

```
1.  ......
2.  statuses: Array[MapStatus]): Iterator[(BlockManagerId, Seq[(BlockId,
    Long)])] = {
3.      assert (statuses != null)
4.  ......
5.  for ((status, mapId) <- statuses.iterator.zipWithIndex) {
6.  ......
7.          val size = status.getSizeForBlock(part)
8.          if (size != 0) {
```

```
9.              splitsByAddress.getOrElseUpdate(status.location, ListBuffer()) +=
10.                 ((ShuffleBlockId(shuffleId, mapId, part), size))
11.           }
12.         }
13.       }
14.     }
15.     splitsByAddress.iterator
16. ......
```

重新回到 BlockStoreShuffleReader，在 ShuffleBlockFetcherIterator 实例化的过程中调用了 initialize()，initialize 首先是 splitLocalRemoteBlocks() 划分本地和远程的 blocks，Utils.randomize(remoteRequests) 把远程请求通过随机的方式添加到的队列中，fetchUpToMaxBytes()发送远程请求获取我们的 block，fetchLocalBlocks()获取本地的 blocks。

下面看一下 splitLocalRemoteBlocks 代码。

- splitLocalRemoteBlocks 返回的类型是 ArrayBuffer[FetchRequest]。
- math.max(maxBytesInFlight / 5, 1L)远程请求最大值为 maxBytesInFlight（单次请求的最大字节数）的 1/5，这里用 5 个并发线程从 5 个节点上进行读取。
- new ArrayBuffer[FetchRequest]缓存远程[FetchRequest]的对象。
- totalBlocks 是 block 的数量。
- for 循环遍历 blocksByAddress，blocksByAddress 是前面的 status location 返回的。
- 最后通过 blockManager 去拿数据，这里先进行一个判断：address.executorId == blockManager.blockManagerId.executorId，从本地拿数据，否则从远程拿数据。
- 从远程拿数据首先获得迭代器 blockInfos.iterator，curRequestSize 是累计获取数据块的大小，curBlocks 是当前提供的块，ArrayBuffer 中的元素数据类型是（BlockId, Long)，如果节点频繁地请求，数据量不是很大，也有可能造成网络故障。
- while 循环累加到 curRequestSize >= targetRequestSize，即请求的数据块大小已经大于远程节点数据块大小的时候，在 remoteRequests 中加入 FetchRequest。

Spark 2.2.1 版本的 ShuffleBlockFetcherIterator.scala 的 splitLocalRemoteBlocks 的代码如下：

```
1.  private[this] def splitLocalRemoteBlocks(): ArrayBuffer[FetchRequest] = {
2.      //远程请求最多 maxBytesInFlight/5 的长度，让它们小于 maxBytesInFlight 的原
        //因是允许并发，并行读取 5 个节点以上，而不是阻塞从一个节点读取输出
3.      val targetRequestSize = math.max(maxBytesInFlight / 5, 1L)
4.      logDebug("maxBytesInFlight: " + maxBytesInFlight + ", targetRequestSize:
        " + targetRequestSize
5.        + ", maxBlocksInFlightPerAddress: " + maxBlocksInFlightPerAddress)
6.
7.      //分割本地和远程块。远程模块进一步切分为 FetchRequests 大小（最大为
8.      //maxBytesInFlight，其为了限制数据传送数量）
9.
10.     val remoteRequests = new ArrayBuffer[FetchRequest]
11.
12.     // 跟踪块总数（包括零大小块）
13.     var totalBlocks = 0
14.     for ((address, blockInfos) <- blocksByAddress) {
15.       totalBlocks += blockInfos.size
16.       if (address.executorId == blockManager.blockManagerId.executorId) {
17.         //滤出零大小的块
18.         localBlocks ++= blockInfos.filter(_._2 != 0).map(_._1)
19.         numBlocksToFetch += localBlocks.size
```

```
20.       } else {
21.         val iterator = blockInfos.iterator
22.         var curRequestSize = 0L
23.         var curBlocks = new ArrayBuffer[(BlockId, Long)]
24.         while (iterator.hasNext) {
25.           val (blockId, size) = iterator.next()
26.           //跳过空块
27.           if (size > 0) {
28.             curBlocks += ((blockId, size))
29.             remoteBlocks += blockId
30.             numBlocksToFetch += 1
31.             curRequestSize += size
32.           } else if (size < 0) {
33.             throw new BlockException(blockId, "Negative block size " + size)
34.           }
35.           if (curRequestSize >= targetRequestSize ||
36.               curBlocks.size >= maxBlocksInFlightPerAddress) {
37.             // 增加 FetchRequest
38.             remoteRequests += new FetchRequest(address, curBlocks)
39.             logDebug(s"Creating fetch request of $curRequestSize at $address "
40.               + s"with ${curBlocks.size} blocks")
41.             curBlocks = new ArrayBuffer[(BlockId, Long)]
42.             curRequestSize = 0
43.           }
44.         }
45.         // 增加最后的请求
46.         if (curBlocks.nonEmpty) {
47.           remoteRequests += new FetchRequest(address, curBlocks)
48.         }
49.       }
50.     }
51.     logInfo(s"Getting $numBlocksToFetch non-empty blocks out of
        $totalBlocks blocks")
52.     remoteRequests
53.   }
```

Spark 2.4.3 版本的 ShuffleBlockFetcherIterator.scala 源码与 Spark 2.2.1 版本相比具有如下特点。

- 删掉上段代码中第 13、15 行代码。
- 上段代码中第 16 行以后新增代码，如果块小于 0，抛出负块大小的异常，如果块等于 0，抛出零块大小的异常。
- 上段代码中第 18 行调整为直接获取 blockInfos 的第一个元素信息 blockId。
- 上段代码中第 25 行以后新增代码，如果块小于 0，抛出负块大小的异常，如果块等于 0，抛出零块大小的异常。
- 删掉上段代码中第 27、32、33 行代码。

```
1.  ……
2.          blockInfos.find(_._2 <= 0) match {
3.           case Some((blockId, size)) if size < 0 =>
4.             throw new BlockException(blockId, "Negative block size " + size)
5.           case Some((blockId, size)) if size == 0 =>
6.             throw new BlockException(blockId, "Zero-sized blocks should be
               excluded.")
7.           case None => // do nothing.
8.          }
```

```
9.         localBlocks ++= blockInfos.map(_._1)
10. ......
11.           if (size < 0) {
12.             throw new BlockException(blockId, "Negative block size " + size)
13.           } else if (size == 0) {
14.             throw new BlockException(blockId, "Zero-sized blocks should be
              excluded.")
15.           } else {
16. ......
```

继续回到 ShuffleBlockFetcherIterator.scala 的 initialize 代码中，下一句关键的代码是 fetchUpToMaxBytes()。fetchUpToMaxBytes()用于获取远程的 block，具体代码如下：

```
1.  private def fetchUpToMaxBytes(): Unit = {
2.      //发送获取请求 maxBytesInFlight，如果无法立即从远程主机获取，将请求延迟到下次、
   //处理
3.      //处理延迟的获取请求
4.      if (deferredFetchRequests.nonEmpty) {
5.        for ((remoteAddress, defReqQueue) <- deferredFetchRequests) {
6.          while (isRemoteBlockFetchable(defReqQueue) &&
7.              !isRemoteAddressMaxedOut(remoteAddress, defReqQueue.front)) {
8.            val request = defReqQueue.dequeue()
9.            logDebug(s"Processing deferred fetch request for $remoteAddress
10.            with" + s"${request.blocks.length} blocks")
11.           send(remoteAddress, request)
12.           if (defReqQueue.isEmpty) {
13.             deferredFetchRequests -= remoteAddress
14.           }
15.         }
16.       }
17.     }
18.
19.     //处理常规的获取请求
20.     while (isRemoteBlockFetchable(fetchRequests)) {
21.       val request = fetchRequests.dequeue()
22.       val remoteAddress = request.address
23.       if (isRemoteAddressMaxedOut(remoteAddress, request)) {
24.         logDebug(s"Deferring fetch request for $remoteAddress with
   ${request.blocks.size} blocks")
25.         val defReqQueue = deferredFetchRequests.getOrElse(remoteAddress,
   new Queue[FetchRequest]())
26.         defReqQueue.enqueue(request)
27.         deferredFetchRequests(remoteAddress) = defReqQueue
28.       } else {
29.         send(remoteAddress, request)
30.       }
31.     }
32.
33.     def send(remoteAddress: BlockManagerId, request: FetchRequest): Unit
   = {
34.       sendRequest(request)
35.       numBlocksInFlightPerAddress(remoteAddress) =
36.         numBlocksInFlightPerAddress.getOrElse(remoteAddress, 0) +
   request.blocks.size
37.     }
38.
39.     def isRemoteBlockFetchable(fetchReqQueue: Queue[FetchRequest]):
   Boolean = {
```

```
40.       fetchReqQueue.nonEmpty &&
41.         (bytesInFlight == 0 ||
42.           (reqsInFlight + 1 <= maxReqsInFlight &&
43.             bytesInFlight + fetchReqQueue.front.size <= maxBytesInFlight))
44.     }
45.
46.     // 检查发送的新请求是否超过从给定远程地址获取的最大块数
47.     def isRemoteAddressMaxedOut(remoteAddress: BlockManagerId, request:
   FetchRequest): Boolean = {
48.       numBlocksInFlightPerAddress.getOrElse(remoteAddress, 0) +
   request.blocks.size >
49.         maxBlocksInFlightPerAddress
50.     }
51.   }
```

fetchUpToMaxBytes 里是一个 while 循环，首先单次的请求小于 maxReqsInFlight，然后 sendRequest。sendRequest 方法是至关重要的代码。

- req.blocks.map 根据 blockId 获得 block 的大小。
- shuffleClient.fetchBlocks 根据 shuffleClient 的 fetchBlocks 获取数据。里面给了 BlockFetchingListener 监听器，请求成功时有自己的处理方式，请求失败时也有自己的处理方式。

下面看一下 ShuffleBlockFetcherIterator.scala 中的 sendRequest 方法中的 fetchBlocks。fetchBlocks 是 ShuffleClient.java 的方法。ShuffleClient 是一个抽象类。ShuffleClient.java 的子类是 BlockTransferService。

Spark 2.2.1 版本的 BlockTransferService.scala 的 fetchBlocks 方法的源码如下：

```
1.  override def fetchBlocks(
2.        host: String,
3.        port: Int,
4.        execId: String,
5.        blockIds: Array[String],
6.        listener: BlockFetchingListener,
7.        tempShuffleFileManager: TempShuffleFileManager): Unit
```

Spark 2.4.3 版本的 BlockTransferService.scala 的 fetchBlocks 方法的源码与 Spark 2.2.1 版本相比具有如下特点。

- tempShuffleFileManager 调整为 DownloadFileManager。

```
1.        ......
2.  tempFileManager: DownloadFileManager): Unit
3.  ......
```

BlockTransferService.scala 中的 fetchBlocks 仍然是抽象的。我们继续看它的子类 NettyBlockTransferService，默认采用 Netty 方式，使用 RPC 通信。

回到 ShuffleBlockFetcherIterator.scala 的 sendRequest 方法如下：

```
1.  private[this] def sendRequest(req: FetchRequest) {
2.     ......
3.      shuffleClient.fetchBlocks(address.host, address.port, address
        .executorId, blockIds.toArray, blockFetchingListener, this)
4.
```

回到调用 sendRequest 的地方，ShuffleBlockFetcherIterator.scala 的 fetchUpToMaxBytes 如下：

```
1.   private def fetchUpToMaxBytes(): Unit = {
2.     ......
3.        send(remoteAddress, request)
4.     ......
```

继续回到调用 fetchUpToMaxBytes 的地方，ShuffleBlockFetcherIterator.scala 的 initialize 方法如下：

```
1.  private[this] def initialize(): Unit = {
2.     ......
3.      fetchUpToMaxBytes()
4.   ......
```

ShuffleBlockFetcherIterator.scala 的 initialize 方法的下一步是 fetchLocalBlocks()。远程获取数据使用的是 Netty 方式，那本地获取数据的方式我们可以看一下 fetchLocalBlocks。

Spark 2.2.1 版本的 ShuffleBlockFetcherIterator.scala 的源码如下：

```
1.    private[this] def fetchLocalBlocks() {
2.      val iter = localBlocks.iterator
3.      while (iter.hasNext) {
4.        val blockId = iter.next()
5.        try {
6.          val buf = blockManager.getBlockData(blockId)
7.          shuffleMetrics.incLocalBlocksFetched(1)
8.          shuffleMetrics.incLocalBytesRead(buf.size)
9.          buf.retain()
10.         results.put(new SuccessFetchResult(blockId, blockManager.
            blockManagerId, 0, buf, false))
11.       } catch {
12.         case e: Exception =>
13.           //如果出现异常，就立即停止
14.           logError(s"Error occurred while fetching local blocks", e)
15.           results.put(new FailureFetchResult(blockId, blockManager.
              blockManagerId, e))
16.           return
17.       }
18.     }
19.   }
```

Spark 2.4.3 版本的 ShuffleBlockFetcherIterator.scala 源码与 Spark 2.2.1 版本相比具有如下特点。

- 上段代码中第 1 行之后新增一行代码，打印开始获取本地块的日志。
- 上段代码中第 10 行 SuccessFetchResult 实例的第三个参数调整为 buf.size()。

```
1.  ......
2.  logDebug(s"Start fetching local blocks: ${localBlocks.mkString(", ")}")
3.  ......
4.    results.put(new SuccessFetchResult(blockId, blockManager. blockManagerId,
      buf.size(), buf, false))
5.
```

ShuffleBlockFetcherIterator.scala 的 fetchLocalBlocks 首先获取 localBlocks.iterator 迭代器，基于迭代器进行循环遍历，数据是被 blockManager 管理的。下面看一下 blockManager 的 getBlockData。

blockManager.scala 的 getBlockData 方法的源码如下：

```
1.  override def getBlockData(blockId: BlockId): ManagedBuffer = {
2.      if (blockId.isShuffle) {
3.        shuffleManager.shuffleBlockResolver.getBlockData(blockId
          .asInstanceOf[ShuffleBlockId])
4.      } else {
5.        getLocalBytes(blockId) match {
6.          case Some(blockData) => new BlockManagerManagedBuffer(blockInfoManager,
            blockId, blockData, true)
7.          case None =>
8.    //如果这个块管理器接收一个它没有的块的请求，那可能是master的块状态已经过时了，因
      //此发送一个RPC，将这个块标记为在这个块管理器中不可用
9.            reportBlockStatus(blockId, BlockStatus.empty)
10.           throw new BlockNotFoundException(blockId.toString)
11.       }
12.     }
13.   }
```

跟进去 ShuffleBlockResolver 的 getBlockData，这里是空方法。

```
1.    def getBlockData(blockId: ShuffleBlockId): ManagedBuffer
```

其子类 IndexShuffleBlockResolver 的 getBlockData 是 sort Shuffle 的方式，其中 DataInputStream 就是文件读取。

Spark 2.2.1 版本的 IndexShuffleBlockResolver.scala 的源码如下：

```
1.     override def getBlockData(blockId: ShuffleBlockId): ManagedBuffer =
       {
2.     //这个块实际上是单个映射输出文件的范围，所以找出合并文件，然后从我们
       //的索引中找到偏移量
3.     val indexFile = getIndexFile(blockId.shuffleId, blockId.mapId)
4.
5.     val in = new DataInputStream(new FileInputStream(indexFile))
6.     try {
7.       ByteStreams.skipFully(in, blockId.reduceId * 8)
8.       val offset = in.readLong()
9.       val nextOffset = in.readLong()
10.      new FileSegmentManagedBuffer(
11.        transportConf,
12.        getDataFile(blockId.shuffleId, blockId.mapId),
13.        offset,
14.        nextOffset - offset)
15.    } finally {
16.      in.close()
17.    }
18.  }
```

Spark 2.4.3 版本的 IndexShuffleBlockResolver.scala 源码与 Spark 2.2.1 版本相比具有如下特点。

- 上段代码中第 3 行之后新增 channel、channel.position 代码。SPARK-22982：如果此文件输入流的位置被另一段代码向前寻址，如果使用的文件描述符不正确，则此代码将获取错误的偏移量（这可能导致一个 reducer 被发送到另一个 reducer 的数据）。在 SPARK-22982 期间，此处添加的明确位置检查是一个有用的调试辅助工具，可能有助于防止此类问题在将来再次发生，这就是为什么即使 SPARK-22982 已修复，而它们仍留在此处的原因。
- 上段代码中第 5 行调整为 Channels.newInputStream(channel)，构造从给定通道读取字节的流。
- 上段代码中第 7～9 行替换为以下第 7～14 行代码。

```
1.  ......
2.    val channel = Files.newByteChannel(indexFile.toPath)
3.     channel.position(blockId.reduceId * 8L)
4.  ......
5.  val in = new DataInputStream(Channels.newInputStream(channel))
6.  ......
7.       val offset = in.readLong()
8.       val nextOffset = in.readLong()
9.       val actualPosition = channel.position()
10.      val expectedPosition = blockId.reduceId * 8L + 16
11.      if (actualPosition != expectedPosition) {
12.        throw new Exception(s"SPARK-22982: Incorrect channel position
           after index file reads: " +
13.          s"expected $expectedPosition but actual position was $actualPosition.")
14.      }
15.......
```

再回到 ShuffledRDD.scala 的 compute。

```
1.   override def compute(split: Partition, context: TaskContext):
     Iterator[(K, C)] = {
2.     val dep = dependencies.head.asInstanceOf[ShuffleDependency[K, V, C]]
3.     SparkEnv.get.shuffleManager.getReader(dep.shuffleHandle, split
       .index, split.index + 1, context)
4.       .read()
5.       .asInstanceOf[Iterator[(K, C)]]
6.   }
```

其中，ShuffleReader.scala 的 read 方法返回的是 Iterator。

```
1.  private[spark] trait ShuffleReader[K, C] {
2.   /**读取此汇聚任务的组合 key-values 键值对。 */
3.   def read(): Iterator[Product2[K, C]]
```

ShuffleReader.scala 的子类 BlockStoreShuffleReader.scala 的 read 方法获得一个 Iterator，获得数据以后进行处理，回到 ShuffleBlockFetcherIterator 的 initialize 方法，构造完成。

第 4 篇　Spark+AI 解密篇

第 32 章　Apache Spark+深度学习实战及内幕解密

本章讲解深度学习实战、Spark+PyTorch 案例实战、Spark+TensorFlow 实战、Spark 上的深度学习内核解密等内容。

32.1　深度学习动手实践

本节讲解人工智能下的深度学习、深度学习数据预处理、单节点深度学习训练、分布式深度学习训练。

32.1.1　人工智能下的深度学习

随着数据量和复杂性的不断增加，深度学习成为提供大数据预测分析解决方案的理想方法，需要增加计算处理能力和更先进的图形处理器。通过深度学习，能够利用非结构化数据（例如图像、文本和语音），应用到图像识别、自动翻译、自然语言处理等领域。图像分类：识别和分类图像，便于排序和更准确的搜索。目标检测：快速的目标检测使汽车自动驾驶和人脸识别成为现实。自然语言处理：准确理解口语，为语音到文本的转换和智能家居提供动力。

深度学习面临的挑战：虽然大数据和人工智能提供了无限的潜力，但从大数据中提取可操作的信息并不是一项普通的任务。隐藏在非结构化数据（图像、声音、文本等）中的大量快速增长的信息，需要先进的技术和跨学科团队（数据工程、数据科学和业务）的密切合作。

基于 Databricks 云平台能够轻松构建、训练和部署深度学习应用程序。

- Databricks 云平台集群提供一个交互式环境，可以轻松地使用深度学习的框架，如 Tensorflow、Keras、Pytorch、Mxnet、Caffe、Cntk 和 Theano。
- Databricks 提供处理数据准备、模型训练和大规模预测的云集群平台。
- Spark 分布式计算进行性能优化，可以在强大的 GPU 硬件上大规模运行。
- 交互式数据科学。Databricks 云平台支持多种编程语言，支持实时数据集的深度学习模型训练。

本章基于 Databricks 云平台进行深度学习与大数据案例实战。

Databricks 云平台是美国 Databricks 公司的公有云服务，Databricks 云平台试用版使用美国亚马逊 AWS 云服务的云主机。

Databricks Cloud TensorFlow 在 Spark 上的部署及测试，包括以下步骤：

- 亚马逊公司 AWS 账号注册。

- ❑ Databricks 公司云平台账号注册，关联配置 AWS 的账户角色信息及存储信息。
- ❑ Databricks 配置云平台集群，配置 Databricks Runtime 5.5 ML 及集群信息。
- ❑ Databricks 云平台集群环境创建 Notebook，检查 Spark 及机器学习库版本。
- ❑ Databricks 云平台集群进行 TensorFlow+Spark 案例实战。

（1）亚马逊公司 AWS 账号注册。

登录亚马逊公司官网（https://amazonaws-china.com/cn/），单击右上角的“注册”按钮进行注册，如图 32-1 所示。

图 32-1　AWS 账号注册

按照官网提示，依次输入电子邮件地址、密码、AWS 账户名称、信用卡等信息，完成 AWS 的账户注册。

（2）Databricks 公司云平台账号注册，关联配置 AWS 的账户角色信息及存储信息。

Databricks 平台提供了免费试用版和社区版，两种版本都提供免费的数据块单元（DBU），数据块单元是每小时基于虚拟机实例类型的 Apache Spark 处理能力单元。Databricks 平台免费试用版较灵活，Databricks 在 AWS 账户中使用计算和 S3 存储资源。在注册过程中，将授予 DataBricks 访问 AWS 账户和 S3 存储桶的权限。注册试用时，可以申请免费的 AWS 资源。Databricks 社区版是完全资源化的，不需要提供任何计算或存储资源。

本案例将使用 Databricks 平台免费试用版。

- ❑ 登录 Databricks 公司官网（https://databricks.com/signup）进行注册，依次输入姓名、公司名称、电子邮件、电话号码、用途、角色、信用卡账单等信息。
- ❑ 注册时，在 Databricks 平台配置 AWS 角色信息。

根据 Databricks 官网的提示（https://docs.databricks.com/administration-guide/account-settings/aws-accounts.html），按照提示内容在 AWS 网站一步步操作，创建 AWS 跨账户角色和访问策略，获取 AWS 角色的全局资源描述符（Amazon Resource Name，ARN）用来指定具体角色，如图 32-2 所示。

Roles > testco-role
Summary
Role ARN
arn:aws:iam::
:role/testco-role

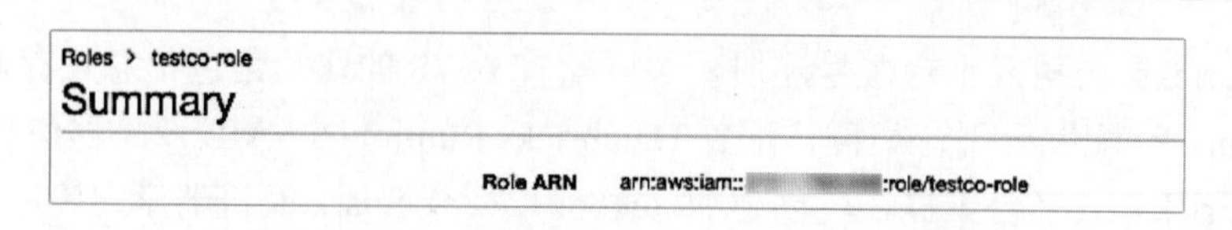

图 32-2　AWS 创建的角色全局资源描述符

将 AWS 云平台的角色全局资源描述符复制到 Databricks 账户页面的 Role ARN 文本栏，并选择 AWS 区域，完成 Databricks 账户与 AWS 角色的关联配置，如图 32-3 所示。

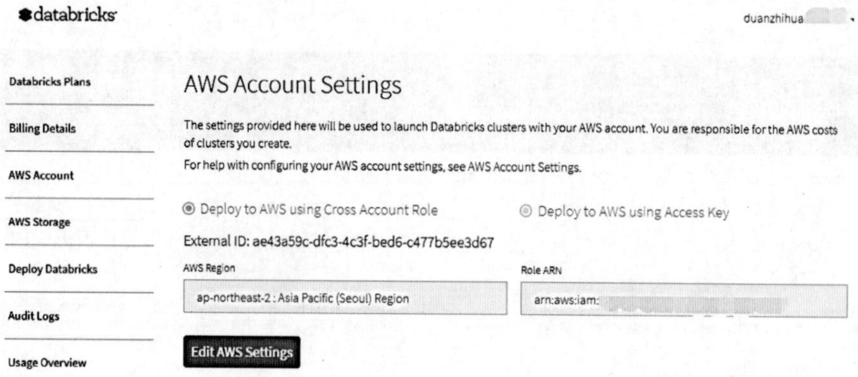

databricks
duanzhihua
Databricks Plans
Billing Details
AWS Account
AWS Storage
Deploy Databricks
Audit Logs
Usage Overview
AWS Account Settings
The settings provided here will be used to launch Databricks clusters with your AWS account. You are responsible for the AWS costs of clusters you create.
For help with configuring your AWS account settings, see AWS Account Settings.
Deploy to AWS using Cross Account Role
Deploy to AWS using Access Key
External ID: ae43a59c-dfc3-4c3f-bed6-c477b5ee3d67
AWS Region
ap-northeast-2 : Asia Pacific (Seoul) Region
Role ARN
arn:aws:iam:
Edit AWS Settings

图 32-3　Databricks 账户关联 AWS 的角色全局资源描述符

- 注册时，在 Databricks 平台配置 AWS 存储信息。

根据 Databricks 官网的提示（https://docs.databricks.com/administration-guide/account-settings/aws-storage.html#aws-storage），按照提示内容在 AWS 网站一步步操作，在 AWS 平台生成 S3 存储桶策略、配置 S3 桶，在 Databricks 平台将 AWS S3 存储桶配置应用于 Databricks 账户，如图 32-4 所示。

https://accounts.cloud.databricks.com/#rootBucket
databricks
duanzhihua
Databricks Plans
Billing Details
AWS Account
AWS Storage
Deploy Databricks
AWS Storage Settings
Databricks stores your account-wide assets, such as libraries, in an Amazon Web Services S3 bucket.
These assets are currently stored in an S3 bucket in your AWS account. If you want to move them to a different S3 bucket in your AWS account, contact support@databricks.com.

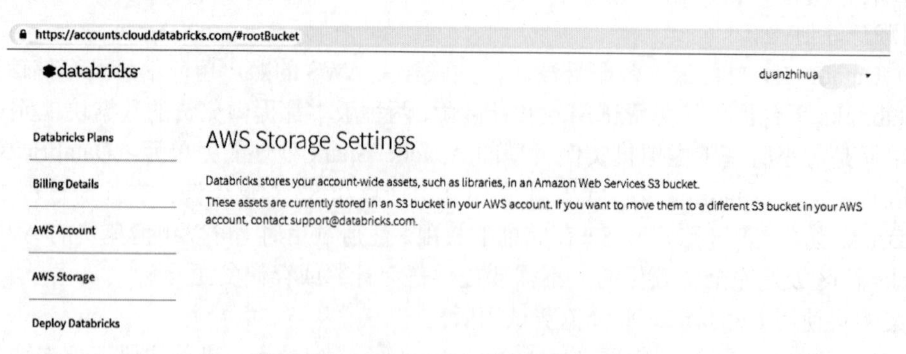

图 32-4　Databricks 平台的 AWS 存储配置

在 Databricks 云平台完成 AWS 角色信息及存储信息的关联配置，根据 Databricks 官网的提示（https://docs.databricks.com/getting-started/try-databricks.html）依次完成其他的配置，将在 Databricks 部署栏目中获取激活的 Databricks 云集群环境登录地址，如图 32-5 所示。

图 32-5　Databricks 云平台集群环境登录地址

单击云平台集群环境页面的链接地址，进入 Databricks 云平台集群环境，如图 32-6 所示。

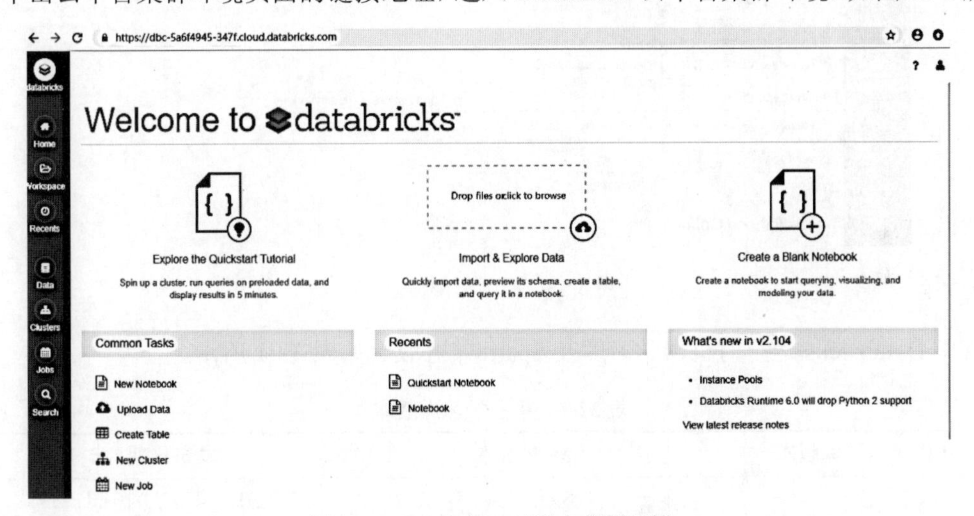

图 32-6　Databricks 云平台集群环境

（3）Databricks 配置云平台集群，配置 Databricks Runtime 5.5 ML 及集群信息。

如图 32-7 所示，在 Databricks 的集群（Clusters）页面单击 Create Cluster 按钮，创建一个新集群。

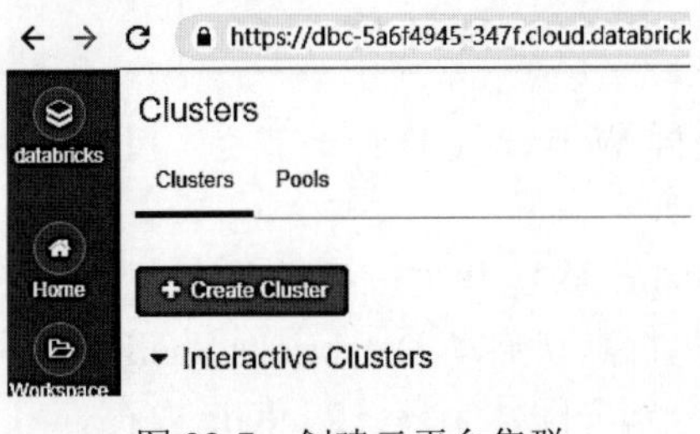

图 32-7　创建云平台集群

如图 32-8 所示，配置新集群的信息：集群名字（DeepLearning）、 Databricks 运行时机器学习库的版本（Databricks Runtime 5.5 LTS ML（includes Apache Spark 2.4.3, Scala 2.11））、Python 的版本（Python 3）、Worker 云节点类型（i3.large）、Driver 云节点的类型（i3.large）等信息。

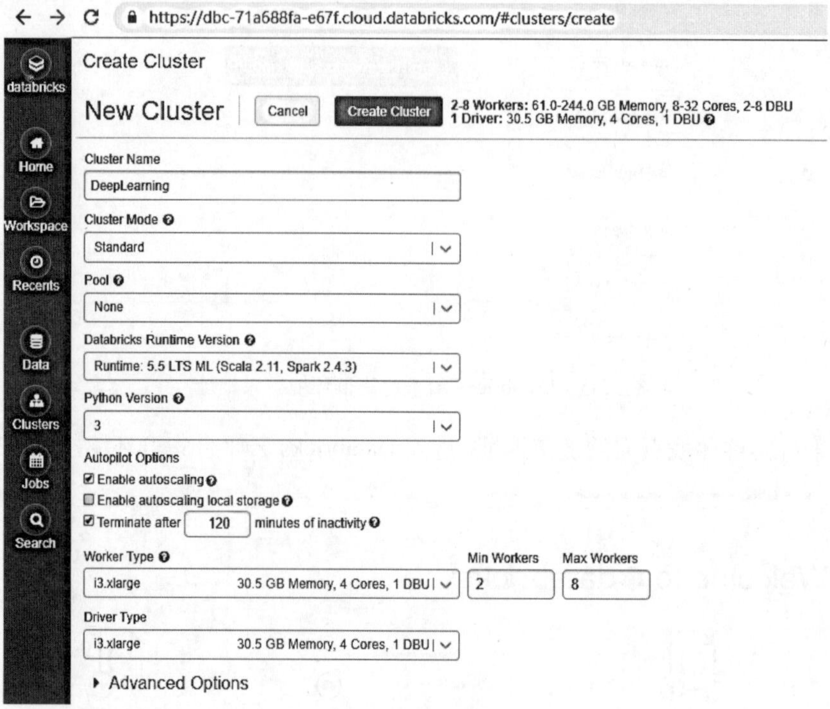

图 32-8　配置集群信息

其中 Databricks 运行时机器学习库及 Spark 兼容版本如表 32-1 所示。

表 32-1　机器学习库版本

机器学习库版本	Spark 版本	发布日期
5.5	Spark 2.4	2019 年 7 月 10 日
5.4	Spark 2.4	2019 年 6 月 3 日
5.3	Spark 2.4	2019 年 4 月 3 日
5.2	Spark 2.4	2019 年 1 月 21 日
5.1	Spark 2.4	2018 年 12 月 18 日
3.5-LTS	Spark 2.2	2017 年 12 月 21 日

其中 Spark 分布式计算集群 Worker 及 Driver 节点，可选配的 AWS 云主机资源（CPU、GPU）部分清单如表 32-2 所示。

如图 32-9 所示，创建 Databricks 云集群以后，单击“三角形”按钮，启动 Spark 集群。

如图 32-10 所示，Spark 集群默认启动 Databricks Shell 应用程序，运行在 2 个 Worker 云节点上，每个 Worker 云节点分别分配 4 个核、24.9GB 内存。

表 32-2　AWS 云主机 CPU/GPU 及内存配置

存储优化实例	vCPUs	内存(GB)
……	……	……
i3.large	2	15
i3.xlarge	4	31
i3.2xlarge	8	61
i3.4xlarge	16	122
i3.8xlarge	32	244
i3.16xlarge (Beta)	64	488
i2.xlarge	4	31
i2.2xlarge	8	61
i2.4xlarge	16	122
i2.8xlarge	32	244
GPU 实例	vCPUs	内存(GB)
p2.xlarge GPU	4	61
p2.8xlarge GPU	32	488
p2.16xlarge GPU	64	732
p3.2xlarge GPU	8	61
p3.8xlarge GPU	32	244
p3.16xlarge (Beta)	64	488
……	……	……

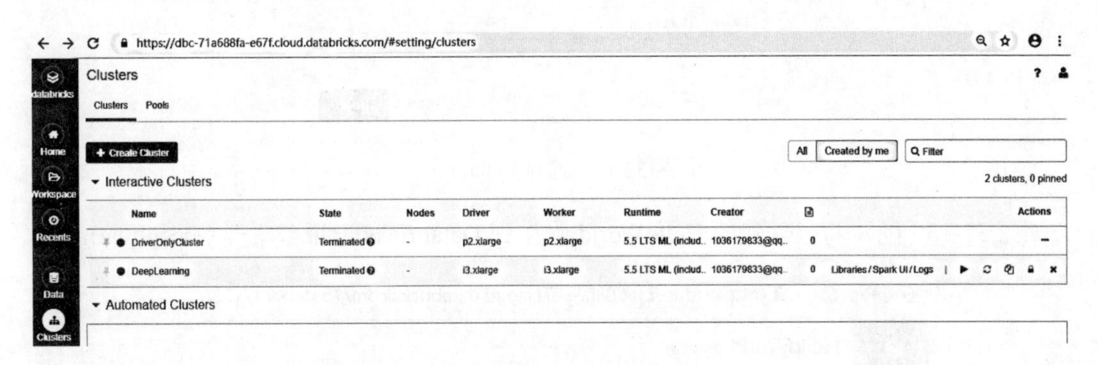

图 32-9　启动 Databricks 云集群页面

（4）在 Databricks 云平台集群环境创建 Notebook，检查 Spark 及机器学习库版本。

根据 Databricks 官网的提示（https://docs.databricks.com/user-guide/notebooks/notebook-manage.html#import-notebook）创建 Notebook。

如图 32-11 所示，单击 Create、Notebook 选项进行创建。

如图 32-12 所示，依次输入名称（HelloWorld）、开发语言（选择 Python 语言）、云平台集群（DeepLearning），单击 Create 按钮创建 Notebook。

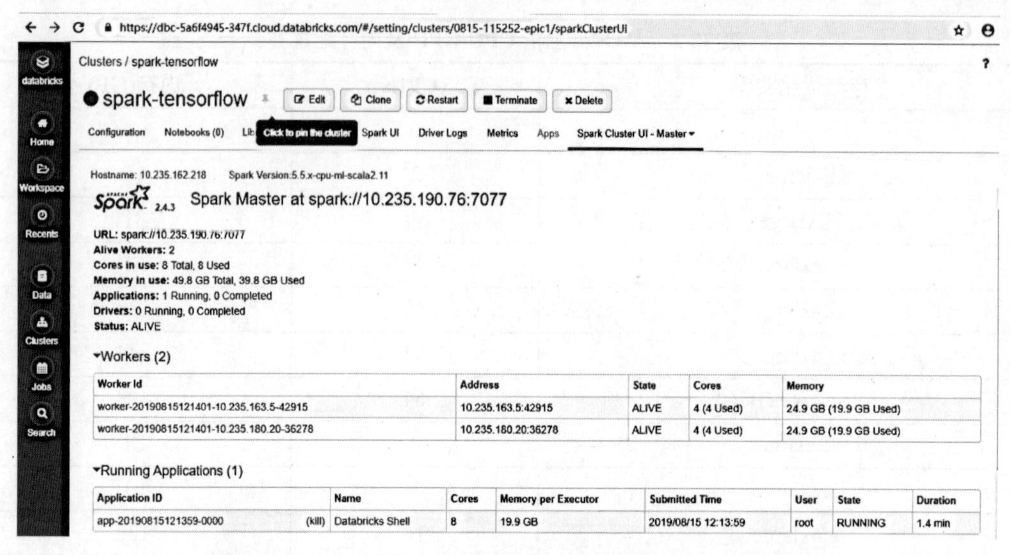

图 32-10　Spark Web UI 页面

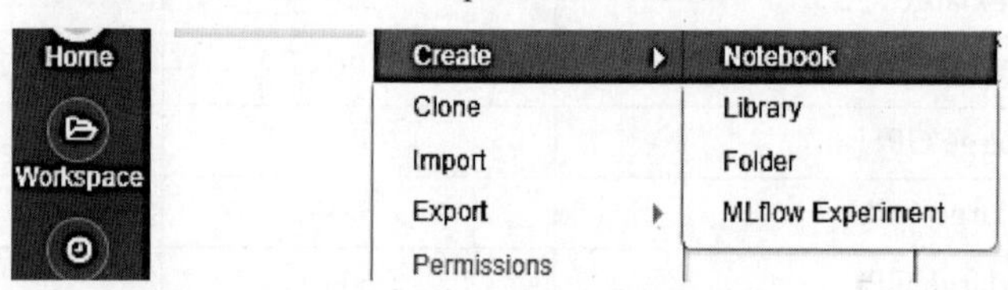

图 32-11　创建 Notebook

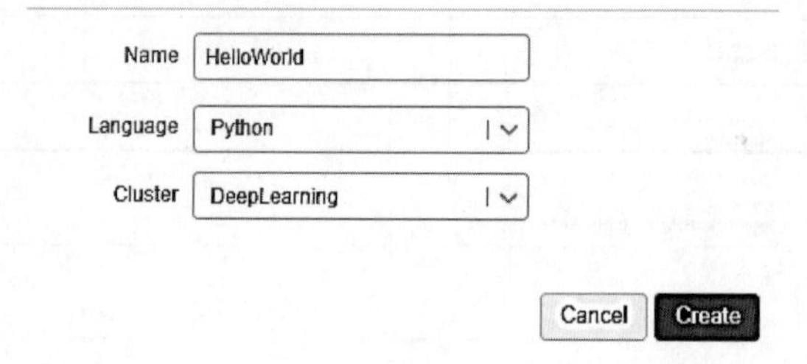

图 32-12　选择 Python

如图 32-13 所示，将已创建的 HelloWorld 附加到 Databricks 云平台集群（DeepLearning）。

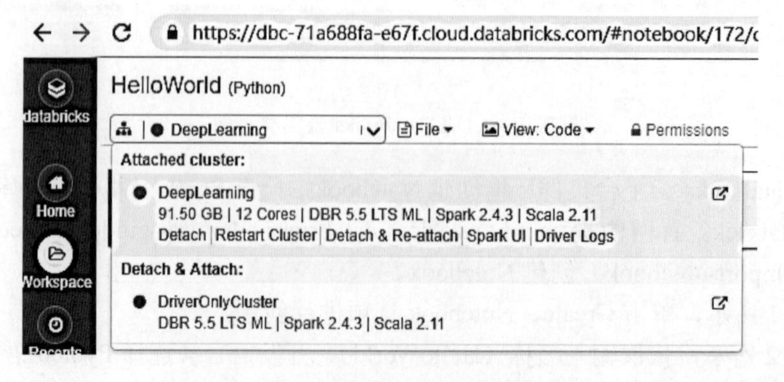

图 32-13　Notebook 附加到云集群

如表 32-3 所示，附加的 HelloWorld 已经定义了以下 Apache Spark 变量。

表 32-3　Apache Spark 变量

类	变 量 名
SparkContext	sc
SQLContext/HiveContext	sqlContext
SparkSession (Spark 2.x)	spark

注意，不要自己创建 SparkSession、SparkContext 或者 SQLContext，这会导致不一致的行为。

在 Databricks Notebook 中检查 Spark 和 Databricks 机器学习库运行时版本，在 Notebook 中运行语句：

```
1.  spark.version
2.  spark.conf.get("spark.databricks.clusterUsageTags.sparkVersion")
```

运行结果如图 32-14 所示。

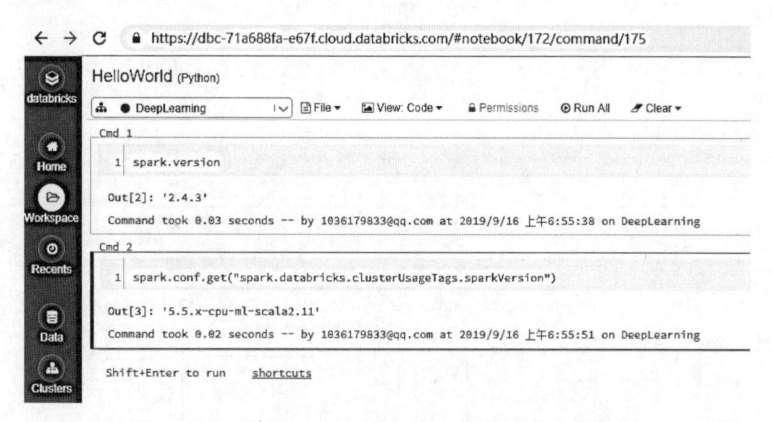

图 32-14　检查 Spark 及 Databricks 机器学习库版本

在 Clusters 栏目中单击 DeepLearning 集群，进入 DeepLearning 云平台集群页面，单击 Spark Cluster UI - Master，查询 Spark 云平台集群的信息，如图 32-15 所示。

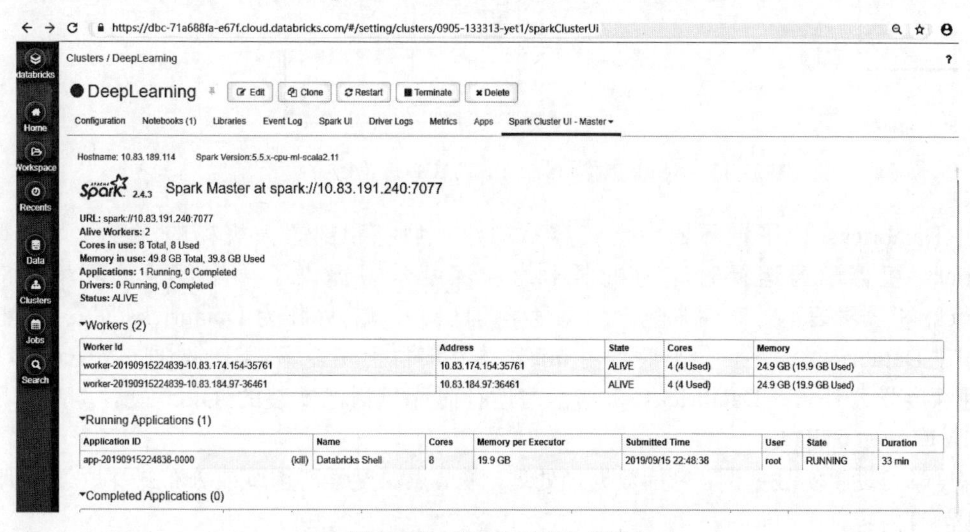

图 32-15　Spark 云平台集群

单击 Metrics 命令，查询 Spark 运行度量的监控信息，如图 32-16 所示。

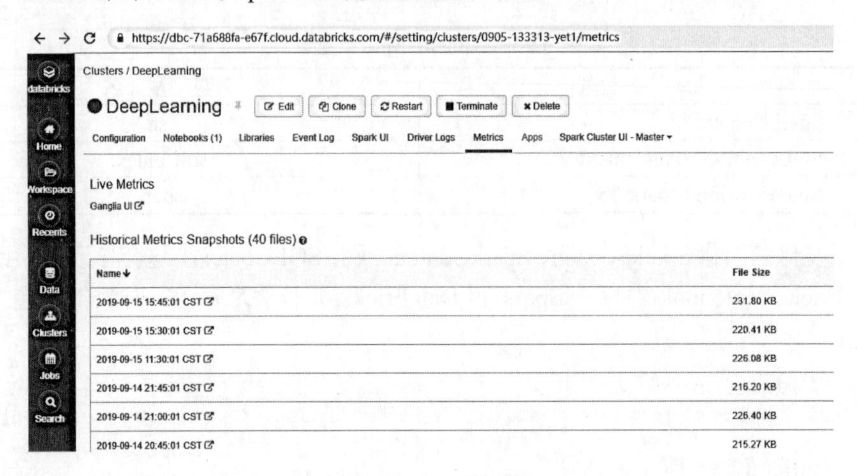

图 32-16　Spark 运行监控

单击 Ganglia UI，查询 Spark 云平台集群工作负载、内存、CPU、网络的运行情况，如图 32-17 所示。

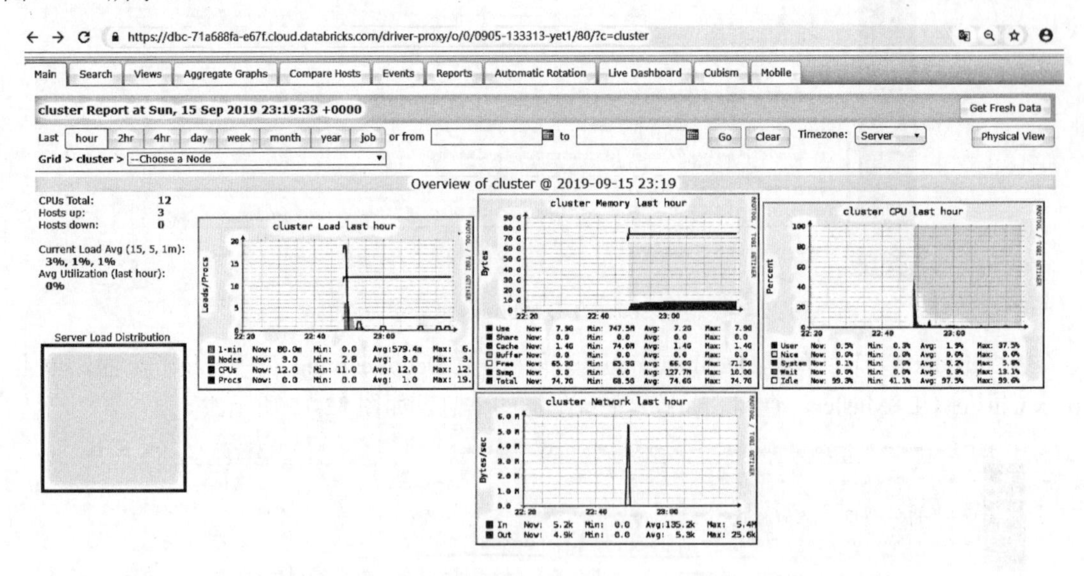

图 32-17　Spark 工作负载、内存、CPU、网络的运行情况

在 Databricks 云平台环境中比较容易构建、训练和部署大规模的深度学习模型，Databricks 机器学习运行时库为机器学习和深度学习提供了一个现成的环境，对于 Databricks 机器学习运行时库未包含的深度学习库，可以将库作为 Databricks 库安装，也可以在创建 Databricks 云平台集群时使用 init 脚本在集群上安装库。图形处理单元（GPU）可以加速深度学习任务，Databricks 机器学习运行时库包括已安装的 GPU 硬件驱动程序和 NVIDIA 库，如 CUDA。

深度学习动手实战案例，包括数据预处理、单节点深度学习训练或分布式深度学习训练、模型预测等内容。

32.1.2　深度学习数据预处理

本节讲解两种深度学习数据预处理的方法。

- ❑ Petastorm 数据预处理。
- ❑ TensorFlow TFRecord 数据格式转换。

32.1.2.1　Petastorm数据预处理方法

Petastorm 是美国 Uber 公司开发的开源数据访问库，Petastorm 支持深度学习模型读取 Apache Parquet 格式的数据，对深度学习模型进行单机或分布式训练及评估。Petastorm 支持基于 Python 的机器学习框架，例如 TensorFlow、PyTorch、PySpark，也可以直接用在 Python 代码中。Petastorm 的 Github 链接地址为 https://github.com/uber/petastorm，Petastorm API 接口文档的链接地址为 https://petastorm.readthedocs.io/en/latest/。

图 32-18 是 Petastorm 架构示意图，通过 Spark 的 PySpark 接口将多个数据源组合成一个数据集，数据集的格式为 Petastorm，Petastorm 数据集将用于 TensorFlow、PyTorch 等深度学习框架模型训练及预测。

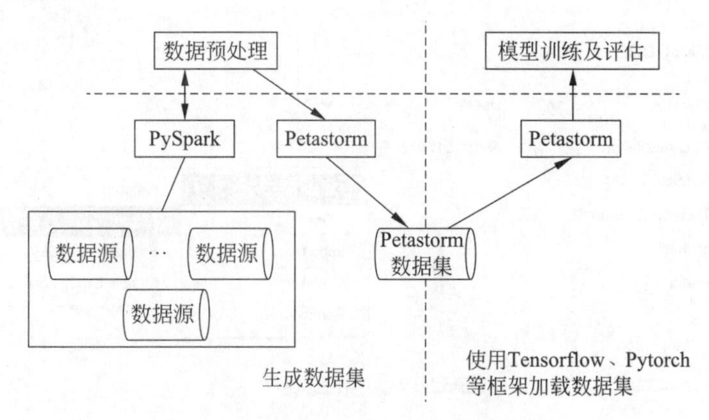

图 32-18　Petastorm 架构示意图

Petastorm 支持直接加载 Apache Parquet 格式的数据，这对于 Databricks 和 Spark 用户很方便，因为 Parquet 是 Spark 推荐使用的数据格式。

接下来使用 Petastorm 从 Parquet 文件加载数据，在 Databrick 云平台中进行 Petastorm 案例实战，实战主要步骤如下。

- ❑ Databricks 云平台安装部署 Petastorm 库。
- ❑ 创建 Databricks 云平台集群 Notebook。
- ❑ 使用 Spark 加载和预处理数据。
- ❑ Spark 使用 Parquet 格式将数据保存到 Databricks dbfs 文件系统。
- ❑ 使用 Petastorm 加载 Parquet 格式的文件数据，深度学习框架加载 Petastorm 数据，进行模型训练或预测。

（1）Databricks 云平台安装部署 Petastorm 库。

Databricks 云平台集群运行的执行环境中如果要使用第三方或本地生成的代码库，可以在 Databricks 集群上安装库。库可以用 Python、Java、Scala、R 编写，可以上传 Java、Scala 和 Python 库，并指向 PyPI、Maven 和 CRAN 存储库中的外部库。在 Databricks 云平台集群中，可以使用 Web UI 页面、命令行模式和通过调用库 API 来管理库。

Databricks 支持三种应用范围的库安装模式。

- 工作区库：工作区库具有与集群安装的库相同的属性及工作区中的相同路径。工作区库是创建集群安装库的模板。要允许工作区中的所有用户共享库，可以在共享文件夹中创建库。若要使其仅对单个用户可用，可以在“用户”文件夹中创建库。
- Databricks 云平台集群安装库：Databricks 云平台集群安装的库仅存在于要安装它的集群的上下文中，拥有集群安装库所需的所有属性：Jar 包的 dbfs 路径、Maven 依赖的坐标、PyPI 包等。
- Notebook 安装库：笔记本范围的库仅存在于安装它的 Notebook 的上下文中。如果在 Notebook 中使用 pip install petastorm 安装，则仅在 Driver 节点上可用。

Databricks 云平台运行的初始执行环境中没有安装 Petastorm 包，可以使用 Databricks 云平台集群安装库模式，通过 Web UI 页面轻松安装 Petastorm 包。

在 Databricks 云平台的工作区的用户处单击右键，在弹出的快捷菜单中选择 Create 及 Library 项，如图 32-19 所示。

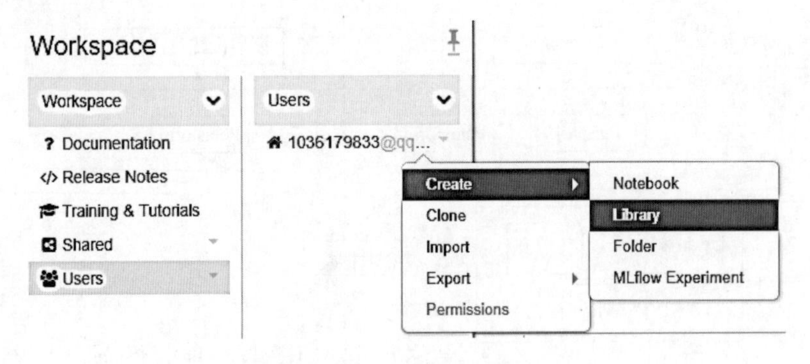

图 32-19　单击创建库

如图 32-20 所示，单击 PyPI 页面栏目，输入要创建的库的名称为 Petastorm，单击 Create 按钮。

Create Library

Library Source

Upload | DBFS/S3 | PyPI | Maven | CRAN

Repository

Optional

Package

petastorm

Create　Cancel

图 32-20　输入要安装的库名称

如图 32-21 所示，在 petastorm 库页面中，选择要安装部署的 Databricks 云平台集群，单击 Install 按钮进行安装。

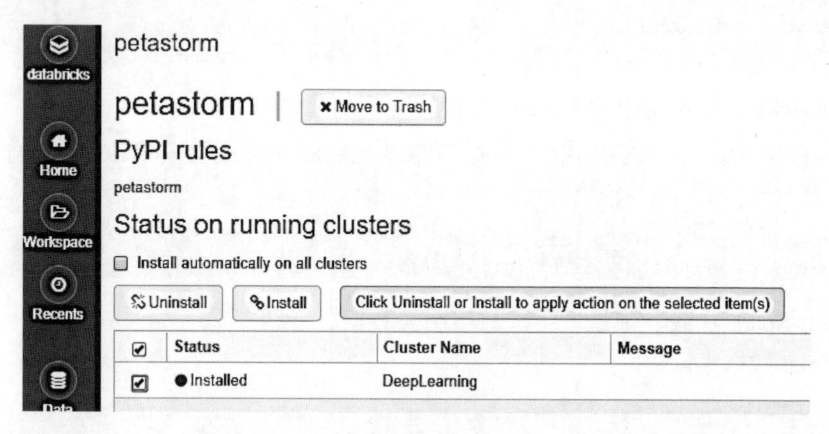

图 32-21　在选择的 Databricks 集群中安装库

如图 32-22 所示，在 Databricks 云平台集群中，单击 Libraries 栏目按钮，查询已经安装的第三方库，显示 petastorm 已经安装成功。

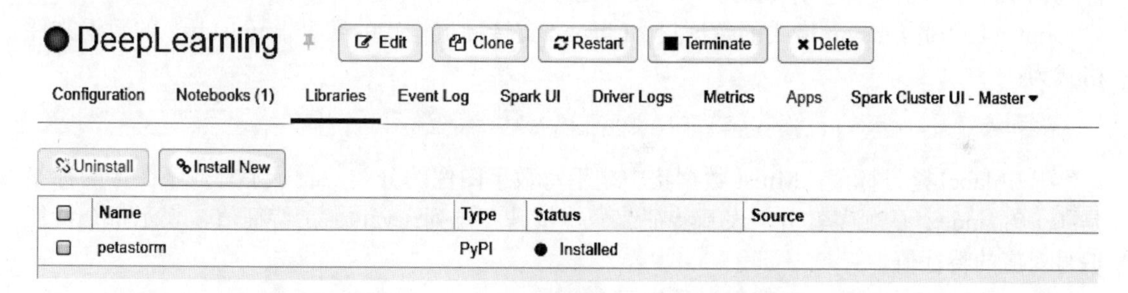

图 32-22　Databricks 集群中查询已经安装的库

（2）创建 Databricks 云平台集群 Notebook。

在 Workspace 栏目中单击右键，在弹出的快捷菜单中选择 Create 选项，依次输入名称（Petastorm）、开发语言（选择 python 语言）、云平台集群，单击 Create 按钮创建 Notebook，如图 32-23 所示。

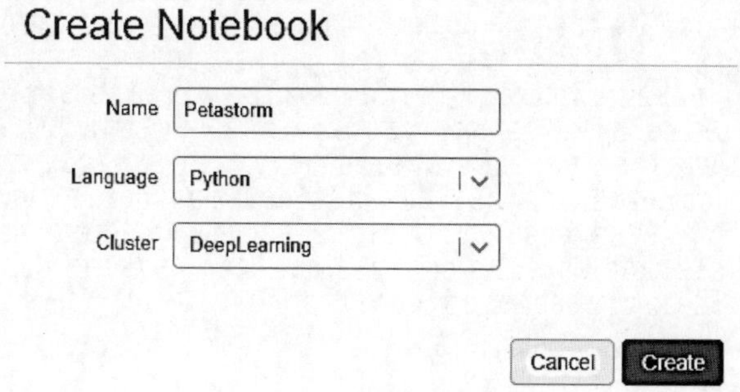

Create Notebook

Name　Petastorm

Language　Python

Cluster　DeepLearning

Cancel　Create

图 32-23　创建 Notebook

（3）导入 os、subprocess、uuid 等库。

```
1.  import os
2.  import subprocess
3.  import uuid
```

在 Databricks 文件系统中为 Notebook 设置唯一的工作目录。

```
1.  work_dir = os.path.join("/ml/tmp/petastorm", str(uuid.uuid4()))
2.  dbutils.fs.mkdirs(work_dir)
3.
4.  def get_local_path(dbfs_path):
5.    return os.path.join("/dbfs", dbfs_path.lstrip("/"))
```

（4）使用 Spark 加载、预处理手写数字图像数据。

下载 Mnist 数据集。

```
1.  data_url  =  "https://www.csie.ntu.edu.tw/~cjlin/libsvmtools/datasets/
    multiclass/mnist.bz2"
2.  libsvm_path = os.path.join(work_dir, "mnist.bz2")
3.  subprocess.check_output(["wget", data_url, "-O", get_local_path(libsvm_path)])
```

其中 mnist.bz2 数据来源是 Yann LeCun, L. Bottou, Y. Bengio 及 P. Haffner 提供的手写数字数据集（http://yann.lecun.com/exdb/mnist/），训练集是 6 万条数据，测试集是 1 万条数据，图像包括 784 个特征，标签分成 0～9 共 10 个分类。

mnist.bz2 是 Libsvm 格式的数据集，Libsvm 支持稀疏格式的数据，mnist.bz2 的数据文件格式为：

```
[label] [index1]:[value1] [index2]:[value2] ...
```

其中 label 是目标值，Mnist 数据集中是手写数字图像的分类，表示手写数字图像的标签是 0～9；index 是顺序索引，表示特征编号，按升序排列；value 是特征值，表示每个特征编号对应的特征值。

mnist.bz2 文件的第一条数据记录如下：

```
5 153:3 154:18 155:18 156:18 157:126 158:136 159:175 160:26 161:166 162:255
163:247 164:127 177:30 178:36 179:94 180:154 181:170 182:253 183:253 184:253
185:253 186:253 187:225 188:172 189:253 190:242 191:195 192:64 204:49 205:238
206:253 207:253 208:253 209:253 210:253 211:253 212:253 213:253 214:251 215:93
216:82 217:82 218:56 219:39 232:18 233:219 234:253 235:253 236:253 237:253
238:253 239:198 240:182 241:247 242:241 261:80 262:156 263:107 264:253 265:253
266:205 267:11 269:43 270:154 290:14 291:1 292:154 293:253 294:90 320:139
321:253 322:190 323:2 348:11 349:190 350:253 351:70 377:35 378:241 379:225
380:160 381:108 382:1 406:81 407:240 408:253 409:253 410:119 411:25 435:45
436:186 437:253 438:253 439:150 440:27 464:16 465:93 466:252 467:253 468:187
494:249 495:253 496:249 497:64 519:46 520:130 521:183 522:253 523:253 524:207
525:2 545:39 546:148 547:229 548:253 549:253 550:253 551:250 552:182 571:24
572:114 573:221 574:253 575:253 576:253 577:253 578:201 579:78 597:23 598:66
599:213 600:253 601:253 602:253 603:253 604:198 605:81 606:2 623:18 624:171
625:219 626:253 627:253 628:253 629:253 630:195 631:80 632:9 649:55 650:172
651:226 652:253 653:253 654:253 655:253 656:244 657:133 658:11 677:136 678:253
679:253 680:253 681:212 682:135 683:132 684:16
……
```

Spark 使用内置的 Libsvm 数据源格式加载 Mnist 数据。

```
1.  df = spark.read.format("libsvm") \
2.    .option("numFeatures", "784") \
3.    .load(libsvm_path)
```

Spark 使用 org.apache.spark.ml.source.libsvm 包实现 SQL 数据源 API，用于将 libsvm 数据加载为数据帧，Spark 加载的数据帧有两列：label 列是存储类型为浮点数的标签，feature 列是存储类型为向量的特征向量。要使用 libsvm 数据源，需要将 libsvm 设置为 DataFrameReader 中的格式，然后指定可选项。

Spark 的 libsvm 数据源支持以下可选项：numfeatures 选项表示特征数，numfeatures 如果未指定或不是一个正数，特征数将自动确定。当数据集已拆分为多个文件并且希望分别加载它们时，numfeatures 比较有用，因为一些文件中可能不存在某些特征，这会导致特征维度不一致。vectorType 选项表示特征向量类型，包括稀疏类型或密集类型，默认值是稀疏类型。

执行 Spark 数据帧的 show 方法，展示数据帧的内容。

```
1.  df.show()
```

在 Databricks 云平台 Notebook 中运行上述代码，运行结果如下：

```
+-----+--------------------+
|label|            features|
+-----+--------------------+
|  5.0|(784,[152,153,154...|
|  0.0|(784,[127,128,129...|
|  4.0|(784,[160,161,162...|
|  1.0|(784,[158,159,160...|
|  9.0|(784,[208,209,210...|
|  2.0|(784,[155,156,157...|
|  1.0|(784,[124,125,126...|
|  3.0|(784,[151,152,153...|
|  1.0|(784,[152,153,154...|
|  4.0|(784,[134,135,161...|
|  3.0|(784,[123,124,125...|
|  5.0|(784,[216,217,218...|
|  3.0|(784,[143,144,145...|
|  6.0|(784,[72,73,74,99...|
|  1.0|(784,[151,152,153...|
|  7.0|(784,[211,212,213...|
|  2.0|(784,[151,152,153...|
|  8.0|(784,[159,160,161...|
|  6.0|(784,[100,101,102...|
|  9.0|(784,[209,210,211...|
+-----+--------------------+
only showing top 20 rows
```

执行 Spark 数据帧的 take(1)方法，获取数据帧中第一条记录的内容。

```
1.  df.take(1)
```

在 Databricks 云平台 Notebook 中运行上述代码，获取的第一条记录是一个 Row 类型的记录，记录的第一个元素是手写数字图像的标签，为数字 5；记录的第二个元素是一个稀疏向量，表示各个维度的特征值，运行结果如下：

```
Out[6]: [Row(label=5.0, features=SparseVector(784, {152: 3.0, 153: 18.0, 154:
18.0, 155: 18.0, 156: 126.0, 157: 136.0, 158: 175.0, 159: 26.0, 160: 166.0,
161: 255.0, 162: 247.0, 163: 127.0, 176: 30.0, 177: 36.0, 178: 94.0, 179: 154.0,
```

```
180: 170.0, 181: 253.0, 182: 253.0, 183: 253.0, 184: 253.0, 185: 253.0, 186:
225.0, 187: 172.0, 188: 253.0, 189: 242.0, 190: 195.0, 191: 64.0, 203: 49.0,
204: 238.0, 205: 253.0, 206: 253.0, 207: 253.0, 208: 253.0, 209: 253.0, 210:
253.0, 211: 253.0, 212: 253.0, 213: 251.0, 214: 93.0, 215: 82.0, 216: 82.0,
217: 56.0, 218: 39.0, 231: 18.0, 232: 219.0, 233: 253.0, 234: 253.0, 235: 253.0,
236: 253.0, 237: 253.0, 238: 198.0, 239: 182.0, 240: 247.0, 241: 241.0, 260:
80.0, 261: 156.0, 262: 107.0, 263: 253.0, 264: 253.0, 265: 205.0, 266: 11.0,
268: 43.0, 269: 154.0, 289: 14.0, 290: 1.0, 291: 154.0, 292: 253.0, 293: 90.0,
319: 139.0, 320: 253.0, 321: 190.0, 322: 2.0, 347: 11.0, 348: 190.0, 349: 253.0,
350: 70.0, 376: 35.0, 377: 241.0, 378: 225.0, 379: 160.0, 380: 108.0, 381:
1.0, 405: 81.0, 406: 240.0, 407: 253.0, 408: 253.0, 409: 119.0, 410: 25.0,
434: 45.0, 435: 186.0, 436: 253.0, 437: 253.0, 438: 150.0, 439: 27.0, 463:
16.0, 464: 93.0, 465: 252.0, 466: 253.0, 467: 187.0, 493: 249.0, 494: 253.0,
495: 249.0, 496: 64.0, 518: 46.0, 519: 130.0, 520: 183.0, 521: 253.0, 522:
253.0, 523: 207.0, 524: 2.0, 544: 39.0, 545: 148.0, 546: 229.0, 547: 253.0,
548: 253.0, 549: 253.0, 550: 250.0, 551: 182.0, 570: 24.0, 571: 114.0, 572:
221.0, 573: 253.0, 574: 253.0, 575: 253.0, 576: 253.0, 577: 201.0, 578: 78.0,
596: 23.0, 597: 66.0, 598: 213.0, 599: 253.0, 600: 253.0, 601: 253.0, 602:
253.0, 603: 198.0, 604: 81.0, 605: 2.0, 622: 18.0, 623: 171.0, 624: 219.0,
625: 253.0, 626: 253.0, 627: 253.0, 628: 253.0, 629: 195.0, 630: 80.0, 631:
9.0, 648: 55.0, 649: 172.0, 650: 226.0, 651: 253.0, 652: 253.0, 653: 253.0,
654: 253.0, 655: 244.0, 656: 133.0, 657: 11.0, 676: 136.0, 677: 253.0, 678:
253.0, 679: 253.0, 680: 212.0, 681: 135.0, 682: 132.0, 683: 16.0})))]
```

（5）Spark 使用 Parquet 格式将数据保存到 Databricks dbfs 文件系统。

在 Databricks 云平台 Notebook 中使用 scala 语言注册一个 UDF 用户自定义函数 toArray，MLlib vector 是一种用户定义类型（UDT），该函数将 MLlib 向量转换为密集数组。

```
1.  %scala
2.
3.  import org.apache.spark.ml.linalg.Vector
4.
5.  val toArray = udf { v: Vector => v.toArray }
6.  spark.sqlContext.udf.register("toArray", toArray)
```

Spark 使用 Parquet 格式将数据保存到 Databricks dbfs 文件系统。

```
1.  parquet_path = os.path.join(work_dir, "parquet")
2.  df.selectExpr("toArray(features) AS features", "int(label) AS label")
3.    .repartition(10) \
4.    .write.mode("overwrite") \
5.    .option("parquet.block.size", 1024 * 1024) \
6.    .parquet(parquet_path)
```

第 2 行代码，Spark 在 selectExpr 语句中查询的第一列，调用用户自定义的 UDF 函数 toArray，将图像的特征数据从稀疏向量转换为密集数组；在 selectExpr 语句中查询的第二列，调用 Spark 的内置函数 int，将标签数据从浮点数类型转换为整型。

第 3 行代码，调用 Spark 的 repartition 方法，将要保存的文件设置为 10 个分区。

第 4 行代码，设置保存文件的方式是覆盖模式（overwrite）。

第 5 行代码，设置 parquet.block.size 大小。因为 Petastorm 将对 Parquet 文件进行分批取样，批处理大小对于 I/O 和计算都很重要，因此使用 parquet.block.size 来控制大小，这里设置为 1024 * 1024。

第 6 行代码，Spark 将数据按照 Parquet 格式写入 Databricks dbfs 文件。

在 Databricks 云平台 Notebook 中运行上述代码，运行结果如图 32-24 所示。

```
1  parquet_path = os.path.join(work_dir, "parquet")
2  df.selectExpr("toArray(features) AS features", "int(label) AS label") \
3    .repartition(10) \
4    .write.mode("overwrite") \
5    .option("parquet.block.size", 1024 * 1024) \
6    .parquet(parquet_path)
```

```
▾ (1) Spark Jobs
    ▾ Job 0    View (Stages: 2/2)
        Stage 0: 4/4
        Stage 1: 10/10
```

图 32-24　Parquet 格式的文件运行结果

单击其中的 View 栏目，查询 Databricks 云平台的 Spark Web UI，如图 32-25 所示。

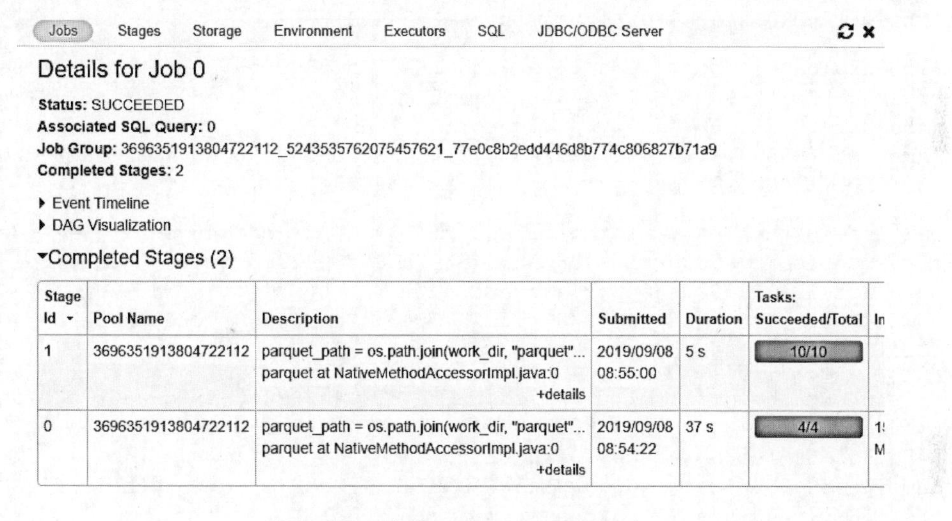

图 32-25　查询 Spark Web UI 页面

查询保存的 Parquet 文件的目录路径。

```
1.   dbutils.fs.ls(parquet_path)
```

在 Databricks 云平台 Notebook 中运行上述代码，运行结果如下：

```
Out[11]:
[FileInfo(path='dbfs:/ml/tmp/petastorm/1873d423-2cd7-469d-aa06-a2ef8ebfc0
7e/parquet/_SUCCESS', name='_SUCCESS', size=0),

FileInfo(path='dbfs:/ml/tmp/petastorm/1873d423-2cd7-469d-aa06-a2ef8ebfc07
e/parquet/_committed_4085920245667218548',
name='_committed_4085920245667218548', size=1010),

FileInfo(path='dbfs:/ml/tmp/petastorm/1873d423-2cd7-469d-aa06-a2ef8ebfc07
e/parquet/_started_4085920245667218548',
name='_started_4085920245667218548', size=0),

FileInfo(path='dbfs:/ml/tmp/petastorm/1873d423-2cd7-469d-aa06-a2ef8ebfc07
e/parquet/part-00000-tid-4085920245667218548-42bb503e-0c70-4b6a-8a18-220d
545a43cf-6-1-c000.snappy.parquet',
name='part-00000-tid-4085920245667218548-42bb503e-0c70-4b6a-8a18-220d545a
43cf-6-1-c000.snappy.parquet', size=1434532),
```

```
FileInfo(path='dbfs:/ml/tmp/petastorm/1873d423-2cd7-469d-aa06-a2ef8ebfc07
e/parquet/part-00001-tid-4085920245667218548-42bb503e-0c70-4b6a-8a18-220d
545a43cf-7-1-c000.snappy.parquet',
name='part-00001-tid-4085920245667218548-42bb503e-0c70-4b6a-8a18-220d545a
43cf-7-1-c000.snappy.parquet', size=1437531),

FileInfo(path='dbfs:/ml/tmp/petastorm/1873d423-2cd7-469d-aa06-a2ef8ebfc07
e/parquet/part-00002-tid-4085920245667218548-42bb503e-0c70-4b6a-8a18-220d
545a43cf-8-1-c000.snappy.parquet',
name='part-00002-tid-4085920245667218548-42bb503e-0c70-4b6a-8a18-220d545a
43cf-8-1-c000.snappy.parquet', size=1438274),

FileInfo(path='dbfs:/ml/tmp/petastorm/1873d423-2cd7-469d-aa06-a2ef8ebfc07
e/parquet/part-00003-tid-4085920245667218548-42bb503e-0c70-4b6a-8a18-220d
545a43cf-9-1-c000.snappy.parquet',
name='part-00003-tid-4085920245667218548-42bb503e-0c70-4b6a-8a18-220d545a
43cf-9-1-c000.snappy.parquet', size=1434409),

FileInfo(path='dbfs:/ml/tmp/petastorm/1873d423-2cd7-469d-aa06-a2ef8ebfc07
e/parquet/part-00004-tid-4085920245667218548-42bb503e-0c70-4b6a-8a18-220d
545a43cf-10-1-c000.snappy.parquet',
name='part-00004-tid-4085920245667218548-42bb503e-0c70-4b6a-8a18-220d545a
43cf-10-1-c000.snappy.parquet', size=1436426),

FileInfo(path='dbfs:/ml/tmp/petastorm/1873d423-2cd7-469d-aa06-a2ef8ebfc07
e/parquet/part-00005-tid-4085920245667218548-42bb503e-0c70-4b6a-8a18-220d
545a43cf-11-1-c000.snappy.parquet',
name='part-00005-tid-4085920245667218548-42bb503e-0c70-4b6a-8a18-220d545a
43cf-11-1-c000.snappy.parquet', size=1437159),

FileInfo(path='dbfs:/ml/tmp/petastorm/1873d423-2cd7-469d-aa06-a2ef8ebfc07
e/parquet/part-00006-tid-4085920245667218548-42bb503e-0c70-4b6a-8a18-220d
545a43cf-12-1-c000.snappy.parquet',
name='part-00006-tid-4085920245667218548-42bb503e-0c70-4b6a-8a18-220d545a
43cf-12-1-c000.snappy.parquet', size=1435083),

FileInfo(path='dbfs:/ml/tmp/petastorm/1873d423-2cd7-469d-aa06-a2ef8ebfc07
e/parquet/part-00007-tid-4085920245667218548-42bb503e-0c70-4b6a-8a18-220d
545a43cf-13-1-c000.snappy.parquet',
name='part-00007-tid-4085920245667218548-42bb503e-0c70-4b6a-8a18-220d545a
43cf-13-1-c000.snappy.parquet', size=1438322),

FileInfo(path='dbfs:/ml/tmp/petastorm/1873d423-2cd7-469d-aa06-a2ef8ebfc07
e/parquet/part-00008-tid-4085920245667218548-42bb503e-0c70-4b6a-8a18-220d
545a43cf-14-1-c000.snappy.parquet',
name='part-00008-tid-4085920245667218548-42bb503e-0c70-4b6a-8a18-220d545a
43cf-14-1-c000.snappy.parquet', size=1437911),

FileInfo(path='dbfs:/ml/tmp/petastorm/1873d423-2cd7-469d-aa06-a2ef8ebfc07
e/parquet/part-00009-tid-4085920245667218548-42bb503e-0c70-4b6a-8a18-220d
545a43cf-15-1-c000.snappy.parquet',
name='part-00009-tid-4085920245667218548-42bb503e-0c70-4b6a-8a18-220d545a
43cf-15-1-c000.snappy.parquet', size=1430914)]
```

执行 Spark 数据帧的 show 方法，展示数据帧将稀疏向量转换为密集数组的内容。

```
1.  df.selectExpr("toArray(features) AS features", "int(label) AS label").show()
```

在 Databricks 云平台 Notebook 中运行上述代码，运行结果如下：

```
+--------------------+------+
|            features|label|
+--------------------+------+
|[0.0, 0.0, 0.0, 0...|     5|
|[0.0, 0.0, 0.0, 0...|     0|
|[0.0, 0.0, 0.0, 0...|     4|
|[0.0, 0.0, 0.0, 0...|     1|
|[0.0, 0.0, 0.0, 0...|     9|
|[0.0, 0.0, 0.0, 0...|     2|
|[0.0, 0.0, 0.0, 0...|     1|
|[0.0, 0.0, 0.0, 0...|     3|
|[0.0, 0.0, 0.0, 0...|     1|
|[0.0, 0.0, 0.0, 0...|     4|
|[0.0, 0.0, 0.0, 0...|     3|
|[0.0, 0.0, 0.0, 0...|     5|
|[0.0, 0.0, 0.0, 0...|     3|
|[0.0, 0.0, 0.0, 0...|     6|
|[0.0, 0.0, 0.0, 0...|     1|
|[0.0, 0.0, 0.0, 0...|     7|
|[0.0, 0.0, 0.0, 0...|     2|
|[0.0, 0.0, 0.0, 0...|     8|
|[0.0, 0.0, 0.0, 0...|     6|
|[0.0, 0.0, 0.0, 0...|     9|
+--------------------+------+
only showing top 20 rows
```

执行 Spark 数据帧的 take(1)方法，获取数据帧将稀疏向量转换为密集数组中第一条记录的内容。

```
2.  df.selectExpr("toArray(features) AS features", "int(label) AS label").take(1)
```

在 Databricks 云平台 Notebook 中运行上述代码，获取的第一条记录是一个 Row 类型的记录，记录的第一个元素是一个大小为 784 的数组，已经将稀疏向量转换为一个密集数组，表示手写数字图像 784 个维度的特征值；记录的第二个元素是手写数字图像的标签，为数字 5，运行结果如下：

```
Out[12]: [Row(features=[0.0, 0.0, 0.0, 0.0, 0.0, 0.0, 0.0, 0.0, 0.0, 0.0, 0.0,
0.0, 0.0, 0.0, 0.0, 0.0, 0.0, 0.0, 0.0, 0.0, 0.0, 0.0, 0.0, 0.0, 0.0, 0.0,
0.0, 0.0, 0.0, 0.0, 0.0, 0.0, 0.0, 0.0, 0.0, 0.0, 0.0, 0.0, 0.0, 0.0, 0.0,
0.0, 0.0, 0.0, 0.0, 0.0, 0.0, 0.0, 0.0, 0.0, 0.0, 0.0, 0.0, 0.0, 0.0, 0.0,
0.0, 0.0, 0.0, 0.0, 0.0, 0.0, 0.0, 0.0, 0.0, 0.0, 0.0, 0.0, 0.0, 0.0, 0.0,
0.0, 0.0, 0.0, 0.0, 0.0, 0.0, 0.0, 0.0, 0.0, 0.0, 0.0, 0.0, 0.0, 0.0, 0.0,
0.0, 0.0, 0.0, 0.0, 0.0, 0.0, 0.0, 0.0, 0.0, 0.0, 0.0, 0.0, 0.0, 0.0, 0.0,
0.0, 0.0, 0.0, 0.0, 0.0, 0.0, 0.0, 0.0, 0.0, 0.0, 0.0, 0.0, 0.0, 0.0, 0.0,
0.0, 0.0, 0.0, 0.0, 0.0, 0.0, 0.0, 0.0, 0.0, 0.0, 0.0, 0.0, 0.0, 0.0, 0.0,
0.0, 0.0, 0.0, 0.0, 0.0, 0.0, 0.0, 0.0, 0.0, 0.0, 0.0, 0.0, 0.0, 0.0, 0.0,
0.0, 0.0, 0.0, 0.0, 0.0, 0.0, 3.0, 18.0, 18.0, 18.0, 126.0, 136.0, 175.0, 26.0,
166.0, 255.0, 247.0, 127.0, 0.0, 0.0, 0.0, 0.0, 0.0, 0.0, 0.0, 0.0, 0.0, 0.0,
0.0, 0.0, 30.0, 36.0, 94.0, 154.0, 170.0, 253.0, 253.0, 253.0, 253.0, 253.0,
225.0, 172.0, 253.0, 242.0, 195.0, 64.0, 0.0, 0.0, 0.0, 0.0, 0.0, 0.0, 0.0,
0.0, 0.0, 0.0, 0.0, 49.0, 238.0, 253.0, 253.0, 253.0, 253.0, 253.0, 253.0,
253.0, 253.0, 251.0, 93.0, 82.0, 82.0, 56.0, 39.0, 0.0, 0.0, 0.0, 0.0, 0.0,
0.0, 0.0, 0.0, 0.0, 0.0, 0.0, 0.0, 18.0, 219.0, 253.0, 253.0, 253.0, 253.0,
253.0, 198.0, 182.0, 247.0, 241.0, 0.0, 0.0, 0.0, 0.0, 0.0, 0.0, 0.0, 0.0,
0.0, 0.0, 0.0, 0.0, 0.0, 0.0, 0.0, 0.0, 0.0, 0.0, 80.0, 156.0, 107.0, 253.0,
253.0, 205.0, 11.0, 0.0, 43.0, 154.0, 0.0, 0.0, 0.0, 0.0, 0.0, 0.0, 0.0, 0.0,
0.0, 0.0, 0.0, 0.0, 0.0, 0.0, 0.0, 0.0, 0.0, 0.0, 0.0, 14.0, 1.0, 154.0, 253.0,
```

```
90.0, 0.0, 0.0, 0.0, 0.0, 0.0, 0.0, 0.0, 0.0, 0.0, 0.0, 0.0, 0.0, 0.0, 0.0,
0.0, 0.0, 0.0, 0.0, 0.0, 0.0, 0.0, 0.0, 0.0, 0.0, 0.0, 139.0, 253.0, 190.0,
2.0, 0.0, 0.0, 0.0, 0.0, 0.0, 0.0, 0.0, 0.0, 0.0, 0.0, 0.0, 0.0, 0.0, 0.0,
0.0, 0.0, 0.0, 0.0, 0.0, 0.0, 0.0, 0.0, 0.0, 0.0, 11.0, 190.0, 253.0, 70.0,
0.0, 0.0, 0.0, 0.0, 0.0, 0.0, 0.0, 0.0, 0.0, 0.0, 0.0, 0.0, 0.0, 0.0, 0.0,
0.0, 0.0, 0.0, 0.0, 0.0, 0.0, 0.0, 0.0, 0.0, 0.0, 35.0, 241.0, 225.0, 160.0,
108.0, 1.0, 0.0, 0.0, 0.0, 0.0, 0.0, 0.0, 0.0, 0.0, 0.0, 0.0, 0.0, 0.0, 0.0,
0.0, 0.0, 0.0, 0.0, 0.0, 0.0, 0.0, 0.0, 0.0, 0.0, 81.0, 240.0, 253.0, 253.0,
119.0, 25.0, 0.0, 0.0, 0.0, 0.0, 0.0, 0.0, 0.0, 0.0, 0.0, 0.0, 0.0, 0.0, 0.0,
0.0, 0.0, 0.0, 0.0, 0.0, 0.0, 0.0, 0.0, 0.0, 0.0, 45.0, 186.0, 253.0, 253.0,
150.0, 27.0, 0.0, 0.0, 0.0, 0.0, 0.0, 0.0, 0.0, 0.0, 0.0, 0.0, 0.0, 0.0, 0.0,
0.0, 0.0, 0.0, 0.0, 0.0, 0.0, 0.0, 0.0, 0.0, 0.0, 16.0, 93.0, 252.0, 253.0,
187.0, 0.0, 0.0, 0.0, 0.0, 0.0, 0.0, 0.0, 0.0, 0.0, 0.0, 0.0, 0.0, 0.0, 0.0,
0.0, 0.0, 0.0, 0.0, 0.0, 0.0, 0.0, 0.0, 0.0, 0.0, 0.0, 249.0, 253.0, 249.0,
64.0, 0.0, 0.0, 0.0, 0.0, 0.0, 0.0, 0.0, 0.0, 0.0, 0.0, 0.0, 0.0, 0.0, 0.0,
0.0, 0.0, 0.0, 0.0, 0.0, 0.0, 0.0, 46.0, 130.0, 183.0, 253.0, 253.0, 207.0,
2.0, 0.0, 0.0, 0.0, 0.0, 0.0, 0.0, 0.0, 0.0, 0.0, 0.0, 0.0, 0.0, 0.0, 0.0,
0.0, 0.0, 0.0, 0.0, 0.0, 39.0, 148.0, 229.0, 253.0, 253.0, 253.0, 250.0, 182.0,
0.0, 0.0, 0.0, 0.0, 0.0, 0.0, 0.0, 0.0, 0.0, 0.0, 0.0, 0.0, 0.0, 0.0, 0.0,
0.0, 0.0, 0.0, 24.0, 114.0, 221.0, 253.0, 253.0, 253.0, 253.0, 201.0, 78.0,
0.0, 0.0, 0.0, 0.0, 0.0, 0.0, 0.0, 0.0, 0.0, 0.0, 0.0, 0.0, 0.0, 0.0, 0.0,
0.0, 0.0, 23.0, 66.0, 213.0, 253.0, 253.0, 253.0, 253.0, 198.0, 81.0, 2.0,
0.0, 0.0, 0.0, 0.0, 0.0, 0.0, 0.0, 0.0, 0.0, 0.0, 0.0, 0.0, 0.0, 0.0, 0.0,
0.0, 18.0, 171.0, 219.0, 253.0, 253.0, 253.0, 253.0, 195.0, 80.0, 9.0, 0.0,
0.0, 0.0, 0.0, 0.0, 0.0, 0.0, 0.0, 0.0, 0.0, 0.0, 0.0, 0.0, 0.0, 0.0, 0.0,
55.0, 172.0, 226.0, 253.0, 253.0, 253.0, 253.0, 244.0, 133.0, 11.0, 0.0, 0.0,
0.0, 0.0, 0.0, 0.0, 0.0, 0.0, 0.0, 0.0, 0.0, 0.0, 0.0, 0.0, 0.0, 0.0, 0.0,
0.0, 136.0, 253.0, 253.0, 253.0, 212.0, 135.0, 132.0, 16.0, 0.0, 0.0, 0.0,
0.0, 0.0, 0.0, 0.0, 0.0, 0.0, 0.0, 0.0, 0.0, 0.0, 0.0, 0.0, 0.0, 0.0, 0.0,
0.0, 0.0, 0.0, 0.0, 0.0, 0.0, 0.0, 0.0, 0.0, 0.0, 0.0, 0.0, 0.0, 0.0, 0.0,
0.0, 0.0, 0.0, 0.0, 0.0, 0.0, 0.0, 0.0, 0.0, 0.0, 0.0, 0.0, 0.0, 0.0, 0.0,
0.0, 0.0, 0.0, 0.0, 0.0, 0.0, 0.0, 0.0, 0.0, 0.0, 0.0, 0.0, 0.0, 0.0, 0.0,
0.0, 0.0, 0.0, 0.0, 0.0, 0.0, 0.0, 0.0, 0.0, 0.0, 0.0, 0.0, 0.0, 0.0, 0.0,
0.0, 0.0, 0.0, 0.0, 0.0, 0.0, 0.0, 0.0, 0.0, 0.0, 0.0, 0.0, 0.0, 0.0, 0.0,
0.0, 0.0, 0.0, 0.0, 0.0, 0.0, 0.0], label=5)]
```

（6）使用 Petastorm 加载 Parquet 格式的文件数据，深度学习框架加载 Petastorm 数据进行模型训练或预测。导入 tensorflow、keras、petastorm 等库。

```
1.  import tensorflow as tf
2.  from tensorflow import keras
3.  from tensorflow.keras import models, layers
4.
5.  from petastorm import make_batch_reader
6.  from petastorm.tf_utils import make_petastorm_dataset
```

使用 tensorflow.keras 构建神经网络模型。

```
1.  def get_model():
2.    model = models.Sequential()
3.    model.add(layers.Conv2D(32, kernel_size=(3, 3),
4.                           activation='relu',
5.                           input_shape=(28, 28, 1)))
6.    model.add(layers.Conv2D(64, (3, 3), activation='relu'))
7.    model.add(layers.MaxPooling2D(pool_size=(2, 2)))
8.    model.add(layers.Dropout(0.25))
9.    model.add(layers.Flatten())
10.   model.add(layers.Dense(128, activation='relu'))
11.   model.add(layers.Dropout(0.5))
```

```
12.   model.add(layers.Dense(10, activation='softmax'))
13.   return model
```

调用 tensorflow.keras 的 model.summary()方法，显示神经网络的模型结构。

```
1.  model = get_model()
2.  model.summary()
```

在 Databricks 云平台 Notebook 中运行上述代码，运行结果如下：

```
_________________________________________________________________
Layer (type)                 Output Shape              Param #
=================================================================
conv2d (Conv2D)              (None, 26, 26, 32)        320
_________________________________________________________________
conv2d_1 (Conv2D)            (None, 24, 24, 64)        18496
_________________________________________________________________
max_pooling2d (MaxPooling2D) (None, 12, 12, 64)        0
_________________________________________________________________
dropout (Dropout)            (None, 12, 12, 64)        0
_________________________________________________________________
flatten (Flatten)            (None, 9216)              0
_________________________________________________________________
dense (Dense)                (None, 128)               1179776
_________________________________________________________________
dropout_1 (Dropout)          (None, 128)               0
_________________________________________________________________
dense_1 (Dense)              (None, 10)                1290
=================================================================
Total params: 1,199,882
Trainable params: 1,199,882
Non-trainable params: 0
_________________________________
```

在 ARROW-4723（https://issues.apache.org/jira/browse/ARROW-4723）解决之前，需要_*的白名单，这是 Databricks 运行时在将数据保存为 Parquet 文件时创建的。

```
1.  import pyarrow.parquet as pq
2.
3.  underscore_files = [f for f in os.listdir(get_local_path(parquet_path))
    if f.startswith("_")]
4.  pq.EXCLUDED_PARQUET_PATHS.update(underscore_files)
```

使用 Petastorm 通过优化 fuse 挂载文件 file:/dbfs/ml 加载 Parquet 数据，并创建 tf.data.dataset，将数据反馈到深度学习框架中，使用 tf.keras 拟合一个简单的神经网络模型。

```
1.  petastorm_dataset_url = "file://" + get_local_path(parquet_path)
2.  with make_batch_reader(petastorm_dataset_url, num_epochs=100) as reader:
3.    dataset = make_petastorm_dataset(reader) \
4.      .map(lambda x: (tf.reshape(x.features, [-1, 28, 28, 1]), tf.one_hot
      (x.label, 10)))
5.    model = get_model()
6.    optimizer = keras.optimizers.Adadelta()
7.    model.compile(optimizer=optimizer,
8.                  loss='categorical_crossentropy',
9.                  metrics=['accuracy'])
10.   model.fit(dataset, steps_per_epoch=10, epochs=10)
```

第 2 行代码，使用 make_batch_reader 方法将 Parquet 文件加载到批次数据，make_batch_reader

方法可以使用 cur_shard 和 shard_count 参数对分布式训练中的数据进行切分，其中 cur_shard 参数表示当前的分片号，读取分片的每个节点都应该传入范围[0，shard_count)内的唯一分片号，shard_count 分片计数也必须提供，cur_shard 默认为 None。shard_count 参数是一个整数，表示要将此数据集分成的分片数，默认为 None。

第 3 行代码使用 make_petastorm_dataset 方法加载 Petastorm 数据集，创建一个 tensorflow.data.dataset。

第 4 行代码将图像输入特征转换为[−1,28,28,1]，将图像标签数据转换为一个大小为 10 的独热编码向量。

第 5 行代码获取 tensorflow.keras 构建的神经网络模型。

第 6 行代码构建 keras.optimizers.Adadelta 优化器。

第 7 行代码设置交叉熵损失函数、优化器、训练性能评估指标。

第 10 行代码对数据集进行拟合训练，设置训练 10 个时代。

在 Databricks 云平台 Notebook 中运行上述代码，运行结果如下：

```
......
Epoch 1/10

 1/10 [==>...........................]-ETA:9s -loss: 14.7244 -acc: 0.0847
 2/10 [=====>........................]-ETA:4s -loss: 14.5767 -acc: 0.0938
 3/10 [========>.....................]-ETA:3s -loss: 14.5303 -acc: 0.0925
 4/10 [===========>..................]-ETA:2s -loss: 14.3506 -acc: 0.1047
 5/10 [==============>...............]-ETA:1s -loss: 14.1715 -acc: 0.1182
 6/10 [=================>............]-ETA:1s -loss: 14.1418 -acc: 0.1188
 7/10 [====================>.........]-ETA:1s -loss: 14.0581 -acc: 0.1239
 8/10 [=======================>......]-ETA:0s -loss: 14.0479 -acc: 0.1244
 9/10 [==========================>...]-ETA:0s -loss: 14.0283 -acc: 0.1262
10/10 [==============================]-3s 312ms/step -loss: 13.9735 -acc: 0.1302
Epoch 2/10

 1/10 [==>...........................]-ETA:1s -loss: 13.9399 -acc: 0.1273
 2/10 [=====>........................]-ETA:1s -loss: 13.9163 -acc: 0.1212
 3/10 [========>.....................]-ETA:1s -loss: 13.6176 -acc: 0.1414
 4/10 [===========>..................]-ETA:1s -loss: 13.4217 -acc: 0.1561
 5/10 [==============>...............]-ETA:1s -loss: 13.4392 -acc: 0.1564
 6/10 [=================>............]-ETA:0s -loss: 13.3645 -acc: 0.1626
 7/10 [====================>.........]-ETA:0s -loss: 13.3282 -acc: 0.1654
 8/10 [=======================>......]-ETA:0s -loss: 13.2618 -acc: 0.1705
 9/10 [==========================>...]-ETA:0s -loss: 13.2318 -acc: 0.1731
10/10 [==============================]-2s 230ms/step -loss: 13.2144 -acc: 0.1739
Epoch 3/10

 1/10 [==>...........................]-ETA:2s -loss: 13.4882 -acc: 0.1576
 2/10 [=====>........................]-ETA:2s -loss: 13.1572 -acc: 0.1788
 3/10 [========>.....................]-ETA:1s -loss: 13.1161 -acc: 0.1818
 4/10 [===========>..................]-ETA:1s -loss: 13.0851 -acc: 0.1848
 5/10 [==============>...............]-ETA:1s -loss: 12.9737 -acc: 0.1903
 6/10 [=================>............]-ETA:1s -loss: 12.9464 -acc: 0.1919
 7/10 [====================>.........]-ETA:0s -loss: 13.0099 -acc: 0.1879
 8/10 [=======================>......]-ETA:0s -loss: 13.0424 -acc: 0.1864
 9/10 [==========================>...]-ETA:0s -loss: 13.0102 -acc: 0.1872
10/10 [==============================]-3s 266ms/step -loss: 13.0377 -acc: 0.1861
Epoch 4/10
```

```
 1/10 [==>...........................]-ETA:1s -loss: 12.3996 -acc: 0.2242
 2/10 [=====>........................]-ETA:1s -loss: 12.5912 -acc: 0.2030
 3/10 [========>.....................]-ETA:1s -loss: 12.7312 -acc: 0.1960
 4/10 [===========>..................]-ETA:1s -loss: 12.7252 -acc: 0.1985
 5/10 [==============>...............]-ETA:1s -loss: 12.6314 -acc: 0.2061
 6/10 [=================>............]-ETA:0s -loss: 12.6049 -acc: 0.2091
 7/10 [====================>.........]-ETA:0s -loss: 12.4860 -acc: 0.2165
 8/10 [=======================>......]-ETA:0s -loss: 12.4042 -acc: 0.2220
 9/10 [==========================>...]-ETA:0s -loss: 12.3595 -acc: 0.2242
10/10 [==============================]-2s 231ms/step -loss: 12.2572 -acc: 0.2315
Epoch 5/10

 1/10 [==>...........................]-ETA:1s -loss: 11.8911 -acc: 0.2606
 2/10 [=====>........................]-ETA:1s -loss: 11.1848 -acc: 0.3000
 3/10 [========>.....................]-ETA:1s -loss: 11.2493 -acc: 0.2949
 4/10 [===========>..................]-ETA:1s -loss: 11.4436 -acc: 0.2833
 5/10 [==============>...............]-ETA:1s -loss: 11.6419 -acc: 0.2715
 6/10 [=================>............]-ETA:0s -loss: 11.5821 -acc: 0.2747
 7/10 [====================>.........]-ETA:0s -loss: 11.6864 -acc: 0.2684
 8/10 [=======================>......]-ETA:0s -loss: 11.7404 -acc: 0.2652
 9/10 [==========================>...]-ETA:0s -loss: 11.6863 -acc: 0.2687
10/10 [==============================]-2s 228ms/step -loss: 11.5537 -acc: 0.2770
Epoch 6/10

 1/10 [==>...........................]-ETA:2s -loss: 11.5167 -acc: 0.2788
 2/10 [=====>........................]-ETA:1s -loss: 11.0847 -acc: 0.3061
 3/10 [========>.....................]-ETA:1s -loss: 10.8250 -acc: 0.3232
 4/10 [===========>..................]-ETA:1s -loss: 10.9759 -acc: 0.3106
 5/10 [==============>...............]-ETA:1s -loss: 10.9387 -acc: 0.3127
 6/10 [=================>............]-ETA:0s -loss: 10.9266 -acc: 0.3121
 7/10 [====================>.........]-ETA:0s -loss: 10.8498 -acc: 0.3160
 8/10 [=======================>......]-ETA:0s -loss: 10.8261 -acc: 0.3167
 9/10 [==========================>...]-ETA:0s -loss: 10.7997 -acc: 0.3192
10/10 [==============================]-2s 221ms/step -loss: 10.7908 -acc: 0.3176
Epoch 7/10

 1/10 [==>...........................]-ETA:2s -loss: 11.3390 -acc: 0.2727
 2/10 [=====>........................]-ETA:1s -loss: 11.3668 -acc: 0.2727
 3/10 [========>.....................]-ETA:1s -loss: 11.1813 -acc: 0.2848
 4/10 [===========>..................]-ETA:1s -loss: 10.7486 -acc: 0.3136
 5/10 [==============>...............]-ETA:1s -loss: 10.6412 -acc: 0.3164
 6/10 [=================>............]-ETA:0s -loss: 10.3897 -acc: 0.3313
 7/10 [====================>.........]-ETA:0s -loss: 10.3950 -acc: 0.3307
 8/10 [=======================>......]-ETA:0s -loss: 10.1370 -acc: 0.3455
 9/10 [==========================>...]-ETA:0s -loss: 9.8825 -acc: 0.3616
10/10 [==============================]-2s 221ms/step -loss: 9.5830 -acc: 0.3800
Epoch 8/10

 1/10 [==>...........................]-ETA:2s -loss: 7.1236 -acc: 0.5152
 2/10 [=====>........................]-ETA:1s -loss: 6.9989 -acc: 0.533
 3/10 [========>.....................]-ETA:1s -loss: 6.7808 -acc: 0.5455
 4/10 [===========>..................]-ETA:1s -loss: 6.8211 -acc: 0.5439
 5/10 [==============>...............]-ETA:1s -loss: 6.4934 -acc: 0.5636
 6/10 [=================>............]-ETA:0s -loss: 6.4202 -acc: 0.5636
 7/10 [====================>.........]-ETA:0s -loss: 6.4668 -acc: 0.5524
 8/10 [=======================>......]-ETA:0s -loss: 6.2038 -acc: 0.5591
 9/10 [==========================>...]-ETA:0s -loss: 5.9697 -acc: 0.5670
10/10 [==============================]-2s 211ms/step -loss: 5.7125 -acc: 0.5727
Epoch 9/10
```

```
 1/10 [==>...........................]-ETA:1s -loss: 2.7693 -acc: 0.6485
 2/10 [=====>........................]-ETA:1s -loss: 2.0824 -acc: 0.6879
 3/10 [========>.....................]-ETA:1s -loss: 1.8513 -acc: 0.6929
 4/10 [===========>..................]-ETA:1s -loss: 1.6466 -acc: 0.7106
 5/10 [==============>...............]-ETA:1s -loss: 1.5040 -acc: 0.7152
 6/10 [=================>............]-ETA:0s -loss: 1.3818 -acc: 0.7242
 7/10 [====================>.........]-ETA:0s -loss: 1.3180 -acc: 0.7290
 8/10 [=======================>......]-ETA:0s -loss: 1.2585 -acc: 0.7280
 9/10 [==========================>...]-ETA:0s -loss: 1.2089 -acc: 0.729
10/10 [==============================]-2s 212ms/step -loss: 1.1646 -acc: 0.7321
Epoch 10/10

 1/10 [==>...........................]-ETA:1s -loss: 0.6752 -acc: 0.7758
 2/10 [=====>........................]-ETA:2s -loss: 0.5980 -acc: 0.7970
 3/10 [========>.....................]-ETA:2s -loss: 0.6095 -acc: 0.7960
 4/10 [===========>..................]-ETA:1s -loss: 0.6318 -acc: 0.7924
 5/10 [==============>...............]-ETA:1s -loss: 0.7044 -acc: 0.7782
 6/10 [=================>............]-ETA:1s -loss: 0.6681 -acc: 0.7970
 7/10 [====================>.........]-ETA:0s -loss: 0.6413 -acc: 0.8009
 8/10 [=======================>......]-ETA:0s -loss: 0.6573 -acc: 0.802
 9/10 [==========================>...]-ETA:0s -loss: 0.6498 -acc: 0.8054
10/10 [==============================]-3s 315ms/step -loss: 0.6449 -acc: 0.8091
```

经过 10 个时代的迭代计算，基于训练集的手写数字图像分类的损失度为 0.6449，精确度为 0.8091。

（7）清理工作目录文件。

在 Databricks 文件系统中查询目录信息。

```
1.  dbutils.fs.ls(work_dir)
```

在 Databricks 云平台 Notebook 中运行上述代码，运行结果如下：

```
Out[14]:
[FileInfo(path='dbfs:/ml/tmp/petastorm/5fc94eba-4ddc-4f2f-9498-0f023de35c
9b/mnist.bz2', name='mnist.bz2', size=15179306),
FileInfo(path='dbfs:/ml/tmp/petastorm/5fc94eba-4ddc-4f2f-9498-0f023de35c9
b/parquet/', name='parquet/', size=0)]
```

使用 dbutils.fs.rm 命令清理工作目录。

```
1.  dbutils.fs.rm(work_dir, recurse=True)
```

在 Databricks 云平台 Notebook 中运行上述代码，运行结果如下：

```
Out[15]: True
```

32.1.2.2　TensorFlow、TFRecord数据格式转换

分布式深度学习框架也可以使用 TFRecord 格式作为数据源。TFRecord 格式是一种简单的面向记录的二进制格式，许多 TensorFlow 应用程序使用 TFRecord 来训练数据，tf.data.TFRecordDataset 是 TensorFlow 数据集，由 TFRecords 文件中的记录组成。

本节讲解 TensorFlow、TFRecord 数据格式转换，包括以下内容。

- 使用 TensorFlow 框架将数据保存为 TFRecord 文件。
- 使用 Spark 将数据帧及数据集保存为 TFRecord 文件。

- 使用 TensorFlow 框架加载 TFRecord 文件。

1. 以Mnist为例演示如何使用TensorFlow框架将数据保存到TFRecord文件

（1）创建 Databricks 云平台集群 Notebook。

在 Workspace 栏目中单击右键，在弹出的快捷菜单中选择 Create，弹出 Create Notebook 对话框依次输入名称（TFRecord）、开发语言（选择 Python 语言）、云平台集群，单击 Create 按钮创建 Notebook，如图 32-26 所示。

Create Notebook

Name　TFRecord

Language　Python

Cluster　Select...

Cancel　Create

图 32-26　创建 Notebook

（2）导入 tensorflow、numpy、urllib 等库。

```
1.  import gzip
2.  import os
3.
4.  import numpy
5.  from six.moves import urllib
6.  import tensorflow as tf
```

（3）定义 Mnist 数据下载位置和保存 TFRecord 文件的位置。

```
1.  params = {}
2.  params['download_data_location'] = '/dbfs/ml/MNISTDemo/mnistData/'
3.  params['tfrecord_location'] = '/dbfs/ml/MNISTDemo/mnistData/'
```

（4）构建函数 download 用于下载 Mnist 数据集。

```
1.  def download(directory, filename):
2.    """从 Mnist 数据集中下载文件"""
3.    filepath = os.path.join(directory, filename)
4.    if tf.gfile.Exists(filepath):
5.      return filepath
6.    if not tf.gfile.Exists(directory):
7.      tf.gfile.MakeDirs(directory)
8.    # Mnist 数据源: http://yann.lecun.com/exdb/mnist/
9.    url = 'https://storage.googleapis.com/cvdf-datasets/mnist/' + filename + '.gz'
10.   temp_file_name, _ = urllib.request.urlretrieve(url)
11.   tf.gfile.Copy(temp_file_name, filepath)
12.   with tf.gfile.GFile(filepath) as f:
13.      size = f.size()
14.   print('Successfully downloaded', filename, size, 'bytes.')
15.   return filepath
```

（5）构建解压 Mnist 图像数据集的函数 extract_images、构建解压 Mnist 标签文件的函数 extract_labels、构建将图像标签转换为独热编码的函数 dense_to_one_hot、构建读取文件字节流的函数_read32，每次读取 4 字节。

Mnist 数据集（http://yann.lecun.com/exdb/mnist/），存储在一个非常简单的文件格式中，用于存储向量和多维矩阵，包括四个文件：

- ❑ train-images-idx3-ubyte：训练集图像（60000 张图片）。
- ❑ train-labels-idx1-ubyte：训练集标签（60000 张图片）。
- ❑ t10k-images-idx3-ubyte：测试集图像（10000 张图片）。
- ❑ t10k-labels-idx1-ubyte：测试集标签（10000 张图片）。

训练集标签文件(train-labels-idx1-ubyte)的格式：从第 0000 个字节读取 4 字节，读取训练集标签文件的魔术数字 2049；从第 0004 个字节读取 4 字节，读取训练集标签文件的记录数（60000）。

```
[offset] [type]            [value]             [description]
0000     32 bit integer   0x00000801(2049)    magic number (MSB first)
0004     32 bit integer   60000               number of items
0008     unsigned byte    ??                  label
0009     unsigned byte    ??                  label
……
xxxx     unsigned byte    ??                  label
```

训练集图像文件(train-images-idx3-ubyte)的格式：从第 0000 个字节读取 4 字节，读取训练集图像文件的魔术数字 2051；从第 0004 个字节读取 4 字节，读取训练集图像文件的记录数（60000）；从第 0008 个字节读取 4 字节，读取训练集图像文件每个图像数据的行数（28）；从第 0012 字节读取 4 字节，读取训练集图像文件每个图像数据的列数（28）。

```
[offset] [type]            [value]             [description]
0000     32 bit integer   0x00000803(2051)    magic number
0004     32 bit integer   60000               number of images
0008     32 bit integer   28                  number of rows
0012     32 bit integer   28                  number of columns
0016     unsigned byte    ??                  pixel
0017     unsigned byte    ??                  pixel
……
xxxx     unsigned byte    ??                  pixel
```

测试集标签文件(t10k-labels-idx1-ubyte)的格式：从第 0000 个字节读取 4 字节，读取测试集标签文件的魔术数字 2049；从第 0004 字节读取 4 字节，读取测试集标签文件的记录数（10000）。

```
[offset] [type]            [value]             [description]
0000     32 bit integer   0x00000801(2049)    magic number (MSB first)
0004     32 bit integer   10000               number of items
0008     unsigned byte    ??                  label
0009     unsigned byte    ??                  label
……
xxxx     unsigned byte    ??                  label
```

测试集图像文件 (t10k-images-idx3-ubyte) 的格式：从第 0000 个字节读取 4 字节，读取测试集图像文件的魔术数字 2051；从第 0004 字节读取 4 字节，读取测试集图像文件的记录数（10000）；从第 0008 个字节读取 4 字节，读取测试集图像文件每个图像数据的行数（28）；从第 0012 字节读取 4 字节，读取测试集图像文件每个图像数据的列数（28）。

```
[offset] [type]          [value]           [description]
0000     32 bit integer  0x00000803(2051)  magic number
0004     32 bit integer  10000             number of images
0008     32 bit integer  28                number of rows
0012     32 bit integer  28                number of columns
0016     unsigned byte   ??                pixel
0017     unsigned byte   ??                pixel
……
xxxx     unsigned byte   ??                pixel
```

在 Databricks 云平台 Notebook 中构建_read32、extract_images、dense_to_one_hot、extract_labels 函数如下：

```
1. def _read32(bytestream):
2.   dt = numpy.dtype(numpy.uint32).newbyteorder('>')
3.   return numpy.frombuffer(bytestream.read(4), dtype=dt)[0]
4.
5. def extract_images(f):
6.   """将图像数据提取到 4 维 uint8 类型的 numpy 数组中，格式为[index, y, x, depth]
7.   参数：
8.     f：可以传递到 gzip 阅读器的文件对象
9.   返回：
10.    data：一个 4 维 uint8 numpy 数组 [index, y, x, depth]
11.  抛出异常：
12.    ValueError：如果 bytestream 不是从 2051 开始的
13.  """
14.  print('Extracting', f.name)
15.  with gzip.GzipFile(fileobj=f) as bytestream:
16.    magic = _read32(bytestream)
17.    if magic != 2051:
18.      raise ValueError('Invalid magic number %d in MNIST image file: %s' %
19.                       (magic, f.name))
20.    num_images = _read32(bytestream)
21.    rows = _read32(bytestream)
22.    cols = _read32(bytestream)
23.    buf = bytestream.read(rows * cols * num_images)
24.    data = numpy.frombuffer(buf, dtype=numpy.uint8)
25.    data = data.reshape(num_images, rows, cols, 1)
26.    return data
27.
28. def dense_to_one_hot(labels_dense, num_classes):
29.  """将类标签从标量转换为一个独热向量"""
30.  num_labels = labels_dense.shape[0]
31.  index_offset = numpy.arange(num_labels) * num_classes
32.  labels_one_hot = numpy.zeros((num_labels, num_classes))
33.  labels_one_hot.flat[index_offset + labels_dense.ravel()] = 1
34.  return labels_one_hot
35.
36. def extract_labels(f, one_hot=False, num_classes=10):
37.  """将标签提取到一维 uint8 类型的 numpy 数组[index]
38.  参数：
39.    f：可以传递到 gzip 阅读器的文件对象
40.    one_hot：进行独热编码
41.    num_classes：独热编码的分类数
42.  返回：
43.    labels：一个一维 uint8 类型的 numpy 数组
44.  抛出异常：
```

```
45.     ValueError: 如果 bytestream 不是从 2049 开始的
46.   """
47.   print('Extracting', f.name)
48.   with gzip.GzipFile(fileobj=f) as bytestream:
49.     magic = _read32(bytestream)
50.     if magic != 2049:
51.       raise ValueError('Invalid magic number %d in MNIST label file: %s' %
52.                        (magic, f.name))
53.     num_items = _read32(bytestream)
54.     buf = bytestream.read(num_items)
55.     labels = numpy.frombuffer(buf, dtype=numpy.uint8)
56.     if one_hot:
57.       return dense_to_one_hot(labels, num_classes)
58.     return labels
```

第 1～3 行，构建一个_read32 函数，其中：

第 3 行代码，每次调用_read32 函数，从文件中读取 4 字节的数据。

第 5～26 行，构建一个 extract_images 函数，其中：

第 16 行代码，调用_read32 函数读取图像数据文件的魔术数字。

第 17 行代码，判断魔术数字是否等于 2051，校验图像数据文件的文件格式。

第 20 行代码，调用_read32 函数读取图像数据文件的记录数，如果是训练集，为 60000 张图片；如果是测试集，为 10000 张图片。

第 21 行代码，调用_read32 函数读取图像数据文件每个图像数据的行数，数值为 28。

第 22 行代码，调用_read32 函数读取图像数据文件每个图像数据的列数，数值为 28。

第 25 行代码，将加载读取的缓冲区（rows * cols * num_images）数据，调整为(num_images, rows, cols, 1)的维度格式。

第 28～34 行，构建一个 dense_to_one_hot 函数，将类标签从标量转换为一个独热向量。

第 36～58 行，构建一个 extract_labels 函数，其中：

第 49 行代码，调用_read32 函数读取图像标签文件的魔术数字。

第 50 行代码，判断标签文件的魔术数字是否等于 2049，校验图像标签文件的文件格式。

第 53 行代码，调用_read32 函数读取图像标签文件的记录数，如果是训练集，为 60000 张图片；如果是测试集，为 10000 张图片。

第 57 行代码，如果设置独热编码参数为 True，则调用 dense_to_one_hot 函数，对于训练集，返回一个 60000 行 10 列的独热编码向量，每一行记录中手写数字标签索引对应的值为 1，其他位置对应的值为 0。

（6）构建 load_dataset 函数，加载解析 Mnist 图像集数据。

```
1.  def load_dataset(directory, images_file, labels_file):
2.    """下载解析 Mnist 数据集"""
3.
4.    images_file = download(directory, images_file)
5.    labels_file = download(directory, labels_file)
6.
7.    with tf.gfile.Open(images_file, 'rb') as f:
8.      images = extract_images(f)
9.
10.   with tf.gfile.Open(labels_file, 'rb') as f:
11.     labels = extract_labels(f)
12.
13.   return images, labels
```

（7）下载 Mnist 数据，从原训练集中划出前 5000 条记录作为验证集，从第 5000～60000 作为新的训练集（共 55000 条记录），分别得到训练集、测试集、验证集。

```
1.  directory = params['download_data_location']
2.  validation_size=5000
3.  train_images, train_labels = load_dataset(directory, 'train-images-idx3-ubyte',
    'train-labels-idx1-ubyte')
4.  test_images, test_labels = load_dataset(directory, 't10k-images-idx3-ubyte',
    't10k-labels-idx1-ubyte')
5.  validation_images = train_images[:validation_size]
6.  validation_labels = train_labels[:validation_size]
7.  train_images = train_images[validation_size:]
8.  train_labels = train_labels[validation_size:]
```

在 Databricks 云平台 Notebook 中运行上述代码，运行结果如下：

```
Successfully downloaded train-images-idx3-ubyte 9912422 bytes.
Successfully downloaded train-labels-idx1-ubyte 28881 bytes.
Extracting /dbfs/ml/MNISTDemo/mnistData/train-images-idx3-ubyte
Extracting /dbfs/ml/MNISTDemo/mnistData/train-labels-idx1-ubyte
Successfully downloaded t10k-images-idx3-ubyte 1648877 bytes.
Successfully downloaded t10k-labels-idx1-ubyte 4542 bytes.
Extracting /dbfs/ml/MNISTDemo/mnistData/t10k-images-idx3-ubyte
Extracting /dbfs/ml/MNISTDemo/mnistData/t10k-labels-idx1-ubyte
```

查询训练集数据的大小：

```
1.  train_images.shape[0]
```

在 Databricks 云平台 Notebook 中运行上述代码，运行结果如下：

```
Out[9]: 55000
```

（8）定义一个 TFRecordWriter，以 TFRecord train.tfrecords 格式打开。

```
1.  name = "train.tfrecords"
2.  filename = os.path.join(params['tfrecord_location'], name)
3.  tfrecord_writer = tf.python_io.TFRecordWriter(filename)
```

（9）定义两个 helper 函数来解析 int 和 bytes 类型。

```
1.  def _int64_feature(value):
2.    return tf.train.Feature(int64_list=tf.train.Int64List(value=[value]))
3.
4.  def _bytes_feature(value):
5.    return tf.train.Feature(bytes_list=tf.train.BytesList(value=[value]))
```

（10）分析 Mnist 训练集数据并将其保存为 TFRecord 文件。

```
1.  num_examples = train_images.shape[0]
2.  images = train_images
3.  labels = train_labels
4.
5.  rows = images.shape[1]
6.  cols = images.shape[2]
7.  depth = images.shape[3]
8.
9.  for index in range(num_examples):
10.   # 1. 将数据转换为 tf.train.Feature 格式
```

```
11.   image_raw = images[index].tostring()
12.   feature = {
13.     'height': _int64_feature(rows),
14.     'width': _int64_feature(cols),
15.     'depth': _int64_feature(depth),
16.     'label': _int64_feature(int(labels[index])),
17.     'image_raw': _bytes_feature(image_raw)
18.   }
19.   # 2. 创建一个 tf.train.Features
20.   features = tf.train.Features(feature=feature)
21.   # 3. 创建一个示例
22.   example = tf.train.Example(features=features)
23.   # 4. 将示例序列化为一个字符串
24.   example_to_string = example.SerializeToString()
25.   # 5. 保存为一个 TFRecord 格式的文件
26.   tfrecord_writer.write(example_to_string)
```

第 11～18 行代码，将 Mnist 训练集的每一行记录转换为 tf.train.Feature 格式。

第 20 行代码，传入 feature 变量，创建一个 tf.train.Features 实例。

第 22 行代码，使用 tf.train.Example 创建一个示例。

第 24 行代码，将 tf.train.Example 示例序列化为一个字符串。

第 26 行代码，将字符串保存为一个 TFRecord 格式的文件。

在 Databricks 文件系统中查询保存的 TFRecord 格式文件。

```
1.  dbutils.fs.ls("file:"+params['tfrecord_location'])
```

在 Databricks 云平台 Notebook 中运行上述代码，训练集保存的 TFRecord 格式文件是 train.tfrecords，运行结果如图 32-27 所示。

```
dbutils.fs.ls("file:"+params['tfrecord_location'] )

Out[56]: [FileInfo(path='file:/dbfs/ml/MNISTDemo/mnistData/t10k-images-idx3-ubyte', name='t10k-images-idx3-ubyte', size=1648877),
 FileInfo(path='file:/dbfs/ml/MNISTDemo/mnistData/t10k-labels-idx1-ubyte', name='t10k-labels-idx1-ubyte', size=4542),
 FileInfo(path='file:/dbfs/ml/MNISTDemo/mnistData/test.tfrecords', name='test.tfrecords', size=8420000),
 FileInfo(path='file:/dbfs/ml/MNISTDemo/mnistData/train-images-idx3-ubyte', name='train-images-idx3-ubyte', size=9912422),
 FileInfo(path='file:/dbfs/ml/MNISTDemo/mnistData/train-labels-idx1-ubyte', name='train-labels-idx1-ubyte', size=28881),
 FileInfo(path='file:/dbfs/ml/MNISTDemo/mnistData/train.tfrecords', name='train.tfrecords', size=46310000)]
```

图 32-27　查询保存的 TFRecord 格式文件

（11）将上述过程封装为一个函数 convert_and_save_to。

```
1.  def convert_and_save_to(images, labels, name, params):
2.    """转换一个 TF dataset 为 tfrecords"""
3.    num_examples = images.shape[0]
4.
5.    rows = images.shape[1]
6.    cols = images.shape[2]
7.    depth = images.shape[3]
8.
9.    filename = os.path.join(params['tfrecord_location'], name + '.tfrecords')
10.   print('Writing', filename)
11.   with tf.python_io.TFRecordWriter(filename) as writer:
12.     for index in range(num_examples):
13.       image_raw = images[index].tostring()
14.       feature={
15.             'label': _int64_feature(int(labels[index])),
```

```
16.              'image_raw': _bytes_feature(image_raw)
17.              }
18.        features=tf.train.Features(feature=feature)
19.        example=tf.train.Example(features=features)
20.        writer.write(example.SerializeToString())
```

调用 convert_and_save_to 函数，加载训练集、测试集数据，将数据集转换为 Examples 实例，并将结果写入 TFRecord 文件。

```
1.  convert_and_save_to(train_images, train_labels, 'train', params)
2.  convert_and_save_to(test_images, test_labels, 'test', params)
```

在 Databricks 云平台 Notebook 中运行上述代码，运行结果如下。

```
Writing /dbfs/ml/MNISTDemo/mnistData/train.tfrecords
Writing /dbfs/ml/MNISTDemo/mnistData/test.tfrecords
```

在 Databricks 文件系统中查询保存的 TFRecord 格式文件。

```
1.  display(dbutils.fs.ls("file:"+params['tfrecord_location']))
```

在 Databricks 云平台 Notebook 中运行上述代码，训练集数据保存为 file:/dbfs/ml/MNISTDemo/mnistData/train.tfrecords 文件，测试集数据保存为 file:/dbfs/ml/MNISTDemo/mnistData/test.tfrecords 文件，运行结果如图 32-28 所示。

```
1  display(dbutils.fs.ls("file:"+params['tfrecord_location']))
```

▸ (3) Spark Jobs

path	name	size
file:/dbfs/ml/MNISTDemo/mnistData/t10k-images-idx3-ubyte	t10k-images-idx3-ubyte	1648877
file:/dbfs/ml/MNISTDemo/mnistData/t10k-labels-idx1-ubyte	t10k-labels-idx1-ubyte	4542
file:/dbfs/ml/MNISTDemo/mnistData/test.tfrecords	test.tfrecords	8420000
file:/dbfs/ml/MNISTDemo/mnistData/train-images-idx3-ubyte	train-images-idx3-ubyte	9912422
file:/dbfs/ml/MNISTDemo/mnistData/train-labels-idx1-ubyte	train-labels-idx1-ubyte	28881
file:/dbfs/ml/MNISTDemo/mnistData/train.tfrecords	train.tfrecords	46310000

图 32-28　查询保存的 TFRecord 格式文件

2. 使用Spark将数据帧及数据集保存为TFRecord文件案例

（1）创建 Databricks 云平台集群 Notebook。

在 Workspace 栏目单击右键，在弹出的快捷菜单中选择 Create，弹出 Create Notebook 对话框，依次输入名称（spark_tfrecord）、开发语言（选择 python 语言）、云平台集群，单击 Create 按钮创建 Notebook，如图 32-29 所示。

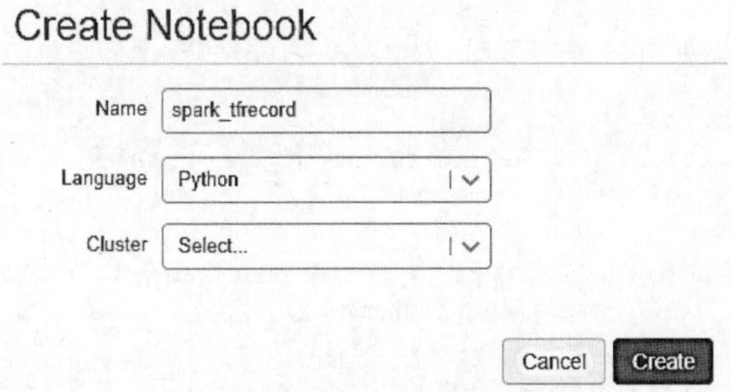

图 32-29　创建 Notebook

（2）导入 numpy、tensorflow 等库。

```
1.  import gzip
2.  import os
3.  import tempfile
4.
5.  import numpy
6.  from six.moves import urllib
7.
8.  import tensorflow as tf
```

（3）执行 Databricks 的 shell 命令，删除/dbfs/ml/MNISTDemo 目录及文件。

```
1.  %sh
2.  rm -rf  /dbfs/ml/MNISTDemo
```

（4）定义 Mnist 数据下载位置和保存 TFRecord 文件的位置。

```
1.  params = {}
2.  params['download_data_location'] = '/dbfs/ml/MNISTDemo/mnistData/'
3.  params['tfrecord_location'] = '/dbfs/ml/MNISTDemo/mnistData/'
```

（5）构建函数 download 用于下载 Mnist 数据集。

```
1.  def download(directory, filename):
2.    """如果还未完成，从 Mnist 数据集下载一个文件"""
3.    filepath = os.path.join(directory, filename)
4.    if tf.gfile.Exists(filepath):
5.      return filepath
6.    if not tf.gfile.Exists(directory):
7.      tf.gfile.MakeDirs(directory)
8.    # CVDF http://yann.lecun.com/exdb/mnist/的镜像
9.    url = 'https://storage.googleapis.com/cvdf-datasets/mnist/' + filename + '.gz'
10.   _, zipped_filepath = tempfile.mkstemp(suffix='.gz')
11.   print('Downloading %s to %s' % (url, zipped_filepath))
12.   urllib.request.urlretrieve(url, zipped_filepath)
13.   tf.gfile.Copy(zipped_filepath, filepath)
14.   os.remove(zipped_filepath)
15.   return filepath
```

（6）构建_read32、extract_images、extract_labels 函数。

```
1.  def _read32(bytestream):
2.    dt = numpy.dtype(numpy.uint32).newbyteorder('>')
3.    return numpy.frombuffer(bytestream.read(4), dtype=dt)[0]
4.
5.  def extract_images(f):
6.    """
7.    Extract the images into a 4D uint8 numpy array.
8.    """
9.    print('Extracting', f.name)
10.   with gzip.GzipFile(fileobj=f) as bytestream:
11.     magic = _read32(bytestream)
12.     if magic != 2051:
13.       raise ValueError('Invalid magic number %d in MNIST image file: %s' %
14.                        (magic, f.name))
15.     num_images = _read32(bytestream)
16.     rows = _read32(bytestream)
17.     cols = _read32(bytestream)
18.     buf = bytestream.read(rows * cols * num_images)
```

```
19.     data = numpy.frombuffer(buf, dtype=numpy.uint8)
20.     data = data.reshape(num_images, rows, cols, 1)
21.     return data
22.
23. def extract_labels(f, one_hot=False, num_classes=10):
24.   """
25.   Extract the labels into a 1D uint8 numpy array.
26.   """
27.   print('Extracting', f.name)
28.   with gzip.GzipFile(fileobj=f) as bytestream:
29.     magic = _read32(bytestream)
30.     if magic != 2049:
31.       raise ValueError('Invalid magic number %d in MNIST label file: %s' %
32.                        (magic, f.name))
33.     num_items = _read32(bytestream)
34.     buf = bytestream.read(num_items)
35.     labels = numpy.frombuffer(buf, dtype=numpy.uint8)
36.     return labels
```

（7）构建数据集下载函数 load_dataset。

```
1.  def load_dataset(directory, images_file, labels_file):
2.    """Download and parse MNIST dataset."""
3.
4.    images_file = download(directory, images_file)
5.    labels_file = download(directory, labels_file)
6.
7.    with tf.gfile.Open(images_file, 'rb') as f:
8.      images = extract_images(f)
9.      images = images.reshape(images.shape[0], images.shape[1] * images.shape[2])
10.     images = images.astype(numpy.float32)
11.     images = numpy.multiply(images, 1.0 / 255.0)
12.
13.   with tf.gfile.Open(labels_file, 'rb') as f:
14.     labels = extract_labels(f)
15.
16.   return images, labels
```

第 8 行代码获取 Mnist 图像数据，数据维度为(num_images, rows, cols, 1)。

第 9 行代码将图像数据转换为新的维度（数据集记录数，图像数据数 784(28*28)）。

第 11 行代码将图像数据进行归一化处理，将[0, 255)范围的数值调整为[0, 1)范围的浮点数。

（8）下载 Mnist 数据， 获取训练集、测试集。

```
1.   directory = params['download_data_location']
2.
3.  train_images, train_labels = load_dataset(directory, 'train-images-idx3-ubyte',
    'train-labels-idx1-ubyte')
4.  test_images, test_labels = load_dataset(directory, 't10k-images-idx3-ubyte',
    't10k-labels-idx1-ubyte')
```

在 Databricks 云平台 Notebook 中运行上述代码，运行结果如下：

```
Downloading  https://storage.googleapis.com/cvdf-datasets/mnist/train-images-
idx3-ubyte.gz to /tmp/tmp4e5hcd0_.gz
Downloading  https://storage.googleapis.com/cvdf-datasets/mnist/train-labels-
idx1-ubyte.gz to /tmp/tmp4w7f1q2c.gz
Extracting /dbfs/ml/MNISTDemo/mnistData/train-images-idx3-ubyte
Extracting /dbfs/ml/MNISTDemo/mnistData/train-labels-idx1-ubyte
```

```
Downloading  https://storage.googleapis.com/cvdf-datasets/mnist/t10k-images-idx3-ubyte.gz to /tmp/tmpwg3cfsbc.gz
Downloading  https://storage.googleapis.com/cvdf-datasets/mnist/t10k-labels-idx1-ubyte.gz to /tmp/tmp2jvr7jzs.gz
Extracting /dbfs/ml/MNISTDemo/mnistData/t10k-images-idx3-ubyte
Extracting /dbfs/ml/MNISTDemo/mnistData/t10k-labels-idx1-ubyte
```

（9）将数据保存到 Spark 数据帧。

```
1.   from pyspark.sql.types import *
2.  data = [(train_images[i].tolist(), int(train_labels[i])) for i in range
    (len(train_images))]
3.  print(data)
4.  schema = StructType([StructField("image", ArrayType(FloatType())),
5.                      StructField("label", LongType())])
6.  df = spark.createDataFrame(data, schema)
7.  df.show()
```

在 Databricks 云平台 Notebook 中运行上述代码，Data 为加载的 Mnist 图像训练集数据，是一个元组列表，元组列表中的每一条记录包含一张图片数据的信息，包含两个元素，元组的第一个元素是 784 大小的图像数据，元组的第二个元素是图像数据的标签，Data 的第一条数据记录如下，其表示的数字标签是 5。

```
[([0.0, 0.0, 0.0, 0.0, 0.0, 0.0, 0.0, 0.0, 0.0, 0.0, 0.0, 0.0, 0.0, 0.0, 0.0,
0.0, 0.0, 0.0, 0.0, 0.0, 0.0, 0.0, 0.0, 0.0, 0.0, 0.0, 0.0, 0.0, 0.0, 0.0,
0.0, 0.0, 0.0, 0.0, 0.0, 0.0, 0.0, 0.0, 0.0, 0.0, 0.0, 0.0, 0.0, 0.0, 0.0,
0.0, 0.0, 0.0, 0.0, 0.0, 0.0, 0.0, 0.0, 0.0, 0.0, 0.0, 0.0, 0.0, 0.0, 0.0,
0.0, 0.0, 0.0, 0.0, 0.0, 0.0, 0.0, 0.0, 0.0, 0.0, 0.0, 0.0, 0.0, 0.0, 0.0,
0.0, 0.0, 0.0, 0.0, 0.0, 0.0, 0.0, 0.0, 0.0, 0.0, 0.0, 0.0, 0.0, 0.0, 0.0,
0.0, 0.0, 0.0, 0.0, 0.0, 0.0, 0.0, 0.0, 0.0, 0.0, 0.0, 0.0, 0.0, 0.0, 0.0,
0.0, 0.0, 0.0, 0.0, 0.0, 0.0, 0.0, 0.0, 0.0, 0.0, 0.0, 0.0, 0.0, 0.0, 0.0,
0.0, 0.0, 0.0, 0.0, 0.0, 0.0, 0.0, 0.0, 0.0, 0.0, 0.0, 0.0, 0.0, 0.0, 0.0,
0.0, 0.0, 0.0, 0.0, 0.0, 0.0, 0.0, 0.0, 0.0, 0.0, 0.0, 0.0, 0.0, 0.0, 0.0,
0.0, 0.0, 0.011764707043766975, 0.07058823853731155, 0.07058823853731155,
0.07058823853731155, 0.4941176772117615, 0.5333333611488342, 0.686274528503418,
0.10196079313755035, 0.6509804129600525, 1.0, 0.9686275124549866, 0.4980392456054687 5,
0.0, 0.0, 0.0, 0.0, 0.0, 0.0, 0.0, 0.0, 0.0, 0.0, 0.0, 0.0, 0.11764706671237946,
0.1411764770746231, 0.3686274588108063, 0.6039215922355652, 0.6666666865348816,
0.9921569228172302, 0.9921569228172302, 0.9921569228172302, 0.9921569228172302,
0.9921569228172302, 0.8823530077934265, 0.6745098233222961, 0.9921569228172302,
0.9490196704864502, 0.7647059559822083, 0.250980406999588, 0.0, 0.0, 0.0, 0.0,
0.0, 0.0, 0.0, 0.0, 0.0, 0.0, 0.0, 0.19215688109397888, 0.9333333969116211,
0.9921569228172302, 0.9921569228172302, 0.9921569228172302, 0.9921569228172302,
0.9921569228172302, 0.9921569228172302, 0.9921569228172302, 0.9921569228172302,
0.9843137860298157, 0.364705890417099, 0.32156863808631897, 0.32156863808631897,
0.2196078598499298, 0.15294118225574493, 0.0, 0.0, 0.0, 0.0, 0.0, 0.0, 0.0,
0.0, 0.0, 0.0, 0.0, 0.0, 0.07058823853731155, 0.8588235974311829, 0.9921569228172302,
0.9921569228172302, 0.9921569228172302, 0.9921569228172302, 0.9921569228172302,
0.7764706611633301, 0.7137255072593689, 0.9686275124549866, 0.945098102092 7429,
0.0, 0.0, 0.0, 0.0, 0.0, 0.0, 0.0, 0.0, 0.0, 0.0, 0.0, 0.0, 0.0, 0.0, 0.0,
0.0, 0.0, 0.0, 0.3137255012989044, 0.6117647290229797, 0.41960787773132324,
0.9921569228172302, 0.9921569228172302, 0.803921639919281, 0.0431372560560703 3,
0.0, 0.16862745583057404, 0.6039215922355652, 0.0, 0.0, 0.0, 0.0, 0.0, 0.0,
0.0, 0.0, 0.0, 0.0, 0.0, 0.0, 0.0, 0.0, 0.0, 0.0, 0.0, 0.0, 0.0, 0.054901964962482 45,
0.003921568859368563, 0.6039215922355652, 0.9921569228172302, 0.3529411852359772,
0.0, 0.0, 0.0, 0.0, 0.0, 0.0, 0.0, 0.0, 0.0, 0.0, 0.0, 0.0, 0.0, 0.0, 0.0,
0.0, 0.0, 0.0, 0.0, 0.0, 0.0, 0.0, 0.0, 0.0, 0.0, 0.545098066329956,
0.9921569228172302, 0.7450980544090271, 0.007843137718737125, 0.0, 0.0, 0.0,
```

```
0.0, 0.0, 0.0, 0.0, 0.0, 0.0, 0.0, 0.0, 0.0, 0.0, 0.0, 0.0, 0.0, 0.0, 0.0,
0.0, 0.0, 0.0, 0.0, 0.0, 0.0, 0.04313725605607033, 0.7450980544090271,
0.9921569228172302, 0.27450981736183167, 0.0, 0.0, 0.0, 0.0, 0.0, 0.0, 0.0,
0.0, 0.0, 0.0, 0.0, 0.0, 0.0, 0.0, 0.0, 0.0, 0.0, 0.0, 0.0, 0.0, 0.0, 0.0,
0.0, 0.0, 0.0, 0.13725490868091583, 0.9450981020927429, 0.8823530077934265,
0.627451002597808, 0.4235294461250305, 0.003921568859368563, 0.0, 0.0, 0.0,
0.0, 0.0, 0.0, 0.0, 0.0, 0.0, 0.0, 0.0, 0.0, 0.0, 0.0, 0.0, 0.0, 0.0, 0.0, 0.0,
0.0, 0.0, 0.0, 0.0, 0.317647069692611 7, 0.9411765336990356, 0.9921569228172302,
0.9921569228172302, 0.4666666984558105 5, 0.09803922474384308, 0.0, 0.0, 0.0,
0.0, 0.0, 0.0, 0.0, 0.0, 0.0, 0.0, 0.0, 0.0, 0.0, 0.0, 0.0, 0.0, 0.0, 0.0, 0.0,
0.0, 0.0, 0.0, 0.0, 0.1764705926179886, 0.729411780834198, 0.9921569228172302,
0.9921569228172302, 0.5882353186607361, 0.10588236153125763, 0.0, 0.0, 0.0,
0.0, 0.0, 0.0, 0.0, 0.0, 0.0, 0.0, 0.0, 0.0, 0.0, 0.0, 0.0, 0.0, 0.0, 0.0, 0.0,
0.0, 0.0, 0.0, 0.0, 0.062745101749897, 0.364705890417099, 0.988235354423523,
0.9921569228172302, 0.7333333492279053, 0.0, 0.0, 0.0, 0.0, 0.0, 0.0, 0.0,
0.0, 0.0, 0.0, 0.0, 0.0, 0.0, 0.0, 0.0, 0.0, 0.0, 0.0, 0.0, 0.0, 0.0, 0.0, 0.0,
0.0, 0.9764706492424011, 0.9921569228172302, 0.9764706492424011, 0.250980406999588,
0.0, 0.0, 0.0, 0.0, 0.0, 0.0, 0.0, 0.0, 0.0, 0.0, 0.0, 0.0, 0.0, 0.0, 0.0, 0.0,
0.0, 0.0, 0.0, 0.0, 0.0, 0.0, 0.1803921610116958 6, 0.5098039507865906, 0.7176470756530762,
0.9921569228172302, 0.9921569228172302, 0.8117647767066956, 0.00784313771873712 5,
0.0, 0.0, 0.0, 0.0, 0.0, 0.0, 0.0, 0.0, 0.0, 0.0, 0.0, 0.0, 0.0, 0.0, 0.0, 0.0,
0.0, 0.0, 0.0, 0.15294118225574493, 0.5803921818733215, 0.8980392813682556,
0.9921569228172302, 0.9921569228172302, 0.9921569228172302, 0.9803922176361084,
0.7137255072593689, 0.0, 0.0, 0.0, 0.0, 0.0, 0.0, 0.0, 0.0, 0.0, 0.0, 0.0,
0.0, 0.0, 0.0, 0.0, 0.0, 0.0, 0.0, 0.0941176563501358, 0.44705885648727417,
0.8666667342185974, 0.9921569228172302, 0.9921569228172302, 0.9921569228172302,
0.9921569228172302, 0.7882353663444519, 0.30588236451148987, 0.0, 0.0, 0.0,
0.0, 0.0, 0.0, 0.0, 0.0, 0.0, 0.0, 0.0, 0.0, 0.0, 0.0, 0.0, 0.0, 0.0, 0.09019608050584793,
0.2588235437870025 6, 0.8352941870689392, 0.9921569228172302, 0.9921569228172302,
0.9921569228172302, 0.9921569228172302, 0.7764706611633301, 0.317647069692611 7,
0.00784313771873712 5, 0.0, 0.0, 0.0, 0.0, 0.0, 0.0, 0.0, 0.0, 0.0, 0.0, 0.0,
0.0, 0.0, 0.0, 0.0, 0.0, 0.07058823853731155, 0.6705882549285889, 0.8588235974311829,
0.9921569228172302, 0.9921569228172302, 0.9921569228172302, 0.9921569228172302,
0.7647059559822083, 0.3137255012989044, 0.03529411926865578, 0.0, 0.0, 0.0,
0.0, 0.0, 0.0, 0.0, 0.0, 0.0, 0.0, 0.0, 0.0, 0.0, 0.0, 0.0, 0.0, 0.2156862914562225 3,
0.6745098233222961, 0.8862745761871338, 0.9921569228172302, 0.9921569228172302,
0.9921569228172302, 0.9921569228172302, 0.9568628072738647, 0.5215686559677124,
0.04313725605607033, 0.0, 0.0, 0.0, 0.0, 0.0, 0.0, 0.0, 0.0, 0.0, 0.0, 0.0, 0.0,
0.0, 0.0, 0.0, 0.0, 0.0, 0.0, 0.5333333611488342, 0.9921569228172302, 0.9921569228172302,
0.9921569228172302, 0.8313726186752319, 0.529411792755127, 0.5176470875740051,
0.062745101749897, 0.0, 0.0, 0.0, 0.0, 0.0, 0.0, 0.0, 0.0, 0.0, 0.0, 0.0, 0.0,
0.0, 0.0, 0.0, 0.0, 0.0, 0.0, 0.0, 0.0, 0.0, 0.0, 0.0, 0.0, 0.0, 0.0, 0.0, 0.0,
0.0, 0.0, 0.0, 0.0, 0.0, 0.0, 0.0, 0.0, 0.0, 0.0, 0.0, 0.0, 0.0, 0.0, 0.0, 0.0,
0.0, 0.0, 0.0, 0.0, 0.0, 0.0, 0.0, 0.0, 0.0, 0.0, 0.0, 0.0, 0.0, 0.0, 0.0, 0.0,
0.0, 0.0, 0.0, 0.0, 0.0, 0.0, 0.0, 0.0, 0.0, 0.0, 0.0, 0.0, 0.0, 0.0, 0.0, 0.0,
0.0, 0.0, 0.0, 0.0, 0.0, 0.0, 0.0, 0.0, 0.0, 0.0, 0.0, 0.0, 0.0, 0.0, 0.0, 0.0,
0.0, 0.0, 0.0, 0.0, 0.0, 0.0, 0.0, 0.0, 0.0, 0.0, 0.0, 0.0, 0.0], 5), ……
```

df.show()的运行结果如下，第 1 列是每张图片的图像数据，第 2 列是每张图片对应的数字标签。

```
+--------------------+-----+
|               image|label|
+--------------------+-----+
|[0.0, 0.0, 0.0, 0...|    5|
|[0.0, 0.0, 0.0, 0...|    0|
|[0.0, 0.0, 0.0, 0...|    4|
|[0.0, 0.0, 0.0, 0...|    1|
|[0.0, 0.0, 0.0, 0...|    9|
```

```
|[0.0, 0.0, 0.0, 0...|    2|
|[0.0, 0.0, 0.0, 0...|    1|
|[0.0, 0.0, 0.0, 0...|    3|
|[0.0, 0.0, 0.0, 0...|    1|
|[0.0, 0.0, 0.0, 0...|    4|
|[0.0, 0.0, 0.0, 0...|    3|
|[0.0, 0.0, 0.0, 0...|    5|
|[0.0, 0.0, 0.0, 0...|    3|
|[0.0, 0.0, 0.0, 0...|    6|
|[0.0, 0.0, 0.0, 0...|    1|
|[0.0, 0.0, 0.0, 0...|    7|
|[0.0, 0.0, 0.0, 0...|    2|
|[0.0, 0.0, 0.0, 0...|    8|
|[0.0, 0.0, 0.0, 0...|    6|
|[0.0, 0.0, 0.0, 0...|    9|
+--------------------+-----+
only showing top 20 rows
```

调用 Spark 的 take(1)方法，从 Spark 数据帧中获取第 1 条记录。

```
1.  df.take(1)
```

在 Databricks 云平台 Notebook 中运行上述代码，运行结果如下，第一条记录是一个 Spark Row 记录，第一列是图像数据，数据大小为 784；第二列是数字标签，表示数字 5。

```
Out[10]: [Row(image=[0.0, 0.0, 0.0, 0.0, 0.0, 0.0, 0.0, 0.0, 0.0, 0.0, 0.0,
0.0, 0.0, 0.0, 0.0, 0.0, 0.0, 0.0, 0.0, 0.0, 0.0, 0.0, 0.0, 0.0, 0.0, 0.0,
0.0, 0.0, 0.0, 0.0, 0.0, 0.0, 0.0, 0.0, 0.0, 0.0, 0.0, 0.0, 0.0, 0.0, 0.0,
0.0, 0.0, 0.0, 0.0, 0.0, 0.0, 0.0, 0.0, 0.0, 0.0, 0.0, 0.0, 0.0, 0.0, 0.0,
0.0, 0.0, 0.0, 0.0, 0.0, 0.0, 0.0, 0.0, 0.0, 0.0, 0.0, 0.0, 0.0, 0.0, 0.0,
0.0, 0.0, 0.0, 0.0, 0.0, 0.0, 0.0, 0.0, 0.0, 0.0, 0.0, 0.0, 0.0, 0.0, 0.0,
0.0, 0.0, 0.0, 0.0, 0.0, 0.0, 0.0, 0.0, 0.0, 0.0, 0.0, 0.0, 0.0, 0.0, 0.0,
0.0, 0.0, 0.0, 0.0, 0.0, 0.0, 0.0, 0.0, 0.0, 0.0, 0.0, 0.0, 0.0, 0.0, 0.0,
0.0, 0.0, 0.0, 0.0, 0.0, 0.0, 0.0, 0.0, 0.0, 0.0, 0.0, 0.0, 0.0, 0.0, 0.0,
0.0, 0.0, 0.0, 0.0, 0.0, 0.0, 0.0, 0.0, 0.0, 0.0, 0.0, 0.0, 0.0, 0.0, 0.0,
0.0, 0.0, 0.0, 0.0, 0.0, 0.0, 0.011764707043766975, 0.07058823853731155,
0.07058823853731155, 0.07058823853731155, 0.4941176772117615, 0.5333333611488342,
0.686274528503418, 0.10196079313755035, 0.6509804129600525, 1.0, 0.9686275124549866,
0.49803924560546875, 0.0, 0.0, 0.0, 0.0, 0.0, 0.0, 0.0, 0.0, 0.0, 0.0, 0.0,
0.11764706671237946, 0.1411764770746231, 0.3686274588108063, 0.6039215922355652,
0.6666666865348816, 0.9921569228172302, 0.9921569228172302, 0.9921569228172302,
0.9921569228172302, 0.9921569228172302, 0.8823530077934265, 0.6745098233222961,
0.9921569228172302, 0.9490196704864502, 0.7647059559822083, 0.250980406999588,
0.0, 0.0, 0.0, 0.0, 0.0, 0.0, 0.0, 0.0, 0.0, 0.0, 0.0, 0.19215688109397888,
0.9333333969116211, 0.9921569228172302, 0.9921569228172302, 0.9921569228172302,
0.9921569228172302, 0.9921569228172302, 0.9921569228172302, 0.9921569228172302,
0.9921569228172302, 0.9843137860298157, 0.364705890417099, 0.32156863808631897,
0.32156863808631897, 0.2196078598499298, 0.15294118225574493, 0.0, 0.0, 0.0,
0.0, 0.0, 0.0, 0.0, 0.0, 0.0, 0.0, 0.0, 0.0, 0.07058823853731155, 0.8588235974311829,
0.9921569228172302, 0.9921569228172302, 0.9921569228172302, 0.9921569228172302,
0.9921569228172302, 0.7764706611633301, 0.7137255072593689, 0.9686275124549866,
0.9450981020927429, 0.0, 0.0, 0.0, 0.0, 0.0, 0.0, 0.0, 0.0, 0.0, 0.0, 0.0,
0.0, 0.0, 0.0, 0.0, 0.0, 0.0, 0.0, 0.3137255012989044, 0.6117647290229797,
0.41960787773132324, 0.9921569228172302, 0.9921569228172302, 0.803921639919281,
0.04313725605607033, 0.0, 0.16862745583057404, 0.6039215922355652, 0.0, 0.0,
0.0, 0.0, 0.0, 0.0, 0.0, 0.0, 0.0, 0.0, 0.0, 0.0, 0.0, 0.0, 0.0, 0.0, 0.0,
0.0, 0.0, 0.05490196496248245, 0.003921568859368563, 0.6039215922355652,
0.9921569228172302, 0.3529411852359772, 0.0, 0.0, 0.0, 0.0, 0.0, 0.0, 0.0,
0.0, 0.0, 0.0, 0.0, 0.0, 0.0, 0.0, 0.0, 0.0, 0.0, 0.0, 0.0, 0.0, 0.0, 0.0,
```

```
0.0, 0.0, 0.0, 0.545098066329956, 0.9921569228172302, 0.7450980544090271,
0.007843137718737125, 0.0, 0.0, 0.0, 0.0, 0.0, 0.0, 0.0, 0.0, 0.0, 0.0, 0.0, 0.0,
0.0, 0.0, 0.0, 0.0, 0.0, 0.0, 0.0, 0.0, 0.0, 0.0, 0.0, 0.0, 0.04313725605607033,
0.7450980544090271, 0.9921569228172302, 0.27450981736183167, 0.0, 0.0, 0.0,
0.0, 0.0, 0.0, 0.0, 0.0, 0.0, 0.0, 0.0, 0.0, 0.0, 0.0, 0.0, 0.0, 0.0, 0.0,
0.0, 0.0, 0.0, 0.0, 0.0, 0.0, 0.0, 0.13725490868091583, 0.9450981020927429,
0.8823530077934265, 0.6274510025978088, 0.4235294461250305, 0.003921568859368563,
0.0, 0.0, 0.0, 0.0, 0.0, 0.0, 0.0, 0.0, 0.0, 0.0, 0.0, 0.0, 0.0, 0.0, 0.0,
0.0, 0.0, 0.0, 0.0, 0.0, 0.0, 0.0, 0.0, 0.3176470696926117, 0.9411765336990356,
0.9921569228172302, 0.9921569228172302, 0.46666669845581055, 0.09803922474384308,
0.0, 0.0, 0.0, 0.0, 0.0, 0.0, 0.0, 0.0, 0.0, 0.0, 0.0, 0.0, 0.0, 0.0, 0.0,
0.0, 0.0, 0.0, 0.0, 0.0, 0.0, 0.0, 0.0, 0.1764705926179886, 0.729411780834198,
0.9921569228172302, 0.9921569228172302, 0.5882353186607361, 0.10588236153125763,
0.0, 0.0, 0.0, 0.0, 0.0, 0.0, 0.0, 0.0, 0.0, 0.0, 0.0, 0.0, 0.0, 0.0, 0.0,
0.0, 0.0, 0.0, 0.0, 0.0, 0.0, 0.0, 0.0, 0.062745101749897, 0.364705890417099,
0.988235354423523, 0.9921569228172302, 0.7333333492279053, 0.0, 0.0, 0.0, 0.0,
0.0, 0.0, 0.0, 0.0, 0.0, 0.0, 0.0, 0.0, 0.0, 0.0, 0.0, 0.0, 0.0, 0.0, 0.0,
0.0, 0.0, 0.0, 0.0, 0.0, 0.0, 0.9764706492424011, 0.9921569228172302,
0.9764706492424011, 0.250980406999588, 0.0, 0.0, 0.0, 0.0, 0.0, 0.0, 0.0,
0.0, 0.0, 0.0, 0.0, 0.0, 0.0, 0.0, 0.0, 0.0, 0.0, 0.0, 0.0, 0.0,
0.18039216101169586, 0.5098039507865906, 0.7176470756530762, 0.9921569228172302,
0.9921569228172302, 0.8117647767066956, 0.007843137718737125, 0.0, 0.0, 0.0,
0.0, 0.0, 0.0, 0.0, 0.0, 0.0, 0.0, 0.0, 0.0, 0.0, 0.0, 0.0, 0.0, 0.0, 0.0,
0.15294118225574493, 0.5803921818733215, 0.8980392813682556, 0.9921569228172302,
0.9921569228172302, 0.9921569228172302, 0.9803922176361084, 0.7137255072593689,
0.0, 0.0, 0.0, 0.0, 0.0, 0.0, 0.0, 0.0, 0.0, 0.0, 0.0, 0.0, 0.0, 0.0, 0.0,
0.0, 0.0, 0.0, 0.0941176563501358, 0.44705885648727417, 0.8666667342185974,
0.9921569228172302, 0.9921569228172302, 0.9921569228172302, 0.9921569228172302,
0.7882353663444519, 0.30588236451148987, 0.0, 0.0, 0.0, 0.0, 0.0, 0.0, 0.0,
0.0, 0.0, 0.0, 0.0, 0.0, 0.0, 0.0, 0.0, 0.0, 0.0, 0.09019608050584793,
0.25882354378700256, 0.8352941870689392, 0.9921569228172302, 0.9921569228172302,
0.9921569228172302, 0.9921569228172302, 0.7764706611633301, 0.3176470696926117,
0.007843137718737125, 0.0, 0.0, 0.0, 0.0, 0.0, 0.0, 0.0, 0.0, 0.0, 0.0, 0.0, 0.0,
0.0, 0.0, 0.0, 0.0, 0.07058823853731155, 0.6705882549285889, 0.8588235974311829,
0.9921569228172302, 0.9921569228172302, 0.9921569228172302, 0.9921569228172302,
0.7647059559822083, 0.3137255012989044, 0.03529411926865578, 0.0, 0.0, 0.0,
0.0, 0.0, 0.0, 0.0, 0.0, 0.0, 0.0, 0.0, 0.0, 0.0, 0.0, 0.0, 0.0, 0.21568629145622253,
0.6745098233222961, 0.8862745761871338, 0.9921569228172302, 0.9921569228172302,
0.9921569228172302, 0.9921569228172302, 0.9568628072738647, 0.5215686559677124,
0.04313725605607033, 0.0, 0.0, 0.0, 0.0, 0.0, 0.0, 0.0, 0.0, 0.0, 0.0, 0.0, 0.0,
0.0, 0.0, 0.0, 0.0, 0.0, 0.0, 0.5333333611488342, 0.9921569228172302, 0.9921569228172302,
0.9921569228172302, 0.8313726186752319, 0.529411792755127, 0.5176470875740051,
0.062745101749897, 0.0, 0.0, 0.0, 0.0, 0.0, 0.0, 0.0, 0.0, 0.0, 0.0, 0.0, 0.0,
0.0, 0.0, 0.0, 0.0, 0.0, 0.0, 0.0, 0.0, 0.0, 0.0, 0.0, 0.0, 0.0, 0.0, 0.0,
0.0, 0.0, 0.0, 0.0, 0.0, 0.0, 0.0, 0.0, 0.0, 0.0, 0.0, 0.0, 0.0, 0.0, 0.0,
0.0, 0.0, 0.0, 0.0, 0.0, 0.0, 0.0, 0.0, 0.0, 0.0, 0.0, 0.0, 0.0, 0.0, 0.0,
0.0, 0.0, 0.0, 0.0, 0.0, 0.0, 0.0, 0.0, 0.0, 0.0, 0.0, 0.0, 0.0, 0.0, 0.0,
0.0, 0.0, 0.0, 0.0, 0.0, 0.0, 0.0, 0.0, 0.0, 0.0, 0.0, 0.0, 0.0, 0.0, 0.0,
0.0, 0.0, 0.0, 0.0, 0.0, 0.0, 0.0, 0.0, 0.0, 0.0, 0.0, 0.0, 0.0], label=5)]
```

（10）将 Spark 数据帧保存为 TFRecord 格式的文件。

```
1.  path = 'file:' + params['tfrecord_location'] + 'df-mnist.tfrecord'
2.  num_partition = 4
3.  df.repartition(num_partition).write.format("tfrecords")
    .mode("overwrite").save(path)
```

第 2 行代码设置分区数为 4。

第 3 行代码将 Spark 数据帧重新分区为 4 个分区，按照 TFRecord 格式覆盖写入 Databricks

文件系统。

（11）在 Databricks 文件系统中查询保存的 TFRecord 格式文件。

```
1.  display(dbutils.fs.ls(path))
```

在 Databricks 云平台 Notebook 中运行上述代码，运行结果如图 32-30 所示，保存为 4 个分区的 TFRcord 格式的文件，分别为 part-r-00000、part-r-00001、part-r-00002、part-r-00003。

```
display(dbutils.fs.ls(path))
```

path	name	size
file:/dbfs/ml/MNISTDemo/mnistData/df-mnist.tfrecord/_SUCCESS	_SUCCESS	0
file:/dbfs/ml/MNISTDemo/mnistData/df-mnist.tfrecord/part-r-00000	part-r-00000	47850000
file:/dbfs/ml/MNISTDemo/mnistData/df-mnist.tfrecord/part-r-00001	part-r-00001	47850000
file:/dbfs/ml/MNISTDemo/mnistData/df-mnist.tfrecord/part-r-00002	part-r-00002	47850000
file:/dbfs/ml/MNISTDemo/mnistData/df-mnist.tfrecord/part-r-00003	part-r-00003	47850000

图 32-30　Spark 保存为 TFRecord 格式文件

3. TensorFlow加载TFRecord文件案例

（1）创建 Databricks 云平台集群 Notebook。

在 Workspace 栏单击右键，在弹出的快捷菜单中选择 Create，弹出 Create Notebook 对话框，依次输入名称（loadTFRecord）、开发语言（选择 Python 语言）、云平台集群，单击 Create 按钮创建 Notebook，如图 32-31 所示。

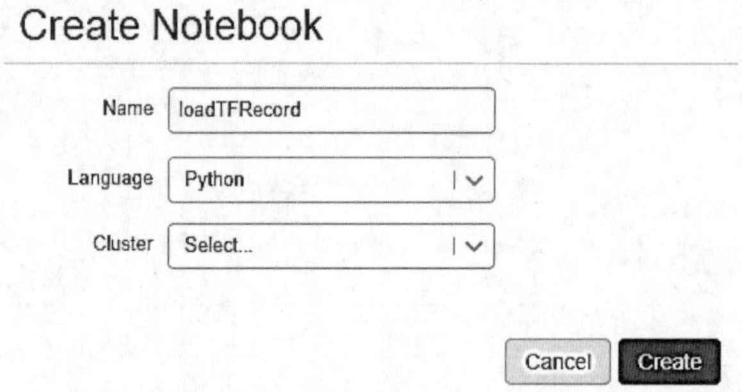

图 32-31　创建 Notebook

（2）导入 os、tensorflow 库。

```
1.  import os
2.  import tensorflow as tf
```

（3）获取 TFRecord 文件的目录及文件路径。

将 Mnist 训练集数据保存在 Databricks 文件系统的 dbfs:/ml 目录，dbfs:/ml 映射到 Driver 节点和 Worker 节点上的文件/dbfs/ml。Databricks 文件系统的 dbfs:/ml 是一个特殊的文件夹，为深度学习工作负载提供高性能的 I/O。

这里使用“TensorFlow 框架将数据保存为 TFRecord 文件”案例中已经保存 TFRecord 文件。

```
1.  tfrecord_location = '/dbfs/ml/MNISTDemo/mnistData/'
2.  name = "train.tfrecords"
3.  filename = os.path.join(tfrecord_location, name)
```

在 Databricks 文件系统中查询文件如下：

```
1.  print(filename)
2.  dbutils.fs.ls("file:"+tfrecord_location)
```

在 Databricks 云平台 Notebook 中运行上述代码，运行结果如下：

```
/dbfs/ml/MNISTDemo/mnistData/train.tfrecords
Out[22]: [FileInfo(path='file:/dbfs/ml/MNISTDemo/mnistData/t10k-images-idx3-ubyte',
name='t10k-images-idx3-ubyte', size=1648877),
 FileInfo(path='file:/dbfs/ml/MNISTDemo/mnistData/t10k-labels-idx1-ubyte',
name='t10k-labels-idx1-ubyte', size=4542),
 FileInfo(path='file:/dbfs/ml/MNISTDemo/mnistData/test.tfrecords', name='test.
tfrecords', size=8420000),
 FileInfo(path='file:/dbfs/ml/MNISTDemo/mnistData/train-images-idx3-ubyte',
name='train-images-idx3-ubyte', size=9912422),
 FileInfo(path='file:/dbfs/ml/MNISTDemo/mnistData/train-labels-idx1-ubyte',
name='train-labels-idx1-ubyte', size=28881),
 FileInfo(path='file:/dbfs/ml/MNISTDemo/mnistData/train.tfrecords',
name='train.tfrecords', size=46310000)]
```

（4）创建一个 TFRecordDataset 作为输入管道。

```
1.  dataset = tf.data.TFRecordDataset(filename)
```

（5）定义一个解码器来读取和解析数据。

```
1.  def decode(serialized_example):
2.    """
3.    分析给定 serialized_example 中的图像和标签，用作 dataset.map 的映射函数
4.    """
5.    IMAGE_SIZE = 28
6.    IMAGE_PIXELS = IMAGE_SIZE * IMAGE_SIZE
7.
8.    # 1. 定义一个分析器
9.    features = tf.parse_single_example(
10.       serialized_example,
11.       #由于两个 key 都是必需的，因此未指定默认值
12.       features={
13.           'image_raw': tf.FixedLenFeature([], tf.string),
14.           'label': tf.FixedLenFeature([], tf.int64),
15.       })
16.
17.   # 2. 转换数据
18.   image = tf.decode_raw(features['image_raw'], tf.uint8)
19.   label = tf.cast(features['label'], tf.int32)
20.   # 3. 调整维度
21.   image.set_shape((IMAGE_PIXELS))
22.   return image, label
```

（6）使用 map 方法把记录解析成 tensors 张量，map 方法接受一个 python 函数解码并将其应用于每个示例。

```
1.  dataset = dataset.map(decode)
```

（7）构建一个函数 normalize 来预处理数据。

```
1.  def normalize(image, label):
2.    """将图像数据从 [0, 255] 调整为[-0.5, 0.5] 浮点数 """
```

```
3.   image = tf.cast(image, tf.float32) * (1. / 255) - 0.5
4.   return image, label
```

（8）构建数据集。

```
1.  dataset = dataset.map(normalize)
2.  batch_size = 1000
3.  dataset = dataset.shuffle(1000 + 3 * batch_size)
4.  dataset = dataset.repeat(2)
5.  dataset = dataset.batch(batch_size)
```

第 1 行代码调用 dataset.map 方法接受一个 python 函数 normalize 并将其应用于每个示例。

第 3 行代码调用 dataset.shuffle 方法对数据集的数据进行洗牌（shuffle）。

第 4 行代码调用 dataset.repeat 方法指定数据集重复的次数。

第 5 行代码调用 dataset.batch 方法从数据集中取出批处理数据的个数。

（9）创建一个迭代器。

```
1.  iterator = dataset.make_one_shot_iterator()
2.  image_batch, label_batch = iterator.get_next()
```

（10）使用 TFRecord 文件作为深度学习框架的训练数据，可以将其用于模型训练。

```
1.  sess = tf.Session()
2.  image_batch, label_batch = sess.run([image_batch, label_batch])
3.  print(image_batch.shape)
4.  print(label_batch.shape)
```

在 Databricks 云平台 Notebook 中运行上述代码， Mnist 数据集每次获取一个批次数据，是一个 1000 行 784 列的图像数据，Mnist 标签数据每次获取一个批次数据，是一个 1000 行 1 列的标签数据，运行结果如下：

```
(1000, 784)
(1000,)
```

查询 Mnist 数据集的图像数据。

```
1.  print(image_batch)
```

在 Databricks 云平台 Notebook 中运行上述代码，运行结果如下：

```
[[-0.5 -0.5 -0.5 ... -0.5 -0.5 -0.5]
 [-0.5 -0.5 -0.5 ... -0.5 -0.5 -0.5]
 [-0.5 -0.5 -0.5 ... -0.5 -0.5 -0.5]
 ...
 [-0.5 -0.5 -0.5 ... -0.5 -0.5 -0.5]
 [-0.5 -0.5 -0.5 ... -0.5 -0.5 -0.5]
 [-0.5 -0.5 -0.5 ... -0.5 -0.5 -0.5]]
```

查询 Mnist 数据集的标签数据。

```
1.  print(label_batch)
```

在 Databricks 云平台 Notebook 中运行上述代码，运行结果如下：

```
[2 3 6 3 3 1 5 0 4 1 5 4 0 2 0 9 5 9 0 4 3 0 6 7 6 6 0 6 3 6 5 9 6 1 2 9 7
1 1 2 1 8 9 3 5 1 7 6 8 7 8 0 1 0 8 5 6 2 1 1 9 7 2 8 2 7 0 1 7 9 4 5 7 2 1
4 2 6 9 3 1 0 8 8 8 2 8 8 3 4 3 7 6 0 6 0 8 7 0 7 4 2 8 7 1 6 9 5 7 7 8 6 6
```

```
1 2 0 7 1 3 4 4 9 8 3 2 6 7 4 1 9 0 4 8 1 4 3 9 6 8 6 4 1 4 9 6 1 2 3 0 3 0
2 2 9 3 0 6 2 0 0 3 4 5 2 5 0 3 0 5 7 2 6 8 7 4 4 7 0 7 5 6 7 5 3 1 2 3 0 4
0 2 3 1 0 5 2 8 3 0 8 1 5 1 8 4 1 9 7 2 2 2 0 7 5 1 5 6 3 8 5 6 5 0 4 4 8 0
9 8 1 3 6 8 0 9 1 3 5 3 3 3 2 4 1 9 1 8 2 3 7 8 7 4 0 9 2 9 9 3 6 3 8 6 6 1
3 1 1 7 8 1 2 2 4 5 8 6 5 1 2 6 3 2 2 8 6 2 8 0 8 7 9 6 0 0 4 8 5 3 5 8 4 8
7 6 5 7 0 0 0 9 2 1 2 1 4 7 9 0 7 9 9 2 7 6 2 5 8 6 7 5 0 3 4 8 9 0 1 5 1 3
5 2 6 0 3 8 0 5 6 1 6 2 8 8 7 6 1 1 3 7 7 7 6 3 4 2 3 3 6 8 3 1 5 3 1 8 7 2
6 5 7 8 8 6 0 1 1 7 3 7 9 4 2 0 7 2 3 4 9 3 0 9 1 5 3 8 1 4 6 9 6 1 6 4 3 2
2 4 6 5 9 6 6 5 5 6 1 3 4 9 6 6 9 9 0 8 4 2 2 3 5 7 0 7 1 3 1 4 5 7 4 8 3 7
7 5 7 5 9 6 0 6 7 8 7 1 1 1 5 5 7 3 2 0 5 2 1 2 4 6 6 2 4 6 2 8 4 3 2 1 6 0
7 0 5 8 6 6 6 5 6 0 6 4 8 2 0 6 9 1 7 2 2 1 2 7 2 9 6 1 5 7 5 9 6 3 9 3 4 1
9 5 7 6 7 5 8 2 6 2 6 7 7 0 3 9 4 0 5 2 8 9 9 9 6 5 3 3 0 2 5 0 0 3 7 6 4 1
7 5 1 2 9 9 0 0 6 0 6 5 2 7 6 8 7 7 5 3 8 2 3 2 6 1 1 7 6 7 1 2 3 0 8 3 3 5
1 0 0 7 8 2 1 8 5 0 4 0 8 8 0 2 0 0 4 4 8 4 6 5 8 2 6 8 3 7 1 1 6 1 0 4 6 9
4 6 3 8 0 1 6 7 6 7 8 1 7 7 7 5 9 8 1 7 2 0 4 8 2 7 3 5 9 1 0 1 0 7 0 4 3 7
0 6 1 8 7 0 7 4 2 4 8 9 8 4 5 9 5 1 6 9 9 7 3 0 5 4 8 6 1 2 8 2 9 1 1 2 8 3
0 6 4 7 0 4 7 7 2 4 1 6 3 9 6 3 8 0 1 2 7 3 3 9 8 0 2 0 2 9 5 7 4 6 2 2 8 3
7 2 2 7 5 8 1 1 1 3 0 3 9 7 0 0 8 7 3 7 7 7 9 4 4 2 6 5 0 9 4 4 1 3 1 7 3 7
5 8 9 9 8 4 7 9 2 9 2 0 8 2 8 2 1 3 5 3 0 1 2 2 1 4 5 3 0 2 4 0 6 0 1 1 1 4
9 2 2 1 2 5 0 4 6 2 6 1 3 8 7 6 8 6 1 1 1 8 7 9 2 3 7 6 4 0 4 5 0 4 8 8 1 7
8 1 0 5 7 7 6 6 7 7 0 4 1 7 4 6 8 4 3 1 1 0 0 9 2 7 6 7 2 5 1 5 0 5 5 3 2 2
7 3 8 1 3 3 3 9 9 8 1 8 1 5 1 7 2 2 9 8 1 3 2 7 6 8 3 3 6 3 0 5 3 8 5 1 7 2
2 6 8 4 2 1 1 4 9 9 9 2 3 9 3 0 8 8 8 9 3 4 1 1 3 8 7 4 7 3 6 6 0 1 0 0 2 3
8 1 4 4 2 5 6 7 4 5 6 6 4]
```

32.1.3　单节点深度学习训练

Databricks 机器学习运行时库支持的深度学习框架包括 TensorFlow、PyTorch、Keras、XGBoost 等，本节讲解 PyTorch 在 Databricks GPU 单节点上实现 Mnist 手写数字识别的案例实战。

（1）创建 Databricks 云平台的 GPU 单节点集群。

在创建 Databricks 云平台单节点 GPU 集群时，可以仅设置 1 个 Spark Driver 节点，0 个 Spark Worker 节点，运行单节点的 Pytorch 程序，这是一种经济高效的方法。注意：Spark 在此设置下不起作用，因为 Spark 集群至少需要一个 Spark Worker 节点来运行 Spark 命令或导入表，在此设置下如果要运行 Spark 命令，Databricks 云平台集群会自动启动 1 个 Spark Worker 节点及 1 个 Spark Driver 节点。

如图 32-32 所示，创建单节点集群（DriverOnlyCluster），选择 Databricks 运行时库(Databricks Runtime Version)为 GPU 的版本（Runtime: 5.5 LTS ML (GPU, Scala 2.11, Spark 2.4.3)），最小 Spark Worker 节点数（Min Workers）为 0，最大 Spark Worker 节点数（Max Workers）为 1，按需确定节点的数量，初始 Workers 节点数设置为 0。

如图 32-33 所示，查询 Databricks 云平台的 Spark UI 页面，集群中仅仅运行一个 Spark Driver 节点。

（2）创建 Databricks 云平台集群 Notebook。

在 Workspace 栏单击右键，在弹出的快捷菜单中选择 Create，弹出“Create Notebook”对话框，依次输入名称（SingleNodeTraining）、开发语言（选择 Python 语言）、云平台集群(DriverOnlyCluster)，单击“Create”按钮创建 Notebook，如图 32-34 所示。

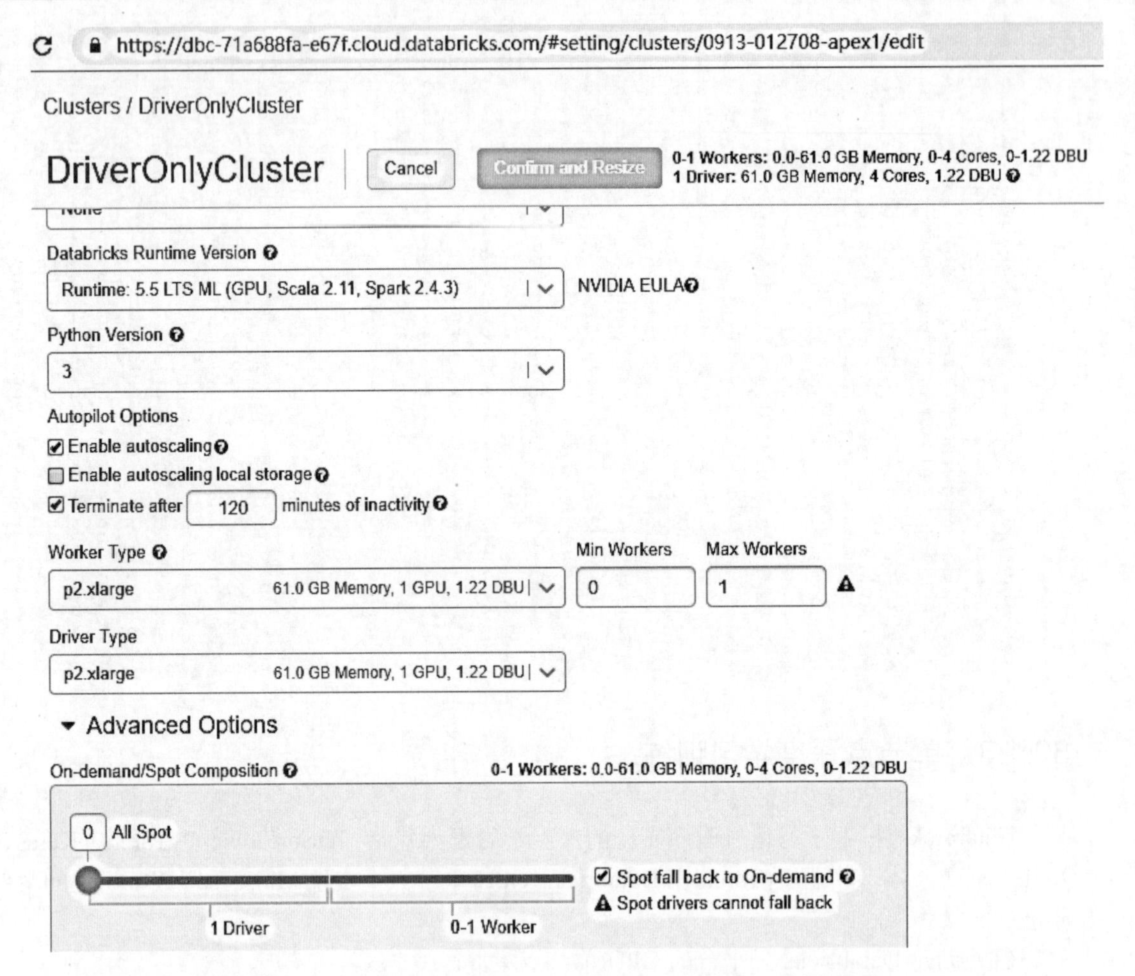

图 32-32　创建 1 个 Driver 节点、0 个 Worker 节点的单节点集群

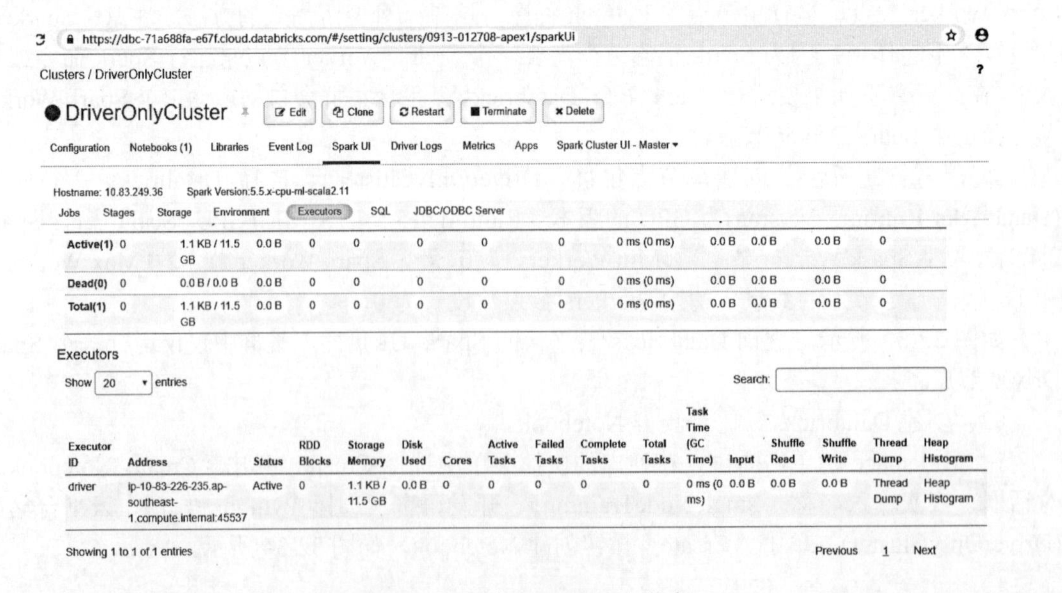

图 32-33　仅一个 Spark Driver 节点

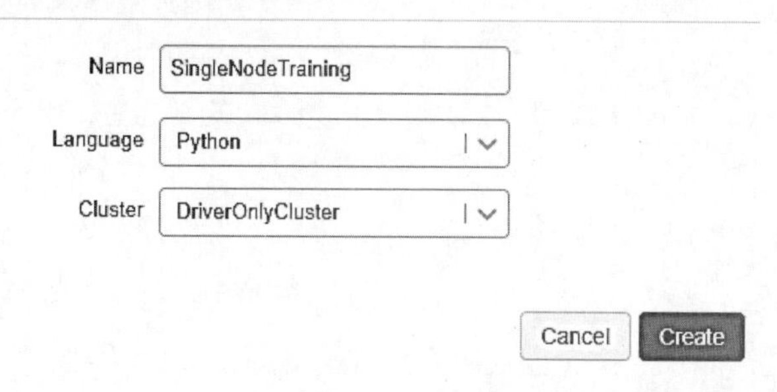

图 32-34　创建 Notebook

（3）导入 PyTorch 的 torch、torchvision 等库。

```
1.  from __future__ import absolute_import
2.  from __future__ import division
3.  from __future__ import print_function
4.  from __future__ import unicode_literals
5.
6.  from collections import namedtuple
7.
8.  import torch
9.  from torch.autograd import Variable
10. import torch.nn as nn
11. import torch.nn.functional as F
12. import torch.optim as optim
13. from torchvision import datasets, transforms
```

（4）设置深度学习模型的批处理大小、学习率、是否使用 GPU 等参数。

```
1.   MNIST_DIR = '/tmp/data/mnist'
2.
3.  Params = namedtuple('Params', ['batch_size', 'test_batch_size', 'epochs',
    'lr', 'momentum', 'seed', 'cuda', 'log_interval'])
4.   args = Params(batch_size=64, test_batch_size=1000, epochs=10, lr=0.01,
    momentum=0.5, seed=1, cuda=True, log_interval=200)
```

第 4 行代码设置深度学习模型的参数。

- batch_size：小批量训练集中的记录数。
- test_batch_size：小批量测试集的记录数。test_batch_size 通常大于 batch_size，因为在测试集预测期间不需要执行反向传播算法。
- epochs：所有训练集数据迭代遍历的次数。
- lr：学习率。
- momentum：动量因子（默认值为 0）。将使用学习率和动量因子动量地随机梯度下降算法。
- cuda：设置为 True，将使用 GPU 资源。

（5）下载 Mnist 数据集，对数据集打散洗牌，划分批处理数据，并对图像特征进行标准化处理。

```
1.  torch.manual_seed(args.seed)
2.
3.  data_transform_fn = transforms.Compose([
4.      transforms.ToTensor(),
5.      transforms.Normalize((0.1307,), (0.3081,))])
6.
7.  train_loader = torch.utils.data.DataLoader(
8.          datasets.MNIST(MNIST_DIR, train=True, download=True,
9.                         transform=data_transform_fn),
10.         batch_size=args.batch_size, shuffle=True, num_workers=1)
11.
12. test_loader = torch.utils.data.DataLoader(
13.         datasets.MNIST(MNIST_DIR, train=False,
14.                        transform=data_transform_fn),
15.         batch_size=args.test_batch_size, shuffle=True, num_workers=1)
```

第 3 行代码使用 transforms.Compose 方法对图像数据进行数据增强，将图像数据转为 Tensor 结构、数据标准化的步骤整合在一起处理。

第 4 行代码将图像数据转换为 Tensor 数据结构。

第 5 行代码将图像数据进行标准化处理。

第 7 行代码构建训练集的批数据迭代器。torch.utils.data.DataLoader 是一个 Pytorch 数据加载器，组合数据集和采样，在给定数据集上提供迭代器。DataLoader 支持 map-style 和 iterable-style 的数据集，支持单进程或多进程加载、自定义加载顺序、可选的自动批处理（排序）和内存固定。

第 12 行代码构建测试集的批次数据迭代器。

在 Databricks 云平台 Notebook 中运行上述代码，运行结果如下：

```
Downloading http://yann.lecun.com/exdb/mnist/train-images-idx3-ubyte.gz to
/tmp/data/mnist/MNIST/raw/train-images-idx3-ubyte.gz

0it [00:00, ?it/s]
  0%|          | 0/9912422 [00:00<?, ?it/s]
  0%|          | 16384/9912422 [00:01<02:50, 58138.16it/s]
  0%|          | 40960/9912422 [00:01<02:29, 66107.33it/s]
  1%|          | 98304/9912422 [00:01<01:56, 84026.60it/s]
  2%|▏         | 155648/9912422 [00:01<01:34, 103557.94it/s]
  3%|▎         | 311296/9912422 [00:02<01:09, 137981.72it/s]
  6%|▋         | 622592/9912422 [00:02<00:49, 188171.96it/s]
  9%|▉         | 868352/9912422 [00:02<00:36, 248055.84it/s]
 19%|█▉        | 1892352/9912422 [00:02<00:23, 345527.55it/s]
 26%|██▌       | 2596864/9912422 [00:03<00:15, 468914.88it/s]
 33%|███▎      | 3317760/9912422 [00:03<00:10, 626092.15it/s]
 41%|████      | 4063232/9912422 [00:03<00:07, 820416.51it/s]
 49%|████▉     | 4841472/9912422 [00:03<00:04, 1052848.12it/s]
 57%|█████▋    | 5636096/9912422 [00:04<00:03, 1316703.96it/s]
 65%|██████▌   | 6463488/9912422 [00:04<00:02, 1606136.44it/s]
 74%|███████▎  | 7299072/9912422 [00:04<00:01, 1901902.93it/s]
 80%|████████  | 7946240/9912422 [00:04<00:00, 2045197.04it/s]
 84%|████████▍ | 8355840/9912422 [00:05<00:00, 1916069.56it/s]
 93%|█████████▎| 9191424/9912422 [00:05<00:00, 2193578.33it/s]
 98%|█████████▊| 9682944/9912422 [00:05<00:00, 2120908.11it/s]
9920512it [00:05, 1733382.46it/s]
Extracting /tmp/data/mnist/MNIST/raw/train-images-idx3-ubyte.gz
```

```
Downloading http://yann.lecun.com/exdb/mnist/train-labels-idx1-ubyte.gz to
/tmp/data/mnist/MNIST/raw/train-labels-idx1-ubyte.gz

0it [00:00, ?it/s]
  0%|          | 0/28881 [00:00<?, ?it/s]
 57%|█████▋    | 16384/28881 [00:00<00:00, 66757.47it/s]
32768it [00:00, 44646.36it/s]
Extracting /tmp/data/mnist/MNIST/raw/train-labels-idx1-ubyte.gz
Downloading http://yann.lecun.com/exdb/mnist/t10k-images-idx3-ubyte.gz to
/tmp/data/mnist/MNIST/raw/t10k-images-idx3-ubyte.gz

0it [00:00, ?it/s]
  0%|          | 0/1648877 [00:00<?, ?it/s]
  1%|          | 16384/1648877 [00:00<00:28, 57955.03it/s]
  2%|▏         | 40960/1648877 [00:01<00:24, 65981.13it/s]
  6%|▌         | 98304/1648877 [00:01<00:18, 83833.18it/s]
  9%|▉         | 155648/1648877 [00:01<00:14, 103353.27it/s]
 14%|█▍        | 229376/1648877 [00:01<00:11, 123385.76it/s]
 19%|█▉        | 319488/1648877 [00:02<00:08, 148314.26it/s]
 25%|██▌       | 417792/1648877 [00:02<00:07, 159861.67it/s]
 31%|███       | 507904/1648877 [00:02<00:05, 192150.97it/s]
 33%|███▎      | 540672/1648877 [00:03<00:06, 167942.76it/s]
 34%|███▍      | 565248/1648877 [00:03<00:07, 138545.73it/s]
 36%|███▋      | 598016/1648877 [00:03<00:07, 135843.39it/s]
 41%|████      | 671744/1648877 [00:03<00:06, 161997.94it/s]
 45%|████▌     | 745472/1648877 [00:04<00:04, 187209.47it/s]
 50%|████▉     | 819200/1648877 [00:04<00:03, 210111.00it/s]
 55%|█████▍    | 901120/1648877 [00:04<00:03, 235269.26it/s]
 60%|█████▉    | 983040/1648877 [00:04<00:02, 256800.01it/s]
 65%|██████▍   | 1064960/1648877 [00:05<00:02, 274355.29it/s]
 70%|██████▉   | 1146880/1648877 [00:05<00:01, 288159.77it/s]
 75%|███████▍  | 1228800/1648877 [00:05<00:01, 298794.58it/s]
 80%|███████▉  | 1318912/1648877 [00:05<00:01, 314548.52it/s]
 85%|████████▍ | 1400832/1648877 [00:06<00:00, 318505.46it/s]
 90%|████████▉ | 1482752/1648877 [00:06<00:00, 321055.85it/s]
 95%|█████████▌| 1572864/1648877 [00:06<00:00, 331799.01it/s]
1654784it [00:06, 244848.04it/s]
Extracting /tmp/data/mnist/MNIST/raw/t10k-images-idx3-ubyte.gz
Downloading http://yann.lecun.com/exdb/mnist/t10k-labels-idx1-ubyte.gz to
/tmp/data/mnist/MNIST/raw/t10k-labels-idx1-ubyte.gz

0it [00:00, ?it/s]
  0%|          | 0/4542 [00:00<?, ?it/s]
8192it [00:00, 16306.28it/s]
Extracting /tmp/data/mnist/MNIST/raw/t10k-labels-idx1-ubyte.gz
Processing...
Done!
```

（6）构建深度学习卷积神经网络（Convolutional Neural Networks，CNN）模型。

创建一个简单的 CNN 模型，其中包含两个卷积层（conv）和两个完全连接层（fc）。在 conv 和 fc 层之间创建一个 dropout 层。

```
1.  class Net(nn.Module):
2.      def __init__(self):
3.          super(Net, self).__init__()
```

```
4.          self.conv1 = nn.Conv2d(1, 10, kernel_size=5)
5.          self.conv2 = nn.Conv2d(10, 20, kernel_size=5)
6.          self.conv2_drop = nn.Dropout2d()
7.          self.fc1 = nn.Linear(320, 50)
8.          self.fc2 = nn.Linear(50, 10)
9.
10.     def forward(self, x):
11.         x = F.relu(F.max_pool2d(self.conv1(x), 2))
12.         x = F.relu(F.max_pool2d(self.conv2_drop(self.conv2(x)), 2))
13.         x = x.view(-1, 320)
14.         x = F.relu(self.fc1(x))
15.         x = F.dropout(x, training=self.training)
16.         x = self.fc2(x)
17.         return F.log_softmax(x)
18.
19. model = Net()
20. model.share_memory() # gradients are allocated lazily, so they are not shared here
```

在 Databricks 云平台 Notebook 中运行上述代码，运行结果如下：

```
Out[4]: Net(
  (conv1): Conv2d(1, 10, kernel_size=(5, 5), stride=(1, 1))
  (conv2): Conv2d(10, 20, kernel_size=(5, 5), stride=(1, 1))
  (conv2_drop): Dropout2d(p=0.5)
  (fc1): Linear(in_features=320, out_features=50, bias=True)
  (fc2): Linear(in_features=50, out_features=10, bias=True)
)
```

（7）模型训练。

为了训练模型，定义了负对数似然损失函数，构建一个带动量的随机梯度下降优化器。然后调用反向传播函数 loss.backward()，调用 optimizer.step()更新模型参数。

```
1.  def train_epoch(epoch, args, model, data_loader, optimizer):
2.      model.train()
3.      for batch_idx, (data, target) in enumerate(data_loader):
4.          if args.cuda:
5.              data, target = data.cuda(), target.cuda()
6.          data, target = Variable(data), Variable(target)
7.          optimizer.zero_grad()
8.          output = model(data)
9.          loss = F.nll_loss(output, target)
10.         loss.backward()
11.         optimizer.step()
12.         if batch_idx % args.log_interval == 0:
13.             print('Train Epoch: {} [{}/{} ({:.0f}%)]\tLoss: {:.6f}'.format(
14.                 epoch, batch_idx * len(data), len(data_loader.dataset),
15.                 100. * batch_idx / len(data_loader), loss.data.item()))
16.
17.
18. def test_epoch(model, data_loader):
19.     model.eval()
20.     test_loss = 0
21.     correct = 0
22.     for data, target in data_loader:
23.         if args.cuda:
24.             data, target = data.cuda(), target.cuda()
25.         data, target = Variable(data, volatile=True), Variable(target)
26.         output = model(data)
```

```
27.         test_loss += F.nll_loss(output, target, size_average=False)
            .data.item() # 批量损失度汇总
28.         pred = output.data.max(1)[1] # 得到对数概率最大值的索引
29.         correct += pred.eq(target.data).cpu().sum()
30.
31.     test_loss /= len(data_loader.dataset)
32.     print('\nTest set: Average loss: {:.4f}, Accuracy: {}/{} ({:.0f}%)\n'.format(
33.         test_loss, correct, len(data_loader.dataset),
34.         100. * correct / len(data_loader.dataset)))
35.
36.
37. # 循环遍历每一个时代，基于训练集进行训练（在每次遍历之后进行评估）
38. if args.cuda:
39.     model = model.cuda()
40. optimizer = optim.SGD(model.parameters(), lr=args.lr, momentum=args.momentum)
41. for epoch in range(1, args.epochs + 1):
42.     train_epoch(epoch, args, model, train_loader, optimizer)
43.     test_epoch(model, test_loader)
```

第 1～15 行代码构建一个函数 train_epoch，基于训练集在每一个时代中进行迭代计算，其中：

第 8 行代码基于小批量训练集数据，获取 CNN 模型的预测概率值。

第 9 行代码使用 F.nll_loss 函数计算训练集数据的实际值及预测概率值的损失度。

第 10 行代码使用 loss.backward()方法实现反向传播算法。

第 11 行代码使用 optimizer.step()更新 CNN 模型的权重参数。

第 18～34 行代码构建一个 test_epoch，基于测试集在每一个时代中进行预测，其中：

第 26 行代码基于小批量测试集数据，获取 CNN 模型的预测概率值。

第 27 行代码使用 F.nll_loss 函数计算测试集预测概率值及实际值的批量损失度的汇总值。

第 28 行代码获取图像预测概率最大值的索引，即手写数字图片的预测值。

第 29 行代码，将图像的预测值与实际值进行比较，如果相等，则将统计值进行累加，获取图像预测准确的统计计数。

第 31 行代码在每一个遍历的时代中，将汇总损失度除以测试集记录数，获取测试集的平均损失度。

第 40 行代码 optim.SGD 方法的第一个参数为 CNN 模型参数，第二个参数是学习率，第三个参数是动量，定义一个随机梯度下降的优化器。

第 42 行代码调用 train_epoch 函数，基于训练集进行模型训练。

第 43 行代码调用 test_epoch 函数，基于测试集进行模型预测。

在 Databricks 云平台 Notebook 中运行上述代码，运行结果如下：

```
/local_disk0/tmp/1568343698268-0/PythonShell.py:17: UserWarning: Implicit
dimension choice for log_softmax has been deprecated. Change the call to include
dim=X as an argument.
  from six import StringIO  # cString does not support unicode well
Train Epoch: 1 [0/60000 (0%)]  Loss: 2.365850
Train Epoch: 1 [12800/60000 (21%)] Loss: 1.327535
Train Epoch: 1 [25600/60000 (43%)] Loss: 0.818490
Train Epoch: 1 [38400/60000 (64%)] Loss: 0.544349
Train Epoch: 1 [51200/60000 (85%)] Loss: 0.685446
/local_disk0/tmp/1568343698268-0/PythonShell.py:25: UserWarning: volatile
was removed and now has no effect. Use `with torch.no_grad():` instead.
```

```
  import matplotlib as mpl
/databricks/python/lib/python3.6/site-packages/torch/nn/_reduction.py:46:
UserWarning: size_average and reduce args will be deprecated, please use
reduction='sum' instead.
  warnings.warn(warning.format(ret))

Test set: Average loss: 0.2061, Accuracy: 9389/10000 (93%)

Train Epoch: 2 [0/60000 (0%)]  Loss: 0.373014
Train Epoch: 2 [12800/60000 (21%)] Loss: 0.389995
Train Epoch: 2 [25600/60000 (43%)] Loss: 0.281899
Train Epoch: 2 [38400/60000 (64%)] Loss: 0.279353
Train Epoch: 2 [51200/60000 (85%)] Loss: 0.384622

Test set: Average loss: 0.1296, Accuracy: 9591/10000 (95%)

Train Epoch: 3 [0/60000 (0%)]  Loss: 0.417406
Train Epoch: 3 [12800/60000 (21%)] Loss: 0.244383
Train Epoch: 3 [25600/60000 (43%)] Loss: 0.344068
Train Epoch: 3 [38400/60000 (64%)] Loss: 0.166291
Train Epoch: 3 [51200/60000 (85%)] Loss: 0.415093

Test set: Average loss: 0.1044, Accuracy: 9679/10000 (96%)

Train Epoch: 4 [0/60000 (0%)]  Loss: 0.361026
Train Epoch: 4 [12800/60000 (21%)] Loss: 0.210600
Train Epoch: 4 [25600/60000 (43%)] Loss: 0.350343
Train Epoch: 4 [38400/60000 (64%)] Loss: 0.156569
Train Epoch: 4 [51200/60000 (85%)] Loss: 0.184896

Test set: Average loss: 0.0843, Accuracy: 9732/10000 (97%)

Train Epoch: 5 [0/60000 (0%)]  Loss: 0.206659
Train Epoch: 5 [12800/60000 (21%)] Loss: 0.149261
Train Epoch: 5 [25600/60000 (43%)] Loss: 0.194065
Train Epoch: 5 [38400/60000 (64%)] Loss: 0.190172
Train Epoch: 5 [51200/60000 (85%)] Loss: 0.294799

Test set: Average loss: 0.0802, Accuracy: 9757/10000 (97%)

Train Epoch: 6 [0/60000 (0%)]  Loss: 0.295716
Train Epoch: 6 [12800/60000 (21%)] Loss: 0.253169
Train Epoch: 6 [25600/60000 (43%)] Loss: 0.289470
Train Epoch: 6 [38400/60000 (64%)] Loss: 0.358406
Train Epoch: 6 [51200/60000 (85%)] Loss: 0.203917

Test set: Average loss: 0.0671, Accuracy: 9787/10000 (97%)

Train Epoch: 7 [0/60000 (0%)]  Loss: 0.149162
Train Epoch: 7 [12800/60000 (21%)] Loss: 0.143309
Train Epoch: 7 [25600/60000 (43%)] Loss: 0.211849
Train Epoch: 7 [38400/60000 (64%)] Loss: 0.120907
Train Epoch: 7 [51200/60000 (85%)] Loss: 0.196015

Test set: Average loss: 0.0642, Accuracy: 9802/10000 (98%)

Train Epoch: 8 [0/60000 (0%)]  Loss: 0.305896
Train Epoch: 8 [12800/60000 (21%)] Loss: 0.182161
Train Epoch: 8 [25600/60000 (43%)] Loss: 0.147750
Train Epoch: 8 [38400/60000 (64%)] Loss: 0.111774
```

```
Train Epoch: 8 [51200/60000 (85%)] Loss: 0.256410

Test set: Average loss: 0.0623, Accuracy: 9810/10000 (98%)

Train Epoch: 9 [0/60000 (0%)]  Loss: 0.136523
Train Epoch: 9 [12800/60000 (21%)] Loss: 0.117827
Train Epoch: 9 [25600/60000 (43%)] Loss: 0.060807
Train Epoch: 9 [38400/60000 (64%)] Loss: 0.193793
Train Epoch: 9 [51200/60000 (85%)] Loss: 0.057915

Test set: Average loss: 0.0556, Accuracy: 9818/10000 (98%)

Train Epoch: 10 [0/60000 (0%)] Loss: 0.117015
Train Epoch: 10 [12800/60000 (21%)]    Loss: 0.313570
Train Epoch: 10 [25600/60000 (43%)]    Loss: 0.195568
Train Epoch: 10 [38400/60000 (64%)]    Loss: 0.254384
Train Epoch: 10 [51200/60000 (85%)]    Loss: 0.176877

Test set: Average loss: 0.0528, Accuracy: 9840/10000 (98%)
```

经过 10 个时代的迭代，训练集的损失度达到了 0.117015；测试集的平均损失度为 0.0528，精确度达到了 98%。

32.1.4　分布式深度学习训练

本节讲解使用 PyTorch 将单节点深度学习代码迁移到 HorovodRunner 分布式节点，利用 PyTorch 和 Horovod 的 HorovodRunner 进行 Mnist 手写数字识别案例的分布式深度学习训练。

Horovod 是美国 Uber 公司开源的一个跨多节点的分布式训练框架，基于 Ring-AllReduce 方法进行深度学习分布式训练，支持多种深度学习架构，包括 TensorFlow、Keras、PyTorch、Mxnet 等。Horovod 的名字来源于俄罗斯传统民间舞蹈，舞蹈者围成一个圈跳舞，类似于 TensorFlow 使用 Horovod 通信的场景。

32.1.4.1　HorovodRunner使用Horovod进行分布式训练

PyTorch 单节点训练迁移到分布式节点训练包括以下步骤。

- ❑ 准备单节点训练代码。
- ❑ 将单节点代码迁移到 HorovodRunner。注意，通过设置 CPU、GPU 参数，Databricks 集群启用 CPU 或 GPU 运行时不需要更改代码，Databricks 云平台集群必须包括 2 个 Spark Worker 节点，使用 Databricks 机器学习运行时库 5.4 以上版本。本案例将使用已创建的 Databricks 云平台集群（DeepLearning），包括 1 个 Spark Driver 节点和 2 个 Spark Worker 节点，如图 32-35 所示。

接下来进行 Mnist 手写数字识别案例的分布式训练实战。

（1）创建 Databricks 云平台集群 Notebook。

在 Workspace 中单击右键，在弹出的快捷菜单中选择 Create，弹出 Create Notebook 对话框，依次输入名称（DistributedTraining）、开发语言（选择 Python 语言）、云平台集群(DeepLearning)，单击 Create 按钮创建 Notebook，如图 32-36 所示。

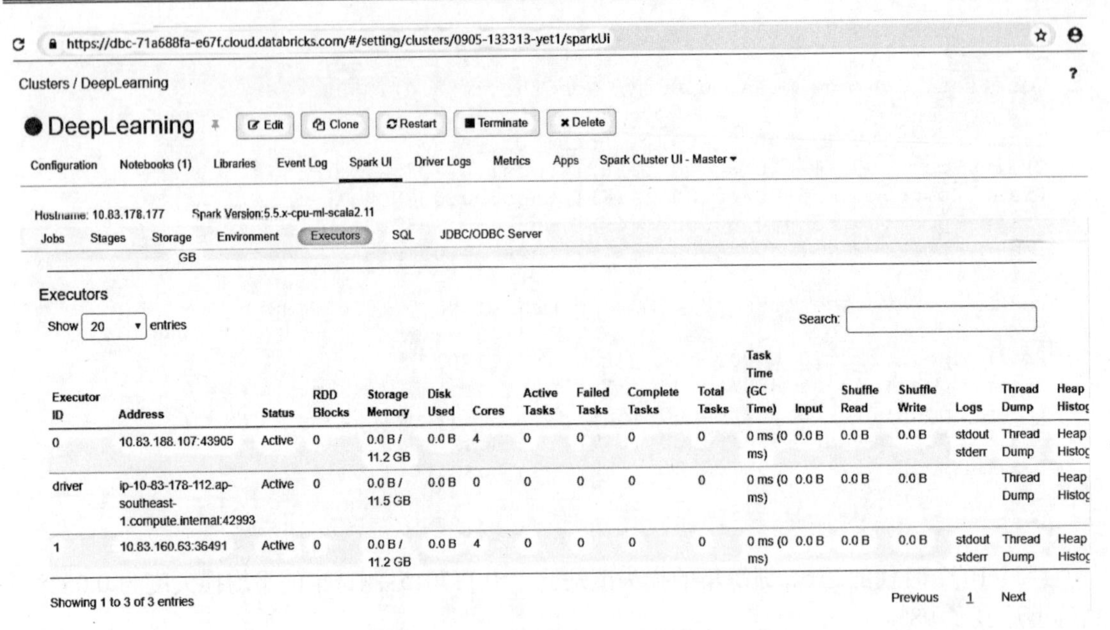

图 32-35 2 个 Spark Worker 节点

Create Notebook

Name: DistributedTraining

Language: Python

Cluster: DeepLearning

Cancel Create

图 32-36 创建 Notebook

（2）导入 torch 库及构建卷积神经网络 CNN 模型。

```
1.  import torch
2.  import torch.nn as nn
3.  import torch.nn.functional as F
4.
5.  class Net(nn.Module):
6.      def __init__(self):
7.          super(Net, self).__init__()
8.          self.conv1 = nn.Conv2d(1, 10, kernel_size=5)
9.          self.conv2 = nn.Conv2d(10, 20, kernel_size=5)
10.         self.conv2_drop = nn.Dropout2d()
11.         self.fc1 = nn.Linear(320, 50)
12.         self.fc2 = nn.Linear(50, 10)
13.
14.     def forward(self, x):
15.         x = F.relu(F.max_pool2d(self.conv1(x), 2))
16.         x = F.relu(F.max_pool2d(self.conv2_drop(self.conv2(x)), 2))
17.         x = x.view(-1, 320)
18.         x = F.relu(self.fc1(x))
19.         x = F.dropout(x, training=self.training)
20.         x = self.fc2(x)
21.         return F.log_softmax(x)
```

查看构建的 CNN 网络模型信息。

```
1.  net =Net()
2.  net
```

在 Databricks 云平台 Notebook 中运行上述代码，运行结果如下：

```
Out[3]: Net(
  (conv1): Conv2d(1, 10, kernel_size=(5, 5), stride=(1, 1))
  (conv2): Conv2d(10, 20, kernel_size=(5, 5), stride=(1, 1))
  (conv2_drop): Dropout2d(p=0.5)
  (fc1): Linear(in_features=320, out_features=50, bias=True)
  (fc2): Linear(in_features=50, out_features=10, bias=True)
)
```

（3）配置单节点训练参数。

```
1.  batch_size = 100
2.  num_epochs = 5
3.  momentum = 0.5
4.  log_interval = 100
```

（4）构建一个函数 train_one_epoch，基于训练集在每一个时代中进行迭代计算。

```
1.   def train_one_epoch(model, device, data_loader, optimizer, epoch):
2.      model.train()
3.      for batch_idx, (data, target) in enumerate(data_loader):
4.         data, target = data.to(device), target.to(device)
5.         optimizer.zero_grad()
6.         output = model(data)
7.         loss = F.nll_loss(output, target)
8.         loss.backward()
9.         optimizer.step()
10.        if batch_idx % log_interval == 0:
11.           print('Train Epoch: {} [{}/{} ({:.0f}%)]\tLoss: {:.6f}'.format(
12.              epoch, batch_idx * len(data), len(data_loader) * len(data),
13.              100. * batch_idx / len(data_loader), loss.item()))
```

（5）定义一个日志目录。

```
1.  from time import time
2.  import os
3.  PYTORCH_DIR = '/dbfs/ml/horovod_pytorch'
4.  LOG_DIR = os.path.join(PYTORCH_DIR, str(time()), 'MNISTDemo')
5.  os.makedirs(LOG_DIR)
```

（6）定义一个函数 save_checkpoint，用于检查点和持久化模型。

```
1.  def save_checkpoint(model, optimizer, epoch):
2.    filepath = LOG_DIR + '/checkpoint-{epoch}.pth.tar'.format(epoch=epoch)
3.    state = {
4.      'model': model.state_dict(),
5.      'optimizer': optimizer.state_dict(),
6.    }
7.    torch.save(state, filepath)
```

（7）使用 PyTorch 进行单节点训练。

```
1.  from torchvision import datasets, transforms
2.
3.  def train(learning_rate):
4.    device = torch.device('cuda' if torch.cuda.is_available() else 'cpu')
```

```
5.
6.    train_dataset = datasets.MNIST(
7.      'data',
8.      train=True,
9.      download=True,
10.     transform=transforms.Compose([transforms.ToTensor(),
        transforms.Normalize((0.1307,), (0.3081,))]))
11.   data_loader = torch.utils.data.DataLoader(train_dataset, batch_size=
      batch_size, shuffle=True)
12.
13.   model = Net().to(device)
14.
15.   optimizer = optim.SGD(model.parameters(), lr=learning_rate,
      momentum=momentum)
16.
17.   for epoch in range(1, num_epochs + 1):
18.     train_one_epoch(model, device, data_loader, optimizer, epoch)
19.     save_checkpoint(model, optimizer, epoch)
```

（8）设置学习率参数为0.001，执行train函数进行模型训练。

```
1.  train(learning_rate = 0.001)
```

在Databricks云平台Notebook中运行上述代码，运行结果如下：

```
Downloading http://yann.lecun.com/exdb/mnist/train-images-idx3-ubyte.gz to
data/MNIST/raw/train-images-idx3-ubyte.gz

0it [00:00, ?it/s]
  0%|          | 0/9912422 [00:00<?, ?it/s]
  0%......
 99%|
 99%|█████████▉| 9854976/9912422 [01:08<00:00, 260747.32it/s]
100%|█████████▉| 9895936/9912422 [01:09<00:00, 271949.73it/s]Extracting
data/MNIST/raw/train-images-idx3-ubyte.gz
Downloading http://yann.lecun.com/exdb/mnist/train-labels-idx1-ubyte.gz to
data/MNIST/raw/train-labels-idx1-ubyte.gz

0it [00:00, ?it/s]
  0%|          | 0/28881 [00:00<?, ?it/s]
 57%|█████▋    | 16384/28881 [00:01<00:00, 66031.91it/s]Extracting data/
MNIST/raw/train-labels-idx1-ubyte.gz
Downloading http://yann.lecun.com/exdb/mnist/t10k-images-idx3-ubyte.gz to
data/MNIST/raw/t10k-images-idx3-ubyte.gz

0it [00:00, ?it/s]
  0%|          | 0/1648877 [00:00<?, ?it/s]
......

100%|█████████▉| 1646592/1648877 [00:07<00:00, 188895.65it/s]

1654784it [00:07, 210883.72it/s] Extracting data/MNIST/raw/t10k-images-idx3-
ubyte.gz
Downloading http://yann.lecun.com/exdb/mnist/t10k-labels-idx1-ubyte.gz to
data/MNIST/raw/t10k-labels-idx1-ubyte.gz

0it [00:00, ?it/s]
  0%|          | 0/4542 [00:00<?, ?it/s]

8192it [00:00, 16648.47it/s] Extracting data/MNIST/raw/t10k-labels-idx1-ubyte.gz
Processing...
```

```
Done!
/local_disk0/tmp/1568240486375-0/PythonShell.py:21: UserWarning: Implicit
dimension choice for log_softmax has been deprecated. Change the call to include
dim=X as an argument.
 # This import has to be done *before* all other imports that could import
MPL. Otherwise,
Train Epoch: 1 [0/60000 (0%)]  Loss: 2.343230

9920512it [01:20, 271949.73it/s] Train Epoch: 1 [10000/60000 (17%)]
Loss: 2.276206
Train Epoch: 1 [20000/60000 (33%)] Loss: 2.294955

32768it [00:20, 66031.91it/s] Train Epoch: 1 [30000/60000 (50%)]
Loss: 2.298310
Train Epoch: 1 [40000/60000 (67%)] Loss: 2.280913
Train Epoch: 1 [50000/60000 (83%)] Loss: 2.258589
Train Epoch: 2 [0/60000 (0%)]      Loss: 2.238731
Train Epoch: 2 [10000/60000 (17%)] Loss: 2.237631
Train Epoch: 2 [20000/60000 (33%)] Loss: 2.240550
Train Epoch: 2 [30000/60000 (50%)] Loss: 2.212544
Train Epoch: 2 [40000/60000 (67%)] Loss: 2.218340
Train Epoch: 2 [50000/60000 (83%)] Loss: 2.109492
Train Epoch: 3 [0/60000 (0%)]      Loss: 2.087557
Train Epoch: 3 [10000/60000 (17%)] Loss: 2.058743
Train Epoch: 3 [20000/60000 (33%)] Loss: 1.904088
Train Epoch: 3 [30000/60000 (50%)] Loss: 1.885665
Train Epoch: 3 [40000/60000 (67%)] Loss: 1.790212
Train Epoch: 3 [50000/60000 (83%)] Loss: 1.661432
Train Epoch: 4 [0/60000 (0%)]      Loss: 1.688230
Train Epoch: 4 [10000/60000 (17%)] Loss: 1.482129
Train Epoch: 4 [20000/60000 (33%)] Loss: 1.372943
Train Epoch: 4 [30000/60000 (50%)] Loss: 1.237064
Train Epoch: 4 [40000/60000 (67%)] Loss: 1.156223
Train Epoch: 4 [50000/60000 (83%)] Loss: 1.230186
Train Epoch: 5 [0/60000 (0%)]      Loss: 1.048005
Train Epoch: 5 [10000/60000 (17%)] Loss: 1.011086
Train Epoch: 5 [20000/60000 (33%)] Loss: 1.001369
Train Epoch: 5 [30000/60000 (50%)] Loss: 1.090772
Train Epoch: 5 [40000/60000 (67%)] Loss: 0.947790
Train Epoch: 5 [50000/60000 (83%)] Loss: 0.977159
```

遍历到第 5 个时代，基于训练集的损失度为 0.977159。

（9）迁移到 HorovodRunner。

HorovodRunner 采用一个 Python 方法，该方法包含带有 Horovod 钩子的深度学习训练代码。这个方法在 Spark Driver 节点上进行序列化并发送给 Spark worker 节点，Horovod MPI 作业使用 Spark 屏障执行模式嵌入到 Spark 作业中。

```
1.  import horovod.torch as hvd
2.  from sparkdl import HorovodRunner
3.
4.  def train_hvd(learning_rate):
5.    hvd.init()  #初始化 Horovod
6.    device = torch.device('cuda' if torch.cuda.is_available() else 'cpu')
7.
8.    if device.type == 'cuda':
9.      # Horovod: 把 GPU 固定到 local rank.
10.     torch.cuda.set_device(hvd.local_rank())
```

```
11.
12.   train_dataset = datasets.MNIST(
13.     root='data-%d'% hvd.rank(), #为每个 worker 使用不同的根目录以避免竞争条件
14.     train=True,
15.     download=True,
16.     transform=transforms.Compose([transforms.ToTensor(),
        transforms.Normalize((0.1307,), (0.3081,))])
17.   )
18.
19.   from torch.utils.data.distributed import DistributedSampler
20.
21.   #配置采样器，使每个 worker 进程获得输入数据集的不同样本
22.   train_sampler = DistributedSampler(train_dataset, num_replicas=hvd.size(),
      rank=hvd.rank())
23.   #使用 train_sampler 在每个 worker 上加载不同的数据样本
24.   train_loader = torch.utils.data.DataLoader(train_dataset, batch_size=
      batch_size, sampler=train_sampler)
25.
26.   model = Net().to(device)
27.
28.   #同步分布式训练中的有效批量大小是按 workers 数量进行缩放的，学习率的提高弥补了批量
      #大小的增加
29.   optimizer = optim.SGD(model.parameters(), lr=learning_rate * hvd.size(),
      momentum=momentum)
30.
31.   #用 Horovod 的 DistributedOptimizer 包装优化器
32.   optimizer = hvd.DistributedOptimizer(optimizer, named_parameters=
      model.named_parameters())
33.
34.   #广播初始参数，以便所有 workers 进程都从相同的参数开始
35.   hvd.broadcast_parameters(model.state_dict(), root_rank=0)
36.
37.   for epoch in range(1, num_epochs + 1):
38.     train_one_epoch(model, device, train_loader, optimizer, epoch)
39.     #只在第一个 worker 进程上保存检查点
40.     if hvd.rank() == 0:
41.       save_checkpoint(model, optimizer, epoch)
```

第 1 行代码导入 horovod.torch 包，别名命名为 hvd。

第 2 行代码从 sparkdl 库中导入 HorovodRunner 包。

第 5 行代码初始化 Horovod。

第 10 行代码将 GPU 或 CPU 固定到本地排序（Local Rank）进程，如图 32-37 所示。

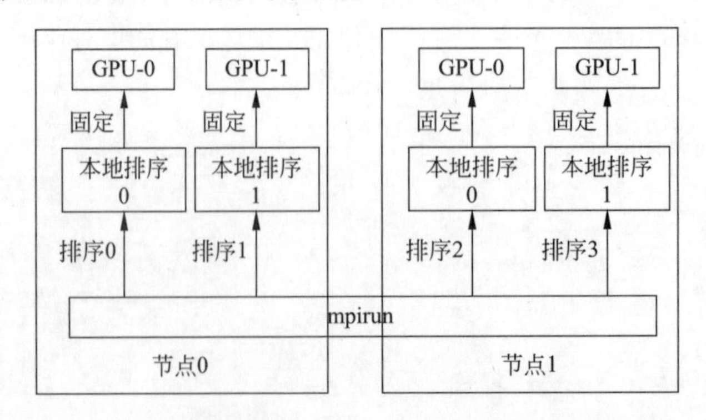

图 32-37　Horovod GPU 固定

第 13 行代码为每个 worker 使用不同的根目录。

第 22 行代码设置样本采样器，在每个 worker 上加载不同的数据样本。

第 29 行代码在随机梯度下降算法中，将学习率乘以 Horovod 的大小，提高了学习率。

第 31 行代码使用 Horovod 封装 DistributedOptimizer 优化器。

第 35 行代码广播模型的初始参数，所有 workers 进程都从相同的参数开始。

第 40 行代码判断是否为第一个 worker 节点，只在第一个 worker 上保存检查点。

（10）使用 Horovod hooks 定义的 train_hvd 函数，可以轻松构建 HorovodRunner 并运行分布式训练。

```
1.  hr = HorovodRunner(np=2) #假设集群有 2 个 workers.
2.  hr.run(train_hvd, learning_rate = 0.001)
```

在 Databricks 云平台 Notebook 中运行上述代码，运行结果如图 32-38 所示。

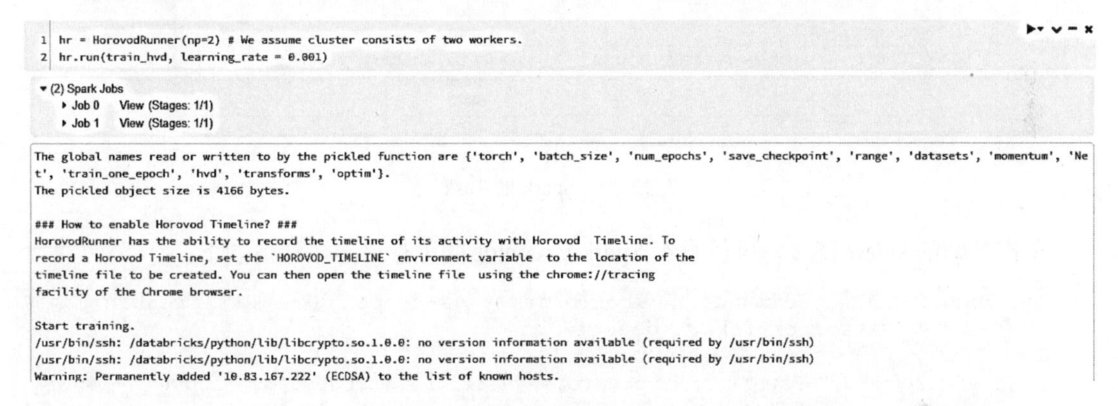

图 32-38　HorovodRunner 运行截图

单击 Spark Jobs 的 View 按钮，弹出 Spark Web UI 页面，可查看 Spark 的运行情况，如图 32-39 所示。

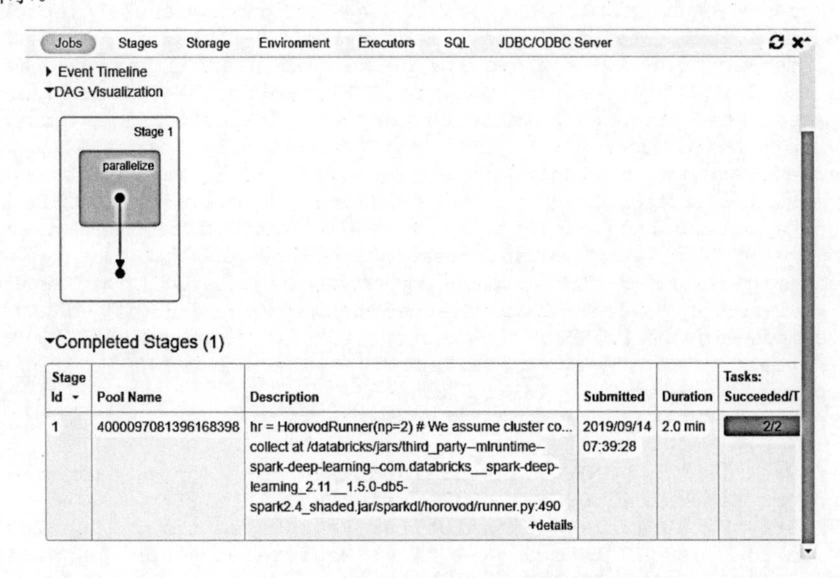

图 32-39　spark 中运行 HorovodRunner

单击其中的 collect 链接，查看 Spark 的 Task，如图 32-40 所示。

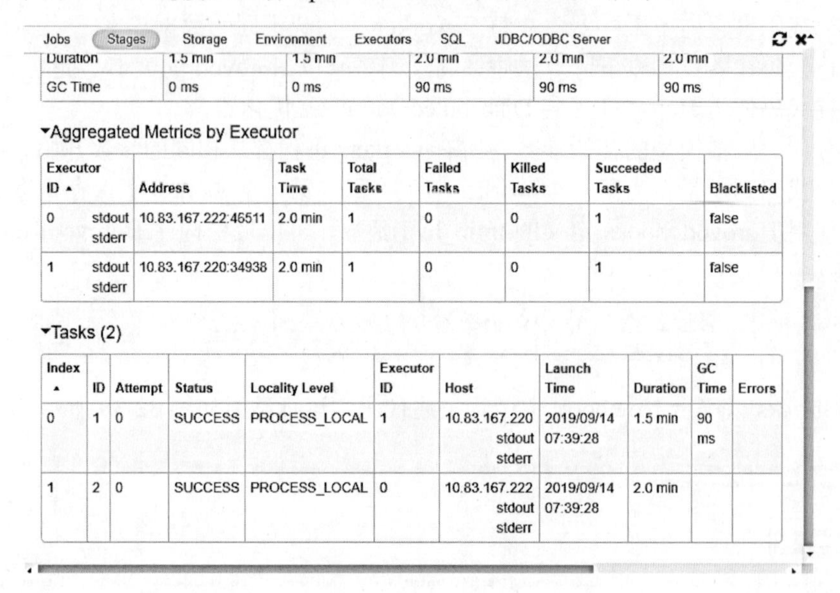

Executor ID ▴		Address	Task Time	Total Tasks	Failed Tasks	Killed Tasks	Succeeded Tasks	Blacklisted
0	stdout stderr	10.83.167.222:46511	2.0 min	1	0	0	1	false
1	stdout stderr	10.83.167.220:34938	2.0 min	1	0	0	1	false

Index ▴	ID	Attempt	Status	Locality Level	Executor ID	Host	Launch Time	Duration	GC Time	Errors
0	1	0	SUCCESS	PROCESS_LOCAL	1	10.83.167.220 stdout stderr	2019/09/14 07:39:28	1.5 min	90 ms	
1	2	0	SUCCESS	PROCESS_LOCAL	0	10.83.167.222 stdout stderr	2019/09/14 07:39:28	2.0 min		

图 32-40　Spark 的 Task

单击其中的 stderr 链接，可以查看 Spark Executor 的运行日志，详情如下：

```
1. Spark Executor Command: "/usr/lib/jvm/java-8-openjdk-amd64/jre/bin/java"
   "-cp" "/databricks/spark/dbconf/log4j
2. ......
3. "-Xmx20396M" "-Dspark.hadoop.hive.server2.thrift.http.port=10000" "-Dspark.
   driver.port=39986" "-Dspark.shuffle.service.port=4048" "-Dspark.ui.port=40943"
   "-Dspark.rpc.message.maxSize=256"  "-Djava.io.tmpdir=/local_disk0/tmp"
   "-XX:ReservedCodeCacheSize=256m"  "-XX:+UseCodeCacheFlushing"  "-Djava.
   security.properties=/databricks/spark/dbconf/java/extra.security" "-XX:
   +PrintFlagsFinal" "-XX:+PrintGCDateStamps" "-verbose:gc" "-XX:+PrintGCDetails"
   "-Xss4m" "-Djavax.xml.datatype.DatatypeFactory=com.sun.org.apache.xerces.
   internal.jaxp.datatype.DatatypeFactoryImpl" "-Djavax.xml.parsers.Document
   BuilderFactory= com.sun.org.apache.xerces.internal.jaxp.DocumentBuilder
   FactoryImpl" "-Djavax.xml.parsers.SAXParserFactory=com.sun.org.apache.
   xerces.internal.jaxp.SAXParserFactoryImpl" "-Djavax.xml.validation.Schema
   Factory:http://www.w3.org/2001/XMLSchema=com.sun.org.apache.xerces.in
   ternal.jaxp.validation.XMLSchemaFactory" "-Dorg.xml.sax.driver=com.sun.
   org.apache.xerces.internal.parsers.SAXParser" "-Dorg.w3c.dom.DOMImplementation
   SourceList=com.sun.org.apache.xerces.internal.dom.DOMXSImplementation
   SourceImpl" "-Djavax.net.ssl.sessionCacheSize=10000" "-Ddatabricks.service
   Name=spark-executor-1" "org.apache.spark.executor.CoarseGrainedExecutorBackend"
   "--driver-url" "spark://CoarseGrainedScheduler@ip-10-83-169-48.ap-southeast-
   1.compute.internal:39986" "--executor-id" "1" "--hostname" "10.83.167.220"
   "--cores"  "4"  "--app-id"  "app-20190914073342-0000"  "--worker-url"
   "spark://Worker@10.83.167.220:35468"
4. ========================================
5.
6. 19/09/14 07:33:48 INFO CoarseGrainedExecutorBackend: Started daemon with
   process name: 2043@0905-133313-yet1-10-83-167-220
7. 19/09/14 07:33:48 INFO SignalUtils: Registered signal handler for TERM
8. 19/09/14 07:33:48 INFO SignalUtils: Registered signal handler for HUP
9. 19/09/14 07:33:48 INFO SignalUtils: Registered signal handler for INT
10. 19/09/14 07:33:49 INFO SecurityManager: Changing view acls to: root
```

```
11. 19/09/14 07:33:49 INFO SecurityManager: Changing modify acls to: root
12. 19/09/14 07:33:49 INFO SecurityManager: Changing view acls groups to:
13. 19/09/14 07:33:49 INFO SecurityManager: Changing modify acls groups to:
14. 19/09/14 07:33:49 INFO SecurityManager: SecurityManager: authentication
    disabled; ui acls disabled; users  with view permissions: Set(root);
    groups with view permissions: Set(); users  with modify permissions:
    Set(root); groups with modify permissions: Set()
15. 19/09/14 07:33:49 INFO TransportClientFactory: Successfully created
    connection to ip-10-83-169-48.ap-southeast-1.compute.internal/10.83.169.48:
    39986 after 90 ms (0 ms spent in bootstraps)
16. 19/09/14 07:33:49 WARN SparkConf: The configuration key 'spark.akka.
    frameSize' has been deprecated as of Spark 1.6 and may be removed in the
    future. Please use the new key 'spark.rpc.message.maxSize' instead.
17. 19/09/14 07:33:49 INFO SecurityManager: Changing view acls to: root
18. 19/09/14 07:33:49 INFO SecurityManager: Changing modify acls to: root
19. 19/09/14 07:33:49 INFO SecurityManager: Changing view acls groups to:
20. 19/09/14 07:33:49 INFO SecurityManager: Changing modify acls groups to:
21. 19/09/14 07:33:49 INFO SecurityManager: SecurityManager: authentication
    disabled; ui acls disabled; users  with view permissions: Set(root);
    groups with view permissions: Set(); users  with modify permissions:
    Set(root); groups with modify permissions: Set()
22. 19/09/14 07:33:50 INFO TransportClientFactory: Successfully created
    connection to ip-10-83-169-48.ap-southeast-1.compute.internal/10.83.169.48:
    39986 after 2 ms (0 ms spent in bootstraps)
23. 19/09/14 07:33:50 INFO DiskBlockManager: Created local directory at
    /local_disk0/spark-37019628-9790-4356-9c7f-6279299cc116/executor-7e99
    c0e1-eb84-4a2e-9af9-a2eabaaf60c0/blockmgr-b9543348-d9e7-4c83-965d-8c6
    b8518877b
24. 19/09/14 07:33:50 INFO MemoryStore: MemoryStore started with capacity 10.4 GB
25. 19/09/14 07:33:50 INFO CoarseGrainedExecutorBackend: Connecting to driver:
    spark://CoarseGrainedScheduler@ip-10-83-169-48.ap-southeast-1.compute
    .internal:39986
26. 19/09/14 07:33:50 INFO WorkerWatcher: Connecting to worker spark://
    Worker@10.83.167.220:35468
27. 19/09/14 07:33:50 INFO TransportClientFactory: Successfully created
    connection to /10.83.167.220:35468 after 2 ms (0 ms spent in bootstraps)
28. 19/09/14 07:33:50 INFO WorkerWatcher: Successfully connected to spark://
    Worker@10.83.167.220:35468
29. 19/09/14 07:33:50 INFO CoarseGrainedExecutorBackend: Successfully registered
    with driver
30. 19/09/14 07:33:50 INFO Executor: Starting executor ID 1 on host 10.83.167.220
31. 19/09/14 07:33:50 INFO Utils: Successfully started service 'org.apache.
    spark.network.netty.NettyBlockTransferService' on port 34938.
32. 19/09/14 07:33:50 INFO NettyBlockTransferService: Server created on
    10.83.167.220:34938
33. 19/09/14 07:33:50 INFO BlockManager: Using org.apache.spark.storage.
    RandomBlockReplicationPolicy for block replication policy
34. 19/09/14 07:33:50 INFO BlockManagerMaster: Registering BlockManager
    BlockManagerId(1, 10.83.167.220, 34938, None)
35. 19/09/14 07:33:50 INFO BlockManagerMaster: Registered BlockManager
    BlockManagerId(1, 10.83.167.220, 34938, None)
36. 19/09/14 07:33:50 INFO BlockManager: external shuffle service port = 4048
37. 19/09/14 07:33:50 INFO BlockManager: Registering executor with local
    external shuffle service.
38. 19/09/14 07:33:50 INFO TransportClientFactory: Successfully created
    connection to /10.83.167.220:4048 after 3 ms (0 ms spent in bootstraps)
39. 19/09/14 07:33:50 INFO BlockManager: Initialized BlockManager: BlockManagerId
```

```
    (1, 10.83.167.220, 34938, None)
40. 19/09/14 07:33:50 INFO Executor: Using REPL class URI: spark://ip-10-83-
    169-48.ap-southeast-1.compute.internal:39986/classes
41. 19/09/14 07:39:28 INFO CoarseGrainedExecutorBackend: Got assigned task 1
42. 19/09/14 07:39:28 INFO Executor: Running task 0.0 in stage 1.0 (TID 1)
43. 19/09/14 07:39:28 INFO Executor: Fetching spark://ip-10-83-169-48.
    ap-southeast-1.compute.internal:39986/files/__clusterWide_lib_d07886c
    2c0c33873a79d861d18598d72.pypi with timestamp 1568446440117
44. 19/09/14 07:39:28 INFO TransportClientFactory: Successfully created
    connection to ip-10-83-169-48.ap-southeast-1.compute.internal/10.83.169.
    48:39986 after 1 ms (0 ms spent in bootstraps)
45. 19/09/14 07:39:28 INFO Utils: Fetching spark://ip-10-83-169-48.ap-
    southeast-1.compute.internal:39986/files/__clusterWide_lib_d07886c2c0
    c33873a79d861d18598d72.pypi to /databricks/spark/work/app-20190914073342-
    0000/1/./fetchFileTemp6355356320783870634.tmp
46. 19/09/14 07:39:28 INFO Utils: Extracting pypi install information from:
    /databricks/spark/work/app-20190914073342-0000/1/./__clusterWide_lib_
    d07886c2c0c33873a79d861d18598d72.pypi
47. 19/09/14 07:39:28 INFO Utils: resolved command to be run: List(/databricks/
    python/bin/pip, install, petastorm, --disable-pip-version-check)
48. 19/09/14 07:39:54 INFO Executor: Fetching spark://ip-10-83-169-48.ap-
    southeast-1.compute.internal:39986/files/__clusterWide_lib_8b1900a170
    7071321949d9ccd001f51d.pypi with timestamp 1568446438304
49. 19/09/14 07:39:54 INFO Utils: Fetching spark://ip-10-83-169-48.ap-
    southeast-1.compute.internal:39986/files/__clusterWide_lib_8b1900a170
    7071321949d9ccd001f51d.pypi to /databricks/spark/work/app-20190914073342-
    0000/1/./fetchFileTemp4457756856657058466.tmp
50. 19/09/14 07:39:54 INFO Utils: Extracting pypi install information from:
    /databricks/spark/work/app-20190914073342-0000/1/./__clusterWide_lib_
    8b1900a1707071321949d9ccd001f51d.pypi
51. 19/09/14 07:39:54 INFO Utils: resolved command to be run: List(/databricks/
    python/bin/pip, install, pydot, --disable-pip-version-check)
52. 19/09/14 07:39:55 INFO Executor: Fetching spark://ip-10-83-169-48.ap-
    southeast-1.compute.internal:39986/jars/__clusterWide_lib_d07886c2c0c
    33873a79d861d18598d72.pypi with timestamp 1568446467702
53. 19/09/14 07:39:55 INFO Utils: Fetching spark://ip-10-83-169-48.ap-southeast-
    1.compute.internal:39986/jars/__clusterWide_lib_d07886c2c0c33873a79d8
    61d18598d72.pypi to /databricks/spark/work/app-20190914073342-0000/1/./
    fetchFileTemp4764250937250260922.tmp
54. 19/09/14 07:39:55 INFO Utils: /databricks/spark/work/app-20190914073342-
    0000/1/./fetchFileTemp4764250937250260922.tmp has been previously copied
    to /databricks/spark/work/app-20190914073342-0000/1/./__clusterWide_lib_
    d07886c2c0c33873a79d861d18598d72.pypi
55. 19/09/14 07:39:55 INFO Executor: Adding file:/databricks/spark/work/
    app-20190914073342-0000/1/./__clusterWide_lib_d07886c2c0c33873a79d861
    d18598d72.pypi to class loader for default
56. 19/09/14 07:39:55 INFO Executor: Fetching spark://ip-10-83-169-48.ap-
    southeast-1.compute.internal:39986/jars/__clusterWide_lib_8b1900a1707
    071321949d9ccd001f51d.pypi with timestamp 1568446440090
57. 19/09/14 07:39:55 INFO Utils: Fetching spark://ip-10-83-169-48.ap-
    southeast-1.compute.internal:39986/jars/__clusterWide_lib_8b1900a1707
    071321949d9ccd001f51d.pypi to /databricks/spark/work/app-20190914073342-
    0000/1/./fetchFileTemp1046645223658009713.tmp
58. 19/09/14 07:39:55 INFO Utils: /databricks/spark/work/app-20190914073342-
    0000/1/./fetchFileTemp1046645223658009713.tmp has been previously copied
    to /databricks/spark/work/app-20190914073342-0000/1/./__clusterWide_lib_
    8b1900a1707071321949d9ccd001f51d.pypi
59. 19/09/14 07:39:55 INFO Executor: Adding file:/databricks/spark/work/app-
    20190914073342-0000/1/./__clusterWide_lib_8b1900a1707071321949d9ccd00
    1f51d.pypi to class loader for default
```

```
60. 19/09/14 07:39:55 INFO TorrentBroadcast: Started reading broadcast
    variable 2
61. 19/09/14 07:39:56 INFO TransportClientFactory: Successfully created
    connection to /10.83.167.222:46511 after 1 ms (0 ms spent in bootstraps)
62. 19/09/14 07:39:56 INFO MemoryStore: Block broadcast_2_piece0 stored as
    bytes in memory (estimated size 8.0 KB, free 10.4 GB)
63. 19/09/14 07:39:56 INFO TorrentBroadcast: Reading broadcast variable 2 took
    124 ms
64. 19/09/14 07:39:56 INFO MemoryStore: Block broadcast_2 stored as values
    in memory (estimated size 10.9 KB, free 10.4 GB)
65. 19/09/14 07:39:56 INFO TorrentBroadcast: Started reading broadcast
    variable 1
66. 19/09/14 07:39:56 INFO TransportClientFactory: Successfully created
    connection to ip-10-83-169-48.ap-southeast-1.compute.internal/10.83.169.
    48:34436 after 2 ms (0 ms spent in bootstraps)
67. 19/09/14 07:39:56 INFO MemoryStore: Block broadcast_1_piece0 stored as
    bytes in memory (estimated size 3.8 KB, free 10.4 GB)
68. 19/09/14 07:39:56 INFO TorrentBroadcast: Reading broadcast variable 1 took
    22 ms
69. 19/09/14 07:39:56 INFO MemoryStore: Block broadcast_1 stored as values
    in memory (estimated size 448.0 B, free 10.4 GB)
70. Using TensorFlow backend.
71. 19/09/14 07:39:58 INFO BarrierTaskContext: Task 1 from Stage 1(Attempt
    0) has entered the global sync, current barrier epoch is 0.
72. 19/09/14 07:39:58 INFO BarrierTaskContext: Task 1 from Stage 1(Attempt
    0) finished global sync successfully, waited for 0 seconds, current barrier
    epoch is 1.
73. Executing command: ['mpirun', '--allow-run-as-root', '-np', '2', '-H',
    '10.83.167.220,10.83.167.222', '--stdin', 'none', '--tag-output', '-mca',
    'rmaps', 'seq', '--bind-to', 'none', '-x', 'NCCL_DEBUG=INFO', '-mca',
    'pml', 'ob1', '-mca', 'btl', '^openib', '-mca', 'plm_rsh_agent', 'ssh -o
    StrictHostKeyChecking=no -i /tmp/HorovodRunner_0b1cf5a5f591/id_rsa',
    'bash', '/tmp/HorovodRunner_0b1cf5a5f591/launch.sh'].
74.
75. /usr/bin/ssh: /databricks/python/lib/libcrypto.so.1.0.0: no version
    information available (required by /usr/bin/ssh)
76. /usr/bin/ssh: /databricks/python/lib/libcrypto.so.1.0.0: no version
    information available (required by /usr/bin/ssh)
77. Warning: Permanently added '10.83.167.222' (ECDSA) to the list of known
    hosts.
78. ……
```

其中第 71、72 行日志显示 Spark 的 BarrierTaskContext 屏障任务上下文信息。

第 73 行日志执行 HorovodRunner 命令。

图 32-41 是 Spark 屏障执行模式的示意图。

在 Databricks 云平台 Notebook 中运行结果的日志如下：

```
The global names read or written to by the pickled function are {'transforms',
'hvd', 'save_checkpoint', 'Net', 'batch_size', 'datasets', 'momentum',
'range', 'num_epochs', 'train_one_epoch', 'optim', 'torch'}.
The pickled object size is 4167 bytes.

### How to enable Horovod Timeline? ###
HorovodRunner has the ability to record the timeline of its activity with
Horovod Timeline. To record a Horovod Timeline, set the `HOROVOD_TIMELINE`
environment variable to the location of the timeline file to be created. You
can then open the timeline file using the chrome://tracingfacility of the
Chrome browser.
```

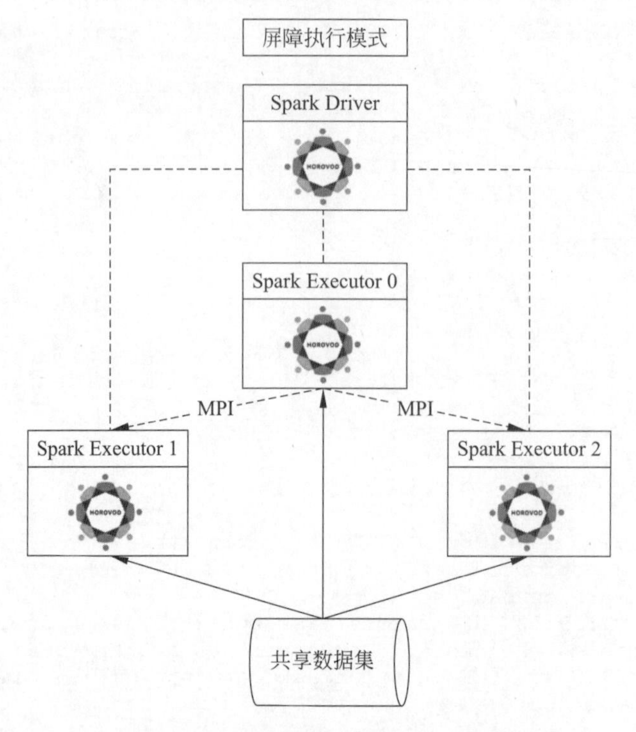

图 32-41　Spark 屏障执行模式与 Horovod

```
Start training.
/usr/bin/ssh: /databricks/python/lib/libcrypto.so.1.0.0: no version information
available (required by /usr/bin/ssh)
/usr/bin/ssh: /databricks/python/lib/libcrypto.so.1.0.0: no version information
available (required by /usr/bin/ssh)
Warning: Permanently added '10.83.168.239' (ECDSA) to the list of known hosts.
[1,0]<stderr>:Using TensorFlow backend.
[1,0]<stderr>:OMP: Info #209: KMP_AFFINITY: decoding x2APIC ids.
[1,0]<stderr>:OMP: Info #207: KMP_AFFINITY: Affinity capable, using global
cpuid leaf 11 info
[1,0]<stderr>:OMP: Info #154: KMP_AFFINITY: Initial OS proc set respected:
{0,1,2,3}
[1,0]<stderr>:OMP: Info #156: KMP_AFFINITY: 4 available OS procs
[1,0]<stderr>:OMP: Info #157: KMP_AFFINITY: Uniform topology
[1,0]<stderr>:OMP: Info #179: KMP_AFFINITY: 1 packages x 2 cores/pkg x 2
threads/core (2 total cores)
[1,0]<stderr>:OMP: Info #211: KMP_AFFINITY: OS proc to physical thread map:
[1,0]<stderr>:OMP: Info #171: KMP_AFFINITY: OS proc 0 maps to package 0 core
0 thread 0
[1,0]<stderr>:OMP: Info #171: KMP_AFFINITY: OS proc 2 maps to package 0 core
0 thread 1
[1,0]<stderr>:OMP: Info #171: KMP_AFFINITY: OS proc 1 maps to package 0 core
1 thread 0
[1,0]<stderr>:OMP: Info #171: KMP_AFFINITY: OS proc 3 maps to package 0 core
1 thread 1
[1,0]<stderr>:OMP: Info #247: KMP_AFFINITY: pid 2242 tid 2242 thread 0 bound
to OS proc set {0}
[1,1]<stderr>:OMP: Info #209: KMP_AFFINITY: decoding x2APIC ids.
[1,1]<stderr>:OMP: Info #207: KMP_AFFINITY: Affinity capable, using global
cpuid leaf 11 info
[1,1]<stderr>:OMP: Info #154: KMP_AFFINITY: Initial OS proc set respected:
```

```
{0,1,2,3}
[1,1]<stderr>:OMP: Info #156: KMP_AFFINITY: 4 available OS procs
[1,1]<stderr>:OMP: Info #157: KMP_AFFINITY: Uniform topology
[1,1]<stderr>:OMP: Info #179: KMP_AFFINITY: 1 packages x 2 cores/pkg x 2
threads/core (2 total cores)
[1,1]<stderr>:OMP: Info #211: KMP_AFFINITY: OS proc to physical thread map:
[1,1]<stderr>:OMP: Info #171: KMP_AFFINITY: OS proc 0 maps to package 0 core
0 thread 0
[1,1]<stderr>:OMP: Info #171: KMP_AFFINITY: OS proc 2 maps to package 0 core
0 thread 1
[1,1]<stderr>:OMP: Info #171: KMP_AFFINITY: OS proc 1 maps to package 0 core
1 thread 0
[1,1]<stderr>:OMP: Info #171: KMP_AFFINITY: OS proc 3 maps to package 0 core
1 thread 1
[1,1]<stderr>:OMP: Info #247: KMP_AFFINITY: pid 2231 tid 2231 thread 0 bound
to OS proc set {0}
[1,0]<stdout>:Downloading
http://yann.lecun.com/exdb/mnist/train-images-idx3-ubyte.gz to data-0/MNIST/
raw/train-images-idx3-ubyte.gz
[1,1]<stdout>:Downloading http://yann.lecun.com/exdb/mnist/train-images-idx3-
ubyte.gz to data-1/MNIST/raw/train-images-idx3-ubyte.gz

......
0it [00:00, ?it/s][1,0]<stderr>:
  0%|          | 0/9912422 [00:00<?, ?it/s][1,1]<stderr>:
  0%|          | 0/9912422 [00:00<?, ?it/s][1,0]<stderr>:
......

100%|██████████| 9895936/9912422 [00:37<00:00, 95397.83it/s][1,0]<stderr>:
 52%|█████     | 5160960/9912422 [00:37<00:31, 149397.39it/s][1,1]<stdout>:
Extracting data-1/MNIST/raw/train-images-idx3-ubyte.gz

*** WARNING: skipped 8652 bytes of output ***
......
100%|██████████| 1646592/1648877 [00:14<00:00, 123471.72it/s][1,1]<stdout>:
Extracting data-1/MNIST/raw/t10k-images-idx3-ubyte.gz
[1,1]<stdout>:Downloading
http://yann.lecun.com/exdb/mnist/t10k-labels-idx1-ubyte.gz to data-1/MNIST/
raw/t10k-labels-idx1-ubyte.gz
......
8192it [00:00, 15741.48it/s] [1,1]<stdout>:Extracting data-1/MNIST/raw/t10k-
labels-idx1-ubyte.gz
[1,1]<stdout>:Processing...
[1,1]<stdout>:Done!
[1,0]<stderr>:
......
100%|██████████| 9895936/9912422 [01:20<00:00, 143040.67it/s][1,0]<stdout>:
Extracting data-0/MNIST/raw/train-images-idx3-ubyte.gz
[1,0]<stdout>:Downloading
http://yann.lecun.com/exdb/mnist/train-labels-idx1-ubyte.gz to data-0/MNIST/raw/
train-labels-idx1-ubyte.gz

......
32768it [00:01, 32229.41it/s] [1,0]<stdout>:Extracting data-0/MNIST/raw/train-
labels-idx1-ubyte.gz
[1,0]<stdout>:Downloading  http://yann.lecun.com/exdb/mnist/t10k-images-idx3-
ubyte.gz to data-0/MNIST/raw/t10k-images-idx3-ubyte.gz
......
```

```
0it [00:00, ?it/s][1,0]<stderr>:[1,0]<stderr>:
......
[1,0]<stderr>:
 99%|██████████████| 1638400/1648877 [00:11<00:00, 172690.45it/s][1,0]<stdout>:
Extracting data-0/MNIST/raw/t10k-images-idx3-ubyte.gz
[1,0]<stdout>:Downloading  http://yann.lecun.com/exdb/mnist/t10k-labels-idx1-
ubyte.gz to data-0/MNIST/raw/t10k-labels-idx1-ubyte.gz
......
0it [00:00, ?it/s][1,0]<stderr>:
......
8192it [00:00, 16461.47it/s] [1,0]<stderr>:[1,0]<stdout>: Extracting data-0/
MNIST/raw/t10k-labels-idx1-ubyte.gz
[1,0]<stdout>:Processing...
[1,0]<stdout>:Done!
[1,1]<stderr>:
[1,1]<stderr>:
1654784it [00:54, 123471.72it/s] [1,1]<stderr>:OMP: Info #247: KMP_AFFINITY:
pid 2231 tid 2241 thread 1 bound to OS proc set {1}
[1,1]<stderr>:OMP: Info #247: KMP_AFFINITY: pid 2231 tid 2240 thread 2 bound
to OS proc set {2}
[1,0]<stderr>:<string>:21:  UserWarning:  Implicit  dimension  choice  for
log_softmax has been deprecated. Change the call to include dim=X as an
argument.
[1,0]<stderr>:OMP: Info #247: KMP_AFFINITY: pid 2242 tid 2252 thread 1 bound
to OS proc set {1}
[1,0]<stderr>:OMP: Info #247: KMP_AFFINITY: pid 2242 tid 2251 thread 2 bound
to OS proc set {2}
[1,0]<stderr>:OMP: Info #247: KMP_AFFINITY: pid 2242 tid 2247 thread 3 bound
to OS proc set {3}
[1,1]<stderr>:OMP: Info #247: KMP_AFFINITY: pid 2231 tid 2236 thread 3 bound
to OS proc set {3}
[1,0]<stdout>:Train Epoch: 1 [0/30000 (0%)]        Loss: 2.294775
[1,1]<stdout>:Train Epoch: 1 [0/30000 (0%)]        Loss: 2.319295
[1,0]<stdout>:Train Epoch: 1 [10000/30000 (33%)]   Loss: 2.298507
[1,1]<stdout>:Train Epoch: 1 [10000/30000 (33%)]   Loss: 2.294906
[1,0]<stderr>:
9920512it [01:40, 143040.67it/s] [1,1]<stdout>:Train Epoch: 1 [20000/30000 (67%)]
Loss: 2.268322
[1,0]<stdout>:Train Epoch: 1 [20000/30000 (67%)]   Loss: 2.277032
[1,0]<stdout>:Train Epoch: 2 [0/30000 (0%)]        Loss: 2.216177
[1,1]<stdout>:Train Epoch: 2 [0/30000 (0%)]        Loss: 2.225912
[1,0]<stderr>:
[1,0]<stderr>:
1654784it [00:28, 172690.45it/s] [1,0]<stdout>:Train Epoch: 2 [10000/30000 (33%)]
Loss: 2.091685
[1,1]<stdout>:Train Epoch: 2 [10000/30000 (33%)]   Loss: 2.096915
[1,0]<stdout>:Train Epoch: 2 [20000/30000 (67%)]   Loss: 2.078182
[1,1]<stdout>:Train Epoch: 2 [20000/30000 (67%)]   Loss: 2.016624
[1,1]<stdout>:Train Epoch: 3 [0/30000 (0%)]        Loss: 1.927540
[1,0]<stdout>:Train Epoch: 3 [0/30000 (0%)]        Loss: 1.839728
[1,0]<stdout>:Train Epoch: 3 [10000/30000 (33%)]   Loss: 1.649760
[1,1]<stdout>:Train Epoch: 3 [10000/30000 (33%)]   Loss: 1.776043
[1,0]<stdout>:Train Epoch: 3 [20000/30000 (67%)]   Loss: 1.668431
[1,1]<stdout>:Train Epoch: 3 [20000/30000 (67%)]   Loss: 1.440771
[1,1]<stdout>:Train Epoch: 4 [0/30000 (0%)]        Loss: 1.303080
[1,0]<stdout>:Train Epoch: 4 [0/30000 (0%)]        Loss: 1.305101
[1,1]<stdout>:Train Epoch: 4 [10000/30000 (33%)]   Loss: 1.169402
[1,0]<stdout>:Train Epoch: 4 [10000/30000 (33%)]   Loss: 1.017144
[1,0]<stdout>:Train Epoch: 4 [20000/30000 (67%)]   Loss: 1.397225
[1,1]<stdout>:Train Epoch: 4 [20000/30000 (67%)]   Loss: 1.152224
[1,0]<stdout>:Train Epoch: 5 [0/30000 (0%)]        Loss: 1.005466
```

```
[1,1]<stdout>:Train Epoch: 5 [0/30000 (0%)]        Loss: 1.233254
[1,0]<stdout>:Train Epoch: 5 [10000/30000 (33%)]   Loss: 0.861171
[1,1]<stdout>:Train Epoch: 5 [10000/30000 (33%)]   Loss: 0.917755
[1,0]<stdout>:Train Epoch: 5 [20000/30000 (67%)]   Loss: 0.952472
[1,1]<stdout>:Train Epoch: 5 [20000/30000 (67%)]   Loss: 0.880607
[1,1]<stderr>:<string>:21: UserWarning: Implicit dimension choice for
log_softmax has been deprecated. Change the call to include dim=X as an
argument.
[1,1]<stderr>:
9920512it [02:51, 57786.38it/s][1,1]<stderr>:
[1,1]<stderr>:
1654784it [02:12, 12458.99it/s] [1,1]<stderr>:
[1,0]<stderr>:
9920512it [02:51, 57726.68it/s]
[1,0]<stderr>:
1654784it [01:30, 18335.52it/s]
```

HorovodRunner 采用了一种 Python 方法，该方法使用 Horovod 钩子包含深入学习的训练代码。这个方法会在 Spark driver 上进行序列化并发送给 Spark workers。Horovod MPI 作业使用 Spark 的屏障执行模式作为 Spark 作业嵌入。第一个执行器使用 BarrierTaskContext 收集所有任务执行器的 IP 地址，并使用 mpirun 触发 Horovod 作业。每个 Python MPI 进程都会加载序列化的用户程序，对其进行反序列化并运行。注意，可以使用 np=–1 在 Driver 节点上生成子流程，以加快开发周期，示例代码如下：

```
hr = HorovodRunner(np=-1)
hr.run(run_training)
```

（11）查询 PyTorch 和 Horovod 分布式训练保存的模型结果文件。

查询 Databricks 文件系统中/ml/horovod_pytorch/的目录。

```
1.   print(dbutils.fs.ls("dbfs:/ml/horovod_pytorch/"))
```

在 Databricks 云平台 Notebook 中运行上述代码，分别显示运行的时间戳和目录信息，运行结果如下：

```
[FileInfo(path='dbfs:/ml/horovod_pytorch/1568241076.7296364/',
name='1568241076.7296364/', size=0), FileInfo(path='dbfs:/ml/horovod_pytorch/
1568446574.306003/', name='1568446574.306003/', size=0), FileInfo(path='dbfs:/ml/
horovod_pytorch/1568465145.6853325/', name='1568465145.6853325/', size=0)]
```

查询时间戳 1568465145.6853325 目录保存的模型文件。

```
1.  print(dbutils.fs.ls("dbfs:/ml/horovod_pytorch/1568465145.6853325/MNISTDemo/"))
```

在 Databricks 云平台 Notebook 中运行上述代码，显示每一个时代分别保存一个模型文件，运行结果如下。

```
[FileInfo(path='dbfs:/ml/horovod_pytorch/1568465145.6853325/MNISTDemo/che
ckpoint-1.pth.tar', name='checkpoint-1.pth.tar', size=177215), FileInfo(path=
'dbfs:/ml/horovod_pytorch/1568465145.6853325/MNISTDemo/checkpoint-2.pth.t
ar', name='checkpoint-2.pth.tar', size=177215), FileInfo(path='dbfs:/ml/
horovod_pytorch/1568465145.6853325/MNISTDemo/checkpoint-3.pth.tar',
name='checkpoint-3.pth.tar', size=177215), FileInfo(path='dbfs:/ml/horovod_
pytorch/1568465145.6853325/MNISTDemo/checkpoint-4.pth.tar', name='checkpoint-4.
pth.tar', size=177215), FileInfo(path='dbfs:/ml/horovod_pytorch/1568465145.
6853325/MNISTDemo/checkpoint-5.pth.tar', name='checkpoint-5.pth.tar', size=177215)]
```

32.1.4.2　HorovodEstimator使用Horovod和Spark MLlib进行分布式训练

HorovodEstimator 是一个 Spark MLlib 估计器 API，利用美国 Uber 公司开发的 Horovod 框架，有助于 Spark 数据帧进行深度神经网络分布式、多 GPU 的训练，简化了 Spark ETL 与 TensorFlow 中模型训练的集成。

HorovodEstimator 可以与 Spark MLlib Pipelines 一起使用，但估计器的持久化还不受支持。HorovodEstimator 拟合模型以后将返回一个 MLlib Transformer（TFTransformer），转换器可用于 Spark 数据帧上的分布式预测，可以存储模型检查点（用于恢复训练）、事件文件（包含训练期间记录的度量）、tf.savedModel 可以存储模型到指定的目录中（也用于 Spark 之外的其他框架进行预测）。HorovodEstimator 不提供容错，如果在训练过程中发生错误，HorovodEstimator 不会尝试恢复，可以重新运行 fit()从最新的检查点恢复训练。

HorovodEstimator 通过以下方式使用 Horovod 进行分布式训练。

- 将训练代码和数据分发到集群中的每个节点。
- 在 Driver 节点和 workers 节点之间启用 SSH 无密登录，通过 MPI 启动训练。
- 编写自定义数据读取和模型导出逻辑。
- 同步运行模式训练与评估。

使用 HorovodEstimator 在 Mnist 数据集进行深层神经网络训练步骤如下。

（1）创建 Databricks 云平台集群 Notebook。

在 Workspace 中单击右键，在弹出的快捷菜单中选择 Create，弹出 Create Notebook 对话框，依次输入名称（HorovodEstimator）、开发语言（选择 Python 语言）、云平台集群（DeepLearning），单击 Create 按钮创建 Notebook，如图 32-42 所示。

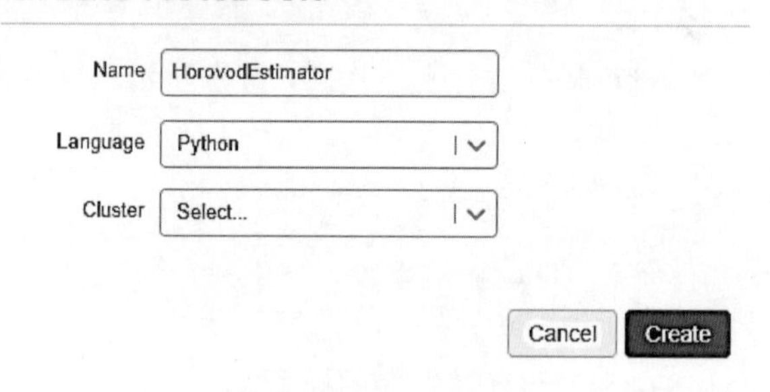

图 32-42　创建 Notebook

（2）导入 numpy、tensorflow、horovod、pyspark、HorovodEstimator 库。

```
1.  import numpy as np
2.  import tensorflow as tf
3.  import horovod.tensorflow as hvd
4.
5.  from pyspark.sql.types import *
6.  from pyspark.sql.functions import rand, when
7.
8.  from sparkdl.estimators.horovod_estimator.estimator import HorovodEstimator
```

（3）加载 Mnist 数据集，将图像表示为浮动数组。

Mnist 数据集是手写数字的大型数据库，如图 32-43 所示。

图 32-43　Mnist 数据集

通过 tf.keras.datasets.mnist 加载 Mnist 数据集。

```
1.  (x_train, y_train), (x_test, y_test) = tf.keras.datasets.mnist.load_
    data("/tmp/mnist")
2.  x_train = x_train.reshape((x_train.shape[0], -1))
3.  data = [(x_train[i].astype(float).tolist(), int(y_train[i])) for i in
    range(len(y_train))]
4.  schema = StructType([StructField("image", ArrayType(FloatType())),
5.                       StructField("label_col", LongType())])
6.  df = spark.createDataFrame(data, schema)
7.  display(df)
```

第 1 行代码通过 tf.keras.datasets.mnist.load_data 加载 Mnist 数据集。

第 4 行代码定义 Spark 数据帧的模式，第一列元素是手写数字的图像数据，第二列元素是手写数字的图像标签。

第 6 行代码将图像转换为 Spark 的数据帧。

在 Databricks 云平台 Notebook 中运行上述代码，数据帧第 1 列（image）是图像数据，第 2 列（label_col）是手写数字标签，运行结果如图 32-44 所示。

▸ (1) Spark Jobs
▾ df: pyspark.sql.dataframe.DataFrame
▾ image: array
element: float
label_col: long

image	label_col
▸ [0,0,0,0,0,0,0,0,0,0,0,0,0,0,0,0,0,0,0,0,0,0,0,0,0,0,0,0,0,0,0,0,0,0,0,0,0,0,0,0,0,0,0,0,0,0,0,0,0,0,0,0,0,0,0,0,0,0,0,0,0,0,0,0,0,0,0,0,0,0,0,0,0,0,0,0,0,0,0,0,0,0,0,0,0,0,0,0,0,0,0,0,0,0,0,0, ··········	5
▸ [0,0,0,0,0,0,0,0,0,0,0,0,0,0,0,0,0,0,0,0,0,0,0,0,0,0,0,0,0,0,0,0,0,0,0,0,0,0,0,0,0,0,0,0,0,0,0,0,0,0,0,0,0,0,0,0,0,0,0,0,0,0,0,0,0,0,0,0,0,0,0,0,0,0,0,0,0,0,0,0,0,0,0,0,0,0,0,0,0,0,0,0,0,0,0,0, ··········	0
▸ [0,0,0,0,0,0,0,0,0,0,0,0,0,0,0,0,0,0,0,0,0,0,0,0,0,0,0,0,0,0,0,0,0,0,0,0,0,0,0,0,0,0,0,0,0,0,0,0,0,0,0,0,0,0,0,0,0,0,0,0,0,0,0,0,0,0,0,0,0,0,0,0,0,0,0,0,0,0,0,0,0,0,0,0,0,0,0,0,0,0,0,0,0,0,0,0, ··········	4
▸ [0,0,0,0,0,0,0,0,0,0,0,0,0,0,0,0,0,0,0,0,0,0,0,0,0,0,0,0,0,0,0,0,0,0,0,0,0,0,0,0,0,0,0,0,0,0,0,0,0,0,0,0,0,0,0,0,0,0,0,0,0,0,0,0,0,0,0,0,0,0,0,0,0,0,0,0,0,0,0,0,0,0,0,0,0,0,0,0,0,0,0,0,0,0,0,0, ··········	1
▸ [0,0,0,0,0,0,0,0,0,0,0,0,0,0,0,0,0,0,0,0,0,0,0,0,0,0,0,0,0,0,0,0,0,0,0,0,0,0,0,0,0,0,0,0,0,0,0,0,0,0,0,0,0,0,0,0,0,0,0,0,0,0,0,0,0,0,0,0,0,0,0,0,0,0,0,0,0,0,0,0,0,0,0,0,0,0,0,0,0,0,0,0,0,0,0,0, ··········	9
▸ [0,0,0,0,0,0,0,0,0,0,0,0,0,0,0,0,0,0,0,0,0,0,0,0,0,0,0,0,0,0,0,0,0,0,0,0,0,0,0,0,0,0,0,0,0,0,0,0,0,0,0,0,0,0,0,0,0,0,0,0,0,0,0,0,0,0,0,0,0,0,0,0,0,0,0,0,0,0,0,0,0,0,0,0,0,0,0,0,0,0,0,0,0,0,0,0, ··········	2
▸ [0,0,0,0,0,0,0,0,0,0,0,0,0,0,0,0,0,0,0,0,0,0,0,0,0,0,0,0,0,0,0,0,0,0,0,0,0,0,0,0,0,0,0,0,0,0,0,0,0,0,0,0,0,0,0,0,0,0,0,0,0,0,0,0,0,0,0,0,0,0,0,0,0,0,0,0,0,0,0,0,0,0,0,0,0,0,0,0,0,0,0,0,0,0,0,0, ··········	1
▸ [0,0,0,0,0,0,0,0,0,0,0,0,0,0,0,0,0,0,0,0,0,0,0,0,0,0,0,0,0,0,0,0,0,0,0,0,0,0,0,0,0,0,0,0,0,0,0,0,0,0,0,0,0,0,0,0,0,0,0,0,0,0,0,0,0,0,0,0,0,0,0,0,0,0,0,0,0,0,0,0,0,0,0,0,0,0,0,0,0,0,0,0,0,0,0,0, ··········	3

图 32-44　Spark DataFrame 展示

（4）查询 HorovodEstimator API 的帮助文档。

```
1.  help(HorovodEstimator)
```

在 Databricks 云平台 Notebook 中运行上述代码，运行结果如下：

```
help(HorovodEstimator)
Help on class HorovodEstimator in module sparkdl.estimators.horovod_estimator.estimator:

class HorovodEstimator(pyspark.ml.base.Estimator,
sparkdl.estimators.horovod_ estimator.params.HasModelFunction,
sparkdl.estimators.horovod_estimator.params.HasFeatureMapping,
sparkdl.estimators.horovod_estimator.params.HasLabelCol,
sparkdl.estimators.horovod_estimator.params.HasModelDir,
sparkdl.estimators.horovod_estimator.params.HasMaxSteps,
sparkdl.estimators.horovod_estimator.params.HasBatchSize,
sparkdl.estimators.horovod_estimator.params.HasIsValidationCol,
sparkdl.estimators.horovod_estimator.params.HasModelFunctionParams,
sparkdl.estimators.horovod_estimator.params.HasSaveCheckpointsSecs,
sparkdl.estimators.horovod_estimator.params.HasValidationInterval)
……
```

（5）构建一个 model_fn 函数。

定义在 Python 函数中训练的深度学习模型。model_fn 函数是 tf.estimator-style 风格，model_fn 函数定义模型的网络结构；在训练、评估和预测（推断）阶段指定模型对单个数据批的输出。

```
1.  def model_fn(features, labels, mode, params):
2.      """
3.      参数：
4.      * features：数据帧输入列名转换为张量的字典（每个张量对应于输入列中的批量数据）
5.      * labels：张量，批量数字标签
6.      * mode：指定是否运行估计器进行训练、评估或预测
7.      * params：超参数的可选指令。将在参数中接收传递给 HorovodEstimator 的内容，
8.                  允许为超参数调整配置估计器
9.      返回：tf.estimator.EstimatorSpec 描述模型
10.     """
11.     from tensorflow.examples.tutorials.mnist import mnist
12.     # HorovodEstimator 将 Spark SQL 标量类型作为张量 [None]进行建模（一批可变大
        #小的标量），Spark SQL array 数组类型(包括 VectorUDT 向量)作为张量[None, None]
        #（一批可变大小的密集可变长度数组）。在这里，图像数据从 ArrayType(FloatType())
        #列作为 [None, None]的浮点张量，因为每个浮点数组的长度是 784，所以把张量调整成
        #[None, 784]的维度
13.     input_layer = features['image']
14.     #input_layer = tf.reshape(input_layer, shape=[-1, 784])
        #在 Spark 数据帧中是[记录数,784]的数据格式，这里无须转换，可以将这行代码注释掉，
        #否则批量输出的分类、预测概率结果将是重复的同一条记录
15.     logits = mnist.inference(input_layer, hidden1_units=params["hidden1_units"],
16.                              hidden2_units=params["hidden2_units"])
17.     serving_key = tf.saved_model.signature_constants.DEFAULT_SERVING_
        SIGNATURE_DEF_KEY
18.     #生成一个预测输出名称到张量的字典（用于预测模式）。通过拟合估计器，TFTransformer
        #产生输出列，张量的输出对应于 DEFAULT_SERVING_SIGNATURE_DEF_KEY
19.     predictions = {
20.         "classes": tf.argmax(input=logits, axis=1, name="classes_tensor"),
```

```
21.         "probabilities": tf.nn.softmax(logits, name="softmax_tensor"),
22.     }
23.     export_outputs = {serving_key: tf.estimator.export.PredictOutput(predictions)}
24.     #如果估计器以预测模式运行，可以在这里停止构建模型图，并简单地返回模型的推断输出
25.     if mode == tf.estimator.ModeKeys.PREDICT:
26.         return tf.estimator.EstimatorSpec(mode=mode, predictions=predictions,
        export_outputs=export_outputs)
27.     # 计算损失（训练和评估模式）
28.     onehot_labels = tf.one_hot(indices=tf.cast(labels, tf.int32), depth=10)
29.     loss = tf.losses.softmax_cross_entropy(onehot_labels=onehot_labels,
    logits=logits)
30.     if mode == tf.estimator.ModeKeys.TRAIN:
31.         #  设置 hooks；每个 worker 节点都要运行
32.         logging_hooks = [tf.train.LoggingTensorHook(tensors={"predictions":
        "classes_tensor"}, every_n_iter=5000)]
33.         # Horovod：按 workers 数量调整学习率，添加分布式优化器
34.         optimizer = tf.train.MomentumOptimizer(
35.             learning_rate=0.001 * hvd.size(), momentum=0.9)
36.         optimizer = hvd.DistributedOptimizer(optimizer)
37.         train_op = optimizer.minimize(
38.             loss=loss,
39.             global_step=tf.train.get_global_step())
40.         return tf.estimator.EstimatorSpec(mode=mode, loss=loss, train_op=
        train_op,export_outputs=export_outputs, training_hooks=logging_hooks)
41.     #如果在评估模式下运行，将模型评估度量（准确性）添加到 EstimatorSpec 中，以便
42.     #模型评估运行时记录它们
43.     eval_metric_ops = {"accuracy": tf.metrics.accuracy(
44.         labels=labels, predictions=predictions["classes"])}
45.     return tf.estimator.EstimatorSpec(
46.         mode=mode, loss=loss, eval_metric_ops=eval_metric_ops, export_outputs=
        export_outputs)
```

第 35 行代码 Horovod 按照 workers 数量调整学习率。

第 36 行代码使用 Horovod 的 DistributedOptimizer 方法封装分布式优化器。

（6）创建一个 HorovodEstimator，用于在 Spark 数据帧上对模型进行分布式训练。

```
1.  # 模型检查点将保存到 Driver 节点的本地文件系统中
2.  model_dir = "/tmp/horovod_estimator"
3.  dbutils.fs.rm(model_dir[5:], recurse=True)
4.  # Create estimator
5.  est = HorovodEstimator(modelFn=model_fn,
6.                         featureMapping={"image": "image"},
7.                         modelDir=model_dir,
8.                         labelCol="label_col",
9.                         batchSize=64,
10.                        maxSteps=5000,
11.                        isValidationCol="isVal",
12.                        modelFnParams={"hidden1_units": 100, "hidden2_units": 50},
13.                        saveCheckpointsSecs=30)
```

（7）在 Mnist 数据帧上拟合估计器，然后使用拟合模型（深度学习管道的 TFTransformer）转换数据。

拟合估计器在 modelDir 目录中会产生以下结果。

- 存储模型的检查点文件，用于恢复训练。
- 事件文件，包含训练和模型评估期间记录的度量日志，可通过 TensorBoard 进行可视化。

❑ 训练结束时：tf.savedModel 将模型文件保存到 modeldir 目录，可用于预测。

新的类别和概率输出列由转换器添加到 Spark 数据帧中。

```
1.  # 增加 1 列指示每一行是否在训练/验证集中；对数据执行随机拆分
2.  df_with_val = df.withColumn("isVal", when(rand() > 0.8, True)
    .otherwise(False))
3.  #拟合评估器获取一个 TFTransformer
4.  transformer = est.fit(df_with_val)
5.  #将 TFTransformer 应用于训练数据并显示结果。注意，预测的“classes”往往与训练集中
    #的标签列相匹配
6.  res = transformer.transform(df)
7.  display(res)
```

在 Databricks 云平台 Notebook 中运行上述代码，运行结果如图 32-45 所示。

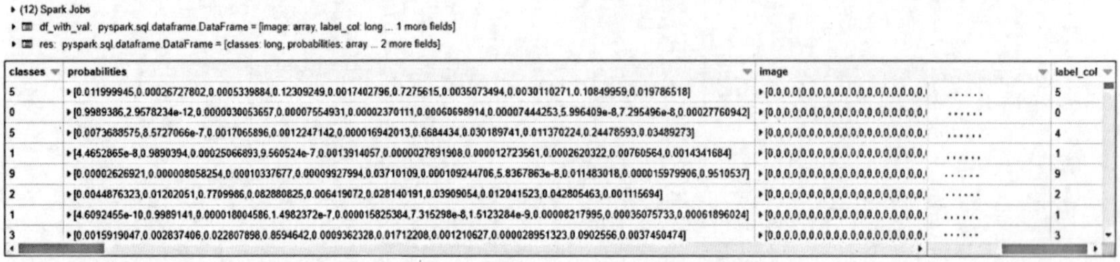

▸ (12) Spark Jobs

▸ df_with_val: pyspark.sql.dataframe.DataFrame = [image: array, label_col: long ... 1 more fields]

▸ res: pyspark.sql.dataframe.DataFrame = [classes: long, probabilities: array ... 2 more fields]

classes	probabilities	image	label_col
5	▸[0.011999945,0.00026727802,0.0005339884,0.12309249,0.0017402796,0.7275615,0.0035073494,0.0030110271,0.10849959,0.019786518]	▸[0,0,0,0,0,0,0,0,0,0,0,0,0,0,0,0,0,0,0, ...	5
0	▸[0.9989386,2.9578234e-12,0.0000030053657,0.00007554931,0.00002370111,0.00060698914,0.00007444253,5.996409e-8,7.295496e-8,0.00027760942]	▸[0,0,0,0,0,0,0,0,0,0,0,0,0,0,0,0,0,0,0, ...	0
5	▸[0.0073688575,8.5727066e-7,0.0017065896,0.0012247142,0.000016942013,0.6684434,0.030189741,0.011370224,0.24478593,0.03489273]	▸[0,0,0,0,0,0,0,0,0,0,0,0,0,0,0,0,0,0,0, ...	4
1	▸[4.4652865e-8,0.9890394,0.00025066893,9.560524e-7,0.0013914057,0.0000027891908,0.000012723561,0.0002620322,0.00760564,0.0014341684]	▸[0,0,0,0,0,0,0,0,0,0,0,0,0,0,0,0,0,0,0, ...	1
9	▸[0.0002626921,0.000008058254,0.00010337677,0.00009927994,0.03710109,0.000109244706,5.8367863e-8,0.011483018,0.000015979906,0.9510537]	▸[0,0,0,0,0,0,0,0,0,0,0,0,0,0,0,0,0,0,0, ...	9
2	▸[0.0044876323,0.01202051,0.7709986,0.082880825,0.006419072,0.028140191,0.03909054,0.012041523,0.042805463,0.001115694]	▸[0,0,0,0,0,0,0,0,0,0,0,0,0,0,0,0,0,0,0, ...	2
1	▸[4.6092455e-10,0.9989141,0.000018004586,1.4982372e-7,0.000015825384,7.315298e-8,1.5123284e-9,0.00008217995,0.00035075733,0.00061896024]	▸[0,0,0,0,0,0,0,0,0,0,0,0,0,0,0,0,0,0,0, ...	1
3	▸[0.0015919047,0.002837406,0.022807898,0.8594642,0.0009362328,0.01712208,0.001210627,0.000028951323,0.0902556,0.0037450474]	▸[0,0,0,0,0,0,0,0,0,0,0,0,0,0,0,0,0,0,0, ...	3

图 32-45　Spark 数据帧的运行结果

图中第一条记录的详细内容如下，第 1 列是模型的预测值，表示预测手写数字标签是数字 5；第 2 列是模型预测的概率值，其中最大的概率值是 0.7275615，它的索引是 5，也就是预测值 5；第 3 列是手写数字的图像数据，是一个 784 大小的数组，第 4 列是手写数字的实际值，为数字 5。

```
5[0.011999945,0.00026727802,0.0005339884,0.12309249,0.0017402796,0.7275615,
0.0035073494,0.0030110271,0.10849959,0.019786518][0,0,0,0,0,0,0,0,0,0,0,
0,0,0,0,0,0,0,0,0,0,0,0,0,0,0,0,0,0,0,0,0,0,0,0,0,0,0,0,0,0,0,0,0,0,0,0,0,
0,0,0,0,0,0,0,0,0,0,0,0,0,0,0,0,0,0,0,0,0,0,0,0,0,0,0,0,0,0,0,0,0,0,0,0,0,
0,0,0,0,0,0,0,0,0,0,0,0,0,0,0,0,0,0,0,0,0,0,0,0,0,0,0,0,0,0,0,0,0,0,0,0,0,0
,0,0,0,0,0,0,0,0,0,0,0,0,0,0,0,0,0,0,0,0,0,0,0,0,0,0,0,0,0,0,0,0,3,18,18,
18,126,136,175,26,166,255,247,127,0,0,0,0,0,0,0,0,0,0,0,0,30,36,94,154,17
0,253,253,253,253,253,225,172,253,242,195,64,0,0,0,0,0,0,0,0,0,0,0,49,238,2
53,253,253,253,253,253,253,253,251,93,82,82,56,39,0,0,0,0,0,0,0,0,0,0,0,0,
0,18,219,253,253,253,253,253,198,182,247,241,0,0,0,0,0,0,0,0,0,0,0,0,0,0,
0,0,0,0,80,156,107,253,253,205,11,0,43,154,0,0,0,0,0,0,0,0,0,0,0,0,0,0,0,
0,0,0,0,14,1,154,253,90,0,0,0,0,0,0,0,0,0,0,0,0,0,0,0,0,0,0,0,0,0,0,0,0,0,
0,139,253,190,2,0,0,0,0,0,0,0,0,0,0,0,0,0,0,0,0,0,0,0,0,0,0,0,11,190,25
3,70,0,0,0,0,0,0,0,0,0,0,0,0,0,0,0,0,0,0,0,0,0,0,0,0,35,241,225,160,108,
1,0,0,0,0,0,0,0,0,0,0,0,0,0,0,0,0,0,0,0,0,0,0,0,81,240,253,253,119,25,0,
0,0,0,0,0,0,0,0,0,0,0,0,0,0,0,0,0,0,0,0,0,0,45,186,253,253,150,27,0,0,0,
0,0,0,0,0,0,0,0,0,0,0,0,0,0,0,0,0,0,0,0,16,93,252,253,187,0,0,0,0,0,0,0,
0,0,0,0,0,0,0,0,0,0,0,0,0,0,0,0,0,0,0,0,249,253,249,64,0,0,0,0,0,0,0,0,0,0,
0,0,0,0,0,0,0,0,0,0,0,0,46,130,183,253,253,207,2,0,0,0,0,0,0,0,0,0,0,0,0,0,
0,0,0,0,0,39,148,229,253,253,253,250,182,0,0,0,0,0,0,0,0,0,0,0,0,0,0,0,0,
0,0,24,114,221,253,253,253,253,201,78,0,0,0,0,0,0,0,0,0,0,0,0,0,0,0,0,0,0,2
3,66,213,253,253,253,253,198,81,2,0,0,0,0,0,0,0,0,0,0,0,0,0,0,0,0,0,18,171,
219,253,253,253,253,195,80,9,0,0,0,0,0,0,0,0,0,0,0,0,0,0,0,0,55,172,226,2
53,253,253,253,244,133,11,0,0,0,0,0,0,0,0,0,0,0,0,0,0,0,0,0,0,136,253,253,
```

```
253,212,135,132,16,0,0,0,0,0,0,0,0,0,0,0,0,0,0,0,0,0,0,0,0,0,0,0,0,0,0,0,0,0,0,
0,0,0,0,0,0,0,0,0,0,0,0,0,0,0,0,0,0,0,0,0,0,0,0,0,0,0,0,0,0,0,0,0,0,0,0,0,0,0,
0,0,0,0,0,0,0,0,0,0,0,0,0,0,0,0,0,0,0,0,0,0,0,0,0,0,0,0,0,0,0,0,0,0,0,0,0,0,0]5
```

（8）可以从检查点文件还原模型以恢复训练。

增加评估器的 MaxSteps 参数，从以前保存的检查点开始，再次进行训练。

```
1.  est.setMaxSteps(10000)
2.  new_transformer = est.fit(df_with_val)
3.  new_res = transformer.transform(df)
4.  display(new_res)
```

在 Databricks 云平台 Notebook 中运行上述代码，第 1 列是模型的预测值；第 2 列是模型预测的概率值；第 3 列是手写数字的图像数据，第 4 列是手写数字的实际值，运行结果如图 32-46 所示。

▸ (12) Spark Jobs

▸ new_res: pyspark.sql.dataframe.DataFrame = [classes: long, probabilities: array ... 2 more fields]

classes	probabilities	image		label_col
5	▸[0.011999945,0.00026727802,0.0005339884,0.12309249,0.0017402796,0.7275615,0.0035073494,0.0030110271,0.10849959,0.019786518]	▸[0,0,0,0,0,0,0,0,0,0,0,0,0,0,0,0,0,0,0,0,0	……	5
0	▸[0.9989386,2.9578234e-12,0.0000030053657,0.00007554931,0.00002370111,0.00060698914,0.00007444253,5.996409e-8,7.295496e-8,0.00027760942]	▸[0,0,0,0,0,0,0,0,0,0,0,0,0,0,0,0,0,0,0,0,0	……	0
5	▸[0.0073688575,8.5727066e-7,0.0017065896,0.0012247142,0.000016942013,0.6684434,0.030189741,0.011370224,0.24478593,0.03489273]	▸[0,0,0,0,0,0,0,0,0,0,0,0,0,0,0,0,0,0,0,0,0	……	4
1	▸[4.4652865e-8,0.9890394,0.00025066893,9.560524e-7,0.0013914057,0.0000027891908,0.000012723561,0.0002620322,0.00760564,0.0014341684]	▸[0,0,0,0,0,0,0,0,0,0,0,0,0,0,0,0,0,0,0,0,0	……	1
9	▸[0.00002626921,0.000008058254,0.00010337677,0.00009927994,0.03710109,0.000109244706,5.8367863e-8,0.011483018,0.000015979906,0.9510537]	▸[0,0,0,0,0,0,0,0,0,0,0,0,0,0,0,0,0,0,0,0,0	……	9
2	▸[0.0044876323,0.01202051,0.7709986,0.082880825,0.006419072,0.028140191,0.03909054,0.012041523,0.042805463,0.001115694]	▸[0,0,0,0,0,0,0,0,0,0,0,0,0,0,0,0,0,0,0,0,0	……	2
1	▸[4.6092455e-10,0.9989141,0.000018004586,1.4982372e-7,0.000015825384,7.315298e-8,1.5123284e-9,0.00008217995,0.00035075733,0.00061896024]	▸[0,0,0,0,0,0,0,0,0,0,0,0,0,0,0,0,0,0,0,0,0	……	1
3	▸[0.0015919047,0.002837406,0.022807898,0.8694642,0.0009362328,0.01712208,0.001210627,0.000028951323,0.0902556,0.0037450474]	▸[0,0,0,0,0,0,0,0,0,0,0,0,0,0,0,0,0,0,0,0,0	……	3

Showing the first 291 rows.

图 32-46　再次进行训练的结果

（9）将模型检查点复制到 Databricks dbfs 文件系统。

将模型检查点和事件文件复制到 Databricks dbfs 文件系统进行持久化，在当前集群终止后也可以访问模型文件。

```
1.  dbutils.fs.cp("file:/tmp/horovod_estimator/", "dbfs:/horovod_estimator/",
    recurse=True)
```

在 Databricks 云平台 Notebook 中运行上述代码，运行结果如下：

```
Out[16]: True
```

检查本地文件系统中的模型文件信息。

```
1.  %sh
2.  ls -ltr /tmp/horovod_estimator
```

在 Databricks 云平台 Notebook 中运行上述代码，运行结果如下：

```
total 4828
drwxr-x--- 3 root root   4096 Sep 15 07:33 1
-rw-r--r-- 1 root root    546 Sep 15 07:33 model.ckpt-0.index
-rw-r--r-- 1 root root 672488 Sep 15 07:33 model.ckpt-0.data-00000-of-00001
-rw-r--r-- 1 root root 107671 Sep 15 07:33 model.ckpt-0.meta
-rw-r--r-- 1 root root    546 Sep 15 07:33 model.ckpt-2752.index
-rw-r--r-- 1 root root 672488 Sep 15 07:33 model.ckpt-2752.data-00000-of-00001
-rw-r--r-- 1 root root 107671 Sep 15 07:33 model.ckpt-2752.meta
-rw-r--r-- 1 root root 352098 Sep 15 07:34 events.out.tfevents.1568532786.
0905-133313-yet1-10-83-169-36
drwxr-xr-x 2 root root   4096 Sep 15 07:35 eval
-rw-r--r-- 1 root root 268815 Sep 15 07:35 graph.pbtxt
```

```
-rw-r--r-- 1 root root    546 Sep 15 07:35 model.ckpt-5000.index
-rw-r--r-- 1 root root 672488 Sep 15 07:35 model.ckpt-5000.data-00000-of-00001
-rw-r--r-- 1 root root 107671 Sep 15 07:35 model.ckpt-5000.meta
-rw-r--r-- 1 root root    546 Sep 15 07:35 model.ckpt-7771.index
-rw-r--r-- 1 root root 672488 Sep 15 07:35 model.ckpt-7771.data-00000-of-00001
-rw-r--r-- 1 root root 107671 Sep 15 07:35 model.ckpt-7771.meta
-rw-r--r-- 1 root root    546 Sep 15 07:35 model.ckpt-10000.index
-rw-r--r-- 1 root root 672488 Sep 15 07:35 model.ckpt-10000.data-00000-of-00001
-rw-r--r-- 1 root root    270 Sep 15 07:35 checkpoint
-rw-r--r-- 1 root root 107671 Sep 15 07:35 model.ckpt-10000.meta
-rw-r--r-- 1 root root 352105 Sep 15 07:35 events.out.tfevents.1568532900.
0905-133313-yet1-10-83-169-36
drwxr-xr-x 4 root root   4096 Sep 15 07:35 saved_model
```

检查 Databricks 文件系统中的模型文件信息。

```
1.  print(dbutils.fs.ls("dbfs:/horovod_estimator/"))
```

在 Databricks 云平台 Notebook 中运行上述代码，运行结果如下：

```
[FileInfo(path='dbfs:/horovod_estimator/1/', name='1/', size=0), FileInfo(path=
'dbfs:/horovod_estimator/checkpoint', name='checkpoint', size=270), FileInfo(path=
'dbfs:/horovod_estimator/eval/', name='eval/', size=0), FileInfo(path='dbfs:/
horovod_estimator/events.out.tfevents.1568532786.0905-133313-yet1-10-83-1
69-36', name='events.out.tfevents.1568532786.0905-133313-yet1-10-83-169-36',
size=352098), FileInfo(path='dbfs:/horovod_estimator/events.out.tfevents.
1568532900.0905-133313-yet1-10-83-169-36', name='events.out.tfevents.1568532900.
0905-133313-yet1-10-83-169-36', size=352105), FileInfo(path='dbfs:/horovod_
estimator/graph.pbtxt', name='graph.pbtxt', size=268815), FileInfo(path='dbfs:/
horovod_estimator/model.ckpt-0.data-00000-of-00001', name='model.ckpt-0.data-
00000-of-00001', size=672488), FileInfo(path='dbfs:/horovod_estimator/model.
ckpt-0.index', name='model.ckpt-0.index', size=546), FileInfo(path='dbfs:/
horovod_estimator/model.ckpt-0.meta', name='model.ckpt-0.meta', size=107671),
FileInfo(path='dbfs:/horovod_estimator/model.ckpt-10000.data-00000-of-000
01', name='model.ckpt-10000.data-00000-of-00001', size=672488), FileInfo(path=
'dbfs:/horovod_estimator/model.ckpt-10000.index', name='model.ckpt-10000.index',
size=546), FileInfo(path='dbfs:/horovod_estimator/model.ckpt-10000.meta',
name='model.ckpt-10000.meta', size=107671), FileInfo(path='dbfs:/horovod_
estimator/model.ckpt-2752.data-00000-of-00001', name='model.ckpt-2752.data-
00000-of-00001', size=672488), FileInfo(path='dbfs:/horovod_estimator/model.
ckpt-2752.index', name='model.ckpt-2752.index', size=546), FileInfo(path=
'dbfs:/horovod_estimator/model.ckpt-2752.meta', name='model.ckpt-2752.meta',
size=107671), FileInfo(path='dbfs:/horovod_estimator/model.ckpt-5000.data-
00000-of-00001', name='model.ckpt-5000.data-00000-of-00001', size=672488),
FileInfo(path='dbfs:/horovod_estimator/model.ckpt-5000.index', name='model.
ckpt-5000.index', size=546), FileInfo(path='dbfs:/horovod_estimator/model.
ckpt-5000.meta', name='model.ckpt-5000.meta', size=107671), FileInfo(path=
'dbfs:/horovod_estimator/model.ckpt-7771.data-00000-of-00001', name='model.
ckpt-7771.data-00000-of-00001', size=672488), FileInfo(path='dbfs:/horovod_
estimator/model.ckpt-7771.index', name='model.ckpt-7771.index', size=546),
FileInfo(path='dbfs:/horovod_estimator/model.ckpt-7771.meta', name='model.ckpt-
7771.meta', size=107671), FileInfo(path='dbfs:/horovod_estimator/saved_
model/', name='saved_model/', size=0)]
```

（10）清理本地文件系统 tmp 目录下的 horovod_estimator 目录及文件。

```
1.  %sh
2.  #rm -rf /tmp/horovod_estimator
```

32.2　Spark+PyTorch 案例实战

本节讲解 PyTorch 在 Spark 上的安装、使用 PyTorch 进行图像识别、PyTorch 性能调优实践等内容。

32.2.1　PyTorch 在 Spark 上的安装

PyTorch 包含在 Databricks 机器学习运行时库 5.1 版本及更高版本中，可以使用 Databricks 机器学习运行时库创建集群并使用 PyTorch。

建议使用 Databricks 机器学习运行时库（已包含 PyTorch）。如果要使用 Databricks 运行时库（无 PyTorch），则可以将 PyTorch 安装为 Databricks PyPI 库。

（1）GPU 集群环境中安装 PyTorch 及 torchvision 版本。

```
torch==1.1.0
torchvision==0.3.0
```

（2）CPU 集群环境中可以使用 wheel 安装 PyTorch 及 torchvision 版本。

Python 3 环境：

```
https://download.pytorch.org/whl/cpu/torch-1.1.0-cp35-cp35m-linux_x86_64.whl
https://download.pytorch.org/whl/cpu/torchvision-0.3.0-cp35-cp35m-linux_x86_64.whl
```

Python 2 环境：

```
https://download.pytorch.org/whl/cpu/torch-1.1.0-cp27-cp27mu-linux_x86_64.whl
https://download.pytorch.org/whl/cpu/torchvision-0.3.0-cp27-cp27mu-linux_x86_64.whl
```

32.2.2　使用 PyTorch 实战图像识别

本节使用 Spark 及 PyTorch，基于 ResNet-50 网络模型，对鲜花图像数据（郁金香、向日葵、玫瑰、蒲公英、菊花）进行分布式图像识别，步骤如下。

（1）准备预训练的模型和鲜花集数据。

- 从 torchvision.models 加载预训练的 ResNet-50 模型。
- 将鲜花数据下载到 Databricks 文件系统空间。

（2）Spark 加载鲜花数据，并转换为 Spark 数据帧。

（3）通过 Pandas UDF 进行模型预测。

图 32-47 是 Spark+PyTorch 案例示意图。

ResNet 网络模型有不同的网络层数，比较常用的是 50-layer、101-layer 和 152-layer，都是由 ResNet 模块堆叠在一起实现的，ResNet 网络结构如图 32-48 所示。

Spark+PyTorch 图像识别步骤如下。

（1）创建 Databricks 云平台集群 Notebook。

在 Workspace 中单击右键，在弹出的快捷菜单中选择 Create，依次输入名称、开发语言（选择 Python 语言）、云平台集群，单击 Create 按钮创建 Notebook，如图 32-49 所示。

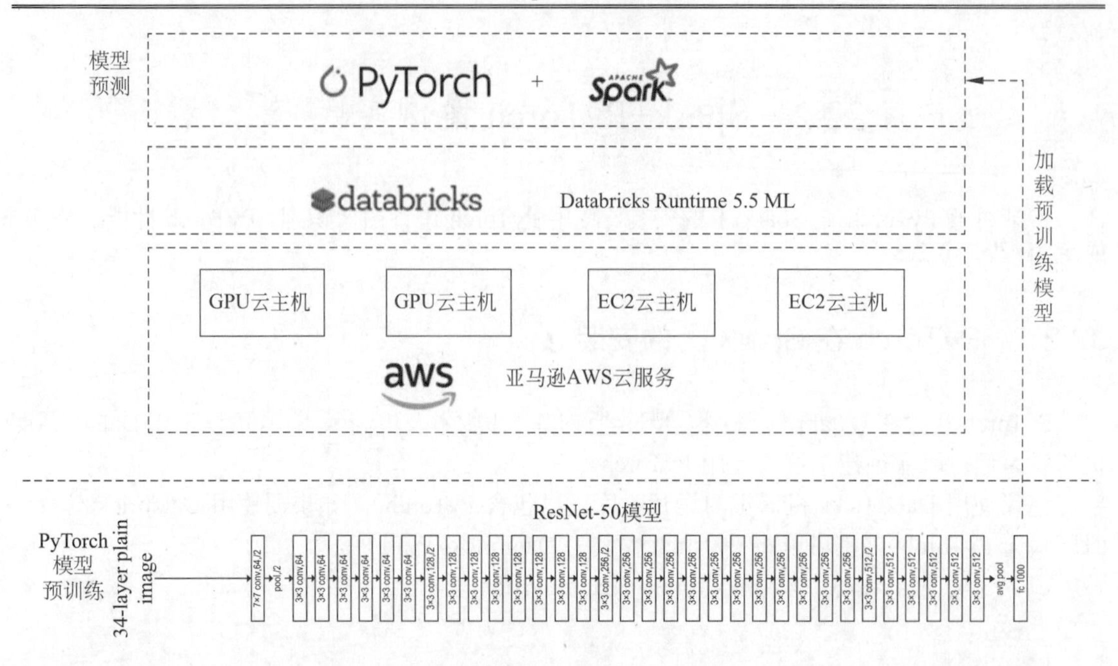

图 32-47　Spark+PyTorch 案例实战

layer name	output size	18-layer	34-layer	50-layer	101-layer	152-layer
conv1	112×112	7×7, 64, stride 2				
		3×3 max pool, stride 2				
conv2_x	56×56	[3×3, 64 3×3, 64] ×2	[3×3, 64 3×3, 64] ×3	[1×1, 64 3×3, 64 1×1, 256] ×3	[1×1, 64 3×3, 64 1×1, 256] ×3	[1×1, 64 3×3, 64 1×1, 256] ×3
conv3_x	28×28	[3×3, 128 3×3, 128] ×2	[3×3, 128 3×3, 128] ×4	[1×1, 128 3×3, 128 1×1, 512] ×4	[1×1, 128 3×3, 128 1×1, 512] ×4	[1×1, 128 3×3, 128 1×1, 512] ×8
conv4_x	14×14	[3×3, 256 3×3, 256] ×2	[3×3, 256 3×3, 256] ×6	[1×1, 256 3×3, 256 1×1, 1024] ×6	[1×1, 256 3×3, 256 1×1, 1024] ×23	[1×1, 256 3×3, 256 1×1, 1024] ×36
conv5_x	7×7	[3×3, 512 3×3, 512] ×2	[3×3, 512 3×3, 512] ×3	[1×1, 512 3×3, 512 1×1, 2048] ×3	[1×1, 512 3×3, 512 1×1, 2048] ×3	[1×1, 512 3×3, 512 1×1, 2048] ×3
	1×1	average pool, 1000-d fc, softmax				
FLOPs		1.8×10^9	3.6×10^9	3.8×10^9	7.6×10^9	11.3×10^9

图 32-48　ResNet 网络结构

Create Notebook

Name　pytorch

Language　Python

Cluster　Select...

Cancel　Create

图 32-49　创建 Notebook

（2）设置启用 CPU 资源。

在启用 CPU 的 Apache Spark 集群上运行 notebook，设置变量 cuda=false。在启用 GPU 的 Apache Spark 集群上运行 notebook，设置变量 cuda=true。

```
1.   cuda = False
```

（3）启动 Arrow 支持。Apache Arrow 是一种内存中的列式数据格式，在 Spark 中用于高效传输 JVM 和 Python 进程之间的数据。将 Spark 数据帧转换为 Pandas 数据帧时，可以使用 Arrow 进行优化。

```
1.  spark.conf.set("spark.sql.execution.arrow.enabled", "true")
2.  spark.conf.set("spark.sql.execution.arrow.maxRecordsPerBatch", "2048")
```

（4）导入 pandas、torch、pyspark 等库。

```
1.  import os
2.  import shutil
3.  import tarfile
4.  import time
5.  import zipfile
6.
7.  try:
8.      from urllib.request import urlretrieve
9.  except ImportError:
10.     from urllib import urlretrieve
11.
12. import pandas as pd
13.
14. import torch
15. from torch.utils.data import Dataset
16. from torchvision import datasets, models, transforms
17. from torchvision.datasets.folder import default_loader  # private API
18.
19. from pyspark.sql.functions import col, pandas_udf, PandasUDFType
20. from pyspark.sql.types import ArrayType, FloatType
```

（5）PyTorch 设置是否使用 GPU。

```
1.   use_cuda = cuda and torch.cuda.is_available()
2.  device = torch.device("cuda" if use_cuda else "cpu")
```

（6）准备预训练模型和数据。

定义输入和输出目录。建议使用 Databricks Runtime 5.4 ML 或更高版本，将训练集数据保存到 Databricks 文件系统 dbfs:/ml 目录，该文件映射到 Driver 及 Worker 节点上的文件 /dbfs/ml。dbfs:/ml 是一个特殊的文件夹，为深度学习工作负载提供高性能的 I/O。

```
1.  URL = "http://download.tensorflow.org/example_images/flower_photos.tgz"
2.  input_local_dir = "/dbfs/ml/tmp/flower/"
3.  output_file_path = "/tmp/predictions"
```

（7）在 Spark Driver 节点上加载 ResNet-50 预训练模型，并广播 ResNet-50 模型的状态，如图 32-50 所示。

```
1.  bc_model_state = sc.broadcast(models.resnet50(pretrained=True).state_dict())
```

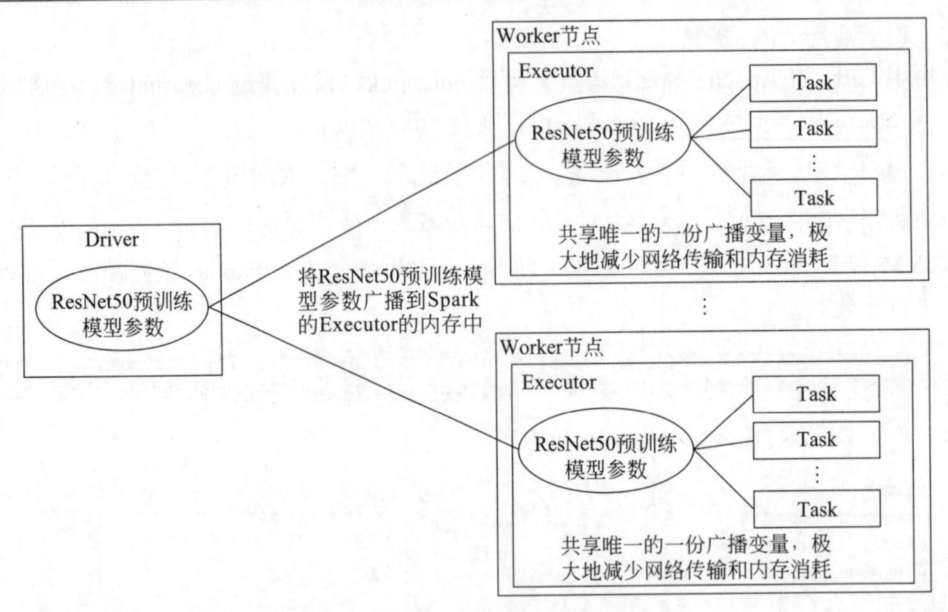

图 32-50　广播 ResNet50 预训练模型参数

（8）定义 get_model_for_eval 方法，返回一个 PyTorch ResNet-50 预训练模型实例，其加载 Spark 广播变量 ResNet50 模型的参数。

```
1.   def get_model_for_eval():
2.    """获取广播变量模型"""
3.    model = models.resnet50(pretrained=True)
4.    model.load_state_dict(bc_model_state.value)
5.    model.eval()
6.    return model
```

（9）定义 maybe_download_and_extract 方法。从 Tensorflow 网站（http://download.tensorflow.org/example_images/flower_photos.tgz）下载鲜花文件并解压缩，解压的文件包括郁金香、向日葵、玫瑰、蒲公英、菊花等图像类型。

```
1.  def maybe_download_and_extract(url, download_dir):
2.      filename = url.split('/')[-1]
3.      file_path = os.path.join(download_dir, filename)
4.      print(file_path)
5.      if not os.path.exists(file_path):
6.          if not os.path.exists(download_dir):
7.              os.makedirs(download_dir)
8.
9.          file_path, _ = urlretrieve(url=url, filename=file_path)
10.         print()
11.         print("Download finished. Extracting files.")
12.
13.         if file_path.endswith(".zip"):
14.             # Unpack the zip-file.
15.             zipfile.ZipFile(file=file_path, mode="r").extractall(download_dir)
16.         elif file_path.endswith((".tar.gz", ".tgz")):
17.             # Unpack the tar-ball.
18.             tarfile.open(name=file_path, mode="r:gz").extractall(download_dir)
19.
20.         print("Done.")
21.     else:
22.         print("Data has apparently already been downloaded and unpacked.")
```

（10）在 Databricks 文件系统中检查已经下载鲜花数据的文件目录。

```
1.   print(dbutils.fs.ls("dbfs:/ml/tmp/flower_photos/"))
```

在 Databricks 云平台 Notebook 中运行上述代码，运行结果如下：

```
[FileInfo(path='dbfs:/ml/tmp/flower_photos/LICENSE.txt', name='LICENSE.txt', size=418049), FileInfo(path='dbfs:/ml/tmp/flower_photos/ daisy/', name='daisy/', size=0),
FileInfo(path='dbfs:/ml/tmp/flower_photos/dandelion/',  name='dandelion/', size=0),
FileInfo(path='dbfs:/ml/tmp/flower_photos/roses/', name='roses/', size=0),
FileInfo(path='dbfs:/ml/tmp/flower_photos/sunflowers/', name='sunflowers/', size=0),
FileInfo(path='dbfs:/ml/tmp/flower_photos/tulips/', name='tulips/', size=0)]
```

查看 Databricks 文件系统中菊花目录的文件信息。

```
1.  print(dbutils.fs.ls("dbfs:/ml/tmp/flower_photos/daisy/"))
```

在 Databricks 云平台 Notebook 中运行上述代码，运行结果如下：

```
[FileInfo(path='dbfs:/ml/tmp/flower_photos/daisy/100080576_f52e8ee070_n.jpg', name='100080576_f52e8ee070_n.jpg', size=26797),
FileInfo(path='dbfs:/ml/tmp/flower_photos/daisy/10140303196_b88d3d6cec.jpg', name='10140303196_b88d3d6cec.jpg', size=117247),
FileInfo(path='dbfs:/ml/tmp/flower_photos/daisy/10172379554_b296050f82_n.jpg', name='10172379554_b296050f82_n.jpg', size=36410), FileInfo
……
```

也可以将鲜花文件下载到本地电脑，鲜花目录如图 32-51 所示。

名称	类型	修改日期
daisy	文件夹	2019/9/3 20:59
dandelion	文件夹	2019/9/3 20:59
roses	文件夹	2019/9/3 20:59
sunflowers	文件夹	2019/9/3 20:59
tulips	文件夹	2019/9/3 20:59
LICENSE	文本文档	2016/2/9 10:59

图 32-51　鲜花目录

单击向日葵的文件目录，向日葵的图片如图 32-52 所示。

图 32-52　向日葵图片

（11）获取鲜花数据集各目录中图像文件的数量。

```
1.  local_dir = input_local_dir + 'flower_photos/'
2.  files = [os.path.join(dp, f) for dp, dn, filenames in os.walk(local_dir)
    for f in filenames if os.path.splitext(f)[1] == '.jpg']
3.  len(files)
```

在 Databricks 云平台 Notebook 中运行上述代码，运行结果如下：

```
Out[44]: 3670
```

（12）将图像文件名加载到 Spark 数据框中。

```
1.  files_df = spark.createDataFrame(
2.    map(lambda path: (path,), files), ["path"]
3.  ).repartition(10)  # 分区数应该是节点总数的倍数 display(files_df.limit(10))
```

第 1 行代码调用 spark.createDataFrame 方法创建数据帧。

第 2 行代码中 createDataFrame 方法的第一个输入参数是 map 函数，在 map 函数中遍历每一个图像的文件名，调用匿名函数将每一个文件名组成(path,)的格式；createDataFrame 方法的第二个参数是数据帧的列名。

第 3 行代码调用 Spark 的 repartition 方法进行重分区，将图像文件名的数据分为 10 个分区。

在 Databricks 云平台 Notebook 中运行上述代码，展示 10 条记录的图像路径及文件名，运行结果如图 32-53 所示。

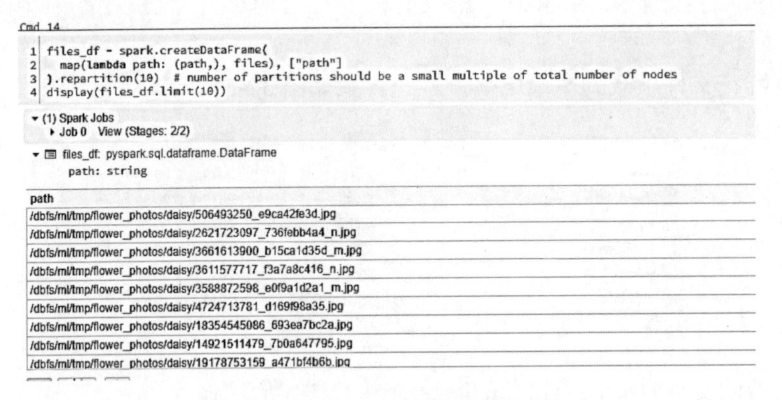

图 32-53　展示 10 条图像路径及文件名

单击图中的 View 文本，可以查询 Databricks 云平台 Spark Jobs 的执行情况，如图 32-54 所示。

（13）创建自定义 PyTorch 的数据集类 ImageDataset。

```
1.  class ImageDataset(Dataset):
2.    def __init__(self, paths, transform=None):
3.      self.paths = paths
4.      self.transform = transform
5.    def __len__(self):
6.      return len(self.paths)
7.    def __getitem__(self, index):
8.      image = default_loader(self.paths[index])
9.      if self.transform is not None:
10.       image = self.transform(image)
11.     return image
```

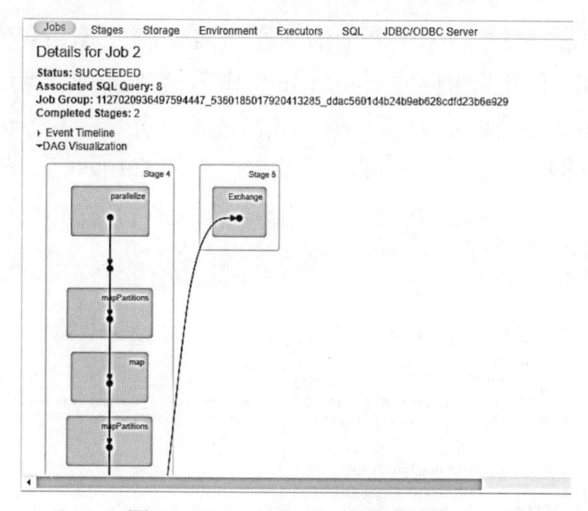

图 32-54　Spark Jobs 的执行情况

（14）定义模型预测的函数。

```
1.  def predict_batch(paths):
2.    transform = transforms.Compose([
3.      transforms.Resize(224),
4.      transforms.CenterCrop(224),
5.      transforms.ToTensor(),
6.      transforms.Normalize(mean=[0.485, 0.456, 0.406],
7.                         std=[0.229, 0.224, 0.225])
8.    ])
9.    images = ImageDataset(paths, transform=transform)
10.   loader = torch.utils.data.DataLoader(images, batch_size=500, num_workers=8)
11.   model = get_model_for_eval()
12.   model.to(device)
13.   all_predictions = []
14.   with torch.no_grad():
15.     for batch in loader:
16.       predictions = list(model(batch.to(device)).cpu().numpy())
17.       for prediction in predictions:
18.         all_predictions.append(prediction)
19.   return pd.Series(all_predictions)
```

第 2 行代码使用 PyTorch 对图像数据进行数据增强，将多个数据转换步骤整合在一起。

第 3 行代码按给定大小进行图像尺寸变化。

第 4 行代码图像中心收缩到给定的大小。

第 5 行代码将图像数据或者数组转换为 Tensor 数据结构。

第 6 行代码对图像数据按通道进行标准化处理。

第 10 行代码调用 PyTorch 的 torch.utils.data.DataLoader 方法加载图像数据，每个批次包含 500 张图像。

第 11 行代码获取 Spark 广播的 ResNet-50 预训练模型参数实例。

第 16 行代码获取每批次图像数据的预测分类。

（15）在本地测试函数。

```
1.  predictions = predict_batch(pd.Series(files[:200]))
```

（16）将函数包装为 Pandas UDF。

```
1.  predict_udf = pandas_udf(ArrayType(FloatType()), PandasUDFType.SCALAR)
    (predict_batch)
```

Pandas_UDF 是在 Spark 2.3 版本中新增的 API，Spark 通过 Arrow 传输数据，使用 Pandas 处理数据。Pandas_UDF 使用关键字 pandas_udf 作为装饰器或声明一个函数进行定义，Pandas_UDF 包括 Scalar（标量映射）和 Grouped Map（分组映射）等类型。

（17）通过 Pandas UDF 进行模型预测，将结果保存到 Parquet 文件中。

```
1.  predictions_df = files_df.select(col('path'), predict_udf(col('path')).
    alias("prediction"))
2.  predictions_df.write.mode("overwrite").parquet(output_file_path)
```

在 Databricks 云平台 Notebook 中运行上述代码，运行结果如图 32-55 所示。

```
1 # Make predictions.
2 predictions_df = files_df.select(col('path'), predict_udf(col('path')).alias("prediction"))
3 predictions_df.write.mode("overwrite").parquet(output_file_path)
```

▸ (1) Spark Jobs
▾ predictions_df: pyspark.sql.dataframe.DataFrame
path: string
▾prediction: array
element: float

图 32-55　结果保存为 Parquet 文件

（18）加载保存的 Parquet 文件并检查预测结果。

```
1.  result_df = spark.read.load(output_file_path)
2.  display(result_df)
```

在 Databricks 云平台 Notebook 中运行上述代码，运行结果如图 32-56 所示。

```
1 result_df = spark.read.load(output_file_path)
2 display(result_df)
```

▸ (3) Spark Jobs
▾ result_df: pyspark.sql.dataframe.DataFrame
path: string
▾prediction: array
element: float

path	prediction
/dbfs/ml/tmp/flower/flower_photos/daisy/16527403771_2391f137c4_n.jpg	▸[-2.2758918,1.0739485,-2.0188758,-3.6599488,-2.0186877,-2.4476438,-3.4474115,1.6215701,0.8783954,-0.6093757,0.9798171,2.6876547,1.8692478,0.
/dbfs/ml/tmp/flower/flower_photos/daisy/174131220_c853df1287.jpg	▸[-3.9154813,1.8891343,-3.2071137,-6.0027475,-2.0269387,-4.3915834,-4.424579,0.41725093,-0.012642965,-1.6194166,1.0658956,3.166872,1.3070763,
/dbfs/ml/tmp/flower/flower_photos/daisy/2001380507_19488ff96a_n.jpg	▸[-1.2487776,2.8638716,-1.7897179,-4.3353453,-1.8561887,-1.8472861,-3.3719778,1.2056043,-1.0880729,-0.6782577,0.17372435,2.8038085,1.5205016,
/dbfs/ml/tmp/flower/flower_photos/daisy/19653086178_28156b7ce4_m.jpg	▸[-3.2210941,2.444324,0.3837378,-2.8573961,-1.5967155,-0.77265745,-1.6578965,1.5971467,0.21213132,0.712795,5.109698,8.589855,4.937847,4.6269
/dbfs/ml/tmp/flower/flower_photos/daisy/15327813273_06cdf42210.jpg	▸[-2.246256,0.7695596,-2.5386806,-5.892159,-1.2902511,-4.4793973,-3.6465945,1.7347807,1.8964009,-1.5031966,1.7116117,3.0810332,1.0332546,0.81
/dbfs/ml/tmp/flower/flower_photos/daisy/1150395827_6f94a5c6e4_n.jpg	▸[-0.22875884,1.0064328,-2.4156303,-5.4791174,-1.5175345,-3.102098,-5.0674195,1.783064,0.94117194,-2.6632545,-0.11523515,1.2735127,-0.1123275
/dbfs/ml/tmp/flower/flower_photos/daisy/158869618_f1a6704236_n.jpg	▸[-2.990019,2.1711266,-1.92595,-5.8859186,-2.793842,-2.616717,-4.6183662,3.0952213,0.33314267,-2.9027998,-0.97704434,2.1988714,0.4714657,-0.37
/dbfs/ml/tmp/flower/flower_photos/daisy/1124324295_503f3a0804.jpg	▸[-0.4325683,1.2019345,-2.4378092,-3.495108,-2.0047512,-2.677111,-2.0389104,1.490081,1.7636256,-0.21664326,3.2460136,4.3464265,2.442171,2.189

Showing the first 95 rows.

Command took 7.64 seconds -- by duanzhihua@189.cn at 2019/9/2 下午9:40:00 on spark-tensorflow

图 32-56　模型预测结果

如图 32-57 所示，其中预测分类值是一个大小为 1000 的数组，根据 ResNet-50 模型预测

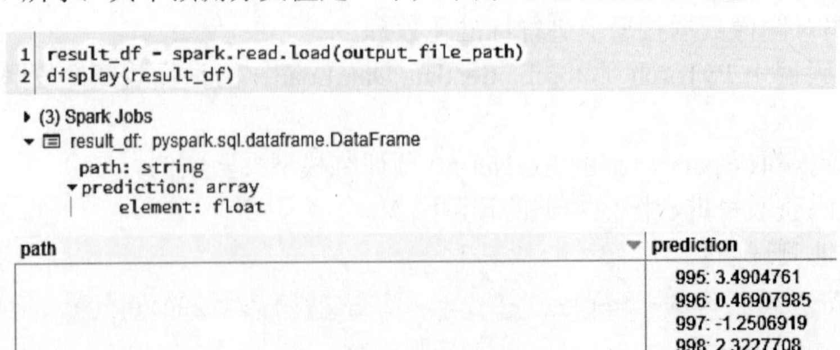

图 32-57　预测分类

1000 个分类的概率。本案例鲜花的分类实际为 5 类：郁金香、向日葵、玫瑰、蒲公英、菊花，感兴趣的读者可以改写 ResNet-50 模型代码进行优化，得到 5 个类别的预测值。

32.2.3　PyTorch 性能调优最佳实践

本节讲解鲜花图像数据集案例基于 ResNet-50 模型的性能调优，鲜花图像数据集包括郁金香、向日葵、玫瑰、蒲公英、菊花等图像数据，使用 Spark 及 PyTorch 进行分布式图像识别。

在 32.2.2 节我们使用 ResNet-50 模型进行鲜花分类预测，分类值是一个大小为 1000 的数组，根据 ResNet-50 模型预测 1000 个分类的概率。本节将调整为预测 5 个分类，得到 5 个类别的预测值。

基于 ResNet-50 模型进行调优实战的两种场景如下。

- 微调 ResNet-50 模型卷积神经网络：自定义全连接层，使用预训练的网络 ResNet-50 模型来初始化自己的网络，而不是随机初始化模型参数，然后微调卷积神经网络的所有层参数。
- 将 ResNet-50 模型卷积神经网络作为固定的图像特征提取器。自定义全连接层，在 ResNet-50 网络模型中，冻结除最后一层以外的所有网络（设置 requires_grad == False，反向传播时不计算梯度），只重新训练最后的全连接层参数（只有这层参数会在反向传播时更新参数）。

接下来进行第一个场景的调优实战：微调 ResNet-50 模型卷积神经网络。

（1）创建 Databricks 云平台集群 Notebook。

在 Workspace 中单击右键，在弹出的快捷菜单中选择 Create，依次输入名称(PytorchFinetuningConvnet)、开发语言（选择 Python 语言）、云平台集群（DeepLearning），单击 Create 按钮创建 Notebook，如图 32-58 所示。

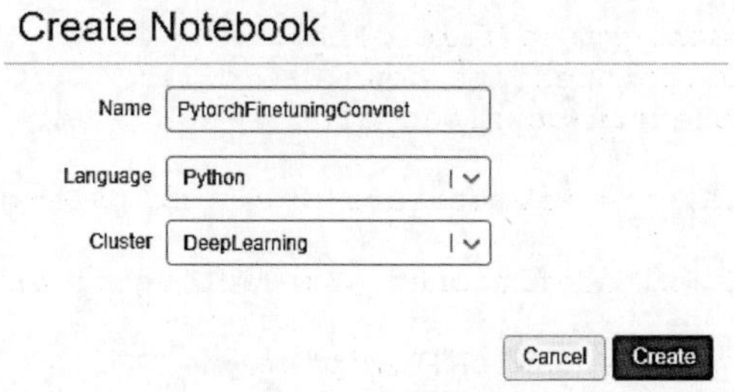

Create Notebook

Name	PytorchFinetuningConvnet
Language	Python
Cluster	DeepLearning

Cancel　Create

图 32-58　创建 Notebook

（2）设置启用 CPU 资源、Arrow 支持，导入 pandas、pytorch、pyspark 等库，设置是否使用 GPU、准备预训练模型和数据等内容。

```
1.  cuda = False
2.
3.  # 启动 Arrow 支持
4.  spark.conf.set("spark.sql.execution.arrow.enabled", "true")
5.  spark.conf.set("spark.sql.execution.arrow.maxRecordsPerBatch", "2048")
6.
7.  import os
```

```
8.  import shutil
9.  import tarfile
10. import time
11. import zipfile
12.
13. try:
14.     from urllib.request import urlretrieve
15. except ImportError:
16.     from urllib import urlretrieve
17.
18. import pandas as pd
19.
20. import torch
21. from torch.utils.data import Dataset
22. from torchvision import datasets, models, transforms, utils
23. from torchvision.datasets.folder import default_loader  # private API
24.
25. from pyspark.sql.functions import col, pandas_udf, PandasUDFType
26. from pyspark.sql.types import ArrayType, FloatType
27. import torch.nn as nn
28. import matplotlib.pyplot as plt
29.
30.
31. use_cuda = cuda and torch.cuda.is_available()
32. device = torch.device("cuda" if use_cuda else "cpu")
33.
34.
35. URL = "http://download.tensorflow.org/example_images/flower_photos.tgz"
36. input_local_dir = "/dbfs/ml/tmp/"
37. output_file_path = "/tmp/predictions"
38.
39.
40.
41. def maybe_download_and_extract(url, download_dir):
42.     filename = url.split('/')[-1]
43.     file_path = os.path.join(download_dir, filename)
44.
45.     if not os.path.exists(file_path):
46.         if not os.path.exists(download_dir):
47.             os.makedirs(download_dir)
48.
49.         file_path, _ = urlretrieve(url=url, filename=file_path)
50.         print()
51.         print("Download finished. Extracting files.")
52.
53.         if file_path.endswith(".zip"):
54.             # Unpack the zip-file.
55.             zipfile.ZipFile(file=file_path, mode="r").extractall(download_dir)
56.         elif file_path.endswith((".tar.gz", ".tgz")):
57.             # Unpack the tar-ball.
58.             tarfile.open(name=file_path, mode="r:gz").extractall(download_dir)
59.
60.         print("Done.")
61.     else:
62.         print("Data has apparently already been downloaded and unpacked.")
63.
64.
65.
66. maybe_download_and_extract(url=URL, download_dir=input_local_dir)
67.
```

```
68.
69.
70. local_dir = input_local_dir + 'flower_photos/'
71. files = [os.path.join(dp, f) for dp, dn, filenames in os.walk(local_dir)
    for f in filenames if os.path.splitext(f)[1] == '.jpg']
72. len(files)
```

在 Databricks 云平台 Notebook 中运行上述代码，查询下载的鲜花数据集包括 3670 张图片，运行结果如下：

```
Out[20]: 3670
```

（3）查看解压缩图像文件的信息。

```
1.  import random
2.  random.shuffle(files)
3.  files
```

在 Databricks 云平台 Notebook 中运行上述代码，运行结果如下：

```
Out[21]: ['/dbfs/ml/tmp/flower_photos/roses/5398569540_7d134c42cb_n.jpg',
 '/dbfs/ml/tmp/flower_photos/dandelion/5651310874_c8be336c2b.jpg',
 '/dbfs/ml/tmp/flower_photos/roses/527513005_41497ca4dc.jpg',
 '/dbfs/ml/tmp/flower_photos/dandelion/16650892835_9228a3ef67_m.jpg',
 '/dbfs/ml/tmp/flower_photos/tulips/13910678178_25e8b1a5e5.jpg',
 '/dbfs/ml/tmp/flower_photos/roses/6803363808_9f9ce98186_m.jpg',
 '/dbfs/ml/tmp/flower_photos/daisy/8938566373_d129e7af75.jpg',
 '/dbfs/ml/tmp/flower_photos/sunflowers/1244774242_25a20d99a9.jpg',
 '/dbfs/ml/tmp/flower_photos/tulips/14861513337_4ef0bfa40d.jpg',
 '/dbfs/ml/tmp/flower_photos/tulips/4508346090_a27b988f79_n.jpg',
 '/dbfs/ml/tmp/flower_photos/sunflowers/10386540696_0a95ee53a8_n.jpg',
 '/dbfs/ml/tmp/flower_photos/sunflowers/2425164088_4a5d2cdf21_n.jpg',
 '/dbfs/ml/tmp/flower_photos/tulips/14017640283_c417141832_n.jpg',
……
```

（4）自定义图像数据集 ImageDataset。

ImageDataset 类继承至 Dataset 类，Dataset 类是 PyTorch 中的一个类，是 PyTorch 中数据集加载类中继承的父类。ImageDataset 类重载 Dataset 父类中的两个成员函数，__len__返回数据集的大小，__getitem__是支持数据集索引的函数，通过 dataset[i]可以得到数据集中的第 i+1 个数据。

```
1.  #定义一个字典 对应编号
2.  dict = {'sunflowers': 0, 'dandelion': 1, 'roses': 2,  'daisy': 3,'tulips': 4}
3.
4.
5.
6.  class ImageDataset(Dataset):
7.    def __init__(self, paths, transform=None):
8.      self.paths = paths
9.      self.transform = transform
10.   def __len__(self):
11.     return len(self.paths)
12.   def __getitem__(self, index):
13.     image = default_loader(self.paths[index])
14.     classname = self.paths[index].split('/')[-2]
15.     target = dict[classname]
16.     print(image,target)
```

```
17.     if self.transform is not None:
18.       image = self.transform(image)
19.     return image,target
20.
```

（5）使用 Dataloader 读取自定义数据集，划分训练集、验证集。

```
1.  ####定义一个 dataloader
2.
3.  transform = transforms.Compose([
4.      transforms.Resize(224),
5.      transforms.CenterCrop(224),
6.      transforms.ToTensor(),
7.      transforms.Normalize(mean=[0.485, 0.456, 0.406],
8.                  std=[0.229, 0.224, 0.225])
9.    ])
10. pdpaths_train=pd.Series(files[:200])
11. pdpaths_val=pd.Series(files[200:300])
12. images_train = ImageDataset(pdpaths_train, transform=transform)
13. images_val = ImageDataset(pdpaths_val, transform=transform)
14. image_datasets ={'train': images_train  ,'val': images_val}
15.
16. dataloaders = {x: torch.utils.data.DataLoader(image_datasets[x],
    batch_size=500,shuffle=True,num_workers=2)   for x in ['train', 'val']}
17. dataset_sizes = {x: len(image_datasets[x]) for x in ['train', 'val']}
18. dataset_sizes
```

第 3～9 行代码使用 transforms.Compose 对图像数据进行数据增强，将多个数据转换步骤整合在一起。

第 10 行代码划分训练集数据，使用鲜花数据集的前 200 张图像作为训练集，在第 12 行代码封装为 ImageDataset 结构。

第 11 行代码划分验证集数据，使用鲜花数据集的第 200～300 张图像作为验证集，在第 13 行代码封装为 ImageDataset 结构。

第 16 行代码将训练集、验证集数据封装进 PyTorch 的 DataLoader 数据结构。

- DataLoader 可以按批次大小分批次读取，这里设置 batch_size=500。批次大小设置过小，可能会导致训练效果不理想，例如设置 batch_size=4，每次训练 4 张图片，准确率仅在 30%左右。
- DataLoader 可以对数据进行随机读取，并对数据进行洗牌操作(shuffle)，打乱数据集内数据分布的顺序。如果不对数据进行洗牌操作，模型训练可能会发生过拟合现象。例如，如果模型都是对同一个分类的鲜花图片进行训练，训练集准确度较高，但测试集要预测的是其他分类的图片，可能会导致每张图片都预测到同一个分类，测试效果不好。因此，需设置 shuffle 参数。
- DataLoader 采用多线程并行加载数据，利用多核处理器加快载入数据的效率。

在 Databricks 云平台 Notebook 中运行上述代码，运行结果如下：

```
Out[23]: {'train': 200, 'val': 100}
```

其中训练集包括 200 张图片，验证集包括 100 张图片。

（6）构建图像可视化函数，在 Databricks Notebook 中显示鲜花图像数据。

```
1.  import torchvision
2.  import numpy as np
```

```
3.
4.  def imshow(inp, title=None):
5.      """Imshow for Tensor."""
6.      inp = inp.numpy().transpose((1, 2, 0))
7.      mean = np.array([0.485, 0.456, 0.406])
8.      std = np.array([0.229, 0.224, 0.225])
9.      inp = std * inp + mean
10.     inp = np.clip(inp, 0, 1)
11.
12.     plt.imshow(inp)
13.     display()
14.     if title is not None:
15.         plt.title(title)
16.     plt.pause(0.001)  # pause a bit so that plots are updated
17.
18.
19. # 获取训练集的批次数据
20. inputs, classes = next(iter(dataloaders['train'] ))
21.
22. # 为每次批次数据生成表格
23. out = torchvision.utils.make_grid(inputs)
24.
25. imshow(out, title=[list(dict.keys())[list(dict.values()).index(x)] for
    x in classes])
```

在 Databricks 云平台 Notebook 中运行上述代码，运行结果如图 32-59 所示。

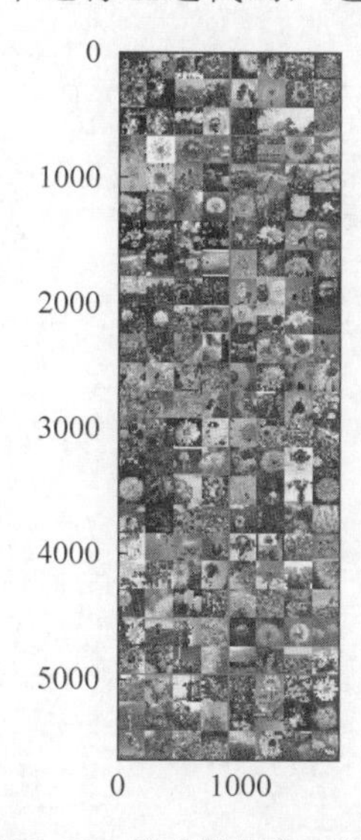

图 32-59　鲜花数据集可视化

（7）构建模型训练函数。

```
1.  def train_model(model, criterion, optimizer, scheduler, num_epochs=25):
2.      since = time.time()
3.
4.      best_model_wts = copy.deepcopy(model.state_dict())
5.      best_acc = 0.0
6.
7.      for epoch in range(num_epochs):
8.          print('Epoch {}/{}'.format(epoch, num_epochs - 1))
9.          print('-' * 10)
```

```
10.
11.         #每个时代包括训练及验证阶段
12.         for phase in ['train', 'val']:
13.             if phase == 'train':
14.                 model.train()  # 设置模型为训练模式
15.             else:
16.                 model.eval()   # 设置模型为验证模式
17.
18.
19.             running_loss = 0.0
20.             running_corrects = 0
21.
22.             # 数据迭代
23.             for inputs, labels in dataloaders[phase]:
24.                 inputs = inputs.to(device)
25.                 labels = labels.to(device)
26.
27.                 # 梯度为 0
28.                 optimizer.zero_grad()
29.
30.                 # 前向传播算法
31.                 # 训练时记录信息
32.                 with torch.set_grad_enabled(phase == 'train'):
33.                     outputs = model(inputs)
34.                     #_, preds = torch.max(outputs, 1)
35.                     p = torch.nn.functional.softmax(outputs, dim=1)
36.                     _, preds = torch.max(p, 1)
37.                     loss = criterion(outputs, labels)
38.
39.                     # 训练阶段进行反向传播及优化
40.                     if phase == 'train':
41.                         loss.backward()
42.                         optimizer.step()
43.
44.                 # 统计
45.                 running_loss += loss.item() * inputs.size(0)
46.                 running_corrects += torch.sum(preds == labels.data)
47.             if phase == 'train':
48.                 scheduler.step()
49.
50.             epoch_loss = running_loss / dataset_sizes[phase]
51.             epoch_acc = running_corrects.double() / dataset_sizes[phase]
52.
53.             print('{} Loss: {:.4f} Acc: {:.4f}'.format(
54.                 phase, epoch_loss, epoch_acc))
55.
56.             # 深拷贝模型
57.             if phase == 'val' and epoch_acc > best_acc:
58.                 best_acc = epoch_acc
59.                 best_model_wts = copy.deepcopy(model.state_dict())
60.
61.         print()
62.
63.     time_elapsed = time.time() - since
64.     print('Training complete in {:.0f}m {:.0f}s'.format(
65.         time_elapsed // 60, time_elapsed % 60))
66.     print('Best val Acc: {:4f}'.format(best_acc))
67.
```

```
68.     # 加载最好模型的权重
69.     model.load_state_dict(best_model_wts)
70.     return model
```

第 14 行代码，如果是模型训练阶段，设置模型为训练模式。

第 16 行代码，如果是模型训练阶段，设置模型为评估模式。

第 28 行代码，对每一次批量数据进行遍历，设置梯度为 0。

第 35 行代码通过 torch.nn.functional.softmax 计算每个图像分类的概率。

第 36 行代码通过 torch.max(p,1) 获取概率的最大值，并获取概率值最大值的索引（即预测的分类）。

第 37 行代码计算模型损失度。

第 41 行代码进行反向传播算法。

第 42 行代码使用优化器更新模型参数。

（8）调整 ResNet-50 模型结构。

ResNet-50 预训练模型的分类值是一个大小为 1000 的数组，将预测 1000 个分类的概率。本案例进行优化，通过自定义全连接层，调整为预测 5 个分类。

```
1.  import torch.optim as optim
2.  from torch.optim import lr_scheduler
3.  import copy
4.
5.
6.  model_ft = models.resnet50(pretrained=True)
7.  num_ftrs = model_ft.fc.in_features
8.  model_ft.fc = nn.Linear(num_ftrs, 5)
9.
10. model_ft = model_ft.to(device)
11.
12. criterion = nn.CrossEntropyLoss()
13.
14. # 模型所有参数都进行优化
15. optimizer_ft = optim.SGD(model_ft.parameters(), lr=0.1, momentum=0.9)
16.
17. # 每 7 个时代衰减 lr 的系数为 0.1
18. exp_lr_scheduler = lr_scheduler.StepLR(optimizer_ft, step_size=7, gamma=0.1)
19.
20.
21. model_ft = train_model(model_ft, criterion, optimizer_ft, exp_lr_scheduler,
    num_epochs=25)
22. model_ft
```

第 6 行代码加载 ResNet-50 预训练模型。

第 8 行代码构建自定义全连接层，全连接层输出结果为 5 个分类的概率值。

第 21 行代码，调用 train_model 函数进行模型训练。

在 Databricks 云平台 Notebook 中运行上述代码，运行结果如下：

```
1.  Downloading: "https://download.pytorch.org/models/resnet50-19c8e357.pth"
    to /root/.cache/torch/checkpoints/resnet50-19c8e357.pth
2.
3.    0%|          | 0/102502400 [00:00<?, ?it/s]
4.    9%|█         | 8732672/102502400 [00:00<00:01, 87264386.87it/s]
5.   17%|██        | 17522688/102502400 [00:00<00:00, 87446770.61it/s]
```

```
6.   24%|███        | 24838144/102502400 [00:00<00:00, 81102966.53it/s]
7.   31%|███▍       | 31924224/102502400 [00:00<00:00, 77730121.64it/s]
8.   40%|████       | 40501248/102502400 [00:00<00:00, 79972978.11it/s]
9.   50%|█████      | 51273728/102502400 [00:00<00:00, 86658908.93it/s]
10.  59%|██████     | 60186624/102502400 [00:00<00:00, 87374117.72it/s]
11.  67%|███████    | 68485120/102502400 [00:00<00:00, 85995381.42it/s]
12.  76%|████████   | 77684736/102502400 [00:00<00:00, 87702170.53it/s]
13.  84%|█████████  | 86491136/102502400 [00:01<00:00, 87794026.87it/s]
14.  93%|██████████ | 95092736/102502400 [00:01<00:00, 82167719.95it/s]
15. 100%|███████████| 102502400/102502400 [00:01<00:00, 84452947.87it/s]
16. Epoch 0/24
17. ----------
18. train Loss: 1.6695 Acc: 0.1450
19. val Loss: 1.7228 Acc: 0.4600
20.
21. Epoch 1/24
22. ----------
23. train Loss: 1.5072 Acc: 0.5400
24. val Loss: 2.9553 Acc: 0.4900
25.
26. Epoch 2/24
27. ----------
28. train Loss: 3.1260 Acc: 0.5650
29. val Loss: 2.5895 Acc: 0.5200
30.
31. Epoch 3/24
32. ----------
33. train Loss: 1.3772 Acc: 0.7450
34. val Loss: 5.2863 Acc: 0.6300
35.
36. Epoch 4/24
37. ----------
38. train Loss: 0.7740 Acc: 0.8000
39. val Loss: 9.0408 Acc: 0.7300
40.
41. Epoch 5/24
42. ----------
43. train Loss: 0.2245 Acc: 0.9400
44. val Loss: 38.6440 Acc: 0.4800
45.
46. Epoch 6/24
47. ----------
48. train Loss: 0.2053 Acc: 0.9350
49. val Loss: 13.1141 Acc: 0.7000
50.
51. Epoch 7/24
52. ----------
53. train Loss: 0.0048 Acc: 1.0000
54. val Loss: 7.6553 Acc: 0.7300
55.
56. Epoch 8/24
57. ----------
58. train Loss: 0.0043 Acc: 1.0000
59. val Loss: 5.1990 Acc: 0.7900
60.
61. Epoch 9/24
62. ----------
63. train Loss: 0.0034 Acc: 1.0000
```

```
64. val Loss: 3.9482 Acc: 0.8000
65.
66. Epoch 10/24
67. ----------
68. train Loss: 0.0027 Acc: 1.0000
69. val Loss: 3.1290 Acc: 0.7900
70.
71. Epoch 11/24
72. ----------
73. train Loss: 0.0021 Acc: 1.0000
74. val Loss: 2.5868 Acc: 0.7900
75.
76. Epoch 12/24
77. ----------
78. train Loss: 0.0018 Acc: 1.0000
79. val Loss: 2.1853 Acc: 0.8000
80.
81. Epoch 13/24
82. ----------
83. train Loss: 0.0016 Acc: 1.0000
84. val Loss: 1.8850 Acc: 0.7900
85.
86. Epoch 14/24
87. ----------
88. train Loss: 0.0014 Acc: 1.0000
89. val Loss: 1.6154 Acc: 0.7900
90.
91. Epoch 15/24
92. ----------
93. train Loss: 0.0014 Acc: 1.0000
94. val Loss: 1.4184 Acc: 0.7900
95.
96. Epoch 16/24
97. ----------
98. train Loss: 0.0014 Acc: 1.0000
99. val Loss: 1.2669 Acc: 0.8000
100.
101.    Epoch 17/24
102.    ----------
103.    train Loss: 0.0014 Acc: 1.0000
104.    val Loss: 1.1489 Acc: 0.7900
105.
106.    Epoch 18/24
107.    ----------
108.    train Loss: 0.0013 Acc: 1.0000
109.    val Loss: 1.0562 Acc: 0.7900
110.
111.    Epoch 19/24
112.    ----------
113.    train Loss: 0.0013 Acc: 1.0000
114.    val Loss: 0.9814 Acc: 0.7900
115.
116.    Epoch 20/24
117.    ----------
118.    train Loss: 0.0013 Acc: 1.0000
119.    val Loss: 0.9210 Acc: 0.7800
120.
121.    Epoch 21/24
122.    ----------
123.    train Loss: 0.0013 Acc: 1.0000
```

```
124.    val Loss: 0.8708 Acc: 0.7800
125.
126.    Epoch 22/24
127.    ----------
128.    train Loss: 0.0013 Acc: 1.0000
129.    val Loss: 0.8302 Acc: 0.7800
130.
131.    Epoch 23/24
132.    ----------
133.    train Loss: 0.0013 Acc: 1.0000
134.    val Loss: 0.7964 Acc: 0.7900
135.
136.    Epoch 24/24
137.    ----------
138.    train Loss: 0.0013 Acc: 1.0000
139.    val Loss: 0.7679 Acc: 0.8000
140.
141.    Training complete in 55m 26s
142.    Best val Acc: 0.800000
143.    Out[26]: ResNet(
144.      (conv1): Conv2d(3, 64, kernel_size=(7, 7), stride=(2, 2), padding=(3,
          3), bias=False)
145.      (bn1): BatchNorm2d(64, eps=1e-05, momentum=0.1, affine=True,
          track_running_stats=True)
146.      (relu): ReLU(inplace)
147.      (maxpool): MaxPool2d(kernel_size=3, stride=2, padding=1,
          dilation=1, ceil_mode=False)
148.      (layer1): Sequential(
149.        (0): Bottleneck(
150.          (conv1):Conv2d(64, 64, kernel_size=(1, 1), stride=(1, 1), bias=False)
151.          (bn1): BatchNorm2d(64, eps=1e-05, momentum=0.1, affine=True,
            track_running_stats=True)
152.          (conv2): Conv2d(64, 64, kernel_size=(3, 3), stride=(1, 1),
            padding=(1, 1), bias=False)
153.          (bn2): BatchNorm2d(64, eps=1e-05, momentum=0.1, affine=True,
            track_running_stats=True)
154.          (conv3):Conv2d(64, 256, kernel_size=(1, 1), stride=(1, 1), bias=False)
155.          (bn3): BatchNorm2d(256, eps=1e-05, momentum=0.1, affine=True,
            track_running_stats=True)
156.          (relu): ReLU(inplace)
157.          (downsample): Sequential(
158.            (0): Conv2d(64, 256, kernel_size=(1, 1), stride=(1, 1), bias=False)
159.            (1): BatchNorm2d(256, eps=1e-05, momentum=0.1, affine=True,
              track_running_stats=True)
160.          )
161.        )
162.        (1): Bottleneck(
163.          (conv1):Conv2d(256, 64, kernel_size=(1, 1), stride=(1, 1), bias=False)
164.          (bn1): BatchNorm2d(64, eps=1e-05, momentum=0.1, affine=True,
            track_running_stats=True)
165.          (conv2): Conv2d(64, 64, kernel_size=(3, 3), stride=(1, 1),
            padding=(1, 1), bias=False)
166.          (bn2): BatchNorm2d(64, eps=1e-05, momentum=0.1, affine=True,
            track_running_stats=True)
167.          (conv3):Conv2d(64, 256, kernel_size=(1, 1), stride=(1, 1), bias=False)
168.          (bn3): BatchNorm2d(256, eps=1e-05, momentum=0.1, affine=True,
            track_running_stats=True)
169.          (relu): ReLU(inplace)
170.        )
171.        (2): Bottleneck(
```

```
172.          (conv1): Conv2d(256, 64, kernel_size=(1, 1), stride=(1, 1), bias=False)
173.          (bn1): BatchNorm2d(64, eps=1e-05, momentum=0.1, affine=True,
          track_running_stats=True)
174.          (conv2): Conv2d(64, 64, kernel_size=(3, 3), stride=(1, 1),
          padding=(1, 1), bias=False)
175.          (bn2): BatchNorm2d(64, eps=1e-05, momentum=0.1, affine=True,
          track_running_stats=True)
176.          (conv3): Conv2d(64, 256, kernel_size=(1, 1), stride=(1, 1), bias=False)
177.          (bn3): BatchNorm2d(256, eps=1e-05, momentum=0.1, affine=True,
          track_running_stats=True)
178.          (relu): ReLU(inplace)
179.        )
180.      )
181.      (layer2): Sequential(
182.        (0): Bottleneck(
183.          (conv1): Conv2d(256, 128, kernel_size=(1, 1), stride=(1, 1), bias=False)
184.          (bn1): BatchNorm2d(128, eps=1e-05, momentum=0.1, affine=True,
          track_running_stats=True)
185.          (conv2): Conv2d(128, 128, kernel_size=(3, 3), stride=(2, 2),
          padding=(1, 1), bias=False)
186.          (bn2): BatchNorm2d(128, eps=1e-05, momentum=0.1, affine=True,
          track_running_stats=True)
187.          (conv3): Conv2d(128, 512, kernel_size=(1, 1), stride=(1, 1),
          bias=False)
188.          (bn3): BatchNorm2d(512, eps=1e-05, momentum=0.1, affine=True,
          track_running_stats=True)
189.          (relu): ReLU(inplace)
190.          (downsample): Sequential(
191.            (0): Conv2d(256, 512, kernel_size=(1, 1), stride=(2, 2),
          bias=False)
192.            (1): BatchNorm2d(512, eps=1e-05, momentum=0.1, affine=True,
          track_running_stats=True)
193.          )
194.        )
195.        (1): Bottleneck(
196.          (conv1): Conv2d(512, 128, kernel_size=(1, 1), stride=(1, 1),
          bias=False)
197.          (bn1): BatchNorm2d(128, eps=1e-05, momentum=0.1, affine=True,
          track_running_stats=True)
198.          (conv2): Conv2d(128, 128, kernel_size=(3, 3), stride=(1, 1),
          padding=(1, 1), bias=False)
199.          (bn2): BatchNorm2d(128, eps=1e-05, momentum=0.1, affine=True,
          track_running_stats=True)
200.          (conv3): Conv2d(128, 512, kernel_size=(1, 1), stride=(1, 1),
          bias=False)
201.          (bn3): BatchNorm2d(512, eps=1e-05, momentum=0.1, affine=True,
          track_running_stats=True)
202.          (relu): ReLU(inplace)
203.        )
204.        (2): Bottleneck(
205.          (conv1): Conv2d(512, 128, kernel_size=(1, 1), stride=(1, 1),
          bias=False)
206.          (bn1): BatchNorm2d(128, eps=1e-05, momentum=0.1, affine=True,
          track_running_stats=True)
207.          (conv2): Conv2d(128, 128, kernel_size=(3, 3), stride=(1, 1),
          padding=(1, 1), bias=False)
208.          (bn2): BatchNorm2d(128, eps=1e-05, momentum=0.1, affine=True,
          track_running_stats=True)
209.          (conv3): Conv2d(128, 512, kernel_size=(1, 1), stride=(1, 1),
          bias=False)
```

```
210.            (bn3): BatchNorm2d(512, eps=1e-05, momentum=0.1, affine=True,
                track_running_stats=True)
211.            (relu): ReLU(inplace)
212.          )
213.          (3): Bottleneck(
214.            (conv1): Conv2d(512, 128, kernel_size=(1, 1), stride=(1, 1),
                bias=False)
215.            (bn1): BatchNorm2d(128, eps=1e-05, momentum=0.1, affine=True,
                track_running_stats=True)
216.            (conv2): Conv2d(128, 128, kernel_size=(3, 3), stride=(1, 1),
                padding=(1, 1), bias=False)
217.            (bn2): BatchNorm2d(128, eps=1e-05, momentum=0.1, affine=True,
                track_running_stats=True)
218.            (conv3): Conv2d(128, 512, kernel_size=(1, 1), stride=(1, 1),
                bias=False)
219.            (bn3): BatchNorm2d(512, eps=1e-05, momentum=0.1, affine=True,
                track_running_stats=True)
220.            (relu): ReLU(inplace)
221.          )
222.        )
223.        (layer3): Sequential(
224.          (0): Bottleneck(
225.            (conv1): Conv2d(512, 256, kernel_size=(1, 1), stride=(1, 1),
                bias=False)
226.            (bn1): BatchNorm2d(256, eps=1e-05, momentum=0.1, affine=True,
                track_running_stats=True)
227.            (conv2): Conv2d(256, 256, kernel_size=(3, 3), stride=(2, 2),
                padding=(1, 1), bias=False)
228.            (bn2): BatchNorm2d(256, eps=1e-05, momentum=0.1, affine=True,
                track_running_stats=True)
229.            (conv3): Conv2d(256, 1024, kernel_size=(1, 1), stride=(1, 1),
                bias=False)
230.            (bn3): BatchNorm2d(1024, eps=1e-05, momentum=0.1, affine=True,
                track_running_stats=True)
231.            (relu): ReLU(inplace)
232.            (downsample): Sequential(
233.              (0): Conv2d(512, 1024, kernel_size=(1, 1), stride=(2, 2),
                bias=False)
234.              (1): BatchNorm2d(1024, eps=1e-05, momentum=0.1, affine=True,
                track_running_stats=True)
235.            )
236.          )
237.          (1): Bottleneck(
238.            (conv1): Conv2d(1024, 256, kernel_size=(1, 1), stride=(1, 1),
                bias=False)
239.            (bn1): BatchNorm2d(256, eps=1e-05, momentum=0.1, affine=True,
                track_running_stats=True)
240.            (conv2): Conv2d(256, 256, kernel_size=(3, 3), stride=(1, 1),
                padding=(1, 1), bias=False)
241.            (bn2): BatchNorm2d(256, eps=1e-05, momentum=0.1, affine=True,
                track_running_stats=True)
242.            (conv3): Conv2d(256, 1024, kernel_size=(1, 1), stride=(1, 1),
                bias=False)
243.            (bn3): BatchNorm2d(1024, eps=1e-05, momentum=0.1, affine=True,
                track_running_stats=True)
244.            (relu): ReLU(inplace)
245.          )
246.          (2): Bottleneck(
247.            (conv1): Conv2d(1024, 256, kernel_size=(1, 1), stride=(1, 1),
                bias=False)
```

```
248.            (bn1): BatchNorm2d(256, eps=1e-05, momentum=0.1, affine=True,
                track_running_stats=True)
249.            (conv2): Conv2d(256, 256, kernel_size=(3, 3), stride=(1, 1),
                padding=(1, 1), bias=False)
250.            (bn2): BatchNorm2d(256, eps=1e-05, momentum=0.1, affine=True,
                track_running_stats=True)
251.            (conv3): Conv2d(256, 1024, kernel_size=(1, 1), stride=(1, 1),
                bias=False)
252.            (bn3): BatchNorm2d(1024, eps=1e-05, momentum=0.1, affine=True,
                track_running_stats=True)
253.            (relu): ReLU(inplace)
254.          )
255.          (3): Bottleneck(
256.            (conv1): Conv2d(1024, 256, kernel_size=(1, 1), stride=(1, 1),
                bias=False)
257.            (bn1): BatchNorm2d(256, eps=1e-05, momentum=0.1, affine=True,
                track_running_stats=True)
258.            (conv2): Conv2d(256, 256, kernel_size=(3, 3), stride=(1, 1),
                padding=(1, 1), bias=False)
259.            (bn2): BatchNorm2d(256, eps=1e-05, momentum=0.1, affine=True,
                track_running_stats=True)
260.            (conv3): Conv2d(256, 1024, kernel_size=(1, 1), stride=(1, 1),
                bias=False)
261.            (bn3): BatchNorm2d(1024, eps=1e-05, momentum=0.1, affine=True,
                track_running_stats=True)
262.            (relu): ReLU(inplace)
263.          )
264.          (4): Bottleneck(
265.            (conv1): Conv2d(1024, 256, kernel_size=(1, 1), stride=(1, 1),
                bias=False)
266.            (bn1): BatchNorm2d(256, eps=1e-05, momentum=0.1, affine=True,
                track_running_stats=True)
267.            (conv2): Conv2d(256, 256, kernel_size=(3, 3), stride=(1, 1),
                padding=(1, 1), bias=False)
268.            (bn2): BatchNorm2d(256, eps=1e-05, momentum=0.1, affine=True,
                track_running_stats=True)
269.            (conv3): Conv2d(256, 1024, kernel_size=(1, 1), stride=(1, 1),
                bias=False)
270.            (bn3): BatchNorm2d(1024, eps=1e-05, momentum=0.1, affine=True,
                track_running_stats=True)
271.            (relu): ReLU(inplace)
272.          )
273.          (5): Bottleneck(
274.            (conv1): Conv2d(1024, 256, kernel_size=(1, 1), stride=(1, 1),
                bias=False)
275.            (bn1): BatchNorm2d(256, eps=1e-05, momentum=0.1, affine=True,
                track_running_stats=True)
276.            (conv2): Conv2d(256, 256, kernel_size=(3, 3), stride=(1, 1),
                padding=(1, 1), bias=False)
277.            (bn2): BatchNorm2d(256, eps=1e-05, momentum=0.1, affine=True,
                track_running_stats=True)
278.            (conv3): Conv2d(256, 1024, kernel_size=(1, 1), stride=(1, 1),
                bias=False)
279.            (bn3): BatchNorm2d(1024, eps=1e-05, momentum=0.1, affine=True,
                track_running_stats=True)
280.            (relu): ReLU(inplace)
281.          )
282.        )
283.        (layer4): Sequential(
284.          (0): Bottleneck(
```

```
285.          (conv1): Conv2d(1024, 512, kernel_size=(1, 1), stride=(1, 1),
              bias=False)
286.          (bn1): BatchNorm2d(512, eps=1e-05, momentum=0.1, affine=True,
              track_running_stats=True)
287.          (conv2): Conv2d(512, 512, kernel_size=(3, 3), stride=(2, 2),
              padding=(1, 1), bias=False)
288.          (bn2): BatchNorm2d(512, eps=1e-05, momentum=0.1, affine=True,
              track_running_stats=True)
289.          (conv3): Conv2d(512, 2048, kernel_size=(1, 1), stride=(1, 1),
              bias=False)
290.          (bn3): BatchNorm2d(2048, eps=1e-05, momentum=0.1, affine=True,
              track_running_stats=True)
291.          (relu): ReLU(inplace)
292.          (downsample): Sequential(
293.            (0): Conv2d(1024, 2048, kernel_size=(1, 1), stride=(2, 2),
              bias=False)
294.            (1): BatchNorm2d(2048, eps=1e-05, momentum=0.1, affine=True,
              track_running_stats=True)
295.          )
296.        )
297.        (1): Bottleneck(
298.          (conv1): Conv2d(2048, 512, kernel_size=(1, 1), stride=(1, 1),
              bias=False)
299.          (bn1): BatchNorm2d(512, eps=1e-05, momentum=0.1, affine=True,
              track_running_stats=True)
300.          (conv2): Conv2d(512, 512, kernel_size=(3, 3), stride=(1, 1),
              padding=(1, 1), bias=False)
301.          (bn2): BatchNorm2d(512, eps=1e-05, momentum=0.1, affine=True,
              track_running_stats=True)
302.          (conv3): Conv2d(512, 2048, kernel_size=(1, 1), stride=(1, 1),
              bias=False)
303.          (bn3): BatchNorm2d(2048, eps=1e-05, momentum=0.1, affine=True,
              track_running_stats=True)
304.          (relu): ReLU(inplace)
305.        )
306.        (2): Bottleneck(
307.          (conv1): Conv2d(2048, 512, kernel_size=(1, 1), stride=(1, 1),
              bias=False)
308.          (bn1): BatchNorm2d(512, eps=1e-05, momentum=0.1, affine=True,
              track_running_stats=True)
309.          (conv2): Conv2d(512, 512, kernel_size=(3, 3), stride=(1, 1),
              padding=(1, 1), bias=False)
310.          (bn2): BatchNorm2d(512, eps=1e-05, momentum=0.1, affine=True,
              track_running_stats=True)
311.          (conv3): Conv2d(512, 2048, kernel_size=(1, 1), stride=(1, 1),
              bias=False)
312.          (bn3): BatchNorm2d(2048, eps=1e-05, momentum=0.1, affine=True,
              track_running_stats=True)
313.          (relu): ReLU(inplace)
314.        )
315.      )
316.      (avgpool): AdaptiveAvgPool2d(output_size=(1, 1))
317.      (fc): Linear(in_features=2048, out_features=5, bias=True)
318.    )
```

第 138～142 行日志显示，在第 24 个时代中，训练集的损失度为 0.0013，训练集的准确度达到了 100%；验证集的损失度为 0.7679，验证集的准确度为 80%，整个模型训练完成以后，验证集预测最好的准确度达到为 80%。

第 144～318 行日志显示了 ResNet 网络模型的结构，其中第 317 行日志显示 ResNet 网络模型最后一层全连接层已经调整，调整为 5 个分类。

（9）使用 Spark 的广播变量广播已经调整的 ResNet 网络模型参数，如图 32-60 所示。

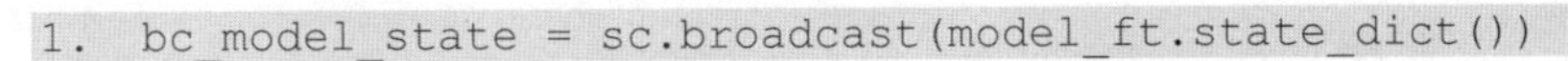

```
1.  bc_model_state = sc.broadcast(model_ft.state_dict())
```

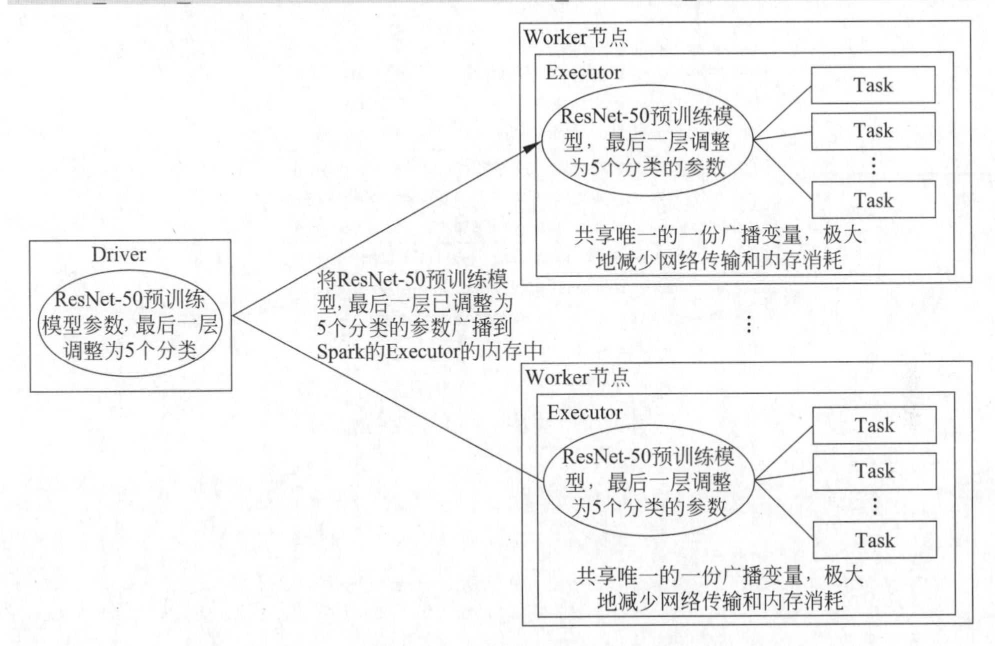

图 32-60　广播已经调整的 ResNet 模型

（10）构建获取模型的函数 get_model_for_eval，用于在每一个 Executor 上加载广播变量的模型参数。

```
1.  def get_model_for_eval():
2.    """Gets the broadcasted model."""
3.    model = models.resnet50(pretrained=True)
4.    num_ftrs = model.fc.in_features
5.    model.fc = nn.Linear(num_ftrs, 5)
6.    model.load_state_dict(bc_model_state.value)
7.    model.eval()
8.    return model
```

（11）读取鲜花图像数据集的文件名信息，构建 Spark 的数据帧。

```
1.  files_df = spark.createDataFrame(
2.    map(lambda path: (path,), files), ["path"]
3.  ).repartition(10)  #  分区数应该是节点总数的一个倍数
4.  display(files_df.limit(10))
```

在 Databricks 云平台 Notebook 中运行上述代码，运行结果如图 32-61 所示。

（12）构建一个预测分类的函数 predict_batch。

```
1.  def predict_batch(paths):
2.    transform = transforms.Compose([
3.      transforms.Resize(224),
4.      transforms.CenterCrop(224),
```

```
1  files_df = spark.createDataFrame(
2    map(lambda path: (path,), files), ["path"]
3  ).repartition(10)  # number of partitions should be a
4  display(files_df.limit(10))
```

▸ (1) Spark Jobs
▸ files_df: pyspark.sql.dataframe.DataFrame = [path: string]

path
/dbfs/ml/tmp/flower_photos/tulips/2834890466_1cf220fba1.jpg
/dbfs/ml/tmp/flower_photos/roses/3872230296_6c477309f3_n.jpg
/dbfs/ml/tmp/flower_photos/daisy/3695826945_9f374e8a00_m.jpg
/dbfs/ml/tmp/flower_photos/roses/5398569540_7d134c42cb_n.jpg
/dbfs/ml/tmp/flower_photos/dandelion/16159487_3a6615a565_n.jpg
/dbfs/ml/tmp/flower_photos/dandelion/5605502523_05acb00ae7_n.jpg
/dbfs/ml/tmp/flower_photos/dandelion/6994931380_a7588c1192_m.jpg
/dbfs/ml/tmp/flower_photos/tulips/16677199221_eab3f22378_n.jpg

图 32-61　创建 Spark 数据帧

```
5.      transforms.ToTensor(),
6.      transforms.Normalize(mean=[0.485, 0.456, 0.406],
7.                           std=[0.229, 0.224, 0.225])
8.    ])
9.    images = ImageDataset(paths, transform=transform)
10.   loader = torch.utils.data.DataLoader(images, batch_size=500, num_workers=2)
11.   model = get_model_for_eval()
12.
13.
14.   model.to(device)
15.   all_predictions = []
16.   with torch.no_grad():
17.     for inputs, _ in loader:
18.       logit = model(inputs.to(device))
19.       p = torch.nn.functional.softmax(logit, dim=1)
20.       predictions = list(p.cpu().numpy())
21.
22.       _, preds = torch.max(p, 1)
23.
24.       for prediction in predictions:
25.         all_predictions.append(prediction)
26.   return pd.Series(all_predictions)
```

第 18 行代码，获取模型的输出值。

第 19 行代码，通过 torch.nn.functional.softmax 计算每个图像分类的概率。

第 22 行代码通过 torch.max(p,1)获取概率最大值的索引，即预测的分类。

（13）将鲜花图像数据集的后 70 张图片作为测试集，执行模型预测函数，基于测试集进行图像分类预测。

```
1.  predictions = predict_batch(pd.Series(files[3600:]))
2.  predictions
```

在 Databricks 云平台 Notebook 中运行上述代码，运行结果如图 32-62 所示。

```
| predictions
t[32]: 0     [0.99999964, 1.9640058e-10, 8.15348e-40, 4.215...
    [3.7912924e-29, 8.577873e-21, 0.0, 1.0, 1.0466...
    [9.36769e-30, 7.5132405e-29, 0.0, 1.0, 1.26556...
    [2.9831755e-20, 1.0921925e-17, 5.98468e-24, 1....
    [5.9661543e-22, 1.2350547e-18, 1.1036354e-23, ...
                            [0.0, 1.0, 0.0, 0.0, 0.0]
                  [1e-45, 9.67784e-40, 0.0, 0.0, 1.0]
    [5.0348404e-24, 1.2443821e-23, 1.0, 2.2628429e...
    [6.9152935e-14, 3.1582024e-12, 3.2131374e-11, ...
    [0.018632956, 0.0013400139, 0.980027, 2.861755...
    [4.7869203e-06, 4.8847604e-10, 0.0, 7.06774e-1...
    [1.0, 1.1293946e-14, 0.0, 6.3618347e-19, 3.501...
    [6.241782e-20, 3.8135187e-19, 1.226e-42, 1.0, ...
    [5.5644087e-11, 0.9686086, 2.0561506e-32, 4.01...
    [3.2120874e-20, 5.4280928e-18, 1.624188e-23, 1...
    [1.0, 2.8063351e-15, 0.0, 1.2857059e-16, 6.914...
    [1.2923027e-11, 0.9199084, 2.9199468e-26, 3.70...
```

图 32-62　预测分类概率

（14）将图片分类的预测结果进行可视化。

```
1.  def visualize_model(model, num_images=6):
2.     was_training = model.training
3.     model.eval()
4.     images_so_far = 0
5.     fig = plt.figure()
6.
7.     with torch.no_grad():
8.         for i, (inputs, labels) in enumerate(dataloaders['val']):
9.             inputs = inputs.to(device)
10.            labels = labels.to(device)
11.
12.            outputs = model(inputs)
13.            #_, preds = torch.max(outputs, 1)
14.            p = torch.nn.functional.softmax(outputs, dim=1)
15.            _, preds = torch.max(p, 1)
16.
17.            for j in range(inputs.size()[0]):
18.                images_so_far += 1
19.                ax = plt.subplot(num_images//2, 2, images_so_far)
20.                ax.axis('off')
21.                #ax.set_title('predicted: {}'.format(   preds[j] ))
22.                ax.set_title('predicted: {}'.format(  list(dict.keys())
                   [list(dict.values()). index(preds[j])] ))
23.                imshow(inputs.cpu().data[j])
24.
25.                if images_so_far == num_images:
26.                    model.train(mode=was_training)
27.                    return
28.        model.train(mode=was_training)
29. visualize_model(get_model_for_eval())
```

在 Databricks 云平台 Notebook 中运行上述代码，运行结果如图 32-63 所示。

（15）加载图像数据集文件信息，构建 Spark 数据帧，对图像分类进行概率预测并展示。

```
30. predict_udf = pandas_udf(ArrayType(FloatType()), PandasUDFType.SCALAR)
    (predict_batch)
31. #进行预测
32. predictions_df = files_df.select(col('path'), predict_udf(col('path'))
    .alias("prediction"))
33. predictions_df.show()
34.
```

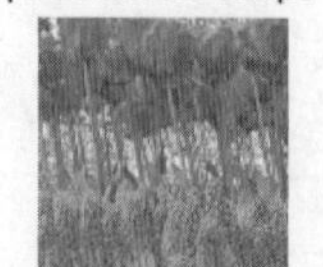

图 32-63　预测鲜花分类

在 Databricks 云平台 Notebook 中运行上述代码，运行结果如图 32-64 所示。

```
▸ (1) Spark Jobs
▸ predictions_df: pyspark.sql.dataframe.DataFrame = [
+--------------------+--------------------+
|                path|          prediction|
+--------------------+--------------------+
|/dbfs/ml/tmp/flow...|[2.1983705E-29, 1...|
|/dbfs/ml/tmp/flow...|[0.0, 0.0, 1.0, 0...|
|/dbfs/ml/tmp/flow...|[1.345232E-29, 7....|
|/dbfs/ml/tmp/flow...|[2.1507191E-19, 1...|
|/dbfs/ml/tmp/flow...|[1.3531507E-5, 0....|
|/dbfs/ml/tmp/flow...|[0.0, 1.0, 0.0, 5...|
|/dbfs/ml/tmp/flow...|[0.0, 1.0, 0.0, 3...|
|/dbfs/ml/tmp/flow...|[1.6270133E-26, 5...|
|/dbfs/ml/tmp/flow...|[0.0, 0.0, 1.0, 0...|
|/dbfs/ml/tmp/flow...|[5.007156E-6, 0.1...|
|/dbfs/ml/tmp/flow...|[9.193572E-25, 3....|
|/dbfs/ml/tmp/flow...|[0.0, 0.0, 1.0, 0...|
|/dbfs/ml/tmp/flow...|[4.2292105E-25, 4...|
|/dbfs/ml/tmp/flow...|[1.0, 1.3509457E-...|
|/dbfs/ml/tmp/flow...|[3.3224828E-26, 7...|
|/dbfs/ml/tmp/flow...|[6.2025663E-29, 6...|
|/dbfs/ml/tmp/flow...|[7.690239E-29, 1....|
|/dbfs/ml/tmp/flow...|[0.0, 1.0, 0.0, 0...|
```

图 32-64　spark 数据帧展示预测概率

（16）获取 Spark 数据帧前 10 个数据。

```
1.   predictions_df.take(10)
```

在 Databricks 云平台 Notebook 中运行上述代码，每一行的第一个元素是图像数据的文件名路径，第二个元素是预测概率的列表，包括 5 个分类的预测概率，运行结果如下：

```
Out[35]:[Row(path='/dbfs/ml/tmp/flower_photos/dandelion/4633792226_80f89c89ec_m.jpg',
prediction=[0.0, 1.0, 0.0, 1.504630213084129e-40, 7.270874286605925e-20]),
 Row(path='/dbfs/ml/tmp/flower_photos/sunflowers/2443095419_17b920d155_m.jpg',
prediction=[1.0, 1.299387108384157e-18, 0.0, 8.739798446333178e-18,
2.2670898112481337e-22]),
```

```
 Row(path='/dbfs/ml/tmp/flower_photos/daisy/1342002397_9503c97b49.jpg',
prediction=[5.18901363578869e-26, 5.378985078989713e-21, 0.0, 1.0,
6.620340397851115e-14]),
 Row(path='/dbfs/ml/tmp/flower_photos/tulips/490541142_c37e2b4191_n.jpg',
prediction=[6.4889165053343e-35, 5.700089186374614e-31, 3.959891495276964e-38,
2.7875349319759924e-38, 1.0]),
 Row(path='/dbfs/ml/tmp/flower_photos/daisy/3494265422_9dba8f2191_n.jpg',
prediction=[3.3079394778923188e-09, 0.04229852929711342, 1.2799433046814156e-26,
0.8811672925949097, 0.07653418183326721]),
 Row(path='/dbfs/ml/tmp/flower_photos/roses/9159362388_c6f4cf3812_n.jpg',
prediction=[0.0, 0.0, 1.0, 0.0, 0.0]),
 Row(path='/dbfs/ml/tmp/flower_photos/tulips/7094415739_6b29e5215c_m.jpg',
prediction=[8.655538091817405e-11, 0.9999998807907104, 8.157689107557431e-11,
1.7209943337093137e-07, 1.198488669418296e-10]),
 Row(path='/dbfs/ml/tmp/flower_photos/tulips/4580206494_9386c81ed8_n.jpg',
prediction=[9.138736913882894e-07, 4.0676513890502974e-06, 8.879409506334923e-06,
8.41552676433821e-08, 0.9999860525131226]),
 Row(path='/dbfs/ml/tmp/flower_photos/tulips/16055807744_000bc07afc_m.jpg',
prediction=[5.566283895024331e-16, 3.243216649016431e-14, 1.30373775994165e-16,
3.8746704917702767e-17, 1.0]),
 Row(path='/dbfs/ml/tmp/flower_photos/daisy/4131565290_0585c4dd5a_n.jpg',
prediction=[9.924455035559276e-09, 1.5768972616569954e-06, 5.1966660796074e-25,
0.999993085861206, 5.319029696693178e-06])]
```

（17）将结果保存为 parquet 格式的文件，然后使用 Spark 重新加载 parquet 文件，并进行展示。

```
1.  predictions_df.write.mode("overwrite").parquet(output_file_path)
2.  result_df = spark.read.load(output_file_path)
3.  display(result_df)
```

在 Databricks 云平台 Notebook 中运行上述代码，运行结果如图 32-65 所示。

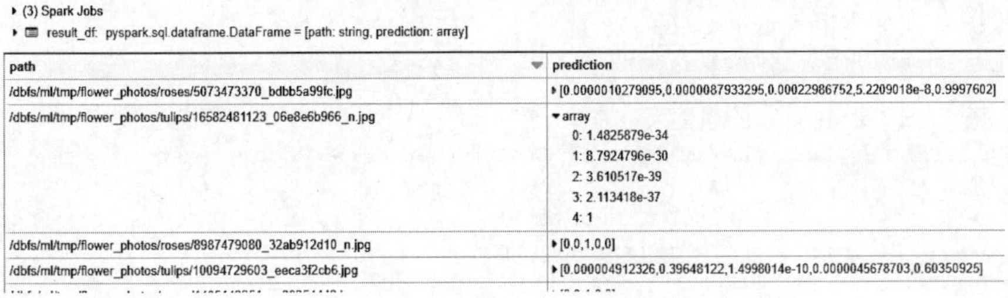

▸ (3) Spark Jobs

▸ result_df: pyspark.sql.dataframe.DataFrame = [path: string, prediction: array]

path	prediction
/dbfs/ml/tmp/flower_photos/roses/5073473370_bdbb5a99fc.jpg	▸[0.0000010279095,0.0000087933295,0.00022986752,5.2209018e-8,0.9997602]
/dbfs/ml/tmp/flower_photos/tulips/16582481123_06e8e6b966_n.jpg	▾array 0: 1.4825879e-34 1: 8.7924796e-30 2: 3.610517e-39 3: 2.113418e-37 4: 1
/dbfs/ml/tmp/flower_photos/roses/8987479080_32ab912d10_n.jpg	▸[0,0,1,0,0]
/dbfs/ml/tmp/flower_photos/tulips/10094729603_eeca3f2cb6.jpg	▸[0.000004912326,0.39648122,1.4998014e-10,0.000045678703,0.60350925]

图 32-65　预测 5 个分类的概率

图中第 2 条记录为：

```
/dbfs/ml/tmp/flower_photos/tulips/16582481123_06e8e6b966_n.jpg  array
 0: 1.4825879e-34
 1: 8.7924796e-30
 2: 3.610517e-39
 3: 2.113418e-37
 4: 1
```

文件名目录中的 tulips 表示这张图片是一朵郁金香，而预测概率的最大值是 1，其索引是 4。查询在代码中定义的鲜花字典编号：

```
dict = {'sunflowers': 0, 'dandelion': 1, 'roses': 2,  'daisy': 3,'tulips': 4}
```

索引 4 对应的字典名称是郁金香，表明图片的预测分类是郁金香，和实际值是一致的，采用 ResNet 模型预测的效果较好。

接下来进行第二个场景的调优：将 ResNet-50 模型卷积神经网络作为固定的图像特征提取器。在 ResNet-50 网络模型中，冻结除最后一层以外的所有网络（设置 requires_grad == False，反向传播时不计算梯度）；自定义全连接层，只重新训练最后的全连接层参数（只有这层参数会在反向传播时更新参数）。

在第一个场景案例代码的基础上进行调整，代码如下：

```
1.  import torch.optim as optim
2.  from torch.optim import lr_scheduler
3.  import copy
4.
5.  model_conv = models.resnet50(pretrained=True)
6.  for param in model_conv.parameters():
7.      param.requires_grad = False
8.
9.  #默认情况下，新构建模块的参数 requires_grad=True
10. num_ftrs = model_conv.fc.in_features
11. model_conv.fc = nn.Linear(num_ftrs, 5)
12.
13. model_conv = model_conv.to(device)
14.
15. # Print model's state_dict
16. print("Model's state_dict:")
17. for param_tensor in model_conv.state_dict():
18.     print(param_tensor, "\t", model_conv.state_dict()[param_tensor].size())
19.
20. criterion = nn.CrossEntropyLoss()
21.
22. #注意，只有最后一层的参数被优化更新
23. #和第一个场景相反
24. optimizer_conv = optim.SGD(model_conv.fc.parameters(), lr=0.001, momentum=0.9)
25.
26. # 每 7 个时代衰减 lr 的系数为 0.1
27. exp_lr_scheduler = lr_scheduler.StepLR(optimizer_conv, step_size=7, gamma=0.1)
28.
29. model_conv = train_model(model_conv, criterion, optimizer_conv,
    exp_lr_scheduler, num_epochs=25)
30. model_conv
```

第 6、7 行代码，冻结 ResNet-50 的网络模型参数，不训练卷积层网络参数。

第 10、11 行代码，构建最后一层全连接层模块，新建模块的梯度参数默认设置为 True，只训练更新全连接层的参数。

第 17、18 行遍历模型的参数字典，打印每一层的模型参数信息。

在 Databricks 云平台 Notebook 中运行上述代码，运行结果如下：

```
Model's state_dict:
conv1.weight     torch.Size([64, 3, 7, 7])
bn1.weight   torch.Size([64])
bn1.bias     torch.Size([64])
bn1.running_mean     torch.Size([64])
bn1.running_var      torch.Size([64])
bn1.num_batches_tracked      torch.Size([])
layer1.0.conv1.weight    torch.Size([64, 64, 1, 1])
```

```
layer1.0.bn1.weight     torch.Size([64])
layer1.0.bn1.bias   torch.Size([64])
layer1.0.bn1.running_mean   torch.Size([64])
layer1.0.bn1.running_var    torch.Size([64])
layer1.0.bn1.num_batches_tracked    torch.Size([])
layer1.0.conv2.weight   torch.Size([64, 64, 3, 3])
layer1.0.bn2.weight     torch.Size([64])
layer1.0.bn2.bias   torch.Size([64])
layer1.0.bn2.running_mean   torch.Size([64])
layer1.0.bn2.running_var    torch.Size([64])
layer1.0.bn2.num_batches_tracked    torch.Size([])
layer1.0.conv3.weight   torch.Size([256, 64, 1, 1])
layer1.0.bn3.weight     torch.Size([256])
layer1.0.bn3.bias   torch.Size([256])
layer1.0.bn3.running_mean   torch.Size([256])
layer1.0.bn3.running_var    torch.Size([256])
layer1.0.bn3.num_batches_tracked    torch.Size([])
layer1.0.downsample.0.weight    torch.Size([256, 64, 1, 1])
layer1.0.downsample.1.weight    torch.Size([256])
layer1.0.downsample.1.bias  torch.Size([256])
layer1.0.downsample.1.running_mean  torch.Size([256])
layer1.0.downsample.1.running_var   torch.Size([256])
layer1.0.downsample.1.num_batches_tracked   torch.Size([])
layer1.1.conv1.weight   torch.Size([64, 256, 1, 1])
layer1.1.bn1.weight     torch.Size([64])
layer1.1.bn1.bias   torch.Size([64])
layer1.1.bn1.running_mean   torch.Size([64])
layer1.1.bn1.running_var    torch.Size([64])
layer1.1.bn1.num_batches_tracked    torch.Size([])
layer1.1.conv2.weight   torch.Size([64, 64, 3, 3])
layer1.1.bn2.weight     torch.Size([64])
layer1.1.bn2.bias   torch.Size([64])
layer1.1.bn2.running_mean   torch.Size([64])
layer1.1.bn2.running_var    torch.Size([64])
layer1.1.bn2.num_batches_tracked    torch.Size([])
layer1.1.conv3.weight   torch.Size([256, 64, 1, 1])
layer1.1.bn3.weight     torch.Size([256])
layer1.1.bn3.bias   torch.Size([256])
layer1.1.bn3.running_mean   torch.Size([256])
layer1.1.bn3.running_var    torch.Size([256])
layer1.1.bn3.num_batches_tracked    torch.Size([])
layer1.2.conv1.weight   torch.Size([64, 256, 1, 1])
layer1.2.bn1.weight     torch.Size([64])
layer1.2.bn1.bias   torch.Size([64])
layer1.2.bn1.running_mean   torch.Size([64])
layer1.2.bn1.running_var    torch.Size([64])
layer1.2.bn1.num_batches_tracked    torch.Size([])
layer1.2.conv2.weight   torch.Size([64, 64, 3, 3])
layer1.2.bn2.weight     torch.Size([64])
layer1.2.bn2.bias   torch.Size([64])
layer1.2.bn2.running_mean   torch.Size([64])
layer1.2.bn2.running_var    torch.Size([64])
layer1.2.bn2.num_batches_tracked    torch.Size([])
layer1.2.conv3.weight   torch.Size([256, 64, 1, 1])
layer1.2.bn3.weight     torch.Size([256])
layer1.2.bn3.bias   torch.Size([256])
layer1.2.bn3.running_mean   torch.Size([256])
layer1.2.bn3.running_var    torch.Size([256])
layer1.2.bn3.num_batches_tracked    torch.Size([])
layer2.0.conv1.weight   torch.Size([128, 256, 1, 1])
```

```
layer2.0.bn1.weight     torch.Size([128])
layer2.0.bn1.bias    torch.Size([128])
layer2.0.bn1.running_mean    torch.Size([128])
layer2.0.bn1.running_var    torch.Size([128])
layer2.0.bn1.num_batches_tracked    torch.Size([])
layer2.0.conv2.weight    torch.Size([128, 128, 3, 3])
layer2.0.bn2.weight     torch.Size([128])
layer2.0.bn2.bias    torch.Size([128])
layer2.0.bn2.running_mean    torch.Size([128])
layer2.0.bn2.running_var    torch.Size([128])
layer2.0.bn2.num_batches_tracked    torch.Size([])
layer2.0.conv3.weight    torch.Size([512, 128, 1, 1])
layer2.0.bn3.weight     torch.Size([512])
layer2.0.bn3.bias    torch.Size([512])
layer2.0.bn3.running_mean    torch.Size([512])
layer2.0.bn3.running_var    torch.Size([512])
layer2.0.bn3.num_batches_tracked    torch.Size([])
layer2.0.downsample.0.weight    torch.Size([512, 256, 1, 1])
layer2.0.downsample.1.weight    torch.Size([512])
layer2.0.downsample.1.bias  torch.Size([512])
layer2.0.downsample.1.running_mean  torch.Size([512])
layer2.0.downsample.1.running_var   torch.Size([512])
layer2.0.downsample.1.num_batches_tracked   torch.Size([])
layer2.1.conv1.weight    torch.Size([128, 512, 1, 1])
layer2.1.bn1.weight     torch.Size([128])
layer2.1.bn1.bias    torch.Size([128])
layer2.1.bn1.running_mean    torch.Size([128])
layer2.1.bn1.running_var    torch.Size([128])
layer2.1.bn1.num_batches_tracked    torch.Size([])
layer2.1.conv2.weight    torch.Size([128, 128, 3, 3])
layer2.1.bn2.weight     torch.Size([128])
layer2.1.bn2.bias    torch.Size([128])
layer2.1.bn2.running_mean    torch.Size([128])
layer2.1.bn2.running_var    torch.Size([128])
layer2.1.bn2.num_batches_tracked    torch.Size([])
layer2.1.conv3.weight    torch.Size([512, 128, 1, 1])
layer2.1.bn3.weight     torch.Size([512])
layer2.1.bn3.bias    torch.Size([512])
layer2.1.bn3.running_mean    torch.Size([512])
layer2.1.bn3.running_var    torch.Size([512])
layer2.1.bn3.num_batches_tracked    torch.Size([])
layer2.2.conv1.weight    torch.Size([128, 512, 1, 1])
layer2.2.bn1.weight     torch.Size([128])
layer2.2.bn1.bias    torch.Size([128])
layer2.2.bn1.running_mean    torch.Size([128])
layer2.2.bn1.running_var    torch.Size([128])
layer2.2.bn1.num_batches_tracked    torch.Size([])
layer2.2.conv2.weight    torch.Size([128, 128, 3, 3])
layer2.2.bn2.weight     torch.Size([128])
layer2.2.bn2.bias    torch.Size([128])
layer2.2.bn2.running_mean    torch.Size([128])
layer2.2.bn2.running_var    torch.Size([128])
layer2.2.bn2.num_batches_tracked    torch.Size([])
layer2.2.conv3.weight    torch.Size([512, 128, 1, 1])
layer2.2.bn3.weight     torch.Size([512])
layer2.2.bn3.bias    torch.Size([512])
layer2.2.bn3.running_mean    torch.Size([512])
layer2.2.bn3.running_var    torch.Size([512])
layer2.2.bn3.num_batches_tracked    torch.Size([])
layer2.3.conv1.weight    torch.Size([128, 512, 1, 1])
```

```
layer2.3.bn1.weight      torch.Size([128])
layer2.3.bn1.bias    torch.Size([128])
layer2.3.bn1.running_mean    torch.Size([128])
layer2.3.bn1.running_var     torch.Size([128])
layer2.3.bn1.num_batches_tracked    torch.Size([])
layer2.3.conv2.weight    torch.Size([128, 128, 3, 3])
layer2.3.bn2.weight      torch.Size([128])
layer2.3.bn2.bias    torch.Size([128])
layer2.3.bn2.running_mean    torch.Size([128])
layer2.3.bn2.running_var     torch.Size([128])
layer2.3.bn2.num_batches_tracked    torch.Size([])
layer2.3.conv3.weight    torch.Size([512, 128, 1, 1])
layer2.3.bn3.weight      torch.Size([512])
layer2.3.bn3.bias    torch.Size([512])
layer2.3.bn3.running_mean    torch.Size([512])
layer2.3.bn3.running_var     torch.Size([512])
layer2.3.bn3.num_batches_tracked    torch.Size([])
layer3.0.conv1.weight    torch.Size([256, 512, 1, 1])
layer3.0.bn1.weight      torch.Size([256])
layer3.0.bn1.bias    torch.Size([256])
layer3.0.bn1.running_mean    torch.Size([256])
layer3.0.bn1.running_var     torch.Size([256])
layer3.0.bn1.num_batches_tracked    torch.Size([])
layer3.0.conv2.weight    torch.Size([256, 256, 3, 3])
layer3.0.bn2.weight      torch.Size([256])
layer3.0.bn2.bias    torch.Size([256])
layer3.0.bn2.running_mean    torch.Size([256])
layer3.0.bn2.running_var     torch.Size([256])
layer3.0.bn2.num_batches_tracked    torch.Size([])
layer3.0.conv3.weight    torch.Size([1024, 256, 1, 1])
layer3.0.bn3.weight      torch.Size([1024])
layer3.0.bn3.bias    torch.Size([1024])
layer3.0.bn3.running_mean    torch.Size([1024])
layer3.0.bn3.running_var     torch.Size([1024])
layer3.0.bn3.num_batches_tracked    torch.Size([])
layer3.0.downsample.0.weight    torch.Size([1024, 512, 1, 1])
layer3.0.downsample.1.weight    torch.Size([1024])
layer3.0.downsample.1.bias  torch.Size([1024])
layer3.0.downsample.1.running_mean  torch.Size([1024])
layer3.0.downsample.1.running_var   torch.Size([1024])
layer3.0.downsample.1.num_batches_tracked   torch.Size([])
layer3.1.conv1.weight    torch.Size([256, 1024, 1, 1])
layer3.1.bn1.weight      torch.Size([256])
layer3.1.bn1.bias    torch.Size([256])
layer3.1.bn1.running_mean    torch.Size([256])
layer3.1.bn1.running_var     torch.Size([256])
layer3.1.bn1.num_batches_tracked    torch.Size([])
layer3.1.conv2.weight    torch.Size([256, 256, 3, 3])
layer3.1.bn2.weight      torch.Size([256])
layer3.1.bn2.bias    torch.Size([256])
layer3.1.bn2.running_mean    torch.Size([256])
layer3.1.bn2.running_var     torch.Size([256])
layer3.1.bn2.num_batches_tracked    torch.Size([])
layer3.1.conv3.weight    torch.Size([1024, 256, 1, 1])
layer3.1.bn3.weight      torch.Size([1024])
layer3.1.bn3.bias    torch.Size([1024])
layer3.1.bn3.running_mean    torch.Size([1024])
layer3.1.bn3.running_var     torch.Size([1024])
layer3.1.bn3.num_batches_tracked    torch.Size([])
layer3.2.conv1.weight    torch.Size([256, 1024, 1, 1])
```

```
layer3.2.bn1.weight     torch.Size([256])
layer3.2.bn1.bias   torch.Size([256])
layer3.2.bn1.running_mean   torch.Size([256])
layer3.2.bn1.running_var    torch.Size([256])
layer3.2.bn1.num_batches_tracked    torch.Size([])
layer3.2.conv2.weight   torch.Size([256, 256, 3, 3])
layer3.2.bn2.weight     torch.Size([256])
layer3.2.bn2.bias   torch.Size([256])
layer3.2.bn2.running_mean   torch.Size([256])
layer3.2.bn2.running_var    torch.Size([256])
layer3.2.bn2.num_batches_tracked    torch.Size([])
layer3.2.conv3.weight   torch.Size([1024, 256, 1, 1])
layer3.2.bn3.weight     torch.Size([1024])
layer3.2.bn3.bias   torch.Size([1024])
layer3.2.bn3.running_mean   torch.Size([1024])
layer3.2.bn3.running_var    torch.Size([1024])
layer3.2.bn3.num_batches_tracked    torch.Size([])
layer3.3.conv1.weight   torch.Size([256, 1024, 1, 1])
layer3.3.bn1.weight     torch.Size([256])
layer3.3.bn1.bias   torch.Size([256])
layer3.3.bn1.running_mean   torch.Size([256])
layer3.3.bn1.running_var    torch.Size([256])
layer3.3.bn1.num_batches_tracked    torch.Size([])
layer3.3.conv2.weight   torch.Size([256, 256, 3, 3])
layer3.3.bn2.weight     torch.Size([256])
layer3.3.bn2.bias   torch.Size([256])
layer3.3.bn2.running_mean   torch.Size([256])
layer3.3.bn2.running_var    torch.Size([256])
layer3.3.bn2.num_batches_tracked    torch.Size([])
layer3.3.conv3.weight   torch.Size([1024, 256, 1, 1])
layer3.3.bn3.weight     torch.Size([1024])
layer3.3.bn3.bias   torch.Size([1024])
layer3.3.bn3.running_mean   torch.Size([1024])
layer3.3.bn3.running_var    torch.Size([1024])
layer3.3.bn3.num_batches_tracked    torch.Size([])
layer3.4.conv1.weight   torch.Size([256, 1024, 1, 1])
layer3.4.bn1.weight     torch.Size([256])
layer3.4.bn1.bias   torch.Size([256])
layer3.4.bn1.running_mean   torch.Size([256])
layer3.4.bn1.running_var    torch.Size([256])
layer3.4.bn1.num_batches_tracked    torch.Size([])
layer3.4.conv2.weight   torch.Size([256, 256, 3, 3])
layer3.4.bn2.weight     torch.Size([256])
layer3.4.bn2.bias   torch.Size([256])
layer3.4.bn2.running_mean   torch.Size([256])
layer3.4.bn2.running_var    torch.Size([256])
layer3.4.bn2.num_batches_tracked    torch.Size([])
layer3.4.conv3.weight   torch.Size([1024, 256, 1, 1])
layer3.4.bn3.weight     torch.Size([1024])
layer3.4.bn3.bias   torch.Size([1024])
layer3.4.bn3.running_mean   torch.Size([1024])
layer3.4.bn3.running_var    torch.Size([1024])
layer3.4.bn3.num_batches_tracked    torch.Size([])
layer3.5.conv1.weight   torch.Size([256, 1024, 1, 1])
layer3.5.bn1.weight     torch.Size([256])
layer3.5.bn1.bias   torch.Size([256])
layer3.5.bn1.running_mean   torch.Size([256])
layer3.5.bn1.running_var    torch.Size([256])
layer3.5.bn1.num_batches_tracked    torch.Size([])
layer3.5.conv2.weight   torch.Size([256, 256, 3, 3])
```

```
layer3.5.bn2.weight      torch.Size([256])
layer3.5.bn2.bias    torch.Size([256])
layer3.5.bn2.running_mean    torch.Size([256])
layer3.5.bn2.running_var     torch.Size([256])
layer3.5.bn2.num_batches_tracked    torch.Size([])
layer3.5.conv3.weight    torch.Size([1024, 256, 1, 1])
layer3.5.bn3.weight      torch.Size([1024])
layer3.5.bn3.bias    torch.Size([1024])
layer3.5.bn3.running_mean    torch.Size([1024])
layer3.5.bn3.running_var     torch.Size([1024])
layer3.5.bn3.num_batches_tracked    torch.Size([])
layer4.0.conv1.weight    torch.Size([512, 1024, 1, 1])
layer4.0.bn1.weight      torch.Size([512])
layer4.0.bn1.bias    torch.Size([512])
layer4.0.bn1.running_mean    torch.Size([512])
layer4.0.bn1.running_var     torch.Size([512])
layer4.0.bn1.num_batches_tracked    torch.Size([])
layer4.0.conv2.weight    torch.Size([512, 512, 3, 3])
layer4.0.bn2.weight      torch.Size([512])
layer4.0.bn2.bias    torch.Size([512])
layer4.0.bn2.running_mean    torch.Size([512])
layer4.0.bn2.running_var     torch.Size([512])
layer4.0.bn2.num_batches_tracked    torch.Size([])
layer4.0.conv3.weight    torch.Size([2048, 512, 1, 1])
layer4.0.bn3.weight      torch.Size([2048])
layer4.0.bn3.bias    torch.Size([2048])
layer4.0.bn3.running_mean    torch.Size([2048])
layer4.0.bn3.running_var     torch.Size([2048])
layer4.0.bn3.num_batches_tracked    torch.Size([])
layer4.0.downsample.0.weight    torch.Size([2048, 1024, 1, 1])
layer4.0.downsample.1.weight    torch.Size([2048])
layer4.0.downsample.1.bias  torch.Size([2048])
layer4.0.downsample.1.running_mean  torch.Size([2048])
layer4.0.downsample.1.running_var   torch.Size([2048])
layer4.0.downsample.1.num_batches_tracked   torch.Size([])
layer4.1.conv1.weight    torch.Size([512, 2048, 1, 1])
layer4.1.bn1.weight      torch.Size([512])
layer4.1.bn1.bias    torch.Size([512])
layer4.1.bn1.running_mean    torch.Size([512])
layer4.1.bn1.running_var     torch.Size([512])
layer4.1.bn1.num_batches_tracked    torch.Size([])
layer4.1.conv2.weight    torch.Size([512, 512, 3, 3])
layer4.1.bn2.weight      torch.Size([512])
layer4.1.bn2.bias    torch.Size([512])
layer4.1.bn2.running_mean    torch.Size([512])
layer4.1.bn2.running_var     torch.Size([512])
layer4.1.bn2.num_batches_tracked    torch.Size([])
layer4.1.conv3.weight    torch.Size([2048, 512, 1, 1])
layer4.1.bn3.weight      torch.Size([2048])
layer4.1.bn3.bias    torch.Size([2048])
layer4.1.bn3.running_mean    torch.Size([2048])
layer4.1.bn3.running_var     torch.Size([2048])
layer4.1.bn3.num_batches_tracked    torch.Size([])
layer4.2.conv1.weight    torch.Size([512, 2048, 1, 1])
layer4.2.bn1.weight      torch.Size([512])
layer4.2.bn1.bias    torch.Size([512])
layer4.2.bn1.running_mean    torch.Size([512])
layer4.2.bn1.running_var     torch.Size([512])
layer4.2.bn1.num_batches_tracked    torch.Size([])
layer4.2.conv2.weight    torch.Size([512, 512, 3, 3])
```

```
layer4.2.bn2.weight     torch.Size([512])
layer4.2.bn2.bias   torch.Size([512])
layer4.2.bn2.running_mean   torch.Size([512])
layer4.2.bn2.running_var    torch.Size([512])
layer4.2.bn2.num_batches_tracked    torch.Size([])
layer4.2.conv3.weight   torch.Size([2048, 512, 1, 1])
layer4.2.bn3.weight     torch.Size([2048])
layer4.2.bn3.bias   torch.Size([2048])
layer4.2.bn3.running_mean   torch.Size([2048])
layer4.2.bn3.running_var    torch.Size([2048])
layer4.2.bn3.num_batches_tracked    torch.Size([])
fc.weight   torch.Size([5, 2048])
fc.bias     torch.Size([5])
Epoch 0/24
----------
train Loss: 1.6752 Acc: 0.1800
val Loss: 1.5340 Acc: 0.3600

Epoch 1/24
----------
train Loss: 1.5418 Acc: 0.2450
val Loss: 1.5069 Acc: 0.3000

Epoch 2/24
----------
train Loss: 1.4638 Acc: 0.3850
val Loss: 1.4089 Acc: 0.4100

Epoch 3/24
----------
train Loss: 1.3377 Acc: 0.5500
val Loss: 1.2249 Acc: 0.5900

Epoch 4/24
----------
train Loss: 1.1427 Acc: 0.6900
val Loss: 1.0607 Acc: 0.7300

Epoch 5/24
----------
train Loss: 0.9560 Acc: 0.8750
val Loss: 0.9918 Acc: 0.6900

Epoch 6/24
----------
train Loss: 0.8328 Acc: 0.8850
val Loss: 0.9178 Acc: 0.6800

Epoch 7/24
----------
train Loss: 0.7157 Acc: 0.8750
val Loss: 0.8904 Acc: 0.6900

Epoch 8/24
----------
train Loss: 0.7008 Acc: 0.8750
val Loss: 0.8616 Acc: 0.7100

Epoch 9/24
----------
```

```
train Loss: 0.6830 Acc: 0.8950
val Loss: 0.8336 Acc: 0.7300

Epoch 10/24
----------
train Loss: 0.6645 Acc: 0.9050
val Loss: 0.8087 Acc: 0.7400

Epoch 11/24
----------
train Loss: 0.6471 Acc: 0.9200
val Loss: 0.7884 Acc: 0.7600

Epoch 12/24
----------
train Loss: 0.6320 Acc: 0.9250
val Loss: 0.7728 Acc: 0.7800

Epoch 13/24
----------
train Loss: 0.6194 Acc: 0.9150
val Loss: 0.7614 Acc: 0.7800

Epoch 14/24
----------
train Loss: 0.6091 Acc: 0.9250
val Loss: 0.7593 Acc: 0.7900

Epoch 15/24
----------
train Loss: 0.6081 Acc: 0.9200
val Loss: 0.7578 Acc: 0.7900

Epoch 16/24
----------
train Loss: 0.6072 Acc: 0.9200
val Loss: 0.7566 Acc: 0.7900

Epoch 17/24
----------
train Loss: 0.6063 Acc: 0.9200
val Loss: 0.7556 Acc: 0.8000

Epoch 18/24
----------
train Loss: 0.6053 Acc: 0.9200
val Loss: 0.7549 Acc: 0.8000

Epoch 19/24
----------
train Loss: 0.6044 Acc: 0.9200
val Loss: 0.7543 Acc: 0.8000

Epoch 20/24
----------
train Loss: 0.6034 Acc: 0.9200
val Loss: 0.7538 Acc: 0.8000

Epoch 21/24
----------
```

```
train Loss: 0.6025 Acc: 0.9200
val Loss: 0.7540 Acc: 0.8000

Epoch 22/24
----------
train Loss: 0.6024 Acc: 0.9200
val Loss: 0.7542 Acc: 0.8100

Epoch 23/24
----------
train Loss: 0.6023 Acc: 0.9200
val Loss: 0.7544 Acc: 0.8100

Epoch 24/24
----------
train Loss: 0.6022 Acc: 0.9200
val Loss: 0.7547 Acc: 0.8100

Training complete in 23m 5s
Best val Acc: 0.810000
……
```

在第 24 个时代中，训练集的损失度为 0.6022，训练集的准确度达到了 92%；验证集的损失度为 0.7547，验证集的准确度为 81%，整个模型训练完成以后，验证集预测最好的准确度达到了 81%。

执行模型预测可视化函数。

```
1.   visualize_model(get_model_for_eval())
```

在 Databricks 云平台 Notebook 中运行上述代码，运行结果如图 32-66 所示。

图 32-66　ResNet-50 模型作为固定图像特征提取器的预测结果

32.3　Spark+TensorFlow 实战

本节讲解 TensorFlow 在 Spark 上的安装、TensorBoard 解密、Spark-TensorFlow 的数据转换等内容。

32.3.1　TensorFlow 在 Spark 上的安装

Databricks 运行时机器学习库中已包括 TensorFlow 和 TensorBoard，可以在不安装任何包的情况下使用这些库。

以下是包含 TensorFlow 的 Databricks 运行时机器学习库版本。

```
Databricks Runtime 5.5 ML: 1.13.1
Databricks Runtime 5.1 - 5.4 ML: 1.12.0
Databricks Runtime 5.0 ML: 1.10.0
```

建议使用 Databricks 机器学习运行时库（已包含 TensorFlow 和 TensorBoard）。如果要使用 Databricks 运行时库（无 TensorFlow），则可以将 TensorFlow 安装为 Databricks PyPI 库。

- GPU 集群环境中安装 TensorFlow 版本。

<tensorflow version>替换为 1.14.0 或 2.0.0-beta1。

```
tensorflow-gpu==<tensorflow version>
```

- CPU 集群环境安装 TensorFlow 版本。

<tensorflow version>替换为 1.14.0 或 2.0.0-beta1。

```
tensorflow==<tensorflow version>
```

32.3.2　TensorBoard 解密

TensorBoard 是用于检查和理解 TensorFlow 运行图的 Web 应用程序。TensorBoard 提供机器学习所需的可视化功能和工具，包括跟踪可视化损失度及准确率等指标、可视化模型图（操作和层）、查看权重、偏差或其他张量随时间变化的直方图、将嵌入投射到较低的维度空间、显示图片、文字和音频数据、剖析 TensorFlow 程序以及更多功能。

接下来讲解在 Databricks Notebook 中进行 Mnist 图像识别的 TensorBoard 案例。

- 上传文件（labels_1024.tsv、sprite_1024.png）到 Databricks 文件系统。TensorBoard 有一个内置的可视化工具 Embedding Projector，将高维的数据按照特定的算法映射到二维或者三维空间进行展示。
- 上传 Mnist.py 代码文件到 Databricks，运行程序，将结果写入到事件文件。
- 运行 TensorBoard 命令，在 Web 页面对图进行可视化。

（1）上传文件（labels_1024.tsv、sprite_1024）到 Databricks 文件系统。

如图 32-67 所示，在 Databricks 云平台首页面中单击 Import & Explore Data 按钮，导入数据文件。

如图 32-68 所示，进入 Create New Table 页面，单击 Select 按钮选择 labels_1024.tsv 和 sprite_1024.png 文件进行上传，上传的文件将保存到 Databricks 文件系统。

其中 labels_1024.tsv 包含 1024 行记录，每行记录是一个数字，表示图像对应的数字标签。

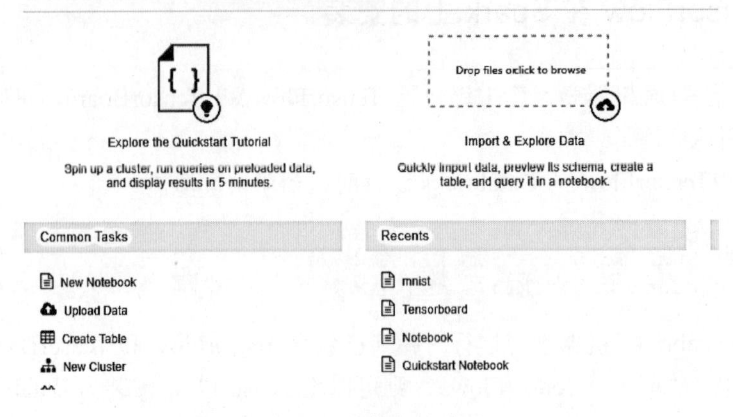

图 32-67　文件上传页面

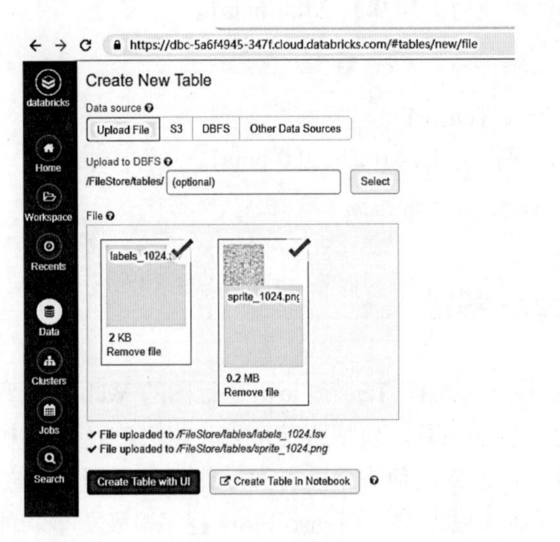

图 32-68　上传文件到 Databricks 文件系统

labels_1024.tsv 部分记录如下：

```
2
2
0
8
9
9
0
2
5
1
……
```

其中 sprite_1024.png 是一张 32×32 的图片，一共包括 1024 个数字，用于在 TensorBoard 中进行可视化展示，如图 32-69 所示。

如图 32-70 所示，在 Databricks 文件系统中，检查已经上传的文件。

72104149590690159734966540740131
34727121174235124463556041957893
74643070291732977627847361369314
17696054992194873974449254767905
85665781016467317182029955156034
46546545144723271818185089250111
09031642361113952945939036557227
12841733887922415987230442419577
28268577918180301994182129759264
15829204002847124027433003196525
92930420711215339786561381051315
56185179462250656372088541140337
61621928619525442838245031775797
19214292049148184598837600302664
93332391268056663882758961841259
19754089910523789406395213136574
22632654897130383193446421825488
40023277087447969098046063548339
33378082170654380963809968685786
02402231975108462679329822927359
18020521376712580377409186774349
19517397691378336728585114431077
07944855408210845040613326726931
46254206217341054311749948402451
16471942415538314568941538032512
83440883317359632613607217142421
79611248177480731310770355276692
83522560829288887493066321322930
05781446029147473988471212237323
91740355863267663279117564951334
78911691445406223151203812671623
90122089902519781041796426813754

图 32-69　Mnist 数据集

```
1 print(dbutils.fs.ls("dbfs:/FileStore/tables/"))

[FileInfo(path='dbfs:/FileStore/tables/labels_1024.tsv', name='labels_1024.tsv', size=2048), FileInfo(path='dbfs:/FileStore/tables/sprite_1024.png', name='sprite_1024.png', size=173020)]
Command took 0.14 seconds -- by duanzhihua@189.cn at 2019/8/18 下午7:17:52 on spark-tensorflow
```

图 32-70　Databricks 文件检查

（2）上传 Mnist.py 代码文件到 Databricks。

如图 32-71 所示，在 Databricks 的 Workspace 中上传 Mnist.py 代码文件，Mnist.py 将实现手写数字图像识别，运行程序，将 Tensorflow 摘要对象的结果保存到事件文件中。

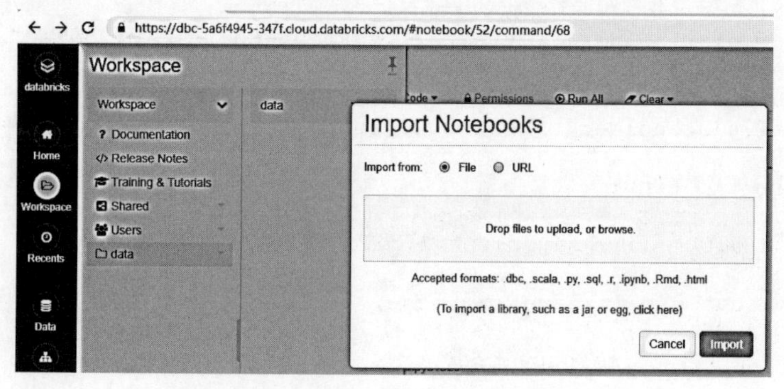

图 32-71　上传代码文件

Mnist.py 代码如下：

```
1.  import os
2.  import os.path
3.  import shutil
```

```
4.  import tensorflow_as_tf
5.
6.  LOGDIR = "/tmp/mnist_tutorial/"
7.  ### MNIST 嵌入 ###
8.  mnist = tf.contrib.learn.datasets.mnist.read_data_sets(train_dir=LOGDIR
    + "data", one_hot=True)
9.
10.
11.
12. def conv_layer(input, size_in, size_out, name="conv"):
13.   with tf.name_scope(name):
14.     w = tf.Variable(tf.truncated_normal([5, 5, size_in, size_out],
        stddev=0.1), name="W")
15.     b = tf.Variable(tf.constant(0.1, shape=[size_out]), name="B")
16.     conv = tf.nn.conv2d(input, w, strides=[1, 1, 1, 1], padding="SAME")
17.     act = tf.nn.relu(conv + b)
18.     tf.summary.histogram("weights", w)
19.     tf.summary.histogram("biases", b)
20.     tf.summary.histogram("activations", act)
21.     return tf.nn.max_pool(act, ksize=[1, 2, 2, 1], strides=[1, 2, 2, 1],
        padding="SAME")
22.
23.
24. def fc_layer(input, size_in, size_out, name="fc"):
25.   with tf.name_scope(name):
26.     w = tf.Variable(tf.truncated_normal([size_in, size_out], stddev=0.1),
        name="W")
27.     b = tf.Variable(tf.constant(0.1, shape=[size_out]), name="B")
28.     act = tf.matmul(input, w) + b
29.     tf.summary.histogram("weights", w)
30.     tf.summary.histogram("biases", b)
31.     tf.summary.histogram("activations", act)
32.     return act
33.
34.
35. def mnist_model(learning_rate, use_two_fc, use_two_conv, hparam):
36.   tf.reset_default_graph()
37.   sess = tf.Session()
38.
39.   # 设置占位符，并重新调整数据的 shape
40.   x = tf.placeholder(tf.float32, shape=[None, 784], name="x")
41.   x_image = tf.reshape(x, [-1, 28, 28, 1])
42.   tf.summary.image('input', x_image, 3)
43.   y = tf.placeholder(tf.float32, shape=[None, 10], name="labels")
44.
45.   if use_two_conv:
46.     conv1 = conv_layer(x_image, 1, 32, "conv1")
47.     conv_out = conv_layer(conv1, 32, 64, "conv2")
48.   else:
49.     conv_out = conv_layer(x_image, 1, 16, "conv")
50.
51.   flattened = tf.reshape(conv_out, [-1, 7 * 7 * 64])
52.
53.
54.   if use_two_fc:
55.     fc1 = fc_layer(flattened, 7 * 7 * 64, 1024, "fc1")
56.     relu = tf.nn.relu(fc1)
57.     embedding_input = relu
58.     tf.summary.histogram("fc1/relu", relu)
```

```
59.      embedding_size = 1024
60.      logits = fc_layer(relu, 1024, 10, "fc2")
61.    else:
62.      embedding_input = flattened
63.      embedding_size = 7*7*64
64.      logits = fc_layer(flattened, 7*7*64, 10, "fc")
65.
66.    with tf.name_scope("xent"):
67.      xent = tf.reduce_mean(
68.          tf.nn.softmax_cross_entropy_with_logits(
69.              logits=logits, labels=y), name="xent")
70.      tf.summary.scalar("xent", xent)
71.
72.    with tf.name_scope("train"):
73.      train_step = tf.train.AdamOptimizer(learning_rate).minimize(xent)
74.
75.    with tf.name_scope("accuracy"):
76.      correct_prediction = tf.equal(tf.argmax(logits, 1), tf.argmax(y, 1))
77.      accuracy = tf.reduce_mean(tf.cast(correct_prediction, tf.float32))
78.      tf.summary.scalar("accuracy", accuracy)
79.
80.    summ = tf.summary.merge_all()
81.
82.
83.    embedding = tf.Variable(tf.zeros([1024, embedding_size]),
       name="test_embedding")
84.    assignment = embedding.assign(embedding_input)
85.    saver = tf.train.Saver()
86.
87.    sess.run(tf.global_variables_initializer())
88.    writer = tf.summary.FileWriter(LOGDIR + hparam)
89.    writer.add_graph(sess.graph)
90.
91.    config = tf.contrib.tensorboard.plugins.projector.ProjectorConfig()
92.    embedding_config = config.embeddings.add()
93.    embedding_config.tensor_name = embedding.name
94.    embedding_config.sprite.image_path = "dbfs:/FileStore/tables/sprite_1024.png"
95.    embedding_config.metadata_path = "dbfs:/FileStore/tables/labels_1024.tsv"
96.
97.    # 指定单个缩略图的宽度和高度
98.    embedding_config.sprite.single_image_dim.extend([28, 28])
99.    tf.contrib.tensorboard.plugins.projector
       .visualize_embeddings(writer, config)
100.
101.       for i in range(100):
102.         batch = mnist.train.next_batch(100)
103.         if i % 5 == 0:
104.           [train_accuracy, s] = sess.run([accuracy, summ], feed_dict={x:
               batch[0], y: batch[1]})
105.           writer.add_summary(s, i)
106.         if i % 10 == 0:
107.           sess.run(assignment, feed_dict={x: mnist.test.images[:1024],
               y: mnist.test.labels[:1024]})
108.           saver.save(sess, os.path.join(LOGDIR, "model.ckpt"), i)
109.         sess.run(train_step, feed_dict={x: batch[0], y: batch[1]})
110.
111.     def make_hparam_string(learning_rate, use_two_fc, use_two_conv):
112.       conv_param = "conv=2" if use_two_conv else "conv=1"
113.       fc_param = "fc=2" if use_two_fc else "fc=1"
```

```
114.       return "lr_%.0E,%s,%s" % (learning_rate, conv_param, fc_param)
115.
116.     def main():
117.       # 可以试图增加一些学习率
118.       for learning_rate in [1E-3, 1E-4]:
119.
120.         #  将“False”作为尝试不同模型架构的值
121.         for use_two_fc in [True]:
122.           for use_two_conv in [False, True]:
123.             # 为每个字符串构造一个超参数字符串（例如："lr_1E-3,fc=2,conv=2"）
124.             hparam = make_hparam_string(learning_rate, use_two_fc, use_two_conv)
125.             print('Starting run for %s' % hparam)
126.
127.             #  实际使用新设置运行
128.             mnist_model(learning_rate, use_two_fc, use_two_conv, hparam)
129.       print('Done training!')
130.       print('Run `tensorboard --logdir=%s` to see the results.' % LOGDIR)
131.       print('Running on mac? If you want to get rid of the dialogue asking
          to give '
132.             'network permissions to TensorBoard, you can provide this flag: '
133.             '--host=localhost')
134.
135.     if __name__ == '__main__':
136.       main()
```

第8行下载 Mnist 训练集数据。

第12～21行构建神经网络卷积层，将权重、偏爱因子、卷积结果保存到 TensorFlow 的直方图中。

第24～32层构建神经网络全连接层，将权重、偏爱因子、全连接层结果保存到 TensorFlow 的直方图中。

第35～109行构建神经网络模型。TensorFlow 使用 tf.name_scope 命名空间在指定的区域中定义各种对象及操作，如定义 xent、train、accuracy 等命名空间；在 TensorFlow 使用 tf.summary.scalar 定义各种标量信息，如定义 xent、accuracy 标量。

第83～99行将高维向量进行可视化，通过 PCA、T-SNE 等方法将高维向量投影到三维坐标系。

第94、95行加载读取 Databricks 文件系统中的 sprite_1024.png、labels_1024.tsv。注意，Databricks 文件系统以“dbfs:”字符串开头。

第88、89、99、105行将 TensorFlow 运行图、摘要对象的结果保存到事件文件中。

第111～114行构建 make_hparam_string 方法，拼接学习率与卷积层名称、全连接层名称的字符串。

第116～136行构建 Mnist 程序运行的主入口 main 函数，在模型参数调优中，尝试不同的学习率、构建不同的网络模型结构，调用 Mnist 模型函数运行，将结果保存到事件文件。

在 Databricks 云平台中运行 Mnist.py，运行结果如下：

```
1.  ……
2.  Successfully downloaded train-images-idx3-ubyte.gz 9912422 bytes.
3.  ……
4.  Extracting /tmp/mnist_tutorial/data/train-images-idx3-ubyte.gz
5.  Successfully downloaded train-labels-idx1-ubyte.gz 28881 bytes.
6.  ……
7.  Extracting /tmp/mnist_tutorial/data/train-labels-idx1-ubyte.gz
8.  ……
```

```
9.  Successfully downloaded t10k-images-idx3-ubyte.gz 1648877 bytes.
10. Extracting /tmp/mnist_tutorial/data/t10k-images-idx3-ubyte.gz
11. Successfully downloaded t10k-labels-idx1-ubyte.gz 4542 bytes.
12. Extracting /tmp/mnist_tutorial/data/t10k-labels-idx1-ubyte.gz
13. ……
14. Starting run for lr_1E-03,conv=1,fc=2
15. ……
16. Starting run for lr_1E-03,conv=2,fc=2
17. Starting run for lr_1E-04,conv=1,fc=2
18. Starting run for lr_1E-04,conv=2,fc=2
19. Done training!
20. Run `tensorboard --logdir=/tmp/mnist_tutorial/` to see the results.
21. Running on mac? If you want to get rid of the dialogue asking to give network
    permissions to TensorBoard, you can provide this flag: --host=localhost
```

在 Databricks 云平台 cmd 中输入命令：

```
%sh ls /tmp/mnist_tutorial/data
```

执行结果如下，Mnist.py 将 Mnist 的训练集图像数据、训练集标签数据及测试集图像数据、测试集标签数据下载到本地 Linux 文件系统。

```
t10k-images-idx3-ubyte.gz
t10k-labels-idx1-ubyte.gz
train-images-idx3-ubyte.gz
train-labels-idx1-ubyte.gz
```

在 Databricks 云平台 cmd 中输入命令：

```
%sh ls /tmp/mnist_tutorial/
```

执行结果如下，Mnist.py 将已训练的神经网络模型数据、不同学习率及卷积层的事件数据保存到本地 Linux 文件系统。

```
checkpoint
data
lr_1E-03,conv=1,fc=2
lr_1E-03,conv=2,fc=2
lr_1E-04,conv=1,fc=2
lr_1E-04,conv=2,fc=2
model.ckpt-50.data-00000-of-00001
model.ckpt-50.index
model.ckpt-50.meta
model.ckpt-60.data-00000-of-00001
model.ckpt-60.index
model.ckpt-60.meta
model.ckpt-70.data-00000-of-00001
model.ckpt-70.index
model.ckpt-70.meta
model.ckpt-80.data-00000-of-00001
model.ckpt-80.index
model.ckpt-80.meta
model.ckpt-90.data-00000-of-00001
model.ckpt-90.index
model.ckpt-90.meta
```

（3）在 Databricks 云平台运行 TensorBoard 命令，启动 Web 页面，对 Tensorflow 运行图可视化。

```
1.  dbutils.tensorboard.start("/tmp/mnist_tutorial/")
```

执行结果如图 32-72 所示。

```
1  dbutils.tensorboard.start("/tmp/mnist_tutorial/")
```

```
Looking for active tensorboard process...
No active process found.
Starting tensorboard process...
Tensorboard process started.
```

TensorBoard log directory set to: /tmp/mnist_tutorial/. View TensorBoard

图 32-72　查看 TensorBoard

单击 View TensorBoard，查询 TensorBoard 的各个页面。

图 32-73 是图像识别神经网络精确度的可视化图。

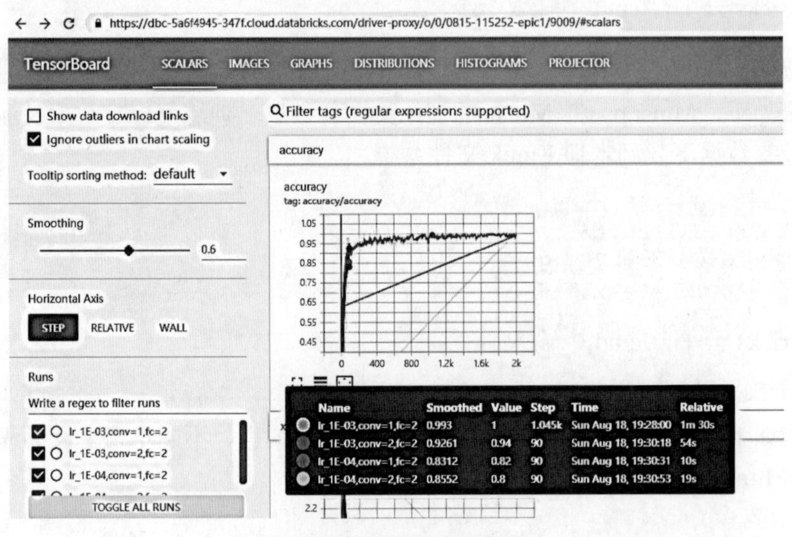

图 32-73　精确度可视化图

图 32-74 是 TensorFlow 中定义的标量的可视化图。

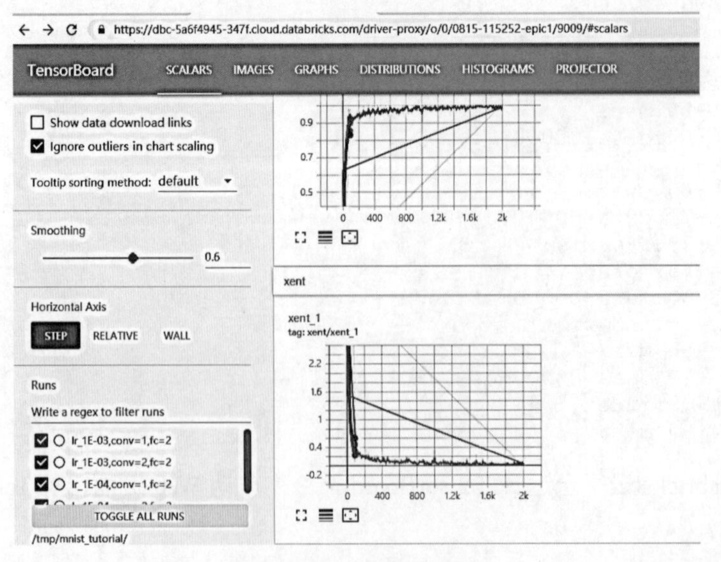

图 32-74　标量示意图

图 32-75 是 TensorFlow 中输入图像数据的可视化图。

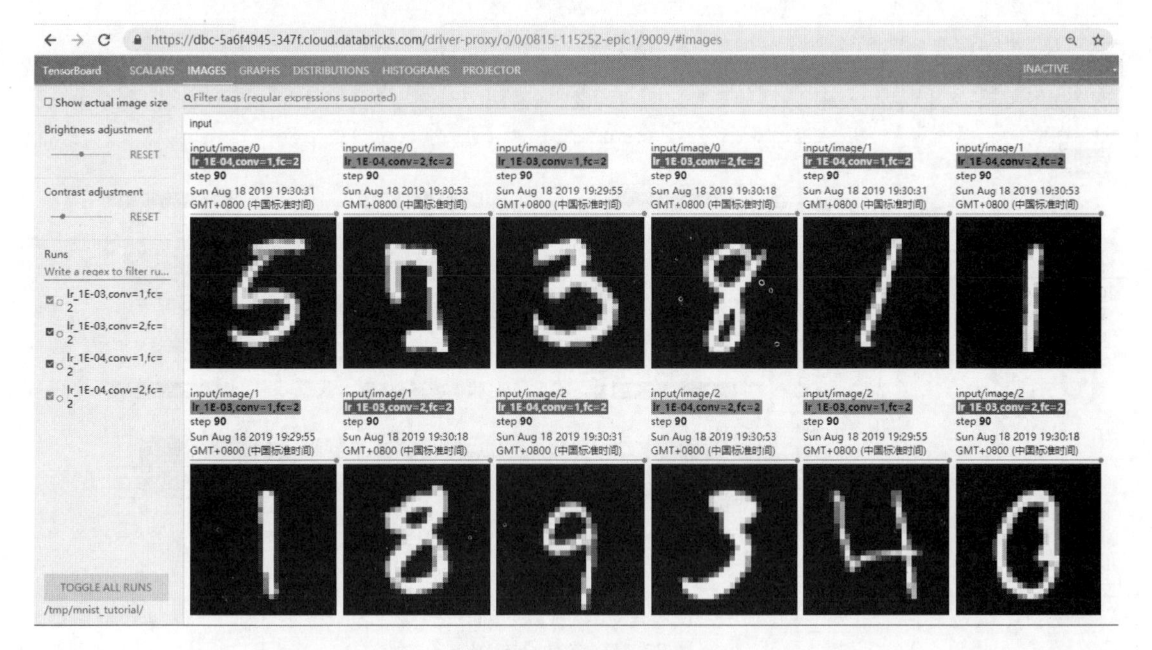

图 32-75　图像数据示意图

图 32-76 是手写数字图像识别神经网络的网络结构图。

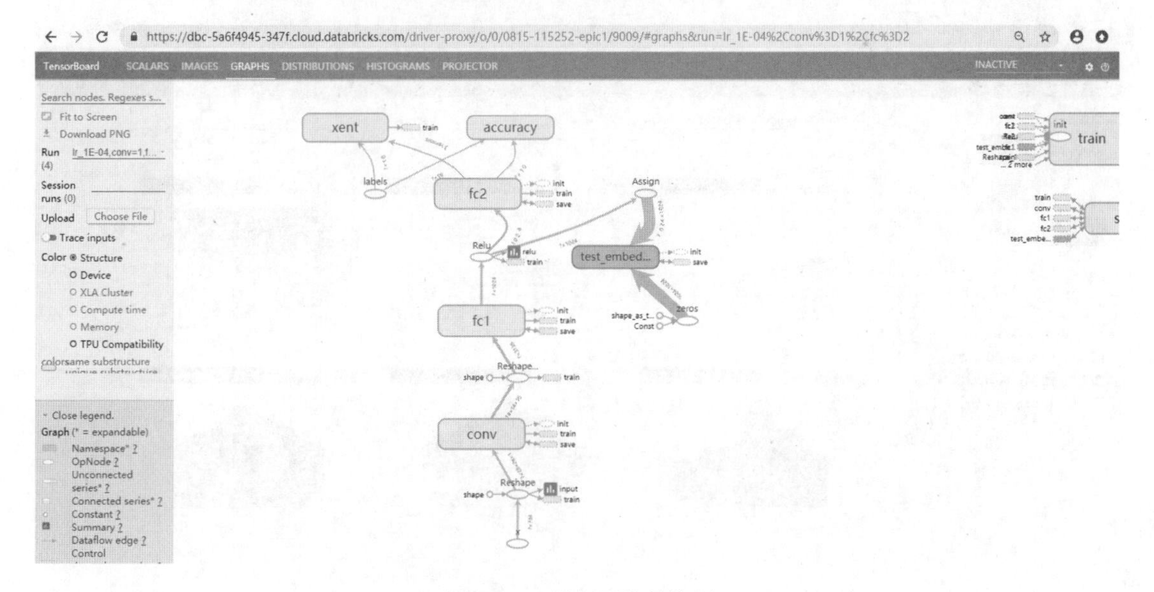

图 32-76　神经网络结构图

图 32-77 是神经网络各卷积层、全连接网络层的参数分布示意图。

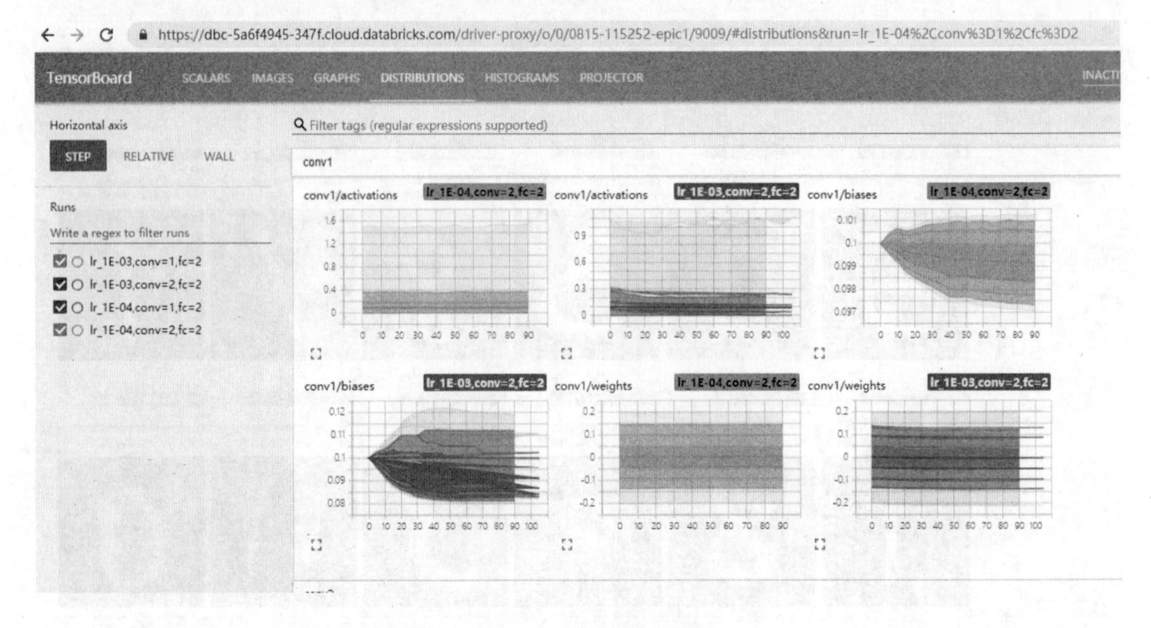

图 32-77　参数分布图

图 32-78 是神经网络各卷积层、全连接网络层的参数直方图。

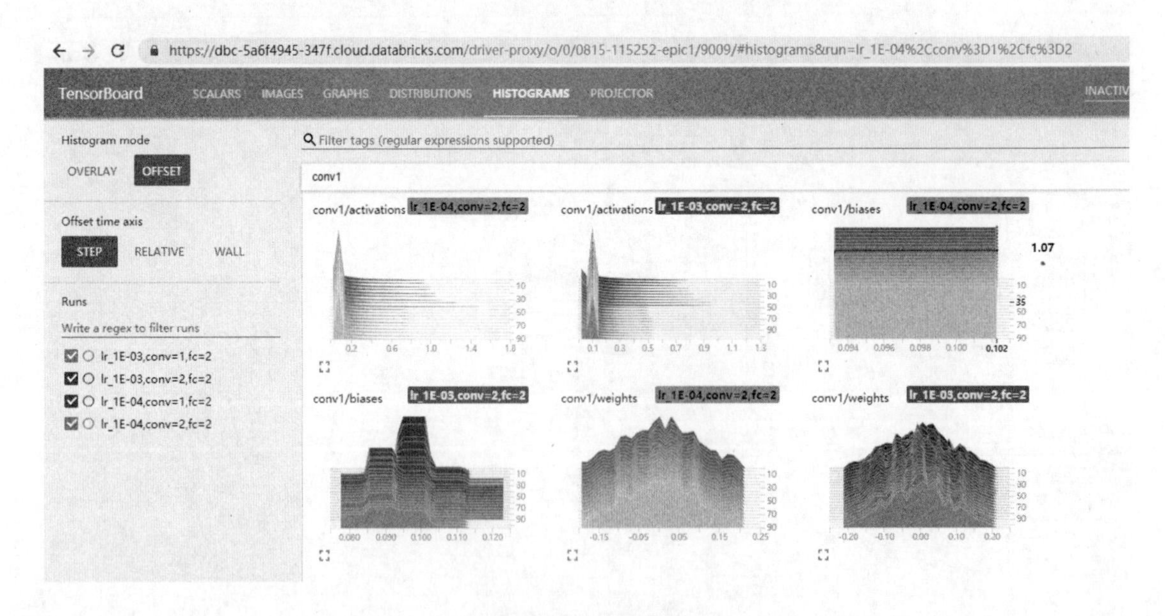

图 32-78　直方图

图 32-79 是将高维向量投影到三维坐标系的示意图。

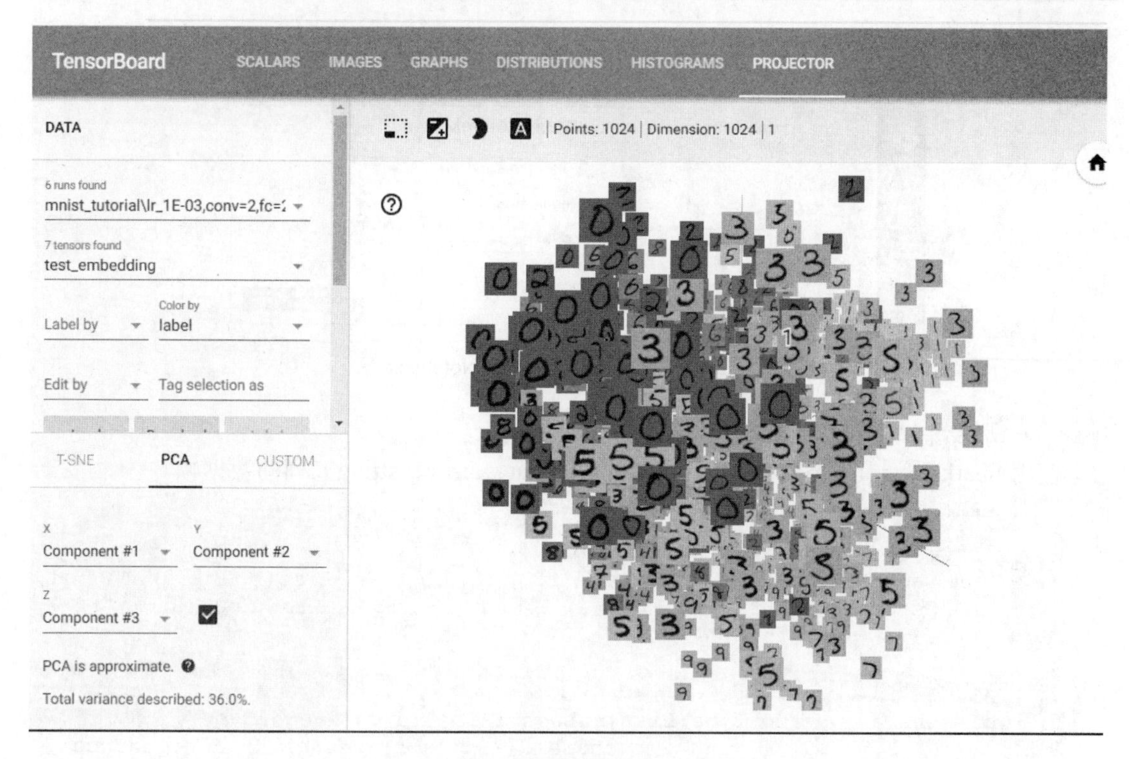

图 32-79　高维到低维示意图

32.3.3　Spark-TensorFlow 的数据转换

Spark TensorFlow Connector 是 TensorFlow 生态系统中的一个库，支持在 Spark DataFrame 数据帧和 TensorFlow TFRecords 之间进行转换（TFRecords 是 TensorFlow 存储数据的常用格式）。使用 Spark TensorFlow 连接器，可以使用 Spark 数据帧 API 将 TFRecords 文件读取到 Spark 的数据帧中，也可以将 Spark 的数据帧保存为 TFRecords 格式。

Spark TensorFlow 连接器已经包含在 Databricks 机器学习运行时库中，为机器学习和 Spark 大数据提供一个方便的环境，可以使用 Databricks 云平台集群的运行时机器学习库对 Spark 与 TensorFlow 的格式进行转换。

接下来讲解使用 Databricks 云平台集群的 Notebook 进行 Spark-TensorFlow 的数据转换案例。

（1）创建 Databricks 云平台集群 Notebook。

在 Workspace 中单击右键，在弹出的快捷菜单中选择 Create，弹出 Create Notebook 对话框，依次输入名称、开发语言（选择 Scala 语言）、云平台集群，单击 Create 按钮创建 Notebook，如图 32-80 所示。

（2）导入 Spark SQL 等库。

```
1.  import org.apache.commons.io.FileUtils
2.  import org.apache.spark.sql.{ DataFrame, Row }
3.  import org.apache.spark.sql.catalyst.expressions.GenericRow
4.  import org.apache.spark.sql.types._
```

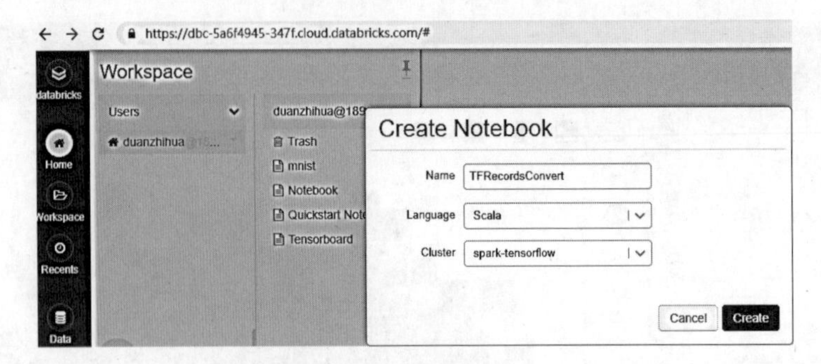

图 32-80　创建基于 Scala 的 Notebook

（3）创建 Spark 数据帧。

使用 Spark 构造具有各种类型（int、long、float、array、string）列的数据帧。

```
1.  //定义 DataFrame 数据
2.  val testRows: Array[Row] = Array(
3.    new GenericRow(Array[Any](11, 1, 23L, 10.0F, 14.0, List(1.0, 2.0), "r1")),
4.    new GenericRow(Array[Any](21, 2, 24L, 12.0F, 15.0, List(2.0, 2.0), "r2"))
5.  )
6.
7.  //数据帧模式
8.  val schema = StructType(List(StructField("id", IntegerType),
9.                          StructField("IntegerTypeLabel", IntegerType),
10.                         StructField("LongTypeLabel", LongType),
11.                         StructField("FloatTypeLabel", FloatType),
12.                         StructField("DoubleTypeLabel", DoubleType),
13.                         StructField("VectorLabel", ArrayType(DoubleType,
                            true)),
14.                         StructField("name", StringType)))
15. //创建数据帧
16. val rdd = spark.sparkContext.parallelize(testRows)
17. val df: DataFrame = spark.createDataFrame(rdd, schema)
```

第 2～4 行代码构建了 2 行数据记录。

第 8 行代码构建数据结构的模式，包括 int、long、float、array、string 等数据类型。

第 16 行代码使用 sparkContext.parallelize 方法将数据记录进行并行化处理，构建为 RDD。

第 17 行代码使用 spark.createDataFrame 方法构建数据帧。

在 Databricks 云平台 Notebook 中运行上述代码，运行结果如下：

```
df:org.apache.spark.sql.DataFrame
    id:integer
    IntegerTypeLabel:integer
    LongTypeLabel:long
    FloatTypeLabel:float
    DoubleTypeLabel:double
    VectorLabel:array
    element:double
    name:string

testRows: Array[org.apache.spark.sql.Row] = Array([11,1,23,10.0,14.0,List(1.0,
2.0),r1], [21,2,24,12.0,15.0,List(2.0, 2.0),r2])
schema: org.apache.spark.sql.types.StructType =
    StructType(StructField(id,IntegerType,true),
```

```
    StructField(IntegerTypeLabel,IntegerType,true),
    StructField(LongTypeLabel,LongType,true),
    StructField(FloatTypeLabel,FloatType,true),
    StructField(DoubleTypeLabel,DoubleType,true),
    StructField(VectorLabel,ArrayType(DoubleType,true),true),
    StructField(name,StringType,true))
rdd: org.apache.spark.rdd.RDD[org.apache.spark.sql.Row] = ParallelCollectionRDD[0]
at parallelize at command-90:16
df: org.apache.spark.sql.DataFrame = [id: int, IntegerTypeLabel: int ... 5
more fields]
```

（4）展示 Spark 数据帧的内容。

```
1.  df.show()
```

在 Databricks 云平台 Notebook 中运行上述代码，运行结果如下：

```
(3) Spark Jobs
+---+----------------+-------------+--------------+---------------+-----------+----+
| id|IntegerTypeLabel|LongTypeLabel|FloatTypeLabel|DoubleTypeLabel|VectorLabel|name|
+---+----------------+-------------+--------------+---------------+-----------+----+
| 11|               1|           23|          10.0|           14.0| [1.0, 2.0]|  r1|
| 21|               2|           24|          12.0|           15.0| [2.0, 2.0]|  r2|
+---+----------------+-------------+--------------+---------------+-----------+----+
```

（5）将 Spark 数据帧导出保存为 TFRecords 格式的文件。注意，使用 overwrite 令将覆盖现有数据。

```
1.  val path = "/tmp/dl/spark-tf-connector/test-output.tfrecord"
2.  df.write.format("tfrecords").option("recordType", "Example")
    .mode("overwrite").save(path)
```

在 Databricks 云平台 Notebook 中运行上述代码，运行结果如下。

```
(1) Spark Jobs
path: String = /tmp/dl/spark-tf-connector/test-output.tfrecord
```

（6）Spark 加载 TFRecords 格式的文件，转换为 Spark 数据帧。

```
1.  // 如果未提供自定义模式，则从 TFRecords 推断数据帧模式
2.  val importedDf1: DataFrame = spark.read.format("tfrecords")
    .option("recordType", "Example").load(path)
3.  importedDf1.show()
4.
5.  //使用自定义模式将 TFRecords 读取到数据帧中
6.  val importedDf2: DataFrame = spark.read.format("tfrecords")
    .schema(schema).load(path)
7.  importedDf2.show()
```

在 Databricks 云平台 Notebook 中运行上述代码，运行结果如下：

```
(7) Spark Jobs
importedDf1:org.apache.spark.sql.DataFrame
    VectorLabel:array
    element:float
    DoubleTypeLabel:float
    name:string
    LongTypeLabel:long
    IntegerTypeLabel:long
    FloatTypeLabel:float
    id:long
```

```
importedDf2:org.apache.spark.sql.DataFrame
    id:integer
    IntegerTypeLabel:integer
    LongTypeLabel:long
    FloatTypeLabel:float
    DoubleTypeLabel:double
    VectorLabel:array
    element:double
    name:string
+-----------+---------------+----+-------------+----------------+--------------+---+
|VectorLabel|DoubleTypeLabel|name|LongTypeLabel|IntegerTypeLabel|FloatTypeLabel| id|
+-----------+---------------+----+-------------+----------------+--------------+---+
| [1.0, 2.0]|           14.0|  r1|           23|               1|          10.0| 11|
| [2.0, 2.0]|           15.0|  r2|           24|               2|          12.0| 21|
+-----------+---------------+----+-------------+----------------+--------------+---+

+---+----------------+-------------+--------------+---------------+-----------+----+
| id|IntegerTypeLabel|LongTypeLabel|FloatTypeLabel|DoubleTypeLabel|VectorLabel|name|
+---+----------------+-------------+--------------+---------------+-----------+----+
| 11|               1|           23|          10.0|           14.0| [1.0, 2.0]|  r1|
| 21|               2|           24|          12.0|           15.0| [2.0, 2.0]|  r2|
+---+----------------+-------------+--------------+---------------+-----------+----+

importedDf1: org.apache.spark.sql.DataFrame = [VectorLabel: array<float>,
DoubleTypeLabel: float ... 5 more fields]
importedDf2: org.apache.spark.sql.DataFrame = [id: int, IntegerTypeLabel:
int ... 5 more fields]
```

查询 df.show()和 importeddf1.show()分别执行的结果，验证 Spark 导入的数据帧与原始的数据帧匹配；importeddf1.show()和 importeddf2.show()分别执行的结果，表明未提供自定义模式的数据帧与提供自定义模式的数据帧在类型上可能不一致，例如，对于 element 字段，Spark 推断其为 float 类型，但在模式中定义的是 double 类型。

在 Databricks 云平台的 Spark Web UI 页面，查询 Spark 的运行情况如图 32-81 所示。

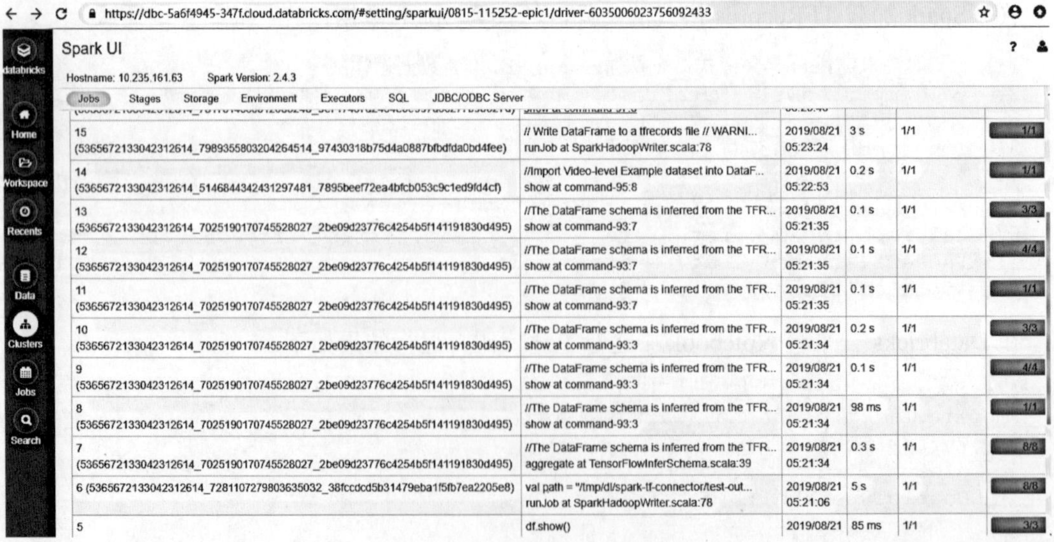

图 32-81　Spark 运行界面

（7）将现有的 TFRecord 数据集加载到 Spark 中。

下载 video_level-train-0.tfrecord 数据源文件，保存到 Databricks 文件系统。

```
%sh curl -s http://us.data.yt8m.org/2/video/train/trainIc.tfrecord >
/dbfs/tmp/dl/spark-tf-connector/video_level-train-0.tfrecord
```

声明模式，并将下载保存的数据文件导入 Spark 数据帧。

```
1.  //将视频级示例数据集导入数据帧
2.  val videoSchema = StructType(List(StructField("id", StringType),
3.              StructField("labels", ArrayType(IntegerType, true)),
4.              StructField("mean_rgb", ArrayType(FloatType, true)),
5.              StructField("mean_audio", ArrayType(FloatType, true))))
6.  val videoDf: DataFrame = spark.read.format("tfrecords")
    .schema(videoSchema).option("recordType", "Example")
7.    .load("dbfs:/tmp/dl/spark-tf-connector/video_level-train-0.tfrecord")
8.  videoDf.show(5)
```

在 Databricks 云平台 Notebook 中运行上述代码，运行结果如下：

```
(1) Spark Jobs
videoDf:org.apache.spark.sql.DataFrame
    id:string
    labels:array
    element:integer
    mean_rgb:array
    element:float
    mean_audio:array
    element:float
+----+--------------------+--------------------+--------------------+
|  id|              labels|            mean_rgb|          mean_audio|
+----+--------------------+--------------------+--------------------+
|rjIc|[2, 26, 45, 130, ...|[0.80833536, 0.73...|[-0.47192606, -0....|
|SUIc|              [1343]|[-0.65343785, 0.7...|[0.41343987, 1.44...|
|MjIc|  [2, 45, 212, 1745]|[-0.17644894, 1.0...|[-1.3482137, 0.72...|
|rzIc|[11, 20, 22, 29, ...|[0.06379097, 0.74...|[-0.35289493, -0....|
|zXIc|         [0, 1, 828]|[0.21688479, -1.2...|[-0.90491796, -0....|
+----+--------------------+--------------------+--------------------+
only showing top 5 rows

videoSchema: org.apache.spark.sql.types.StructType =
    StructType(StructField(id,StringType,true),
    StructField(labels,ArrayType(IntegerType,true),true),
    StructField(mean_rgb,ArrayType(FloatType,true),true),
    StructField(mean_audio,ArrayType(FloatType,true),true))
videoDf: org.apache.spark.sql.DataFrame = [id: string, labels: array<int> ...
2 more fields]
```

（8）将 Spark 数据帧数据导出保存为 TFRecords 格式的文件，然后再读取加载，将其导回 Spark 数据帧。

将数据帧写入 TFRecords 文件，此命令将覆盖现有数据。

```
1.  videoDf.write.format("tfrecords").option("recordType", "Example")
    .mode("overwrite").save("dbfs:/tmp/dl/spark-tf-connector/youtube-8m-
    video.tfrecords")
```

Spark 读取加载的 TFRecords 文件，导入数据帧，验证数据帧与原始数据帧匹配。

```
1.  val importedDf1: DataFrame = spark.read.format("tfrecords").option
    ("recordType", "Example").schema(videoSchema).load("dbfs:/tmp/dl/spark-
    tf-connector/youtube-8m-video.tfrecords")
2.  importedDf1.show(5)
```

在 Databricks 云平台 Notebook 中运行上述代码，运行结果如下：

```
(1) Spark Jobs
importedDf1:org.apache.spark.sql.DataFrame
    id:string
    labels:array
    element:integer
    mean_rgb:array
    element:float
    mean_audio:array
    element:float
+----+--------------------+--------------------+--------------------+
|  id|              labels|            mean_rgb|          mean_audio|
+----+--------------------+--------------------+--------------------+
|rjIc|[2, 26, 45, 130, ...|[0.80833536, 0.73...|[-0.47192606, -0... |
|SUIc|              [1343]|[-0.65343785, 0.7...|[0.41343987, 1.44...|
|MjIc|  [2, 45, 212, 1745]|[-0.17644894, 1.0...|[-1.3482137, 0.72...|
|rzIc|[11, 20, 22, 29, ...|[0.06379097, 0.74...|[-0.35289493, -0... |
|zXIc|        [0, 1, 828]|[0.21688479, -1.2...|[-0.90491796, -0... |
+----+--------------------+--------------------+--------------------+
only showing top 5 rows

importedDf1: org.apache.spark.sql.DataFrame = [id: string, labels: array<int> ...
2 more fields]
```

（9）删除下载的数据文件。

查询下载的数据文件信息。

```
print(dbutils.fs.ls("dbfs:/tmp/dl/spark-tf-connector/youtube-8m-video
.tfrecords"))
```

在 Databricks 云平台 Notebook 中运行上述代码，运行结果如下：

```
WrappedArray(FileInfo(dbfs:/tmp/dl/spark-tf-connector/youtube-8m-video
.tfrecords/_SUCCESS, _SUCCESS, 0), FileInfo(dbfs:/tmp/dl/spark-tf-connector/
youtube-8m-video.tfrecords/part-r-00000, part-r-00000, 4530186))
```

在 Databricks 云平台 Notebook 中执行以下命令，删除下载的文件。

```
%fs rm -r /tmp/dl/spark-tf-connector/
```

32.4　Spark 上的深度学习内核解密

本节讲解使用 TensorFlow 进行图片的分布式处理、数据模型源码剖析、逻辑节点源码剖析、构建索引源码剖析、深度学习下 Spark 作业源码剖析、性能调优最佳实践等内容。

32.4.1　使用 TensorFlow 进行图片的分布式处理

基于 Databricks 云平台提供的运行时机器学习库，我们在 Databricks 云平台创建自己的

Spark 集群，调用 TensorFlow 深度学习框架进行模型训练及预测。

图 32-82 是 TensorFlow 与 Spark 集成架构图。

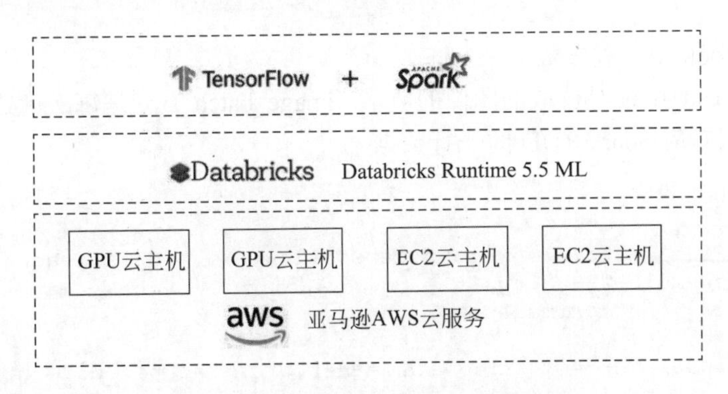

图 32-82 TensorFlow 与 Spark 集成架构图

案例使用 Google 公司的 Inception 3 网络模型对图像数据进行分类预测。例如，输入一张图片，预测图片是珊瑚礁、潜水者、鸭嘴兽、虎鲨等。Inception 3 是基于大型图像数据库 ImageNet 数据的预训练模型，本案例将使用 Spark 分布式计算框架，在 Spark 的每一个任务中利用 TensorFlow Inception 3 模型对图像分类进行预测。

Inception 3 网络模型如表 32-4 所示。

表 32-4 Inception 3 网络模型

类 型	感受野大小/步幅或说明	输入尺寸
conv	3×3/2	299×299×3
conv	3×3/1	149×149×32
conv padded	3×3/1	147×147×32
pool	3×3/2	147×147×64
conv	3×3/1	73×73×64
conv	3×3/2	71×71×80
conv	3×3/1	35×35×192
3×Inception	论文图 5	35×35×288
5×Inception	论文图 6	17×17×768
2×Inception	论文图 7	8×8×1280
pool	8×8	8×8×2048
linear	logits	1×1×2048
softmax	classifier	1×1×1000

其中 Inception 3 模型的论文链接为 https://arxiv.org/pdf/1512.00567.pdf。

32.4.2　数据模型源码剖析

（1）在 Notebook 中设置参数。

其中 images_read_limit 是读取的图像的数量。image_batch_size 是每次批处理的图像数量，一个批处理的数据对应 Spark RDD 的一行数据。

```
1.  # notebook 设置
2.
3.  MODEL_URL = 'http://download.tensorflow.org/models/image/imagenet/
    inception-2015-12-05.tgz'
4.  model_dir = '/tmp/imagenet'
5.
6.  IMAGES_INDEX_URL = 'http://image-net.org/imagenet_data/urls/imagenet_
    fall11_urls.tgz'
7.  images_read_limit = 1000L  # Increase this to read more images
8.
9.  #图像的批处理数
10. # 一个批处理对应一个 RDD 行
11. image_batch_size = 3
12.
13. num_top_predictions = 5
```

（2）在 Databricks Notebook 中导入 Tensorflow 库、Numpy 等库。

```
1.  import numpy as np
2.  import tensorflow as tf
3.  import os
4.  from tensorflow.python.platform import gfile
5.  import os.path
6.  import re
7.  import sys
8.  import tarfile
9.  from subprocess import Popen, PIPE, STDOUT
```

（3）下载预训练模型。

在 Databricks Notebook 中下载预训练模型。MODEL_URL 是 Tensorflow 预训练模型（Inception 3）的下载地址（http://download.tensorflow.org/models/image/imagenet/inception-2015-12-05.tgz）。

```
1.  def maybe_download_and_extract():
2.    """下载并提取模型 tar 文件."""
3.    import urllib.request
4.    dest_directory = model_dir
5.    if not os.path.exists(dest_directory):
6.      os.makedirs(dest_directory)
7.    filename = MODEL_URL.split('/')[-1]
8.    filepath = os.path.join(dest_directory, filename)
9.    if not os.path.exists(filepath):
10.     filepath2, _ = urllib.request.urlretrieve(MODEL_URL, filepath)
11.     print("filepath2", filepath2)
12.     statinfo = os.stat(filepath)
13.     print('Succesfully downloaded', filename, statinfo.st_size, 'bytes.')
14.     tarfile.open(filepath, 'r:gz').extractall(dest_directory)
```

```
15.   else:
16.     print('Data already downloaded:', filepath, os.stat(filepath))
17. maybe_download_and_extract()
```

在 Databricks Notebook 中运行上述代码，将预训练模型下载到 Databricks AWS 云主机中，运行结果如下：

```
1.  filepath2  /tmp/imagenet/inception-2015-12-05.tgz
2.  Succesfully  downloaded inception-2015-12-05.tgz 88931400 bytes
```

我们将 Inception 3 预训练模型下载到本地，解压缩以后查询文件如图 32-83 所示。

名称	类型	大小
classify_image_graph_def.pb	PB 文件	93,432 KB
cropped_panda	JPG 文件	3 KB
imagenet_2012_challenge_label_map_proto.pbtxt	PBTXT 文件	64 KB
imagenet_synset_to_human_label_map	文本文档	725 KB
LICENSE	文件	12 KB

图 32-83 Inception 3 预训练模型

其中 classify_image_graph_def.pb 文件为 Inception 3 网络模型文件。

其中 imagenet_2012_challenge_label_map_proto.pbtxt 文件是 ImageNet 数据集的标签映射文件，设置 UID 与分类标签的对应关系，target_class 是分类代码，从 1 到 1000，共 1000 个类别，记为 Node_ID；target_class_string 是 n 开头的字符串，记为 UID。部分记录格式如下：

```
# -*- protobuffer -*-
# LabelMap from ImageNet 2012 full data set UID to int32 target class.
entry {
  target_class: 449
  target_class_string: "n01440764"
}
entry {
  target_class: 450
  target_class_string: "n01443537"
}
entry {
  target_class: 442
  target_class_string: "n01484850"
}
......
```

其中 imagenet_synset_to_human_label_map.txt 文件是 UID 与类别名称（人类可读字符串）的对应关系。部分记录格式如下：

```
n00004475   organism, being
n00005787   benthos
n00006024   heterotroph
n00006484   cell
n00007846   person, individual, someone, somebody, mortal, soul
n00015388   animal, animate being, beast, brute, creature, fauna
n00017222   plant, flora, plant life
n00021265   food, nutrient
n00021939   artifact, artefact
......
```

（4）在 Databricks Notebook 中加载 Inception 3 模型数据，并广播给 Spark Workers。

```
1.  model_path = os.path.join(model_dir, 'classify_image_graph_def.pb')
2.  with gfile.FastGFile(model_path, 'rb') as f:
3.    model_data = f.read()
4.  model_data_bc = sc.broadcast(model_data)
```

第 4 行代码使用 SparkContext 的 Broadcast 广播变量，在 Spark 分布式集群环境下，将 Inception 3 模型数据从 Driver 节点发送到 Worker 各个节点，如图 32-84 所示。

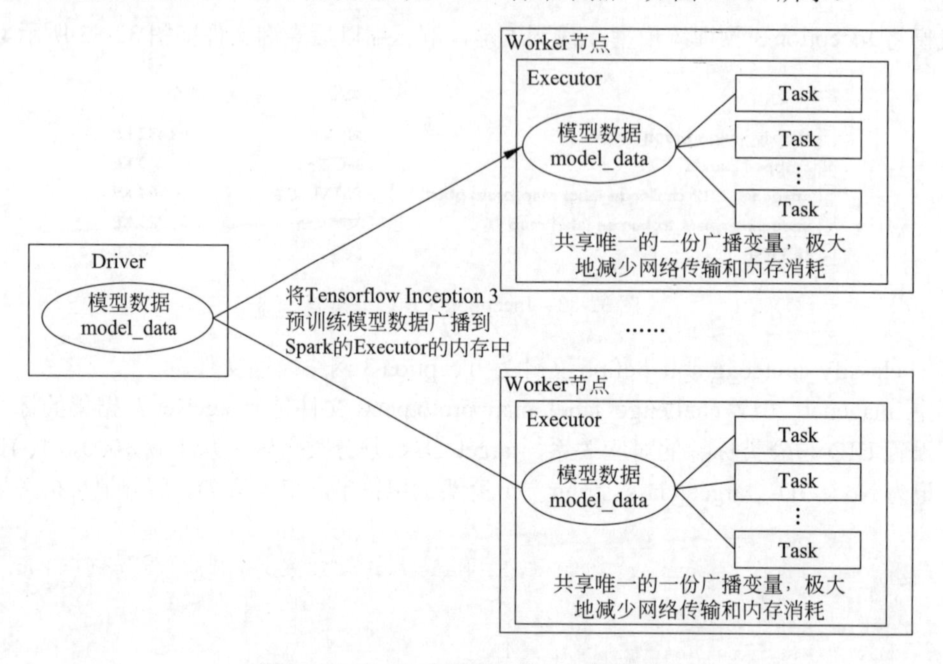

图 32-84　Spark 广播 Inception 3 模型数据

32.4.3　逻辑节点源码剖析

根据分类编号查询分类的名称。创建 NodeLookup 类，NodeLookup 类提供了 load 方法，load 方法的第一个参数是 label_lookup_path 文件路径，根据分类编码 Node_ID 查询 UID；第二个参数是 uid_lookup_path 文件路径，根据 UID 查找分类的可读字符串。

NodeLookup 类提供了 id_to_string 方法，输入一个分类编码 Node_ID，将返回人类的可读字符串。

```
1.  class NodeLookup(object):
2.    """ 将整数节点 ID 转换为人类可读标签"""
3.
4.    def __init__(self,
5.                 label_lookup_path=None,
6.                 uid_lookup_path=None):
7.      if not label_lookup_path:
8.        label_lookup_path = os.path.join(
9.            model_dir, 'imagenet_2012_challenge_label_map_proto.pbtxt')
10.     if not uid_lookup_path:
11.       uid_lookup_path = os.path.join(
```

```
12.         model_dir, 'imagenet_synset_to_human_label_map.txt')
13.     self.node_lookup = self.load(label_lookup_path, uid_lookup_path)
14.
15.   def load(self, label_lookup_path, uid_lookup_path):
16.     """为每个 SoftMax 节点加载一个可读的英文名称
17.
18.     参数:
19.       label_lookup_path: 字符串 UID 到整数节点 ID
20.       uid_lookup_path: 字符串 UID 到可读字符串
21.
22.     Returns:
23.       从整数节点 ID 到人类可读字符串的字典
24.     """
25.     if not gfile.Exists(uid_lookup_path):
26.       tf.logging.fatal('File does not exist %s', uid_lookup_path)
27.     if not gfile.Exists(label_lookup_path):
28.       tf.logging.fatal('File does not exist %s', label_lookup_path)
29.
30.     # 加载从字符串 uid 到人类可读字符串的映射
31.     proto_as_ascii_lines = gfile.GFile(uid_lookup_path).readlines()
32.     uid_to_human = {}
33.     p = re.compile(r'[n\d]*[ \S,]*')
34.     for line in proto_as_ascii_lines:
35.       parsed_items = p.findall(line)
36.       uid = parsed_items[0]
37.       human_string = parsed_items[2]
38.       uid_to_human[uid] = human_string
39.
40.     # 加载从字符串 uid 到整数节点 id 的映射
41.     node_id_to_uid = {}
42.     proto_as_ascii = gfile.GFile(label_lookup_path).readlines()
43.     for line in proto_as_ascii:
44.       if line.startswith('  target_class:'):
45.         target_class = int(line.split(': ')[1])
46.       if line.startswith('  target_class_string:'):
47.         target_class_string = line.split(': ')[1]
48.         node_id_to_uid[target_class] = target_class_string[1:-2]
49.
50.     #  加载整数节点 ID 到可读字符串的最终映射
51.     node_id_to_name = {}
52.     for key, val in node_id_to_uid.items():
53.       if val not in uid_to_human:
54.         tf.logging.fatal('Failed to locate: %s', val)
55.       name = uid_to_human[val]
56.       node_id_to_name[key] = name
57.
58.     return node_id_to_name
59.
60.   def id_to_string(self, node_id):
61.     if node_id not in self.node_lookup:
62.       return ''
63.     return self.node_lookup[node_id]
64.
65. node_lookup = NodeLookup().node_lookup
66. # 广播变量节点查找表
67. node_lookup_bc = sc.broadcast(node_lookup)
```

第 65 行代码构建一个 NodeLookup().node_lookup 实例 node_lookup。

第 67 行代码使用 SparkContext 的 Broadcast 广播变量，在 Spark 分布式集群环境下，将 node_lookup 实例从 Driver 节点发送到 Worker 各个节点，如图 32-85 所示。

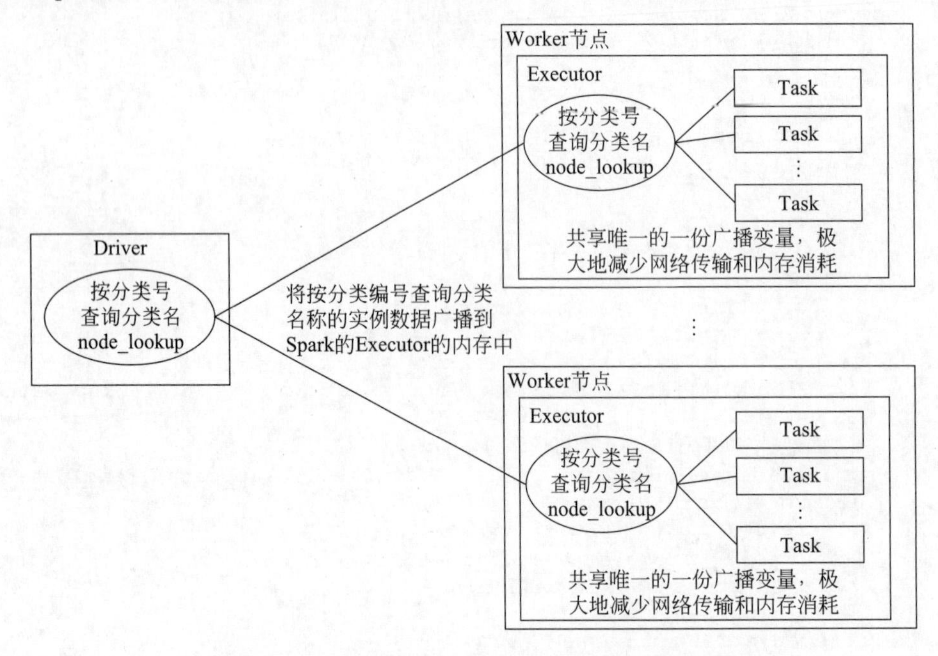

图 32-85　Spark 广播分类编号查询分类名称的数据

32.4.4　构建索引源码剖析

读取图像文件索引，分批加载图像文件（http://image-net.org/imagenet_data/urls/imagenet_fall11_urls.tgz）。

```
1.  # 读取图像文件
2.
3.  def run(cmd):
4.    p = Popen(cmd, shell=True, stdin=PIPE, stdout=PIPE, stderr=STDOUT,
      close_fds=True)
5.    return p.stdout.read()
6.
7.  def read_file_index():
8.    import urllib.request
9.    content = urllib.request.urlopen(IMAGES_INDEX_URL)
10.   data = content.read(images_read_limit)
11.   tmpfile = "/tmp/imagenet.tgz"
12.   with open(tmpfile, 'wb') as f:
13.     f.write(data)
14.   run("tar -xOzf %s > /tmp/imagenet.txt" % tmpfile)
15.   with open("/tmp/imagenet.txt", 'r') as f:
16.     lines = [l.split() for l in f]
17.     input_data = [tuple(elts) for elts in lines if len(elts) == 2]
18.     return [input_data[i:i+image_batch_size] for i in range(0,len(input_data),
        image_batch_size)]
19. batched_data = read_file_index()
20. print("There are %d batches" % len(batched_data))
```

在 Databricks Notebook 中运行上述代码，将图像文件分成 7 批，每批 3 张图片，运行结果如下：

```
There are 7 batches
```

32.4.5 深度学习下的 Spark 作业源码剖析

使用 sc.parallelize 算子把批图像数据并行化，Spark Workers 将并行处理多个 URL 下载图像。定义在 Spark Workers 上任务运行的方法 apply_inference_on_batch，Spark 在图像数据集中并行执行这些方法，采用 TensorFlow Inception 3 模型对图像类别进行预测。

```
1.  def run_inference_on_image(sess, img_id, img_url, node_lookup):
2.    """下载图像并对其进行预测
3.
4.    参数:
5.      image: 图像文件的 URL
6.
7.    返回:
8.      (image ID, image URL, scores),
9.      其 scores 是(human-readable node names, score)元组对的列表
10.   """
11.   import urllib.request
12.
13.   try:
14.     image_data = urllib.request.urlopen(img_url, timeout=1.0).read()
15.   except:
16.     return (img_id, img_url, None)
17.   # 一些有用的张量:
18.   #SoftMax:0: 包含归一化预测的张量，包括 1000 个标签
19.   #pool_3:0: 包含图像 2048 个浮点型描述的倒数第二层的张量
20.   #DecodeJpeg/contents:0:包含提供图像 JPEG 编码的字符串的张量。通过将 image_data
    作为输入提供给图来运行 SoftMax 张量
21.   softmax_tensor = sess.graph.get_tensor_by_name('softmax:0')
22.   try:
23.     predictions = sess.run(softmax_tensor,
24.                            {'DecodeJpeg/contents:0': image_data})
25.   except:
26.     # 处理格式错误的 JPEG 文件的问题
27.     return (img_id, img_url, None)
28.   predictions = np.squeeze(predictions)
29.   top_k = predictions.argsort()[-num_top_predictions:][::-1]
30.   scores = []
31.   for node_id in top_k:
32.     if node_id not in node_lookup:
33.       human_string = ''
34.     else:
35.       human_string = node_lookup[node_id]
36.     score = predictions[node_id]
37.     scores.append((human_string, score))
38.   return (img_id, img_url, scores)
39.
40. def apply_inference_on_batch(batch):
41.   """ 对一批图像进行预测，没有明确告诉 TensorFlow 使用 GPU，它能够根据哪个更快，在
```

```
42.    CPU 和 GPU 之间自动进行选择
43.    """
44.    with tf.Graph().as_default() as g:
45.      graph_def = tf.GraphDef()
46.      graph_def.ParseFromString(model_data_bc.value)
47.      tf.import_graph_def(graph_def, name='')
48.      with tf.Session() as sess:
49.        labeled = [run_inference_on_image(sess, img_id, img_url, node_lookup_bc
           .value) for (img_id, img_url) in batch]
50.        return [tup for tup in labeled if tup[2] is not None]
51. urls = sc.parallelize(batched_data)
52. labeled_images = urls.flatMap(apply_inference_on_batch)
```

第 21 行代码获取 TensorFlow 的 SoftMax 归一化预测张量，包括 1000 个分类标签。

第 23 行代码中输入图像数据，TensorFlow 在 Inception 3 模型中执行 SoftMax 预测张量，得到 1000 个类别的预测概率。

第 29 行代码使用 argsort 方法获取分类预测概率前 5 名的类别索引，即分类编码（Node_ID）。

第 35 行代码根据分类编码（Node_ID）从 Spark 的广播变量 node_lookup 中查询对应的类别名称。

第 38 行代码返回每个 Spark 任务的预测结果（img_id, img_url, scores）。

第 51 行代码执行 SparkContext 的 parallelize 方法，并行化处理图像的批次数据。

第 52 行代码调用 Spark 的 flatMap 转换算子方法，将在 Spark 上运行 TensorFlow，但实际上不会运行 Spark 作业，因为 flatMap 不是 RDD 的行动算子。

调用 SparkContext 的 collect()行动算子方法，将 Spark 集群提交作业，运行 Spark 作业来预测图像分类，如图 32-86 所示。

```
1.  local_labeled_images = labeled_images.collect()
```

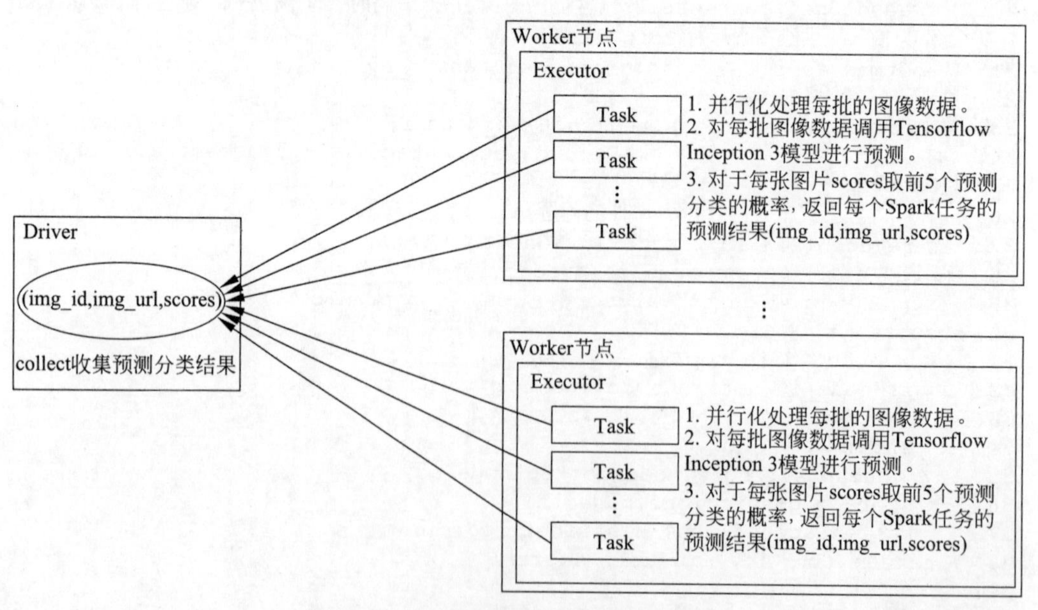

图 32-86　Spark collect 收集预测分类结果

在 Databricks Notebook 中运行上述代码，运行结果如下：

```
Out[36]: [('n00004475_6590',
  'http://farm4.static.flickr.com/3175/2737866473_7958dc8760.jpg',
  [('Band Aid', 0.1705848),
   ("pajama, pyjama, pj's, jammies", 0.080555655),
   ('diaper, nappy, napkin', 0.054097027),
   ('maraca', 0.034667443),
   ('remote control, remote', 0.027286513)]),
 ('n00004475_15899',
  'http://farm4.static.flickr.com/3276/2875184020_9944005d0d.jpg',
  [('bathtub, bathing tub, bath, tub', 0.25855055),
   ('tub, vat', 0.064664155),
   ('weasel', 0.05953106),
   ('black-footed ferret, ferret, Mustela nigripes', 0.04664932),
   ('hair spray', 0.02649683)]),
 ('n00004475_32312',
  'http://farm3.static.flickr.com/2531/4094333885_e8462a8338.jpg',
  [('orangutan, orang, orangutang, Pongo pygmaeus', 0.69150686),
   ('chimpanzee, chimp, Pan troglodytes', 0.09066241),
   ('siamang, Hylobates syndactylus, Symphalangus syndactylus', 0.01035614),
   ('gorilla, Gorilla gorilla', 0.0061912835),
   ('patas, hussar monkey, Erythrocebus patas', 0.0018502222)]),
 ('n00004475_35466',
  'http://farm4.static.flickr.com/3289/2809605169_8efe2b8f27.jpg',
  [('tiger cat', 0.14697573),
   ('tabby, tabby cat', 0.13278556),
   ('Egyptian cat', 0.06901597),
   ('washbasin, handbasin, washbowl, lavabo, wash-hand basin', 0.047761083),
   ('washer, automatic washer, washing machine', 0.029126083)]),
 ('n00004475_39382',

'http://2.bp.blogspot.com/_SrRTF97Kbfo/SUqT9y-qTVI/AAAAAAAABmg/saRXhruwS6
M/s400/bARADEI.jpg',
  [('Windsor tie', 0.46045992),
   ('suit, suit of clothes', 0.45713198),
   ('groom, bridegroom', 0.0056763305),
   ('bow tie, bow-tie, bowtie', 0.003930869),
   ('Loafer', 0.0030105752)]),
 ('n00004475_42770',
  'http://farm4.static.flickr.com/3488/4051378654_238ca94313.jpg',
  [('robin, American robin, Turdus migratorius', 0.5495006),
   ('water ouzel, dipper', 0.20509002),
   ('brambling, Fringilla montifringilla', 0.017242523),
   ('house finch, linnet, Carpodacus mexicanus', 0.0068506193),
   ('chickadee', 0.006212114)]),
 ('n00004475_54295',
  'http://farm4.static.flickr.com/3368/3198142470_6eb0be5f32.jpg',
  [('howler monkey, howler', 0.25775447),
   ('titi, titi monkey', 0.07344895),
   ('orangutan, orang, orangutang, Pongo pygmaeus', 0.05589796),
   ('patas, hussar monkey, Erythrocebus patas', 0.04289965),
   ('brown bear, bruin, Ursus arctos', 0.027977169)]),
 ('n00005787_66',
  'http://ib.berkeley.edu/labs/koehl/images/hannah.jpg',
```

```
[('wreck', 0.42330453),
  ('scuba diver', 0.1708945),
  ('coral reef', 0.16631156),
  ('gar, garfish, garpike, billfish, Lepisosteus osseus', 0.05191869),
  ('coho, cohoe, coho salmon, blue jack, silver salmon, Oncorhynchus kisutch',
   0.044225577)]),
('n00005787_97',
 'http://farm1.static.flickr.com/45/139488995_bd06578562.jpg',
 [('platypus, duckbill, duckbilled platypus, duck-billed platypus,
Ornithorhynchus anatinus',
   0.23488732),
  ('stingray', 0.17786986),
  ('tiger shark, Galeocerdo cuvieri', 0.0740401),
  ('electric ray, crampfish, numbfish, torpedo', 0.062862106),
  ('sturgeon', 0.05023549)]),
('n00005787_105',
 'http://farm3.static.flickr.com/2285/2658605078_f409b25597.jpg',
 [('coral reef', 0.5244134),
  ('coral fungus', 0.21099082),
  ('scuba diver', 0.009617118),
  ('sea slug, nudibranch', 0.009392346),
  ('sea anemone, anemone', 0.007957744)])]
```

32.4.6　性能调优最佳实践

本节讲述调试和性能调整的技巧，以便在数据块上进行模型推理。典型的模型推理有以下两部分。

- 数据输入管道。数据输入管道对数据 I/O 要求较高。
- 模型推理。模型推理对计算要求较高。

如何确定深度学习端到端工作流的瓶颈，下面提供了一些方法。

- 将模型简化为一个简单的模型，并测算每秒的计算度量。如果最佳模型和普通模型之间端到端时间相差较小，那么数据输入管道很可能是瓶颈，否则模型推理就是瓶颈。
- 如果使用 GPU 进行模型推断，需检查 GPU 利用率。如果 GPU 利用率一直不是很高，那么数据输入管道可能是瓶颈。

数据输入管道的性能调优方法。

利用 GPU 可以有效地优化模型推理的运行速度。随着 GPU 和其他加速器的速度越来越快，数据输入管道必须跟上需求。数据输入管道可以将数据读入 Spark 数据帧，对其进行转换，并将其加载为模型推理的输入。

如果数据输入是瓶颈，下面是一些提高 I/O 吞吐量的技巧。

（1）设置每批的最大记录数。只要记录可以放入内存，最大记录数的增加可以减少调用 UDF 函数的 I/O 开销。

Spark 批大小的配置语句如下：

```
spark.conf.set("spark.sql.execution.arrow.maxRecordsPerBatch", "5000")
```

（2）批量加载数据，对输入数据通过 Pandas UDF 进行预处理。

对于 TensorFlow、Keras、Databricks，建议使用 tf.data API。通过在 map 函数中设置

num_parallel_calls，调用 prefetch 和 batch 方法，进行数据获取和批处理，以及并行分析转换。

```
dataset.map(parse_example, num_parallel_calls=num_process)
.prefetch(prefetch_size).batch(batch_size)
```

（3）对于 PyTorch，建议使用 DataLoader 类。可以为批处理设置批大小，为并行数据加载设置 num_workers。

```
torch.utils.data.DataLoader(images, batch_size=batch_size, num_workers=num_process)
```